NEW AND DYNAMIC ELECTRONIC MEDIA

Learning about the human body is a fascinating, but challenging pursuit. Different students will tackle the challenges in different ways. With that in mind, we offer a variety of supplemental materials to complement this textbook and help you reach your goals for success in this course. Print supplements are described in the preface. Below are descriptions of materials that take advantage of contemporary mediums such as CD-ROMS and the Internet, much of which is available to you through your purchase of this new textbook.

Companion CD-ROM

Attached to the inside of the front cover is an envelope containing a companion CD-ROM for the text. It contains two sets of quizzes for each chapter, which will help you review material and determine your understanding. In addition, you can create Mastery tests in order to practice for exams. The mastery test allows you to choose how many questions you want to attempt, and to pick questions from multiple chapters. The questions will come up randomly so you can test your knowledge as many times as you would like. A full glossary with audio pronunciations is also included on the CD.

Dedicated Website

A dedicated companion website to accompany this text is found at the Wiley home for anatomy and physiology – **APCental** (**www.wiley.com/college/apcentral**). Clicking on the book's cover will take you to the site dedicated to *Principles of Anatomy and Physiology 10e*. Click on the Student Companion Site to access the many resources available to support and enhance your study. The site requires that you register with the access code provided with your new textbook. This access code is enclosed with directions for implementing it in the envelope with the companion CD on the inside front cover. Access to the website can be purchased online for those users who have not purchased a new text.

Resources on the website include an interactive survey that will help you determine your learning style and tips for effective study. In addition, you will find chapter self-quizzes for each chapter, interactive illustrated activities to help you learn anatomical terms, a pronunciation dictionary with terms by chapter with self-quizzing on terminology and flashcard options, web-based activities called Insights and Explorations, search engines for investigating homeostatic disorders, and instructor-reviewed web links. We also include some related material to enhance the relevance of your study to your lives and future careers such as Essays on Wellness and a review of Medical Tests and Procedures.

Interactions: Exploring the Functions of the Human Body
► Foundations (0-471-40162-5)

Foundations is the first CD in the Interactions Series and has been bundled at no charge with the purchase of a new copy of *Principles of Anatomy and Physiology 10e*. This dynamic new CD is a rich resource of animations and interactive activities that provide you with a basic knowledge of the levels of organization within the human body and how they interrelate. It lays the groundwork for understanding of the core principles of homeostasis, communication, energy flow, fluid flow, and boundaries important to the study of anatomy and physiology. On each of the chapter opening pages you will find suggested activities on the *Foundations* CD that relate to that chapter. Enjoy this dynamic new tool.

In addition to *Foundations*, there are eight more CDs in the *Interactions* Series. Each CD-ROM focuses on the physiological processes of one or two body systems. These are listed below. Please contact your Wiley sales representative for additional information about **Interactions** and purchase options.

- ► *Regulation: The Nervous and Endocrine Systems* (0-471-20782-9)
- ► *Support and Movement: The Skeletal and Muscular Systems* (0-471-20801-9)
- ► *Distribution: The Cardiovascular and Lymphatic Systems* (0-471-20802-7)
- ► *Disease Resistance: The Lymphatic and Immune Systems* (0-471-26523-3)
- ► *Energy Acquisition and Use: The Digestive System and Metabolism* (0-471-26524-1)
- ► *Gas Exchange and pH Balance: The Respiratory System* (0-471-26525-X)
- ► *Balancing Fluids and pH: The Urinary System* (0-471-26526-8)
- ► *Continuity: The Reproductive System* (0-471-26527-6)

PRINCIPLES OF ANATOMY AND PHYSIOLOGY

Tenth Edition

Gerard J. Tortora

Bergen Community College

Sandra Reynolds Grabowski

Purdue University

John Wiley & Sons, Inc.

Senior Editor	Bonnie Roesch
Marketing Manager	Clay Stone
Developmental Editor	Ellen Ford
Production Director	Pamela Kennedy
Associate Production Manager	Kelly Tavares
Text and Cover Designer	Karin Gerdes Kincheloe
Art Coordinator	Claudia Durrell
Photo Editor	Hilary Newman
Chapter Opener Illustrations	Keith Kasnot

Cover photos: Photo of woman: ©PhotoDisc, Inc.
Photo of man: ©Ray Massey/Stone
Illustration and photo credits follow the Glossary.

This book was typeset by Progressive Information Technologies. It was printed and bound by Von Hoffmann Press, Inc. The cover was also printed by Von Hoffman Press, Inc.

The paper in this book was manufactured by a mill whose forest management programs include sustained yield harvesting of its timberlands. Sustained yield harvesting principles ensure that the number of trees cut each year does not exceed the amount of new growth.

This book is printed on acid-free paper. ∞

USA ISBN: 0-471-41501-4
WIE ISBN: 0-471-22472-3

Printed in the United States of America.
10 9 8 7 6 5 4

About The Cover

In choosing the cover illustration for the tenth edition of *Principles of Anatomy and Physiology,* we wanted to reflect the underlying principle in the textbook as well as our approach to presentation of the content. In a word, we wanted to suggest the idea of balance. Successfully practicing yoga requires finding the balance between the physical, mental, and spiritual aspects of the human body. Achieving this balance produces flexibility and strength, as demonstrated by the figures on the cover.

Similarly, balance has always been at the heart of the study of anatomy and physiology. Homeostasis, the underlying principle discussed throughout this textbook, is the state of equilibrium of the internal environment of the human body. During your study of anatomy and physiology you will learn about the structures and functions needed to regulate and maintain this balanced state and its importance in promoting optimal well being.

Balance also applies to the way we present the content in the tenth edition of *Principles of Anatomy & Physiology*. We offer a balanced coverage of structure and function; between normal and abnormal anatomy and physiology; between concepts and applications; between narrative and illustrations; between print and media; and among supplemental materials developed to meet the variety of teaching and learning styles. We hope that this balanced approach will provide you with the flexibility and strength to succeed in this course and in your future careers.

About the Authors

Gerard J. Tortora is Professor of Biology and former Coordinator at Bergen Community College in Paramus, New Jersey, where he teaches human anatomy and physiology as well as microbiology. He received his bachelor's degree in biology from Fairleigh Dickinson University and his master's degree in science education from Montclair State College. He is a member of many professional organizations, such as the Human Anatomy and Physiology Society (HAPS), the American Society of Microbiology (ASM), American Association for the Advancement of Science (AAAS), National Education Association (NEA), and the Metropolitan Association of College and University Biologists (MACUB).

Above all, Jerry is devoted to his students and their aspirations. In recognition of this commitment, Jerry was the recipient of MACUB's 1992 President's Memorial Award. In 1996, he received a National Institute for Staff and Organizational Development (NISOD) excellence award from the University of Texas and was selected to represent Bergen Community College in a campaign to increase awareness of the contributions of community colleges to higher education.

Jerry is the author of several best-selling science textbooks and laboratory manuals, a calling that often requires an additional 40 hours per week beyond his teaching responsibilities. Nevertheless, he still makes time for four or five weekly aerobic workouts that include biking and running. He also enjoys attending college basketball and professional hockey games and performances at the Metropolitan Opera House.

To my children, Lynne, Gerard, Kenneth, Anthony, and Andrew, who make it all worthwhile. — **G.J.T.**

Sandra Reynolds Grabowski is an instructor in the Department of Biological Sciences at Purdue University in West Lafayette, Indiana. For 25 years, she has taught human anatomy and physiology to students in a wide range of academic programs. In 1992 students selected her as one of the top ten teachers in the School of Science at Purdue.

Sandy received her BS in biology and her PhD in neurophysiology from Purdue. She is a founding member of the Human Anatomy and Physiology Society (HAPS) and has served HAPS as President and as editor of *HAPS News*. In addition, she is a member of the American Anatomy Association, the Association for Women in Science (AWIS), the National Science Teachers Association (NSTA), and the Society for College Science Teachers (SCST).

To students around the world, whose questions and comments continue to inspire the fine-tuning of this textbook. — **S.R.G.**

Preface

An anatomy and physiology course can be the gateway to a gratifying career in a host of health-related professions. As active teachers of the course, we recognize both the rewards and challenges in providing a strong foundation for understanding the complexities of the human body to an increasingly diverse population of students. Building on the unprecedented success of previous editions, the tenth edition of *Principles of Anatomy and Physiology* continues to offer a balanced presentation of content under the umbrella of our primary and unifying theme of homeostasis, supported by relevant discussions of disruptions to homeostasis. In addition, years of student feedback have convinced us that readers learn anatomy and physiology more readily when they remain mindful of the relationship between structure and function. As a writing team—an anatomist and a physiologist—our very different specializations offer practical advantages in fine-tuning the balance between anatomy and physiology.

Most importantly, our students continue to remind us of their needs for—and of the power of—simplicity, directness, and clarity. To meet these needs each chapter has been written and revised to include:

- clear, compelling, and up-to-date discussions of anatomy and physiology
- expertly executed and generously sized art
- classroom-tested pedagogy
- outstanding student study support.

As we revised the content for this edition we kept our focus on these important criteria for success in the anatomy and physiology classroom and have refined or added new elements to enhance the teaching and learning process.

Homeostasis: A Unifying Theme

The dynamic physiological constancy known as homeostasis is the prime theme in *Principles of Anatomy and Physiology*. We immediately introduce this unifying concept in Chapter 1 and describe how various feedback mechanisms work to maintain physiological processes within the narrow range that is compatible with life. Homeostatic mechanisms are discussed throughout the book, and homeostatic processes are clarified and reinforced through our well-received series of homeostasis feedback illustrations.

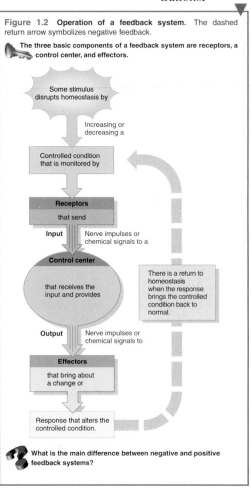

Figure 1.2 Operation of a feedback system. The dashed return arrow symbolizes negative feedback.

The three basic components of a feedback system are receptors, a control center, and effectors.

Some stimulus disrupts homeostasis by

Increasing or decreasing a

Controlled condition that is monitored by

Receptors that send

Input — Nerve impulses or chemical signals to a

Control center that receives the input and provides

There is a return to homeostasis when the response brings the controlled condition back to normal.

Output — Nerve impulses or chemical signals to

Effectors that bring about a change or

Response that alters the controlled condition.

What is the main difference between negative and positive feedback systems?

Figure 21.14 Negative feedback regulation of blood pressure via baroreceptor reflexes.

When blood pressure decreases, heart rate increases.

Some stimulus disrupts homeostasis by

Decreasing

Blood pressure

Receptors
Baroreceptors in arch of aorta and carotid sinus are stretched less

Input — Decreased rate of nerve impulses

Control centers
CV center in medulla oblongata
and adrenal medulla

Return to homeostasis when increased cardiac output and increased vascular resistance bring blood pressure back to normal

Output — Increased sympathetic, decreased parasympathetic stimulation
Increased secretion of epinephrine and norepinephrine from adrenal medulla

Effectors

| Increased stroke volume and heart rate lead to increased cardiac output (CO) | Constriction of blood vessels increases systemic vascular resistance (SVR) |

Increased blood pressure

Does this negative feedback cycle represent the changes that occur when you lie down or when you stand up?

New to this edition are ten *Focus on Homeostasis* pages, one each for the integumentary, skeletal, muscular, nervous, endocrine, cardiovascular, lymphatic and immune, respiratory, digestive, and urinary systems. Incorporating both graphic and narrative elements, these pages explain, clearly and succinctly, how the system under consideration contributes to the homeostasis of each of the other body systems. Use of this feature will enhance student understanding of the links between body systems and how interactions among systems contribute to the homeostasis of the body as a whole

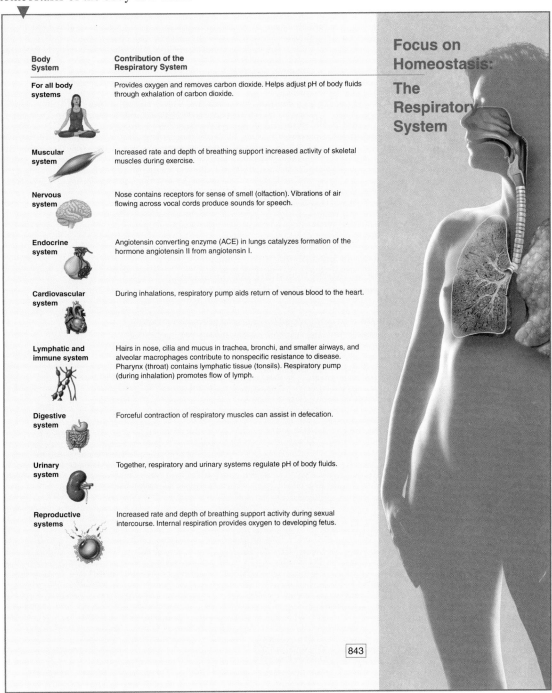

Focus on Homeostasis: The Respiratory System

Body System	Contribution of the Respiratory System
For all body systems	Provides oxygen and removes carbon dioxide. Helps adjust pH of body fluids through exhalation of carbon dioxide.
Muscular system	Increased rate and depth of breathing support increased activity of skeletal muscles during exercise.
Nervous system	Nose contains receptors for sense of smell (olfaction). Vibrations of air flowing across vocal cords produce sounds for speech.
Endocrine system	Angiotensin converting enzyme (ACE) in lungs catalyzes formation of the hormone angiotensin II from angiotensin I.
Cardiovascular system	During inhalations, respiratory pump aids return of venous blood to the heart.
Lymphatic and immune system	Hairs in nose, cilia and mucus in trachea, bronchi, and smaller airways, and alveolar macrophages contribute to nonspecific resistance to disease. Pharynx (throat) contains lymphatic tissue (tonsils). Respiratory pump (during inhalation) promotes flow of lymph.
Digestive system	Forceful contraction of respiratory muscles can assist in defecation.
Urinary system	Together, respiratory and urinary systems regulate pH of body fluids.
Reproductive systems	Increased rate and depth of breathing support activity during sexual intercourse. Internal respiration provides oxygen to developing fetus.

843

In addition, we believe students can better understand normal physiological processes by examining situations in which diseases or disorders impair those processes. We offer three features that highlight these disruptions to homeostasis.

A perennial favorite among students, the intriguing **Clinical Applications** in every chapter explore the clinical, professional, or everyday relevance of a particular anatomical structure or its related function. Many are new to this edition, and all have been reviewed for accuracy and relevance. Each application directly follows the discussion to which it relates.

Aspirin and Thrombolytic Agents

In patients with heart and blood vessel disease, the events of hemostasis may occur even without external injury to a blood vessel. At low doses, **aspirin** inhibits vasoconstriction and platelet aggregation by blocking synthesis of thromboxane A2. It also reduces the chance of thrombus formation. Due to these effects, aspirin reduces the risk of transient ischemic attacks (TIA), strokes, myocardial infarction, and blockage of peripheral arteries.

Thrombolytic agents are chemical substances that are injected into the body to dissolve blood clots that have already formed to restore circulation. They either directly or indirectly activate plasminogen. The first thrombolytic agent, approved in 1982 for dissolving clots in the coronary arteries of the heart, was **streptokinase,** which is produced by streptococcal bacteria. A genetically engineered version of human **tissue plasminogen activator (t-PA)** is now used to treat both heart attacks and brain attacks (strokes) that are caused by blood clots. ■

The **Disorders: Homeostatic Imbalances** sections at the end of most chapters include concise discussions of major diseases and disorders that illustrate departures from normal homeostasis. They provide answers to many questions that students ask about medical problems.

DISORDERS: HOMEOSTATIC IMBALANCES

Disorders of epithelial tissues are mainly specific to individual organs, for example, peptic ulcer disease (PUD), which erodes the epithelial lining of the stomach or small intestine. For this reason, epithelial disorders are described together with the relevant body system throughout the text. The most prevalent disorders of connective tissues are **autoimmune diseases**—diseases in which antibodies produced by the immune system fail to distinguish what is foreign from what is self and attack the body's own tissues. The most common autoimmune disorder is rheumatoid arthritis, which attacks the synovial membranes of joints. Because connective tissue is one of the most abundant and widely distributed of the four main types of tissues, its disorders often affect multiple body systems. Common disorders of muscle tissue and nervous tissue are described at the ends of Chapters 10 and 12, respectively.

Sjögren's Syndrome

Sjögren's syndrome (SHŌ-grenz) is a common autoimmune disorder that causes inflammation and destruction of exocrine glands, especially the lacrimal (tear) glands and salivary glands. Signs include dryness of the mucous membranes in the eyes and mouth and salivary gland enlargement. Systemic effects include arthritis, difficulty swallowing, pancreatitis (inflammation of the pancreas), pleuritis (inflammation of the pleurae of the lungs), and migraine headaches. The disorder affects more females than males by a ratio of 9 to 1. About 20% of older adults experience some signs of Sjögren's. Treatment is supportive, including using artificial tears to moisten the eyes, sipping fluids, chewing sugarless gum, and using a saliva substitute to moisten the mouth.

Systemic Lupus Erythematosus

Systemic lupus erythematosus (er-i-thē-ma-TŌ-sus), **SLE,** or simply lupus, is a chronic inflammatory disease of connective tissue occurring mostly in nonwhite women during their childbearing years. It is an autoimmune disease that can cause tissue damage in every body system. The disease, which can range from a mild condition in most patients to a rapidly fatal disease, is marked by periods of exacerbation and remission. The prevalence of SLE is about 1 in 2000 persons, with females more likely to be afflicted than males by a ratio of 8 or 9 to 1.

Although the cause of SLE is not known, genetic, environmental, and hormonal factors are implicated. The genetic component is suggested by studies of twins and family history. Environmental factors include viruses, bacteria, chemicals, drugs, exposure to excessive sunlight, and emotional stress. With regard to hormones, sex hormones, such as estrogens, may trigger SLE.

Signs and symptoms of SLE include painful joints, low-grade fever, fatigue, mouth ulcers, weight loss, enlarged lymph nodes and spleen, sensitivity to sunlight, rapid loss of large amounts of scalp hair, and anorexia. A distinguishing feature of lupus is an eruption across the bridge of the nose and cheeks called a "butterfly rash." Other skin lesions may occur, including blistering and ulceration. The erosive nature of some SLE skin lesions was thought to resemble the damage inflicted by the bite of a wolf—thus, the term *lupus* (= wolf). The most serious complications of the disease involve inflammation of the kidneys, liver, spleen, lungs, heart, brain, and gastrointestinal tract. Because there is no cure for SLE, treatment is supportive, including anti-inflammatory drugs, such as aspirin, and immunosuppressive drugs.

MEDICAL TERMINOLOGY

Vocabulary-building glossaries of selected **Medical Terminology** and conditions also appear at the end of appropriate chapters. This feature has been expanded and updated for this edition.

Atrophy (AT-rō-fē; *a-* = without; *-trophy* = nourishment) A decrease in the size of cells, with a subsequent decrease in the size of the affected tissue or organ.

Hypertrophy (hī-PER-trō-fē; *hyper-* = above or excessive) Increase in the size of a tissue because its cells enlarge without undergoing cell division.

Tissue rejection An immune response of the body directed at foreign proteins in a transplanted tissue or organ; immunosuppressive drugs, such as cyclosporine, have largely overcome tissue rejection in heart-, kidney-, and liver-transplant patients.

Tissue transplantation The replacement of a diseased or injured tissue or organ. The most successful transplants involve use of a person's own tissues or those from an identical twin.

Xenotransplantation (zen′-ō-trans-plan-TĀ-shun; *xeno-* = strange, foreign) The replacement of a diseased or injured tissue or organ with cells or tissues from an animal. Porcine (from pigs) and bovine (from cows) heart valves are used for some heart-valve replacement surgeries.

Organization, Special Topics, and Content Improvements

The book follows the same unit and topic sequence as its nine earlier editions. It is divided into five principal sections: Unit 1, "Organization of the Human Body," provides an understanding of the structural and functional levels of the body, from molecules to organ systems. Unit 2, "Principles of Support and Movement," analyzes the anatomy and physiology of bones, joints, and muscles. Unit 3, "Control Systems of the Human Body," emphasizes the importance of neural communication in the immediate maintenance of homeostasis, the role of sensory receptors in providing information about the internal and external environment, and the significance of hormones in maintaining long-term homeostasis. Unit 4, "Maintenance of the Human Body," explains how body systems function to maintain homeostasis on a moment-to-moment basis through the processes of circulation, respiration, digestion, cellular

metabolism, urinary functions, and buffer systems. Unit 5, "Continuity," covers the anatomy and physiology of the reproductive systems, development, and the basic concepts of genetics and inheritance.

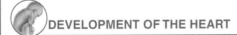

DEVELOPMENT OF THE HEART

▶ O B J E C T I V E

• Describe the development of the heart.

The *heart,* a derivative of **mesoderm,** begins to develop early in the third week after fertilization. In the ventral region of the embryo inferior to the foregut, the heart develops from a group of mesodermal cells called the **cardiogenic area.** The next step in its development is the formation of a pair of tubes, the **endocardial tubes,** which develop from the cardiogenic area (Figure 20.18). These tubes then unite to form a common tube, referred to as the **primitive heart tube.** Next, the primitive heart tube develops into five distinct regions: (1) **truncus arteriosus,** (2) **bulbus cordis,** (3) **ventricle,** (4) **atrium,** and (5) **sinus venosus.** Because the bulbus cordis and ventricle grow most rapidly, and because the heart enlarges more rapidly than its superior and inferior attachments, the heart first assumes a U-shape and then an S-shape. The flexures of the heart reorient the regions so that the atrium and sinus venosus eventually come to lie superior to the bulbus cordis, ventricle, and truncus arteriosus. Contractions

Development We often tell our students that they can better appreciate the "logic" of human anatomy by becoming aware of how various structures developed in the first place. As in previous editions, illustrated discussions of development are found near the conclusion of most body system chapters. Placing this coverage at the end of chapters enables students to master the anatomical terminology they need before attempting to learn about embryonic and fetal structures. The fetus icon designates the start of each development discussion.

AGING AND THE IMMUNE SYSTEM

▶ O B J E C T I V E

• Describe the effects of aging on the immune system.

With advancing age, most people become more susceptible to all types of infections and malignancies. Their response to vaccines is decreased, and they tend to produce more autoantibodies (antibodies against their body's own molecules). In addition, the immune system exhibits lowered levels of function. For example, T cells become less responsive to antigens, and fewer T cells respond to infections. This may result from age-related atrophy of the thymus or decreased production of thymic hormones. Because the T cell population decreases with age, B cells are also less responsive. Consequently, antibody levels do not increase as rapidly in response to a challenge by an antigen, resulting in increased susceptibility to various infections. It is for this key reason that elderly individuals are encouraged to get influenza (flu) vaccinations each year.

Aging Students need to be reminded from time to time that anatomy and physiology is not static. As the body ages, its structures and related functions subtly change. Moreover, aging is a professionally relevant topic for the majority of this book's readers, who will go on to careers in *health-related* fields in which the average age of the client population is steadily advancing. For these reasons, age-related changes in anatomy and physiology are discussed at the end of sixteen chapters.

EXERCISE AND BONE TISSUE

▶ O B J E C T I V E

• Describe how exercise and mechanical stress affect bone tissue.

Within limits, bone has the ability to alter its strength in response to changes in mechanical stress. When placed under stress, bone tissue adapts by becoming stronger through increased deposition of mineral salts and production of collagen fibers. Another effect of stress is to increase the production of calcitonin, a hormone that inhibits bone resorption. Without mechanical stress, bone does not remodel normally because bone resorption outstrips bone formation. Removal of mechanical stress weakens bone through demineralization (loss of bone minerals) and decreased numbers of collagen fibers.

The main mechanical stresses on bone are those that result from the pull of skeletal muscles and the pull of gravity. If a person is bedridden or has a fractured bone in a cast, the strength of the unstressed bones diminishes. Astronauts subjected to the microgravity of space also lose bone mass. In both cases, bone loss can be dramatic—as much as 1% per week. Bones of athletes, which are repetitively and highly stressed, become notably thicker and stronger than those of nonathletes. Weight-bearing activities, such as walking or moderate weight lifting, help build and retain bone mass. Adolescents and young adults should engage in regular weight-bearing exercise prior to the closure of the epiphyseal plates to help build total mass prior to its inevitable reduction with aging. Even elderly people can strengthen their bones by engaging in weight-bearing exercise.

Exercise Physical exercise can produce favorable changes in some anatomical structures and enhance many physiological functions, most notably those associated with the muscular, skeletal, and cardiovascular systems. This information is especially relevant to readers embarking on careers in physical education, sports training, and dance. Hence, key chapters include brief discussions of exercise-related considerations, which are signaled by a distinctive running shoe icon.

Every chapter in the tenth edition of *Principles of Anatomy and Physiology* incorporates a host of improvements to both the text and the art, many suggested by reviewers, educators, and students. In addition, most chapters offer new Clinical Applications. Here are some of the more noteworthy changes.

Chapter 2 / The Chemical Level of Organization
A new Clinical Application on "Saturated Fats, Cholesterol, and Atherosclerosis." Some sections are more concise, and the complete discussion of oxidation–reduction reactions is now in chapter 25.

Chapter 3 / The Cellular Level of Organization
Description of a newly discovered cellular organelle, the proteasome. New Clinical Applications on Medical Uses of Isotonic, Hypertonic, and Hypotonic Solutions; Proteasomes and Disease; Genomics; and Antisense Therapy. Transport through the Golgi complex, tumor-suppressor genes, and causes of cancer updated and revised.

Chapter 4 / The Tissue Level of Organization
New illustrations of cell junctions and types of multicellular exocrine glands; several new photomicrographs. Revisions in sections on cell junctions, the basement membrane, and ground substance of connective tissue.

Chapter 5 / The Integumentary System
New section on types of hairs. New Clinical Applications on Photosensitivity, Hair Removal, and Hair and Hormones; New illustrations of the rule of nines, types of epidermal cells, epidermal wound healing, and deep wound healing. Structure of epidermis and development of the integumentary system revised.

Chapter 6 / The Skeletal System: Bone Tissue
New illustrations of the histology of bone, endochondral ossification, and fracture repair.

Chapter 8 / The Skeletal System: The Appendicular Skeleton
New illustration of the hip bone. Addition of Hip Fracture to the disorders section at the end of the chapter.

Chapter 9 / Joints
New Clinical Applications on Rotator Cuff Injuries, Separated Shoulder, Tennis Elbow, Little League Elbow, Dislocated Radius, Swollen Knee, Dislocated Knee, and Lyme Disease.

Chapter 10 / Muscle Tissue
New coverage of exercise and skeletal muscles, fibromyalgia, and benefits of stretching. New illustrations on microscopic organization of skeletal muscle tissue, the neuromuscular junction, and development of the muscular system. Proteins in skeletal muscle and uses of Botox revised.

Chapter 11 / The Muscular System
New illustration of a newly discovered skeletal muscle, the sphenomandibularis muscle (see page 325).

Chapter 12 / Nervous Tissue
New Medical Terminology list. New illustrations of the structure of a neuron, myelinated and unmyelinated axons, and signal transmission at synapses. Updated definitions of sensory neuron, interneuron, and motor neuron.

Chapter 13 / The Spinal Cord and Spinal Nerves
New Medical Terminology list. New illustration of the connective tissue coverings of a spinal nerve.

Chapter 14 / The Brain and Cranial Nerves
New illustrations of the parts of the brain, protective coverings of the brain, pathways of cerebrospinal fluid flow, thalamus, development of the brain, seven figures on cranial nerves. Revised discussions of the thalamus and cranial nerves.

Chapter 15 / Sensory, Motor and Integrative Systems
New illustrations of muscle spindles, tendon organs, somatic sensory pathways, and motor pathways. New Clinical Applications on Phantom Limb Sensation, Amyotrophic Lateral Sclerosis, Damage to Basal Ganglia, and Damage to the Cerebellum. Section on Muscle Spindles rewritten to simplify terminology but more clearly explain function; sections on somatic motor pathways, memory, and sleep updated.

Chapter 16 / The Special Senses
New illustrations of taste buds, effects of tastants on taste neurons, and development of the eyes and ears. New Clinical Applications on Hyposmia, Macular Degeneration, Taste Aversion, Presbyopia, and Color and Night Blindness. New section on Development of the Eyes and Ears. Expanded Medical Terminology list.

Chapter 17 / The Autonomic Nervous System
New Table comparing sympathetic and parasympathetic divisions of the autonomic nervous system; new illustrations of sympathetic and parasympathetic divisions.

Chapter 18 / The Endocrine System
New illustrations of the location of endocrine glands, thyroid gland, parathyroid glands, adrenal glands, pancreas, and development of the endocrine system. New Clinical Applications on Administering Hormones, Oxytocin and Childbirth, Congenital Adrenal Hyperplasia, and Posttraumatic Stress Disorder. Photographs of people with endocrine disorders, including an interesting case of a twin with giantism (see page 625). New Medical Terminology list.

Chapter 19 / The Blood
New description of red bone marrow. New Clinical Applications on Iron Overload and Tissue Damage, Induced Polycythemia in Athletes, and Aspirin and Thrombolytic Agents.

Chapter 20 / The Heart
New section on ATP production in cardiac muscle fibers. New Clinical Applications on Heart Valve Disorders, Reperfusion Damage, Ectopic Pacemaker, Regeneration of Cardiac Muscle Tissue, Artificial Hearts. New illustrations of the serous pericardium, structure of the heart, coronary circulation, pacemaker potentials, route of action potential depolarization and repolarization, and development of the heart. Arrhythmias added to disorders section. Expanded Medical Terminology list.

Chapter 21 / Blood Vessels and Hemodynamics
New illustrations of the structure of blood vessels, types of capillaries, action of skeletal muscles, ANS innervation of the heart,

circulatory routes, principal branches of the aorta, ascending aorta, branches of the brachiocephalic trunk in the neck, cerebral arterial circle, branches of the abdominal aorta, arteries of the pelvis and lower limbs, pulmonary circulation, fetal circulation, and development of blood vessels and blood cells.

Chapter 22 / The Lymphatic and Immune System and Resistance to Disease

New section on stress and immunity. New illustrations of lymphatic tissue, routes for lymph drainage, lymph node structure, development of the lymphatic system, phagocytosis, and stages of inflammation. New Clinical Applications on Ruptured Spleen, Edema and Lymph Flow, and Microbial Evasion of Phagocytosis. Autoimmune diseases, severe combined immunodeficiency disease, SLE, and lymphomas added to the disorders section. Descriptions of thymus and lymph nodes revised.

Chapter 23 / The Respiratory System

New illustrations of the structures of the respiratory system, location of peripheral chemoreceptors that help regulate respiration, and development of the bronchial tubes and lungs. New Clinical Applications on Laryngitis, Cancer of the Larynx, and Nebulization. New discussion and explanation of ventilation-perfusion coupling, nitrogen narcosis, and decompression sickness. Addition of ARDS to disorders section. Updated percentages of carbon dioxide dissolved in blood plasma and carried by bicarbonate ions and hemoglobin.

Chapter 24 / The Digestive System

New illustrations of organs of the gastrointestinal tract, peritoneal folds, salivary glands, histology of the stomach, secretion of HCl by stomach cells, histology of the small intestine, anatomy of the large intestine, and histology of the large intestine. New Clinical Applications on Root Canal Therapy, Jaundice, Occult Blood, and Absorption of Alcohol.

Chapter 25 / Metabolism

New section on energy homeostasis. New Clinical Application on Emotional Eating and new Medical Terminology list. Updated sections on regulation of food intake and obesity.

Chapter 26 / The Urinary System

New illustrations of a sagittal section through the kidney, blood supply of the kidneys, anatomy of the ureter, urinary bladder, and urethra, and development of the urinary system. More concise sections on reabsorption and secretion. New Clinical Applications on Nephroptosis and Loss of Plasma Proteins. Addition of Urinary Bladder Cancer to the disorders section.

Chapter 27 / Fluid, Electrolyte, and Acid–Base Homeostasis

New Summary table of factors that maintain body water balance. Discussion and related figure describing how kidneys contribute to acid–base balance by secreting hydrogen ions moved from Chapter 26 to this chapter.

Chapter 28 / The Reproductive Systems

New illustrations to compare mitosis and meiosis and to show development of internal reproductive organs. New table reviewing oogenesis and development of ovarian follicles; new explanation of mittelschmerz and the role of leptin in puberty. New Clinical Applications on Uterine Prolapse, Episiotomy, and Female Athlete Triad. Addition of genital warts and premenstrual dysphoric disorder to the disorders section.

Chapter 29 / Development and Inheritance

New illustrations of fertilization, cleavage, formation of primary germ layers, gastrulation, development of the notochordal process, neurulation, development of chorionic villi, embryo folding, and development of pharyngeal arches. Beautiful new photographs of embryonic and fetal development. New Clinical Applications on Anencephaly and Stem Cell Research. Addition of Trinucleotide Repeat Diseases to the disorders section. Updated and revised sections on embryonic and fetal development, noninvasive prenatal tests, and inheritance.

Enhancement to the Illustration Program

New Design A textbook with beautiful illustrations or photographs on most pages requires a carefully crafted and functional design. The new design for the tenth edition has been transformed to assist students in making the most of the text's many features and outstanding art. A larger trim size provides even more space for the already large and highly praised illustrations and accommodates the larger font size for easier reading. Each page is carefully laid out to place related text, figures, and tables near one another, minimizing the need for page turning while reading a topic. New to this edition is the red print used to indicate the first mention of a figure or table. Not only is the reader alerted to refer to the figure or table, but the color print also serves as a place locator for easy return to the narrative.

Distinctive icons incorporated throughout the chapters signal special features and make them easy to find during review. These include the **key** with Key Concepts Statements; the **ques-** tion mark with the applicable questions that enhance every figure; the **stethoscope** indicating a clinical application within the chapter narrative; the **fetus icon** announcing the developmental anatomy section; the **running shoe** highlighting content relevant to exercise and the icons that indicate the study outline and distinctive types of **chapter-ending questions.**

Figure 12.3 **Structure of a typical neuron.** Arrows indicate the direction of information flow: dendrites → cell body → axon → axon terminals.

The basic parts of a neuron are dendrites, a cell body, and an axon.

DENDRITES
CELL BODY
Axon collateral
Initial segment
Axon hillock
Mitochondrion
AXON
Nucleus of Schwann cell
Cytoplasm
Nissl bodies
Neurofibril
Nucleus

Schwann cell:
Cytoplasm
Myelin sheath
Neurolemma
Node of Ranvier
Neurofibril
Axon:
Axoplasm
Axolemma

Axon terminal
Synaptic end bulb

(a) Parts of a motor neuron

(b) Motor neuron

What roles do the dendrites, cell body, and axon play in communication of signals?

New Art Studying human anatomy and physiology is both a visual and a descriptive enterprise. Countless students have now benefited from the detailed and generously proportioned bone and muscle art that characterizes our book. Continuing in this tradition, you will find exciting new three-dimensional illustrations gracing the pages of virtually every chapter in the book. More than 90 figures are new to this edition, and nearly every figure has been revised or improved in some way.

Figure 6.3 **Histology of bone.** Sections through the diaphysis of a long bone, from the surrounding periosteum on the right, to compact bone in the middle, to spongy bone and the medullary cavity on the left. The inset at the upper right shows an osteocyte in a lacuna. See Table 4.3J on page 126 for a photomicrograph of compact bone tissue and Figure 6.14a for a scanning electron micrograph of spongy bone tissue.

In compact bone, osteocytes lie in lacunae arranged in concentric circles around a central canal.

Compact bone
Spongy bone
Periosteum
Medullary cavity

See Figure 6.4b for details
Medullary cavity
Trabeculae
Inner circumferential lamellae
Spongy bone

Interstitial lamellae
Outer circumferential lamellae
Concentric lamellae
Blood vessels
Lymphatic vessel
Lacuna
Canaliculi
Osteon
Osteocyte
Periosteal vein
Periosteal artery
Periosteum:
Outer fibrous layer
Inner osteogenic layer
Central canal
Perforating canal

Lymphatic vessel
Compact bone

Osteons (Haversian systems) in compact bone and trabeculae in spongy bone

The central (Haversian) canals may become blocked. What effect would this have on osteocytes?

Figure 20.8 **The coronary circulation.** The views of the heart from the anterior aspect in (a) and (b) are drawn as if the heart were transparent to reveal blood vessels on the posterior aspect. (See Tortora, *A Photographic Atlas of the Human Body,* Figures 6.8 and 6.9.)

The right and left coronary arteries deliver blood to the heart; the coronary veins drain blood from the heart into the coronary sinus.

Ascending aorta
Pulmonary trunk
RIGHT CORONARY
Right atrium
MARGINAL BRANCH
Right ventricle

Arch of aorta
LEFT CORONARY
Left auricle
CIRCUMFLEX BRANCH
ANTERIOR INTER-VENTRICULAR BRANCH
POSTERIOR INTER-VENTRICULAR BRANCH
Left ventricle

(a) Anterior view of coronary arteries

Figure 22.9 **Phagocytosis of a microbe.**

The major types of phagocytes are neutrophils and macrophages.

1 CHEMOTAXIS
Microbe
2 ADHERENCE
Pseudopod
Lysosome
Plasma membrane
Digestive enzymes
3 INGESTION
4 DIGESTION
5 KILLING
Phagocyte
Digested microbe in phagolysosome
Residual body (indigestible material)

(a) Phases of phagocytosis

What chemicals are responsible for killing ingested microbes?

New Histology-Based Art As part of our goal for continuous improvement, many of the anatomical illustrations based on histological preparations have been revised and redrawn for this edition.

Figure 24.23 **Histology of the small intestine.**

Circular folds, villi and microvilli increase the surface area of the small intestine for digestion and absorption.

Lumen of small intestine

Villus   Blood   Lacteal
capillary

Simple columnar epithelium
Lamina propria

Opening of intestinal gland
Lymphatic nodule
Muscularis mucosae
Arteriole
Venule
Lymphatic vessel
Circular layer of muscle
Myenteric plexus
Longitudinal layer of muscle

MUCOSA

SUBMUCOSA

MUSCULARIS

SEROSA

(a) Three-dimensional view of layers of the small intestine showing villi

Microvilli
Absorptive cell (digests and absorbs nutrients)

Goblet cell (secretes mucus)

Enteroendocrine cell (secretes the hormones secretin, cholecystokinin, or GIP)

Paneth cell (secretes lysozyme and is capable of phagocytosis)

(b) Enlarged villus showing lacteal, capillaries, intestinal glands, and cell types

What is the functional significance of the blood capillary network and lacteal in the center of each villus?

New Photomicrographs Dr. Michael Ross of the University of Florida has again provided us with more beautiful, customized photomicrographs of various tissues of the body. We have always considered Dr. Ross' photos among the best available and their inclusion in this edition greatly enhances the illustration program.

Platelet

White blood cell (leukocyte)

Red blood cell (erythrocyte)

Blood plasma

Red blood cells

White blood cells

Platelets

LM 1230x

Blood in blood vessels

Blood smear

Cadaver Photos As before, we provide an assortment of large, clear cadaver photos at strategic points in many chapters. What is more, many anatomy illustrations are keyed to the large cadaver photos available in a companion text, *A Photographic Atlas of the Human Body*, by Gerard J. Tortora.

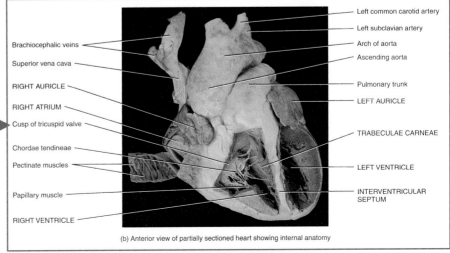

Brachiocephalic veins
Superior vena cava
RIGHT AURICLE
RIGHT ATRIUM
Cusp of tricuspid valve
Chordae tendineae
Pectinate muscles
Papillary muscle
RIGHT VENTRICLE

Left common carotid artery
Left subclavian artery
Arch of aorta
Ascending aorta
Pulmonary trunk
LEFT AURICLE

TRABECULAE CARNEAE

LEFT VENTRICLE

INTERVENTRICULAR SEPTUM

(b) Anterior view of partially sectioned heart showing internal anatomy

Helpful Orientation Diagrams Students sometimes need help figuring out the plane of view of anatomy illustrations — descriptions alone do not always suffice. An orientation diagram that depicts and explains the perspective of the view represented in the figure accompanies every major anatomy illustration. There are three types of diagrams: (1) planes used to indicate where certain sections are made when a part of the body is cut; (2) diagrams containing a directional arrow and the word "View" to indicate the direction from which the body part is viewed, and (3) diagrams with arrows leading from or to them that direct attention to enlarged and detailed parts of illustrations.

Frontal plane

View

Key Concept Statements This art-related feature summarizes an idea that is discussed in the text and demonstrated in a figure. Each Key Concept Statement is positioned adjacent to its figure and is denoted by a distinctive key icon.

Figure 21.1 Comparative structure of blood vessels. The relative size of the capillary in (c) is enlarged.

Arteries carry blood from the heart to tissues; veins carry blood from tissues to the heart.

TUNICA INTERNA:
Endothelium
Basement membrane
Internal elastic lamina

TUNICA MEDIA:
Smooth muscle

External elastic lamina

TUNICA EXTERNA

Valve

Lumen
(a) Artery

Lumen
(b) Vein

Lumen
Basement membrane
Endothelium

(c) Capillary

Internal elastic lamina
External elastic lamina
Tunica externa
Lumen with blood cells
Tunica interna
Tunica media
Connective tissue

LM 200x
(d) Transverse section through an artery

Connective tissue
Red blood cell
Capillary endothelial cells

LM 600x
(e) Red blood cells passing through a capillary

Figure Questions This highly applauded feature asks readers to synthesize verbal and visual information, think critically, or draw conclusions about what they see in a figure. Each Figure Question appears adjacent to its illustration and is highlighted in this edition by the purple question mark icon. Answers are located at the end of each chapter.

Which vessel—the femoral artery or the femoral vein—has a thicker wall? Which has a wider lumen?

Correlation of Sequential Processes Correlation of sequential processes in text and art is achieved through the use of numbered lists in the narrative that correspond to numbered segments in the accompanying figure. This approach is used extensively throughout the book to lend clarity to the flow of complex processes.

Figure 20.10 **The conduction system of the heart.** Autorhythmic fibers in the SA node, located in the right atrial wall (a), act as the heart's pacemaker, initiating cardiac action potentials (b) that cause contraction of the heart's chambers.

The conduction system ensures that the chambers of the heart contract in a coordinated manner.

Frontal plane

Left atrium

Right atrium

1 SINOATRIAL (SA) NODE

2 ATRIOVENTRICULAR (AV) NODE

3 ATRIOVENTRICULAR (AV) BUNDLE (BUNDLE OF HIS)

4 RIGHT AND LEFT BUNDLE BRANCHES

Left ventricle

Right ventricle

5 PURKINJE FIBERS

(a) Anterior view of frontal section

+ 10 mV

Action potential

Membrane potential

Threshold

− 60 mV

Pacemaker potential

0   0.8   1.6   2.4

Time (sec)

(b) Pacemaker potentials and action potentials in autorhythmic fibers of SA node

Which component of the conduction system provides the only electrical connection between the atria and the ventricles?

Functions Overview This feature succinctly lists the functions of an anatomical structure or body system depicted within a figure. The juxtaposition of text and art further reinforces the connection between structure and function.

OVERVIEW OF KIDNEY FUNCTIONS

OBJECTIVE

• **List the functions of the kidneys.**

The kidneys do the major work of the urinary system. The other parts of the system are mainly passageways and storage areas. Functions of the kidneys include:

• ***Excreting wastes and foreign s...*** urine, the kidneys help excrete w... have no useful function in the bod... in urine result from metabolic reac... include ammonia and urea from the deamination of amino acids; bilirubin from the catabolism of hemoglobin; creatinine from the breakdown of creatine phosphate in muscle fibers; and uric acid from the catabolism of nucleic acids.

Figure 26.1 **Organs of the urinary system in a female.** (See Tortora, *A Photographic Atlas of the Human Body,* Figure 13.2.)

Urine formed by the kidneys passes first into the ureters, then to the urinary bladder for storage, and finally through the urethra for elimination from the body.

Diaphragm
Esophagus
Left adrenal (suprarenal) gland
RIGHT KIDNEY
Left renal vein
LEFT KIDNEY
Right renal artery
Abdominal aorta
Inferior vena cava
RIGHT URETER
LEFT URETER
Rectum
Left ovary
URINARY BLADDER
Uterus
URETHRA

Anterior view

Functions of the Urinary System

1. The kidneys excrete wastes in urine, regulate blood volume and composition, help regulate blood pressure, synthesize glucose, release erythropoietin, and participate in vitamin D synthesis.
2. The ureters transport urine from the kidneys to the urinary bladder.
3. The urinary bladder stores urine.
4. The urethra discharges urine from the body.

Which organs constitute the urinary system?

19

Hallmark Features

The tenth edition of *Principles of Anatomy and Physiology* builds on the legacy of thoughtfully designed and class-tested pedagogical features that provide a complete learning system for students as they navigate their way through the text and course. All have been revised to reflect the enhancements to the text.

Chapter-opening Pages Each chapter now begins with a chapter-opening page that includes a beautiful new piece of art related to the body system under consideration. A perspective on the chapter's content from either a recent student of the course or a practitioner in a health related field introduces current users to the relevance of the material or particularly interesting portions of the chapter. Suggestions for energizing their study with dynamic multimedia activities on the Foundations CD included with the text and specially developed internet activities on the companion website are highlighted.

Helpful Exhibits Students of anatomy and physiology need extra help learning the many structures that constitute certain body systems—most notably skeletal muscles, articulations, blood vessels, and nerves. As in previous editions, the chapters that present these topics are organized around **Exhibits,** each of which consists of an overview, a tabular summary of the relevant anatomy, and an associated suite of illustrations or photographs. Each Exhibit is prefaced by an Objective and closes with a Checkpoint activity. We trust you will agree that our spaciously designed Exhibits are ideal study vehicles for learning anatomically complex body systems.

Student Objectives and Checkpoints **Objectives** are found at the beginning of major sections throughout each chapter. Complementing this format, **Checkpoint** questions appear at strategic intervals within chapters to give students the chance to validate their understanding as they read.

Tools for Mastering Vocabulary Students—even the best ones—generally find it difficult at first to read and pronounce anatomical and physiological terms. Moreover, as teachers we are sympathetic to the needs of the growing ranks of college students who speak English as a second language. For these reasons, we have endeavored to ensure that this book has a strong and helpful vocabulary component. The key terms in every chapter are emphasized by use of **boldface type.** We include **pronunciation guides** when major, or especially hard-to-pronounce, structures and functions are introduced in the discussions, Tables, or Exhibits. **Word roots** citing the Greek or Latin derivations of anatomical terms are offered as an additional aid. As a further service to readers, we provide a list of **Medical Terminology** at the conclusion of most chapters, and a comprehensive **Glossary** at the back of the book. Also included at the end of the book is a list of the basic building blocks of medical terminology—**Combining Forms, Word Roots, Prefixes, and Suffixes.**

Study Outline As always, readers will benefit from the popular end-of-chapter **Study Outline** that is page referenced to the chapter discussions.

End-of-chapter Questions The **Self-Quiz Questions** are written in a variety of styles that are calculated to appeal to readers' different testing preferences. **Critical Thinking Questions** challenge readers to apply concepts to real-life situations. The style of these questions ought to make students smile on occasion as well as think! Answers to the Self-Quiz and Critical Thinking Questions are located in Appendix E.

Complete Teaching And Learning Package

Continuing the tradition of providing a complete teaching and learning package, the tenth edition of *Principles of Anatomy and Physiology* is available with a host of carefully planned supplementary materials that will help you and your students attain the maximal benefit from our textbook. Please contact your Wiley sales representative for additional information about any of these resources.

Media

Dynamic CD-ROMs and an enhanced Companion Website have been developed to complement the tenth edition of *Principles of Anatomy and Physiology*. These are fully described on the inside front cover and include:

- **Student Companion CD-ROM,** provided free with the purchase of a new textbook

- **Dedicated Book Companion Website** free to students who purchase a new textbook

- **Interactions: Exploring the Functions of the Human Body.** The first CD in this dramatic new series—Foundations— is provided free to adopters of the text and to students purchasing a new text.

For Instructors

- **Full-color Overhead Transparencies** This set of full-color acetates includes over 650 figures, including histology micrographs, from the text — complete with figure numbers and captions. Transparencies have enlarged labels for effective use as overhead projections in the classroom. (0-471- 25425-8)

- **Instructor's Resource CD-ROM** For lecture presentation purposes, this cross-platform CD-ROM includes all of the line art from the text in labeled and unlabeled formats. In addition, a pre-designed set of Powerpoint Slides is included with text and images. The slides are easily edited to customize your presentations to your specific needs. (0-471- 25350-2)

• **Professor's Resource Manual** by Lee Famiano of Cuyahoga CC. This on-line manual provides instructors with tools to enhance their lectures. Key features include: Lecture Outlines, What's New and Different in this Chapter, Critical Thinking Questions, Teaching Tips, and a guide to other educational resources. The entire manual is found on the text's dedicated companion website and is fully downloadable. The electronic format allows you to customize lecture outlines or activities for delivery either in print or electronically to your students. (0-471-25355-3)

• **Printed Test Bank** by P. James Nielsen of Western Illinois University. A testbank of nearly 3,000 questions, many new to the tenth edition, is available. A variety of formats—multiple choice, short answer, matching and essay—are provided to accommodate different testing preferences. (0-471-25421-5)

• **Computerized Test Bank** An electronic version of the printed test bank is available on a cross-platform CD-ROM. (0-471-25146-1)

• **Interactions: Exploring the Functions of the Human Body**

• **WebCT** or **Blackboard** Course Management Systems with content prepared by Sharon Simpson are available.

• **Faculty Resource Network –** New from Wiley is a support structure to help instructors implement the dynamic new media that supports this text into their classrooms, laboratories, or on-line courses. Consult with your Wiley representative for details about this program.

For Students

• **Student Companion CD-ROM**

• **Interactions: Exploring the Functions of the Human Body**

• **Book Companion Website**

• **Learning Guide (0-471-43447-7)** by Kathleen Schmidt Prezbindowski, College of Mount St. Joseph. Designed specifically to fit the needs of students with different learning styles, this well-received guide helps students to more closely examine important concepts through a variety of activities and exercises. The 29 chapters in the *Learning Guide* parallel those of the textbook and include many activities, quizzes and tests for review and study.

• **Illustrated Notebook (0-471-25150-X)** A true companion to the text, this unique notebook is a tool for organized note taking in class and for review during study. Following the sequence in the textbook, each left-handed page displays an unlabeled black and white copy of every text figure. Students can fill in the labels during lecture or lab at the instructor's directions and take additional notes on the lined right-handed pages.

• **Anatomy and Physiology: A Companion Coloring Book (0-471-39515-3)** This helpful study aid features over 500 striking, original illustrations that, by actively coloring them, give students a clear understanding of key anatomical structures and physiological processes.

For the Laboratory

• **A Brief Atlas of the Human Skeleton** This brief photographic review of the human skeleton is provided free with every new copy of the text.

• **Laboratory Manual for Anatomy and Physiology** by Connie Allen and Valerie Harper, Edison Community College **(0-471-39464-5)** This new laboratory manual presents material covered in the 2-semester undergraduate anatomy & physiology laboratory course in a clear and concise way, while maintaining a student-friendly tone. The manual is very interactive and contains activities and experiments that enhance students' ability to both visualize anatomical structures and understand physiological topics.

• **Cat Dissection Manual** by Connie Allen and Valerie Harper, Edison Community College **(0-471-26457-1)** This manual includes photographs and illustrations of the cat along with guidelines for dissection. It is available independently as well as bundled with the main manual depending upon your an adoption needs.

• **Fetal Pig Dissection Manual** by Connie Allen and Valerie Harper, Edison Community College **(0-471-26458-X)** This manual includes photographs and illustrations of the fetal pig along with guidelines for dissection. It is available independently as well as bundled with the main manual depending upon your an adoption needs.

• **Photographic Atlas of the Human Body with Selected Cat, Sheep, and Cow Dissections** by Gerard Tortora **(0-471-37487-3)** This four-colored atlas is designed to support both study and laboratory experiences. Organized by body systems, the clearly labeled photographs provide a stunning visual reference to gross anatomy. Histological micrographs are also included. Many of the illustrations within *Principles of Anatomy and Physiology,* 10e are cross-referenced to this atlas.

Like each of our students, this book has a life of its own. The structure, content, and production values of *Principles of Anatomy and Physiology* are shaped as much by its relationship with educators and readers as by the vision that gave birth to the book ten editions ago. Today you, our readers, are the "heart" of this book. We invite you to continue the tradition of sending in your suggestions to us so that we can include them in the eleventh edition.

Gerard J. Tortora
Department of Science and Health, S229
Bergen Community College
400 Paramus Road
Paramus, NJ 07652

Sandra Reynolds Grabowski
Department of Biological Sciences
1392 Lilly Hall of Life Sciences
Purdue University
West Lafayette, IN 47907-1392
email: Sgrabows@bilbo.bio.purdue.edu

Acknowledgements

For the tenth edition of *Principles of Anatomy and Physiology*, we have enjoyed the opportunity of collaborating with a group of dedicated and talented professionals. Accordingly, we would like to recognize and thank the members of our book team, who often worked evenings and weekends, as well as days, to bring this book to you. At John Wiley & Sons, Inc., our Editor Bonnie Roesch again illuminates the path toward ever better books with her creative ideas and dedication. Bonnie was our valued editor during the seventh and eighth editions—welcome back and thanks! Karin Kincheloe is the Wiley designer whose vision for the tenth edition is a larger, more colorful, more user-friendly new style. Moreover, Karin laid out each page of the book to achieve the best possible placement of text, figures, and other elements. Both instructors and students will appreciate and benefit from the pedagogically effective and visually pleasing design elements that augment the content changes made to this edition. Danke sehr! Claudia Durrell, our Art Coordinator, has collaborated on this text since its seventh edition. Her artistic ability, organizational skills, attention to detail, and understanding of our illustration preferences greatly enhance the visual appeal and style of the figures. She remains a cornerstone of our projects— thank you, Claudia, for all your contributions. Kelly Tavares, Senior Production Editor, demonstrated her untiring expertise during each step of the production process. She coordinated all aspects of actually making and manufacturing the book. Kelly also was "on press" as the book was being printed to ensure the highest possible quality. Muito obrigada, Kelly for all the extra hours you spent to implement book-improving changes! Ellen Ford, Developmental Editor, shepherded the manuscript and electronic files during the revision process. She also contacted many reviewers for their comments and sought out former students to obtain their perspectives on learning anatomy and physiology for the new chapter-opening pages. Hillary Newman, Photo Editor, provided us with all of the photos we requested and did it with efficiency, accuracy, and professionalism. Mary O'Sullivan, Assistant Editor, coordinated the development of many of the supplements that support this text. We are most appreciative! Wiley Editorial Assistants Justin Bow and Kelli Coaxum helped with various aspects of the project and took care of many details. Thanks to all of you!

Jerri K. Lindsey, Tarrant County Junior College, Northeast and Caryl Tickner, Stark State College wrote the end-of-chapter Self-Quiz Questions. Joan Barber of Delaware Technical and Community College contributed the Critical Thinking Questions. Thanks to all three for writing questions that students will appreciate. And a special thank you to Kathleen Prezbindowski, who has authored the *Learning Guide* for many editions. The high quality of her study activities ensures student success.

Outstanding illustrations and photographs have always been a signature feature of *Principles of Anatomy and Physiology*. Respected scientific and medical illustrators on our team of exceptional artists include Mollie Borman, Leonard Dank, Sharon Ellis, Wendy Hiller Gee, Jean Jackson, Keith Kasnot, Lauren Keswick, Steve Oh, Lynn O'Kelley, Hilda Muinos, Tomo Narashima, Nadine Sokol, and Kevin Somerville. Mark Nielsen of the University of Utah provided many of the cadaver photos that appear in this edition. Artists at Imagineering created the amazing computer graphics images and provided all labeling of figures.

Reviewers

We are extremely grateful to our colleagues who reviewed the manuscript and offered insightful suggestions for improvement. The contributions of all these people, who generously provided their time and expertise to help us maintain the book's accuracy and clarity, are acknowledged in the list that follows.

Patricia Ahanotu, Georgia Perimeter College
Cynthia L. Beck, George Mason University
Clinton L. Benjamin, Lower Columbia College
Anna Berkovitz, Purdue University
Charles J. Biggers, University of Memphis
Mark Bloom, Tyler Junior College
Michele Boiani, University of Pennsylvania
Bruce M. Carlson, University of Michigan
Barbara Janson Cohen, Delaware County Community College
Matthew Jarvis Cohen, Delaware County Community College
Marcia Carol Coss, George Mason University
Victor P. Eroschenko, University of Idaho
Lorraine Findlay, Nassau Community College
Candice Francis, Palomar College
Christina Gan, Rogue Community College
Gregory Garman, Centralia College

Alan Gillen, Pensacola Christian College

Chaya Gopalan, St. Louis Community College

Janet Haynes, Long Island University

Clare Hayes, Metropolitan State College of Denver

James Junker, Campbell University

Gerald Karp, University of Florida

William Kleinelp, Middlesex County College

John Langdon, University of Indianapolis

John Lepri, University of North Carolina at Greensboro

Jerri K. Lindsey, Tarrant County College

Mary Katherine K. Lockwood, University of New Hampshire

Jennifer Lundmark, California State University, Sacramento

Paul Malven, Purdue University

G. K. Maravelas, Bristol Community College

Jane Marks, Paradise Valley Community College

Lee Meserve, Bowling Green State University

Javanika Mody, Anne Arundel Community College

Robert L. Moskowitz, Community College of Philadelphia

Shigihiro Nakajima, University of Illinois at Chicago

Jerry D. Norton, Georgia State University

Justicia Opoku, University of Maryland

Weston Opitz, Kansas Wesleyan University

Joann Otto, Purdue University

David Parker, Northern Virginia Community College

Karla Pouillon, Everett Community College

Linda Powell, Community College of Philadelphia

C. Lee Rocket, Bowling Green State University

Esmond J. Sanders, University of Alberta

Louisa Schmid, Tyler Junior College

Hans Schöler, University of Pennsylvania

Charles Sinclair, University of Indianapolis

Dianne Snyder, Augusta State University

Dennis Strete, McLennan Community College

Eric Sun, Macon State College

Antonietto Tan, Worcester State College

Jim Van Brunt, Rogue Community College

Jyoti Wagle, Houston Community College

Curt Walker, Dixie State College

DeLoris M. Wenzel, The University of Georgia

David Westmoreland, US Air Force Academy

Frederick E. Williams, University of Toledo

In addition, every chapter was read and reviewed by either an Anatomy & Physiology student or a health care professional now working in his or her chosen career. Their comments on the relevance and interest of the chapter topics is presented on each chapter opening page. We appreciate their time and effort, and especially their enthusiasm for the subject matter. We think that enthusiasm is catching! Thanks go out to the following:

Tamatha Adkins, RN

Molly Causby, Georgia Perimeter College

Margaret Chambers, George Mason University

RoNell Coco, Clark College

John Curra, Respiratory Technologist

Stephanie Hall Ford, Barry University

Helen Hart Ford, Radiological Technologist

Elizabeth Garrison, Modesto Junior College

Emily Gordon, Edison Community College

Kim Green, Edison Community College

Mike Grosse, Licensed Physical Therapy Assistant

Caroline Guerra, Broward Community College

Jill Haan, Cardiographic Technician

Dorie Hart, RN

Lisa P. Hubbard, Macon State College

Mark Johnson, Edison Community College

Wendy Lawrence, Troy State University

Candice Machado, Modesto Junior College

Susan Mahoney, Physical Therapist

Christine McGrellis, Mohawk Valley Community College

Jacqueline Opera, Medical Laboratory Technologist

Denise Pacheco, Modesto Junior College

Joan Petrokovsky, Licensed Massage Therapist

Toni Sheridan, Pharmacist

Barbara Simone, RN, FNP

Tiffany Smith, Stark State College of Technology

Sabrina von Brueckwitz, Edison Community College

To the Student

Your book has a variety of special features that will make your time studying anatomy and physiology a more rewarding experience. These have been developed based on feedback from students – like you – who have used previous editions of the text. A review of the preface will give you insight, both visually and in narrative, to all of the text's distinctive features.

Our experience in the classroom has taught us that student's appreciate a hint – both visually and verbally – at the beginning of each chapter about what to expect from its contents. Each chapter of your book begins with a stunning illustration depicting the system or main content being covered in the chapter. In addition, a short introduction to the chapter contents – usually written by a student who has recently completed the course, but occasionally by a practitioner in an allied health field – offers you an insight into some of the most intriguing or relevant aspects of the chapter. Links to activities on your **Foundations** CD, and special web-based activities called **Insights and Explorations** are suggested to make your study time worthwhile and interesting.

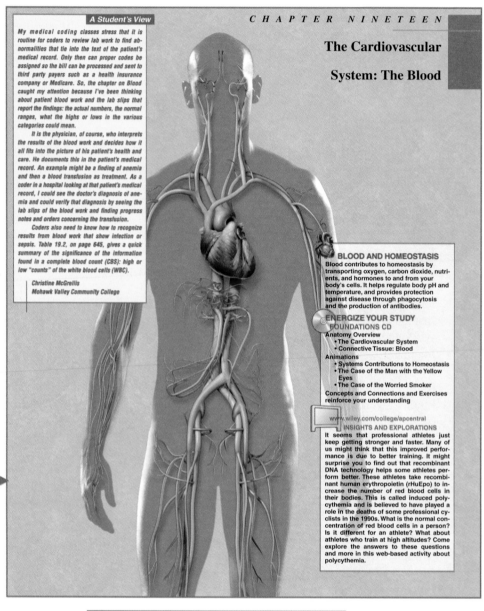

CHAPTER NINETEEN

The Cardiovascular System: The Blood

A Student's View

My medical coding classes stress that it is routine for coders to review lab work to find abnormalities that tie into the text of the patient's medical record. Only then can proper codes be assigned so the bill can be processed and sent to third party payers such as a health insurance company or Medicare. So, the chapter on Blood caught my attention because I've been thinking about patient blood work and the lab slips that report the findings: the actual numbers, the normal ranges, what the highs or lows in the various categories could mean.

It is the physician, of course, who interprets the results of the blood work and decides how it all fits into the picture of his patient's health and care. He documents this in the patient's medical record. An example might be a finding of anemia and then a blood transfusion as treatment. As a coder in a hospital looking at that patient's medical record, I could see the doctor's diagnosis of anemia and could verify that diagnosis by seeing the lab slips of the blood work and finding progress notes and orders concerning the transfusion.

Coders also need to know how to recognize results from blood work that show infection or sepsis. Table 19.2, on page 645, gives a quick summary of the significance of the information found in a complete blood count (CBS): high or low "counts" of the white blood cells (WBC).

Christine McGrellis
Mohawk Valley Community College

BLOOD AND HOMEOSTASIS
Blood contributes to homeostasis by transporting oxygen, carbon dioxide, nutrients, and hormones to and from your body's cells. It helps regulate body pH and temperature, and provides protection against disease through phagocytosis and the production of antibodies.

ENERGIZE YOUR STUDY
FOUNDATIONS CD
Anatomy Overview
• The Cardiovascular System
• Connective Tissue: Blood
Animations
• Systems Contributions to Homeostasis
• The Case of the Man with the Yellow Eyes
• The Case of the Worried Smoker
Concepts and Connections and Exercises reinforce your understanding

www.wiley.com/college/apcentral
INSIGHTS AND EXPLORATIONS
It seems that professional athletes just keep getting stronger and faster. Many of us might think that this improved performance is due to better training. It might surprise you to find out that recombinant DNA technology helps some athletes perform better. These athletes take recombinant human erythropoietin (rHuEpo) to increase the number of red blood cells in their bodies. This is called induced polycythemia and is believed to have played a role in the deaths of some professional cyclists in the 1990s. What is the normal concentration of red blood cells in a person? Is it different for an athlete? What about athletes who train at high altitudes? Come explore the answers to these questions and more in this web-based activity about polycythemia.

FUNCTIONS AND PROPERTIES OF BLOOD

▶ OBJECTIVES

• Describe the functions of blood.
• Describe the physical characteristics and principal components of blood.

Most cells of a multicellular organism cannot move around to obtain oxygen and nutrients or eliminate carbon dioxide and other wastes. Instead, these needs are met by two fluids: blood and interstitial fluid. **Blood** is a connective tissue composed of a liquid matrix ca

As you begin each narrative section of the chapter, be sure to take note of the **objectives** at the beginning of the section to help you focus on what is important as you read it.

▶ CHECKPOINT
1. In what ways is blood plasma similar to and different from interstitial fluid?
2. How many kilograms or pounds of blood is present in your body?
3. How does the volume of blood plasma in your body compare to the volume of fluid in a two-liter bottle of Coke?
4. What are the major solutes in blood plasma? What does each do?
5. What is the significance of lower-than-normal or higher-than-normal hematocrit?

At the end of the section, take time to try and answer the **Checkpoint** questions placed there. If you can, then you are ready to move on to the next section. If you experience difficulty answering the questions, you may want to re-read the section before continuing.

Studying the figures (illustrations that include artwork and photographs) in this book is as important as reading the text. To get the most out of the visual parts of this book, use the tools we have added to the figures to help you understand the concepts being presented. Start by reading the **legend**, which explains what the figure is about. Next, study the **key concept statement**, which reveals a basic idea portrayed in the figure. Added to many figures you will also find an **orientation diagram** to help you understand the perspective from which you are viewing a particular piece of anatomical art. Finally, at the bottom of each figure you will find a **figure question**. If you try to answer these questions as you go along, they will serve as self-checks to help you understand the material. Often it will be possible to answer a question by examining the figure itself. Other questions will encourage you to integrate the knowledge you've gained by carefully reading the text associated with the figure. Still other questions may prompt you to think critically about the topic at hand or predict a consequence in advance of its description in the text.

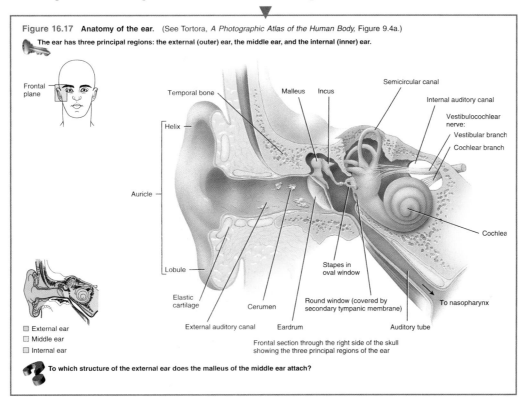

Figure 16.17 Anatomy of the ear. (See Tortora, *A Photographic Atlas of the Human Body,* Figure 9.4a.)

The ear has three principal regions: the external (outer) ear, the middle ear, and the internal (inner) ear.

Frontal section through the right side of the skull showing the three principal regions of the ear

☐ External ear
☐ Middle ear
☐ Internal ear

To which structure of the external ear does the malleus of the middle ear attach?

At the end of each chapter are other resources that you will find useful. **The Study Outline** is a concise statement of important topics discussed in the chapter. Page numbers are listed next to key concepts so you can easily refer to the specific passages in the text for clarification or amplification.

STUDY OUTLINE

SPINAL CORD ANATOMY
(p. 420)

1. The spinal cord is protected by the vertebral column, the meninges, cerebrospinal fluid, and denticulate ligaments.
2. The three meninges are coverings that run continuously around the spinal cord and brain. They are the dura mater, arachnoid mater, and pia mater.
3. The spinal cord begins as a continuation of the medulla oblongata and ends at about the second lumbar vertebra in an adult.
4. The spinal cord contains cervical and lumbar enlargements that serve as points of origin for nerves to the limbs.
5. The tapered inferior portion of the spinal cord is the conus medullaris, from which arise the filum terminale and cauda equina.
6. Spinal nerves connect to each segment of the spinal cord by two roots. The posterior or dorsal root contains sensory axons, and the anterior or ventral root contains motor neuron axons.
7. The anterior median fissure and the posterior median sulcus partially divide the spinal cord into right and left sides.
8. The gray matter in the spinal cord is divided into horns, and the white matter into columns. In the center of the spinal cord is the central canal, which runs the length of the spinal cord.
9. Parts of the spinal cord observed in transverse section are the gray commissure; central canal; anterior, posterior, and lateral gray horns; and anterior, posterior, and lateral white columns, which contain ascending and descending tracts. Each part has specific functions.
10. The spinal cord conveys sensory and motor information by way of ascending and descending tracts, respectively.

The **Self-quiz Questions** are designed to help you evaluate your understanding of the chapter contents. **Critical Thinking Questions** are word problems that allow you to apply the concepts you have studied in the chapter to specific situations.

SELF-QUIZ QUESTIONS

Fill in the blanks in the following statements.

1. _____ are fast, predictable, automatic responses to changes in the environment.
2. Because they contain both sensory and motor axons, spinal nerves are considered to be _____ nerves.

Indicate whether the following statements are true or false.

3. Reflexes permit the body to make exceedingly rapid adjustments to homeostatic imbalances.
4. Autonomic reflexes involve responses of smooth muscle, cardiac muscle, and glands.

Choose the one best answer to the following questions.

(a) endoneurium, (b) fascicle, (c) perineurium, (d) epineurium, (e) neurolemma.

6. Which of the following is *not* a function of the spinal cord? (a) reflex center, (b) integration of EPSPs and IPSPs, (c) conduction pathway for sensory impulses, (d) conduction pathway for motor impulses, (e) interpretation of sensory stimuli.
7. Which of the following are true? (1) The anterior (ventral) gray horns contain cell bodies of neurons that cause skeletal muscle contraction. (2) The gray commissure connects the white matter of the right and left sides of the spinal cord. (3) Cell bodies of autonomic motor neurons are located in the lateral gray horns. (4) Sensory (ascending) tracts conduct motor impulses down the spinal ... (5) Gray matter in the spinal cord consists of cell bodies of

CRITICAL THINKING QUESTIONS

1. Pearl, a lifeguard at the local beach, stepped on a discarded lit cigarette with her bare foot. Trace the reflex arcs set in motion by her accident. Name the reflex types.
   HINT *How will she remain standing when she picks up her foot?*
2. Why doesn't your spinal cord creep up toward your he ... time you bend over? Why doesn't it get all twisted out o ... when you exercise?
   HINT *How would you prevent a boat from floating aw ... dock?*

3. Jose's severe headaches and other symptoms were suggestive of meningitis, so his physician ordered a spinal tap. List the structures that the needle will pierce from the most superficial to the deepest. Why would the physican order a test in the spinal region to check a problem in Jose's head?

You will also find the **Answers to Figure Questions** at the end of chapters.

ANSWERS TO FIGURE QUESTIONS

13.1 The superior boundary of the spinal dura mater is the foramen magnum of the occipital bone. The inferior boundary is the second sacral vertebra.

13.2 The cervical enlargement connects with sensory and motor nerves of the upper limbs.

13.3 A horn is an area of gray matter, and a column is a region of white matter in the spinal cord.

13.4 The anterior corticospinal tract is located on the anterior side of the spinal cord, originates in the cortex of the cerebrum, and

the motor impulses leave the spinal cord on the side opposite the entry of sensory impulses.

13.10 All spinal nerves are mixed (have sensory and motor components) because the posterior root containing sensory axons and the anterior root containing motor axons unite to form the spinal nerve.

13.11 The anterior rami serve the upper and lower limbs.

13.12 Severing the spinal cord at level C2 causes respiratory arrest because it prevents descending nerve impulses from reaching the

---

Learning the language of anatomy and physiology can be one the more challenging aspects of taking this course. Throughout the text we have included pronunciations, and sometimes, Word Roots, for many terms that may be new to you. These appear in parentheses immediately following the new words, and the pronunciations are repeated in the glossary at the back of the book. You companion CD includes the complete glossary and offers you the opportunity to hear the words pronounced. In addition, the Companion Website offers you a review of these terms by chapter, pronounces them for you, and allows you the opportunity to create flash cards or quiz yourself on the many new terms.

Look at the words carefully and say them out loud several times. Learning to pronounce a new word will help you remember it and make it a useful part of your medical vocabulary. Take a few minutes to review the following pronunciation key, so it will be familiar to you when you encounter new words. The key is repeated at the beginning of the Glossary, as well.

Pronounciation Key

1. The most strongly accented syllable appears in capital letters, for example, bilateral (bī-LAT-er-al) and diagnosis (dī-ag-NŌ-sis).
2. If there is a secondary accent, it is noted by a prime ('), for example, constitution (kon'-sti-TOO-shun) and physiology (fiz'-ē-OL-ō-jē). Any additional secondary accents are also noted by a prime, for example, decarboxylation (dē'-kar-bok'-si-LĀ-shun).
3. Vowels marked by a line above the letter are pronounced with the long sound, as in the following common words:

   ā as in māke   ō as in pōle
   ē as in bē   ū as in cute
   ī as in īvy

4. Vowels not marked by a line above the letter are pronounced with the short sound, as in the following words:

   a as in above or at   o as in not
   e as in bet   u as in bud
   i as in sip

5. Other vowel sounds are indicated as follows:

   oy as in oil
   oo as in root

6. Consonant sounds are pronounced as in the following words:

   b as in bat   m as in mother
   ch as in chair   n as in no
   d as in dog   p as in pick
   f as in father   r as in rib
   g as in get   s as in so
   h as in hat   t as in tea
   j as in jump   v as in very
   k as in can   w as in welcome
   ks as in tax   z as in zero
   kw as in quit   zh as in lesion
   l as in let

Brief Table of Contents

Contents

Clinical Applications

Chapter 17 The Autonomic Nervous System 565

 Clinical Applications

**FOCUS ON HOMEOSTASIS:
THE NERVOUS SYSTEM 582**

Chapter 18 The Endocrine System 586

 Clinical Applications

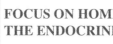 **FOCUS ON HOMEOSTASIS:
THE ENDOCRINE SYSTEM 624**

An Introduction to the Human Body

A Student's View

W h a t makes the difference in anatomy and physiology is being able to understand what you are reading. There are many aspects of this chapter that make it easy to comprehend and understand. The checkpoints at the end of each section are to the point and focus on the essential topics in the section that you might want to review more. Exhibits also get right to the point, simplifying the material and making it easy to memorize the data provided. Tables provide art and I find that essential in learning the material. Without good visual aids it is harder to understand the purpose of things and their functions (ex. table 1.2).

An aspect in this chapter that is the most useful in my career is the medical section, which explains different medical procedures. In studying to be a nurse providing this information is very helpful to me.

Denice Pacheco
Modesto Junior College

THE HUMAN BODY AND HOMEOSTASIS

Humans have many ways to maintain homeostasis, the state of relative stability of the body's internal environment. Disruptions to homeostasis often set in motion corrective cycles, called negative feedback systems, that help restore the conditions needed for health and life.

ENERGIZE YOUR STUDY
FOUNDATIONS CD

Homeostasis Folder
- **Explore Animations on Homeostasis and Feedback Cycles**
- **Concepts and Connections and Exercises reinforce your understanding**

Systems Folder
- **Take a tour of the first page of each Body System Overview to introduce yourself with structures and function of your body**

www.wiley.com/college/apcentral

INSIGHTS AND EXPLORATIONS

Would you like to be able to see inside the human body and explore the location of various organs using detailed computer images of real people? The National Library of Medicine (NLM) began the Visible Human Project in 1986. They obtained transverse MRI, CT, and anatomical images of a man and a woman who had donated their bodies for the study. The anatomical images were produced by making transverse cryosections at 1 mm or 0.33 mm intervals. At the NLM web site scientists and students, including you, can download samples of full scale images for study and use in research.

The fascinating journey that begins the exploration of the human body starts with an introduction to the disciplines of anatomy and physiology. Then, we will consider the organization of living things and explore the properties that all living things share. Next, you will discover how the body regulates its own internal environment, a ceaseless process called homeostasis that is a major theme in every chapter of this book. Finally, we introduce the basic vocabulary that will help you speak about the body in a way that is understood by scientists and health-care professionals alike.

ANATOMY AND PHYSIOLOGY DEFINED

▶ OBJECTIVE

- **Define anatomy and physiology, and name several subdisciplines of these sciences.**

Two branches of science—anatomy and physiology—provide the foundation for understanding the body's parts and functions. **Anatomy** (a-NAT-ō-mē; *ana-* = up; *-tomy* = process of cutting) is the science of body structures and the relationships among structures. It was first studied by **dissection** (dis-SEK-shun; *dis-* = apart; *-section* = act of cutting), the careful cutting apart of body structures to study their relationships. Today, a variety of imaging techniques also contribute to the advancement of anatomical knowledge. We will describe and compare some common imaging techniques in Table 1.4. Whereas anatomy deals with structures of the body, **physiology** (fiz′-ē-OL-o-jē; *physio-* = nature; *-logy* = study of) is the science of body functions—how the body parts work. Table 1.1 describes several subdisciplines of anatomy and physiology.

You will learn about the human body by studying its anatomy and physiology together. Always keep in mind that structure and function are intertwined. The structure of a part of the body allows performance of certain functions. For example, the bones of the skull join tightly to form a rigid case that protects the brain. The bones of the fingers, by contrast, are more loosely joined, which allows a variety of movements. The walls of the air sacs in the lungs are very thin, permitting rapid movement of inhaled oxygen into the blood. By contrast, the lining of the urinary bladder is much thicker. Yet its construction allows for considerable stretching as the urinary bladder fills with urine.

▶ CHECKPOINT

1. What body function might a respiratory therapist strive to improve?
2. Give an example of how the structure of a part of the body is related to its function.

Table 1.1	Selected Subdisciplines of Anatomy and Physiology		
Subdisciplines of Anatomy	**Study of**	**Subdisciplines of Physiology**	**Study of**
Embryology (em′-brē-OL-ō-jē; *embry-* = embryo; *-logy* = study of)	Structures that emerge from the time of the fertilized egg through the eighth week in utero.	**Neurophysiology** (NOOR-ō-fiz-ē-ol′-ō-jē; *neuro-* = nerve)	Functional properties of nerve cells.
Developmental biology	Structures that emerge from the time of the fertilized egg to the adult form.	**Endocrinology** (en′-dō-kri-NOL-ō-jē; *endo-* = within; *-crin* = secretion)	Hormones (chemical regulators in the blood) and how they control body functions.
Histology (hiss′-TOL-ō-jē; *hist-* = tissue)	Microscopic structure of tissues.	**Cardiovascular physiology** (kar-dē-ō-VAS-kū-lar; *cardi-* = heart; *-vascular* = blood vessels)	Functions of the heart and blood vessels.
Surface anatomy	Anatomical landmarks on the surface of the body through visualization and palpation.	**Immunology** (im′-ū-NOL-ō-jē; *immun-* = not susceptible)	How the body defends itself against disease-causing agents.
Gross anatomy	Structures that can be examined without using a microscope.	**Respiratory physiology** (RES-pir-a-to′-rē; *respira-* = to breathe)	Functions of the air passageways and lungs.
Systemic anatomy	Structure of specific systems of the body such as the nervous or respiratory systems.	**Renal physiology** (RĒ-nal; *ren-* = kidney)	Functions of the kidneys.
Regional anatomy	Specific regions of the body such as the head or chest.	**Exercise physiology**	Changes in cell and organ functions as a result of muscular activity.
Radiographic anatomy (rā′-dē-ō-GRAF-ik; *radio-* = ray; *-graphic* = to write)	Body structures that can be visualized with x rays.		
Pathological anatomy (path′-ō-LOJ-i-kal; *path-* = disease)	Structural changes (from gross to microscopic) associated with disease.	**Pathophysiology** (PATH-ō-fiz-ē-ol′-ō-jē)	Functional changes associated with disease and aging.

LEVELS OF BODY ORGANIZATION

OBJECTIVES

- **Describe the levels of structural organization that make up the human body.**
- **List the eleven systems of the human body, the organs present in each, and their general functions.**

The levels of organization of a language—letters of the alphabet, words, sentences, paragraphs, and so on—provide a useful comparison to the levels of organization of the human body. Your exploration of the human body will extend from some of the smallest body structures and their functions to the largest structure—an entire person. From the smallest to the largest size of their components, six levels of organization are relevant to understanding anatomy and physiology: the chemical, cellular, tissue, organ, system, and organismal levels of organization (Figure 1.1).

1 The *chemical level* includes **atoms,** the smallest units of matter that participate in chemical reactions, and **molecules,** two or more atoms joined together. Certain atoms, such as carbon (C), hydrogen (H), oxygen (O), nitrogen (N),

Figure 1.1 Levels of structural organization in the human body.

🔑 The levels of structural organization are chemical, cellular, tissue, organ, system, and organismal.

1 CHEMICAL LEVEL

Atoms
(C, H, O, N, P)

Molecule
(DNA)

2 CELLULAR LEVEL

Smooth muscle cell

3 TISSUE LEVEL

Smooth muscle tissue

Serous membrane

4 ORGAN LEVEL

Smooth muscle tissue layers

Stomach

Epithelial tissue

5 SYSTEM LEVEL

Esophagus
Liver
Stomach
Pancreas
Gallbladder
Small intestine
Large intestine

Digestive system

6 ORGANISMAL LEVEL

Which level of structural organization is composed of two or more different types of tissues that work together to perform a specific function?

phosphorus (P), calcium (Ca), and sulfur (S), are essential for maintaining life. Two familiar examples of molecules found in the body are deoxyribonucleic acid (DNA), the genetic material passed from one generation to the next, and glucose, commonly known as blood sugar. Chapters 2 and 25 focus on the chemical level of organization.

2 At the *cellular level,* molecules combine to form **cells,** the basic structural and functional units of an organism. Cells are the smallest living units in the human body. Among the many kinds of cells in your body are smooth muscle cells, nerve cells, and epithelial cells. The focus of Chapter 3 is the cellular level of organization.

3 The next level of structural organization is the *tissue level.* **Tissues** are groups of cells and the materials surrounding them that work together to perform a particular function. There are just four basic types of tissue in your body:

epithelial tissue, connective tissue, muscle tissue, and *nervous tissue.* Chapter 4 describes the tissue level of organization. Shown here is smooth muscle tissue, which consists of tightly packed smooth muscle cells.

4 At the *organ level,* different kinds of tissues are joined together. **Organs** are structures that are composed of two or more different types of tissues; they have specific functions and usually have recognizable shapes. Examples of organs are the stomach, heart, liver, lungs, and brain. Figure 1.1 shows how several tissues make up the stomach. The stomach's outer covering is a layer of epithelial tissue and connective tissue that reduces friction when the stomach moves and rubs against other organs. Underneath are the *smooth muscle tissue layers,* which contract to churn and mix food and then push it into the next digestive organ, the small intestine. The innermost lining is an *epithelial tissue layer*

Table 1.2	The Eleven Systems of the Human Body		

Integumentary System

- Hair
- Skin and associated glands
- Fingernails (and toenails)

Components: Skin, and structures derived from it, such as hair, nails, sweat glands, and oil glands.

Functions: Protects the body; helps regulate body temperature; eliminates some wastes; helps make vitamin D; and detects sensations such as touch, pain, warmth, and cold.

② Muscular System

- Skeletal muscle
- Tendon

Components: Muscles composed of skeletal muscle tissue, so-named because it is usually attached to bones.

Functions: Produces body movements, such as walking; stabilizes body position (posture); generates heat.

Skeletal System

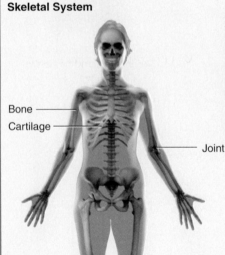

- Bone
- Cartilage
- Joint

Components: Bones and joints of the body and their associated cartilages.

Functions: Supports and protects the body; aids body movements; houses cells that produce blood cells; stores minerals and lipids (fats).

④ Nervous System

- Brain
- Spinal cord
- Nerve

Components: Brain, spinal cord, nerves, and special sense organs, such as the eye and ear.

Functions: Generates action potentials (nerve impulses) to regulate body activities; detects changes in the body's internal and external environment, interprets the changes, and responds by causing muscular contractions or glandular secretions.

that produces fluid and chemicals responsible for digestion in the stomach.

5 The next level of structural organization in the body is the *system level*. A **system** consists of related organs that have a common function. An example is the digestive system, which breaks down and absorbs food. Its organs include the mouth, salivary glands, pharynx (throat), esophagus, stomach, small intestine, large intestine, liver, gallbladder, and pancreas. Sometimes an organ is part of more than one system. The pancreas, for example, is part of both the digestive system and the hormone-producing endocrine system.

6 The largest organizational level is the *organismal level*. An **organism** is any living individual. All the parts of the human body functioning together constitute the total organism—one living person.

In the chapters that follow, you will study the anatomy and physiology of the major body systems. Table 1.2 lists the components and introduces the functions of these systems. You will also discover that all body systems influence one another. As an example, consider how just two body systems—the integumentary and skeletal systems—cooperate. The integumentary system, which includes the skin, hair, and nails, protects all other body systems, including the skeletal system, which is composed of all the bones and joints of the body. The skin serves as a barrier between the outside environment and internal tissues and organs. It also participates in the production of vitamin D, which is needed for proper deposition of calcium and other minerals into bone. The skeletal system, in turn, provides support for the integumentary system. It also serves as a reservoir for calcium, storing calcium in times of plenty and releasing it for other tissues in times of need. In addition, red bone marrow, present within some bones, generates the white blood cells that

---

5 Endocrine System

Pituitary gland
Pineal gland
Thyroid gland
Thymus
Adrenal gland
Pancreas
Ovary
Testis

Components: Hormone-producing glands (pineal gland, hypothalamus, pituitary gland, thymus, thyroid gland, parathyroid glands, adrenal glands, pancreas, ovaries, and testes) and hormone-producing cells in several other organs.

Functions: Regulates body activities by releasing hormones, which are chemical messengers transported in blood from an endocrine gland to a target organ.

6 Lymphatic and Immune System

Tonsil
Thymus
Thoracic duct
Spleen
Lymph node
Lymphatic vessel

Components: Lymphatic fluid and vessels; also includes spleen, thymus, lymph nodes, and tonsils.

Functions: Returns proteins and fluid to blood; carries lipids from gastrointestinal tract to blood; includes structures where lymphocytes that protect against disease-causing organisms mature and proliferate.

7 Cardiovascular System

Blood vessels:
Artery
Vein
Heart

Components: Blood, heart, and blood vessels.

Functions: Heart pumps blood through blood vessels; blood carries oxygen and nutrients to cells and carbon dioxide and wastes away from cells and helps regulate acid-base balance, temperature, and water content of body fluids; blood components help defend against disease and mend damaged blood vessels.

8 Respiratory System

Pharynx
Larynx (voice box)
Trachea (windpipe)
Bronchus
Lung

Components: Lungs and air passageways such as the pharynx (throat), larynx (voice box), trachea (windpipe), and bronchial tubes leading into and out of them.

Functions: Transfers oxygen from inhaled air to blood and carbon dioxide from blood to exhaled air; helps regulate acid–base balance of body fluids; air flowing out of lungs through vocal cords produces sounds.

(continues)

Table 1.2   The Eleven Systems of the Human Body (*continued*)

Digestive System

- Mouth
- Salivary gland
- Esophagus
- Liver
- Gallbladder
- Large intestine
- Small intestine
- Anus
- Stomach
- Pancreas (posterior to stomach)

Components: Organs of gastrointestinal tract, a long tube that includes the mouth, esophagus, stomach, small and large intestines, and anus; also includes accessory organs that assist in digestive processes, such as the salivary glands, liver, gallbladder, and pancreas.

Functions: Achieves physical and chemical breakdown of food; absorbs nutrients; eliminates solid wastes.

Urinary System

- Kidney
- Ureter
- Urinary bladder
- Urethra

Components: Kidneys, ureters, urinary bladder, and urethra.

Functions: Produces, stores, and eliminates urine; eliminates wastes and regulates volume and chemical composition of blood; helps maintain the acid–base balance of body fluids; maintains body's mineral balance; helps regulate production of red blood cells.

- Uterine (Fallopian) tube
- Uterus
- Vagina
- Mammary gland
- Ovary

- Penis
- Testis
- Ductus (vas) deferens
- Seminal vesicle
- Prostate

Reproductive Systems

Components: Gonads (testes in males and ovaries in females) and associated organs (uterine tubes, uterus, and vagina in females and epididymis, ductus deferens, and penis in males).

Functions: Gonads produce gametes (sperm or oocytes) that unite to form a new organism; gonads also release hormones that regulate reproduction and other body processes; associated organs transport and store gametes.

---

help the skin resist invasion by disease-causing organisms. As you study each of the body systems in more detail, you will discover how they work together to maintain health, provide protection from disease, and allow for reproduction of the human species.

Palpation, Auscultation, and Percussion

Health-care professionals and students of anatomy and physiology commonly use three noninvasive techniques to assess certain aspects of body structure and function. In **palpation** (pal-PĀY-shun; *palp-* = gently touching) the examiner feels body surfaces with the hands. An example is palpating an artery to find the pulse and measure the heart rate. In **auscultation** (aus-cul-TĀY-shun; *auscult-* = listening) the examiner listens to body sounds to evaluate the functioning of certain organs, often using

a stethoscope to amplify the sounds. An example is auscultation of the lungs during breathing to check for crackling sounds associated with abnormal fluid accumulation in the lungs. In **percussion** (pur-KUSH-un; *percus-* = beat through) the examiner taps on the body surface with the fingertips and listens to the resulting echo. For example, percussion may reveal the abnormal presence of fluid in the lungs or air in the intestines. It may also provide information about the size, consistency, and position of an underlying structure. ■

▶ C H E C K P O I N T

3. Define the following terms: atom, molecule, cell, tissue, organ, system, and organism.

4. At what levels of organization would an exercise physiologist study the human body? (*Hint: Refer to Table 1.1.*)

5. Referring to Table 1.2, which body systems help eliminate wastes?

CHARACTERISTICS OF THE LIVING HUMAN ORGANISM

▶ **O B J E C T I V E S**

- **Define the important life processes of the human body.**
- **Define homeostasis and explain its relationship to interstitial fluid.**

Basic Life Processes

* 6 most important life process of the human body

Organisms carry on certain processes that distinguish them from nonliving things. Following are the six most important life processes of the human body:

① **Metabolism** (me-TAB-ō-lizm) is the sum of all the chemical processes that occur in the body. One phase of metabolism is **catabolism** (ka-TAB-ō-lizm; *catabol-* = throwing down; *-ism* = a condition), the breaking down of complex chemical substances into simpler ones. The other phase of metabolism is **anabolism** (a-NAB-ō-lizm; *anabol-* = a raising up), the building up of complex chemical substances from smaller, simpler ones. In metabolic processes, oxygen taken in by the respiratory system and nutrients broken down in the digestive system provide the chemical energy to power cellular activities. Metabolism also includes breaking down large, complex molecules into smaller, simpler ones and using the resulting building blocks to assemble the body's structural components. For example, digestive processes split proteins in food into amino acids, which are the building blocks of proteins. Amino acids then are used to build new proteins that make up body structures such as muscles and bones.

② **Responsiveness** is the body's ability to detect and respond to changes in its internal or external environment. A decrease in body temperature would be an example of a change in the internal environment. Turning your head toward the sound of squealing brakes is an example of responsiveness to a change in the external environment. Different cells in the body respond to environmental changes in characteristic ways. Nerve cells respond by generating electrical signals known as nerve impulses (action potentials). Muscle cells respond by contracting, which generates force to move body parts.

③ **Movement** includes motion of the whole body, individual organs, single cells, and even tiny structures inside cells. For example, the coordinated action of leg muscles moves your whole body from one place to another when you walk or run. After you eat a meal that contains fats, your gallbladder contracts and squirts bile into the gastrointestinal tract. The bile aids in the digestion of fats. When a body tissue is damaged or infected, certain white blood cells move from the blood into the affected tissue to help clean up and repair the area. Even inside individual cells, various cellular parts move from one position to another to carry out their functions.

④ **Growth** is an increase in body size that results from an increase in the size of existing cells, the number of cells, or both. In addition, a tissue sometimes increases in size because the amount of material between cells increases. In a growing bone, for example, mineral deposits accumulate around the bone cells, causing the bone to enlarge in length and width.

⑤ **Differentiation** is the process a cell undergoes to develop from an unspecialized to a specialized state. As you will see later, each type of cell in the body has a specialized structure and function. Specialized cells differ in structure and function from the ancestor cells that gave rise to them. For example, red blood cells and several types of white blood cells differentiate from the same unspecialized ancestor cells in red bone marrow. Such ancestor cells, which can divide and give rise to progeny that undergo differentiation, are known as **stem cells.** Also through differentiation, a fertilized egg (ovum) develops into an embryo, and then into a fetus, an infant, a child, and finally an adult.

⑥ **Reproduction** refers either to the formation of new cells for tissue growth, repair, or replacement or to the production of a new individual. Through the fertilization of an ovum by a sperm cell, life continues from one generation to the next.

When the life processes cease to occur properly, the result is death of cells and tissues, which may lead to death of the organism. Clinically, loss of the heartbeat, absence of spontaneous breathing, and loss of brain functions indicate death in the human body.

 Autopsy

An **autopsy** (AW-top-sē = seeing with one's own eyes) is a postmortem (after death) examination of the body and dissection of its internal organs to confirm or determine the cause of death. An autopsy can uncover the existence of diseases not detected during life, determine the extent of injuries, and explain how those injuries may have contributed to a person's death. It also may provide more information about a disease, assist in the accumulation of statistical data, and educate health-care students. Moreover, an autopsy can reveal conditions that may affect offspring or siblings (such as congenital heart defects). Sometimes, an autopsy is legally required, such as during a criminal investigation. It may also be useful in resolving disputes between beneficiaries and insurance companies about the cause of death. ■

Homeostasis

The French physiologist Claude Bernard (1813–1878) first proposed that the cells of many-celled organisms flourish because they live in the relative constancy of *"le milieu interieur"* —the internal environment—despite continual changes in the organisms' external environment. The American physiologist Walter B. Cannon (1871–1945) coined the term *homeostasis* to describe this dynamic constancy. **Homeostasis** (hō′mē-ō-STĀ-sis;

homeo- = sameness; *-stasis* = standing still) is the condition of equilibrium in the body's internal environment. It occurs due to the ceaseless interplay of all the body's regulatory processes. Homeostasis is a dynamic condition. In response to changing conditions, the body's equilibrium point can change over a narrow range that is compatible with maintaining life. For example, the level of glucose in blood normally stays between 70 and 110 milligrams of glucose per 100 milliliters of blood.* Each body structure, from the cellular level to the systemic level, contributes in some way to keeping the internal environment within normal limits.

Body Fluids

An important aspect of homeostasis is maintaining the volume and composition of **body fluids,** which are dilute, watery solutions found inside cells as well as surrounding them. The fluid within cells is **intracellular fluid** (*intra-* = inside), abbreviated **ICF.** The fluid outside body cells is **extracellular fluid** (*extra-* = outside), abbreviated **ECF.** Dissolved in the water of ICF and ECF are oxygen, nutrients, proteins, and a variety of *ions* (electrically charged chemical particles). All these substances are needed to maintain life. The ECF that fills the narrow spaces between cells of tissues is known as **interstitial fluid** (in´-ter-STISH-al; *inter-* = between). As you progress with your studies, you will learn that the ECF within blood vessels is termed **blood plasma,** that within lymphatic vessels is called **lymph,** that in and around the brain and spinal cord is called **cerebrospinal fluid,** that in joints is called **synovial fluid,** and that in the eyes is called **aqueous humor** and **vitreous body.**

As Bernard predicted, the proper functioning of body cells depends on precise regulation of the composition of their surrounding fluid. Because interstitial fluid surrounds all body cells, it is often called the body's *internal environment.* The composition of interstitial fluid changes as substances move back and forth between it and blood plasma. Such exchange of materials occurs across the thin walls of the smallest blood vessels in the body, the *blood capillaries.* This movement in both directions across capillary walls provides needed materials, such as glucose, oxygen, ions, and so on, to tissue cells. It also removes wastes, such as carbon dioxide, from interstitial fluid.

▶ C H E C K P O I N T

6. Which life process in the human body sustains all the others?

7. Describe the locations of intracellular fluid, extracellular fluid, interstitial fluid, and blood plasma.

8. Why is interstitial fluid called the internal environment of the body?

_____
*Appendix A describes metric measurements.

CONTROL OF HOMEOSTASIS

▶ O B J E C T I V E S

• **Describe the components of a feedback system.**
• **Contrast the operation of negative and positive feedback systems.**
• **Explain why homeostatic imbalances cause disorders.**

Homeostasis in the human body is continually being disturbed. Some disruptions come from the external environment (outside the body) in the form of physical insults such as intense heat or lack of oxygen. Other disruptions originate in the internal environment (within the body), such as a blood glucose level that is too low. Homeostatic imbalances may also occur due to psychological stresses in our social environment—the demands of work and school, for example. In most cases the disruption of homeostasis is mild and temporary, and the responses of body cells quickly restore balance in the internal environment. In some cases the disruption of homeostasis may be intense and prolonged, as in poisoning, overexposure to temperature extremes, or severe infection. Under these circumstances, regulation of homeostasis may fail.

Fortunately, the body has many regulating systems that usually bring the internal environment back into balance. Most often, the nervous system and the endocrine system, working together or independently, provide the needed corrective measures. The nervous system regulates homeostasis by sending messages in the form of *nerve impulses* to organs that can counteract deviations from the balanced state. The endocrine system includes many glands that secrete messenger molecules called *hormones* into the blood. Whereas nerve impulses typically cause rapid changes, hormones usually work more slowly. Both means of regulation, however, work toward the same end, namely maintaining homeostasis. As you will see, both systems carry out their mission mainly through negative feedback systems.

Feedback Systems

The body can regulate its internal environment through a multitude of feedback systems. A **feedback system** is a cycle of events in which the status of a body condition is continually monitored, evaluated, changed, remonitored, reevaluated, and so on. Each monitored variable, such as body temperature, blood pressure, or blood glucose level, is termed a *controlled condition.* Any disruption that changes a controlled condition is called a *stimulus.* Three basic components make up a feedback system—a receptor, a control center, and an effector (Figure 1.2).

• A **receptor** is a body structure that monitors changes in a controlled condition and sends input to a control center. Typically, the input is in the form of nerve impulses or chemical signals. For example, nerve endings in the skin that sense temperature are one of many different types of receptors in the body.

Figure 1.2 Operation of a feedback system. The dashed return arrow symbolizes negative feedback.

🔑 The three basic components of a feedback system are receptors, a control center, and effectors.

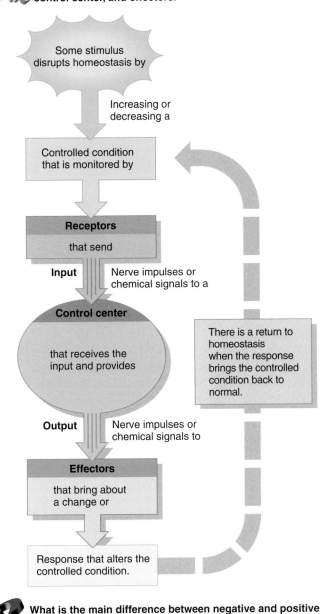

❓ What is the main difference between negative and positive feedback systems?

- A **control center** in the body sets the range of values within which a controlled condition should be maintained, evaluates the input it receives from receptors, and generates output commands when they are needed. Output from the control center typically occurs as nerve impulses, hormones, or other chemical signals.

- An **effector** is a body structure that receives output from the control center and produces a *response* or effect that changes the controlled condition. Nearly every organ or tis-

sue in the body can behave as an effector. For example, when your body temperature drops sharply, your brain (control center) sends nerve impulses (output) to your skeletal muscles (effectors). The result is shivering, which generates heat and raises your body temperature.

A group of receptors and effectors communicating with their control center forms a feedback system that can regulate a controlled condition in the body's internal environment. In a feedback system, or *feedback loop,* the response of the system "feeds back" to change the controlled condition in some way. Feedback systems can produce either negative feedback or positive feedback. If the response reverses the original stimulus, as in the body temperature regulation example, the system is operating by *negative feedback.* If the response enhances or intensifies the original stimulus, the system is operating by *positive feedback.*

Negative Feedback Systems

A **negative feedback system** reverses a change in a controlled condition. First, a stimulus disrupts homeostasis by altering the controlled condition. The receptors that are part of the feedback system detect the change and send input to a control center. The control center evaluates the input and, if necessary, issues output commands to an effector. The effector produces a physiological response that is able to return the controlled condition to its normal state.

Consider one negative feedback system that helps regulate blood pressure. Blood pressure (BP) is the force exerted by blood as it presses against the walls of blood vessels. When the heart beats faster or harder, BP increases. If some internal or external stimulus causes blood pressure (controlled condition) to rise, the following sequence of events occurs (Figure 1.3). *Baroreceptors,* which are pressure-sensitive nerve cells (the receptors) that are located in the walls of certain blood vessels, detect the higher pressure. The baroreceptors send nerve impulses (input) to the brain (control center), which interprets the impulses and responds by sending nerve impulses (output) to the heart (the effector). Heart rate decreases, which causes BP to decrease (response). This sequence of events quickly returns the controlled condition—blood pressure—to normal, and homeostasis is restored. Notice that the activity of the effector causes BP to drop, a result that is opposite to the original stimulus (an increase in BP). Thus, this is a negative feedback system.

Positive Feedback Systems

A **positive feedback system** tends to strengthen or reinforce a change in one of the body's controlled conditions. A positive feedback system operates similarly to a negative feedback system, except for the way the response affects the controlled condition. A stimulus alters a controlled condition, which is monitored by receptors that send input to a control center. The control center provides commands to an effector, but this time

Figure 1.3 Homeostatic regulation of blood pressure by a negative feedback system. Note that the response is fed back into the system, and the system continues to lower blood pressure until there is a return to normal blood pressure (homeostasis).

 If the response reverses the stimulus, a system is operating by negative feedback.

 What would happen to heart rate if some stimulus caused blood pressure to decrease? Would this occur by way of positive or negative feedback?

the effector produces a physiological response that *reinforces* the initial change in the controlled condition. The action of a positive feedback system continues until it is interrupted by some mechanism outside the system.

Normal childbirth provides a good example of a positive feedback system (Figure 1.4). The first contractions of labor (stimulus) push part of the baby into the cervix, the lowest part of the uterus, which opens into the vagina. Stretch-sensitive nerve cells (receptors) monitor the amount of stretching of the cervix (controlled condition). As stretching increases, they send more nerve impulses (input) to the brain (control center), which in turn releases the hormone oxytocin (output) into the blood. Oxytocin causes muscles in the wall of the uterus (effector) to contract even more forcefully. The contractions push the baby farther down the uterus, which stretches the cervix even more. The cycle of stretching, hormone release, and ever-stronger contractions is interrupted only by the birth of the baby. Then, stretching of the cervix ceases and oxytocin is no longer released. As you will see later, blood clotting is also an example of a positive feedback system.

These examples suggest some important differences between positive and negative feedback systems. Because a positive feedback system continually reinforces a change in a controlled condition, some event outside the system must shut it off. If the action of a positive feedback system is not stopped, it can "run away" because it cannot be controlled. It may even produce life-threatening conditions in the body. The action of a negative feedback system, by contrast, slows and then stops as the controlled condition returns to its normal state. Usually, positive feedback systems reinforce conditions that do not happen very often whereas negative feedback systems regulate conditions in the body that remain fairly stable over long periods.

Homeostatic Imbalances

As long as all the body's controlled conditions remain within certain narrow limits, body cells function efficiently, negative feedback loops maintain homeostasis, and the body stays healthy. Should one or more components of the body lose their ability to contribute to homeostasis, however, the normal equilibrium among body processes may be disturbed. If the homeostatic imbalance is moderate, a disorder or disease may occur; if it is severe, death may result.

A **disorder** is any derangement or abnormality of function. **Disease** is a more specific term for an illness characterized by a recognizable set of signs and symptoms. A *local disease* affects one part or a limited region of the body; a *systemic disease* affects either the entire body or several parts of it. Diseases alter body structures and functions in characteristic ways. A person with a disease may experience **symptoms,** which are subjective changes in body functions that are not apparent to an observer. Examples of symptoms are headache, nausea, and anxiety. Objective changes that a clinician can observe and measure, on the other hand, are called **signs.** Signs of disease can be either

Figure 1.4 Positive feedback control of labor contractions during birth of a baby. The solid return arrow symbolizes positive feedback.

🔑 **If the response enhances or intensifies the stimulus, a system is operating by positive feedback.**

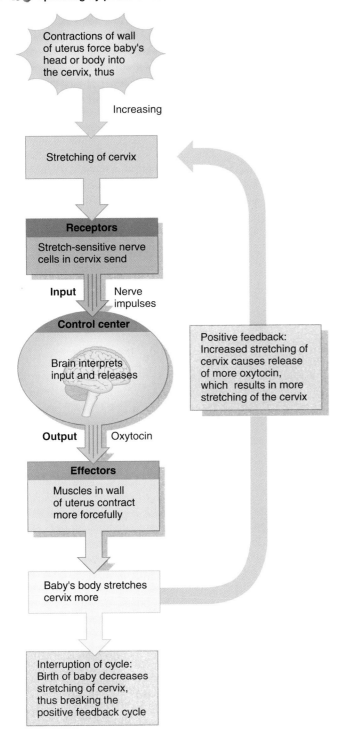

❓ **Why do positive feedback systems that are part of a normal physiological response include a termination mechanism?**

anatomical, such as swelling or a rash, or physiological, such as fever, high blood pressure, or paralysis.

The science that deals with why, when, and where diseases occur and how they are transmitted among individuals in a community is known as **epidemiology** (ep′-i-dē-mē-OL-ō-jē; *epi-* = upon; *-demi* = people). **Pharmacology** (far′-ma-KOL-ō-jē; *pharmac-* = drug) is the science that deals with the effects and uses of drugs in the treatment of disease.

🩺 Diagnosis of Disease

Diagnosis (dī′ag-NŌ-sis; *dia-* = through; *-gnosis* = knowledge) is the science and skill of distinguishing one disorder or disease from another. The patient's signs and symptoms, his or her medical history, a physical exam, and laboratory tests provide the basis for making a diagnosis. Taking a *medical history* consists of collecting information about events that might be related to a patient's illness. These include the chief complaint, history of present illness, past medical problems, family medical problems, social history, and review of symptoms. A *physical examination* is an orderly evaluation of the body and its functions. This process includes inspection (looking at or into the body using various instruments), palpation, auscultation, percussion, measuring vital signs (temperature, pulse, respiratory rate, and blood pressure), and sometimes laboratory tests. ■

▶ C H E C K P O I N T

9. What types of disturbances can act as stimuli that initiate a feedback system?

10. How are negative and positive feedback systems similar? How are they different?

11. Contrast and give examples of symptoms and signs of a disease.

AGING AND HOMEOSTASIS

▶ O B J E C T I V E

• **Describe some of the effects of aging.**

As you will see later, aging is a normal process characterized by a progressive decline in the body's responses to restore homeostasis. Aging produces observable changes in structure and function and increases vulnerability to stress and disease. The changes associated with aging are apparent in all body systems. Examples include wrinkled skin, gray hair, loss of bone mass, decreased muscle mass and strength, diminished reflexes, decreased sensory perception and motor responses, decreased production of some hormones, increased incidence of coronary artery disease and congestive heart failure, increased susceptibility to infections and malignancies, decreased lung capacity, less efficient functioning of the digestive system, decreased kidney function, menopause, and enlarged prostate. These and other effects of aging will be discussed in detail in later chapters.

ANATOMICAL TERMINOLOGY

OBJECTIVES

- **Describe the orientation of the body in the anatomical position.**
- **Relate the common names to the corresponding anatomical descriptive terms for various regions of the human body.**
- **Define the anatomical planes and sections used to describe the human body.**
- **Describe the major body cavities, the organs they contain, and their associated linings.**

Scientists and health-care professionals use a common language of special terms when referring to body structures and their functions. The language of anatomy has precisely defined meanings that allow us to communicate without using unneeded or ambiguous words. For example, is it correct to say, "The wrist is above the fingers"? This might be true if your arms are at your sides. But if you hold your hands up above your head, your fingers would be above your wrists. To prevent this kind of confusion, anatomists developed a standard anatomical position and a special vocabulary for relating body parts to one another.

Body Positions

Descriptions of any region or part of the human body assume that it is in a specific stance called the **anatomical position.** In the anatomical position, the subject stands erect facing the observer, with the head level and the eyes facing directly forward. The feet are flat on the floor and directed forward, and the arms are at the sides with the palms turned forward (Figure 1.5). In the anatomical position, the body is upright. Two terms describe a reclining body. If the body is lying face down, it is in the **prone** position. If the body is lying face up, it is in the **supine** position.

Regional Names

The human body is divided into several major regions that can be identified externally. The principal regions are the head, neck, trunk, upper limbs, and lower limbs (Figure 1.5). The **head** consists of the skull and face. Whereas the *skull* encloses and protects the brain, the *face* is the anterior (front) portion of the head that includes the eyes, nose, mouth, forehead, cheeks, and chin. The **neck** supports the head and attaches it to the trunk. The **trunk** consists of the chest, abdomen, and pelvis. Each **upper limb** attaches to the trunk and consists of the shoulder, armpit, arm (portion of the limb from the shoulder to the elbow), forearm (portion of the limb from the elbow to the wrist), wrist, and hand. Each **lower limb** also attaches to the trunk and consists of the buttock, thigh (portion of the limb from the buttock to the knee), leg (portion of the limb from the knee to the ankle), ankle, and foot. The *groin* is the area on the anterior surface of the body marked by a crease on each side, where the trunk attaches to the thighs.

Figure 1.5 shows the common names of major parts of the body. The corresponding anatomical descriptive form (adjective) for each part appears in parentheses next to the common name. For example, if you receive a tetanus shot in your *buttock,* it is a *gluteal* injection. Because the descriptive form of a body part usually is based on a Greek or Latin word, it may look different from the common name for the same part or area. For example, the Latin word for armpit is *axilla* (ak-SIL-a). Thus, one of the nerves passing within the armpit is named the axillary nerve. You will learn more about the Greek and Latin word roots of anatomical and physiological terms as you read this book.

Directional Terms

To locate various body structures, anatomists use specific **directional terms,** words that describe the position of one body part relative to another. Several directional terms are grouped in pairs that have opposite meanings, such as anterior (front) and posterior (back). Exhibit 1.1 and Figure 1.6 on pages 14–15 present the main directional terms.

Planes and Sections

You will also study parts of the body relative to planes, imaginary flat surfaces that pass through the body parts (Figure 1.7 on page 16). A **sagittal plane** (SAJ-i-tal; *sagitt-* = arrow) is a vertical plane that divides the body or an organ into right and left sides. More specifically, when such a plane passes through the midline of the body or an organ and divides it into *equal* right and left sides, it is called a **midsagittal plane** or a **median plane.** If the sagittal plane does not pass through the midline but instead divides the body or an organ into *unequal* right and left sides, it is called a **parasagittal plane** (*para-* = near). A **frontal** or **coronal plane** (kō-RŌ-nal; *corona* = crown) divides the body or an organ into anterior (front) and posterior (back) portions. A **transverse plane** divides the body or an organ into superior (upper) and inferior (lower) portions. Other names for a transverse plane are a cross-sectional or horizontal plane. Sagittal, frontal, and transverse planes are all at right angles to one another. An **oblique plane,** by contrast, passes through the body or an organ at an angle between the transverse plane and either a sagittal or frontal plane.

When you study a body region, you often view it in section. A **section** is one flat surface of a three-dimensional structure. It is important to know the plane of the section so you can understand the anatomical relationship of one part to another. Figure 1.8 on page 16 indicates how three different sections—*transverse, frontal,* and *midsagittal*—provide different views of the brain.

Body Cavities

Body cavities are spaces within the body that help protect, separate, and support internal organs. Bones, muscles, and ligaments

Figure 1.5 The anatomical position. The common names and corresponding anatomical terms (in parentheses) are indicated for specific body regions. For example, the head is the cephalic region.

In the anatomical position, the subject stands erect facing the observer with the head level and the eyes facing forward. The feet are flat on the floor and directed forward, and the arms are at the sides with the palms facing forward.

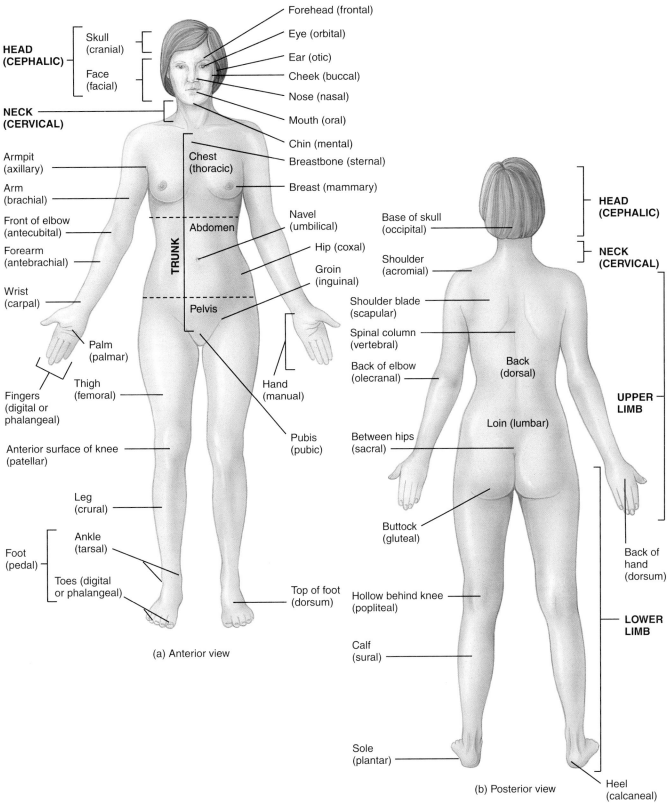

(a) Anterior view

(b) Posterior view

What is the usefulness of defining one standard anatomical position?

Exhibit 1.1   Directional Terms (Figure 1.6)

► **OBJECTIVE**

• Define each directional term used to describe the human body.

Overview

Most of the directional terms used to describe the human body can be grouped into pairs that have opposite meanings. For example, **superior** means toward the upper part of the body, whereas **inferior** means toward the lower part of the body. Moreover, it is important to understand that directional terms have relative meanings; they only make sense when used to describe the position of one structure relative to another.

For example, your knee is superior to your ankle, even though both are located in the inferior half of the body. Study the directional terms below and the example of how each is used. As you read the examples, look at Figure 1.6 to see the location of each structure.

► **CHECKPOINT**

Which directional terms can be used to specify the relationships between (1) the elbow and the shoulder, (2) the left and right shoulders, (3) the sternum and the humerus, and (4) the heart and the diaphragm?

Directional Term	Definition	Example of Use
Superior (soo'-PEER-ē-or) (cephalic or cranial)	Toward the head, or the upper part of a structure.	The heart is superior to the liver.
Inferior (in'-FEER-ē-or) (caudal)	Away from the head, or the lower part of a structure.	The stomach is inferior to the lungs.
Anterior (an-TEER-ē-or) (ventral)*	Nearer to or at the front of the body.	The sternum (breastbone) is anterior to the heart.
Posterior (pos-TEER-ē-or) (dorsal)	Nearer to or at the back of the body.	The esophagus is posterior to the trachea (windpipe).
Medial (MĒ-dē-al)	Nearer to the midline.†	The ulna is medial to the radius.
Lateral (LAT-er-al)	Farther from the midline.	The lungs are lateral to the heart.
Intermediate (in'-ter-MĒ-dē-at)	Between two structures.	The transverse colon is intermediate between the ascending and descending colons.
Ipsilateral (ip-si-LAT-er-al)	On the same side of the body as another structure.	The gallbladder and ascending colon are ipsilateral.
Contralateral (CON-tra-lat-er-al)	On the opposite side of the body from another structure.	The ascending and descending colons are contralateral.
Proximal (PROK-si-mal)	Nearer to the attachment of a limb to the trunk; nearer to the origination of a structure.	The humerus is proximal to the radius.
Distal (DIS-tal)	Farther from the attachment of a limb to the trunk; farther from the origination of a structure.	The phalanges are distal to the carpals.
Superficial (soo'-per-FISH-al)	Toward or on the surface of the body.	The ribs are superficial to the lungs.
Deep (DĒP)	Away from the surface of the body.	The ribs are deep to the skin of the chest and back.

*Ventral refers to the belly side, whereas dorsal refers to the back side. In four-legged animals, anterior = cephalic (toward the head), ventral = inferior, posterior = caudal (toward the tail), and dorsal = superior.

†The midline is an imaginary vertical line that divides the body into equal right and left sides.

Figure 1.6   Directional terms.

Directional terms precisely locate various parts of the body relative to one another.

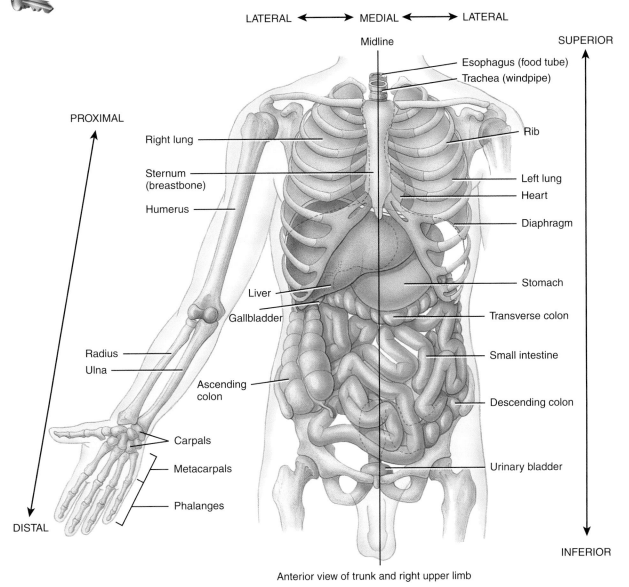

Anterior view of trunk and right upper limb

Is the radius proximal to the humerus? Is the esophagus anterior to the trachea? Are the ribs superficial to the lungs? Is the urinary bladder medial to the ascending colon? Is the sternum lateral to the descending colon?

Figure 1.7   Planes through the human body.

Frontal, transverse, sagittal, and oblique planes divide the body in specific ways.

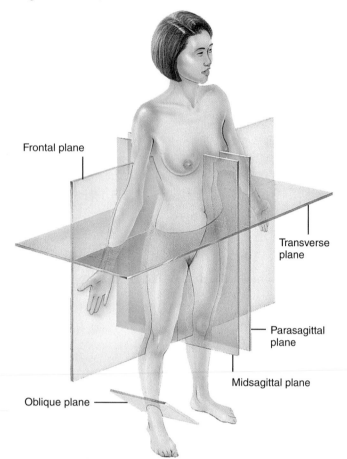

Frontal plane

Transverse plane

Parasagittal plane

Midsagittal plane

Oblique plane

Right anterolateral view

Which plane divides the heart into anterior and posterior portions?

Figure 1.8   Planes and sections through different parts of the brain.   The diagrams (left) show the planes, and the photographs (right) show the resulting sections. Note: The arrows in the diagrams indicate the direction from which each section is viewed. This aid is used throughout the book to indicate viewing perspectives.

Planes divide the body in various ways to produce sections.

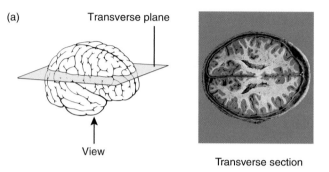

(a)   Transverse plane

View

Transverse section

(b)   Frontal plane

View

Frontal section

(c)   Midsagittal plane

View

Midsagittal section

Which plane divides the brain into unequal right and left portions?

separate the various body cavities from one another. The two major cavities are the dorsal and ventral body cavities (Figure 1.9).

Dorsal Body Cavity

The **dorsal body cavity** is located near the dorsal (posterior) surface of the body and has two subdivisions, the cranial cavity and the vertebral canal. The cranial bones form the **cranial cavity,** which contains the brain. The bones of the vertebral column (backbone) form the **vertebral (spinal) canal,** which contains the spinal cord. Three layers of protective tissue, the **meninges** (me-NIN-jēz), line the dorsal body cavity.

Ventral Body Cavity

The other major body cavity—the **ventral body cavity**—is located on the ventral (anterior) aspect of the body. The ventral

body cavity also has two main subdivisions, the superior thoracic and the inferior abdominopelvic cavities. The **diaphragm** (DĪ-a-fram = partition or wall) is a dome-shaped muscle that separates the thoracic cavity from the abdominopelvic cavity. The organs inside the ventral body cavity are termed the **viscera** (VIS-er-a).

The superior part of the ventral body cavity is the **thoracic cavity** (thor-AS-ik; *thorac-* = chest) or chest cavity (Figure 1.10). The ribs, the muscles of the chest, the sternum (breastbone), and the thoracic portion of the vertebral column (backbone) encircle

Figure 1.9 **Body cavities.** The dashed lines in (a) and (b) indicate the border between the abdominal and pelvic cavities. (See Tortora, *A Photographic Atlas of the Human Body*, Figures 6.5, 6.11, and 11.12.)

🔑 **The two major body cavities are the dorsal and ventral cavities.**

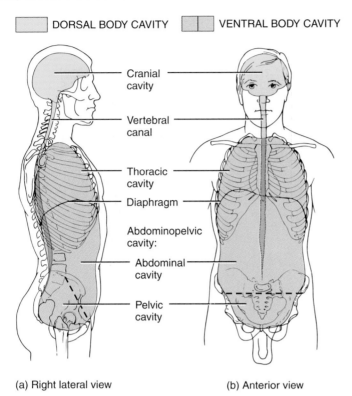

DORSAL BODY CAVITY   VENTRAL BODY CAVITY

- Cranial cavity
- Vertebral canal
- Thoracic cavity
- Diaphragm
- Abdominopelvic cavity:
  - Abdominal cavity
  - Pelvic cavity

(a) Right lateral view   (b) Anterior view

🔎 **In which cavities are the following organs located: urinary bladder, stomach, heart, small intestine, lungs, internal female reproductive organs, thymus, spleen, liver? Use the following symbols for your response: T = thoracic cavity, A = abdominal cavity, or P = pelvic cavity.**

Figure 1.10 **The thoracic cavity.** The dashed lines indicate the borders of the mediastinum. Note: When transverse sections are viewed inferiorly (from below), the anterior aspect of the body appears on top and the left side of the body appears on the right side of the illustration. (See Tortora, *A Photographic Atlas of the Human Body*, Figure 6.6.)

🔑 **The thoracic cavity is the superior part of the ventral body cavity.**

Transverse plane

View

ANTERIOR

- Muscle
- Heart
- PERICARDIAL CAVITY
- Right lung
- Aorta
- RIGHT PLEURAL CAVITY
- Rib
- Sternum (breastbone)
- Thymus
- Left lung
- Esophagus
- Vertebral column (backbone)
- LEFT PLEURAL CAVITY

POSTERIOR

Inferior view of transverse section of thoracic cavity

 What is the name of the cavity that surrounds the heart? Which cavities surround the lungs?

Figure 1.11 **The mediastinum.** The dashed lines indicate the borders of the mediastinum.

 The mediastinum is medial to the lungs. It extends from the sternum to the vertebral column and from the neck to the diaphragm.

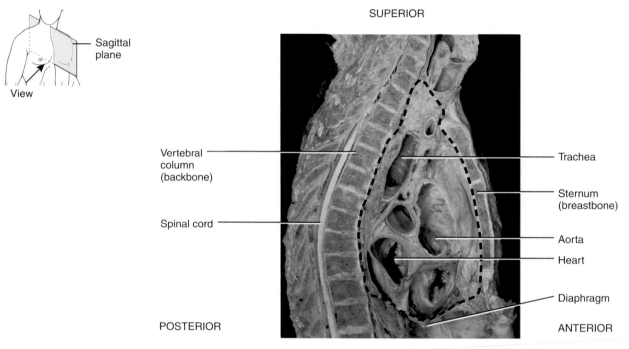

SUPERIOR

Vertebral column (backbone)

Spinal cord

Trachea

Sternum (breastbone)

Aorta

Heart

Diaphragm

POSTERIOR

ANTERIOR

INFERIOR

Sagittal section of thoracic cavity

? Which of the following structures are contained in the mediastinum: right lung, heart, esophagus, spinal cord, trachea, rib, thymus, left pleural cavity?

the thoracic cavity. The **pericardial cavity** (per′-i-KAR-dē-al; *peri-* = around; *-cardial* = heart), a fluid-filled space that surrounds the heart, and two **pleural cavities** (PLOOR-al; *pleur-* = rib or side) lie within the thoracic cavity. Each pleural cavity surrounds one lung and contains a small amount of fluid. The central part of the thoracic cavity is called the **mediastinum** (mē′-dē-as-TĪ-num; *media-* = middle; *-stinum* = partition). It is between the lungs, extending from the sternum to the vertebral column and from the neck to the diaphragm (Figure 1.11). The mediastinum contains all thoracic viscera except the lungs themselves. Among the structures in the mediastinum are the heart, esophagus, trachea, thymus, and several large blood vessels.

The inferior part of the ventral body cavity is the **abdominopelvic cavity** (ab-dom′-i-nō-PEL-vik; see Figure 1.8), which extends from the diaphragm to the groin and is encircled by the abdominal wall and the bones and muscles of the pelvis. As the name suggests, the abdominopelvic cavity is divided into two portions, even though no wall separates them (Figure 1.12). The superior portion, the **abdominal cavity** (*abdomin-* = belly), contains the stomach, spleen, liver, gallbladder, small intestine, and most of the large intestine. The inferior portion, the **pelvic**

cavity (*pelv-* = basin), contains the urinary bladder, portions of the large intestine, and internal organs of the reproductive system.

Thoracic and Abdominal Cavity Membranes

A thin, slippery **serous membrane** covers the viscera within the thoracic and abdominal cavities and also lines the walls of the thorax and abdomen. The parts of a serous membrane are (1) the *parietal layer* (pa-RĪ-e-tal), which lines the walls of the cavities, and (2) the *visceral layer,* which covers and adheres to the viscera within the cavities. Serous fluid between the two layers reduces friction, allowing the viscera to slide somewhat during movements, for example, when the lungs inflate and deflate during breathing.

The serous membrane of the pleural cavities is called the **pleura** (PLOO-ra). The *visceral pleura* clings to the surface of the lungs, whereas the *parietal pleura* lines the chest wall. In between is the pleural cavity, filled with a small volume of serous fluid. The serous membrane of the pericardial cavity is the **pericardium.** The *visceral pericardium* covers the surface of

Figure 1.12 **The abdominopelvic cavity.** The dashed line shows the approximate boundary between the abdominal and pelvic cavities. (See Tortora, *A Photographic Atlas of the Human Body,* Figure 12.2.)

🔑 The abdominopelvic cavity extends from the diaphragm to the groin.

Anterior view

❓ To which body systems do the organs shown here within the abdominal and pelvic cavities belong? (*Hint: Refer to Table 1.2.*)

the heart, whereas the *parietal pericardium* lines the chest wall. Between them is the pericardial cavity. The **peritoneum** (per-i-tō-NĒ-um) is the serous membrane of the abdominal cavity. The *visceral peritoneum* covers the abdominal viscera, whereas the

parietal peritoneum lines the abdominal wall. Between them is the *peritoneal cavity.* Most abdominal organs are located in the peritoneal cavity. Some are located behind the parietal peritoneum between it and the posterior abdominal wall. Such organs are said to be *retroperitoneal* (reʹ-trō-per-i-tō-NĒ-al; *retro* = behind). The kidneys, adrenal glands, pancreas, duodenum of the small intestine, ascending and descending colons of the large intestine, and portions of the abdominal aorta and inferior vena cava are retroperitoneal.

Table 1.3 presents a summary of body cavities and their membranes.

Abdominopelvic Regions and Quadrants

To describe the location of the many abdominal and pelvic organs more easily, anatomists and clinicians use two methods of dividing the abdominopelvic cavity into smaller compartments. In the first method, two horizontal and two vertical lines, aligned like a tick-tack-toe grid, partition this cavity into nine **abdominopelvic regions** (Figure 1.13a). The top horizontal line, the *subcostal line,* is drawn just inferior to the rib cage, across the inferior portion of the stomach; the bottom horizontal line, the *transtubercular line,* is drawn just inferior to the tops of the hip bones. Two vertical lines, the left and right *midclavicular lines,* are drawn through the midpoints of the clavicles (collar bones), just medial to the nipples. The four lines divide the abdominopelvic cavity into a larger middle section and smaller left and right sections. The names of the nine abdominopelvic regions are right hypochondriac, epigastric, left hypochondriac, right lumbar, umbilical, left lumbar, right inguinal (iliac), hypogastric (pubic), and left inguinal (iliac).

Table 1.3	Summary of Body Cavities and Their Membranes
Cavity	**Comments**
Dorsal cavity	
Cranial cavity	Formed by cranial bones and contains brain.
Vertebral cavity	Formed by vertebral column and contains spinal cord.
Ventral cavity	
Thoracic cavity	Superior portion of ventral body cavity; contains pleural and pericardial cavities and the mediastinum.
Pleural cavity	Each surrounds a lung; the serous membrane of the pleural cavities is the pleura.
Pericardial cavity	Surrounds the heart; the serous membrane of the pericardial cavity is the pericardium.
Mediastinum	Central portion of thoracic cavity between the lungs; extends from sternum to vertebral column and from neck to diaphragm; contains heart, thymus, esophagus, trachea, and several large blood vessels.
Abdominopelvic cavity	Inferior portion of ventral body cavity; subdivided into abdominal and pelvic cavities.
Abdominal cavity	Contains stomach, spleen, liver, gallbladder, small intestine, and most of large intestine; the serous membrane of the abdominal cavity is the peritoneum.
Pelvic cavity	Contains urinary bladder, portions of large intestine, and internal organs of reproduction.

Figure 1.13   Regions and quadrants of the abdominopelvic cavity.

 The nine-region designation is used for anatomical studies, whereas the quadrant designation is used to locate the site of pain, tumor, or some other abnormality.

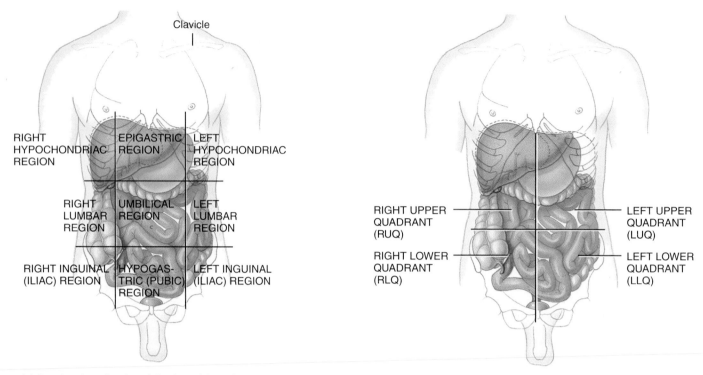

(a) Anterior view showing abdominopelvic regions

(b) Anterior view showing abdominopelvic quadrants

In which abdominopelvic region is each of the following found: most of the liver, transverse colon, urinary bladder, spleen? In which abdominopelvic quadrant would pain from appendicitis (inflammation of the appendix) be felt?

The second method is simpler and divides the abdominopelvic cavity into **quadrants** (KWOD-rantz; *quad-* = one-fourth), as shown in Figure 1.13b. In this method, a vertical line and a horizontal line are passed through the **umbilicus** (um-bi-LĪ-kus; *umbilic-* = navel) or *belly button*. The names of the abdominopelvic quadrants are right upper quadrant (RUQ), left upper quadrant (LUQ), right lower quadrant (RLQ), and left lower quadrant (LLQ). Whereas the nine-region division is more widely used for anatomical studies, quadrants are more commonly used by clinicians for describing the site of abdominopelvic pain, tumor, or other abnormality.

▶ C H E C K P O I N T

12. Locate each region shown in Figure 1.5 on your own body, and then identify it by its common name and the corresponding anatomical descriptive form.

13. What structures separate the various body cavities from one another?

14. Locate the nine abdominopelvic regions and the four abdominopelvic quadrants on yourself, and list some of the organs found in each.

MEDICAL IMAGING

▶ O B J E C T I V E

• Describe the principles and importance of medical imaging procedures in the evaluation of organ functions and the diagnosis of disease.

Various kinds of **medical imaging** procedures allow visualization of structures inside our bodies and are increasingly helpful for precise diagnosis of a wide range of anatomical and physiological disorders. The grandparent of all medical imaging techniques is conventional radiography, in medical use since the late 1940s. The newer imaging technologies not only contribute to diagnosis of disease, but they also are advancing our understanding of normal physiology. Table 1.4 describes some commonly used medical imaging techniques. Other imaging methods, such as cardiac catheterization, will be discussed in later chapters.

▶ C H E C K P O I N T

15. Which forms of medical imaging use radiation? Which do not?

16. Which imaging modality best reveals the physiology of a structure?

Table 1.4 Common Medical Imaging Procedures

Radiography

Procedure: A single barrage of x rays passes through the body, producing an image of interior structures on x-ray–sensitive film. The resulting two-dimensional image is a *radiograph* (RĀ-dē-ō-graf′), commonly called an *x ray.*

Comments: Produces clear images of bony or dense structures, which appear bright, but poor images of soft tissues or organs, which appear hazy or dark.

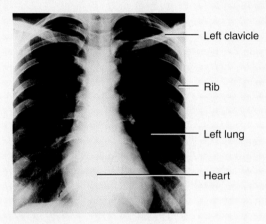

Left clavicle

Rib

Left lung

Heart

Anterior view of thorax

Computed tomography (CT) [formerly called computerized axial tomography (CAT) scanning]

Procedure: Computer-assisted radiography in which an x-ray beam traces an arc at multiple angles around a section of the body. The resulting transverse section of the body, called a *CT scan,* is reproduced on a video monitor.

Comments: Visualizes soft tissues and organs with much more detail than conventional radiographs. Differing tissue densities show up as various shades of gray. Multiple scans can be assembled to build three-dimensional views of structures.

ANTERIOR

Heart

Aorta

Backbone
Left rib
Left scapula

POSTERIOR

Inferior view of transverse section of the thorax

Magnetic resonance imaging (MRI)

Procedure: The body is exposed to a high-energy magnetic field, which causes protons (small positive particles within atoms, such as hydrogen) in body fluids and tissues to arrange themselves in relation to the field. Then a pulse of radiowaves "reads" these ion patterns, and a color-coded image is assembled on a video monitor. The resulting image is a two- or three-dimensional blueprint of cellular chemistry.

Comments: Relatively safe, but can't be used on patients with metal in their bodies. Shows fine details for soft tissues but not for bones. Most useful for differentiating between normal and abnormal tissues. Used to detect tumors and artery-clogging fatty plaques, reveal brain abnormalities, measure blood flow, and detect a variety of musculoskeletal, liver, and kidney disorders.

Sagittal section of brain

Sonography

Procedure: High-frequency sound waves produced by a handheld wand reflect off body tissues and are detected by the same instrument. The image, which may be still or moving, is called a *sonogram* (SON-ō-gram) and is reproduced on a video monitor.

Comments: Safe, noninvasive, painless, and uses no dyes. Most commonly used to visualize the fetus during pregnancy. Also used to observe the size, location, and actions of organs and blood flow through blood vessels.

Forehead

Eye

Hand

Courtesy of Andrew Joseph Tortora and Damaris Soler

(continues)

Table 1.4   Common Medical Imaging Procedures *(continued)*

Positron emission tomography (PET)

Procedure: A substance that emits positrons (positively charged particles) is injected into the body, where it is taken up by tissues. The collision of positrons with negatively charged electrons in body tissues produces gamma rays (similar to x rays) that are detected by gamma cameras positioned around the subject. A computer receives signals from the gamma cameras and constructs a *PET scan* image, displayed in color on a video monitor. The PET scan shows where the injected substance is being used in the body. In the PET scan image shown here, the black and blue colors indicate minimal activity, whereas the red, orange, yellow, and white colors indicate areas of increasingly greater activity.

Comments: Used to study the physiology of body structures, such as metabolism in the brain or heart.

ANTERIOR

POSTERIOR

Transverse section showing blood flow through brain (darkened area at upper left indicates where a stroke has occurred)

STUDY OUTLINE

ANATOMY AND PHYSIOLOGY DEFINED (p. 2)

1. Anatomy is the science of body structures and the relationships among structures; physiology is the science of body functions.
2. Dissection is the careful cutting apart of body structures to study their relationships.
3. Some subdisciplines of anatomy are embryology, developmental biology, histology, surface anatomy, gross anatomy, systemic anatomy, regional anatomy, radiographic anatomy, and pathological anatomy (see Table 1.1 on page 2).
4. Some subdisciplines of physiology are neurophysiology, endocrinology, cardiovascular physiology, immunology, respiratory physiology, renal physiology, exercise physiology, and pathophysiology (see Table 1.1 on page 2).

LEVELS OF BODY ORGANIZATION (p. 3)

1. The human body consists of six levels of structural organization: chemical, cellular, tissue, organ, system, and organismal levels.
2. Cells are the basic structural and functional living units of an organism and the smallest living units in the human body.
3. Tissues are groups of cells and the materials surrounding them that work together to perform a particular function.
4. Organs are composed of two or more different types of tissues; they have specific functions and usually have recognizable shapes.
5. Systems consist of related organs that have a common function.
6. An organism is any living individual.
7. Table 1.2 on pages 4–6 introduces the eleven systems of the human organism: the integumentary, skeletal, muscular, nervous, endocrine, cardiovascular, lymphatic and immune, respiratory, digestive, urinary, and reproductive systems.

CHARACTERISTICS OF THE LIVING HUMAN ORGANISM (p. 7)

1. All organisms carry on certain processes that distinguish them from nonliving things.
2. The most important life processes of the human body are metabolism, responsiveness, movement, growth, differentiation, and reproduction.
3. Homeostasis is a condition of equilibrium in the body's internal environment produced by the interplay of all the body's regulatory processes.
4. Body fluids are dilute, watery solutions. Intracellular fluid (ICF) is inside cells, and extracellular fluid (ECF) is outside cells. Interstitial fluid is the ECF that fills spaces between tissue cells, whereas plasma is the ECF within blood vessels.
5. Because it surrounds all body cells, interstitial fluid is called the body's internal environment.

CONTROL OF HOMEOSTASIS (p. 8)

1. Disruptions of homeostasis come from external and internal stimuli and psychological stresses.
2. When disruption of homeostasis is mild and temporary, responses of body cells quickly restore balance in the internal environment. If disruption is extreme, regulation of homeostasis may fail.
3. Most often, the nervous and endocrine systems acting together or separately regulate homeostasis. The nervous system detects body changes and sends nerve impulses to counteract changes in controlled conditions. The endocrine system regulates homeostasis by secreting hormones.
4. Feedback systems include three components. (1) Receptors moni-

tor changes in a controlled condition and send input to a control center. (2) The control center sets the value at which a controlled condition should be maintained, evaluates the input it receives from receptors, and generates output commands when they are needed. (3) Effectors receive output from the control center and produce a response (effect) that alters the controlled condition.

5. If a response reverses the original stimulus, the system is operating by negative feedback. If a response enhances the original stimulus, the system is operating by positive feedback.

6. One example of negative feedback is a system that helps regulate blood pressure. If a stimulus causes blood pressure (controlled condition) to rise, baroreceptors (pressure-sensitive nerve cells, the receptors) in blood vessels send impulses (input) to the brain (control center). The brain sends impulses (output) to the heart (effector). As a result, heart rate decreases (response) and blood pressure decreases to normal (restoration of homeostasis).

7. One example of positive feedback occurs during the birth of a baby. When labor begins, the cervix of the uterus is stretched (stimulus), and stretch-sensitive nerve cells in the cervix (receptors) send nerve impulses (input) to the brain (control center). The brain responds by releasing oxytocin (output), which stimulates the uterus (effector) to contract more forcefully (response). Movement of the baby further stretches the cervix, more oxytocin is released, and even more forceful contractions occur. The cycle is broken with the birth of the baby.

8. Disruptions of homeostasis—homeostatic imbalances—can lead to disorders, diseases, and even death.

9. Disorder is a general term for any derangement or abnormality of function. A disease is an illness with a definite set of signs and symptoms.

10. Symptoms are subjective changes in body functions that are not apparent to an observer, whereas signs are objective changes that can be observed and measured.

AGING AND HOMEOSTASIS (p. 11)

1. Aging produces observable changes in structure and function and increases vulnerability to stress and disease.

2. Changes associated with aging occur in all body systems.

ANATOMICAL TERMINOLOGY (p. 12)

1. Descriptions of any region of the body assume the body is in the anatomical position, in which the subject stands erect facing the observer, with the head level and the eyes facing directly forward. The feet are flat on the floor and directed forward, and the arms are at the sides, with the palms turned forward.

2. A body lying face down is prone, whereas a body lying face up is supine.

3. Regional names are terms given to specific regions of the body. The principal regions are the head, neck, trunk, upper limbs, and lower limbs.

4. Within the regions, specific body parts have common names and are specified by corresponding anatomical terms. Examples are chest (thoracic), nose (nasal), and wrist (carpal).

5. Directional terms indicate the relationship of one part of the body to another. Exhibit 1.1 on page 14 summarizes commonly used directional terms.

6. Planes are imaginary flat surfaces that are used to divide the body or organs to visualize interior structures. A midsagittal plane

divides the body or an organ into equal right and left sides. A parasagittal plane divides the body or an organ into unequal right and left sides. A frontal plane divides the body or an organ into anterior and posterior portions. A transverse plane divides the body or an organ into superior and inferior portions. An oblique plane passes through the body or an organ at an angle between a transverse plane and either a midsagittal, parasagittal, or frontal plane.

7. Sections are flat surfaces resulting from cuts through body structures. They are named according to the plane on which the cut is made and include transverse, frontal, and sagittal sections.

8. Body cavities are spaces in the body that help protect, separate, and support internal organs.

9. The dorsal and ventral cavities are the two major body cavities.

10. The dorsal cavity is subdivided into the cranial cavity, which contains the brain, and the vertebral canal, which contains the spinal cord. The meninges are protective tissues that line the dorsal cavity.

11. The diaphragm subdivides the ventral body cavity into a superior thoracic cavity and an inferior abdominopelvic cavity. The viscera are organs within the ventral body cavity. A serous membrane lines the wall of the cavity and adheres to the viscera.

12. The thoracic cavity is subdivided into three smaller cavities: a pericardial cavity, which contains the heart, and two pleural cavities, each of which contains a lung.

13. The central part of the thoracic cavity is the mediastinum. It is located between the pleural cavities, extending from the sternum to the vertebral column and from the neck to the diaphragm. It contains all thoracic viscera except the lungs.

14. The abdominopelvic cavity is divided into a superior abdominal and an inferior pelvic cavity.

15. Viscera of the abdominal cavity include the stomach, spleen, liver, gallbladder, small intestine, and most of the large intestine.

16. Viscera of the pelvic cavity include the urinary bladder, portions of the large intestine, and internal organs of the reproductive system.

17. Serous membranes line the walls of the thoracic and abdominal cavities and cover the organs within them. They include the pleura, associated with the lungs; the pericardium, associated with the heart; and the peritoneum, associated with the abdominal cavity.

18. Table 1.3 on page 19 summarizes body cavities and their membranes.

19. To describe the location of organs more easily, the abdominopelvic cavity is divided into abdominopelvic regions: right hypochondriac, epigastric, left hypochondriac, right lumbar, umbilical, left lumbar, right inguinal (iliac), hypogastric (pubic), and left inguinal (iliac).

20. To locate the site of an abdominopelvic abnormality in clinical studies, the abdominopelvic cavity is divided into quadrants: right upper quadrant (RUQ), left upper quadrant (LUQ), right lower quadrant (RLQ), and left lower quadrant (LLQ).

MEDICAL IMAGING (p. 20)

1. Medical imaging techniques allow visualization of internal structures to diagnose abnormal anatomy and deviations from normal physiology.

2. Table 1.4 on pages 21–22 summarizes several medical imaging techniques.

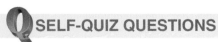 **SELF-QUIZ QUESTIONS**

Fill in the blanks in the following statements.

1. The basic structural and functional unit of the human organism is the _____.

2. The phase of metabolism that involves breaking down large, complex molecules into smaller, simpler ones is called _____.

3. The condition in which the body's internal environment remains within certain physiological limits is termed _____.

Indicate whether the following statements are true or false.

4. In a negative feedback system, the response enhances or intensifies the original stimulus.

5. The ventral body cavity contains the heart, lungs, and abdominal viscera.

6. The parietal layer is the part of a serous membrane that lines the walls of the thoracic and abdominal cavities.

Choose the one best answer to the following questions.

7. Which of the following statements are *true* of a feedback system? (1) A feedback system consists of a control center, a receptor, and an effector. (2) The control center receives input from the effector. (3) The receptor monitors environmental changes that affect the body. (4) The effector produces a response to return the controlled condition to normal. (5) A feedback system involves a cycle of events in which information about the status of a condition is monitored and fed back to a central control region. (a) 1 and 2, (b) 1, 2, and 3, (c) 1, 3, and 5, (d) 1, 3, 4, and 5, (e) 2 and 4.

8. Which of the following statements are *true* of a serous membrane? (1) It lines body cavities that do not open directly to the exterior. (2) It lines body cavities that open directly to the exterior. (3) It is a double-layered membrane. (4) Examples include the pleura, pericardium, and peritoneum. (5) It lines the cavity only and does not cover the organs within the cavity. (a) 1, 3, 4, and 5, (b) 1, 3, and 4, (c) 1, 3 and 5, (d) 1 and 3, (e) 1 and 5.

9. A vertical plane that divides the body or an organ into right and left sides is termed a (an) (a) frontal plane, (b) sagittal plane, (c) transverse plane, (d) oblique plane, (e) coronal plane.

10. The two systems that regulate most homeostatic responses of the body are the (a) nervous and cardiovascular systems, (b) respiratory and cardiovascular systems, (c) endocrine and cardiovascular systems, (d) cardiovascular and urinary systems, (e) endocrine and nervous systems.

11. The fluid found within body cells is (a) blood plasma, (b) extracellular fluid, (c) interstitial fluid, (d) cerebrospinal fluid, (e) intracellular fluid.

12. Match the following life processes to their definitions:
_____ (a) responsiveness
_____ (b) reproduction
_____ (c) growth
_____ (d) metabolism
_____ (e) differentiation
_____ (f) movement

(1) development of a cell from an unspecialized to specialized state
(2) motion of body or body parts
(3) ability to detect and respond to change
(4) formation of new cells or a new individual
(5) sum of all of the body's chemical processes
(6) increase in size

13. Match the following:
_____ (a) nervous system
_____ (b) endocrine system
_____ (c) urinary system
_____ (d) cardiovascular system
_____ (e) muscular system
_____ (f) respiratory system
_____ (g) digestive system
_____ (h) skeletal system
_____ (i) integumentary system
_____ (j) lymphatic & immune system
_____ (k) reproductive system

(1) regulates body activities through chemicals transported in blood to various target organs of the body
(2) produces gametes; releases hormones from gonads
(3) protects against disease; returns fluids to blood
(4) protects body by forming a barrier between the outside environment and internal organs
(5) transports oxygen and nurtrients to cells; protects against disease; carries wastes away from cells
(6) regulates body activities through action potentials; receives sensory information, interprets the information, and responds to the information
(7) carries out the physical and chemical breakdown of food and absorption of nutrients
(8) transfers oxygen and carbon dioxide between air and blood
(9) supports and protects the body; provides internal framework
(10) powers movements of the body and stabilizes body position
(11) eliminates wastes; regulates the volume and chemical composition of blood

14. Match the following common names and anatomical descriptive adjectives:
_____ (a) axillary
_____ (b) inguinal
_____ (c) cervical
_____ (d) cranial
_____ (e) brachial
_____ (f) orbital
_____ (g) gluteal
_____ (h) buccal

(1) skull
(2) eye
(3) cheek
(4) armpit
(5) arm
(6) groin
(7) buttock
(8) neck

15. Match each of the following directional terms to its definition:
_____ (a) at the front of the body
_____ (b) closer to the trunk
_____ (c) toward the upper part of a structure
_____ (d) nearer to the midline of the body
_____ (e) farther from the midline of the body
_____ (f) at the back of the body
_____ (g) farther from the trunk
_____ (h) toward the lower part of a structure

(1) superior
(2) inferior
(3) anterior (ventral)
(4) posterior (dorsal)
(5) medial
(6) lateral
(7) proximal
(8) distal

CRITICAL THINKING QUESTIONS

1. On her first anatomy and physiology exam, Heather defined homeostasis as "the condition in which the body approaches room temperature and stays there." Do you agree with Heather's definition?
   HINT *Does an oral thermometer measure down to 72°F (25°C)?*

2. Elena complained of numbness and tingling sensations in both of her hands. Her physician recommended bilateral splints for carpal tunnel syndrome. Where would she wear these splints?
   HINT *The splints somewhat restrict movement of the hand.*

3. Eighty-year-old Harold fell off a ladder and may have broken his arm. The emergency room physician ordered some x rays, but Harold said that he "wants one of those fancy, new x rays—an MRI. And while you're taking my picture," he continued, "check out that pacemaker near my heart, too." The physician refused to order an MRI. Why?
   HINT *Pacemakers contain metal.*

 ANSWERS TO FIGURE QUESTIONS

1.1 Organs have two or more different types of tissues that work together to perform a specific function.

1.2 The difference between negative and positive feedback systems is that in negative feedback systems, the response reverses the original stimulus, whereas in positive feedback systems, the response enhances the original stimulus.

1.3 When something causes blood pressure to decrease, then heart rate increases due to operation of this negative feedback system.

1.4 Because positive feedback systems continually intensify or reinforce the original stimulus, a termination mechanism is needed to end the response.

1.5 Having one standard anatomical position allows directional terms to be clear, and any body part can be described in relation to any other part.

1.6 No, No, Yes, Yes, No.

1.7 The frontal plane divides the heart into anterior and posterior portions.

1.8 The parasagittal plane divides the brain into unequal right and left portions.

1.9 P, A, T, A, T, P, T, A, A.

1.10 The pericardial cavity surrounds the heart, and the pleural cavities surround the lungs.

1.11 Structures in the mediastinum are the heart, esophagus, trachea, thymus, and several large blood vessels.

1.12 The illustrated abdominal cavity organs all belong to the digestive system (liver, gallbladder, stomach, appendix, small intestine, and most of the large intestine). Illustrated pelvic cavity organs belong to the urinary system (the urinary bladder) and the digestive system (part of the large intestine).

1.13 The liver is mostly in the epigastric region; the transverse colon is in the umbilical region; the urinary bladder is in the hypogastric region; the spleen is in the left hypochondriac region. The pain associated with appendicitis would be felt in the right lower quadrant (RLQ).

The Chemical Level
of Organization

CHEMISTRY AND HOMEOSTASIS

Maintaining the proper assortment and quantity of thousands of different chemicals in your body, and interactions of these chemicals with one another, are two important aspects of homeostasis.

ENERGIZE YOUR STUDY
FOUNDATIONS CD

Chemistry Folder

- Explore Animations on Enzyme Functions or Water and Fluid Flow, reinforce your understanding of Bonding and Polarity and much more
- Support you knowledge of Common Biomolecules with a dynamic Overview
- Concepts and Connections and Exercises reinforce your understanding and provide some fun

www.wiley.com/college/apcentral

INSIGHTS AND EXPLORATIONS

Oxygen is something we all require to live. Heart attacks, if they do not cause death, can cause severe damage by depriving the brain of oxygen. Scientists have actually had mice survive by breathing a liquid chemical that can supply oxygen to their cells. How is it that a liquid could provide oxygen? Think about how oxygen must get from our lungs to the cells in our fingers tips or the tips of our toes. It is dissolved in a liquid. Come along and explore how scientists believe liquid ventilation can be used to save the lives of heart attack patients, patients with respiratory failure and premature infants with lungs too immature to inflate properly.

You learned in Chapter 1 that the chemical level of organization, the lowest level of structural organization, consists of atoms and molecules. These small particles combine to form body structures and systems of astonishing size and complexity. In this chapter, we consider how atoms bond together to form molecules and how atoms and molecules release or store energy in processes known as chemical reactions. Water, which accounts for nearly two-thirds of body weight, is vitally important for chemical reactions and homeostasis in the body. Finally, you will learn about five families of molecules whose unique proper-ties contribute to assembly of the body's structures or to power-ing the processes that characterize life.

Chemistry (KEM-is-trē) is the science of the structure and interactions of matter. All living and nonliving things consist of **matter,** which is anything that occupies space and has mass. The amount of matter in any object is its **mass.** On Earth, *weight* is the force of gravity acting on matter. Things weigh less when they are farther from Earth because the pull of gravity is weaker. In outer space, weight is close to zero but mass remains the same as it is on Earth.

HOW MATTER IS ORGANIZED

▶ O B J E C T I V E S

- **Identify the main chemical elements of the human body.**
- **Describe the structures of atoms, ions, molecules, free radicals, and compounds.**

Chemical Elements

All forms of matter—both living and nonliving—are made up of a limited number of building blocks called **chemical elements.** Each element is a substance that cannot be split into a simpler substance by ordinary chemical means. Scientists now recognize 112 elements. Of these, 92 occur naturally on Earth. The rest have been produced from the natural elements using particle accelerators or nuclear reactors. Each element is designated by a **chemical symbol,** one or two letters of the element's name in English, Latin, or another language. Examples of chemical symbols are H for hydrogen, C for carbon, O for oxygen, N for nitrogen, Ca for calcium, and Na for sodium (*natrium* = sodium).*

Twenty-six different elements normally are present in your body. Just four elements—oxygen, carbon, hydrogen, and nitrogen—constitute about 96% of the body's mass. Eight others—calcium, phosphorus (P), potassium (K), sulfur (S), sodium, chlorine (Cl), magnesium (Mg), and iron (Fe)—contribute 3.8% of the body's mass. An additional 14 elements—the *trace elements*—are present in tiny amounts. Together, they account for the remaining 0.2% of the body's mass. Several trace elements have important functions in the body. For example, iodine is needed to make thyroid hormones. The functions of some trace elements are unknown. Table 2.1 lists the main chemical elements that compose the human body.

Structure of Atoms

Each element is made up of **atoms,** the smallest units of matter that retain the properties and characteristics of an element. Atoms are extremely small. Two hundred thousand of the largest atoms would fit on the period at the end of this sentence. Hydrogen atoms, the smallest atoms, have a diameter less than 0.1 nanometer $(0.1 \times 10^{-9}$ m $= 0.0000000001$ m), and the largest atoms are only five times larger.

Dozens of different **subatomic particles** compose individual atoms. However, only three types of subatomic particles are important for understanding chemical reactions: protons, neutrons, and electrons (Figure 2.1). The dense central core of an atom is its **nucleus.** Within the nucleus are positively charged **protons (p^+)** and uncharged (neutral) **neutrons (n^0).** The tiny negatively charged **electrons (e^-)** move about in a large space surrounding the nucleus. They do not follow a fixed path or orbit but instead form a negatively charged "cloud" that envelopes the nucleus (Figure 2.1a).

Even though their exact positions cannot be predicted, specific groups of electrons are most likely to move about within certain regions around the nucleus. These regions are called **electron shells.** We depict the electron shells as simple circles around the nucleus even though some of their shapes are not spherical. Because each electron shell can hold a specific number of electrons, the electron shell model best conveys this aspect of atomic structure (Figure 2.1b). For instance, the first electron shell (nearest the nucleus) never holds more than 2 electrons. The second shell holds a maximum of 8 electrons, whereas the third can hold up to 18 electrons. The electron shells fill with electrons in a specific order, beginning with the first shell. For example, notice in Figure 2.2 that sodium (Na), which has 11 electrons total, contains 2 electrons in the first shell, 8 electrons in the second shell, and 1 electron in the third shell. The most massive element present in the human body is iodine, which has a total of 53 electrons: 2 in the first shell, 8 in the second shell, 18 in the third shell, 18 in the fourth shell, and 7 in the fifth shell.

---

*The periodic table of elements, which lists all of the known chemical elements, can be found in Appendix B.

Table 2.1 Main Chemical Elements in the Body

Chemical Element (Symbol)	% Of Total Body Mass	Significance
Oxygen (O)	65.0	Part of water and many organic (carbon-containing molecules); used to generate ATP, a molecule used by cells to temporarily store chemical energy.
Carbon (C)	18.5	Forms backbone chains and rings of all organic molecules: carbohydrates, lipids (fats), proteins, and nucleic acids (DNA and RNA).
Hydrogen (H)	9.5	Constituent of water and most organic molecules; ionized form (H^+) makes body fluids more acidic.
Nitrogen (N)	3.2	Component of all proteins and nucleic acids.
Calcium (Ca)	1.5	Contributes to hardness of bones and teeth; ionized form (Ca^{2+}) needed for blood clotting, release of hormones, contraction of muscle, and many other processes.
Phosphorus (P)	1.0	Component of nucleic acids and ATP; required for normal bone and tooth structure.
Potassium (K)	0.35	Ionized form (K^+) is the most plentiful cation (positively charged particle) in intracellular fluid; needed to generate action potentials.
Sulfur (S)	0.25	Component of some vitamins and many proteins.
Sodium (Na)	0.2	Ionized form (Na^+) is the most plentiful cation in extracellular fluid; essential for maintaining water balance; needed to generate action potentials.
Chlorine (Cl)	0.2	Ionized form (Cl^-) is the most plentiful anion (negatively charged particle) in extracellular fluid; essential for maintaining water balance.
Magnesium (Mg)	0.1	Ionized form (Mg^{2+}) needed for action of many enzymes, molecules that increase the rate of chemical reactions in organisms.
Iron (Fe)	0.005	Ionized forms (Fe^{2+} and Fe^{3+}) are part of hemoglobin (oxygen-carrying protein in red blood cells) and some enzymes (proteins that catalyze chemical reactions in living cells).
Trace elements	0.2	Aluminum (Al), Boron (B), Chromium (Cr), Cobalt (Co), Copper (Cu), Fluorine (F), Iodine (I), Manganese (Mn), Molybdenum (Mo), Selenium (Se), Silicon (Si), Tin (Sn), Vanadium (V), and Zinc (Zn).

Figure 2.1 **Two representations of the structure of an atom.** Electrons move about the nucleus, which contains neutrons and protons. (a) In the electron cloud model of an atom, the shading represents the chance of finding an electron in regions outside the nucleus. (b) In the electron shell model, filled circles represent individual electrons, which are grouped into concentric circles according to the shells they occupy. Both models depict a carbon atom, with six protons, six neutrons, and six electrons.

 An atom is the smallest unit of matter that retains the properties and characteristics of its element.

Protons (p^+) ⎤
⎟ Nucleus
Neutrons (n^0) ⎦

Electrons (e^-)

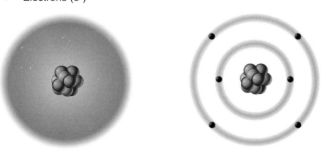

(a) Electron cloud model   (b) Electron shell model

How are the electrons of carbon distributed between the first and second electron shells?

The number of electrons in an atom of an element always equals the number of protons. Because each electron and proton carries one charge, the negatively charged electrons and the positively charged protons balance each other. Thus, each atom is electrically neutral; its total charge is zero.

Atomic Number and Mass Number

The *number of protons* in the nucleus of an atom, designated by the atom's **atomic number,** distinguishes the atoms of one element from those of another. Figure 2.2 shows that atoms of different elements have different atomic numbers because they have different numbers of protons. For example, oxygen has an atomic number of 8 because its nucleus has eight protons, whereas sodium has an atomic number of 11 because its nucleus has 11 protons.

The **mass number** of an atom is the sum of its protons and neutrons. For sodium, which has 11 protons and 12 neutrons, the mass number is 23 (Figure 2.2). Although all atoms of one element have the same number of protons, they may have different numbers of neutrons and thus different mass numbers. **Isotopes** are atoms of an element that have different numbers of neutrons and therefore different mass numbers. In a sample of oxygen, for example, most atoms have 8 neutrons, but a few have 9 or 10, even though all have 8 protons and 8 electrons. Most isotopes are stable, which means that their nuclear structure does not

Figure 2.2   Atomic structures of several stable atoms.

 The atoms of different elements have different atomic numbers because they have different numbers of protons.

Hydrogen (H)
Atomic number =1
Mass number =**1** or 2
Atomic mass =1.01

Carbon (C)
Atomic number = 6
Mass number = **12** or 13
Atomic mass = 12.01

Nitrogen (N)
Atomic number = 7
Mass number = **14** or 15
Atomic mass = 14.01

Oxygen (O)
Atomic number = 8
Mass number = **16**, 17, or 18
Atomic mass = 16.00

Sodium (Na)
Atomic number = 11
Mass number = **23**
Atomic mass = 22.99

Chlorine (Cl)
Atomic number = 17
Mass number = **35** or 37
Atomic mass = 35.45

Potassium (K)
Atomic number = 19
Mass number = **39**, 40, or 41
Atomic mass = 39.10

Iodine (I)
Atomic number = 53
Mass number = **127**
Atomic mass = 126.90

Atomic number = number of protons in an atom
Mass number = number of protons and neutrons in an atom (boldface indicates most common isotope)
Atomic mass = average mass of all stable atoms of a given element in daltons

Which four of these elements are present most abundantly in living organisms?

change over time. The stable isotopes of oxygen are designated ^{16}O, ^{17}O, and ^{18}O (or O-16, O-17, and O-18). The numbers indicate the mass number (total number of protons and neutrons) of each isotope. As you will discover, the electrons of an atom determine the atom's chemical properties. Although the isotopes of an element have different numbers of neutrons, they have the same number of electrons and thus identical chemical properties.

Certain isotopes called **radioactive isotopes** are unstable; their nuclei decay (spontaneously change) into a stable configuration. Examples are H-3, C-14, O-15, and O-19. As they decay, these atoms emit radiation—either subatomic particles or packets of energy—and in the process often transform into a different element. For example, the radioactive isotope of carbon, C-14, decays to N-14. The decay of a radioisotope may be fast, occurring in a fraction of a second, or slow, taking millions of years. The **half-life** of an isotope is the time required for half of the radioactive atoms in a sample of that isotope to decay into a more stable form. For instance, the half-life of C-14 is 5600 years, whereas the half-life of I-131 is 8 days.

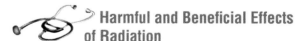

Harmful and Beneficial Effects of Radiation

Radioactive isotopes have both harmful and helpful effects. They can pose a serious threat to the human body because their radiations can break apart molecules, which may produce tissue damage as well as cause various types of cancer. Although the decay of naturally occurring radioactive isotopes typically releases just a small amount of radiation into the environment, localized accumulations can occur. One example of this is radon-222, a colorless and odorless gas that is a naturally occurring radioactive breakdown product of uranium. Radon may seep out of the soil and accumulate in buildings. Radon exposure not only greatly increases the risk of lung cancer in smokers but also causes many cases of lung cancer in nonsmokers. Beneficial uses of certain radioisotopes include their use in medical imaging procedures, for example, to diagnose heart disease, and to treat cancer by killing cancerous cells. ■

Atomic Mass

The standard unit for measuring the mass of atoms and their subatomic particles is a **dalton,** also known as an *atomic mass unit (amu).* A neutron has a mass of 1.008 daltons, and a proton has a mass of 1.007 daltons. The mass of an electron, however, is 0.0005 dalton, almost 2000 times smaller than the mass of a neutron or proton. The **atomic mass** (also called the *atomic weight*) of an element is the average mass of all its naturally occurring isotopes. The atomic mass of chlorine, for example, is 35.45 daltons. About 76% of all chlorine atoms have 18 neutrons (mass number = 35), whereas 24% have 20 neutrons (mass number = 37). Typically, the atomic mass of an element is close to the mass number of its most abundant isotope.

The mass of a single atom is slightly less than the sum of the masses of its neutrons, protons, and electrons because some mass (less than 1%) was lost when the atom's subatomic particles came together to form an atom. This aspect of atom formation explains why an element's atomic mass can be slightly less than the mass number of its smallest stable isotope. For example, the atomic mass of sodium is 22.99 daltons although its mass number is 23.

Ions, Molecules, Free Radicals, and Compounds

As we discussed, atoms of the same element have the same number of protons. In addition, the atoms of each element have a characteristic way of losing, gaining, or sharing their electrons when interacting with other atoms to achieve stability. The way that electrons behave enables atoms in the body to exist in electrically charged forms called ions, or to join with each other into the complex combinations called molecules. If an atom either *gives up* or *gains* electrons, it becomes an ion. An **ion** is an atom that has a positive or negative charge because it has unequal numbers of protons and electrons. *Ionization* is the process of giving up or gaining electrons. An ion of an atom is symbolized by writing its chemical symbol followed by the number of its positive (+) or negative (−) charges. Thus, Ca^{2+} stands for a calcium ion that has two positive charges because it has lost two electrons.

By contrast, when two or more atoms *share* electrons, the resulting combination is called a **molecule** (MOL-e-kūl). A molecule may consist of two atoms of the same kind, such as an oxygen molecule (Figure 2.3a). A *molecular formula* indicates the elements and the number of atoms of each element that make up a molecule. The molecular formula for a molecule of oxygen is O_2. The subscript 2 indicates that the molecule contains two atoms. Two or more different kinds of atoms may also form a molecule, as in a water molecule (H_2O). In H_2O one atom of oxygen shares electrons with two atoms of hydrogen.

A **free radical** is an electrically charged atom or group of atoms with an unpaired electron in its outermost shell. A common example is superoxide, which is formed by the addition of

Figure 2.3 Atomic structures of an oxygen molecule and a superoxide free radical.

A free radical has an unpaired electron in its outermost electron shell.

(a) Oxygen molecule (O_2)   (b) Superoxide free radical (O_2^-)

Unpaired electron

What substances in the body can inactivate oxygen-derived free radicals?

an electron to an oxygen molecule (Figure 2.3b). Having an unpaired electron makes a free radical unstable, highly reactive, and destructive to nearby molecules. Free radicals become stable by either giving up their unpaired electron to, or taking on an electron from, another molecule. In so doing, free radicals may break apart important body molecules.

A **compound** is a substance that contains atoms of two or more different elements. Most of the atoms in the body are joined into compounds. Water (H_2O) and sodium chloride (NaCl), common table salt, are compounds. A molecule of oxygen (O_2), however, is not a compound because it consists of atoms of only one element. Thus, while all compounds are molecules, not all molecules are compounds.

Free Radicals and Their Effects on Health

In our bodies, several processes can generate free radicals. They may result from exposure to ultraviolet radiation in sunlight or to x rays. Some reactions that occur during normal metabolic processes produce free radicals. Moreover, certain harmful substances, such as carbon tetrachloride (a solvent used in dry cleaning), give rise to free radicals when they participate in metabolic reactions in the body. Among the many disorders and diseases linked to oxygen-derived free radicals are cancer, atherosclerosis, Alzheimer disease, emphysema, diabetes mellitus, cataracts, macular degeneration, rheumatoid arthritis, and deterioration associated with aging. Consuming more *antioxidants*— substances that inactivate oxygen-derived free radicals—is thought to slow the pace of damage caused by free radicals. Important dietary antioxidants include selenium, beta-carotene, and vitamins C and E. ■

▶ **CHECKPOINT**

1. What are the names and chemical symbols of the 12 most abundant chemical elements in the human body?

2. Compare the meanings of atomic number, mass number, and atomic mass.

3. What are isotopes and radioactive isotopes of chemical elements?

CHEMICAL BONDS

* **Describe how valence electrons form chemical bonds.**
* **Distinguish among ionic, covalent, and hydrogen bonds.**

The forces that hold together the atoms of a molecule or a compound are **chemical bonds.** The likelihood that an atom will form a chemical bond with another atom depends on the number of electrons in its outermost shell, also called the **valence shell.** An atom with a valence shell holding eight electrons is *chemically stable,* which means it is unlikely to form chemical bonds with other atoms. Neon, for example, has eight electrons in its valence shell, and for this reason it does not easily bond with other atoms. Hydrogen and helium are exceptions; their valence shell is the first electron shell, which holds a maximum of two electrons. Because helium has two valence electrons (see Figure 2.2), it too is stable and seldom bonds with other atoms.

The atoms of most biologically important elements do not have eight electrons in their valence shells. Under the right conditions, two or more atoms can interact in ways that produce a chemically stable arrangement of eight valence electrons for each atom. This chemical principle, called the **octet rule** (*octet = set of eight*), helps explain why atoms interact in predictable ways. One atom is more likely to interact with another atom if doing so will leave both with eight valence electrons. For this to happen, an atom either empties its partially filled valence shell, fills it with donated electrons, or shares electrons with other atoms. The way that valence electrons are distributed determines what kind of chemical bond results. We will consider three kinds of chemical bonds: ionic bonds, covalent bonds, and hydrogen bonds.

Ionic Bonds

When atoms lose or gain one or more valence electrons, ions are formed. Positively and negatively charged ions are attracted to one another—opposites attract. The force of attraction that holds ions having opposite charges together is an **ionic bond.** Consider sodium and chlorine atoms to see how this happens. Sodium has one valence electron (Figure 2.4a). If sodium *loses* this electron, it is left with the eight electrons in its second shell, a complete octet. As a result, however, the total number of protons (11) exceeds the number of electrons (10). Thus, the sodium atom has become a **cation,** or positively charged ion. A sodium ion has a charge of 1 + and is written Na^+. By contrast, chlorine has seven valence electrons (Figure 2.4b). If chlorine *gains* an electron from a neighboring atom, it will have a complete octet in its third electron shell. After gaining an electron, the total number of electrons (18) exceeds the number of protons (17), and the chlorine atom has become an **anion,** a negatively charged ion. The ionic form of chlorine is called a chloride ion. It has a charge of 1 − and is written Cl^-. When an atom of

Figure 2.4 Ions and ionic bond formation. (a) A sodium atom can have a complete octet of electrons in its outermost shell by losing one electron. (b) A chlorine atom can have a complete octet by gaining one electron. (c) An ionic bond may form between oppositely charged ions. (d) In a crystal of NaCl, each Na^+ is surrounded by six Cl^-. In (a), (b), and (c), the electron that is lost or accepted is colored red.

🔑 **An ionic bond is the force of attraction that holds together oppositely charged ions.**

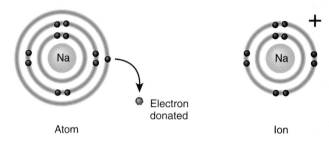

(a) Sodium: 1 valence electron

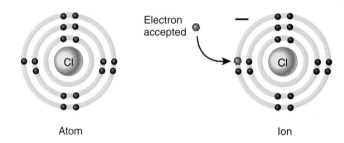

(b) Chlorine: 7 valence electrons

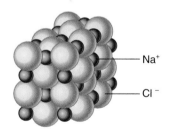

(c) Ionic bond in sodium chloride (NaCl)

(d) Packing of ions in a crystal of sodium chloride

❓ **What are cations and anions?**

sodium donates its sole valence electron to an atom of chlorine, the resulting positive and negative charges pull both ions tightly together into an ionic bond (Figure 2.4c). The resulting compound is sodium chloride, written NaCl.

In general, ionic compounds exist as solids, with an orderly, repeating arrangement of the ions, as in a crystal of NaCl (Figure 2.4d). A crystal of NaCl may be large or small—the total number of ions can vary—but the ratio of Na^+ to Cl^- is always 1:1. In the body, ionic bonds are found mainly in teeth and bones, where they give great strength to the tissue. Most other ions in the body are dissolved in body fluids. An ionic compound that dissociates (breaks apart) into positive and negative ions in solution is called an **electrolyte** (e-LEK-trō-līt). Electrolytes are so-named because their solutions can conduct an electric current. (In Chapter 27 we will discuss the chemistry and importance of electrolytes.)

When a single positively charged ion of sodium and a single negatively charged ion of chlorine form an ionic bond, the net charge of the resulting compound, NaCl, is zero. The same is true for every compound with ionic bonds—the net charge of the compound is zero. In your study of the human body, however, you will discover many compounds that are positively or negatively charged. For example, ammonium (NH_4^+) and hydroxide (OH^-) are very common ionic compounds found in the body. The reason these compounds can carry a charge is that they are the result of covalent bonds (discussed next), not ionic bonds. Table 2.2 lists the names and symbols of the most common ions in the body.

Covalent Bonds

When a **covalent bond** forms, two or more atoms *share* electrons rather than gaining or losing whole electrons. Atoms form a covalently bonded molecule by sharing one, two, or three pairs of their valence electrons. The larger the number of electron pairs shared between two atoms, the stronger the covalent bond. The electrons in each pair are said to be shared because they spend most of their time in the region between the nuclei of the two atoms. Covalent bonds may form between atoms of the same element or between atoms of different elements. They are the most common chemical bonds in the body, and the compounds that result from them form most of the body's structures.

A **single covalent bond** results when two atoms share one electron pair. For example, a molecule of hydrogen forms when two hydrogen atoms share their single valence electrons (Figure 2.5a), which allows both atoms to have a full valence shell at least part of the time. A **double covalent bond** results when two atoms share two pairs of electrons, as happens in an oxygen molecule (Figure 2.5b). A **triple covalent bond** occurs when two atoms share three pairs of electrons, as in a molecule of nitrogen (Figure 2.5c). Notice in the *structural formulas* for covalently bonded molecules in Figure 2.5 that the number of lines between the chemical symbols for two atoms indicates whether the bond is a single ($-$), double ($=$), or triple ($\equiv$) covalent bond.

The same principles of covalent bonding that apply to atoms of the same element also apply to covalent bonds between atoms of different elements. The gas methane (CH_4) contains covalent bonds formed between the atoms of two different elements (Figure 2.5d). The valence shell of the carbon atom can hold eight electrons but has only four of its own. The single electron shell of a hydrogen atom can hold two electrons, but each hydrogen atom has only one of its own. A methane molecule contains four separate single covalent bonds. Each hydrogen atom shares one pair of electrons with the carbon atom.

In some covalent bonds, two atoms share the electrons equally—one atom does not attract the shared electrons more strongly than the other atom. This type of bond is a **nonpolar covalent bond.** The bonds between two identical atoms are always nonpolar covalent bonds (Figure 2.5a–c). Also, the bonds between carbon and hydrogen atoms are nonpolar, for example, each of the four C–H bonds in a methane molecule (Figure 2.5d).

In a **polar covalent bond,** the sharing of electrons between two atoms is unequal—one atom attracts the shared electrons more strongly than the other. When polar covalent bonds form, the resulting molecule has a partial negative charge near the atom that attracts electrons more strongly. This atom has greater **electronegativity,** the power to attract electrons to itself. At least one other atom in the molecule then will have a partial positive charge. The partial charges are indicated by a lowercase Greek delta with a minus or plus sign: δ^- and δ^+. A very important example of a polar covalent bond in living systems is the bond between oxygen and hydrogen, for example, in a molecule of water (Figure 2.6 on page 34). Later in the chapter, we will see how polar covalent bonds allow water to dissolve many molecules that are important to life. The N–H and O–C bonds are also polar bonds.

Table 2.2	Common Ions and Ionic Compounds in the Body		
Cations		**Anions**	
Name	Symbol	Name	Symbol
Hydrogen ion	H^+	Fluoride ion	F^-
Sodium ion	Na^+	Chloride ion	Cl^-
Potassium ion	K^+	Iodide ion	I^-
Ammonium ion	NH_4^+	Hydroxide ion	OH^-
Hydronium ion	H_3O^+	Nitrate ion	NO_3^-
Magnesium ion	Mg^{2+}	Bicarbonate ion	HCO_3^-
Calcium ion	Ca^{2+}	Oxide ion	O^{2-}
Iron (II) ion	Fe^{2+}	Sulfide ion	S^{2-}
Iron (III) ion	Fe^{3+}	Phosphate ion	PO_4^{3-}

Figure 2.5 Covalent bond formation. The red electrons are shared equally. In writing the structural formula of a covalently bonded molecule, each straight line between the chemical symbols for two atoms denotes a pair of shared electrons. In molecular formulas, the number of atoms in each molecule is noted by subscripts.

🔑 **In a covalent bond, two atoms share one, two, or three pairs of valence electrons.**

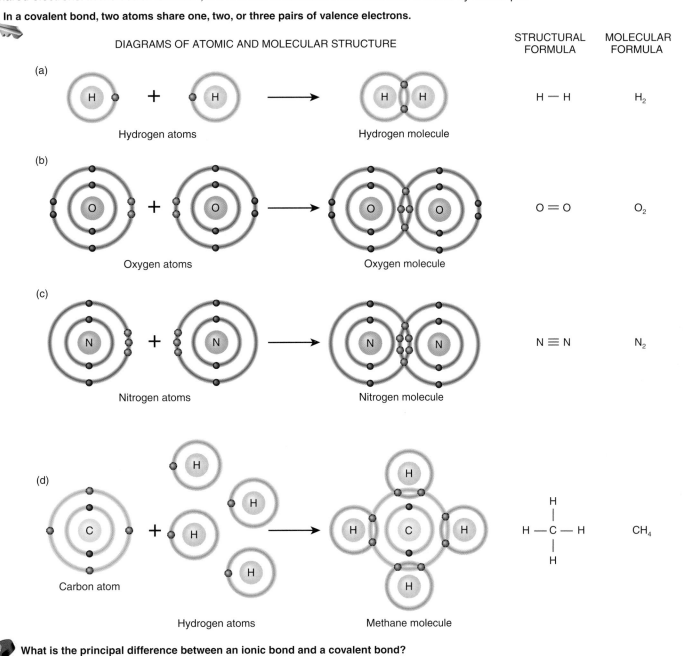

DIAGRAMS OF ATOMIC AND MOLECULAR STRUCTURE | STRUCTURAL FORMULA | MOLECULAR FORMULA

(a) Hydrogen atoms → Hydrogen molecule   H — H H_2

(b) Oxygen atoms → Oxygen molecule   O = O O_2

(c) Nitrogen atoms → Nitrogen molecule   N ≡ N N_2

(d) Carbon atom + Hydrogen atoms → Methane molecule   H — C — H (with H above and H below) CH_4

❓ **What is the principal difference between an ionic bond and a covalent bond?**

Hydrogen Bonds

The polar covalent bonds that form between hydrogen atoms and other atoms can give rise to a third type of chemical bond, a hydrogen bond (Figure 2.7). A **hydrogen bond** forms when a hydrogen atom with a partial positive charge (δ^+) attracts the partial negative charge (δ^-) of neighboring electronegative atoms, most often oxygen or nitrogen. Thus, hydrogen bonds result from attraction of oppositely charged parts of molecules rather than from sharing of electrons as in covalent bonds. Hydrogen bonds are weak—only about 5% as strong as covalent bonds. Thus, they cannot bind atoms into molecules. However, hydrogen bonds do establish important links between molecules or between different parts of a large molecule, such as a protein or nucleic acid (both discussed later in this chapter).

Figure 2.6 **Polar covalent bonds between oxygen and hydrogen atoms in a water molecule.** The red electrons are shared unequally. Because the oxygen nucleus attracts the shared electrons more strongly, the oxygen end of a water molecule has a partial negative charge, written δ^-, and the hydrogen ends have partial positive charges, written δ^+.

🔑 A polar covalent bond occurs when one atomic nucleus attracts the shared electrons more strongly than does the nucleus of another atom in the molecule.

Oxygen atom   Hydrogen atoms   Water molecule

🔑 **Which atom in a water molecule has greater electronegativity?**

The hydrogen bonds that link neighboring water molecules give water considerable *cohesion,* the tendency of like particles to stay together. The cohesion of water molecules creates a very high **surface tension,** a measure of the difficulty of stretching or breaking the surface of a liquid. At the boundary between water and air, water's surface tension is very high because the water molecules are much more attracted to one another than they are attracted to molecules in the air. The influence of water's surface tension on the body can be seen in the way it increases the work required for breathing. A thin film of watery fluid coats the air sacs of the lungs. So, each inhalation must have enough force to

overcome the opposing effect of surface tension as the air sacs stretch and enlarge when taking in air.

Even though single hydrogen bonds are weak, very large molecules may contain hundreds of these bonds. Acting collectively, hydrogen bonds provide considerable strength and stability and help determine the three-dimensional shape of large molecules. As you will see later in this chapter, a large molecule's shape determines how it functions.

▶ CHECKPOINT

4. Which electron shell is the valence shell of an atom, and what is its significance?

5. How does valence relate to the octet rule?

6. What information is conveyed when you write the molecular or structural formula for a molecule?

Figure 2.7 **Hydrogen bonding among water molecules.** Each water molecule forms hydrogen bonds, indicated by dotted lines, with three to four neighboring water molecules.

🔑 Hydrogen bonds occur because hydrogen atoms in one water molecule are attracted to the partial negative charge of the oxygen atom in another water molecule.

Hydrogen bonds

🔑 **Why would you expect ammonia (NH₃) to form hydrogen bonds with water molecules?**

CHEMICAL REACTIONS

▶ OBJECTIVES

- **Define a chemical reaction.**
- **Describe the various forms of energy.**
- **Compare exergonic and endergonic chemical reactions.**
- **Describe the role of activation energy and catalysts in chemical reactions.**
- **Describe synthesis, decomposition, exchange, and reversible reactions.**

A **chemical reaction** occurs when new bonds form or old bonds break between atoms. Chemical reactions are the foundation of all life processes, and as we have seen, the interactions of valence electrons are the basis of all chemical reactions. Consider how hydrogen and oxygen molecules react to form water molecules (Figure 2.8). The starting substances—two H_2 and one O_2—are known as the **reactants.** The ending substances—two molecules of H_2O—are the **products.** The arrow in the figure indicates the direction in which the reaction proceeds. In a chemical reaction, the total mass of the reactants equals the total mass of the products, a relationship known as the **law of conservation of mass.** Because of this law, the number of atoms of each element is the same before and after the reaction. However, because the atoms are rearranged, the reactants and products have different chemical properties. Through thousands of different chemical reactions, body structures are built and body functions are carried out. The term **metabolism** refers to all the chemical reactions occurring in the body.

Forms of Energy and Chemical Reactions

Each chemical reaction involves energy changes. **Energy** (*en-* = in; *-ergy* = work) is the capacity to do work. Two principal forms of energy are **potential energy,** energy stored by matter due to its position, and **kinetic energy,** the energy associated with matter in motion. For example, the energy stored in a battery, in water behind a dam, or in a person poised to jump down some steps is potential energy. When the battery is used to run a clock, or the gates of the dam are opened and the falling water turns a generator, or the person jumps, potential energy is converted into kinetic energy. **Chemical energy** is a form of potential energy that is stored in the bonds of compounds and molecules. The total amount of energy present at the beginning and end of a chemical reaction is the same. Although energy can be neither created nor destroyed, it may be converted from one form to another. This principle is known as the **law of conservation of energy.** For example, some of the chemical energy in the foods we eat is eventually converted into various forms of kinetic energy, such as mechanical energy used to walk and talk. Conversion of energy from one form to another generally releases heat, some of which is used to maintain normal body temperature.

Energy Transfer in Chemical Reactions

In chemical reactions, breaking old bonds requires energy, and forming new bonds releases energy. Because most chemical reactions involve both breaking bonds in the reactants and forming bonds in the products, the *overall reaction* may either release energy or absorb energy. **Exergonic reactions** (*ex-* = out) release more energy than they absorb. In an exergonic reaction, the energy released when new bonds form is *greater* than the energy needed to break apart old bonds. As a result, surplus energy is released as the reaction occurs. At the conclusion of an exergonic reaction, the products have *less* potential energy than the reactants. By contrast, **endergonic reactions** (*end-* = within) absorb more energy than they release. In an endergonic reaction, the energy released when new bonds form is *less* than the energy needed to break apart old bonds, so energy must be absorbed

Figure 2.8 The chemical reaction between two hydrogen molecules (H_2) and one oxygen molecule (O_2) to form two molecules of water (H_2O). Note that the reaction occurs by breaking old bonds and making new bonds.

The number of atoms of each element is the same before and after a chemical reaction.

2 H_2	O_2	2 H_2O
Reactants		Products

 Why does this reaction require two molecules of H_2?

from sources outside the reactants for the reaction to take place. At the conclusion of an endergonic reaction, the products have *more* potential energy than the reactants.

A key feature of the body's metabolism is the coupling of exergonic reactions and endergonic reactions. Energy released from an exergonic reaction often is used to drive an endergonic one. In general, exergonic reactions occur as nutrients, such as glucose, are broken down. Some of the energy released may be trapped in the covalent bonds of *adenosine triphosphate (ATP),* which we describe more fully later in this chapter. If a molecule of glucose is completely broken down, the chemical energy in its bonds can be used to produce as many as 38 molecules of ATP. The energy transferred to the ATP molecules is then used to drive endergonic reactions needed to build body structures, such as muscles and bones. The energy in ATP is also used to do mechanical work, such as occurs during the contraction of muscle or the movement of substances into or out of cells.

Activation Energy

Because particles of matter such as atoms, ions, and molecules have kinetic energy, they are continuously moving and colliding with one another. A sufficiently forceful collision can disrupt the movement of valence electrons, causing an existing chemical bond to break or a new one to form. The collision energy needed to break chemical bonds in the reactants is called the **activation energy** (Figure 2.9). This initial energy "investment" is needed to start a reaction. The reactants must absorb enough energy for their chemical bonds to become unstable and their valence electrons to form new combinations. Then, as new bonds form, energy is released to the surroundings.

Both the concentration of particles and the temperature influence the chance that a collision will occur and cause a chemical reaction.

- *Concentration.* The more particles of matter present in a confined space, the greater the chance that they will collide. The concentration of particles increases when more are added to a given space or when the pressure on the space

Figure 2.9   Activation energy.

Activation energy is the energy needed to break chemical bonds in the reactant molecules so a reaction can start.

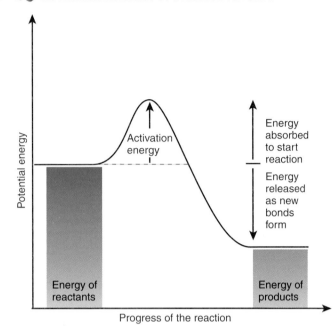

Why is the reaction illustrated here exergonic?

Figure 2.10   Comparison of energy needed for a chemical reaction to proceed with a catalyst (green curve) and without a catalyst (red curve).

Catalysts speed up chemical reactions by lowering the activation energy.

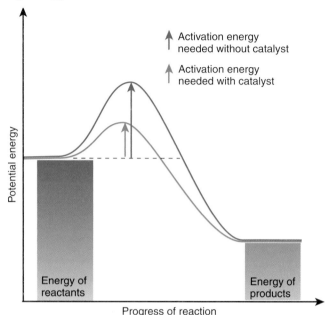

Does a catalyst change the potential energies of the products and reactants?

increases, which forces the particles closer together so that they collide more often.

- **Temperature.** As temperature rises, particles of matter move about more rapidly. Thus, the higher the temperature of matter, the more forcefully particles will collide, and the greater the chance that a collision will produce a reaction.

Catalysts

As we have seen, chemical reactions occur when chemical bonds break or form after atoms, ions, or molecules collide with one another. Body temperature and the concentrations of molecules in body fluids, however, are far too low for most chemical reactions to occur rapidly enough to maintain life. Raising the temperature and the number of reacting particles of matter in the body could increase the frequency of collisions and thus increase the rate of chemical reactions, but doing so could also damage or kill the body's cells.

Substances called catalysts solve this problem. **Catalysts** are chemical compounds that speed up chemical reactions by lowering the activation energy needed for a reaction to occur (Figure 2.10). A catalyst does not alter the difference in potential energy between the reactants and the products. Rather, it lowers the amount of energy needed to start the reaction.

For chemical reactions to occur, some particles of matter—especially large molecules—must not only collide with sufficient force, but they must "hit" one another at precise spots. A catalyst helps to properly orient the colliding particles. Thus,

they interact at the spots that make the reaction happen. Although the action of a catalyst helps to speed up a chemical reaction, the catalyst itself is unchanged at the end of the reaction. A single catalyst molecule can repeatedly assist one chemical reaction after another. The most important catalysts in the body are enzymes, which we will discuss later in this chapter.

Types of Chemical Reactions

After a chemical reaction takes place, the atoms of the reactants are rearranged to yield products with new chemical properties. In this section we will look at the types of chemical reactions common to all living cells. Once you have learned them, you will be able to understand the chemical reactions discussed later in the book.

Synthesis Reactions—Anabolism

When two or more atoms, ions, or molecules combine to form new and larger molecules, the processes are **synthesis reactions.** The word *synthesis* means "to put together." A synthesis reaction can be expressed as follows:

$$A \quad + \quad B \quad \xrightarrow{\text{Combine to form}} \quad AB$$

Atom, ion,   Atom, ion,   New molecule AB
or molecule A   or molecule B

One example of a synthesis reaction is the reaction between two hydrogen molecules and one oxygen molecule to form two molecules of water (see Figure 2.8). Another example of a synthesis reaction is:

$$N_2 \quad + \quad 3\,H_2 \quad \xrightarrow{\text{Combine to form}} \quad 2\,NH_3$$

One nitrogen molecule   Three hydrogen molecules   Two ammonia molecules

All the synthesis reactions that occur in your body are collectively referred to as **anabolism** (a-NAB-ō-lizm). Overall, anabolic reactions are usually endergonic because they absorb more energy than they release. Combining simple molecules like amino acids (discussed shortly) to form large molecules such as proteins is an example of anabolism.

Decomposition Reactions—Catabolism

Decomposition reactions split up large molecules into smaller atoms, ions, or molecules. A decomposition reaction is expressed as follows:

$$AB \quad \xrightarrow{\text{Breaks down into}} \quad A \quad + \quad B$$

Molecule AB   Atom, ion, or molecule A   Atom, ion, or molecule B

For example, under the proper conditions, a molecule of methane can decompose into a carbon atom and two hydrogen molecules:

$$CH_4 \quad \xrightarrow{\text{Breaks down into}} \quad C \quad + \quad 2\,H_2$$

One methane molecule   One carbon atom   Two hydrogen molecules

The decomposition reactions that occur in your body are collectively referred to as **catabolism** (ka-TAB-ō-lizm). Overall, catabolic reactions are usually exergonic because they release more energy than they absorb. For instance, the series of reactions that break down glucose to pyruvic acid, with the net production of two molecules of ATP, are important catabolic reactions in the body. These reactions will be discussed in Chapter 25.

Exchange Reactions

Many reactions in the body are **exchange reactions;** they consist of both synthesis and decomposition reactions. One type of exchange reaction works like this:

$$AB + CD \longrightarrow AD + BC$$

The bonds between A and B and between C and D break (decomposition), and new bonds then form (synthesis) between A and D and between B and C. An example of an exchange reaction is:

$$HCl \quad + \quad NaHCO_3 \quad \longrightarrow \quad H_2CO_3 \quad + \quad NaCl$$

Hydrochloric acid   Sodium bicarbonate   Carbonic acid   Sodium chloride

Notice that the ions in both compounds have "switched partners": The hydrogen ion (H^+) from HCl has combined with the bicarbonate ion (HCO_3^-) from $NaHCO_3$, and the sodium ion (Na^+) from $NaHCO_3$ has combined with the chloride ion (Cl^-) from HCl.

Reversible Reactions

Some chemical reactions proceed in only one direction, from reactants to products, as previously indicated by the single arrows. Other chemical reactions may be reversible. In a **reversible reaction,** the products can revert to the original reactants. A reversible reaction is indicated by two half arrows pointing in opposite directions:

$$AB \underset{\text{Combine to form}}{\overset{\text{Breaks down into}}{\rightleftarrows}} A + B$$

Some reactions are reversible only under special conditions:

$$AB \underset{\text{Heat}}{\overset{\text{Water}}{\rightleftarrows}} A + B$$

In that case, whatever is written above or below the arrows indicates the condition needed for the reaction to occur. In these reactions, AB breaks down into A and B only when water is added, and A and B react to produce AB only when heat is applied. Many reversible reactions in the body require enzymes. Often, different enzymes guide the reactions in opposite directions.

▶ CHECKPOINT

7. What is the relationship between reactants and products in a chemical reaction?
8. Compare potential energy and kinetic energy.
9. Describe the law of conservation of energy.
10. How do catalysts affect activation energy?
11. How are anabolism and catabolism related to synthesis and decomposition reactions, respectively?

INORGANIC COMPOUNDS AND SOLUTIONS

▶ OBJECTIVES

• **Describe the properties of water and inorganic acids, bases, and salts.**
• **Distinguish among solutions, colloids, and suspensions.**
• **Define pH and explain the role of buffer systems in homeostasis.**

Most of the chemicals in your body exist in the form of compounds. Biologists and chemists divide these compounds into two principal classes: inorganic compounds and organic compounds. In general, **inorganic compounds** lack carbon and are structurally simple. They include water and many salts, acids, and bases. Inorganic compounds may have either ionic or covalent bonds. Water makes up 55–60% of a lean adult's total body mass; all other inorganic compounds add an additional 1–2%. **Organic compounds,** by contrast, always contain carbon, usually contain hydrogen, and always have covalent bonds. Two carbon-containing compounds—carbon dioxide (CO_2) and bicarbonate ion (HCO_3^-)—are classified as inorganic.

Water

Water is the most important and abundant inorganic compound in all living systems. Although you might be able to survive for weeks without food, without water you would die in a matter of days. Nearly all the body's chemical reactions occur in a watery medium. Water has many properties that make it such an indispensable compound for life. The most important property of water is its polarity—the uneven sharing of valence electrons that confers a partial negative charge near the one oxygen atom and two partial positive charges near the two hydrogen atoms in a water molecule (see Figure 2.6). This property alone makes water an excellent solvent for other ionic or polar substances, gives water molecules cohesion (the tendency to stick together), and allows water to minimize temperature changes.

Water as a Solvent

In medieval times people searched in vain for a "universal solvent," a substance that would dissolve all other materials. They found nothing that worked as well as water. Although it is the most versatile solvent known, however, it is not a universal solvent. If it were, no container could hold it because the container would dissolve! In a **solution,** a substance called the **solvent** dissolves another substance called the **solute.** Usually there is more solvent than solute in a solution. For example, your sweat is a dilute solution of water (the solvent) plus small amounts of salts (the solutes).

The versatility of water as a solvent for ionized or polar substances is due to its polar covalent bonds and its bent shape, which allows each water molecule to interact with several neighboring ions or molecules. Solutes that contain polar covalent bonds are **hydrophilic** (*hydro-* = water; *-philic* = loving), which means they dissolve easily in water. Common examples of hydrophilic solutes are sugar and salt. Molecules that contain mainly nonpolar covalent bonds, by contrast, are **hydrophobic** (*-phobic* = fearing). They are not very water soluble. Examples of hydrophobic compounds include animal fats and vegetable oils.

To understand the dissolving power of water, consider what happens when a crystal of a salt such as sodium chloride (NaCl) is placed in water (Figure 2.11). The electronegative oxygen atom in water molecules attracts the sodium ions (Na^+), and the electropositive hydrogen atoms in water molecules attracts the chloride ions (Cl^-). Soon, water molecules surround and separate some Na^+ and Cl^- from each other at the surface of the crystal, breaking the ionic bonds that held NaCl together. The

Figure 2.11 How polar water molecules dissolve salts and polar substances. When a crystal of sodium chloride is placed in water, the slightly negative oxygen end (red) of water molecules is attracted to the positive sodium ions (Na^+), and the slightly positive hydrogen portions (gray) of water molecules are attracted to the negative chloride ions (Cl^-).

 Water is a versatile solvent because its polar covalent bonds, in which electrons are shared unequally, create positive and negative regions.

Hydrated sodium ion

Water molecule

Crystal of NaCl

Hydrated chloride ion

Table sugar (sucrose) easily dissolves in water but is not an electrolyte. Is it likely that all the covalent bonds between atoms in table sugar are nonpolar bonds?

water molecules surrounding the ions also lessen the chance that ions with opposite charges will come together and reform an ionic bond.

The ability of water to form solutions and suspensions is essential to health and survival. Because water can dissolve or suspend so many different substances, it is an ideal medium for metabolic reactions. Water enables dissolved reactants to collide and form products. Water also dissolves waste products, which allows them to be flushed out of the body in the urine.

Water in Chemical Reactions

Water serves as the medium for most chemical reactions in the body, and it also participates as a reactant or product in certain reactions. During digestion, for example, decomposition reactions break down large nutrient molecules into smaller molecules by the addition of water molecules. This type of reaction is called **hydrolysis** (hī-DROL-i-sis; *-lysis* = to loosen or break apart).

Hydrolysis reactions enable dietary nutrients to be absorbed into the body. When two smaller molecules join to form a larger molecule in a **dehydration synthesis reaction** (*de-* = from, down, or out; *hydra-* = water), by contrast, a water molecule is formed and removed from the reactants. As you will see later in the chapter, such reactions occur during synthesis of proteins and other large molecules (for example, see Figure 2.21).

High Heat Capacity of Water

In comparison to most substances, water can absorb or release a relatively large amount of heat with only a modest change in its own temperature. For this reason, water is said to have a high *heat capacity*. The reason for this property is the large number of hydrogen bonds in water. As water absorbs heat energy, some of the energy is used to break hydrogen bonds. Less energy is then left over to increase the motion of water molecules, which would increase the water's temperature. The high heat capacity of water is the reason it is used in automobile radiators; it cools the engine by absorbing heat without its own temperature rising to an unacceptably high level. The large amount of water in the body has a similar effect: It lessens the impact of environmental temperature changes, helping to maintain the homeostasis of body temperature.

Water also requires a large amount of heat to change from a liquid to a gas. Its *heat of vaporization* is high. As water evaporates from the surface of the skin, it removes a large quantity of heat, providing an important cooling mechanism.

Water as a Lubricant

Water is a major part of mucus and other lubricating fluids. Lubrication is especially necessary in the chest (pleural and pericardial cavities) and abdomen (peritoneal cavity), where internal organs touch and slide over one another. It is also needed at joints, where bones, ligaments, and tendons rub against one another. Inside the gastrointestinal tract, mucus and other watery secretions moisten foods, which aids their smooth passage.

Solutions, Colloids, and Suspensions

A **mixture** is a combination of elements or compounds that are physically blended together but not bound by chemical bonds. For example, the air you are breathing is a mixture of gases that includes nitrogen, oxygen, argon, and carbon dioxide. Three common liquid mixtures are solutions, colloids, and suspensions.

Once mixed together, solutes in a solution remain evenly dispersed among the solvent molecules. Because the solute particles in a solution are very small, a solution looks clear and transparent.

A **colloid** differs from a solution mainly because of the size of its particles. The solute particles in a colloid are large enough

to scatter light, just as water droplets in fog scatter light from a car's headlight beams. For this reason, colloids usually look translucent or opaque. Milk is an example of a liquid that is both a colloid and a solution: The large milk proteins make it a colloid, whereas calcium salts, milk sugar (lactose), ions, and other small particles are in solution.

The solutes in both solutions and colloids do not settle out and accumulate on the bottom of the container. In a **suspension,** by contrast, the suspended material may mix with the liquid or suspending medium for some time, but eventually it will settle out. Blood is an example of a suspension. When freshly drawn from the body, blood has an even, reddish color. After blood sits for a while in a test tube, red blood cells settle out of the suspension and drift to the bottom of the tube (see Figure 19.1a on page 635). The upper layer, the liquid portion of blood, appears pale yellow and is called blood plasma. Blood plasma is both a solution of ions and other small solutes and a colloid due to the presence of plasma proteins.

The **concentration** of a solution may be expressed in several ways. One common way is by a mass per volume **percentage,** which gives the relative mass of a solute found in a given volume of solution. Another way expresses concentration in units of **moles per liter (mol/L),** which relate to the total number of molecules in a given volume of solution. A **mole** is the amount of any substance that has a mass in grams equal to the sum of the atomic masses of all its atoms. For example, one mole of the element chlorine (atomic mass = 35.45) is 35.45 grams and one mole of the salt sodium chloride (NaCl) is 58.44 grams (22.99 for Na + 35.45 for Cl). Just as a dozen always means 12 of something, a mole of anything has the same number of particles: 6.023×10^{23}. This huge number is called *Avogadro's number.* Thus, measurements of substances that are stated in moles tell us about the numbers of atoms, ions, or molecules present. This is important when chemical reactions are occurring because each reaction requires a set number of atoms of specific elements. Table 2.3 describes these ways of expressing concentration.

Table 2.3	Percentage and Molarity
Definition	**Example**
Percentage (mass per volume) Number of grams of a substance per 100 milliliters (mL) of solution.	To make a 10% NaCl solution, take 10 gm of NaCl and add enough water to make a total of 100 mL of solution.
Molarity = moles (mol) per liter A 1 molar (1 M) solution = 1 mole of a solute in 1 liter of solution.	To make a 1 molar (1 M) solution of NaCl, Dissolve 1 mole of NaCl (58.44 gm) in enough water to make a total of 1 liter of solution.

Inorganic Acids, Bases, and Salts

When inorganic acids, bases, or salts dissolve in water, they **dissociate** (dis′-sō-sē-ĀT); they separate into ions and become surrounded by water molecules. An **acid** (Figure 2.12a) is a substance that dissociates into one or more **hydrogen ions (H^+)** and one or more anions. Because H^+ is a single proton with one positive charge, an acid also is a **proton donor.** A **base,** by contrast (Figure 2.12b), dissociates into one or more **hydroxide ions (OH^-)** and one or more cations. Because hydroxide ions have a strong attraction for protons, a base is a **proton acceptor.**

A **salt,** when dissolved in water, dissociates into cations and anions, neither of which is H^+ or OH^- (Figure 2.12c). In the body, salts are electrolytes that are important for carrying electrical currents (ions flowing from one place to another), especially in nerve and muscle tissues. The ions of salts also provide many essential chemical elements in intracellular and extracellular fluids such as blood, lymph, and the interstitial fluid of tissues.

Acids and bases react with one another to form salts. For example, the reaction of hydrochloric acid (HCl) and potassium hydroxide (KOH), a base, produces the salt potassium chloride (KCl) and water (H_2O). This exchange reaction can be written as follows:

$$HCl + KOH \longrightarrow H^+ + Cl^- + K^+ + OH^- \longrightarrow KCl + H_2O$$

| Acid | Base | | Dissociated ions | | Salt | Water |

Acid–Base Balance: The Concept of pH

To ensure homeostasis, intracellular and extracellular fluids must contain almost balanced quantities of acids and bases. The more hydrogen ions (H^+) dissolved in a solution, the more acidic the solution; conversely, the more hydroxide ions (OH^-), the more basic (alkaline) the solution. The chemical reactions that take place in the body are very sensitive to even small changes in the acidity or alkalinity of the body fluids in which they occur. Any departure from the narrow limits of normal H^+ and OH^- concentrations greatly disrupts body functions.

A solution's acidity or alkalinity is expressed on the **pH scale,** which extends from 0 to 14 (Figure 2.13). This scale is based on the concentration of H^+ in moles per liter. A pH of 7 means that a solution contains one ten-millionth (0.0000001) of a mole of hydrogen ions per liter. The number 0.0000001 is written as 1×10^{-7} in scientific notation, which indicates that the number is 1 with the decimal point moved seven places to the left. To convert this value to pH, the negative exponent (-7) is changed to a positive number (7). A solution with a H^+ concentration of 0.0001 (10^{-4}) moles per liter has a pH of 4; a solution with a H^+ concentration of 0.000000001 (10^{-9}) moles per liter has a pH of 9; and so on.

The midpoint of the pH scale is 7, where the concentrations of H^+ and OH^- are equal. A substance with a pH of 7, such as

Figure 2.12   Dissociation of inorganic acids, bases, and salts.

 Dissociation is the separation of inorganic acids, bases, and salts into ions in a solution.

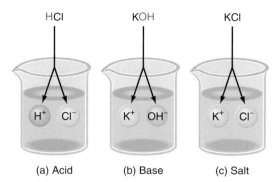

(a) Acid   (b) Base   (c) Salt

The compound $CaCO_3$ (calcium carbonate) dissociates into a calcium ion Ca^{2+} and a carbonate ion CO_3^{2-}. Is it an acid, a base, or a salt? What about H_2SO_4, which dissociates into two H^+ and one SO_4^{2-}?

pure water, is neutral. A solution that has more H^+ than OH^- is an **acidic solution** and has a pH below 7. A solution that has more OH^- than H^+ is a **basic (alkaline) solution** and has a pH above 7. It is important to realize that a change of one whole number on the pH scale represents a *tenfold* change in the number of H^+. A pH of 6 denotes 10 times more H^+ than a pH of 7, whereas a pH of 8 indicates 10 times fewer H^+ than a pH of 7 and 100 times fewer H^+ than a pH of 6.

Maintaining pH: Buffer Systems

Although the pH of body fluids may differ, the normal limits for each fluid are quite narrow. Table 2.4 shows the pH values for certain body fluids together with those of some common substances. Homeostatic mechanisms maintain the pH of blood between 7.35 and 7.45, which is slightly more basic than pure water. Saliva is slightly acidic, and semen is slightly basic. Because the kidneys help remove excess acid from the body, urine can be quite acidic.

Even though strong acids and bases are continually taken into and formed by the body, the pH of fluids inside and outside cells remains almost constant. One important reason is the presence of **buffer systems,** which function to convert strong acids or bases into weak acids or bases. Strong acids (or bases) ionize easily and contribute many H^+ or (OH^-) to a solution. Therefore, they can change pH drastically, which can disrupt the body's metabolism. Weak acids (or bases) do not ionize as much and contribute fewer H^+ (or OH^-). Hence, they have less effect on the pH. The chemical compounds that can convert strong acids or bases into weak ones are called **buffers.**

One important buffer system in the body is the **carbonic acid–bicarbonate buffer system.** Carbonic acid (H_2CO_3) can

Figure 2.13 **The pH scale.** A pH below 7 indicates an acidic solution — more H^+ than OH^-. $[H^+]$ = hydrogen ion concentration; $[OH^-]$ = hydroxide ion concentration.

The lower the numerical value of the pH, the more acidic is the solution because the H^+ concentration becomes progressively greater. A pH above 7 indicates a basic (alkaline) solution; that is, there are more OH^- than H^+. The higher the pH, the more basic the solution.

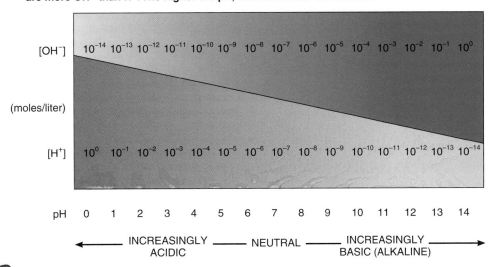

At pH 7 (neutrality), the concentrations of H^+ and OH^- are equal (10^{-7} mol/liter). What is the concentration of H^+ and OH^- at pH 6? Which pH is more acidic, 6.82 or 6.91? Which pH is closer to neutral, 8.41 or 5.59?

Table 2.4 pH Values of Selected Substances	
Substance	**pH Value**
• Gastric juice (found in the stomach)	1.2–3.0
Lemon juice	2.3
Vinegar	3.0
Carbonated soft drink	3.0–3.5
Orange juice	3.5
• Vaginal fluid	3.5–4.5
Tomato juice	4.2
Coffee	5.0
• Urine	4.6–8.0
• Saliva	6.35–6.85
Milk	6.8
Distilled (pure) water	7.0
• Blood	7.35–7.45
• Semen (fluid containing sperm)	7.20–7.60
• Cerebrospinal fluid (fluid associated with nervous system)	7.4
• Pancreatic juice (digestive juice of the pancreas)	7.1–8.2
• Bile (liver secretion that aids fat digestion)	7.6–8.6
Milk of magnesia	10.5
Lye	14.0

• Denotes substances in the human body.

act as a weak acid, and the bicarbonate ion (HCO_3^-) can act as a weak base. Hence, this buffer system can compensate for either an excess or a shortage of H^+. For example, if there is an excess of H^+ (an acidic condition), HCO_3^- can function as a weak base and remove the excess H^+, as follows:

$$H^+ \;+\; HCO_3^- \;\longrightarrow\; H_2CO_3 \;\longrightarrow\; H_2O \;+\; CO_2$$

Hydrogen ion   Bicarbonate ion (weak base)   Carbonic acid   Water   Carbon dioxide

If there is a shortage of H^+ (an alkaline condition), by contrast, H_2CO_3 can function as a weak acid and provide needed H^+ as follows:

$$H_2CO_3 \;\longrightarrow\; H^+ \;+\; HCO_3^-$$

Carbonic acid (weak acid)   Hydrogen ion   Bicarbonate ion

Chapter 27 describes buffers and their roles in maintaining acid–base balance in more detail.

▶ **CHECKPOINT**

12. How do inorganic compounds differ from organic compounds?

13. Describe two ways to express the concentration of a solution.

14. What functions does water perform in the body?

15. What is the chemical reaction whereby bicarbonate ions prevent buildup of excess H^+?

ORGANIC COMPOUNDS

▶ O B J E C T I V E S

- **Describe the functional groups of organic molecules.**
- **Identify the building blocks and functions of carbohydrates, lipids, proteins, enzymes, deoxyribonucleic acid (DNA), ribonucleic acid (RNA), and adenosine triphosphate (ATP).**

Inorganic compounds are relatively simple. Their molecules have only a few atoms and cannot be used by cells to perform complicated biological functions. Many organic molecules, by contrast, are relatively large and have unique characteristics that allow them to carry out complex functions. Important categories of organic compounds, which make up about 40% of body mass, include carbohydrates, lipids, proteins, nucleic acids, and adenosine triphosphate (ATP).

Carbon and Its Functional Groups

Carbon has several properties that make it particularly useful to living organisms. For one thing, it can form bonds with one to thousands of other carbon atoms to produce large molecules that can have many different shapes. Due to this property of carbon, the body can build many different organic compounds, each of which has a unique structure and function. Moreover, the large size of most carbon-containing molecules and the fact that some do not dissolve easily in water make them useful materials for building body structures.

Organic compounds are held together mostly or entirely by covalent bonds. Carbon has four electrons in its outermost (valence) shell. It can bond covalently with a variety of atoms, including other carbon atoms, to form rings and straight or branched chains. Other elements that most often bond with carbon in organic compounds are hydrogen (one bond), oxygen (two bonds), and nitrogen (three bonds). Sulfur (two bonds) and phosphorus (five bonds) are also present in organic compounds. The other elements listed in Table 2.1 are present in a smaller number of organic compounds.

The chain of carbon atoms in an organic molecule is called the **carbon skeleton.** Many of the carbons are bonded to hydrogen atoms. Also attached to the carbon skeleton are distinctive **functional groups,** in which other elements form bonds with carbon and hydrogen atoms. Each type of functional group has a specific arrangement of atoms that confers characteristic chemical properties upon organic molecules. Table 2.5 lists the most common functional groups of organic molecules and describes

Table 2.5	**Major Functional Groups**		
Name and Structural Formula*	**Occurrence and Significance**	**Name and Structural Formula***	**Occurrence and Significance**
Hydroxyl R—O—H	*Alcohols* contain an —OH group, which is polar and hydrophilic due to its electronegative O atom. Molecules with many —OH groups dissolve easily in water.	**Ester** R—C(=O)—O—R	*Esters* predominate in dietary fats and oils and also occur in our body triglycerides. Aspirin is an ester of salicylic acid, a pain-relieving molecule found in bark of the willow tree.
Sulfhydryl R—S—H	*Thiols* have an —SH group, which is polar and hydrophilic due to its electronegative S atom. Certain amino acids, the building blocks of proteins, contain —SH groups, which help stabilize the shape of proteins. An example is the amino acid cysteine.	**Phosphate** R—O—P(=O)(O⁻)—O⁻	*Phosphates* contain a phosphate group ($-PO_4^{2-}$), which is very hydrophilic due to the dual negative charges. An important example is adenosine triphosphate (ATP), which transfers chemical energy between organic molecules during chemical reactions.
Carbonyl R—C(=O)—R or R—C(=O)—H	*Ketones* contain a carbonyl group within the carbon skeleton. The carbonyl group is polar and hydrophilic due to its electronegative O atom. *Aldehydes* have a carbonyl group at the end of the carbon skeleton.	**Amino** R—N(H)(H) or R—N⁺(H)(H)—H	*Amines* have an —NH₂ group, which can act as a base and pick up a hydrogen ion, giving the amino group a positive charge. At the pH of body fluids, most amino groups have a charge of 1+. All amino acids have an amino group at one end.
Carboxyl R—C(=O)—OH or R—C(=O)—O⁻	*Carboxylic acids* contain a carboxyl group at the end of the carbon skeleton. All amino acids have a —COOH group at one end. The negatively charged form predominates at the pH of body cells and is hydrophilic.		

*The letter R represents the carbon skeleton of the molecule.

some of their properties. Because organic molecules often are big, there are shorthand methods for representing their structural formulas. Figure 2.14 shows two ways to indicate the structure of the sugar glucose, a molecule with a ring-shaped carbon skeleton that has several hydroxyl groups attached.

Small organic molecules can combine into very large molecules that are called **macromolecules** (*macro-* = large). Macromolecules are usually **polymers** (*poly-* = many; *-mers* = parts). A polymer is a large molecule formed by the covalent bonding of many identical or similar small building-block molecules called **monomers** (*mono-* = one). Usually, the reaction that joins two monomers is a dehydration synthesis. In this type of reaction, a hydrogen atom is removed from one monomer and a hydroxyl group is removed from the other to form a molecule of water (see Figure 2.15). Macromolecules such as carbohydrates, lipids, proteins, and nucleic acids are assembled in cells via dehydration synthesis reactions.

Molecules that have the same molecular formula but different structures are called **isomers** (Ī-so-merz; *iso-* = equal or the same). For example, the molecular formulas for the sugars glucose and fructose are both $C_6H_{12}O_6$. The individual atoms, however, are positioned differently along the carbon skeleton (see Figure 2.15).

Carbohydrates

Carbohydrates include sugars, starches, glycogen, and cellulose. Even though they are a large and diverse group of organic compounds and have several functions, carbohydrates represent only 2 – 3% of your total body mass. Plants store carbohydrate as starch and use the carbohydrate cellulose to build plant cell walls. Although humans eat cellulose, the most plentiful organic substance on Earth, they cannot digest it. It does, however, create bulk, which helps to move food and wastes along the gastrointestinal tract. In humans and animals, carbohydrates function mainly as a source of chemical energy for generating ATP needed to drive metabolic reactions. Only a few carbohydrates are used for building structural units. One example is deoxyribose, a type of sugar that is a building block of deoxyribonucleic acid (DNA), the molecule that carries inherited genetic information.

Carbon, hydrogen, and oxygen are the elements found in carbohydrates. The ratio of hydrogen to oxygen atoms is usually 2:1, the same as in water. Although there are exceptions, carbohydrates generally contain one water molecule for each carbon atom. This is the reason they are called carbohydrates, which means "watered carbon." The three major groups of carbohydrates, based on their sizes, are monosaccharides, disaccharides, and polysaccharides (Table 2.6).

Monosaccharides and Disaccharides: The Simple Sugars

Monosaccharides and disaccharides are known as **simple sugars.** The monomers of carbohydrates, **monosaccharides** (mon′-ō-SAK-a-rīds; *sacchar-* = sugar), contain from three to seven carbon atoms. They are designated by names ending in "-ose" with a prefix that indicates the number of carbon atoms. For example, monosaccharides with three carbons are called *trioses* (*tri-* = three). There are also *tetroses* (four-carbon sugars), *pentoses* (five-carbon sugars), *hexoses* (six-carbon sugars), and *heptoses* (seven-carbon sugars). Cells throughout the body break down the hexose glucose to produce ATP.

Two monosaccharide molecules can combine by dehydration synthesis to form one **disaccharide** (dī-SAK-a-rīd; *di-* = two) molecule and a molecule of water. For example, molecules of the monosaccharides glucose and fructose combine to form a molecule of the disaccharide sucrose (table sugar), as shown in

Figure 2.14 Alternate ways to write the structural formula for glucose.

🔑 In standard shorthand, carbon atoms are understood to be at locations where two bond lines intersect, and single hydrogen atoms are not indicated.

All atoms written out   Standard shorthand

❓ How many hydroxyl groups does a molecule of glucose have? How many carbon atoms are part of glucose's carbon skeleton?

Table 2.6	Major Carbohydrate Groups
Type of Carbohydrate	**Examples**
Monosaccharides	Glucose (the main blood sugar) Fructose (found in fruits) Galactose (in milk sugar) Deoxyribose (in DNA) Ribose (in RNA)
Disaccharides	Sucrose (table sugar) = glucose + fructose Lactose (milk sugar) = glucose + galactose Maltose = glucose + glucose
Polysaccharides	Glycogen, the stored form of carbohydrate in animals Starch, the stored form of carbohydrate in plants and main carbohydrate in food Cellulose, part of cell walls in plants; not digested by humans but aids movement of food through intestines

Figure 2.15   Structural and molecular formulas for the monosaccharides glucose and fructose and the disaccharide sucrose. In dehydration synthesis (read from left to right), two smaller molecules, glucose and fructose, are joined to form a larger molecule of sucrose. Note the loss of a water molecule. In hydrolysis (read from right to left), the addition of a water molecule to the larger sucrose molecule breaks the disaccharide into two smaller molecules, glucose and fructose.

🔑 **Monosaccharides are the monomers used to build carbohydrates.**

Dehydration synthesis

Hydrolysis

Glucose ($C_6H_{12}O_6$)   Fructose ($C_6H_{12}O_6$)   Sucrose ($C_{12}H_{22}O_{11}$)   Water

❓ **How many carbon atoms can you count in fructose? In sucrose?**

Figure 2.15. Glucose and fructose are isomers. Although they have the same molecular formula, they are different monosaccharides because the relative positions of the oxygen and carbon atoms are different. Notice that the formula for sucrose is $C_{12}H_{22}O_{11}$, not $C_{12}H_{24}O_{12}$, because a molecule of water is removed as the two monosaccharides are joined.

Disaccharides can also be split into smaller, simpler molecules by hydrolysis. A molecule of sucrose, for example, may be hydrolyzed into its components, glucose and fructose, by the addition of water. Figure 2.15 also illustrates this reaction.

Polysaccharides

The third major group of carbohydrates are the **polysaccharides** (pol′-ē-SAK-a-rīds). Each polysaccharide molecule contains tens or hundreds of monosaccharides joined through dehydration synthesis reactions. Unlike simple sugars, polysaccharides usually are not soluble in water and do not taste sweet. The main polysaccharide in the human body is glycogen, which is made entirely of glucose monomers linked to one another in branching chains (Figure 2.16). A small amount of carbohydrates are stored as glycogen in the liver and skeletal muscles. Like disaccharides, polysaccharides can be broken down into monosaccharides through hydrolysis reactions. For example, when the blood glucose level falls, liver cells have the ability to break down glycogen into glucose and release it into the blood. In this way, glucose is provided for body cells, which break it down to synthesize ATP.

Lipids

A second important group of organic compounds is **lipids** (*lip-* = fat). Lipids make up 18–25% of body mass in lean adults. Like carbohydrates, lipids contain carbon, hydrogen, and oxygen. Unlike carbohydrates, they do not have a 2:1 ratio of hydrogen to oxygen. The proportion of electronegative oxygen atoms in lipids is usually smaller than in carbohydrates, so there are fewer polar covalent bonds. As a result, most lipids are insoluble in polar solvents such as water; they are *hydrophobic*. Non-

Figure 2.16   Part of a glycogen molecule, the main polysaccharide in the human body.

🔑 **Glycogen is made up of glucose monomers and is the storage form of carbohydrate in the human body.**

Glucose monomer

❓ **Which body cells store glycogen?**

polar solvents such as chloroform and ether, however, readily dissolve lipids. Because they are hydrophobic, only the smallest lipids (some fatty acids) can dissolve in watery blood plasma. To become more soluble in blood plasma, other lipid molecules form complexes with hydrophilic protein molecules. The resulting lipid and protein particles are termed **lipoproteins.**

The diverse lipid family includes triglycerides (fats and oils), phospholipids (lipids that contain phosphorus), steroids (lipids that contain rings of carbon atoms), eicosanoids (20-carbon lipids), and a variety of other lipids, including fatty acids, fat-soluble vitamins (vitamins A, D, E, and K), and lipoproteins. Table 2.7 introduces the various types of lipids and highlights their roles in the human body.

Triglycerides

The most plentiful lipids in your body and in your diet are the **triglycerides** (trī-GLI-cer-īdes; *tri-* = three), also known as **triacylglycerols.** At room temperature, triglycerides may be either solids (fats) or liquids (oils), and they are the body's most highly concentrated form of chemical energy. Triglycerides provide more than twice as much energy per gram as do carbohydrates and proteins. Our capacity to store triglycerides in adipose (fat) tissue is unlimited for all practical purposes. Excess dietary carbohydrates, proteins, fats, and oils all have the same fate: They are deposited in adipose tissue as triglycerides.

A triglyceride consists of two types of building blocks, a single glycerol molecule and three fatty acid molecules. A three-carbon **glycerol** molecule forms the backbone of a triglyceride (Figure 2.17). Three **fatty acids** are attached by dehydration synthesis reactions, one to each carbon of the glycerol backbone. The chemical bond formed where each water molecule is removed is an *ester linkage* (see Table 2.5). The reverse reaction, hydrolysis, breaks down a single molecule of a triglyceride into three fatty acids and glycerol.

Saturated fats are triglycerides that contain only *single covalent bonds* between fatty acid carbon atoms. Because they lack double bonds, each carbon atom is *saturated with hydrogen atoms* (see, for example, palmitic acid and stearic acid in Figure 2.17c). Triglycerides with mainly saturated fatty acids usually are solid at room temperature. Although saturated fats occur mostly in animal tissues, they also are found in a few plant products, such as cocoa butter, palm oil, and coconut oil.

Monounsaturated fats contain fatty acids with *one double covalent bond* between two fatty acid carbon atoms. Thus, they are not completely saturated with hydrogen atoms (see, for example, oleic acid in Figure 2.17c). Olive oil and peanut oil are rich in triglycerides with monounsaturated fatty acids.

Polyunsaturated fats contain *more than one double covalent bond* between fatty acid carbon atoms. An example is linoleic acid. Canola oil, corn oil, safflower oil, sunflower oil, and soybean oil contain a high percentage of polyunsaturated fatty acids.

Table 2.7	Types of Lipids in the Body			
Type of Lipid	**Functions**		**Type of Lipid**	**Functions**
Triglycerides *(fats and oils)*	Protection, insulation, energy storage.		**Other Lipids**	
Phospholipids	Major lipid component of cell membranes.		Fatty acids	Catabolized to generate adenosine triphosphate (ATP) or used to synthesize triglycerides and phospholipids.
Steroids			Carotenes	Needed for synthesis of vitamin A, which is used to make visual pigments in the eyes.
Cholesterol	Minor component of all animal cell membranes; precursor of bile salts, vitamin D, and steroid hormones.		Vitamin E	Promotes wound healing, prevents tissue scarring, contributes to the normal structure and function of the nervous system, and functions as an antioxidant.
Bile salts	Needed for absorption of dietary lipids.			
Vitamin D	Helps regulate calcium level in the body; needed for bone growth and repair.		Vitamin K	Required for synthesis of blood-clotting proteins.
Adrenocortical hormones	Help regulate metabolism, resistance to stress, and salt and water balance.		Lipoproteins	Transport lipids in the blood, carry triglycerides and cholesterol to tissues, and remove excess cholesterol from the blood.
Sex hormones	Stimulate reproductive functions and sexual characteristics.			
Eicosanoids	Have diverse effects on blood clotting, inflammation, immunity, stomach acid secretion, airway diameter, lipid breakdown, and smooth muscle contraction.			

Figure 2.17 The formation of a triglyceride (triacylglycerol) from a glycerol and three fatty acid molecules. Each time a glycerol (a) and a fatty acid (b) are joined in dehydration synthesis, a molecule of water is removed. An ester linkage joins the glycerol to each of the three molecules of fatty acids, which vary in length and the number and location of double bonds between carbon atoms ($C = C$). Shown here (c) is a triglyceride molecule that contains two saturated fatty acids and a monounsaturated fatty acid. The kink (bend) in the oleic acid occurs at the double bond.

🔑 **One glycerol and three fatty acids are the building blocks of triglycerides.**

Palmitic acid ($C_{15}H_{31}COOH$)

(b) Fatty acid molecule

(a) Glycerol molecule

Ester linkage

Palmitic acid ($C_{15}H_{31}COOH$) + H_2O
(Saturated)

Stearic acid ($C_{17}H_{35}COOH$) + H_2O
(Saturated)

(c) Triglyceride (fat) molecule

Oleic acid ($C_{17}H_{33}COOH$) + H_2O
(Monounsaturated)

❓ **Does the oxygen in the water molecule removed during dehydration synthesis come from the glycerol or from a fatty acid?**

Phospholipids

Like triglycerides, **phospholipids** have a glycerol backbone and two fatty acid chains attached to the first two carbons. In the third position, however, a phosphate group (PO_4^{3-}) links a small charged group that usually contains nitrogen (N) to the backbone (Figure 2.18). This portion of the molecule (the "head") is polar and can form hydrogen bonds with water molecules. The two fatty acids (the "tails"), by contrast, are nonpolar and can interact only with other lipids. Molecules that have both polar and nonpolar parts are said to be **amphipathic** (am-fi-PATH-ic; *amphi-* = on both sides; *-pathic* = feeling). Amphipathic phos-

pholipids line up tails-to-tails in a double row to make up much of the membrane that surrounds each cell (Figure 2.18c).

Steroids

The structure of **steroids** differs considerably from that of the triglycerides. Steroids have four rings of carbon atoms (colored gold in Figure 2.19). Body cells synthesize other steroids from cholesterol (Figure 2.19a), which has a large nonpolar region consisting of the four rings and a hydrocarbon tail. In the body, the commonly encountered steroids, such as cholesterol, estrogens, testosterone, cortisol, bile salts, and vitamin D, are further

Figure 2.18 Phospholipids. (a) In the synthesis of phospholipids, two fatty acids attach to the first two carbons of the glycerol backbone. A phosphate group links a small charged group to the third carbon in glycerol. In (b), the circle represents the polar head region, and the two wavy lines represent the two nonpolar tails. Double bonds in the fatty acid hydrocarbon chain often form kinks in the tail.

 Phospholipids are amphipathic molecules, having both polar and nonpolar regions.

(a) Chemical structure of a phospholipid

(b) Simplified way to draw a phospholipid

(c) Arrangement of phospholipids in a portion of a cell membrane

 Which portion of a phospholipid is hydrophilic, and which portion is hydrophobic?

classified as **sterols** because they also have at least one hydroxyl (alcohol) group (−OH). The polar hydroxyl groups make sterols weakly amphipathic.

Saturated Fats, Cholesterol, and Atherosclerosis

Atherosclerosis is a progressively worsening disorder in which fatty plaques (deposits) form in the walls of arteries. Plaque buildup narrows the passageway and severely limits blood flow in advanced cases. Although many factors contribute, people who eat a diet high in saturated fats and cholesterol run a greater risk of developing atherosclerosis than do people who eat a diet lower in these lipids. To reduce dietary intake of saturated fats and cholesterol, eat less red meat (beef, pork, and lamb) and fat-laden dairy products (whole milk, butter, and cheese). ■

Other Lipids

Eicosanoids (ī-KŌ-sa-noids; *eicosan-* = twenty) are lipids derived from a 20-carbon fatty acid called arachidonic acid. The two principal subclasses of eicosanoids are the **prostaglandins** (pros'-ta-GLAN-dins) and the **leukotrienes** (loo'-kō-TRĪ-ēnz).

Figure 2.19 Steroids. All steroids have four rings of carbon atoms.

Cholesterol, which is synthesized in the liver, is the starting material for synthesis of other steroids in the body.

(a) Cholesterol

(b) Estradiol (an estrogen or female sex hormone)

(c) Testosterone (a male sex hormone)

(d) Cortisol

How is the structure of estradiol different from that of testosterone?

Prostaglandins have a wide variety of functions. They modify responses to hormones, contribute to the inflammatory response (Chapter 22), prevent stomach ulcers, dilate (enlarge) airways to the lungs, regulate body temperature, and influence formation of blood clots, to name just a few effects. Leukotrienes participate in allergic and inflammatory responses.

Body lipids also include fatty acids (which can undergo either hydrolysis to provide ATP or dehydration synthesis to build triglycerides and phospholipids); fat-soluble vitamins such as beta-carotenes (the yellow-orange pigments in egg yolk, carrots, and tomatoes that are converted to vitamin A); vitamins D, E, and K; and lipoproteins.

Proteins

Proteins are large molecules that contain carbon, hydrogen, oxygen, and nitrogen. Some proteins also contain sulfur. A normal, lean adult body is 12 – 18% protein. Much more complex in structure than carbohydrates or lipids, proteins have many roles in the body and are largely responsible for the structure of body tissues. Enzymes are proteins that speed up most biochemical reactions. Other proteins work as "motors" to drive muscle contraction. Antibodies are proteins that defend against invading microbes. Some hormones that regulate homeostasis also are proteins. Table 2.8 describes several important functions of proteins.

Amino Acids and Polypeptides

The monomers of proteins are **amino acids** (a-MĒ-nō). Each of the 20 different amino acids has three important functional

Table 2.8	Functions of Proteins		
Type of Protein	**Functions**	**Type of Protein**	**Functions**
Structural	Form structural framework of various parts of the body. *Examples:* collagen in bone and other connective tissues, and keratin in skin, hair, and fingernails.	**Immunological**	Aid responses that protect body against foreign substances and invading pathogens. *Examples:* antibodies and interleukins.
Regulatory	Function as hormones that regulate various physiological processes; control growth and development; as neurotransmitters, mediate responses of the nervous system. *Examples:* the hormone insulin, which regulates blood glucose level, and a neurotransmitter known as substance P, which mediates sensation of pain in the nervous system.	**Transport**	Carry vital substances throughout body. *Example:* hemoglobin, which transports most oxygen and some carbon dioxide in the blood.
		Catalytic	Act as enzymes that regulate biochemical reactions. *Examples:* salivary amylase, sucrase, and ATPase.
Contractile	Allow shortening of muscle cells, which produces movement. *Examples:* myosin and actin.		

groups attached to a central carbon atom (Figure 2.20a): (1) an amino group ($-NH_2$), (2) an acidic carboxyl group ($-COOH$), and (3) a side chain (R group). At the normal pH of body fluids, both the amino group and the carboxyl group are ionized (Figure 2.20b). The distinctive side chain gives each amino acid its individual chemical identity (Figure 2.20c).

Synthesis of a protein takes place in stepwise fashion—one amino acid is joined to a second, a third is then added to the first two, and so on. The covalent bond joining each pair of amino acids is a **peptide bond.** It always forms between the carbon of the carboxyl group ($-COOH$) of one amino acid and the nitrogen of the amino group ($-NH_2$) of another. At the site of formation of a peptide bond, a molecule of water is removed (Figure 2.21). Thus, this is a dehydration synthesis reaction. Breaking a peptide bond, as occurs during digestion of dietary proteins, is a hydrolysis reaction (Figure 2.21).

When two amino acids combine, a **dipeptide** results. Adding another amino acid to a dipeptide produces a **tripeptide.** Further additions of amino acids result in the formation of a chainlike **peptide** (4–9 amino acids) or **polypeptide** (10–2000 or more amino acids). Small proteins may consist of a single polypeptide chain with as few as 50 amino acids. Other proteins have hundreds or thousands of amino acids and may consist of two or more polypeptide chains folded together.

A great variety of proteins is possible because each variation in the number or sequence of amino acids can produce a different protein. The situation is similar to using an alphabet of 20 letters to form words. Each different amino acid is like a letter, and their various combinations (peptides, polypeptides, or proteins) are like the words.

Levels of Structural Organization in Proteins

Proteins exhibit four levels of structural organization. The **primary structure** is the unique sequence of amino acids that are linked by covalent peptide bonds to form a polypeptide chain

Figure 2.20 Amino acids. (a) In keeping with their name, amino acids have an amino group (shaded blue) and a carboxyl (acid) group (shaded red). The side chain (R group) is different in each amino acid. (b) At pH close to 7, both the amino group and the carboxyl group are ionized. (c) Glycine is the simplest amino acid; the side chain is a single H atom. Cysteine is one of two amino acids that contain sulfur (S). The side chain in tyrosine contains a six-carbon ring. Lysine has a second amino group at the end of its side chain.

🔑 **Body proteins contain 20 different amino acids, each of which has a unique side chain.**

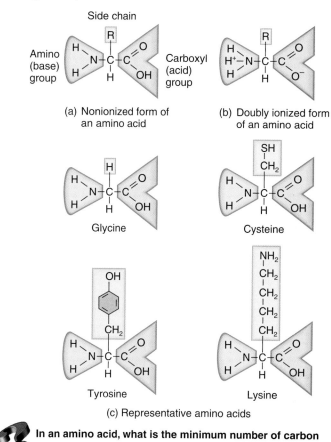

(a) Nonionized form of an amino acid

(b) Doubly ionized form of an amino acid

Glycine

Cysteine

Tyrosine

Lysine

(c) Representative amino acids

❓ **In an amino acid, what is the minimum number of carbon atoms? Of nitrogen atoms?**

Figure 2.21 Formation of a peptide bond between two amino acids during dehydration synthesis. In this example, glycine is joined to alanine, forming a dipeptide (read from left to right). Breaking a peptide bond occurs via hydrolysis (read from right to left).

🔑 **Amino acids are the monomers used to build proteins.**

Glycine   Alanine

Dehydration synthesis

Hydrolysis

Peptide bond

Glycylalanine (a dipeptide)

Water

❓ **What type of reaction takes place during catabolism of proteins?**

(Figure 2.22a). A protein's primary structure is genetically determined. Any changes in a protein's amino acid sequence can have serious consequences for body cells. In sickle-cell disease, for example, a nonpolar amino acid (valine) replaces a polar amino acid (glutamate) at just two locations in the oxygen-carrying protein hemoglobin. This change of amino acids diminishes hemoglobin's water solubility. As a result, the altered hemoglobin tends to form crystals inside red blood cells, producing deformed, sickle-shaped cells that cannot properly squeeze through narrow blood vessels.

The **secondary structure** of a protein is the repeated twisting or folding of neighboring amino acids in the polypeptide chain (Figure 2.22b). Two common secondary structures are *alpha helixes* (clockwise spirals) and *pleated sheets*. The secondary structure of a protein is stabilized by hydrogen bonds, which form at regular intervals along the polypeptide backbone.

The **tertiary structure** (TUR-shē-er′-ē) refers to the three-dimensional shape of a polypeptide chain. Each protein has a unique tertiary structure that determines how it will function. The tertiary folding pattern may allow amino acids at opposite ends of the chain to be close neighbors (Figure 2.22c). Several types of bonds can contribute to a protein's tertiary structure. The strongest but least common bonds are S−S covalent bonds, called *disulfide bridges*. These bonds form between the sulfhydryl groups of two monomers of the amino acid cysteine. Many weak bonds—hydrogen bonds, ionic bonds, and hydrophobic interactions—also help determine the folding pattern. Some parts of a polypeptide are attracted to water (hydrophilic), and other parts are repelled by it (hydrophobic). Because most proteins in our body exist in watery surroundings, the folding process places most amino acids that have hydrophobic side chains in the central core of the protein, away from the protein's surface. Often, helper molecules known as *chaperones* aid the folding process.

When a protein contains more than one polypeptide chain, the arrangement of the individual polypeptide chains relative to one another is the **quaternary structure** (KWA-ter-ner′-ē; Figure 2.22d). The bonds that hold polypeptide chains together are similar to those that maintain the tertiary structure.

Proteins vary tremendously in structure. Different proteins have different architectures and different three-dimensional shapes. This variation in structure and shape is directly related to their diverse functions. In practically every case, the function of a protein depends on its ability to recognize and bind to some other molecule, much as a key fits in a lock. Thus, a hormone binds to a specific protein on a cell whose function it will alter, and an antibody protein binds to a foreign substance (antigen) that has invaded the body. A protein's unique shape permits it to interact with other molecules to carry out a specific function.

Homeostatic mechanisms maintain the temperature and chemical composition of body fluids, which allow body proteins to keep their proper three-dimensional shapes. If a protein encounters an altered environment, it may unravel and lose its characteristic shape (secondary, tertiary, and quaternary struc-

ture). This process is called **denaturation.** Denatured proteins are no longer functional. A common example of permanent denaturation is seen in frying an egg. In a raw egg the soluble egg-white protein (albumin) is a clear, viscous fluid. When heat is applied to the egg, however, the protein denatures, becomes insoluble, and turns white.

Enzymes

In living cells, most catalysts are protein molecules called **enzymes** (EN-zīms). Some enzymes consist of two parts—a protein portion, called the **apoenzyme** (ā′-pō-EN-zīm), and a nonprotein portion, called a **cofactor.** The cofactor may be a metal ion (such as iron, magnesium, zinc, or calcium) or an organic molecule called a *coenzyme.* Coenzymes often are derived from vitamins. The names of enzymes usually end in the suffix *-ase.* All enzymes can be grouped according to the types of chemical reactions they catalyze. For example, *oxidases* add oxygen, *kinases* add phosphate, *dehydrogenases* remove hydrogen, *ATPases* split ATP, *anhydrases* remove water, *proteases* break down proteins, and *lipases* break down triglycerides.

Enzymes catalyze specific reactions. They do so with great efficiency and with many built-in controls. Three important properties of enzymes are as follows

1. *Enzymes are highly specific.* Each particular enzyme binds only to specific **substrates**—the reactant molecules on which the enzyme acts. Of the more than 1000 known enzymes in your body, each has a characteristic three-dimensional shape with a specific surface configuration, which allows it to recognize and bind to certain substrates. In some cases, the part of the enzyme that catalyzes the reaction, called the **active site,** is thought to "fit" the substrate like a key fits in a lock. In other cases the active site changes its shape to fit snugly around the substrate once the substrate enters the active site. This change in shape is known as an *induced fit.*

Not only is an enzyme matched to a particular substrate, it also catalyzes a specific reaction. From among the large number of diverse molecules in a cell, an enzyme must recognize the correct substrate and then take it apart or merge it with another substrate to form one or more specific products.

2. *Enzymes are very efficient.* Under optimal conditions, enzymes can catalyze reactions at rates that are from 100 million to 10 billion times more rapid than those of similar reactions occurring without enzymes. The number of substrate molecules that a single enzyme molecule can convert to product molecules in one second is generally between 1 and 10,000 and can be as high as 600,000.

3. *Enzymes are subject to a variety of cellular controls.* Their rate of synthesis and their concentration at any given time are under the control of a cell's genes. Substances within the cell may either enhance or inhibit the activity of a given enzyme. Many enzymes have both active and inactive forms in cells. The rate at which the inactive form becomes active or vice versa is determined by the chemical environment inside the cell.

Figure 2.22 **Levels of structural organization in proteins.** (a) The primary structure is the sequence of amino acids in the polypeptide. (b) Common secondary structures include alpha helixes and pleated sheets. For simplicity, the amino acid side groups are not shown here. (c) The tertiary structure is the overall folding pattern that produces a distinctive, three-dimensional shape. (d) The quaternary structure in a protein is the arrangement of two or more polypeptide chains relative to one another.

🔑 **The unique shape of each protein permits it to carry out specific functions.**

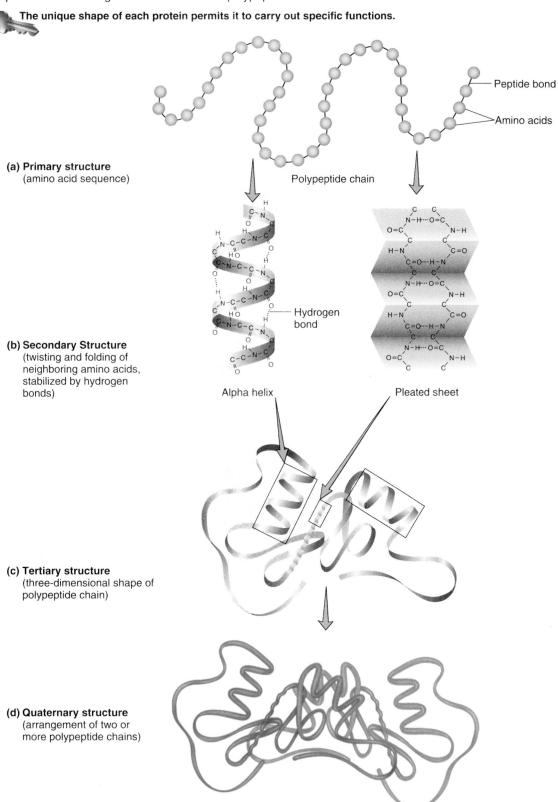

(a) Primary structure
(amino acid sequence)

Peptide bond

Amino acids

Polypeptide chain

(b) Secondary Structure
(twisting and folding of neighboring amino acids, stabilized by hydrogen bonds)

Hydrogen bond

Alpha helix

Pleated sheet

(c) Tertiary structure
(three-dimensional shape of polypeptide chain)

(d) Quaternary structure
(arrangement of two or more polypeptide chains)

❓ **Do all proteins have a quaternary structure?**

Enzymes lower the activation energy of a chemical reaction by decreasing the "randomness" of the collisions between molecules. They also help bring the substrates together in the proper orientation so that the reaction can occur. Figure 2.23 depicts how an enzyme works:

➊ The substrates make contact with the active site on the surface of the enzyme molecule, forming a temporary intermediate compound called the **enzyme-substrate complex.** In this reaction the two substrate molecules are sucrose (a disaccharide) and water.

➋ The substrate molecules are transformed by either the rearrangement of existing atoms, the breakdown of the substrate molecule, or the combination of several substrate molecules into the products of the reaction. Here the products are two monosaccharides: glucose and fructose.

➌ After the reaction is completed and the reaction products move away from the enzyme, the unchanged enzyme is free to attach to other substrate molecules.

Sometimes a single enzyme may catalyze a reversible reaction in either direction, depending on the relative amounts of the substrates and products. For example, the enzyme *carbonic anhydrase* catalyzes the following reversible reaction:

$$CO_2 + H_2O \underset{}{\overset{\textit{Carbonic anhydrase}}{\rightleftharpoons}} H_2CO_3$$

Carbon   Water   Carbonic acid
dioxide

During exercise, when more CO_2 is produced and released into the blood, the reaction flows to the right, increasing the amount of carbonic acid in the blood. Then, as you exhale CO_2, its level in the blood falls and the reaction flows to the left, converting carbonic acid to CO_2 and H_2O.

Galactosemia

Galactosemia (ga-lak-tō-SĒ-mē-a; *galactos-* = milk; *-emia-* = in the blood) is an inherited disorder in which the monosaccharide galactose builds up in the blood. Galactose level increases because the enzyme that converts galactose to glucose (galactose-1-phosphate uridylyl transferase) is absent. An infant with the disorder will fail to thrive within a week after birth due to anorexia (loss of appetite), vomiting, and diarrhea. The main source of galactose in the infant's diet is the disaccharide lactose (milk sugar), which undergoes hydrolysis into glucose and galactose. Hence, treatment of galactosemia consists of eliminating milk from the diet. If treatment is delayed, the infant will remain physically small and become mentally retarded. ■

Nucleic Acids: Deoxyribonucleic Acid (DNA) and Ribonucleic Acid (RNA)

Nucleic acids (nū-KLĒ-ic), so named because they were first discovered in the nuclei of cells, are huge organic molecules that contain carbon, hydrogen, oxygen, nitrogen, and phosphorus. Nucleic acids are of two varieties. The first, **deoxyribonucleic acid (DNA)** (dē-ok′-sē-rī-bō-nū-KLĒ-ik), forms the inherited genetic material inside each cell. In humans, each **gene** is a segment of a DNA molecule. Our genes determine the traits we inherit, and by controlling protein synthesis they regulate most of the activities that take place in body cells throughout a lifetime. When a cell divides, its hereditary information passes on to the next generation of cells. **Ribonucleic acid (RNA),** the second type of nucleic acid, relays instructions from the genes to guide each cell's synthesis of proteins from amino acids.

The monomers of nucleic acids are **nucleotides.** A nucleic acid is a chain of repeating nucleotides. Each nucleotide of DNA consists of three parts (Figure 2.24a):

1. *Nitrogenous base.* DNA contains four different nitrogenous bases, which contain atoms of C, H, O, and N. In DNA the four **nitrogenous bases** are adenine (A), thymine (T), cytosine (C), and guanine (G). Whereas adenine and guanine are larger, double-ring bases called **purines** (PŪR-ēnz), thymine and cytosine are smaller, single-ring bases called **pyrimidines** (pī-RIM-i-dēnz). The nucleotides are named according to the base that is present. For instance, a nucleotide containing thymine is called a thymine nucleotide, one containing adenine is called an adenine nucleotide, and so on.

Figure 2.23   How an enzyme works.

🔑 **An enzyme speeds up a chemical reaction without being altered or consumed.**

H₂O

Substrates
Sucrose and
Water

Enzyme
Sucrase

Active site
of enzyme

➊ Enzyme and substrate come together at active site of enzyme, forming an enzyme–substrate complex

Products
Glucose
Fructose

➌ When reaction is complete, enzyme is unchanged and free to catalyze same reaction again on a new substrate

➋ Enzyme catalyzes reaction and transforms substrate into products

❓ **Why is it that sucrase cannot catalyze the formation of sucrose from glucose and fructose?**

Figure 2.24 DNA molecule. (a) A nucleotide consists of a base, a pentose sugar, and a phosphate group. (b) The paired bases project toward the center of the double helix. The structure is stabilized by hydrogen bonds (dotted lines) between each base pair. There are two hydrogen bonds between adenine and thymine and three between cytosine and guanine.

 Nucleotides are the monomers of nucleic acids.

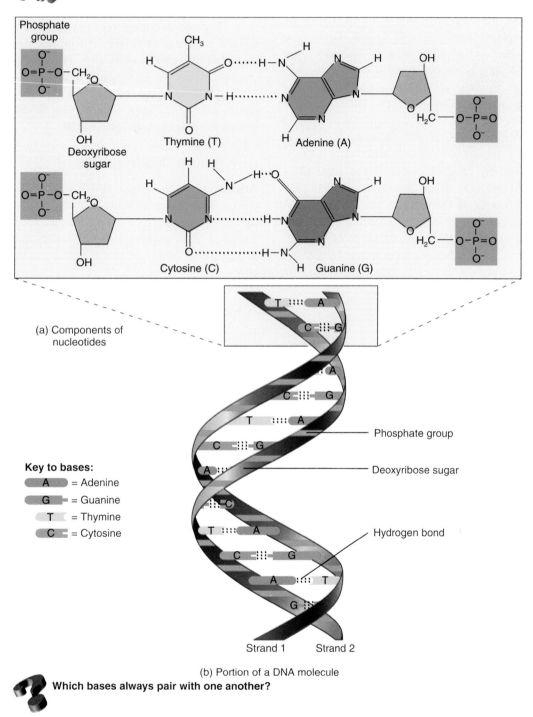

(a) Components of nucleotides

Key to bases:

A = Adenine
G = Guanine
T = Thymine
C = Cytosine

Strand 1   Strand 2

(b) Portion of a DNA molecule

Which bases always pair with one another?

2. Pentose sugar. A five-carbon sugar called **deoxyribose** attaches to each base in DNA.

3. Phosphate group. Alternating phosphate groups (PO_4^{3-}) and pentoses form the "backbone" of a DNA strand; the bases project inward from the backbone chain (Figure 2.24b).

In 1953, F.H.C. Crick of Great Britain and J.D. Watson, a young American scientist, published a brief paper describing how these three components might be arranged in DNA. Their insights into data gathered by others led them to construct a model so elegant and simple that the scientific world immediately knew it was correct! In the Watson–Crick **double helix** model, DNA resembles a spiral ladder (Figure 2.24b). Two strands of alternating phosphate groups and deoxyribose sugars form the uprights of the ladder. Paired bases, held together by hydrogen bonds, form the rungs. Adenine always pairs with thymine, and cytosine always pairs with guanine. If you know the sequence of bases in one strand of DNA, you can predict the sequence on the complementary (second) strand. Each time DNA is copied, as when living cells divide to increase their number, the two strands unwind. Each strand serves as the template or mold on which to construct a new second strand. Any change that occurs in the base sequence of a DNA strand is called a *mutation.* Some mutations can result in the death of a cell, cause cancer, or produce genetic defects in future generations.

RNA, the second variety of nucleic acid, differs from DNA in several respects. In humans, RNA is single-stranded. The sugar in the RNA nucleotide is the pentose **ribose,** and RNA contains the pyrimidine base uracil (U) instead of thymine. Cells contain three different kinds of RNA: messenger RNA, ribosomal RNA, and transfer RNA. Each has a specific role to perform in carrying out the instructions coded in DNA (described on page 87).

DNA Fingerprinting

A technique called DNA fingerprinting is used in research and in courts of law to ascertain whether a person's DNA matches the DNA obtained from samples or pieces of legal evidence such as blood stains or hairs. In each person, certain DNA segments contain base sequences that are repeated several times. Both the number of repeat copies in one region and the number of regions subject to repeat are different from one person to another. DNA fingerprinting can be done with minute quantities of DNA—for example, from a single strand of hair, a drop of semen, or a spot of blood. It also can be used to identify a crime victim or a child's biological parents and even to determine whether two people have a common ancestor. ■

Adenosine Triphosphate

Adenosine triphosphate (a-DEN-ō-sēn) or **ATP** is the "energy currency" of living systems (Figure 2.25). It functions to transfer the energy liberated in exergonic catabolic reactions to cellular activities that require energy (endergonic reactions). Among these cellular activities are muscular contractions, movement of chromosomes during cell division, movement of structures within cells, transport of substances across cell membranes, and

Figure 2.25 Structures of ATP and ADP. "Squiggles" (~) indicate the two phosphate bonds that can be used to transfer energy. Energy transfer typically involves hydrolysis of the last phosphate bond of ATP.

ATP transfers chemical energy to power cellular activities.

 What are some cellular activities that depend on energy supplied by ATP?

synthesis of larger molecules from smaller ones. Structurally, ATP consists of three phosphate groups attached to an adenosine unit composed of adenine and the five-carbon sugar ribose.

When the third phosphate group (PO_4^{3-}), symbolized by Ⓟ in the following discussion, is hydrolyzed by the addition of a water molecule, the overall reaction liberates energy. This energy is used by the cell to power its activities. The enzyme that catalyzes the hydrolysis of ATP is called *ATPase*. Removal of the third phosphate group leaves a molecule called **adenosine diphosphate (ADP)**. This reaction is as follows:

$$\text{ATP} + \text{H}_2\text{O} \xrightarrow{\textit{ATPase}} \text{ADP} + \text{Ⓟ} + \text{E}$$

| Adenosine triphosphate | Water | | Adenosine diphosphate | Phosphate group | Energy |

The energy supplied by the catabolism of ATP into ADP is constantly being used by the cell. As the supply of ATP at any given time is limited, a mechanism exists to replenish it: The enzyme *ATP synthase* catalyzes the addition of a phosphate group to ADP. The reaction is as follows:

$$\text{ADP} + \text{Ⓟ} + \text{E} \xrightarrow{\textit{ATP synthase}} \text{ATP} + \text{H}_2\text{O}$$

| Adenosine diphosphate | Phosphate group | Energy | | Adenosine triphosphate | Water |

As you can see from this reaction, energy is required to produce ATP. The energy needed to attach a phosphate group to

ADP is supplied mainly by the catabolism of glucose in a process called cellular respiration. Cellular respiration has two phases, anaerobic and aerobic:

1. ***Anaerobic phase.*** In the absence of oxygen, glucose is partially broken down by a series of catabolic reactions into pyruvic acid. Each glucose molecule that is converted into a pyruvic acid molecule yields two molecules of ATP.

2. ***Aerobic phase.*** In the presence of oxygen, glucose is completely broken down into carbon dioxide and water. These reactions generate heat and 36 to 38 ATP molecules.

Chapters 10 and 25 cover the details of cellular respiration.

▶ CHECKPOINT

16. How are carbohydrates classified?
17. How are dehydration synthesis and hydrolysis reactions related?
18. What is the importance to the body of triglycerides, phospholipids, steroids, lipoproteins, and eicosanoids?
19. Distinguish among saturated, monounsaturated, and polyunsaturated fats.
20. Define a protein. What is a peptide bond?
21. What are the levels of structural organization in proteins?
22. How do DNA and RNA differ?
23. In the reaction catalyzed by ATP synthase, what are the substrates and products? Is this an exergonic or endergonic reaction?

STUDY OUTLINE

HOW MATTER IS ORGANIZED (p. 27)

1. All forms of matter are composed of chemical elements.
2. Oxygen, carbon, hydrogen, and nitrogen make up about 96% of body mass.
3. Each element is made up of small units called atoms.
4. Atoms consist of a nucleus, which contains protons and neutrons, plus electrons that move about the nucleus in regions called electron shells.
5. The number of protons (the atomic number) distinguishes the atoms of one element from those of another element.
6. The mass number of an atom is the sum of its protons and neutrons.
7. Different atoms of an element that have the same number of protons but different numbers of neutrons are called isotopes. Radioactive isotopes are unstable and decay.
8. The atomic mass of an element is the average mass of all naturally occurring isotopes of that element.
9. An atom that *gives up* or *gains* electrons becomes an ion—an atom that has a positive or negative charge because it has unequal numbers of protons and electrons. Positively charged ions are cations; negatively charged ions are anions.

10. If two atoms share electrons, a molecule is formed. Compounds contain atoms of two or more elements.
11. A free radical is an electrically charged atom or group of atoms with an unpaired electron in its outermost shell. A common example is superoxide, which is formed by the addition of an electron to an oxygen molecule.

CHEMICAL BONDS (p. 31)

1. Forces of attraction called chemical bonds hold atoms together. These bonds result from gaining, losing, or sharing electrons in the valence shell.
2. Most atoms become stable when they have an octet of eight electrons in their valence (outermost) electron shell.
3. When the force of attraction between ions of opposite charge holds them together, an ionic bond has formed.
4. In a covalent bond, atoms share pairs of valence electrons. Covalent bonds may be single, double, or triple and either nonpolar or polar.
5. An atom of hydrogen that forms a polar covalent bond with an oxygen atom or a nitrogen atom may also form a weaker bond,

called a hydrogen bond, with an electronegative atom. The polar covalent bond causes the hydrogen atom to have a partial positive charge (δ^+) that attracts the partial negative charge (δ^-) of neighboring electronegative atoms, often oxygen or nitrogen.

CHEMICAL REACTIONS (p. 34)

1. When atoms combine with or break apart from other atoms, a chemical reaction occurs. The starting substances are the reactants, and the ending ones are the products.
2. Energy, the capacity to do work, is of two principal kinds: potential (stored) energy and kinetic energy (energy of motion).
3. Endergonic reactions require energy, whereas exergonic reactions release energy. ATP couples endergonic and exergonic reactions.
4. The initial energy investment needed to start a reaction is the activation energy. Reactions are more likely when the concentrations and the temperatures of the reacting particles are higher.
5. Catalysts accelerate chemical reactions by lowering the activation energy. Most catalysts in living organisms are protein molecules called enzymes.
6. Synthesis reactions involve the combination of reactants to produce larger molecules. The reactions are anabolic and usually endergonic.
7. In decomposition reactions, a substance is broken down into smaller molecules. The reactions are catabolic and usually exergonic.
8. Exchange reactions involve the replacement of one atom or atoms by another atom or atoms.
9. In reversible reactions, end products can revert to the original reactants.

INORGANIC COMPOUNDS AND SOLUTIONS (p. 37)

1. Inorganic compounds usually are small and usually lack carbon. Organic substances always contain carbon, usually contain hydrogen, and always have covalent bonds.
2. Water is the most abundant substance in the body. It is an excellent solvent and suspending medium, participates in hydrolysis and dehydration synthesis reactions, and serves as a lubricant. Because of its many hydrogen bonds, water molecules are cohesive, which causes a high surface tension. Water also has a high capacity for absorbing heat and a high heat of vaporization.
3. Inorganic acids, bases, and salts dissociate into ions in water. An acid ionizes into anions and hydrogen ions (H^+); a base ionizes into cations and hydroxide ions (OH^-). A salt ionizes into neither H^+ nor OH^-.
4. Mixtures are combinations of elements or compounds that are physically blended together but are not bound by chemical bonds. Solutions, colloids, and suspensions are mixtures with different properties.

5. Two ways to express the concentration of a solution are percentage (mass per volume) and moles per liter. A mole (abbreviated mol) is the amount in grams of any substance that has a mass equal to the combined atomic mass of all its atoms.
6. The pH of body fluids must remain fairly constant for the body to maintain homeostasis. On the pH scale, 7 represents neutrality. Values below 7 indicate acidic solutions, and values above 7 indicate alkaline solutions.
7. Buffer systems usually consist of a weak acid and a weak base. They react with excess H^+ and excess OH^-, which helps maintain pH homeostasis.
8. One important buffer system is the carbonic acid – bicarbonate buffer system. The bicarbonate ion (HCO_3^-) acts as a weak base, and carbonic acid (H_2CO_3) acts as a weak acid.

ORGANIC COMPOUNDS (p. 42)

1. Carbon, with its four valence electrons, bonds covalently with other carbon atoms to form large molecules of many different shapes. Attached to the carbon skeletons of organic molecules are functional groups that confer distinctive chemical properties.
2. Small organic molecules are joined together to form larger molecules by dehydration synthesis reactions in which a molecule of water is removed. In the reverse process, called hydrolysis, large molecules are broken down into smaller ones by adding water.
3. Carbohydrates provide most of the chemical energy needed to generate ATP. They may be monosaccharides, disaccharides, or polysaccharides.
4. Lipids are a diverse group of compounds that include triglycerides (fats and oils), phospholipids, steroids, and eicosanoids. Triglycerides protect, insulate, provide energy, and are stored. Phospholipids are important cell membrane components. Eicosanoids (prostaglandins and leukotrienes) modify hormone responses, contribute to inflammation, dilate airways, and regulate body temperature.
5. Proteins are constructed from amino acids. They give structure to the body, regulate processes, provide protection, help muscles contract, transport substances, and serve as enzymes. Levels of structural organization among proteins are primary, secondary, tertiary, and quaternary.
6. Deoxyribonucleic acid (DNA) and ribonucleic acid (RNA) are nucleic acids consisting of nitrogenous bases, five-carbon (pentose) sugars, and phosphate groups. DNA is a double helix and is the primary chemical in genes. RNA takes part in protein synthesis.
7. Adenosine triphosphate (ATP) is the principal energy-transferring molecule in living systems. When it transfers energy to an endergonic reaction, it is decomposed to adenosine diphosphate (ADP) and Ⓟ. ATP is synthesized from ADP and Ⓟ using the energy supplied by various decomposition reactions, particularly those of glucose.

Q SELF-QUIZ QUESTIONS

Fill in the blanks in the following statements.

1. Atoms consist of _____, _____, and _____.

2. To be chemically stable, the atoms of all elements except hydrogen and helium must hold _____ electrons in the valence shell.

3. The number of protons located in an atom's nucleus is known as its _____ .

4. The factors that influence the chance that a collision will occur and cause a chemical reaction are _____ and _____.

Indicate whether the following statements are true or false.

5. The four elements which compose most of the body's mass are carbon, hydrogen, oxygen, and nitrogen.

6. Ionic bonds occur when atoms lose, gain, and share electrons in their valence shell.

Choose the one best answer to the following questions.

7. Which of the following statements are *true* of covalent bonds? (1) The atoms form a molecule by sharing one, two, or three pairs of their valence electrons. (2) Covalent bonds can form only between atoms of different elements. (3) The greater the number of electron pairs shared, the stronger the covalent bond. (4) Covalent bonds are the least common chemical bonds in the body. (5) Covalent bonds can be either polar or nonpolar. (a) 1, 2, and 3, (b) 1, 3, and 5, (c) 2, 4, and 5, (d) 1, 2, and 5, (e) 1, 3, and 4.

8. Organic compounds that function primarily to provide a readily available source of chemical energy to generate the ATP that drives metabolic reactions are (a) lipids, (b) nucleic acids, (c) water, (d) carbohydrates, (e) proteins.

9. The body's most highly concentrated form of chemical energy is (a) a triglyceride, (b) a lipoprotein, (c) glycogen, (d) a steroid, (e) an eicosanoid.

10. The organic compounds that serve in structural, regulatory, contractile, immunological, transport, and catalytic capacities are (a) lipids, (b) carbohydrates, (c) proteins, (d) nucleic acids, (e) buffers.

11. Which of the following statements is *not true* concerning DNA? (a) It contains the five-carbon sugar deoxyribose. (b) It composed of a single strand of nucelotides. (c) DNA has a double helix construction. (d) DNA contains the nitrogenous bases adenine, guanine, cytosine and thymine. (e) It is the cell's genetic material.

12. Match the following reactions with the term that describes them.
    (1) synthesis reaction   (2) exchange reaction
    (3) decomposition reaction   (4) reversible reaction
    ____ (a) $H_2 + Cl_2 \longrightarrow 2\ HCl$
    ____ (b) $3\ NaOH + H_3PO_4 \longrightarrow Na_3PO_4 + 3\ H_2O$
    ____ (c) $CaCO_3 + CO_2 + H_2O \longrightarrow Ca(HCO_3)_2$
    ____ (d) $NH_3 + H_2O \rightleftharpoons NH_4^+ + OH^-$

13. Which of the following statements regarding ATP are *true?* (1) ATP is the energy currency for the cell. (2) The energy supplied by the hydrolysis of ATP is constantly being used by the cell. (3) Energy is required to produce ATP. (4) The production of ATP involves both aerobic and anaerobic phases. (5) The process of producing energy in the form of ATP is termed the law of conservation of energy. (a) 1, 2, 3, and 4, (b) 1, 2, 3, and 5, (c) 2, 4, and 5, (d) 1, 2, and 4, (e) 3, 4, and 5.

14. Match the following.
    (1) acid   (2) base   (3) buffer   (4) enzyme
    (5) pH   (6) ATP   (7) salt   (8) water
    ____ (a) a polar covalent molecule that serves as a solvent, has a high heat capacity, creates a high surface tension, and serves as a lubricant
    ____ (b) a substance that dissociates into one or more hydrogen ions and one or more anions
    ____ (c) a substance that dissociates into cations and anions, neither of which is a hydrogen ion or a hydroxyl ion
    ____ (d) a proton acceptor
    ____ (e) a measure of hydrogen ion concentration
    ____ (f) a chemical compound that can convert strong acids and bases into weak ones
    ____ (g) a catalyst for chemical reactions that is specific, efficient, and under cellular control
    ____ (h) a compound that functions to temporarily store and then transfer energy liberated in exergonic reactions to cellular activities that require energy

15. Match the following.
    (1) chemical element   (2) matter   (3) atom
    (4) compound   (5) molecule   (6) isotope
    (7) free radical   (8) ion
    ____ (a) an electrically charged atom or group of atoms with an unpaired electron in its outermost shell
    ____ (b) a combination resulting when two or more atoms share electrons
    ____ (c) a building block of matter that cannot be split into a simpler chemical substance by ordinary chemical reactions
    ____ (d) anything that occupies space and has mass
    ____ (e) an atom of an element that differs in the number of neutrons, and therefore mass number, from other atoms of the same element
    ____ (f) the smallest unit of matter that retains the properties and characteristics of an element
    ____ (g) an atom that has given up or gained electrons
    ____ (h) a substance that can be broken down into two or more different elements by ordinary chemical means

CRITICAL THINKING QUESTIONS

1. When Selma joined the "Be Kind to Invertebrates Society" (their slogan is "Have you hugged a bug today?"), she decided to eat a diet free of sugars and fatty acids. How likely is it that Selma will stick to her diet?
   HINT   *Selma's diet makes even bread and water look good.*

2. Ming read an article in the *Daily News* describing a DNA fingerprinting technique being used as evidence in a criminal trial. The article explained that DNA is formed from 20 unique combinations of amino acids that are different in every individual. What should Ming's letter to the editor say?
   HINT   *What do the initials "DNA" stand for?*

3. When asked the pH of blood, math major Sean guessed "the same as water." Kareem told him that he was way off! Explain in terms that Sean will understand.
   HINT   *Express the pH of blood mathematically.*

ANSWERS TO FIGURE QUESTIONS

2.1 In carbon, the first shell contains two electrons and the second shell contains four electrons.

2.2 The four most plentiful elements in living organisms are oxygen, carbon, hydrogen, and nitrogen.

2.3 Antioxidants such as vitamins C and E can inactivate oxygen free radicals.

2.4 A cation is a positively charged ion; an anion is a negatively charged ion.

2.5 An ionic bond involves the *loss* and *gain* of electrons; a covalent bond involves the *sharing* of pairs of electrons.

2.6 The oxygen atom in a water molecule has greater electronegativity than the hydrogen atoms.

2.7 The N atom in ammonia is electronegative. Because it attracts electrons more strongly than do the H atoms, the nitrogen end of ammonia acquires a slight negative charge. H atoms in water molecules (or in other ammonia molecules) can form hydrogen bonds with the nitrogen. Likewise, O atoms in water molecules can form hydrogen bonds with H atoms in ammonia molecules.

2.8 The law of conservation of mass stipulates that the number of hydrogen atoms in the reactants equals the number in the products—in this case, four hydrogen atoms total. Put another way, two molecules of H_2 are needed so that the number of H atoms and O atoms in the reactants is the same as the number of H atoms and O atoms in the products.

2.9 This reaction is exergonic because the reactants have more potential energy than the products.

2.10 No. A catalyst does not change the potential energies of the products and reactants; it only lowers the activation energy needed to get the reaction going.

2.11 Because sugar easily dissolves in a polar solvent (water), you can correctly predict that it has several polar covalent bonds.

2.12 $CaCO_3$ is a salt, and H_2SO_4 is an acid.

2.13 At pH = 6, $[H^+] = 10^{-6}$ mol/liter and $[OH^-] = 10^{-8}$ mol/liter. A pH of 6.82 is more acidic than a pH of 6.91. Both pH = 8.41 and pH = 5.59 are 1.41 pH units from neutral (pH = 7).

2.14 Glucose has five $-OH$ groups and 6 carbon atoms.

2.15 There are 6 carbons in fructose and 12 in sucrose.

2.16 Cells in the liver and in skeletal muscle store glycogen.

2.17 The oxygen in the water molecule comes from a fatty acid.

2.18 The polar head is hydrophilic, and the nonpolar tails are hydrophobic.

2.19 The only differences are the number of double bonds in and the types of functional groups attached to ring A.

2.20 An amino acid has a minimum of two carbon atoms and one nitrogen atom. See the structure of glycine in Figure 2.20c.

2.21 Hydrolysis occurs during catabolism of proteins.

2.22 Proteins consisting of a single polypeptide chain do not have a quaternary structure.

2.23 Sucrase has specificity for the sucrose molecule and thus would not "recognize" glucose and fructose.

2.24 Thymine always pairs with adenine, and cytosine always pairs with guanine.

2.25 A few cellular activities that depend on energy supplied by ATP are muscular contractions, movement of chromosomes, transport of substances across cell membranes, and synthesis (anabolic) reactions.

The Cellular Level of Organization

CELLS AND HOMEOSTASIS

About 200 different types of specialized cells carry out a multitude of functions that help each body system contribute to the homeostasis of the whole body. At the same time, all cells share key structures and functions that support their intense activity.

ENERGIZE YOUR STUDY

FOUNDATIONS CD

Cell Folder

- Explore Animations that overview Membrane Functions, provide details on Transport Mechanisms, or illuminate the process of Protein Synthesis
- Discover the amazing diversity of cellular structures that provide for so many different functions
- Exercises and Concepts and Connections give you opportunities to reinforce your knowledge

www.wiley.com/college/apcentral

INSIGHTS AND EXPLORATIONS

KARTAGENERS SYNDROME

One in every 68,000 Caucasians has a genetic disease called Kartegeners syndrome. In these individuals the cilia and flagella on their cells do not move normally. This means cells with cilia, such as those in the trachea, bronchi, and uterine tubes of females or the sperm cells of males with flagella do not function normally. One of the most common outcomes is infertility. How is it that cilia and flagella work to cause cells to move? Explore this and more in this web-based activity.

About 200 different types of cells are the basic units that compose your body. Each **cell** is a living structural and functional unit that is enclosed by a membrane. All cells arise from existing cells by the process of **cell division,** in which one "parent" cell divides into two new "daughter" cells. In your body, different types of cells fulfill unique roles that support homeostasis and contribute to the many functional capabilities of the human organism. **Cell biology** is the study of cellular structure and function. As you study the various parts of a cell and their relationships to each other, you will learn that cell structure and function are intimately related. Cells carry out a dazzling array of chemical reactions to create and maintain life processes. They do so, in part, by isolating specific types of chemical reactions within specialized structures inside the cell. In this chapter, you will learn about cellular structures and how they carry out their functions.

PARTS OF A CELL

▶ O B J E C T I V E

• **Name and describe the three main parts of a cell.**

Figure 3.1 provides an overview of the typical structures found in body cells. Whereas most cells have many of the structures shown in this diagram, no one cell has all of them. For ease of study, we can divide a cell into three main parts: plasma membrane, cytoplasm, and nucleus.

• The **plasma membrane** forms the cell's flexible outer surface, separating the cell's internal environment from the external environment outside the cell. It is a selective barrier that regulates the flow of materials into and out of a cell. This selectivity helps establish and maintain the appropriate environment for normal cellular activities. The plasma membrane also plays a key role in communication, both among cells and between cells and their external environment.

• The **cytoplasm** (SĪ-tō-plasm; -*plasm* = formed or molded) consists of all the cellular contents between the plasma

Figure 3.1 Typical structures found in body cells.

The cell is the basic living, structural and functional unit of the body.

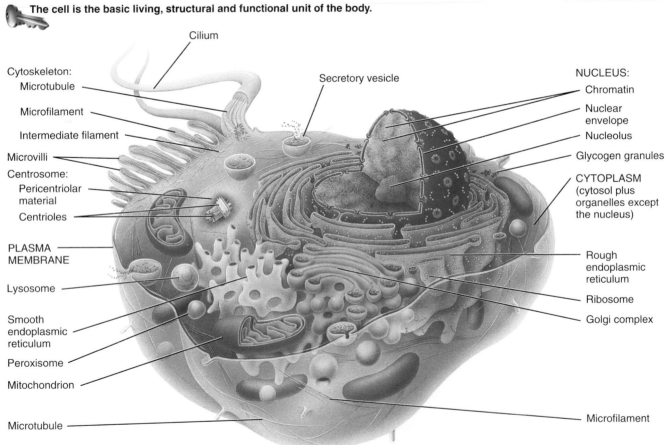

Cilium

Cytoskeleton:
    Microtubule
    Microfilament
    Intermediate filament
Microvilli
Centrosome:
    Pericentriolar material
    Centrioles

PLASMA MEMBRANE

Lysosome

Smooth endoplasmic reticulum

Peroxisome

Mitochondrion

Microtubule

Secretory vesicle

NUCLEUS:
    Chromatin
    Nuclear envelope
    Nucleolus
Glycogen granules
CYTOPLASM (cytosol plus organelles except the nucleus)

Rough endoplasmic reticulum

Ribosome

Golgi complex

Microfilament

Sectional view

 What are the three principal parts of a cell?

60

membrane and the nucleus. This compartment has two components: cytosol and organelles. **Cytosol** (SĪ-tō-sol) is the fluid portion of cytoplasm. It contains water, dissolved solutes, and suspended particles. Surrounded by cytosol are several different types of **organelles** (or-ga-NELZ = little organs). Each type of organelle has a characteristic shape and specific functions. Examples are the cytoskeleton, ribosomes, endoplasmic reticulum, Golgi complex, lysosomes, peroxisomes, and mitochondria.

- The **nucleus** (NOO-klē-us = nut kernel) is a large organelle that houses most of a cell's DNA. Within the nucleus are the **chromosomes** (*chromo-* = colored), each of which consists of a single molecule of DNA associated with several proteins. A chromosome contains thousands of hereditary units called **genes** that control most aspects of cellular structure and function.

THE PLASMA MEMBRANE

OBJECTIVE

- **Describe the structure and functions of the plasma membrane.**

The **plasma membrane** is a flexible yet sturdy barrier that surrounds and contains the cytoplasm of a cell. The *fluid mosaic model* describes its structure. According to this model, the molecular arrangement of the plasma membrane resembles an ever-moving sea of fluid lipids that contains a mosaic of many different proteins (Figure 3.2). Some proteins float freely like icebergs in the lipid sea whereas others are anchored at specific locations like boats at a dock. The membrane lipids allow passage of several types of lipid-soluble molecules but act as a barrier to the entry or exit of charged or polar substances. By contrast, some of the proteins in the plasma membrane allow movement of polar molecules and ions into and out of the cell.

The Lipid Bilayer

The basic structural framework of the plasma membrane is the **lipid bilayer.** It consists of two back-to-back layers made up of three types of lipid molecules—phospholipids, cholesterol, and glycolipids (Figure 3.2). About 75% of the membrane lipids are **phospholipids,** lipids that contain phosphate groups. Present in smaller amounts are **cholesterol,** a steroid with an attached −OH (hydroxyl) group, and various **glycolipids,** lipids with attached carbohydrate groups.

The bilayer arrangement occurs because the lipids are **amphipathic** (am-fē-PATH-ik) molecules, which means that they have both polar and nonpolar parts. In phospholipids (see Figure 2.18 on page 47), the polar part is the phosphate-containing

Figure 3.2 **The fluid mosaic arrangement of lipids and proteins in the plasma membrane.**

Membranes are fluid structures because the lipids and many of the proteins are free to rotate and move sideways in their own half of the bilayer.

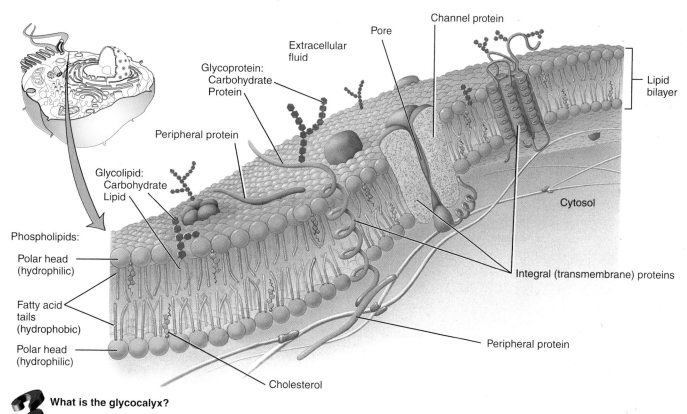

What is the glycocalyx?

"head," which is *hydrophilic* (*hydro-* = water; *-philic* = loving). The nonpolar parts are the two long fatty acid "tails," which are *hydrophobic* (*-phobic* = fearing) hydrocarbon chains. Because "like seeks like," the phospholipid molecules orient themselves in the bilayer with their hydrophilic heads facing outward. In this way, the heads face a watery fluid on either side—cytosol on the inside and extracellular fluid on the outside. The hydrophobic fatty acid tails in each half of the bilayer point toward the center of the membrane, forming a nonpolar, hydrophobic region in the membrane's interior.

Glycolipids account for about 5% of membrane lipids. Their carbohydrate groups form a polar "head," whereas their fatty acid "tails" are nonpolar. Glycolipids appear only in the membrane layer that faces the extracellular fluid, which is one reason the bilayer is asymmetric, or different, on its two sides. The remaining 20% of plasma membrane lipids are weakly amphipathic cholesterol molecules (see Figure 2.19a on page 48), which are interspersed among the other lipids in both layers of the membrane. The tiny −OH group is the only polar region of cholesterol, and it forms hydrogen bonds with the polar heads of phospholipids and glycolipids. The stiff steroid rings and hydrocarbon tail of cholesterol are nonpolar; they fit among the fatty acid tails of the phospholipids and glycolipids.

Arrangement of Membrane Proteins

Membrane proteins are divided into two categories—integral and peripheral—according to whether or not they are firmly embedded in the membrane (Figure 3.2). **Integral proteins** extend into or through the lipid bilayer among the fatty acid tails and can be removed only by methods that disrupt membrane structure. Most integral proteins are **transmembrane proteins,** which span the entire lipid bilayer and protrude into both the cytosol and extracellular fluid. A few integral proteins are tightly attached to one side of the bilayer by covalent bonding to fatty acids in the lipid bilayer. **Peripheral proteins,** by contrast, associate more loosely with the polar heads of membrane lipids or with integral proteins at the inner or outer surface of the membrane. In addition, peripheral proteins can be removed without disrupting membrane integrity. Like membrane lipids, integral membrane proteins are amphipathic. Their hydrophilic regions protrude into either the watery extracellular fluid or the cytosol, and their hydrophobic regions extend among the fatty acid tails.

Many membrane proteins are **glycoproteins,** proteins with carbohydrate groups attached to the ends that protrude into the extracellular fluid. The carbohydrates are *oligosaccharides* (*oligo-* = few; *saccharides* = sugars), chains of 2 to 60 monosaccharides that may be straight or branched. The carbohydrate portions of glycolipids and glycoproteins form an extensive sugary coat called the **glycocalyx** (glī-kō-KAL-iks). The composition of the glycocalyx acts like a molecular "signature" that enables cells to recognize one another. For example, a white blood cell's ability to detect a "foreign" glycocalyx is one basis

of the immune response that helps us destroy invading organisms. In addition, the glycocalyx enables cells to adhere to one another in some tissues, and it protects cells from being digested by enzymes in the extracellular fluid. The hydrophilic properties of the glycocalyx attract a film of fluid to the surface of many cells. This action makes red blood cells slippery as they flow through narrow blood vessels and protects cells that line the airways and the gastrointestinal tract from drying out.

Functions of Membrane Proteins

Generally, the types of lipids in cellular membranes vary only slightly from one membrane to another. In contrast, the membranes of different cells and various intracellular organelles have remarkably different assortments of proteins. Their proteins determine many of the functions that membranes can perform (Figure 3.3). Some integral membrane proteins are **ion channels** that have a *pore* or hole through which specific ions, such as potassium ions (K^+), can flow into or out of the cell. Most ion channels are *selective;* they allow only a single type of ion to pass through. Other membrane proteins act as **transporters,** which selectively move a polar substance from one side of the membrane to the other. Integral proteins called **receptors** serve as cellular recognition sites. Each type of receptor recognizes and binds a specific type of molecule. For instance, insulin receptors bind the hormone insulin. A specific molecule that binds to a receptor is called a **ligand** (LĪ-gand; *liga* = tied) of that receptor. Some integral and peripheral proteins are **enzymes** that catalyze specific chemical reactions at the inside or outside surface of the cell. Membrane glycoproteins and glycolipids often are **cell-identity markers.** They may enable a cell to recognize other cells of the same kind during tissue formation or to recognize and respond to potentially dangerous foreign cells. The ABO blood type markers are one example of cell identity markers. When you receive a blood transfusion, the blood type must be compatible with your own. Integral and peripheral proteins may serve as **linkers,** which anchor proteins in the plasma membranes of neighboring cells to one another or to protein filaments inside and outside the cell.

Membrane Fluidity

Membranes are fluid structures, rather like cooking oil, because most of the membrane lipids and many of the membrane proteins easily rotate and move sideways in their own half of the bilayer. Neighboring lipid molecules exchange places about 10 million times per second and may wander completely around a cell in only a few minutes! Membrane fluidity depends both on the number of double bonds in the fatty acid tails of the lipids that make up the bilayer and on the amount of cholesterol present. Each double bond puts a "kink" in the fatty acid tail (see Figure 2.18 on page 47), which increases membrane fluidity because it prevents lipid molecules from packing tightly in the

Figure 3.3 Functions of membrane proteins.

 Membrane proteins largely reflect the functions a cell can perform.

☐ Extracellular fluid   ▦ Plasma membrane   ☐ Cytosol

Ion channel
Allows specific ion (○) to move through water-filled pore. Most plasma membranes include specific channels for several common ions.

Pore

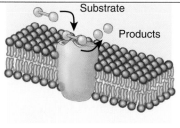

Transporter
Transports specific substances (○) across membrane by changing shape. For example, amino acids, needed to synthesize new proteins, enter body cells via transporters.

Receptor
Ligand
Recognizes specific ligand (▽) and alters cell's function in some way. For example, antidiuretic hormone binds to receptors in the kidneys and changes the water permeability of certain plasma membranes.

Substrate
Enzyme
Products
Catalyzes reaction inside or outside cell (depending on which direction the active site faces). For example, lactase protruding from epithelial cells lining your small intestine splits the disaccharide lactose in the milk you drink.

MHC protein
Cell Identity Marker
Distinguishes your cells from anyone else's (unless you are an identical twin). An important class of such markers are the major histocompatability (MHC) proteins.

Linker
Anchors filaments inside and outside to the plasma membrane, providing structural stability and shape for the cell. May also participate in movement of the cell or link two cells together.

? When stimulating a cell, the hormone insulin first binds to a protein in the plasma membrane. This action best represents which membrane protein function?

membrane. Due to the fluidity of membrane lipids, the lipid bilayer self-seals if it is torn or punctured. When a needle is pushed through a plasma membrane and pulled out, the puncture site seals spontaneously, and the cell does not burst. This property of lipid bilayers allows scientists to fertilize an oocyte by injecting a sperm cell through a tiny syringe. It also permits removal and replacement of a cell's nucleus in cloning experiments, such as was done to create the cloned sheep named Dolly.

Despite the great mobility of membrane lipids and proteins in their own half of the bilayer, they seldom flip-flop from one half of the bilayer to the other. Flip-flopping is rare because it is difficult for hydrophilic parts of membrane molecules to pass through the hydrophobic core of the membrane. Thus, the halves of the membrane bilayer remain asymmetric. The outside half contains some lipids and proteins that are different from those in the inside half.

Because of the way it forms hydrogen bonds with neighboring phospholipid and glycolipid heads and fills the space between bent fatty acid tails, cholesterol makes the lipid bilayer stronger but less fluid at normal body temperature. At low temperatures, cholesterol has the opposite effect—it increases membrane fluidity. This effect of cholesterol is significant in cold-blooded animals but is not important in warm-blooded humans. In atherosclerosis (one form of arteriosclerosis), commonly called "hardening of the arteries," smooth muscle cells and cholesterol build up to form atherosclerotic plaques inside blood vessels. The added cholesterol stiffens the plaque, thus contributing to the decreased flexibility of blood vessels that characterizes this disease.

Membrane Permeability

A membrane is said to be *permeable* to substances that can pass through it and *impermeable* to those that cannot. Although plasma membranes are not completely permeable to any substance, they do permit some substances to pass more readily than others. This property of membranes is termed **selective permeability.**

The lipid bilayer portion of the membrane is permeable to nonpolar, uncharged molecules, such as oxygen, carbon dioxide, and steroids, but is impermeable to ions and charged or polar molecules, such as glucose. It is also slightly permeable to water and urea, a waste product from the breakdown of amino acids. The slight permeability to water and urea is an unexpected property given that water and urea are polar molecules. These two small molecules are thought to pass through the lipid bilayer in the following way. As the fatty acid tails of membrane phospholipids and glycolipids randomly move about, small gaps briefly appear in the hydrophobic environment of the membrane's interior. Water and urea molecules are small enough to move from one gap to another until they have crossed the membrane.

Transmembrane proteins that act as channels and transporters increase the plasma membrane's permeability to a variety of small- and medium-sized polar and charged substances (including ions) that cannot cross the lipid bilayer. Channels and transporters are very selective. Each one helps only a specific molecule or ion to cross the membrane. Macromolecules, such as proteins, are unable to pass across the plasma membrane except by endocytosis and exocytosis (discussed later in this chapter).

Gradients Across the Plasma Membrane

Because the plasma membrane of a cell is selectively permeable, it admits some substances into the cytosol while excluding others. This property allows a living cell to maintain different concentrations of selected substances on either side of the plasma membrane. A **concentration gradient** is a difference in the concentration of a chemical from one place to another, such as from the inside to the outside of the plasma membrane. Many ions and molecules are more concentrated in either the cytosol or the extracellular fluid. For instance, oxygen molecules and sodium ions (Na^+) are more concentrated in the extracellular fluid than in the cytosol, whereas the opposite is true for carbon dioxide molecules and potassium ions (K^+) (Figure 3.4a).

The plasma membrane also creates a difference in the distribution of positively and negatively charged ions between one side of the plasma membrane and the other. Typically, the inner surface of the plasma membrane is more negatively charged and the outer surface is more positively charged (Figure 3.4b). A difference in electrical charges between two regions constitutes an *electrical potential*. Because it occurs across the plasma membrane, this charge difference is termed the *membrane potential*.

As you will see shortly, the concentration gradient and membrane potential are important because they help move substances across the plasma membrane. In many cases a substance will move across a plasma membrane *down its concentration gradient*. That is to say, a substance will move "downhill," from where it is more concentrated to where it is less concentrated. Similarly, a positively charged substance will tend to move toward a negatively charged area, and a negatively charged substance will tend to move toward a positively charged area. Because the concentration gradient and the membrane potential both affect ion movements, the combined influence is termed an ion's **electrochemical gradient.**

▶ C H E C K P O I N T

1. How do hydrophobic and hydrophilic regions govern the arrangement of membrane lipids in a bilayer?

2. What substances can and cannot diffuse through the lipid bilayer?

3. Defend this statement: The proteins present in a plasma membrane determine the functions that a membrane can perform.

4. How does cholesterol affect membrane fluidity?

5. Why are membranes said to have selective permeability?

6. What factors contribute to an electrochemical gradient?

Figure 3.4   Gradients across the plasma membrane. (a) Sodium ions and oxygen molecules are more concentrated in the extracellular fluid, whereas potassium ions and carbon dioxide are more concentrated in cytosol. (b) Because the inner surface of the plasma membrane of most cells is negative relative to the outer surface, an electrical gradient exists across the membrane.

🗝 **Each substance moves across the membrane down its own electrochemical gradient.**

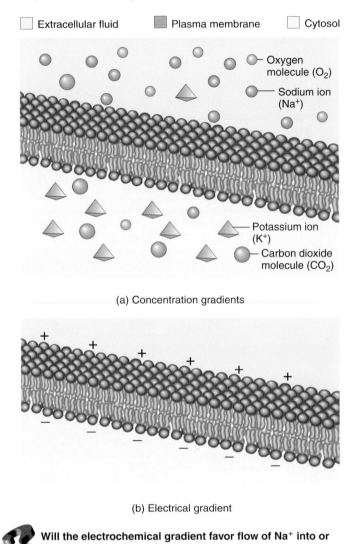

☐ Extracellular fluid   ▨ Plasma membrane   ☐ Cytosol

— Oxygen molecule (O_2)
— Sodium ion (Na^+)
— Potassium ion (K^+)
— Carbon dioxide molecule (CO_2)

(a) Concentration gradients

(b) Electrical gradient

❓ **Will the electrochemical gradient favor flow of Na^+ into or out of cells?**

TRANSPORT ACROSS THE PLASMA MEMBRANE

▶ O B J E C T I V E

• **Describe the processes that transport substances across the plasma membrane.**

Transport of materials across the plasma membrane is essential to the life of a cell. Certain substances must move into the cell to support metabolic reactions. Others that have been produced by the cell for export or cellular waste products must move out. As you have seen, some substances can cross the lipid bilayer part of the plasma membrane, whereas others use ion channels or transporters formed by membrane proteins.

Substances move across cellular membranes via transport processes that are classified as active or passive depending on whether they require cellular energy (Figure 3.5). In *passive transport,* a substance moves down its concentration or electrochemical gradient across the membrane using only its own kinetic energy. The three types of passive transport are diffusion through the lipid bilayer, diffusion through channels, and facilitated diffusion. In *active transport,* cellular energy is used to drive the substance "uphill" against its concentration gradient.

Besides passive and active transport, materials can enter and leave cells by endocytosis and exocytosis. In these processes, the material becomes enclosed within a tiny membrane-surrounded sac called a vesicle. In *endocytosis,* vesicles detach from the plasma membrane while bringing materials into the cell. In *exocytosis,* vesicles merge with the plasma membrane to release materials from the cell. Even large particles, such as whole bacteria and red blood cells, and macromolecules, such as polysaccharides and proteins, may enter cells via endocytosis and leave cells via exocytosis.

Principles of Diffusion

Learning why materials diffuse across membranes requires an understanding of how diffusion occurs in a solution. **Diffusion** is the random mixing of particles that occurs in a solution because of the particles' kinetic energy. Both the *solutes,* the dissolved substances, and the *solvent,* the liquid that does the dissolving, undergo diffusion. If a particular solute is present in high concentration in one area of a solution and in low concentration in another area, solute molecules will diffuse toward the area where they are less concentrated. They move *down their concentration gradient.* After some time, an equilibrium is reached and the particles are evenly distributed throughout the solution. At equilibrium, the particles continue to move about randomly, due to their kinetic energy, without further change in their concentrations.

An example of diffusion occurs if you place a crystal of dye in a water-filled container (Figure 3.6). Just next to the dye, the color is intense because the dye concentration is higher there. At increasing distances, the color is lighter and lighter because the dye concentration is lower. Some time later, at equilibrium, the solution of water and dye has a uniform color. The dye molecules and water molecules have diffused down their concentration gradients until they are evenly mixed in solution.

In the example of dye diffusion, no membrane was involved. Substances may also diffuse through a membrane, if the membrane is permeable to them. Several factors influence the diffusion rate of substances across plasma membranes:

1. ***Steepness of the concentration gradient.*** The greater the difference in concentration between the two sides of the membrane, the higher the rate of diffusion. When charged particles are diffusing, it is the steepness of the electrochemical gradient that determines diffusion rate across the membrane.

Figure 3.5 Processes for transport of materials across the plasma membrane.

🔑 In passive transport, a substance moves down its electrochemical gradient, whereas in active transport, cellular energy is used to drive the substance "uphill," against its electrochemical gradient.

- Extracellular fluid
- Plasma membrane
- Cytosol

Concentration gradient

Diffusion through the lipid bilayer | Diffusion through a channel | Facilitated diffusion | ATP → ADP

PASSIVE TRANSPORT | ACTIVE TRANSPORT

❓ In which transport processes do transmembrane proteins take part?

Figure 3.6 Diffusion. A crystal of dye placed in a cylinder of water dissolves (beginning) and then diffuses from the region of higher dye concentration to regions of lower dye concentration (intermediate). At equilibrium, dye concentration is uniform throughout although random movement continues.

🔑 In diffusion, a substance moves down its concentration gradient.

Beginning   Intermediate   Equilibrium

❓ How would having a fever affect body processes that involve diffusion?

2.   *Temperature.* The higher the temperature, the faster the rate of diffusion. In a person who has a fever, all of the body's diffusion processes occur more rapidly.

3.   *Mass of the diffusing substance.* The larger the mass of the diffusing particle, the slower its diffusion rate. Molecules with smaller molecular weights diffuse more rapidly than larger ones.

4.   *Surface area.* The larger the membrane surface area available for diffusion, the faster the diffusion rate. For example, the air sacs of the lungs have a large surface area available for diffusion of oxygen from the air into the blood. Some lung diseases, such as emphysema, reduce the surface area. This slows the rate of diffusion and makes breathing more difficult.

5.   *Diffusion distance.* The greater the distance over which diffusion must occur, the longer it takes. Diffusion across a plasma membrane takes only a fraction of a second because the membrane is so thin. In pneumonia, fluid collects in the lungs and increases diffusion distance, thus slowing diffusion of oxygen into the blood.

We will begin to explore membrane transport by seeing how the solvent water crosses membranes and then continue by looking at the various categories of solutes.

Osmosis

Osmosis (oz-MŌ-sis) is the net movement of a solvent through a selectively permeable membrane. In living systems, the solvent is water, which moves by osmosis across plasma membranes from an area of *higher water concentration* to an area of *lower water concentration*. Another way to understand this idea is to consider the solute concentration: In osmosis, water moves through a selectively permeable membrane from an area of *lower solute concentration* to an area of *higher solute concentration*. During osmosis, water molecules pass through a plasma membrane in two ways: (1) by moving through the lipid bilayer, as previously described, and (2) by moving through **aquaporins** (*aqua-* = water), integral membrane proteins that function as water channels.

Osmosis occurs only when a membrane is permeable to water but is not permeable to certain solutes. A simple experiment can demonstrate osmosis. Consider a U-shaped tube in which a selectively permeable membrane separates the left and right arms of the tube. A volume of pure water is poured into the left arm, and the same volume of a solution containing an impermeable solute is poured into the right arm (Figure 3.7a). Because

Figure 3.7   Principle of osmosis. Water molecules move through the selectively permeable membrane; the solute molecules in the right arm cannot pass through the membrane. (a) As the experiment starts, water molecules move from the left arm into the right arm, down the water concentration gradient. (b) After some time, the volume of water in the left arm has decreased and the volume of solution in the right arm has increased. At equilibrium, net osmosis has stopped. Hydrostatic pressure forces just as many water molecules to move from right to left as osmotic pressure forces water molecules to move from left to right. (c) If pressure is applied to the solution in the right arm, the starting conditions can be restored. This pressure, which stops osmosis, is equal to the osmotic pressure.

🔑 **Osmosis is the movement of water molecules through a selectively permeable membrane.**

(a) Starting conditions   (b) Equilibrium   (c) Restoring starting conditions

❓ **Will the fluid level in the right arm rise until the water concentrations are the same in both arms?**

the *water* concentration is higher on the left and lower on the right, net movement of water molecules—osmosis—occurs from left to right. Note that water is moving down its concentration gradient. At the same time, the membrane prevents diffusion of the solute from the right arm into the left arm. As a result, the volume of water in the left arm decreases, and the volume of solution in the right arm increases (Figure 3.7b).

You might think that osmosis would continue until no water remained on the left side, but this is *not* what happens. In this experiment, the higher the column of solution in the right arm becomes, the more pressure it exerts on its side of the membrane. Pressure exerted in this way by a liquid is known as *hydrostatic pressure* and it forces water molecules to move back into the left arm. Equilibrium is reached when just as many water molecules move from right to left due to the hydrostatic pressure as move from left to right due to osmosis (Figure 3.7b).

A solution containing solute particles that cannot cross the membrane exerts a force, called the **osmotic pressure.** The osmotic pressure of a solution is proportional to the concentration of the solute particles that cannot cross the membrane—the higher the solute concentration, the higher the solution's osmotic pressure. Consider what would happen if a piston were used to apply more pressure to the fluid in the right arm. With enough pressure, the volume of fluid in each arm could be restored to the starting volume, and the concentration of solute in the right arm would be the same as it was in the beginning (Figure 3.7c). The amount of pressure needed to restore the starting condition equals the osmotic pressure. So, osmotic pressure is the pressure needed to stop the movement of water into a solution containing solutes when a membrane permeable only to the water separates the two solutions. Notice that the osmotic pressure of a solution does not produce the movement of water during osmosis. Rather it is the pressure that would *prevent* such water movement.

Normally, the osmotic pressure of the cytosol is the same as the osmotic pressure of the interstitial fluid outside cells. Because the osmotic pressure on both sides of the plasma membrane (which is selectively permeable) is the same, cell volume remains relatively constant. When body cells are placed in a solution having a different osmotic pressure than their cytosol, however, the shape and volume of the cells changes. As water moves by osmosis into or out of the cells, their volume increases or decreases. A solution's **tonicity** (*tonic* = tension) is a measure of the solution's ability to change the volume of cells by altering their water content.

Any solution in which a cell—for example, a red blood cell (RBC)—maintains its normal shape and volume is an **isotonic solution** (*iso-* = same) (Figure 3.8a). The concentrations of solutes that cannot cross the plasma membrane are the same on both sides of the membrane in this solution. For instance, a 0.9% NaCl solution (0.9 grams of sodium chloride in 100 mL of solution), called a *normal (physiological) saline solution,* is isotonic for RBCs. Whereas the RBC plasma membrane permits the water to move back and forth, it behaves as though it is impermeable to Na^+ and Cl^-, the solutes. (Any Na^+ or Cl^- ions that

Figure 3.8 Tonicity and its effects on red blood cells (RBCs). One example of an isotonic solution for RBCs is 0.9% NaCl.

 Cells placed in an isotonic solution maintain their shape because there is no net water movement into or out of the cell.

| Isotonic solution | Hypotonic solution | Hypertonic solution |

(a) Normal RBC shape · (b) RBC undergoes hemolysis · (c) RBC undergoes crenation

 Will a 2% solution of NaCl cause hemolysis or crenation of RBCs? Why?

enter the cell through channels or transporters are immediately moved back out by active transport or other means. These ions thus behave as though they cannot penetrate the membrane.) When RBCs are bathed in 0.9% NaCl, water molecules enter and exit at the same rate, allowing the RBCs to keep their normal shape and volume.

A different situation results if RBCs are placed in a **hypotonic solution** (*hypo-* = less than), a solution that has a *lower* concentration of solutes than the cytosol inside the RBCs (Figure 3.8b). In this case, water molecules enter the cells faster than they leave, causing the RBCs to swell and eventually to burst. The rupture of RBCs in this manner is called **hemolysis** (hē-MOL-i-sis; *hemo-* = blood; *-lysis* = to loosen or split apart). Pure water is very hypotonic and causes rapid hemolysis.

A **hypertonic solution** (*hyper-* = greater than) has a *higher* concentration of solutes than does the cytosol inside RBCs (Figure 3.8c). One example of a hypertonic solution is a 2% NaCl solution. In such a solution, water molecules move out of the cells faster than they enter, causing the cells to shrink. The shrinkage of RBCs in this manner is called **crenation** (kre-NĀ-shun).

Medical Uses of Isotonic, Hypertonic, and Hypotonic Solutions

RBCs and other body cells may be damaged or destroyed if exposed to hypertonic or hypotonic solutions. For this reason, most **intravenous (IV) solutions,** liquids infused into the blood of a vein, are isotonic. Examples are isotonic saline (0.9% NaCl) and D5W, which stands for dextrose 5% in water. Sometimes infusion of a hypertonic solution is useful to treat patients who have *cerebral edema,* excess interstitial fluid in the brain. Infusion of a hypertonic solution, such as mannitol, relieves such fluid overload by causing osmosis of water from interstitial fluid into the blood. Then, the kidneys excrete excess water from the blood into the urine. Hypotonic solutions, either IV or oral, can be used to treat people who are dehydrated. The water in the hypotonic solution moves from blood into interstitial fluid and then into body cells to rehydrate them. The sports drink or water that you consume to "rehydrate" after a workout is hypotonic. ■

Diffusion Through the Lipid Bilayer

Nonpolar, hydrophobic molecules diffuse through the lipid bilayer of the plasma membrane, into and out of cells. Such molecules include oxygen, carbon dioxide, and nitrogen gases; fatty acids, steroids, and fat-soluble vitamins (A, E, D, and K); small alcohols; and ammonia. Two polar molecules—water and urea—also can diffuse through the lipid bilayer. Because the plasma membrane is somewhat permeable to all these substances, they do not contribute to the tonicity of body fluids. Diffusion through the lipid bilayer is important in the movement of oxygen and carbon dioxide between blood and body cells, and between blood and air within the lungs during breathing. It also is the route for absorption of some nutrients and excretion of some wastes by body cells.

Diffusion Through Membrane Channels

Most membrane channels are *ion channels;* they allow passage of small, inorganic ions, which are too hydrophilic to penetrate the nonpolar interior of the lipid bilayer. Each ion can diffuse across the membrane only at sites containing channels that allow that specific ion to pass. In typical plasma membranes, the most numerous ion channels are selective for K^+ (potassium ions) or Cl^- (chloride ions); fewer channels are available for Na^+ (sodium ions) or Ca^{2+} (calcium ions). Diffusion of solutes through channels is generally slower than diffusion through the lipid bilayer because channels occupy a smaller fraction of the membrane's total surface area. Still, diffusion through channels is a very fast process: More than a million potassium ions can flow through a K^+ channel in one second!

A channel is said to be "gated" when part of the channel protein acts as a "plug" or "gate," changing shape in one way to open the pore and in another way to close it (Figure 3.9). Some gated channels randomly alternate between the open and closed positions, whereas others are regulated by chemical or electrical changes inside and outside the cell. When the gates of a channel are open, ions diffuse into or out of cells, down their electrochemical gradients. The plasma membranes of different types of cells may have fewer or more ion channels and thus display different permeabilities to various ions.

Facilitated Diffusion

Several solutes that are too polar or highly charged to diffuse through the lipid bilayer and are too big to diffuse through membrane channels can cross the plasma membrane by **facilitated**

Figure 3.9 Diffusion of K^+ through a gated membrane channel. Most membranes have channels that are selective for potassium ions (K^+), sodium ions (Na^+), calcium ions (Ca^{2+}), and chloride ions (Cl^-).

An ion channel is an integral transmembrane protein that permits specific small, inorganic ions to diffuse through its pore.

☐ Extracellular fluid   ▨ Plasma membrane   ☐ Cytosol

(a) Gate open   (b) Gate closed

How does a gated channel work?

diffusion. This process originally was envisioned as one in which a "carrier" protein bound to a solute and diffused with it across the membrane. The name is unfortunate, however, because the process is not truly diffusion. Instead, a solute binds to a specific transporter on one side of the membrane and is released on the other side after the transporter undergoes a change in shape. For this reason, some experts advocate changing the name to *facilitated transport.* Also, the proteins that mediate facilitated diffusion are no longer termed "carriers." We now know that they don't move across the membrane like a ferry boat carrying passengers across a lake.

Like diffusion, nonetheless, the net result of facilitated diffusion is movement down a concentration gradient. This happens because the solute binds more often to the transporter on the side of the membrane where the solute is more concentrated before being moved to the side where it is less concentrated. Once the concentration is the same on both sides of the membrane, solute molecules bind to the transporter on the cytosolic side and move out to the extracellular fluid as rapidly as they bind to the transporter on the extracellular side and move into the cytosol. Thus, the rate of facilitated diffusion is determined by the steepness of the concentration gradient between the two sides of the membrane.

The number of facilitated diffusion transporters available in a plasma membrane places an upper limit, called the *transport maximum,* on the rate at which facilitated diffusion can occur. Once all the transporters are occupied, the transport maximum is reached, and a further increase in the concentration gradient does not produce faster facilitated diffusion. Thus, the process of facilitated diffusion exhibits *saturation.*

Solutes that can move across plasma membranes by facilitated diffusion include glucose, fructose, galactose, and some vitamins. Glucose enters many body cells by facilitated diffusion, as follows (Figure 3.10):

1 Glucose binds to a glucose transporter protein called GluT at the extracellular surface of the membrane.

2 As the GluT transporter changes shape, glucose passes through the membrane.

3 GluT releases glucose into the cytosol.

After glucose enters a cell by facilitated diffusion, an enzyme called *hexokinase* attaches a phosphate group to produce a different molecule, namely glucose 6-phosphate. This reaction keeps the intracellular concentration of glucose very low, so that the glucose concentration gradient always favors facilitated diffusion of glucose into, not out of, cells.

The selective permeability of the plasma membrane is often regulated to achieve homeostasis. For instance, the hormone insulin promotes the insertion of many copies of a specific type of glucose transporter into the plasma membranes of certain cells. Thus, the effect of insulin is to increase the transport maximum for facilitated diffusion of glucose into cells. With more transporters available, body cells can pick up glucose from the blood more rapidly.

Figure 3.10 Facilitated diffusion of glucose across a plasma membrane. The transporter (GluT) binds to glucose in the extracellular fluid, changes its shape, and releases glucose into the cytosol.

 Facilitated diffusion requires a transporter but does not use ATP.

What factors affect the rate of facilitated diffusion?

Active Transport

Some polar or charged solutes that must enter or leave body cells cannot cross the plasma membrane through any form of passive transport because they would need to move "uphill," *against* their concentration gradient. Such solutes may be able to cross the membrane by **active transport.** Active transport is an energy-requiring process in which transporter proteins move solutes across the membrane against a concentration gradient. Two sources of cellular energy can be used to drive active transport: (1) Energy obtained from hydrolysis of ATP is the source in *primary active transport;* (2) Energy stored in an ionic concentration gradient is the source in *secondary active transport.* Like facilitated diffusion, active transport processes exhibit a transport maximum and saturation. Solutes actively transported across the plasma membrane include several ions, such as Na^+, K^+, H^+ (hydrogen ions), Ca^{2+}, I^- (iodide ions), and Cl^-; amino acids; and monosaccharides. (Note that some of these substances also cross the membrane via channels or facilitated diffusion when the proper channel proteins or transporters are present.)

Primary Active Transport

In **primary active transport,** energy derived from hydrolysis of ATP changes the shape of a transporter protein, which "pumps"

a substance across a plasma membrane against its concentration gradient. Indeed, transporter proteins that carry out primary active transport are often called **pumps.** A typical body cell expends about 40% of the ATP it generates on primary active transport. Chemicals that turn off ATP production—for example, the poison cyanide—are lethal because they shut down active transport in cells throughout the body.

The most prevalent primary active transport mechanism expels sodium ions (Na^+) from cells and brings potassium ions (K^+) in. Because of the specific ions it moves, this transporter is called the **sodium-potassium pump.** Because a part of the sodium-potassium pump acts as an *ATPase,* an enzyme that hydrolyzes ATP, another name for this pump is **Na^+/K^+ ATPase.** All cells have thousands of sodium-potassium pumps in their plasma membranes. These sodium-potassium pumps maintain a low concentration of Na^+ in the cytosol by pumping them into the extracellular fluid against the Na^+ concentration gradient. At the same time, the pumps move K^+ into cells against the K^+ concentration gradient. Because K^+ and Na^+ slowly leak back across the plasma membrane down their electrochemical gradients—through passive transport or secondary active transport—the sodium-potassium pumps must work nonstop to maintain a low concentration of Na^+ and a high concentration of K^+ in the cytosol.

Figure 3.11 depicts the operation of the sodium-potassium pump.

1 Three Na$^+$ in the cytosol bind to the pump protein.

2 Binding of Na$^+$ triggers the hydrolysis of ATP into ADP, a reaction that also attaches a phosphate group (Ⓟ) to the pump protein. This chemical reaction changes the shape of the pump protein, expelling the three Na$^+$ into the extracellular fluid. Now the shape of the pump protein favors binding of two K$^+$ in the extracellular fluid to the pump protein.

3 The binding of K$^+$ triggers release of the phosphate group from the pump protein. This reaction again causes the shape of the pump protein to change.

4 As the pump protein reverts to its original shape, it releases K$^+$ into the cytosol. At this point, the pump is again ready to bind three Na$^+$, and the cycle repeats.

The different concentrations of Na$^+$ and K$^+$ in cytosol and extracellular fluid are crucial for maintaining normal cell volume and also for the ability of some cells to generate electrical signals such as action potentials. The tonicity of a solution is proportional to the concentration of its solute particles that cannot penetrate the membrane. Because sodium ions that diffuse into a cell or enter through secondary active transport are immediately pumped out, it is as if they never entered. In effect, sodium ions behave as if they cannot penetrate the membrane. Thus, sodium ions are an important contributor to the tonicity of the extracellular fluid. A similar condition holds for K$^+$ in the cytosol. By helping to maintain normal tonicity on each side of the plasma membrane, the sodium-potassium pumps ensure that cells neither shrink nor swell due to osmotic movement of water out of or into cells by osmosis.

Secondary Active Transport

In **secondary active transport,** the energy stored in a Na$^+$ or H$^+$ concentration gradient is used to drive other substances across the membrane against their own concentration gradients. Because a Na$^+$ or H$^+$ gradient is established by primary active transport, secondary active transport *indirectly* uses energy obtained from the hydrolysis of ATP.

The sodium-potassium pump maintains a steep concentration gradient of Na$^+$ across the plasma membrane. As a result, the sodium ions have stored or potential energy, just as water behind a dam has potential energy. Accordingly, if there is a route for Na$^+$ to leak back in, some of the stored energy can be converted to kinetic energy (energy of motion) and used to transport other substances *against their concentration gradients.* In essence, secondary active transport proteins harness the energy in the Na$^+$ concentration gradient by providing routes for Na$^+$ to leak into cells. In secondary active transport, a transporter protein simultaneously binds to Na$^+$ and another substance and then changes its shape so that both substances cross the membrane at the same time. These transporters are either *symporters,* which move two substances in the same direction, or *antiporters,* which move two substances in opposite directions across the membrane.

Plasma membranes contain several antiporters and symporters that are powered by the Na$^+$ gradient (Figure 3.12a). For instance, the concentration of calcium ions (Ca^{2+}) is low in the cytosol because Na$^+$/Ca^{2+} antiporters eject calcium ions. Like-

Figure 3.11 The sodium-potassium pump (Na$^+$/K$^+$ ATPase) expels sodium ions (Na$^+$) and brings potassium ions (K$^+$) into the cell.

🔑 Sodium-potassium pumps maintain a low intracellular concentration of sodium ions.

What is the role of ATP in the operation of this pump?

Figure 3.12 Secondary active transport mechanisms. (a) Antiporters carry two substances across the membrane in opposite directions. (b) Symporters carry two substances across the membrane in the same direction.

Secondary active transport mechanisms use the energy stored in an ionic concentration gradient (here, for Na^+). Because primary active transport pumps that hydrolyze ATP maintain the gradient, secondary active transport mechanisms consume ATP indirectly.

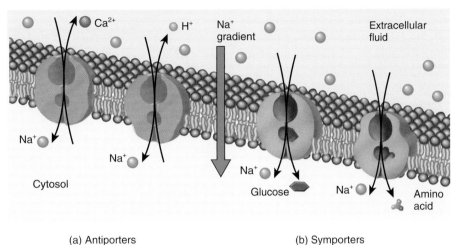

(a) Antiporters (b) Symporters

What is the main difference between primary and secondary active transport mechanisms?

wise, Na^+/H^+ antiporters help regulate the cytosol's pH (H^+ concentration) by expelling excess H^+. By contrast, dietary glucose and amino acids are absorbed into cells that line the small intestine by Na^+-glucose and Na^+-amino acid symporters (Figure 3.12b). In each case, sodium ions are moving down their concentration gradient while the other solutes move "uphill," against their concentration gradients. Keep in mind that all these symporters and antiporters can do their job because the sodium-potassium pumps maintain a low concentration of Na^+ in the cytosol.

Digitalis Increases Ca^{2+} in Heart Muscle Cells

Because it strengthens the heartbeat, digitalis often is given to patients with *heart failure,* a condition of weakened pumping action by the heart. Digitalis exerts its effect by slowing the sodium-potassium pumps, which lets more Na^+ accumulate inside heart muscle cells. The result is a smaller Na^+ concentration gradient across the plasma membrane, which causes the Na^+/Ca^{2+} antiporters to slow down. As a result, more Ca^{2+} remains inside heart muscle cells. The slight increase in the level of Ca^{2+} in the cytosol of heart muscle cells increases the force of their contractions and thus strengthens the heartbeat. ■

Transport in Vesicles

A **vesicle,** as noted earlier, is a small, spherical sac that has budded off from an existing membrane. Vesicles transport a variety of substances from one structure to another within cells, as you will see later in this chapter. Vesicles also import materials from and release materials into extracellular fluid. During **endocytosis** (*endo-* = within), materials move into a cell in a vesicle formed from the plasma membrane. In **exocytosis** (*exo-* = out), materials move out of a cell by the fusion of vesicles formed inside the cell with the plasma membrane. Both endocytosis and exocytosis require energy supplied by ATP.

Endocytosis

Here we consider three types of endocytosis: receptor-mediated endocytosis, phagocytosis, and pinocytosis.

Receptor-mediated endocytosis is a highly selective type of endocytosis by which cells take up specific ligands. (Recall that ligands are molecules that bind to specific receptors.) A vesicle forms after a receptor protein in the plasma membrane recognizes and binds to a particular particle in the extracellular fluid. For instance, cells take up cholesterol contained in low-density lipoproteins (LDLs), transferrin (an iron-transporting protein in the blood), some vitamins, antibodies, and certain hormones by receptor-mediated endocytosis. Receptor-mediated endocytosis of LDLs (and other ligands) occurs as follows (Figure 3.13):

1 *Binding.* On the extracellular side of the plasma membrane, an LDL particle that contains cholesterol binds to a specific receptor in the plasma membrane to form a receptor-LDL complex. The receptors are integral membrane proteins that are concentrated in regions of the plasma membrane, called *clathrin-coated pits.* Here, a protein called clathrin attaches to the membrane on its cytoplasmic side. Many clathrin molecules come together, forming a basketlike structure around the receptor-LDL complexes that causes the membrane to invaginate (fold inward).

2 *Vesicle formation.* The invaginated edges of the membrane around the clathrin-coated pit fuse and a small piece of the membrane pinches off. The resulting vesicle, known as a *clathrin-coated vesicle,* contains the receptor-LDL complexes.

3 *Uncoating.* Almost immediately after it is formed, the clathrin-coated vesicle loses its clathrin coat to become an *uncoated vesicle.* Clathrin molecules either return to the inner surface of the plasma membrane or help form coats on other vesicles inside the cell.

4 *Fusion with endosome.* The uncoated vesicle quickly fuses with a vesicle known as an *endosome.* Within an endosome, the LDL particles separate from their receptors.

5 *Recycling of receptors to plasma membrane.* Most of the receptors accumulate in elongated protrusions of the endosome. These pinch off, forming a *transport vesicle* that contains the receptors and returns them to the plasma membrane. An LDL receptor is returned to the plasma membrane about 10 minutes after it enters a cell.

6 *Degradation in lysosome.* Other transport vesicles, which contain the LDL particles, bud off the endosome and soon fuse with a *lysosome.* Within lysosomes are many digestive enzymes. Certain enzymes break down the large protein and lipid molecules of the LDL particle into amino acids, fatty acids, and cholesterol. These smaller molecules then leave the lysosome. The cell uses cholesterol for rebuilding its membranes and for synthesis of steroids, such as estrogen. Fatty acids and amino acids can be used for ATP production or to build other molecules needed by the cell.

Viruses and Receptor-mediated Endocytosis

Although receptor-mediated endocytosis normally imports needed materials, some viruses are able to use this mechanism to enter and infect body cells. For example, the human immunodeficiency virus (HIV), which causes acquired immunodeficiency syndrome (AIDS), can attach to a receptor called CD4. This receptor is present in the plasma membrane of white blood cells called helper T cells. After binding to CD4, HIV enters the helper T cell via receptor-mediated endocytosis. ■

Figure 3.13 Receptor-mediated endocytosis of a low-density lipoprotein (LDL) particle.

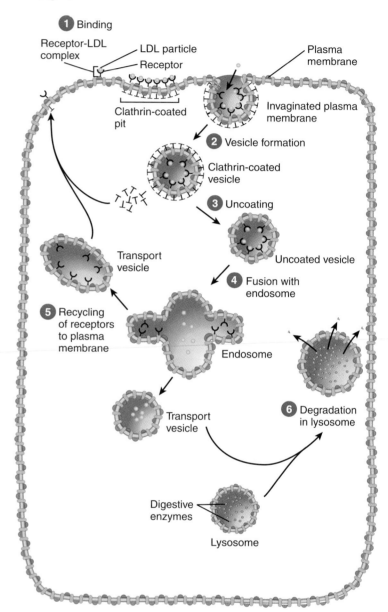

Receptor-mediated endocytosis imports materials that are needed by cells.

? Besides LDL particles, what are several examples of ligands that can undergo receptor-mediated endocytosis?

Phagocytosis (fag′-ō-sī-TŌ-sis; *phago-* = to eat) is a form of endocytosis in which the cell engulfs large solid particles, such as worn-out cells, whole bacteria, or viruses (Figure 3.14). Only a few body cells, termed **phagocytes,** are able to carry out phagocytosis. Two main types of phagocytes are *macrophages,* located in many body tissues, and *neutrophils,* a type of white blood cell. Phagocytosis begins when the particle binds to a plasma membrane receptor, causing the cell to extend **pseudopods** (SOO-dō-pods; *pseudo-* = false; *-pods* = feet), which are

Figure 3.14 Phagocytosis. Pseudopods surround a particle and the membranes fuse to form a phagosome.

🔑 **Phagocytosis is a vital defense mechanism that helps protect the body from disease.**

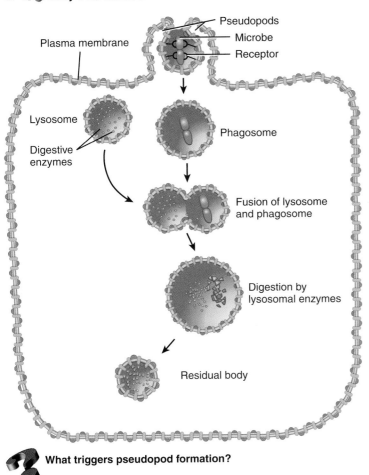

What triggers pseudopod formation?

Figure 3.15 Pinocytosis. The plasma membrane folds inward, forming a pinocytic vesicle.

🔑 **Most body cells carry out pinocytosis, the nonselective uptake of tiny droplets of extracellular fluid.**

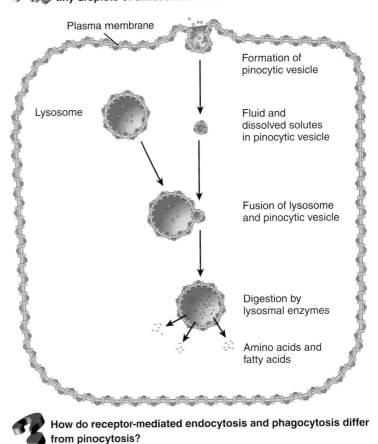

How do receptor-mediated endocytosis and phagocytosis differ from pinocytosis?

projections of its plasma membrane and cytoplasm. Pseudopods surround the particle outside the cell, and the membranes fuse to form a vesicle called a *phagosome,* which enters the cytoplasm. The phagosome fuses with one or more lysosomes, and lysosomal enzymes break down the ingested material. In most cases, any undigested materials in the phagosome remain indefinitely in a vesicle called a *residual body.* The process of phagocytosis is a vital defense mechanism that helps protect the body from disease. Through phagocytosis, macrophages dispose of billions of aged, worn-out red blood cells every day and neutrophils help rid the body of invading microorganisms.

Most body cells carry out **pinocytosis** (pi-nō-sī-TŌ-sis; *pino-* = to drink), a form of endocytosis in which tiny droplets of extracellular fluid are taken up (Figure 3.15). No receptor proteins are involved; all solutes dissolved in the extracellular fluid are brought into the cell. During pinocytosis, the plasma membrane folds inward and forms a *pinocytic vesicle* containing a droplet of extracellular fluid. The pinocytic vesicle detaches or "pinches off" from the plasma membrane and enters the cytosol. Within the cell, the pinocytic vesicle fuses with a lysosome,

where enzymes degrade the engulfed solutes. The resulting smaller molecules, such as amino acids and fatty acids, leave the lysosome to be used elsewhere in the cell.

Exocytosis

In contrast with endocytosis, which brings materials into a cell, **exocytosis** results in *secretion,* the liberation of materials from a cell. All cells carry out exocytosis, but it is especially important in two types of cells: (1) Secretory cells liberate digestive enzymes, hormones, mucus, or other secretions; (2) Nerve cells release substances called *neurotransmitters* via exocytosis (see Figure 12.14 on page 405). During exocytosis, membrane-enclosed vesicles called *secretory vesicles* form inside the cell, fuse with the plasma membrane, and release their contents into the extracellular fluid.

Segments of the plasma membrane lost through endocytosis are recovered or recycled by exocytosis. The balance between endocytosis and exocytosis keeps the surface area of a cell's plasma membrane relatively constant. Membrane exchange is quite extensive in certain cells. In your pancreas, for example, the cells that secrete digestive enzymes can recycle an amount of plasma membrane equal to the cell's entire surface area in 90 minutes.

Transcytosis

In **transcytosis,** vesicles undergo endocytosis on one side of a cell, move across the cell, and then undergo exocytosis on the opposite side. As the vesicles fuse with the plasma membrane, the vesicular contents are released into the extracellular fluid. Transcytosis occurs most often across the endothelial cells that line blood vessels and is a means for materials to move between blood plasma and interstitial fluid. For instance, when a woman is pregnant, some of her antibodies cross the placenta into the fetal circulation via transcytosis.

Table 3.1 summarizes the processes by which materials are transported into and out of cells.

Table 3.1	**Transport of Materials Into and Out of Cells**	
Transport Process	**Description**	**Substances Transported**
Osmosis	Movement of water molecules across a selectively permeable membrane from an area of higher water concentration to an area of lower water concentration.	Solvent: water in living systems.
Diffusion	Random mixing of molecules or ions due to their kinetic energy. A substance diffuses down a concentration gradient until it reaches equilibrium.	
Diffusion through the lipid bilayer	Passive diffusion of a substance through the lipid bilayer of the plasma membrane.	Nonpolar, hydrophobic solutes: oxygen, carbon dioxide, and nitrogen; fatty acids, steroids, and fat-soluble vitamins; glycerol, small alcohols; ammonia. Polar molecules: water and urea.
Diffusion through membrane channels	Passive diffusion of a substance down its electrochemical gradient through channels that span a lipid bilayer; some channels are gated.	Small inorganic solutes, mainly ions: K^+, Cl^-, Na^+, and Ca^{2+}. Water.
Facilitated Diffusion	Passive movement of a substance down its concentration gradient via transmembrane proteins that act as transporters; maximum diffusion rate is limited by number of available transporters.	Polar or charged solutes: glucose, fructose, galactose, and some vitamins.
Active Transport	Transport in which cell expends energy to move a substance across the membrane against its concentration gradient through transmembrane proteins that act as transporters; maximum transport rate is limited by number of available transporters.	Polar or charged solutes.
Primary active transport	Transport of a substance across the membrane against its concentration gradient by pumps; transmembrane proteins that use energy supplied by hydrolysis of ATP.	Na^+, K^+, Ca^{2+}, H^+, I^-, Cl^-, and other ions.
Secondary active transport	Coupled transport of two substances across the membrane using energy supplied by a Na^+ or H^+ concentration gradient maintained by primary active transport pumps. Antiporters move Na^+ (or H^+) and another substance in opposite directions across the membrane; symporters move Na^+ (or H^+) and another substance in the same direction across the membrane.	Antiport: Ca^{2+}, H^+ out of cells. Symport: glucose, amino acids into cells.
Transport In Vesicles	Movement of substances into or out of a cell in vesicles that bud from the plasma membrane; requires energy supplied by ATP.	
Endocytosis	Movement of substances into a cell in vesicles.	
Receptor-mediated endocytosis	Ligand-receptor complexes trigger infolding of a clathrin-coated pit that forms a vesicle containing ligands.	Ligands: transferrin, low-density lipoproteins (LDLs), some vitamins, certain hormones, and antibodies.
Phagocytosis	"Cell eating"; movement of a solid particle into a cell after pseudopods engulf it to form a phagosome.	Bacteria, viruses, and aged or dead cells.
Pinocytosis	"Cell drinking"; movement of extracellular fluid into a cell by infolding of plasma membrane to form a pinocytic vesicle.	Solutes in extracellular fluid.
Exocytosis	Movement of substances out of a cell in secretory vesicles that fuse with the plasma membrane and release their contents into the extracellular fluid.	Neurotransmitters, hormones, and digestive enzymes.

▶ C H E C K P O I N T

7. What is the key difference between passive and active transport?

8. How do symporters and antiporters carry out their functions?

9. What factors can increase the rate of diffusion?

10. What is osmotic pressure?

11. Which substances can diffuse directly through the lipid bilayer?

12. How does diffusion through membrane channels compare to facilitated diffusion?

13. What is the difference between primary and secondary active transport?

14. In what ways are endocytosis and exocytosis similar and different?

CYTOPLASM

▶ O B J E C T I V E

- **Describe the structure and function of cytoplasm, cytosol, and organelles.**

Cytoplasm has two components: (1) the cytosol and (2) a variety of organelles, tiny structures that perform different functions in the cell.

Cytosol

The **cytosol** is the fluid portion of the cytoplasm that surrounds organelles (see Figure 3.1) and constitutes about 55% of total cell volume. Although it varies in composition and consistency from one part of a cell to another, cytosol is 75–90% water plus various dissolved and suspended components. Among these are different types of ions, glucose, amino acids, fatty acids, proteins, lipids, ATP, and waste products. Also present in some cells are various organic molecules that aggregate into masses for storage. These aggregations may appear and disappear at different times in the life of a cell. Examples include *lipid droplets* that contain triglycerides and clusters of glycogen molecules called *glycogen granules* (see Figure 3.1).

The cytosol is the site of many chemical reactions required for a cell's existence. For example, enzymes in cytosol catalyze *glycolysis,* a series of 10 chemical reactions that produces two molecules of ATP from one molecule of glucose. Other types of cytosolic reactions provide the building blocks for maintaining cell structures and for growth.

Organelles

As noted earlier, **organelles** are specialized structures that have characteristic shapes and that perform specific functions in cellular growth, maintenance, and reproduction. Despite the many chemical reactions occurring in a cell at the same time, there is little interference between one type of reaction and another be-

cause they occur in different organelles. Each type of organelle has its own set of enzymes that carry out specific reactions, and each is a functional compartment where specific biochemical processes take place. The numbers and types of organelles vary in different cells, depending on the cell's function. Moreover, organelles often cooperate with each other to maintain homeostasis. Although the nucleus is a large organelle, it is discussed separately because of its special importance in directing the life of a cell.

The Cytoskeleton

The **cytoskeleton** is a network of several different kinds of protein filaments that extend throughout the cytosol (see Figure 3.1). The cytoskeleton provides a structural framework for the cell, serving as a scaffold that helps to determine a cell's shape and to organize the cellular contents. The cytoskeleton also aids movement of organelles within the cell, of chromosomes during cell division, and of whole cells such as phagocytes. Although its name implies rigidity, the cytoskeleton is continually reorganizing as a cell moves and changes shape, as, for example, during cell division. Three types of filamentous proteins contribute to the cytoskeleton. In the order of their increasing diameter, these structures are microfilaments, intermediate filaments, and microtubules.

Microfilaments, the thinnest elements of the cytoskeleton, are composed of the protein *actin* and are most prevalent at the periphery of a cell (Figure 3.16a). Microfilaments have two general functions: helping generate movement and providing mechanical support. With respect to movement, microfilaments are involved in muscle contraction, cell division, and cell locomotion, such as occurs during the migration of embryonic cells during development, the invasion of tissues by white blood cells to fight infection, or the migration of skin cells during wound healing.

Microfilaments provide much of the mechanical support that is responsible for the basic strength and shapes of cells. They anchor the cytoskeleton to integral proteins in the plasma membrane. Microfilaments also provide mechanical support for cell extensions called **microvilli** (*micro-* = small; *-villi* = tufts of hair), which are nonmotile, microscopic fingerlike projections of the plasma membrane. Within each microvillus is a core of parallel microfilaments. Because they greatly increase the surface area of the cell, microvilli are abundant on cells involved in absorption, such as the epithelial cells that line the small intestine.

As their name suggests, **intermediate filaments** are thicker than microfilaments but thinner than microtubules (Figure 3.16b). Several different proteins can compose intermediate filaments, which are exceptionally strong. They are found in parts of cells subject to mechanical stress and also help stabilize the position of organelles such as the nucleus.

The largest of the cytoskeletal components, **microtubules** are long, unbranched hollow tubes composed mainly of the protein *tubulin.* An organelle called the centrosome (discussed

Figure 3.16   Cytoskeleton.

The cytoskeleton is a network of three types of protein filaments that extend throughout the cytoplasm: microfilaments, intermediate filaments, and microtubules.

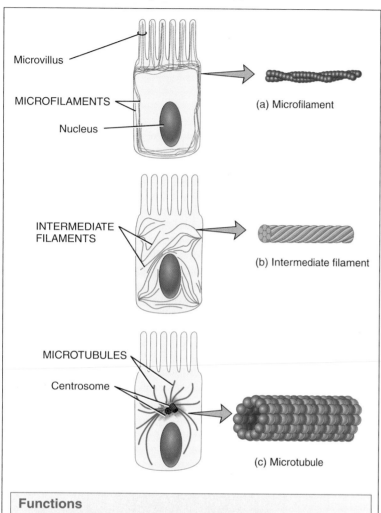

(a) Microfilament

(b) Intermediate filament

(c) Microtubule

Functions

1. **The cytoskeleton serves as a scaffold that helps to determine a cell's shape and to organize the cellular contents.**
2. **The cytoskeleton aids movement of organelles within the cell, of chromosomes during cell division, and of whole cells such as phagocytes.**

 Which cytoskeletal component helps form the structure of centrioles, cilia, and flagella?

shortly) serves as the site where assembly of microtubules begins. The microtubules grow outward from the centrosome toward the periphery of the cell (Figure 3.16c). Microtubules help determine cell shape and function in the movement of organelles, such as secretory vesicles. They also participate in the movement of chromosomes during cell division and in the movement of specialized cell projections, such as cilia and flagella. *Motor proteins* called *kinesins* and *dyneins* are responsible for powering the movements in which microtubules participate.

These proteins act like miniature engines that propel substances and organelles along a microtubule, much as a train engine pushes or pulls cars along a railroad track.

Centrosome

The **centrosome,** located near the nucleus, consists of two components: a pair of centrioles and pericentriolar material (Figure 3.17a). The two **centrioles** are cylindrical structures, each composed of nine clusters of three microtubules (triplets) arranged in a circular pattern (Figure 3.17b). The long axis of one centriole is at a right angle to the long axis of the other (Figure 3.17c). Surrounding the centrioles is **pericentriolar material,** which contains hundreds of ring-shaped complexes composed of the protein *tubulin.* These tubulin complexes are the organizing centers for growth of the mitotic spindle, which plays a critical role in cell division, and for microtubule formation in nondividing cells.

Cilia and Flagella

Microtubules are the dominant components of cilia and flagella, both of which are motile projections of the cell surface. **Cilia** (SIL-ē-a = eyelashes; singular is *cilium*) are numerous, short, hairlike projections that extend from the surface of the cell (see Figure 3.1). Each cilium contains a core of 20 microtubules surrounded by plasma membrane (Figure 3.18a). The microtubules are arranged such that one pair in the center is surrounded by nine clusters of two fused microtubules (doublets). Each cilium is anchored to a *basal body* just below the surface of the plasma membrane. A basal body is similar in structure to a centriole and functions in initiating the assembly of cilia and flagella.

A cilium displays an oarlike pattern of beating in which it is relatively stiff during the power stroke, but relatively flexible during the recovery stroke (Figure 3.18b). The coordinated movement of many cilia on the surface of a cell causes the steady movement of fluid along the cell's surface. Many cells of the respiratory tract, for example, have hundreds of cilia that help sweep foreign particles trapped in mucus away from the lungs. Their movement is paralyzed by nicotine in cigarette smoke. For this reason, smokers cough often to remove foreign particles from their airways. Cells that line the uterine (Fallopian) tubes also have cilia that sweep oocytes (egg cells) toward the uterus.

Flagella (fla-JEL-a = whip; singular is *flagellum*) are similar in structure to cilia but are typically much longer. Flagella usually move an entire cell. A flagellum generates forward motion along its axis by rapidly wiggling in a wavelike pattern (Figure 3.18c). The only example of a flagellum in the human body is a sperm cell's tail, which propels the sperm toward its rendezvous with an oocyte.

Ribosomes

Ribosomes (RĪ-bō-sōms) are the sites of protein synthesis. The name of these tiny organelles reflects their high content of one

Figure 3.17 Centrosome.

Located near the nucleus, the centrosome consists of a pair of centrioles and pericentriolar material.

(a) Details of a centrosome

(b) Arrangement of microtubules in centrosome

(c) Centrioles

TEM 76,000x

Function
The pericentriolar material contains tubulins that build microtubules in nondividing cells and form the mitotic spindle during cell division.

If you observed that a cell did not have a centrosome, what could you predict about its capacity for cell division?

Figure 3.18 Cilia and flagella.

A cilium contains a core of microtubules with one pair in the center surrounded by nine clusters of doublet microtubules.

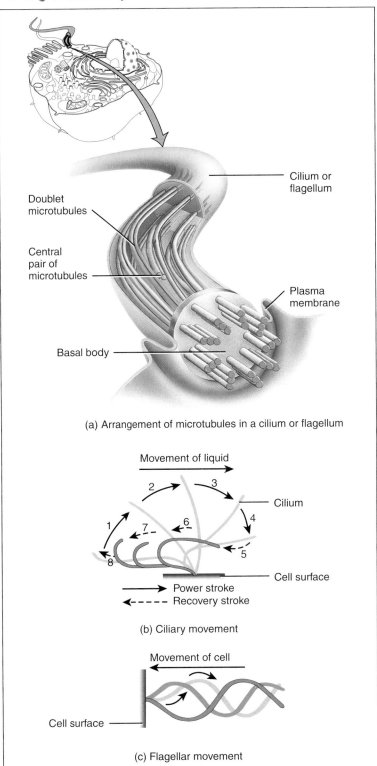

(a) Arrangement of microtubules in a cilium or flagellum

(b) Ciliary movement

(c) Flagellar movement

Functions
1. Cilia move fluids along a cell's surface.
2. A flagellum moves an entire cell.

 What is the functional difference between cilia and flagella?

type of ribonucleic acid, **ribosomal RNA (rRNA),** but each one also includes more than 50 proteins. Structurally, a ribosome consists of two subunits, one about half the size of the other (Figure 3.19). The large and small subunits are made separately in the nucleolus, a spherical body inside the nucleus. Once produced, the large and small subunits exit the nucleus and then come together in the cytoplasm.

Some ribosomes are attached to the outer surface of the nuclear membrane and to an extensively folded membrane called the endoplasmic reticulum. These ribosomes synthesize proteins destined for specific organelles, for insertion in the plasma membrane, or for export from the cell. Other ribosomes are "free" or unattached to other cytoplasmic structures. Free ribosomes synthesize proteins used in the cytosol. Ribosomes are also located within mitochondria, where they synthesize mitochondrial proteins.

Endoplasmic Reticulum

The **endoplasmic reticulum** (en′-dō-PLAS-mik re-TIK-ū-lum; *-plasmic* = cytoplasm; *reticulum* = network) or **ER** is an extensive network of folded membranes that extends from the nuclear envelope (the membrane around the nucleus), to which it is connected, throughout the cytoplasm (Figure 3.20). ER membranes are shaped like flattened sacs or tubules. The ER is so extensive that it constitutes more than half of the membranous surfaces within most cells. Two distinct, but interrelated, forms of ER exist in cells. Ribosomes adhere to the surface of *rough* ER whereas *smooth* ER looks "smooth" because it lacks ribosomes.

Figure 3.20 Endoplasmic reticulum.

 The endoplasmic reticulum is a network of membrane-enclosed sacs or tubules that extend throughout the cytoplasm and connect to the nuclear envelope.

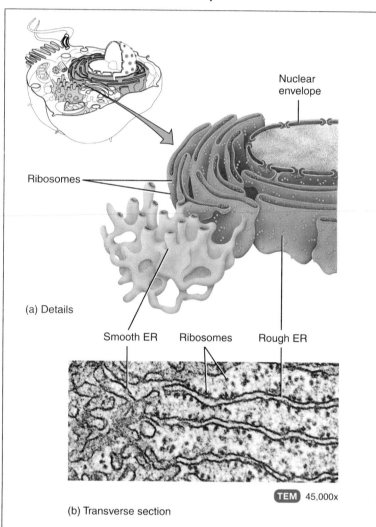

Nuclear envelope

Ribosomes

(a) Details

Smooth ER   Ribosomes   Rough ER

TEM 45,000x

(b) Transverse section

Functions
1. Rough ER synthesizes glycoproteins and phospholipids that are transferred into cellular organelles, inserted into the plasma membrane, or secreted during exocytosis.
2. Smooth ER synthesizes fatty acids and steroids, such as estrogens and testosterone; inactivates or detoxifies drugs and other potentially harmful substances; removes the phosphate group from glucose 6-phosphate; and stores and releases calcium ions that trigger contraction in muscle cells.

Figure 3.19 Ribosomes.

 Ribosomes are the sites of protein synthesis.

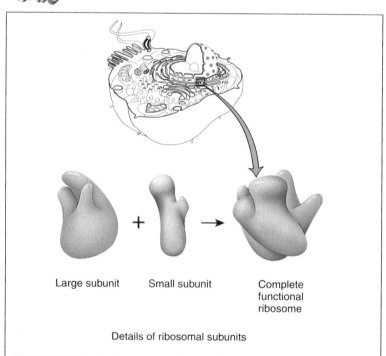

Large subunit   Small subunit   Complete functional ribosome

Details of ribosomal subunits

Functions
1. Ribosomes associated with endoplasmic reticulum synthesize proteins destined for insertion in the plasma membrane or secretion from the cell.
2. Free ribosomes synthesize proteins used in the cytosol.

 Where are subunits of ribosomes synthesized and assembled?

 What are the structural and functional differences between rough and smooth ER?

Proteins synthesized by ribosomes attached to **rough ER** enter the space within the ER for processing and sorting. In some cases, enzymes attach carbohydrates to the proteins to form glycoproteins. In other cases, enzymes attach the proteins to phospholipids, also synthesized by the rough ER. These molecules may be incorporated into organelle membranes, inserted into the plasma membrane, or secreted via exocytosis.

Smooth ER extends from the rough ER to form a network of membranous tubules (Figure 3.20). Synthesis of fatty acids and steroids, such as estrogens and testosterone, occurs in the smooth ER. In liver cells, enzymes located in the smooth ER inactivate or detoxify drugs and other potentially harmful substances, for example, alcohol. In liver, kidney, and intestinal cells a smooth ER enzyme removes the phosphate group from glucose 6-phosphate, which allows the "free" glucose to enter the bloodstream. In muscle cells, calcium ions that trigger con-traction are released from the sarcoplasmic reticulum, a form of smooth ER.

Golgi Complex

Most of the proteins synthesized by ribosomes attached to rough ER are ultimately transported to other regions of the cell. The first step in the transport pathway is through an organelle called the **Golgi complex** (GOL-jē). It consists of 3 to 20 **Golgi cisternae** (sis-TER-nē = cavities; singular is *cisterna*). These small, flattened membranous sacs with bulging edges resemble a stack of pita bread (Figure 3.21). The cisternae are often curved, giving the Golgi complex a cuplike shape. Most cells have several Golgi complexes, and Golgi complexes are more extensive in cells that secrete proteins. As you will see, this fact is a clue to the organelle's role in the cell.

Figure 3.21 Golgi complex.

The opposite faces of a Golgi complex differ in size, shape, content, and enzymatic activities.

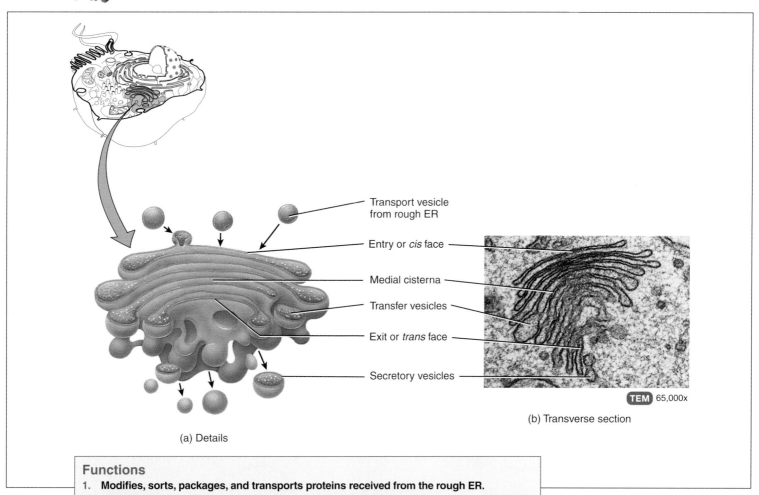

Transport vesicle from rough ER

Entry or *cis* face

Medial cisterna

Transfer vesicles

Exit or *trans* face

Secretory vesicles

TEM 65,000x

(b) Transverse section

(a) Details

Functions
1. Modifies, sorts, packages, and transports proteins received from the rough ER.
2. Forms secretory vesicles that discharge processed proteins via exocytosis into extracellular fluid; forms membrane vesicles that ferry new molecules to the plasma membrane; forms transport vesicles that carry molecules to other organelles, such as lysosomes.

 How do the entry and exit faces differ in function?

The cisternae at the opposite ends of a Golgi complex differ from each other in size, shape, and enzymatic activity. The convex **entry** or *cis* **face** is a cisterna that faces the rough ER. The concave **exit** or *trans* **face** is a cisterna that faces the plasma membrane. Sacs between the entry and exit faces are called **medial cisternae.** Transport vesicles from the ER merge to form the entry face. From the entry face, the cisterna are thought to mature, in turn becoming medial and then exit cisternae.

The entry, medial, and exit regions of the Golgi complex contain different enzymes that permit each to modify, sort, and package proteins for transport to different destinations. The entry face receives and modifies proteins produced by the rough ER. The medial cisternae add carbohydrates to proteins to form glycoproteins and lipids to proteins to form lipoproteins. The exit face modifies the molecules further and then sorts and packages them for transport to their destinations.

Proteins arriving at, passing through, and exiting the Golgi complex do so through maturation of the cisternae and exchanges that occur via transfer vesicles (Figure 3.22):

1 Proteins synthesized by ribosomes on the rough ER are surrounded by a piece of the ER membrane, which eventually buds from the membrane surface in the form of transport vesicles.

2 Transport vesicles move toward the entry face of the Golgi complex.

3 Fusion of several transport vesicles creates the entry face of the Golgi complex and releases proteins into its lumen.

4 The proteins move from the entry face into one or more medial cisternae. Enzymes in the medial cisternae modify the proteins to form glycoproteins, glycolipids, and lipoproteins. Transfer vesicles that bud from the edges of the cisternae move specific enzymes back toward the entry face and also move some partially modified proteins forward toward the exit face.

5 The products of the medial cisternae move into the lumen of the exit face.

6 Within the exit face cisterna, the products are further modified and are sorted and packaged.

7 Some of the processed proteins leave the exit face in **secretory vesicles.** These vesicles deliver the proteins to the plasma membrane, where they are discharged by exocytosis into the extracellular fluid. For example, certain pancreatic cells release the hormone insulin in this way.

8 Other processed proteins leave the exit face in **membrane vesicles** that deliver their contents to the plasma membrane

Figure 3.22   Processing and packaging of proteins by the Golgi complex.

All proteins exported from the cell are processed in the Golgi complex.

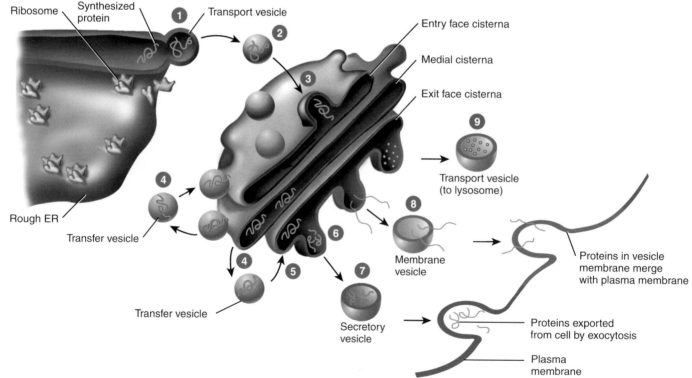

What are the three general destinations for proteins that leave the Golgi complex?

for incorporation into the membrane. In doing so, the Golgi complex adds new segments of plasma membrane as existing segments are lost and modifies the number and distribution of membrane molecules.

9 Finally, some processed proteins leave the exit face in **transport vesicles** that will carry the proteins to another cellular destination. For instance, transport vesicles ferry digestive enzymes to lysosomes, whose structure and functions are discussed next.

Lysosomes

Lysosomes (LĪ-sō-sōms; *lyso-* = dissolving; *-somes* = bodies) are membrane-enclosed vesicles that form from the Golgi complex (Figure 3.23). Inside are as many as 60 kinds of powerful digestive and hydrolytic enzymes that can break down a wide variety of molecules. Lysosomal enzymes work best at an acidic pH, and the lysosomal membrane includes active transport pumps that move hydrogen ions (H^+) into the lysosome. Thus, the lysosomal interior has a pH of 5, which is 100 times more acidic than the cytosolic pH of 7. The lysosomal membrane also includes transporters that move the final products of digestion, such as glucose, fatty acids, and amino acids, into the cytosol.

Lysosomal enzymes also help recycle the cell's own structures. A lysosome can engulf another organelle, digest it, and return the digested components to the cytosol for reuse. The process by which worn-out organelles are digested is known as **autophagy** (aw-TOF-a-jē; *auto-* = self; *-phagy* = eating). During autophagy, the organelle to be digested is enclosed by a membrane derived from the ER to create a vesicle called an *autophagosome,* which then fuses with a lysosome. In this way, a human liver cell, for example, recycles about half of its cytoplasmic contents every week. Lysosomal enzymes may also destroy their own cell, a process known as **autolysis** (aw-TOL-i-sis). Autolysis occurs in some pathological conditions and is also responsible for the tissue deterioration that occurs just after death.

Although most lysosomal enzymes act within a cell, there are some instances in which the enzymes operate in extracellular digestion. One example occurs during fertilization. The head of a sperm cell releases lysosomal enzymes that aid its penetration of the oocyte.

Tay-Sachs Disease

Some disorders are caused by faulty or absent lysosomal enzymes. For instance, **Tay-Sachs disease,** which most often affects children of Ashkenazi (eastern European Jewish) descent, is an inherited condition characterized by the absence of a single lysosomal enzyme called Hex A. This enzyme normally breaks down a membrane glycolipid called ganglioside G_{M2} that is especially prevalent in nerve cells. As ganglioside G_{M2} accumulates, because it is not broken down, the nerve cells function less efficiently. Children with Tay-Sachs disease typically experience

Figure 3.23 Lysosomes.

🔑 Lysosomes contain several types of powerful digestive enzymes.

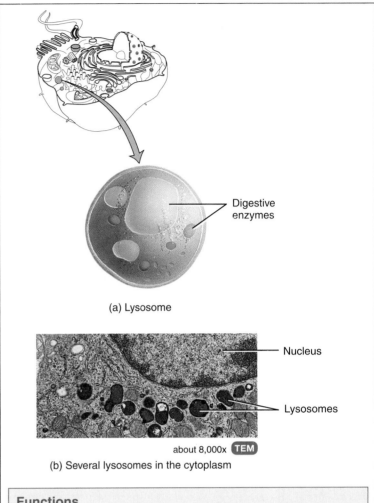

(a) Lysosome

Digestive enzymes

Nucleus

Lysosomes

about 8,000x **TEM**

(b) Several lysosomes in the cytoplasm

Functions
1. Digest substances that enter a cell via endocytosis and transport final products of digestion into cytosol.
2. Carry out autophagy, the digestion of worn-out organelles.
3. Carry out autolysis, the digestion of entire cell.
4. Carry out extracellular digestion.

 What is the name of the process by which worn-out organelles are digested by lysosomes?

seizures and muscle rigidity. They gradually become blind, demented, and uncoordinated and usually die before the age of 5. Tests can now reveal whether an adult is a carrier of the defective gene. ■

Peroxisomes

Another group of organelles similar in structure to lysosomes, but smaller, are the **peroxisomes** (pe-ROKS-i-sōms; *peroxi-* = peroxide; *somes* = bodies; see Figure 3.1). Peroxisomes contain several *oxidases,* enzymes that can oxidize (remove hydrogen atoms from) various organic substances. For instance, amino acids and fatty acids are oxidized in peroxisomes as part of normal metabolism. In addition, enzymes in peroxisomes oxidize toxic substances, such as alcohol. A byproduct of the oxidation reactions is hydrogen peroxide (H_2O_2), a potentially toxic compound. However, peroxisomes also contain the enzyme *catalase,* which decomposes H_2O_2. Because production and degradation of H_2O_2 occurs within the same organelle, peroxisomes protect other parts of the cell from the toxic effects of H_2O_2. New peroxisomes form by budding off from preexisting ones.

Proteasomes

Although lysosomes degrade proteins delivered to them in vesicles, cytosolic proteins also require disposal at certain times in the life of a cell. Continuous destruction of unneeded, damaged, or faulty proteins is the function of tiny structures called **proteasomes** (PRŌ-tē-a-sōmes = protein bodies). For example, proteins that are part of metabolic pathways are degraded after they have accomplished their function. Such protein destruction halts the pathway once the appropriate response has been achieved. A typical body cell contains many thousands of proteasomes, in both the cytosol and the nucleus. They were discovered only recently because they are far too small to discern under the light microscope and do not show up well in electron micrographs. Proteasomes were so-named because they contain myriad *proteases,* enzymes that cut proteins into small peptides. Once the enzymes of a proteasome have chopped up a protein into smaller chunks, other enzymes then break down the peptides into amino acids, which can be recycled into new proteins.

Proteasomes and Disease

Proteasomes are thought to be a factor in several diseases. For instance, people who have **cystic fibrosis** produce a misshapen membrane transporter protein whose mission normally is to help pump chloride ions (Cl^-) out of certain cells. Proteosomes degrade the defective transporter before it can reach the plasma membrane. The result is an imbalance in the transport of fluid and ions across the plasma membrane that causes the buildup of thick mucus outside certain types of cells. The accumulated mucus clogs the airways in the lungs, causing breathing difficulty, and prevents proper secretion of digestive enzymes by the pancreas, causing digestive problems. Disease could also result from failure of proteasomes to degrade abnormal proteins. For example, clumps of misfolded proteins accumulate in brain cells of people with Parkinson disease and Alzheimer disease. Discovering why the proteasomes fail to clear these abnormal proteins is a goal of ongoing research. ■

Mitochondria

Because they generate most of the ATP, the "powerhouses" of the cell are its **mitochondria** (mī-tō-KON-drē-a; *mito-* = thread; *-chondria* = granules; singular is *mitochondrion*). A cell may have as few as a hundred or as many as several thousand mitochondria. Active cells, such as those found in the muscles, liver, and kidneys, have a large number of mitochondria because they use ATP at a high rate. Mitochondria are usually located in a cell where the energy need is greatest, such as in among the contractile proteins in muscle cells.

A mitochondrion consists of an **outer mitochondrial membrane** and an **inner mitochondrial membrane** with a small fluid-filled space between them (Figure 3.24). Both membranes are similar in structure to the plasma membrane. The inner mitochondrial membrane contains a series of folds called **cristae** (KRIS-tē = ridges). The large central fluid-filled cavity of a mitochondrion, enclosed by the inner mitochondrial membrane, is the **matrix.** The elaborate folds of the cristae provide an enormous surface area for the chemical reactions that are part of the aerobic phase of *cellular respiration.* These reactions produce most of a cell's ATP and the enzymes that catalyze them are located on the cristae and in the matrix.

Like peroxisomes, mitochondria self-replicate, a process that occurs during times of increased cellular energy demand or before cell division. Because ribosomes are also present in the mitochondrial matrix, synthesis of some proteins needed for mitochondrial functions occurs there. Mitochondria even have their own DNA, in the form of multiple copies of a circular DNA molecule that contains 37 genes. These mitochondrial genes control the synthesis of 2 ribosomal RNAs, 22 transfer RNAs, and 13 proteins that build mitochondrial components.

Although the nucleus of each somatic cell contains genes from both your mother and father, mitochondrial genes are inherited only from your mother. The head of a sperm (the part that penetrates and fertilizes an oocyte) normally lacks most organelles, such as mitochondria, ribosomes, endoplasmic reticulum, and the Golgi complex. Also, any sperm mitochondria that do enter the oocyte are soon destroyed.

▶ CHECKPOINT

15. What does cytoplasm have that cytosol lacks?

16. Which organelles are surrounded by a membrane and which are not?

17. Which organelles contribute to synthesizing protein hormones and packaging them into secretory vesicles?

18. What happens on the cristae and in the matrix of mitochondria?

NUCLEUS

▶ OBJECTIVE

• **Describe the structure and function of the nucleus.**

The **nucleus** is a spherical or oval-shaped structure that usually is the most prominent feature of a cell (Figure 3.25). Most body

Figure 3.24 Mitochondria.

🔑 Within mitochondria, chemical reactions called cellular respiration generate ATP.

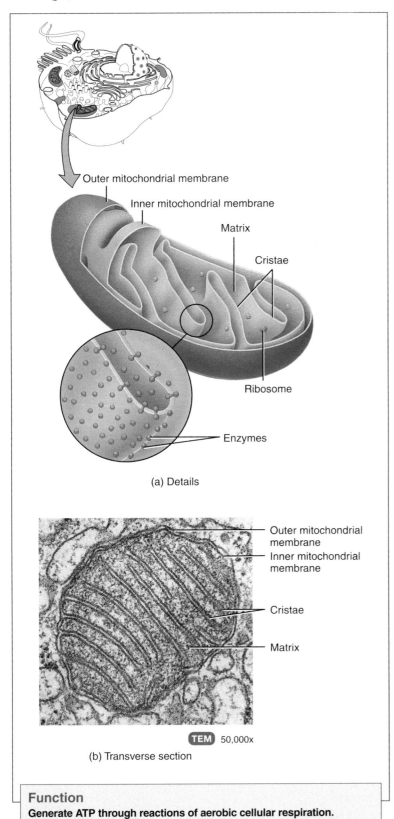

Outer mitochondrial membrane
Inner mitochondrial membrane
Matrix
Cristae
Ribosome
Enzymes

(a) Details

Outer mitochondrial membrane
Inner mitochondrial membrane
Cristae
Matrix

TEM 50,000x

(b) Transverse section

Function
Generate ATP through reactions of aerobic cellular respiration.

❓ How do the cristae of a mitochondrion contribute to its ATP-producing function?

cells have a single nucleus, although some, such as mature red blood cells, have none. In contrast, skeletal muscle cells and a few other types of cells have more than one nucleus. A double membrane called the **nuclear envelope** separates the nucleus from the cytoplasm. Both layers of the nuclear envelope are lipid bilayers similar to the plasma membrane. The outer membrane of the nuclear envelope is continuous with rough ER and resembles it in structure. Many openings called **nuclear pores** extend through the nuclear envelope. Each nuclear pore consists of a circular arrangement of proteins that surrounds a large central opening that is about 10 times wider than the pore of a channel protein in the plasma membrane.

Nuclear pores control the movement of substances between the nucleus and the cytoplasm. Small molecules and ions diffuse passively through the pores. Most large molecules, however, such as RNAs and proteins, cannot pass through the nuclear pores by diffusion. Instead, their passage involves an active transport process in which the molecules are recognized and selectively transported through the nuclear pore into or out of the nucleus. For example, proteins needed for nuclear functions move from the cytosol into the nucleus whereas RNA molecules move from the nucleus into the cytosol.

Inside the nucleus are one or more spherical bodies called **nucleoli** (noo′-KLĒ-ō-lī; singular is *nucleolus*) that function in producing ribosomes. Each nucleolus is a cluster of protein, DNA, and RNA that is not enclosed by a membrane. Nucleoli are the sites of synthesis of rRNA and assembly of rRNA and proteins into ribosomal subunits. Nucleoli are quite prominent in cells that synthesize large amounts of protein, such as muscle and liver cells. Nucleoli disperse and disappear during cell division and reorganize once new cells are formed.

Within the nucleus are most of the cell's hereditary units, called **genes,** which control cellular structure and direct cellular activities. Genes are arranged in single file along **chromosomes** (*chromo-* = colored). Human somatic (body) cells have 46 chromosomes, 23 inherited from each parent. Each chromosome is a long molecule of DNA that is coiled together with several proteins (Figure 3.26 on page 85). This complex of DNA, proteins, and some RNA is called **chromatin.** The total genetic information carried in a cell or an organism is its **genome.**

In cells that are not dividing, the chromatin appears as a diffuse, granular mass. Electron micrographs reveal that chromatin has a "beads-on-a-string" structure. Each "bead" is a **nucleosome** and consists of double-stranded DNA wrapped twice around a core of eight proteins called **histones,** which help organize the coiling and folding of DNA. The "string" between the "beads" is **linker DNA,** which holds adjacent nucleosomes together. Another histone promotes coiling of nucleosomes into a larger diameter **chromatin fiber,** which then folds into large loops. In cells that are not dividing, this is how DNA is packed. Just before cell division takes place, however, the DNA replicates (duplicates) and the loops condense even more, forming a pair of **chromatids.** As you will see shortly, during cell division a pair of chromatids constitutes a chromosome.

Figure 3.25   Nucleus.

The nucleus contains most of the cell's genes, which are located on chromosomes.

(a) Details of the nucleus

(b) Details of the nuclear envelope

about 8,000x **TEM**

(c) Transverse section of the nucleus

Functions
1. **Controls cellular structure.**
2. **Directs cellular activities.**
3. **Produces ribosomes in nucleoli.**

What is chromatin?

The main parts of a cell and their functions are summarized in Table 3.2 on page 86.

Genomics

In the last decade of the 20th century, the genomes of humans, mice, fruit flies, and more than 50 microbes were sequenced. As a result, research in the field of **genomics,** the study of the relationships between the genome and the biological functions of an organism, has flourished. The Human Genome Project began in 1990 as an effort to sequence all of the nearly 3.2 billion nu-

cleotides of our genome. By June 2000, scientists had produced a "working draft" of the human genome, having sequenced about 90% of the nucleotides. More than 99.9% of the nucleotide bases are identical in everyone. Less than 0.1% of our DNA (1 in each 1000 bases) accounts for inherited differences among humans. Surprisingly, at least half of the human genome consists of repeated sequences that do not code for proteins, so-called "junk" DNA. The average gene consists of 3,000 nucleotides, but sizes vary greatly. The largest known human gene codes for the protein dystrophin and has 2.4 million nucleotides. Scientists now estimate the total number of genes in

Figure 3.26 Packing of DNA into a chromosome in a dividing cell. When packing is complete, two identical DNA molecules and their histones form a pair of chromatids, which are held together by a centromere.

 A chromosome is a highly coiled and folded DNA molecule that is combined with protein molecules.

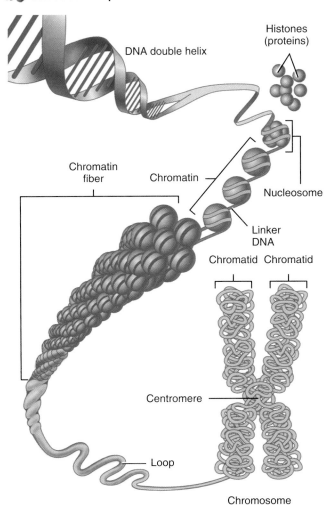

What are the components of a nucleosome?

the human genome at about 35,000 to 45,000, far fewer than the 100,000 previously predicted to exist. With an increasingly complete understanding of the human genome, genomic medicine aims to detect and treat genetic diseases at an early stage. Soon physicians may be able to provide counseling and treatment that is more effective for disorders that have significant genetic components, for instance, hypertension (high blood pressure), obesity, diabetes, and cancer. ■

▶ C H E C K P O I N T
19. How is DNA packed in the nucleus?

PROTEIN SYNTHESIS

▶ O B J E C T I V E

- **Describe the sequence of events that take place during protein synthesis.**

Although cells synthesize many chemicals to maintain homeostasis, much of the cellular machinery is devoted to synthesizing large numbers of diverse proteins. The proteins, in turn, determine the physical and chemical characteristics of cells and, therefore, of organisms. Just as genome means all of the genes in an organism, **proteome** means all of an organism's proteins. Some proteins are used to assemble cellular structures such as the plasma membrane, the cytoskeleton, and other organelles. Other proteins serve as hormones, antibodies, and contractile elements in muscle tissue. Still others are enzymes that regulate the rates of the numerous chemical reactions that occur in cells or transporters that carry various materials in the blood, such as transport of oxygen by hemoglobin.

Gene expression is the process whereby a gene's DNA is used to direct synthesis of a specific protein. First, the information encoded in a specific region of DNA is *transcribed* (copied) to produce a specific molecule of RNA (ribonucleic acid). Then, the RNA attaches to a ribosome, where the information contained in RNA is *translated* into a corresponding sequence of amino acids to form a new protein molecule (Figure 3.27).

Figure 3.27 Overview of gene expression. Synthesis of a specific protein requires transcription of a gene's DNA into RNA and translation of RNA into a corresponding sequence of amino acids.

 Whereas transcription occurs in the nucleus, translation occurs in the cytoplasm.

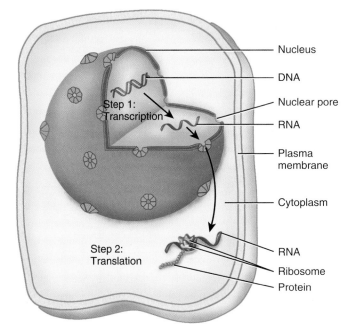

Why are proteins important in the life of a cell?

Table 3.2 Cell Parts and Their Functions

Part	Structure	Functions
Plasma Membrane	Fluid-mosaic lipid bilayer (phospholipids, cholesterol, and glycolipids) studded with proteins; surrounds cytoplasm.	Protects cellular contents; makes contact with other cells; contains channels, transporters, receptors, enzymes, cell-identity markers, and linker proteins; mediates the entry and exit of substances.
Cytoplasm	Cellular contents between the plasma membrane and nucleus—cytosol and organelles.	Site of all intracellular activities except those occurring in the nucleus.
Cytosol	Composed of water, solutes, suspended particles, lipid droplets, and glycogen granules.	Medium in which many of cell's metabolic reactions occur.
Organelles	Specialized structures with characteristic shapes.	Each organelle has specific functions.
Cytoskeleton	Network of three types of protein filaments: microfilaments, intermediate filaments, and microtubules.	Maintains shape and general organization of cellular contents; responsible for cellular movements.
Centrosome	A pair of centrioles plus pericentriolar material.	The pericentriolar material contains tubulins, which are used for growth of the mitotic spindle and microtubule formation.
Cilia and flagella	Motile cell surface projections that contain 20 microtubules and a basal body.	Cilia move fluids over a cell's surface; flagella move an entire cell.
Ribosome	Composed of two subunits containing ribosomal RNA and proteins; may be free in cytosol or attached to rough ER.	Protein synthesis.
Endoplasmic reticulum (ER)	Membranous network of flattened sacs or tubules. Rough ER is covered by ribosomes and is attached to nuclear envelope; smooth ER lacks ribosomes.	Rough ER synthesizes glycoproteins and phospholipids that are transferred to cellular organelles, inserted into the plasma membrane or secreted during exocytosis. Smooth ER synthesizes fatty acids and steroids; inactivates or detoxifies drugs; removes phosphate group from glucose 6-phosphate; and stores and releases calcium ions in muscle cells.
Golgi complex	Consists of 3–20 flattened membranous sacs called cisternae; structurally and functionally divided into entry *(cis)* face, medial cisternae, and exit *(trans)* face.	Entry *(cis)* face accepts proteins from rough ER; medial cisternae form glycoproteins, glycolipids, and lipoproteins; exit *(trans)* face modifies the molecules further, then sorts and packages them for transport to their destinations.
Lysosome	Vesicle formed from Golgi complex; contains digestive enzymes.	Fuses with and digests contents of endosomes, pinocytic vesicles, and phagosomes and transports final products of digestion into cytosol; digests worn-out organelles (autophagy), entire cells (autolysis), and extracellular materials.
Peroxisome	Vesicle containing oxidases (oxidative enzymes) and catalase (decomposes hydrogen peroxide); new peroxisomes bud from preexisting ones.	Oxidizes amino acids and fatty acids; detoxifies harmful substances, such as alcohol; produces hydrogen peroxide.
Proteasome	Tiny structure that contains proteases (proteolytic enzymes).	Degrades unneeded, damaged, or faulty proteins by cutting them into small peptides.
Mitochondrion	Consists of outer and inner mitochondrial membranes, cristae, and matrix; new mitochondria form from preexisting ones.	Site of aerobic cellular respiration reactions that produce most of a cell's ATP.
Nucleus	Consists of nuclear envelope with pores, nucleoli, and chromosomes, which exist as a tangled mass of chromatin in interphase cells.	Contains genes, which control cellular structure and direct cellular functions.

DNA and RNA store genetic information as sets of three nucleotides. A sequence of three such nucleotides in DNA is called a **base triplet.** Each DNA base triplet is transcribed as a complementary sequence of three RNA nucleotides, called a **codon.** A given codon specifies a particular amino acid. The **genetic code** is the set of rules that relate the base triplet sequence of DNA to the corresponding codons of RNA and the amino acids they specify.

The first step in protein synthesis is transcription, which occurs inside the nucleus.

Transcription

During **transcription,** the genetic information represented by the sequence of base triplets in DNA serves as a template for copying the information into a complementary sequence of codons in a strand of RNA. Three kinds of RNA are made from the DNA template:

- **Messenger RNA (mRNA)** directs the synthesis of a protein.
- **Ribosomal RNA (rRNA)** joins with ribosomal proteins to make ribosomes.
- **Transfer RNA (tRNA)** binds to an amino acid and holds it in place on a ribosome until it is incorporated into a protein during translation. One end of the tRNA carries a specific amino acid, and the opposite end consists of a triplet of nucleotides called an **anticodon.** By pairing between complementary bases, the tRNA anticodon attaches to the mRNA codon. Each of the more than 20 different types of tRNA binds to only one of the 20 different amino acids.

The enzyme **RNA polymerase** catalyzes transcription of DNA. However, the enzyme must be instructed where to start the transcription process and where to end it. Only one of the two DNA strands serves as a template for RNA synthesis. The segment of DNA where transcription begins is a special nucleotide sequence called a **promoter,** which is located near the beginning of a gene (Figure 3.28a). This is where RNA polymerase attaches to the DNA. During transcription, bases pair in a complementary manner: The bases cytosine (C), guanine (G), and thymine (T) in the DNA template pair with guanine, cytosine, and adenine (A), respectively, in the RNA strand (Figure 3.28b). However, adenine in the DNA template pairs with uracil (U), not thymine, in RNA:

A		U
T		A
G		C
C	→	G
A		U
T		A
Template DNA base sequence		Complementary RNA base sequence

Transcription of the DNA strand ends at another special nucleotide sequence called a **terminator,** which specifies the end of the gene (Figure 3.28a). When RNA polymerase reaches the

Figure 3.28 Transcription. DNA transcription begins at a promoter and ends at a terminator.

🔑 **During transcription, the genetic information in DNA is copied to RNA.**

(a) Overview

(b) Details

If the DNA template had the base sequence AGCT, what would be the mRNA base sequence? What enzyme catalyzes transcription of DNA?

terminator, the enzyme detaches from the transcribed RNA molecule and the DNA strand.

Not all parts of a gene actually code for parts of a protein. Within a gene are regions called **introns** that *do not* code for parts of proteins. They are located between regions called **exons** that *do* code for segments of a protein. Immediately after transcription, the transcript includes information from both introns and exons and is called **pre-mRNA.** The introns are removed from pre-mRNA by **small nuclear ribonucleoproteins** (snRNPs, pronounced "snurps"; Figure 3.28b). The snRNPs are enzymes that cut out the introns and splice together the exons. The resulting product is a functional mRNA molecule that passes through a pore in the nuclear envelope to reach the cytoplasm, where translation takes place.

Although the human genome contains 35,000 to 45,000 genes, there are probably 500,000 to 1 million human proteins. How can so many proteins be coded for by so few genes? Part of the answer lies in **alternative splicing** of mRNA, a process in which the pre-mRNA transcribed from a gene is spliced in different ways to produce several different mRNAs. The different mRNAs are then translated into different proteins. In this way, one gene may code for ten or more different proteins. In addition, chemical modifications are made to proteins after translation, for example, as proteins pass through the Golgi complex. Such chemical alterations can produce two or more different proteins from a single translation.

Antisense Therapy

In 1998 the U.S. Food and Drug Administration (FDA) approved Fomivirsen (Vitravene), the first drug designed to shut down gene expression by blocking the action of mRNA. The drug is an "antisense" DNA oligonucleotide (a few DNA nucleotides strung together) that is complementary to the "sense" of a specific mRNA. The DNA oligonucleotide base pairs with the mRNA and thereby blocks translation of the mRNA into a key protein needed for replication of the cytomegalovirus retinitis virus. The drug is injected into the eye to inhibit infection by this virus, which can rapidly cause blindness. Other antisense DNA oligonucleotides are being tested in clinical trials for treatment of certain cancers. ■

Translation

Translation is the process whereby the nucleotide sequence in an mRNA molecule specifies the amino acid sequence of a protein. Ribosomes in the cytoplasm carry out translation. The small subunit of a ribosome has a *binding site* for mRNA; the large subunit has two binding sites for tRNA molecules, a *P site* and an *A site* (Figure 3.29). The first tRNA molecule bearing its specific amino acid attaches to mRNA at the P site. The A site holds the next tRNA molecule bearing its amino acid. Translation occurs in the following way (Figure 3.30):

Figure 3.29 Translation. During translation, an mRNA molecule binds to a ribosome. Then, the mRNA nucleotide sequence specifies the amino acid sequence of a protein.

Ribosomes have a binding site for mRNA and a P site and an A site for attachment of tRNA.

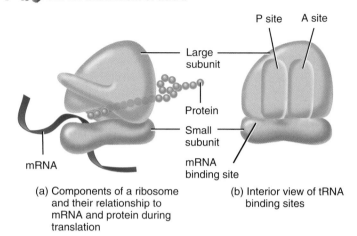

(a) Components of a ribosome and their relationship to mRNA and protein during translation

(b) Interior view of tRNA binding sites

What roles do the P and A sites serve?

1 An mRNA molecule binds to the small ribosomal subunit at the mRNA binding site. A special tRNA, called *initiator tRNA,* binds to the start codon (AUG) on mRNA, where translation begins. The tRNA anticodon (UAC) attaches to the mRNA codon (AUG) by pairing between the complementary bases. Besides being the start codon, AUG is also the codon for the amino acid methionine. Thus, methionine is always the first amino acid in a growing polypeptide.

2 Next, the large ribosomal subunit attaches to the small ribosomal subunit–mRNA complex, creating a functional ribosome. The initiator tRNA, with its amino acid (methionine), fits into the P site of the ribosome.

3 The anticodon of another tRNA with its attached amino acid pairs with the second mRNA codon at the A site of the ribosome.

4 A component of the large ribosomal subunit catalyzes the formation of a peptide bond between methionine, which separates from its tRNA at the P site, and the amino acid carried by the tRNA at the A site.

5 After peptide bond formation, the tRNA at the P site detaches from the ribosome, and the ribosome shifts the mRNA strand by one codon. The tRNA in the A site bearing the dipeptide shifts into the P site, allowing another tRNA with its amino acid to bind to a newly exposed codon at the A site. Steps **3** through **5** occur repeatedly, and the polypeptide progressively lengthens.

6 Protein synthesis ends when the ribosome reaches a stop codon at the A site, at which time the completed protein detaches from the final tRNA. When the tRNA vacates the A site, the ribosome splits into its large and small subunits.

Figure 3.30   Protein elongation and termination of protein synthesis during translation.

During protein synthesis the small and large ribosomal subunits join. When the process is complete, they separate.

1 Initiator tRNA attaches to a start codon.

2 Large and small ribosomal subunits join to form a functional ribosome and initiator tRNA fits into P site.

3 Anticodon of incoming tRNA pairs with next mRNA codon at A site.

4 Amino acid on tRNA at P site forms a peptide bond with amino acid at A site.

5 tRNA at P site leaves ribosome, ribosome shifts by one codon; tRNA previously at A site is now at the P site.

6 Protein synthesis stops when the ribosome reaches stop codon on mRNA.

Key:

= Adenine

= Guanine

= Cytosine

= Uracil

Summary of movement of ribosome along mRNA

What is the function of a stop codon?

Protein synthesis progresses at a rate of about 15 peptide bonds per second. As the ribosome moves along the mRNA and before it completes synthesis of the whole protein, another ribosome may attach behind it and begin translation of the same mRNA strand. Several ribosomes attached to the same mRNA constitute a **polyribosome.** The simultaneous movement of several ribosomes along the same mRNA molecule permits the translation of one mRNA into several identical proteins at the same time.

Recombinant DNA

Beginning in 1973, scientists developed techniques for inserting genes from other organisms into a variety of host cells. This manipulation can cause the host organism to produce proteins it normally does not synthesize. Organisms so altered are called **recombinants,** and their DNA—a combination of DNA from different sources—is called **recombinant DNA.** When recombinant DNA functions properly, the host will synthesize the protein specified by the new gene it has acquired. The technology that has arisen from manipulating genetic material is **genetic engineering.**

The practical applications of recombinant DNA technology are enormous. Strains of recombinant bacteria are now producing large quantities of many important therapeutic substances. Just a few of these are *human growth hormone (hGH),* required for normal growth and metabolism; *insulin,* a hormone that helps regulate blood glucose level and is used by diabetics; *interferon (IFN),* an antiviral (and possibly anticancer) substance; and *erythropoietin (EPO),* a hormone that stimulates production of red blood cells. ■

▶ C H E C K P O I N T
20. What is the difference between transcription and translation?

CELL DIVISION

▶ O B J E C T I V E S

• **Discuss the stages, events, and significance of somatic cell division.**
• **Describe the signals that induce cell division.**

As cells become damaged, diseased, or worn out, they are replaced by cell division. Cell division also occurs during growth of tissues. The two types of cell division—somatic cell division and reproductive cell division—accomplish different goals for the organism.

In **somatic cell division,** a cell undergoes a nuclear division called **mitosis** and a cytoplasmic division called **cytokinesis** to produce two identical **daughter cells.** Each daughter cell has the same number and kind of chromosomes as the original cell. Somatic cell division replaces dead or injured cells and adds new ones for tissue growth.

Reproductive cell division is the mechanism that produces gametes—sperm and oocytes. This process consists of a special two-step division called **meiosis,** in which the number of chromosomes in the nucleus is reduced by half, followed by cytokinesis. Meiosis is described in Chapter 28; here we focus on somatic cell division.

The Cell Cycle in Somatic Cells

The **cell cycle** is an orderly sequence of events by which a somatic cell duplicates its contents and divides in two. Human cells, except for gametes, contain 23 pairs of chromosomes. The two chromosomes of each pair—one originally derived from the mother and the other from the father—are called **homologous chromosomes** or **homologs.** Each homolog carries one copy of each gene located on that chromosome. The two members of the same gene pair are located at identical sites on the two homologs. When a cell reproduces, it must replicate (duplicate) all its chromosomes so the genes may be passed on to the next generation of cells. The cell cycle consists of two major periods: interphase, when a cell is not dividing, and the mitotic phase, when a cell is dividing (Figure 3.31).

Figure 3.31   The cell cycle. Not illustrated is cytokinesis, division of the cytoplasm, which occurs during late anaphase or early telophase of the mitotic phase.

In a complete cell cycle, a cell duplicates its contents and divides into two daughter cells.

In which phase of the cell cycle does DNA replication occur?

Interphase

During **interphase** the cell replicates its DNA. It also produces additional organelles and cytosolic components in anticipation of cell division. Interphase is a state of high metabolic activity, and during this time the cell does most of its growing. Interphase consists of three phases: G_1, S, and G_2 (Figure 3.31). The S stands for synthesis of DNA. Because the G-phases are periods when there is no activity related to DNA duplication, they are thought of as gaps or interruptions in DNA duplication.

The **G_1 phase** is the interval between the mitotic phase and the S phase. During G_1, the cell is metabolically active; it duplicates most of its organelles and cytosolic components but not its DNA. Virtually all the cellular activities described in this chapter happen during G_1. For a cell with a total cell cycle time of 24 hours, G_1 lasts 8 to 10 hours. The duration of this phase, however, is quite variable. It is very short in many embryonic cells or cancer cells.

The **S phase** is the interval between G_1 and G_2 and lasts about 8 hours. During the S phase, DNA replication occurs. As a result, the two daughter cells formed during cell division will have identical genetic material. The centrosome also replicates during the S phase. The **G_2 phase** is the interval between the S phase and the mitotic phase. It lasts 4 to 6 hours. During G_2, cell growth continues and enzymes and other proteins are synthesized in preparation for cell division. Cells that remain in G_1 for a very long time, perhaps destined never to divide again, are said to be in the **G_0 state.** For example, most nerve cells are in this state. Once a cell enters the S phase, however, it is committed to go through cell division.

When DNA replicates during the S phase, its helical structure partially uncoils, and the two strands separate at the points where hydrogen bonds connect base pairs (Figure 3.32). Each exposed base of the old DNA strand then pairs with the complementary base of a newly synthesized nucleotide. A new DNA strand takes shape as chemical bonds form between neighboring nucleotides. The uncoiling and complementary base pairing continues until each of the two original DNA strands is joined with a newly formed complementary DNA strand. The original DNA molecule has become two identical DNA molecules.

A microscopic view of a cell during interphase shows a clearly defined nuclear envelope, a nucleolus, and a tangled mass of chromatin (Figure 3.33a). Once a cell completes its activities during the G_1, S, and G_2 phases of interphase, the mitotic phase begins.

Mitotic Phase

The **mitotic (M) phase** of the cell cycle consists of nuclear division, or mitosis, and cytoplasmic division, or cytokinesis. The events that occur during mitosis and cytokinesis are plainly visible under a microscope because chromatin condenses into discrete chromosomes.

NUCLEAR DIVISION: MITOSIS Distribution of the two sets of chromosomes, one set into each of two separate nuclei, occurs during **mitosis.** The process results in the *exact* partitioning of genetic information. For convenience, biologists divide the process into four stages: prophase, metaphase, anaphase, and telophase.

1. **Prophase.** During early prophase, the chromatin fibers condense and shorten into chromosomes that are visible under the light microscope (Figure 3.33b). The condensation process may

Figure 3.32 Replication of DNA. The two strands of the double helix separate by breaking the hydrogen bonds (shown as dotted lines) between nucleotides. New, complementary nucleotides attach at the proper sites, and a new strand of DNA is synthesized alongside each of the original strands. Arrows indicate hydrogen bonds forming again between pairs of bases.

🔑 **Replication doubles the amount of DNA.**

Key:
- A = Adenine
- G = Guanine
- T = Thymine
- C = Cytosine

Hydrogen bonds

Old strand   New strand   New strand   Old strand

Why is it crucial that DNA replication occurs before cytokinesis in somatic cell division?

Figure 3.33   Cell division: mitosis and cytokinesis.   Begin the sequence at (a) at the top of the figure and read clockwise until you complete the process.

🔑 **In somatic cell division, a single diploid cell divides to produce two identical diploid daughter cells.**

(a) INTERPHASE

LM   all at 325x

Centrosome:
— Centrioles
— Pericentriolar material
Nucleolus
Nuclear envelope
Chromatin
Plasma membrane
Cytosol

(f) DAUGHTER CELLS
IN INTERPHASE

Centromere
Chromosome
(two chromatids
joined at centromere)

Kinetochore

Mitotic spindle
(microtubules)

Fragments of
nuclear envelope

Early   Late
(b) PROPHASE

Metaphase plate

(c) METAPHASE

Cleavage furrow

(e) TELOPHASE

Cleavage furrow

Chromosome

Late   Early
(d) ANAPHASE

❓ **When does cytokinesis begin?**

prevent entangling of the long DNA strands as they move during mitosis. Because DNA replication took place during the S phase of interphase, each prophase chromosome consists of a pair of identical, double-stranded *chromatids.* A constricted region called a *centromere* holds the chromatid pair together. At the outside of each centromere is a protein complex known as the **kinetochore** (ki-NET-ō-kor). Later in prophase, tubulins in the pericentriolar material of the centrosomes start to form the **mitotic spindle,** a football-shaped assembly of microtubules that attach to the kinetochore (Figure 3.33b). As the microtubules lengthen, they push the centrosomes to the poles (ends) of the cell so that the spindle extends from pole to pole. The spindle is responsible for the separation of chromatids to opposite poles of the cell. Then, the nucleolus disappears and the nuclear envelope breaks down.

2. **Metaphase.** During metaphase, the microtubules align the centromeres of the chromatid pairs at the exact center of the mitotic spindle (Figure 3.33c). This midpoint region is called the **equatorial plane.**

3. **Anaphase.** During anaphase, the centromeres split, separating the two members of each chromatid pair, which move toward opposite poles of the cell (Figure 3.33d). Once separated, the chromatids are termed chromosomes. As the chromosomes are pulled by the microtubules during anaphase, they appear V-shaped because the centromeres lead the way, dragging the trailing arms of the chromosomes toward the pole.

4. **Telophase.** The final stage of mitosis, telophase, begins after chromosomal movement stops (Figure 3.33e). The identical sets of chromosomes now at opposite poles of the cell, uncoil and revert to the threadlike chromatin form. A nuclear envelope forms around each chromatin mass, nucleoli reappear in the daughter nuclei, and the mitotic spindle disappears.

CYTOPLASMIC DIVISION: CYTOKINESIS Division of a parent cell's cytoplasm and organelles into two daughter cells is called **cytokinesis** (sī′-tō-ki-NĒ-sis; *-kinesis* = motion). This process begins in late anaphase or early telophase with formation of a **cleavage furrow,** a slight indentation of the plasma membrane. The cleavage furrow usually appears midway between the centrosomes and extends around the periphery of the cell (see Figure 3.33d and e). Actin microfilaments that lie just inside the plasma membrane form a *contractile ring* that pulls the plasma membrane progressively inward. The ring constricts the center of the cell, like tightening a belt around the waist, and ultimately pinches it in two. Because the plane of the cleavage furrow is always perpendicular to the mitotic spindle, the two sets of chromosomes end up in separate daughter cells. When cytokinesis is complete, interphase begins (Figure 33.3f).

Considering the cell cycle in its entirety, the sequence of events is

$G_1 \longrightarrow$ S phase $\longrightarrow G_2$ phase $\longrightarrow$ mitosis $\longrightarrow$ cytokinesis

Table 3.3 summarizes the events of the cell cycle in somatic cells.

Table 3.3	Events of the Somatic Cell Cycle
Phase	**Activity**
INTERPHASE	Period between cell divisions; chromosomes not visible under light microscope.
G_1 **phase**	Metabolically active cell duplicates organelles and cytosolic components.
S phase	Replication of DNA and centrosomes.
G_2 **phase**	Cell growth, enzyme and protein synthesis continues.
MITOTIC PHASE	Parent cell produces daughter cells with identical chromosomes; chromosomes visible under light microscope.
Mitosis	Nuclear division; distribution of two sets of chromosomes into separate nuclei.
Prophase	Chromatin fibers condense into paired chromatids; nucleolus and nuclear envelope disappear; each centrosome moves to an opposite pole of the cell.
Metaphase	Centromeres of chromatid pairs line up at metaphase plate.
Anaphase	Centromeres split; identical sets of chromosomes move to opposite poles of cell.
Telophase	Nuclear envelopes and nucleoli reappear; chromosomes resume chromatin form; mitotic spindle disappears.
Cytokinesis	Cytoplasmic division; contractile ring forms cleavage furrow around center of cell, dividing cytoplasm into separate and equal portions.

Control of Cell Destiny

A cell has three possible destinies—to remain alive and functioning without dividing, to grow and divide, or to die. Homeostasis is maintained when there is a balance between cell proliferation and cell death. The signals that tell a cell when to exist in the G_0 phase, when to divide, and when to die have been the subjects of intense and fruitful research during the past decade. A key signal that induces cell division is **maturation promoting factor (MPF).** One component of MPF is a group of enzymes called **cdc2 proteins,** so named because they participate in the cell division cycle (cdc). Another component of MPF is a protein called **cyclin,** so named because its level rises and falls during the cell cycle. Cyclin builds up in the cell during interphase and activates cdc2 proteins and thus MPF. As a result, the cell undergoes mitosis.

Cellular death is also regulated. Throughout the lifetime of an organism, certain cells undergo **apoptosis** (ap-ō-TŌ-sis = a falling off), an orderly, genetically programmed death. In apoptosis, a triggering agent from either outside or inside the cell causes "cell-suicide" genes to produce enzymes that damage the cell in several ways, including disrupting its cytoskeleton and nucleus. As a result, the cell shrinks and pulls away from neighboring cells. The DNA within the nucleus fragments, and the cytoplasm shrinks, although the plasma membrane remains intact. Phagocytes in the vicinity then ingest the dying cell. Apoptosis is especially useful because it removes unneeded cells during development before birth, for example, the separation of webbed digits during fetal development. It continues to occur after birth to regulate the number of cells in a tissue and eliminate potentially dangerous cells such as cancer cells.

Apoptosis, a normal type of cell death, contrasts with **necrosis** (ne-KRŌ-sis = death), a pathological type of cell death that results from tissue injury. In necrosis, many adjacent cells swell, burst, and spill their cytoplasm into the interstitial fluid. The cellular debris usually stimulates an inflammatory response by the immune system, which does not occur in apoptosis.

Tumor-suppressor Genes

Abnormalities in genes that regulate the cell cycle or apoptosis are associated with many diseases. For example, some cancers are caused by damage to genes called **tumor-suppressor genes,** which produce proteins that normally inhibit cell division. Loss or alteration of a tumor-suppressor gene called *p53* on chromosome 17 is the most common genetic change leading to a wide variety of tumors, including breast and colon cancers. The normal p53 protein arrests cells in the G_1 phase, which prevents cell division. Normal p53 protein also assists in repair of damaged DNA and induces apoptosis in the cells where DNA repair was not successful. For this reason, the *p53* gene is nicknamed "the guardian angel of the genome." ■

▶ CHECKPOINT

21. Distinguish between the somatic and reproductive types of cell division. Why is each important?

22. Define interphase. When does DNA replicate?

23. What are the major events of each stage of the mitotic phase?

24. How are apoptosis and necrosis similar and different?

Figure 3.34   Diverse shapes and sizes of human cells.   The relative difference in size between the smallest and largest cells is actually much greater than shown here.

The nearly 100 trillion cells in an average adult can be classified into about 200 different cell types.

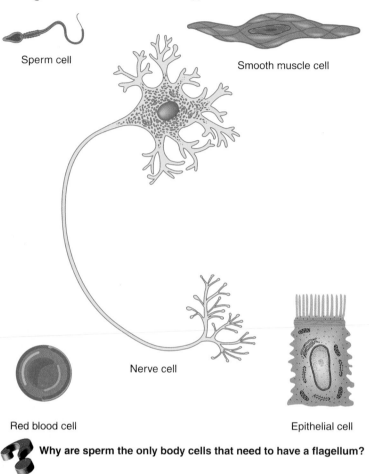

Sperm cell

Smooth muscle cell

Nerve cell

Red blood cell

Epithelial cell

Why are sperm the only body cells that need to have a flagellum?

CELLULAR DIVERSITY

The body of an average human adult is composed of nearly 100 trillion cells. All of these cells can be classified into about 200 different cell types. Cells vary considerably in size. High-powered microscopes are needed to see the smallest cells of the body. The largest cell, a single oocyte, is barely visible to the unaided eye. The sizes of cells are measured in units called *micrometers*. One micrometer (μm) is equal to 1 one-millionth of a meter, or 10^{-6} m (1/25,000 of an inch). Whereas a red blood cell has a diameter of 8 μm, an oocyte has a diameter of about 140 μm.

The shapes of cells also vary considerably (Figure 3.34). They may be round, oval, flat, cuboidal, columnar, elongated, star-shaped, cylindrical, or disc-shaped. A cell's shape is related to its function in the body. For example, a sperm cell has a long whiplike tail (flagellum) that it uses for locomotion. The disc shape of a red blood cell gives it a large surface area that enhances its ability to pass oxygen to other cells. The long, spindle shape of a relaxed smooth muscle cell shortens as it contracts. This shape change allows groups of smooth muscle cells to narrow or widen the passage for blood flowing through blood vessels. In this way, they regulate blood flow through various tissues. Some cells contain microvilli, which greatly increase their surface area. Microvilli are common in the epithelial cells that line the small intestine, where the large surface area speeds the absorption of digested food. Nerve cells have long extensions that permit them to conduct nerve impulses over great distances. As you will see, cellular diversity also permits organization of cells into more complex tissues and organs.

AGING AND CELLS

► O B J E C T I V E

• **Describe the cellular changes that occur with aging.**

Aging is a normal process accompanied by a progressive alteration of the body's homeostatic adaptive responses. It produces observable changes in structure and function and increases vulnerability to environmental stress and disease. The specialized branch of medicine that deals with the medical problems and care of elderly persons is **geriatrics** (jer′-ē-AT-riks; *ger-* = old age; *-iatrics* = medicine).

Although many millions of new cells normally are produced each minute, several kinds of cells in the body—heart muscle cells, skeletal muscle cells, and nerve cells—do not divide because they are arrested permanently in the G_0 phase. Experiments have shown that many other cell types have only a limited capability to divide. Normal cells grown outside the body divide only a certain number of times and then stop. These observations suggest that cessation of mitosis is a normal, genetically programmed event. According to this view, "aging genes" are part of the genetic blueprint at birth. These genes have an important function in normal cells but their activities slow over time. They bring about aging by slowing down or halting processes vital to life.

Another aspect of aging involves **telomeres** (TĒ-lō-merz), specific DNA sequences found only at the tips of each chromosome. These pieces of DNA protect the tips of chromosomes from erosion and from sticking to one another. However, in most normal body cells each cycle of cell division shortens the telomeres. Eventually, after many cycles of cell division, the telomeres can be completely gone and even some of the functional chromosomal material may be lost. These observations suggest that erosion of DNA from the tips of our chromosomes contributes greatly to aging and death of cells.

Glucose, the most abundant sugar in the body, plays a role in the aging process. It is haphazardly added to proteins inside and outside cells, forming irreversible cross-links between adjacent protein molecules. With advancing age, more cross-links form, which contributes to the stiffening and loss of elasticity that occur in aging tissues.

Free radicals produce oxidative damage in lipids, proteins, or nucleic acids by "stealing" an electron to accompany their unpaired electrons. Some effects are wrinkled skin, stiff joints, and hardened arteries. Normal cellular metabolism—for example, aerobic cellular respiration in mitochondria—produces some free radicals. Others are present in air pollution, radiation, and certain foods we eat. Naturally occurring enzymes in peroxisomes and in the cytosol normally dispose of free radicals. Certain dietary substances, such as vitamin E, vitamin C, beta-carotene, and selenium, are antioxidants that inhibit free radical formation.

Whereas some theories of aging explain the process at the cellular level, others concentrate on regulatory mechanisms operating within the entire organism. For example, the immune system may start to attack the body's own cells. This *autoimmune response* might be caused by changes in cell-identity markers at the surface of cells that cause antibodies to attach to and mark the cell for destruction. As changes in the proteins on the plasma membrane of cells increase, the autoimmune response intensifies, producing the well-known signs of aging. In the chapters that follow, we will discuss the effects of aging on each body system.

Progeria and Werner Syndrome

Progeria (prō-JER-ē-a) is a noninherited disease characterized by normal development in the first year of life followed by rapid aging. The condition is expressed by dry and wrinkled skin, total baldness, and birdlike facial features. Death usually occurs around age 13.

Werner syndrome is a rare, inherited disease that causes a rapid acceleration of aging, usually while the person is only in his or her twenties. It is characterized by wrinkling of the skin, graying of the hair and baldness, cataracts, muscular atrophy, and a tendency to develop diabetes mellitus, cancer, and cardiovascular disease. Most afflicted individuals die before age 50. Recently, the gene that causes Werner syndrome has been identified. Researchers hope to use the information to gain insight into the mechanisms of aging. ■

► C H E C K P O I N T

25. What is one reason that some tissues become stiffer as they age?

DISORDERS: HOMEOSTATIC IMBALANCES

Cancer

Cancer is a group of diseases characterized by uncontrolled cell proliferation. When cells in a part of the body divide without control, the excess tissue that develops is called a **tumor** or **neoplasm** (NĒ-ō-plazm; *neo-* = new). The study of tumors is called **oncology** (on-KOL-ō-jē; *onco-* = swelling or mass). Tumors may be cancerous and often fatal, or they may be harmless. A cancerous neoplasm is called a **malignant tumor** or **malignancy.** One property of most malignant tumors is their ability to undergo **metastasis** (me-TAS-ta-sis), the spread of cancerous cells to other parts of the body. A **benign tumor** is a neoplasm that does not metastasize. An example is a wart. A benign tumor may be surgically removed if it interferes with normal body function or becomes disfiguring.

Types of Cancer

The name of a cancer is derived from the type of tissue in which it develops. Most human cancers are **carcinomas** (kar-si-NŌ-maz; *carcin-* = cancer; *-omas* = tumors), malignant tumors that arise from epithelial cells. **Melanomas** (mel-a-NŌ-maz; *melan-* = black), for example, are cancerous growths of melanocytes, skin epithelial cells that produce the pigment melanin. **Sarcoma** (sar-KŌ-ma; *sarc-* = flesh) is a general term for any cancer arising from muscle cells or connective tissues. For example, **osteogenic sarcoma** (*osteo-* = bone; *-genic* = origin), the most frequent type of childhood cancer, destroys normal bone tissue. **Leukemia** (loo-KĒ-mē-a; *leuk-* = white; *-emia* = blood) is a cancer of blood-forming organs characterized by rapid growth of abnormal leukocytes (white blood cells). **Lymphoma** (lim-FŌ-ma) is a malignant disease of lymphatic tissue—for example, of lymph nodes.

Growth and Spread of Cancer

Cells of malignant tumors duplicate rapidly and continuously. As malignant cells invade surrounding tissues, they often trigger **angiogenesis,** the growth of new networks of blood vessels. Proteins that stimulate angiogenesis in tumors are called **tumor angiogenesis factors (TAFs).** However, inhibitors of angiogenesis may also be present. Thus, the formation of new blood vessels can occur either by overproduction of TAFs or by the lack of naturally occurring angiogenesis inhibitors. As the cancer grows, it begins to compete with normal tissues for space and nutrients. Eventually, the normal tissue decreases in size and dies. Some malignant cells may detach from the initial (primary) tumor and invade a body cavity or enter the blood or lymph, then circulate to and invade other body tissues, establishing secondary tumors. Malignant cells resist the antitumor defenses of the body. The pain associated with cancer develops when the tumor presses on nerves or blocks a passageway in an organ so that secretions build up pressure.

Causes of Cancer

Several factors may trigger a normal cell to lose control and become cancerous. One cause is environmental agents: substances in the air we breathe, the water we drink, and the food we eat. A chemical agent or radiation that produces cancer is called a **carcinogen** (car-SIN-ō-jen). Carcinogens induce **mutations,** permanent structural changes in the DNA base sequence of a gene. The World Health Organization estimates that carcinogens are associated with 60–90% of all human cancers. Examples of carcinogens are hydrocarbons found in cigarette tar, radon gas from the earth, and ultraviolet (UV) radiation in sunlight.

Intensive research efforts are now directed toward studying cancer-causing genes, or **oncogenes** (ON-kō-jēnz). These genes, when inappropriately activated, have the ability to transform a normal cell into a cancerous cell. Most oncogenes derive from normal genes called **proto-oncogenes** that regulate growth and development. The proto-oncogene undergoes some change that either causes it to produce an abnormal product or disrupts its control. It may be expressed inappropriately or make its products in excessive amounts or at the wrong time. Some oncogenes cause excessive production of growth factors, chemicals that stimulate cell growth. Others may trigger changes in a cell-surface receptor, causing it to send signals as though it were being activated by a growth factor. As a result, the growth pattern of the cell becomes abnormal.

Proto-oncogenes in every cell carry out normal cellular functions until a malignant change occurs. It appears that some proto-oncogenes are activated to oncogenes by mutations in which the DNA of the proto-oncogene is altered. Other proto-oncogenes are activated by a rearrangement of the chromosomes so that segments of DNA are exchanged. Rearrangement activates proto-oncogenes by placing them near genes that enhance their activity.

Some cancers have a viral origin. Viruses are tiny packages of nucleic acids, either RNA or DNA, that can reproduce only while inside the cells they infect. Some viruses, termed **oncogenic viruses,** cause cancer by stimulating abnormal proliferation of cells. For instance, the *human papillomavirus (HPV)* causes virtually all cervical cancers in women. The virus produces a protein that causes proteasomes to destroy the p53 protein. (Recall that p53 protein normally suppresses cell division.) In the absence of this suppressor protein, cells proliferate without control.

Carcinogenesis: A Multistep Process

Carcinogenesis (kar'-si-nō-JEN-e-sis), the process by which cancer develops, is a multistep process in which as many as 10 distinct mutations may have to accumulate in a cell before it becomes cancerous. The progression of genetic changes leading to cancer is best understood for colon (colorectal) cancer. Such cancers, as well as lung and breast cancer, take years or decades to develop. In colon cancer, the tumor begins as an area of increased cell proliferation that results from one mutation. This growth then progresses to abnormal, but noncancerous, growths called adenomas. After two or three additional mutations, a mutation of the tumor-suppressor gene *p53* occurs and a carcinoma develops. Recall that tumor-suppressor genes normally suppress unregulated cell growth. The fact that so many mutations are needed for a cancer to develop indicates that cell growth is normally controlled with many sets of checks and balances.

Treatment of Cancer

Many cancers are removed surgically. However, when cancer is widely distributed throughout the body or exists in organs such as the brain whose functioning would be greatly harmed by surgery, chemotherapy and radiation therapy may be used instead. Sometimes surgery, chemotherapy, and radiation therapy are used in combination. Chemotherapy involves administering drugs that cause death of cancerous cells. Radiation therapy breaks chromosomes, thus blocking cell division. Because cancerous cells divide rapidly, they are more vulnerable to the destructive effects of chemotherapy and radiation therapy than are normal cells. Unfortunately for the patients, hair follicle cells, red bone marrow cells, and cells lining the gastrointestinal tract also are rapidly dividing. Hence, the side effects of chemotherapy and radiation therapy include hair loss due to death of hair follicle cells, vomiting and nausea due to death of cells lining the stomach and intestines, and susceptibility to infection due to slowed production of white blood cells in red bone marrow.

Treating cancer is difficult because it is not a single disease and because the cells in a single tumor population rarely behave all in the same way. Although most cancers are thought to derive from a single abnormal cell, by the time a tumor reaches a clinically detectable size, it may contain a diverse population of abnormal cells. For example, some cancerous cells metastasize readily, and others do not. Some are sensitive to chemotherapy drugs and some are drug resistant. Because of differences in drug resistance, a single chemotherapeutic agent may destroy susceptible cells but permit resistant cells to proliferate.

MEDICAL TERMINOLOGY

Anaplasia (an′-a-PLĀ-zē-a; *an* = not; *plasia* = to shape) The loss of tissue differentiation and function that is characteristic of most malignancies.

Atrophy (AT-rō-fē; *a* = without; *-trophy* = nourishment) A decrease in the size of cells, with a subsequent decrease in the size of the affected tissue or organ; wasting away.

Dysplasia (dis-PLĀ-zē-a; *dys-* = abnormal) Alteration in the size, shape, and organization of cells due to chronic irritation or inflammation; may progress to neoplasia (tumor formation, usually malignant) or revert to normal if the irritation is removed.

Hyperplasia (hī-per-PLĀ-zē-a; *hyper-* = over) Increase in the number of cells of a tissue due to an increase in the frequency of cell division.

Hypertrophy (hī-PER-trō-fē) Increase in the size of cells without cell division.

Metaplasia (met′-a-PLĀ-zē-a; *meta-* = change) The transformation of one type of cell into another.

Progeny (PROJ-e-nē; *pro-* = forward; *-geny* = production) Offspring or descendants.

STUDY OUTLINE

INTRODUCTION (p. 60)

1. A cell is the basic, living, structural and functional unit of the body.
2. Cell biology is the scientific study of cellular structure and function.

PARTS OF A CELL (p. 60)

1. Figure 3.1 on page 60 provides an overview of the typical structures in body cells.
2. The principal parts of a cell are the plasma membrane; the cytoplasm, which consists of cytosol and organelles; and the nucleus.

THE PLASMA MEMBRANE (p. 61)

1. The plasma membrane surrounds and contains the cytoplasm of a cell.
2. The membrane is composed of a 50:50 mix by weight of proteins and lipids that are held together by noncovalent forces.
3. According to the fluid mosaic model, the membrane is a mosaic of proteins floating like icebergs in a bilayer sea of lipids.
4. The lipid bilayer consists of two back-to-back layers of phospholipids, cholesterol, and glycolipids. The bilayer arrangement occurs because the lipids are amphipathic, having both polar and nonpolar parts.
5. Integral proteins extend into or through the lipid bilayer, whereas peripheral proteins associate with membrane lipids or integral proteins at the inner or outer surface of the membrane.
6. Many integral proteins are glycoproteins, with sugar groups attached to the ends that face the extracellular fluid. Together with glycolipids, the glycoproteins form a glycocalyx on the extracellular surface of cells.
7. Membrane proteins have a variety of functions. Ion channels and transporters are membrane proteins that help specific solutes across the membrane; receptors serve as cellular recognition sites; and linkers anchor proteins in the plasma membranes to protein filaments inside and outside the cell. Some membrane proteins are enzymes and others are cell-identity markers.

8. Membrane fluidity is greater when there are more double bonds in the fatty acid tails of the lipids that make up the bilayer. Cholesterol makes the lipid bilayer stronger but less fluid at normal body temperature. Because of its fluidity, the lipid bilayer self-seals when torn or punctured.
9. The membrane's selective permeability permits some substances to pass more readily than others. The lipid bilayer is permeable to water, urea, and most nonpolar, uncharged molecules. It is impermeable to ions and charged or polar molecules other than water and urea. Channels and transporters increase the plasma membrane's permeability to small- and medium-sized polar and charged substances, including ions, that cannot cross the lipid bilayer.
10. The selective permeability of the plasma membrane supports the existence of concentration gradients, differences in the concentrations of chemicals between one side of the membrane and the other.

TRANSPORT ACROSS THE PLASMA MEMBRANE (p. 64)

1. In passive transport, a substance moves down its concentration gradient across the membrane using its own kinetic energy of motion. In active transport, cellular energy is used to drive the substance "uphill" against its concentration gradient.
2. In endocytosis and exocytosis, tiny vesicles either detach from or merge with the plasma membrane to move materials across the membrane, into or out of a cell.
3. In diffusion, molecules or ions move from an area of higher concentration to an area of lower concentration until an equilibrium is reached.
4. The rate of diffusion across a plasma membrane is affected by the steepness of the concentration gradient, temperature, mass of the diffusing substance, surface area available for diffusion, and the distance over which diffusion must occur.
5. Osmosis is the net movement of water through a selectively permeable membrane from an area of higher water concentration to an area of lower water concentration.
6. In an isotonic solution, red blood cells maintain their normal

shape; in a hypotonic solution, they undergo hemolysis; in a hypertonic solution, they undergo crenation.

7. Nonpolar, hydrophobic molecules, such as oxygen, carbon dioxide, nitrogen, steroids, fat-soluble vitamins (A, E, D, and K), small alcohols, and ammonia plus water and urea diffuse through the lipid bilayer of the plasma membrane.

8. Ion channels selective for K^+, Cl^-, Na^+, and Ca^{2+} allow these small, inorganic ions (which are too hydrophilic to penetrate the membrane's nonpolar interior) to diffuse across the plasma membrane.

9. In facilitated diffusion, a solute such as glucose binds to a specific transporter on one side of the membrane and is released on the other side after the transporter undergoes a change in shape.

10. Substances can cross the membrane against their concentration gradient by active transport. Actively transported substances include several ions, such as Na^+, K^+, H^+, Ca^{2+}, I^-, and Cl^-; amino acids; and monosaccharides.

11. Two sources of energy are used to drive active transport: Energy obtained from hydrolysis of ATP is the source in primary active transport, and energy stored in a Na^+ or H^+ concentration gradient is the source in secondary active transport.

12. The most prevalent primary active transport pump is the sodium-potassium pump, also known as Na^+/K^+ ATPase.

13. Secondary active transport mechanisms include both symporters and antiporters that are powered by either a Na^+ or H^+ concentration gradient. Symporters move two substances in the same direction, whereas antiporters move two substances in opposite directions across the membrane.

14. Receptor-mediated endocytosis is the selective uptake of large molecules and particles (ligands) that bind to specific receptors in membrane areas called clathrin-coated pits.

15. Phagocytosis is the ingestion of solid particles. Some white blood cells destroy microbes that enter the body in this way.

16. Pinocytosis is the ingestion of extracellular fluid. In this process, a pinocytic vesicle surrounds the fluid.

17. Exocytosis involves movement of secretory or waste products out of a cell by fusion of vesicles with the plasma membrane.

18. In transcytosis, vesicles undergo endocytosis on one side of a cell, move across the cell, and then undergo exocytosis on the opposite side.

CYTOPLASM (p. 75)

1. Cytoplasm is all the cellular contents between the plasma membrane and the nucleus. It consists of cytosol and organelles.

2. Cytosol is the fluid portion of cytoplasm, containing water, ions, glucose, amino acids, fatty acids, proteins, lipids, ATP, and waste products. It is the site of many chemical reactions required for a cell's existence.

3. Organelles are specialized structures with characteristic shapes that have specific functions.

4. The cytoskeleton is a network of several kinds of protein filaments that extend throughout the cytoplasm. Components of the cytoskeleton are microfilaments, intermediate filaments, and microtubules. The cytoskeleton provides a structural framework for the cell and is responsible for cell movements.

5. The centrosome consists of a pair of centrioles and pericentriolar material. The pericentriolar material organizes microtubules in nondividing cells and the mitotic spindle in dividing cells.

6. Cilia and flagella are formed by basal bodies and are motile pro-

jections of the cell surface. Cilia move fluid along the cell surface whereas flagella move an entire cell.

7. Ribosomes, composed of ribosomal RNA and ribosomal proteins, consist of two subunits made in the nucleus. They are sites of protein synthesis.

8. Endoplasmic reticulum (ER) is a network of membranes that form flattened sacs or tubules; it extends from the nuclear envelope throughout the cytoplasm.

9. Rough ER is studded with ribosomes. Proteins synthesized by ribosomes attached to rough ER enter the space within the ER for processing and sorting. It produces secretory proteins, membrane proteins, and organelle proteins. It also forms glycoproteins and attaches proteins to phospholipids, which it also synthesizes.

10. Smooth ER lacks ribosomes. It synthesizes fatty acids and steroids; inactivates or detoxifies drugs and other potentially harmful substances; removes phosphate from glucose 6-phosphate; and releases calcium ions that trigger contraction in muscle cells.

11. The Golgi complex consists of flattened sacs called cisternae. The entry, medial, and exit regions of the Golgi complex contain different enzymes that permit each to modify, sort, and package proteins for transport to different destinations.

12. Processed proteins leave the Golgi complex in secretory vesicles, membrane vesicles, or transport vesicles that will carry the proteins to another cellular destination.

13. Lysosomes are membrane-enclosed vesicles that contain digestive enzymes. Late endosomes, phagosomes, and pinocytic vesicles deliver materials to lysosomes for degradation. Lysosomes function in digestion of worn-out organelles (autophagy), digestion of a host cell (autolysis), and extracellular digestion.

14. Peroxisomes contain oxidases that oxidize amino acids, fatty acids, and toxic substances and, in the process, produce hydrogen peroxide. Catalase, also present in peroxisomes, destroys hydrogen peroxide.

15. Proteasomes continually degrade unneeded, damaged, or faulty proteins. Their proteases cut proteins into small peptides.

16. Mitochondria consist of a smooth outer membrane, an inner membrane containing cristae, and a fluid-filled cavity called the matrix. They are called "powerhouses" of the cell because they produce most of a cell's ATP.

NUCLEUS (p. 82)

1. The nucleus consists of a double nuclear envelope; nuclear pores, which control the movement of substances between the nucleus and cytoplasm; nucleoli, which produce ribosomes; and genes arranged on chromosomes.

2. Human somatic cells have 46 chromosomes, 23 inherited from each parent. The total genetic information carried in a cell or an organism is its genome.

PROTEIN SYNTHESIS (p. 85)

1. Cells make proteins by transcribing and translating the genetic information contained in DNA.

2. The genetic code is the set of rules that relate the base triplet sequences of DNA to the corresponding codons of RNA and the amino acids they specify.

3. In transcription, the genetic information in the sequence of base triplets in DNA serves as a template for copying the information into a complementary sequence of codons in messenger RNA.

Transcription begins on DNA in a region called a promoter. Regions of DNA that code for protein synthesis are called exons; those that do not are called introns.

4. Newly synthesized pre-mRNA is modified before leaving the nucleus.

5. Translation is the process by which the nucleotide sequence of mRNA specifies the amino acid sequence of a protein. In translation, mRNA binds to a ribosome, specific amino acids attach to tRNA, and anticodons of tRNA bind to codons of mRNA, thus bringing specific amino acids into position on a growing polypeptide. Translation begins at the start codon and ends at the stop codon.

CELL DIVISION (p. 90)

1. Cell division is the process by which cells reproduce themselves. It consists of nuclear division (mitosis or meiosis) and cytoplasmic division (cytokinesis).

2. Cell division that results in an increase in the number of body cells is called somatic cell division and involves a nuclear division called mitosis plus cytokinesis.

3. Cell division that results in the production of sperm and oocytes is called reproductive cell division and consists of a nuclear division called meiosis plus cytokinesis.

4. The cell cycle is an orderly sequence of events in which a somatic cell duplicates its contents and divides in two. It consists of interphase and a mitotic phase.

5. Before the mitotic phase, the chromosomes replicate so that identical chromosomes can be passed on to the next generation of cells.

6. A cell that is between divisions and is carrying on every life process except division is in interphase, which consists of three phases: G_1, S, and G_2. During the G_1 phase, the cell duplicates its organelles and cytosolic components; during the S phase, DNA and centrosome replication occur; during the G_2 phase, enzymes and other proteins are synthesized.

7. Mitosis is the replication of the chromosomes and the distribution of two identical sets of chromosomes into separate and equal nuclei; it consists of prophase, metaphase, anaphase, and telophase.

8. Cytokinesis usually begins in late anaphase and ends in telophase. A cleavage furrow forms at the cell's equatorial plane and progresses inward, pinching through the cell to form two separate portions of cytoplasm.

9. A cell can either remain alive and functioning without dividing, grow and divide, or die. Maturation promoting factor (MPF) induces cell division (both mitosis and meiosis). MPF consists of cyclin, which builds up during interphase, and cdc2 proteins, which are activated by cyclin.

10. Apoptosis is programmed cell death, a normal type of cell death. It first occurs during embryological development and continues for the lifetime of an organism.

11. Certain genes regulate both cell division and apoptosis. Abnormalities in these genes are associated with a wide variety of diseases and disorders.

CELLULAR DIVERSITY (p. 94)

1. The almost 200 different types of cells in the body vary considerably in size and shape.

2. The sizes of cells are measured in micrometers. One micrometer (μm) equals 10^{-6} m (1/25,000 of an inch).

3. A cell's shape is related to its function.

AGING AND CELLS (p. 95)

1. Aging is a normal process accompanied by progressive alteration of the body's homeostatic adaptive responses.

2. Many theories of aging have been proposed, including genetically programmed cessation of cell division, shortening of telomeres, addition of glucose to proteins, buildup of free radicals, and an intensified autoimmune response.

Q SELF-QUIZ QUESTIONS

Fill in the blanks in the following statements.

1. The three principal parts of the cell are the _____, _____, and _____.

2. _____ refers to programmed cell death, whereas _____ refers to cell death resulting from tissue injury.

3. The fluid portion of the cytoplasm is the _____.

4. List three causes of cellular aging.

Indicate whether the following statements are true or false.

5. Duplicated chromosomes held together by centromeres are called chromatids.

6. The sodium pump is an example of primary active transport.

Choose the one best answer to the following questions.

7. The basic structural unit of the plasma membrane is the (a) lipid bilayer, (b) integral protein, (c) cholesterol molecule, (d) peripheral protein, (e) glycoprotein-glycolipid complex.

8. Integral proteins can function in the cell membrane in all the following ways *except* (a) as a channel, (b) as a transporter, (c) as a receptor, (d) as an exocytosis vesicle, (e) as a cell-identity marker.

9. Which of the following factors influence the diffusion rate of substances through a plasma membrane? (1) concentration gradient, (2) diffusion distance, (3) surface area, (4) size of diffusing substance, (5) temperature. (a) 1, 2, 3, and 5, (b) 1, 2, 3, 4, and 5, (c) 2, 3, 4, and 5, (d) 1, 2, and 5, (e) 2, 3, and 5.

10. A cell would lose water volume and shrink if placed in (a) a hypertonic solution, (b) a hypotonic solution, (c) an isotonic solution, (d) a hydrophobic solution, (e) an ionic solution.

11. Which of the following statements regarding the nucleus are *true*? (1) Nucleoli within the nucleus are the sites of ribosome synthesis. (2) Most of the cell's hereditary units, called genes, are located within the nucleus. (3) The nuclear membrane is a solid, impermeable membrane. (4) Protein synthesis occurs within the nucleus. (5) In nondividing cells, DNA is found in the nucleus in the form of chromatin. (a) 1, 2, and 3, (b) 1, 2, and 4, (c) 1, 2, and 5, (d) 2, 4, and 5, (e) 2, 3, and 4.

12. Match the following:

_____ (a) mitosis
_____ (b) meiosis
_____ (c) prophase
_____ (d) metaphase
_____ (e) anaphase
_____ (f) telophase
_____ (g) cytokinesis
_____ (h) interphase

(1) cytoplasmic division
(2) somatic cell division resulting in identical daughter cells
(3) reproductive cell division that reduces the number of chromosomes by half
(4) stage of cell division when replication of DNA occurs
(5) stage when chromatin fibers condense and shorten to form chromosomes
(6) stage when centromeres split and sister chromatids move to opposite poles of the cell
(7) stage when centromeres of chromatid pairs line up at the center of the mitotic spindle
(8) stage when chromosomes uncoil and revert to chromatin

13. Match the following:

_____ (a) codon
_____ (b) RNA polymerase
_____ (c) intron
_____ (d) exon
_____ (e) transcription
_____ (f) translation
_____ (g) messenger RNA
_____ (h) transfer RNA
_____ (i) ribosomal RNA
_____ (j) snRNP
_____ (k) promotor
_____ (l) terminator

(1) DNA region that does not code for synthesis of a part of a protein
(2) DNA region that codes for synthesis of a part of a protein
(3) enzyme that removes all introns and joins remaining exons
(4) nucleotide sequence, located near the beginning of a gene, that indicates where transcription begins; RNA polymerase attachment site
(5) the copying of the DNA message onto messenger RNA
(6) joins with ribosomal proteins to make ribosomes
(7) binds to amino acids and holds them in place on a ribosome to be incorporated into a protein
(8) nucleotide sequence near the end of a gene that indicates where transcription ends; releases RNA polymerase
(9) reading of messenger RNA for protein synthesis
(10) enzyme that catalyzes transcription of DNA
(11) a transcribed sequence of three RNA nucleotides
(12) directs synthesis of a protein

14. Match the following:

_____ (a) cytoskeleton
_____ (b) centrosome
_____ (c) ribosomes
_____ (d) rough ER
_____ (e) smooth ER
_____ (f) Golgi complex
_____ (g) lysosomes
_____ (h) peroxisomes
_____ (i) mitochondria
_____ (j) cilia
_____ (k) proteasomes

(1) membrane-enclosed vesicles formed in the Golgi complex that contain strongly hydrolytic enzymes
(2) network of protein filaments that extend throughout the cytoplasm, providing cellular shape and organization
(3) sites of protein synthesis
(4) contain enzymes that break apart unneeded, damaged, or faulty proteins into small peptides
(5) site where secretory proteins and membrane molecules are synthesized
(6) membrane-enclosed vesicles that contain enzymes that oxidize various organic substances
(7) microtubular structures extending from the plasma membrane and involved in movement of materials along the cell's surface
(8) modifies, sorts, packages, and transports molecules synthesized in the rough ER
(9) an organizing center for growth of the mitotic spindle
(10) function in ATP generation
(11) functions in synthesizing fatty acids and steroids, helping liver cells release glucose into the bloodstream, and detoxification

15. Match the following:

_____ (a) diffusion
_____ (b) osmosis
_____ (c) facilitated diffusion
_____ (d) primary active transport
_____ (e) secondary active transport
_____ (f) vesicular transport
_____ (g) phagocytosis
_____ (h) pinocytosis
_____ (i) exocytosis
_____ (j) receptor-mediated endocytosis
_____ (k) trancytosis

(1) passive transport in which a solute binds to a specific transporter on one side of the membrane and is released on the other side
(2) movement of materials out of the cell by fusing of secretory vesicles with the plasma membrane
(3) the random mixing of particles in a solution due to the kinetic energy of the particles
(4) transport of substances either into or out of the cell by means of a small, spherical membranous sac formed by budding off from existing membranes

(5) uses energy derived from hydrolysis of ATP to change the shape of a transporter protein, which "pumps" a substance across a cellular membrane against its concentration gradient
(6) vesicular movement involving endocytosis on one side of a cell and subsequent exocytosis on the opposite side of the cell
(7) type of endocytosis that involves the nonselective uptake of tiny droplets of extracellular fluid
(8) type of endocytosis in which large solid particles are taken in

(9) movement of water from an area of higher to an area of lower water concentration through a selectively permeable membrane
(10) process that allows a cell to take specific ligands from the ECF by forming vesicles
(11) indirectly uses energy obtained from the breakdown of ATP; involves symporters and antiporters

CRITICAL THINKING QUESTIONS

1. In old detective novels, a dead body is sometimes called "a stiff." But the corpse doesn't remain stiff forever. After a time, it begins to soften. What causes this change in the tissues?
HINT *Worn out organelles suffer a similar fate.*

2. Mucin is a protein present in saliva and other secretions. When mixed with water, it becomes the slippery substance known as mucus. Trace the route taken by mucin through the cell, from its synthesis to its secretion, listing all the organelles and processes involved.
HINT *All proteins are produced initially by the same organelle.*

3. If you were trying to locate the DNA sequence that coded for an mRNA, would you look at a gene's introns or exons?
HINT *The mRNA is complementary to specific parts of the DNA.*

ANSWERS TO FIGURE QUESTIONS

3.1 The three main parts of a cell are the plasma membrane, cytoplasm, and nucleus.

3.2 The glycocalyx is the sugary coat on the extracellular surface of the plasma membrane. It is composed of the carbohydrate portions of membrane glycolipids and glycoproteins.

3.3 The membrane protein that binds to insulin is a receptor.

3.4 Both the concentration gradient and the electrical gradient will favor flow of Na$^+$ into cells.

3.5 Transmembrane proteins allow diffusion through channels, facilitated diffusion, and active transport.

3.6 Because fever involves an increase in body temperature, the rates of all diffusion processes would increase.

3.7 No, the water concentrations can never be the same in the two arms because the left arm contains pure water and the right arm contains a solution that is less than 100% water.

3.8 A 2% solution of NaCl will cause crenation of RBCs because it is hypertonic.

3.9 A gated channel opens or closes in response to chemical or electrical changes inside or outside the cell. When the gate is open, ions can diffuse through the channel.

3.10 The rate of facilitated diffusion depends on the concentration gradient and the number of available transporters.

3.11 ATP adds a phosphate group to the pump protein, which changes the pump's three-dimensional shape.

3.12 In secondary active transport, hydrolysis of ATP is used only indirectly to drive the activity of symporter or antiporter proteins, whereas this reaction directly powers the pump protein in primary active transport.

3.13 Cholesterol, iron, vitamins, and hormones are examples of ligands that can undergo receptor-mediated endocytosis.

3.14 The binding of particles to a plasma membrane receptor triggers pseudopod formation.

3.15 Receptor-mediated endocytosis and phagocytosis involve receptor proteins, pinocytosis does not.

3.16 Centrioles, cilia, and flagella all contain microtubules.

3.17 A cell without a centrosome probably would not be able to undergo cell division.

3.18 Cilia move fluids across cell surfaces, whereas flagella move an entire cell.

3.19 Large and small ribosomal subunits are synthesized in the nucleolus in the nucleus and then join in the cytoplasm.

3.20 Rough ER has attached ribosomes, whereas smooth ER does not.

3.21 The entry face receives and modifies proteins from rough ER while the exit face modifies, sorts, and packages molecules for transport to other destinations.

3.22 Some proteins are secreted from the cell by exocytosis, some are incorporated into the plasma membrane, and some enter transfer vesicles that deliver their cargo to other destinations, such as lysosomes.

3.23 Digestion of worn-out organelles by lysosomes is called autophagy.

3.24 Mitochondrial cristae increase the surface area available for chemical reactions and contain some of the enzymes needed for ATP production.

3.25 Chromatin is a complex of DNA, proteins, and some RNA.

3.26 A nucleosome is a double-stranded DNA wrapped twice around a core of eight histones (proteins).

3.27 Proteins determine the physical and chemical characteristics of cells.

3.28 The DNA base sequence AGCT would be transcribed into the RNA base sequence UCGA. RNA polymerase catalyzes transcription of DNA.

3.29 The P site holds the tRNA attached to the growing polypeptide. The A site holds the tRNA carrying the next amino acid to be added to the growing polypeptide.

3.30 When a ribosome encounters a stop codon at the A site, the completed protein detaches from the final tRNA.

3.31 Chromosomes replicate during the S phase.

3.32 DNA replication occurs before cytokinesis so that each of the new daughter cells will have a complete genome.

3.33 Cytokinesis usually starts in late anaphase or early telophase.

3.34 Sperm, which use the flagella for locomotion, are the only body cells that need to move considerable distances.

The Tissue Level
of Organization

My favorite learning tool in Chapter 4 is the photomicrographs. Have you ever glanced at a friend's vacation photo and asked, "Where was that taken? But with a closer look you see palm trees and a sandy beach so you know it's not the North Pole. Then you look closer and notice them mimicking the movements of a hula dancer standing next to them and say, "When did you go to Hawaii? That's the exact method I use for identifying tissues. I cover up the text and focus solely on the photomicrographs. I look at the shapes and arrangements of cells and ask myself "Where was that taken?" This time instead of traveling the world I am traveling the human body. Absorbing the visual characteristics of each photomicrograph I try not to predict or assume. Once I feel comfortable with the cells, structures, and overall design I uncover the text and read. With my eyes shifting from the words to the photomicrograph and back again I see how the design of the tissue fits the description, location, and function. Pretty soon I realize the palm trees of the human body are neurons with reaching leaves (dendrites), a long trunk (axon), and a coconut (nucleus). The sandy beaches of the body are stratified squamous epithelium with the waves of dermis pushing their way up onto them. Our bodies even have hula dancers (columnar epithelium) with grass skirts (cilia) moving in rhythm to expel foreign particles.

Learning anatomy and physiology through this visual diagnostic journey allows me to hone my detective skills. Skills that I will need to become a good ultrasound technician.

RoNell Coco
Clark College

TISSUES AND HOMEOSTASIS

The four basic types of tissues in the human body contribute to homeostasis by providing diverse functions including protection, support, communication among cells, and resistance to disease, to name just a few.

ENERGIZE YOUR STUDY
FOUNDATIONS CD
Tissue Folder
- **Learn how cells form tissues with an Animation on Intercellular Junctions**
- **Explore the details of the major tissue types and see where they are found throughout your body**
- **Discover how much you have learned using Exercises like Name That Tissue or Getting it Together**

Systems Folder
- **Concept and Connections: Integumentary System and Tissues**

www.wiley.com/college/apcentral

INSIGHTS AND EXPLORATIONS

Americans are fat! Obesity is a common health risk factor among Americans. While checking out at the grocery store the racks are full of magazines blaring headlines about how to lose weight in time for the holidays. Or how this celebrity gets and keeps the weight off. One popular method is not to diet but to have the fat tissue (adipose tissue) removed by liposuction. In this web based exercise we will explore how liposuction can be used to remove a specific tissue in the body.

A cell is a complex collection of compartments, each of which carries out a host of biochemical reactions that make life possible. However, a cell seldom functions as an isolated unit in the body. Instead, cells usually work together in groups called tissues. A **tissue** is a group of similar cells that usually has a common embryonic origin and functions together to carry out specialized activities. The structure and properties of a specific tissue are influenced by factors such as the nature of the extracellular material that surrounds the tissue cells and the connections between the cells that compose the tissue. Tissues may be hard, semisolid, or even liquid in their consistency, a range exemplified by bone, fat, and blood. In addition, tissues vary greatly with respect to the kinds of cells present, how the cells are arranged, and the types of materials found between the cells. **Histology** (hiss'-TOL-ō-jē; *histo-* = tissue; *-logy* = study of) is the science that deals with the study of tissues. A **pathologist** (pa-THOL-ō-gist; *patho-* = disease) is a physician who specializes in laboratory studies of cells and tissues to help other physicians make accurate diagnoses. One of the principal functions of a pathologist is to examine tissues for any changes that might indicate disease.

TYPES OF TISSUES AND THEIR ORIGINS

▶ OBJECTIVE

- **Name the four basic types of tissues that make up the human body and describe the characteristics of each.**

Body tissues can be classified into four basic types according to their function and structure:

1. Epithelial tissue covers body surfaces and lines hollow organs, body cavities, and ducts. It also forms glands.

2. Connective tissue protects and supports the body and its organs. Various types of connective tissue bind organs together, store energy reserves as fat, and help provide immunity to disease-causing organisms.

3. Muscle tissue generates the physical force needed to make body structures move.

4. Nervous tissue detects changes in a variety of conditions inside and outside the body and responds by generating nerve impulses that help maintain homeostasis.

Epithelial tissue and connective tissue, except for bone tissue and blood, are discussed in detail in this chapter. The general features of bone tissue and blood will be introduced here, but a detailed discussion of their characteristics and functions is presented in Chapters 6 and 19, respectively. Similarly, the structure and function of muscle tissue and nervous tissue are examined in detail in Chapters 10 and 12, respectively.

All tissues of the body develop from three **primary germ layers,** the first tissues that form in a human embryo: **ectoderm, mesoderm,** and **endoderm** (see Figure 29.7b on page 1071). All three primary germ layers contribute to epithelial tissues. Mesoderm gives rise to all connective tissues and most muscle tissues. Ectoderm develops into nervous tissue. (Table 29.1 on page 1072 provides a list of structures derived from each layer.)

Normally, most cells within a tissue remain anchored to other cells, to basement membranes (described shortly), and to connective tissues. A few cells, such as phagocytes, move freely through the body, searching for invaders. Before birth, however, many cells migrate extensively as part of the growth and development process.

Biopsy

A **biopsy** (BĪ-op-sē; *bio-* = life, *-opsy* = to view) is the removal of a sample of living tissue for microscopic examination. This procedure is used to help diagnose many disorders, especially cancers, and to discover the cause of unexplained infections and inflammations. Both normal and potentially diseased tissues are removed for purposes of comparison. Once the tissue sample is removed, either surgically or through a needle and syringe, it may be preserved, stained to highlight special properties, or cut into thin sections for microscopic observation. Sometimes a biopsy is conducted while a patient is anesthetized during surgery to help a physician determine the most appropriate treatment. For example, if a biopsy of thyroid tissue reveals malignant cells, the surgeon can proceed with the most appropriate procedure right away. ■

▶ CHECKPOINT

1. Define a tissue.

2. What are the four basic types of body tissues?

CELL JUNCTIONS

▶ OBJECTIVE

- **Describe the structure and functions of the five main types of cell junctions.**

Most epithelial cells and some muscle and nerve cells are tightly joined into functional units. **Cell junctions** are contact points between the plasma membranes of tissue cells. Depending on their structure, cell junctions may serve one of three functions: (1) forming seals between cells, like a "zip-lock" at the top of a plastic storage bag, (2) anchoring cells to one another or to extracellular material, or (3) providing channels that allow ions and molecules to pass from cell to cell within a tissue. Here we consider the five most important types of cell junctions: tight junctions, adherens junctions, desmosomes, hemidesmosomes, and gap junctions (Figure 4.1):

branes together (Figure 4.1a). Cells of epithelial tissues that line the stomach, intestines, and urinary bladder have many tight junctions. They retard the passage of substances between cells and thus prevent the contents of these organs from leaking into the blood or surrounding tissues.

Tight Junctions

Tight junctions consist of weblike strands of transmembrane proteins that fuse the outer surfaces of adjacent plasma mem-

Adherens Junctions

Adherens junctions (ad-HER-ens) contain *plaque,* a dense layer of proteins on the inside of the plasma membrane that

Figure 4.1 Cell junctions.

Most epithelial cells and some muscle and nerve cells contain cell junctions.

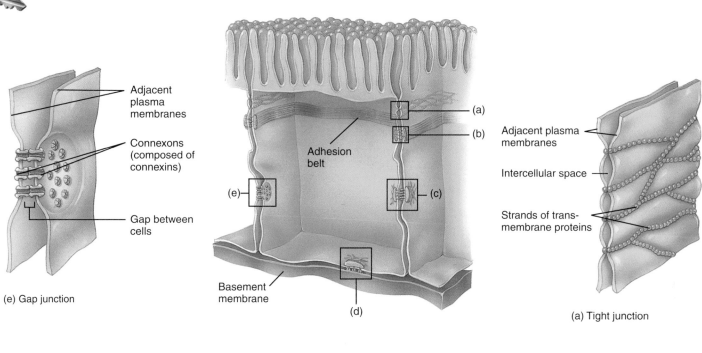

(e) Gap junction

(a) Tight junction

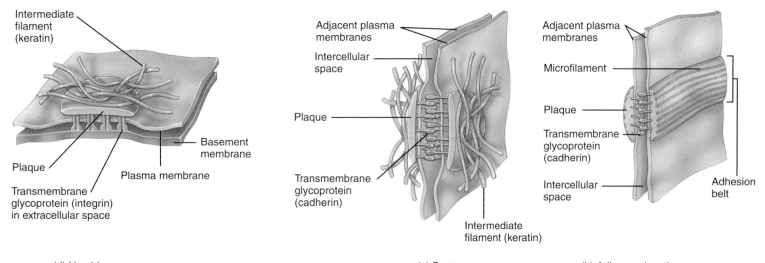

(d) Hemidesmosome

(c) Desmosome

(b) Adherens junction

 Which type of cell junction functions in communication between adjacent cells?

attaches to both cytoskeleton proteins and membrane proteins (Figure 4.1b). Actin microfilaments extend from the plaque into the cell's cytosol. Transmembrane glycoproteins called **cadherins** insert into the plaque from the opposite side. Each cadherin partially crosses the intercellular space (the space between the cells) and connects to cadherins of the adjacent cell, thus attaching the two cells. In epithelial cells, adherens junctions often form extensive zones called *adhesion belts* that encircle the cell. Adherens junctions help epithelial surfaces resist separation.

Desmosomes

Like adherens junctions, **desmosomes** (DEZ-mō-sōms; *desmo-* = band) contain plaque and have transmembrane glycoproteins (cadherins) that extend into the intercellular space between adjacent cell membranes and attach cells to one another (Figure 4.1c). Intermediate filaments (made of keratin) extend from desmosomes on one side of the cell across the cytosol to desmosomes on the opposite side of the cell. This structural arrangement contributes to the stability of the cells and tissue. Desmosomes are common among the cells that make up the epidermis (the outermost layer of the skin) and between cardiac muscle cells in the heart. Desmosomes prevent epidermal cells from separating under tension and cardiac muscle cells from pulling apart during contraction.

Hemidesmosomes

Hemidesmosomes (*hemi-* = half) resemble desmosomes but they lack links to adjacent cells. They look like half of a desmosome, thus the name (Figure 4.1d). However, the transmembrane glycoproteins in hemidesmosomes are **integrins** rather than cadherins. On the inside of the plasma membrane, integrins attach to intermediate filaments made of the protein keratin. On the outside of the plasma membrane, the integrins attach to the protein *laminin*, which is present in the basement membrane (discussed shortly). Thus, hemidesmosomes anchor cells to the basement membrane.

Gap Junctions

At **gap junctions**, membrane proteins called **connexins** form tiny fluid-filled tunnels called **connexons** that connect neighboring cells (Figure 4.1e). The plasma membranes of gap junctions are not fused together as in tight junctions but are separated by a very narrow intercellular gap (space). Ions and small molecules can diffuse through connexons from the cytosol of one cell to another. Gap junctions allow the cells in a tissue to communicate with each other. In a developing embryo, some of the chemical and electrical signals that regulate growth and cell differentiation travel via gap junctions. Moreover, gap junctions enable nerve or muscle impulses to spread rapidly among cells, a process that is crucial for the normal operation of some parts of the nervous

system and for the contraction of muscle in the heart and gastrointestinal tract.

▶ CHECKPOINT

3. Which type of cell junctions allow cellular communication?
4. Which types of cell junctions are found in epithelial tissues?

EPITHELIAL TISSUE

▶ OBJECTIVES

- **Describe the general features of epithelial tissues.**
- **For each different type of epithelium, list its location, structure, and function.**

An **epithelial tissue** (ep-i-THĒ-lē-al), or **epithelium** (plural is *epithelia*), consists of cells arranged in continuous sheets, in either single or multiple layers. Because the cells are closely packed and are held tightly together by many cell junctions, there is little intercellular space between adjacent plasma membranes.

The various surfaces of epithelial cells often differ in structure and have specialized functions. The **apical (free) surface** of an epithelial cell faces the body surface, a body cavity, the lumen (interior space) of an internal organ, or a tubular duct that receives secretions from the cells (Figure 4.2). Apical surfaces may contain cilia or microvilli. The **lateral surfaces** of an epithelial cell face the adjacent cells on either side. As you have seen in Figure 4.1, lateral surfaces may contain tight junctions, adherens junctions, desmosomes, and gap junctions. The **basal surface** of an epithelial cell is opposite the apical surface and adheres to extracellular materials. Hemidesmosomes in the basal surfaces of epithelial cells anchor the epithelium to the basement membrane. In subsequent discussions about epithelia with multiple layers the term *apical layer* refers to the most superficial layer of cells whereas the *basal layer* is the deepest layer of cells.

The **basement membrane** is a thin extracellular layer that commonly consists of two layers, the basal lamina and reticular lamina. The *basal lamina* (*lamina* = thin layer) is closer to the epithelial cells and is secreted by them. It contains proteins such as collagen (described shortly) and laminin, as well as glycoproteins and proteoglycans (also described shortly). The laminin molecules in the basal lamina adhere to integrins in hemidesmosomes and thus attach epithelial cells to the basement membrane (see Figure 4.1d). The *reticular lamina* is closer to the underlying connective tissue and contains fibrous proteins produced by connective tissue cells called fibroblasts. Besides their function of attaching to and supporting the overlying epithelial tissue, basement membranes have other roles. They form a surface along which epithelial cells migrate during growth or wound healing, they restrict passage of larger molecules between epithelium and connective tissue, and they participate in filtration of blood in the kidneys.

Epithelial tissue is **avascular** (*a-* = without; *vascular* = vessel); that is, it lacks its own blood supply. The blood vessels that bring in nutrients and remove wastes are located in the

Figure 4.2 Surfaces of epithelial cells and the structure and location of the basement membrane.

The basement membrane is found between epithelium and connective tissue.

Apical (free) surface
Lateral surfaces
Epithelium
Basal surface
Basal lamina
Reticular lamina
Basement membrane
Connective tissue
Nerve
Blood vessel

What are the functions of the basement membrane?

adjacent connective tissue. Exchange of substances between connective tissue and epithelium occurs by diffusion. Although epithelial tissue is avascular, it has a nerve supply.

Because epithelial tissue forms boundaries between the body's organs, or between the body and the external environment, it is repeatedly subjected to physical stress and injury. A high rate of cell division, however, allows epithelial tissue to constantly renew and repair itself by sloughing off dead or injured cells and replacing them with new ones. Epithelial tissue plays many different roles in the body, the most important of which are protection, filtration, secretion, absorption, and excretion. In addition, epithelial tissue combines with nervous tissue to form special organs for smell, hearing, vision, and touch.

Epithelial tissue may be divided into two types. **Covering and lining epithelium** forms the outer covering of the skin and some internal organs. It also forms the inner lining of blood vessels, ducts, and body cavities, and the interior of the respiratory, digestive, urinary, and reproductive systems. **Glandular epithelium** constitutes the secreting portion of glands, such as the thyroid gland, adrenal glands, and sweat glands.

Covering and Lining Epithelium

The types of covering and lining epithelial tissue are classified according to two characteristics: the arrangement of cells into layers and the shapes of the cells.

1. *Arrangement of cells in layers.* The cells of covering and lining epithelia are arranged in one or more layers depending on the function the epithelium performs:

   a. *Simple epithelium* is a single layer of cells that functions in diffusion, osmosis, filtration, *secretion* (the production and release of substances such as mucus, sweat, or enzymes), and *absorption* (the intake of fluids or other substances by cells).

   b. *Stratified epithelium* consists of two or more layers of cells that protect underlying tissues in locations where there is considerable wear and tear.

   c. *Pseudostratified epithelium* contains only a single layer of cells, but it appears to have multiple layers because the cell nuclei lie at different levels and not all cells reach the apical surface. Cells that do extend to the apical surface are either ciliated or secrete mucus.

2. *Cell shapes.*

   a. *Squamous* cells (SKWĀ-mus = flat) are arranged like floor tiles and are thin, which allows for the rapid movement of substances through them.

   b. *Cuboidal* cells are as tall as they are wide and are shaped like cubes or hexagons. They may have microvilli at their apical surface and function in either secretion or absorption.

   c. *Columnar* cells are much taller than they are wide and protect underlying tissues. Their apical surface may have cilia or microvilli, and they often are specialized for secretion and absorption.

   d. *Transitional* cells change shape, from cuboidal to flat and back, as organs stretch to a larger size then collapse to a smaller size.

   Considering the arrangements of layers and the cell shapes, the types of covering and lining epithelia are:

   I. Simple epithelium
      A. Simple squamous epithelium
      B. Simple cuboidal epithelium
      C. Simple columnar epithelium (nonciliated and ciliated)

   II. Stratified epithelium
      A. Stratified squamous epithelium (keratinized and nonkeratinized)*
      B. Stratified cuboidal epithelium*
      C. Stratified columnar epithelium*
      D. Transitional epithelium

   III. Pseudostratified columnar epithelium (nonciliated and ciliated)

Each of these covering and lining epithelial tissues is described in the following sections and illustrated in (Table 4.1). The tables throughout this chapter contain figures consisting of a

---

*This classification is based on the shape of the cells in the apical layer.

photomicrograph, a corresponding diagram, and an inset that identifies a major location of the tissue in the body. Along with the illustrations are descriptions, locations, and functions of the tissues.

Simple Epithelium

SIMPLE SQUAMOUS EPITHELIUM This tissue consists of a single layer of flat cells that resembles a tiled floor when viewed from their apical surface (Table 4.1A). The nucleus of each cell is a flattened oval or sphere and is centrally located. Simple squamous epithelium is present at sites where the processes of filtration, such as blood filtration in the kidneys, or diffusion, for instance diffusion of oxygen into blood vessels of the lungs, are occurring. It is not found in body areas that are subject to wear and tear.

The simple squamous epithelium that lines the heart, blood vessels, and lymphatic vessels is known as **endothelium** (*endo-* = within; *-thelium* = covering); the type that forms the epithelial layer of serous membranes is called **mesothelium** (*meso-* = middle). Unlike other epithelial tissues, which arise from embryonic ectoderm or endoderm, endothelium and mesothelium both are derived from embryonic mesoderm.

SIMPLE CUBOIDAL EPITHELIUM The cuboidal shape of the cells in this tissue (Table 4.1B) is obvious when the tissue is sectioned and viewed from the side. Cell nuclei are usually round and centrally located. Simple cuboidal epithelium performs the functions of secretion and absorption.

SIMPLE COLUMNAR EPITHELIUM When viewed from the side, the cells appear rectangular with oval nuclei near the base of the cells. Simple columnar epithelium exists in two forms: nonciliated simple columnar epithelium and ciliated simple columnar epithelium.

Nonciliated simple columnar epithelium contains two types of cells—columnar epithelial cells with microvilli at their apical surface and goblet cells (Table 4.1C). **Microvilli** are microscopic fingerlike cytoplasmic projections that increase the surface area of the plasma membrane (see Figure 3.1 on page 60). Their presence increases the rate of absorption by the cell. **Goblet cells** are modified columnar epithelial cells that secrete mucus, a slightly sticky fluid, at their apical surfaces. Before it is released, mucus accumulates in the upper portion of the cell, causing that area to bulge out. The whole cell then resembles a goblet or wine glass. Secreted mucus serves as a lubricant for the linings of the digestive, respiratory, reproductive, and most of the urinary tracts. Mucus also helps prevent destruction of the stomach lining by acidic gastric juice.

Ciliated simple columnar epithelium contains columnar epithelial cells with cilia at their apical surface (Table 4.1D). In the airways of the upper respiratory tract, goblet cells are interspersed among ciliated columnar cells. Mucus secreted by the goblet cells forms a film over the airway surface that traps inhaled foreign particles. The cilia beat in unison, moving the mucus and any foreign particles toward the throat, where it can be coughed up and swallowed or spit out. Cilia also help move

oocytes expelled from the ovaries through the uterine (Fallopian) tubes into the uterus.

Stratified Epithelium

In contrast to simple epithelium, stratified epithelium has at least two layers of cells. Thus, it is more durable and can better protect underlying tissues. Some cells of stratified epithelia also produce secretions. The name of the specific kind of stratified epithelium depends on the shape of the cells in the apical layer.

STRATIFIED SQUAMOUS EPITHELIUM Cells in the apical layer of this type of epithelium are flat, whereas in the deep layers, cells vary in shape from cuboidal to columnar (Table 4.1E). The deepest cells continually undergo cell division. As new cells grow, the cells of the basal layer are pushed upward toward the apical layer. As they move farther from the deeper layers and from their blood supply in the underlying connective tissue, they become dehydrated, shrunken, and harder and then die. At the apical layer, the cells lose their cell junctions and are sloughed off, but they are replaced as new cells continually emerge from the basal cells.

Stratified squamous epithelium exists in both keratinized and nonkeratinized forms. In *keratinized stratified squamous epithelium*, the apical layer and several layers deep to it are partially dehydrated and contain a layer of **keratin**, a tough, fibrous protein that helps protect the skin and underlying tissues from heat, microbes, and chemicals. *Nonkeratinized stratified squamous epithelium* does not contain keratin in the apical layer and several layers deep to it and remains moist.

Papanicolaou Test

A **Papanicolaou test** (pa-pa-NI-kō-lō), **Pap test,** or **Pap smear** involves collecting and microscopically examining epithelial cells that have sloughed off the apical layer of a tissue. A very common type of Pap test involves examining the cells from the nonkeratinized stratified squamous epithelium of the cervix and vagina. This type of Pap test is performed mainly to detect early changes in the cells of the female reproductive system that may indicate cancer or a precancerous condition. An annual Pap test is recommended for all women as part of a routine pelvic exam. ■

STRATIFIED CUBOIDAL EPITHELIUM This fairly rare type of epithelium sometimes consists of more than two layers of cells (Table 4.1F). Its function is mainly protective but it also functions in limited secretion and absorption.

STRATIFIED COLUMNAR EPITHELIUM Like stratified cuboidal epithelium, this type of tissue also is uncommon. Usually the basal layer consists of shortened, irregularly shaped cells; only the apical layer has cells that are columnar in form (Table 4.1G). This type of epithelium functions in protection and secretion.

(text continues on page 115)

Table 4.1 Epithelial Tissues

Covering and Lining Epithelium

A. Simple squamous epithelium

Description: Single layer of flat cells; centrally located nucleus.

Location: Lines heart, blood vessels, lymphatic vessels, air sacs of lungs, glomerular (Bowman's) capsule of kidneys, and inner surface of the tympanic membrane (eardrum); forms epithelial layer of serous membranes, such as the peritoneum.

Function: Filtration, diffusion, osmosis, and secretion in serous membranes.

Plasma membrane

Cytoplasm

Nucleus

Peritoneum

LM 243x

Surface view of simple squamous epithelium of mesothelial lining of peritoneum

Simple squamous cell

Basement membrane

Connective tissue

Simple squamous epithelium

B. Simple cuboidal epithelium

Description: Single layer of cube-shaped cells; centrally located nucleus.

Location: Covers surface of ovary, lines anterior surface of capsule of the lens of the eye, forms the pigmented epithelium at the back of the eye, lines kidney tubules and smaller ducts of many glands, and makes up the secreting portion of some glands such as the thyroid gland and the ducts of some glands such as the pancreas.

Function: Secretion and absorption.

Duodenum

Pancreas

Simple cuboidal epithelium

Nucleus of simple cuboidal cell

Simple cuboidal cell

Lumen of duct

Basement membrane

Connective tissue

Connective tissue

LM 330x

Sectional view of simple cuboidal epithelium of intralobular duct of pancreas

Simple cuboidal epithelium

(continues)

Table 4.1   Epithelial Tissues *(continued)*

Covering and Lining Epithelium

C. Nonciliated simple columnar epithelium

Description: Single layer of nonciliated rectangular cells with nuclei near base of cells; contains goblet cells and cells with microvilli in some locations.

Location: Lines the gastrointestinal tract from the stomach to the anus, ducts of many glands, and gallbladder.

Function: Secretion and absorption.

Sectional view of nonciliated simple columnar epithelium of lining of jejunum of small intestine

Nonciliated simple columnar epithelium

D. Ciliated simple columnar epithelium

Description: Single layer of ciliated rectangular cells with nuclei near base of cells; contains goblet cells in some locations.

Location: Lines a few portions of upper respiratory tract, uterine (Fallopian) tubes, uterus, some paranasal sinuses, and central canal of spinal cord.

Function: Moves mucus and other substances by ciliary action.

Sectional view of ciliated simple columnar epithelium of uterine tube

Ciliated simple columnar epithelium

Covering and Lining Epithelium

E. Stratified squamous epithelium

Description: Several layers of cells; cuboidal to columnar shape in deep layers; squamous cells form the apical layer and several layers deep to it; cells from the basal layer replace surface cells as they are lost.

Location: Keratinized variety forms superficial layer of skin; nonkeratinized variety lines wet surfaces, such as lining of the mouth, esophagus, part of epiglottis, and vagina, and covers the tongue.

Function: Protection.

Vagina

Stratified squamous epithelium

Connective tissue

Flattened squamous cell at apical surface

Basement membrane

Connective tissue

LM 200x

Sectional view of stratified squamous epithelium of vagina

Stratified squamous epithelium

F. Stratified cuboidal epithelium

Description: Two or more layers of cells in which the cells in the apical layer are cube-shaped.

Location: Ducts of adult sweat glands and esophageal glands and part of male urethra.

Function: Protection and limited secretion and absorption.

Esophagus

Nucleus of stratified cuboidal cell

Lumen of duct

Stratified cuboidal epithelium

Connective tissue

Apical surface

Basement membrane

Connective tissue

LM 380x

Sectional view of stratified cuboidal epithelium of the duct of an esophageal gland

Stratified cuboidal epithelium

(continues)

Table 4.1 Epithelial Tissues (continued)

Covering and Lining Epithelium

G. Stratified columnar epithelium

Description: Several layers of irregularly shaped cells; columnar cells are only in the apical layer.

Location: Lines part of urethra, large excretory ducts of some glands, such as esophageal glands, small areas in anal mucous membrane, and part of the conjunctiva of the eye.

Function: Protection and secretion.

Esophagus

Stratified columnar epithelium

Lumen of duct

Nucleus of stratified columnar cell

Connective tissue

Apical surface

Basement membrane

Connective tissue

LM 300x

Sectional view of stratified columnar epithelium of the duct of an esophageal gland

Stratified columnar epithelium

H. Transitional epithelium

Description: Appearance is variable (transitional); shape of cells in apical layer ranges from squamous (when stretched) to cuboidal (when relaxed).

Location: Lines urinary bladder and portions of ureters and urethra.

Function: Permits distention.

Urinary bladder

Lumen of urinary bladder

Nucleus of transitional cell

Connective tissue

Transitional epithelium

Apical surface

Basement membrane

Connective tissue

LM 500x

Sectional view of transitional epithelium of urinary bladder in relaxed state

Relaxed transitional epithelium

Covering and Lining Epithelium

I. Pseudostratified columnar epithelium

Description: Not a true stratified tissue; nuclei of cells are at different levels; all cells are attached to basement membrane, but not all reach the surface.

Location: Pseudostratified ciliated columnar epithelium lines the airways of most of upper respiratory tract; pseudostratified nonciliated columnar epithelium lines larger ducts of many glands, epididymis, and part of male urethra.

Function: Secretion and movement of mucus by ciliary action.

Trachea

Mucus in goblet cell | Cilia | Lumen of trachea

Nucleus of ciliated columnar cell

Nucleus of goblet cell

Nucleus of basal cell

Connective tissue

LM 300x

Sectional view of pseudostratified ciliated columnar epithelium of trachea

Pseudostratified ciliated columnar epithelium

Mucus in goblet cell | Ciliated columnar cell

Cilia

Basement membrane

Basal cell

Connective tissue

Pseudostratified ciliated columnar epithelium

(continues)

Table 4.1	Epithelial Tissues *(continued)*

Glandular Epithelium

J. Endocrine glands

Description: Secretory products (hormones) diffuse into blood after passing through interstitial fluid.

Location: Examples include pituitary gland at base of brain, pineal gland in brain, thyroid and parathyroid glands near larynx (voice box), adrenal glands superior to kidneys, pancreas near stomach, ovaries in pelvic cavity, testes in scrotum, and thymus in thoracic cavity.

Function: Produce hormones that regulate various body activities.

LM 500x

Sectional view of endocrine gland (thyroid gland)

Thyroid gland

Thyroid follicle

Blood vessel

Hormone-producing (epithelial) cell

Stored precursor of hormone

Thyroid follicle

Endocrine gland (thyroid gland)

K. Exocrine glands

Description: Secretory products released into ducts.

Location: Sweat, oil, and earwax glands of the skin; digestive glands such as salivary glands, which secrete into mouth cavity, and pancreas, which secretes into the small intestine.

Function: Produce mucus, sweat, oil, earwax, saliva, or digestive enzymes.

Skin

Secretory portion of sweat gland

LM 300x

Sectional view of the secretory portion of an exocrine gland (sweat gland)

Lumen of duct of sweat gland

Nucleus of secretory cell of sweat gland

Basement membrane

Exocrine gland (sweat gland)

TRANSITIONAL EPITHELIUM This kind of stratified epithelium is present only in the urinary system and has a variable appearance. In its relaxed or unstretched state (Table 4.1H), transitional epithelium looks similar to stratified cuboidal epithelium, except that the cells in the apical layer tend to be large and rounded. As the tissue is stretched, its cells become flatter, giving the appearance of stratified squamous epithelium. Because of its elasticity, transitional epithelium is ideal for lining hollow structures that are subjected to expansion from within, such as the urinary bladder. It functions to accommodate a variable volume of fluid without rupturing.

Pseudostratified Columnar Epithelium

The third category of covering and lining epithelium is pseudostratified columnar epithelium (Table 4.1I). The nuclei of the cells are at various depths. Even though all the cells are attached to the basement membrane in a single layer, some cells do not extend to the surface. When viewed from the side, these features give the false impression of a multilayered tissue—thus the name *pseudo*stratified epithelium (*pseudo-* = false). In *pseudostratified ciliated columnar epithelium,* the cells that extend to the surface either secrete mucus (goblet cells) or bear cilia. The mucus traps foreign particles and the cilia sweep away mucus for eventual elimination from the body. *Pseudostratified nonciliated columnar epithelium* contains cells without cilia and lacks goblet cells.

Glandular Epithelium

The function of glandular epithelium is secretion, which is accomplished by glandular cells that often lie in clusters deep to the covering and lining epithelium. A **gland** may consist of a single cell or a group of cells that secrete substances into ducts (tubes), onto a surface, or into the blood. All glands of the body are classified as either endocrine or exocrine.

The secretions of **endocrine glands** (Table 4.1J) enter the interstitial fluid and then diffuse directly into the bloodstream without flowing through a duct. These secretions, called *hormones,* regulate many metabolic and physiological activities to maintain homeostasis. The pituitary, thyroid, and adrenal glands are examples of endocrine glands. Endocrine glands will be described in detail in Chapter 18.

Exocrine glands (*exo-* = outside; *-crine* = secretion; (Table 4.1K) secrete their products into ducts that empty onto the surface of a covering and lining epithelium. Thus, the product of an exocrine gland is released at the skin surface or into the lumen of a hollow organ. The secretions of exocrine glands include mucus, sweat, oil, earwax, saliva, and digestive enzymes. Examples of exocrine glands are sudoriferous (sweat) glands, which produce sweat to help lower body temperature, and salivary glands, which secrete saliva. Saliva contains mucus and digestive enzymes among other substances. As you will see later, some glands of the body, such as the pancreas, ovaries, and testes, are mixed glands that contain both endocrine and exocrine tissue.

Structural Classification of Exocrine Glands

Exocrine glands are classified into unicellular and multicellular types. **Unicellular glands** are single-celled. Goblet cells, an important unicellular exocrine gland, secrete mucus directly onto the apical surface of a lining epithelium rather than into ducts. Most glands are **multicellular glands,** composed of many cells that form a distinctive microscopic structure or macroscopic organ. Examples are sudoriferous, sebaceous (oil), and salivary glands.

Multicellular glands are categorized according to two criteria: by whether the ducts of the glands are branched or unbranched and by the shape of the secretory portions of the gland (Figure 4.3). If the duct of the gland does not branch, it is a **simple gland.** If the duct branches, it is a **compound gland.** Glands with tubular secretory parts are **tubular glands;** those with more rounded secretory portions are **acinar glands** (AS-i-nar; *acin-* = berry). **Tubuloacinar glands** have both tubular and more rounded secretory parts.

Combinations of the degree of duct branching and the shape of the secretory part are the criteria for the following structural classification scheme for multicellular exocrine glands:

I. Simple glands
   A. **Simple tubular.** Tubular secretory part is straight and attaches to a single unbranched duct. Example: glands in the large intestine.
   B. **Simple branched tubular.** Tubular secretory part is branched and attaches to a single unbranched duct. Example: gastric glands.
   C. **Simple coiled tubular.** Tubular secretory part is coiled and attaches to a single unbranched duct. Example: sweat glands.
   D. **Simple acinar.** Secretory portion is rounded and attaches to a single unbranched duct. Example: glands of penile urethra.
   E. **Simple branched acinar.** Rounded secretory part is branched and attaches to a single unbranched duct. Example: sebaceous glands.
II. Compound glands
   A. **Compound tubular.** Secretory portion is tubular and attaches to a branched duct. Example: bulbourethral (Cowper's) glands.
   B. **Compound acinar.** Secretory portion is rounded and attaches to a branched duct. Example: mammary glands.
   C. **Compound tubuloacinar.** Secretory portion is both tubular and rounded and attaches to a branched duct. Example: acinar glands of the pancreas.

Functional Classification of Exocrine Glands

The functional classification of exocrine glands is based on how their secretion is released. In **merocrine glands** (MER-ō-krin;

Figure 4.3 Multicellular exocrine glands. Pink represents the secretory portion; lavender represents the duct.

🔑 **Structural classification of multicellular exocrine glands is based on the branching pattern of the duct and the shape of the secreting portion.**

Simple tubular   Simple branched tubular   Simple coiled tubular   Simple acinar   Simple branched acinar

Compound tubular   Compound acinar   Compound tubuloacinar

❓ **How do simple multicellular glands differ from compound ones?**

mero- = a part), the secretion is synthesized on ribosomes attached to rough ER; processed, sorted, and packaged by the Golgi complex; and released from the cell in secretory vesicles via exocytosis (Figure 4.4a). Most exocrine glands of the body are merocrine glands. Examples are the salivary glands and pancreas. **Apocrine glands** (AP-ō-krin; apo- = from) accumulate their secretory product at the apical surface of the secreting cell. Then, that portion of the cell pinches off from the rest of the cell to release the secretion (Figure 4.4b). The remaining part of the cell repairs itself and repeats the process. Electron micrographic studies have called into question whether humans have apocrine glands. What were once thought to be apocrine glands—for example, mammary glands that secrete milk—are probably merocrine glands. The cells of **holocrine glands** (HŌ-lō-krin; holo- = entire) accumulate a secretory product in their cytosol. As the secretory cell matures, it ruptures and becomes the secretory prod-

uct (Figure 4.4c). The sloughed off cell is replaced by a new cell. One example of a holocrine gland is a sebaceous gland of the skin.

▶ CHECKPOINT

5. Describe the various layering arrangements and cell shapes of epithelium.

6. What characteristics are common to all epithelial tissues?

7. How is the structure of the following kinds of epithelium related to their functions: simple squamous, simple cuboidal, simple columnar (nonciliated and ciliated), stratified squamous (keratinized and nonkeratinized), stratified cuboidal, stratified columnar, transitional, and pseudostratified columnar (ciliated and nonciliated)?

8. Where are endothelium and mesothelium located?

9. What distinguishes endocrine glands and exocrine glands? Name and give examples of the three functional classes of exocrine glands.

Figure 4.4 Functional classification of multicellular exocrine glands.

 The functional classification of exocrine glands is based on whether a secretion is a product of a cell or consists of an entire or a partial glandular cell.

Secretion

Secretory vesicle

Golgi complex

Rough ER

Nucleus

(a) Merocrine secretion

Pinched off portion of cell is secretion

(b) Apocrine secretion

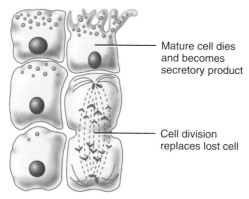

Mature cell dies and becomes secretory product

Cell division replaces lost cell

(c) Holocrine secretion

 What class of glands are sebaceous (oil) glands? Salivary glands?

CONNECTIVE TISSUE

▶ O B J E C T I V E S

- Describe the general features of connective tissue.
- Describe the structure, location, and function of the various types of connective tissues.

Connective tissue is one of the most abundant and widely distributed tissues in the body. In its various forms, connective tissue has a variety of functions. It binds together, supports, and strengthens other body tissues; protects and insulates internal organs; compartmentalizes structures such as skeletal muscles; is the major transport system within the body (blood, a fluid connective tissue); is the major site of stored energy reserves (adipose, or fat, tissue); and is the main site of immune responses.

General Features of Connective Tissue

Connective tissue consists of two basic elements: cells and matrix. A connective tissue's **matrix** fills the wide spaces between its cells. The matrix consists of protein-based fibers and ground substance, the material between the cells and the fibers. It is usually secreted by the connective tissue cells and determines the tissue's qualities. For instance, in cartilage, the matrix is firm but pliable. In bone, by contrast, the matrix is hard and not pliable.

In contrast to epithelia, connective tissues do not usually occur on body surfaces, such as the covering or lining of internal organs, the lining of body cavities, or the external surface of the body. However, a type of connective tissue called areolar connective tissue lines joint cavities. Also, unlike epithelia, connective tissues usually are highly vascular; that is, they have a rich blood supply. Exceptions include cartilage, which is avascular, and tendons, which have a scanty blood supply. Except for cartilage, connective tissues, like epithelia, have a nerve supply.

Connective Tissue Cells

Mesodermal embryonic cells called mesenchymal cells give rise to the cells of connective tissue. Each major type of connective tissue contains an immature class of cells whose name ends in *-blast,* which means "to bud or sprout." These immature cells are called *fibroblasts* in loose and dense connective tissue, *chondroblasts* in cartilage, and *osteoblasts* in bone. Blast cells retain the capacity for cell division and secrete the matrix that is characteristic of the tissue. In cartilage and bone, once the matrix is produced, the fibroblasts differentiate into mature cells whose names end in *-cyte,* namely chondrocytes and osteocytes. Mature cells have reduced capacity for cell division and matrix formation and are mostly involved in maintaining the matrix.

The types of cells present in connective tissues vary according to the type of tissue and include the following (Figure 4.5):

1.   Fibroblasts (FĪ-brō-blasts; *fibro-* = fibers) are large, flat cells with branching processes. They are present in several connective tissues, where they usually are the most numerous connective tissue cells. Fibroblasts migrate through the connective tissue, secreting the fibers and ground substance of the matrix.

2.   Macrophages (MAK-rō-fā-jez; *macro-* = large; *-phages* = eaters) develop from monocytes, a type of white blood cell. Macrophages have an irregular shape with short branching projections and are capable of engulfing bacteria and cellular debris by phagocytosis. Some are *fixed macrophages,* which means they reside in a particular tissue. Examples are alveolar macrophages in the lungs or spleen macrophages in the spleen. Others are *wandering macrophages,* which roam the tissues and gather at sites of infection or inflammation.

3.   Plasma cells are small cells that develop from a type of white blood cell called a B lymphocyte. Plasma cells secrete antibodies, proteins that attack or neutralize foreign substances in the body. Thus, plasma cells are an important part of the body's immune system. Although they are found in many places in the body, most plasma cells reside in connective tissues, especially in the gastrointestinal and respiratory tracts. They are also abundant in the salivary glands, lymph nodes, and red bone marrow.

4.   Mast cells are abundant alongside the blood vessels that supply connective tissue. They produce histamine, a chemical that dilates small blood vessels as part of the inflammatory response, the body's reaction to injury or infection.

5.   Adipocytes, also called fat cells or adipose cells, are connective tissue cells that store triglycerides (fats). They are found below the skin and around organs such as the heart and kidneys.

6.   White blood cells are not found in significant numbers in normal connective tissue. However, in response to certain conditions they migrate from blood into connective tissues. For example, *neutrophils* gather at sites of infection, and *eosinophils* migrate to sites of parasitic invasions and allergic responses.

Connective Tissue Matrix

Each type of connective tissue has unique properties, due to accumulation of specific matrix materials between the cells. The matrix consists of fluid, semifluid, gelatinous, or calcified ground substance and protein fibers.

Ground Substance

As mentioned earlier, the **ground substance** is the component of a connective tissue between the cells and fibers. The ground substance supports cells, binds them together, and provides a medium through which substances are exchanged between the blood and cells. It plays an active role in how tissues develop, migrate, proliferate, and change shape, and in how they carry out their metabolic functions.

Ground substance contains water and an assortment of large molecules, many of which are complex combinations of polysaccharides and proteins. The polysaccharides include

Figure 4.5   Representative cells and fibers in connective tissues.

Fibroblasts are usually the most numerous connective tissue cells.

What is the function of fibroblasts?

hyaluronic acid, chondroitin sulfate, dermatan sulfate, and keratan sulfate. Collectively, they are referred to as **glycosaminoglycans** (glī-kos-a-mē′-nō-GLĪ-kans) or **GAGs.** Except for hyaluronic acid, the GAGs are associated with proteins called **proteoglycans** (prō′-tē-ō-GLĪ-kans). The proteoglycans form a core protein and the GAGs project from the protein like the bristles of a brush. One of the most important properties of GAGs is that they trap water, making the ground substance more jellylike.

Hyaluronic acid (hī′-a-loo-RON-ik) is a viscous, slippery substance that binds cells together, lubricates joints, and helps maintain the shape of the eyeballs. White blood cells, sperm cells, and some bacteria produce *hyaluronidase,* an enzyme that breaks apart hyaluronic acid, thus causing the ground substance of connective tissue to become more liquid. The ability to produce hyaluronidase helps white blood cells move more easily through connective tissues to reach sites of infection and aids penetration of an oocyte by a sperm cell during fertilization. It may also allow bacteria to spread more rapidly through connective tissues. **Chondroitin sulfate** (kon-DROY-tin) provides support and adhesiveness in cartilage, bone, skin, and blood vessels. The skin, tendons, blood vessels, and heart valves contain **dermatan sulfate,** whereas bone, cartilage, and the cornea of the eye contain **keratan sulfate.** Also present in the ground substance are **adhesion proteins,** which are responsible for linking components of the ground substance to each other and to the surfaces of cells. The main adhesion protein of connective tissue is **fibronectin,** which binds to both collagen fibers (discussed shortly) and ground substance thereby linking them together. It also attaches cells to the ground substance.

Fibers

Three types of **fibers** are embedded in the matrix between the cells: collagen fibers, elastic fibers, and reticular fibers. They function to strengthen and support connective tissues.

Collagen fibers (*colla* = glue) are very strong and resist pulling forces, but they are not stiff, which promotes tissue flexibility. Different types of collagen fibers in various tissues have slightly different properties. For example, the collagen fibers found in cartilage attract more water molecules than do the collagen fibers in bone, which gives cartilage a more cushioning consistency. Collagen fibers often occur in bundles lying parallel to one another (Figure 4.5). The bundle arrangement affords great strength. Chemically, collagen fibers consist of the protein *collagen.* This is the most abundant protein in your body, representing about 25% of the total protein. Collagen fibers are found in most types of connective tissues, especially bone, cartilage, tendons, and ligaments.

Elastic fibers, which are smaller in diameter than collagen fibers, branch and join together to form a network within a tissue. An elastic fiber consists of molecules of the protein *elastin* surrounded by a glycoprotein named *fibrillin,* which strengthens and stabilizes elastic fibers. Because of their unique molecular structure, elastic fibers are strong but can be stretched up to 150% of their relaxed length without breaking. Equally important, elastic fibers have the ability to return to their original shape after being stretched, a property called *elasticity.* Elastic fibers are plentiful in skin, blood vessel walls, and lung tissue.

Reticular fibers (*reticul-* = net), consisting of *collagen* arranged in fine bundles and a coating of glycoprotein, provide support in the walls of blood vessels and form a network around the cells in some tissues, for instance, areolar tissue, adipose tissue, and smooth muscle tissue. Produced by fibroblasts, reticular fibers are much thinner than collagen fibers and form branching networks. Like collagen fibers, reticular fibers provide support and strength. Reticular fibers are plentiful in reticular connective tissue, which forms the **stroma** (= bed or covering) or supporting framework of many soft organs, such as the spleen and lymph nodes. These fibers also help form the basement membrane.

Marfan Syndrome

Marfan syndrome (MAR-fan) is an inherited disorder caused by a defective fibrillin gene. The result is abnormal development of elastic fibers. Tissues rich in elastic fibers are malformed or weakened. Structures affected most seriously are the covering layer of bones (periosteum), the ligament that suspends the lens of the eye, and the walls of the large arteries. People with Marfan syndrome tend to be tall and have disproportionately long arms, legs, fingers, and toes. A common symptom is blurred vision caused by displacement of the lens of the eye. The most life-threatening complication of Marfan syndrome is weakening of the aorta (the main artery that emerges from the heart), which can suddenly burst. ■

Classification of Connective Tissues

Classifying connective tissues can be challenging because of the diversity of cells and matrix present, and the differences in their relative proportions. Thus, the grouping of connective tissues into categories is not always clear-cut. We will classify them as follows:

I. Embryonic connective tissue
   A. Mesenchyme
   B. Mucous connective tissue
II. Mature connective tissue
   A. Loose connective tissue
      1. Areolar connective tissue
      2. Adipose tissue
      3. Reticular connective tissue

B. Dense connective tissue
  1. Dense regular connective tissue
  2. Dense irregular connective tissue
  3. Elastic connective tissue
C. Cartilage
  1. Hyaline cartilage
  2. Fibrocartilage
  3. Elastic cartilage
D. Bone tissue
E. Blood tissue
F. Lymph

Note that our classification scheme has two major subclasses of connective tissue: embryonic and mature. **Embryonic connective tissue** is present primarily in the *embryo,* the developing human from fertilization through the first two months of pregnancy, and in the *fetus,* the developing human from the third month of pregnancy to birth.

One example of embryonic connective tissue found almost exclusively in the embryo is **mesenchyme** (MEZ-en-kīm), the tissue from which all other connective tissues eventually arise (Table 4.2A). Mesenchyme is composed of irregularly shaped cells, a semifluid ground substance, and delicate reticular fibers. Another kind of embryonic tissue is **mucous connective tissue (Wharton's jelly),** found mainly in the umbilical cord of the fetus. Mucous connective tissue is a form of mesenchyme that contains widely scattered fibroblasts, a more viscous jellylike ground substance, and collagen fibers (Table 4.2B).

The second major subclass of connective tissue, **mature connective tissue,** is present in the newborn. Its cells arise from mesenchyme. Mature connective tissue is of several types, which we explore next.

Types of Mature Connective Tissue

The six types of mature connective tissue are (1) loose connective tissue, (2) dense connective tissue, (3) cartilage, (4) bone tissue, (5) blood tissue, and (6) lymph. We now examine each in detail.

Loose Connective Tissue

In **loose connective tissue** the fibers are loosely intertwined and many cells are present. The types of loose connective tissue are areolar connective tissue, adipose tissue, and reticular connective tissue.

AREOLAR CONNECTIVE TISSUE   One of the most widely distributed connective tissues in the body is **areolar connective tissue** (a-RĒ-ō-lar; *areol-* = a small space). It contains several kinds of cells, including fibroblasts, macrophages, plasma cells, mast cells, adipocytes, and a few white blood cells (Table 4.3A). All three types of fibers—collagen, elastic, and reticular—are

arranged randomly throughout the tissue. The ground substance contains hyaluronic acid, chondroitin sulfate, dermatan sulfate, and keratan sulfate. Combined with adipose tissue, areolar connective tissue forms the *subcutaneous layer,* the layer of tissue that attaches the skin to underlying tissues and organs.

ADIPOSE TISSUE   Adipose tissue is a loose connective tissue in which the cells, called **adipocytes** (*adipo-* = fat), are specialized for storage of triglycerides (Table 4.3B). Adipocytes are derived from fibroblasts. Because the cell fills up with a single, large triglyceride droplet, the cytoplasm and nucleus are pushed to the periphery of the cell. Adipose tissue is found wherever areolar connective tissue is located. Adipose tissue is a good insulator and can therefore reduce heat loss through the skin. It is a major energy reserve and generally supports and protects various organs. As the amount of adipose tissue increases with weight gain, new blood vessels form. Thus, an obese person has many more miles of blood vessels than does a lean person, a situation that can cause high blood pressure.

Most adipose tissue in adults is **white adipose tissue,** the type just described. Another type, called **brown adipose tissue (BAT),** obtains its darker color from a very rich blood supply and numerous mitochondria, which contain colored pigments that participate in aerobic cellular respiration. Although BAT is widespread in the fetus and infant, in adults only small amounts are present. BAT generates considerable heat and probably helps to maintain body temperature in the newborn. The heat generated by the many mitochondria is carried away to other body tissues by the extensive blood supply.

Liposuction

A surgical procedure, called **liposuction** (*lip-* = fat) or **suction lipectomy** (*-ectomy* = to cut out), involves suctioning out small amounts of adipose tissue from various areas of the body. The technique can be used as a body-contouring procedure in regions such as the thighs, buttocks, arms, breasts, and abdomen. Postsurgical complications that may develop include fat emboli (clots), infection, fluid depletion, injury to internal structures, and severe postoperative pain. ■

RETICULAR CONNECTIVE TISSUE   Reticular connective tissue consists of fine interlacing reticular fibers and reticular cells (Table 4.3C). Reticular connective tissue forms the stroma (supporting framework) of the liver, spleen, and lymph nodes and helps bind together smooth muscle cells. Additionally, reticular fibers in the spleen filter blood and remove worn-out blood cells and reticular fibers in lymph nodes filter lymph and remove bacteria.

Dense Connective Tissue

Dense connective tissue contains more numerous, thicker, and denser fibers but considerably fewer cells than loose connective

Table 4.2 Embryonic Connective Tissues

A. Mesenchyme

Description: Consists of irregularly shaped mesenchymal cells embedded in a semifluid ground substance that contains reticular fibers.

Location: Under skin and along developing bones of embryo; some mesenchymal cells are found in adult connective tissue, especially along blood vessels.

Function: Forms all other kinds of connective tissue.

Embryo

Ground substance

Nucleus of mesenchymal cell

Reticular fiber

LM 300x

Sectional view of mesenchyme of a developing embryo

Mesenchyme

B. Mucous connective tissue

Description: Consists of widely scattered fibroblasts embedded in a viscous, jellylike ground substance that contains fine collagen fibers.

Location: Umbilical cord of fetus.

Function: Support.

Umbilical cord

Fetus

Epithelial surface cell of umbilical cord

Ground substance

Collagen fiber

Nucleus of fibroblast

LM 275x

Sectional view of mucous connective tissue of the umbilical cord

Mucous connective tissue

Table 4.3   Mature Connective Tissues

Loose Connective Tissue

A. Areolar connective tissue

Description: Consists of fibers (collagen, elastic, and reticular) and several kinds of cells (fibroblasts, macrophages, plasma cells, adipocytes, and mast cells) embedded in a semifluid ground substance.

Location: Subcutaneous layer deep to skin; papillary (superficial) region of dermis of skin; lamina propria of mucous membranes; and around blood vessels, nerves, and body organs.

Function: Strength, elasticity, and support.

Skin

Subcutaneous layer

Mast cell
Collagen fibers
Elastic fibers

LM 300x

Sectional view of subcutaneous areolar connective tissue

Areolar connective tissue

B. Adipose tissue

Description: Consists of adipocytes, cells specialized to store triglycerides (fats) as a large centrally located droplet; nucleus and cytoplasm are peripherally located.

Location: Subcutaneous layer deep to skin, around heart and kidneys, yellow bone marrow, and padding around joints and behind eyeball in eye socket.

Function: Reduces heat loss through skin, serves as an energy reserve, supports, and protects. In newborns, brown adipose tissue generates considerable heat that helps maintain proper body temperature.

Heart

Fat

Nucleus of adipocyte
Cytoplasm
Fat-storage area of adipocyte
Blood vessel
Plasma membrane

Adipose tissue

LM 300x

Sectional view of adipose tissue showing adipocytes of white fat

tissue. There are three types: dense regular connective tissue, dense irregular connective tissue, and elastic connective tissue.

DENSE REGULAR CONNECTIVE TISSUE   In this tissue, bundles of collagen fibers are *regularly* arranged in parallel patterns that confer great strength (Table 4.3D). The tissue withstands pulling along the axis of the fibers. Fibroblasts, which produce the fibers and ground substance, appear in rows between the fibers. The tissue is silvery white and tough, yet somewhat pliable. Examples are tendons and most ligaments.

Loose Connective Tissue

C. Reticular connective tissue

Description: Consists of a network of interlacing reticular fibers and reticular cells.

Location: Stroma (supporting framework) of liver, spleen, lymph nodes; red bone marrow, which gives rise to blood cells; reticular lamina of the basement membrane; and around blood vessels and muscles.

Function: Forms stroma of organs; binds together smooth muscle tissue cells; filters and removes worn-out blood cells in the spleen and microbes in lymph nodes.

Lymph node

Nucleus of reticular cell

Reticular fiber

LM 225x

Sectional view of reticular connective tissue of a lymph node

Reticular connective tissue

Dense Connective Tissue

D. Dense regular connective tissue

Description: Matrix looks shiny white; consists mainly of collagen fibers arranged in bundles; fibroblasts present in rows between bundles.

Location: Forms tendons (attach muscle to bone), most ligaments (attach bone to bone), and aponeuroses (sheetlike tendons that attach muscle to muscle or muscle to bone).

Function: Provides strong attachment between various structures.

Tendon

Skeletal muscle

Nucleus of fibroblast

Collagen fiber

LM 250x

Sectional view of dense regular connective tissue of a tendon

Dense regular connective tissue

(continues)

DENSE IRREGULAR CONNECTIVE TISSUE This tissue contains collagen fibers that are packed more closely together than in loose connective tissue and are usually *irregularly* arranged (Table 4.3E). It is found in parts of the body where pulling forces are exerted in various directions. The tissue often occurs in sheets, such as in the dermis of the skin, which underlies the epidermis, or the pericardium around the heart. Heart valves, the perichondrium (the membrane surrounding cartilage), and the periosteum (the membrane surrounding bone) are dense irregular connective tissues, although they have a fairly orderly arrangement of their collagen fibers.

Table 4.3	Mature Connective Tissues *(continued)*

Dense Connective Tissue

E. Dense irregular connective tissue

Description: Consists predominantly of collagen fibers, randomly arranged, and a few fibroblasts.

Location: Fasciae (tissue beneath skin and around muscles and other organs), reticular (deeper) region of dermis of skin, periosteum of bone, perichondrium of cartilage, joint capsules, membrane capsules around various organs (kidneys, liver, testes, lymph nodes), pericardium of the heart, and heart valves.

Function: Provides strength.

Sectional view of dense irregular connective tissue of reticular region of dermis

LM 275x

Dense irregular connective tissue

F. Elastic connective tissue

Description: Consists of predominantly freely branching elastic fibers; fibroblasts present in spaces between fibers.

Location: Lung tissue, walls of elastic arteries, trachea, bronchial tubes, true vocal cords, suspensory ligament of penis, and ligaments between vertebrae.

Function: Allows stretching of various organs.

Sectional view of elastic connective tissue of aorta

LM 435x

Elastic connective tissue

ELASTIC CONNECTIVE TISSUE Branching elastic fibers predominate in elastic connective tissue (Table 4.3F), giving the unstained tissue a yellowish color. Fibroblasts are present in the spaces between the fibers. Elastic connective tissue is quite strong and can recoil to its original shape after being stretched. Elasticity is important to the normal functioning of lung tissue, which recoils as you exhale, and elastic arteries, whose recoil between heart beats helps maintain blood flow.

Cartilage

G. Hyaline cartilage   *Description:* Consists of a bluish-white, shiny ground substance with fine collagen fibers and many chondrocytes; most abundant type of cartilage.

Location: Ends of long bones, anterior ends of ribs, nose, parts of larynx, trachea, bronchi, bronchial tubes, and embryonic and fetal skeleton.

Function: Provides smooth surfaces for movement at joints, as well as flexibility and support.

Skeleton

Fetus

Perichondrium

Lacuna containing chondrocyte

Nucleus of chondrocyte

Ground substance

LM 450x

Sectional view of hyaline cartilage of a developing fetal bone

Hyaline cartilage

H. Fibrocartilage   *Description:* Consists of chondrocytes scattered among bundles of collagen fibers within the matrix.

Location: Pubic symphysis (point where hip bones join anteriorly), intervertebral discs (discs between vertebrae), menisci (cartilage pads) of knee, and portions of tendons that insert into cartilage.

Function: Support and fusion.

Tendon of quadriceps femoris muscle

Patella (knee cap)

Nucleus of chondrocyte

Collagen fibers in ground substance

Lacuna containing chondrocyte

LM 300x

Sectional view of fibrocartilage of tendon

Fibrocartilage

Portion of right lower limb

(continues)

Cartilage

Cartilage consists of a dense network of collagen fibers and elastic fibers firmly embedded in chondroitin sulfate, a gel-like component of the ground substance. Cartilage can endure considerably more stress than loose and dense connective tissues. Whereas the strength of cartilage is due to its collagen fibers, its resilience (ability to assume its original shape after deformation) is due to chondroitin sulfate.

Table 4.3   Mature Connective Tissues *(continued)*

Cartilage

I. Elastic cartilage   *Description:* Consists of chondrocytes located in a threadlike network of elastic fibers within the matrix.

Location: Lid on top of larynx (epiglottis), part of external ear (auricle), and auditory (Eustachian) tubes.

Function: Gives support and maintains shape.

Auricle of ear

Perichondrium

Nucleus of chondrocyte in lacuna

Elastic fiber in ground substance

LM 250x

Sectional view of elastic cartilage of auricle of ear

Elastic cartilage

Bone Tissue

J. Compact bone   *Description:* Compact bone tissue consists of osteons (Haversian systems) that contain lamellae, lacunae, osteocytes, canaliculi, and central (Haversian) canals. By contrast, spongy bone tissue (see Figure 6.3 on page 167) consists of thin columns called trabeculae; spaces between trabeculae are filled with red bone marrow.

Location: Both compact and spongy bone tissue make up the various parts of bones of the body.

Function: Support, protection, storage; houses blood-forming tissue; serves as levers that act together with muscle tissue to enable movement.

Femur

Canaliculi

Central (Haversian) canal

Lacuna

Lamellae

LM 550x

Sectional view of an osteon (Haversian system) of femur (thigh bone)

Osteocyte   Calcified matrix

Canaliculi   Lacuna

Details of an osteocyte

The cells of mature cartilage, called **chondrocytes** (KON-drō-sīts; *chondro-* = cartilage), occur singly or in groups within spaces called **lacunae** (la-KOO-nē = little lakes; singular is *lacuna*) in the matrix. A membrane of dense irregular connective tissue called the **perichondrium** (per′-i-KON-drē-um; *peri-* = around) covers the surface of most cartilage. Unlike other connective tissues, cartilage has no blood vessels or nerves, except in the perichondrium. Three kinds of cartilage are recognized: hyaline cartilage, fibrocartilage, and elastic cartilage.

HYALINE CARTILAGE   This type of cartilage contains a resilient gel as its ground substance and appears in the body as a bluish-

Blood Tissue

K. Blood

Description: Consists of blood plasma and formed elements: red blood cells (erythrocytes), white blood cells (leukocytes), and platelets (thrombocytes).

Location: Within blood vessels (arteries, arterioles, capillaries, venules, and veins) and within the chambers of the heart.

Function: Red blood cells transport oxygen and carbon dioxide; white blood cells carry on phagocytosis and are involved in allergic reactions and immune system responses; platelets are essential for the clotting of blood.

Blood in blood vessels   Blood smear

LM 1230x

Platelet

White blood cell (leukocyte)

Red blood cell (erythrocyte)

Blood plasma

Red blood cells

White blood cells

Platelets

white, shiny substance. The fine collagen fibers are not visible with ordinary staining techniques, and prominent chondrocytes are found in lacunae (Table 4.3G). Most hyaline cartilage is surrounded by a perichondrium. The exceptions are the articular cartilage in joints and at the epiphyseal plates, the regions where bones lengthen as a person grows. Hyaline cartilage is the most abundant cartilage in the body. It affords flexibility and support and, at joints, reduces friction and absorbs shock. Hyaline cartilage is the weakest of the three types of cartilage.

FIBROCARTILAGE Chondrocytes are scattered among clearly visible bundles of collagen fibers within the matrix of this type of cartilage (Table 4.3H). Fibrocartilage lacks a perichondrium. This tissue combines strength and rigidity and is the strongest of the three types of cartilage. One location of fibrocartilage is the intervertebral discs, the disk-shaped material between the backbones.

ELASTIC CARTILAGE   In this tissue, chondrocytes are located within a threadlike network of elastic fibers within the matrix (Table 4.3I). A perichondrium is present. Elastic cartilage provides strength and elasticity and maintains the shape of certain structures, such as the external ear.

REPAIR AND GROWTH OF CARTILAGE   Metabolically, cartilage is a relatively inactive tissue that grows slowly. When injured or inflamed, cartilage repair proceeds slowly, in large part because cartilage is avascular. Substances needed for repair and blood cells that participate in tissue repair must diffuse or migrate into the cartilage. The growth of cartilage follows two basic patterns: interstitial growth and appositional growth.

In **interstitial growth,** the cartilage increases rapidly in size due to the division of existing chondrocytes and the continuous deposition of increasing amounts of matrix by the chondrocytes. As the chondrocytes synthesize new matrix, they are pushed away from each other. These events cause the cartilage to expand from within, which is the reason for the term *inter*stitial. This growth pattern occurs while the cartilage is young and pliable, during childhood and adolescence.

In **appositional growth,** activity of cells in the inner chondrogenic layer of the perichondrium leads to growth. The deeper cells of the perichondrium, the fibroblasts, divide; some differentiate into chondroblasts. As differentiation continues, the chondroblasts surround themselves with matrix and become chondrocytes. As a result, matrix accumulates beneath the perichondrium on the outer surface of the cartilage, causing it to grow in width. Appositional growth starts later than interstitial growth and continues through adolescence.

Bone Tissue

Cartilage, joints, and bones make up the skeletal system. The skeletal system supports soft tissues, protects delicate structures, and works with skeletal muscles to generate movement. Bones store calcium and phosphorus; house red bone marrow, which produces blood cells; and contain yellow bone marrow, a storage site for triglycerides. Bones are organs composed of several different connective tissues, including **bone** or **osseous tissue** (OS-ē-us), the periosteum, red and yellow bone marrow, and the endosteum (a membrane that lines a space within bone that stores yellow bone marrow). Bone tissue is classified as either compact or spongy, depending on how its matrix and cells are organized.

The basic unit of **compact bone** is an **osteon** or **Haversian system** (Table 4.3J). Each osteon has four parts:

1.   The **lamellae** (la-MEL-lē = little plates) are concentric rings of matrix that consist of mineral salts (mostly calcium and phosphates), which give bone its hardness, and collagen fibers,

which give bone its strength. The lamellae are responsible for the compact nature of this type of bone tissue.

2.   **Lacunae** are small spaces between lamellae that contain mature bone cells called **osteocytes.**

3.   Projecting from the lacunae are **canaliculi** (KAN-a-lik-ū-lī = little canals), networks of minute canals containing the processes of osteocytes. Canaliculi provide routes for nutrients to reach osteocytes and for wastes to leave them.

4.   A **central (Haversian) canal** contains blood vessels and nerves.

Spongy bone lacks osteons. Rather, it consists of columns of bone called **trabeculae** (tra-BEK-ū-lē = little beams), which contain lamellae, osteocytes, lacunae, and canaliculi. Spaces between lamellae are filled with red bone marrow. Chapter 6 presents bone tissue histology in more detail.

Tissue Engineering

The technology of **tissue engineering** has allowed scientists to grow new tissues in the laboratory for replacement of damaged tissues in the body. Tissue engineers have already developed laboratory-grown versions of skin and cartilage. In the procedure, scaffolding beds of biodegradable synthetic materials or collagen are used as substrates that permit body cells such as skin cells or cartilage cells to be cultured. As the cells divide and assemble, the scaffolding degrades, and the new, permanent tissue is then implanted in the patient. Other structures being developed by tissue engineers include bones, tendons, heart valves, bone marrow, and intestines. Work is also underway to develop insulin-producing cells for diabetics, dopamine-producing cells for Parkinson disease patients, and even entire livers and kidneys. ■

Blood Tissue

Blood tissue (or simply blood) is a connective tissue with a liquid matrix called **blood plasma,** a pale yellow fluid that consists mostly of water with a wide variety of dissolved substances—nutrients, wastes, enzymes, plasma proteins, hormones, respiratory gases, and ions (Table 4.3K). Suspended in the blood plasma are formed elements—red blood cells (erythrocytes), white blood cells (leukocytes), and platelets (thrombocytes). **Red blood cells** transport oxygen to body cells and remove carbon dioxide from them. **White blood cells** are involved in phagocytosis, immunity, and allergic reactions. **Platelets** participate in blood clotting. The details of blood are considered in Chapter 19.

Lymph

Lymph is the extracellular fluid that flows in lymphatic vessels. It is a connective tissue that consists of a clear fluid similar to blood plasma but with much less protein. In the lymph are various cells and chemicals, whose composition varies from one part of the body to another. For example, lymph leaving lymph nodes includes many lymphocytes, a type of white blood cell, whereas lymph from the small intestine has a high content of newly absorbed dietary lipids. The details of lymph are considered in Chapter 22.

10. In what ways do connective tissues differ from epithelia?

11. What are the features of the cells, ground substance, and fibers that make up connective tissue?

12. How are connective tissues classified? List the various types.

13. How are the structures of the following connective tissues related to their functions: areolar connective tissue, adipose tissue, reticular connective tissue, dense regular connective tissue, dense irregular connective tissue, elastic connective tissue, hyaline cartilage, fibrocartilage, elastic cartilage, and bone tissue?

14. How are interstitial and appositional growth of cartilage different?

MEMBRANES

OBJECTIVES

• Define a membrane.
• Describe the classification of membranes.

Membranes are flat sheets of pliable tissue that cover or line a part of the body. The combination of an epithelial layer and an underlying connective tissue layer constitutes an **epithelial membrane.** The principal epithelial membranes of the body are mucous membranes, serous membranes, and the cutaneous membrane, or skin. (Skin is discussed in detail in Chapter 5, and will not be discussed here.) Another kind of membrane, a **synovial membrane,** lines joints and contains connective tissue but no epithelium.

Epithelial Membranes

Mucous Membranes

A **mucous membrane** or **mucosa** lines a body cavity that opens directly to the exterior. Mucous membranes line the entire digestive, respiratory, and reproductive tracts, and much of the urinary tract. They consist of both a lining layer of epithelium and an underlying layer of connective tissue.

The epithelial layer of a mucous membrane is an important feature of the body's defense mechanisms because it is a barrier that microbes and other pathogens have difficulty penetrating. Usually, tight junctions connect the cells, so materials cannot leak in between them. Goblet cells and other cells of the epithelial layer of a mucous membrane secrete mucus, and this slippery fluid prevents the cavities from drying out. It also traps particles in the respiratory passageways and lubricates food as it moves through the gastrointestinal tract. In addition, the epithelial layer secretes some of the enzymes needed for digestion and is the site of food and fluid absorption in the gastrointestinal tract. The epithelia of mucous membranes vary greatly in different parts of the body. For example, the epithelium of the mucous membrane of the small intestine is nonciliated simple columnar (see Table 4.1C), whereas the epithelium

of the large airways to the lungs is pseudostratified ciliated columnar (see Table 4.1I).

The connective tissue layer of a mucous membrane is areolar connective tissue and is called the **lamina propria** (LAM-i-na PRŌ-prē-a). The lamina propria is so named because it belongs to the mucous membrane (propria = one's own). The lamina propria supports the epithelium, binds it to the underlying structures, and allows some flexibility of the membrane. It also holds blood vessels in place and protects underlying muscles from abrasion or puncture. Oxygen and nutrients diffuse from the lamina propria to the epithelium covering it whereas carbon dioxide and wastes diffuse in the opposite direction.

Serous Membranes

A **serous membrane** (serous = watery) or **serosa** lines a body cavity that does not open directly to the exterior, and it covers the organs that lie within the cavity. Serous membranes consist of areolar connective tissue covered by mesothelium (simple squamous epithelium). Serous membranes have two layers: The layer attached to the cavity wall is called the **parietal layer** (pa-RĪ-e-tal; pariet- = wall); the layer that covers and attaches to the organs inside the cavity is the **visceral layer** (viscer- = body organ). The mesothelium of a serous membrane secretes **serous fluid,** a watery lubricating fluid that allows organs to glide easily over one another or to slide against the walls of cavities.

The serous membrane lining the thoracic cavity and covering the lungs is the **pleura.** The serous membrane lining the heart cavity and covering the heart is the **pericardium.** The serous membrane lining the abdominal cavity and covering the abdominal organs is the **peritoneum.**

Synovial Membranes

Synovial membranes (sin-Ō-vē-al; syn- = together, referring here to a place where bones come together) line the cavities of freely movable joints. Like serous membranes, synovial membranes line structures that do not open to the exterior. Unlike mucous, serous, and cutaneous membranes, they lack epithelium and are therefore not epithelial membranes. Synovial membranes are composed of areolar connective tissue with elastic fibers and varying numbers of adipocytes. Synovial membranes secrete **synovial fluid,** which lubricates and nourishes the cartilage covering the bones at movable joints; these membranes are called articular synovial membranes. Other synovial membranes line cushioning sacs, called bursae, and tendon sheaths in the hands and feet, thus easing the movement of muscle tendons.

► CHECKPOINT

15. Define the following kinds of membranes: mucous, serous, cutaneous, and synovial.

16. Where is each type of membrane located in the body? What are their functions?

MUSCLE TISSUE

- **Describe the general features of muscle tissue.**
- **Contrast the structure, location, and mode of control of skeletal, cardiac, and smooth muscle tissue.**

Muscle tissue consists of elongated cells called *muscle fibers* that can use ATP to generate force. As a result, muscle tissue produces body movements, maintains posture, and generates heat. Based on its location and certain structural and functional features, muscle tissue is classified into three types: skeletal, cardiac, and smooth (Table 4.4).

Skeletal muscle tissue is named for its location—usually attached to the bones of the skeleton (Table 4.4A). It is also *striated;* the fibers contain alternating light and dark bands called *striations* that are visible under a light microscope. Skeletal muscle is *voluntary* because it can be made to contract or relax by conscious control. A single skeletal muscle fiber is very long, roughly cylindrical in shape, and has many nuclei located at the

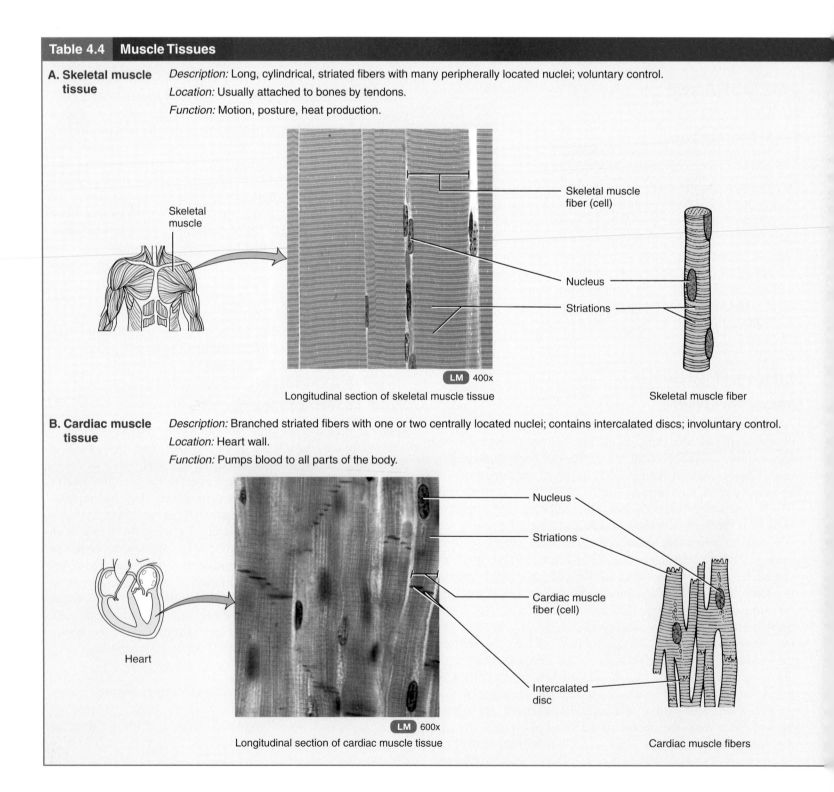

| Table 4.4 | Muscle Tissues |

A. Skeletal muscle tissue

Description: Long, cylindrical, striated fibers with many peripherally located nuclei; voluntary control.
Location: Usually attached to bones by tendons.
Function: Motion, posture, heat production.

Skeletal muscle

Skeletal muscle fiber (cell)

Nucleus

Striations

LM 400x

Longitudinal section of skeletal muscle tissue

Skeletal muscle fiber

B. Cardiac muscle tissue

Description: Branched striated fibers with one or two centrally located nuclei; contains intercalated discs; involuntary control.
Location: Heart wall.
Function: Pumps blood to all parts of the body.

Heart

Nucleus

Striations

Cardiac muscle fiber (cell)

Intercalated disc

LM 600x

Longitudinal section of cardiac muscle tissue

Cardiac muscle fibers

periphery of the cell. Within a whole muscle, the individual muscle fibers are parallel to each other.

Cardiac muscle tissue forms most of the wall of the heart (Table 4.4B). Like skeletal muscle, it is striated. However, unlike skeletal muscle tissue, it is *involuntary;* its contraction is not consciously controlled. Cardiac muscle fibers are branched and usually have only one centrally located nucleus, although an occasional cell has two nuclei. They attach end to end to each other by transverse thickenings of the plasma membrane called **intercalated discs** (*intercalat-* = to insert between), which con-

tain both desmosomes and gap junctions. Intercalated discs are unique to cardiac muscle. The desmosomes strengthen the tissue and hold the fibers together during their vigorous contractions. The gap junctions provide a route for quick conduction of muscle action potentials throughout the heart.

Smooth muscle tissue is located in the walls of hollow internal structures such as blood vessels, airways to the lungs, the stomach, intestines, gallbladder, and urinary bladder (Table 4.4C). Its contraction helps constrict or narrow the lumen of blood vessels, physically break down and move food along

C. Smooth muscle tissue

Description: Spindle-shaped (thickest in middle and tapering at both ends), nonstriated fibers with one centrally located nucleus; involuntary control.

Location: Walls of hollow internal structures such as blood vessels, airways to the lungs, stomach, intestines, gallbladder, urinary bladder, and uterus.

Function: Motion (constriction of blood vessels and airways, propulsion of foods through gastrointestinal tract, contraction of urinary bladder and gallbladder).

Artery

Smooth muscle

Smooth muscle fiber (cell)

Nucleus of smooth muscle fiber

LM 410x

Longitudinal section of smooth muscle tissue

Smooth muscle fiber

the gastrointestinal tract, move fluids through the body, and eliminate wastes. Smooth muscle fibers are usually *involuntary,* and they are nonstriated (lack striations), hence the term *smooth.* A smooth muscle fiber is small, thickest in the middle, and tapering at each end. It contains a single, centrally located nucleus. Gap junctions connect many individual fibers in some smooth muscle tissues, for example, in the wall of the intestines. Such muscle tissues can produce powerful contractions as many muscle fibers contract in unison. In other locations such as the iris of the eye, smooth muscle fibers contract individually, like skeletal muscle fibers, because gap junctions are absent. Chapter 10 provides a detailed discussion of muscle tissue.

▶ CHECKPOINT

17. Which muscles are striated and which are smooth?

18. Which types of muscle tissue have gap junctions?

NERVOUS TISSUE

▶ OBJECTIVE

• Describe the structural features and functions of nervous tissue.

Despite the awesome complexity of the nervous system, it consists of only two types of cells: neurons and neuroglia. **Neurons,** or nerve cells, are sensitive to various stimuli. They convert stimuli into nerve impulses (action potentials) and conduct these impulses to other neurons, to muscle tissue, or to glands. Most neurons consist of three basic parts: a cell body and two kinds of cell processes—dendrites and axons (Table 4.5). The **cell body** contains the nucleus and other organelles. **Dendrites** (*dendr-* = tree) are tapering, highly branched, and usually short

cell processes. They are the major receiving or input portion of a neuron. The **axon** (*axo-* = axis) of a neuron is a single, thin, cylindrical process that may be very long. It is the output portion of a neuron, conducting nerve impulses toward another neuron or to some other tissue.

Even though **neuroglia** (noo-RŌG-lē-a; *-glia* = glue) do not generate or conduct nerve impulses, these cells do have many important supportive functions (see Table 12.1 on page 392). The detailed structure and function of neurons and neuroglia are considered in Chapter 12.

▶ CHECKPOINT

19. What are the functions of the dendrites, cell body, and axon of a neuron?

TISSUE REPAIR: RESTORING HOMEOSTASIS

▶ OBJECTIVE

• Describe the role of tissue repair in restoring homeostasis.

Tissue repair is the process that replaces worn-out, damaged, or dead cells. New cells originate by cell division from the **stroma,** the supporting connective tissue, or from the **parenchyma,** cells that constitute the functioning part of the tissue or organ. In adults, each of the four basic tissue types (epithelial, connective, muscle, and nervous) has a different capacity for replenishing parenchymal cells lost by damage, disease, or other processes.

Epithelial cells, which endure considerable wear and tear (and even injury) in some locations, have a continuous capacity for renewal. In some cases, immature, undifferentiated cells called **stem cells** divide to replace lost or damaged cells. For example,

Table 4.5	Nervous Tissue

Description: Consists of neurons (nerve cells) and neuroglia. Neurons consist of a cell body and processes extending from the cell body (multiple dendrites and a single axon). Neuroglia do not generate or conduct nerve impulses but have other important functions.

Location: Nervous system.

Function: Exhibits sensitivity to various types of stimuli, converts stimuli into nerve impulses (action potentials), and conducts nerve impulses to other neurons, muscle fibers, or glands.

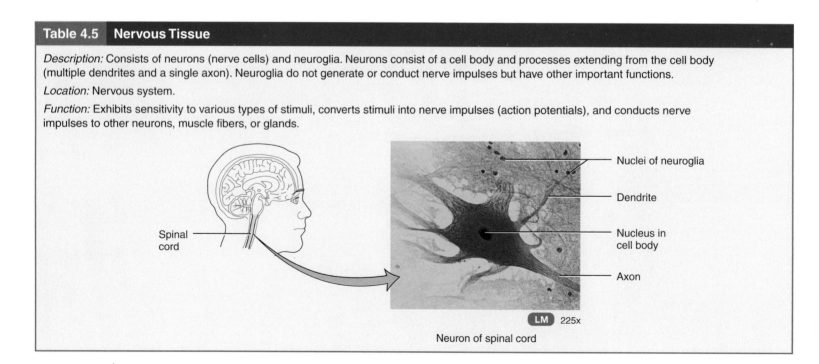

Spinal cord

Nuclei of neuroglia

Dendrite

Nucleus in cell body

Axon

LM 225x

Neuron of spinal cord

stem cells reside in protected locations in the epithelia of the skin and gastrointestinal tract to replenish cells sloughed from the apical layer, and stem cells in red bone marrow continually provide new red and white blood cells and platelets. In other cases, mature, differentiated cells can undergo cell division; examples are hepatocytes (liver cells) and endothelial cells in blood vessels.

Some connective tissues also have a continuous capacity for renewal. One example is bone, which has an ample blood supply. Other connective tissues such as cartilage can replenish cells less readily in part because of a smaller blood supply.

Muscle tissue has a relatively poor capacity for renewal of lost cells. Even though skeletal muscle tissue contains stem cells called *satellite cells,* they do not divide rapidly enough to replace extensively damaged muscle fibers. Cardiac muscle tissue lacks satellite cells, and existing cardiac muscle fibers do not undergo mitosis to form new cells. Recent evidence suggests, however, that stem cells do migrate into the heart from the blood. There, they can differentiate and replace a limited number of thyroid cardiac muscle fibers and endothelial cells in heart blood vessels. Smooth muscle fibers can proliferate to some extent, but they do so much more slowly than the cells of epithelial or connective tissues.

Nervous tissue has the poorest capacity for renewal. Although experiments have revealed the presence of some stem cells in the brain, they normally do not undergo mitosis to replace damaged neurons. Discovering why this is so is a major goal of researchers who seek ways to encourage repair of damaged or diseased nervous tissue.

The restoration of an injured tissue or organ to normal structure and function depends entirely on whether parenchymal cells are active in the repair process. If parenchymal cells accomplish the repair, **tissue regeneration** is possible, and a near-perfect reconstruction of the injured tissue may occur. However, if fibroblasts of the stroma are active in the repair, the replacement tissue will be a new connective tissue. The fibroblasts synthesize collagen and other matrix materials that aggregate to form scar tissue, a process known as **fibrosis.** Because scar tissue is not specialized to perform the functions of the parenchymal tissue, the original function of the tissue or organ is impaired.

When tissue damage is extensive, as in large, open wounds, both the connective tissue stroma and the parenchymal cells are active in repair. This repair involves the rapid cell division of many fibroblasts and the manufacture of new collagen fibers to provide structural strength. Blood capillaries also sprout new buds to supply the healing tissue the materials it needs. All these processes create an actively growing connective tissue called **granulation tissue.** This new tissue forms across a wound or surgical incision to provide a framework (stroma) that supports the epithelial cells that migrate into the open area and fill it. The newly formed granulation tissue also secretes a fluid that kills bacteria.

Three factors affect tissue repair: nutrition, blood circulation, and age. Nutrition is vital because the healing process places a great demand on the body's store of nutrients. Adequate protein in the diet is important because most of the structural components of a tissue are proteins. Several vitamins also play a direct role in wound healing and tissue repair. For example, vita-min C directly affects the normal production and maintenance of matrix materials, especially collagen. Vitamin C also strengthens and promotes the formation of new blood vessels. With vitamin C deficiency, even superficial wounds fail to heal, and the walls of the blood vessels become fragile and are easily ruptured.

In tissue repair, proper blood circulation is essential to transport oxygen, nutrients, antibodies, and many defensive cells to the injured site. The blood also plays an important role in the removal of tissue fluid, bacteria, foreign bodies, and debris, elements that would otherwise interfere with healing.

 Adhesions

Scar tissue can form **adhesions,** abnormal joining of tissues. Adhesions commonly form in the abdomen around a site of previous inflammation such as an inflamed appendix, and they can develop after surgery. Although adhesions do not always cause problems, they can decrease tissue flexibility, cause obstruction (such as in the intestine), and make a subsequent operation more difficult. An *adhesiotomy,* the surgical release of adhesions, may be required. ■

AGING AND TISSUES

▶ O B J E C T I V E

• **Describe the effects of aging on tissues.**

Generally, tissues heal faster and leave less obvious scars in the young than in the aged. In fact, surgery performed on fetuses leaves no scars. The younger body is generally in a better nutritional state, its tissues have a better blood supply, and its cells have a higher metabolic rate. Thus, cells can synthesize needed materials and divide more quickly. The extracellular components of tissues also change with age. Glucose, the most abundant sugar in the body, plays a role in the aging process. Glucose is haphazardly added to proteins inside and outside cells, forming irreversible cross-links between adjacent protein molecules. With advancing age, more cross-links form, which contributes to the stiffening and loss of elasticity that occur in aging tissues. Collagen fibers, responsible for the strength of tendons, increase in number and change in quality with aging. These changes in the collagen of arterial walls are as much responsible for their loss of extensibility as are the deposits associated with atherosclerosis, the deposition of fatty materials in arterial walls. Elastin, another extracellular component, is responsible for the elasticity of blood vessels and skin. It thickens, fragments, and acquires a greater affinity for calcium with age—changes that may also be associated with the development of atherosclerosis.

▶ C H E C K P O I N T

20. How are stromal and parenchymal repair of a tissue different?

21. What is the importance of granulation tissue?

22. What common changes occur in epithelial and connective tissues with aging?

DISORDERS: HOMEOSTATIC IMBALANCES

Disorders of epithelial tissues are mainly specific to individual organs, for example, peptic ulcer disease (PUD), which erodes the epithelial lining of the stomach or small intestine. For this reason, epithelial disorders are described together with the relevant body system throughout the text. The most prevalent disorders of connective tissues are **autoimmune diseases**—diseases in which antibodies produced by the immune system fail to distinguish what is foreign from what is self and attack the body's own tissues. The most common autoimmune disorder is rheumatoid arthritis, which attacks the synovial membranes of joints. Because connective tissue is one of the most abundant and widely distributed of the four main types of tissues, its disorders often affect multiple body systems. Common disorders of muscle tissue and nervous tissue are described at the ends of Chapters 10 and 12, respectively.

Sjögren's Syndrome

Sjögren's syndrome (SHŌ-grenz) is a common autoimmune disorder that causes inflammation and destruction of exocrine glands, especially the lacrimal (tear) glands and salivary glands. Signs include dryness of the mucous membranes in the eyes and mouth and salivary gland enlargement. Systemic effects include arthritis, difficulty swallowing, pancreatitis (inflammation of the pancreas), pleuritis (inflammation of the pleurae of the lungs), and migraine headaches. The disorder affects more females than males by a ratio of 9 to 1. About 20% of older adults experience some signs of Sjögren's. Treatment is supportive, including using artificial tears to moisten the eyes, sipping fluids, chewing sugarless gum, and using a saliva substitute to moisten the mouth.

Systemic Lupus Erythematosus

Systemic lupus erythematosus (er-i-thē-ma-TŌ-sus), **SLE,** or simply lupus, is a chronic inflammatory disease of connective tissue occurring mostly in nonwhite women during their childbearing years. It is an autoimmune disease that can cause tissue damage in every body system. The disease, which can range from a mild condition in most patients to a rapidly fatal disease, is marked by periods of exacerbation and remission. The prevalence of SLE is about 1 in 2000 persons, with females more likely to be afflicted than males by a ratio of 8 or 9 to 1.

Although the cause of SLE is not known, genetic, environmental, and hormonal factors are implicated. The genetic component is suggested by studies of twins and family history. Environmental factors include viruses, bacteria, chemicals, drugs, exposure to excessive sunlight, and emotional stress. With regard to hormones, sex hormones, such as estrogens, may trigger SLE.

Signs and symptoms of SLE include painful joints, low-grade fever, fatigue, mouth ulcers, weight loss, enlarged lymph nodes and spleen, sensitivity to sunlight, rapid loss of large amounts of scalp hair, and anorexia. A distinguishing feature of lupus is an eruption across the bridge of the nose and cheeks called a "butterfly rash." Other skin lesions may occur, including blistering and ulceration. The erosive nature of some SLE skin lesions was thought to resemble the damage inflicted by the bite of a wolf—thus, the term *lupus* (= wolf). The most serious complications of the disease involve inflammation of the kidneys, liver, spleen, lungs, heart, brain, and gastrointestinal tract. Because there is no cure for SLE, treatment is supportive, including anti-inflammatory drugs, such as aspirin, and immunosuppressive drugs.

MEDICAL TERMINOLOGY

Atrophy (AT-rō-fē; *a-* = without; *-trophy* = nourishment) A decrease in the size of cells, with a subsequent decrease in the size of the affected tissue or organ.

Hypertrophy (hī-PER-trō-fē; *hyper-* = above or excessive) Increase in the size of a tissue because its cells enlarge without undergoing cell division.

Tissue rejection An immune response of the body directed at foreign proteins in a transplanted tissue or organ; immunosuppressive drugs, such as cyclosporine, have largely overcome tissue rejection in heart-, kidney-, and liver-transplant patients.

Tissue transplantation The replacement of a diseased or injured tissue or organ. The most successful transplants involve use of a person's own tissues or those from an identical twin.

Xenotransplantation (zen'-ō-trans-plan-TĀ-shun; *xeno-* = strange, foreign) The replacement of a diseased or injured tissue or organ with cells or tissues from an animal. Porcine (from pigs) and bovine (from cows) heart valves are used for some heart-valve replacement surgeries.

STUDY OUTLINE

TYPES OF TISSUES AND THEIR ORIGINS (p. 104)

1. A tissue is a group of similar cells that usually has a similar embryological origin and is specialized for a particular function.
2. The various tissues of the body are classified into four basic types: epithelial, connective, muscle, and nervous.
3. All tissues of the body develop from three primary germ layers,

the first tissues that form in a human embryo: ectoderm, mesoderm, and endoderm.

CELL JUNCTIONS (p. 104)

1. Cell junctions are points of contact between adjacent plasma membranes.

2. Tight junctions form fluid-tight seals between cells; adherens junctions, desmosomes, and hemidesmosomes anchor cells to one another or to the basement membrane; and gap junctions permit electrical and chemical signals to pass between cells.

EPITHELIAL TISSUE (p. 106)

1. The subtypes of epithelia include covering and lining epithelia and glandular epithelia.
2. An epithelium consists mostly of cells with little extracellular material between adjacent plasma membranes. The apical, lateral, and basal surfaces of epithelial cells are modified in various ways to carry out specific functions. Epithelium is arranged in sheets and attached to a basement membrane. Although it is avascular, it has a nerve supply. Epithelia are derived from all three primary germ layers and have a high capacity for renewal.
3. Epithelial layers are simple (one layer), stratified (several layers), or pseudostratified (one layer that appears to be several). The cell shapes may be squamous (flat), cuboidal (cubelike), columnar (rectangular), or transitional (variable).
4. Simple squamous epithelium consists of a single layer of flat cells (Table 4.1A). It is found in parts of the body where filtration or diffusion are priority processes. One type, endothelium, lines the heart and blood vessels. Another type, mesothelium, forms the serous membranes that line the thoracic and abdominopelvic cavities and cover the organs within them.
5. Simple cuboidal epithelium consists of a single layer of cube-shaped cells that function in secretion and absorption (Table 4.1B). It is found covering the ovaries, in the kidneys and eyes, and lining some glandular ducts.
6. Nonciliated simple columnar epithelium consists of a single layer of nonciliated rectangular cells (Table 4.1C). It lines most of the gastrointestinal tract. Specialized cells containing microvilli perform absorption. Goblet cells secrete mucus.
7. Ciliated simple columnar epithelium consists of a single layer of ciliated rectangular cells (Table 4.1D). It is found in a few portions of the upper respiratory tract, where it moves foreign particles trapped in mucus out of the respiratory tract.
8. Stratified squamous epithelium consists of several layers of cells; cells of the apical layer and several layers deep to it are flat (Table 4.1E). A nonkeratinized variety lines the mouth; a keratinized variety forms the epidermis, the most superficial layer of the skin.
9. Stratified cuboidal epithelium consists of several layers of cells; cells at the apical layer are cube-shaped (Table 4.1F). It is found in adult sweat glands and a portion of the male urethra.
10. Stratified columnar epithelium consists of several layers of cells; cells of the apical layer have a columnar shape (Table 4.1G). It is found in a portion of the male urethra and large excretory ducts of some glands.
11. Transitional epithelium consists of several layers of cells whose appearance varies with the degree of stretching (Table 4.1H). It lines the urinary bladder.
12. Pseudostratified columnar epithelium has only one layer but gives the appearance of many (Table 4.1I). A ciliated variety contains goblet cells and lines most of the upper respiratory tract; a nonciliated variety has no goblet cells and lines ducts of many glands, the epididymis, and part of the male urethra.
13. A gland is a single cell or a group of epithelial cells adapted for secretion.
14. Endocrine glands secrete hormones into the blood (Table 4.1J).

15. Exocrine glands (mucous, sweat, oil, and digestive glands) secrete into ducts or directly onto a free surface (Table 4.1K).
16. Structural classification of exocrine glands includes unicellular and multicellular glands.
17. Functional classification of exocrine glands includes holocrine, apocrine, and merocrine glands.

CONNECTIVE TISSUE (p. 117)

1. Connective tissue is one of the most abundant body tissues.
2. Connective tissue consists of cells and a matrix of ground substance and fibers; it has abundant matrix with relatively few cells. It does not usually occur on free surfaces, has a nerve supply (except for cartilage), and is highly vascular (except for cartilage, tendons, and ligaments).
3. Cells in connective tissue are derived from mesenchymal cells.
4. Cell types include fibroblasts (secrete matrix), macrophages (perform phagocytosis), plasma cells (secrete antibodies), mast cells (produce histamine), adipocytes (store fat), and white blood cells (migrate from blood in response to infections).
5. The ground substance and fibers make up the matrix.
6. The ground substance supports and binds cells together, provides a medium for the exchange of materials, and is active in influencing cell functions.
7. Substances found in the ground substance include water and polysaccharides such as hyaluronic acid, chondroitin sulfate, dermatan sulfate, and keratan sulfate (glycosaminoglycans). Also present are proteoglycans and adhesion proteins.
8. The fibers in the matrix provide strength and support and are of three types. (a) Collagen fibers (composed of collagen) are found in large amounts in bone, tendons, and ligaments. (b) Elastic fibers (composed of elastin, fibrillin, and other glycoprotein) are found in skin, blood vessel walls, and lungs. (c) Reticular fibers (composed of collagen and glycoprotein) are found around fat cells, nerve fibers, and skeletal and smooth muscle cells.
9. The two major subclasses of connective tissue are embryonic connective tissue (found in the embryo and fetus) and mature connective tissue (present in the newborn).
10. The embryonic connective tissues are mesenchyme, which forms all other connective tissues (Table 4.2A), and mucous connective tissue, found in the umbilical cord of the fetus, where it gives support (Table 4.2B).
11. Mature connective tissue differentiates from mesenchyme. It is subdivided into several kinds: loose or dense connective tissue, cartilage, bone tissue, blood tissue, and lymph.
12. Loose connective tissue includes areolar connective tissue, adipose tissue, and reticular connective tissue.
13. Areolar connective tissue consists of the three types of fibers, several types of cells, and a semifluid ground substance (Table 4.3A). It is found in the subcutaneous layer, in mucous membranes, and around blood vessels, nerves, and body organs.
14. Adipose tissue consists of adipocytes, which store triglycerides (Table 4.3B). It is found in the subcutaneous layer, around organs, and in yellow bone marrow. Brown adipose tissue (BAT) generates heat.
15. Reticular connective tissue consists of reticular fibers and reticular cells and is found in the liver, spleen, and lymph nodes (Table 4.3C).
16. Dense connective tissue includes dense regular connective tissue, dense irregular connective tissue, and elastic connective tissue.

17. Dense regular connective tissue consists of parallel bundles of collagen fibers and fibroblasts (Table 4.3D). It forms tendons, most ligaments, and aponeuroses.

18. Dense irregular connective tissue consists of usually randomly arranged collagen fibers and a few fibroblasts (Table 4.3E). It is found in fasciae, the dermis of skin, and membrane capsules around organs.

19. Elastic connective tissue consists of branching elastic fibers and fibroblasts (Table 4.3F). It is found in the walls of large arteries, lungs, trachea, and bronchial tubes.

20. Cartilage contains chondrocytes and has a rubbery matrix (chondroitin sulfate) containing collagen and elastic fibers.

21. Hyaline cartilage is found in the embryonic skeleton, at the ends of bones, in the nose, and in respiratory structures (Table 4.3G). It is flexible, allows movement, and provides support.

22. Fibrocartilage is found in the pubic symphysis, intervertebral discs, and menisci (cartilage pads) of the knee joint (Table 4.3H).

23. Elastic cartilage maintains the shape of organs such as the epiglottis of the larynx, auditory (Eustachian) tubes, and external ear (Table 4.3I).

24. Cartilage enlarges by interstitial growth (from within) and appositional growth (from without).

25. Bone or osseous tissue consists of a matrix of mineral salts and collagen fibers that contribute to the hardness of bone, and osteocytes that are located in lacunae (Table 4.3J). It supports, protects, helps provide movement, stores minerals, and houses blood-forming tissue.

26. Blood tissue consists of blood plasma and formed elements—red blood cells, white blood cells, and platelets (Table 4.3K). Its cells function to transport oxygen and carbon dioxide, carry on phagocytosis, participate in allergic reactions, provide immunity, and bring about blood clotting.

27. Lymph is the extracellular fluid that flows in lymphatic vessels. It is a clear fluid similar to blood plasma but with less protein.

MEMBRANES (p. 129)

1. An epithelial membrane consists of an epithelial layer overlying a connective tissue layer. Examples are mucous, serous, and cutaneous membranes.

2. Mucous membranes line cavities that open to the exterior, such as the gastrointestinal tract.

3. Serous membranes line closed cavities (pleura, pericardium, peritoneum) and cover the organs in the cavities. These membranes consist of parietal and visceral layers.

4. Synovial membranes line joint cavities, bursae, and tendon sheaths and consist of areolar connective tissue instead of epithelium.

MUSCLE TISSUE (p. 130)

1. Muscle tissue consists of fibers that are specialized for contraction. It provides motion, maintenance of posture, and heat production.

2. Skeletal muscle tissue is attached to bones and is striated, and its action is voluntary (Table 4.4A).

3. Cardiac muscle tissue forms most of the heart wall and is striated, and its action is involuntary (Table 4.4B).

4. Smooth muscle tissue is found in the walls of hollow internal structures (blood vessels and viscera) and is nonstriated, and its action is involuntary (Table 4.4C).

NERVOUS TISSUE (p. 132)

1. The nervous system is composed of neurons (nerve cells) and neuroglia (protective and supporting cells) (Table 4.5).

2. Neurons are sensitive to stimuli, convert stimuli into nerve impulses, and conduct nerve impulses.

3. Most neurons consist of a cell body and two types of processes, dendrites and axons.

TISSUE REPAIR: RESTORING HOMEOSTASIS (p. 132)

1. Tissue repair is the replacement of worn-out, damaged, or dead cells by healthy ones.

2. Stem cells may divide to replace lost or damaged cells.

3. If the injury is superficial, tissue repair involves parenchymal regeneration; if damage is extensive, granulation tissue is involved.

4. Good nutrition and blood circulation are vital to tissue repair.

AGING AND TISSUES (p. 133)

1. Tissues heal faster and leave less obvious scars in the young than in the aged; surgery performed on fetuses leaves no scars.

2. The extracellular components of tissues, such as collagen and elastic fibers, also change with age.

Q SELF-QUIZ QUESTIONS

Fill in the blanks in the following statements.

1. A group of cells that usually have a common origin and a specialized function is called a _____.

2. Secretions of _____ glands enter the interstitial fluid, and then diffuse into the bloodstream.

3. The three components that can be used to classify connective tissue are _____, _____, and _____.

Indicate whether the following statements are true or false.

4. Connective tissue fibers that are arranged in bundles and lend strength and flexibility to a tissue are collagen fibers.

5. Due to its ample blood supply, damaged cartilage is quickly repaired.

Choose the one best answer to the following questions.

6. The tissue type that can detect changes in the internal and external environments and respond to the changes is (a) nervous tissue, (b) muscle tissue, (c) connective tissue, (d) epithelial tissue.

7. Which of the following statements are *true* of epithelium? (1) The cells are arranged in continuous single- or multiple-layer sheets. (2) The attachment between the basal layer and the connective tissue is called the basement membrane. (3) This tissue has an

extensive blood supply. (4) This tissue has a high rate of cell division. (5) This tissue functions in protection, secretion, absorption, and excretion. (a) 1, 2, 3, and 4, (b) 2, 3, 4, and 5, (c) 1, 2, 4, and 5, (d) 1, 3, and 5, (e) 2, 4, and 5.

8. The type of exocrine gland that forms its secretory product and simply releases it from the cell by exocytosis is the (a) apocrine gland, (b) merocrine gland, (c) holocrine gland, (d) endocrine gland, (e) apical gland.

9. The connective tissue cells responsible for secreting fibers and ground substance are (a) macrophages, (b) mast cells, (c) fibroblasts, (d) adipocytes, (e) plasma cells.

10. The membrane lining a body cavity that opens directly to the exterior is a (a) serous membrane, (b) mucous membrane, (c) synovial membrane, (d) plasma membrane, (e) basement membrane.

11. The muscle tissue that forms the bulk of the wall of the heart is (a) skeletal muscle, (b) smooth muscle, (c) involuntary, nonstriated muscle, (d) cardiac muscle, (e) striated, voluntary muscle.

12. Match the following:
_____ (a) contains a single layer of flat cells; found in the body where filtration (kidney) or diffusion (lungs) are priority processes
_____ (b) found in the superficial part of skin; provides protection from heat, microbes, and chemicals
_____ (c) contains cube-shaped cells; found in the kidney that function in secretion and absorption
_____ (d) lines the upper respiratory tract and uterine tubes; wavelike motion of cilia propel material through the lumen
_____ (e) found in the urinary bladder; contains cells that can change shape (stretch or relax)
_____ (f) contains cells with microvilli and goblet cells; found in linings of the digestive, reproductive and urinary tracts
_____ (g) contains cells that are all attached to the basement membrane, although some do not reach the surface; those cells that do extend to the surface secrete mucus or bear cilia at their apical surface
_____ (h) a fairly rare type of epithelium that has a mainly protective function

(1) pseudostratified ciliated columnar epithelium
(2) ciliated simple columnar epithelium
(3) transitional epithelium
(4) simple squamous epithelium
(5) simple cuboidal epithelium
(6) nonciliated simple columnar epithelium
(7) stratified cuboidal epithelium
(8) keratinized stratified squamous epithelium

13. Match the following:
_____ (a) the tissue from which all other connective tissues eventually arise
_____ (b) connective tissue with a clear, liquid matrix that flows in lymphatic vessels
_____ (c) connective tissue consisting of several kinds of cells, containing all three fiber types randomly arranged, and found in the subcutaneous layer of the skin
_____ (d) a loose connective tissue specialized for triglyceride storage
_____ (e) tissue that contains reticular fibers and reticular cells and forms the stroma of certain organs such as the spleen
_____ (f) tissue with irregularly arranged collagen fibers found in the dermis of the skin
_____ (g) tissue found in the lungs that is strong and can recoil back to its original shape after being stretched
_____ (h) tissue that affords flexibility at joints and reduces joint friction
_____ (i) tissue that provides strength and rigidity and is the strongest of the three types of cartilage
_____ (j) bundles of collagen arranged in parallel patterns; compose tendons and ligaments
_____ (k) tissue that forms the internal framework of the body and works with skeletal muscle to generate movement
_____ (l) connective tissue with formed elements suspended in a liquid matrix called plasma

(1) blood
(2) fibrocartilage
(3) mesenchyme
(4) dense regular connective tissue
(5) lymph
(6) hyaline cartilage
(7) dense irregular connective tissue
(8) areolar connective tissue
(9) reticular connective tissue
(10) bone (osseous tissue)
(11) elastic connective tissue
(12) adipose tissue

14. Match the following:

_____ (a) prevents organ contents from leaking into the blood or surrounding tissues

_____ (b) forms adhesion belts that help epithelial surfaces resist separation

_____ (c) makes tissues stable by linking the cytoskeletons of cells together

_____ (d) anchors cells to basement membrane

_____ (e) allows epithelial cells in a tissue to communicate; enables nerve or muscle impulses to spread rapidly between cells

(1) gap junction
(2) tight junction
(3) desmosome
(4) hemidesmosome
(5) adherens junction

15. Match the following:

_____ (a) result of repair of extensive tissue damage by both fibroblasts and parenchymal cells; provides a supporting framework for new tissue

_____ (b) near perfect repair of damaged tissue by parenchymal cells

_____ (c) replacement of damaged tissue by collagen fibers and other matrix materials, forming scar tissue

(1) fibrosis
(2) tissue regeneration
(3) granulation tissue

CRITICAL THINKING QUESTIONS

1. Imagine that you live 50 years in the future, and you can custom-design a human to suit the environment. Your assignment is to customize the tissue makeup for life on a large planet with strong gravity; a cold, dry climate; and a thin atmosphere. What adaptations would you incorporate into the structure and/or amount of your tissues, and why?

HINT *Think of the adaptations seen in people and animals that live in the Arctic versus those that live near the equator.*

2. The neighborhood kids are walking around with straight pins and sewing needles stuck into their fingertips. There is no visible bleeding. What type of tissue have they pierced? What if they stuck a needle straight into the fingertip?

HINT *If you pierced an earlobe, it would bleed.*

3. Kendra was reading her *Principles of Anatomy and Physiology* text at the college cafeteria's all-you-can-eat salad bar. She was having trouble visualizing tissues, until her roommate plopped down a bowl of gelatin salad. It was pale pink with grapes, shredded carrots, and coconut. As the salad quivered in the bowl, Kendra suddenly understood connective tissue. Relate the structure of the salad to the structure of connective tissue.

HINT *The fruit is suspended in the jellylike gelatin.*

ANSWERS TO FIGURE QUESTIONS

4.1 Gap junctions allow the spread of electrical and chemical signals between adjacent cells.

4.2 The basement membrane provides physical support of epithelium, serves as a filter in the kidneys, and guides cell migration during development and tissue repair.

4.3 Simple multicellular exocrine glands have a nonbranched duct; compound multicellular exocrine glands have a branched duct.

4.4 Sebaceous (oil) glands are holocrine glands, and salivary glands are merocrine glands.

4.5 Fibroblasts secrete the fibers and ground substance of the matrix.

The Integumentary System

A Practitioner's View

A *strong* knowledge base of the integumentary system is essential in my work as a certified Wound Ostomy Continence Nurse. The first section of the chapter, which discusses the structure and function of the skin presents a thorough overview with nice supporting figures to enhance one's understanding. This is the foundation to understanding wound healing concepts. I was also glad to see age related skin changes discussed. Many of us encounter a large population of aged in our practices and an understanding of this helps to gear our interventions to their special needs.

My favorite section of the chapter was Maintaining Homeostasis: Skin Wound Healing. Wound care products prescribed today heal based on these principles. A good working knowledge of wound healing phases is very helpful and often misunderstood by health care providers.

Dorri Hart
Wound Ostomy and Continence Nurse
Westchester Medical Center

THE INTEGUMENTARY SYSTEM AND HOMEOSTASIS

The integumentary system contributes to homeostasis by protecting the body and helping regulate body temperature. It also allows you to sense pleasurable, painful, and other stimuli in your external environment.

ENERGIZE YOUR STUDY

FOUNDATIONS CD

Anatomy Overview
- The Integumentary System

Animations
- Negative Feedback Control of Body Temperature
- Systems Contributions to Homeostasis

Concept and Connections
- The Integumentary System and Tissues

www.wiley.com/college/apcentral

INSIGHTS AND EXPLORATIONS

We have seen the pictures of people being rescued from burning buildings, trucks, cars and other accidents. According to the American Burn Association, 1.25 million people seek medical attention for burns every year in the United States. The standard treatment of third-degree burns (burns that extend into the deepest layers of the skin) is to remove and replace the damaged skin with new skin, ideally from other areas of the burned patient. For patients who are burned over 80 to 90 percent of their body this is not possible. Come explore how scientists are growing skin from burn patients to use in skin grafts.

Skin and its accessory structures—hair and nails, along with various glands, muscles, and nerves—make up the **integumentary system** (in-teg-ū-MEN-tar-ē; *inte-* = whole; *-gument* = body covering). The integumentary system helps protect the body, helps maintain a constant body temperature, and provides sensory information about the surrounding environment. Of all the body's organs, none is more easily inspected or more exposed to infection, disease, and injury than the skin. Although the skin's location makes it vulnerable to damage from trauma, sunlight, microbes, and pollutants in the environment, it has protective features that ward off such damage. Because of its visibility, skin reflects our emotions and some aspects of normal physiology, as evidenced by frowning, blushing, and sweating. Changes in skin color may indicate homeostatic imbalances in the body; for example, a bluish skin color is one sign of heart failure. Abnormal skin eruptions or rashes such as chickenpox, cold sores, or measles may reveal systemic infections or diseases of internal organs. Other conditions may involve just the skin itself, such as warts, age spots, or pimples. So important is the skin to one's image that many people spend much time and money to restore it to a more normal or youthful appearance. **Dermatology** (der′-ma-TOL-ō-jē; *dermato-* = skin; *-logy* = study of) is the medical specialty for diagnosis and treatment of integumentary system disorders.

STRUCTURE OF THE SKIN

▶ O B J E C T I V E S

- **Describe the layers of the epidermis and the cells that compose them.**
- **Compare the composition of the papillary and reticular regions of the dermis.**
- **Explain the basis for different skin colors.**

The **skin** or **cutaneous membrane** covers the external surface of the body. It is the largest organ of the body in surface area and weight. In adults, the skin covers an area of about 2 square meters (22 square feet) and weighs 4.5–5 kg (10–11 lb), about 16% of total body weight. It ranges in thickness from 0.5 mm (0.02 in.) on the eyelids to 4.0 mm (0.16 in.) on the heels. However, over most of the body it is 1–2 mm (0.04–0.08 in.) thick. Structurally, the skin consists of two main parts (Figure 5.1). The superficial, thinner portion, which is composed of *epithelial tissue,* is the **epidermis** (ep′-i-DERM-is; *epi-* = above). The deeper, thicker *connective tissue* part is the **dermis.**

Deep to the dermis, but not part of the skin, is the **subcutaneous (subQ) layer.** Also called the **hypodermis** (*hypo-* = below), this layer consists of areolar and adipose tissues. Fibers that extend from the dermis anchor the skin to the subcutaneous layer, which, in turn, attaches to underlying tissues and organs. The subcutaneous layer serves as a storage depot for fat and contains large blood vessels that supply the skin. This region (and sometimes the dermis) also contains nerve endings called **lamellated (Pacinian) corpuscles** (pa-SIN-ē-an) that are sensitive to pressure (Figure 5.1).

Epidermis

The **epidermis** is keratinized stratified squamous epithelium. It contains four principal types of cells: keratinocytes, melanocytes, Langerhans cells, and Merkel cells (Figure 5.2 on page 142). About 90% of epidermal cells are **keratinocytes** (ker-a-TIN-ō-sīts; *keratino-* = hornlike; *-cytes* = cells), which are arranged in four or five layers and produce the protein **keratin** (Figure 5.2a). Recall from Chapter 4 that keratin is a tough, fibrous protein that helps protect the skin and underlying tissues from heat, microbes, and chemicals. Keratinocytes also produce lamellar granules, which release a water-repellent sealant.

About 8% of the epidermal cells are **melanocytes** (MEL-a-nō-sīts; *melano-* = black), which produce the pigment melanin (Figure 5.2b). Their long, slender projections extend between the keratinocytes and transfer melanin granules to them. **Melanin** is a brown-black pigment that contributes to skin color and absorbs damaging ultraviolet (UV) light. Once inside keratinocytes, the melanin granules cluster to form a protective veil over the nucleus, on the side toward the skin surface. In this way, they shield the nuclear DNA from being damaged by UV light. Although keratinocytes gain some protection from melanin granules, melanocytes themselves are particularly susceptible to damage by UV light.

Langerhans cells (LANG-er-hans) arise from red bone marrow and migrate to the epidermis (Figure 5.2c), where they constitute a small fraction of the epidermal cells. They participate in immune responses mounted against microbes that invade the skin, and are easily damaged by UV light.

Merkel cells are the least numerous of the epidermal cells. They are located in the deepest layer of the epidermis, where they contact the flattened process of a sensory neuron (nerve cell), a structure called a **tactile (Merkel) disc** (Figure 5.2d). Merkel cells and tactile discs detect different aspects of touch sensations (see Table 15.2 on page 507).

Several distinct layers of keratinocytes in various stages of development form the epidermis (Figure 5.3 on page 143). In most regions of the body the epidermis has four strata or layers—stratum basale, stratum spinosum, stratum granulosum, and a thin stratum corneum. Where exposure to friction is greatest, such as in the fingertips, palms, and soles, the epidermis has five layers—stratum basale, stratum spinosum, stratum granulosum, stratum lucidum, and a thick stratum corneum.

Stratum Basale

The deepest layer of the epidermis is the **stratum basale** (*basal-* = base), composed of a single row of cuboidal or columnar keratinocytes. Some cells in this layer are *stem cells* that un-

Figure 5.1 Components of the integumentary system. The skin consists of a thin, superficial epidermis and a deep, thicker dermis. Deep to the skin is the subcutaneous layer, which attaches the dermis to underlying organs and tissues.

The integumentary system includes the skin and its accessory structures—hair, nails, and skin glands—along with associated smooth muscles and nerves.

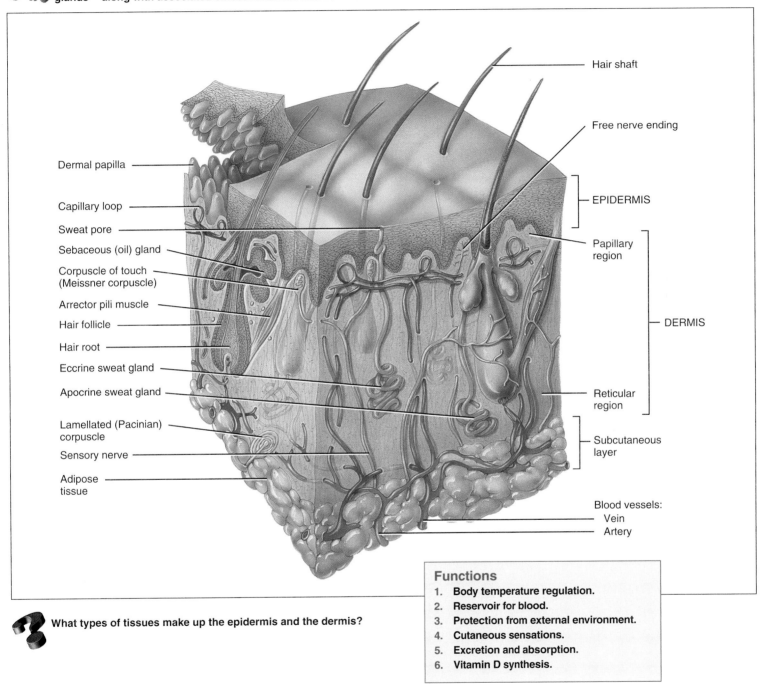

Hair shaft

Free nerve ending

Dermal papilla

EPIDERMIS

Capillary loop

Sweat pore

Papillary region

Sebaceous (oil) gland

Corpuscle of touch (Meissner corpuscle)

Arrector pili muscle

Hair follicle

DERMIS

Hair root

Eccrine sweat gland

Apocrine sweat gland

Reticular region

Lamellated (Pacinian) corpuscle

Sensory nerve

Subcutaneous layer

Adipose tissue

Blood vessels:
Vein
Artery

Functions
1. **Body temperature regulation.**
2. **Reservoir for blood.**
3. **Protection from external environment.**
4. **Cutaneous sensations.**
5. **Excretion and absorption.**
6. **Vitamin D synthesis.**

What types of tissues make up the epidermis and the dermis?

dergo cell division to continually produce new keratinocytes. The nuclei of keratinocytes in the stratum basale are large, and their cytoplasm contains many ribosomes, a small Golgi complex, a few mitochondria, and some rough endoplasmic reticulum. The cytoskeleton within keratinocytes of the stratum basale includes scattered intermediate filaments, called tonofilaments.

The tonofilaments are composed of a protein that will later form the keratin in more superficial epidermal layers. Tonofilaments attach to desmosomes, which bind cells of the stratum basale to each other and to the cells of the adjacent stratum spinosum, and to hemidesmosomes, which bind the keratinocytes to the basement membrane between the epidermis and the dermis. Keratin

Figure 5.2 Types of cells in the epidermis. Besides keratinocytes, the epidermis contains melanocytes, which produce the pigment melanin; Langerhans cells, which participate in immune responses; and Merkel cells, which function in the sensation of touch.

Most of the epidermis consists of keratinocytes, which produce the protein keratin (protects underlying tissues) and lamellar granules (contain a waterproof sealant).

(a) Keratinocyte

(b) Melanocyte

(c) Langerhans cell

(d) Merkel cell

What is the function of melanin?

also protects deeper layers from injury. Melanocytes, Langerhans cells, and Merkel cells with their associated tactile discs are scattered among the keratinocytes of the basal layer. The stratum basale is also known as the **stratum germinativum** (jer′-mi-na-TĒ-vum; *germ-* = sprout), a name that emphasizes its role in forming new cells.

Skin Grafts

New skin cannot regenerate if an injury destroys the stratum basale and its stem cells. Skin wounds of this magnitude require skin grafts in order to heal. A **skin graft** involves covering the wound with a patch of healthy skin taken from a donor site. To avoid tissue rejection, the transplanted skin is usually taken from the same individual *(autograft)* or an identical twin *(isograft)*. If skin damage is so extensive that an autograft would cause harm, a self-donation procedure called *autologous skin transplantation* (aw-TOL-ō-gus) may be used. In this procedure, performed most often for severely burned patients, small amounts of an individual's epidermis are removed, and the keratinocytes are cultured in the laboratory to produce thin sheets of skin. The new skin is transplanted back to the patient so that it covers the burn wound and generates a permanent skin. Also available as skin grafts for wound coverage are products developed from the foreskins of circumsized infants (Apligraft and Transite) that are grown in the laboratory. ■

Stratum Spinosum

Superficial to the stratum basale is the **stratum spinosum** (*spinos-* = thornlike), where 8 to 10 layers of many-sided keratinocytes fit closely together. Cells in the more superficial portions of this layer become somewhat flattened. These keratinocytes have the same organelles as cells of the stratum basale, and some cells in this layer may retain their ability to undergo cell division. When cells of the stratum spinosum are prepared for microscopic examination, they shrink and pull apart such that they appear to be covered with thornlike spines (see Figure 5.2a), although they are rounded and larger in living tissue. Each spiny projection in a prepared tissue section is a point where bundles of tonofilaments are inserting into a desmosome, tightly joining the cells to one another. This arrangement provides both strength and flexibility to the skin. Projections of both Langerhans cells and melanocytes also appear in this stratum.

Stratum Granulosum

At the middle of the epidermis, the **stratum granulosum** (*granulos-* = little grains) consists of three to five layers of flattened keratinocytes that are undergoing apoptosis. (Recall from Chapter 3 that apoptosis is an orderly, genetically programmed cell death in which the nucleus fragments before the cells die.) The nuclei and other organelles of these cells begin to degenerate, and tonofilaments become more apparent. A distinctive feature of cells in this layer is the presence of darkly staining granules of a protein called **keratohyalin** (ker′-a-tō-HĪ-a-lin), which converts the tonofilaments into keratin. Also present in the keratinocytes are membrane-enclosed **lamellar granules,** which release a lipid-rich secretion. This secretion fills the spaces between cells of the stratum granulosum, stratum lucidum, and stratum corneum. The lipid-rich secretion acts as a water-repellent sealant, retarding loss of body fluids and entry of foreign materials. As their nuclei break down during apoptosis, the keratinocytes of the stratum granulosum can no longer carry on vital metabolic reactions, and they die. Thus, the stratum granulosum marks the transition between the deeper, metabolically active strata and the dead cells of the more superficial strata.

Stratum Lucidum

The **stratum lucidum** (*lucid-* = clear) is present only in the thick skin of the fingertips, palms, and soles. It consists of three to five layers of flattened clear, dead keratinocytes that contain large amounts of keratin and thickened plasma membranes.

Stratum Corneum

The **stratum corneum** (*corne-* = horn or horny) consists of 25 to 30 layers of flattened dead keratinocytes. These cells are con-

Figure 5.3   Layers of the epidermis.

The epidermis consists of keratinized stratified squamous epithelium.

Stratum corneum
Stratum lucidum
Stratum granulosum
Stratum spinosum
Stratum basale

Dead keratinocytes
Lamellar granules
Keratinocyte
Langerhans cell
Merkel cell
Tactile disc
Sensory neuron
Melanocyte
Dermis

(a) Four principal cell types in epidermis

Superficial

Deep

Stratum corneum
Stratum lucidum
Stratum granulosum
Stratum spinosum
Stratum basale
Dermal papilla
Dermis

LM 480x

(b) Photomicrograph of a portion of skin

Which epidermal layer includes stem cells that continually undergo cell division?

tinuously shed and replaced by cells from the deeper strata. The interior of the cells contains mostly keratin. Between the cells are lipids from lamellar granules that help make this layer an effective water-repellent barrier. Its multiple layers of dead cells also help to protect deeper layers from injury and microbial invasion. Constant exposure of skin to friction stimulates the formation of a *callus,* an abnormal thickening of the stratum corneum.

Keratinization and Growth of the Epidermis

Newly formed cells in the stratum basale are slowly pushed to the surface. As the cells move from one epidermal layer to the next, they accumulate more and more keratin, a process called **keratinization** (ker′-a-tin-i-ZĀ-shun). Then they undergo apoptosis. Eventually the keratinized cells slough off and are replaced by underlying cells that, in turn, become keratinized. The whole process by which cells form in the stratum basale, rise to the surface, become keratinized, and slough off takes about four

weeks in an average epidermis of 0.1 mm (0.004 in.) thickness. The rate of cell division in the stratum basale increases when the outer layers of the epidermis are stripped away, as occurs in abrasions and burns. The mechanisms that regulate this remarkable growth are not well understood, but hormonelike proteins such as **epidermal growth factor (EGF)** play a role.

Table 5.1 summarizes the distinctive features of the epidermal strata.

Psoriasis

Psoriasis is a common and chronic skin disorder in which keratinocytes divide and move more quickly than normal from the stratum basale to the stratum corneum. They are shed prematurely in as little as 7 to 10 days. The immature keratinocytes make an abnormal keratin, which forms flaky, silvery scales at the skin surface, most often on the knees, elbows, and scalp (dandruff). Effective treatments — various topical ointments and UV phototherapy — suppress cell division, decrease the rate of cell growth, or inhibit keratinization. ∎

Table 5.1	Summary of Epidermal Strata
Stratum	**Description**
Basale	Deepest layer, composed of a single row of cuboidal or columnar keratinocytes that contain scattered tonofilaments (intermediate filaments); stem cells undergo cell division to produce new keratinocytes; melanocytes, Langerhans cells, and Merkel cells associated with tactile discs are scattered among the keratinocytes.
Spinosum	Eight to ten rows of many-sided keratinocytes with bundles of tonofilaments; includes projections of melanocytes and Langerhans cells.
Granulosum	Three to five rows of flattened keratinocytes, in which organelles are beginning to degenerate; cells contain the protein keratohyalin, which converts tonofilaments into keratin, and lamellar granules, which release a lipid-rich, water-repellent secretion.
Lucidum	Present only in skin of fingertips, palms, and soles; consists of three to five rows of clear, flat, dead keratinocytes with large amounts of keratin.
Corneum	Twenty-five to thirty rows of dead, flat keratinocytes that contain mostly keratin.

Dermis

The second, deeper part of the skin, the **dermis,** is composed mainly of connective tissue containing collagen and elastic fibers. The few cells present in the dermis include fibroblasts, macrophages, and some adipocytes. Blood vessels, nerves, glands, and hair follicles are embedded in dermal tissue. Based on its tissue structure, the dermis can be divided into a papillary region and a reticular region.

The **papillary region** is the superficial part of the dermis, about one-fifth of the thickness of the total layer (see Figure 5.1). It consists of areolar connective tissue containing fine elastic fibers. Its surface area is greatly increased by small, fingerlike projections called **dermal papillae** (pa-PIL-ē = nipples). These nipple-shaped structures indent the epidermis and some contain capillary loops. Other dermal papillae also contain tactile receptors called **corpuscles of touch** or **Meissner corpuscles,** nerve endings that are sensitive to touch. Also present in the dermal papillae are **free nerve endings,** dendrites that lack any apparent structural specialization. Different free nerve endings initiate signals that give rise to sensations of warmth, coolness, pain, tickling, and itching.

The deeper part of the dermis is called the **reticular region** (*reticul-* = netlike). It consists of dense irregular connective tissue containing bundles of collagen and some coarse elastic fibers. The bundles of collagen fibers in the reticular region interlace in a netlike manner. A few adipose cells, hair follicles, nerves, sebaceous (oil) glands, and sudoriferous (sweat) glands occupy the spaces between fibers.

The combination of collagen and elastic fibers in the reticular region provides the skin with strength, *extensibility* (ability to stretch), and *elasticity* (ability to return to original shape after stretching). The extensibility of skin can readily be seen in pregnancy and obesity. Extreme stretching, however, may produce small tears in the dermis, causing *striae* (STRĪ-ē = streaks), or stretch marks, that are visible as red or silvery white streaks on the skin surface.

The surfaces of the palms, fingers, soles, and toes have a series of ridges and grooves. They appear either as straight lines or as a pattern of loops and whorls, as on the tips of the digits. These **epidermal ridges** develop during the third and fourth fetal months as the epidermis conforms to the contours of the underlying dermal papillae of the papillary region. The ridges increase the surface area of the epidermis and thus increase the grip of the hand or foot by increasing friction. Because the ducts of sweat glands open on the tops of the epidermal ridges as sweat pores, the sweat and ridges form fingerprints (or footprints) upon touching a smooth object. The epidermal ridge pattern is genetically determined and is unique for each individual. Normally, the ridge pattern does not change during life, except to enlarge, and thus can serve as the basis for identification through fingerprints or footprints. The study of the pattern of epidermal ridges is called **dermatoglyphics** (der′-ma-tō-GLIF-iks; *glyphe* = carved work).

Table 5.2 summarizes the structural features of the papillary and reticular regions of the dermis.

Photodamage and Photosensitivity Reactions

Although basking in the warmth of the sun may feel good, it is not a healthy practice. Both the longer-wavelength ultraviolet A (UVA) rays and the shorter-wavelength UVB rays cause **photodamage** of the skin. Light-skinned and dark-skinned individuals alike experience the effect of acute overexposure to UVB rays—sunburn. Even if sunburn does not occur, the UVB rays can damage DNA, producing genetic mutations in epidermal cells. In turn, the mutations can cause skin cancer. As UVA rays penetrate

Table 5.2	Summary of Papillary and Reticular Regions of the Dermis
Region	**Description**
Papillary	The superficial portion of the dermis (about one-fifth); consists of areolar connective tissue with elastic fibers; contains dermal papillae that house capillaries, corpuscles of touch, and free nerve endings.
Reticular	The deeper portion of the dermis (about four-fifths); consists of dense irregular connective tissue with bundles of collagen and some coarse elastic fibers. Spaces between fibers contain some adipose cells, hair follicles, nerves, sebaceous glands, and sudoriferous glands.

to the dermis, they produce oxygen free radicals that disrupt collagen and elastic fibers in the extracellular matrix. This is the main reason for the severe wrinkling that occurs in those who spend a great deal of time in the sun without protection.

Besides causing photodamage to the skin, exposure to direct sunlight may also produce **photosensitivity reactions** in some individuals taking certain oral medications, such as erythromycin, or herbal supplements, such as St. John's wort. The reactions may include redness, peeling, hives, blisters, and even shock. ■

The Structural Basis of Skin Color

Melanin, carotene, and hemoglobin are three pigments that impart a wide variety of colors to skin. The amount of **melanin** causes the skin's color to vary from pale yellow to tan to black. Melanocytes are most plentiful in the epidermis of the penis, nipples of the breasts, area just around the nipples (areolae), face, and limbs. They are also present in mucous membranes. Because the *number* of melanocytes is about the same in all people, differences in skin color are due mainly to the *amount of pigment* the melanocytes produce and disperse to keratinocytes. In some people, melanin accumulates in patches called *freckles*. As one grows older, *age (liver) spots* may develop. These flat blemishes look like freckles and range in color from light brown to black. Like freckles, age spots are accumulations of melanin.

Melanocytes synthesize melanin from the amino acid *tyrosine* in the presence of an enzyme called *tyrosinase*. Synthesis occurs in an organelle called a **melanosome.** Exposure to UV light increases the enzymatic activity within melanosomes and thus stimulates melanin production. Both the amount and darkness of melanin increase, which gives the skin a tanned appearance and further protects the body against UV radiation. Thus, within limits, melanin serves a protective function. Nevertheless, repeatedly exposing the skin to UV light causes skin cancer. A tan is lost when the melanin-containing keratinocytes are shed from the stratum corneum.

Carotene (KAR-ō-tēn; *carot* = carrot), a yellow-orange pigment, is the precursor of vitamin A, which is used to synthesize pigments needed for vision. Carotene is found in the stratum corneum and fatty areas of the dermis and subcutaneous layer.

When little melanin or carotene is present, the epidermis appears translucent. Thus, the skin of white people appears pink to red, depending on the amount and oxygen content of the blood moving through capillaries in the dermis. The red color is due to **hemoglobin,** the oxygen-carrying pigment inside red blood cells.

Albinism (AL-bin-izm; *albin-* = white) is the inherited inability of an individual to produce melanin. Most **albinos** (al-BĪ-nōs), people affected by albinism, have melanocytes that are unable to synthesize tyrosinase. Melanin is missing from their hair, eyes, and skin. In another condition, called **vitiligo** (vit-i-LĪ-gō), the partial or complete loss of melanocytes from patches of skin produces irregular white spots. The loss of melanocytes may be related to an immune system malfunction in which antibodies attack the melanocytes.

 Skin Color Clues

The color of skin and mucous membranes can provide clues for diagnosing certain conditions. When blood is not picking up an adequate amount of oxygen in the lungs, such as in someone who has stopped breathing, the mucous membranes, nail beds, and skin appear bluish or **cyanotic** (sī-a-NOT-ik; *cyan-* = blue). This occurs because hemoglobin that is depleted of oxygen looks deep, purplish blue. **Jaundice** (JON-dis; *jaund-* = yellow) is due to a buildup of the yellow pigment bilirubin in the blood. This condition gives a yellowish appearance to the whites of the eyes and the skin. Jaundice usually indicates liver disease. **Erythema** (er-i-THĒ-ma; *eryth-* = red), redness of the skin, is caused by engorgement of capillaries in the dermis with blood due to skin injury, exposure to heat, infection, inflammation, or allergic reactions. All skin color changes are observed most readily in people with lighter-colored skin and may be more difficult to discern in people with darker skin. ■

Types of Skin

Although the skin over the entire body is similar in structure, there are quite a few local variations related to thickness of the epidermis, strength, flexibility, degree of keratinization, distribution and type of hair, density and types of glands, pigmentation, vascularity (blood supply), and innervation (nerve supply). Based on structural and functional properties, we recognize two major types of skin: thin (hairy) and thick (hairless).

Thin skin covers all parts of the body except for the palms, palmar surfaces of the digits, and the soles. Its epidermis is thin, just 0.10–0.15 mm (0.004–0.006 in.). A distinct stratum lucidum is lacking, and the strata spinosum and corneum are relatively thin. Thin skin has lower, broader, and fewer dermal papillae, and thus lacks epidermal ridges. Additionally, even though thin skin has hair follicles, arrector pili muscles, and sebaceous (oil) glands, it has fewer sweat glands than thick skin. Finally, thin skin has a sparser distribution of sensory receptors than thick skin.

Thick skin covers the palms, palmar surfaces of the fingers, and soles. Its epidermis is relatively thick, 0.6–4.5 mm (0.024–0.18 in.), and features a distinct stratum lucidum as well as thicker strata spinosum and corneum. The dermal papillae of thick skin are higher, narrower, and more numerous than those in thin skin, and thus thick skin has epidermal ridges. Thick skin lacks hair follicles, arrector pili muscles, and sebaceous glands, and has more sweat glands than thin skin. Sensory receptors are more densely clustered in thick skin.

Table 5.3 summarizes the features of thin and thick skin.

▶ CHECKPOINT

1. What structures are included in the integumentary system?
2. How does the subcutaneous layer relate to the skin?
3. How does the process of keratinization occur?
4. What are the structural and functional differences between the epidermis and dermis?
5. How are epidermal ridges formed?
6. In what ways are thin and thick skin similar and different?

Table 5.3	Comparison of Thin and Thick Skin	
Feature	**Thin Skin**	**Thick Skin**
Distribution	All parts of the body except palms, palmar surface of digits, and soles.	Palms, palmar surface of digits, and soles.
Epidermal thickness	0.10–0.15 mm (0.004–0.006 in.).	0.6–4.5 mm (0.024–0.18 in.).
Epidermal strata	Stratum lucidum essentially lacking; thinner strata spinosum and corneum.	Thick strata lucidum, spinosum, and corneum.
Epidermal ridges	Lacking due to poorly developed and fewer dermal papillae.	Present due to well-developed and more numerous dermal papillae.
Hair follicles and arrector pili muscles	Present.	Absent.
Sebaceous glands	Present.	Absent.
Sudoriferous glands	Fewer.	More numerous.
Sensory receptors	Sparser.	Denser.

ACCESSORY STRUCTURES OF THE SKIN

▶ O B J E C T I V E

• **Contrast the structure, distribution, and functions of hair, skin glands, and nails.**

Accessory structures of the skin—hair, skin glands, and nails—develop from the embryonic epidermis. They have a host of important functions. For example, hair and nails protect the body, and sweat glands help regulate body temperature.

Hair

Hairs, or *pili* (PI-lē), are present on most skin surfaces except the palms, palmar surfaces of the fingers, soles, and plantar surfaces of the feet. In adults, hair usually is most heavily distributed across the scalp, in the eyebrows, in the axillae (armpits), and around the external genitalia. Genetic and hormonal influences largely determine hair thickness and the pattern of distribution.

Anatomy of a Hair

Each hair is composed of columns of dead, keratinized cells bonded together by extracellular proteins. The **shaft** is the superficial portion of the hair, most of which projects from the surface of the skin (Figure 5.4a). The shaft of straight hair is round in transverse section, that of wavy hair is oval, and that of curly hair is kidney-shaped. The **root** is the portion of the hair deep to the shaft that penetrates into the dermis, and sometimes into the subcutaneous layer. The shaft and root both consist of three concen-

tric layers (Figure 5.4c, d). The inner *medulla,* which may be lacking in thinner hair, is composed of two or three rows of irregularly shaped cells containing pigment granules and air spaces. The middle *cortex* consists of elongated cells that contain pigment granules in dark hair but mostly air in gray or white hair. The *cuticle of the hair,* the outermost layer, consists of a single layer of cells. In the deeper part of the hair root, the cells of the cuticle of the hair are nucleated. In the upper part of the root and in the shaft they are scale-like and lack nuclei. Cuticle cells on the shaft are arranged like shingles on the side of a house, with their free edges pointing toward the end of the hair, and they are the most heavily keratinized of hair cells (Figure 5.4b).

Surrounding the root of the hair is the **hair follicle,** which is made up of an external root sheath and an internal root sheath (Figure 5.4c d). The *external root sheath* is a downward continuation of the epidermis. Near the surface, it contains all the epidermal layers. At the base of the hair follicle, the external root sheath contains only the stratum basale. The *internal root sheath* is produced by the matrix (described shortly) and forms a cellular tubular sheath of epithelium between the external root sheath and the hair.

The base of each hair follicle is an onion-shaped structure, the **bulb** (Figure 5.4c). This structure houses a nipple-shaped indentation, the **papilla of the hair,** which contains areolar connective tissue and many blood vessels that nourish the growing hair follicle. The bulb also contains a germinal layer of cells called the **matrix.** The matrix cells arise from the stratum basale, the site of cell division. Hence, matrix cells are responsible for the growth of existing hairs, and they produce new hairs when old hairs are shed. This replacement process occurs within the same follicle. Matrix cells also give rise to the cells of the internal root sheath.

Sebaceous (oil) glands (discussed shortly) and a bundle of smooth muscle cells are also associated with hairs (Figure 5.4a). The smooth muscle is the **arrector pili** (a-REK-tor PI-lē; *arrect* = to raise). It extends from the superficial dermis of the skin to the side of the hair follicle. In its normal position, hair emerges at an angle to the surface of the skin. Under physiologic or emotional stress, such as cold or fright, autonomic nerve endings stimulate the arrector pili muscles to contract, which pulls the hair shafts perpendicular to the skin surface. This action causes "goose bumps" or "gooseflesh" because the skin around the shaft forms slight elevations.

Surrounding each hair follicle are dendrites of neurons, called **hair root plexuses,** that are sensitive to touch (Figure 5.4a). The hair root plexuses generate nerve impulses if their hair shaft is moved.

Hair Removal

A substance that removes hair is called a **depilatory.** It dissolves the protein in the hair shaft, turning it into a gelatinous mass than can be wiped away. Because the hair root is not affected, regrowth of the hair occurs. In **electrolysis,** an electric current is used to destroy the hair matrix so the hair cannot regrow. ■

Figure 5.4   Hair.

Hairs are growths of epidermis composed of dead, keratinized cells.

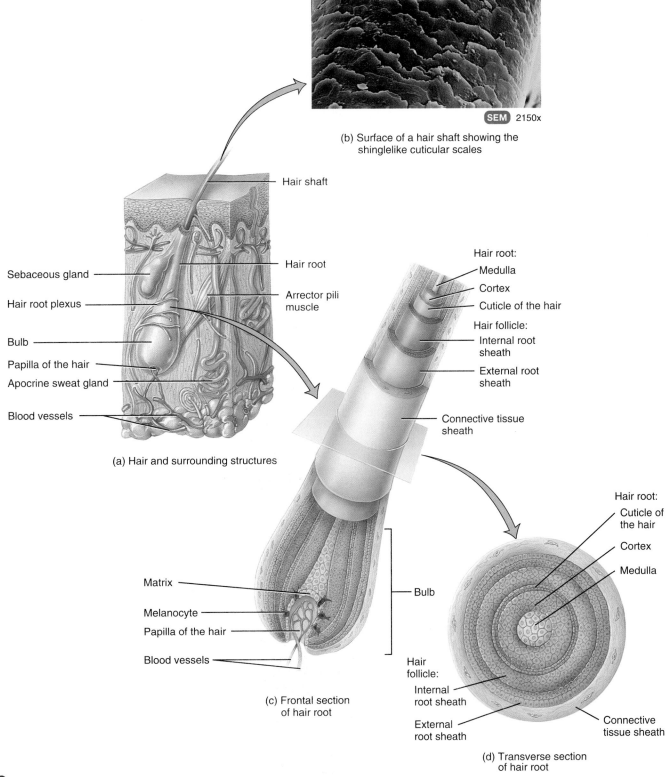

(b) Surface of a hair shaft showing the shinglelike cuticular scales

SEM 2150x

Hair shaft

Sebaceous gland

Hair root plexus

Bulb

Papilla of the hair

Apocrine sweat gland

Blood vessels

Hair root

Arrector pili muscle

(a) Hair and surrounding structures

Hair root:
- Medulla
- Cortex
- Cuticle of the hair

Hair follicle:
- Internal root sheath
- External root sheath

Connective tissue sheath

Matrix

Melanocyte

Papilla of the hair

Blood vessels

Bulb

Hair follicle:
- Internal root sheath
- External root sheath

(c) Frontal section of hair root

Hair root:
- Cuticle of the hair
- Cortex
- Medulla

Connective tissue sheath

(d) Transverse section of hair root

Why does it hurt when you pluck a hair out but not when you have a haircut?

Hair Growth

Each hair follicle goes through a growth cycle, which consists of a growth stage and a resting stage. During the **growth stage,** cells of the matrix differentiate, keratinize, and die. This process forms the root sheath and hair shaft. As new cells are added at the base of the hair root, the hair grows longer. In time, the growth of the hair stops and the **resting stage** begins. After the resting stage, a new growth cycle begins. The old hair root falls out or is pushed out of the hair follicle, and a new hair begins to grow in its place. Scalp hair grows for 2 to 6 years and rests for about 3 months. At any time, about 85% of scalp hairs are in the growth stage. Visible hair is dead, but until the hair is pushed out of its follicle by a new hair, portions of its root within the scalp are alive.

Normal hair loss in the adult scalp is about 100 hairs per day. Both the rate of growth and the replacement cycle may be altered by illness, radiation therapy, chemotherapy, age, genetics, gender, and severe emotional stress. Rapid weight-loss diets that severely restrict calories or protein increase hair loss. The rate of shedding also increases for three to four months after childbirth. **Alopecia** (al′-o-PĒ-shē-a), the partial or complete lack of hair, may result from genetic factors, aging, endocrine disorders, chemotherapy for cancer, or skin disease.

Types of Hairs

Hair follicles develop between the ninth and twelfth weeks after fertilization as downgrowths of the stratum basale of the epidermis into the dermis. Usually by the fifth month of development, the follicles produce very fine, nonpigmented hairs called **lanugo** (la-NOO-gō = wool or down) that cover the body of the fetus. This hair is shed before birth, except in the scalp, eyebrows, and eyelashes. A few months after birth, slightly thicker hairs replace these downy hairs. Over the remainder of the body of an infant, a new growth of short, fine hair occurs. These hairs are known as **vellus hairs** (VEL-us = fleece), commonly called "peach fuzz." In response to hormones (androgens) secreted at puberty, coarse pigmented and frequently curly hair develops in the axillae (armpits) and pubic region. In males, these hairs also appear on the face and other parts of the body. The hairs that develop at puberty, together with those of the head, eyebrows, and eyelashes are called terminal hairs. About 95% of body hair on males is terminal hair (5% vellus hair), whereas about 35% of body hair on females is terminal (65% vellus hair).

Hair Color

The color of hair is due primarily to the amount and type of melanin in its keratinized cells. Melanin is synthesized by melanocytes scattered in the matrix of the bulb and passes into cells of the cortex and medulla (Figure 5.4c). Dark-colored hair contains mostly true melanin, whereas blond and red hair contain variants of melanin in which there is iron and more sulfur. Graying hair occurs because of a progressive decline in tyrosinase, whereas white hair results from accumulation of air bubbles in the medullary shaft.

Functions of Hair

Although the protection it offers is limited, hair on the head guards the scalp from injury and the sun's rays. It also decreases heat loss from the scalp. Eyebrows and eyelashes protect the eyes from foreign particles, as does hair in the nostrils and in the external ear canal. Touch receptors associated with hair follicles (hair root plexuses) are activated whenever a hair is even slightly moved. Thus, hairs function in sensing light touch.

Hair and Hormones

At puberty, when the testes begin secreting significant quantities of androgens (masculinizing sex hormones), males develop the typical male pattern of hair growth, including a beard and a hairy chest. In females at puberty, the ovaries and the adrenal glands produce small quantities of androgens, which promote hair growth in the axillae and pubic region. Occasionally, a tumor of the adrenal glands, testes, or ovaries produces an excessive amount of androgens. The result in females or prepubertal males is **hirsutism** (HER-soo-tizm; *hirsut-* = shaggy), a condition of excessive body hair.

Surprisingly, androgens also must be present for occurrence of the most common form of baldness, **androgenic alopecia** or **male-pattern baldness.** In genetically predisposed adults, androgens inhibit hair growth. On men, hair loss is most obvious at the temples and crown. Women are more likely to have thinning of hair on top of the head. The first drug approved for enhancing scalp hair growth was minoxidil (Rogaine). It causes vasodilation (widening of blood vessels), thus increasing circulation. In about a third of the people who try it, minoxidil improves hair growth, causing scalp follicles to enlarge and lengthening the growth cycle. For many, however, the hair growth is meager. Minoxidil does not help people who already are bald. ■

Skin Glands

Several kinds of exocrine glands are associated with the skin: sebaceous (oil) glands, sudoriferous (sweat) glands, and ceruminous glands. Mammary glands, which are specialized sudoriferous glands that secrete milk, are discussed in Chapter 28 along with the female reproductive system.

Sebaceous Glands

Sebaceous glands (se-BĀ-shus; *sebace-* = greasy) or **oil glands** are simple, branched acinar glands. With few exceptions, they are connected to hair follicles (see Figures 5.1 and 5.4a). The secreting portion of a sebaceous gland lies in the dermis and usually opens into the neck of a hair follicle. In some locations, such as the lips, glans penis, labia minora, and tarsal glands of the eyelids, sebaceous glands open directly onto the surface of the skin. Absent in the palms and soles, sebaceous glands are small in most areas of the trunk and limbs, but large in the skin of the breasts, face, neck, and upper chest.

Sebaceous glands secrete an oily substance called **sebum** (SĒ-bum), which is a mixture of triglycerides, cholesterol, proteins, and inorganic salts. Sebum coats the surface of hairs and helps keep them from drying and becoming brittle. Sebum also prevents excessive evaporation of water from the skin, keeps the skin soft and pliable, and inhibits the growth of certain bacteria.

 Acne

Acne is an inflammation of sebaceous glands that usually begins at puberty, when the sebaceous glands grow in size and increase their production of sebum. Androgens from the testes, ovaries, and adrenal glands play the greatest role in stimulating sebaceous glands. Acne occurs predominantly in sebaceous follicles that have been colonized by bacteria, some of which thrive in the lipid-rich sebum. The infection may cause a cyst or sac of connective tissue cells to form, which can destroy and displace epidermal cells. This condition, called **cystic acne,** can permanently scar the epidermis. ■

Sudoriferous Glands

There are three to four million **sweat glands,** or **sudoriferous glands** (soo′-dor-IF-er-us; *sudori-* = sweat; *-ferous* = bearing). The cells of sweat glands release their secretions by exocytosis and empty them into hair follicles or onto the skin surface through pores. They are divided into two main types, eccrine and apocrine, based on their structure, location, and type of secretion.

Eccrine sweat glands (*eccrine* = secreting outwardly) are simple, coiled tubular glands and are much more common than apocrine sweat glands. They are distributed throughout the skin, except for the margins of the lips, nail beds of the fingers and toes, glans penis, glans clitoris, labia minora, and eardrums. Eccrine sweat glands are most numerous in the skin of the forehead, palms, and soles; their density can be as high as 450 per square centimeter (3000 per square inch) in the palms. The secretory portion of eccrine sweat glands is located mostly in the deep dermis (sometimes in the upper subcutaneous layer). The excretory duct projects through the dermis and epidermis and ends as a pore at the surface of the epidermis (see Figure 5.1).

The sweat produced by eccrine sweat glands (about 600 mL per day) consists of water, ions (mostly Na^+ and Cl^-), urea, uric acid, ammonia, amino acids, glucose, and lactic acid. The main function of eccrine sweat glands is to help regulate body temperature through evaporation. As sweat evaporates, large quantities of heat energy leave the body surface. Sweat, or perspiration, usually occurs first on the forehead and scalp, extends to the face and the rest of the body, and occurs last on the palms and soles. Under conditions of emotional stress, however, the palms, soles, and axillae are the first surfaces to sweat. Eccrine sweat also plays a small role in eliminating wastes such as urea, uric acid, and ammonia. Sweat that evaporates from the skin before it is perceived as moisture is termed **insensible perspiration.** Sweat that is excreted in larger amounts and is seen as moisture on the skin is called **sensible perspiration.**

Apocrine sweat glands are also simple, coiled tubular glands. They are found mainly in the skin of the axilla (armpit), groin, areolae (pigmented areas around the nipples) of the breasts, and bearded regions of the face in adult males. These glands were once thought to release their secretions in an apocrine manner—by pinching off a portion of the cell. We now know, however, that their secretion is released by exocytosis, which is characteristic of the release of secretions by merocrine glands (see Figure 4.4a on page 117). Although in this sense they are merocrine glands, the term *apocrine* is still used. The secretory portion of these sweat glands is located mostly in the subcutaneous layer, and the excretory duct opens into hair follicles (see Figure 5.1). Their secretory product is slightly viscous compared to eccrine secretions and contains the same components as eccrine sweat plus lipids and proteins. In women, cells of apocrine sweat glands enlarge about the time of ovulation and shrink during menstruation. Whereas eccrine sweat glands start to function soon after birth, apocrine sweat glands do not begin to function until puberty. Apocrine sweat glands are stimulated during emotional stress and sexual excitement; these secretions are commonly known as a "cold sweat."

Table 5.4 presents a comparison of eccrine and apocrine sweat glands.

Ceruminous Glands

Modified sweat glands in the external ear, called **ceruminous glands** (se-ROO-mi-nus; *cer-* = wax), produce a waxy secretion. The secretory portions of ceruminous glands lie in the subcutaneous layer, deep to sebaceous glands. Their excretory ducts

Table 5.4	Comparison of Eccrine and Apocrine Sweat Glands	
Feature	**Eccrine Sweat Glands**	**Apocrine Sweat Glands**
Distribution	Throughout skin, except for margins of lips, nail beds, glans penis and clitoris, labia minora, and eardrums.	Skin of the axilla, groin, areolae, and bearded regions of the face.
Location of secretory portion	Mostly in deep dermis.	Mostly in subcutaneous layer.
Termination of excretory duct	Surface of epidermis.	Hair follicle.
Secretion	Less viscous; consists of water, ions (Na^+, Cl^-), urea, uric acid, ammonia, amino acids, glucose, and lactic acid.	More viscous; consists of the same components as eccrine sweat glands plus lipids and proteins.
Functions	Regulation of body temperature and waste removal.	Stimulated during emotional stress and sexual excitement.
Onset of function	Soon after birth.	Puberty.

open either directly onto the surface of the external auditory canal (ear canal) or into ducts of sebaceous glands. The combined secretion of the ceruminous and sebaceous glands is called **cerumen,** or earwax. Cerumen, together with hairs in the external auditory canal, provides a sticky barrier that impedes the entrance of foreign bodies.

Impacted Cerumen

Some people produce an abnormally large amount of cerumen in the external auditory canal. The cerumen may accumulate until it becomes impacted (firmly wedged), which prevents sound waves from reaching the eardrum. The treatment for **impacted cerumen** is usually periodic ear irrigation with enzymes to dissolve the wax or removal of wax with a blunt instrument by trained medical personnel. The use of cotton-tipped swabs or sharp objects is not recommended for this purpose because they may push the cerumen farther into the external auditory canal and damage the eardrum. ■

Nails

Nails are plates of tightly packed, hard, keratinized epidermal cells. The cells form a clear, solid covering over the dorsal surfaces of the distal portions of the digits. Each nail consists of a nail body, a free edge, and a nail root (Figure 5.5). The **nail body** is the portion of the nail that is visible, the **free edge** is the part that may extend past the distal end of the digit, and the **nail root** is the portion that is buried in a fold of skin. Most of the nail body appears pink because of blood flowing through underlying capillaries. The free edge is white because there are no

underlying capillaries. The whitish, crescent-shaped area of the proximal end of the nail body is called the **lunula** (LOO-noo-la = little moon). It appears whitish because the vascular tissue underneath does not show through due to the thickened stratum basale in the area. Beneath the free edge is a thickened region of stratum corneum called the **hyponychium** (hī′-pō-NIK-ē-um; *hypo-* = below), which secures the nail to the fingertip. The **eponychium** (ep′-ō-NIK-ē-um; *ep-* = above; *-onych* = nail) or **cuticle** is a narrow band of epidermis that extends from and adheres to the margin (lateral border) of the nail wall. It occupies the proximal border of the nail and consists of stratum corneum.

The epithelium deep to the nail root is the **nail matrix,** where cells divide by mitosis to produce growth. Nail growth occurs by the transformation of superficial cells of the matrix into nail cells. In the process, the harder outer layer is pushed forward over the stratum basale. The growth rate of nails is determined by the rate of mitosis in matrix cells, which is influenced by factors such as a person's age, health, and nutritional status. Nail growth also varies according to the season, the time of day, and environmental temperature. The average growth in the length of fingernails is about 1 mm (0.04 in.) per week. The growth rate is somewhat slower in toenails. The longer the digit, the faster the nail grows.

Functionally, nails help us grasp and manipulate small objects in various ways, provide protection against trauma to the ends of the digits, and allow us to scratch various parts of the body.

► CHECKPOINT

7. Describe the structure of a hair. What causes "goose bumps"?

8. Contrast the locations and functions of sebaceous (oil) glands, sudoriferous (sweat) glands, and ceruminous glands.

9. Describe the parts of a nail.

Figure 5.5 Nails.   Shown is a fingernail.

Nail cells arise by transformation of superficial cells of the nail matrix.

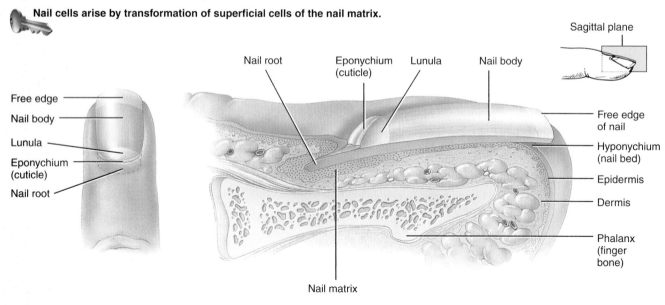

(a) Dorsal view

(b) Sagittal section showing internal detail

Why are nails so hard?

FUNCTIONS OF THE SKIN

OBJECTIVE

• Describe how the skin contributes to regulation of body temperature, protection, sensation, excretion and absorption, and synthesis of vitamin D.

The skin helps regulate body temperature, serves as a water-repellent and protective barrier between the external environment and internal tissues, contains sensory nerve endings, excretes a small amount of salts and several organic compounds and has some capacity to absorb substances, and helps to synthesize the active form of vitamin D.

Thermoregulation

The skin contributes to **thermoregulation,** the homeostatic regulation of body temperature, in two ways: by liberating sweat at its surface and by adjusting the flow of blood in the dermis. In response to high environmental temperature or heat produced by exercise, the evaporation of sweat from the skin surface helps lower body temperature. In response to low environmental temperature, production of sweat is decreased, which helps conserve heat. During moderate exercise, the flow of blood through the skin increases, which increases the amount of heat radiated from the body.

The dermis houses an extensive network of blood vessels that carry 8–10% of the total blood flow in a resting adult. For this reason, the skin acts as a *blood reservoir.* During very strenuous exercise, skin blood vessels constrict (narrow) somewhat, diverting more blood to contracting skeletal muscles and the heart. Because of this shunting of blood away from the skin, however, less heat is lost from the skin, and body temperature may rise.

Protection

The skin covers the body and provides protection in various ways. Keratin in the skin protects underlying tissues from microbes, abrasion, heat, and chemicals and the tightly interlocked keratinocytes resist invasion by microbes. Lipids released by lamellar granules retard evaporation of water from the skin surface, thus protecting the body from dehydration, and they also retard entry of water across the skin surface when we take a shower or go swimming. The oily sebum from the sebaceous glands also protects skin and hairs from drying out and contains bactericidal chemicals that kill surface bacteria. The acidic pH of perspiration retards the growth of some microbes. The pigment melanin provides some protection against the damaging effects of UV light. Two types of cells carry out protective functions that are immunological in nature. Epidermal Langerhans cells alert the immune system to the presence of potentially harmful microbial invaders by recognizing and processing them, and macrophages in the dermis phagocytize bacteria and viruses that manage to penetrate the skin surface.

Cutaneous Sensations

Cutaneous sensations are those that arise in the skin. These include tactile sensations—touch, pressure, vibration, and tickling—as well as thermal sensations such as warmth and coolness. Another cutaneous sensation, pain, usually is an indication of impending or actual tissue damage. Some of the wide variety of abundantly distributed nerve endings and receptors in the skin include the tactile discs of the epidermis, the corpuscles of touch in the dermis, and hair root plexuses around each hair follicle. Chapter 15 provides more details on the topic of cutaneous sensations.

Excretion and Absorption

The skin normally has a small role in *excretion,* the elimination of substances from the body, and *absorption,* the passage of materials from the external environment into body cells. Despite the almost waterproof nature of the stratum corneum, about 400 mL of water evaporates through it daily. A sedentary person loses an additional 200 mL per day of water as sweat, whereas a physically active person loses much more. Besides removing water and heat (by evaporation), sweat also is the vehicle for excretion of small amounts of salts, carbon dioxide, and two organic molecules that result from the breakdown of proteins—ammonia and urea.

The absorption of water-soluble substances through the skin is negligible, but certain lipid-soluble materials do penetrate the skin. These include fat-soluble vitamins (A, D, E, and K), certain drugs, and the gases oxygen and carbon dioxide. Toxic materials that can be absorbed through the skin include organic solvents such as acetone (in some nail polish removers) and carbon tetrachloride (dry-cleaning fluid); salts of heavy metals such as lead, mercury, and arsenic; and the toxins in poison ivy and poison oak. Since topical (applied to the skin) steroids, such as cortisone, are lipid-soluble, they move easily into the papillary region of the dermis. Here, they exert their anti-inflammatory properties by inhibiting histamine production by mast cells. Recall that histamine contributes to inflammation.

Synthesis of Vitamin D

Synthesis of vitamin D requires activation of a precursor molecule in the skin by UV rays in sunlight. Enzymes in the liver and kidneys then modify the activated molecule, finally producing *calcitriol,* the most active form of vitamin D. Calcitriol is a hormone that aids in the absorption of calcium in foods from the gastrointestinal tract into the blood.

Transdermal Drug Administration

Most drugs are either absorbed into the body through the digestive system or injected into subcutaneous tissue or muscle. An alternative route, **transdermal drug administration,** enables a drug contained within an adhesive skin patch to pass across the epidermis and into the blood vessels of the dermis. The drug is released at a controlled rate over one to several days. Because the major barrier to penetration of most drugs is the stratum corneum, transdermal absorption is most rapid in regions of the skin where this layer is thin, such as the scrotum, face, and scalp. A growing number of drugs are available for transdermal administration. These include nitroglycerin, for prevention of angina pectoris (chest pain associated with heart disease); scopolamine, for motion sickness; estradiol, used for estrogen-replacement therapy during menopause; and nicotine, used to help people stop smoking. ■

▶ CHECKPOINT

10. In what two ways does the skin help regulate body temperature?
11. In what ways does the skin serve as a protective barrier?
12. What sensations arise from stimulation of neurons in the skin?
13. What types of molecules can penetrate the stratum corneum?

MAINTAINING HOMEOSTASIS: SKIN WOUND HEALING

▶ OBJECTIVE
• **Explain how epidermal wounds and deep wounds heal.**

Skin damage sets in motion a sequence of events that repairs the skin to its normal (or near-normal) structure and function. Two kinds of wound-healing processes can occur, depending on the depth of the injury. Epidermal wound healing occurs following wounds that affect only the epidermis; deep wound healing occurs following wounds that penetrate the dermis.

Epidermal Wound Healing

Even though the central portion of an epidermal wound may extend to the dermis, the edges of the wound usually involve only slight damage to superficial epidermal cells. Common types of epidermal wounds include abrasions, in which a portion of skin has been scraped away, and minor burns.

In response to an epidermal injury, basal cells of the epidermis surrounding the wound break contact with the basement membrane. The cells then enlarge and migrate across the wound (Figure 5.6a). The cells appear to migrate as a sheet until advancing cells from opposite sides of the wound meet. When epidermal cells encounter each other, they stop migrating due to a cellular response called **contact inhibition.** Migration of the epidermal cells stops completely when each is finally in contact with other epidermal cells on all sides.

While some basal epidermal cells migrate, epidermal growth factor, a hormone, stimulates basal stem cells to divide and replace the ones that have moved into the wound. Replacement of migrating basal cells continues until the wound is resurfaced (Figure 5.6b). Following this, the relocated cells divide to build new strata, thus thickening the new epidermis.

Deep Wound Healing

Deep wound healing occurs when an injury extends to the dermis and subcutaneous layer. Because multiple tissue layers must be repaired, the healing process is more complex than in epidermal wound healing. In addition, scar tissue is formed, and the healed tissue loses some of its normal function. Deep wound healing occurs in four phases: an inflammatory phase, a migratory phase, a proliferative phase, and a maturation phase.

During the **inflammatory phase,** a blood clot forms in the wound and loosely unites the wound edges (Figure 5.7a). This phase of deep wound healing involves **inflammation,** a vascular and cellular response that helps eliminate microbes, foreign material, and dying tissue in preparation for repair. The vasodilation and increased permeability of blood vessels associated with inflammation enhance delivery of helpful cells. These include phagocytic white blood cells called neutrophils, monocytes, which develop into macrophages that phagocytize microbes, and mesenchymal cells, which develop into fibroblasts (Figure 5.7a).

Figure 5.6 Epidermal wound healing.

🔑 In an epidermal wound, the injury does not extend into the dermis.

(a) Division of basal epithelial cells and migration across wound

(b) Thickening of epidermis

 Would you expect an epidermal wound to bleed? Why or why not?

Figure 5.7 Deep wound healing. The initial inflammatory phase (a) is followed by a migratory phase, a proliferative phase, and finally (b) a maturation phase.

In a deep wound, the injury extends deep to the epidermis.

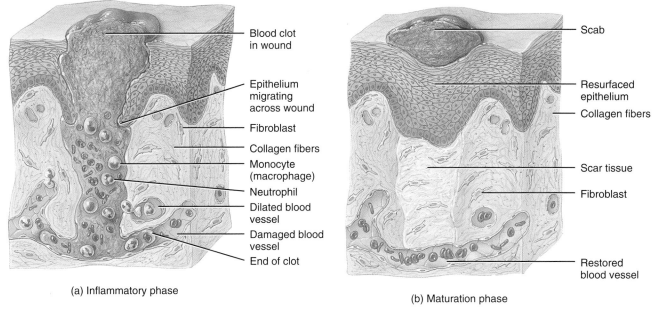

(a) Inflammatory phase

(b) Maturation phase

Deep wound healing

 Why is the arrival of phagocytic white blood cells (neutrophils and macrophages) at the scene of a wound helpful?

The three phases that follow do the work of repairing the wound. In the **migratory phase,** the clot becomes a scab, and epithelial cells migrate beneath the scab to bridge the wound. Fibroblasts migrate along fibrin threads and begin synthesizing scar tissue (collagen fibers and glycoproteins), and damaged blood vessels begin to regrow. During this phase, the tissue filling the wound is called **granulation tissue.** The **proliferative phase** is characterized by extensive growth of epithelial cells beneath the scab, deposition by fibroblasts of collagen fibers in random patterns, and continued growth of blood vessels. Finally, during the **maturation phase,** the scab sloughs off once the epidermis has been restored to normal thickness. Collagen fibers become more organized, fibroblasts decrease in number, and blood vessels are restored to normal (Figure 5.7b).

The process of scar tissue formation is called **fibrosis.** Sometimes, so much scar tissue is formed during deep wound healing that a raised scar—one that is elevated above the normal epidermal surface—results. If such a scar remains within the boundaries of the original wound, it is a **hypertrophic scar.** If it extends beyond the boundaries into normal surrounding tissues, it is a **keloid scar.** Scar tissue differs from normal skin in that its collagen fibers are more densely arranged. Scar tissue also has fewer blood vessels and might not contain the same number of hairs, skin glands, or sensory structures as undamaged skin. Because of the arrangement of collagen fibers and the scarcity of blood vessels, scars usually are lighter in color than normal skin.

▶ CHECKPOINT

14. Why doesn't epidermal wound healing result in scar formation?

 ## DEVELOPMENT OF THE INTEGUMENTARY SYSTEM

▶ OBJECTIVE

• **Describe the development of the epidermis, its accessory structures, and the dermis.**

At the end of many chapters throughout this book, we will discuss the development of body systems. About 15 days after the fertilization of an oocyte, a portion of the developing embryo differentiates into three **primary germ layers: ectoderm** (*ecto-* = outside), **mesoderm** (*meso-* = middle), and **endoderm** (*endo-* = within). They are the embryonic tissues from which all tissues and organs of the body eventually develop (see Table 29.1 on page 1072).

The *epidermis* is derived from the **ectoderm.** At the beginning of the eighth week after fertilization the ectoderm consists of simple cuboidal epithelium. These cells become flattened and are known as the **periderm.** By the fourth month all layers of the epidermis are formed and each layer assumes its characteristic structure.

The *dermis* is derived from **mesodermal cells** in a zone beneath the ectoderm. There they undergo a process that changes them into the connective tissues that begin to form the dermis at about 11 weeks.

Nails are developed at about 10 weeks. Initially they consist of a thick layer of epithelium called the **primary nail field.** The nail itself is keratinized epithelium and grows distally from its base. It is not until the ninth month that the nails actually reach the tips of the digits.

As noted earlier, *hair follicles* develop between the ninth and twelfth weeks as downgrowths of the stratum basale of the epidermis into the deeper dermis. The downgrowths soon differentiate into the bulb (dividing epithelial cells) and the papilla (connective tissue) of the hair, beginnings of the epithelial portions of sebaceous glands, and other structures associated with hair follicles. By the fifth month, the follicles produce **lanugo** (delicate fetal hair). It is produced first on the head and then on other parts of the body. The lanugo is usually shed prior to birth.

The epithelial (secretory) portions of *sebaceous glands* develop from the sides of the hair follicles at about four months and remain connected to the follicles. The epithelial portions of *sudoriferous glands* are also derived from downgrowths of the stratum basale of the epidermis into the dermis. They appear at about five months on the palms and soles and a little later in other regions. The connective tissue and blood vessels associated with the glands develop from **mesoderm.**

By the fifth month of development, secretions from sebaceous glands mix with sloughed off epidermal cells and hairs to form a fatty substance called **vernix caseosa** (VER-niks KĀ-sē-ō-sa; *vernix* = varnish, *caseosa* = cheese). This substance covers and protects the skin of the fetus from the constant exposure to the amniotic fluid in which it is bathed. In addition, the vernix caseosa facilitates the birth of the fetus because of its slippery nature and protects the skin from being damaged by the nails.

▶ **C H E C K P O I N T**
15. Which structures develop as downgrowths of the stratum basale?

AGING AND THE INTEGUMENTARY SYSTEM

▶ O B J E C T I V E

• **Describe the effects of aging on the integumentary system.**

The pronounced effects of skin aging do not become noticeable until people reach their late forties. Most of the age-related changes occur in the dermis. Collagen fibers in the dermis begin to decrease in number, stiffen, break apart, and disorganize into a shapeless, matted tangle. Elastic fibers lose some of their elasticity, thicken into clumps, and fray, an effect that is greatly accelerated in the skin of smokers. Fibroblasts, which produce both collagen and elastic fibers, decrease in number. As a result, the skin forms the characteristic crevices and furrows known as wrinkles.

With further aging, Langerhans cells dwindle in number and macrophages become less-efficient phagocytes, thus decreasing the skin's immune responsiveness. Moreover, decreased size of sebaceous glands leads to dry and broken skin that is more susceptible to infection. Production of sweat diminishes, which probably contributes to the increased incidence of heat stroke in the elderly. There is a decrease in the number of functioning melanocytes, resulting in gray hair and atypical skin pigmentation. An increase in the size of some melanocytes produces pigmented blotching (age spots). Walls of blood vessels in the dermis become thicker and less permeable, and subcutaneous adipose tissue is lost. Aged skin (especially the dermis) is thinner than young skin, and the migration of cells from the basal layer to the epidermal surface slows considerably. With the onset of old age, skin heals poorly and becomes more susceptible to pathological conditions such as skin cancer and pressure ulcers.

Growth of nails and hair slows during the second and third decades of life. The nails also may become more brittle with age, often due to dehydration or repeated use of cuticle remover or nail polish.

• • •

To appreciate the many ways that skin contributes to homeostasis of other body systems, examine *Focus on Homeostasis: The Integumentary System.* This focus box is the first of ten, found at the end of selected chapters, that explain the how the body system under consideration contributes to the homeostasis of all the other body systems. The *Focus on Homeostasis* feature will help you understand how the individual body systems interact to contribute to the homeostasis of the entire body. Next, in Chapter 6, we will explore the how bone tissue is formed and how bones are assembled into the skeletal system, which protects many of our internal organs.

DISORDERS: HOMEOSTATIC IMBALANCES

Skin Cancer

Excessive exposure to the sun has caused virtually all of the one million cases of **skin cancer** diagnosed annually in the United States. There are three common forms of skin cancer. **Basal cell carcinomas** account for about 78% of all skin cancers. The tumors arise from cells in the stratum basale of the epidermis and rarely metastasize. **Squamous cell carcinomas,** which account for about 20% of all skin cancers, arise from squamous cells of the epidermis, and they have a variable tendency to metastasize. Most arise from preexisting lesions of damaged tissue on sun-exposed skin. Basal and squamous cell carcinomas are together known as *nonmelanoma skin cancer.* They are 50% more common in males than in females.

Malignant melanomas arise from melanocytes and account for about 2% of all skin cancers. They are the most prevalent life-threatening cancer in young women. The estimated lifetime risk of developing melanoma is now 1 in 75, double the risk only 15 years ago. In part, this increase is due to depletion of the ozone layer, which absorbs some UV light high in the atmosphere. But the main reason for the increase is that more people are spending more time in the sun and in tanning beds.

(text continues on page 156)

Body System	Contribution of the Integumentary System
For all body systems	Skin and hair provide barriers that protect all internal organs from damaging agents in the external environment; sweat glands and skin blood vessels help regulate body temperature, needed for proper functioning of other body systems.
Skeletal system	Skin helps activate vitamin D, needed for proper absorption of dietary calcium and phosphorus to build and maintain bones.
Muscular system	Skin helps provide calcium ions, needed for muscle contraction.
Nervous system	Nerve endings in skin and subcutaneous tissue provide input to the brain for touch pressure, thermal, and pain sensations.
Endocrine system	Keratinocytes in skin help activate vitamin D to calcitriol, a hormone that aids absorption of dietary calcium and phosphorus.
Cardiovascular system	Local chemical changes in dermis cause widening and narrowing of skin blood vessels, which help adjust blood flow to the skin.
Lymphatic and immune system	Skin is "first line of defense" in immunity, providing mechanical barriers and chemical secretions that discourage penetration and growth of microbes; Langerhans cells in epidermis participate in immune responses by recognizing and processing foreign antigens; macrophages in the dermis phagocytize microbes that penetrate the skin surface.
Respiratory system	Hairs in nose filter dust particles from inhaled air; stimulation of pain nerve endings in skin may alter breathing rate.
Digestive system	Skin helps activate vitamin D to the hormone calcitriol, which promotes absorption of dietary calcium and phosphorus in the small intestine.
Urinary system	Kidney cells receive partially activated vitamin D hormone from skin and convert it to calcitriol; some waste products are excreted from body in sweat, contributing to excretion by urinary system.
Reproductive systems	Nerve endings in skin and subcutaneous tissue respond to erotic stimuli, thereby contributing to sexual pleasure; suckling of a baby stimulates nerve endings in skin, leading to milk ejection; mammary glands (modified sweat glands) produce milk; skin stretches during pregnancy as fetus enlarges.

Malignant melanomas metastasize rapidly and can kill a person within months of diagnosis.

The key to successful treatment of malignant melanoma is early detection. The early warning signs of malignant melanoma are identified by the acronym ABCD (Figure 5.8). *A* is for *asymmetry;* malignant melanomas tend to lack symmetry. *B* is for *border;* malignant melanomas have irregular—notched, indented, scalloped, or indistinct—borders. *C* is for *color;* malignant melanomas have uneven coloration and may contain several colors. *D* is for *diameter;* ordinary moles typically are smaller than 6 mm (0.25 in.), about the size of a pencil eraser. Once a malignant melanoma has the characteristics of A, B, and C, it is usually larger than 6 mm.

Among the risk factors for skin cancer are the following:

1. **Skin type.** Individuals with light-colored skin who never tan but always burn are at high risk.

2. **Sun exposure.** People who live in areas with many days of sunlight per year and at high altitudes (where ultraviolet light is more intense) have a higher risk of developing skin cancer. Likewise, people who engage in outdoor occupations and those who have suffered three or more severe sunburns have a higher risk.

3. **Family history.** Skin cancer rates are higher in some families than in others.

4. **Age.** Older people are more prone to skin cancer owing to longer total exposure to sunlight.

5. **Immunological status.** Individuals who are immunosuppressed have a higher incidence of skin cancer.

Burns

A **burn** is tissue damage caused by excessive heat, electricity, radioactivity, or corrosive chemicals that denature the proteins in the skin cells. Burns destroy some of the skin's important contributions to homeostasis—protection against microbial invasion and desiccation, and thermoregulation.

Burns are graded according to their severity. A *first-degree burn* involves only the epidermis. It is characterized by mild pain and erythema (redness) but no blisters. Skin functions remain intact. Immediate flushing with cold water may lessen the pain and damage caused by a first-degree burn. Generally, healing of a first-degree burn will occur in 3 to 6 days and may be accompanied by flaking or peeling. One example of a first-degree burn is mild sunburn.

A *second-degree burn* destroys a portion of the epidermis and possibly parts of the dermis. Some skin functions are lost. In a second-degree burn, redness, blister formation, edema, and pain result. In a blister the epidermis has separated from the dermis due to the accumulation of tissue fluid between them. Associated structures, such as hair follicles, sebaceous glands, and sweat glands, usually are not injured. If there is no infection, second-degree burns heal without skin grafting in about 3 to 4 weeks, but scarring may result. First- and second-degree burns are collectively referred to as *partial-thickness burns.*

A *third-degree burn* or *full-thickness burn* destroys a portion of the epidermis, the underlying dermis, and associated structures. Most skin functions are lost. Such burns vary in appearance from marble-white to mahogany colored to charred, dry wounds. There is marked edema, and the burned region is numb because sensory nerve endings have been destroyed. Regeneration occurs slowly, and much granulation tissue forms before being covered by epithelium. Skin grafting may be required to promote healing and to minimize scarring.

The injury to the skin tissues directly in contact with the damaging agent is the *local effect* of a burn. Generally, however, the *systemic effects* of a major burn are a greater threat to life. The systemic effects of a burn may include (1) a large loss of water, plasma, and plasma proteins, which causes shock; (2) bacterial infection; (3) reduced circulation of blood; (4) decreased production of urine; and (5) diminished immune responses.

The seriousness of a burn is determined by its depth and extent of area involved, as well as the person's age and general health. According to the American Burn Association's classification of burn injury, a major burn includes third-degree burns over 10% of body surface area; or second-degree burns over 25% of body surface area; or any third-degree burns on the face, hands, feet, or perineum (per-i-NĒ-um, which includes the anal and urogenital regions). When the burn area exceeds 70%, more than half the victims die. A quick means for estimating the surface area affected by a burn in an adult is the **rule of nines** (Figure 5.9):

Figure 5.9 Rule of nines method for determining the extent of a burn. The percentages are the approximate proportions of the body surface area.

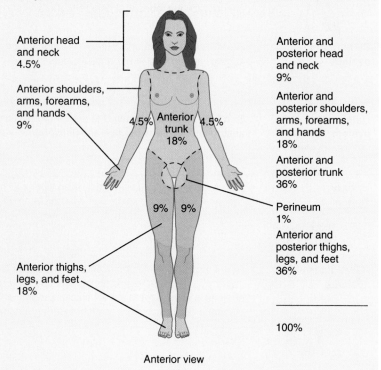

Anterior head and neck 4.5%

Anterior shoulders, arms, forearms, and hands 9%

4.5% | Anterior trunk 18% | 4.5%

9% | 9%

Anterior thighs, legs, and feet 18%

Anterior and posterior head and neck 9%

Anterior and posterior shoulders, arms, forearms, and hands 18%

Anterior and posterior trunk 36%

Perineum 1%

Anterior and posterior thighs, legs, and feet 36%

100%

Anterior view

Figure 5.8 Comparison of a normal nevus (mole) and a malignant melanoma.

(a) Normal nevus (mole)

(b) Malignant melanoma

1. Count 9% if the anterior and posterior surfaces of the head and neck are affected.

2. Count 9% for the anterior and posterior surfaces of each upper limb (total of 18% for both upper limbs).

3. Count four times nine or 36% for the anterior and posterior surfaces of the trunk, including the buttocks.

4. Count 9% for the anterior and 9% for the posterior surfaces of each lower limb as far up as the buttocks (total of 36% for both lower limbs).

5. Count 1% for the perineum.

A more accurate way to estimate the amount of surface area affected by a burn is the **Lund–Browder method,** which estimates the extent by comparing the areas affected to the percentage of total surface area for body parts (Table 5.5). Because most of the proportions of the body change with growth, some percentages vary for different ages.

Pressure Ulcers

Pressure ulcers, also known as *decubitus ulcers* (dē-KŪ-bi-tus) or *bedsores,* are caused by a constant deficiency of blood flow to tissues. Typically the affected tissue overlies a bony projection that has been subjected to prolonged pressure against an object such as a bed, cast, or splint. If the pressure is relieved in a few hours, redness occurs but no lasting tissue damage results. Blistering of the affected area may indicate superficial damage, whereas a reddish-blue discoloration may indicate deep tissue damage. Prolonged pressure causes tissue ulceration. Small breaks in the epidermis become infected, and the sensitive subcutaneous layer and deeper tissues are damaged. Eventually, the tissue dies. Pressure ulcers occur most often in bedridden patients. With proper care, pressure ulcers are preventable.

Table 5.5	Lund–Browder Burn Assessment Chart				
Area	**Age (years)**				
	0–1	**1–4**	**5–9**	**10–15**	**Adult**
Head	19	17	13	10	7
Neck	2	2	2	2	2
Anterior trunk	13	13	13	13	13
Posterior trunk	13	13	13	13	13
Right buttock	$2\frac{1}{2}$	$2\frac{1}{2}$	$2\frac{1}{2}$	$2\frac{1}{2}$	$2\frac{1}{2}$
Left buttock	$2\frac{1}{2}$	$2\frac{1}{2}$	$2\frac{1}{2}$	$2\frac{1}{2}$	$2\frac{1}{2}$
Genitalia	1	1	1	1	1
Right arm	4	4	4	4	4
Left arm	4	4	4	4	4
Right forearm	3	3	3	3	3
Left forearm	3	3	3	3	3
Right hand	$2\frac{1}{2}$	$2\frac{1}{2}$	$2\frac{1}{2}$	$2\frac{1}{2}$	$2\frac{1}{2}$
Left hand	$2\frac{1}{2}$	$2\frac{1}{2}$	$2\frac{1}{2}$	$2\frac{1}{2}$	$2\frac{1}{2}$
Right thigh	$5\frac{1}{2}$	$6\frac{1}{2}$	$8\frac{1}{2}$	$8\frac{1}{2}$	$9\frac{1}{2}$
Left thigh	$5\frac{1}{2}$	$6\frac{1}{2}$	$8\frac{1}{2}$	$8\frac{1}{2}$	$9\frac{1}{2}$
Right leg	5	5	$5\frac{1}{2}$	6	7
Left leg	5	5	$5\frac{1}{2}$	6	7
Right foot	$3\frac{1}{2}$	$3\frac{1}{2}$	$3\frac{1}{2}$	$3\frac{1}{2}$	$3\frac{1}{2}$
Left foot	$3\frac{1}{2}$	$3\frac{1}{2}$	$3\frac{1}{2}$	$3\frac{1}{2}$	$3\frac{1}{2}$

MEDICAL TERMINOLOGY

Cold sore A lesion, usually in oral mucous membrane, caused by Type 1 herpes simplex virus (HSV) transmitted by oral or respiratory routes. The virus remains dormant until triggered by factors such as ultraviolet light, hormonal changes, and emotional stress. Also called a *fever blister.*

Contact dermatitis (der-ma-TĪ-tis; *dermat-* = skin; *-itis* = inflammation of) Inflammation of the skin characterized by redness, itching, and swelling and caused by exposure of the skin to chemicals that bring about an allergic reaction, such as poison ivy toxin.

Corn A painful conical thickening of the stratum corneum of the epidermis found principally over toe joints and between the toes, often caused by friction or pressure. Corns may be hard or soft, depending on their location. Hard corns are usually found over toe joints, and soft corns are usually found between the fourth and fifth toes.

Cutaneous anthrax One of the three forms of anthrax that accounts for 95% of all naturally occurring anthrax. Endospores (spores) introduced at the site of a cut or abrasion produce a lesion that turns into a black scab and falls off in a week or two. The disease is characterized by low-grade fever, malaise (general feeling of discomfort), and enlarged lymph nodes. Without antibiotic treatment the mortality rate is as high as 20%.

Hemangioma (he-man′-jē-Ō-ma; *hem-* = blood; *-angi-* = blood vessel; *-oma* = tumor) Localized tumor of the skin and subcutaneous layer that results from an abnormal increase in blood vessels. One type is a **portwine stain,** a flat, and pink, red, or purple lesion present at birth, usually at the nape of the neck.

Hives Condition of the skin marked by reddened elevated patches that are often itchy. Most commonly caused by infections, physical trauma, medications, emotional stress, food additives, and certain food allergies. Also called *urticaria* (yur-ti- KAR-ē-a).

Impetigo (im′-pe-TĪ-gō) Superficial skin infection caused by *Staphylococcus* bacteria; most common in children.

Intradermal (in-tra-DER-mal; *intra-* = within) Within the skin. Also called *intracutaneous.*

Laceration (las-er-Ā-shun; *lacer-* = torn) An irregular tear of the skin.

Nevus (NĒ-vus) A round, flat, or raised area of pigmented skin that may be present at birth or may develop later. Varies in color from yellow-brown to black. Also called a *mole* or *birthmark.*

Pruritus (proo-RĪ-tus; *pruri-* = to itch) Itching, one of the most common dermatological disorders. It may be caused by skin disorders (infections), systemic disorders (cancer, kidney failure), psychogenic factors (emotional stress), or allergic reactions.

Topical In reference to a medication, applied to the skin surface rather than ingested or injected.

Wart Mass produced by uncontrolled growth of epithelial skin cells; caused by a papillomavirus. Most warts are noncancerous.

STUDY OUTLINE

STRUCTURE OF THE SKIN (p. 140)

1. The integumentary system consists of the skin and its accessory structures—hair, nails, glands, muscles, and nerves.
2. The skin is the largest organ of the body in surface area and weight. The principal parts of the skin are the epidermis (superficial) and dermis (deep).
3. The subcutaneous layer (hypodermis) is deep to the dermis and not part of the skin. It anchors the dermis to underlying tissues and organs, and it contains lamellated (Pacinian) corpuscles.
4. The types of cells in the epidermis are keratinocytes, melanocytes, Langerhans cells, and Merkel cells.
5. The epidermal layers, from deep to superficial, are the stratum basale, stratum spinosum, stratum granulosum, stratum lucidum (in thick skin only), and stratum corneum. Stem cells in the stratum basale undergo continuous cell division, producing keratinocytes for the other layers.
6. The dermis consists of papillary and reticular regions. The papillary region is composed of areolar connective tissue containing fine elastic fibers, dermal papillae, and Meissner corpuscles. The reticular region is composed of dense irregular connective tissue containing interlaced collagen and coarse elastic fibers, adipose tissue, hair follicles, nerves, sebaceous (oil) glands, and ducts of sudoriferous (sweat) glands.
7. Epidermal ridges provide the basis for fingerprints and footprints.
8. Thin skin covers all parts of the body except for the palms, palmar surfaces of the digits, and the soles. Thick skin covers the palms, palmar surfaces of the digits, and soles.
9. The color of skin is due to melanin, carotene, and hemoglobin.

ACCESSORY STRUCTURES OF THE SKIN (p. 146)

1. Accessory structures of the skin—hair, skin glands, and nails—develop from the embryonic epidermis.
2. A hair consists of a shaft, most of which is superficial to the surface, a root that penetrates the dermis and sometimes the subcutaneous layer, and a hair follicle.
3. Associated with each hair follicle is a sebaceous (oil) gland, an arrector pili muscle, and a hair root plexus.
4. New hairs develop from division of matrix cells in the bulb; hair replacement and growth occur in a cyclic pattern consisting of alternating growth and resting stages.
5. Hairs offer a limited amount of protection—from the sun, heat loss, and entry of foreign particles into the eyes, nose, and ears. They also function in sensing light touch.
6. Sebaceous (oil) glands are usually connected to hair follicles; they are absent in the palms and soles. Sebaceous glands produce sebum, which moistens hairs and waterproofs the skin. Clogged sebaceous glands may produce acne.
7. There are two types of sudoriferous (sweat) glands: eccrine and apocrine. Eccrine sweat glands have an extensive distribution; their ducts terminate at pores at the surface of the epidermis. Apocrine sweat glands are limited to the skin of the axillae, groin, and areolae; their ducts open into hair follicles. They begin functioning at puberty and are stimulated during emotional stress and sexual excitement.

8. Ceruminous glands are modified sudoriferous glands that secrete cerumen. They are found in the external auditory canal (ear canal).
9. Lanugo of the fetus is shed before birth, except in the scalp, eyebrows, and eyelashes. Most body hair on males is terminal (coarse, pigmented); most body hair on females is vellus (fine).
10. Nails are hard, keratinized epidermal cells over the dorsal surfaces of the distal portions of the digits.
11. The principal parts of a nail are the nail body, free edge, nail root, lunula, eponychium, and matrix. Cell division of the matrix cells produces new nails.

FUNCTIONS OF THE SKIN (p. 151)

1. Skin functions include body temperature regulation, protection, sensation, excretion and absorption, and synthesis of vitamin D.
2. The skin participates in thermoregulation by liberating sweat at its surface and by adjusting the flow of blood in the dermis.
3. The skin provides physical, chemical, and biological barriers that help protect the body.
4. Cutaneous sensations include tactile sensations, thermal sensations, and pain.

MAINTAINING HOMEOSTASIS: SKIN WOUND HEALING (p. 152)

1. In an epidermal wound, the central portion of the wound usually extends down to the dermis, whereas the wound edges involve only superficial damage to the epidermal cells.
2. Epidermal wounds are repaired by enlargement and migration of basal cells, contact inhibition, and division of migrating and stationary basal cells.
3. During the inflammatory phase of deep wound healing, a blood clot unites the wound edges, epithelial cells migrate across the wound, vasodilation and increased permeability of blood vessels enhance delivery of phagocytes, and mesenchymal cells develop into fibroblasts.
4. During the migratory phase, fibroblasts migrate along fibrin threads and begin synthesizing collagen fibers and glycoproteins.
5. During the proliferative phase, epithelial cells grow extensively.
6. During the maturation phase, the scab sloughs off, the epidermis is restored to normal thickness, collagen fibers become more organized, fibroblasts begin to disappear, and blood vessels are restored to normal.

DEVELOPMENT OF THE INTEGUMENTARY SYSTEM (p. 154)

1. The epidermis develops from the embryonic ectoderm, and the accessory structures of the skin (hair, nails, and skin glands) are epidermal derivatives.
2. The dermis is derived from mesodermal cells.

AGING AND THE INTEGUMENTARY SYSTEM (p. 154)

1. Most effects of aging begin to occur when people reach their late forties.
2. Among the effects of aging are wrinkling, loss of subcutaneous adipose tissue, atrophy of sebaceous glands, and decrease in the number of melanocytes and Langerhans cells.

Q SELF-QUIZ QUESTIONS

Fill in the blanks in the following statements.

1. Thick skin can be found on the _____, _____ and _____.

2. The combination of collagen and elastic fibers in the reticular region of the dermis provides the skin with _____, _____, and _____.

3. Modified sweat glands in the ear are called _____ glands.

4. The pigments that give skin a wide variety of colors are _____, _____, and _____.

Choose the one best answer to the following questions.

5. The substance that prevents excessive evaporation of water from the skin, keeps the skin soft and pliable, and inhibits the growth of bacteria is (a) sebum, (b) sweat, (c) cerumen, (d) carotene, (e) melatonin.

6. Epidermal ridges (a) indicate the predominant direction of underlying collagen fiber bundles, (b) increase the grip of the hand or foot, (c) provide the pigments of skin color, (d) synthesize vitamin D in the epidermis of the skin.

7. To permanently remove a hair, you must destroy the (a) matrix, (b) shaft, (c) cuticle, (d) arrector pili, (e) external root sheath.

8. Which of the following statements are *true*? (1) Nails are composed of tightly packed, hard, keratinized cells of the epidermis that form a clear, solid covering over the dorsal surface of the terminal end of digits. (2) The free edge of the nail is white due to absence of capillaries. (3) Nails help us grasp and manipulate small objects. (4) Nails protect the ends of digits from trauma. (5) Nail color is due to a combination of melanin and carotene. (a) 1, 2, and 3, (b) 1, 3, and 4, (c) 1, 2, 3, and 4, (d) 2, 3, and 4, (e) 1, 3, and 5.

9. When performing surgery, the physician's scalpel would first cut through which layer of the epidermis? (a) stratum granulosum, (b) stratum basale, (c) stratum corneum, (d) stratum spinosum, (e) stratum lucidum.

10. Which of the following statements are *true*? (1) A first-degree burn involves only the surface epidermis. (2) In a second-degree burn, no skin function is lost. (3) First- and second-degree burns are collectively referred to as partial thickness burns. (4) A third-degree burn destroys the epidermis, dermis, and epidermal derivatives. (5) When burns exceed 20% of body surface area, more than half the victims die. (a) 1, 2, and 3, (b) 2, 3, and 4, (c) 3, 4, and 5, (d) 1, 2, and 4, (e) 1, 3, and 4.

11. Match the following:
    ____ (a) deep region of the dermis composed primarily of dense irregular connective tissue
    ____ (b) composed of keratinized stratified squamous epithelial tissue
    ____ (c) not considered part of the skin, it contains areolar and adipose tissues and blood vessels; attaches skin to underlying tissues and organs
    ____ (d) superficial region of the dermis; composed of areolar connective tissue

    (1) subcutaneous layer (hypodermis)
    (2) papillary region
    (3) reticular region
    (4) epidermis

12. Match the following:
    ____ (a) a skin disorder in which keratinocytes divide and move more quickly to the skin surface, are shed prematurely, and make an abnormal keratin
    ____ (b) an inherited inability to produce melanin
    ____ (c) a condition in which a partial or complete loss of melanocytes from patches of skin produce irregular white spots
    ____ (d) a yellowed appearance of the whites of the eyes and of light colored skin, usually indicating liver disease
    ____ (e) a bluish appearance of mucous membranes, nail beds, and light-colored skin due to oxygen depletion
    ____ (f) redness of the skin caused by engorgement of capillaries in the dermis with blood
    ____ (g) partial or complete lack of hair
    ____ (h) an inflammation of sebaceous glands
    ____ (i) a precancerous skin lesion induced by sunlight
    ____ (j) the most common form of skin cancer
    ____ (k) a skin sore caused by long-standing, pressure-induced deficiency of blood flow to tissues overlying a bony projection
    ____ (l) the process of scar-tissue formation
    ____ (m) a rapidly metastasizing, potentially fatal form of skin cancer.

    (1) fibrosis
    (2) erythema
    (3) vitiligo
    (4) cyanosis
    (5) psoriasis
    (6) alopecia
    (7) jaundice
    (8) acne
    (9) malignant melanoma
    (10) solar keratosis
    (11) decubitus ulcer
    (12) basal cell carcinoma
    (13) albinism

13. Match the following:

_____ (a) produce the protein that helps protect the skin and underlying tissues from light, heat, microbes, and many chemicals

_____ (b) produce the pigment that contributes to skin color and absorbs ultraviolet light

_____ (c) cells that arise from red bone marrow, migrate to the epidermis, and participate in immune responses

_____ (d) cells thought to function in the sensation of touch

_____ (e) an abnormal thickening of the epidermis

_____ (f) release a lipid-rich secretion that functions as a water-repellent sealant in the stratum granulosum

_____ (g) pressure-sensitive cells found mostly in the subcutaneous layer

_____ (h) a fatty substance that covers and protects the skin of the fetus from the constant exposure to amniotic fluid

(1) Merkel cells
(2) callus
(3) keratinocytes
(4) Langerhans cells
(5) melanocytes
(6) lamellar granules
(7) lamellated (Pacinian) corpuscles
(8) vernix caseosa

14. Match the following:

_____ (a) liberation of sweat at the surface and adjustment of blood flow in the dermis

_____ (b) provision of a chemical barrier and resistance to microbial invasion

_____ (c) input of touch, pressure, pain, and heat and cold information

_____ (d) elimination of usually unneeded substances and passage of materials from the external environment into body cells

_____ (e) production of calcitriol

(1) protection
(2) sensation
(3) thermoregulation
(4) vitamin D synthesis
(5) excretion and absorption

15. Match the following and place the phases in the correct order:

_____ (a) epithelial cells migrate under scab to bridge the wound; formation of granulation tissue

_____ (b) sloughing of scab; reorganization of collagen fibers; blood vessels return to normal

_____ (c) vasodilation and increased permeability of blood vessels to deliver cells involved in phagocytosis; clot formation

_____ (d) extensive growth of epithelial cells beneath scab; random deposition of collagen fibers; continued growth of blood vessels

(1) proliferative phase
(2) inflammatory phase
(3) maturation phase
(4) migratory phase

Correct order of phases:

1) _____, 2) _____, 3) _____, 4) _____,

CRITICAL THINKING QUESTIONS

1. The amount of dust that collects in a house with an assortment of dogs, cats, and people is truly amazing. A lot of these dust particles had a previous "life" as part of the home's living occupants. Where did the dust originate on the human body?
   HINT *Dogs, cats, and humans all shed in their own way.*

2. Kiko is fastidious about personal hygiene. She's convinced that body secretions on the skin are unsanitary and wants to have all her exocrine glands removed. Is this wise?
   HINT *There are about 3 to 4 million sudoriferous glands alone.*

3. Your nephew has been learning about cells in science class, and now he refuses to take a bath. He asks, "If all cells have a semipermeable membrane, and the skin is made out of cells, then won't I swell up and pop if I take a bath?" Explain the relevant anatomy to your nephew before he starts attracting flies.
   HINT *Semipermeable membranes function in living cells.*

ANSWERS TO FIGURE QUESTIONS

5.1 The epidermis is epithelial tissue, whereas the dermis is connective tissue.

5.2 Melanin protects the DNA of the nucleus of keratinocytes from damage from UV light.

5.3 The stratum basale is the layer of the epidermis that contains stem cells.

5.4 Plucking a hair stimulates hair root plexuses in the dermis, some of which are sensitive to pain. Because the cells of a hair shaft are already dead and the hair shaft lacks nerves, cutting hair is not painful.

5.5 Nails are hard because they are composed of tightly packed, hard, keratinized epidermal cells.

5.6 Epidermal wounds do not bleed because there are no blood vessels in the epidermis.

5.7 In addition to phagocytizing microbes, neutrophils and macrophages can help clean up cellular debris that results from the wound.

The Skeletal System:

Bone Tissue

BONE TISSUE AND HOMEOSTASIS

Bone tissues are continuously growing, remodeling, and repairing themselves. They contribute to homeostasis of the body by providing support, protection, the production of blood cells, and the storage of minerals and triglycerides.

ENERGIZE YOUR STUDY

FOUNDATIONS CD
Anatomy Overview
  • Connective Tissue: Bone Tissue

www.wiley.com/college/apcentral

INSIGHTS AND EXPLORATIONS
Have you seen the advertisements on television by a famous actress talking about how she lost an inch in height in one year? The actress goes on to urge women who have reached menopause to use hormone replacement therapy to prevent a bone disease called osteoporosis. Explore how women can get shorter as they get older and how to prevent it in this web-based activity.

A bone is made up of several different tissues working together: bone or osseous tissue, cartilage, dense connective tissues, epithelium, adipose tissue, and nervous tissue. For this reason, each individual bone is an organ. Bone tissue is a complex and dynamic living tissue. It continually engages in a process called remodeling—building new bone tissue and breaking down old bone tissue. The entire framework of bones and their cartilages together constitute the **skeletal system.** This chapter will survey the various components of bones so you can understand how bones form, how they age, and how exercise affects the density and strength of bones. The study of bone structure and the treatment of bone disorders is called **osteology** (os-tē-OL-ō-jē; *osteo-* = bone; *-logy* = study of).

FUNCTIONS OF BONE AND THE SKELETAL SYSTEM

▶ OBJECTIVE

• **Discuss the functions of the skeletal system.**

Bone tissue and the skeletal system perform several basic functions:

1. **Support.** The skeleton serves as the structural framework for the body by supporting soft tissues and providing attachment points for the tendons of most skeletal muscles.

2. **Protection.** The skeleton protects many internal organs from injury. For example, cranial bones protect the brain, vertebrae protect the spinal cord, and the rib cage protects the heart and lungs.

3. **Assistance in movement.** Because skeletal muscles attach to bones, when muscles contract, they pull on bones. Together, bones and muscles produce movement. This function is discussed in detail in Chapter 10.

4. **Mineral homeostasis.** Bone tissue stores several minerals, especially calcium and phosphorus. On demand, bone releases minerals into the blood to maintain critical mineral balances (homeostasis) and to distribute the minerals to other parts of the body.

5. **Blood cell production.** Within certain bones, a connective tissue called **red bone marrow** produces red blood cells, white blood cells, and platelets, a process called **hemopoiesis** (hēm-ō-poy-Ē-sis; *hemo-* = blood; *poiesis* = making). Red bone marrow consists of developing blood cells, adipocytes, fibroblasts, and macrophages within a network of reticular fibers. It is present in developing bones of the fetus and in some adult bones, such as the pelvis, ribs, breastbone, backbones, skull, and ends of the arm bones and thighbones.

6. **Triglyceride storage.** Triglycerides stored in the adipose cells of **yellow bone marrow** are an important chemical energy reserve. Yellow bone marrow consists mainly of adipose cells, which store triglycerides, and a few blood cells. In the newborn, all bone marrow is red and is involved in hemopoiesis. With increasing age, much of the bone marrow changes from red to yellow.

▶ CHECKPOINT

1. What kinds of tissues make up the skeletal system?
2. How do red and yellow bone marrow differ in composition, location, and function?

STRUCTURE OF BONE

▶ OBJECTIVE

• **Describe the parts of a long bone.**

The structure of a bone may be analyzed by considering the parts of a long bone, for instance, the humerus (the arm bone), as shown in Figure 6.1. A long bone is one that has greater length than width. A typical long bone consists of the following parts:

1. The **diaphysis** (dī-AF-i-sis = growing between) is the bone's shaft or body—the long, cylindrical, main portion of the bone.

2. The **epiphyses** (e-PIF-i-sēz = growing over; singular is *epiphysis*) are the distal and proximal ends of the bone.

3. The **metaphyses** (me-TAF-i-sēz; *meta-* = between; singular is *metaphysis*) are the regions in a mature bone where the diaphysis joins the epiphyses. In a growing bone, each metaphysis includes an **epiphyseal plate** (ep′-i-FIZ-ē-al), a layer of hyaline cartilage that allows the diaphysis of the bone to grow in length (described later in the chapter). When bone growth in length stops, the cartilage in the epiphyseal plate is replaced by bone and the resulting bony structure is known as the **epiphyseal line.**

4. The **articular cartilage** is a thin layer of hyaline cartilage covering the epiphysis where the bone forms an articulation (joint) with another bone. Articular cartilage reduces friction and absorbs shock at freely movable joints. Because articular cartilage lacks a perichondrium, repair of damage is limited.

5. The **periosteum** (per′-ē-OS-tē-um; *peri-* = around) is a tough sheath of dense irregular connective tissue that surrounds the bone surface wherever it is not covered by articular cartilage. The periosteum contains bone-forming cells that enable bone to grow in diameter or thickness, but not in length. It also protects the bone, assists in fracture repair, helps nourish bone tissue, and serves as an attachment point for ligaments and tendons.

Figure 6.1 Parts of a long bone. The spongy bone tissue of the epiphysis and metaphysis contains red bone marrow, whereas the medullary cavity of the diaphysis contains yellow bone marrow (in adults).

A long bone is covered by articular cartilage at its proximal and distal epiphyses and by periosteum around the diaphysis.

(a) Partially sectioned humerus (arm bone)

(b) Partially sectioned femur (thigh bone)

Functions of Bone Tissue

1. **Supports soft tissues and provides attachment for skeletal muscles.**
2. **Protects internal organs.**
3. **Assists in movement together with skeletal muscles.**
4. **Stores and releases minerals.**
5. **Contains red bone marrow, which produces blood cells.**
6. **Contains yellow bone marrow, which stores triglycerides (fats).**

What is the functional significance of the periosteum?

6. The **medullary cavity** (MED-ū-lar′-ē; *medulla-* = marrow, pith) or **marrow cavity** is the space within the diaphysis that contains fatty yellow bone marrow in adults.

7. The **endosteum** (end-OS-tē-um; *endo-* = within) is a thin membrane that lines the medullary cavity. It contains a single layer of bone-forming cells and a small amount of connective tissue.

▶ C H E C K P O I N T

3. Diagram the parts of a long bone, and list the functions of each part.

HISTOLOGY OF BONE TISSUE

OBJECTIVE

• **Describe the histological features of bone tissue.**

Like other connective tissues, **bone,** or **osseous tissue** (OS-ē -us), contains an abundant matrix of intercellular materials that surround widely separated cells. The matrix is about 25% water, 25% collagen fibers, and 50% crystallized mineral salts. The abundant inorganic mineral salts are mainly *hydroxyapatite* (calcium phosphate and calcium carbonate). In addition, bone matrix includes small amounts of magnesium hydroxide, fluoride, and sulfate. As these mineral salts are deposited in the framework formed by the collagen fibers of the matrix, they crystallize and the tissue hardens. This process of **calcification** is initiated by osteoblasts, the bone-building cells.

It was once thought that calcification simply occurred when enough mineral salts were present to form crystals. Now, however, we know that the process occurs only in the presence of collagen fibers. Mineral salts begin to crystallize in the microscopic spaces between collagen fibers. After the spaces are filled, mineral crystals accumulate around the collagen fibers. The combination of crystallized salts and collagen fibers is responsible for the hardness that is characteristic of bone.

Although a bone's *hardness* depends on the crystallized inorganic mineral salts, a bone's *flexibility* depends on its collagen fibers. Like reinforcing metal rods in concrete, collagen fibers and other organic molecules provide *tensile strength,* which is resistance to being stretched or torn apart. Soaking a bone in an acidic solution, such as vinegar, dissolves its mineral salts, causing the bone to become rubbery and flexible. As you will see shortly, bone cells called osteoclasts secrete enzymes and acids that break down bone matrix.

Four types of cells are present in bone tissue: osteogenic cells, osteoblasts, osteocytes, and osteoclasts (Figure 6.2).

1. **Osteogenic cells** (os′-tē-ō-JEN-ik; *-genic* = producing) are unspecialized stem cells derived from mesenchyme, the tissue from which all connective tissues are formed. They are the only bone cells to undergo cell division; the resulting daughter cells develop into osteoblasts. Osteogenic cells are found along the inner portion of the periosteum, in the endosteum, and in the canals within bone that contain blood vessels.

2. **Osteoblasts** (OS-tē-ō-blasts′; *-blasts* = buds or sprouts) are bone-building cells. They synthesize and secrete collagen fibers and other organic components needed to build the matrix of bone tissue, and they initiate calcification (described shortly). As osteoblasts surround themselves with matrix, they become trapped in their secretions and become osteocytes. (Note: *Blasts* in bone or any other connective tissue secrete matrix.)

3. **Osteocytes** (OS-tē-ō-sīts′; *-cytes* = cells), mature bone cells, are the main cells in bone tissue and maintain its daily metabolism, such as the exchange of nutrients and wastes with the blood. Like osteoblasts, osteocytes do not undergo cell division. (Note: *Cytes* in bone or any other tissue maintain the tissue.)

4. **Osteoclasts** (OS-tē-ō-clasts′; *-clast* = break) are huge cells derived from the fusion of as many as 50 monocytes (a type of white blood cell) and are concentrated in the endosteum. On the side of the cell that faces the bone surface, the osteoclast's plasma membrane is deeply folded into a *ruffled border.* Here the cell releases powerful lysosomal enzymes and acids that digest the protein and mineral components of the underlying bone matrix. This breakdown of bone matrix, termed **resorption,** is part of the normal development, growth, maintenance, and repair of bone. (Note: *Clasts* in bone break down matrix.)

Bone is not completely solid but has many small spaces between its cells and matrix components. Some spaces are channels for blood vessels that supply bone cells with nutrients. Other spaces are storage areas for red bone marrow. Depending on the size and distribution of the spaces, the regions of a bone may be

Figure 6.2   Types of cells in bone tissue.

Osteogenic cells undergo cell division and develop into osteoblasts, which secrete bone matrix.

Osteogenic cell (develops into an osteoblast)

Osteoblast (forms bone matrix)

Osteocyte (maintains bone tissue)

Osteoclast (functions in resorption, the breakdown of bone matrix)

Ruffled border

Why is bone resorption important?

categorized as compact or spongy (see Figure 6.1). Overall, about 80% of the skeleton is compact bone and 20% is spongy bone.

Compact Bone Tissue

Compact bone tissue contains few spaces. It forms the external layer of all bones and makes up the bulk of the diaphyses of long bones. Compact bone tissue provides protection and support and resists the stresses produced by weight and movement.

Compact bone tissue is arranged in units called **osteons** or **Haversian systems** (Figure 6.3). Blood vessels, lymphatic ves-

sels, and nerves from the periosteum penetrate the compact bone through transverse **perforating (Volkmann's) canals.** The vessels and nerves of the perforating canals connect with those of the medullary cavity, periosteum, and **central (Haversian) canals.** The central canals run longitudinally through the bone. Around the canals are **concentric lamellae** (la-MEL-ē)—rings of hard, calcified matrix. Between the lamellae are small spaces called **lacunae** (la-KOO-nē = little lakes; singular is *lacuna*), which contain osteocytes. Radiating in all directions from the lacunae are tiny **canaliculi** (kan'-a-LIK-ū-lī = small channels), which are filled with extracellular fluid. Inside the canaliculi are slender fingerlike processes of osteocytes (see inset at right

Figure 6.3 Histology of bone.   Sections through the diaphysis of a long bone, from the surrounding periosteum on the right, to compact bone in the middle, to spongy bone and the medullary cavity on the left. The inset at the upper right shows an osteocyte in a lacuna. See Table 4.3J on page 126 for a photomicrograph of compact bone tissue and Figure 6.14a for a scanning electron micrograph of spongy bone tissue.

In compact bone, osteocytes lie in lacunae arranged in concentric circles around a central canal.

Osteons (Haversian systems) in compact bone and trabeculae in spongy bone

As people age, some central (Haversian) canals may become blocked. What effect would this have on the surrounding osteocytes?

of Figure 6.3). Neighboring osteocytes communicate via gap junctions. The canaliculi connect lacunae with one another and with the central canals. Thus, an intricate, miniature canal system throughout the bone provides many routes for nutrients and oxygen to reach the osteocytes and for wastes to diffuse away. This is very important because diffusion through the lamellae is extremely slow.

Osteons in compact bone tissue are aligned in the same direction along lines of stress. In the shaft, for example, they are parallel to the long axis of the bone. As a result, the shaft of a long bone resists bending or fracturing even when considerable force is applied from either end. Compact bone tissue tends to be thickest in those parts of a bone where stresses are applied in relatively few directions. The lines of stress in a bone are not static. They change as a person learns to walk and in response to repeated strenuous physical activity, such as occurs when a person undertakes weight training. The lines of stress in a bone also can change because of fractures or physical deformity. Thus, the organization of osteons changes over time in response to the physical demands placed on the skeleton.

The areas between osteons contain **interstitial lamellae** (in′-ter-STISH-al), which also have lacunae with osteocytes and canaliculi. Interstitial lamellae are fragments of older osteons that have been partially destroyed during bone rebuilding or growth. Lamellae that encircle the bone just beneath the periosteum are called **outer circumferential lamellae;** those that encircle the medullary cavity are called **inner circumferential lamellae.**

Spongy Bone Tissue

In contrast to compact bone tissue, **spongy bone tissue** does not contain osteons. As shown in Figure 6.4, it consists of **trabeculae** (tra-BEK-ū-lē = little beams; singular is *trabecula*), an irregular latticework of thin columns of bone. The macroscopic spaces between the trabeculae of some bones are filled with red bone marrow. Within each trabecula are osteocytes that lie in lacunae. Radiating from the lacunae are canaliculi. Because osteocytes are located on the superficial surfaces of trabeculae, they receive nourishment directly from the blood circulating through the medullary cavities.

Spongy bone tissue makes up most of the bone tissue of short, flat, and irregularly shaped bones. It also forms most of the epiphyses of long bones and a narrow rim around the medullary cavity of the diaphysis of long bones.

At first glance, the structure of the osteons of compact bone tissue may appear to be highly organized, whereas the trabeculae of spongy bone tissue may appear to be far less well organized. However, the trabeculae in spongy bone tissue are precisely oriented along lines of stress, a characteristic that helps bones resist stresses and transfer force without breaking. Spongy bone tissue tends to be located where bones are not heavily stressed or where stresses are applied from many directions.

Spongy bone tissue is different from compact bone tissue in two respects. First, spongy bone tissue is light, which reduces the overall weight of a bone so that it moves more readily when pulled by a skeletal muscle. Second, the trabeculae of spongy

Figure 6.4 Spongy bone. See Figure 6.14a for a scanning electron micrograph of spongy bone tissue.

🔑 **In spongy bone, osteocytes lie in lacunae arranged irregularly in trabeculae.**

Space for red bone marrow

Trabeculae

(a) Enlarged aspect of spongy bone trabeculae

Lacuna

Interstitial lamellae

Canaliculi

Osteocyte

Osteoclast

Osteoblasts aligned along trabeculae of new bone

(b) Details of a section of a trabecula

❓ **Where is spongy bone found?**

bone tissue support and protect the red bone marrow. The spongy bone tissue in the hip bones, ribs, breastbone, backbones, and the ends of long bones is the only site of red bone marrow and, thus, of hemopoiesis in adults.

Bone Scan

A **bone scan** is a diagnostic procedure that takes advantage of the fact that bone is living tissue. A small amount of a radioactive tracer compound that is readily absorbed by bone is injected intravenously. The degree of uptake of the tracer is related to the amount of blood flow to the bone. A scanning device measures the radiation emitted from the bones, and the information is translated into a photograph or diagram that can be read like an x ray. Normal bone tissue is identified by a consistent gray color throughout because of its uniform uptake of the radioactive tracer. Darker or lighter areas, however, may indicate bone abnormalities. Darker areas are called "hot spots," areas of increased metabolism that absorb more of the radioactive tracer. Hot spots may indicate bone cancer, abnormal healing of fractures, or abnormal bone growth. Lighter areas are called "cold spots," areas of decreased metabolism that absorb less of the radioactive tracer. Cold spots may indicate problems such as

degenerative bone disease, decalcified bone, fractures, bone infections, Paget's disease, and rheumatoid arthritis. A bone scan detects abnormalities 3 to 6 months sooner than standard x-ray procedures and exposes the patient to less radiation. ■

▶ C H E C K P O I N T

4. Why is bone considered a connective tissue?
5. What are the four types of cells in bone tissue?
6. What is the composition of the matrix of bone tissue?
7. How are spongy and compact bone tissue different in terms of their microscopic appearance, location, and function?

BLOOD AND NERVE SUPPLY OF BONE

▶ O B J E C T I V E

• **Describe the blood and nerve supply of bone.**

Bone is richly supplied with blood. Blood vessels, which are especially abundant in portions of bone containing red bone marrow, pass into bones from the periosteum. We will consider the blood supply to a long bone such as the mature tibia (shin bone) shown in Figure 6.5.

Figure 6.5   Blood supply of a mature long bone, the tibia (shin bone).

🔑 **Bone is richly supplied with blood vessels.**

- Articular cartilage
- Epiphyseal vein
- Epiphyseal artery
- Epiphyseal line
- Metaphyseal vein
- Metaphyseal artery
- Periosteum
- Periosteal artery
- Periosteal vein
- Medullary cavity
- Compact bone
- Nutrient foramen
- Nutrient vein
- Nutrient artery

Epiphysis — Metaphysis — Diaphysis

 Where do periosteal arteries enter bone tissue?

Periosteal arteries accompanied by nerves enter the diaphysis through many perforating (Volkmann's) canals and supply the periosteum and outer part of the compact bone (see Figure 6.3). Near the center of the diaphysis, a large **nutrient artery** passes through a hole in compact bone called the **nutrient foramen.** On entering the medullary cavity, the nutrient artery divides into proximal and distal branches that supply both the inner part of compact bone tissue of the diaphysis and the spongy bone tissue and red marrow as far as the epiphyseal plates (or lines). Some bones, like the tibia, have only one nutrient artery; others like the femur (thigh bone) have several. The ends of long bones are supplied by the metaphyseal and epiphyseal arteries, which arise from arteries that supply the associated joint. The **metaphyseal arteries** enter the metaphyses of a long bone and, together with the nutrient artery, supply the red bone marrow and bone tissue of the metaphyses. The **epiphyseal arteries** enter the epiphyses of a long bone and supply the red bone marrow and bone tissue of the epiphyses.

Veins that carry blood away from long bones are evident in three places: (1) One or two **nutrient veins** accompany the nutrient artery in the diaphysis; (2) numerous **epiphyseal veins** and **metaphyseal veins** exit with their respective arteries in the epiphyses; and (3) many small **periosteal veins** exit with their respective arteries in the periosteum.

Nerves accompany the blood vessels that supply bones. The periosteum is rich in sensory nerves, some of which carry pain sensations. These nerves are especially sensitive to tearing or tension, which explains the severe pain resulting from a fracture or a bone tumor.

▶ C H E C K P O I N T

8. Explain the location and roles of the nutrient arteries, nutrient foramina, epiphyseal arteries, and periosteal arteries.

BONE FORMATION

▶ O B J E C T I V E

- **Describe the steps of intramembranous and endochondral ossification.**

The process by which bone forms is called **ossification** (os′-i-fi-KĀ-shun; *ossi-* = bone; *-fication* = making) or **osteogenesis.** The "skeleton" of a human embryo is composed of loose fibrous connective tissue membranes and pieces of hyaline cartilage, which are shaped like bones and are the sites where ossification occurs. These embryonic tissues provide the template for subsequent ossification, which begins during the sixth or seventh week of embryonic development and follows one of two patterns.

The two methods of bone formation involve the replacement of a preexisting connective tissue with bone. These two methods of ossification do not lead to differences in the structure of mature bones, but are simply different methods of bone development. In the first type of ossification, called **intramembranous ossification** (in′-tra-MEM-bra-nus; *intra-* = within; *membran-* = membrane), bone forms directly on or within loose fibrous connective tissue membranes. In the second type, **endochondral ossification** (en′-dō-KON-dral; *endo-* = within; *-chondral* = cartilage), bone forms within hyaline cartilage.

Intramembranous Ossification

Intramembranous ossification is the simpler of the two methods of bone formation. The flat bones of the skull and mandible (lower jawbone) are formed in this way. Also, the "soft spots" that help the fetal skull pass through the birth canal later harden as they undergo intramembranous ossification, which occurs as follows (Figure 6.6):

1 *Development of the center of ossification.* At the site where the bone will develop, mesenchymal cells cluster together and differentiate, first into osteogenic cells and then into osteoblasts. (Recall that *mesenchyme* is the tissue from which all other connective tissues arise.) The site of such a cluster is called a **center of ossification.** Osteoblasts secrete the organic matrix of bone until they are surrounded by it.

2 *Calcification.* Then secretion of matrix stops and the cells, now called osteocytes, lie in lacunae and extend their narrow cytoplasmic processes into canaliculi that radiate in all directions. Within a few days, calcium and other mineral salts are deposited and the matrix hardens or calcifies.

3 *Formation of trabeculae.* As the bone matrix forms, it develops into trabeculae that fuse with one another to form spongy bone. Blood vessels grow into the spaces between the trabeculae and the mesenchyme along the surface of the newly formed bone. Connective tissue that is associated with the blood vessels in the trabeculae differentiates into red bone marrow.

4 *Development of the periosteum.* At the periphery of the bone, the mesenchyme condenses and develops into the periosteum. Eventually, a thin layer of compact bone replaces the surface layers of the spongy bone, but spongy bone remains in the center. Much of the newly formed bone is remodeled (destroyed and reformed) as the bone is transformed into its adult size and shape.

Figure 6.6   Intramembranous ossification. Follow along on this figure as you read the corresponding numbered paragraphs in the text. Parts ❶ and ❷ show a smaller field of view at higher magnification than parts ❸ and ❹.

🔑 **Intramembranous ossification involves the formation of bone directly on or within fibrous connective tissue membranes formed by clusters of mesenchymal cells.**

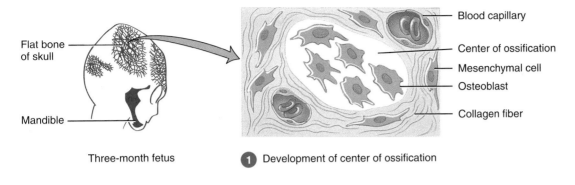

Three-month fetus

Flat bone of skull

Mandible

Blood capillary
Center of ossification
Mesenchymal cell
Osteoblast
Collagen fiber

❶ Development of center of ossification

Osteocyte in lacuna
Canaliculus
Osteoblast
Newly calcified bone matrix

❷ Calcification

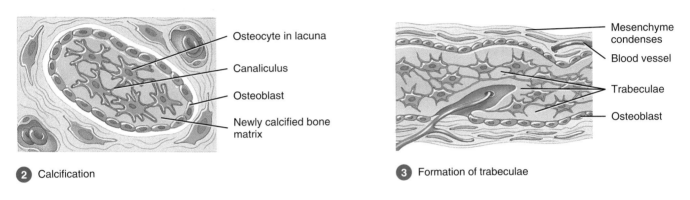

Mesenchyme condenses
Blood vessel
Trabeculae
Osteoblast

❸ Formation of trabeculae

Periosteum
Spongy bone tissue
Compact bone tissue

❹ Development of the periosteum

 Which bones of the body develop by intramembranous ossification?

Endochondral Ossification

The replacement of cartilage by bone is called **endochondral ossification.** Although most bones of the body are formed in this way, the process is best observed in a long bone. It proceeds as follows (Figure 6.7):

1 *Development of the cartilage model.* At the site where the bone is going to form, mesenchymal cells crowd together in the shape of the future bone, and then develop into chondroblasts. The chondroblasts secrete cartilage matrix, producing a **cartilage model** consisting of hyaline cartilage. A membrane called the **perichondrium** (per-i-KON-drē-um) develops around the cartilage model.

2 *Growth of the cartilage model.* Once chondroblasts become deeply buried in cartilage matrix, they are called chondrocytes. The cartilage model grows in length by continual cell division of chondrocytes accompanied by further secretion of the cartilage matrix. This type of growth is termed **interstitial growth** and results in an increase in length. In contrast, growth of the cartilage in thickness is due mainly to the addition of more matrix material to the periphery of the model by new chondroblasts that develop from the perichondrium. This growth pattern in which matrix is deposited on the cartilage surface is called **appositional growth** (a-pō-ZISH-i-nal).

As the cartilage model continues to grow, chondrocytes in its mid-region hypertrophy (increase in size). Some hypertrophied cells burst and release their contents, which increases the pH of the surrounding matrix. This change in pH triggers calcification. Other chondrocytes within the calcifying cartilage die because nutrients can no longer diffuse quickly enough through the matrix. As chondrocytes die, lacunae form and eventually merge into small cavities.

3 *Development of the primary ossification center.* Primary ossification proceeds *inward* from the external surface of the bone. A nutrient artery penetrates the perichondrium and the calcifying cartilage model through a nutrient foramen in the mid-region of the cartilage model, stimulating osteogenic cells in the perichondrium to differentiate into osteoblasts. Deep to the perichondrium the osteoblasts secrete a thin shell of compact bone called the **periosteal bone collar.** Once the perichondrium starts to form bone, it is known as the **periosteum.** Near the middle of the model, periosteal capillaries grow into the disintegrating calcified cartilage. Upon growing into the cartilage model, the capillaries induce growth of a **primary ossification center,** a region where bone tissue will replace most of the cartilage. Osteoblasts then begin to deposit bone matrix over the remnants of calcified cartilage, forming spongy bone trabeculae. As the ossification center grows toward the ends of the bone, osteoclasts break down some of the newly formed spongy bone trabeculae. This activity leaves a cavity, the medullary (marrow) cavity, in the core of the model. The medullary cavity then fills with red bone marrow.

4 *Development of the secondary ossification centers.* The diaphysis (shaft), which was once a solid mass of hyaline cartilage, is replaced by compact bone, the core of which contains a red bone marrow–filled medullary cavity. When branches of the epiphyseal artery enter the epiphyses, **secondary ossification centers** develop, usually around the time of birth. Bone formation is similar to that in primary ossification centers. One difference, however, is that spongy bone remains in the interior of the epiphyses (no medullary cavities are formed there). Secondary ossification proceeds *outward* from the center of the epiphysis toward the outer surface of the bone.

5 *Formation of articular cartilage and the epiphyseal plate.* The hyaline cartilage that covers the epiphyses becomes the articular cartilage. Prior to adulthood, hyaline cartilage remains between the diaphysis and epiphysis as the **epiphyseal plate,** which is responsible for the lengthwise growth of long bones.

▶ CHECKPOINT

9. What are the major events of intramembranous and endochondral ossification and how are they different?

BONE GROWTH

▶ OBJECTIVES

- **Describe how bone grows in length and thickness.**
- **Explain the role of nutrients and hormones in regulating bone growth.**

During childhood, bones throughout the body grow in thickness by appositional growth, and long bones lengthen by the addition of bone material on the diaphyseal side of the epiphyseal plate. Bones stop growing in length at about age 25, although they may continue to thicken.

Growth in Length

To understand how a bone grows in length, you need to know some of the details of the structure of the epiphyseal plate (Figure 6.8 on page 172). The **epiphyseal plate** is a layer of hyaline cartilage in the metaphysis of a growing bone that consists of four zones (Figure 6.8b).

1. *Zone of resting cartilage.* This layer is nearest the epiphysis and consists of small, scattered chondrocytes. The term "resting" is used because the cells do not function in bone growth. Rather, they anchor the epiphyseal plate to the bone of the epiphysis.

Figure 6.7   Endochondral ossification.

During endochondral ossification, bone gradually replaces a cartilage model.

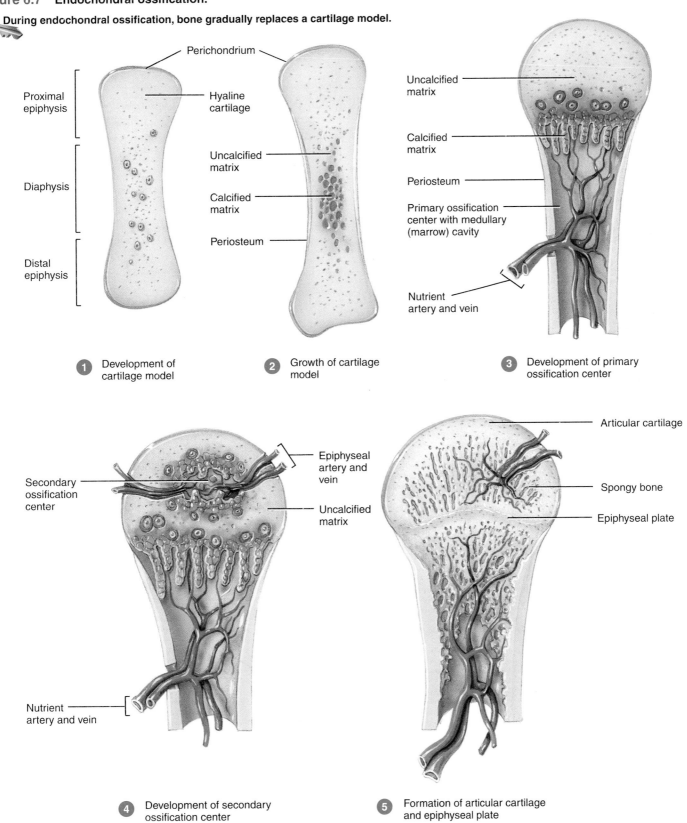

① Development of cartilage model

② Growth of cartilage model

③ Development of primary ossification center

④ Development of secondary ossification center

⑤ Formation of articular cartilage and epiphyseal plate

 If radiographs of an 18-year-old basketball player show clear epiphyseal plates but no epiphyseal lines, is she likely to grow taller?

Figure 6.8   The epiphyseal plate is a layer of hyaline cartilage in the metaphysis of a growing bone. The epiphyseal plate appears as a dark band between whiter calcified areas in the radiograph.

🔑 **The epiphyseal plate allows the diaphysis of a bone to increase in length.**

(a) Radiograph showing the epiphyseal plate of the femur of a 3-year-old

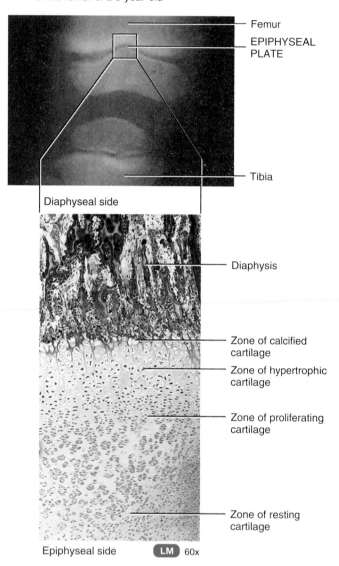

Diaphyseal side

— Femur
— EPIPHYSEAL PLATE
— Tibia

— Diaphysis

— Zone of calcified cartilage
— Zone of hypertrophic cartilage

— Zone of proliferating cartilage

— Zone of resting cartilage

Epiphyseal side   60x

(b) Histology of the epiphyseal plate

❓ **What activities of the epiphyseal plate account for the lengthwise growth of the diaphysis?**

2. *Zone of proliferating cartilage.* Slightly larger chondrocytes in this zone are arranged like stacks of coins. The chondrocytes divide to replace those that die at the diaphyseal side of the epiphyseal plate.

3. *Zone of hypertrophic cartilage.* In this layer, the chondrocytes are even larger and remain arranged in columns. The chondrocytes accumulate glycogen in their cytoplasm, and the matrix between lacunae narrows in response to the growth of lacunae. The lengthening of the diaphysis is the result of cell divisions in the zone of proliferating cartilage and maturation of the cells in the zone of hypertrophic cartilage.

4. *Zone of calcified cartilage.* The final zone of the epiphyseal plate is only a few cells thick and consists mostly of dead chrondrocytes because the matrix around them has calcified. Osteoclasts dissolve the calcified cartilage, and osteoblasts and capillaries from the diaphysis invade the area. The osteoblasts lay down bone matrix, replacing the calcified cartilage. As a result, the diaphyseal border of the epiphyseal plate is firmly cemented to the bone of the diaphysis.

The activity of the epiphyseal plate is the only way that the diaphysis can increase in length. As a bone grows, chondrocytes proliferate on the epiphyseal side of the plate. New chondrocytes cover older ones, which are then destroyed by the process of calcification. Thus, the cartilage is replaced by bone on the diaphyseal side of the plate. In this way the thickness of the epiphyseal plate remains relatively constant, but the bone on the diaphyseal side increases in length.

Between the ages of 18 and 25, the epiphyseal plates close; that is, the epiphyseal cartilage cells stop dividing, and bone replaces all the cartilage. The epiphyseal plate fades, leaving a bony structure called the **epiphyseal line.** The appearance of the epiphyseal line signifies that the bone has stopped growing in length. The clavicle is the last bone to stop growing. If a bone fracture damages the epiphyseal plate, the fractured bone may be shorter than normal once adult stature is reached. This is because damage to cartilage, which is avascular, accelerates closure of the epiphyseal plate, thus inhibiting lengthwise growth of the bone.

Growth in Thickness

Unlike cartilage, which can thicken by both interstitial and appositional growth, bone can grow in thickness or diameter only by **appositional growth** (Figure 6.9):

① At the bone surface, periosteal cells differentiate into osteoblasts, which secrete the collagen fibers and other organic molecules that form bone matrix. The osteoblasts become surrounded by matrix and develop into osteocytes. This process forms bone ridges on either side of a periosteal blood vessel. The ridges slowly enlarge and create a groove for the periosteal blood vessel.

Figure 6.9   Bone growth in diameter: appositional growth.

🔑 Whereas cartilage can grow by both interstitial and appositional growth, bone can grow in diameter only by appositional growth.

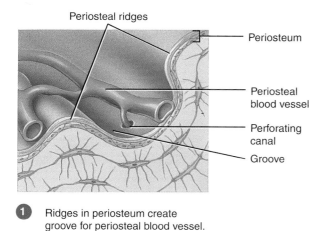

Periosteal ridges
Periosteum
Periosteal blood vessel
Perforating canal
Groove

1 Ridges in periosteum create groove for periosteal blood vessel.

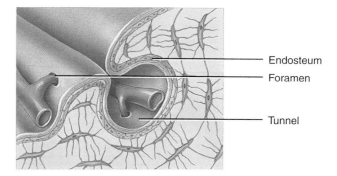

Endosteum
Foramen
Tunnel

2 Periosteal ridges fuse, forming an endosteum-lined tunnel.

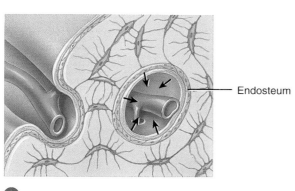

Endosteum

3 Osteoblasts in endosteum build new concentric lamellae inward toward center of tunnel, forming a new osteon.

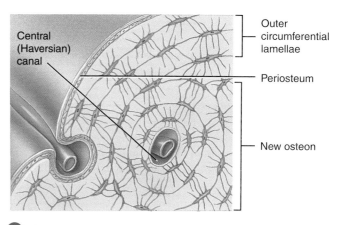

Central (Haversian) canal
Outer circumferential lamellae
Periosteum
New osteon

4 Bone grows outward as osteoblasts in periosteum build new outer circumferential lamellae. Osteon formation repeats as new periosteal ridges fold over blood vessels.

 How does the medullary cavity enlarge during growth in diameter?

2 Eventually, the ridges fold together and fuse, and the groove becomes a tunnel that encloses the blood vessel. The former periosteum now becomes the endosteum that lines the tunnel.

3 Osteoblasts in the endosteum deposit bone matrix, forming new concentric lamellae. The formation of additional concentric lamellae proceeds inward toward the periosteal blood vessel. In this way, the tunnel fills in, and a new osteon is created.

4 As an osteon is forming, osteoblasts under the periosteum deposit new outer circumferential lamellae, further increasing the thickness of the bone. As additional periosteal blood vessels become enclosed as in step **1**, the growth process continues.

Recall that as new bone tissue is being deposited on the outer surface of bone, the bone tissue lining the medullary cavity is destroyed by osteoclasts in the endosteum. In this way, the medullary cavity enlarges as the bone increases in diameter.

Factors Affecting Bone Growth

Growth and maintenance of bone depend on adequate dietary intake of minerals and vitamins, as well as on sufficient levels of several hormones. Large amounts of calcium and phosphorus are needed while bones are growing, as are smaller amount of fluoride, magnesium, iron, and manganese. Vitamin C is needed for synthesis of collagen, the main bone protein, and also for differentiation of osteoblasts into osteocytes. Vitamins K and B_{12} also are needed for protein synthesis, whereas vitamin A stimulates activity of osteoblasts.

During childhood, the most important hormones that stimulate bone growth are the insulinlike growth factors (IGFs), which are produced by bone tissue and the liver. IGFs promote cell division at the epiphyseal plate and in the periosteum and enhance synthesis of the proteins needed to build new bone. Production of IGFs, in turn, is stimulated by human growth hormone (hGH) from the anterior lobe of the pituitary gland. Thyroid hormones (T_3 and T_4), from the thyroid gland, and insulin, from the pancreas, also are needed for normal bone growth.

At puberty, hormones known as sex steroids start to be released in larger quantities. The **sex steroids** include estrogens (produced by the ovaries) and androgens (produced by the testes in males). Although females have much higher levels of estrogens and males have higher levels of androgens, females also have low levels of androgens, and males have low levels of estrogens. The adrenal glands of both sexes produce androgens, and other tissues, such as adipose tissue, can convert androgens to estrogens. These hormones are responsible for increased osteoblast activity and synthesis of bone matrix and the sudden "growth spurt" that occurs during the teenage years. Estrogens also promote changes in the skeleton that are typical of females, for example, widening of the pelvis. Ultimately, sex steroids, especially estrogens in both sexes, shut down growth at the epiphyseal plates. At this point, elongation of the bones ceases. Lengthwise growth of bones typically ends earlier in females than in males due to the higher levels of estrogens in females.

Hormonal Abnormalities that Affect Height

Excessive or deficient secretion of hormones that normally govern bone growth causes a person to be abnormally tall or short. During childhood, oversecretion of hGH produces giantism, in which a person becomes much taller and heavier than normal, whereas undersecretion of hGH or thyroid hormones produces short stature. Because estrogens terminate growth at the epiphyseal plates, both men and women who lack estrogens or receptors for estrogens grow taller than normal. ■

▶ CHECKPOINT

10. Describe the zones of the epiphyseal plate and their functions.

11. What is the significance of the epiphyseal line?

12. How is bone growth in length different from bone growth in thickness?

BONES AND HOMEOSTASIS

▶ O B J E C T I V E S

- **Describe the processes involved in bone remodeling.**
- **Describe the sequence of events in repair of a fracture.**
- **Describe the role of bone in calcium homeostasis.**

Bone Remodeling

Even after bones have reached their adult shapes and sizes, they continue to be renewed. **Bone remodeling** is an ongoing process whereby osteoclasts first carve out small tunnels in old bone tissue and then osteoblasts rebuild it anew. In different parts of the skeleton, a full cycle of bone remodeling may take as little as 2 to 3 months or last much longer. For example, remodeling of the distal portion of the femur takes about four months. Remodeling normally serves two purposes: It renews bone tissue before deterioration sets in, and it redistributes bone matrix along lines of mechanical stress. Remodeling also is the way that injured bone heals.

The breakdown of matrix by osteoclasts is called **bone resorption.** In this process, an osteoclast attaches tightly to the bone surface at the endosteum or periosteum and forms a leakproof seal at the edges of its ruffled border (see Figure 6.2). Then it releases protein-digesting lysosomal enzymes and several acids into the sealed pocket. The enzymes digest collagen fibers and other organic substances while the acids dissolve the bone minerals. Working together, several osteoclasts carve out a small tunnel in the old bone. The degraded bone proteins and matrix minerals, mainly calcium and phosphorus, enter an osteoclast by endocytosis, cross the cell in vesicles, and undergo exocytosis on the side opposite the ruffled border. Now in the interstitial fluid, the products of bone resorption diffuse into nearby blood capillaries. Once a small area of bone has been resorbed, osteoclasts depart and osteoblasts move in to rebuild the bone in that area.

To achieve homeostasis in bone, the bone-resorbing actions of osteoclasts must balance the bone-making actions of osteoblasts. If osteoblasts form too much new bone tissue, the bones become abnormally thick and heavy. If bone tissue becomes overly calcified, thick bumps called *spurs* may appear and may interfere with movement at joints. Alternatively, a loss of too much calcium or inadequate formation of new tissue weakens bone tissue.

The control of bone growth and bone remodeling is complex and not completely understood. Several circulating hormones as well as locally produced substances are involved. Sex steroids slow resorption of old bone and promote deposition of new bone. One way that estrogens slow resorption is by promoting apoptosis (programmed death) of osteoclasts. Other hormones that govern remodeling include parathyroid hormone and calcitriol, the active form of vitamin D, both of which also increase blood calcium level (see Figure 6.12).

Fracture and Repair of Bone

A **fracture** is any break in a bone. Fractures are named according to their severity, the shape or position of the fracture line, or even the physician who first described them. Among the common kinds of fractures are the following (Figure 6.10):

- **Open (compound) fracture:** The broken ends of the bone protrude through the skin (Figure 6.10a). Conversely, a **closed (simple) fracture** does not break the skin.

- **Comminuted fracture** (KOM-i-noo-ted; *com-* = together; *-minuted* = crumbled): The bone splinters at the site of impact, and smaller bone fragments lie between the two main fragments (Figure 6.10b).

- **Greenstick fracture:** A partial fracture in which one side of the bone is broken and the other side bends; occurs only in children, whose bones are not yet fully ossified and contain more organic material than inorganic material (Figure 6.10c).

- **Impacted fracture:** One end of the fractured bone is forcefully driven into the interior of the other (Figure 6.10d).

- **Pott's fracture:** A fracture of the distal end of the lateral leg bone (fibula), with serious injury of the distal tibial articulation (Figure 6.10e).

- **Colles' fracture** (KOL-ez): A fracture of the distal end of the lateral forearm bone (radius) in which the distal fragment is displaced posteriorly (Figure 6.10f).

In some cases, a bone may fracture without visibly breaking. For example, a **stress fracture** is a series of microscopic fissures in bone that forms without any evidence of injury to other tissues. In healthy adults, stress fractures result from repeated, strenuous activities such as running, jumping, or aerobic dancing. Stress fractures also result from disease processes that disrupt normal bone calcification, such as osteoporosis (discussed later). About 25% of stress fractures involve the tibia. Although standard x-ray images often fail to reveal the presence of stress fractures, they show up clearly in a bone scan.

The following steps occur in the repair of a bone fracture (Figure 6.11):

1 *Formation of fracture hematoma.* Because of the fracture, blood vessels crossing the fracture line are broken. These include vessels in the periosteum, osteons (Haversian systems), medullary cavity, and perforating canals. As blood leaks from the torn ends of the vessels, it forms a clot around the site of the fracture. This clot, called a **fracture hematoma** (hē′-ma-TŌ-ma; *hemat-* = blood; *-oma* = tumor), usually forms 6 to 8 hours after the injury. Because the circulation of blood stops at the site where the fracture hematoma forms, nearby bone cells die. Swelling and inflammation occur in response to dead bone cells, producing additional cellular debris. Blood capillaries grow into the blood clot, and phagocytes (neutrophils and macrophages)

Figure 6.10   Types of bone fractures.

A fracture is any break in a bone.

(a) Open fracture

(b) Comminuted fracture

(c) Greenstick fracture

(d) Impacted fracture

(e) Pott's fracture

(f) Colles' fracture

 What is the difference between an open fracture and a closed fracture?

and osteoclasts begin to remove the dead or damaged tissue in and around the fracture hematoma. This stage may last up to several weeks.

2 *Fibrocartilaginous callus formation.* The infiltration of new blood capillaries into the fracture hematoma helps organize it into an actively growing connective tissue called a **procallus.** Next, fibroblasts from the periosteum and osteogenic cells from the periosteum, endosteum, and red bone marrow invade the procallus. The fibroblasts produce collagen fibers, which help connect the broken ends of the bone. Phagocytes continue to remove cellular debris. Osteogenic cells develop

Figure 6.11   Steps in repair of a bone fracture.

Bone heals more rapidly than cartilage because its blood supply is more plentiful.

1 Formation of fracture hematoma

2 Fibrocartilaginous callus fomation

3 Bony callus formation

4 Bone remodeling

Why does it sometimes take months for a fracture to heal?

into chondroblasts in areas of avascular healthy bone tissue and begin to produce fibrocartilage. Eventually, the procallus is transformed into a **fibrocartilaginous callus,** a mass of repair tissue that bridges the broken ends of the bone. Formation of the fibrocartilaginous callus takes about 3 weeks.

3 **Bony callus formation.** In areas closer to well-vascularized healthy bone tissue, osteogenic cells develop into osteoblasts, which begin to produce spongy bone trabeculae. The trabeculae join living and dead portions of the original bone fragments. In time, the fibrocartilage is converted to spongy bone, and the callus is then referred to as a **bony callus.** The bony callus lasts about 3 to 4 months.

4 **Bone remodeling.** The final phase of fracture repair is **bone remodeling** of the callus. Dead portions of the original fragments of broken bone are gradually resorbed by osteoclasts. Compact bone replaces spongy bone around the periphery of the fracture. Sometimes, the repair process is so thorough that the fracture line is undetectable, even in a radiograph (x ray). Often, a thickened area on the surface of the bone remains as evidence of a healed fracture, and a healed bone may eventually be stronger than it was before the break.

Although bone has a generous blood supply, healing sometimes takes months. The calcium and phosphorus needed to strengthen and harden new bone are deposited only gradually, and bone cells generally grow and reproduce slowly. Moreover, the temporary disruption in its blood supply helps explain the slowness of healing of severely fractured bones.

Treatments for Fractures

Treatments for fractures vary according to the person's age, the type of fracture, and the bone involved. The ultimate goals of fracture treatment are realignment of the bone fragments, immobilization to maintain realignment, and restoration of function. For bones to unite, the fractured ends must be brought into alignment, a process called **reduction.** In **closed reduction,** the fractured ends of a bone are brought into alignment by manual manipulation, and the skin remains intact. In **open reduction,** the fractured ends of a bone are brought into alignment by a surgical procedure in which internal fixation devices such as screws, plates, pins, rods, and wires are used. Following reduction, a fractured bone may be kept immobilized by a cast, sling, splint, elastic bandage, external fixation device, or a combination of these devices. ■

Bone's Role in Calcium Homeostasis

Bone is the body's major calcium reservoir, storing 99% of total body calcium. The level of calcium in the blood can be regulated by controlling the rates of calcium resorption from bone into blood and of calcium deposition from blood into bone. Most functions of nerve cells depend on having a stable level of calcium ions (Ca^{2+}) in extracellular fluid. Also, many enzymes require Ca^{2+} as a cofactor (an additional substance needed for an enzymatic reaction to occur), and blood clotting requires Ca^{2+}. For this reason, the blood plasma level of Ca^{2+} is very closely regulated between 9 and 11 mg/100 mL. Even small changes in Ca^{2+} concentration outside this range may prove fatal—the heart may stop (cardiac arrest) if the concentration goes too high, or breathing may cease (respiratory arrest) if the level falls too low. The role of bone in calcium homeostasis is to "buffer" the blood Ca^{2+} level, releasing Ca^{2+} into blood plasma when the level decreases, and taking Ca^{2+} back when the level rises. Hormones regulate these exchanges.

The most important hormone that regulates Ca^{2+} exchange between bone and blood is **parathyroid hormone (PTH),** secreted by the parathyroid glands (see Figure 18.13 on page 607). PTH secretion operates via a negative feedback system (Figure 6.12). If some stimulus causes blood Ca^{2+} level to decrease, parathyroid gland cells (receptors) detect this change and increase their production of a molecule known as cyclic adenosine monophosphate (cyclic AMP). The gene for PTH within the nucleus of a parathyroid gland cell, which acts as the control center, detects the increased production of cyclic AMP (the input). As a result, PTH synthesis speeds up, and more PTH (the output) is released into the blood. The presence of higher levels of PTH increases the number and activity of osteoclasts (effectors), which step up the pace of bone resorption. The resulting release of Ca^{2+} from bone into blood returns the blood Ca^{2+} level to normal.

PTH also acts on the kidneys (effectors) to decrease loss of Ca^{2+} in the urine, so more is retained in the blood, and stimulates formation of **calcitriol,** a hormone that promotes absorption of calcium from the gastrointestinal tract. Both of these actions also help elevate blood Ca^{2+} level.

Another hormone makes a potential contribution to the homeostasis of blood Ca^{2+} by its influence on bone. When blood Ca^{2+} rises above normal, *parafollicular cells* in the thyroid gland secrete **calcitonin (CT).** CT inhibits activity of osteoclasts, speeds blood Ca^{2+} uptake by bone, and accelerates Ca^{2+} deposition into bones. The net result is that CT promotes bone formation and decreases blood Ca^{2+} level. Despite these effects, the role of CT in normal calcium homeostasis is uncertain because it can be completely absent without causing symptoms. Nevertheless, calcitonin harvested from salmon (Miacalcin) is an effective drug for treating osteoporosis because it slows bone resorption.

Figure 18.14 on page 608 summarizes the roles of parathyroid hormone, calcitriol, and calcitonin in regulation of blood Ca^{2+} level.

▶ CHECKPOINT

13. Define remodeling, and describe the roles of osteoblasts and osteoclasts in the process.

14. Define a fracture and outline the four steps involved in fracture repair.

15. How do hormones act on bone to regulate calcium homeostasis?

Figure 6.12   Negative feedback system for the regulation of blood calcium (Ca²⁺) concentration.   PTH = parathyroid hormone.

 Release of calcium from bone matrix and retention of calcium by the kidneys are the two main ways that blood calcium level can be increased.

What body functions depend on proper levels of Ca²⁺?

EXERCISE AND BONE TISSUE

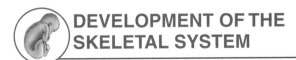

▶ O B J E C T I V E

- **Describe how exercise and mechanical stress affect bone tissue.**

Within limits, bone has the ability to alter its strength in response to changes in mechanical stress. When placed under stress, bone tissue adapts by becoming stronger through increased deposition of mineral salts and production of collagen fibers. Another effect of stress is to increase the production of calcitonin, a hormone that inhibits bone resorption. Without mechanical stress, bone does not remodel normally because bone resorption outstrips bone formation. Removal of mechanical stress weakens bone through demineralization (loss of bone minerals) and decreased numbers of collagen fibers.

The main mechanical stresses on bone are those that result from the pull of skeletal muscles and the pull of gravity. If a person is bedridden or has a fractured bone in a cast, the strength of the unstressed bones diminishes. Astronauts subjected to the microgravity of space also lose bone mass. In both cases, bone loss can be dramatic—as much as 1% per week. Bones of athletes, which are repetitively and highly stressed, become notably thicker and stronger than those of nonathletes. Weight-bearing activities, such as walking or moderate weight lifting, help build and retain bone mass. Adolescents and young adults should engage in regular weight-bearing exercise prior to the closure of the epiphyseal plates to help build total mass prior to its inevitable reduction with aging. Even elderly people can strengthen their bones by engaging in weight-bearing exercise.

▶ C H E C K P O I N T

16. What types of mechanical stresses may be used to strengthen bone tissue?

DEVELOPMENT OF THE SKELETAL SYSTEM

▶ O B J E C T I V E

- **Describe the development of the skeletal system and the limbs.**

Both intramembranous and endochondral ossification begin when *mesenchymal cells,* connective tissue cells derived from **mesoderm,** migrate into the area where bone formation will occur. In some skeletal structures, mesenchymal cells develop into chondroblasts that form cartilage. In other skeletal structures, mesenchymal cells develop into osteoblasts that form *bone tissue* by intramembranous or endochondral ossification.

Discussion of the development of the skeletal system provides an excellent opportunity to trace limb development. During the middle of the fourth week, the upper limbs appear as small elevations at the sides of the trunk called **upper limb buds** (Figure 6.13a). By the end of the fourth week, the **lower**

Figure 6.13 **Features of a human embryo during weeks four through eight of development.** The liver and heart prominences are surface bulges resulting from the underlying growth of these organs. The embryo receives oxygen and nutrients from its mother through blood vessels in the umbilical cord.

 After the limb buds develop, endochondral ossification of the limb bones begins during the seventh embryonic week.

(a) Four-week embryo showing development of limb buds

(b) Six-week embryo showing development of hand and foot plates

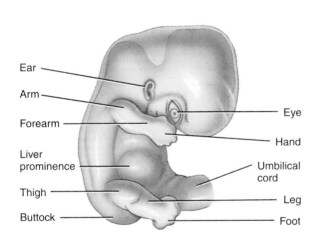

(c) Seven-week embryo showing development of arm, forearm, and hand in upper limb bud and thigh, leg, and foot in lower limb bud

(d) Eight-week embryo in which limb buds have developed into upper and lower limbs

Which of the three basic embryonic tissues—ectoderm, mesoderm, and endoderm—gives rise to the skeletal system?

limb buds appear. The limb buds consist of masses of general **mesoderm** covered by **ectoderm.** At this point, a mesenchymal skeleton exists in the limbs; some of the masses of mesoderm surrounding the developing bones will become the skeletal muscles of the limbs.

By the sixth week, the limb buds develop a constriction around the middle portion. The constriction produces distal segments of the upper buds called **hand plates** and distal segments of the lower buds called **foot plates** (Figure 6.13b). These plates represent the beginnings of the hands and feet, respectively. At this stage of limb development, a cartilaginous skeleton is pre-

sent. By the seventh week (Figure 6.13c), the arm, forearm, and hand are evident in the upper limb bud, and the thigh, leg, and foot appear in the lower limb bud. Endochondral ossification has begun. By the eighth week (Figure 6.13d), as the shoulder, elbow, and wrist areas become apparent, the upper limb bud is appropriately called the upper limb, and the lower limb bud is now the lower limb.

About 22 to 24 days after fertilization, mesodermal cells form a solid cylinder of cells called the **notochord** (nō-tō-KORD; *noto-* = the back; *chord* = a string). See Figure 29.9a on page 1074. This structure plays an extremely important role in inducing, through chemical signals, unspecialized embryonic cells to develop into specialized tissues and organs. For example, the notochord induces certain mesodermal cells to develop into the vertebral bodies and forms the nucleus pulposus of the intervertebral discs (see Figure 7.16 on pages 203–204).

▶ CHECKPOINT

17. When and how do the limbs develop?

AGING AND BONE TISSUE

▶ OBJECTIVE

• **Describe the effects of aging on bone tissue.**

From birth through adolescence, more bone tissue is produced than is lost during bone remodeling. In young adults the rates of bone deposition and resorption are about the same. As the level of sex steroids diminishes during middle age, especially in women after menopause, a decrease in bone mass occurs because bone resorption outpaces bone deposition. In old age, loss of bone through resorption occurs more rapidly than bone gain. Because women's bones generally are smaller and less massive than men's bones to begin with, loss of bone mass in old age typically has a greater adverse effect in women.

There are two principal effects of aging on bone tissue: loss of bone mass and brittleness. Loss of bone mass results from **demineralization,** the loss of calcium and other minerals from bone matrix. This loss usually begins after age 30 in females, accelerates greatly around age 45 as levels of estrogens decrease, and continues until as much as 30% of the calcium in bones is lost by age 70. Once bone loss begins in females, about 8% of bone mass is lost every ten years. In males, calcium loss typically does not begin until after age 60, and about 3% of bone mass is lost every ten years. The loss of calcium from bones is one of the problems in osteoporosis (described shortly).

The second principal effect of aging on the skeletal system, brittleness, results from a decreased rate of protein synthesis. Recall that the organic part of bone matrix, mainly collagen fibers, gives bone its tensile strength. The loss of tensile strength causes the bones to become very brittle and susceptible to fracture. In some elderly people, collagen fiber synthesis slows, in part, due to diminished production of human growth hormone. In addition to increasing the susceptibility to fractures, loss of bone mass also leads to deformity, pain, stiffness, loss of height, and loss of teeth.

▶ CHECKPOINT

18. What is demineralization, and how does it affect the functioning of bone?

19. What changes occur in the organic part of bone matrix with aging?

DISORDERS: HOMEOSTATIC IMBALANCES

Osteoporosis

Osteoporosis (os′-tē-ō-pō-RŌ-sis; *por-* = passageway; *-osis* = condition) is literally a condition of porous bones (Figure 6.14). The basic problem is that bone resorption outpaces bone deposition. In large part this is due to depletion of calcium from the body—more calcium is lost in urine, feces, and sweat than is absorbed from the diet. Bone mass becomes so depleted that bones fracture, often spontaneously, under the mechanical stresses of everyday living. For example, a hip fracture might result from simply sitting down too quickly. In the United States, osteoporosis causes more than a million fractures a year, mainly in the hip, wrist, and vertebrae. Osteoporosis afflicts the entire skeletal system. In addition to fractures, osteoporosis causes shrinkage of vertebrae, height loss, hunched backs, and bone pain.

Thirty million people in the United States suffer from osteoporosis. The disorder primarily affects middle-aged and elderly people, 80% of them women. Older women suffer from osteoporosis more often than men for two reasons: Women's bones are less massive than men's bones, and production of estrogens in women declines dramatically at menopause, whereas production of the main androgen, testosterone in older men wanes gradually and only slightly. Estrogens and testosterone stimulate osteoblast activity and synthesis of bone matrix. Besides gender, risk factors for developing osteoporosis include a family history of the disease, European or Asian ancestry, thin or small body build, an inactive lifestyle, cigarette smoking, a diet low in calcium and vitamin D, more than two alcoholic drinks a day, and the use of certain medications.

In postmenopausal women, treatment of osteoporosis may include estrogen replacement therapy (ERT; low doses of estrogens) or hor-

Figure 6.14 Comparison of spongy bone tissue from (a) a normal young adult and (b) a person with osteoporosis. Notice the weakened trabeculae in (b). Compact bone tissue is similarly affected by osteoporosis.

(a) Normal bone   (b) Osteoporotic bone

mone replacement therapy (HRT; a combination of estrogens and progesterone, another sex steroid). Although such treatments help combat osteoporosis, they increase a woman's risk of breast cancer. The drug Raloxifene (Evista) mimics the beneficial effects of estrogens on bone without increasing the risk of breast cancer. Another drug that may be used is the nonhormone drug Alendronate (Fosamax), which blocks resorption of bone by osteoclasts.

Perhaps more important than treatment is prevention. Adequate calcium intake and weight-bearing exercise in her early years may be more beneficial to a woman than drugs and calcium supplements when she is older.

Rickets and Osteomalacia

Rickets and **osteomalacia** (os'-tē-ō-ma-LĀ-shē-a; *-malacia* = softness) are disorders in which bones fail to calcify. Although the organic matrix is still produced, calcium salts are not deposited, and the bones become "soft" or rubbery and easily deformed. Rickets affects the growing bones of children. Because new bone formed at the epiphyseal plates fails to ossify, bowed legs and deformities of the skull, rib cage, and pelvis are common. In osteomalacia, sometimes called "adult rickets," new bone formed during remodeling fails to calcify. This disorder causes varying degrees of pain and tenderness in bones, especially in the hip and leg. Bone fractures also result from minor trauma.

MEDICAL TERMINOLOGY

Osteoarthritis (os'-tē-ō-ar-THRĪ-tis; *arthr* = joint) The degeneration of articular cartilage such that the bony ends touch; the resulting friction of bone against bone worsens the condition. Usually associated with the elderly.

Osteogenic sarcoma (os'-tē-Ō-JEN-ik sar-KŌ-ma; *sarcoma* = connective tissue tumor) Bone cancer that primarily affects osteoblasts and occurs most often in teenagers during their growth spurt; the most common sites are the metaphyses of the thigh bone (femur), shin bone (tibia), and arm bone (humerus). Metastases occur most often in lungs; treatment consists of multidrug chemotherapy and removal of the malignant growth, or amputation of the limb.

Osteomyelitis (os'-tē-ō-mī-e-LĪ-tis) An infection of bone characterized by high fever, sweating, chills, pain, and nausea, pus formation, edema, and warmth over the affected bone and rigid overlying muscles. Bacteria, usually *Staphylococcus aureus,* often cause it. The bacteria may reach the bone from outside the body (through open fractures, penetrating wounds, or orthopedic surgical procedures); from other sites of infection in the body (abscessed teeth, burn infections, urinary tract infections, or upper respiratory infections) via the blood; and from adjacent soft tissue infections (as occurs in diabetes mellitus).

Osteopenia (os'-tē-ō-PĒ-nē-a; *penia* = poverty) Reduced bone mass due to a decrease in the rate of bone synthesis to a level insufficient to compensate for normal bone resorption; any decrease in bone mass below normal. An example is osteoporosis.

STUDY OUTLINE

INTRODUCTION (p. 162)

1. A bone is made up of several different tissues: bone or osseous tissue, cartilage, dense connective tissues, epithelium, various blood-forming tissues, adipose tissue, and nervous tissue.
2. The entire framework of bones and their cartilages constitute the skeletal system.

FUNCTIONS OF BONE AND THE SKELETAL SYSTEM (p. 162)

1. The skeletal system functions in support, protection, movement, mineral homeostasis, blood cell production, and triglyceride storage.

STRUCTURE OF BONE (p. 162)

1. Parts of a typical long bone are the diaphysis (shaft), proximal and distal epiphyses (ends), metaphyses, articular cartilage, periosteum, medullary (marrow) cavity, and endosteum.

HISTOLOGY OF BONE TISSUE (p. 164)

1. Bone tissue consists of widely separated cells surrounded by large amounts of matrix.
2. The four principal types of cells in bone tissue are osteogenic cells, osteoblasts, osteocytes, and osteoclasts.
3. The matrix of bone contains abundant mineral salts (mostly hydroxyapatite) and collagen fibers.
4. Compact bone tissue consists of osteons (Haversian systems) with little space between them.
5. Compact bone tissue lies over spongy bone tissue in the epiphyses and makes up most of the bone tissue of the diaphysis. Functionally, compact bone tissue protects, supports, and resists stress.
6. Spongy bone tissue does not contain osteons. It consists of trabeculae surrounding many red bone marrow–filled spaces.
7. Spongy bone tissue forms most of the structure of short, flat, and irregular bones, and the epiphyses of long bones. Functionally, spongy bone tissue trabeculae offer resistance along lines of stress, support and protect red bone marrow, and make bones lighter for easier movement.

BLOOD AND NERVE SUPPLY OF BONE (p. 167)

1. Long bones are supplied by periosteal, nutrient, and epiphyseal arteries; veins accompany the arteries.
2. Nerves accompany blood vessels in bone; the periosteum is rich in sensory neurons.

BONE FORMATION (p. 168)

1. Bone forms by a process called ossification (osteogenesis), which begins when mesenchymal cells become transformed into osteogenic cells. These undergo cell division and give rise to cells that differentiate into osteoblasts and osteoclasts.
2. Ossification begins during the sixth or seventh week of embryonic life. The two types of ossification, intramembranous and endochondral, involve the replacement of a preexisting connective tissue with bone.
3. Intramembranous ossification occurs within fibrous connective tissue membranes.
4. Endochondral ossification occurs within a hyaline cartilage model. The primary ossification center of a long bone is in the diaphysis. Cartilage degenerates, leaving cavities that merge to form the medullary cavity. Osteoblasts lay down bone. Next, ossification occurs in the epiphyses, where bone replaces cartilage, except for the epiphyseal plate.

BONE GROWTH (p. 170)

1. The epiphyseal plate consists of four zones: zones of resting cartilage, proliferating cartilage, hypertrophic cartilage, and calcified cartilage.
2. Because of the activity of the epiphyseal plate, the diaphysis of a bone increases in length.
3. Bone grows in thickness or diameter due to the addition of new bone tissue by periosteal osteoblasts around the outer surface of the bone (appositional growth).
4. Dietary minerals (especially calcium and phosphorus) and vitamins (C, K, and B_{12}) are needed for bone growth and maintenance. Insulinlike growth factors (IGFs), human growth hormone, thyroid hormones, insulin, estrogens, and androgens stimulate bone growth. The sex steroids, especially estrogens, also shut down growth at the epiphyseal plates.

BONES AND HOMEOSTASIS (p. 174)

1. Bone remodeling is an ongoing process whereby osteoclasts carve out small tunnels in old bone tissue and then osteoblasts rebuild it anew.
2. In bone resorption, osteoclasts release enzymes and acids that degrade collagen fibers and dissolve mineral salts.
3. Sex steroids slow resorption of old bone and promote new bone deposition. Estrogens slow resorption by promoting apoptosis of osteoclasts.
4. A fracture is any break in a bone.
5. Fracture repair involves formation of a fracture hematoma, of fibrocartilaginous callus, of bony callus, and bone remodeling.
6. Types of fractures include closed (simple), open (compound), comminuted, greenstick, impacted, stress, Pott's, and Colles'.
7. Bone is the major reservoir for calcium in the body.
8. Parathyroid hormone (PTH) secreted by the parathyroid gland increases blood Ca^{2+} level, whereas calcitonin (CT) from the thyroid gland has the potential to decrease blood Ca^{2+} level.

EXERCISE AND BONE TISSUE (p. 178)

1. Mechanical stress increases bone strength by increasing deposition of mineral salts and production of collagen fibers.
2. Removal of mechanical stress weakens bone through demineralization and collagen fiber reduction.

DEVELOPMENT OF THE SKELETAL SYSTEM (p. 178)

1. Bone forms from mesoderm by intramembranous or endochondral ossification.
2. Limbs develop from limb buds, which consist of mesoderm and ectoderm.

AGING AND BONE TISSUE (p. 180)

1. The principal effect of aging is demineralization, a loss of calcium from bones, which may result in osteoporosis.
2. Another effect is a decreased production of matrix proteins (mostly collagen fibers), which makes bones more brittle and thus more susceptible to fracture.

Q SELF-QUIZ QUESTIONS

Fill in the blanks in the following statements.

1. The crystallized mineral salts in bone contribute to bone's _____, while the collagen fibers and other organic molecules provide bone with _____.
2. Compact bone tissue is composed of _____; spongy bone tissue is composed of _____.
3. Endochondral ossification refers to the formation of bone within _____; intramembranous ossification refers to the formation of bone directly from _____.

Indicate whether the following statements are true or false.

4. The activity of the epiphyseal plate is the only mechanism by which the diaphysis can increase in length.
5. Homeostasis in bone occurs when there is a balance between formation of bone by osteocytes and resorption of bone by osteoclasts.

Choose the one best answer to the following questions.

6. Chrondocytes are actively dividing in the (a) zone of hypertrophic cartilage, (b) zone of proliferative cartilage, (c) epiphyseal line, (d) zone of resting cartilage, (e) zone of calcified cartilage.

7. Which of the following statements are *true*? (1) Osteogenic cells develop directly into osteoclasts. (2) Osteogenic cells are unspecialized stem cells. (3) Osteoblasts form bone. (4) Osteocytes are the principal cells of bone tissue. (5) Osteoclasts function in bone resorption. (6) Osteoblasts maintain daily cellular activities of bone tissue. (a) 1, 2, 4, 5, and 6, (b) 2, 3, 4, and 5, (c) 1, 3, 5, and 6, (d) 2, 4, and 6, (e) 1, 2, 4, 5, and 6.

8. Which of the following statements is *true*? (a) The osteons in compact bone are arranged parallel to the lines of stress. (b) Spongy bone tissue composes the shaft of the long bones of the limbs. (c) Yellow bone marrow is found only in embryonic bone. (d) The sex steroids are the most important hormones regulating blood calcium levels. (e) In adult bones, hemopoiesis occurs in compact bone tissue only.

9. Place in order the steps involved in intramembranous ossification. (1) Fusion of bony matrices to form trabeculae. (2) Clusters of osteoblasts form a center of ossification that secretes the organic matric. (3) Replacement of spongy bone with compact bone on the bone's surface. (4) Development of periosteum on the bone's periphery. (5) Hardening of the matrix by deposition of calcium and mineral salts. (a) 2, 4, 5, 1, 3, (b) 4, 3, 5, 1, 2, (c) 1, 2, 5, 4, 3, (d) 2, 5, 1, 4, 3, (e) 5, 1, 3, 4, 2.

10. Which of the following statements are *true*? (1) Weight-bearing activities help build and retain bone mass. (2) Removal of mechanical stress weakens bone. (3) In the absence of mechanical stress, resorption outstrips bone formation. (4) Mechanical stress decreases the production of calcitonin, a hormone that inhibits bone resorption. (5) The main mechanical stresses on bone are those resulting from the contraction of skeletal muscles and the pull of gravity. (a) 1, 2, 3, and 5, (b) 2, 3, 4, and 5, (c) 1, 3, and 5, (d) 2, 4, and 5, (e) 2, 3, and 4.

11. Which of the following are functions of bone tissue and the skeletal system? (1) support, (2) excretion, (3) assistance in movement, (4) mineral homeostasis, (5) blood cell production. (a) 1, 2, and 3, (b) 2, 3, and 4, (c) 3, 4, and 5, (d) 1, 3, 4, and 5, (e) 2, 3, 4, and 5.

12. Match the following.
____ (a) space within the shaft of a bone that contains red or yellow bone marrow
____ (b) triglyceride storage tissue
____ (c) hemopoietic tissue
____ (d) thin layer of hyaline cartilage covering the ends of bones where they form a joint
____ (e) distal and proximal ends of bones
____ (f) the long, cylindrical main portion of the bone; the shaft
____ (g) in a growing bone, the region that contains the growth plate
____ (h) the tough membrane that surrounds the bone surface wherever cartilage is not present
____ (i) membrane lining the medullary cavity
____ (j) a remnant of the active epiphyseal plate; a sign that the bone has stopped growing in length

(1) articular cartilage
(2) endosteum
(3) medullary cavity
(4) diaphysis
(5) epiphyses
(6) metaphysis
(7) periosteum
(8) red bone marrow
(9) yellow bone marrow
(10) epiphyseal line

13. Match the following.
____ (a) small spaces between lamellae that contain osteocytes
____ (b) perforating canals that penetrate compact bone; carry blood vessels, lymphatic vessels, and nerves from the periosteum
____ (c) areas between osteons
____ (d) microscopic unit of compact bone tissue
____ (e) tiny canals filled with extracellular fluid; connect lacunae to each other and to the central canal
____ (f) canals that extend longitudinally through the bone and connect blood vessels and nerves to the osteocytes
____ (g) irregular lattice work of thin columns of bone found in spongy bone tissue
____ (h) rings of hard calcified matrix found around the central canals
____ (i) an opening in the shaft of the bone allowing an artery to pass into the bone
____ (j) a layer of hyaline cartilage in the area between the shaft and end of a growing bone

(1) osteon
(2) Volkmann's canals
(3) Haversian canals
(4) concentric lamellae
(5) epiphyseal plate
(6) trabeculae
(7) interstitial lamellae
(8) canaliculi
(9) nutrient foramen
(10) lacunae

14. Match the following.
____ (a) blood clot that forms around the fracture site
____ (b) a condition of porous bones characterized by decreased bone mass and increased susceptibility to fractures
____ (c) splintered bone, with smaller fragments lying between main fragments
____ (d) a broken bone that does not break through the skin
____ (e) a partial break in a bone in which one side of the bone is broken and the other side bends
____ (f) a broken bone that protrudes through the skin
____ (g) microscopic bone breaks resulting from inability to withstand repeated stressful impact
____ (h) connective tissue mass that forms a bridge between the broken ends of bones
____ (i) condition in which new bone formed by remodeling fails to calcify in adults
____ (j) an infection of bone

(1) closed (simple) fracture
(2) open (compound) fracture
(3) fibrocartiagenous callus
(4) greenstick fracture
(5) stress fracture
(6) comminuted fracture
(7) osteoporosis
(8) osteomalacia
(9) fracture hematoma
(10) osteomyelitis

15. Match the following.

_____ (a) decreases blood calcium levels by accelerating calcium deposition in bones

_____ (b) required for collagen synthesis

_____ (c) during childhood, it promotes growth at epiphyseal plate; production stimulated by human growth hormone

_____ (d) involved in bone growth by increasing osteoblast activity; causes long bones to stop growing in length

_____ (e) raises blood calcium level by increasing bone resorption

(1) PTH
(2) CT
(3) insulinlike growth factors
(4) sex steroids
(5) vitamin C

CRITICAL THINKING QUESTIONS

1. A father brings his young daughter Lynne Marie to the emergency room after she fell off her bicycle. The attending physician tells the father that his daughter has suffered a greenstick fracture in her forearm. The father is confused—there weren't any sticks where she fell in the street. What should the physician explain to the father?
HINT *Only children suffer this type of fracture.*

2. Aunt Edith is 95 years old today. She claims that she's been getting shorter every year, and that soon she'll disappear altogether.

What's happening to Aunt Edith?
HINT *Her mind is as sharp as ever; think about her bones.*

3. Astronauts in space exercise as part of their daily routine, yet they still have problems with bone weakness after prolonged stays in space. Why does this happen?
HINT *What is missing in space that you have to work against on Earth?*

ANSWERS TO FIGURE QUESTIONS

6.1 The periosteum is essential for growth in bone diameter, bone repair, and bone nutrition. It also serves as a point of attachment for ligaments and tendons.

6.2 Bone resorption is necessary for the development, growth, maintenance, and repair of bone.

6.3 Because the central (Haversian) canals are the main blood supply to the osteocytes of an osteon, their blockage would lead to death of the osteocytes.

6.4 Spongy bone makes up most of short, flat, and irregularly shaped bones; most of the epiphyses of long bones; and a narrow rim around the medullary cavity of the diaphysis of long bones.

6.5 Periosteal arteries enter bone tissue through perforations (Volkmann's canals).

6.6 Flat bones of the skull and mandible (lower jawbone) develop by intramembranous ossification.

6.7 Yes, she probably will grow taller. Epiphyseal lines are indications of growth zones that have ceased to function. The absence of epiphyseal lines indicates that the bone is still lengthening.

6.8 The lengthwise growth of the diaphysis is caused by cell divisions in the zone of proliferating cartilage and maturation of the cells in the zone of hypertrophic cartilage.

6.9 The medullary cavity enlarges by activity of the osteoclasts in the endosteum.

6.10 An open fracture breaks through the skin whereas a closed fracture does not.

6.11 Healing of bone fractures can take months because calcium and phosphorus deposition is a slow process, and bone cells generally grow and reproduce slowly.

6.12 Heartbeat, respiration, nerve cell functioning, enzyme functioning, and blood clotting all depend on proper levels of calcium.

6.13 The skeletal system develops from mesoderm.

The Skeletal System:

The Axial Skeleton

While growing up I always referred to the "vertebral column" as the spine or backbone and never gave it much more thought other than it was in the middle of my back. As I got older I understood that the spine contained nerves and if damaged could cause paralysis. Boy, was I surprised when I began reading the section on the spine—26 vertebrae (actually 33, but as it develops some fuse). I never fathomed that it consisted of that many bones. Like I said, I always referred to it as my backbone, singular, not backbones.

Through my reading I discovered that the spine was broken down into regions. At first I had a difficult time differentiating exactly where one region ended and the next began on an actual skeleton, because I couldn't remember how many vertebrae were in each region. So, I referred back to the book and that's when it stood out, as plain as the nose on my face—seven, twelve, and five—breakfast, lunch, and dinner.

Not only is the vertebral column broken down into regions, but also each vertebra has a name. Knowing the regions made it easy, but the cervical region had two vertebrae that differed considerably from all the others and so were named accordingly. Unfortunately, this just gave me more to memorize, as if I didn't have enough to learn already. Atlas and axis, how was I going to remember which was which? The book actually supplied a great anecdote, "named for the mythological Atlas who supported the world on his shoulders." Sometimes it's hard to create our own little anecdotes for remembering difficult things, so when the book supplies them it's great.

| Lisa Hubbard
| Macon State College

THE AXIAL SKELETON AND HOMEOSTASIS

The bones of the axial skeleton contribute to homeostasis by protecting many of the body's organs such as the brain, spinal cord, heart, and lungs. They are also important in calcium storage and release.

ENERGIZE YOUR STUDY
FOUNDATIONS CD
Anatomy Overview
  • The Skeletal System
  • Connective Tissue: Bone Tissue
Animation
  • Systems Contributions to Homeostasis
Concepts and Connections and Exercises reinforce your understanding

www.wiley.com/college/apcentral
INSIGHTS AND EXPLORATIONS
Almost all of us have had the experience of our jaws catching and clicking at some time. Now imagine being in a romantic restaurant where the dinner music isn't violins playing, but the clicking of your companion's jaw. This is called temporomandibular joint syndrome or TMJ syndrome, and the symptoms don't just include a clicking jaw, but also facial pain, swelling of the face, the jaw actually locking open or closed, and swallowing difficulties. Over 10 million people in the United States, 90 percent of whom are women of childbearing age suffer from this problem. Explore more about this syndrome in a web-based activity.

Because the skeletal system forms the framework of the body, a familiarity with the names, shapes, and positions of individual bones will help you locate other organs. For example, the radial artery, the site where the pulse is usually taken, is named for its closeness to the radius, the lateral bone of the forearm. The ulnar nerve is named for its closeness to the ulna, the medial bone of the forearm. The frontal lobe of the brain lies deep to the frontal (forehead) bone. The tibialis anterior muscle lies along the anterior surface of the tibia (shin bone).

Movements such as throwing a ball, biking, and walking require an interaction between bones and muscles. To understand how muscles produce different movements, you will learn where the muscles attach on individual bones and the types of joints acted on by the contracting muscles. The bones, muscles, and joints together form an integrated system called the **musculoskeletal system.** The branch of medical science concerned with the prevention or correction of disorders of the musculoskeletal system is called **orthopedics** (or′-thō-PĒ-diks; *ortho-* = correct; *pedi* = child).

DIVISIONS OF THE SKELETAL SYSTEM

▶ O B J E C T I V E

• **Describe how the skeleton is divided into axial and appendicular divisions.**

The adult human skeleton consists of 206 named bones, most of which are paired on the right and left sides of the body. The skeletons of infants and children have more than 206 bones because some of their bones fuse later in life. Examples are the hip bones and some bones of the backbone.

Bones of the adult skeleton are grouped into two principal divisions: the 80 bones of the **axial skeleton** and the 126 bones of the **appendicular skeleton** (*appendic-* = to hang onto). Table 7.1 presents the 80 bones of the axial skeleton and the 126 bones of the appendicular skeleton. Figure 7.1 shows how both divisions join to form the complete skeleton. The longitudinal **axis,** or center, of the human body is a vertical line that runs through the body's center of gravity, extending down through the head to the space between the feet. The axial skeleton consists of the bones arranged along the axis: skull bones, auditory ossicles (ear bones), hyoid bone (see Figure 7.4), ribs, breastbone, and bones of the backbone. The appendicular skeleton consists of the bones of the **upper** and **lower limbs (extremities),** plus the bones forming the **girdles** that connect the limbs to the axial skeleton. Functionally, the auditory ossicles in the middle ear are not part of either the axial or appendicular skeleton, but they are grouped with the axial skeleton for convenience. The auditory ossicles vibrate in response to sound waves that strike the eardrum and have a key role in hearing (see Chapter 16).

We will organize our study of the skeletal system around the two divisions of the skeleton, with emphasis on how the many bones of the body are interrelated. In this chapter we focus on the axial skeleton, looking first at the skull and then at the bones of the backbone and the chest. In Chapter 8 we explore the appendicular skeleton, examining in turn the bones of the pectoral (shoulder) girdle and upper limbs, and then the pelvic (hip) girdle and the lower limbs. But before we examine the skull, we direct our attention to some general characteristics of bones.

▶ C H E C K P O I N T

1. Which bones make up the axial and appendicular divisions of the skeleton?

Table 7.1	The Bones of the Adult Skeletal System	
Division of the Skeleton	**Structure**	**Number of Bones**
Axial Skeleton	**Skull**	
	Cranium	8
	Face	14
	Hyoid	1
	Auditory ossicles	6
	Vertebral column	26
	Thorax	
	Sternum	1
	Ribs	24
		Subtotal = 80
Appendicular Skeleton	**Pectoral (shoulder) girdles**	
	Clavicle	2
	Scapula	2
	Upper limbs (extremities)	
	Humerus	2
	Ulna	2
	Radius	2
	Carpals	16
	Metacarpals	10
	Phalanges	28
	Pelvic (hip) girdle	
	Hip, pelvic, or coxal bone	2
	Lower limbs (extremities)	
	Femur	2
	Fibula	2
	Tibia	2
	Patella	2
	Tarsals	14
	Metatarsals	10
	Phalanges	28
		Subtotal = 126
		Total = 206

Figure 7.1   Divisions of the skeletal system.   The axial skeleton is colored blue, the appendicular skeleton tan. (See Tortora, *A Photographic Atlas of the Human Body,* Figure 3.1.)

 The adult human skeleton consists of 206 bones grouped into axial and appendicular divisions.

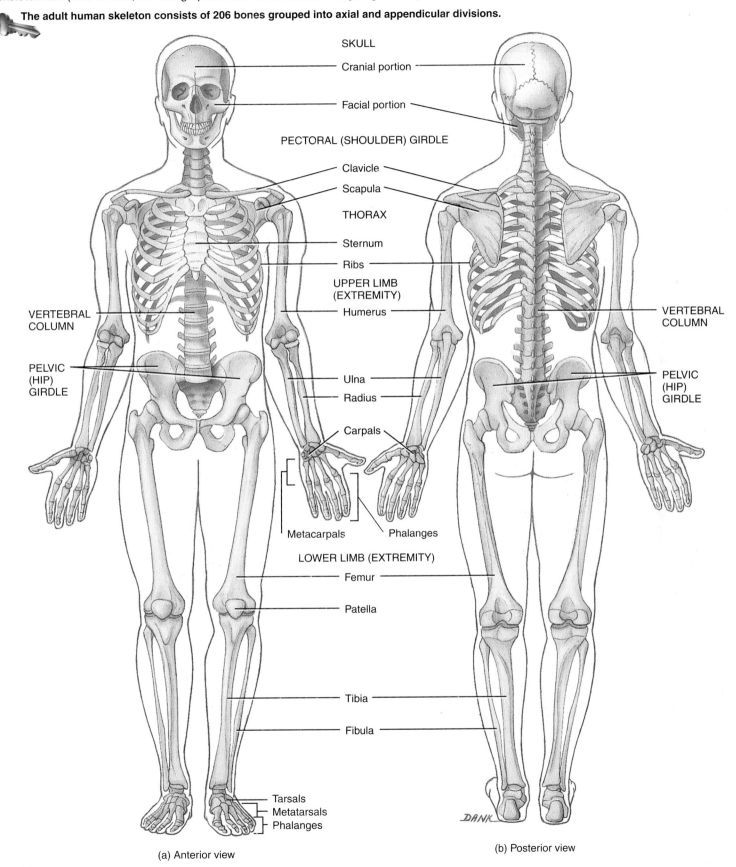

SKULL
Cranial portion
Facial portion

PECTORAL (SHOULDER) GIRDLE
Clavicle
Scapula

THORAX
Sternum
Ribs

UPPER LIMB (EXTREMITY)
Humerus

VERTEBRAL COLUMN

PELVIC (HIP) GIRDLE

Ulna
Radius

Carpals

Metacarpals   Phalanges

LOWER LIMB (EXTREMITY)
Femur
Patella

VERTEBRAL COLUMN

PELVIC (HIP) GIRDLE

Tibia
Fibula

Tarsals
Metatarsals
Phalanges

(a) Anterior view   (b) Posterior view

DANK

Which of the following structures are part of the axial skeleton, and which are part of the appendicular skeleton? Skull, clavicle, vertebral column, shoulder girdle, humerus, pelvic girdle, and femur.

TYPES OF BONES

▶ O B J E C T I V E

• **Classify bones based on their shape and location.**

Most bones of the body can be classified into five main types based on shape: long, short, flat, irregular, and sesamoid (Figure 7.2). **Long bones** have greater length than width and consist of a shaft and a variable number of extremities (ends). They are usually somewhat curved for strength. A slightly curved bone absorbs the stress of the body's weight at several different points, so that it is evenly distributed. If such bones were straight, the weight of the body would be unevenly distributed, and the bone would fracture more easily. Long bones consist mostly of compact bone tissue in their diaphyses but they also contain considerable amounts of spongy bone tissue in their epiphyses. Long bones include those in the thigh (femur), leg (tibia and fibula), arm (humerus), forearm (ulna and radius), and fingers and toes (phalanges).

Short bones are somewhat cube-shaped because they are nearly equal in length and width. They consist of spongy bone tissue except at the surface, where there is a thin layer of compact bone tissue. Examples of short bones are the wrist or carpal bones (except for the pisiform, which is a sesamoid bone) and the ankle or tarsal bones (except for the calcaneus, which is an irregular bone).

Flat bones are generally thin and composed of two nearly parallel plates of compact bone tissue enclosing a layer of spongy bone tissue. Flat bones afford considerable protection and provide extensive areas for muscle attachment. Flat bones include the cranial bones, which protect the brain; the breastbone (sternum) and ribs, which protect organs in the thorax; and the shoulder blades (scapulae).

Irregular bones have complex shapes and cannot be grouped into any of the previous categories. They vary in the amount of spongy and compact bone present. Such bones include the vertebrae of the backbone and some facial bones.

Sesamoid bones (= shaped like a sesame seed) develop in certain tendons where there is considerable friction, tension, and physical stress, such as the palms and soles. They may vary in number from person to person, are not always completely ossified, and typically measure only a few millimeters in diameter. Notable exceptions are the two patellae (kneecaps), which are large sesamoid bones that are normally present in everyone. Functionally, sesamoid bones protect tendons from excessive wear and tear, and they often change the direction of pull of a tendon, which improves the mechanical advantage at a joint.

An additional type of bone is not included in this classification by shape, but instead is classified by location. **Sutural bones** (SOO-chur-al; *sutur-* = seam) are small bones located within joints, called sutures, between certain cranial bones (see Figure 7.6). Their number varies greatly from person to person.

Figure 7.2 **Types of bones based on shape.** The bones are not drawn to scale.

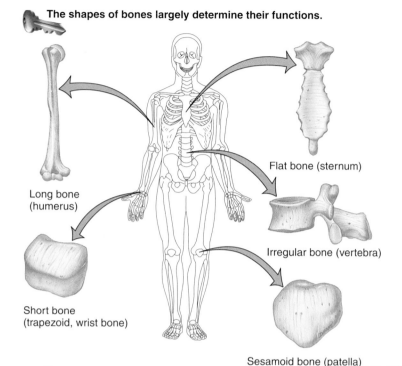

🔑 The shapes of bones largely determine their functions.

Long bone (humerus)

Flat bone (sternum)

Irregular bone (vertebra)

Short bone (trapezoid, wrist bone)

Sesamoid bone (patella)

Which type of bone primarily protects and provides a large surface area for muscle attachment?

▶ C H E C K P O I N T

2. Give examples of long, short, flat, and irregular bones.

BONE SURFACE MARKINGS

▶ O B J E C T I V E

• **Describe the principal surface markings on bones and the functions of each.**

Bones have characteristic **surface markings,** structural features adapted for specific functions. Sometimes they are not present at birth but develop later in response to certain forces and are most prominent during adult life. In response to tension on a bone surface where tendons, ligaments, aponeuroses, and fasciae pull on the periosteum of bone, new bone is deposited, resulting in raised or roughened areas. Conversely, compression on a bone surface results in a depression.

There are two major types of surface markings: (1) *depressions and openings,* which form joints or allow the passage of soft tissues (such as blood vessels and nerves), and (2) *processes,* which are projections or outgrowths that either help form joints or serve as attachment points for connective tissue (such as ligaments and tendons). Table 7.2 describes the various surface markings and provides examples of each.

Table 7.2	Bone Surface Markings	
Marking	**Description**	**Example**
Depressions and Openings: Sites allowing the passage of soft tissue (nerves, blood vessels, ligaments, tendons) or formation of joints		
Fissure (FISH-ur)	Narrow slit between adjacent parts of bones through which blood vessels or nerves pass.	Superior orbital fissure of the sphenoid bone (Figure 7.13).
Foramen (fō-RĀ-men); plural is **foramina**	Opening (*foramen* = hole) through which blood vessels, nerves, or ligaments pass.	Optic foramen of the sphenoid bone (Figure 7.13).
Fossa (FOS-a)	Shallow depression (*fossa* = trench).	Coronoid fossa of the humerus (Figure 8.5d).
Sulcus (SUL-kus)	Furrow (*sulcus* = groove) along a bone surface that accommodates a blood vessel, nerve, or tendon.	Intertubercular sulcus of the humerus (Figure 8.5d).
Meatus (mē-Ā-tus)	Tubelike opening (*meatus* = passageway).	External auditory meatus of the temporal bone (Figure 7.4).
Processes: Projections or outgrowths on bone that form joints or attachment points for connective tissue, such as ligaments and tendons.		
Processes that form joints:		
Condyle (KON-dīl)	Large, round protuberance (*condylus* = knuckle) at the end of a bone.	Lateral condyle of the femur (Figure 8.12a).
Facet	Smooth flat articular surface.	Superior articular facet of a vertebra (Figure 7.18d).
Head	Rounded articular projection supported on the neck (constricted portion) of a bone.	Head of the femur (Figure 8.12a).
Processes that form attachment points for connective tissue:		
Crest	Prominent ridge or elongated projection.	Iliac crest of the hip bone (Figure 8.9b).
Epicondyle	Projection above (*epi-* = above) a condyle.	Medial epicondyle of the femur (Figure 8.12a).
Line	Long, narrow ridge or border (less prominent than a crest).	Linea aspera of the femur (Figure 8.12b).
Spinous process	Sharp, slender projection.	Spinous process of a vertebra (Figure 7.17).
Trochanter (trō-KAN-ter)	Very large projection.	Greater trochanter of the femur (Figure 8.12b).
Tubercle (TOO-ber-kul)	Small, rounded projection (*tuber-* = knob).	Greater tubercle of the humerus (Figure 8.5a).
Tuberosity	Large, rounded, usually roughened projection.	Ischial tuberosity of the hip bone (Figure 8.9b).

▶ CHECKPOINT

3. List and describe several bone surface markings, and give an example of each. Check your list against Table 7.2.

SKULL

▶ OBJECTIVES

- **Name the cranial and facial bones and indicate the number of each.**
- **Describe the following unique features of the skull: sutures, paranasal sinuses, and fontanels.**

The **skull,** which contains 22 bones, rests on the superior end of the vertebral column. It includes two sets of bones: cranial bones and facial bones. The **cranial bones** (*crani-* = brain case) form the cranial cavity, which encloses and protects the brain. The eight cranial bones are the frontal bone, two parietal bones, two temporal bones, the occipital bone, the sphenoid bone, and the ethmoid bone. Fourteen **facial bones** form the face: two nasal bones, two maxillae (or maxillas), two zygomatic bones, the mandible, two lacrimal bones, two palatine bones, two inferior nasal conchae, and the vomer. Figures 7.3 through 7.8 illustrate these bones from different viewing directions.

General Features

Besides forming the large cranial cavity, the skull also forms several smaller cavities, including the nasal cavity and orbits (eye sockets), which open to the exterior. Certain skull bones also contain cavities that are lined with mucous membranes and are called paranasal sinuses. The sinuses open into the nasal cavity. Also within the skull are small cavities that house the structures involved in hearing and equilibrium.

Other than the auditory ossicles, which are involved in hearing and are located within the temporal bones, the mandible is the only movable bone of the skull. Immovable joints called sutures, which are especially noticeable on the outer surface of the skull, hold most of the skull bones together.

The skull has many surface markings, such as foramina and fissures through which blood vessels and nerves pass. You will learn the names of important skull bone surface markings as we describe each bone.

In addition to protecting the brain, the cranial bones also have other functions. Their inner surfaces attach to membranes (meninges) that stabilize the positions of the brain, blood vessels, and nerves. The outer surfaces of cranial bones provide large areas of attachment for muscles that move various parts of the head. The bones also provide attachment for some muscles that produce facial expressions. Besides forming the framework of the face, the facial bones protect and provide support for the entrances to the digestive and respiratory systems. Together, the cranial and facial bones protect and support the delicate special sense organs for vision, taste, smell, hearing, and equilibrium.

Cranial Bones

Frontal Bone

The **frontal bone** forms the forehead (the anterior part of the cranium), the roofs of the orbits, and most of the anterior part of

the cranial floor (Figure 7.3). Soon after birth, the left and right sides of the frontal bone are united by the *metopic suture*, which usually disappears by age six to eight.

If you examine the anterior view of the skull in Figure 7.3, you will note the *frontal squama,* a scalelike plate of bone that forms the forehead. It gradually slopes inferiorly from the coronal suture, on the top of the skull, then angles abruptly and becomes almost vertical. Superior to the orbits the frontal bone thickens, forming the *supraorbital margin* (*supra =* above; *-orbi =* circle). From this margin, the frontal bone extends posteriorly to form the roof of the orbit, which is part of the floor of the cranial cavity. Within the supraorbital margin, slightly medial to its midpoint, is a hole called the *supraorbital foramen.* As you read about each foramen associated with a cranial bone, refer to Table 7.3 to note which structures pass through it. The *frontal sinuses* lie deep to the frontal squama. Sinuses, or more technically paranasal sinuses, are mucous membrane–lined cavities in certain skull bones (discussed later).

Figure 7.3 Skull. (See Tortora, *A Photographic Atlas of the Human Body,* Figure 3.2.)

🔑 **The skull consists of 8 cranial bones and 14 facial bones.**

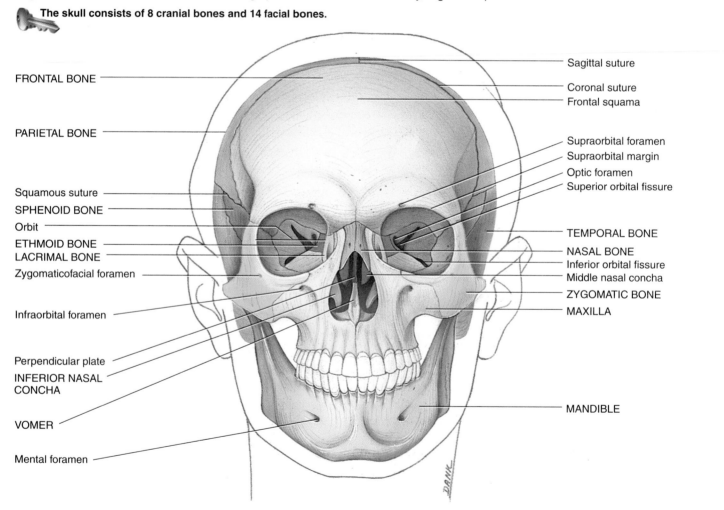

FRONTAL BONE

PARIETAL BONE

Squamous suture
SPHENOID BONE
Orbit
ETHMOID BONE
LACRIMAL BONE
Zygomaticofacial foramen

Infraorbital foramen

Perpendicular plate
INFERIOR NASAL CONCHA

VOMER

Mental foramen

Sagittal suture
Coronal suture
Frontal squama

Supraorbital foramen
Supraorbital margin
Optic foramen
Superior orbital fissure

TEMPORAL BONE
NASAL BONE
Inferior orbital fissure
Middle nasal concha
ZYGOMATIC BONE
MAXILLA

MANDIBLE

Anterior view

❓ **Which of the bones shown here are cranial bones?**

Black Eyes

Just superior to the supraorbital margin is a sharp ridge. A blow to the ridge often fractures the bone or lacerates the skin over it, resulting in bleeding. Bruising of the skin over the ridge causes tissue fluid and blood to accumulate in the surrounding connective tissue. The resulting swelling and discoloration is called a black eye. ■

Parietal Bones

The two **parietal bones** (pa-RĪ-e-tal; *pariet-* = wall) form the greater portion of the sides and roof of the cranial cavity (Figure 7.4). The internal surfaces of the parietal bones contain many protrusions and depressions that accommodate the blood vessels supplying the dura mater, the superficial connective tissue covering of the brain. There are no foramina in the parietal bones.

Temporal Bones

The two **temporal bones** (*tempor-* = temple) form the inferior lateral aspects of the cranium and part of the cranial floor. In the lateral view of the skull (Figure 7.4), note the *temporal squama*, the thin, flat part of the temporal bone that forms the anterior and superior part of the temple. Projecting from the inferior portion of the temporal squama is the *zygomatic process*, which articulates (forms a joint) with the temporal process of the zygomatic (cheek) bone. Together, the zygomatic process of the temporal bone and the temporal process of the zygomatic bone form the *zygomatic arch*.

A socket called the *mandibular fossa* is located on the inferior posterior surface of the zygomatic process of the temporal bone. Anterior to the mandibular fossa is a rounded elevation, the *articular tubercle* (Figure 7.4). The mandibular fossa and

Figure 7.4   Skull. Although the hyoid bone is not part of the skull, it is included in the illustration for reference. (See Tortora, *A Photographic Atlas of the Human Body,* Figure 3.3.)

Together the zygomatic process of the temporal bone and the temporal process of the zygomatic bone form the zygomatic arch.

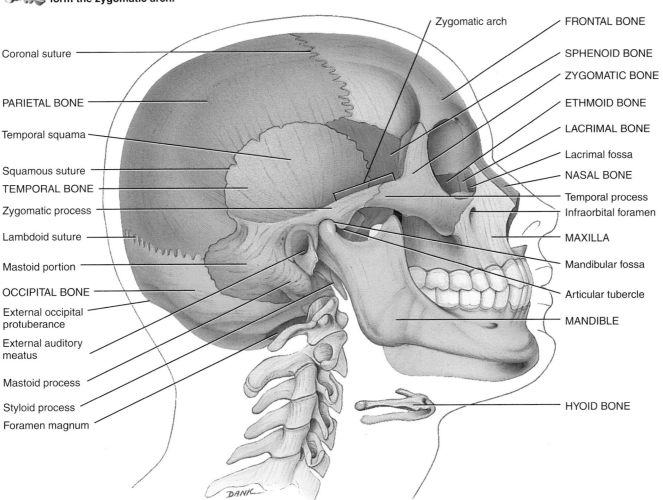

Right lateral view

What are the major bones on either side of the squamous suture, the lambdoid suture, and the coronal suture?

articular tubercle articulate with the mandible (lower jawbone) to form the *temporomandibular joint (TMJ).*

Located posteriorly on the temporal bone is the *mastoid portion (mastoid = breast-shaped).* See Figure 7.4. It is located posterior and inferior to the *external auditory meatus (meatus = passageway),* or ear canal, which directs sound waves into the ear. In an adult, this portion of the bone contains several *mastoid "air cells."* These tiny air-filled compartments are separated from the brain by thin bony partitions. In cases of **mastoiditis** (inflammation of the mastoid air cells, for example, from a middle-ear infection), the infection may spread to the brain.

The *mastoid process* is a rounded projection of the mastoid portion of the temporal bone posterior to the external auditory meatus. It is the point of attachment for several neck muscles.

The *internal auditory meatus* (Figure 7.5) is the opening through which the facial nerve (cranial nerve VII) and vestibulocochlear nerve (cranial nerve VIII) pass. The *styloid process (styl- = stake or pole)* projects inferiorly from the inferior surface of the temporal bone and serves as a point of attachment for muscles and ligaments of the tongue and neck (see Figure 7.4). Between the styloid process and the mastoid process is the *stylomastoid foramen* (see Figure 7.7).

At the floor of the cranial cavity (see Figure 7.8a) is the *petrous portion (petrous = rock)* of the temporal bone. This part is triangular and is located at the base of the skull between the sphenoid and occipital bones. The petrous portion houses the internal ear and the middle ear, structures involved in hearing and equilibrium. It also contains the *carotid foramen,* through which

Figure 7.5   Skull.   (See Tortora, *A Photographic Atlas of the Human Body,* Figure 3.4.)

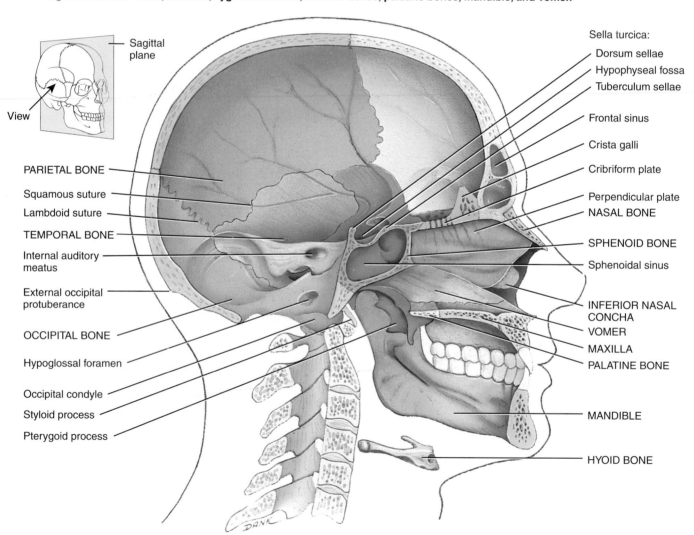

The cranial bones are the frontal, parietal, temporal, occipital, sphenoid, and ethmoid bones. The facial bones are the nasal bone, maxillae, zygomatic bones, lacrimal bones, palatine bones, mandible, and vomer.

Medial view of sagittal section

 With which bones does the temporal bone articulate?

the carotid artery passes (see Figure 7.7). Posterior to the carotid foramen and anterior to the occipital bone is the *jugular foramen,* a passageway for the jugular vein.

Occipital Bone

The **occipital bone** (ok-SIP-i-tal; *occipit-* = back of head) forms the posterior part and most of the base of the cranium (Figure 7.6; also see Figure 7.4). Also view the occipital bone and surrounding structures in the inferior view of the skull in Figure 7.7. The *foramen magnum* (= large hole) is in the inferior part of the bone. Within this foramen, the medulla oblongata (inferior part of the brain) connects with the spinal cord. The vertebral and spinal arteries also pass through this foramen. The *occipital condyles* are oval processes with convex surfaces, one on either side of the foramen magnum (Figure 7.7 on page 195). They articulate with depressions on the first cervical vertebra (atlas) to form the *atlanto-occipital joint.* Superior to each occipital

condyle on the inferior surface of the skull is the *hypoglossal foramen* (*hypo-* = under; *-glossal* = tongue). (See Figure 7.5.)

The *external occipital protuberance* is a prominent midline projection on the posterior surface of the bone just above the foramen magnum. You may be able to feel this structure as a bump on the back of your head, just above your neck. (See Figure 7.4.) A large fibrous, elastic ligament, the *ligamentum nuchae* (*nucha-* = nape of neck), which helps support the head, extends from the external occipital protuberance to the seventh cervical vertebra. Extending laterally from the protuberance are two curved ridges, the *superior nuchal lines,* and below these are two *inferior nuchal lines,* which are areas of muscle attachment (Figure 7.7).

Sphenoid Bone

The **sphenoid bone** (SFĒ-noyd = wedge-shaped) lies at the middle part of the base of the skull (Figures 7.7 and 7.8). This

Figure 7.6   Skull.   The sutures are exaggerated for emphasis. (See Tortora, *A Photographic Atlas of the Human Body,* Figure 3.5.)

🔑 **The occipital bone forms most of the posterior and inferior portions of the cranium.**

Posterior view

 Which bones form most of the lateral aspect of the cranium posterior to the ears?

Figure 7.7 Skull.   The mandible (lower jawbone) has been removed. (See Tortora, *A Photographic Atlas of the Human Body,* Figure 3.7.)

 The occipital condyles of the occipital bone articulate with the first cervical vertebra to form the atlanto-occipital joints.

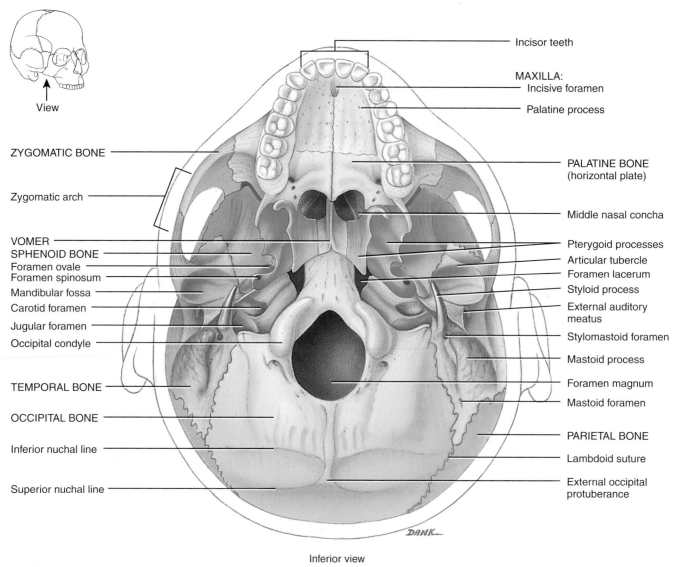

Inferior view

What parts of the nervous system join within the foramen magnum?

bone is the keystone of the cranial floor because it articulates with all the other cranial bones, holding them together. Viewing the floor of the cranium from a superior perspective (Figure 7.8a), note the sphenoid articulations. It joins anteriorly with the frontal bone, laterally with the temporal bones, and posteriorly with the occipital bone. The sphenoid lies posterior and slightly superior to the nasal cavity and forms part of the floor, sidewalls, and rear wall of the orbit (see Figure 7.12).

The shape of the sphenoid resembles a bat with outstretched wings (see Figure 7.8b). The *body* of the sphenoid is the cube-like medial portion between the ethmoid and occipital bones. It contains the *sphenoidal sinuses,* which drain into the nasal cav-

ity (see Figure 7.13). The *sella turcica* (SEL-a TUR-si-ka; *sella* = saddle; *turcica* = Turkish) is a bony saddle-shaped structure on the superior surface of the body of the sphenoid (Figure 7.8a). The anterior part of the sella turcica, which forms the horn of the saddle, is a ridge called the **tuberculum sellae.** The seat of the saddle is a depression, the **hypophyseal fossa** (hī-pō-FIZ-ē-al), which contains the pituitary gland. The posterior part of the sella turcica, which forms the back of the saddle, is another ridge called the **dorsum sellae.**

The *greater wings* of the sphenoid project laterally from the body and form the anterolateral floor of the cranium. The greater wings also form part of the lateral wall of the skull just anterior

Figure 7.8 **Sphenoid bone.** (See Tortora, *A Photographic Atlas of the Human Body,* Figures 3.8. and 3.9.)

The sphenoid bone is called the keystone of the cranial floor because it articulates with all other cranial bones, holding them together.

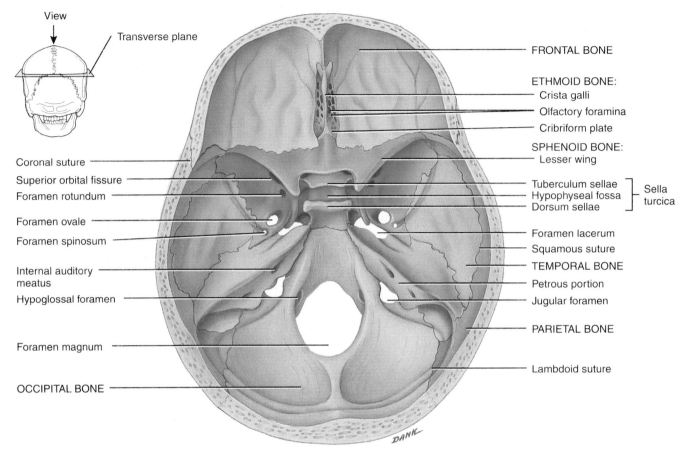

View

Transverse plane

FRONTAL BONE

ETHMOID BONE:
Crista galli
Olfactory foramina
Cribriform plate

SPHENOID BONE:
Lesser wing

Coronal suture

Superior orbital fissure

Foramen rotundum

Tuberculum sellae
Hypophyseal fossa } Sella turcica
Dorsum sellae

Foramen ovale

Foramen spinosum

Foramen lacerum
Squamous suture
TEMPORAL BONE

Internal auditory meatus

Hypoglossal foramen

Petrous portion
Jugular foramen

PARIETAL BONE

Foramen magnum

Lambdoid suture

OCCIPITAL BONE

DANK

(a) Superior view of sphenoid bone in floor of cranium

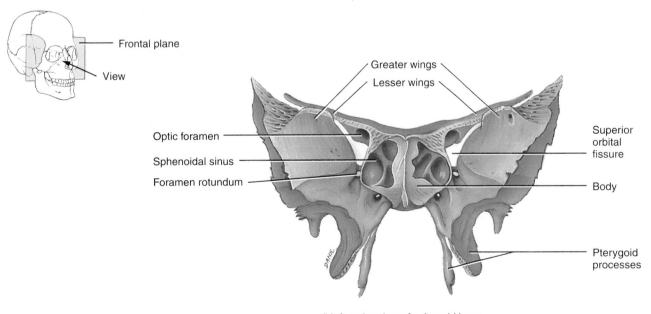

Frontal plane

View

Greater wings

Lesser wings

Optic foramen

Sphenoidal sinus

Foramen rotundum

Superior orbital fissure

Body

Pterygoid processes

DANK

(b) Anterior view of sphenoid bone

 Starting at the crista galli of the ethmoid bone and going in a clockwise direction, what are the names of the bones that articulate with the sphenoid bone?

to the temporal bone and can be viewed externally. The *lesser wings,* which are smaller, form a ridge of bone anterior and superior to the greater wings. They form part of the floor of the cranium and the posterior part of the orbit of the eye.

Between the body and lesser wing just anterior to the sella turcica is the *optic foramen* (*optic* = eye). Lateral to the body between the greater and lesser wings is a triangular slit called the *superior orbital fissure.* This fissure may also be seen in the anterior view of the orbit in Figure 7.12.

In Figures 7.7 and 7.8b, you can see the *pterygoid processes* (TER-i-goyd = winglike) by looking at the inferior part of the sphenoid bone. These structures project inferiorly from the points where the body and greater wings unite and form the lateral posterior region of the nasal cavity. Some of the muscles that move the mandible attach to the pterygoid processes. At the base of the lateral pterygoid process in the greater wing is the *foramen ovale* (= oval hole). The *foramen lacerum* (= lacerated) is bounded anteriorly by the sphenoid bone and medially by the sphenoid and occipital bones. The foramen is covered in part by a layer of fibrocartilage in living subjects. Another foramen associated with the sphenoid bone is the *foramen rotundum* (= round hole) located at the junction of the anterior and medial parts of the sphenoid bone.

Ethmoid Bone

The **ethmoid bone** (ETH-moyd = like a sieve) is a light, sponge-like bone located on the midline in the anterior part of the cranial floor medial to the orbits (Figure 7.9). It is anterior to the sphenoid and posterior to the nasal bones. The ethmoid bone forms (1) part of the anterior portion of the cranial floor; (2) the medial wall of the orbits; (3) the superior portions of the nasal septum, a partition that divides the nasal cavity into right and left sides; and (4) most of the superior sidewalls of the nasal cavity.

The *lateral masses* of the ethmoid bone compose most of the wall between the nasal cavity and the orbits. They contain 3 to 18 air spaces, or "cells." The ethmoidal cells together form the *ethmoidal sinuses* (see Figure 7.13). The *perpendicular plate* forms the superior portion of the nasal septum (Figure 7.9). The *cribriform plate* (*cribri-* = sieve) lies in the anterior floor of the cranium and forms the roof of the nasal cavity. The cribriform plate contains the *olfactory foramina* (*olfact-* = to smell) through which axons of the olfactory nerve pass. Projecting upward from the cribriform plate is a triangular process called the *crista galli* (*crista* = crest; *galli* = cock). This structure serves as a point of attachment for the membranes that cover the brain.

The lateral masses of the ethmoid bone contain two thin, scroll-shaped projections lateral to the nasal septum. These are the *superior nasal concha* (KONG-ka = shell) and the *middle nasal concha.* The plural term is *conchae* (KONG-kē). A third pair of conchae, the inferior nasal conchae, are separate bones (discussed shortly). The conchae cause turbulence in inhaled air, which results in many inhaled particles striking and becoming trapped in the mucus that lines the nasal passageways. This tur-

bulence thus cleanses the inhaled air before it passes into the rest of the respiratory tract. Turbulent airflow around the superior nasal conchae also aids in the distribution of olfactory stimulants for the sensation of smell. Air striking the mucous lining of the conchae is also warmed and moistened.

Facial Bones

The shape of the face changes dramatically during the first two years after birth. The brain and cranial bones expand, the teeth form and erupt (emerge), and the paranasal sinuses increase in size. Growth of the face ceases at about 16 years of age.

Nasal Bones

The paired **nasal bones** meet at the midline (see Figure 7.3) and form part of the bridge of the nose. The rest of the supporting tissue of the nose consists of cartilage.

Maxillae

The paired **maxillae** (mak-SIL-ē = jawbones; singular is *maxilla*) unite to form the upper jawbone. They articulate with every bone of the face except the mandible, or lower jawbone (see Figures 7.4 and 7.7). The maxillae form part of the floors of the orbits, part of the lateral walls and floor of the nasal cavity, and most of the hard palate. The hard palate is a bony partition formed by the palatine processes of the maxillae and horizontal plates of the palatine bones that forms the roof of the mouth.

Each maxilla contains a large *maxillary sinus* that empties into the nasal cavity (see Figure 7.13). The *alveolar process* (al-VĒ-ō-lar; *alveol-* = small cavity) of the maxilla is an arch that contains the *alveoli* (sockets) for the maxillary (upper) teeth. The *palatine process* is a horizontal projection of the maxilla that forms the anterior three-quarters of the hard palate. The union and fusion of the maxillary bones normally is completed before birth.

The *infraorbital foramen* (*infra-* = below), which can be seen in the anterior view of the skull in Figure 7.3, is an opening in the maxilla below the orbit. A final structure associated with the maxilla and sphenoid bone is the *inferior orbital fissure*. It is located between the greater wing of the sphenoid and the maxilla (see Figure 7.12).

Cleft Palate and Cleft Lip

Usually the palatine processes of the maxillary bones unite during weeks 10 to 12 of embryonic development. Failure to do so can result in one type of **cleft palate.** The condition may also involve incomplete fusion of the horizontal plates of the palatine bones (see Figures 7.6 and 7.7). Another form of this condition, called **cleft lip,** involves a split in the upper lip. Cleft lip and cleft palate often occur together. Depending on the extent and position of the cleft, speech and swallowing may be affected. In addition, children with cleft palate tend to have many ear infec-

Figure 7.9 **Ethmoid bone.** (See Tortora, *A Photographic Atlas of the Human Body,* Figure 3.10.)

The ethmoid bone forms part of the anterior portion of the cranial floor, the medial wall of the orbits, the superior portions of the nasal septum, and most of the sidewalls of the nasal cavity.

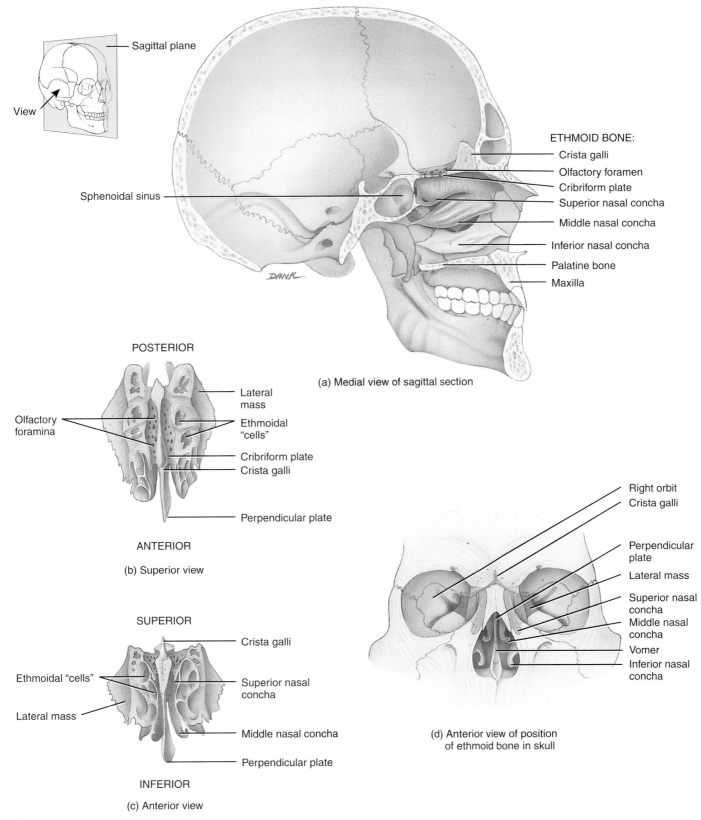

Sagittal plane

View

Sphenoidal sinus

ETHMOID BONE:
Crista galli
Olfactory foramen
Cribriform plate
Superior nasal concha
Middle nasal concha
Inferior nasal concha
Palatine bone
Maxilla

(a) Medial view of sagittal section

POSTERIOR

Lateral mass
Olfactory foramina
Ethmoidal "cells"
Cribriform plate
Crista galli
Perpendicular plate

ANTERIOR

(b) Superior view

SUPERIOR

Crista galli
Ethmoidal "cells"
Superior nasal concha
Lateral mass
Middle nasal concha
Perpendicular plate

INFERIOR

(c) Anterior view

Right orbit
Crista galli
Perpendicular plate
Lateral mass
Superior nasal concha
Middle nasal concha
Vomer
Inferior nasal concha

(d) Anterior view of position of ethmoid bone in skull

What part of the ethmoid bone forms the superior part of the nasal septum? The medial walls of the orbits?

tions that can lead to hearing loss. Facial and oral surgeons recommend closure of cleft lip during the first few weeks following birth, and surgical results are excellent. Repair of cleft palate typically is done between 12 and 18 months of age, ideally before the child begins to talk. Speech therapy may be needed, because the palate is important for pronouncing consonants, and orthodontic therapy may be needed to align the teeth. Again, results are usually excellent. ■

Zygomatic Bones

The two **zygomatic bones** (*zygo-* = like a yoke), commonly called cheekbones, form the prominences of the cheeks and part of the lateral wall and floor of each orbit (see Figure 7.12). They articulate with the maxilla and the frontal, sphenoid, and temporal bones.

Lacrimal Bones

The paired **lacrimal bones** (LAK-ri-mal; *lacrim-* = teardrops), the smallest bones of the face, are thin and resemble a fingernail in size and shape (see Figures 7.3 and 7.4). These bones are posterior and lateral to the nasal bones and form a part of the medial wall of each orbit. The lacrimal bones each contain a *lacrimal fossa,* a vertical tunnel formed with the maxilla, that houses the lacrimal sac, a structure that gathers tears and passes them into the nasal cavity (see Figure 7.12).

Palatine Bones

The two **palatine bones** (PAL-a-tīn) are L-shaped (see Figure 7.7). They form the posterior portion of the hard palate, part of the floor and lateral wall of the nasal cavity, and a small portion of the floors of the orbits. The *horizontal plates* of the palatine bones form the posterior portion of the hard palate, which separates the nasal cavity from the oral cavity (see Figures 7.6 and 7.7).

Inferior Nasal Conchae

The two **inferior nasal concha** are inferior to the middle nasal conchae of the ethmoid bone (see Figures 7.3 and 7.9a). They are scroll-like bones that form a part of the inferior lateral wall of the nasal cavity and project into the nasal cavity. The inferior nasal conchae are separate bones; they are not part of the ethmoid bone. All three pairs of nasal conchae help swirl and filter air before it passes into the lungs. However, only the superior nasal conchae are involved in the sense of smell.

Vomer

The **vomer** (VŌ-mer = plowshare) is a roughly triangular bone on the floor of the nasal cavity that articulates superiorly with the perpendicular plate of the ethmoid bone and inferiorly with both the maxillae and palatine bones along the midline (see Figures 7.3 and 7.7). It is part of the nasal septum, the partition that divides the nasal cavity into right and left sides.

Mandible

The **mandible** (*mand-* = to chew), or lower jawbone, is the largest, strongest facial bone (Figure 7.10). Other than the auditory ossicles, it is the only movable skull bone. Notice that the mandible consists of a curved, horizontal portion, the *body,* and two perpendicular portions, the *rami* (RĀ-mī = branches; singular is *ramus*). The *angle* of the mandible is the area where each ramus meets the body. Each ramus has a posterior *condylar process* (KON-di-lar) that articulates with the mandibular fossa and articular tubercle of the temporal bone (see Figure 7.4). This articulation is the **temporomandibular joint (TMJ).** The mandible also has an anterior *coronoid process* (KOR-ō-noyd) to which the temporalis muscle attaches. The depression between the coronoid and condylar processes is called the *mandibular notch.* The *alveolar process* is an arch containing the *alveoli* (sockets) for the mandibular (lower) teeth.

The *mental foramen* (*ment-* = chin) is located below the mandibular second premolar tooth. Another foramen in the mandible is the *mandibular foramen* on the medial surface of each ramus. Dentists use both foramina to reach nerves while injecting anesthetics. The mandibular foramen is the beginning of the *mandibular canal,* which runs obliquely in the ramus and anteriorly to the body deep to the roots of the teeth. The inferior alveolar nerves and blood vessels, which are distributed to the mandibular teeth, pass through the canal.

Temporomandibular Joint Syndrome

One problem associated with the temporomandibular joint (TMJ) is **temporomandibular joint (TMJ) syndrome.** It is

Figure 7.10   Mandible.

🔑 **The mandible is the largest and strongest facial bone.**

Condylar process — Coronoid process — Mandibular foramen — Mandibular notch — Ramus — Body — Angle — Alveolar process — Mental foramen

DANK

Right lateral view

❓ **What is the distinctive functional feature of the mandible among all the skull bones?**

characterized by dull pain around the ear, tenderness of the jaw muscles, a clicking or popping noise when opening or closing the mouth, limited or abnormal opening of the mouth, headache, tooth sensitivity, and abnormal wearing of the teeth. TMJ syndrome can be caused by improperly aligned teeth, grinding or clenching the teeth, trauma to the head and neck, or arthritis. Treatments include applying moist heat or ice, eating a soft diet, taking pain relievers such as aspirin, muscle retraining, adjusting or reshaping the teeth, orthodontic treatment, or surgery. ■

Nasal Septum

The inside of the nose, called the nasal cavity, is divided into right and left sides by a vertical partition called the **nasal septum.** The three components of the nasal septum are the vomer, septal cartilage, and the perpendicular plate of the ethmoid bone (Figure 7.11). The anterior border of the vomer articulates with the septal cartilage, which is hyaline cartilage, to form the anterior portion of the septum. The superior border of the vomer articulates with the perpendicular plate of the ethmoid bone to form the remainder of the nasal septum.

Deviated Nasal Septum

A **deviated nasal septum** is one that is deflected laterally from the midline of the nose. The deviation usually occurs at the junction of the vomer bone with the septal cartilage. Septal deviations may occur due to a developmental abnormality or trauma. If the deviation is severe, it may entirely block the nasal passageway. Even a partial blockage may lead to infection. If inflammation occurs, it may cause nasal congestion, blockage of the paranasal sinus openings, chronic sinusitis, headache, and nosebleeds. The condition usually can be corrected surgically. ■

Orbits

Seven bones of the skull join to form each **orbit** (eye socket), which contains the eyeball and associated structures (Figure 7.12). The three cranial bones of the orbit are the frontal, sphenoid, and ethmoid; the four facial bones are the palatine, zygomatic, lacrimal, and maxilla. Each pyramid-shaped orbit has four regions that converge posteriorly:

1. Parts of the frontal and sphenoid bones comprise the roof of the orbit.

2. Parts of the zygomatic and sphenoid bones form the lateral wall of the orbit.

3. Parts of the maxilla, zygomatic, and palatine bones make up the floor of the orbit.

4. Parts of the maxilla, lacrimal, ethmoid, and sphenoid bones form the medial wall of the orbit.

Associated with each orbit are five openings:

1. The *optic foramen* is at the junction of the roof and medial wall.

2. The *superior orbital fissure* is at the superior lateral angle of the apex.

3. The *inferior orbital fissure* is at the junction of the lateral wall and floor.

4. The *supraorbital foramen* is on the medial side of the supraorbital margin of the frontal bone.

5. The *lacrimal fossa* is in the lacrimal bone.

Foramina

We mentioned most of the **foramina** of the skull in the descriptions of the cranial and facial bones that they penetrate.

Figure 7.11   Nasal septum.   (See Tortora, *A Photographic Atlas of the Human Body,* Figure 3.4.)

The structures that form the nasal septum are the perpendicular plate of the ethmoid bone, the vomer, and septal cartilage.

Sagittal section

 What is the function of the nasal septum?

Figure 7.12 **Details of the orbit (eye socket).** (See Tortora, *A Photographic Atlas of the Human Body,* Figure 3.11.)

🔑 **The orbit is a pyramid-shaped structure that contains the eyeball and associated structures.**

FRONTAL BONE

Supraorbital margin

SPHENOID BONE

Optic foramen

Superior orbital fissure

PALATINE BONE

Zygomaticofacial foramen

ZYGOMATIC BONE

Infraorbital foramen

Supraorbital foramen

NASAL BONE

LACRIMAL BONE

ETHMOID BONE

Lacrimal fossa

MAXILLA

Inferior orbital fissure

Anterior view showing the bones of the right orbit

❓ **Which seven bones form the orbit?**

As preparation for studying other systems of the body, especially the nervous and cardiovascular systems, these foramina and the structures passing through them are listed in Table 7.3. For your convenience and for future reference, the foramina are listed alphabetically.

Unique Features of the Skull

The skull exhibits several unique features not seen in other bones of the body. These include sutures, paranasal sinuses, and fontanels.

Sutures

A **suture** (SOO-chur = seam) is an immovable joint in an adult that is found only between skull bones and that holds most skull bones together. Sutures in the skulls of infants and children often are movable. The names of many sutures reflect the bones they unite. For example, the frontozygomatic suture is between the frontal bone and the zygomatic bone. Similarly, the sphenoparietal suture is between the sphenoid bone and the parietal bone. In other cases, however, the names of sutures are not so obvious. Of the many sutures found in the skull, we will identify only four prominent ones:

1. The **coronal suture** (kō-RŌ-nal; *coron-* = crown) unites the frontal bone and both parietal bones (see Figure 7.4).

2. The **sagittal suture** (SAJ-i-tal; *sagitt-* = arrow) unites the two parietal bones on the superior midline of the skull (see Figure 7.6). The sagittal suture is so named because in the infant,

before the bones of the skull are firmly united, the suture and the fontanels (soft spots) associated with it resemble an arrow.

3. The **lambdoid suture** (LAM-doyd) unites the two parietal bones to the occipital bone. This suture is so named because of its resemblance to the Greek letter lambda (Λ), as can be seen in Figure 7.6. Sutural bones may occur within the sagittal and lambdoid sutures.

4. The **squamous sutures** (SKWĀ-mus; *squam-* = flat) unite the parietal and temporal bones on the lateral aspects of the skull (see Figure 7.4).

Paranasal Sinuses

The **paranasal sinuses** (*para-* = beside) are paired cavities in certain cranial and facial bones near the nasal cavity. They are most evident in a sagittal section of the skull (Figure 7.13 on page 202). The paranasal sinuses are lined with mucous membranes that are continuous with the lining of the nasal cavity. Skull bones containing the paranasal sinuses are the frontal, sphenoid, ethmoid, and maxillary. Besides producing mucus, the paranasal sinuses serve as resonating chambers for sound as we speak or sing.

Sinusitis

Secretions produced by the mucous membranes of the paranasal sinuses drain into the nasal cavity. An inflammation of the membranes due to an allergic reaction or infection is called **sinusitis.** If the membranes swell enough to block drainage into the nasal cavity, fluid pressure builds up in the paranasal sinuses, and a si-

Table 7.3 Principal Foramina of the Skull

Foramen	Location	Structures Passing Through*
Carotid (relating to carotid artery in neck)	Petrous portion of temporal bone (Figure 7.7).	Internal carotid artery and sympathetic nerves for eyes.
Hypoglossal (*hypo-* = under; *glossus* = tongue)	Superior to base of occipital condyles (Figure 7.8).	Cranial nerve XII (hypoglossal) and branch of ascending pharyngeal artery.
Infraorbital (*infra-* = below)	Inferior to orbit in maxilla (Figure 7.13).	Infraorbital nerve and blood vessels and a branch of the maxillary division of cranial nerve V (trigeminal).
Jugular (*jugul-* = the throat)	Posterior to carotid canal between petrous portion of temporal bone and occipital bone (Figure 7.8).	Internal jugular vein, cranial nerves IX (glossopharyngeal), X (vagus), and XI (accessory).
Lacerum (*lacerum* = lacerated)	Bounded anteriorly by sphenoid bone, posteriorly by petrous portion of temporal bone, and medially by sphenoid and occipital bones (Figure 7.8a).	Branch of ascending pharyngeal artery.
Magnum (= large)	Occipital bone (Figure 7.7).	Medulla oblongata and its membranes (meninges), cranial nerve XI (accessory), and vertebral and spinal arteries.
Mandibular (*mand-* = to chew)	Medial surface of ramus of mandible (Figure 7.10).	Inferior alveolar nerve and blood vessels.
Mastoid (= breast-shaped)	Posterior border of mastoid process of temporal bone (Figure 7.7).	Emissary vein to transverse sinus and branch of occipital artery to dura mater.
Mental (*ment-* = chin)	Inferior to second premolar tooth in mandible (Figure 7.10).	Mental nerve and vessels.
Olfactory (*olfact* = to smell)	Cribriform plate of ethmoid bone (Figure 7.8).	Cranial nerve I (olfactory).
Optic (= eye)	Between superior and inferior portions of small wing of sphenoid bone (Figure 7.13).	Cranial nerve II (optic) and ophthalmic artery.
Ovale (= oval)	Greater wing of sphenoid bone (Figure 7.8).	Mandibular branch of cranial nerve V (trigeminal).
Rotundum (= round)	Junction of anterior and medial parts of sphenoid bone (Figure 7.8).	Maxillary branch of cranial nerve V (trigeminal).
Stylomastoid (*stylo-* = stake or pole)	Between styloid and mastoid processes of temporal bone (Figure 7.7).	Cranial nerve VII (facial) and stylomastoid artery.
Supraorbital (*supra-* = above)	Supraorbital margin of orbit in frontal bone (Figure 7.13).	Supraorbital nerve and artery.

* The cranial nerves listed here are described in Table 14.3 on pages 485–489.

nus headache results. A severely deviated nasal septum or nasal polyps, growths that can be removed surgically, may also cause chronic sinusitis. ■

Fontanels

The skeleton of a newly formed embryo consists of cartilage and fibrous connective tissue membrane structures shaped like bones. Gradually, ossification occurs—bone replaces the cartilage and fibrous connective tissue membranes. At birth, membrane-filled spaces called **fontanels** (fon-ta-NELZ = little fountains) are present between the cranial bones (Figure 7.14). Commonly called "soft spots," fontanels are areas of fibrous connective tissue membranes. Eventually, they will be replaced with bone by intramembranous ossification and become sutures. Functionally, the fontanels provide some flexibility to the fetal skull. They allow the skull to change shape as it passes through the birth canal and permit rapid growth of the brain during infancy. Although an infant may have many fontanels at birth, the form and location of six are fairly constant:

- The unpaired **anterior fontanel,** located at the midline between the two parietal bones and the frontal bone, is roughly diamond-shaped and is the largest fontanel. It usually closes 18 to 24 months after birth.

- The unpaired **posterior fontanel** is located at the midline between the two parietal bones and the occipital bone. Because it is much smaller than the anterior fontanel, it generally closes about 2 months after birth.

- The paired **anterolateral fontanels,** located laterally between the frontal, parietal, temporal, and sphenoid bones, are small and irregular in shape. Normally, they close about 3 months after birth.

Figure 7.13   Paranasal sinuses. (See Tortora, *A Photographic Atlas of the Human Body,* Figure 3.4.)

🔑 Paranasal sinuses are mucous membrane-lined spaces in the frontal, sphenoid, ethmoid, and maxillary bones that connect to the nasal cavity.

Sagittal plane

View

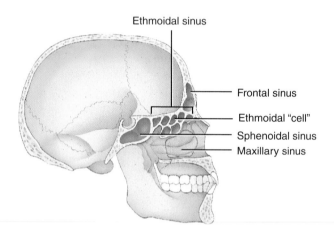

Ethmoidal sinus

Frontal sinus

Ethmoidal "cell"

Sphenoidal sinus

Maxillary sinus

Sagittal section

❓ What are the functions of the paranasal sinuses?

Figure 7.14   Fontanels at birth. (See Tortora, *A Photographic Atlas of the Human Body,* Figure 3.12.)

🔑 Fontanels are fibrous connective tissue membrane-filled spaces between cranial bones that are present at birth.

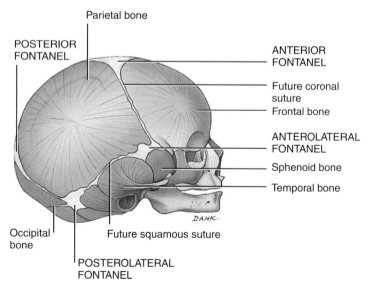

Parietal bone

POSTERIOR FONTANEL

ANTERIOR FONTANEL

Future coronal suture

Frontal bone

ANTEROLATERAL FONTANEL

Sphenoid bone

Temporal bone

DANK

Occipital bone

Future squamous suture

POSTEROLATERAL FONTANEL

Right lateral view

❓ Which fontanel do four different skull bones border?

- The paired **posterolateral fontanels,** located laterally between the parietal, occipital, and temporal bones, are irregularly shaped. They begin to close 1 to 2 months after birth, but closure is generally not complete until 12 months.

The amount of closure in fontanels helps a physician gauge the degree of brain development. In addition, the anterior fontanel serves as a landmark for withdrawal of blood for analysis from the superior sagittal sinus (a large vein on the midline surface of the brain).

HYOID BONE

▶ O B J E C T I V E

- **Describe the relationship of the hyoid bone to the skull.**

The single **hyoid bone** (HĪ-oyd = U-shaped) is a unique component of the axial skeleton because it does not articulate with any other bone (see Figure 7.4). Rather, it is suspended from the styloid processes of the temporal bones by ligaments and muscles. Located in the anterior neck between the mandible and larynx, the hyoid bone supports the tongue, providing attachment sites for some tongue muscles and for muscles of the neck and pharynx. The hyoid bone consists of a horizontal *body* and paired projections called the *lesser horns* and the *greater horns* (Figure 7.15). Muscles and ligaments attach to these paired projections.

The hyoid bone and the cartilages of the larynx and trachea are often fractured during strangulation. As a result, they are carefully examined at autopsy when strangulation is suspected.

▶ C H E C K P O I N T

4. Describe the general features of the skull.

5. What bones constitute the orbit?

6. What structures make up the nasal septum?

7. Define the following: foramen, suture, paranasal sinus, and fontanel.

8. What are the functions of the hyoid bone?

Figure 7.15   Hyoid bone. (See Tortora, *A Photographic Atlas of the Human Body,* Figure 3.13.)

🔑 The hyoid bone supports the tongue, providing attachment sites for muscles of the tongue, neck, and pharynx.

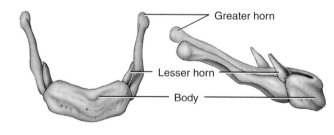

Greater horn

Lesser horn

Body

(a) Anterior view   (b) Right lateral view

❓ In what way is the hyoid bone different from all the other bones of the axial skeleton?

VERTEBRAL COLUMN

- Identify the regions and normal curves of the vertebral column and describe its structural and functional features.

The **vertebral column,** also called the *spine* or *backbone,* makes up about two-fifths of the total height of the body and is composed of a series of bones called **vertebrae** (VER-te-brē; singular is *vertebra*). Together with the sternum and ribs, the vertebral column forms the skeleton of the trunk of the body. Whereas the vertebral column consists of bone and connective tissue, the spinal cord consists of nervous tissue. The length of the column is about 71 cm (28 in.) in an average adult male and about 61 cm (24 in.) in an average adult female. The vertebral column functions as a strong, flexible rod that can move forward, backward, sideways, and rotate. It encloses and protects the spinal cord, supports the head, and serves as a point of attachment for the ribs, pelvic girdle, and muscles of the back.

The total number of vertebrae during early development is 33. Then, several vertebrae in the sacral and coccygeal regions fuse. As a result, the adult vertebral column, also called the spinal column, typically contains 26 vertebrae (Figure 7.16). These are distributed as follows:

- 7 **cervical vertebrae** (*cervic-* = neck) are in the neck region.
- 12 **thoracic vertebrae** (*thorax* = chest) are posterior to the thoracic cavity.
- 5 **lumbar vertebrae** (*lumb-* = loin) support the lower back.
- 1 **sacrum** (SĀ-krum = sacred bone) consists of five fused *sacral vertebrae.*
- 1 **coccyx** (KOK-siks = cuckoo, because the shape resembles the bill of a cuckoo bird), consists of four fused *coccygeal vertebrae* (kok-SIJ-ē-al).

Whereas the cervical, thoracic, and lumbar vertebrae are movable, the sacrum and coccyx are not. We will discuss each of these regions in detail shortly.

Figure 7.16 Vertebral column. The numbers in parentheses in (a) indicate the number of vertebrae in each region. In (d), the relative size of the disc has been enlarged for emphasis. A "window" has been cut in the annulus fibrosus to view the nucleus pulposus. (See Tortora, *A Photographic Atlas of the Human Body,* Figure 3.15.)

🔑 **The adult vertebral column typically contains 26 vertebrae.**

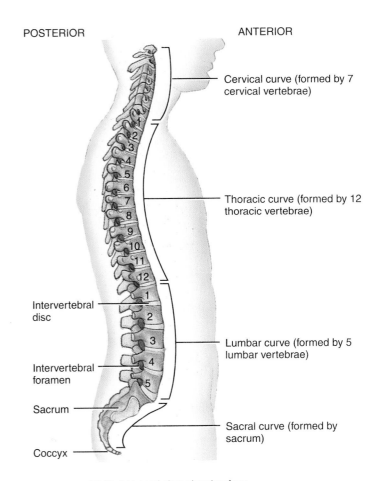

Cervical vertebrae (7)

Thoracic vertebrae (12)

Intervertebral disc

Lumbar vertebrae (5)

Sacrum (1)

Coccyx (1)

(a) Anterior view showing regions of the vertebral column

POSTERIOR ANTERIOR

Cervical curve (formed by 7 cervical vertebrae)

Thoracic curve (formed by 12 thoracic vertebrae)

Intervertebral disc

Lumbar curve (formed by 5 lumbar vertebrae)

Intervertebral foramen

Sacrum

Sacral curve (formed by sacrum)

Coccyx

(b) Right lateral view showing four normal curves

(continues)

Figure 7.16 *(continued)*

Intervertebral foramen

Vertebral body

Nucleus pulposus

Annulus fibrosus

Single curve in fetus   Four curves in adult

Normal intervertebral disc

Compressed intervertebral disc in a weight-bearing situation

(c) Fetal and adult curves

(d) Intervertebral disc

Which curves of the adult vertebral column are concave (relative to the anterior side of the body)?

Normal Curves of the Vertebral Column

When viewed from the side, the vertebral column shows four slight bends called **normal curves** (Figure 7.16b). Relative to the front of the body, the *cervical* and *lumbar curves* are convex (bulging out), whereas the *thoracic* and *sacral curves* are concave (cupping in). The curves of the vertebral column increase its strength, help maintain balance in the upright position, absorb shocks during walking, and help protect the vertebrae from fracture.

In the fetus, there is only a single anteriorly concave curve (Figure 7.16c). At about the third month after birth, when an infant begins to hold its head erect, the cervical curve develops. Later, when the child sits up, stands, and walks, the lumbar curve develops. The thoracic and sacral curves are called *primary curves* because they form first during fetal development. The cervical and lumbar curves are known as *secondary curves* because they begin to form later, several months after birth. All curves are fully developed by age 10. However, secondary curves may be progressively lost in old age.

Various conditions may exaggerate the normal curves of the vertebral column, or the column may acquire a lateral bend, resulting in **abnormal curves** of the vertebral column. Three such abnormal curves—kyphosis, lordosis, and scoliosis—are described in the Medical Terminology section on page 214.

Intervertebral Discs

Between the bodies of adjacent vertebrae from the second cervical vertebra to the sacrum are **intervertebral discs** (Figure 7.16d). Each disc has an outer fibrous ring consisting of fibrocartilage called the *annulus fibrosus* (*annulus* = ringlike) and an inner soft, pulpy, highly elastic substance called the *nucleus pulposus* (*pulposus* = pulplike). The discs form strong joints, permit various movements of the vertebral column, and absorb vertical shock. Under compression, they flatten, broaden, and

bulge from their intervertebral spaces. Superior to the sacrum, the intervertebral discs constitute about one-fourth the length of the vertebral column.

Parts of a Typical Vertebra

Vertebrae in different regions of the spinal column vary in size, shape, and detail, but they are similar enough that we can discuss the structure and functions of a typical vertebra (Figure 7.17). Vertebrae typically consist of a body, a vertebral arch, and several processes.

Body

The **body** is the thick, disk-shaped anterior portion that is the weight-bearing part of a vertebra. Its superior and inferior surfaces are roughened for the attachment of cartilaginous intervertebral discs. The anterior and lateral surfaces contain nutrient foramina for blood vessels.

Vertebral Arch

The **vertebral arch** extends posteriorly from the body of the vertebra and together with the body of the vertebra surrounds the spinal cord. Two short, thick processes, the *pedicles* (PED-i-kuls = little feet), form the vertebral arch. The pedicles project posteriorly from the body to unite with the laminae. The *laminae* (LAM-i-nē = thin layers) are the flat parts that join to form the posterior portion of the vertebral arch. The *vertebral foramen* lies between the vertebral arch and body and contains the spinal cord, adipose tissue, areolar connective tissue, and blood vessels. Collectively, the vertebral foramina of all vertebrae form the **vertebral (spinal) canal,** which is the inferior part of the dorsal body cavity. The pedicles exhibit superior and inferior notches called *vertebral notches.* When the vertebral notches are stacked on top of one another, they form an opening between adjoining

Figure 7.17 Structure of a typical vertebra, as illustrated by a thoracic vertebra. In (b), only one spinal nerve has been included, and it has been extended beyond the intervertebral foramen for clarity. The sympathetic chain is part of the autonomic nervous system (see Figure 17.2 on page 569). (See Tortora, *A Photographic Atlas of the Human Body,* Figure 3.16.)

🔑 **A vertebra consists of a body, a vertebral arch, and several processes.**

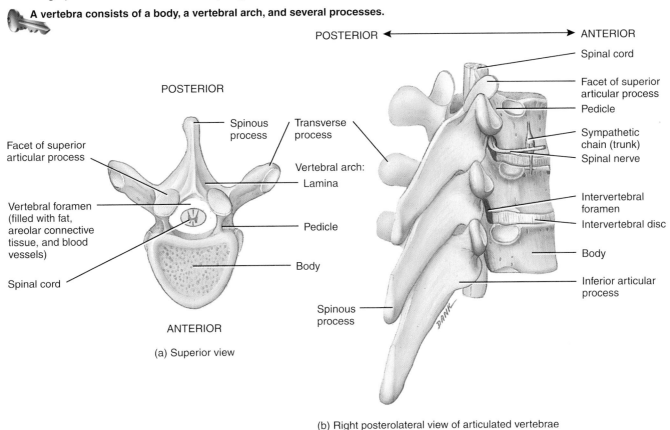

(a) Superior view

(b) Right posterolateral view of articulated vertebrae

❓ **What are the functions of the vertebral and intervertebral foramina?**

vertebrae on both sides of the column. Each opening, called an *intervertebral foramen,* permits the passage of a single spinal nerve. Thoracic and lumbar spinal nerves pass through the intervertebral foramina.

Processes

Seven *processes* arise from the vertebral arch. At the point where a lamina and pedicle join, a *transverse process* extends laterally on each side. A single *spinous process* (spine) projects posteriorly from the junction of the laminae. These three processes serve as points of attachment for muscles. The remaining four processes form joints with other vertebrae above or below. The two *superior articular processes* of a vertebra articulate with the two inferior *articular processes* of the vertebra immediately above them. The two *inferior articular processes* of a vertebra articulate with the two superior articular processes of the vertebra immediately below them. The articulating surfaces of the articular processes are called *facets* (= little faces). The articulations formed between the bodies and articular facets of successive vertebrae are termed *intervertebral joints.*

Regions of the Vertebral Column

We turn now to the five regions of the vertebral column, beginning at the superior end. Note that vertebrae in each region are numbered in sequence, from superior to inferior.

Cervical Region

The bodies of **cervical vertebrae** (C1–C7) are smaller than those of thoracic vertebrae (Figure 7.18a). Their vertebral arches, however, are larger. All cervical vertebrae have three foramina: one vertebral foramen and two transverse foramina (Figure 7.18b). The vertebral foramina of cervical vertebrae are the largest in the spinal column because they house the cervical enlargement of the spinal cord. Each cervical transverse process contains a *transverse foramen* through which the vertebral artery and its accompanying vein and nerve pass. The spinous processes of C2 through C6 are often *bifid*—that is, split into two parts (Figure 7.18c, d).

The first two cervical vertebrae differ considerably from the others. The first cervical vertebra (C1), the **atlas,** supports

Figure 7.18   Cervical vertebrae.   (See Tortora, *A Photographic Atlas of the Human Body*, Figure 3.17.)

🔑 **The cervical vertebrae are found in the neck region.**

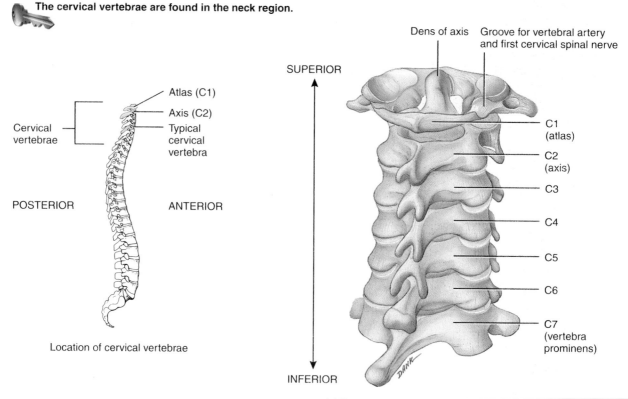

SUPERIOR

Cervical
vertebrae

Atlas (C1)

Axis (C2)

Typical
cervical
vertebra

POSTERIOR   ANTERIOR

Location of cervical vertebrae

INFERIOR

Dens of axis   Groove for vertebral artery
and first cervical spinal nerve

C1
(atlas)

C2
(axis)

C3

C4

C5

C6

C7
(vertebra
prominens)

(a) Posterior view of articulated cervical vertebrae

the head and is named for the mythological Atlas who supported the world on his shoulders (Figure 7.18a, b). The atlas is a ring of bone with *anterior* and *posterior arches* and large *lateral masses*. It lacks a body and a spinous process. The superior surfaces of the lateral masses, called *superior articular facets*, are concave. They articulate with the occipital condyles of the occipital bone to form the *atlanto-occipital joints*. These articulations permit the movement seen when moving the head to signify "yes." The inferior surfaces of the lateral masses, the *inferior articular facets*, articulate with the second cervical vertebra. The transverse processes and transverse foramina of the atlas are quite large.

The second cervical vertebra (C2), the **axis** (see Figure 7.18a, c), does have a body. A peglike process called the *dens* (= tooth) or *odontoid process* projects up through the anterior portion of the vertebral foramen of the atlas. The dens makes a pivot on which the atlas and head rotate, as in moving the head to signify "no." This arrangement permits side-to-side rotation of the head. The articulation formed between the anterior arch of the atlas and dens of the axis, and between their articular facets, is called the *atlanto-axial joint*. In some instances of trauma, the dens of the axis may be driven into the medulla oblongata of the brain. When whiplash injuries result in death, this type of injury is the usual cause.

The third through sixth cervical vertebrae (C3–C6), represented by the vertebra in Figure 7.18d, correspond to the struc-

tural pattern of the typical cervical vertebra previously described. The seventh cervical vertebra (C7), called the *vertebra prominens*, is somewhat different (see Figure 7.18a). It has a single large spinous process that can be seen and felt at the base of the neck.

Thoracic Region

Thoracic vertebrae (T1–T12; Figure 7.19 on page 208) are considerably larger and stronger than cervical vertebrae. In addition, the spinous processes on T1 and T2 are long, laterally flattened, and directed inferiorly. In contrast, the spinous processes on T11 and T12 are shorter, broader, and directed more posteriorly. Compared to cervical vertebrae, thoracic vertebrae also have longer and larger transverse processes.

The most distinctive feature of the thoracic vertebrae is that they articulate with the ribs. The articulating surfaces of the vertebrae are called *facets* and *demifacets* (= half-facets). Except for T11 and T12, the transverse processes have facets for articulating with the *tubercles* of the ribs. The bodies of thoracic vertebrae also have facets or demifacets for articulation with the *heads* of the ribs. The articulations between the thoracic vertebrae and ribs are called *vertebrocostal joints*. As you can see in Figure 7.19a, T1 has a superior facet and an inferior demifacet, one on each side of the vertebral body. T2 through T8 have a superior and inferior demifacet, one on each side of the vertebral

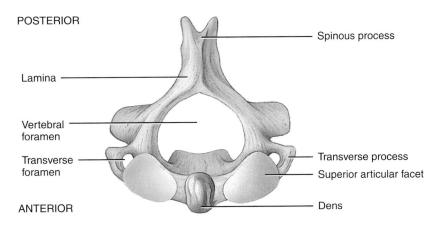

POSTERIOR

Vertebral foramen

Superior articular facet

Transverse foramen

ANTERIOR

Posterior arch

Groove for vertebral artery and first cervical spinal nerve

Lateral mass

Transverse process

Articular surface for dens of axis

Anterior arch

(b) Superior view of the atlas (C1)

POSTERIOR

Lamina

Vertebral foramen

Transverse foramen

ANTERIOR

Spinous process

Transverse process

Superior articular facet

Dens

(c) Superior view of the axis (C2)

POSTERIOR

Lamina

Vertebral foramen

Transverse foramen

Transverse process

ANTERIOR

Bifid spinous process

Superior articular facet

Pedicle

Body

(d) Superior view of a typical cervical vertebra

 Which bones permit the movement of the head to signify no?

body. T9 has a superior demifacet on each side of the vertebral body, and T10 through T12 have a superior facet on each side of the vertebral body. Movements of the thoracic region are limited by thin intervertebral discs and by the attachment of the ribs to the sternum.

Lumbar Region

The **lumbar vertebrae** (L1–L5) are the largest and strongest in the vertebral column (Figure 7.20 on page 209) because the amount of body weight supported by the vertebrae increases toward the inferior end of the backbone. Their processes are short and thick. The superior articular processes are directed medially instead of superiorly and the inferior articular processes are directed laterally instead of inferiorly. The spinous processes are thick and broad and project posteriorly. The spinous processes are well adapted for the attachment of the large back muscles.

Table 7.4 on page 210 summarizes the major structural differences among cervical, thoracic, and lumbar vertebrae.

Figure 7.19 **Thoracic vertebrae.** (See Tortora, *A Photographic Atlas of the Human Body,* Figure 3.16.)

🔑 The thoracic vertebrae are found in the chest region and articulate with the ribs.

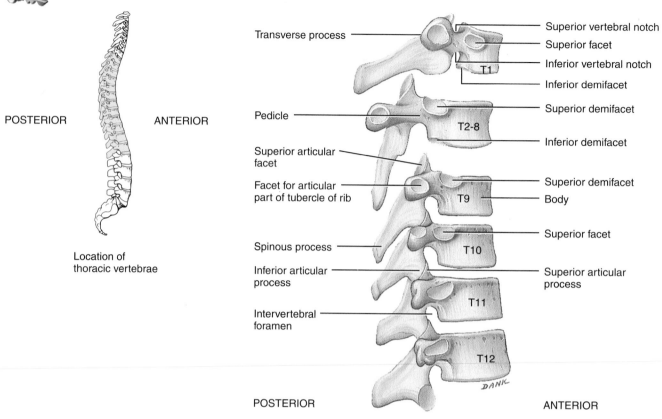

Location of thoracic vertebrae

(a) Right lateral view of several articulated thoracic vertebrae

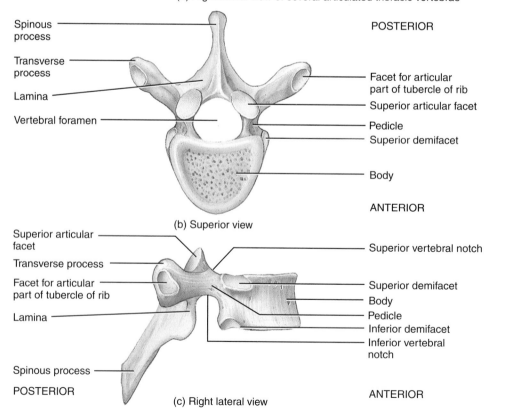

(b) Superior view

(c) Right lateral view

❓ Which parts of thoracic vertebrae articulate with the ribs?

Figure 7.20 Lumbar vertebrae. (See Tortora, *A Photographic Atlas of the Human Body,* Figure 3.18.)

 Lumbar vertebrae are found in the lower back.

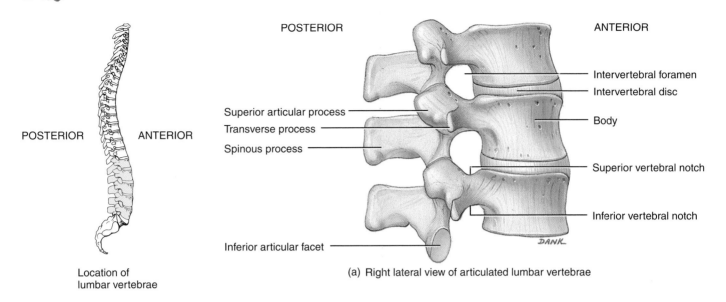

Location of lumbar vertebrae

(a) Right lateral view of articulated lumbar vertebrae

(b) Superior view

(c) Right lateral view

Why do the lumbar vertebrae need to be the largest and strongest in the vertebral column?

Sacrum

The **sacrum** is a triangular bone formed by the union of five sacral vertebrae, indicated in Figure 7.21a as S1–S5. The fusion of the sacral vertebrae begins between ages 16 and 18 and is usually completed by age 30. The sacrum serves as a strong foundation for the pelvic girdle. It is positioned at the posterior portion of the pelvic cavity medial to the two hip bones. The female sacrum is shorter, wider, and more curved

Table 7.4	Comparison of Major Structural Features of Cervical, Thoracic, and Lumbar Vetebrae		
Characteristic	**Cervical**	**Thoracic**	**Lumbar**
Overall structure	See Figure 7.18d	See Figure 7.19b	See Figure 7.20b
Size	Small	Larger	Largest
Foramina	One vertebral and two transverse	One vertebral	One vertebral
Spinous processes	Slender and often bifid (C2–C6)	Long and fairly thick (most project inferiorly)	Short and blunt (project posteriorly rather than inferiorly)
Transverse processes	Small	Fairly large	Large and blunt
Articular facets for ribs	Absent	Present	Absent
Direction of articular facets			
Superior	Posterosuperior	Posterolateral	Medial
Inferior	Anteroinferior	Anteromedial	Lateral
Size of intervertebral discs	Thick relative to size of vertebral bodies	Thin relative to vertebral bodies	Massive

between S2 and S3 than the male sacrum (see Table 8.1 on page 231).

The concave anterior side of the sacrum faces the pelvic cavity. It is smooth and contains four *transverse lines (ridges)* that mark the joining of the sacral vertebral bodies (Figure 7.21a). At the ends of these lines are four pairs of *anterior sacral foramina.* The lateral portion of the superior surface contains a smooth surface called the *sacral ala* (= wing), which is formed by the fused transverse processes of the first sacral vertebra (S1).

The convex, posterior surface of the sacrum contains a *median sacral crest,* which is the fused spinous processes of the upper sacral vertebrae, and a *lateral sacral crest,* which is the

Figure 7.21   Sacrum and coccyx.   (See Tortora, *A Photographic Atlas of the Human Body,* Figure 3.19.)

🔑 **The union of five sacral vertebrae forms the sacrum, and the union of usually four coccygeal vertebrae forms the coccyx.**

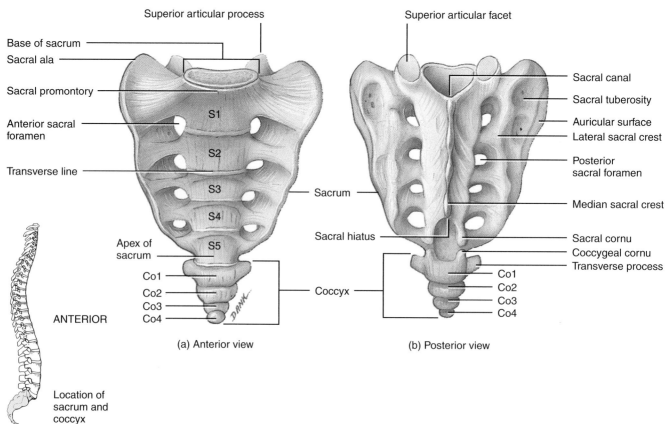

(a) Anterior view

(b) Posterior view

🔎 **How many foramina pierce the sacrum, and what is their function?**

fused transverse processes of the sacral vertebrae. Four pairs of *posterior sacral foramina* (Figure 7.21b) communicate with the anterior sacral foramina through which nerves and blood vessels pass. The *sacral canal* is a continuation of the vertebral canal. The laminae of the fifth sacral vertebra, and sometimes the fourth, fail to meet. This leaves an inferior entrance to the vertebral canal called the *sacral hiatus* (hī-Ā-tus = opening). On either side of the sacral hiatus are the *sacral cornua,* the inferior articular processes of the fifth sacral vertebra. Ligaments connect them to the coccyx.

The narrow inferior portion of the sacrum is known as the *apex.* The broad superior portion of the sacrum is the *base.* The anteriorly projecting border of the base, called the *sacral promontory* (PROM-on-tō-rē), is one of the points used for measurements of the pelvis. On both lateral surfaces, the sacrum has a large *auricular surface* (*auricular* = ear) that articulates with the ilium of each hipbone to form the *sacroiliac joint.* Posterior to the auricular surface is a roughened surface, the *sacral tuberosity,* that contains depressions for the attachment of ligaments. The sacral tuberosity is another surface of the sacrum that unites with the hip bones to form the sacroiliac joints. The *superior articular processes* of the sacrum articulate with the fifth lumbar vertebra, and the base of the sacrum articulates with the body of the fifth lumbar vertebra, to form the *lumbosacral joint.*

Coccyx

The **coccyx** is also triangular and is formed by the fusion of usually four coccygeal vertebrae, indicated in Figure 7.21 as Co1–Co4. The coccygeal vertebrae fuse when a person is between 20 and 30 years of age. The dorsal surface of the body of the coccyx contains two long *coccygeal cornua* that are connected by ligaments to the sacral cornua. The coccygeal cornua are the pedicles and superior articular processes of the first coccygeal vertebra. On the lateral surfaces of the coccyx are a series of *transverse processes,* the first pair being the largest. The coccyx articulates superiorly with the apex of the sacrum. In females, the coccyx points inferiorly; in males, it points anteriorly (see Table 8.1 on page 231).

Caudal Anesthesia

Anesthetic agents that act on the sacral and coccygeal nerves are sometimes injected through the sacral hiatus, a procedure called **caudal anesthesia** or **epidural block.** The procedure is used most often to relieve pain during labor and to provide anesthesia to the perineal area. Because the sacral hiatus is between the sacral cornua, the cornua are important bony landmarks for locating the hiatus. Anesthetic agents may also be injected through the posterior sacral foramina. ■

▶ C H E C K P O I N T

9. What are the functions of the vertebral column?

10. When do the secondary vertebral curves develop?

11. What are the main distinguishing characteristics of the bones of the various regions of the vertebral column?

THORAX

O B J E C T I V E

• **Identify the bones of the thorax.**

The term **thorax** refers to the entire chest. The skeletal part of the thorax, the **thoracic cage,** is a bony enclosure formed by the sternum, costal cartilages, ribs, and the bodies of the thoracic vertebrae (Figure 7.22). The thoracic cage is narrower at its superior end and broader at its inferior end and is flattened from front to back. It encloses and protects the organs in the thoracic and superior abdominal cavities and provides support for the bones of the shoulder girdle and upper limbs.

Sternum

The **sternum,** or breastbone, is a flat, narrow bone located in the center of the anterior thoracic wall that measures about 15 cm (6 in.) in length and consists of three parts (Figure 7.22). The superior part is the **manubrium** (ma-NOO-brē-um = handlelike); the middle and largest part is the **body;** and the inferior, smallest part is the **xiphoid process** (ZĪ-foyd = sword-shaped).

The junction of the manubrium and body forms the *sternal angle.* The manubrium has a depression on its superior surface, the *suprasternal notch.* Lateral to the suprasternal notch are *clavicular notches* that articulate with the medial ends of the clavicles to form the *sternoclavicular joints.* The manubrium also articulates with the costal cartilages of the first and second ribs to form the *sternocostal joints.*

The body of the sternum articulates directly or indirectly with the costal cartilages of the second through tenth ribs. The xiphoid process consists of hyaline cartilage during infancy and childhood and does not ossify completely until about age 40. No ribs are attached to it, but the xiphoid process provides attachment for some abdominal muscles. Incorrect positioning of the hands of a rescuer during cardiopulmonary resuscitation (CPR) may fracture the xiphoid process, driving it into internal organs. During thoracic surgery, the sternum may be split along the midline to allow surgeons access to structures in the thoracic cavity such as the thymus, heart, and great vessels of the heart.

Ribs

Twelve pairs of **ribs** give structural support to the sides of the thoracic cavity (see Figure 7.22). The ribs increase in length from the first through seventh, and then decrease in length to the twelfth rib. Each rib articulates posteriorly with its corresponding thoracic vertebra.

The first through seventh pairs of ribs have a direct anterior attachment to the sternum by a strip of hyaline cartilage called *costal cartilage* (*cost-* = rib). These ribs are called *true (vertebrosternal) ribs.* The remaining five pairs of ribs are termed *false ribs* because their costal cartilages either attach indirectly to the sternum or do not attach to the sternum at all. The cartilages of

Figure 7.22 **Skeleton of the thorax.** (See Tortora, *A Photographic Atlas of the Human Body,* Figure 3.20.)

The bones of the thorax enclose and protect organs in the thoracic cavity and in the superior abdominal cavity.

Anterior view

Which ribs are true ribs? Which are false ribs?

the eighth, ninth, and tenth pairs of ribs attach to each other and then to the cartilages of the seventh pair of ribs. These false ribs are called *vertebrochondral ribs.* The eleventh and twelfth false ribs are also known as *floating ribs* because the costal cartilage at their anterior ends does not attach to the sternum at all. Floating ribs attach only posteriorly to the thoracic vertebrae. Inflammation of one or more costal cartilages, called *costochondritis,* is characterized by local tenderness and pain in the anterior chest wall that may radiate. The symptoms mimic the chest pain associated with a heart attack (angina pectoris).

Figure 7.23a shows the parts of a typical (third through ninth) rib. The *head* is a projection at the posterior end of the rib. It consists of one or two *facets* that articulate with facets on the bodies of adjacent thoracic vertebrae to form *vertebrocostal joints.* The *neck* is a constricted portion just lateral to the head. A knoblike structure on the posterior surface where the neck joins the body is called a *tubercle* (TOO- ber-kul). The *nonarticular part* of the tubercle affords attachment to the ligament of the tubercle. The *articular part* of the tubercle articulates with the facet of a transverse process of the inferior of the two vertebrae to which the head of the rib is connected. These articulations also form vertebrocostal joints. The *body* is the main part of the rib. A short distance beyond the tubercle, an abrupt change in the curvature of the body occurs. This point is called the *costal angle.* The inner surface of the rib has a *costal groove* that protects blood vessels and a small nerve.

In summary, the posterior portion of the rib connects to a thoracic vertebra by its head and the articular part of a tubercle.

The facet of the head fits into a facet on the body of one vertebra or into the demifacets of two adjoining vertebrae. The articular part of the tubercle articulates with the facet of the transverse process of the vertebra.

Intercostal muscles, blood vessels, and nerves occupy spaces between ribs, called *intercostal spaces.* Surgical access to the lungs or other structures in the thoracic cavity is commonly obtained through an intercostal space. Special rib retractors are used to create a wide separation between ribs. The costal cartilages are sufficiently elastic in younger persons to permit considerable bending without breaking.

Rib Fractures

Rib fractures are the most common chest injuries, and they usually result from direct blows, most often from impact with a steering wheel, falls, and crushing injuries to the chest. Ribs tend to break at the point where the greatest force is applied, but they may also break at their weakest point—the site of greatest curvature, which is just anterior to the costal angle. In some cases, fractured ribs may puncture the heart, great vessels of the heart, lungs, trachea, bronchi, esophagus, spleen, liver, and kidneys. Rib fractures are usually quite painful. ■

▶ CHECKPOINT

12. What are the functions of the bones of the thorax?

13. What are the parts of the sternum?

14. How are ribs classified based on their attachment to the sternum?

Figure 7.23 The structure of ribs. Each rib has a head, a neck, and a body. The facets and the articular part of the tubercle are where the rib articulates with a vertebra. (See Tortora, *A Photographic Atlas of the Human Body,* Figure 3.21.)

Each rib articulates posteriorly with its corresponding thoracic vertebra.

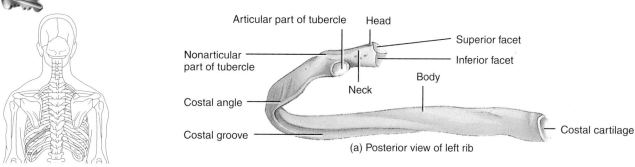

Articular part of tubercle Head
Nonarticular part of tubercle
Superior facet
Inferior facet
Neck
Body
Costal angle
Costal groove
Costal cartilage

(a) Posterior view of left rib

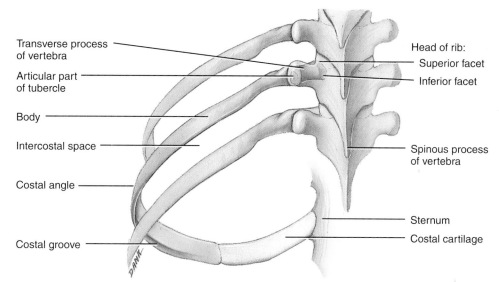

Transverse process of vertebra
Articular part of tubercle
Body
Intercostal space
Costal angle
Costal groove

Head of rib:
Superior facet
Inferior facet
Spinous process of vertebra
Sternum
Costal cartilage

(b) Posterior view of left ribs articulated with thoracic vertebrae and sternum

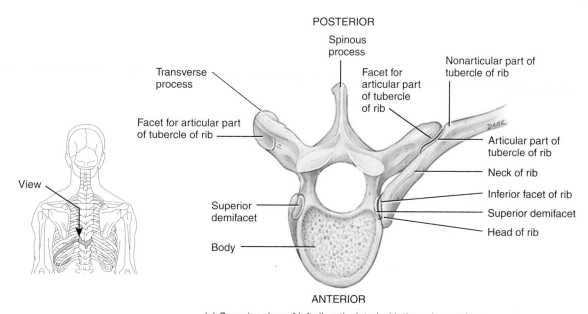

POSTERIOR

Spinous process
Transverse process
Facet for articular part of tubercle of rib
Superior demifacet
Body

Nonarticular part of tubercle of rib
Facet for articular part of tubercle of rib
Articular part of tubercle of rib
Neck of rib
Inferior facet of rib
Superior demifacet
Head of rib

View

ANTERIOR

(c) Superior view of left rib articulated with thoracic vertebra

How does a rib articulate with a thoracic vertebra?

DISORDERS: HOMEOSTATIC IMBALANCES

Herniated (Slipped) Disc

In their function as shock absorbers, intervertebral discs are constantly compressed. If the anterior and posterior ligaments of the discs become injured or weakened, the pressure developed in the nucleus pulposus may be great enough to rupture the surrounding fibrocartilage (annulus fibrosus). Then, the nucleus pulposus may herniate (protrude) posteriorly or into one of the adjacent vertebral bodies. This condition is called a **herniated (slipped) disc.** It occurs most often in the lumbar region because that part of the vertebral column bears much of the weight of the body, and it is the region of the most flexing and bending.

Most often, the nucleus pulposus slips posteriorly toward the spinal cord and spinal nerves (Figure 7.24). This movement exerts pressure on the spinal nerves, causing acute pain. If the roots of the sciatic nerve, which passes from the spinal cord to the foot, are compressed, the pain radiates down the posterior thigh, through the calf, and occasionally into the foot. If pressure is exerted on the spinal cord itself, some of its neurons may be destroyed. Treatment of a herniated disc may involve a *laminectomy,* a procedure in which parts of the laminae of the vertebra and intervertebral disc are removed to relieve pressure on the nerves.

Spina Bifida

Spina bifida (SPĪ-na BIF-i-da) is a congenital defect of the vertebral column in which laminae fail to unite at the midline. In serious cases, the membranes (meninges) around the spinal cord or the spinal cord itself protrude through the opening and produce serious problems, such

Figure 7.24   Herniated (slipped) disc.

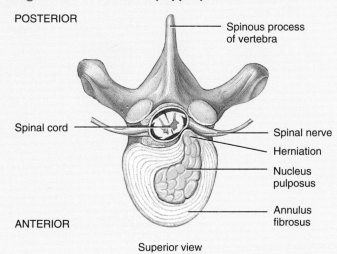

Superior view

as partial or complete paralysis, partial or complete loss of urinary bladder control, and the absence of reflexes. An increased risk of spina bifida is associated with low levels of folic acid, one of the B vitamins, during pregnancy. Spina bifida may be diagnosed before birth by a test of the mother's blood for alphafetoprotein, a protein produced by the fetus; by sonography; or by amniocentesis (withdrawal of amniotic fluid for analysis).

MEDICAL TERMINOLOGY

Kyphosis (kī-FŌ-sis; *kyphos-* = hump; *-osis* = condition). An exaggeration of the thoracic curve of the vertebral column. In tuberculosis of the spine, vertebral bodies may partially collapse, causing an acute angular bending of the vertebral column. In the elderly, degeneration of the intervertebral discs leads to kyphosis. Rickets or poor posture may also cause kyphosis. It is also common in females with advanced osteoporosis. The term *round-shouldered* refers to mild kyphosis.

Lordosis (lor-DŌ-sis; *lord-* = bent backward), sometimes called *swayback.* An exaggeration of the lumbar curve of the vertebral column. It may result from increased weight of the abdomen as in pregnancy or extreme obesity, poor posture, rickets, or tuberculosis of the spine.

Lumbar spine stenosis (*sten-* = narrowed). Narrowing of the spinal canal in the lumbar part of the vertebral column, due to hypertrophy of surrounding bone or soft tissues. It may be caused by arthritic changes in the intervertebral discs and is a common cause of back and leg pain.

Scoliosis (skō-lē-Ō-sis; *scolio* = crooked). Lateral bending of the vertebral column, usually in the thoracic region. It may result from congenitally (present at birth) malformed vertebrae, chronic sciatica, paralysis of muscles on one side of the vertebral column, poor posture, or one leg being shorter than the other.

STUDY OUTLINE

INTRODUCTION (p. 186)

1. Bones, joints, and muscles working together form the musculoskeletal system.

DIVISIONS OF THE SKELETAL SYSTEM (p. 186)

1. The axial skeleton consists of bones arranged along the longitudinal axis. The parts of the axial skeleton are the skull, auditory os-

sicles (ear bones), hyoid bone, vertebral column, sternum, and ribs.

2. The appendicular skeleton consists of the bones of the pectoral (shoulder) girdles, bones of the upper limbs, pelvic (hip) girdles, and bones of the lower limbs.

TYPES OF BONES (p. 188)

1. On the basis of shape, bones are classified as long, short, flat, irregular, or sesamoid. Sesamoid bones develop in tendons or ligaments.
2. Sutural bones are found within the sutures of some cranial bones.

BONE SURFACE MARKINGS (p. 188)

1. Surface markings are structural features visible on the surfaces of bones.
2. Each marking—whether a depression, an opening, or a process—is structured for a specific function, such as joint formation, muscle attachment, or passage of nerves and blood vessels (Table 7.2 on page 189).

SKULL (p. 189)

1. The skull consists of 22 cranial and facial bones.
2. The eight cranial bones are the frontal, parietal (2), temporal (2), occipital, sphenoid, and ethmoid.
3. The 14 facial bones are the nasal (2), maxillae (2), zygomatic (2), lacrimal (2), palatine (2), inferior nasal conchae (2), vomer, and mandible.
4. The nasal septum consists of the vomer, perpendicular plate of the ethmoid, and septal cartilage. It divides the nasal cavity into left and right sides.
5. Seven skull bones form each of the orbits (eye sockets).
6. The foramina of the skull bones provide passages for nerves and blood vessels (Table 7.3 on page 201).
7. Sutures are immovable joints that connect most bones of the skull.

Examples are the coronal, sagittal, lambdoid, and squamous sutures.

8. Paranasal sinuses are cavities in bones of the skull that communicate with the nasal cavity. They are lined by mucous membranes. The cranial bones containing paranasal sinuses are the frontal, sphenoid, ethmoid, and maxillae.
9. Fontanels are fibrous, connective tissue membrane-filled spaces between the cranial bones of fetuses and infants. The major fontanels are the anterior, posterior, anterolaterals, and posterolaterals. After birth, they fill in with bone and become sutures.

HYOID BONE (p. 202)

1. The hyoid bone is a U-shaped bone that does not articulate with any other bone.
2. It supports the tongue and provides attachment for some tongue muscles and for some muscles of the pharynx and neck.

VERTEBRAL COLUMN (p. 203)

1. The vertebral column, sternum, and ribs constitute the skeleton of the trunk of the body.
2. The 26 bones of the adult vertebral column are the cervical vertebrae (7), the thoracic vertebrae (12), the lumbar vertebrae (5), the sacrum (5 fused vertebrae), and the coccyx (usually 4 fused vertebrae).
3. The vertebral column contains normal curves (cervical, thoracic, lumbar, and sacral) that give strength, support, and balance.
4. The vertebrae are similar in structure, each usually consisting of a body, vertebral arch, and seven processes. Vertebrae in the different regions of the column vary in size, shape, and detail.

THORAX (p. 211)

1. The thoracic skeleton consists of the sternum, ribs, costal cartilages, and thoracic vertebrae.
2. The thoracic cage protects vital organs in the chest area and upper abdomen.

Q SELF-QUIZ QUESTIONS

Fill in the blanks in the following statements.

1. Membrane-filled spaces between cranial bones that enable the fetal skull to change size and shape during passage through the birth canal are called _____.
2. The ribs articulate posteriorly with the _____ vertebrae.
3. _____ separate adjacent vertebrae, from the second cervical vertebra to the sacrum.
4. The hypophyseal fossa of the sella turcica of the sphenoid bone contains the _____.

Indicate whether the following statements are true or false.

5. The axial skeleton consists of the bones that lie along the axis: skull bones, ear bones, hyoid bone, ribs, breastbone, and pelvic and pectoral girdles.
6. The first cervical vertebra is the axis, and the second is the atlas.

Choose the one best answer to the following questions.

7. Which of the following are true for the cranial and/or facial bones? (1) Their inner surfaces attach to membranes that stabilize positions of the brain, blood vessels, and nerves. (2) Their outer surfaces provide areas of attachment for muscles that move various parts of the head. (3) They protect and provide support for the entrances to the digestive, respiratory, and integumentary systems. (4) They provide attachment for the muscles of facial expression. (5) They protect and support the delicate special sense organs. (a) 1, 2, 4, and 5, (b) 2, 3, 4, and 5, (c) 1, 3, 4, and 5, (d) 1, 2, 3, and 5, (e) 1, 2, 3, 4, and 5.
8. In which of the following bones are paranasal sinuses *not* found? (a) frontal bone, (b) sphenoid bone, (c) lacrimal bones, (d) ethmoid bone, (e) maxillae.
9. Which of the following are functions of the vertebral curves? (1) protection of the heart and lungs, (2) increase in strength, (3) maintenance of balance in the upright position, (4) absorption of shock during walking, (5) protection of vertebrae against fracture. (a) 1, 2, 3, and 4, (b) 2, 3, 4, and 5, (c) 1, 3, 4, and 5, (d) 1, 2, 3, and 5, (e) 1, 2, 3, 4, and 5.

10. The partition that divides the nasal cavity into right and left sides in the (a) pterygoid process, (b) petrous portion of the temporal bone, (c) hypophyseal fossa, (d) hard palate, (e) nasal septum.

11. The suture that unites the two parietal bones is the (a) coronal suture, (b) lambdoid suture, (c) squamous suture, (d) sagittal suture, (e) frontal suture.

12. Match the following.
___ (a) supraorbital foramen
___ (b) temporomandibular joint
___ (c) external auditory meatus
___ (d) foramen magnum
___ (e) optic foramen
___ (f) cribriform plate
___ (g) palatine process
___ (h) ramus, body, and condylar process
___ (i) transverse foramen, bifid processes
___ (j) dens
___ (k) promontory
___ (l) costal cartilages
___ (m) xiphoid process

(1) temporal bone
(2) sphenoid bone
(3) cervical vertebrae
(4) ethmoid bone
(5) articulation of mandibular fossa and articular tubercle of the temporal bone to the mandible
(6) occipital bone
(7) frontal bone
(8) maxillae
(9) mandible
(10) axis
(11) sacrum
(12) sternum
(13) ribs

13. Match the following (answers may be used more than once).
___ (a) bones that have greater length than width and consist of a shaft and a variable number of extremities
___ (b) cube-shaped bones that are nearly equal in length and width
___ (c) bones that develop in certain tendons where there is considerable friction, tension, and physical stress
___ (d) small bones located within joints between certain cranial bones
___ (e) thin bones composed of two nearly parallel plates of compact bone enclosing a layer of spongy bone
___ (f) bones with complex shapes, including the vertebrae and some facial bones
___ (g) patellae are examples
___ (h) bones that provide considerable protection and extensive areas for muscle attachment
___ (i) include femur, tibia, fibula, humerus, ulna, and radius
___ (j) include cranial bones, sternum, and ribs
___ (k) include almost all of the carpal (wrist) and tarsal (ankle) bones

(1) irregular bones
(2) long bones
(3) short bones
(4) flat bones
(5) sesamoid bones
(6) sutural bones

14. Match the following.
___ (a) prominent ridge or elongated projection
___ (b) tubelike opening
___ (c) large round protuberance used as attachment point for tendons or ligaments
___ (d) smooth, flat articular surface
___ (e) sharp, slender projection
___ (f) hole for passage of blood vessels, nerves or ligaments
___ (g) large, rounded, rough projection

(1) foramen
(2) tuberosity
(3) spinous process
(4) crest
(5) facet
(6) condyle
(7) meatus

15. Match the following.
___ (a) forms the forehead
___ (b) form the inferior lateral aspects of the cranium and part of the cranial floor, contain zygomatic process and mastoid process
___ (c) forms part of the anterior portion of the cranial floor, medial wall of the orbits, superior portions of nasal septum, and most of the sidewalls of the nasal cavity
___ (d) form the prominence of the cheek and part of the lateral wall and floor of each orbit
___ (e) the largest, strongest facial bone; is the only movable skull bone
___ (f) a roughly triangular bone on the floor of the nasal cavity; one of the components of the nasal septum
___ (g) form greater portion of the sides and roof of the cranial cavity
___ (h) forms the posterior part and most of the base of the cranium; contains the foramen magnum
___ (i) called the keystone of the cranial floor; contains the sella turcica, optic foramen, and pterygoid processes
___ (j) form the bridge of the nose
___ (k) the smallest bones of the face; contain a vertical groove that houses a structure that gathers tears and passes them into the nasal cavity
___ (l) does not articulate with any other bone
___ (m) unite to form the upper jawbone and articulate with every bone of the face except the lower jawbone
___ (n) form the posterior part of the hard palate, part of the floor and lateral wall of the nasal cavity, and a small portion of the floors of the orbits
___ (o) scroll-like bones that form a part of the lateral walls of the nasal cavity; functions in the turbulent circulation and filtration of air

(1) temporal bones
(2) parietal bones
(3) frontal bone
(4) occipital bone
(5) sphenoid bone
(6) ethmoid bone
(7) nasal bones
(8) maxillae
(9) zygomatic bones
(10) lacrimal bones
(11) palatine bones
(12) vomer
(13) mandible
(14) inferior nasal conchae
(15) hyoid bone

CRITICAL THINKING QUESTIONS

1. While investigating her new baby brother, 4-year-old Latisha found a soft spot on the baby's skull and announced that the baby needed to go back because "it's not finished yet." Explain the presence of soft spots in the infant's skull.
   HINT *An infant should be "soft in the head," even if an adult should not.*

2. Thirty-five-year-old Barbara (old enough to know better) was bouncing down the stairs in her stocking feet when she slipped and landed hard on her buttocks. She felt a pain sharp enough to bring tears to her eyes, and she needed to sit very gingerly on the way to and at the emergency room. The attending physician said that she had broken a bone but that she wouldn't be given a cast. What bone do you think she broke?
   HINT *Your buttocks usually keep this bone from being banged when you sit.*

3. The advertisement reads "New Posture Perfect Mattress! Keeps spine perfectly straight—just like when you were born! A straight spine equals a great sleep!" Would you buy a mattress from this company? Explain.
   HINT *Take a sideways view of the vertebral column.*

ANSWERS TO FIGURE QUESTIONS

7.1 Axial skeleton: skull and vertebral column. Appendicular skeleton: clavicle, shoulder girdle, humerus, pelvic girdle, and femur.

7.2 Flat bones protect and provide a large surface area for muscle attachment.

7.3 The frontal, parietal, sphenoid, ethmoid, and temporal bones are cranial bones.

7.4 Squamous suture: parietal and temporal bones. Lambdoid suture: parietal and occipital bones. Coronal suture: parietal and frontal bones.

7.5 The temporal bone articulates with the parietal, sphenoid, zygomatic, and occipital bones.

7.6 The parietal bones form the posterior, lateral portion of the cranium.

7.7 The medulla oblongata of the brain connects with the spinal cord in the foramen magnum.

7.8 Crista galli of ethmoid bone, frontal, parietal, temporal, occipital, temporal, parietal, frontal, crista galli of ethmoid bone.

7.9 The perpendicular plate of the ethmoid bone forms the superior part of the nasal septum, and the lateral masses compose most of the medial walls of the orbits.

7.10 The mandible is the only movable skull bone, other than the auditory ossicles.

7.11 The nasal septum divides the nasal cavity into right and left sides.

7.12 Bones forming the orbit are the frontal, sphenoid, zygomatic, maxilla, lacrimal, ethmoid, and palatine.

7.13 The paranasal sinuses produce mucus and serve as resonating chambers for vocalization.

7.14 Four different skull bones border the anterolateral fontanel.

7.15 The hyoid bone does not articulate with any other bone.

7.16 The thoracic and sacral curves are concave.

7.17 The vertebral foramina enclose the spinal cord, whereas the intervertebral foramina provide spaces for spinal nerves to exit the vertebral column.

7.18 The atlas moving on the axis permits movement of the head to signify "no."

7.19 The facets and demifacets on the bodies of the thoracic vertebrae articulate with the facets on the head of the ribs, and the facets on the transverse processes of these vertebrae articulate with the tubercles of the ribs.

7.20 The lumbar vertebrae are stout because the amount of body weight supported by vertebrae increases toward the inferior end of the vertebral column.

7.21 There are four pairs of sacral foramina, for a total of eight. Each anterior sacral foramen joins a posterior sacral foramen at the intervertebral foramen. Nerves and blood vessels pass through these tunnels in the bone.

7.22 True ribs: pairs 1–7; false ribs: pairs 8–12.

7.23 The facet on the head of a rib fits into a facet on the body of a vertebra, and the articular part of the tubercle of a rib articulates with the facet of the transverse process of a vertebra.

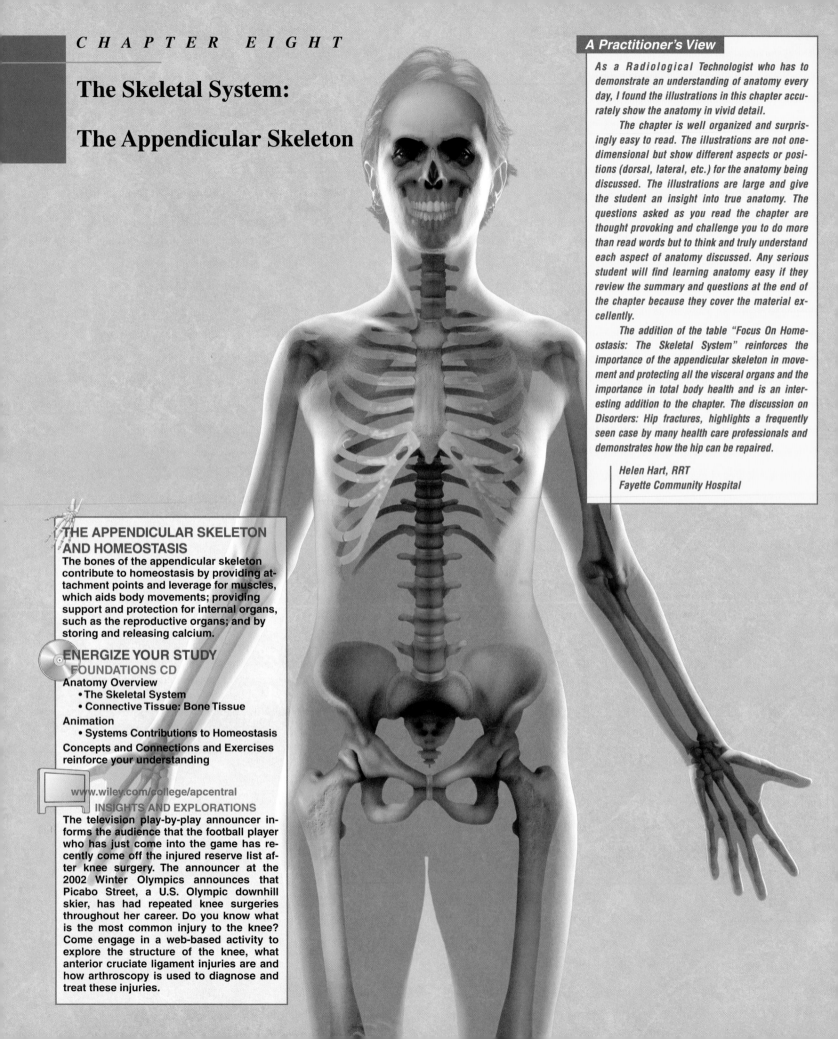

The Skeletal System:

The Appendicular Skeleton

THE APPENDICULAR SKELETON AND HOMEOSTASIS

The bones of the appendicular skeleton contribute to homeostasis by providing attachment points and leverage for muscles, which aids body movements; providing support and protection for internal organs, such as the reproductive organs; and by storing and releasing calcium.

ENERGIZE YOUR STUDY
FOUNDATIONS CD
Anatomy Overview
   • The Skeletal System
   • Connective Tissue: Bone Tissue
Animation
   • Systems Contributions to Homeostasis
Concepts and Connections and Exercises reinforce your understanding

www.wiley.com/college/apcentral

INSIGHTS AND EXPLORATIONS
The television play-by-play announcer informs the audience that the football player who has just come into the game has recently come off the injured reserve list after knee surgery. The announcer at the 2002 Winter Olympics announces that Picabo Street, a U.S. Olympic downhill skier, has had repeated knee surgeries throughout her career. Do you know what is the most common injury to the knee? Come engage in a web-based activity to explore the structure of the knee, what anterior cruciate ligament injuries are and how arthroscopy is used to diagnose and treat these injuries.

The two main divisions of the skeletal system are the axial skeleton and the appendicular skeleton. Whereas the axial skeleton serves mainly to protect and support internal organs, the appendicular skeleton, the focus of this chapter, mainly facilitates movement. The appendicular skeleton includes bones that make up the upper and lower limbs as well as bones, called girdles, that attach the limbs to the axial skeleton. As you study this chapter, you will see how the bones of the appendicular skeleton act with each other and with skeletal muscles to bring about a variety of movements.

PECTORAL (SHOULDER) GIRDLE

OBJECTIVE

• **Identify the bones of the pectoral (shoulder) girdle and their principal markings.**

The **pectoral** (PEK-tō-ral) or **shoulder girdles** attach the bones of the upper limbs to the axial skeleton (Figure 8.1). Each of the two pectoral girdles consists of two bones, the *clavicle* on the anterior side and the *scapula* on the posterior side. The clavicle articulates with the manubrium of the sternum at the *sternoclavicular joint*. The scapula articulates with the clavicle at the *acromioclavicular joint* and with the humerus at the *gleno-humeral (shoulder) joint*. The pectoral girdles do not articulate with the vertebral column. They are held in position by complex muscle attachments.

Clavicle

Each slender, S-shaped **clavicle** (KLAV-i-kul = key) or *collarbone* lies horizontally in the superior and anterior part of the thorax superior to the first rib (Figure 8.2). The medial half of the clavicle is convex anteriorly, whereas the lateral half is concave anteriorly. The medial end of the clavicle, the *sternal extremity,* is rounded and articulates with the manubrium of the sternum to form the *sternoclavicular joint.* The broad, flat, lateral end, the

Figure 8.1   Right pectoral (shoulder) girdle.

The clavicle is the anterior bone of the pectoral girdle, and the scapula is the posterior bone.

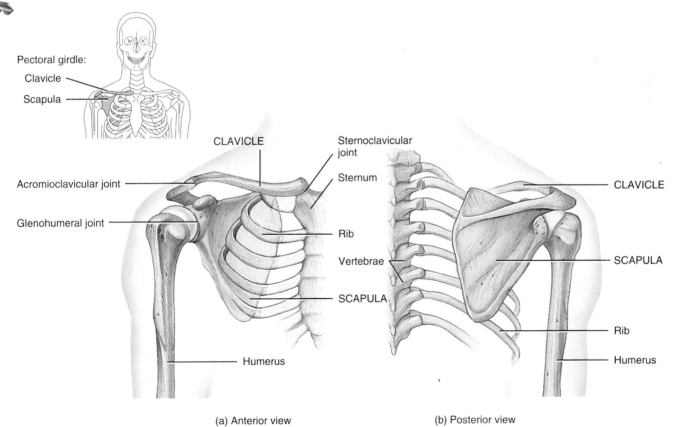

Pectoral girdle:
Clavicle
Scapula

CLAVICLE
Acromioclavicular joint
Glenohumeral joint
Humerus

Sternoclavicular joint
Sternum
Rib
Vertebrae
SCAPULA

(a) Anterior view

CLAVICLE
SCAPULA
Rib
Humerus

(b) Posterior view

What is the function of the pectoral girdles?

219

Figure 8.2   Right clavicle.

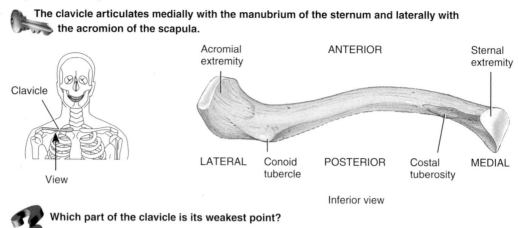

The clavicle articulates medially with the manubrium of the sternum and laterally with the acromion of the scapula.

Inferior view

Which part of the clavicle is its weakest point?

acromial extremity (a-KRŌ-mē-al), articulates with the acromion of the scapula. This joint is called the *acromioclavicular joint.* (See Figure 8.1.) The *conoid tubercle* (KŌ-noyd = cone-like) on the inferior surface of the lateral end of the bone is a point of attachment for the conoid ligament. The *costal tuberosity* on the inferior surface of the medial end is a point of attachment for the costoclavicular ligament.

Fractured Clavicle

The clavicle transmits mechanical force from the upper limb to the trunk. If the force transmitted to the clavicle is excessive, as in falling on one's outstretched arm, a **fractured clavicle** may result. The clavicle is one of the most frequently broken bones in the body. Because the junction of the clavicle's two curves is its weakest point, the clavicular midregion is the most frequent fracture site. As a result of automobile accidents involving people who wear shoulder harness seatbelts, compression of the clavicle, even in the absence of fracture, often causes damage to the median nerve, which lies between the clavicle and the second rib. ■

Scapula

Each **scapula** (SCAP-ū-la; plural is *scapulae*), or *shoulder blade,* is a large, triangular, flat bone situated in the superior part of the posterior thorax between the levels of the second and seventh ribs (Figure 8.3). The medial borders of the scapulae lie about 5 cm (2 in.) from the vertebral column.

A sharp ridge, the *spine,* runs diagonally across the posterior surface of the flattened, triangular *body* of the scapula. The lat-

eral end of the spine projects as a flattened, expanded process called the *acromion* (a-KRŌ-mē-on; *acrom-* = topmost), easily felt as the high point of the shoulder. Tailors measure the length of the upper limb from the acromion. The acromion articulates with the acromial extremity of the clavicle to form the *acromioclavicular joint.* Inferior to the acromion is a shallow depression called the *glenoid cavity.* This cavity accepts the head of the humerus (arm bone) to form the *glenohumeral joint* (see Figure 8.1).

The thin edge of the bone near the vertebral column is the *medial (vertebral) border.* The thick edge closer to the arm is the *lateral (axillary) border.* The medial and lateral borders join at the *inferior angle.* The superior edge of the scapula, called the *superior border,* joins the vertebral border at the *superior angle.* The *scapular notch* is a prominent indentation along the superior border through which the suprascapular nerve passes.

At the lateral end of the superior border of the scapula is a projection of the anterior surface called the *coracoid process* (KOR-a-koyd = like a crow's beak), to which the tendons of muscles attach. Superior and inferior to the spine are two fossae: the *supraspinous fossa* (sū-pra-SPĪ-nus) and the *infraspinous fossa* (in-fra-SPĪ-nus), respectively. Both serve as surfaces of attachment for the tendons of shoulder muscles called the supraspinatus and infraspinatus muscles. On the anterior surface is a slightly hollowed-out area called the *subscapular fossa,* also a surface of attachment for the tendons of shoulder muscles.

▶ C H E C K P O I N T

1.   Which bones or parts of bones of the pectoral girdle form the sternoclavicular, acromioclavicular, and glenohumeral joints?

Figure 8.3   Right scapula (shoulder blade).   (See Tortora, *A Photographic Atlas of the Human Body,* Figure 3.22.)

 The glenoid cavity of the scapula articulates with the head of the humerus to form the shoulder joint.

(a) Anterior view

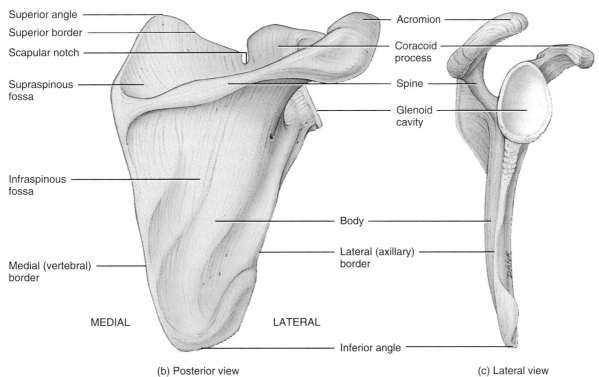

(b) Posterior view

(c) Lateral view

Which part of the scapula forms the high point of the shoulder?

UPPER LIMB

• **Identify the bones of the upper limb and their principal markings.**

Each **upper limb (upper extremity)** has 30 bones in three locations—(1) the humerus in the arm; (2) the ulna and radius in the forearm; and (3) the 8 carpals in the carpus (wrist), the 5 metacarpals in the metacarpus (palm), and the 14 phalanges (bones of the digits) in the hand (Figure 8.4).

Humerus

The **humerus** (HŪ-mer-us), or arm bone, is the longest and largest bone of the upper limb (Figure 8.5). It articulates proximally with the scapula and distally at the elbow with both the ulna and the radius.

The proximal end of the humerus features a rounded *head* that articulates with the glenoid cavity of the scapula to form the *glenohumeral joint.* Distal to the head is the *anatomical neck,* the site of the epiphyseal line, which is visible as an oblique groove. The *greater tubercle* is a lateral projection distal to the anatomical neck. It is the most laterally palpable bony landmark of the shoulder region. The *lesser tubercle* projects anteriorly. Between both tubercles runs an *intertubercular sulcus.* The *surgical neck* is a constriction in the humerus just distal to the tubercles, where the head tapers to the shaft; it is so named because fractures often occur here.

The *body (shaft)* of the humerus is roughly cylindrical at its proximal end, but it gradually becomes triangular until it is flattened and broad at its distal end. Laterally, at the middle portion of the shaft, there is a roughened, V-shaped area called the *deltoid tuberosity.* This area serves as a point of attachment for the tendons of the deltoid muscle.

Several prominent features are evident at the distal end of the humerus. The *capitulum* (ka-PIT-ū-lum; *capit-* = head) is a rounded knob on the lateral aspect of the bone that articulates with the head of the radius. The *radial fossa* is an anterior depression that receives the head of the radius when the forearm is flexed (bent). The *trochlea* (TRŌK-lē-a), located medial to the capitulum, is a spool-shaped surface that articulates with the ulna. The *coronoid fossa* (KŌR-o-noyd = crown-shaped) is an anterior depression that receives the coronoid process of the ulna when the forearm is flexed. The *olecranon fossa* (ō-LEK-ra-non = elbow) is a posterior depression that receives the olecranon of the ulna when the forearm is extended (straightened). The *medial epicondyle* and *lateral epicondyle* are rough projections on either side of the distal end to which the tendons of most muscles of the forearm are attached. The ulnar nerve lies on the posterior surface of the medial epicondyle and may easily be palpated by rolling a finger over the skin surface above the medial epicondyle.

Figure 8.4 Right upper limb.

Each upper limb includes a humerus, ulna, radius, carpals, metacarpals, and phalanges.

Clavicle

Scapula

HUMERUS

ULNA

RADIUS

CARPALS

METACARPALS

PHALANGES

Anterior view

How many bones make up each upper limb?

Figure 8.5 **Right humerus in relation to the scapula, ulna, and radius.** (See Tortora, *A Photographic Atlas of the Human Body,* Figure 3.23.)

🔑 The humerus is the longest and largest bone of the upper limb.

(a) Anterior view   (b) Posterior view

Which parts of the humerus articulate with the radius at the elbow? With the ulna at the elbow?

Ulna and Radius

The **ulna** is located on the medial aspect (the little-finger side) of the forearm and is longer than the radius (Figure 8.6). At the proximal end of the ulna (Figure 8.6b) is the *olecranon,* which forms the prominence of the elbow. The *coronoid process* (Figure 8.6a) is an anterior projection that, together with the olecranon, receives the trochlea of the humerus. The *trochlear notch* is a large curved area between the olecranon and coronoid process that forms part of the elbow joint (see Figure 8.7b). Just inferior to the coronoid process is the *ulnar tuberosity.* The distal end of the ulna consists of a *head* that is separated from the wrist by a

fibrocartilage disc. A *styloid process* is on the posterior side of the distal end.

The **radius** is located on the lateral aspect (thumb side) of the forearm (see Figure 8.6). The proximal end of the radius has a disc-shaped *head* that articulates with the capitulum of the humerus and the radial notch of the ulna. Inferior to the head is the constricted *neck.* A roughened area inferior to the neck on the medial side, called the *radial tuberosity,* is a point of attachment for the tendons of the biceps brachii muscle. The shaft of the radius widens distally to form a *styloid process* on the lateral side. Fracture of the distal end of the radius is the most common fracture in adults older than 50 years.

Figure 8.6 **Right ulna and radius in relation to the humerus and carpals.** (See Tortora, *A Photographic Atlas of the Human Body,* Figure 3.24.)

In the forearm, the longer ulna is on the medial side, whereas the shorter radius is on the lateral side.

(a) Anterior view

(b) Posterior view

What part of the ulna is called the "elbow"?

The ulna and radius articulate with the humerus at the *elbow joint*. The articulation occurs in two places: where the head of the radius articulates with the capitulum of the humerus (Figure 8.7a), and where the trochlear notch of the ulna receives the trochlea of the humerus (Figure 8.7b).

The ulna and the radius connect with one another at three sites. First, a broad, flat, fibrous connective tissue called the *interosseous membrane* (in'-ter-OS-ē-us; *inter-* = between, *osse-* = bone) joins the shafts of the two bones. This membrane also provides a site of attachment for some tendons of deep

Figure 8.7 **Articulations formed by the ulna and radius.** (a) Elbow joint. (b) Joint surfaces at proximal end of the ulna. (c) Joint surfaces at distal ends of radius and ulna. The interosseous membrane also connects the ulna and the radius.

🔑 The elbow joint is formed by two articulations: (1) the trochlear notch of the ulna with the trochlea of the humerus and (2) the head of the radius with the capitulum of the humerus.

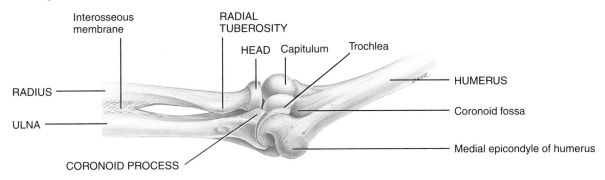

(a) Medial view in relation to humerus

(b) Lateral view of proximal end of ulna

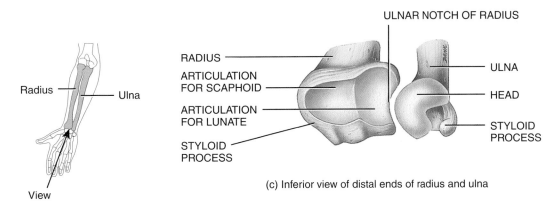

(c) Inferior view of distal ends of radius and ulna

❓ How many joints are formed between the radius and ulna? What are they called?

skeletal muscles of the forearm. The ulna and radius also articulate at their proximal and distal ends. Proximally, the head of the radius articulates with the ulna's *radial notch,* a depression that is lateral and inferior to the trochlear notch (Figure 8.7b). This articulation is the *proximal radioulnar joint.* Distally, the head of the ulna articulates with the *ulnar notch* of the radius (Figure 8.7c). This articulation is the *distal radioulnar joint.* Finally, the distal end of the radius articulates with three bones of the wrist—the lunate, the scaphoid, and the triquetrum—to form the *radiocarpal (wrist) joint.*

Carpals, Metacarpals, and Phalanges

The **carpus** (wrist) is the proximal part of the hand and includes eight small bones, the **carpal bones (carpals),** joined to one another by ligaments (Figure 8.8). Articulations between carpal bones are called *intercarpal joints*. The carpals are arranged in two transverse rows of four bones each. Their names reflect their shapes. The carpals in the proximal row, from lateral to medial, are the **scaphoid** (SKAF-oyd = boatlike), **lunate** (LOO-nāt = moon-shaped), **triquetrum** (trī-KWĒ-trum = three-cornered), and **pisiform** (PIS-i-form = pea-shaped). The carpals in the distal row, from lateral to medial, are the **trapezium** (tra-PĒ-zē-um = four-sided figure with no two sides parallel), **trapezoid** (TRAP-e-zoyd), **capitate** (KAP-i-tāt = head-shaped), and **hamate** (HAM-āt = hooked). The capitate is the largest carpal bone;

its rounded projection, the head, articulates with the lunate. The hamate is named for a large hook-shaped projection on its anterior surface. In about 70% of carpal fractures, only the scaphoid is broken. This is because the force of a fall on an outstretched hand is transmitted from the capitate through the scaphoid to the radius.

The concave space formed by the pisiform and hamate (on the ulnar side), and the scaphoid and trapezium (on the radial side), plus the *flexor retinaculum* (deep fascia) is the **carpal tunnel.** The long flexor tendons of the fingers and thumb and the median nerve pass through the carpal tunnel. Narrowing of the carpal tunnel may give rise to a condition called carpal tunnel syndrome (see page 356).

The **metacarpus** (*meta-* = beyond), or palm, is the intermediate region of the hand and consists of five bones called **metacarpal bones (metacarpals).** Each metacarpal bone con-

Figure 8.8   Right wrist and hand in relation to the ulna and radius.   (See Tortora, *A Photographic Atlas of the Human Body,* Figure 3.25.)

The skeleton of the hand consists of the proximal carpals, the intermediate metacarpals, and the distal phalanges.

(a) Anterior view

(b) Posterior view

Which is the most frequently fractured wrist bone?

sists of a proximal *base,* an intermediate *shaft,* and a distal *head* (Figure 8.8b). The metacarpals are numbered I to V (or 1–5), starting with the thumb, which is lateral. The bases articulate with the distal row of carpal bones to form the *carpometacarpal joints.* The heads articulate with the proximal phalanges to form the *metacarpophalangeal joints.* The heads of the metacarpals are commonly called "knuckles" and are readily visible in a clenched fist.

The **phalanges** (fa-LAN-jēz; *phalan-* = a battle line), or bones of the digits, make up the distal part of the hand. There are 14 phalanges in each hand and, like the metacarpals, the phalanges are numbered I to V (or 1–5), beginning with the thumb, which is lateral. A single bone of a digit is referred to as a **phalanx** (FĀ-lanks). Each phalanx consists of a proximal *base,* an intermediate *shaft,* and a distal *head.* There are two phalanges in the thumb (*pollex*), and three phalanges in each of the other four digits. In order from the thumb, the other four digits are commonly called the index finger, middle finger, ring finger, and little finger. The first row of phalanges, the *proximal row,* articulates with the metacarpal bones and second row of phalanges. The second row of phalanges, the *middle row,* articulates with the proximal row and the third row. The third row of phalanges, the *distal row,* articulates with the middle row. The thumb has no middle phalanx. Joints between phalanges are called *interphalangeal joints.*

▶ C H E C K P O I N T

2. Which bones form the upper limb, from proximal to distal?

3. What joints do the bones of the upper limb form?

PELVIC (HIP) GIRDLE

▶ O B J E C T I V E

- **Identify the bones of the pelvic girdle and their principal markings.**

The **pelvic (hip) girdle** consists of the two **hip bones,** also called **coxal bones** (KOK-sal; *cox-* = hip) (Figure 8.9). The hip bones unite anteriorly at a joint called the **pubic symphysis** (PŪ-bik SIM-fi-sis). They unite posteriorly with the sacrum at the *sacroiliac joints.* The complete ring composed of the hip bones, pubic symphysis, and sacrum forms a deep, basinlike structure called the **bony pelvis** (*pelv-* = basin). The plural is *pelves* (PEL-vēz) or *pelvises.* Functionally, the bony pelvis provides a strong and stable support for the vertebral column and pelvic organs. The pelvic girdle of the bony pelvis also accepts the bones of the lower limbs, connecting them to the axial skeleton.

Each of the two hip bones of a newborn consists of three bones separated by cartilage: a superior *ilium,* an inferior and anterior *pubis,* and an inferior and posterior *ischium* (IS-kē-um). Eventually, the three separate bones fuse together and become one bone (Figure 8.10a). Although the hip bones function as single bones, anatomists commonly discuss them as if they still consisted of three bones.

Ilium

The **ilium** (IL-ē-um = flank) is the largest of the three components of the hip bone (Figure 8.10b, c). A superior *ala* (= wing)

Figure 8.9 Bony pelvis.   Shown here is the female bony pelvis. (See Tortora, *A Photographic Atlas of the Human Body,* Figure 3.27.)

🔑 **The hip bones unite anteriorly at the pubic symphysis and posteriorly at the sacrum to form the bony pelvis.**

Pelvic (hip) girdle

Hip bone

Sacrum

Coccyx

Pubic symphysis

Sacroiliac joint

Sacral promontory

Pelvic brim

Acetabulum

Obturator foramen

Anterior view

 What are the functions of the bony pelvis?

Figure 8.10 **Right hip bone.** The lines of fusion of the ilium, ischium, and pubis depicted in (a) are not always visible in an adult. (See Tortora, *A Photographic Atlas of the Human Body,* Figure 3.26.)

🔑 **The acetabulum is the socket formed where the three parts of the hip bone converge.**

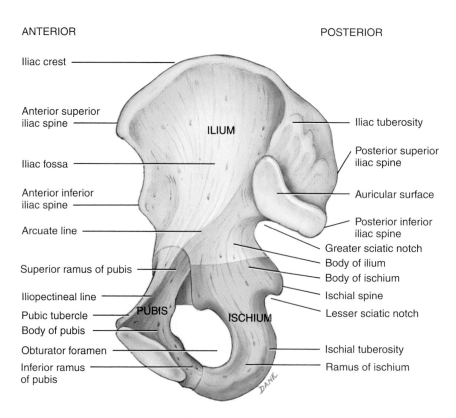

POSTERIOR

ANTERIOR

Gluteal lines
Anterior
Inferior
Posterior

ILIUM

Iliac crest

Anterior superior
iliac spine

Ala

Anterior inferior
iliac spine

Posterior
superior
iliac spine

Posterior
inferior
iliac spine

Body of ilium
Acetabulum

Greater sciatic notch

Acetabular notch

Body of ischium

Superior ramus
of pubis

Ischial spine

Pubic tubercle

Lesser sciatic notch

Pubic crest

Ischial tuberosity

PUBIS

Obturator foramen

Ramus of ischium

ISCHIUM

Inferior ramus
of pubic

(b) Detailed lateral view

SUPERIOR

ILIUM

ISCHIUM

PUBIS

POSTERIOR

ANTERIOR

**(a) Lateral view showing
parts of hip bone**

ANTERIOR

POSTERIOR

Iliac crest

Anterior superior
iliac spine

ILIUM

Iliac tuberosity

Iliac fossa

Posterior superior
iliac spine

Anterior inferior
iliac spine

Auricular surface

Arcuate line

Posterior inferior
iliac spine

Greater sciatic notch

Superior ramus of pubis

Body of ilium

Body of ischium

Iliopectineal line

Ischial spine

Pubic tubercle

PUBIS

ISCHIUM

Lesser sciatic notch

Body of pubis

Obturator foramen

Ischial tuberosity

Inferior ramus
of pubis

Ramus of ischium

(c) Detailed medial view

❓ **Which part of the hip bone articulates with the femur? With the sacrum?**

and an inferior *body,* which enters into the formation of the *acetabulum,* the socket for the head of the femur, comprise the ilium. Its superior border, the *iliac crest,* ends anteriorly in a blunt *anterior superior iliac spine.* Below this spine is the *anterior inferior iliac spine.* Posteriorly, the iliac crest ends in a sharp *posterior superior iliac spine.* Below this spine is the *posterior inferior iliac spine.* The spines serve as points of attachment for the tendons of the muscles of the trunk, hip, and thighs. Below the posterior inferior iliac spine is the *greater sciatic notch* (sī-AT-ik), through which the sciatic nerve, the longest nerve in the body, passes.

The medial surface of the ilium contains the *iliac fossa,* a concavity where the tendon of the iliacus muscle attaches. Posterior to this fossa are the *iliac tuberosity,* a point of attachment for the sacroiliac ligament, and the *auricular surface (auric-* = ear-shaped), which articulates with the sacrum to form the *sacroiliac joint* (see Figure 8.9). Projecting anteriorly and inferiorly from the auricular surface is a ridge called the *arcuate line* (AR-kū–at; *arc-* = bow).

The other conspicuous markings of the ilium are three arched lines on its lateral surface called the *posterior gluteal line (glut-* = buttock), the *anterior gluteal line,* and the *inferior gluteal line.* The tendons of the gluteal muscles attach to the ilium between these lines.

Ischium

The **ischium** (IS-kē-um = hip) is the inferior, posterior portion of the hip bone (see Figure 8.10b, c). A superior *body* and an inferior *ramus (ram-* = branch; plural is *rami),* which joins the pubis, comprise the ischium. Its features include the prominent *ischial spine,* a *lesser sciatic notch* below the spine, and a rough and thickened *ischial tuberosity.* This prominent tuberosity may hurt someone's thigh when you sit on their lap. Together, the ramus and the pubis surround the *obturator foramen* (OB-too-rā-tōr; *obtur-* = closed up), the largest foramen in the skeleton. The foramen is so-named because, even though blood vessels and nerves pass through it, it is nearly completely closed by the fibrous *obturator membrane.*

Pubis

The **pubis** or **os pubis,** meaning pubic bone, is the anterior and inferior part of the hip bone (Figure 8.10b, c). A *superior ramus,* an *inferior ramus,* and a *body* between the rami comprise the pubis. The anterior border of the body is the *pubic crest,* and at its lateral end is a projection, the *pubic tubercle.* This tubercle is the beginning of a raised line, the *iliopectineal line* (il-ē-ō-pek-TIN-ē-al), which extends superiorly and laterally along the superior ramus to merge with the arcuate line of the ilium. These lines, as you will see shortly, are important landmarks for distinguishing the superior and inferior portions of the bony pelvis.

The *pubic symphysis* is the joint between the two hip bones (see Figure 8.9). It consists of a disc of fibrocartilage. The inferior rami of the two pubic bones converge to form the pubic arch. In the later stages of pregnancy, the hormone relaxin (produced by the ovaries and placenta) increases the flexibility of the pubic symphysis to ease delivery of the baby. Weakening of the joint, together with an already compromised center of gravity due to an enlarged uterus, also alters the gait during pregnancy.

The *acetabulum* (as-e-TAB-ū-lum = vinegar cup) is a deep fossa formed by the ilium, ischium, and pubis. It functions as the socket that accepts the rounded head of the femur. Together, the acetabulum and the femoral head form the *hip (coxal) joint.* On the inferior side of the acetabulum is a deep indentation, the *acetabular notch.* It forms a foramen through which blood vessels and nerves pass, and it serves as a point of attachment for ligaments of the femur (for example, the ligament of the head of the femur).

True and False Pelves

The bony pelvis is divided into superior and inferior portions by a boundary called the *pelvic brim* (Figure 8.11a). You can trace the pelvic brim by following the landmarks around parts of the hip bones. Beginning posteriorly at the *sacral promontory* of the sacrum, trace laterally and inferiorly along the *arcuate lines* of the ilium of the hip bones. Continue inferiorly along the *iliopectineal lines* of the pubis. Finally, trace anteriorly to the superior portion of the pubic symphysis. Together, these points form an oblique plane that is higher in the back than in the front. The circumference of this plane is the pelvic brim.

The portion of the bony pelvis superior to the pelvic brim is the **false (greater) pelvis** (Figure 8.11b). It is bordered by the lumbar vertebrae posteriorly, the upper portions of the hip bones laterally, and the abdominal wall anteriorly. The space enclosed by the false pelvis is part of the abdomen. It does not contain pelvic organs, except for the urinary bladder (when it is full) and the uterus during pregnancy.

The **true (lesser) pelvis** is the part of the bony pelvis inferior to the pelvic brim (Figure 8.11b). It is bounded by the sacrum and coccyx posteriorly, inferior portions of the ilium and ischium laterally, and the pubic bones anteriorly. The true pelvis surrounds the pelvic cavity (see Figure 1.9 on page 17). The superior opening of the true pelvis is the pelvic brim, also called the *pelvic inlet;* the inferior opening of the true pelvis is the *pelvic outlet.* The *pelvic axis* is an imaginary line that curves through the true pelvis and joins the central points of the planes of the pelvic inlet and outlet. During childbirth, the pelvic axis is the route taken by the baby's head as it descends through the pelvis.

▶ CHECKPOINT

4. Where are the true and false pelves?

Figure 8.11   True and false pelves.   Shown here is the female pelvis. For simplicity, the landmarks of the pelvic brim are shown only on the left side of the body, and the outline of the pelvic brim is shown only on the right side. The entire pelvic brim is shown in Figure 8.9.

🔑 **The pelvic brim separates the true and false pelves.**

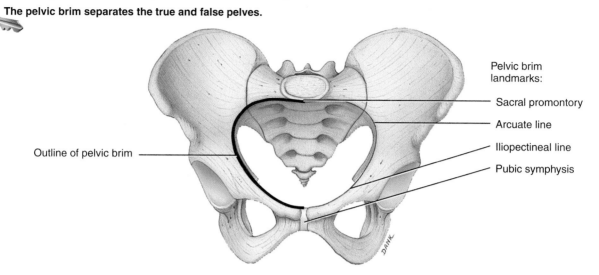

(a) Anterior view of borders of pelvic brim

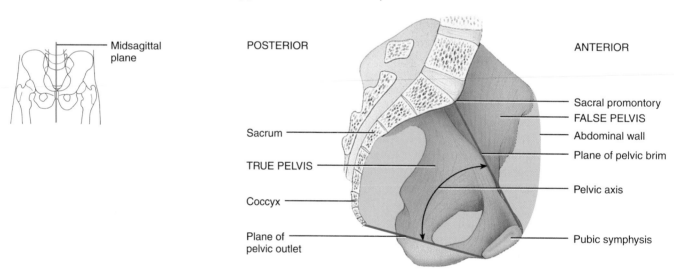

(b) Midsagittal section indicating locations of true and false pelves

❓ **What is the significance of the pelvic axis?**

COMPARISON OF FEMALE AND MALE PELVES

OBJECTIVE

• **Compare the principal structural differences between female and male pelves.**

In the following discussion, it is assumed that the male and female are comparable in age and physical stature. Generally, the bones of males are larger and heavier than those of females and possess larger surface markings. Gender-related differences in the features of skeletal bones are readily apparent when comparing the female and male pelves. Most of the structural differences in the pelves are adaptations to the requirements of pregnancy and childbirth. The female's pelvis is wider and shallower than the male's (Table 8.1). As a result, there is more space in the true pelvis of the female, especially in the pelvic inlet and pelvic outlet, which accommodate the passage of the infant's head at birth. Table 8.1 lists and illustrates the structural differences between the female and male pelves.

▶ CHECKPOINT

5.   Why are structural differences between female and male pelves important?

Table 8.1	Comparison of Female and Male Pelves	
Point of Comparison	**Female**	**Male**
General structure	Light and thin.	Heavy and thick.
False (greater) pelvis	Shallow.	Deep.
Pelvic brim (inlet)	Larger and more oval.	Smaller and heart-shaped.
Acetabulum	Small and faces anteriorly.	Large and faces laterally.
Obturator foramen	Oval.	Round.
Pubic arch	Greater than 90° angle.	Less than 90° angle.

Iliac crest	Less curved.	More curved.
Ilium	Less vertical.	More vertical.
Greater sciatic notch	Wide.	Narrow.
Coccyx	More movable and more curved anteriorly.	Less movable and less curved anteriorly.
Sacrum	Shorter, wider (see anterior views), and more curved anteriorly.	Longer, narrower (see anterior views), and less curved anteriorly.

Right lateral views

(continues)

Table 8.1	Comparison of Female and Male Pelves *(continued)*	
Point of Comparison	**Female**	**Male**
Pelvic outlet	Wider.	Narrower.
Ischial tuberosity	Shorter, farther apart, and more laterally projecting.	Longer, closer together, and more medially projecting.

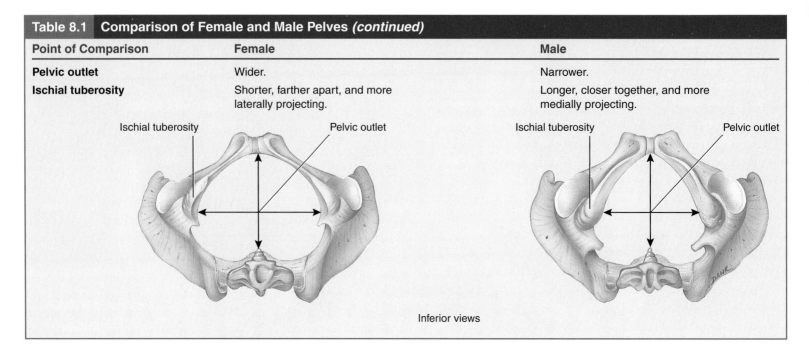

Inferior views

COMPARISON OF PECTORAL AND PELVIC GIRDLES

▶ **OBJECTIVE**

• **Describe the differences in the pectoral and pelvic girdles.**

Now that we've studied the structures of the pectoral and pelvic girdles, we can note some of their significant differences. The pectoral girdle does not directly articulate with the vertebral column, whereas the pelvic girdle directly articulates with the vertebral column via the sacroiliac joint. The sockets (glenoid fossae) for the upper limbs in the pectoral girdle are shallow and maximize movement, whereas the sockets (acetabula) for the lower limbs in the pelvic girdle are deep and allow less movement. Overall, the structure of the pectoral girdle offers more mobility than strength, whereas that of the pelvic girdle offers more strength and less mobility.

LOWER LIMB

▶ **OBJECTIVE**

• **Identify the bones of the lower limb and their principal markings.**

Each **lower limb (lower extremity)** has 30 bones in four locations—(1) the femur in the thigh; (2) the patella (kneecap); (3) the tibia and fibula in the leg; (4) and the 7 tarsals in the tarsus (ankle), the 5 metatarsals in the metatarsus, and the 14 phalanges (bones of the digits) in the foot (Figure 8.12).

Femur

The **femur,** or thigh bone, is the longest, heaviest, and strongest bone in the body (Figure 8.13 on page 234). Its proximal end articulates with the acetabulum of the hip bone. Its distal end articulates with the tibia and patella. The *body (shaft)* of the femur angles medially and, as a result, the knee joints are closer to the midline. The angle of convergence is greater in females because the female pelvis is broader.

The proximal end of the femur consists of a rounded *head* that articulates with the acetabulum of the hip bone to form the *hip (coxal) joint.* The head contains a small centered depression (pit) called the *fovea capitis* (FŌ-vē-a CAP-i-tis; *fovea* = pit; *capitis* = of the head). The ligament of the head of the femur connects the fovea capitis of the femur to the acetabulum of the hip bone. The *neck* of the femur is a constricted region distal to the head. The *greater trochanter* (trō-KAN-ter) and *lesser trochanter* are projections that serve as points of attachment for the tendons of some of the thigh and buttock muscles. The greater trochanter is the prominence felt and seen anterior to the hollow on the side of the hip. It is a landmark commonly used to locate the site for intramuscular injections into the lateral surface of the thigh. The lesser trochanter is inferior and medial to the greater trochanter. Between the anterior surface of the trochanters is a narrow *intertrochanteric line* (Figure 8.13a).

Figure 8.12 Right lower limb.

Each lower limb includes a femur, patella (kneecap), tibia, fibula, tarsals (ankle bones), metatarsals, and phalanges (bones of the digits).

— Hip bone

— Sacrum

— FEMUR

— PATELLA

— TIBIA

— FIBULA

— TARSALS

— METATARSALS

— PHALANGES

Anterior view

How many bones make up each lower limb?

Between the posterior surface of the trochanters is the *intertrochanteric crest* (Figure 8.13b).

Inferior to the intertrochanteric crest on the posterior surface of the body of the femur is a vertical ridge called the *gluteal tuberosity*. It blends into another vertical ridge called the *linea aspera* (LIN-ē-a AS-per-a; *asper-* = rough). Both ridges serve as attachment points for the tendons of several thigh muscles.

The distal end of the femur is expanded and includes the *medial condyle* and the *lateral condyle*. These articulate with the medial and lateral condyles of the tibia. Superior to the condyles are the *medial epicondyle* and the *lateral epicondyle*. A depressed area between the condyles on the posterior surface is called the *intercondylar fossa* (in-ter-KON-di-lar). The *patellar surface* is located between the condyles on the anterior surface.

Patella

The **patella** (= little dish), or kneecap, is a small, triangular bone located anterior to the knee joint (Figure 8.14 on page 235). It is a sesamoid bone that develops in the tendon of the quadriceps femoris muscle. The broad superior end of the patella is called the *base.* The pointed inferior end is the *apex.* The posterior surface contains two *articular facets,* one for the medial condyle and the other for the lateral condyle of the femur. The patellar ligament attaches the patella to the tibial tuberosity. The *patellofemoral joint,* between the posterior surface of the patella and the patellar surface of the femur, is the intermediate component of the *tibiofemoral (knee) joint.* The patella functions to increase the leverage of the tendon of the quadriceps femoris muscle, to maintain the position of the tendon when the knee is bent (flexed), and to protect the knee joint.

 Patellofemoral Stress Syndrome

Patellofemoral stress syndrome ("runner's knee") is one of the most common problems in runners. During normal flexion and extension of the knee, the patella tracks (glides) up and down in the groove between the femoral condyles. In patellofemoral stress syndrome, normal tracking does not occur. Instead, the patella also tracks laterally, and the increased pressure on the joint causes aching or tenderness around or under the patella. The pain typically occurs after a person has been sitting for a while, especially after exercise. Squatting or walking down stairs worsens it. One cause of runner's knee is constantly walking, running, or jogging on the same side of the road. Because roads slope down on the sides, the knee that is closer to the center of the road endures greater mechanical stress because it does not fully extend during a stride. Other predisposing factors include having knock-knees, running on hills, and running long distances. ∎

Figure 8.13   Right femur in relation to the hip bone, patella, tibia, and fibula.   (See Tortora, *A Photographic Atlas of the Human Body,* Figure 3.28.)

The acetabulum of the hip bone and head of the femur articulate to form the hip joint.

Femur

Hip bone

HEAD

GREATER TROCHANTER

NECK

INTERTROCHANTERIC LINE   CREST

LESSER TROCHANTER

GREATER TROCHANTER

GLUTEAL TUBEROSITY

BODY

FEMUR

LINEA ASPERA

LATERAL EPICONDYLE

MEDIAL EPICONDYLE

MEDIAL CONDYLE

LATERAL CONDYLE

Patella

LATERAL EPICONDYLE

INTERCONDYLAR FOSSA

LATERAL CONDYLE

Fibula

Tibia

Fibula

DANK

(a) Anterior view

(b) Posterior view

Tibia and Fibula

The **tibia,** or shin bone, is the larger, medial, weight-bearing bone of the leg (Figure 8.15 on page 236). The tibia articulates at its proximal end with the femur and fibula, and at its distal end with the fibula and the talus bone of the ankle. An interosseous membrane connects the tibia and fibula.

The proximal end of the tibia is expanded into a *lateral condyle* and a *medial condyle*. These articulate with the condyles of the femur to form the lateral and medial *tibiofemoral (knee)* *joints.* The inferior surface of the lateral condyle articulates with the head of the fibula. The slightly concave condyles are separated by an upward projection called the *intercondylar eminence* (Figure 8.15b). The *tibial tuberosity* on the anterior surface is a point of attachment for the patellar ligament. Inferior to and continuous with the tibial tuberosity is a sharp ridge that can be felt below the skin. This ridge is the *anterior border (crest)* or *shin.*

The medial surface of the distal end of the tibia forms the *medial malleolus* (mal-LĒ-ō-lus = hammer). This structure articulates with the talus bone of the ankle and forms the protru-

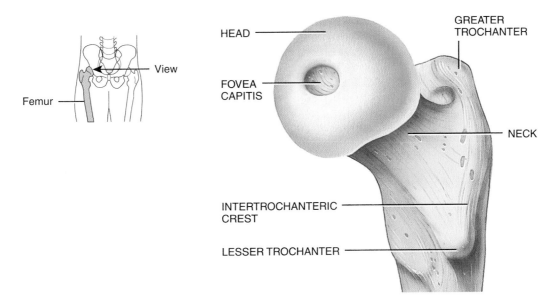

HEAD

GREATER TROCHANTER

FOVEA CAPITIS

NECK

View

Femur

INTERTROCHANTERIC CREST

LESSER TROCHANTER

(c) Medial view of proximal end of femur

 Why is the angle of convergence of the femurs greater in females than males?

sion on the medial surface of the ankle. The *fibular notch* (Figure 8.15c) articulates with the distal end of the fibula to form the *distal tibiofibular joint.*

The **fibula** is parallel and lateral to the tibia, but it is considerably smaller than the tibia. The proximal end, the *head* of the fibula, articulates with the inferior surface of the lateral condyle

of the tibia below the level of the knee joint to form the *proximal tibiofibular joint.* The distal end has a projection called the *lateral malleolus* that articulates with the talus bone of the ankle. The lateral malleolus is the prominence on the lateral surface of the ankle. As noted, the fibula also articulates with the tibia at the fibular notch.

Figure 8.14   Right patella.

The patella articulates with the lateral and medial condyles of the femur.

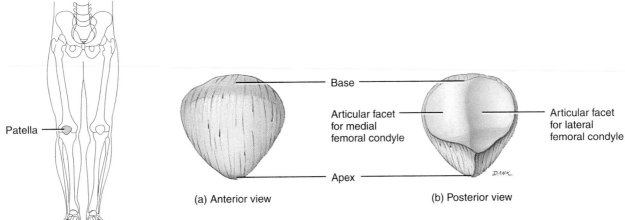

Patella

Base

Articular facet for medial femoral condyle

Articular facet for lateral femoral condyle

Apex

(a) Anterior view

(b) Posterior view

 Because the patella develops in the tendon of a muscle, it is classified as which type of bone?

Figure 8.15   Right tibia and fibula in relation to the femur, patella, and talus.   (See Tortora, *A Photographic Atlas of the Human Body,* Figure 3.30.)

The tibia articulates with the femur and fibula proximally, and with the fibula and talus distally.

(a) Anterior view

(b) Posterior view

(c) Lateral view of distal end of tibia

 Which leg bone bears the weight of the body?

Tarsals, Metatarsals, and Phalanges

The **tarsus** (ankle) is the proximal region of the foot and consists of seven **tarsal bones** (Figure 8.16). They include the **talus** (TĀ-lus = ankle bone) and **calcaneus** (kal-KĀ-nē-us = heel), located in the posterior part of the foot. The calcaneus is the largest and strongest tarsal bone. The anterior tarsal bones are the **cuboid** (= cube-shaped), **navicular** (= like a little boat), and three **cuneiform bones** (= wedge-shaped) called the **first (medial), second (intermediate),** and **third (lateral) cuneiforms.** Joints between tarsal bones are called *intertarsal joints.* The talus, the uppermost tarsal bone, is the only bone of the foot that articulates with the fibula and tibia. It articulates on one side

with the medial malleolus of the tibia and on the other side with the lateral malleolus of the fibula. These articulations form the *talocrural (ankle) joint.* During walking, the talus transmits about half the weight of the body to the calcaneus and half to other tarsal bones.

The **metatarsus** is the intermediate region of the foot and consists of five **metatarsal bones** numbered I to V (or 1–5) from the medial to lateral position (Figure 8.16). Like the metacarpals of the palm of the hand, each metatarsal consists of a proximal *base,* an intermediate *shaft,* and a distal *head.* The metatarsals articulate proximally with the first, second, and third cuneiform bones and with the cuboid to form the *tarsometatarsal joints.* Distally, they articulate with the proximal

Figure 8.16   Right foot.   (See Tortora, *A Photographic Atlas of the Human Body,* Figure 3.31.)

The skeleton of the foot consists of the proximal tarsals, the intermediate metatarsals, and the distal phalanges.

(a) Superior view

(b) Inferior view

Which tarsal bone articulates with the tibia and fibula?

row of phalanges to form the *metatarsophalangeal joints.* The first metatarsal is thicker than the others because it bears more weight.

The **phalanges** comprise the distal component of the foot and resemble those of the hand both in number and arrangement. They are numbered I to V (or 1–5) beginning with the great toe, which is medial. Each also consists of a proximal *base,* an intermediate *shaft,* and a distal *head.* The great or big toe (*hallux*) has two large, heavy phalanges called proximal and distal phalanges. The other four toes each have three phalanges—proximal, middle, and distal. Joints between phalanges of the foot, like those of the hand, are called *interphalangeal joints.*

Arches of the Foot

The bones of the foot are arranged in two **arches** (Figure 8.17). The arches enable the foot to support the weight of the body, provide an ideal distribution of body weight over the hard and soft tissues of the foot, and provide leverage while walking. The arches are not rigid; they yield as weight is applied and spring back when the weight is lifted, thus helping to absorb shocks. Usually, the arches are fully developed by age 12 or 13.

The *longitudinal arch* has two parts, both of which consist of tarsal and metatarsal bones arranged to form an arch from the anterior to the posterior part of the foot. The *medial part* of the longitudinal arch originates at the calcaneus. It rises to the talus and descends through the navicular, the three cuneiforms, and the heads of the three medial metatarsals. The *lateral part* of the longitudinal arch also begins at the calcaneus. It rises at the cuboid and descends to the heads of the two lateral metatarsals.

The *transverse arch* is found between the medial and lateral aspects of the foot and is formed by the navicular, three cuneiforms, and the bases of the five metatarsals.

 Flatfoot and Clawfoot

The bones composing the arches are held in position by ligaments and tendons. If these ligaments and tendons are weakened, the height of the medial longitudinal arch may decrease or "fall." The result is **flatfoot,** the causes of which include excessive weight, postural abnormalities, weakened supporting tissues, and genetic predisposition. A custom-designed arch support often is prescribed to treat flatfoot.

Clawfoot is a condition in which the medial longitudinal arch is abnormally elevated. It is often caused by muscle deformities, such as may occur in diabetics whose neurological lesions lead to atrophy of muscles of the foot. ■

▶ C H E C K P O I N T

6. Which bones form the lower limb, from proximal to distal?

7. What joints do the bones of the lower limb form with one another?

8. What are the functions of the arches of the foot?

• • •

To appreciate the skeletal system's contributions to homeostasis of other body systems, examine *Focus on Homeostasis: The Skeletal System.* Next, in Chapter 9, we will see how joints both hold the skeleton together and permit it to participate in movements.

Figure 8.17   Arches of the right foot.

Arches help the foot support and distribute the weight of the body and provide leverage during walking.

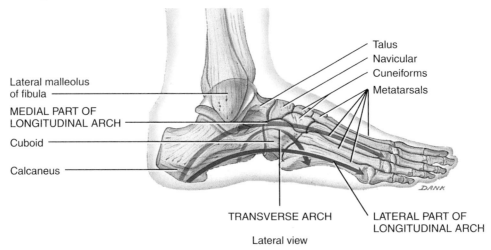

What structural feature of the arches allows them to absorb shocks?

Body System	Contribution of the Skeletal System
For all body systems	Bones provide support and protection for internal organs; bones store and release calcium, which is needed for proper functioning of most body tissues.
Integumentary system	Bones provide strong support for overlying muscles and skin while joints provide flexibility that allows skin to bend.
Muscular system	Bones provide attachment points for muscles and leverage for muscles to bring about body movements; contraction of skeletal muscle requires calcium ions.
Nervous system	Skull and vertebrae protect brain and spinal cord; normal blood level of calcium is needed for normal functioning of neurons and neuroglia.
Endocrine system	Bones store and release calcium, needed during exocytosis of hormone-filled vesicles and for normal actions of many hormones.
Cardiovascular system	Red bone marrow carries out hemopoiesis (blood cell formation); rhythmical beating of the heart requires calcium ions.
Lymphatic and immune system	Red bone marrow produces lymphocytes, white blood cells that are involved in immune responses.
Respiratory system	Axial skeleton of thorax protects lungs; rib movements assist breathing; some muscles used for breathing attach to bones via tendons.
Digestive system	Teeth masticate (chew) food; rib cage protects esophagus, stomach, and liver; pelvis protects portions of the intestines.
Urinary system	Ribs partially protect kidneys and pelvis protects urinary bladder and urethra.
Reproductive systems	Pelvis protects ovaries, uterine (Fallopian) tubes, and uterus in females and part of ductus (vas) deferens and accessory glands in males; bones are an important source of calcium needed for milk synthesis during lactation.

239

Focus on Homeostasis:

The Skeletal System

DISORDERS: HOMEOSTATIC IMBALANCES

Hip Fracture

Although any region of the hip girdle may fracture, the term **hip fracture** most commonly applies to a break in the bones associated with the hip joint—the head, neck, or trochanteric regions of the femur, or the bones that form the acetabulum. In the United States, 300,000 to 500,000 people sustain hip fractures each year. The incidence of hip fractures is increasing, in part due to longer life spans. Decreases in bone mass due to osteoporosis and an increased tendency to fall predispose elderly people to hip fractures.

Hip fractures often require surgical treatment, the goal of which is to repair and stabilize the fracture, increase mobility, and decrease pain.

Sometimes the repair is accomplished by using surgical pins, screws, nails, and plates to secure the head of the femur. In severe hip fractures, the femoral head or the acetabulum of the hip bone may be replaced by prostheses (artificial devices). The procedure of replacing either the femoral head or the acetabulum is *hemiarthroplasty* (hem-ē-AR-thrō-plas-tē; *hemi-* = one half; *-arthro-* = joint; *-plasty* = molding). Replacement of both the femoral head and acetabulum is *total hip arthroplasty*. The acetabular prosthesis is made of plastic, whereas the femoral prosthesis is metal; both are designed to withstand a high degree of stress. The prostheses are attached to healthy portions of bone with acrylic cement and screws.

MEDICAL TERMINOLOGY

Clubfoot or ***talipes equinovarus*** (*-pes* = foot; *equino-* = horse) An inherited deformity in which the foot is twisted inferiorly and medially, and the angle of the arch is increased; occurs in 1 of every 1000 births. Treatment consists of manipulating the arch to a normal curvature by casts or adhesive tape, usually soon after birth. Corrective shoes or surgery may also be required.

Genu valgum (JĒ-noo VAL-gum; *genu-* = knee; *valgum* = bent outward) A decreased space between the knees and increased space between the ankles due to a medial bending of the legs in relation to the thigh. Also called **knock-knee.**

Genu varum (JĒ-noo VAR-um; *varum* = bent toward the midline)

An increased space between the knees due to a lateral bending of the legs relative to the thigh. Also called **bowleg.**

Hallux valgus (HAL-uks VAL-gus; *hallux* = great toe) Angulation of the great toe away from the midline of the body, typically caused by wearing tightly fitting shoes. Involves lateral deviation of the proximal phalanx of the great toe and medial displacement of metatarsal I. Also called a **bunion.**

Pelvimetry (pel-VIM-e-trē) Measurement of the size of the inlet and outlet of the birth canal; may be done by ultrasound or physical examination.

STUDY OUTLINE

PECTORAL (SHOULDER) GIRDLE (p. 219)

1. Each pectoral (shoulder) girdle consists of a clavicle and scapula.
2. Each pectoral girdle attaches an upper limb to the axial skeleton.

UPPER LIMB (p. 222)

1. There are 60 bones in the two upper limbs (extremities).
2. The bones of each upper limb include the humerus, the ulna, the radius, the carpals, the metacarpals, and the phalanges.

PELVIC (HIP) GIRDLE (p. 227)

1. The pelvic (hip) girdle consists of two hip bones.
2. Each hip bone consists of three fused components: ilium, pubis, and ischium.
3. The hip bones, sacrum, and pubic symphysis form the bony pelvis. It supports the vertebral column and pelvic viscera and attaches the lower limbs to the axial skeleton.
4. The true pelvis is separated from the false pelvis by the pelvic brim.

COMPARISON OF FEMALE AND MALE PELVES (p. 230)

1. Bones of males are generally larger and heavier than bones of

females. They also have more prominent markings for muscle attachments.
2. The female pelvis is adapted for pregnancy and childbirth. Gender-related differences in pelvic structure are listed and illustrated in Table 8.1 on pages 231–232.

COMPARISON OF PECTORAL AND PELVIC GIRDLES (p. 232)

1. The pectoral girdle does not directly articulate with the vertebral column; the pelvic girdle does.
2. The glenoid fossae of the scapulae are shallow and maximize movement; the acetabula of the hip bones are deep and allow less movement.

LOWER LIMB (p. 232)

1. There are 60 bones in the two lower limbs (extremities).
2. The bones of each lower limb include the femur, the patella, the tibia, the fibula, the tarsals, the metatarsals, and the phalanges.
3. The bones of the foot are arranged in two arches, the longitudinal arch and the transverse arch, to provide support and leverage.

SELF-QUIZ QUESTIONS

Fill in the blanks in the following statements.

1. Functions of the axial skeleton are related mainly to _____ of internal organs, whereas functions of the appendicular skeleton are related more to _____.

2. The bones that comprise the palm are called the _____.

Indicate whether the following statements are true or false.

3. The arches of the foot enable the foot to support the weight of the body, provide an ideal distribution of body weight over the hard and soft tissues of the foot, and provide leverage for walking.

4. Functionally, the pelvis provides a strong and stable support for the vertebral column and pelvic viscera.

Choose the one best answer to the following questions.

5. Which of the following statements are *true*? (1) The pectoral girdle consists of the scapula, the clavicle, and the sternum. (2) Although the joints of the pectoral girdle are not very stable, they allow free movement in many directions. (3) The anterior component of the pectoral girdle is the scapula. (4) The pectoral girdle articulates directly with the vertebral column. (5) The posterior component of the pectoral girdle is the sternum. (a) 1, 2, and 3, (b) only 2, (c) only 4, (d) 2, 3, and 5, (e) 3, 4, and 5.

6. Which of the following bones are carpal bones? (1) phalanges, (2) scaphoid, lunate, and pisiform, (3) trapezium, trapezoid, capitate, (4) hamate and triquetrum, (5) calcaneus and talus. (a) 1, 2, and 3, (b) 2, 3, and 4, (c) 1, 2, 3, and 4, (d) 2, 3, 4, and 5, (e) 1, 2, 3, 4, and 5.

7. Which of the following statements is *true*? (a) The ilium is the smallest component of the hip bone. (b) The greater sciatic notch is found on the ischium. (c) The ischium is the anterior, superior portion of the hip bone. (d) The pubic symphysis is the joint between the two hip bones. (e) The obturator foramen is the smallest foramen in the body.

8. Which of the following articulates with the distal end of the tibia? (a) calcaneous, (b) navicular, (c) cuboid, (d) cuneiform, (e) talus.

9. Which of the following statements are *true*? (1) Bones of males are generally larger and heavier than those of females. (2) The female pelvis is wider than the male pelvis. (3) The true pelvis of males contains more space than that of the females. (4) Skeletal bones of males generally possess larger surface features than those of females. (5) Most structural differences between male and female pelves are related to pregnancy and childbirth. (a) 1, 2, and 3, (b) 2, 3, and 4, (c) 3, 4, and 5, (d) 1, 2, 3, and 5, (e) 1, 2, 4, and 5.

10. The acetabulum is part of the (a) fibula, (b) humerus, (c) hip bone, (d) femur, (e) tibia.

11. Which of the following is *not* a tarsal bone? (a) patella, (b) navicular, (c) calcaneus, (d) talus, (e) cuneiform.

12. Which of the following are *true* concerning the elbow joint? (1) When the forearm is extended, the olecranon fossa receives the olecranon. (2) When the forearm is flexed, the radial fossa receives the coronoid process. (3) The head of the radius articulates with the capitulum. (4) The trochlea articulates with the trochlear notch. (5) The head of the ulna atriculates with the ulnar notch of the radius. (a) 1, 2, 3, 4, and 5, (b) 1, 3, and 4, (c) 1, 3, 4, and 5, (d) 1, 2, 3, and 4, (e) 2, 3, and 4.

13. Which of the following is mismatched? (a) narrowing of carpal tunnel: carpal tunnel syndrome, (b) abnormal elevation of medial longitudinal arch: flat foot, (c) inherited foot deformity causing the foot to twist inferiorly and medially: clubfoot, (d) medial bending of legs resulting in decreased space between the knees and increased space between the ankle: genu valgum, (e) condition where great toes angle away from the midline of the body; often caused by tight-fitting shoes: bunion.

14. Match the following:
    _____ (a) olecranon process
    _____ (b) olecranon fossa
    _____ (c) trochlea
    _____ (d) greater trochanter
    _____ (e) medial malleolus
    _____ (f) acromial extremity
    _____ (g) capitulum
    _____ (h) acromion
    _____ (i) radial tuberosity
    _____ (j) acetabulum
    _____ (k) lateral malleolus
    _____ (l) glenoid cavity
    _____ (m) coronoid process
    _____ (n) linea aspera
    _____ (o) anterior border

    (1) clavicle
    (2) scapula
    (3) humerus
    (4) ulna
    (5) radius
    (6) femur
    (7) tibia
    (8) fibula
    (9) coxal bone

15. Match the following:
    _____ (a) a large, triangular, flat bone found in the posterior part of the thorax
    _____ (b) an S-shaped bone lying horizontally in the superior and anterior part of the thorax
    _____ (c) articulates proximally with the scapula and distally with the radius and ulna
    _____ (d) located on the medial aspect of the forearm
    _____ (e) located on the lateral aspect of the forearm
    _____ (f) the longest, heaviest, and strongest bone of the body
    _____ (g) the larger, medial bone of the leg
    _____ (h) the smaller, lateral bone of the leg
    _____ (i) heel bone
    _____ (j) carpal bone

    (1) calcaneus
    (2) scapula
    (3) scaphoid
    (4) radius
    (5) femur
    (6) clavicle
    (7) ulna
    (8) tibia
    (9) humerus
    (10) fibula

CRITICAL THINKING QUESTIONS

1. Young Tommy watches way too much television, especially cartoons and monster movies. So when he heard that his great uncle had "clawfoot," he couldn't decide if he should be thrilled or terrified! Explain "clawfoot" to Tommy.
   HINT *His great uncle is not changing his species—it's a human condition.*

2. The local newspaper reported that Farmer White caught his hand in a piece of machinery last Tuesday. He lost the lateral two fingers of his left hand. Science reporter Kent Clark, really a high school junior, reports that Farmer White has three remaining phalanges. Is Kent correct, or does he need a refresher course in anatomy?
   HINT *How many phalanges are in each finger?*

3. Given that the hips and shoulders are located in the trunk of the body, why aren't they part of the axial skeleton like the rest of the bones that make up the trunk?
   HINT *Have you ever seen a snake with shoulders?*

ANSWERS TO FIGURE QUESTIONS

8.1 The pectoral girdles attach the upper limbs to the axial skeleton.

8.2 The weakest part of the clavicle is its midregion at the junction of the two curves.

8.3 The highest point of the shoulder is the acromion.

8.4 Each upper limb has 30 bones.

8.5 The radius articulates at the elbow with the capitulum and radial fossa of the humerus. The ulna articulates at the elbow with the trochlea, coronoid fossa, and olecranon fossa of the humerus. The interosseous membrane connects their shafts.

8.6 The "elbow" part of the ulna is the olecranon.

8.7 The radius and ulna form three joints, the connection at the interosseus membrane and the proximal and distal radioulnar joints.

8.8 The most frequently fractured wrist bone is the scaphoid.

8.9 The bony pelvis attaches the lower limbs to the axial skeleton and supports the backbone and pelvic viscera.

8.10 The femur articulates with the acetabulum of the hip bone; the sacrum articulates with the auricular surface of the hip bone.

8.11 The pelvic axis is the course taken by a baby's head as it descends through the pelvis during childbirth.

8.12 Each lower limb has 30 bones.

8.13 The angle of convergence of the femurs is greater in females than males because the female pelvis is broader.

8.14 The patella is a sesamoid bone, the largest one in the body.

8.15 The tibia is the weight-bearing bone of the leg.

8.16 The talus articulates with the tibia and the fibula.

8.17 The arches are not rigid; they yield when weight is applied and spring back when weight is lifted.

Joints

A Student's View

As a current EMT student and future medical school candidate, I cannot stress enough the importance of this topic in Anatomy and Physiology. The chapter demonstrates the proper way joints should work and look in a healthy individual. This material will most likely show up on tests, whether it is in another related class, or a test in the practical world with a trauma or medical patient. I personally found the images of Figure 9.4 helpful during a test for an EMT class.

When I think back to Anatomy and Physiology, I remember my professor acting out the movements in Figures 9.5–9.10 and this made the material much more interesting when it came time to study. The class got a good laugh acting out the movements, and this helped us remember the topic better. So even though it takes time, dedication, and perseverance to do well in Anatomy and Physiology; a little fun is helpful too. As I continue to embark on the exciting field of medicine, the material in this chapter provides me an excellent foundation to learn and build upon.

Mark Johnson
Edison Community College

JOINTS AND HOMEOSTASIS

The joints of the skeletal system contribute to homeostasis by holding bones together in ways that allow for movement and flexibility.

ENERGIZE YOUR STUDY

FOUNDATIONS CD

Anatomy Overview
- The Skeletal System
- Connective Tissue: Bone Tissue

Animation
- Systems Contributions to Homeostasis

Concepts and Connections and Exercises reinforce your understanding

www.wiley.com/college/apcentral

INSIGHTS AND EXPLORATIONS

Many of us wake up in the morning with aches and pains in our joints. But we know that when we just get up and get going we will feel better. For individuals suffering from osteoarthritis, the promise of a better day does not come. In fact, for these people, as the day wears on, their joints, such as knees and fingers, become even more stiff and painful. You've probably even seen some people whose hands have very large joints that look misshapen and swollen. Come explore more about this chronic bone disease in a web-based activity.

ones are too rigid to bend without being damaged. Fortunately, flexible connective tissues form joints that hold bones together while still permitting, in most cases, some degree of movement. A **joint,** also called an **articulation** or **arthrosis,** is a point of contact between two bones, between bone and cartilage, or between bone and teeth. When we say one bone *articulates* with another bone, we mean that the bones form a joint. Because most movements of the body occur at joints, you can appreciate their importance if you imagine how a cast over your knee joint makes walking difficult, or how a splint on a finger limits your ability to manipulate small objects. The scientific study of joints is termed **arthrology** (ar-THROL-ō-jē; *arthr-* = joint; *-logy* = study of). The study of motion of the human body is called **kinesiology** (ki-nē-sē′-OL-ō-jē; *kinesi* = movement).

JOINT CLASSIFICATIONS

▶ O B J E C T I V E

• **Describe the structural and functional classifications of joints.**

Joints are classified structurally, based on their anatomical characteristics, and functionally, based on the type of movement they permit.

The structural classification of joints is based on two criteria: (1) the presence or absence of a space between the articulating bones, called a synovial cavity, and (2) the type of connective tissue that binds the bones together. Structurally, joints are classified as one of the following types:

• **Fibrous joints:** The bones are held together by fibrous connective tissue that is rich in collagen fibers; they lack a synovial cavity.

• **Cartilaginous joints:** The bones are held together by cartilage; they lack a synovial cavity.

• **Synovial joints:** The bones forming the joint have a synovial cavity and are united by the dense irregular connective tissue of an articular capsule, and often by accessory ligaments.

The functional classification of joints relates to the degree of movement they permit. Functionally, joints are classified as one of the following types:

• **Synarthrosis** (sin′-ar-THRŌ-sis; *syn-* = together): An immovable joint. The plural is *synarthroses.*

• **Amphiarthrosis** (am′-fē-ar-THRŌ-sis; *amphi-* = on both sides): A slightly movable joint. The plural is *amphiarthroses.*

• **Diarthrosis** (dī-ar-THRŌ-sis = movable joint): A freely movable joint. The plural is *diarthroses.* All diarthroses are synovial joints. They have a variety of shapes and permit several different types of movements.

The following sections present the joints of the body according to their structural classification. As we examine the structure of each type of joint, we will also explore its functional attributes.

FIBROUS JOINTS

▶ O B J E C T I V E

• **Describe the structure and functions of the three types of fibrous joints.**

Fibrous joints lack a synovial cavity, and the articulating bones are held very closely together by fibrous connective tissue. They permit little or no movement. The three types of fibrous joints are sutures, syndesmoses, and gomphoses.

Sutures

A **suture** (SOO-chur; *sutur-* = seam) is a fibrous joint composed of a thin layer of dense fibrous connective tissue that unites only bones of the skull. An example is the coronal suture between the parietal and frontal bones (Figure 9.1a). The irregular, interlocking edges of sutures give them added strength and decrease their chance of fracturing. Because a suture is immovable, it is classified functionally as a synarthrosis.

Some sutures, although present during childhood, are replaced by bone in the adult. Such a suture is called a **synostosis** (sin′-os-TŌ-sis; *os-* = bone), or bony joint—a joint in which there is a complete fusion of bone across the suture line. An example is the *metopic suture* between the left and right sides of the frontal bone that begins to fuse during infancy.

Syndesmoses

A **syndesmosis** (sin′-dez-MŌ-sis; *syndesmo-* = band or ligament) is a fibrous joint in which, compared to a suture, there is a greater distance between the articulating bones and more fibrous connective tissue. The fibrous connective tissue is arranged either as a bundle (ligament) or as a sheet (interosseous membrane) (Figure 9.1b). One example of a syndesmosis is the distal tibiofibular joint, where the anterior tibiofibular ligament connects the tibia and fibula. Another example is the interosseous membrane between the parallel borders of the tibia and fibula. Because it permits slight movement, a syndesmosis is classified functionally as an amphiarthrosis.

Figure 9.1   Fibrous joints.

At a fibrous joint the bones are held together by fibrous connective tissue.

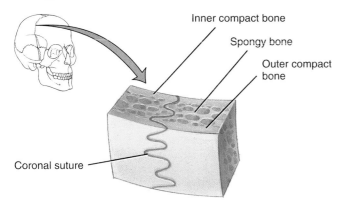

- Inner compact bone
- Spongy bone
- Outer compact bone
- Coronal suture

(a) Suture between skull bones

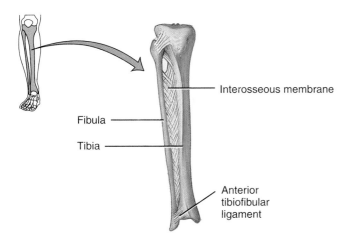

- Interosseous membrane
- Fibula
- Tibia
- Anterior tibiofibular ligament

(b) Syndesmoses between tibia and fibula

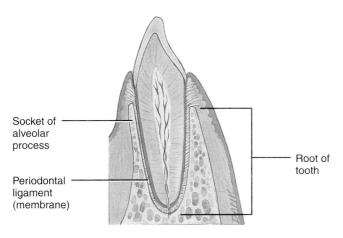

- Socket of alveolar process
- Periodontal ligament (membrane)
- Root of tooth

(c) Gomphosis between tooth and socket of alveolar process

Functionally, why are sutures classified as synarthroses, and syndesmoses as amphiarthroses?

Gomphoses

A **gomphosis** (gom-FŌ-sis; *gompho-* = a bolt or nail) or *dento-alveolar joint* is a type of fibrous joint in which a cone-shaped peg fits into a socket. The only examples of gomphoses are the articulations of the roots of the teeth with the sockets (alveoli) of the alveolar processes of the maxillae and mandible (Figure 9.1c). The dense fibrous connective tissue between a tooth and its socket is the periodontal ligament (membrane). A gomphosis is classified functionally as a synarthrosis, an immovable joint.

CARTILAGINOUS JOINTS

> **OBJECTIVE**
>
> • **Describe the structure and functions of the two types of cartilaginous joints.**

Like a fibrous joint, a **cartilaginous joint** lacks a synovial cavity and allows little or no movement. Here the articulating bones are tightly connected by either hyaline cartilage or fibrocartilage. The two types of cartilaginous joints are synchondroses and symphyses.

Synchondroses

A **synchondrosis** (sin'-kon-DRŌ-sis; *chondro-* = cartilage) is a cartilaginous joint in which the connecting material is hyaline cartilage. An example of a synchondrosis is the epiphyseal plate that connects the epiphysis and diaphysis of a growing bone (Figure 9.2a). A photomicrograph of the epiphyseal plate is shown in Figure 6.8 on page 174. Functionally, a synchondrosis is a synarthrosis. When bone elongation ceases, bone replaces the hyaline cartilage, and the synchondrosis becomes a synostosis, a bony joint. Another example of a synchondrosis is the joint between the first rib and the manubrium of the sternum, which also ossifies during adult life and becomes an immovable synostosis.

Symphyses

A **symphysis** (SIM-fi-sis = growing together) is a cartilaginous joint in which the ends of the articulating bones are covered with hyaline cartilage, but a broad, flat disc of fibrocartilage connects the bones. All symphyses occur in the midline of the body. The pubic symphysis between the anterior surfaces of the hip bones is one example of a symphysis (Figure 9.2b). This type of joint is also found at the intervertebral joints between the bodies of vertebrae (see Figure 7.17 on page 205). A portion of the intervertebral disc is fibrocartilage. A symphysis is an amphiarthrosis, a slightly movable joint.

Figure 9.2 Cartilaginous joints.

 At a cartilaginous joint the bones are held together by cartilage.

(a) Synchondrosis

(b) Symphysis

What is the structural difference between a synchondrosis and a symphysis?

Figure 9.3 Structure of a typical synovial joint. Note the two layers of the articular capsule—the fibrous capsule and the synovial membrane. Synovial fluid fills the joint cavity between the synovial membrane and the articular cartilage.

The distinguishing feature of a synovial joint is the synovial cavity between the articulating bones.

Frontal section

What is the functional classification of synovial joints?

► C H E C K P O I N T

1. Where in the body can each type of fibrous and cartilaginous joint be found?

2. What type of tissue holds the articulating bones together in fibrous and cartilaginous joints?

3. Which of the fibrous and cartilaginous joints are classified as amphiarthroses?

SYNOVIAL JOINTS

► O B J E C T I V E S

• **Describe the structure of synovial joints.**
• **Describe the six subtypes of synovial joints.**
• **Describe the structure and function of bursae and tendon sheaths.**

Structure of Synovial Joints

Synovial joints (si-NŌ-vē-al) have certain characteristics that distinguish them from other joints. The unique characteristic of a synovial joint is the presence of a space called a **synovial (joint) cavity** between the articulating bones (Figure 9.3). The synovial cavity allows a joint to be freely moveable. Hence, all synovial joints are classified functionally as diarthroses. The bones at a synovial joint are covered by **articular cartilage,** which is hyaline cartilage. The cartilage covers the articulating surface of the bones with a smooth, slippery surface but does not bind them together. Articular cartilage reduces friction between bones in the joint during movement and helps to absorb shock.

Articular Capsule

A sleevelike **articular capsule** surrounds a synovial joint, encloses the synovial cavity, and unites the articulating bones. The articular capsule is composed of two layers, an outer fibrous capsule and an inner synovial membrane (Figure 9.3). The outer layer, the **fibrous capsule,** usually consists of dense irregular connective tissue that attaches to the periosteum of the articulating bones. The flexibility of the fibrous capsule permits considerable movement at a joint while its great tensile strength (resistance to stretching) helps prevent the bones from dislocating. The fibers of some fibrous capsules are arranged in parallel bundles that are highly adapted for resisting strains. Such fiber bundles are called **ligaments** (*liga-* = bound or tied) and are often designated by individual names. The strength of ligaments is one of the main mechanical factors that hold bones close together in a synovial joint. The inner layer of the articular capsule, the **synovial membrane,** is composed of areolar connective tissue with elastic fibers. At many synovial joints the synovial membrane includes accumulations of adipose tissue, called **articular fat pads.** An example is the infrapatellar fat pad in the knee (see Figure 9.14c).

Synovial Fluid

The synovial membrane secretes **synovial fluid** (*ov-* = egg), which forms a thin film over the surfaces within the articular capsule. This viscous, clear or pale yellow fluid was named for its similarity in appearance and consistency to uncooked egg white (albumin). Synovial fluid consists of hyaluronic acid, secreted by fibroblast-like cells in the synovial membrane, and interstitial fluid filtered from blood plasma. Its several functions include reducing friction by lubricating the joint, and supplying nutrients to and removing metabolic wastes from the chondrocytes within articular cartilage. (Recall that cartilage is an avascular tissue.) Synovial fluid also contains phagocytic cells that remove microbes and the debris that results from normal wear and tear in the joint. When a synovial joint is immobile for a time, the fluid is quite viscous (gel-like), but as joint movement increases, the fluid becomes less viscous. One of the benefits of a warm-up before exercise is that it stimulates the production and secretion of synovial fluid.

Accessory Ligaments and Articular Discs

Many synovial joints also contain **accessory ligaments** called extracapsular ligaments and intracapsular ligaments. *Extracapsular ligaments* lie outside the articular capsule. Examples are the fibular and tibial collateral ligaments of the knee joint (see Figure 9.14d). *Intracapsular ligaments* occur within the articular capsule but are excluded from the synovial cavity by folds of the synovial membrane. Examples are the anterior and posterior cruciate ligaments of the knee joint (see Figure 9.14d).

Inside some synovial joints, such as the knee, are pads of fibrocartilage that lie between the articular surfaces of the bones and are attached to the fibrous capsule. These pads are called **articular discs** or **menisci** (me-NIS-sī; singular is *meniscus*). Figure 9.14d depicts the lateral and medial menisci in the knee joint. The discs usually subdivide the synovial cavity into two separate spaces as in the temporomandibular joint (TMJ), thus permitting separate movements in the respective spaces. By modifying the shape of the joint surfaces of the articulating bones, articular discs allow two bones of different shapes to fit more tightly. Articular discs also help to maintain the stability of the joint and direct the flow of synovial fluid to the areas of greatest friction.

Torn Cartilage and Arthroscopy

The tearing of articular discs (menisci) in the knee, commonly called **torn cartilage,** occurs often among athletes. Such dam-

aged cartilage will begin to wear and may precipitate arthritis unless surgically removed (meniscectomy). Surgical repair of the torn cartilage may be assisted by **arthroscopy** (ar-THROS-kō-pē; -*scopy* = observation). This procedure involves examination of the interior of a joint, usually the knee, with an arthroscope, a lighted, pencil-thin instrument used for visualization. Arthroscopy is used to determine the nature and extent of damage following knee injury; to help remove torn cartilage and repair damaged cruciate ligaments in the knee; to help obtain tissue samples for analysis; to monitor the progression of disease and the effects of therapy; and to help perform surgery on other joints, such as the shoulder, elbow, ankle, and wrist. ■

Nerve and Blood Supply

The nerves that supply a joint are the same as those that supply the skeletal muscles that move the joint. Synovial joints contain many nerve endings that are distributed to the articular capsule and associated ligaments. Some of the nerve endings convey information about pain from the joint to the spinal cord and brain for processing. Other nerve endings are responsive to the degree of movement and stretch at a joint. This information, too, is relayed to the spinal cord and brain, which may respond by sending impulses through different nerves to the muscles to adjust body movements.

Arteries in the vicinity of a synovial joint send out many branches that penetrate the ligaments and articular capsule to deliver oxygen and nutrients. Veins remove carbon dioxide and wastes from the joints. The arterial branches from several different arteries typically join together around a joint before penetrating the articular capsule. The articulating portions of a synovial joint receive nourishment from synovial fluid, whereas all other joint tissues are supplied by blood capillaries.

Sprain and Strain

A **sprain** is the forcible wrenching or twisting of a joint that stretches or tears its ligaments but does not dislocate the bones. It occurs when the ligaments are stressed beyond their normal capacity. Sprains also may damage surrounding blood vessels, muscles, tendons, or nerves. Severe sprains may be so painful that the joint cannot be moved. There is considerable swelling, which results from hemorrhage of ruptured blood vessels. The ankle joint is most often sprained; the lower back is another frequent location for sprains. A **strain** is a stretched or partially torn muscle. It often occurs when a muscle contracts suddenly and powerfully—for example, in sprinters when they accelerate too quickly. ■

Bursae and Tendon Sheaths

The various movements of the body create friction between moving parts. Saclike structures called **bursae** (BER-sē = purses; singular is *bursa*) are strategically situated to reduce friction in some synovial joints, such as the shoulder and knee joints (see Figures 9.11 and 9.14c). Bursae are not strictly part of synovial joints, but do resemble joint capsules because their walls consist of connective tissue lined by a synovial membrane. They are also filled with a fluid similar to synovial fluid. Bursae are located between the skin and bone in places where skin rubs over bone. They are also found between tendons and bones, muscles and bones, and ligaments and bones. The fluid-filled bursal sacs cushion the movement of one body part over another.

In addition to bursae, structures called tendon sheaths also reduce friction at joints. **Tendon sheaths** are tubelike bursae that wrap around tendons where there is considerable friction. This occurs where tendons pass through synovial cavities, such as the tendon of the biceps brachii muscle at the shoulder joint (see Figure 9.11c). Other areas of considerable friction where tendon sheaths are found are at the wrist and ankle, where many tendons come together in a confined space (see Figure 11.22 on page 374), and in the fingers and toes, where there is a great deal of movement (see Figure 11.18a on page 358).

 Bursitis

An acute or chronic inflammation of a bursa is called **bursitis.** The condition may be caused by trauma, by an acute or chronic infection (including syphilis and tuberculosis), or by rheumatoid arthritis (described on page 267). Repeated, excessive exertion of a joint often results in bursitis, with local inflammation and the accumulation of fluid. Symptoms include pain, swelling, tenderness, and limited movement. Treatment may include oral anti-inflammatory agents and injections of cortisol-like steroids. ■

Types of Synovial Joints

Although all synovial joints have a generally similar structure, the shapes of the articulating surfaces vary and thus various types of movements are possible. Accordingly, synovial joints are divided into six subtypes: planar, hinge, pivot, condyloid, saddle, and ball-and-socket joints.

Planar Joints

The articulating surfaces of bones in a **planar joint** are flat or slightly curved (Figure 9.4a). Planar joints primarily permit side-to-side and back-and-forth gliding movements (described shortly). These joints are said to be *nonaxial* because the motion they allow does not occur around an axis. However, x-ray films made during wrist and ankle movements reveal some rotation of the small carpal and tarsal bones in addition to their predominant gliding movements. Examples of planar joints are the intercarpal joints (between carpal bones at the wrist), intertarsal joints (between tarsal bones at the ankle), sternoclavicular joints (between the manubrium of the sternum and the clavicle), acromioclavicular joints (between the acromion of the scapula and the clavicle),

sternocostal joints (between the sternum and ends of the costal cartilages at the tips of the second through seventh pairs of ribs), and vertebrocostal joints (between the heads and tubercles of ribs and transverse processes of thoracic vertebrae).

Hinge Joints

In a **hinge joint,** the convex surface of one bone fits into the concave surface of another bone (Figure 9.4b). As the name implies, hinge joints produce an angular, opening-and-closing motion like that of a hinged door. Hinge joints are *monaxial (uniaxial)* because they typically allow motion around a single axis. Examples of hinge joints are the knee, elbow, ankle, and interphalangeal joints.

Pivot Joints

In a **pivot joint,** the rounded or pointed surface of one bone articulates with a ring formed partly by another bone and partly by a ligament (Figure 9.4c). A pivot joint is monaxial because it allows rotation around its own longitudinal axis only. Examples of pivot joints are the atlanto-axial joint, in which the atlas rotates around the axis and permits the head to turn from side to side as in signifying "no" (see Figure 9.9a), and the radioulnar joints that enable the palms to turn anteriorly and posteriorly (see Figure 9.10h).

Condyloid Joints

In a **condyloid joint** (KON-di-loyd; *condyl-* = knuckle) or *ellipsoidal* joint, the convex oval-shaped projection of one bone fits into the oval-shaped depression of another bone (Figure 9.4d). A condyloid joint is *biaxial* because the movement it permits is around two axes. Notice that you can move your index finger both up-and-down and from side-to-side. Examples are the wrist and metacarpophalangeal joints for the second through fifth digits.

Saddle Joints

In a **saddle joint,** the articular surface of one bone is saddle-shaped, and the articular surface of the other bone fits into the "saddle" as a sitting rider would sit (Figure 9.4e). A saddle joint is a modified condyloid joint in which the movement is somewhat freer. Saddle joints are *biaxial,* producing side-to-side and up-and-down movements. An example of a saddle joint is the carpometacarpal joint between the trapezium of the carpus and metacarpal of the thumb.

Ball-and-Socket Joints

A **ball-and-socket joint** consists of the ball-like surface of one bone fitting into a cuplike depression of another bone (Figure 9.4f). Such joints are *multiaxial (polyaxial)* because they permit movement around three axes plus all directions in between. Examples of functional ball-and-socket joints are the shoulder

Figure 9.4 Subtypes of synovial joints. For each subtype, a drawing of the actual joint and a simplified diagram are shown.

Synovial joints are classified into subtypes based on the shapes of the articulating bone surfaces.

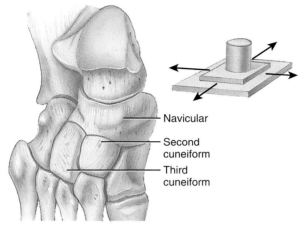

Navicular

Second cuneiform

Third cuneiform

(a) Planar joint between the navicular and second and third cuneiforms of the tarsus in the foot

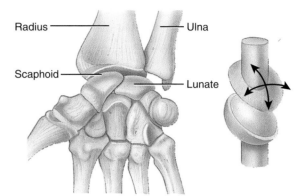

Radius

Ulna

Scaphoid

Lunate

(d) Condyloid joint between radius and scaphoid and lunate bones of the carpus (wrist)

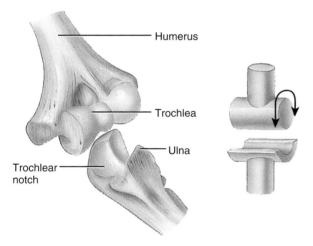

Humerus

Trochlea

Ulna

Trochlear notch

(b) Hinge joint between trochlea of humerus and trochlear notch of ulna at the elbow

Radius

Ulna

Trapezium

Metacarpal of thumb

(e) Saddle joint between trapezium of carpus (wrist) and metacarpal of thumb

Head of radius

Radial notch

Annular ligament

Ulna

Radius

(c) Pivot joint between head of radius and radial notch of ulna

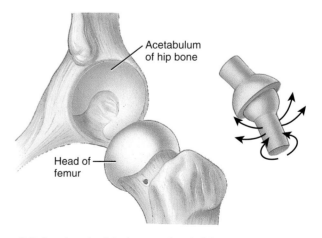

Acetabulum of hip bone

Head of femur

(f) Ball-and-socket joint between head of the femur and acetabulum of the hip bone

 Which of the joints shown here are biaxial?

and hip joints. At the shoulder joint, the head of the humerus fits into the glenoid cavity of the scapula. At the hip joint, the head of the femur fits into the acetabulum of the hip bone.

Table 9.1 summarizes the structural and functional categories of joints.

▶ CHECKPOINT

4. How does the structure of synovial joints classify them as diarthroses?

5. What are the functions of articular cartilage, synovial fluid, and articular discs?

6. What types of sensations are perceived at joints, and from what sources do joint components receive their nourishment?

7. In what ways are bursae similar to joint capsules? How do they differ?

8. Which types of joints are nonaxial, monaxial, biaxial, and multiaxial?

Table 9.1	Summary of Structural and Functional Classifications of Joints		
Structural Classification	**Description**	**Functional Classification**	**Example**
Fibrous	Articulating bones held together by fibrous connective tissue; no synovial cavity.		
Suture	Articulating bones united by a thin layer of dense fibrous connective tissue, found between bones of the skull. With age, some sutures are replaced by a synostosis, in which the bones fuse across the former suture.	Synarthrosis (immovable).	Frontal suture.
Syndesmosis	Articulating bones united by dense fibrous connective tissue, either a ligament or an interosseous membrane.	Amphiarthrosis (slightly movable).	Distal tibiofibular joint.
Gomphosis	Articulating bones united by periodontal ligament; cone-shaped peg fits into a socket.	Synarthrosis.	At roots of teeth in alveoli (sockets) of maxillae and mandible.
Cartilaginous	Articulating bones united by cartilage; no synovial cavity.		
Synchondrosis	Connecting material is hyaline cartilage; becomes a synostosis when bone elongation ceases.	Synarthrosis.	Epiphyseal plate at between the joint diaphysis and epiphysis of a long bone.
Symphysis	Connecting material is a broad, flat disc of fibrocartilage.	Amphiarthrosis.	Intervertebral joints and pubic symphysis.
Synovial	Characterized by a synovial cavity, articular cartilage, and an articular capsule; may contain accessory ligaments, articular discs, and bursae.		
Planar	Articulated surfaces are flat or slightly curved.	Nonaxial diarthrosis (freely movable); gliding motion.	Intercarpal, intertarsal, sternocostal (between sternum and the 2nd–7th pairs of ribs), and vertebrocostal joints.
Hinge	Convex surface fits into a concave surface.	Monaxial diarthrosis; angular motion.	Elbow, ankle, and interphalangeal joints.
Pivot	Rounded or pointed surface fits into a ring formed partly by bone and partly by a ligament.	Monaxial diarthrosis; rotation.	Atlanto-axial and radioulnar joints.
Condyloid	Oval-shaped projection fits into an oval-shaped depression.	Biaxial diarthrosis; angular motion.	Radiocarpal and metacarpophalangeal joints.
Saddle	Articular surface of one bone is saddle-shaped, and the articular surface of the other bone "sits" in the saddle.	Biaxial diarthrosis; angular motion.	Carpometarcarpal joint between trapezium and thumb.
Ball-and-socket	Ball-like surface fits into a cuplike depression.	Multiaxial diarthrosis; angular motion and rotation.	Shoulder and hip joints.

TYPES OF MOVEMENTS AT SYNOVIAL JOINTS

> **OBJECTIVE**
>
> • **Describe the types of movements that can occur at synovial joints.**

Anatomists, physical therapists, and kinesiologists use specific terminology to designate the movements that can occur at synovial joints. These precise terms may indicate the form of motion, the direction of movement, or the relationship of one body part to another during movement. Movements at synovial joints are grouped into four main categories: (1) gliding, (2) angular movements, (3) rotation, and (4) special movements. This last category includes movements that occur only at certain joints.

Gliding

Gliding is a simple movement in which relatively flat bone surfaces move back-and-forth and from side-to-side with respect to one another (Figure 9.5). There is no significant alteration of the angle between the bones. Gliding movements are limited in range due to the structure of the articular capsule and associated ligaments and bones. Gliding occurs at planar joints.

Angular Movements

In **angular movements,** there is an increase or a decrease in the angle between articulating bones. The major angular movements are flexion, extension, lateral extension, hyperextension, abduction, adduction, and circumduction. These movements are discussed with respect to the body in the anatomical position.

Figure 9.5   Gliding movements at synovial joints.

 Gliding motion consists of side-to-side and back-and-forth movements.

Intercarpal joints

 Which subtype of synovial joints exhibits gliding movements?

Flexion, Extension, Lateral Flexion, and Hyperextension

Flexion and extension are opposite movements. In **flexion** (FLEK-shun; *flex-* = to bend) there is a decrease in the angle between articulating bones; in **extension** (eks-TEN-shun; *exten-* = to stretch out) there is an increase in the angle between articulating bones, often to restore a part of the body to the anatomical position after it has been flexed (Figure 9.6). Both movements usually occur in the sagittal plane. Hinge, pivot, condyloid, saddle, and ball-and-socket joints all permit flexion and extension. All of the following are examples of flexion:

• Bending the head toward the chest at the atlanto-occipital joint between the atlas (the first vertebra) and the occipital bone of the skull, and at the cervical intervertebral joints between the cervical vertebrae (Figure 9.6a)

• Bending the trunk forward at the intervertebral joints

• Moving the humerus forward at the shoulder joint, as in swinging the arms forward while walking (Figure 9.6b)

• Moving the forearm toward the arm at the elbow joint between the humerus, ulna, and radius (Figure 9.6c)

• Moving the palm toward the forearm at the wrist or radiocarpal joint between the radius and carpals (Figure 9.6d)

• Bending the digits of the hand or feet at the interphalangeal joints between phalanges

• Moving the femur forward at the hip joint between the femur and hip bone, as in walking (Figure 9.6e)

• Moving the leg toward the thigh at the knee or tibiofemoral joint between the tibia, femur, and patella, as occurs when bending the knee (Figure 9.6f)

Although flexion and extension usually occur in the sagittal plane, there are a few exceptions. For example, flexion of the thumb involves movement of the thumb medially across the palm at the carpometacarpal joint between the trapezium and metacarpal of the thumb, as in touching the thumb to the opposite side of the palm. Another example is movement of the trunk sideways to the right or left at the waist. This movement, which occurs in the frontal plane and involves the intervertebral joints, is called **lateral flexion** (Figure 9.6g).

Continuation of extension beyond the anatomical position is called **hyperextension** (*hyper-* = beyond or excessive). Examples of hyperextension include:

• Bending the head backward at the atlanto-occipital and cervical intervertebral joints (Figure 9.6a)

• Bending the trunk backward at the intervertebral joints

• Moving the humerus backward at the shoulder joint, as in swinging the arms backward while walking (Figure 9.6b)

• Moving the palm backward at the wrist joint (Figure 9.6d)

• Moving the femur backward at the hip joint, as in walking (Figure 9.6e)

Figure 9.6 Angular movements at synovial joints—flexion, extension, hyperextension, and lateral flexion.

In angular movements, there is an increase or decrease in the angle between articulating bones.

(a) Atlanto-occipital and cervical intervertebral joints

(b) Shoulder joint

(c) Elbow joint

(d) Wrist joint

(e) Hip joint

(f) Knee joint

(g) Intervertebral joints

What are two examples of flexion that do not occur in the sagittal plane?

Hyperextension of hinge joints, such as the elbow, interphalangeal, and knee joints, is usually prevented by the arrangement of ligaments and the anatomical alignment of the bones.

Abduction, Adduction, and Circumduction

Abduction (ab-DUK-shun; *ab-* = away; *-duct* = to lead) is the movement of a bone away from the midline, whereas **adduction** (ad-DUK-shun; *ad-* = toward) is the movement of a bone toward the midline. Both movements usually occur in the frontal plane. Condyloid, saddle, and ball-and-socket joints permit abduction and adduction. Examples of abduction include moving the humerus laterally at the shoulder joint, moving the palm laterally at the wrist joint, and moving the femur laterally at the hip joint (Figure 9.7a–c). The movement that returns each of these body parts to the anatomical position is adduction (Figure 9.7a–c).

With respect to the digits, the midline of the body is not used as a point of reference for abduction and adduction. In abduction of the fingers (but not the thumb), an imaginary line is drawn through the longitudinal axis of the middle (longest)

Figure 9.7 **Angular movements at synovial joints—abduction and adduction.**

 Abduction and adduction usually occur in the frontal plane.

| (a) Shoulder joint | (b) Wrist joint | (c) Hip joint |

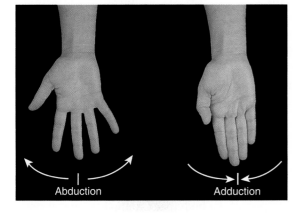

(d) Metacarpophalangeal joints of the fingers (not the thumb)

Is considering adduction as "adding your limb to your trunk" an effective learning device?

finger, and the fingers move away (spread out) from the middle finger (Figure 9.7d). In abduction of the thumb, the thumb moves away from the palm in the sagittal plane (see Figure 11.18c on page 358). Abduction of the toes is relative to an imaginary line drawn through the second toe. Adduction of the fingers and toes returns them to the anatomical position. Adduction of the thumb moves the thumb toward the palm in the sagittal plane (see Figure 11.18c).

Circumduction (ser-kum-DUK-shun; *circ-* = circle) is movement of the distal end of a body part in a circle (Figure 9.8). Circumduction occurs as a result of a continuous sequence of flexion, abduction, extension, and adduction. Examples of circumduction are moving the humerus in a circle at the shoulder joint (Figure 9.8a), moving the hand in a circle at the wrist joint,

Figure 9.8 **Angular movements at synovial joints—circumduction.**

 Circumduction is the movement of the distal end of a body part in a circle.

| (a) Shoulder joint | (b) Hip joint |

 Which movements in continuous sequence produce circumduction?

moving the thumb in a circle at the carpometacarpal joint, moving the fingers in a circle at the metacarpophalangeal joints (between the metacarpals and phalanges), and moving the femur in a circle at the hip joint (Figure 9.8b). Both the shoulder and hip joints permit circumduction. Flexion, abduction, extension, and adduction are more limited in the hip joints than in the shoulder joints due to the tension on certain ligaments and muscles (see Exhibits 9.1 and 9.3).

Rotation

In **rotation** (rō-TĀ-shun; *rota-* = revolve), a bone revolves around its own longitudinal axis. Pivot and ball-and-socket joints permit rotation. One example is turning the head from side-to-side at the atlanto-axial joint (between the atlas and axis), as in signifying "no" (Figure 9.9a). Another is turning the trunk from side-to-side at the intervertebral joints while keeping the hips and lower limbs in the anatomical position. In the limbs, rotation is defined relative to the midline, and specific qualifying terms are used. If the anterior surface of a bone of the limb is turned toward the midline, the movement is called *medial (internal) rotation*. You can medially rotate the humerus at the shoulder joint as follows: Starting in the anatomical position, flex your elbow and then draw your palm across the chest (Figure 9.9b). Medial rotation of the forearm at the radioulnar joints (between the radius and ulna) involves turning the palm medially from the anatomical position (see Figure 9.10h). You can medially rotate the femur at the hip joint as follows: Lie on your back, bend your knee, and then move your leg and foot laterally from the midline. Although you are moving your leg and foot laterally, the femur is rotating medially (Figure 9.9c). Medial rotation of the leg at the knee joint can be produced by sitting on a chair, bending your knee, raising your lower limb off the floor, and turning your toes medially. If the anterior surface of the bone of a limb is turned away from the midline, the movement is called *lateral (external) rotation* (see Figure 9.9b, c).

Special Movements

As noted previously, **special movements** occur only at certain joints. They include elevation, depression, protraction, retraction, inversion, eversion, dorsiflexion, plantar flexion, supination, pronation, and opposition (Figure 9.10):

- **Elevation** (el-e-VĀ-shun = to lift up) is an upward movement of a part of the body, such as closing the mouth at the temporomandibular joint (between the mandible and temporal bone) to elevate the mandible (Figure 9.10a) or shrugging the shoulders at the acromioclavicular joint to elevate the scapula.

- **Depression** (de-PRESH-un = to press down) is a downward movement of a part of the body, such as opening the mouth to depress the mandible (Figure 9.10b) or returning shrugged shoulders to the anatomical position to depress the scapula.

- **Protraction** (prō-TRAK-shun = to draw forth) is a movement of a part of the body anteriorly in the transverse plane. You can protract your mandible at the temporomandibular joint by thrusting it outward (Figure 9.10c) or protract your clavicles at the acromioclavicular and sternoclavicular joints by crossing your arms.

- **Retraction** (rē-TRAK-shun = to draw back) is a movement of a protracted part of the body back to the anatomical position (Figure 9.10d).

Figure 9.9 Rotation at synovial joints.

 In rotation, a bone revolves around its own longitudinal axis.

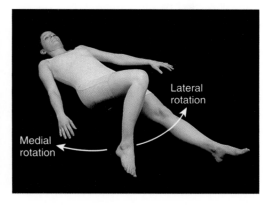

(a) Atlanto-axial joint   (b) Shoulder joint   (c) Hip joint

How do medial and lateral rotation differ?

Figure 9.10 Special movements at synovial joints.

 Special movements occur only at certain synovial joints.

(a)   Temporomandibular joint   (b)   (c)   Temporomandibular joint   (d)

(e)   Intertarsal joints   (f)   (g) Ankle joint

(h) Radioulnar joint

What movement of the shoulder girdle occurs while bringing the arms forward until the elbows touch?

- **Inversion** (in-VER-zhun = to turn inward) is movement of the soles medially at the intertarsal joints (between the tarsals) so that the soles face each other (Figure 9.10e).

- **Eversion** (ē-VER-zhun = to turn outward) is a movement of the soles laterally at the intertarsal joints so that the soles face away from each other (Figure 9.10f).

- **Dorsiflexion** (dor-si-FLEK-shun) refers to bending of the foot at the ankle or talocrural joint (between the tibia, fibula, and talus) in the direction of the dorsum (superior surface) (Figure 9.10g). Dorsiflexion occurs when you stand on your heels.

- **Plantar flexion** involves bending of the foot at the ankle joint in the direction of the plantar or inferior surface (see Figure 9.10g), as when standing on your toes.

- **Supination** (soo-pi-NĀ-shun) is a movement of the forearm at the proximal and distal radioulnar joints in which the palm is turned anteriorly or superiorly (Figure 9.10h). This position of the palms is one of the defining features of the anatomical position.

- **Pronation** (prō-NĀ-shun) is a movement of the forearm at the proximal and distal radioulnar joints in which the distal end of the radius crosses over the distal end of the ulna and the palm is turned posteriorly or inferiorly (Figure 9.10h).

- **Opposition** (op-ō-ZISH-un) is the movement of the thumb at the carpometacarpal joint (between the trapezium and metacarpal of the thumb) in which the thumb moves across the palm to touch the tips of the fingers on the same hand (see Figure 11.18c on page 358). This is the distinctive digital movement that gives humans and other primates the ability to grasp and manipulate objects very precisely.

A summary of the movements that occur at synovial joints is presented in Table 9.2.

Table 9.2   Summary of Movements at Synovial Joints

Movement	Description	Movement	Description
Gliding	Movement of relatively flat bone surfaces back-and-forth and from side-to-side over one another; little change in the angle between bones.	*Special*	Occurs at specific joints.
		Elevation	Superior movement of a body part.
Angular	Increase or decrease in the angle between bones.	Depression	Inferior movement of a body part.
		Protraction	Anterior movement of a body part in the transverse plane.
Flexion	Decrease in the angle between articulating bones, usually in the sagittal plane.	Retraction	Posterior movement of a body part in the transverse plane.
Lateral flexion	Movement of the trunk in the frontal plane.		
Extension	Increase in the angle between articulating bones, usually in the sagittal plane.	Inversion	Medial movement of the soles so that they face each other.
Hyperextension	Extension beyond the anatomical position.	Eversion	Lateral movement of the soles so that they face away from each other.
Abduction	Movement of a bone away from the midline, usually in the frontal plane.	Dorsiflexion	Bending the foot in the direction of the dorsum (superior surface).
Adduction	Movement of a bone toward the midline, usually in the frontal plane.	Plantar flexion	Bending the foot in the direction of the plantar surface (sole).
Circumduction	Flexion, abduction, extension, and adduction in succession, in which the distal end of a body part moves in a circle.	Supination	Movement of the forearm that turns the palm anteriorly or superiorly.
		Pronation	Movement of the forearm that turns the palm posteriorly or inferiorly.
Rotation	Movement of a bone around its longitudinal axis; in the limbs, it may be medial (toward midline) or lateral (away from midline).	Opposition	Movement of the thumb across the palm to touch fingertips on the same hand.

Joint Dislocation

Dislocation (dis′-lō-KĀ-shun; *dis-* = apart) or **luxation** (luks-Ā-shun; *luxatio* = dislocation) is the displacement of a bone from a joint, with resultant tearing of ligaments, tendons, and articular capsules. It is usually caused by a blow or fall. A partial or incomplete dislocation is termed a **subluxation** (sub-luks-Ā-shun). The joint most commonly dislocated in adults is the shoulder joint because its socket is quite shallow. Usually, the head of the humerus becomes displaced inferiorly, where the articular capsule is least protected. Dislocations of the mandible, elbow, fingers, knee, or hip are less common. In children, the most common upper limb dislocation involves the elbow joint (see dislocation of the radial head on page 267). ■

▶ C H E C K P O I N T

9. On yourself or with a partner, demonstrate each movement listed in Table 9.2.

SELECTED JOINTS OF THE BODY

In previous chapters we discussed the major bones and their markings. In this chapter we have examined how joints are clas-sified according to their structure and function, and we have introduced the movements that occur at joints. Table 9.3 (se-lected joints of the axial skeleton) and Table 9.4 (selected joints of the appendicular skeleton) will help you integrate the infor-mation you learned about bones with the information about joint classifications and movements. These tables list some of the major joints of the body according to their articular components (the bones that enter into their formation), their structural and functional classification, and the type(s) of movement that occurs at each joint.

Next we examine in detail four selected joints of the body in a series of exhibits. Each exhibit considers a specific synovial joint and contains (1) a definition—a description of the type of joint and the bones that form the joint; (2) the anatomical com-ponents—a description of the major connecting ligaments, articular disc, articular capsule, and other distinguishing features of the joint; and (3) the joint's possible movements. Each exhibit also refers you to a figure that illustrates the joint. The joints described are the shoulder (humeroscapular or glenohumeral) joint, elbow joint, hip (coxal) joint, and knee (tibiofemoral) joint. Because the shoulder, elbow, hip, and knee joints are described in Exhibits 9.1 through 9.4, they are not included in Tables 9.3 and 9.4.

Table 9.3   Selected Joints of the Axial Skeleton

Joint	Articular Components	Classification	Movements
Suture	Between skull bones.	*Structural:* fibrous. *Functional:* synarthrosis.	None.
Temporomandibular (TMJ)	Between condylar process of mandible and mandibular fossa and articular tubercle of temporal bone.	*Structural:* synovial (combined hinge and planar joint). *Functional:* diarthrosis.	Depression, elevation, protraction, retraction, lateral displacement, and slight rotation of mandible.
Atlanto-occipital	Between superior articular facets of atlas and occipital condyles of occipital bone.	*Structural:* synovial (condyloid). *Functional:* diarthrosis.	Flexion and extension of head and slight lateral flexion of head to either side.
Atlanto-axial	(1) Between dens of axis and anterior arch of atlas and (2) between lateral masses of atlas and axis.	*Structural:* synovial (pivot) between dens and anterior arch, and synovial (planar) between lateral masses. *Functional:* diarthrosis.	Rotation of head.
Intervertebral	(1) Between vertebral bodies and (2) between vertebral arches.	*Structural:* cartilaginous (symphysis) between vertebral bodies, and synovial (planar) between vertebral arches. *Functional:* amphiarthrosis between vertebral bodies, and diarthrosis between vertebral arches.	Flexion, extension, lateral flexion, and rotation of vertebral column.
Vertebrocostal	(1) Between facets of heads of ribs and facets of bodies of adjacent thoracic vertebrae and intervertebral discs between them and (2) between articular part of tubercles of ribs and facets of transverse processes of thoracic vertebrae.	*Structural:* synovial (planar). *Functional:* diarthrosis.	Slight gliding.
Sternocostal	Between sternum and first seven pairs of ribs.	*Structural:* cartilaginous (synchondrosis) between sternum and first pair of ribs, and synovial (planar) between sternum and second through seventh pairs of ribs. *Functional:* synarthrosis between sternum and first pair of ribs, and diarthrosis between sternum and second through seventh pairs of ribs.	None between sternum and first pair of ribs; slight gliding between sternum and second through seventh pairs of ribs.
Lumbosacral	(1) Between body of fifth lumbar vertebra and base of sacrum and (2) between inferior articular facets of fifth lumbar vertebra and superior articular facets of first vertebra of sacrum.	*Structural:* cartilaginous (symphysis) between body and base, and synovial (planar) between articular facets. *Functional:* amphiarthrosis between body and base, and diarthrosis between articular facets.	Flexion, extension, lateral flexion, and rotation of vertebral column.

Table 9.4	Selected Joints of the Appendicular Skeleton		
Joint	**Articular Components**	**Classification**	**Movements**
Sternoclavicular	Between sternal end of clavicle, manubrium of sternum, and first costal cartilage.	*Structural:* synovial (planar and pivot). *Functional:* diarthrosis.	Gliding, with limited movements in nearly every direction.
Acromioclavicular	Between acromion of scapula and acromial end of clavicle.	*Structural:* synovial (planar). *Functional:* diarthrosis.	Gliding and rotation of scapula on clavicle.
Radioulnar	Proximal radioulnar joint between head of radius and radial notch of ulna; distal radioulnar joint between ulnar notch of radius and head of ulna.	*Structural:* synovial (pivot). *Functional:* diarthrosis.	Rotation of forearm.
Wrist (radiocarpal)	Between distal end of radius and scaphoid, lunate, and triquetrum of carpus.	*Structural:* synovial (condyloid). *Functional:* diarthrosis.	Flexion, extension, abduction, adduction, circumduction, and slight hyperextension of wrist.
Intercarpal	Between proximal row of carpal bones, distal row of carpal bones, and between both rows of carpal bones (midcarpal joints).	*Structural:* synovial (planar), except for hamate, scaphoid, and lunate (midcarpal) joint, which is synovial (saddle). *Functional:* diarthrosis.	Gliding plus flexion and abduction at midcarpal joints.
Carpometacarpal	Carpometacarpal joint of thumb between trapezium of carpus and first metacarpal; carpometacarpal joints of remaining digits formed between carpus and second through fifth metacarpals.	*Structural:* synovial (saddle) at thumb and synovial (planar) at remaining digits. *Functional:* diarthrosis.	Flexion, extension, abduction, adduction, and circumduction at thumb, and gliding at remaining digits.
Metacarpophalangeal and metatarsophalangeal	Between heads of metacarpals (or metatarsals) and bases of proximal phalanges.	*Structural:* synovial (condyloid). *Functional:* diarthrosis.	Flexion, extension, abduction, adduction, and circumduction of phalanges.
Interphalangeal	Between heads of phalanges and bases of more distal phalanges.	*Structural:* synovial (hinge). *Functional:* diarthrosis.	Flexion and extension of phalanges.
Sacroiliac	Between auricular surfaces of sacrum and ilia of hip bones.	*Structural:* synovial (planar). *Functional:* diarthrosis.	Slight gliding (even more so during pregnancy).
Pubic symphysis	Between anterior surfaces of hip bones.	*Structural:* cartilaginous (symphysis). *Functional:* amphiarthrosis.	Slight movements (even more so during pregnancy).
Tibiofibular	Proximal tibiofibular joint between lateral condyle of tibia and head of fibula; distal tibiofibular joint between distal end of fibula and fibular notch of tibia.	*Structural:* synovial (planar) at proximal joint, and fibrous (syndesmosis) at distal joint. *Functional:* diarthrosis at proximal joint, and amphiarthrosis at distal joint.	Slight gliding at proximal joint, and slight rotation of fibula during dorsiflexion of foot.
Ankle (talocrural)	(1) Between distal end of tibia and its medial malleolus and talus and (2) between lateral malleolus of fibula and talus.	*Structural:* synovial (hinge). *Functional:* diarthrosis.	Dorsiflexion and plantar flexion of foot.
Intertarsal	Subtalar joint between talus and calcaneus of tarsus; talocalcaneonavicular joint between talus and calcaneus and navicular of tarsus; calcaneocuboid joint between calcaneus and cuboid of tarsus.	*Structural:* synovial (planar) at subtalar and calcaneocuboid joints, and synovial at talocalcaneonavicular joint. *Functional:* diarthrosis.	Inversion and eversion of foot.
Tarsometatarsal	Between three cuneiforms of tarsus and bases of five metatarsal bones.	*Structural:* synovial (planar). *Functional:* diarthrosis.	Slight gliding.

Exhibit 9.1 Shoulder Joint (Figure 9.11)

► **O B J E C T I V E S**

- Describe the anatomical components of the shoulder joint.
- Explain the movements that can occur at this joint.

Definition

The **shoulder joint** is a ball-and-socket joint formed by the head of the humerus and the glenoid cavity of the scapula. It also is referred to as the *humeroscapular* or *glenohumeral joint.*

Anatomical Components

1. ***Articular capsule.*** Thin, loose sac that completely envelops the joint and extends from the glenoid cavity to the anatomical neck of the humerus. The inferior part of the capsule is its weakest area.

2. ***Coracohumeral ligament.*** Strong, broad ligament that strengthens the superior part of the articular capsule and extends from the coracoid process of the scapula to the greater tubercle of the humerus.

3. ***Glenohumeral ligaments.*** Three thickenings of the articular capsule over the anterior surface of the joint. They extend from the glenoid cavity to the lesser tubercle and anatomical neck of the humerus. These ligaments are often indistinct or absent and provide only minimal strength.

4. ***Transverse humeral ligament.*** Narrow sheet extending from the greater tubercle to the lesser tubercle of the humerus.

5. ***Glenoid labrum.*** Narrow rim of fibrocartilage around the edge of the glenoid cavity. It slightly deepens and enlarges the glenoid cavity.

6. ***Bursae.*** Four *bursae* are associated with the shoulder joint. They are the *subscapular bursa, subdeltoid bursa, subacromial bursa,* and *subcoracoid bursa.*

Movements

The shoulder joint allows flexion, extension, abduction, adduction, medial rotation, lateral rotation, and circumduction of the arm (see Figures 9.6–9.9). It has more freedom of movement than any other joint of the body. This freedom results from the looseness of the articular capsule and shallowness of the glenoid cavity in relation to the large size of the head of the humerus.

Although the ligaments of the shoulder joint strengthen it to some extent, most of the strength results from the muscles that surround the joint, especially the *rotator cuff muscles.* These muscles (supraspinatus, infraspinatus, teres minor, and subscapularis) join the scapula to the humerus (see also Figure 11.15 on pages 346–347). The tendons of the muscles, together called the **rotator cuff,** encircle the joint (except for the inferior portion) and fuse with the articular capsule. The rotator cuff muscles work as a group to hold the head of the humerus in the glenoid cavity.

► **C H E C K P O I N T**

Which tendons at the shoulder joint of a baseball pitcher are most likely to be torn due to excessive circumduction?

Figure 9.11 Right shoulder (humeroscapular or glenohumeral) joint.
(See Tortora, *A Photographic Atlas of the Human Body,* Figures 4.1 and 4.2.)

🔑 Most of the stability of the shoulder joint results from the arrangement of the rotator cuff muscles.

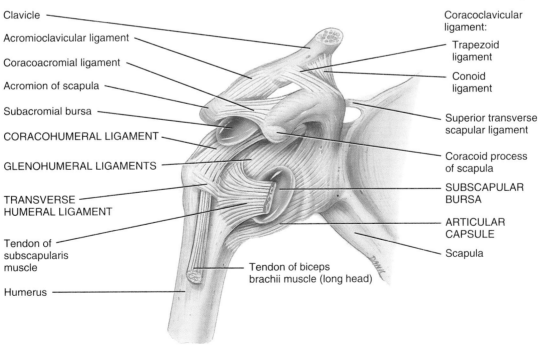

Clavicle

Acromioclavicular ligament

Coracoacromial ligament

Acromion of scapula

Subacromial bursa

CORACOHUMERAL LIGAMENT

GLENOHUMERAL LIGAMENTS

TRANSVERSE HUMERAL LIGAMENT

Tendon of subscapularis muscle

Humerus

Coracoclavicular ligament:
Trapezoid ligament
Conoid ligament

Superior transverse scapular ligament

Coracoid process of scapula

SUBSCAPULAR BURSA

ARTICULAR CAPSULE

Scapula

Tendon of biceps brachii muscle (long head)

(a) Anterior view

(continues)

Exhibit 9.1 Shoulder Joint (Figure 9.11) (continued)

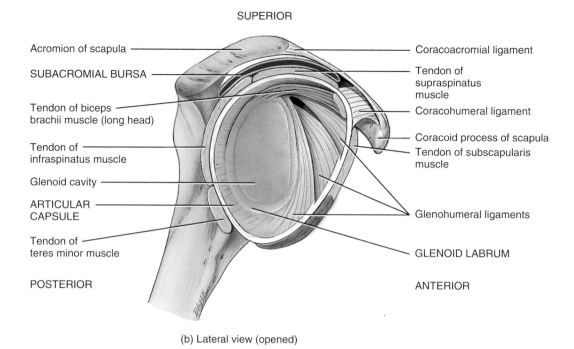

SUPERIOR

Acromion of scapula

SUBACROMIAL BURSA

Tendon of biceps brachii muscle (long head)

Tendon of infraspinatus muscle

Glenoid cavity

ARTICULAR CAPSULE

Tendon of teres minor muscle

POSTERIOR

View

Coracoacromial ligament

Tendon of supraspinatus muscle

Coracohumeral ligament

Coracoid process of scapula
Tendon of subscapularis muscle

Glenohumeral ligaments

GLENOID LABRUM

ANTERIOR

(b) Lateral view (opened)

Frontal plane

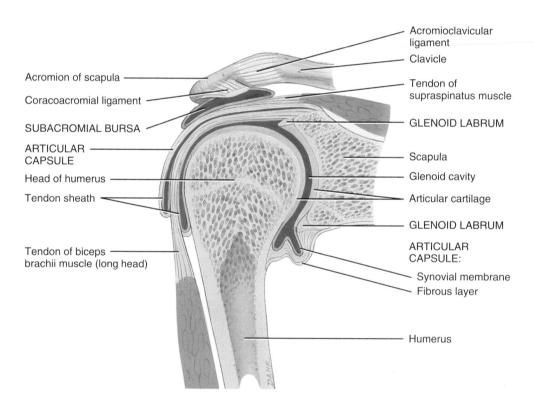

Acromion of scapula

Coracoacromial ligament

SUBACROMIAL BURSA

ARTICULAR CAPSULE

Head of humerus

Tendon sheath

Tendon of biceps brachii muscle (long head)

Acromioclavicular ligament

Clavicle

Tendon of supraspinatus muscle

GLENOID LABRUM

Scapula

Glenoid cavity

Articular cartilage

GLENOID LABRUM

ARTICULAR CAPSULE:

Synovial membrane
Fibrous layer

Humerus

(c) Frontal section

 Why does the shoulder joint have more freedom of movement than any other joint of the body?

Exhibit 9.2   Elbow Joint (Figure 9.12)

► **O B J E C T I V E S**

• Describe the anatomical components of the elbow joint.
• Explain the movements that can occur at this joint.

Definition

The **elbow joint** is a hinge joint formed by the trochlea of the humerus, the trochlear notch of the ulna, and the head of the radius.

Anatomical Components

1. ***Articular capsule.*** The anterior part covers the anterior part of the joint, from the radial and coronoid fossae of the humerus to the coronoid process of the ulna and the annular ligament of the radius. The posterior part extends from the capitulum, olecranon fossa, and lateral epicondyle of the humerus to the annular ligament of the radius, the olecranon of the ulna, and the ulna posterior to the radial notch.

2. ***Ulnar collateral ligament.*** Thick, triangular ligament that extends from the medial epicondyle of the humerus to the coronoid process and olecranon of the ulna.

3. ***Radial collateral ligament.*** Strong, triangular ligament that extends from the lateral epicondyle of the humerus to the annular ligament of the radius and the radial notch of the ulna.

Movements

The elbow joint allows flexion and extension of the forearm (see Figure 9.6c).

► C H E C K P O I N T

At the elbow joint, which ligaments connect (a) the humerus and the ulna, and (b) the humerus and the radius?

Figure 9.12   Right elbow joint.   (See Tortora, *A Photographic Atlas of the Human Body,* Figures 4.3 and 4.4.)

🔑 **The elbow joint is formed by parts of three bones: humerus, ulna, and radius.**

(a) Medial aspect

(b) Lateral aspect

 Which movements are possible at a hinge joint?

Exhibit 9.3 Hip Joint (Figure 9.13)

▶ OBJECTIVES

- Describe the anatomical components of the hip joint.
- Explain the movements that can occur at this joint.

Definition

The **hip joint** *(coxal joint)* is a ball-and-socket joint formed by the head of the femur and the acetabulum of the hip bone.

Anatomical Components

1. **Articular capsule.** Very dense and strong capsule that extends from the rim of the acetabulum to the neck of the femur. One of the strongest structures of the body, the capsule consists of circular and longitudinal fibers. The circular fibers, called the *zona orbicularis,* form a collar around the neck of the femur. Accessory ligaments known as the iliofemoral ligament, pubofemoral ligament, and ischiofemoral ligament reinforce the longitudinal fibers.

2. **Iliofemoral ligament.** Thickened portion of the articular capsule that extends from the anterior inferior iliac spine of the hip bone to the intertrochanteric line of the femur.

3. **Pubofemoral ligament.** Thickened portion of the articular capsule that extends from the pubic part of the rim of the acetabulum to the neck of the femur.

4. **Ischiofemoral ligament.** Thickened portion of the articular capsule that extends from the ischial wall of the acetabulum to the neck of the femur.

5. **Ligament of the head of the femur.** Flat, triangular band that extends from the fossa of the acetabulum to the fovea capitis of the head of the femur.

6. **Acetabular labrum.** Fibrocartilage rim attached to the margin of the acetabulum that enhances the depth of the acetabulum. Because the diameter of the acetabular rim is smaller than that of the head of the femur, dislocation of the femur is rare.

7. **Transverse ligament of the acetabulum.** Strong ligament that crosses over the acetabular notch. It supports part of the acetabular labrum and is connected with the ligament of the head of the femur and the articular capsule.

Movements

The hip joint allows flexion, extension, abduction, adduction, circumduction, medial rotation, and lateral rotation of thigh (see Figures 9.6–9.9). The extreme stability of the hip joint is related to the very strong articular capsule and its accessory ligaments, the manner in which the femur fits into the acetabulum, and the muscles surrounding the joint. Although the shoulder and hip joints are both ball-and-socket joints, the movements at the hip joints do not have as wide a range of motion. Flexion is limited by the anterior surface of the thigh coming into contact with the anterior abdominal wall when the knee is flexed and by tension of the hamstring muscles when the knee is extended. Extension is limited by tension of the iliofemoral, pubofemoral, and ischiofemoral ligaments. Abduction is limited by the tension of the pubofemoral ligament, and adduction is limited by contact with the opposite limb and tension in the ligament of the head of the femur. Medial rotation is limited by the tension in the ischiofemoral ligament, and lateral rotation is limited by tension in the iliofemoral and pubofemoral ligaments.

▶ CHECKPOINT

What factors limit the degree of flexion and abduction at the hip joint?

Figure 9.13 Right hip (coxal) joint. (See Tortora, *A Photographic Atlas of the Human Body,* Figure 4.5.)

The articular capsule of the hip joint is one of the strongest structures in the body.

- Rectus femoris tendon
- Greater trochanter of femur
- PUBOFEMORAL LIGAMENT
- ILIOFEMORAL LIGAMENT
- Obturator canal
- Obturator membrane
- Lesser trochanter of femur
- Hip bone
- Femur

(a) Anterior view

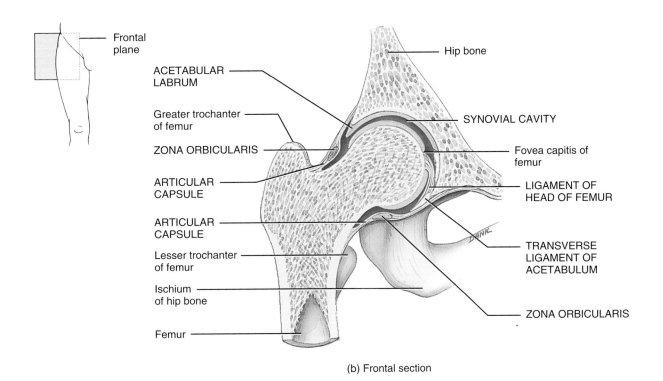

Frontal plane

ACETABULAR LABRUM

Greater trochanter of femur

ZONA ORBICULARIS

ARTICULAR CAPSULE

ARTICULAR CAPSULE

Lesser trochanter of femur

Ischium of hip bone

Femur

Hip bone

SYNOVIAL CAVITY

Fovea capitis of femur

LIGAMENT OF HEAD OF FEMUR

TRANSVERSE LIGAMENT OF ACETABULUM

ZONA ORBICULARIS

(b) Frontal section

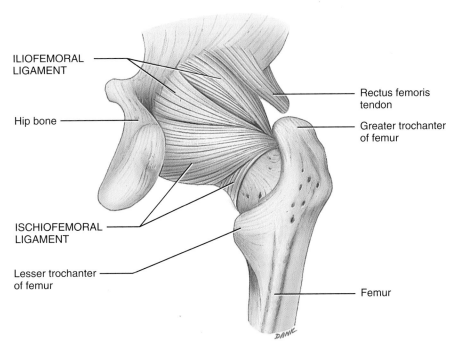

ILIOFEMORAL LIGAMENT

Hip bone

ISCHIOFEMORAL LIGAMENT

Lesser trochanter of femur

Rectus femoris tendon

Greater trochanter of femur

Femur

(c) Posterior view

 Which ligaments limit the degree of extension that is possible at the hip joint?

Exhibit 9.4   Knee Joint (Figure 9.14)

▶ **OBJECTIVES**

• Describe the anatomical components of the knee joint.
• Explain the movements that can occur at this joint.

Definition

The **knee joint** *(tibiofemoral joint)* is the largest and most complex joint of the body, actually consisting of three joints within a single synovial cavity:
• Laterally is a tibiofemoral joint, between the lateral condyle of the femur, lateral meniscus, and lateral condyle of the tibia. It is a modified hinge joint.
• Medially is a second tibiofemoral joint, between the medial condyle of the femur, medial meniscus, and medial condyle of the tibia. It also a modified hinge joint.
• An intermediate patellofemoral joint, between the patella and the patellar surface of the femur, is a planar joint.

Anatomical Components

1. **Articular capsule.** No complete, independent capsule unites the bones. The ligamentous sheath surrounding the joint consists mostly of muscle tendons or expansions of them. There are, however, some capsular fibers connecting the articulating bones.

2. **Medial and lateral patellar retinacula.** Fused tendons of insertion of the quadriceps femoris muscle and the fascia lata (deep fascia of thigh) that strengthen the anterior surface of the joint.

3. **Patellar ligament.** Continuation of the common tendon of insertion of the quadriceps femoris muscle that extends from the patella to the tibial tuberosity. This ligament also strengthens the anterior surface of the joint. The posterior surface of the ligament is separated from the synovial membrane of the joint by an infrapatellar fat pad.

4. **Oblique popliteal ligament.** Broad, flat ligament that extends from the intercondylar fossa of the femur to the head of the tibia. The tendon of the semimembranosus muscle is superficial to the ligament and passes from the medial condyle of the tibia to the lateral condyle of the femur. The ligament and tendon strengthen the posterior surface of the joint.

5. **Arcuate popliteal ligament.** Extends from the lateral condyle of the femur to the styloid process of the head of the fibula. It strengthens the lower lateral part of the posterior surface of the joint.

6. **Tibial collateral ligament.** Broad, flat ligament on the medial surface of the joint that extends from the medial condyle of the femur to the medial condyle of the tibia. Tendons of the sartorius, gracilis, and semitendinosus muscles, all of which strengthen the medial aspect of the joint, cross the ligament. Because the tibial collateral ligament is firmly attached to the medial meniscus, tearing of the ligament frequently results in tearing of the meniscus and damage to the anterior cruciate ligament, described under 8a.

7. **Fibular collateral ligament.** Strong, rounded ligament on the lateral surface of the joint that extends from the lateral condyle of the femur to the lateral side of the head of the fibula. It strengthens the lateral aspect of the joint. The ligament is covered by the tendon of the biceps femoris muscle. The tendon of the popliteal muscle is deep to the ligament.

8. **Intracapsular ligaments.** Ligaments within the capsule that connect the tibia and femur. The anterior and posterior cruciate ligaments (KROO-shē-āt = shaped like a cross) are named based on their origins relative to the intercondylar area of the tibia. Following their originations, they cross on their way to their destinations on the femur.

    a. **Anterior cruciate ligament (ACL).** Extends posteriorly and laterally from a point *anterior* to the intercondylar area of the tibia to the posterior part of the medial surface of the lateral condyle of the femur. This ligament is stretched or torn in about 70% of all serious knee injuries. The ACL limits hyperextension of the knee and prevents the anterior sliding of the tibia on the femur.

    b. **Posterior cruciate ligament (PCL).** Extends anteriorly and medially from a depression on the *posterior* intercondylar area of the tibia and lateral meniscus to the anterior part of the lateral surface of the medial condyle of the femur. The PCL prevents the posterior sliding of the tibia (and anterior sliding of the femur) when the knee is flexed. This is very important when walking down stairs or a steep incline.

9. **Articular discs (menisci).** Two fibrocartilage discs between the tibial and femoral condyles that help compensate for the irregular shapes of the bones and circulate synovial fluid.

    a. **Medial meniscus.** Semicircular piece of fibrocartilage (C-shaped). Its anterior end is attached to the anterior intercondylar fossa of the tibia, anterior to the anterior cruciate ligament. Its posterior end is attached to the posterior intercondylar fossa of the tibia between the attachments of the posterior cruciate ligament and lateral meniscus.

    b. **Lateral meniscus.** Nearly circular piece of fibrocartilage (approaches an incomplete O in shape). Its anterior end is attached anterior to the intercondylar eminence of the tibia, and lateral and posterior to the anterior cruciate ligament. Its posterior end is attached posterior to the intercondylar eminence of the tibia, and anterior to the posterior end of the medial meniscus. The medial and lateral menisci are connected to each other by the *transverse ligament* and to the margins of the head of the tibia by the *coronary ligaments* (not illustrated).

10. The more important *bursae* of the knee include the following:

    a. *Prepatellar bursa* between the patella and skin.

    b. *Intrapatellar bursa* between superior part of tibia and patellar ligament.

    c. *Suprapatellar bursa* between inferior part of femur and deep surface of quadriceps femoris muscle.

Movements

The knee joint allows flexion, extension, slight medial rotation, and lateral rotation of leg in flexed position (see Figures 9.6f and 9.9c).

▶ **CHECKPOINT**

Why is the knee joint said to be "three joints in one"?

Figure 9.14 **Right knee (tibiofemoral) joint.** (See Tortora, *A Photographic Atlas of the Human Body,* Figures 4.6 through 4.8.)

🔑 **The knee joint is the largest and most complex joint in the body.**

Femur

SUPRAPATELLAR BURSA

Quadriceps femoris tendon

Vastus lateralis muscle

Vastus medialis muscle

Patella

MEDIAL PATELLAR RETINACULUM

LATERAL PATELLAR RETINACULUM

INFRAPATELLAR FAT PAD

FIBULAR COLLATERAL LIGAMENT

TIBIAL COLLATERAL LIGAMENT

Head of fibula

ARTICULAR CAPSULE

INFRAPATELLAR BURSA

PATELLAR LIGAMENT

Fibula

Tibia

(a) Anterior superficial view

Femur

ARTICULAR CAPSULE

Adductor magnus tendon

Lateral head of gastrocnemius muscle

Medial head of gastrocnemius muscle

OBLIQUE POPLITEAL LIGAMENT

TIBIAL COLLATERAL LIGAMENT

ARCUATE POPLITEAL LIGAMENT

Popliteus muscle

FIBULAR COLLATERAL LIGAMENT

Semimembranosus tendon

Tibia

Posterior ligament of head of fibula

Fibula

(b) Posterior deep view

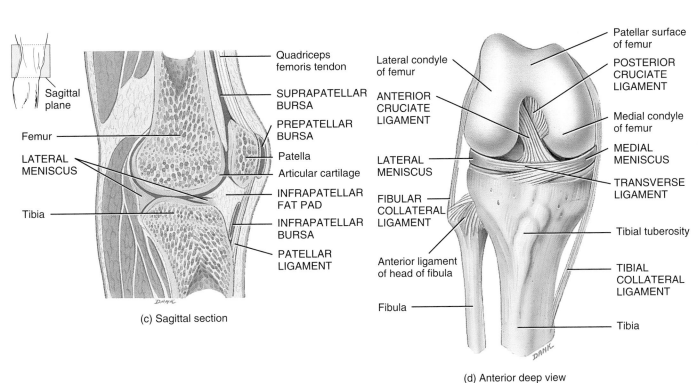

Sagittal plane

Femur

Quadriceps femoris tendon

SUPRAPATELLAR BURSA

PREPATELLAR BURSA

LATERAL MENISCUS

Patella

Articular cartilage

Tibia

INFRAPATELLAR FAT PAD

INFRAPATELLAR BURSA

PATELLAR LIGAMENT

(c) Sagittal section

Lateral condyle of femur

Patellar surface of femur

POSTERIOR CRUCIATE LIGAMENT

ANTERIOR CRUCIATE LIGAMENT

Medial condyle of femur

LATERAL MENISCUS

MEDIAL MENISCUS

FIBULAR COLLATERAL LIGAMENT

TRANSVERSE LIGAMENT

Tibial tuberosity

Anterior ligament of head of fibula

TIBIAL COLLATERAL LIGAMENT

Fibula

Tibia

(d) Anterior deep view

🦵 **What movement occurs at the knee joint when the quadriceps femoris (anterior thigh) muscles contract?**

265

FACTORS AFFECTING CONTACT AND RANGE OF MOTION AT SYNOVIAL JOINTS

▶ O B J E C T I V E

• **Describe five factors that influence the type of movement and range of motion possible at a synovial joint.**

The articular surfaces of synovial joints contact one another and determine the type and range of motion that is possible. **Range of motion (ROM)** is the range, measured in degrees of a circle, through which the bones of a joint can be moved. The following factors contribute to keeping the articular surfaces in contact and affect range of motion:

1. *Structure or shape of the articulating bones.* The structure or shape of the articulating bones determines how closely they can fit together. The articular surfaces of some bones interlock with one another. This spatial relationship is very obvious at the hip joint, where the head of the femur articulates with the acetabulum of the hip bone. An interlocking fit allows rotational movement.

2. *Strength and tension of the joint ligaments.* The different components of a fibrous capsule are tense or taut only when the joint is in certain positions. Tense ligaments restrict the range of motion and guide the movement of the articulating bones. In the knee joint, for example, the anterior cruciate ligament is taut and the posterior cruciate ligament is loose when the knee is straightened, and the reverse occurs when the knee is bent.

3. *Arrangement and tension of the muscles.* Muscle tension reinforces the restraint placed on a joint by its ligaments, and thus restricts movement. A good example of the effect of muscle tension on a joint is seen at the hip joint. When the thigh is raised with the knee straight, the movement is restricted by the tension of the hamstring muscles on the posterior surface of the thigh. But if the knee is bent, the tension on the hamstring muscles is lessened, and the thigh can be raised farther.

4. *Apposition of soft parts.* The point at which one body surface contacts another may limit mobility. For example, if you bend your arm at the elbow, it can move no farther after the anterior surface of the forearm meets with and presses against the biceps brachii muscle of the arm.

5. *Disuse.* Movement at a joint may be restricted if a joint has not been used for an extended period. For example, if an elbow joint is immobilized by a cast, only a limited range of motion at the joint may be possible for a time after the cast is removed. Restricted movement due to disuse may result from decreased amounts of synovial fluid, from diminished flexibility of ligaments and tendons, and from muscular atrophy, a reduction in size or wasting of a muscle.

▶ C H E C K P O I N T

10. Besides the hip joint, which joints have bones that interlock and allow rotational movement?

AGING AND JOINTS

▶ O B J E C T I V E

• **Explain the effects of aging on joints.**

Aging usually results in decreased production of synovial fluid in joints. In addition, the articular cartilage becomes thinner with age, and ligaments shorten and lose some of their flexibility. The effects of aging on joints, which vary considerably from one person to another, are affected by genetic factors and by wear and tear. Although degenerative changes in joints may begin in those as young as 20 years of age, most changes do not occur until much later. By age 80, almost everyone develops some type of degeneration in the knees, elbows, hips, and shoulders. Males also commonly develop degenerative changes in the vertebral column, resulting in a hunched-over posture and pressure on nerve roots. One type of arthritis, called osteoarthritis, is at least partially age-related. Nearly everyone over age 70 has evidence of some osteoarthritic changes.

▶ C H E C K P O I N T

11. Which joints show evidence of degeneration in nearly all individuals as aging progresses?

DISORDERS: HOMEOSTATIC IMBALANCES

Common Joint Injuries

Rotator cuff injury is a common injury among baseball pitchers and volleyball players due to shoulder movements that involve vigorous circumduction. Most often, there is tearing of the supraspinatus muscle tendon of the rotator cuff. This tendon is especially predisposed to wear-and-tear changes because of its location between the head of the humerus and acromion of the scapula, which compresses the tendon during shoulder movements. A **separated shoulder** is an injury of the acromioclavicular joint, the joint formed by the acromion of the scapula and the acromial end of the clavicle. It most often happens with forceful trauma, as may happen when the shoulder strikes the ground in a fall.

Tennis elbow most commonly refers to pain at or near the lateral epicondyle of the humerus, usually caused by an improperly executed backhand. The extensor muscles strain or sprain, resulting in pain. **Little-league elbow** typically develops because of a heavy pitching schedule or throwing many curve balls, especially in youngsters. In this injury, the elbow may enlarge, fragment, or separate.

A **dislocation of the radial head** is the most common upper limb dislocation in children. In this injury, the head of the radius slides past or ruptures the radial annular ligament, a ligament that forms a collar around the head of the radius at the proximal radioulnar joint. Dislocation is most apt to occur when a strong pull is applied to the forearm while it is extended and supinated, for instance while swinging a child around with outstretched arms.

The knee joint is the joint most vulnerable to damage because of the stresses to which it is subjected and because there is no interlocking of the articulating bones. A **swollen knee** may occur immediately or hours after an injury. Immediate swelling is due to escape of blood from damaged blood vessels adjacent to areas involving rupture of the anterior cruciate ligament, damage to synovial membranes, torn menisci, fractures, or collateral ligament sprains. Delayed swelling is due to excessive production of synovial fluid, a condition commonly referred to as "water on the knee." A common type of knee injury in football is **rupture of the tibial collateral ligaments,** often associated with tearing of the anterior cruciate ligament and medial meniscus (torn cartilage). Usually, a hard blow to the lateral side of the knee while the foot is fixed on the ground causes the damage. A **dislocated knee** refers to the displacement of the tibia relative to the femur. The most common type is dislocation anteriorly, resulting from hyperextension of the knee. A frequent consequence of a dislocated knee is damage to the popliteal artery.

Rheumatism and Arthritis

Rheumatism (ROO-ma-tizm) is any painful disorder of the supporting structures of the body—bones, ligaments, tendons, or muscles—that is not caused by infection or injury. **Arthritis** is a form of rheumatism in which the joints are swollen, stiff, and painful. It afflicts about 40 million people in the United States.

Rheumatoid Arthritis

Rheumatoid arthritis (RA) is an autoimmune disease in which the immune system of the body attacks its own tissues—in this case, its own cartilage and joint linings. RA is characterized by inflammation of the joint, which causes swelling, pain, and loss of function. Usually, this form of arthritis occurs bilaterally: If one wrist is affected, the other is also likely to be affected, although often not to the same degree.

The primary symptom of rheumatoid arthritis is inflammation of the synovial membrane. If untreated, the membrane thickens, and synovial fluid accumulates. The resulting pressure causes pain and tenderness. The membrane then produces an abnormal granulation tissue, called *pannus,* that adheres to the surface of the articular cartilage and sometimes erodes the cartilage completely. When the cartilage is destroyed, fibrous tissue joins the exposed bone ends. The fibrous tissue ossifies and fuses the joint so that it becomes immovable—the ultimate crippling effect of rheumatoid arthritis. The growth of the granulation tissue causes the distortion of the fingers that characterizes hands of RA sufferers.

Osteoarthritis

Osteoarthritis (OA) (os'-tē-ō-ar-THRĪ-tis) is a degenerative joint disease in which joint cartilage is gradually lost. It results from a combination of aging, irritation of the joints, and wear and abrasion. Commonly known as "wear-and-tear" arthritis, osteoarthritis is the leading cause of disability in older persons.

Osteoarthritis is a progressive disorder of synovial joints, particularly weight-bearing joints. Articular cartilage deteriorates and new bone forms in the subchondral areas and at the margins of the joint. The cartilage slowly degenerates, and as the bone ends become exposed, spurs (small bumps) of new osseous tissue are deposited on them. These spurs decrease the space of the joint cavity and restrict joint movement. Unlike rheumatoid arthritis, osteoarthritis affects mainly the articular cartilage, although the synovial membrane often becomes inflamed late in the disease. A major distinction between osteoarthritis and rheumatoid arthritis is that osteoarthritis first afflicts the larger joints (knees, hips), whereas rheumatoid arthritis first strikes smaller joints.

Gouty Arthritis

Uric acid (a substance that gives urine its name) is a waste product produced during the metabolism of nucleic acid (DNA and RNA) subunits. A person who suffers from **gout** (GOWT) either produces excessive amounts of uric acid or is not able to excrete as much as normal. The result is a buildup of uric acid in the blood. This excess acid then reacts with sodium to form a salt called sodium urate. Crystals of this salt accumulate in soft tissues such as the kidneys and in the cartilage of the ears and joints.

In **gouty arthritis,** sodium urate crystals are deposited in the soft tissues of the joints. The crystals irritate and erode the cartilage, causing inflammation, swelling, and acute pain. Eventually, the crystals destroy all joint tissues. If the disorder is untreated, the ends of the articulating bones fuse, and the joint becomes immovable.

Lyme Disease

A spiral-shaped bacterium called *Borrelia burgdorferi* causes **Lyme disease,** named for the town of Lyme, Connecticut, where it was first reported in 1975. The bacteria are transmitted to humans mainly by deer ticks *(Ixodes dammini).* These ticks are so small that their bites often go unnoticed. Within a few weeks of the tick bite, a rash may appear at the site. Although the rash often resembles a bull's eye target, there are many variations, and some people never develop a rash. Other symptoms include joint stiffness, fever and chills, headache, stiff neck, nausea, and low back pain. In advanced stages of the disease, arthritis is the main complication. It usually afflicts the larger joints such as the knee, ankle, hip, elbow, or wrist. Antibiotics are generally effective against Lyme disease, especially if they are given promptly. However, some symptoms may linger for years.

MEDICAL TERMINOLOGY

Arthralgia   (ar-THRAL-jē-a; *arthr-* = joint; *-algia* = pain) Pain in a joint.

Bursectomy   (bur-SEK-tō-mē; *-ectomy* = removal of) Removal of a bursa.

Chondritis   (kon-DRĪ-tis; *chondr-* = cartilage)   Inflammation of cartilage.

Synovitis   (sin′-ō-VĪ-tis) Inflammation of a synovial membrane in a joint.

 # STUDY OUTLINE

INTRODUCTION (p. 244)

1. A joint (articulation or arthrosis) is a point of contact between two bones, between bone and cartilage, or between bone and teeth.
2. A joint's structure may permit no movement, slight movement, or free movement.

JOINT CLASSIFICATIONS (p. 244)

1. Structural classification is based on the presence or absence of a synovial cavity and the type of connecting tissue. Structurally, joints are classified as fibrous, cartilaginous, or synovial.
2. Functional classification of joints is based on the degree of movement permitted. Joints may be synarthroses (immovable), amphiarthroses (slightly movable), or diarthroses (freely movable).

FIBROUS JOINTS (p. 244)

1. Bones held together by fibrous connective tissue are fibrous joints.
2. These joints include immovable sutures (found between skull bones), slightly movable syndesmoses (such as the distal tibiofibular joint), and immovable gomphoses (roots of teeth in alveoli of mandible and maxilla).

CARTILAGINOUS JOINTS (p. 245)

1. Bones held together by cartilage are cartilaginous joints.
2. These joints include immovable synchondroses united by hyaline cartilage (epiphyseal plates between diaphyses and epiphyses) and slightly movable symphyses united by fibrocartilage (pubic symphysis).

SYNOVIAL JOINTS (p. 246)

1. Synovial joints contain a space between bones called the synovial cavity. All synovial joints are diarthroses.
2. Other characteristics of synovial joints are the presence of articular cartilage and an articular capsule, made up of a fibrous capsule and a synovial membrane.
3. The synovial membrane secretes synovial fluid, which forms a thin, viscous film over the surfaces within the articular capsule.

4. Many synovial joints also contain accessory ligaments (extracapsular and intracapsular) and articular discs (menisci).

5. Synovial joints contain an extensive nerve and blood supply. The nerves convey information about pain, joint movements, and the degree of stretch at a joint. Blood vessels penetrate the articular capsule and ligaments.

6. Bursae are saclike structures, similar in structure to joint capsules, that reduce friction in joints such as the shoulder and knee joints.

7. Tendon sheaths are tubelike bursae that wrap around tendons where there is considerable friction.

8. Subtypes of synovial joints are planar, hinge, pivot, condyloid, saddle, and ball-and-socket.

9. Planar joint: articulating surfaces are flat; bone glides back and forth and side to side (nonaxial); found between carpals and tarsals.

10. Hinge joint: convex surface of one bone fits into concave surface of another; motion is angular around one axis (monaxial); examples are the elbow, knee, and ankle joints.

11. Pivot joint: a round or pointed surface of one bone fits into a ring formed by another bone and a ligament; movement is rotation (monaxial); examples are the atlanto-axial and radioulnar joints.

12. Condyloid joint: an oval-shaped projection of one bone fits into an oval cavity of another; motion is angular around two axes (biaxial); examples are the wrist joint and metacarpophalangeal joints for the second through fifth digits.

13. Saddle joint: articular surface of one bone is shaped like a saddle and the other bone fits into the "saddle" like a sitting rider; motion is angular around two axes (biaxial); an example is the carpometacarpal joint between the trapezium and the metacarpal of the thumb.

14. Ball-and-socket joint: ball-shaped surface of one bone fits into cuplike depression of another; motion is angular and rotational around three axes and all directions in between (multiaxial); examples are the shoulder and hip joints.

15. Table 9.1 on page 250 summarizes the structural and functional categories of joints.

TYPES OF MOVEMENTS AT SYNOVIAL JOINTS (p. 251)

1. In a gliding movement, the nearly flat surfaces of bones move back-and-forth and from side-to-side. Gliding movements occur at planar joints.

2. In angular movements, a change in the angle between bones occurs. Examples are flexion–extension, lateral flexion, hyperextension, and abduction–adduction. Circumduction refers to flexion, abduction, extension, and adduction in succession. Angular movements occur at hinge, condyloid, saddle, and ball-and-socket joints.

3. In rotation, a bone moves around its own longitudinal axis. Rotation can occur at pivot and ball-and-socket joints.

4. Special movements occur at specific synovial joints. Examples are elevation–depression, protraction–retraction, inversion–eversion, dorsiflexion–plantar flexion, supination–pronation, and opposition.

5. Table 9.2 on page 256 summarizes the various types of movements at synovial joints.

SELECTED JOINTS OF THE BODY (p. 256)

1. Tables 9.3 and 9.4 on pages 257–258 provide a summary of selected joints of the body, including articular components, structural and functional classifications, and movements.

2. The shoulder (humeroscapular or glenohumeral) joint is between the head of the humerus and glenoid cavity of the scapula (Exhibit 9.1 on page 259).

3. The elbow joint is between the trochlea of the humerus, the trochlear notch of the ulna, and the head of the radius (Exhibit 9.2 on page 261).

4. The hip (coxal) joint is between the head of the femur and acetabulum of the hip bone (Exhibit 9.3 on page 262).

5. The knee (tibiofemoral) joint is between the patella and patellar surface of femur; lateral condyle of femur, lateral meniscus, and lateral condyle of tibia; and medial condyle of femur, medial meniscus, and medial condyle of tibia (Exhibit 9.4 on page 264).

FACTORS AFFECTING CONTACT AND RANGE OF MOTION AT SYNOVIAL JOINTS (p. 266)

1. The ways that articular surfaces of synovial joints contact one another determines the type of movement that is possible.

2. Factors that contribute to keeping the surfaces in contact and affect range of motion are structure or shape of the articulating bones, strength and tension of the joint ligaments, arrangement and tension of the muscles, apposition of soft parts, hormones, and disuse.

AGING AND JOINTS (p. 266)

1. With aging, a decrease in synovial fluid, thinning of articular cartilage, and decreased flexibility of ligaments occur.

2. Most individuals experience some degeneration in the knees, elbows, hips, and shoulders due to the aging process.

SELF-QUIZ QUESTIONS

Fill in the blanks in the following statements.

1. A point of contact between two bones, between bone and cartilage, or between bone and teeth is called a(n) _____.

2. _____ lubricates, reduces friction, and supplies nutrients to and removes wastes from joints.

3. A degenerative joint disease that is a leading cause of disability in the elderly is _____.

Choose the one best answer to the following questions.

4. A joint where bones are held together by collagen-rich fibrous connective tissue is a (a) diarthrotic joint, (b) synovial joint, (c) cartilaginous joint, (d) fibrous joint, (e) functional joint.

5. A slightly movable joint is a(n) (a) synarthrosis, (b) amphiarthrosis, (c) diarthrosis, (d) suture, (e) synchondrosis.

6. Which of the following is a cartilaginous joint? (a) symphysis, (b) gomphosis, (c) suture, (d) syndesmosis, (e) synovial joint.

7. Which of the following statements are *true*? (1) The bones at a synovial joint are covered by a mucous membrane. (2) The articular capsule surrounds a synovial joint, encloses the synovial cavity, and unites the articulating bones. (3) The fibrous capsule of the articular capsule permits considerable movement at a joint. (4) The tensile strength of the fibrous capsule helps prevent bones from disarticulating. (5) All joints contain a fibrous capsule. (a) 1, 2, 3, and 4, (b) 2, 3, 4, and 5, (c) 2, 3, and 4, (d) 1, 2, and 3, (e) 2, 4, and 5.

8. Which of the following keep the articular surfaces of synovial joints in contact and affect range of motion? (1) structure or shape of the articulating bones, (2) strength and tension of the joint ligaments, (3) arrangement and tension of muscles, (4) lack of use, (5) apposition of soft parts. (a) 1, 2, 3, and 5, (b) 2, 3, 4, and 5, (c) 1, 3, 4, and 5, (d) 1, 3, and 5, (e) 1, 2, 3, 4, and 5.

9. Which of the following does *not* reduce friction in joints? (a) bursae, (b) articular cartilage, (c) tendon sheaths, (d) synovial fluid, (e) accessory ligaments.

Indicate whether the following statements are true or false.

10. The fibers of some fibrous capsules are arranged in parallel bundles called ligaments.

11. Movement of the trunk sideways at the waist is an example of hyperextension.

12. Match the following:
    ___ (a) movement in which relatively flat bone surfaces move back-and-forth and side-to-side with respect to one another
    ___ (b) decrease in angle between bones
    ___ (c) increase in angle between bones
    ___ (d) movement of bone away from midline
    ___ (e) movement of bone toward midline
    ___ (f) movement of distal end of a part of the body in a circle
    ___ (g) a bone revolves around its own longitudinal axis

    (1) abduction
    (2) rotation
    (3) flexion
    (4) adduction
    (5) gliding
    (6) extension
    (7) circumduction

13. Match the following:
    ___ (a) a fibrous joint that unites the bones of the skull; a synarthrosis
    ___ (b) a fibrous joint between the tibia and fibula; an amphiarthrosis
    ___ (c) the articulation between bone and teeth
    ___ (d) the epiphyseal plate
    ___ (e) joint between the two pubic bones
    ___ (f) a bony joint

    (1) synostosis
    (2) synchondrosis
    (3) syndesmosis
    (4) suture
    (5) symphysis
    (6) gomphosis

14. Match the following:

____ (a) rounded or pointed surface of one bone articulates with a ring formed by another bone and a ligament; allows rotation around its own axis

____ (b) articulating bone surfaces are flat or slightly curved; permit gliding movement

____ (c) convex, oval projection of one bone fits into oval depression of another bone; permits movement in two axes

____ (d) convex surface of one bone articulates with concave surface of another bone; permits flexion and extension

____ (e) ball-shaped surface of one bone articulates with cuplike depression of another bone; permits largest degree of movement

____ (f) modified condyloid joint where articulating bones resemble a rider sitting in a saddle

(1) hinge joint
(2) saddle joint
(3) ball-and-socket joint
(4) planar joint
(5) condyloid joint
(6) pivot joint

15. Match the following:

____ (a) upward movement of a body part

____ (b) downward movement of a body part

____ (c) movement of a body part anteriorly in the transverse plane

____ (d) movement of an anteriorly projected body part back to the anatomical position

____ (e) movement of the soles medially at the intertarsal joints

____ (f) movement of the soles laterally

____ (g) action that occurs when you stand on your heels

____ (h) action that occurs when you stand on your toes

____ (i) movement of the forearm to turn the palm anteriorly

____ (j) movement of the forearm to turn the palm posteriorly

____ (k) movement of thumb across the palm to touch the tips of the fingers of the same hand

(1) pronation
(2) plantar flexion
(3) eversion
(4) retraction
(5) opposition
(6) elevation
(7) depression
(8) inversion
(9) protraction
(10) dorsiflexion
(11) supination

CRITICAL THINKING QUESTIONS

1. Katie loves pretending that she's a human cannonball. As she jumps off the diving board, she assumes the proper position before she pounds into the water: head and thighs tucked against her chest; back rounded; arms pressed against her sides while her forearms, crossed in front of her shins, hold her legs tightly folded against her chest. Use the proper anatomical terms to describe the position of Katie's back, head, and limbs.
   HINT *She is about as far as she can get from the anatomical position.*

2. After the cast was removed from his arm, 85-year-old Mr. Singh was referred to physical therapy for a frozen elbow joint. When the therapist said she would be working to free up his "hinge," Mr. Singh complained, "You're supposed to be fixing my elbow, not my back door!" Explain the situation to Mr. Singh.
   HINT *Think about how a door moves.*

3. Kendal was sitting in the cafeteria about 30 minutes before his big anatomy and physiology exam. "I'm not worried," he told his lab partner while he drank his third cup of coffee. "I've got it all figured out. There are three types of joints: Cartilaginous joints are made out of hyaline cartilage, like the intervertebral discs and the suture; fibrous joints have an articular capsule; and the serous joints have cavities that do not open to the outside of the body." Kendal's lab partner was speechless. Can you explain the three types better?
   HINT *Synovial fluid lubricates diarthroses.*

ANSWERS TO FIGURE QUESTIONS

9.1 Functionally, sutures are classified as synarthroses because they are immovable; syndesmoses are classified as amphiarthroses because they are slightly movable.

9.2 The difference between a synchondrosis and a symphysis is the type of cartilage that holds the joint together: hyaline cartilage in a synchondrosis and fibrocartilage in a symphysis.

9.3 Functionally, synovial joints are diarthroses, freely movable joints.

9.4 Condyloid and saddle joints are biaxial joints.

9.5 Gliding movements occur at planar joints.

9.6 Two examples of flexion that do not occur in the sagittal plane are flexion of the thumb and lateral flexion of the trunk.

9.7 When you adduct your arm or leg, you bring it closer to the midline of the body, thus "adding" it to the trunk.

9.8 Circumduction involves flexion, abduction, extension, and adduction in continuous sequence.

9.9 The anterior surface of a bone or limb rotates toward the midline in medial rotation, and away from the midline in lateral rotation.

9.10 Bringing the arms forward until the elbows touch is an example of protraction.

9.11 The shoulder joint is so freely movable because of the looseness of its articular capsule and the shallowness of the glenoid cavity in relation to the size of the head of the humerus.

9.12 A hinge joint permits flexion and extension.

9.13 Tension in three ligaments—iliofemoral, pubofemoral, and ischiofemoral—limits the degree of extension at the hip joint.

9.14 Contraction of the quadriceps causes extension at the knee joint.

Muscle Tissue

A Practitioner's View

As a rehabilitation specialist, I need to understand and apply the concepts of the various different types of contractions when working to restore clients to their full functional status. Therapists must understand how a muscle or muscle groups function during activities. For example, if a client complains of knee buckling when going down stairs, the clinician should consider that the quadriceps muscle group may be weak when performing an eccentric contraction. Therefore, the client's therapeutic program should include strengthening eccentric exercises for the knee extensors (quadriceps).

A significant part of my practice as a physical therapist is on the effects of muscle atrophy. Atrophy frequently occurs after immobilization resulting from a casting of a limb due to injury, fracture, or after surgical repair. Or, an individual may favor a joint or muscle due to prolonged pain, which also could result in atrophy. My role as the physical therapist is to restore the range of motion, strength, and endurance of all muscle and muscle groups involved. Exercise has the beneficial effect of restoring the physiology of the muscle fibers assisting with muscle metabolism and restoration of the contractile elements within the muscle.

Susan Mahoney, Physical Therapist
North Shore Education Consortium

MUSCLE TISSUE AND HOMEOSTASIS

Muscle tissue contributes to homeostasis by producing heat to maintain normal body temperature, moving substances through the body, and producing body movements.

ENERGIZE YOUR STUDY
FOUNDATIONS CD

Anatomy Overview
- The Muscular System
- Muscle Tissue

Animations
- Systems Contributions to Homeostasis
- Negative Feedback Control of Temperature

Exercises
- Warming Up

www.wiley.com/college/apcentral

INSIGHTS AND EXPLORATIONS

Have you ever thought about traveling in space to a far away planet, or simply spending a vacation aboard a space station? If so, you would experience reduced gravity, so your muscles would not have to work as hard to keep you upright or to move about. As a result, your muscles would start to atrophy or get smaller and weaker. At the end of your vacation you might have to check into a planet Earth gym before resuming your normal activities on Earth. This is a problem for astronauts who stay in space for extended periods of time. Find out more about muscle atrophy in space and how research about astronauts may help patients here on Earth who suffer from muscular disorders in this web-based activity.

Although bones provide leverage and form the framework of the body, they cannot move body parts by themselves. Motion results from the alternating contraction and relaxation of muscles, which constitute 40–50% of total body weight in adults. Your muscular strength reflects the prime function of muscle—changing chemical energy into mechanical energy to generate force, perform work, and produce movement. In addition, muscle tissues stabilize the body's position, regulate organ volume, generate heat, and propel fluids and food matter through various body systems. The scientific study of muscles is known as **myology** (mī-OL-ō-jē; *myo-* = muscle; *-logy* = study of).

OVERVIEW OF MUSCLE TISSUE

▶ O B J E C T I V E

• **Correlate the three types of muscle tissue with their functions and special properties.**

Types of Muscle Tissue

The three types of muscle tissue are skeletal, cardiac, and smooth (see the comparison in Table 4.4 on pages 130–131). Although the different types of muscle tissue share some properties, they differ from one another in their microscopic anatomy, location, and control by the nervous and endocrine systems.

Skeletal muscle tissue is so-named because most skeletal muscles function to move bones of the skeleton. (A few skeletal muscles attach to and move the skin or other skeletal muscles.) Skeletal muscle tissue is *striated:* Alternating light and dark bands *(striations)* are seen when the tissue is examined with a microscope (see Figure 10.5). Skeletal muscle tissue works mainly in a *voluntary* manner. Its activity can be consciously controlled by neurons (nerve cells) that are part of the somatic (voluntary) division of the nervous system. (Figure 12.1 on page 387 depicts the divisions of the nervous system.) Most skeletal muscles also are controlled subconsciously to some extent. For instance, the diaphragm continues to alternately contract and relax while we are asleep to power breathing. Also, we do not need to consciously think about contracting the skeletal muscles that maintain posture or stabilize body positions.

Only the heart contains **cardiac muscle tissue,** which forms most of the heart wall. Cardiac muscle is also *striated*, but its action is *involuntary.* The alternating contraction and relaxation of the heart is not consciously controlled. Rather, the heart beats because it has a pacemaker that initiates each contraction. This built-in or intrinsic rhythm is termed **autorhythmicity.** Several hormones and neurotransmitters can adjust heart rate by speeding or slowing the pacemaker.

Smooth muscle tissue is located in the walls of hollow internal structures, such as blood vessels, airways, and most organs in the abdominopelvic cavity. It is also found in the skin, attached to hair follicles. Under a microscope, this tissue lacks the striations of skeletal and cardiac muscle tissue. For this reason, it looks *nonstriated* or *smooth.* The action of smooth muscle is usually *involuntary,* and some smooth muscle tissue has autorhythmicity. Both cardiac muscle and smooth muscle are regulated by neurons that are part of the autonomic (involuntary) division of the nervous system and by hormones released by endocrine glands.

Functions of Muscle Tissue

Through sustained contraction or alternating contraction and relaxation, muscle tissue has four key functions. It produces body movements, stabilizes body positions, stores and moves substances within the body, and generates heat.

1. *Producing body movements.* Movements of the whole body such as walking and running, and localized movements such as grasping a pencil or nodding the head, rely on the integrated functioning of bones, joints, and skeletal muscles.

2. *Stabilizing body positions.* Skeletal muscle contractions stabilize joints and help maintain body positions, such as standing or sitting. Postural muscles contract continuously when a person is awake; for example, sustained contractions in neck muscles hold the head upright.

3. *Storing and moving substances within the body.* Sustained contractions of ringlike bands of smooth muscle called *sphincters* may prevent outflow of the contents of a hollow organ. Temporary storage of food in the stomach or urine in the urinary bladder is possible because smooth muscle sphincters close off the outlets of these organs. Cardiac muscle contractions pump blood through the body's blood vessels. Contraction and relaxation of smooth muscle in the walls of blood vessels help adjust their diameter and thus regulate the rate of blood flow. Smooth muscle contractions also move food and substances such as bile and enzymes through the gastrointestinal tract, push gametes (sperm and oocytes) through the passageways of the reproductive systems, and propel urine through the urinary system. Skeletal muscle contractions promote the flow of lymph and aid the return of blood to the heart.

4. *Generating heat.* As muscle tissue contracts, it produces heat, a process known as **thermogenesis.** Much of the heat generated by muscle is used to maintain normal body temperature. Involuntary contractions of skeletal muscle, known as shivering, can increase the rate of heat production severalfold.

Properties of Muscle Tissue

Muscle tissue has four special properties that enable it to function and contribute to homeostasis:

1. **Electrical excitability,** a property of both muscle and nerve cells, is the ability to respond to certain stimuli by producing electrical signals such as *action potentials*. (Chapter 12 describes how action potentials arise; see page 398.) Action potentials can propagate (travel) along a cell's plasma membrane due to the presence of specific voltage-gated channels. For muscle cells, two main types of stimuli trigger action potentials. One is autorhythmic electrical signals arising in the muscle tissue itself, such as occurs in the heart's pacemaker. The other is chemical stimuli, such as neurotransmitters released by neurons, hormones distributed by the blood, or even local changes in pH.

2. **Contractility** is the ability of muscle tissue to contract forcefully when stimulated by an action potential. When muscle contracts, it generates tension (force of contraction) while pulling on its attachment points. In an **isometric contraction** (*iso-* = equal; *-metric* = measure or length), the muscle develops tension but does not shorten. An example is holding a book in an outstretched hand. If the tension generated is great enough to overcome the resistance of the object to being moved, the muscle shortens and movement occurs. In an **isotonic contraction** (*-tonic* = tension), the tension developed by the muscle remains almost constant while the muscle shortens. An example is lifting a book off a table.

3. **Extensibility** is the ability of muscle to stretch without being damaged. Extensibility allows a muscle to contract forcefully even if it is already stretched. Normally, smooth muscle is subject to the greatest amount of stretching. For example, each time the stomach fills with food, the muscle in its wall is stretched. Cardiac muscle also is stretched each time the heart fills with blood. During normal activities, the stretch on skeletal muscle remains almost constant.

4. **Elasticity** is the ability of muscle tissue to return to its original length and shape after contraction or extension.

This chapter focuses mainly on the structure and function of skeletal muscle tissue. Cardiac muscle and smooth muscle are examined in detail later (in discussions of the autonomic nervous system in Chapter 17, of the heart in Chapter 20, and of the various organs that contain smooth muscle in several chapters).

▶ C H E C K P O I N T
1. What features distinguish the three types of muscle tissue?
2. What are the general functions of muscle tissue?

SKELETAL MUSCLE TISSUE

▶ O B J E C T I V E S

- **Explain the relation of connective tissue components, blood vessels, and nerves to skeletal muscles.**
- **Describe the microscopic anatomy of a skeletal muscle fiber.**

Each skeletal muscle is a separate organ composed of hundreds to thousands of cells called **muscle fibers** because of their elongated shapes. Thus, the terms *muscle cell* and *muscle fiber* are synonymous. Connective tissues surround muscle fibers and whole muscles, and blood vessels and nerves penetrate into muscle (Figure 10.1). To understand how contraction of skeletal muscle can generate tension, one first needs to understand its gross and microscopic anatomy.

Connective Tissue Components

Connective tissue surrounds and protects muscle tissue. A **fascia** (FASH-ē-a = bandage) is a sheet or broad band of fibrous connective tissue that supports and surrounds muscles and other organs of the body. The **superficial fascia** (or **subcutaneous layer**) separates muscle from skin (see Figure 11.21 on page 371). It is composed of areolar connective tissue and adipose tissue and provides a pathway for nerves, blood vessels, and lymphatic vessels to enter and exit muscles. The adipose tissue of superficial fascia stores most of the body's triglycerides, serves as an insulating layer that reduces heat loss, and protects muscles from physical trauma. **Deep fascia** is dense irregular connective tissue that lines the body wall and limbs and holds muscles with similar functions together. Deep fascia allows free movement of muscles, carries nerves, blood vessels, and lymphatic vessels, and fills spaces between muscles.

Three layers of connective tissue extend from the deep fascia to further protect and strengthen skeletal muscle (Figure 10.1). The outermost layer, encircling the whole muscle, is the **epimysium** (ep-i-MĪZ-ē-um; *epi-* = upon). **Perimysium** (per-i-MĪZ-ē-um; *peri-* = around) surrounds groups of 10 to 100 or more individual muscle fibers, separating them into bundles called **fascicles** (FAS-i-kuls = little bundles). Many fascicles are large enough to be seen with the naked eye. They give a cut of meat its characteristic "grain," and if you tear a piece of meat, it rips apart along the fascicles. Both epimysium and perimysium are dense irregular connective tissue. Penetrating the interior of each fascicle and separating individual muscle fibers (muscle cells) from one another is **endomysium** (en′-dō-MĪZ-ē-um; *endo-* = within), a thin sheath of areolar connective tissue.

The epimysium, perimysium, and endomysium all are continuous with the connective tissue that attaches skeletal muscle to other structures, such as bone or another muscle. All three connective tissue layers may extend beyond the muscle fibers to form a **tendon**—a cord of dense regular connective tissue that attaches a muscle to the periosteum of a bone. An example is the calcaneal (Achilles) tendon of the gastrocnemius (calf) muscle, which attaches the muscle to the calcaneus (shown in Figure 11.22 on page 374). When the connective tissue elements extend as a broad, flat layer, the tendon is called an **aponeurosis** (*apo-* = from; *neur-* = a sinew). An example is the epicranial aponeurosis on top of the skull between the frontal and occipital bellies of the occipitofrontalis muscle (shown in Figure 11.4c on page 321).

Figure 10.1 Organization of skeletal muscle and its connective tissue coverings.

A skeletal muscle consists of individual muscle fibers (cells) bundled into fascicles and surrounded by three connective tissue layers that are extensions of the deep fascia.

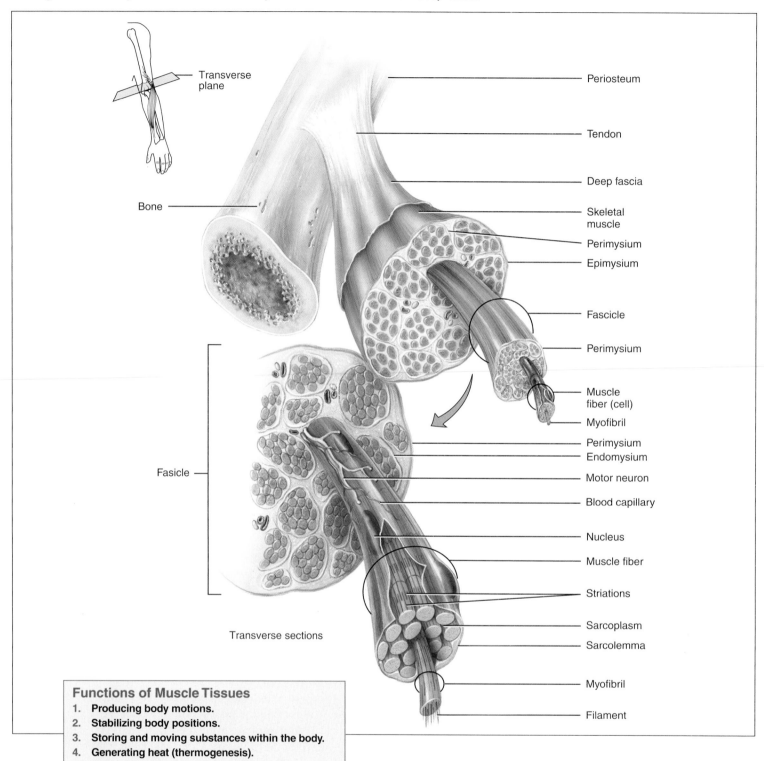

Transverse plane

Bone

Fasicle

Transverse sections

Periosteum

Tendon

Deep fascia

Skeletal muscle

Perimysium

Epimysium

Fascicle

Perimysium

Muscle fiber (cell)

Myofibril

Perimysium

Endomysium

Motor neuron

Blood capillary

Nucleus

Muscle fiber

Striations

Sarcoplasm

Sarcolemma

Myofibril

Filament

Functions of Muscle Tissue
1. **Producing body motions.**
2. **Stabilizing body positions.**
3. **Storing and moving substances within the body.**
4. **Generating heat (thermogenesis).**

 Which connective tissue coat surrounds groups of muscle fibers, separating them into fascicles?

Nerve and Blood Supply

Skeletal muscles are well supplied with nerves and blood vessels. Generally, an artery and one or two veins accompany each nerve that penetrates a skeletal muscle. The neurons that stimulate skeletal muscle to contract are *somatic motor neurons*. Each somatic motor neuron has a threadlike axon that extends from the brain or spinal cord to a group of skeletal muscle fibers (Figure 10.2). The axon is wrapped with *myelin,* which is provided by nearby Schwann cells. The axon of a somatic motor neuron typically branches many times. Each axon collateral (branch) extends to a different skeletal muscle fiber. The structural point of contact and functional site of communication between the motor neuron and the muscle fiber is termed the **neuromuscular junction (NMJ).** At the NMJ, several axon terminals expand into clusters of *synaptic end bulbs.*

Microscopic blood vessels called capillaries are plentiful in muscle tissue; each muscle fiber is in close contact with one or more capillaries (Figure 10.2). The blood capillaries bring in oxygen and nutrients and remove heat and the waste products of muscle metabolism. Especially during contraction, a muscle fiber synthesizes and uses considerable ATP (adenosine triphosphate). These reactions require oxygen, glucose, fatty acids, and other substances that are supplied in the blood.

Microscopic Anatomy of a Skeletal Muscle Fiber

The most important components of a skeletal muscle are the muscle fibers themselves. During embryonic development, each skeletal muscle fiber arises from the fusion of a hundred or more small mesodermal cells called *myoblasts* (Figure 10.3a). Hence, each mature skeletal muscle fiber has a hundred or more nuclei. Once fusion has occurred, the muscle fiber loses its ability to undergo mitosis. Thus, the number of skeletal muscle fibers is set before birth, and most of these cells last a lifetime. The dramatic muscle growth that occurs after birth occurs mainly by enlargement of existing muscle fibers. During childhood, human growth hormone and other hormones stimulate an increase in the size of skeletal muscle fibers. The hormone testosterone (from the testes in males and in small amounts from other tissues in females) promotes further enlargement of muscle fibers. A few myoblasts do persist in mature skeletal muscle as *satellite cells* (Figure 10.3b). These cells retain the capacity to fuse with one another or with damaged muscle fibers to regenerate functional muscle fibers. Mature muscle fibers range from 10 to 100 μm* in diameter. In humans, although a typical length is about 10 cm (4 in.), some are up to 30 cm (12 in.) long.

Sarcolemma, T Tubules, and Sarcoplasm

The multiple nuclei of a skeletal muscle fiber are located just beneath the **sarcolemma** (*sarc-* = flesh; *-lemma* = sheath), the plasma membrane of a muscle cell (Figure 10.3b). Thousands of tiny invaginations of the sarcolemma, called **T (transverse) tubules,** tunnel in from the surface toward the center of each muscle fiber. T tubules are open to the outside of the fiber and

_____

*One micrometer (μm) is 10^{-6} meter (1/25,000 in.).

Figure 10.2 Blood and nerve supply to skeletal muscle. The site of synaptic contact between a somatic motor neuron and a skeletal muscle fiber is the neuromuscular junction (NMJ). Two NMJs are present in this view.

🔑 **Branches of one motor neuron typically contact many muscle fibers.**

Axon terminal

Blood capillary

Synaptic end bulbs

Axon collateral (branch)

Somatic motor neuron

Skeletal muscle fiber

Axon collateral

Schwann cell

Synaptic end bulbs

SEM 1650x

 What are the functions of blood capillaries in muscle?

Figure 10.3   Microscopic organization of skeletal muscle.   (a) During embryonic development, many myoblasts fuse to form one skeletal muscle fiber. Once fusion has occurred, a skeletal muscle fiber loses the ability to undergo cell division, but satellite cells retain this ability. (b) The
sarcolemma of the fiber encloses sarcoplasm and myofibrils, which are striated. Sarcoplasmic
reticulum (SR) wraps around each myofibril. Thousands of T tubules, filled with interstitial fluid,
invaginate from the sarcolemma toward the center of the muscle fiber. A triad is a T tubule and
the two terminal cisterns of the SR on either side of it.

🔑 **The contractile elements of muscle fibers are the myofibrils, which contain overlapping thick
and thin filaments.**

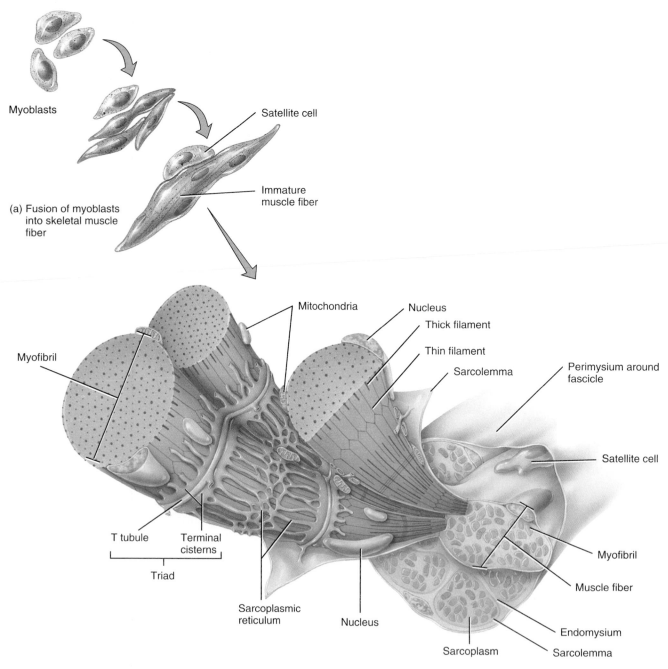

Myoblasts

Satellite cell

Immature
muscle fiber

(a) Fusion of myoblasts
into skeletal muscle
fiber

Mitochondria

Nucleus

Thick filament

Thin filament

Myofibril

Sarcolemma

Perimysium around
fascicle

Satellite cell

T tubule

Terminal
cisterns

Triad

Sarcoplasmic
reticulum

Nucleus

Myofibril

Muscle fiber

Endomysium

Sarcoplasm

Sarcolemma

(b) Details of several myofibrils

 Which structure shown here releases calcium ions to trigger muscle contraction?

thus are filled with interstitial fluid. Muscle action potentials propagate along the sarcolemma and through the T tubules, quickly spreading throughout the muscle fiber. This arrangement ensures that an action potential excites all parts of the muscle fiber at almost the same instant.

Within the sarcolemma is the **sarcoplasm,** the cytoplasm of a muscle fiber. Sarcoplasm includes a substantial amount of glycogen, which can be split into glucose that is used for synthesis of ATP. In addition, the sarcoplasm contains a red-colored protein called **myoglobin** (mī-ō-GLŌB-in). This protein, found only in muscle binds oxygen molecules that diffuse into muscle fibers from interstitial fluid. Myoglobin releases oxygen when mitochondria need it for ATP production. The mitochondria lie in rows throughout the muscle fiber, strategically close to the muscle proteins that use ATP during contraction.

Myofibrils and Sarcoplasmic Reticulum

At high magnification, the sarcoplasm appears stuffed with little threads. These small structures are the **myofibrils,** the contractile elements of skeletal muscle (Figure 10.3b). Myofibrils are about 2 μm in diameter and extend the entire length of a muscle fiber. Their prominent striations make an entire muscle fiber look striated.

A fluid-filled system of membranous sacs called the **sarcoplasmic reticulum (SR)** encircles each myofibril. This elaborate system is similar to smooth endoplasmic reticulum in nonmuscle cells. Dilated end sacs of the sarcoplasmic reticulum called *terminal cisterns* (= reservoirs) butt against the T tubule from both sides. A transverse tubule and the two terminal cisterns on either side of it form a *triad* (*tri-* = three). In a relaxed muscle fiber, the sarcoplasmic reticulum stores calcium ions (Ca^{2+}). Release of Ca^{2+} from the terminal cisterns of the sarcoplasmic reticulum triggers muscle contraction.

Muscular Atrophy and Hypertrophy

Muscular atrophy (A-trō-fē; *a-* = without, *-trophy* = nourishment) is a wasting away of muscles. Individual muscle fibers decrease in size because of progressive loss of myofibrils. The atrophy that occurs if muscles are not used is termed *disuse atrophy.* Bedridden individuals and people with casts experience disuse atrophy because the number of nerve impulses to inactive muscle is greatly reduced. If the nerve supply to a muscle is disrupted or cut, the muscle undergoes *denervation atrophy.* In about 6 months to 2 years, the muscle will be one-quarter of its original size, and the muscle fibers will be replaced by fibrous connective tissue. The transition to connective tissue, when complete, cannot be reversed.

Muscular hypertrophy (hī-PER-trō-fē; *hyper-* = above or excessive) is an increase in the diameter of muscle fibers owing to the production of more myofibrils, mitochondria, sarcoplasmic reticulum, and so forth. It results from very forceful, repetitive muscular activity, such as strength training. Because hypertrophied muscles contain more myofibrils, they are capable of contractions that are more forceful. ■

Filaments and the Sarcomere

Within myofibrils are two types of even smaller structures called **filaments,** which are only 1–2 μm long (Figure 10.3b). The diameter of the *thin filaments* is about 8 nm,* whereas that of the *thick filaments* is about 16 nm. Overall, there are two thin filaments for every thick filament. The filaments inside a myofibril do not extend the entire length of a muscle fiber. Instead, they are arranged in compartments called **sarcomeres** (*-mere* = part), which are the basic functional units of a myofibril (Figure 10.4a). Narrow, plate-shaped regions of dense material called **Z discs** separate one sarcomere from the next.

The thick and thin filaments overlap one another to a greater or lesser extent, depending on whether the muscle is contracted, relaxed, or stretched. The pattern of their overlap, consisting of a variety of zones and bands (Figure 10.4b), creates the striations that can be seen both in single myofibrils and in whole muscle fibers. The darker middle part of the sarcomere is the **A band,** which extends the entire length of the thick filaments (Figure 10.4b). Toward each end of the A band is a *zone of overlap,* where the thick and thin filaments lie side by side. The **I band** is a lighter, less dense area that contains the rest of the thin filaments but no thick filaments (Figure 10.4b). A Z disc passes through the center of each I band. A narrow **H zone** in the center of each A band contains thick but not thin filaments. Supporting proteins that hold the thick filaments together at the center of the H zone form the **M line,** so-named because it is at the *middle* of the sarcomere. Figure 10.5 shows the relations of the zones, bands, and lines as seen in a transmission electron micrograph.

Exercise-Induced Muscle Damage

Comparison of electron micrographs of muscle tissue taken from athletes before and after intense exercise reveal considerable exercise-induced muscle damage, including torn sarcolemmas in some muscle fibers, damaged myofibrils, and disrupted Z discs. Microscopic muscle damage after exercise also is indicated by increases in blood levels of proteins, such as myoglobin and the enzyme creatine kinase, that are normally confined within muscle fibers. From 12 to 48 hours after a period of strenuous exercise, skeletal muscles often become sore. Such **delayed onset muscle soreness (DOMS)** is accompanied by stiffness, tenderness, and swelling. Although the causes of DOMS are not completely understood, microscopic muscle damage appears to be a major factor. ■

*One nanometer (nm) is 10^{-9} meter (0.001 μm).

Figure 10.4 **The arrangement of filaments within a sarcomere.** A sarcomere extends from one Z disc to the next.

🔑 **Myofibrils contain two types of filaments: thick filaments and thin filaments.**

(a) Myofibril

(b) Filaments

❓ **Among the following, which is smallest: muscle fiber, thick filament, or myofibril? Which is largest?**

Figure 10.5 **Characteristic zones and bands of a sarcomere.**

🔑 **The striations of skeletal muscle are alternating darker A bands and lighter I bands.**

TEM 21,600x

❓ **What is the ratio between thick and thin filaments in skeletal muscle?**

Muscle Proteins

Myofibrils are built from three kinds of proteins: (1) contractile proteins, which generate force during contraction; (2) regulatory proteins, which help switch the contraction process on and off; and (3) structural proteins, which keep the thick and thin filaments in the proper alignment, give the myofibril elasticity and extensibility, and link the myofibrils to the sarcolemma and extracellular matrix.

The two *contractile proteins* in muscle are myosin and actin, which are the main components of thick and thin filaments, respectively. **Myosin** functions as a *motor protein* in all three types of muscle tissue. Motor proteins push or pull their cargo to achieve movement by converting the chemical energy in ATP to the mechanical energy of motion or force production. In skeletal muscle, about 300 molecules of myosin form a single thick filament. Each myosin molecule is shaped like two golf clubs twisted together (Figure 10.6a). The *myosin tail* (twisted golf club handles) points toward the M line in the center of the sarcomere. Tails of neighboring myosin molecules lie parallel to one another, forming the shaft of the thick filament. The two

projections of each myosin molecule (golf club heads) are called *myosin heads* or *crossbridges.* The heads project outward from the shaft in a spiraling fashion, each extending toward one of the six thin filaments that surround each thick filament.

Thin filaments extend from anchoring points within the Z discs (see Figure 10.4b). Their main component is the protein **actin.** Individual actin molecules join to form an actin filament that is twisted into a helix (Figure 10.6b). On each actin molecule is a *myosin-binding site,* where a myosin head can attach. Smaller amounts of two *regulatory proteins*—**tropomyosin** and **troponin**—are also part of the thin filament. In relaxed muscle, myosin is blocked from binding to actin because strands of tropomyosin cover the *myosin-binding sites* on actin. The tropomyosin strands, in turn, are held in place by troponin molecules.

Besides contractile and regulatory proteins, muscle contains about a dozen *structural proteins,* which contribute to the alignment, stability, elasticity, and extensibility of myofibrils. Several key structural proteins are titin, myomesin, and dystrophin. **Titin** *(titan* = gigantic) is the third most plentiful protein in skeletal muscle (after actin and myosin). This molecule's name reflects its huge size. With a molecular weight of about 3 million daltons, titin is 50 times larger than an average-sized protein. Each titin molecule spans half a sarcomere, from a Z disc to an M line (see Figure 10.4b), a distance of 1 to 1.2 μm in relaxed muscle. Titin anchors a thick filament to both a Z disc and the M line, thereby helping stabilize the position of the thick filament. The part of the titin molecule that extends from the Z disc to the beginning of the thick filament is very elastic. Because it can stretch to at least four times its resting length and then spring back unharmed, titin accounts for much of the elasticity and extensibility of myofibrils. Titin probably helps the sarcomere return to its resting length after a muscle has contracted or been stretched.

Molecules of the protein *myomesin* form the M line. The M line proteins bind to titin and connect adjacent thick filaments to one another. *Dystrophin* is a cytoskeletal protein that links thin filaments of the sarcomere to integral membrane proteins of the sarcolemma. In turn, the membrane proteins attach to proteins in the connective tissue matrix that surrounds muscle fibers. Hence, dystrophin and its associated proteins are thought to reinforce the sarcolemma and help transmit the tension generated by the sarcomeres to the tendons.

Figure 10.6 **Structure of thick and thin filaments.** (a) A thick filament contains about 300 myosin molecules, one of which is enlarged. The myosin tails form the shaft of the thick filament, whereas the myosin heads project outward toward the surrounding thin filaments. (b) Thin filaments contain actin, troponin, and tropomyosin.

Contractile proteins (myosin and actin) generate force during contraction, whereas regulatory proteins (troponin and tropomyosin) help switch contraction on and off.

Thick filament

Myosin tail

Myosin heads (crossbridges)

(a) One thick filament (above) and a myosin molecule (below)

Actin   Troponin   Tropomyosin

Myosin-binding site (covered by tropomyosin)

(b) Portion of a thin filament

Which proteins connect into the Z disc? Which proteins are present in the A band? In the I band?

▶ C H E C K P O I N T

3. What are the types of fascia that cover skeletal muscles?

4. Why is a rich blood supply important for muscle contraction?

5. How are the structures of thin and thick filaments different?

CONTRACTION AND RELAXATION OF SKELETAL MUSCLE FIBERS

▶ O B J E C T I V E S

- **Outline the steps involved in the sliding filament mechanism of muscle contraction.**
- **Describe how muscle action potentials arise at the neuromuscular junction.**

When scientists examined the first electron micrographs of skeletal muscle in the mid-1950s, they were surprised to see that the lengths of the thick and thin filaments were the same in both relaxed and contracted muscle. It had been thought that muscle contraction must be a folding process, somewhat like closing an accordion. However, researchers discovered that skeletal muscle shortens during contraction because the thick and thin filaments slide past one another. The model describing the contraction of muscle is known as the **sliding filament mechanism.**

The Sliding Filament Mechanism

Muscle contraction occurs because myosin heads attach to and "walk" along the thin filaments at both ends of a sarcomere, progressively pulling the thin filaments toward the M line (Figure 10.7). As a result, the thin filaments slide inward and meet at the center of a sarcomere. They may even move so far inward that their ends overlap (Figure 10.7c). As the thin filaments slide inward, the Z discs come closer together, and the sarcomere shortens. However, the lengths of the individual thick and thin filaments do not change. Shortening of the sarcomeres causes shortening of the whole muscle fiber, which leads to shortening of the entire muscle.

The Contraction Cycle

At the onset of contraction, the sarcoplasmic reticulum releases calcium ions (Ca^{2+}) into the cytosol. There, they bind to troponin and cause the troponin–tropomyosin complexes to move away from the myosin-binding sites on actin. Once the binding sites are "free," the **contraction cycle**—the repeating sequence of events that causes the filaments to slide—begins. The contraction cycle consists of four steps (Figure 10.8):

❶ **ATP hydrolysis.** The myosin head includes an ATP-binding pocket and an ATPase, an enzyme that hydrolyzes ATP into ADP (adenosine diphosphate) and a phosphate group. This hydrolysis reaction energizes the myosin head. Notice that the products of ATP hydrolysis—ADP and a phosphate group—are still attached to the myosin head.

❷ **Attachment of myosin to actin to form crossbridges.** The energized myosin head attaches to the myosin-binding site on actin and then releases the previously hydrolyzed phosphate group.

❸ **Power stroke.** The release of the phosphate group triggers the power stroke of contraction. During the power stroke, the pocket on the myosin head where ADP is still bound opens. As a result, the myosin head rotates and releases the ADP. The myosin head generates force as it rotates toward the center of the sarcomere, sliding the thin filament past the thick filament toward the M line.

❹ **Detachment of myosin from actin.** At the end of the power stroke, the myosin head remains firmly attached to actin until it binds another molecule of ATP. As ATP binds to the ATP-binding pocket on the myosin head, the myosin head detaches from actin.

The contraction cycle repeats as the myosin ATPase again hydrolyzes ATP. The reaction reorients the myosin head and transfers energy from ATP to the myosin head. Then, the myosin head is again energized and ready to combine with the next myosin-binding site on the thin filament. The contraction cycle repeats for as long as ATP is available and the Ca^{2+} level near the thin filament is sufficiently high. The myosin heads keep

Figure 10.7 Sliding filament mechanism of muscle contraction, as it occurs in two adjacent sarcomeres.

During muscle contractions, thin filaments move toward the M line of each sarcomere.

(a) Relaxed muscle

(b) Partially contracted muscle

(c) Maximally contracted muscle

 What happens to the I band and H zone as muscle contracts? Do the lengths of the thick and thin filaments change?

Figure 10.8 The contraction cycle. Sarcomeres exert force and shorten through repeated cycles during which the myosin heads (crossbridges) attach to actin, rotate, and detach.

During the power stroke of contraction, myosin heads rotate and move the thin filaments past the thick filaments toward the center of the sarcomere.

1 Myosin heads hydrolyze ATP and become reoriented and energized

2 Myosin heads bind to actin, forming crossbridges

Contraction cycle continues if ATP is available and Ca²⁺ level in the sarcoplasm is high

4 As myosin heads bind ATP, the crossbridges detach from actin

3 Myosin heads rotate toward center of the sarcomere (power stroke)

ATP ADP P

What would happen if ATP suddenly were not available after the sarcomere had started to shorten?

rotating back and forth with each power stroke, pulling the thin filaments toward the M line. Each of the 600 myosin heads in one thick filament attaches and detaches about five times per second. At any one instant, some of the myosin heads are attached to actin and generating force, whereas others are detached and getting ready to bind again.

Contraction is analogous to running on a foot-powered treadmill. One foot (myosin head) strikes the belt (thin filament) and pushes it backward (toward the M line). Then the other foot comes down and imparts a second push. The belt (thin filament) moves smoothly while the runner (thick filament) remains stationary. Each myosin head progressively "walks" along a thin filament, coming closer to the Z disc with each "step," while the thin filament moves toward the M line. And like the legs of a runner, the myosin head needs a constant supply of energy to keep going—one molecule of ATP for each contraction cycle!

As the contraction cycle continues, movement of myosin heads applies the force that draws the Z discs toward each other, and the sarcomere shortens. During a maximal muscle contraction, the distance between two Z discs can decrease to half the resting length. The Z discs, in turn, pull on neighboring sarcomeres, and the whole muscle fiber shortens. Some of the compo-

nents of muscle are elastic: They stretch slightly before they transfer the tension generated by the sliding filaments. The elastic components include titin molecules, connective tissue around the muscle fibers (endomysium, perimysium, and epimysium), and tendons that attach muscle to bone. As the cells of a skeletal muscle start to shorten, they first pull on their connective tissue coverings and tendons. The coverings and tendons stretch and then become taut, and the tension passed through the tendons pulls on the bones to which they are attached. The result is movement of a part of the body. However, the contraction cycle does not always result in shortening of the muscle fibers and the whole muscle. In isometric contractions, the myosin heads rotate and generate tension, but the thin filaments cannot slide inward because the tension they generate is not large enough to move the load on the muscle (see Figure 10.17c).

Excitation–Contraction Coupling

An increase in Ca²⁺ concentration in the cytosol starts muscle contraction, whereas a decrease stops it. When a muscle fiber is relaxed, the concentration of Ca²⁺ in its cytosol is very low, only about 0.1 micromole per liter (0.1 μm/L). A huge amount

Figure 10.9 The role of Ca²⁺ in the regulation of contraction by troponin and tropomyosin. (a) During relaxation, the level of Ca²⁺ in the sarcoplasm is low, only 0.1 μM (0.001 mM), because calcium ions are pumped into the sarcoplasmic reticulum (SR) by Ca²⁺ active transport pumps. (b) A muscle action potential propagating along a transverse tubule opens Ca²⁺ release channels in the SR, calcium ions flow into the cytosol, and contraction begins.

An increase in the Ca²⁺ level in the sarcoplasm starts the sliding of thin filaments. When the level of Ca²⁺ in the sarcoplasm declines, sliding stops.

Sarcolemma   T tubules   Terminal cistern of SR   Muscle action potential

Ca²⁺ release channels closed   Ca²⁺ release channels open

Thin filament

Troponin   Myosin-binding site on actin   Myosin

Tropomyosin

Key:
- = Ca²⁺
- ⊗ = Ca²⁺ active transport pumps
- < = Ca²⁺ release channels

Troponin holds tropomyosin in position to block myosin-binding sites on actin.

Ca²⁺ binds to troponin, which changes the shape of the troponin–tropomyosin complex and uncovers the myosin-binding sites on actin.

(a) Relaxation   (b) Contraction

What are three functions of ATP in muscle contraction?

---

of Ca²⁺, however, is stored inside the sarcoplasmic reticulum (Figure 10.9a). As a muscle action potential propagates along the sarcolemma and into the T tubules, it causes **Ca²⁺ release channels** in the SR membrane to open (Figure 10.9b). When these channels are open, Ca²⁺ flows out of the SR into the cytosol around the thick and thin filaments. As a result, the Ca²⁺ concentration in the cytosol rises tenfold or more. The released calcium ions combine with troponin, causing it to change shape. This conformational change moves the troponin–tropomyosin complex away from the myosin-binding sites on actin. Once these binding sites are free, myosin heads bind to them, and the contraction cycle begins. The events just described constitute **excitation–contraction coupling,** the steps that connect excitation (a muscle action potential propagating along the sarcolemma and into the T tubules) to contraction (sliding of the filaments).

The sarcoplasmic reticulum membrane also contains **Ca²⁺**

active transport pumps that use ATP to constantly move Ca²⁺ from the cytosol into the SR (Figure 10.9). While muscle action potentials continue to propagate through the T tubules, the Ca²⁺ release channels are open. Calcium ions flow into the cytosol more rapidly than they are transported back by the pumps. After the last action potential has propagated throughout the T tubules, the Ca²⁺ release channels close. As the pumps move Ca²⁺ back into the SR, the concentration of calcium ions in the cytosol quickly decreases. Inside the SR, molecules of a calcium-binding protein, appropriately called **calsequestrin,** bind to the Ca²⁺, enabling even more Ca²⁺ to be sequestered within the SR. As a result, the concentration of Ca²⁺ inside the SR is 10,000 times higher than in the cytosol in a relaxed muscle fiber. As the Ca²⁺ level drops in the cytosol, the troponin–tropomyosin complexes slide back over and cover the myosin-binding sites, and the muscle fiber relaxes.

Rigor Mortis

After death, cellular membranes start to become leaky. Calcium ions leak out of the sarcoplasmic reticulum into the cytosol and allow myosin heads to bind to actin. ATP synthesis has ceased, however, so the crossbridges cannot detach from actin. The resulting condition, in which muscles are in a state of rigidity (cannot contract or stretch), is called **rigor mortis** (rigidity of death). Rigor mortis begins 3–4 hours after death and lasts about 24 hours; then it disappears as proteolytic enzymes from lysosomes digest the crossbridges. ■

Length–Tension Relationship

Figure 10.10 plots the **length–tension relationship** for skeletal muscle, which shows how the forcefulness of muscle contraction depends on the length of the sarcomeres within a muscle *before contraction begins.* At a sarcomere length of about 2.0–2.4 μm, the zone of overlap in each sarcomere is optimal, and the muscle fiber can develop maximum tension. Notice in Figure 10.10 that maximum tension (100%) occurs when the zone of overlap between a thick and thin filament extends from the edge of the H zone to one end of a thick filament.

As the sarcomeres of a muscle fiber are stretched to a longer length, the zone of overlap shortens, and fewer myosin heads can make contact with thin filaments. So, the tension the fiber can produce decreases. When a skeletal muscle fiber is stretched to 170% of its optimal length, there is no overlap between the thick and thin filaments. Because none of the myosin heads can bind to thin filaments, the muscle fiber cannot contract, and tension is zero. As sarcomere lengths become increasingly shorter than the optimum, the tension that can develop again decreases. This is because thick filaments crumple as they are compressed by the Z discs, resulting in fewer myosin heads making contact with thin filaments. Normally, resting muscle fiber length is held very close to the optimum by firm attachments of skeletal muscle to bones (via their tendons) and to other inelastic tissues, so that overstretching does not occur.

The Neuromuscular Junction

A muscle fiber contracts in response to one or more action potentials propagating along its sarcolemma and through its T tubule system. Muscle action potentials arise at the **neuromuscular junction (NMJ),** the region of synaptic contact between a somatic motor neuron and a skeletal muscle fiber (Figure 10.11a). A **synapse** is a region where communication occurs between two neurons, or between a neuron and a target cell—in this case, between a motor neuron and a skeletal muscle fiber. At most synapses a small gap, called the *synaptic cleft,* separates the two cells. Because the cells do not physically touch, the action potential from one cell cannot "jump the gap" to directly excite the next cell. Instead, the first cell communicates with the second indirectly, by releasing a chemical called a **neurotransmitter.**

Figure 10.10 Length–tension relationship in a skeletal muscle fiber. Maximum tension during contraction occurs when the resting sarcomere length is 2.0–2.4 μm.

A muscle fiber develops its greatest tension when there is an optimal zone of overlap between thick and thin filaments.

Why is tension maximal at a sarcomere length of 2.2 μm?

As you saw in Figure 10.2, each motor neuron axon terminal divides into a cluster of synaptic end bulbs (Figure 10.11a). Suspended in the cytosol within each end bulb are hundreds of membrane-enclosed sacs called **synaptic vesicles.** Inside each synaptic vesicle are thousands of molecules of **acetylcholine** (as'-ē-til-KŌ-lēn), abbreviated **ACh,** the neurotransmitter released at the NMJ.

The muscle fiber part of the NMJ is the sarcolemma opposite the synaptic end bulbs. This region of the sarcolemma is called the **motor end plate.** Within each motor end plate are 30 to 40 million **acetylcholine receptors,** which are integral transmembrane proteins that bind specifically to ACh. As you will see, the ACh receptors also are ligand-gated ion channels. A neuromuscular junction thus includes all the synaptic end bulbs on one side of the synaptic cleft, plus the motor end plate of the muscle fiber on the other side.

A nerve impulse elicits a muscle action potential in the following way (Figure 10.11c).

1 *Release of acetylcholine.* Arrival of the nerve impulse at the synaptic end bulbs causes many synaptic vesicles to undergo exocytosis. During exocytosis, the synaptic vesicles fuse with the motor neuron's plasma membrane, which liberates ACh into the synaptic cleft. The ACh then diffuses across the synaptic cleft between the motor neuron and the motor end plate.

2 *Activation of ACh receptors.* Binding of two molecules of ACh to the receptor opens the ion channel part of the ACh receptor. Once the channel is open, small cations, most importantly Na^+, can flow across the membrane.

3 *Production of muscle action potential.* The inflow of Na^+ (down its electrochemical gradient) makes the inside of the muscle fiber more positively charged. This change in the membrane potential is the signal that triggers a muscle action potential. Each nerve impulse normally elicits one muscle action potential. The muscle action potential then propagates along the sarcolemma and into the T tubule system.

4 *Termination of ACh activity.* The effect of ACh binding lasts only briefly because ACh is rapidly broken down by an enzyme called **acetylcholinesterase (AChE).** This enzyme is attached to collagen fibers in the extracellular matrix of the synaptic cleft. AChE breaks down ACh into acetyl and choline, products that cannot activate the ACh receptor.

If another nerve impulse releases more acetylcholine, steps 2 and 3 repeat. When action potentials cease in the motor neuron, ACh release stops, and AChE rapidly breaks down the ACh already present in the synaptic cleft. This ends the production of muscle action potentials, and the Ca^{2+} release channels in the sarcoplasmic reticulum membrane close.

The NMJ usually is near the midpoint of a skeletal muscle fiber. Muscle action potentials arise at the NMJ and then propagate toward both ends of the fiber. This arrangement permits nearly simultaneous activation (and thus contraction) of all parts of the muscle fiber.

Figure 10.12 summarizes the events that occur during contraction and relaxation of a skeletal muscle fiber.

Pharmacology of the NMJ

Several plant products and drugs selectively block certain events at the NMJ. *Botulinum toxin,* produced by the bacterium *Clostridium botulinum,* blocks exocytosis of synaptic vesicles at the NMJ. As a result, ACh is not released, and muscle contraction does not occur. The bacteria proliferate in improperly canned foods and their toxin is one of the most lethal chemicals known. A tiny amount can cause death by paralyzing skeletal muscles. Breathing stops due to paralysis of respiratory muscles, including the diaphragm. Yet, it is also the first bacterial toxin to be used as a medicine (Botox). Injections of Botox into the affected muscles can help patients who have strabismus (crossed eyes) or blepharospasm (uncontrollable blinking). It is also used as a cosmetic treatment to relax muscles that cause facial wrinkles and to alleviate chronic back pain due to muscle spasms in the lumbar region.

The plant derivative *curare,* a poison used by South American Indians on arrows and blowgun darts, causes muscle paralysis by binding to and blocking ACh receptors. In the presence of curare, the ion channels do not open. Curare-like drugs are often used during surgery to relax skeletal muscles. A family of chemicals called *anticholinesterase agents* have the property of slowing the enzymatic activity of acetylcholinesterase, thus slowing removal of ACh from the synaptic cleft. At low doses, these agents can strengthen weak muscle contractions. One example is neostigmine, which is used to treat patients with myasthenia gravis (see page 302). Neostigmine is also used as an antidote for curare poisoning and to terminate the effects of curare after surgery. ∎

► C H E C K P O I N T

6. What roles do contractile, regulatory, and structural proteins play in muscle contraction and relaxation?

7. How do calcium ions and ATP contribute to muscle contraction and relaxation?

8. How does sarcomere length influence the maximum tension that is possible during muscle contraction?

9. How is the motor end plate different from other parts of the sarcolemma?

Figure 10.11 **Structure of the neuromuscular junction (NMJ), the synapse between a somatic motor neuron and a skeletal muscle fiber.**

Synaptic end bulbs at the tips of axon terminals contain synaptic vesicles filled with acetylcholine (ACh).

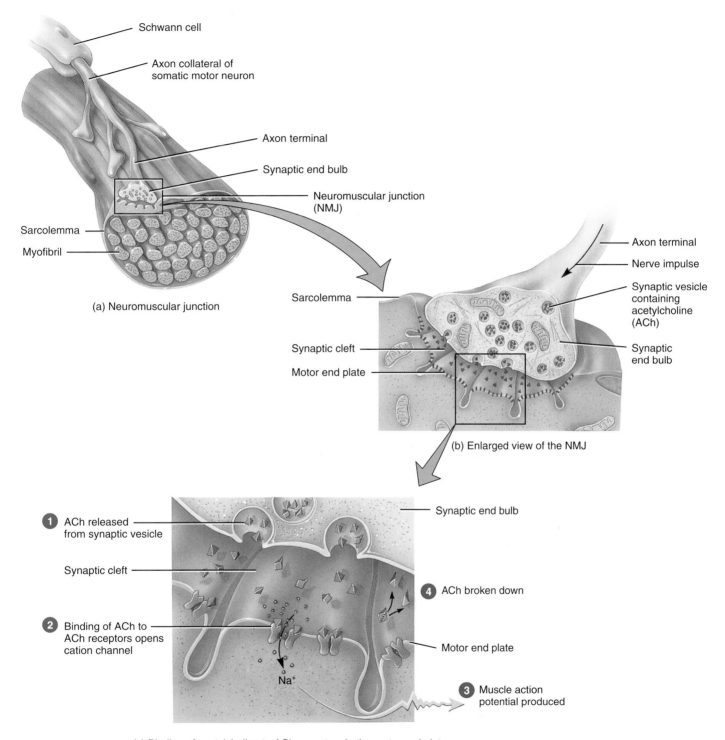

(a) Neuromuscular junction

(b) Enlarged view of the NMJ

1 ACh released from synaptic vesicle

Synaptic cleft

2 Binding of ACh to ACh receptors opens cation channel

Na^+

4 ACh broken down

Motor end plate

3 Muscle action potential produced

Synaptic end bulb

(c) Binding of acetylcholine to ACh receptors in the motor end plate

 What part of the sarcolemma contains acetylcholine receptors?

Figure 10.12   Summary of the events of contraction and relaxation in a skeletal muscle fiber.

 Acetylcholine released at the neuromuscular junction triggers a muscle action potential, which leads to muscle contraction.

1. Nerve impulse arrives at axon terminal of motor neuron and triggers release of acetylcholine (ACh).

Nerve impulse

Muscle action potential (AP)

Synaptic vesicle filled with ACh

T tubule

2. ACh diffuses across synaptic cleft, binds to its receptors in the motor end plate, and triggers a muscle action potential (AP).

ACh receptor

4. Muscle AP travelling along T tubule opens Ca^{2+} release channels in the sarcoplasmic reticulum (SR) membrane, which allows calcium ions to flood into the sarcoplasm.

3. Acetylcholinesterase in synaptic cleft destroys ACh so another muscle action potential does not arise unless more ACh is released from motor neuron.

SR

Ca^{2+}

Elevated Ca^{2+}

5. Ca^{2+} binds to troponin on the thin filament, exposing the binding sites for myosin.

6. Contraction: power strokes use ATP; myosin heads bind to actin, swivel, and release; thin filaments are pulled toward center of sarcomere.

Ca^{2+} active transport pumps

8. Troponin–tropomyosin complex slides back into position where it blocks the myosin binding sites on actin.

7. Ca^{2+} release channels in SR close and Ca^{2+} active transport pumps use ATP to restore low level of calcium ions in sarcoplasm.

9. Muscle relaxes.

Which numbered steps in this figure are part of excitation–contraction coupling?

MUSCLE METABOLISM

▶ **OBJECTIVE**

• **Describe the reactions by which muscle fibers produce ATP.**

Production of ATP in Muscle Fibers

Unlike most cells of the body, skeletal muscle fibers often switch between virtual inactivity, when they are relaxed and using only a modest amount of ATP, and great activity, when they are contracting and using ATP at a rapid pace. Contraction of muscle requires a huge amount of ATP for powering the contraction cycle, for pumping Ca^{2+} into the sarcoplasmic reticulum to achieve muscle relaxation, and for other metabolic reactions. However, the ATP present inside muscle fibers is enough to power contraction for only a few seconds. If strenuous exercise continues for more than a few seconds, more ATP is made. Muscle fibers have three ways to produce ATP: (1) from creatine phosphate, (2) by anaerobic cellular respiration, and (3) by aerobic cellular respiration (Figure 10.13). Whereas using creatine phosphate for ATP production is unique to muscle fibers, all body cells make ATP by the reactions of anaerobic and aerobic cellular respiration. We consider the events of cellular respiration briefly here and then in detail in Chapter 25.

Creatine Phosphate

While they are relaxed, muscle fibers produce more ATP than they need for resting metabolism. The excess ATP is used to synthesize **creatine phosphate,** an energy-rich molecule that is

Figure 10.13 Production of ATP for muscle contraction. (a) Creatine phosphate, formed from ATP while the muscle is relaxed, transfers a high-energy phosphate group to ADP, forming ATP, during muscle contraction. (b) Breakdown of muscle glycogen into glucose and production of pyruvic acid from glucose via glycolysis produce both ATP and lactic acid. Because no oxygen is needed, this is an anaerobic pathway. (c) Within mitochondria, pyruvic acid, fatty acids, and amino acids are used to produce ATP via aerobic cellular respiration, an oxygen-requiring set of reactions.

🔑 **During a long-term event such as a marathon race, most ATP is produced aerobically.**

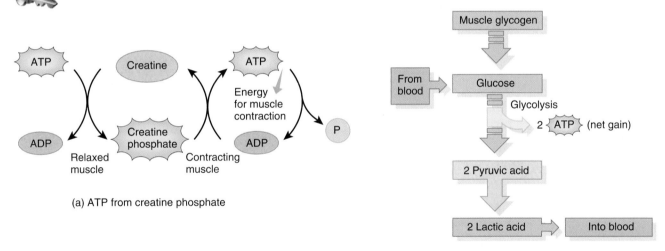

(a) ATP from creatine phosphate

(b) ATP from anaerobic respiration

(c) ATP from aerobic cellular respiration

 Where inside a skeletal muscle fiber are the events shown here occurring?

found only in muscle fibers (Figure 10.13a). The enzyme *creatine kinase (CK)* catalyzes the transfer of one of the high-energy phosphate groups from ATP to creatine, forming creatine phosphate and ADP. Creatine phosphate is three to six times more plentiful than ATP in the sarcoplasm of a relaxed muscle fiber. When contraction begins and the ADP level starts to rise, CK catalyzes the transfer of a high-energy phosphate group from creatine phosphate back to ADP. This direct phosphorylation reaction quickly forms new ATP molecules. Together, creatine phosphate and ATP provide enough energy for muscles to contract maximally for about 15 seconds. This amount of energy is sufficient for maximal short bursts of activity—for example, to run a 100-meter dash.

Creatine Supplementation

Creatine is a small, amino acid-like molecule that is both synthesized in the body and derived from foods. Adults need to synthesize and ingest a total of about 2 grams of creatine daily to make up for the urinary loss of creatinine, the breakdown product of creatine. Some studies have demonstrated improved performance during intense exercise in subjects who had ingested creatine supplements. For example, college football players who received supplements of 15 grams per day for 28 days gained more muscle mass and had larger gains in lifting power and sprinting performance than the control subjects. Other studies, however, have failed to find a performance-enhancing effect of creatine supplementation. Moreover, ingesting extra creatine decreases the body's own synthesis of creatine, and it is not known whether natural synthesis recovers after long-term creatine supplementation. Further research is needed to determine both the long-term safety and the value of creatine supplementation. ∎

Anaerobic Cellular Respiration

Anaerobic cellular respiration is a series of ATP-producing reactions that do not require oxygen. When muscle activity continues and the supply of creatine phosphate is depleted, glucose is catabolized to generate ATP. Glucose easily passes from the blood into contracting muscle fibers via facilitated diffusion, and it is also produced by the breakdown of glycogen within muscle fibers (Figure 10.13b). Then, a series of ten reactions known as *glycolysis* quickly breaks down each glucose molecule into two molecules of pyruvic acid. (Figure 25.4 on page 912 shows the reactions of glycolysis.) These reactions use two ATP but form four ATP for a net gain of two.

Ordinarily, the pyruvic acid formed by glycolysis in the cytosol enters mitochondria. There, it enters a series of oxygen-requiring reactions that produce a large amount of ATP. During some activities, however, not enough oxygen is available. Then, anaerobic reactions convert most of the pyruvic acid to lactic acid in the cytosol. About 80% of the lactic acid produced in this way diffuses out of the skeletal muscle fibers into the blood. Liver cells can convert some of the lactic acid back to glucose. This conversion has two benefits: providing new glucose molecules and reducing acidity. In these ways, anaerobic cellular respiration can provide enough energy for about 30 to 40 seconds of maximal muscle activity. Together, creatine phosphate and glycolysis can provide enough ATP to run a 400-meter race.

Aerobic Cellular Respiration

Muscle activity that lasts longer than half a minute depends increasingly on **aerobic cellular respiration,** a series of oxygen-requiring reactions that produce ATP in mitochondria. If sufficient oxygen is present, pyruvic acid enters the mitochondria, where it is completely oxidized in reactions that generate ATP, carbon dioxide, water, and heat (Figure 10.13c). Although aerobic cellular respiration is slower than glycolysis, it yields much more ATP. Each glucose molecule yields about 36 molecules of ATP and a typical fatty acid molecule yields more than 100 molecules of ATP via aerobic cellular respiration.

Muscle tissue has two sources of oxygen: (1) oxygen that diffuses into muscle fibers from the blood and (2) oxygen released by myoglobin within muscle fibers. Both myoglobin (found only in muscle cells) and hemoglobin (found only in red blood cells) are oxygen-binding proteins. They bind oxygen when it is plentiful and release oxygen when it is scarce.

Aerobic cellular respiration provides enough ATP for prolonged activity so long as sufficient oxygen and nutrients are available. Besides pyruvic acid obtained from glycolysis of glucose, these nutrients include fatty acids (from the breakdown of triglycerides in adipose cells) and amino acids (from the breakdown of proteins). In activities that last more than 10 minutes, the aerobic system provides more than 90% of the needed ATP. At the end of an endurance event, such as a marathon race, nearly 100% of the ATP is being produced by aerobic cellular respiration.

Muscle Fatigue

The inability of a muscle to contract forcefully after prolonged activity is called **muscle fatigue.** Fatigue results mainly from changes within muscle fibers. Even before actual muscle fatigue occurs, a person may have feelings of tiredness and the desire to cease activity. This response, termed *central fatigue,* may be a protective mechanism in that it might cause a person to stop exercising before muscle becomes too damaged. As you will see, certain types of skeletal muscle fibers fatigue more quickly than others.

Although the precise mechanisms that cause muscle fatigue are still not clear, several factors are thought to contribute. One important factor is inadequate release of calcium ions from the SR, resulting in a decline of Ca^{2+} concentration in the sarcoplasm. Depletion of creatine phosphate also is associated with fatigue. Surprisingly, however, the ATP levels in fatigued muscle often are not much lower than those in resting muscle. Other factors that contribute to muscle fatigue include insufficient oxygen, depletion of glycogen and other nutrients, buildup of lactic acid and ADP, and failure of action potentials in the motor neuron to release enough acetylcholine.

Oxygen Consumption After Exercise

During prolonged periods of muscle contraction, increases in breathing effort and blood flow enhance oxygen delivery to muscle tissue. After muscle contraction has stopped, heavy breathing continues for a while, and oxygen consumption remains above the resting level. Depending on the intensity of the exercise, the recovery period may be just a few minutes or several hours. In 1922, A. V. Hill coined the term **oxygen debt** for the added oxygen, over and above the resting oxygen consumption, that is taken into the body after exercise. He proposed that this extra oxygen was used to "pay back" or restore metabolic conditions to the resting level in three ways: (1) to convert lactic acid back into glycogen stores in the liver, (2) to resynthesize creatine phosphate and ATP in muscle fibers, and (3) to replace the oxygen removed from myoglobin.

The metabolic changes that occur *during exercise,* however, can account for only some of the extra oxygen used *after exercise.* Only a small amount of glycogen resynthesis occurs from lactic acid. Instead, most glycogen is made much later from dietary carbohydrates. Much of the lactic acid that remains after exercise is converted back to pyruvic acid and used for ATP production via aerobic cellular respiration in the heart, liver, kidneys, and skeletal muscle. Postexercise oxygen use also is boosted by ongoing changes. First, the elevated body temperature after strenuous exercise increases the rate of chemical reactions throughout the body. Faster reactions use ATP more rapidly, and more oxygen is needed to produce the ATP. Second, the heart and muscles used in breathing are still working harder than they were at rest, and thus they consume more ATP. Third, tissue repair processes are occurring at an increased pace. For these reasons, **recovery oxygen uptake** is a better term than oxygen debt for the elevated use of oxygen after exercise.

▶ CHECKPOINT

10. Which ATP-producing reactions are aerobic and which are anaerobic?

11. Which sources provide ATP during a 1000-meter run?

12. What factors contribute to muscle fatigue?

13. Why is the term *recovery oxygen uptake* more accurate than *oxygen debt?*

CONTROL OF MUSCLE TENSION

OBJECTIVES

• **Describe the structure and function of a motor unit.**

• **Explain the phases of a twitch contraction.**

• **Describe how frequency of stimulation affects muscle tension.**

A single nerve impulse in a motor neuron elicits a single muscle action potential in all the muscle fibers with which it forms synapses. In contrast to action potentials, which always have the same size in a given neuron or muscle fiber, the contraction that results from a single muscle action potential has significantly smaller force than the maximum force the fiber is capable of producing. The total force or tension that a single fiber can produce depends mainly on the rate at which nerve impulses arrive at the neuromuscular junction. The number of impulses per second is the *frequency of stimulation.* In addition, as we saw in Figure 10.10, the amount of stretch before contraction influences the maximum tension that is possible during contraction. Finally, conditions such as nutrient and oxygen availability can influence the tension a single fiber can generate. When considering the contraction of a whole muscle, the total tension it can produce depends on the number of fibers that are contracting in unison.

Motor Units

Even though each skeletal muscle fiber has only a single neuromuscular junction, the axon of a motor neuron branches out and forms neuromuscular junctions with many different muscle fibers. A **motor unit** consists of a somatic motor neuron plus all the skeletal muscle fibers it stimulates (Figure 10.14). A single motor neuron makes contact with an average of 150 muscle fibers and all muscle fibers in one motor unit contract in unison. Typically, the muscle fibers of a motor unit are dispersed throughout a muscle rather than clustered together.

Muscles that control precise movements consist of many small motor units. For instance, muscles of the larynx (voice box) that control voice production have as few as two or three muscle fibers per motor unit, and muscles controlling eye movements may have 10 to 20 muscle fibers per motor unit. In contrast, skeletal muscles responsible for large-scale and powerful movements, such as the biceps brachii muscle in the arm and the gastrocnemius muscle in the calf of the leg, have as many as 2000 to 3000 muscle fibers in some motor units. Remember, all the muscle fibers of a motor unit contract and relax together. Accordingly, the total strength of a contraction depends, in part, on how large the motor units are and how many motor units are activated at the same time.

Figure 10.14 Motor units. Shown are two somatic motor neurons, one in purple and one in green, each supplying the muscle fibers of its motor unit.

 A motor unit consists of a somatic motor neuron plus all the muscle fibers it stimulates.

Neuromuscular junction

Spinal cord

Motor neurons

Muscle fibers (cells)

What is the effect of the size of a motor unit on its strength of contraction? (Assume that each muscle fiber can generate about the same amount of tension.)

Twitch Contraction

A **twitch contraction** is the brief contraction of all the muscle fibers in a motor unit in response to a single action potential in its motor neuron. In the laboratory, a twitch also can be produced by direct electrical stimulation of a motor neuron or its muscle fibers. The record of a muscle contraction, called a **myogram,** is shown in Figure 10.15. Twitches of skeletal muscle fibers last anywhere from 20 to 200 msec. This duration is very long compared to the brief 1–2 msec* duration of an action potential.

Note that a brief delay occurs between application of the stimulus (time zero on the graph) and the beginning of contraction. The delay, which lasts about two milliseconds, is termed the **latent period.** During the latent period, calcium ions are being released from the sarcoplasmic reticulum, the filaments start to exert tension, the elastic components stretch, and finally shortening begins. The second phase, the **contraction period,** lasts 10–100 msec. During the third phase, the **relaxation period,** also lasting 10–100 msec, the active transport of Ca^{2+} back into the sarcoplasmic reticulum causes relaxation. The actual duration of these periods depends on the type of muscle fiber. Some fibers, such as those that move the eyes, are fast-twitch fibers (described shortly); they have a contraction period as brief as 10 msec and an equally brief relaxation period. Others, such as those that move the legs, are slow-twitch fibers, with contraction and relaxation periods of about 100 msec each.

If two stimuli are applied, one immediately after the other, the muscle will respond to the first stimulus but not to the second. When a muscle fiber receives enough stimulation to contract, it temporarily loses its excitability and cannot respond for a time. The period of lost excitability, called the **refractory period,** is a characteristic of all muscle and nerve cells. The duration of the refractory period varies with the muscle involved. Skeletal muscle has a short refractory period of about five milliseconds, whereas cardiac muscle has a long refractory period of about 300 milliseconds.

Frequency of Stimulation

When the second of two stimuli occurs after the refractory period is over, the skeletal muscle will respond to both stimuli (Figure 10.16). If the second stimulus occurs after the refractory period but before the muscle fiber has relaxed, the second contraction will actually be stronger than the first (Figure 10.16b). This phenomenon, in which stimuli arriving at different times cause larger contractions, is called **wave summation.**

When a skeletal muscle is stimulated at a rate of 20 to 30 times per second, it can only partially relax between stimuli. The result is a sustained but wavering contraction called **unfused tetanus** (*tetan-* = rigid, tense; Figure 10.16c). Stimula-

---

Figure 10.15 Myogram of a twitch contraction. The arrow indicates the time at which the stimulus occurred.

 A myogram is a record of a muscle contraction.

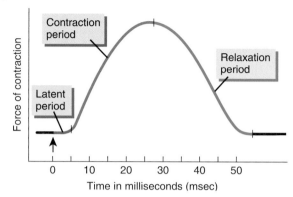

What events occur during the latent period?

tion at a higher rate of 80 to 100 stimuli per second causes **fused tetanus,** a sustained contraction in which individual twitches cannot be discerned (Figure 10.16d). Wave summation and both kinds of tetanus result from the addition of Ca^{2+} released from the sarcoplasmic reticulum by the second, and subsequent, stimuli to the Ca^{2+} still in the sarcoplasm from the first stimulus. Because the Ca^{2+} level builds up, the peak tension generated during fused tetanus is 5 to 10 times larger than the peak tension produced during a single twitch. Even so, smooth, sustained voluntary muscle contractions are achieved mainly by out-of-synchrony unfused tetanus in different motor units.

The stretch of elastic components is also related to wave summation. During wave summation, elastic components are not given much time to spring back between contractions, and thus they remain taut. While in this state, the elastic components do not require very much stretching before the beginning of the next muscular contraction. The combination of the tautness of the elastic components and the partially contracted state of the filaments enables the force of another contraction to be added more quickly to the one before.

Motor Unit Recruitment

The process in which the number of active motor units increases is called **motor unit recruitment.** Typically, the different motor units in a whole muscle are not stimulated to contract in unison. While some motor units are contracting, others are relaxed. This pattern of motor unit activity delays muscle fatigue by allowing alternately contracting motor units to relieve one another. In this way, contraction of a whole muscle can be sustained for long periods. The weakest motor units are recruited first, with progressively stronger motor units being added if the task requires more force.

Recruitment is one factor responsible for producing smooth movements rather than a series of jerks. As mentioned, the num-

Figure 10.16 Myograms showing the effects of different frequencies of stimulation. (a) Single twitch. (b) When a second stimulus occurs before the muscle has relaxed, the second contraction is stronger than the first, a phenomenon called wave summation. (The dashed line indicates the force of contraction expected in a single twitch.) (c) In unfused tetanus, the curve looks jagged due to partial relaxation of the muscle between stimuli. (d) In fused tetanus, which occurs when there are 80–100 stimuli per second, the contraction force is steady and sustained.

🔑 Due to wave summation, the tension produced during a sustained contraction is greater than during a single twitch.

(a) Single twitch   (b) Wave summation   (c) Unfused tetanus   (d) Fused tetanus

❓ Would the peak force of the second contraction in (b) be larger or smaller if the second stimulus were applied a few milliseconds later?

ber of muscle fibers innervated by one motor neuron varies greatly. Precise movements are brought about by small changes in muscle contraction. Therefore, the muscles that produce precise movements are made up of small motor units. For this reason, when a motor unit is recruited or turned off, only slight changes occur in muscle tension. By contrast, large motor units are active where large tension is needed and precision is less important.

🩺 Aerobic Training versus Strength Training

Regular, repeated activities such as jogging or aerobic dancing increase the supply of oxygen-rich blood available to skeletal muscles for aerobic cellular respiration. By contrast, activities such as weight lifting rely more on anaerobic production of ATP through glycolysis. Such anaerobic activities stimulate synthesis of muscle proteins and result, over time, in increased muscle size (muscle hypertrophy). As a result, aerobic training builds endurance for prolonged activities, whereas anaerobic training builds muscle strength for short-term feats. **Interval training** is a workout regimen that incorporates both types of training—for example, alternating sprints with jogging. ∎

Muscle Tone

Even at rest, a skeletal muscle exhibits **muscle tone** (*tonos* = tension), a small amount of tautness or tension in the muscle due to weak, involuntary contractions of its motor units. Recall that skeletal muscle contracts only after it is activated by acetylcholine released by nerve impulses in its motor neurons. Hence, muscle tone is established by neurons in the brain and spinal cord that excite the muscle's motor neurons. When the motor neurons serving a skeletal muscle are damaged or cut, the muscle becomes **flaccid** (FLAS-id = flabby), a state of limpness in which muscle tone is lost. To sustain muscle tone, small groups of motor units are alternately active and inactive in a constantly shifting pattern. Muscle tone keeps skeletal muscles firm, but it does not result in a force strong enough to produce movement. For example, when the muscles in the back of the neck are in normal tonic contraction, they keep the head upright and prevent it from slumping forward on the chest. Muscle tone also is important in smooth muscle tissues, such as those found in the gastrointestinal tract, where the walls of the digestive organs maintain a steady pressure on their contents. The tone of smooth muscle fibers in the walls of blood vessels plays a crucial role in regulating blood pressure.

Isotonic and Isometric Contractions

Isotonic contractions are used for body movements and for moving objects. The two types of isotonic contractions are concentric and eccentric. In a **concentric isotonic contraction,** a muscle shortens and pulls on another structure, such as a tendon, to produce movement and to reduce the angle at a joint. Picking a book up off a table involves concentric isotonic contractions of the biceps brachii muscle in the arm (Figure 10.17a). By contrast, as you lower the book to place it back on the table, the previously shortened biceps lengthens in a controlled manner while it continues to contract. When the length of a muscle increases during a contraction, the contraction is an **eccentric isotonic contraction** (Figure 10.17b). During an eccentric contraction, the tension exerted by the myosin crossbridges resists movement of a load (the book, in this case) and slows the lengthening process. For reasons that are not well understood, repeated eccentric isotonic contractions produce more muscle damage and more delayed-onset muscle soreness than do concentric isotonic contractions.

In **isometric contractions,** the myosin crossbridges generate considerable tension but the muscle doesn't shorten because the force of the load equals the muscle tension. An example would be holding a book steady using an outstretched arm (Figure 10.17c). These contractions are important for maintaining posture and for supporting objects in a fixed position. Although isometric contractions do not result in body movement, energy is still expended. The book pulls the arm downward, stretching the shoulder and arm muscles. The isometric contraction of the shoulder and arm muscles counteracts the stretch. Isometric contractions are important because they stabilize some joints as others are moved. Most activities include both isotonic and isometric contractions.

▶ CHECKPOINT

14. How are the sizes of motor units related to the degree of muscular control they allow?

15. What is motor unit recruitment?

16. Why is muscle tone important?

17. What do each of following terms mean: concentric isotonic contraction, eccentric isotonic contraction, and isometric contraction?

TYPES OF SKELETAL MUSCLE FIBERS

▶ OBJECTIVE

• **Compare the structure and function of the three types of skeletal muscle fibers.**

Skeletal muscle fibers are not all alike in composition and function. For example, muscle fibers vary in their content of myoglobin, the red-colored protein that binds oxygen in muscle fibers.

Figure 10.17 Comparison between isotonic (concentric and eccentric) and isometric contractions. Parts (a) and (b) show isotonic contraction of the biceps brachii muscle in the arm; part (c) shows isometric contraction of shoulder and arm muscles.

 In an isotonic contraction, tension remains constant as muscle length decreases or increases; in an isometric contraction, tension increases greatly without a change in muscle length.

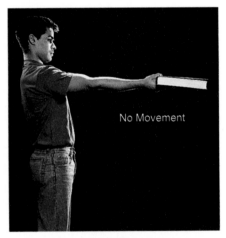

(a) Concentric contraction while picking up a book

(b) Eccentric contraction while lowering a book

(c) Isometric contraction while holding a book steady

What type of contraction is occurring in your neck muscles while you are walking?

Skeletal muscle fibers that have a high myoglobin content are termed *red muscle fibers;* those that have a low content of myoglobin are called *white muscle fibers.* Red muscle fibers also contain more mitochondria and are supplied by more blood capillaries than are white muscle fibers.

Skeletal muscle fibers also contract and relax at different speeds. A fiber is categorized as either slow or fast depending on how rapidly the ATPase in its myosin heads hydrolyzes ATP. In addition, skeletal muscle fibers vary in which metabolic reactions they use to generate ATP and in how quickly they fatigue. Based on these structural and functional characteristics, skeletal muscle fibers are classified into three main types: (1) slow oxidative fibers, (2) fast oxidative-glycolytic fibers, and (3) fast glycolytic fibers.

Slow Oxidative Fibers

Slow oxidative (SO) fibers are smallest in diameter and thus are the least powerful type of muscle fibers. They appear dark red because they contain large amounts of myoglobin and many blood capillaries. Because they have many large mitochondria, SO fibers generate ATP mainly by aerobic cellular respiration, which is why they are called oxidative fibers. These fibers are said to be "slow" because the ATPase in the myosin heads hydrolyzes ATP relatively slowly and the contraction cycle proceeds at a slower pace than in "fast" fibers. As a result, SO fibers have a slow speed of contraction. Their twitch contractions last from 100 to 200 msec, and they take longer to reach peak tension. However, slow fibers are very resistant to fatigue and are capable of prolonged, sustained contractions for many hours. These slow-twitch, fatigue-resistant fibers are adapted for maintaining posture and for aerobic, endurance-type activities such as running a marathon.

Fast Oxidative-Glycolytic Fibers

Fast oxidative-glycolytic (FOG) fibers are intermediate in diameter between the other two types of fibers. Like slow oxidative fibers, they contain large amounts of myoglobin and many blood capillaries. Thus, they have a dark red appearance. FOG fibers can generate considerable ATP by aerobic cellular respiration, which gives them a moderately high resistance to fatigue. Because their intracellular glycogen level is high, they also generate ATP by anaerobic glycolysis. FOG fibers are "fast" because the ATPase in their myosin heads hydrolyzes ATP three to five times faster than the myosin ATPase in SO fibers, which makes their speed of contraction faster. Thus, twitches of FOG fibers reach peak tension more quickly than those of SO fibers but are briefer in duration—less than 100 msec. FOG fibers contribute to activities such as walking and sprinting.

Fast Glycolytic Fibers

Fast glycolytic (FG) fibers are largest in diameter and contain the most myofibrils. Hence, they can generate the most powerful contractions. FG fibers have low myoglobin content, relatively few blood capillaries, few mitochondria, and appear white in color. They contain large amounts of glycogen and generate ATP mainly by glycolysis. Due to their large size and their ability to hydrolyze ATP rapidly, FG fibers contract strongly and quickly. These fast-twitch fibers are adapted for intense anaerobic movements of short duration, such as weight lifting or throwing a ball, but they fatigue quickly. Strength training programs that engage a person in activities requiring great strength for short times produce increases in the size, strength, and glycogen content of fast glycolytic fibers. The FG fibers of a weight lifter may be 50% larger than those of a sedentary person or endurance athlete. The increase in size is due to increased synthesis of muscle proteins. The overall result is muscle enlargement due to hypertrophy of the FG fibers.

Distribution and Recruitment of Different Types of Fibers

Most skeletal muscles are a mixture of all three types of skeletal muscle fibers, about half of which are SO fibers. The proportions vary somewhat, depending on the action of the muscle, the person's training regimen, and genetic factors. For example, the continually active postural muscles of the neck, back, and legs have a high proportion of SO fibers. Muscles of the shoulders and arms, in contrast, are not constantly active but are used briefly now and then to produce large amounts of tension, such as in lifting and throwing. These muscles have a high proportion of FG fibers. Leg muscles, which not only support the body but are also used for walking and running, have large numbers of both SO and FOG fibers.

Even though most skeletal muscles are a mixture of all three types of skeletal muscle fibers, the skeletal muscle fibers of any given motor unit are all of the same type. However, the different motor units in a muscle are recruited in a specific order, depending on need. For example, if weak contractions suffice to perform a task, only SO motor units are activated. If more force is needed, the motor units of FOG fibers are also recruited. Finally, if maximal force is required, motor units of FG fibers are also called into action. Activation of various motor units is controlled by the brain and spinal cord.

Table 10.1 summarizes the characteristics of the three types of skeletal muscle fibers.

▶ C H E C K P O I N T

18. Why are some skeletal muscle fibers classified as "fast" whereas others are said to be "slow"?

Table 10.1 Characteristics of the Three Types of Skeletal Muscle Fibers

- Slow oxidative fiber
- Fast glycolytic fiber
- Fast oxidative-glycolytic fiber

LM 440x

Transverse section of three types of skeletal muscle fibers

	Slow Oxidative (SO) Fibers	Fast Oxidative-Glycolytic (FOG) Fibers	Fast Glycolytic (FG) Fibers
Structural Characteristic			
Fiber diameter	Smallest.	Intermediate.	Largest.
Myoglobin content	Large amount.	Large amount.	Small amount.
Mitochondria	Many.	Many.	Few.
Capillaries	Many.	Many.	Few.
Color	Red.	Red-pink.	White (pale).
Functional Characteristic			
Capacity for generating ATP and method used	High capacity, by aerobic (oxygen-requiring) cellular respiration.	Intermediate capacity, by both aerobic (oxygen-requiring) cellular respiration and glycolysis (anaerobic).	Low capacity, by anaerobic cellular respiration (glycolysis).
Rate of ATP hydrolysis by myosin ATPase	Slow.	Fast.	Fast.
Contraction velocity	Slow.	Fast.	Fast.
Fatigue resistance	High.	Intermediate.	Low.
Creatine kinase	Lowest amount.	Intermediate amount.	Highest amount.
Glycogen stores	Low.	Intermediate.	High.
Order of recruitment	First.	Second.	Third.
Location where fibers are abundant	Postural muscles such as those of the neck.	Leg muscles.	Arm muscles.
Primary functions of fibers	Maintaining posture and aerobic endurance activities.	Walking, sprinting.	Rapid, intense movements of short duration.

EXERCISE AND SKELETAL MUSCLE TISSUE

▶ OBJECTIVE

- **Describe the effects of exercise on different types of skeletal muscle fibers.**

The relative ratio of fast-twitch and slow-twitch fibers in each muscle is genetically determined and helps account for individual differences in physical performance. For example, people with a higher proportion of fast glycolytic (FG) fibers often excel in activities that require periods of intense activity, such as weight lifting or sprinting. People with higher percentages of slow oxidative (SO) fibers are better at activities that require endurance, such as long-distance running.

Although the total number of skeletal muscle fibers usually does not increase, the characteristics of those present can change to some extent. Various type of exercises can induce changes in the fibers in a skeletal muscle. Endurance-type (aerobic) exercises, such as running or swimming, cause a gradual transformation of some fast glycolytic (FG) fibers into fast oxidative-glycolytic (FOG) fibers. The transformed muscle fibers show slight increases in diameter, number of mitochondria, blood supply, and strength. Endurance exercises also result in cardiovascular and respiratory changes that cause skeletal muscles to receive better supplies of oxygen and nutrients but do not increase muscle mass. By contrast, exercises that require great strength for short periods produce an increase in the size and strength of fast glycolytic (FG) fibers. The increase in size is due to increased synthesis of thick and thin filaments. The overall result is muscle enlargement (hypertrophy), as evidenced by the bulging muscles of body builders.

Anabolic Steroids

The illegal use of **anabolic steroids** by athletes has received widespread attention. These steroid hormones, similar to testosterone, are taken to increase muscle size and thus strength during athletic contests. The large doses needed to produce an effect, however, have damaging, sometimes even devastating side effects, including liver cancer, kidney damage, increased risk of heart disease, stunted growth, wide mood swings, and increased irritability and aggression. Additionally, females who take anabolic steroids may experience atrophy of the breasts and uterus, menstrual irregularities, sterility, facial hair growth, and deepening of the voice. Males may experience diminished testosterone secretion, atrophy of the testes, and baldness. ■

▶ C H E C K P O I N T
19. On a cellular level, what causes muscle hypertrophy?

CARDIAC MUSCLE TISSUE

▶ O B J E C T I V E

• **Describe the main structural and functional characteristics of cardiac muscle tissue.**

The principal tissue in the heart wall is **cardiac muscle tissue** (described in more detail in Chapter 20 and illustrated in Figure 20.9 on page 673). Between the layers of **cardiac muscle fibers,** the contractile cells of the heart, are sheets of connective tissue that contain blood vessels, nerves, and the conduction system of the heart. Cardiac muscle fibers have the same arrangement of actin and myosin and the same bands, zones, and Z discs as skeletal muscle fibers. A unique feature of cardiac muscle fibers are their *intercalated discs* (in-TER-ka-lāt-ed; *intercal-* = to insert between). These microscopic structures are irregular transverse thickenings of the sarcolemma that connect the ends of cardiac muscle fibers to each other. The discs contain *desmosomes,* which hold the fibers together, and *gap junctions,* which allow muscle action potentials to spread from one cardiac muscle fiber to another (see Figure 4.1e on page 105).

In response to a single action potential, cardiac muscle tissue remains contracted 10 to 15 times longer than skeletal muscle tissue (see Figure 20.11 on page 676). The long contraction is due to prolonged delivery of Ca²⁺ into the sarcoplasm. In cardiac muscle fibers, calcium ions enter the sarcoplasm both from the sarcoplasmic reticulum (as in skeletal muscle fibers) and from the interstitial fluid that bathes the fibers. Because the channels that allow inflow of Ca²⁺ from interstitial fluid stay open for a relatively long time, a cardiac muscle contraction lasts much longer than a skeletal muscle twitch.

We have seen that skeletal muscle tissue contracts only when stimulated by acetylcholine released by a nerve impulse in a motor neuron. In contrast, cardiac muscle tissue contracts when stimulated by its own autorhythmic muscle fibers. Under normal resting conditions, cardiac muscle tissue contracts and relaxes about 75 times a minute. This continuous, rhythmic

activity is a major physiological difference between cardiac and skeletal muscle tissue. Accordingly, cardiac muscle tissue requires a constant supply of oxygen, and the mitochondria in cardiac muscle fibers are larger and more numerous than in skeletal muscle fibers. This structural feature correctly suggests that cardiac muscle depends greatly on aerobic cellular respiration to generate ATP. Moreover, cardiac muscle fibers can use lactic acid produced by skeletal muscle fibers to make ATP, a benefit during exercise.

▶ C H E C K P O I N T
20. What are the similarities and differences between skeletal and cardiac muscle?

SMOOTH MUSCLE TISSUE

▶ O B J E C T I V E

• **Describe the main structural and functional characteristics of smooth muscle tissue.**

Like cardiac muscle tissue, **smooth muscle tissue** is usually activated involuntarily. Of the two types of smooth muscle tissue, the more common type is **visceral (single-unit) smooth muscle tissue** (Figure 10.18a). It is found in wraparound sheets that form

Figure 10.18 Two types of smooth muscle tissue. In (a), one autonomic motor neuron synapses with several visceral smooth muscle fibers, and action potentials spread to neighboring fibers through gap junctions. In (b), three autonomic motor neurons synapse with individual multiunit smooth muscle fibers. Stimulation of one multiunit fiber causes contraction of that fiber only.

🔑 **Visceral smooth muscle fibers connect to one another by gap junctions and contract as a single unit. Multiunit smooth muscle fibers lack gap junctions and contract independently.**

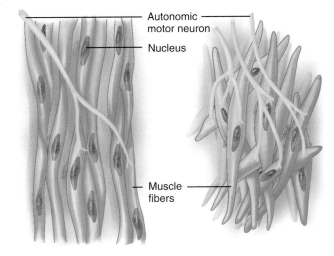

(a) Visceral (single-unit) smooth muscle tissue

(b) Multiunit smooth muscle tissue

 Which type of smooth muscle is more like cardiac muscle than skeletal muscle, with respect to both its structure and function?

part of the walls of small arteries and veins and of hollow organs such as the stomach, intestines, uterus, and urinary bladder. Like cardiac muscle, visceral smooth muscle is autorhythmic. Because the fibers connect to one another by gap junctions, muscle action potentials spread throughout the network. When a neurotransmitter, hormone, or autorhythmic signal stimulates one fiber, the muscle action potential spreads to neighboring fibers, which then contract in unison, as a single unit.

The second type of smooth muscle tissue is **multiunit smooth muscle tissue** (Figure 10.18b). It consists of individual fibers, each with its own motor neuron terminals and with few gap junctions between neighboring fibers. Whereas stimulation of one visceral muscle fiber causes contraction of many adjacent fibers, stimulation of one multiunit fiber causes contraction of that fiber only. Multiunit smooth muscle tissue is found in the walls of large arteries, in airways to the lungs, in the arrector pili muscles that attach to hair follicles, in the muscles of the iris that adjust pupil diameter, and in the ciliary body that adjusts focus of the lens in the eye.

Microscopic Anatomy of Smooth Muscle

A single relaxed smooth muscle fiber is 30–200 μm long. It is thickest in the middle (3–8 μm) and tapers at each end (Figure 10.19). Within each fiber is a single, oval, centrally located nucleus.

Figure 10.19 Microscopic anatomy of a smooth muscle fiber. A photomicrograph of smooth muscle is shown in Table 4.4 on page 131.

Smooth muscle fibers have thick and thin filaments but no transverse tubules and scanty sarcoplasmic reticulum.

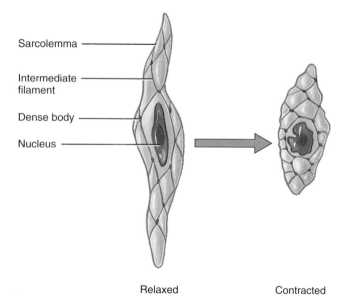

Sarcolemma

Intermediate filament

Dense body

Nucleus

Relaxed   Contracted

How does the speed of onset and duration of contraction in a smooth muscle fiber compare with that in a skeletal muscle fiber?

The sarcoplasm of smooth muscle fibers contains both *thick filaments* and *thin filaments,* in ratios between 1:10 and 1:15, but they are not arranged in orderly sarcomeres as in striated muscle. Smooth muscle fibers also contain *intermediate filaments.* Because the various filaments have no regular pattern of overlap, smooth muscle fibers do not exhibit striations (see Table 4.4 on page 131). This is the reason for the name *smooth.* Smooth muscle fibers also lack transverse tubules and have only scanty sarcoplasmic reticulum for storage of Ca^{2+}.

In smooth muscle fibers, intermediate filaments attach to structures called **dense bodies,** which are functionally similar to Z discs in striated muscle fibers. Some dense bodies are dispersed throughout the sarcoplasm; others are attached to the sarcolemma. Bundles of intermediate filaments stretch from one dense body to another (Figure 10.19). During contraction, the sliding filament mechanism involving thick and thin filaments generates tension that is transmitted to intermediate filaments. These, in turn, pull on the dense bodies attached to the sarcolemma, causing a lengthwise shortening of the muscle fiber. As a smooth muscle fiber contracts, it rotates as a corkscrew turns. The fiber twists in a helix as it contracts and rotates in the opposite direction as it relaxes.

Physiology of Smooth Muscle

Although the principles of contraction are similar in all three types of muscle tissue, smooth muscle tissue exhibits some important physiological differences. Compared with contraction in a skeletal muscle fiber, contraction in smooth muscle fiber starts more slowly and lasts much longer. Moreover, smooth muscle can both shorten and stretch to a greater extent than other muscle types.

An increase in the concentration of Ca^{2+} in smooth muscle cytosol initiates contraction, just as in striated muscle. Sarcoplasmic reticulum (the reservoir for Ca^{2+} in striated muscle) is scanty in smooth muscle. Calcium ions flow into smooth muscle cytosol from both the interstitial fluid and sarcoplasmic reticulum. Because there are no transverse tubules in smooth muscle fibers, however, it takes longer for Ca^{2+} to reach the filaments in the center of the fiber and trigger the contractile process. This accounts, in part, for the slow onset and prolonged contraction of smooth muscle.

Several mechanisms regulate contraction and relaxation of smooth muscle cells. In one, a regulatory protein called **calmodulin** (cal-MOD-ū-lin) binds to Ca^{2+} in the cytosol. (Recall that troponin has this role in striated muscle fibers.) After binding to Ca^{2+}, calmodulin activates an enzyme called *myosin light chain kinase.* This enzyme uses ATP to phosphorylate (add a phosphate group to) a portion of the myosin head. Once the phosphate group is attached, the myosin head can bind to actin, and contraction can occur. Myosin light chain kinase works rather slowly, also contributing to the slowness of smooth muscle contraction.

Not only do calcium ions enter smooth muscle fibers slowly, they also move slowly out of the muscle fiber when excitation declines, which delays relaxation. The prolonged presence of Ca^{2+} in the cytosol provides for **smooth muscle tone,** a state of continued partial contraction. Smooth muscle tissue can thus sustain long-term tone, which is important in the gastrointestinal tract, where the walls maintain a steady pressure on the contents of the tract, and in the walls of blood vessels called arterioles, which maintain a steady pressure on blood.

Most smooth muscle fibers contract or relax in response to action potentials from the autonomic nervous system. In addition, many smooth muscle fibers contract or relax in response to stretching, hormones, or local factors such as changes in pH, oxygen and carbon dioxide levels, temperature, and ion concentrations. For example, the hormone epinephrine, released by the adrenal medulla, causes relaxation of smooth muscle in the airways and in some blood vessel walls (those that have so-called β_2 receptors; see Table 17.4 on page 580).

Unlike striated muscle fibers, smooth muscle fibers can stretch considerably and still maintain their contractile function. When smooth muscle fibers are stretched, they initially contract, developing increased tension. Within a minute or so, the tension decreases. This phenomenon, which is called the **stress-relaxation response,** allows smooth muscle to undergo great changes in length while retaining the ability to contract effectively. Thus, even though smooth muscle in the walls of blood vessels and hollow organs such as the stomach, intestines, and urinary bladder can stretch, the pressure on the contents within them changes very little. After the organ empties, however, the smooth muscle in the wall rebounds, and the wall retains its firmness.

▶ C H E C K P O I N T

21. How do visceral and multiunit smooth muscle differ?

22. In what ways are the properties of skeletal and smooth muscle similar and different?

REGENERATION OF MUSCLE TISSUE

▶ O B J E C T I V E

• **Explain how muscle fibers regenerate.**

Because mature skeletal muscle fibers have lost the ability to undergo cell division, growth of skeletal muscle after birth is due mainly to hypertrophy, the enlargement of existing cells, rather than to **hyperplasia,** an increase in the number of fibers. Satellite cells divide slowly and fuse with existing fibers to assist both in muscle growth and in repair of damaged fibers. The number of new skeletal muscle fibers formed, however, is not enough to compensate for significant skeletal muscle damage or degenera-

tion. In such cases, skeletal muscle tissue undergoes **fibrosis,** the replacement of muscle fibers by fibrous scar tissue. For this reason, skeletal muscle tissue can regenerate only to a limited extent.

Until recently it was believed that damaged cardiac muscle fibers could not be replaced and that healing took place exclusively by fibrosis. New research described in Chapter 20 indicates that, under certain circumstances, cardiac muscle tissue can regenerate. In addition, cardiac muscle fibers can undergo hypertrophy in response to increased workload. Hence, many athletes have enlarged hearts.

Smooth muscle tissue, like skeletal and cardiac muscle tissue, can undergo hypertrophy. In addition, certain smooth muscle fibers, such as those in the uterus, retain their capacity for division and thus can grow by hyperplasia. Also, new smooth muscle fibers can arise from cells called *pericytes,* stem cells found in association with blood capillaries and small veins. Smooth muscle fibers can also proliferate in certain pathological conditions, such as occur in the development of atherosclerosis (see page 689). Compared with the other two types of muscle tissue, smooth muscle tissue has considerably greater powers of regeneration. Such powers are still limited when compared with other tissues, such as epithelium.

Table 10.2 summarizes the major characteristics of the three types of muscle tissue.

DEVELOPMENT OF MUSCLE

▶ O B J E C T I V E

• **Describe the development of muscles.**

Except for the muscles of the iris of the eyes and the arrector pili muscles attached to hairs, all muscles of the body are derived from **mesoderm.** As the mesoderm develops, part of it becomes arranged in dense columns on either side of the developing nervous system. These columns of mesoderm undergo segmentation into a series of blocks of cells called **somites** (Figure 10.20a on page 301). The first pair of somites appears on the 20th day of embryonic development. Eventually, 42 to 44 pairs of somites are formed by the end of the fifth week.

With the exception of the skeletal muscles of the head and limbs, *skeletal muscles* develop from the **mesoderm of somites.** Because there are very few somites in the head region of the embryo, most of the skeletal muscles there develop from the **general mesoderm** in the head region. The skeletal muscles of the limbs develop from masses of general mesoderm around developing bones in embryonic limb buds (origins of future limbs; see Figure 6.13a on page 181).

The cells of a somite differentiate into three regions: (1) a **myotome,** which forms the skeletal muscles of the head, neck, and limbs; (2) a **dermatome,** which forms the connective

Table 10.2   Summary of the Major Features of the Three Types of Muscle Tissue

Characteristic	Skeletal Muscle	Cardiac Muscle	Smooth Muscle
Microscopic appearance and features	Long cylindrical fiber with many peripherally located nuclei; striated.	Branched cylindrical fiber with one centrally located nucleus; intercalated discs join neighboring fibers; striated.	Fiber is thickest in the middle, tapered at each end, and has one centrally positioned nucleus; not striated.
Location	Attached primarily to bones by tendons.	Heart.	Walls of hollow viscera, airways, blood vessels, iris and ciliary body of eye, arrector pili of hair follicles.
Fiber diameter	Very large (10–100 μm).	Large (14 μm).	Small (3–8 μm).
Connective tissue components	Endomysium, perimysium, and epimysium.	Endomysium.	Endomysium.
Fiber length	100 μm–30 cm.	50–100 μm.	30–200 μm.
Contractile proteins organized into sarcomeres	Yes.	Yes.	No.
Sarcoplasmic reticulum	Abundant.	Some.	Scanty.
Transverse tubules present	Yes, aligned with each A–I band junction.	Yes, aligned with each Z disc.	No.
Junctions between fibers	None.	Intercalated discs contain gap junctions and desmosomes.	Gap junctions in visceral smooth muscle; none in multiunit smooth muscle.
Autorhythmicity	No.	Yes.	Yes, in visceral smooth muscle.
Source of Ca^{2+} for contraction	Sarcoplasmic reticulum.	Sarcoplasmic reticulum and interstitial fluid.	Sarcoplasmic reticulum and interstitial fluid.
Regulator proteins for contraction	Troponin and tropomyosin.	Troponin and tropomyosin.	Calmodulin and myosin light chain kinase.
Speed of contraction	Fast.	Moderate.	Slow.
Nervous control	Voluntary (somatic nervous system).	Involuntary (autonomic nervous system).	Involuntary (autonomic nervous system).
Contraction regulated by	Acetylcholine released by somatic motor neurons.	Acetylcholine and norepinephrine released by autonomic motor neurons; several hormones.	Acetylcholine and norepinephrine released by autonomic motor neurons; several hormones, local chemical changes; stretching.
Capacity for regeneration	Limited, via satellite cells.	Limited, under certain conditions.	Considerable, via pericytes (compared with other muscle tissues, but limited compared with epithelium).

tissues, including the dermis; and (3) a **sclerotome,** which gives rise to the vertebrae (Figure 10.20b).

Cardiac muscle develops from **mesodermal cells** that migrate to and envelop the developing heart while it is still in the form of primitive heart tubes (see in Figure 20.18 on page 688).

Smooth muscle develops from **mesodermal cells** that migrate to and envelop the developing gastrointestinal tract and viscera.

Figure 10.20 Location and structure of somites, key structures in the development of the muscular system.

🔑 **Most muscles are derived from mesoderm.**

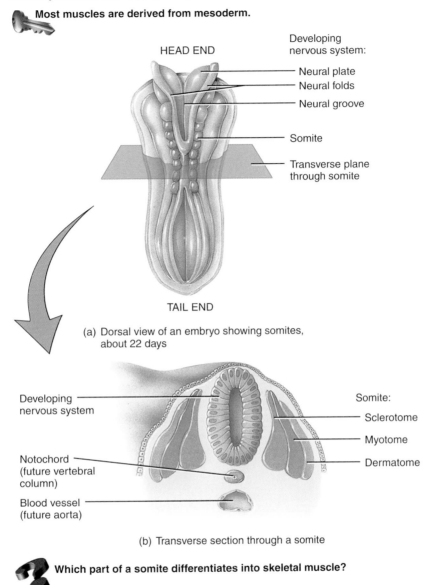

(a) Dorsal view of an embryo showing somites, about 22 days

(b) Transverse section through a somite

❓ **Which part of a somite differentiates into skeletal muscle?**

AGING AND MUSCLE TISSUE

▶ OBJECTIVE

• **Explain how aging affects skeletal muscle.**

Beginning at about 30 years of age, humans undergo a slow, progressive loss of skeletal muscle mass that is replaced largely by fibrous connective tissue and adipose tissue. In part, this decline is due to increasing inactivity. Accompanying the loss of muscle mass is a decrease in maximal strength, a slowing of muscle reflexes, and a loss of flexibility. In some muscles, a selective loss of muscle fibers of a given type may occur. With aging, the relative number of slow oxidative fibers appears to increase. This could be due either to atrophy of the other fiber types or their conversion into slow oxidative fibers. Whether this is an effect of aging itself or mainly reflects the more limited physical activity of older people is still an unresolved question. Nevertheless, aerobic activities and strength training programs are effective in older people and can slow or even reverse the age-associated decline in muscular performance.

▶ CHECKPOINT

23. Which type of muscle fibers can regenerate most effectively after damage?

24. Why does muscle strength decrease with aging?

DISORDERS: HOMEOSTATIC IMBALANCES

Skeletal muscle function may be abnormal due to disease or damage of any of the components of a motor unit: somatic motor neuron, neuromuscular junctions, or muscle fibers. The term **neuromuscular disease** encompasses problems at all three sites, whereas the term **myopathy** (mī-OP-a-thē; -*pathy* = disease) signifies a disease or disorder of the skeletal muscle tissue itself.

Myasthenia Gravis

Myasthenia gravis (mī-as-THĒ-nē-a GRAV-is) is an autoimmune disease that causes chronic, progressive damage of the neuromuscular junction. In people with myasthenia gravis, the immune system inappropriately produces antibodies that bind to and block some ACh receptors, thereby decreasing the number of functional ACh receptors at the motor end plates of skeletal muscles (see Figure 10.12). Because 75% of patients with myasthenia gravis have hyperplasia or tumors of the thymus, it is possible that thymic abnormalities cause the disorder. As the disease progresses, more ACh receptors are lost. Thus, muscles become increasingly weaker, fatigue more easily, and may eventually cease to function.

Myasthenia gravis occurs in about 1 in 10,000 people and is more common in women, who typically are ages 20 to 40 at onset, than in men, who usually are ages 50 to 60 at onset. The muscles of the face and neck are most often affected. Initial symptoms include weakness of the eye muscles, which may produce double vision, and difficulty in swallowing. Later, the person has difficulty chewing and talking. Eventually the muscles of the limbs may become involved. Death may result from paralysis of the respiratory muscles, but often the disorder does not progress to this stage.

Anticholinesterase drugs such as pyridostigmine (Mestinon) or neostigmine are the first line of treatment for myasthenia gravis. They act as inhibitors of acetylcholinesterase, the enzyme that breaks down ACh. Thus, the inhibitors raise the level of ACh that is available to bind with still-functional receptors. More recently, steroid drugs, such as prednisone, have been used with success to reduce antibody levels. Another treatment is plasmapheresis, a procedure that removes the antibodies from the blood. Often, surgical removal of the thymus (thymectomy) is helpful.

Muscular Dystrophy

The term **muscular dystrophy** refers to a group of inherited muscle-destroying diseases that cause progressive degeneration of skeletal muscle fibers. The most common form of muscular dystrophy is *DMD—Duchenne muscular dystrophy* (doo-SHAN). Because the mutated gene is on the X chromosome, which males have only one of, DMD strikes boys almost exclusively. (Sex-linked inheritance is described in Chapter 29.) Worldwide, about 1 in every 3500 male babies—21,000 in all—are born with DMD each year. The disorder usually becomes apparent between the ages of 2 and 5, when parents notice the child falls often and has difficulty running, jumping, and hopping. By age 12 most boys with DMD are unable to walk. Respiratory or cardiac failure usually causes death between the ages of 20 and 30.

In DMD, the gene that codes for the protein dystrophin is mutated, and little or no dystrophin is present. Without the reinforcing effect of dystrophin, the sarcolemma easily tears during muscle contraction. Because their plasma membrane is damaged, muscle fibers slowly rupture and die. The dystrophin gene was discovered in 1987, and by 1990 the first attempts were made to treat DMD patients with gene therapy. The muscles of three boys with DMD were injected with myoblasts bearing functional dystrophin genes, but only a few muscle fibers gained the ability to produce dystrophin. Similar clinical trials with additional patients have also failed. An alternate approach to the problem is to find a way to induce muscle fibers to produce the protein utrophin, which is similar to dystrophin. Experiments with dystrophin-deficient mice suggest this approach may work.

Fibromyalgia

Fibromyalgia (*algia* = painful condition) is a painful, nonarticular rheumatic disorder that usually appears between the ages of 25 and 50. An estimated 3 million people in the United States suffer from fibromyalgia, which is 15 times more common in women than in men. The disorder affects the fibrous connective tissue components of muscles, tendons, and ligaments. A striking sign is pain that results from gentle pressure at specific "tender points." Even without pressure, there is pain, tenderness, and stiffness of muscles, tendons, and surrounding soft tissues. Besides muscle pain, those with fibromyalgia report severe fatigue, poor sleep, headaches, depression, and inability to carry out their daily activities. Often, a gentle aerobic fitness program is beneficial.

Abnormal Contractions of Skeletal Muscle

One kind of abnormal muscular contraction is a **spasm,** a sudden involuntary contraction of a single muscle in a large group of muscles. A painful spasmodic contraction is known as a **cramp.** A **tic** is a spasmodic twitching made involuntarily by muscles that are ordinarily under voluntary control. Twitching of the eyelid and facial muscles are examples of tics. A **tremor** is a rhythmic, involuntary, purposeless contraction that produces a quivering or shaking movement. A **fasciculation** is an involuntary, brief twitch of an entire motor unit that is visible under the skin; it occurs irregularly and is not associated with movement of the affected muscle. Fasciculations may be seen in multiple sclerosis (see page 413) or in amyotrophic lateral sclerosis (Lou Gehrig's disease, see page 514). A **fibrillation** is a spontaneous contraction of a single muscle fiber that is not visible under the skin but can be recorded by electromyography. Fibrillations may signal destruction of motor neurons.

MEDICAL TERMINOLOGY

Hypertonia Increased muscle tone, either rigidity (muscle stiffness) or spasticity (muscle stiffness associated with an increase in tendon reflexes).

Hypotonia (*hypo-* = below or deficient) Decreased or lost muscle tone, usually due to damage to somatic motor neurons.

Muscle strain Tearing of a muscle because of forceful impact, accompanied by bleeding and severe pain. Also known as a Charley horse or pulled muscle. It often occurs in contact sports and typically affects the quadriceps femoris muscle on the anterior surface of the thigh. The condition is treated by RICE therapy: rest

(R), ice immediately after the injury (I), compression via a supportive wrap (C), and elevation of the limb (E).

Myalgia (mī-AL-jē-a; *-algia* = painful condition) Pain in or associated with muscles.

Myoma (mī-Ō-ma; *-oma* = tumor) A tumor consisting of muscle tissue.

Myomalacia (mī′-ō-ma-LĀ-shē-a; *-malacia* = soft) Pathological softening of muscle tissue.

Myositis (mī′-ō-SĪ-tis; *-itis* = inflammation of) Inflammation of muscle fibers (cells).

Myotonia (mī′-ō-TŌ-nē-a; *-tonia* = tension) Increased muscular excitability and contractility, with decreased power of relaxation; tonic spasm of the muscle.

Volkmann's contracture (FŌLK-manz kon-TRAK-tur; *contra-* = against) Permanent shortening (contracture) of a muscle due to replacement of destroyed muscle fibers by fibrous connective tissue, which lacks extensibility. Destruction of muscle fibers may occur from interference with circulation caused by a tight bandage, a piece of elastic, or a cast.

 ## STUDY OUTLINE

INTRODUCTION (p. 274)

1. Motion results from alternating contraction and relaxation of muscles, which constitute 40–50% of total body weight.
2. The prime function of muscle is changing chemical energy into mechanical energy to perform work.

OVERVIEW OF MUSCLE TISSUE (p. 274)

1. The three types of muscle tissue are skeletal, cardiac, and smooth. Skeletal muscle tissue is primarily attached to bones; it is striated and voluntary. Cardiac muscle tissue forms the wall of the heart; it is striated and involuntary. Smooth muscle tissue is located primarily in internal organs; it is nonstriated (smooth) and involuntary.
2. Through contraction and relaxation, muscle tissue performs four important functions: producing body movements; stabilizing body positions; moving substances within the body and regulating organ volume; and producing heat.
3. Four special properties of muscle tissues are electrical excitability, the property of responding to stimuli by producing action potentials; contractility, the ability to generate tension to do work; extensibility, the ability to be extended (stretched); and elasticity, the ability to return to original shape after contraction or extension.
4. An isotonic contraction occurs when a muscle shortens and moves a constant load; tension remains almost constant. An isometric contraction occurs without shortening of the muscle; tension increases greatly.

SKELETAL MUSCLE TISSUE (p. 275)

1. Connective tissues surrounding muscle are the epimysium, covering the entire muscle; perimysium, covering fascicles; and the endomysium, covering muscle fibers. Superficial fascia separates muscle from skin.
2. Tendons and aponeuroses are extensions of connective tissue beyond muscle fibers that attach the muscle to bone or to other muscle.
3. Skeletal muscles are well supplied with nerves and blood vessels. Generally, an artery and one or two veins accompany each nerve that penetrates a skeletal muscle.
4. Somatic motor neurons provide the nerve impulses that stimulate skeletal muscle to contract.
5. Blood capillaries bring in oxygen and nutrients and remove heat and waste products of muscle metabolism.
6. The major cells of skeletal muscle tissue are termed skeletal muscle fibers. Each muscle fiber has 100 or more nuclei because it arises from fusion of many myoblasts. Satellite cells are myoblasts that persist after birth. The sarcolemma is a muscle fiber's plasma membrane; it surrounds the sarcoplasm. T tubules are invaginations of the sarcolemma.
7. Each muscle fiber (cell) contains hundreds of myofibrils, the contractile elements of skeletal muscle. Sarcoplasmic reticulum surrounds each myofibril. Within a myofibril are thin and thick filaments, arranged in compartments called sarcomeres.
8. The overlapping of thick and thin filaments produces striations. Darker A bands alternate with lighter I bands.
9. Myofibrils are built from three types of proteins: contractile, regulatory, and structural. The contractile proteins are myosin (thick filament) and actin (thin filament). Regulatory proteins are tropomyosin and troponin (both are part of the thin filament). Structural proteins include titin (links Z disc to M line and stabilizes thick filament), myomesin (forms M line), and dystrophin (links thin filaments to sarcolemma).
10. Projecting myosin heads (crossbridges) contain actin-binding and ATP-binding sites and are the motor proteins that power muscle contraction.

CONTRACTION AND RELAXATION OF SKELETAL MUSCLE FIBERS (p. 282)

1. Muscle contraction occurs because myosin heads attach to and "walk" along the thin filaments at both ends of a sarcomere, progressively pulling the thin filaments toward the center of a sarcomere. As the thin filaments slide inward, the Z discs come closer together, and the sarcomere shortens.
2. The contraction cycle is the repeating sequence of events that causes sliding of the filaments: (1) myosin ATPase hydrolyzes ATP and becomes energized, (2) the myosin head attaches to actin, (3) the myosin head generates force as it rotates toward the center

of the sarcomere (power stroke), and (4) binding of ATP to myosin detaches myosin from actin. The myosin head again hydrolyzes the ATP, returns to its original position, and binds to a new site on actin as the cycle continues.

3. An increase in Ca^{2+} concentration in the cytosol starts filament sliding, whereas a decrease turns off the sliding process.

4. The muscle action potential propagating into the T tubule system causes opening of Ca^{2+} release channels in the SR membrane. Calcium ions diffuse from the SR into the cytosol and combine with troponin. This binding causes the troponin–tropomyosin complex to move away from the myosin-binding sites on actin.

5. Ca^{2+} active transport pumps continually remove Ca^{2+} from the sarcoplasm into the SR. When the concentration of calcium ions in the cytosol decreases, the troponin–tropomyosin complexes slide back over and block the myosin-binding sites, and the muscle fiber relaxes.

6. A muscle fiber develops its greatest tension when there is an optimal zone of overlap between thick and thin filaments. This dependence is the length–tension relationship.

7. The neuromuscular junction (NMJ) is the synapse between a somatic motor neuron and a skeletal muscle fiber. The NMJ includes the axon terminals and synaptic end bulbs of a motor neuron, plus the adjacent motor end plate of the muscle fiber sarcolemma.

8. When a nerve impulse reaches the synaptic end bulbs of a somatic motor neuron, it triggers exocytosis of the synaptic vesicles, which releases acetylcholine (ACh). ACh diffuses across the synaptic cleft and binds to ACh receptors, initiating a muscle action potential. Acetylcholinesterase then quickly destroys ACh.

MUSCLE METABOLISM (p. 289)

1. Muscle fibers have three sources for ATP production: creatine, anaerobic cellular respiration, and aerobic cellular respiration.

2. Creatine kinase catalyzes the transfer of a high-energy phosphate group from creatine phosphate to ADP to form new ATP molecules. Together, creatine phosphate and ATP provide enough energy for muscles to contract maximally for about 15 seconds.

3. Glucose is converted to pyruvic acid in the reactions of glycolysis, which yield two ATPs without using oxygen. Such anaerobic cellular respiration can provide enough energy for 30–40 seconds of maximal muscle activity.

4. Muscular activity that lasts longer than half a minute depends on aerobic cellular respiration, mitochondrial reactions that require oxygen to produce ATP.

5. The inability of a muscle to contract forcefully after prolonged activity is muscle fatigue.

6. Elevated oxygen use after exercise is called recovery oxygen uptake.

CONTROL OF MUSCLE TENSION (p. 291)

1. A motor neuron and the muscle fibers it stimulates form a motor unit. A single motor unit may contain as few as two or as many as 3000 muscle fibers.

2. Recruitment is the process of increasing the number of active motor units.

3. A twitch contraction is a brief contraction of all the muscle fibers in a motor unit in response to a single action potential.

4. A record of a contraction is called a myogram. It consists of a latent period, a contraction period, and a relaxation period. The

refractory period is the time when a muscle fiber has temporarily lost excitability. Skeletal muscles have a short refractory period, whereas cardiac muscle has a long refractory period.

5. Wave summation is the increased strength of a contraction that occurs when a second stimulus arrives before the muscle has completely relaxed after a previous stimulus.

6. Repeated stimuli can produce unfused tetanus, a sustained muscle contraction with partial relaxation between stimuli. More rapidly repeating stimuli produce fused tetanus, a sustained contraction without partial relaxation between stimuli.

7. Continual involuntary activation of a small number of motor units produces muscle tone, which is essential for maintaining posture.

8. In a concentric isotonic contraction, the muscle shortens to produce movement and to reduce the angle at a joint. During an eccentric isotonic contraction, the muscle lengthens.

9. Isometric contractions, in which tension is generated without muscle shortening, are important because they stabilize some joints as others are moved.

TYPES OF SKELETAL MUSCLE FIBERS (p. 295)

1. On the basis of their structure and function, skeletal muscle fibers are classified as slow oxidative (SO), fast oxidative-glycolytic (FOG), and fast glycolytic (FG) fibers.

2. Most skeletal muscles contain a mixture of all three fiber types. Their proportions vary with the typical action of the muscle.

3. The motor units of a muscle are recruited in the following order: first SO fibers, then FOG fibers, and finally FG fibers.

4. Table 10.1 on page 296 summarizes the three types of skeletal muscle fibers.

EXERCISE AND SKELETAL MUSCLE TISSUE (p. 296)

1. Various type of exercises can induce changes in the fibers in a skeletal muscle. Endurance-type (aerobic) exercises cause a gradual transformation of some fast glycolytic (FG) fibers into fast oxidative-glycolytic (FOG) fibers.

2. Exercises that require great strength for short periods produce an increase in the size and strength of fast-glycolytic (FG) fibers. The increase in size is due to increased synthesis of thick and thin filaments.

CARDIAC MUSCLE TISSUE (p. 297)

1. Cardiac muscle is found only in the heart. Cardiac muscle fibers have the same arrangement of actin and myosin and the same bands, zones, and Z discs as skeletal muscle fibers. The fibers connect to one another through intercalated discs, which contain both desmosomes and gap junctions.

2. Cardiac muscle tissue remains contracted 10 to 15 times longer than skeletal muscle tissue due to prolonged delivery of Ca^{2+} into the sarcoplasm.

3. Cardiac muscle tissue contracts when stimulated by its own autorhythmic fibers. Due to its continuous, rhythmic activity, cardiac muscle depends greatly on aerobic cellular respiration to generate ATP.

SMOOTH MUSCLE TISSUE (p. 297)

1. Smooth muscle is nonstriated and involuntary.

2. Smooth muscle fibers contain intermediate filaments and dense

bodies (whose function is similar to that of the Z discs in striated muscle).

3. Visceral (single-unit) smooth muscle is found in the walls of hollow viscera and of small blood vessels. Many fibers form a network that contracts in unison.

4. Multiunit smooth muscle is found in large blood vessels, large airways to the lungs, arrector pili muscles, and the eye (for adjusting pupil diameter and lens focus). The fibers operate independently rather than in unison.

5. The duration of contraction and relaxation of smooth muscle is longer than in skeletal muscle.

6. Smooth muscle fibers contract in response to nerve impulses, hormones, and local factors.

7. Smooth muscle fibers can stretch considerably and still maintain their contractile function.

REGENERATION OF MUSCLE TISSUE (p. 299)

1. Skeletal muscle fibers cannot divide and have limited powers of regeneration; cardiac muscle fibers can neither divide nor regenerate under certain circumstances; smooth muscle fibers have limited capacity for division and regeneration.

2. Table 10.2 on page 300 summarizes the major characteristics of the three types of muscle tissue.

DEVELOPMENT OF MUSCLE (p. 299)

1. With few exceptions, muscles develop from mesoderm.

2. Skeletal muscles of the head and limbs develop from general mesoderm. Other skeletal muscles develop from the mesoderm of somites.

AGING AND MUSCLE TISSUE (p. 301)

1. Beginning at about 30 years of age, there is a slow, progressive loss of skeletal muscle, which is replaced by fibrous connective tissue and fat.

2. Aging also results in a decrease in muscle strength, slower muscle reflexes, and loss of flexibility.

SELF-QUIZ QUESTIONS

Fill in the blanks in the following statements.

1. The properties of muscle that enable it to carry out its functions and contribute to homeostasis are _____, _____, _____, and _____.

2. The contractile proteins in muscles are _____ and _____.

3. A single somatic motor neuron and all of the muscle fibers it stimulates is known as a _____.

Indicate whether the following statements are true or false.

4. The key functions of muscle are motion of the body, movement of substances within the body, body position stability, organ volume regulation, heat production, and energy formation.

5. The sequence of events resulting in skeletal muscle contraction are (a) generation of a nerve impulse, (b) release of the neurotransmitter acetylcholine, (c) generation of a muscle action potential, (d) release of calcium ions from the sarcoplasmic reticulum, (e) calcium ion binding to the troponin–tropomyosin complex, (f) power stroke with actin and myosin binding and release.

Choose the one best answer to the following questions.

6. In muscle physiology, the latent period refers to (a) the period of lost excitability that occurs when two stimuli are applied immediately one after the other, (b) the brief contraction of a motor unit, (c) the period of elevated oxygen use after exercise, (d) an inability of a muscle to contract forcefully after prolonged activity, (e) a brief delay that occurs between application of a stimulus and the beginning of contraction.

7. In muscle contraction, what starts filament sliding? (a) an increase in Ca^{2+} concentration in the cytosol, (b) a decrease in Na^+ concentration in the cytosol, (c) the bending of the myosin head, (d) the release of energy from the breakdown of ATP, (e) the power stroke.

8. What would happen if ATP were suddenly unavailable after the sarcomere had begun to shorten? (a) Nothing. The contraction would proceed normally. (b) The myosin heads would be unable to detach from actin. (c) Troponin would bind with the myosin heads. (d) Actin and myosin filaments would separate completely and be unable to recombine. (e) The myosin heads would detach completely from actin and bind to the troponin–tropomyosin complex.

9. The calcium-binding protein in the sarcoplasmic reticulum of skeletal muscle is (a) calmodulin, (b) acetylcholinesterase, (c) troponin, (d) titin, (e) calsequestrin.

10. Which of the following are sources of ATP for muscle contraction? (1) creatine phosphate, (2) glycolysis, (3) anaerobic cellular respiration, (4) aerobic cellular respiration, (5) acetylcholine. (a) 1, 2, and 3, (b) 2, 3, and 4, (c) 2, 3, and 5, (d) 1, 2, 3, and 4, (e) 2, 3, 4, and 5.

11. Which of the following statements are true? (1) Skeletal muscles with a high myoglobin content are termed white muscle fibers. (2) Slow oxidative (SO) fibers generate ATP primarily by aerobic cellular respiration, are resistant to fatigue, and are capable of prolonged, sustained contractions. (3) Fast glycolytic (FG) fibers generate the most powerful contractions, generate ATP mainly by glycolysis, and are adapted for intense anaerobic movements. (4) Fast oxidative-glycolytic (FOG) fibers generate considerable ATP by aerobic cellular respiration but can also generate ATP by anaerobic glycolysis. (5) Most skeletal muscles are a mixture of SO, FOG, and FG fibers in various proportions. (6) Skeletal muscle fibers of any one motor unit can vary in muscle fiber types. (a) 1, 2, 3, and 4, (b) 2, 3, 4, 5, and 6, (c) 2, 3, 4, and 5, (d) 3, 4, 5, and 6, (e) 2, 4, and 6.

12. Match the following:

_____ (a) a sheath of areolar connective tissue that wraps around individual skeletal muscle fibers

_____ (b) dense irregular connective tissue that separates a muscle into groups of individual muscle fibers

_____ (c) bundles of muscle fibers

_____ (d) the outermost connective tissue layer that encircles an entire skeletal muscle

_____ (e) dense irregular connective tissue that lines the body wall and limbs and holds muscles with similar function together

_____ (f) a cord of dense regular connective tissue that attaches muscle to the periosteum of bone

_____ (g) areolar and adipose connective tissue that separates muscle from skin

_____ (h) connective tissue elements extended as a broad, flat layer

(1) aponeurosis
(2) deep fascia
(3) superficial fascia
(4) tendon
(5) endomysium
(6) perimysium
(7) epimysium
(8) fascicles

13. Match the following:

_____ (a) point of contact between a motor neuron and a muscle fiber

_____ (b) invaginations of the sarcolemma from the surface toward the center of the muscle fiber

_____ (c) myoblasts that persist in mature skeletal muscle

_____ (d) plasma membrane of a muscle fiber

_____ (e) oxygen-binding protein found only in muscle fibers

_____ (f) Ca^{2+}-storing tubular system similar to smooth endoplasmic reticulum

_____ (g) the contracting unit of a skeletal muscle fiber

_____ (h) middle area in the sarcomere where thick and thin filaments are found

_____ (i) area in the sarcomere where thin filaments are present but thick filaments are not

_____ (j) separates the sarcomeres from each other

(1) A band
(2) I band
(3) Z disc
(4) sarcomere
(5) neuromuscular junction
(6) myoglobin
(7) satellite cells
(8) T tubules
(9) sarcoplasmic reticulum
(10) sarcolemma

14. Match the following (some questions have more than one answer):

_____ (a) has fibers joined by intercalated discs

_____ (b) thick and thin filaments are not arranged as orderly sarcomeres

_____ (c) uses satellite cells to repair damaged muscle fibers

_____ (d) striated

_____ (e) contraction begins slowly but lasts for long periods

_____ (f) has a prolonged contraction due to prolonged calcium delivery from both the sarcoplasmic reticulum and the interstitial fluid

_____ (g) does not exhibit autorhythmicity

_____ (h) uses pericytes to repair damaged muscle fibers

_____ (i) uses troponin as a regulatory protein

_____ (j) can be classified as single-unit or multiunit

(1) skeletal muscle
(2) cardiac muscle
(3) smooth muscle

15. Match the following:

_____ (a) the smooth muscle action that allows the fibers to maintain their contractile function even when stretched

_____ (b) a brief contraction of all the muscle fibers in a motor unit of a muscle in response to a single action potential in its motor neuron

_____ (c) sustained contraction of a muscle

_____ (d) larger contractions resulting from stimuli arriving at different times

_____ (e) process of increasing the number of activated motor units

_____ (f) contraction in which the muscle shortens

_____ (g) inability of a muscle to maintain its strength of contraction or tension during prolonged activity

_____ (h) produced by the continual involuntary activation of a small number of skeletal muscle motor units; results in firmness in skeletal muscle

_____ (i) contraction in which muscle tension is generated without shortening of the muscle

_____ (j) contraction in which a muscle lengthens

(1) muscle fatigue
(2) twitch contraction
(3) wave summation
(4) tetanus
(5) concentric isotonic contraction
(6) motor unit recruitment
(7) muscle tone
(8) eccentric isotonic contraction
(9) isometric contraction
(10) stress-relaxation response

CRITICAL THINKING QUESTIONS

1. How can smooth muscle contract if it has no striations or sarcomeres?
   HINT *Look for similarities in the structure of smooth and striated muscle.*

2. Peng is an amateur body builder with a huge collection of weight-lifting equipment in his basement. His little sister Ming tried to lift the 100-pound barbell—it didn't budge. Peng lifted the same barbell over his head, held it there, and then lowered it back to the ground. Ming was very impressed with her big brother. Describe the types of contractions performed by Ming and her brother.
   HINT *Ming's contraction is the same type used by her brother to hold the barbell up.*

3. As the students from Mr. Klopfer's A&P class filed out of the room, they were massaging their writing hands. "Man, he was on a roll today! Two hours of notes with no break. My hand is killing me!" What's causing the muscle pain?
   HINT *After two hours of A&P, the brain is not the only part of the anatomy that is fatigued.*

ANSWERS TO FIGURE QUESTIONS

10.1 Perimysium bundles groups of muscle fibers into fascicles.

10.2 Blood capillaries in muscle deliver oxygen and nutrients and remove heat and waste products of muscle metabolism.

10.3 The sarcoplasmic reticulum releases calcium ions to trigger muscle contraction.

10.4 Size, from smallest to largest: thick filament, myofibril, and muscle fiber.

10.5 There are two thin filaments for each thick filament in skeletal muscle.

10.6 Actin, titin, and nebulin connect into the Z disc. A bands contain myosin, actin, troponin, tropomyosin, titin, and nebulin; I bands contain actin, troponin, tropomyosin, titin, and nebulin.

10.7 The I bands and H zones disappear. The lengths of the filaments do not change.

10.8 If ATP were not available, the myosin heads would not be able to detach from actin.

10.9 Three functions of ATP in muscle contraction are (1) its hydrolysis by an ATPase activates the myosin head so it can bind to actin and rotate; (2) its binding to myosin causes detachment from actin after the power stroke; and (3) it powers the pumps that transport Ca^{2+} from the cytosol back into the sarcoplasmic reticulum.

10.10 A sarcomere length of 2.2 μm gives a generous zone of overlap between the parts of the thick filaments that have myosin heads and the thin filaments without the overlap being so extensive that sarcomere shortening is limited.

10.11 The portion of the sarcolemma that contains acetylcholine receptors is the motor end plate.

10.12 Steps ④ through ⑥ are part of excitation–contraction coupling (muscle action potential through binding of myosin heads to actin).

10.13 Glycolysis, exchange of phosphate between creatine phosphate and ADP, and glycogen breakdown occur in the cytosol. Oxidation of pyruvic acid, amino acids, and fatty acids (aerobic cellular respiration) occurs in mitochondria.

10.14 Motor units having many muscle fibers are capable of more forceful contractions than those having only a few fibers.

10.15 Events occurring during the latent period are the events of excitation–contraction coupling: Release of Ca^{2+} from SR and binding to troponin result in attachment of myosin heads to actin and the onset of rotation.

10.16 If the second stimulus were applied a little later, the second contraction would be smaller than the one illustrated in (b).

10.17 Holding your head upright without movement involves mainly isometric contractions.

10.18 Visceral smooth muscle and cardiac muscle are similar in that both contain gap junctions, which allow action potentials to spread from one cell to its neighbors.

10.19 Contraction in a smooth muscle fiber starts more slowly and lasts much longer than contraction in a skeletal muscle fiber.

10.20 The myotome of a somite differentiates into skeletal muscle.

The Muscular System

THE MUSCULAR SYSTEM AND HOMEOSTASIS

The muscular system and muscle tissues of your body contribute to homeostasis by stabilizing body position, producing movements, regulating organ volume, moving substances within the body, and producing heat.

ENERGIZE YOUR STUDY
FOUNDATIONS CD

Anatomy Overview
• The Muscular System

Animation
• Systems Contributions to Homeostasis
• Negative Feedback Control of Temperature

Exercises
• Warming Up

www.wiley.com/college/apcentral

INSIGHTS AND EXPLORATIONS

You have probably seen the airplane passenger who has to get up during a long or evening flight and walk about the cabin. Or maybe you have sat next to someone who has to jiggle his legs during a movie or theater event. These are individuals who have restless leg syndrome. They have an overwhelming urge to move their legs. Some of these people are awakened at night by creepy-crawly sensation in their legs. Come learn more about the restless leg syndrome in this web-based activity.

ogether, the voluntarily controlled muscles of your body comprise the **muscular system.** The nearly 700 individual muscles that make up the muscular system, for instance, the deltoid muscle, include both skeletal muscle tissue and connective tissue. Most muscles have important functions in producing movements of body parts. A few muscles function mainly to stabilize bones so that other skeletal muscles can execute a movement more effectively. This chapter presents many of the major muscles in the body. Most of them are found on both the right and left sides. For each muscle, we'll identify its attachment sites and its innervation—the nerve or nerves that stimulate it to contract. Developing a working knowledge of these key aspects of skeletal muscle anatomy will enable you to understand how normal movements occur. This knowledge is especially crucial for allied health and physical rehabilitation professionals who work with people whose normal patterns of movement and physical mobility have been disrupted by physical trauma, surgery, or muscular paralysis.

HOW SKELETAL MUSCLES PRODUCE MOVEMENT

► O B J E C T I V E S

- **Describe the relationship between bones and skeletal muscles in producing body movements.**
- **Define lever and fulcrum, and compare the three types of levers based on location of the fulcrum, effort, and load.**
- **Identify the types of fascicle arrangements in a skeletal muscle, and relate the arrangements to strength of contraction and range of motion.**
- **Explain how the prime mover, antagonist, synergist, and fixator in a muscle group work together to produce movement.**

Muscle Attachment Sites: Origin and Insertion

Skeletal muscles that produce movement do so by exerting force on tendons, which in turn pull on bones or other structures, for instance, skin. Most muscles cross at least one joint and typically attach to the articulating bones that form the joint (Figure 11.1a).

When a muscle contracts, it pulls one of the articulating bones toward the other. Typically, the two bones do not move equally in response to contraction. Rather, one bone remains stationary or near its original position. Either other muscles stabilize that bone by contracting and pulling it in the opposite direction or the bone's structure makes it less movable. The site of a muscle's attachment to the more stationary bone is its **origin.** The site of a muscle's attachment to the more movable bone is its **insertion.** Usually, the insertion moves toward the origin as a muscle contracts. A good analogy is a spring on a screen door. The spring has its origin on the doorframe, the stationary attachment, and its insertion on the door, the moveable attachment. Often, the origin is proximal to the insertion, especially for muscles in the limbs. The fleshy part of the muscle between the tendons of the origin and insertion is the **belly.**

Muscles that move a body part often do not lie over the moving part. For example, one of the functions of the biceps brachii muscle is to move the forearm (Figure 11.1b). Note, however, that the belly of the muscle is over the humerus, not the forearm. You will also see that muscles crossing two joints, such as the rectus femoris and sartorius, have more complex actions than muscles that cross only one joint.

Tenosynovitis

Tenosynovitis (ten′-ō-sin-ō-VĪ-tis), commonly known as **tendinitis,** is a painful inflammation of the tendons, tendon sheaths, and synovial membranes of joints. The tendons most often affected are at the wrists, shoulders, elbows (resulting in *tennis elbow*), finger joints (resulting in *trigger finger*), ankles, and feet. The affected sheaths sometimes become visibly swollen due to fluid accumulation. The joint is tender and movement of the body part often causes pain. Trauma, strain, or excessive exercise may cause tenosynovitis. For instance, tying shoelaces too tightly may cause tenosynovitis of the dorsum of the foot. Also, gymnasts are prone to developing the condition because of chronic, repetitive, and maximum hyperextension at the wrists. ■

Lever Systems and Leverage

In producing movement, bones act as levers, and joints function as the fulcrums of these levers. A **lever** is a rigid structure that can move around a fixed point called a **fulcrum,** symbolized by △. A lever is acted on at two different points by two different forces: the **effort** [E], which causes movement, and the **load** [L] or **resistance,** which opposes movement. The effort is the force exerted by muscular contraction, whereas the load is typically the weight of the body part that is moved. Motion occurs when the effort applied to the bone at the insertion exceeds the load. Consider the biceps brachii flexing the forearm at the elbow as an object is lifted (Figure 11.1b). When the forearm is raised, the elbow is the fulcrum. The weight of the forearm plus the weight of the object in the hand is the load. The force of contraction of the biceps brachii pulling the forearm up is the effort.

Levers produce trade-offs between effort and the speed and range of motion. In one situation, a lever operates at a *mechanical advantage*—has **leverage**—when a smaller effort can move a heavier load. Here the trade-off is that the effort must move a greater distance (must have a longer range of motion) and faster

Figure 11.1   Relationship of skeletal muscles to bones.   (a) Muscles are attached to bones by tendons known as the origin and insertion. (b) Skeletal muscles produce movements by pulling on bones. Bones serve as levers, and joints act as fulcrums for the levers. Here the movement of the forearm illustrates the lever-fulcrum principle. Note where the load (resistance) and effort are applied in this example.

In the limbs, the origin of a muscle is usually proximal and the insertion is usually distal.

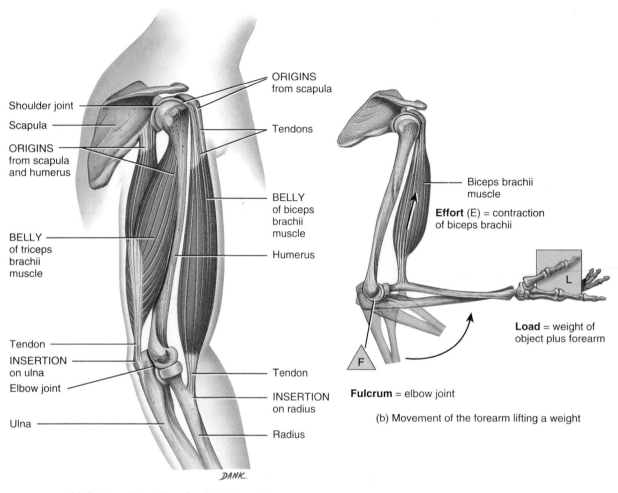

(a) Origin and insertion of a skeletal muscle

Where is the belly of the muscle that extends the forearm located?

than the load. Recall that range of motion refers to the range, measured in degrees of a circle, through which the bones of a joint can be moved. The lever formed by the mandible at the temporomandibular joints (fulcrums) and the effort provided by contraction of the jaw muscles produce a powerful mechanical advantage that crushes food. In another situation, a lever operates at a *mechanical disadvantage* when a larger effort moves a lighter load. In this case the trade-off is that the effort must move a shorter distance and slower than the load. The lever formed by the humerus at the shoulder joint (fulcrum) and the effort provided by the back and shoulder muscles produces a

mechanical "disadvantage" that enables a major-league pitcher to hurl a baseball at nearly 100 miles per hour!

The positions of the effort, load, and fulcrum on the lever determine whether the lever operates at a mechanical advantage or disadvantage. When the load is close to the fulcrum and the effort is applied farther away, the lever operates at a mechanical advantage. When you chew food, the load (the food) is positioned close to the fulcrums (your temporomandibular joints) while your jaw muscles exert effort farther out from the joints. By contrast, when the effort is applied close to the fulcrum and the load is farther away, the lever operates at a mechanical disadvantage. When a pitcher throws a baseball, the back and shoulder muscles apply intense effort very close to the fulcrum (the shoulder joint) while the lighter load (the ball) is propelled at the far end of the lever (the arm bone).

Levers are categorized into three types according to the positions of the fulcrum, the effort, and the load:

1. The fulcrum is between the effort and the load in **first-class levers** (Figure 11.2a). (Think **EFL**.) Scissors and seesaws are examples of first-class levers. A first-class lever can produce either a mechanical advantage or disadvantage depending on whether the effort or the load is closer to the fulcrum. As we've seen in the preceding examples, if the effort is farther from the fulcrum than the load, a heavy load can be moved, but not very far or fast. If the effort is closer to the fulcrum than the load, only a lighter load can be moved, but it moves far and fast.

There are few first-class levers in the body. One example is the lever formed by the head resting on the vertebral column (Figure 11s.2sa). When the head is raised, the weight of the an-

terior portion of the skull is the load. The joint between the atlas and the occipital bone (atlanto-occipital joint) forms the fulcrum. The contraction of the posterior neck muscles provides the effort.

2. The load is between the fulcrum and the effort in **second-class levers** (Figure 11.2b). (Think **FLE**.) They operate like a wheelbarrow. Second-class levers always produce a mechanical advantage because the load is always closer to the fulcrum than the effort. This arrangement sacrifices speed and range of motion for force. Most authorities believe that there are no second-class levers in the body.

3. The effort is between the fulcrum and the load in **third-class levers** (Figure 11.2c). (Think **FEL**.) These levers operate like a pair of forceps and are the most common levers in the body. Third-class levers always produce a mechanical disadvantage because the effort is always closer to the fulcrum than the load. In the body, this arrangement favors speed and range of motion over force. The elbow joint, the bones of the arm and

Figure 11.2 Types of levers.

Levers are divided into three types based on the locations of the fulcrum, effort, and load (resistance).

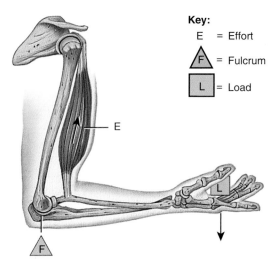

Key:
E = Effort
F = Fulcrum
L = Load

(a) First-class lever

(b) Second-class lever

(c) Third-class lever

In which class of lever is the effort applied between the fulcrum and the load?

forearm, and the biceps brachii muscle are one example of a third-class lever (Figure 11.2c). As we have seen, in flexing the forearm at the elbow, the weight of the hand and forearm is the load, the elbow joint is the fulcrum, and the contraction of the biceps brachii muscle provides the effort. Another example of the action of a third-class lever is adduction of the thigh, in which the thigh is the load, the hip joint is the fulcrum, and the contraction of the adductor muscles is the effort.

Effects of Fascicle Arrangement

Recall from Chapter 10 that the skeletal muscle fibers (cells) within a muscle are arranged in bundles known as **fascicles.** Within a fascicle, all muscle fibers are parallel to one another. The fascicles, however, may form one of five patterns with respect to the tendons: parallel, fusiform (shaped like a cigar), circular, triangular, or pennate (shaped like a feather) (Table 11.1).

Fascicular arrangement affects a muscle's power and range of motion. As a muscle fiber contracts, it shortens to about 70%

of its resting length. Thus, the longer the fibers in a muscle, the greater the range of motion it can produce. By contrast, the power of a muscle depends on its total cross-sectional area, because a short fiber can contract as forcefully as a long one. Fascicular arrangement often represents a compromise between power and range of motion. Pennate muscles, for instance, have a large number of fascicles distributed over their tendons, giving them greater power but a smaller range of motion. Parallel muscles, in contrast, have comparatively few fascicles that extend the length of the muscle. Thus, they have a greater range of motion but less power.

Coordination Within Muscle Groups

Movements often are the result of several skeletal muscles acting as a group rather than acting alone. Most skeletal muscles are arranged in opposing (antagonistic) pairs at joints—that is, flexors–extensors, abductors–adductors, and so on. Within opposing pairs, one muscle, called the **prime mover** or **agonist**

Table 11.1	Arrangement of Fascicles	
Parallel		**Fusiform**
Fascicles parallel to longitudinal axis of muscle; terminate at either end end in flat tendons.		Fascicles nearly parallel to longitudinal axis of muscle; terminate in flat tendons; muscle tapers toward tendons, where diameter is less than at belly.
Example: Stylohyoid muscle (see Figure 11.8)		*Example:* Digastric muscle (see Figure 11.8)
Circular		**Triangular**
Fascicles in concentric circular arrangements form sphincter muscles that enclose an orifice (opening).		Fascicles spread over broad area converge at thick central tendon; gives muscle a triangular appearance.
Example: Orbicularis oculi muscle (see Figure 11.4)		*Example:* Pectoralis major muscle (see Figure 11.3a)
Pennate		

Short fascicles in relation to total muscle length; tendon extends nearly entire length of muscle.

Unipennate	**Bipennate**	**Multipennate**
Fascicles are arranged on only one side of tendon.	Fascicles are arranged on both sides of centrally positioned tendons.	Fascicles attach obliquely from many directions to several tendons.
Example: Extensor digitorum longus muscle (see Figure 11.22b)	*Example:* Rectus femoris muscle (see Figure 11.20a)	*Example:* Deltoid muscle (see Figure 11.10a

(= leader), contracts to cause an action while the other muscle, the **antagonist** (*anti-* = against), stretches and yields to the effects of the prime mover. In the process of flexing the forearm at the elbow, for instance, the biceps brachii is the prime mover, and the triceps brachii is the antagonist (see Figure 11.1). The antagonist and prime mover are usually located on opposite sides of the bone or joint, as is the case in this example.

With an opposing pair of muscles, the roles of the prime mover and antagonist can switch for different movements. For example, while extending the forearm at the elbow, the triceps brachii is the prime mover, and the biceps brachii is the antagonist. The roles of the two muscles reverse during flexion of the elbow. If a prime mover and its antagonist contract at the same time with equal force, there will be no movement.

Sometimes a prime mover crosses other joints before it reaches the joint at which its primary action occurs. The biceps brachii, for example, spans both the shoulder and elbow joints, and its primary action is on the forearm. To prevent unwanted movements at intermediate joints or to otherwise aid the movement of the prime mover, muscles called **synergists** (SIN-er-gists; *syn* = together; *ergon* = work) contract and stabilize the intermediate joints. As an example, muscles that flex the fingers (prime movers) cross the intercarpal and radiocarpal joints (intermediate joints). If movement at these intermediate joints was unrestrained, you would not be able to flex your fingers without flexing the wrist at the same time. Synergistic contraction of the wrist extensors stabilizes the wrist joint and prevents it from moving (unwanted movement) while the flexor muscles of the fingers contract to bring about efficient flexion of the fingers (primary action). Synergists are usually located close to the prime mover.

Some muscles in a group also act as **fixators,** which stabilize the origin of the prime mover so that the prime mover can act more efficiently. Fixators steady the proximal end of a limb while movements occur at the distal end. For example, the scapula in the pectoral (shoulder) girdle is a freely movable bone that serves as the origin for several muscles that move the arm. However, when the arm muscles contract, the scapula must be held steady. Fixator muscles accomplish this by holding the scapula firmly against the back of the chest. In abduction of the arm, the deltoid muscle serves as the prime mover, whereas fixators (pectoralis minor, trapezius, subclavius, serratus anterior muscles, and others) hold the scapula firmly (see Figure 11.14). These fixators stabilize the scapula, which serves as the attachment site for the origin of the deltoid muscle, whereas the insertion of the muscle pulls on the humerus to abduct the arm. Under different conditions—that is, for different movements—many muscles may act, at various times, as prime movers, antagonists, synergists, or fixators.

In the limbs, a **compartment** is a group of skeletal muscles, and their associated blood vessels and nerves, that have a common function. In the upper limbs, for example, flexor compartment muscles are on the anterior surface whereas extensor compartment muscles are on the posterior surface.

Benefits of Stretching

The overall goal of **stretching** is to achieve normal range of motion of joints and mobility of soft tissues surrounding the joints. For most individuals, the best stretching routine involves *static stretching,* that is, slow sustained stretching that holds a muscle in a lengthened position. The muscles should be stretched to the point of slight discomfort (not pain) and held for about 15–30 seconds. Stretching should be done after warming up as this will increase range of motion. Among the benefits of stretching are the following:

1. *Improved physical performance.* A flexible joint has the ability to move through a greater range of motion, which improves physical performance.

2. *Decreased risk of injury.* Stretching decreases resistance in various soft tissues and thus there is less likelihood of exceeding maximum tissue extensibility during an activity. This decreases the risk of injury.

3. *Reduced muscle soreness.* Stretching can reduce some of the muscle soreness that results from delayed onset mucle soreness (DOMS) after exercise (see page 279).

4. *Improved posture.* Poor posture results from improper posture and gravity over a number of years. Stretching can help realign soft tissues to improve and maintain good posture. ■

▶ C H E C K P O I N T

1. With respect to muscles, what do the terms origin, insertion, and belly mean?

2. When you flex your knee, where are the fulcrum and load? Where is the muscle that provides the effort located?

3. Why can a parallel muscle have a greater range of motion than a pennate muscle?

HOW SKELETAL MUSCLES ARE NAMED

▶ O B J E C T I V E

• **Explain seven features used in naming skeletal muscles.**

Several features of skeletal muscles provide descriptive ways to name muscles. The names of most of the nearly 700 skeletal muscles contain combinations of word roots for their distinctive features. Learning the terms that refer to these features will help you remember the names of muscles. Such muscle features include the pattern of the muscle's fascicles; the size, shape, action, number of origins, and location of the muscle; and the sites of origin and insertion of the muscle. Study Table 11.2 to become familiar with the terms used in muscle names.

▶ C H E C K P O I N T

4. Select ten muscles in Figure 11.3 on pages 316–317 and identify the features on which their names are based. (Hint: *Use the prefix, suffix, and root of each muscle's name as a guide.*)

Table 11.2 Characteristics Used to Name Muscles

Name	Meaning	Example	Figure
DIRECTION: Orientation of muscle fascicles relative to the body's midline.			
Rectus	Parallel to midline	Rectus abdominis	11.10b
Transverse	Perpendicular to midline	Transverse abdominis	11.10b
Oblique	Diagonal to midline	External oblique	11.10a
SIZE: Relative size of the muscle.			
Maximus	Largest	Gluteus maximus	11.20c
Minimus	Smallest	Gluteus minimus	11.20c
Longus	Longest	Adductor longus	11.20a
Brevis	Shortest	Adductor brevis	11.20b
Latissimus	Widest	Latissimus dorsi	11.15b
Longissimus	Longest	Longissimus capitis	11.19a
Magnus	Large	Adductor magnus	11.20a
Major	Larger	Pectoralis major	11.15a
Minor	Smaller	Pectoralis minor	11.14a
Vastus	Huge	Vastus lateralis	11.20a
SHAPE: Relative shape of the muscle.			
Deltoid	Triangular	Deltoid	11.10a
Trapezius	Trapezoid	Trapezius	11.3b
Serratus	Saw-toothed	Serratus anterior	11.14b
Rhomboid	Diamond-shaped	Rhomboid major	11.15c
Orbicularis	Circular	Orbicularis oculi	11.4a
Pectinate	Comblike	Pectineus	11.20a
Piriformis	Pear-shaped	Piriformis	11.20c
Platys	Flat	Platysma	11.4a
Quadratus	Square, four-sided	Quadratus femoris	11.20c
Gracilis	Slender	Gracilis	11.20a
ACTION: Principal action of the muscle.			
Flexor	Decreases a joint angle	Flexor carpi radialis	11.17a
Extensor	Increases a joint angle	Extensor carpi ulnaris	11.17c
Abductor	Moves a bone away from the midline	Abductor pollicis longus	11.17c
Adductor	Moves a bone closer to the midline	Adductor longus	11.20a
Levator	Raises or elevates a body part	Levator scapulae	11.14a
Depressor	Lowers or depresses a body part	Depressor labii inferioris	11.4b
Supinator	Turns palm superiorly or anteriorly	Supinator	11.17b
Pronator	Turns palm inferiorly or posteriorly	Pronator teres	11.17a
Sphincter	Decreases the size of an opening	External anal sphincter	11.12
Tensor	Makes a body part rigid	Tensor fasciae latae	11.20a
Rotator	Rotates a bone around its longitudinal axis	Obturator externus	11.20b
NUMBER OF ORIGINS: Number of tendons of origin.			
Biceps	Two origins	Biceps brachii	11.16a
Triceps	Three origins	Triceps brachii	11.16b
Quadriceps	Four origins	Quadriceps femoris	11.20a
LOCATION: Structure near which a muscle is found. *Example:* Temporalis, a muscle near the temporal bone.			11.4c
ORIGIN AND INSERTION: Sites where muscle originates and inserts. *Example:* Sternocleidomastoid, originating on the sternum and clavicle and inserting on mastoid process of temporal bone.			11.3a

PRINCIPAL SKELETAL MUSCLES

Exhibits 11.1 through 11.20 will assist you in learning the names of the principal skeletal muscles in various regions of the body. The muscles in the exhibits are divided into groups according to the part of the body on which they act. As you study groups of muscles in the exhibits, refer to Figure 11.3 to see how each group is related to the others.

The exhibits contain the following elements:

- *Objective.* This statement describes the main outcomes to expect after studying the exhibit.

- *Overview.* These paragraphs provide a general orientation to the muscles under consideration and emphasize how the muscles are organized within various regions. The discussion also highlights any distinguishing or interesting features about the muscles.

- *Muscle names.* The word roots indicate how the muscles are named. Once you have mastered the naming of the muscles, their actions will have more meaning.

- *Origins, insertions, and actions.* For each muscle, you are also given its origin, insertion, and actions, the main movements that occur when the muscle contracts.

- *Relating muscles to movements.* These exercises will help you organize the muscles in a body region under consideration according to the actions they produce.

- *Innervation.* This section lists the nerve or nerves that cause contraction of each muscle. In general, cranial nerves, which arise from the lower parts of the brain, serve muscles in the head region. By contrast, spinal nerves, which arise from the spinal cord within the vertebral column, innervate muscles in the rest of the body. Cranial nerves are designated by both a name and a Roman numeral—for example, the facial nerve (cranial nerve VII). Spinal nerves are numbered in groups according to the part of the spinal cord from which they arise: C = cervical (neck region), T = thoracic (chest region), L = lumbar (lower back region), and S = sacral (buttocks region). An example is T1, the first thoracic spinal nerve.

- *Questions.* These review, thought-provoking, and application of knowledge questions relate specifically to information in each exhibit.

- *Figures.* The figures in the exhibits present superficial and deep, anterior and posterior, or medial and lateral views to show each muscle's position as clearly as possible. The muscle names in all capital letters are specifically referred to in the tabular part of the exhibit.

Here is a list of the exhibits and accompanying figures that describe the principal skeletal muscles:

• • •

To appreciate the many ways the muscular system contributes to homeostasis of other body systems, examine *Focus on Homeostasis: The Muscular System* on page 379. Next, in Chapter 12, you will see how the nervous system is organized, how neurons generate nerve impulses that activate muscle tissues as well as other neurons, and how synapses function.

Figure 11.3 Principal superficial skeletal muscles.

Most movements require several skeletal muscles acting in groups.

Epicranial aponeurosis

Occipitofrontalis (frontal belly)

Temporalis

Orbicularis oculi

Nasalis

Masseter

Orbicularis oris

Depressor anguli oris

Platysma

Omohyoid

Sternocleidomastoid

Scalenes

Sternohyoid

Trapezius

Latissimus dorsi

Deltoid

Pectoralis major

Serratus anterior

Rectus abdominis

Biceps brachii

External oblique

Brachialis

Brachioradialis

Triceps brachii

Extensor carpi radialis
longus

Extensor carpi radialis
longus and brevis

Extensor digitorum

Brachioradialis

Tensor fasciae latae

Flexor carpi radialis

Iliacus

Palmaris longus

Psoas major

Flexor carpi ulnaris

Extensor pollicis
longus

Abductor pollicis longus

Pectineus

Thenar muscles

Adductor longus

Hypothenar muscles

Sartorius

Adductor magnus

Gracilis

Vastus lateralis

Rectus femoris

Iliotibial tract

Vastus medialis

Tendon of
quadriceps
femoris

Patellar ligament

Tibialis anterior

Patella

Gastrocnemius

Fibularis longus

Soleus

Tibia

Tibia

Flexor digitorum longus

Calcaneal
(Achilles)
tendon

DANK

(a) Anterior view

Figure 11.3

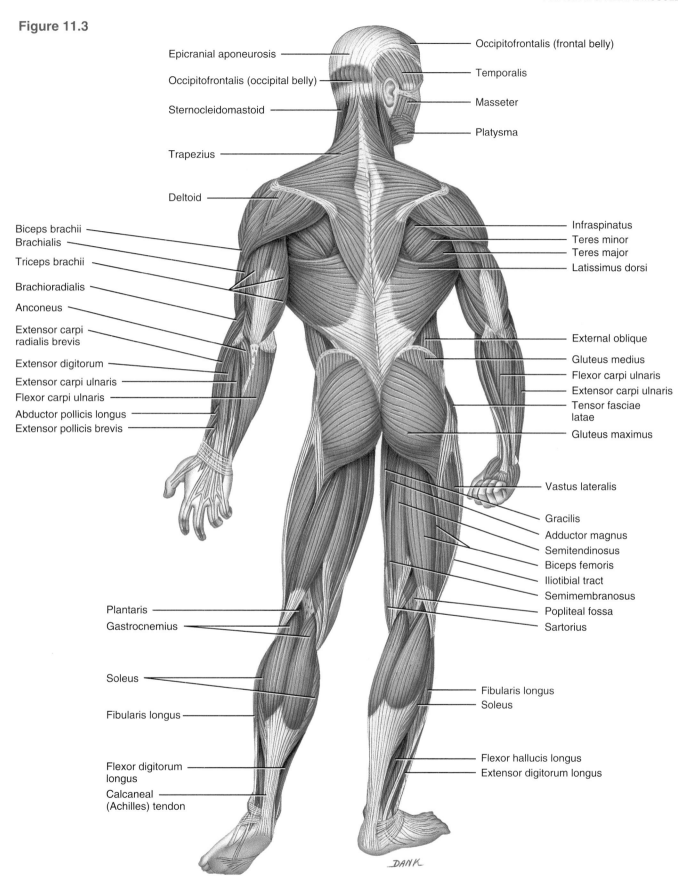

Epicranial aponeurosis

Occipitofrontalis (occipital belly)

Sternocleidomastoid

Trapezius

Deltoid

Biceps brachii
Brachialis
Triceps brachii

Brachioradialis

Anconeus

Extensor carpi
radialis brevis

Extensor digitorum
Extensor carpi ulnaris
Flexor carpi ulnaris
Abductor pollicis longus
Extensor pollicis brevis

Occipitofrontalis (frontal belly)

Temporalis

Masseter

Platysma

Infraspinatus
Teres minor
Teres major
Latissimus dorsi

External oblique

Gluteus medius
Flexor carpi ulnaris
Extensor carpi ulnaris
Tensor fasciae
latae

Gluteus maximus

Vastus lateralis

Gracilis
Adductor magnus
Semitendinosus
Biceps femoris
Iliotibial tract
Semimembranosus
Popliteal fossa
Sartorius

Plantaris
Gastrocnemius

Soleus

Fibularis longus

Flexor digitorum
longus

Calcaneal
(Achilles) tendon

Fibularis longus
Soleus

Flexor hallucis longus
Extensor digitorum longus

DANK

(b) Posterior view

 Give an example of a muscle named for each of the following characteristics: direction of fascicles, muscle shape, muscle action, muscle size, origin, insertion, location, and number of tendons of origin.

Exhibit 11.1   Muscles of Facial Expression (Figure 11.4)

▶ **OBJECTIVE**

• Describe the origin, insertion, action, and innervation of the muscles of facial expression.

The muscles of facial expression provide humans with the ability to express a wide variety of emotions. The muscles themselves lie within the layers of superficial fascia. They usually originate in the fascia or bones of the skull and insert into the skin. Because of their insertions, the muscles of facial expression move the skin rather than a joint when they contract.

Among the noteworthy muscles in this group are those surrounding the orifices (openings) of the head such as the eyes, nose, and mouth. These muscles function as *sphincters* (SFINGK-ters), which close the orifices, and *dilators,* which open the orifices. For example, the **orbicularis oculi** muscle closes the eye, whereas the **levator palpebrae superioris** muscle opens the eye. The **occipitofrontalis** is an unusual muscle in this group because it is made up of two parts: an anterior part called the **frontal belly,** which is superficial to the frontal bone, and a posterior part called the **occipital belly,** which is superficial to the occipital bone. The two muscular portions are held together by a strong aponeurosis (sheetlike tendon), the **epicranial aponeurosis** (ep-i-KRĀ-nē-al ap-ō-noo-RŌ-sis) or **galea aponeurotica** (GĀ-lē-a ap-ō-noo′-RŌ-ti-ka), which covers the superior and lateral surfaces of the skull. The **buccinator** muscle forms the major muscular portion of the cheek. It functions in whistling, blowing, and sucking and assists in chewing. The duct of the parotid gland (a salivary gland) pierces the buccinator muscle to reach the oral cavity. The buccinator muscle is so-named because it compresses the cheeks (*bucc-* = cheek) during blowing—for example, when a musician plays a wind instrument such as a trumpet.

Innervation

• All muscles of facial expression except levator palpebrae superioris: facial nerve (cranial nerve VII)

• Levator palpebrae superioris: oculomotor nerve (cranial nerve III)

 Bell's Palsy

Bell's palsy, also known as **facial paralysis,** is a unilateral paralysis of the muscles of facial expression. It is due to damage or disease of the facial nerve (cranial nerve VII). Although the cause is unknown, inflammation of the facial nerve or infection by the herpes simplex virus have been suggested. The paralysis causes the entire side of the face to droop in severe cases. The person cannot wrinkle the forehead, close the eye, or pucker the lips on the affected side. Difficulty in swallowing and drooling also occur. Eighty percent of patients recover completely within a few weeks to a few months. For others, paralysis is permanent. ■

Relating Muscles to Movements

Arrange the muscles in this exhibit into two groups: (1) those that act on the mouth and (2) those that act on the eyes.

▶ CHECKPOINT

What muscles would you use to do the following: show surprise, express sadness, bare your upper teeth, pucker your lips, squint, and blow up a balloon?

Muscle	Origin	Insertion	Action
Scalp Muscles			
Occipitofrontalis (ok-sip′-i-tō-frun-TĀ-lis)			
Frontal belly	Epicranial aponeurosis.	Skin superior to supraorbital margin.	Draws scalp anteriorly, raises eyebrows, and wrinkles skin of forehead horizontally.
Occipital belly (*occipit-* = back of the head)	Occipital bone and mastoid process of temporal bone.	Epicranial aponeurosis.	Draws scalp posteriorly.

Muscle	Origin	Insertion	Action
Mouth Muscles			
Orbicularis oris (or-bi'-kū-LAR-ioos OR-is; *orb-* = circular; *oris* = of the mouth)	Muscle fibers surrounding opening of mouth.	Skin at corner of mouth.	Closes and protrudes lips, compresses lips against teeth, and shapes lips during speech.
Zygomaticus major (zī-gō-MA-ti-kus; *zygomatic* = cheek bone; *major* = greater)	Zygomatic bone.	Skin at angle of mouth and orbicularis oris.	Draws angle of mouth superiorly and laterally, as in smiling or laughing.
Zygomaticus minor (*minor* = lesser)	Zygomatic bone.	Upper lip.	Raises (elevates) upper lip, exposing maxillary teeth.
Levator labii superioris (le-VĀ-tor LĀ-bē-ī soo-per'-ē-OR-is; *levator* = raises or elevates; *labii* = lip; *superioris* = upper)	Superior to infraorbital foramen of maxilla.	Skin at angle of mouth and orbicularis oris.	Raises upper lip.
Depressor labii inferioris (de-PRE-sor LĀ-bē-ī in-fer'-ē-OR-is; *depressor* = depresses or lowers; *inferioris* = lower)	Mandible.	Skin of lower lip.	Depresses (lowers) lower lip.
Depressor anguli oris (*angul* = angle or corner)	Mandible.	Angle of mouth.	Draws angle of mouth laterally and inferiorly, as in opening mouth.
Buccinator (BUK-si-nā'-tor; *bucc-* = cheek)	Alveolar processes of maxilla and mandible and pterygomandibular raphe (fibrous band extending from the pterygoid process to the mandible).	Orbicularis oris.	Presses cheeks against teeth and lips, as in whistling, blowing, and sucking; draws corner of mouth laterally; and assists in mastication (chewing) by keeping food between the teeth (and not between teeth and cheeks).
Risorius (ri-ZOR-ē-us; *risor* = laughter)	Fascia over parotid (salivary) gland.	Skin at angle of mouth.	Draws angle of mouth laterally, as in tenseness.
Mentalis (men-TĀ-lis; *ment-* = the chin)	Mandible.	Skin of chin.	Elevates and protrudes lower lip and pulls skin of chin up, as in pouting.
Neck Muscle			
Platysma (pla-TIZ-ma; *platy* = flat, broad)	Fascia over deltoid and pectoralis major muscles.	Mandible, muscles around angle of mouth, and skin of lower face.	Draws outer part of lower lip inferiorly and posteriorly as in pouting; depresses mandible.
Orbit and Eyebrow Muscles			
Orbicularis oculi (or-bi'-kū-LAR-is OK-ū-lī; *oculi* = of the eye)	Medial wall of orbit.	Circular path around orbit.	Closes eye.
Corrugator supercilii (KOR-a- gā'-tor soo-per-SI-lē-ī; *corrugat* = wrinkle; *supercilii* = of the eyebrow)	Medial end of superciliary arch of frontal bone.	Skin of eyebrow.	Draws eyebrow inferiorly and wrinkles skin of forehead vertically as in frowning.
Levator palpebrae superioris (le-VĀ-tor PAL-pe-brē soo-per'-ē-OR-is; *palpebrae* = eyelids) (see also Figure 11.5a)	Roof of orbit (lesser wing of sphenoid bone).	Skin of upper eyelid.	Elevates upper eyelid (opens eye).

(continues)

Exhibit 11.1   Muscles of Facial Expression (Figure 11.4) *(continued)*

Figure 11.4   Muscles of facial expression.   (See Tortora, *A Photographic Atlas of the Human Body,* Figures 5.2–5.4.)

When they contract, muscles of facial expression move the skin rather than a joint.

(a) Anterior superficial view   (b) Anterior deep view

Epicranial aponeurosis

TEMPORALIS

OCCIPITOFRONTALIS
(OCCIPITAL BELLY)

Posterior auricular

Zygomatic arch

Mandible

MASSETER

Sternocleidomastoid

Splenius capitis

Trapezius

Levator scapulae

Middle
scalene

OCCIPITOFRONTALIS
(FRONTAL BELLY)

ORBICULARIS OCULI

Nasalis
LEVATOR LABII
SUPERIORIS

ZYGOMATICUS MINOR

ZYGOMATICUS MAJOR

BUCCINATOR

RISORIUS

ORBICULARIS ORIS

DEPRESSOR ANGULI ORIS

DEPRESSOR LABII INFERIORIS

MENTALIS

PLATYSMA

DANK

(c) Right lateral superficial view

 Which muscles of facial expression cause frowning, smiling, pouting, and squinting?

Exhibit 11.2   Muscles that Move the Eyeballs—Extrinsic Eye Muscles (Figure 11.5)

▶ **OBJECTIVE**

• Describe the origin, insertion, action, and innervation of the extrinsic eye muscles.

Muscles that move the eyeballs are called **extrinsic eye muscles** because they originate outside the eyeballs (in the orbit) and insert on the outer surface of the sclera ("white of the eye"). The extrinsic eye muscles are some of the fastest contracting and most precisely controlled skeletal muscles in the body.

Three pairs of extrinsic eye muscles control movements of the eyeballs: (1) superior and inferior recti, (2) lateral and medial recti, and (3) superior and inferior oblique. The four recti muscles (superior, inferior, lateral, and medial) arise from a tendinous ring in the orbit and insert into the sclera of the eye. Whereas the **superior** and **inferior recti** lie in the same vertical plane, the **medial** and **lateral recti** lie in the same horizontal plane. The actions of the recti muscles can be deduced from their insertions on the sclera. The superior and inferior recti move the eyeballs superiorly and inferiorly, respectively; the lateral and medial recti move the eyeballs laterally and medially, respectively. It should be noted that neither the superior nor the inferior rectus muscle pulls directly parallel to the long axis of the eyeballs, and as a result, both muscles move the eyeballs medially.

It is not as easy to deduce the actions of the oblique muscles (superior and inferior) because of their paths through the orbits. For example, the **superior oblique** muscle originates posteriorly near the tendinous ring, then passes anteriorly, and ends in a round tendon. The tendon extends through a pulleylike loop called the *trochlea* (= pulley) in the anterior and medial part of the roof of the orbit. Finally, the tendon turns and inserts on the posterolateral aspect of the eyeballs. Accordingly, the superior oblique muscle moves the eyeballs inferiorly and laterally. The **inferior oblique** muscle originates on the maxilla at the anteromedial aspect of the floor of the orbit. It then passes posteriorly and laterally and inserts on the posterolateral aspect of the eyeballs. Because of this arrangement, the inferior oblique muscle moves the eyeballs superiorly and laterally.

Innervation

• Superior rectus, inferior rectus, medial rectus, and inferior oblique: oculomotor nerve (cranial nerve III)
• Lateral rectus: abducens nerve (cranial nerve VI)
• Superior oblique: trochlear nerve (cranial nerve IV)

 Strabismus

Strabismus is a condition in which the two eyes are not properly aligned. A lesion of the oculomotor nerve (cranial nerve III), which controls the superior, inferior, and medial recti and the inferior oblique muscles, causes the eyeball to move laterally when at rest. The person cannot move the eyeball medially and inferiorly. A lesion in the abducens nerve (cranial nerve VI), which innervates the lateral rectus muscle, causes the eyeball to move medially when at rest with inability to move the eyeball laterally. ■

Relating Muscles to Movements

Arrange the muscles in this exhibit according to their actions on the eyeballs: (1) elevation, (2) depression, (3) abduction, (4) adduction, (5) medial rotation, and (6) lateral rotation. The same muscle may be mentioned more than once.

▶   **CHECKPOINT**

Which muscles contract and relax in each eye as you gaze to your left without moving your head?

Muscle	Origin	Insertion	Action
Superior rectus (*rectus* = fascicles parallel to midline)	Common tendinous ring (attached to orbit around optic foramen).	Superior and central part of eyeball.	Moves eyeball superiorly (elevation) and medially (adduction), and rotates it medially.
Inferior rectus	Same as above.	Inferior and central part of eyeball.	Moves eyeball inferiorly (depression) and medially (adduction), and rotates it laterally.
Lateral rectus	Same as above.	Lateral side of eyeball.	Moves eyeball laterally (abduction).
Medial rectus	Same as above.	Medial side of eyeball.	Moves eyeball medially (adduction).
Superior oblique (*oblique* = fascicles diagonal to midline)	Sphenoid bone, superior and medial to the tendinous ring in the orbit.	Eyeball between superior and lateral recti. The muscle inserts into the superior and lateral surfaces of the eyeball via a tendon that passes through the trochlea.	Moves eyeball inferiorly (depression) and laterally (abduction), and rotates it medially.
Inferior oblique	Maxilla in floor of orbit.	Eyeball between inferior and lateral recti.	Moves eyeball superiorly (elevation) and laterally (abduction), and rotates it laterally.

Figure 11.5 Extrinsic muscles of the eyeball. (See Tortora, *A Photographic Atlas of the Human Body,* Figures 5.4 and 5.5.)

 The extrinsic muscles of the eyeball are some of the fastest contracting and most precisely controlled skeletal muscles in the body.

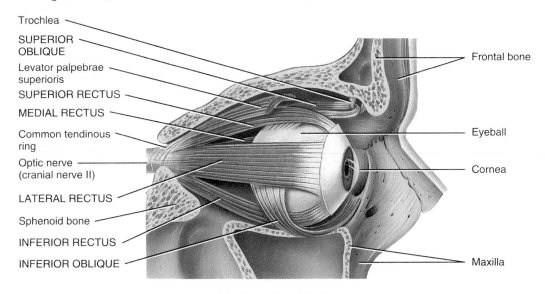

Trochlea
SUPERIOR OBLIQUE
Levator palpebrae superioris
SUPERIOR RECTUS
MEDIAL RECTUS
Common tendinous ring
Optic nerve (cranial nerve II)
LATERAL RECTUS
Sphenoid bone
INFERIOR RECTUS
INFERIOR OBLIQUE

Frontal bone
Eyeball
Cornea
Maxilla

(a) Lateral view of right eyeball

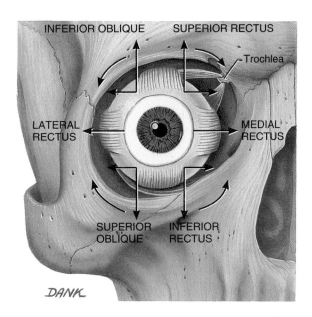

INFERIOR OBLIQUE   SUPERIOR RECTUS
Trochlea
LATERAL RECTUS
MEDIAL RECTUS
SUPERIOR OBLIQUE   INFERIOR RECTUS

DANK

(b) Movements of right eyeball in response to contraction of extrinsic muscles

Why does the inferior oblique muscle move the eyeball superiorly and laterally?

Exhibit 11.3   Muscles that Move the Mandible (Lower Jaw Bone) (Figure 11.6)

▶ O B J E C T I V E

• Describe the origin, insertion, action, and innervation of the muscles that move the mandible.

The muscles that move the mandible (lower jaw) at the temporomandibular joint (TMJ) are known as the muscles of mastication because they are involved in chewing (mastication). Of the four pairs of muscles involved in mastication, three are powerful closers of the jaw and account for the strength of the bite: **masseter, temporalis,** and **medial pterygoid.** Of these, the masseter is the strongest muscle of mastication. The medial and **lateral pterygoid** muscles assist in mastication by moving the mandible from side to side to help grind food. Additionally, these muscles protrude the mandible.

In 1996, researchers at the University of Maryland identified a new muscle in the skull and named it the **sphenomandibularis muscle.** It extends from the lateral surface of the sphenoid bone to the medial surface of the condylar process and ramus of the mandible (Figure 11.6b). The muscle is believed to be either a fifth muscle of mastication or a previously unidentified component of an already identified muscle (temporalis or medial pterygoid).

Innervation

• Masseter, temporalis, medial pterygoid, lateral pterygoid: mandibular division of the trigeminal nerve (cranial nerve V)

• Sphenomandibularis: maxillary branch of the trigeminal nerve (cranial nerve V)

Relating Muscles to Movements

Arrange the muscles in this exhibit according to their actions on the mandible: (1) elevation, (2) depression, (3) retraction, (4) protraction, and (5) side-to-side movement. The same muscle may be mentioned more than once.

▶   C H E C K P O I N T

What would happen if you lost tone in the masseter and temporalis muscles?

Figure 11.6   Muscles that move the mandible (lower jaw bone).

🔑 The muscles that move the mandible are also known as muscles of mastication.

Parietal bone

TEMPORALIS

Occipital bone

Zygomatic arch (cut)

Temporomandibular joint (TMJ)

MEDIAL PTERYGOID

Ramus of mandible (cut)

Frontal bone

Nasal bone

Zygomatic bone (cut)

LATERAL PTERYGOID

Maxilla

Buccinator

Orbicularis oris

Body of mandible

DANK

(a) Right lateral deep view

Muscle	Origin	Insertion	Action
Masseter (MA-se-ter = a chewer) (see Figure 11.4c)	Maxilla and zygomatic arch.	Angle and ramus of mandible.	Elevates mandible, as in closing mouth, and retracts (draws back) mandible.
Temporalis (tem'-pō-RĀ-lis; tempor- = temples)	Temporal bone.	Coronoid process and ramus of mandible.	Elevates and retracts mandible.
Medial pterygoyd (TER-i-goyd; medial = closer to midline; pterygoid = like a wing)	Medial surface of lateral portion of pterygoid process of sphenoid bone; maxilla.	Angle and ramus of mandible.	Elevates and protracts (protrudes) mandible and moves mandible from side to side.
Lateral pterygoid (TER-i-goyd; lateral = farther from midline)	Greater wing and lateral surface of lateral portion of pterygoid process of sphenoid bone.	Condyle of mandible; temporomandibular joint (TMJ).	Protracts mandible, depresses mandible as in opening mouth, and moves mandible from side to side.

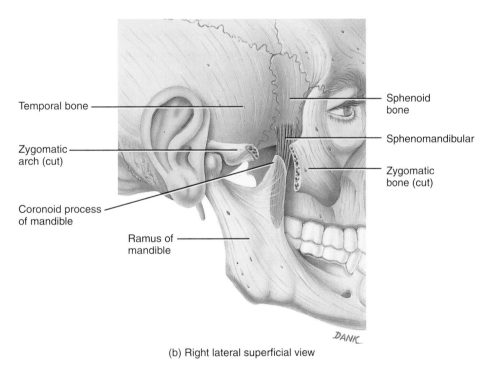

(b) Right lateral superficial view

 Which is the strongest muscle of mastication?

Exhibit 11.4   Muscles that Move the Tongue—Extrinsic Tongue Muscles (Figure 11.7)

▶ OBJECTIVE

• Describe the origin, insertion, action, and innervation of the extrinsic muscles of the tongue.

The tongue is a highly mobile structure that is vital to digestive functions such as mastication, perception of taste, and deglutition (swallowing). It is also important in speech. The tongue's mobility is greatly aided by its suspension from the mandible, styloid process of the temporal bone, and hyoid bone.

The tongue is divided into lateral halves by a median fibrous septum. The septum extends throughout the length of the tongue. Inferiorly, the septum attaches to the hyoid bone. Muscles of the tongue are of two principal types: extrinsic and intrinsic. **Extrinsic tongue muscles** originate outside the tongue and insert into it. They move the entire tongue in various directions, such as anteriorly, posteriorly, and laterally. **Intrinsic tongue muscles** originate and insert within the tongue. These muscles alter the shape of the tongue rather than moving the entire tongue. The extrinsic and intrinsic muscles of the tongue insert into both lateral halves of the tongue.

When you study the extrinsic tongue muscles, you will notice that all of their names end in *glossus,* meaning tongue. You will also notice that the actions of the muscles are obvious, considering the positions of the mandible, styloid process, hyoid bone, and soft palate, which serve as origins for these muscles. For example, the **genioglossus** (originates on the mandible) pulls the tongue downward and forward, the **styloglossus** (originates on the styloid process) pulls the tongue upward and backward, the **hyoglossus** (originates on the hyoid bone) pulls the tongue downward and flattens it, and the **palatoglossus** (originates on the soft palate) raises the back portion of the tongue.

Innervation

• Genioglossus, styloglossus, and hyoglossus: hypoglossal nerve (cranial nerve XII)
• Palatoglossus: the pharyngeal plexus, which contains axons from both the vagus nerve (cranial nerve X) and the accessory nerve (cranial nerve XI)

 Intubation During Anesthesia

When general anesthesia is administered during surgery, a total relaxation of the genioglossus muscle results. This causes the tongue to fall posteriorly, which may obstruct the airway to the lungs. To avoid this, the mandible is either manually thrust forward and held in place, or a tube is inserted from the lips through the laryngopharynx (inferior portion of the throat) into the trachea (endotracheal intubation). ■

Relating Muscles to Movements

Arrange the muscles in this exhibit according to the following actions on the tongue: (1) depression, (2) elevation, (3) protraction, and (4) retraction. The same muscle may be mentioned more than once.

▶ CHECKPOINT

When your physician says, "Open your mouth, stick out your tongue, and say *ahh*," so she can examine the inside of your mouth for possible signs of infection, which muscles do you contract?

Muscle	Origin	Insertion	Action
Genioglossus (jē′-nē-ō-GLOS-us; *genio-* = the chin; *glossus* = tongue)	Mandible.	Undersurface of tongue and hyoid bone.	Depresses tongue and thrusts it anteriorly (protraction).
Styloglossus (stī′-lō-GLOS-us; *stylo* = stake or pole; styloid process of temporal bone)	Styloid process of temporal bone.	Side and undersurface of tongue.	Elevates tongue and draws it posteriorly (retraction).
Palatoglossus (pal′-a-tō-GLOS- us; *palato-* = the roof of the mouth or palate)	Anterior surface of soft palate.	Side of tongue.	Elevates posterior portion of tongue and draws soft palate down on tongue.
Hyoglossus (hī′-ō-GLOS-us)	Greater horn and body of hyoid bone.	Side of tongue.	Depresses tongue and draws down its sides.

Figure 11.7 Muscles that move the tongue.

The extrinsic and intrinsic muscles of the tongue are arranged in both lateral halves of the tongue.

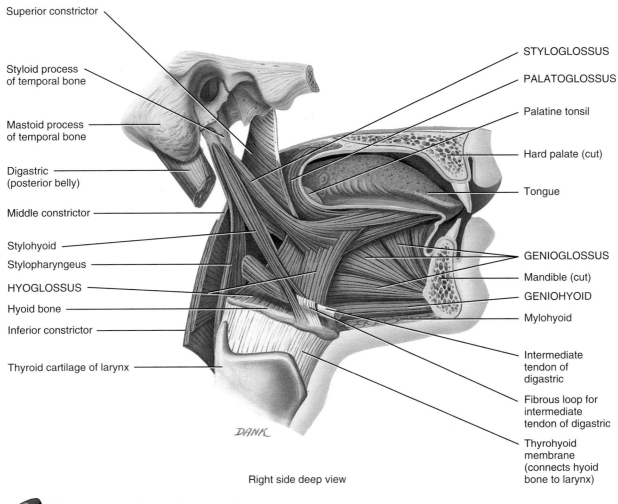

Superior constrictor

Styloid process of temporal bone

Mastoid process of temporal bone

Digastric (posterior belly)

Middle constrictor

Stylohyoid

Stylopharyngeus

HYOGLOSSUS

Hyoid bone

Inferior constrictor

Thyroid cartilage of larynx

STYLOGLOSSUS

PALATOGLOSSUS

Palatine tonsil

Hard palate (cut)

Tongue

GENIOGLOSSUS

Mandible (cut)

GENIOHYOID

Mylohyoid

Intermediate tendon of digastric

Fibrous loop for intermediate tendon of digastric

Thyrohyoid membrane (connects hyoid bone to larynx)

DANK

Right side deep view

What are the functions of the tongue?

Exhibit 11.5 **Muscles of the Anterior Neck (Figure 11.8)**

► **O B J E C T I V E**

• Describe the origin, insertion, action, and innervation of the muscles of the anterior neck.

Two groups of muscles are associated with the anterior aspect of the neck: (1) the **suprahyoid muscles,** so-called because they are located superior to the hyoid bone, and (2) the **infrahyoid muscles,** so-named because they lie inferior to the hyoid bone. Acting with the infrahyoid muscles, the suprahyoid muscles stabilize the hyoid bone. Thus, the hyoid bone serves as a firm base upon which the tongue can move.

As a group, the suprahyoid muscles elevate the hyoid bone, floor of the oral cavity, and tongue during swallowing. As its name suggests, the **digastric** muscle has two bellies, anterior and posterior, united by an intermediate tendon that is held in position by a fibrous loop (see Figure 11.7). This muscle elevates the hyoid bone and larynx (voice box) during swallowing and speech and depresses the mandible. The **stylohyoid** muscle elevates and draws the hyoid bone posteriorly, thus elongating the floor of the oral cavity during swallowing. The **mylohyoid** muscle elevates the hyoid bone and helps press the tongue against the roof of the oral cavity during swallowing to move food from the oral cavity into the throat. The **geniohyoid** muscle (see Figure 11.7) elevates and draws the hyoid bone anteriorly to shorten the floor of the oral cavity and to widen the throat to receive food that is being swallowed. It also depresses the mandible.

The infrahyoid muscles are sometimes called "strap" muscles because of their ribbonlike appearance. Most of the infrahyoid muscles depress the hyoid bone and some move the larynx during swallowing and speech. The **omohyoid** muscle, like the digastric muscle, is composed of two bellies connected by an intermediate tendon. In this case, however, the two bellies are referred to as *superior* and *inferior,* rather than anterior and posterior. Together, the omohyoid, **sternohyoid,** and **thyrohyoid** muscles depress the hyoid bone. In addition, the **sternothyroid** muscle depresses the thyroid cartilage (Adam's apple) of the larynx, whereas the thyrohyoid muscle elevates the thyroid cartilage. The actions are necessary during the production of low and high tones, respectively, during phonation.

Innervation

• Anterior belly of the digastric and the mylohyoid muscles: mandibular division of the trigeminal nerve (cranial nerve V)

• Posterior belly of the digastric and the stylohyoid: facial nerve (cranial nerve VII)

• Geniohyoid muscle: first cervical spinal nerve

• Omohyoid, sternohyoid, and sternothyroid: branches of ansa cervicalis (spinal nerves C1–C3)

• Thyrohyoid: branches of ansa cervicalis (spinal nerves C1–C2) and descending hypoglossal nerve (cranial nerve XII)

Relating Muscles to Movements

Arrange the muscles in this exhibit according to the following actions on the hyoid bone: (1) elevating it, (2) drawing it anteriorly, (3) drawing it posteriorly, and (4) depressing it; and on the thyroid cartilage: (1) elevating it and (2) depressing it. The same muscle may be mentioned more than once.

► **C H E C K P O I N T**

Which tongue, facial, and mandibular muscles do you use for chewing?

Muscle	Origin	Insertion	Action
Suprahyoid muscles			
Digastric (dī′-GAS-trik; *di-* = two; *gastr-* = belly)	Anterior belly from inner side of inferior border of mandible; posterior belly from mastoid process of temporal bone.	Body of hyoid bone via an intermediate tendon.	Elevates hyoid bone and depresses mandible, as in opening the mouth.
Stylohyoid (stī′-lō-HĪ-oid; *stylo* = stake or pole, styloid process of temporal bone; *hyo-* = U-shaped, pertaining to hyoid bone)	Styloid process of temporal bone.	Body of hyoid bone.	Elevates hyoid bone and draws it posteriorly.
Mylohyoid (mī′-lō-HĪ-oid) (*mylo-* = mill)	Inner surface of mandible.	Body of hyoid bone.	Elevates hyoid bone and floor of mouth and depresses mandible.
Geniohyoid (jē′-nē-ō-HĪ-oid; *genio* = chin) (see also Figure 11.7)	Inner surface of mandible.	Body of hyoid bone.	Elevates hyoid bone, draws hyoid bone and tongue anteriorly, and depresses mandible.
Infrahyoid muscles			
Omohyoid (ō-mō-HĪ-oid; *omo-* = relationship to the shoulder)	Superior border of scapula and superior transverse ligament.	Body of hyoid bone.	Depresses hyoid bone.
Sternohyoid (ster′-no-HĪ-oid; *sterno-* = sternum)	Medial end of clavicle and manubrium of sternum.	Body of hyoid bone.	Depresses hyoid bone.
Sternothyroid (ster′-nō-THĪ-roid; *thyro-* = thyroid gland)	Manubrium of sternum.	Thyroid cartilage of larynx.	Depresses thyroid cartilage of larynx.
Thyrohyoid (thī′-rō-HĪ-oid)	Thyroid cartilage of larynx.	Greater horn of hyoid bone.	Elevates thyroid cartilage and depresses hyoid bone.

Figure 11.8 Muscles of the anterior neck.

The suprahyoid muscles elevate the hyoid bone, the floor of the oral cavity, and the tongue during swallowing.

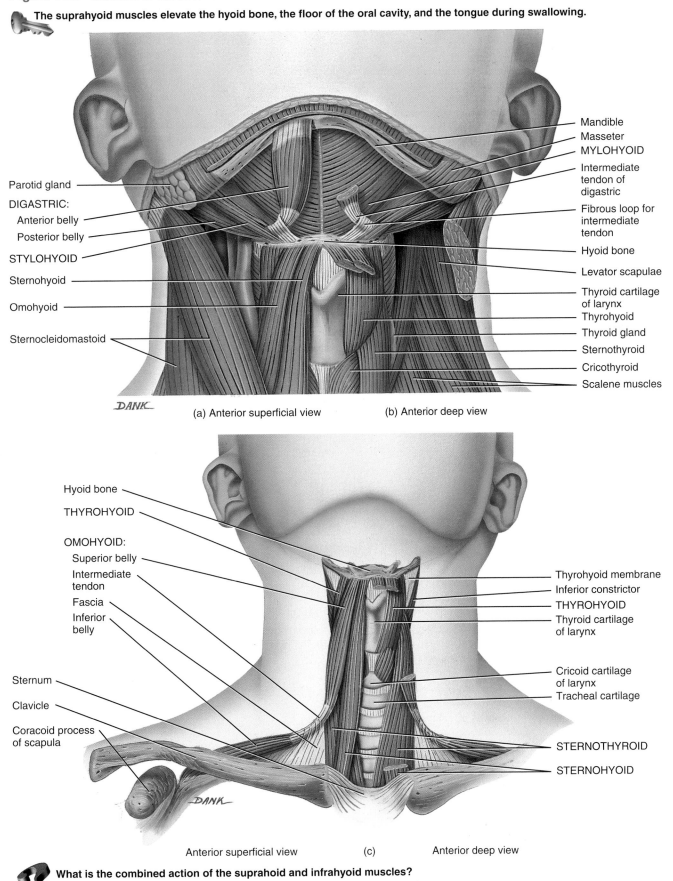

Parotid gland

DIGASTRIC:
  Anterior belly
  Posterior belly

STYLOHYOID

Sternohyoid

Omohyoid

Sternocleidomastoid

Mandible
Masseter
MYLOHYOID
Intermediate tendon of digastric
Fibrous loop for intermediate tendon
Hyoid bone
Levator scapulae
Thyroid cartilage of larynx
Thyrohyoid
Thyroid gland
Sternothyroid
Cricothyroid
Scalene muscles

DANK

(a) Anterior superficial view   (b) Anterior deep view

Hyoid bone

THYROHYOID

OMOHYOID:
  Superior belly
  Intermediate tendon
  Fascia
  Inferior belly

Sternum

Clavicle

Coracoid process of scapula

Thyrohyoid membrane
Inferior constrictor
THYROHYOID
Thyroid cartilage of larynx
Cricoid cartilage of larynx
Tracheal cartilage
STERNOTHYROID
STERNOHYOID

DANK

Anterior superficial view   (c)   Anterior deep view

What is the combined action of the suprahoid and infrahyoid muscles?

329

Exhibit 11.6 Muscles that Move the Head (Figure 11.9)

▶ **OBJECTIVE**

• Describe the origin, insertion, action, and innervation of the muscles that move the head.

The head is attached to the vertebral column at the atlanto-occipital joint formed by the atlas and occipital bone. Balance and movement of the head on the vertebral column involves the action of several neck muscles. For example, contraction of the two **sternocleidomastoid** muscles together (bilaterally) flexes the cervical portion of the vertebral column and extends the head. Acting singly (unilaterally), the sternocleidomastoid muscle laterally flexes and rotates the head. Bilateral contraction of the **semispinalis capitis, splenius capitis,** and **longissimus capitis** muscles extends the head. However, when these same muscles contract unilaterally, their actions are quite different, involving primarily rotation of the head.

The sternocleidomastoid muscle is an important landmark that divides the neck into two major triangles: anterior and posterior. The triangles are important because of the structures that lie within their boundaries.

The **anterior triangle** is bordered superiorly by the mandible, inferiorly by the sternum, medially by the cervical midline, and laterally by the anterior border of the sternocleidomastoid muscle. The anterior triangle is subdivided into an unpaired submental triangle and three paired triangles: submandibular, carotid, and muscular. The anterior triangle contains submental, submandibular, and deep cervical lymph nodes; the submandibular salivary gland and a portion of the parotid salivary gland;

the facial artery and vein; carotid arteries and internal jugular vein; and cranial nerves IX (glossopharyngeal), X (vagus), XI (accessory), and XII (hypoglossal).

The **posterior triangle** is bordered inferiorly by the clavicle, anteriorly by the posterior border of the sternocleidomastoid muscle, and posteriorly by the anterior border of the trapezius muscle. The posterior triangle is subdivided into two triangles, occipital and supraclavicular, by the inferior belly of the omohyoid muscle. The posterior triangle contains part of the subclavian artery, external jugular vein, cervical lymph nodes, brachial plexus, and the accessory nerve (cranial nerve XI).

Innervation

• Sternocleidomastoid: accessory nerve (cranial nerve XI)
• All capitis muscles: cervical spinal nerves

Relating Muscles to Movements

Arrange the muscles in this exhibit according to the following actions on the head: (1) flexion, (2) lateral flexion, (3) extension, (4) rotation to side opposite contracting muscle, and (5) rotation to same side as contracting muscle. The same muscle may be mentioned more than once.

▶ **CHECKPOINT**
What muscles do you contract to signify "yes" and "no"?

Muscle	Origin	Insertion	Action
Sternocleidomastoid (ster′-nō-klī′-dō-MAS-toid; *sterno-* = breastbone; *cleido-* = clavicle; *mastoid* = mastoid process of temporal bone)	Sternum and clavicle.	Mastoid process of temporal bone.	Acting together (bilaterally), flex cervical portion of vertebral column, extend head, and elevate sternum during forced inhalation; acting singly (unilaterally), laterally flex and rotate head to side opposite contracting muscle.
Semispinalis capitis (se′-mē-spi-NA-lis KAP-i-tis; *semi-* = half; *spine* = spinous process; *capit-* = head) (see Figure 11.19a)	Transverse processes of first six or seven thoracic vertebrae and seventh cervical vertebra, and articular processes of fourth, fifth, and sixth cervical vertebrae.	Occipital bone between superior and inferior nuchal lines.	Acting together, extend head; acting singly, rotate head to side opposite contracting muscle.
Splenius capitis (SPLE-nē-us KAP-i-tis; *splen-* = bandage) (see Figure 11.19a)	Ligamentum nuchae and spinous processes of seventh cervical vertebra and first three or four thoracic vertebrae.	Occipital bone and mastoid process of temporal bone.	Acting together, extend head; acting singly, laterally flex and rotate head to same side as contracting muscle.
Longissimus capitis (lon-JIS-i-mus KAP-i-tis; *longissimus* = longest) (see Figure 11.19a)	Transverse processes of upper four thoracic vertebrae and articular processes of last four cervical vertebrae.	Mastoid process of temporal bone.	Acting together, extend head; acting singly, laterally flex and rotate head to same side as contracting muscle.

Figure 11.9   Triangles of the neck.

 The sternocleidomastoid muscle divides the neck into two principal triangles: anterior and posterior.

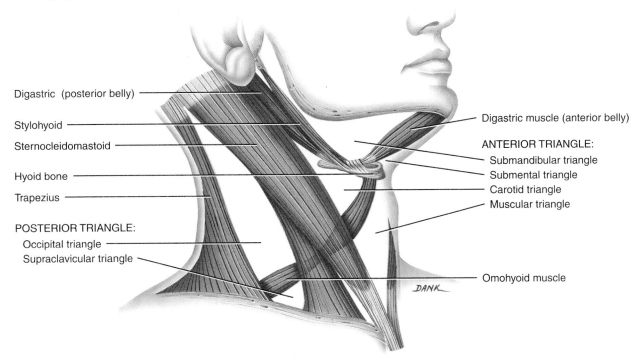

Right lateral view

Why are triangles important?

Exhibit 11.7   Muscles that Act on the Abdominal Wall (Figure 11.10)

► OBJECTIVE

• Describe the origin, insertion, action, and innervation of the muscles that act on the abdominal wall.

The anterolateral abdominal wall is composed of skin, fascia, and four pairs of muscles: the external oblique, internal oblique, transversus abdominis, and rectus abdominis. The first three muscles are arranged from superficial to deep. The **external oblique** is the superficial muscle. Its fascicles extend inferiorly and medially. The **internal oblique** is the intermediate flat muscle. Its fascicles extend at right angles to those of the external oblique. The **transversus abdominis** is the deep muscle, with most of its fascicles directed transversely around the abdominal wall. Together, the external oblique, internal oblique, and transversus abdominis form three layers of muscle around the abdomen. In each layer, the muscle fascicles extend in a different direction. This is a structural arrangement that affords considerable protection to the abdominal viscera, especially when the muscles have good tone.

The **rectus abdominis** muscle is a long muscle that extends the entire length of the anterior abdominal wall, from the pubic crest and pubic symphysis to the cartilages of ribs 5–7 and the xiphoid process of the sternum. The anterior surface of the muscle is interrupted by three transverse fibrous bands of tissue called **tendinous intersections,** believed to be remnants of septa that separated myotomes during embryological development (see Figure 10.20b on page 301).

As a group, the muscles of the anterolateral abdominal wall help contain and protect the abdominal viscera; flex, laterally flex, and rotate the vertebral column at the intervertebral joints; compress the abdomen during forced exhalation; and produce the force required for defecation, urination, and childbirth.

The aponeuroses of the external oblique, internal oblique, and transversus abdominis muscles form the **rectus sheaths,** which enclose the rectus abdominis muscles. The sheaths meet at the midline to form the **linea alba** (= white line), a tough, fibrous band that extends from the xiphoid process of the sternum to the pubic symphysis. In the latter stages of pregnancy, the linea alba stretches to increase the distance between the rectus abdominis muscles. The inferior free border of the external oblique aponeurosis, plus some collagen fibers, forms the **inguinal ligament,** which runs from the anterior superior iliac spine to the pubic tubercle (see Figure 11.20a). Just superior to the medial end of the inguinal ligament is a triangular slit in the aponeurosis referred to

as the **superficial inguinal ring,** the outer opening of the **inguinal canal** (see Figure 28.2). The canal contains the spermatic cord and ilioinguinal nerve in males, and the round ligament of the uterus and ilioinguinal nerve in females.

The posterior abdominal wall is formed by the lumbar vertebrae, parts of the ilia of the hip bones, psoas major and iliacus muscles (described in Exhibit 11.17), and quadratus lumborum muscle. Whereas the anterolateral abdominal wall can contract and distend, the posterior abdominal wall is bulky and stable by comparison.

Innervation

• Rectus abdominis: thoracic spinal nerves T7–T12
• External oblique: thoracic spinal nerves T7–T12 and the iliohypogastric nerve
• Internal oblique and transversus abdominis: thoracic spinal nerves T8–T12, the iliohypogastric nerve, and the ilioinguinal nerve
• Quadratus lumborum: thoracic spinal nerve T12 and lumbar spinal nerves L1–L3 or L1–L4

 Inguinal Hernia

The inguinal region is a weak area in the abdominal wall. It is often the site of an **inguinal hernia,** a rupture or separation of a portion of the inguinal area of the abdominal wall resulting in the protrusion of a part of the small intestine. Hernia is much more common in males than in females because the inguinal canals in males are larger to accommodate the spermatic cord and ilioinguinal nerve. They are especially weak points in the abdominal wall. ■

Relating Muscles to Movements

Arrange the muscles in this exhibit according to the following actions on the vertebral column: (1) flexion, (2) lateral flexion, (3) extension, and (4) rotation. The same muscle may be mentioned more than once.

► CHECKPOINT

Which muscles do you contract when you "suck in your tummy," thereby compressing the anterior abdominal wall?

Muscle	Origin	Insertion	Action
Rectus abdominis (REK-tus ab-DOM-in-is; *rectus* = fascicles parallel to midline; *abdomin* = abdomen)	Pubic crest and pubic symphysis.	Cartilage of fifth to seventh ribs and xiphoid process.	Flexes vertebral column, especially lumbar portion, and compresses abdomen to aid in defecation, urination, forced exhalation, and childbirth.
External oblique (ō-BLĒK; *external* = closer to surface; *oblique* = fascicles diagonal to midline)	Inferior eight ribs.	Iliac crest and linea alba.	Acting together (bilaterally), compress abdomen and flex vertebral column; acting singly (unilaterally), laterally flex vertebral column, especially lumbar portion, and rotate vertebral column.
Internal oblique (ō-BLĒK; *internal* = farther from surface)	Iliac crest, inguinal ligament, and thoracolumbar fascia.	Cartilage of last three or four ribs and linea alba.	Acting together, compress abdomen and flex vertebral column; acting singly, laterally flex vertebral column, especially lumbar portion, and rotate vertebral column.
Transversus abdominis (tranz-VER-sus ab-DOM-in-is; *transverse* = fascicles perpendicular to midline)	Iliac crest, inguinal ligament, lumbar fascia, and cartilages of inferior six ribs.	Xiphoid process, linea alba, and pubis.	Compresses abdomen.
Quadratus lumborum (kwod-RĀ-tus lum-BOR-um; *quad* = four; *lumbo-* = lumbar region) (see Figure 11.11)	Iliac crest and iliolumbar ligament.	Inferior border of twelfth rib and transverse processes of first four lumbar vertebrae.	Acting together, pull twelfth ribs inferiorly during forced exhalation, fix twelfth ribs to prevent their elevation during deep inhalation, and help extend lumbar portion of vertebral column; acting singly, laterally flex vertebral column, especially lumbar portion.

(continues)

Exhibit 11.7   Muscles that Act on the Abdominal Wall (Figure 11.10) (continued)

Figure 11.10   Muscles of the male anterolateral abdominal wall.   (See Tortora, *A Photographic Atlas of the Human Body*, Figure 5.7.)

The anterolateral abdominal muscles protect the abdominal viscera, move the vertebral column, and assist in forced exhalation, defecation, urination, and childbirth.

(a) Anterior superficial view   (b) Anterior deep view

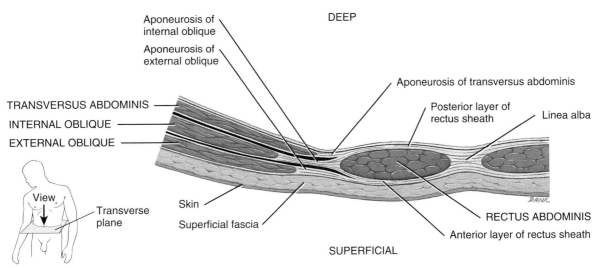

Aponeurosis of internal oblique

Aponeurosis of external oblique

DEEP

Aponeurosis of transversus abdominis

Posterior layer of rectus sheath

Linea alba

TRANSVERSUS ABDOMINIS

INTERNAL OBLIQUE

EXTERNAL OBLIQUE

View

Transverse plane

Skin

Superficial fascia

SUPERFICIAL

RECTUS ABDOMINIS

Anterior layer of rectus sheath

(c) Transverse section of anterior abdominal wall superior to umbilicus (navel)

 Which abdominal muscle aids in urination?

Exhibit 11.8 Muscles Used in Breathing (Figure 11.11)

▶ OBJECTIVE

• Describe the origin, insertion, action, and innervation of the muscles used in breathing.

The muscles described here alter the size of the thoracic cavity so that breathing can occur. Inhalation (breathing in) occurs when the thoracic cavity increases in size, and exhalation (breathing out) occurs when the thoracic cavity decreases in size.

The dome-shaped **diaphragm** is the most important muscle that powers breathing. It also separates the thoracic and abdominal cavities. The diaphragm is composed of a peripheral muscular portion and a central portion called the central tendon. The **central tendon** is a strong aponeurosis that serves as the tendon of insertion for all peripheral muscular fibers of the diaphragm. It fuses with the inferior surface of the fibrous pericardium (external covering of the heart) and the parietal pleurae (external coverings of the lungs).

Movements of the diaphragm also help return venous blood passing through the abdomen to the heart. Together with the anterolateral abdominal muscles, the diaphragm helps to increase intra-abdominal pressure to evacuate the pelvic contents during defecation, urination, and childbirth. This mechanism is further assisted when you take a deep breath and close the rima glottidis (the space between vocal folds). The trapped air in the respiratory system prevents the diaphragm from elevating. The increase in intra-abdominal pressure as just described will also help support the vertebral column and prevent flexion during weight lifting. This greatly assists the back muscles in lifting a heavy weight.

The diaphragm has three major openings through which various structures pass between the thorax and abdomen. These structures include the aorta, along with the thoracic duct and azygos vein, which pass through the **aortic hiatus;** the esophagus with accompanying vagus (X) cranial nerves, which pass through the **esophageal hiatus;** and the inferior vena cava, which passes through the **caval opening (foramen for the vena cava).** In a condition called a hiatus hernia, the stomach protrudes superiorly through the esophageal hiatus.

Other muscles involved in breathing are called **intercostal** muscles. They span the intercostal spaces, the spaces between ribs. These muscles are arranged in three layers. The 11 **external intercostal muscles** occupy the superficial layer, and their fibers run obliquely inferiorly and anteriorly from the rib above to the rib below. They elevate the ribs during inhalation to help expand the thoracic cavity. The 11 **internal intercostal muscles** occupy the intermediate layer of the intercostal spaces. The fibers of these muscles run obliquely inferiorly and posteriorly from the inferior border of the rib above to the superior border of the rib below. They draw adjacent ribs together during forced exhalation to help decrease the size of the thoracic cavity. The deepest muscle layer is made up of the **transversus thoracis.** These poorly developed muscles (not illustrated) extend in the same direction as the internal intercostals and may have the same role.

As you will see in Chapter 23, the diaphragm and external intercostal muscles are used during quiet inhalation and exhalation. However, during deep, forceful inhalation, the sternocleidomastoid, scalene, and pectoralis minor muscles are also used and during deep, forceful exhalation, the external oblique, internal oblique, transversus abdominis, rectus abdominis, and internal intercostals are also used.

Innervation

• Diaphragm: phrenic nerve, which contains axons from cervical spinal nerves C3–C5
• External and internal intercostals: thoracic spinal nerves T2–T12

Relating Muscles to Movements

Arrange the muscles in this exhibit according to the following actions on the size of the thorax: (1) increase in vertical length, (2) increase in lateral and anteroposterior dimensions, and (3) decrease in lateral and anteroposterior dimensions.

▶ CHECKPOINT

What are the names of the three openings in the diaphragm, and which structures pass through each?

Muscle	Origin	Insertion	Action
Diaphragm (DĪ-a-fram; *dia* = across; *-phragm* = wall)	Xiphoid process of the sternum, costal cartilages of inferior six ribs, and lumbar vertebrae.	Central tendon.	Contraction of the diaphragm causes it to flatten and increases the vertical dimension of the thoracic cavity, resulting in inhalation; relaxation of the diaphragm causes it to move superiorly and decreases the vertical dimension of the thoracic cavity, resulting in exhalation.
External intercostals (in′-ter-KOS-tals; *external* = closer to surface; *inter-* = between; *costa* = rib)	Inferior border of rib above.	Superior border of rib below.	Contraction elevates the ribs and increases the anteroposterior and lateral dimensions of the thoracic cavity, resulting in inhalation; relaxation depresses the ribs and decreases the anteroposterior and lateral dimensions of the thoracic cavity, resulting in exhalation.
Internal intercostals (in′-ter-KOS-tals; *internal* = father from surface)	Superior border of rib below.	Inferior border of rib above.	Contraction draws adjacent ribs together to further decrease the anteroposterior and lateral dimensions of the thoracic cavity during forced exhalation.

Figure 11.11 Muscles used in breathing, as seen in a male.

 Openings in the diaphragm permit passage of the aorta, esophagus, and inferior vena cava.

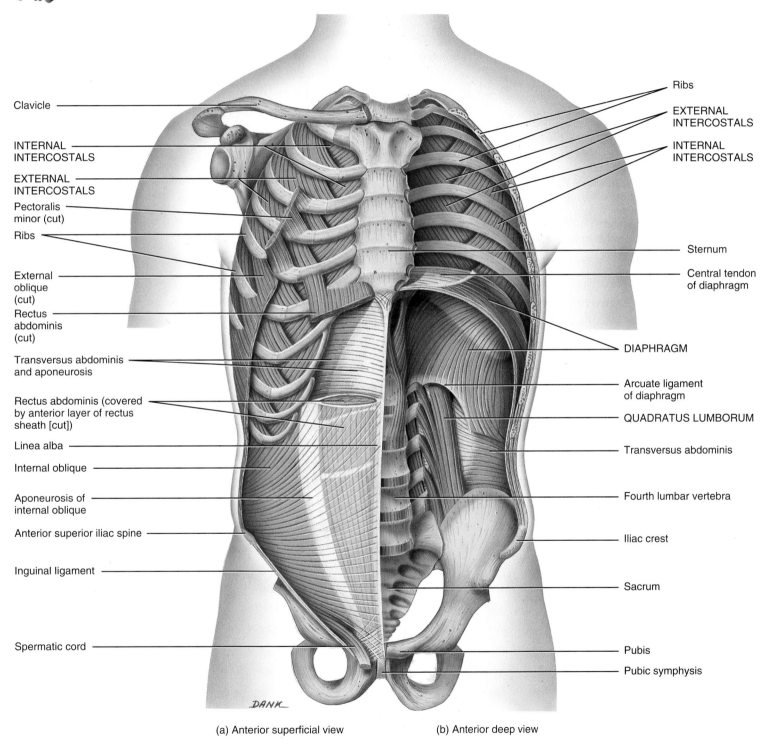

(a) Anterior superficial view   (b) Anterior deep view

Which muscle associated with breathing is innervated by the phrenic nerve?

Exhibit 11.9 Muscles of the Pelvic Floor (Figure 11.12)

▶ OBJECTIVE

- Describe the origin, insertion, action, and innervation of the muscles of the pelvic floor.

The muscles of the pelvic floor are the levator ani and coccygeus. Together with the fascia covering their internal and external surfaces, these muscles are referred to as the **pelvic diaphragm,** which stretches from the pubis anteriorly to the coccyx posteriorly, and from one lateral wall of the pelvis to the other. This arrangement gives the pelvic diaphragm the appearance of a funnel suspended from its attachments. The anal canal and urethra pierce the pelvic diaphragm in both sexes, and the vagina goes through it in females.

The two components of the **levator ani** muscle are the **pubococcygeus** and **iliococcygeus.** Figure 11.12 shows these muscles in the female and Figure 11.13 in the male. The levator ani is the largest and most important muscle of the pelvic floor. It supports the pelvic viscera and resists the inferior thrust that accompanies increases in intra-abdominal pressure during functions such as forced exhalation, coughing, vomiting, urination, and defecation. The muscle also functions as a sphincter at the anorectal junction, urethra, and vagina. In addition to assisting the levator ani, the **coccygeus** pulls the coccyx anteriorly after it has been pushed posteriorly during defecation or childbirth.

Innervation

- Pubococcygeus and iliococcygeus: sacral spinal nerves S2–S4
- Coccygeus: sacral spinal nerves S4–S5

 Injury of Levator Ani and Stress Incontinence

During childbirth, the levator ani muscle supports the head of the fetus, and the muscle may be injured during a difficult childbirth or traumatized during an episiotomy (a cut made with surgical scissors to prevent or direct tearing of the perineum during the birth of a baby). The consequence of such injury may be urinary stress incontinence, leakage of urine whenever intra-abdominal pressure is increased—for example, during coughing. One way to treat urinary stress incontinence is to strengthen and tighten the muscles that support the pelvic viscera. This is accomplished by *Kegel exercises,* the alternate contraction and relaxation of muscles of the pelvic floor. ■

Relating Muscles to Movements

Arrange the muscles in this exhibit according to the following actions on the pelvic viscera: supporting and maintaining their position, and resisting an increase in intra-abdominal pressure; and according to the following action on the anus, urethra, and vagina: constriction. The same muscle may be mentioned more than once.

▶ CHECKPOINT

Which muscles are strengthened by Kegel exercises?

Muscle	Origin	Insertion	Action
Levator ani (le-VĀ-tor Ā-nē; *levator* = raises; *ani* = anus)	This muscle is divisible into two parts, the pubococcygeus muscle and the iliococcygeus muscle.		
Pubococcygeus (pū′-bō-kok-SIJ-ē-us; *pubo-* = pubis; *coccygeus* = coccyx)	Pubis.	Coccyx, urethra, anal canal, central tendon of perineum, and anococcygeal raphe (narrow fibrous band that extends from anus to coccyx).	Supports and maintains position of pelvic viscera; resists increase in intra-abdominal pressure during forced exhalation, coughing, vomiting, urination, and defecation; constricts anus, urethra, and vagina; and supports fetal head during childbirth.
Iliococcygeus (il′-ē-ō-kok-SIJ-ē-us; *ilio-* = ilium)	Ischial spine.	Coccyx.	As above.
Coccygeus (kok-SIJ-ē-us)	Ischial spine.	Lower sacrum and upper coccyx.	Supports and maintains position of pelvic viscera; resists increase in intra-abdominal pressure during forced exhalation, coughing, vomiting, urination, and defecation; and pulls coccyx anteriorly following defecation or childbirth.

Figure 11.12 **Muscles of the pelvic floor, as seen in the female perineum.**

 The pelvic diaphragm supports the pelvic viscera.

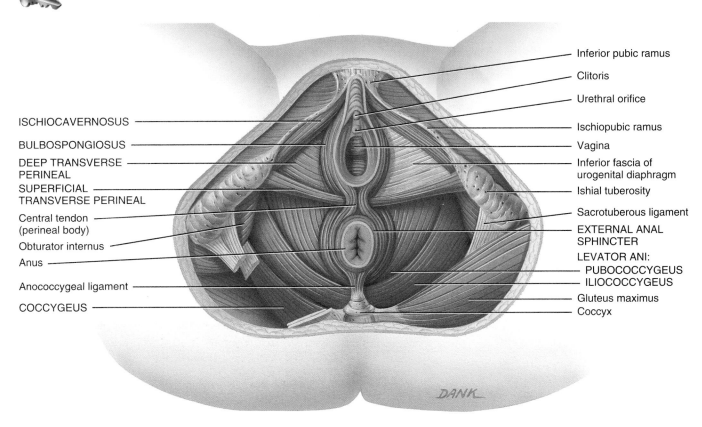

ISCHIOCAVERNOSUS

BULBOSPONGIOSUS

DEEP TRANSVERSE PERINEAL

SUPERFICIAL TRANSVERSE PERINEAL

Central tendon (perineal body)

Obturator internus

Anus

Anococcygeal ligament

COCCYGEUS

Inferior pubic ramus

Clitoris

Urethral orifice

Ischiopubic ramus

Vagina

Inferior fascia of urogenital diaphragm

Ishial tuberosity

Sacrotuberous ligament

EXTERNAL ANAL SPHINCTER

LEVATOR ANI:
  PUBOCOCCYGEUS
  ILIOCOCCYGEUS

Gluteus maximus

Coccyx

Inferior superficial view

What are the borders of the pelvic diaphragm?

Exhibit 11.10   Muscles of the Perineum (Figures 11.12 and 11.13)

▶ OBJECTIVE

• Describe the origin, insertion, action, and innervation of the muscles of the perineum.

The **perineum** is the region of the trunk inferior to the pelvic diaphragm. A diamond-shaped area extends from the pubic symphysis anteriorly, to the coccyx posteriorly, and to the ischial tuberosities laterally. The female and the male perineums may be compared in Figures 11.12 and 11.13, respectively. A transverse line drawn between the ischial tuberosities divides the perineum into an anterior **urogenital triangle** that contains the external genitals and a posterior **anal triangle** that contains the anus (see Figure 28.23 on page 1039). In the center of the perineum is a wedge-shaped mass of fibrous tissue called the **central tendon (perineal body).** Several perineal muscles insert into this strong tendon.

The muscles of the perineum are arranged in two layers: **superficial** and **deep.** The muscles of the superficial layer are the **superficial transverse perineal muscle,** the **bulbospongiosus,** and the **ischiocavernosus.** The deep muscles of the perineum are the **deep transverse perineal muscle** and the **external urethral sphincter.** The deep transverse perineal muscle, external urethral sphincter, and their fascia are known as the **urogenital diaphragm.** The muscles of this diaphragm assist in urination and ejaculation in males and urination in

females. The **external anal sphincter** closely adheres to the skin around the margin of the anus and keeps the anal canal and anus closed except during defecation.

Innervation

• All muscles of the perineum except external anal sphincter: perineal branch of the pudendal nerve of the sacral plexus (shown in Exhibit 13.4 on page 444).

• External anal sphincter: sacral spinal nerve S4 and the inferior rectal branch of the pudendal nerve.

Relating Muscles to Movements

Arrange the muscles in this exhibit according to the following actions: (1) expulsion of urine and semen, (2) erection of the clitoris and penis, (3) closing the anal orifice, and (4) constricting the vaginal orifice. The same muscle may be mentioned more than once.

▶ CHECKPOINT

What are the borders and contents of the urogenital triangle and the anal triangle?

Muscle	Origin	Insertion	Action
Superficial Perineal Muscles			
Superficial transverse perineal (per-i-NĒ-al; *superficial* = closer to surface; *transverse* = across; *perineus* = perineum)	Ischial tuberosity.	Central tendon of perineum.	Helps stabilize central tendon of perineum.
Bulbospongiosus (bul′-bō-spon′-jē-Ō-sus; *bulb* = a bulb; *spongio-* = sponge)	Central tendon of perineum.	Inferior fascia of urogenital diaphragm, corpus spongiosum of penis, and deep fascia on dorsum of penis in male; pubic arch and root and dorsum of clitoris in female.	Helps expel urine during urination, helps propel semen along urethra, assists in erection of the penis in male; constricts vaginal orifice and assists in erection of clitoris in female.
Ischiocavernosus (is′-kē-ō-ka′-ver-NŌ-sus; *ischio-* = the hip)	Ischial tuberosity and ischial and pubic rami.	Corpus cavernosum of penis in male and clitoris in female.	Maintains erection of penis in male and clitoris in female.
Deep Perineal Muscles			
Deep transverse perineal (per-i-NĒ-al; *deep* = farther from surface)	Ischial rami.	Central tendon of perineum.	Helps expel last drops of urine and semen in male and urine in female.
External urethral sphincter (ū-RĒ-thral SFINGK-ter)	Ischial and pubic rami.	Median raphe in male and vaginal wall in female.	Helps expel last drops of urine and semen in male and urine in female.
External anal sphincter (Ā-nal)	Anococcygeal ligament.	Central tendon of perineum.	Keeps anal canal and anus closed.

Figure 11.13 Muscles of the male perineum.

 The urogenital diaphragm assists in urination in females and males, ejaculation in males, and helps strengthen the pelvic floor.

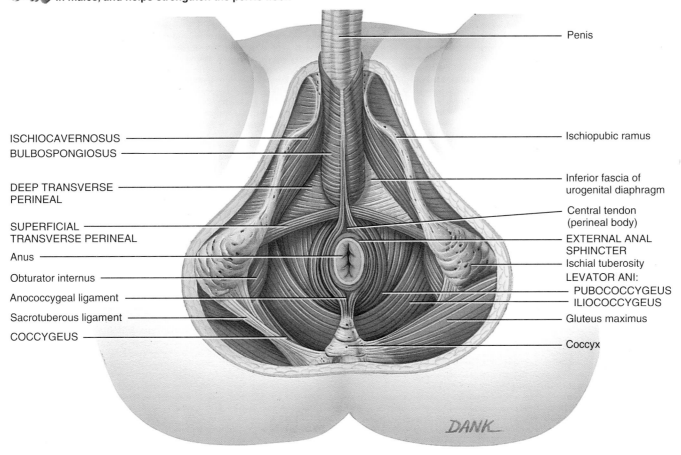

ISCHIOCAVERNOSUS

BULBOSPONGIOSUS

DEEP TRANSVERSE PERINEAL

SUPERFICIAL TRANSVERSE PERINEAL

Anus

Obturator internus

Anococcygeal ligament

Sacrotuberous ligament

COCCYGEUS

Penis

Ischiopubic ramus

Inferior fascia of urogenital diaphragm

Central tendon (perineal body)

EXTERNAL ANAL SPHINCTER

Ischial tuberosity

LEVATOR ANI:
PUBOCOCCYGEUS
ILIOCOCCYGEUS

Gluteus maximus

Coccyx

DANK

Inferior superficial view

What are the borders of the perineum?

Exhibit 11.11 Muscles that Move the Pectoral (Shoulder) Girdle (Figure 11.14)

► O B J E C T I V E

• Describe the origin, insertion, action, and innervation of the muscles that move the pectoral girdle.

The main action of the muscles that move the pectoral girdle is to stabilize the scapula so it can function as a steady origin for most of the muscles that move the humerus. Because scapular movements usually accompany humeral movements in the same direction, the muscles also move the scapula to increase the range of motion of the humerus. For example, it would not be possible to abduct the humerus past the horizontal if the scapula did not move with the humerus. During abduction, the scapula follows the humerus by rotating upward.

Muscles that move the pectoral girdle can be classified into two groups based on their location in the thorax: **anterior** and **posterior thoracic muscles.** The anterior thoracic muscles are the subclavius, pectoralis minor, and serratus anterior. The **subclavius** is a small, cylindrical muscle under the clavicle that extends from the clavicle to the first rib. It steadies the clavicle during movements of the pectoral girdle. The **pectoralis minor** is a thin, flat, triangular muscle that is deep to the pectoralis major. Besides its role in movements of the scapula, the pectoralis minor muscle assists in forced inhalation. The **serratus anterior** is a large, flat, fan-shaped muscle between the ribs and scapula. It is named because of the saw-toothed appearance of its origins on the ribs.

The posterior thoracic muscles are the trapezius, levator scapulae, rhomboid major, and rhomboid minor. The **trapezius** is a large, flat, triangular sheet of muscle extending from the skull and vertebral column medially to the pectoral girdle laterally. It is the most superficial back muscle and covers the posterior neck region and superior portion of the trunk. The two trapezius muscles form a trapezium (diamond-shaped quadrangle)—hence its name. The **levator scapulae** is a narrow, elongated muscle in the posterior portion of the neck. It is deep to the sternocleidomastoid and trapezoid muscles. As its name suggests, one of its actions is to elevate the scapula. The **rhomboid major** and **rhomboid minor** lie deep to the trapezius and are not always distinct from each other. They appear as parallel bands that pass inferiolaterally from the vertebrae to the scapula. They are named based on their shape—that is, a rhomboid (an oblique parallelogram). The rhomboid major is about two times wider than the rhomboid minor. Both muscles are used when forcibly lowering the raised upper limbs, as in driving a stake with a sledgehammer.

To understand the actions of muscles that move the scapula, it is first helpful to describe the various movements of the scapula:

• **Elevation:** superior movement of the scapula, such as shrugging the shoulders or lifting a weight over the head.
• **Depression:** inferior movement of the scapula, as in doing a "pull-up."
• **Abduction (protraction):** movement of the scapula laterally and anteriorly, as in doing a "push-up" or punching.
• **Adduction (retraction):** movement of the scapula medially and posteriorly, as in pulling the oars in a rowboat.
• **Upward rotation:** movement of the inferior angle of the scapula laterally so that the glenoid cavity is moved upward. This movement is required to abduct the humerus past the horizontal.
• **Downward rotation:** movement of the inferior angle of the scapula medially so that the glenoid cavity is moved downward. This movement is seen when a gymnast on parallel bars supports the weight of the body on the hands.

Innervation

Muscles that move the shoulder receive their innervation mainly from nerves that emerge from the cervical and brachial plexuses (shown in Exhibits 13.1 and 13.2 on pages 436 and 438).

• Subclavius: subclavian nerve
• Pectoralis minor: medial pectoral nerve
• Serratus anterior: long thoracic nerve
• Trapezius: accessory nerve (cranial nerve XI) and cervical spinal nerves C3–C5
• Levator scapulae: dorsal scapular nerve and cervical spinal nerves C3–C5
• Rhomboid major and rhomboid minor: dorsal scapular nerve

Relating Muscles to Movements

Arrange the muscles in this exhibit according to the following actions on the scapula: (1) depression, (2) elevation, (3) abduction, (4) adduction, (5) upward rotation, and (6) downward rotation. The same muscle may be mentioned more than once.

► C H E C K P O I N T

What muscles in this exhibit are used to raise your shoulders, lower your shoulders, join your hands behind your back, and join your hands in front of your chest?

Muscle	Origin	Insertion	Action
Anterior Thoracic Muscles			
Subclavius (sub-KLĀ-vē-us; *sub-* = under; *clavius* = clavicle)	First rib.	Clavicle.	Depresses and moves clavicle anteriorly and helps stabilize pectoral girdle.
Pectoralis minor (pek'-tō-RĀ-lis; *pector-* = the breast, chest, thorax; *minor* = lesser)	Second through fifth, third through fifth, or second through fourth ribs.	Coracoid process of scapula.	Abducts scapula and rotates it downward; elevates third through fifth ribs during forced inhalation when scapula is fixed.
Serratus anterior (ser-Ā-tus; *serratus* = saw-toothed; *anterior* = front)	Superior eight or nine ribs.	Vertebral border and inferior angle of scapula.	Abducts scapula and rotates it upward; elevates ribs when scapula is stabilized; known as "boxer's muscle."

Posterior Thoracic Muscles

Trapezius (tra-PĒ-zē-us; *trapezi-* = trapezoid-shaped)	Superior nuchal line of occipital bone, ligamentum nuchae, and spines of seventh cervical and all thoracic vertebrae.	Clavicle and acromion and spine of scapula.	Superior fibers elevate scapula and can help extend head; middle fibers adduct scapula; inferior fibers depress scapula; superior and inferior fibers together rotate scapula upward; stabilizes scapula.
Levator scapulae (le-VĀ-tor SKA-pū-lē; *levator* = raises; *scapulae* = of the scapula)	Superior four or five cervical vertebrae.	Superior vertebral border of scapula.	Elevates scapula and rotates it downward.
Rhomboid major (rom-BOYD; *rhomboid* = rhomboid or diamond-shaped) (see Figure 11.15c)	Spines of second to fifth thoracic vertebrae.	Vertebral border of scapula inferior to spine.	Elevates and adducts scapula and rotates it downward; stabilizes scapula.
Rhomboid minor (rom-BOYD) (see Figure 11.15c)	Spines of seventh cervical and first thoracic vertebrae.	Vertebral border of scapula superior to spine.	Elevates and adducts scapula and rotates it downward; stabilizes scapula.

Figure 11.14 **Muscles that move the pectoral (shoulder) girdle.** (See Tortora, *A Photographic Atlas of the Human Body,* Figure 5.8.)

🔑 Muscles that move the pectoral girdle originate on the axial skeleton and insert on the clavicle or scapula.

(a) Anterior deep view (b) Anterior deeper view

❓ What is the main action of the muscles that move the pectoral girdle?

Exhibit 11.12 Muscles that Move the Humerus (Arm Bone) (Figure 11.15)

▶ **O B J E C T I V E**

• Describe the origin, insertion, action, and innervation of the muscles that move the humerus.

Of the nine muscles that cross the shoulder joint, all except the pectoralis major and latissimus dorsi originate on the scapula. The pectoralis major and latissimus dorsi thus are called **axial muscles** because they originate on the axial skeleton. The remaining seven muscles, the **scapular muscles,** arise from the scapula.

Of the two axial muscles that move the humerus, the **pectoralis major** is a large, thick, fan-shaped muscle that covers the superior part of the thorax. It has two origins: a smaller clavicular head and a larger sternocostal head. The **latissimus dorsi** is a broad, triangular muscle located on the inferior part of the back. It is commonly called the "swimmer's muscle" because its many actions are used while swimming.

Among the scapular muscles, the **deltoid** is a thick, powerful shoulder muscle that covers the shoulder joint and forms the rounded contour of the shoulder. This muscle is a frequent site of intramuscular injections. As you study the deltoid, note that its fascicles originate from three different points and that each group of fascicles moves the humerus differently. The **subscapularis** is a large triangular muscle that fills the subscapular fossa of the scapula and forms part of the posterior wall of the axilla. The **supraspinatus** is a rounded muscle, named for its location in the supraspinous fossa of the scapula. It lies deep to the trapezius. The **infraspinatus** is a triangular muscle, also named for its location in the infraspinous fossa of the scapula. The **teres major** is a thick, flattened muscle inferior to the teres minor that also helps form part of the posterior wall of the axilla. The **teres minor** is a cylindrical, elongated muscle, often inseparable from the infraspinatus, which lies along its superior border. The **coracobrachialis** is an elongated, narrow muscle in the arm.

The strength and stability of the shoulder joint are not provided by the shape of the articulating bones or its ligaments. Instead, four deep muscles of the shoulder—subscapularis, supraspinatus, infraspinatus, and teres minor—strengthen and stabilize the shoulder joint. These muscles join the scapula to the humerus. Their flat tendons fuse together to form the **rotator (musculotendinous) cuff,** a nearly complete circle of tendons around the shoulder joint, like the cuff on a shirtsleeve. The supraspinatus muscle is especially predisposed to wear and tear because of its location between the head of the humerus and acromion of the scapula, which compress its tendon during shoulder movements, especially abduction of the arm.

Innervation

Nerves that emerge from the brachial plexus (shown in Exhibit 13.2 on page 438) innervate muscles that move the arm.

• Pectoralis major: medial and lateral pectoral nerves
• Latissimus dorsi: thoracodorsal nerve
• Deltoid and teres minor: axillary nerve
• Subscapularis: upper and lower subscapular nerves
• Supraspinatus and infraspinatus: suprascapular nerve
• Teres major: lower subscapular nerve
• Coracobrachialis: musculocutaneous nerve

 Impingement Syndrome

One of the most common causes of shoulder pain and dysfunction in athletes is known as **impingement syndrome.** The repetitive movement of the arm over the head that is common in baseball, overhead racquet sports, lifting weights over the head, spiking a volleyball, and swimming puts these athletes at risk for developing this syndrome. It may also be caused by a direct blow or stretch injury. Continual pinching of the supraspinatus tendon as a result of overhead motions causes it to become inflamed and results in pain. If movement is continued despite the pain, the tendon may degenerate near the attachment to the humerus and ultimately may tear away from the bone (rotator cuff injury). Treatment consists of resting the injured tendons, strengthening the shoulder through exercise, and surgery if the injury is particularly severe. ■

Relating Muscles to Movements

Arrange the muscles in this exhibit according to the following actions on the humerus at the shoulder joint: (1) flexion, (2) extension, (3) abduction, (4) adduction, (5) medial rotation, and (6) lateral rotation. The same muscle may be mentioned more than once.

▶ **C H E C K P O I N T**

Why are two muscles that cross the shoulder joint called axial muscles, and the seven others called scapular muscles?

Muscle	Origin	Insertion	Action
Axial Muscles that Move the Humerus			
Pectoralis major (pek′-tō-RĀ-lis; *pector-* = chest; *major* = larger) (see also Figure 11.10a)	Clavicle (clavicular head), sternum, and costal cartilages of second to sixth ribs and sometimes first to seventh ribs (sternocostal head).	Greater tubercle and intertubercular sulcus of humerus.	As a whole, adducts and medially rotates arm at shoulder joint; clavicular head alone flexes arm, and sternocostal head alone extends arm at shoulder joint.
Latissimus dorsi (la-TIS-i-mus DOR-sī; *latissimus* = widest; *dorsi* = of the back)	Spines of inferior six thoracic vertebrae, lumbar vertebrae, crests of sacrum and ilium, inferior four ribs.	Intertubercular sulcus of humerus.	Extends, adducts, and medially rotates arm at shoulder joint; draws arm inferiorly and posteriorly.
Scapular Muscles that Move the Humerus			
Deltoid (DEL-toyd = triangularly shaped)	Acromial extremity of clavicle (anterior fibers), acromion of scapula (lateral fibers), and spine of scapula (posterior fibers).	Deltoid tuberosity of humerus.	Lateral fibers abduct arm at shoulder joint; anterior fibers flex and medially rotate arm at shoulder joint; posterior fibers extend and laterally rotate arm at shoulder joint.
Subscapularis (sub-scap′-ū-LA-ris; *sub-* = below; *scapularis* = scapula).	Subscapular fossa of scapula.	Lesser tubercle of humerus.	Medially rotates arm at shoulder joint.
Supraspinatus (soo-pra-spī-NĀ-tus; *supra-* = above; *spina-* = spine [of the scapula])	Supraspinous fossa of scapula.	Greater tubercle of humerus.	Assists deltoid muscle in abducting arm at shoulder joint.
Infraspinatus (in′-fra-spī-NĀ-tus; *infra-* = below)	Infraspinous fossa of scapula.	Greater tubercle of humerus.	Laterally rotates and adducts arm at shoulder joint.
Teres major (TE-rēz; *teres* = long and round)	Inferior angle of scapula.	Intertubercular sulcus of humerus.	Extends arm at shoulder joint and assists in adduction and medial rotation of arm at shoulder joint.
Teres minor (TE-rēz)	Inferior lateral border of scapula.	Greater tubercle of humerus.	Laterally rotates, extends, and adducts arm at shoulder joint.
Coracobrachialis (kor′-a-kō-brā-kē-Ā′-lis; *coraco-* = coracoid process [of the scapula]; *brachi-* = arm)	Coracoid process of scapula.	Middle of medial surface of shaft of humerus.	Flexes and adducts arm at shoulder joint.

(continues)

Exhibit 11.12 **Muscles that Move the Humerus (Arm Bone) (Figure 11.15)** *(continued)*

Figure 11.15 **Muscles that move the humerus (arm bone).** (See Tortora, *A Photographic Atlas of the Human Body,* Figures 5.9 and 5.10.)

🔑 The tendons that form the rotator cuff provide the strength and stability of the shoulder joint.

(a) Anterior deep view (the intact pectoralis major muscle is shown in Figure 11.10a)

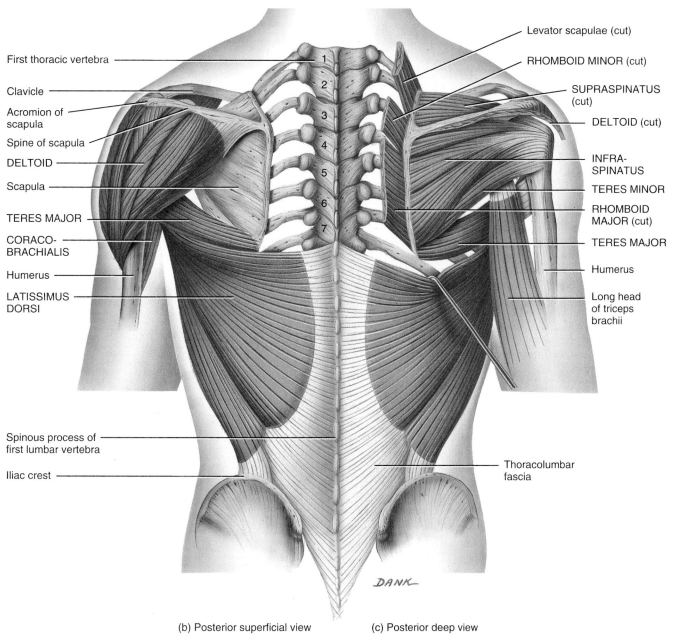

Levator scapulae (cut)

RHOMBOID MINOR (cut)

SUPRASPINATUS (cut)

DELTOID (cut)

INFRA-SPINATUS

TERES MINOR

RHOMBOID MAJOR (cut)

TERES MAJOR

Humerus

Long head of triceps brachii

First thoracic vertebra

Clavicle

Acromion of scapula

Spine of scapula

DELTOID

Scapula

TERES MAJOR

CORACO-BRACHIALIS

Humerus

LATISSIMUS DORSI

Spinous process of first lumbar vertebra

Iliac crest

Thoracolumbar fascia

DANK

(b) Posterior superficial view   (c) Posterior deep view

 Which tendons make up the rotator cuff?

Exhibit 11.13   Muscles that Move the Radius and Ulna (Forearm Bones) (Figure 11.16)

▶ OBJECTIVE

• Describe the origin, insertion, action, and innervation of the muscles that move the radius and ulna.

Most of the muscles that move the radius and ulna (forearm bones) cause flexion and extension at the elbow, which is a hinge joint. The biceps brachii, brachialis, and brachioradialis muscles are the flexor muscles. The extensor muscles are the triceps brachii and the anconeus.

The **biceps brachii** is the large muscle located on the anterior surface of the arm. As indicated by its name, it has two heads of origin (long and short), both from the scapula. The muscle spans both the shoulder and elbow joints. In addition to its role in flexing the forearm at the elbow joint, it also supinates the forearm at the radioulnar joints and flexes the arm at the shoulder joint. The **brachialis** is deep to the biceps brachii muscle. It is the most powerful flexor of the forearm at the elbow joint. For this reason, it is called the "workhorse" of the elbow flexors. The **brachioradialis** flexes the forearm at the elbow joint, especially when a quick movement is required or when a weight is lifted slowly during flexion of the forearm.

The **triceps brachii** is the large muscle located on the posterior surface of the arm. It is the more powerful of the extensors of the forearm at the elbow joint. As its name implies, it has three heads of origin, one from the scapula (long head) and two from the humerus (lateral and medial heads). The long head crosses the shoulder joint; the other heads do not. The **anconeus** is a small muscle located on the lateral part of the posterior aspect of the elbow that assists the triceps brachii in extending the forearm at the elbow joint.

Some muscles that move the radius and ulna are involved in pronation and supination at the radioulnar joints. The pronators, as suggested by their names, are the **pronator teres** and **pronator quadratus** muscles. The supinator of the forearm is the aptly named **supinator** muscle.

You use the powerful action of the supinator when you twist a corkscrew or turn a screw with a screwdriver.

In the limbs, functionally related skeletal muscles and their associated blood vessels and nerves are grouped together by fascia into regions called **compartments.** In the arm, the biceps brachii, brachialis, and coracobrachialis muscles comprise the **anterior (flexor) compartment.** The triceps brachii muscle forms the **posterior (extensor) compartment.**

Innervation

Muscles that move the radius and ulna are innervated by nerves derived from the brachial plexus (shown in Exhibit 13.2 on page 438).

• Biceps brachii: musculocutaneous nerve
• Brachialis: musculocutaneous and radial nerves
• Brachioradialis, triceps brachii, and anconeus: radial nerve
• Pronator teres and pronator quadratus: median nerve
• Supinator: deep radial nerve

Relating Muscles to Movements

Arrange the muscles in this exhibit according to the following actions on the elbow joint: (1) flexion and (2) extension; the following actions on the forearm at the radioulnar joints: (1) supination and (2) pronation; and the following actions on the humerus at the shoulder joint: (1) flexion and (2) extension. The same muscle may be mentioned more than once.

▶ CHECKPOINT

Which muscles are in the anterior and posterior compartments of the arm?

Muscle	Origin	Insertion	Action
Forearm Flexors			
Biceps brachii (BĪ-ceps BRĀ-kē-ī ; *biceps* = two heads [of origin]; *brachii* = arm)	*Long head* originates from tubercle above glenoid cavity of scapula (supraglenoid tubercle); *short head* originates from coracoid process of scapula.	Radial tuberosity of radius and bicipital aponeurosis.*	Flexes forearm at elbow joint, supinates forearm at radioulnar joints, and flexes arm at shoulder joint.
Brachialis (brā-kē-Ā-lis)	Distal, anterior surface of humerus.	Ulnar tuberosity and coronoid process of ulna.	Flexes forearm at elbow joint.
Brachioradialis (brā′-kē-ō-rā-dē-Ā-lis; *radi-* = radius) (see Figure 11.17a)	Lateral border of distal end of humerus.	Superior to styloid process of radius.	Flexes forearm at elbow joint; supinates and pronates forearm at radioulnar joints to neutral position.

Forearm Extensors

Triceps brachii (TRĪ-ceps BRĀ-kē-ī; *triceps* = three heads [of origin])	*Long head:* infraglenoid tubercle, a projection inferior to glenoid cavity of scapula.	Olecranon of ulna.	Extends forearm at elbow joint and extends arm at shoulder joint.
	Lateral head: lateral and posterior surface of humerus superior to radial groove.		
	Medial head: entire posterior surface of humerus inferior to a groove for the radial nerve.		
Anconeus (an-KŌ-nē-us; *ancon* = the elbow) (see Figure 11.17c)	Lateral epicondyle of humerus.	Olecranon and superior portion of shaft of ulna.	Extends forearm at elbow joint.

Forearm Pronators

Pronator teres (PRŌ-nā-tor TE-rēz; *pronator* = turns palm downward or posteriorly) (see Figure 11.17a)	Medial epicondyle of humerus and coronoid process of ulna.	Midlateral surface of radius.	Pronates forearm at radioulnar joints and weakly flexes forearm at elbow joint.
Pronator quadratus (PRŌ-nā-tor kwod-RĀ-tus; *quadratus* = square, four-sided) (see Figure 11.17a)	Distal portion of shaft of ulna.	Distal portion of shaft of radius.	Pronates forearm at radioulnar joints.

Forearm Supinator

Supinator (SOO-pi-nā-tor; *supinator* = turns palm upward or anteriorly) (see Figure 11.17b)	Lateral epicondyle of humerus and ridge near radial notch of ulna (supinator crest).	Lateral surface of proximal one-third of radius.	Supinates forearm at radioulnar joints.

*The *bicipital aponeurosis* is a broad aponeurosis from the tendon of insertion of the biceps brachii muscle that descends medially across the brachial artery and fuses with deep fascia over the forearm flexor muscles.

(continues)

Exhibit 11.13   Muscles that Move the Radius and Ulna (Forearm Bones) (Figure 11.16) *(continued)*

Figure 11.16   Muscles that move the radius and ulna (forearm bones).   (See Tortora, *A Photographic Atlas of the Human Body,* Figure 5.11.)

Whereas the anterior arm muscles flex the forearm, the posterior arm muscles extend it.

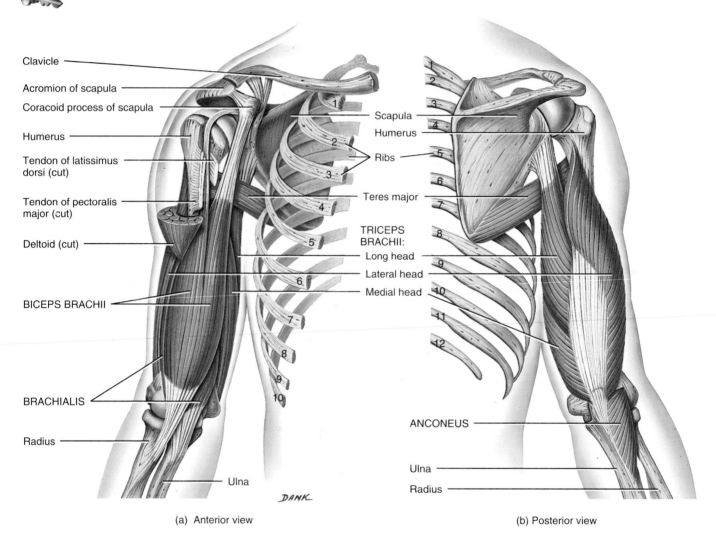

(a) Anterior view

(b) Posterior view

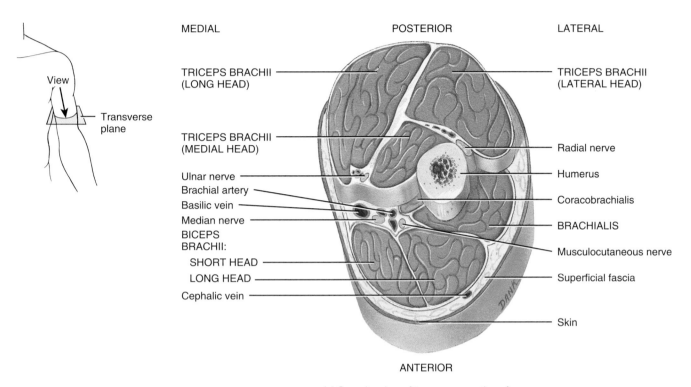

MEDIAL   POSTERIOR   LATERAL

View

Transverse plane

TRICEPS BRACHII (LONG HEAD)

TRICEPS BRACHII (LATERAL HEAD)

TRICEPS BRACHII (MEDIAL HEAD)

Radial nerve

Ulnar nerve

Humerus

Brachial artery

Basilic vein

Coracobrachialis

Median nerve

BRACHIALIS

BICEPS BRACHII:

SHORT HEAD

Musculocutaneous nerve

LONG HEAD

Superficial fascia

Cephalic vein

Skin

ANTERIOR

(c) Superior view of transverse section of arm

Which muscles are the most powerful flexor and the most powerful extensor of the forearm?

Exhibit 11.14   Muscles that Move the Wrist, Hand, Thumb, and Fingers (Figure 11.17)

▶ OBJECTIVE

- Describe the origin, insertion, action, and innervation of the muscles that move the wrist, hand, thumb, and fingers.

Muscles of the forearm that move the wrist, hand, thumb, and fingers are many and varied. Those in this group that act on the digits are known as **extrinsic hand muscles** because they originate *outside* the hand and insert within it. As you will see, the names for the muscles that move the wrist, hand, and digits give some indication of their origin, insertion, or action. Based on location and function, the muscles of the forearm are divided into two groups: (1) anterior compartment muscles and (2) posterior compartment muscles. The **anterior (flexor) compartment** muscles of the forearm originate on the humerus, typically insert on the carpals, metacarpals, and phalanges, and function as flexors. The bellies of these muscles form the bulk of the forearm. The **posterior (extensor) compartment** muscles of the forearm originate on the humerus, insert on the metacarpals and phalanges, and function as extensors. Within each compartment, the muscles are grouped as superficial or deep.

The **superficial anterior compartment** muscles are arranged in the following order from lateral to medial: **flexor carpi radialis, palmaris longus** (absent in about 10% of the population), and **flexor carpi ulnaris** (the ulnar nerve and artery are just lateral to the tendon of this muscle at the wrist). The **flexor digitorum superficialis** muscle is deep to the other three muscles and is the largest superficial muscle in the forearm.

The **deep anterior compartment** muscles are arranged in the following order from lateral to medial: **flexor pollicis longus** (the only flexor of the distal phalanx of the thumb) and **flexor digitorum profundus** (ends in four tendons that insert into the distal phalanges of the fingers).

The **superficial posterior compartment** muscles are arranged in the following order from lateral to medial: **extensor carpi radialis longus, extensor carpi radialis brevis, extensor digitorum** (occupies most of the posterior surface of the forearm and divides into four tendons that insert into the middle and distal phalanges of the fingers), **extensor digiti minimi** (a slender muscle generally connected to the extensor digitorum), and the **extensor carpiulnaris.**

The **deep posterior compartment** muscles are arranged in the following order from lateral to medial: **abductor pollicis longus, extensor pollicis brevis, extensor pollicis longus,** and **extensor indicis.**

The tendons of the muscles of the forearm that attach to the wrist or continue into the hand, along with blood vessels and nerves, are held close to bones by strong fasciae. The tendons are also surrounded by tendon sheaths. At the wrist, the deep fascia is thickened into fibrous bands called **retinacula** (*retinacul* = a holdfast). The **flexor retinaculum** is located over the palmar surface of the carpal bones. The long flexor tendons of the digits and wrist and the median nerve pass through the flexor retinaculum. The **extensor retinaculum** is located over the dorsal surface of the carpal bones. The extensor tendons of the wrist and digits pass through it.

Innervation

Nerves derived from the brachial plexus (shown in Exhibit 13.2 on page 438) innervate muscles that move the wrist, hand, thumb, and fingers.

- Flexor carpi radialis, palmaris longus, flexor digitorum superficialis, and flexor pollicis longus: median nerve
- Flexor carpi ulnaris: ulnar nerve
- Flexor digitorum profundus: median and ulnar nerves
- Extensor carpi radialis longus, extensor carpi radialis brevis, and extensor digitorum: radial nerve
- Extensor digiti minimi, extensor carpi ulnaris, and all of the deep muscles of the posterior compartment: deep radial nerve

Relating Muscles to Movements

Arrange the muscles in this exhibit according to the following actions on the wrist joint: (1) flexion, (2) extension, (3) abduction, and (4) adduction; the following actions on the fingers at the metacarpophalangeal joints: (1) flexion and (2) extension; the following actions on the fingers at the interphalangeal joints: (1) flexion and (2) extension; the following actions on the thumb at the carpometacarpal, metacarpophalangeal, and interphalangeal joints: (1) extension and (2) abduction; and the following action on the thumb at the interphalangeal joint: flexion. The same muscle may be mentioned more than once.

▶ CHECKPOINT

Which muscles and actions of the wrist, hand, and digits are used when writing?

Muscle	Origin	Insertion	Action
Superficial Anterior (Flexor) Compartment of the Forearm			
Flexor carpi radialis (FLEK-sor KAR-pē rā′-dē -Ā-lis; *flexor* = decreases angle at joint; *carpi* = of the wrist; *radi-* = radius)	Medial epicondyle of humerus.	Second and third metacarpals.	Flexes and abducts hand (radial deviation) at wrist joint.
Palmaris longus (pal-MA-ris LON-gus; *palma* = palm; *longus* = long)	Medial epicondyle of humerus.	Flexor retinaculum and palmar aponeurosis (deep fascia in center of palm).	Weakly flexes hand at wrist joint.
Flexor carpi ulnaris FLEK-sor KAR-pē ul-NAR-is; *ulnar-* = ulna)	Medial epicondyle of humerus and superior posterior border of ulna.	Pisiform, hamate, and base of fifth metacarpal.	Flexes and adducts hand (ulnar deviation) at wrist joint.
Flexor digitorum superficialis FLEK-sor di-ji-TOR-um soo′- per-fish′-ē -Ā-lis; *digit* = finger or toe; *superficialis* = closer to surface)	Medial epicondyle of humerus, coronoid process of ulna, and a ridge along lateral margin of anterior surface (anterior oblique line) of radius.	Middle phalanx of each finger.*	Flexes middle phalanx of each finger at proximal interphalangeal joint, proximal phalanx of each finger at metacarpophalangeal joint, and hand at wrist joint.

Muscle	Origin	Insertion	Action
Deep Anterior (Flexor) Compartment of the Forearm			
Flexor pollicis longus (FLEK-sor POL-li-sis LON-gus; *pollic-* = thumb)	Anterior surface of radius and interosseous membrane (sheet of fibrous tissue that holds shafts of ulna and radius together).	Base of distal phalanx of thumb.	Flexes distal phalanx of thumb at interphalangeal joint.
Flexor digitorum profundus (FLEK-sor di'-ji-TOR-um prō-FUN-dus; *profundus* = deep)	Anterior medial surface of body of ulna.	Base of distal phalanx of each finger.	Flexes distal and middle phalanges of each finger at interphalangeal joints, proximal phalanx of each finger at metacarpophalangeal joint, and hand at wrist joint.
Superficial Posterior (Extensor) Compartment of the Forearm			
Extensor carpi radialis longus (eks-TEN-sor KAR-pē rā'-dē-Ā-lis LON-gus; *extensor* = increases angle at joint)	Lateral supracondylar ridge of humerus.	Second metacarpal.	Extends and abducts hand at wrist joint.
Extensor carpi radialis brevis (eks-TEN-sor KAR-pē rā'-dē-Ā-lis BREV-is; *brevis* = short)	Lateral epicondyle of humerus.	Third metacarpal.	Extends and abducts hand at wrist joint.
Extensor digitorum (eks-TEN-sor di'-ji-TOR-um)	Lateral epicondyle of humerus.	Distal and middle phalanges of each finger.	Extends distal and middle phalanges of each finger at interphalangeal joints, proximal phalanx of each finger at metacarpophalangeal joint, and hand at wrist joint.
Extensor digiti minimi (eks-TEN-sor DIJ-i-tē MIN-i-mē; *digit* = finger or toe; *minimi* = little finger)	Lateral epicondyle of humerus.	Tendon of extensor digitorum on fifth phalanx.	Extends proximal phalanx of little finger at metacarpophalangeal joint and hand at wrist joint.
Extensor carpi ulnaris (eks-TEN-sor KAR-pē ul-NAR-is)	Lateral epicondyle of humerus and posterior border of ulna.	Fifth metacarpal.	Extends and adducts hand at wrist joint.
Deep Posterior (Extensor) Compartment of the Forearm			
Abductor pollicis longus (ab-DUK-tor POL-li-sis LON-gus; *abductor* = moves part away from midline)	Posterior surface of middle of radius and ulna and interosseous membrane.	First metacarpal.	Abducts and extends thumb at carpometacarpal joint and abducts hand at wrist joint.
Extensor pollicis brevis (eks-TEN-sor POL-li-sis BREV-is)	Posterior surface of middle of radius and interosseous membrane.	Base of proximal phalanx of thumb.	Extends proximal phalanx of thumb at metacarpophalangeal joint, first metacarpal of thumb at carpometacarpal joint, and hand at wrist joint.
Extensor pollicis longus (eks-TEN-sor POL-li-sis LON-gus)	Posterior surface of middle of ulna and interosseous membrane.	Base of distal phalanx of thumb.	Extends distal phalanx of thumb at interphalangeal joint, first metacarpal of thumb at carpometacarpal joint, and abducts hand at wrist joint.
Extensor indicis (eks-TEN-sor IN-di-kis; *indicis* = index)	Posterior surface of ulna.	Tendon of extensor digitorum of index finger.	Extends distal and middle phalanges of index finger at interphalangeal joints, proximal phalanx of index finger at metacarpophalangeal joint, and hand at wrist joint.

*Reminder: The thumb or pollex is the first digit and has two phalanges: proximal and distal. The remaining digits, the fingers, are numbered 2–5, and each has three phalanges: proximal, middle, and distal.

(continues)

Figure 11.17 **Muscles that move the wrist, hand, thumb, and fingers.** (See Tortora, *A Photographic Atlas of the Human Body,* Figures 5.12 and 5.13.)

The anterior compartment muscles function as flexors, and the posterior compartment muscles function as extensors.

Biceps brachii
Brachialis
Brachial artery
Median nerve
Medial epicondyle of humerus
Tendon of biceps brachii
PRONATOR TERES
BRACHIORADIALIS
SUPINATOR
PALMARIS LONGUS
FLEXOR CARPI RADIALIS
FLEXOR CARPI ULNARIS
FLEXOR DIGITORUM PROFUNDUS
PRONATOR TERES (cut)
FLEXOR DIGITORUM SUPERFICIALIS
FLEXOR POLLICIS LONGUS
ABDUCTOR POLLICIS LONGUS
PRONATOR QUADRATUS
Flexor retinaculum
Metacarpals
Tendon of flexor digitorum superficialis
Tendon of flexor digitorum profundus

PL
PT
FCR
FDS
FCU

Ulna

Key to abbreviations in (b)

PL = Palmaris longus
PT = Pronator teres
FCR = Flexor carpi radialis
FDS = Flexor digitorum superficialis
FCU = Flexor carpi ulnaris

(a) Anterior superficial view

(b) Anterior deep view

Triceps brachii

Humerus

BRACHIORADIALIS

EXTENSOR CARPI RADIALIS LONGUS

Medial epicondyle of humerus
Lateral epicondyle of humerus
Olecranon of ulna
ANCONEUS
EXTENSOR CARPI ULNARIS
EXTENSOR DIGITORUM
EXTENSOR CARPI RADIALIS BREVIS
EXTENSOR DIGITI MINIMI

SUPINATOR

FLEXOR CARPI ULNARIS

FLEXOR DIGITORUM PROFUNDUS

Tendon of pronator teres

ABDUCTOR POLLICIS LONGUS

EXTENSOR POLLICIS LONGUS

EXTENSOR POLLICIS BREVIS

EXTENSOR INDICIS

Tendon of extensor carpi ulnaris
Extensor retinaculum

Carpals

Tendon of extensor digiti minimi

Tendon of extensor indicis

Tendons of extensor digitorum

Dorsal interossei

DANK

(c) Posterior superficial view

(d) Posterior deep view

What structures pass through the flexor retinaculum?

355

Exhibit 11.15   Intrinsic Muscles of the Hand (Figure 11.18)

▶ OBJECTIVE

• Describe the origin, insertion, action, and innervation of the intrinsic muscles of the hand.

Several of the muscles discussed in Exhibit 11.14 are extrinsic muscles of the hand. They originate outside the hand and insert within the hand. The extrinsic muscles of the hand produce powerful but crude movements of the digits. The **intrinsic hand muscles** are located in the palm. They produce the less powerful but more intricate and precise movements of the digits. The muscles in this group are so-named because their origins and insertions are *within* the hand. The functional importance of the hand muscles is readily apparent when one considers that certain hand injuries can cause permanent disability.

The intrinsic muscles of the hand are divided into three groups: (1) **thenar,** (2) **hypothenar,** and (3) **intermediate.** The four thenar muscles act on the thumb and form the **thenar eminence,** the lateral rounded contour on the palm. The thenar muscles include the **abductor pollicis brevis, opponens pollicis, flexor pollicis brevis,** and **adductor pollicis.**

The three hypothenar muscles act on the little finger and form the **hypothenar eminence,** the rounded contour on the medial aspect of the palm. The hypothenar muscles are the **abductor digiti minimi, flexor digiti minimi brevis,** and **opponens digiti minimi.**

The 12 intermediate (midpalmar) muscles act on all the digits except the thumb. The intermediate muscles include the **lumbricals, palmar interossei,** and **dorsal interossei.** Both sets of interossei muscles are located between the metacarpals and are important in abduction, adduction, flexion, and extension of the fingers, and in movements in skilled activities such as writing, typing, and playing a piano.

Besides free motion, the hand performs several special movements. *Power grip* is the forcible movement of the fingers and thumb against the palm, as in hand-shaking. *Precision handling* is a change in position of a handled object that requires exact control of finger and thumb positions, as in winding a watch or threading a needle. *Pinching* is compression between the thumb and index finger or between the thumb and first two fingers.

Movements of the thumb are very important in the precise activities of the hand. They are defined in different planes from comparable movements of the fingers because the thumb can be positioned at a right angle to the fingers. Figure 11.18c shows the five main movements of the thumb. They include *flexion* (movement of the thumb medially across the palm), *extension* (movement of the thumb laterally away from the palm), *abduction* (movement of the thumb in an anteroposterior plane away from the palm), *adduction* (movement of the thumb in an antero-posterior plane toward the palm), and *opposition* (movement of the thumb across the palm so that the tip of the thumb meets the tip of a finger). Opposition is the most distinctive digital movement. It gives humans and other primates the ability to precisely grasp and manipulate objects.

Innervation

Nerves derived from the brachial plexus (shown in Exhibit 13.2 on page 438) innervate intrinsic muscles of the hand.

• Abductor pollicis brevis and opponens pollicis: median nerve
• Adductor pollicis, abductor digiti minimi, flexor digiti minimi brevis, opponens digiti minimi, dorsal interossei, and palmar interossei: ulnar nerve
• Flexor pollicis brevis and lumbricals: median and ulnar nerves

 Carpal Tunnel Syndrome

The **carpal tunnel** is a narrow passageway formed anteriorly by the flexor retinaculum and posteriorly by the carpal bones. Through this tunnel pass the median nerve, the most superficial structure, and the long flexor tendons for the digits. Structures within the carpal tunnel, especially the median nerve, are vulnerable to compression, and the resulting condition is called **carpal tunnel syndrome.** Compression of the median nerve leads to sensory changes over the lateral side of the hand and muscle weakness in the thenar eminence. This results in pain, numbness, and tingling of the fingers. The condition may be caused by inflammation of the digital tendon sheaths, fluid retention, excessive exercise, infection, trauma, and repetitive activities that involve flexion of the wrist, such as keyboarding, cutting hair, and playing a piano. ■

Relating Muscles to Movements

Arrange the muscles in this exhibit according to the following actions on the thumb at the carpometacarpal and metaphalangeal joints: (1) abduction, (2) adduction, (3) flexion, and (4) opposition; and the following actions on the fingers at the metacarpophalangeal and interphalangeal joints: (1) abduction, (2) adduction, (3) flexion, and (4) extension. The same muscle may be mentioned more than once.

▶ CHECKPOINT

How do the actions of the extrinsic and intrinsic muscles of the hand differ?

Muscle	Origin	Insertion	Action
Thenar (lateral aspect of palm)			
Abductor pollicis brevis (ab-DUK-tor POL-li-sis BREV-is; *abductor* = moves part away from middle; *pollic-* = the thumb; *brevis* = short)	Flexor retinaculum, scaphoid, and trapezium.	Lateral side of proximal phalanx of thumb.	Abducts thumb at carpometacarpal joint.
Opponens pollicis (op-PŌ-nenz POL-li-sis; *opponens* = opposes)	Flexor retinaculum and trapezium.	Lateral side of first metacarpal (thumb).	Moves thumb across palm to meet little finger (opposition) at the carpometacarpal joint.
Flexor pollicis brevis (FLEK-sor POL-li-sis BREV-is; *flexor* = decreases angle at joint)	Flexor retinaculum, trapezium, capitate, and trapezoid.	Lateral side of proximal phalanx of thumb.	Flexes thumb at carpometacarpal and metacarpophalangeal joints.
Adductor pollicis (ad-DUK-tor POL-li-sis; *adductor* = moves part toward midline)	Oblique head: capitate and second and third metacarpals; transverse head: third metacarpal.	Medial side of proximal phalanx of thumb by a tendon containing a sesamoid bone.	Adducts thumb at carpometacarpal and metacarpophalangeal joints.
Hypothenar (medial aspect of palm)			
Abductor digiti minimi (ab-DUK-tor DIJ-i-tē MIN-i-mē; *digit* = finger or toe; *minimi* = little)	Pisiform and tendon of flexor carpi ulnaris.	Medial side of proximal phalanx of little finger.	Abducts and flexes little finger at metacarpophalangeal joint.
Flexor digiti minimi brevis (FLEK-sor DIJ-i-tē MIN-i-mē BREV-is)	Flexor retinaculum and hamate.	Medial side of proximal phalanx of little finger.	Flexes little finger at carpometacarpal and metacar-pophalangeal joints.
Opponens digiti minimi (op-PŌ-nenz DIJ-i-tē MIN-i-mē)	Flexor retinaculum and hamate.	Medial side of fifth metacarpal (little finger).	Moves little finger across palm to meet thumb (opposition) at the carpometacarpal joint.
Intermediate (Midpalmar)			
Lumbricals (LUM-bri-kals; *lumbric-* = earthworm) (four muscles)	Lateral sides of tendons and flexor digitorum profundus of each finger.	Lateral sides of tendons of extensor digitorum on proximal phalanges of each finger.	Flex each finger at metacarpo-phalangeal joints and extend each finger at interphalangeal joints.
Palmar interossei (in´-ter-OS-ē-ī (*palmar* = palm; *inter-* = between; *ossei* = bones) (four muscles)	Sides of shafts of metacarpals of all digits (except the middle one)	Sides of bases of proximal phalanges of all digits (except the middle one).	Adduct each finger at metacarpo-phalangeal joints; flex each finger at metacarpophalangeal joints.
Dorsal interossei (in´-ter-OS-ē-ī; *dorsal* = back surface) (four muscles)	Adjacent sides of metacarpals.	Proximal phalanx of each finger.	Abduct fingers 2–4 at metacarpo-phalangeal joints; flex fingers 2–4 at metacarpophalangeal joints; and extend each finger at interphalangeal joints.

(continues)

Exhibit 11.15   Intrinsic Muscles of the Hand (Figure 11.18) *(continued)*

Figure 11.18   Intrinsic muscles of the hand.   (See Tortora, *A Photographic Atlas of the Human Body,* Figures 5.14 and 5.15.)

 The intrinsic muscles of the hand produce intricate and precise movements of the thumb and fingers.

(a) Anterior superficial view

(b) Anterior deep view

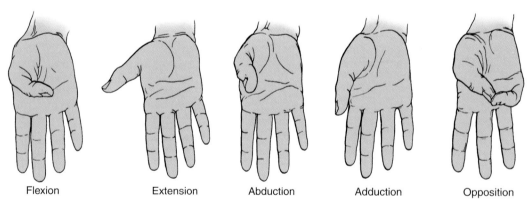

Flexion   Extension   Abduction   Adduction   Opposition

(c) Movements of the thumb

Which structure moves when muscles of the thenar eminence contract?

Exhibit 11.16 Muscles that Move the Vertebral Column (Backbone) (Figure 11.19)

► OBJECTIVE

- Describe the origin, insertion, action, and innervation of the muscles that move the vertebral column.

The muscles that move the vertebral column (backbone) are quite complex because they have multiple origins and insertions and there is considerable overlap among them. One way to group the muscles is on the basis of the general direction of the muscle bundles and their approximate lengths. For example, the splenius muscles arise from the midline and extend laterally and superiorly to their insertions (Figure 11.19a). The erector spinae muscle arises from either the midline or more laterally but usually runs almost longitudinally, with neither a significant lateral nor medial direction as it is traced superiorly. The transversospinalis arise laterally but extend toward the midline as they are traced superiorly. Deep to these three muscle groups are small segmental muscles that extend between spinous processes or transverse processes of vertebrae. Because the scalene muscles also assist in moving the vertebral column, they are included in this exhibit. Note in Exhibit 11.17 that the rectus abdominis, external oblique, internal oblique, and quadratus lumborum muscles also play a role in moving the vertebral column.

The bandage-like **splenius** muscles are attached to the sides and back of the neck. The two muscles in this group are named on the basis of their superior attachments (insertions): **splenius capitis** (head region) and **splenius cervicis** (cervical region). They extend the head and laterally flex and rotate the head.

The **erector spinae** is the largest muscle mass of the back, forming a prominent bulge on either side of the vertebral column. It is the chief extensor of the vertebral column. It is also important in controlling flexion, lateral flexion, and rotation of the vertebral column and in maintaining the lumbar curve, because the main mass of the muscle is in the lumbar region. It consists of three groups: iliocostalis (laterally placed), longissimus (intermediately placed), and spinalis (medially placed). These groups, in turn, consist of a series of overlapping muscles, and the muscles within the groups are named according to the regions of the body with which they are associated. The **iliocostalis group** consists of three muscles: the **iliocostalis cervicis** (cervical region), **iliocostalis thoracis** (thoracic region), and **iliocostalis lumborum** (lumbar region). The **longissimus group** resembles a herring bone and consists of three muscles: the **longissimus capitis** (head region), **longissimus cervicis** (cervical region), and **longissimus thoracic** (thoracic region). The **spinalis group** also consists of three muscles: the **spinalis capitis, spinalis cervicis,** and **spinalis thoracis.**

The **transversospinales** are named because their fibers run from the transverse processes to the spinous processes of the vertebrae. The semispinalis muscles in this group are also named according to the region of the body with which they are associated: **semispinalis capitis** (head region), **semispinalis cervicis** (cervical region), and **semispinalis thoracic** (thoracic region). These muscles extend the vertebral column and rotate the head. The **multifidus** muscle in this group, as its name implies, is segmented into several bundles. It extends and laterally flexes the vertebral column and rotates the head. The **rotatores** muscles of this group are short and are found along the entire length of the vertebral column. They extend and rotate the vertebral column.

Within the **segmental** muscle group (Figure 11.17b) are the **interspinales** and **intertransversarii** muscles, which unite the spinous and transverse processes of consecutive vertebrae. They function primarily in stabilizing the vertebral column during its movements.

Within the **scalene** group (Figure 11.17c), the **anterior scalene** muscle is anterior to the middle scalene muscle, the **middle scalene** muscle is intermediate in placement and is the longest and largest of the scalene muscles, and the **posterior scalene** muscle is posterior to the middle scalene muscle and is the smallest of the scalene muscles. These muscles flex, laterally flex, and rotate the head and assist in deep inhalation.

Innervation

- Splenius capitis: middle cervical nerves
- Splenius cervicis: inferior cervical nerves
- Iliocostalis cervicis and semispinalis capitis: cervical and thoracic nerves
- Iliocostalis thoracis: thoracic nerves
- Iliocostalis lumborum: lumbar nerves
- Longissimus capitis: middle and inferior cervical nerves
- Longissimus cervicis, longissimus thoracis, all spinalis muscles, multifidus, rotatores, and interspinales: cervical, thoracic, and lumbar nerves
- Semispinalis cervicis and semispinalis thoracis: cervical and thoracic nerves
- Intertransversarii: cervical, thoracic and lumbar spinal nerves
- Anterior scalene: cervical spinal nerves C5–C6
- Middle scalene: cervical spinal nerves C3–C8
- Posterior scalene: cervical spinal nerves C6–C8.

 Back Injuries and Heavy Lifting

Full flexion at the waist, as in touching your toes, overstretches the erector spinae muscles. Because of the length-tension relationship (see page 285), muscles that are overstretched cannot contract effectively. Straightening up from such a position is therefore initiated by the hamstring muscles on the back of the thigh and the gluteus maximus muscles of the buttocks. The erector spinae muscles join in as the degree of flexion decreases. Improperly lifting a heavy weight, however, can strain the erector spinae muscles. The result can be painful muscle spasms, tearing of tendons and ligaments of the lower back, and rupturing of intervertebral discs. The lumbar muscles are adapted for maintaining posture, not for lifting. This is why it is important to kneel and use the powerful extensor muscles of the thighs and buttocks while lifting a heavy load. ■

Relating Muscles to Movements

Arrange the muscles in this exhibit according to the following actions on the head at the atlanto-occipital and intervertebral joints: (1) flexion, (2) extension, (3) lateral flexion, (4) rotation to same side as contracting muscle, and (5) rotation to opposite side as contracting muscle; the following actions on the vertebral column at the intervertebral joints: (1) flexion, (2) extension, (3) lateral flexion, (4) rotation, and (5) stabilization; and the following action on the ribs: elevation during deep inhalation. The same muscle may be mentioned more than once.

► CHECKPOINT

What are the four major groups of muscles that move the vertebral column?

(continues)

Exhibit 11.16 Muscles that Move the Vertebral Column (Backbone) (Figure 11.19) (continued)

Muscle	Origin	Insertion	Action
Splenius (SPLĒ-nē-us)			
Splenius capitis (KAP-i-tis; *splenium* = bandage; *capit-* = head)	Ligamentum nuchae and spinous processes of seventh cervical vertebra and first three or four thoracic vertebrae.	Occipital bone and mastoid process of temporal bone.	Acting together (bilaterally), extend head; acting singly (unilaterally), laterally flex and rotate head to same side as contracting muscle.
Splenius cervicis (SER-vi-kis; *cervic-* = neck)	Spinous processes of third through sixth thoracic vertebrae.	Transverse processes of first two or four cervical vertebrae.	Acting together, extend head; acting singly, laterally flex and rotate head to same side as contracting muscle.
Erector Spinae (e-REK-tor SPI-nē) Consists of iliocostalis muscles (lateral), longissimus muscles (intermediate), and spinalis muscles (medial).			
Iliocostalis Group (Lateral)			
Iliocostalis cervicis (il'-ē-ō-kos-TAL-is SER-vi-kis; *ilio-* = flank; *costa-* = rib)	Superior six ribs.	Transverse processes of fourth to sixth cervical vertebrae.	Acting together, muscles of each region (cervical, thoracic, and lumbar) extend and maintain erect posture of vertebral column of their respective regions; acting singly, laterally flex vertebral column of their respective regions.
Iliocostalis thoracis (il'-ē-ō-kos-TAL-is thō-RA-sis; *thorac-* = chest)	Inferior six ribs.	Superior six ribs.	
Iliocostalis lumborum (il'-ē-ō-kos-TAL-is lum-BOR-um)	Iliac crest.	Inferior six ribs.	
Longissimus Group (Intermediate)			
Longissimus capitis (lon-JIS-i-mus KAP-i-tis; *longissimus* = longest)	Transverse processes of superior four thoracic vertebrae and articular processes of inferior four cervical vertebrae.	Mastoid process of temporal bone.	Acting together, both longissimus capitis muscles extend head; acting singly, rotate head to same side as contracting muscle. Acting together, longissimus cervicis and both longissimus thoracis muscles extend vertebral column of their respective regions; acting singly, laterally flex vertebral column of their respective regions.
Longissimus cervicis (lon-JIS-i-mus SER-vi-kis)	Transverse processes of fourth and fifth thoracic vertebrae.	Transverse processes of second to sixth cervical vertebrae.	
Longissimus thoracis (lon-JIS-i-mus thō-RA-sis)	Transverse processes of lumbar vertebrae.	Transverse processes of all thoracic and superior lumbar vertebrae and ninth and tenth ribs.	
Spinalis Group (Medial)			
Spinalis capitis (spi-NĀ-lis KAP-i-tis; *spinal-* = vertebral column)	Arises with semispinalis capitis.	Occipital bone.	Acting together, muscles of each region (cervical, thoracic, and lumbar) extend vertebral column of their respective regions.
Spinalis cervicis (spi-NĀ-lis SER-vi-kis)	Ligamentum nuchae and spinous process of seventh cervical vertebra.	Spinous process of axis.	
Spinalis thoracis (spi-NĀ-lis thō-RA-sis)	Spinous processes of superior lumbar and inferior thoracic vertebrae.	Spinous processes of superior thoracic vertebrae.	

Muscle	Origin	Insertion	Action
Transversospinales (trans-ver-sō-spi-NA-lēz)			
Semispinalis capitis (sem′-ē-spi-NĀ-lis KAP-i-tis; *semi-* = partially or one-half)	Transverse processes of first six or seven thoracic vertebrae and seventh cervical vertebra, and articular processes of fourth, fifth, and sixth cervical vertebrae.	Occipital bone.	Acting together, extend head; acting singly, rotate head to side opposite contracting muscle.
Semispinalis cervicis (sem′-ē-spi-NĀ-lis SER-vi-kis)	Transverse processes of superior five or six thoracic vertebrae.	Spinous processes of first to fifth cervical vertebrae.	Acting together, both semispinalis cervicis and both semispinalis thoracis muscles extend vertebral column of their respective regions; acting singly, rotate head to side opposite contracting muscle.
Semispinalis thoracis (sem′-ē-spi-NĀ-lis thō-RA-sis)	Transverse processes of sixth to tenth thoracic vertebrae.	Spinous processes of superior four thoracic and last two cervical vertebrae.	
Multifidus (mul-TIF-i-dus; *multi* = many; *fid-* = segmented)	Sacrum, ilium, transverse processes of lumbar, thoracic, and inferior four cervical vertebrae.	Spinous process of a more superior vertebra.	Acting together, extend vertebral column; acting singly, laterally flex vertebral column and rotate head to side opposite contracting muscle.
Rotatores (rō′-ta-TŌ-rez; singular is **rotatore;** *rotatore* = to rotate)	Transverse processes of all vertebrae.	Spinous process of vertebra superior to the one of origin.	Acting together, extend vertebral column; acting singly, rotate vertebral column to side opposite contracting muscle.
Segmental (seg-MEN-tal)			
Interspinales (in-ter-spī-NĀ-lēz; *inter-* = SPĪ-nāl-ez; *inter-* = between)	Superior surface of all spinous processes.	Inferior surface of spinous process of vertebra superior to the one of origin.	Acting together, extend vertebral column; acting singly, stabilize vertebral column during movement.
Intertransversarii (in′-ter- trans-vers-AR-ē-ī ; singular is **intertransversarius**)	Transverse processes of all vertebrae.	Transverse process of vertebra superior to the one of origin.	Acting together, extend vertebral column; acting singly, laterally flex vertebral column and stabilize it during movements.
Scalenes (SKĀ-lēnz)			
Anterior scalene (SKĀ-lēn; *anterior* = front; *scalene* = uneven)	Transverse processes of third through sixth cervical vertebrae.	First rib.	Acting together, right and left anterior scalene and middle scalene muscles flex head and elevate first ribs during deep inhalation; acting singly, laterally flex head and rotate head to side opposite contracting muscle.
Middle scalene (SKĀ-lēn)	Transverse processes of inferior six cervical vertebrae.	First rib.	
Posterior scalene (SKĀ-lēn)	Transverse processes of fourth through sixth cervical vertebrae.	Second rib.	Acting together, flex head and elevate second ribs during deep inhalation; acting singly, laterally flex head and rotate head to side opposite contracting muscle.

(continues)

Exhibit 11.16 Muscles that Move the Vertebral Column (Backbone) (Figure 11.19) *(continued)*

Figure 11.19 Muscles that move the vertebral column (backbone).

The erector spinae group is the largest muscular mass of the body and is the chief extensor of the vertebral column.

LONGISSIMUS CAPITIS

SPINALIS CERVICIS

LONGISSIMUS CERVICIS

ILIOCOSTALIS THORACIS

SPINALIS THORACIS

ILIOCOSTALIS LUMBORUM

SEMISPINALIS CAPITIS
Ligamentum nuchae
SPINALIS CAPITIS

SPLENIUS CAPITIS

SPLENIUS CERVICIS

ILIOCOSTALIS CERVICIS

SEMISPINALIS CERVICIS

LONGISSIMUS THORACIS

SEMISPINALIS THORACIS

INTERTRANSVERSARIUS

ROTATORE

MULTIFIDUS

DANK

(a) Posterior view

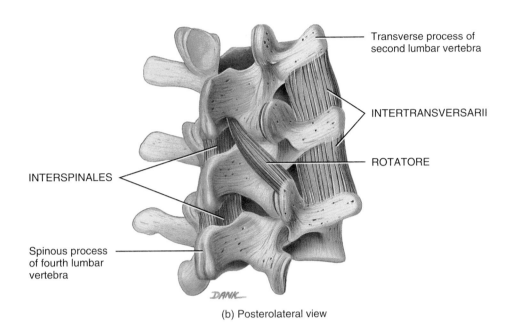

Transverse process of second lumbar vertebra

INTERTRANSVERSARII

INTERSPINALES

ROTATORE

Spinous process of fourth lumbar vertebra

DANK

(b) Posterolateral view

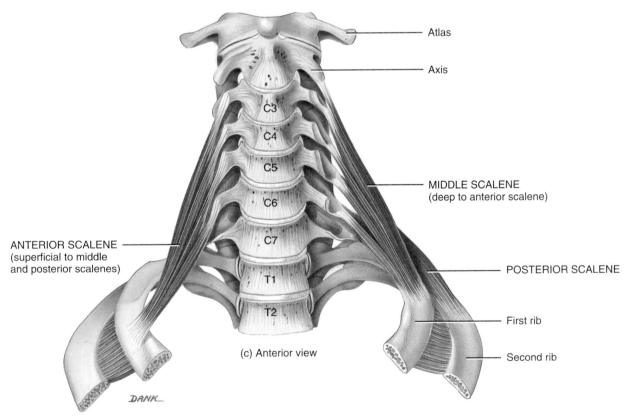

Atlas

Axis

C3

C4

C5

C6

MIDDLE SCALENE
(deep to anterior scalene)

C7

ANTERIOR SCALENE
(superficial to middle
and posterior scalenes)

POSTERIOR SCALENE

T1

T2

First rib

(c) Anterior view

Second rib

DANK

Which muscles originate at the midline and extend laterally and upward to their insertion?

Exhibit 11.17   Muscles that Move the Femur (Thigh Bone) (Figure 11.20)

▶ OBJECTIVE

- Describe the origin, insertion, action, and innervation of the muscles that move the femur.

As you will see, muscles of the lower limbs are larger and more powerful than those of the upper limbs because lower limb muscles function in stability, locomotion, and maintenance of posture. Upper limb muscles are characterized by versatility of movement. In addition, muscles of the lower limbs often cross two joints and act equally on both.

The majority of muscles that move the femur originate on the pelvic girdle and insert on the femur. The **psoas major** and **iliacus** muscles have a common insertion. Together, they are known as the **iliopsoas** (il'-ē-ō-SŌ-as) muscle because they share a common insertion (lesser trochanter of femur). Sometimes people "cheat" when doing situps by using their iliopsoas muscle instead of the abdominal muscles. By doing so, this can potentially produce back pain due to the origin of the iliopsoas on the backbone. There are three gluteal muscles: gluteus maximus, gluteus medius, and gluteus minimus. The **gluteus maximus** is the largest and heaviest of the three muscles and is one of the largest muscles in the body. It is the chief extensor of the femur. The **gluteus medius** is mostly deep to the gluteus maximus and is a powerful abductor of the femur at the hip joint. It is a common site for an intramuscular injection. The **gluteus minimus** is the smallest of the gluteal muscles and lies deep to the gluteus medius.

The **tensor fasciae latae** muscle is located on the lateral surface of the thigh. The *fascia lata* is a layer of deep fascia, composed of dense connective tissue, that encircles the entire thigh. It is well developed laterally where, together with the tendons of the tensor fasciae and gluteus maximus muscles, it forms a structure called the **iliotibial tract.** The tract inserts into the lateral condyle of the tibia.

The **piriformis, obturator internus, obturator externus, superior gemellus, inferior gemellus,** and **quadratus femoris** muscles are all deep to the gluteus maximus muscle and function as lateral rotators of the femur at the hip joint.

Three muscles on the medial aspect of the thigh are the **adductor longus, adductor brevis,** and **adductor magnus.** They originate on the pubic bone and insert on the femur. All three muscles adduct, flex, and medially rotate the femur at the hip joint. The **pectineus** muscle also adducts and flexes the femur at the hip joint.

Technically, the adductor muscles and pectineus muscles are components of the medial compartment of the thigh and could be included in

Exhibit 11.18. However, they are included here because they act on the femur.

Innervation

Nerves derived from the lumbar and sacral plexuses (shown in Exhibits 13.3 and 13.4 on pages 442 and 444) innervate the muscles that move the femur.

- Psoas major: lumbar nerves L2–L3
- Iliacus and pectineus: femoral nerve
- Gluteus maximus: inferior gluteal nerve
- Gluteus medius and minimus and tensor fasciae latae: superior gluteal nerve
- Piriformis: sacral spinal nerves S1 or S2, mainly S1
- Obturator internus and superior gemellus: nerve to obturator internus
- Obturator externus, adductor longus, and adductor brevis: obturator nerve
- Inferior gemellus and quadratus femoris: nerve to quadratus femoris
- Adductor magnus: obturator and sciatic nerves

 Pulled Groin

Certain athletic and other activities may cause a **pulled groin,** a strain, stretching, or tearing of the distal attachments of the medial muscles of the thigh, iliopsoas, and/or adductor muscles. Generally, a pulled groin results from activities that involve quick sprints, as might occur during soccer, tennis, football, and running. ■

Relating Muscles to Movements

Arrange the muscles in this exhibit according to the following actions on the thigh at the hip joint: (1) flexion, (2) extension, (3) abduction, (4) adduction, (5) medial rotation, and (6) lateral rotation. The same muscle may be mentioned more than once.

▶ CHECKPOINT
What forms the iliotibial tract?

Muscle	Origin	Insertion	Action
Psoas major (SŌ-as; *psoa* = a muscle of the loin)	Transverse processes and bodies of lumbar vertebrae.	With iliacus into lesser trochanter of femur.	Psoas major and iliacus muscles acting together flex thigh at hip joint, rotate thigh laterally, and flex trunk on the hip as in sitting up from the supine position
Iliacus (il'-ē-AK-us; *iliac-* = ilium)	Iliac fossa and sacrum.	With psoas major into lesser trochanter of femur.	
Gluteus maximus (GLOO-tē-us MAK-si-mus; *glute-* = rump or buttock; *maximus* = largest)	Iliac crest, sacrum, coccyx, and aponeurosis of sacrospinalis.	Iliotibial tract of fascia lata and lateral part of linea aspera under greater trochanter (gluteal tuberosity) of femur.	Extends thigh at hip joint and laterally rotates thigh.
Gluteus medius (GLOO-tē-us MĒ-dē-us; *medi-* = middle)	Ilium.	Greater trochanter of femur.	Abducts thigh at hip joint and medially rotates thigh.
Gluteus minimus (GLOO-tē-us MIN-i-mus; *minim-* = smallest)	Ilium.	Greater trochanter of femur.	Abducts thigh at hip joint and medially rotates thigh.
Tensor fasciae latae (TEN-sor FA-shē-ē LĀ-tē; *tensor* = makes tense; *fasciae* = of the band; *lat-* = wide)	Iliac crest.	Tibia by way of the iliotibial tract.	Flexes and abducts thigh at hip joint.
Piriformis (pir-i-FOR-mis; *piri-* = pear; *form-* = shape)	Anterior sacrum.	Superior border of greater trochanter of femur.	Laterally rotates and abducts thigh at hip joint.
Obturator internus (OB-too-rā'-tor in-TER-nus; *obturator* = obturator foramen; *intern-* = inside)	Inner surface of obturator foramen, pubis, and ischium.	Greater trochanter of femur.	Laterally rotates and abducts thigh at hip joint.
Obturator externus (OB-too-rā'-tor ex-TER-nus; *extern-* = outside)	Outer surface of obturator membrane.	Deep depression inferior to greater trochanter (trochanteric fossa) of femur.	Laterally rotates and abducts thigh at hip joint.
Superior gemellus (jem-EL-lus; *superior* = above; *gemell-* = twins)	Ischial spine.	Greater trochanter of femur.	Laterally rotates and abducts thigh at hip joint.
Inferior gemellus (jem-EL-lus; *inferior* = below)	Ischial tuberosity.	Greater trochanter of femur.	Laterally rotates and abducts thigh at hip joint.
Quadratus femoris (kwod-RĀ-tus FEM-or-is; *quad-* = square, four-sided; *femoris* = femur)	Ischial tuberosity.	Elevation superior to mid-portion of intertrochanteric crest (quadrate tubercle) on posterior femur.	Laterally rotates and stabilizes hip joint.
Adductor longus (LONG-us; *adductor* = moves part closer to midline; *longus* = long)	Pubic crest and pubic symphysis.	Linea aspera of femur.	Adducts and flexes thigh at hip joint and medially rotates thigh.
Adductor brevis (BREV-is; *brevis* = short)	Inferior ramus of pubis.	Superior half of linea aspera of femur.	Adducts and flexes thigh at hip joint and medially rotates thigh.
Adductor magnus (MAG-nus; *magnus* = large)	Inferior ramus of pubis and ischium to ischial tuberosity.	Linea aspera of femur.	Adducts thigh at hip joint and medially rotates thigh; anterior part flexes thigh at hip joint, and posterior part extends thigh at hip joint.
Pectineus (pek-TIN-ē-us; *pectin-* = a comb)	Superior ramus of pubis.	Pectineal line of femur, between lesser trochanter and linea aspera.	Flexes and adducts thigh at hip joint.

(continues)

Exhibit 11.17 Muscles that Move the Femur (Thigh Bone) (Figure 11.20) (continued)

Figure 11.20 Muscles that move the femur (thigh bone). (See Tortora, *A Photographic Atlas of the Human Body,* Figures 5.16 and 5.17.)

Most muscles that move the femur originate on the pelvic (hip) girdle and insert on the femur.

(a) Anterior superficial view

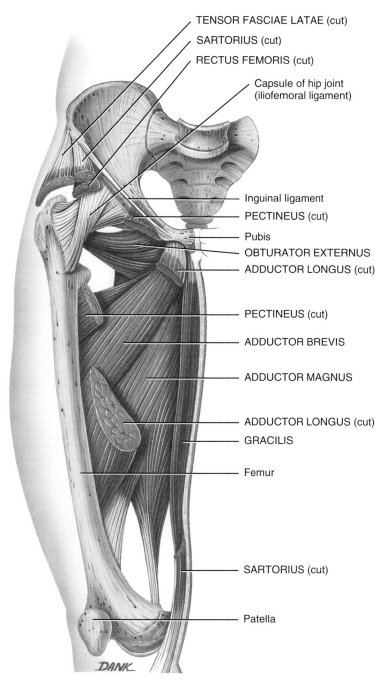

TENSOR FASCIAE LATAE (cut)

SARTORIUS (cut)

RECTUS FEMORIS (cut)

Capsule of hip joint
(iliofemoral ligament)

Inguinal ligament

PECTINEUS (cut)

Pubis

OBTURATOR EXTERNUS

ADDUCTOR LONGUS (cut)

PECTINEUS (cut)

ADDUCTOR BREVIS

ADDUCTOR MAGNUS

ADDUCTOR LONGUS (cut)

GRACILIS

Femur

SARTORIUS (cut)

Patella

(b) Anterior deep view (femur rotated laterally)

(continues)

Exhibit 11.17 **Muscles that Move the Femur (Thigh Bone) (Figure 11.20)** *(continued)*

Iliac crest

GLUTEUS MAXIMUS (cut)

Sacrum

Coccyx

OBTURATOR INTERNUS

Ischial tuberosity

Sciatic nerve

GRACILIS

SARTORIUS

GLUTEUS MEDIUS (cut)

GLUTEUS MINIMUS

PIRIFORMIS

SUPERIOR GEMELLUS

Greater trochanter

INFERIOR GEMELLUS

OBTURATOR EXTERNUS

QUADRATUS FEMORIS

GLUTEUS MAXIMUS (cut)

Femur

ADDUCTOR MAGNUS

HAMSTRINGS:

SEMITENDINOSUS

BICEPS FEMORIS

SEMIMEMBRANOSUS

Vastus lateralis

Femur deep to popliteal fossa

Plantaris

Gastrocnemius

Tendon of biceps femoris

(c) Posterior superficial view

 What are the main differences between the muscles of the upper and lower limbs?

Exhibit 11.18 Muscles that Act on the Femur (Thigh Bone) and Tibia and Fibula (Leg Bones) (Figures 11.20 and 11.21)

▶ OBJECTIVE

• Describe the origin, insertion, action, and innervation of the muscles that act on the femur, tibia, and fibula.

Deep fascia separate the muscles that act on the femur (thigh bone) and tibia and fibula (leg bones) into medial, anterior, and posterior compartments. The muscles of the **medial (adductor) compartment of the thigh** adduct the femur at the hip joint. (See the adductor magnus, adductor longus, adductor brevis, and pectineus, which are components of the medial compartment, in Exhibit 11.17.) The **gracilis,** the other muscle in the medial compartment, not only adducts the thigh, but also flexes the leg at the knee joint. For this reason, it is discussed here. The gracilis is a long, straplike muscle on the medial aspect of the thigh and knee.

The muscles of the **anterior (extensor) compartment of the thigh** extend the leg (and flex the thigh). This compartment contains the quadriceps femoris and sartorius muscles. The **quadriceps femoris** muscle is the biggest muscle in the body, covering most of the anterior surface and sides of the thigh. The muscle is actually a composite muscle, usually described as four separate muscles: (1) **rectus femoris,** on the anterior aspect of the thigh; (2) **vastus lateralis,** on the lateral aspect of the thigh; (3) **vastus medialis,** on the medial aspect of the thigh; and (4) **vastus intermedius,** located deep to the rectus femoris between the vastus lateralis and vastus medialis. The common tendon for the four muscles is known as the **quadriceps tendon,** which inserts into the patella. The tendon continues below the patella as the **patellar ligament,** which attaches to the tibial tuberosity. The quadriceps femoris muscle is the great extensor muscle of the leg. The **sartorius** is a long, narrow muscle that forms a band across the thigh from the ilium of the hip bone to the medial side of the tibia. The various movements it produces help effect the cross-legged sitting position in which the heel of one limb is placed on the knee of the opposite limb. It is known as the tailor's muscle because tailors often assume this cross-legged sitting position. (Because the major action of the sartorius muscle is to move the thigh rather than the leg, it could have been included in Exhibit 11.17.)

The muscles of the **posterior (flexor) compartment of the thigh** flex the leg (and extend the thigh). This compartment is composed of three muscles collectively called the **hamstrings:** (1) **biceps femoris,** (2) **semitendinosus,** and (3) **semimembranosus.** The hamstrings are so-named because their tendons are long and stringlike in the popliteal area and from an old practice of butchers in which they hung hams for smoking by these long tendons. Because the hamstrings span two joints (hip and knee), they are both extensors of the thigh and flexors of the

leg. The **popliteal fossa** is a diamond-shaped space on the posterior aspect of the knee bordered laterally by the tendons of the biceps femoris muscle and medially by the tendons of the semitendinosus and semimembranosus muscles.

Innervation

The obturator and femoral nerves, derived from the lumbar plexus (shown in Exhibit 13.3 on page 442), and the sciatic nerve, derived from the sacral plexus (shown in Exhibit 13.4 on page 444), innervate muscles that act on the femur, tibia, and fibula.

• Gracilis: obturator nerve
• Quadriceps femoris and sartorius: femoral nerve
• Biceps femoris: tibial and common peroneal nerves from the sciatic nerve
• Semimembranosus and semitendinosus: tibial nerve from the sciatic nerve

 Pulled Hamstrings

A strain or partial tear of the proximal hamstring muscles is referred to as **pulled hamstrings** or **hamstring strains.** They are common sports injuries in individuals who run very hard and/or are required to perform quick starts and stops. Sometimes the violent muscular exertion required to perform a feat tears off part of the tendinous origins of the hamstrings, especially the biceps femoris, from the ischial tuberosity. This is usually accompanied by a contusion (bruising), tearing of some of the muscle fibers, and rupture of blood vessels, producing a hematoma (collection of blood) and pain. Adequate training with good balance between the quadriceps femoris and hamstrings and stretching exercises before running or competing are important in preventing this injury. ■

Relating Muscles to Movements

Arrange the muscles in this exhibit according to the following actions on the thigh at the hip joint: (1) abduction, (2) adduction, (3) lateral rotation, (4) flexion, and (5) extension; and according to the following actions on the leg at the knee joint: (1) flexion and (2) extension. The same muscle may be mentioned more than once.

▶ CHECKPOINT

Which muscles are part of the medial, anterior, and posterior compartments of the thigh?

(continues)

Exhibit 11.18 | **Muscles that Act on the Femur (Thigh Bone) and Tibia and Fibula (Leg Bones) (Figures 11.20 and 11.21)**

Muscle	Origin	Insertion	Action
Medial (adductor) compartment of the thigh			
Adductor magnus (MAG-nus)			
Adductor longus (LONG-us)	See Exhibit 11.17.		
Adductor brevis (BREV-is)			
Pectineus (pek-TIN-ē-us)			
Gracilis (gra-SIL-is; *gracilis* = slender)	Body and inferior ramus of pubis.	Medial surface of body of tibia.	Adducts thigh at hip joint, medially rotates thigh, and flexes leg at knee joint.
Anterior (extensor) compartment of the thigh			
Quadriceps femoris (KWOD-ri-ceps FEM-or-is; *quadriceps* = four heads [of origin]; *femoris* = femur)			
Rectus femoris (REK-tus FEM-or-is; *rectus* = fascicles parallel to midline)	Anterior inferior iliac spine.	Patella via quadriceps tendon and then tibial tuberosity via patellar ligament.	All four heads extend leg at knee joint; rectus femoris muscle acting alone also flexes thigh at hip joint.
Vastus lateralis (VAS-tus lat'-e-RĀ-lis; *vast-* = huge; *lateralis* = lateral)	Greater trochanter and linea aspera of femur.		
Vastus medialis (VAS-tus mē'-dē-A-lis; *medialis* = medial)	Linea aspera of femur.		
Vastus intermedius (VAS-tus in'-ter-MĒ-dē-us; *intermedius* = middle)	Anterior and lateral surfaces of body of femur.		
Sartorius (sar-TOR-ē-us; *sartor* = tailor; longest muscle in body)	Anterior superior iliac spine.	Medial surface of body of tibia.	Flexes leg at knee joint; flexes, abducts, and laterally rotates thigh at hip joint.
Posterior (flexor) compartment of the thigh			
Hamstrings A collective designation for three separate muscles.			
Biceps femoris (BĪ-ceps FEM-or-is; *biceps* = two heads of origin)	Long head arises from ischial tuberosity; short head arises from linea aspera of femur.	Head of fibula and lateral condyle of tibia.	Flexes leg at knee joint and extends thigh at hip joint.
Semitendinosus (sem'-ē-ten-di-NŌ-sus; *semi-* = half; *tendo* = tendon)	Ischial tuberosity.	Proximal part of medial surface of shaft of tibia.	Flexes leg at knee joint and extends thigh at hip joint.
Semimembranosus (sem'-ē-mem-bra-NŌ-sus; *membran-* = membrane)	Ischial tuberosity.	Medial condyle of tibia.	Flexes leg at knee joint and extends thigh at hip joint.

(continued)

Figure 11.21 Muscles that act on the femur (thigh bone) and tibia and fibula (leg bones).

 Muscles that act on the leg originate in the hip and thigh; deep fascia separates them into compartments.

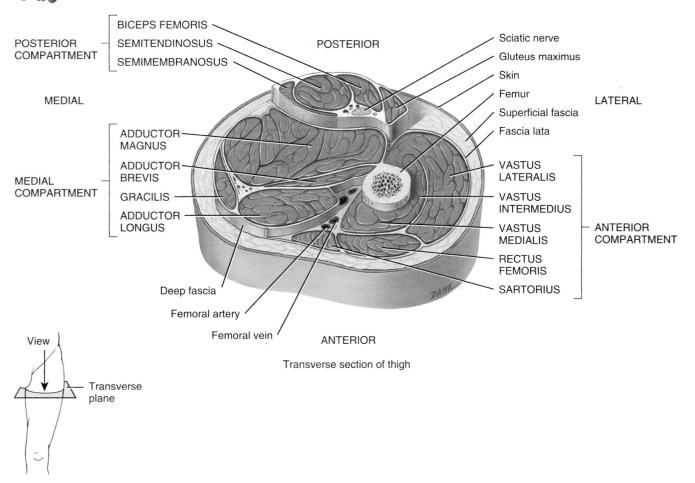

Transverse section of thigh

Which muscles comprise the quadriceps femoris and hamstring muscles?

Exhibit 11.19 Muscles that Move the Foot and Toes (Figure 11.22)

▶ OBJECTIVE

• Describe the origin, insertion, action, and innervation of the muscles that move the foot and toes.

Muscles that move the foot and toes are located in the leg. The muscles of the leg, like those of the thigh, are divided by deep fascia into three compartments: anterior, lateral, and posterior. The **anterior compartment of the leg** consists of muscles that dorsiflex the foot. In a situation analogous to the wrist, the tendons of the muscles of the anterior compartment are held firmly to the ankle by thickenings of deep fascia called the **superior extensor retinaculum** *(transverse ligament of the ankle)* and **inferior extensor retinaculum** *(cruciate ligament of the ankle).*

Within the anterior compartment, the **tibialis anterior** is a long, thick muscle against the lateral surface of the tibia, where it is easy to palpate. The **extensor hallucis longus** is a thin muscle between and partly deep to the tibialis anterior and **extensor digitorum longus** muscles. This featherlike muscle is lateral to the tibialis anterior muscle, where it can easily be palpated. The **fibularis (peroneus) tertius** muscle is part of the extensor digitorum longus, with which it shares a common origin.

The **lateral (fibular) compartment of the leg** contains two muscles that plantar flex and evert the foot: the **fibularis (peroneus) longus** and **fibularis (peroneus) brevis.**

The **posterior compartment of the leg** consists of muscles in superficial and deep groups. The superficial muscles share a common tendon of insertion, the **calcaneal (Achilles) tendon,** the strongest tendon of the body. It inserts into the calcaneal bone of the ankle. The superficial and most of the deep muscles plantar flex the foot at the ankle joint. The superficial muscles of the posterior compartment are the gastrocnemius, soleus, and plantaris—the so-called calf muscles. The large size of these muscles is directly related to the characteristic upright stance of humans. The **gastrocnemius** is the most superficial muscle and forms the prominence of the calf. The **soleus,** which lies deep to the gastrocnemius, is broad and flat. It derives its name from its resemblance to a flat fish (sole). The **plantaris** is a small muscle that may be absent; sometimes there are two of them in each leg. It runs obliquely between the gastrocnemius and soleus muscles.

The deep muscles of the posterior compartment are the popliteus, tibialis posterior, flexor digitorum longus, and flexor hallucis longus. The **popliteus** is a triangular muscle that forms the floor of the popliteal fossa. The **tibialis posterior** is the deepest muscle in the posterior compartment. It lies between the flexor digitorum longus and flexor hallucis longus muscles. The **flexor digitorum longus** is smaller than the **flexor hallucis longus,** even though the former flexes four toes, whereas the latter flexes only the great toe at the interphalangeal joint.

Innervation

Nerves derived from the sacral plexuses (shown in Exhibit 13.4 on page 444) innervate muscles that move the foot and toes.

• Muscles of the anterior compartment: deep fibular (perineal) nerve
• Muscles of the lateral compartment: superficial fibular (perineal) nerve
• Muscles of the posterior compartment: tibial nerve

 Shinsplint Syndrome

Shinsplint syndrome, or simply **shinsplints,** refers to pain or soreness along the tibia, specifically the medial, distal two-thirds. It may be caused by tendinitis of the anterior compartment muscles, especially the tibialis anterior muscle, inflammation of the periosteum (periostitis) around the tibia, or stress fractures of the tibia. The tendinitis usually occurs when poorly conditioned runners run on hard or banked surfaces with poorly supportive running shoes. The condition may also occur with vigorous activity of the legs following a period of relative inactivity. The muscles in the anterior compartment (mainly the tibialis anterior) can be strengthened to balance the stronger posterior compartment muscles. ■

Relating Muscles to Movements

Arrange the muscles in this exhibit according to the following actions on the foot at the ankle joint: (1) dorsiflexion and (2) plantar flexion; according to the following actions on the foot at the intertarsal joints: (1) inversion and (2) eversion; and according to the following actions on the toes at the metatarsophalangeal and intertarsal joints: (1) flexion and (2) extension. The same muscle may be mentioned more than once.

▶ CHECKPOINT

What are the superior extensor retinaculum and inferior extensor retinaculum?

Muscle	Origin	Insertion	Action
Anterior Compartment of the Leg			
Tibialis anterior (tib′-ē-Ā-lis = tibia; *anterior* = front)	Lateral condyle and body of tibia and interosseous membrane (sheet of fibrous tissue that holds shafts of tibia and fibula together).	First metatarsal and first (medial) cuneiform.	Dorsiflexes foot at ankle joint and inverts foot at intertarsal joints.
Extensor hallucis longus (HAL-ū-sis LON-gus; *extensor* = increases angle at joint; *halluc-* = hallux or great toe; *longus* = long)	Anterior surface of fibula and interosseous membrane.	Distal phalanx of great toe.	Dorsiflexes foot at ankle joint and extends proximal phalanx of great toe at metatarsophalangeal joint.
Extensor digitorum longus (di′-ji-TOR-um LON-gus)	Lateral condyle of tibia, anterior surface of fibula, and interosseous membrane.	Middle and distal phalanges of toes 2–5.*	Dorsiflexes foot at ankle joint and extends distal and middle phalanges of each toe at interphalangeal joints and proximal phalanx of each toe at metatarsophalangeal joint.
Fibularis (Peroneus) tertius (fib-ū-LAR-is TER-shus; *peron-* = fibula; *tertius* = third)	Distal third of fibula and interosseous membrane.	Base of fifth metatarsal.	Dorsiflexes foot at ankle joint and everts foot at intertarsal joints.
Lateral (Fibular) Compartment of the Leg			
Fibularis (Peroneus) longus (fib-ū-LAR-is LON-gus)	Head and body of fibula and lateral condyle of tibia.	First metatarsal and first cuneiform.	Plantar flexes foot at ankle joint and everts foot at intertarsal joints.
Fibularis (Peroneus) brevis (fib-ū-LAR-is BREV-is; *brevis* = short)	Body of fibula.	Base of fifth metatarsal.	Plantar flexes foot at ankle joint and everts foot at intertarsal joints.
Superficial Posterior Compartment of the Leg			
Gastrocnemius (gas′-trok-NĒ-mē-us; *gastro-* = belly; *cnem-* = leg)	Lateral and medial condyles of femur and capsule of knee.	Calcaneus by way of calcaneal (Achilles) tendon.	Plantar flexes foot at ankle joint and flexes leg at knee joint.
Soleus (SŌ-lē-us; *sole* = a type of flat fish)	Head of fibula and medial border of tibia.	Calcaneus by way of calcaneal (Achilles) tendon.	Plantar flexes foot at ankle joint.
Plantaris (plan-TĀR-is; *plantar-* = sole of foot)	Femur superior to lateral condyle.	Calcaneus by way of calcaneal (Achilles) tendon.	Plantar flexes foot at ankle joint and flexes leg at knee joint.
Deep Posterior Compartment of the Leg			
Popliteus (pop-LIT-ē-us; *poplit-* = the back of the knee)	Lateral condyle of femur.	Proximal tibia.	Flexes leg at knee joint and medially rotates tibia to unlock the extended knee.
Tibialis posterior (tib′-ē-Ā-lis; *posterior* = back)	Tibia, fibula, and interosseous membrane.	Second, third, and fourth metatarsals; navicular; all three cuneiforms; and cuboid.	Plantar flexes foot at ankle joint and inverts foot at intertarsal joints.
Flexor digitorum longus (di′-ji-TOR-um LON-gus; *digit* = finger or toe)	Posterior surface of tibia.	Distal phalanges of toes 2–5.	Plantar flexes foot at ankle joint; flexes distal and middle phalanges of each toe at interphalangeal joints and proximal phalanx of each toe at metatarsophalangeal joint.
Flexor hallucis longus (HAL-ū-sis LON-gus; *flexor* = decreases angle at joint)	Inferior two-thirds of fibula.	Distal phalanx of great toe.	Plantar flexes foot at ankle joint; flexes distal phalanx of great toe at interphalangeal joint and proximal phalanx of great toe at metatarsophalangeal joint.

*Reminder: The great toe or hallux is the first toe and has two phalanges: proximal and distal. The remaining toes are numbered 2–5, and each has three phalanges: proximal, middle, and distal.

(continues)

Exhibit 11.19   Muscles that Move the Foot and Toes (Figure 11.22) *(continued)*

Figure 11.22   Muscles that move the foot and toes.   (See Tortora, *A Photographic Atlas of the Human Body,* Figures 5.18 and 5.19.)

The superficial muscles of the posterior compartment share a common tendon of insertion, the calcaneal (Achilles) tendon, which inserts into the calcaneal bone of the ankle.

Quadriceps femoris
Tendon of quadriceps femoris
Iliotibial tract
Biceps femoris
Patella
PLANTARIS
Head of fibula
Patellar ligament
Tibia
TIBIALIS ANTERIOR
GASTROCNEMIUS
FIBULARIS LONGUS
SOLEUS
EXTENSOR DIGITORUM LONGUS
FLEXOR DIGITORUM LONGUS
FIBULARIS BREVIS
FIBULARIS TERTIUS
EXTENSOR HALLUCIS LONGUS
Calcaneal (Achilles) tendon
Fibula
EXTENSOR HALLUCIS BREVIS
EXTENSOR DIGITORUM BREVIS
Metatarsals

Superior extensor retinaculum
Inferior extensor retinaculum

DANK

(a) Anterior superficial view

(b) Right lateral superficial view

Gracilis

Sartorius

Biceps femoris

Semitendinosus

Semimembranosus

Femur

Popliteal fossa

PLANTARIS

GASTROCNEMIUS (cut)

Tendon of biceps femoris (cut)

Tibia

POPLITEUS

GASTROCNEMIUS

SOLEUS (cut)

Fibula

TIBIALIS POSTERIOR

SOLEUS

FIBULARIS LONGUS

FLEXOR DIGITORUM LONGUS

FLEXOR HALLUCIS LONGUS

FIBULARIS BREVIS

Tibia

Tendon of tibialis posterior

Fibula

Calcaneal (Achilles) tendon (cut)

(c) Posterior superficial view

(d) Posterior deep view

 What structures firmly hold the tendons of the anterior compartment muscles to the ankle?

Exhibit 11.20   Intrinsic Muscles of the Foot (Figure 11.23)

▶ **OBJECTIVE**

• Describe the origin, insertion, action, and innervation of the intrinsic muscles of the foot.

The muscles in this exhibit are termed **intrinsic muscles** because they originate and insert *within* the foot. The intrinsic muscles of the foot are, for the most part, comparable to those in the hand. Whereas the muscles of the hand are specialized for precise and intricate movements, those of the foot are limited to support and locomotion. The deep fascia of the foot forms the **plantar aponeurosis (fascia)** that extends from the calcaneus bone to the phalanges of the toes. The aponeurosis supports the longitudinal arch of the foot and encloses the flexor tendons of the foot.

The intrinsic muscles of the foot are divided into two groups: **dorsal** and **plantar.** There is only one dorsal muscle, the **extensor digitorum brevis,** a four-part muscle deep to the tendons of the extensor digitorum longus muscle, which extends toes 2−5 at the metatarsophalangeal joints.

The plantar muscles are arranged in four layers, the most superficial layer being termed the first layer. Three muscles are in the first layer. The **abductor hallucis,** which lies along the medial border of the sole, is comparable to the abductor pollicis brevis in the hand, and abducts the great toe at the metatarsophalangeal joint. The **flexor digitorum brevis,** which lies in the middle of the sole, flexes toes 2−5 at the interphalangeal and metatarsophalangeal joints. The **abductor digiti minimi,** which lies along the lateral border of the sole, is comparable to the same muscle in the hand, and abducts the little toe.

The second layer consists of the **quadratus plantae,** a rectangular muscle that arises by two heads and flexes toes 2−5 at the metatarsophalangeal joints, and the **lumbricals,** four small muscles that are similar to the lumbricals in the hands. They flex the proximal phalanges and extend the distal phalanges of toes 2−5.

Three muscles comprise the third layer. The **flexor hallucis brevis,** which lies adjacent to the plantar surface of the metatarsal of the great toe, is comparable to the same muscle in the hand, and flexes the great toe. The **adductor hallucis,** which has an oblique and transverse head like the adductor pollicis in the hand, adducts the great toe. The **flexor digiti minimi brevis,** which lies superficial to the metatarsal of the little toe, is comparable to the same muscle in the hand, and flexes the little toe.

The fourth layer is the deepest and consists of two muscle groups. The **dorsal interossei** are four muscles that adduct toes 2−4, flex the

proximal phalanges, and extend the distal phalanges. The three **plantar interossei** adduct toes 3−5, flex the proximal phalanges, and extend the distal phalanges. The interossei of the feet are similar to those of the hand. However, their actions are relative to the midline of the second digit rather than the third digit as in the hand.

Innervation

Nerves derived from the sacral plexus (shown in Exhibit 13.4 on page 444) innervate intrinsic muscles of the foot.

• Extensor digitorum brevis: deep fibular (perineal) nerve
• Abductor hallucis and flexor digitorum brevis: medial plantar nerve
• Abductor digiti minimi, quadratus plantae, adductor hallucis, flexor digiti minimi brevis, and dorsal and plantar interossei: lateral plantar nerve
• Lumbricals: medial and lateral plantar nerves

 Plantar Fasciitis

Plantar fasciitis (fas-ē-ĪE-tis) or **painful heel syndrome** is an inflammatory reaction due to chronic irritation of the plantar aponeurosis (fascia) at its origin on the calcaneus (heel bone). The aponeurosis becomes less elastic with age. This condition is also related to weight-bearing activities (walking, jogging, lifting heavy objects), improperly constructed or fitting shoes, excess weight (puts pressure on the feet), and poor biomechanics (flat feet, high arches, and abnormalities in gait may cause uneven distribution of weight on the feet). Plantar fasciitis is the most common cause of heel pain in runners and arises in response to the repeated impact of running. ■

Relating Muscles to Movements

Arrange the muscles in this exhibit according to the following actions on the great toe at the metatarsophalangeal joint: (1) flexion, (2) extension, (3) abduction, and (4) adduction; and according to the following actions on toes 2−5 at the metatarsophalangeal and interphalangeal joints: (1) flexion, (2) extension, (3) abduction, and (4) adduction. The same muscle may be mentioned more than once.

▶ **CHECKPOINT**

How do the intrinsic muscles of the hand and foot differ in function?

Muscle	Origin	Insertion	Action
Dorsal			
Extensor digitorum brevis (*extensor* = increases angle at joint; *digit* = finger or toe; *brevis* = short) (see Figure 11.22a, b)	Calcaneus and inferior extensor retinaculum.	Tendons of extensor digitorum longus on toes 2–4 and proximal phalanx of great toe.*	Extensor hallucis brevis extends great toe at metatarsophalangeal joint and extensor digitorum brevis extends toes 2–4 at interphalangeal joints.
Plantar			
First Layer (most superficial)			
Abductor hallucis (*abductor* = moves part away from midline; *hallucis* = hallux or great toe)	Calcaneus, plantar aponeurosis, and flexor retinaculum.	Medial side of proximal phalanx of great toe with the tendon of the flexor hallucis brevis.	Abducts and flexes great toe at metatarsophalangeal joint.
Flexor digitorum brevis (*flexor* = decreases angle at joint)	Calcaneus and plantar aponeurosis.	Sides of middle phalanx of toes 2–5.	Flexes toes 2–5 at proximal interphalangeal and metatarsophalangeal joints.
Abductor digiti minimi (*minimi* = little)	Calcaneus and plantar aponeurosis.	Lateral side of proximal phalanx of little toe with the tendon of the flexor digiti minimi brevis.	Abducts and flexes little toe at metatarsophalangeal joint.
Second layer			
Quadratus plantae (kwod-RĀ-tus; *quad-* = square, four-sided; *planta* = the sole of the foot)	Calcaneus.	Tendon of flexor digitorum longus.	Assists flexor digitorum longus to flex toes 2–5 at interphalangeal and metatarsophalangeal joints.
Lumbricals (LUM-bri-kals; *lumbric-* = earthworm)	Tendons of flexor digitorum longus.	Tendons of extensor digitorum longus on proximal phalanges of toes 2–5.	Extend toes 2–5 at interphalangeal joints and flex toes 2–5 at metatarsophalangeal joints.
Third Layer			
Flexor hallucis brevis	Cuboid and third (lateral) cuneiform.	Medial and lateral sides of proximal phalanx of great toe via a tendon containing a sesamoid bone.	Flexes great toe at metatarsophalangeal joint.
Adductor hallucis	Metatarsals 2–4, ligaments of 3–5 metatarsophalangeal joints, and tendon of peroneus longus.	Lateral side of proximal phalanx of great toe.	Adducts and flexes great toe at metatarsophalangeal joint.
Flexor digiti minimi brevis	Metatarsal 5 and tendon of peroneus longus.	Lateral side of proximal phalanx of little toe.	Flexes little toe at metatarsophalangeal joint.
Fourth Layer (deepest)			
Dorsal interossei (in-ter-OS-ē-ī) (not illustrated)	Adjacent side of metatarsals.	Proximal phalanges: both sides of toe 2 and lateral side of toes 3 and 4.	Abduct and flex toes 2–4 at metatarsophalangeal joints and extend toes at interphalangeal joints.
Plantar interossei	Metatarsals 3–5.	Medial side of proximal phalanges of toes 3–5.	Adduct and flex proximal metatarsophalangeal joints and extend toes at interphalangeal joints.

* The tendon that inserts into the proximal phalanx of the great toe, together with its belly, is often described as a separate muscle, the extensor hallucis brevis.

Exhibit 11.20   Intrinsic Muscles of the Foot (Figure 11.23) (continued)

Figure 11.23   Intrinsic muscles of the foot.

Whereas the intrinsic muscles of the hand are specialized for precise and intricate movements, those of the foot are limited to support and movement.

(a) Plantar superficial and deep view

(b) Plantar deep view

 What structure supports the longitudinal arch and encloses the flexor tendons of the foot?

Body System	Contribution of the Muscular System
For all body systems	The muscular system and muscle tissues produce body movements, stabilize body positions, move substances within the body, and produce heat that helps maintain normal body temperature.
Integumentary system	Pull of skeletal muscles on attachments to skin of face cause facial expressions; muscular exercise increases skin blood flow.
Skeletal system	Skeletal muscle causes movement of body parts by pulling on attachments to bones; skeletal muscle provides stability for bones and joints.
Nervous system	Smooth, cardiac, and skeletal muscles carry out commands for the nervous system; shivering—involuntary contraction of skeletal muscles that is regulated by the brain—generates heat to raise body temperature.
Endocrine system	Regular activity of skeletal muscles (exercise) improves action of some hormones, such as insulin; muscles protect some endocrine glands.
Cardiovascular system	Cardiac muscle powers pumping action of heart; contraction and relaxation of smooth muscle in blood vessel walls help adjust the amount of blood flowing through various body tissues; contraction of skeletal muscles in the legs assists return of blood to the heart; regular exercise causes cardiac hypertrophy (enlargement) and increases heart's pumping efficiency; lactic acid produced by active skeletal muscles may be used for ATP production by the heart.
Lymphatic and immune system	Skeletal muscles protect some lymph nodes and lymphatic vessels and promote flow of lymph inside lymphatic vessels; exercise may increase or decrease some immune responses.
Respiratory system	Skeletal muscles involved with breathing cause air to flow into and out of the lungs; smooth muscle fibers adjust size of airways; vibrations in skeletal muscles of larynx control air flowing past vocal cords, regulating voice production; coughing and sneezing, due to skeletal muscle contractions, help clear airways; regular exercise improves efficiency of breathing.
Digestive system	Skeletal muscles protect and support organs in the abdominal cavity; alternating contraction and relaxation of skeletal muscles power chewing and initiate swallowing; smooth muscle sphincters control volume of organs of the gastrointestinal (GI) tract; smooth muscles in walls of GI tract mix and move its contents through the tract.
Urinary system	Skeletal and smooth muscle sphincters and smooth muscle in wall of urinary bladder control whether urine is stored in the bladder or voided (urination).
Reproductive systems	Skeletal and smooth muscle contractions eject semen from male; smooth muscle contractions propel oocyte along uterine tube, help regulate flow of menstrual blood from uterus, and force baby from uterus during childbirth; during intercourse skeletal muscle contractions are associated with orgasm and pleasurable sensations in both sexes.

379

DISORDERS: HOMEOSTATIC IMBALANCES

Running Injuries

Nearly 70% of those who jog or run sustain some type of running-related injury. Even though most such injuries are minor, some are quite serious. Moreover, untreated or inappropriately treated minor injuries may become chronic. Among runners, common sites of injury include the ankle, knee, calcaneal (Achilles) tendon, hip, groin, foot, and back. Of these, the knee often is the most severely injured area.

Running injuries are frequently related to faulty training techniques. This may involve improper (or lack of) warm-up routines, running too much, or running too soon after an injury. Or it might involve extended running on hard and/or uneven surfaces. Poorly constructed or worn-out running shoes can also contribute to injury, as can any biomechanical problem (such as a fallen arch) aggravated by running.

Most sports injuries should be treated initially with RICE therapy, which stands for Rest, Ice, Compression, and Elevation. Immediately apply ice, and rest and elevate the injured part. Then apply an elastic bandage, if possible, to compress the injured tissue. Continue using RICE for 2 to 3 days, and resist the temptation to apply heat, which may worsen the swelling. Follow-up treatment may include alternating moist heat and ice massage to enhance blood flow in the injured area. Sometimes it is helpful to take nonsteroidal anti-inflammatory drugs (NSAIDs) or to have local injections of corticosteroids. During the recovery period, it is important to keep active using an alternative fitness program that does not worsen the original injury. This activity should be determined in consultation with a physician. Finally, careful exercise is needed to rehabilitate the injured area itself.

Compartment Syndrome

In several exhibits in this chapter, we noted that skeletal muscles in the limbs are organized into units called *compartments*. In a disorder called **compartment syndrome,** some external or internal pressure constricts the structures within a compartment, resulting in damaged blood vessels and subsequent reduction of the blood supply (ischemia) to the structures within the compartment. Common causes of compartment syndrome include crushing and penetrating injuries, contusion (damage to subcutaneous tissues without the skin being broken), muscle strain (overstretching of a muscle), or an improperly fitted cast. As a result of hemorrhage, tissue injury, and edema (buildup of interstitial fluid), the pressure increase in the compartment can have serious consequences. Because the fascia that enclose the compartments are very strong, accumulated blood and interstitial fluid cannot escape, and the increased pressure can literally choke off the blood flow and deprive nearby muscles and nerves of oxygen. One treatment option is **fasciotomy** (fash-ē-OT-ō-mē), a surgical procedure in which muscle fascia is cut to relieve the pressure. Without intervention, nerves suffer damage, and muscles develop scar tissue that results in permanent shortening of the muscles, a condition called *contracture.*

STUDY OUTLINE

HOW SKELETAL MUSCLES PRODUCE MOVEMENT (p. 309)

1. Skeletal muscles that produce movement do so by pulling on bones.
2. The attachment to the more stationary bone is the origin; the attachment to the more movable bone is the insertion.
3. Bones serve as levers, and joints serve as fulcrums. Two different forces act on the lever: load (resistance) and effort.
4. Levers are categorized into three types—first-class, second-class, and third-class (most common)—according to the positions of the fulcrum, the effort, and the load on the lever.
5. Fascicular arrangements include parallel, fusiform, circular, triangular, and pennate. Fascicular arrangement affects a muscle's power and range of motion.
6. A prime mover produces the desired action; an antagonist produces an opposite action. Synergists assist a prime mover by reducing unnecessary movement. Fixators stabilize the origin of a prime mover so that it can act more efficiently.

HOW SKELETAL MUSCLES ARE NAMED (p. 313)

1. Distinctive features of different skeletal muscles include direction of muscle fascicles; size, shape, action, number of origins (or heads), and location of the muscle; and sites of origin and insertion of the muscle.
2. Most skeletal muscles are named based on combinations of these features.

PRINCIPAL SKELETAL MUSCLES (p. 313)

1. Muscles of facial expression move the skin rather than a joint when they contract, and they permit us to express a wide variety of emotions.
2. The extrinsic muscles that move the eyeballs are among the fastest contracting and most precisely controlled skeletal muscles in the body. They permit us to elevate, depress, abduct, adduct, and medially and laterally rotate the eyeballs.
3. Muscles that move the mandible (lower jaw) are also known as the muscles of mastication because they are involved in chewing.
4. The extrinsic muscles that move the tongue are important in chewing, swallowing, and speech.
5. Suprahyoid muscles of the anterior neck are located above the hyoid bone. They elevate the hyoid bone, oral cavity, and tongue during swallowing. Infrahyoid muscles of the anterior neck depress the hyoid bone and some move the larynx during swallowing and speech.

6. Muscles that move the head alter the position of the head and help balance the head on the vertebral column.
7. Muscles that act on the abdominal wall help contain and protect the abdominal viscera, move the vertebral column, compress the abdomen, and produce the force required for defecation, urination, vomiting, and childbirth.
8. Muscles used in breathing alter the size of the thoracic cavity so that inhalation and exhalation can occur.
9. Muscles of the pelvic floor support the pelvic viscera, resist the thrust that accompanies increases in intra-abdominal pressure, and function as sphincters at the anus, urethra, and vagina.
10. Muscles of the perineum assist in urination, erection of the penis and clitoris, ejaculation, and defecation.
11. Muscles that move the pectoral (shoulder) girdle stabilize the scapula so it can function as a stable point of origin for most of the muscles that move the humerus.
12. Muscles that move the humerus (arm bone) originate for the most part on the scapula (scapular muscles); the remaining muscles originate on the axial skeleton (axial muscles).
13. Muscles that move the radius and ulna (forearm bones) are involved in flexion and extension at the elbow joint and are organized into flexor and extensor compartments.
14. Muscles that move the wrist, hand, thumb, and fingers are many and varied. Those muscles that act on the thumb and fingers are called extrinsic muscles.
15. The intrinsic muscles of the hand are important in skilled activities and provide humans with the ability to grasp precisely and manipulate objects.
16. Muscles that move the vertebral column are quite complex because they have multiple origins and insertions and because there is considerable overlap among them.
17. Muscles that move the femur (thigh bone) originate for the most part on the pelvic girdle and insert on the femur. These muscles are larger and more powerful than comparable muscles in the upper limb.
18. Muscles that move the femur (thigh bone) and tibia and fibula (leg bones) are separated into medial (adductor), anterior (extensor), and posterior (flexor) compartments.
19. Muscles that move the foot and toes are divided into anterior, lateral, and posterior compartments.
20. Intrinsic muscles of the foot, unlike those of the hand, normally are limited to the functions of support and locomotion.

SELF-QUIZ QUESTIONS

Fill in the blanks in the following statements.

1. The less movable end of the muscle is the _____; the more movable end is the _____.

2. The muscle that forms the major portion of the cheek is the _____.

Indicate whether the following statements are true or false.

3. Because of their insertions, the muscles of facial expression move the skin rather than a joint when they contract.

4. Opposition is the single most distinctive digital movement that gives humans and other primates the ability to precisely grasp and manipulate objects.

Choose the one best answer to the following questions.

5. Which of the following statements is *false*? (a) Fixators steady the proximal end of a limb whereas movements occur at the distal end. (b) Synergists prevent unwanted movements in intermediate joints. (c) The prime mover, or agonist, causes a desired action. (d) The antagonist supports and aids the agonist's movement. (e) Fascicular arrangement affects a muscle's power and range of motion.

6. Which of the following are important characteristics used in the naming of muscles? (1) muscle shape, (2) muscle size, (3) direction of muscle of fibers, (4) muscle action, (5) muscle location, (6) sites of origin and insertion, (7) number of insertions. (a) 1, 2, 3, 4, 5, and 6, (b) 2, 3, 4, 5, 6, and 7, (c) 2, 4, 6, and 7, (d) 1, 3, 5, and 7, (e) all are correct.

7. The facial nerve (cranial nerve VII) innervates (a) most muscles of facial expression, (b) the muscles that move the head, (c) extrinsic muscles that move the tongue, (d) muscles of mastication, (e) muscles that move the mandible.

8. The muscle that closes the mouth is the (a) risorius, (b) levator labii superioris, (c) platysma, (d) mentalis, (e) masseter.

9. Which of the following are muscle-generated actions on the mandible? (1) elevation, (2) depression, (3) retraction, (4) protraction, (5) abduction, (6) adduction, (7) side-to-side movement. (a) 1, 2, 3, and 5, (b) 2, 3, 4, and 6, (c) 1, 2, 3, 4, and 7, (d) 3, 4, 5, and 7, (e) 1, 3, 5, and 7.

10. Which of the following are functions of the anterolateral abdominal muscle group? (1) helps contain and protect the abdominal viscera, (2) helps contain and protect the lower area of the heart and lungs, (3) flexes, laterally flexes, and rotates the vertebral column, (4) expands the abdomen during forced expiration, (5) produces the force required for defecation, urination, and childbirth. (a) 1, 3, and 5, (b) 2, 4, and 5, (c) 1, 2, and 4, (d) 1 and 5, (e) 1, 2, and 5.

11. In order for movement to occur (a) both the prime mover and antagonist must contract simultaneously, (b) muscles that move a body part cannot lie over the moving part, (c) a muscle must cross at least two joints, (d) the muscle fascicles must have a circular arrangement, (e) a muscle must be innervated by a spinal nerve.

12. Match the following:

_____ (a) compression of median nerve resulting in pain and numbness and tingling in the fingers

_____ (b) tendinitis of the anterior compartment muscles of the leg; inflammation of the tibial periosteum

_____ (c) improperly aligned eyeballs due to lesions in either the oculomotor or abducens nerves

_____ (d) stretching or tearing of distal attachments of adductor muscles and/or iliopsoas

_____ (e) rupture of a portion of the inguinal area of the abdominal wall resulting in protrusion of part of the small intestine

_____ (f) caused by repetitive movement of the arm over the head which results in inflammation of the supraspinatus tendon

_____ (g) inflammation due to chronic irritation of the plantar aponeurosis at its origin on the calcaneous; most common cause of heel pain in runners

_____ (h) painful inflammation of tendons, tendon sheaths, and the synovial membranes of joints

_____ (i) paralysis of facial muscles as a result of damage to the facial nerve

_____ (j) common condition in individuals who perform quick starts and stops; tearing away of part of the tendinous origins from the ischial tuberosity

_____ (k) may occur as a result of injury to levator ani muscle

_____ (l) external or internal pressure constricts structures in a compartment, causing a reduction of blood supply to the structures

(1) tenosynovitis (tendinitis)
(2) Bell's palsy
(3) inguinal hernia
(4) urinary stress incontinence
(5) compartment syndrome
(6) pulled groin
(7) pulled hamstrings
(8) strabismus
(9) shinsplints
(10) plantar fasciitis
(11) impingement syndrome
(12) carpal tunnel syndrome

13. Match the following:

_____ (a) most common lever in the body

_____ (b) lever formed by the head resting on the vertebral column

_____ (c) always produces a mechanical advantage

_____ (d) EFL

_____ (e) FLE

_____ (f) FEL

_____ (g) adduction of the thigh

(1) first-class lever

(2) second-class lever

(3) third-class lever

14. Match the following:

_____ (a) rectus femoris, vastus lateralis, vastus medialis, vastus intermedius

_____ (b) biceps femoris, semitendinosus, semimembranosus

_____ (c) erector spinae; includes iliocostalis, longissimus, and spinalis groups

_____ (d) thenar, hypothenar, intermediate

_____ (e) biceps brachii, brachialis, coracobrachialis

_____ (f) latissimus dorsi

_____ (g) subscapularis, supraspinatus, infraspinatus, teres minor

_____ (h) diaphragm, external intercostals, internal intercostals

(1) breathing muscles

(2) constitute flexor compartment of the arm

(3) hamstrings

(4) intrinsic muscle groups of the hand

(5) muscles that strengthen and stabilize the shoulder joint; the rotator cuff

(6) quadriceps femoris muscle

(7) largest muscle mass of the back

(8) swimmer's muscle

15. Match the following (some answers may be used more than once):

_____ (a) trapezius

_____ (b) orbicularis oculi

_____ (c) levator ani

_____ (d) rectus abdominis

_____ (e) triceps brachii

_____ (f) gastrocnemius

_____ (g) temporalis

_____ (h) external anal sphincter

_____ (i) external oblique

_____ (j) splenius capitis

_____ (k) digastric

_____ (l) styloglossus

_____(m) masseter

_____ (n) adductor longus

_____ (o) zygomaticus major

_____ (p) latissimus dorsi

_____ (q) flexor carpi radialis

_____ (r) pronator teres

_____ (s) sternocleidomastoid

_____ (t) quadriceps femoris

_____ (u) deltoid

_____ (v) tibialis anterior

_____(w) sartorius

_____ (x) gluteus maximus

_____ (y) superior rectus

(1) muscle of facial expression

(2) muscle of mastication

(3) muscle that moves the eyeballs

(4) extrinsic muscle that moves the tongue

(5) suprahyoid muscle

(6) muscle of the perineum

(7) muscle that moves the head

(8) abdominal wall muscle

(9) pelvic floor muscle

(10) pectoral girdle muscle

(11) muscle that moves the humerus

(12) muscle that moves the radius and ulna

(13) muscle that moves the wrist, hand, and digits

(14) muscle that moves the vertebral column

(15) muscle that moves the femur

(16) muscle that acts on the femur, tibia, and fibula

(17) muscle that moves the foot and toes

 CRITICAL THINKING QUESTIONS

1. While you proceed through your daily activities, your mucles take turns being the agonist or antagonist—that is, being contracted or relaxed. Name some antagonistic pairs of muscles in the arm, thigh, torso, and leg.

HINT *Antagonistic pairs are often located on opposite sides of a limb or body region.*

2. Melody was trying to lift a heavy package using the proper back-saving position that she had been taught at work. She squatted down with her knees bent and her back straight. Then she lifted the package using her leg mucles and keeping her back straight. Identify the resistance, fulcrum, and effort involved in lifting the package. What type of lever is this?

HINT *Determine the relative order of the effort, the fulcrum, and the resistance.*

3. When Carlos read the first question on his take-home exam, he raised one eyebrow, whistled, then closed his eyes and shook his head. Name the muscles involved in these actions.

HINT *Try it yourself.*

ANSWERS TO FIGURE QUESTIONS

11.1 The belly of the muscle that extends the forearm, the triceps brachii, is located posterior to the humerus.

11.2 In a third-class lever, the effort is applied between the fulcrum and the load.

11.3 Possible correct responses: direction of fibers: external oblique; shape: deltoid; action: extensor digitorum; size: gluteus maximus; origin and insertion: sternocleidomastoid; location: tibialis anterior; number of tendons of origin: biceps brachii.

11.4 Frowning: frontalis; smiling: zygomaticus major; pouting: platysma; squinting: orbicularis oculi.

11.5 The inferior oblique muscle moves the eyeball superiorly and laterally because it originates at the anteromedial aspect of the floor of the orbit and inserts on the posterolateral aspect of the eyeball.

11.6 The masseter is the strongest of the chewing muscles.

11.7 Functions of the tongue include chewing, perception of taste, swallowing, and speech.

11.8 The suprahyoid and infrahyoid muscles fix the hyoid bone to assist in tongue movements.

11.9 Triangles are important because of the structures that lie within their boundaries.

11.10 The rectus abdominis muscle aids in urination.

11.11 The phrenic nerve innervates the diaphragm.

11.12 The borders of the pelvic diaphragm are the pubic symphysis anteriorly, the coccyx posteriorly, and the walls of the pelvis laterally.

11.13 The borders of the perineum are the pubic symphysis anteriorly, the coccyx posteriorly, and the ischial tuberosities laterally.

11.14 The main action of the muscles that move the pectoral girdle is to stabilize the scapula to assist in movements of the humerus.

11.15 The rotator cuff consists of the flat tendons of the subscapularis, supraspinatus, infraspinatus, and teres minor muscles that form a nearly complete circle around the shoulder joint.

11.16 The most powerful forearm flexor is the brachialis. The most powerful forearm extensor is the triceps brachii.

11.17 Flexor tendons of the digits and wrist and the median nerve pass through the flexor retinaculum.

11.18 Muscles of the thenar eminence act on the thumb.

11.19 The splenius muscles extend laterally and upward from the ligamentum nuchae at the midline.

11.20 Upper limb muscles exhibit diversity of movement; lower limb muscles function in stability, locomotion, and maintenance of posture. Moreover, lower limb muscles usually cross two joints and act equally on both.

11.21 The quadriceps femoris consists of the rectus femoris, vastus lateralis, vastus medialis, and vastus intermedius. The hamstrings consist of the biceps femoris, semitendinosus, and semimembranosus.

11.22 The superior and inferior extensor retinacula firmly hold the tendons of the anterior compartment muscles to the ankle.

11.23 The plantar aponeurosis supports the longitudinal arch and encloses the flexor tendons of the foot.

Nervous Tissue

I was a housewife and mother pregnant with our third child. When she was delivered, we found out she had spina bifida—a neuro tube defect. I was educated quickly and basically in the nervous system and how it affects many different areas in the body. As a result, my interest was peeked and I returned to school to study Nursing. I am now one year away from graduation with my Associates Degree, and am still learning about how the Nervous system works. In Anatomy we learned the basics of this science and in Physiology we studied the chemistry of how the nervous system works. I can now interact in knowledgeable discussions with my child's doctors about her condition.

In nursing classes, the nervous system is introduced all over again while studying pain, inflammation, fluid electrolyte imbalances, oxygenation, digestion, urinary function, regulatory mechanisms, movement and coordination and nursing care in specialized settings. Understanding pain is a large part of our studies. So knowing the communication through propagation of nerve impulses and how ions are involved with nerve conduction and fluid and electrolyte balance is a very important part of Nursing. Neurotransmitters are very interesting and several pain medications mimic their action. For example, endorphins and encephalin have analgesic effects like morphine.

What I have learned in Anatomy & Physiology will follow me not only in the rest of my education as I continue on through my Associates, Bachelors, and Masters of Nursing, but also in my career as a nurse and my career as a mom of three beautiful girls — one of whom has a spinal cord defect.

Elizabeth Garrison
Modesto Junior College

NERVOUS TISSUE AND HOMEOSTASIS

The excitable characteristic of nervous tissue allows for the generation of nerve impulses (action potentials) that provide communication and regulation of most body tissues.

ENERGIZE YOUR STUDY

FOUNDATIONS CD

Anatomy Overview
• The Nervous System
• Nervous Tissue

Animations
• Systems Contributions to Homeostasis
• The Case of the Girl with the Fruity Breath

Concepts and Connections and Exercises reinforce your understanding

www.wiley.com/college/apcentral

INSIGHTS AND EXPLORATIONS

A movie and TV actor announce that they are taking time off from their work to deal with a recently diagnosed disease. A political personality announces that having Parkinson disease will not interfere with her ability to be the next governor of a state. You may have seen a famous athlete or movie actress with tremors in their hands or a head that shakes. What is Parkinson disease? How does it affect these individuals? What can be done to treat this disease? Come explore some of the answers to these questions in this web-based activity.

With a mass of only 2 kg (4.5 lb), about 3% of total body weight, the nervous system is one of the smallest and yet the most complex of the 11 body systems. The two main subdivisions of the nervous system are the **central nervous system (CNS),** which consists of the brain and spinal cord, and the **peripheral** (pe-RIF-er-al) **nervous system (PNS),** which includes all nervous tissue outside the CNS.

Together, the nervous system and the endocrine system share responsibility for maintaining homeostasis. Their objective is the same—to keep controlled conditions within limits that maintain life—but their means of achieving that objective differ. Whereas the nervous system regulates body activities by responding rapidly using nerve impulses (action potentials), the endocrine system responds more slowly, though no less effectively, by releasing hormones. The roles of the nervous and endocrine systems in maintaining homeostasis are compared in Chapter 18, on page 587.

Besides helping maintain homeostasis, the nervous system is responsible for our perceptions, behaviors, and memories. It also initiates all voluntary movements. Because the nervous system is quite complex, we will consider different aspects of its structure and function in several related chapters. This chapter focuses on the organization of the nervous system and the properties of the cells that make up nervous tissue—neurons (nerve cells) and neuroglia. In chapters that follow, we will examine the structure and functions of the spinal cord and spinal nerves, and of the brain and cranial nerves. Then we will discuss the somatic senses—touch, pressure, warmth, cold, pain, and others—and the sensory and motor pathways to understand how nerve impulses pass into the spinal cord and brain or from the spinal cord and brain to muscles and glands. Our exploration continues with a discussion of the special senses: smell, taste, vision, hearing, and equilibrium. Finally, we conclude our study of the nervous system by examining the autonomic nervous system, the part of the nervous system that operates without voluntary control.

The branch of medical science that deals with the normal functioning and disorders of the nervous system is **neurology** (noo-ROL-ō-jē; *neuro-* = nerve or nervous system; *-logy* = study of).

OVERVIEW OF THE NERVOUS SYSTEM

▶ O B J E C T I V E S

• **List the structures and basic functions of the nervous system.**
• **Describe the organization of the nervous system.**

Structures of the Nervous System

The **nervous system** is a complex, highly organized network of billions of neurons and even more neuroglia. The structures that make up the nervous system include the brain, cranial nerves and their branches, the spinal cord, spinal nerves and their branches, ganglia, enteric plexuses, and sensory receptors (Figure 12.1). The skull encloses the **brain,** which contains about 100 billion (10^{11}) neurons. Twelve pairs (right and left) of **cranial nerves,** numbered I through XII, emerge from the base of the brain. A **nerve** is a bundle of hundreds to thousands of axons plus associated connective tissue and blood vessels that lies outside the brain and spinal cord. Each nerve follows a defined path and serves a specific region of the body. For example, the right cranial nerve I carries signals for the sense of smell from the right side of the nose to the brain.

The **spinal cord** connects to the brain through the foramen magnum of the skull and is encircled by the bones of the vertebral column. It contains about 100 million neurons. Thirty-one pairs of **spinal nerves** emerge from the spinal cord, each serving a specific region on the right or left side of the body. **Ganglia** (GANG-lē-a = swelling or knot) are small masses of nervous tissue, containing primarily cell bodies of neurons, that are located outside the brain and spinal cord. Ganglia are closely associated with cranial and spinal nerves. In the walls of organs of the gastrointestinal tract are extensive networks of neurons, called **enteric plexuses,** that help regulate the digestive system. **Sensory receptors** are either the dendrites of sensory neurons (described shortly) or separate, specialized cells that monitor changes in the internal or external environment, such as photoreceptors in the retina of the eye.

Functions of the Nervous System

The nervous system carries out a complex array of tasks, such as sensing various smells, producing speech, remembering, providing signals that control body movements, and regulating the operation of internal organs. These diverse activities can be grouped into three basic functions: sensory, integrative, and motor.

• **Sensory function.** Sensory receptors *detect* internal stimuli, such as an increase in blood acidity, and external stimuli, such as a raindrop landing on your arm. The neurons that carry sensory information from cranial and spinal nerves into the brain and spinal cord or from a lower to a higher level in the spinal cord and brain are **sensory** or **afferent neurons** (AF-er-ent NOO-ronz; *af-* = toward; *-ferrent* = carried).

• **Integrative function.** The nervous system *integrates* (processes) sensory information by analyzing and storing some of it and by making decisions for appropriate

Figure 12.1 Major structures of the nervous system.

 The nervous system includes the brain, cranial nerves, spinal cord, spinal nerves, ganglia enteric plexuses, and sensory receptors.

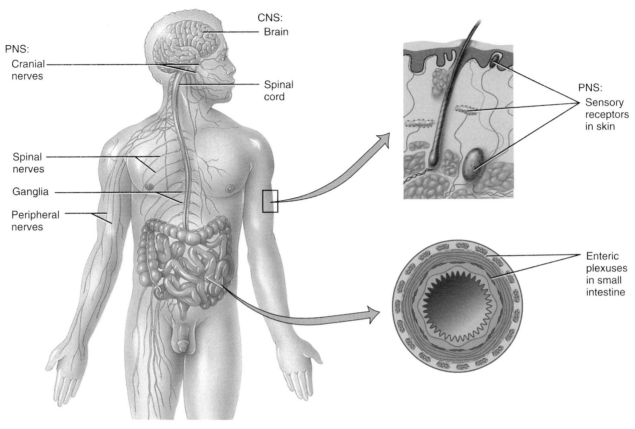

CNS:
Brain

PNS:
Cranial nerves

Spinal cord

Spinal nerves

Ganglia

Peripheral nerves

PNS:
Sensory receptors in skin

Enteric plexuses in small intestine

What is the total number of cranial and spinal nerves in the human body?

responses. Many of the neurons that participate in integration are **interneurons,** whose axons extend only for a short distance and contact nearby neurons in the brain, spinal cord, or a ganglion. Interneurons comprise the vast majority of neurons in the body.

- *Motor function.* The nervous system's motor function involves *responding* to integration decisions. The neurons that serve this function are **motor** or **efferent neurons** (EF-er-ent; *ef-* = away from). Motor neurons carry information from the brain toward the spinal cord or out of the brain and spinal cord into cranial or spinal nerves. The cells and organs contacted by motor neurons in cranial and spinal nerves are termed **effectors.** Muscle fibers and glandular cells are examples of effectors.

Organization of the Nervous System

The CNS (brain and spinal cord) integrates and correlates many different kinds of incoming sensory information. The CNS is

also the source of thoughts, emotions, and memories. Most nerve impulses that stimulate muscles to contract and glands to secrete originate in the CNS. Components of the PNS are cranial nerves and their branches, spinal nerves and their branches, ganglia, and sensory receptors. The PNS may be subdivided further into a **somatic nervous system (SNS)** (*somat-* = body), an **autonomic nervous system (ANS)** (*auto-* = self; *-nomic* = law), and an **enteric nervous system (ENS)** (*enter-* = intestines) (Figure 12.2). The SNS consists of (1) sensory neurons that convey information from somatic receptors in the head, body wall, and limbs and from receptors for the special senses of vision, hearing, taste, and smell to the CNS and (2) motor neurons that conduct impulses from the CNS to *skeletal muscles* only. Because these motor responses can be consciously controlled, the action of this part of the PNS is *voluntary.*

The ANS consists of (1) sensory neurons that convey information from autonomic sensory receptors, located primarily in visceral organs such as the stomach and lungs, to the CNS, and (2) motor neurons that conduct nerve impulses from the CNS to

Figure 12.2 Organization of the nervous system.   Subdivisions of the PNS are the somatic nervous system (SNS), the autonomic nervous system (ANS), and the enteric nervous system (ENS).

🔑 The two main subsystems of the nervous system are (1) the central nervous system (CNS), consisting of the brain and spinal cord, and (2) the peripheral nervous system (PNS), consisting of all nervous tissue outside the CNS.

What terms are given to neurons that carry input to the CNS? That carry output from the CNS?

smooth muscle, cardiac muscle, and *glands.* Because its motor responses are not normally under conscious control, the action of the ANS is *involuntary.* The motor part of the ANS consists of two branches, the **sympathetic division** and the **parasympathetic division.** With a few exceptions, effectors are innervated by both divisions, and usually the two divisions have opposing actions. For example, sympathetic neurons increase heart rate, whereas parasympathetic neurons slow it down.

The ENS is the "brain of the gut," and its operation is involuntary. Once considered part of the ANS, the ENS consists of approximately 100 million neurons in enteric plexuses that extend the entire length of the gastrointestinal (GI) tract. Many of the neurons of the enteric plexuses function independently of the ANS and CNS to some extent, although they also communicate with the CNS via sympathetic and parasympathetic neurons. Sensory neurons of the ENS monitor chemical changes within the GI tract and the stretching of its walls. Enteric motor neurons govern contraction of GI tract smooth muscle, secretions of the GI tract organs such as acid secretion by the stomach, and activity of GI tract endocrine cells.

▶ CHECKPOINT

1. What are the components of the CNS and PNS?

2. What kinds of problems would result from damage of sensory neurons, interneurons, and motor neurons?

3. What are the components and functions of the SNS, ANS, and ENS? Which subdivisions have voluntary versus involuntary actions?

HISTOLOGY OF NERVOUS TISSUE

▶ OBJECTIVES

• **Contrast the histological characteristics and the functions of neurons and neuroglia.**

• **Distinguish between gray matter and white matter.**

Nervous tissue consists of two types of cells: neurons and neuroglia. Neurons provide most of the unique functions of the nervous system, such as sensing, thinking, remembering, controlling muscle activity, and regulating glandular secretions. Neuroglia support, nourish, and protect the neurons and maintain homeostasis in the interstitial fluid that bathes neurons.

Neurons

Like muscle cells, **neurons** have the property of **electrical excitability,** the ability to produce *action potentials* or *impulses* in response to stimuli. Once they arise, action potentials propagate from one point to the next along the plasma membrane due to the presence of specific types of ion channels.

Parts of a Neuron

Most neurons have three parts: (1) a cell body, (2) dendrites, and (3) an axon (Figure 12.3). The **cell body** contains a nucleus surrounded by cytoplasm that includes typical organelles such as lysosomes, mitochondria, and a Golgi complex. Neuronal cell

Figure 12.3 Structure of a typical neuron. Arrows indicate the direction of information flow: dendrites → cell body → axon → axon terminals.

🔑 **The basic parts of a neuron are dendrites, a cell body, and an axon.**

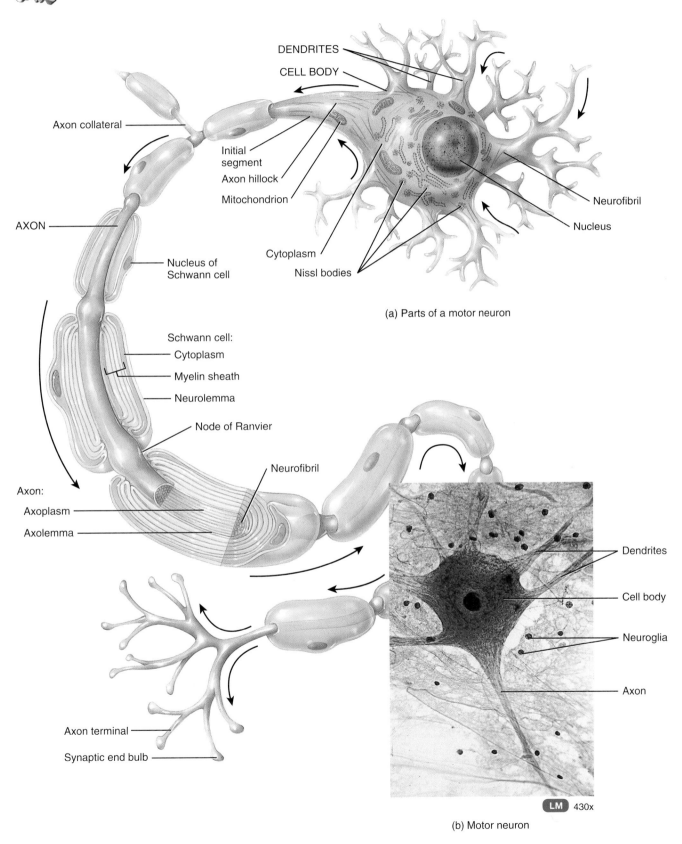

DENDRITES

CELL BODY

Axon collateral

Initial segment

Axon hillock

Mitochondrion

AXON

Nucleus of Schwann cell

Cytoplasm

Nissl bodies

Neurofibril

Nucleus

(a) Parts of a motor neuron

Schwann cell:
Cytoplasm
Myelin sheath
Neurolemma
Node of Ranvier

Neurofibril

Axon:
Axoplasm
Axolemma

Axon terminal

Synaptic end bulb

Dendrites

Cell body

Neuroglia

Axon

LM 430x

(b) Motor neuron

❓ **What roles do the dendrites, cell body, and axon play in communication of signals?**

bodies also contain prominent clusters of rough endoplasmic reticulum, termed *Nissl bodies,* the sites of protein synthesis in neurons. The German histologist Franz Nissl first described this unique microscopic feature of neurons. Newly synthesized proteins produced by Nissl bodies are used to replace cellular components, as material for growth of neurons, and to regenerate damaged axons in the PNS. The cytoskeleton includes both *neurofibrils,* composed of bundles of intermediate filaments that provide the cell shape and support, and *microtubules,* which assist in moving materials between the cell body and axon. Many neurons also contain *lipofuscin,* a pigment that occurs as clumps of yellowish brown granules in the cytoplasm. Lipofuscin is a product of neuronal lysosomes that accumulates as the neuron ages, but it does not seem to harm the neuron.

Two kinds of processes or extensions emerge from the cell body of a neuron: multiple dendrites and a single axon. *Nerve fiber* is a general term for any neuronal process (dendrite or axon). **Dendrites** (= little trees) are the receiving or input portions of a neuron. They usually are short, tapering, and highly branched. In many neurons the dendrites form a tree-shaped array of processes extending from the cell body. Their cytoplasm contains Nissl bodies, mitochondria, and other organelles.

The single **axon** (= axis) of a neuron propagates nerve impulses toward another neuron, a muscle fiber, or a gland cell. An axon is a long, thin, cylindrical projection that often joins the cell body at a cone-shaped elevation called the **axon hillock** (= small hill). The first part of the axon is the **initial segment.** In most neurons, impulses arise at the junction of the axon hillock and the initial segment, which is called the **trigger zone,** and then conduct along the axon. An axon contains mitochondria, microtubules, and neurofibrils. Because rough endoplasmic reticulum is not present, protein synthesis does not occur in the axon. The cytoplasm of an axon, called **axoplasm,** is surrounded by a plasma membrane known as the **axolemma** (*lemma* = sheath or husk). Along the length of an axon, side branches called **axon collaterals** may branch off, typically at a right angle to the axon. The axon and its collaterals end by dividing into many fine processes called **axon terminals.**

The site of communication between two neurons or between a neuron and an effector cell is called a **synapse.** The tips of some axon terminals swell into bulb-shaped structures called **synaptic end bulbs,** whereas others exhibit a string of swollen bumps called **varicosities.** Both synaptic end bulbs and varicosities contain many tiny membrane-enclosed sacs called **synaptic vesicles** that store a chemical **neurotransmitter.** Neurons were long thought to liberate just one type of neurotransmitter at all synaptic end bulbs. We now know that many neurons contain two or even three neurotransmitters. When neurotransmitter molecules are released from synaptic vesicles, they excite or inhibit other neurons, muscle fibers, or gland cells.

The cell body is the portion of a neuron that synthesizes new cell products or recycles old ones. However, because some substances are needed in the axon or at the axon terminals, two types of transport systems carry materials from the cell body to the axon terminals and back. The slower system, which moves materials about 1–5 mm per day, is called **slow axonal transport.** It conveys axoplasm in one direction only—from the cell body toward the axon terminals. It supplies new axoplasm for developing or regenerating axons and replenishes axoplasm in growing and mature axons.

The faster system, which is capable of moving materials a distance of 200–400 mm per day, is called **fast axonal transport.** It uses proteins that function as "motors" to move materials in both directions—away from and toward the cell body—along the surfaces of microtubules. Fast axonal transport moves various organelles and materials that form the membranes of the axolemma, synaptic end bulbs, and synaptic vesicles. Some materials transported back to the cell body are degraded or recycled; other influence neuronal growth.

Tetanus

Fast axonal transport is the route by which some toxins and disease-causing viruses make their way from axon terminals near skin cuts to cell bodies of neurons, where they can cause damage. For example, the toxin produced by *Clostridium tetani* bacteria is carried by fast axonal transport to the CNS. There it disrupts the actions of motor neurons, causing prolonged painful muscle spasms, a condition called **tetanus.** The delay between the release of the toxin and the first appearance of symptoms is due, in part, to the time required for transport of the toxin to the cell body. For this reason, a laceration or puncture injury to the head or neck is a more serious matter than a similar injury in the leg. The closer the site of injury is to the brain, the shorter the transit time. Thus, treatment must begin quickly to prevent symptoms of tetanus. ■

Structural Diversity in Neurons

Neurons display great diversity in size and shape. For example, their cell bodies range in diameter from 5 micrometers (μm) (slightly smaller than a red blood cell) up to 135 μm (barely large enough to see with the unaided eye). The pattern of dendritic branching is varied and distinctive for neurons in different parts of the nervous system. A few small neurons lack an axon, and many others have very short axons. The longest axons, however, are almost as long as a person is tall, extending from the toes to the lowest part of the brain.

Both structural and functional features are used to classify the various neurons in the body. Structurally, neurons are classified according to the number of processes extending from the cell body (Figure 12.4).

1. **Multipolar neurons** usually have several dendrites and one axon (see also Figure 12.3). Most neurons in the brain and spinal cord are of this type.

2. **Bipolar neurons** have one main dendrite and one axon. They are found in the retina of the eye, in the inner ear, and in the olfactory area of the brain.

3. **Unipolar neurons** are sensory neurons that begin in the embryo as bipolar neurons. During development, the axon and

Figure 12.4 **Structural classification of neurons.** Breaks indicate that axons are longer than shown.

 A multipolar neuron has many processes extending from the cell body, whereas a bipolar neuron has two and a unipolar neuron has one.

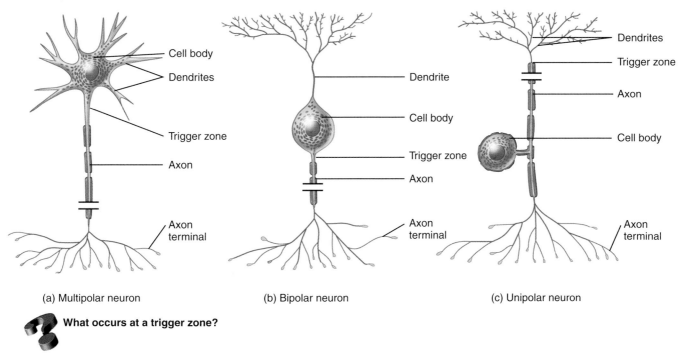

(a) Multipolar neuron   (b) Bipolar neuron   (c) Unipolar neuron

What occurs at a trigger zone?

dendrite fuse into a single process that divides into two branches a short distance from the cell body. Both branches have the characteristic structure and function of an axon. They are long, cylindrical processes that propagate action potentials. However, the axon branch that extends into the periphery has dendrites at its distal tip, whereas the axon branch that extends into the CNS ends in synaptic end bulbs. The dendrites monitor a sensory stimulus such as touch or stretching. The trigger zone for nerve impulses in a unipolar neuron is at the junction of the dendrites and axon (Figure 12.4c). The impulses then propagate toward the synaptic end bulbs.

Some neurons are named for the histologist who first described them; examples are **Purkinje cells** (pur-KIN-jē) in the cerebellum (Figure 12.5a) and **Renshaw cells** in the spinal cord. Other neurons are named for an aspect of their shape or appearance. For example, **pyramidal cells** (pi-RAM-i-dal), found in the brain, have a cell body shaped like a pyramid (Figure 12.5b). Often, a distinctive pattern of dendritic branching allows identification of a particular type of neuron in the CNS.

Neuroglia

Neuroglia (noo-RŌG-lē-a; -*glia* = glue) or **glia** constitute about half the volume of the CNS. Their name derives from the idea of early histologists that they were the "glue" that held nervous tissue together. We now know that neuroglia are not merely passive bystanders but rather active participants in the operation of nervous tissue. Generally, neuroglia are smaller than neurons,

Figure 12.5 **Two examples of CNS neurons.** Arrows indicate the direction of information flow.

 The dendritic branching pattern often is distinctive for a particular type of neuron.

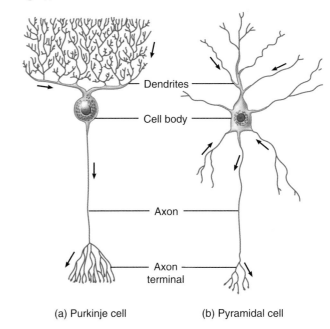

(a) Purkinje cell   (b) Pyramidal cell

 Why are pyramidal cells so-named?

and they are 5 to 50 times more numerous. In contrast to neurons, glia do not generate or propagate action potentials, and they can multiply and divide in the mature nervous system. In cases of injury or disease, neuroglia multiply to fill in the spaces formerly occupied by neurons. Brain tumors derived from glia, called **gliomas,** tend to be highly malignant and to grow rapidly. Of the six types of neuroglia, four—astrocytes, oligodendrocytes, microglia, and ependymal cells—are found only in the CNS. The remaining two types—Schwann cells and satellite cells—are present in the PNS. Table 12.1 describes the structure and functions of neuroglia.

Myelination

The axons of most mammalian neurons are surrounded by a multilayered lipid and protein covering, called the **myelin sheath,** that is produced by neuroglia. The sheath electrically insulates the axon of a neuron and increases the speed of nerve impulse conduction. Axons without such a covering are said to be **unmyelinated** (Figure 12.6a,c on page 394), whereas those with it are **myelinated** (Figure 12.6b,d). Electron micrographs reveal that even unmyelinated axons are surrounded by a thin coat of neuroglial plasma membrane.

Two types of neuroglia produce myelin sheaths: Schwann cells (in the PNS) and oligodendrocytes (in the CNS). In the PNS, **Schwann cells** begin to form myelin sheaths around axons during fetal development. Each Schwann cell wraps about 1 millimeter (1 mm = 0.04 in.) of a single axon's length by spiraling many times around the axon (Figure 12.6a). Eventually, multiple layers of glial plasma membrane surround the axon, with the Schwann cell's cytoplasm and nucleus forming the outermost layer. The inner portion, consisting of up to 100 layers of

Table 12.1	Neuroglia	
Type	**Appearance**	**Functions**
Central Nervous System		
Astrocytes (AS-trō-sīts; *astro-* = star; *-cyte* = cell)	Star-shaped, with many processes.	Help maintain appropriate chemical environment for generation of nerve impulses; provide nutrients to neurons; take up excess neurotransmitters; assist with migration of neurons during brain development; help form the blood–brain barrier.
Oligodendrocytes (OL-i-gō-den′-drō-sīts; *oligo-* = few; *dendro-* = tree)	Smaller than astrocytes, with fewer processes; round or oval cell body.	Form supporting network around CNS neurons; produce myelin sheath around several adjacent axons of CNS neurons.
Microglia (mī-KROG-lē-a; *micro-* = small)	Small cells with few processes; derived from mesodermal cells that also give rise to monocytes and macrophages.	Protect CNS cells from disease by engulfing invading microbes; migrate to areas of injured nerve tissue where they clear away debris of dead cells (but may also kill healthy cells).

Schwann cell membrane, is the myelin sheath. The outer nucleated cytoplasmic layer of the Schwann cell, which encloses the myelin sheath, is the **neurolemma (sheath of Schwann).** A neurolemma is found only around axons in the PNS. When an axon is injured, the neurolemma aids regeneration by forming a regeneration tube that guides and stimulates regrowth of the axon. Gaps in the myelin sheath, called **nodes of Ranvier** (RON-vē-ā), appear at intervals along the axon (see Figure 12.3). Each Schwann cell wraps one axon segment between two nodes.

In the CNS, an **oligodendrocyte** myelinates parts of several axons. Each oligodendrocyte puts forth about 15 broad, flat processes that spiral around CNS axons, forming a myelin sheath. A neurolemma is not present, however, because the oligodendrocyte cell body and nucleus do not envelop the axon. Nodes of Ranvier are present, but they are fewer in number.

Axons in the CNS display little regrowth after injury. This is thought to be due, in part, to the absence of a neurolemma, and in part to an inhibitory influence exerted by the oligodendrocytes on axon regrowth.

The amount of myelin increases from birth to maturity, and its presence greatly increases the speed of nerve impulse conduction. An infant's responses to stimuli are neither as rapid nor as coordinated as those of an older child or an adult, in part because myelination is still in progress during infancy. Certain diseases, such as multiple sclerosis (see page 413) and Tay-Sachs disease (see page 81), destroy myelin sheaths.

Gray and White Matter

In a freshly dissected section of the brain or spinal cord, some regions look white and glistening, whereas others appear gray

Type	Appearance	Functions
Ependymal cells (ep-EN-di-mal; *epen-* = above; *dym-* = garment)	Epithelial cells arranged in a single layer; range in shape from cuboidal to columnar; many are ciliated.	Line ventricles of the brain and central canal of the spinal cord (spaces filled with cerebrospinal fluid); form cerebrospinal fluid and assist in its circulation.

Cilia — Ependymal cell

Peripheral Nervous System

Schwann cells	Flattened cells that encircle PNS axons.	Each cell surrounds multiple unmyelinated axons with a single layer of its plasma membrane or produces part of the myelin sheath around a single axon of a PNS neuron; participate in regeneration of PNS axons.

Unmyelinated axons — Schwann cell

Schwann cell — Myelin sheath — Axon

Myelinated axon

Satellite cells (SAT-i-līt)	Flattened cells arranged around the cell bodies of neurons in ganglia.	Support neurons in PNS ganglia.

Satellite cell — Neuron cell body in a ganglion

Figure 12.6   Unmyelinated and myelinated axons.   Notice that one layer of Schwann cell plasma membrane surrounds unmyelinated axons.

Axons of most mammalian neurons are surrounded by a myelin sheath produced by Schwann cells in the PNS and by oligodendrocytes in the CNS.

Schwann cell:
Cytoplasm
Nucleus

Unmyelinated axons

(a) Transverse section of unmyelinated axons

Schwann cell:
Nucleus
Cytoplasm

Axolemma of axon

Neurolemma

Myelin sheath

(b) Transverse sections of stages in the formation of a myelin sheath

Schwann cell:
Cytoplasm
Neurolemma
Nucleus
Unmyelinated axons

TEM   2700x

(c) Transverse section of unmyelinated axons

Schwann cell:
Nucleus
Cytoplasm
Neurolemma
Myelin sheath
Myelinated axon

TEM   5000x

(d) Transverse section of myelinated axon

What is the functional advantage of myelination?

(Figure 12.7). The **white matter** is aggregations of myelinated and unmyelinated axons of many neurons. The whitish color of myelin gives white matter its name. The **gray matter** of the nervous system contains neuronal cell bodies, dendrites, unmyeli-nated axons, axon terminals, and neuroglia. It looks grayish, rather than white, because the Nissl bodies impart a gray color and there is little or no myelin in these areas. Blood vessels are present in both white and gray matter.

Figure 12.7 Distribution of gray and white matter in the spinal cord and brain.

White matter consists of myelinated and unmyelinated axons of many neurons. Gray matter consists of neuron cell bodies, dendrites, axon terminals, unmyelinated axons, and neuroglia.

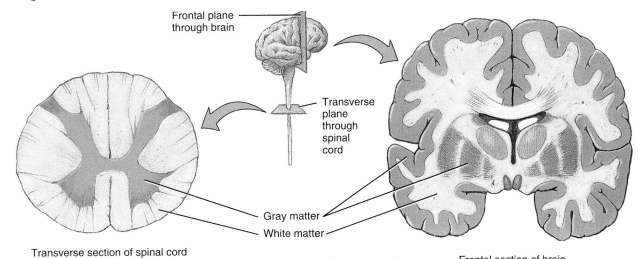

Frontal plane through brain

Transverse plane through spinal cord

Gray matter
White matter

Transverse section of spinal cord

Frontal section of brain

 What is responsible for the white appearance of white matter?

In the spinal cord, the white matter surrounds an inner core of gray matter shaped like a butterfly or the letter H; in the brain, a thin shell of gray matter covers the surface of the largest portions of the brain, the cerebrum and cerebellum (Figure 12.7). When used to describe nervous tissue, a **nucleus** is a cluster of neuronal cell bodies within the CNS. Many nuclei of gray matter also lie deep within the brain. The arrangements of gray and white matter in the spinal cord and brain are discussed more extensively in Chapters 13 and 14, respectively.

► CHECKPOINT
4. Describe the parts of a neuron and the functions of each.
5. Give several examples of the structural diversity of neurons.
6. Define neurolemma. Why is it important?
7. With reference to the nervous system, what is a nucleus?

ELECTRICAL SIGNALS IN NEURONS

OBJECTIVES
• **Describe the cellular properties that permit communication among neurons and effectors.**
• **Compare the basic types of ion channels, and explain how they relate to action potentials and graded potentials.**
• **Describe the factors that maintain a resting membrane potential.**
• **List the sequence of events that generate an action potential.**

Like muscle fibers, neurons are electrically excitable. They communicate with one another using two types of electrical signals: (1) *Action potentials* allow communication over both short and long distances within the body. (2) *Graded potentials* are used for short-distance communication only. Production of both types of signal depends on two basic features of the plasma membrane of excitable cells: the existence of a resting membrane potential and the presence of specific types of ion channels.

Like most other cells in the body, the plasma membrane of excitable cells exhibits a **membrane potential,** an electrical voltage difference across the membrane. In excitable cells, this voltage is termed the **resting membrane potential.** The membrane potential is like voltage stored in a battery. If you connect the positive and negative terminals of a battery with a piece of wire, electrons will flow along the wire. This flow of charged particles is called **current.** In living cells, the flow of ions (rather than electrons) constitutes the electrical current.

Graded potentials and action potentials occur because the plasma membranes of neurons contain many different kinds of ion channels that open or close in response to specific stimuli. Because the lipid bilayer of the plasma membrane is a good electrical insulator, the main paths for current to flow across the membrane are through the ion channels.

Ion Channels

The presence of various types of ion channels in the plasma membranes of neurons and muscle fibers gives these cells their property of electrical excitability. When ion channels are open, they allow specific ions to move across the plasma membrane, down their electrochemical gradient. Ions move from where they are more concentrated to where they are less concentrated. Also, positively charged cations move toward a negatively charged area, and negatively charged anions move toward a positively charged area. As ions move, they constitute a flow of electrical current that can change the membrane potential.

Ion channels open and close due to the presence of "gates." The gate is a part of the channel protein that can seal the channel pore shut or move aside to open the pore (see Figure 3.9 on page 68). The electrical signals produced by neurons and muscle fibers rely on four types of ion channels: leakage channels, voltage-gated channels, ligand-gated channels, and mechanically gated channels.

1. In **leakage channels,** the gates randomly alternate between open and closed positions. Typically, plasma membranes have many more potassium ion (K^+) leakage channels than sodium ion (Na^+) leakage channels. Thus, the membrane's permeability to K^+ is much higher than its permeability to Na^+.

2. A **voltage-gated channel** opens in response to a change in membrane potential (voltage) (Figure 12.8a). Voltage-gated channels participate in the generation and conduction of action potentials.

3. A **ligand-gated channel** opens and closes in response to a specific chemical stimulus. A wide variety of chemical ligands—including neurotransmitters, hormones, and particular ions—can open or close ligand-gated channels. The neurotransmitter acetylcholine, for example, opens cation channels that allow Na^+ and Ca^{2+} to diffuse inward and K^+ to diffuse outward (Figure 12.8b). Ligand-gated channels operate in two basic ways. The ligand molecule itself may open or close the channel by binding to a portion of the channel protein, as in the case of acetylcholine. Or the ligand may act *indirectly* via a type of membrane protein called a G protein that activates another molecule, a "second messenger," in the cytosol, which in turn operates the channel's gate. Some hormones and neurotransmitters work by such second-messenger systems (see Figure 18.4 on page 593).

4. A **mechanically gated channel** opens or closes in response to mechanical stimulation in the form of vibration (such as sound waves), pressure (such as touch), or tissue stretching. The force distorts the channel from its resting position, opening the gate. Examples of mechanically gated channels are those found in auditory receptors in the ears, in receptors that monitor stretching of internal organs, and in touch receptors in the skin.

Resting Membrane Potential

The resting membrane potential exists because of a small buildup of negative ions in the cytosol along the inside of the membrane, and an equal buildup of positive ions in the extracel-

Figure 12.8 Voltage-gated and ligand-gated channels in the plasma membrane. (a) A change in membrane potential opens voltage-gated K^+ channels during an action potential. (b) A chemical stimulus—here, the neurotransmitter acetylcholine—opens a ligand-gated channel.

 Gated channels open and close in response to a particular type of stimulus.

(a) Voltage-gated channel

(b) Ligand-gated channel

 What type of gated channel (not shown here) is activated by a touch on the arm?

lular fluid along the outside surface of the membrane (Figure 12.9a). Such a separation of positive and negative electrical charges is a form of potential energy, which is measured in volts or millivolts (1 mV = 0.001V). The greater the difference in charge across the membrane, the larger the membrane potential (voltage). Notice in Figure 12.9a that the buildup of charge occurs only very close to the membrane. The cytosol or extracellular fluid elsewhere contains equal numbers of positive and negative charges and is electrically neutral.

In neurons, the resting membrane potential ranges from −40 to −90 mV. A typical value is −70 mV. The minus sign indicates that the inside is negative relative to the outside. A cell that exhibits a membrane potential is said to be **polarized.** Most body cells are polarized; the membrane potential varies from +5 mV to −100 mV in different types of cells.

Two conditions allow maintenance of the resting membrane potential in excitable cells:

1. ***Unequal distribution of ions across the plasma membrane.*** Recall that the concentrations of major anions and cations are different outside and inside cells (Figure 12.9b). Extracellular fluid is rich in Na⁺ and chloride ions (Cl^-). In cytosol, however, the main cation is K⁺, and the two dominant anions are organic phosphates and amino acids in proteins.

2. ***Relative permeability of the plasma membrane to Na⁺ and K⁺.*** In a resting neuron or muscle fiber, the permeability of the plasma membrane is 50 to 100 times greater to K⁺ than to Na⁺.

To understand how these conditions maintain the resting membrane potential, first consider what would happen if the membrane were permeable only to K⁺. This cation would tend to leak out of the cell into the extracellular fluid down its concentration gradient. But, as more and more positive potassium ions exited, the interior of the membrane would become increasingly negative. The resulting electrical difference would then start to attract the positively charged K⁺ back into the cell. Hence, an **electrochemical gradient**—an electrical difference plus a concentration difference—exists across the membrane. Eventually, just as many K⁺ would be entering because of the electrical difference as would be exiting because of the concentration (chemical) gradient. The membrane potential that just balances the K⁺ concentration difference is −90 mV and is called the *potassium equilibrium potential.* The resting membrane potential (−70 mV) is close to, but not exactly equal to, the potassium equilibrium potential, which means that the membrane must be slightly permeable to other ions as well.

Actually, the membrane is moderately permeable to K⁺ and Cl^- and very slightly permeable to Na⁺. In theory, the electrical effect of K⁺ outflow might be canceled if Na⁺ could flow inward, in effect exchanging one positive ion for another. However, because the membrane permeability to Na⁺ is so low, inward leakage of Na⁺ is far too slow to keep pace with outward leakage of K⁺. A second way to cancel the electrical effect of K⁺ outflow might be the simultaneous exit of anions. Most anions in the cell, however, are not free to leave; they are attached either to large proteins or to other organic molecules, such as phosphates in ATP. Finally, inward leakage of Cl^- down its concentration gradient cannot cancel the electrical effect of K⁺ outflow. Any chloride ions that enter the cell can only make

Figure 12.9 Distributions of (a) charges and (b) ions that produce the resting membrane potential.

The resting membrane potential is due to a small buildup of anions, mainly organic phosphates (PO_4^{3-}) and proteins, in the cytosol just inside the membrane and an equal buildup of cations, mainly sodium ions (Na⁺), in the extracellular fluid just outside the membrane.

(a) Distribution of charges

(b) Distribution of ions

 What is a typical value for the resting membrane potential of a neuron?

the inside more negative. Because of the low permeability of the membrane to Na^+ and to anions inside the cell and the higher permeability to K^+, the fluid along the inner surface of the plasma membrane becomes increasingly more negatively charged as K^+ leaves.

Note that both the electrical and concentration gradients promote inflow of Na^+. The negatively charged interior of the membrane attracts cations, and the concentration of Na^+ is higher outside the cell. Even though membrane permeability to Na^+ is very low, a slow leak would eventually destroy the electrochemical gradient. The small inward Na^+ leak and outward K^+ leak are counteracted by the sodium-potassium pumps (Na^+/K^+ ATPase; see Figure 3.11 on page 70). These pumps help maintain the resting membrane potential by pumping out Na^+ as fast as it leaks in. At the same time, the sodium-potassium pumps bring in K^+. However, the potassium ions also redistribute according to electrical and chemical gradients, as previously described.

Because the sodium-potassium pumps expel three Na^+ for each two K^+ imported, they are *electrogenic*, which means they contribute to the negativity of the resting membrane potential. Their total contribution, however, is very small, only -3 mV of the total -70 mV resting membrane potential in a typical neuron.

Graded Potentials

When a stimulus causes ligand-gated or mechanically gated channels to open or close in an excitable cell's plasma membrane, a **graded potential** arises. A graded potential is a small deviation from the membrane potential that makes the membrane either more polarized (inside more negative) or less polarized (inside less negative). When the response makes the membrane even more polarized (a greater difference in the charge between inside and outside), it is termed a **hyperpolarizing graded potential** (Figure 12.10a). When the response makes the membrane less polarized, it is termed a **depolarizing graded potential** (Figure 12.10b).

To say that these electrical signals are *graded* means that they vary in amplitude (size), depending on the strength of the stimulus. They are larger or smaller depending on how many ion channels have opened (or closed) and how long each remains open. The opening or closing of ion channels alters the flow of specific ions across the membrane, producing a flow of current that is *localized*, which means that it spreads along the plasma membrane for a short distance and then dies out. For this reason, graded potentials are useful for communication only when the distance is less than a few hundred micrometers.

Typically, ligand-gated and mechanically gated channels are present in the dendrites of sensory neurons, and ligand-gated channels are most numerous in the dendrites and cell bodies of interneurons and motor neurons. Ligand-gated channels are occasionally present in axons. Hence, graded potentials occur most often in the dendrites and cell body of a neuron, and less often in the axon. Graded potentials have different names

depending on which type of stimulus causes them and where they occur. For example, when a neurotransmitter binds to its receptors at a ligand-gated channel, it produces a graded potential called a *postsynaptic potential* (explained shortly). On the other hand, the graded potentials that occur in sensory receptors and sensory neurons are termed *receptor potentials* and *generator potentials* (explained in Chapter 15).

Generation of Action Potentials

An **action potential (AP)** or **impulse** is a sequence of rapidly occurring events that take place in two phases (Figure 12.11).

Figure 12.10 Graded potentials. Most graded potentials occur in the dendrites and cell body (areas colored blue in the inset).

During a hyperpolarizing graded potential, the membrane polarization is more negative than the resting level. During a depolarizing graded potential the membrane polarization is less negative than the resting level.

(a) Hyperpolarizing graded potential

(b) Depolarizing graded potential

What types of ion channels produce graded potentials when they open or close?

During the **depolarizing phase,** the negative membrane potential decreases toward zero and eventually becomes positive. Then, the **repolarizing phase** restores the membrane potential to the resting state of −70mV. During an action potential, two types of voltage-gated channels open and then close. These channels are present mainly in the plasma membrane of the axon and axon terminals. The first channels that open, the Na⁺ channels, allow Na⁺ to rush into the cell, which causes the depolarizing phase. Then K⁺ channels open, allowing K⁺ to flow out, which produces the repolarizing phase. Together, the depolarization and repolarization phases last about 1 msec (0.001 sec) in a typical neuron.

Action potentials arise according to the **all-or-none principle.** When depolarization reaches a certain level termed the **threshold** (about −55 mV in many neurons), the voltage-gated channels open, and an action potential that is always the same size (amplitude) occurs. This situation is similar to pushing on the first domino in a long row of standing dominos. When the push on the first domino is strong enough (reaches threshold), that domino falls against the second domino, and the *entire* row

topples (an action potential occurs). Stronger pushes on the first domino produce the identical effect—toppling of the entire row. Thus, pushing on the first domino produces an all-or-none event: The dominoes all fall or none fall. Because they can travel long distances without dying out, APs function in communication over both short and long distances. Different neurons may have different thresholds for generation of an action potential, but the threshold in any one neuron usually is constant.

Depolarizing Phase

When a depolarizing graded potential or some other stimulus causes the membrane to depolarize to threshold, voltage-gated Na⁺ channels open rapidly. Both the electrical and the chemical gradients favor inward movement of Na⁺, and the resulting inrush of Na⁺ causes the depolarizing phase of the action potential (Figure 12.11). The inflow of Na⁺ changes the membrane potential from −55 mV to +30 mV. At the peak of the action potential, the inside of the membrane is 30 mV more positive than the outside.

Figure 12.11 Action potential (AP) or impulse. When a stimulus depolarizes the membrane to threshold (−55 mV), an AP is generated. The action potential arises at the trigger zone (here, at the junction of the axon hillock and the initial segment) and then propagates along the axon to the axon terminals. The green colored regions in the inset indicate the parts of a neuron that typically have voltage-gated Na⁺ and K⁺ channels (axon and axon terminals).

🔑 **An action potential consists of depolarizing and repolarizing phases.**

 Which channels are open during depolarization? During repolarization?

Each voltage-gated Na⁺ channel has two separate gates, an *activation gate* and an *inactivation gate*. In the *resting state* of a voltage-gated Na⁺ channel, the inactivation gate is open, but the activation gate is closed (step 1 in Figure 12.12). As a result, Na⁺ cannot move into the cell through these channels. At threshold, voltage-gated Na⁺ channels are activated. In the *activated state* of a voltage-gated Na⁺ channel, both the activation and inactivation gates in the channel are open and Na⁺ inflow begins (step 2 in Figure 12.12). As more channels open, Na⁺ inflow increases, the membrane depolarizes further, more Na⁺ channels open, and so on. This is an example of a positive feedback mechanism. Depolarization causes inflow of Na⁺, further depolarizing the membrane and opening more Na⁺ channels.

Shortly after the activation gates open the inactivation gates close (step 3 in Figure 12.12). Now the channel is in an *inactivated state*. While a voltage-gated Na⁺ channel is open for a few ten-thousandths of a second, about 20,000 Na⁺ flow across the membrane and change the membrane potential considerably. By contrast, the Na⁺ concentration hardly changes because millions of Na⁺ are present nearby in the extracellular fluid. Because only a few thousand Na⁺ enter during a single action potential, the sodium-potassium pumps easily bail them out and maintain the low concentration of Na⁺ inside the cell.

Repolarizing Phase

Besides opening voltage-gated Na⁺ channels, a threshold-level depolarization also opens voltage-gated K⁺ channels (steps 3 and 4 in Figure 12.12). The voltage-gated K⁺ channels open more slowly, however, so their opening occurs at about the same time the voltage-gated Na⁺ channels are closing. The slower opening of voltage-gated K⁺ channels and the closing of previously open Na⁺ channels produce the repolarizing phase of the action potential. As the Na⁺ channels are inactivated, Na⁺ inflow slows. At the same time, the K⁺ channels are opening, and K⁺ outflow accelerates. Slowing of Na⁺ inflow and acceleration of K⁺ outflow causes the membrane potential to change from +30 mV to −70 mV. Repolarization also allows inactivated Na⁺ channels to revert to the resting state.

While the voltage-gated K⁺ channels are open, outflow of K⁺ may be large enough to cause an **after-hyperpolarizing phase** of the action potential (see Figure 12.11). During this phase, the membrane is even more permeable to K⁺ than in the resting state, and the membrane potential drifts toward the potassium equilibrium potential (about −90 mV). As the voltage-gated K⁺ channels close, however, the membrane potential returns to the resting level of −70 mV. In contrast to voltage-gated Na⁺ channels, most voltage-gated K⁺ channels do not exhibit an inactivated state. Instead, they alternate between closed (resting) and open (activated) states.

Refractory Period

The period of time after an action potential begins during which an excitable cell cannot generate another action potential is called the **refractory period** (see key in Figure 12.11). During the **absolute refractory period,** a second action potential cannot be initiated, even with a very strong stimulus. This period coincides with the period of Na⁺ channel activation and inactivation (steps 2–4 in Figure 12.12). Inactivated Na⁺ channels cannot reopen; they first must return to the resting state (step 1 in Figure 12.12). In contrast to action potentials, graded potentials do not exhibit a refractory period.

Large-diameter axons have a brief absolute refractory period of about 0.4 msec. Because a second nerve impulse can arise very quickly, up to 1000 impulses per second are possible. Small-diameter axons have absolute refractory periods as long as 4 msec. Thus, they can transmit a maximum of 250 impulses per second. Under normal body conditions, the maximum frequency of nerve impulses in different axons ranges between 10 and 1000 per second.

The **relative refractory period** is the period of time during which a second action potential can be initiated, but only by a larger-than-normal stimulus. It coincides with the period when the voltage-gated K⁺ channels are still open after inactivated Na⁺ channels have returned to their resting state (see Figure 12.11).

Propagation of Nerve Impulses

To communicate information from one part of the body to another, nerve impulses must travel from where they arise at a trigger zone, often the axon hillock, to the axon terminals. This mode of travel is called **propagation** or **conduction,** and it depends on positive feedback. As sodium ions flow in, they cause depolarization that opens voltage-gated Na⁺ channels in adjacent segments of the membrane. Thus, the nerve impulse self-propagates along the membrane, rather like toppling that long row of dominoes by pushing over the first one in the line. Also, because the membrane is refractory behind the leading edge of a nerve impulse, it normally moves in one direction only—from where it arises at the trigger zone toward the axon terminals.

Neurotoxins and Local Anesthetics

Certain shellfish and other creatures contain **neurotoxins,** substances that produce their poisonous effects by acting on the nervous system. One particularly lethal neurotoxin is tetrodotoxin (TTX), present in the viscera of Japanese pufferfish. TTX effectively blocks production of action potentials by inserting itself into voltage-gated Na⁺ channels so they cannot open.

Local anesthetics are drugs that block pain and other somatic sensations. Examples include procaine (Novocaine) and Lidocaine, which may be used to produce anesthesia in the skin during suturing of a gash, in the mouth during dental work, or in the lower body during childbirth. Like TTX, these drugs act by blocking the opening of voltage-gated Na⁺ channels. Nerve impulses cannot propagate past the obstructed region, so pain signals do not reach the CNS. Smaller-diameter axons, such

Figure 12.12 Changes in ion flow through voltage-gated channels during the depolarizing and repolarizing phases of an action potential. Leakage channels and sodium-potassium pumps are not shown.

Inflow of sodium ions (Na$^+$) causes the depolarizing phase and outflow of potassium ions (K$^+$) causes the repolarizing phase of an action potential.

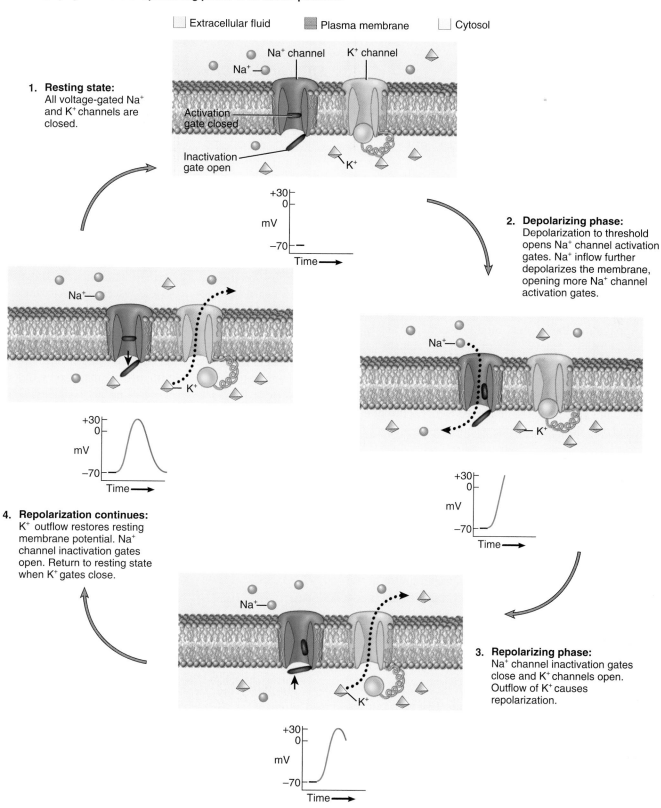

Extracellular fluid   Plasma membrane   Cytosol

1. Resting state: All voltage-gated Na$^+$ and K$^+$ channels are closed.

Na$^+$ channel   K$^+$ channel
Na$^+$
Activation gate closed
Inactivation gate open
K$^+$

2. Depolarizing phase: Depolarization to threshold opens Na$^+$ channel activation gates. Na$^+$ inflow further depolarizes the membrane, opening more Na$^+$ channel activation gates.

Na$^+$
K$^+$

4. Repolarization continues: K$^+$ outflow restores resting membrane potential. Na$^+$ channel inactivation gates open. Return to resting state when K$^+$ gates close.

3. Repolarizing phase: Na$^+$ channel inactivation gates close and K$^+$ channels open. Outflow of K$^+$ causes repolarization.

Na$^+$
K$^+$

Given the existence of leakage channels for both K$^+$ and Na$^+$, could the membrane repolarize if the voltage-gated K$^+$ channels did not exist?

as those carrying pain signals, are more sensitive than larger-diameter axons to low doses of anesthetic drugs.

In addition, localized cooling of a nerve produces an anesthetic effect. It retards impulse conduction because axons propagate impulses at lower speeds when cooled. Pain resulting from injured tissue can be reduced by the application of ice because propagation of the pain sensations along axons is partially blocked. ■

Continuous and Saltatory Conduction

The type of impulse propagation described so far occurs in muscle fibers and unmyelinated axons. Such step-by-step depolarization and repolarization of each adjacent segment of the plasma membrane is called **continuous conduction** (Figure 12.13a). In continuous conduction, ions flow through their voltage-gated channels in each adjacent segment of the membrane. Note that the impulse propagates only a relatively short distance in a few milliseconds.

Nerve impulses propagate more rapidly along myelinated axons than along unmyelinated axons. Compare parts a and b in Figure 12.13 to see that the impulse propagates much farther along the myelinated axon in a few milliseconds. **Saltatory conduction** (SAL-ta-tō-rē; *saltat-* = leaping) is the special mode of impulse propagation that occurs along myelinated axons. It happens because of the uneven distribution of voltage-gated channels. Few voltage-gated channels are present in regions where a myelin sheath covers the axolemma. By contrast, at the nodes of Ranvier (where there is no myelin sheath), the axolemma has many voltage-gated channels. Hence, current carried by Na^+ and K^+ flows across the membrane mainly at the nodes.

When a nerve impulse propagates along a myelinated axon, an electric current (carried by ions) flows through the extracellular fluid surrounding the myelin sheath and through the cytosol from one node to the next. The nerve impulse at the first node generates ionic currents in the cytosol and extracellular fluid that depolarize the membrane to threshold, opening voltage-gated Na^+ channels at the second node. The resulting ionic flow through the opened channels constitutes a nerve impulse at the second node. Then, the nerve impulse at the second node generates an ionic current that opens voltage-gated Na^+ channels at the third node, and so on. Each node repolarizes after it depolarizes.

The flow of current across the membrane only at the nodes of Ranvier has two consequences.

- The impulse appears to leap from node to node as each nodal area depolarizes to threshold, thus the name "saltatory." Because an impulse "leaps" long intervals as current flows from one node to the next in saltatory conduction, it travels much faster than it would by continuous conduction in an unmyelinated axon of the same diameter.

- Opening a smaller number of channels only at the nodes, rather than many channels in each adjacent segment of membrane, represents a more energy-efficient mode of conduction. Because only small regions of the membrane depolarize and repolarize, minimal inflow of Na^+ and outflow of K^+ occurs each time a nerve impulse passes by. Thus, less ATP is used by sodium-potassium pumps to maintain the low intracellular concentration of Na^+ and the low extracellular concentration of K^+.

Effect of Axon Diameter

Larger-diameter axons propagate impulses faster than smaller ones. The largest-diameter axons (about $5-20$ μm) are called **A fibers,** and all are myelinated. A fibers have a brief absolute refractory period and conduct impulses at speeds of 12 to 130 m/sec ($27-280$ mi/hr). The axons of sensory neurons that propagate impulses associated with touch, pressure, position of joints, and some thermal sensations are A fibers, as are the axons of motor neurons that conduct impulses to skeletal muscles.

B fibers are axons with diameters of $2-3$ μm. Like A fibers, B fibers are myelinated and exhibit saltatory conduction at speeds up to 15 m/sec (32 mi/hr). B fibers have a somewhat longer absolute refractory period than A fibers. B fibers conduct sensory nerve impulses from the viscera to the brain and spinal cord. They also constitute all the axons of the autonomic motor neurons that extend from the brain and spinal cord to the ANS relay stations called autonomic ganglia.

C fibers are the smallest diameter axons ($0.5-1.5$ μm) and all are unmyelinated. Nerve impulse propagation along a C fiber ranges from 0.5 to 2 m/sec ($1-4$ mi/hr). C fibers exhibit the longest absolute refractory periods. These unmyelinated axons conduct some sensory impulses for pain, touch, pressure, heat, and cold from the skin, and pain impulses from the viscera. Autonomic motor fibers that extend from autonomic ganglia to stimulate the heart, smooth muscle, and glands are C fibers. Examples of motor functions of B and C fibers are constricting and dilating the pupils, increasing and decreasing the heart rate, and contracting and relaxing the urinary bladder—actions of the autonomic nervous system.

Encoding of Stimulus Intensity

How can your sensory systems detect stimuli of differing intensities if all nerve impulses are the same size? Why does a light touch feel different from firmer pressure? The main answer to this question is the *frequency of impulses*—how often they are generated at the trigger zone. Thus, a light touch generates a low frequency of widely spaced nerve impulses. By contrast, a firmer pressure elicits nerve impulses that pass down the axon closer together at a higher frequency. In addition to the "frequency code" for stimulus intensity, a second factor is the number of sensory neurons recruited (activated) by the stimulus. A firm pressure stimulates more pressure-sensitive neurons than does a light touch.

Figure 12.13 **Conduction (propagation) of a nerve impulse after it arises at the trigger zone.** Dotted lines indicate ionic current flow. The insets show the path of current flow. (a) In continuous conduction along an unmyelinated axon, ionic currents flow across each adjacent segment of the membrane. (b) In saltatory conduction along a myelinated axon, the nerve impulse at the first node generates ionic currents in the cytosol and interstitial fluid that open voltage-gated Na^+ channels at the second node, and so on at each subsequent node.

🔑 **Unmyelinated axons exhibit continuous conduction, whereas myelinated axons exhibit saltatory conduction.**

(a) Continuous conduction

(b) Saltatory conduction

❓ **What factors determine the speed of propagation of a nerve impulse?**

Comparison of Electrical Signals Produced by Excitable Cells

We have seen that excitable cells—neurons and muscle fibers—produce two types of electrical signals, graded potentials and action potentials (impulses). One obvious difference between them is that action potentials permit communication over long distances because they are propagated, whereas graded potentials function only in short-distance communication because they are not propagated. Table 12.2 presents a summary of the differences between graded potentials and action potentials.

As we discussed in Chapter 10, propagation of a muscle action potential along the sarcolemma and into the T tubule system initiates the events of muscle contraction. Although action potentials in muscle fibers and in neurons are similar, there are some notable differences. Whereas the typical resting membrane potential of a neuron is −70 mV, it is closer to −90 mV in

skeletal and cardiac muscle fibers. The duration of a nerve impulse is 0.5–2 msec, but a muscle action potential is considerably longer—about 1.0–5.0 msec for skeletal muscle fibers and 10–300 msec for cardiac and smooth muscle fibers. Finally, the conduction speed of action potentials along the largest-diameter, myelinated axons is about 18 times faster than the conduction speed along the sarcolemma of a skeletal muscle fiber.

▶ CHECKPOINT

8. What are the meanings of the terms *resting membrane potential, depolarization, repolarization, nerve impulse,* and *refractory period*?

9. How is saltatory conduction different from continuous conduction?

10. What factors determine the speed of propagation of nerve impulses?

11. How is the intensity of a stimulus encoded in the nervous system?

Table 12.2	Comparison of Graded Potentials and Action Potentials	
Characteristic	**Graded Potentials**	**Action Potentials**
Origin	Arise mainly in dendrites and cell body (some arise in axons).	Arise at trigger zones and propagate along the axon.
Types of channels	Ligand-gated or mechanically gated ion channels.	Voltage-gated channels for Na^+ and K^+.
Conduction	Not propagated; localized and thus permit communication over a few micrometers.	Propagate and thus permit communication over longer distances.
Amplitude	Depending on strength of stimulus, varies from less than 1 mV to more than 50 mV.	All-or-none; typically about 100 mV.
Duration	Typically longer, ranging from several msec to several min.	Shorter, ranging from 0.5 to 2 msec.
Polarity	May be hyperpolarizing (inhibitory to generation of an action potential) or depolarizing (excitatory to generation of an action potential).	Always consist of depolarizing phase followed by repolarizing phase and return to resting membrane potential.
Refractory period	Not present, thus spatial and temporal summation can occur.	Present, thus summation cannot occur.

SIGNAL TRANSMISSION AT SYNAPSES

▶ **OBJECTIVES**

- **Explain the events of signal transmission at a chemical synapse.**
- **Distinguish between spatial and temporal summation.**
- **Give examples of excitatory and inhibitory neurotransmitters, and describe how they act.**

In Chapter 10 we described the events occurring at one type of synapse, the neuromuscular junction. Our focus here is synaptic communication among the billions of neurons in the nervous system. Synapses are essential for homeostasis because they allow information to be filtered and integrated. During learning, the structure and function of particular synapses change. The changes may allow some signals to be transmitted while others are blocked. For instance, the extent to which some of your synapses change because of studying will determine your performance on anatomy and physiology tests! Some diseases and neurological disorders result from disruptions of synaptic communication. Synapses also are the sites of action for many therapeutic and addictive chemicals.

At a synapse between neurons, the neuron sending the signal is called the **presynaptic neuron,** and the neuron receiving the message is called the **postsynaptic neuron.** Most synapses are either **axodendritic** (from axon to dendrite), **axosomatic** (from axon to soma), or **axoaxonic** (from axon to axon). The two types of synapses—electrical and chemical—differ both structurally and functionally.

Electrical Synapses

At an **electrical synapse,** action potentials (impulses) conduct directly between adjacent cells through **gap junctions.** Each gap junction contains a hundred or so tubular *connexons,* which act like tunnels to connect the cytosol of the two cells (see Figure 4.1e on page 105). As ions flow from one cell to the next through the connexons, the action potential spreads from cell to cell. Gap junctions are common in visceral smooth muscle, cardiac muscle, and the developing embryo. They also occur in the CNS.

Electrical synapses have two main advantages:

1. *Faster communication.* Because action potentials conduct directly through gap junctions, electrical synapses are faster than chemical synapses. At an electrical synapse, the action potential passes directly from the presynaptic cell to the postsynaptic cell. The events that occur at a chemical synapse take some time and impose a slight delay in communication.

2. *Synchronization.* Electrical synapses can synchronize the activity of a group of neurons or muscle fibers. In other words, a large number of neurons or muscle fibers can produce action potentials in unison if they are connected by gap junctions. The value of synchronized action potentials in the heart or in visceral smooth muscle is coordinated contraction of these fibers.

Chemical Synapses

Although the plasma membranes of presynaptic and postsynaptic neurons in a **chemical synapse** are close, they do not touch. The **synaptic cleft,** a space of 20–50 nm that is filled with interstitial fluid, separates the two neurons. Because nerve impulses cannot conduct across the synaptic cleft, an alternate, indirect form of communication occurs across this space. In response to a nerve impulse, the presynaptic neuron releases a neurotransmitter that diffuses through the fluid in the synaptic cleft and binds to receptors in the plasma membrane of the postsynaptic neuron. The postsynaptic neuron receives the chemical signal and, in turn, produces a **postsynaptic potential,** a type of graded potential. Thus, the presynaptic neuron converts an electrical signal (nerve impulse) into a chemical signal (released neurotransmitter). The postsynaptic neuron receives the chemical signal and, in turn, generates an electrical signal (postsynaptic potential). The time required for these processes at a chemical synapse, the **synaptic delay,** is about 0.5 msec. This delay is the reason that chemical synapses relay signals more slowly than electrical synapses.

A typical chemical synapse transmits a signal as follows (Figure 12.14):

1 A nerve impulse arrives at a synaptic end bulb (or at a varicosity) of a presynaptic axon.

2 The depolarizing phase of the nerve impulse opens **voltage-gated Ca²⁺ channels,** which are present in the membrane of synaptic end bulbs. Because calcium ions are more concentrated in the extracellular fluid, Ca^{2+} flows inward through the opened channels.

3 An increase in the concentration of Ca^{2+} inside the presynaptic neuron is the signal that triggers exocytosis of some of the synaptic vesicles. As vesicle membranes merge with the plasma membrane, neurotransmitter molecules within the vesicles are released into the synaptic cleft. Each synaptic vesicle contains several thousand molecules of neurotransmitter.

4 The neurotransmitter molecules diffuse across the synaptic cleft and bind to **neurotransmitter receptors** in the postsynaptic neuron's plasma membrane. The receptor shown in Figure 12.14 is part of a ligand-gated channel; in other cases the receptor may be a separate protein in the membrane.

5 Binding of neurotransmitter molecules to their receptors on ligand-gated channels opens the channels and allows particular ions to flow across the membrane.

6 As ions flow through the opened channels, the voltage across the membrane changes. This change in membrane voltage is a **postsynaptic potential.** Depending on which ions the channels admit, the postsynaptic potential may be a depolarization or hyperpolarization. For example, opening of Na^+ channels allows inflow of Na^+, which causes depolarization. However, opening of Cl^- channels allows inflow of Cl^-, which causes hyperpolarization.

7 When a depolarizing postsynaptic potential reaches threshold, it triggers one or more nerve impulses.

At most chemical synapses, only *one-way information transfer* can occur—from a presynaptic neuron to a postsynaptic neuron or an effector, such as a muscle fiber or a gland cell. For example, synaptic transmission at a neuromuscular junction (NMJ) proceeds from a somatic motor neuron to a skeletal muscle

Figure 12.14 Signal transmission at a chemical synapse. Through exocytosis of synaptic vesicles, a presynaptic neuron releases neurotransmitter molecules. After diffusing across the synaptic cleft, the neurotransmitter binds to receptors in the plasma membrane of the postsynaptic neuron and produces a postsynaptic potential.

At a chemical synapse, a presynaptic neuron converts an electrical signal (nerve impulse) into a chemical signal (neurotransmitter release). The postsynaptic neuron then converts the chemical signal back into an electrical signal (postsynaptic potential).

 Why may electrical synapses work in two directions, but chemical synapses can transmit a signal in one direction only?

fiber (but not in the opposite direction). Only synaptic end bulbs of presynaptic neurons can release neurotransmitter, and only the postsynaptic neuron's membrane has the correct receptor proteins to recognize and bind that neurotransmitter. As a result, action potentials move along their pathways in one direction.

Excitatory and Inhibitory Postsynaptic Potentials

A neurotransmitter causes either an excitatory or an inhibitory graded potential. If it *depolarizes* the postsynaptic membrane, it is excitatory because it brings the membrane closer to threshold (see Figure 12.10b). A depolarizing postsynaptic potential is called an **excitatory postsynaptic potential (EPSP).** Often, EPSPs result from opening of *cation* channels. These channels allow the three most plentiful cations (Na^+, K^+, and Ca^{2+}) to pass through, but Na^+ inflow is greater than either Ca^{2+} inflow or K^+ outflow. Although a single EPSP normally does not initiate a nerve impulse, the postsynaptic neuron does become more excitable. It is partially depolarized and thus more likely to reach threshold when the next EPSP occurs.

If, in contrast, a neurotransmitter causes *hyperpolarization* of the postsynaptic membrane (see Figure 12.10a), it is inhibitory. During hyperpolarization, generation of a nerve impulse is more difficult than usual. The membrane potential is more negative and thus even farther from threshold than it was in its resting state. A hyperpolarizing postsynaptic potential is termed an **inhibitory postsynaptic potential (IPSP).** IPSPs often result from the opening of ligand-gated Cl^- or K^+ channels. When Cl^- channels open, a larger number of chloride ions diffuse inward. When K^+ channels open, a larger number of potassium ions diffuse outward. In both cases, the ionic flow causes the inside to become more negative (hyperpolarized).

Removal of Neurotransmitter

Removal of the neurotransmitter from the synaptic cleft is essential for normal synaptic function. If a neurotransmitter could linger in the synaptic cleft, it would influence the postsynaptic neuron, muscle fiber, or gland cell indefinitely. Neurotransmitter is removed in three ways:

1. *Diffusion.* Some of the released neurotransmitter molecules diffuse away from the synaptic cleft. Once a neurotransmitter molecule is out of reach of its receptors, it can no longer exert an effect.

2. *Enzymatic degradation.* Certain neurotransmitters are inactivated through enzymatic degradation. For example, the enzyme acetylcholinesterase breaks down acetylcholine in the synaptic cleft.

3. *Uptake by cells.* Many neurotransmitters are actively transported back into the neuron that released them (reuptake). Others are transported into neighboring neuroglia (uptake). The

neurons that release norepinephrine, for example, rapidly take up the norepinephrine and recycle it into new synaptic vesicles. The membrane proteins that accomplish such uptake are called *neurotransmitter transporters.* Several therapeutically important drugs selectively block reuptake of specific neurotransmitters by interfering with these transporters. For example, the drug fluoxetine (Prozac) is a **selective serotonin reuptake inhibitor (SSRI).** By inhibiting serotonin transporters, Prozac prolongs the activity of this neurotransmitter at synapses in the brain. SSRIs provide relief for those suffering from some forms of depression.

Spatial and Temporal Summation of Postsynaptic Potentials

A typical neuron in the CNS receives input from 1000 to 10,000 synapses. Integration of these inputs, which is known as **summation,** occurs at the trigger zone. The greater the summation of EPSPs, the greater the chance that threshold will be reached. At threshold, one or more nerve impulses (action potentials) arise.

When summation results from buildup of neurotransmitter released simultaneously by *several* presynaptic end bulbs, it is called **spatial summation** (Figure 12.15a). When summation results from buildup of neurotransmitter released by a *single* presynaptic end bulb two or more times in rapid succession, it is called **temporal summation** (Figure 12.15b). Because a typical EPSP lasts about 15 msec, the second (and subsequent) release of neurotransmitter must occur soon after the first one if temporal summation is to occur. Summation is rather like a vote on the Internet. If many people are voting "yes" or "no" on an issue, the situation is like spatial summation. If one person is voting repeatedly, the situation is like temporal summation. Most of the time, spatial and temporal summations are acting together to influence the chance that a neuron fires an impulse.

A single postsynaptic neuron receives input from many presynaptic neurons, some of which release excitatory neurotransmitters and some of which release inhibitory neurotransmitters. The sum of all the excitatory and inhibitory effects at any given time determines the effect on the postsynaptic neuron, which may respond in the following ways:

1. *EPSP.* If the total excitatory effects are greater than the total inhibitory effects but less than the threshold level of stimulation, the result is an EPSP that does not reach threshold. Subsequent stimuli can more easily generate a nerve impulse through summation because the neuron is partially depolarized.

2. *Nerve impulse(s).* If the total excitatory effects are greater than the total inhibitory effects and the threshold level of stimulation is reached or surpassed, the EPSP spreads to the initial segment of the axon and triggers one or more nerve impulses. Impulses continue to be generated as long as the EPSP is above the threshold level.

3. *IPSP.* If the total inhibitory effects are greater than the excitatory effects, the membrane hyperpolarizes (IPSP). The result is

Figure 12.15 Spatial and temporal summation. (a) When presynaptic neurons a and b separately cause EPSPs (arrows) in postsynaptic neuron c, the threshold level is not reached in neuron c. Spatial summation occurs only when neurons a and b act simultaneously on neuron c; their EPSPs sum to reach the threshold level and trigger a nerve impulse. (b) Temporal summation occurs when stimuli applied to the same axon in rapid succession (arrows) cause overlapping EPSPs that sum. When depolarization reaches the threshold level, a nerve impulse is triggered.

 The sum of all excitatory and inhibitory postsynaptic potentials determines whether an action potential is generated.

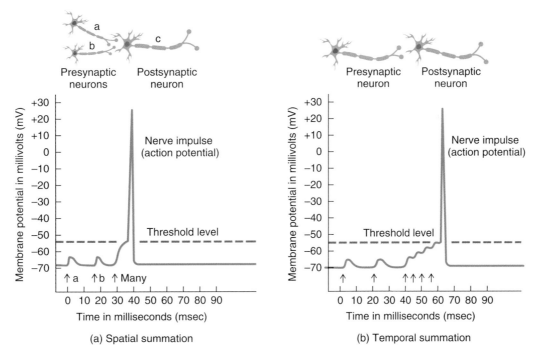

(a) Spatial summation (b) Temporal summation

What would be the result if in addition to the four EPSPs indicated by arrows in part (b) an IPSP occurred at time 55 msec?

inhibition of the postsynaptic neuron and an inability to generate a nerve impulse.

Table 12.3 summarizes the structural and functional elements of a neuron.

Strychnine Poisoning

The importance of inhibitory neurons can be appreciated by observing what happens when their activity is blocked. Normally, inhibitory neurons in the spinal cord called *Renshaw cells* release the neurotransmitter glycine (described shortly) at inhibitory synapses with somatic motor neurons. This inhibitory input to their motor neurons prevents excessive contraction of skeletal muscles. Strychnine is a lethal poison that binds to and blocks glycine receptors. Then, the normal, delicate balance between excitation and inhibition in the CNS is disturbed, and motor neurons generate nerve impulses without restraint. All skeletal muscles, including the diaphragm, contract fully and remain contracted. Because the diaphragm cannot relax, the victim cannot breathe. ■

▶ CHECKPOINT

12. How is neurotransmitter removed from the synaptic cleft?

13. How are excitatory and inhibitory postsynaptic potentials similar and different?

14. Why are action potentials said to be "all-or-none," whereas EPSPs and IPSPs are described as "graded"?

NEUROTRANSMITTERS

▶ OBJECTIVE

• **Describe the classes and functions of neurotransmitters.**

About 100 substances are either known or suspected neurotransmitters. Some neurotransmitters bind to their receptors and act quickly to open or close ion channels in the membrane. Others act more slowly via second-messenger systems to influence chemical reactions inside cells. The result of either process can be excitation or inhibition of postsynaptic neurons. Many

Table 12.3	Summary of Neuronal Structure and Function	
	Structure	**Functions**
	Dendrites	Receive stimuli through activation of ligand-gated or mechanically gated ion channels; in sensory neurons, produce generator or receptor potentials; in motor neurons and interneurons, produce excitatory and inhibitory postsynaptic potentials (EPSPs and IPSPs).
	Cell body	Receives stimuli and produces EPSPs and IPSPs through activation of ligand- or mechanically gated ion channels.
	Junction of axon hillock and initial segment of axon	Trigger zone in many neurons; integrates EPSPs and IPSPs and, if sum is a depolarization that reaches threshold, initiates action potential (nerve impulse).
	Axon	Propagates (conducts) nerve impulses from initial segment (or from dendrites of sensory neurons) to axon terminals in a self-reinforcing manner; impulse amplitude does not change as it propagates along the axon.
	Axon terminals and synaptic end bulbs (or varicosities)	Inflow of Ca^{2+} caused by depolarizing phase of nerve impulse triggers neurotransmitter release by exocytosis of synaptic vesicles.

Plasma membrane includes chemically gated channels
Plasma membrane includes voltage-gated Na^+ and K^+ channels
Plasma membrane includes voltage-gated Ca^{2+} channels

neurotransmitters are also hormones released into the bloodstream by endocrine cells in organs throughout the body. Within the brain, certain neurons, called **neurosecretory cells,** also secrete hormones. Neurotransmitters can be divided into two classes based on size: small-molecule neurotransmitters and neuropeptides.

Small-Molecule Neurotransmitters

The small-molecule neurotransmitters include acetylcholine, amino acids, biogenic amines, ATP and other purines, and nitric oxide.

Acetylcholine

The best-studied neurotransmitter is **acetylcholine (ACh),** which is released by many PNS neurons and by some CNS neurons. ACh is an excitatory neurotransmitter at some synapses, such as the neuromuscular junction, where it acts directly to open ligand-gated cation channels. It is also an inhibitory neurotransmitter at other synapses, where its effects on ion channels occur indirectly via receptors that link to a G protein. For example, ACh slows heart rate at inhibitory synapses made by

parasympathetic neurons of the vagus nerve (cranial nerve X). The enzyme *acetylcholinesterase (AChE)* inactivates ACh by splitting it into acetate and choline fragments.

Amino Acids

Several amino acids are neurotransmitters in the CNS. **Glutamate** (glutamic acid) and **aspartate** (aspartic acid) have powerful excitatory effects. Most excitatory neurons in the CNS and perhaps half of the synapses in the brain communicate via glutamate. At some glutamate synapses, binding of the neurotransmitter to its receptors opens Ca^{2+} channels. The consequent inflow of calcium ions produces an EPSP. Inactivation of glutamate occurs via reuptake. Glutamate transporters actively transport glutamate back into the synaptic end bulbs and neighboring neuroglia.

Gamma aminobutyric (GAM-ma am-i-nō-bū-TIR-ik) **acid (GABA)** and **glycine** are important inhibitory neurotransmitters. Both cause IPSPs by opening Cl^- channels. GABA is found only in the CNS, where it is the most common inhibitory neurotransmitter. As many as one-third of all brain synapses use GABA. Antianxiety drugs such as diazepam (Valium) enhance the action of GABA. In the spinal cord, about half of the inhibitory synapses use the amino acid glycine and half use GABA.

Excitotoxicity

A high level of glutamate in the interstitial fluid of the CNS causes **excitotoxicity**—destruction of neurons through prolonged activation of excitatory synaptic transmission. The most common cause of excitotoxicity is oxygen deprivation of the brain due to ischemia (inadequate blood flow), as happens during a "brain attack" or stroke. Lack of oxygen causes the glutamate transporters to fail, and glutamate accumulates in the interstitial spaces between neurons and glia, literally stimulating the neurons to death. Clinical trials are underway to see if antiglutamate drugs can offer some protection from excitotoxicity after a person suffers a stroke. ■

Biogenic Amines

Certain amino acids are modified and decarboxylated (carboxyl group removed) to produce biogenic amines. Those that are prevalent in the nervous system include norepinephrine, epinephrine, dopamine, and serotonin. Depending on the type of receptor (there are three or more different types for each biogenic amine), biogenic amines may cause either excitation or inhibition.

Norepinephrine (NE) plays roles in arousal (awakening from deep sleep), dreaming, and regulating mood. A smaller number of neurons in the brain use **epinephrine** as a neurotransmitter. Both epinephrine and norepinephrine also serve as hormones. Cells of the adrenal medulla, the inner portion of the adrenal gland, release them into the blood.

Brain neurons containing the neurotransmitter **dopamine (DA)** are active during emotional responses, addictive behaviors, and pleasurable experiences. In addition, dopamine-releasing neurons help regulate skeletal muscle tone and some aspects of movement due to contraction of skeletal muscles. The muscular stiffness that occurs in Parkinson disease is due to degeneration of neurons that release dopamine (see page 521). One form of schizophrenia is due to accumulation of excess dopamine.

Norepinephrine, dopamine, and epinephrine are **catecholamines** (cat-e-KŌL-a-mēns). They all include an amino group (—NH₂) and a catechol ring composed of six carbons and two adjacent hydroxyl (—OH) groups. Catecholamines are synthesized from the amino acid tyrosine. Inactivation of catecholamines occurs via reuptake into synaptic end bulbs. Then they are either recycled back into the synaptic vesicles or destroyed by the enzymes. The two enzymes that break down catecholamines are **catechol-*O*-methyltransferase** (kat′-e-kōl-ō-meth-il-TRANS-fer-ās), or **COMT,** and **monoamine oxidase** (mon-ō-AM-īn OK-si-dās), or **MAO.**

Serotonin, also known as **5-hydroxytryptamine (5-HT),** is concentrated in the neurons in a part of the brain called the raphe nucleus. It is thought to be involved in sensory perception, temperature regulation, control of mood, appetite, and the induction of sleep.

ATP and Other Purines

The characteristic ring structure of the adenosine portion of ATP (shown in Figure 2.25 on page 54) is called a purine ring. Adenosine itself, as well as its triphosphate, diphosphate, and monophosphate derivatives (ATP, ADP, and AMP), is an excitatory neurotransmitter in both the CNS and the PNS. Most of the synaptic vesicles that contain ATP also contain another neurotransmitter. In the PNS, ATP and norepinephrine are released together from some sympathetic neurons, whereas ATP and acetylcholine are released together from some parasympathetic neurons.

Nitric Oxide

The simple gas **nitric oxide (NO)** is an important neurotransmitter that has widespread effects throughout the body. NO includes a single nitrogen atom, by contrast to nitrous oxide (N₂O), or laughing gas, which has two nitrogen atoms. N₂O is sometimes used as an anesthetic during dental procedures.

The enzyme **nitric oxide synthase (NOS)** catalyzes formation of NO from the amino acid arginine. Based on the presence of NOS, it is estimated that more than 2% of the neurons in the brain produce NO. Unlike all previously known neurotransmitters, NO is not synthesized in advance and packaged into synaptic vesicles. Rather, it is formed on demand and acts immediately. Its action is brief because NO is a highly reactive free radical. It exists for less than 10 seconds before it combines with oxygen and water to form inactive nitrates and nitrites. Because NO is lipid soluble, it diffuses out of cells that produce it and into neighboring cells, where it activates an enzyme for production of a second messenger called cyclic GMP. Some research suggests that NO plays a role in memory and learning.

The first recognition of NO as a regulatory molecule was the discovery in 1987 that EDRF (endothelium-derived relaxing factor) was actually NO. Endothelial cells in blood vessel walls release NO, which diffuses into neighboring smooth muscle cells and causes relaxation. The result is vasodilation, an increase in blood vessel diameter. The effects of such vasodilation range from a lowering of blood pressure to erection of the penis in males. Sildenafil (Viagra) alleviates erectile dysfunction (impotence) by enhancing the effect of NO. In larger quantities, NO is highly toxic. Phagocytic cells, such as macrophages and certain white blood cells, produce NO to kill microbes and tumor cells.

Neuropeptides

Neurotransmitters consisting of 3 to 40 amino acids linked by peptide bonds are called **neuropeptides** (noor-ō-PEP-tīds). Numerous and widespread in both the CNS and the PNS, neuropeptides have both excitatory and inhibitory actions. Neuropeptides are formed in the neuron cell body, packaged into vesicles, and transported to axon terminals. Besides their role as neurotransmitters, many neuropeptides serve as hormones that regulate physiological responses elsewhere in the body.

In 1974, scientists discovered that certain brain neurons have plasma membrane receptors for opiate drugs such as morphine and heroin. The quest to find the naturally occurring substances that use these receptors brought to light the first neuropeptides: two molecules, each a chain of five amino acids, named **enkephalins** (en-KEF-a-lins). Their potent analgesic (pain-relieving) effect is 200 times stronger than morphine. Besides the enkephalins, other so-called *opioid peptides* include the **endorphins** (en-DOR-fins) and **dynorphins** (dī-NOR-fins). It is thought that opioid peptides are the body's natural painkillers. Acupuncture may produce analgesia (loss of pain sensation) by increasing the release of opioids. They have also been linked to improved memory and learning; feelings of pleasure or euphoria; control of body temperature; regulation of hormones that affect the onset of puberty, sexual drive, and reproduction; and mental illnesses such as depression and schizophrenia.

Another neuropeptide, **substance P,** is released by neurons that transmit pain-related input from peripheral pain receptors into the central nervous system. Enkephalin suppresses the release of substance P, thus decreasing the number of nerve impulses being relayed to the brain for pain sensations. Substance P has also been shown to counter the effects of certain nerve-damaging chemicals, prompting speculation that it might prove useful as a treatment for nerve degeneration.

Table 12.4 provides brief portraits of these neuropeptides, as well as others that are discussed in later chapters.

Table 12.4	Neuropeptides
Substance	**Comment**
Substance P	Found in sensory neurons, spinal cord pathways, and parts of brain associated with pain; enhances perception of pain.
Enkephalins	Inhibit pain impulses by suppressing release of substance P.
Endorphins	Inhibit pain by blocking release of substance P; may have a role in memory and learning, sexual activity, and control of body temperature.
Dynorphins	May be related to controlling pain and registering emotions.
Hypothalamic releasing and inhibiting hormones	Produced by the hypothalamus; regulate the release of hormones by the anterior pituitary.
Angiotensin II	Stimulates thirst; may regulate blood pressure in the brain. As a hormone causes vasoconstriction and promotes release of aldosterone, which increases the rate of salt and water reabsorption by the kidneys.
Cholecystokinin (CCK)	Found in the brain and small intestine; may regulate feeding as a "stop eating" signal. As a hormone, regulates pancreatic enzyme secretion during digestion, and contraction of smooth muscle in the gastrointestinal tract.

Mimicking and Blocking the Effects of Neurotransmitters

Substances naturally present in the body as well as drugs and toxins can modify the effects of neurotransmitters in several ways:

1. Neurotransmitter synthesis can be stimulated or inhibited. For instance, many patients with Parkinson disease receive benefit from the drug L-DOPA because it is a precursor of dopamine. For a while, taking L-DOPA boosts dopamine production in affected brain areas.

2. Neurotransmitter release can be enhanced or blocked. Amphetamines promote release of dopamine and norepinephrine. Botulinum toxin causes paralysis by blocking release of acetylcholine from somatic motor neurons.

3. The neurotransmitter receptors can be activated or blocked. An agent that binds to receptors and enhances or mimics the effect of a natural neurotransmitter is an **agonist.** Isoproterenol (Isuprel) is a powerful agonist of epinephrine and norepinephrine. It can be used to dilate the airways during an asthma attack. An agent that binds to and blocks neurotransmitter receptors is an **antagonist.** Zyprexa, a drug prescribed for schizophrenia, is an antagonist of serotonin and dopamine.

4. Neurotransmitter removal can be stimulated or inhibited.

For example, cocaine produces euphoria—intensely pleasurable feelings—by blocking transporters for dopamine reuptake. This action allows dopamine to linger longer in synaptic clefts, producing excessive stimulation of certain brain regions. ■

▶ CHECKPOINT

15. Which neurotransmitters are excitatory and which are inhibitory?

16. In what ways is nitric oxide different from all previously known neurotransmitters?

NEURAL CIRCUITS

▶ OBJECTIVE

• **Identify the various types of neural circuits in the nervous system.**

The CNS contains billions of neurons organized into complicated networks called **neural circuits,** each of which is a functional group of neurons that processes a specific kind of information. In a **simple series circuit,** a presynaptic neuron stimulates only a single postsynaptic neuron. The second neuron then stimulates another, and so on. Most neural circuits, however, are more complex.

A single presynaptic neuron may synapse with several postsynaptic neurons. Such an arrangement, called **divergence,** permits one presynaptic neuron to influence several postsynaptic neurons (or several muscle fibers or gland cells) at the same time. In a **diverging circuit,** the nerve impulse from a single presynaptic neuron causes the stimulation of increasing numbers of cells along the circuit (Figure 12.16a). For example, a small number of neurons in the brain that govern a particular body movement stimulate a much larger number of neurons in the spinal cord. Sensory signals also spread into diverging circuits and are often relayed to several regions of the brain. This arrangement has the effect of amplifying the signal.

In another arrangement, called **convergence,** several presynaptic neurons synapse with a single postsynaptic neuron. This arrangement permits more effective stimulation or inhibition of the postsynaptic neuron. In a **converging circuit** (Figure 12.16b), the postsynaptic neuron receives nerve impulses from several different sources. For example, a single motor neuron that synapses with skeletal muscle fibers at neuromuscular junctions receives input from several pathways that originate in different brain regions.

Some circuits are constructed so that once the presynaptic cell is stimulated, it will cause the postsynaptic cell to transmit a series of nerve impulses. One such circuit is called a **reverber-** **ating circuit** (Figure 12.16c). In this pattern, the incoming impulse stimulates the first neuron, which stimulates the second, which stimulates the third, and so on. Branches from later neurons synapse with earlier ones. This arrangement sends impulses back through the circuit again and again. The output signal may last from a few seconds to many hours, depending on the number of synapses and the arrangement of neurons in the circuit. Inhibitory neurons may turn off a reverberating circuit after a period of time. Among the body responses thought to be the result of output signals from reverberating circuits are breathing, coordinated muscular activities, waking up, sleeping (when reverberation stops), and short-term memory.

A fourth type of circuit is the **parallel after-discharge circuit** (Figure 12.16d). In this circuit, a single presynaptic cell stimulates a group of neurons, each of which synapses with a common postsynaptic cell. A differing number of synapses between the first and last neurons imposes varying synaptic delays, so that the last neuron exhibits multiple EPSPs or IPSPs. If the input is excitatory, the postsynaptic neuron then can send out a stream of impulses in quick succession. Parallel after-discharge circuits may be involved in precise activities such as mathematical calculations.

▶ **CHECKPOINT**

17. What is a neural circuit?

18. What are the functions of diverging, converging, reverberating, and parallel after-discharge circuits?

Figure 12.16 Examples of neural circuits.

🔑 A neural circuit is a functional group of neurons that processes a specific kind of information.

(a) Diverging circuit   (b) Converging circuit   (c) Reverberating circuit   (d) Parallel after-discharge circuit

 A motor neuron in the spinal cord typically receives input from neurons that originate in several different regions of the brain. Is this an example of convergence or divergence?

REGENERATION AND REPAIR OF NERVOUS TISSUE

▶ O B J E C T I V E S

- **Define plasticity and neurogenesis.**
- **Describe the events involved in damage and repair of peripheral nerves.**

Throughout one's life, the nervous system exhibits **plasticity,** the capability to change based on experience. At the level of individual neurons, the changes that can occur include the sprouting of new dendrites, synthesis of new proteins, and changes in synaptic contacts with other neurons. Undoubtedly, both chemical and electrical signals drive the changes that occur. Despite plasticity, however, mammalian neurons have very limited powers of **regeneration,** the capability to replicate or repair themselves. In the PNS, damage to dendrites and myelinated axons may be repaired if the cell body remains intact and if the Schwann cells that produce myelination remain active. In the CNS, little or no repair of damage to neurons occurs. Even when the cell body remains intact, a severed axon cannot be repaired or regrown.

Neurogenesis in the CNS

Neurogenesis—the birth of new neurons from undifferentiated stem cells—occurs regularly in some animals. For example, new neurons appear and disappear every year in some songbirds. Until recently, the dogma in humans and other primates was "no new neurons" in the adult brain. Then, in 1992, Canadian researchers published their unexpected finding that **epidermal growth factor (EGF)** stimulated cells taken from the brains of adult mice to proliferate into both neurons and astrocytes. Previously, EGF was known to trigger mitosis in a variety of non-neuronal cells and to promote wound healing and tissue regeneration. In 1998 scientists discovered that significant numbers of new neurons do arise in the adult human hippocampus, an area of the brain that is crucial for learning.

The nearly complete lack of neurogenesis in other regions of the brain and spinal cord, however, seems to result from two factors: (1) inhibitory influences from neuroglia, particularly oligodendrocytes, and (2) absence of growth-stimulating cues that were present during fetal development. Axons in the CNS are myelinated by oligodendrocytes that do not form neurolemmas (sheaths of Schwann). In addition, CNS myelin is one of the factors inhibiting regeneration of neurons. Perhaps this same mechanism stops axonal growth once a target region has been reached during development. Also, after axonal damage, nearby astrocytes proliferate rapidly, forming a type of scar tissue and constituting a physical barrier to regeneration. Thus, injury of the brain or spinal cord usually is permanent. Nevertheless, ongoing research seeks ways to improve the environment for existing spinal cord axons to bridge the injury gap. Scientists also are trying to find ways to stimulate dormant stem cells to replace neurons lost through damage or disease and to develop tissue-cultured neurons that can be used for transplantation purposes.

Damage and Repair in the PNS

Axons and dendrites that are associated with a neurolemma may undergo repair if the cell body is intact, if the Schwann cells are functional, and if scar tissue formation does not occur too rapidly (Figure 12.17). Most nerves in the PNS consist of processes that are covered with a neurolemma. A person who injures axons of a nerve in an upper limb, for example, has a good chance of regaining nerve function.

Figure 12.17   Damage and repair of a neuron in the PNS.

 Myelinated axons in the peripheral nervous system may be repaired if the cell body remains intact and if Schwann cells remain active.

(a) Normal neuron

(b) Chromatolysis and Wallerian degeneration

(c) Regeneration

What is the role of the neurolemma in regeneration?

When there is damage to an axon, changes usually occur both in the cell body of the affected neuron and in the portion of the axon distal to the site of injury. Changes also may occur in the portion of the axon proximal to the site of injury.

About 24 to 48 hours after injury to a process of a normal central or peripheral neuron (Figure 12.17a), the Nissl bodies break up into fine granular masses. This alteration is called **chromatolysis** (krō′-ma-TOL-i-sis; *chromato-* = color; *-lysis* = destruction). By the third to fifth day, the part of the axon distal to the damaged region becomes slightly swollen and then breaks up into fragments; the myelin sheath also deteriorates (Figure 12.17b). Even though the axon and myelin sheath degenerate, the neurolemma remains. Degeneration of the distal portion of the axon and myelin sheath is called **Wallerian degeneration.**

Following chromatolysis, signs of recovery in the cell body become evident. Macrophages phagocytize the degeneration debris. Synthesis of RNA and protein accelerates, which favors rebuilding or **regeneration** of the axon. Recovery often takes several months. The Schwann cells on either side of the injured site multiply by mitosis, grow toward each other, and may form a **regeneration tube** across the injured area (Figure 12.17c). The tube guides growth of a new axon from the proximal area across the injured area into the distal area previously occupied by the original axon. However, new axons cannot grow if the gap at the site of injury is too large or if the gap becomes filled with collagen fibers.

During the first few days following damage, buds of regenerating axons begin to invade the tube formed by the Schwann cells (Figure 12.17b). Axons from the proximal area grow at a rate of about 1.5 mm (0.06 in.) per day across the area of damage, find their way into the distal regeneration tubes, and grow toward the distally located receptors and effectors. Thus, some sensory and motor connections are reestablished and some functions restored. In time, the Schwann cells form a new myelin sheath.

▶ C H E C K P O I N T

19. What factors contribute to a lack of neurogenesis in most parts of the brain?

20. What is the function of the regeneration tube in repair of neurons?

DISORDERS: HOMEOSTATIC IMBALANCES

Multiple Sclerosis

Multiple sclerosis (MS) is a disease that causes a progressive destruction of myelin sheaths of neurons in the CNS. It afflicts about 350,000 people in the United States and 2 million people worldwide. It usually appears between the ages of 20 and 40, affecting females twice as often as males. MS is most common in whites, less common in blacks, and rare in Asians. MS is an autoimmune disease—the body's own immune system spearheads the attack. The condition's name describes the anatomical pathology: In *multiple* regions the myelin sheaths deteriorate to *scleroses,* which are hardened scars or plaques. Magnetic resonance imaging (MRI) studies reveal numerous plaques in the white matter of the brain and spinal cord. The destruction of myelin sheaths slows and then short-circuits propagation of nerve impulses.

The most common form of the condition is relapsing-remitting MS, which usually appears in early adulthood. The first symptoms may include a feeling of heaviness or weakness in the muscles, abnormal sensations, or double vision. An attack is followed by a period of remission during which the symptoms temporarily disappear. One attack follows another over the years, usually every year or two. The result is a progressive loss of function interspersed with remission periods, during which symptoms abate.

Although the cause of MS is unclear, both genetic susceptibility and exposure to some environmental factor (perhaps a herpesvirus) appear to contribute. Since 1993 many patients with relapsing-remitting MS have been treated with injections of beta interferon. This treatment lengthens the time between relapses, decreases the severity of relapses, and slows formation of new lesions in some cases. Unfortunately, not all MS patients can tolerate beta interferon, and therapy becomes less effective as the disease progresses.

Epilepsy

Epilepsy is characterized by short, recurrent, periodic attacks of motor, sensory, or psychological malfunction, although it almost never affects intelligence. The attacks, called *epileptic seizures,* afflict about 1% of the world's population. They are initiated by abnormal, synchronous electrical discharges from millions of neurons in the brain, perhaps resulting from abnormal reverberating circuits. The discharges stimulate many of the neurons to send nerve impulses over their conduction pathways. As a result, lights, noise, or smells may be sensed when the eyes, ears, and nose have not been stimulated. Moreover, the skeletal muscles of a person having a seizure may contract involuntarily. *Partial seizures* begin in a small focus on one side of the brain and produce milder symptoms, whereas *generalized seizures* involve larger areas on both sides of the brain and loss of consciousness.

Epilepsy has many causes, including brain damage at birth (the most common cause); metabolic disturbances (hypoglycemia, hypocalcemia, uremia, hypoxia); infections (encephalitis or meningitis); toxins (alcohol, tranquilizers, hallucinogens); vascular disturbances (hemorrhage, hypotension); head injuries; and tumors and abscesses of the brain. However, most epileptic seizures have no demonstrable cause.

Epileptic seizures often can be eliminated or alleviated by antiepileptic drugs, such as phenytoin, carbamazepine, and valproate sodium. An implantable device that stimulates the vagus nerve (cranial nerve X) also has produced dramatic results in reducing seizures in some patients whose epilepsy was not well-controlled by drugs.

MEDICAL TERMINOLOGY

Demyelination   (dē-mī-e-li-NĀ-shun) Loss or destruction of myelin sheaths around axons in the CNS or PNS.

Guillain-Barré Syndrome (GBS)   An acute demyelinating disorder in which macrophages strip myelin from axons in the PNS. It is the most common cause of acute paralysis in North America and Europe and may result from the immune system's response to a bacterial infection. Most patients recover completely or partially, but about 15% remain paralyzed.

Neuropathy   (noo-ROP-a-thē; *neuro-* = a nerve; *-pathy* = disease) Any disorder that affects the nervous system but particularly a

disorder of a cranial or spinal nerve. An example is **facial neuropathy** (Bell's palsy), a disorder of the facial nerve (cranial nerve VII).

Rabies   (RĀ-bēz; *rabi-* = mad, raving) A fatal disease caused by a virus that reaches the CNS via fast axonal transport. It is usually transmitted by the bite of a dog or other meat-eating animal. The symptoms are excitement, aggressiveness, and madness, followed by paralysis and death.

STUDY OUTLINE

OVERVIEW OF THE NERVOUS SYSTEM (p. 386)

1. The central nervous system (CNS) consists of the brain and spinal cord. The peripheral nervous system (PNS) consists of all nervous tissue outside the CNS.
2. Structures that make up the nervous system are the brain, 12 pairs of cranial nerves and their branches, the spinal cord, 31 pairs of spinal nerves and their branches, ganglia, enteric plexuses, and sensory receptors.
3. The nervous system helps maintain homeostasis and integrates all body activities by sensing changes (sensory function), interpreting them (integrative function), and reacting to them (motor function).
4. Sensory (afferent) neurons carry sensory information from cranial and spinal nerves into the brain and spinal cord or from a lower to a higher level in the spinal cord and brain. Interneurons have short axons that contact nearby neurons in the brain, spinal cord, or a ganglion. Motor (efferent) neurons carry information from the brain toward the spinal cord or out of the brain and spinal cord into cranial or spinal nerves.
5. Components of the PNS include the somatic nervous system (SNS), autonomic nervous system (ANS), and enteric nervous system (ENS).
6. The SNS consists of neurons that conduct impulses from somatic and special sense receptors to the CNS and motor neurons from the CNS to skeletal muscles.
7. The ANS contains sensory neurons from visceral organs and motor neurons that convey impulses from the CNS to smooth muscle tissue, cardiac muscle tissue, and glands.
8. The ENS consists of neurons in enteric plexuses in the gastrointestinal (GI) tract that function independently of the ANS and CNS to some extent. The ENS monitors sensory changes in and controls operation of the GI tract.

HISTOLOGY OF NERVOUS TISSUE (p. 388)

1. Nervous tissue consists of neurons (nerve cells) and neuroglia. Neurons have the property of electrical excitability and are responsible for most unique functions of the nervous system: sensing,

thinking, remembering, controlling muscle activity, and regulating glandular secretions. Neuroglia support, nurture, and protect the neurons and maintain homeostasis in the interstitial fluid that bathes neurons.

2. Most neurons have three parts. The dendrites are the main receiving or input region. Integration occurs in the cell body, which includes typical cellular organelles. The output part typically is a single axon, which propagates nerve impulses toward another neuron, a muscle fiber, or a gland cell.
3. Synapses are the site of functional contact between two excitable cells. Axon terminals contain synaptic vesicles filled with neurotransmitter molecules.
4. Slow axonal transport and fast axonal transport are systems for conveying materials to and from the cell body and axon terminals.
5. On the basis of their structure, neurons are either multipolar, bipolar, or unipolar.
6. Neuroglia include astrocytes, oligodendrocytes, microglia, ependymal cells, Schwann cells, and satellite cells (summarized in Table 12.1 on pages 392–393).
7. Two types of neuroglia produce myelin sheaths: oligodendrocytes myelinate axons in the CNS, and Schwann cells myelinate axons in the PNS.
8. White matter consists of aggregates of myelinated processes, whereas gray matter contains cell bodies, dendrites, and axon terminals of neurons, unmyelinated axons, and neuroglia.
9. In the spinal cord, gray matter forms an H-shaped inner core that is surrounded by white matter. In the brain, a thin, superficial shell of gray matter covers the cerebral and cerebellar hemispheres.

ELECTRICAL SIGNALS IN NEURONS (p. 395)

1. Neurons communicate with one another using action potentials, which allow communication over both short and long distances within the body, and graded potentials, which are used for short-distance communication only.
2. The electrical signals produced by neurons and muscle fibers rely on four kinds of ion channels: leakage channels, voltage-gated channels, ligand-gated channels, and mechanically gated channels.

3. A typical value for the resting membrane potential is −70 mV. A cell that exhibits a membrane potential is polarized.

4. A graded potential is a small deviation from the resting membrane potential that occurs because ligand-gated or mechanically gated channels open or close. A graded potential makes the membrane potential either more negative (more polarization), termed hyperpolarization, or less negative (less polarization), termed depolarization.

5. The amplitude of a graded potential varies, depending on the strength of the stimulus.

6. According to the all-or-none principle, if a stimulus is strong enough to generate an action potential, the impulse generated is of a constant size. A stronger stimulus does not generate a larger impulse.

7. During an action potential, voltage-gated Na^+ and K^+ channels open and close in sequence. This results first in depolarization, the loss, and then the reversal of membrane polarization (from −70 mV to +30 mV). Then repolarization, the recovery of the resting membrane potential (from +30 mV to −70 mV), occurs.

8. During the first part of the refractory period (RP), another impulse cannot be generated at all (absolute RP); a little later, it can be triggered only by a larger-than-normal stimulus (relative RP).

9. Because an action potential conducts or propagates (travels) from point to point along the membrane without getting smaller, it is useful for long-distance communication. Table 12.2 on page 404 compares graded potentials and action potentials.

10. Nerve impulse propagation in which the impulse "leaps" from one node of Ranvier to the next along a myelinated axon is saltatory conduction. Saltatory conduction is faster than continuous conduction.

11. Axons with larger diameters conduct impulses at higher speeds than do axons with smaller diameters.

12. The intensity of a stimulus is encoded in the frequency of action potentials and in the number of sensory neurons that are recruited.

SIGNAL TRANSMISSION AT SYNAPSES (p. 404)

1. A synapse is the functional junction between one neuron and another, or between a neuron and an effector such as a muscle or a gland. The two types of synapses are electrical and chemical.

2. A chemical synapse produces only one-way information transfer—from a presynaptic neuron to a postsynaptic neuron.

3. An excitatory neurotransmitter is one that can depolarize the postsynaptic neuron's membrane, bringing the membrane potential closer to threshold. An inhibitory neurotransmitter hyperpolarizes the membrane of the postsynaptic neuron.

4. Neurotransmitter is removed from the synaptic cleft in three ways: diffusion, enzymatic degradation, and uptake by cells (neurons and neuroglia).

5. If several presynaptic end bulbs release their neurotransmitter at about the same time, the combined effect may generate a nerve impulse, due to summation. Summation may be spatial or temporal.

6. The postsynaptic neuron is an integrator. It receives excitatory and inhibitory signals, integrates them, and then responds accordingly.

7. Table 12.3 on page 408 summarizes the structural and functional elements of a neuron.

NEUROTRANSMITTERS (p. 407)

1. Both excitatory and inhibitory neurotransmitters are present in the CNS and the PNS. A given neurotransmitter may be excitatory in some locations and inhibitory in others.

2. Neurotransmitters can be divided into two classes based on size: (1) small-molecule neurotransmitters (acetylcholine, amino acids, biogenic amines, ATP and other purines, and nitric oxide), and (2) neuropeptides, which are composed of 3 to 40 amino acids.

3. Chemical synaptic transmission may be modified by affecting synthesis, release, or removal of a neurotransmitter or by blocking or stimulating neurotransmitter receptors.

4. Table 12.4 on page 410 describes several important neuropeptides.

NEURAL CIRCUITS (p. 410)

1. Neurons in the central nervous system are organized into networks called neural circuits.

2. Neural circuits include simple series, diverging, converging, reverberating (oscillatory), and parallel after-discharge circuits.

REGENERATION AND REPAIR OF NERVOUS TISSUE (p. 412)

1. The nervous system exhibits plasticity (the capability to change based on experience), but it has very limited powers of regeneration (the capability to replicate or repair damaged neurons).

2. Neurogenesis, the birth of new neurons from undifferentiated stem cells, is normally very limited. Repair of damaged axons does not occur in most regions of the CNS.

3. Axons and dendrites that are associated with a neurolemma in the PNS may undergo repair if the cell body is intact, the Schwann cells are functional, and scar tissue formation does not occur too rapidly.

SELF-QUIZ QUESTIONS

Fill in the blanks in the following statements.

1. The subdivisions of the PNS are the _____, _____, and _____.

2. A cell that exhibits a membrane potential is said to be _____.

Indicate whether the following statements are true or false.

3. The nearly complete lack of neurogenesis and regenerative success in the intact CNS seems to result from inhibitory influences from neuroglia and absence of growth cues that were present during development.

4. The two main factors contributing to the resting membrane potential are the equal distribution of ions across the plasma membrane and the relative impermeability of the plasma membrane to sodium and potassium ions.

Choose the one best answer to the following questions.

5. Which of the following statements are *true*? (1) The sensory function of the nervous system involves sensory receptors sensing certain changes in the internal and external environments. (2) Sensory neurons receive electrical signals from sensory receptors. (3)

The integrative function of the nervous system involves analyzing sensory information, storing some aspects of it, and making decisions regarding appropriate behaviors. (4) Interneurons carry nerve impulses to effectors. (5) Motor function involves responding to integration decisions. (a) 1, 2, 3, and 4, (b) 2, 4, and 5, (c) 1, 2, 3, and 5, (d) 1, 2, and 4, (e) 2, 3, 4, and 5.

6. Which of the following is *not* a neuronal circuit in the body? (a) diverging circuit, (b) parallel predischarge circuit, (c) converging circuit, (d) reverberating circuit, (e) parallel after-discharge circuit.

7. Which of the following statements are *true*? (1) Two or more neurotransmitters may be present in many neurons. (2) Some amino acids, including glutamate and aspartate, are neurotransmitters. (3) The catecholamine neurotransmitter acetylcholine is synthesized from the amino acid tyrosine. (4) The simple gas nitric oxide functions as a neurotransmitter. (5) An excitatory neurotransmitter can never be inhibitory regardless of the neuron that produces it. (a) 1, 2, and 4, (b) 2, 3, and 5, (c) 1, 3, and 5, (d) 1, 3, and 4, (e) 2, 4, and 5.

8. Which of the following statements are *true*? (1) If the excitatory effect is greater than the inhibitory effect but less than the threshold of stimulation, the result is a subthreshold EPSP. (2) If the excitatory effect is greater than the inhibitory effect and reaches or surpasses the threshold level of stimulation, the result is a threshold or suprathreshold EPSP and one or more nerve impulses. (3) If the inhibitory effect is greater than the excitatory effect, the membrane hyperpolarizes, resulting in inhibition of the postsynaptic neuron and the inability of the neuron to generate a nerve impulse. (4) The greater the summation of hyperpolarizations, the more likely a nerve impulse will be initiated. (5) Inhibitory postsynaptic potentials often result from the opening of ligand-gated Cl^- or K^+ channels. (a) 1, 4, and 5, (b) 2, 4, and 5, (c) 1, 3, and 5, (d) 2, 3, and 4, (e) 1, 2, 3, and 5.

9. Which of the following statements are *true*? (1) The basic types of ion channels are gated, leakage, and electrical. (2) Ion channels allow for the generation of graded potentials and action potentials. (3) The major stimuli that operate gated channels are voltage changes, ligands (chemicals), and mechanical pressure. (4) Excitable cells that possess ligand-gated or mechanically gated channels in the plasma membrane produce action potentials when a stimulus causes those channels to open or close. (5) A graded potential is a small deviation from the resting potential that makes the membrane either more polarized or less polarized. (a) 1, 2, and 3, (b) 2, 3, and 4, (c) 2, 3, and 5, (d) 2, 3, 4, and 5, (e) 1, 3, and 5.

10. Which of the following statements are *true*? (1) The frequency of impulses and the number of activated sensory neurons encodes differences in stimuli intensity. (2) Larger-diameter axons conduct nerve impulses faster than smaller-diameter ones. (3) Continuous conduction is faster than saltatory conduction. (4) The diameter of an axon and the presence or absence of a myelin sheath are the most important factors that determine the speed of nerve impulse propagation. (5) Action potentials are localized, whereas graded potentials are propagated. (a) 1, 3, and 5, (b) 3 and 4, (c) 2, 4, and 5, (d) 2 and 4, (e) 1, 2, and 4.

11. Neurotransmitters are removed from the synaptic cleft by (1) axonal transport, (2) diffusion, (3) neurosecretory cells, (4) enzymatic breakdown, (5) cellular uptake. (a) 1, 2, 3, and 4, (b) 2, 4, and 5, (c) 2, 3, and 4, (d) 1, 4, and 5, (e) 1, 2, 3, 4, and 5.

12. Which of the following is *not* considered part of the PNS? (1) cranial nerves, (2) spinal cord, (3) ganglia, (4) spinal nerves, (5) enteric plexuses. (a) 2 only, (b) 1, 3, and 5, (c) 2, 3, and 5, (d) 5 only, (e) 2 and 5.

13. Match the following:

_____ (a) neurons with just one process extending from the cell body; are always sensory neurons

_____ (b) small phagocytic neuroglia; serve as the brain's immune cells

_____ (c) help maintain an appropriate chemical environment for generation of action potentials by neurons; part of the blood–brain barrier

_____ (d) provide myelin sheath for CNS axons

_____ (e) contains neuronal cell bodies, dendrites, axon terminals, unmyelinated axons, and neuroglia

_____ (f) a cluster of cell bodies and dendrites of neurons within the CNS

_____ (g) form CSF and assist its circulation

_____ (h) neurons having several dendrites and one axon; most common neuronal type

_____ (i) neurons with one main dendrite and one axon; found in the retina of the eye

_____ (j) provide myelin sheath for PNS axons

_____ (k) support neurons in PNS ganglia

_____ (l) a cluster of neuronal cell bodies located outside the brain and spinal cord

_____ (n) aggregation of myelinated processes from many neurons

(1) astrocytes
(2) oligodendrocytes
(3) ganglia
(4) ependymal cells
(5) satellite cells
(6) unipolar neurons
(7) bipolar neurons
(8) multipolar neurons
(9) gray matter
(10) white matter
(11) microglia
(12) Schwann cells
(13) nucleus

14. Match the following:

_____ (a) a sequence of rapidly occurring events that decreases and eventually reverses the membrane potential and then restores it to the resting state; a nerve impulse

_____ (b) a small deviation from the resting membrane potential that makes the membrane either more or less negative

_____ (c) period of time when a second action potential can be initiated with a very strong stimulus

_____ (d) the minimum level of depolarization required for a nerve impulse to be generated

_____ (e) the recovery of the resting membrane potential

_____ (f) a neurotransmitter-caused depolarization of the postsynaptic membrane

_____ (g) a neurotransmitter-caused hyperpolarization of the postsynaptic membrane

_____ (h) time during which a neuron cannot produce an action potential even with a very strong stimulus

_____ (i) polarization that is less negative than the resting level

_____ (j) the hyperpolarization that occurs after the repolarizing phase of an action potential

(1) graded potential
(2) action potential
(3) excitatory postsynaptic potential
(4) inhibitory postsynaptic potential
(5) absolute refractory period
(6) repolarization
(7) after hyperpolarizing phase
(8) threshold
(9) relative refractory period
(10) depolarizing graded potential

15. Match the following:

_____ (a) the part of the neuron that contains the nucleus

_____ (b) rough endoplasmic reticulum in neurons

_____ (c) store neurotransmitter

_____ (d) the process that propagates nerve impulses toward another neuron, muscle fiber, or gland cell

_____ (e) the receiving or input portions of a neuron

_____ (f) a multilayered lipid and protein covering for axons produced by neuroglia

_____ (g) the outer nucleated cytoplasmic layer of a Schwann cell

_____ (h) first portion of the axon

_____ (i) site of communication between two neurons or between a neuron and an effector cell

_____ (j) form the cytoskeleton of a neuron

_____ (k) gaps in the myelin sheath of an axon

_____ (l) general term for any neuronal process

_____ (m) area where the axon joins the cell body

_____ (n) area where nerve impulses arise

(1) myelin sheath
(2) neurolemma
(3) nodes of Ranvier
(4) cell body
(5) Nissl bodies
(6) neurofibrils
(7) dendrites
(8) axon
(9) axon hillock
(10) initial segment
(11) trigger zone
(12) nerve fiber
(13) synapse
(14) synaptic vesicles

CRITICAL THINKING QUESTIONS

1. Elena skipped her anatomy and physiology class (never again!) and borrowed her friend's notes. Unfortunately, her friend doesn't draw very well, and the notes are out of order. The diagram of the motor neuron looks a lot like the one of the converging circuit, and the bipolar neuron resembles a simple circuit. What do you tell Elena to help her straighten out these notes?
HINT *Neuronal circuits involve more than one neuron.*

2. When asked to define "gray matter," Ho answered that it's white matter from a really old person. Explain gray matter to Ho.
HINT *At least he's correct that there is a colored substance which collects in neurons with age.*

3. The buzzing of the alarm clock woke Carrie. She stretched, yawned, and started to salivate as she smelled the brewing coffee. She could feel her stomach rumble. List the divisions of the nervous system that are involved in each of these actions.
HINT *Which activities are voluntary?*

?ANSWERS TO FIGURE QUESTIONS

12.1 The total number of cranial and spinal nerves in your body is $(12 \times 2) + (31 \times 2) = 86$.

12.2 Sensory or afferent neurons carry input to the CNS; motor or efferent neurons carry output from the CNS.

12.3 Dendrites receive (motor neurons or interneurons) or generate (sensory neurons) inputs; the cell body also receives input signals; the axon conducts nerve impulses (action potentials) and transmits the message to another neuron or effector cell by releasing a neurotransmitter at its synaptic end bulbs.

12.4 Nerve impulses arise at the trigger zone.

12.5 The cell body of a pyramidal cell is shaped like a pyramid.

12.6 Myelination increases the speed of nerve impulse conduction.

12.7 Myelin makes white matter look shiny and white.

12.8 A touch on the arm activates mechanically gated channels.

12.9 A typical value for the resting membrane potential in a neuron is -70 mV.

12.10 Graded potentials occur when ligand- or mechanically gated channels open or close.

12.11 Voltage-gated Na^+ channels are open during the depolarizing phase, and voltage-gated K^+ channels are open during the repolarizing phase.

12.12 Yes, because the leakage channels would still allow K^+ to exit more rapidly than Na^+ could enter the axon. Some mammalian myelinated axons have only a few voltage-gated K^+ channels.

12.13 The diameter of an axon, presence or absence of a myelin sheath, and temperature determine propagation speed of a nerve impulse.

12.14 In some electrical synapses (gap junctions), ions may flow equally well in either direction, so either neuron may be the presynaptic one. At a chemical synapse, one neuron releases neurotransmitter and the other neuron has receptors that bind this chemical. Thus, the signal can proceed in only one direction.

12.15 If an IPSP occurred at 55 msec, threshold depolarization would likely not be reached, and a nerve impulse would not be generated.

12.16 A motor neuron receiving input from several other neurons is an example of convergence.

12.17 The neurolemma provides a regeneration tube that guides regrowth of a severed axon.

The Spinal Cord and Spinal Nerves

A Student's View

The spinal cord and its nerves are simply amazing! Just as a computer monitor receives lots of information at incredible speeds, similarly, a brain receives information from the spinal cord and the peripheral nervous system. But, unlike a computer and its monitor, most of the time the nerves send out messages to the brain without you realizing you're doing anything!

From atlas (which carries the weight of the head) to cauda equina (otherwise known as the horse's tail), the spinal cord is protected by a firm yet flexible casing called the spinal column. Most information carriers are protected somehow, like telephone and cable wires, PBX cable (used for electricity), or even cords to your everyday appliances! And most of those information carriers lead to a "brain", like a television or a coffee maker.

When those other appliances are off, they are off. But with our spinal cord and spinal nerves, even while we sleep those information carriers promote homeostasis for our bodies, allowing responses to our internal and external environment.

So whether you're thinking about what you are doing, or not, the spinal cord is at work and thankfully has a great protector (the spinal column), so it can keep on doing its job for "boss", the brain!

Kim Green
Edison Community College

THE SPINAL CORD NERVES AND HOMEOSTASIS

The spinal cord and nerves contribute to homeostasis by providing quick, reflexive responses to many stimuli. The spinal cord is the pathway for sensory impulses to the brain and motor impulses to spinal nerves.

ENERGIZE YOUR STUDY

FOUNDATIONS CD

Anatomy Overview
- The Nervous System

Animation
- Systems Contributions to Homeostasis

Concepts and Connections and Exercises reinforce your understanding

www.wiley.com/college/apcentral

INSIGHTS AND EXPLORATIONS

In movies he was one of the strongest people alive — Superman. In real life, Christopher Reeve, while riding his horse in a jumping competition, was thrown head first to the ground. The first two vertebrae of his spinal column were crushed and his spinal cord damaged leading to paralysis and loss of sensation from his neck down. The National Institutes of Health (NIH) reported that for 1996, more than 10,000 Americans had spinal cord injuries and over 200,000 individuals were living with paralysis of their legs and/or arms. At one time there was little hope that individuals paralyzed by spinal cord damage could regain function, but recently scientists have been able to observe some axon regeneration and recovery in rats with severe spinal cord damage. Come explore how some nerves, those in the peripheral nervous system, often re-grow and recover and what scientists are discovering that may help spinal nerves regenerate.

The spinal cord and spinal nerves contain neural circuits that mediate some of your most rapid reactions to environmental changes. If you pick up something hot, for instance, the grasping muscles may relax and you may drop the hot object even before the sensation of extreme heat or pain reaches your conscious perception. This is an example of a spinal cord reflex—a quick, automatic response to certain kinds of stimuli that involves neurons only in the spinal nerves and spinal cord. Besides processing reflexes, the spinal cord gray matter also is a site for integration (summing) of excitatory postsynaptic potentials (EPSPs) and inhibitory postsynaptic potentials (IPSPs). These graded potentials arise as neurotransmitter molecules interact with their receptors at synapses in the spinal cord. The white matter of the spinal cord contains a dozen major sensory and motor tracts, the "highways" for conduction of sensory nerve impulses toward the brain and motor nerve impulses from the brain toward effector tissues. Recall that the spinal cord is continuous with the brain. Together, they comprise the central nervous system (CNS).

SPINAL CORD ANATOMY

▶ OBJECTIVE

• **Describe the protective structures and the gross anatomical features of the spinal cord.**

Protective Structures

Two types of connective tissue coverings—bony vertebrae and tough, connective tissue meninges—plus a cushion of cerebrospinal fluid (produced in the brain) surround and protect the delicate nervous tissue of the spinal cord (and the brain as well).

Vertebral Column

The spinal cord is located within the vertebral canal of the vertebral column. The vertebral foramina of all the vertebrae, stacked one on top of the other, form the canal. The surrounding vertebrae provide a sturdy shelter for the enclosed spinal cord (see Figure 13.1b). The vertebral ligaments, meninges, and cerebrospinal fluid provide additional protection.

Meninges

The **meninges** (me-NIN-jēz) are three connective tissue coverings that encircle the spinal cord and brain. The **spinal meninges** surround the spinal cord (Figure 13.1a) and are continuous with the **cranial meninges,** which encircle the brain (shown in Figure 14.4a on page 456). The most superficial of the three spinal meninges is the **dura mater** (DOO-ra MĀ-ter = tough mother), which is composed of dense, irregular connective tissue. It forms a sac from the level of the foramen magnum in the occipital bone, where it is continuous with the dura mater of the brain, to the second sacral vertebra. The spinal cord is also protected by a cushion of fat and connective tissue located in the **epidural space,** a space between the dura mater and the wall of the vertebral canal (Figure 13.1b).

The middle **meninx** (MĒ-ninks, singular form of meninges) is an avascular covering called the **arachnoid mater** (a-RAK-noyd; *arachn-* = spider; *-oid* = similar to) because of its spider's web arrangement of delicate collagen fibers and some elastic fibers. It is deep to the dura mater and is continuous with the arachnoid mater of the brain. Between the dura mater and the arachnoid mater is a thin **subdural space,** which contains interstitial fluid.

The innermost meninx is the **pia mater** (PĒ-a MĀ-ter; *pia* = delicate), a thin transparent connective tissue layer that adheres to the surface of the spinal cord and brain. It consists of interlacing bundles of collagen fibers and some fine elastic fibers. Within the pia mater are many blood vessels that supply oxygen and nutrients to the spinal cord. Between the arachnoid mater and the pia mater is the **subarachnoid space,** which contains cerebrospinal fluid.

All three spinal meninges cover the spinal nerves up to the point of exit from the spinal column through the intervertebral foramina. Triangular-shaped membranous extensions of the pia mater suspend the spinal cord in the middle of its dural sheath. These extensions, called **denticulate ligaments** (den-TIK-ū-lāt = small tooth), are thickenings of the pia mater. They project laterally and fuse with the arachnoid mater and inner surface of the dura mater between the anterior and posterior nerve roots of spinal nerves on either side. Extending all along the length of the spinal cord, the denticulate ligaments protect the spinal cord against shock and sudden displacement.

External Anatomy of the Spinal Cord

The **spinal cord,** although roughly cylindrical, is flattened slightly in its anterior–posterior dimension. In adults, it extends from the medulla oblongata, the inferior part of the brain, to the superior border of the second lumbar vertebra (Figure 13.2). In newborn infants, it extends to the third or fourth lumbar vertbra. During early childhood, both the spinal cord and the vertebral column grow longer as part of overall body growth. Around age 4 or 5, however, elongation of the spinal cord stops. Because the vertebral column continues to elongate, the spinal cord does not extend the entire length of the vertebral column in

Figure 13.1   Gross anatomy of the spinal cord.   The spinal meninges are evident in both views.

Meninges are connective tissue coverings that surround the spinal cord and brain.

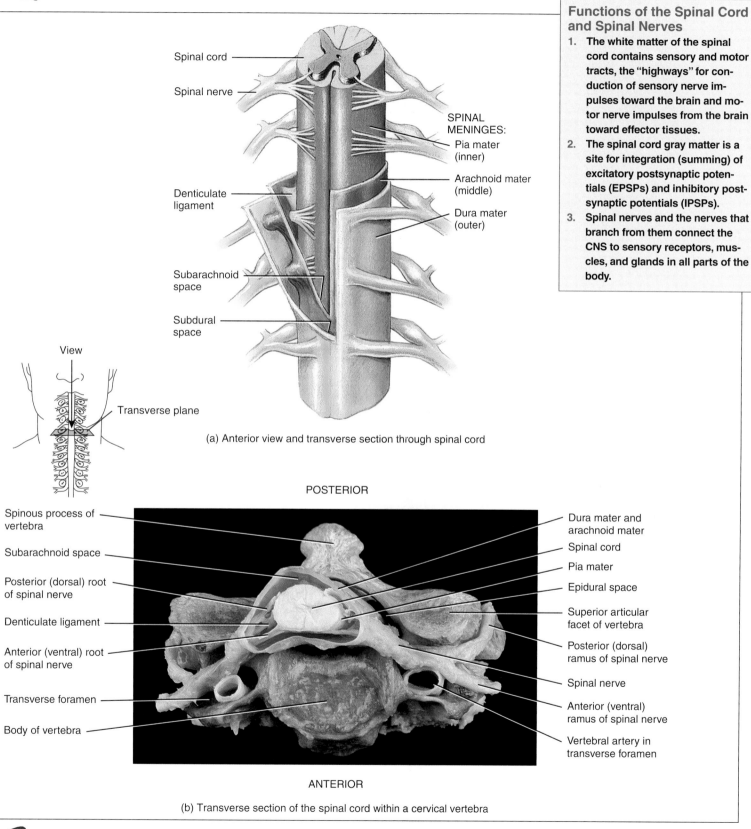

Spinal cord

Spinal nerve

SPINAL MENINGES:

Pia mater (inner)

Arachnoid mater (middle)

Denticulate ligament

Dura mater (outer)

Subarachnoid space

Subdural space

View

Transverse plane

(a) Anterior view and transverse section through spinal cord

POSTERIOR

Spinous process of vertebra

Subarachnoid space

Posterior (dorsal) root of spinal nerve

Denticulate ligament

Anterior (ventral) root of spinal nerve

Transverse foramen

Body of vertebra

Dura mater and arachnoid mater

Spinal cord

Pia mater

Epidural space

Superior articular facet of vertebra

Posterior (dorsal) ramus of spinal nerve

Spinal nerve

Anterior (ventral) ramus of spinal nerve

Vertebral artery in transverse foramen

ANTERIOR

(b) Transverse section of the spinal cord within a cervical vertebra

Functions of the Spinal Cord and Spinal Nerves

1. The white matter of the spinal cord contains sensory and motor tracts, the "highways" for conduction of sensory nerve impulses toward the brain and motor nerve impulses from the brain toward effector tissues.
2. The spinal cord gray matter is a site for integration (summing) of excitatory postsynaptic potentials (EPSPs) and inhibitory postsynaptic potentials (IPSPs).
3. Spinal nerves and the nerves that branch from them connect the CNS to sensory receptors, muscles, and glands in all parts of the body.

What are the superior and inferior boundaries of the spinal dura mater?

Figure 13.2 **External anatomy of the spinal cord and the spinal nerves.** (See Tortora, *A Photographic Atlas of the Human Body,* Figures 8.2, 8.3.)

The spinal cord extends from the medulla oblongata of the brain to the superior border of the second lumbar vertebra.

Posterior view of entire spinal cord and portions of spinal nerves

What portion of the spinal cord connects with nerves of the upper limbs?

adults. The length of the adult spinal cord ranges from 42 to 45 cm (16–18 in.). Its diameter is about 2 cm (0.75 in.) in the midthoracic region, somewhat larger in the lower cervical and midlumbar regions, and smallest at the inferior tip.

When the spinal cord is viewed externally, two conspicuous enlargements can be seen. The superior enlargement, the **cervical enlargement,** extends from the fourth cervical vertebra to the first thoracic vertebra. Nerves to and from the upper limbs arise from the cervical enlargement. The inferior enlargement, called the **lumbar enlargement,** extends from the ninth to the twelfth thoracic vertebra. Nerves to and from the lower limbs arise from the lumbar enlargement.

Inferior to the lumbar enlargement, the spinal cord tapers to a conical portion referred to as the **conus medullaris** (KŌ-nus med-ū-LAR-is; *conus* = cone), which ends at the level of the intervertebral disc between the first and second lumbar vertebrae in adults. Arising from the conus medullaris is the **filum terminale** (FĪ-lum ter-mi-NAL-ē = terminal filament), an extension of the pia mater that extends inferiorly and anchors the spinal cord to the coccyx.

Nerves that arise from the inferior part of the spinal cord do not leave the vertebral column at the same level as they exit from the cord. The roots (points of attachment to the spinal cord) of these nerves angle inferiorly in the vertebral canal from the end of the spinal cord like wisps of hair. Appropriately, the roots of these nerves are collectively named the **cauda equina** (KAW-da ē-KWĪ-na), meaning "horse's tail."

Spinal cord organization appears to be segmented because the 31 pairs of spinal nerves emerge at regular intervals from intervertebral foramina (Figure 13.2). Indeed, each pair of spinal nerves is said to arise from a *spinal segment.* Within the spinal cord, however, there is no obvious segmentation dividing the white matter or the gray matter. The naming of spinal nerves and spinal segments is based on their location. There are eight pairs of *cervical nerves* (represented as C1–C8), twelve pairs of *thoracic nerves* (T1–T12), five pairs of *lumbar nerves* (L1–L5), five pairs of *sacral nerves* (S1–S5), and one pair of *coccygeal nerves* (Co1).

Spinal nerves are the paths of communication between the spinal cord and the nerves innervating specific regions of the body. Two bundles of axons, called **roots,** connect each spinal nerve to a segment of the cord (see Figure 13.3a). The **posterior** or **dorsal root** contains only sensory axons, which conduct nerve impulses from sensory receptors in the skin, muscles, and internal organs into the central nervous system. Each posterior root also has a swelling, the **posterior** or **dorsal root ganglion,** which contains the cell bodies of sensory neurons. The **anterior** or **ventral root** contains axons of motor neurons, which conduct nerve impulses from the CNS to effector organs and cells.

Spinal Tap

In a **spinal tap (lumbar puncture),** a local anesthetic is given, and a long needle is inserted into the subarachnoid space. The procedure is used to withdraw cerebrospinal fluid (CSF) for diagnostic purposes; to introduce antibiotics, contrast media for myelography, or anesthetics; to administer chemotherapy; to measure CSF pressure; and to evaluate the effects of treatment. In adults, a spinal tap is normally performed between the third and fourth or fourth and fifth lumbar vertebrae. Because this region is inferior to the lowest portion of the spinal cord, it provides relatively safe access. (A line drawn across the highest points of the iliac crests, called the *supracristal line,* passes through the spinous process of the fourth lumbar vertebra.) ■

Internal Anatomy of the Spinal Cord

Two grooves penetrate the white matter of the spinal cord and divide it into right and left sides (Figure 13.3). The **anterior median fissure** is a deep, wide groove on the anterior (ventral) side. The **posterior median sulcus** is a shallower, narrow furrow on the posterior (dorsal) side. The gray matter of the spinal cord is shaped like the letter H or a butterfly and is surrounded by white matter. The gray matter consists primarily of the cell bodies of neurons, neuroglia, unmyelinated axons, and the dendrites of interneurons and motor neurons. The white matter consists of bundles of myelinated and unmyelinated axons of sensory neurons, interneurons, and motor neurons. The **gray commissure** (KOM-mi-shur) forms the crossbar of the H. In the center of the gray commissure is a small space called the **central canal;** it extends the entire length of the spinal cord. At its superior end, the central canal is continuous with the fourth ventricle (a space that contains cerebrospinal fluid) in the medulla oblongata of the brain. Anterior to the gray commissure is the **anterior (ventral) white commissure,** which connects the white matter of the right and left sides of the spinal cord.

In the gray matter of the spinal cord and brain, clusters of neuronal cell bodies form functional groups called nuclei. *Sensory nuclei* receive input from receptors via sensory neurons, whereas *motor nuclei* provide output to effector tissues via motor neurons. The gray matter on each side of the spinal cord is subdivided into regions called **horns.** The **anterior (ventral) gray horns** contain cell bodies of somatic motor neurons and motor nuclei, which provide nerve impulses for contraction of skeletal muscles. The **posterior (dorsal) gray horns** contain somatic and autonomic sensory nuclei. Between the anterior and posterior gray horns are the **lateral gray horns,** which are present only in the thoracic, upper lumbar, and sacral segments of the spinal cord. The lateral horns contain the cell bodies of autonomic motor neurons that regulate activity of smooth muscle, cardiac muscle, and glands.

The white matter, like the gray matter, also is organized into regions. The anterior and posterior gray horns divide the white matter on each side into three broad areas called **columns:** (1) **anterior (ventral) white columns,** (2) **posterior (dorsal) white columns,** and (3) **lateral white columns.** Each column, in

Figure 13.3   Internal anatomy of the spinal cord: the organization of gray matter and white matter.   For simplicity, dendrites are not shown in this and several other illustrations of transverse sections of the spinal cord. Blue and red arrows in (a) indicate the direction of nerve impulse propagation.

 In the spinal cord, white matter surrounds the gray matter.

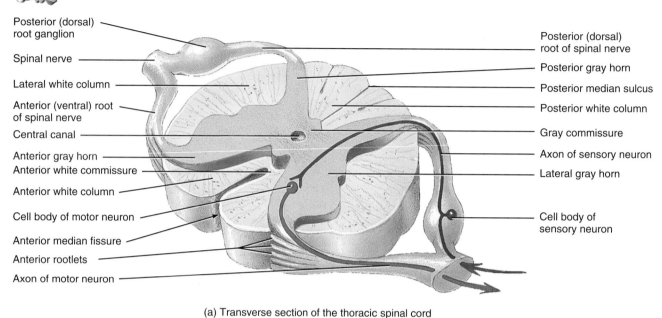

Posterior (dorsal) root ganglion

Spinal nerve

Lateral white column

Anterior (ventral) root of spinal nerve

Central canal

Anterior gray horn

Anterior white commissure

Anterior white column

Cell body of motor neuron

Anterior median fissure

Anterior rootlets

Axon of motor neuron

Posterior (dorsal) root of spinal nerve

Posterior gray horn

Posterior median sulcus

Posterior white column

Gray commissure

Axon of sensory neuron

Lateral gray horn

Cell body of sensory neuron

(a) Transverse section of the thoracic spinal cord

View

Transverse plane

Posterior median sulcus

Posterior white column

Posterior gray horn

Gray commissure

Lateral white column

Central canal

Anterior gray horn

Anterior white column

Anterior median fissure

LM   5x

(b) Transverse section of the thoracic spinal cord

 What is the difference between a horn and a column in the spinal cord?

turn, contains distinct bundles of axons having a common origin or destination and carrying similar information. These bundles, which may extend long distances up or down the spinal cord, are called **tracts**. **Sensory (ascending) tracts** consist of axons that conduct nerve impulses toward the brain. Tracts consisting of axons that carry nerve impulses down the spinal cord are called **motor (descending) tracts**. Sensory and motor tracts of the spinal cord are continuous with sensory and motor tracts in the brain.

► CHECKPOINT

1. Where are the spinal meninges located? Where are the epidural, subdural, and subarachnoid spaces located?

2. What are the cervical and lumbar enlargements?

3. Define conus medullaris, filum terminale, and cauda equina. What is a spinal segment? How is the spinal cord partially divided into right and left sides?

4. What does each of the following terms mean? Gray commissure, central canal, anterior gray horn, lateral gray horn, posterior gray horn, anterior white column, lateral white column, posterior white column, ascending tract, and descending tract.

SPINAL CORD PHYSIOLOGY

► OBJECTIVES

• **Describe the functions of the major sensory and motor tracts of the spinal cord.**

• **Describe the functional components of a reflex arc and the ways reflexes maintain homeostasis.**

The spinal cord has two principal functions in maintaining homeostasis: nerve impulse propagation and information integration. The *white matter tracts* in the spinal cord are highways for nerve impulse propagation. Along these tracts, sensory impulses flow toward the brain, and motor impulses flow from the brain toward skeletal muscles and other effector tissues. The *gray matter* of the spinal cord receives and integrates incoming and outgoing information.

Sensory and Motor Tracts

The first way the spinal cord promotes homeostasis is by conducting nerve impulses along tracts. Often, the name of a tract indicates its position in the white matter and where it begins and ends. For example, the anterior spinothalamic tract is located in the *anterior* white column; it begins in the *spinal cord* and ends in the *thalamus* (a region of the brain). Notice that the location of the axon terminals comes last in the name. This regularity in naming allows you to determine the direction of information flow along any tract named according to this convention. Thus, because the anterior spinothalamic tract conveys nerve impulses

from the spinal cord toward the brain, it is a sensory (ascending) tract. Figure 13.4 highlights the major sensory and motor tracts in the spinal cord. These tracts are described in detail in Chapter 15 and summarized in Tables 15.3 and 15.4 on pages 511 and 515.

Nerve impulses from sensory receptors propagate up the spinal cord to the brain along two main routes on each side: the spinothalamic tracts and the posterior columns. The **lateral** and **anterior spinothalamic tracts** convey nerve impulses for sensing pain, warmth, coolness, itching, tickling, deep pressure, and a crude, poorly localized sense of touch. The right and left **posterior columns** carry nerve impulses for several kinds of sensation. These include (1) proprioception, awareness of the positions and movements of muscles, tendons, and joints; (2) discriminative touch, the ability to feel exactly what part of the body is touched; (3) two-point discrimination, the ability to distinguish the touching of two different points on the skin, even though they are close together; (4) light pressure sensations; and (5) vibration sensations.

The sensory systems keep the CNS informed of changes in the external and internal environments. Responses to this information are brought about by motor systems, which enable us to move about and change our physical relationship to the world around us. As sensory information is conveyed to the CNS, it becomes part of a large pool of sensory input. Each piece of incoming information is integrated with all the other information arriving from activated sensory neurons.

Through the activity of interneurons, integration occurs in several regions of the spinal cord and brain. As a result, motor responses to make a muscle contract or a gland secrete can be initiated at several levels. Most regulation of involuntary activities of smooth muscle, cardiac muscle, and glands by the autonomic nervous system (ANS) originates in the brain stem (the lower part of the brain that is continuous with the spinal cord) and in a nearby brain region called the hypothalamus.

The cerebral cortex (superficial gray matter of the cerebrum) plays a major role in controlling precise, voluntary muscular movements. Other brain regions provide important integration for regulation of automatic movements, such as arm swinging during walking. Motor output to skeletal muscles travels down the spinal cord in two types of descending pathways: direct and indirect. The **direct pathways** include the *lateral corticospinal, anterior corticospinal,* and *corticobulbar tracts*. They convey nerve impulses that originate in the cerebral cortex and are destined to cause precise, *voluntary* movements of skeletal muscles. **Indirect pathways** include the *rubrospinal, tectospinal,* and *vestibulospinal tracts*. They convey nerve impulses from the brain stem and other parts of the brain that govern automatic movements and help coordinate body movements with visual stimuli. Indirect pathways also maintain skeletal muscle tone, maintain contraction of postural muscles, and play a major role in equilibrium by regulating muscle tone in response to movements of the head.

Figure 13.4 **Locations of major sensory and motor tracts, shown in a transverse section of the spinal cord.** Sensory tracts are indicated on one half and motor tracts on the other half of the cord, but actually all tracts are present on both sides.

🔑 **The name of a tract often indicates its location in the white matter and where it begins and ends.**

Based on its name, what are the position in the spinal cord, origin, and destination of the anterior corticospinal tract? Is this a sensory or a motor tract?

Reflexes and Reflex Arcs

The second way the spinal cord promotes homeostasis is by serving as an integrating center for some reflexes. A **reflex** is a fast, involuntary, unplanned sequence of actions that occurs in response to a particular stimulus. Some reflexes are inborn, such as pulling your hand away from a hot surface before you even feel that it is hot. Other reflexes are learned or acquired. For instance, you learn many reflexes while acquiring driving expertise. Slamming on the brakes in an emergency is one example. When integration takes place in the spinal cord gray matter, the reflex is a **spinal reflex.** By contrast, if integration occurs in the brain stem rather than the spinal cord, the reflex is a **cranial reflex.** An example is the tracking movements of your eyes as you read this sentence. You are probably most aware of **somatic reflexes,** which involve contraction of skeletal muscles. Equally important, however, are the **autonomic (visceral) reflexes,** which generally are not consciously perceived. They involve responses of smooth muscle, cardiac muscle, and glands. As you will see in Chapter 17, body functions such as heart rate, digestion, urination, and defecation are controlled by the autonomic nervous system through autonomic reflexes.

Nerve impulses propagating into, through, and out of the CNS follow specific pathways, depending on the kind of information, its origin, and its destination. The pathway followed by nerve impulses that produce a reflex is a **reflex arc (reflex circuit).** A reflex arc includes the following five functional components (Figure 13.5):

1 **Sensory receptor.** The distal end of a sensory neuron (dendrite) or an associated sensory structure serves as a sensory receptor. It responds to a specific **stimulus**—a change in the internal or external environment—by producing a graded potential called a generator (or receptor) potential (described on page 501). If a generator potential reaches the threshold level of depolarization, it will trigger one or more nerve impulses in the sensory neuron.

2 **Sensory neuron.** The nerve impulses propagate from the sensory receptor along the axon of the sensory neuron to the axon terminals, which are located in the gray matter of the spinal cord or brain stem.

3 **Integrating center.** One or more regions of gray matter within the CNS act as an integrating center. In the simplest type of reflex, the integrating center is a single synapse between a sensory neuron and a motor neuron. A reflex pathway having only one synapse in the CNS is termed a

Figure 13.5 General components of a reflex arc. The arrows show the direction of nerve impulse propagation.

🔑 A reflex is a fast, predictable sequence of involuntary actions that occur in response to certain changes in the environment.

2 SENSORY NEURON
(axon conducts impulses from receptor to integrating center)

1 SENSORY RECEPTOR
(responds to a stimulus by producing a generator or receptor potential)

Interneuron

3 INTEGRATING CENTER
(one or more regions within the CNS that relay impulses from sensory to motor neurons)

4 MOTOR NEURON
(axon conducts impulses from integrating center to effector)

5 EFFECTOR
(muscle or gland that responds to motor nerve impulses)

 What initiates a nerve impulse in a sensory neuron? Which branch of the nervous system includes all integrating centers for reflexes?

monosynaptic reflex arc (*mono-* = one). More often, the integrating center consists of one or more interneurons, which may relay impulses to other interneurons as well as to a motor neuron. A **polysynaptic reflex arc** (*poly-* = many) involves more than two types of neurons and more than one CNS synapse.

4 Motor neuron. Impulses triggered by the integrating center propagate out of the CNS along a motor neuron to the part of the body that will respond.

5 Effector. The part of the body that responds to the motor nerve impulse, such as a muscle or gland, is the effector. Its action is called a reflex. If the effector is skeletal muscle, the reflex is a **somatic reflex.** If the effector is smooth muscle, cardiac muscle, or a gland, the reflex is an **autonomic (visceral) reflex.**

Because reflexes are normally so predictable, they provide useful information about the health of the nervous system and can greatly aid diagnosis of disease. Damage or disease anywhere along its reflex arc can cause a reflex to be absent or abnormal. For example, tapping the patellar ligament normally causes reflex extension of the knee joint. Absence of the patellar reflex could indicate damage of the sensory or motor neurons, or a spinal cord injury in the lumbar region. Somatic reflexes generally can be tested simply by tapping or stroking the body surface. Most autonomic reflexes, by contrast, are not practical diagnostic tools because it is difficult to stimulate visceral receptors, which are deep inside the body. An exception is the pupillary light reflex, in which the pupils of both eyes decrease in diameter when either eye is exposed to light. Because the reflex arc includes synapses in the brain stem and the nearby midbrain, the absence of a normal pupillary light reflex may indicate damage or injury of these brain regions.

Next, we examine four important somatic spinal reflexes: the stretch reflex, the tendon reflex, the flexor (withdrawal) reflex, and the crossed extensor reflex.

The Stretch Reflex

A **stretch reflex** causes contraction of a skeletal muscle (the effector) in response to stretching of the muscle. This type of reflex occurs via a monosynaptic reflex arc. The reflex can occur by activation of a single sensory neuron that forms one synapse in the CNS with a single motor neuron. Stretch reflexes can be elicited by tapping on tendons attached to muscles at the elbow, wrist, knee, and ankle joints.

A stretch reflex operates as follows (Figure 13.6):

1 Slight stretching of a muscle stimulates sensory receptors in the muscle called **muscle spindles** (shown in more detail

in Figure 15.4 on page 506). The spindles monitor changes in the length of the muscle.

2 In response to being stretched, a muscle spindle generates one or more nerve impulses that propagate along a somatic sensory neuron through the posterior root of the spinal nerve and into the spinal cord.

3 In the spinal cord (integrating center), the sensory neuron makes an excitatory synapse with and thereby activates a motor neuron in the anterior gray horn.

4 If the excitation is strong enough, one or more nerve impulses arise in the motor neuron and propagate along its

Figure 13.6 Stretch reflex. This monosynaptic reflex arc has only one synapse in the CNS—between a single sensory neuron and a single motor neuron. A polysynaptic reflex arc to antagonistic muscles that includes two synapses in the CNS and one interneuron is also illustrated. Plus signs (+) indicate excitatory synapses; the minus sign (−) indicates an inhibitory synapse.

The stretch reflex causes contraction of a muscle that has been stretched.

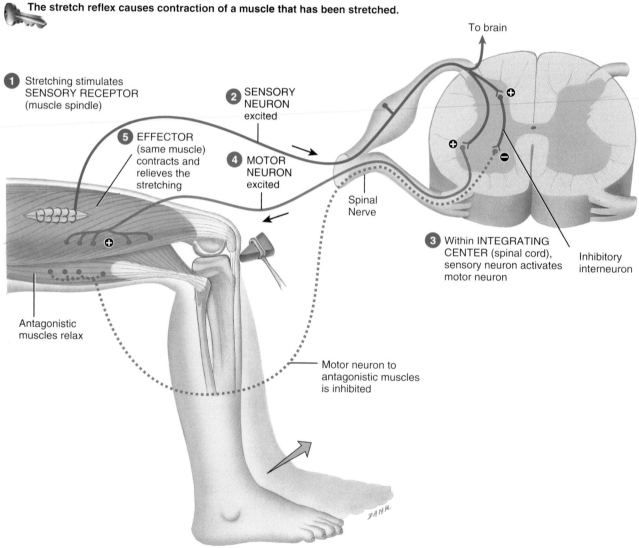

What makes this an ipsilateral reflex?

axon, which extends from the spinal cord into the anterior root and through peripheral nerves to the stimulated muscle. The axon terminals of the motor neuron form neuromuscular junctions (NMJs) with skeletal muscle fibers of the stretched muscle.

5 Acetylcholine released by nerve impulses at the NMJs triggers one or more muscle action potentials in the stretched muscle (effector), and the muscle contracts. Thus, muscle stretch is followed by muscle contraction, which relieves the stretching.

In the reflex arc just described, sensory nerve impulses enter the spinal cord on the same side from which motor nerve impulses leave it. This arrangement is called an **ipsilateral reflex** (ip′-si-LAT-er-al = same side). All monosynaptic reflexes are ipsilateral.

In addition to the large-diameter motor neurons that innervate typical skeletal muscle fibers, smaller-diameter motor neurons innervate smaller, specialized muscle fibers within the muscle spindles themselves. The brain regulates muscle spindle sensitivity through pathways to these smaller motor neurons. This regulation ensures proper muscle spindle signaling over a wide range of muscle lengths during voluntary and reflex contractions. By adjusting how vigorously a muscle spindle responds to stretching, the brain sets an overall level of **muscle tone,** which is the small degree of contraction present while the muscle is at rest. Because the stimulus for the stretch reflex is stretching of muscle, this reflex helps avert injury by preventing overstretching of muscles.

Although the stretch reflex pathway itself is monosynaptic (just two neurons and one synapse), a polysynaptic reflex arc to the antagonistic muscles operates at the same time. This arc involves three neurons and two synapses. An axon collateral (branch) from the muscle spindle sensory neuron also synapses with an inhibitory interneuron in the integrating center. In turn, the interneuron synapses with and inhibits a motor neuron that normally excites the antagonistic muscles (Figure 13.6). Thus, when the stretched muscle contracts during a stretch reflex, antagonistic muscles that oppose the contraction relax. This type of neural circuit, which simultaneously causes contraction of one muscle and relaxation of its antagonists, is termed **reciprocal innervation.** Reciprocal innervation prevents conflict between opposing muscles and is vital in coordinating body movements.

Axon collaterals of the muscle spindle sensory neuron also relay nerve impulses to the brain over specific ascending pathways. In this way, the brain receives input about the state of stretch or contraction of skeletal muscles, enabling it to coordinate muscular movements and posture. The nerve impulses that pass to the brain also allow conscious perception that the reflex has occurred.

The Tendon Reflex

The stretch reflex operates as a feedback mechanism to control muscle *length* by causing muscle contraction. In contrast, the **tendon reflex** operates as a feedback mechanism to control muscle *tension* by causing muscle relaxation before muscle force becomes so great that tendons might be torn. Although the tendon reflex is less sensitive than the stretch reflex, it can override the stretch reflex when tension is great, making you drop a very heavy weight, for example. Like the stretch reflex, the tendon reflex is ipsilateral. The sensory receptors for this reflex are called **tendon (Golgi tendon) organs** (shown in more detail in Figure 15.4 on page 506), which lie within a tendon near its junction with a muscle. Whereas muscle spindles are sensitive to changes in muscle length, tendon organs detect and respond to changes in muscle tension that are caused by passive stretch or muscular contraction.

A tendon reflex operates as follows (Figure 13.7):

1 As the tension applied to a tendon increases, the tendon organ (sensory receptor) is stimulated (depolarized to threshold).

2 Nerve impulses arise and propagate into the spinal cord along a sensory neuron.

3 Within the spinal cord (integrating center), the sensory neuron activates an inhibitory interneuron that synapses with a motor neuron.

4 The inhibitory neurotransmitter inhibits (hyperpolarizes) the motor neuron, which then generates fewer nerve impulses.

5 The muscle attached to the same tendon relaxes and relieves excess tension.

Thus, as tension on the tendon organ increases, the frequency of inhibitory impulses increases, and inhibition of the motor neurons to the muscle developing excess tension (effector) causes relaxation of the muscle. In this way, the tendon reflex protects the tendon and muscle from damage due to excessive tension.

Note in Figure 13.7 that the sensory neuron from the tendon organ also synapses with an excitatory interneuron in the spinal cord. The excitatory interneuron, in turn, synapses with motor neurons controlling antagonistic muscles. Thus, while the tendon reflex brings about relaxation of the muscle attached to the tendon organ, it also triggers contraction of antagonists. Here we have another example of reciprocal innervation. The sensory neuron also relays nerve impulses to the brain by way of sensory tracts, thus informing the brain about the state of muscle tension throughout the body.

The Flexor and Crossed Extensor Reflexes

Another reflex involving a polysynaptic reflex arc results when, for instance, you step on a tack. In response to such a painful stimulus, you immediately withdraw your leg. This reflex, called

Figure 13.7 Tendon reflex. This reflex arc is polysynaptic—more than one CNS synapse and more than two different neurons are involved in the pathway. The sensory neuron synapses with two interneurons. An inhibitory interneuron causes relaxation of the effector, and a stimulatory interneuron causes contraction of the antagonistic muscle. Plus signs (+) indicate excitatory synapses; the minus sign (−) indicates an inhibitory synapse.

🔑 **The tendon reflex causes relaxation of the muscle attached to the stimulated tendon organ.**

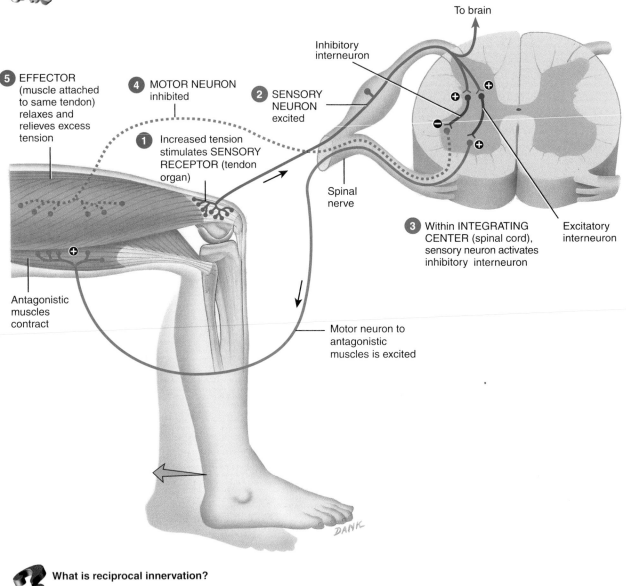

❓ What is reciprocal innervation?

the **flexor** or **withdrawal reflex,** operates as follows (Figure 13.8):

1. Stepping on a tack stimulates the dendrites (sensory receptor) of a pain-sensitive neuron.

2. This sensory neuron then generates nerve impulses, which propagate into the spinal cord.

3. Within the spinal cord (integrating center), the sensory neuron activates interneurons that extend to several spinal cord segments.

4. The interneurons activate motor neurons in several spinal cord segments. As a result, the motor neurons generate nerve impulses, which propagate toward the axon terminals.

5. Acetylcholine released by the motor neurons causes the flexor muscles in the thigh (effectors) to contract, producing withdrawal of the leg. This reflex is protective because contraction of flexor muscles moves a limb away from the source of a possibly damaging stimulus.

Figure 13.8 **Flexor (withdrawal) reflex.** This reflex arc is polysynaptic and ipsilateral. Plus signs (+) indicate excitatory synapses.

The flexor reflex causes withdrawal of a part of the body in response to a painful stimulus.

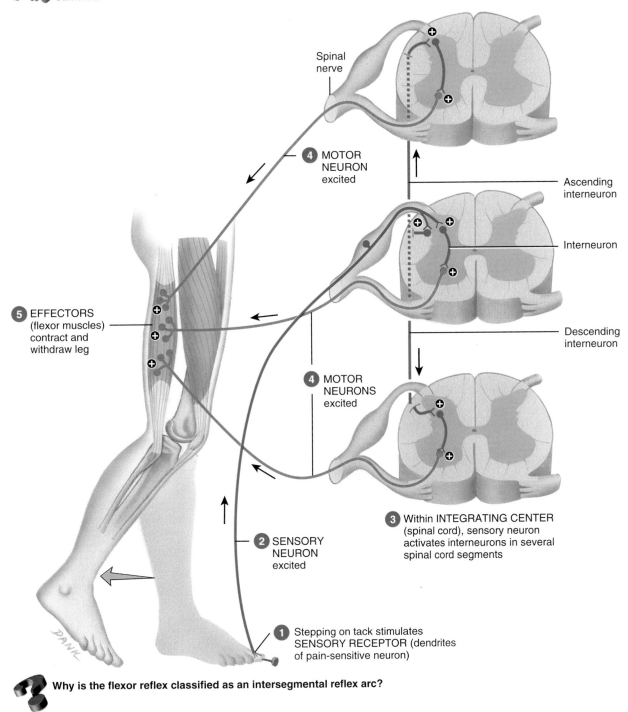

Why is the flexor reflex classified as an intersegmental reflex arc?

The flexor reflex, like the stretch reflex, is ipsilateral—the incoming and outgoing impulses propagate into and out of the same side of the spinal cord. The flexor reflex also illustrates another feature of polysynaptic reflex arcs. Moving your entire lower or upper limb away from a painful stimulus involves con- traction of more than one muscle group. Hence, several motor neurons must simultaneously convey impulses to several limb muscles. Because nerve impulses from one sensory neuron ascend and descend in the spinal cord and activate interneurons in several segments of the spinal cord, this type of reflex is called an **intersegmental reflex arc** (*inter-* = between). Through intersegmental reflex arcs, a single sensory neuron can activate several motor neurons, thereby stimulating more than one effector. The monosynaptic stretch reflex, in contrast, involves muscles receiving nerve impulses from one spinal cord segment only.

Something else may happen when you step on a tack: You may start to lose your balance as your body weight shifts to the other foot. Besides initiating the flexor reflex that causes you to withdraw the limb, the pain impulses from stepping on the tack also initiate a balance-maintaining **crossed extensor reflex,** which operates as follows (Figure 13.9):

Figure 13.9 Crossed extensor reflex. The flexor reflex arc is shown (at left) to enable comparison with the crossed extensor reflex arc. Plus signs (+) indicate excitatory synapses.

A crossed extensor reflex causes contraction of muscles that extend joints in the limb opposite a painful stimulus.

Why is the crossed extensor reflex classified as a contralateral reflex arc?

1. Stepping on a tack stimulates the sensory receptor of a pain-sensitive neuron in the right foot.

2. This sensory neuron then generates nerve impulses, which propagate into the spinal cord.

3. Within the spinal cord (integrating center), the sensory neuron activates several interneurons that synapse with motor neurons on the left side of the spinal cord in several spinal cord segments. Thus, incoming pain signals cross to the opposite side through interneurons at that level, and at several levels above and below the point of entry into the spinal cord.

4. The interneurons excite motor neurons that innervate extensor muscles in several spinal cord segments. The motor neurons, in turn, generate more nerve impulses, which propagate toward the axon terminals.

5. Acetylcholine released by the motor neurons causes extensor muscles in the thigh (effectors) of the unstimulated left limb to contract, producing extension of the left leg. In this way, weight can be placed on the foot of the limb that must now support the entire body. A comparable reflex occurs with painful stimulation of the left lower limb or either upper limb.

Unlike the flexor reflex, which is an ipsilateral reflex, the crossed extensor reflex involves a **contralateral reflex arc** (kon′-tra-LAT-er-al = opposite side): Sensory impulses enter one side of the spinal cord and motor impulses exit on the opposite side. Thus, a crossed extensor reflex synchronizes the extension of the contralateral limb with the withdrawal (flexion) of the stimulated limb. Reciprocal innervation also occurs in both the flexor reflex and the crossed extensor reflex. In the flexor reflex, when the flexor muscles of a painfully stimulated lower limb are contracting, the extensor muscles of the same limb are relaxing to some degree. If both sets of muscles contracted at the same time, the two sets of muscles would pull on the bones in opposite directions, which might immobilize the limb. Because of reciprocal innervation, however, one set of muscles contracts while the other relaxes.

Plantar Flexion Reflex and the Babinski Sign

Several reflexes have clinical significance. Among the most important is the **plantar flexion reflex,** elicited by gently stroking the lateral outer margin of the sole. The normal response is a curling under of all the toes. Damage to descending motor pathways, such as the corticospinal tract, alters this reflex. Then, stroking the sole causes the great toe to extend, with or without fanning of the other toes, a response called the **Babinski sign.** In children under 18 months of age, the Babinski sign is normal due to incomplete myelination of axons in the corticospinal tract. The normal response after 18 months of age is the plantar flexion reflex. ■

5. Which spinal cord tracts are ascending tracts? Which are descending tracts?

6. How are somatic and autonomic reflexes similar and different?

7. Describe the mechanism and function of a stretch reflex, tendon reflex, flexor (withdrawal) reflex, and crossed extensor reflex.

8. What does each of the following terms mean in relation to reflex arcs? Monosynaptic, ipsilateral, polysynaptic, intersegmental, contralateral, and reciprocal innervation.

SPINAL NERVES

▶ OBJECTIVES

- Describe the components, connective tissue coverings, and branching of a spinal nerve.
- Define plexus, and identify the distribution of nerves of the cervical, brachial, lumbar, and sacral plexuses.
- Describe the clinical significance of dermatomes.

Spinal nerves and the nerves that branch from them are part of the peripheral nervous system (PNS). They connect the CNS to sensory receptors, muscles, and glands in all parts of the body. The 31 pairs of spinal nerves are named and numbered according to the region and level of the vertebral column from which they emerge (see Figure 13.2). The first cervical pair emerges between the atlas (first cervical vertebra) and the occipital bone. All other spinal nerves emerge from the vertebral column through the intervertebral foramina between adjoining vertebrae.

Not all spinal cord segments are aligned with their corresponding vertebrae. Recall that the spinal cord ends near the level of the superior border of the second lumbar vertebra, and that the roots of the lower lumbar, sacral, and coccygeal nerves descend at an angle to reach their respective foramina before emerging from the vertebral column. This arrangement constitutes the cauda equina (see Figure 13.2).

As noted earlier, a typical **spinal nerve** has two connections to the cord: a posterior root and an anterior root (see Figure 13.3a). The posterior and anterior roots unite to form a spinal nerve at the intervertebral foramen. Because the posterior root contains sensory axons and the anterior root contains motor axons, a spinal nerve is a **mixed nerve.** The posterior root contains a posterior root ganglion in which cell bodies of sensory neurons are located.

Connective Tissue Coverings of Spinal Nerves

Each cranial nerve and spinal nerve contains layers of protective connective tissue coverings (Figure 13.10). Individual axons,

Figure 13.10 **Organization and connective tissue coverings of a spinal nerve.** Part b: From Richard G. Kessel and Randy H. Kardon, *Tissues and Organs: A Text-Atlas of Scanning Electron Microscopy.* Copyright © 1979 by W. H. Freeman and Company. Reprinted by perimission.

🔑 **Three layers of connective tissue wrappings protect axons: endoneurium surrounds individual axons, perineurium surrounds bundles of axons (fascicles), and epineurium surrounds an entire nerve.**

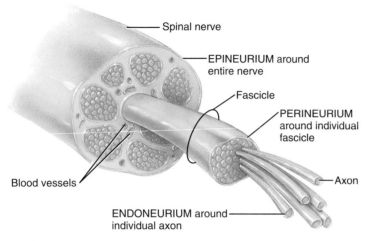

(a) Transverse section showing the coverings of a spinal nerve

(b) Transverse section of 12 nerve fascicles

 Why are all spinal nerves classified as mixed nerves?

whether myelinated or unmyelinated, are wrapped in **endoneurium** (en′-dō-NOO-rē-um; *endo-* = within or inner). Groups of axons with their endoneurium are arranged in bundles called **fascicles,** each of which is wrapped in **perineurium** (per′-i-NOO-rē-um; *peri-* = around). The superficial covering over the entire nerve is the **epineurium** (ep′-i-NOO-rē-um; *epi-* = over). The dura mater of the spinal meninges fuses with the epineurium as the nerve passes through the intervertebral foramen. Note the presence of many blood vessels, which nourish nerves, within the perineurium and epineurium (Figure 13.10b). You may recall from Chapter 10 that the connective tissue coverings of skeletal muscles—endomysium, perimysium, and epimysium—are similar in organization to those of nerves.

Distribution of Spinal Nerves

Branches

A short distance after passing through its intervertebral foramen, a spinal nerve divides into several branches (Figure 13.11).

These branches are known as **rami** (RĀ-mī = branches). The **posterior (dorsal) ramus** (RĀ-mus; singular form) serves the deep muscles and skin of the dorsal surface of the trunk. The **anterior (ventral) ramus** serves the muscles and structures of the upper and lower limbs and the skin of the lateral and ventral surfaces of the trunk. In addition to posterior and anterior rami, spinal nerves also give off a **meningeal branch.** This branch reenters the vertebral canal through the intervertebral foramen and supplies the vertebrae, vertebral ligaments, blood vessels of the spinal cord, and meninges. Other branches of a spinal nerve are the **rami communicantes** (kō-mū-ni-KAN-tēz), components of the autonomic nervous system whose structure and function are discussed in Chapter 17.

Plexuses

The anterior rami of spinal nerves, except for thoracic nerves T2–T12, do not go directly to the body structures they supply. Instead, they form networks on both the left and right sides of

Figure 13.11 **Branches of a typical spinal nerve, shown in transverse section through the thoracic portion of the spinal cord.** (See Tortora, *A Photographic Atlas of the Human Body*, Figure 8.4.)

 The branches of a spinal nerve are the posterior ramus, the anterior ramus, the meningeal branch, and the rami communicantes.

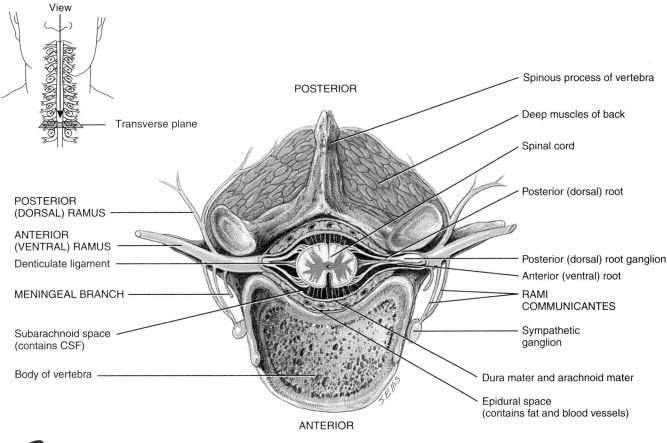

Which spinal nerve branches serve the upper and lower limbs?

the body by joining with various numbers of axons from anterior rami of adjacent nerves. Such a network of axons is called a **plexus** (= braid or network). The principal plexuses are the **cervical plexus, brachial plexus, lumbar plexus,** and **sacral plexus.** A smaller **coccygeal plexus** is also present. Refer to Figure 13.2 to see their relationships to one another. Emerging from the plexuses are nerves bearing names that are often descriptive of the general regions they serve or the course they take. Each of the nerves, in turn, may have several branches named for the specific structures they innervate.

Exhibits 13.1–13.4 summarize the principal plexuses. The anterior rami of spinal nerves T2–T12 are called intercostal nerves and will be discussed next.

Intercostal Nerves

The anterior rami of spinal nerves T2–T12 do not enter into the formation of plexuses and are known as **intercostal** or **thoracic nerves.** These nerves directly innervate the structures they sup-

ply in the intercostal spaces. After leaving its intervertebral foramen, the anterior ramus of nerve T2 innervates the intercostal muscles of the second intercostal space and supplies the skin of the axilla and posteromedial aspect of the arm. Nerves T3–T6 extend along the costal grooves of the ribs and then to the intercostal muscles and skin of the anterior and lateral chest wall. Nerves T7–T12 supply the intercostal muscles and abdominal muscles, and the overlying skin. The posterior rami of the intercostal nerves supply the deep back muscles and skin of the posterior aspect of the thorax.

Exhibit 13.1 Cervical Plexus (Figure 13.12)

▶ **OBJECTIVE**

• Describe the origin and distribution of the cervical plexus.

The **cervical plexus** (SER-vi-kul) is formed by the anterior (ventral) rami of the first four cervical nerves (C1–C4), with contributions from C5. There is one on each side of the neck alongside the first four cervical vertebrae. The roots of the plexus indicated in Figure 13.12 are the anterior rami.

The cervical plexus supplies the skin and muscles of the head, neck, and superior part of the shoulders and chest. The phrenic nerves arise from the cervical plexuses and supply motor fibers to the diaphragm. Branches of the cervical plexus also run parallel to the accessory nerve (cranial nerve XI) and hypoglossal nerve (cranial nerve XII).

 Injuries to the Phrenic Nerves

Complete severing of the spinal cord above the origin of the phrenic nerves (C3, C4, and C5) causes respiratory arrest. Breathing stops because the phrenic nerves no longer send impulses to the diaphragm. ■

▶ CHECKPOINT

Which nerve that arises from the cervical plexus causes contraction of the diaphragm?

Nerve	Origin	Distribution
Superficial (Sensory) Branches		
Lesser occipital	C2	Skin of scalp posterior and superior to ear.
Great auricular (aw-RIK-ū-lar)	C2–C3	Skin anterior, inferior, and over ear, and over parotid glands.
Transverse cervical	C2–C3	Skin over anterior aspect of neck.
Supraclavicular	C3–C4	Skin over superior portion of chest and shoulder.
Deep (Largely Motor) Branches		
Ansa cervicalis (AN-sa ser-vi-KAL-is)		This nerve divides into superior and inferior roots.
Superior root	C1	Infrahyoid and geniohyoid muscles of neck.
Inferior root	C2–C3	Infrahyoid muscles of neck.
Phrenic (FREN-ik)	C3–C5	Diaphragm.
Segmental branches	C1–C5	Prevertebral (deep) muscles of neck, levator scapulae, and middle scalene muscles.

Figure 13.12 Cervical plexus in anterior view. (See Tortora, *A Photographic Atlas of the Human Body,* Figure 8.7.)

The cervical plexus supplies the skin and muscles of the head, neck, superior portion of the shoulders and chest, and diaphragm.

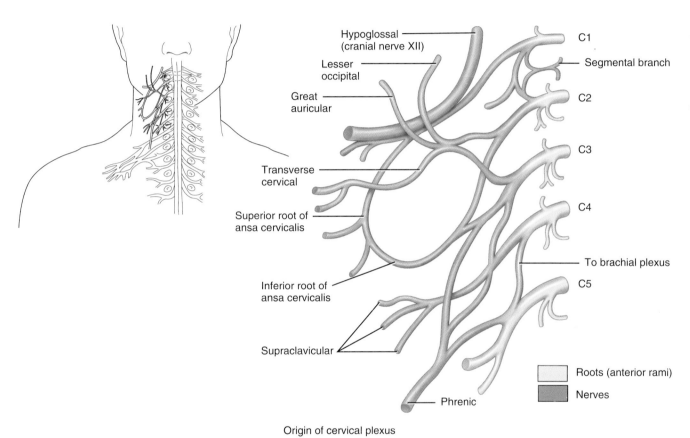

Hypoglossal (cranial nerve XII)

Lesser occipital

Great auricular

Transverse cervical

Superior root of ansa cervicalis

Inferior root of ansa cervicalis

Supraclavicular

Phrenic

C1

Segmental branch

C2

C3

C4

To brachial plexus

C5

	Roots (anterior rami)
	Nerves

Origin of cervical plexus

Why does complete severing of the spinal cord at level C2 cause respiratory arrest?

Exhibit 13.2   Brachial Plexus (Figures 13.13 and 13.14)

► **OBJECTIVE**

• Describe the origin, distribution, and effects of damage to the brachial plexus.

The anterior (ventral) rami of spinal nerves C5–C8 and T1 form the **brachial plexus** (BRĀ-kē-al), which extends inferiorly and laterally on either side of the last four cervical and first thoracic vertebrae (Figure 13.13a). It passes above the first rib posterior to the clavicle and then enters the axilla.

The brachial plexus provides the entire nerve supply of the shoulders and upper limbs (Figure 13.13b). Five important nerves arise from the brachial plexus: (1) The **axillary nerve** supplies the deltoid and teres minor muscles. (2) The **musculocutaneous nerve** supplies the flexors of the arm. (3) The **radial nerve** supplies the muscles on the posterior aspect of the arm and forearm. (4) The **median nerve** supplies most of the muscles of the anterior forearm and some of the muscles of the hand. (5) The **ulnar nerve** supplies the anteromedial muscles of the forearm and most of the muscles of the hand.

Injuries to Nerves Emerging from the Brachial Plexus

Injury to the superior roots of the brachial plexus (C5–C6) may result from forceful pulling away of the head from the shoulder, as might occur from a heavy fall on the shoulder or during childbirth in which the infant's head is excessively stretched. The presentation of this injury is characterized by an upper limb in which the shoulder is adducted, the arm is medially rotated, the elbow is extended, the forearm is pronated, and the wrist is flexed (Figure 13.14a). This condition is called **Erb-Duchene palsy** or **waiter's tip position.** There is loss of sensation along the lateral side of the arm.

Radial (and axillary) **nerve injury** can be caused by improperly administered intramuscular injections into the deltoid muscle. The radial nerve may also be injured when a cast is applied too tightly around the mid-humerus. Radial nerve injury is indicated by **wrist drop,** the inability to extend the wrist and fingers (Figure 13.14b). Sensory loss is minimal due to the overlap of sensory innervation by adjacent nerves.

Median nerve injury is indicated by numbness, tingling, and pain in the palm and fingers. There is also inability to pronate the forearm and flex the proximal interphalangeal joints of all digits and the distal interphalangeal joints of the second and third digits (Figure 13.14c). In addition, wrist flexion is weak and is accompanied by adduction, and thumb movements are weak.

Ulnar nerve injury is indicated by an inability to abduct or adduct the fingers, atrophy of the interosseus muscles of the hand, hyperextension of the metacarpophalangeal joints, and flexion of the interphalangeal joints, a condition called **clawhand** (Figure 13.14d). There is loss of sensation over the little finger.

Long thoracic nerve injury results in paralysis of the serratus anterior muscle. The medial border of the scapula protrudes, giving it the appearance of a wing. When the arm is raised, the vertebral border and inferior angle of the scapula pull away from the thoracic wall and protrude outward, a condition called **winged scapula** (Figure 13.14e). The arm cannot be abducted beyond the horizontal position. ■

► **CHECKPOINT**

Injury of which nerve could cause paralysis of the serratus anterior muscle?

Nerve	Origin	Distribution
Dorsal scapular (SKAP-ū-lar)	C5	Levator scapulae, rhomboid major, and rhomboid minor muscles.
Long thoracic (thor-RAS-ik)	C5–C7	Serratus anterior muscle.
Subclavian (sub-KLĀ-vē-an)	C5–C6	Subclavius muscle.
Suprascapular	C5–C6	Supraspinatus and infraspinatus muscles.
Musculocutaneous (mus'-kū-lo-kū-TĀN-ē-us)	C5–C7	Coracobrachialis, biceps brachii, and brachialis muscles.
Lateral pectoral (PEK-to-ral)	C5–C7	Pectoralis major muscle.
Upper subscapular	C5–C6	Subscapularis muscle.
Thoracodorsal (tho-RA-sō-dor-sal)	C6–C8	Latissimus dorsi muscle.
Lower subscapular	C5–C6	Subscapularis and teres major muscles.
Axillary (AK-si-lar-ē)	C5–C6	Deltoid and teres minor muscles; skin over deltoid and superior posterior aspect of arm.
Median	C5–T1	Flexors of forearm, except flexor carpi ulnaris and some muscles of the hand (lateral palm); skin of lateral two-thirds of palm of hand and fingers.
Radial		Triceps brachii and other extensor muscles of arm and extensor muscles of forearm; skin of posterior arm and forearm, lateral two-thirds of dorsum of hand, and fingers over proximal and middle phalanges.
Medial pectoral	C8–T1	Pectoralis major and pectoralis minor muscles.
Medial cutaneous nerve of arm (kū'-TĀ-nē-us)	C8–T1	Skin of medial and posterior aspects of distal third of arm.
Medial cutaneous nerve of forearm	C8–T1	Skin of medial and posterior aspects of forearm.
Ulnar	C8–T1	Flexor carpi ulnaris, flexor digitorum profundus, and most muscles of the hand; skin of medial side of hand, little finger, and medial half of ring finger.

Figure 13.13 Brachial plexus in anterior view. (See Tortora, *A Photographic Atlas of the Human Body,* Figures 8.8, 8.9.)

The brachial plexus supplies the shoulders and upper limbs.

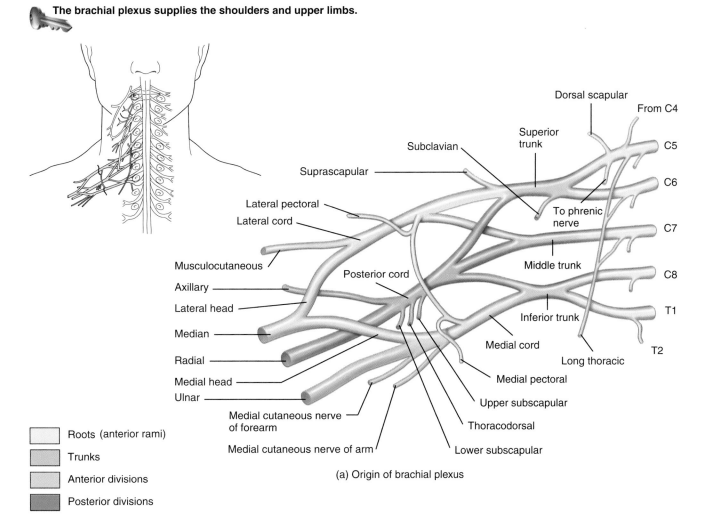

Dorsal scapular

From C4

C5

Superior trunk

Subclavian

Suprascapular

C6

Lateral pectoral

To phrenic nerve

Lateral cord

C7

Musculocutaneous

Posterior cord

Middle trunk

C8

Axillary

Lateral head

Inferior trunk

T1

Median

Medial cord

Long thoracic

T2

Radial

Medial pectoral

Medial head

Upper subscapular

Ulnar

Thoracodorsal

Medial cutaneous nerve
of forearm

Lower subscapular

Medial cutaneous nerve of arm

(a) Origin of brachial plexus

Roots (anterior rami)

Trunks

Anterior divisions

Posterior divisions

(continues)

Exhibit 13.2   Brachial Plexus (Figures 13.13 and 13.14) *(continued)*

Figure 13.13   *(continued)*

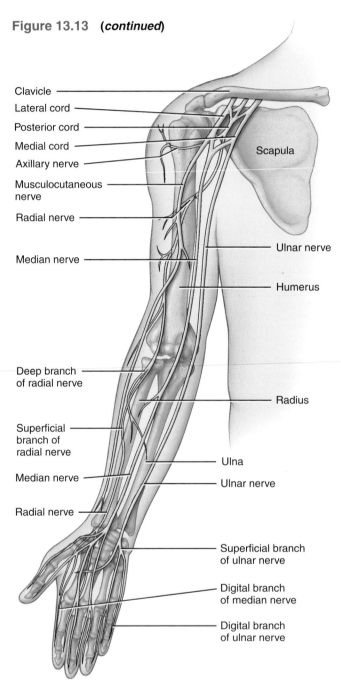

Clavicle

Lateral cord

Posterior cord

Medial cord

Axillary nerve

Musculocutaneous nerve

Radial nerve

Median nerve

Deep branch of radial nerve

Superficial branch of radial nerve

Median nerve

Radial nerve

Scapula

Ulnar nerve

Humerus

Radius

Ulna

Ulnar nerve

Superficial branch of ulnar nerve

Digital branch of median nerve

Digital branch of ulnar nerve

(b) Distribution of nerves from the brachial plexus

 What five important nerves arise from the brachial plexus?

Figure 13.14 Injuries to the brachial plexus.

Injuries to the brachial plexus affect the sensations and movements of the upper limbs.

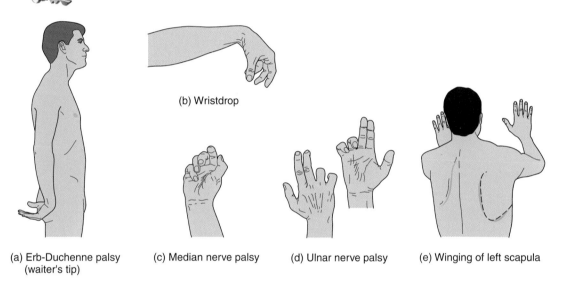

(a) Erb-Duchenne palsy (waiter's tip)

(b) Wristdrop

(c) Median nerve palsy

(d) Ulnar nerve palsy

(e) Winging of left scapula

 Injury to which nerve of the brachial plexus affects sensations on the palm and fingers?

Exhibit 13.3   Lumbar Plexus (Figure 13.15)

► O B J E C T I V E

• Describe the origin and distribution of the lumbar plexus.

Anterior (ventral) rami of spinal nerves L1–L4 form the **lumbar plexus** (LUM-bar) (Figure 13.15). Unlike the brachial plexus, there is no intricate intermingling of fibers in the lumbar plexus. On either side of the first four lumbar vertebrae, the lumbar plexus passes obliquely outward, posterior to the psoas major muscle and anterior to the quadratus lumborum muscle. It then gives rise to its peripheral nerves.

The lumbar plexus supplies the anterolateral abdominal wall, external genitals, and part of the lower limbs.

 Lumbar Plexus Injuries

The largest nerve arising from the lumbar plexus is the femoral nerve. **Femoral nerve injury,** which can occur in stab or gunshot wounds, is indicated by an inability to extend the leg and by loss of sensation in the skin over the anteromedial aspect of the thigh.

Obturator nerve injury, a common complication of childbirth, results in paralysis of the adductor muscles of the leg and loss of sensation over the medial aspect of the thigh. ■

Figure 13.15   Lumbar plexus in anterior view.   (See Tortora, *A Photographic Atlas of the Human Body,* Figure 8.10.)

🔑 **The lumbar plexus supplies the anterolateral abdominal wall, external genitals, and part of the lower limbs.**

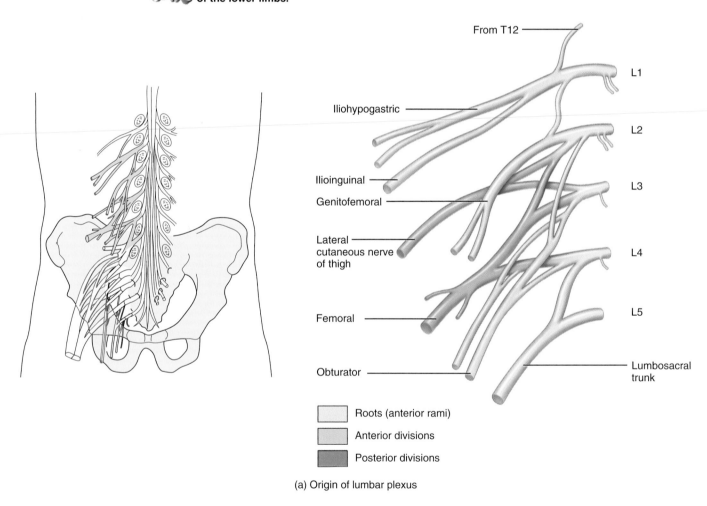

(a) Origin of lumbar plexus

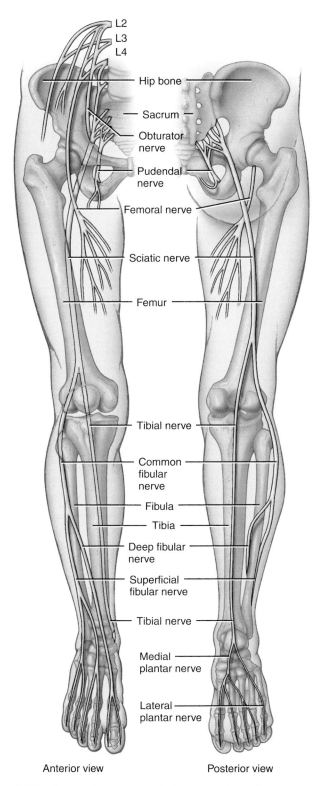

Anterior view   Posterior view

(b) Distribution of nerves from the lumbar and sacral plexuses

 What are the signs of femoral nerve injury?

Nerve	Origin	Distribution
Iliohypogastric (il′-ē-ō-hī-pō-GAS-trik)	L1	Muscles of anterolateral abdominal wall; skin of inferior abdomen and buttock.
Ilioinguinal (il′-ē-ō-IN-gwi-nal)	L1	Muscles of anterolateral abdominal wall; skin of superior medial aspect of thigh, root of penis and scrotum in male, and labia majora and mons pubis in female.
Genitofemoral (jen′-i-tō-FEM-or-al)	L1–L2	Cremaster muscle; skin over middle anterior surface of thigh, scrotum in male, and labia majora in female.
Lateral cutaneous nerve of thigh	L2–L3	Skin over lateral, anterior, and posterior aspects of thigh.
Femoral	L2–L4	Flexor muscles of thigh and extensor muscles of leg; skin over anterior and medial aspect of thigh and medial side of leg and foot.
Obturator (OB-too-rā-tor)	L2–L4	Adductor muscles of leg; skin over medial aspect of thigh.

Exhibit 13.4   Sacral and Coccygeal Plexuses (Figure 13.16)

▶ **OBJECTIVE**

• Describe the origin and distribution of the sacral plexus.

The anterior (ventral) rami of spinal nerves L4–L5 and S1–S4 form the **sacral plexus** (SĀ-kral) (Figure 13.16). This plexus is situated largely anterior to the sacrum. The sacral plexus supplies the buttocks, perineum, and lower limbs. The largest nerve in the body—the sciatic nerve—arises from the sacral plexus.

The anterior (ventral) rami of spinal nerves S4–S5 and the coccygeal nerves form a small **coccygeal plexus,** which supplies a small area of skin in the coccygeal region.

Sciatic Nerve Injury

Injury to the sciatic nerve and its branches results in **sciatica,** pain that may extend from the buttock down the posterior and lateral aspect of the leg and the lateral aspect of the foot. The sciatic nerve may be injured because of a herniated (slipped) disc, dislocated hip, osteoarthritis of the lumbosacral spine, pressure from the uterus during pregnancy, or an improperly administered gluteal intramuscular injection.

In the majority of sciatic nerve injuries, the common fibular portion is the most affected, frequently from fractures of the fibula or by pressure from casts or splints. Damage to the common fibular nerve causes the foot to be plantar flexed, a condition called **footdrop,** and inverted, a condition called **equinovarus.** There is also loss of function along the anterolateral aspects of the leg and dorsum of the foot and toes. Injury to the tibial portion of the sciatic nerve results in dorsiflexion of the foot plus eversion, a condition called **calcaneovalgus.** Loss of sensation on the sole also occurs. ■

▶ **CHECKPOINT**
Injury of which nerve causes footdrop?

Nerve	Origin	Distribution
Superior gluteal (GLOO-tē-al)	L4–L5 and S1	Gluteus minimus and gluteus medius muscles and tensor fasciae latae.
Inferior gluteal	L5–S2	Gluteus maximus muscle.
Nerve to piriformis (pir-i-FORM-is)	S1–S2	Piriformis muscle.
Nerve to quadratus femoris (quad-RĀ-tus FEM-or-is) **and inferior gemellus** (jem-EL-us)	L4–L5 and S1	Quadratus femoris and inferior gemellus muscles.
Nerve to obturator internus (OB-too-rā′-tor in-TER-nus) **and superior gemellus**	L5–S2	Obturator internus and superior gemellus muscles.
Perforating cutaneous (kū′-TĀ-ne-us)	S2–S3	Skin over inferior medial aspect of buttock.
Posterior cutaneous nerve of thigh	S1–S3	Skin over anal region, inferior lateral aspect of buttock, superior posterior aspect of thigh, superior part of calf, scrotum in male, and labia majora in female.
Sciatic (sī-AT-ik)	L4–S3	Actually two nerves—tibial and common fibular—bound together by common sheath of connective tissue. It splits into its two divisions, usually at the knee. (See below for distributions.) As sciatic nerve descends through the thigh, it sends branches to hamstring muscles and adductor magnus.
Tibial (TIB-ē-al)	L4–S3	Gastrocnemius, plantaris, soleus, popliteus, tibialis posterior, flexor digitorum longus, and flexor hallucis longus muscles. Branches of tibial nerve in foot are medial plantar nerve and lateral plantar nerve.
Medial plantar (PLAN-tar)		Abductor hallucis, flexor digitorum brevis, and flexor hallucis brevis muscles; skin over medial two-thirds of plantar surface of foot.
Lateral plantar		Remaining muscles of foot not supplied by medial plantar nerve; skin over lateral third of plantar surface of foot.
Common fibular (FIB-ū-lar)	L4–S2	Divides into a superficial fibular and a deep fibular branch.
Superficial fibular		Fibularis longus and fibularis brevis muscles; skin over distal third of anterior aspect of leg and dorsum of foot.
Deep fibular		Tibialis anterior, extensor hallucis longus, fibularis tertius, and extensor digitorum longus and brevis muscles; skin on adjacent sides of great and second toes.
Pudendal (pū-DEN-dal)	S2–S4	Muscles of perineum; skin of penis and scrotum in male and clitoris, labia majora, labia minora, and vagina in female.

Figure 13.16 Sacral and coccygeal plexuses in anterior view. The distribution of the nerves of the sacral plexus is shown in Figure 13.15b. (See Tortora, *A Photographic Atlas of the Human Body,* Figure 8.11.)

 The sacral plexus supplies the buttocks, perineum, and lower limbs.

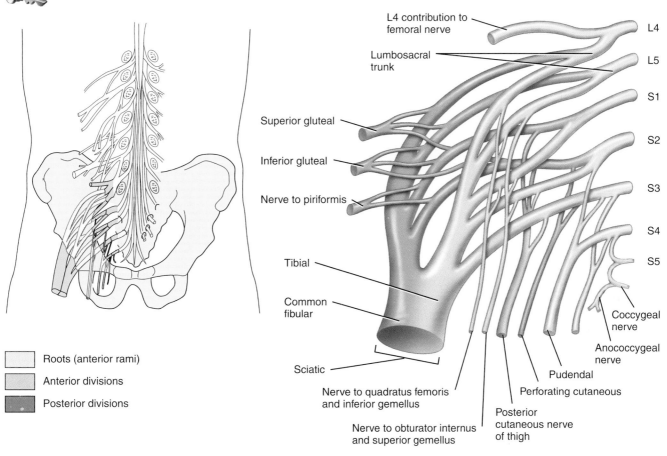

Roots (anterior rami)

Anterior divisions

Posterior divisions

L4 contribution to femoral nerve

Lumbosacral trunk

Superior gluteal

Inferior gluteal

Nerve to piriformis

Tibial

Common fibular

Sciatic

Nerve to quadratus femoris and inferior gemellus

Nerve to obturator internus and superior gemellus

Posterior cutaneous nerve of thigh

Perforating cutaneous

Pudendal

Anococcygeal nerve

Coccygeal nerve

L4

L5

S1

S2

S3

S4

S5

Origin of sacral plexus

What is the origin of the sacral plexus?

Dermatomes

The skin over the entire body is supplied by somatic sensory neurons that carry nerve impulses from the skin into the spinal cord and brain stem. Likewise, somatic motor neurons that carry impulses out of the spinal cord innervate the underlying skeletal muscles. Each spinal nerve contains sensory neurons that serve a specific, predictable segment of the body. Cranial nerve V (the trigeminal nerve) serves most of the skin of the face and scalp. The area of the skin that provides sensory input to the CNS via one pair of spinal nerves or cranial nerve V is called a **dermatome** (*derma-* = skin; *-tome* = thin segment) (Figure 13.17). The nerve supply in adjacent dermatomes overlaps somewhat. In regions where the overlap is considerable, little loss of sensation may result if only one of the nerves supplying the dermatome is damaged.

Knowing which spinal cord segments supply each dermatome makes it possible to locate damaged regions of the spinal cord. If the skin in a particular region is stimulated but the sensation is not perceived, the nerves supplying that dermatome are probably damaged. Information about the innervation patterns of spinal nerves can also be used therapeutically. Cutting posterior roots or infusing local anesthetics can block pain either permanently or transiently. Because dermatomes overlap, deliberate production of a region of complete anesthesia may require that at least three adjacent spinal nerves be cut or blocked by an anesthetic drug.

Spinal Cord Transection and Muscle Function

An injury that entirely severs the spinal cord is said to cause complete **transection** (tran-SEK-shun; *trans-* = across; *-section* = a cut). After transection, a person will have permanent loss of sensations in dermatomes below the injury because ascending nerve impulses cannot propagate past the transection to reach the brain. At the same time, voluntary muscle contractions will be lost below the transection because nerve impulses descending from the brain also cannot pass. The extent of paralysis of skeletal muscles depends on the level of injury. The following list outlines which muscle functions may be *retained* at progressively lower levels of spinal cord transection.

- C1–C3: no function maintained from the neck down; ventilator needed for breathing
- C4–C5: diaphragm, which allows breathing
- C6–C7: some arm and chest muscles, which allows feeding, some dressing, and propelling wheelchair
- T1–T3: intact arm function
- T4–T9: control of trunk above the umbilicus
- T10–L1: most thigh muscles, which allows walking with long leg braces
- L1–L2: most leg muscles, which allows walking with short leg braces ■

Figure 13.17 Distribution of dermatomes.

🔑 A dermatome is an area of skin that provides sensory input to the CNS via the posterior roots of one pair of spinal nerves or via cranial nerve V (trigeminal nerve).

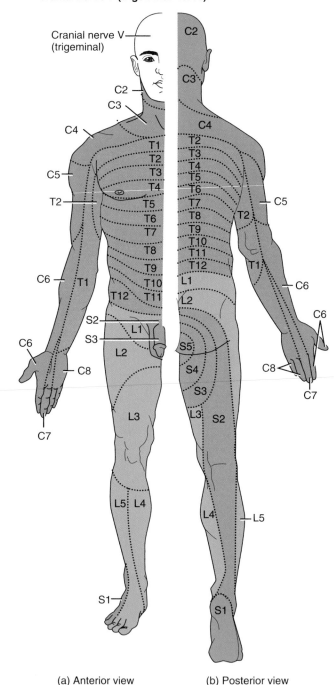

(a) Anterior view   (b) Posterior view

❓ Which is the only spinal nerve that does not have a corresponding dermatome?

▶ CHECKPOINT

9. How are spinal nerves named and numbered? Why are all spinal nerves classified as mixed nerves?

10. How do spinal nerves connect to the spinal cord?

11. Which regions of the body are supplied by plexuses and by intercostal nerves?

DISORDERS: HOMEOSTATIC IMBALANCES

Shingles

Shingles is an acute infection of the peripheral nervous system caused by herpes zoster (HER-pēz ZOS-ter), the virus that also causes chickenpox. After a person recovers from chickenpox, the virus retreats to a posterior root ganglion. If the virus is reactivated, the immune system usually prevents it from spreading. From time to time, however, the reactivated virus overcomes a weakened immune system, leaves the ganglion, and travels down sensory neurons of the skin by fast axonal transport (described on page 390). The result is pain, discoloration of the skin, and a characteristic line of skin blisters. The line of blisters marks the distribution (dermatome) of the particular cutaneous sensory nerve belonging to the infected posterior root ganglion.

Poliomyelitis

Poliomyelitis, or simply **polio,** is caused by a virus called poliovirus. The onset of the disease is marked by fever, severe headache, a stiff neck and back, deep muscle pain and weakness, and loss of certain somatic reflexes. In its most serious form, the virus produces paralysis by destroying cell bodies of motor neurons, specifically those in the anterior horns of the spinal cord and in the nuclei of the cranial nerves. Polio can cause death from respiratory or heart failure if the virus invades neurons in vital centers that control breathing and heart functions in the brain stem. Even though polio vaccines have virtually eradicated polio in the United States, outbreaks of polio continue throughout the world. Due to international travel, polio could easily be reintroduced into North America if individuals are not vaccinated appropriately.

Several decades after suffering a severe attack of polio and following their recovery from it, some individuals develop a condition called **post-polio syndrome.** This neurological disorder is characterized by progressive muscle weakness, extreme fatigue, loss of function, and pain, especially in muscles and joints. Post-polio syndrome seems to involve a slow degeneration of motor neurons that innervate muscle fibers. Triggering factors appear to be a fall, a minor accident, surgery, or prolonged bed rest. Possible causes include overuse of surviving motor neurons over time, smaller motor neurons because of the initial infection by the virus, reactivation of dormant polio viral particles, immune-mediated responses, hormone deficiencies, and environmental toxins. Treatment consists of muscle-strengthening exercises, administration of pyridostigmine to enhance the action of acetylcholine in stimulating muscle contraction, and administration of nerve growth factors to stimulate both nerve and muscle growth.

MEDICAL TERMINOLOGY

Meningitis (men-in-JĪ-tis; *-itis* = inflammation) Inflammation of the meninges.

Nerve block Loss of sensation in a region due to injection of a local anesthetic; an example is local dental anesthesia.

Neuralgia (noo-RAL-jē-a; *neur-* = nerve; *-algia* = pain) Attacks of pain along the entire course or a branch of a sensory nerve.

Neuritis (*neur-* = nerve; *-itis* = inflammation) Inflammation of one or several nerves that may result from irritation to the nerve produced by direct blows, bone fractures, contusions, or penetrating injuries. Additional causes include infections, vitamin deficiency (usually thiamine), and poisons such as carbon monoxide, carbon tetrachloride, heavy metals, and some drugs.

Paresthesia (par-es-THĒ-zē-a; *par-* = departure from normal; *-esthesia* = sensation) An abnormal sensation such as burning, pricking, tickling, or tingling resulting from a disorder of a sensory nerve.

STUDY OUTLINE

SPINAL CORD ANATOMY (p. 420)

1. The spinal cord is protected by the vertebral column, the meninges, cerebrospinal fluid, and denticulate ligaments.
2. The three meninges are coverings that run continuously around the spinal cord and brain. They are the dura mater, arachnoid mater, and pia mater.
3. The spinal cord begins as a continuation of the medulla oblongata and ends at about the second lumbar vertebra in an adult.
4. The spinal cord contains cervical and lumbar enlargements that serve as points of origin for nerves to the limbs.
5. The tapered inferior portion of the spinal cord is the conus medullaris, from which arise the filum terminale and cauda equina.
6. Spinal nerves connect to each segment of the spinal cord by two roots. The posterior or dorsal root contains sensory axons, and the anterior or ventral root contains motor neuron axons.
7. The anterior median fissure and the posterior median sulcus partially divide the spinal cord into right and left sides.
8. The gray matter in the spinal cord is divided into horns, and the white matter into columns. In the center of the spinal cord is the central canal, which runs the length of the spinal cord.
9. Parts of the spinal cord observed in transverse section are the gray commissure; central canal; anterior, posterior, and lateral gray horns; and anterior, posterior, and lateral white columns, which contain ascending and descending tracts. Each part has specific functions.
10. The spinal cord conveys sensory and motor information by way of ascending and descending tracts, respectively.

SPINAL CORD PHYSIOLOGY (p. 425)

1. The white matter tracts in the spinal cord are highways for nerve impulse propagation. Along these tracts, sensory impulses flow toward brain, and motor impulses flow from the brain toward skeletal muscles and other effector tissues.
2. Sensory impulses travel along two main routes in the white matter of the spinal cord: the posterior columns and the spinothalamic tracts.
3. Motor impulses propagate along two main routes in the white matter of the spinal cord: direct pathways and indirect pathways.
4. A second major function of the spinal cord is to serve as an integrating center for spinal reflexes. This integration occurs in the gray matter.
5. A reflex is a fast, predictable sequence of involuntary actions, such as muscle contractions or glandular secretions, which occurs in response to certain changes in the environment.
6. Reflexes may be spinal or cranial and somatic or autonomic (visceral).
7. The components of a reflex arc are sensory receptor, sensory neuron, integrating center, motor neuron, and effector.
8. Somatic spinal reflexes include the stretch reflex, the tendon reflex, the flexor (withdrawal) reflex, and the crossed extensor reflex; all exhibit reciprocal innervation.
9. A two-neuron or monosynaptic reflex arc consists of one sensory neuron and one motor neuron. A stretch reflex, such as the patellar reflex, is an example.
10. The stretch reflex is ipsilateral and is important in maintaining muscle tone.
11. A polysynaptic reflex arc contains sensory neurons, interneurons, and motor neurons. The tendon reflex, flexor (withdrawal) reflex, and crossed extensor reflexes are examples.
12. The tendon reflex is ipsilateral and prevents damage to muscles and tendons when muscle force becomes too extreme. The flexor reflex is ipsilateral and moves a limb away from the source of a painful stimulus. The crossed extensor reflex extends the limb contralateral to a painfully stimulated limb, allowing the weight of the body to shift when a supporting limb is withdrawn.
13. A clinically important somatic reflex is the plantar flexion reflex, which is expressed as the Babinski sign when descending motor pathways are damaged.

SPINAL NERVES (p. 433)

1. The 31 pairs of spinal nerves are named and numbered according to the region and level of the spinal cord from which they emerge.
2. There are 8 pairs of cervical, 12 pairs of thoracic, 5 pairs of lumbar, 5 pairs of sacral, and 1 pair of coccygeal nerves.
3. Spinal nerves typically are connected with the spinal cord by a posterior root and an anterior root. All spinal nerves contain both sensory and motor axons (are mixed nerves).
4. Three connective tissue coverings associated with spinal nerves are the endoneurium, perineurium, and epineurium.
5. Branches of a spinal nerve include the posterior ramus, anterior ramus, meningeal branch, and rami communicantes.
6. The anterior rami of spinal nerves, except for T2–T12, form networks of nerves called plexuses.
7. Emerging from the plexuses are nerves bearing names that typically describe the general regions they supply or the route they follow.
8. Nerves of the cervical plexus supply the skin and muscles of the head, neck, and upper part of the shoulders; they connect with some cranial nerves and innervate the diaphragm.
9. Nerves of the brachial plexus supply the upper limbs and several neck and shoulder muscles.
10. Nerves of the lumbar plexus supply the anterolateral abdominal wall, external genitals, and part of the lower limbs.
11. Nerves of the sacral plexus supply the buttocks, perineum, and part of the lower limbs.
12. Nerves of the coccygeal plexus supply the skin of the coccygeal region.
13. Anterior rami of nerves T2–T12 do not form plexuses and are called intercostal (thoracic) nerves. They are distributed directly to the structures they supply in intercostal spaces.
14. Sensory neurons within spinal nerves and cranial nerve V (trigeminal nerve) serve specific, constant segments of the skin called dermatomes.
15. Knowledge of dermatomes helps a physician determine which segment of the spinal cord or which spinal nerve is damaged.

SELF-QUIZ QUESTIONS

Fill in the blanks in the following statements.

1. _____ are fast, predictable, automatic responses to changes in the environment.
2. Because they contain both sensory and motor axons, spinal nerves are considered to be _____ nerves.

Indicate whether the following statements are true or false.

3. Reflexes permit the body to make exceedingly rapid adjustments to homeostatic imbalances.
4. Autonomic reflexes involve responses of smooth muscle, cardiac muscle, and glands.

Choose the one best answer to the following questions.

5. The connective tissue covering around an entire nerve is the (a) endoneurium, (b) fascicle, (c) perineurium, (d) epineurium, (e) neurolemma.
6. Which of the following is *not* a function of the spinal cord? (a) reflex center, (b) integration of EPSPs and IPSPs, (c) conduction pathway for sensory impulses, (d) conduction pathway for motor impulses, (e) interpretation of sensory stimuli.
7. Which of the following are true? (1) The anterior (ventral) gray horns contain cell bodies of neurons that cause skeletal muscle contraction. (2) The gray commissure connects the white matter of the right and left sides of the spinal cord. (3) Cell bodies of autonomic motor neurons are located in the lateral gray horns. (4) Sensory (ascending) tracts conduct motor impulses down the spinal cord. (5) Gray matter in the spinal cord consists of cell bodies of

neurons, neuroglia, unmyelinated axons and dendrites of interneurons and motor neurons. (a) 1, 2, 3, and 5, (b) 2 and 4, (c) 2, 3, 4, and 5, (d) 1, 3, and 5, (e) 1, 2, 3, and 4.

8. Place the events of a reflex arc in the correct order. (1) Impulses received by one or more integrating regions within the CNS. (2) Sensory receptor activated by a stimulus. (3) Nerve impulse travels into spinal cord through posterior root of spinal cord. (4) An effector responds to the motor nerve impulses. (5) Motor nerve impulses travel through anterior root of spinal nerve. (a) 2, 3, 1, 5, 4, (b) 3, 2, 1, 5, 4, (c) 2, 5, 1, 3, 4, (d) 4, 2, 3, 1, 5, (e) 2, 5, 3, 1, 4.

9. Which of the following are functions of the nerves of the posterior column? (1) proprioception, (2) discriminative touch, (3) pain, (4) thermal sensations, (5) pressure, (6) vibration, (7) two-point discrimination. (a) 1, 2, 4, and 5, (b) 2, 4, 6, and 7, (c) 1, 2, 5, 6 and 7, (d) 3, 4, 5, 6, and 7, (e) 1, 3, 5, 6, and 7.

10. Match the following (some answers may be used more than once):

_____ (a) a reflex resulting in the contraction of a muscle when it is stretched
_____ (b) receptors that monitor changes in muscle length
_____ (c) a balance-maintaining reflex
_____ (d) operates as a feedback mechanism to control muscle tension by causing muscle relaxation when muscle force becomes too extreme
_____ (e) reflex arc that consists of one sensory and one motor neuron
_____ (f) acts as a feedback mechanism to control muscle length by causing muscle contraction
_____ (g) sensory impulses enter on one side of the spinal cord and motor impulses exit on the opposite side
_____ (h) occurs when sensory nerve impulse travels up and down the spinal cord, thereby activating several motor neurons and more than one effector
_____ (i) withdrawal reflex
_____ (j) receptors that monitor changes in muscle tension
_____ (k) maintains proper muscle tone
_____ (l) reflex pathway that contains sensory neurons, interneurons, and motor neurons
_____ (m) motor nerve impulses exit the spinal cord on the same side that sensory impulses entered the spinal cord
_____ (n) protects the tendon and muscle from damage due to excessive tension

(1) stretch reflex
(2) tendon reflex
(3) flexor reflex
(4) crossed extensor reflex
(5) intersegmental reflex arc
(6) contralateral reflex arc
(7) ipsilateral reflex arc
(8) muscle spindles
(9) tendon organs
(10) monosynaptic reflex
(11) polysynaptic reflex

11. Which of the following statements is *false*? (a) The two main spinal cord sensory pathways are the spinothalamic and anterior columns. (b) The spinothalamic tracts convey impulses for sensing pain, temperature, touch, and deep pressure. (c) Direct pathways convey nerve impulses destined to cause precise, voluntary movements of skeletal muscles. (d) Indirect pathways convey nerve impulses that program automatic movements, help coordinate body movements with visual stimuli, maintain skeletal muscle tone and posture, and contribute to equilibrium. (e) The direct pathways are motor pathways.

12. The nerves that do *not* enter into the formation of a plexus are (a) C1–C5, (b) T2–T12, (c) C4–T5, (d) L1–L3, (e) T6–L2.

13. Which of the following statements is *false*? (a) The most superficial of the spinal meninges is the dura mater. (b) The epidural space contains fat and connective tissue. (c) The middle spinal meninx is the arachnoid mater. (d) The innermost spinal meninx is the pia mater. (e) The subarachnoid space contains interstitial fluid and blood cells.

14. Match the following:

_____ (a) the joining together of the anterior rami of adjacent nerves
_____ (b) spinal nerve branches that serve the deep muscles and skin of the posterior surface of the trunk
_____ (c) spinal nerve branches that serve the muscles and structures of the upper and lower limbs and the lateral and ventral trunk
_____ (d) area of the spinal cord from which nerves to and from the upper limbs arise
_____ (e) area of the spinal cord from which nerves to and from the lower limbs arise
_____ (f) the roots form the nerves that arise from the inferior part of the spinal cord but do not leave the vertebral column at the same level as they exit the cord
_____ (g) contains motor neuron axons that conduct impulses from the spinal cord to the periphery
_____ (h) contains sensory neuron axons that conduct impulses from the periphery into the spinal cord
_____ (i) an extension of the pia mater that anchors the spinal cord to the coccyx
_____ (j) extending the length of the spinal cord, these pia mater thickenings fuse with the arachnoid mater and dura mater and help protect the spinal cord from shock and sudden displacement
_____ (k) space within the spinal cord filled with cerebrospinal fluid

(1) cervical enlargement
(2) lumbar enlargement
(3) central canal
(4) denticulate ligaments
(5) cauda equina
(6) posterior (dorsal) root
(7) anterior (ventral) root
(8) posterior ramus
(9) anterior ramus
(10) plexus
(11) filum terminale

15. Match the following:
    (1) cervical plexus   (2) brachial plexus   (3) lumbar plexus   (4) sacral plexus   (5) coccygeal plexus

    ____ (a) provides the entire nerve supply of the shoulders and upper limbs
    ____ (b) provides the nerve supply of the skin and muscles of the head, neck, and superior part of the shoulders and chest
    ____ (c) provides the nerve supply of the anterolateral abdominal wall, external genitals, and part of the lower limbs
    ____ (d) innervates the buttocks, perineum, and lower limbs

    ____ (e) phrenic nerve arises from this plexus
    ____ (f) median nerve arises from this plexus
    ____ (g) sciatic nerve arises from this plexus
    ____ (h) femoral nerve arises from this plexus
    ____ (i) supplies a small area of skin in coccygeal region
    ____ (j) injury to this plexus can affect breathing

CRITICAL THINKING QUESTIONS

1. Pearl, a lifeguard at the local beach, stepped on a discarded lit cigarette with her bare foot. Trace the reflex arcs set in motion by her accident. Name the reflex types.
   HINT   *How will she remain standing when she picks up her foot?*

2. Why doesn't your spinal cord creep up toward your head every time you bend over? Why doesn't it get all twisted out of position when you exercise?
   HINT   *How would you prevent a boat from floating away from a dock?*

3. Jose's severe headaches and other symptoms were suggestive of meningitis, so his physician ordered a spinal tap. List the structures that the needle will pierce from the most superficial to the deepest. Why would the physican order a test in the spinal region to check a problem in Jose's head?
   HINT   *The central nervous system needs protection in both its locations.*

ANSWERS TO FIGURE QUESTIONS

13.1 The superior boundary of the spinal dura mater is the foramen magnum of the occipital bone. The inferior boundary is the second sacral vertebra.

13.2 The cervical enlargement connects with sensory and motor nerves of the upper limbs.

13.3 A horn is an area of gray matter, and a column is a region of white matter in the spinal cord.

13.4 The anterior corticospinal tract is located on the anterior side of the spinal cord, originates in the cortex of the cerebrum, and ends in the spinal cord. It contains descending axons and thus is a motor tract.

13.5 A sensory receptor produces a generator potential, which triggers a nerve impulse if the generator potential reaches threshold. Reflex integrating centers are in the CNS.

13.6 In an ipsilateral reflex, the sensory and motor neurons are on the same side of the spinal cord.

13.7 Reciprocal innervation is a type of neural circuit involving simultaneous contraction of one muscle and relaxation of its antagonist.

13.8 The flexor reflex is intersegmental because impulses go out over motor neurons located in several spinal nerves, each arising from a different segment of the spinal cord.

13.9 The crossed extensor reflex is a contralateral reflex arc because the motor impulses leave the spinal cord on the side opposite the entry of sensory impulses.

13.10 All spinal nerves are mixed (have sensory and motor components) because the posterior root containing sensory axons and the anterior root containing motor axons unite to form the spinal nerve.

13.11 The anterior rami serve the upper and lower limbs.

13.12 Severing the spinal cord at level C2 causes respiratory arrest because it prevents descending nerve impulses from reaching the phrenic nerve, which stimulates contraction of the diaphragm, the main muscle needed for breathing.

13.13 The axillary, musculocutaneous, radial, median, and ulnar nerves are five important nerves that arise from the brachial plexus.

13.14 Injury to the median nerve affects sensations on the palm and fingers.

13.15 Signs of femoral nerve injury include inability to extend the leg and loss of sensation in the skin over the anterolateral aspect of the thigh.

13.16 The origin of the sacral plexus is the anterior rami of spinal nerves L4–L5 and S1–S4.

13.17 The only spinal nerve without a corresponding dermatome is C1.

The Brain and
Cranial Nerves

A Student's View

One of the most complex and fascinating structures in the human body is the brain. The body is capable of almost anything, but nothing is possible without the brain receiving information, analyzing the data, and taking action. From the exterior, the brain is protected only by bone, meninges, and fluid. Internally, the blood-brain barrier acts as a gatekeeper by identifying substances and either granting or denying entrance to the brain. Each section of the brain is highly specialized in its duty, yet the organ works as a united machine. Although the brain has amazing potential in thoughts and actions, just minutes of deprivation in blood or oxygen can shut the machine down.

The brain is aware of its surroundings via input from the spinal cord and cranial nerves. Cranial nerves with sensory functions allow us to smell and see. Nerves with both motor and sensory functions are responsible for everything from tasting and chewing, to breathing and the beating of the heart. Simple tasks often taken for granted, such as smiling or swallowing, are all made possible by cranial nerves.

As the daughter of two surgical intensive care nurses, I have seen and heard of the horrific results of trauma to the brain and cranial nerves. Trauma to the brain or cranial nerves devastates more than trauma to any other organ or body system. Head injuries and severed nerves are injuries that can't always be repaired. An intensive care unit (ICU) can have drastic effects on its staff. I would bet that after working with patients with head injuries in an (ICU), more nurses faithfully wear seatbelts and helmets.

Emily J. Gordon
Edison Community College

THE BRAIN, CRANIAL NERVES AND HOMEOSTASIS

Your brain contributes to homeostasis by mediating complex functions through sensory input, integration of new and stored information, making decisions, and signaling actions.

ENERGIZE YOUR STUDY

FOUNDATIONS CD

Anatomy Overview
• The Nervous System

Animation
• Systems Contributions to Homeostasis
• The Case of the Man with Yellow Eyes
• The Case of the Girl with Fruity Breath

Concepts and Connections and Exercises reinforce your understanding

www.wiley.com/college/apcentral

INSIGHTS AND EXPLORATIONS

Have you considered becoming a vegetarian lately? Recently, there have been many reports of people getting sick and dying from the always fatal Mad Cow Disease (also called bovine spongiform encephalopathy), which has found its way to humans in Europe. Doctors doing autopsies on individuals who have died from this disease even wear chain-metal gloves to protect themselves from accidental exposure! The brains of the victims of this disease are so altered that the brain tissue has a sponge-like appearance when viewed through a microscope. Hence, the name of the disease. Come on a web-based adventure to learn about the agents that can cause this dreaded disease.

olving an equation, feeling hungry, laughing—the neural processes needed for each of these activities occurs in a different region of the **brain,** that portion of the central nervous system contained within the cranium. About 100 billion neurons and 10–50 trillion neuroglia make up the brain, which has a mass of about 1300 g (almost 3 lb) in adults. On average, each neuron forms 1000 synapses with other neurons. Thus, the total number of synapses, about a thousand trillion or 10^{15}, is larger than the number of stars in the galaxy.

The brain is the center for registering sensations, correlating them with one another and with stored information, making decisions, and taking actions. It also is the center for the intellect, emotions, behavior, and memory. But the brain encompasses yet a larger domain: It directs our behavior toward others. With ideas that excite, artistry that dazzles, or rhetoric that mesmerizes, one person's thoughts and actions may influence and shape the lives of many others. As you will see, different regions of the brain are specialized for different functions. As well, different parts of the brain work together to accomplish certain functions. This chapter explores how the brain is protected and nourished, what functions occur in the major regions of the brain, and how the spinal cord and the 12 pairs of cranial nerves connect with the brain.

BRAIN ORGANIZATION, PROTECTION, AND NOURISHMENT

> ### OBJECTIVES
>
> - Identify the major parts of the brain.
> - Describe how the brain is protected.
> - Explain the formation and circulation of cerebrospinal fluid.

Major Parts of the Brain

The brain consists of four major parts: brain stem, cerebellum, diencephalon, and cerebrum (Figure 14.1). The **brain stem** is continuous with the spinal cord and consists of the medulla oblongata, pons, and midbrain. Posterior to the brain stem is the **cerebellum** (ser′-e-BEL-um = little brain). Superior to the brain stem is the **diencephalon** (dī′-en-SEF-a-lon; *di-* = through; *-encephalon* = brain), consisting mainly of the thalamus and hypothalamus and including the epithalamus and the subthalamus. Supported on the diencephalon and brain stem, the **cerebrum** (se-RĒ-brum = brain) is the largest part of the brain.

Brain Blood Flow and the Blood–Brain Barrier

Blood flows to the brain mainly via the internal carotid and vertebral arteries (see Figure 21.20 on page 726); the internal jugular veins return blood from the head to the heart (see Figure 21.25 on page 740).

In an adult, the brain represents only 2% of total body weight, but it consumes about 20% of the oxygen and glucose used at rest. Neurons synthesize ATP almost exclusively from glucose via reactions that use oxygen (oxidative phosphorylation in mitochondria). When activity of neurons and neuroglia increases in a region of the brain, blood flow to that area also increases. Even a brief slowing of brain blood flow may cause unconsciousness. Typically, an interruption in blood flow for 1 or 2 minutes impairs neuronal function, and total deprivation of oxygen for about 4 minutes causes permanent injury. Because virtually no glucose is stored in the brain, the supply of glucose also must be continuous. If blood entering the brain has a low level of glucose, mental confusion, dizziness, convulsions, and loss of consciousness may occur.

The existence of a **blood–brain barrier (BBB)** protects brain cells from harmful substances and pathogens by preventing passage of many substances from blood into brain tissue. Tight junctions seal together the endothelial cells of brain capillaries, which also are surrounded by a thick basement membrane. In addition, the processes of many astrocytes (one type of neuroglia) press up against the capillaries. The astrocyte processes selectively pass some substances from the blood to neurons but inhibit the passage of others. A few water-soluble substances (for example, glucose) cross the BBB by active transport. Other substances, such as creatinine, urea, and most ions, cross the BBB very slowly. Still other substances—proteins and most antibiotic drugs—do not pass at all from the blood into brain tissue. However, the BBB does not prevent passage of lipid-soluble substances, such as oxygen, carbon dioxide, alcohol, and most anesthetic agents, into brain tissue. Trauma, certain toxins, and inflammation can cause a breakdown of the blood–brain barrier.

Breaching the Blood–Brain Barrier

We have seen how the BBB prevents the passage into brain tissue of potentially harmful substances. But another consequence of the BBB's efficient protection is that it also prevents the passage of certain drugs that could be therapeutic for brain cancer or other CNS disorders. Researchers are finding ways to move drugs past the BBB. In one method, the drug is injected in a concentrated sugar solution. The high osmotic pressure of the sugar solution causes the endothelial cells of the capillaries to shrink, which opens gaps between their tight junctions and makes the BBB more leaky. As a result, the drug can enter the brain tissue. ■

Figure 14.1 The brain. The pituitary gland is discussed with the endocrine system in Chapter 18. (See Tortora, *A Photographic Atlas of the Human Body,* Figures 8.12, 8.13, and 8.15.)

🔑 **The four principal parts of the brain are the brain stem, cerebellum, diencephalon, and cerebrum.**

Sagittal plane

View

CEREBRUM

DIENCEPHALON:
  Thalamus
  Epithalamus
  Hypothalamus
  Pineal gland

BRAIN STEM:
  Midbrain
  Pons
  Medulla oblongata

Pituitary gland

CEREBELLUM

Spinal cord

POSTERIOR

ANTERIOR

(a) Sagittal section, medial view

(continues)

Protective Coverings of the Brain

The cranium (see Figure 7.4 on page 192) and the cranial meninges surround and protect the brain. The **cranial meninges** are continuous with the spinal meninges, have the same basic structure, and bear the same names: the outer **dura mater,** the middle **arachnoid mater,** and the inner **pia mater** (Figure 14.2). Blood vessels that enter brain tissue pass along the surface of the brain, and as they penetrate inward, they are sheathed by a loose-fitting sleeve of pia mater. Three extensions of the dura mater separate parts of the brain. (1) The **falx cerebri** (FALKS CER-e-brē; *falx* = sickle-shaped) separates the two hemispheres (sides) of the cerebrum. (2) The **falx cerebelli** (cer-e-BEL-ī)

separates the two hemispheres of the cerebellum. (3) The **tentorium cerebelli** (ten-TŌ-rē-um = tent) separates the cerebrum from the cerebellum.

Cerebrospinal Fluid

Cerebrospinal fluid (CSF) is a clear, colorless liquid that protects the brain and spinal cord against chemical and physical injuries. It also carries oxygen, glucose, and other needed chemicals from the blood to neurons and neuroglia. CSF continuously circulates through cavities in the brain and spinal cord and around the brain and spinal cord in the subarachnoid space (between the arachnoid mater and pia mater).

Figure 14.1 *(continued)*

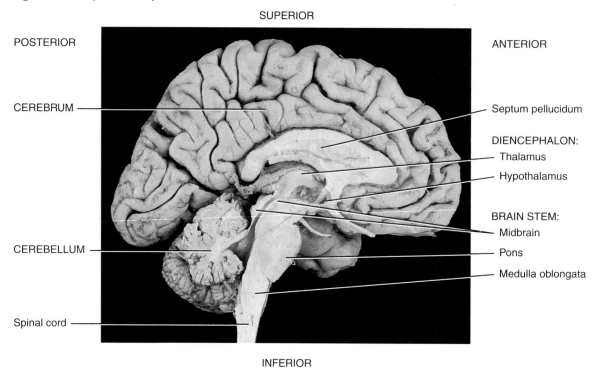

SUPERIOR

POSTERIOR

ANTERIOR

CEREBRUM

Septum pellucidum

DIENCEPHALON:
  Thalamus
  Hypothalamus

BRAIN STEM:
  Midbrain
  Pons
  Medulla oblongata

CEREBELLUM

Spinal cord

INFERIOR

(b) Sagittal section, medial view

 Which of the four principal parts of the brain is the largest?

Figure 14.2   The protective coverings of the brain.

 Cranial bones and the cranial meninges protect the brain.

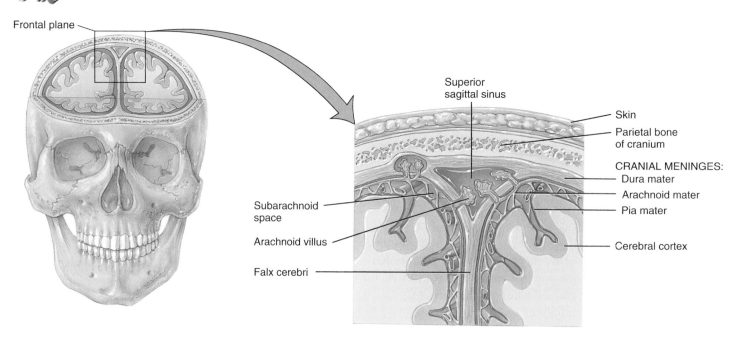

Frontal plane

Superior
sagittal sinus

Skin

Parietal bone
of cranium

CRANIAL MENINGES:
  Dura mater
  Arachnoid mater
  Pia mater

Subarachnoid
space

Arachnoid villus

Falx cerebri

Cerebral cortex

Frontal section through skull showing
the cranial meninges

 What are the three layers of the cranial meninges, from superficial to deep?

Formation of CSF in the Ventricles

Figure 14.3 shows the four CSF-filled cavities within the brain, which are called **ventricles** (VEN-tri-kuls = little cavities). A **lateral ventricle** is located in each hemisphere of the cerebrum. Anteriorly, the lateral ventricles are separated by a thin membrane, the **septum pellucidum** (SEP-tum pe-LOO-si-dum; *pellucid* = transparent). The **third ventricle** is a narrow cavity along the midline superior to the hypothalamus and between the right and left halves of the thalamus. The **fourth ventricle** lies between the brain stem and the cerebellum.

The total volume of CSF is 80 to 150 mL (3 to 5 oz) in an adult. CSF contains glucose, proteins, lactic acid, urea, cations (Na^+, K^+, Ca^{2+}, Mg^{2+}), and anions (Cl^- and HCO_3^-); it also contains some white blood cells. The CSF contributes to homeostasis in three main ways:

1. **Mechanical protection.** CSF serves as a shock-absorbing medium that protects the delicate tissue of the brain and spinal cord from jolts that would otherwise cause them to hit the bony walls of the cranial and vertebral cavities. The fluid also buoys the brain so that it "floats" in the cranial cavity.

2. **Chemical protection.** CSF provides an optimal chemical environment for accurate neuronal signaling. Even slight changes in the ionic composition of CSF within the brain can seriously disrupt production of action potentials and postsynaptic potentials.

3. **Circulation.** CSF is a medium for exchange of nutrients and waste products between the blood and nervous tissue.

The sites of CSF production are the **choroid plexuses** (KŌ-royd = membranelike), which are networks of capillaries (microscopic blood vessels) in the walls of the ventricles. The capillaries are covered by ependymal cells that form cerebrospinal fluid from blood plasma by filtration and secretion. Because the ependymal cells are joined by tight junctions, materials entering CSF from choroid capillaries cannot leak between these cells; instead, they must pass through the ependymal cells. This **blood–cerebrospinal fluid barrier** permits certain substances to enter the CSF but excludes others, protecting the brain and spinal cord from potentially harmful blood-borne substances.

Circulation of CSF

The CSF formed in the choroid plexuses of each lateral ventricle flows into the third ventricle through two narrow, oval openings, the **interventricular foramina (foramina of Monro)** (Figure 14.4a). More CSF is added by the choroid plexus in the roof of the third ventricle. The fluid then flows through the **cerebral aqueduct (aqueduct of Sylvius),** which passes through the midbrain, into the fourth ventricle. The choroid plexus of the fourth ventricle contributes more fluid. CSF enters the subarachnoid space through three openings in the roof of the fourth ventricle: a **median aperture (of Magendie)** and the paired **lateral**

Figure 14.3   Locations of ventricles within a "transparent" brain.   One interventricular foramen on each side connects a lateral ventricle to the third ventricle, and the cerebral aqueduct connects the third ventricle to the fourth ventricle.

Ventricles are cavities within the brain that are filled with cerebrospinal fluid.

POSTERIOR

ANTERIOR

Cerebrum

LATERAL VENTRICLES

INTERVENTRICULAR FORAMEN

FOURTH VENTRICLE

Cerebellum

THIRD VENTRICLE

LATERAL APERTURE

CEREBRAL AQUEDUCT

MEDIAN APERTURE

Pons

Medulla oblongata

CENTRAL CANAL

Spinal cord

Right lateral view of brain

Which brain region is anterior to the fourth ventricle? Which is posterior to it?

Figure 14.4   Pathways of circulating cerebrospinal fluid.   (See Tortora, *A Photographic Atlas of the Human Body,* Figures 8.15 and 8.18.)

🔑 **CSF is formed by ependymal cells that cover the choroid plexuses of the ventricles.**

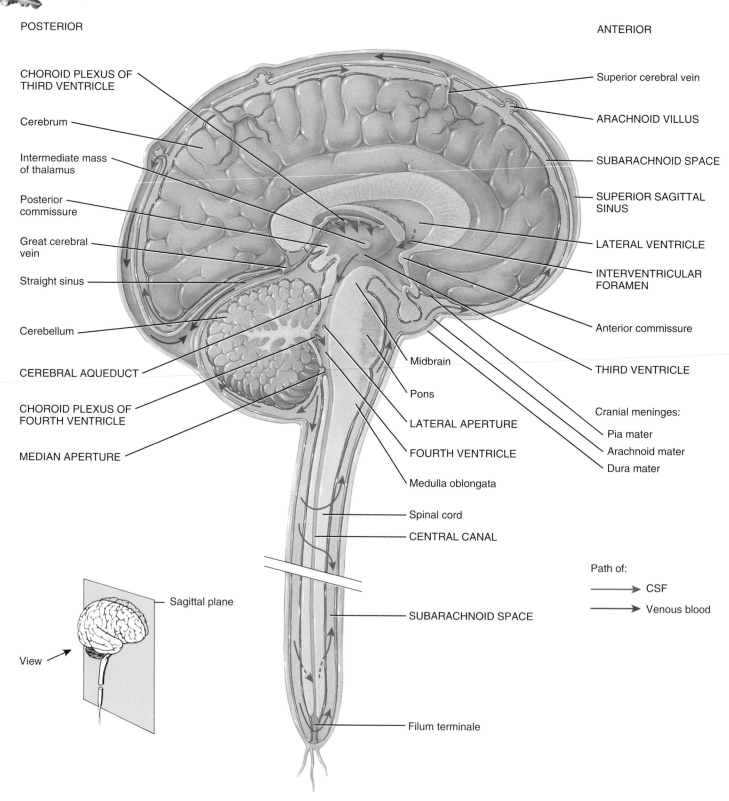

(a) Sagittal section of brain and spinal cord

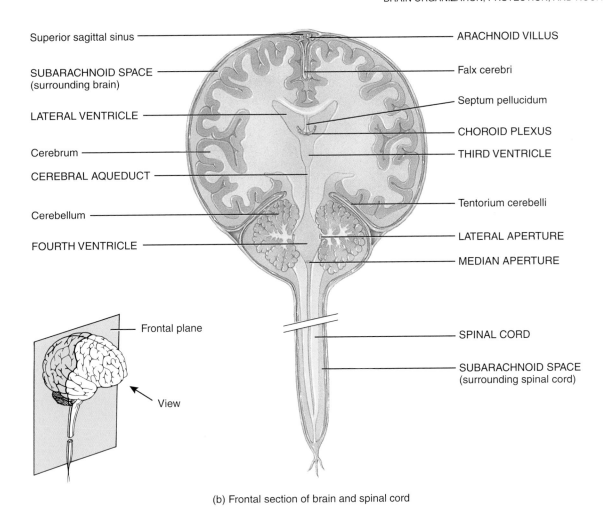

Superior sagittal sinus — ARACHNOID VILLUS

SUBARACHNOID SPACE (surrounding brain) — Falx cerebri

LATERAL VENTRICLE — Septum pellucidum

— CHOROID PLEXUS

Cerebrum — THIRD VENTRICLE

CEREBRAL AQUEDUCT —

Cerebellum — Tentorium cerebelli

FOURTH VENTRICLE — LATERAL APERTURE

— MEDIAN APERTURE

Frontal plane

View — SPINAL CORD

SUBARACHNOID SPACE (surrounding spinal cord)

(b) Frontal section of brain and spinal cord

(continues)

apertures (of Luschka), one on each side. CSF then circulates in the central canal of the spinal cord and in the subarachnoid space around the surface of the brain and spinal cord.

CSF is gradually reabsorbed into the blood through **arachnoid villi,** fingerlike extensions of the arachnoid that project into the dural venous sinuses, especially the **superior sagittal sinus** (see Figure 14.2). (A sinus is like a vein but has thinner walls.) Normally, CSF is reabsorbed as rapidly as it is formed by the choroid plexuses, at a rate of about 20 mL/hr (480 mL/day). Because the rates of formation and reabsorption are the same, the pressure of CSF normally is constant. Figure 14.4c summarizes the production and flow of CSF.

Hydrocephalus

Abnormalities in the brain—tumors, inflammation, or developmental malformations—can interfere with the drainage of CSF from the ventricles into the subarachnoid space. When excess CSF accumulates in the ventricles, the CSF pressure rises. Elevated CSF pressure causes a condition called **hydrocephalus**

(hī′-drō-SEF-a-lus; *hydro-* = water; *cephal-* = head). In a baby in which the fontanels have not yet closed, the head bulges due to the increased pressure. If the condition persists, the fluid buildup compresses and damages the delicate nervous tissue. Hydrocephalus is relieved by draining the excess CSF. A neurosurgeon may implant a drain line, called a shunt, into the lateral ventricle to divert CSF into the superior vena cava or abdominal cavity. In adults, hydrocephalus may occur after head injury, meningitis, or subarachnoid hemorrhage. ■

▶ **CHECKPOINT**

1. How do the sizes and locations of the cerebrum and cerebellum compare?

2. Where are the cranial meninges located?

3. What is the significance of the blood–brain barrier?

4. What structures are the sites of CSF production, and where are they located?

5. What is the difference between the blood–brain barrier and the blood–cerebrospinal fluid barrier?

Figure 14.4 *(continued)*

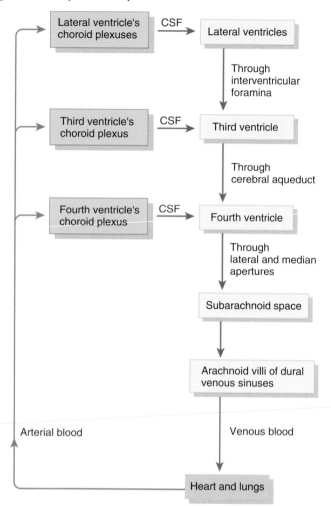

(c) Summary of the formation, circulation, and absorption of cerebrospinal fluid (CSF)

 Where is CSF reabsorbed?

THE BRAIN STEM

OBJECTIVE

• **Describe the structures and functions of the brain stem.**

The brain stem is the part of the brain between the spinal cord and the diencephalon. It consists of three structurally and functionally connected regions: (1) the medulla oblongata, (2) pons, and (3) midbrain. Extending through the brain stem is the reticular formation, a netlike region of interspersed gray and white matter.

Medulla Oblongata

The **medulla oblongata** (me-DOOL-la ob'-long-GA-ta), or more simply the **medulla,** is a continuation of the spinal cord. It

forms the inferior part of the brain stem (Figure 14.5; also see Figure 14.1). The medulla begins at the foramen magnum and extends to the inferior border of the pons, a distance of about 3 cm (1.2 in.).

Within the medulla's white matter are all sensory (ascending) and motor (descending) tracts extending between the spinal cord and other parts of the brain. Some of the white matter forms bulges on the anterior aspect of the medulla. These protrusions are the **pyramids** (Figure 14.6 on page 460, see also Figure 14.5), formed by the largest motor tracts that pass from the cerebrum to the spinal cord. Just superior to the junction of the medulla with the spinal cord, 90% of the axons in the left pyramid cross to the right side, and 90% of the axons in the right pyramid cross to the left side. This crossing is called the **decussation of pyramids** (dē'-ku-SĀ-shun; *decuss-* = crossing) and explains why one side of the brain controls movements on the opposite side of the body.

The medulla also contains several nuclei, masses of gray matter where neurons form synapses with one another. Several of these nuclei control vital body functions. The **cardiovascular center** regulates the rate and force of the heartbeat and the diameter of blood vessels. The **medullary rhythmicity area** of the respiratory center adjusts the basic rhythm of breathing. Other nuclei in the medulla control reflexes for vomiting, coughing, and sneezing.

Just lateral to each pyramid is an oval-shaped swelling called an **olive** (see Figures 14.5 and 14.6). Within the olive is the **inferior olivary nucleus.** Neurons here relay impulses from proprioceptors (monitoring joint and muscle positions) to the cerebellum.

Nuclei associated with sensations of touch, conscious proprioception, pressure, and vibration are located in the posterior part of the medulla. These nuclei are the right and left **gracile nucleus** (GRAS-il = slender) and **cuneate nucleus** (kū-nē-āt = wedge). Many ascending sensory axons form synapses in these nuclei, and postsynaptic neurons then relay the sensory information to the thalamus on the opposite side of the brain (see Figure 15.6a on page 510). The axons ascend to the thalamus in a band of white matter called the **medial lemniscus** (lemNIS-kus = ribbon), which extends through the medulla, pons, and midbrain (see Figure 14.7b). Finally, the medulla contains nuclei that receive sensory input from or provide motor output to cranial nerves VIII (vestibulocochlear), IX (glossopharyngeal), X (vagus), XI (accessory), and XII (hypoglossal).

 Injury of the Medulla

Given the many vital activities controlled by the medulla, it is not surprising that a hard blow to the back of the head or upper neck can be fatal. Damage to the medullary rhythmicity area is particularly serious and can rapidly lead to death. Symptoms of nonfatal injury to the medulla may include cranial nerve malfunctions on the same side of the body as the injury, paralysis and loss of sensation on the opposite side of the body, and irregularities in breathing or heart rhythm. ■

Figure 14.5 **Medulla oblongata in relation the rest of the brain stem.**
(See Tortora, *A Photographic Atlas of the Human Body,* Figure 8.19.)

🔑 The brain stem consists of the medulla oblongata, pons, and midbrain.

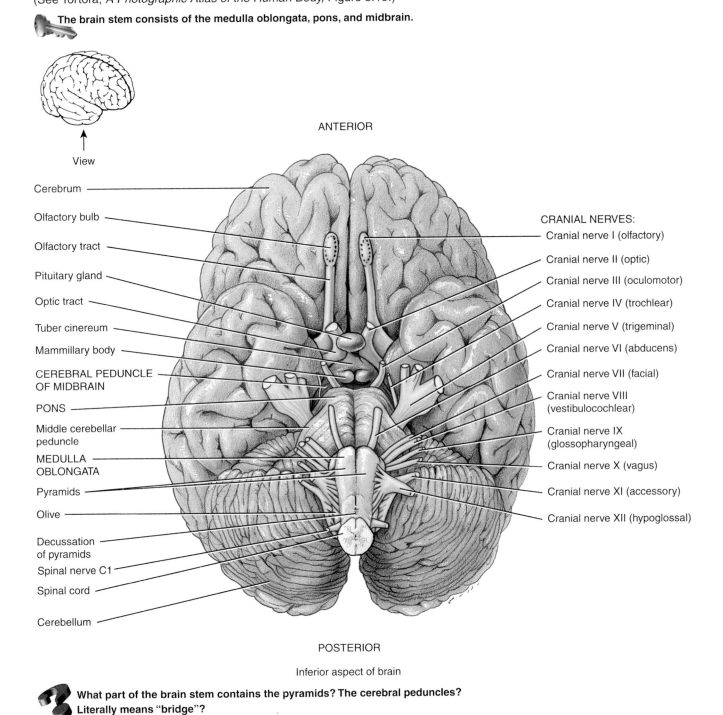

ANTERIOR

View

Cerebrum

Olfactory bulb

Olfactory tract

Pituitary gland

Optic tract

Tuber cinereum

Mammillary body

CEREBRAL PEDUNCLE
OF MIDBRAIN

PONS

Middle cerebellar
peduncle

MEDULLA
OBLONGATA

Pyramids

Olive

Decussation
of pyramids

Spinal nerve C1

Spinal cord

Cerebellum

CRANIAL NERVES:
Cranial nerve I (olfactory)

Cranial nerve II (optic)

Cranial nerve III (oculomotor)

Cranial nerve IV (trochlear)

Cranial nerve V (trigeminal)

Cranial nerve VI (abducens)

Cranial nerve VII (facial)

Cranial nerve VIII
(vestibulocochlear)

Cranial nerve IX
(glossopharyngeal)

Cranial nerve X (vagus)

Cranial nerve XI (accessory)

Cranial nerve XII (hypoglossal)

POSTERIOR

Inferior aspect of brain

❓ What part of the brain stem contains the pyramids? The cerebral peduncles?
Literally means "bridge"?

Pons

The **pons** (= bridge) lies directly superior to the medulla and anterior to the cerebellum and is about 2.5 cm (1 in.) long (see Figures 14.1 and 14.5). Like the medulla, the pons consists of both nuclei and tracts. As its name implies, the pons is a bridge that connects parts of the brain with one another. These connections are provided by bundles of axons. Some axons of the pons connect the right and left sides of the cerebellum. Others are part of ascending sensory tracts and descending motor tracts.

Several **pontine nuclei** are the sites at which signals for voluntary movements that originate in the cerebral cortex are relayed into the cerebellum. Other nuclei in the pons are the **pneumotaxic area** (noo-mō-TAK-sik) and the **apneustic area** (ap-NOO-stik), shown in Figure 23.25 on page 837. Together

Figure 14.6 Internal anatomy of the medulla oblongata.

 The pyramids of the medulla contain the largest motor tracks that run from the cerebrum to the spinal cord.

View

Transverse plane

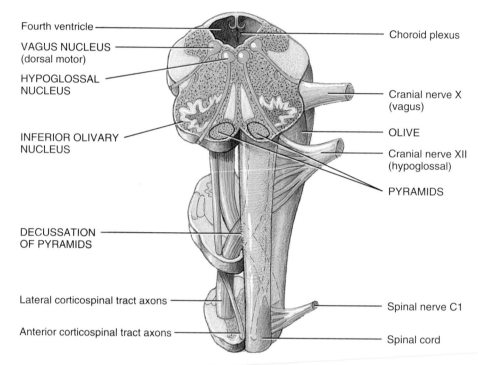

Fourth ventricle

VAGUS NUCLEUS
(dorsal motor)

HYPOGLOSSAL
NUCLEUS

INFERIOR OLIVARY
NUCLEUS

DECUSSATION
OF PYRAMIDS

Lateral corticospinal tract axons

Anterior corticospinal tract axons

Choroid plexus

Cranial nerve X
(vagus)

OLIVE

Cranial nerve XII
(hypoglossal)

PYRAMIDS

Spinal nerve C1

Spinal cord

Transverse section and anterior surface of medulla oblongata

What does decussation mean? What is the functional consequence of decussation of the pyramids?

with the medullary rhythmicity area, the pneumotaxic and apneustic areas help control breathing. The pons also contains nuclei associated with cranial nerves V (trigeminal), VI (abducens), VII (facial), and VIII (vestibulocochlear).

Midbrain

The **midbrain** or **mesencephalon** extends from the pons to the diencephalon (see Figures 14.1 and 14.5) and is about 2.5 cm (1 in.) long. The cerebral aqueduct passes through the midbrain, connecting the third ventricle above with the fourth ventricle below. Like the medulla and the pons, the midbrain contains both tracts and nuclei.

The anterior part of the midbrain contains a pair of tracts called **cerebral peduncles** (pe-DUNK-kuls or PĒ-dung-kuls = little feet; Figure 14.5 and Figure 14.7b). They contain axons of corticospinal, corticopontine, and corticobulbar motor neurons, which conduct nerve impulses from the cerebrum to the spinal cord, medulla, and pons. The cerebral peduncles also contain axons of sensory neurons that extend from the medulla to the thalamus.

The posterior part of the midbrain, called the **tectum** (= roof), contains four rounded elevations (Figure 14.7a). The two superior elevations are known as the **superior colliculi**

(ko-LIK-ū-lī = little hills, singular is *colliculus*). They serve as reflex centers for certain visual activities. Through neural circuits from the retina of the eye to the superior colliculi to the extrinsic eye muscles, visual stimuli elicit eye movements for tracking moving images and scanning stationary images as you are doing to read this sentence. Other superior colliculi reflexes are the pupillary reflex (adjusts size of the pupil), the accommodation reflex (adjusts shape of lens for close versus far vision), and reflexes that govern movements of the eyes, head, and neck in response to visual stimuli. The two inferior elevations, the **inferior colliculi,** are part of the auditory pathway, relaying impulses from the receptors for hearing in the ear to the thalamus. They also are reflex centers for the startle reflex, sudden movements of the head and body that occur when you are surprised by a loud noise.

The midbrain contains several nuclei, including the left and right **substantia nigra** (sub-STAN-shē-a = substance; NĪ-gra = black), which are large, darkly pigmented nuclei. Neurons that release dopamine extend from the substantia nigra to the basal ganglia and help control subconscious muscle activities (Figure 14.7b). Loss of these neurons is associated with Parkinson disease (see page 521). Also present are the left and right **red nuclei** (Figure 14.7b), which look reddish due to their rich blood supply and an iron-containing pigment in their neuronal

Figure 14.7   Midbrain.   (See Tortora, *A Photographic Atlas of the Human Body,* Figure 8.23.)

 The midbrain connects the pons to the diencephalon.

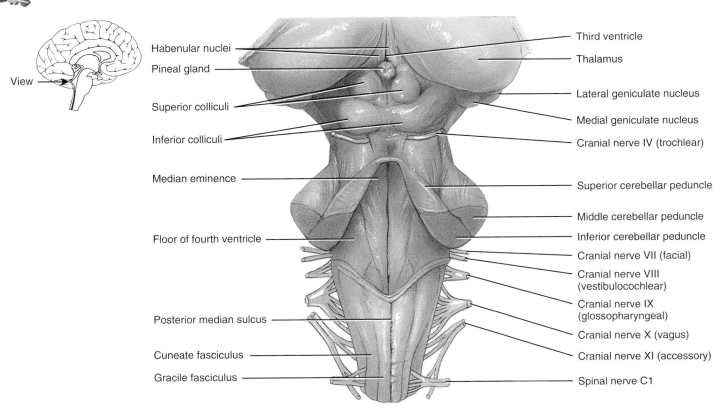

Third ventricle

Thalamus

Habenular nuclei

Pineal gland

View

Superior colliculi

Inferior colliculi

Median eminence

Floor of fourth ventricle

Posterior median sulcus

Cuneate fasciculus

Gracile fasciculus

Lateral geniculate nucleus

Medial geniculate nucleus

Cranial nerve IV (trochlear)

Superior cerebellar peduncle

Middle cerebellar peduncle

Inferior cerebellar peduncle

Cranial nerve VII (facial)

Cranial nerve VIII (vestibulocochlear)

Cranial nerve IX (glossopharyngeal)

Cranial nerve X (vagus)

Cranial nerve XI (accessory)

Spinal nerve C1

(a) Posterior view of midbrain in relation to brain stem

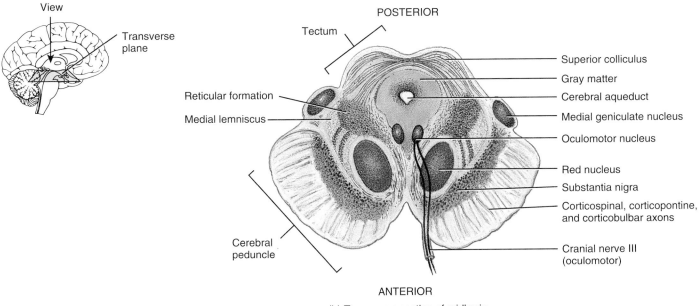

View

Transverse plane

POSTERIOR

Tectum

Reticular formation

Medial lemniscus

Cerebral peduncle

Superior colliculus

Gray matter

Cerebral aqueduct

Medial geniculate nucleus

Oculomotor nucleus

Red nucleus

Substantia nigra

Corticospinal, corticopontine, and corticobulbar axons

Cranial nerve III (oculomotor)

ANTERIOR

(b) Transverse section of midbrain

What is the importance of the cerebral peduncles?

cell bodies. Axons from the cerebellum and cerebral cortex form synapses in the red nuclei, which function with the cerebellum to coordinate muscular movements. Finally, nuclei in the midbrain are associated with cranial nerves III (oculomotor) and IV (trochlear).

Reticular Formation

In addition to the well-defined nuclei already described, much of the brainstem consists of small clusters of neuronal cell bodies (gray matter) interspersed among small bundles of myelinated axons (white matter). The broad region where white matter and gray matter exhibit a netlike arrangement is known as the **reticular formation** (*ret-* = net; Figure 14.7b). It extends from the upper part of the spinal cord, throughout the brainstem, and into the lower part of the diencephalon. Neurons within the reticular formation have both ascending (sensory) and descending (motor) functions. Part of the reticular formation, called the **reticular activating system (RAS),** consists of sensory axons that project to the cerebral cortex (see Figure 15.10 on page 518). The RAS helps maintain consciousness and is active during awakening from sleep. For example, we awaken to the sound of an alarm clock, to a flash of lightning, or to a painful pinch because of RAS activity that arouses the cerebral cortex. The reticular formation's main descending function is to help regulate muscle tone, which is the slight degree of contraction in normal resting muscles.

The functions of the brain stem are summarized in Table 14.1 on page 472.

▶ C H E C K P O I N T

6. Where are the medulla, pons, and midbrain located relative to one another?

7. Define decussation of pyramids. Why is it important?

8. What body functions are governed by nuclei in the brain stem?

9. What are two important functions of the reticular formation?

▎ THE CEREBELLUM

▶ O B J E C T I V E

• **Describe the structure and functions of the cerebellum.**

The **cerebellum,** the second-largest part of the brain, occupies the inferior and posterior aspects of the cranial cavity. It accounts for about a tenth of the brain mass yet contains nearly half of the neurons in the brain. The cerebellum is posterior to the medulla and pons and inferior to the posterior portion of the cerebrum (see Figure 14.1). A deep groove known as the **transverse fissure,** and the **tentorium cerebelli,** which supports the posterior part of the cerebrum, separate the cerebellum from the cerebrum (see Figure 14.4b).

In superior or inferior views, the shape of the cerebellum is somewhat like a butterfly. The central constricted area is the **ver-**

mis (= worm), and the lateral "wings" or lobes are the **cerebellar hemispheres** (Figure 14.8a, b). Each hemisphere consists of lobes separated by deep and distinct fissures. The **anterior lobe** and **posterior lobe** govern subconscious aspects of skeletal muscle movements. The **flocculonodular lobe** (*flocculo-* = wool-like tuft) on the inferior surface contributes to equilibrium and balance.

The superficial layer of the cerebellum, called the **cerebellar cortex,** consists of gray matter in a series of slender, parallel ridges called **folia** (= leaves). Deep to the gray matter are tracts called **arbor vitae** (= tree of life) that resemble branches of a tree. Even deeper, within the white matter, are the **cerebellar nuclei,** regions of gray matter that give rise to axons carrying impulses from the cerebellum to other brain centers and to the spinal cord.

Three paired **cerebellar peduncles** attach the cerebellum to the brain stem (see Figure 14.7a). These bundles of white matter consist of axons that conduct impulses between the cerebellum and other parts of the brain. The **inferior cerebellar peduncles** carry sensory information from the vestibular apparatus of the inner ear and from proprioceptors throughout the body into the cerebellum. The axons extend from the inferior olivary nucleus of the medulla and from the spinocerebellar tracts of the spinal cord into the cerebellum. The **middle cerebellar peduncles** are the largest peduncles. Their axons carry commands for voluntary movements (that originate in motor areas of the cerebral cortex) from the pontine nuclei into the cerebellum. The **superior cerebellar peduncles** contain axons that extend from the cerebellum to the red nuclei of the midbrain and to several nuclei of the thalamus.

A main function of the cerebellum is to evaluate how well movements initiated by motor areas in the cerebrum are actually being carried out. When movements initiated by the cerebral motor areas are not being carried out correctly, the cerebellum detects the discrepancies. It then sends feedback signals to motor areas of the cerebral cortex, via its connections to the red nucleus and thalamus. The feedback signals help correct the errors, smooth the movements, and coordinate complex sequences of skeletal muscle contractions. Besides coordinating skilled movements, the cerebellum is the main brain region that regulates posture and balance. These aspects of cerebellar function make possible all skilled muscular activities, from catching a baseball to dancing to speaking.

The presence of reciprocal connections between the cerebellum and association areas of the cerebral cortex suggest that the cerebellum may also have nonmotor functions such as cognition (acquisition of knowledge) and language processing. This view is supported by imaging studies like MRI and PET.

The functions of the cerebellum are summarized in Table 14.1 on page 472.

▶ C H E C K P O I N T

10. For each of the three pairs of cerebellar peduncles, where do their axons begin and end?

Figure 14.8   Cerebellum.   (See Tortora, *A Photographic Atlas of the Human Body,* Figure 8.25.)

 The cerebellum coordinates skilled movements and regulates posture and balance.

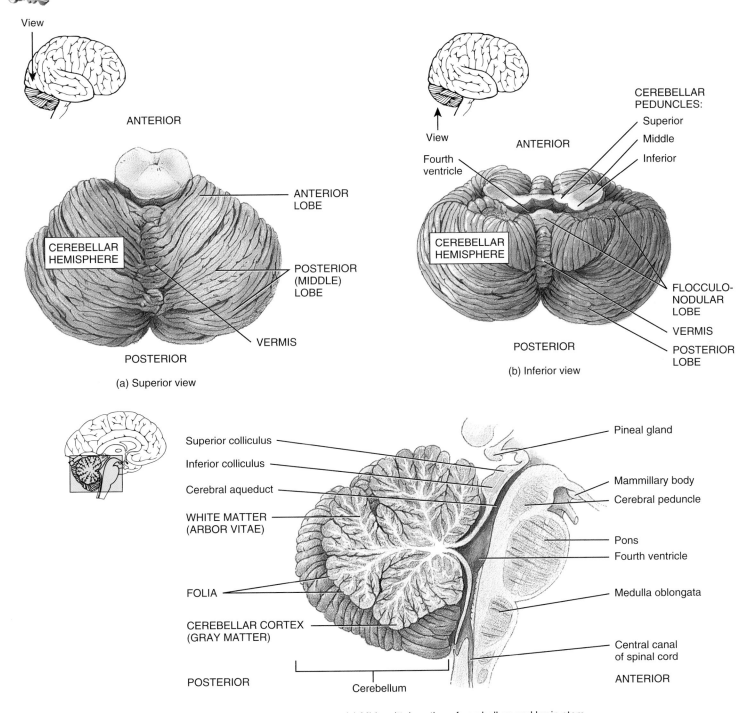

View

ANTERIOR

ANTERIOR LOBE

CEREBELLAR HEMISPHERE

POSTERIOR (MIDDLE) LOBE

VERMIS

POSTERIOR

(a) Superior view

View

ANTERIOR

CEREBELLAR PEDUNCLES:
Superior
Middle
Inferior

Fourth ventricle

CEREBELLAR HEMISPHERE

FLOCCULO-NODULAR LOBE

VERMIS

POSTERIOR

POSTERIOR LOBE

(b) Inferior view

Superior colliculus

Inferior colliculus

Cerebral aqueduct

WHITE MATTER (ARBOR VITAE)

FOLIA

CEREBELLAR CORTEX (GRAY MATTER)

POSTERIOR

Cerebellum

Pineal gland

Mammillary body

Cerebral peduncle

Pons

Fourth ventricle

Medulla oblongata

Central canal of spinal cord

ANTERIOR

(c) Midsagittal section of cerebellum and brain stem

Which structures contain the axons that carry information into and out of the cerebellum?

THE DIENCEPHALON

OBJECTIVE

• Describe the components and functions of the diencephalon.

The **diencephalon** extends from the brain stem to the cerebrum and surrounds the third ventricle. It includes the thalamus, hypothalamus, epithalamus, and subthalamus.

Thalamus

The **thalamus** (THAL-a-mus = inner chamber) measures about 3 cm (1.2 in.) in length and makes up 80% of the diencephalon.

It consists of paired oval masses of gray matter, organized into nuclei, with interspersed tracts of white matter (Figure 14.9). In most people, a bridge of gray matter called the **intermediate mass** crosses the third ventricle, joining the right and left halves of the thalamus (see Figure 14.10).

The thalamus is the major relay station for most sensory impulses that reach the primary sensory areas of the cerebral cortex from the spinal cord, brain stem, and midbrain. Although crude perception of painful, thermal, and pressure sensations arises at the level of the thalamus, precise localization of these sensations depends on nerve impulses arriving at the cerebral cortex.

Figure 14.9 Thalamus. Note the position of the thalamus in (a), the lateral view, and in (b), the medial view. The various thalamic nuclei shown in (c) and (d) are correlated by color to the cortical regions to which they project in (a) and (b). (See Tortora, *A Photographic Atlas of the Human Body,* Figures 8.18 and 8.22.)

🔑 The thalamus is the main relay station for sensory impulses that reach the cerebral cortex from other parts of the brain and the spinal cord.

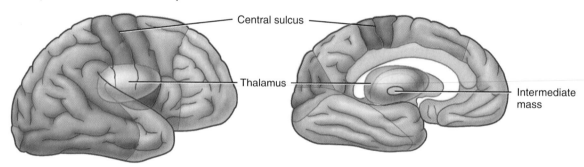

(a) Lateral view of right cerebral hemisphere

(b) Medial view of left cerebral hemisphere

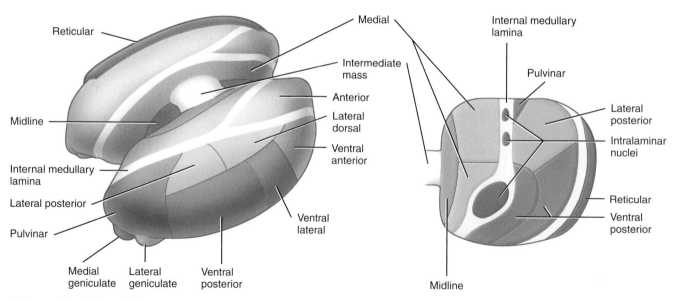

(c) Superolateral view of thalamus showing locations of thalamic nuclei (reticular nucleus is shown on the left side only; all other nuclei are shown on the right side)

(d) Transverse section of right side of thalamus showing locations of thalamic nuclei

Oblique plane

View

Falx cerebri

Cerebrum

Corpus callosum

THALAMUS

Cerebellum

ANTERIOR

Skin

Caudate nucleus

Putamen

Globus pallidus

Third ventricle

Tentorium cerebelli

POSTERIOR

(e) Oblique section of brain

 What structure usually connects the right and left halves of the thalamus?

The thalamus contributes to motor functions by transmitting information from the cerebellum and basal ganglia to the primary motor area of the cerebral cortex. It also relays nerve impulses between different areas of the cerebrum. The thalamus contributes to regulation of autonomic activities and maintenance of consciousness. Axons that connect the thalamus and cerebral cortex pass through the **internal capsule,** a thick band of white matter lateral to the thalamus (see Figure 14.13).

A vertical Y-shaped sheet of white matter called the **internal medullary lamina** divides the gray matter of the right and left sides of the thalamus (Figure 14.9c). It consists of myelinated axons that enter and leave the various thalamic nuclei.

Based on their positions and functions, there are seven major groups of thalamic nuclei on each side (Figure 14.9c,d).

1. The **anterior nucleus** connects to the hypothalamus and limbic system (described on page 471). It functions in emotions, regulation of alertness, and memory.

2. The **medial nuclei** connect to the cerebral cortex, limbic system, and basal ganglia. They function in emotions, learning, memory, awareness, and cognition (thinking and knowing).

3. Nuclei in the **lateral group** connect to the superior colliculi, limbic system, and cortex in all lobes of the cerebrum. The **lateral dorsal nucleus** functions in expression of emotions. The **lateral posterior nucleus** and **pulvinar nucleus** help integrate sensory information.

4. Five nuclei are part of the **ventral group.** The **ventral anterior nucleus** contributes to motor functions, possibly movement planning. The **ventral lateral nucleus** connects to the cerebellum and motor parts of the cerebral cortex. Its neurons are active during movements on the opposite side of the body. The **ventral posterior nucleus** relays impulses for somatic sensations such as touch, pressure, proprioception, vibration, heat, cold, and pain from the face and body to the cerebral cortex. The **lateral geniculate nucleus** (je-NIK-ū-lat = bent like a knee) relays visual impulses for sight from the retina to the primary visual area of the cerebral cortex. The **medial geniculate nucleus** relays auditory impulses for hearing from the ear to the primary auditory area of the cerebral cortex.

5. **Intralaminar nuclei** lie within the internal medullary lamina and make connections with the reticular formation, cerebellum, basal ganglia, and wide areas of the cerebral cortex. They function in pain perception, integration of sensory and motor information, and arousal (activation of the cerebral cortex from the brainstem reticular formation).

6. The **midline nucleus** forms a thin band adjacent to the third ventricle and has a presumed function in memory and olfaction.

7. The **reticular nucleus** surrounds the lateral aspect of the thalamus, next to the internal capsule. This nucleus monitors, filters, and integrates activities of other thalamic nuclei.

Hypothalamus

The **hypothalamus** (*hypo-* = under) is a small part of the diencephalon located inferior to the thalamus. It is composed of a dozen or so nuclei in four major regions:

- The *mammillary region* (*mammill-* = nipple-shaped), adjacent to the midbrain, is the most posterior part of the hypothalamus. It includes the mammillary bodies and posterior hypothalamic nucleus (Figure 14.10). The **mammillary bodies** are two, small, rounded projections that serve as relay stations for reflexes related to the sense of smell (see also Figure 14.5).

- The *tuberal region,* the widest part of the hypothalamus, includes the dorsomedial, ventromedial, and arcuate nuclei, plus the stalklike **infundibulum,** which connects the pituitary gland to the hypothalamus (Figure 14.10). The **median eminence** is a slightly raised region that encircles the infundibulum.

- The *supraoptic region* (*supra-* = above; *-optic* = eye) lies superior to the optic chiasm (point of crossing of optic nerves) and contains the paraventricular nucleus, supraoptic nucleus, anterior hypothalamic nucleus, and suprachiasmatic nucleus (Figure 14.10). Axons from the paraventricular and supraoptic nuclei form the hypothalamohypophyseal tract, which extends through the infundibulum to the posterior lobe of the pituitary.

- The *preoptic region* anterior to the supraoptic region is usually considered part of the hypothalamus because it participates with the hypothalamus in regulating certain autonomic activities. The preoptic region contains the medial and lateral preoptic nuclei (Figure 14.10).

The hypothalamus controls many body activities and is one of the major regulators of homeostasis. Sensory impulses related to both somatic and visceral senses arrive at the hypothalamus, as do impulses from receptors for vision, taste, and smell. Other receptors within the hypothalamus itself continually monitor osmotic pressure, glucose level, certain hormone concentrations, and the temperature of blood. The hypothalamus has several very important connections with the pituitary gland and produces a variety of hormones, which are described in more detail in Chapter 18. Whereas some functions can be attributed to specific hypothalamic nuclei, others are not so precisely localized. Important functions of the hypothalamus are as follows:

1. ***Control of the ANS.*** The hypothalamus controls and integrates activities of the autonomic nervous system, which reg-

Figure 14.10 Hypothalamus.   Selected portions of the hypothalamus and a three-dimensional representation of hypothalamic nuclei are shown (after Netter).

 The hypothalamus controls many body activities and is an important regulator of homeostasis.

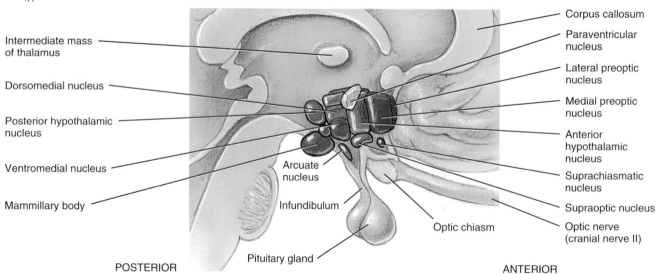

Sagittal section of brain showing hypothalamic nuclei

 What are the four major regions of the hypothalamus, from posterior to anterior?

ulates contraction of smooth and cardiac muscle and the secretions of many glands. Axons extend from the hypothalamus to sympathetic and parasympathetic nuclei in the brain stem and spinal cord. Through the ANS, the hypothalamus is a major regulator of visceral activities, including regulation of heart rate, movement of food through the gastrointestinal tract, and contraction of the urinary bladder.

2. Production of hormones. The hypothalamus produces several hormones and has two types of important connections with the pituitary gland, an endocrine gland located inferior to the hypothalamus (see Figure 14.1). First, hypothalamic hormones are released into capillary networks in the median eminence. The bloodstream carries these hormones directly to the anterior lobe of the pituitary, where they stimulate or inhibit secretion of anterior pituitary hormones. Second, axons extend from the paraventricular and supraoptic nuclei through the infundibulum into the posterior lobe of the pituitary. The cell bodies of these neurons make one of two hormones (oxytocin or antidiuretic hormone). Their axons transport the hormones to the posterior pituitary, where they are released.

3. Regulation of emotional and behavioral patterns. Together with the limbic system (described shortly), the hypothalamus participates in expressions of rage, aggression, pain, and pleasure, and the behavioral patterns related to sexual arousal.

4. Regulation of eating and drinking. The hypothalamus regulates food intake through the arcuate and paraventricular nuclei. It also contains a **thirst center.** When certain cells in the hypothalamus are stimulated by rising osmotic pressure of the extracellular fluid, they cause the sensation of thirst. The intake of water by drinking restores the osmotic pressure to normal, removing the stimulation and relieving the thirst.

5. Control of body temperature. If the temperature of blood flowing through the hypothalamus is above normal, the hypothalamus directs the autonomic nervous system to stimulate activities that promote heat loss. When blood temperature is below normal, by contrast, the hypothalamus generates impulses that promote heat production and retention.

6. Regulation of circadian rhythms and states of consciousness. The suprachiasmatic nucleus establishes patterns of awakening and sleep that occur on a circadian (daily) schedule.

Epithalamus

The **epithalamus** (*epi-* = above), a small region superior and posterior to the thalamus, consists of the pineal gland and habenular nuclei. The **pineal gland** (PĪN-ē-al = pinecone-like) is about the size of a small pea and protrudes from the posterior midline of the third ventricle (see Figure 14.1). The pineal gland is part of the endocrine system because it secretes the hormone **melatonin.** As more melatonin is liberated during darkness than in light, this hormone is thought to promote sleepiness. Melatonin also appears to contribute to the setting of the body's biological clock. The **habenular nuclei** (ha-BEN-ū-lar), shown in Figure 14.7a, are involved in olfaction, especially emotional responses to odors.

Subthalamus

The **subthalamus,** a small area immediately below the thalamus, includes tracts and the paired **subthalamic nuclei,** which connect to motor areas of the cerebrum. Parts of two pairs of midbrain nuclei—the red nucleus and the substantia nigra—also extend into the subthalamus. The subthalamic nuclei, red nuclei, and substantia nigra work together with the basal ganglia, cerebellum, and cerebrum in the control of body movements.

The functions of the four parts of the diencephalon are summarized in Table 14.1 on page 472.

Circumventricular Organs

Parts of the diencephalon, called **circumventricular** (ser'-kum-ven-TRIK-ū-lar) **organs (CVOs)** because they lie in the walls of the third and fourth ventricles, can monitor chemical changes in the blood because they lack a blood–brain barrier. CVOs include part of the hypothalamus, the pineal gland, the pituitary gland, and a few other nearby structures. Functionally, these regions coordinate homeostatic activities of the endocrine and nervous systems, such as the regulation of blood pressure, fluid balance, hunger, and thirst. CVOs are also thought to be the sites of entry into the brain of HIV, the virus that causes AIDS. Once in the brain, HIV may cause dementia (irreversible deterioration of mental state) and other neurological disorders.

▶ CHECKPOINT

11. Why is the thalamus considered a "relay station" in the brain?
12. Why is the hypothalamus part of both the nervous system and the endocrine system?

THE CEREBRUM

▶ OBJECTIVES

- **Describe the cortex, convolutions, fissures, and sulci of the cerebrum.**
- **List and locate the lobes of the cerebrum.**
- **Describe the nuclei that comprise the basal ganglia.**
- **List the structures and describe the functions of the limbic system.**

The cerebrum is the "seat of intelligence." It provides us with the ability to read, write, and speak; to make calculations and compose music; and to remember the past, plan for the future, and imagine things that have never existed before.

The right and left halves of the cerebrum, called **cerebral hemispheres,** consist of an outer rim of gray matter, an internal region of cerebral white matter, and gray matter nuclei deep within the white matter. The outer rim of gray matter is the **cerebral cortex** (*cortex* = rind or bark) (Figure 14.11a). Although only 2–4 mm (0.08–0.16 in.) thick, the cerebral cortex contains billions of neurons. Deep to the cerebral cortex lies the cerebral white matter.

During embryonic development, when brain size increases rapidly, the gray matter of the cortex enlarges much faster than the deeper white matter. As a result, the cortical region rolls and folds upon itself. The folds are called **gyri** (JĪ-rī = circles) or **convolutions** (Figure 14.11a,b). The deepest grooves between folds are known as **fissures;** the shallower grooves between folds are termed **sulci** (SUL-sī = grooves). (The singular terms are *gyrus* and *sulcus.*) The most prominent fissure, the **longitudinal fissure,** separates the cerebrum into right and left halves called **cerebral hemispheres.** The hemispheres are connected internally by the **corpus callosum** (kal-LŌ-sum; *corpus* = body; *callosum* = hard), a broad band of white matter containing axons that extend between the hemispheres (see Figure 14.9b).

Figure 14.11 Cerebrum. Because the insula cannot be seen externally, it has been projected to the surface in (b). (See Tortora, *A Photographic Atlas of the Human Body,* Figure 8.14.)

 The cerebrum is the "seat of intelligence"; it provides us with the ability to read, write, and speak; to make calculations and compose music; to remember the past and plan for the future; and to create.

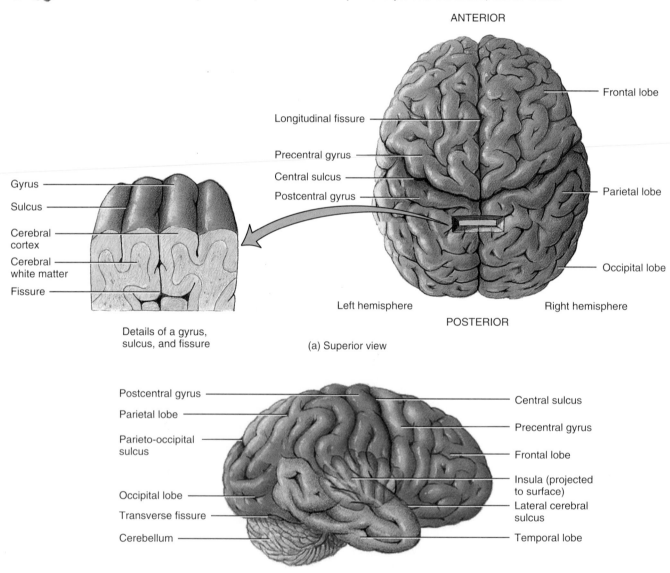

Details of a gyrus, sulcus, and fissure

(a) Superior view

(b) Right lateral view

 During development, which part of the brain—gray matter or white matter—enlarges more rapidly? What are the brain folds, shallow grooves, and deep grooves called?

Lobes of the Cerebrum

Each cerebral hemisphere can be further subdivided into four lobes. The lobes are named after the bones that cover them: frontal, parietal, temporal, and occipital lobes (see Figure 14.11a,b). The **central sulcus** (SUL-kus) separates the **frontal lobe** from the **parietal lobe.** A major gyrus, the **precentral gyrus**—located immediately anterior to the central sulcus—contains the primary motor area of the cerebral cortex. Another major gyrus, the **postcentral gyrus,** which is located immediately posterior to the central sulcus, contains the primary somatosensory area of the cerebral cortex. The **lateral cerebral sulcus** separates the **frontal lobe** from the **temporal lobe.** The **parieto-occipital sulcus** separates the **parietal lobe** from the **occipital lobe.** A fifth part of the cerebrum, the **insula,** cannot be seen at the surface of the brain because it lies within the lateral cerebral sulcus, deep to the parietal, frontal, and temporal lobes (Figure 14.11b).

Cerebral White Matter

The **cerebral white matter** consists of myelinated and unmyelinated axons in three types of tracts (Figure 14.12):

1. **Association tracts** contain axons that conduct nerve impulses between gyri in the same hemisphere.

2. **Commissural tracts** contain axons that conduct nerve impulses from gyri in one cerebral hemisphere to corresponding gyri in the other cerebral hemisphere. Three important groups of commissural tracts are the **corpus callosum, anterior commissure,** and **posterior commissure.**

3. **Projection tracts** contain axons that conduct nerve impulses from the cerebrum to lower parts of the CNS (thalamus, brainstem, or spinal cord) or from lower parts of the CNS to the cerebrum. An example is the **internal capsule,** a thick band of white matter that contains both ascending and descending axons (see Figure 14.13b).

Basal Ganglia

Deep within each cerebral hemisphere are three nuclei (masses of gray matter) that are collectively termed the **basal ganglia** (Figure 14.13). (Recall that "ganglion" usually means a collection of neuronal cells bodies *outside* the CNS. The name here is the one exception to that general rule. An alternative term used in some textbooks—*basal nuclei*—is not used by most neuroscientists. It can be confused with the names of other brain regions, such as the nucleus basalis, which deteriorates in people who suffer from Alzheimer disease.)

Two of the basal ganglia nuclei are side-by-side, just lateral to the thalamus. They are the **globus pallidus** (*globus* = ball; *pallidus* = pale), which is closer to the thalamus, and the **putamen** (pū-TĀ-men = shell), which is closer to the cerebral cortex. The third member of the basal ganglia is the **caudate nucleus** (*caud-* = tail), which has a large "head" connected to a

Figure 14.12   Organization of white matter tracts of the left cerebral hemisphere.

Association tracts, commissural tracts, and projection tracts form white matter areas in the cerebral hemispheres.

Medial view of tracts revealed by removing gray matter from a midsagittal section

Which tracts carry impulses between gyri of the same hemisphere? Between gyri in opposite hemispheres? Between the cerebrum, thalamus, brain stem, and spinal cord?

Figure 14.13 **Basal ganglia.** In (a) the basal ganglia have been projected to the surface. (See Tortora, *A Photographic Atlas of the Human Body,* Figures 8.17, 8.18, and 8.22.)

The basal ganglia control automatic movements of skeletal muscles and muscle tone.

Lateral ventricle

Thalamus

Tail of caudate nucleus

Occipital lobe of cerebrum

Body of caudate nucleus

Frontal lobe of cerebrum

Putamen

Head of caudate nucleus

Amygdala

POSTERIOR

ANTERIOR

(a) Lateral view of right side of brain

Frontal plane

View

Longitudinal fissure

Septum pellucidum

Internal capsule

Insula

Thalamus

Subthalamic nucleus

Hypothalamus

Cerebrum

Corpus callosum

Lateral ventricle

Caudate nucleus

Putamen — Corpus striatum

Globus pallidus

Third ventricle

Optic tract

(b) Anterior view of frontal section

 Where are the basal ganglia located relative to the thalamus?

smaller "tail" by a long comma-shaped "body." Together, the caudate nucleus and putamen are known as the **corpus striatum.** Nearby structures that are functionally linked to the basal ganglia are the *substantia nigra* of the midbrain (see Figure

14.7b) and the *subthalamic nuclei* of the diencephalon (see Figure 14.13b). Axons from the substantia nigra terminate in the caudate nucleus and putamen. The subthalamic nuclei interconnect with the globus pallidus.

The basal ganglia receive input from the cerebral cortex and provide output back to motor parts of the cortex via medial and ventral group nuclei of the thalamus. In addition, the nuclei of the basal ganglia have extensive connections with one another. A major function of the basal ganglia is to help regulate initiation and termination of movements. Activity of neurons in the putamen precedes or anticipates body movements whereas activity of neurons in the caudate nucleus occurs prior to eye movements. The globus pallidus helps regulate the muscle tone required for specific body movements. The basal ganglia also control subconscious contractions of skeletal muscles. Examples include automatic arm swings while walking and laughing in response to a joke. Movement disruptions are a hallmark of Parkinson disease (see page 521). In this disorder, neurons that extend from the substantia nigra to the putamen and caudate nucleus degenerate.

Besides these influences on motor functions, the basal ganglia have other roles. They help initiate and terminate some cognitive processes, such as attention, memory, and planning, and may act with the limbic system to regulate emotional behaviors. Some psychiatric disorders, such as obsessive-compulsive disorder, schizophrenia, and chronic anxiety, are thought to involve dysfunction of circuits between the basal ganglia and the limbic system.

The Limbic System

Encircling the upper part of the brain stem and the corpus callosum is a ring of structures on the inner border of the cerebrum and floor of the diencephalon that constitutes the **limbic system** (*limbic* = border). The main components of the limbic system are as follows (Figure 14.14):

1. The so-called *limbic lobe* is a rim of cerebral cortex on the medial surface of each hemisphere. It includes the **cingulate gyrus** (*cingul-* = belt), which lies above the corpus callosum, and the **parahippocampal gyrus,** which is in the temporal lobe below. The **hippocampus** (= seahorse) is a portion of the parahippocampal gyrus that extends into the floor of the lateral ventricle.

2. The **dentate gyrus** (*dentate* = toothed) lies between the hippocampus and parahippocampal gyrus.

3. The **amygdala** (*amygda-* = almond-shaped) is composed of several groups of neurons located close to the tail of the caudate nucleus.

4. The **septal nuclei** are located within the septal area formed by the regions under the corpus callosum and the paraterminal gyrus (a cerebral gyrus).

5. The **mammillary bodies of the hypothalamus** are two round masses close to the midline near the cerebral peduncles.

6. Two nuclei of the thalamus, the **anterior nucleus** and the **medial nucleus,** participate in limbic circuits.

7. The **olfactory bulbs** are flattened bodies of the olfactory pathway that rest on the cribriform plate.

8. The **fornix, stria terminalis, stria medullaris, medial forebrain bundle,** and **mammillothalamic tract** are linked by bundles of interconnecting myelinated axons.

Figure 14.14 Components of the limbic system and surrounding structures.

The limbic system governs emotional aspects of behavior.

Sagittal plane

View

Anterior nucleus of thalamus
Mammillothalamic tract
Corpus callosum
Cingulate gyrus (in frontal lobe)
Anterior commissure
Mammillary body in hypothalamus
Septal nuclei
Olfactory bulb
Amygdala
Parahippocampal gyrus (in temporal lobe)

Fornix
Stria medullaris
Stria terminalis
Hippocampus (in temporal lobe)
Dentate gyrus

POSTERIOR

Sagittal section

ANTERIOR

Which part of the limbic system functions with the cerebrum in memory?

The limbic system is sometimes called the "emotional brain" because it plays a primary role in a range of emotions, including pain, pleasure, docility, affection, and anger. It also is involved in olfaction (smelling) and memory. Experiments have shown that when different areas of animals' limbic system are stimulated, the animals' reactions indicate that they are experiencing intense pain or extreme pleasure. Stimulation of other limbic system areas in animals produces tameness and signs of affection. Stimulation of a cat's amygdala or certain nuclei of the hypothalamus produces a behavioral pattern called rage—the cat extends its claws, raises its tail, opens its eyes wide, hisses, and spits. By contrast, removal of the amygdala produces an animal that lacks fear and aggression. Likewise, a person whose amygdala is damaged fails to recognize fearful expressions in others or to express fear herself in situations where this emotion would normally be appropriate.

The hippocampus, together with other parts of the cerebrum, functions in memory. People with damage to certain limbic system structures forget recent events and cannot commit anything to memory.

Table 14.1 summarizes the functions of various parts of the brain.

Brain Injuries

Brain injuries are commonly associated with head trauma and result in part from displacement and distortion of neuronal tissue at the moment of impact. Additional tissue damage occurs when

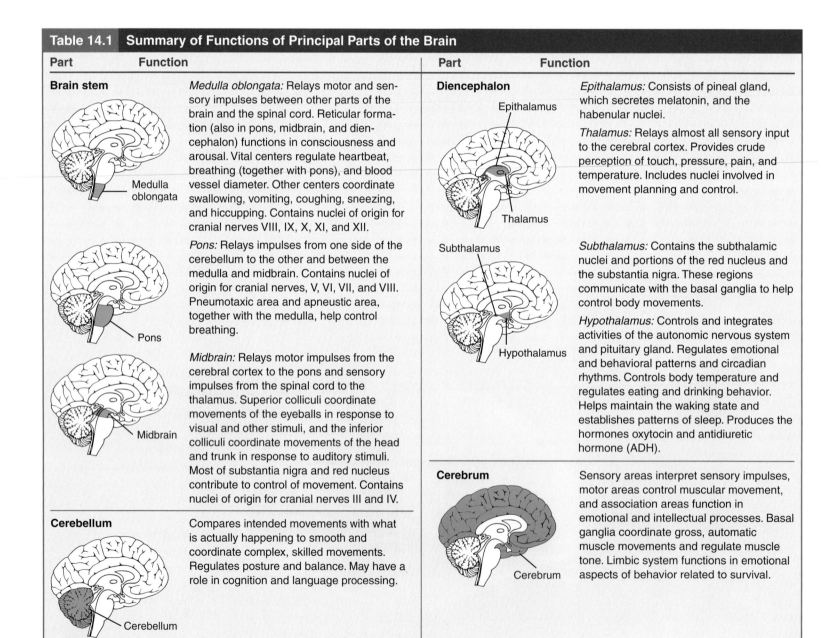

Table 14.1   Summary of Functions of Principal Parts of the Brain

Part	Function	Part	Function
Brain stem	*Medulla oblongata:* Relays motor and sensory impulses between other parts of the brain and the spinal cord. Reticular formation (also in pons, midbrain, and diencephalon) functions in consciousness and arousal. Vital centers regulate heartbeat, breathing (together with pons), and blood vessel diameter. Other centers coordinate swallowing, vomiting, coughing, sneezing, and hiccupping. Contains nuclei of origin for cranial nerves VIII, IX, X, XI, and XII.	**Diencephalon**	*Epithalamus:* Consists of pineal gland, which secretes melatonin, and the habenular nuclei.
	Pons: Relays impulses from one side of the cerebellum to the other and between the medulla and midbrain. Contains nuclei of origin for cranial nerves, V, VI, VII, and VIII. Pneumotaxic area and apneustic area, together with the medulla, help control breathing.		*Thalamus:* Relays almost all sensory input to the cerebral cortex. Provides crude perception of touch, pressure, pain, and temperature. Includes nuclei involved in movement planning and control.
	Midbrain: Relays motor impulses from the cerebral cortex to the pons and sensory impulses from the spinal cord to the thalamus. Superior colliculi coordinate movements of the eyeballs in response to visual and other stimuli, and the inferior colliculi coordinate movements of the head and trunk in response to auditory stimuli. Most of substantia nigra and red nucleus contribute to control of movement. Contains nuclei of origin for cranial nerves III and IV.		*Subthalamus:* Contains the subthalamic nuclei and portions of the red nucleus and the substantia nigra. These regions communicate with the basal ganglia to help control body movements.
			Hypothalamus: Controls and integrates activities of the autonomic nervous system and pituitary gland. Regulates emotional and behavioral patterns and circadian rhythms. Controls body temperature and regulates eating and drinking behavior. Helps maintain the waking state and establishes patterns of sleep. Produces the hormones oxytocin and antidiuretic hormone (ADH).
Cerebellum	Compares intended movements with what is actually happening to smooth and coordinate complex, skilled movements. Regulates posture and balance. May have a role in cognition and language processing.	**Cerebrum**	Sensory areas interpret sensory impulses, motor areas control muscular movement, and association areas function in emotional and intellectual processes. Basal ganglia coordinate gross, automatic muscle movements and regulate muscle tone. Limbic system functions in emotional aspects of behavior related to survival.

normal blood flow is restored after a period of ischemia (reduced blood flow). The sudden increase in oxygen level produces large numbers of oxygen free radicals (charged oxygen molecules with an unpaired electron). Brain cells recovering from the effects of a stroke or cardiac arrest also release free radicals. Free radicals cause damage by disrupting cellular DNA and enzymes and by altering plasma membrane permeability.

Various degrees of brain injury are described by specific terms. A **concussion** is an abrupt, but temporary, loss of consciousness (from seconds to hours) following a blow to the head or the sudden stopping of a moving head. It is the most common brain injury. A concussion produces no obvious bruising of the brain. Signs of a concussion are headache, drowsiness, lack of concentration, confusion, or post-traumatic amnesia (memory loss).

A **contusion** is bruising of the brain due to trauma and includes the leakage of blood from microscopic vessels. It is usually associated with a concussion. In a contusion, the pia mater may be torn, allowing blood to enter the subarachnoid space. The area most commonly affected is the frontal lobe. A contusion usually results in an immediate loss of consciousness (generally lasting no longer than 5 minutes), loss of reflexes, transient cessation of respiration, and decreased blood pressure. Vital signs typically stabilize in a few seconds.

A **laceration** is a tear of the brain, usually from a skull fracture or a gunshot wound. A laceration results in rupture of large blood vessels, with bleeding into the brain and subarachnoid space. Consequences include cerebral hematoma (localized pool of blood, usually clotted, that swells against the brain tissue), edema, and increased intracranial pressure. ■

▶ C H E C K P O I N T

13. How are the lobes of the cerebrum separated from one another? What is the insula?

14. How is cerebral white matter organized?

15. What functions do the basal ganglia have?

16. What system is the amygdala part of and what is its function?

CEREBRAL CORTEX AREAS AND FUNCTIONS

▶ O B J E C T I V E

- **Describe the locations and functions of the sensory, association, and motor areas of the cerebral cortex.**

Specific types of sensory, motor, and integrative signals are processed in certain regions of the cerebral cortex (Figure 14.15). Generally, **sensory areas** receive and interpret sensory impulses, **motor areas** initiate movements, and **association areas** deal with more complex integrative functions such as memory, emotions, reasoning, will, judgment, personality traits, and intelligence.

Sensory Areas

Sensory impulses arrive mainly in the posterior half of both cerebral hemispheres, in regions behind the central sulci. In the cortex, primary sensory areas have the most direct connections with peripheral sensory receptors.

Secondary sensory areas and sensory association areas often are adjacent to the primary areas. They usually receive input both from the primary areas and from other brain regions. Secondary sensory areas and sensory association areas integrate sensory experiences to generate meaningful patterns of recognition and awareness. Whereas a person with damage in the *primary* visual area would be blind in at least part of his visual field, a person with damage to a visual *association* area might see normally yet be unable to recognize a friend.

The following are some important sensory areas (Figure 14.15):

- The **primary somatosensory area** (areas 1, 2, and 3) is located directly posterior to the central sulcus of each cerebral hemisphere in the postcentral gyrus of each parietal lobe. It extends from the lateral cerebral sulcus, along the lateral surface of the parietal lobe to the longitudinal fissure, and then along the medial surface of the parietal lobe within the longitudinal fissure.

    The primary somatosensory area receives nerve impulses for touch, proprioception (joint and muscle position), pain, itching, tickle, and thermal sensations. A "map" of the entire body is present in the primary somatosensory area: Each point within the area receives impulses from a specific part of the body (see Figure 15.5a on page 508). The size of the cortical area receiving impulses from a particular part of the body depends on the number of receptors present there rather than on the size of the body part. For example, a larger region of the somatosensory area receives impulses from the lips and fingertips than from the thorax or hip. The major function of the primary somatosensory area is to localize exactly the points of the body where sensations originate.

- The **primary visual area** (area 17) is located at the posterior tip of the occipital lobe mainly on the medial surface (next to the longitudinal fissure). It receives impulses that convey information for vision. Axons of neurons with cell bodies in the eye form the optic nerve (cranial nerve II), which ends in the lateral geniculate nucleus of the thalamus. From the thalamus, neurons project to the primary visual area, carrying information concerning shape, color, and movement of visual stimuli.

- The **primary auditory area** (areas 41 and 42) is located in the superior part of the temporal lobe near the lateral cerebral sulcus. It interprets the basic characteristics of sound such as pitch and rhythm.

- The **primary gustatory area** (area 43) is located at the base of the postcentral gyrus superior to the lateral

Figure 14.15 **Functional areas of the cerebrum.** Broca's speech area is in the left cerebral hemisphere of most people; it is shown here to indicate its relative location. The numbers, still used today, are from K. Brodmann's map of the cerebral cortex, first published in 1909.

🔑 Particular areas of the cerebral cortex process sensory, motor, and integrative signals.

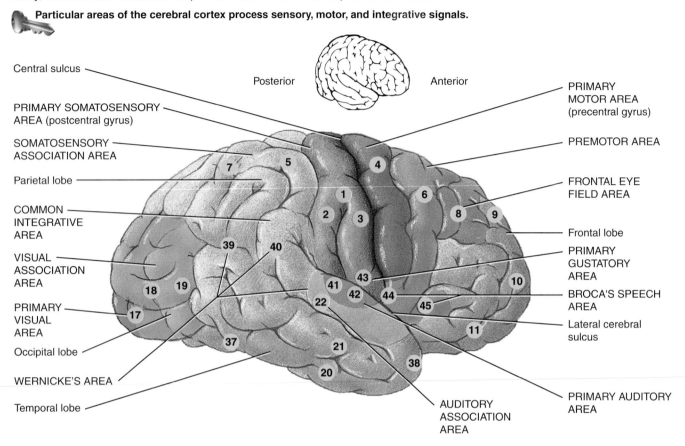

Lateral view of right cerebral hemisphere

 What area(s) of the cerebrum integrate(s) interpretation of visual, auditory, and somatic sensations? Translates thoughts into speech? Controls skilled muscular movements? Interprets sensations related to taste? Interprets pitch and rhythm? Interprets shape, color, and movement of objects? Controls voluntary scanning movements of the eyes?

cerebral sulcus in the parietal cortex. It receives impulses for taste.

- The **primary olfactory area** (area 28) is located in the temporal lobe on the medial aspect (and thus not visible in Figure 14.15). It receives impulses for smell.

Motor Areas

Motor output from the cerebral cortex flows mainly from the anterior part of each hemisphere. Among the most important motor areas are the following (Figure 14.15):

- The **primary motor area** (area 4) is located in the precentral gyrus of the frontal lobe. Each region in the primary motor area controls voluntary contractions of specific muscles or groups of muscles (see Figure 15.5b on page 508).

Electrical stimulation of any point in the primary motor area causes contraction of specific skeletal muscle fibers on the opposite side of the body. As is true for the primary somatosensory area, body parts do not "map" to the primary motor area in proportion to their size. More cortical area is devoted to those muscles involved in skilled, complex, or delicate movement. For instance, the cortical region devoted to muscles that move the fingers is much larger than the region for muscles that move the toes.

- **Broca's speech area** (areas 44 and 45) is located in the frontal lobe close to the lateral cerebral sulcus. Speaking and understanding language are complex activities that involve several sensory, association, and motor areas of the cortex. In 97% of the population, these language areas are localized in the *left* hemisphere. The planning and production of speech occurs in Broca's (BRŌ-kaz) speech area, located in one frontal lobe—the *left* frontal lobe in most people. From Broca's speech area, nerve impulses pass to the premotor regions that control the muscles of the larynx, pharynx, and mouth. The impulses from the premotor area result in specific, coordinated muscle contractions that enable you to speak. Simultaneously, impulses propagate from Broca's speech area to the primary motor area. From here, impulses also control the breathing muscles to regulate the proper flow of air past the vocal cords. The coordinated contractions of your speech and breathing muscles enable you to speak your thoughts.

Association Areas

The association areas of the cerebrum consist of some motor and sensory areas, plus large areas on the lateral surfaces of the occipital, parietal, and temporal lobes and on the frontal lobes anterior to the motor areas. Association areas are connected with one another by association tracts and include the following (Figure 14.15):

- The **somatosensory association area** (areas 5 and 7) is just posterior to and receives input from the primary somatosensory area, as well as from the thalamus and other parts of the brain. Its role is to integrate and interpret sensations. This area permits you to determine the exact shape and texture of an object without looking at it, to determine the orientation of one object with respect to another as they are felt, and to sense the relationship of one body part to another. Another role of the somatosensory association area is the storage of memories of past sensory experiences, enabling you to compare current sensations with previous experiences.

- The **visual association area** (areas 18 and 19) is located in the occipital lobe. It receives sensory impulses from the primary visual area and the thalamus. It relates present and past visual experiences and is essential for recognizing and evaluating what is seen.

- The **auditory association area** (area 22) is located inferior and posterior to the primary auditory area in the temporal cortex. It allows one to realize whether a sound is speech, music, or noise.

- **Wernicke's (posterior language) area** (area 22, and possibly areas 39 and 40) is a broad region in the temporal and parietal lobes. This area interprets the meaning of speech by recognizing spoken words. It is active as you translate words into thoughts. The regions in the *right* hemisphere that correspond to Broca's and Wernicke's areas in the left hemisphere also contribute to verbal communication by adding emotional content, for instance anger or joy, to spoken words.

- The **common integrative area** (areas 5, 7, 39, and 40) is bordered by somatosensory, visual, and auditory association areas. It receives nerve impulses from these areas and from the primary gustatory area, primary olfactory area, the thalamus, and parts of the brain stem. This area integrates sensory interpretations from the association areas and impulses from other areas, allowing formation of thoughts based on a variety of sensory inputs. It then transmits signals to other parts of the brain to cause the appropriate response to the interpretation of the sensory signals.

- The **premotor area** (area 6) is a motor association area that is immediately anterior to the primary motor area. Neurons in this area communicate with the primary motor cortex, the sensory association areas in the parietal lobe, the basal ganglia, and the thalamus. The premotor area deals with learned motor activities of a complex and sequential nature. It generates nerve impulses that cause specific groups of muscles to contract in a specific sequence, for instance to write a word. The premotor area controls learned skilled movements and serves as a memory bank for such movements.

- The **frontal eye field area** (area 8) in the frontal cortex is sometimes included in the premotor area. It controls voluntary scanning movements of the eyes—reading this sentence, for instance.

 Aphasia

Much of what we know about language areas comes from studies of patients with language or speech disturbances that have resulted from brain damage. Broca's speech area, the auditory association area, and other language areas are located in the left cerebral hemisphere of most people, regardless of whether they are left-handed or right-handed. Injury to language areas of the cerebral cortex results in **aphasia** (a-FĀ-zē-a; *a-* = without; *-phasia* = speech), an inability to use or comprehend words. Damage to Broca's speech area results in *nonfluent aphasia,* an inability to properly articulate or form words; people with nonfluent aphasia know what they wish to say but cannot speak. Damage to the common integrative area or auditory association area (areas 39 and 22) results in *fluent aphasia,* characterized by faulty understanding of spoken or written words. A person experiencing this type of aphasia may fluently produce strings of words that have no meaning. The underlying deficit may be **word deafness,** an inability to understand spoken words; or **word blindness,** an inability to understand written words; or both. ■

Hemispheric Lateralization

Although the brain is almost symmetrical on its right and left sides, subtle anatomical differences between the two hemispheres exist. For example, in about two-thirds of the population, the planum temporale, a region of the temporal lobe that includes Wernicke's area, is 50% larger on the left side than on the right side. This asymmetry appears in the human fetus at about 30 weeks of gestation. Moreover, although the two hemispheres share performance of many functions, each hemisphere also specializes in performing certain unique functions. This functional asymmetry is termed **hemispheric lateralization.**

The most obvious example of hemispheric lateralization is this: The left hemisphere receives somatic sensory signals from and controls muscles on the right side of the body, whereas the right hemisphere receives sensory signals from and controls the left side of the body. Beyond these differences, however, in most people the left hemisphere is more important for reasoning, numerical and scientific skills, spoken and written language, and the ability to use and understand sign language. Patients with damage in the left hemisphere, for example, often exhibit aphasia. Conversely, the right hemisphere is more specialized for musical and artistic awareness; spatial and pattern perception; recognition of faces and emotional content of language; discrimination of different smells; and generating mental images of sight, sound, touch, taste, and smell to compare relationships among them. Patients with damage in the right hemisphere regions that correspond to Broca's and Wernicke's areas in the left hemisphere speak in a monotonous voice, having lost the ability to impart emotional inflection to what they say.

An example of hemispheric lateralization is shown in Figure 14.16. The photograph shows superimposed MRI and PET images, averaged from six men and five women, as several pleasant scents were presented to both nostrils. The MRI image depicts brain regions whereas the PET image demonstrates areas of increased blood flow, corresponding to areas of increased neuronal activity. In both hemispheres, an area termed the *piriform cortex,* near the junction of the frontal and temporal lobes, "lights up" in the PET images. This cortical area receives input from neurons in the olfactory bulb on the same side. In addition, activation occurs mainly on the right side in a region termed the *orbitofrontal cortex,* corresponding roughly to Brodmann's area 11 (see Figure 14.15). Indeed, people who suffer damage to the right orbitofrontal cortex have great difficulty identifying odors and discriminating different odors.

Despite some dramatic differences in functions of the two hemispheres, there is considerable variation from one person to another. Also, lateralization seems less pronounced in females than in males, both for language (left hemisphere) and for visual and spatial skills (right hemisphere). For instance, females are less likely than males to suffer aphasia after damage to the left hemisphere. A possibly related observation is that females have, on average, a 12% larger anterior commissure and a broader posterior portion of the corpus callosum than males. Recall that both the anterior commissure and the corpus callosum are commissural tracts that provide communication between the two hemispheres.

Table 14.2 summarizes some of the distinctive functions that are more likely to reside in the left or the right hemisphere.

Brain Waves

At any instant, brain neurons are generating millions of impulses (action potentials) and graded potentials (excitatory and

Figure 14.16 Cortical areas activated by olfactory stimuli. This is a composite picture in which multiple MRI (magnetic resonance image) and PET (positron emission tomography) scans were superimposed. The MRI scans reveal brain tissue and an outline of the skull whereas the PET scans "light up" regions of increased blood flow, an indicator of increased activity of neurons.

 Hemispheric lateralization means that each cerebral hemisphere performs unique functions.

Transverse section through head

Which area of the cerebrum contains neurons that demonstrate hemispheric lateralization for olfaction?

inhibitory postsynaptic potentials). Taken together, these neuronal electrical signals are called **brain waves.** Brain waves generated by neurons close to the brain surface, mainly neurons in the cerebral cortex, can be detected by sensors called electrodes placed on the forehead and scalp. A record of such waves is called an **electroencephalogram** (e-lek'-trō-en-SEF-a-lō-gram) or **EEG.** Electroencephalograms are useful for studying normal brain functions, such as changes that occur during sleep. They also help neurologists diagnose a variety of brain disorders, such as epilepsy, tumors, metabolic abnormalities, sites of trauma, and degenerative diseases.

Patterns of activation of brain neurons produce four types of brain waves (Figure 14.17):

1. **Alpha waves.** These rhythmic waves occur at a frequency of about 8–13 cycles per second. (The unit commonly used to express frequency is the hertz [Hz]. One hertz is one cycle per second.) Alpha waves are present in the EEGs of nearly all normal individuals when they are awake and resting with their eyes closed. These waves disappear entirely during sleep.

Table 14.2	Functional Differences Between the Two Cerebral Hemispheres
Left Hemisphere Functions	**Right Hemisphere Functions**
Receives somatic sensory signals from and controls muscles on right side of body. Also receives images from right half of visual field.	Receives somatic sensory signals from and controls muscles on left side of body. Also receives images from left half of visual field.
Reasoning.	Musical and artistic awareness.
Numerical and scientific skills.	Space and pattern perception.
Ability to use and understand sign language.	Recognition of faces and emotional content of facial expressions.
Spoken and written language.	Generating emotional content of language.
	Generating mental images to compare spatial relationships.
	Identifying and discriminating among odors.

2. **Beta waves.** The frequency of these waves is between 14 and 30 Hz. Beta waves generally appear when the nervous system is active—that is, during periods of sensory input and mental activity.

3. **Theta waves.** These waves have frequencies of 4–7 Hz. Theta waves normally occur in children and adults experiencing emotional stress. They also occur in many disorders of the brain.

4. **Delta waves.** The frequency of these waves is 1–5 Hz. Whereas delta waves occur during deep sleep in adults, they are normal in awake infants. When produced by an awake adult, they indicate brain damage.

▶ CHECKPOINT

17. What areas of the cerebral cortex are needed for normal language abilities?

18. What does the term "hemispheric lateralization" mean?

19. What is the diagnostic value of an EEG?

Figure 14.17 Types of brain waves recorded in an electroencephalogram (EEG).

🔑 Brain waves indicate electrical activity of the cerebral cortex.

Alpha

Beta

Theta

Delta

|← 1 sec →|

 Which type of brain wave indicates emotional stress?

CRANIAL NERVES

▶ OBJECTIVE

• **Identify the cranial nerves by name and number, and give the functions of each.**

The twelve pairs of **cranial nerves** are so-named because they pass through various foramina in the cranial bones. Like the 31 pairs of spinal nerves, they are part of the peripheral nervous system (PNS). Each cranial nerve has both a number, designated by a roman numeral, and a name (see Figure 14.5). The numbers indicate the order, from anterior to posterior, in which the nerves arise from the brain. The names designate a nerve's distribution or function.

Cranial nerves emerge from the nose (cranial nerve I), the eyes (cranial nerve II), the inner ear (cranial nerve VIII), the brainstem (cranial nerves III–XII), and the spinal cord (part of cranial nerve XI). Two cranial nerves, I (olfactory) and II (optic), contain only sensory axons and thus are called **sensory nerves.** The rest are **mixed nerves** because they contain axons of both sensory and motor neurons. The cell bodies of sensory neurons are located in ganglia outside the brain, whereas the cell bodies of motor neurons lie in nuclei within the brain. Cranial nerves III (oculomotor), IV (trochlear), VI (abducens), XI (accessory), and XII (hypoglossal) are mainly motor. Whereas a few of their axons are sensory axons from muscle proprioceptors, most of their axons are motor neurons that innervate skeletal muscles. Cranial nerves III, VII, IX, and X include both somatic and autonomic motor axons. The somatic axons innervate skeletal muscles whereas the autonomic axons, which are part of parasympathetic division, innervate glands, smooth muscle, and cardiac muscle.

Cranial Nerve I: Olfactory Nerve

The **olfactory nerve** (ol-FAK-tō-rē; *olfact-* = to smell), **cranial nerve I,** is entirely sensory; it contains axons that conduct nerve impulses for the sense of smell or olfaction. The olfactory mucosa occupies the superior part of the nasal cavity, covering the inferior surface of the cribriform plate and extending down along the superior nasal concha and upper part of the middle nasal concha (see Figure 16.1a on page 528). In the olfactory mucosa are the olfactory receptors. These bipolar neurons have a single odor-sensitive dendrite projecting from one side of the cell body and an unmyelinated axon extending from the other side. Bundles of axons of olfactory receptors extend through about 20 olfactory foramina in the cribriform plate of the ethmoid bone on each side of the nose (see Figure 16.1b on page 528). These 40 or so bundles of axons are the olfactory nerves, and collectively they form the right and left cranial nerve I.

Olfactory nerves end in the brain in paired masses of gray matter called the **olfactory bulbs,** two extensions of the brain that rest on the cribriform plate. Within the olfactory bulbs, the axon terminals of olfactory receptors form synapses with the dendrites and cell bodies of the next neurons in the olfactory pathway. The axons of these neurons make up the **olfactory tracts,** which extend posteriorly from the olfactory bulbs (see Figure 14.5). Axons in the olfactory tracts end in the primary olfactory area in the temporal lobe of the cerebral cortex.

Cranial Nerve II: Optic Nerve

The **optic nerve** (OP-tik; *opti-* = the eye, vision, eye) is entirely sensory; it contains axons that conduct nerve impulses for vision. In the retina, rods and cones initiate visual signals and relay them to bipolar cells, which pass the signals on to ganglion cells (see Figure 16.9 on page 537). Axons of all the ganglion cells in one retina join to form an optic nerve, which passes through the optic foramen. About 10 mm (0.4 in.) posterior to the eyeball, the two optic nerves merge to form the **optic chiasm** (KĪ-azm = a crossover, as in the letter X). Within the chiasm, axons from the medial half of each eye cross to the opposite side; axons from the lateral half remain on the same side. Posterior to the chiasm, the regrouped axons, some from each eye, form the **optic tracts** (see Figure 16.16 on page 545). Most axons in the optic tracts end in the lateral geniculate nucleus of the thalamus. There they synapse with neurons whose axons extend to the primary visual area of the cerebral cortex (area 17 in Figure 14.15). A few axons pass through the optic chiasm and then extend to the superior colliculi of the midbrain. They synapse with motor neurons that control the extrinsic and intrinsic eye muscles.

Cranial Nerve III: Oculomotor Nerve

The **oculomotor nerve** (ok′-ū-lō-MŌ-tor; *oculo-* = eye; *-motor* = a mover), **cranial nerve III,** is a mixed but mainly motor cranial nerve. Its motor nucleus is in the ventral part of the midbrain (Figure 14.18a). The oculomotor nerve extends anteriorly and divides into superior and inferior branches, both of which pass through the superior orbital fissure into the orbit. Axons in the superior branch innervate the superior rectus (an extrinsic eyeball muscle) and the levator palpebrae superioris (the muscle of the upper eyelid). Axons in the inferior branch supply the medial rectus, inferior rectus, and inferior oblique muscles — all extrinsic eyeball muscles. These somatic motor neurons control movements of the eyeball and upper eyelid.

The inferior branch of the oculomotor nerve also provides parasympathetic innervation to intrinsic eyeball muscles, which are smooth muscle. They include the ciliary muscle of the eyeball and the circular muscles (sphincter pupillae) of the iris. Parasympathetic impulses propagate from the oculomotor nucleus in the midbrain to the **ciliary ganglion,** a relay center of the autonomic nervous system. From the ciliary ganglion, parasympathetic axons extend to the ciliary muscle, which adjusts the lens for near vision. Other parasympathetic axons stimulate the circular muscles of the iris to contract when bright light stimulates the eye, causing a decrease in the size of the pupil (constriction).

The sensory portion of the oculomotor nerve consists of afferent axons extending from proprioceptors in the extrinsic eyeball muscles supplied by the nerve to the midbrain. These axons convey nerve impulses for **proprioception,** the nonvisual perception of the movements and position of the body.

Cranial Nerve IV: Trochlear Nerve

The **trochlear nerve** (TRŌK-lē-ar; *trochle-* = a pulley), **cranial nerve IV,** is a mixed but mainly motor cranial nerve. It is the smallest of the 12 cranial nerves and is the only cranial nerve that arises from the posterior aspect of the brain stem.

The motor portion originates in a nucleus in the midbrain, and axons from the nucleus pass through the superior orbital fissure of the orbit (Figure 14.18b). These somatic motor axons innervate the superior oblique muscle of the eyeball, another extrinsic eyeball muscle that controls movement of the eyeball.

The sensory portion of the trochlear nerve consists of axons that extend from proprioceptors in the superior oblique muscle to a nucleus of the nerve in the midbrain. These axons convey nerve impulses for proprioception.

Cranial Nerve V: Trigeminal Nerve

The **trigeminal nerve** (trī-JEM-i-nal = triple, for its three branches), **cranial nerve V,** is a mixed cranial nerve and the largest of the cranial nerves. The trigeminal nerve emerges from two roots on the ventrolateral surface of the pons. The large sensory root has a swelling called the **trigeminal ganglion,** which is located in a fossa on the inner surface of the petrous portion of the temporal bone. The ganglion contains cell bodies of most of the primary sensory neurons. The smaller motor root originates in a nucleus in the pons.

As indicated by its name, the trigeminal nerve has three branches: ophthalmic, maxillary, and mandibular (Figure 14.19 on page 480). The **ophthalmic nerve** (of-THAL-mik; *ophthalm-* = the eye), the smallest branch, enters the orbit via the superior

Figure 14.18 Oculomotor, trochlear, and abducens nerves (cranial nerves III, IV, and VI).

The oculomotor nerve has the widest distribution to extrinsic eye muscles.

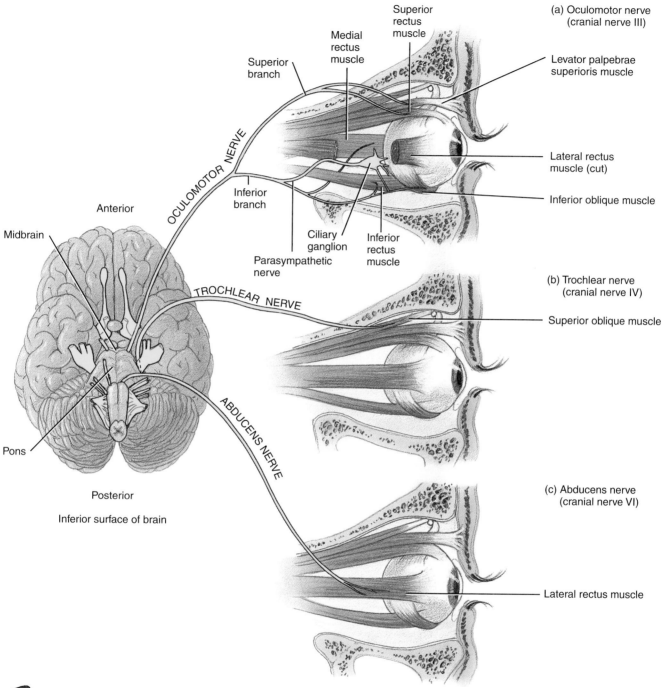

(a) Oculomotor nerve
(cranial nerve III)

(b) Trochlear nerve
(cranial nerve IV)

(c) Abducens nerve
(cranial nerve VI)

Which branch of the oculomotor nerve is distributed to the superior rectus muscle? Which is the smallest cranial nerve?

orbital fissure. The **maxillary nerve** (*maxilla* = upper jaw bone) is intermediate in size between the ophthalmic and mandibular nerves and enters the foramen rotundum. The **mandibular nerve** (*mandibula* = lower jaw bone), the largest branch, exits through the foramen ovale.

Sensory axons in the trigeminal nerve carry nerve impulses for touch, pain, and thermal sensations. The ophthalmic nerve contains sensory axons from the skin over the upper eyelid, eyeball, lacrimal glands, upper part of the nasal cavity, side of the nose, forehead, and anterior half of the scalp. The maxillary

Figure 14.19 Trigeminal nerve (cranial nerve V).

The three branches of the trigeminal nerve leave the cranium through the superior orbital fissure, foramen rotundum, and foramen ovale.

 How does the trigeminal nerve compare in size with the other cranial nerves?

nerve includes sensory axons from the mucosa of the nose, palate, part of the pharynx, upper teeth, upper lip, and lower eyelid. The mandibular nerve contains sensory axons from the anterior two-thirds of the tongue (not taste), cheek and mucosa deep to it, lower teeth, skin over the mandible and side of the head anterior to the ear, and mucosa of the floor of the mouth. Sensory axons from the three branches of the trigeminal nerve enter the semilunar ganglion and terminate in nuclei in the pons. There are also sensory fibers from proprioceptors in the muscles of mastication.

Somatic motor axons of the trigeminal nerve are part of the mandibular nerve and supply muscles of mastication (masseter, temporalis, medical pterygoid, lateral pterygoid, anterior belly of digastric, and mylohyoid muscles). These motor neurons control chewing movements.

Dental Anesthesia

The inferior alveolar nerve, a branch of the mandibular nerve, supplies all the teeth in one-half of the mandible; it is often anesthetized in dental procedures. The same procedure will anesthetize the lower lip because the mental nerve is a branch of the inferior alveolar nerve. Because the lingual nerve runs very close to the inferior alveolar nerve near the mental foramen, it too is often anesthetized at the same time. For anesthesia to the upper teeth, the superior alveolar nerve endings, which are branches of the maxillary nerve, are blocked by inserting the needle beneath the mucous membrane. The anesthetic solution is then infiltrated slowly throughout the area of the roots of the teeth to be treated. ∎

Cranial Nerve VI: Abducens Nerve

The **abducens nerve** (ab-DOO-senz; *ab-* = away; *-ducens* = to lead), **cranial nerve VI,** is a mixed but mainly motor cranial nerve that originates from a nucleus in the pons (see Figure 14.18c). Somatic motor axons extend from the nucleus to the lateral rectus muscle of the eyeball, an extrinsic eyeball muscle, through the superior orbital fissure of the orbit. Nerve impulses cause abduction of the eyeball (lateral rotation), the reason for the name of the nerve. The sensory axons extend from proprioceptors in the lateral rectus muscle to the pons.

Cranial Nerve VII: Facial Nerve

The **facial nerve** (FĀ-shal = face), **cranial nerve VII,** is a mixed cranial nerve. Its sensory axons extend from the taste buds of the anterior two-thirds of the tongue through the **geniculate ganglion,** a cluster of cell bodies of sensory neurons that lie beside the facial nerve, and end in the pons (Figure 14.20). The sensory portion of the facial nerve also contains axons from proprioceptors in muscles of the face and scalp.

Axons of somatic motor neurons arise from a nucleus in the pons, enter the petrous portion of the temporal bone, and innervate facial, scalp, and neck muscles. Nerve impulses propagating along these axons cause contraction of muscles of facial expression muscles plus the stylohyoid muscle and the posterior belly of the digastric muscle.

Axons of parasympathetic neurons that are part of the facial nerve end in two parasympathetic ganglia: the **pterygopalatine**

Figure 14.20 Facial nerve (cranial nerve VII).

The facial nerve causes contraction of the muscles of facial expression.

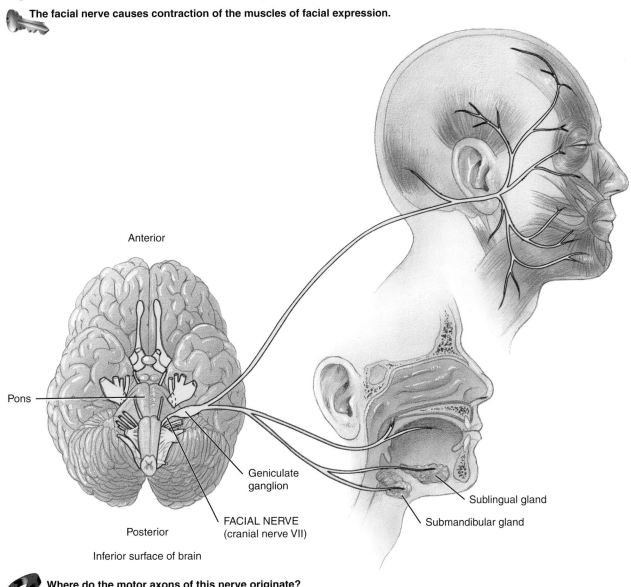

Anterior

Pons

Geniculate ganglion

FACIAL NERVE (cranial nerve VII)

Posterior

Inferior surface of brain

Sublingual gland

Submandibular gland

Where do the motor axons of this nerve originate?

ganglion and the **submandibular ganglion.** From the two ganglia, other parasympathetic axons extend to lacrimal glands (which secrete tears), nasal glands, palatine glands, and saliva-producing sublingual and submandibular glands.

Cranial Nerve VIII: Vestibulocochlear Nerve

The **vestibulocochlear nerve** (vest-tib-ū-lō-KOK-lē-ar; *vestibulo-* = small cavity; *-cochlear* = a spiral, snail-like), **cranial nerve VIII,** was formerly known as the **acoustic** or **auditory nerve.** It is a mixed but mainly sensory cranial nerve and has two branches, the vestibular nerve and the cochlear nerve. The **vestibular nerve** carries impulses for equilibrium whereas the **cochlear nerve** carries impulses for hearing.

Sensory axons in the vestibular nerve extend from the semicircular canals, the saccule, and the utricle of the inner ear to the **vestibular ganglion,** where their cell bodies are located (see Figure 16.20b on page 550), and end in vestibular nuclei in the pons. Some sensory axons also enter the cerebellum via the inferior cerebellar peduncle. Axons of motor neurons in the vestibular nerve project from the pons to hair cells of the semicircular canals, saccule, and utricle.

Sensory axons in the cochlear nerve arise in the spiral organ (organ of Corti) in the cochlea of the internal ear. The cell bodies of cochlear branch sensory neurons are located in the **spiral ganglion** of the cochlea. From there, axons extend to nuclei in the medulla oblongata. Axons of motor neurons in the cochlear nerve project from the pons to hair cells of the spiral organ.

Cranial Nerve IX: Glossopharyngeal Nerve

The **glossopharyngeal nerve** (glos'-ō-fa-RIN-jē-al; *glosso-* = tongue; *-pharyngeal* = throat), **cranial nerve IX,** is a mixed cranial nerve. Sensory axons of the glossopharyngeal nerve arise from taste buds and somatic sensory receptors on the posterior one-third of the tongue, from proprioceptors in swallowing muscles supplied by the motor portion, from stretch receptors in the carotid sinus, and from chemoreceptors in the carotid body near the carotid arteries (Figure 14.21). The cell bodies of these sensory neurons are located in the superior and inferior ganglia. From the ganglia, sensory axons pass through the jugular foramen and end in the medulla.

Axons of motor neurons in the glossopharyngeal nerve arise in nuclei of the medulla and exit the skull through the jugular

Figure 14.21   Glossopharyngeal nerve (cranial nerve IX).

Sensory axons in the glossopharyngeal nerve carry signals from the taste buds.

Anterior

Parotid gland

Otic ganglion

Stylopharyngeus muscle

Soft palate

Palatine tonsil

Inferior ganglion

Tongue

Superior ganglion

Carotid sinus and carotid body

Medulla oblongata

GLOSSOPHARYNGEAL NERVE (cranial nerve IX)

Posterior

Inferior surface of brain

Through which foramen does the glossopharyngeal nerve exit the skull?

foramen. Somatic motor neurons innervate the stylopharyngeus muscle, and autonomic motor neurons (parasympathetic) stimulate the parotid gland to secrete saliva. Some of the cell bodies of parasympathetic motor neurons are located in the **otic ganglion.**

Cranial Nerve X: Vagus Nerve

The **vagus nerve** (VĀ-gus = vagrant or wandering), **cranial nerve X,** is a mixed cranial nerve that is distributed from the head and neck into the thorax and abdomen (Figure 14.22). The nerve derives its name from its wide distribution. In the neck, it lies medial and posterior to the internal jugular vein and common carotid artery.

Sensory axons in the vagus nerve arise from the skin of the external ear, a few taste buds in the epiglottis and pharynx, and proprioceptors in muscles of the neck and throat. Also, sensory axons come from baroreceptors (stretch receptors) and chemoreceptors in the carotid sinus and carotid body near the carotid arteries, chemoreceptors in the aortic body near the arch of the aorta, and visceral sensory receptors in most organs of the thoracic and abdominal cavities. These axons pass through the jugular foramen and end in the medulla and pons.

Axons of autonomic motor neurons (parasympathetic) in the vagus nerve originate in nuclei of the medulla and end in the lungs and heart. Vagal parasympathetic axons also supply glands of the gastrointestinal (GI) tract and smooth muscle of the respiratory passageways, esophagus, stomach, gallbladder, small

Figure 14.22 **Vagus nerve (cranial nerve X).**

The vagus nerve is widely distributed in the head, neck, thorax, and abdomen.

Carotid sinus and carotid body

Anterior

Lungs

Superior ganglion

Heart

Medulla oblongata

Liver and gallbladder

Inferior ganglion

Stomach

VAGUS NERVE (cranial nerve X)

Pancreas

Posterior

Inferior surface of brain

Large intestine

Small intestine

Where is the vagus nerve located in the neck region?

intestine, and most of the large intestine (see Figure 17.2 on page 569).

Cranial Nerve XI: Accessory Nerve

The **accessory nerve** (ak-SES-ō-rē = assisting), **cranial nerve XI,** is a mixed cranial nerve. It differs from all other cranial nerves because it originates from *both* the brain stem and the spinal cord (Figure 14.23). The **cranial root** is motor and arises from nuclei in the medulla oblongata, passes through the jugular foramen, and supplies the voluntary muscles of the pharynx, larynx, and soft palate that are used in swallowing. The **spinal root** is mixed but mainly motor. Its motor axons arise in the anterior gray horn of the first five segments of the cervical portion of the spinal cord. The axons from the segments come together, pass through the foramen magnum, and then exit through the jugular foramen along with axons in the cranial part. The spinal root conveys motor impulses to the sternocleidomastoid and trapezius muscles to coordinate head movements. Sensory axons in the spinal root originate from proprioceptors in the muscles supplied by its motor neurons and end in the medulla oblongata.

Cranial Nerve XII: Hypoglossal Nerve

The **hypoglossal (XII) nerve** (hī′-pō-GLOS-al; *hypo-* = below; *-glossal* = tongue), **cranial nerve XII,** is a mixed cranial nerve. The sensory portion of the hypoglossal nerve consists of axons that originate from proprioceptors in the tongue muscles and end in the medulla oblongata (Figure 14.24). The sensory fibers conduct nerve impulses for proprioception. The somatic motor axons originate in a nucleus in the medulla oblongata, pass through the hypoglossal canal, and supply the muscles of the tongue. These axons conduct nerve impulses for speech and swallowing.

Table 14.3 presents a summary of cranial nerves, including clinical applications related to their dysfunction.

▶ CHECKPOINT

20. How are cranial nerves named and numbered?

21. What is the difference between a mixed cranial nerve and a sensory cranial nerve?

22. For each cranial nerve, what sort of test could reveal if it is damaged?

Figure 14.23   Accessory nerve (cranial nerve XI).

🔑 The accessory nerve exits the cranium through the jugular foramen.

Anterior

Medulla oblongata

Spinal cord

Posterior

Inferior surface of brain

ACCESSORY NERVE (cranial nerve XI)

Trapezius muscle

Sternocleidomastoid muscle

How does the accessory nerve differ from the other cranial nerves?

Figure 14.24 **Hypoglossal nerve (cranial nerve XII).**

 The hypoglossal nerve exits the cranium through the hypoglossal canal.

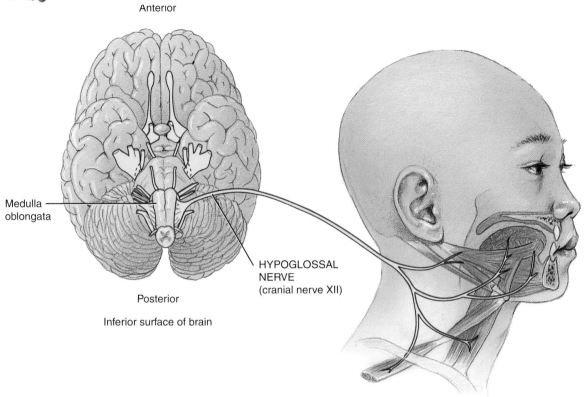

Anterior

Medulla oblongata

HYPOGLOSSAL NERVE (cranial nerve XII)

Posterior

Inferior surface of brain

? What important motor functions does the hypoglossal nerve mediate?

Table 14.3	**Summary of Cranial Nerves**	
Number and Name*	**Type and Location**	**Function and Clinical Application**
Cranial Nerve I: **Olfactory nerve** (ol-FAK-tō-rē; *olfact-* = to smell) — Olfactory bulb — Olfactory nerve — Olfactory tract	**Sensory** Arises in olfactory mucosa, passes through foramina in the cribriform plate of the ethmoid bone, and ends in the olfactory bulb. The olfactory tract extends via two pathways to olfactory areas of cerebral cortex.	*Function:* Smell. *Clinical application:* Loss of the sense of smell, called *anosmia,* may result from head injuries in which the cribriform plate of the ethmoid bone is fractured and from lesions along the olfactory pathway.

*A mnemonic device that can be used to remember the names of the nerves is: "**O**h, **o**h, **o**h, **t**o **t**ouch **a**nd **f**eel **v**ery **g**reen **v**egetables—**AH**!" Each boldfaced letter corresponds to the first letter of a pair of cranial nerves.

(continues)

Table 14.3   Summary of Cranial Nerves *(continued)*

Number and Name*	Type and Location	Function and Clinical Application
Cranial Nerve II: **Optic nerve** (OP-tik; *opti-* = the eye, vision) 	**Sensory** Arises in the retina of the eye, passes through the optic foramen, forms the optic chiasm and then the optic tracts, and terminates in the lateral geniculate nuclei of thalamus. From the thalamus, axons extend to the primary visual area (area 17) of the cerebral cortex.	*Function:* Vision *Clinical application:* Fractures in the orbit, damage along the visual pathway, and diseases of the nervous system may result in visual field defects and loss of visual acuity. Blindness due to a defect in or loss of one or both eyes is called *anopia*.
Cranial Nerve III: **Oculomotor nerve** (ok′-ū-lō-MŌ-tor; *oculo-* = eye; *-motor* = a mover) 	**Mixed (mainly motor)** *Sensory portion:* Consists of axons from proprioceptors in eyeball muscles that pass through the superior orbital fissure and terminate in the midbrain. *Motor portion:* Originates in the midbrain and passes through the superior orbital fissure. Axons of somatic motor neurons innervate the levator palpebrae superioris muscle of the upper eyelid and four extrinsic eyeball muscles (superior rectus, medial rectus, inferior rectus, and inferior oblique). Parasympathetic axons innervate the ciliary muscle of eyeball and the circular muscles (sphincter pupillae) of the iris.	*Sensory function:* Proprioception. *Somatic motor function:* Movement of upper eyelid and eyeball. *Autonomic motor function (parasympathetic):* Accommodation of lens for near vision and constriction of pupil. *Clinical application:* Nerve damage causes *strabismus* (a deviation of the eye in which both eyes do not fix on the same object), *ptosis* (drooping) of the upper eyelid, dilation of the pupil, movement of the eyeball downward and outward on the damaged side, loss of accommodation for near vision, and *diplopia* (double vision).
Cranial Nerve IV: **Trochlear nerve** (TRŌK-lē-ar; *trochle-* = a pulley) 	**Mixed (mainly motor)** *Sensory portion:* Consists of axons from proprioceptors in the superior oblique muscles, which pass through the superior orbital fissure and terminate in the midbrain. *Motor portion:* Originates in the midbrain and passes through the superior orbital fissure. Innervates the superior oblique muscle, an extrinsic eyeball muscle.	*Sensory function:* Proprioception. *Somatic motor function:* Movement of the eyeball. *Clinical application:* In trochlear nerve paralysis, diplopia and strabismus occur.

Image labels:
- Optic nerve
- Optic tract
- Oculomotor nerve
- Trochlear nerve

Number and Name*	Type and Location	Function and Clinical Application
Cranial Nerve V: **Trigeminal nerve** (trī-JEM-i-nal = triple, for its three branches) Trigeminal nerve	**Mixed** *Sensory portion:* Consists of three branches, all of which end in the pons. (1) The **ophthalamic nerve** (*ophthalm-* = the eye) contains axons from the skin over the upper eyelid, eyeball, lacrimal glands, nasal cavity, side of nose, forehead, and anterior half of scalp that pass through superior orbital fissure. (2) The **maxillary nerve** (*maxilla* = upper jaw bone) contains axons from the mucosa of the nose, palate, parts of the pharynx, upper teeth, upper lip, and lower eyelid that pass through the foramen rotundum. (3) The **mandibullar nerve** (*mandibula* = lower jaw bone) contains axons from the anterior two-thirds of the tongue (somatic sensory axons but not axons for the special sense of taste), the lower teeth, skin over mandible, cheek and mucosa deep to it, and side of head in front of ear that pass through the foramen ovale. *Motor portion:* Is part of the mandibular branch, which originates in the pons, passes through the foramen ovale, and innervates muscles of mastication (masseter, temporalis, medical pterygoid, lateral pterygoid, anterior belly of digastric, and mylohyoid muscles).	*Sensory function:* Conveys impulses for touch, pain, and temperature sensations and proprioception. *Somatic motor function:* Chewing. *Clinical application:* Neuralgia (pain) of one or more branches of the trigeminal nerve is called *trigeminal neuralgia (tic douloureux).* Injury of the mandibullar nerve may cause paralysis of the chewing muscles and a loss of the sensations of touch, temperature, and proprioception in the lower part of the face. Dentists apply anesthetic drugs to branches of the maxillary nerve for anesthesia of upper teeth and to branches of the mandibular nerve for anesthesia of lower teeth.
Cranial Nerve VI: **Abducens nerve** (ab-DOO-senz; *ab-* = away; *-ducens* = to lead) Abducens nerve	**Mixed (mainly motor)** *Sensory portion:* Consists of axons from proprioceptors in the lateral rectus muscle, which pass through the superior orbital fissure and end in the pons. *Motor portion:* Originates in the pons, passes through the superior orbital fissure, and innervates the lateral rectus muscle, an extrinsic eyeball muscle.	*Sensory function:* Proprioception. *Somatic motor function:* Movement of the eyeball. *Clinical application:* With damage to this nerve, the affected eyeball cannot move laterally beyond the midpoint, and the eye usually is directed medially.

(continues)

Table 14.3	Summary of Cranial Nerves *(continued)*	
Number and Name*	**Type and Location**	**Function and Clinical Application**
Cranial Nerve VII: **Facial nerve** (FĀ-shal = face) — Facial nerve	**Mixed** *Sensory portion:* Arises from taste buds on the anterior two-thirds of the tongue, passes through the stylomastoid foramen and geniculate ganglion (located beside the facial nerve), and ends in the pons. From there, axons extend to the thalamus, and then to the gustatory areas of the cerebral cortex. Also contains axons from proprioceptors in muscles of the face and scalp. *Motor portion:* Originates in the pons and passes through the stylomastoid foramen. Axons of somatic motor neurons innervate facial, scalp, and neck muscles. Parasympathetic axons innervate lacrimal, sublingual, submandibular, nasal, and palatine glands.	*Sensory function:* Proprioception and taste. *Somatic motor function:* Facial expression. *Autonomic motor function (parasympathetic):* Secretion of saliva and tears. *Clinical application:* Injury produces *Bell's palsy* (paralysis of the facial muscles), loss of taste, decreased salivation, and loss of ability to close the eyes, even during sleep.
Cranial Nerve VIII: **Vestibulocochlear nerve** (vest-tib-ū-lō-KOK-lē-ar; *vestibulo-* = small cavity; *-cochlear* = a spiral, snail-like) — Vestibulocochlear nerve	**Mixed (mainly sensory)** *Vestibular nerve, sensory portion:* Arises in the semicircular canals, saccule, and utricle and forms the vestibular ganglion. Axons end in the pons and cerebellum. *Vestibular nerve, motor portion:* Originates in the pons and terminates on hair cells of the semicircular canals, saccule, and utricle. *Cochlear nerve, sensory portion:* Arises in the spiral organ (organ of Corti), forms the spiral ganglion, passes through nuclei in the medulla, and ends in the thalamus. Axons synapse with thalamic neurons that relay impulses to the primary auditory area (areas 41 and 42) of the cerebral cortex. *Cochlear nerve, motor portion:* Originates in the pons and terminates on hair cells of the spiral organ.	*Vestibular branch sensory function:* Conveys impulses related to equilibrium. *Vestibular branch motor function:* Adjusts sensitivity of hair cells. *Cochlear branch sensory function:* Conveys impulses for hearing. *Cochlear branch motor function:* Modifies function of hair cells by altering their response to sound waves. *Clinical application:* Injury to the vestibular branch may cause *vertigo* (a subjective feeling of rotation), *ataxia* (muscular incoordination), and *nystagmus* (involuntary rapid movement of the eyeball). Injury to the cochlear branch may cause *tinnitus* (ringing in the ears) or deafness.
Cranial Nerve IX: **Glossopharyngeal nerve** (glos'-ō-fa-RIN-jē-al; *glosso-* = tongue; *-pharyngeal* = throat) — Glossopharyngeal nerve	**Mixed** *Sensory portion:* Consists of axons from taste buds and somatic sensory receptors on posterior one-third of the tongue, from proprioceptors in swallowing muscles supplied by the motor portion, and from stretch receptors in carotid sinus and chemoreceptors in carotid body near the carotid arteries. Axons pass through the jugular foramen and end in the medulla. *Motor portion:* Originates in the medulla and passes through the jugular foramen. Axons of somatic motor neurons innervate the stylopharyngeus muscle, a muscle of pharynx that elevates the larynx during swallowing. Parasympathetic axons innervate the parotid (salivary) gland.	*Sensory function:* Taste and somatic sensations (touch, pain, temperature) from posterior third of tongue; proprioception in swallowing muscles; monitoring of blood pressure; monitoring of O_2 and CO_2 in blood for regulation of breathing rate and depth. *Somatic motor function:* Elevates the pharynx during swallowing and speech. *Autonomic motor function (parasympathetic):* Stimulates secretion of saliva. *Clinical application:* Injury causes difficulty in swallowing, reduced secretion of saliva, loss of sensation in the throat, and loss of taste sensation.

Number and Name*	Type and Location	Function and Clinical Application

Cranial Nerve X:
Vagus nerve
(VĀ-gus = vagrant or wandering)

Vagus nerve

Mixed

Sensory portion: Consists of axons from small number of taste buds in the epiglottis and pharynx, proprioceptors in muscles of the neck and throat, stretch receptors and chemoreceptors in carotid sinus and carotid body near the carotid arteries, chemoreceptors in aortic body near arch of the aorta, and visceral sensory receptors in most organs of the thoracic and abdominal cavities. Axons pass through the jugular foramen and end in the medulla and pons.

Motor portion: Originates in medulla and passes through the jugular foramen. Axons of somatic motor neurons innervate skeletal muscles in the throat and neck. Parasympathetic axons innervate smooth muscle in the airways, esophagus, stomach, small intestine, most of large intestine, and gallbladder; cardiac muscle in the heart; and glands of the gastrointestinal (GI) tract.

Sensory function: Taste and somatic sensations (touch, pain, temperature, and proprioception) from epiglottis and pharynx; monitoring of blood pressure; monitoring of O_2 and CO_2 in blood for regulation of breathing rate and depth; sensations from visceral organs in thorax and abdomen.

Somatic motor function: Swallowing, coughing, and voice production.

Autonomic motor function (parasympathetic): Smooth muscle contraction and relaxation in organs of the GI tract; slowing of the heart rate; secretion of digestive fluids.

Clinical application: Injury interrupts sensations from many organs in the thoracic and abdominal cavities, interferes with swallowing, paralyzes vocal cords, and causes heart rate to increase.

Cranial Nerve XI:
Accessory nerve
(ak-SES-ō-rē = assisting)

Accessory nerve

Mixed (mainly motor)

Sensory portion: Consists of axons from proprioceptors in muscles of the pharynx, larynx, and soft palate that pass through the jugular foramen and end in the medulla.

Motor portion: Consists of a cranial root and a spinal root. *Cranial root* arises in the medulla, passes through the jugular foramen, and supplies muscles of the pharynx, larynx, and soft palate. *Spinal root* originates in the anterior gray horn of the first five cervical segments of the spinal cord, passes through the jugular foramen, and supplies the sternocleidomastoid and trapezius muscles.

Sensory function: Proprioception.

Somatic motor function: Cranial part mediates swallowing movements; spinal part mediates movement of head and shoulders.

Clinical application: If nerves are damaged, the sternocleidomastoid and trapezius muscles become paralyzed, with resulting inability to raise the shoulders and difficulty in turning the head.

Cranial Nerve XII:
Hypoglossal nerve
(hī'-pō-GLOS-al; hypo- = below; -glossal = tongue)

Hypoglossal nerve

Mixed (mainly motor)

Sensory portion: Consists of axons from proprioceptors in tongue muscles that pass through the hypoglossal canal and end in the medulla.

Motor portion: Originates in the medulla, passes through the hypoglossal canal, and supplies muscles of the tongue.

Sensory function: Proprioception.

Motor function: Movement of tongue during speech and swallowing.

Clinical application: Injury results in difficulty in chewing, speaking, and swallowing. The tongue, when protruded, curls toward the affected side, and the affected side atrophies.

DEVELOPMENT OF THE NERVOUS SYSTEM

▶ O B J E C T I V E

• Describe how the parts of the brain develop.

Development of the nervous system begins in the third week with a thickening of the **ectoderm** called the **neural plate** (Figure 14.25). The plate folds inward and forms a longitudinal groove, the **neural groove.** The raised edges of the neural plate are called **neural folds.** As development continues, the neural folds increase in height and meet to form a tube called the **neural tube.**

Three layers of cells differentiate from the wall that encloses the neural tube. The outer or **marginal layer** cells develop into the *white matter* of the nervous system. The middle or **mantle layer** cells develop into the *gray matter.* The inner or **ependymal layer** cells eventually form the *lining of the central canal of the spinal cord* and *ventricles of the brain.*

The **neural crest** is a mass of tissue between the neural tube and the skin ectoderm (Figure 14.25b). It differentiates and eventually forms the *posterior (dorsal) root ganglia of spinal nerves, spinal nerves, ganglia of cranial nerves, cranial nerves, ganglia of the autonomic nervous system, adrenal medulla,* and *meninges.*

During the fourth week of embryonic development, the anterior part of the neural tube develops into three enlarged areas called **primary brain vesicles.** These are the **prosencephalon** (prōs'-en-SEF-a-lon; *pros-* = before) or forebrain, **mesencephalon** (mes'-en-SEF-a-lon; *mes-* = middle) or midbrain, and **rhombencephalon** (rom'-ben-SEF-a-lon; *rhomb-* = behind) or hindbrain (Figure 14.26a). During the fifth week of development, the prosencephalon develops into two **secondary brain vesicles** called the **telencephalon** (tel'-en-SEF-a-lon; *tel-* = distant) and the **diencephalon** (dī-en-SEF-a-lon; *di-* = through) (Figure 14.26b). The rhombencephalon also develops into two secondary brain vesicles called the **metencephalon** (met'-en-SEF-a-lon; *met-* = after) and the **myelencephalon** (mī-el-en-SEF-a-lon; *myel-* = marrow). The area of the neural tube inferior to the myelencephalon gives rise to the *spinal cord.*

The brain vesicles continue to develop as follows:

• The telencephalon develops into the *cerebral hemispheres,* including the *basal ganglia,* and houses the paired *lateral ventricles.*

• The diencephalon develops into the *epithalamus, thalamus, subthalamus, hypothalamus,* and *pineal gland* and houses the *third ventricle.*

Figure 14.25 Origin of the nervous system.   (a) Dorsal view of an embryo in which the neural folds have partially united, forming the early neural tube. (b) Transverse sections through the embryo showing the formation of the neural tube.

The nervous system begins developing in the third week from a thickening of ectoderm called the neural plate.

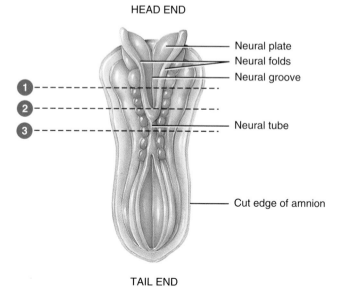

HEAD END

Neural plate
Neural folds
Neural groove
Neural tube

Cut edge of amnion

TAIL END

(a) Dorsal view

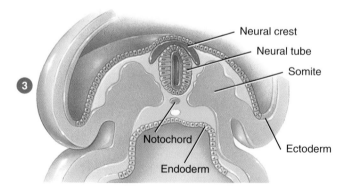

Future neural crest
Neural plate
Ectoderm
Notochord
Endoderm
Mesoderm

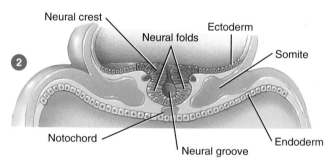

Neural crest
Ectoderm
Neural folds
Somite
Notochord
Endoderm
Neural groove

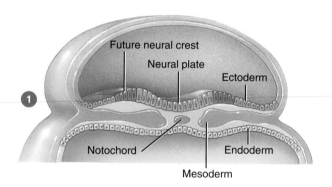

Neural crest
Neural tube
Somite
Notochord
Ectoderm
Endoderm

(b) Transverse sections

 What is the origin of the gray matter of the nervous system?

The various parts of the brain develop from the primary brain vesicles.

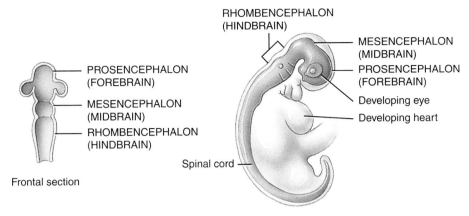

(a) Three–four-week embryo showing primary brain vesicles

(b) Five-week embryo showing secondary brain vesicles

Which primary brain vesicle does not develop into a secondary brain vesicle?

• The mesencephalon develops into the *midbrain,* which surrounds the *cerebral aqueduct.*

• The metencephalon becomes the *pons* and *cerebellum* and houses part of the *fourth ventricle.*

• The myelencephalon develops into the *medulla oblongata* and houses the remainder of the *fourth ventricle.*

Two neural tube defects—spina bifida (see page 214) and anencephaly (absence of the skull and cerebral hemispheres, discussed on page 1073)—are associated with low levels of folic acid (folate), one of the B vitamins. Although many foods, especially grain products such as cereals and bread, are fortified with folic acid, the incidence of both disorders is greatly reduced when women who may become pregnant are taking folic acid supplements.

AGING AND THE NERVOUS SYSTEM

▶ O B J E C T I V E

• **Describe the effects of aging on the nervous system.**

The brain grows rapidly during the first few years of life. Growth is due mainly to an increase in the size of neurons already present, the proliferation and growth of neuroglia, the development of dendritic branches and synaptic contacts, and continuing myelination of axons. From early adulthood onward, brain mass declines. By the time a person reaches 80, the brain weighs about 7% less than it did in young adulthood. Although the number of neurons present does not decrease very much, the number of synaptic contacts declines. Associated with the decrease in brain mass is a decreased capacity for sending nerve impulses to and from the brain. As a result, processing of information diminishes. Conduction velocity decreases, voluntary motor movements slow down, and reflex times increase.

▶ C H E C K P O I N T

23. What parts of the brain develop from each primary brain vesicle?

24. How is brain mass related to age?

DISORDERS: HOMEOSTATIC IMBALANCES

Cerebrovascular Accident

The most common brain disorder is a **cerebrovascular accident (CVA),** also called a **stroke** or **brain attack.** CVAs affect 500,000 people a year in the United States and represent the third leading cause of death, behind heart attacks and cancer. A CVA is characterized by abrupt onset of persisting neurological symptoms, such as paralysis or loss of sensation, that arise from destruction of brain tissue. Common causes of CVAs are intracerebral hemorrhage (from a blood vessel in the pia mater or brain), emboli (blood clots), and atherosclerosis (formation of cholesterol-containing plaques that block blood flow) of the cerebral arteries.

Among the risk factors implicated in CVAs are high blood pressure, high blood cholesterol, heart disease, narrowed carotid arteries, transient ischemic attacks (TIAs; discussed next), diabetes, smoking, obesity, and excessive alcohol intake.

A clot-dissolving drug called tissue plasminogen activator (t-PA) can open up blocked blood vessels in the brain. The drug is most effective when administered within three hours of the onset of the CVA, and is helpful only for CVAs due to a blood clot. Use of t-PA can decrease the permanent disability associated with CVAs by 50%.

Transient Ischemic Attacks

A **transient ischemic attack (TIA)** is an episode of temporary cerebral dysfunction caused by impaired blood flow to the brain. Symptoms include dizziness, weakness, numbness, or paralysis in a limb or in one side of the body; drooping of one side of the face; headache; slurred speech or difficulty understanding speech; and a partial loss of vision or double vision. Sometimes nausea or vomiting also occurs. The onset of symptoms is sudden and reaches maximum intensity almost immediately. A TIA usually persists for 5 to 10 minutes and only rarely lasts as long as 24 hours. It leaves no permanent neurological deficits. The causes of the impaired blood flow that lead to TIAs are blood clots, atherosclerosis, and certain blood disorders. About one-third of patients who experience a TIA will have a CVA within 5 years. Therapy for TIAs includes drugs such as aspirin, which blocks the aggregation of blood platelets, and anticoagulants; cerebral artery bypass grafting; and carotid endarterectomy (removal of the cholesterol-containing plaques and inner lining of an artery).

Alzheimer Disease

Alzheimer disease (ALTZ-hī -mer) or **AD** is a disabling senile dementia, the loss of reasoning and ability to care for oneself, that afflicts about 11% of the population over age 65. In the United States, AD afflicts about 4 million people and claims over 100,000 lives a year. Thus, AD is the fourth leading cause of death among the elderly, after heart disease, cancer, and stroke. The cause of most AD cases is still unknown, but evidence suggests it is due to a combination of genetic factors, environmental or lifestyle factors, and the aging process. Mutations in three different genes (coding for presenilin-1, presenilin-2, and amyloid precursor protein) lead to early-onset forms of AD in afflicted families but account for less than 1% of all cases. Another gene, called *APOE,* codes for *apolipoprotein E,* a protein that helps transport cholesterol in the blood. People who have one or two copies of the e4 allele (form) of *APOE* have a much higher risk of developing AD and an earlier age of onset when compared with people who have either the e2 or e3 alleles. (A person has two alleles of most genes, one inherited from the mother and one from the father.) An environmental risk factor for developing AD is a history of head injury. A similar dementia occurs in boxers, probably caused by repeated blows to the head.

Individuals with AD initially have trouble remembering recent events. They then become confused and forgetful, often repeating questions or getting lost while traveling to familiar places. Disorientation grows and memories of past events disappear, and episodes of paranoia, hallucination, or violent changes in mood may occur. As their minds continue to deteriorate, they lose their ability to read, write, talk, eat, or walk. The disease culminates in dementia. A person with AD usually dies of some complication that afflicts bedridden patients, such as pneumonia.

At autopsy, brains of AD victims show three distinct structural abnormalities:

1. ***Loss of neurons that liberate acetylcholine.*** A major center of neurons that liberate ACh is the nucleus basalis, which is below the globus pallidus. Axons of these neurons project widely throughout the cerebral cortex and limbic system. Their destruction is a hallmark of Alzheimer disease.

2. ***Beta-amyloid plaques,*** clusters of abnormal proteins deposited outside neurons.

3. ***Neurofibrillary tangles,*** abnormal bundles of protein filaments inside neurons in affected brain regions.

Drugs that inhibit acetylcholinesterase (AChE), the enzyme that inactivates ACh, improve alertness and behavior in about 5% of AD patients. Tacrine, the first anticholinesterase inhibitor approved for treatment of AD in the United States, has significant side effects and requires dosing four times a day. Donepezil, approved in 1998, is less toxic to the liver and has the advantage of once-a-day dosing. Some evidence suggests that vitamin E (an antioxidant), estrogen, ibuprofen, and ginko biloba extract may have slight beneficial effects in AD patients.

MEDICAL TERMINOLOGY

Agnosia (ag-NŌ-zē-a; *a-* = without; *-gnosia* = knowledge) Inability to recognize the significance of sensory stimuli such as sounds, sights, smells, tastes, and touch.

Apraxia (a-PRAK-sē-a; *-praxia* = coordinated) Inability to carry out purposeful movements in the absence of paralysis.

Delirium (de-LIR-ē-um = off the track) A transient disorder of abnormal cognition and disordered attention accompanied by disturbances of the sleep-wake cycle and psychomotor behavior (hyperactivity or hypoactivity of movements and speech). Also called **acute confusional state (ACS).**

Dementia (de-MEN-shē-a; *de-* = away from; *-mentia* = mind) Permanent or progressive general loss of intellectual abilities, including impairment of memory, judgment, and abstract thinking and changes in personality.

Encephalitis (en′-sef-a-LĪ-tis) An acute inflammation of the brain caused by either a direct attack by any of several viruses or an allergic reaction to any of the many viruses that are normally harmless to the central nervous system. If the virus affects the spinal cord as well, the condition is called **encephalomyelitis.**

Lethargy (LETH-ar-jē) A condition of functional sluggishness.

Stupor (STŪ-por) Unresponsiveness from which a patient can be aroused only briefly and only by vigorous and repeated stimulation.

STUDY OUTLINE

BRAIN ORGANIZATION, PROTECTION, AND NOURISHMENT (p. 452)

1. The major parts of the brain are the brain stem, cerebellum, diencephalon, and cerebrum.
2. Blood flow to the brain is mainly via the internal carotid and vertebral arteries.
3. Any interruption of the oxygen or glucose supply to the brain can result in weakening of, permanent damage to, or death of brain cells.
4. The blood–brain barrier (BBB) causes different substances to move between the blood and the brain tissue at different rates and prevents the movement of some substances from blood into the brain.
5. Cranial bones and the cranial meninges protect the brain. The cranial meninges are continuous with the spinal meninges. From superficial to deep they are the dura mater, arachnoid mater, and pia mater.
6. Cerebrospinal fluid (CSF) provides mechanical protection, chemical protection, and circulation of nutrients. It forms in the choroid plexuses and circulates through the lateral ventricles, third ventricle, fourth ventricle, subarachnoid space, and central canal. Most of the fluid is absorbed into the blood across the arachnoid villi of the superior sagittal blood sinus.

THE BRAIN STEM (p. 458)

1. The medulla oblongata attaches to the spinal cord and contains both motor and sensory tracts. It contains nuclei that are reflex centers for regulation of heart rate, respiratory rate, vasoconstriction, swallowing, coughing, vomiting, and sneezing. It also contains nuclei associated with cranial nerves VIII through XII.
2. The pons is superior to the medulla. It connects the spinal cord with the brain and links parts of the brain with one another by way of tracts. Pontine nuclei relay nerve impulses related to voluntary skeletal movements from the cerebral cortex to the cerebellum. The pons contains the pneumotaxic and apneustic centers, which

help control breathing. It contains nuclei associated with cranial nerves V–VII and the vestibular branch of cranial nerve VIII.
3. The midbrain is between the pons and diencephalon and surrounds the cerebral aqueduct. It conveys motor impulses from the cerebrum to the cerebellum and spinal cord, sends sensory impulses from the spinal cord to the thalamus, and mediates auditory and visual reflexes. It also contains nuclei associated with cranial nerves III and IV.
4. A large part of the brain stem consists of small areas of gray matter and white matter called the reticular formation. It helps maintain consciousness, causes awakening from sleep, and contributes to regulating muscle tone.

THE CEREBELLUM (p. 462)

1. The cerebellum occupies the inferior and posterior aspects of the cranial cavity. It consists of two lateral hemispheres and a medial, constricted vermis.
2. It connects to the brain stem by three pairs of cerebellar peduncles.
3. The cerebellum functions to coordinate movements and to maintain normal muscle tone, posture, and balance.

THE DIENCEPHALON (p. 464)

1. The diencephalon surrounds the third ventricle and consists of the thalamus, hypothalamus, epithalamus, and subthalamus.
2. The thalamus is superior to the midbrain and contains nuclei that relay sensory impulses to the cerebral cortex. It also allows crude appreciation of painful, thermal, and pressure sensations and mediates some motor activities.
3. The hypothalamus is inferior to the thalamus. It controls the autonomic nervous system, secretes hormones, functions in rage and aggression, governs body temperature, regulates food and fluid intake, and establishes circadian rhythms.
4. The epithalamus consists of the pineal gland and the habenular nuclei. The pineal gland secretes melatonin, which promotes sleep and helps set the body's biological clock.

5. The subthalamus connects to motor areas of the cerebrum.
6. Circumventricular organs (CVOs) can monitor chemical changes in the blood because they lack the blood–brain barrier.

THE CEREBRUM (p. 467)

1. The cerebrum is the largest part of the brain. Its cortex contains gyri (convolutions), fissures, and sulci.
2. The cerebral lobes are named the frontal, parietal, temporal, and occipital.
3. The white matter is deep to the cortex and consists of myelinated and unmyelinated axons extending to other regions as association, commissural, and projection tracts.
4. The basal ganglia are several groups of nuclei in each cerebral hemisphere. They help control automatic movements of skeletal muscles and help regulate muscle tone.
5. The limbic system encircles the upper part of the brain stem and the corpus callosum. It functions in emotional aspects of behavior and memory.
6. Table 14.1 on page 472 summarizes the functions of various parts of the brain.

CEREBRAL CORTEX AREAS AND FUNCTIONS (p. 473)

1. The sensory areas of the cerebral cortex allow perception of sensory impulses. The motor areas are the regions that govern muscular movement. The association areas are concerned with more complex integrative functions.
2. The primary somatosensory area (areas 1, 2, and 3) receives nerve impulses from somatic sensory receptors for touch, proprioception, pain, and temperature. Each point within the area receives impulses from a specific part of the face or body.
3. The primary visual area (area 17) receives impulses that convey visual information. The primary auditory area (areas 41 and 42) interprets the basic characteristics of sound such as pitch and rhythm. The primary gustatory area (area 43) receives impulses for taste. The primary olfactory area (area 28) receives impulses for smell.
4. Motor areas include the primary motor area (area 4), which controls voluntary contractions of specific muscles or groups of muscles, and Broca's speech area (areas 44 and 45), which controls production of speech.
5. The somatosensory association area (areas 5 and 7) permits you to determine the exact shape and texture of an object without looking at it and to sense the relationship of one body part to another. The visual association area (areas 18 and 19) relates present to past visual experiences and is essential for recognizing and evaluating what is seen. The auditory association area (area 22) deals with the meanings of sounds.

6. Wernicke's area (area 22 and possibly 39 and 40) interprets the meaning of speech by translating words into thoughts. The common integrative area (areas 5, 7, 39, and 40) integrates sensory interpretations from the association areas and impulses from other areas, allowing thoughts based on sensory inputs.
7. The premotor area (area 6) generates nerve impulses that cause specific groups of muscles to contract in specific sequences. The frontal eye field area (area 8) controls voluntary scanning movements of the eyes.
8. Subtle anatomical differences exist between the two hemispheres, and each has unique functions. Each hemisphere receives sensory signals from and controls the opposite side of the body. The left hemisphere is more important for language, numerical and scientific skills, and reasoning. The right hemisphere is more important for musical and artistic awareness, spatial and pattern perception, recognition of faces, emotional content of language, identifying odors, and generating mental images of sight, sound, touch, taste, and smell.
9. Brain waves generated by the cerebral cortex are recorded from the surface of the head in an electroencephalogram (EEG). The EEG may be used to diagnose epilepsy, infections, and tumors.

CRANIAL NERVES (p. 477)

1. Twelve pairs of cranial nerves originate from the nose, eyes, inner ear, brainstem, and spinal cord.
2. They are named primarily based on their distribution and are numbered I–XII in order of attachment to the brain. Table 14.3 on pages 485–489 summarizes the cranial nerves.

DEVELOPMENT OF THE NERVOUS SYSTEM (p. 490)

1. The development of the nervous system begins with a thickening of a region of the ectoderm called the neural plate.
2. During embryological development, primary brain vesicles form from the neural tube and serve as forerunners of various parts of the brain.
3. The telencephalon forms the cerebrum, the diencephalon develops into the thalamus and hypothalamus, the mesencephalon develops into the midbrain, the metencephalon develops into the pons and cerebellum, and the myelencephalon forms the medulla.

AGING AND THE NERVOUS SYSTEM (p. 491)

1. The brain grows rapidly during the first few years of life.
2. Age-related effects involve loss of brain mass and decreased capacity for sending nerve impulses.

SELF-QUIZ QUESTIONS

Fill in the blanks in the following statements.

1. The sites of CSF production are the _____.

2. The _____ separates the cerebrum into right and left halves.

Indicate whether the following statements are true or false.

3. The hypothalamus controls many body activities and is one of the major regulators of homeostasis.

4. Delta waves in an EEG produced by an awake adult indicate periods of sensory input and mental activity.

Choose the one best answer to the following questions.

5. Which of the following statements is *false*? (a) The blood supply to the brain is provided mainly by the internal carotid and vertebral arteries. (b) Neurons in the brain rely almost exclusively on aerobic respiration to produce ATP. (c) An interruption of blood flow to the brain for even 30 seconds may impair brain function. (d) Glucose supply to the brain must be continuous. (e) Even a brief interruption of blood flow to the brain may cause unconsciousness.

6. In which of the following ways does cerebrospinal fluid contribute to homeostasis? (1) mechanical protection, (2) chemical protection, (3) electrical protection, (4) circulation, (5) osmolarity. (a) 1, 2, and 3, (b) 2, 3, and 4, (c) 3, 4, and 5, (d) 1, 2, and 4, (e) 2, 4, and 5.

7. Which of the following are functions of the hypothalamus? (1) control of the ANS, (2) control of the pituitary gland, (3) regulation of emotional and behavioral patterns, (4) regulation of eating and drinking, (5) control of body temperature, (6) regulation of circadian rhythms and states of consciousness. (a) 1, 2, 4, and 6, (b) 2, 3, 5, and 6, (c) 1, 3, 5, and 6, (d) 1, 4, 5, and 6, (e) 1, 2, 3, 4, 5, and 6.

8. Which of the following statements is *false*? (a) Association tracts transmit nerve impulses between gyri in the same hemisphere. (b) Commissural tracts transmit impulses from the gyri in one cerebral hemisphere to the corresponding gyri in the other hemisphere. (c) Projection tracts form descending and ascending tracts that transmit impulses from the cerebrum and other parts of the brain to the spinal cord, or from the spinal cord to the brain. (d) The internal capsule is an example of commissural tracts. (e) The corpus callosum is an example of commissural tracts.

9. Which of the following statements is *true*? (a) The right and left hemispheres of the cerebrum are completely symmetrical. (b) The left hemisphere controls the left side of the body. (c) The right hemisphere is more important for spoken and written language. (d) The left hemisphere is more important for musical and artistic awareness. (e) Hemispheric lateralization is more pronounced in males than in females.

10. Which of the following is *not* part of the diencephalon? (a) pons, (b) thalamus, (c) mammillary bodies, (d) hypothalamus, (e) pineal gland.

11. The blood–brain barrier (1) allows passage of lipid soluble substances into the brain, (2) involves astrocytes selectively allowing passage of certain substances into the brain, (3) allows passage of proteins into the brain, (4) is formed by the meninges of the brain, (5) can interfere with the treatment of brain disorders. (a) 2 and 5, (b) 1, 3, and 5, (c) 2, 3, and 4, (d) 1, 2, and 5, (e) 1, 2, 3, and 5.

12. Match the following:

(1) cranial nerve I   (2) cranial nerve II   (3) cranial nerve III
(4) cranial nerve IV   (5) cranial nerve V   (6) cranial nerve VI
(7) cranial nerve VII   (8) cranial nerve VIII   (9) cranial nerve IX
(10) cranial nerve X   (11) cranial nerve XI   (12) cranial nerve XII

____ (a) oculomotor
____ (b) trigeminal
____ (c) abducens
____ (d) vestibulocochlear
____ (e) accessory
____ (f) vagus
____ (g) facial
____ (h) glossopharyngeal
____ (i) olfactory
____ (j) trochlear
____ (k) optic
____ (l) hypoglossal
____ (m) functions in sense of smell
____ (n) functions in hearing and equilibrium
____ (o) functions in chewing
____ (p) functions in facial expression and secretion of saliva and tears
____ (q) functions in movement of tongue during speech and swallowing
____ (r) functions in secretion of digestive fluids
____ (s) functions in secretion of saliva, taste, regulation of blood pressure, and muscle sense

13. Match the following:

____ (a) protrusions in the medulla formed by large motor tracts
____ (b) dura mater extension that separates the two cerebral hemispheres
____ (c) fingerlike extensions of the arachnoid mater where CSF is reabsorbed
____ (d) dura mater extension that separates the two cerebellar hemispheres
____ (e) located in the hypothalamus; relay stations for reflexes related to smell
____ (f) folds in the cerebral cortex
____ (g) shallow grooves in the cerebral cortex
____ (h) bundles of white matter that relay information between the cerebellum and other parts of the brain
____ (i) a thick band of sensory and motor tracts that connect the cerebral cortex with the brain stem and spinal cord
____ (j) dura mater extension that separates the cerebrum from the cerebellum
____ (k) thin membranous partition between the lateral ventricles

(1) gyri
(2) internal capsule
(3) mammillary bodies
(4) tentorium cerebelli
(5) pyramids
(6) falx cerebelli
(7) septum pellucidum
(8) cerebellar peduncles
(9) falx cerebri
(10) sulci
(11) arachnoid villi

14. Match the following (some answers may be used more than once):

____ (a) emotional brain

____ (b) bridge connecting the spinal cord with the brain and parts of the brain with each other

____ (c) sensory relay area

____ (d) alerts the cerebral cortex to incoming sensory signals and helps regulate muscle tone

____ (e) the motor command center; regulates posture and balance

____ (f) lack the blood–brain barrier; can monitor chemical changes in the blood

____ (g) site of decussation of pyramids

____ (h) site of pneumotaxic and apneustic areas

____ (i) secretes melatonin

____ (j) contains sensory, motor, and association areas

____ (k) responsible for maintaining consciousness and awakening from sleep

____ (l) controls ANS

____ (m) contains reflex centers for movements of the eyes, head, and neck in response to visual and other stimuli, and reflex center for movements of the head and trunk in response to auditory stimuli

____ (n) plays an essential role in awareness and in the acquisition of knowledge

____ (o) several groups of nuclei that control large autonomic movements of skeletal muscles and help regulate muscle tone required for specific body movements

____ (p) produces hormones that regulate endocrine gland functions

____ (q) contains the vital cardiovascular center and medullary rhythmicity area

(1) medulla oblongata
(2) pons
(3) midbrain
(4) cerebellum
(5) pineal gland
(6) thalamus
(7) hypothalamus
(8) cerebrum
(9) limbic system
(10) reticular formation
(11) circumventricular organs
(12) reticular activating system
(13) basal ganglia

15. Match the following:

____ (a) allows planning and production of speech

____ (b) interprets pitch and rhythm

____ (c) controls voluntary contraction of muscles

____ (d) allows recognition and evaluation of visual experiences

____ (e) integration and interpretation of somatic sensations; comparison of past to present sensations

____ (f) receives impulses for touch, proprioception, pain and temperature

____ (g) interprets taste

____ (h) interpretation of sounds as speech, music or noise

____ (i) receives impulses from many sensory and association areas as well as the thalamus and brain stem; allows formation of thoughts so appropriate action can occur

____ (j) translates words into thoughts

____ (k) receives impulses for smell

____ (l) allows interpretation of shape, color and movement

(1) primary visual area
(2) primary auditory area
(3) primary gustatory area
(4) primary olfactory area
(5) primary somatosensory area
(6) primary motor area
(7) somatosensory association area
(8) visual association area
(9) Broca's area
(10) auditory association area
(11) Wernicke's area
(12) common integrative area

CRITICAL THINKING QUESTIONS

1. An elderly relative suffered a CVA (stroke) and now has difficulty moving her right arm, and she also has speech problems. What areas of the brain were damaged by the stroke?
   HINT *What results from decussation of the pyramids of the medulla oblongata?*

2. Casey complained to her swim coach that "my bathing cap is so tight, it'll squeeze my brains out my ears!" Her coach is majoring in physical therapy and tells her that that's anatomically impossible. Explain the coach's position.
   HINT *Put your hands on your head and squeeze. What do you feel?*

3. Ando's first trip to the dentist after a ten-year absence resulted in extensive dental work, so he received numbing injections of anesthetic in several locations during the session. While having lunch right after the appointment, soup dribbles out of his mouth because he has no feeling in his left upper lip, his right lower lip, and the tip of his tongue. What happened to Ando?
   HINT *The dentist wanted to block the sensation of pain from the teeth.*

ANSWERS TO FIGURE QUESTIONS

14.1 The largest part of the brain is the cerebrum.

14.2 From superficial to deep, the three cranial meninges are the dura mater, arachnoid, and pia mater.

14.3 The brain stem is anterior to the fourth ventricle, and the cerebellum is posterior to it.

14.4 The arachnoid villi that project into the dural venous sinuses reabsorb CSF.

14.5 The medulla oblongata contains the pyramids; the midbrain contains the cerebral peduncles; "pons" means "bridge."

14.6 Decussation means crossing to the opposite side. The pyramids contain motor tracts that extend from the cortex into the spinal cord and convey impulses for contraction of skeletal muscle. Hence, the functional consequence of decussation of the pyramids is that one side of the cerebrum controls muscles on the opposite side of the body.

14.7 The cerebral peduncles are the main sites where tracts extend between the superior parts of the brain and the inferior parts of the brain and the spinal cord.

14.8 The cerebellar peduncles contain the axons that carry information into and out of the cerebellum.

14.9 The intermediate mass connects the right and left halves of the thalamus.

14.10 From posterior to anterior, the four major regions of the hypothalamus are the mammillary, tuberal, supraoptic, and preoptic regions.

14.11 The gray matter enlarges more rapidly, in the process producing convolutions or gyri (folds), sulci (shallow grooves), and fissures (deep grooves).

14.12 Association tracts connect gyri of the same hemisphere; commissural tracts connect gyri in opposite hemispheres; projection tracts connect the cerebrum with the thalamus, brain stem, and spinal cord.

14.13 The basal ganglia are lateral, superior, and inferior to the thalamus.

14.14 The hippocampus functions in memory.

14.15 Common integrative area; motor speech area; premotor area; gustatory areas; auditory areas; visual areas; frontal eye field area.

14.16 The orbitofrontal cortex in the right hemisphere.

14.17 In an EEG, theta waves indicate emotional stress.

14.18 Superior branch; trochlear nerve.

14.19 The trigeminal nerve is the largest cranial nerve.

14.20 Motor axons of the facial nerve originate in the pons.

14.21 Jugular foramen.

14.22 Between and behind the internal jugular vein and common carotid artery.

14.23 It originates from both the brain and spinal cord.

14.24 Speech and swallowing.

14.25 Gray matter derives from the mantle layer cells of the neural tube.

14.19 The mesencephalon does not develop into a secondary brain vesicle.

CHAPTER FIFTEEN

Sensory, Motor, and
Integrative Systems

A Student's View

As a physical therapy assistant, I am often asked, "What is causing this pain?" and "what can I do to treat it?" This is a common question posed to any of us who work in the health care field. A person who suffers from low back problems want to know why his toes are always numb. A stroke patient has lost all muscle strength and control on his left side and wants to get it back. A person whose facial muscles are left flaccid and useless due to a virus has come to me for answers.

These are the types of situations I deal with everyday while working at a privately owned out-patient physical therapy practice; situations which can be explained through motor and sensory tracts; cases which utilize stimulation to facilitate muscle contraction; or posture and exercise to decrease pain.

Education plays an extremely important role in any rehabilitation process and can often be the one thing patients need most. A picture is worth a thousand words and with the clear, concise illustrations in this chapter, I often utilize this book at our facility as a vital tool to help patients understand the rehabilitation process.

Mike Grosse,
Licensed Physical Therapy Assistant
Physical Therapy of Clifton

SENSORY, MOTOR, AND INTEGRATIVE SYSTEMS AND HOMEOSTASIS

The sensory and motor pathways of the body provide routes for input into the brain and spinal cord and output to targeted organs for response, such as muscle contraction.

ENERGIZE YOUR STUDY
FOUNDATIONS CD

Anatomy Overview
• The Nervous System

Animation
• Systems Contributions to Homeostasis

Concepts and Connections and Exercises reinforce your understanding

www.wiley.com/college/apcentral

INSIGHTS AND EXPLORATIONS

You have probably heard about people who lose a limb but still sense that it is still there — a phantom limb. Imagine having lost your arm but still feel the sensation of reaching out to pick up perhaps a glass of water or even worse to feel pain in a limb that you no longer have. Come explore some theories about this phenomenon in a web-based exercise.

The previous three chapters described the organization of the nervous system. In this chapter, we explore the levels and components of sensation. We also examine the pathways that convey somatic sensory nerve impulses from the body to the brain, and the pathways that carry impulses from the brain to skeletal muscles to produce movements. As sensory impulses reach the CNS, they become part of a large pool of sensory input. However, not every bit of input to the CNS elicits a response.

Rather, each piece of incoming information is integrated with other arriving and previously stored information. Integration occurs at many places along pathways in the CNS, such as the spinal cord, brain stem, cerebellum, basal ganglia, and cerebral cortex. The motor responses that govern muscle contraction also are modified at several of these levels. To conclude this chapter, we introduce two of the brain's integrative functions: (1) wakefulness and sleep and (2) learning and memory.

SENSATION

OBJECTIVES

• Define sensation, and discuss the components of sensation.
• Describe the different ways to classify sensory receptors.

In its broadest definition, **sensation** is the conscious or subconscious awareness of external or internal stimuli. The nature of the sensation and the type of reaction generated vary according to the ultimate CNS destination of nerve impulses that convey sensory information. In the spinal cord, sensory impulses are the input part of spinal reflexes such as the stretch reflex. Sensory impulses that reach the lower brain stem elicit more complex reflexes, such as changes in heart rate or breathing rate. Sensory impulses reaching the thalamus can provide a crude awareness of the location of a stimulus and the type of sensation, such as touch, pain, hearing, or taste. When sensory impulses reach the cerebral cortex, we can precisely locate and identify specific sensations. In addition, memories of previous sensations are stored in the cerebral cortex.

Perception is the conscious awareness and the interpretation of meaning of sensations. We have no perception of some sensory impulses because they never reach the thalamus and cerebral cortex. For example, certain sensory receptors constantly monitor the pressure of blood in blood vessels. Because the nerve impulses conveying blood pressure information propagate to the cardiovascular center in the medulla oblongata rather than to the cerebral cortex, however, blood pressure is not consciously perceived.

Sensory Modalities

Each unique type of sensation—such as touch, pain, vision, or hearing—is called a **sensory modality.** A given sensory neuron carries information for only one sensory modality. Neurons relaying impulses for touch to the somatosensory area of the cerebral cortex, for example, do not also transmit impulses for pain. Likewise, nerve impulses from the eyes are perceived as sight, whereas those from the ears are perceived as sounds.

The different sensory modalities can be grouped into two classes: general senses and special senses.

1. The **general senses** include both **somatic senses** (*somat-* = of the body) and **visceral senses.** Somatic sensory modalities include tactile sensations (touch, pressure, and vibration), thermal sensations (warm and cold), pain sensations, and proprioceptive sensations, which allow perception of both the static (nonmoving) positions of limbs and body parts (joint and muscle position sense) and movements of the limbs and head. Visceral sensations provide information about conditions within internal organs.

2. The **special senses** include the sensory modalities of smell, taste, vision, hearing, and equilibrium or balance.

In this chapter we discuss the somatic senses and visceral pain. The special senses are the focus of Chapter 16. Visceral senses are discussed further in Chapter 17 and in association with individual organs in later chapters.

The Process of Sensation

The process of sensation begins in a **sensory receptor,** which is either a specialized cell or the dendrites of a sensory neuron. Each sensory receptor monitors a particular condition in the internal or external environment but is sensitive to stimuli for one sensory modality only. Thus, a given sensory receptor responds vigorously to one particular kind of **stimulus,** a change in the environment that can activate certain sensory receptors. A sensory receptor responds only weakly or not at all to other kinds of stimuli. This characteristic of sensory receptors is known as **selectivity.**

For a sensation to arise, the following four events typically occur:

1. *Stimulation of the sensory receptor.* An appropriate stimulus must occur within the sensory receptor's *receptive field,* that is, the body region where stimulation elicits a response.

2. *Transduction of the stimulus.* A sensory receptor *transduces* (converts) energy in a stimulus into a graded potential. Recall that graded potentials vary in amplitude (size), depending on the strength of the stimulus that evokes them, and are not propagated. (See page 398 to review the differences between action potentials and graded potentials.) Each type of sensory receptor exhibits selectivity: It can transduce only one kind of stimulus. For example, odorant molecules in the air stimulate olfactory (smell) receptors in the nose, which transduce the molecules' chemical energy into electrical energy in the form of a graded potential.

3. *Generation of impulses.* When a graded potential in a sensory neuron reaches threshold, it triggers one or more nerve impulses, which then propagate toward the CNS. Sensory neurons that conduct impulses from the PNS into the CNS are called **first-order neurons.**

4. *Integration of sensory input.* A particular region of the CNS receives and integrates the sensory nerve impulses. Conscious sensations or perceptions are integrated in the cerebral cortex. You seem to see with your eyes, hear with your ears, and feel pain in an injured part of your body because sensory impulses from each part of the body arrive in a specific region of the cerebral cortex, which interprets the sensation as coming from the stimulated sensory receptors.

Sensory Receptors

Types of Sensory Receptors

Several structural and functional characteristics of sensory receptors can be used to group them into different classes. On a microscopic level, sensory receptors may be (1) free nerve endings, (2) encapsulated nerve endings of first-order sensory neurons, or (3) separate cells that synapse with first-order sensory neurons (Figure 15.1).

Free nerve endings are bare dendrites; they lack any structural specializations that can be seen under a light microscope (Figure 15.1a). Receptors for pain, thermal, tickle, itch, and some touch sensations are free nerve endings. Receptors for

Figure 15.1   Types of sensory receptors and their relationship to first-order sensory neurons. (a) Free nerve endings, in this case, a cold-sensitive receptor. These endings are bare dendrites (no apparent structural specialization) of first-order neurons. (b) An encapsulated nerve ending, in this case a pressure-sensitive receptor. Encapsulated nerve endings are dendrites of first-order neurons. (c) A separate receptor cell—here, a gustatory (taste) receptor—and its synapse with a first-order neuron.

Free nerve endings and encapsulated nerve endings produce generator potentials that trigger nerve impulses in first-order neurons. Separate sensory receptors produce a receptor potential that causes release of neurotransmitter. The neurotransmitter then triggers nerve impulses in a first-order neuron.

Which senses are served by receptors that are separate cells?

other somatic and visceral sensations, such as touch, pressure, and vibration, are **encapsulated nerve endings.** Their dendrites are enclosed in a connective tissue capsule that has a distinctive microscopic structure—for example, lamellated (pacinian) corpuscles (Figure 15.1b). The different types of capsules enhance the sensitivity or specificity of the receptor.

Sensory receptors for some special senses are specialized, **separate cells** that synapse with first-order sensory neurons. These include *hair cells* for hearing and equilibrium in the inner ear and *gustatory receptors* in taste buds (Figure 15.1c). *Photoreceptors* in the retina of the eye for vision also are specialized, separate cells.

Sensory receptors produce two different kinds of graded potentials—generator potentials and receptor potentials—in response to a stimulus. When stimulated, the dendrites of free nerve endings, encapsulated nerve endings, and the receptive part of olfactory receptors produce **generator potentials** (Figure 15.1a, b). When a generator potential is large enough to reach threshold, it triggers one or more nerve impulses in the axon of a first-order sensory neuron. The resulting nerve impulse propagates along the axon into the CNS. Thus, generator potentials generate action potentials.

Hair cells of the inner ear, gustatory receptors, and photoreceptors, by contrast, do not generate action potentials. Their graded potentials, termed **receptor potentials,** trigger release of neurotransmitter through exocytosis of synaptic vesicles (Figure 15.1c). The neurotransmitter molecules liberated from synaptic vesicles diffuse across the synaptic cleft and produce a postsynaptic potential (PSP) in the first-order neuron. In turn, the PSP may trigger one or more nerve impulses, which propagate along the axon into the CNS.

The amplitude of both generator potentials and receptor potentials varies with the intensity of the stimulus. An intense stimulus produces a large generator potential or receptor potential, whereas a weak stimulus elicits a small one. Similarly, large generator potentials or receptor potentials trigger impulses at high frequencies in the first-order neuron, whereas small generator potentials or receptor potentials trigger impulses at lower frequencies.

Another way to group sensory receptors is based on the location of the receptors and the origin of the stimuli that activate them. **Exteroceptors** (EKS-ter-ō-sep′-tors) are located at or near the external surface of the body; they are sensitive to stimuli originating outside the body and provide information about the external environment. The sensations of hearing, vision, smell, taste, touch, pressure, vibration, temperature, and pain are conveyed by exteroceptors. **Interoceptors** (IN-ter-ō-sep′-tors) are located in blood vessels, visceral organs, muscles, and the nervous system and monitor conditions in the *internal* environment. The nerve impulses produced by interoceptors usually are not consciously perceived; occasionally, however, activation of interoceptors by strong stimuli may be felt as pain or pressure. **Proprioceptors** (PRŌ-prē-ō-sep′-tors; *proprio-* = one's own) are located in muscles, tendons, joints, and the inner ear. They provide information about body position, muscle length and tension, and the position and movement of our joints.

A third way to group sensory receptors is according to the type of stimulus they detect. Most stimuli are in the form of electromagnetic energy, such as light or heat; mechanical energy, such as sound waves or pressure changes; or chemical energy, such as in a molecule of glucose. Table 15.1 summarizes the classification of sensory receptors and lists different types of sensory receptors according to the type of stimulus they detect.

Table 15.1	Classification of Sensory Receptors
Basis of Classification	**Description**
Microscopic Features	
Free nerve endings	Bare dendrites associated with pain, thermal, tickle, itch, and some touch sensations.
Encapsulated nerve endings	Dendrites enclosed in a connective tissue capsule, such as a corpuscle of touch.
Separate cells	Receptor cell synapses with first-order sensory neurons; located in the retina of the eye (photoreceptors), inner ear (hair cells), and taste buds of the tongue (gustatory receptor cells).
Receptor Location and Activating Stimuli	
Exteroceptors	Located at or near body surface; sensitive to stimuli originating outside body; provide information about external environment; convey visual, smell, taste, touch, pressure, vibration, thermal, and pain sensations.
Interoceptors	Located in blood vessels, visceral organs, and nervous system; provide information about internal environment; impulses produced usually are not consciously perceived but occasionally may be felt as pain or pressure.
Proprioceptors	Located in muscles, tendons, joints, and inner ear; provide information about body position, muscle length and tension, position and motion of joints, and equilibrium (balance).
Type of Stimulus Detected	
Mechanoreceptors	Detect mechanical pressure; provide sensations of touch, pressure, vibration, proprioception, and hearing and equilibrium; also monitor stretching of blood vessels and internal organs.
Thermoreceptors	Detect changes in temperature.
Nociceptors	Respond to stimuli resulting from physical or chemical damage to tissue.
Photoreceptors	Detect light that strikes the retina of the eye.
Chemoreceptors	Detect chemicals in mouth (taste), nose (smell), and body fluids.
Osmoreceptors	Sense the osmotic pressure of body fluids.

Adaptation in Sensory Receptors

A characteristic of most sensory receptors is **adaptation,** in which the generator potential or receptor potential decreases in amplitude during a maintained, constant stimulus. As a result, the frequency of nerve impulses in the first-order neuron also decreases during a prolonged stimulus. Because of adaptation, the perception of a sensation may fade or disappear even though the stimulus persists. For example, when you first step into a hot shower, the water may feel very hot, but soon the sensation decreases to one of comfortable warmth even though the stimulus (the high temperature of the water) does not change.

Receptors vary in how quickly they adapt. **Rapidly adapting receptors** adapt very quickly. They are specialized for signaling *changes* in a stimulus. Receptors associated with pressure, touch, and smell are rapidly adapting. **Slowly adapting receptors,** by contrast, adapt slowly and continue to trigger nerve impulses as long as the stimulus persists. Slowly adapting receptors monitor stimuli associated with pain, body position, and chemical composition of the blood.

▶ CHECKPOINT

1. How is a sensation different from a perception?

2. What is a sensory modality?

3. How are generator potentials and receptor potentials similar and different?

4. What is the difference between rapidly adapting and slowly adapting receptors?

SOMATIC SENSATIONS

▶ OBJECTIVES

- **Describe the location and function of the receptors for tactile, thermal, and pain sensations.**
- **Identify the receptors for proprioception and describe their functions.**

Somatic sensations arise from stimulation of sensory receptors embedded in the skin or subcutaneous layer; in mucous membranes of the mouth, vagina, and anus; in muscles, tendons, and joints; and in the inner ear. The sensory receptors for somatic sensations are distributed unevenly—some parts of the body surface are densely populated with receptors whereas other parts contain only a few. The areas with the highest density of somatic sensory receptors are the tip of the tongue, the lips, and the fingertips. Somatic sensations that arise from stimulating the skin surface are **cutaneous sensations** (kū-TĀ-nē-us; *cutane-* = skin). Somatic sensations are of four modalities: tactile, thermal, pain, and proprioceptive.

Tactile Sensations

The **tactile sensations** (TAK-tīl; *tact-* = touch) are those of touch, pressure, vibration, itch, and tickle. Although we perceive differences among these sensations, they arise by activation of some of the same types of receptors. Several types of encapsulated mechanoreceptors attached to large-diameter, myelinated A fibers mediate sensations of touch, pressure, and vibration. Other touch sensations, as well as itch and tickle sensations, are detected by free nerve endings attached to small-diameter, unmyelinated C fibers. Recall that larger diameter, myelinated axons propagate nerve impulses more rapidly than do smaller diameter, unmyelinated axons. Tactile receptors in the skin or subcutaneous layer include corpuscles of touch, hair root plexuses, type I and II cutaneous mechanoreceptors, lamellated corpuscles, and free nerve endings (Figure 15.2).

Sensations of **touch** generally result from stimulation of tactile receptors in the skin or subcutaneous layer. **Crude touch** is the ability to perceive that something has contacted the skin, even though its exact location, shape, size, or texture cannot be determined. **Fine touch** provides specific information about a touch sensation, such as exactly what point on the body is touched plus the shape, size, and texture of the source of stimulation.

There are two types of rapidly adapting touch receptors. **Corpuscles of touch,** or **Meissner corpuscles** (MĪS-ner), are receptors for fine touch that are located in the dermal papillae of hairless skin. Each corpuscle is an egg-shaped mass of dendrites enclosed by a capsule of connective tissue. Because corpuscles of touch are rapidly adapting receptors, they generate nerve impulses mainly at the onset of a touch. They are abundant in the hands, eyelids, tip of the tongue, lips, nipples, soles, clitoris, and tip of the penis. **Hair root plexuses** are rapidly adapting touch receptors found in hairy skin; they consist of free nerve endings wrapped around hair follicles. Hair root plexuses detect movements on the skin surface that disturb hairs. For example, a flea landing on a hair causes movement of the hair shaft that stimulates the free nerve endings.

There also are two types of slowly adapting touch receptors. **Type I cutaneous mechanoreceptors,** also known as **Merkel discs,** function in fine touch. Merkel discs are saucer-shaped, flattened free nerve endings that contact Merkel cells of the stratum basale (see Figure 5.2 on page 142). These mechanoreceptors are plentiful in the fingertips, hands, lips, and external genitalia. **Type II cutaneous mechanoreceptors,** or **Ruffini corpuscles,** are elongated, encapsulated receptors located deep in the dermis, and in ligaments and tendons. Present in the hands and abundant on the soles, they are most sensitive to stretching that occurs as digits or limbs are moved.

Pressure is a sustained sensation that is felt over a larger area than is touch. It occurs with deformation of deeper tissues. Receptors that contribute to sensations of pressure include corpuscles of touch, type I mechanoreceptors, and lamellated corpuscles. **Lamellated,** or **Pacinian, corpuscles** (pa-SIN-ē-an) are large oval structures composed of a multilayered connective tissue capsule that encloses a dendrite. Like corpuscles of touch, lamellated corpuscles adapt rapidly. They are widely distributed in the body: in the dermis and subcutaneous layer; in submucosal tissues that underlie mucous and serous membranes;

Figure 15.2 Structure and location of sensory receptors in the skin and subcutaneous layer.

The somatic sensations of touch, pressure, vibration, warmth, cold, and pain arise from sensory receptors in the skin, subcutaneous layer, and mucous membranes.

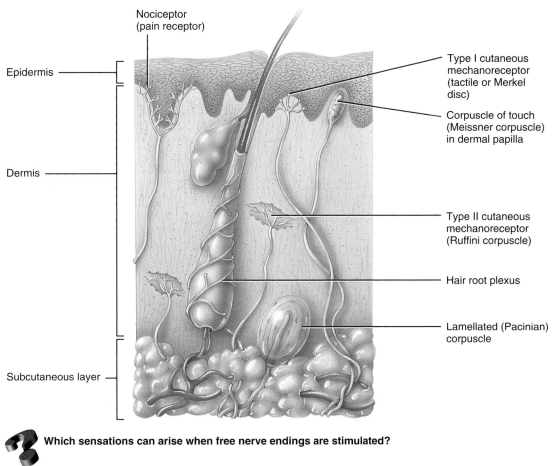

Which sensations can arise when free nerve endings are stimulated?

around joints, tendons, and muscles; in the periosteum; and in the mammary glands, external genitalia, and certain viscera, such as the pancreas and urinary bladder.

Sensations of **vibration** result from rapidly repetitive sensory signals from tactile receptors. The receptors for vibration sensations are corpuscles of touch and lamellated corpuscles. Whereas corpuscles of touch can detect lower-frequency vibrations, lamellated corpuscles detect higher-frequency vibrations.

The **itch** sensation results from stimulation of free nerve endings by certain chemicals, such as bradykinin, often because of a local inflammatory response. Free nerve endings and lamellated corpuscles are thought to mediate the **tickle** sensation. This intriguing sensation typically arises only when someone else touches you, not when you touch yourself. The explanation of this puzzle seems to lie in the impulses that conduct to and from the cerebellum when you are moving your fingers and touching yourself that don't occur when someone else is tickling you.

 Phantom Limb Sensation

Patients who have had a limb amputated may still experience sensations such as itching, pressure, tingling, or pain as if the limb were still there. This phenomenon is called **phantom limb sensation.** One explanation for phantom limb sensations is that the cerebral cortex interprets impulses arising in the proximal portions of sensory neurons that previously carried impulses from the limb as coming from the nonexistent (phantom) limb. Another explanation for phantom limb sensations is that the brain itself contains networks of neurons that generate sensations of body awareness. In this view, neurons in the brain that previously received sensory impulses from the missing limb are still active, giving rise to false sensory perceptions. ∎

Thermal Sensations

Thermoreceptors are free nerve endings that have receptive fields about 1 mm in diameter on the skin surface. Two distinct **thermal sensations**—coldness and warmth—are mediated by different receptors. **Cold receptors** are located in the stratum basale of the epidermis and are attached to medium-diameter, myelinated A fibers, although a few connect to small-diameter, unmyelinated C fibers. Temperatures between 10° and 40°C (50–105°F) activate cold receptors. Because skin temperature is about 34°C, the sensation of coolness that occurs when the receptive field of a cold receptor is touched by a small probe having a *higher* temperature is termed *paradoxical cold*. **Warm receptors** are located in the dermis and are attached to small-diameter, unmyelinated C fibers; they are activated by temperatures between 32° and 48°C (90–118°F). Cold and warm receptors both adapt rapidly at the onset of a stimulus but continue to generate impulses at a lower frequency throughout a prolonged stimulus. Temperatures below 10°C and above 48°C stimulate mainly nociceptors, rather than thermoreceptors, producing painful sensations.

Pain Sensations

Pain is indispensable for survival. It serves a protective function by signaling the presence of noxious, tissue-damaging conditions. From a medical standpoint, the subjective description and indication of the location of pain may help pinpoint the underlying cause of disease.

Nociceptors (*noci-* = harmful), the receptors for pain, are free nerve endings found in every tissue of the body except the brain (Figure 15.2). Intense thermal, mechanical, or chemical stimuli can activate nociceptors. Tissue irritation or injury releases chemicals such as prostaglandins, kinins, and potassium ions (K^+) that stimulate nociceptors. Pain may persist even after a pain-producing stimulus is removed because pain-mediating chemicals linger, and because nociceptors exhibit very little adaptation. Conditions that elicit pain include excessive distention (stretching) of a structure, prolonged muscular contractions, muscle spasms, or ischemia (inadequate blood flow to an organ).

Types of Pain

There are two types of pain: fast and slow. The perception of **fast pain** occurs very rapidly, usually within 0.1 second after a stimulus is applied, because the nerve impulses propagate along medium-diameter, myelinated A-delta fibers. This type of pain is also known as acute, sharp, or pricking pain. The pain felt from a needle puncture or knife cut to the skin are examples of fast pain. Fast pain is not felt in deeper tissues of the body. The perception of **slow pain,** by contrast, begins a second or more after a stimulus is applied. It then gradually increases in intensity over a period of several seconds or minutes. Impulses for slow pain conduct along small-diameter, unmyelinated C fibers. This type of pain, which may be excruciating, is also referred to as chronic, burning, aching, or throbbing pain. Slow pain can occur both in the skin and in deeper tissues or internal organs. An example is the pain associated with a toothache. You can perceive the difference in onset of these two types of pain best when you injure a body part that is far from the brain because the conduction distance is long. When you stub a toe, for example, you first feel the sharp sensation of fast pain and then feel the slower, aching sensation of slow pain.

Pain that arises from stimulation of receptors in the skin is called **superficial somatic pain,** whereas stimulation of recep-

tors in skeletal muscles, joints, tendons, and fascia causes **deep somatic pain. Visceral pain** results from stimulation of nociceptors in visceral organs. If stimulation is *diffuse* (involves large areas), visceral pain can be severe. Diffuse stimulation of visceral nociceptors might result from distention or ischemia of an internal organ. For example, a kidney stone or a gallstone might cause severe pain by obstructing and distending a ureter or bile duct.

Localization of Pain

Fast pain is very precisely localized to the stimulated area. For example, if someone pricks you with a pin, you know exactly which part of your body was stimulated. Slow somatic pain also is well localized but more diffuse; it usually appears to come from a larger area of the skin. In some instances of visceral slow pain, the affected area is where the pain is felt. If the pleural membranes around the lungs are inflamed, for example, you experience chest pain.

In many instances of visceral pain, the pain is felt in or just deep to the skin that overlies the stimulated organ, or in a surface area far from the stimulated organ. This phenomenon is called **referred pain.** Figure 15.3 shows skin regions to which visceral pain may be referred. In general, the visceral organ involved and the area to which the pain is referred are served by the same segment of the spinal cord. For example, sensory fibers from the heart, the skin over the heart, and the skin along the medial aspect of the left arm enter spinal cord segments T1 to T5. Thus, the pain of a heart attack typically is felt in the skin over the heart and along the left arm.

⚕ Analgesia: Relief from Pain

Pain sensations sometimes occur out of proportion to minor damage, persist chronically due to an injury, or even appear for no obvious reason rather than warning of actual or impending damage. In such cases, **analgesia** (*an-* = without; *-algesia* = pain) or pain relief is needed. Analgesic drugs such as aspirin and ibuprofen (for example, Advil) block formation of prostaglandins, which stimulate nociceptors. Local anesthetics, such as Novocaine, provide short-term pain relief by blocking conduction of nerve impulses along the axons of first-order pain neurons. Morphine and other opiate drugs alter the quality of pain perception in the brain; pain is still sensed, but it is no longer perceived as being so noxious. ∎

Proprioceptive Sensations

Proprioceptive sensations allow us to know where our head and limbs are located and how they are moving even if we are not looking at them. These sensations arise in receptors termed **proprioceptors.** Those proprioceptors imbedded in muscles and tendons inform us of the degree to which muscles are contracted, the amount of tension on tendons, and the positions of

Figure 15.3 Distribution of referred pain. The colored parts of the diagrams indicate skin areas to which visceral pain is referred.

Nociceptors are present in almost every tissue of the body.

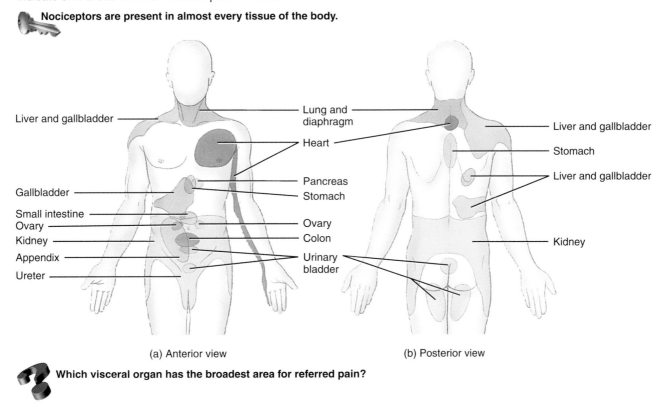

(a) Anterior view (b) Posterior view

Which visceral organ has the broadest area for referred pain?

joints. Hair cells of the vestibular apparatus in the inner ear are proprioceptors that monitor the orientation of the head relative to the ground and head position during movements. They provide information for maintaining balance and equilibrium (described in Chapter 16). Because proprioceptors adapt slowly and only slightly, the brain continually receives nerve impulses related to the position of different body parts and makes adjustments to ensure coordination.

Proprioceptors also convey the rate of movement of one body part in relation to others, so that we can walk, type, or dress without using our eyes. **Kinesthesia** (kin′-es-THĒ-zē-a; *kin-* = motion; *-esthesia* = perception) is the perception of body movements. Proprioceptive sensations also allow us to estimate the weight of objects and determine the muscular effort necessary to perform a task. For example, as you pick up a bag you quickly realize whether it contains popcorn or books, and you then exert the correct amount of effort needed to lift it. Here we discuss three types of proprioceptors: muscle spindles within skeletal muscles, tendon organs within tendons, and joint kinesthetic receptors within synovial joint capsules.

Muscle Spindles

Muscle spindles are the proprioceptors in skeletal muscles that participate in stretch reflexes (shown in Figure 13.6 on page 428).

By adjusting how vigorously a muscle spindle responds to stretching, the brain sets an overall level of **muscle tone,** the small degree of contraction that is present while the muscle is at rest.

Each **muscle spindle** consists of a several slowly adapting sensory nerve endings that wrap around 3 to 10 specialized muscle fibers, called **intrafusal muscle fibers** (*intrafusal* = within a spindle). A connective tissue capsule encloses the assembly of sensory nerve endings and intrafusal fibers and anchors the spindle to endomysium and perimysium (Figure 15.4). Muscle spindles are interspersed among ordinary skeletal muscle fibers and aligned parallel to them. In muscles that produce finely controlled movements, such as movements of the fingers or eyes, muscle spindles are plentiful. Muscles involved in coarser but more forceful movements, for instance, the quadriceps femoris and hamstrings muscles of the thigh, have fewer muscle spindles. The only skeletal muscles that lack spindles are the tiny muscles of the middle ear.

The main function of muscle spindles is to measure *muscle length*—how much a muscle is being stretched. Either sudden or prolonged stretching of the central areas of the intrafusal muscle fibers stimulates the sensory nerve endings. The resulting nerve impulses propagate into the CNS. Quickly, information from muscle spindles arrives at the somatic sensory areas of the cerebral cortex, which allows conscious perception of limb positions and movements. At the same time, impulses from muscle

Figure 15.4 Two types of proprioceptors: a muscle spindle and a tendon organ. In muscle spindles, which monitor changes in skeletal muscle length, sensory nerve endings wrap around the central portion of intrafusal muscle fibers. In tendon organs, which monitor the force of muscle contraction, sensory nerve endings are activated by increasing tension on a tendon.

🔑 **Proprioceptors provide information about body position and movement.**

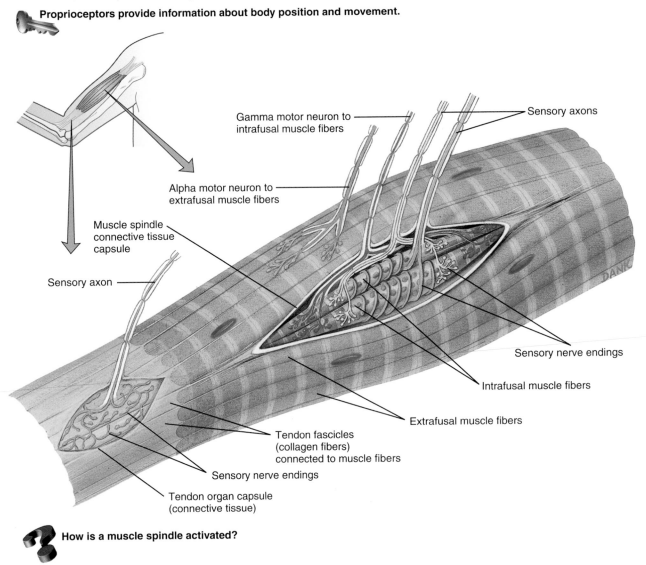

❓ **How is a muscle spindle activated?**

spindles also pass to the cerebellum, where the input is used to coordinate muscle contractions.

In addition to their sensory nerve endings near the middle of intrafusal fibers, muscle spindles contain motor neurons called **gamma motor neurons.** These motor neurons terminate near both ends of the intrafusal fibers. As the two ends of an intrafusal fiber both contract, they exert a "pull" on the sensory nerve endings in the middle. The brain regulates sensitivity of muscle spindles through pathways to the gamma motor neurons. As the frequency of impulses in its gamma motor neuron increases, a muscle spindle becomes more sensitive to stretching of its midregion.

Surrounding muscle spindles are ordinary skeletal muscle fibers, called **extrafusal muscle fibers** (*extrafusal* = outside a

spindle), which are innervated by large-diameter A fibers called **alpha motor neurons.** The cell bodies of both gamma and alpha motor neurons are located in the anterior gray horn of the spinal cord (or in the brainstem for muscles in the head). During the stretch reflex, impulses in muscle spindle sensory axons propagate into the spinal cord and brain stem and activate alpha motor neurons that innervate extrafusal muscle fibers in the same muscle. In this way, activation of its muscle spindles causes contraction of a skeletal muscle.

Tendon Organs

Tendon organs are located at the junction of a tendon and a muscle. By initiating tendon reflexes (see Figure 13.7 on page 430), tendon organs protect tendons and their associated muscles from damage due to excessive tension. Each **tendon organ** consists of a thin capsule of connective tissue that encloses a few tendon fascicles (bundles of collagen fibers) (Figure 15.4). Penetrating the capsule are one or more sensory nerve endings that entwine among and around the collagen fibers. When tension is applied to a tendon, the tendon organs generate nerve impulses that propagate into the CNS, providing information about changes in *muscle force*. Tendon reflexes decrease muscle force by causing muscle relaxation.

Joint Kinesthetic Receptors

Several types of **joint kinesthetic receptors** are present within and around the articular capsules of synovial joints. Free nerve endings and type II cutaneous mechanoreceptors (Ruffini corpuscles) in the capsules of joints respond to pressure. Small lamellated (pacinian) corpuscles in the connective tissue outside articular capsules respond to acceleration and deceleration of joints during movement. Articular ligaments contain receptors similar to tendon organs that adjust reflex inhibition of the adjacent muscles when excessive strain is placed on the joint.

Table 15.2 summarizes the somatic sensory receptors and the sensations they convey.

▶ C H E C K P O I N T

5. Which somatic sensory receptors are encapsulated?
6. Which somatic sensory receptors adapt slowly, and which adapt rapidly?
7. Which somatic sensory receptors mediate fine touch sensations?
8. How does fast pain differ from slow pain?
9. What is referred pain, and how is it useful in diagnosing internal disorders?
10. What aspects of muscle function are monitored by muscle spindles and tendon organs?

Table 15.2 Summary of Receptors for Somatic Sensations

Receptor Type	Receptor Structure and Location	Sensations	Adaptation Rate
Tactile Receptors			
Corpuscles of touch (Meissner corpuscles)	Capsule surrounds mass of dendrites in dermal papillae of hairless skin.	Fine touch, pressure, and slow vibrations.	Rapid.
Hair root plexuses	Free nerve endings wrapped around hair follicles in skin.	Touch.	Rapid.
Type I cutaneous mechanoreceptors (tactile or Merkel disc)	Saucer-shaped free nerve endings make contact with Merkel cells in epidermis.	Touch and pressure.	Slow.
Type II cutaneous mechanoreceptors (Ruffini corpuscles)	Elongated capsule surrounds dendrites deep in dermis and in ligaments and tendons.	Stretching of skin.	Slow.
Lamellated (pacinian) corpuscles	Oval, layered capsule surrounds dendrites; present in dermis and subcutaneous layer, submucosal tissues, joints, periosteum, and some viscera.	Pressure, fast vibrations, and tickling.	Rapid.
Itch and tickle receptors	Free nerve endings and lamellated corpuscles in skin and mucous membranes.	Itching and tickling.	Both slow and rapid.
Thermoreceptors			
Warm receptors and **cold receptors**	Free nerve endings in skin and mucous membranes of mouth, vagina, and anus.	Warmth or cold.	Rapid initially, then slow.
Pain Receptors			
Nociceptors	Free nerve endings in every tissue of the body except the brain.	Pain.	Slow.
Proprioceptors			
Muscle spindles	Sensory nerve endings wrap around central area of encapsulated intrafusal muscle fibers within most skeletal muscles.	Muscle length.	Slow.
Tendon organs	Capsule encloses collagen fibers and sensory nerve endings at junction of tendon and muscle.	Muscle tension.	Slow.
Joint kinesthetic receptors	Lamellated corpuscles, Ruffini corpuscles, tendon organs, and free nerve endings.	Joint position and movement.	Rapid.

SOMATIC SENSORY AND MOTOR MAPS IN THE CEREBRAL CORTEX

OBJECTIVE

• **Describe the location and functions of the primary somatosensory areas and primary motor areas in the cerebral cortex.**

Specific areas of the cerebral cortex receive somatic sensory input from particular parts of the body and other areas of the cerebral cortex provide output instructions for movement of particular parts of the body. The somatic sensory map and the somatic motor map relate body parts to these cortical areas.

Precise localization of somatic sensations occurs when nerve impulses arrive at the **primary somatosensory area**

(areas 1, 2, and 3 in Figure 14.15 on page 474), which occupies the postcentral gyri of the parietal lobes of the cerebral cortex. Each region in this area receives sensory input from a different part of the body. Figure 15.5a maps the destination of somatic sensory signals from different parts of the left side of the body in the somatosensory area of the right cerebral hemisphere. The left cerebral hemisphere has a similar primary somatosensory area that receives sensory input from the right side of the body.

Note that some parts of the body—chiefly the lips, face, tongue, and thumb—provide input to large regions in the somatosensory area. Other parts of the body, such as the trunk and lower limbs, project to much smaller cortical regions. The relative sizes of these regions in the somatosensory area are proportional to the number of specialized sensory receptors

Figure 15.5 Somatic sensory and somatic motor maps in the cerebral cortex. (a) Primary somatosensory area (postcentral gyrus) and (b) primary motor area (precentral gyrus) of the right cerebral hemisphere. The left hemisphere has similar representation. (After Penfield and Rasmussen.)

Each point on the body surface maps to a specific region in both the primary somatosensory area and the primary motor area.

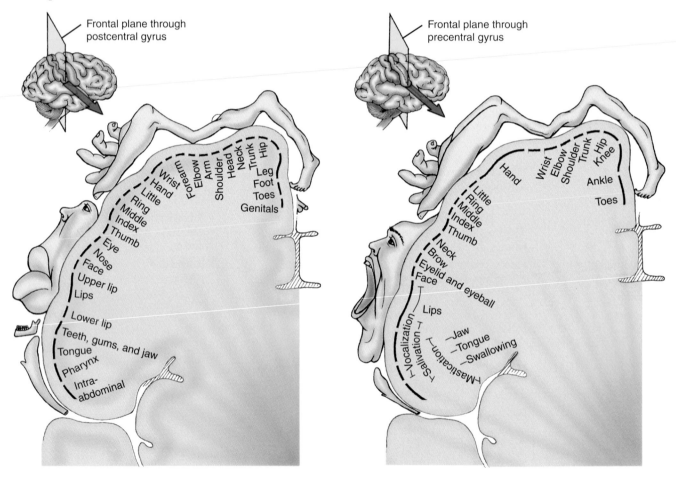

(a) Frontal section of primary somatosensory area in right cerebral hemisphere

(b) Frontal section of primary motor area in right cerebral hemisphere

How do the somatosensory and motor representations compare for the hand, and what does this difference imply?

within the corresponding part of the body. For example, there are many sensory receptors in the skin of the lips but few in the skin of the trunk. The size of the cortical region that represents a body part may expand or shrink somewhat, depending on the quantity of sensory impulses received from that body part. For example, people who learn to read Braille eventually have a larger representation of the fingertips in the somatosensory area.

Control of body movements occurs via neural circuits in several regions of the brain. The **primary motor area** (area 4 in Figure 14.15 on page 474), located in the precentral gyrus of the frontal lobe (Figure 15.5b), is a major control region for planning and initiating voluntary movements. The adjacent **premotor area** (area 6) and even the primary somatosensory area in the postcentral gyrus also contribute axons to the descending motor pathways. As is true for somatic sensory representation in the somatosensory area, different muscles are represented unequally in the primary motor area. The cortical area devoted to a muscle is proportional to the number of motor units in that muscle. Muscles in the thumb, fingers, lips, tongue, and vocal cords have large representations, whereas the trunk has a much smaller representation. By comparing Figures 15.5a and 15.5b, you can see that somatosensory and somatic motor representations are similar but not identical for most parts of the body.

▶ C H E C K P O I N T

11. How are various parts of the body represented in the somatosensory cortex? Which body parts have the largest representation?

12. What is the significance of the genitals being represented in the primary somatosensory area but not in the primary motor area of the cortex? (*Hint: Genital tissues contain smooth muscle but no skeletal muscle.*)

SOMATIC SENSORY PATHWAYS

▶ O B J E C T I V E

• **Describe the neuronal components and functions of the posterior column–medial lemniscus, the anterolateral, and the spinocerebellar pathways.**

Somatic sensory pathways relay information from somatic sensory receptors to the primary somatosensory area in the cerebral cortex and to the cerebellum. The pathways to the cerebral cortex consist of thousands of sets of three neurons: a first-order neuron, a second-order neuron, and a third-order neuron.

1. **First-order neurons** conduct impulses from somatic receptors into the brain stem or spinal cord. From the face, mouth, teeth, and eyes, somatic sensory impulses propagate along *cranial nerves* into the brain stem. From the neck, body, and posterior aspect of the head, somatic sensory impulses propagate along *spinal nerves* into the spinal cord.

2. **Second-order neurons** conduct impulses from the brain stem and spinal cord to the thalamus. Axons of second-order neurons *decussate* (cross over to the opposite side) in the brain stem or spinal cord before ascending to the ventral posterior nucleus of the thalamus. Thus, all somatic sensory information from one side of the body reaches the thalamus on the opposite side.

3. **Third-order neurons** conduct impulses from the thalamus to the primary somatosensory area of the cortex on the same side.

Somatic sensory impulses entering the spinal cord ascend to the cerebral cortex via two general pathways: (1) the posterior column–medial lemniscus pathway and (2) the anterolateral (spinothalamic) pathways.

Posterior Column–Medial Lemniscus Pathway to the Cortex

Nerve impulses for conscious proprioception and most tactile sensations ascend to the cerebral cortex along the **posterior column–medial lemniscus pathway** (Figure 15.6a). The name of the pathway comes from the names of two white-matter tracts that convey the impulses: the posterior column of the spinal cord and the medial lemniscus of the brain stem.

First-order neurons extend from sensory receptors in the trunk and limbs into the spinal cord and ascend to the medulla oblongata on the same side of the body. The cell bodies of these first-order neurons are in the posterior (dorsal) root ganglia of spinal nerves. In the spinal cord, their axons form the **posterior (dorsal) columns,** which consist of two parts: the **gracile fasciculus** (GRAS-īl fa-SIK-ū-lus) and the **cuneate fasciculus** (KŪ-nē-āt). (See Table 15.3 on page 511.) The axon terminals synapse with second-order neurons whose cell bodies are located in the gracile nucleus or cuneate nucleus of the medulla. Impulses from the neck, upper limbs, and upper chest propagate along axons in the cuneate fasciculus and arrive at the cuneate nucleus. Impulses from the trunk and lower limbs propagate along axons in the gracile fasciculus and arrive at the gracile nucleus. Impulses for somatic sensations from the head arrive in the brain stem via sensory axons that are part of all cranial nerves except the olfactory nerve (cranial nerve I) and optic nerve (cranial nerve II). Most somatic sensory axons carrying impulses from the face are part of the trigeminal nerve (cranial nerve V).

The axon of the second-order neuron crosses to the opposite side of the medulla and enters the **medial lemniscus** (lem-NIS-kus = ribbon), a thin ribbonlike projection tract that extends from the medulla to the ventral posterior nucleus of the thalamus. In the thalamus, the axon terminals of second-order neurons synapse with third-order neurons, which project their axons to the primary somatosensory area of the cerebral cortex.

Impulses conducted along the posterior column–medial lemniscus pathway give rise to several highly evolved and refined sensations:

• **Fine touch** is the ability to recognize specific information about a touch sensation, such as what point on the body is touched plus the shape, size, and texture of the source of stimulation.

Figure 15.6 Somatic sensory pathways.

Nerve impulses propagate along sets of first-order, second-order, and third-order neurons to the primary somatosensory area (postcentral gyrus) of the cerebral cortex.

(a) Posterior column-medial lemniscus pathway

(b) Anterolateral (spinothalamic) pathways

What sorts of sensory deficits could be produced by damage to the right lateral spinothalamic tract?

- **Stereognosis** is the ability to recognize the size, shape, and texture of an object by feeling it. Examples are reading Braille or identifying a paperclip by feeling it.
- **Proprioception** is the awareness of the precise position of body parts, and **kinesthesia** is the awareness of directions of movement. Proprioceptors also allow **weight discrimination,** the ability to assess the weight of an object.
- **Vibratory sensations** arise when rapidly fluctuating touch stimuli are present.

Anterolateral Pathways to the Cortex

The **anterolateral** or **spinothalamic pathways** (spī-nō-tha-LAM-ik) relay impulses for pain, thermal, tickle, and itch sensations. In addition, they carry some impulses that give rise to a very crude, poorly localized touch or pressure sensation. Impulses for vibratory sensations travel by way of both the posterior column–medial lemniscus pathway and the anterolateral pathway.

Like the posterior column–medial lemniscus pathway, the anterolateral pathways are composed of three-neuron sets (Figure 15.6b). The first-order neuron connects a receptor of the neck, trunk, or limbs with the spinal cord. The cell body of the first-order neuron is in the posterior root ganglion. The axon terminals of the first-order neuron synapse with the second-order neuron, whose cell body is located in the posterior gray horn of the spinal cord.

The axon of the second-order neuron crosses to the opposite side of the spinal cord. Then, it passes upward to the brain stem in either the **lateral spinothalamic tract** or the **anterior spinothalamic tract.** The lateral spinothalamic tract conveys sensory impulses for pain and temperature, whereas the anterior spinothalamic tract conveys impulses for tickle, itch, crude touch, pressure, and vibrations. The axon of the second-order neuron ends in the ventral posterior nucleus of the thalamus,

where it synapses with the third-order neuron. The axon of the third-order neuron projects to the primary somatosensory area of the cerebral cortex on the same side as the thalamus.

Somatic Sensory Pathways to the Cerebellum

Two tracts in the spinal cord—the **posterior spinocerebellar tract** (spī-nō-ser-e-BEL-ar) and the **anterior spinocerebellar tract**—are the major routes whereby proprioceptive impulses reach the cerebellum. Although not consciously perceived, sensory impulses conveyed to the cerebellum along these two pathways are critical for posture, balance, and coordination of skilled movements. Table 15.3 summarizes the major somatic sensory tracts in the spinal cord and pathways in the brain.

Table 15.3 Major Somatic Sensory Tracts in the Spinal Cord and Pathways in the Brain	
Tract and Location	**Functions and Pathways**
Posterior column: Gracile fasciculus Cuneate fasciculus	**Posterior column:** Conveys nerve impulses for the sensations of fine touch, stereognosis, conscious proprioception, kinesthesia, weight discrimination, and vibration. Axons of first-order neurons from one side of the body form the posterior column on the same side and end in the medulla, where they synapse with dendrites and cell bodies of second-order neurons. Axons of second-order neurons decussate, enter the medial lemniscus on the opposite side, and extend to the thalamus. Third-order neurons transmit nerve impulses from the thalamus to the primary somatosensory cortex on the side opposite the site of stimulation.
Lateral spinothalamic tract Anterior spinothalamic tract	**Lateral spinothalamic:** Conveys nerve impulses for pain and thermal sensations. Axons of first-order neurons from one side of the body synapse with dendrites and cell bodies of second-order neurons in the posterior gray horn on the same side. Axons of second-order neurons decussate, enter the lateral spinothalamic tract on the opposite side, and extend to the thalamus. Third-order neurons transmit nerve impulses from the thalamus to the primary somatosensory cortex on the side opposite the site of stimulation. **Anterior spinothalamic:** Conveys nerve impulses for itch, tickle, pressure, vibrations, and crude, poorly localized touch sensations. Axons of first-order neurons from one side of the body synapse with dendrites and cell bodies of second-order neurons in the posterior gray horn on the same side. Axons of second-order neurons decussate, enter the anterior spinothalamic tract on the opposite side, and extend to the thalamus. Third-order neurons transmit nerve impulses from the thalamus to the primary somatosensory cortex on the side opposite the site of stimulation.
Posterior spinocerebellar tract Anterior spinocerebellar tract	**Anterior and posterior spinocerebellar:** Convey nerve impulses from proprioceptors in the trunk and lower limb of one side of the body to the same side of the cerebellum. The proprioceptive input informs the cerebellum of actual movements, allowing it to coordinate, smooth, and refine skilled movements and maintain posture and balance.

Syphilis

Syphilis is a sexually transmitted disease caused by the bacterium *Treponema pallidum*. If the infection is not treated, the third stage of syphilis typically causes debilitating neurological symptoms. A common outcome is progressive degeneration of the posterior portions of the spinal cord, including the posterior columns, posterior spinocerebellar tracts, and posterior roots. Somatic sensations are lost, and the person's gait becomes uncoordinated and jerky because proprioceptive impulses fail to reach the cerebellum. ■

▶ C H E C K P O I N T

13. What are three differences between the posterior column–medial lemniscus pathway and the anterolateral pathways?

14. What type of sensory information is carried in the spinocerebellar tracts, and what is its usefulness?

SOMATIC MOTOR PATHWAYS

▶ O B J E C T I V E S

• **Identify the locations and functions of lower motor neurons.**
• **Describe the four neural circuits that provide input to lower motor neurons.**
• **Compare the locations and functions of the direct and indirect motor pathways.**
• **Explain how the basal ganglia and cerebellum modulate movements.**

Neural circuits in the brain and spinal cord orchestrate all voluntary and involuntary movements. Ultimately, all excitatory and inhibitory signals that control movement converge on the alpha motor neurons that extend out of the brain stem and spinal cord to innervate skeletal muscles in the head and body. These neurons, also known as **lower motor neurons (LMNs),** have their cell bodies in the brain stem and spinal cord. Their axons extend from the motor nuclei of cranial nerves to skeletal muscles of the face and head and from all levels of the spinal cord to skeletal muscles of the limbs and trunk. Only lower motor neurons provide output from the CNS to skeletal muscle fibers. For this reason, they are also called the *final common pathway.*

Neurons in four distinct but highly interactive neural circuits, collectively termed the **somatic motor pathways,** participate in control of movement by providing input to lower motor neurons (Figure 15.7):

① *Local circuit neurons.* Both excitatory and inhibitory inputs arrive at lower motor neurons from nearby interneurons. These **local circuit neurons** are located close to the lower motor neuron cell bodies in the brain stem and spinal cord. Local circuit neurons receive input from somatic sensory receptors, such as nociceptors and muscle spindles, as well as from higher centers in the brain. They help coordinate rhythmic activity in specific muscle groups, such as alternating flexion and extension of the lower limbs during walking.

② *Upper motor neurons.* Local circuit neurons and lower motor neurons receive input from **upper motor neurons (UMNs).** Most upper motor neurons synapse with interneurons, which, in turn, synapse with lower motor neurons. (A few upper motor neurons synapse directly with lower motor neurons.) UMNs from the cerebral cortex are essential for planning, initiating, and directing sequences of voluntary movements. Other UMNs originate in motor centers of the brain stem: the red nucleus, the vestibular nucleus, the superior colliculus, and the reticular formation. UMNs from the brain stem regulate muscle tone, control postural muscles, and help maintain balance and orientation of the head and body.

③ *Basal ganglia neurons.* Neurons of the basal ganglia assist movement by providing input to upper motor neurons. Neural circuits interconnect the basal ganglia with motor areas of the cerebral cortex, thalamus, subthalamic nucleus, and substantia nigra. These circuits help initiate and terminate movements, suppress unwanted movements, and establish a normal level of muscle tone.

④ *Cerebellar neurons.* Neurons of the cerebellum also aid movement by controlling the activity of upper motor neurons. Neural circuits interconnect the cerebellum with motor areas of the cerebral cortex (via the thalamus) and the brain stem. A prime function of the cerebellum is to monitor differences between intended movements and movements actually performed. Then, it issues commands to upper motor neurons to reduce errors in movement. The cerebellum thus coordinates body movements and helps maintain normal posture and balance.

Paralysis

Damage or disease of *lower* motor neurons produces **flaccid paralysis** of muscles on the same side of the body. There is neither voluntary nor reflex action of the innervated muscle fibers, muscle tone is decreased or lost, and the muscle remains limp or flaccid. Injury or disease of *upper* motor neurons in the cerebral cortex causes **spastic paralysis** of muscles on the opposite side of the body. In this condition muscle tone is increased, reflexes are exaggerated, and pathological reflexes such as the Babinski sign (see page 433) appear. ■

Organization of Upper Motor Neuron Pathways

The axons of upper motor neurons extend from the brain to lower motor neurons via two types of somatic motor pathways—direct and indirect. **Direct motor pathways** provide input to lower motor neurons via axons that extend directly from the cerebral cortex. **Indirect motor pathways** provide input to lower motor neurons from motor centers in the brain stem. These brain stem centers, in turn, receive signals from neurons in the basal ganglia, cerebellum, and cerebral cortex. Direct and

Figure 15.7 Somatic motor pathways for coordination and control of movement. Lower motor neurons receive input directly from (❶) local circuit neurons (purple arrow) and (❷) upper motor neurons in the cerebral cortex and brainstem (green arrows). Neural circuits involving basal ganglia neurons (❸) and cerebellar neurons (❹) regulate activity of upper motor neurons (red arrows).

🔑 Because lower motor neurons provide all output to skeletal muscles, they are called the final common pathway.

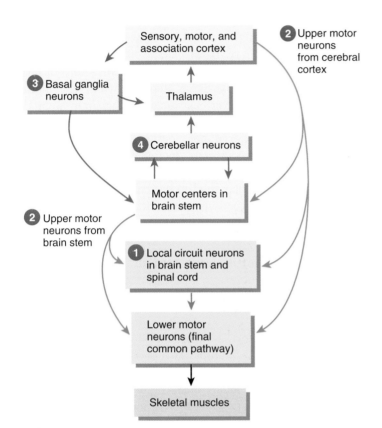

How do the functions of upper motor neurons from the cerebral cortex and from the brain stem differ?

Figure 15.8 Direct motor pathways whereby signals initiated by the primary motor area in the right hemisphere control skeletal muscles on the left side of the body. Spinal cord tracts carrying impulses of direct motor pathways are the lateral corticospinal tract and anterior corticospinal tract.

🔑 Direct pathways convey impulses that result in precise, voluntary movements.

 What two other tracts (not shown in the figure) convey impulses resulting in precise, voluntary movements?

indirect pathways both govern generation of nerve impulses in the lower motor neurons, the neurons that stimulate contraction of skeletal muscles.

Direct Motor Pathways

Nerve impulses for voluntary movements propagate from the cerebral cortex to lower motor neurons via the direct motor pathways (Figure 15.8), also known as the *pyramidal pathways.* Areas of the cerebral cortex that contain large, pyramid-shaped cell bodies of upper motor neurons include not only the primary motor area in the precentral gyrus (area 4 in Figure 14.15 on page 474) but also the premotor area (area 6) and even the primary somatosensory area in the postcentral gyrus (areas 1, 2, and 3). Axons of these cortical UMNs descend through the internal capsule of the cerebrum. In the medulla oblongata, the axon bundles form the ventral bulges known as the pyramids.

About 90% of the axons of upper motor neurons *decussate* (cross over) to the *contralateral* (opposite) side in the medulla oblongata. The 10% that remain on the *ipsilateral* (same) side eventually decussate at the spinal cord levels where they synapse with an interneuron or lower motor neuron. Thus, the right cerebral cortex controls muscles on the left side of the body, and the left cerebral cortex controls muscles on the right side of the body.

Three tracts contain axons of upper motor neurons belonging to the direct motor pathways.

1. *Lateral corticospinal tracts.* Axons of UMNs that decussate in the medulla form the **lateral corticospinal tracts** (kor'-ti-kō-SPĪ-nal) in the right and left lateral white columns of the spinal cord (Figure 15.8 and Table 15.4). These motor neurons control muscles located in distal parts of the limbs. The distal muscles are responsible for precise, agile, and highly skilled movements of the limbs, hands, and feet. Examples include the movements needed to button a shirt or play the piano.

2. *Anterior corticospinal tracts.* Axons of cortical UMNs that do not decussate in the medulla form the **anterior corticospinal tracts** in the right and left anterior white columns (Figure 15.8 and Table 15.4). At each spinal cord level, some of these axons decussate via the anterior white commissure. Then, they synapse with interneurons or lower motor neurons in the anterior gray horn. Axons of these lower motor neurons exit the cervical and upper thoracic segments of the cord in the anterior roots of spinal nerves. They terminate in skeletal muscles that control movements of the neck and part of the trunk, thus coordinating movements of the axial skeleton.

3. *Corticobulbar tracts.* Some axons of upper motor neurons that conduct impulses for the control of skeletal muscles in the head extend through the internal capsule to the midbrain, where they join the **corticobulbar tracts** (kor'-ti-kō-BUL-bar) in the right and left cerebral peduncles (see Table 15.4). Some of the axons in the corticobulbar tracts have decussated, whereas others have not. The axons terminate in the motor nuclei of nine pairs of cranial nerves in the pons and medulla oblongata: cranial nerves III (oculomotor), IV (trochlear), V (trigeminal), VI (abducens), VII (facial), IX (glossopharyngeal), X (vagus), XI (accessory), and XII (hypoglossal). The lower motor neurons of cranial nerves convey impulses that control precise, voluntary movements of the eyes, tongue, and neck, plus chewing, facial expression, and speech.

Table 15.4 summarizes the functions and pathways of the tracts in the direct motor pathways.

Amyotrophic Lateral Sclerosis

Amyotrophic lateral sclerosis (ALS) (ā'-mī-ō-TROF-ik; *a-* = without; *myo-* = muscle; *trophic* = nourishment) is a progressive degenerative disease of unknown cause. The disease attacks motor areas of the cerebral cortex, axons of upper motor neurons in the lateral white columns (corticospinal and rubrospinal tracts), and lower motor neuron cell bodies. It causes progressing muscle weakness and atrophy. ALS often begins in sections of the spinal cord that serve the hands and arms but rapidly spreads to involve the whole body and face, without affecting intellect or sensations. Death typically occurs in 2 to 5 years. ALS is commonly known as *Lou Gehrig's disease* after the New York Yankees baseball player who died of it at age 37 in 1941. ■

Indirect Motor Pathways

The **indirect motor pathways** or **extrapyramidal pathways** include all somatic motor tracts other than the corticospinal and corticobulbar tracts. Nerve impulses conducted along the indirect pathways follow complex, polysynaptic circuits that involve the motor cortex, basal ganglia, thalamus, cerebellum, reticular formation, and nuclei in the brain stem. Axons of upper motor neurons that carry nerve impulses from the indirect pathways descend from various nuclei of the brain stem into five major tracts of the spinal cord and terminate on local circuit neurons or lower motor neurons. These tracts are the **rubrospinal** (ROO-brō-spī-nal), **tectospinal** (TEK-tō-spī-nal), **vestibulospinal** (ves-TIB-ū-lō-spī-nal), **lateral reticulospinal** (re-TIK-ū-lō-spī-nal), and **medial reticulospinal tracts.** Table 15.4 summarizes the functions and pathways of the tracts in the indirect motor pathways.

Modulation of Movement by the Basal Ganglia

In contrast to upper motor neurons, the basal ganglia and cerebellum do not project directly to local circuit neurons or lower motor neurons. Rather, they modulate movement through their effects on upper motor neurons. Two parts of the basal ganglia, the caudate nucleus and the putamen, receive input from sensory, association, and motor areas of the cerebral cortex and from the substantia nigra. Output from the basal ganglia comes from the globus pallidus and substantia nigra, which send feedback signals to the motor cortex by way of the thalamus. (Figure 14.13b on page 470 shows these parts of the basal ganglia.) This feedback circuit—from cortex to basal ganglia to thalamus to cortex—appears to function in initiating and terminating movements. Neurons in the putamen generate impulses just before body movements occur whereas neurons in the caudate nucleus generate impulses just before eye movements occur.

The basal ganglia also suppress unwanted movements by their inhibitory effects on the thalamus and superior colliculus. This function of the basal ganglia is revealed in certain types of basal ganglia damage, such as occurs in Huntington disease. Abnormal, uncontrollable movements or tremors arise with loss of basal ganglia inhibition. The basal ganglia also influence muscle tone. The globus pallidus sends impulses that reduce muscle tone into the reticular formation. Damage or destruction of some basal ganglia connections causes a generalized increase in muscle tone.

In addition to their motor functions, the basal ganglia also influence many aspects of cortical function, including sensory, limbic, cognitive, and linguistic functions.

Table 15.4	Major Somatic Motor Pathways in the Brain and Tracts in the Midbrain and Spinal Cord
Tract and Location	**Functions and Pathways**

Direct (pyramidal) tracts

Lateral corticospinal tract

Anterior corticospinal tract

Spinal cord

Lateral corticospinal: Conveys nerve impulses from the motor cortex to skeletal muscles on opposite side of body for precise, voluntary movements of the limbs, hands, and feet. Axons of upper motor neurons (UMNs) descend from the precentral and postcentral gyri of the cortex into the medulla. Here 90% decussate (cross over to the opposite side) and then enter the contralateral side of the spinal cord to form this tract. At their level of termination, these UMNs end in the anterior gray horn on the same side. They provide input to lower motor neurons, which innervate skeletal muscles.

Anterior corticospinal: Conveys nerve impulses from the motor cortex to skeletal muscles on opposite side of body for movements of the axial skeleton. Axons of UMNs descend from the cortex into the medulla. Here the 10% that do not decussate enter the spinal cord and form this tract. At their level of termination, these UMNs decussate and end in the anterior gray horn on the opposite side. They provide input to lower motor neurons, which innervate skeletal muscles.

Cerebral peduncle

Corticobulbar tract

Midbrain of brain stem

Corticobulbar: Conveys nerve impulses from the motor cortex to skeletal muscles of the head and neck to coordinate precise, voluntary movements. Axons of UMNs descend from the cortex into the brain stem, where some decussate and others do not. They provide input to lower motor neurons in the nuclei of cranial nerves III, IV, V, VI, VII, IX, X, XI, and XII, which control voluntary movements of the eyes, tongue and neck; chewing; facial expression; and speech.

Indirect (extrapyramidal) tracts

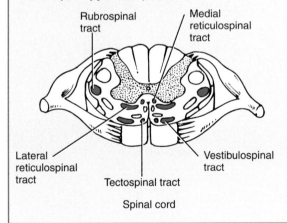

Rubrospinal tract

Medial reticulospinal tract

Lateral reticulospinal tract

Vestibulospinal tract

Tectospinal tract

Spinal cord

Rubrospinal: Conveys nerve impulses from the red nucleus (which receives input from the cerebral cortex and cerebellum) to contralateral skeletal muscles that govern precise movements of the distal parts of the limbs.

Tectospinal: Conveys nerve impulses from the superior colliculus to contralateral skeletal muscles that move the head and eyes in response to visual stimuli.

Vestibulospinal: Conveys nerve impulses from the vestibular nucleus (which receives input about head movements from the vestibular apparatus in the inner ear) to regulate ipsilateral muscle tone for maintaining balance in response to head movements.

Lateral reticulospinal: Conveys nerve impulses from the reticular formation to facilitate flexor reflexes, inhibit extensor reflexes, and decrease muscle tone in muscles of the axial skeleton and proximal parts of the limbs.

Medial reticulospinal: Conveys nerve impulses from the reticular formation to facilitate extensor reflexes, inhibit flexor reflexes, and increase muscle tone in muscles of the axial skeleton and proximal parts of the limbs.

Damage to the Basal Ganglia

Damage to the basal ganglia results in uncontrollable, abnormal body movements, often accompanied by muscle rigidity and tremors (shaking) while at rest. In **Parkinson disease (PD),** neurons that extend from the substantia nigra to the caudate nucleus and putamen degenerate. Loss of these dopamine-releasing neurons causes a generalized increase in muscle tone and stiffness of the face, arms, and legs. **Huntington disease (HD)** is a hereditary disorder in which the caudate nucleus and putamen degenerate, with loss of neurons that normally release GABA or acetylcholine. A key sign of HD is **chorea** (KŌ-rē-a = a dance) in which rapid, jerky movements occur involuntarily and without purpose. Although symptoms of HD often do not appear until age 30 or 40, the person also experiences progressive mental deterioration. Death occurs in 10 to 20 years. ■

Modulation of Movement by the Cerebellum

The cerebellum is active in both learning and performing rapid, coordinated, highly skilled movements such as hitting a golf ball, speaking, and swimming. It also functions to maintain proper posture and equilibrium (balance). Cerebellar function involves four activities (Figure 15.9):

1 The cerebellum *monitors intentions for movement* by receiving impulses from the motor cortex and basal ganglia via the pontine nuclei in the pons regarding what movements are planned (red lines).

2 The cerebellum *monitors actual movement* by receiving input from proprioceptors in joints and muscles that reveals what actually is happening (blue lines). These nerve impulses travel in the anterior and posterior spinocerebellar tracts. Nerve impulses from the vestibular (equilibrium-sensing) apparatus in the inner ear and from the eyes also enter the cerebellum.

3 The cerebellum *compares the command signals* (intentions for movement) *with sensory information* (actual movement performed).

4 If there is a discrepancy between intended and actual movement, the cerebellum *sends out corrective feedback* to upper motor neurons. Corrective feedback travels via the thalamus to UMNs in the cerebral cortex but goes directly to UMNs in brain stem motor centers (green lines). As movements occur, the cerebellum continuously provides error corrections to upper motor neurons, which decreases errors and smoothes the motion. It also contributes over longer periods to the learning of new motor skills.

Skilled activities such as tennis or volleyball provide good examples of the contribution of the cerebellum to movement. To make a good stop volley or to block a spike, you must bring your racket or arms forward just far enough to make solid contact. How do you stop at exactly the right point? Before you even hit

Figure 15.9   Input to and output from the cerebellum.

The cerebellum coordinates and smoothes contractions of skeletal muscles during skilled movements and helps maintain posture and balance.

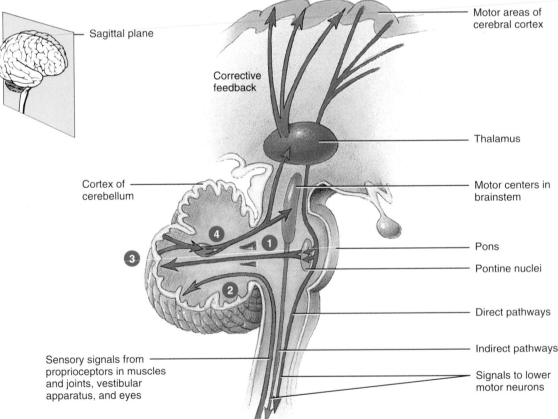

Sagittal section through brain and spinal cord

 Which tracts carry information from proprioceptors in joints and muscles to the cerebellum?

the ball, the cerebellum has sent nerve impulses to the cerebral cortex and basal ganglia informing them where your swing must stop. In response to impulses from the cerebellum, the cortex and basal ganglia transmit motor impulses to opposing body muscles to stop the swing.

🩺 Damage to the Cerebellum

A hallmark of cerebellar trauma or disease is **ataxia** (a-TAK-sē-a; *a-* = without; *-taxia* = coordination), an abnormal condition in which movements are jerky and uncoordinated. People with ataxia have a staggering gait and, if blindfolded, cannot touch the tip of their nose with a finger because they cannot coordinate movements with their sense of where a body part is located. Often, speech is slurred due to a lack of coordination of speech muscles. Cerebellar damage also produces **intention tremor,** a shaking that occurs during deliberate voluntary movement. Chronic alcoholics often suffer degeneration of the anterior part of the cerebellum, which controls the lower limbs. Such people have an impaired walking gait with relatively normal control of arm movements. ∎

▶ CHECKPOINT

15. Why are the two main somatic motor pathways named "direct" and "indirect"?

16. Given the normal functions of the basal ganglia and the cerebellum, what differences would you expect to see with damage of the basal ganglia versus damage to the cerebellum?

INTEGRATIVE FUNCTIONS OF THE CEREBRUM

▶ OBJECTIVE

• **Compare the integrative cerebral functions of wakefulness and sleep, and learning and memory.**

We turn now to a fascinating, though incompletely understood, function of the cerebrum: integration. The **integrative functions** include cerebral activities such as sleep and wakefulness, learning and memory, and emotional responses. (The role of the limbic system in emotional behavior was discussed in Chapter 14.)

Wakefulness and Sleep

Humans sleep and awaken in a 24-hour cycle called a **circadian rhythm** (ser-KĀ-dē-an; *circa-* = about; *-dia* = a day) that is established by the suprachiasmatic nucleus of the hypothalamus (see Figure 14.10 on page 466). A person who is awake is in a state of readiness and is able to react consciously to various stimuli. Moreover, EEG recordings show that the cerebral cortex is very active during wakefulness, whereas fewer impulses arise during most stages of sleep.

The Role of the Reticular Activating System in Awakening

Stimulation of some parts of the reticular formation increases activity of the cerebral cortex. Thus, a portion of the reticular formation is known as the **reticular activating system (RAS)** (Figure 15.10). When this area is active, many nerve impulses pass upward to widespread areas of the cerebral cortex, both directly and via the thalamus. The effect is a generalized increase in cortical activity.

Arousal, or awakening from sleep, also involves increased activity in the RAS. For arousal to occur, the RAS must be stimulated. Many sensory stimuli can activate the RAS: painful stimuli detected by nociceptors, touch and pressure on the skin, movement of the limbs, bright light, or the buzz of an alarm clock. Once the RAS is activated, the cerebral cortex is also activated, and arousal occurs. The result is a state of wakefulness called **consciousness.** Notice in Figure 15.10 that even though the RAS receives input from somatic sensory receptors, the eyes, and the ears, there is no input from olfactory receptors; even strong odors may fail to cause arousal. People who die in house fires usually succumb to smoke inhalation without awakening. For this reason, all sleeping areas should have a nearby smoke detector that emits a loud alarm. A vibrating pillow or flashing light can serve the same purpose for those who are hearing-impaired.

Sleep

Sleep is a state of altered consciousness or partial unconsciousness from which an individual can be aroused. Although it is essential, the exact functions of sleep are still unclear. Sleep deprivation impairs attention, learning, and performance. Normal sleep consists of two components: non-rapid eye movement (NREM) sleep and rapid eye movement (REM) sleep. **NREM sleep** consists of four gradually merging stages, each of which is characterized by a different kind of EEG activity (Figure 15.11a on page 519):

1. *Stage 1* is a transition stage between wakefulness and sleep that normally lasts 1–7 minutes. The person is relaxed with eyes closed and has fleeting thoughts. Alpha waves, which are present in people awake with eyes closed, diminish. People awakened during this stage often say they have not been sleeping.

2. *Stage 2* or *light sleep* is the first stage of true sleep. In it, a person is a little more difficult to awaken. Fragments of dreams may be experienced, and the eyes may slowly roll from side to side. The EEG shows *sleep spindles*—bursts of sharply pointed waves that occur at 12–14 Hz (hertz = cycles per second) and last 1–2 seconds.

3. *Stage 3* is a period of moderately deep sleep. Body temperature and blood pressure decrease. It is difficult to awaken the person, and the EEG shows a mixture of sleep spindles and larger, lower-frequency waves. This stage occurs about 20 minutes after falling asleep.

Figure 15.10   The reticular activating system (RAS) consists of neurons whose axons project from the reticular formation through the thalamus to the cerebral cortex.

 Increased activity of the RAS causes awakening from sleep (arousal).

Sagittal section through brain and spinal cord

? Why should every sleeping room have a smoke detector?

4. *Stage 4* or *slow-wave sleep* is the deepest level of sleep. Slow, large-amplitude delta waves dominate the EEG. Although brain metabolism decreases significantly during slow-wave sleep and body temperature drops slightly, most reflexes are intact, and muscle tone is decreased only slightly. When sleepwalking occurs, it does so during this stage.

Typically, a person goes from stage 1 to stage 4 of NREM sleep in less than an hour (Figure 15.11b). During a typical 7- or 8-hour sleep period, there are three to five episodes of **REM sleep**, during which the eyes move rapidly back and forth under closed eyelids. The person may rapidly ascend through stages 3 and 2 before entering REM sleep. The first episode of REM sleep lasts 10–20 minutes. Then, another interval of NREM sleep follows.

REM and NREM sleep alternate throughout the night. REM periods, which occur approximately every 90 minutes, gradually lengthen, until the final one lasts about 50 minutes. In adults, REM sleep totals 90–120 minutes during a typical sleep period. As a person ages, the average total time spent sleeping decreases; moreover, the percentage of REM sleep declines. As

much as 50% of an infant's sleep is REM sleep, as opposed to 35% for 2-year-olds and 25% for adults. Sleep spindles appear at about 3 months of age. Although we do not yet understand the function of REM sleep, the high percentage of REM sleep in infants and children is thought to be important for the maturation of the brain. Neuronal activity is high during REM sleep—brain blood flow and oxygen use is higher during REM sleep than during intense mental or physical activity while awake. Elderly people often have lower-amplitude delta waves and early morning awakenings.

Different parts of the brain mediate NREM and REM sleep. Neurons in the preoptic area of the hypothalamus, the basal forebrain, and the medulla oblongata govern NREM sleep, whereas neurons in the pons and midbrain turn REM sleep on and off. Sleep spindles are thought to arise in the reticular nucleus of the thalamus. Several lines of evidence suggest the existence of sleep-inducing chemicals in the brain. One apparent sleep-inducer is adenosine, which accumulates during periods of high ATP (adenosine triphosphate) use by the nervous system. Adenosine binds to specific receptors, called A1 receptors, and inhibits certain cholinergic (acetylcholine-releasing) neurons of

Figure 15.11 The stages of sleep. (a) EEG recordings during sleep stages. In an awake person whose eyes are closed, alpha waves predominate; slow, high-amplitude delta waves characterize stage 4 sleep. (b) Alternating intervals of NREM and REM sleep during a typical sleep period.

🔑 **As non-rapid eye movement (NREM) sleep progresses from stage 1 to stage 4, the EEG waves become slower (lower frequency) and larger (higher amplitude).**

(a) EEG waves during sleep stages

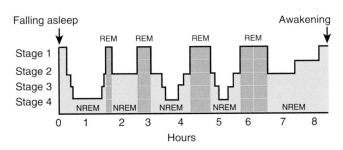

(b) Pattern of NREM and REM sleep over one sleep period

❓ **In which phase of sleep do dreaming and paralysis of most skeletal muscles occur?**

the RAS that participate in arousal. Thus, activity in the RAS during sleep is low due to the inhibitory effect of adenosine. Caffeine (in coffee) and theophylline (in tea)—substances known for their ability to maintain wakefulness—bind to and block the A1 receptors, preventing adenosine from binding and inducing sleep.

Several physiological changes occur during sleep. Most dreaming occurs during REM sleep, and the EEG readings are similar to those of a person who is awake. With the exception of motor neurons that govern breathing and eye movements, most somatic motor neurons are inhibited during REM sleep, which decreases muscle tone and even paralyzes the skeletal muscles. Many people experience a momentary feeling of paralysis if they are awakened during REM sleep. During sleep, activity in the parasympathetic division of the autonomic nervous system (ANS) increases while sympathetic activity decreases. Heart rate and blood pressure decrease during NREM sleep and decrease further during REM sleep. Periodically during REM sleep, however, sympathetic activity increases. In men, this often causes erection of the penis, even when dream content is not sexual. Presence of penile erections during REM sleep in a man with erectile dysfunction (inability to attain an erection while awake) indicates a psychological, rather than a physical, cause of his problem.

Learning and Memory

Without memory, we would repeat mistakes and be unable to learn. Similarly, we would not be able to repeat our successes or accomplishments, except by chance. Although both learning and memory have been extensively studied, we still have no completely satisfactory explanation for how we recall information or how we remember events. However, we do know something about how information is acquired and stored, and it is clear that there are different categories of memory.

Learning is the ability to acquire new information or skills through instruction or experience. **Memory** is the process by which information acquired through learning is stored and retrieved. For an experience to become part of memory, it must produce persistent structural and functional changes that represent the experience in the brain. This capability for change associated with learning is termed **plasticity.** Nervous system plasticity underlies our ability to change our behavior in response to stimuli from the external and internal environments. It involves changes in individual neurons—for example, synthesis of different proteins or sprouting of new dendrites—as well as changes in the strengths of synaptic connections among neurons. The parts of the brain known to be involved with memory include the association cortex of the frontal, parietal, occipital, and temporal lobes; parts of the limbic system, especially the hippocampus and amygdala; and the diencephalon. The primary somatosensory and primary motor areas in the brain also exhibit plasticity. If a particular body part is used more intensively or in a newly learned activity, such as reading Braille, the cortical areas devoted to that body part gradually expand.

Memory occurs in stages over a period of time. **Immediate memory** is the ability to recall ongoing experiences for a few seconds. It provides a perspective to the present time that allows us to know where we are and what we are doing. **Short-term memory** is the temporary ability to recall a few pieces of information for seconds to minutes. One example is when you look up an unfamiliar telephone number, cross the room to the phone, and then dial the new number. If the number has no special significance, it is usually forgotten within a few seconds. Brain areas involved in immediate and short-term memory include the hippocampus, the mammillary bodies, and two nuclei of the thalamus (anterior and medial nuclei). Some evidence supports the notion that short-term memory depends more on electrical and chemical events in the brain than on structural changes, such as the formation of new synapses.

Information in short-term memory may later be transformed into a more permanent type of memory, called **long-term memory,** which lasts from days to years. If you use that new telephone number often enough, it becomes part of long-term memory. Information in long-term memory usually can be retrieved for use whenever needed. The reinforcement that results from the frequent retrieval of a piece of information is called **memory consolidation.** Long-term memories for information that can be expressed by language, such as a telephone number, apparently are stored in wide regions of the cerebral cortex. Memories for motor skills, such as how to serve a tennis ball, are stored in the basal ganglia and cerebellum as well as in the cerebral cortex.

Although the brain receives many stimuli, we pay attention to only a few of them at a time. It has been estimated that only 1% of all the information that comes to our consciousness is stored as long-term memory. Moreover, much of what goes into long-term memory is eventually forgotten. Memory does not record every detail as if it were magnetic tape. Even when details are lost, we can often explain the idea or concept using our own words and ways of viewing things.

Several conditions that inhibit the electrical activity of the brain, such as anesthesia, coma, electroconvulsive therapy (ECT), and ischemia of the brain, disrupt retention of recently acquired information without altering previously established long-term memories. People who suffer retrograde amnesia (loss of past memories) cannot remember anything that occurred during the 30 minutes or so before the amnesia developed. As a person recovers from amnesia, the most recent memories return last.

Anatomical changes occur in neurons when they are stimulated. For example, electron micrographs of neurons subjected to prolonged, intense activity reveal an increase in the number of presynaptic terminals and enlargement of synaptic end bulbs in presynaptic neurons, as well as an increase in the number of dendritic branches in postsynaptic neurons. Moreover, neurons grow new synaptic end bulbs with increasing age, presumably because of increased use. Opposite changes occur when neurons are inactive. For example, in animals that have lost their eyesight, the cerebral cortex in the visual area becomes thinner.

Believed to underlie some aspects of memory is a phenomenon called long-term potentiation (LTP), in which transmission at some synapses within the hippocampus is enhanced (potentiated) for hours or weeks after a brief period of high-frequency stimulation. The neurotransmitter released is glutamate, which acts on NMDA* glutamate receptors on the postsynaptic neurons. In some cases, induction of LTP depends on the release of nitric oxide (NO) from the postsynaptic neurons after they have been activated by glutamate. The NO, in turn, diffuses into the presynaptic neurons and causes LTP.

▶ C H E C K P O I N T

17. Describe how sleep and wakefulness are related to the reticular activating system (RAS).

18. What are the four stages of non-rapid eye movement (NREM) sleep? How is NREM sleep distinguished from rapid eye movement (REM) sleep?

19. Define memory. What are the three kinds of memory? What is memory consolidation?

---

* Named after the chemical *N*-methyl D-aspartate, which is used to detect this type of glutamate receptor.

DISORDERS: HOMEOSTATIC IMBALANCES

Spinal Cord Injury

The spinal cord may be damaged by a tumor either within or adjacent to the spinal cord, herniated intervertebral discs, blood clots, penetrating wounds caused by projectile fragments, or other trauma. In many cases of traumatic injury of the spinal cord, the patient has an improved outcome if an anti-inflammatory corticosteroid drug called methylprednisolone is given within 8 hours of the injury.

Depending on the location and extent of spinal cord damage, paralysis may occur. **Monoplegia** (*mono-* = one; *-plegia* = blow or strike) is paralysis of one limb only. **Diplegia** (*di-* = two) is paralysis of both upper limbs or both lower limbs. **Paraplegia** (*para-* = beyond) is paralysis of both lower limbs. **Hemiplegia** (*hemi-* = half) is paralysis of the upper limb, trunk, and lower limb on one side of the body, and **quadriplegia** (*quad-* = four) is paralysis of all four limbs.

Complete transection of the spinal cord means that the cord is severed from one side to the other, thus cutting all sensory and motor tracts. It results in a loss of all sensations and voluntary movement *below* the level of the transection. **Hemisection** is a partial transection of the cord on either the right or the left side. After hemisection, three main symptoms, known together as *Brown-Séquard syndrome,* occur below the level of the injury: (1) Damage of the posterior column causes loss of proprioception and fine touch sensations on the ipsilateral (same) side as the injury. (2) Damage of the lateral corticospinal tract causes ipsilateral paralysis. (3) Damage of the spinothalamic tracts causes loss of pain and temperature sensations on the contralateral (opposite) side.

Following complete transection, and to varying degrees after hemisection, spinal shock occurs. **Spinal shock** is an immediate response to spinal cord injury characterized by temporary **areflexia** (a'-rē-FLEK-sē-a), loss of reflex function. The areflexia occurs in parts of the body served by spinal nerves below the level of the injury. Signs of acute spinal shock include slow heart rate, low blood pressure, flaccid paralysis of skeletal muscles, loss of somatic sensations, and urinary bladder dysfunction. Spinal shock may begin within 1 hour after injury and may last from several minutes to several months, after which reflex activity gradually returns.

Parkinson Disease

Parkinson disease (PD) is a progressive disorder of the CNS that typically affects its victims around age 60. Neurons that extend from the substantia nigra to the putamen and caudate nucleus, where they release the neurotransmitter dopamine (DA), degenerate in PD. In the caudate nucleus of the basal ganglia are neurons that liberate the neurotransmitter acetylcholine (ACh). Although the level of ACh does not change as the level of DA declines, the imbalance of neurotransmitter activity—too little DA and too much ACh—is thought to cause most of the symptoms. The cause of PD is unknown, but toxic environmental chemicals, such as pesticides, herbicides, and carbon monoxide, are suspected contributing agents. Only 5% of PD patients have a family history of the disease.

In PD patients, involuntary skeletal muscle contractions often interfere with voluntary movement. For instance, the muscles of the upper limb may alternately contract and relax, causing the hand to shake. This shaking, called **tremor,** is the most common symptom of PD. Also, muscle tone may increase greatly, causing rigidity of the involved body part. Rigidity of the facial muscles gives the face a masklike appearance. The expression is characterized by a wide-eyed, unblinking stare and a slightly open mouth with uncontrolled drooling.

Motor performance is also impaired by **bradykinesia** (*brady-* = slow), slowness of movements. Activities such as shaving, cutting food, and buttoning a blouse take longer and become increasingly more difficult as the disease progresses. Muscular movements also exhibit **hypokinesia** (*hypo-* = under), decreasing range of motion. For example, words are written smaller, letters are poorly formed, and eventually handwriting becomes illegible. Often, walking is impaired; steps become shorter and shuffling, and arm swing diminishes. Even speech may be affected.

Treatment of PD is directed toward increasing levels of DA and decreasing levels of ACh. Although people with PD do not manufacture enough dopamine, taking it orally is useless because DA cannot cross the blood–brain barrier. Even though symptoms are partially relieved by a drug developed in the 1960s called levodopa (L-dopa), a precursor of DA, the drug does not slow the progression of the disease. As more and more affected brain cells die, the drug becomes useless. Another drug called selegiline (deprenyl) is used to inhibit monoamine oxidase, an enzyme that degrades catecholamine neurotransmitters such as dopamine. This drug slows progression of PD and may be used together with levodopa.

For more than a decade, surgeons have sought to reverse the effects of Parkinson disease by transplanting dopamine-rich fetal nerve tissue into the basal ganglia (usually the putamen) of patients with severe PD. Only a few postsurgical patients have shown any degree of improvement, such as less rigidity and improved quickness of motion. Another surgical technique that has produced improvement for some patients is *pallidotomy,* in which a part of the globus pallidus that generates tremors and produces muscle rigidity is destroyed.

MEDICAL TERMINOLOGY

Cerebral palsy (CP) A motor disorder that causes loss of muscle control and coordination; caused by damage of the motor areas of the brain during fetal life, birth, or infancy. Radiation during fetal life, temporary lack of oxygen during birth, and hydrocephalus during infancy may also cause cerebral palsy.

Coma (KŌ-ma) A state of deep unconsciousness from which a person cannot be aroused. A comatose patient lies in a sleeplike state, usually with eyes closed. In the lightest stages of coma, brain stem and spinal cord reflexes persist, but in the deepest stages, even reflexes are lost. If respiratory and cardiovascular controls are damaged, the patient dies.

Insomnia (in-SOM-nē-a; *in-* = not; *-somnia* = sleep) Difficulty in falling asleep and staying asleep.

Narcolepsy (NAR-kō-lep-sē; *narco-* = numbness; *-lepsy* = a seizure) A condition in which REM sleep cannot be inhibited during waking periods. As a result, involuntary periods of sleep that last about 15 minutes occur throughout the day.

Sleep apnea (AP-nē-a; *a-* = without; *-pnea* = breath) A disorder in which a person repeatedly stops breathing for 10 or more seconds while sleeping. Most often, it occurs because loss of muscle tone in pharyngeal muscles allows the airway to collapse.

STUDY OUTLINE

SENSATION (p. 499)

1. Sensation is a conscious or subconscious awareness of external or internal stimuli.
2. The nature of a sensation and the type of reaction generated vary according to the destination of sensory impulses in the CNS.
3. Each different type of sensation is a sensory modality; usually, a given sensory neuron serves only one modality.
4. General senses include somatic senses (touch, pressure, vibration, warmth, cold, pain, itch, tickle, and proprioception) and visceral

senses; special senses include the modalities of smell, taste, vision, hearing, and equilibrium.

5. For a sensation to arise, four events typically occur: stimulation, transduction, generation of impulses, and integration.

6. Simple receptors, consisting of free nerve endings and encapsulated nerve endings, are associated with the general senses; complex receptors are associated with the special senses.

7. Sensory receptors respond to stimuli by producing receptor or generator potentials.

8. Table 15.1 on page 501 summarizes the classification of sensory receptors.

9. Adaptation is a decrease in sensitivity during a long-lasting stimulus. Receptors are either rapidly adapting or slowly adapting.

SOMATIC SENSATIONS (p. 502)

1. Somatic sensations include tactile sensations (touch, pressure, vibration, itch, and tickle), thermal sensations (warmth and cold), pain, and proprioception.

2. Receptors for tactile, thermal, and pain sensations are located in the skin, subcutaneous layer, and mucous membranes of the mouth, vagina, and anus.

3. Receptors for proprioceptive sensations (position and movement of body parts) are located in muscles, tendons, joints, and the inner ear.

4. Receptors for touch are (a) hair root plexuses and corpuscles of touch (Meissner corpuscles), which are rapidly adapting, and (b) slowly adapting type I cutaneous mechanoreceptors (tactile or Merkel discs). Type II cutaneous mechanoreceptors (Ruffini corpuscles), which are slowly adapting, are sensitive to stretching. Receptors for pressure include corpuscles of touch, type I mechanoreceptors, and lamellated (Pacinian) corpuscles. Receptors for vibration are corpuscles of touch and lamellated corpuscles. Itch receptors are free nerve endings, whereas both free nerve endings and lamellated corpuscles mediate the tickle sensation.

5. Thermoreceptors are free nerve endings. Cold receptors are located in the stratum basale of the epidermis; warm receptors are located in the dermis.

6. Pain receptors (nociceptors) are free nerve endings that are located in nearly every body tissue.

7. Nerve impulses for fast pain propagate along medium-diameter, myelinated A-delta fibers, whereas those for slow pain conduct along small-diameter unmyelinated C fibers.

8. Proprioceptors include muscle spindles, tendon organs, joint kinesthetic receptors, and hair cells of the inner ear.

9. Table 15.2 on page 507 summarizes the somatic sensory receptors and the sensations they convey.

SOMATIC SENSORY AND MOTOR MAPS IN THE CEREBRAL CORTEX (p. 508)

1. Specific regions of the primary somatosensory area (postcentral gyrus) of the cerebral cortex receive somatic sensory input from different parts of the body.

2. The primary motor area (precentral gyrus) of the cortex is the major control region for planning and initiating voluntary movements.

SOMATIC SENSORY PATHWAYS (p. 509)

1. Somatic sensory pathways from receptors to the cerebral cortex involve three-neuron sets: first-order, second-order, and third-order neurons.

2. Axon collaterals (branches) of somatic sensory neurons simultaneously carry signals into the cerebellum and the reticular formation of the brain stem.

3. Impulses propagating along the posterior column-medial lemniscus pathway relay fine touch, stereognosis, proprioception, and vibratory sensations.

4. The neural pathway for pain and thermal sensations is the lateral spinothalamic tract.

5. The neural pathway for tickle, itch, crude touch, and pressure sensations is the anterior spinothalamic pathway.

6. The pathways to the cerebellum are the anterior and posterior spinocerebellar tracts, which transmit impulses for subconscious muscle and joint position sense from the trunk and lower limbs.

7. Table 15.3 on page 511 summarizes the major somatic sensory pathways.

SOMATIC MOTOR PATHWAYS (p. 512)

1. All excitatory and inhibitory signals that control movement converge on the alpha motor neurons, also known as lower motor neurons (LMNs) or the final common pathway.

2. Neurons in four neural circuits, collectively termed the somatic motor pathways, participate in control of movement by providing input to lower motor neurons: local circuit neurons, upper motor neurons, basal ganglia neurons, and cerebellar neurons.

3. The axons of upper motor neurons extend from the brain to lower motor neurons via direct and indirect motor pathways. The direct (pyramidal) pathways include the lateral and anterior corticospinal tracts and corticobulbar tracts. Indirect (extrapyramidal) pathways extend from several motor centers of the brain stem into the spinal cord.

4. Neurons of the basal ganglia assist movement by providing input to the upper motor neurons. They help initiate and terminate movements, suppress unwanted movements, and establish a normal level of muscle tone.

5. The cerebellum is active in learning and performing rapid, coordinated, highly skilled movements. It also contributes to maintaining balance and posture.

6. Table 15.4 on page 514 summarizes the major somatic motor pathways.

INTEGRATIVE FUNCTIONS OF THE CEREBRUM (p. 517)

1. Sleep and wakefulness are integrative functions that are controlled by the suprachiasmatic nucleus and the reticular activating system (RAS).

2. Non-rapid eye movement (NREM) sleep consists of four stages identified by characteristic waves in EEG recordings.

3. Most dreaming occurs during rapid eye movement (REM) sleep.

4. Memory, the ability to store and recall thoughts, involves persistent changes in the brain, a capability called plasticity. Three types are immediate, short-term, and long-term memory.

SELF-QUIZ QUESTIONS

Fill in the blanks in the following statements.

1. _____ is the conscious or subconscious awareness of external or internal stimuli, whereas _____ is the conscious awareness and interpretation of sensory impulses.

2. _____ is a decline in sensitivity during a maintained, constant stimulus.

Indicate whether the following statements are true or false.

3. Awakening from sleep involves increased activity in the reticular activating system.

4. Lower motor neurons terminate in skeletal muscle fibers.

Choose the one best answer to the following questions.

5. Which of the following are *true*? (1) Slow pain is a result of impulse propagation along myelinated A-delta nerve fibers. (2) Visceral pain occurs when nociceptors in the skin are stimulated. (3) Referred pain is pain felt in an area far from the stimulated organ. (4) Nociceptors exhibit very little adaptation. (5) Nociceptors are located in every body tissue. (a) 1, 3, 4, and 5, (b) 2, 3, and 5, (c) 1 and 5, (d) 3 and 4, (e) 3, 4, and 5.

6. You cannot "hear" with your eyes because (a) hearing is a special sense and vision is a general sense, (b) the sensory neurons for sight carry information only for the modality of vision, (c) the impulses for hearing are transmitted to the somatosensory area of the cerebral cortex, (d) hearing receptors are selective and vision receptors are not, (e) hearing receptors produce a generator potential and vision receptors produce a receptor potential.

7. Which of the following events typically occur if a sensation is to arise? (1) stimulation of the sensory receptor, (2) generation of a nerve impulse, (3) transduction of the stimulus, (4) integration of sensory input, (5) effector response to the stimulus. (a) 1, 2, 3, and 4, (b) 2, 3, 4, and 5, (c) 1, 3, 4, and 5, (d) 1, 4, and 5, (e) 2, 4, and 5.

8. Which of the following statements is *false*? (a) First-order sensory neurons carry signals from the somatic sensory receptors into either the brain stem or spinal cord. (b) Second-order neurons carry signals from the spinal cord and brain stem to the thalamus. (c) Third-order neurons project to the primary somatosensory area of the cortex, where conscious perception of the sensation results. (d) The somatic sensory pathways to the cerebellum are the posterior column-medial lemniscus pathway and the anterolateral pathway. (e) Axons of second-order neurons decussate in the spinal cord or brain stem before ascending to the thalamus.

9. Which of the following is *not* an aspect of cerebellar function? (a) monitoring movement intention, (b) monitoring actual movement, (c) comparing intent with actual performance, (d) sending out corrective signals, (e) directing sensory input to effectors.

10. During REM sleep (1) neuronal activity in the pons and midbrain is high, (2) most somatic motor neurons are inhibited, (3) most dreaming occurs, (4) sleepwalking can occur, (5) there is an increase in heart rate and blood pressure. (a) 1, 2, 4 and 5, (b) 2, 3, and 5, (c) 1, 2, 3, 4, and 5, (d) 2, 3, and 4, (e) 1, 2, and 3.

11. Which of the following statements is *false*? (a) Learning is the ability to acquire new knowledge or skills through instruction or experience. (b) Memory is the process of storing and retrieving information acquired by learning. (c) For an experience to become a memory, it must produce persistent functional and structural changes in the brain that represent the experience. (d) Short-term memory involves memory consolidation. (e) The capacity for change associated with learning is termed plasticity.

12. Which of the following statements is *incorrect*? (a) The graded potentials produced by receptors that serve the senses of touch, pressure, stretching, vibration, pain proprioception, and smell are generator potentials. (b) The graded potentials produced by receptors that serve the special senses of vision, hearing, equilibrium, and taste are receptor potentials. (c) When a generator potential is large enough to reach threshold, it generates one or more nerve impulses in a first-order sensory neuron. (d) A receptor potential generates nerve impulses in second-order neuron. (e) The amplitude of both generator and receptor potentials vary with the intensity of the stimulus.

13. Match the following:

_____ (a) located in the precentral gyrus, this is the major control region of the cerebral cortex for initiation of voluntary movements

_____ (b) direct pathways conveying impulses from the cortex to the spinal cord that result in precise, voluntary movements

_____ (c) contain motor neurons that control skilled movements of the hands and feet

_____ (d) tracts include rubrospinal, tectospinal, vestibulospinal, lateral reticulospinal, and medial reticulospinal

_____ (e) contain neurons that help initiate and terminate movements; can suppress unwanted movements; influence muscle tone

_____ (f) mainly carry pain and temperature impulses

_____ (g) the major routes relaying proprioceptive input to the cerebellum; critical for posture, balance and coordination of skilled movements

_____ (h) include the gracile fasciculus and cuneate fasciculus

_____ (i) contain motor neurons that coordinate movements of the axial skeleton

_____ (j) contain axons that convey impulses for precise, voluntary movements of the eyes, tongue, and neck, plus chewing, facial expression, and speech

_____ (k) convey sensations of fine touch, stereognosis, proprioception, and weight discrimination to the cerebral cortex

(1) posterior column

(2) anterolateral (spinothalamic) pathways

(3) spinocerebellar pathways

(4) lateral corticospinal tracts

(5) anterior corticospinal tracts

(6) corticobulbar pathways

(7) extrapyramidal pathways

(8) pyramidal pathways

(9) primary motor area

(10) basal ganglia

(11) posterior column–medial lemniscus pathways

14. Match the following:

_____ (a) specialized groupings of muscle fibers interspersed among regular skeletal muscle fibers and oriented parallel to them; monitor changes in the length of a skeletal muscle

_____ (b) inform the CNS about changes in muscle tension

_____ (c) free nerve ending receptors for pain

_____ (d) encapsulated receptors for touch located in the dermal papillae; found in hairless skin, eyelids, tip of the tongue, and lips

_____ (e) lamellated corpuscles that detect pressure

_____ (f) type II cutaneous mechanoreceptors; most sensitive to stretching that occurs as digits or limbs are moved

_____ (g) located in the stratum basale and activated by low temperatures

_____ (h) located in the dermis and activated by high temperatures

_____ (i) found within and around the articular capsules of synovial joints; respond to pressure and acceleration and deceleration of joints

_____ (j) type I cutaneous mechanoreceptors that function in fine touch

(1) Meissner corpuscles
(2) Merkel discs
(3) Ruffini corpuscles
(4) Pacinian corpuscles
(5) cold receptors
(6) warm receptors
(7) nociceptors
(8) tendon organs
(9) joint kinesthetic receptors
(10) muscle spindles

15. Match the following:

(1) exteroceptors   (2) interoceptors   (3) proprioceptors
(4) mechanoreceptors (5) thermoreceptors   (6) nociceptors
(7) photoreceptors   (8) chemoreceptors

_____ (a) receptors located in muscles, tendons, joints, and the inner ear

_____ (b) receptors located in blood vessels, visceral organs, muscles and the nervous system

_____ (c) receptors that detect temperature changes

_____ (d) receptors that detect light that strikes the retina of the eye

_____ (e) receptors located at or near the external surface of the body

_____ (f) receptors that provide information about body position, muscle tension, and position and activity of joints

_____ (g) receptors that detect chemicals in the mouth, nose, and body fluids

_____ (h) receptors that detect mechanical pressure or stretching

_____ (i) receptors that respond to stimuli resulting from physical or chemical damage to tissues

CRITICAL THINKING QUESTIONS

1. When Joni first stepped onto the sailboat, she smelled the tangy sea air and felt the motion of water beneath her feet. After a few minutes, she no longer noticed the smell, but unfortunately she was aware of the rolling motion for hours. What types of receptors are involved in smell and detection of motion? Why did her sensation of smell fade but the rolling sensation remain?
HINT *One way to classify receptors is by the type of stimulus that they detect.*

2. Damara was laughing uncontrollably; her brother had her by the leg and was tickling her foot. How does her brain know that her foot is being tickled?
HINT *If you blocked the pathway for tickle, you'd still be able to feel pain.*

3. Yoshio used to predict the weather by how much the bunion (abnormally swollen joint on big toe) on his left foot was bothering him. Last year, Yoshio's left foot was amputated due to complications from diabetes, but sometimes he still thinks he feels that bunion. Explain his experience.
HINT *How did Yoshio's toe communicate with his brain before the amputation?*

?ANSWERS TO FIGURE QUESTIONS

15.1 The special senses of vision, taste, hearing, and equilibrium are served by separate sensory cells.

15.2 Pain, thermal sensations, and tickle and itch arise with activation of different free nerve endings.

15.3 The kidneys have the broadest area for referred pain.

15.4 Muscle spindles are activated when the central areas of the intrafusal fibers are stretched.

15.5 The hand has a larger representation in the motor area than in the somatosensory area, which implies greater precision in the hand's movement control than fine ability in its sensation.

15.6 Damage to the right lateral spinothalamic tract could result in loss of pain and thermal sensations on the left side of the body.

15.7 Cerebral cortex UMNs are active in planning, initiating, and directing sequences of voluntary movements. Brain stem UMNs regulate muscle tone, control postural muslces, and help maintain balance and orientation of the head and body.

15.8 The corticobulbar and rubrospinal tracts convey impulses that result in precise, voluntary movements (see Table 15.4 on page 515).

15.9 The anterior and posterior spinocerebellar tracts carry information from proprioceptors in joints and muscles to the cerebellum.

15.10 A smoke detector responds to smoke by sounding a loud bell or buzzer, which awakens sleepers by providing auditory input that stimulates the RAS.

15.11 Dreaming and paralysis of skeletal muscles occur during REM sleep.

The Special Senses

I found the Special Senses chapter to be very understandable and I was pleased to be able to apply knowledge already learned. This chapter reinforces chapters about the muscular system and the nervous system by explaining which muscles and nerves control which senses and how. The everyday life examples and medical relationships helped me understand the information better.

I really like the way each sense is broken down into anatomy, physiology, and then diseases. This makes it so much easier to understand. I found the embryonic development of the eyes and ears especially fascinating.

The whole chapter allows you to use previous knowledge and apply that to the special senses. The Self-Quiz Questions are helpful in reviewing. The pronunciation guides, checkpoints, and questions that accompany the figures are a big help in thinking about what you have read. The Study Outline at the end of the chapter gives readers a base to elaborate on using their knowledge gained from the chapter.

Molly Causby
Georgia Perimeter College

THE SPECIAL SENSES AND HOMEOSTASIS
Sensory organs have special receptors that allow us to smell, taste, see, hear, and maintain equilibrium or balance.

ENERGIZE YOUR STUDY
FOUNDATIONS CD
Anatomy Overview
• The Nervous System
Animation
• Systems Contributions to Homeostasis
• The Case of the Girl with Fruity Breath
• The Case of the Man with Yellow Eyes
Concepts and Connections and Exercises
reinforce your understanding

www.wiley.com/college/apcentral
INSIGHTS AND EXPLORATIONS
If you are reading this, it must mean that you can see. You may not have thought about losing your eyesight, or if you did, it was probably because of some accident, perhaps during a sport activity when a bat or ball came close to your eye. For individuals with macular degeneration there is no accident, no specific event that leads to blindness. In this chronic progressive disease, vision is lost gradually, day by day, month by month, year by year. Individuals with this disease can go for long periods without any change in their vision then suddenly experience a significant loss. Come explore in this web-based activity the anatomy of the eye and this disease called macular degeneration.

Receptors for the special senses—smell, taste, vision, hearing, and equilibrium—are housed in complex sensory organs such as the eyes and ears. In this chapter we examine the structure and function of the special sense organs, and the pathways involved in conveying information from them to the central nervous system. **Ophthalmology** (of-thal-MOL-ō-jē; *ophthalmo-* = eye; *-logy* = study of) is the science that deals with the eye and its disorders. The other special senses are, in large part, the concern of **otorhinolaryngology** (ō-tō-rī′nō-lar-in-GOL-ō-jē; *oto-* = ear; *rhino-* = nose; *laryngo-* = larynx).

OLFACTION: SENSE OF SMELL

▶ O B J E C T I V E

- **Describe the olfactory receptors and the neural pathway for olfaction.**

Both smell and taste are chemical senses; the sensations arise from the interaction of molecules with smell or taste receptors. Because impulses for smell and taste propagate to the limbic system (and to higher cortical areas as well), certain odors and tastes can evoke strong emotional responses or a flood of memories.

Anatomy of Olfactory Receptors

The nose contains 10–100 million receptors for the sense of smell or **olfaction** (ol-FAK-shun; *olfact-* = smell). The total area of the olfactory epithelium is 5 cm^2 (a little less than 1 in.^{2}). It occupies the superior part of the nasal cavity, covering the inferior surface of the cribriform plate and extending along the superior nasal concha and upper part of the middle nasal concha (Figure 16.1a). The olfactory epithelium consists of three kinds of cells: olfactory receptors, supporting cells, and basal stem cells (Figure 16.1b).

Olfactory receptors are the first-order neurons of the olfactory pathway. They are bipolar neurons whose exposed tip is a knob-shaped dendrite and whose axon ends in the olfactory bulb. The sites of olfactory transduction are the **olfactory hairs,** which are cilia that project from the dendrite. (Recall that transduction is the conversion of stimulus energy into a graded potential in a sensory receptor.) Chemicals that have an odor and can therefore stimulate the olfactory hairs are termed **odorants.** Olfactory receptors respond to the chemical stimulation of an odorant molecule by producing a generator potential, thus initiating the olfactory response. From the base of each olfactory receptor, a single axon projects through the cribriform plate and into the olfactory bulb.

Supporting cells are columnar epithelial cells of the mucous membrane lining the nose. They provide physical support, nourishment, and electrical insulation for the olfactory receptors, and they help detoxify chemicals that come in contact with the olfactory epithelium. **Basal stem cells** lie between the bases of the supporting cells and continually undergo cell division to produce new olfactory receptors, which live for only a month or so before being replaced. This process is remarkable because olfactory receptors are neurons, and in general, mature neurons are not replaced.

Within the connective tissue that supports the olfactory epithelium are **olfactory (Bowman's) glands,** which produce mucus that is carried to the surface of the epithelium by ducts. The secretion moistens the surface of the olfactory epithelium and dissolves odorants. Both supporting cells of the nasal epithelium and olfactory glands are innervated by branches of the facial nerve (cranial nerve VII), which can be stimulated by certain chemicals. Impulses in these nerves, in turn, stimulate the lacrimal glands in the eyes and nasal mucous glands. The result is tears and a runny nose after inhaling substances such as pepper or the vapors of household ammonia.

Physiology of Olfaction

Many attempts have been made to distinguish among and classify "primary" sensations of smell. Genetic evidence now suggests the existence of hundreds of primary odors. Our ability to recognize about 10,000 different odors probably depends on patterns of activity in the brain that arise from activation of many different combinations of olfactory receptors.

Olfactory receptors react to odorant molecules in the same way that most sensory receptors react to their specific stimuli: A generator potential (depolarization) develops and triggers one or more nerve impulses. How the generator potential arises in olfactory receptors is known in some cases. Many different odorants bind to receptors that are linked to G proteins in the plasma membrane and activate the enzyme adenylate cyclase (see page 592). The result is the following chain of events: production of cyclic adenosine monophosphate (cAMP) → opening of sodium ion (Na$^+$) channels → inflow of Na$^+$ → depolarizing generator potential → nerve impulses arise and propagate along axon of olfactory receptor.

Odor Thresholds and Adaptation

Olfaction, like all the special senses, has a low threshold. Only a few molecules of certain substances need be present in air to be perceived as an odor. A good example is the chemical methyl mercaptan, which smells like rotten cabbage and can be detected in concentrations as low as 1/25 billionth of a milligram per milliliter of air. Because the natural gas used for cooking and heating is odorless but lethal and potentially explosive if it accumulates, a small amount of methyl mercaptan is added to provide olfactory warning of gas leaks.

Adaptation (decreasing sensitivity) to odors occurs rapidly. Olfactory receptors adapt by about 50% in the first second or so after stimulation but adapt very slowly thereafter. Still, complete

527

Figure 16.1 Olfactory epithelium and olfactory receptors. (a) Location of olfactory epithelium in nasal cavity. (b) Anatomy of olfactory receptors, consisting of first-order neurons whose axons extend through the cribriform plate and terminate in the olfactory bulb. (See Tortora, *A Photographic Atlas of the Human Body,* Figures 9.1b and 9.1c.)

 The olfactory epithelium consists of olfactory receptors, supporting cells, and basal stem cells.

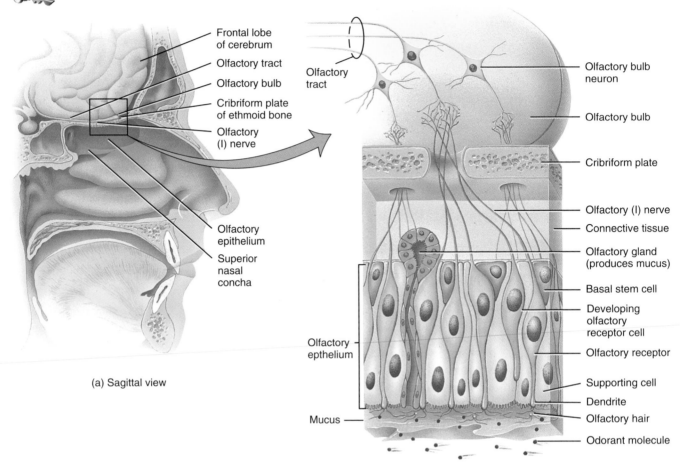

(a) Sagittal view

(b) Enlarged aspect of olfactory receptors

Which part of an olfactory receptor detects an odorant molecule?

insensitivity to certain strong odors occurs in about a minute after exposure. Apparently, reduced sensitivity involves an adaptation process in the central nervous system as well.

The Olfactory Pathway

On each side of the nose, bundles of the slender, unmyelinated axons of olfactory receptors extend through about 20 olfactory foramina in the cribriform plate of the ethmoid bone (Figure 16.1b). These 40 or so bundles of axons collectively form the right and left **olfactory nerves** (cranial nerve I). The olfactory nerves terminate in the brain in paired masses of gray matter called the **olfactory bulbs,** which are located below the frontal lobes of the cerebrum and lateral to the crista galli of the ethmoid bone. Within the olfactory bulbs, the axon terminals of

olfactory receptors—the first-order neurons—form synapses with the dendrites and cell bodies of second-order neurons in the olfactory pathway.

Axons of olfactory bulb neurons extend posteriorly and form the **olfactory tract** (Figure 16.1b). These axons project to the lateral olfactory area, which is located at the inferior and medial surface of the temporal lobe. This cortical region is part of the limbic system and includes some of the amygdala (see Figure 14.14 on page 471). Because many olfactory tract axons end in the lateral olfactory area, it is probably the primary olfactory area, where conscious awareness of smells begins. Connections to other limbic system regions and to the hypothalamus account for our emotional and memory-evoked responses to odors. Examples include sexual excitement upon smelling a certain perfume, nausea upon smelling a food that once made you violently ill, or an odor-evoked memory of a childhood experience.

From the lateral olfactory area, pathways also extend to the frontal lobe, both directly and indirectly via the thalamus. An important region for odor identification and discrimination is the orbitofrontal area (area 11 in Figure 14.15 on page 474). People who suffer damage in this area have difficulty identifying different odors. Positron emission tomography (PET) studies suggest some degree of hemispheric lateralization: The orbitofrontal area of the *right* hemisphere exhibits greater activity during olfactory processing.

Hyposmia

Women often have a keener sense of smell than men do, especially at the time of ovulation. Smoking seriously impairs the sense of smell in the short term and may cause long-term damage to olfactory receptors. With aging the sense of smell deteriorates. **Hyposmia** (hī-POZ-mē-a; *osmi* = smell, odor), a reduced ability to smell, affects half of those over age 65 and 75% of those over age 80. Hyposmia also can be caused by neurological changes, such as a head injury, Alzheimer disease, or Parkinson disease; certain drugs, such as antihistamines, analgesics, or steroids; and the damaging effects of smoking. ■

▶ CHECKPOINT

1. How do basal stem cells contribute to olfaction?

2. What is the sequence of events from the binding of an odorant molecule to an olfactory hair to the arrival of a nerve impulse in an olfactory bulb?

GUSTATION: SENSE OF TASTE

▶ OBJECTIVE

- **Describe the gustatory receptors and the neural pathway for gustation.**

Like olfaction, taste is a chemical sense; to be detected by either sense, stimulating molecules must be dissolved. Taste or **gustation** (gus-TĀ-shun; *gust-* = taste) is much simpler than olfaction in that only five primary tastes can be distinguished: *sour, sweet, bitter, salty,* and *umami* (ū-MAM-ē). The umami taste, recently described by Japanese scientists, is described as "meaty" or "savory." All other flavors, such as chocolate, pepper, and coffee, are combinations of the five primary tastes, plus accompanying olfactory and tactile (touch) sensations. Odors from food can pass upward from the mouth into the nasal cavity, where they stimulate olfactory receptors. Because olfaction is much more sensitive than taste, a given concentration of a food substance may stimulate the olfactory system thousands of times more strongly than it stimulates the gustatory system. When persons with colds or allergies complain that they cannot taste their food, they are reporting blockage of olfaction, not of taste.

Anatomy of Taste Buds and Papillae

The receptors for sensations of taste are located in the taste buds (Figure 16.2). The nearly 10,000 taste buds of a young adult are mainly on the tongue, but they are also found on the soft palate (posterior portion of roof of mouth), pharynx (throat), and epiglottis (cartilage lid over voice box). The number of taste buds declines with age. Each **taste bud** is an oval body consisting of three kinds of epithelial cells: supporting cells, gustatory receptor cells, and basal cells (see Figure 16.2c). The **supporting cells** surround about 50 **gustatory receptor cells.** A single, long microvillus, called a **gustatory hair,** projects from each gustatory receptor cell to the external surface through the **taste pore,** an opening in the taste bud. **Basal cells,** found at the periphery of the taste bud near the connective tissue layer, produce supporting cells, which then develop into gustatory receptor cells with a life span of about 10 days. At their base, the gustatory receptor cells synapse with dendrites of the first-order neurons that form the first part of the gustatory pathway. The dendrites of each first-order neuron branch profusely and contact many gustatory receptor cells in several taste buds.

Taste buds are found in elevations on the tongue called **papillae** (pa-PIL-ē), which provide a rough texture to the upper surface of the tongue (Figure 16.2a, b). Three types of papillae contain taste buds.

1. About 12 very large, circular **vallate (circumvallate) papillae** (VAL-āt = wall-like) form an inverted V-shaped row at the back of the tongue. Each of these papillae houses 100–300 taste buds.

2. **Fungiform papillae** (FUN-ji-form = mushroomlike) are mushroom-shaped elevations scattered over the entire surface of the tongue that contain about five taste buds each.

3. **Foliate papillae** (FŌ-lē-āt = leaflike) are located in small trenches on the lateral margins of the tongue but most of their taste buds degenerate in early childhood.

In addition, the entire surface of the tongue has **filiform papillae** (FIL-i-form = threadlike). These pointed, threadlike structures contain tactile receptors but no taste buds. They increase friction between the tongue and food, making it easier for the tongue to move food in the oral cavity.

Physiology of Gustation

Chemicals that stimulate gustatory receptor cells are known as **tastants.** Once a tastant is dissolved in saliva, it can make contact with the plasma membrane of the gustatory hairs, which are the sites of taste transduction. The result is a receptor potential that stimulates exocytosis of synaptic vesicles from the gustatory receptor cell. In turn, the liberated neurotransmitter molecules trigger nerve impulses in the first-order sensory neurons that synapse with gustatory receptor cells.

How the receptor potential arises is different for various tastants. The sodium ions (Na^+) in a salty food enter gustatory

Figure 16.2   The relationship of gustatory receptor cells in taste buds to tongue papillae.   (See Tortora, *A Photographic Atlas of the Human Body,* Figure 9.2.)

Gustatory (taste) receptor cells are located in taste buds.

(a) Dorsum of tongue showing location of papillae

(b) Details of papillae

(c) Structure of a taste bud

What role do supporting cells in taste buds play?

receptor cells via Na$^+$ channels in the plasma membrane. The accumulation of Na$^+$ inside causes depolarization, which opens Ca^{2+} channels. In turn, inflow of Ca^{2+} triggers exocytosis of synaptic vesicles and liberation of neurotransmitter. The hydrogen ions (H$^+$) in sour tastants may flow into gustatory receptor cells via H$^+$ channels. They also influence opening and closing of other types of ion channels. Again, the result is a depolarization that leads to release of neurotransmitter.

Other tastants, responsible for stimulating sweet, bitter, and umami tastes, do not themselves enter gustatory receptor cells. Rather, they bind to receptors on the plasma membrane that are linked to G proteins. The G-proteins then activate several different chemicals known as second messengers inside the gustatory receptor cell. Different second messengers cause depolarization in different ways, but the result is the same—release of neurotransmitter.

If all tastants cause release of neurotransmitter from many gustatory receptor cells, why do foods taste different? The answer to this question is thought to lie in the patterns of nerve impulses in groups of first-order taste neurons that synapse with the gustatory receptor cells. Different tastes arise from activation of different groups of taste neurons. In addition, although each individual gustatory receptor cell responds to more than one of the five primary tastes, it may respond more strongly to some tastants than to others. Figure 16.3 shows the responses of three groups of taste neurons to sweet, salty, and sour chemicals.

Taste Thresholds and Adaptation

The threshold for taste varies for each of the primary tastes. The threshold for bitter substances, as measured by using quinine, is

Figure 16.3 Effects of different tastants on three groups of taste neurons. The height of each column is the number of nerve impulses generated in five seconds.

Each gustatory receptor cell and its first-order neuron can respond to several different types of tastants.

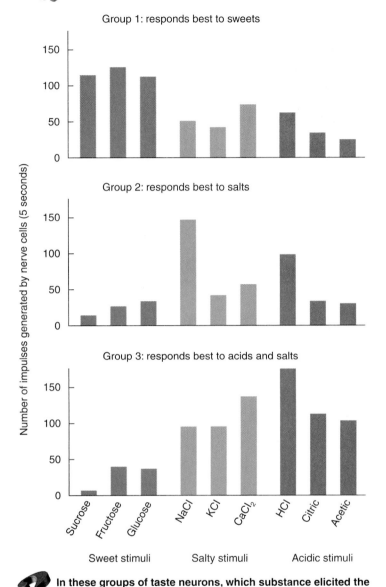

In these groups of taste neurons, which substance elicited the strongest sweet response?

lowest. Because poisonous substances often are bitter, the low threshold (or high sensitivity) may have a protective function. The threshold for sour substances, as measured by using hydrochloric acid, is somewhat higher. The thresholds for salty substances, represented by sodium chloride, and for sweet substances, as measured by using sucrose, are about the same and are higher than those for bitter or sour substances.

Complete adaptation to a specific taste can occur in 1–5 minutes of continuous stimulation. Taste adaptation is due to changes that occur in the taste receptors, in olfactory receptors, and in neurons of the gustatory pathway in the CNS.

The Gustatory Pathway

Three cranial nerves include axons of first-order gustatory neurons from taste buds. The facial nerve (cranial nerve VII) serves the anterior two-thirds of the tongue, the glossopharyngeal nerve (cranial nerve IX) serves the posterior one-third of the tongue, and the vagus nerve (cranial nerve X) serves the throat and epiglottis. From taste buds, impulses propagate along these cranial nerves to the medulla oblongata. From the medulla, some axons carrying taste signals project to the limbic system and the hypothalamus, whereas others project to the thalamus. Taste signals that project from the thalamus to the primary gustatory area in the parietal lobe of the cerebral cortex (see area 43 in Figure 14.15 on page 474) give rise to the conscious perception of taste.

Taste Aversion

Probably because of taste projections to the hypothalamus and limbic system, there is a strong link between taste and pleasant or unpleasant emotions. Sweet foods evoke reactions of pleasure while bitter ones cause expressions of disgust even in newborn babies. This phenomenon is the basis for **taste aversion,** in which people and animals quickly learn to avoid a food if it upsets the digestive system. The advantage of avoiding foods that cause such illness is longer survival. However, the drugs and radiation treatments used to combat cancer often cause nausea and gastrointestinal upset regardless of what foods are consumed. Thus, cancer patients may lose their appetite because they develop taste aversions for most foods. ■

▶ CHECKPOINT

3. How do olfactory receptors and gustatory receptor cells differ in structure and function?

4. Compare the olfactory and gustatory pathways.

VISION

▶ OBJECTIVES

- **List and describe the accessory structures of the eye and the structural components of the eyeball.**
- **Discuss image formation by describing refraction, accommodation, and constriction of the pupil.**
- **Describe the processing of visual signals in the retina and the neural pathway for vision.**

More than half the sensory receptors in the human body are located in the eyes, and a large part of the cerebral cortex is devoted to processing visual information. In this section of the chapter, we examine the accessory structures of the eye, the eyeball, how visual images are formed, the physiology of vision, and the visual pathway.

Accessory Structures of the Eye

The **accessory structures** of the eye are the eyelids, eyelashes, eyebrows, the lacrimal (tearing) apparatus, and extrinsic eye muscles.

Eyelids

The upper and lower **eyelids,** or **palpebrae** (PAL-pe-brē), shade the eyes during sleep, protect the eyes from excessive light and foreign objects, and spread lubricating secretions over the eyeballs (Figure 16.4). The upper eyelid is more movable than the lower and contains in its superior region the **levator palpebrae superioris muscle.** The space between the upper and lower eyelids that exposes the eyeball is the **palpebral fissure.** Its angles are known as the **lateral commissure** (KOM-i-shur), which is narrower and closer to the temporal bone, and the **medial commissure,** which is broader and nearer the nasal bone. In the medial commissure is a small, reddish elevation, the **lacrimal caruncle** (KAR-ung-kul), which contains sebaceous (oil) glands and sudoriferous (sweat) glands. The whitish material that sometimes collects in the medial commissure comes from these glands.

From superficial to deep, each eyelid consists of epidermis, dermis, subcutaneous tissue, fibers of the orbicularis oculi muscle, a tarsal plate, tarsal glands, and conjunctiva (Figure 16.5a). The **tarsal plate** is a thick fold of connective tissue that gives form and support to the eyelids. Embedded in each tarsal plate is a row of elongated modified sebaceous glands, known as **tarsal** or **Meibomian glands** (mī-BŌ-mē-an), that secrete a fluid that helps keep the eyelids from adhering to each other. Infection of the tarsal glands produces a tumor or cyst on the eyelid called a **chalazion** (ka-LĀ-zē-on). The **conjunctiva** (kon′-junk-TĪ-va) is a thin, protective mucous membrane composed of stratified columnar epithelium with numerous goblet cells that is supported by areolar connective tissue. The **palpebral conjunctiva** lines the inner aspect of the eyelids, and the **bulbar conjunctiva** passes from the eyelids onto the anterior surface of the eyeball. Dilation and congestion of the blood vessels of the bulbar conjunctiva due to local irritation or infection are the cause of bloodshot eyes.

Eyelashes and Eyebrows

The **eyelashes,** which project from the border of each eyelid, and the **eyebrows,** which arch transversely above the upper eyelids, help protect the eyeballs from foreign objects, perspiration, and the direct rays of the sun. Sebaceous glands at the base of the hair follicles of the eyelashes, called **sebaceous ciliary glands,** release a lubricating fluid into the follicles. Infection of these glands is called a **sty.**

The Lacrimal Apparatus

The **lacrimal apparatus** (*lacrim-* = tears) is a group of structures that produces and drains **lacrimal fluid** or **tears.** The

Figure 16.4 Surface anatomy of the right eye.

The palpebral fissure is the space between the upper and lower eyelids that exposes the eyeball.

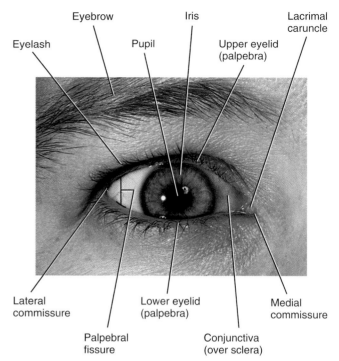

Eyebrow   Iris   Lacrimal caruncle

Eyelash   Pupil   Upper eyelid (palpebra)

Lateral commissure   Lower eyelid (palpebra)   Medial commissure

Palpebral fissure   Conjunctiva (over sclera)

Which structure shown here is continuous with the inner lining of the eyelids?

lacrimal glands, each about the size and shape of an almond, secrete lacrimal fluid, which drains into 6–12 **excretory lacrimal ducts** that empty tears onto the surface of the conjunctiva of the upper lid (Figure 16.5b). From here the tears pass medially over the anterior surface of the eyeball to enter two small openings called **lacrimal puncta.** Tears then pass into two ducts, the **lacrimal canals,** which lead into the **lacrimal sac** and then into the **nasolacrimal duct.** This duct carries the lacrimal fluid into the nasal cavity just inferior to the inferior nasal concha.

Lacrimal fluid is a watery solution containing salts, some mucus, and **lysozyme,** a protective bactericidal enzyme. The fluid protects, cleans, lubricates, and moistens the eyeball. After being secreted, lacrimal fluid is spread medially over the surface of the eyeball by the blinking of the eyelids. Each gland produces about 1 mL of lacrimal fluid per day.

Normally, tears are cleared away as fast as they are produced, either by evaporation or by passing into the lacrimal canals and then into the nasal cavity. If an irritating substance makes contact with the conjunctiva, however, the lacrimal glands are stimulated to oversecrete, and tears accumulate (watery eyes). Lacrimation is a protective mechanism, as the tears dilute and wash away the irritating substance. Watery eyes also occur when an inflammation of the nasal mucosa, such as occurs with a cold, obstructs the nasolacrimal ducts and blocks drainage of

Figure 16.5 Accessory structures of the eye.

Accessory structures of the eye are the eyelids, eyelashes, eyebrows, the lacrimal apparatus, and extrinsic eye muscles.

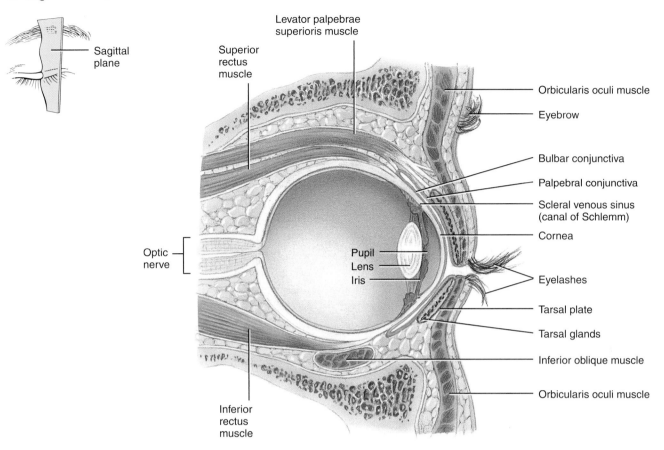

(a) Sagittal section of eye and its accessory structures

(b) Anterior view of the lacrimal apparatus

FLOW OF TEARS

Lacrimal gland
↓
Excretory lacrimal ducts
↓
Superior or inferior lacrimal canal
↓
Lacrimal sac
↓
Nasolacrimal duct
↓
Nasal cavity

 What is lacrimal fluid, and what are its functions?

tears. Humans are unique in expressing emotions, both happiness and sadness, by **crying.** In response to parasympathetic stimulation, the lacrimal glands produce excessive lacrimal fluid that may spill over the edges of the eyelids and even fill the nasal cavity with fluid. Thus, crying produces a runny nose.

Extrinsic Eye Muscles

Six extrinsic eye muscles move each eye: the **superior rectus, inferior rectus, lateral rectus, medial rectus, superior oblique,** and **inferior oblique** (Figures 16.5a and 16.6). They receive innervation from cranial nerves III, IV, or VI. In general, the motor units in these muscles are small. Some motor neurons serve only two or three muscle fibers—fewer than in any other part of the body except the larynx (voice box). Such small motor units permit smooth, precise, and rapid movement of the eyes. As indicated in Exhibit 11.2 on page 322, the extrinsic eye muscles move the eyeball laterally, medially, superiorly, and inferiorly. Looking to the right, for example, requires simultaneous contraction of the right lateral rectus and left medial rectus muscles and relaxation of the left lateral rectus and right medial rectus. The oblique muscles preserve rotational stability of the eyeball. Circuits in the brain stem and cerebellum coordinate and synchronize the movements of the eyes.

Anatomy of the Eyeball

The adult **eyeball** measures about 2.5 cm (1 in.) in diameter. Of its total surface area, only the anterior one-sixth is exposed; the remainder is recessed and protected by the orbit, into which it fits. Anatomically, the wall of the eyeball consists of three layers: fibrous tunic, vascular tunic, and retina.

Fibrous Tunic

The **fibrous tunic,** the superficial coat of the eyeball, is avascular and consists of the anterior cornea and posterior sclera (Figure 16.6). The **cornea** (KOR-nē-a) is a transparent coat that covers the colored iris. Because it is curved, the cornea helps focus light onto the retina. Its outer surface consists of nonkeratinized stratified squamous epithelium. The middle coat of the cornea consists of collagen fibers and fibroblasts, and the inner surface is simple squamous epithelium. The **sclera** (SKLE-ra; *scler-* = hard), the "white" of the eye, is a layer of dense connective tissue made up mostly of collagen fibers and fibroblasts. The sclera covers the entire eyeball except the cornea; it gives shape to the eyeball, makes it more rigid, and protects its inner parts. At the junction of the sclera and cornea is an opening known as the **scleral venous sinus (canal of Schlemm).** A fluid called aqueous humor drains into this sinus (see Figure 16.5a).

 Corneal Transplants

Corneal transplants are the most common organ transplant operation—and the most successful type of transplant as well.

Because the cornea is avascular, antibodies in the blood that might cause rejection do not enter the transplanted tissue, and rejection rarely occurs. The defective cornea is removed and a donor cornea of similar diameter is sewn in. The shortage of donor corneas has been partially overcome by the development of artificial corneas made of plastic. ■

Vascular Tunic

The **vascular tunic** or **uvea** (Ū-vē-a) is the middle layer of the eyeball and has three parts: choroid, ciliary body, and iris (Figure 16.6). The highly vascularized **choroid** (KŌ-royd), which is the posterior portion of the vascular tunic, lines most of the internal surface of the sclera. It provides nutrients to the posterior surface of the retina.

In the anterior portion of the vascular tunic, the choroid becomes the **ciliary body** (SIL-ē-ar′-ē). It extends from the **ora serrata** (Ō-ra ser-RĀ-ta), the jagged anterior margin of the retina, to a point just posterior to the junction of the sclera and cornea. The ciliary body consists of the ciliary processes and the ciliary muscle. The **ciliary processes** are protrusions or folds on the internal surface of the ciliary body. They contain blood capillaries that secrete aqueous humor. Extending from the ciliary process are **zonular fibers (suspensory ligaments)** that attach to the lens. The **ciliary muscle** is a circular band of smooth muscle that alters the shape of the lens, adapting it for near or far vision. When the muscle contracts, the ciliary muscle is pulled anteriorly and inward toward the pupil, the tension in the zonular fibers is reduced, and the shape of the lens is altered into a more spherical shape.

The **iris,** the colored portion of the eyeball, is shaped like a flattened donut. It is suspended between the cornea and the lens and is attached at its outer margin to the ciliary processes. It consists of circular and radial smooth muscle fibers. A principal function of the iris is to regulate the amount of light entering the vitreous chamber of the eyeball through the **pupil,** the hole in the center of the iris. Autonomic reflexes regulate pupil diameter in response to light levels (Figure 16.7 on page 536). When bright light stimulates the eye, parasympathetic neurons stimulate the **circular muscles (sphincter pupillae)** of the iris to contract, causing a decrease in the size of the pupil (constriction). In dim light, sympathetic neurons stimulate the **radial muscles (dilator pupillae)** of the iris to contract, causing an increase in the pupil's size (dilation).

Retina

The third and inner coat of the eyeball, the **retina,** lines the posterior three-quarters of the eyeball and is the beginning of the visual pathway (see Figure 16.6). An ophthalmoscope allows an observer to peer through the pupil, providing a magnified image of the retina and the blood vessels that course across the retina's anterior surface (Figure 16.8 on page 537). The surface of the retina is the only place in the body where blood vessels can be viewed directly and examined for pathological changes, such as

Figure 16.6 **Anatomy of the eyeball.** (See Tortora, *A Photographic Atlas of the Human Body,* Figure 9.3b.)

 The wall of the eyeball consists of three layers: the fibrous tunic, the vascular tunic, and the retina.

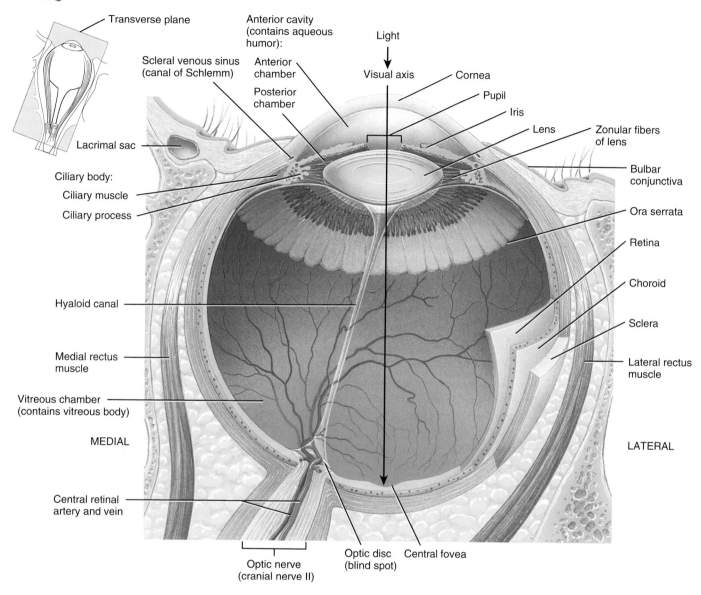

Superior view of transverse section of right eyeball

What are the components of the fibrous tunic and vascular tunic?

those that occur with hypertension or diabetes mellitus. Several landmarks are visible. The **optic disc** is the site where the optic nerve exits the eyeball. Bundled together with the optic nerve are the **central retinal artery,** a branch of the ophthalmic artery, and the **central retinal vein.** Branches of the central retinal artery fan out to nourish the anterior surface of the retina; the central retinal vein drains blood from the retina through the optic disc.

The optic part of the retina consists of a pigmented layer and a neural layer. The **pigmented layer** is a sheet of melanin-containing epithelial cells located between the choroid and the neural part of the retina. Melanin in the choroid and in the pigmented layer absorbs stray light rays, which prevents reflection and scattering of light within the eyeball. As a result, the image cast on the retina by the cornea and lens remains sharp and clear. Albinos lack melanin in all parts of the body, including the eye. They often need to wear sunglasses, even indoors, because even moderately bright light is perceived as bright glare due to light scattering.

Figure 16.7 Responses of the pupil to light of varying brightness.

Contraction of the circular muscles causes constriction of the pupil; contraction of the radial muscles causes dilation of the pupil.

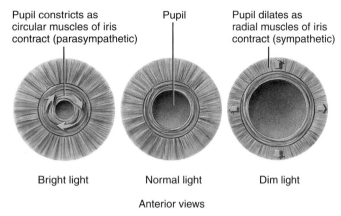

Pupil constricts as circular muscles of iris contract (parasympathetic)

Pupil

Pupil dilates as radial muscles of iris contract (sympathetic)

Bright light   Normal light   Dim light

Anterior views

Which division of the autonomic nervous system causes pupillary constriction? Which causes pupillary dilation?

Figure 16.8 A normal retina, as seen through an ophthalmoscope. Blood vessels in the retina can be viewed directly and examined for pathological changes.

The optic disc is the site where the optic nerve exits the eyeball. The central fovea is the area of highest visual acuity.

TEMPORAL SIDE

NASAL SIDE

Central fovea

Macula lutea

Optic disc

Retinal blood vessels

Right eye

For which diseases may evidence be seen through an ophthalmoscope?

The **neural layer** of the retina is a multilayered outgrowth of the brain that extensively processes visual data before sending nerve impulses into axons that form the optic nerve. Three distinct layers of retinal neurons—the **photoreceptor layer,** the **bipolar cell layer,** and the **ganglion cell layer**—are separated by two zones, the outer and inner synaptic layers, where synaptic contacts are made (Figure 16.9). Note that light passes through the ganglion and bipolar cell layers and both synaptic layers before it reaches the photoreceptor layer. Two other types of cells present in the retina are called **horizontal cells** and **amacrine cells.** These cells form laterally directed neural circuits that modify the signals being transmitted along the pathway from photoreceptors to bipolar cells to ganglion cells.

Two types of photoreceptors transduce light rays into receptor potentials: rods and cones. Each retina has about 6 million cones and 120 million rods. **Rods** have a low light threshold, allowing us to see in dim light, such as moonlight. Because they do not provide color vision, in dim light we see only shades of gray. Brighter lights stimulate the **cones,** which have a higher threshold and produce color vision. Most of our visual experiences are mediated by the cone system, the loss of which produces legal blindness. In contrast, a person who loses rod vision mainly has difficulty seeing in dim light and thus should not drive at night.

The **macula lutea** (MAK-ū-la LOO-tē-a; *macula* = a small, flat spot; *lute-* = yellowish) is in the exact center of the posterior portion of the retina, at the visual axis of the eye. The **central fovea** (see Figure 16.6), a small depression in the center of the macula lutea, contains only cones. In addition, the layers of bipolar and ganglion cells, which scatter light to some extent, do not cover the cones here; these layers are displaced to the periphery of the central fovea. As a result, the central fovea is the area of highest **visual acuity** or **resolution** (sharpness of vision). A main reason that you move your head and eyes while looking at something is to place images of interest on your central fovea—as you do to read the words in this sentence! Rods are absent in the central fovea and are more plentiful toward the periphery of the retina. Because rod vision is more sensitive than cone vision, you can see a faint object (such as a dim star) better if you gaze slightly to one side rather than looking directly at it.

From photoreceptors, information flows to bipolar cells through the outer synaptic layer, and then from bipolar cells through the inner synaptic layer to ganglion cells. The axons of ganglion cells extend posteriorly to the optic disc and exit the eyeball as the optic nerve. The optic disc is also called the **blind spot.** Because it contains no rods or cones, we cannot see an image that strikes the blind spot. Normally, you are not aware of having a blind spot, but you can easily demonstrate its presence. Cover your left eye and gaze directly at the cross below. Then increase or decrease the distance between the book and your eye. At some point the square will disappear as its image falls on the blind spot.

Lens

Behind the pupil and iris, within the cavity of the eyeball, is the **lens** (see Figure 16.5). Proteins called **crystallins,** arranged like the layers of an onion, make up the lens, which normally is perfectly transparent and lacks blood vessels. It is enclosed by a clear connective tissue capsule and held in position by encir-

Figure 16.9 Microscopic structure of the retina. The downward blue arrow at left indicates the direction of the signals passing through the neural layer of the retina. Eventually, nerve impulses arise in ganglion cells and propagate along their axons, which make up the optic nerve (cranial nerve II). (See Tortora, *A Photographic Atlas of the Human Body,* Figure 9.3c.)

🔑 In the retina, visual signals pass from photoreceptors to bipolar cells to ganglion cells.

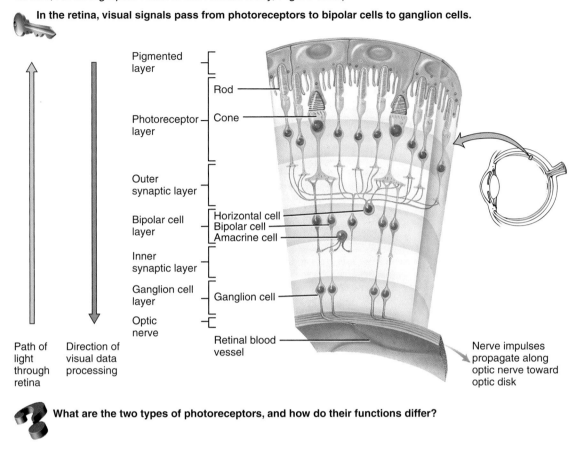

What are the two types of photoreceptors, and how do their functions differ?

Interior of the Eyeball

cling zonular fibers, which attach to the ciliary processes. The lens helps focus images on the retina to facilitate clear vision.

The lens divides the interior of the eyeball into two cavities: the anterior cavity and vitreous chamber. The **anterior cavity**—the space anterior to the lens—consists of two chambers. The **anterior chamber** lies between the cornea and the iris, whereas the **posterior chamber** lies behind the iris and in front of the zonular fibers and lens (Figure 16.10). Both chambers of the anterior cavity are filled with **aqueous humor** (*aqua* = water), a watery fluid that nourishes the lens and cornea. Aqueous humor continually filters out of blood capillaries in the ciliary processes and enters the posterior chamber. It then flows forward between the iris and the lens, through the pupil, and into the anterior chamber. From the anterior chamber, aqueous humor drains into the scleral venous sinus (canal of Schlemm) and then into the blood. Normally, aqueous humor is completely replaced about every 90 minutes.

The pressure in the eye, called **intraocular pressure,** is produced mainly by the aqueous humor and partly by the vitreous body (described shortly); normally it is about 16 mmHg (millimeters of mercury). The intraocular pressure maintains the shape of the eyeball and prevents the eyeball from collapsing.

The second, and larger, cavity of the eyeball is the **vitreous chamber,** which lies between the lens and the retina. Within the vitreous chamber is the **vitreous body,** a jellylike substance that contributes to intraocular pressure. It holds the retina flush against the choroid, so that the retina provides an even surface for the reception of clear images. Unlike the aqueous humor, the vitreous body does not undergo constant replacement. It is formed during embryonic life and is not replaced thereafter. The vitreous body also contains phagocytic cells that remove debris, keeping this part of the eye clear for unobstructed vision. Occasionally, collections of debris may cast a shadow on the retina and create the appearance of specks that dart in and out of the field of vision. These "vitreal floaters," which are more common in older individuals, are usually harmless and do not require treatment. The **hyaloid canal** is a narrow channel that runs through the vitreous body from the optic disc to the posterior aspect of the lens. In the fetus, it was occupied by the hyaloid artery (see Figure 16.24d).

Figure 16.10 **The iris separates the anterior and posterior chambers of the eye.** The section passes through the anterior portion of the eyeball at the junction of the cornea and sclera. Arrows indicate the flow of aqueous humor.

 The lens separates the posterior chamber of the anterior cavity from the vitreous chamber.

 Where is aqueous humor produced, what is its circulation path, and where does it drain from the eyeball?

Table 16.1 summarizes the structures associated with the eyeball.

Age-related Macular Disease

Age-related macular disease (AMD) is a degenerative disorder of the retina and pigmented layer in persons 50 years of age and older. In AMD, abnormalities occur in the region of the macula lutea, which is ordinarily the area of most acute vision. Victims of advanced AMD retain their peripheral vision but lose the ability to see straight ahead. For instance, they cannot see facial features to identify a person in front of them. AMD is the leading cause of blindness in those over age 75, afflicting 13 million Americans, and is 2.5 times more common in pack-a-day smokers than in nonsmokers. Initially, a person may experience blurring and distortion at the center of the visual field. In "dry" AMD, central vision gradually diminishes because the pig-mented layer atrophies and degenerates. There is no effective treatment. In about 10% of cases, dry AMD progresses to "wet" AMD, in which new blood vessels form in the choroid and leak plasma or blood under the retina. Vision loss can be slowed by using laser surgery to destroy leaking blood vessels. ■

Image Formation

In some ways the eye is like a camera: Its optical elements focus an image of some object on a light-sensitive "film"—the retina—while ensuring the correct amount of light to make the proper "exposure." To understand how the eye forms clear images of objects on the retina, we must examine three processes: (1) the refraction or bending of light by the lens and cornea; (2) accommodation, the change in shape of the lens; and (3) constriction or narrowing of the pupil.

Table 16.1	Summary of Structures of the Eyeball
Structure	**Function**
Fibrous tunic Cornea Sclera	*Cornea:* Admits and refracts (bends) light. *Sclera:* Provides shape and protects inner parts.
Vascular tunic Iris — Ciliary body Choroid	*Iris:* Regulates amount of light that enters eyeball. *Ciliary body:* Secretes aqueous humor and alters shape of lens for near or far vision (accommodation). *Choroid:* Provides blood supply and absorbs scattered light.
Retina Retina	Receives light and converts it into receptor potentials and nerve impulses. Output to brain is via axons of ganglion cells, which form the optic nerve (cranial nerve II).
Lens Lens	Refracts light.
Anterior cavity Anterior cavity	Contains aqueous humor that helps maintain shape of eyeball and supplies oxygen and nutrients to lens and cornea.
Vitreous chamber Vitreous chamber	Contains vitreous body that helps maintain shape of eyeball and keeps the retina attached to the choroid.

Refraction of Light Rays

When light rays traveling through a transparent substance (such as air) pass into a second transparent substance with a different density (such as water), they bend at the junction between the two. This bending is called **refraction** (Figure 16.11a). As light rays enter the eye, they are refracted at the anterior and posterior surfaces of the cornea. Both surfaces of the lens of the eye further refract the light rays so they come into exact focus on the retina.

Images focused on the retina are inverted; they are upside down (Figure 16.11b, c). They also undergo right to left reversal;

Figure 16.11 Refraction of light rays. (a) Refraction is the bending of light rays at the junction of two transparent substances with different densities. (b) The cornea and lens refract light rays from distant objects so the image is focused on the retina. (c) In accommodation, the lens becomes more spherical, which increases the refraction of light.

Images focused on the retina are inverted and left-to-right reversed.

Light ray before refraction

Air

Water

Light ray after refraction

(a) Refraction of light rays

Nearly parallel rays from distant object

Lens

(b) Viewing distant object

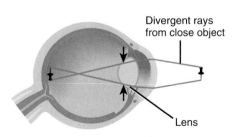

Divergent rays from close object

Lens

(c) Accommodation

What sequence of events occurs during accommodation?

that is, light from the right side of an object strikes the left side of the retina, and vice versa. The reason the world does not look inverted and reversed is that the brain "learns" early in life to coordinate visual images with the orientations of objects. The brain stores the inverted and reversed images we acquired when we first reached for and touched objects and interprets those visual images as being correctly oriented in space.

About 75% of the total refraction of light occurs at the cornea. The lens provides the remaining 25% of focusing power and also changes the focus to view near or distant objects. When an object is 6 m (20 ft) or more away from the viewer, the light rays reflected from the object are nearly parallel to one another (Figure 16.11b). These parallel rays must be bent enough to fall exactly focused on the central fovea, where vision is sharpest. However, because light rays that are reflected from objects closer than 6 m (20 ft) are divergent rather than parallel (Figure 16.11c), the rays must be refracted more if they are to be focused on the retina. This additional refraction is accomplished in the process called accommodation.

Accommodation and the Near Point of Vision

A surface that curves outward, like the surface of a ball, is said to be *convex*. When the surface of a lens is convex, that lens will refract incoming light rays toward each other, such that they eventually intersect. If the surface of a lens curves inward, like the inside of a hollow ball, the lens is said to be *concave* and causes light rays to refract away from each other. The lens of the eye is convex on both its anterior and posterior surfaces, and its focusing power increases as its curvature becomes greater. When the eye is focusing on a close object, the lens becomes more curved and refracts the light rays more. This increase in the curvature of the lens for near vision is called **accommodation** (Figure 16.11c). The **near point of vision** is the minimum distance from the eye that an object can be clearly focused with maximum accommodation. This distance is about 10 cm (4 in.) in a young adult.

How does accommodation occur? When you are viewing distant objects, the ciliary muscle is relaxed and the lens is flatter because it is stretched in all directions by taut zonular fibers. When you view a close object, the ciliary muscle contracts, which pulls the ciliary process and choroid forward toward the lens. This action releases tension on the lens and zonular fibers. Because it is elastic, the lens becomes more spherical (more convex), which increases its focusing power and causes greater convergence of the light rays.

 Presbyopia

With aging, the lens loses elasticity and thus its ability to accommodate. Therefore, older people cannot read print at the same close range as can youngsters. This condition is called **presbyopia** (prez-bē-Ō-pē-a; *presby-* = old; *-opia* = pertaining to the eye or vision). By age 40 the near point of vision may have increased to 20 cm (8 in.), and at age 60 to 80 cm (31 in.). Presbyopia usually begins in the mid-forties. At about that age, peo-

ple who have not previously worn glasses begin to need them for reading. Those who already wear glasses typically start to need bifocals, lenses that have two different refractive powers for distant and close vision. ■

Refraction Abnormalities

The normal eye, known as an **emmetropic eye** (em′-e-TROP-ik), can sufficiently refract light rays from an object 6 m (20 ft) away so that a clear image is focused on the retina. Many people, however, lack this ability because of refraction abnormalities. Among these abnormalities are **myopia** (mī-Ō-pē-a), or nearsightedness, which occurs when the eyeball is too long relative to the focusing power of the cornea and lens. In **hypermetropia** (hī′-per-me-TRŌ-pē-a), or farsightedness (also known as **hyperopia**), the eyeball length is short relative to the focusing power of the cornea and lens. Figure 16.12 illustrates these conditions and explains how they are corrected. Another refraction abnormality is **astigmatism** (a-STIG-ma-tizm), in which either the cornea or the lens has an irregular curvature. As a result, parts of the image are out of focus, and thus vision is blurred or distorted.

Most errors of vision can be corrected by eyeglasses, contact lenses, or surgical procedures. A contact lens floats on a film of tears over the cornea. The anterior outer surface of the contact lens corrects the visual defect, whereas its posterior surface matches the curvature of the cornea. LASIK and PRK (photorefractive keratectomy) are two computer-guided laser surgery techniques that involve reshaping the cornea to permanently correct refraction abnormalities.

Constriction of the Pupil

The circular muscle fibers of the iris also have a role in the formation of clear retinal images. Part of the accommodation mechanism consists of the contraction of the circular muscles of the iris to constrict the pupil. **Constriction of the pupil** is a narrowing of the diameter of the hole through which light enters the eye. This autonomic reflex occurs simultaneously with accommodation and prevents light rays from entering the eye through the periphery of the lens. Light rays entering at the periphery would not be brought to focus on the retina and would result in blurred vision. The pupil, as noted earlier, also constricts in bright light.

Convergence

Because of the position of their eyes in their heads, many animals see one set of objects off to the left through one eye, and an entirely different set of objects off to the right through the other. In humans, both eyes focus on only one set of objects—a characteristic called **binocular vision.** This feature of our visual system allows the perception of depth and an appreciation of the three-dimensional nature of objects.

Binocular vision occurs when light rays from an object strike corresponding points on the two retinas. When we stare straight ahead at a distant object, the incoming light rays are aimed directly at both pupils and are refracted to comparable spots on the retinas of both eyes. As we move closer to an object, however, the eyes must rotate medially if the light rays from the object are to strike the same points on both retinas. The term **convergence** refers to this medial movement of the two eyeballs so that both are directed toward the object being viewed. The nearer the object, the greater the degree of convergence needed to maintain binocular vision. The coordinated action of the extrinsic eye muscles brings about convergence.

Physiology of Vision

Photoreceptors and Photopigments

Rods and cones were named for the different appearance of each's *outer segment*—the distal end of the photoreceptor next to the pigmented layer. The outer segments of rods are cylindrical or rod-shaped, whereas those of cones are tapered or cone-shaped (Figure 16.13). Transduction of light energy into

Figure 16.12 Refraction abnormalities in the eyeball and their correction. (a) Normal (emmetropic) eye. (b) In the nearsighted or myopic eye, the image is focused in front of the retina. The condition may result from an elongated eyeball or thickened lens. (c) Correction of myopia is by use of a concave lens that diverges entering light rays so that they come into focus directly on the retina. (d) In the farsighted or hypermetropic eye, the image is focused behind the retina. The condition results from a shortened eyeball or a thin lens. (e) Correction of hypermetropia is by a convex lens that converges entering light rays so that they focus directly on the retina.

Figure 16.13 Structure of rod and cone photoreceptors. The inner segments contain the metabolic machinery for synthesis of photopigments and production of ATP. The photopigments are embedded in the membrane discs or folds of the outer segments. New discs in rods and new folds in cones form at the junction of the outer and inner segments. Pigment epithelial cells phagocytize the old discs and folds that slough off the distal tip of the outer segments.

🔑 Transduction of light energy into a receptor potential occurs in the outer segments of rods and cones.

🔑 In myopia (nearsightedness), distant objects can't be seen clearly; in hypermetropia (farsightedness), nearby objects can't be seen clearly.

(a) Normal (emmetropic) eye

(b) Nearsighted (myopic) eye, uncorrected

(c) Nearsighted (myopic) eye, corrected

(d) Farsighted (hypermetropic) eye, uncorrected

(e) Farsighted (hypermetropic) eye, corrected

 What is presbyopia?

ROD   CONE

LIGHT DIRECTION

 What are the functional similarities between rods and cones?

a receptor potential occurs in the outer segment. The photopigments are integral proteins in the plasma membrane of the outer segment. In cones the plasma membrane folds back and forth in a pleated fashion, whereas in rods the pleats pinch off from the plasma membrane to form discs. The outer segment of each rod contains a stack of about 1000 discs, piled up like coins inside a wrapper.

Photoreceptor outer segments are renewed at an astonishingly rapid pace. In rods, one to three new discs are added to the base of the outer segment every hour while old discs slough off at the tip and are phagocytized by pigment epithelial cells. The *inner segment* contains the cell nucleus, Golgi complex, and many mitochondria. At its proximal end, the photoreceptor expands into bulblike synaptic terminals filled with synaptic vesicles.

The first step in visual transduction is absorption of light by a **photopigment,** a colored protein that undergoes structural changes when it absorbs light. Light absorption initiates the events that lead to the production of a receptor potential. The single type of photopigment in rods is **rhodopsin** (*rhod-* = rose; *opsin-* = related to vision). Three different **cone photopigments** are present in the retina, one in each of three types of cones. Color vision results from different colors of light selectively activating the different cone photopigments.

All photopigments associated with vision contain two parts: a glycoprotein known as **opsin** and a derivative of vitamin A called **retinal.** Vitamin A derivatives are formed from carotenoids, the plant pigments that give carrots their orange color. Good vision depends on adequate dietary intake of carotenoid-rich vegetables such as carrots, spinach, broccoli, and yellow squash, or foods that contain vitamin A, such as liver.

Retinal is the light-absorbing part of all visual photopigments. In the human retina, there are four different opsins, one for each cone photopigment and another for rhodopsin. Small variations in the amino acid sequences of the different opsins permit the rods and cones to absorb different colors (wavelengths) of incoming light. Rhodopsin absorbs blue to green light most effectively, whereas the three different cone photopigments most effectively absorb blue, green, or yellow-orange light.

Photopigments respond to light in the following cyclical process (Figure 16.14):

1 In darkness, retinal has a bent shape, called *cis*-retinal, which fits snugly into the opsin portion of the photopigment. When *cis*-retinal absorbs a photon of light, it straightens out to a shape called *trans*-retinal. This *cis*-to-*trans* conversion is called **isomerization** and is the first step in visual transduction. After retinal isomerizes, several unstable chemical intermediates form and disappear. These chemical changes lead to production of a receptor potential (see Figure 16.15).

2 In about a minute, *trans*-retinal completely separates from opsin. The final products look colorless, so this part of the cycle is termed **bleaching** of photopigment.

3 An enzyme called **retinal isomerase** converts *trans*-retinal back to *cis*-retinal.

4 The *cis*-retinal then can bind to opsin, reforming a functional photopigment. This part of the cycle—resynthesis of a photopigment—is called **regeneration.**

The pigmented layer adjacent to the photoreceptors stores a large quantity of vitamin A and contributes to the regeneration process in rods. The extent of rhodopsin regeneration decreases drastically if the retina detaches from the pigmented layer. Cone photopigments regenerate much more quickly than does rhodopsin and are less dependent on the pigmented layer. After complete bleaching, regeneration of half of the rhodopsin takes 5 minutes, whereas half of the cone photopigments regenerate in only 90 seconds. Full regeneration of bleached rhodopsin takes 30 to 40 minutes.

Light and Dark Adaptation

When you emerge from dark surroundings (say, a tunnel) into the sunshine, **light adaptation** occurs—your visual system adjusts in seconds to the brighter environment by decreasing its sensitivity. On the other hand, when you enter a darkened room such as a theater, your visual system undergoes **dark adaptation**—its sensitivity increases slowly over many minutes.

Figure 16.14 The cyclical bleaching and regeneration of photopigment. Blue arrows indicate bleaching steps; black arrows indicate regeneration steps.

🔑 **Retinal, a derivative of vitamin A, is the light-absorbing part of all visual photopigments.**

Rod disc in outer segment

Rhodopsin molecule

Light

Disc membrane

cis-retinal

4 *Cis*-retinal binds to opsin

opsin

1 Isomerization of retinal

Colored photopigment (rhodopsin)

cis-retinal

trans-retinal

opsin

trans-retinal

opsin

trans-retinal

opsin

3 Retinal isomerase converts *trans*- to *cis*-retinal

opsin

Colorless products

2 *Trans*-retinal separates from opsin

What is the conversion of *cis*-retinal to *trans*-retinal called?

Bleaching and regeneration of the photopigments account for some (but not all) of the sensitivity changes during light and dark adaptation.

As the light level increases, more and more photopigment is bleached. While light is bleaching some photopigment molecules, however, others are being regenerated. In daylight, regeneration of rhodopsin cannot keep up with the bleaching process, so rods contribute little to daylight vision. In contrast, cone photopigments regenerate rapidly enough that some of the *cis* form is always present, even in very bright light.

If the light level decreases abruptly, sensitivity increases rapidly at first and then more slowly. In complete darkness, full regeneration of cone photopigments occurs during the first 8 minutes of dark adaptation. During this time, a threshold (barely perceptible) light flash is seen as having color. Rhodopsin regenerates more slowly, and our visual sensitivity increases until even a single photon (the smallest unit of light) can be detected. In that situation, although much dimmer light can be detected, threshold flashes appear gray-white, regardless of their color. At very low light levels, such as starlight, objects appear as shades of gray because only the rods are functioning.

Release of Neurotransmitter by Photoreceptors

The absorption of light and isomerization of retinal initiates chemical changes in the photoreceptor outer segment that lead to production of a receptor potential. To understand how the receptor potential arises, however, we first need to examine the operation of photoreceptors in the absence of light. In darkness, sodium ions (Na^+) flow into photoreceptor outer segments through ligand-gated Na^+ channels (Figure 16.15a). The ligand that holds these channels open is **cyclic GMP (guanosine monophosphate)** or **cGMP.** The inflow of Na^+, called the "dark current," partially depolarizes the photoreceptor. As a result, in darkness the membrane potential of a photoreceptor is about -30 mV. This is much closer to zero than a typical neuron's resting membrane potential of -70 mV. The partial depolarization during darkness triggers continual release of neurotransmitter at the synaptic terminals. The neurotransmitter in rods, and perhaps in cones, is the amino acid glutamate (glutamic acid). At synapses between rods and some bipolar cells, glutamate is an inhibitory neurotransmitter: It triggers inhibitory postsynaptic potentials (IPSPs) that hyperpolarize the bipolar cells and prevent them from sending signals on to the ganglion cells.

When light strikes the retina and *cis*-retinal undergoes isomerization, enzymes are activated that break down cGMP. As a result, some cGMP-gated Na^+ channels close, Na^+ inflow decreases, and the membrane potential becomes more negative, approaching -70 mV (Figure 16.15b). This sequence of events produces a hyperpolarizing receptor potential that decreases the release of glutamate. Dim lights cause small and brief receptor potentials that partially turn off glutamate release; brighter lights elicit larger and longer receptor potentials that more completely shut down neurotransmitter release. Thus, light excites the bipolar cells that synapse with rods by turning off the release of an inhibitory neurotransmitter!

Figure 16.15 Operation of rod photoreceptors.

Light causes a hyperpolarizing receptor potential in photoreceptors, which decreases release of an inhibitory neurotransmitter (glutamate).

cGMP

cGMP-gated
Na+ channels open

Na+

Inflow of Na+
(dark current)

Membrane potential
of −30 mV

Glutamate released
at synaptic terminals
inhibits bipolar cell

Glutamate

(a) In darkness

Isomerization of retinal
activates enzyme that
breaks down cGMP

cGMP-gated
Na+ channels close

Inflow of Na+ slows

Hyperpolarizing
receptor potential

Glutamate release
turned off, which excites
bipolar cell

(b) In light

What is the function of cGMP in photoreceptors?

Color Blindness and Night Blindness

Most forms of **color blindness,** an inherited inability to distinguish between certain colors, result from the absence or a deficiency of one of the three cone photopigments. The most common type is red-green color blindness, in which a photopigment sensitive to orange-red light or green light is missing. As a result, the person cannot distinguish between red and green. Prolonged vitamin A deficiency and the consequent below normal amount of rhodopsin may cause **night blindness** or **nyctalopia** (nik′-ta-LŌ-pē-a), an inability to see well at low light levels. ■

The Visual Pathway

After considerable processing of visual signals in the retina at synapses among the various types of neurons (see Figure 16.9), the axons of retinal ganglion cells provide output from the retina to the brain. They exit the eyeball as the **optic nerve (cranial nerve II).**

Processing of Visual Input in the Retina

Within the retina, certain features of visual input are enhanced while other features may be discarded. Input from several cells may either converge upon a smaller number of postsynaptic neurons or diverge to a large number. Overall, however, convergence predominates because there are only 1 million ganglion cells but 126 million photoreceptors.

Once receptor potentials arise in rods and cones, they spread through the inner segments to the synaptic terminals. Neurotransmitter molecules released by rods and cones induce local graded potentials in both bipolar cells and horizontal cells. Between 6 and 600 rods synapse with a single bipolar cell in the outer synaptic layer, whereas a cone more often synapses with just one bipolar cell. The convergence of many rods onto a single bipolar cell increases the light sensitivity of rod vision but slightly blurs the image that is perceived. Cone vision, although less sensitive, has higher acuity because of the one-to-one synapses between cones and their bipolar cells. Stimulation of rods by light excites their bipolar cells, whereas cone bipolar cells may be either excited or inhibited when a light is turned on.

Horizontal cells transmit inhibitory signals to bipolar cells in the areas lateral to excited rods and cones. This lateral inhibition enhances contrasts in the visual scene between areas of the retina that are strongly stimulated and adjacent areas that are more weakly stimulated. Horizontal cells also assist in the differentiation of various colors. Amacrine cells, which are excited by bipolar cells, synapse with ganglion cells and transmit information to them that signals a change in the level of illumination of the retina. When bipolar or amacrine cells transmit excitatory signals to ganglion cells, the ganglion cells become depolarized and initiate nerve impulses.

Brain Pathway and Visual Fields

The axons within the optic nerve pass through the **optic chiasm** (kī-AZ-m = a crossover, as in the letter X), a crossing point of

Figure 16.16 The visual pathway. (a) Partial dissection of the brain reveals the optic radiations (axons extending from the thalamus to the occipital lobe). (b) An object in the binocular visual field can be seen with both eyes. In (c) and (d), note that information from the right side of the visual field of each eye projects to the left side of the brain, and information from the left side of the visual field of each eye projects to the right side of the brain.

 The axons of ganglion cells in the temporal half of each retina extend to the thalamus on the same side; the axons of ganglion cells in the nasal half of each retina extend to the thalamus on the opposite side.

the optic nerves (Figure 16.16a, b). Some axons cross to the opposite side, whereas others remain uncrossed. After passing through the optic chiasm, the axons, now part of the **optic tract,** enter the brain and terminate in the lateral geniculate nucleus of the thalamus. Here they synapse with neurons whose axons form the **optic radiations,** which project to the primary visual areas in the occipital lobes of the cerebral cortex (area 17 in Figure 14.15 on page 474).

Everything that can be seen by one eye is that eye's **visual field.** Because our eyes are located anteriorly in the head, the visual fields of the two eyes overlap considerably (Figure 16.16b). We have binocular vision due to the large region where the visual fields of the two eyes overlap—the **binocular visual field.** The visual field of each eye is divided into two regions: the **nasal** or **central half** and the **temporal** or **peripheral half** (Figure 16.16c, d). For each eye, light rays from an object in the nasal half of the visual field fall on the temporal half of the retina, and light rays from an object in the temporal half of the visual field fall on the nasal half of the retina. Moreover, visual information from the *right* half of each visual field is conveyed to the *left* side of the brain, whereas visual information from the *left* half of each visual field is conveyed to the *right* side of the brain, as follows (Figure 16.16c, d):

1. The axons of all retinal ganglion cells in one eye exit the eyeball at the optic disc and form the optic nerve on that side.

2. At the optic chiasm, axons from the temporal half of each retina do not cross but continue directly to the lateral geniculate nucleus of the thalamus on the same side.

3. In contrast, axons from the nasal half of each retina cross and continue to the opposite thalamus.

4. Each optic tract consists of crossed and uncrossed axons that project from the optic chiasm to the thalamus on one side.

5. Axon collaterals (branches) of the retinal ganglion cells project to the midbrain, where they participate in circuits that govern constriction of the pupils in response to light and coordination of head and eye movements. Collaterals also extend to the suprachiasmatic nucleus of the hypothalamus, which establishes patterns of sleep and other activities that occur on a circadian or daily schedule in response to intervals of light and darkness.

6. The axons of thalamic neurons form the optic radiations as they project from the thalamus to the primary visual area of the cortex on the same side.

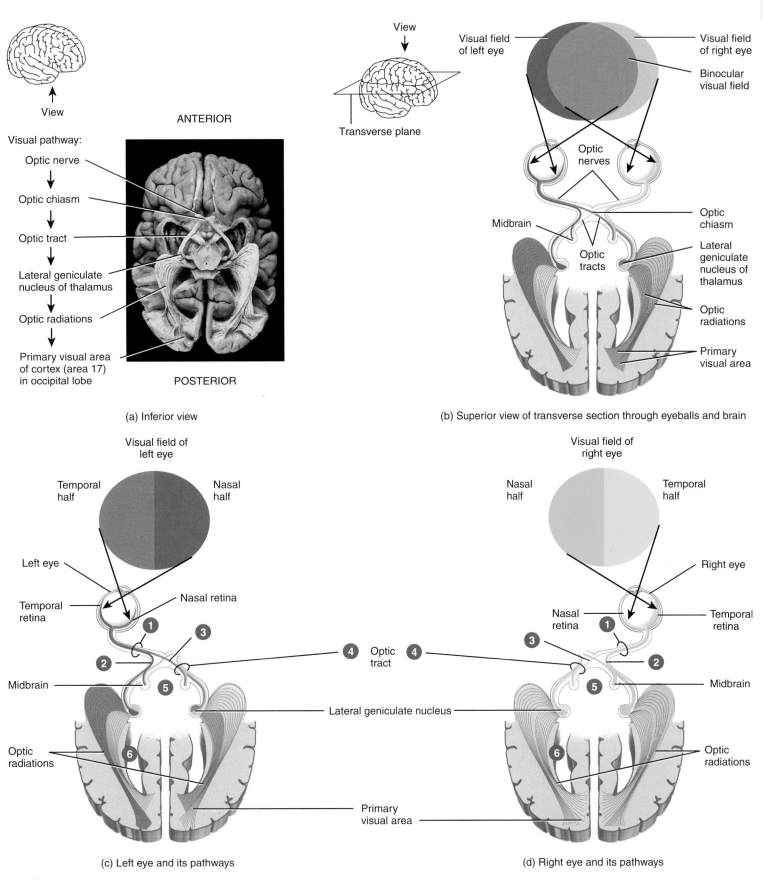

View

ANTERIOR

Visual pathway:

Optic nerve
↓
Optic chiasm
↓
Optic tract
↓
Lateral geniculate
nucleus of thalamus
↓
Optic radiations
↓
Primary visual area
of cortex (area 17)
in occipital lobe

POSTERIOR

(a) Inferior view

View

Transverse plane

Visual field
of left eye

Visual field
of right eye

Binocular
visual field

Optic
nerves

Optic
chiasm

Midbrain

Optic
tracts

Lateral
geniculate
nucleus of
thalamus

Optic
radiations

Primary
visual area

(b) Superior view of transverse section through eyeballs and brain

Visual field of
left eye

Temporal
half

Nasal
half

Left eye

Temporal
retina

Nasal retina

Midbrain

Optic
radiations

Optic
tract

Lateral geniculate nucleus

Primary
visual area

(c) Left eye and its pathways

Visual field of
right eye

Nasal
half

Temporal
half

Right eye

Nasal
retina

Temporal
retina

Midbrain

Optic
radiations

(d) Right eye and its pathways

Light rays from an object in the temporal half of the visual field strike which half of the retina?

Although we have just described the visual pathway as a single pathway, visual signals are thought to be processed by at least three separate systems in the cerebral cortex, each with its own function. One system processes information related to the shape of objects, another system processes information regarding color of objects, and a third system processes information about movement, location, and spatial organization.

▶ C H E C K P O I N T

5. What is the function of the lacrimal apparatus?

6. What types of cells make up the neural layer and the pigmented layer of the retina?

7. How do photopigments respond to light and recover in darkness?

8. How do receptor potentials arise in photoreceptors?

9. What is the pathway whereby nerve impulses triggered by an object in the nasal half of the visual field of the left eye reach the primary visual area of the cortex?

HEARING AND EQUILIBRIUM

▶ O B J E C T I V E S

• **Describe the anatomy of the structures in the three main regions of the ear.**

• **List the major events in the physiology of hearing.**

• **Identify the receptor organs for equilibrium, and describe how they function.**

The ear is an engineering marvel because its sensory receptors can transduce sound vibrations with amplitudes as small as the diameter of an atom of gold (0.3 nm) into electrical signals 1000 times faster than photoreceptors can respond to light. Besides receptors for sound waves, the ear also contains receptors for equilibrium.

Anatomy of the Ear

The ear is divided into three main regions: the external ear, which collects sound waves and channels them inward; the middle ear, which conveys sound vibrations to the oval window; and the internal ear, which houses the receptors for hearing and equilibrium.

External (Outer) Ear

The **external (outer) ear** consists of the auricle, external auditory canal, and eardrum (Figure 16.17). The **auricle (pinna)** is a flap of elastic cartilage shaped like the flared end of a trumpet and covered by skin. The rim of the auricle is the **helix;** the inferior portion is the **lobule.** Ligaments and muscles attach the auricle to the head. The **external auditory canal** (*audit-* = hearing) is a curved tube about 2.5 cm (1 in.) long that lies in the temporal bone and leads from the auricle to the eardrum. The **eardrum** or **tympanic membrane** (tim-PAN-ik; *tympan-* = a drum) is a thin, semitransparent partition between the external auditory canal and middle ear. The eardrum is covered by epidermis and lined by simple cuboidal epithelium. Between the epithelial layers is connective tissue composed of collagen, elastic fibers, and fibroblasts.

Near the exterior opening, the external auditory canal contains a few hairs and specialized sebaceous (oil) glands called **ceruminous glands** (se-ROO-mi-nus) that secrete earwax or **cerumen** (se-ROO-men). The combination of hairs and cerumen helps prevent dust and foreign objects from entering the ear. Cerumen usually dries up and falls out of the ear canal. Some people, however, produce a large amount of cerumen, which can become impacted and can muffle incoming sounds.

Middle Ear

The **middle ear** is a small, air-filled cavity in the temporal bone that is lined by epithelium (Figure 16.18). It is separated from the external ear by the eardrum and from the internal ear by a thin bony partition that contains two small membrane-covered openings: the oval window and the round window. Extending across the middle ear and attached to it by ligaments are the three smallest bones in the body, the **auditory ossicles** (OS-si-kuls), which are connected by synovial joints. The bones, named for their shapes, are the malleus, incus, and stapes—commonly called the hammer, anvil, and stirrup, respectively. The "handle" of the **malleus** (MAL-ē-us) attaches to the internal surface of the eardrum. The head of the malleus articulates with the body of the incus. The **incus** (ING-kus), the middle bone in the series, articulates with the head of the stapes. The base or footplate of the **stapes** (STĀ-pēz) fits into the **oval window.** Directly below the oval window is another opening, the **round window,** which is enclosed by a membrane called the **secondary tympanic membrane.**

Besides the ligaments, two tiny skeletal muscles also attach to the ossicles (Figure 16.18 on page 548). The **tensor tympani muscle,** which is innervated by the mandibular branch of the trigeminal nerve (cranial nerve V), limits movement and increases tension on the eardrum to prevent damage to the inner ear from loud noises. The **stapedius muscle,** which is innervated by the facial nerve (cranial nerve VII), is the smallest of all skeletal muscles. By dampening large vibrations of the stapes due to loud noises, it protects the oval window, but it also decreases the sensitivity of hearing. For this reason, paralysis of the stapedius muscle is associated with **hyperacusia** (abnormally sensitive hearing). Because it takes a fraction of a second for the tensor tympani and stapedius muscles to contract, they can protect the inner ear from prolonged loud noises, but not from brief ones such as a gunshot.

The anterior wall of the middle ear contains an opening that leads directly into the **auditory (pharyngotympanic) tube,** commonly known as the **Eustachian tube.** The auditory tube, which consists of both bone and hyaline cartilage, connects the middle ear with the nasopharynx (upper portion of the throat). It is normally closed at its medial (pharyngeal) end. During swallowing and yawning, it opens, allowing air to enter or leave the middle ear until the pressure in the middle ear equals the atmospheric pressure. When the pressures are balanced, the eardrum

Figure 16.17 Anatomy of the ear. (See Tortora, *A Photographic Atlas of the Human Body,* Figure 9.4a.)

The ear has three principal regions: the external (outer) ear, the middle ear, and the internal (inner) ear.

Frontal plane

Temporal bone
Malleus
Incus
Semicircular canal
Internal auditory canal
Vestibulocochlear nerve:
Vestibular branch
Cochlear branch
Helix
Auricle
Cochlea
Lobule
Stapes in oval window
Elastic cartilage
Cerumen
Round window (covered by secondary tympanic membrane)
To nasopharynx
External auditory canal
Eardrum
Auditory tube

☐ External ear
☐ Middle ear
☐ Internal ear

Frontal section through the right side of the skull showing the three principal regions of the ear

To which structure of the external ear does the malleus of the middle ear attach?

vibrates freely as sound waves strike it. If the pressure is not equalized, intense pain, hearing impairment, ringing in the ears, and vertigo could develop. The auditory tube also is a route whereby pathogens may travel from the nose and throat to the middle ear.

Internal (Inner) Ear

The **internal (inner) ear** is also called the **labyrinth** (LAB-i-rinth) because of its complicated series of canals (Figure 16.19 on page 549). Structurally, it consists of two main divisions: an outer bony labyrinth that encloses an inner membranous labyrinth. The **bony labyrinth** is a series of cavities in the temporal bone divided into three areas: (1) the semicircular canals and (2) the vestibule, both of which contain receptors for equilibrium, and (3) the cochlea, which contains receptors for hearing. The bony labyrinth is lined with periosteum and contains **perilymph.** This fluid, which is chemically similar to cerebrospinal fluid, surrounds the **membranous labyrinth,** a series of sacs and tubes inside the bony labyrinth and having the same general form. The membra-

nous labyrinth is lined by epithelium and contains **endolymph.** The level of potassium ions (K^+) in endolymph is unusually high for an extracellular fluid, and potassium ions play a role in the generation of auditory signals (described shortly).

The **vestibule** (VES-ti-būl) is the oval central portion of the bony labyrinth. The membranous labyrinth in the vestibule consists of two sacs called the **utricle** (Ū-tri-kl = little bag) and the **saccule** (SAK-ūl = little sac), which are connected by a small duct. Projecting superiorly and posteriorly from the vestibule are the three bony **semicircular canals,** each of which lies at approximately right angles to the other two. Based on their positions, they are named the anterior, posterior, and lateral semicircular canals. The anterior and posterior semicircular canals are vertically oriented; the lateral one is horizontally oriented. At one end of each canal is a swollen enlargement called the **ampulla** (am-PUL-la = saclike duct). The portions of the membranous labyrinth that lie inside the bony semicircular canals are called the **semicircular ducts.** These structures communicate with the utricle of the vestibule.

Figure 16.18   The right middle ear containing the auditory ossicles.   (See Tortora, *A Photographic Atlas of the Human Body,* Figure 3.14.)

 Common names for the malleus, incus, and stapes are the hammer, anvil, and stirrup, respectively.

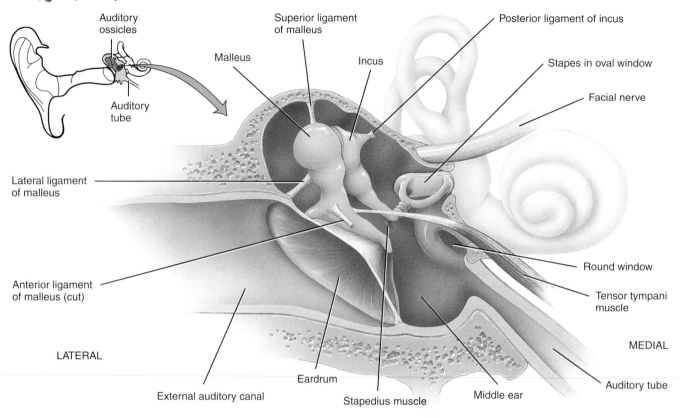

Frontal section showing location of auditory ossicles

What structures separate the middle ear from the external ear and from the internal ear?

The vestibular branch of the vestibulocochlear nerve (cranial nerve VIII) consists of *ampullary, utricular,* and *saccular nerves.* These nerves contain both first-order sensory neurons and motor neurons that synapse with receptors for equilibrium. The first-order sensory neurons carry sensory information from the receptors, and the motor neurons carry feedback signals to the receptors, apparently to modify their sensitivity. Cell bodies of the sensory neurons are located in the **vestibular ganglia** (see Figure 16.20b).

Anterior to the vestibule is the **cochlea** (KOK-lē-a = snail-shaped), a bony spiral canal (Figure 16.20a) that resembles a snail's shell and makes almost three turns around a central bony core called the **modiolus** (mō-DĪ-ō-lus; Figure 16.20b). Sections through the cochlea reveal that it is divided into three channels (Figure 16.20a–c). Together, the partitions that separate the channels are shaped like the letter Y. The stem of the Y is a bony shelf that protrudes into the canal; the wings of the Y are composed mainly of membranous labyrinth. The channel above the bony partition is the **scala vestibuli,** which ends at the oval win-

dow; the channel below is the **scala tympani,** which ends at the round window.

The scala vestibuli and scala tympani both contain peri-lymph and are completely separated, except for an opening at the apex of the cochlea, the **helicotrema** (hel-i-kō-TRĒ-ma). See Figure 16.21. The cochlea adjoins the wall of the vestibule, into which the scala vestibuli opens. The perilymph in the vestibule is continuous with that of the scala vestibuli. The third channel (between the wings of the Y) is the **cochlear duct** or **scala media.** The **vestibular membrane** separates the cochlear duct from the scala vestibuli, and the **basilar membrane** separates the cochlear duct from the scala tympani.

Resting on the basilar membrane is the **spiral organ** or **organ of Corti** (Figure 16.20c, d). The spiral organ is a coiled sheet of epithelial cells, including supporting cells and about 16,000 **hair cells,** which are the receptors for hearing. There are two groups of hair cells: The *inner hair cells* are arranged in a single row whereas the *outer hair cells* are arranged in three rows. At the apical tip of each hair cell is a **hair bundle,** consisting

Figure 16.19 **The right internal ear.** The outer, cream-colored area is part of the bony labyrinth; the inner, pink-colored area is the membranous labyrinth.

The bony labyrinth contains perilymph, and the membranous labyrinth contains endolymph.

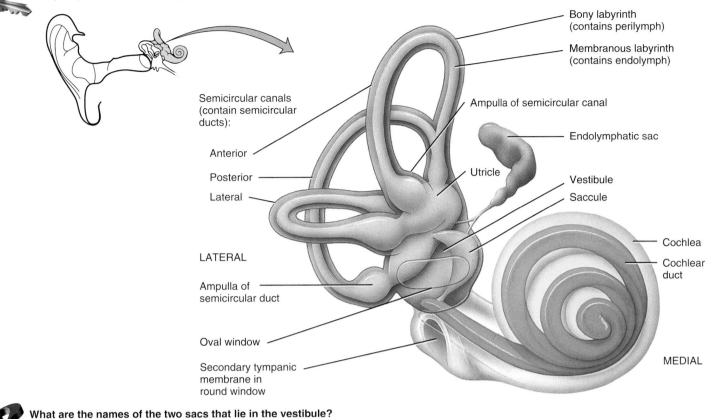

What are the names of the two sacs that lie in the vestibule?

Figure 16.20 **Semicircular canals, vestibule, and cochlea of the right ear.** Note that the cochlea makes nearly three complete turns. (See Tortora, *A Photographic Atlas of the Human Body,* Figure 9.4b.)

The three channels in the cochlea are the scala vestibuli, the scala tympani, and the cochlear duct.

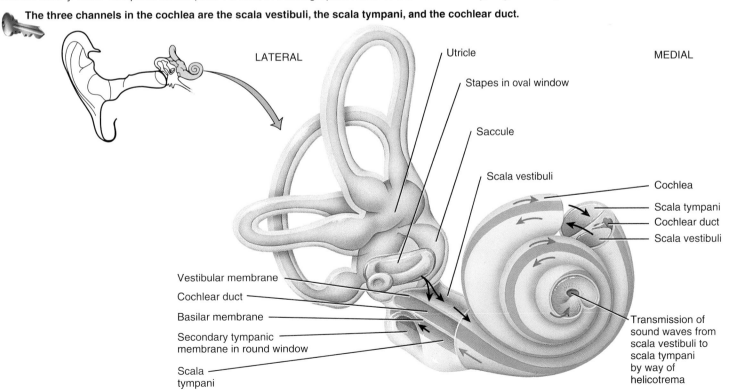

(a) Sections through the cochlea

(continues)

Figure 16.20 *(continued)*

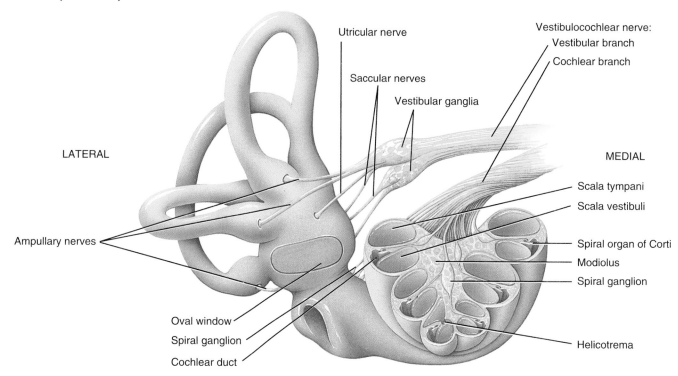

LATERAL

Utricular nerve

Saccular nerves

Vestibular ganglia

Vestibulocochlear nerve:
Vestibular branch
Cochlear branch

MEDIAL

Ampullary nerves

Scala tympani

Scala vestibuli

Spiral organ of Corti

Modiolus

Spiral ganglion

Oval window

Spiral ganglion

Cochlear duct

Helicotrema

(b) Components of the vestibulocochlear nerve (cranial nerve VIII)

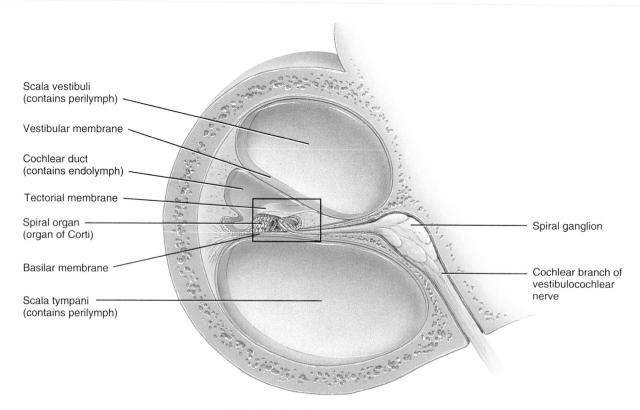

Scala vestibuli
(contains perilymph)

Vestibular membrane

Cochlear duct
(contains endolymph)

Tectorial membrane

Spiral organ
(organ of Corti)

Basilar membrane

Scala tympani
(contains perilymph)

Spiral ganglion

Cochlear branch of
vestibulocochlear
nerve

(c) Section through one turn of the cochlea

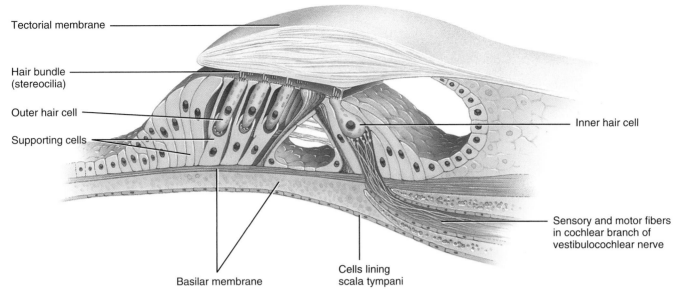

Tectorial membrane

Hair bundle (stereocilia)

Outer hair cell

Supporting cells

Inner hair cell

Sensory and motor fibers in cochlear branch of vestibulocochlear nerve

Basilar membrane

Cells lining scala tympani

(d) Enlargement of spiral organ (organ of Corti)

 What are the three subdivisions of the bony labyrinth?

of 30–100 *stereocilia* that extend into the endolymph of the cochlear duct. Despite their name, stereocilia are actually long, hairlike microvilli arranged in several rows of graded height.

At their basal ends, inner and outer hair cells synapse both with first-order sensory neurons and with motor neurons from the cochlear branch of the vestibulocochlear nerve (cranial nerve VIII). Cell bodies of the sensory neurons are located in the **spiral ganglion** (Figure 16.20b, c). Although outer hair cells outnumber them by 3 to 1, the inner hair cells synapse with 90–95% of the first-order sensory neurons in the cochlear nerve that relay auditory information to the brain. By contrast, 90% of the motor neurons in the cochlear nerve synapse with outer hair cells. Projecting over and in contact with the hair cells of the spiral organ is the **tectorial membrane** (*tector-* = covering), a flexible gelatinous membrane.

The Nature of Sound Waves

Sound waves are a series of alternating high- and low-pressure regions traveling in the same direction through some medium (such as air). They originate from a vibrating object in much the same way that ripples arise and travel over the surface of a pond when you toss a stone into it. The sounds heard most acutely by the human ear are those from sources that vibrate at frequencies between 500 and 5000 hertz (Hz; 1 Hz = 1 cycle per second). The entire audible range extends from 20 to 20,000 Hz. Sounds of speech contain frequencies mainly between 100 and 3000 Hz, and the "high C" sung by a soprano has a dominant frequency at 1048 Hz. The sounds from a jet plane several miles away range from 20 to 100 Hz.

The *frequency* of a sound vibration is its *pitch.* The higher the frequency of vibration, the higher is the pitch. Also, the larger the *intensity* (size or amplitude) of the vibration, the *louder* is the sound. Sound intensity is measured in units called **decibels (dB).** Each ten decibels represents a tenfold increase in sound intensity. The hearing threshold—the point at which an average young adult can just distinguish sound from silence—is defined as 0 dB at 1000 Hz. Rustling leaves have a decibel level of 15; whispered speech, 30; normal conversation, 60; a vacuum cleaner, 75; shouting, 80; and a nearby motorcycle or jackhammer, 90. Sound becomes uncomfortable to a normal ear at about 120 dB, and painful above 140 dB.

Loud Sounds and Hair Cell Damage

Exposure to loud music and the engine roar of jet planes, revved-up motorcycles, lawn mowers, and vacuum cleaners damages hair cells of the cochlea. Because prolonged noise exposure causes hearing loss, employers in the United States must require workers to use hearing protectors when occupational noise levels exceed 90 dB. Rock concerts and even inexpensive headphones can easily produce sounds over 110 dB. Continued exposure to high-intensity sounds is one cause of **deafness,** a significant or total hearing loss. The louder the sounds, the more rapid is the hearing loss. Deafness usually begins with loss of sensitivity for high-pitched sounds. If you are listening to music through headphones and bystanders can hear it, the dB level is in the damaging range. Most people fail to notice their progressive hearing loss until destruction is extensive and they begin having difficulty understanding speech. Wearing earplugs with a noise-reduction rating of 30 dB while engaging in noisy activities can protect the sensitivity of your ears. ■

Physiology of Hearing

The following events are involved in hearing (Figure 16.21):

1 The auricle directs sound waves into the external auditory canal.

2 When sound waves strike the eardrum, the alternating high- and low-pressure of the air cause the eardrum to vibrate back and forth. The distance it moves, which is very small, depends on the intensity and frequency of the sound waves. The eardrum vibrates slowly in response to low-frequency (low-pitched) sounds and rapidly in response to high-frequency (high-pitched) sounds.

3 The central area of the eardrum connects to the malleus, which also starts to vibrate. The vibration is transmitted from the malleus to the incus and then to the stapes.

4 As the stapes moves back and forth, it pushes the membrane of the oval window in and out. The oval window vibrates about 20 times more vigorously than the eardrum because the ossicles efficiently transmit small vibrations spread over a large surface area (eardrum) into larger vibrations of a smaller surface (oval window).

5 The movement of the oval window sets up fluid pressure waves in the perilymph of the cochlea. As the oval window bulges inward, it pushes on the perilymph of the scala vestibuli.

6 Pressure waves are transmitted from the scala vestibuli to the scala tympani and eventually to the round window, causing it to bulge outward into the middle ear. (See **9** in the figure.)

7 As the pressure waves deform the walls of the scala vestibuli and scala tympani, they also push the vestibular membrane back and forth, creating pressure waves in the endolymph inside the cochlear duct.

8 The pressure waves in the endolymph cause the basilar membrane to vibrate, which moves the hair cells of the spiral organ against the tectorial membrane. Bending of the stereocilia produces receptor potentials that ultimately lead to the generation of nerve impulses.

Figure 16.21 Events in the stimulation of auditory receptors in the right ear. The numbers correspond to the events listed in the text. The cochlea has been uncoiled to more easily visualize the transmission of sound waves and their distortion of the vestibular and basilar membranes of the cochlear duct.

Hair cells of the spiral organ (organ of Corti) convert a mechanical vibration (stimulus) into an electrical signal (receptor potential).

 Which part of the basilar membrane vibrates most vigorously in response to high-frequency (high-pitched) sounds?

Sound waves of various frequencies cause certain regions of the basilar membrane to vibrate more intensely than other regions. Each segment of the basilar membrane is "tuned" for a particular pitch. Because the membrane is narrower and stiffer at the base of the cochlea (portion closer to the oval window), high-frequency (high-pitched) sounds near 20,000 Hz induce maximal vibrations in this region. Toward the apex of the cochlea near the helicotrema, the basilar membrane is wider and more flexible; low-frequency (low-pitched) sounds near 20 Hz cause maximal vibration of the basilar membrane there. Loudness is determined by the intensity of sound waves. High-intensity sound waves cause larger vibrations of the basilar membrane, which leads to a higher frequency of nerve impulses reaching the brain. Louder sounds also may stimulate a larger number of hair cells.

The hair cells transduce mechanical vibrations into electrical signals. As the basilar membrane vibrates, the hair bundles at the apex of the hair cell bend back and forth and slide against one another. A *tip link* protein connects the tip of each stereocilium to a mechanically gated ion channel called the **transduction channel** in its taller stereocilium neighbor. As the stereocilia bend in the direction of the taller stereocilia, the tip links tug on the transduction channels and open them. These channels allow cations in the endolymph, primarily K^+, to enter the hair cell cytosol. As cations enter, they produce a depolarizing receptor potential. Depolarization quickly spreads along the plasma membrane and opens voltage-gated Ca^{2+} channels in the base of the hair cell. The resulting inflow of Ca^{2+} triggers exocytosis of synaptic vesicles containing a neurotransmitter, which is probably glutamate. As more neurotransmitter is released, the frequency of nerve impulses in the first-order sensory neurons that synapse with the base of the hair cell increases. Bending of the stereocilia in the opposite direction closes the transduction channels, allows repolarization or even hyperpolarization to occur, and reduces neurotransmitter release from the hair cells. This decreases the frequency of nerve impulses in the sensory neurons.

Besides its role in detecting sounds, the cochlea has the surprising ability to produce sounds. These usually inaudible sounds, called **otoacoustic emissions,** can be picked up by placing a sensitive microphone next to the eardrum. They are caused by vibrations of the outer hair cells that occur in response to sound waves and to signals from motor neurons. As they depolarize and repolarize, the outer hair cells rapidly shorten and lengthen. This vibratory behavior appears to change the stiffness of the tectorial membrane and is thought to enhance the movement of the basilar membrane, which amplifies the responses of the inner hair cells. At the same time, the outer hair cell vibrations set up a traveling wave that goes back toward the stapes and leaves the ear as an otoacoustic emission. Detection of these inner ear–produced sounds is a fast, inexpensive, and noninvasive way to screen newborns for hearing defects. In deaf babies, otoacoustic emissions are not produced or are greatly reduced in size.

The Auditory Pathway

First-order sensory neurons in the cochlear branch of each vestibulocochlear nerve (cranial nerve VIII) terminate in the cochlear nuclei of the medulla oblongata on the same side. From there, axons carrying auditory signals project to the superior olivary nuclei in the pons on both sides. Slight differences in the timing of impulses arriving from the two ears at the olivary nuclei allow us to locate the source of a sound. From both the cochlear nuclei and the olivary nuclei, axons ascend to the inferior colliculus in the midbrain, and then to the medial geniculate body of the thalamus. From the thalamus, auditory signals project to the primary auditory area in the superior temporal gyrus of the cerebral cortex (Brodmann's areas 41 and 42 in Figure 14.15 on page 474). Because many auditory axons decussate (cross over) in the medulla while others remain on the same side, the right and left primary auditory areas receive nerve impulses from both ears.

Cochlear Implants

Cochlear implants translate sounds into electronic signals that can be interpreted by the brain. They are useful for people with deafness that is caused by damage to hair cells in the cochlea. Sound waves are picked up by a tiny microphone and converted into electrical signals. The signals then travel to electrodes implanted in the cochlea, where they trigger nerve impulses in sensory neurons in the cochlear branch of the vestibulocochlear nerve. These artifically induced nerve impulses propagate over their normal pathways to the brain. The sounds perceived are crude compared to normal hearing, but they provide a sense of rhythm and loudness; information about certain noises, such as those made by telephones and automobiles; and the pitch and cadence of speech. Some patients hear well enough with a cochlear implant to use the telephone. ■

Physiology of Equilibrium

There are two kinds of **equilibrium** (balance). One kind, called **static equilibrium,** refers to the maintenance of the position of the body (mainly the head) relative to the force of gravity. The second kind, **dynamic equilibrium,** is the maintenance of body position (mainly the head) in response to sudden movements such as rotation, acceleration, and deceleration. Collectively, the receptor organs for equilibrium are called the **vestibular apparatus** (ves-TIB-ū-lar), which includes the saccule, utricle, and semicircular ducts.

Otolithic Organs: Saccule and Utricle

The walls of both the utricle and the saccule contain a small, thickened region called a **macula** (MAK-ū-la; Figure 16.22). The two maculae (plural), which are perpendicular to one

Figure 16.22 Location and structure of receptors in the maculae of the right ear.
Both first-order sensory neurons (blue) and motor neurons (red) synapse with the hair cells.

 The movement of stereocilia initiates depolarizing receptor potentials.

(a) Overall structure of a section of the macula

(b) Details of two hair cells

(c) Position of macula with head upright (left) and tilted forward (right)

With which type of equilibrium are the maculae mainly concerned?

another, are the receptors for static equilibrium. They also contribute to some aspects of dynamic equilibrium. For static equilibrium, they provide sensory information on the position of the head in space and are essential for maintaining appropriate posture and balance. For dynamic equilibrium, they detect linear acceleration and deceleration—for example, the sensations you feel while in an elevator or a car that is speeding up or slowing down.

The two maculae consist of two kinds of cells: **hair cells,** which are the sensory receptors, and **supporting cells.** Hair cells feature **hair bundles** that consist of 70 or more *stereocilia,* which are actually microvilli, plus one *kinocilium,* a conventional cilium anchored firmly to its basal body and extending beyond the longest stereocilia. As in the cochlea, the stereocilia are connected by tip links. Scattered among the hair cells are columnar supporting cells that probably secrete the thick, gelatinous, glycoprotein layer, called the **otolithic membrane,** that rests on the hair cells. A layer of dense calcium carbonate crystals, called **otoliths** (*oto-* = ear; *-liths* = stones) extends over the entire surface of the otolithic membrane.

Because the otolithic membrane sits on top of the macula, if you tilt your head forward, the otolithic membrane (and the otoliths as well) is pulled by gravity. It slides "downhill" over the hair cells in the direction of the tilt, bending the hair bundles. However, if you are sitting upright in a car that suddenly jerks forward, the otolithic membrane, due to its inertia, lags behind the head movement. As the otolithic membrane lags behind, it pulls on the hair bundles and makes them bend in the other direction. Bending of the hair bundles in one direction stretches the tip links, which pull open transduction channels thereby producing depolarizing receptor potentials; bending in the opposite direction closes the transduction channels and produces repolarization.

As the hair cells depolarize and repolarize, they release neurotransmitter at a faster or slower rate. The hair cells synapse with first-order sensory neurons in the vestibular branch of the vestibulocochlear nerve (cranial nerve VIII) (see Figure 16.21a). These neurons fire impulses at a slow or rapid pace depending on how much neurotransmitter is present. Motor neurons also synapse with the hair cells and sensory neurons. Evidently, they regulate the sensitivity of the hair cells and sensory neurons.

Semicircular Ducts

The three semicircular ducts, together with the saccule and the utricle, function in dynamic equilibrium. The ducts lie at right angles to one another in three planes (Figure 16.23): The two vertical ducts are the anterior and posterior semicircular ducts, and the horizontal one is the lateral semicircular duct (see also Figure 16.19). This positioning permits detection of rotational acceleration or deceleration. In the **ampulla,** the dilated portion of each duct, is a small elevation called the **crista.** Each crista contains a group of **hair cells** and **supporting cells** covered by a mass of gelatinous material called the **cupula** (KŪ-pū-la). When the head moves, the attached semicircular ducts and hair cells move with it. The endolymph within the ampulla, however, is not attached and lags behind due to its inertia. As the moving hair cells drag along the stationary fluid, the hair bundles bend. Bending of the hair bundles produces receptor potentials. In turn, the receptor potentials lead to nerve impulses that pass along the vestibular branch of the vestibulocochlear nerve (cranial nerve VIII).

Equilibrium Pathways

Most of the vestibular branch axons of the vestibulocochlear nerve (cranial nerve VIII) enter the brain stem and terminate in several vestibular nuclei in the medulla and pons. The remaining axons enter the cerebellum through the inferior cerebellar peduncle (see Figure 14.5b). Bidirectional pathways connect the vestibular nuclei and cerebellum.

Axons from all the vestibular nuclei extend to the nuclei of cranial nerves that control eye movements—oculomotor (cranial nerve III), trochlear (cranial nerve IV), and abducens (cranial nerve VI). Other axons from vestibular nuclei extend to the nucleus of the accessory nerve (cranial nerve XI), which helps control head and neck movements. In addition, axons from the lateral vestibular nucleus form the vestibulospinal tract, which conveys impulses to skeletal muscles that regulate muscle tone in response to head movements.

Various pathways between the vestibular nuclei, cerebellum, and cerebrum enable the cerebellum to play a key role in maintaining static and dynamic equilibrium. The cerebellum continuously receives updated sensory information from the utricle and saccule. It monitors this information and makes corrective adjustments. Essentially, in response to input from the utricle, saccule, and semicircular ducts, the cerebellum continuously sends nerve impulses to the motor areas of the cerebrum. This feedback allows correction of signals from the motor cortex to specific skeletal muscles to maintain equilibrium.

Table 16.2 on page 557 summarizes the structures of the ear related to hearing and equilibrium.

▶ C H E C K P O I N T

10. How are sound waves transmitted from the auricle to the spiral organ of Corti?

11. How do hair cells in the cochlea and vestibular apparatus transduce mechanical vibrations into electrical signals?

12. What is the pathway for auditory impulses from the cochlea to the cerebral cortex?

13. Compare the function of the maculae in maintaining static equilibrium with the role of the cristae in maintaining dynamic equilibrium.

14. What is the role of vestibular input to the cerebellum?

Figure 16.23   Location and structure of the semicircular ducts of the right ear.   Both first-order sensory neurons (blue) and motor neurons (red) synapse with the hair cells. The ampullary nerves are branches of the vestibular division of the vestibulocochlear nerve (cranial nerve VIII).

 The positions of the semicircular ducts permit detection of rotational movements.

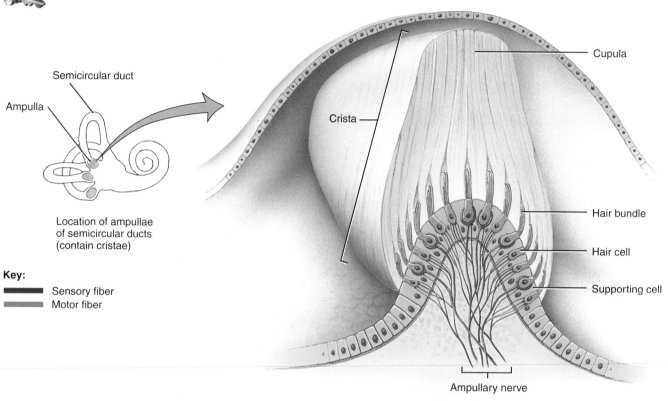

(a) Details of a crista

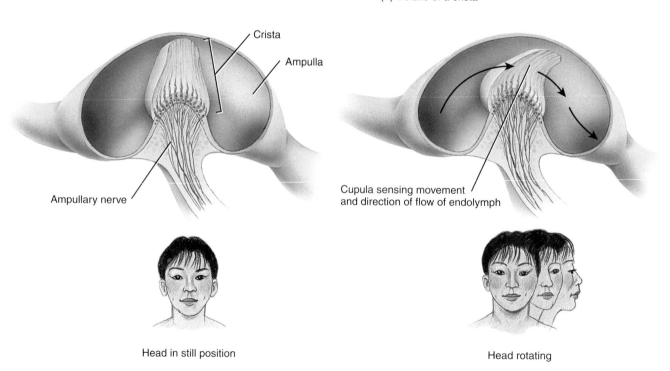

(b) Position of a crista with the head in the still position (left)
and when the head rotates (right)

With which type of equilibrium are the semicircular ducts, the utricle, and the saccule associated?

Table 16.2 Summary of Structures of the Ear

Regions of the Ear and Key Structures	Function
External (outer) ear	*Auricle (pinna):* Collects sound waves.
External auditory canal, Auricle, Eardrum	*External auditory canal (meatus):* Directs sound waves to the eardrum.
	Eardrum (tympanic membrane): Sound waves cause it to vibrate, which, in turn, causes the malleus to vibrate.
Middle ear	*Auditory ossicles:* Transmit and amplify vibrations from tympanic membrane to oval window.
Auditory ossicles, Auditory tube	*Auditory tube:* Equalizes air pressure on both sides of the tympanic membrane.
Internal (inner) ear	*Cochlea:* Contains a series of fluids, channels, and membranes that transmit vibrations to the spiral organ (organ of Corti), the organ of hearing; hair cells in the spiral organ produce receptor potentials, which elicit nerve impulses in the cochlear branch of the vestibulocochlear nerve (cranial nerve VIII).
Semicircular ducts, Utricle, Cochlea, Saccule	*Vestibular apparatus:* Includes semicircular ducts, utricle, and saccule, which generate nerve impulses that propagate along the vestibular branch of the vestibulocochlear nerve (cranial nerve VIII).
	Semicircular ducts: Contain cristae, site of hair cells for dynamic equilibrium.
	Utricle: Contains macula, site of hair cells for static and dynamic equilibrium.
	Saccule: Contains macula, site of hair cells for static and dynamic equilibrium.

DEVELOPMENT OF THE EYES AND EARS

▶ OBJECTIVE

• **Describe the development of the eyes and the ears.**

Eyes

The *eyes* begin to develop about 22 days after fertilization when the **ectoderm** of the lateral walls of the prosencephalon (forebrain) bulges out to form a pair of shallow grooves called the **optic grooves** (Figure 16.24a). Within a few days, as the neural tube is closing, the optic grooves enlarge and grow toward the surface ectoderm and become known as the **optic vesicles** (Figure 16.24b). When the optic vesicles reach the surface ectoderm, the surface ectoderm thickens to form the **lens placodes.** In addition, the distal portions of the optic vesicles invaginate (Figure 16.24c), forming the **optic cups**; they remain attached to the prosencephalon by narrow, hollow proximal structures called **optic stalks** (Figure 16.24d).

The lens placodes also invaginate and develop into lens vesicles that sit in the optic cups. The lens vesicles eventually develop into the *lenses.* Blood is supplied to the developing lenses (and retina) by the hyaloid arteries. These arteries gain access to the developing eyes through a groove on the inferior surface of the optic cup and optic stalk called the **choroid fissure.** As the lenses mature, part of the hyaloid arteries that pass through the vitreous chamber degenerate; the remaining portions of the hyaloid arteries become the *central retinal arteries.*

The inner wall of the optic cup forms the *neural layer* of the optic part of the retina, while the outer layer forms the *pigmented layer* of the optic part of the retina. Axons from the neural layer grow through the optic stalk to the brain, converting the optic stalk to the *optic nerve* (cranial nerve II). Although myelination of the optic nerves begins late in fetal life, it is not completed until the tenth week after birth.

Figure 16.24   Development of the eyes.

🔑 **The eyes begin to develop about 22 days after fertilization from ectoderm of the prosencephalon.**

External view, about 28-day embryo

(a) About 22 days

(b) About 28 days

(c) About 31 days

(d) About 32 days

❓ **Which structure gives rise to the neural and pigmented layers of the optic part of the retina?**

The anterior portion of the optic cup forms the epithelium of the *ciliary body, iris,* and *circular and radial muscles* of the iris. The connective tissue of the ciliary body, *ciliary muscle,* and *zonular fibers* of the lens develop from **mesenchyme** around the anterior portion of the optic cup.

Mesenchyme surrounding the optic cup and optic stalk differentiates into an inner layer that gives rise to the *choroid* and an outer layer that develops into the *sclera* and part of the cornea. The remainder of the cornea is derived from surface **ectoderm.**

The *anterior chamber* develops from a cavity that forms in the **mesenchyme** between the iris and cornea; the *posterior chamber* develops from a cavity that forms in the **mesenchyme** between the iris and lens.

Some mesenchyme around the developing eye enters the optic cup through the choroid fissure. This mesenchyme occupies

Figure 16.25 Development of the ears.

The first part of the ears to develop are the internal ears, which begin to form about 22 days after fertilization as thickenings of surface ectoderm.

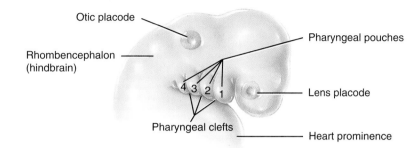

External view, about 28-day embryo

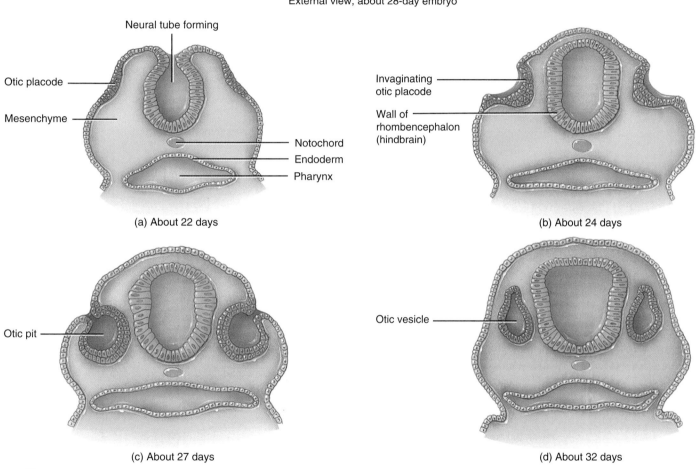

(a) About 22 days

(b) About 24 days

(c) About 27 days

(d) About 32 days

How do the three parts of the ear differ in origin?

the space between the lens and retina and differentiates into a delicate network of fibers. Later the spaces between the fibers fill with a jellylike substance, thus forming the *vitreous body* in the vitreous chamber.

The *eyelids* form from surface **ectoderm** and **mesenchyme.** The upper and lower eyelids meet and fuse at about eight weeks of development and remain closed until about 26 weeks of development.

Ears

The first portion of the ear to develop is the *internal ear.* It begins to form about 22 days after fertilization as a thickening of the surface ectoderm, called **otic placodes** (Figure 16.25a), that appear on either side of the rhombencephalon (hindbrain). The otic placodes invaginate quickly (Figure 16.25b) to form the **otic pits** (Figure 16.25c). Next, the otic pits pinch off from the surface ectoderm to form the **otic vesicles** within the mesenchyme of the head (Figure 16.25d). During later development, the otic vesicles will form the structures associated with the *membranous labyrinth* of the internal ear. **Mesenchyme** around the otic vesicles produces cartilage that later ossifies to form the bone associated with the *bony labyrinth* of the internal ear.

The *middle ear* develops from a structure called the first **pharyngeal (branchial) pouch,** an **endoderm**-lined outgrowth of the primitive pharynx (see the inset in Figure 16.25). The pharyngeal pouches are discussed in detail in Chapter 29 on page 1079. The *auditory ossicles* develop from the first and second pharyngeal pouches.

The *external ear* develops from the first **pharyngeal cleft,** an **endoderm**-lined groove between the first and second pharyngeal pouches (see the inset in Figure 16.25). The pharyngeal clefts are discussed in detail in Chapter 29 on page 1079.

DISORDERS: HOMEOSTATIC IMBALANCES

Cataracts

A common cause of blindness is a loss of transparency of the lens known as a **cataract.** The lens becomes cloudy (less transparent) due to changes in the structure of the lens proteins. Cataracts often occur with aging but may also be caused by injury, excessive exposure to ultraviolet rays, certain medications (such as long-term use of steroids), or complications of other diseases (for example, diabetes). People who smoke also have increased risk of developing cataracts. Fortunately, sight can usually be restored by surgical removal of the old lens and implantation of a new artificial one.

Glaucoma

Glaucoma is the most common cause of blindness in the United States, afflicting about 2% of the population over age 40. Glaucoma is an abnormally high intraocular pressure due to a buildup of aqueous humor within the anterior cavity. The fluid compresses the lens into the vitreous body and puts pressure on the neurons of the retina. Persistent pressure results in a progression from mild visual impairment to irreversible destruction of neurons of the retina, damage to the optic nerve, and blindness. Glaucoma is painless, and the other eye compensates largely, so a person may experience considerable retinal damage and loss of vision before the condition is diagnosed. Because glaucoma occurs more often with advancing age, regular measurement of intraocular pressure is an increasingly important part of an eye exam as people grow older. Risk factors include race (blacks are more susceptible), increasing age, family history, and past eye injuries and disorders.

Deafness

Deafness is significant or total hearing loss. **Sensorineural deafness** is caused by either impairment of hair cells in the cochlea or damage of the cochlear branch of the vestibulocochlear nerve (cranial nerve VIII). This type of deafness may be caused by atherosclerosis, which reduces blood supply to the ears; by repeated exposure to loud noise, which destroys hair cells of the spiral organ; and by certain drugs such as aspirin and streptomycin. **Conduction deafness** is caused by impairment of the external and middle ear mechanisms for transmitting sounds to the cochlea. Causes of conduction deafness include otosclerosis, the deposition of new bone around the oval window; impacted cerumen; injury to the eardrum; and aging, which often results in thickening of the eardrum and stiffening of the joints of the auditory ossicles.

Ménière's Disease

Ménière's disease (men′-ē-ĀRZ) results from an increased amount of endolymph that enlarges the membranous labyrinth. Among the symptoms are fluctuating hearing loss (caused by distortion of the basilar membrane of the cochlea) and roaring tinnitus (ringing). Spinning or whirling vertigo (dizziness) is characteristic of Ménière's disease. Almost total destruction of hearing may occur over a period of years.

Otitis Media

Otitis media is an acute infection of the middle ear caused mainly by bacteria and associated with infections of the nose and throat. Symptoms include pain, malaise, fever, and a reddening and outward bulging of the eardrum, which may rupture unless prompt treatment is received. (This may involve draining pus from the middle ear.) Bacteria passing into the auditory tube from the nasopharynx are the primary cause of all middle ear infections. Children are more susceptible than adults are to middle ear infections because their auditory tubes are almost horizontal, which decreases drainage.

MEDICAL TERMINOLOGY

Amblyopia   (am′-blē-Ō-pē-a; *ambly-* = dull or dim) Term used to describe the loss of vision in an otherwise normal eye that, because of muscle imbalance, cannot focus in synchrony with the other eye. Sometimes called "wandering eyeball."

Anosmia   (an-OZ-mē-a; *a-* = without; *osmi* = smell, odor) Total lack of the sense of smell.

Blepharitis   (blef-a-RĪ-tis; *blephar-* = eyelid; *-itis* = inflammation of) An inflammation of the eyelid.

Conjunctivitis (pinkeye)   An inflammation of the conjunctiva; when caused by bacteria such as pneumococci, staphylococci, or *Hemophilus influenzae,* it is very contagious and more common in children. Conjunctivitis may also be caused by irritants, such as dust, smoke, or pollutants in the air, in which case it is not contagious.

Keratitis   (ker′-a-TĪ-tis; *kerat-* = cornea) An inflammation or infection of the cornea.

Miosis   (mī-Ō-sis) Constriction of the pupil.

Mydriasis (mi-DRĪ-a-sis) Dilation of the pupil.

Nystagmus (nis-TAG-mus; *nystagm-* = nodding or drowsy) A rapid involuntary movement of the eyeballs, possibly caused by a disease of the central nervous system. It is associated with conditions that cause vertigo.

Otalgia (ō-TAL-jē-a; *oto-* = ear; *-algia* = pain) Earache.

Ptosis (TŌ-sis = fall) Falling or drooping of the eyelid (or slippage of any organ below its normal position).

Retinoblastoma (ret-i-nō-blas-TŌ-ma; *-oma* = tumor) A tumor arising from immature retinal cells; it accounts for 2% of childhood cancers.

Scotoma (skō-TŌ-ma = darkness) An area of reduced or lost vision in the visual field.

Strabismus (stra-BIZ-mus) An imbalance in the extrinsic eye muscles that causes a misalignment of one eye so that its line of vision is not parallel with that of the other eye (cross-eyes) and both eyes are not pointed at the same object at the same time; the condition produces a squint.

Tinnitus (ti-NĪ-tus) A ringing, roaring, or clicking in the ears.

Trachoma (tra-KŌ-ma) A serious form of conjunctivitis and the greatest single cause of blindness in the world. It is caused by the bacterium *Chlamydia trachomatis*. The disease produces an excessive growth of subconjunctival tissue and invasion of blood vessels into the cornea, which progresses until the entire cornea is opaque, causing blindness.

Vertigo (VER-ti-gō = dizziness) A sensation of spinning or movement in which the world seems to revolve or the person seems to revolve in space.

STUDY OUTLINE

OLFACTION: SENSE OF SMELL (p. 527)

1. The receptors for olfaction, which are bipolar neurons, are in the nasal epithelium.
2. In olfactory reception, a generator potential develops and triggers one or more nerve impulses.
3. The threshold of smell is low, and adaptation to odors occurs quickly.
4. Axons of olfactory receptors form the olfactory nerves (collectively, cranial nerve I), which convey nerve impulses to the olfactory bulbs, olfactory tracts, limbic system, and cerebral cortex (temporal and frontal lobes).

GUSTATION: SENSATION OF TASTE (p. 529)

1. The receptors for gustation, the gustatory receptor cells, are located in taste buds.
2. Dissolved chemicals, called tastants, stimulate gustatory receptor cells by flowing through ion channels in the plasma membrane or by binding to receptors attached to G-proteins in the membrane.
3. Receptor potentials developed in gustatory receptor cells cause the release of neurotransmitter, which can generate nerve impulses in first-order sensory neurons.
4. The threshold varies with the taste involved, and adaptation to taste occurs quickly.
5. Gustatory receptor cells trigger nerve impulses in cranial nerves VII, IX, and X. Taste signals then pass to the medulla oblongata, thalamus, and cerebral cortex (parietal lobe).

VISION (p. 531)

1. Accessory structures of the eyes include the eyebrows, eyelids, eyelashes, lacrimal apparatus, and extrinsic eye muscles.
2. The lacrimal apparatus consists of structures that produce and drain tears.
3. The eye is constructed of three layers: (a) fibrous tunic (sclera and cornea), (b) vascular tunic (choroid, ciliary body, and iris), and (c) retina.

4. The retina consists of a pigmented layer and a neural layer that includes a photoreceptor layer, bipolar cell layer, ganglion cell layer, horizontal cells, and amacrine cells.
5. The anterior cavity contains aqueous humor; the vitreous chamber contains the vitreous body.
6. Image formation on the retina involves refraction of light rays by the cornea and lens, which focus an inverted image on the central fovea of the retina.
7. For viewing close objects, the lens increases its curvature (accommodation) and the pupil constricts to prevent light rays from entering the eye through the periphery of the lens.
8. The near point of vision is the minimum distance from the eye at which an object can be clearly focused with maximum accommodation.
9. In convergence, the eyeballs move medially so they are both directed toward an object being viewed.
10. The first step in vision is the absorption of light by photopigments in rods and cones and isomerization of *cis*-retinal. Receptor potentials in rods and cones decrease the release of inhibitory neurotransmitter, which induces graded potentials in bipolar cells and horizontal cells.
11. Horizontal cells transmit inhibitory signals to bipolar cells; bipolar or amacrine cells transmit excitatory signals to ganglion cells, which depolarize and initiate nerve impulses.
12. Impulses from ganglion cells are conveyed into the optic nerve (cranial nerve II), through the optic chiasm and optic tract, to the thalamus. From the thalamus, impulses for vision propagate to the cerebral cortex (occipital lobe). Axon collaterals of retinal ganglion cells extend to the midbrain and hypothalamus.

HEARING AND EQUILIBRIUM (p. 546)

1. The external (outer) ear consists of the auricle, external auditory canal, and eardrum (tympanic membrane).
2. The middle ear consists of the auditory tube, ossicles, oval window, and round window.
3. The internal (inner) ear consists of the bony labyrinth and mem-

branous labyrinth. The internal ear contains the spiral organ (organ of Corti), the organ of hearing.

4. Sound waves enter the external auditory canal, strike the eardrum, pass through the ossicles, strike the oval window, set up waves in the perilymph, strike the vestibular membrane and scala tympani, increase pressure in the endolymph, vibrate the basilar membrane, and stimulate hair bundles on the spiral organ (organ of Corti).

5. Hair cells convert mechanical vibrations into a receptor potential, which releases neurotransmitter that can initiate nerve impulses in first-order sensory neurons.

6. Sensory axons in the cochlear branch of the vestibulocochlear nerve (cranial nerve VIII) terminate in the medulla oblongata. Auditory signals then pass to the inferior colliculus, thalamus, and temporal lobes of the cerebral cortex.

7. Static equilibrium is the orientation of the body relative to the pull of gravity. The maculae of the utricle and saccule are the sense organs of static equilibrium.

8. Dynamic equilibrium is the maintenance of body position in response to movement. The cristae in the semicircular ducts are the main sense organs of dynamic equilibrium.

9. Most vestibular branch axons of the vestibulocochlear nerve enter the brain stem and terminate in the medulla and pons; other axons enter the cerebellum.

DEVELOPMENT OF THE EYES AND EARS (p. 557)

1. The eyes begin their development about 22 days after fertilization from ectoderm of the lateral walls of the prosencephalon (forebrain).

2. The ears begin their development about 22 days after fertilization from a thickening of ectoderm on either side of the rhombencephalon (hindbrain). The sequence of development of the ear is internal ear, middle ear, and external ear.

SELF-QUIZ QUESTIONS

Fill in the blanks in the following statements.

1. The five primary taste sensations are _____, _____, _____, _____, and _____.

2. ____ equilibrium refers to the maintenance of the position of the body relative to the force gravity; ____ equilibrium refers to the maintenance of body position in response to sudden movements such as rotation, acceleration, and deceleration.

Indicate whether the following statements are true or false.

3. Of all of the special senses, only smell and taste sensations project both to higher cortical areas and to the limbic system.

4. Rods are most important for color vision, whereas cones are most important for seeing shades of gray in dim light.

Choose the one best answer to the following questions.

5. Which of the following statements is *incorrect*? (a) Olfactory receptors respond to the chemical stimulation of an odorant molecule by producing a receptor potential. (b) Basal stem cells continually produce new olfactory receptors. (c) Adaptation to odors is rapid and occurs in both olfactory receptors and the CNS. (d) Olfaction has a low threshold. (e) The lateral olfactory area is considered to be the primary olfaction area, where conscious awareness of smells begins.

6. Which of the following statements is *incorrect*? (a) Taste is a chemical sense. (b) The receptors for taste sensations are found in taste buds located on the tongue, the soft palate, the pharynx, and the larynx. (c) Gustatory hairs are the sites of taste transduction. (d) The threshold for bitter substances is the highest. (e) Complete adaptation to taste can occur in 1 to 5 minutes.

7. Which of the following are functions of the visual processing systems in the cerebral cortex? (1) information processing regarding shapes of objects, (2) information processing regarding weights of objects, (3) information processing regarding colors of objects, (4) information processing regarding ages of objects, (5) information processing regarding movement, location, and spatial organization, (a) 1, 2, and 3, (b) 2, 3, and 4, (c) 1, 3, and 5, (d) 1, 3, and 4, (e) 2, 4, and 5.

8. Which of the following statements is *incorrect*? (a) Retinal is the light-absorbing portion of all visual photopigments. (b) Whereas the photopigment in rods is rhodopsin, three different cone opsins are present in the retina. (c) Retinal is a derivative of vitamin C. (d) Rhodopsin absorbs blue to green light most effectively; the cone opsins absorb blue, green, or yellow-orange light. (e) Bleaching and regeneration of the photopigments account for much but not all of the sensitivity changes during light and dark adaptation.

9. The ability to change the curvature of the lens for near vision is called (a) refraction, (b) adaptation, (c) accommodation, (d) transduction, (e) convergence.

10. Which of the following sets of terms are *mismatched*? (a) emmetropia: normal vision, (b) myopia: eyeball is too long, (c) hypermetropia: nearsightedness, (d) astigmatism: irregular curvature of lens or cornea, (e) presbyopia: loss of lens elasticity.

11. Which of the following is the *correct* sequence for the visual pathway? (a) cornea, pupil, iris, photoreceptors, ganglion cells, bipolar cells, thalamus, visual cortex, (b) cornea, pupil, photoreceptors, bipolar cells, ganglion cells, thalamus, visual cortex, (c) sclera, iris, photoreceptors, bipolar cells, ganglion cells, thalamus, visual cortex, (d) cornea, vitreous body, aqueous humor, ciliary body, photoreceptors, thalamus, visual cortex, (e) cornea, aqueous humor, pupil, vitreous body, photoreceptors, bipolar cells, ganglion cells, thalamus, visual cortex.

12. Which of the following is the *correct* sequence for the auditory pathway? (a) external auditory canal, tympanic membrane, auditory ossicles, oval window, cochlea and organ of Corti, (b) tympanic membrane, external auditory canal, auditory ossicles, cochlea and organ of Corti, round window, (c) auditory ossicles, tympanic membrane, cochlea and organ of Corti, round window, oval window, external auditory canal, (d) auricle, tympanic membrane, round window, cochlea and organ of Corti, oval window, (e) external auditory canal, tympanic membrane, auditory ossicles, internal auditory canal, organ of Corti, oval window.

13. Match the following:

____ (a) upper and lower eyelids

____ (b) produces and drains tears

____ (c) arch transversely above the eyeballs and help protect the eyeballs from foreign objects, perspiration, and the direct rays of the sun

____ (d) move the eyeball medially, laterally, superiorly, or inferiorly

____ (e) shade the eyes during sleep, spread lubricating secretions over the eyeballs

____ (f) a thick fold of connective tissue that gives form and support to the eyelids

____ (g) modified sebaceous glands; secretion helps keep eyelids from adhering to each other

____ (h) project from the border of each eyelid; help protect the eyeballs from foreign objects, perspiration, and direct rays of the sun

____ (i) a thin, protective mucous membrane that lines the inner aspect of the eyelids and passes from the eyelids onto the anterior surface of the eyeball

(1) palpebrae
(2) tarsal or Meibomian glands
(3) conjunctiva
(4) eyelashes
(5) lacrimal apparatus
(6) extrinsic eye muscles
(7) eyebrows
(8) tarsal plate

14. Match the following:

____ (a) posterior portion of the vascular tunic; provides nutrients to the posterior surface of the retina

____ (b) colored portion of the eyeball; regulates the amount of light entering the posterior part of the eyeball

____ (c) beginning of the visual pathway; contains rods and cones

____ (d) biconvex transparent structure that fine tunes focusing of light rays for clear vision

____ (e) transparent part of fibrous tunic; helps focus light

____ (f) circular band of smooth muscle that alters the shape of the lens for near or far vision

____ (g) site where the optic nerve exits the eyeball; the blind spot

____ (h) fluid in the anterior cavity that helps nourish the lens and cornea; helps maintain shape of the eyeball

____ (i) the hole in the center of the iris

____ (j) jellylike substance in the vitreous chamber that helps prevent the eyeball from collapsing and holds the retina flush against the internal portions of the eyeball

____ (k) white of the eye; gives shape to the eyeball, makes it more rigid, protects its inner parts; a part of the fibrous tunic

____ (l) small depression in the center of the macula lutea that contains only cone photoreceptors and is the area of highest visual acuity

____ (m) contain blood capillaries that secrete aqueous humor; attach to suspensory ligaments of lens

(1) cornea
(2) sclera
(3) choroid
(4) ciliary processes
(5) ciliary muscle
(6) iris
(7) pupil
(8) retina
(9) optic disc
(10) central fovea
(11) aqueous humor
(12) lens
(13) vitreous body

15. Match the following:

____ (a) partition between external auditory canal and middle ear; eardrum

____ (b) oval central portion of the bony labyrinth

____ (c) receptor for static equilibrium; also contributes to some aspects of dynamic equilibrium; consists of hair cells and supporting cells

____ (d) organ of Corti; organ for hearing

____ (e) ear bones: anvil, malleus, stapes

____ (f) the pressure equalization tube that connects the middle ear to the nasopharynx

____ (g) contains the organ of Corti

____ (h) fluid found within the membranous labyrinth; pressure waves in this fluid cause vibration of the basilar membrane

____ (i) receptor organs for equilibrium; the saccule, utricle, and semicircular canals

____ (j) swollen enlargement in semicircular canals; contains structures involved in dynamic equilibrium

____ (k) opening between the middle ear and internal ear; is enclosed by a membrane called the secondary tympanic membrane

____ (l) the flap of elastic cartilage covered by skin that captures sound waves; the pinna

____ (m) fluid found inside bony labyrinth; bulging of the oval window causes pressure waves in this fluid

____ (n) opening between the middle and inner ear; receives base of stapes

(1) auricle
(2) tympanic membrane
(3) auditory ossicles
(4) vestibular apparatus
(5) ampulla
(6) cochlea
(7) perilymph
(8) oval window
(9) round window
(10) auditory or Eustachian tube
(11) vestibule
(12) endolymph
(13) spiral organ
(14) macula

CRITICAL THINKING QUESTIONS

1. Mother held a dish of steaming food under Brenna's nose. "How can you say it tastes terrible? You haven't even tried it!" Brenna's lips remained clamped shut. How can Brenna be so sure she won't like the food? Trace the pathway of the food's smell to the brain.
HINT *How would Brenna know if she had tried this food before?*

2. Fred will be 81 years old on his next birthday. He started working on the fishing boat when he was 15, but lately he's had to let his grandson do the piloting because his vision has gotten so cloudy. Although he's a smoker, his health is generally good, and he's had his blood pressure and intraocular pressure checked yearly. What could be the problem with his eyes?
HINT *Fishermen are constantly exposed to bright sunlight and glare off the water.*

3. Mari's mother always says to her, "Anything I say goes in one ear and out the other!" (At least that's what Mari thinks her mom says.) Follow the path of sound from the external ear to the organ of hearing.
HINT *The inner ear on one side of the head does not connect to the ear on the other side.*

ANSWERS TO FIGURE QUESTIONS

16.1 The olfactory hairs detect odorant molecules.

16.2 Supporting cells develop into gustatory receptor cells.

16.3 Fructose elicited the largest number of nerve impulses in Group 1 neurons.

16.4 The conjunctiva is continuous with the inner lining of the eyelids.

16.5 Lacrimal fluid, or tears, is a watery solution containing salts, some mucus, and lysozyme that protects, cleans, lubricates, and moistens the eyeball.

16.6 The fibrous tunic consists of the cornea and sclera; the vascular tunic consists of the choroid, ciliary body, and lens.

16.7 The parasympathetic division of the ANS causes pupillary constriction, whereas the sympathetic division causes pupillary dilation.

16.8 An ophthalmoscopic examination can reveal evidence of hypertension, diabetes mellitus, cataract, and macular degeneration.

16.9 The two types of photoreceptors are rods and cones. Rods provide black-and-white vision in dim light, whereas cones provide high visual acuity and color vision in bright light.

16.10 After its secretion by the ciliary process, aqueous humor flows into the posterior chamber, around the iris, into the anterior chamber, and out of the eyeball through the scleral venous sinus.

16.11 During accommodation the ciliary muscle contracts, causing the zonular fibers to slacken. The lens then becomes more convex, increasing its focusing power.

16.12 Presbyopia is the loss of lens elasticity that occurs with aging.

16.13 Both rods and cones transduce light into receptor potentials, use a photopigment embedded in outer segment membrane discs or folds, and release neurotransmitter at synapses with bipolar cells and horizontal cells.

16.14 The conversion of *cis*-retinal to *trans*-retinal is called isomerization.

16.15 Cyclic GMP is the ligand that opens Na^+ channels in photoreceptors.

16.16 Light rays from an object in the temporal half of the visual field fall on the nasal half of the retina.

16.17 The malleus is attached to the eardrum.

16.18 The eardrum separates the middle ear from the external ear; the oval and round windows separate the middle ear from the internal ear.

16.19 The two sacs in the vestibule are the utricle and saccule.

16.20 The bony labyrinth consists of semicircular canals, vestibule, and cochlea.

16.21 The region of the basilar membrane close to the oval and round windows vibrates most vigorously in response to high-frequency sounds.

16.22 Maculae mainly operate in static equilibrium.

16.23 The semicircular ducts, utricle, and saccule are mainly associated with dynamic equilibrium.

16.24 The optic cup forms the neural and pigmented layers of the optic part of the retina.

16.25 The internal ear develops from surface ectoderm, the middle ear develops from pharyngeal pouches, and the external ear develops from a pharyngeal cleft.

The Autonomic

Nervous System

Two outstanding features in this chapter greatly contributed to my study of the autonomic nervous system. The chapter review test (along with the answer key allows me to test myself and then pinpoint with great accuracy the specific concepts that I am not yet grasping. Because I can identify these "weak" areas so quickly, I am able to maximize study time and also pat myself on the back for the understanding I did acquire on the first read through. The chapter review test answer key was also incredibly helpful because the "wrong" answer did not "stay" in my thoughts long, as I was able to correct it almost immediately. I believe the brevity of its stay and the fewer times I have to erase an inaccurate neuronal pathway, the better.

The artwork (for example, figure 17.2) is a terrific visual aid. I found my copy of this figure and table 17.4, folded and smudged, in my pharmacology book. I used this figure when I studied pharmacology last year in order to have "at a glance" the adrenergic and cholinergic systems delineated. Being able to visually locate the positions of the alpha and beta receptors. Increased my understanding of the pharmacology of various medications. Since I am pursuing a degree as an Advanced Registered Nurse Practitioner, which includes prescribing medication, this is extremely important to me.

Stephanie Hall Ford
Barry University

THE AUTONOMIC NERVOUS SYSTEM AND HOMEOSTASIS

The ANS contributes to homeostasis by responding to subconsciously perceived visceral sensations and exciting or inhibiting smooth muscle, cardiac muscle, glands, and adipose tissue.

ENERGIZE YOUR STUDY

FOUNDATIONS CD

Anatomy Overview
• The Nervous System

Animation
• Systems Contributions to Homeostasis

Concepts and Connections and Exercises reinforce your understanding

www.wiley.com/college/apcentral

INSIGHTS AND EXPLORATIONS

Bet you know someone, maybe even yourself, who is always complaining of cold fingers, toes, nose, or ears, especially during the winter months. The National Institute of Arthritis and Musculoskeletal Skin Diseases estimates that 5 to 10% of the general population, women more often than men, has Raynaud's phenomenon. For these individuals when they are exposed to cold temperatures the blood vessels in their fingers toes, nose, and ears constrict and dilate inappropriately leading to color changes and pain. They may have patriotic digits that first become white, followed by blue and finally red. We will explore this phenomenon in a web-based activity about the autonomic nervous system.

Like the somatic nervous system, the **autonomic nervous system (ANS)** operates via reflex arcs. A continual flow of nerve impulses from *autonomic sensory neurons* in visceral organs and blood vessels propagate into *integrating centers* in the central nervous system (CNS). Then, impulses in *autonomic motor neurons* propagate to various effector tissues, thereby regulating the activity of smooth muscle, cardiac muscle, and many glands. Structurally, the ANS includes autonomic sensory neurons, integrating centers in the CNS, and autonomic motor neurons. Functionally, the ANS usually operates without conscious control. The system was originally named *autonomic* because it was thought to function autonomously or in a self-governing manner, without control by the CNS. However, centers in the hypothalamus and brain stem do regulate ANS reflexes.

In this chapter, we compare structural and functional features of the somatic and autonomic nervous systems. Then we discuss the anatomy of the motor portion of the ANS and compare the organization and actions of its two major parts, the sympathetic and parasympathetic divisions.

COMPARISON OF SOMATIC AND AUTONOMIC NERVOUS SYSTEMS

▶ O B J E C T I V E

- **Compare the structural and functional differences between the somatic and autonomic parts of the nervous system.**

The somatic nervous system includes both sensory and motor neurons. Sensory neurons convey input from receptors for the special senses (vision, hearing, taste, smell, and equilibrium) and from receptors for somatic senses (pain, thermal, tactile, and proprioceptive sensations). All these sensations normally are consciously perceived. In turn, somatic motor neurons innervate skeletal muscle—the effector tissue of the somatic nervous system—and produce voluntary movements. When a somatic motor neuron stimulates a skeletal muscle, the muscle contracts; the effect always is excitation. If somatic motor neurons cease to stimulate a muscle, the result is a paralyzed, limp muscle that has no muscle tone. In addition, even though we are generally not conscious of breathing, the muscles that generate respiratory movements are skeletal muscles controlled by somatic motor neurons. If the respiratory motor neurons become inactive, breathing stops. A few skeletal muscles, such as those in the middle ear, are controlled by reflexes and cannot be contracted voluntarily.

The main input to the ANS comes from **autonomic sensory neurons.** Mostly, these neurons are associated with interoceptors, such as chemoreceptors that monitor blood CO_2 level and mechanoreceptors that detect the degree of stretch in the walls of organs or blood vessels. These sensory signals are not consciously perceived most of the time, although intense activation of interoceptors may produce conscious sensations. Two examples of visceral sensations that are perceived are sensations of pain from damaged viscera and angina pectoris (chest pain) from inadequate blood flow to the heart. Input that influences the ANS also includes some sensations monitored by somatic sensory and special sensory neurons. For example, somatic pain can produce dramatic changes in some autonomic activities.

Autonomic motor neurons regulate visceral activities by either increasing (exciting) or decreasing (inhibiting) ongoing activities in their effector tissues, which are cardiac muscle, smooth muscle, and glands. Changes in the diameter of the pupils, dilation and constriction of blood vessels, and adjustment of the rate and force of the heartbeat are examples of autonomic motor responses. Unlike skeletal muscle, tissues innervated by the ANS often function to some extent even if their nerve supply is damaged. The heart continues to beat, for instance, when it is removed for transplantation into another person. Single-unit smooth muscle, such as in the lining of the gastrointestinal tract, contracts rhythmically on its own, and glands produce some secretions in the absence of ANS control.

Most autonomic responses cannot be consciously altered or suppressed to any great degree. You probably cannot voluntarily slow your heartbeat to half its normal rate. For this reason, some autonomic responses are the basis for polygraph ("lie detector") tests. Nevertheless, practitioners of yoga or other techniques of meditation may learn how to modulate at least some of their autonomic activities through long practice. **Biofeedback,** in which monitoring devices display information about a body function such as heart rate or blood pressure, enhances the ability to learn such conscious control. Signals from the general somatic and special senses, acting via the limbic system, also influence responses of autonomic motor neurons. Seeing a bike about to hit you, hearing squealing brakes of a nearby car, or being grabbed by an attacker, for example, would increase the rate and force of your heartbeat.

Recall from Chapter 10 that the axon of a single, myelinated somatic motor neuron extends from the CNS all the way to the skeletal muscle fibers in its motor unit (Figure 17.1a). By contrast, the autonomic motor pathways consist of two motor neurons in series, one following the other (Figure 17.1b). The first neuron has its cell body in the CNS; its myelinated axon extends from the CNS to an **autonomic ganglion.** (A ganglion is a collection of neuronal cell bodies outside the CNS.) The cell body of the second neuron is also in that autonomic ganglion; its unmyelinated axon extends directly from the ganglion to the effector (smooth muscle, cardiac muscle, or a gland). In addition, whereas all somatic motor neurons release only acetylcholine (ACh) as their neurotransmitter, autonomic motor neurons release either ACh or norepinephrine (NE).

The output (motor) part of the ANS has two principal branches: the **sympathetic division** and the **parasympathetic**

Figure 17.1 **Motor neuron pathways in the (a) somatic nervous system and (b) autonomic nervous system (ANS).** Note that autonomic motor neurons release either acetylcholine (ACh) or norepinephrine (NE); somatic motor neurons release ACh.

Somatic nervous system stimulation always excites its effectors (skeletal muscle fibers); stimulation by the autonomic nervous system either excites or inhibits visceral effectors.

(a) Somatic nervous system

(b) Autonomic nervous system

 What does dual innervation mean?

division. Most organs have **dual innervation:** they receive impulses from both sympathetic and parasympathetic neurons. In general, nerve impulses from one division of the ANS stimulate the organ to increase its activity (excitation), whereas impulses from the other division decrease the organ's activity (inhibition).

For example, an increased rate of nerve impulses from the sympathetic division increases heart rate, whereas an increased rate of nerve impulses from the parasympathetic division decreases heart rate. Table 17.1 summarizes the similarities and differences between the somatic and autonomic nervous systems.

Table 17.1	Summary of Somatic and Autonomic Nervous Systems	
	Somatic Nervous System	**Autonomic Nervous System**
Sensory input	Special senses and somatic senses.	Mainly from interoceptors; some from special senses and somatic senses.
Control of motor output	Voluntary control from cerebral cortex, with contributions from basal ganglia, cerebellum, brain stem, and spinal cord.	Involuntary control from limbic system, hypothalamus, brain stem, and spinal cord; limited control from cerebral cortex.
Motor neuron pathway	One-neuron pathway: somatic motor neurons extending from CNS synapse directly with effector.	Two-neuron pathway: preganglionic neurons extending from CNS synapse with postganglionic neurons in an autonomic ganglion, and postganglionic neurons extending from ganglion synapse with a visceral effector. Also, preganglionic neurons extending from CNS synapse with cells of adrenal medulla.
Neurotransmitters and hormones	All somatic motor neurons release ACh.	Preganglionic axons release acetylcholine (ACh); postganglionic axons release ACh (parasympathetic division and sympathetic fibers to sweat glands) or norepinephrine (NE; remainder of sympathetic division); adrenal medulla releases epinephrine and norepinephrine.
Effectors	Skeletal muscle.	Smooth muscle, cardiac muscle, and glands.
Responses	Contraction of skeletal muscle.	Contraction or relaxation of smooth muscle; increased or decreased rate and force of contraction of cardiac muscle; increased or decreased secretions of glands.

1. Why is the autonomic nervous system so-named?
2. What are the main input and output components of the autonomic nervous system?

ANATOMY OF AUTONOMIC MOTOR PATHWAYS

► O B J E C T I V E S

- **Describe preganglionic and postganglionic neurons of the autonomic nervous system.**
- **Compare the anatomical components of the sympathetic and parasympathetic divisions of the autonomic nervous system.**

Anatomical Components

The first of the two motor neurons in any autonomic motor pathway is called a **preganglionic neuron** (Figure 17.1b). Its cell body is in the brain or spinal cord, and its axon exits the CNS as part of a cranial or spinal nerve. The axon of a preganglionic neuron is a small-diameter, myelinated type B fiber that extends to an autonomic ganglion. There it synapses with a **postganglionic neuron,** the second neuron in the autonomic motor pathway (Figure 17.1b). Notice that the postganglionic neuron lies entirely outside the CNS. Its cell body and dendrites are located in an autonomic ganglion, where it forms synapses with one or more preganglionic axons. The axon of a postganglionic neuron is a small-diameter, unmyelinated type C fiber that terminates in a visceral effector. Thus, preganglionic neurons convey nerve impulses from the CNS to autonomic ganglia, and postganglionic neurons relay the impulses from autonomic ganglia to visceral effectors.

Preganglionic Neurons

In the sympathetic division, the preganglionic neurons have their cell bodies in the lateral horns of the gray matter in the 12 thoracic segments and the first two lumbar segments of the spinal cord (Figure 17.2). For this reason, the sympathetic division is also called the **thoracolumbar division** (thōr′-a-kō-LUM-bar), and the axons of the sympathetic preganglionic neurons are known as the **thoracolumbar outflow.**

Cell bodies of preganglionic neurons of the parasympathetic division are located in the nuclei of four cranial nerves in the brain stem (III, VII, IX, and X) and in the lateral gray horns of the second through fourth sacral segments of the spinal cord (see Figure 17.3). Hence, the parasympathetic division is also known as the **craniosacral division,** and the axons of the parasympathetic preganglionic neurons are referred to as the **craniosacral outflow.**

Autonomic Ganglia

The autonomic ganglia may be divided into three general groups: Two of the groups are components of the sympathetic

Figure 17.2 **Structure of the sympathetic division of the autonomic nervous system.** Solid lines represent preganglionic axons; dashed lines represent postganglionic axons. Although the innervated structures are shown only for one side of the body for diagrammatic purposes, the sympathetic division actually innervates tissues and organs on both sides.

 Cell bodies of sympathetic preganglionic neurons are located in the lateral horns of gray matter in the 12 thoracic and first two lumbar segments of the spinal cord.

division, and one group is a component of the parasympathetic division.

SYMPATHETIC GANGLIA The sympathetic ganglia are the sites of synapses between sympathetic preganglionic and postganglionic neurons. The two groups of sympathetic ganglia are sympathetic trunk ganglia and prevertebral ganglia. **Sympathetic trunk ganglia** (also called *vertebral chain ganglia* or *paravertebral ganglia*) lie in a vertical row on either side of the vertebral column. These ganglia extend from the base of the skull to the coccyx (Figure 17.2). Because the sympathetic trunk ganglia are near the spinal cord, most sympathetic preganglionic axons are short. Postganglionic axons from sympathetic trunk ganglia mostly innervate organs above the diaphragm. Examples of sympathetic trunk ganglia are the **superior, middle,** and **inferior cervical ganglia** (Figure 17.2).

The second group of sympathetic ganglia, the **prevertebral** (*collateral*) **ganglia,** lies anterior to the vertebral column and close to the large abdominal arteries. In general, postganglionic axons from prevertebral ganglia innervate organs below the diaphragm. There are three major prevertebral ganglia: (1) The **celiac ganglion** (SĒ-lē-ak) is on either side of the celiac artery just inferior to the diaphragm. (2) The **superior mesenteric ganglion** is near the beginning of the superior mesenteric artery in the upper abdomen. (3) The **inferior mesenteric ganglion** is near the beginning of the inferior mesenteric artery in the middle of the abdomen (Figure 17.2; see also Figure 17.4).

PARASYMPATHETIC GANGLIA Preganglionic axons of the parasympathetic division synapse with postganglionic neurons in **terminal** (*intramural*) **ganglia.** Most of these ganglia are located close to or actually within the wall of a visceral organ. Because the axons of parasympathetic preganglionic neurons extend from the CNS to a terminal ganglion in an innervated organ, they are longer than most of the axons of sympathetic preganglionic neurons. Examples of terminal ganglia include the ciliary ganglion, pterygopalatine ganglion, submandibular ganglion, and otic ganglion (Figure 17.3 on page 570).

Autonomic Plexuses

In the thorax, abdomen, and pelvis, axons of both sympathetic and parasympathetic neurons form tangled networks called **autonomic plexuses,** many of which lie along major arteries.

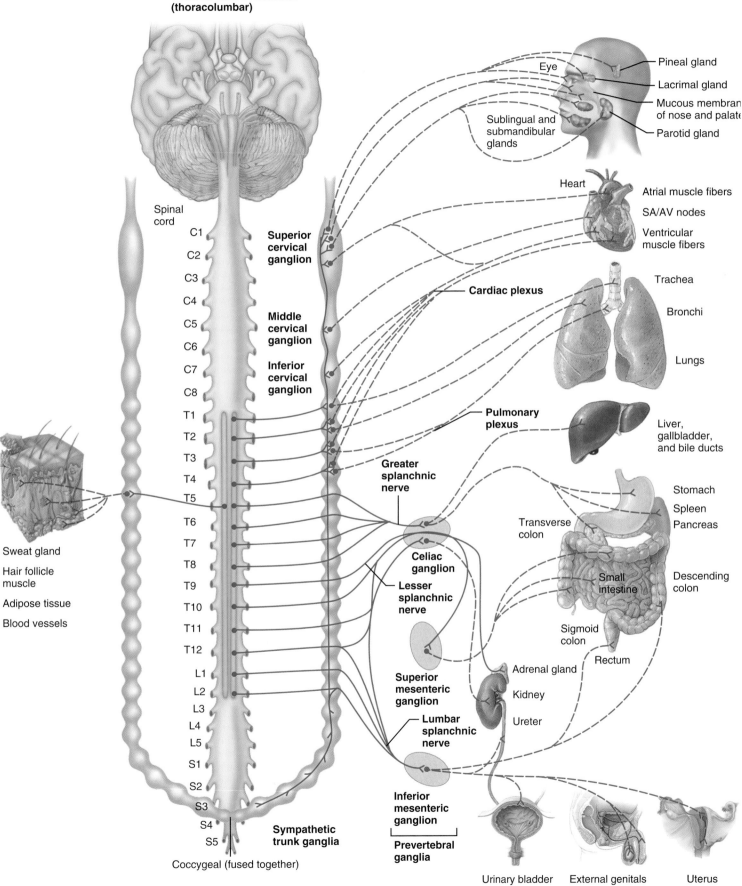

**SYMPATHETIC DIVISION
(thoracolumbar)**

Pineal gland

Eye

Lacrimal gland

Mucous membran
of nose and palat

Sublingual and
submandibular
glands

Parotid gland

Spinal
cord

C1

C2

C3

C4

C5

C6

C7

C8

T1

T2

T3

T4

T5

T6

T7

T8

T9

T10

T11

T12

L1

L2

L3

L4

L5

S1

S2

S3

S4

S5

**Superior
cervical
ganglion**

**Middle
cervical
ganglion**

**Inferior
cervical
ganglion**

Cardiac plexus

**Pulmonary
plexus**

**Greater
splanchnic
nerve**

**Celiac
ganglion**

**Lesser
splanchnic
nerve**

**Superior
mesenteric
ganglion**

**Lumbar
splanchnic
nerve**

**Inferior
mesenteric
ganglion**

**Sympathetic
trunk ganglia**

Coccygeal (fused together)

**Prevertebral
ganglia**

Heart

Atrial muscle fibers

SA/AV nodes

Ventricular
muscle fibers

Trachea

Bronchi

Lungs

Liver,
gallbladder,
and bile ducts

Stomach

Spleen

Pancreas

Transverse
colon

Descending
colon

Small
intestine

Sigmoid
colon

Rectum

Adrenal gland

Kidney

Ureter

Sweat gland

Hair follicle
muscle

Adipose tissue

Blood vessels

Urinary bladder   External genitals   Uterus

Which division, sympathetic or parasympathetic, has longer preganglionic axons? Why?

PARASYMPATHETIC DIVISION
(craniosacral)

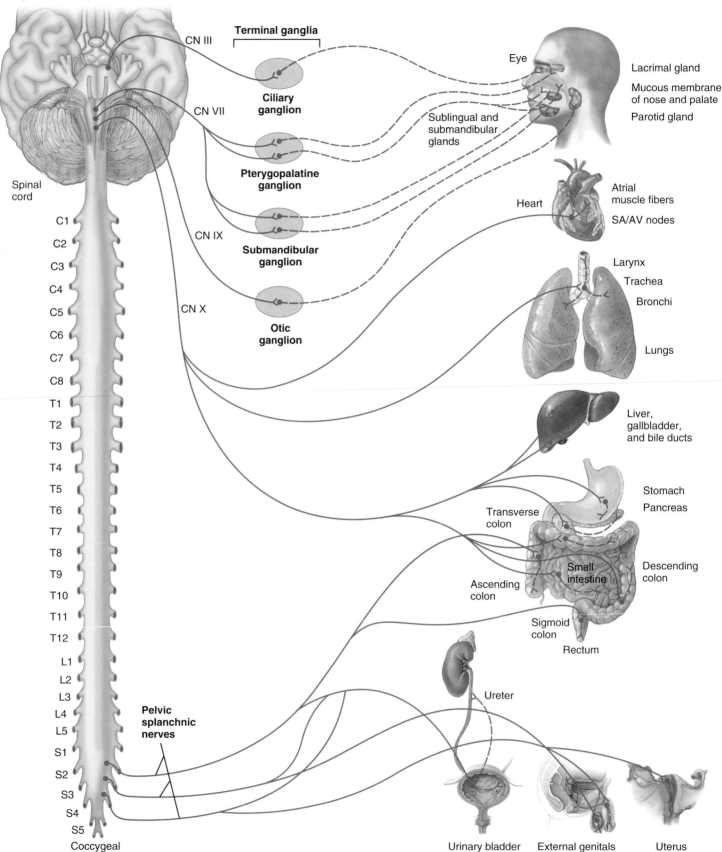

Terminal ganglia

CN III

Ciliary ganglion

CN VII

Pterygopalatine ganglion

Spinal cord

Eye

Lacrimal gland

Mucous membrane of nose and palate

Parotid gland

Sublingual and submandibular glands

C1
C2
C3
C4
C5

CN IX

Submandibular ganglion

CN X

Otic ganglion

Heart

Atrial muscle fibers

SA/AV nodes

C6
C7
C8
T1

Larynx

Trachea

Bronchi

Lungs

T2
T3
T4
T5
T6

Liver, gallbladder, and bile ducts

Stomach

Pancreas

T7
T8
T9

Transverse colon

Ascending colon

Small intestine

Descending colon

T10
T11
T12
L1

Sigmoid colon

Rectum

L2
L3
L4
L5

Pelvic splanchnic nerves

Ureter

S1
S2
S3
S4
S5
Coccygeal

Urinary bladder   External genitals   Uterus

Which ganglia are associated with the parasympathetic division? Sympathetic division?

Figure 17.3 Structure of the parasympathetic division of the autonomic nervous system. Solid lines represent preganglionic axons; dashed lines represent postganglionic axons. Although the innervated structures are shown only for one side of the body for diagrammatic purposes, the parasympathetic division actually innervates tissues and organs on both sides.

🔑 **Cell bodies of parasympathetic preganglionic neurons are located in brainstem nuclei and in the lateral horns of gray matter in the second through fourth sacral segments of the spinal cord.**

⟵————————————————————

The autonomic plexuses also may contain sympathetic ganglia and axons of autonomic sensory neurons. Often plexuses are named after the artery along which they are distributed. The major plexuses in the thorax are the **cardiac plexus,** which is at

the base of the heart and surrounds the large blood vessels emerging from the heart, and the **pulmonary plexus,** which is located mostly posterior to each lung (Figure 17.4). See also Figure 17.2.

The abdomen and pelvis also contain major autonomic plexuses (Figure 17.4). The **celiac** *(solar)* **plexus** is found at the level of the last thoracic and first lumbar vertebrae. It is the largest autonomic plexus and surrounds the celiac and superior mesenteric arteries. It contains two large celiac ganglia and a dense network of autonomic axons. Secondary plexuses that arise from the celiac plexus are distributed to the liver, gallbladder, stomach, pancreas, spleen, kidneys, medulla (inner region) of the adrenal gland, testes, and ovaries. The **superior mesenteric plexus,** a secondary plexus of the celiac plexus, contains the superior mesenteric ganglion and supplies the small and

Figure 17.4 Autonomic plexuses in the thorax, abdomen, and pelvis.

🔑 **An autonomic plexus is a network of sympathetic and parasympathetic axons that sometimes also includes autonomic sensory axons and sympathetic ganglia.**

Right vagus nerve (cranial nerve X)

Right primary bronchus

Right sympathetic trunk ganglion

Greater splanchnic nerve

Lesser splanchnic nerve

Inferior vena cava (cut)

Celiac trunk (artery)

Right kidney

Superior mesenteric artery

Inferior mesenteric artery

Right sympathetic trunk ganglion

Trachea

Left vagus nerve

Arch of aorta

Cardiac plexus

Pulmonary plexus

Esophagus

Thoracic aorta

Esophageal plexus

Diaphragm

Celiac ganglion and plexus

Superior mesenteric ganglion and plexus

Inferior mesenteric ganglion and plexus

Hypogastric plexus

 Which is the largest autonomic plexus?

large intestine. The **inferior mesenteric plexus,** also a secondary plexus of the celiac plexus, contains the inferior mesenteric ganglion, which innervates the large intestine. The **hypogastric plexus** is anterior to the fifth lumbar vertebra and supplies pelvic viscera. The **renal plexuses,** located near the kidneys, contain the renal ganglion and supply the renal arteries within the kidneys and the ureters.

Postganglionic Neurons

Once axons of sympathetic preganglionic neurons pass to sympathetic trunk ganglia, they may connect with postganglionic neurons in one of the following ways (Figure 17.5):

1 An axon may synapse with postganglionic neurons in the ganglion it first reaches.

2 An axon may ascend or descend to a higher or lower ganglion before synapsing with postganglionic neurons.

3 An axon may continue, without synapsing, through the sympathetic trunk ganglion to end at a prevertebral ganglion and synapse with postganglionic neurons there.

In addition, some preganglionic sympathetic axons extend to and terminate in the adrenal medulla.

A single sympathetic preganglionic fiber has many axon collaterals (branches) and may synapse with 20 or more postganglionic neurons. This pattern of projection is an example of divergence and helps explain why many sympathetic responses affect almost the entire body simultaneously. After exiting their ganglia, the postganglionic axons typically terminate in several visceral effectors (see Figure 17.2).

Axons of preganglionic neurons of the parasympathetic division pass to terminal ganglia near or within a visceral effector (see Figure 17.3). In the ganglion, the presynaptic neuron usually synapses with only four or five postsynaptic neurons, all of which supply a single visceral effector. Thus, parasympathetic responses can be localized to a single effector. With this background in mind, we can now examine some specific structural features of the sympathetic and parasympathetic divisions of the ANS.

Structure of the Sympathetic Division

Cell bodies of sympathetic preganglionic neurons are part of the lateral horns of all thoracic segments and of the first two lumbar segments of the spinal cord (see Figure 17.2). The preganglionic axons leave the spinal cord through the anterior root of a spinal nerve along with the somatic motor neurons at the same segmental level. After exiting through the intervertebral foramina, the myelinated preganglionic sympathetic axons enter a short pathway called a **white ramus** before passing to the nearest sympathetic trunk ganglion on the same side (Figure 17.5). Collectively, the white rami are called the **white rami communicantes** (kō-mū-ni-KAN-tēz; singular is **ramus communicans**). The "white" in their name indicates that they contain myelinated axons. Only the thoracic and first two or three lum-

bar nerves have white rami communicantes. The white rami communicantes connect the anterior ramus of the spinal nerve with the ganglia of the sympathetic trunk. The paired sympathetic trunk ganglia are arranged anterior and lateral to the vertebral column, one on either side. Typically, there are 3 cervical, 11 or 12 thoracic, 4 or 5 lumbar, 4 or 5 sacral sympathetic trunk ganglia, and 1 coccygeal ganglion. The right and left coccygeal ganglia are fused together and usually lie at the midline. Although the sympathetic trunk ganglia extend inferiorly from the neck, chest, and abdomen to the coccyx, they receive preganglionic axons only from the thoracic and lumbar segments of the spinal cord (see Figure 17.2).

The cervical portion of each sympathetic trunk is located in the neck and is subdivided into superior, middle, and inferior ganglia (see Figure 17.2). The **superior cervical ganglion** is anterior to a transverse process of the second cervical vertebra. Postganglionic neurons leaving the superior cervical ganglion serve the head and heart. They are distributed to sweat glands, smooth muscle of the eye, blood vessels of the face, lacrimal glands, nasal mucosa, and the submandibular, sublingual, parotid salivary glands, and the heart. Gray rami communicantes (described shortly) from the ganglion also pass to the upper two to four cervical spinal nerves. The **middle cervical ganglion** lies near the sixth cervical vertebra, and the **inferior cervical ganglion** is located near the first rib, anterior to the transverse processes of the seventh cervical vertebra. Postganglionic neurons from the middle and inferior cervical ganglia innervate the heart.

The thoracic portion of each sympathetic trunk lies anterior to the necks of the corresponding ribs. This region of the sympathetic trunk receives most of the sympathetic preganglionic axons. Postganglionic neurons from the thoracic sympathetic trunk innervate the heart, lungs, bronchi, and other thoracic viscera. In the skin, these neurons also innervate sweat glands, blood vessels, and arrector pili muscles of hair follicles.

The lumbar portion of each sympathetic trunk lies lateral to the corresponding lumbar vertebrae. The sacral region of the sympathetic trunk lies in the pelvic cavity on the medial side of the sacral foramina. Unmyelinated postganglionic axons from the lumbar and sacral sympathetic trunk ganglia enter a short pathway called a **gray ramus** and then merge with a spinal nerve or join the hypogastric plexus via direct visceral branches. The **gray rami communicantes** are structures containing the postganglionic axons that connect the ganglia of the sympathetic trunk to spinal nerves (Figure 17.5). The axons are unmyelinated. Gray rami communicantes outnumber the white rami because there is a gray ramus leading to each of the 31 pairs of spinal nerves.

As preganglionic axons extend from a white ramus communicans into the sympathetic trunk, they give off several axon collaterals (branches). These collaterals terminate and synapse in several ways. Some synapse in the first ganglion at the level of entry. Others pass up or down the sympathetic trunk for a variable distance to form the **sympathetic chains,** the fibers on

Figure 17.5 **Types of connections between ganglia and postganglionic neurons in the sympathetic division of the ANS.** Numbers correspond to descriptions in the text. Also illustrated are the gray and white rami communicantes.

Sympathetic ganglia lie in two chains on either side of the vertebral column (sympathetic trunk ganglia) and near large abdominal arteries anterior to the vertebral column (prevertebral ganglia).

—— Preganglionic neuron
----- Postganglionic neuron

Anterior view

What substance gives the white rami their white appearance?

which the ganglia are strung (Figure 17.5). Many postganglionic axons rejoin the spinal nerves through gray rami and supply peripheral visceral effectors such as sweat glands, smooth muscle in blood vessels, and arrector pili muscles of hair follicles.

Some preganglionic axons pass through the sympathetic trunk without terminating in it. Beyond the trunk, they form nerves known as **splanchnic nerves** (SPLANK-nik; see Figure 17.2), which extend to and terminate in the outlying prevertebral ganglia. Splanchnic nerves from the thoracic area terminate in the **celiac ganglion,** where the preganglionic neurons synapse with postganglionic cell bodies. Preganglionic axons from the fifth through ninth or tenth thoracic ganglia (T5–T9 or T10) form the *greater splanchnic nerve*. It pierces the diaphragm, and enters the celiac ganglion of the celiac plexus. From there, postganglionic neurons extend to the stomach, spleen, liver, kidney, and small intestine. Preganglionic axons from the tenth and eleventh thoracic ganglia (T10–T11) form the *lesser splanchnic nerve*. It pierces the diaphragm, and passes through the celiac plexus to enter the superior mesenteric ganglion of the superior mesenteric plexus. Postganglionic neurons from this ganglion innervate the small intestine and colon. The *lowest splanchnic nerve,* not always present, is formed by preganglionic axons from the twelfth thoracic ganglia (T12) or a branch of the lesser splanchnic nerve. It passes through the diaphragm, and enters the renal plexus near the kidney. Postganglionic neurons from the renal plexus supply kidney arterioles and the ureter.

Preganglionic axons that form the *lumbar splanchnic nerve* from the first through fourth lumbar ganglia (L1–L4) enter the inferior mesenteric plexus and terminate in the inferior mesenteric ganglion, where they synapse with postganglionic neurons. Axons of postganglionic neurons extend through the hypogastric plexus and supply the distal colon and rectum, urinary bladder, and genital organs. Postganglionic axons leaving the prevertebral ganglia follow the course of various arteries to abdominal and pelvic visceral effectors.

Sympathetic *preganglionic neurons* also extend to the adrenal medullae. Developmentally, the adrenal medullae and sympathetic ganglia are derived from the same tissue, the neural crest (see Figure 14.25 on page 490). The adrenal medullae are modified sympathetic ganglia, and their cells are similar to sympathetic postganglionic neurons. Rather than extending to another organ, however, these cells release hormones into the blood. Upon stimulation by sympathetic preganglionic neurons, the adrenal medullae release a mixture of catecholamine hormones—about 80% **epinephrine,** 20% **norepinephrine,** and a trace amount of **dopamine.**

Horner's Syndrome

In **Horner's syndrome,** the sympathetic innervation to one side of the face is lost due to an inherited mutation, an injury, or a disease that affects sympathetic outflow through the superior cervical ganglion. Symptoms occur on the affected side and include ptosis (drooping of the upper eyelid), miosis (constricted pupil), and anhidrosis (lack of sweating). ■

Structure of the Parasympathetic Division

Cell bodies of parasympathetic preganglionic neurons are found in nuclei in the brain stem and in the lateral horns of the second through fourth sacral segments of the spinal cord (see Figure 17.3). Their axons emerge as part of a cranial nerve or as part of the anterior root of a spinal nerve. The **cranial parasympathetic outflow** consists of preganglionic axons that extend from the brain stem in four cranial nerves. The **sacral parasympathetic outflow** consists of preganglionic axons in anterior roots of the second through fourth sacral nerves. The preganglionic axons of both the cranial and sacral outflows end in terminal ganglia, where they synapse with postganglionic neurons.

The cranial outflow has five components: four pairs of ganglia and the plexuses associated with the vagus nerve (cranial nerve X). The four pairs of cranial parasympathetic ganglia

innervate structures in the head and are located close to the organs they innervate (see Figure 17.3).

1. The **ciliary ganglia** lie lateral to each optic nerve (cranial nerve II) near the posterior aspect of the orbit. Preganglionic axons pass with the oculomotor nerves (cranial nerve III) to the ciliary ganglia. Postganglionic axons from the ganglia innervate smooth muscle fibers in the eyeball.

2. The **pterygopalatine ganglia** (ter′-i-gō-PAL-a-tīn) are located laterally to the sphenopalatine foramen, between the sphenoid and palatine bones. They receive preganglionic axons from the facial nerve (cranial nerve VII) and send postganglionic axons to the nasal mucosa, palate, pharynx, and lacrimal glands.

3. The **submandibular ganglia** are found near the ducts of the submandibular salivary glands. They receive preganglionic axons from the facial nerves and send postganglionic axons to the submandibular and sublingual salivary glands.

4. The **otic ganglia** are situated just inferior to each foramen ovale. They receive preganglionic axons from the glossopharyngeal nerves (cranial nerve IX) and send postganglionic axons to the parotid salivary glands.

Preganglionic axons that leave the brain as part of the vagus nerves (cranial nerve X) carry nearly 80% of the total craniosacral outflow. Vagal axons extend to many terminal ganglia in the thorax and abdomen. Because the terminal ganglia are close to or in the walls of their visceral effectors, postganglionic parasympathetic axons are very short. As the vagus nerve passes through the thorax, it sends axons to the heart and the airways of the lungs. In the abdomen, it supplies the liver, gallbladder, stomach, pancreas, small intestine, and part of the large intestine.

The sacral parasympathetic outflow consists of preganglionic axons from the anterior roots of the second through fourth sacral nerves (S2–S4) and they form the **pelvic splanchnic nerves** (see Figure 17.3). These nerves synapse with parasympathetic postganglionic neurons located in terminal ganglia in the walls of the innervated viscera. From the ganglia, parasympathetic postganglionic axons innervate smooth muscle and glands in the walls of the colon, ureters, urinary bladder, and reproductive organs.

▶ CHECKPOINT

3. Why is the sympathetic division called the thoracolumbar division even though its ganglia extend from the cervical to the sacral region?

4. Prepare a list of the organs served by each sympathetic and parasympathetic ganglion.

5. Describe the locations of sympathetic trunk ganglia, prevertebral ganglia, and terminal ganglia. Which types of autonomic neurons synapse in each type of ganglion?

6. Why may the sympathetic division produce simultaneous effects throughout the body, whereas parasympathetic effects typically are localized to specific organs?

ANS NEUROTRANSMITTERS AND RECEPTORS

OBJECTIVE

• **Describe the neurotransmitters and receptors involved in autonomic responses.**

Based on the neurotransmitter they produce and release, autonomic neurons are classified as either cholinergic or adrenergic. The receptors for the neurotransmitters are integral membrane proteins that are located in the plasma membrane of the postsynaptic neuron or effector cell.

Cholinergic Neurons and Receptors

Cholinergic neurons (kō′-lin-ER-jik) release the neurotransmitter **acetylcholine (ACh).** In the ANS, the cholinergic neurons include (1) all sympathetic and parasympathetic preganglionic neurons, (2) sympathetic postganglionic neurons that innervate most sweat glands, and (3) all parasympathetic postganglionic neurons (Figure 17.6).

ACh is stored in synaptic vesicles and released by exocytosis. It then diffuses across the synaptic cleft and binds with specific **cholinergic receptors,** which are integral membrane proteins in the *postsynaptic* plasma membrane. The two types of cholinergic receptors, both of which bind ACh, are nicotinic receptors and muscarinic receptors. **Nicotinic receptors** are present in the plasma membrane of dendrites and cell bodies of both sympathetic and parasympathetic postganglionic neurons (Figure 17.6a, b) and in the motor end plate at the neuromuscular junction. They are so-named because nicotine mimics the action of ACh by binding to these receptors. (Nicotine, a natural substance in tobacco leaves, is not normally present in nonsmokers.) **Muscarinic receptors** are present in the plasma membranes of all effectors (smooth muscle, cardiac muscle, and glands) innervated by parasympathetic postganglionic axons. In addition, most sweat glands receive their innervation from *cholinergic* sympathetic postganglionic neurons and possess muscarinic receptors (see Figure 17.6b). These receptors are so-named because a mushroom poison called muscarine mimics the actions of ACh by binding to them. Nicotine does not activate muscarinic receptors, and muscarine does not activate nicotinic receptors, but ACh does activate both types of cholinergic receptor.

Activation of nicotinic receptors by ACh causes depolarization and thus excitation of the postsynaptic cell—either a postganglionic neuron, an autonomic effector, or a skeletal muscle fiber. Activation of muscarinic receptors by ACh sometimes causes depolarization (excitation) and sometimes causes hyperpolarization (inhibition), depending on which particular cell bears the muscarinic receptors. For example, binding of ACh to muscarinic receptors inhibits (relaxes) smooth muscle sphincters

Figure 17.6 Cholinergic neurons (aqua) and adrenergic neurons (orange) in the sympathetic and parasympathetic divisions. Cholinergic neurons release acetylcholine, whereas adrenergic neurons release norepinephrine. Cholinergic and adrenergic receptors all are integral membrane proteins located in the plasma membrane of a postsynaptic neuron or an effector cell.

🔑 **Most sympathetic postganglionic neurons are adrenergic; other autonomic neurons are cholinergic.**

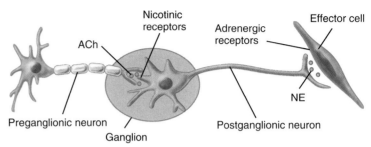

(a) Sympathetic division–innervation to most effector tissues

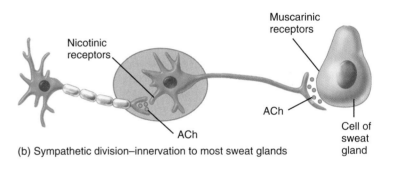

(b) Sympathetic division–innervation to most sweat glands

(c) Parasympathetic division

 Which neurons are cholinergic and possess nicotinic ACh receptors? What type of receptors for ACh do the effector tissues innervated by these neurons possess?

in the gastrointestinal tract. By contrast, ACh excites smooth muscle fibers in the circular muscles of the iris of the eye, causing them to contract. Because acetylcholine is quickly inactivated by the enzyme **acetylcholinesterase (AChE),** effects triggered by cholinergic neurons are brief.

Adrenergic Neurons and Receptors

In the ANS, **adrenergic neurons** (ad′-ren-ER-jik) release **norepinephrine (NE),** also known as **noradrenalin** (Figure 17.6a). Most sympathetic postganglionic neurons are adrenergic. Like ACh, NE is synthesized and stored in synaptic vesicles and released by exocytosis. Molecules of NE diffuse across the synaptic cleft and bind to specific adrenergic receptors on the postsynaptic membrane, causing either excitation or inhibition of the effector cell.

Adrenergic receptors bind both norepinephrine and epinephrine. Moreover, the receptors are activated by both norepinephrine released as a neurotransmitter by sympathetic postganglionic neurons and by epinephrine and norepinephrine released as hormones into the blood by the adrenal medullae. The two main types of adrenergic receptors are **alpha (α) receptors** and **beta (β) receptors,** which are found on visceral effectors innervated by most sympathetic postganglionic axons. These receptors are further classified into subtypes—α_1, α_2, β_1, β_2, and β_3—based on the specific responses they elicit and by their selective binding of drugs that activate or block them. Although there are some exceptions, activation of α_1 and β_1 receptors generally produces excitation, whereas activation of α_2 and β_2 receptors causes inhibition of effector tissues. β_3 receptors are present only on cells of brown adipose tissue, where their activation causes thermogenesis (heat production). Cells of most effectors contain either alpha or beta receptors; some visceral effector cells contain both. Norepinephrine stimulates alpha receptors more strongly than beta receptors, whereas epinephrine is a potent stimulator of both alpha and beta receptors.

The activity of norepinephrine at a synapse is terminated either when the NE is taken up by the axon that released it or when the NE is enzymatically inactivated by either **catechol-O-methyltransferase (COMT)** or **monoamine oxidase (MAO).** Compared to ACh, norepinephrine lingers in the synaptic cleft for a longer time. Thus, effects triggered by adrenergic neurons typically are longer lasting than those triggered by cholinergic neurons.

Table 17.2 describes the locations of cholinergic and adrenergic receptors and summarizes the responses that occur when each type of receptor is activated.

Receptor Agonists and Antagonists

A large variety of drugs and natural products can selectively activate or block specific cholinergic or adrenergic receptors. An **agonist** is a substance that binds to and activates a receptor, in the process mimicking the effect of a natural neurotransmitter or hormone. Phenylephrine, an adrenergic agonist at α_1 receptors, is a common ingredient in cold and sinus medications. Because it constricts blood vessels in the nasal mucosa, phenylephrine reduces production of mucus, thus relieving nasal congestion. An **antagonist** is a substance that binds to and blocks a receptor, thereby preventing a natural neurotransmitter or hormone from exerting its effect. For example, atropine, which blocks mus-

carinic ACh receptors, dilates the pupils, reduces glandular secretions, and relaxes smooth muscle in the gastrointestinal tract. As a result, it is used to dilate the pupils during eye examinations, in the treatment of smooth muscle disorders such as iritis and intestinal hypermotility, and as an antidote for chemical warfare agents that inactivate acetylcholinesterase.

Propranolol (Inderal) often is prescribed for patients with hypertension (high blood pressure). It is a nonselective beta blocker, meaning it binds to all types of beta receptors and prevents their activation by epinephrine and norepinephrine. The desired effects of propranolol are due to its *blockade* of β_1 receptors—namely, decreased heart rate and force of contraction and a consequent decrease in blood pressure. Undesired effects due to blockade of β_2 receptors may include hypoglycemia (low blood glucose), resulting from decreased glycogen breakdown and decreased gluconeogenesis (the conversion of a noncarbohydrate into glucose in the liver), and mild bronchoconstriction (narrowing of the airways). If these side effects pose a threat to the patient, a selective β_1 blocker such as metoprolol (Lopressor) can be prescribed instead of propranolol.

▶ C H E C K P O I N T

7. Why are cholinergic and adrenergic neurons so named?

8. What neurotransmitters and hormones bind to adrenergic receptors?

9. What do the terms *agonist* and *antagonist* mean?

PHYSIOLOGICAL EFFECTS OF THE ANS

▶ O B J E C T I V E

*   **Describe the major responses of the body to stimulation by the sympathetic and parasympathetic divisions of the ANS.**

As noted earlier, most body organs receive innervation from both divisions of the ANS, which typically work in opposition to one another. The balance between sympathetic and parasympathetic activity or "tone" is regulated by the hypothalamus. Typically, the hypothalamus turns up sympathetic tone at the same time it turns down parasympathetic tone, and vice versa. The two divisions can affect body organs differently because their postganglionic neurons release different neurotransmitters and because the effector organs possess different adrenergic and cholinergic receptors. A few structures receive only sympathetic innervation—sweat glands, arrector pili muscles attached to hair follicles in the skin, the kidneys, the spleen, most blood vessels, and the adrenal medullae (see Figure 17.2). In these structures there is no opposition from the parasympathetic division. Still, an increase in sympathetic tone has one effect, whereas a decrease in sympathetic tone produces the opposite effect.

Sympathetic Responses

During physical or emotional stress, the sympathetic division dominates the parasympathetic division. High sympathetic tone

Table 17.2	Locations and Responses of Adrenergic and Cholinergic Receptors	
Type of Receptor	**Major Locations**	**Effects of Receptor Activation**
Cholinergic	Integral proteins in postsynaptic plasma membranes; activated by the neurotransmitter acetylcholine.	
Nicotinic	Plasma membrane of postganglionic sympathetic and parasympathetic neurons.	Excitation → impulses in postaganglionic neurons.
	Cells of adrenal medullae.	Epinephrine and norepinephrine secretion.
	Sarcolemma of skeletal muscle fibers (motor end plate).	Excitation → contraction.
Muscarinic	Effectors innervated by parasympathetic postganglionic neurons.	In some receptors, excitation; in others, inhibition.
	Sweat glands innervated by cholinergic sympathetic postganglionic neurons.	Increased sweating.
	Skeletal muscle blood vessels innervated by cholinergic sympathetic postganglionic neurons.	Inhibition → relaxation → vasodilation.
Adrenergic	Integral proteins in postsynaptic plasma membranes; activated by the neurotransmitter norepinephrine, and by hormonal norepinephrine and epinephrine.	
α_1	Smooth muscle fibers in blood vessels that serve salivary glands, skin, mucosal membranes, kidneys, and abdominal viscera; radial muscle in iris of eye; sphincter muscles of stomach and urinary bladder.	Excitation → contraction, which causes vasoconstriction, dilation of pupil, and closing of sphincters.
	Salivary gland cells.	Secretion of K^+ and water.
	Sweat glands on palms and soles.	Increased sweating.
α_2	Smooth muscle fibers in some blood vessels.	Inhibition → relaxation → vasodilation.
	Cells of pancreatic islets that secrete the hormone insulin (beta cells).	Decreased insulin secretion.
	Pancreatic acinar cells.	Inhibition of digestive enzyme secretion.
	Platelets in blood.	Aggregation to form platelet plug.
β_1	Cardiac muscle fibers.	Excitation → increased force and rate of contraction.
	Juxtaglomerular cells of kidneys.	Renin secretion.
	Posterior pituitary.	Secretion of antidiuretic hormone.
	Adipose cells.	Breakdown of triglycerides → release of fatty acids into blood.
β_2	Smooth muscle in walls of airways; in blood vessels that serve the heart, skeletal muscle, adipose tissue, and liver; and in walls of visceral organs, such as the urinary bladder.	Inhibition → relaxation, which causes dilation of airways, vasodilation, and relaxation of organ walls.
	Ciliary muscle in eye.	Inhibition → relaxation.
	Hepatocytes in liver.	Glycogenolysis (breakdown of glycogen into glucose).
β_3	Brown adipose tissue.	Thermogenesis (heat production).

favors body functions that can support vigorous physical activity and rapid production of ATP. At the same time, the sympathetic division reduces body functions that favor the storage of energy. Besides physical exertion, a variety of emotions—such as fear, embarrassment, or rage—stimulate the sympathetic division. Visualizing body changes that occur during "E situations" such as exercise, emergency, excitement, and embarrassment will help you remember most of the sympathetic responses. Activation of the sympathetic division and release of hormones by the adrenal medullae set in motion a series of physiological re-

sponses collectively called the **fight-or-flight response,** which includes the following effects:

1. The pupils of the eyes dilate.

2. Heart rate, force of heart contraction, and blood pressure increase.

3. The airways dilate, allowing faster movement of air into and out of the lungs.

4. The blood vessels that supply the kidneys and gastrointestinal tract constrict, which decreases blood flow through these tissues. The result is a slowing of urine formation and digestive activities, which are not essential during exercise.

5. Blood vessels that supply organs involved in exercise or fighting off danger—skeletal muscles, cardiac muscle, liver, and adipose tissue—dilate, allowing greater blood flow through these tissues.

6. Liver cells perform glycogenolysis (breakdown of glycogen to glucose), and adipose tissue cells perform lipolysis (breakdown of triglycerides to fatty acids and glycerol).

7. Release of glucose by the liver increases blood glucose level.

8. Processes that are not essential for meeting the stressful situation are inhibited. For example, muscular movements of the gastrointestinal tract and digestive secretions slow down or even stop.

The effects of sympathetic stimulation are longer lasting and more widespread than the effects of parasympathetic stimulation for three reasons: (1) Sympathetic postganglionic axons diverge more extensively; as a result, many tissues are activated simultaneously. (2) Acetylcholinesterase quickly inactivates acetylcholine, whereas norepinephrine lingers in the synaptic cleft for a longer period. (3) Epinephrine and norepinephrine secreted into the blood from the adrenal medulla intensify and prolong the responses caused by NE liberated from sympathetic postganglionic axons. These blood-borne hormones circulate throughout the body, affecting all tissues that have alpha and beta receptors. In time, blood-borne NE and epinephrine are inactivated by enzymatic destruction in the liver.

Raynaud Disease

In **Raynaud disease,** the digits (fingers and toes) become ischemic (lack blood) after exposure to cold or with emotional stress. The condition is due to excessive sympathetic stimulation of smooth muscle in the arterioles of the digits and a heightened response to stimuli that cause vasoconstriction. When arterioles in the digits vasoconstrict in response to sympathetic stimulation, blood flow is greatly diminished. As a result, the digits may blanch (look white due to blockage of blood flow) or become cyanotic (look blue due to deoxygenated blood in capillaries). In extreme cases, the digits may become necrotic from lack of oxygen and nutrients. With rewarming after cold exposure, the arterioles may dilate, causing the fingers and toes to look red. Many patients with Raynaud disease have low blood pressure. Some have increased numbers of alpha-adrenergic receptors. The disease is most common in young women and occurs more often in cold climates. Patients with Raynaud disease should avoid exposure to cold, wear warm clothing, and keep the hands and feet warm. Drugs used to treat the condition include nifedipine, a calcium channel blocker that relaxes vascular smooth muscle, and prazosin, which relaxes smooth muscle by blocking alpha receptors. ■

Parasympathetic Responses

In contrast to the "fight-or-flight" activities of the sympathetic division, the parasympathetic division enhances "rest-and-digest" activities. Parasympathetic responses support body functions that conserve and restore body energy during times of rest and recovery. In the quiet intervals between periods of exercise, parasympathetic impulses to the digestive glands and the smooth muscle of the gastrointestinal tract predominate over sympathetic impulses. This allows energy-supplying food to be digested and absorbed. At the same time, parasympathetic responses reduce body functions that support physical activity.

The acronym "SLUDD" can be helpful in remembering five parasympathetic responses. It stands for salivation (S), lacrimation (L), urination (U), digestion (D), and defecation (D). All of these activities are stimulated mainly by the parasympathetic division. Besides the increasing SLUDD responses, other important parasympathetic responses are "three decreases": decreased heart rate, decreased diameter of airways (bronchoconstriction), and decreased diameter (constriction) of the pupils.

Table 17.3 compares the structural and functional features of the sympathetic and parasympathetic divisions of the ANS. Table 17.4 on pages 580–581 lists the responses of glands, cardiac muscle, and smooth muscle to stimulation by the sympathetic and parasympathetic divisions of the ANS.

▶ C H E C K P O I N T

10. Define cholinergic and adrenergic neurons and receptors.

11. What are some examples of the antagonistic effects of the sympathetic and parasympathetic divisions of the autonomic nervous system?

12. What happens during the fight-or-flight response?

13. Why is the parasympathetic division of the ANS called an energy conservation/restoration system?

14. Describe the sympathetic response in a frightening situation for each of the following body parts: hair follicles, iris of eye, lungs, spleen, adrenal medullae, urinary bladder, stomach, intestines, gallbladder, liver, heart, arterioles of the abdominal viscera, and arterioles of skeletal muscles.

INTEGRATION AND CONTROL OF AUTONOMIC FUNCTIONS

▶ O B J E C T I V E S

• **Describe the components of an autonomic reflex.**

• **Explain the relationship of the hypothalamus to the ANS.**

Autonomic Reflexes

Autonomic reflexes are responses that occur when nerve impulses pass over an autonomic reflex arc. These reflexes play a key role in regulating controlled conditions in the body, such

Table 17.3	Comparison of Sympathetic and Parasympathetic Divisions of the ANS	
	Sympathetic	**Parasympathetic**
Distribution	Wide regions of the body: skin, sweat glands, arrector pili muscles of hair follicles, adipose tissue, smooth muscle of blood vessels.	Limited mainly to head and to viscera of thorax, abdomen, and pelvis; some blood vessels.
Location of preganglionic neuron cell bodies and site of outflow	Cell bodies of preganglionic neurons are located in lateral gray horns of spinal cord segments T1–L2. Axons of preganglionic neurons constitute thoracolumbar outflow.	Cell bodies of preganglionic neurons are located in the nuclei of cranial nerves III, VII, IX, and X and and the lateral gray horns of spinal cord segments S2–S4. Axons of preganglionic neurons constitute craniosacral outflow.
Associated ganglia	Two types: sympathetic trunk ganglia and prevertebral ganglia.	One type: terminal ganglia.
Ganglia locations	Close to CNS and distant from visceral effectors.	Typically near or within wall of visceral effectors.
Axon length and divergence	Preganglionic neurons with short axons synapse with many postganglionic neurons with long axons that pass to many visceral effectors.	Preganglionic neurons with long axons usually synapse with four to five postganglionic neurons with short axons that pass to a single visceral effector.
Rami communicantes	Both present; white rami communicantes contain myelinated preganglionic axons, and gray rami communicantes contain unmyelinated postganglionic axons.	Neither present.
Neurotransmitters	Preganglionic neurons release acetylcholine (ACh), which is excitatory and stimulates postganglionic neurons; most postganglionic neurons release norepinephrine (NE); postganglionic neurons that innervate most sweat glands and some blood vessels in skeletal muscle release ACh.	Preganglionic neurons release acetylcholine (ACh), which is excitatory and stimulates postganglionic neurons; postganglionic neurons release ACh.
Physiological effects	Fight-or-flight responses.	"Rest-and-digest" activities.

as *blood pressure,* by adjusting heart rate, force of ventricular contraction, and blood vessel diameter; *digestion,* by adjusting the motility (movement) and muscle tone of the gastrointestinal tract; and *defecation* and *urination,* by regulating the opening and closing of sphincters.

The components of an autonomic reflex arc are as follows:

1. *Receptor.* Like the receptor in a somatic reflex arc (see Figure 13.5 on page 427), the receptor in an autonomic reflex arc is the distal end of a sensory neuron, which responds to a stimulus and produces a change that will ultimately trigger nerve impulses. Autonomic sensory receptors are mostly associated with interoceptors.

2. *Sensory neuron.* Conducts nerve impulses from receptors to the CNS.

3. *Integrating center.* Interneurons within the CNS relay signals from sensory neurons to motor neurons. The main integrating centers for most autonomic reflexes are located in the hypothalamus and brain stem. Some autonomic reflexes, such as those for urination and defecation, have integrating centers in the spinal cord.

4. *Motor neurons.* Nerve impulses triggered by the integrating center propagate out of the CNS along motor neurons to an effector. In an autonomic reflex arc, two motor neurons connect the CNS to an effector: The preganglionic neuron conducts motor impulses from the CNS to an autonomic ganglion, and the postganglionic neuron conducts motor impulses from an autonomic ganglion to an effector (see Figure 17.1).

5. *Effector.* In an autonomic reflex arc, the effectors are smooth muscle, cardiac muscle, and glands, and the reflex is called an autonomic reflex.

Autonomic Control by Higher Centers

Normally, we are not aware of muscular contractions of the digestive organs, heartbeat, changes in the diameter of blood vessels, and pupil dilation and constriction because the integrating centers for these autonomic responses are in the spinal cord or the lower regions of the brain. Somatic or autonomic sensory neurons deliver input to these centers, and autonomic motor neurons provide output that adjusts activity in the visceral effector, usually without our conscious perception.

The hypothalamus is the major control and integration center of the ANS. The hypothalamus receives sensory input related to visceral functions, olfaction (smell), and gustation (taste), as well as input related to changes in temperature, osmolarity, and

Table 17.4	Effects of Sympathetic and Parasympathetic Divisions of the ANS	
Visceral Effector	**Effect of Sympathetic Stimulation** (α or β adrenergic receptors, except as noted)*	**Effect of Parasympathetic Stimulation** (muscarinic ACh receptors)
Glands		
Adrenal medullae	Secretion of epinephrine and norepinephrine (nicotinic ACh receptors).	No known effect.
Lacrimal (tear)	Slight secretion of tears (α).	Secretion of tears.
Pancreas	Inhibits secretion of digestive enzymes and the hormone insulin (α_2); promotes secretion of the hormone glucagon (β_2).	Secretion of digestive enzymes and the hormone insulin.
Posterior pituitary	Secretion of antidiuretic hormone (ADH) (β_1).	No known effect.
Pineal	Increases synthesis and release of melatonin (β).	No known effect.
Sweat	Increases sweating in most body regions (muscarinic ACh receptors); sweating on palms of hands and soles of feet (α_1).	No known effect.
Adipose tissue[†]	Lipolysis (breakdown of triglycerides into fatty acids and glycerol) (β_1); release of fatty acids into blood (β_1 and β_3).	No known effect.
Liver[†]	Glycogenolysis (conversion of glycogen into glucose); gluconeogenesis (conversion of noncarbohydrates into glucose); decreased bile secretion (α and β_2).	Glycogen synthesis; increased bile secretion.
Kidney, juxtaglomerular cells[†]	Secretion of renin (β_1).	No known effect.
Cardiac (Heart) Muscle	Increased heart rate and force of atrial and ventricular contractions (β_1).	Decreased heart rate; decreased force of atrial contraction.
Smooth Muscle		
Iris, radial muscle	Contraction → dilation of pupil (α_1).	No known effect.
Iris, circular muscle	No known effect.	Contraction → constriction of pupil.
Ciliary muscle of eye	Relaxation for distant vision (β_2).	Contraction for close vision.
Lungs, bronchial muscle	Relaxation → airway dilation (β_2).	Contraction → airway constriction.
Gallbladder and ducts	Relaxation (β_2).	Contraction → increased release of bile into small intestine.
Stomach and intestines	Decreased motility and tone (α_1, α_2, β_2); contraction of sphincters (α_1).	Increased motility and tone; relaxation of sphincters.

*Subcategories of α and β receptors are listed if known.
[†]Grouped with glands because they release substances into the blood.

levels of various substances in blood. In addition, it receives input relating to emotions from the limbic system. Output from the hypothalamus influences autonomic centers both in the brain stem (such as the cardiovascular, salivation, swallowing, and vomiting centers) and in the spinal cord (such as the defecation and urination reflex centers in the sacral spinal cord).

Anatomically, the hypothalamus is connected to both the sympathetic and parasympathetic divisions of the ANS by axons of neurons whose dendrites and cell bodies are in various hypothalamic nuclei. The axons form tracts from the hypothalamus to sympathetic and parasympathetic nuclei in the brain stem and spinal cord through relays in the reticular formation. The posterior and lateral parts of the hypothalamus control the sympathetic division. Stimulation of these areas produces an increase in heart rate and force of contraction, a rise in blood pressure

due to constriction of blood vessels, an increase in body temperature, dilation of the pupils, and inhibition of the gastrointestinal tract. In contrast, the anterior and medial parts of the hypothalamus control the parasympathetic division. Stimulation of these areas results in a decrease in heart rate, lowering of blood pressure, constriction of the pupils, and increased secretion and motility of the gastrointestinal tract.

 Autonomic Dysreflexia

Autonomic dysreflexia is an exaggerated response of the sympathetic division of the ANS that occurs in about 85% of individuals with spinal cord injury at or above the level of T6. The condition is seen after recovery from spinal shock (see page

Visceral Effector	Effect of Sympathetic Stimulation (α or β adrenergic receptors, except as noted) *	Effect of Parasympathetic Stimulation (muscarinic ACh receptors)
Smooth Muscle (continued)		
Spleen	Contraction and discharge of stored blood into general circulation (α_1).	No known effect.
Ureter	Increases motility (α_1).	Increases motility (?).
Urinary bladder	Relaxation of muscular wall (β_2); contraction of sphincter (α_1).	Contraction of muscular wall; relaxation of sphincter.
Uterus	Inhibits contraction in nonpregnant woman (β_2); promotes contraction in pregnant woman (α_1).	Minimal effect.
Sex organs	In males: contraction of smooth muscle of ductus (vas) deferens, seminal vesicle, prostate $\rightarrow$ ejaculation of semen (α_1).	Vasodilation; erection of clitoris (females) and penis (males).
Hair follicles, arrector pili muscle	Contraction $\rightarrow$ erection of hairs (α_1).	No known effect.
Vascular Smooth Muscle		
Salivary gland arterioles	Vasoconstriction, which decreases secretion (α_1).	Vasodilation, which increases K^+ and water secretion.
Gastric gland arterioles	Vasoconstriction, which inhibits secretion (α_1).	Secretion of gastric juice.
Intestinal gland arterioles	Vasoconstriction, which inhibits secretion (α_1).	Secretion of intestinal juice.
Coronary (heart) arterioles	Relaxation $\rightarrow$ vasodilation (β_2); contraction $\rightarrow$ vasoconstriction (α_1, α_2); contraction $\rightarrow$ vasoconstriction (muscarinic ACh receptors).	Contraction $\rightarrow$ vasoconstriction.
Skin and mucosal arterioles	Contraction $\rightarrow$ vasoconstriction (α_1).	Vasodilation, which may not be physiologically significant.
Skeletal muscle arterioles	Contraction $\rightarrow$ vasoconstriction (α_1); relaxation $\rightarrow$ vasodilation (β_2); relaxation $\rightarrow$ vasodilation (muscarinic ACh receptors).	No known effect.
Abdominal viscera arterioles	Contraction $\rightarrow$ vasoconstriction (α_1, β_2).	No known effect.
Brain arterioles	Slight contraction $\rightarrow$ vasoconstriction (α_1).	No known effect.
Kidney arterioles	Constriction of blood vessels $\rightarrow$ decreased urine volume (α_1).	No known effect.
Systemic veins	Contraction $\rightarrow$ constriction (α_1); relaxation $\rightarrow$ dilation (β_2).	No known effect.

520) and occurs due to interruption of the control of ANS neurons by higher centers. When certain sensory impulses, such as those resulting from stretching of a full urinary bladder, are unable to ascend the spinal cord, mass stimulation of the sympathetic nerves inferior to the level of injury occurs. Other triggers include stimulation of pain receptors and visceral contractions. Among the effects of increased sympathetic activity is severe vasoconstriction, which elevates blood pressure. In response, the cardiovascular center in the medulla oblongata (1) increases parasympathetic output via the vagus nerve, which decreases heart rate, and (2) decreases sympathetic output, which causes dilation of blood vessels superior to the level of the injury.

Autonomic dysreflexia is characterized by a pounding headache; hypertension; flushed, warm skin with profuse sweating above the injury level; pale, cold, and dry skin below the injury level; and anxiety. It is an emergency condition that requires immediate intervention. If untreated, autonomic dysreflexia can cause seizures, stroke, or heart attack. ∎

▶ C H E C K P O I N T

15. Give three examples of controlled conditions in the body that are kept in homeostatic balance by autonomic reflexes.

16. How does an autonomic reflex arc differ from a somatic reflex arc?

• • •

Now that our exploration of the nervous system is complete, you can appreciate the many ways that this system contributes to homeostasis of other body systems by examining *Focus on Homeostasis: The Nervous System* on page 582. Next, in Chapter 18, we will see how the hormones released by the endocrine system also help maintain homeostasis of many body processes.

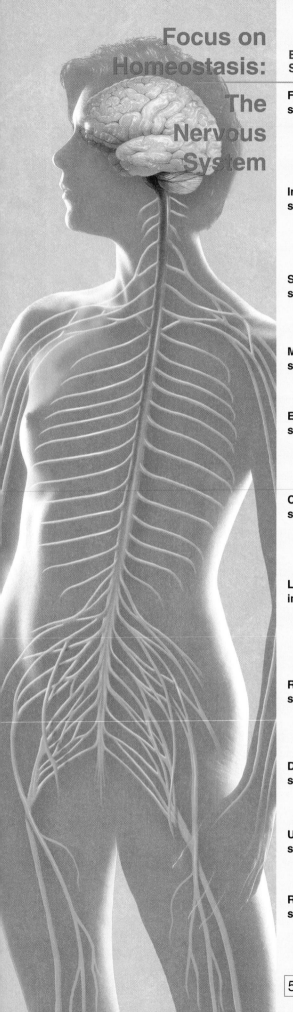

Focus on Homeostasis: The Nervous System

Body System	Contribution of the Nervous System
For all body systems	Together with hormones from the endocrine system, nerve impulses provide communication and regulation of most body tissues.
Integumentary system	Sympathetic nerves of the autonomic nervous system (ANS) control contraction of smooth muscles attached to hair follicles and secretion of perspiration from sweat glands.
Skeletal system	Nociceptors (pain receptors) in bone tissue warn of bone trauma or damage.
Muscular system	Somatic motor neurons receive instructions from motor areas of the brain and stimulate contraction of skeletal muscles to bring about body movements; basal ganglia and reticular formation set level of muscle tone; cerebellum coordinates skilled movements.
Endocrine system	Hypothalamus regulates secretion of hormones from anterior and posterior pituitary; ANS regulates secretion of hormones from adrenal medulla and pancreas.
Cardiovascular system	Cardiovascular center in the medulla oblongata provides nerve impulses to ANS that govern heart rate and the forcefulness of the heartbeat; nerve impulses from ANS also regulate blood pressure and blood flow through blood vessels.
Lymphatic and immune system	Certain neurotransmitters help regulate immune responses; activity in nervous system may increase or decrease immune responses.
Respiratory system	Respiratory areas in brain stem control breathing rate and depth; ANS helps regulate diameter of airways.
Digestive system	ANS and enteric nervous system (ENS) help regulate digestion; parasympathetic division of ANS stimulates many digestive processes.
Urinary system	ANS helps regulate blood flow to kidneys, thereby influencing the rate of urine formation; brain and spinal cord centers govern emptying of the urinary bladder.
Reproductive systems	Hypothalamus and limbic system govern a variety of sexual behaviors; ANS brings about erection of penis in males and clitoris in females and ejaculation of semen in males; hypothalamus regulates release of anterior pituitary hormones that control gonads (ovaries and testes); nerve impulses elicited by touch stimuli from suckling infant cause release of oxytocin and milk ejection in nursing mothers.

STUDY OUTLINE

COMPARISON OF SOMATIC AND AUTONOMIC NERVOUS SYSTEMS (p. 566)

1. The somatic nervous system operates under conscious control; the ANS usually operates without conscious control.
2. Sensory input for the somatic nervous system is mainly from the special senses and somatic senses; sensory input for the ANS is from interoceptors, in addition to special senses and somatic senses.
3. The axons of somatic motor neurons extend from the CNS and synapse directly with an effector. Autonomic motor pathways consist of two motor neurons in series. The axon of the first motor neuron extends from the CNS and synapses in a ganglion with the second motor neuron; the second neuron synapses with an effector.
4. The output (motor) portion of the ANS has two divisions: sympathetic and parasympathetic. Most body organs receive dual innervation; usually one ANS division causes excitation and the other causes inhibition.
5. Somatic nervous system effectors are skeletal muscles; ANS effectors include cardiac muscle, smooth muscle, and glands.
6. Table 17.1 on page 567 compares the somatic and autonomic nervous systems.

ANATOMY OF AUTONOMIC MOTOR PATHWAYS (p. 568)

1. Preganglionic neurons are myelinated; postganglionic neurons are unmyelinated.
2. The cell bodies of sympathetic preganglionic neurons are in the lateral gray horns of the 12 thoracic and the first two or three lumbar segments of the spinal cord; the cell bodies of parasympathetic preganglionic neurons are in four cranial nerve nuclei (III, VII, IX, and X) in the brain stem and lateral gray horns of the second through fourth sacral segments of the spinal cord.
3. Autonomic ganglia are classified as sympathetic trunk ganglia (on both sides of vertebral column), prevertebral ganglia (anterior to vertebral column), and terminal ganglia (near or inside visceral effectors).
4. Sympathetic preganglionic neurons synapse with postganglionic neurons in ganglia of the sympathetic trunk or in prevertebral ganglia; parasympathetic preganglionic neurons synapse with postganglionic neurons in terminal ganglia.

ANS NEUROTRANSMITTERS AND RECEPTORS (p. 575)

1. Cholinergic neurons release acetylcholine, which binds to nicotinic or muscarinic cholinergic receptors.

2. In the ANS, the cholinergic neurons include all sympathetic and parasympathetic preganglionic neurons, all parasympathetic postganglionic neurons, and sympathetic postganglionic neurons that innervate most sweat glands.
3. In the ANS, adrenergic neurons release norepinephrine. Both epinephrine and norepinephrine bind to alpha and beta adrenergic receptors.
4. Most sympathetic postganglionic neurons are adrenergic.
5. Table 17.2 on page 577 summarizes the types of cholinergic and adrenergic receptors.
6. An agonist is a substance that binds to and activates a receptor, mimicking the effect of a natural neurotransmitter or hormone. An antagonist is a substance that binds to and blocks a receptor, thereby preventing a natural neurotransmitter or hormone from exerting its effect.

PHYSIOLOGICAL EFFECTS OF THE ANS (p. 576)

1. The sympathetic division favors body functions that can support vigorous physical activity and rapid production of ATP (fight-or-flight response); the parasympathetic division regulates activities that conserve and restore body energy.
2. The effects of sympathetic stimulation are longer lasting and more widespread than the effects of parasympathetic stimulation.
3. Table 17.3 on page 579 compares structural and functional features of the sympathetic and parasympathetic divisions.
4. Table 17.4 on pages 580–581 lists sympathetic and parasympathetic responses.

INTEGRATION AND CONTROL OF AUTONOMIC FUNCTIONS (p. 578)

1. An autonomic reflex adjusts the activities of smooth muscle, cardiac muscle, and glands.
2. An autonomic reflex arc consists of a receptor, a sensory neuron, an integrating center, two autonomic motor neurons, and a visceral effector.
3. The hypothalamus is the major control and integration center of the ANS. It is connected to both the sympathetic and the parasympathetic divisions.

SELF-QUIZ QUESTIONS

Fill in the blanks in the following statements.

1. Cholinergic neurons release _____ and adrenergic neurons release _____.

2. Because of the location of the preganglionic cell bodies, the sympathetic division of the ANS is also called the _____ division, whereas the parasympathetic division is also called the _____ division.

Indicate whether the following statements are true or false.

3. Structurally, the ANS includes autonomic sensory neurons and autonomic motor neurons.

4. Organs that receive both sympathetic and parasympathetic motor impulses are said to have dual innervation.

Choose the one best answer to the following questions.

5. Which of the following statements are *true*? (1) The somatic nervous system and the ANS both include sensory and motor neurons. (2) Somatic sensory receptors are interoceptors. (3) The effect of an autonomic motor neuron is either excitation or inhibition, whereas that of a somatic motor neuron is always excitation. (4) Autonomic sensory neurons are mostly associated with interoceptors and mechanoreceptors. (5) Autonomic motor pathways always consist of two motor neurons in series. (6) Somatic motor pathways may or may not consist of two motor neurons in series. (a) 1, 2, 3, 4, and 5, (b) 1, 3, 4, and 5, (c) 2, 3, 5, and 6, (d) 1, 3, 5, and 6, (e) 2, 4, 5, and 6.

6. Which of the following statements is *false*? (a) The first neuron in an autonomic pathway is the preganglionic neuron. (b) The preganglionic neuron's cell body is within the CNS. (c) The postganglionic neuron's cell body is within the CNS. (d) Postganglionic neurons relay impulses from autonomic ganglia to visceral effectors. (e) All somatic motor neurons release acetylcholine.

7. Which of the following statements is *false*? (a) A single sympathetic preganglionic fiber may synapse with 20 or more postganglionic fibers, which partly explains why sympathetic responses are widespread throughout the body. (b) Parasympathetic effects tend to be localized because parasympathetic neurons usually synapse in the terminal ganglia with only four or five postsynaptic neurons (all of which supply a single effector). (c) Only the thoracic and first lumbar nerves have white rami communicantes. (d) The white rami communicantes connect the posterior ramus of the spinal nerve with the ganglia of the sympathetic trunk. (e) White rami communicantes contain myelinated fibers.

8. Which of the following are cholinergic neurons? (1) all sympathetic preganglionic neurons, (2) all parasympathetic preganglionic neurons, (3) all parasympathetic postganglionic neurons, (4) all sympathetic postganglionic neurons, (5) some sympathetic postganglionic neurons. (a) 1, 2, 3, and 5, (b) 1, 2, 3, and 4, (c) 2, 3, and 5, (d) 2 and 5, (e) 1, 3, and 5.

9. Which of the following statements are *true*? (1) Most sympathetic postganglionic axons are adrenergic. (2) Cholinergic receptors include nicotinic and muscarinic receptors. (3) Adrenergic receptors include alpha and beta receptors. (4) Muscarinic receptors are present on all effectors innervated by parasympathetic postganglionic axons. (5) In general, norepinephrine stimulates alpha receptors more vigorously than beta receptors, whereas epinephrine is a potent stimulator of both alpha and beta receptors. (a) 1, 2, 3, 4, and 5, (b) 2, 3, 4 and 5, (c) 1, 3, 4, and 5, (d) 3, 4, and 5, (e) 1, 2, 3, and 4.

10. Which of the following are reasons why the effects of sympathetic stimulation are longer lasting and more widespread than those of parasympathetic stimulation? (1) There is greater divergence of sympathetic postganglionic fibers. (2) There is less divergence of sympathetic postganglionic fibers. (3) Acetylcholinesterase quickly inactivates ACh, whereas norepinephrine lingers in the synaptic cleft for a longer time. (4) Norepinephrine and epinephrine secreted into the blood by the adrenal medullae intensify the actions of the sympathetic division. (5) ACh remains in the synaptic cleft until norepinephrine is produced. (a) 1 and 3, (b) 1, 3, and 5, (c) 1, 3, and 4, (d) 2, 3, and 4, (e) 2, 3, and 5.

11. A beta blocker works (a) by activating β receptors, (b) as an agonist, (c) by inactivating epinephrine and norepinephrine, (d) by preventing activation of β receptors by epinephrine and norepinephrine, (e) by preventing activation of muscarinic receptors by ACh.

12. Which of the following are components of an autonomic reflex arc? (1) receptor, (2) sensory neuron, (3) integrating center, (4) two motor neurons, (5) effector. (a) 3, 4, and 5, (b) 1, 3, and 4, (c) 2, 4, and 5, (d) 1, 4, and 5, (e) 1, 2, 3, 4, and 5.

13. Match the following:

_____ (a) plexus located mostly posterior to each lung

_____ (b) plexus containing the inferior mesenteric ganglion and supplying the large intestine

_____ (c) plexus located anterior to the fifth lumbar vertebra and the superior portion of the sternum; supplies pelvic viscera

_____ (d) plexus at the base of the heart surrounding the large blood vessels emerging from the heart

_____ (e) the largest autonomic plexus; found at the level of the last thoracic and first lumbar vertebrae

_____ (f) plexus containing the superior mesenteric ganglion and supplying the small and large intestine

(1) cardiac plexus
(2) pulmonary plexus
(3) celiac (solar) plexus
(4) superior mesenteric plexus
(5) inferior mesenteric plexus
(6) hypogastric plexus

14. Match the following:
    (1) sympathetic trunk ganglia
    (2) prevertebral ganglia
    (3) terminal ganglia
    ____ (a) also known as intramural ganglia
    ____ (b) includes the celiac, superior mesenteric, and inferior mesenteric ganglia
    ____ (c) also called vertebral chain or paravertebral ganglia
    ____ (d) lie in a vertical row on either side of the vertebral column
    ____ (e) postganglionic axons, in general, innervate organs below the diaphragm
    ____ (f) ganglia located at the end of an autonomic motor pathway close to or actually within the wall of a visceral organ
    ____ (g) includes ciliary, pterygopalatine, submandibular, and otic ganglia

15. Match the following:
    (1) increased activity of the sympathetic division
    (2) increased activity of the parasympathetic division
    ____ (a) stimulates urination and defecation
    ____ (b) prepares the body for emergency situations
    ____ (c) fight-or-flight response
    ____ (d) promotes digestion and absorption of food
    ____ (e) concerned primarily with processes involving the expenditure of energy
    ____ (f) controlled by the posterior and lateral portions of the hypothalamus
    ____ (g) controlled by the anterior and medial portions of the hypothalamus
    ____ (h) causes a decrease in heart rate

CRITICAL THINKING QUESTIONS

1. It's Thanksgiving and you've just eaten a huge turkey dinner with all the trimmings. Now you're going to watch the football game on the television. Which division of the nervous system will be handling your body's after-dinner activities? List several organs involved and the effects of the nervous system on their functions.
   HINT *The most strenuous activity of the afternoon will be reaching for the TV's remote control.*

2. After his lunch, baby Bobby was looking uncomfortable, squirming and fretting. Suddenly, a pungent odor wafted from his diaper and he looked very content. Trace the reflex arc responsible for Bobby's present condition.
   HINT *Baby Bobby's body responds to the stimuli automatically—say that three times fast.*

3. The advertisements made cigarette smoking sound cool and relaxing, so Jamal decided to try it. After smoking half a pack, Jamal felt nervous and edgy, his muscles felt "twitchy," and his hands were shaking. What's happened to Jamal?
   HINT *How does your nervous system cause a muscle twitch?*

ANSWERS TO FIGURE QUESTIONS

17.1 Dual innervation means that a body organ innervated by the ANS receives both sympathetic and parasympathetic neurons.

17.2 Most parasympathetic preganglionic axons are longer than most sympathetic preganglionic axons because most parasympathetic ganglia are in the walls of visceral organs, whereas most sympathetic ganglia are close to the spinal cord in the sympathetic trunk.

17.3 Terminal ganglia are associated with the parasympathetic division; sympathetic trunk and prevertebral ganglia are associated with the sympathetic division.

17.4 The largest autonomic plexus is the celiac plexus.

17.5 White rami look white due to the presence of myelin.

17.6 Cholinergic neurons that have nicotinic ACh receptors include sympathetic postganglionic neurons innervating sweat glands and all parasympathetic postganglionic neurons. The effectors innervated by these cholinergic neurons possess muscarinic receptors.

The Endocrine System

THE ENDOCRINE SYSTEM AND HOMEOSTASIS

Circulating or local hormones of the endocrine system contribute to homeostasis by regulating activity and growth of target cells in your body. Hormones also regulate your metabolism.

ENERGIZE YOUR STUDY

FOUNDATIONS CD

Anatomy Overview
- The Endocrine System
- Epithelial Tissue: Glandular Epithelium

Animations
- Systems Contributions to Homeostasis
- The Case of the Girl with Fruity Breath

Concepts and Connections and Exercises reinforce your understanding

www.wiley.com/college/apcentral

INSIGHTS AND EXPLORATIONS

Do you find yourself getting depressed as the days grow shorter with the onset of winter? For some people this depression is chronic and is known as Seasonal Affective Disorder (SAD). It is estimated that 25% of the population experiences a mild form of SAD, while 5% suffer from a more severe form of the disorder. Come explore via the web why light can help individuals suffering from this disorder.

A Student's View

Sometimes reading can become so monotonous that the information just becomes black characters on a white page, and your mind goes on autopilot — making it harder to interpret. The Clinical Applications help to answer the question, "Why am I doing this?"

A few of the clinical applications in this chapter have even helped remind me why I chose the medical field; to help people and family members. As often as I have seen my father take his insulin shots, it never dawned on me why he had to take the insulin in the form of an injection and not an oral medicine. Even after reading and learning about water-soluble and lipid-soluable hormones I didn't put two and two together until after reading the Clinical Application, which in turn led to thoughts of where else this would apply. Getting my mind buzzing with more ideas and ultimately getting the response of, "So, that's why", is a great feeling.

The further I get in my studies of anatomy and physiology the more information I am able to pull together and apply when reading a Clinical Application. I don't believe knowledge is power unless you know how to apply that knowledge. The various clinical applications show the big picture and turn the text into images that allow me to fully understand what I have just read. Seeing my hard work come together by being able to understand all aspects of the clinical application inspires me to continue learning. In a perfect world, all my textbooks would include applications like these.

Lisa P. Hubbard
Macon State College

A round age twelve, as boys and girls enter puberty, they start to develop striking differences in physical appearance and behavior. Perhaps no other period in life so dramatically shows the impact of the endocrine system in directing development and regulating body functions. In girls, estrogens promote accumulation of adipose tissue in the breasts and hips, sculpting a feminine shape. At the same time, or a little later in boys, testosterone begins to help build muscle mass and enlarge the vocal cords, producing a lower-pitched voice. These changes are just a few examples of the powerful influence of endocrine secretions. Less dramatically, perhaps, multitudes of hormones help maintain homeostasis on a daily basis. They regulate the activity of smooth muscle, cardiac muscle, and some glands; alter metabolism; spur growth and development; influence reproductive processes; and participate in circadian (daily) rhythms established by the suprachiasmatic nucleus of the hypothalamus.

COMPARISON OF CONTROL BY THE NERVOUS AND ENDOCRINE SYSTEMS

▶ OBJECTIVE

- **Compare control of body functions by the nervous system and endocrine system.**

Although the nervous and endocrine systems act together to co-ordinate functions of all body systems, their means of control are different. The nervous system acts through nerve impulses conducted along axons of neurons. At synapses, nerve impulses trigger the release of mediator (messenger) molecules called *neurotransmitters* (shown in Figure 12.14 on page 405). The endocrine system also controls body activities by releasing mediators, which are called *hormones*.

A **hormone** (*hormon* = to excite or get moving) is a mediator molecule that is released in one part of the body but regulates activity of cells in other parts of the body. Most hormones enter interstitial fluid and then the bloodstream. The circulating blood delivers hormones to cells throughout the body. Both neurotransmitters and hormones exert their effects by binding to receptors on or in their "target" cells. Several neurotransmitters are also hormones. One familiar example is norepinephrine, which is released as a neurotransmitter by sympathetic postganglionic neurons and as a hormone by cells of the adrenal medullae.

Responses of the endocrine system often are slower than responses of the nervous system. Although some hormones act within seconds, most take several minutes or more to cause a response. Moreover, the effects attained by activating the nervous system are generally briefer than the effects produced by the endocrine system. The nervous system causes muscles to contract or relax and glands to secrete either more or less of their product. The influence of the endocrine system is much broader. It helps regulate virtually all types of body cells.

We will also have several opportunities to see how the nervous and endocrine systems function together as an interlocking "supersystem." Certain parts of the nervous system stimulate or inhibit the release of hormones, which in turn may promote or inhibit the generation of nerve impulses. Table 18.1 compares the characteristics of the nervous and endocrine systems. In this chapter, we focus on the major endocrine glands and hormone-producing tissues and examine how their hormones govern body activities.

▶ CHECKPOINT

1. How many similarities and differences can you list between the nervous and endocrine systems with regard to the control of homeostasis?

ENDOCRINE GLANDS

▶ OBJECTIVE

- **Distinguish between exocrine and endocrine glands.**

Recall from Chapter 4 that the body contains two kinds of glands: exocrine glands and endocrine glands. **Exocrine glands** (*exo-* = outside) secrete their products into ducts that carry the secretions into body cavities, into the lumen of an organ, or to the

Table 18.1	Comparison of Control by the Nervous and Endocrine Systems	
Characteristic	**Nervous System**	**Endocrine System**
Mediator molecules	Neurotransmitters released locally in response to nerve impulses.	Hormones delivered to tissues throughout the body by the blood.
Site of mediator action	Close to site of release, at a synapse; binds to receptors in postsynaptic membrane.	Far from site of release (usually); binds to receptors on or in target cells.
Types of target cells	Muscle (smooth, cardiac, and skeletal) cells, gland cells, other neurons.	Virtually all body cells.
Time to onset of action	Typically within milliseconds (thousandths of a second).	Seconds to hours or days.
Duration of action	Generally briefer.	Generally longer.

outer surface of the body. Exocrine glands include sudoriferous (sweat), sebaceous (oil), mucous, and digestive glands. **Endocrine glands,** by contrast, secrete their products (hormones) into the interstitial fluid surrounding the secretory cells, rather than into ducts. Then, hormones diffuse into capillaries and blood carries them away. Because most hormones are required in very small amounts, circulating levels typically are low.

The endocrine glands include the pituitary, thyroid, parathyroid, adrenal, and pineal glands (Figure 18.1). In addition, several organs and tissues are not endocrine glands exclusively but contain cells that secrete hormones. These include the hypothalamus, thymus, pancreas, ovaries, testes, kidneys, stomach, liver, small intestine, skin, heart, adipose tissue, and placenta. Taken together, all endocrine glands and hormone-secreting cells constitute the **endocrine system.** The science of the structure and function of the endocrine glands and the diagnosis and treatment of disorders of the endocrine system is **endocrinology** (en′-dō-kri-NOL-ō-jē; *endo-* = within; *-crino* = to secrete; *-logy* = study of).

HORMONE ACTIVITY

▶ **O B J E C T I V E S**

- **Describe how hormones interact with target-cell receptors.**
- **Compare the two chemical classes of hormones based on their solubility.**

The Role of Hormone Receptors

Although a given hormone travels throughout the body in the blood, it affects only specific target cells. Hormones, like neurotransmitters, influence their target cells by chemically binding to specific protein or glycoprotein **receptors.** Only the target cells for a given hormone have receptors that bind and recognize that hormone. For example, thyroid-stimulating hormone (TSH) binds to receptors on cells of the thyroid gland, but it does not bind to cells of the ovaries because ovarian cells do not have TSH receptors.

Figure 18.1 Location of many endocrine glands. Also shown are other organs that contain endocrine cells and associated structures.

🔑 Endocrine glands secrete hormones, which circulating blood delivers to target tissues.

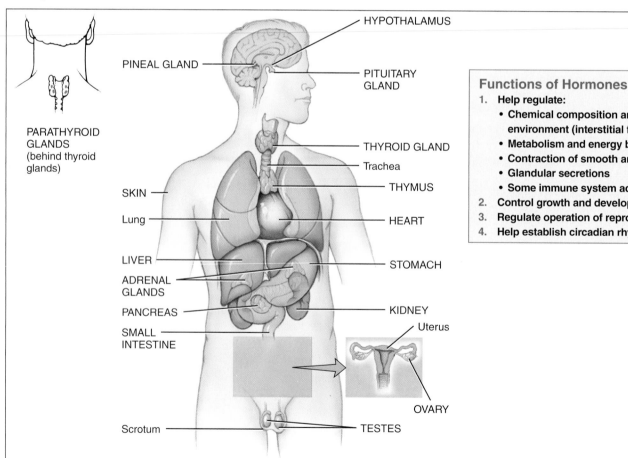

PINEAL GLAND

PARATHYROID GLANDS (behind thyroid glands)

SKIN

Lung

LIVER

ADRENAL GLANDS

PANCREAS

SMALL INTESTINE

Scrotum

HYPOTHALAMUS

PITUITARY GLAND

THYROID GLAND

Trachea

THYMUS

HEART

STOMACH

KIDNEY

Uterus

OVARY

TESTES

Functions of Hormones
1. **Help regulate:**
   - **Chemical composition and volume of internal environment (interstitial fluid)**
   - **Metabolism and energy balance**
   - **Contraction of smooth and cardiac muscle fibers**
   - **Glandular secretions**
   - **Some immune system activities**
2. **Control growth and development.**
3. **Regulate operation of reproductive systems.**
4. **Help establish circadian rhythms.**

❓ What is the basic difference between endocrine glands and exocrine glands?

Receptors, like other cellular proteins, are constantly being synthesized and broken down. Generally, a target cell has 2000 to 100,000 receptors for a particular hormone. If a hormone is present in excess, the number of target-cell receptors may decrease—an effect called **down-regulation.** For example, when certain cells of the testes are exposed to a high concentration of luteinizing hormone (LH), the number of LH receptors decreases. Down-regulation decreases the responsiveness of target cells to the hormone. In contrast, when a hormone (or neurotransmitter) is deficient, the number of receptors may increase. This phenomenon, known as **up-regulation,** makes a target tissue more sensitive to a hormone.

Blocking Hormone Receptors

Synthetic hormones that block the receptors for some naturally occurring hormones are available as drugs. For example, RU486 (mifepristone), which is used to induce abortion, binds to the receptors for progesterone (a female sex hormone) and prevents progesterone from exerting its normal effect. When RU486 is given to a pregnant woman, the uterine conditions needed for nurturing an embryo are not maintained, embryonic development stops, and the embryo is sloughed off along with the uterine lining. This example illustrates an important endocrine principle: If a hormone is prevented from interacting with its receptors, the hormone cannot perform its normal functions. ■

Circulating and Local Hormones

Most hormones are **circulating hormones**—they pass from the secretory cells that make them into interstitial fluid and then into the blood (Figure 18.2a). Other hormones, termed **local hormones,** act locally on neighboring cells or on the same cell that secreted them without first entering the bloodstream (Figure 18.2b). Among local hormones are those that act on neighboring cells, called **paracrines** (*para-* = beside or near), and those that act on the same cell that secreted them, termed **autocrines** (*auto-* = self). One example of a local hormone is interleukin 2 (IL-2), which is released by helper T cells during immune responses. IL-2 helps activate other nearby immune cells, a paracrine effect. But, it also acts as an autocrine by stimulating the same cell that released it to proliferate. This action generates more helper T cells that can secrete even more IL-2 and thus strengthen the immune response. Another example of a local hormone is the gas nitric oxide (NO), which is released by endothelial cells lining blood vessels. NO causes relaxation of nearby vascular smooth muscle fibers, which in turn causes vasodilation and increases the flow of blood in that region.

Local hormones usually are inactivated quickly; circulating hormones may linger in the blood and exert their effects for a few minutes or occasionally for a few hours. In time, circulating hormones are inactivated by the liver and excreted by the kid-

Figure 18.2 Comparison between circulating hormones and local hormones (autocrines and paracrines).

🔑 Circulating hormones act on distant target cells. Paracrines act on neighboring cells and autocrines act on the same cell that produced them.

(a) Circulating hormones

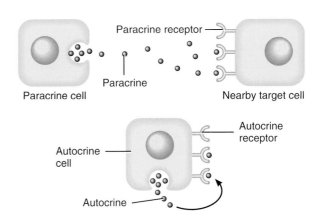

(b) Local hormones (paracrines and autocrines)

❓ In the stomach, one stimulus for secretion of hydrochloric acid by parietal cells is the release of histamine by neighboring mast cells. Is histamine an autocrine or a paracrine in this situation?

neys. In cases of kidney or liver failure, excessive levels of hormones may build up in the blood.

Chemical Classes of Hormones

Chemically, hormones can be divided into two broad classes: those that are soluble in lipids, and those that are soluble in water. This chemical classification is also useful functionally because the ways in which the two classes exert their effects are different.

Lipid-soluble Hormones

The lipid-soluble hormones include steroid hormones, thyroid hormones, and nitric oxide.

1. **Steroid hormones** are derived from cholesterol. Each steroid hormone is unique due to the presence of different functional groups attached at various sites on the four rings at the core of its structure. These small differences allow for a large diversity of functions.

2. Two **thyroid hormones** (T_3 and T_4) are synthesized by attaching iodine to the amino acid tyrosine. The benzene ring of tyrosine plus the attached iodines make T_3 and T_4 very lipid soluble.

3. The gas **nitric oxide (NO)** is both a hormone and a neurotransmitter. Its synthesis is catalyzed by the enzyme nitric oxide synthase.

Water-soluble Hormones

The water-soluble hormones include amine hormones, peptide and protein hormones, and eicosanoid hormones.

1. **Amine hormones** are synthesized by decarboxylating (removing a molecule of CO_2) and modifying certain amino acids. They are called amines because they retain an amino group ($-NH_3^+$). The catecholamines—epinephrine, norepinephrine, and dopamine—are synthesized by modifying the amino acid tyrosine. Histamine is synthesized from the amino acid histidine by mast cells and platelets. Serotonin and melatonin are derived from tryptophan.

2. **Peptide hormones** and **protein hormones** are amino acid polymers. The smaller peptide hormones consist of chains of 3 to 49 amino acids whereas the larger protein hormones include 50 to 200 amino acids. Examples of peptide hormones are antidiuretic hormone and oxytocin; protein hormones include human growth hormone and insulin. Several of the protein hormones—for example, thyroid-stimulating hormone (TSH)—have attached carbohydrate groups and thus are **glycoprotein hormones.**

3. The **eicosanoid hormones** (ī-KŌ-sa-noid; *eicos-* = twenty forms; *-oid* = resembling) are derived from arachidonic acid, a 20-carbon fatty acid. The two major types of eicosanoids are **prostaglandins** and **leukotrienes**. The eicosanoids are important local hormones, and they may act as circulating hormones as well.

Table 18.2 summarizes the classes of lipid-soluble and water-soluble hormones. It also provides an overview of the major hormones and their sites of secretion.

Hormone Transport in the Blood

Most water-soluble hormone molecules circulate in the watery blood plasma in a "free" form (not attached to plasma proteins), whereas most lipid-soluble hormone molecules are bound to **transport proteins.** The transport proteins, which are synthesized by cells in the liver, have three functions:

1. They make lipid-soluble hormones temporarily water soluble, thus increasing their solubility in blood.

2. They retard passage of small hormone molecules through the filtering mechanism in the kidneys, thus slowing the rate of hormone loss in the urine.

3. They provide a ready reserve of hormone, already present in the bloodstream.

In general, 0.1–10% of the molecules of a lipid-soluble hormone are not bound to a transport protein. This **free fraction** diffuses out of capillaries, binds to receptors, and triggers responses. As free hormone molecules leave the blood and bind to their receptors, transport proteins release new ones to replenish the free fraction.

Administering Hormones

Both steroid hormones and thyroid hormones are effective when taken by mouth. They are not split apart during digestion and easily cross the intestinal lining because they are lipid soluble. By contrast, peptide and protein hormones, such as insulin, are not effective oral medications because digestive enzymes destroy them by breaking their peptide bonds. This is why people who need insulin must take it by injection. ∎

▶ CHECKPOINT

2. What is the difference between down-regulation and up-regulation?

3. Identify the chemical classes of hormones, and give an example of each.

4. How are hormones transported in the blood?

MECHANISMS OF HORMONE ACTION

▶ OBJECTIVE

• **Describe the two general mechanisms of hormone action.**

The response to a hormone depends on both the hormone and the target cell. Various target cells respond differently to the same hormone. Insulin, for example, stimulates synthesis of glycogen in liver cells but synthesis of triglycerides in adipose cells.

The response to a hormone is not always the synthesis of new molecules, as is the case for insulin. Other hormonal effects include changing the permeability of the plasma membrane, stimulating transport of a substance into or out of the target cells, altering the rate of specific metabolic reactions, or causing contraction of smooth muscle or cardiac muscle. In part, these varied effects of hormones are possible because a single hormone can set in motion several different cellular responses. First, however, a hormone must "announce its arrival" to a target cell by binding to its receptors. For lipid-soluble hormones, the receptors are located inside target cells. For water-soluble hormones, the receptors are part of the plasma membrane of target cells.

Table 18.2 Summary of Hormones by Chemical Class

Chemical Class	Hormones	Site of Secretion
Lipid-soluble		
Steroid hormones	Aldosterone, cortisol, and androgens.	Adrenal cortex.
	Calcitriol.	Kidneys.
	Testosterone.	Testes.
	Estrogens and progesterone.	Ovaries.
Aldosterone		
Thyroid hormones	T_3 (triiodothyronine) and T_4 (thyroxine).	Thyroid gland (follicular cells).
Triiodothyronine (T_3)		
Gas	Nitric oxide (NO).	Endothelial cells lining blood vessels.
Water-soluble		
Amines	Epinephrine and norepinephrine (catecholamines).	Adrenal medulla.
	Melatonin.	Pineal gland.
	Histamine.	Mast cells in connective tissues.
Norepinephrine	Serotonin.	Platelets in blood.
Peptides and proteins	All hypothalamic releasing and inhibiting hormones.	Hypothalamus.
	Oxytocin, antidiuretic hormone.	Posterior pituitary.
	Human growth hormone, thyroid-stimulating hormone, adrenocorticotropic hormone, follicle-stimulating hormone, luteinizing hormone, prolactin, melanocyte-stimulating hormone.	Anterior pituitary.
	Insulin, glucagon, somatostatin, pancreatic polypeptide.	Pancreas.
	Parathyroid hormone.	Parathyroid glands.
	Calcitonin.	Thyroid gland (parafollicular cells).
Oxytocin	Gastrin, secretin, cholecystokinin, GIP (glucose-dependent insulinotropic peptide).	Stomach and small intestine (enteroendocrine cells).
	Erythropoietin.	Kidneys.
	Leptin.	Adipose tissue.
Eicosanoids	Prostaglandins, leukotrienes.	All cells except red blood cells.
A leukotriene (LTB_4)		

Peptides and proteins structure:

Glutamine ——— Isoleucine
Asparagine   Tyrosine
Cysteine —S—S— Cysteine
Proline
Leucine
Glycine
NH₂

Action of Lipid-soluble Hormones

Lipid-soluble hormones, including steroid hormones and thyroid hormones, bind to receptors within target cells. Their mechanism of action is as follows (Figure 18.3):

1 A free lipid-soluble hormone molecule diffuses from the blood, through interstitial fluid, and through the lipid bilayer of the plasma membrane into a cell.

2 If the cell is a target cell, the hormone binds to and activates receptors located within the cytosol or nucleus. The activated receptor-hormone complex then alters gene expression: It turns specific genes of the nuclear DNA on or off.

3 As the DNA is transcribed, new messenger RNA (mRNA) forms, leaves the nucleus, and enters the cytosol. There, it directs synthesis of a new protein, often an enzyme, on the ribosomes.

4 The new proteins alter the cell's activity and cause the responses typical of that hormone.

Action of Water-soluble Hormones

Because amine, peptide, protein, and eicosanoid hormones are not lipid soluble, they cannot diffuse through the lipid bilayer of the plasma membrane and bind to receptors inside target cells. Instead, water-soluble hormones bind to receptors that protrude from the target cell surface. The receptors are integral transmembrane proteins in the plasma membrane. When a water-soluble hormone binds to its receptor at the outer surface of the plasma membrane, it acts as the **first messenger.** The first messenger (the hormone) then causes production of a **second messenger** inside the cell, where specific hormone-stimulated responses take place. One common second messenger is **cyclic AMP (cAMP).** Neurotransmitters, neuropeptides, and several sensory transduction mechanisms (for example, vision; see Figure 16.14 on page 542) also act via second-messenger systems.

The action of a typical water-soluble hormone occurs as follows (Figure 18.4):

1 A water-soluble hormone (the first messenger) diffuses from the blood through interstitial fluid and then binds to its receptor at the exterior surface of a target cell's plasma membrane. The hormone-receptor complex activates another membrane protein, called a **G protein.** The activated G protein in turn activates **adenylate cyclase.**

2 Adenylate cyclase converts ATP into cyclic AMP (cAMP). Because the enzyme's active site is on the inner surface of the plasma membrane, this reaction occurs in the cytosol of the cell.

3 Cyclic AMP (the second messenger) activates one or more protein kinases, which may be free in the cytosol or bound to the plasma membrane. A protein kinase is an enzyme that phosphorylates (adds a phosphate group to) cellular proteins. The donor of the phosphate group is ATP, which is converted to ADP.

4 Activated protein kinases phosphorylate one or several other enzymes. Phosphorylation activates some enzymes and inactivates others, rather like turning a switch on or off.

5 Phosphorylated enzymes, in turn, catalyze reactions that produce physiological responses. Different protein kinases exist within different target cells and within different organelles of the same target cell. Thus, one protein kinase might trigger glycogen synthesis, a second triglyceride breakdown, a third protein synthesis, and so forth. Moreover, phosphorylation by a protein kinase can inhibit certain enzymes. For example, some of the kinases unleashed when epinephrine binds to liver cells inactivate an enzyme needed for glycogen synthesis.

6 After a brief period, an enzyme called **phosphodiesterase** inactivates cAMP. Thus, the cell's response is turned off unless new hormone molecules continue to bind to their receptors in the plasma membrane.

The binding of a hormone to its receptor activates many G-protein molecules, which in turn activate molecules of adenylate cyclase. Unless further stimulated by the binding of more hormone molecules to receptors, G proteins slowly inactivate, thus helping to stop the hormone response. G proteins are a common feature of most second-messenger systems.

Figure 18.3 Mechanism of action of the lipid-soluble steroid hormones and thyroid hormones.

Lipid-soluble hormones bind to receptors inside target cells.

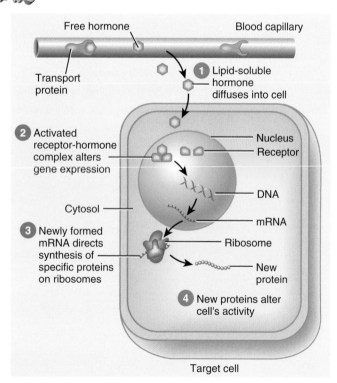

- Free hormone
- Blood capillary
- Transport protein
- **1** Lipid-soluble hormone diffuses into cell
- **2** Activated receptor-hormone complex alters gene expression
- Nucleus
- Receptor
- Cytosol
- DNA
- **3** Newly formed mRNA directs synthesis of specific proteins on ribosomes
- mRNA
- Ribosome
- New protein
- **4** New proteins alter cell's activity
- Target cell

How quickly do you think a steroid hormone can begin to exert its effects?

Figure 18.4 Mechanism of action of the water-soluble hormones (amines, peptides, proteins, and eicosanoids).

Water-soluble hormones bind to receptors embedded in the plasma membrane of target cells.

Target cell

Why is cAMP called a "second messenger"?

Cyclic AMP and other second messengers alter a cell's function in specific ways. For example, an increase in cAMP causes adipose cells to break down triglycerides and release fatty acids more rapidly, but cAMP stimulates thyroid cells to secrete more thyroid hormone. Many hormones exert at least some of their physiological effects through the *increased* synthesis of cAMP. Examples include antidiuretic hormone (ADH), thyroid-stimulating hormone (TSH), adrenocorticotropic hormone (ACTH), glucagon, epinephrine, and hypothalamic-releasing hormones. In other cases, such as growth hormone-inhibiting hormone (GHIH), the level of cyclic AMP *decreases* in response to the binding of a hormone to its receptor.

Besides cAMP, other second messengers include calcium ions (Ca^{2+}), cGMP (cyclic guanosine monophosphate, a cyclic nucleotide similar to cAMP), inositol trisphosphate (IP_3), and diacylglycerol (DAG). Although it is a lipid-soluble hormone, nitric oxide exerts its effect inside smooth muscle fibers by activating guanylyl cyclase. This enzyme, in turn, converts guanosine triphosphate (GTP) to cGMP, which causes calcium ions to enter storage areas of the smooth muscle fiber. The lowered Ca^{2+} level in the cytosol then causes muscle relaxation. A given hormone (or neurotransmitter) may use different second messengers in different target cells.

Hormones that bind to plasma membrane receptors can induce their effects at very low concentrations because they initiate a cascade or chain reaction, each step of which multiplies or amplifies the initial effect. For example, the binding of a single molecule of epinephrine to its receptor on a liver cell may activate a hundred or so G proteins, each of which activates an adenylate cyclase molecule. If each adenylate cyclase produces even 1000 cAMP, then 100,000 of these second messengers will be liberated inside the cell. Each cAMP may activate a protein kinase, which in turn can act on hundreds or thousands of substrate molecules. Some of the kinases phosphorylate and activate a key enzyme needed for glycogen breakdown. The end result of the binding of epinephrine to its receptor is the breakdown of millions of glycogen molecules into glucose monomers.

Cholera Toxin and G Proteins

The toxin produced by cholera bacteria is deadly. It produces massive watery diarrhea, and an infected person can rapidly die from the resulting dehydration. The cholera toxin modifies the G proteins in intestinal epithelial cells such that they become locked in an activated state. As a result, the intracellular cAMP concentration skyrockets. One of the effects of cAMP in these cells is to stimulate an active transport pump that ejects chloride ions (Cl^-) from the cells into the lumen of the intestines. Water follows the Cl^- by osmosis, and positively charged sodium ions follow the negatively charged Cl^-. Thus, cholera toxin causes a huge outflow of Na^+, Cl^-, and water into the feces. Treatment consists of replacement of lost fluids, either intravenously or by mouth (oral rehydration therapy), plus antibiotic therapy with tetracycline. ■

Hormone Interactions

The responsiveness of a target cell to a hormone depends on (1) the hormone's concentration, (2) the abundance of the target cell's hormone receptors, and (3) influences exerted by other hormones. A target cell responds more vigorously when the level of a hormone rises or when it has more receptors (up-regulation). In addition, the actions of some hormones on target cells require a simultaneous or recent exposure to a second hormone. In such cases, the second hormone is said to have a **permissive effect.** For example, epinephrine alone only weakly stimulates lipolysis (the breakdown of triglycerides), but when small amounts of thyroid hormones (T_3 and T_4) are also present, the same amount of epinephrine stimulates lipolysis much more powerfully. Sometimes the permissive hormone up-regulates receptors for the other hormone, and sometimes it promotes the synthesis of an enzyme required for the expression of the other hormone's effects.

When the effect of two hormones acting together is greater or more extensive than the sum of each hormone acting alone, the two hormones are said to have a **synergistic effect.** For example, normal development of oocytes in the ovaries requires both FSH from the anterior pituitary and estrogens from the ovaries. Neither hormone alone is sufficient.

When one hormone opposes the actions of another hormone, the two hormones are said to have **antagonistic effects.** An example of an antagonistic pair of hormones is insulin, which promotes synthesis of glycogen by liver cells, and glucagon, which stimulates breakdown of glycogen in the liver.

CONTROL OF HORMONE SECRETION

▶ O B J E C T I V E

• **Describe the three types of signals that can control hormone secretion.**

The release of most hormones occurs in short bursts, with little or no secretion between bursts. When stimulated, an endocrine gland will release its hormone in more frequent bursts, increasing the concentration of the hormone in the blood. In the absence of stimulation, the blood level of the hormone decreases. Regulation of secretion normally prevents overproduction or underproduction of any given hormone.

Hormone secretion is regulated by (1) signals from the nervous system, (2) chemical changes in the blood, and (3) other hormones. For example, nerve impulses to the adrenal medullae regulate the release of epinephrine; blood Ca^{2+} level regulates the secretion of parathyroid hormone; and a hormone from the anterior pituitary (ACTH) stimulates the release of cortisol by the adrenal cortex. Most hormonal regulatory systems work via negative feedback, but a few operate via positive feedback. For example, during childbirth, the hormone oxytocin stimulates contractions of the uterus, and uterine contractions, in turn, stimulate more oxytocin release, a positive feedback effect.

▶ C H E C K P O I N T

5. What factors determine the responsiveness of a target cell to a hormone?

6. What are the differences among permissive effects, synergistic effects, and antagonistic effects of hormones?

7. How do negative and positive feedback systems differ from one another?

Now that we have an overview of the roles of hormones in the endocrine system, we turn to discussions of the various endocrine glands and the hormones they secrete.

HYPOTHALAMUS AND PITUITARY GLAND

▶ O B J E C T I V E S

• **Describe the locations of and relationships between the hypothalamus and pituitary gland.**

• **Describe the location, histology, hormones, and functions of the anterior and posterior pituitary.**

For many years, the pituitary gland was called the "master" endocrine gland because it secretes several hormones that control other endocrine glands. We now know that the pituitary gland itself has a master—the **hypothalamus.** This small region of the brain below the thalamus is the major integrating link between the nervous and endocrine systems. It receives input from the limbic system, cerebral cortex, thalamus, and reticular activating system. It also receives sensory signals from internal organs and from the retina. Painful, stressful, and emotional experiences all cause changes in hypothalamic activity. In turn, the hypothalamus controls the autonomic nervous system and regulates body temperature, thirst, hunger, sexual behavior, and defensive reactions such as fear and rage.

The hypothalamus is an important regulatory center in the nervous system and is also a crucial endocrine gland. Cells in the hypothalamus synthesize at least nine different hormones, and the pituitary gland secretes seven. Together, these 16 hormones play important roles in the regulation of virtually all aspects of growth, development, metabolism, and homeostasis.

The **pituitary gland** (pi-TOO-i-tār-ē) or **hypophysis** (hī-POF-i-sis) lies in the hypophyseal fossa of the sella turcica of the sphenoid bone and attaches to the hypothalamus by a stalk, the **infundibulum** (= a funnel; Figure 18.5). The pituitary gland is a pea-shaped structure measuring 1–1.5 cm (0.5 in.) in diameter and has two anatomically and functionally separate lobes. The **anterior pituitary (anterior lobe)** accounts for about 75% of the total weight of the gland. It develops from an outgrowth of ectoderm called the hypophyseal pouch in the roof of the mouth (see Figure 18.21b). The **posterior pituitary (posterior lobe)** also develops from an ectodermal outgrowth, called the neurohypophyseal bud (see Figure 18.21b). It contains axons and axon terminals of more than 10,000 neurons whose cell bodies are located in the supraoptic and paraventricular nuclei of the hypothalamus (see Figure 18.8). The axon terminals in the posterior pituitary are associated with specialized neuroglia called **pituicytes** (pi-TOO-i-sītz).

A third region called the **pars intermedia** atrophies during human fetal development and ceases to exist as a separate lobe in adults (see Figure 18.21b). However, some of its cells migrate into adjacent parts of the anterior pituitary, where they persist.

Anterior Pituitary

The **anterior pituitary** or **adenohypophysis** (ad′-e-nō-hī-POF-i-sis; *adeno-* = gland; *-hypophysis* = undergrowth) secretes hormones that regulate a wide range of bodily activities, from

Figure 18.5 Hypothalamus and pituitary gland, and their blood supply. The small figure at right indicates that releasing and inhibiting hormones synthesized by hypothalamic neurosecretory cells are transported within axons and released at the axon terminals. The hormones diffuse into capillaries of the primary plexus of the hypophyseal portal system and are carried by the hypophyseal portal veins to the secondary plexus of the hypophyseal portal system for distribution to target cells in the anterior pituitary.

🔑 **Hypothalamic hormones are an important link between the nervous and endocrine systems.**

Posterior pituitary — Infundibulum

Anterior pituitary

Sagittal section of pituitary gland

Hypothalamic neurosecretory cell

Hypophyseal portal veins

Hypothalamus

Pituitary gland

HYPOTHALAMUS

Primary plexus of the hypophyseal portal system

Infundibulum

Superior hypophyseal artery

Posterior hypophyseal veins

Hypophyseal portal veins

Sphenoid bone

POSTERIOR PITUITARY

ANTERIOR PITUITARY

Capillary plexus of the infundibular process

Secondary plexus of the hypophyseal portal system

Hypophyseal fossa

Anterior hypophyseal veins

POSTERIOR

ANTERIOR

Inferior hypophyseal artery

 What is the functional importance of the hypophyseal portal veins?

growth to reproduction. Release of anterior pituitary hormones is stimulated by **releasing hormones** and suppressed by **inhibiting hormones** from the hypothalamus. Thus, the hypothalamic hormones are an important link between the nervous and endocrine systems.

Hypophyseal Portal System

Hypothalamic hormones reach the anterior pituitary through a portal system. Usually, blood passes from the heart through an artery to a capillary to a vein and back to the heart. In a portal system, blood flows from one capillary network into a portal vein, and then into a second capillary network without passing through the heart. The name of the portal system gives the location of the second capillary network. In the **hypophyseal portal system** (hī′-pō-FIZ-ē-al), blood flows from capillaries in the hypothalamus into portal veins that carry blood to capillaries of the anterior pituitary.

The **superior hypophyseal arteries**, branches of the internal carotid arteries, bring blood into the hypothalamus (Figure 18.5). At the juncture of the median eminence of the hypothalamus and the infundibulum, these arteries divide into a capillary network called the **primary plexus of the hypophyseal portal system.** From the primary plexus, blood drains into the **hypophyseal portal veins** that pass down the outside of the infundibulum. In the anterior pituitary, the hypophyseal portal veins divide again and form another capillary network called the **secondary plexus of the hypophyseal portal system.**

Near the median eminence and above the optic chiasm are clusters of specialized neurons, called **neurosecretory cells.** They synthesize the hypothalamic releasing and inhibiting hormones in their cell body and package the hormones inside vesicles, which reach the axon terminals by axonal transport. When nerve impulses reach the axon terminals, they stimulate the vesicles to undergo exocytosis. The hormones then diffuse into the primary plexus of the hypophyseal portal system. Quickly, the hypothalamic hormones flow with the blood through the portal veins and into the secondary plexus. This direct route permits hypothalamic hormones to act immediately on anterior pituitary cells, before the hormones are diluted or destroyed in the general circulation. Hormones secreted by anterior pituitary cells pass into the secondary plexus capillaries, which drain into the anterior hypophyseal veins. Now in the general circulation, anterior pituitary hormones travel to target tissues throughout the body.

Types of Anterior Pituitary Cells

Five types of anterior pituitary cells—somatotrophs, thyrotrophs, gonadotrophs, lactotrophs, and corticotrophs—secrete seven hormones (Table 18.3):

1. **Somatotrophs** secrete **human growth hormone (hGH)** or **somatotropin** (sō′-ma-tō-TRŌ-pin; *somato-* = body; *-tropin* = change). Human growth hormone in turn stimulates several tissues to secrete **insulinlike growth factors**, hormones that stimulate general body growth and regulate aspects of metabolism.

2. **Thyrotrophs** secrete **thyroid-stimulating hormone (TSH)** or **thyrotropin** (thī-rō-TRŌ-pin; *thyro-* = pertaining to the thyroid gland). TSH controls the secretions and other activities of the thyroid gland.

3. **Gonadotrophs** (*gonado-* = seed) secrete two hormones: **follicle-stimulating hormone (FSH)** and **luteinizing hormone (LH)** (LOO-tē-in′-īz-ing). FSH and LH both act on the gonads. They stimulate secretion of estrogens and progesterone and the maturation of oocytes in the ovaries, and they stimulate secretion of testosterone and sperm production in the testes.

4. **Lactotrophs** (*lacto-* = milk) secrete **prolactin (PRL),** which initiates milk production in the mammary glands.

5. **Corticotrophs** secrete **adrenocorticotropic hormone (ACTH)** or **corticotropin** (kor′-ti-kō-TRŌ-pin; *cortico-* = rind or bark), which stimulates the adrenal cortex to secrete glucocorticoids. Some corticotrophs, remnants of the pars intermedia, also secrete **melanocyte-stimulating hormone (MSH).**

Hormones that influence another endocrine gland are called **tropic hormones** or **tropins.** Several of the anterior pituitary hormones are tropins. The two gonadotropins, follicle stimulating hormone and luteinizing hormone, regulate the functions of the gonads (ovaries and testes). Thyrotropin stimulates the thy-

Table 18.3	Hormones of the Anterior Pituitary		
Hormone	**Secreted by**	**Releasing Hormone (Stimulates Secretion)**	**Inhibiting Hormone (Suppresses Secretion)**
Human growth hormone (hGH) or **somatotropin**	Somatotrophs.	Growth hormone-releasing hormone (GHRH), also known as somatocrinin.	Growth hormone-inhibiting hormone (GHIH), also known as somatostatin.
Thyroid-stimulating hormone (TSH) or **thyrotropin**	Thyrotrophs.	Thyrotropin-releasing hormone (TRH).	Growth hormone-inhibiting hormone (GHIH).
Follicle-stimulating hormone (FSH)	Gonadotrophs.	Gonadotropic-releasing hormone (GnRH).	—
Luteinizing hormone (LH)	Gonadotrophs.	Gonadotropic-releasing hormone (GnRH).	—
Prolactin (PRL)	Lactotrophs.	Prolactin-releasing hormone (PRH); TRH.	Prolactin-inhibiting hormone (PIH), which is dopamine.
Adrenocorticotropic hormone (ACTH) or **corticotropin**	Corticotrophs.	Corticotropin-releasing hormone (CRH).	—
Melanocyte-stimulating hormone	Corticotrophs.	Corticotropin-releasing hormone (CRH).	Dopamine.

roid gland, whereas corticotropin acts on the cortex of the adrenal gland.

Control of Secretion by the Anterior Pituitary

Secretion of anterior pituitary hormones is regulated in two ways. First, neurosecretory cells in the hypothalamus secrete five releasing hormones, which stimulate secretion of anterior pituitary hormones, and two inhibiting hormones, which suppress secretion of anterior pituitary hormones (Table 18.3). Second, negative feedback in the form of hormones released by target glands decreases secretions of three types of anterior pituitary cells (Figure 18.6). In such negative feedback loops, the secretory activity of thyrotrophs, gonadotrophs, and corticotrophs decreases when blood levels of their target gland hormones rise. For example, corticotropin (ACTH) stimulates the cortex of the adrenal gland to secrete glucocorticoids, mainly cortisol. In turn, an elevated blood level of cortisol decreases secretion of both corticotropin and corticotropin-releasing hormone (CRH) due to negative feedback suppression of the anterior pituitary corticotrophs and hypothalamic neurosecretory cells.

Human Growth Hormone and Insulinlike Growth Factors

Somatotrophs are the most numerous cells in the anterior pituitary, and human growth hormone (hGH) is the most plentiful anterior pituitary hormone. The main function of hGH is to promote synthesis and secretion of small protein hormones called **insulinlike growth factors (IGFs).** In response to human growth hormone, cells in the liver, skeletal muscle, cartilage, bone, and other tissues secrete IGFs, which may either enter the bloodstream from the liver or act locally in other tissues as autocrine or paracrine substances. IGFs cause cells to grow and multiply by increasing uptake of amino acids into cells and accelerating protein synthesis. IGFs also decrease the breakdown of proteins and the use of amino acids for ATP production. Due to these effects of the IGFs, human growth hormone increases the growth rate of the skeleton and skeletal muscles during childhood and the teenage years. In adults, human growth hormone and IGFs help maintain the mass of muscles and bones and promote healing of injuries and tissue repair.

Insulinlike growth factors also enhance lipolysis in adipose tissue, which results in increased use of the released fatty acids for ATP production by body cells. Besides affecting protein and lipid metabolism, human growth hormone and IGFs influence carbohydrate metabolism by decreasing glucose uptake, thereby decreasing the use of glucose for ATP production by most body cells. This action spares glucose so that neurons may continue to use it for ATP production in times of glucose scarcity. IGFs and human growth hormone may also stimulate liver cells to release glucose into the blood.

Figure 18.6 Negative feedback regulation of hypothalamic neurosecretory cells and anterior pituitary corticotrophs. Solid green arrows show stimulation of secretions whereas dashed red lines show inhibition of secretion via negative feedback.

 Cortisol secreted by the adrenal cortex suppresses secretion of CRH and ACTH.

Which other target gland hormones suppress secretion of hypothalamic and anterior pituitary hormones by negative feedback?

Somatotrophs in the anterior pituitary release bursts of human growth hormone every few hours, especially during sleep. Their secretory activity is controlled mainly by two hypothalamic hormones: (1) growth hormone-releasing hormone (GHRH) promotes secretion of human growth hormone, and (2) growth hormone-inhibiting hormone (GHIH) suppresses it. A major regulator of GHRH and GHIH secretion is blood glucose level (Figure 18.7):

① **Hypoglycemia** (hī′-po-glī-SĒ-mē-a), an abnormally low blood glucose concentration, stimulates the hypothalamus to secrete GHRH, which flows toward the anterior pituitary in the hypophyseal portal veins.

2 Upon reaching the anterior pituitary, GHRH stimulates somatotrophs to release human growth hormone.

3 Human growth hormone stimulates secretion of insulinlike growth factors, which speed up breakdown of liver glycogen into glucose. Thus, glucose enters the blood more rapidly.

4 As a result, blood glucose rises to the normal level (about 90 mg/100 mL of blood plasma).

5 An increase in blood glucose above the normal level inhibits release of GHRH.

6 **Hyperglycemia** (hī′-per-glī-SĒ-mē-a), an abnormally high blood glucose concentration, stimulates the hypothalamus to secrete GHIH (while inhibiting the secretion of GHRH).

7 Upon reaching the anterior pituitary in portal blood, GHIH inhibits secretion of human growth hormone by somatotrophs.

8 A low level of human growth hormone and IGFs slows breakdown of glycogen in the liver, and glucose is released into the blood more slowly.

9 Blood glucose falls to the normal level.

10 A decrease in blood glucose below the normal level (hypoglycemia) inhibits release of GHIH.

Other stimuli that promote secretion of human growth hormone include decreased fatty acids and increased amino acids in the blood; deep sleep (stages 3 and 4 of non-rapid eye movement sleep); increased activity of the sympathetic division of the autonomic nervous system, such as might occur with stress or vigorous physical exercise; and other hormones, including glucagon, estrogens, cortisol, and insulin. Factors that inhibit human growth hormone secretion are increased fatty acids and decreased amino acids in the blood; rapid eye movement sleep; emotional deprivation; obesity; low levels of thyroid hormones; and human growth hormone itself (through negative feedback).

Diabetogenic Effect of hGH

One symptom of excess human growth hormone (hGH) is hyperglycemia. Persistent hyperglycemia, in turn, stimulates the pancreas to secrete insulin continually. Such excessive stimulation, if it lasts for weeks or months, may cause "beta-cell burnout," a greatly decreased capacity of pancreatic beta cells to synthesize and secrete insulin. Thus, excess secretion of human growth hormone may have a **diabetogenic effect;** that is, it causes diabetes mellitus (lack of insulin activity). ■

Thyroid-stimulating Hormone

Thyroid-stimulating hormone (TSH) stimulates the synthesis and secretion of the two thyroid hormones, triiodothyronine (T_3) and thyroxine (T_4), both produced by the thyroid gland.

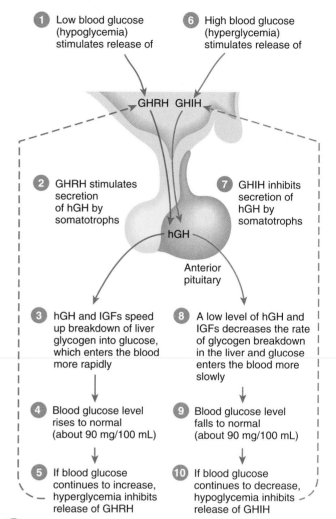

Figure 18.7 Effects of human growth hormone (hGH) and insulinlike growth factors (IGFs). Dashed lines indicate inhibition.

Secretion of hGH is stimulated by growth hormone-releasing hormone (GHRH) and inhibited by growth hormone-inhibiting hormone (GHIH).

1 Low blood glucose (hypoglycemia) stimulates release of

6 High blood glucose (hyperglycemia) stimulates release of

GHRH GHIH

2 GHRH stimulates secretion of hGH by somatotrophs

7 GHIH inhibits secretion of hGH by somatotrophs

hGH

Anterior pituitary

3 hGH and IGFs speed up breakdown of liver glycogen into glucose, which enters the blood more rapidly

8 A low level of hGH and IGFs decreases the rate of glycogen breakdown in the liver and glucose enters the blood more slowly

4 Blood glucose level rises to normal (about 90 mg/100 mL)

9 Blood glucose level falls to normal (about 90 mg/100 mL)

5 If blood glucose continues to increase, hyperglycemia inhibits release of GHRH

10 If blood glucose continues to decrease, hypoglycemia inhibits release of GHIH

If a person has a pituitary tumor that secretes a large amount of hGH and the tumor cells are not responsive to regulation by GHRH and GHIH, will hyperglycemia or hypoglycemia be more likely?

Thyrotropin-releasing hormone (TRH) from the hypothalamus controls TSH secretion. Release of TRH, in turn, depends on blood levels of T_3 and T_4, which inhibit secretion of TRH via negative feedback. There is no thyrotropin-inhibiting hormone. The release of TRH is explained later in the chapter (see Figure 18.12).

Follicle-stimulating Hormone

In females, the ovaries are the targets for follicle-stimulating hormone (FSH). Each month FSH initiates the development of several ovarian follicles, saclike arrangements of secretory cells that surround a developing oocyte. FSH also stimulates follicular cells to secrete estrogens (female sex hormones). In males, FSH stimulates sperm production in the testes. Gonadotropin-releasing hormone (GnRH) from the hypothalamus stimulates FSH release. Release of GnRH and FSH is suppressed by estrogens in females and by testosterone (the principal male sex hormone) in males through negative feedback systems. There is no gonadotropin-inhibiting hormone.

Luteinizing Hormone

In females, luteinizing hormone (LH) triggers **ovulation,** the release of a secondary oocyte (future ovum) by an ovary. LH stimulates formation of the corpus luteum (structure formed after ovulation) in the ovary and the secretion of progesterone (another female sex hormone) by the corpus luteum. Together, FSH and LH also stimulate secretion of estrogen by ovarian cells. Estrogens and progesterone prepare the uterus for implantation of a fertilized ovum and help prepare the mammary glands for milk secretion. In males, LH stimulates the interstitial cells in the testes to secrete testosterone. Secretion of LH, like that of FSH, is controlled by gonadotropin-releasing hormone (GnRH).

Prolactin

Prolactin (PRL), together with other hormones, initiates and maintains milk secretion by the mammary glands. By itself, prolactin has only a weak effect. Only after the mammary glands have been primed by estrogens, progesterone, glucocorticoids, human growth hormone, thyroxine, and insulin, which exert permissive effects, does PRL bring about milk secretion. Ejection of milk from the mammary glands depends on the hormone oxytocin, which is released from the posterior pituitary. Together, milk secretion and ejection constitute *lactation.*

The hypothalamus secretes both inhibitory and excitatory hormones that regulate prolactin secretion. Prolactin-inhibiting hormone (PIH), which is dopamine, inhibits the release of prolactin from the anterior pituitary. As the levels of estrogens and progesterone fall just before menstruation begins, the secretion of PIH diminishes and the blood level of prolactin rises. Breast tenderness just before menstruation may be caused by elevated prolactin. Because the prolactin level is high for only a short period, however, milk production does not start. As the menstrual cycle begins anew and the level of estrogens rises, PIH is again secreted and the prolactin level drops. Prolactin level rises during pregnancy, stimulated by prolactin-releasing hormone (PRH) from the hypothalamus. The sucking action of a nursing infant causes a reduction in hypothalamic secretion of PIH.

The function of prolactin is not known in males, but its hypersecretion causes erectile dysfunction (impotence, the inability to have an erection of the penis). In females, hypersecretion of prolactin causes galactorrhea (inappropriate lactation) and amenorrhea (absence of menstrual cycles).

Adrenocorticotropic Hormone

Corticotrophs secrete mainly adrenocorticotropic hormone (ACTH) or corticotropin. ACTH controls the production and secretion of cortisol and other glucocorticoids by the cortex (outer portion) of the adrenal glands. Corticotropin-releasing hormone (CRH) from the hypothalamus stimulates secretion of ACTH by corticotrophs. Stress-related stimuli, such as low blood glucose or physical trauma, and interleukin-1, a substance produced by macrophages, also stimulate release of ACTH. Glucocorticoids cause negative feedback inhibition of both CRH and ACTH release.

Melanocyte-stimulating Hormone

Melanocyte-stimulating hormone (MSH) increases skin pigmentation in amphibians by stimulating the dispersion of melanin granules in melanocytes. Its exact role in humans is unknown, but the presence of MSH receptors in the brain suggests it may influence brain activity. There is little circulating MSH in humans. However, continued administration of MSH for several days does produce a darkening of the skin. Excessive corticotropin-releasing hormone (CRH) can stimulate MSH release, whereas dopamine inhibits MSH release.

Table 18.4 summarizes the principal actions of the anterior pituitary hormones.

Posterior Pituitary

Although the **posterior pituitary** or **neurohypophysis** does not *synthesize* hormones, it does *store* and *release* two hormones. As noted earlier, it consists of pituicytes and axon terminals of hypothalamic neurosecretory cells. The cell bodies of the neurosecretory cells are in the paraventricular and supraoptic nuclei of the hypothalamus; their axons form the **hypothalamohypophyseal tract** (hī′-pō-thal′-a-mō-hī-pō-FIZ-ē-al). This tract begins in the hypothalamus and ends near blood capillaries in the posterior pituitary (Figure 18.8 on page 601). Different neurosecretory cells produce one of two hormones: **oxytocin (OT)** ok′-sē-TŌ-sin; *okytokos* = swift birth) and **antidiuretic hormone (ADH),** also called **vasopressin.**

After their production in the cell bodies of neurosecretory cells, oxytocin and antidiuretic hormone are packaged into secretory vesicles, which move by fast axonal transport (described on page 390) to the axon terminals in the posterior pituitary. Nerve impulses that propagate along the axon and reach the axon terminals trigger exocytosis of these vesicles. As the vesicles undergo exocytosis, the hormone within them is released.

The **inferior hypophyseal arteries** (see Figure 18.5), which branch from the internal carotid arteries, supply blood to the posterior pituitary. In the posterior pituitary, the inferior hypophyseal arteries drain into the **capillary plexus of the infundibular process,** a capillary network that receives secreted oxytocin and antidiuretic hormone (see Figure 18.5). From this

Table 18.4 Summary of the Principal Actions of Anterior Pituitary Hormones

Hormone and Target Tissues	Principal Actions	Hormone and Target Tissues	Principal Actions
Human growth hormone (hGH) or **somatotropin** Liver	Stimulates liver, muscle, cartilage, bone, and other tissues to synthesize and secrete insulinlike growth factors (IGFs); IGFs promote growth of body cells, protein synthesis, tissue repair, lipolysis, and elevation of blood glucose concentration.	**Luteinizing hormone (LH)** Ovaries Testes	In females, stimulates secretion of estrogens and progesterone, ovulation, and formation of corpus luteum. In males, stimulates interstitial cells in testes to develop and produce testosterone.
Thyroid-stimulating hormone (TSH) or **thyrotropin** Thyroid gland	Stimulates synthesis and secretion of thyroid hormones by thyroid gland.	**Prolactin (PRL)** Mammary glands	Together with other hormones, promotes milk secretion by the mammary glands.
Follicle-stimulating hormone (FSH) Ovaries Testes	In females, initiates development of oocytes and induces ovarian secretion of estrogens. In males, stimulates testes to produce sperm.	**Adrenocorticotropic hormone (ACTH)** or **corticotropin** Adrenal cortex	Stimulates secretion of glucocorticoids (mainly cortisol) by adrenal cortex.
		Melanocyte-stimulating hormone (MSH) Brain	Exact role in humans is unknown but may influence brain activity; when present in excess, can cause darkening of skin.

plexus, hormones pass into the **posterior hypophyseal veins** for distribution to target cells in other tissues.

Oxytocin

During and after delivery of a baby, oxytocin affects two target tissues: the mother's uterus and breasts. During delivery, oxytocin enhances contraction of smooth muscle cells in the wall of the uterus; after delivery, it stimulates milk ejection ("letdown") from the mammary glands in response to the mechanical stimulus provided by a suckling infant. The function of oxytocin in males and in nonpregnant females is not clear. Experiments with animals have suggested that it has actions within the brain that foster parental caretaking behavior toward young offspring. It may also be responsible, in part, for the feelings of sexual pleasure during and after intercourse.

Oxytocin and Childbirth

Years before oxytocin was discovered, it was common practice in midwifery to let a first-born twin nurse at the mother's breast to speed the birth of the second child. Now we know why this practice is helpful—it stimulates release of oxytocin. Even after a single birth, nursing promotes expulsion of the placenta (afterbirth) and helps the uterus regain its smaller size. Synthetic OT (Pitocin) often is given to induce labor or to increase uterine tone and control hemorrhage just after giving birth. ■

Antidiuretic Hormone

An **antidiuretic** is a substance that decreases urine production. ADH causes the kidneys to return more water to the blood, thus decreasing urine volume. In the absence of ADH, urine output increases more than tenfold, from the normal 1 to 2 liters to about 20 liters a day. ADH also decreases the water lost through sweating and causes constriction of arterioles, which increases blood pressure.

The amount of ADH secreted varies with blood osmotic pressure and blood volume. Figure 18.9 on page 602 shows regulation of ADH secretion and the actions of ADH.

1 High blood osmotic pressure—due to dehydration or a decline in blood volume because of hemorrhage, diarrhea, or excessive sweating—stimulates **osmoreceptors,** neu-

Figure 18.8 **Axons of hypothalamic neurosecretory cells form the hypothalamohypophyseal tract, which extends from the paraventricular and supraoptic nuclei to the posterior pituitary.** Hormone molecules synthesized in the cell body of a neurosecretory cell are packaged into secretory vesicles that move down to the axon terminals. Nerve impulses trigger exocytosis of the vesicles thereby releasing the hormone.

 Oxytocin and antidiuretic hormone are synthesized in the hypothalamus and released into the capillary plexus of the infundibular process in the posterior pituitary.

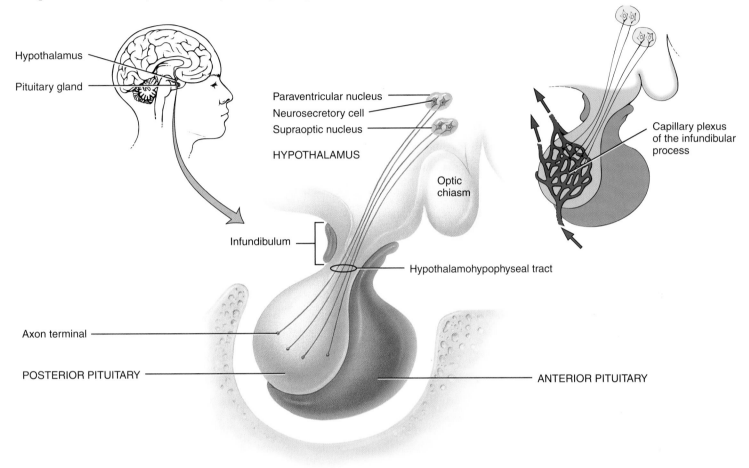

Functionally, how are the hypothalamohypophyseal tract and the hypophyseal portal veins similar? Structurally, how are they different?

rons in the hypothalamus that monitor blood osmotic pressure. Elevated blood osmotic pressure activates the osmoreceptors directly; they also receive excitatory input from other brain areas when blood volume decreases.

2 Osmoreceptors activate the hypothalamic neurosecretory cells that synthesize and release ADH.

3 When neurosecretory cells receive excitatory input from the osmoreceptors, they generate nerve impulses that cause exocytosis of ADH-containing vesicles from their axon terminals in the posterior pituitary. This liberates ADH, which diffuses into blood capillaries of the posterior pituitary.

4 The blood carries ADH to three target tissues: the kidneys,

sudoriferous (sweat) glands, and smooth muscle in blood vessel walls. The kidneys respond by retaining more water, which decreases urine output. Secretory activity of sweat glands decreases, which lowers the rate of water loss by perspiration from the skin. Smooth muscle in the walls of arterioles (small arteries) contracts in response to high levels of ADH, which constricts (narrows) the lumen of these blood vessels and increases blood pressure. The other name of ADH, vasopressin, reflects its effect of raising blood pressure.

5 Low osmotic pressure of blood or increased blood volume inhibits the osmoreceptors.

Figure 18.9 Regulation of secretion and actions of antidiuretic hormone (ADH).

ADH acts to retain body water and increase blood pressure.

① High blood osmotic pressure stimulates hypothalamic osmoreceptors

② Osmoreceptors activate the neurosecretory cells that synthesize and release ADH

③ Nerve impulses liberate ADH from axon terminals in the posterior pituitary into the bloodstream

④ Kidneys retain more water, which decreases urine output

Sudoriferous (sweat) glands decrease water loss by perspiration from the skin

Arterioles constrict, which increases blood pressure

⑤ Low blood osmotic pressure inhibits hypothalamic osmoreceptors

⑥ Inhibition of osmoreceptors reduces or stops ADH secretion

Osmoreceptors

Hypothalamus

ADH

Target tissues

 If you drank a liter of water, what effect would this have on the osmotic pressure of your blood, and how would the level of ADH change in your blood?

⑥ Inhibition of osmoreceptors reduces or stops ADH secretion. The kidneys then retain less water by forming a larger volume of urine, secretory activity of sweat glands increases, and arterioles dilate. The blood volume and osmotic pressure of body fluids return to normal.

Secretion of ADH can also be altered in other ways. Pain, stress, trauma, anxiety, acetylcholine, nicotine, and drugs such as morphine, tranquilizers, and some anesthetics stimulate ADH secretion. Alcohol inhibits ADH secretion, thereby increasing urine output. The resulting dehydration may cause both the thirst and the headache typical of a hangover. Hyposecretion of ADH or nonfunctional ADH receptors causes diabetes insipidus (see page 626).

Table 18.5 lists the posterior pituitary hormones, control of their secretion, and their principal actions.

▶ **CHECKPOINT**

8. In what respect is the pituitary gland actually two glands?

9. How do hypothalamic releasing and inhibiting hormones influence secretions of the anterior pituitary?

10. Describe the structure and importance of the hypothalamohypophyseal tract.

11. Explain how blood levels of T_3/T_4, TSH, and TRH would change in a laboratory animal that has undergone a thyroidectomy (complete removal of its thyroid gland).

Table 18.5	**Summary of Posterior Pituitary Hormones**	
Hormone and Target Tissues	**Control of Secretion**	**Principal Actions**
Oxytocin (OT) Uterus Mammary glands	Neurosecretory cells of hypothalamus secrete OT in response to uterine distention and stimulation of nipples.	Stimulates contraction of smooth muscle cells of uterus during childbirth; stimulates contraction of myoepithelial cells in mammary glands to cause milk ejection.
Antidiuretic hormone (ADH) or **vasopressin** Kidneys Sudoriferous (sweat) glands Arterioles	Neurosecretory cells of hypothalamus secrete ADH in response to elevated blood osmotic pressure, dehydration, loss of blood volume, pain, or stress; low blood osmotic pressure, high blood volume, and alcohol inhibit ADH secretion.	Conserves body water by decreasing urine volume; decreases water loss through perspiration; raises blood pressure by constricting arterioles.

Figure 18.10 Location, blood supply, and histology of the thyroid gland.
(See Tortora, *A Photographic Atlas of the Human Body,* Figure 10.3a.)

🔑 Thyroid hormones regulate (1) oxygen use and basal metabolic rate, (2) cellular metabolism, and (3) growth and development.

(a) Anterior view of thyroid gland

(b) Thyroid follicles

LM 500x

THYROID GLAND

▶ OBJECTIVE

- **Describe the location, histology, hormones, and functions of the thyroid gland.**

The butterfly-shaped **thyroid gland** is located just inferior to the larynx (voice box). It is composed of right and left **lateral lobes,** one on either side of the trachea, that are connected by an **isthmus** (IS-mus = a narrow passage) anterior to the trachea (Figure 18.10a). A small, pyramidal-shaped lobe sometimes extends upward from the isthmus. The normal mass of the thyroid is about 30 g (1 oz). It is highly vascularized and receives 80–120 mL of blood per minute.

Microscopic spherical sacs called **thyroid follicles** (Figure 18.10b; also see Figure 18.13c) make up most of the thyroid gland. The wall of each follicle consists primarily of cells called **follicular cells,** most of which extend to the lumen (internal space) of the follicle. A **basement membrane** surrounds each

 Which cells secrete T_3 and T_4? Which secrete calcitonin? Which of these hormones are also called thyroid hormones?

follicle. When the follicular cells are inactive, their shape is low cuboidal to squamous, but under the influence of TSH they become cuboidal or low columnar and actively secretory. The follicular cells produce two hormones: **thyroxine** (thī-ROK-sēn), which is also called **tetraiodothyronine** (tet-ra-ī-ō-dō-THĪ-rō-nēn) or **T₄** because it contains four atoms of iodine, and **triiodothyronine** (trī-ī′-ō-dō-THĪ-rō-nēn) or **T₃,** which contains three atoms of iodine. T₃ and T₄ are also known as **thyroid hormones.** A few cells called **parafollicular cells** or **C cells** lie between follicles. They produce the hormone **calcitonin** (kal-si-TŌ-nin), which helps regulate calcium homeostasis.

Formation, Storage, and Release of Thyroid Hormones

The thyroid gland is the only endocrine gland that stores its secretory product in large quantity—normally about a 100-day supply. Synthesis and secretion of T₃ and T₄ occurs as follows (Figure 18.11):

1 *Iodide trapping.* Thyroid follicular cells trap iodide ions (I⁻) by actively transporting them from the blood into the cytosol. As a result, the thyroid gland normally contains most of the iodide in the body.

2 *Synthesis of thyroglobulin.* While the follicular cells are trapping I⁻, they are also synthesizing **thyroglobulin (TGB).** TGB is a large glycoprotein that is produced in the rough endoplasmic reticulum, modified in the Golgi complex, and packaged into secretory vesicles. The vesicles then undergo exocytosis, which releases TGB into the lumen of the follicle. The material that accumulates in the lumen is termed **colloid.**

3 *Oxidation of iodide.* Some of the amino acids in TGB are tyrosines that will become iodinated. However, negatively charged iodide ions cannot bind to tyrosine until they undergo oxidation (removal of electrons) to iodine: $2I^- \rightarrow I_2$. The enzyme that catalyzes this reaction is peroxidase. As the iodide ions are being oxidized, they pass through the membrane into the colloid.

4 *Iodination of tyrosine.* As iodine molecules (I₂) form, they react with tyrosines that are part of thyroglobulin molecules in the colloid. Binding of one iodine atom yields monoiodotyrosine (T₁), and a second iodination produces diiodotyrosine (T₂).

5 *Coupling of T₁ and T₂.* During the last step in the synthesis of thyroid hormone, two T₂ molecules join to form T₄ or one T₁ and one T₂ join to form T₃.

6 *Pinocytosis and digestion of colloid.* Droplets of colloid reenter follicular cells by pinocytosis and merge with lysosomes. Digestive enzymes break down TGB, cleaving off molecules of T₃ and T₄.

7 *Secretion of thyroid hormones.* Because T₃ and T₄ are lipid soluble, they diffuse through the plasma membrane into

Figure 18.11 **Steps in the synthesis and secretion of thyroid hormones.**

Thyroid hormones are synthesized by attaching iodine atoms to the amino acid tyrosine.

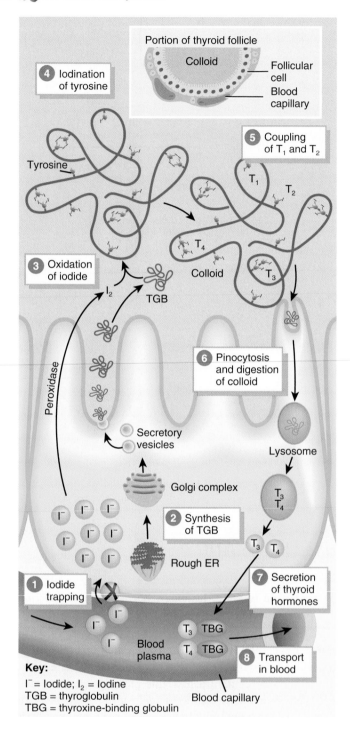

Key:
I⁻ = Iodide; I₂ = Iodine
TGB = thyroglobulin
TBG = thyroxine-binding globulin

What is the storage form of thyroid hormones?

interstitial fluid and then into the blood. T_4 normally is secreted in greater quantity than T_3, but T_3 is several times more potent. Moreover, after T_4 enters a body cell, most of it is converted to T_3 by removal of one iodine.

8 **Transport in the blood.** More than 99% of both the T_3 and the T_4 combine with transport proteins in the blood, mainly **thyroxine-binding globulin (TGB).**

Control of Thyroid Hormone Secretion

Two main factors control the size and the secretory activity of the thyroid gland. First, although iodine is needed for synthesis of thyroid hormones, an abnormally high blood iodine concentration suppresses release of thyroid hormones. Second, thyrotropin-releasing hormone (TRH) from the hypothalamus and thyroid-stimulating hormone (TSH) from the anterior pituitary stimulate synthesis and release of thyroid hormones, as shown in Figure 18.12:

1 Low blood levels of T_3 and T_4 or low metabolic rate stimulate the hypothalamus to secrete TRH.

2 TRH enters the hypophyseal portal veins and flows to the anterior pituitary, where it stimulates thyrotrophs to secrete TSH.

3 TSH stimulates virtually all aspects of thyroid follicular cell activity, including iodide trapping, hormone synthesis and secretion, and growth of the follicular cells.

4 The thyroid follicular cells release T_3 and T_4 into the blood until the metabolic rate returns to normal.

5 An elevated level of T_3 inhibits release of TRH and TSH (negative feedback inhibition).

Conditions that increase ATP demand—a cold environment, hypoglycemia, high altitude, and pregnancy—also increase the secretion of the thyroid hormones.

Actions of Thyroid Hormones

Because most body cells have receptors for thyroid hormones, T_3 and T_4 exert their effects throughout the body. They regulate oxygen use and basal metabolic rate, cellular metabolism, and growth and development (6 in Figure 18.12).

Thyroid hormones increase **basal metabolic rate (BMR),** the rate of oxygen consumption at rest after an overnight fast, by stimulating the use of cellular oxygen to produce ATP. The sodium-potassium pumps (Na^+/K^+ ATPase) that continually eject sodium ions (Na^+) from the cytosol into the extracellular fluid use a large part of the ATP produced by most cells.

A second major effect of thyroid hormones is to stimulate synthesis of additional sodium-potassium pumps, which then use more ATP. As cells produce and use more ATP, more heat is given off, and body temperature rises. This phenomenon is called the **calorigenic effect** of the thyroid hormones. In this way, they play an important role in the maintenance of normal

Figure 18.12 Regulation of secretion and actions of thyroid hormones. TRH = thyrotropin-releasing hormone, TSH = thyroid-stimulating hormone, T_3 = triiodothyronine, and T_4 = thyroxine (tetraiodothyronine).

🔑 **TSH promotes release of thyroid hormones (T_3 and T_4) by the thyroid gland.**

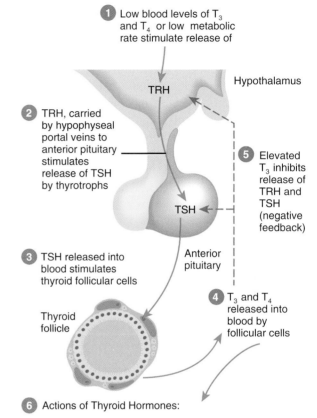

1 Low blood levels of T_3 and T_4 or low metabolic rate stimulate release of

2 TRH, carried by hypophyseal portal veins to anterior pituitary stimulates release of TSH by thyrotrophs

Hypothalamus

TRH

5 Elevated T_3 inhibits release of TRH and TSH (negative feedback)

TSH

3 TSH released into blood stimulates thyroid follicular cells

Anterior pituitary

Thyroid follicle

4 T_3 and T_4 released into blood by follicular cells

6 Actions of Thyroid Hormones:

Increase basal metabolic rate
Increase use of glucose and oxygen for ATP production
Stimulate synthesis of Na^+/K^+ ATPase
Increase body temperature (calorigenic effect)
Stimulate protein synthesis and accelerate tissue growth
Stimulate lipolysis and cholesterol excretion

 How could an iodine-deficient diet lead to goiter, which is an enlargement of the thyroid gland?

body temperature. Normal mammals can survive in freezing temperatures, but those whose thyroid glands have been removed cannot.

In the regulation of metabolism, the thyroid hormones stimulate protein synthesis and increase the use of glucose and fatty acids for ATP production. They also increase lipolysis and enhance cholesterol excretion, thus reducing blood cholesterol level.

The thyroid hormones enhance some actions of the catecholamines (norepinephrine and epinephrine) because they up-regulate beta (β) receptors. For this reason, symptoms of hyperthyroidism include increased heart rate, more forceful heartbeats, and increased blood pressure.

Together with human growth hormone and insulin, thyroid hormones accelerate body growth, particularly the growth of nervous tissue. Deficiency of thyroid hormones during fetal development, infancy, or childhood causes stunted growth and severe mental retardation.

Calcitonin

The hormone produced by the **parafollicular cells** of the thyroid gland (see Figure 18.10b) is **calcitonin (CT)** (kal-si-TŌ-nin). CT can decrease the level of calcium in the blood by inhibiting the action of osteoclasts, the cells that break down bone matrix. The secretion of CT is controlled by a negative feedback system (see Figure 18.14). Calcitonin's importance in normal physiology is unclear because it can be present in excess or completely absent without causing clinical symptoms.

When its blood level is high, calcitonin lowers the amount of blood calcium and phosphates by inhibiting bone resorption (breakdown of bone matrix) by osteoclasts and by accelerating uptake of calcium and phosphates into bone matrix. Miacalcin, a calcitonin extract derived from salmon that is ten times more potent than human calcitonin, is prescribed to treat osteoporosis.

Table 18.6 summarizes the hormones produced by the thyroid gland, control of their secretion, and their principal actions.

▶ C H E C K P O I N T

12. How are the thyroid hormones synthesized, stored, and secreted?

13. How is the secretion of T_3 and T_4 regulated?

14. What are the physiological effects of the thyroid hormones?

PARATHYROID GLANDS

▶ O B J E C T I V E

- **Describe the location, histology, hormone, and functions of the parathyroid glands.**

Partially embedded in the posterior surface of the lateral lobes of the thyroid gland are several small, round masses of tissue called the **parathyroid glands** (*para-* = beside). Each parathyroid gland has a mass of about 40 mg (0.04 g). Usually, one superior and one inferior parathyroid gland are attached to each lateral thyroid lobe (Figure 18.13a).

Microscopically, the parathyroid glands contain two kinds of epithelial cells (Figure 18.13b, c). The more numerous cells, called **chief (principal) cells,** produce **parathyroid hormone (PTH),** or **parathormone.** The function of the other kind of cell, called an *oxyphil cell,* is not known.

Parathyroid Hormone

Parathyroid hormone is the major regulator of the levels of calcium (Ca^{2+}), magnesium (Mg^{2+}), and phosphate (HPO_4^{2-}) ions in the blood. The blood calcium level directly controls the secretion of both calcitonin and parathyroid hormone via negative feedback loops that do not involve the pituitary gland (Figure 18.14 on page 608):

1 A higher-than-normal level of calcium ions (Ca^{2+}) in the blood stimulates parafollicular cells of the thyroid gland to release more calcitonin.

Table 18.6	Summary of Thyroid Gland Hormones	
Hormone and Source	**Control of Secretion**	**Principal Actions**
T₃ (triiodothyronine) and **T₄ (thyroxine)** or **thyroid hormones** from follicular cells	Secretion is increased by thyrotropin-releasing hormone (TRH), which stimulates release of thyroid-stimulating hormone (TSH) in response to low thyroid hormone levels, low metabolic rate, cold, pregnancy, and high ltitudes; TRH and TSH secretion are inhibited in response to high thyroid hormone levels; high iodine level suppresses T_3/T_4 secretion.	Increase basal metabolic rate, stimulate synthesis of proteins, increase use of glucose and fatty acids for ATP production, increase lipolysis, enhance cholesterol excretion, accelerate body growth, and contribute to development of the nervous system.
Calcitonin (CT) from parafollicular cells accelerating	High blood Ca^{2+} levels stimulate secretion; low blood Ca^{2+} levels inhibit secretion.	Lowers blood levels of ionic calcium and phosphates by inhibiting bone resorption by osteoclasts and uptake of calcium and phosphates into bone matrix.

Figure 18.13 Location, blood supply, and histology of the parathyroid glands. (See Tortora, *A Photographic Atlas of the Human Body,* Figure 10.4a.)

The parathyroid glands, normally four in number, are embedded in the posterior surface of the thyroid gland.

(a) Posterior view

(b) Parathyroid gland

LM 325x

(c) Portion of the thyroid gland (left) and parathyroid gland (right)

What are the secretory products of (1) parafollicular cells of the thyroid gland and (2) chief cells of the parathyroid glands?

Figure 18.14 The roles of calcitonin (green arrows), parathyroid hormone (blue arrows), and calcitriol (orange arrows) in calcium homeostasis.

With respect to regulation of blood Ca^{2+} level, calcitonin and PTH are antagonists.

1 High level of Ca^{2+} in blood stimulates thyroid gland parafollicular cells to release more CT.

3 Low level of Ca^{2+} in blood stimulates parathyroid gland chief cells to release more PTH.

6 CALCITRIOL stimulates increased absorption of Ca^{2+} from foods, which increases blood Ca^{2+} level.

5 PTH also stimulates the kidneys to release CALCITRIOL.

4 PARATHYROID HORMONE (PTH) promotes resorption of Ca^{2+} from bone matrix into blood and retards loss of Ca^{2+} in urine, thus increasing blood Ca^{2+} level.

2 CALCITONIN inhibits osteoclasts, thus decreasing blood Ca^{2+} level.

What are the primary target tissues for PTH, CT, and calcitriol?

2 Calcitonin inhibits the activity of osteoclasts, thereby decreasing blood Ca^{2+} level.

3 A lower-than-normal level of Ca^{2+} in the blood stimulates chief cells of the parathyroid gland to release more PTH.

4 PTH promotes resorption of bone matrix, which releases Ca^{2+} into the blood, and slows loss of Ca^{2+} in the urine, raising the blood level of Ca^{2+}.

5 PTH also stimulates the kidneys to synthesize calcitriol, the active form of vitamin D.

6 Calcitriol stimulates increased absorption of Ca^{2+} from foods in the gastrointestinal tract, which helps increase the blood level of Ca^{2+}.

PTH increases the number and activity of osteoclasts. The result is elevated bone *resorption,* which releases ionic calcium (Ca^{2+}) and phosphates (HPO_4^{2-}) into the blood. PTH also acts on the kidneys. First, it slows the rate at which Ca^{2+} and Mg^{2+} are lost from blood into the urine. Second, it increases loss of HPO_4^{2-} from blood into the urine. Because more HPO_4^{2-} is lost in the urine than is gained from the bones, PTH decreases blood HPO_4^{2-} level and increases blood Ca^{2+} and Mg^{2+} levels. A third

effect of PTH on the kidneys is to promote formation of the hormone **calcitriol,** the active form of vitamin D. Calcitriol, also known as *1,25-dihydroxy vitamin D_3,* increases the rate of Ca^{2+}, HPO_4^{2-}, and Mg^{2+} *absorption* from the gastrointestinal tract into the blood. Table 18.7 summarizes control of secretion and the principal actions of parathyroid hormone.

▶ CHECKPOINT

15. How is secretion of parathyroid hormone regulated?

16. In what ways are the actions of PTH and calcitriol similar and different?

ADRENAL GLANDS

▶ OBJECTIVE

• **Describe the location, histology, hormones, and functions of the adrenal glands.**

The paired **adrenal (suprarenal) glands,** one of which lies superior to each kidney (Figure 18.15a), have a flattened pyrami-

Table 18.7	Summary of Parathyroid Gland Hormone	
Hormone and Source	**Control of Secretion**	**Principal Actions**
Parathyroid hormone (PTH) from chief cells Chief cell 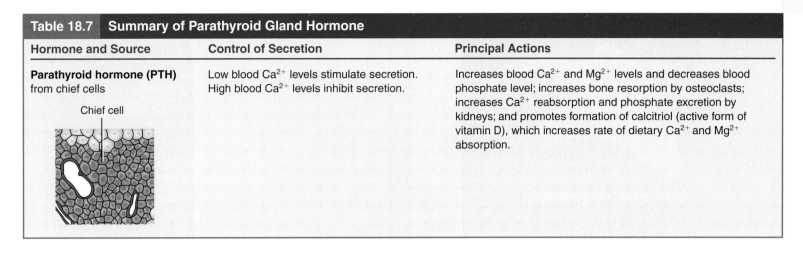	Low blood Ca^{2+} levels stimulate secretion. High blood Ca^{2+} levels inhibit secretion.	Increases blood Ca^{2+} and Mg^{2+} levels and decreases blood phosphate level; increases bone resorption by osteoclasts; increases Ca^{2+} reabsorption and phosphate excretion by kidneys; and promotes formation of calcitriol (active form of vitamin D), which increases rate of dietary Ca^{2+} and Mg^{2+} absorption.

dal shape. In an adult, each adrenal gland is 3–5 cm in height, 2–3 cm in width, and a little less than 1 cm thick. Its adult mass is 3.5–5 g, only half the size at birth. During embryonic development, the adrenal glands differentiate into two structurally and functionally distinct regions: A large, peripherally located **adrenal cortex**, comprising 80–90% of the gland, develops from mesoderm; a small, centrally located **adrenal medulla** develops from ectoderm (Figure 18.15b). A connective tissue

Figure 18.15 Location, blood supply, and histology of the adrenal (suprarenal) glands. (See Tortora, *A Photographic Atlas of the Human Body,* Figure 10.5a.)

🔑 The adrenal cortex secretes steroid hormones that are essential for life; the adrenal medulla secretes norepinephrine and epinephrine.

(a) Anterior view

(continues)

Figure 18.15 *(continued)*

Capsule

Adrenal cortex

Adrenal medulla

Section through left adrenal gland

Capsule

Adrenal cortex:

Zona glomerulosa secretes mineralocorticoids, mainly aldosterone

Zona fasciculata secretes glucocorticoids, mainly cortisol

Zona reticularis secretes androgens

Adrenal medulla chromaffin cells secrete epinephrine and norepinephrine (NE)

LM 50x

(b) Subdivisions of the adrenal gland

 What is the position of the adrenal glands relative to the pancreas and kidneys?

capsule covers the gland. The adrenal glands, like the thyroid gland, are highly vascularized.

The adrenal cortex produces steroid hormones that are essential for life. Complete loss of adrenocortical hormones leads to death due to dehydration and electrolyte imbalances in a few days to a week, unless hormone replacement therapy begins promptly. The adrenal medulla produces three catecholamine hormones—norepinephrine, epinephrine, and a small amount of dopamine.

Adrenal Cortex

The adrenal cortex is subdivided into three zones, each of which secretes different hormones (Figure 18.15b). The outer zone, just deep to the connective tissue capsule, is the **zona glomerulosa** (*zona* = belt; *glomerul-* = little ball). Its cells, which are closely packed and arranged in spherical clusters and arched columns, secrete hormones called **mineralocorticoids** (min′-er-al-ō-KOR-ti-koyds) because they affect mineral homeostasis. The middle zone, or **zona fasciculata** (*fascicul-* = little bundle), is the widest of the three zones and consists of cells arranged in long, straight columns. The cells of the zona fasciculata secrete mainly **glucocorticoids** (gloo′-kō-KOR-ti-koyds), so named because they affect glucose homeostasis. The cells of the inner zone, the **zona reticularis** (*reticul-* = network), are arranged in branching cords. They synthesize small amounts of weak **andro-**

gens (*andro-* = a man), steroid hormones that have masculinizing effects.

Mineralocorticoids

Aldosterone (al-DA-ster-ōn) is the major mineralocorticoid. It regulates homeostasis of two mineral ions, namely sodium ions (Na^+) and potassium ions (K^+), and helps adjust blood pressure and blood volume. Aldosterone also promotes excretion of H^+ in the urine; this removal of acids from the body can help prevent acidosis (blood pH below 7.35).

The **renin–angiotensin–aldosterone** or **RAA pathway** (RĒ-nin an′-jē-ō-TEN-sin) controls secretion of aldosterone (Figure 18.16):

1 Stimuli that initiate the renin–angiotensin–aldosterone pathway include dehydration, Na^+ deficiency, or hemorrhage.

2 These conditions cause a decrease in blood volume.

3 Decreased blood volume leads to decreased blood pressure.

4 Lowered blood pressure stimulates certain cells of the kidneys, called juxtaglomerular cells, to secrete the enzyme **renin.**

5 The level of renin in the blood increases.

6 Renin converts **angiotensinogen,** a plasma protein produced by the liver, into **angiotensin I.**

7 Blood containing increased levels of angiotensin I circulates in the body.

8 As blood flows through capillaries, particularly those of the lungs, the enzyme **angiotensin-converting enzyme (ACE)** converts angiotensin I into the hormone **angiotensin II.**

9 Blood level of angiotensin II increases.

10 Angiotensin II stimulates the adrenal cortex to secrete aldosterone.

11 Blood containing increased levels of aldosterone circulates to the kidneys.

12 In the kidneys, aldosterone increases reabsorption of Na^+ and water so that less is lost in the urine. Aldosterone also stimulates the kidneys to increase secretion of K^+ and H^+ into the urine.

13 With increased water reabsorption by the kidneys, blood volume increases.

14 As blood volume increases, blood pressure increases to normal.

15 Angiotensin II also stimulates contraction of smooth muscle in the walls of arterioles. The resulting vasoconstriction of the arterioles increases blood pressure and thus helps raise blood pressure to normal.

16 Besides angiotensin II, a second stimulator of aldosterone secretion is an increase in the K^+ concentration of blood (or interstitial fluid). A decrease in the blood K^+ level has the opposite effect.

Glucocorticoids

The glucocorticoids, which regulate metabolism and resistance to stress, include **cortisol (hydrocortisone), corticosterone,** and **cortisone.** Of the three, cortisol is the most abundant, accounting for about 95% of glucocorticoid activity.

Control of glucocorticoid secretion is via a typical negative feedback system (Figure 18.17). Low blood levels of glucocorticoids, mainly cortisol, stimulate neurosecretory cells in the hypothalamus to secrete **corticotropin-releasing hormone (CRH).** CRH (together with a low level of cortisol) promotes the

Figure 18.16 **Regulation of aldosterone secretion by the renin–angiotensin–aldosterone (RAA) pathway.**

Aldosterone helps regulate blood volume, blood pressure, and levels of Na^+, K^+, and H^+ in the blood.

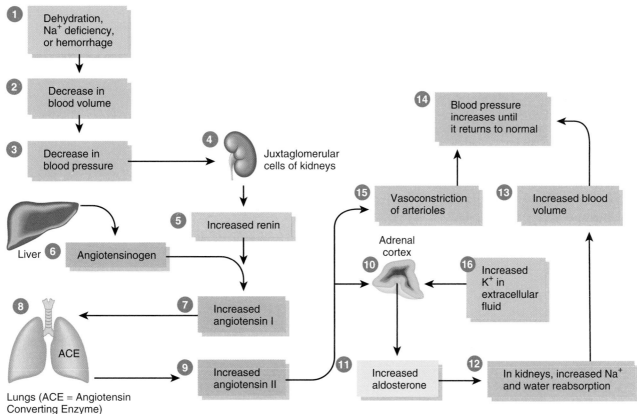

In what two ways can angiotensin II increase blood pressure, and what are its target tissues in each case?

release of ACTH from the anterior pituitary. ACTH flows in the blood to the adrenal cortex, where it stimulates glucocorticoid secretion. To a much smaller extent, ACTH also stimulates secretion of aldosterone. The discussion of stress at the end of the chapter describes how the hypothalamus also increases CRH release in response to a variety of physical and emotional stresses.

Glucocorticoids have the following effects:

1. ***Protein breakdown.*** Glucocorticoids increase the rate of protein breakdown, mainly in muscle fibers, and thus increase the liberation of amino acids into the bloodstream. The amino acids may be used by body cells for synthesis of new proteins or for ATP production.

2. ***Glucose formation.*** Upon stimulation by glucocorticoids, liver cells may convert certain amino acids or lactic acid to glucose, which neurons and other cells can use for ATP production. Such conversion of a substance other than glycogen or another monosaccharide into glucose is called **gluconeogenesis** (gloo′-ko-nē′-ō-JEN-e-sis).

3. ***Lipolysis.*** Glucocorticoids stimulate **lipolysis,** the breakdown of triglycerides and release of fatty acids from adipose tissue into the blood.

4. ***Resistance to stress.*** Glucocorticoids work in many ways to provide resistance to stress. The additional glucose supplied by the liver cells provides tissues with a ready source of ATP to combat a range of stresses, including exercise, fasting, fright, temperature extremes, high altitude, bleeding, infection, surgery, trauma, and disease. Because glucocorticoids make blood vessels more sensitive to other hormones that cause vasoconstriction, they raise blood pressure. This effect would be an advantage in cases of severe blood loss, which causes blood pressure to drop.

5. ***Anti-inflammatory effects.*** Glucocorticoids inhibit white blood cells that participate in inflammatory responses. Unfortunately, glucocorticoids also retard tissue repair, and as a result, they slow wound healing. Although high doses can cause severe mental disturbances, glucocorticoids are very useful in the treatment of chronic inflammatory disorders such as rheumatoid arthritis.

6. ***Depression of immune responses.*** High doses of glucocorticoids depress immune responses. For this reason, glucocorticoids are prescribed for organ transplant recipients to retard tissue rejection by the immune system.

Androgens

In both males and females, the adrenal cortex secretes small amounts of weak androgens. The major androgen secreted by the adrenal gland is **dehydroepiandrosterone (DHEA)** (dē-hī-drō-ep′-ē-an-DROS-ter-ōn). After puberty in males, androgens are also released in much greater quantity by the testes. Thus, the amount of androgens secreted by the adrenal gland is usually so low that their effects are insignificant. In females, however, adrenal androgens play important roles. They promote libido (sex drive) and are converted into estrogens (feminizing sex steroids) by other body tissues. After menopause, when ovarian secretion of estrogens ceases, all female estrogens come from

Figure 18.17 Negative feedback regulation of glucocorticoid secretion.

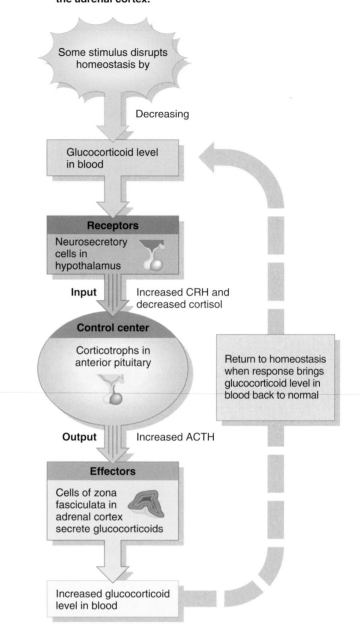

A high level of CRH and a low level of glucocorticoids promote the release of ACTH, which stimulates glucocorticoid secretion by the adrenal cortex.

 If a heart transplant patient receives prednisone (a glucocorticoid) to help prevent rejection of the transplanted tissue, will blood levels of ACTH and CRH be high or low? Explain.

conversion of adrenal androgens. Adrenal androgens also stimulate growth of axillary and pubic hair in boys and girls and contribute to the prepubertal growth spurt. Although control of adrenal androgen secretion is not fully understood, the main hormone that stimulates its secretion is ACTH.

 Congenital Adrenal Hyperplasia

Congenital adrenal hyperplasia (CAH) is a genetic disorder in which one or more enzymes needed for synthesis of cortisol are absent. Because the cortisol level is low, secretion of ACTH by the anterior pituitary is high due to lack of negative feedback inhibition. ACTH, in turn, stimulates growth and secretory activity of the adrenal cortex. As a result, both adrenal glands are enlarged. However, certain steps leading to synthesis of cortisol are blocked. Thus, precursor molecules accumulate, and some of these are weak androgens that can undergo conversion to testosterone. The result is **virilism,** or masculinization. In a female, virile characteristics include growth of a beard, development of a much deeper voice and a masculine distribution of body hair, growth of the clitoris so it may resemble a penis, atrophy of the breasts, and increased muscularity that produces a masculine physique. In prepubertal males, the syndrome causes the same characteristics as in females, plus rapid development of the male sexual organs and emergence of male sexual desires. In adult males, the virilizing effects due to CAH are usually completely obscured by the normal virilizing effects of the testosterone secreted by the testes. As a result, CAH is often difficult to diagnose in adult males. Treatment involves cortisol therapy, which inhibits ACTH secretion and thus reduces production of adrenal androgens. ■

Adrenal Medulla

The inner region of the adrenal gland, the **adrenal medulla,** is a modified sympathetic ganglion of the autonomic nervous system (ANS). It develops from the same embryonic tissue as all other sympathetic ganglia, but its cells lack axons and form clusters around large blood vessels. Rather than releasing a neurotransmitter, the cells of the adrenal medulla secrete hormones. The hormone-producing cells, called **chromaffin cells** (KRŌ-maf-in; *chrom-* = color; *-affin* = affinity for; see Figure 18.15b), are innervated by sympathetic preganglionic neurons in the splanchnic nerve. Because the ANS controls the chromaffin cells directly, hormone release can occur very quickly.

The two major hormones synthesized by the adrenal medulla are **epinephrine** and **norepinephrine (NE),** also called adrenaline and noradrenaline, respectively. Cortisol from the adrenal cortex induces synthesis of the enzyme needed to convert NE to epinephrine. Because the adrenal cortex surrounds the adrenal medulla, the cortisol level of blood in the medulla is normally quite high. Hence, about 80% of the medullary cells secrete epinephrine. The remaining 20% secrete norepinephrine because they lack the converting enzyme. Unlike the hormones of the adrenal cortex, the medullary hormones are not essential for life due to the presence of sympathetic neurons in other parts of the body.

In stressful situations and during exercise, impulses from the hypothalamus stimulate sympathetic preganglionic neurons, which in turn stimulate the chromaffin cells to secrete epinephrine and norepinephrine. These two hormones greatly augment the fight-or-flight response. By increasing heart rate and force of contraction, epinephrine and norepinephrine increase the pumping output of the heart, which increases blood pressure. They also increase blood flow to the heart, liver, skeletal muscles, and adipose tissue; dilate airways to the lungs; and increase blood levels of glucose and fatty acids.

Table 18.8 summarizes the hormones produced by the adrenal glands, control of their secretion, and their principal actions.

Table 18.8	Summary of Adrenal Gland Hormones	
Hormone and Source	**Control of Secretion**	**Principal Actions**
Adrenal cortex hormones		
Mineralocorticoids (mainly aldosterone) from zona glomerulosa cells	Increased blood K$^+$ level and angiotensin II stimulate secretion.	Increase blood levels of Na$^+$ and water and decrease blood level of K$^+$.
Glucocorticoids (mainly cortisol) from zona fasciculata cells	ACTH stimulates release; corticotropin-releasing hormone (CRH) promotes ACTH secretion in response to stress and low blood levels of glucocorticoids.	Increase protein breakdown (except in liver), stimulate gluconeogenesis and lipolysis, provide resistance to stress, dampen inflammation, and depress immune responses.
Androgens (mainly dehydroepiandrosterone or **DHEA)** from zona reticularis cells	ACTH stimulates secretion.	Assist in early growth of axillary and pubic hair in both sexes; in females, contribute to libido and are source of estrogens after menopause.
Adrenal cortex		
Adrenal medulla hormones		
Epinephrine and **norepinephrine** from chromaffin cells	Sympathetic preganglionic neurons release acetylcholine, which stimulates secretion.	Produce effects that enhance those of the sympathetic division of the autonomic nervous system (ANS) during stress.
Adrenal medulla		

► CHECKPOINT

17. How do the adrenal cortex and adrenal medulla compare with regard to location and histology?

18. How is secretion of adrenal cortex hormones regulated?

19. How is the adrenal medulla related to the autonomic nervous system?

PANCREATIC ISLETS

► OBJECTIVE

• **Describe the location, histology, hormones, and functions of the pancreatic islets.**

The **pancreas** (*pan-* = all; *-creas* = flesh) is both an endocrine gland and an exocrine gland. We discuss its endocrine functions here and its exocrine functions in Chapter 24 together with the digestive system. The pancreas is a flattened organ that measures about 12.5–15 cm (4.5–6 in.) in length. It is located in the curve of the duodenum, the first part of the small intestine, and consists of a head, a body, and a tail (Figure 18.18a). Roughly 99% of the pancreatic cells are arranged in clusters called *acini*. The acini produce digestive enzymes, which flow into the gastrointestinal tract through a network of ducts. Scattered among the exocrine acini are 1–2 million tiny clusters of endocrine tissue called **pancreatic islets** or **islets of Langerhans** (LAHNG-er-hanz; Figure 18.18b, c). Abundant capillaries serve both the exocrine and endocrine portions of the pancreas.

Cell Types in the Pancreatic Islets

Each pancreatic islet includes four types of hormone-secreting cells:

1. **Alpha** or **A cells** constitute about 15% of pancreatic islet cells and secrete **glucagon** (GLOO-ka-gon).

2. **Beta** or **B cells** constitute about 80% of pancreatic islet cells and secrete **insulin** (IN-soo-lin).

3. **Delta** or **D cells** constitute about 5% of pancreatic islet cells and secrete somatostatin (identical to growth hormone-inhibiting hormone secreted by the hypothalamus).

4. **F cells** constitute the remainder of pancreatic islet cells and secrete **pancreatic polypeptide.**

The interactions of the four pancreatic hormones are complex and not completely understood. Whereas glucagon raises blood glucose level, insulin lowers it. Somatostatin acts in a paracrine manner to inhibit both insulin and glucagon release from neighboring beta and alpha cells. It may also act as a circulating hormone to slow absorption of nutrients from the gastrointestinal tract. Pancreatic polypeptide inhibits somatostatin secretion, gallbladder contraction, and secretion of digestive enzymes by the pancreas.

Regulation of Glucagon and Insulin Secretion

The principal action of glucagon is to increase blood glucose level when it falls below normal. Insulin, on the other hand, helps lower blood glucose level when it is too high. The level of blood glucose controls secretion of glucagon and insulin via negative feedback (Figure 18.19 on page 616):

1 Low blood glucose level (hypoglycemia) stimulates secretion of glucagon from alpha cells of the pancreatic islets.

2 Glucagon acts on hepatocytes (liver cells) to accelerate the conversion of glycogen into glucose (glycogenolysis) and to promote formation of glucose from lactic acid and certain amino acids (gluconeogenesis).

3 As a result, hepatocytes release glucose into the blood more rapidly, and blood glucose level rises.

4 If blood glucose continues to rise, high blood glucose level (hyperglycemia) inhibits release of glucagon (negative feedback).

5 At the same time, high blood glucose (hyperglycemia) stimulates secretion of insulin by beta cells of the pancreatic islets.

6 Insulin acts on various cells in the body to accelerate facilitated diffusion of glucose into cells, especially skeletal muscle fibers; to speed conversion of glucose into glycogen (glycogenesis); to increase uptake of amino acids by cells and to increase protein synthesis; to speed synthesis of fatty acids (lipogenesis); to slow glycogenolysis; and to slow gluconeogenesis.

7 As a result, blood glucose level falls.

8 If blood glucose level drops below normal, low blood glucose inhibits release of insulin (negative feedback).

Although high blood glucose level is the most important stimulator of insulin release, several hormones and neurotransmitters also regulate the release of insulin and glucagon. Whereas glucagon stimulates insulin release, insulin suppresses glucagon secretion. As blood glucose level declines and less insulin is secreted, the alpha cells are released from the inhibitory effect of insulin. Then, they secrete more glucagon. Indirectly, human growth hormone (hGH) and adrenocorticotropic hormone (ACTH) stimulate secretion of insulin because they act to elevate blood glucose.

Insulin secretion is stimulated by:

• Acetylcholine, the neurotransmitter liberated from axon terminals of parasympathetic vagus nerve axons that innervate the pancreatic islets,

• The amino acids arginine and leucine, which would be present in the blood at higher levels after a protein-containing meal, and

Figure 18.18 Location, blood supply, and histology of the pancreas.
(See Tortora, *A Photographic Atlas of the Human Body,* Figure 10.6a.)

Pancreatic hormones regulate blood glucose level.

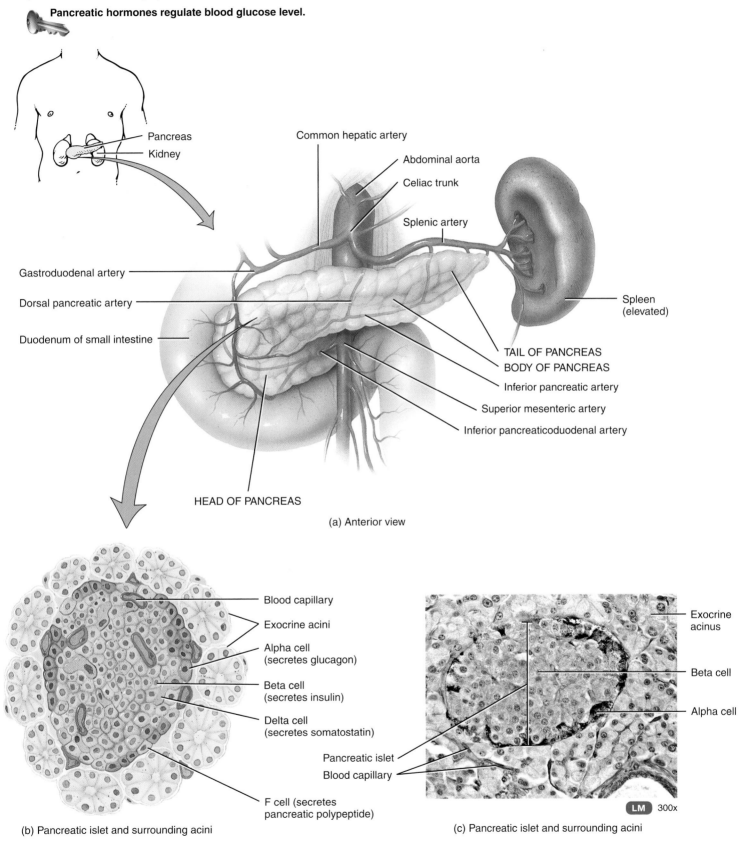

(a) Anterior view

(b) Pancreatic islet and surrounding acini

(c) Pancreatic islet and surrounding acini

 Is the pancreas an exocrine gland or an endocrine gland?

Figure 18.19   Negative feedback regulation of the secretion of glucagon (blue arrows) and insulin (orange arrows).

🔑 **Low blood glucose stimulates release of glucagon, whereas high blood glucose stimulates secretion of insulin.**

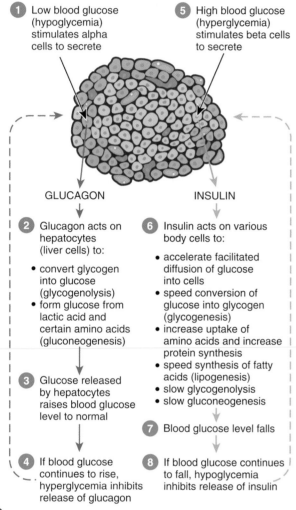

① Low blood glucose (hypoglycemia) stimulates alpha cells to secrete

⑤ High blood glucose (hyperglycemia) stimulates beta cells to secrete

GLUCAGON   INSULIN

② Glucagon acts on hepatocytes (liver cells) to:
- convert glycogen into glucose (glycogenolysis)
- form glucose from lactic acid and certain amino acids (gluconeogenesis)

③ Glucose released by hepatocytes raises blood glucose level to normal

④ If blood glucose continues to rise, hyperglycemia inhibits release of glucagon

⑥ Insulin acts on various body cells to:
- accelerate facilitated diffusion of glucose into cells
- speed conversion of glucose into glycogen (glycogenesis)
- increase uptake of amino acids and increase protein synthesis
- speed synthesis of fatty acids (lipogenesis)
- slow glycogenolysis
- slow gluconeogenesis

⑦ Blood glucose level falls

⑧ If blood glucose continues to fall, hypoglycemia inhibits release of insulin

❓ **Why is glucagon sometimes called an "anti-insulin" hormone?**

- Glucose-dependent insulinotropic peptide (GIP),* a hormone released by enteroendocrine cells of the small intestine in response to the presence of glucose in the gastrointestinal tract.

Thus, digestion and absorption of food containing both carbohydrates and proteins provide strong stimulation for insulin release. Glucagon secretion is stimulated by:

- Increased activity of the sympathetic division of the ANS, as occurs during exercise, and

_____
* GIP—previously called gastric inhibitory peptide—was renamed because at physiological concentration its inhibitory effect on stomach function is negligible.

- A rise in blood amino acids if blood glucose level is low, which could occur after a meal that contained mainly protein.

Table 18.9 summarizes the hormones produced by the pancreas, control of their secretion, and their principal actions.

▶ **CHECKPOINT**

20. How are blood levels of glucagon and insulin controlled?

21. What are the effects on secretion of insulin and glucagon of exercise versus eating a carbohydrate- and protein-rich meal?

OVARIES AND TESTES

▶ **OBJECTIVE**

- **Describe the location, hormones, and functions of the male and female gonads.**

Gonads are the organs that produce gametes—sperm in males and oocytes in females. The female gonads, the **ovaries,** are paired oval bodies located in the pelvic cavity. They produce several steroid hormones including two **estrogens** (estradiol and estrone) and **progesterone.** These female sex hormones, along with FSH and LH from the anterior pituitary, regulate the menstrual cycle, maintain pregnancy, and prepare the mammary glands for lactation. They also help establish and maintain feminine secondary sex characteristics, such as big breasts and wide hips.

The ovaries also produce **inhibin,** a protein hormone that inhibits secretion of follicle-stimulating hormone (FSH). During pregnancy, the ovaries and placenta produce a peptide hormone called **relaxin,** which increases the flexibility of the pubic symphysis during pregnancy and helps dilate the uterine cervix during labor and delivery. These actions help ease the baby's passage by enlarging the birth canal.

The male gonads, the **testes,** are oval glands that lie in the scrotum. The main hormone produced and secreted by the testes is **testosterone,** an **androgen** or male sex hormone. Testosterone regulates production of sperm and stimulates the development and maintenance of masculine secondary sex characteristics, such as beard growth and deepening of the voice. The testes also produce inhibin, which inhibits secretion of FSH. The detailed structure of the ovaries and testes and the specific roles of sex hormones are discussed in Chapter 28.

Table 18.10 summarizes the hormones produced by the ovaries and testes and their principal actions.

▶ **CHECKPOINT**

22. Why are the ovaries and testes classified as endocrine glands as well as organs where gametes are formed?

Table 18.9 Summary of Pancreatic Islet Hormones

Hormone and Source	Control of Secretion	Principal Actions
Glucagon from alpha cells of pancreatic islets Alpha cell	Decreased blood level of glucose, exercise and mainly protein meals stimulate secretion; somatostatin and insulin inhibit secretion.	Raises blood glucose level by accelerating breakdown of glycogen into glucose in liver (glycogenolysis), converting other nutrients into glucose in liver (gluconeogenesis), and releasing glucose into the blood.
Insulin from beta cells of pancreatic islets Beta cell	Increased blood level of glucose, acetylcholine (released by parasympathetic vagus nerve fibers), arginine and leucine (two amino acids), glucagon, GIP, hGH, and ACTH stimulate secretion; somatostatin inhibits secretion.	Lowers blood glucose level by accelerating transport of glucose into cells, converting glucose into glycogen (glycogenesis), and decreasing glycogenolysis and gluconeogenesis; also increases lipogenesis and stimulates protein synthesis.
Somatostatin from delta cells of pancreatic islets Delta cell	Pancreatic polypeptide inhibits secretion.	Inhibits secretion of insulin and glucagon and slows absorption of nutrients from the gastrointestinal tract.
Pancreatic polypeptide from F cells of pancreatic islets F cell	Meals containing protein, fasting, exercise, and acute hypoglycemia stimulate secretion; somatostatin and elevated blood glucose level inhibit secretion.	Inhibits somatostatin secretion, gallbladder contraction, and secretion of pancreatic digestive enzymes.

Table 18.10 Summary of Hormones of the Ovaries and Testes

Hormone	Principal Actions	Hormone	Principal Actions
Ovarian hormones		***Testicular hormones***	
Estrogens and **progesterone** Ovaries	Together with gonadotropic hormones of the anterior pituitary gland, regulate the female reproductive cycle, regulate oogenesis, maintain pregnancy, prepare the mammary glands for lactation, and promote development and maintenance of feminine secondary sex characteristics.	**Testosterone** Testes	Stimulates descent of testes before birth, regulates spermatogenesis, and promotes development and maintenance of masculine secondary sex characteristics.
Relaxin	Increases flexibility of pubic symphysis during pregnancy and helps dilate uterine cervix during labor and delivery.	**Inhibin**	Inhibits secretion of FSH from anterior pituitary.
Inhibin	Inhibits secretion of FSH from anterior pituitary.		

PINEAL GLAND

• **Describe the location, histology, hormone, and functions of the pineal gland.**

The **pineal gland** (PĪN-ē-al = pinecone shape) is a small endocrine gland attached to the roof of the third ventricle of the brain at the midline (see Figure 18.1). It is part of the epithalamus, positioned between the two superior colliculi, and has a mass of 0.1–0.2 g. The gland, which is covered by a capsule formed by the pia mater, consists of masses of neuroglia and secretory cells called **pinealocytes** (pin-ē-AL-ō-sītz). Sympathetic postganglionic axons from the superior cervical ganglion terminate in the pineal gland.

Although many anatomical features of the pineal gland have been known for years, its physiological roles are still unclear. One hormone secreted by the pineal gland is **melatonin,** an amine hormone derived from serotonin. More melatonin is released in darkness, and less melatonin is liberated in strong sunlight. Sympathetic postganglionic axons from the superior cervical ganglion extend to the pineal gland and form synaptic contacts with pinealocytes. In darkness, norepinephrine released by the sympathetic fibers stimulates synthesis and secretion of melatonin, which may promote sleepiness.

Melatonin is thought to contribute to setting the body's biological clock, which is controlled from the suprachiasmatic nucleus of the hypothalamus. During sleep, plasma levels of melatonin increase tenfold and then decline to a low level again before awakening. Small doses of melatonin given orally can induce sleep and reset daily rhythms, which might benefit workers whose shifts alternate between daylight and nighttime hours. Melatonin also is a potent antioxidant that may provide some protection against damaging oxygen free radicals.

In animals that breed during specific seasons, melatonin inhibits reproductive functions. Whether melatonin influences human reproductive function, however, is still unclear. Melatonin levels are higher in children and decline with age into adulthood, but there is no evidence that changes in melatonin secretion correlate with the onset of puberty and sexual maturation. Nevertheless, because melatonin causes atrophy of the gonads in several animal species, the possibility of adverse effects on human reproduction must be studied before its use to reset daily rhythms can be recommended.

Seasonal Affective Disorder and Jet Lag

Seasonal affective disorder (SAD) is a type of depression that afflicts some people during the winter months, when day length is short. It is thought to be due, in part, to overproduction of melatonin. Full-spectrum bright-light therapy—repeated doses of several hours of exposure to artificial light as bright as sunlight—provides relief for some people. Three to six hours of exposure to bright light also appears to speed recovery from jet lag, the fatigue suffered by travelers who quickly cross several time zones. ∎

THYMUS

The **thymus** is located behind the sternum between the lungs. Because of its role in immunity, the details of the structure and functions of the thymus are discussed in Chapter 22. The hormones produced by the thymus—**thymosin, thymic humoral factor (THF), thymic factor (TF),** and **thymopoietin**—promote the maturation of T cells (a type of white blood cell that destroys microbes and foreign substances) and may retard the aging process.

OTHER HORMONES AND GROWTH FACTORS

• **List the hormones and growth factors secreted by cells in tissues and organs other than endocrine glands, and describe their functions.**

Hormones from Other Endocrine Cells

Some tissues and organs other than those described already contain endocrine cells that secrete hormones. Table 18.11 provides an overview of these hormones and their actions.

Eicosanoids

Two families of eicosanoid molecules—the **prostaglandins** (pros′-ta-GLAN-dins), or **PGs,** and the **leukotrienes** (loo-kō-TRĪ-ēns), or **LTs**—act as local hormones in most tissues of the body. They are synthesized by clipping a 20-carbon fatty acid called **arachidonic acid** from membrane phospholipid molecules. From arachidonic acid, different enzymatic reactions produce PGs or LTs. **Thromboxane (TX)** is a modified PG that constricts blood vessels and promotes platelet activation. In response to chemical and mechanical stimuli, virtually all body cells except red blood cells release these local hormones, which act mainly as potent paracrines and autocrines and appear in the blood only in minute quantities. They are present only briefly due to rapid inactivation.

To exert their effects, eicosanoids bind to receptors on target-cell plasma membranes and stimulate or inhibit the synthesis of second messengers such as cyclic AMP. Leukotrienes stimulate chemotaxis of white blood cells and mediate inflammation. The prostaglandins alter smooth muscle contraction, glandular secretions, blood flow, reproductive processes, platelet function, respiration, nerve impulse transmission, lipid metabolism, and immune responses. They also have roles in inflammation, in promoting fever, and in intensifying pain.

Nonsteroidal Anti-inflammatory Drugs

In 1971, scientists solved the long-standing puzzle of how aspirin works. Aspirin and related **nonsteroidal anti-inflamma-**

tory drugs (NSAIDs), such as ibuprofen (Motrin), inhibit a key enzyme in prostaglandin synthesis without affecting synthesis of leukotrienes. NSAIDS are used to treat a wide variety of inflammatory disorders, from rheumatoid arthritis to tennis elbow. The success of NSAIDs in reducing fever, pain, and inflammation points to prostaglandins as contributing to these woes. ■

Growth Factors

Several of the hormones we've described—insulinlike growth factor, thymosin, insulin, thyroid hormones, human growth hormone, prolactin, and erythropoietin—stimulate cell growth and division. In addition, several more recently discovered hormones called **growth factors** play important roles in tissue development, growth, and repair. Growth factors are *mitogenic* substances—they cause growth by stimulating cell division. Many growth factors act locally, as autocrines or paracrines. A summary of sources and actions of six important growth factors is presented in Table 18.12.

▶ **CHECKPOINT**

23. What is the relationship between melatonin secretion and sleep?

24. What hormones are secreted by the gastrointestinal tract, placenta, kidneys, skin, adipose tissue, and heart?

25. What are some functions of prostaglandins, leukotrienes, and growth factors?

Table 18.11	Summary of Hormones Produced by Organs and Tissues that Contain Endocrine Cells
Hormone	**Principal Actions**
Gastrointestinal tract	
Gastrin	Promotes secretion of gastric juice and increases motility of the stomach.
Glucose-dependent insulinotropic peptide (GIP)	Stimulates release of insulin by pancreatic beta cells.
Secretin	Stimulates secretion of pancreatic juice and bile.
Cholecystokinin (CCK)	Stimulates secretion of pancreatic juice, regulates release of bile from the gallbladder, and brings about a feeling of fullness after eating.
Placenta	
Human chorionic gonadotropin (hCG)	Stimulates the corpus luteum in the ovary to continue the production of estrogens and progesterone to maintain pregnancy.
Estrogens and **progesterone**	Maintain pregnancy and help prepare mammary glands to secrete milk.
Human chorionic somatomammotropin (hCS)	Stimulates the development of the mammary glands for lactation.
Kidneys	
Erythropoietin (EPO)	Increases rate of red blood cell formation.
Calcitriol* (active form of vitamin D)	Aids in the absorption of dietary calcium and phosphorus.
Heart	
Atrial natriuretic peptide (ANP)	Decreases blood pressure.
Adipose tissue	
Leptin	Suppresses appetite and may have a permissive effect on activity of GnRH and gonadotropins.

*Synthesis begins in the skin, continues in the liver, and ends in the kidneys.

Table 18.12	Summary of Selected Growth Factors
Growth Factor	**Comment**
Epidermal growth factor (EGF)	Produced in submaxillary (salivary) glands; stimulates proliferation of epithelial cells, fibroblasts, neurons, and astrocytes; suppresses some cancer cells and secretion of gastric juice by the stomach.
Platelet-derived growth factor (PDGF)	Produced in blood platelets; stimulates proliferation of neuroglia, smooth muscle fibers, and fibroblasts; appears to have a role in wound healing; may contribute to the development of atherosclerosis.
Fibroblast growth factor (FGF)	Found in pituitary gland and brain; stimulates proliferation of many cells derived from embryonic mesoderm (fibroblasts, adrenocortical cells, smooth muscle fibers, chondrocytes, and endothelial cells); also stimulates formation of new blood vessels (angiogenesis).
Nerve growth factor (NGF)	Produced in submaxillary (salivary) glands and hippocampus of brain; stimulates the growth of ganglia in embryonic life, maintains sympathetic nervous system; stimulates hypertrophy and differentiation of neurons.
Tumor angiogenesis factors (TAFs)	Produced by normal and tumor cells; stimulate growth of new capillaries, organ regeneration, and wound healing.
Transforming growth factors (TGFs)	Produced by various cells as separate molecules called TGF-alpha and TGF-beta. TGF-alpha has activities similar to epidermal growth factor, and TGF-beta inhibits proliferation of many cell types.

THE STRESS RESPONSE

OBJECTIVE

• **Describe how the body responds to stress.**

It is impossible to remove all stress from our everyday lives. Some stress, called **eustress,** prepares us to meet certain challenges and thus is helpful. Other stress, called **distress,** is harmful. Any stimulus that produces a stress response is called a **stressor.** A stressor may be almost any disturbance—heat or cold, environmental poisons, toxins given off by bacteria, heavy bleeding from a wound or surgery, or a strong emotional reaction. Stressors may be pleasant or unpleasant, and they vary among people and even within the same person at different times.

Homeostatic mechanisms attempt to counteract stress. When they are successful, the internal environment remains within normal physiological limits. If stress is extreme, unusual, or long lasting, the normal mechanisms may not be enough. In 1936, Hans Selye, a pioneer in stress research, showed that a variety of stressful conditions or noxious agents elicit a similar sequence of bodily changes. These changes, called the **stress response** or **general adaptation syndrome (GAS),** are controlled mainly by the hypothalamus. The stress response occurs in three stages: (1) an initial fight-or-flight response, (2) a slower resistance reaction, and eventually (3) exhaustion.

The Fight-or-Flight Response

The **fight-or-flight response,** initiated by nerve impulses from the hypothalamus to the sympathetic division of the autonomic nervous system (ANS), including the adrenal medulla, quickly mobilizes the body's resources for immediate physical activity (Figure 18.20a). It brings huge amounts of glucose and oxygen to the organs that are most active in warding off danger: the brain, which must become highly alert; the skeletal muscles, which may have to fight off an attacker or flee; and the heart, which must work vigorously to pump enough blood to the brain and muscles. During the fight-or-flight response, nonessential body functions such as digestive, urinary, and reproductive activities are inhibited. Reduction of blood flow to the kidneys, however, promotes release of renin, which sets into the motion the renin–angiotensin–aldosterone pathway (see Figure 18.16). Aldosterone causes the kidneys to retain Na$^+$, which leads to water retention and elevated blood pressure. Water retention also helps preserve body fluid volume in the case of severe bleeding.

The Resistance Reaction

The second stage in the stress response is the **resistance reaction** (Figure 18.20b). Unlike the short-lived fight-or-flight response, which is initiated by nerve impulses from the hypothalamus, the resistance reaction is initiated in large part by hypothalamic releasing hormones and is a longer-lasting response. The hormones involved are corticotropin-releasing hormone (CRH), growth hormone-releasing hormone (GHRH), and thyrotropin-releasing hormone (TRH).

CRH stimulates the anterior pituitary to secrete ACTH, which in turn stimulates the adrenal cortex to release more cortisol. Cortisol then stimulates gluconeogenesis by liver cells, breakdown of triglycerides into fatty acids (lipolysis), and catabolism of proteins into amino acids. Tissues throughout the body can use the resulting glucose, fatty acids, and amino acids to produce ATP or to repair damaged cells. Cortisol also reduces inflammation.

A second hypothalamic releasing hormone, GHRH, causes the anterior pituitary to secrete human growth hormone (hGH). Acting via insulinlike growth factors, hGH stimulates lipolysis and glycogenolysis, the breakdown of glycogen to glucose. A third hypothalamic releasing hormone, TRH, stimulates the anterior pituitary to secrete thyroid-stimulating hormone (TSH). TSH promotes secretion of thyroid hormones, which stimulate the increased use of glucose for ATP production. The combined actions of hGH and TSH thereby supply additional ATP for metabolically active cells.

The resistance stage helps the body continue fighting a stressor long after the fight-or-flight response dissipates. Generally, it is successful in seeing us through a stressful episode, and our bodies then return to normal. Occasionally, however, the resistance stage fails to combat the stressor, and the body moves into the state of exhaustion.

Exhaustion

The resources of the body may eventually become so depleted that they cannot sustain the resistance stage, and **exhaustion** ensues. Prolonged exposure to high levels of cortisol and other hormones involved in the resistance reaction causes wasting of muscle, suppression of the immune system, ulceration of the gastrointestinal tract, and failure of pancreatic beta cells. In addition, pathological changes may occur because resistance reactions persist after the stressor has been removed.

Stress and Disease

Although the exact role of stress in human diseases is not known, it is clear that stress can lead to particular diseases by temporarily inhibiting certain components of the immune system. Stress-related disorders include gastritis, ulcerative colitis, irritable bowel syndrome, hypertension, asthma, rheumatoid arthritis (RA), migraine headaches, anxiety, and depression. People under stress are at a greater risk of developing chronic disease or dying prematurely.

Interleukin-1, a cytokine secreted by macrophages of the immune system (see page 784), is an important link between stress and immunity. One action of interleukin-1 is to stimulate secretion of ACTH, which in turn stimulates the production of cortisol. Not only does cortisol provide resistance to stress and inflammation, but it also suppresses further production of interleukin-1. Thus, the immune system turns on the stress response, and the resulting cortisol then turns off one immune system mediator. This negative feedback system keeps the immune response in check once it has accomplished its goal. Because

Figure 18.20 **Responses to stressors during the stress response.** Red arrows (hormonal responses) and green arrows (neural responses) in (a) indicate immediate "fight-or-flight" reactions; black arrows in (b) indicate long-term resistance reactions.

🔑 **Stressors stimulate the hypothalamus to initiate the stress response through the fight-or-flight response and the resistance reaction.**

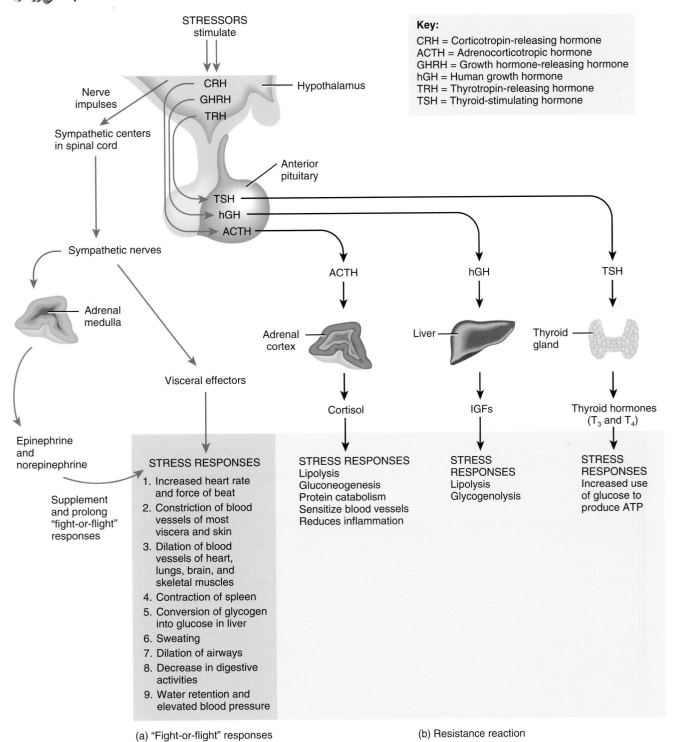

Key:
CRH = Corticotropin-releasing hormone
ACTH = Adrenocorticotropic hormone
GHRH = Growth hormone-releasing hormone
hGH = Human growth hormone
TRH = Thyrotropin-releasing hormone
TSH = Thyroid-stimulating hormone

(a) "Fight-or-flight" responses

(b) Resistance reaction

 What is the basic difference between the stress response and homeostasis?

glucocorticoids, such as cortisol, suppress certain aspects of the immune response, they are used as immunosuppressive drugs for organ transplant recipients.

Posttraumatic Stress Disorder

Posttraumatic stress disorder (PTSD) is an anxiety disorder that may develop in an individual who has experienced, witnessed, or learned about a physically or psychologically distressing event. The immediate cause of PTSD appears to be the specific stressors associated with the events. Among the stressors are terrorism, hostage taking, imprisonment, military duty, serious accidents, torture, sexual or physical abuse, violent crimes, school shootings, massacres, and natural disasters. In the United States, PTSD affects 10% of females and 5% of males. Symptoms of PTSD include reexperiencing the event through nightmares or flashbacks; avoidance of any activity, person, place, or event associated with the stressors; loss of interest and lack of motivation; poor concentration; irritability; and insomnia. Treatment may include the use of antidepressants, mood stabilizers, and antianxiety and antipsychotic agents. ■

▶ C H E C K P O I N T

26. What is the central role of the hypothalamus during stress?

27. What body reactions occur during the fight-or-flight response, the resistance reaction, and exhaustion?

28. How are stress and immunity related?

DEVELOPMENT OF THE ENDOCRINE SYSTEM

▶ O B J E C T I V E

• **Describe the development of endocrine glands.**

The development of the endocrine system is not as localized as the development of other systems because endocrine organs develop in widely separated parts of the embryo.

About three weeks after fertilization, the *pituitary gland (hypophysis)* begins to develop from two different regions of the ectoderm. The *posterior pituitary (neurohypophysis)* is derived from an outgrowth of ectoderm called the **neurohypophyseal bud,** located on the floor of the hypothalamus (Figure 18.21). The *infundibulum,* also an outgrowth of the neurohypophyseal bud, connects the posterior pituitary to the hypothalamus. The *anterior pituitary (adenohypophysis)* is derived from an outgrowth of ectoderm from the roof of the mouth called the **hypophyseal (Rathke's) pouch.** The pouch grows toward the neurohypophyseal bud and eventually loses its connection with the roof of the mouth.

The *thyroid gland* develops during the fourth week as a midventral outgrowth of endoderm, called the **thyroid diverticulum,** from the floor of the pharynx at the level of the second pair of pharyngeal pouches. The outgrowth projects inferiorly and differentiates into the right and left lateral lobes and the isthmus of the gland.

The *parathyroid glands* develop during the fourth week from endoderm as outgrowths from the third and fourth **pharyngeal pouches,** which help to form structures of the head and neck.

The adrenal cortex and adrenal medulla develop during the fifth week and have completely different embryological origins. The *adrenal cortex* is derived from intermediate mesoderm from the same region that produces the gonads. Endocrine tissues that secrete steroid hormones all are derived from mesoderm. The *adrenal medulla* is ectodermal in origin and derives from **neural crest** cells that migrate to the superior pole of the kidney. Recall that neural crest cells also give rise to sympathetic ganglia and other structures of the nervous system (see Figure 14.25b on page 490).

The *pancreas* develops during the fifth through seventh weeks from two outgrowths of endoderm from the part of the **foregut** that later becomes the duodenum (see Figure 29.12c on page 1078). The two outgrowths eventually fuse to form the pancreas. The origin of the ovaries and testes is discussed in the section on the reproductive system.

The *pineal gland* arises during the seventh week as an outgrowth between the thalamus and colliculi of the midbrain from ectoderm associated with the **diencephalon** (see Figure 14.26 on page 491).

The *thymus* arises during the fifth week from endoderm of the third pharyngeal pouches.

AGING AND THE ENDOCRINE SYSTEM

▶ O B J E C T I V E

• **Describe the effects of aging on the endocrine system.**

Although some endocrine glands shrink as we get older, their performance may or may not be compromised. Production of human growth hormone by the anterior pituitary decreases, which is one cause of muscle atrophy as aging proceeds. The thyroid gland often decreases its output of thyroid hormones with age, causing a decrease in metabolic rate, an increase in body fat, and hypothyroidism, which is seen more often in older people. Due to less negative feedback (lower levels of thyroid hormones), thyroid-stimulating hormone level increases with age.

With aging, the blood level of PTH rises, perhaps due to inadequate dietary intake of calcium. In a study of older women who took 2,400 mg/day of supplemental calcium, blood levels of PTH were lower, as in younger women. Both calcitriol and calcitonin levels are lower in older persons. Together, the rise in PTH and the fall in calcitonin level heighten the age-related decrease in bone mass that leads to osteoporosis and increased risk of fractures.

The adrenal glands contain increasingly more fibrous tissue and produce less cortisol and aldosterone with advancing age. However, production of epinephrine and norepinephrine remains

Figure 18.21 Development of the endocrine system.

 Glands of the endocrine system develop from all three primary germ layers: ectoderm, mesoderm, and endoderm.

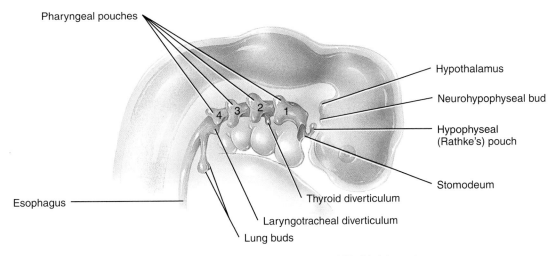

(a) Location of the neurohypophyseal bud, hypophyseal (Rathke's) pouch, thyroid diverticulum and pharyngeal pouches in a 28-day embryo

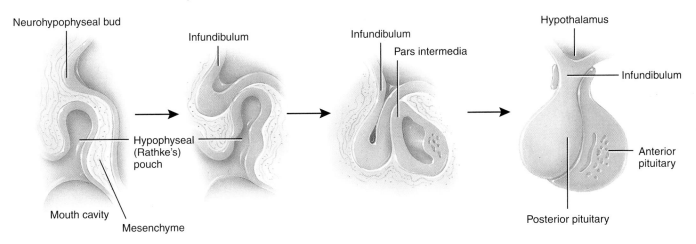

(b) Development of the pituitary gland between five and sixteen weeks

Which two endocrine glands develop from tissues with two different embryological origins?

normal. The pancreas releases insulin more slowly with age, and receptor sensitivity to insulin declines. As a result, blood glucose levels in older people increase faster and return to normal more slowly than in younger individuals.

The thymus is largest in infancy. After puberty, its size begins to decrease, and thymic tissue is replaced by adipose and areolar connective tissue. In older adults, the thymus has atrophied significantly. However, it still produces new T cells for immune responses.

The ovaries decrease in size with age, and they no longer respond to gonadotropins. The resultant decreased output of estrogens leads to conditions such as osteoporosis, high blood cholesterol, and atherosclerosis. FSH and LH levels are high due to less negative feedback inhibition of estrogens. Although

testosterone production by the testes decreases with age, the effects are not usually apparent until very old age, and many elderly males can still produce active sperm in normal numbers.

▶ CHECKPOINT

29. Which hormone is related to the muscle atrophy that occurs with aging?

• • •

To appreciate the many ways the endocrine system contributes to homeostasis of other body systems, examine *Focus on Homeostasis: The Endocrine System* on page 624. Next, in Chapter 19, we will begin to explore the cardiovascular system, starting with a description of the composition and functions of blood.

Body System	Contribution of the Endocrine System
For all body systems	Together with the nervous system, circulating and local hormones of the endocrine system regulate activity and growth of target cells throughout the body; several hormones regulate metabolism, uptake of glucose, and molecules used for ATP production by body cells.
Integumentary system	Androgens stimulate growth of axillary and pubic hair and activation of sebaceous glands; excess melanocyte-stimulating hormone (MSH) causes darkening of skin.
Skeletal system	Human growth hormone (hGH) and insulinlike growth factors (IGFs) stimulate bone growth; estrogens cause closure of the epiphyseal plates at the end of puberty and help maintain bone mass in adults; parathyroid hormone (PTH) and calcitonin regulate release versus uptake of calcium and other minerals between bone matrix and blood; thyroid hormones are needed for normal development and growth of the skeleton.
Muscular system	Epinephrine and norepinephrine help increase blood flow to exercising muscle; PTH maintains proper level of Ca^{2+}, needed for muscle contraction; glucagon, insulin, and other hormones regulate metabolism in muscle fibers; hGH, IGFs, and thyroid hormones help maintain muscle mass.
Nervous system	Several hormones, especially thyroid hormones, insulin, and growth hormone, influence growth and development of the nervous system; PTH maintains proper level of Ca^{2+}, needed for generation and conduction of nerve impulses.
Cardiovascular system	Erythropoietin (EPO) promotes formation of red blood cells; aldosterone and antidiuretic hormone (ADH) increase blood volume; epinephrine and norepinephrine increase heart rate and force of contraction; several hormones elevate blood pressure during exercise and other stresses.
Lymphatic and immune system	Glucocorticoids such as cortisol depress inflammation and immune responses; thymic hormones promote maturation of T cells (a type of white blood cell).
Respiratory system	Epinephrine and norepinephrine dilate (widen) airways during exercise and other stresses; erythropoietin regulates amount of oxygen carried in blood by adjusting number of red blood cells.
Digestive system	Epinephrine and norepinephrine depress activity of the digestive system; gastrin, cholecystokinin, secretin, and GIP help regulate digestion; calcitriol promotes absorption of dietary calcium; leptin suppresses appetite.
Urinary system	ADH, aldosterone, and atrial natriuretic peptide (ANP) adjust the rate of loss of water and ions in the urine thereby regulating blood volume and ion content of the blood.
Reproductive systems	Hypothalamic releasing and inhibiting hormones, follicle-stimulating hormone (FSH), and luteinizing hormone (LH) regulate development, growth, and secretions of the gonads (ovaries and testes); estrogens and testosterone contribute to development of oocytes and sperm and stimulate development of secondary sex characteristics; prolactin promotes milk secretion in mammary glands; oxytocin causes contraction of the uterus and ejection of milk from the mammary glands.

DISORDERS: HOMEOSTATIC IMBALANCES

Disorders of the endocrine system often involve either **hyposecretion** (*hypo-* = too little or under), inadequate release of a hormone, or **hypersecretion** (*hyper-* = too much or above), excessive release of a hormone. In other cases, the problem is faulty hormone receptors, an inadequate number of receptors, or defects in second-messenger systems. Because hormones are distributed in the blood to target tissues throughout the body, problems associated with endocrine dysfunction may also be widespread.

Pituitary Gland Disorders

Pituitary Dwarfism, Giantism, and Acromegaly

Several disorders of the anterior pituitary involve human growth hormone (hGH). Hyposecretion of hGH during the growth years slows bone growth, and the epiphyseal plates close before normal height is reached. This condition is called **pituitary dwarfism.** Other organs of the body also fail to grow, and the body proportions are childlike. Treatment requires administration of hGH during childhood, before the epiphyseal plates close.

Hypersecretion of hGH during childhood causes **giantism,** an abnormal increase in the length of long bones. The person grows to be very tall, but body proportions are about normal. Figure 18.22a shows identical twins; one brother developed giantism due to a pituitary tumor. Hypersecretion of hGH during adulthood is called **acromegaly** (ak′-rō-MEG-a-lē). Although hGH cannot produce further lengthening of the long bones because the epiphyseal plates are already closed, the bones of the hands, feet, cheeks, and jaws thicken and other tissues enlarge. In addition, the eyelids, lips, tongue, and nose enlarge, and the skin thickens and develops furrows, especially on the forehead and soles (Figure 18.22b).

Figure 18.22 Photographs of people with endocrine disorders.

Disorders of the endocrine system often involve hyposecretion or hypersecretion of hormones.

(a) A 22-year old man with pituitary giantism shown beside his identical twin

(b) Acromegaly (excess hGH during adulthood)

(c) Exophthalmos (excess thyroid hormones, as in Graves' disease)

(d) Goiter (enlargement of thyroid gland)

(e) Cushing's syndrome (excess glucocorticoids)

Which endocrine disorder is due to antibodies that mimic the action of TSH?

Diabetes Insipidus

The most common abnormality associated with dysfunction of the posterior pituitary is **diabetes insipidus** (dī-a-BĒ-tēs in-SIP-i-dus; *diabetes* = overflow; *insipidus* = tasteless) or **DI**. This disorder is due to defects in antidiuretic hormone (ADH) receptors or an inability to secrete ADH. *Neurogenic diabetes insipidus* results from hyposecretion of ADH, usually caused by a brain tumor, head trauma, or brain surgery that damages the posterior pituitary or the hypothalamus. In *nephrogenic diabetes insipidus,* the kidneys do not respond to ADH. The ADH receptors may be nonfunctional, or the kidneys may be damaged. A common symptom of both forms of DI is excretion of large volumes of urine, with resulting dehydration and thirst. Bed-wetting is common in afflicted children. Because so much water is lost in the urine, a person with DI may die of dehydration if deprived of water for only a day or so.

Treatment of neurogenic diabetes insipidus involves hormone replacement, usually for life. Either subcutaneous injection or nasal spray application of ADH analogs is effective. Treatment of nephrogenic diabetes insipidus is more complex and depends on the nature of the kidney dysfunction. Restriction of salt in the diet and, paradoxically, the use of certain diuretic drugs, are helpful.

Thyroid Gland Disorders

Thyroid gland disorders affect all major body systems and are among the most common endocrine disorders. **Congenital hypothyroidism,** hyposecretion of thyroid hormones that is present at birth, has devastating consequences if not treated promptly. Previously termed *cretinism,* this condition causes severe mental retardation. At birth, the baby typically is normal because lipid-soluble maternal thyroid hormones crossed the placenta during pregnancy and allowed normal development. Most states require testing of all newborns to ensure adequate thyroid function. If congenital hypothyroidism exists, oral thyroid hormone treatment must be started soon after birth and continued for life.

Hypothyroidism during the adult years produces **myxedema** (mix-e-DĒ-ma), which occurs about five times more often in females than in males. A hallmark of this disorder is edema (accumulation of interstitial fluid) that causes the facial tissues to swell and look puffy. A person with myxedema has a slow heart rate, low body temperature, sensitivity to cold, dry hair and skin, muscular weakness, general lethargy, and a tendency to gain weight easily. Because the brain has already reached maturity, mental retardation does not occur, but the person may be less alert. Oral thyroid hormones reduce the symptoms.

The most common form of hyperthyroidism is **Graves disease,** which also occurs seven to ten times more often in females than in males, usually before age 40. Graves disease is an autoimmune disorder in which the person produces antibodies that mimic the action of thyroid-stimulating hormone (TSH). The antibodies continually stimulate the thyroid gland to grow and produce thyroid hormones. A primary sign is an enlarged thyroid, which may be two to three times its normal size. Graves patients often have a peculiar edema behind the eyes, called **exophthalmos** (ek'-sof-THAL-mos), which causes the eyes to protrude (Figure 18.22c). Treatment may include surgically removing part or all of the thyroid gland (thyroidectomy), using radioactive iodine (^{131}I) to selectively destroy thyroid tissue, and using antithyroid drugs to block synthesis of thyroid hormones.

A **goiter** (GOY-ter; *guttur* = throat) is simply an enlarged thyroid gland. It may be associated with hyperthyroidism, hypothyroidism, or

euthyroidism (*eu-* = good), which means normal secretion of thyroid hormone. In some places in the world, dietary iodine intake is inadequate; the resultant low level of thyroid hormone in the blood stimulates secretion of TSH, which causes thyroid gland enlargement (Figure 18.22d).

Parathyroid Gland Disorders

Hypoparathyroidism—too little parathyroid hormone—leads to a deficiency of Ca^{2+}, which causes neurons and muscle fibers to depolarize and produce action potentials spontaneously. This leads to twitches, spasms, and **tetany** (maintained contraction) of skeletal muscle. The leading cause of hypoparathyroidism is accidental damage to the parathyroid glands or to their blood supply during thyroidectomy surgery.

Hyperparathyroidism, an elevated level of parathyroid hormone, most often is due to a tumor of one of the parathyroid glands. An elevated level of PTH causes excessive resorption of bone matrix, raising the blood levels of calcium and phosphate ions and causing bones to become soft and easily fractured. High blood calcium level promotes formation of kidney stones. Fatigue, personality changes, and lethargy are also seen in patients with hyperparathyroidism.

Adrenal Gland Disorders

Cushing's Syndrome

Hypersecretion of cortisol by the adrenal cortex produces **Cushing's syndrome** (Figure 18.22e). Causes include a tumor of the adrenal gland that secretes cortisol, or a tumor elsewhere that secretes adrenocorticotropic hormone (ACTH), which in turn stimulates excessive secretion of cortisol. The condition is characterized by breakdown of muscle proteins and redistribution of body fat, resulting in spindly arms and legs accompanied by a rounded "moon face," "buffalo hump" on the back, and pendulous (hanging) abdomen. Facial skin is flushed, and the skin covering the abdomen develops stretch marks. The person also bruises easily, and wound healing is poor. The elevated level of cortisol causes hyperglycemia, osteoporosis, weakness, hypertension, increased susceptibility to infection, decreased resistance to stress, and mood swings. People who need long-term glucocorticoid therapy—for instance, to prevent rejection of a transplanted organ—may develop a cushinoid appearance.

Addison's Disease

Undersecretion of glucocorticoids and aldosterone causes **Addison's disease (chronic adrenocortical insufficiency).** The majority of cases are autoimmune disorders in which antibodies cause adrenal cortex destruction or block binding of ACTH to its receptors. Pathogens, such as the bacterium that causes tuberculosis, also may trigger adrenal cortex destruction. Symptoms, which typically do not appear until 90% of the adrenal cortex has been destroyed, include mental lethargy, anorexia, nausea and vomiting, weight loss, hypoglycemia, and muscular weakness. Loss of aldosterone leads to elevated potassium and decreased sodium in the blood, low blood pressure, dehydration, decreased cardiac output, arrhythmias, and even cardiac arrest. The skin may have a "bronzed" appearance that often is mistaken for a suntan. Such was true in the case of President John F. Kennedy, whose Addison's disease was known to only a few while he was alive. Treatment consists of replacing glucocorticoids and mineralocorticoids and increasing sodium in the diet.

Pheochromocytomas

Usually benign tumors of the chromaffin cells of the adrenal medulla, called **pheochromocytomas** (fē-ō-krō′-mō-si-TŌ-mas; *pheo-* = dusky; *chromo-* = color; *cyto-* = cell), cause hypersecretion of epinephrine and norepinephrine. The result is a prolonged version of the fight-or-flight response: rapid heart rate, high blood pressure, high levels of glucose in blood and urine, an elevated basal metabolic rate (BMR), flushed face, nervousness, sweating, and decreased gastrointestinal motility. Treatment is surgical removal of the tumor.

Pancreatic Islet Disorders

The most common endocrine disorder is **diabetes mellitus** (MEL-i-tus; *melli-* = honey sweetened), caused by an inability to produce or use insulin. Diabetes mellitus is the fourth leading cause of death by disease in the United States, primarily because of its damage to the cardiovascular system. Because insulin is unavailable to aid transport of glucose into body cells, blood glucose level is high and glucose "spills" into the urine (glucosuria). Hallmarks of diabetes mellitus are the three "polys": *polyuria,* excessive urine production due to an inability of the kidneys to reabsorb water; *polydipsia,* excessive thirst; and *polyphagia,* excessive eating.

Both genetic and environmental factors contribute to onset of the two types of diabetes mellitus—type 1 and type 2—but the exact mechanisms are still unknown. In **type 1 diabetes** insulin level is low because the person's immune system destroys the pancreatic beta cells. It is also called **insulin-dependent diabetes mellitus (IDDM)** because insulin injections are required to prevent death. Most commonly, IDDM develops in people younger than age 20, though it persists throughout life. By the time symptoms of IDDM arise, 80–90% of the islet beta cells have been destroyed. IDDM is most common in northern Europe, especially in Finland where nearly 1% of the population develops IDDM by 15 years of age. In the United States, IDDM is 1.5–2.0 times more common in whites than in African American or Asian populations.

The cellular metabolism of an untreated type 1 diabetic is similar to that of a starving person. Because insulin is not present to aid the entry of glucose into body cells, most cells use fatty acids to produce ATP. Stores of triglycerides in adipose tissue are catabolized to yield fatty acids and glycerol. The byproducts of fatty acid breakdown— organic acids called ketones or ketone bodies—accumulate. Buildup of ketones causes blood pH to fall, a condition known as **ketoacidosis.** Unless treated quickly, ketoacidosis can cause death.

The breakdown of stored triglycerides also causes weight loss. As lipids are transported by the blood from storage depots to cells, lipid particles are deposited on the walls of blood vessels, leading to atherosclerosis and a multitude of cardiovascular problems, including cerebrovascular insufficiency, ischemic heart disease, peripheral vascular disease, and gangrene. A major complication of diabetes is loss of vision due either to cataracts (excessive glucose attaches to lens proteins, causing cloudiness) or to damage to blood vessels of the retina. Severe kidney problems also may result from damage to renal blood vessels.

Type 1 diabetes is treated through self-monitoring of blood glucose level (up to 7 times daily), regular meals containing 45–50% carbohydrates and less than 30% fats, exercise, and periodic insulin injections (up to 3 times a day). Several implantable pumps are available to provide insulin without the need for repeated injections. Because they lack a reliable glucose sensor, however, the person must self-monitor blood glucose level to determine insulin doses. It is also possible to successfully transplant a pancreas, but immunosuppressive drugs must then be taken for life. Most pancreatic transplants are done in people who also need a kidney transplant because of renal failure. Another promising approach under investigation is transplantation of isolated islets in semipermeable hollow tubes. The tubes allow glucose and insulin to enter and leave but prevent entry of immune system cells that might attack the islet cells.

Type 2 diabetes, also called **non-insulin-dependent diabetes mellitus (NIDDM),** is much more common than type 1, representing more than 90% of all cases. Type 2 diabetes most often occurs in obese people who are over age 35. Clinical symptoms are mild, and the high glucose levels in the blood often can be controlled by diet, exercise, and weight loss. Sometimes, a drug such as *glyburide* (DiaBeta) is used to stimulate secretion of insulin by pancreatic beta cells. Although some type 2 diabetics need insulin, many have a sufficient amount (or even a surplus) of insulin in the blood. For these people, diabetes arises not from a shortage of insulin but because target cells become less sensitive to it due to down-regulation of insulin receptors.

Hyperinsulinism most often results when a diabetic injects too much insulin. The main symptom is **hypoglycemia,** decreased blood glucose level, which occurs because the excess insulin stimulates too much uptake of glucose by body cells. The resulting hypoglycemia stimulates the secretion of epinephrine, glucagon, and human growth hormone. As a consequence, anxiety, sweating, tremor, increased heart rate, hunger, and weakness occur. When blood glucose falls, brain cells are deprived of the steady supply of glucose they need to function effectively. Severe hypoglycemia leads to mental disorientation, convulsions, unconsciousness, and shock. Shock due to an insulin overdose is termed **insulin shock.** Death can occur quickly unless blood glucose level is raised.

MEDICAL TERMINOLOGY

Gynecomastia (gī-ne′-kō-MAS-tē-a; *gyneco-* = woman; *mast-* = breast) Excessive development of mammary glands in a male. Sometimes a tumor of the adrenal gland may secrete sufficient amounts of estrogen to cause the condition.

Hirsutism (HER-soo-tizm; *hirsut-* = shaggy) Presence of excessive bodily and facial hair in a male pattern, especially in women; may be due to excess androgen production due to tumors or drugs.

Thyroid crisis (storm) A severe state of hyperthyroidism that can be life threatening. It is characterized by high body temperature, rapid heart rate, high blood pressure, gastrointestinal symptoms (abdominal pain, vomiting, diarrhea), agitation, tremors, confusion, seizures, and possibly coma.

Virilizing adenoma (*aden* = gland; *oma* = tumor) Tumor of the adrenal gland that liberates excessive androgens, causing virilism (masculinization) in females. Occasionally, adrenal tumor cells liberate estrogens to the extent that a male patient develops gynecomastia. Such a tumor is called a **feminizing adenoma.**

STUDY OUTLINE

INTRODUCTION (p. 587)

1. Hormones regulate the activity of smooth muscle, cardiac muscle, and some glands; alter metabolism; spur growth and development; influence reproductive processes; and participate in circadian (daily) rhythms.

COMPARISON OF CONTROL BY THE NERVOUS AND ENDOCRINE SYSTEMS (p. 587)

1. The nervous system controls homeostasis through nerve impulses and neurotransmitters, which act locally and quickly. The endocrine system uses hormones, which more slowly act in distant parts of the body. (See Table 18.1 on page 587.)
2. The nervous system controls neurons, muscle cells, and glandular cells; the endocrine system regulates virtually all body cells.

ENDOCRINE GLANDS (p. 587)

1. Exocrine (sudoriferous, sebaceous, mucous, and digestive) glands secrete their products through ducts into body cavities or onto body surfaces. Endocrine glands secrete hormones into interstitial fluid. Then, the hormones diffuse into the blood.
2. The endocrine system consists of endocrine glands (pituitary, thyroid, parathyroid, adrenal, and pineal glands) and other hormone-secreting tissues (hypothalamus, thymus, pancreas, ovaries, testes, kidneys, stomach, liver, small intestine, skin, heart, adipose tissue, and placenta).

HORMONE ACTIVITY (p. 588)

1. Hormones affect only specific target cells that have receptors to recognize (bind) a given hormone. The number of hormone receptors may decrease (down-regulation) or increase (up-regulation).
2. Circulating hormones enter the bloodstream; local hormones (paracrines and autocrines) act locally on neighboring cells.
3. Chemically, hormones are either lipid soluble (steroids, thyroid hormones, and nitric oxide) or water soluble (amines; peptides, proteins, and glycoproteins; and eicosanoids). (See Table 18.2 on page 591.)
4. Water-soluble hormone molecules circulate in the watery blood plasma in a "free" form (not attached to plasma proteins), whereas most lipid-soluble hormone molecules are bound to transport proteins synthesized by the liver.

MECHANISMS OF HORMONE ACTION (p. 590)

1. Lipid-soluble steroid hormones and thyroid hormones affect cell function by altering gene expression.
2. Water-soluble hormones alter cell function by activating plasma membrane receptors, which elicit production of a second messenger that activates various enzymes inside the cell.
3. Three hormonal interactions are a permissive effect, a synergistic effect, and an antagonistic effect.

CONTROL OF HORMONE SECRETION (p. 594)

1. Hormone secretion is controlled by signals from the nervous system, chemical changes in blood, and other hormones.
2. Negative feedback systems regulate hormonal secretions.

HYPOTHALAMUS AND PITUITARY GLAND (p. 594)

1. The hypothalamus is the major integrating link between the nervous and endocrine systems.
2. The hypothalamus and pituitary gland regulate virtually all aspects of growth, development, metabolism, and homeostasis.
3. The pituitary gland is attached to the hypothalamus. It consists of two lobes—the anterior pituitary and the posterior pituitary.
4. Secretion of anterior pituitary hormones is stimulated by releasing hormones and suppressed by inhibiting hormones from the hypothalamus.
5. The blood supply to the anterior pituitary is from the superior hypophyseal arteries. Hypothalamic releasing and inhibiting hormones enter the primary plexus and flow to the secondary plexus in the anterior pituitary by the hypophyseal portal veins.
6. The anterior pituitary consists of somatotrophs that produce human growth hormone (hGH); lactotrophs that produce prolactin (PRL); corticotrophs that secrete adrenocorticotropic hormone (ACTH) and melanocyte-stimulating hormone (MSH); thyrotrophs that secrete thyroid-stimulating hormone (TSH); and gonadotrophs that synthesize follicle-stimulating hormone (FSH) and luteinizing hormone (LH). (See Tables 18.3 and 18.4 on pages 596 and 600.)
7. Human growth hormone (hGH) stimulates body growth through insulinlike growth factors (IGFs). Secretion of hGH is inhibited by GHIH (growth hormone-inhibiting hormone, or somatostatin) and promoted by GHRH (growth hormone-releasing hormone).
8. TSH regulates thyroid gland activities. Its secretion is stimulated by TRH (thyrotropin-releasing hormone) and suppressed by GHIH.
9. FSH and LH regulate the activities of the gonads—ovaries and testes. Their secretion is controlled by GnRH (gonadotropin-releasing hormone).
10. Prolactin (PRL) helps initiate milk secretion. Prolactin-inhibiting hormone (PIH) suppresses secretion of PRL, whereas prolactin-releasing hormone (PRH) and TRH stimulate PRL secretion.
11. ACTH regulates the activities of the adrenal cortex and is controlled by CRH (corticotropin-releasing hormone).
12. Dopamine inhibits secretion of MSH.
13. The posterior pituitary contains axon terminals of neurosecretory cells whose cell bodies are in the hypothalamus.
14. Hormones made by the hypothalamus and stored in the posterior pituitary are oxytocin (OT), which stimulates contraction of the uterus and ejection of milk from the breasts, and antidiuretic hormone (ADH), which stimulates water reabsorption by the kidneys and constriction of arterioles. ADH is also known as vasopressin. (See Table 18.5 on page 602.)
15. Oxytocin secretion is stimulated by uterine stretching and suckling during nursing; ADH secretion is controlled by osmotic pressure of the blood and blood volume.

THYROID GLAND (p. 603)

1. The thyroid gland is located inferior to the larynx.
2. It consists of thyroid follicles composed of follicular cells, which secrete the thyroid hormones thyroxine (T_4) and triiodothyronine (T_3), and parafollicular cells, which secrete calcitonin (CT).
3. Thyroid hormones are synthesized from iodine and tyrosine within thyroglobulin (TGB). They are transported in the blood bound to plasma proteins, mostly thyroxine-binding globulin (TBG).
4. Secretion is controlled by TRH from the hypothalamus and thyroid-stimulating hormone (TSH) from the anterior pituitary.
5. Thyroid hormones regulate oxygen use and metabolic rate, cellular metabolism, and growth and development.
6. Calcitonin (CT) can lower the blood level of calcium ions (Ca^{2+}) and promote deposition of Ca^{2+} into bone matrix. Secretion of CT is controlled by Ca^{2+} level in the blood. (See Table 18.6 on page 606.)

PARATHYROID GLANDS (p. 606)

1. The parathyroid glands are embedded in the posterior surfaces of the lateral lobes of the thyroid gland. They consist of chief cells and oxyphil cells.
2. Parathyroid hormone (PTH) regulates the homeostasis of calcium, magnesium, and phosphate ions by increasing blood calcium and magnesium levels and decreasing blood phosphate level. PTH secretion is controlled by the level of calcium in the blood. (See Table 18.7 on page 609.)

ADRENAL GLANDS (p. 608)

1. The adrenal glands are located superior to the kidneys. They consist of an outer adrenal cortex and inner adrenal medulla.
2. The adrenal cortex is divided into a zona glomerulosa, a zona fasciculata, and a zona reticularis; the adrenal medulla consists of chromaffin cells and large blood vessels.
3. Cortical secretions are mineralocorticoids, glucocorticoids, and androgens.
4. Mineralocorticoids (mainly aldosterone) increase sodium and water reabsorption and decrease potassium reabsorption. Secretion is controlled by the renin–angiotensin–aldosterone (RAA) pathway and by K^+ level in the blood.
5. Glucocorticoids (mainly cortisol) promote protein breakdown, gluconeogenesis, and lipolysis; help resist stress; and serve as anti-inflammatory substances. Their secretion is controlled by ACTH.
6. Androgens secreted by the adrenal cortex stimulate growth of axillary and pubic hair, aid the prepubertal growth spurt, and contribute to libido.
7. The adrenal medulla secretes epinephrine and norepinephrine (NE), which produce effects similar to sympathetic responses. They are released during stress. (See Table 18.8 on page 613.)

PANCREATIC ISLETS (p. 614)

1. The pancreas lies in the curve of the duodenum. It has both endocrine and exocrine functions.
2. The endocrine portion consists of pancreatic islets or islets of Langerhans, made up of four types of cells: alpha, beta, delta, and F cells.
3. Alpha cells secrete glucagon, beta cells secrete insulin, delta cells secrete somatostatin, and F cells secrete pancreatic polypeptide.
4. Glucagon increases blood glucose level, whereas insulin decreases blood glucose level. Secretion of both hormones is controlled by the level of glucose in the blood. (See Table 18.9 on page 617.)

OVARIES AND TESTES (p. 616)

1. The ovaries are located in the pelvic cavity and produce estrogens, progesterone, and inhibin. These sex hormones govern the development and maintenance of feminine secondary sex characteristics, reproductive cycles, pregnancy, lactation, and normal female reproductive functions. (See Table 18.10 on page 616.)
2. The testes lie inside the scrotum and produce testosterone and inhibin. These sex hormones govern the development and maintenance of masculine secondary sex characteristics and normal male reproductive functions. (See Table 18.10 on page 616.)

PINEAL GLAND (p. 618)

1. The pineal gland is attached to the roof of the third ventricle of the brain. It consists of secretory cells called pinealocytes, neuroglia, and endings of sympathetic postganglionic axons.
2. The pineal gland secretes melatonin, which contributes to setting the body's biological clock (controlled in the suprachiasmatic nucleus). During sleep, plasma levels of melatonin increase.

THYMUS (p. 618)

1. The thymus secretes several hormones related to immunity.
2. Thymosin, thymic humoral factor (THF), thymic factor (TF), and thymopoietin promote the maturation of T cells.

OTHER HORMONES AND GROWTH FACTORS (p. 618)

1. Body tissues other than those normally classified as endocrine glands contain endocrine tissue and secrete hormones.
2. These include the gastrointestinal tract, placenta, kidneys, skin, and heart. (See Table 18.11 on page 619.)
3. Prostaglandins and leukotrienes are eicosanoids that act as local hormones in most body tissues.
4. Growth factors are local hormones that stimulate cell growth and division. (See Table 18.12 on page 619.)

THE STRESS RESPONSE (p. 620)

1. Productive stress is termed eustress, and harmful stress is termed distress.
2. If stress is extreme, it triggers the stress response (general adaptation syndrome), which occurs in three stages: the fight-or-flight response, resistance reaction, and exhaustion.
3. The stimuli that produce the stress response are called stressors. Stressors include surgical operations, poisons, infections, fever, and strong emotional responses.
4. The fight-or-flight response is initiated by nerve impulses from the hypothalamus to the sympathetic division of the autonomic nervous system and the adrenal medulla. This response rapidly increases circulation, promotes ATP production, and decreases nonessential activities.

5. The resistance reaction is initiated by releasing hormones secreted by the hypothalamus, most importantly CRH, TRH, and GHRH. Resistance reactions are longer lasting and accelerate breakdown reactions to provide ATP for counteracting stress.

6. Exhaustion results from depletion of body resources during the resistance stage.

7. Stress may trigger certain diseases by inhibiting the immune system. An important link between stress and immunity is interleukin-l, produced by macrophages; it stimulates secretion of ACTH.

DEVELOPMENT OF THE ENDOCRINE SYSTEM (p. 622)

1. The development of the endocrine system is not as localized as in other systems because endocrine organs develop in widely separated parts of the embryo.

2. The pituitary gland, adrenal medulla, and pineal gland develop from ectoderm; the adrenal cortex develops from mesoderm; and the thyroid gland, parathyroid glands, pancreas, and thymus develop from endoderm.

AGING AND THE ENDOCRINE SYSTEM (p. 622)

1. Although some endocrine glands shrink as we get older, their performance may or may not be compromised.

2. Production of human growth hormone, thyroid hormones, cortisol, aldosterone, and estrogens decrease with advancing age.

3. With aging, the blood levels of TSH, LH, FSH, and PTH rise.

4. The pancreas releases insulin more slowly with age, and receptor sensitivity to glucose declines.

5. After puberty, thymus size begins to decrease, and thymic tissue is replaced by adipose and areolar connective tissue.

 SELF-QUIZ QUESTIONS

Fill in the blanks in the following statements.

1. The three stages of the stress response or general adaptation syndrome, in order of occurrence, are _____, _____, and _____.

2. The _____ is the major integrating link between the nervous and endocrine systems and acts as an endocrine gland itself.

3. _____ glands secrete their products into ducts that carry the product to the destination site. _____ glands secrete their products into the interstitial fluid, where they diffuse into capillaries.

Indicate whether the following statements are true or false.

4. If the effect of two or more hormones acting together is greater than the sum of each acting alone, then the two hormones are said to have a permissive effect.

5. In the direct gene activation method of hormone action, the hormone enters the target cell and binds to an intracellular receptor. The activated receptor-hormone complex then alters gene expression to produce the protein that causes the physiological responses characteristic of the hormone.

Choose the one best answer to the following questions.

6. Which of the following comparisons are *true*? (1) Nerve impulses produce their effects quickly, whereas hormonal responses generally are slower. (2) Nervous system effects are brief, whereas endocrine system effects are longer lasting. (3) The nervous system controls homeostasis through nerve impulses; the endocrine system works through hormones. (4) The nervous system can stimulate or inhibit the release of hormones by the endocrine system. (5) The nervous system includes organs scattered throughout the body; the endocrine system consists of only the pituitary gland, thyroid gland, adrenal gland, and pancreas. (a) 1, 2, 3, 4, and 5, (b) 1, 2, 3, and 4, (c) 2, 3, 4, and 5, (d) 2, 4, and 5, (e) 1, 4, and 5.

7. Which of the following is *not* a class of hormones? (a) steroids, (b) amines, (c) alcohols, (d) peptides and proteins, (e) eicosanoids.

8. Which of the following factors affect the responsiveness of a target cell to a hormone? (1) hormone concentration, (2) target cell type, (3) abundance of hormone receptors, (4) influences of other hormones, (5) age of the target cell. (a) 1, 2, 3, and 4, (b) 2, 3, 4, and 5, (c) 1, 2, and 3, (d) 1, 3, and 4, (e) 2, 4, and 5.

9. The class of adrenal gland hormones that provide resistance to stress, produce anti-inflammatory effects, and promote normal metabolism to ensure adequate quantities of ATP is (a) glucocorticoids, (b) mineralocorticoids, (c) androgens, (d) catecholamines, (e) sympathomimetics.

10. Hormones (1) generally utilize negative feedback mechanisms to regulate their secretion, (2) will only affect target cells far removed from the hormone-producing secretory cells, (3) must bind to transport proteins in order to circulate in the blood, (4) will only cause an effect on cells with receptors for the hormone, (5) can regulate the responsiveness of the target tissue by controlling the number of receptors for the hormone. (a) 1, 2, and 3, (b) 1, 2, 4, and 5, (c) 2, 3, and 4, (d) 2, 3, 4, and 5, (e) 1, 4, and 5.

11. Place in correct order the action of a water-soluble hormone on its target cell. (1) Adenylate cyclase activated, catalyzing the conversion of ATP to cAMP. (2) Enzymes catalyze reactions that produce physiological response attributed to the hormone. (3) Hormone binds to plasma membrane receptor. (4) Activated protein kinases phosphorylate cellular proteins. (5) Hormone-receptor complex activates G proteins. (6) Cyclic AMP activates protein kinases. (a) 3, 5, 1, 6, 4, 2, (b) 3, 1, 5, 6, 4, 2, (c) 5, 1, 4, 2, 3, 6, (d) 3, 4, 5, 1, 6, 2, (e) 6, 3, 5, 1, 4, 2.

12. The stress response is controlled primarily by the (a) kidneys, (b) adrenal glands, (c) hypothalamus, (d) thyroid gland, (e) heart.

13. Match the following:

_____ (a) increases blood Ca²⁺ level

_____ (b) increases blood glucose level

_____ (c) decreases blood Ca²⁺ level

_____ (d) decreases blood glucose level

_____ (e) initiates and maintains milk secretion by the mammary glands

_____ (f) regulates body's biological clock

_____ (g) stimulates sex hormone production; triggers ovulation

_____ (h) augment fight-or-flight response

_____ (i) regulates metabolism and resistance to stress

_____ (j) helps control water and electrolyte homeostasis

_____ (k) suppresses release of FSH

_____ (l) stimulate growth of axillary and pubic hair

_____ (m) promotes T cell maturation

_____ (n) regulates oxygen use, basal metabolic rate, cellular metabolism, and growth and development

_____ (o) stimulates protein synthesis, inhibits protein breakdown, stimulates lipolysis, and retards use of glucose for ATP production

_____ (p) inhibits water loss through the kidneys

_____ (q) stimulates gamete (egg and sperm) formation

_____ (r) stimulates milk ejection

_____ (s) stimulate and inhibit secretion of anterior pituitary hormones

_____ (t) increases skin pigmentation when present in excess

(1) insulin
(2) glucagon
(3) inhibin
(4) FSH
(5) LH
(6) thyroxine and triiodothyronine
(7) calcitonin
(8) parathormone
(9) melanocyte-stimulating hormone
(10) oxytocin
(11) ADH
(12) prolactin
(13) hGH
(14) hypothalamic regulating hormones
(15) aldosterone
(16) androgens
(17) epinephrine and norepinephrine
(18) melatonin
(19) thymosin
(20) cortisol

14. Match the following hormone-secreting cells to the hormone(s) they release:

_____ (a) ACTH

_____ (b) TSH

_____ (c) glucagon

_____ (d) calcitonin

_____ (e) insulin

_____ (f) progesterone

_____ (g) FSH and LH

_____ (h) hGH

_____ (i) testosterone

_____ (j) thyroxine and triiodothyronine

(1) beta cells of pancreatic islets
(2) alpha cells of pancreatic islets
(3) follicular cells of thyroid gland
(4) parafollicular cells of thyroid gland
(5) testes
(6) ovary
(7) somatotrophs
(8) thyrotrophs
(9) gonadotrophs
(10) corticotrophs

15. Match the endocrine disorder to the problem that produced the disorder:

_____ (a) hyposecretion of insulin or down-regulation of insulin receptors

_____ (b) hypersecretion of hGH before closure of epiphyseal plates

_____ (c) hyposecretion of thyroid hormone that is present at birth

_____ (d) hypersecretion of glucocorticoids

_____ (e) hyposecretion of hGH before closure of epiphyseal plates

_____ (f) hypersecretion of epinephrine and norepinephrine

_____ (g) hypersecretion of hGH after closure of epiphyseal plates

_____ (h) hyposecretion of glucocorticoids and aldosterone

_____ (i) hyposecretion of ADH

_____ (j) hypersecretion of melatonin

_____ (k) hyposecretion of thyroid hormone in adults

_____ (l) hyperthyroidism, an autoimmune disease

(1) giantism
(2) acromegaly
(3) pituitary dwarfism
(4) diabetes insipidus
(5) myxedema
(6) Graves disease
(7) Cushing's syndrome
(8) seasonal affective disorder
(9) Addison's disease
(10) pheochromocytomas
(11) congenital hypothyroidism
(12) diabetes mellitus

CRITICAL THINKING QUESTIONS

1. A possible treatment being investigated for type 1 diabetes is the transplantation of only one type of cell into the patient, instead of the entire source organ. What cell type could this be, and from what organ? How could just one cell type provide a cure for diabetes?

HINT *The organ has both exocrine and endocrine functions.*

2. Amanda hates her new student ID photo. Her hair looks dry, the extra weight that she's gained shows, and her neck looks fat. In fact, there's an odd butterfly shaped swelling across the front of

her neck, under her chin. Amanda's also been feeling very tired and mentally "dull" lately, but she figures all new A&P students feel that way. Should she visit the clinic or just wear turtlenecks?

HINT *She also feels constantly cold.*

3. Jamal is a 15-year-old computer whiz with an interest in medicine. He likes to practice virtual surgery using the computer program he wrote himself. Today he's transplanting an adrenal gland. Describe the location and structure of the adrenal glands.

HINT *It's really two glands in one.*

ANSWERS TO FIGURE QUESTIONS

18.1 Secretions of endocrine glands diffuse into interstitial fluid and then into the blood; exocrine secretions flow into ducts that lead into body cavities or to the body surface.

18.2 In the stomach, histamine is a paracrine because it acts on nearby parietal cells without entering the blood.

18.3 Synthesis of new proteins takes several minutes to half an hour, due to the time needed for gene expression. The effects of steroid and thyroid hormones thus appear slowly.

18.4 Cyclic AMP is termed a second messenger because it translates the presence of the first messenger, the water-soluble hormone, into a response inside the cell.

18.5 The hypophyseal portal veins carry blood from the median eminence of the hypothalamus, where hypothalamic releasing and inhibiting hormones are secreted, to the anterior pituitary, where these hormones act.

18.6 Thyroid hormones suppress secretion of TSH by thyrotrophs and of TRH by hypothalamic neurosecretory cells; gonadal hormones suppress secretion of FSH and LH by gonadotrophs and of GnRH by hypothalamic neurosecretory cells.

18.7 Excess hGH causes hyperglycemia.

18.8 Functionally, both the hypothalamohypophyseal tract and the hypophyseal portal veins carry hypothalamic hormones to the pituitary gland. Structurally, the tract is composed of axons of neurons that extend from the hypothalamus to the posterior pituitary, whereas the portal veins are blood vessels that extend from the hypothalamus to the anterior pituitary.

18.9 Absorption of a liter of water in the intestines would decrease the osmotic pressure of your blood plasma, turning off secretion of ADH and decreasing the ADH level in your blood.

18.10 Follicular cells secrete T_3 and T_4, also known as thyroid hormones. Parafollicular cells secrete calcitonin.

18.11 The storage form of thyroid hormones is thyroglobulin.

18.12 Lack of iodine in the diet → diminished production of T_3 and T_4 → increased release of TSH → growth (enlargement) of the thyroid gland → goiter.

18.13 Parafollicular cells of the thyroid gland secrete calcitonin; chief cells of the parathyroid gland secrete PTH.

18.14 Target tissues for PTH are bone and kidneys; target tissue for CT is bone; target tissue for calcitriol is the GI tract.

18.15 The adrenal glands are superior to the kidneys in the retroperitoneal space.

18.16 Angiotensin II acts to constrict blood vessels by causing contraction of vascular smooth muscle, and it stimulates secretion of aldosterone (by zona glomerulosa cells of the adrenal cortex), which in turn causes the kidneys to conserve water and thereby increase blood volume.

18.17 A transplant recipient who takes prednisone will have low blood levels of ACTH and CRH due to negative feedback suppression of the anterior pituitary and hypothalamus by the prednisone.

18.18 The pancreas is both an endocrine and an exocrine gland.

18.19 Glucagon is considered an anti-insulin hormone because it has several effects that are opposite to those of insulin.

18.20 Homeostasis maintains controlled conditions typical of a normal internal environment; the stress response resets controlled conditions at a different level to cope with various stressors.

18.21 The pituitary gland and the adrenal glands both include tissues having two different embryological origins.

18.22 Antibodies that mimic the action of TSH are produced in Graves disease.

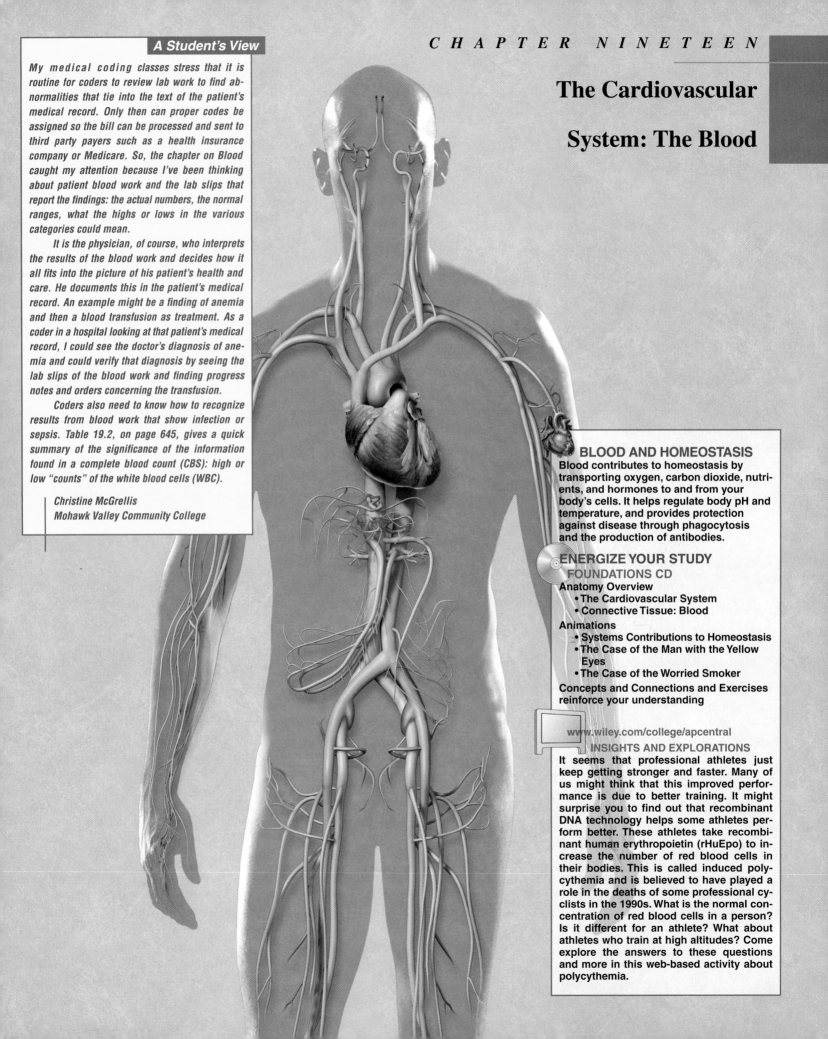

The Cardiovascular System: The Blood

BLOOD AND HOMEOSTASIS

Blood contributes to homeostasis by transporting oxygen, carbon dioxide, nutrients, and hormones to and from your body's cells. It helps regulate body pH and temperature, and provides protection against disease through phagocytosis and the production of antibodies.

ENERGIZE YOUR STUDY

FOUNDATIONS CD

Anatomy Overview
- The Cardiovascular System
- Connective Tissue: Blood

Animations
- Systems Contributions to Homeostasis
- The Case of the Man with the Yellow Eyes
- The Case of the Worried Smoker

Concepts and Connections and Exercises reinforce your understanding

www.wiley.com/college/apcentral

INSIGHTS AND EXPLORATIONS

It seems that professional athletes just keep getting stronger and faster. Many of us might think that this improved performance is due to better training. It might surprise you to find out that recombinant DNA technology helps some athletes perform better. These athletes take recombinant human erythropoietin (rHuEpo) to increase the number of red blood cells in their bodies. This is called induced polycythemia and is believed to have played a role in the deaths of some professional cyclists in the 1990s. What is the normal concentration of red blood cells in a person? Is it different for an athlete? What about athletes who train at high altitudes? Come explore the answers to these questions and more in this web-based activity about polycythemia.

The **cardiovascular system** (*cardio-* = heart; *vascular* = blood vessels) consists of three interrelated components: blood, the heart, and blood vessels. The focus of this chapter is blood; the next two chapters will examine the heart and blood vessels, respectively. Blood transports various substances, helps regulate several life processes, and affords protection against disease. For all of its similarities in origin, composition, and functions, blood is as unique from one person to another as are skin, bone, and hair. Health-care professionals routinely examine and analyze its differences through various blood tests when trying to determine the cause of different diseases. The branch of science concerned with the study of blood, blood-forming tissues, and the disorders associated with them is **hematology** (hēm-a-TOL-ō -jē ; *hema-* or *hemato-* = blood; *-logy* = study of).

FUNCTIONS AND PROPERTIES OF BLOOD

▶ O B J E C T I V E S

- **Describe the functions of blood.**
- **Describe the physical characteristics and principal components of blood.**

Most cells of a multicellular organism cannot move around to obtain oxygen and nutrients or eliminate carbon dioxide and other wastes. Instead, these needs are met by two fluids: blood and interstitial fluid. **Blood** is a connective tissue composed of a liquid matrix called **plasma** that dissolves and suspends various cells and cell fragments. **Interstitial fluid** is the fluid that bathes body cells (see Figure 27.1 on page 992). Blood transports oxygen from the lungs and nutrients from the gastrointestinal tract to body cells. The oxygen and nutrients diffuse from the blood into the interstitial fluid and then into body cells. Carbon dioxide and other wastes move in the reverse direction, from body cells into the interstitial fluid and then into the blood.

Functions of Blood

Blood, the only liquid connective tissue, has three general functions:

1. *Transportation.* Blood transports oxygen from the lungs to the cells of the body and carbon dioxide from the body cells to the lungs for exhalation. It carries nutrients from the gastrointestinal tract to body cells and hormones from endocrine glands to other body cells. Blood also transports heat and waste products to the lungs, kidneys, and skin for elimination from the body.

2. *Regulation.* Circulating blood helps maintain homeostasis in all body fluids. Blood helps regulate pH through buffers. It also helps adjust body temperature through the heat-absorbing and coolant properties of the water (see page 39) in plasma and its variable rate of flow through the skin, where excess heat can be lost from the blood to the environment. Blood osmotic pressure also influences the water content of cells, mainly through interactions of dissolved ions and proteins.

3. *Protection.* Blood can clot, which protects against its excessive loss from the cardiovascular system after an injury. In addition, white blood cells protect against disease by carrying on phagocytosis. Several types of blood proteins, antibodies, such as interferons, and complement, help protect against disease in a variety of ways.

Physical Characteristics of Blood

Blood is denser and more viscous than water and feels slightly sticky. The temperature of blood is 38°C (100.4°F), about 1°C higher than oral or rectal body temperature, and it has a slightly alkaline pH ranging from 7.35 to 7.45. Blood constitutes about 20% of extracellular fluid, amounting to 8% of the total body mass. The blood volume is 5 to 6 liters (1.5 gal) in an average-sized adult male and 4 to 5 liters (1.2 gal) in an average-sized adult female. Several hormonal negative feedback systems ensure that blood volume and osmotic pressure remain relatively constant. Especially important systems are those involving aldosterone, antidiuretic hormone, and atrial natriuretic peptide, which regulate how much water is excreted in the urine (see pages 970–972).

Withdrawing Blood

Blood samples for laboratory testing may be obtained in several ways. The most common procedure is **venipuncture,** withdrawal of blood from a vein using a hypodermic needle and syringe. A tourniquet is wrapped around the arm above the venipuncture site, which causes blood to accumulate in the vein. This increased blood volume makes the vein stand out. Clenching the fist further causes it to stand out, making the venipuncture more successful. A common site for venipuncture is the median cubital vein anterior to the elbow (see Figure 21.26b on page 743). Another method of withdrawing blood is through a **finger** or **heel stick.** Diabetic patients who monitor their daily blood sugar typically perform a finger stick, and it is often used for drawing blood from infants and children. In an **arterial stick,** blood is withdrawn from an artery, for example, when it is important to know the level of oxygen in oxygenated blood. ■

Components of Blood

Whole blood has two components: (1) blood plasma, a watery liquid matrix that contains dissolved substances, and (2) formed elements, which are cells and cell fragments. If a sample of blood is centrifuged (spun) in a small glass tube, the cells sink to the bottom of the tube while the lighter-weight plasma forms a layer on top (Figure 19.1a). Blood is about 45% formed

Figure 19.1 **Components of blood in a normal adult.**

Blood is a connective tissue that consists of blood plasma (liquid) plus formed elements (red blood cells, white blood cells, and platelets).

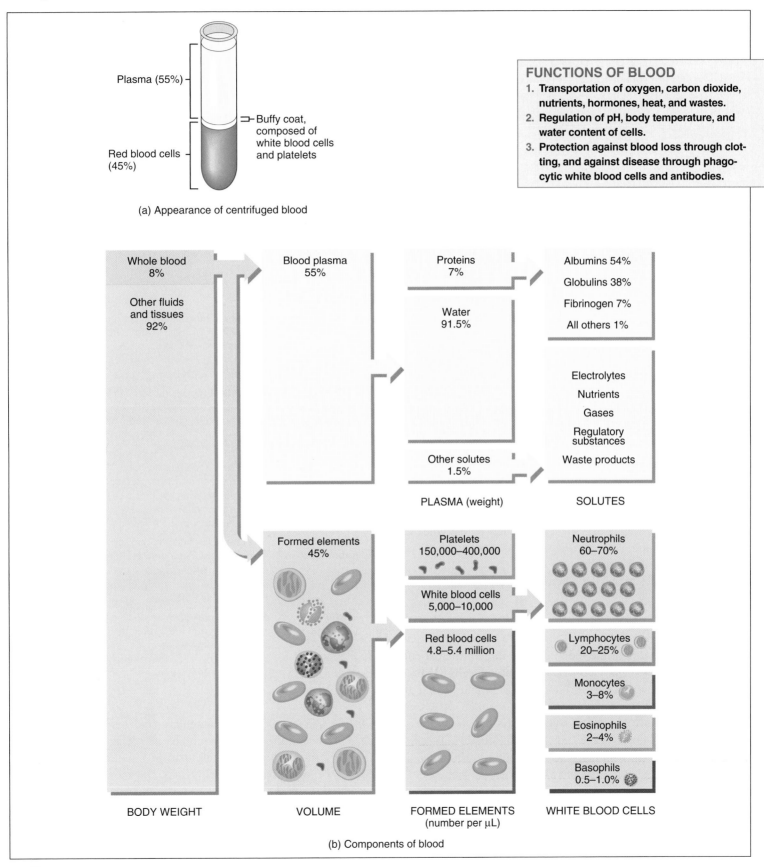

FUNCTIONS OF BLOOD
1. **Transportation of oxygen, carbon dioxide, nutrients, hormones, heat, and wastes.**
2. **Regulation of pH, body temperature, and water content of cells.**
3. **Protection against blood loss through clotting, and against disease through phagocytic white blood cells and antibodies.**

Plasma (55%)

Buffy coat, composed of white blood cells and platelets

Red blood cells (45%)

(a) Appearance of centrifuged blood

Whole blood 8%

Other fluids and tissues 92%

Blood plasma 55%

Proteins 7%

Water 91.5%

Other solutes 1.5%

Albumins 54%
Globulins 38%
Fibrinogen 7%
All others 1%

Electrolytes
Nutrients
Gases
Regulatory substances
Waste products

PLASMA (weight)   SOLUTES

Formed elements 45%

Platelets 150,000–400,000

White blood cells 5,000–10,000

Red blood cells 4.8–5.4 million

Neutrophils 60–70%

Lymphocytes 20–25%

Monocytes 3–8%

Eosinophils 2–4%

Basophils 0.5–1.0%

BODY WEIGHT   VOLUME   FORMED ELEMENTS (number per μL)   WHITE BLOOD CELLS

(b) Components of blood

What is the approximate volume of blood in your body? (*Hint*: Each liter of blood has a mass of one kilogram.)

elements and 55% plasma. Normally, more than 99% of the formed elements are red-colored red blood cells (RBCs). Pale, colorless white blood cells (WBCs) and platelets occupy less than 1% of total blood volume. They form a very thin **buffy coat** layer between the packed RBCs and plasma in centrifuged blood. Figure 19.1b shows the composition of blood plasma and the numbers of the various types of formed elements in blood.

Blood Plasma

When the formed elements are removed from blood, a straw-colored liquid called **blood plasma** (or simply **plasma**) is left. Plasma is about 91.5% water and 8.5% solutes, most of which (7% by weight) are proteins. Some of the proteins in plasma are also found elsewhere in the body, but those confined to blood are called **plasma proteins.** Among other functions, these proteins play a role in maintaining proper blood osmotic pressure, which is an important factor in the exchange of fluids across capillary walls (discussed in Chapter 21).

Hepatocytes (liver cells) synthesize most of the plasma proteins, which include the **albumins** (54% of plasma proteins), **globulins** (38%), and **fibrinogen** (7%). Certain blood cells develop into cells that produce gamma globulins, an important type of globulin. These plasma proteins are also called **antibodies** or **immunoglobulins** because they are produced during certain immune responses. Foreign invaders such as bacteria and viruses stimulate production of millions of different antibodies. An antibody binds specifically to the foreign substance, called an **antigen,** that provoked its production. When an antigen and antibody come together, they form an **antigen–antibody complex,** which disables the invading antigen (see Figure 22.18 on page 791).

Besides proteins, other solutes in plasma include electrolytes, nutrients, regulatory substances such as enzymes and hormones, gases, and waste products such as urea, uric acid, creatinine, ammonia, and bilirubin.

Table 19.1 describes the chemical composition of blood plasma.

Formed Elements

The **formed elements** of the blood include three principal components: **red blood cells (RBCs), white blood cells (WBCs),** and **platelets** (Figure 19.2). Although RBCs and WBCs are whole cells, platelets are cell fragments. Unlike RBCs and platelets, which have just a few roles, WBCs have a number of specialized functions. There are several distinct types of WBCs—neutrophils, lymphocytes, monocytes, eosinophils, and basophils—each having a unique microscopic appearance. The roles of each type of WBC are discussed later in this chapter.

The percentage of total blood volume occupied by RBCs is called the **hematocrit** (he-MAT-ō-krit). For example, a hematocrit of 40 means that 40% of the volume of blood is composed of RBCs. The normal range of hematocrit for adult females is

Table 19.1	Substances in Blood Plasma
Constituent	**Description**
Water (91.5%)	Liquid portion of blood. Acts as solvent and suspending medium for components of blood; absorbs, transports, and releases heat.
Proteins (7.0)	Exert colloid osmotic pressure, which helps maintain water balance between blood and tissues and regulates blood volume.
Albumins	Smallest and most numerous plasma proteins; produced by liver. Function as transport proteins for several steroid hormones and for fatty acids.
Globulins	Produced by liver and by plasma cells, which develop from B lymphocytes. Antibodies (immunoglobulins) help attack viruses and bacteria. Alpha and beta globulins transport iron, lipids, and fat-soluble vitamins.
Fibrinogen	Produced by liver. Plays essential role in blood clotting.
Other Solutes (1.5%)	
Electrolytes	Inorganic salts. Positively charged ions (cations) include Na^+, K^+, Ca^{2+}, Mg^{2+}; negatively charged ions (anions) include Cl^-, HPO_4^{2-}, SO_4^{2-}, and HCO_3^-. Help maintain osmotic pressure and play essential roles in the function of cells.
Nutrients	Products of digestion pass into blood for distribution to all body cells. Include amino acids (from proteins), glucose (from carbohydrates), fatty acids and glycerol (from triglycerides), vitamins, and minerals.
Gases	Oxygen (O_2), carbon dioxide (CO_2), and nitrogen (N_2). Whereas more O_2 is associated with hemoglobin inside red blood cells, more CO_2 is dissolved in plasma. N_2 is present but has no known function in the body.
Regulatory substances	Enzymes, produced by body cells, catalyze chemical reactions. Hormones, produced by endocrine glands, regulate metabolism, growth, and development in body. Vitamins are cofactors for enzymatic reactions.
Waste products	Most are breakdown products of protein metabolism and are carried by blood to organs of excretion. Include urea, uric acid, creatine, creatinine, bilirubin, and ammonia.

38–46% (average = 42); for adult males, it is 40–54% (average = 47). The hormone testosterone, which is present in much higher concentration in males than in females, stimulates synthesis of erythropoietin (EPO), the hormone that stimulates production of RBCs. Thus, testosterone contributes to higher hematocrits in males. Lower values in women during their reproductive years also may be due to excessive loss of blood during menstruation. A significant drop in hematocrit indicates *anemia*, a lower-than-normal number of RBCs. In *polycythemia* the per-

Figure 19.2 Scanning electron micrograph of the formed elements of blood.

The formed elements of blood are red blood cells (RBCs), white blood cells (WBCs), and platelets.

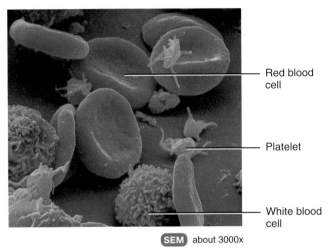

Red blood cell

Platelet

White blood cell

SEM about 3000x

Which formed elements of the blood are cell fragments?

centage of RBCs is abnormally high, and the hematocrit may be 65% or higher. Causes of polycythemia include abnormal increases in RBC production, tissue hypoxia, dehydration, and blood doping or use of EPO by athletes.

Induced Polycythemia in Athletes

Delivery of oxygen to muscles is a limiting factor in muscular feats. As a result, increasing the oxygen-carrying capacity of the blood enhances athletic performance, especially in endurance events. Because red blood cells are the main transport vehicle for oxygen, athletes have tried several means of increasing their hematocrit, causing **induced polycythemia,** to gain a competitive edge. Training at higher altitudes produces a natural increase in hematocrit because the lower amount of oxygen in inhaled air triggers release of more EPO. In past years, some athletes performed *blood doping* to elevate their hematocrit. In this procedure, blood cells are removed from the body and stored for a month or so, during which time the hematocrit returns to normal. Then, a few days before an athletic event, the RBCs are reinjected to increase hematocrit. More recently, athletes have enhanced their RBC production by injecting *Epoetin alfa (Procrit* or *Epogen),* a drug that is used to treat anemia. Epoetin alfa is identical to naturally occurring EPO because it is manufactured by recombinant DNA technology using the human gene that codes for erythropoietin. Practices that increase hematocrit are dangerous, however, because they increase the workload of the heart. The increased numbers of RBCs raise the viscosity of the blood, which increases the resistance to blood flow and makes the blood more difficult for the heart to pump. Increased viscosity also contributes to high blood pressure and increased risk of stroke. During the 1980s, at least 15 competitive cyclists died from heart attacks or strokes linked to suspected use of Epoetin alfa. Although the International Olympics Committee bans Epoetin alfa use, enforcement is difficult because the drug is identical to naturally occurring EPO. ■

▶ C H E C K P O I N T

1. In what ways is blood plasma similar to and different from interstitial fluid?
2. How many kilograms or pounds of blood is present in your body?
3. How does the volume of blood plasma in your body compare to the volume of fluid in a two-liter bottle of Coke?
4. What are the major solutes in blood plasma? What does each do?
5. What is the significance of lower-than-normal or higher-than-normal hematocrit?

FORMATION OF BLOOD CELLS

▶ O B J E C T I V E

- **Explain the origin of blood cells.**

Although some lymphocytes have a lifetime measured in years, most formed elements of the blood are continually dying and being replaced within hours, days, or weeks. Negative feedback systems regulate the total number of RBCs and platelets in circulation, and their numbers normally remain steady. The abundance of the different types of WBCs, however, varies in response to challenges by invading pathogens and other foreign antigens.

The process by which the formed elements of blood develop is called **hemopoiesis** (hē-mō-poy-Ē-sis; *-poiesis* = making) or *hematopoiesis*. Before birth, hemopoiesis first occurs in the yolk sac of an embryo and later in the liver, spleen, thymus, and lymph nodes of a fetus. In the last three months before birth, red bone marrow becomes the primary site of hemopoiesis and continues as the main source of blood cells after birth and throughout life.

Red bone marrow is a highly vascularized connective tissue located in the microscopic spaces between trabeculae of spongy bone tissue. It is present chiefly in bones of the axial skeleton, pectoral and pelvic girdles, and the proximal epiphyses of the humerus and femur (see Figure 6.1a on page 165). About 0.05–0.1% of red bone marrow cells are derived from mesenchymal cells called **pluripotent stem cells** (ploo-RIP-ō-tent; *pluri-* = several). Pluripotent stem cells are cells that have the capacity to develop into several different types of cells (Figure 19.3).

Stem cells in red bone marrow reproduce themselves, proliferate, and differentiate into cells that give rise to blood cells,

Figure 19.3   Origin, development, and structure of blood cells. Some of the generations of some cell lines have been omitted.

Blood cell production is called hemopoiesis and occurs only in red bone marrow after birth.

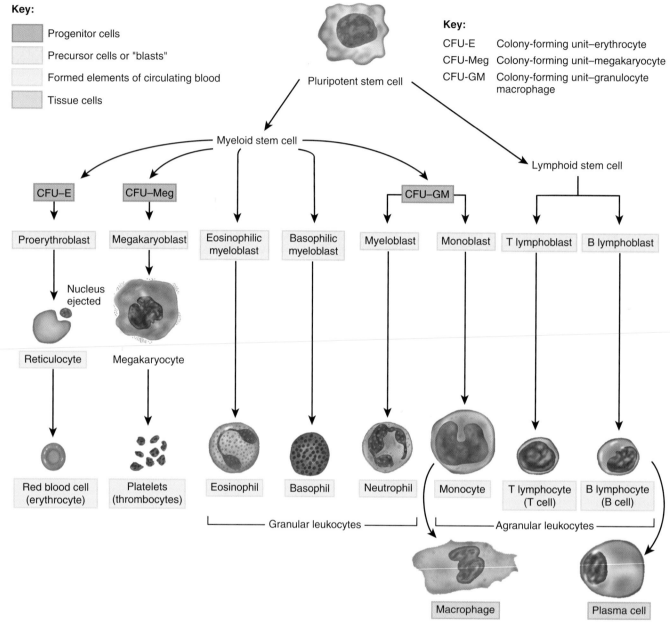

Key:

■ Progenitor cells

□ Precursor cells or "blasts"

□ Formed elements of circulating blood

□ Tissue cells

Key:

CFU-E   Colony-forming unit–erythrocyte

CFU-Meg   Colony-forming unit–megakaryocyte

CFU-GM   Colony-forming unit–granulocyte macrophage

Pluripotent stem cell

Myeloid stem cell   Lymphoid stem cell

CFU–E   CFU–Meg   CFU–GM

Proerythroblast   Megakaryoblast   Eosinophilic myeloblast   Basophilic myeloblast   Myeloblast   Monoblast   T lymphoblast   B lymphoblast

Nucleus ejected

Reticulocyte   Megakaryocyte

Red blood cell (erythrocyte)   Platelets (thrombocytes)   Eosinophil   Basophil   Neutrophil   Monocyte   T lymphocyte (T cell)   B lymphocyte (B cell)

———— Granular leukocytes ————   ———— Agranular leukocytes ————

Macrophage   Plasma cell

From which connective tissue cells do pluripotent stem cells develop?

macrophages, reticular cells, mast cells, and adipocytes. The reticular cells produce reticular fibers, which form the stroma (framework) that supports red bone marrow cells. Blood from nutrient and metaphyseal arteries (see Figure 6.5 on page 169) enters a bone and passes into the enlarged and leaky capillaries, called *sinuses,* that surround red bone marrow cells and fibers. After blood cells form, they enter the sinuses and other blood vessels and leave the bone through nutrient and periosteal veins

(see Figure 6.5 on page 169). With the exception of lymphocytes, formed elements do not divide once they leave red bone marrow.

Pluripotent stem cells produce two further types of stem cells, called *myeloid stem cells* and *lymphoid stem cells.* Myeloid stem cells begin their development in red bone marrow and give rise to red blood cells, platelets, monocytes, neutrophils, eosinophils, and basophils. Lymphoid stem cells begin their development in red bone marrow but complete it in lymphatic tissues; they give rise to lymphocytes. Although the various stem cells have distinctive cell identity markers in their plasma membranes, they cannot be distinguished histologically and resemble lymphocytes.

During hemopoiesis, the myeloid stem cells differentiate into **progenitor cells** (pro-JEN-i-tor). Progenitor cells are no longer capable of reproducing themselves and are committed to giving rise to more specific elements of blood. Some progenitor cells are known as *colony-forming units (CFUs)*. After the CFU designation is an abbreviation that designates the mature elements in blood that they will produce: CFU–E ultimately produces erythrocytes (red blood cells), CFU–Meg produces megakaryocytes, the source of platelets, and CFU–GM ultimately produces granulocytes (specifically, neutrophils) and monocytes. Progenitor cells, like stem cells, resemble lymphocytes and cannot be distinguished by their microscopic appearance alone. Other myeloid stem cells develop directly into cells called precursor cells (described next). Lymphoid stem cells differentiate into T lymphoblasts and B lymphoblasts, which ultimately develop into B lymphocytes (B cells) and T lymphocytes (T cells), respectively.

In the next generation, the cells are called **precursor cells,** also known as **blasts.** Over several cell divisions they develop into the actual formed elements of blood. For example, monoblasts develop into monocytes, eosinophilic myeloblasts develop into eosinophils, and so on. Precursor cells have recognizable microscopic appearances.

Several hormones called **hemopoietic growth factors** regulate the differentiation and proliferation of particular progenitor cells. **Erythropoietin** (e-rith′-rō-POY-e-tin) or **EPO** increases the number of red blood cell precursors. The main producers of EPO are cells in the kidneys that lie between the kidney tubules (peritubular interstitial cells). With renal failure, EPO release slows and RBC production is inadequate. **Thrombopoietin** (throm′-bō-POY-e-tin) or **TPO** is a hormone produced by the liver that stimulates formation of platelets (thrombocytes) from megakaryocytes. Several different cytokines regulate development of different blood cell types. **Cytokines** are small glycoproteins that are produced by red bone marrow cells, leukocytes, macrophages, fibroblasts, and endothelial cells. They typically act as local hormones (autocrines or paracrines). Cytokines stimulate proliferation of progenitor cells in red bone marrow and regulate the activities of cells involved in nonspecific defenses (such as phagocytes) and immune responses (such as B cells and T cells). Two important families of cytokines that stimulate white blood cell formation are **colony-stimulating factors (CSFs)** and **interleukins.**

Medical Uses of Hemopoietic Growth Factors

Hemopoietic growth factors made available through recombinant DNA technology hold tremendous potential for medical uses when a person's natural ability to form new blood cells is diminished or defective. Recombinant erythropoietin (Epoetin alfa) is very effective in treating the diminished red blood cell production that accompanies end-stage kidney disease. Granulocyte-macrophage colony-stimulating factor and granulocyte CSF are given to stimulate white blood cell formation in cancer patients who are undergoing chemotherapy, which kills some of their red bone marrow cells as well as the cancer cells because both cell types are undergoing mitosis. Thrombopoietin shows great promise for preventing platelet depletion during chemotherapy. CSFs and thrombopoietin also improve the outcome of patients who receive bone marrow transplants. ■

▶ CHECKPOINT

6. Which hemopoietic growth factors regulate differentiation and proliferation of CFU–E and formation of platelets from megakaryocytes?

RED BLOOD CELLS

▶ OBJECTIVE

• **Describe the structure, functions, life cycle, and production of red blood cells.**

Red blood cells (RBCs) or **erythrocytes** (e-RITH-rō-sīts; *erythro-* = red; *-cyte* = cell) contain the oxygen-carrying protein **hemoglobin,** which is a pigment that gives whole blood its red color. A healthy adult male has about 5.4 million red blood cells per microliter (μL) of blood,* and a healthy adult female has about 4.8 million. (One drop of blood is about 50 μL.) To maintain normal numbers of RBCs, new mature cells must enter the circulation at the astonishing rate of at least 2 million per second, a pace that balances the equally high rate of RBC destruction.

RBC Anatomy

RBCs are biconcave discs with a diameter of 7–8 μm (Figure 19.4a). Mature red blood cells have a simple structure. Their plasma membrane is both strong and flexible, which allows them to deform without rupturing as they squeeze through narrow capillaries. As you will see later, certain glycolipids in the plasma membrane of RBCs are antigens that account for the various blood groups such as the ABO and Rh groups. RBCs lack a nucleus and other organelles and can neither reproduce nor carry on extensive metabolic activities. The cytosol of RBCs contains hemoglobin molecules, which were synthesized before loss of the nucleus during RBC production and which constitute about 33% of the cell's weight.

---

*1 μL = 1 mm^3 = 10^{-6} liter.

Figure 19.4 The shapes of a red blood cell (RBC) and a hemoglobin molecule, and the structure of a heme group.   In (b), note that each of the four polypeptide chains (blue) of a hemoglobin molecule has one heme group (gold), which contains an iron ion Fe^{2+} (red).

 The iron portion of a heme group binds oxygen for transport by hemoglobin.

(a) RBC shape   (b) Hemoglobin molecule   (c) Iron-containing heme

How many molecules of O_2 can one hemoglobin molecule transport?

RBC Physiology

Red blood cells are highly specialized for their oxygen transport function. Because mature RBCs have no nucleus, all their internal space is available for oxygen transport. Moreover, RBCs lack mitochondria and generate ATP anaerobically (without oxygen). Hence, they do not use up any of the oxygen they transport. Even the shape of an RBC facilitates its function. A biconcave disc has a much greater surface area for its volume than, say, a sphere or a cube. This shape provides a large surface area for the diffusion of gas molecules into and out of the RBC.

Each RBC contains about 280 million hemoglobin molecules and each hemoglobin can carry up to four oxygen molecules. A hemoglobin molecule consists of a protein called **globin,** composed of four polypeptide chains (two alpha and two beta chains), plus four nonprotein pigments called **hemes** (Figure 19.4b). One ringlike heme binds to each polypeptide chain. At the center of the heme ring is an iron ion (Fe^{2+}) that can combine reversibly with one oxygen molecule (Figure 19.4c). The oxygen picked up in the lungs is transported bound to the iron of the heme group. As blood flows through tissue capillaries, the iron-oxygen reaction reverses. Hemoglobin releases oxygen, which diffuses first into the interstitial fluid and then into cells.

Hemoglobin also transports about 13% of the total carbon dioxide, a waste product of metabolism. Blood flowing through tissue capillaries picks up carbon dioxide, some of which combines with amino acids in the globin part of hemoglobin. As blood flows through the lungs, the carbon dioxide is released from hemoglobin and then exhaled.

In addition to its key role in transporting oxygen and carbon dioxide, hemoglobin also plays a role in regulation of blood flow and blood pressure. The gaseous hormone **nitric oxide (NO),** produced by the endothelial cells that line blood vessels, binds to hemoglobin. Under some circumstances, hemoglobin releases NO. The released NO causes *vasodilation,* an increase in blood vessel diameter that occurs when the smooth muscle in the vessel wall relaxes. Vasodilation improves blood flow and enhances oxygen delivery to cells near the site of NO release.

RBC Life Cycle

Red blood cells live only about 120 days because of the wear and tear their plasma membranes undergo as they squeeze through blood capillaries. Without a nucleus and other organelles, RBCs cannot synthesize new components to replace damaged ones. The plasma membrane becomes more fragile with age, and the cells are more likely to burst, especially as they squeeze through narrow channels in the spleen. Ruptured red blood cells are removed from circulation and destroyed by fixed phagocytic macrophages in the spleen and liver, and the breakdown products are recycled, as follows (Figure 19.5):

1 Macrophages in the spleen, liver, or red bone marrow phagocytize ruptured and worn-out red blood cells.

2 The globin and heme portions of hemoglobin are split apart.

3 Globin is broken down into amino acids, which can be reused to synthesize other proteins.

Figure 19.5 Formation and destruction of red blood cells, and the recycling of hemoglobin components. RBCs circulate for about 120 days after leaving red bone marrow before they are phagocytized by macrophages.

The rate of RBC formation by red bone marrow equals the rate of RBC destruction by macrophages.

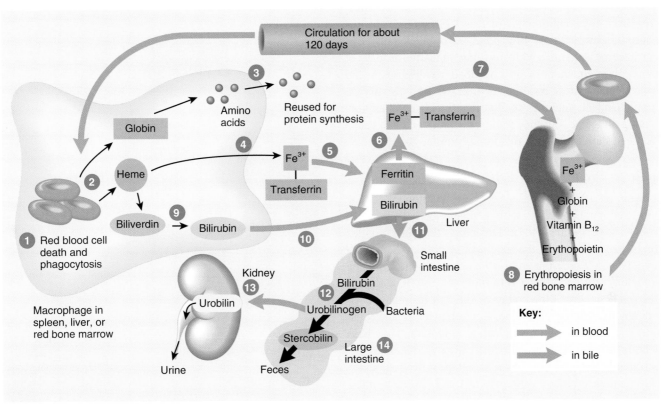

What is the function of transferrin?

4 Iron removed from the heme portion is in the form of Fe^{3+}, which associates with the plasma protein **transferrin** (trans-FER-in; *trans-* = across; *ferr-* = iron), a transporter for Fe^{3+} in the bloodstream.

5 In muscle fibers, liver cells, and macrophages of the spleen and liver, Fe^{3+} detaches from transferrin and attaches to an iron-storage protein called **ferritin.**

6 Upon release from a storage site or absorption from the gastrointestinal tract, Fe^{3+} reattaches to transferrin.

7 The Fe^{3+}-transferrin complex is then carried to bone marrow, where RBC precursor cells take it up through receptor-mediated endocytosis (see Figure 3.13 on page 72) for use in hemoglobin synthesis. Iron is needed for the heme portion of the hemoglobin molecule, whereas amino acids are needed for the globin portion. Vitamin B_{12} is also needed for synthesis of hemoglobin.

8 Erythropoiesis in red bone marrow results in the production of red blood cells, which enter the circulation.

9 When iron is removed from heme, the non-iron portion of heme is converted to **biliverdin** (bil'-i-VER-din), a green

pigment, and then into **bilirubin** (bil'-i-ROO-bin), a yellow-orange pigment.

10 Bilirubin enters the blood and is transported to the liver.

11 Within the liver, bilirubin is secreted by liver cells into bile, which passes into the small intestine and then into the large intestine.

12 In the large intestine, bacteria convert bilirubin into **urobilinogen** (ūr-ō-bī-LIN-ō-jen).

13 Some urobilinogen is absorbed back into the blood, converted to a yellow pigment called **urobilin** (ūr-ō-BĪ-lin), and excreted in urine.

14 Most urobilinogen is eliminated in feces in the form of a brown pigment called **stercobilin** (ster'-kō-BĪ-lin), which gives feces its characteristic color.

Iron Overload and Tissue Damage

Because free iron ions (Fe^{2+} and Fe^{3+}) bind to and damage molecules in cells or in the blood, transferrin and ferritin act as protective "protein escorts" during transport and storage of iron ions. As a result, plasma contains virtually no free iron. Moreover, only small amounts are available inside body cells for use in synthesis of iron-containing molecules such as the cytochrome pigments needed for ATP production in mitochondria (see Figure 25.7 on page 915). In cases of **iron overload,** the amount of iron present in the body builds up. Because we have no method for eliminating excess iron, any condition that increases dietary iron absorption can cause iron overload. At some point, the proteins transferrin and ferritin become saturated with iron ions, and free iron level rises. Common consequences of iron overload are diseases of the liver, heart, pancreatic islets, and gonads. Iron overload also allows certain iron-dependent microbes to flourish. Such microbes normally are not pathogenic, but they multiply rapidly and can cause lethal effects in a short time when free iron is present. ■

Erythropoiesis: Production of RBCs

Erythropoiesis (e-rith′-rō-poy-Ē-sis), the production of RBCs, starts in the red bone marrow with a precursor cell called a proerythroblast (see Figure 19.3). The proerythroblast divides several times, producing cells that begin to synthesize hemoglobin. Ultimately, a cell near the end of the development sequence ejects its nucleus and becomes a **reticulocyte** (re-TIK-ū-lō-sīt). Loss of the nucleus causes the center of the cell to indent, producing a distinctive biconcave shape. Reticulocytes retain some mitochondria, ribosomes, and endoplasmic reticulum. They pass from red bone marrow into the bloodstream by squeezing between the endothelial cells of blood capillaries. Reticulocytes develop into mature red blood cells, within 1 to 2 days after their release from red bone marrow.

Normally, erythropoiesis and red blood cell destruction proceed at roughly the same pace. If the oxygen-carrying capacity of the blood falls because erythropoiesis is not keeping up with RBC destruction, a negative feedback system steps up RBC production (Figure 19.6). The controlled condition is the amount of oxygen delivered to body tissues. Cellular oxygen deficiency, called **hypoxia** (hī-POKS-ē-a), may occur if too little oxygen enters the blood. For example, the lower oxygen content of air at high altitudes reduces the amount of oxygen in the blood. Oxygen delivery may also fall due to anemia, which has many causes: Lack of iron, lack of certain amino acids, and lack of vitamin B_{12} are but a few (see page 654). Circulatory problems that reduce blood flow to tissues may also reduce oxygen delivery. Whatever the cause, hypoxia stimulates the kidneys to step up the release of erythropoietin. This hormone circulates through the blood to the red bone marrow, where it speeds the development of proerythroblasts into reticulocytes. When the number of circulating RBCs increases, more oxygen can be delivered to body tissues.

Premature newborns often exhibit anemia, due in part to inadequate production of erythropoietin. During the first weeks after birth, the liver, not the kidneys, produces most EPO. Because the liver is less sensitive than the kidneys to hypoxia, newborns have a smaller EPO response to anemia than do adults.

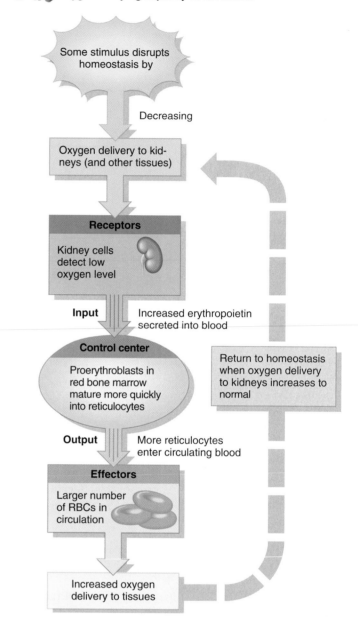

Figure 19.6 Negative feedback regulation of erythropoiesis (red blood cell formation). Lower oxygen content of air at high altitudes, anemia, and circulatory problems may reduce oxygen delivery to body tissues.

The main stimulus for erythropoiesis is hypoxia, a decrease in the oxygen-carrying capacity of the blood.

Some stimulus disrupts homeostasis by

Decreasing

Oxygen delivery to kidneys (and other tissues)

Receptors
Kidney cells detect low oxygen level

Input   Increased erythropoietin secreted into blood

Control center
Proerythroblasts in red bone marrow mature more quickly into reticulocytes

Return to homeostasis when oxygen delivery to kidneys increases to normal

Output   More reticulocytes enter circulating blood

Effectors
Larger number of RBCs in circulation

Increased oxygen delivery to tissues

How might your hematocrit change if you moved from a town at sea level to a high mountain village?

Reticulocyte Count

The rate of erythropoiesis is measured by a **reticulocyte count.** Normally, a little less than 1% of the oldest RBCs are replaced by newcomer reticulocytes on any given day. It then takes 1 to 2 days for the reticulocytes to lose the last vestiges of endoplasmic reticulum and become mature RBCs. Thus, reticulocytes account for about 0.5–1.5% of all RBCs in a normal blood sample. A low "retic" count in a person who is anemic might indicate shortage of erythropoietin or an inability of the red bone marrow to respond to EPO, perhaps because of a nutritional deficiency or leukemia. A high "retic" count might indicate a good red bone marrow response to previous blood loss or to iron therapy in someone who had been iron deficient. It could also point to illegal use of Epoetin alfa by an athlete. ■

▶ C H E C K P O I N T

7. How large are RBCs and what is their microscopic appearance?

8. How is hemoglobin recycled?

9. What is erythropoiesis? How does erythropoiesis affect hematocrit? What factors speed up and slow down erythropoiesis?

WHITE BLOOD CELLS

▶ O B J E C T I V E

• **Describe the structure, functions, and production of white blood cells.**

Types of WBCs

Unlike red blood cells, white blood cells or **leukocytes** (LOO-kō -sīts; *leuko-* = white) have a nucleus and do not contain hemoglobin (Figure 19.7). WBCs are classified as either granular or agranular, depending on whether they contain conspicuous chemical-filled cytoplasmic vesicles (originally called granules) that are made visible by staining. *Granular leukocytes* include

Figure 19.7 Types of white blood cells.

 The shape of their nuclei and the staining properties of their cytoplasmic granules distinguish white blood cells from one another.

(a) Eosinophil (b) Basophil (c) Neutrophil

 LM all 1420x

(d) Small lymphocyte (e) Monocyte

 Which WBCs are called granular leukocytes? Why?

neutrophils, eosinophils, and basophils; *agranular leukocytes* include lymphocytes and monocytes. As shown in Figure 19.3, monocytes and granular leukocytes develop from a myeloid stem cell. In contrast, lymphocytes develop from a lymphoid stem cell.

Granular Leukocytes

After staining, each of the three types of granular leukocytes displays conspicuous granules with distinctive coloration that can be recognized under a light microscope. The large, uniform-sized granules of an **eosinophil** (ē-ō-SIN-ō-fil) are *eosinophilic* (= eosin-loving)—they stain red-orange with acidic dyes (Figure 19.7a). The granules usually do not cover or obscure the nucleus, which most often has two or three lobes connected by either a thin strand or a thick strand. The round, variable-sized granules of a **basophil** (BĀ-sō-fil) are *basophilic* (= basic loving)—they stain blue-purple with basic dyes (Figure 19.7b). The granules commonly obscure the nucleus, which has two lobes. The granules of a **neutrophil** (NOO-trō-fil) are smaller, evenly distributed, and pale lilac in color (Figure 19.7c); the nucleus has two to five lobes, connected by very thin strands of chromatin. As the cells age, the number of nuclear lobes increases. Because older neutrophils thus have several differently shaped nuclear lobes, they are often called *polymorphonuclear leukocytes (PMNs),* polymorphs, or "polys." Younger neutrophils are often called *bands* because their nucleus is more rod-shaped.

Agranular Leukocytes

Even though so-called agranular leukocytes possess cytoplasmic granules, the granules are not visible under a light microscope because of their small size and poor staining qualities.

The nucleus of a **lymphocyte** (LIM-fō-sīt) is round or slightly indented and stains dark. The cytoplasm stains sky blue and forms a rim around the nucleus. The larger the cell, the more cytoplasm is visible. Lymphocytes are classified as large or small based on cell diameter: 6–9 μm in small lymphocytes and 10–14 μm in large lymphocytes (Figure 19.7d). (Although the functional significance of the size difference between small and large lymphocytes is unclear, the distinction is still useful clinically because an increase in the number of large lymphocytes has diagnostic significance in acute viral infections and in some immunodeficiency diseases.)

Monocytes (MON-ō-sīts) are 12–20 μm in diameter (Figure 19.7e). The nucleus of a monocyte is usually kidney shaped or horseshoe shaped, and the cytoplasm is blue-gray and has a foamy appearance. The color and appearance are due to very fine *azurophilic granules* (az′-ū-rō-FIL-ik; *azur-* = blue; *-philic* = loving), which are lysosomes. The blood is merely a conduit for monocytes, which migrate from the blood into the tissues, where they enlarge and differentiate into **macrophages** (= large eaters). Some become **fixed macrophages,** which means they reside in a particular tissue; examples are alveolar macrophages in the lungs, macrophages in the spleen, and stellate reticuloendothelial (Kupffer) cells in the liver. Others become **wandering**

macrophages, which roam the tissues and gather at sites of infection or inflammation.

White blood cells and other nucleated body cells have proteins, called *major histocompatibility (MHC) antigens,* protruding from their plasma membrane into the extracellular fluid. These "cell identity markers" are unique for each person (except identical twins). Although RBCs possess blood group antigens, they lack the MHC antigens.

Functions of WBCs

In a healthy body, some WBCs, especially lymphocytes, can live for several months or years, but most live only a few days. During a period of infection, phagocytic WBCs may live only a few hours. WBCs are far less numerous than red blood cells, about 5000–10,000 cells per μL of blood. RBCs therefore outnumber white blood cells by about 700:1. **Leukocytosis** (loo′-kō-sī-TŌ-sis), an increase in the number of WBCs above 10,000/μm, is a normal, protective response to stresses such as invading microbes, strenuous exercise, anesthesia, and surgery. An abnormally low level of white blood cells (below 5000/μL) is termed **leukopenia** (loo′-kō-PĒ-nē-a). It is never beneficial and may be caused by radiation, shock, and certain chemotherapeutic agents.

The skin and mucous membranes of the body are continuously exposed to microbes and their toxins. Some of these microbes can invade deeper tissues to cause disease. Once pathogens enter the body, the general function of white blood cells is to combat them by phagocytosis or immune responses. To accomplish these tasks, many WBCs leave the bloodstream and collect at points of pathogen invasion or inflammation. Once granulocytes and monocytes leave the bloodstream to fight injury or infection, they never return to it. Lymphocytes, on the other hand, continually recirculate—from blood to interstitial spaces of tissues to lymphatic fluid and back to blood. Only 2% of the total lymphocyte population is circulating in the blood at any given time; the rest are in lymphatic fluid and organs such as the skin, lungs, lymph nodes, and spleen.

WBCs leave the bloodstream by a process termed **emigration** (em′-i-GRĀ-shun; *e-* = out; *migra-* = wander) in which they roll along the endothelium, stick to it, and then squeeze between endothelial cells (Figure 19.8). The precise signals that stimulate emigration through a particular blood vessel vary for the different types of WBCs. Molecules known as **adhesion molecules** help WBCs stick to the endothelium. For example, endothelial cells display adhesion molecules called *selectins* in response to nearby injury and inflammation. Selectins stick to carbohydrates on the surface of neutrophils, causing them to slow down and roll along the endothelial surface. On the neutrophil surface are other adhesion molecules called *integrins,* which tether neutrophils to the endothelium and assist their movement through the blood vessel wall and into the interstitial fluid of the injured tissue.

Neutrophils and macrophages are active in **phagocytosis;** they can ingest bacteria and dispose of dead matter (see Figure 3.14 on page 73). Several different chemicals released by

Figure 19.8 Emigration of white blood cells.

🔑 Adhesion molecules (selectins and integrins) assist the emigration of WBCs from the bloodstream into interstitial fluid.

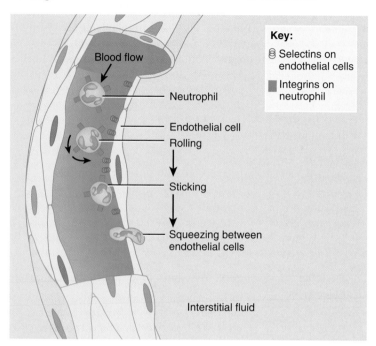

Key:
- 🅢 Selectins on endothelial cells
- ■ Integrins on neutrophil

Blood flow

Neutrophil

Endothelial cell
Rolling

Sticking

Squeezing between endothelial cells

Interstitial fluid

❓ In what way is the "traffic pattern" of lymphocytes in the body different from those of other WBCs?

microbes and inflamed tissues attract phagocytes, a phenomenon called **chemotaxis.** Among the substances that provide stimuli for chemotaxis are toxins produced by microbes; kinins, which are specialized products of damaged tissues; and some of the colony-stimulating factors. The CSFs also enhance the phagocytic activity of neutrophils and macrophages.

Among WBCs, neutrophils respond most quickly to tissue destruction by bacteria. After engulfing a pathogen during phagocytosis, a neutrophil unleashes several chemicals to destroy the pathogen. These chemicals include the enzyme **lysozyme,** which destroys certain bacteria, and **strong oxidants,** such as the superoxide anion (O_2^-), hydrogen peroxide (H_2O_2), and the hypochlorite anion (OCl^-), which is similar to household bleach. Neutrophils also contain **defensins,** proteins that exhibit a broad range of antibiotic activity against bacteria and fungi. Within a neutrophil, vesicles containing defensins merge with phagosomes containing microbes. Defensins form peptide "spears" that poke holes in microbe membranes; the resulting loss of cellular contents kills the invader.

Monocytes take longer to reach a site of infection than do neutrophils, but they arrive in larger numbers and destroy more microbes. Upon arrival they enlarge and differentiate into wandering macrophages, which clean up cellular debris and microbes by phagocytosis after an infection.

Eosinophils leave the capillaries and enter tissue fluid. They are believed to release enzymes, such as histaminase, that combat the effects of histamine and other mediators of inflammation in allergic reactions. Eosinophils also phagocytize antigen–antibody complexes and are effective against certain parasitic worms. A high eosinophil count often indicates an allergic condition or a parasitic infection.

At sites of inflammation, basophils leave capillaries, enter tissues, and liberate heparin, histamine, and serotonin. These substances intensify the inflammatory reaction and are involved in hypersensitivity (allergic) reactions. Basophils are similar in function to mast cells, connective tissue cells that originate from pluripotent stem cells in red bone marrow. Like basophils, mast cells liberate mediators in inflammation, including heparin, histamine, and proteases. Mast cells are widely dispersed in the body, particularly in connective tissues of the skin and mucous membranes of the respiratory and gastrointestinal tracts.

Lymphocytes are the major soldiers in immune system battles (described in detail in Chapter 22). Three main types of lymphocytes are B cells, T cells, and natural killer cells. B cells are particularly effective in destroying bacteria and inactivating their toxins. T cells attack viruses, fungi, transplanted cells, cancer cells, and some bacteria. Immune responses carried out by B cells and T cells help combat infection and provide protection against some diseases. T cells also are responsible for transfusion reactions, allergies, and rejection of transplanted organs. Natural killer cells attack a wide variety of infectious microbes and certain spontaneously arising tumor cells.

An increase in the number of circulating WBCs usually indicates inflammation or infection. A physician may order a **differential white blood cell count** to detect infection or inflammation, determine the effects of possible poisoning by chemicals or drugs, monitor blood disorders (for example, leukemia) and effects of chemotherapy, or detect allergic reactions and parasitic infections. Because each type of white blood cell plays a different role, determining the *percentage* of each type in the blood assists in diagnosing the condition. Table 19.2 lists the significance of both elevated and depressed WBC counts.

Bone Marrow Transplant

A **bone marrow transplant** is the intravenous transfer of red bone marrow from a healthy donor to a recipient, with the goal of establishing normal hemopoiesis and consequently normal blood cell counts in the recipient. In patients with cancer or certain genetic diseases, the defective red bone marrow first must be destroyed by high doses of chemotherapy and whole body radiation just before the transplant takes place. The donor marrow must be very closely matched to that of the recipient, or rejection occurs. When a transplant is successful, stem cells from the donor marrow reseed and grow in the recipient's red bone marrow cavities. Bone marrow transplants have been used to treat aplastic anemia, certain types of leukemia, severe combined immunodeficiency disease (SCID), Hodgkin's disease, non-Hodgkin's lymphoma, multiple myeloma, thalassemia, sickle-cell disease, breast cancer, ovarian cancer, testicular cancer, and hemolytic anemia. ■

WBC Type	High Count May Indicate	Low Count May Indicate
	Table 19.2 Significance of High and Low White Blood Cell Counts	
Neutrophils	Bacterial infection, burns, stress, inflammation.	Radiation exposure, drug toxicity, vitamin B_{12} deficiency, and systemic lupus erythematosus (SLE).
Lymphocytes	Viral infections, some leukemias.	Prolonged illness, immunosuppression, and treatment with cortisol.
Monocytes	Viral or fungal infections, tuberculosis, some leukemias, other chronic diseases.	Bone marrow suppression, treatment with cortisol.
Eosinophils	Allergic reactions, parasitic infections, autoimmune diseases.	Drug toxicity, stress.
Basophils	Allergic reactions, leukemias, cancers, hypothyroidism.	Pregnancy, ovulation, stress, and hyperthyroidism.

▶ C H E C K P O I N T

10. What is the importance of emigration, chemotaxis, and phagocytosis in fighting bacterial invaders?

11. How are leukocytosis and leukopenia different?

12. What is a differential white blood cell count?

13. The first letter of each word in the phrase "*N*ever *l*et *m*onkeys *e*at *b*ananas!" gives the order of WBCs from most numerous to least numerous. Which WBCs do n, l, m, e, and b stand for?

14. What functions do B cells, T cells, and natural killer cells perform?

▌PLATELETS

▶ O B J E C T I V E

- **Describe the structure, function, and origin of platelets.**

Besides the immature cell types that develop into erythrocytes and leukocytes, hemopoietic stem cells also differentiate into cells that produce platelets. Under the influence of the hormone **thrombopoietin,** myeloid stem cells develop into megakaryocyte-colony-forming cells that, in turn, develop into precursor cells called megakaryoblasts (see Figure 19.3). Megakaryoblasts transform into megakaryocytes, huge cells that splinter into 2000 to 3000 fragments. Each fragment, enclosed by a piece of the plasma membrane, is a **platelet (thrombocyte).** Platelets break off from the megakaryocytes in red bone marrow and then enter the blood circulation. Between 150,000 and 400,000 platelets are present in each μL of blood. They are disc-shaped, 2–4 μm in diameter, and have many vesicles but no nucleus.

Platelets help stop blood loss from damaged blood vessels by forming a platelet plug. Their granules also contain chemicals that, once released, promote blood clotting. Platelets have a short life span, normally just 5 to 9 days. Aged and dead platelets are removed by fixed macrophages in the spleen and liver.

Table 19.3 summarizes the formed elements in blood.

Complete Blood Count

A **complete blood count (CBC)** is a valuable test that screens for anemia and various infections. Usually included are counts of RBCs, WBCs, and platelets per μL of whole blood; hematocrit; and differential white blood cell count. The amount of hemoglobin in grams per milliliter of blood also is determined.

Table 19.3	Summary of Formed Elements in Blood		
Name and Appearance	**Number**	**Characterictics***	**Functions**
Red blood cells (RBCs) or *erythrocytes*	4.8 million/μL in females; 5.4 million/μL in males.	7–8 μm diameter, biconcave discs, without a nucleus; live for about 120 days.	Hemoglobin within RBCs transports most of the oxygen and part of the carbon dioxide in the blood.
White blood cells (WBCs) or *leukocytes*	5000–10,000/μL.	Most live for a few hours to a few days.[†]	Combat pathogens and other foreign substances that enter the body.
Granular leukocytes			
Neutrophils	60–70% of all WBCs.	10–12 μm diameter; nucleus has 2–5 lobes connected by thin strands of chromatin; cytoplasm has very fine, pale lilac granules.	Phagocytosis. Destruction of bacteria with lysozyme, defensins, and strong oxidants, such as superoxide anion, hydrogen peroxide, and hypochlorite anion.
Eosinophils	2–4% of all WBCs.	10–12 μm diameter; nucleus has 2 or 3 lobes; large, red-orange granules fill the cytoplasm.	Combat the effects of histamine in allergic reactions, phagocytize antigen–antibody complexes, and destroy certain parasitic worms.
Basophils	0.5–1% of all WBCs.	8–10 μm diameter; nucleus has 2 lobes; large cytoplasmic granules appear deep blue-purple.	Liberate heparin, histamine, and serotonin in allergic reactions that intensify the overall inflammatory response.
Agranular leukocytes			
Lymphocytes (T cells, B cells, and **natural killer cells)**	20–25% of all WBCs.	Small lymphocytes are 6–9 μm in diameter; large lymphocytes are 10–14 μm in diameter; nucleus is round or slightly indented; cytoplasm forms a rim around the nucleus that looks sky blue; the larger the cell, the more cytoplasm is visible.	Mediate immune responses, including antigen–antibody reactions. B cells develop into plasma cells, which secrete antibodies. T cells attack invading viruses, cancer cells, and transplanted tissue cells. Natural killer cells attack a wide variety of infectious microbes and certain spontaneously arising tumor cells.
Monocytes	3–8% of all WBCs.	12–20 μm diameter; nucleus is kidney shaped or horseshoe shaped; cytoplasm is blue-gray and has foamy appearance.	Phagocytosis (after transforming into fixed or wandering macrophages).
Platelets (thrombocytes)	150,000–400,000/μL.	2–4 μm diameter cell fragments that live for 5–9 days; contain many vesicles but no nucleus.	Form platelet plug in hemostasis; release chemicals that promote vascular spasm and blood clotting.

*Colors are those seen when using Wright's stain.
[†]Some lymphocytes, called T and B memory cells, can live for many years once they are established.

Normal hemoglobin ranges are: infants, 14–20 g/100 mL of blood; adult females, 12–16 g/100 mL of blood; and adult males, 13.5–18 g/100 mL of blood. ■

▶ CHECKPOINT

15. How do RBCs, WBCs, and platelets compare with respect to size, number per μL of blood, and life span?

HEMOSTASIS

▶ OBJECTIVES

- **Describe the three mechanisms that contribute to hemostasis.**
- **Identify the stages of blood clotting and explain the various factors that promote and inhibit blood clotting.**

Hemostasis (hē-mō-STĀ-sis) is a sequence of responses that stops bleeding. When blood vessels are damaged or ruptured, the hemostatic response must be quick, localized to the region of damage, and carefully controlled. Three mechanisms reduce blood loss: (1) vascular spasm, (2) platelet plug formation, and (3) blood clotting (coagulation). When successful, hemostasis prevents **hemorrhage** (HEM-o-rij; *-rhage* = burst forth), the loss of a large amount of blood from the vessels. Hemostatic mechanisms can prevent hemorrhage from smaller blood vessels, but extensive hemorrhage from larger vessels usually requires medical intervention.

Vascular Spasm

When arteries or arterioles are damaged, the circularly arranged smooth muscle in their walls contracts immediately. Such a **vascular spasm** reduces blood loss for several minutes to several hours, during which time the other hemostatic mechanisms go into operation. The spasm is probably caused by damage to the smooth muscle, by substances released from activated platelets, and by reflexes initiated by pain receptors.

Platelet Plug Formation

Considering their small size, platelets store an impressive array of chemicals. Within many vesicles are clotting factors, ADP, ATP, Ca^{2+}, and serotonin. Also present are enzymes that produce thromboxane A2, a prostaglandin; *fibrin-stabilizing factor,* which helps to strengthen a blood clot; lysosomes; some mitochondria; membrane systems that take up and store calcium and provide channels for release of the contents of granules; and glycogen. Also within platelets is **platelet-derived growth factor (PDGF),** a hormone that can cause proliferation of vascular endothelial cells, vascular smooth muscle fibers, and fibroblasts to help repair damaged blood vessel walls.

Platelet plug formation occurs as follows (Figure 19.9):

1 Initially, platelets contact and stick to parts of a damaged blood vessel, such as collagen fibers of the connective tissue underlying the damaged endothelial cells. This process is called **platelet adhesion.**

Figure 19.9 Platelet plug formation.

 A platelet plug can stop blood loss completely if the hole in a blood vessel is small enough.

1 Platelet adhesion

2 Platelet release reaction

3 Platelet aggregation

Besides platelet plug formation, which two mechanisms also contribute to hemostasis?

2 Due to adhesion, the platelets become activated, and their characteristics change dramatically. They extend many projections that enable them to contact and interact with one another, and they begin to liberate the contents of their vesicles. This phase is called the **platelet release reaction.** Liberated ADP and thromboxane A2 play a major role by activating nearby platelets. Serotonin and thromboxane A2 function as vasoconstrictors, causing and sustaining contraction of vascular smooth muscle, which decreases blood flow through the injured vessel.

3 The release of ADP makes other platelets in the area sticky, and the stickiness of the newly recruited and activated platelets causes them to adhere to the originally activated platelets. This gathering of platelets is called **platelet aggregation.** Eventually, the accumulation and attachment of large numbers of platelets form a mass called a **platelet plug.**

A platelet plug is very effective in preventing blood loss in a small vessel. Although initially the platelet plug is loose, it becomes quite tight when reinforced by fibrin threads formed during clotting (see Figure 19.10). A platelet plug can stop blood loss completely if the hole in a blood vessel is not too large.

Blood Clotting

Normally, blood remains in its liquid form as long as it stays within its vessels. If it is drawn from the body, however, it thickens and forms a gel. Eventually, the gel separates from the liquid. The straw-colored liquid, called **serum,** is simply plasma minus the clotting proteins. The gel is called a **clot.** It consists of a network of insoluble protein fibers called fibrin in which the formed elements of blood are trapped (Figure 19.10).

Figure 19.10 Part of a blood clot. Notice the platelet and red blood cells entrapped in fibrin threads.

🔑 **A blood clot is a gel that contains formed elements of the blood entangled in fibrin threads.**

Platelet

Red blood cell

Fibrin thread

SEM 15,000x

 What is serum?

The process of gel formation, called **clotting** or **coagulation,** is a series of chemical reactions that culminates in formation of fibrin threads. If blood clots too easily, the result can be **thrombosis**—clotting in an undamaged blood vessel. If the blood takes too long to clot, hemorrhage can occur.

Clotting involves several substances known as **clotting (coagulation) factors.** These factors include calcium ions (Ca^{2+}), several inactive enzymes that are synthesized by hepatocytes (liver cells) and released into the bloodstream, and various molecules associated with platelets or released by damaged tissues. Most clotting factors are identified by Roman numerals that indicate the order of their discovery, and not necessarily the order of their participation in the clotting process.

Clotting is a complex cascade of enzymatic reactions in which each clotting factor activates many molecules of the next one in a fixed sequence. Finally, a large quantity of product (the insoluble protein fibrin) is formed. Clotting can be divided into three stages (Figure 19.11):

1 First, several reactions of the clotting sequence end in formation of prothrombinase. Two pathways, called the extrinsic pathway (Figure 19.11a) and the intrinsic pathway (Figure 19.11b), lead to prothrombinase. Once prothrombinase is formed, the steps involved in the next two stages of clotting are the same for both the extrinsic and intrinsic pathways, and together these two stages are referred to as the common pathway.

2 Prothrombinase converts prothrombin (a plasma protein formed by the liver) into the enzyme thrombin.

3 Thrombin converts soluble fibrinogen (another plasma protein formed by the liver) into insoluble fibrin. Fibrin forms the threads of the clot.

The Extrinsic Pathway

The **extrinsic pathway** of blood clotting has fewer steps than the intrinsic pathway and occurs rapidly—within a matter of seconds if trauma is severe. It is so named because a tissue protein called **tissue factor (TF),** also known as **thromboplastin,** leaks into the blood from cells *outside (extrinsic to)* blood vessels and initiates the formation of prothrombinase. TF is a complex mixture of lipoproteins and phospholipids released from the surfaces of damaged cells. In the presence of Ca^{2+}, TF begins a sequence of reactions that ultimately activates clotting factor X (Figure 19.11a). Once factor X is activated, it combines with factor V in the presence of Ca^{2+} to form the active enzyme prothrombinase, completing the extrinsic pathway.

The Intrinsic Pathway

The **intrinsic pathway** of blood clotting is more complex than the extrinsic pathway, and it occurs more slowly, usually requiring several minutes. The intrinsic pathway is so-named because its activators are either in direct contact with blood or contained *within (intrinsic to)* the blood; outside tissue damage is not needed. If endothelial cells become roughened or damaged,

Figure 19.11 The blood-clotting cascade.

In clotting, coagulation factors are activated in sequence, resulting in a cascade of reactions that includes positive feedback cycles.

What is the outcome of the first stage of clotting?

blood can come in contact with collagen fibers in the connective tissue around the endothelium of the blood vessel. In addition, trauma to endothelial cells causes damage to platelets, resulting in the release of phospholipids by the platelets. Contact with col-

lagen fibers (or with the glass sides of a blood collection tube) activates clotting factor XII (Figure 19.11b), which begins a sequence of reactions that eventually activates clotting factor X. Platelet phospholipids and Ca^{2+} can also participate in the activation of factor X. Once factor X is activated, it combines with factor V to form the active enzyme prothrombinase (just as occurs in the extrinsic pathway), completing the intrinsic pathway.

The Common Pathway

Once prothrombinase is formed, the common pathway follows. In the second stage of blood clotting (Figure 19.11c), prothrombinase and Ca^{2+} catalyze the conversion of prothrombin to thrombin. In the third stage, thrombin, in the presence of Ca^{2+}, converts fibrinogen, which is soluble, to loose fibrin threads, which are insoluble. Thrombin also activates factor XIII (fibrin stabilizing factor), which strengthens and stabilizes the fibrin threads into a sturdy clot. Plasma contains some factor XIII, which is also released by platelets trapped in the clot.

Thrombin has two positive feedback effects. In the first positive feedback loop, which involves factor V, it accelerates the formation of prothrombinase. Prothrombinase, in turn, accelerates the production of more thrombin, and so on. In the second positive feedback loop, thrombin activates platelets, which reinforces their aggregation and the release of platelet phospholipids.

Clot Retraction

Once a clot is formed, it plugs the ruptured area of the blood vessel and thus stops blood loss. **Clot retraction** is the consolidation or tightening of the fibrin clot. The fibrin threads attached to the damaged surfaces of the blood vessel gradually contract as platelets pull on them. As the clot retracts, it pulls the edges of the damaged vessel closer together, decreasing the risk of further damage. During retraction, some serum can escape between the fibrin threads, but the formed elements in blood cannot. Normal retraction depends on an adequate number of platelets in the clot, which release factor XIII and other factors, thereby strengthening and stabilizing the clot. Permanent repair of the blood vessel can then take place. In time, fibroblasts form connective tissue in the ruptured area, and new endothelial cells repair the vessel lining.

Role of Vitamin K in Clotting

Normal clotting depends on adequate vitamin K in the body. Although vitamin K is not involved in actual clot formation, it is required for the synthesis of four clotting factors by hepatocytes: factors II (prothrombin), VII, IX, and X. Vitamin K, which is normally produced by bacteria that inhabit the large intestine, is a fat-soluble vitamin, and it can be absorbed through the lining of the intestine and into the blood only if absorption of lipids is normal. People suffering from disorders that slow absorption of lipids (for example, inadequate release of bile into the small intestine) often experience uncontrolled bleeding as a consequence of vitamin K deficiency.

The various clotting factors, their sources, and the pathways of activation are summarized in Table 19.4.

Hemostatic Control Mechanisms

Many times a day little clots start to form, often at a site of minor roughness or at a developing atherosclerotic plaque inside a blood vessel. Because blood clotting involves amplification and positive feedback cycles, a clot has a tendency to enlarge, creating the potential for impairment of blood flow through undamaged vessels. The **fibrinolytic system** (fī-bri-nō-LIT-ik) dissolves small, inappropriate clots; it also dissolves clots at a site of damage once the damage is repaired. Dissolution of a clot is called **fibrinolysis** (fī-bri-NOL-i-sis). When a clot is formed, an inactive plasma enzyme called **plasminogen** is incorporated into the clot. Both body tissues and blood contain substances that can activate plasminogen to **plasmin (fibrinolysin),** an active plasma enzyme. Among these substances are thrombin, activated factor XII, and tissue plasminogen activator (t-PA), which is synthesized in endothelial cells of most tissues and liberated into the blood. Once plasmin is formed, it can dissolve the clot by digesting fibrin threads and inactivating substances such as fibrinogen, prothrombin, and factors V, VIII, and XII.

Even though thrombin has a positive feedback effect on blood clotting, clot formation normally remains localized at the site of damage. A clot does not extend beyond a wound site into the general circulation, in part because fibrin absorbs thrombin into the clot. Another reason for localized clot formation is that because of the dispersal of some of the clotting factors by the blood, their concentrations are not high enough to bring about widespread clotting.

Several other mechanisms also operate to control blood clotting. For example, endothelial cells and white blood cells produce a prostaglandin called **prostacyclin** (pros-ta-SĪ-klin) that opposes the actions of thromboxane A2. Prostacyclin is a powerful inhibitor of platelet adhesion and release.

Moreover, substances that inhibit clotting, called **anticoagulants,** are present in blood. These include **antithrombin,** which blocks the action of factors XII, XI, IX, X, and II (thrombin). **Heparin,** an anticoagulant that is produced by mast cells and basophils, combines with antithrombin and increases its effectiveness in blocking thrombin. Another anticoagulant, **activated protein C (APC),** inactivates factors V and VIII, the two major clotting factors not blocked by antithrombin, and enhances activity of plasminogen activators. Babies that lack the ability to produce APC due to a genetic mutation usually die of blood clots in infancy.

Anticoagulants

Patients who are at increased risk of forming blood clots may receive an **anticoagulant drug,** a substance that delays, suppresses, or prevents blood clotting. Examples are heparin or warfarin. **Heparin** is often administered during hemodialysis and open-heart surgery. **Warfarin (Coumadin)** acts as an antagonist to vitamin K and thus blocks synthesis of four clotting factors (II, VII, IX, and X). Warfarin is slower acting than heparin. To prevent clotting in donated blood, blood banks and laboratories

Table 19.4	Clotting (Coagulation) Factors		
Number*	**Name(s)**	**Source**	**Pathway(s) of Activation**
I	Fibrinogen.	Liver.	Common.
II	Prothrombin.	Liver.	Common.
III	Tissue factor (thromboplastin).	Damaged tissues and activated platelets.	Extrinsic.
IV	Calcium ions (Ca^{2+}).	Diet, bones, and platelets.	All.
V	Proaccelerin, labile factor, or accelerator globulin (AcG).	Liver and platelets.	Extrinsic and intrinsic.
VII	Serum prothrombin conversion accelerator (SPCA), stable factor, or proconvertin.	Liver.	Extrinsic.
VIII	Antihemophilic factor (AHF), antihemophilic factor A, or antihemophilic globulin (AHG).	Platelets and endothelial cells.	Intrinsic.
IX	Christmas factor, plasma thromboplastin component (PTC), or antihemophilic factor B.	Liver.	Intrinsic.
X	Stuart factor, Prower factor, or thrombokinase.	Liver.	Extrinsic and intrinsic.
XI	Plasma thromboplastin antecedent (PTA) or antihemophilic factor C.	Liver.	Intrinsic.
XII	Hageman factor, glass factor, contact factor, or antihemophilic factor D.	Liver.	Intrinsic.
XIII	Fibrin-stabilizing factor (FSF).	Liver and platelets.	Common.

*There is no factor VI. Prothrombinase (prothrombin activator) is a combination of activated factors V and X.

often add a substance that removes Ca^{2+}; examples are EDTA (ethylene diamine tetraacetic acid) and CPD (citrate phosphate dextrose). ■

Intravascular Clotting

Despite the anticoagulating and fibrinolytic mechanisms, blood clots sometimes form within the cardiovascular system. Such clots may be initiated by roughened endothelial surfaces of a blood vessel resulting from atherosclerosis, trauma, or infection. These conditions induce adhesion of platelets. Intravascular clots may also form when blood flows too slowly (stasis), allowing clotting factors to accumulate locally in high enough concentrations to initiate coagulation. Clotting in an unbroken blood vessel (usually a vein) is called **thrombosis** (throm-BŌ-sis; *thromb-* = clot; *-osis* = a condition of). The clot itself is a **thrombus,** and it may dissolve spontaneously. If it remains intact, however, the thrombus may become dislodged and be swept away in the blood. A blood clot, bubble of air, fat from broken bones, or a piece of debris transported by the bloodstream is called an **embolus** (*em-* = in; *bolus* = a mass). An embolus that breaks away from an arterial wall may lodge in a smaller-diameter artery downstream and block blood flow to a vital organ. When an embolus lodges in the lungs, the condition is called **pulmonary embolism.**

Aspirin and Thrombolytic Agents

In patients with heart and blood vessel disease, the events of hemostasis may occur even without external injury to a blood vessel. At low doses, **aspirin** inhibits vasoconstriction and platelet aggregation by blocking synthesis of thromboxane A2. It also reduces the chance of thrombus formation. Due to these effects, aspirin reduces the risk of transient ischemic attacks (TIA), strokes, myocardial infarction, and blockage of peripheral arteries.

Thrombolytic agents are chemical substances that are injected into the body to dissolve blood clots that have already formed to restore circulation. They either directly or indirectly activate plasminogen. The first thrombolytic agent, approved in 1982 for dissolving clots in the coronary arteries of the heart, was **streptokinase,** which is produced by streptococcal bacteria. A genetically engineered version of human **tissue plasminogen activator (t-PA)** is now used to treat both heart attacks and brain attacks (strokes) that are caused by blood clots. ■

▶ CHECKPOINT

16. What is hemostasis?

17. How do vascular spasm and platelet plug formation occur?

18. What is fibrinolysis? Why does blood rarely remain clotted inside blood vessels?

19. How do the extrinsic and intrinsic pathways of blood clotting differ?

20. What does each term mean: anticoagulant, thrombus, embolus, and thrombolytic agent?

BLOOD GROUPS AND BLOOD TYPES

▶ OBJECTIVE

• **Explain the ABO and Rh blood groups.**

The surfaces of erythrocytes contain a genetically determined assortment of glycoproteins and glycolipids that can act as antigens. Called **isoantigens** or **agglutinogens** (a-gloo-TIN-ō-jens), these antigens occur in characteristic combinations. Based on the presence or absence of various isoantigens, blood is categorized into different **blood groups.** Within a given blood group, there may be two or more different **blood types.** There are at least 24 blood groups and more than 100 isoantigens that can be detected on the surface of red blood cells. Here we discuss two major blood groups—ABO and Rh. The incidence of ABO and Rh blood types varies among different population groups, as indicated in Table 19.5. Among other blood groups are the Lewis, Kell, Kidd, and Duffy systems.

ABO Blood Group

The **ABO blood group** is based on two glycolipid isoantigens called A and B (Figure 19.12). People whose RBCs display *only antigen A* have **type A** blood. Those who have *only antigen B* are **type B.** Individuals who have *both A and B antigens* are **type AB,** whereas those who have *neither antigen A nor B* are **type O.**

In addition to isoantigens on RBCs, blood plasma usually contains **isoantibodies** or **agglutinins** (a-GLOO-ti-nins) that react with the A or B antigens if the two are mixed. These are **anti-A antibody,** which reacts with antigen A, and **anti-B antibody,** which reacts with antigen B. The antibodies present in each of the four blood types are shown in Figure 19.12. You do not have antibodies that react with the antigens of your own RBCs, but most likely you do have antibodies for any antigens that your RBCs lack. Although isoantibodies start to appear in the blood within a few months after birth, the reason for their presence is not clear. Perhaps they are formed in response to bacteria that normally inhabit the gastrointestinal tract. Because the isoantibodies are large IgM-type antibodies (see Table 22.3 on page 790) that do not cross the placenta, ABO incompatibility between a mother and her fetus rarely causes problems.

Table 19.5	Blood Types in the United States				
Population Group	O	A	B	AB	Rh⁺
White	45	40	11	4	85
Black	49	27	20	4	95
Korean	32	28	30	10	100
Japanese	31	38	21	10	100
Chinese	42	27	25	6	100
Native American	79	16	4	1	100

Figure 19.12   Antigens and antibodies of the ABO blood types.

The antibodies in your plasma do not react with the antigens on your red blood cells.

BLOOD TYPE	TYPE A	TYPE B	TYPE AB	TYPE O
Red blood cells	A antigen	B antigen	Both A and B antigens	Neither A nor B antigen
Plasma	Anti-B antibody	Anti-A antibody	Neither antibody	Both anti-A and anti-B antibodies

Which antibodies are usually present in type O blood?

Rh Blood Group

The **Rh blood group** is so-named because the antigen was discovered in the blood of the *Rhesus* monkey. The alleles of three genes may code for the Rh antigen. People whose RBCs have Rh antigens are designated Rh^+ (Rh positive); those who lack Rh antigens are designated Rh^- (Rh negative). Table 19.5 shows the incidence of Rh^+ and Rh^- in various populations. Normally, plasma does not contain anti-Rh antibodies. If an Rh^- person receives an Rh^+ blood transfusion, however, the immune system starts to make anti-Rh antibodies that will remain in the blood. If a second transfusion of Rh^+ blood is given later, the previously formed anti-Rh antibodies will cause **hemolysis** (rupture) of the RBCs in the donated blood, and a severe reaction may occur.

Hemolytic Disease of the Newborn

The most common problem with Rh incompatibility, **hemolytic disease of the newborn (HDN),** may arise during pregnancy (Figure 19.13). Normally, no direct contact occurs between maternal and fetal blood while a woman is pregnant. However, if a small amount of Rh^+ blood leaks from the fetus through the placenta into the bloodstream of an Rh^- mother, the mother will start to make anti-Rh antibodies. Because the greatest possibility of fetal blood leakage into the maternal circulation occurs at delivery, the firstborn baby usually is not affected. If the mother becomes pregnant again, however, her anti-Rh antibodies can cross the placenta and enter the bloodstream of the fetus. If the fetus is Rh^-, there is no problem, because Rh^- blood does not have the Rh antigen. If the fetus is Rh^+, however, hemolysis brought on by fetal–maternal incompatibility may occur in the fetal blood.

Figure 19.13   Development of hemolytic disease of the newborn (HDN).   (a) At birth, a small quantity of fetal blood usually leaks across the placenta into the maternal bloodstream. A problem can arise when the mother is Rh^- and the baby is Rh^+, having inherited an allele for one of the Rh antigens from the father. (b) Upon exposure to Rh antigen, the mother's immune system responds by making anti-Rh antibodies. (c) During a subsequent pregnancy, the maternal antibodies cross the placenta into the fetal blood. If the second fetus is Rh^+, the ensuing antigen–antibody reaction causes hemolysis of fetal RBCs. The result is HDN.

HDN occurs when maternal anti-Rh antibodies cross the placenta and cause hemolysis of fetal RBCs.

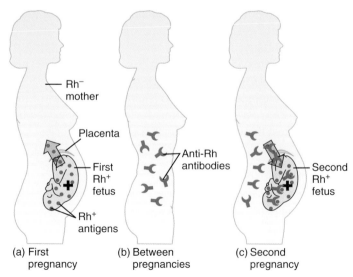

(a) First pregnancy   (b) Between pregnancies   (c) Second pregnancy

Why is the firstborn baby unlikely to have HDN?

An injection of anti-Rh antibodies called *anti-Rh gamma globulin (RhoGAM)* can be given to prevent HDN. All Rh⁻ women should receive RhoGAM soon after every delivery, miscarriage, or abortion. These antibodies bind to and inactivate the fetal Rh antigens, if they are present. Thus, the mother's immune system does not respond to the foreign antigens by producing antibodies. ■

Transfusions

Despite the differences in RBC antigens reflected in the blood group systems, blood is the most easily shared of human tissues, saving many thousands of lives every year through transfusions. A **transfusion** (trans-FŪ-shun) is the transfer of whole blood or blood components (red blood cells only or plasma only) into the bloodstream. A transfusion is most often given to alleviate anemia or when blood volume is low—for example, after a severe hemorrhage. However, the normal components of one person's RBC plasma membrane can trigger damaging antigen–antibody responses in a transfusion recipient. In an incompatible blood transfusion, isoantibodies in the recipient's plasma bind to the isoantigens on the donated RBCs. When these antigen–antibody complexes form, they activate plasma proteins of the complement family (described on page 790). In essence, complement molecules make the plasma membrane of the donated RBCs leaky, causing them to hemolyze (burst) and release hemoglobin into the plasma. The liberated hemoglobin may cause kidney damage by clogging the filtration membranes.

As an example of an incompatible blood transfusion, consider what happens if a person with type A blood receives a transfusion of type B blood. The recipient's blood (type A) contains A antigens on the red blood cells and anti-B antibodies in the plasma. The donor's blood (type B) contains B antigens and anti-A antibodies. In this situation, two things can happen. First, the anti-B antibodies in the recipient's plasma can bind to the B antigens on the donor's erythrocytes, causing hemolysis of the red blood cells. Second, the anti-A antibodies in the donor's plasma can bind to the A antigens on the recipient's red blood cells. However, the second reaction is usually not serious because the donor's anti-A antibodies become so diluted in the recipient's plasma that they do not cause significant hemolysis of the recipient's RBCs. Table 19.6 summarizes the interactions of the four blood types of the ABO system.

People with type AB blood do not have anti-A or anti-B antibodies in their plasma. They are sometimes called *universal recipients* because theoretically they can receive blood from donors of all four blood types. They have no antibodies to attack antigens on donated RBCs (Table 19.6). People with type O blood have neither A nor B antigens on their RBCs and are sometimes called *universal donors* because theoretically they can donate blood to all four ABO blood types. Type O persons requiring blood may receive only type O blood (Table 19.6). In practice, use of the terms universal recipient and universal donor is misleading and dangerous. Blood contains antigens and anti-

bodies other than those associated with the ABO system that can cause transfusion problems. Thus, blood should be carefully cross-matched or screened before transfusion. In about 80% of the population, soluble antigens of the ABO type appear in saliva and other body fluids, in which case blood type can be identified from a sample of saliva.

Typing and Cross-Matching Blood for Transfusion

To avoid blood-type mismatches, laboratory technicians type the patient's blood and then either cross-match it to potential donor blood or screen it for the presence of antibodies. Outside the body, at room temperature, mixing of incompatible blood causes **agglutination** (a-gloo-ti-NĀ-shun), clumping of RBCs that is visible to the naked eye, rather than hemolysis. Agglutination is an antigen–antibody response, whereby the cells become cross-linked to one another. (Note that agglutination is not the same as blood clotting.)

In the procedure for ABO blood typing, single drops of blood are mixed with different antisera, solutions that contain antibodies. One drop of blood is mixed with anti-A serum, which contains anti-A antibodies that will agglutinate red blood cells that possess A antigens. Another drop is mixed with anti-B serum, which contains anti-B antibodies that will agglutinate red blood cells that possess B antigens. If the red blood cells agglutinate only when mixed with anti-A serum, the blood is type A. If the red blood cells agglutinate only when mixed with anti-B serum, the blood is type B. The blood is type AB if both drops agglutinate; if neither drop agglutinates, the blood is type O.

In the procedure for determining Rh factor, a drop of blood is mixed with antiserum containing antibodies that will agglutinate RBCs displaying Rh antigens. If the blood agglutinates, it is Rh⁺; no agglutination indicates Rh⁻.

Once the patient's blood type is known, donor blood of the same ABO and Rh type is selected. In a **cross-match,** the possible

Table 19.6	Summary of ABO Blood Group Interactions			
	Blood Type			
Characteristic	**A**	**B**	**AB**	**O**
Antigen on RBCs	A	B	Both A and B	Neither A nor B
Antibody in plasma	anti-B	anti-A	Neither anti-A nor anti-B	Both anti-A and anti-B
Compatible donor blood types (no hemolysis)	A, O	B, O	A, B, AB, O	O
Incompatible donor blood types (hemolysis)	B, AB	A, AB	—	A, B, AB

donor RBCs are mixed with the recipient's serum. If agglutination does not occur, the recipient does not have antibodies that will attack the donor RBCs. Alternatively, the recipient's serum can be **screened** against a test panel of RBCs having antigens known to cause blood transfusion reactions to detect any antibodies that may be present.

▶ CHECKPOINT

21. What precautions must be taken before giving a blood transfusion?
22. What is hemolysis and how can it occur after a mismatched blood transfusion?

DISORDERS: HOMEOSTATIC IMBALANCES

Anemia

Anemia is a condition in which the oxygen-carrying capacity of blood is reduced. All of the many types of anemia are characterized by reduced numbers of RBCs or a decreased amount of hemoglobin in the blood. The person feels fatigued and is intolerant of cold, both of which are related to lack of oxygen needed for ATP and heat production. Also, the skin appears pale, due to the low content of red-colored hemoglobin circulating in skin blood vessels. Among the most important causes and types of anemia are the following:

* *Inadequate absorption of iron, excessive loss of iron, increased iron requirement, or insufficient intake of iron* causes **iron-deficiency anemia,** the most common type of anemia. Women are at greater risk for iron-deficiency anemia due to menstrual blood losses and increased iron demands of the growing fetus during pregnancy. Gastrointestinal losses, such as occurs with malignancy or ulceration, also contribute to this type of anemia.
* *Insufficient hemopoiesis* resulting from an inability of the stomach to produce intrinsic factor, which is needed for absorption of vitamin B_{12} in the small intestine, causes **pernicious anemia.**
* *Excessive loss of RBCs* through bleeding resulting from large wounds, stomach ulcers, or especially heavy menstruation leads to **hemorrhagic anemia.**
* *RBC plasma membranes rupture prematurely* in **hemolytic anemia.** The released hemoglobin pours into the plasma and may damage the filtering units (glomeruli) in the kidneys. The condition may result from inherited defects such as abnormal red blood cell enzymes, or from outside agents such as parasites, toxins, or antibodies from incompatible transfused blood.
* *Deficient synthesis of hemoglobin* occurs in **thalassemia** (thal'-a-SĒ-mē-a), a group of hereditary hemolytic anemias. The RBCs are small (microcytic), pale (hypochromic), and short-lived. Thalassemia occurs primarily in populations from countries bordering the Mediterranean Sea.
* *Destruction of red bone marrow* results in **aplastic anemia.** Toxins, gamma radiation, and certain medications that inhibit enzymes needed for hemopoiesis are causes.

Sickle-Cell Disease

The RBCs of a person with **sickle-cell disease (SCD)** contain Hb-S, an abnormal kind of hemoglobin. When Hb-S gives up oxygen to the interstitial fluid, it forms long, stiff, rodlike structures that bend the erythrocyte into a sickle shape (Figure 19.14). The sickled cells rupture easily. Even though erythropoiesis is stimulated by the loss of the cells, it cannot keep pace with hemolysis. Prolonged oxygen reduction may eventually cause extensive tissue damage.

Sickle-cell disease is inherited. People with two sickle-cell genes have severe anemia, whereas those with only one defective gene have minor problems. Sickle-cell genes are found primarily among populations, or descendants of populations, that live in the malaria belt around the world, including parts of Mediterranean Europe, sub-Saharan Africa, and tropical Asia. The gene responsible for the tendency of the RBCs to sickle also alters the permeability of the plasma membranes of sickled cells, causing potassium ions to leak out. Low levels of potassium kill the malaria parasites that infect sickled cells. Because of this effect, a person with one normal gene and one sickle-cell gene has higher-than-average resistance to malaria. The possession of a single sickle-cell gene thus confers a survival advantage.

Treatment of SCD consists of administration of analgesics to relieve pain, fluid therapy to maintain hydration, oxygen to reduce the stimulus for the crisis, antibiotics to counter infections, and blood transfusions. People who suffer SCD have normal fetal hemoglobin (Hb-F), a slightly different form of hemoglobin that predominates at birth and is present in small amounts thereafter. In some patients with sickle-cell disease, a drug called hydroxyurea promotes transcription of the normal Hb-F gene, elevates the level of Hb-F, and reduces the chance that the RBCs will sickle. Unfortunately, this drug also has toxic effects on the bone marrow; thus, its safety for long-term use is questionable.

Figure 19.14 Red blood cells from a person with sickle-cell disease.

Red blood cells

Hemophilia

Hemophilia (hē-mō-FIL-ē-a; *-philia* = loving) is an inherited deficiency of clotting in which bleeding may occur spontaneously or after only minor trauma. Different types of hemophilia are due to deficiencies of different blood clotting factors and exhibit varying degrees of severity, ranging from mild to severe bleeding tendencies. The most common type (classic hemophilia) is hemophilia A, in which factor VIII is absent. People with hemophilia B lack factor IX. Hemophilia A and B occur primarily among males because these are sex-linked recessive disorders, whereas hemophilia C affects both males and females. Hemophilia C is caused by a lack of factor XI (which in turn activates factor IX) and is much less severe than hemophilia A or B. This is because an alternate activator of factor IX is present—namely, factor VII. Hemophilia is characterized by spontaneous or traumatic subcutaneous and intramuscular hemorrhaging, nosebleeds, blood in the urine, and hemorrhages in joints that produce pain and tissue damage. Treatment involves transfusions of fresh plasma or concentrates of the deficient clotting factor to relieve the tendency to bleed.

Disseminated Intravascular Clotting

Disseminated intravascular clotting (DIC) is a disorder of hemostasis characterized by simultaneous and unregulated blood clotting and hemorrhage throughout the body. DIC is associated primarily with endothelial damage, tissue damage, and direct activation of clotting factor X, all of which activate the blood-clotting sequence. Common causes of DIC are infections, hypoxia, low blood flow rates, trauma, tumors, hypotension, and hemolysis. DIC may be severe and life threatening. The clots decrease blood flow and eventually cause ischemia, infarction, and necrosis, resulting in multisystem organ dysfunction. A peculiar feature of DIC is that the patient often begins to bleed despite forming clots. Because so many clotting factors are removed by the widespread clotting, too few remain to allow normal clotting of the remaining blood.

Leukemia

Acute leukemia is a malignant disease of blood-forming tissues characterized by uncontrolled production and accumulation of immature leukocytes. In **chronic leukemia,** mature leukocytes accumulate in the bloodstream because they do not die at the end of their normal life span. The *human T cell leukemia-lymphoma virus-1 (HTLV-1)* is strongly associated with some types of leukemia. The abnormal accumulation of immature leukocytes may be reduced by treatment with x rays and antileukemic drugs. In some cases, a bone marrow transplant can cure the leukemia.

MEDICAL TERMINOLOGY

Acute normovolemic hemodilution (nor-mō-vō-LĒ-mik hē-mō-di-LOO-shun) Removal of blood immediately before surgery and its replacement with a cell-free solution to maintain sufficient blood volume for adequate circulation. At the end of surgery, once bleeding has been controlled, the collected blood is returned to the body.

Autologous preoperative transfusion (aw-TOL-o-gus trans-FŪ-zhun; *auto-* = self) Donating one's own blood; can be done up to 6 weeks before elective surgery. Also called *predonation.*

Blood bank A facility that collects and stores a supply of blood for future use by the donor or others. Because blood banks have now assumed added and diverse functions (immunohematology reference work, continuing medical education, bone and tissue storage, and clinical consultation), they are more appropriately referred to as **centers of transfusion medicine.**

Cyanosis (sī-a-NŌ-sis; *cyano-* = blue) Slightly bluish/dark-purple skin discoloration, most easily seen in the nail beds and mucous membranes, due to an increased quantity of reduced hemoglobin (hemoglobin not combined with oxygen) in systemic blood.

Hemochromatosis (hē-mō-krō-ma-TŌ-sis; *chroma* = color) Disorder of iron metabolism characterized by excessive absorption of ingested iron and excess deposits of iron in tissues (especially the liver, heart, pituitary gland, gonads, and pancreas) that result in bronze discoloration of the skin, cirrhosis, diabetes mellitus, and bone and joint abnormalities.

Jaundice (*jaund-* = yellow) An abnormal yellowish discoloration of the sclerae of the eyes, skin, and mucous membranes due to excess bilirubin (yellow-orange pigment) in the blood. *Prehepatic jaundice* is due to excess production of bilirubin; *hepatic jaundice* is due to abnormal bilirubin processing by the liver caused by congenital liver disease, cirrhosis (scar tissue formation) of the liver, or hepatitis (liver inflammation); and *extrahepatic jaundice* is due to blockage of bile drainage by gallstones or cancer of the bowel or pancreas.

Phlebotomist (fle-BOT-ō-mist; *phlebo-* = vein; *-tom* = cut) A technician who specializes in withdrawing blood.

Septicemia (sep′-ti-SĒ-mē-a; *septic-* = decay; *-emia* = condition of blood) Toxins or disease-causing bacteria in the blood. Also called "blood poisoning."

Thrombocytopenia (throm′-bō-sī-tō-PĒ-nē-a; *-penia* = poverty) Very low platelet count that results in a tendency to bleed from capillaries.

Venesection (vē′-ne-SEK-shun; *ven-* = vein) Opening of a vein for withdrawal of blood. Although **phlebotomy** (fle-BOT-ō-mē) is a synonym for venesection, in clinical practice phlebotomy refers to therapeutic bloodletting, such as the removal of some blood to lower its viscosity in a patient with polycythemia.

Whole blood Blood containing all formed elements, plasma, and plasma solutes in natural concentrations.

STUDY OUTLINE

INTRODUCTION (p. 634)

1. The cardiovascular system consists of the blood, heart, and blood vessels.
2. Blood is a connective tissue composed of plasma (liquid portion) and formed elements (cells and cell fragments).

FUNCTIONS AND PROPERTIES OF BLOOD (p. 634)

1. Blood transports oxygen, carbon dioxide, nutrients, wastes, and hormones.
2. It helps regulate pH, body temperature, and water content of cells.
3. It provides protection through clotting and by combating toxins and microbes through certain phagocytic white blood cells or specialized plasma proteins.
4. Physical characteristics of blood include a viscosity greater than that of water; a temperature of 38°C (100.4°F); and a pH of 7.35–7.45.
5. Blood constitutes about 8% of body weight, and its volume is 4–6 liters in adults.
6. Blood is about 55% plasma and 45% formed elements.
7. The hematocrit is the percentage of total blood volume occupied by red blood cells.
8. Plasma consists of 91.5% water and 8.5% solutes. Principal solutes include proteins (albumins, globulins, fibrinogen), nutrients, vitamins, hormones, respiratory gases, electrolytes, and waste products.
9. The formed elements in blood include red blood cells (erythrocytes), white blood cells (leukocytes), and platelets.

FORMATION OF BLOOD CELLS (p. 637)

1. Hemopoiesis is the formation of blood cells from hemopoietic stem cells in red bone marrow.
2. Myeloid stem cells form RBCs, platelets, granulocytes, and monocytes. Lymphoid stem cells give rise to lymphocytes.
3. Several hemopoietic growth factors stimulate differentiation and proliferation of the various blood cells.

RED BLOOD CELLS (p. 639)

1. Mature RBCs are biconcave discs that lack nuclei and contain hemoglobin.
2. The function of the hemoglobin in red blood cells is to transport oxygen and some carbon dioxide.
3. RBCs live about 120 days. A healthy male has about 5.4 million RBCs/μL of blood; a healthy female has about 4.8 million/μL.
4. After phagocytosis of aged RBCs by macrophages, hemoglobin is recycled.
5. RBC formation, called erythropoiesis, occurs in adult red bone marrow of certain bones. It is stimulated by hypoxia, which stimulates release of erythropoietin by the kidneys.
6. A reticulocyte count is a diagnostic test that indicates the rate of erythropoiesis.

WHITE BLOOD CELLS (p. 643)

1. WBCs are nucleated cells. The two principal types are granulo-cytes (neutrophils, eosinophils, and basophils) and agranulocytes (lymphocytes and monocytes).
2. The general function of WBCs is to combat inflammation and infection. Neutrophils and macrophages (which develop from monocytes) do so through phagocytosis.
3. Eosinophils combat the effects of histamine in allergic reactions, phagocytize antigen–antibody complexes, and combat parasitic worms. Basophils liberate heparin, histamine, and serotonin in allergic reactions that intensify the inflammatory response.
4. B lymphocytes, in response to the presence of foreign substances called antigens, differentiate into plasma cells that produce antibodies. Antibodies attach to the antigens and render them harmless. This antigen–antibody response combats infection and provides immunity. T lymphocytes destroy foreign invaders directly. Natural killer cells attack infectious microbes and tumor cells.
5. Except for lymphocytes, which may live for years, WBCs usually live for only a few hours or a few days. Normal blood contains 5000–10,000 WBCs/μL.

PLATELETS (p. 645)

1. Platelets (thrombocytes) are disc-shaped cell fragments that splinter from megakaryocytes. Normal blood contains 150,000–400,000 platelets/μL.
2. Platelets help stop blood loss from damaged blood vessels by forming a platelet plug.

HEMOSTASIS (p. 647)

1. Hemostasis refers to the stoppage of bleeding.
2. It involves vascular spasm, platelet plug formation, and blood clotting (coagulation).
3. In vascular spasm, the smooth muscle of a blood vessel wall contracts, which slows blood loss.
4. Platelet plug formation involves the aggregation of platelets to stop bleeding.
5. A clot is a network of insoluble protein fibers (fibrin) in which formed elements of blood are trapped.
6. The chemicals involved in clotting are known as clotting (coagulation) factors.
7. Blood clotting involves a cascade of reactions that may be divided into three stages: formation of prothrombinase, conversion of prothrombin into thrombin, and conversion of soluble fibrinogen into insoluble fibrin.
8. Clotting is initiated by the interplay of the extrinsic and intrinsic pathways of blood clotting.
9. Normal coagulation requires vitamin K and is followed by clot retraction (tightening of the clot) and ultimately fibrinolysis (dissolution of the clot).
10. Clotting in an unbroken blood vessel is called thrombosis. A thrombus that moves from its site of origin is called an embolus.

BLOOD GROUPS AND BLOOD TYPES (p. 651)

1. ABO and Rh blood groups are genetically determined and based on antigen–antibody responses.

2. In the ABO blood group, the presence or absence of A and B antigens on the surface of RBCs determines blood type.

3. In the Rh system, individuals whose RBCs have Rh antigens are classified as Rh$^+$; those who lack the antigen are Rh$^-$.

4. Hemolytic disease of the newborn (HDN) can occur when an Rh$^-$ mother is pregnant with an Rh$^+$ fetus.

5. Before blood is transfused, a recipient's blood is typed and then either cross-matched to potential donor blood or screened for the presence of antibodies.

SELF-QUIZ QUESTIONS

Fill in the blanks in the following statements.

1. The components of blood are _____ and _____, which include _____, _____, and _____.

2. Plasma minus its clotting proteins is termed _____.

3. Cell fragments enclosed by a piece of the plasma membrane of megakaryocytes that contain serotonin and clotting factors are called _____ or _____.

Indicate whether the following statements are true or false.

4. Hemoglobin functions in transporting both oxygen and carbon dioxide and in regulating blood pressure.

5. When incompatible blood is mixed at room temperature outside of the body, it causes hemolysis of the red blood cells.

Choose the one best answer to the following questions.

6. Which of the following are functions of blood? (1) transportation, (2) excretion, (3) regulation, (4) protection, (5) secretion. (a) 1, 2, 3, and 4, (b) 1,2,4, and 5, (c) 1, 2, and 4, (d) 2, 4, and 5, (e) 1, 3, and 4.

7. The plasma proteins called antibodies are (a) globulins, (b) albumins, (c) fibrinogen, (d) thrombin, (e) fibrin.

8. Which of the following are *not* required for clot formation? (1) vitamin K, (2) calcium, (3) prostacyclin, (4) plasmin, (5) fibrinogen. (a) 1, 2, and 5, (b) 3, 4, and 5, (c) 4 and 5, (d) 1, 2, and 3, (e) 3 and 4.

9. Place the steps involved in hemostasis in the correct order. (1) conversion of fibrinogen into fibrin, (2) conversion of prothrombin into thrombin, (3) adhesion and aggregation of platelets on damaged vessel, (4) prothrombinase formed by extrinsic or intrinsic pathway, (5) reduction of blood loss by initiation of a vascular spasm. (a) 5, 3, 4, 2, 1, (b) 5, 4, 3, 1, 2, (c) 3, 5, 4, 2, 1, (d) 5, 3, 2, 1, 4, (e) 5, 3, 2, 4, 1.

10. Which of the following statements explain why red blood cells (RBCs) are highly specialized for oxygen transport? (1) RBCs contain hemoglobin. (2) RBCs lack a nucleus. (3) RBCs have many mitochondria and thus generate ATP aerobically. (4) The biconcave shape of RBCs provides a large surface area for the inward and outward diffusion of gas molecules. (5) RBCs can carry up to four oxygen molecules for each hemoglobin molecule. (a) 1, 2, 3, and 5, (b) 1, 2, 4, and 5, (c) 2, 3, 4, and 5, (d) 1, 3, and 5, (e) 2, 4, and 5.

11. Which of the following are *true* statements? (1) White blood cells leave the bloodstream by emigration. (2) Adhesion molecules help white blood cells stick to the endothelium, which aids emigration. (3) Neutrophils and macrophages are active in phagocytosis. (4) The attraction of phagocytes to microbes and inflamed tissue is termed chemotaxis. (5) Leucopenia is an increase in white blood cell count that occurs during infection. (a) 1, 2, 4, and 5, (b) 2, 3, 4, and 5, (c) 1, 2, 3, and 4, (d) 1, 3, and 5, (e) 1, 2, and 4.

12. A person with type A Rh$^-$ blood can receive a transfusion with blood of which of the following types? (1) A Rh$^+$, (2) B Rh$^-$, (3) AB Rh$^-$, (4) O Rh$^-$, (5) A Rh$^-$. (a) 1 only, (b) 3 only, (c) 4 only, (d) 4 and 5, (e) 1 and 5.

13. Match the following:

_____ (a) contain hemoglobin and function in gas transport

_____ (b) young neutrophils showing a rod-shaped nucleus

_____ (c) white blood cell showing a kidney-shaped nucleus; capable of phagocytosis

_____ (d) monocytes that roam the tissues and gather at sites of infection or inflammation

_____ (e) occur as B cells, T cells, and natural killer cells

_____ (f) combat the effects of histamine and other mediators of inflammation in allergic reactions; also phagocytize antigen-antibody complexes

_____ (g) respond to tissue destruction by bacteria; release lysozyme, strong oxidants, and defensins

_____ (h) older neutrophils with several differently shaped nuclear lobes

_____ (i) monocytes that leave the blood and reside in a particular tissue, such as Kupffer cells in the liver

_____ (j) involved in inflammatory and allergic reactions; are involved in hypersensitivity reactions

(1) neutrophils
(2) lymphocytes
(3) monocytes
(4) eosinophils
(5) basophils
(6) red blood cells
(7) polymorphs
(8) bands
(9) fixed macrophages
(10) wandering macrophages

14. Match the following:

____(a) individual forms of progenitor cells; named on the basis of the mature elements in blood they will ultimately produce

____(b) cells that give rise to all the formed elements of blood; derived from mesenchyme

____(c) give rise to red blood cells and platelets but not to B cells

____(d) give rise to lymphocytes but not to neutrophils

____(e) hormone that increases the numbers of red blood cell precursors

____(f) cells no longer capable of replenishing themselves; can only give rise to more specific formed elements of blood

____(g) stimulate white blood cell formation

____(h) released from the red bone marrow, they develop into mature red blood cells

____(i) hormone that stimulates formation of platelets

(1) pluripotent stem cells
(2) myeloid stem cells
(3) lymphoid stem cells
(4) reticulocytes
(5) colony-forming units
(6) erythropoietin
(7) thrombopoietin
(8) progenitor cell
(9) cytokines

15. Match the following:

____(a) tissue protein that leaks into the blood from cells outside blood vessels and initiates the formation of prothrombinase

____(b) an anticoagulant

____(c) required for synthesis of blood clotting factors

____(d) its formation is initiated by either the extrinsic or intrinsic pathway or both; catalyzes the conversion of prothrombin to thrombin

____(e) glycoproteins and glycolipids on the surfaces of red blood cells that can act as antigens

____(f) forms the threads of a clot; produced from fibrinogen

____(g) can dissolve a clot by digesting fibrin threads

____(h) serves as the catalyst to form fibrin; formed from prothrombin

(1) prothrombinase
(2) thrombin
(3) fibrin
(4) thromboplastin
(5) plasmin
(6) heparin
(7) agglutinogens
(8) vitamin K

CRITICAL THINKING QUESTIONS

1. During his anatomy class, Josef accidently cut himself while dissecting a sheep heart. Describe the composition of the red fluid flowing from the cut.

HINT *It's a liquid connective tissue.*

2. Last year, prior to donating blood, it was determined that Josef's blood type was B negative. Explain how they determined Josef's blood type.

HINT *The "B" and the "negative" refer to two different antigens.*

3. When Josef removed the bandage from his hand later that day, the bleeding had stopped. How did this happen?

HINT *A cell fragment plays a major role in this process.*

ANSWERS TO FIGURE QUESTIONS

19.1 Blood volume is about 8% of your body mass, roughly 5–6 liters in males and 4–5 liters in females. For instance, a 70 kg person has a blood volume of 5.6 liters (70 kg × 8% × 1 liter/kg).

19.2 Platelets are cell fragments.

19.3 Pluripotent stem cells develop from mesenchyme.

19.4 One hemoglobin molecule can transport a maximum of four O_2 molecules, one O_2 bound to each heme group.

19.5 Transferrin is plasma protein that transports iron in the blood.

19.6 Once you moved to high altitude, your hematocrit would increase due to increased secretion of erythropoietin.

19.7 Neutrophils, eosinophils, and basophils are called granulocytes because all have cytoplasmic granules that are visible through a light microscope when stained.

19.8 Lymphocytes recirculate from blood to tissues and back to blood. After leaving the blood, other WBCs remain in the tissues until they die.

19.9 Besides platelet plug formation, vascular spasm and blood clotting contribute to hemostasis.

19.10 Serum is blood plasma minus the clotting proteins.

19.11 The outcome of the first stage of clotting is the formation of prothrombinase.

19.12 Type O blood usually contains both anti-A and anti-B antibodies.

19.13 Because the mother is most likely to start making anti-Rh antibodies after the first baby is already born, that baby suffers no damage.

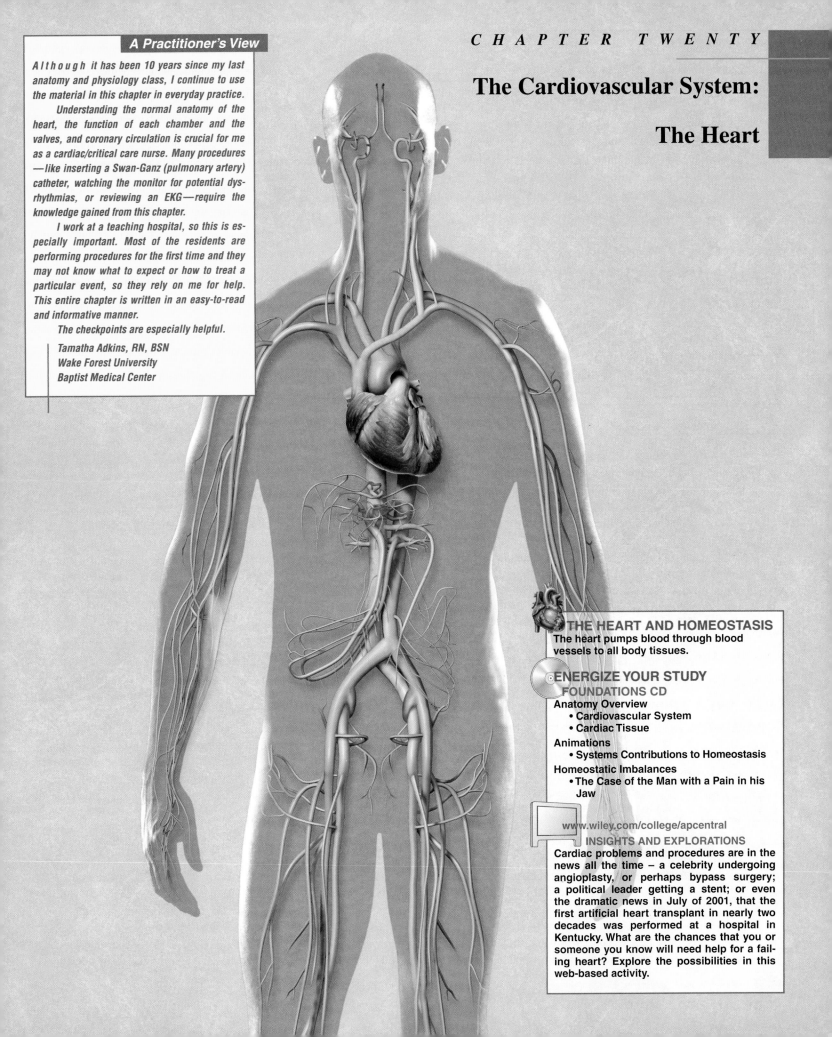

CHAPTER TWENTY

The Cardiovascular System:

The Heart

A Practitioner's View

Although it has been 10 years since my last anatomy and physiology class, I continue to use the material in this chapter in everyday practice.

Understanding the normal anatomy of the heart, the function of each chamber and the valves, and coronary circulation is crucial for me as a cardiac/critical care nurse. Many procedures —like inserting a Swan-Ganz (pulmonary artery) catheter, watching the monitor for potential dysrhythmias, or reviewing an EKG—require the knowledge gained from this chapter.

I work at a teaching hospital, so this is especially important. Most of the residents are performing procedures for the first time and they may not know what to expect or how to treat a particular event, so they rely on me for help. This entire chapter is written in an easy-to-read and informative manner.

The checkpoints are especially helpful.

Tamatha Adkins, RN, BSN
Wake Forest University
Baptist Medical Center

THE HEART AND HOMEOSTASIS

The heart pumps blood through blood vessels to all body tissues.

ENERGIZE YOUR STUDY

FOUNDATIONS CD

Anatomy Overview
- Cardiovascular System
- Cardiac Tissue

Animations
- Systems Contributions to Homeostasis

Homeostatic Imbalances
- The Case of the Man with a Pain in his Jaw

www.wiley.com/college/apcentral

INSIGHTS AND EXPLORATIONS

Cardiac problems and procedures are in the news all the time – a celebrity undergoing angioplasty, or perhaps bypass surgery; a political leader getting a stent; or even the dramatic news in July of 2001, that the first artificial heart transplant in nearly two decades was performed at a hospital in Kentucky. What are the chances that you or someone you know will need help for a failing heart? Explore the possibilities in this web-based activity.

The **cardiovascular system** consists of the blood, the heart, and blood vessels. In the last chapter we examined the composition and functions of blood. For blood to reach body cells and exchange materials with them, it must be constantly propelled through blood vessels. The heart is the pump that circulates the blood through an estimated 100,000 km (60,000 mi) of blood vessels. It beats about 100,000 times every day, which adds up to 35 million beats in a year. Even while you are sleeping, your heart pumps 30 times its own weight each minute, about 5 liters (5.3 qt) to the lungs and the same volume to the rest of the body. At this rate, the heart pumps more than 14,000 liters (3,600 gal) of blood in a day. You don't spend all your time sleeping, however, and your heart pumps even more vigorously when you are active. The study of the normal heart and the diseases associated with it is **cardiology** (kar-dē-OL-ō-jē; *cardio-* = heart; *-logy* = study of). This chapter explores the design of the heart and the unique properties that permit it to pump for a lifetime without rest.

ANATOMY OF THE HEART

> O B J E C T I V E S

- **Describe the location of the heart, and trace its outline on the surface of the chest.**
- **Describe the structure of the pericardium and the heart wall.**
- **Discuss the external and internal anatomy of the chambers of the heart.**

Location of the Heart

For all its might, the cone-shaped heart is relatively small, roughly the same size as a closed fist—about 12 cm (5 in.) long, 9 cm (3.5 in.) wide at its broadest point, and 6 cm (2.5 in.) thick. Its mass averages 250 g (8 oz) in adult females and 300 g (10 oz) in adult males. The heart rests on the diaphragm, near the midline of the thoracic cavity. It lies in the **mediastinum** (mē-dē-a-STĪ-num), a mass of tissue that extends from the sternum to the vertebral column between the lungs (Figure 20.1a). About two-thirds of the mass of the heart lies to the left of the body's midline (Figure 20.1b). Visualize the heart as a cone lying on its side. The pointed end of the heart is the **apex**, which is directed anteriorly, inferiorly, and to the left. The broad portion of the heart opposite the apex is the **base**, which is directed posteriorly, superiorly, and to the right.

In addition to the apex and base, the heart has several surfaces and borders (margins). The **anterior surface** is deep to the sternum and ribs. The **inferior surface** is the part of the heart between the apex and right border and rests mostly on the diaphragm (Figure 20.1b). The **right border** faces the right lung and extends from the inferior surface to the base. The **left border,** also called the *pulmonary border,* faces the left lung and extends from the base to the apex.

Cardiopulmonary Resuscitation

Because the heart lies between two rigid structures—the vertebral column and the sternum (Figure 20.1a)—external pressure (compression) on the chest can be used to force blood out of the heart and into the circulation. In cases in which the heart suddenly stops beating, **cardiopulmonary resuscitation (CPR)**—properly applied cardiac compressions, performed together with artificial ventilation of the lungs—saves lives. It keeps oxygenated blood circulating until the heart can be restarted. Amazingly, it may be possible to perform CPR on yourself. Periodic very forceful coughs appear to squeeze enough blood out of the heart to sustain circulation until emergency help arrives. ■

Pericardium

The membrane that surrounds and protects the heart is the **pericardium** (*peri-* = around). It confines the heart to its position in the mediastinum, while allowing sufficient freedom of movement for vigorous and rapid contraction. The pericardium consists of two main parts: the fibrous pericardium and the serous pericardium (Figure 20.2a on page 662). The superficial **fibrous pericardium** is a tough, inelastic, dense irregular connective tissue. It resembles a bag that rests on and attaches to the diaphragm; its open end is fused to the connective tissues of the blood vessels entering and leaving the heart. The fibrous pericardium prevents overstretching of the heart, provides protection, and anchors the heart in the mediastinum.

The deeper **serous pericardium** is a thinner, more delicate membrane that forms a double layer around the heart (Figure 20.2a). The outer **parietal layer** of the serous pericardium is fused to the fibrous pericardium. The inner **visceral layer** of the serous pericardium, also called the **epicardium** (*epi-* = on top of), adheres tightly to the surface of the heart. Between the parietal and visceral layers of the serous pericardium is a thin film of serous fluid. This fluid, known as **pericardial fluid,** is a slippery secretion of the pericardial cells that reduces friction between the membranes as the heart moves. The space that contains the few milliliters of pericardial fluid is called the **pericardial cavity.**

Pericarditis and Cardiac Tamponade

Inflammation of the pericardium is known as **pericarditis.** If production of pericardial fluid diminishes, painful rubbing together of the parietal and visceral serous pericardial layers may result. A buildup of pericardial fluid (which may also occur in pericarditis) or extensive bleeding into the pericardium are life-threatening conditions. Because the pericardium cannot

Figure 20.1 **Position of the heart and associated structures in the mediastinum (dashed outline).** (See Tortora, *A Photographic Atlas of the Human Body,* Figures 6.5 and 6.6.)

The heart is located in the mediastinum, with two-thirds of its mass to the left of the midline.

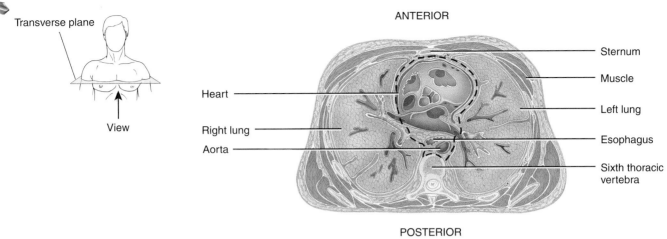

(a) Inferior view of transverse section of thoracic cavity showing the heart in the mediastinum

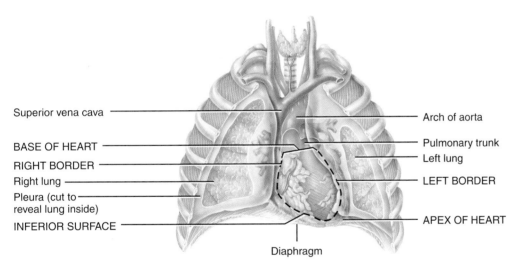

(b) Anterior view of the heart in the mediastinum

What is the mediastinum?

stretch, the buildup of fluid or blood compresses the heart. This compression, known as **cardiac tamponade** (tam′-pon-ĀD), can stop the beating of the heart. ■

Layers of the Heart Wall

The wall of the heart consists of three layers (Figure 20.2a): the epicardium (external layer), the myocardium (middle layer), and the endocardium (inner layer). The outermost **epicardium,** also called the *visceral layer of the serous pericardium,* is the thin, transparent outer layer of the wall. It is composed of mesothelium and delicate connective tissue that imparts a smooth, slip-pery texture to the outermost surface of the heart. The middle **myocardium** (*myo-* = muscle), which is cardiac muscle tissue, makes up the bulk of the heart and is responsible for its pumping action. Although it is striated like skeletal muscle, cardiac muscle is involuntary like smooth muscle. The cardiac muscle fibers swirl diagonally around the heart in interlacing bundles (Figure 20.2c). The innermost **endocardium** (*endo-* = within) is a thin layer of endothelium overlying a thin layer of connective tissue. It provides a smooth lining for the chambers of the heart and covers the valves of the heart. The endocardium is continuous with the endothelial lining of the large blood vessels attached to the heart.

Figure 20.2   Pericardium and heart wall.

 The pericardium is a triple-layered sac that surrounds and protects the heart.

Pericardium
Epicardium
Myocardium
Endocardium

PERICARDIUM

Heart wall

FIBROUS
PERICARDIUM

PARIETAL
LAYER OF
SEROUS
PERICARDIUM

Pericardial
cavity

VISCERAL
LAYER OF
SEROUS
PERICARDIUM
(EPICARDIUM)

ENDOCARDIUM

Trabeculae
carneae

Coronary blood
vessels

MYOCARDIUM
(CARDIAC MUSCLE)

(a) Portion of pericardium and right ventricular heart wall
showing the divisions of the pericardium and layers of
the heart wall

Heart

Parietal layer
of serous
pericardium

Pericardial
cavity

Serous pericardium

Pericardial
cavity

Visceral layer
of serous
pericardium

(b) Simplified relationship of the serous pericardium to the heart

Superior
vena cava

Aorta

Pulmonary
trunk

Superficial
muscle
bundles
in atria

Deep
muscle
bundle in
ventricle

Superficial
muscle
bundles in
ventricles

(c) Cardiac muscle bundles of the myocardium

Which layer is both a part of the pericardium and a part of the heart wall?

Chambers of the Heart

The heart has four chambers. The two superior chambers are the **atria** (= entry halls or chambers), and the two inferior chambers are the **ventricles** (= little bellies). On the anterior surface of each atrium is a wrinkled pouchlike structure called an **auricle** (OR-i-kul; *auri-* = ear), so named because of its resemblance to a dog's ear (Figure 20.3). Each auricle slightly increases the capacity of an atrium so that it can hold a greater volume of blood. Also on the surface of the heart are a series of grooves, called **sulci** (SUL-sē), that contain coronary blood vessels and a variable amount of fat. Each sulcus (SUL-kus) marks the external boundary between two chambers of the heart. The deep **coronary sulcus** (*coron-* = resembling a crown) encircles most of the heart and marks the boundary between the superior atria and inferior ventricles. The **anterior interventricular sulcus** is a shallow groove on the anterior surface of the heart that marks the boundary between the right and left ventricles. This sulcus

continues around to the posterior surface of the heart as the **posterior interventricular sulcus,** which marks the boundary between the ventricles on the posterior aspect of the heart (Figure 20.3c).

Right Atrium

The **right atrium** receives blood from three veins: the *superior vena cava, inferior vena cava,* and *coronary sinus* (Figure 20.4a on page 665). The anterior and posterior walls within the right atrium differ considerably. Whereas the posterior wall is smooth, the anterior wall is rough due to the presence of muscular ridges called **pectinate muscles** (PEK-tin-āt; *pectin* = comb), which also extend into the auricle (Figure 20.4b). Between the right atrium and left atrium is a thin partition called the **interatrial septum** (*inter-* = between; *septum* = a dividing wall or partition). A prominent feature of this septum is an oval depression called the **fossa ovalis,** which is

Figure 20.3   Structure of the heart: surface features.

Sulci and grooves that contain blood vessels and fat and mark the boundaries between the various chambers.

(a) Anterior external view showing surface features

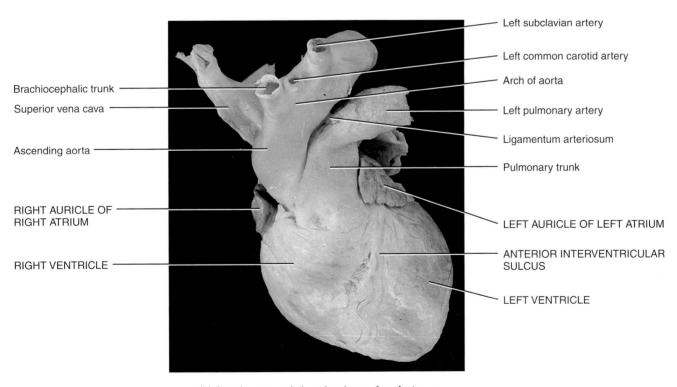

(b) Anterior external view showing surface features

(continues)

(Figure 20.3 continued)

Left common carotid artery

Left subclavian artery

Arch of aorta

Descending aorta

Left pulmonary artery

AURICLE OF LEFT ATRIUM

Left pulmonary veins

LEFT ATRIUM

Coronary sinus

LEFT VENTRICLE

POSTERIOR
INTERVENTRICULAR SULCUS

Brachiocephalic trunk

Superior vena cava

Ascending aorta

Right pulmonary artery

Right pulmonary veins

RIGHT ATRIUM

Right coronary artery

Inferior vena cava

Middle cardiac vein

RIGHT VENTRICLE

(c) Posterior external view showing surface features

The coronary sulcus forms a boundary between which chambers of the heart?

the remnant of the foramen ovale, an opening in the inter-atrial septum of the fetal heart that normally closes soon after birth (see Figure 21.31 on page 754). Blood passes from the right atrium into the right ventricle through a valve called the **tricuspid valve** (trī-KUS-pid; *tri-* = three; *cuspid* = point) because it consists of three leaflets or cusps (Figure 20.4a). The valves of the heart are composed of dense connective tissue covered by endocardium.

Right Ventricle

The **right ventricle** forms most of the anterior surface of the heart. The inside of the right ventricle contains a series of ridges formed by raised bundles of cardiac muscle fibers called **trabeculae carneae** (tra-BEK-ū-lē KAR-nē-ē; *trabeculae* = little beams; *carneae* = fleshy). Some of the trabeculae carneae convey part of the conduction system of the heart (see page 674). The cusps of the tricuspid valve are connected to tendonlike cords, the **chordae tendineae** (KOR-dē ten-DIN-ē-ē; *chord-* = cord; *tend-* = tendon), which, in turn, are connected to cone-shaped trabeculae carneae called **papillary muscles** (*papill-* = nipple). The right ventricle is separated from the left ventricle by a partition called the **interventricular septum.** Blood passes from the right ventricle through the

pulmonary valve into a large artery called the *pulmonary trunk,* which divides into right and left *pulmonary arteries.*

Left Atrium

The **left atrium** forms most of the base of the heart. It receives blood from the lungs through four *pulmonary veins.* Like the right atrium, the inside of the left atrium has a smooth posterior wall. Because pectinate muscles are confined to the auricle of the left atrium, the anterior wall of the left atrium also is smooth. Blood passes from the left atrium into the left ventricle through the **bicuspid (mitral) valve** (*bi-* = two), which has two cusps. The term *mitral* refers to its resemblance to a bishop's miter (hat), which is two-sided.

Left Ventricle

The **left ventricle** forms the apex of the heart. Like the right ventricle, the left ventricle also contains trabeculae carneae and has chordae tendinae that anchor the cusps of the bicuspid valve to papillary muscles. Blood passes from the left ventricle through the **aortic valve** into the largest artery of the body, the *ascending aorta* (*aorte* = to suspend, because the aorta once was believed to lift up the heart). Some of the blood in the aorta flows into the *coronary arteries,* which branch from the

Figure 20.4 **Structure of the heart: internal anatomy.**

Blood flows into the right atrium through the superior vena cava, inferior vena cava, and coronary sinus and into the left atrium through four pulmonary veins.

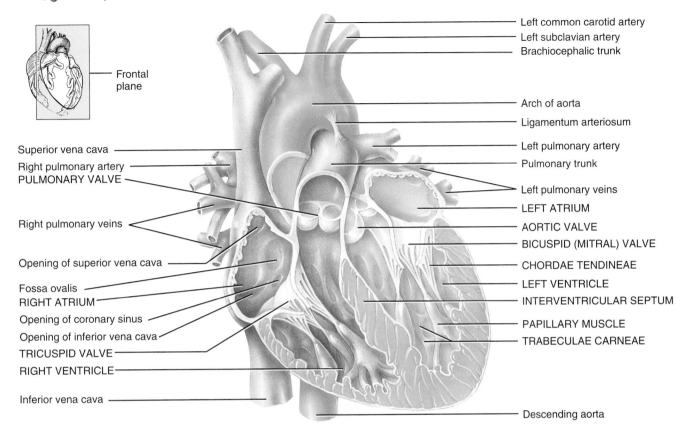

Frontal plane

Left common carotid artery
Left subclavian artery
Brachiocephalic trunk

Arch of aorta
Ligamentum arteriosum

Superior vena cava
Right pulmonary artery
PULMONARY VALVE

Left pulmonary artery
Pulmonary trunk

Left pulmonary veins
LEFT ATRIUM
AORTIC VALVE
BICUSPID (MITRAL) VALVE

Right pulmonary veins

CHORDAE TENDINEAE

Opening of superior vena cava

LEFT VENTRICLE
INTERVENTRICULAR SEPTUM

Fossa ovalis
RIGHT ATRIUM
Opening of coronary sinus
Opening of inferior vena cava
TRICUSPID VALVE
RIGHT VENTRICLE

PAPILLARY MUSCLE
TRABECULAE CARNEAE

Inferior vena cava

Descending aorta

(a) Anterior view of frontal section showing internal anatomy

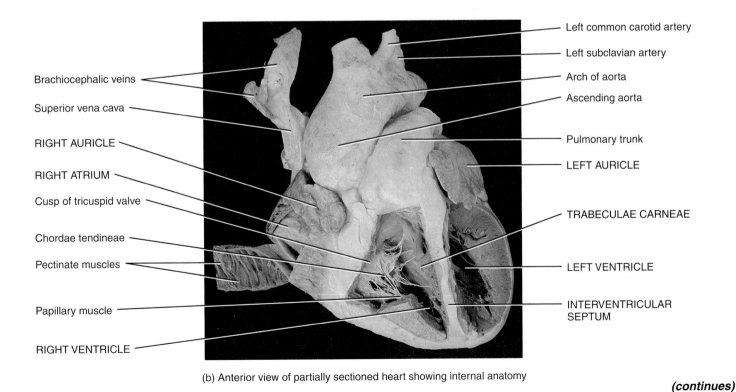

Left common carotid artery

Left subclavian artery

Brachiocephalic veins

Arch of aorta

Ascending aorta

Superior vena cava

RIGHT AURICLE

Pulmonary trunk

RIGHT ATRIUM

LEFT AURICLE

Cusp of tricuspid valve

Chordae tendineae

TRABECULAE CARNEAE

Pectinate muscles

Papillary muscle

LEFT VENTRICLE

INTERVENTRICULAR SEPTUM

RIGHT VENTRICLE

(b) Anterior view of partially sectioned heart showing internal anatomy

(continues)

(Figure 20.4 continued)

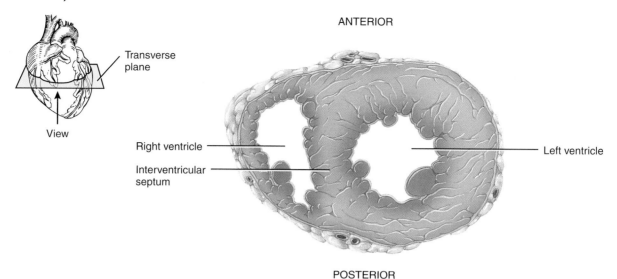

ANTERIOR

Transverse plane

View

Right ventricle

Interventricular septum

Left ventricle

POSTERIOR

(c) Inferior view of transverse section showing differences in thickness of ventricular walls

 How does thickness of the myocardium relate to the workload of a cardiac chamber?

ascending aorta and carry blood to the heart wall. The remainder of the blood passes into the *arch of the aorta* and *descending aorta* (*thoracic aorta* and *abdominal aorta*). Branches of the arch of the aorta and descending aorta carry blood throughout the body.

During fetal life, a temporary blood vessel, called the *ductus arteriosus,* shunts blood from the pulmonary trunk into the aorta. Hence, only a small amount of blood enters the nonfunctioning fetal lungs (see Figure 21.31). The ductus arteriosus normally closes shortly after birth, leaving a remnant known as the **ligamentum arteriosum,** which connects the arch of the aorta and pulmonary trunk (Figure 20.4a).

Myocardial Thickness and Function

The thickness of the myocardium of the four chambers varies according to each chamber's function. The thin-walled atria deliver blood into the adjacent ventricles. Because the ventricles pump blood greater distances, their walls are thicker (Figure 20.4a). Although the right and left ventricles act as two separate pumps that simultaneously eject equal volumes of blood, the right side has a much smaller workload. It pumps blood a short distance to the lungs at lower pressure and the resistance to blood flow is small. The left ventricle pumps blood great distances to all other parts of the body at higher pressure, and the resistance to blood flow is larger. Therefore, the left ventricle works much harder than the right ventricle to maintain the same rate of blood flow. The anatomy of the two ventricles confirms this functional difference—the muscular wall of the left ventricle is considerably thicker than the wall of the right ventricle (Figure 20.4c). Note also that the perimeter of the lumen (space)

of the left ventricle is roughly circular, whereas that of the right ventricle is somewhat crescent shaped.

Fibrous Skeleton of the Heart

In addition to cardiac muscle tissue, the heart wall also contains dense connective tissue that forms the **fibrous skeleton of the heart** (Figure 20.5). Essentially, the fibrous skeleton consists of dense connective tissue rings that surround the valves of the heart, fuse with one another, and merge with the interventricular septum. Four fibrous rings support the four valves of the heart and are fused to each other. The fibrous skeleton forms the foundation to which the heart valves attach and prevents overstretching of the valves as blood passes through them. It also serves as a point of insertion for bundles of cardiac muscle fibers and acts as an electrical insulator between the atria and ventricles.

▶ C H E C K P O I N T

1. Where are the apex and base of the heart?

2. Define each of the following external features of the heart: auricle, coronary sulcus, anterior interventricular sulcus, and posterior interventricular sulcus.

3. What are the characteristic internal features of each chamber of the heart?

4. Which blood vessels deliver blood to the right and left atria?

5. What is the relationship between wall thickness and function for the chambers of the heart?

6. What type of tissue composes the fibrous skeleton of the heart? What functions does this tissue perform?

Figure 20.5 Fibrous skeleton of the heart (depicted in blue).

 Fibrous rings support the four valves of the heart and are fused to each other.

Pulmonary valve — PULMONARY FIBROUS RING
Left coronary artery — CONUS TENDON
Aortic valve — AORTIC FIBROUS RING
LEFT FIBROUS TRIGONE — Right coronary artery
RIGHT FIBROUS TRIGONE —
Bicuspid valve — Tricuspid valve
LEFT ATRIOVENTRICULAR FIBROUS RING — RIGHT ATRIOVENTRICULAR FIBROUS RING

Superior view (the atria have been removed)

In what two ways does the fibrous skeleton contribute to the functioning of heart valves?

HEART VALVES AND CIRCULATION OF BLOOD

▶ O B J E C T I V E S

- **Describe the structure and function of the valves of the heart.**
- **Describe the flow of blood through the chambers of the heart and through the systemic and pulmonary circulations.**
- **Discuss the coronary circulation.**

As each chamber of the heart contracts, it pushes a volume of blood into a ventricle or out of the heart into an artery. Valves open and close in response to *pressure changes* as the heart contracts and relaxes. Each of the four valves helps ensure the one-way flow of blood by opening to let blood through and then closing to prevent the backflow of blood.

Operation of the Atrioventricular Valves

Because they are located between an atrium and a ventricle, the tricuspid and bicuspid valves are termed **atrioventricular (AV) valves.** When an AV valve is open, the pointed ends of the cusps project into the ventricle. Blood moves from the atria into the ventricles through open AV valves when atrial pressure is higher than ventricular pressure (Figure 20.6a, c). At this time, the papillary muscles are relaxed, and the chordae tendineae are slack. When the ventricles contract, the pressure of the blood drives the

Figure 20.6 Responses of the valves to the pumping of the heart.

Heart valves prevent the backflow of blood.

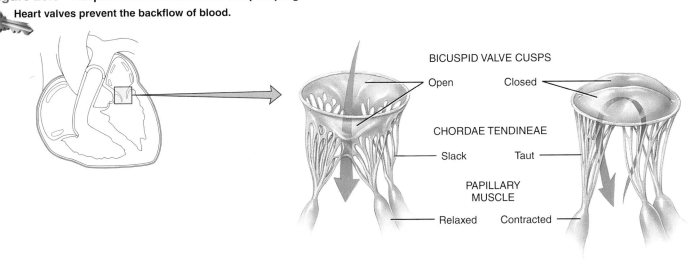

BICUSPID VALVE CUSPS
Open   Closed

CHORDAE TENDINEAE
Slack   Taut

PAPILLARY MUSCLE
Relaxed   Contracted

(a) Bicuspid valve open   (b) Bicuspid valve closed

(continues)

(Figure 20.6 continued)

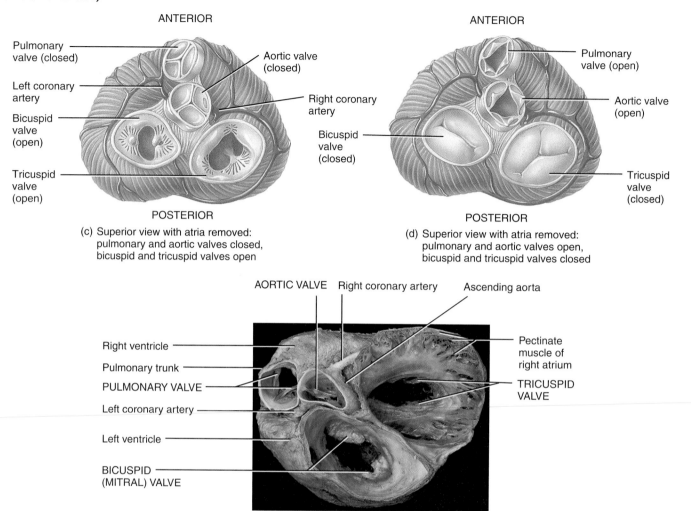

(c) Superior view with atria removed: pulmonary and aortic valves closed, bicuspid and tricuspid valves open

(d) Superior view with atria removed: pulmonary and aortic valves open, bicuspid and tricuspid valves closed

(e) Superior view of atrioventricular and semilunar valves

 How do papillary muscles prevent AV valve cusps from everting or swinging upward into the atria?

cusps upward until their edges meet and close the opening (Figure 20.6b, d). At the same time, the papillary muscles are also contracting, which pulls on and tightens the chordae tendineae. This prevents the valve cusps from everting (being forced to open in the opposite direction into the atria due to the high ventricular pressure). If the AV valves or chordae tendineae are damaged, blood may regurgitate (flow back) into the atria when the ventricles contract.

Operation of the Semilunar Valves

The aortic and pulmonary valves are known as the **semilunar (SL) valves** (*semi-* = half; *lunar* = moon-shaped) because they are made up of three crescent moon–shaped cusps (Figure 20.6c). Each cusp attaches to the artery wall by its convex outer margin. The SL valves allow ejection of blood from the heart into arteries but prevent backflow of blood into the ventricles. The free borders of the cusps project into the lumen of the artery. When the ventricles contract, pressure builds up within the chambers. The semilunar valves open when pressure in the ventricles exceeds the pressure in the arteries, permitting ejection of blood from the ventricles into the pulmonary trunk and aorta (Figure 20.6d). As the ventricles relax, blood starts to flow back toward the heart. As the back-flowing blood fills the cusps, the semilunar valves close (Figure 20.6c).

Surprisingly perhaps, valves do not guard the entrance of the venae cavae into the right atrium or the pulmonary veins into the left atrium. As the atria contract, a small amount of blood does flow backward into these vessels. However, as the atrial muscle contracts, it compresses and nearly collapses the venous entry points thus minimizing backflow of blood.

 Heart Valve Disorders

When heart valves operate normally, they open fully and close completely at the proper times. Failure of a heart valve to open fully is known as **stenosis** (= a narrowing) whereas failure of a valve to close completely is termed **insufficiency** or **incompetence.** In **mitral stenosis,** scar formation or a congenital defect causes narrowing of the mitral valve. **Mitral insufficiency** allows backflow of blood from the left ventricle into the left atrium. One cause is **mitral valve prolapse (MVP),** in which one or both cusps of the mitral valve protrude into the left atrium during ventricular contraction. Although a small volume of blood may flow back into the left atrium during ventricular contraction, mitral valve prolapse does not always pose a serious threat. Mitral insufficiency also can be due to a damaged mitral valve or ruptured chordae tendineae. In **aortic stenosis** the aortic valve is narrowed, and in **aortic insufficiency** there is backflow of blood from the aorta into the left ventricle.

Certain infectious diseases can damage or destroy the heart valves. One example is **rheumatic fever,** an acute systemic inflammatory disease that usually occurs after a streptococcal infection of the throat. The bacteria trigger an immune response in which antibodies that are produced to destroy the bacteria attack and inflame the connective tissues in joints, heart valves, and other organs. Even though rheumatic fever may weaken the entire heart wall, most often it damages the bicuspid (mitral) and aortic valves. ■

Systemic and Pulmonary Circulations

With each beat, the heart pumps blood into two closed circuits—the **systemic circulation** and the **pulmonary circulation** (*pulmon-* = lung). The two circuits are arranged in series: the output of one becomes the input of the other, as would happen if you attached two garden hoses together. The left side of the heart is the pump for the systemic circulation; it receives bright red, oxygen-rich blood from the lungs. The left ventricle ejects blood into the *aorta* (Figure 20.7a). From the aorta, the blood divides into separate streams, entering progressively smaller *systemic arteries* that carry it to all organs throughout the body—except for the air sacs (alveoli) of the lungs, which are supplied by the pulmonary circulation. In systemic tissues, arteries give rise to smaller-diameter *arterioles,* which finally lead into extensive beds of *systemic capillaries.* Exchange of nutrients and gases occurs across the thin capillary walls. Blood unloads O_2 (oxygen) and picks up CO_2 (carbon dioxide). In most cases, blood flows through only one capillary and then enters a *systemic venule.* Venules carry deoxygenated (oxygen-poor) blood away from tissues and merge to form larger *systemic veins.* Ultimately the blood flows back to the right atrium.

Figure 20.7 Systemic and pulmonary circulations.   Throughout this book, blood vessels that carry oxygenated blood (which looks bright red) are colored red, whereas those that carry deoxygenated blood (which looks dark red) are colored blue.

The left side of the heart pumps oxygenated blood into the systemic circulation to all tissues of the body except the air sacs (alveoli) of the lungs. The right side of the heart pumps deoxygenated blood into the pulmonary circulation to the air sacs (alveoli) of the lungs.

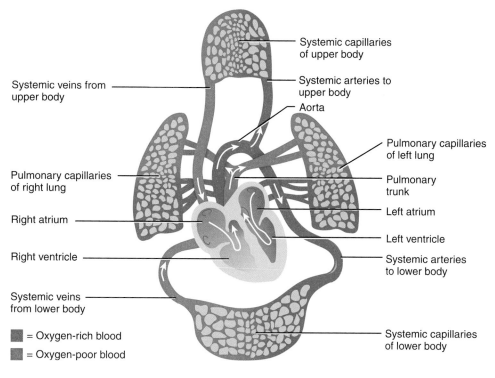

(a) Systemic and pulmonary circulations

(continues)

(Figure 20.7 continued)

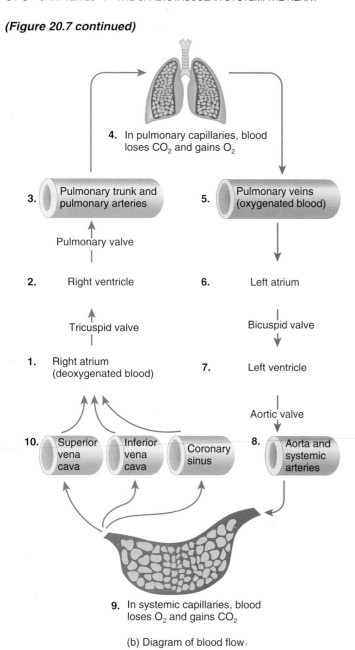

4. In pulmonary capillaries, blood loses CO_2 and gains O_2

3. Pulmonary trunk and pulmonary arteries

5. Pulmonary veins (oxygenated blood)

Pulmonary valve

2. Right ventricle

6. Left atrium

Tricuspid valve

Bicuspid valve

1. Right atrium (deoxygenated blood)

7. Left ventricle

Aortic valve

10. Superior vena cava | Inferior vena cava | Coronary sinus

8. Aorta and systemic arteries

9. In systemic capillaries, blood loses O_2 and gains CO_2

(b) Diagram of blood flow

In part (b), which numbers constitute the pulmonary circulation? Which constitute the systemic circulation?

The right side of the heart is the pump for the pulmonary circulation; it receives all the dark red, deoxygenated blood returning from the systemic circulation. Blood ejected from the right ventricle flows into the *pulmonary trunk,* which branches into *pulmonary arteries* that carry blood to the right and left lungs. In pulmonary capillaries, then, blood unloads CO_2, which is exhaled, and picks up O_2. The freshly oxygenated blood then flows into pulmonary veins and returns to the left atrium. Figure 20.7b reviews the route of blood flow through the chambers and valves of the heart and the pulmonary and systemic circulations.

Coronary Circulation

Nutrients could not diffuse quickly enough from blood in the chambers of the heart to supply all the layers of cells that make up the heart wall. For this reason, the myocardium has its own blood vessels, the **coronary circulation** (*coron-* = crown). The **coronary arteries** branch from the ascending aorta and encircle the heart like a crown encircles the head (Figure 20.8a). While the heart is contracting, little blood flows in the coronary arteries because they are squeezed shut. When the heart relaxes, however, the high pressure of blood in the aorta propels blood through the coronary arteries, into capillaries, and then into **coronary veins** (Figure 20.8b).

Coronary Arteries

Two coronary arteries, the right and left coronary arteries, branch from the ascending aorta and supply oxygenated blood to the myocardium (Figure 20.8a). The **left coronary artery** passes inferior to the left auricle and divides into the anterior interventricular and circumflex branches. The **anterior interventricular branch** or **left anterior descending (LAD) artery** extends along the anterior interventricular sulcus and supplies oxygenated blood to the walls of both ventricles. The **circumflex branch** lies in the coronary sulcus and distributes oxygenated blood to the walls of the left ventricle and left atrium.

The **right coronary artery** gives off small atrial branches that supply the right atrium. It continues inferior to the right auricle and divides into the posterior interventricular and marginal branches. The **posterior interventricular branch** follows the posterior interventricular sulcus and supplies the walls of the two ventricles with oxygenated blood. The **right marginal branch** in the coronary sulcus carries oxygenated blood to the myocardium of the right ventricle.

Most parts of the body receive blood from branches of more than one artery, and where two or more arteries supply the same region, they usually connect. These connections, called **anastomoses** (a-nas'-tō-MŌ-sēs), provide alternate routes for blood to reach a particular organ or tissue. The myocardium contains many anastomoses that connect branches of a given coronary artery or extend between branches of different coronary arteries. They provide detours for arterial blood if a main route becomes obstructed. Thus, heart muscle may receive sufficient oxygen even if one of its coronary arteries is partially blocked.

Coronary Veins

After blood passes through the coronary arteries, it flows into capillaries, where it delivers oxygen and nutrients and collects carbon dioxide and wastes, and then into veins. The deoxygenated blood then drains into a large vascular sinus on the posterior surface of the heart, called the **coronary sinus** (Figure 20.8b), which empties into the right atrium. A vascular sinus is a thin-walled vein that has no smooth muscle to alter its diameter. The main tributaries carrying blood into the coronary sinus are the **great cardiac vein,** which drains the anterior aspect of the heart, and the **middle cardiac vein,** which drains the posterior aspect of the heart.

Reperfusion Damage

When blockage of a coronary artery deprives the heart muscle of oxygen, **reperfusion,** reestablishing the blood flow, may damage the tissue further. This surprising effect is due to the formation of oxygen **free radicals** from the reintroduced oxygen. Free radicals are electrically charged molecules that have an unpaired electron (see Figure 2.3b on page 30). Such molecules are unstable and highly reactive. They cause chain reactions that lead to cellular damage and death. To counter the effects of oxygen free radicals, body cells produce enzymes that convert free radicals to less reactive substances. Two such enzymes are *superoxide dismutase* and *catalase.* In addition, some nutrients, such as vitamin E, vitamin C, beta-carotene, and selenium are antioxidants, which remove oxygen free radicals. Drugs that lessen reperfusion damage after a heart attack or stroke are being developed. ■

▶ C H E C K P O I N T

7. What causes the heart valves to open and to close?

8. In correct sequence, which heart chambers, heart valves, and blood vessels would a drop of blood encounter as it flows out of the right atrium until it reaches the aorta?

9. Which arteries deliver oxygenated blood to the myocardium of the left and right ventricles?

Figure 20.8 **The coronary circulation.** The views of the heart from the anterior aspect in (a) and (b) are drawn as if the heart were transparent to reveal blood vessels on the posterior aspect. (See Tortora, *A Photographic Atlas of the Human Body,* Figures 6.8 and 6.9.)

ARTERIES DELIVER BLOOD TO THE HEART
- VEINS DRAIN BLOOD.

🔑 The right and left coronary arteries deliver blood to the heart; the coronary veins drain blood from the heart into the coronary sinus.

(a) Anterior view of coronary arteries

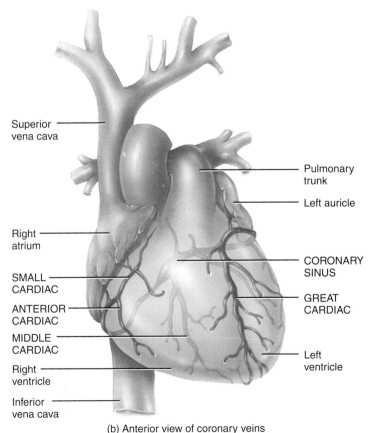

(b) Anterior view of coronary veins

(continues)

(Figure 20.8 continued)

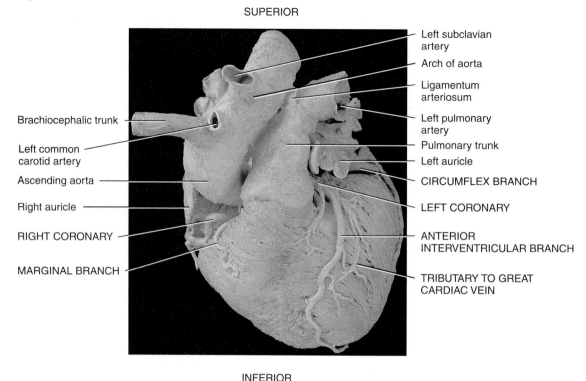

SUPERIOR

Brachiocephalic trunk

Left common
carotid artery

Ascending aorta

Right auricle

RIGHT CORONARY

MARGINAL BRANCH

Left subclavian
artery

Arch of aorta

Ligamentum
arteriosum

Left pulmonary
artery

Pulmonary trunk

Left auricle

CIRCUMFLEX BRANCH

LEFT CORONARY

ANTERIOR
INTERVENTRICULAR BRANCH

TRIBUTARY TO GREAT
CARDIAC VEIN

INFERIOR

(c) Anterior view of coronary blood vessels

 Which coronary blood vessel delivers oxygenated blood to the left atrium and left ventricle?

CARDIAC MUSCLE TISSUE AND THE CARDIAC CONDUCTION SYSTEM

OBJECTIVES

- **Describe the structural and functional characteristics of cardiac muscle tissue and the conduction system of the heart.**
- **Describe how an action potential occurs in cardiac contractile fibers.**
- **Describe the electrical events of a normal electrocardiogram (ECG).**

Histology of Cardiac Muscle Tissue

Compared with skeletal muscle fibers, cardiac muscle fibers are shorter in length and less circular in transverse section (Figure 20.9). They also exhibit branching, which gives individual cardiac muscle fibers a "stair-step" appearance (see Table 4.4B). A typical cardiac muscle fiber is 50–100 μm long and has a diameter of about 14 μm. Usually one centrally located nucleus is present, although an occasional cell may have two nuclei. The ends of cardiac muscle fibers connect to neighboring fibers by irregular transverse thickenings of the sarcolemma called **intercalated discs** (in-TER-kā-lāt-ed; *intercalat-* = to insert between). The discs contain **desmosomes,** which hold the fibers together, and

gap junctions, which allow muscle action potentials to conduct from one muscle fiber to its neighbors.

Mitochondria are larger and more numerous in cardiac than in skeletal muscle fibers. In a cardiac muscle fiber, they take up 25% of the cytosolic space as compared with only 2% of the cytosolic space in a skeletal muscle fiber. Cardiac muscle fibers have the same arrangement of actin and myosin, and the same bands, zones, and Z discs, as skeletal muscle fibers. The transverse tubules of cardiac muscle are wider but less abundant than those of skeletal muscle; the one transverse tubule per sarcomere is located at the Z disc. The sarcoplasmic reticulum of cardiac muscle fibers is somewhat smaller than the SR of skeletal muscle fibers. As a result, cardiac muscle has a smaller intracellular reserve of Ca^{2+}.

Regeneration of Heart Cells

The heart of a heart attack survivor often has regions of infarcted (dead) cardiac muscle tissue that typically are replaced with noncontractile fibrous scar tissue over time. Our inability to repair damage from a heart attack has been attributed to a lack of stem cells in cardiac muscle and to the absence of mitosis in mature cardiac muscle fibers. A recent study of heart transplant recipients by American and Italian scientists, however, provides evidence for significant replacement of heart cells. The researchers

Figure 20.9 Histology of cardiac muscle tissue. (See Table 4.4B for a light microscopic view of cardiac muscle.)

Cardiac muscle fibers connect to neighboring fibers by intercalated discs, which contain desmosomes and gap junctions.

(a) Cardiac muscle fibers

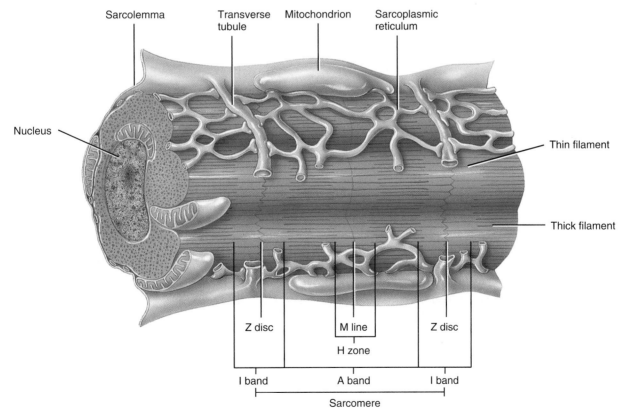

(b) Arrangement of components in a cardiac muscle fiber

 What are the functions of intercalated discs in cardiac muscle fibers?

studied men who had received a heart from a female, and then looked for the presence of a Y chromosome in heart cells. (All female cells except gametes have two X chromosomes and lack the Y chromosome.) Several years after the transplant surgery, between 7% and 16% of the heart cells in the transplanted tissue, including cardiac muscle fibers and endothelial cells in coronary arterioles and capillaries, had been replaced by the recipient's own cells, as evidenced by the presence of a Y chromosome. The study also revealed cells with some of the characteristics of stem cells in both transplanted hearts and control hearts. Evidently, stem cells can migrate from the blood into the heart and differentiate into functional muscle and endothelial cells. The hope is that researchers can learn how to "turn on" such regeneration of heart cells to treat people with heart failure or cardiomyopathy (diseased heart). ■

Autorhythmic Fibers: The Conduction System

An inherent and rhythmical electrical activity is the reason for the heart's continuous beating. The source of this electrical activity is a network of specialized cardiac muscle fibers. These cells are called **autorhythmic fibers** (*auto-* = self) because they are self-excitable. Autorhythmic fibers repeatedly generate action potentials that trigger heart contractions. Autorhythmic fibers continue to stimulate a heart to beat even after it is removed from the body—for example, to be transplanted into another person—although all of its nerves have been cut. (Note: Surgeons do not attempt to reattach heart nerves during heart transplant operations. For this reason, it has been said that heart surgeons are better "plumbers" than they are "electricians.")

During embryonic development, about 1% of the cardiac muscle fibers become autorhythmic fibers, which have two important functions.

1. They act as a **pacemaker,** setting the rhythm of electrical excitation that causes contraction of the heart.

2. They form the **conduction system,** a network of specialized cardiac muscle fibers that provide a path for each cycle of cardiac excitation to progress through the heart. The conduction system ensures that cardiac chambers become stimulated to contract in a coordinated manner, which makes the heart an effective pump.

Cardiac action potentials propagate through the conduction system in the following sequence (Figure 20.10a):

① Cardiac excitation normally begins in the **sinoatrial (SA) node,** located in the right atrial wall just inferior to the opening of the superior vena cava. SA node cells do not have a stable resting potential. Rather, they repeatedly depolarize to threshold spontaneously. The spontaneous depolarization is a **pacemaker potential.** When the pacemaker potential reaches threshold, it triggers an action potential (Figure 20.10b). Each action potential from the SA node propagates throughout both atria via gap junctions in the intercalated discs of atrial muscle fibers. Following the action potential, the atria contract.

② By conducting along atrial muscle fibers, the action potential reaches the **atrioventricular (AV) node,** located in the septum between the two atria, just anterior to the opening of the coronary sinus (Figure 20.10a).

③ From the AV node, the action potential enters the **atrioventricular (AV) bundle** (also known as the **bundle of His**). This bundle is the only site where action potentials can conduct from the atria to the ventricles. (Elsewhere, the fibrous skeleton of the heart electrically insulates the atria from the ventricles.)

④ After propagating along the AV bundle, the action potential enters both the **right** and **left bundle branches.** The bundle branches extend through the interventricular septum toward the apex of the heart.

⑤ Finally, the large-diameter **Purkinje fibers** rapidly conduct the action potential from the apex of the heart upward to the remainder of the ventricular myocardium. Then, the ventricles contract, pushing the blood upward toward the semilunar valves.

On their own, autorhythmic fibers in the SA node initiate an action potential about every 0.6 second, or 100 times per minute. Normally, this rate is faster than that of any other autorhythmic fibers. So, action potentials from the SA node spread through the conduction system and stimulate other areas before the other areas are able to generate an action potential at their own, slower rate. Thus, the SA node is the normal pacemaker of the heart. Nerve impulses from the autonomic nervous system (ANS) and blood-borne hormones (such as epinephrine) *modify the timing and strength* of each heartbeat, but they *do not establish the fundamental rhythm.* In a person at rest, for example, acetylcholine released by the parasympathetic division of the ANS slows SA node pacing to about 75 action potentials per minute or one every 0.8 sec (Figure 20.10b).

If the SA node becomes damaged or diseased, the slower AV node can pick up the pacemaking task. Its rate of spontaneous depolarization is 40 to 60 times per minute. If the activity of both nodes is suppressed, the heartbeat may still be maintained by autorhythmic fibers in the ventricles—the AV bundle, a bundle branch, or Purkinje fibers. However, the pacing rate is so slow, only 20 to 35 beats per minute, that blood flow to the brain is inadequate. When this condition occurs, normal heart rhythm can be restored and maintained by surgically implanting an **artificial pacemaker,** a device that sends out small electrical currents to stimulate the heart to contract. Many of the newer pacemakers, called activity-adjusted pacemakers, automatically speed up the heartbeat during exercise.

Ectopic Pacemaker

Sometimes, a site other than the SA node becomes the pacemaker because it develops abnormal self-excitability. Such a site is called an **ectopic pacemaker** (ek-TOP-ik = displaced). An

Figure 20.10 The conduction system of the heart. Autorhythmic fibers in the SA node, located in the right atrial wall (a), act as the heart's pacemaker, initiating cardiac action potentials (b) that cause contraction of the heart's chambers.

🔑 **The conduction system ensures that the chambers of the heart contract in a coordinated manner.**

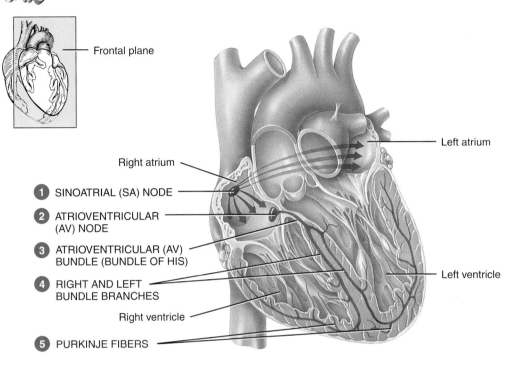

Frontal plane

Left atrium

Right atrium

1 SINOATRIAL (SA) NODE

2 ATRIOVENTRICULAR (AV) NODE

3 ATRIOVENTRICULAR (AV) BUNDLE (BUNDLE OF HIS)

4 RIGHT AND LEFT BUNDLE BRANCHES

Left ventricle

Right ventricle

5 PURKINJE FIBERS

(a) Anterior view of frontal section

+ 10 mV

Action potential

Membrane potential

Threshold

− 60 mV

Pacemaker potential

0   0.8   1.6   2.4

Time (sec) ⟶

(b) Pacemaker potentials and action potentials in autorhythmic fibers of SA node

 Which component of the conduction system provides the only electrical connection between the atria and the ventricles?

ectopic pacemaker may operate only occasionally, producing extra beats, or it may pace the heart for some period. Triggers of ectopic activity include caffeine and nicotine, electrolyte imbalances, hypoxia, and toxic reactions to drugs such as digitalis. Students recording their own electrocardiograms (described shortly) in an anatomy and physiology lab may see an occasional ectopic beat if they have consumed too much coffee, tea, or Coke! ■

Action Potential and Contraction of Contractile Fibers

The action potential initiated by the SA node travels along the conduction system and spreads out to excite the "working" atrial and ventricular muscle fibers, which are called **contractile fibers.** An action potential occurs in a contractile fiber as follows (Figure 20.11):

1 *Depolarization.* Unlike autorhythmic fibers, contractile fibers have a stable resting membrane potential that is close to −90 mV. When a contractile fiber is brought to threshold by an action potential in neighboring fibers, its **voltage-gated fast Na⁺ channels** open. These sodium ion channels are "fast" because they open very rapidly in response to a threshold-level depolarization. Opening of these channels allows Na⁺ inflow because the cytosol is electrically more negative than interstitial fluid and Na⁺ concentration is higher in interstitial fluid. Inflow of Na⁺ down the electrochemical gradient produces a **rapid depolarization.** Within a few milliseconds, the fast Na⁺ channels automatically inactivate and Na⁺ inflow decreases.

2 *Plateau.* The next phase of an action potential in a contractile fiber is the **plateau,** a period of maintained depolarization. It is due in part to opening of **voltage-gated slow Ca²⁺ channels** in the sarcolemma and sarcoplasmic reticulum membrane. To trigger contraction, some calcium ions pass through the sarcolemma from the interstitial fluid (which has a higher Ca²⁺ concentration) into the cytosol. The small inflow of Ca²⁺ from interstitial fluid causes much more Ca²⁺ to pour out of the sarcoplasmic reticulum within the fiber. At the same time, the membrane permeability to potassium ions decreases due to closing of some K⁺ channels. For about 0.25 sec the membrane is depolarized, close to 0 mV, as a small outflow of K⁺ just balances the inflow of Ca²⁺. By comparison, depolarization in a neuron or skeletal muscle fiber is much briefer, about 1 msec (0.001 sec), because it lacks a plateau phase.

3 *Repolarization.* The recovery of the resting membrane potential during the **repolarization** phase of a cardiac action potential resembles that in other excitable cells. After a delay (which is particularly prolonged in cardiac muscle), **voltage-gated K⁺ channels** open, thereby increasing the membrane permeability to potassium ions. Outflow of K⁺ restores the negative resting membrane potential (−90 mV). At the same time, the calcium channels are closing, which also contributes to repolarization.

The mechanism of contraction is similar in cardiac and skeletal muscle: The electrical activity (action potential) leads to the mechanical response (contraction) after a short delay. As Ca²⁺ concentration rises inside a contractile fiber, Ca²⁺ binds to the regulatory protein troponin, which allows the actin and myosin filaments to begin sliding past one another, and tension starts to develop. Substances that alter the movement of Ca²⁺ through slow Ca²⁺ channels influence the strength of heart contractions. Epinephrine, for example, increases contraction force by enhancing Ca²⁺ flow into the cytosol.

In muscle, the **refractory period** is the time interval during which a second contraction cannot be triggered. The refractory period of a cardiac muscle fiber lasts longer than the contraction itself (Figure 20.11). As a result, another contraction cannot begin until relaxation is well underway. For this reason, tetanus (maintained contraction) cannot occur in cardiac muscle as it can in skeletal muscle. The advantage is apparent if you consider how the ventricles work. Their pumping function depends on alternating contraction (when they eject blood) and relaxation (when they refill). If heart muscle could undergo tetanus, blood flow would cease.

Figure 20.11 Action potential in a ventricular contractile fiber. The resting membrane potential is about −90 mV.

A long refractory period prevents tetanus in cardiac muscle fibers.

2 Plateau (maintained depolarization) due to Ca²⁺ inflow when voltage-gated slow Ca²⁺ channels open and some K⁺ channels close

3 Repolarization due to K⁺ outflow when voltage-gated K⁺ channels open

1 Rapid depolarization due to Na⁺ inflow when voltage-gated fast Na⁺ channels open

Membrane potential (mV)

+20, 0, −20, −40, −60, −80, −100

0.3 sec

Depolarization | Repolarization

Refractory period

Contraction

How does the duration of an action potential in a ventricular contractile fiber compare with that in a skeletal muscle fiber?

ATP Production in Cardiac Muscle

In contrast to skeletal muscle, cardiac muscle produces little of the ATP it needs by anaerobic cellular respiration (see Figure 10.13 on page 289). Instead, it relies almost exclusively on aerobic cellular respiration in its numerous mitochondria. The needed oxygen diffuses from blood in the coronary circulation and is released from myoglobin inside cardiac muscle fibers. Cardiac muscle fibers use several fuels to power mitochondrial ATP production. In a person at rest, the heart's ATP comes mainly from oxidation of fatty acids (60%) and glucose (35%), with smaller contributions from lactic acid, amino acids, and ketone bodies. During exercise, the heart's use of lactic acid, produced by actively contracting skeletal muscles, rises.

Like skeletal muscle, cardiac muscle also produces some ATP from creatine phosphate. One sign that a myocardial infarction (heart attack, see page 691) has occurred is the presence in blood of creatine kinase (CK), the enzyme that catalyzes transfer of a phosphate group from creatine phosphate to ADP to make ATP. Normally, CK and other enzymes are confined within cells. Injured or dying cardiac or skeletal muscle fibers release CK into the blood.

Electrocardiogram

As action potentials propagate through the heart, they generate electrical currents that can be detected at the surface of the body. An **electrocardiogram** (e-lek′-trō-KAR-dē-ō-gram), abbreviated either **ECG** or **EKG** (from the German word *Elektrokardiogram*) is a recording of these electrical signals. The ECG is a composite record of action potentials produced by all the heart muscle fibers during each heartbeat. The instrument used to record the changes is an **electrocardiograph.**

In clinical practice, electrodes are positioned on the arms and legs (limb leads) and at six positions on the chest (chest leads) to record the ECG. The electrocardiograph amplifies the heart's electrical signals and produces 12 different tracings from different combinations of limb and chest leads. Each limb and chest electrode records slightly different electrical activity because it is in a different position relative to the heart. By comparing these records with one another and with normal records, it is possible to determine (1) if the conducting pathway is abnormal, (2) if the heart is enlarged, and (3) if certain regions of the heart are damaged.

In a typical Lead II record (right arm to left leg), three clearly recognizable waves appear with each heartbeat (Figure 20.12). The first, called the **P wave,** is a small upward deflection on the ECG. The P wave represents **atrial depolarization,** which spreads from the SA node through contractile fibers in both atria. The second wave, called the **QRS complex,** begins as a downward deflection, continues as a large, upright, triangular wave, and ends as a downward wave. The QRS complex represents **rapid ventricular depolarization,** as the action potential spreads through ventricular contractile fibers. The third wave is a dome-shaped upward deflection called the **T wave.** It indicates

Figure 20.12 Normal electrocardiogram or ECG (Lead II). P wave = atrial depolarization; QRS complex = onset of ventricular depolarization; T wave = ventricular repolarization.

An ECG is a recording of the electrical activity that initiates each heartbeat.

Key:

Atrial contraction
Ventricular contraction

What is the significance of an enlarged Q wave?

ventricular repolarization and occurs just as the ventricles are starting to relax. The T wave is smaller and wider than the QRS complex because repolarization occurs more slowly than depolarization. During the plateau period of steady depolarization, the ECG tracing is flat.

In reading an ECG, the size of the waves can provide clues to abnormalities. Larger P waves, for example, indicate enlargement of an atrium; an enlarged Q wave may indicate a myocardial infarction; and an enlarged R wave generally indicates enlarged ventricles. The T wave is flatter than normal when the heart muscle is receiving insufficient oxygen—as, for example, in coronary artery disease. The T wave may be elevated in hyperkalemia (high blood K⁺ level).

Sometimes it is helpful to evaluate the heart's response to the stress of physical exercise. Such a test is called a **stress electrocardiogram,** or **stress test.** Although narrowed coronary arteries may carry adequate oxygenated blood while a person is at rest, they will not be able to meet the heart's increased need for oxygen during strenuous exercise. This situation creates changes that can be seen on an electrocardiogram.

Analysis of an ECG also involves measuring the time spans between waves, which are called **intervals** or **segments.** For example, the **P-Q interval** is the time from the beginning of the P wave to the beginning of the QRS complex. It represents the conduction time from the beginning of atrial excitation to the beginning of ventricular excitation. Put another way, the P-Q interval is the time required for the action potential to travel through the atria, atrioventricular node, and the remaining fibers of the conduction system. In coronary artery disease and rheumatic fever, scar tissue may form in the heart. As the action potential detours around scar tissue, the P-Q interval lengthens.

The **S-T segment** begins at the end of the S wave and ends at the beginning of the T wave. It represents the time when the ventricular contractile fibers are depolarized during the plateau phase of the action potential. The S-T segment is elevated (above the baseline) in acute myocardial infarction and depressed (below the baseline) when the heart muscle receives insufficient oxygen. The **Q-T interval** extends from the start of the QRS complex to the end of the T wave. It is the time from the beginning of ventricular depolarization to the end of ventricular repolarization. The Q-T interval may be lengthened by myocardial damage, myocardial ischemia (decreased blood flow), or conduction abnormalities.

Correlation of ECG Waves with Atrial and Ventricular Systole

As we have seen, the atria and ventricles depolarize and then contract at different times because the conduction system routes cardiac action potentials along a specific pathway. The term **systole** (SIS-tō-lē = contraction) refers to the phase of contraction; the phase of relaxation is **diastole** (dī-AS-tō-lē = dilation or expansion). The ECG waves predict the timing of atrial and ventricular systole and diastole. At a heart rate of 75 beats per minute, the timing is as follows (Figure 20.13):

1 A cardiac action potential arises in the SA node. It propagates throughout the atrial muscle and down to the AV node in about 0.03 sec. As the atrial contractile fibers depolarize, the P wave appears in the ECG.

2 After the P wave begins, the atria contract (atrial systole). Conduction of the action potential slows at the AV node because the fibers there have much smaller diameters and fewer gap junctions. (Traffic slows in a similar way where a four-lane highway narrows to one lane in a construction zone!) The resulting 0.1-sec delay gives the atria time to contract, thus adding to the volume of blood in the ventricles, before ventricular diastole begins.

3 The action potential propagates rapidly again after entering the AV bundle. About 0.2 sec after onset of the P wave, it has propagated through the bundle branches, Purkinje fibers, and the entire ventricular myocardium. Depolarization progresses down the septum, upward from the apex, and outward from the endocardial surface, producing the QRS complex. At the same time, atrial repolarization is occurring, but it is not usually evident in an ECG because the larger QRS complex masks it. Atrial systole follows atrial repolarization.

4 Contraction of ventricular contractile fibers, that is ventricular systole, begins shortly after the QRS appears and continues during the S-T segment. As contraction proceeds from the apex toward the base of the heart, blood is squeezed upward toward the semilunar valves.

5 Repolarization of ventricular contractile fibers begins at the apex and spreads throughout the ventricular myocardium. This produces the T wave in the ECG about 0.4 sec after the P wave onset.

6 Shortly after the T wave begins, the ventricles start to relax (ventricular diastole). By 0.6 sec, ventricular repolarization is complete and ventricular contractile fibers are relaxed.

During the next 0.2 sec, contractile fibers in both the atria and ventricles are relaxed. At 0.8 sec, the P wave appears again in the ECG, the atria begin to contract, and the cycle repeats. As you can see, events in the heart occur in cycles that repeat for as long as you live. Next, we will see how the pressure changes associated with relaxation and contraction of the heart chambers allow the heart to alternately fill with blood and then eject blood into the aorta and pulmonary trunk.

▶ CHECKPOINT

10. How do cardiac muscle fibers differ structurally and functionally from skeletal muscle fibers?

11. In what ways are autorhythmic fibers similar to and different from contractile fibers?

12. What happens during each of the three phases of an action potential in ventricular contractile fibers?

13. In what ways are ECGs helpful in diagnosing cardiac problems?

14. How does each ECG wave, interval, and segment relate to contraction (systole) and relaxation (diastole) of the atria and ventricles?

THE CARDIAC CYCLE

▶ OBJECTIVES

- **Describe the pressure and volume changes that occur during a cardiac cycle.**
- **Relate the timing of heart sounds to the ECG waves and pressure changes during systole and diastole.**

A single **cardiac cycle** includes all the events associated with one heartbeat. Thus, a cardiac cycle consists of systole and diastole of the atria plus systole and diastole of the ventricles.

Pressure and Volume Changes During the Cardiac Cycle

In each cardiac cycle, the atria and ventricles alternately contract and relax, forcing blood from areas of higher pressure to areas of lower pressure. As a chamber of the heart contracts, blood pressure within it increases. Figure 20.14 shows the relation between the heart's electrical signals (ECG) and changes in atrial

Figure 20.13 Timing and route of action potential depolarization and repolarization through the conduction system and myocardium. Green indicates depolarization and red indicates repolarization. Arrows show the direction of propagation of the action potential.

🔑 Depolarization causes contraction and repolarization causes relaxation of cardiac muscle fibers.

1 Depolarization of atrial contractile fibers produces P wave

P

Action potential in SA node

0   0.2
Seconds

6 Ventricular diastole (relaxation)

P

0   0.2   0.4   0.6   0.8
Seconds

2 Atrial systole (contraction)

P

0   0.2
Seconds

5 Repolarization of ventricular contractile fibers produces T wave

P   T

0   0.2   0.4   0.6
Seconds

3 Depolarization of ventricular contractile fibers produces QRS complex

R

P

Q   S

0   0.2   0.4
Seconds

4 Ventricular systole (contraction)

P

0   0.2   0.4
Seconds

 Where in the conduction system do action potentials propagate most slowly?

pressure, ventricular pressure, aortic pressure, and ventricular volume during the cardiac cycle. The pressures given in Figure 20.14 apply to the left side of the heart; pressures on the right side are considerably lower. Each ventricle, however, expels the same volume of blood per beat, and the same pattern exists for both pumping chambers. When heart rate is 75 beats/min, a cardiac cycle lasts 0.8 sec. To examine and correlate the events taking place during a cardiac cycle, we will begin with atrial systole.

Atrial Systole

During **atrial systole,** which lasts about 0.1 sec, the atria are contracting. At the same time, the ventricles are relaxed.

1 Depolarization of the SA node causes atrial depolarization, marked by the P wave in the ECG.

2 Atrial depolarization causes atrial systole. As the atria contract, they exert pressure on the blood within, which forces blood through the open AV valves into the ventricles.

3 Atrial systole contributes a final 25 mL of blood to the volume already in each ventricle (about 105 mL). The end of atrial systole is also the end of ventricular diastole (relaxation). Thus, each ventricle contains about 130 mL at the end of its relaxation period (diastole). This blood volume is the **end-diastolic volume (EDV).**

4 The QRS complex in the ECG marks the onset of ventricular depolarization.

Ventricular Systole

During **ventricular systole,** which lasts about 0.3 sec, the ventricles are contracting. At the same time, the atria are relaxed, in **atrial diastole.**

5 Ventricular depolarization causes ventricular systole. As ventricular systole begins, pressure rises inside the ventricles and pushes blood up against the atrioventricular (AV) valves, forcing them shut. For about 0.05 seconds, both the SL (semilunar) and AV valves are closed. This is the period of **isovolumetric contraction** (*iso-* = same). During this interval, cardiac muscle fibers are contracting and exerting force but are not yet shortening. Thus, the muscle contraction is isometric (same length). Moreover, because all four valves are closed, ventricular volume remains the same (isovolumic).

6 Continued contraction of the ventricles causes pressure inside the chambers to rise sharply. When left ventricular pressure surpasses aortic pressure at about 80 millimeters of mercury (mmHg) and right ventricular pressure rises above the pressure in the pulmonary trunk (about 20 mmHg), both SL valves open. At this point, ejection of blood from the heart begins. The period when the SL valves are open is **ventricular ejection** and lasts for about 0.25 sec. The pressure in the left ventricle continues to rise to about 120 mmHg,

whereas the pressure in the right ventricle climbs to about 25–30 mmHg.

7 The left ventricle ejects about 70 mL of blood into the aorta and the right ventricle ejects the same volume of blood into the pulmonary trunk. The volume remaining in each ventricle at the end of systole, about 60 mL, is the **end-systolic volume (ESV). Stroke volume,** the volume ejected per beat from each ventricle, equals end-diastolic volume minus end-systolic volume: SV = EDV − ESV. At rest, the stroke volume is about 130 mL – 60 mL = 70 mL (a little more than 2 oz).

8 The T wave in the ECG marks the onset of ventricular repolarization.

Relaxation Period

During the **relaxation period,** which lasts about 0.4 sec, the atria and the ventricles are both relaxed. As the heart beats faster and faster, the relaxation period becomes shorter and shorter, whereas the durations of atrial systole and ventricular systole shorten only slightly.

9 Ventricular repolarization causes **ventricular diastole.** As the ventricles relax, pressure within the chambers falls, and blood in the aorta and pulmonary trunk begins to flow backward toward the regions of lower pressure in the ventricles. Backflowing blood catches in the valve cusps and closes the SL valves. The aortic valve closes at a pressure of about 100 mmHg. Rebound of blood off the closed cusps of the aortic valve produces the **dicrotic wave** on the aortic pressure curve. After the SL valves close, there is a brief interval when ventricular blood volume does not change because all four valves are closed. This is the period of **isovolumetric relaxation.**

10 As the ventricles continue to relax, the pressure falls quickly. When ventricular pressure drops below atrial pressure, the AV valves open, and **ventricular filling** begins. The major part of ventricular filling occurs just after the AV valves open. Blood that has been flowing into and building up in the atria during ventricular systole then rushes rapidly into the ventricles. At the end of the relaxation period, the ventricles are about three-quarters full. The P wave appears in the ECG, signaling the start of another cardiac cycle.

Heart Sounds

Auscultation (aws-kul-TĀ-shun; *ausculta-* = listening) is the act of listening to sounds within the body, and it is usually done with a stethoscope. The sound of the heartbeat comes primarily from blood turbulence caused by the closing of the heart valves. Smoothly flowing blood is silent. Recall the sounds made by white-water rapids or a waterfall as compared with the silence of a smoothly flowing river. During each cardiac cycle, there are four **heart sounds,** but in a normal heart only the first and second heart sounds (S1 and S2) are loud enough to be heard by

Figure 20.14   Cardiac cycle.   (a) ECG. (b) Changes in left atrial pressure (green line), left ventricular pressure (blue line), and aortic pressure (red line) as they relate to the opening and closing of heart valves. (c) Changes in left ventricular volume. (d) Phases of the cardiac cycle.

 A cardiac cycle is composed of all the events associated with one heartbeat.

(a) ECG

0.1 sec	0.3 sec	0.4 sec
Atrial systole	Ventricular systole	Relaxation period

(b) Pressure (mmHg)

⑨ Aortic valve closes

Dicrotic wave

Aortic pressure

⑥ Aortic valve opens

Left ventricular pressure

⑤ Bicuspid valve closes

②

⑩ Bicuspid valve opens

Left atrial pressure

(c) Volume in ventricle (mL)

③ End-diastolic volume

Stroke volume

⑦ End-systolic volume

(d) Phases of the cardiac cycle

| Atrial contraction | Isovolumetric contraction | Ventricular ejection | Isovolumetric relaxation | Ventricular filling |

How much blood remains in each ventricle at the end of ventricular diastole in a resting person? What is this volume called?

listening through a stethoscope. Figure 20.15a shows the timing of heart sounds relative to other events in the cardiac cycle and where each sound is best heard.

The first sound (S1), which can be described as a **lubb** sound, is louder and a bit longer than the second sound. S1 is caused by blood turbulence associated with closure of the AV valves soon after ventricular systole begins. The second sound (S2), which is shorter and not as loud as the first, can be described as a **dupp** sound. S2 is caused by blood turbulence associated with closure of the SL valves at the beginning of ventricular diastole. Although S1 and S2 are due to blood turbulence associated with the closure of valves, they are best heard at the surface of the chest in locations that are slightly different from the locations of the valves (Figure 20.15b). Normally not loud enough to be heard, S3 is due to blood turbulence during rapid ventricular filling, and S4 is due to blood turbulence during atrial systole.

Heart Murmurs

Heart sounds provide valuable information about the mechanical operation of the heart. A **heart murmur** is an abnormal sound consisting of a clicking, rushing, or gurgling noise that is heard before, between, or after the normal heart sounds, or that may mask the normal heart sounds. Although some heart murmurs are "innocent," meaning they are not associated with a significant heart problem, most often a murmur indicates a valve disorder. When a heart valve exhibits stenosis, the heart murmur is heard while the valve should be fully open but is not. For example, mitral stenosis produces a murmur during the relaxation period, between S2 and the next S1. An incompetent heart valve, by contrast, causes a murmur to appear when the valve should be fully closed but is not. So, a murmur due to mitral incompetence occurs during ventricular systole, between S1 and S2. ■

▶ C H E C K P O I N T

15. Why must left ventricular pressure be greater than aortic pressure during ventricular ejection?

16. Does more blood flow through the coronary arteries during ventricular diastole or ventricular systole?

17. During which two periods of the cardiac cycle do the heart muscle fibers exhibit isometric contractions?

18. What events produce the four normal heart sounds? Which ones can usually be heard through a stethoscope?

Figure 20.15   Heart sounds.   (a) Correlation of heart sounds with ECG waves, pressure changes, and valve opening and closing. (b) Location of valves (purple) and auscultation sites (red) for heart sounds.

🔑 **Listening to sounds within the body is called auscultation; it is usually done with a stethoscope.**

(a) Timing of heart sounds

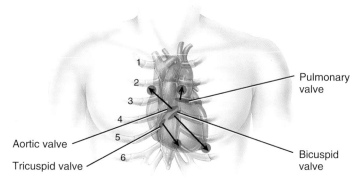

(b) Anterior view of heart valve locations and auscultation sites

 Which heart sound is related to blood turbulence associated with closure of the AV valves?

CARDIAC OUTPUT

O B J E C T I V E

• **Define cardiac output, and describe the factors that affect it.**

Although the heart has autorhythmic fibers that enable it to beat independently, its operation is governed by events occurring in the rest of the body. All body cells must receive a certain amount of oxygenated blood each minute to maintain health and life. When cells are metabolically active, as during exercise, they take up even more oxygen from the blood. During rest periods, cellular metabolic need is reduced, and the workload of the heart decreases.

Cardiac output (CO) is the volume of blood ejected from the left ventricle (or the right ventricle) into the aorta (or pulmonary trunk) each minute. Cardiac output equals the **stroke volume (SV),** the volume of blood ejected by the ventricle during each contraction, multiplied by the **heart rate (HR),** the number of heartbeats per minute:

$$CO = SV \times HR$$
$$(mL/min) \quad (mL/beat) \quad (beats/min)$$

In a typical resting adult male, stroke volume averages 70 mL/beat, and heart rate is about 75 beats/min. Thus, average cardiac output is

$$
\begin{aligned}
CO &= 70 \text{ mL/beat} \times 75 \text{ beats/min} \\
&= 5250 \text{ mL/min} \\
&= 5.25 \text{ L/min}
\end{aligned}
$$

This volume is close to the total blood volume, which is about 5 liters in a typical adult male. Thus, the entire blood volume flows through the pulmonary and systemic circulations each minute. When body tissues use more or less oxygen, cardiac output changes to meet the need. Factors that increase stroke volume or heart rate normally increase CO. During mild exercise, for example, stroke volume may increase to 100 mL/beat, and heart rate to 100 beats/min. Cardiac output then would be 10 L/min. During intense (but still not maximal) exercise, the heart rate may accelerate to 150 beats/min, and stroke volume may rise to 130 mL/beat, providing a cardiac output of 19.5 L/min.

Cardiac reserve is the difference between a person's maximum cardiac output and cardiac output at rest. A cardiac reserve that is four or five times the resting value is average. Top endurance athletes may have a cardiac reserve seven or eight times their resting CO. People with severe heart disease may have little or no cardiac reserve, which limits their ability to carry out even the simple tasks of daily living.

Regulation of Stroke Volume

A healthy heart will pump out all the blood that entered its chambers during the previous diastole. In other words, if more blood returns to the heart during diastole, then more blood is ejected during the next systole. At rest, the stroke volume is 50–60% of the end-diastolic volume because 40–50% remains

in the ventricles after each contraction (end-systolic volume). Three factors regulate stroke volume and ensure that the left and right ventricles pump equal volumes of blood: (1) **preload,** the degree of stretch on the heart before it contracts; (2) **contractility,** the forcefulness of contraction of individual ventricular muscle fibers; and (3) **afterload,** the pressure that must be exceeded before ejection of blood from the ventricles can occur.

Preload: Effect of Stretching

A greater preload (stretch) on cardiac muscle fibers prior to contraction increases their force of contraction. Within limits, the more the heart fills with blood during diastole, the greater the force of contraction during systole. This relationship is known as the **Frank–Starling law of the heart.** The preload is proportional to the volume of blood that fills the ventricles at the end of diastole, the end-diastolic volume (EDV). Normally, the greater the EDV, the more forceful the next contraction.

Two key factors determine EDV: (1) the duration of ventricular diastole and (2) **venous return,** the volume of blood returning to the right ventricle. When heart rate increases, the duration of diastole is shorter. Less filling time means a smaller EDV, and the ventricles may contract before they are adequately filled. By contrast, when venous return increases, a greater volume of blood flows into the ventricles, and the EDV is increased.

When heart rate exceeds about 160 beats/min, stroke volume usually declines due to the short filling time. At such rapid heart rates, EDV is less, and the preload is lower. People who have slow resting heart rates, in contrast, usually have large resting stroke volumes because filling time is prolonged and preload is larger.

The Frank–Starling law of the heart equalizes the output of the right and left ventricles and keeps the same volume of blood flowing to both the systemic and pulmonary circulations. If the left side of the heart pumps a little more blood than the right side, for example, the volume of blood returning to the right ventricle (venous return) increases. With increased EDV, then, the right ventricle contracts more forcefully on the next beat, and the two sides are again in balance.

Contractility

The second factor that influences stroke volume is myocardial **contractility,** which is the strength of contraction at any given preload. Substances that increase contractility are **positive inotropic agents,** whereas those that decrease contractility are **negative inotropic agents.** Thus, for a constant preload, the stroke volume increases when a positive inotropic substance is present. Positive inotropic agents often promote Ca^{2+} inflow during cardiac action potentials, which strengthens the force of the next contraction. Stimulation of the sympathetic division of the autonomic nervous system (ANS), hormones such as epinephrine and norepinephrine, increased Ca^{2+} level in the interstitial fluid, and the drug digitalis all have positive inotropic effects. In contrast, inhibition of the sympathetic division of the ANS, anoxia, acidosis, some anesthetics (for example, halothane), and increased K$^+$ level in the interstitial fluid have negative inotropic effects. *Calcium channel blockers* are drugs that can have a negative inotropic effect by reducing Ca^{2+} inflow, thereby decreasing the strength of the heartbeat.

Afterload

Ejection of blood from the heart begins when pressure in the right ventricle exceeds the pressure in the pulmonary trunk (about 20 mmHg), and when the pressure in the left ventricle exceeds the pressure in the aorta (about 80 mmHg). At that point, the higher pressure in the ventricles causes blood to push the semilunar valves open. The pressure that must be overcome before a semilunar valve can open is termed the **afterload.** At any given preload, an increase in afterload causes stroke volume to decrease, and more blood remains in the ventricles at the end of systole. Conditions that can increase afterload include hypertension (elevated blood pressure) and narrowing of arteries by atherosclerosis (see page 447).

Congestive Heart Failure

In **congestive heart failure (CHF),** the heart is a failing pump. Causes of CHF include coronary artery disease (see page 000), congenital defects, long-term high blood pressure (which increases the afterload), myocardial infarctions (regions of dead heart tissue due to a previous heart attack), and valve disorders. As the pump becomes less effective, more blood remains in the ventricles at the end of each cycle, and gradually the end-diastolic volume (preload) increases. Initially, increased preload may promote increased force of contraction (the Frank–Starling law of the heart), but as the preload increases further, the heart is overstretched and contracts less forcefully. The result is a potentially lethal positive feedback loop: Less-effective pumping leads to even lower pumping capability.

Often, one side of the heart starts to fail before the other. If the left ventricle fails first, it can't pump out all the blood it receives. As a result, blood backs up in the lungs and causes *pulmonary edema,* fluid accumulation in the lungs that can suffocate an untreated person. If the right ventricle fails first, blood backs up in the systemic veins and over time, the kidneys cause an increase in blood volume. In this case, the resulting *peripheral edema* usually is most noticeable in the feet and ankles. ■

Regulation of Heart Rate

As we have seen, cardiac output depends on heart rate as well as stroke volume. Adjustments in heart rate are important in the short-term control of cardiac output and blood pressure. The sinoatrial (SA) node initiates contraction and, if left to itself, would set a constant heart rate of about 100 beats/min. However, tissues require a different volume of blood flow under different conditions. During exercise, for example, cardiac output rises to supply working tissues with increased amounts of oxygen and nutrients. Stroke volume may fall if the ventricular myocardium

is damaged or if blood volume is reduced by bleeding. In these cases, homeostatic mechanisms act to maintain adequate cardiac output by increasing the heart rate and contractility. Among the several factors that contribute to regulation of heart rate, the most important are the autonomic nervous system and hormones released by the adrenal medullae (epinephrine and norepinephrine).

Autonomic Regulation of Heart Rate

Nervous system regulation of the heart originates in the **cardiovascular center** in the medulla oblongata. This region of the brain stem receives input from a variety of sensory receptors and from higher brain centers, such as the limbic system and cerebral cortex. The cardiovascular center then directs appropriate output by increasing or decreasing the frequency of nerve impulses in both the sympathetic and parasympathetic branches of the ANS (Figure 20.16).

Even before physical activity begins, especially in competitive situations, heart rate may climb. This anticipatory increase occurs because the limbic system sends nerve impulses to the cardiovascular center in the medulla. Then, as physical activity begins, **proprioceptors** that are monitoring the position of limbs and muscles send an increased frequency of nerve impulses to the cardiovascular center. Proprioceptor input is a major stimulus for the quick rise in heart rate that occurs at the onset of physical activity. Other sensory receptors that provide input to the cardiovascular center include **chemoreceptors,** which monitor chemical changes in the blood, and **baroreceptors,** which monitor stretching caused by blood pressure in major arteries and veins. Important baroreceptors located in the arch of the aorta and in the carotid arteries (see Figure 21.13 on page 711) detect changes in blood pressure. They provide input to the cardiovascular center when blood pressure changes. Baroreceptor reflexes, which are also important for the regulation of blood pressure, are discussed in detail in Chapter 21. Here we focus on the innervation of the heart by the sympathetic and parasympathetic branches of the ANS.

Sympathetic neurons extend from the medulla oblongata into the spinal cord. From the thoracic region of the spinal cord, sympathetic **cardiac accelerator nerves** extend out to the SA node, AV node, and most portions of the myocardium. Impulses in the cardiac accelerator nerves trigger the release of norepinephrine, which binds to beta-1 receptors on cardiac muscle fibers. This interaction has two separate effects: (1) In SA (and AV) node fibers, norepinephrine speeds the rate of spontaneous depolarization so that these pacemakers fire impulses more rapidly and heart rate increases. (2) In contractile fibers throughout the atria and ventricles, norepinephrine enhances Ca^{2+} entry through the voltage-gated slow Ca^{2+} channels, thereby increasing contractility. The result is a greater ejection of blood during systole. With a moderate increase in heart rate, stroke volume does not decline because the increased contractility offsets the decreased preload. With maximal sympathetic stimulation, however, heart rate may reach 200 beats/min in a 20-year-old person. At such a high heart rate, stroke volume is lower than at rest due

Figure 20.16 Nervous system control of the heart.

The cardiovascular center in the medulla oblongata controls both sympathetic and parasympathetic nerves that innervate the heart.

INPUT TO CARDIOVASCULAR CENTER (nerve impulses)

From higher brain centers: cerebral cortex, limbic system, and hypothalamus

From sensory receptors:
Proprioceptors–monitor movements
Chemoreceptors–monitor blood chemistry
Baroreceptors–monitor blood pressure

OUTPUT TO HEART
(increased frequency of nerve impulses)

Increased rate of spontaneous depolarization in SA node (and AV node) increases heart rate

Increased contractility of atria and ventricles increases stroke volume

Decreased rate of spontaneous depolarization in SA node (and AV node) decreases heart rate

Cardiac accelerator nerves (sympathetic)

Vagus nerves (cranial nerve X, parasympathetic)

Cardiovascular (CV) center

What region of the heart is innervated by the sympathetic division but not by the parasympathetic division?

to the very short filling time. The maximal heart rate declines with age; as a rule, subtracting one's age from 220 provides a good estimate of maximal heart rate in beats per minute.

Parasympathetic nerve impulses reach the heart via the right and left **vagus (X) nerves.** Vagal axons terminate in the SA node, AV node, and atrial myocardium. They release acetylcholine, which decreases heart rate by slowing the rate of spontaneous depolarization in autorhythmic fibers. As only a few vagal fibers innervate ventricular muscle, changes in parasympathetic activity have little effect on contractility of the ventricles.

A continually shifting balance exists between sympathetic and parasympathetic stimulation of the heart. At rest, parasympathetic stimulation predominates. The resting heart rate—about 75 beats/min—is usually lower than the autorhythmic rate of the SA node (about 100 beats/min). With maximal stimulation by the parasympathetic division, the heart can slow to 20 or 30 beats/min, or can even stop momentarily.

Chemical Regulation of Heart Rate

Certain chemicals influence both the basic physiology of cardiac muscle and the heart rate. For example, hypoxia (lowered oxygen level), acidosis (low pH), and alkalosis (high pH) all depress cardiac activity. Several hormones and cations have major effects on the heart:

1. Hormones. Epinephrine and norepinephrine (from the adrenal medullae) enhance the heart's pumping effectiveness. These hormones affect cardiac muscle fibers in much the same way as does norepinephrine released by cardiac accelerator nerves—they increase both heart rate and contractility. Exercise, stress, and excitement cause the adrenal medullae to release more hormones. Thyroid hormones also enhance cardiac contractility and increase heart rate. One sign of hyperthyroidism (excessive thyroid hormone) is **tachycardia,** an elevated resting heart rate.

2. Cations. Given that differences between intracellular and extracellular concentrations of several cations (for example, Na^+ and K^+) are crucial for the production of action potentials in all nerve and muscle fibers, it is not surprising that ionic imbalances can quickly compromise the pumping effectiveness of the heart. In particular, the relative concentrations of three cations—K^+, Ca^{2+}, and Na^+—have a large effect on cardiac function. Elevated blood levels of K^+ or Na^+ decrease heart rate and contractility. Excess Na^+ blocks Ca^{2+} inflow during cardiac action potentials, thereby decreasing the force of contraction, whereas excess K^+ blocks generation of action potentials. A moderate increase in interstitial (and thus intracellular) Ca^{2+} level speeds heart rate and strengthens the heartbeat.

Other Factors in Heart Rate Regulation

Age, gender, physical fitness, and body temperature also influence resting heart rate. A newborn baby is likely to have a resting heart rate over 120 beats/min; the rate then declines through childhood. Senior citizens may develop a more rapid heartbeat, however. Adult females often have slightly higher resting heart rates than adult males, although regular exercise tends to bring resting heart rate down in both sexes. A physically fit person may even exhibit **bradycardia,** a resting heart rate under 60 beats/min. This is a beneficial effect of endurance-type training because a slowly beating heart is more energy efficient than one that beats more rapidly.

Increased body temperature, as occurs during a fever or strenuous exercise, causes the SA node to discharge impulses more quickly, thereby increasing heart rate. Decreased body temperature decreases heart rate and strength of contraction.

During surgical repair of certain heart abnormalities, it is helpful to slow a patient's heart rate by **hypothermia** (hī-pō-THER-mē-a), in which the person's body is deliberately cooled to a low core temperature. Hypothermia also slows metabolism, which reduces the oxygen needs of the tissues, allowing the heart and brain to withstand short periods of interrupted or reduced blood flow during the procedure.

Figure 20.17 summarizes the factors that can increase stroke volume and heart rate to achieve an increase in cardiac output.

▶ C H E C K P O I N T

19. How is cardiac output calculated?

20. Define stroke volume (SV), and explain the factors that regulate it.

Figure 20.17   Factors that increase cardiac output.

🔑 **Cardiac output equals stroke volume multiplied by heart rate.**

Increased end diastolic volume (stretches the heart)

Positive inotropic agents such as increased sympathetic stimulation; catecholamines, glucagon, or thyroid hormones in the blood; increased Ca^{2+} in extracellular fluid

Decreased arterial blood pressure during diastole

↓

| Increased PRELOAD | Increased CONTRACTILITY | Decreased AFTERLOAD |

↓

Within limits, cardiac muscle fibers contract more forcefully with stretching (Frank–Starling law of the heart)

Positive inotropic agents increase force of contraction at all physiological levels of stretch

Semilunar valves open sooner when blood pressure in aorta and pulmonary artery is lower

Increased STROKE VOLUME

Increased CARDIAC OUTPUT

Increased HEART RATE

Increased sympathetic stimulation and decreased parasympathetic stimulation

Catecholamine or thyroid hormones in the blood; moderate increase in extracellular Ca^{2+}

Infants and senior citizens, females, low physical fitness, increased body temperature

| NERVOUS SYSTEM |
| Cardiovascular center in medulla oblongata receives input from cerebral cortex, limbic system, proprioceptors, baroreceptors, and chemoreceptors |

| CHEMICALS |

| OTHER FACTORS |

 When you are exercising, contraction of skeletal muscles helps return blood to the heart more rapidly. Would this effect tend to increase or decrease stroke volume?

21. What is the Frank–Starling law of the heart? What is its significance?

22. Define cardiac reserve. How does it change with training or with heart failure?

23. How do the sympathetic and parasympathetic divisions of the autonomic nervous system adjust heart rate?

EXERCISE AND THE HEART

▶ **O B J E C T I V E**

• **Explain the relationship between exercise and the heart.**

No matter what a person's level of fitness, it can be improved at any age with regular exercise. Of the various types of exercise, some are more effective than others for improving the health of the cardiovascular system. **Aerobics,** any activity that works large body muscles for at least 20 minutes, elevates cardiac output and accelerates metabolic rate. Three to five such sessions a week are usually recommended for improving the health of the cardiovascular system. Brisk walking, running, bicycling, cross-country skiing, and swimming are examples of aerobic activities.

Sustained exercise increases the oxygen demand of the muscles. Whether the demand is met depends mainly on the adequacy of cardiac output and proper functioning of the respiratory system. After several weeks of training, a healthy person increases maximal cardiac output, thereby increasing the maximal rate of oxygen delivery to the tissues. Oxygen delivery also rises because skeletal muscles develop more capillary networks in response to long-term training.

During strenuous activity, a well-trained athlete can achieve a cardiac output double that of a sedentary person, in part because training causes hypertrophy (enlargement) of the heart. Even though the heart of a well-trained athlete is larger, *resting* cardiac output is about the same as in a healthy untrained person, because stroke volume is increased while heart rate is decreased. The resting heart rate of a trained athlete often is only 40 to 60 beats per minute (resting bradycardia).

Help for Failing Hearts

As the heart fails, a person has decreasing ability to exercise or even to move around. A variety of surgical techniques and medical devices exist to aid a failing heart. For some patients, even a 10% increase in the volume of blood ejected from the ventricles can mean the difference between being bedridden and having limited mobility. **Heart transplants** are common today and produce good results, but the availability of donor hearts is very limited. For instance, there are 50 potential candidates for each of the 2500 donor hearts available each year in the United States. Another approach is to use **cardiac assist devices** and surgical procedures that augment heart function without removing the heart. Table 20.1 describes several of these.

Finally, scientists continue to develop and refine **artificial hearts,** mechanical devices that completely replace the functions of the natural heart. During the 1980s several patients received a Jarvik-7 artificial heart, which used an external power source to drive an internal pump by compressed air. In

Table 20.1	Cardiac Assist Devices and Procedures
Device	**Description**
Intra-aortic balloon pump (IABP)	A 40-mL polyurethane balloon mounted on a catheter is inserted into an artery in the groin and threaded into the thoracic aorta. An external pump inflates the balloon with gas at the beginning of ventricular diastole. As the balloon inflates, it pushes blood both backward toward the heart, which improves coronary blood flow, and forward toward peripheral tissues. The balloon then is rapidly deflated just before the next ventricular systole, making it easier for the left ventricle to eject blood. Because the balloon is inflated between heartbeats, this technique is called intra-aortic balloon counterpulsation.
Hemopump	This propeller-like pump is threaded through an artery in the groin and then into the left ventricle. There, the blades of the pump whirl at about 25,000 revolutions per minute, pulling blood out of the left ventricle and pushing it into the aorta.
Left ventricular assist device (LVAD)	The LVAD is a completely portable assist device. It is implanted within the abdomen and powered by a battery pack worn in a shoulder holster. The LVAD is connected to the patient's weakened left ventricle and pumps blood into the aorta. The pumping rate increases automatically during exercise.
Cardiomyoplasty	A large piece of the patient's own skeletal muscle (left latissimus dorsi) is partially freed from its connective tissue attachments and wrapped around the heart, leaving the blood and nerve supply intact. An implanted pacemaker stimulates the skeletal muscle's motor neurons to cause contraction 10–20 times per minute, in synchrony with some of the heartbeats.
Skeletal muscle assist device	A piece of the patient's own skeletal muscle is used to fashion a pouch that is inserted between the heart and the aorta, functioning as a booster heart. A pacemaker stimulates the muscle's motor neurons to elicit contraction.

1990, the U.S. Food and Drug Administration (FDA) banned use of the device because persistent problems with blood clotting caused strokes, and the chest tube for compressed air led to infections. More than a decade later, in July 2001, the first person received a fully self-contained artificial heart, the AbioCor Implantable Replacement Heart. Made of titanium, plastic, and epoxy, the 2-pound AbioCor heart is powered by a battery pack worn outside but with no wires penetrating through the skin. It alternately pumps blood from the left and then the right side of the heart. Because a permanent opening through the chest is not needed, the risk of infection should be much lower than with the Jarvik-7. Without the artificial heart, the first recipient was given little chance to survive more than a month due to his congestive heart failure, kidney disease, and diabetes. After surgery, he lived for 151 days, recovering enough to give several interviews and to enjoy a fishing trip. However, internal bleeding and organ failure unrelated to the AbioCor heart caused his death. By the end of 2001 several more people had received an AbioCor heart as part of the clinical trials needed to test any new device. Their progress provides encouragement that AbioCor hearts will be available on the market within the next few years. ■

DEVELOPMENT OF THE HEART

▶ OBJECTIVE

• **Describe the development of the heart.**

The *heart,* a derivative of **mesoderm,** begins to develop early in the third week after fertilization. In the ventral region of the embryo inferior to the foregut, the heart develops from a group of mesodermal cells called the **cardiogenic area.** The next step in its development is the formation of a pair of tubes, the **endocardial tubes,** which develop from the cardiogenic area (Figure 20.18). These tubes then unite to form a common tube, referred to as the **primitive heart tube.** Next, the primitive heart tube develops into five distinct regions: (1) **truncus arteriosus,** (2) **bulbus cordis,** (3) **ventricle,** (4) **atrium,** and (5) **sinus venosus.** Because the bulbus cordis and ventricle grow most rapidly, and because the heart enlarges more rapidly than its superior and inferior attachments, the heart first assumes a U-shape and then an S-shape. The flexures of the heart reorient the regions so that the atrium and sinus venosus eventually come to lie superior to the bulbus cordis, ventricle, and truncus arteriosus. Contractions

Figure 20.18 Development of the heart.   Arrows within the structures indicate the direction of blood flow.

The heart begins its development from a group of mesodermal cells called the cardiogenic area during the third week after fertilization.

 When during embryonic development does the primitive heart begin to contract?

of the primitive heart begin by the end of the third week of development; they originate in the sinus venosus and force blood through the tubular heart.

At about the seventh week of development, a partition called the **interatrial septum** forms in the atrial region. This septum divides the atrial region into a *right atrium* and *left atrium*. The opening in the partition is the **foramen ovale,** which normally closes at birth or soon thereafter and later forms a depression called the *fossa ovalis* (see Figure 20.4a). An **interventricular septum** also develops; it partitions the ventricular region into a *right ventri-*

cle and *left ventricle.* The bulbus cordis and truncus arteriosus divide into two vessels, the *aorta* (arising from the left ventricle) and the *pulmonary trunk* (arising from the right ventricle). The great veins of the heart, the *superior vena cava* and the *inferior vena cava*, develop from the venous end of the primitive heart tube.

▶ C H E C K P O I N T

24. What are some of the cardiovascular benefits of regular exercise?

25. What structures develop from the bulbus cordis and truncus arteriosus?

DISORDERS: HOMEOSTATIC IMBALANCES

Coronary Artery Disease

In **coronary artery disease (CAD),** atherosclerotic plaques (described shortly) in coronary arteries reduce blood flow to the myocardium. CAD is the number-one killer of men and women in all ethnic groups, causing 750,000 deaths in the United States every year. Some individuals have no signs or symptoms; others experience angina pectoris (chest pain), and still others suffer a heart attack.

Risk Factors for CAD

People who possess combinations of certain risk factors are more likely to develop CAD. *Risk factors* are characteristics, symptoms, or signs present in a disease-free person that are statistically associated with a greater chance of developing a disease. Some risk factors for developing CAD can be minimized by lifestyle changes or by drugs. Among such risk factors for CAD are high blood cholesterol level, high blood pressure, cigarette smoking, obesity, diabetes mellitus, "type A" personality, and sedentary lifestyle. It has been found, for example, that moderate exercise and reducing fat in the diet can not only stop the progression of CAD, but actually reverse the process. Other risk factors are beyond our control. These include genetic predisposition (family history of CAD at an early age), age, and gender. Adult males are more likely than adult females to develop CAD until age 70 when the risks become roughly equal.

Development of Atherosclerotic Plaques

Although the following discussion of atherosclerosis applies to coronary arteries, the process can also occur in arteries outside the heart. Thickening of the walls of arteries and loss of elasticity are the main characteristics of a group of diseases called **arteriosclerosis** (ar-tē-rē-ō-skle-RŌ-sis; *sclero-* = hardening). One form of arteriosclerosis is **atherosclerosis** (ath-er-ō-skle-RŌ-sis), a progressive disease characterized by the formation of lesions called **atherosclerotic plaques** in the walls of large and medium-sized arteries (Figure 20.19). Atherosclerosis is initiated by one or more unknown factors that cause damage to the endothelial lining of the arterial wall. Factors that may initiate the process include high circulating LDL levels, cytomegalovirus (a common herpes virus), prolonged high blood pressure, carbon monoxide in cigarette smoke, and high blood glucose levels in diabetes mellitus. Atherosclerosis is thought to begin when one of these factors injures the endothelium of an artery, promoting the aggregation of platelets and attracting phagocytes.

At the injury site, cholesterol and triglycerides collect in the inner layer of the arterial wall. Contact with platelets, lipids, and other components of blood stimulates smooth muscle cells in the arterial wall to proliferate abnormally. As an atherosclerotic plaque develops and enlarges, it progressively obstructs blood flow. An additional danger is that a plaque may provide a roughened surface that attracts platelets,

Figure 20.19   Photomicrographs of a transverse section of (a) a normal artery and (b) an artery partially obstructed by an atherosclerotic plaque.

Partially obstructed lumen (space through which blood flows)

Atherosclerotic plaque

LM 20x

LM 20x

(a) Normal artery

(b) Obstructed artery

initiating clot formation, and further obstructing blood flow. Moreover, a thrombus or piece of a thrombus may dislodge to become an embolus and obstruct blood flow in other vessels.

Diagnosis of CAD

Cardiac catheterization (kath′-e-ter-i-ZĀ-shun) is an invasive procedure used to visualize the heart's coronary arteries, chambers, valves, and great vessels. It may also be used to measure pressure in the heart and blood vessels; to assess cardiac output; to measure the flow of blood through the heart and blood vessels; and to identify the location of septal and valvular defects. The basic procedure involves inserting a long, flexible, radiopaque **catheter** (plastic tube) into a peripheral vein (for right heart catheterization) or a peripheral artery (for left heart catheterization) and guiding it under fluoroscopy (x-ray observation).

Cardiac angiography (an′-jē-OG-ra-fē) is another invasive procedure in which a cardiac catheter is used to inject a radiopaque contrast medium into blood vessels or heart chambers. The procedure may be used to visualize coronary arteries, the aorta, pulmonary blood vessels, and the ventricles to assess structural abnormalities in blood vessels (such as atherosclerotic plaques and emboli), ventricular volume, wall thickness, and wall motion. Angiography can also be used to inject clot-dissolving drugs, such as streptokinase or tissue plasminogen activator (t-PA), into a coronary artery to dissolve an obstructing thrombus.

Treatment of CAD

Treatment options for CAD include drugs (nitroglycerine, beta blockers, cholesterol-lowering drugs, and clot-dissolving agents) and various surgical and nonsurgical procedures designed to increase the blood supply to or decrease the metabolic demands of the heart.

Coronary artery bypass grafting (CABG) is a surgical procedure in which a blood vessel from another part of the body is attached ("grafted") to a coronary artery to bypass an area of blockage. A piece of the grafted blood vessel is sutured between the aorta and the unblocked portion of the coronary artery (Figure 20.20a).

A nonsurgical procedure used to treat CAD is termed **percutaneous transluminal coronary angioplasty (PTCA)** (*percutaneous* = through the skin; *trans-* = across; *lumen* = an opening or channel in a tube; *angio-* = blood vessel; *-plasty* = to mold or to shape). In this procedure, a balloon catheter is inserted into an artery of an arm or leg and gently guided into a coronary artery (Figure 20.20b). Then, while dye is released, angiograms are taken to locate the plaques. Next, the catheter is advanced to the point of obstruction, and a balloonlike device is inflated with air to squash the plaque against the blood vessel wall. Because 30–50% of PTCA-opened arteries fail due to restenosis (renarrowing) within six months after the procedure is done, a stent may be inserted via a catheter. A *stent* is a stainless steel device, resembling a spring coil. It is permanently placed in an artery to keep the artery open (Figure 20.20c).

Myocardial Ischemia and Infarction

Partial obstruction of blood flow in the coronary arteries may cause **myocardial ischemia** (is-KĒ-mē-a; *ische-* = to obstruct; *-emia* = in the blood), a condition of reduced blood flow to the myocardium. Usually, ischemia causes **hypoxia** (reduced oxygen supply), which may weaken cells without killing them. **Angina pectoris** (an-JĪ-na, or AN-ji-na, PEK-to-ris), literally meaning "strangled chest," is a severe pain

Figure 20.20   Three procedures for reestablishing blood flow in occluded coronary arteries.

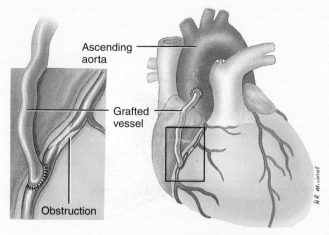

(a) Coronary artery bypass grafting (CABG)

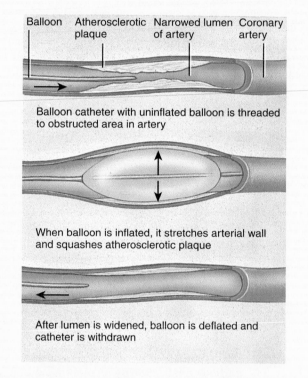

Balloon catheter with uninflated balloon is threaded to obstructed area in artery

When balloon is inflated, it stretches arterial wall and squashes atherosclerotic plaque

After lumen is widened, balloon is deflated and catheter is withdrawn

(b) Percutaneous transluminal coronary angioplasty (PTCA)

(c) Stent in an artery

that usually accompanies myocardial ischemia. Typically, sufferers describe it as a tightness or squeezing sensation, as though the chest were in a vise. Angina pectoris often occurs during exertion, when the heart demands more oxygen, and then disappears with rest. The pain associated with angina pectoris is often referred to the neck, chin, or down the left arm to the elbow. In some people, ischemic episodes occur without producing pain. This condition is known as **silent myocardial ischemia** and is particularly dangerous because the person has no forewarning of an impending heart attack.

A complete obstruction to blood flow in a coronary artery may result in a **myocardial infarction** (in-FARK-shun), or **MI,** commonly called a heart attack. *Infarction* means the death of an area of tissue because of interrupted blood supply. Because the heart tissue distal to the obstruction dies and is replaced by noncontractile scar tissue, the heart muscle loses some of its strength. The aftereffects depend partly on the size and location of the infarcted (dead) area. Besides killing normal heart tissue, an infarction may disrupt the conduction system of the heart and cause sudden death by triggering ventricular fibrillation. Treatment for a myocardial infarction may involve injection of a thrombolytic (clot-dissolving) agent such as streptokinase or t-PA, plus heparin (an anticoagulant), or performing coronary angioplasty or coronary artery bypass grafting. Fortunately, heart muscle can remain alive in a resting person if it receives as little as 10–15% of its normal blood supply. Furthermore, the extensive anastomoses of coronary blood vessels are a significant factor in helping some people survive myocardial infarctions.

Congenital Heart Defects

A defect that is present at birth, and usually before, is called a **congenital defect.** Many such defects are not serious and may go unnoticed for a lifetime. Others are life threatening and must be surgically repaired. Among the several congenital defects that affect the heart are the following:

- **Coarctation** (kō′-ark-TĀ-shun) **of the aorta.** In this condition, a segment of the aorta is too narrow, and thus the flow of oxygenated blood to the body is reduced, the left ventricle is forced to pump harder, and high blood pressure develops.

- **Patent ductus arteriosus.** In some babies, the ductus arteriosus, a temporary blood vessel between the aorta and the pulmonary trunk, remains open rather than closing shortly after birth. As a result, aortic blood flows into the lower-pressure pulmonary trunk, thus increasing the pulmonary trunk blood pressure and overworking both ventricles.

- **Septal defect.** A septal defect is an opening in the septum that separates the interior of the heart into left and right sides. In an **atrial septal defect** the fetal foramen ovale between the two atria fails to close after birth. A **ventricular septal defect** is caused by incomplete development of the interventricular septum. In such cases, oxygenated blood flows directly from the left ventricle into the right ventricle, where it mixes with deoxygenated blood.

- **Tetralogy of Fallot** (tet-RAL-ō-jē of fal-Ō). This condition is a combination of four developmental defects: an interventricular septal defect, an aorta that emerges from both ventricles instead of from the left ventricle only, a stenosed pulmonary valve, and an enlarged right ventricle. Deoxygenated blood from the right ventricle enters the left ventricle through the interventricular septum, mixes with oxygenated blood, and is pumped into the systemic cir-

culation. Moreover, because the aorta emerges from the right ventricle and the pulmonary trunk is stenosed, very little blood reaches the pulmonary circulation. This causes cyanosis, the bluish discoloration most easily seen in nail beds and mucous membranes when the level of deoxygenated hemoglobin is high. For this reason, tetralogy of Fallot is one of the conditions that cause a "blue baby."

Arrhythmias

The usual rhythm of heartbeats, established by the SA node, is **normal sinus rhythm**. An **arrhythmia** (a-RITH-mē-a) or **dysrhythmia** is an irregularity in heart rhythm resulting from a defect in the conduction system of the heart. Arrhythmias can be caused by caffeine, nicotine, alcohol, and certain other drugs; anxiety; hyperthyroidism; potassium deficiency; and certain heart diseases.

Heart Block

One serious arrhythmia is **heart block,** in which propagation of action potentials through the conduction system is either slowed or blocked at some point. The most common site of blockage is the atrioventricular node, a condition called **atrioventricular (AV) block.** In *first-degree AV block,* the P-Q interval is prolonged, usually because conduction through the AV node is slower than normal. In *second-degree AV block,* some of the action potentials from the SA node are not conducted through the AV node. The result is "dropped" beats because excitation doesn't always reach the ventricles. In *third-degree (complete) AV block,* no SA node action potentials get through the AV node. Autorhythmic fibers in the atria and ventricles pace the upper and lower chambers separately. With complete AV block, the ventricular contraction rate is less than 40 beats/min.

Flutter and Fibrillation

Two other abnormal rhythms are **flutter** and **fibrillation. Atrial flutter** consists of rapid atrial contractions (240–360 beats/min) accompanied by a second-degree AV block. Flutter may result from rheumatic heart disease, coronary artery disease, or certain congenital heart diseases. In **atrial fibrillation** contraction of atrial fibers is asynchronous (not in unison) so that atrial pumping ceases altogether. Atrial fibrillation may occur in myocardial infarction, acute and chronic rheumatic heart disease, and hyperthyroidism. In an otherwise strong heart, atrial fibrillation reduces the pumping effectiveness of the heart by 20–30%.

Ventricular fibrillation is the most deadly arrhythmia, in which contractions of ventricular muscle fibers are completely asynchronous. Ventricular pumping stops, blood ejection ceases, and circulatory failure and death occur unless the situation is quickly corrected. A procedure known as **defibrillation,** in which a strong, brief electrical current is passed through the heart, often can stop ventricular fibrillation. The electric shock is applied via large paddle-shaped electrodes pressed against the skin of the chest. Patients who face a high risk of dying from heart rhythm disorders now can receive an **automatic implantable cardioverter defibrillator (AICD),** an implanted device that monitors their heart rhythm and delivers a small shock directly to the heart when a life-threatening rhythm disturbance occurs. Thousands of patients around the world have AICDs, including Dick Cheney, Vice President of the United States, who received a combination pacemaker-defibrillator in 2001.

Ventricular Premature Contraction

Another form of arrhythmia arises when an ectopic focus, a region of the heart other than the conduction system, becomes more excitable than normal and causes an occasional abnormal action potential to occur. As a wave of depolarization spreads outward from the ectopic focus, it causes a **ventricular premature contraction.** The contrac- tion occurs early in diastole before the SA node is normally scheduled to discharge its action potential. Ventricular premature contractions may be relatively benign and may be caused by emotional stress, excessive intake of stimulants such as caffeine or nicotine, and lack of sleep. In other cases, the premature beats may reflect an underlying pathology.

MEDICAL TERMINOLOGY

Angiocardiography (an′-jē-ō-kar′-dē-OG-ra-fē; *angio-* = vessel; *cardio-* = heart) X-ray examination of the heart and great blood vessels after injection of a radiopaque dye into the bloodstream.

Cardiac arrest (KAR-dē-ak a-REST) A clinical term meaning cessation of an effective heartbeat. The heart may be completely stopped or in ventricular fibrillation.

Cardiomegaly (kar′-dē-ō-MEG-a-lē; *mega* = large) Heart enlargement.

Cor pulmonale (CP) (kor pul-mōn-ALE; *cor-* = heart; *pulmon-* = lung) A term referring to right ventricular hypertrophy from disorders that bring about hypertension (high blood pressure) in the pulmonary circulation.

Ejection fraction The fraction of the end-diastolic volume (EDV) that is ejected during an average heartbeat. Equal to stroke volume (SV) divided by EDV.

Palpitation (pal′-pi-TĀ-shun) A fluttering of the heart or an abnormal rate or rhythm of the heart.

Paroxysmal tachycardia (par′-ok-SIZ-mal tak′-e-KAR-dē-a; *tachy-* = swift or fast) A period of rapid heartbeats that begins and ends suddenly.

Sudden cardiac death The unexpected cessation of circulation and breathing due to an underlying heart disease such as ischemia, myocardial infarction, or a disturbance in cardiac rhythm.

STUDY OUTLINE

ANATOMY OF THE HEART (p. 660)

1. The heart is located in the mediastinum; about two-thirds of its mass is to the left of the midline.
2. The heart is shaped like a cone lying on its side; its apex is the pointed, inferior part, whereas its base is the broad, superior part.
3. The pericardium is the membrane that surrounds and protects the heart; it consists of an outer fibrous layer and an inner serous pericardium, which is composed of a parietal and a visceral layer.
4. Between the parietal and visceral layers of the serous pericardium is the pericardial cavity, a potential space filled with a few milliliters of pericardial fluid that reduces friction between the two membranes.
5. Three layers make up the wall of the heart: epicardium (visceral layer of the serous pericardium), myocardium, and endocardium.
6. The epicardium consists of mesothelium and connective tissue, the myocardium is composed of cardiac muscle tissue, and the endocardium consists of endothelium and connective tissue.
7. The heart chambers include two superior chambers, the right and left atria, and two inferior chambers, the right and left ventricles.
8. External features of the heart include the auricles (flaps on each atrium that slightly increase their volume), the coronary sulcus between the atria and ventricles, and the anterior and posterior sulci between the ventricles on the anterior and posterior surfaces of the heart, respectively.
9. The right atrium receives blood from the superior vena cava, inferior vena cava, and coronary sinus. It is separated from the left atrium by the interatrial septum, which contains the fossa ovalis. Blood exits the right atrium through the tricuspid valve.

10. The right ventricle receives blood from the right atrium. It is separated from the left ventricle by the interventricular septum and pumps blood through the pulmonary valve into the pulmonary trunk.
11. Oxygenated blood enters the left atrium from the pulmonary veins and exits through the bicuspid (mitral) valve.
12. The left ventricle pumps oxygenated blood through the aortic valve into the aorta.
13. The thickness of the myocardium of the four chambers varies according to the chamber's function. The left ventricle, with the highest workload, has the thickest wall.
14. The fibrous skeleton of the heart is dense connective tissue that surrounds and supports the valves of the heart.

HEART VALVES AND CIRCULATION OF BLOOD (p. 667)

1. Heart valves prevent backflow of blood within the heart. The atrioventricular (AV) valves, which lie between atria and ventricles, are the tricuspid valve on the right side of the heart and the bicuspid (mitral) valve on the left. The semilunar (SL) valves are the aortic valve, at the entrance to the aorta, and the pulmonary valve, at the entrance to the pulmonary trunk.
2. The left side of the heart is the pump for the systemic circulation, the circulation of blood throughout the body except for the air sacs of the lungs. The left ventricle ejects blood into the aorta, and blood then flows into systemic arteries, arterioles, capillaries, venules, and veins, which carry it back to the right atrium.
3. The right side of the heart is the pump for pulmonary circulation, the circulation of blood through the lungs. The right ventricle ejects blood into the pulmonary trunk, and blood then flows into

pulmonary arteries, pulmonary capillaries, and pulmonary veins, which carry it back to the left atrium.

4. The coronary circulation provides blood flow to the myocardium. The main arteries of the coronary circulation are left and right coronary arteries; the main veins are the cardiac vein and the coronary sinus.

CARDIAC MUSCLE TISSUE AND THE CARDIAC CONDUCTION SYSTEM (p. 672)

1. Cardiac muscle fibers usually contain a single centrally located nucleus. Compared to skeletal muscle fibers, cardiac muscle fibers have more and larger mitochondria, slightly smaller sarcoplasmic reticulum, and wider transverse tubules, which are located at Z discs.
2. Cardiac muscle fibers are connected via end-to-end intercalated discs. Desmosomes in the discs provide strength and gap junctions allow muscle action potentials to conduct from one muscle fiber to its neighbors.
3. Autorhythmic fibers form the conduction system, cardiac muscle fibers that spontaneously depolarize and generate action potentials.
4. Components of the conduction system are the sinoatrial (SA) node (pacemaker), atrioventricular (AV) node, atrioventricular (AV) bundle (bundle of His), bundle branches, and Purkinje fibers.
5. Phases of an action potential in a ventricular contractile fiber include rapid depolarization, a long plateau, and repolarization.
6. Cardiac muscle tissue has a long refractory period, which prevents tetanus.
7. The record of electrical changes during each cardiac cycle is called an electrocardiogram (ECG). A normal ECG consists of a P wave (atrial depolarization), a QRS complex (onset of ventricular depolarization), and a T wave (ventricular repolarization).
8. The P–Q interval represents the conduction time from the beginning of atrial excitation to the beginning of ventricular excitation. The S–T segment represents the time when ventricular contractile fibers are fully depolarized.

THE CARDIAC CYCLE (p. 678)

1. A cardiac cycle consists of the systole (contraction) and diastole (relaxation) of both atria, plus the systole and diastole of both ventricles. With an average heartbeat of 75 beats/min, a complete cardiac cycle requires 0.8 seconds.

2. The phases of the cardiac cycle are (a) atrial systole, (b) ventricular systole, and (c) relaxation period.
3. S1, the first heart sound (lubb), is caused by blood turbulence associated with the closing of the atrioventricular valves. S2, the second sound (dupp), is caused by blood turbulence associated with the closing of semilunar valves.

CARDIAC OUTPUT (P. 683)

1. Cardiac output (CO) is the amount of blood ejected per minute by the left ventricle into the aorta (or by the right ventricle into the pulmonary trunk). It is calculated as follows: CO (mL/min) = stroke volume (SV) in mL/beat × heart rate (HR) in beats per minute.
2. Stroke volume (SV) is the amount of blood ejected by a ventricle during each systole.
3. Cardiac reserve is the difference between a person's maximum cardiac output and his or her cardiac output at rest.
4. Stroke volume is related to preload (stretch on the heart before it contracts), contractility (forcefulness of contraction), and afterload (pressure that must be exceeded before ventricular ejection can begin).
5. According to the Frank–Starling law of the heart, a greater preload (end-diastolic volume) stretching cardiac muscle fibers just before they contract increases their force of contraction until the stretching becomes excessive.
6. Nervous control of the cardiovascular system stems from the cardiovascular center in the medulla oblongata.
7. Sympathetic impulses increase heart rate and force of contraction; parasympathetic impulses decrease heart rate.
8. Heart rate is affected by hormones (epinephrine, norepinephrine, thyroid hormones), ions (Na^+, K^+, Ca^{2+}), age, gender, physical fitness, and body temperature.

EXERCISE AND THE HEART (p. 687)

1. Sustained exercise increases oxygen demand on muscles.
2. Among the benefits of aerobic exercise are increased cardiac output, decreased blood pressure, weight control, and increased fibrinolytic activity.

DEVELOPMENT OF THE HEART (p. 688)

1. The heart develops from mesoderm.
2. The endocardial tubes develop into the four-chambered heart and great vessels of the heart.

SELF-QUIZ QUESTIONS

Fill in the blanks in the following statements.

1. The chamber of the heart with the thickest myocardium is the _____.

2. The phase of heart contraction is called _____; the phase of heart relaxation is called _____.

3. Cardiac output equals _____ multiplied by heart rate.

Indicate whether the following statements are true or false.

4. In auscultation, the lubb represents closing of the semilunar valves and the dupp represents closing of the atrioventricular valves.

5. Positive inotropic agents decrease myocardial contractility.

6. The Frank–Starling law of the heart equalizes the output of the right and left ventricles and keeps the same volume of blood flowing to both the systemic and pulmonary circulations.

Choose the one best answer to the following questions.

7. Which of the following is the correct route of blood through the heart from the systemic circulation to the pulmonary circulation and back to the systemic circulation? (a) right atrium, tricuspid valve, right ventricle, pulmonary semilunar valve, left atrium, mitral valve, left ventricle, aortic semilunar valve, (b) left atrium, tricuspid valve, left ventricle, pulmonary semilunar valve, right atrium, mitral valve, right

ventricle, aortic semilunar valve, (c) left atrium, pulmonary semilunar valve, right atrium, tricuspid valve, left ventricle, aortic semilunar valve, right ventricle, mitral valve, (d) left ventricle, mitral valve, left atrium, pulmonary semilunar valve, right ventricle, tricuspid valve, right atrium, aortic semilunar valve, (e) right atrium, mitral valve, right ventricle, pulmonary semilunar valve, left atrium, tricuspid valve, left ventricle, aortic semilunar valve.

8. Which of the following represents the correct pathway for conduction of an action potential through the heart? (a) AV node, AV bundle, SA node, Purkinje fibers, bundle branches, (b) AV node, bundle branches, AV bundle, SA node, Purkinje fibers, (c) SA node, AV node, AV bundle, bundle branches, Purkinje fibers, (d) SA node, AV bundle, bundle branches, AV node, Purkinje fibers, (e) SA node, AV node, Purkinje fibers, bundle branches, AV bundle.

9. A softball player is found to have a resting cardiac ouput of 5.0 liters per minute and a heart rate of 50 beats per minute. What is her stroke volume? (a) 10 mL; (b) 100 mL; (c) 1000 mL; (d) 6000 mL; (e) The information given is insufficient to calculate stroke volume.

10. When would a murmur due to stenosis of a semilunar valve be heard? (a) between S1 and S2, when the valve should be fully open but is not; (b) between S1 and S2, when the valve should be fully closed but is not; (c) between S2 and S1, when the valve should be fully open but is not; (d) between S2 and S1, when the valve should be fully closed but is not.

11. Which of the following are *true*? (1) ANS regulation of heart rate originates in the cardiovascular center of the medulla oblongata. (2) Proprioceptor input is a major stimulus that accounts for the rapid rise in the heart rate at the onset of physical activity. (3) The vagus nerves release norepinephrine, causing the heart rate to increase. (4) Hormones from the adrenal medulla and the thyroid gland can increase the heart rate. (5) Hypothermia increases the heart rate. (a) 1, 2, 3, and 4, (b) 1, 2, and 4, (c) 2, 3, 4, and 5, (d) 3, 5, and 6, (e) 1, 2, 4, and 5

12. Match the following:
_____ (a) collects oxygenated blood from the pulmonary circulation
_____ (b) pumps deoxygenated blood to the lungs for oxygenation
_____ (c) their contraction pulls on and tightens the chordae tendineae, preventing the valve cusps from everting
_____ (d) increase blood-holding capacity of the atria
_____ (e) tendonlike cords connected to the atrioventricular valve cusps which, along with the papillary muscles, prevent valve eversion
_____ (f) pumps oxygenated blood to all body cells, except the air sacs of the lungs
_____ (g) collects deoxygenated blood from the systemic circulation
_____ (h) prevents backflow of blood from the right ventricle into the right atrium
_____ (i) left atrioventricular valve

(1) right atrium
(2) right ventricle
(3) left atrium
(4) left ventricle
(5) tricuspid valve
(6) bicuspid (mitral) valve
(7) chordae tendineae
(8) auricles
(9) papillary muscles

13. Match the following:
_____ (a) cardiac muscle tissue
_____ (b) blood vessels that pierce the heart muscle and supply blood to the cardiac muscle fibers
_____ (c) separate the upper and lower heart chambers, preventing backflow of blood from the inferior chambers back into the superior chambers
_____ (d) endothelial cells lining the interior of the heart; are continuous with the endothelium of the blood vessels
_____ (e) inner visceral layer of the serous pericardium; adheres tightly to the surface of the heart
_____ (f) the remnant of the foramen ovale, an opening in the interatrial septum of the fetal heart
_____ (g) outer layer of the serous pericardium; is fused to the fibrous pericardium
_____ (h) the gap junction and desmosome connections between individual cardiac muscle fibers
_____ (i) the superficial dense irregular connective tissue covering of the heart
_____ (j) prevent backflow of blood from the arteries into the inferior heart chambers

(1) fibrous pericardium
(2) parietal pericardium
(3) epicardium
(4) myocardium
(5) endocardium
(6) atrioventricular valves
(7) semilunar valves
(8) intercalated discs
(9) fossa ovalis
(10) coronary circulation

14. Match the following:
_____ (a) amount of blood contained in the ventricles at the end of ventricular relaxation
_____ (b) period of time when cardiac muscle fibers are contracting and exerting force but not shortening
_____ (c) amount of blood ejected per beat by each ventricle
_____ (d) amount of blood remaining in the ventricles following ventricular contraction
_____ (e) difference between a person's maximum cardiac output and cardiac output at rest
_____ (f) period of time when semilunar valves are open and blood flows out of the inferior chambers
_____ (g) period when all four valves are closed and ventricular blood volume does not change

(1) cardiac reserve
(2) stroke volume
(3) end-diastolic volume (EDV)
(4) isovolumetric relaxation
(5) end-systolic volume (ESV)
(6) ventricular ejection
(7) isovolumetric contraction

15. Match the following:
_____ (a) indicates ventricular repolarization
_____ (b) represents the time from the beginning of ventricular depolarization to the end of ventricular repolarization
_____ (c) represents atrial depolarization
_____ (d) represents the time when the ventricular contractile fibers are fully depolarized; occurs during the plateau phase of the action potential
_____ (e) represents the onset of ventricular depolarization
_____ (f) represents the conduction time from the beginning of atrial excitation to the beginning of ventricular excitation

(1) P wave
(2) QRS complex
(3) T wave
(4) P–Q interval
(5) S–T segment
(6) Q–T interval

CRITICAL THINKING QUESTIONS

1. Arian, a member of the college track team, volunteered for a study that the exercise physiology class was conducting on the cardiovascular system. His resting heart rate was measured at 55 beats per minute and he had a normal cardiac output (CO). After strenuous exercise, his cardiac reserve was calculated to be six times the resting value. Arian is in excellent physical condition. Predict Arian's resting CO and stroke volume. What would his CO be during strenuous exercise?
HINT *At rest, each ventricle pumps a volume of blood per minute that is about equal to the body's total blood volume.*

2. At family gatherings, Aunt Mercy likes to tell the story of when she had rheumatic fever as a child and how her mother treated her with home remedies. Now Aunt Mercy complains of a "weak heart." Could the childhood illness be related to her heart problem?
HINT *The immune system may mistakenly attack healthy body tissue along with the bacteria.*

3. Mr. Perkins is a large, 62-year-old man with a weakness for sweets and fried foods. His idea of exercise is to get up to change the channel while watching sports on television. Lately, he's been troubled by chest pains when he walks up stairs. His doctor told him to quit smoking and scheduled a cardiac angiography for next week. What is involved in performing this procedure? Why did the doctor order this test?
HINT *Mr Perkins' stress electrocardiogram revealed changes that suggest myocardial ischemia.*

ANSWERS TO FIGURE QUESTIONS

20.1 The mediastinum is the mass of tissue that extends from the sternum to the vertebral column between the lungs.

20.2 The visceral layer of the serous pericardium (epicardium) is both a part of the pericardium and a part of the heart wall.

20.3 The coronary sulcus forms a boundary between the atria and ventricles.

20.4 The greater the workload of a heart chamber, the thicker its myocardium.

20.5 The fibrous skeleton attaches to the heart valves and prevents overstretching of the valves as blood passes through them.

20.6 The papillary muscles contract, which pulls on the chordae tendineae and prevents valve cusps from everting.

20.7 Numbers 2 (right ventricle) through 6 depict the pulmonary circulation, whereas numbers 7 (left ventricle) through 10 and 1 (right atrium) depict the systemic circulation.

20.8 The circumflex artery delivers oxygenated blood to the left atrium and left ventricle.

20.9 The intercalated discs hold the cardiac muscle fibers together and enable action potentials to propagate from one muscle fiber to another.

20.10 The only electrical connection between the atria and the ventricles is the atrioventricular bundle.

20.11 The duration of an action potential is much longer in a ventricular contractile fiber (0.3 sec = 300 msec) than in a skeletal muscle fiber (1–2 msec).

20.12 An enlarged Q wave may indicate a myocardial infarction (heart attack).

20.13 Action potentials propagate most slowly through the AV node.

20.14 The amount of blood in each ventricle at the end of ventricular diastole—the end-diastolic volume—is about 130 mL in a resting person.

20.15 The first sound (S1), or lubb, is associated with closure of the AV valves.

20.16 The ventricular myocardium receives innervation from the sympathetic division only.

20.17 The skeletal muscle "pump" increases stroke volume by increasing preload (end-diastolic volume).

20.18 The heart begins to contract by the end of the third week of development.

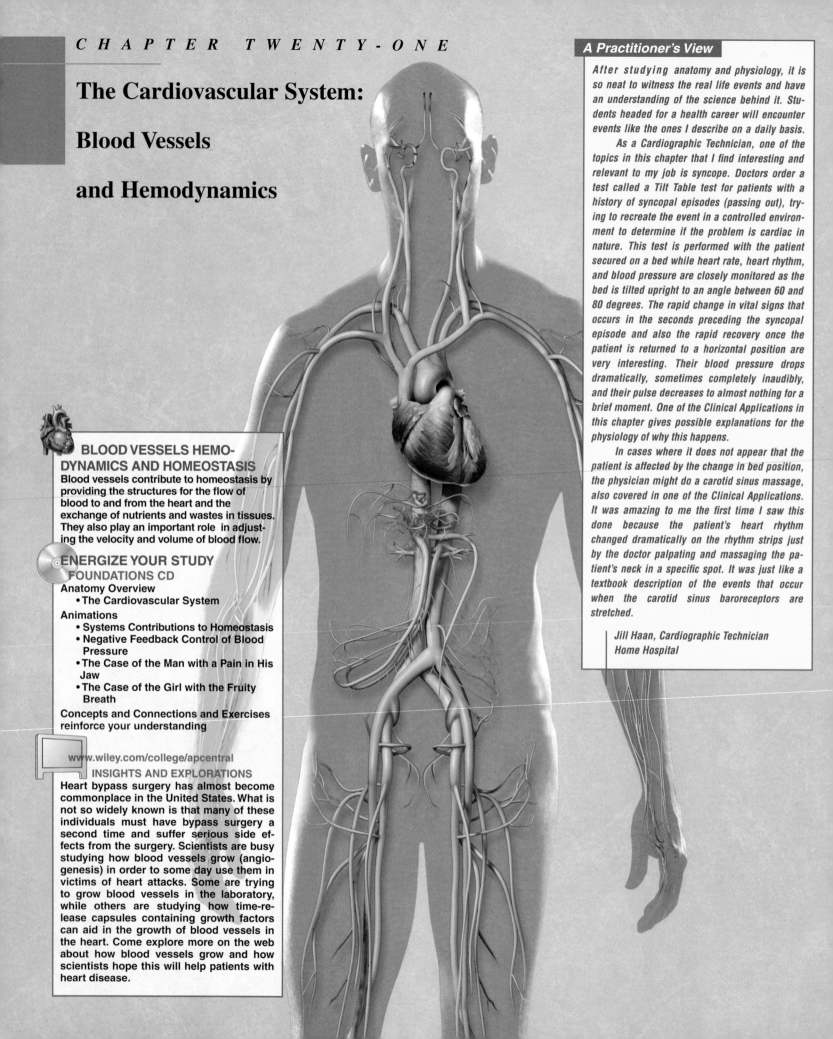

The Cardiovascular System:

Blood Vessels

and Hemodynamics

BLOOD VESSELS HEMO-DYNAMICS AND HOMEOSTASIS

Blood vessels contribute to homeostasis by providing the structures for the flow of blood to and from the heart and the exchange of nutrients and wastes in tissues. They also play an important role in adjusting the velocity and volume of blood flow.

ENERGIZE YOUR STUDY
FOUNDATIONS CD

Anatomy Overview
- The Cardiovascular System

Animations
- Systems Contributions to Homeostasis
- Negative Feedback Control of Blood Pressure
- The Case of the Man with a Pain in His Jaw
- The Case of the Girl with the Fruity Breath

Concepts and Connections and Exercises reinforce your understanding

www.wiley.com/college/apcentral

INSIGHTS AND EXPLORATIONS

Heart bypass surgery has almost become commonplace in the United States. What is not so widely known is that many of these individuals must have bypass surgery a second time and suffer serious side effects from the surgery. Scientists are busy studying how blood vessels grow (angiogenesis) in order to some day use them in victims of heart attacks. Some are trying to grow blood vessels in the laboratory, while others are studying how time-release capsules containing growth factors can aid in the growth of blood vessels in the heart. Come explore more on the web about how blood vessels grow and how scientists hope this will help patients with heart disease.

The cardiovascular system contributes to homeostasis of other body systems by transporting and distributing blood throughout the body to deliver materials such as oxygen, nutrients, and hormones and to carry away wastes. This transport occurs via blood vessels, which form a closed system of tubes that carries blood away from the heart, transports it to the tissues of the body, and then returns it to the heart. Chapters 19 and 20 described the composition and functions of blood and the structure and function of the heart. In this chapter, the focus is on the structure and functions of the various types of blood vessels; on **hemodynamics** (hē-mō-dī-NAM-iks; *hemo-* = blood; *dynamics* = power), the forces involved in circulating blood throughout the body; and on the blood vessels that constitute the major circulatory routes.

STRUCTURE AND FUNCTION OF BLOOD VESSELS

> O B J E C T I V E

- **Contrast the structure and function of arteries, arterioles, capillaries, venules, and veins.**

The five main types of blood vessels are arteries, arterioles, capillaries, venules, and veins. **Arteries** (AR-ter-ēz) carry blood *away from the heart* to other organs. Large, elastic arteries leave the heart and divide into medium-sized, muscular arteries that branch out into the various regions of the body. Medium-sized arteries then divide into small arteries, which, in turn, divide into still smaller arteries called **arterioles** (ar-TER-ē-ōls). As the arterioles enter a tissue, they branch into myriad tiny vessels called **capillaries** (KAP-i-lar′-ēz = hairlike). The thin walls of capillaries allow exchange of substances between the blood and body tissues. Groups of capillaries within a tissue reunite to form small veins called **venules** (VEN-ūls). These, in turn, merge to form progressively larger blood vessels called veins. **Veins** (VĀNZ) are the blood vessels that convey blood from the tissues *back to the heart.* Because blood vessels require oxygen (O_2) and nutrients just like other tissues of the body, larger blood vessels are served by their own blood vessels, called **vasa vasorum** (literally, vasculature of vessels), located within their walls.

Arteries

In ancient times, **arteries** (*ar-* = air; *ter-* = to carry) were found empty at death and thus were thought to contain only air. The wall of an artery has three coats or tunics: (1) tunica interna, (2) tunica media, and (3) tunica externa (Figure 21.1). The innermost coat, the **tunica interna** or **intima,** contains a lining of simple squamous epithelium called *endothelium, a basement membrane,* and a layer of elastic tissue called the *internal elastic lamina.* The endothelium is a continuous layer of cells that line the inner surface of the entire cardiovascular system (the heart and all blood vessels). Normally, endothelium is the only tissue that makes contact with blood. The tunica interna is closest to the **lumen,** the hollow center through which blood flows. The middle coat, or **tunica media,** is usually the thickest layer. It consists of elastic fibers and smooth muscle fibers (cells) that extend circularly (in rings) around the lumen. Due to their plentiful elastic fibers, arteries normally have high *compliance,* which means that their walls easily stretch or expand without tearing in response to a small increase in pressure. The outer coat, the **tunica externa,** is composed mainly of elastic and collagen fibers. In muscular arteries (described shortly), an *external elastic lamina* composed of elastic tissue separates the tunica externa from the tunica media.

Sympathetic neurons of the autonomic nervous system innervate vascular smooth muscle. An increase in sympathetic stimulation typically stimulates the smooth muscle to contract, squeezing the vessel wall and narrowing the lumen. Such a decrease in the diameter of the lumen of a blood vessel is called **vasoconstriction.** In contrast, smooth muscle fibers relax when sympathetic stimulation decreases or when certain chemicals, such as nitric oxide, K^+, H^+, and lactic acid, are present. The resulting increase in lumen diameter is called **vasodilation.** Additionally, when an artery or arteriole is damaged, its smooth muscle contracts, producing vascular spasm of the vessel. Such a vasospasm limits blood flow through the damaged vessel and helps reduce blood loss if the vessel is small.

Elastic Arteries

The largest-diameter arteries, termed **elastic arteries** because the tunica media contains a high proportion of elastic fibers, have walls that are relatively thin in proportion to their overall diameter. Elastic arteries perform an important function: They help propel blood onward while the ventricles are relaxing. As blood is ejected from the heart into elastic arteries, their highly elastic walls stretch, accommodating the surge of blood. By stretching, the elastic fibers store mechanical energy for a short time, functioning as a **pressure reservoir** (Figure 21.2a on page 699). Then, the elastic fibers recoil and convert stored (potential) energy in the vessel into kinetic energy of the blood. Thus, blood continues to move through the arteries even while the ventricles are relaxed (Figure 21.2b). Because they conduct blood from the heart to medium-sized, more-muscular arteries, elastic arteries also are called *conducting arteries.* The aorta and the brachiocephalic, common carotid, subclavian, vertebral, pulmonary, and common iliac arteries are elastic arteries (see Figure 21.18).

Figure 21.1 Comparative structure of blood vessels. The relative size of the capillary in (c) is enlarged.

🔑 Arteries carry blood from the heart to tissues; veins carry blood from tissues to the heart.

TUNICA INTERNA:
— Endothelium

— Basement membrane

— Internal elastic lamina

TUNICA MEDIA:
— Smooth muscle

— External elastic lamina

— TUNICA EXTERNA —

Valve

Lumen
(a) Artery

Lumen
(b) Vein

Lumen

Endothelium

Basement membrane

(c) Capillary

Internal elastic lamina

External elastic lamina

Tunica externa

Lumen with blood cells

Tunica interna

Tunica media

Connective tissue

LM 200x

(d) Transverse section through an artery

Connective tissue

Red blood cell

LM 600x

Capillary endothelial cells

(e) Red blood cells passing through a capillary

❓ Which vessel—the femoral artery or the femoral vein—has a thicker wall? Which has a wider lumen?

Figure 21.2 **Pressure reservoir function of elastic arteries.**

 Recoil of elastic arteries keeps blood flowing during ventricular relaxation (diastole).

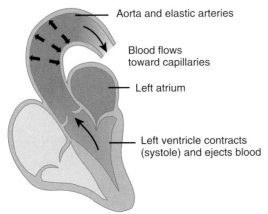

Aorta and elastic arteries

Blood flows toward capillaries

Left atrium

Left ventricle contracts (systole) and ejects blood

(a) Elastic aorta and arteries stretch during ventricular contraction

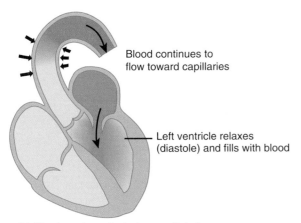

Blood continues to flow toward capillaries

Left ventricle relaxes (diastole) and fills with blood

(b) Elastic aorta and arteries recoil during ventricular relaxation

In atherosclerosis, the walls of elastic arteries become less compliant (stiffer). What effect does reduced compliance have on the pressure reservoir function of arteries?

Muscular Arteries

Medium-sized arteries are called **muscular arteries** because their tunica media contains more smooth muscle and fewer elastic fibers than elastic arteries. Thus, muscular arteries are capable of greater vasoconstriction and vasodilation to adjust the rate of blood flow. The large amount of smooth muscle makes the walls of muscular arteries relatively thick. Muscular arteries also are called *distributing arteries* because they distribute blood to various parts of the body. Examples include the brachial artery in the arm and radial artery in the forearm (see Figure 21.18).

Arterioles

An **arteriole** (= small artery) is a very small (almost microscopic) artery that delivers blood to capillaries (Figure 21.3). Arterioles near the arteries from which they branch have a tunica interna like that of arteries, a tunica media composed of smooth

muscle and very few elastic fibers, and a tunica externa composed mostly of elastic and collagen fibers. In the smallest-diameter arterioles, which are closest to capillaries, the tunics consist of little more than a ring of endothelial cells surrounded by a few scattered smooth muscle fibers.

Arterioles play a key role in regulating blood flow from arteries into capillaries by regulating **resistance,** the opposition to blood flow. In a blood vessel, resistance is due mainly to friction between blood and the inner walls of blood vessels. When the blood vessel diameter is smaller, the friction is greater. Because contraction and relaxation of the smooth muscle in arteriole walls can change their diameter, arterioles are known as *resistance vessels.* Contraction of arteriolar smooth muscle causes vasoconstriction, which increases resistance and decreases blood flow into capillaries supplied by that arteriole. By contrast, relaxation of arteriolar smooth muscle causes vasodilation, which decreases resistance and increases blood flow into capillaries. A change in diameter of arterioles can also significantly affect blood pressure.

Capillaries

Capillaries are microscopic vessels that connect arterioles to venules (Figure 21.3). The flow of blood from arterioles to venules through capillaries is called the **microcirculation.** Capillaries are found near almost every cell in the body, but their number varies with the metabolic activity of the tissue they serve. Body tissues with high metabolic requirements, such as muscles, the liver, the kidneys, and the nervous system, use more O_2 and nutrients and thus have extensive capillary networks. Tissues with lower metabolic requirements, such as tendons and ligaments, contain fewer capillaries. Capillaries are absent in a few tissues, for instance, all covering and lining epithelia, the cornea and lens of the eye, and cartilage.

Capillaries are known as *exchange vessels* because their prime function is the exchange of nutrients and wastes between the blood and tissue cells through the interstitial fluid. The structure of capillaries is well suited to this function. Capillary walls are composed of only a single layer of endothelial cells and a basement membrane (see Figure 21.1e). They have no tunica media or tunica externa. Thus, a substance in the blood must pass through just one cell layer to reach the interstitial fluid and tissue cells. Exchange of materials occurs only through the walls of capillaries and the beginning of venules; the walls of arteries, arterioles, most venules, and veins present too thick a barrier. Capillaries form extensive branching networks that increase the surface area available for rapid exchange of materials. In most tissues, blood flows through only a small part of the capillary network when metabolic needs are low. However, when a tissue is active, such as contracting muscle, the entire capillary network fills with blood.

A **metarteriole** (*met-* = beyond) is a vessel that emerges from an arteriole and supplies a group of 10–100 capillaries that constitute a **capillary bed** (Figure 21.3a). The proximal end of

Figure 21.3 **Arteriole, capillaries, and venule.** Precapillary sphincters regulate the flow of blood through capillary beds.

🔑 In capillaries, nutrients, gases, and wastes are exchanged between the blood and interstitial fluid.

(a) Sphincters relaxed: blood flowing through capillary bed

(b) Sphincters contracted: blood flowing through thoroughfare channel

❓ Why do metabolically active tissues have extensive capillary networks?

a metarteriole is surrounded by scattered smooth muscle fibers whose contraction and relaxation help regulate blood flow through the capillary bed. The distal end of a metarteriole, which empties into a venule, has no smooth muscle fibers and is called a **thoroughfare channel.** Blood flowing through a thoroughfare channel bypasses the capillary bed.

True capillaries emerge from arterioles or metarterioles. At their sites of origin, a ring of smooth muscle fibers called a **precapillary sphincter** controls the flow of blood into a true capillary. When the precapillary sphincters are relaxed (open), blood flows into the capillary bed (Figure 21.3b); when precapillary sphincters contract (close or partially close), blood flow through the capillary bed ceases or decreases (Figure 21.3b). Typically, blood flows intermittently through a capillary bed due to alternating contraction and relaxation of the smooth muscle of metarterioles and the precapillary sphincters. This intermittent contraction and relaxation, which may occur 5 to 10 times per minute, is called **vasomotion.** In part, vasomotion is due to chemicals released by the endothelial cells; nitric oxide is one

example. At any given time, blood flows through only about 25% of a capillary bed.

The body contains three different types of capillaries: continuous capillaries, fenestrated capillaries, and sinusoids (Figure 21.4). Many capillaries are **continuous capillaries,** in which the plasma membranes of endothelial cells form a continuous tube that is interrupted only by **intercellular clefts,** which are gaps between neighboring endothelial cells (Figure 21.4a). Continuous capillaries are found in skeletal and smooth muscle, connective tissues, and the lungs.

Other capillaries of the body are **fenestrated capillaries** (*fenestr-* = window). The plasma membranes of the endothelial cells in these capillaries have many **fenestrations,** small pores (holes) ranging from 70 to 100 nm in diameter (Figure 21.4b). Fenestrated capillaries are found in the kidneys, villi of the small intestine, choroid plexuses of the ventricles in the brain, ciliary processes of the eyes, and endocrine glands.

Sinusoids are wider and more winding than other capillaries. Their endothelial cells may have unusually large fenestrations. In

Figure 21.4 Types of capillaries.

🔑 Capillaries are microscopic blood vessels that connect arterioles and venules.

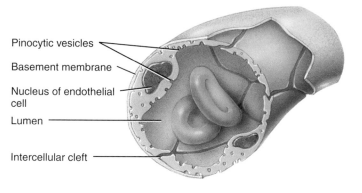

- Pinocytic vesicles
- Basement membrane
- Nucleus of endothelial cell
- Lumen
- Intercellular cleft

(a) Continuous capillary formed by endothelial cells

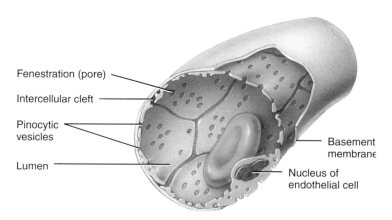

- Fenestration (pore)
- Intercellular cleft
- Pinocytic vesicles
- Lumen
- Basement membrane
- Nucleus of endothelial cell

(b) Fenestrated capillary

- Incomplete basement membrane
- Lumen
- Nucleus of endothelial cell
- Intercellular cleft

(c) Sinusoid

❓ How do materials move across capillary walls?

addition to an incomplete or absent basement membrane (Figure 21.4c), sinusoids have very large intercellular clefts that allow proteins and in some cases even blood cells to pass from a tissue into the bloodstream. For example, newly formed blood cells enter the bloodstream through the sinusoids of red bone marrow. In addition, sinusoids contain specialized lining cells that are adapted to the function of the tissue. Sinusoids in the liver, for example, contain phagocytic cells that remove bacteria and other debris from the blood. The spleen, anterior pituitary, and parathyroid glands also have sinusoids.

Venules

When several capillaries unite, they form small veins called **venules** (= little veins). Venules collect blood from capillaries and drain into veins. The smallest venules, those closest to the capillaries, consist of a tunica interna of endothelium and a tunica media that has only a few scattered smooth muscle fibers and fibroblasts. Like capillaries, the walls of the smallest venules are very porous and are the site where many phagocytic white blood cells emigrate from the bloodstream into an inflamed or infected tissue. As venules become larger and converge to form veins, they contain the tunica externa characteristic of veins.

Veins

Although **veins** are composed of essentially the same three coats as arteries, the relative thicknesses of the layers are different. The tunica interna of veins is thinner than that of arteries; the tunica media of veins is much thinner than in arteries, with relatively little smooth muscle and elastic fibers. The tunica externa of veins is the thickest layer and consists of collagen and elastic fibers; the tunica externa of the inferior vena cava also contains longitudinal fibers of smooth muscle. Veins lack the external or internal elastic laminae found in arteries (see Figure 21.1b). Still, veins are distensible enough to adapt to variations in the volume and pressure of blood passing through them, although they are not designed to withstand high pressure. Furthermore, the lumen of a vein is larger than that of a comparable artery, and veins often appear collapsed (flattened) when sectioned.

Many veins, especially those in the limbs, also feature **valves,** which are thin folds of tunica interna that form flaplike cusps. The valve cusps project into the lumen, pointing toward the heart (Figure 21.5). Because the low blood pressure in veins allows blood returning to the heart to slow and even back up, the valves aid in venous return by preventing the backflow of blood.

A **vascular (venous) sinus** is a vein with a thin endothelial wall that has no smooth muscle to alter its diameter. In a vascular sinus, the surrounding dense connective tissue replaces the tunica media and tunica externa in providing support. For example, dural venous sinuses, which are supported by the dura mater, convey deoxygenated blood from the brain to the heart. Another example of a vascular sinus is the coronary sinus of the heart (see Figure 20.3c on page 664).

Figure 21.5 Venous valves.

Valves in veins allow blood to flow in one direction only—toward the heart.

Transverse plane

Cusps of valve

Transverse section

Frontal plane

Cusps of valve

Longitudinally cut

Photograph of a valve in a vein

Why are valves more important in arm veins and leg veins than in neck veins?

Varicose Veins

Leaky venous valves can cause veins to become dilated and twisted in appearance, a condition called **varicose veins** (*varic-* = a swollen vein). The condition may occur in the veins of almost any body part, but it is most common in the esophagus and in superficial veins of the lower limbs. The valvular defect may be congenital or may result from mechanical stress (prolonged standing or pregnancy) or aging. The leaking venous valves allow the backflow of blood, which causes pooling of blood. This, in turn, creates pressure that distends the vein and allows fluid to leak into the surrounding tissue. As a result, the affected vein and the tissue around it may become inflamed and painfully tender. Veins close to the surface, especially the saphenous vein, are highly susceptible to varicosities, whereas deeper veins are not as vulnerable because surrounding skeletal muscles prevent their walls from stretching excessively. Varicose veins in the anal canal are termed *hemorrhoids*. ■

Anastomoses

Most tissues of the body receive blood from more than one artery. The union of the branches of two or more arteries supply-ing the same body region is called an **anastomosis** (a-nas-tō-MŌ-sis = connecting; plural is **anastomoses**). Anastomoses between arteries provide alternate routes for blood to reach a tissue or organ. If blood flow stops for a short time when normal movements compress a vessel, or if a vessel is blocked by disease, injury, or surgery, then circulation to a part of the body is not necessarily stopped. The alternate route of blood flow to a body part through an anastomosis is known as **collateral circulation**. Anastomoses may also occur between veins and between arterioles and venules. Arteries that do not anastomose are known as **end arteries**. Obstruction of an end artery interrupts the blood supply to a whole segment of an organ, producing necrosis (death) of that segment. Alternate blood routes may also be provided by nonanastomosing vessels that supply the same region of the body.

Blood Distribution

The largest portion of your blood volume at rest—about 60%— is in systemic veins and venules (Figure 21.6). Systemic capillaries hold only about 5% of the blood volume, and arteries and arterioles about 15%. Because systemic veins and venules contain a large percentage of the blood volume, they function as **blood reservoirs** from which blood can be diverted quickly if the need arises. For example, during increased muscular activity, the cardiovascular center in the brain stem sends a larger number of sympathetic impulses to veins. The result is *venoconstriction,* constriction of veins, which reduces the volume of blood in

Figure 21.6 Blood distribution in the cardiovascular system at rest.

Because systemic veins and venules contain more than half the total blood volume, they are called blood reservoirs.

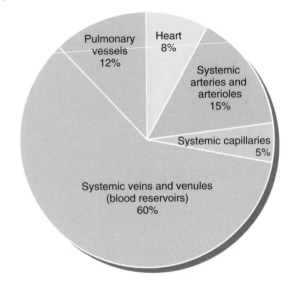

Pulmonary vessels 12%

Heart 8%

Systemic arteries and arterioles 15%

Systemic capillaries 5%

Systemic veins and venules (blood reservoirs) 60%

If your total blood volume is 5 liters, what volume is in your venules and veins right now? In your capillaries?

reservoirs and allows a greater blood volume to flow to skeletal muscles, where it is needed most. A similar mechanism operates in cases of hemorrhage, when blood volume and pressure decrease; in this case, venoconstriction helps counteract the drop in blood pressure. Among the principal blood reservoirs are the veins of the abdominal organs (especially the liver and spleen) and the veins of the skin.

▶ CHECKPOINT

1. What is the function of elastic fibers and smooth muscle in the tunica media of arteries?

2. How are elastic arteries and muscular arteries different?

3. What structural features of capillaries allow the exchange of materials between blood and body cells?

4. Distinguish between pressure reservoirs and blood reservoirs. Why is each important?

5. What is the relationship between anastomoses and collateral circulation?

CAPILLARY EXCHANGE

▶ OBJECTIVE

- **Discuss the pressures that cause movement of fluids between capillaries and interstitial spaces.**

The mission of the entire cardiovascular system is to keep blood flowing through capillaries to allow **capillary exchange,** the movement of substances between blood plasma and interstitial fluid. The 5% of the blood in systemic capillaries at any given time is continually exchanging materials with interstitial fluid. Substances enter and leave capillaries by three basic mechanisms: diffusion, transcytosis, and bulk flow.

Diffusion

The most important method of capillary exchange is simple diffusion. Many substances, such as oxygen (O_2), carbon dioxide (CO_2), glucose, amino acids, and hormones, enter and leave capillaries by simple diffusion. Because O_2 and nutrients normally are present in higher concentrations in blood, they diffuse down their concentration gradients into interstitial fluid and then into body cells. CO_2 and other wastes released by body cells are present in higher concentrations in interstitial fluid, so they diffuse into blood.

All plasma solutes except proteins pass easily across most capillary walls. Except in sinusoids, the gaps between capillary endothelial cells are usually too small to allow proteins through. Thus, most plasma proteins remain in the blood. Smaller water-soluble substances, such as glucose and amino acids, pass through either fenestrations or intercellular clefts (see Figure 21.4). Lipid-soluble materials, such as O_2, CO_2, and steroid hormones, may pass directly through the lipid bilayer of endothelial cell plasma membranes. The gaps between the endothelial cells in liver sinusoids are so large that they do allow proteins to enter the bloodstream. In fact, hepatocytes (liver cells) synthesize and release many proteins, such as fibrinogen (the main clotting protein) and albumin, which then diffuse into the sinusoids through the large intercellular clefts.

Capillaries in most areas of the brain, by contrast, are very "tight;" they allow only a few substances to enter and leave. The endothelial cells of most brain capillaries are nonfenestrated and are sealed together by tight junctions. The resulting blockade to movement of materials into and out of brain capillaries is known as the *blood–brain barrier* (see page 452). In brain areas that lack the blood–brain barrier, for example, the hypothalamus, pineal gland, and pituitary gland, materials undergo capillary exchange more freely.

Transcytosis

A small quantity of material crosses capillary walls by **transcytosis** (*trans-* = across; *cyt-* = cell; *-osis* = process). In this process, substances in blood plasma become enclosed within tiny pinocytic vesicles that first enter endothelial cells by endocytosis, then move across the cell and exit on the other side by exocytosis. This method of transport is important mainly for large, lipid-insoluble molecules that cannot cross capillary walls in any other way. For example, the hormone insulin (a small protein) enters the bloodstream by transcytosis, and certain antibodies (also proteins) pass from the maternal circulation into the fetal circulation by transcytosis.

Bulk Flow: Filtration and Reabsorption

Bulk flow is a passive process by which *large* numbers of ions, molecules, or particles in a fluid move together in the same direction. The substances move in unison, and they move at rates far greater than can be accounted for by diffusion alone. Bulk flow occurs from an area of higher pressure to an area of lower pressure, and it continues as long as a pressure difference exists. Whereas diffusion is more important for *solute exchange* between plasma and interstitial fluid, bulk flow is more important for regulation of the *relative volumes of blood and interstitial fluid.* Pressure-driven movement of fluid and solutes *from* blood capillaries *into* interstitial fluid is termed **filtration.** Pressure-driven movement *from* interstitial fluid *into* blood capillaries is called **reabsorption.**

Two pressures promote filtration: **blood hydrostatic pressure (BHP),** the pressure generated by the pumping action of the heart, and **interstitial fluid osmotic pressure.** The main pressure promoting reabsorption of fluid is **blood colloid osmotic pressure.** The balance of these pressures, called **net filtration pressure (NFP),** determines whether blood volume and interstitial fluid remain steady or change. Overall, the volume of fluid and solutes reabsorbed normally is almost as large as the volume filtered. This near equilibrium is known as **Starling's law of**

the capillaries. Let's see how these hydrostatic and osmotic pressures balance.

Within vessels, the hydrostatic pressure is due to the pressure that water in blood plasma exerts against blood vessel walls. The blood hydrostatic pressure (BHP) is about 35 millimeters of mercury (mmHg) at the arterial end of a capillary, and about 16 mmHg at the capillary's venous end (Figure 21.7). BHP "pushes" fluid out of capillaries into interstitial fluid. The opposing pressure of the interstitial fluid, called **interstitial fluid hydrostatic pressure (IFHP),** "pushes" fluid from interstitial spaces back into capillaries. However, IFHP is close to zero.

(IFHP is difficult to measure, and its reported values vary from small positive values to small negative values.) For our discussion we assume that IFHP equals 0 mmHg all along the capillaries.

The difference in osmotic pressure across a capillary wall is due almost entirely to the presence in blood of plasma proteins, which are too large to pass through either fenestrations or gaps between endothelial cells. **Blood colloid osmotic pressure (BCOP)** is a force caused by the colloidal suspension of these large proteins in plasma that averages 26 mmHg in most capillaries. The effect of BCOP is to "pull" fluid from interstitial

Figure 21.7   Dynamics of capillary exchange (Starling's law of the capillaries). Excess filtered fluid drains into lymphatic capillaries.

🔑 **Blood hydrostatic pressure pushes fluid out of capillaries (filtration), whereas blood colloid osmotic pressure pulls fluid into capillaries (reabsorption).**

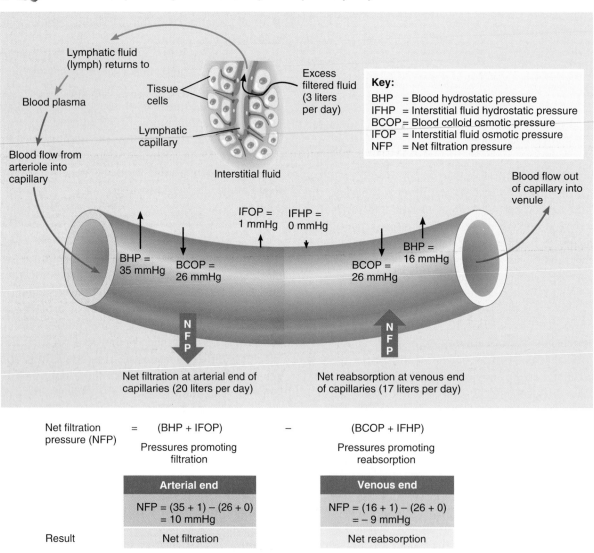

Net filtration at arterial end of capillaries (20 liters per day)

Net reabsorption at venous end of capillaries (17 liters per day)

Net filtration pressure (NFP)	=	(BHP + IFOP)	−	(BCOP + IFHP)
		Pressures promoting filtration		Pressures promoting reabsorption

	Arterial end	**Venous end**
	NFP = (35 + 1) − (26 + 0) = 10 mmHg	NFP = (16 + 1) − (26 + 0) = −9 mmHg
Result	Net filtration	Net reabsorption

 A person who has liver failure cannot synthesize the normal amount of plasma proteins. How does a deficit of plasma proteins affect blood colloid osmotic pressure, and what is the effect on capillary filtration and reabsorption?

spaces into capillaries. Opposing BCOP is **interstitial fluid osmotic pressure (IFOP),** which "pulls" fluid out of capillaries into interstitial fluid. Normally, IFOP is very small—0.1–5 mmHg—because only tiny amounts of protein are present in interstitial fluid. The small amount of protein that leaks from blood plasma into interstitial fluid does not accumulate there because it enters lymphatic fluid and is returned to the blood. For discussion, we can use a value of 1 mmHg for IFOP.

Whether fluids leave or enter capillaries depends on the balance of pressures. If the pressures that push fluid out of capillaries exceed the pressures that pull fluid into capillaries, fluid will move from capillaries into interstitial spaces (filtration). If, however, the pressures that push fluid out of interstitial spaces into capillaries exceed the pressures that pull fluid out of capillaries, then fluid will move from interstitial spaces into capillaries (reabsorption).

The net filtration pressure (NFP), which indicates the direction of fluid movement, is calculated as follows:

$$NFP = (BHP + IFOP) - (BCOP + IFHP)$$

Pressures that   Pressures that
promote filtration   promote reabsorption

At the arterial end of a capillary,

$$NFP = (35 + 1)\ mmHg - (26 + 0)\ mmHg$$
$$= 36 - 26\ mmHg = 10\ mmHg$$

Thus, at the arterial end of a capillary, there is a *net outward pressure* of 10 mmHg, and fluid moves out of the capillary into interstitial spaces (filtration).

At the venous end of a capillary,

$$NFP = (16 + 1)\ mmHg - (26 + 0)\ mmHg$$
$$= 17 - 26\ mmHg = -9\ mmHg$$

At the venous end of a capillary, the negative value (−9 mmHg) represents a *net inward pressure,* and fluid moves into the capillary from tissue spaces (reabsorption).

On average, about 85% of the fluid filtered out of capillaries is reabsorbed. The excess filtered fluid and the few plasma proteins that do escape from blood into interstitial fluid enter lymphatic capillaries (see Figure 22.2 on page 767). As lymph drains into the subclavian veins in the upper thorax (see Figure 22.3 on page 768), these materials return to the blood. Every day about 20 liters of fluid filter out of capillaries in tissues throughout the body. Of this fluid, 17 liters are reabsorbed and 3 liters enter lymphatic capillaries. (Excluded from these numbers are the 180 liters of fluid that are filtered from capillaries in the kidneys and the 178 liters that are reabsorbed during the formation of urine each day.)

Edema

If filtration greatly exceeds reabsorption, the result is **edema** (= swelling), an abnormal increase in interstitial fluid volume.

Edema is not usually detectable in tissues until interstitial fluid volume has risen to 30% above normal. Edema can result from either excess filtration or inadequate reabsorption.

Two situations may cause excess filtration:

- *Increased capillary blood pressure* causes more fluid to be filtered from capillaries.

- *Increased permeability of capillaries* raises interstitial fluid osmotic pressure by allowing some plasma proteins to escape. Such leakiness may be caused by the destructive effects of chemical, bacterial, thermal, or mechanical agents on capillary walls.

One situation commonly causes inadequate reabsorption:

- *Decreased concentration of plasma proteins* lowers the blood colloid osmotic pressure. Inadequate synthesis or loss of plasma proteins is associated with liver disease, burns, malnutrition, and kidney disease. ■

▶ C H E C K P O I N T

6. How can substances enter and leave blood plasma?

7. How do hydrostatic and osmotic pressures determine fluid movement across the walls of capillaries?

8. Write an equation to describe the pressures involved in Starling's law of the capillaries.

HEMODYNAMICS: FACTORS AFFECTING BLOOD FLOW

▶ O B J E C T I V E S

- **Explain the factors that regulate the volume of blood flow.**
- **Explain how blood pressure changes throughout the cardiovascular system.**
- **Describe the factors that determine mean arterial blood pressure and systemic vascular resistance.**
- **Describe the relation between cross-sectional area and velocity of blood flow.**

Blood flow is the volume of blood that flows through any tissue in a given time period (in mL/min). Total blood flow is cardiac output (CO), the volume of blood that circulates through systemic (or pulmonary) blood vessels each minute. In Chapter 20 we saw that cardiac output depends on heart rate and stroke volume: Cardiac output (CO) = heart rate (HR) × stroke volume (SV). How the cardiac output becomes distributed into circulatory routes that serve various body tissues depends on two more factors: (1) the *pressure difference* that drives the blood flow through a tissue and (2) the *resistance* to blood flow in specific blood vessels. Blood flows from regions of higher to lower pressure; the greater the pressure difference, the greater the blood flow. The higher the resistance, by contrast, the smaller the blood flow.

Blood Pressure

As you have just learned, blood flows from regions of higher pressure to regions of lower pressure; the greater the pressure difference, the greater the blood flow. Contraction of the ventricles generates **blood pressure (BP),** the hydrostatic pressure exerted by blood on the walls of a blood vessel. BP is highest in the aorta and large systemic arteries, where in a resting, young adult, BP rises to about 120 mmHg during systole (ventricular contraction) and drops to about 80 mmHg during diastole (ventricular relaxation). **Systolic blood pressure** is the highest pressure attained in arteries during systole, and **diastolic blood pressure** is the lowest arterial pressure during diastole (Figure 21.8). As blood leaves the aorta and flows through the systemic circulation, its pressure falls progressively as the distance from the left ventricle increases. Blood pressure decreases to about 35 mmHg as blood passes from systemic arteries through systemic arterioles and into capillaries, where the pressure fluctuations disappear. At the venous end of capillaries, blood pressure has dropped to about 16 mmHg. Blood pressure continues to drop as blood enters systemic venules and then veins because these vessels are farthest from the left ventricle. Finally, blood pressure reaches 0 mmHg as blood flows into the right ventricle.

Mean arterial blood pressure (MABP), the average pressure in arteries, is roughly one-third of the way between the diastolic and systolic pressures. It can be estimated as follows:

$$MABP = \text{diastolic BP} + 1/3 \text{ (systolic BP} - \text{diastolic BP)}$$

Thus, in a person whose BP is 120/80 mmHg, MABP is about 93 mmHg.

We have already seen that cardiac output equals heart rate multiplied by stroke volume. Another way to calculate cardiac output is to divide mean arterial blood pressure (MABP) by resistance (R): CO = MABP ÷ R. By rearranging the terms of this equation, you can see that MABP = CO × R. If cardiac output rises due to an increase in stroke volume or heart rate, then the mean arterial blood pressure rises so long as resistance remains steady. Likewise, a decrease in cardiac output causes a decrease in blood pressure if resistance does not change.

Blood pressure also depends on the total volume of blood in the cardiovascular system. The normal volume of blood in an adult is about 5 liters (5.3 qt). Any decrease in this volume, as from hemorrhage, decreases the amount of blood that is circulated through the arteries each minute. A modest decrease can be compensated for by homeostatic mechanisms that help maintain blood pressure (described on page 709), but if the decrease in blood volume is greater than 10% of the total, blood pressure drops. Conversely, anything that increases blood volume, such as water retention in the body, tends to increase blood pressure.

Resistance

As noted earlier, **vascular resistance** is the opposition to blood flow due to friction between blood and the walls of blood ves-

Figure 21.8 Blood pressures in various parts of the cardiovascular system. The dashed line is the mean (average) blood pressure in the aorta, arteries, and arterioles.

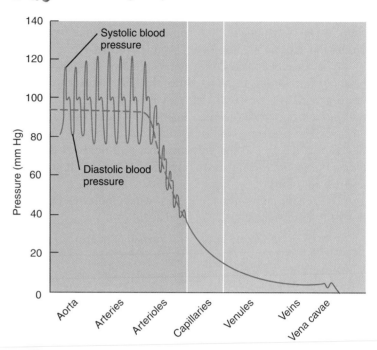

Blood pressure rises and falls with each heartbeat in blood vessels leading to capillaries.

Is the mean blood pressure in the aorta closer to systolic or to diastolic pressure?

sels. Vascular resistance depends on (1) size of the blood vessel lumen, (2) blood viscosity, and (3) total blood vessel length.

1. **Size of the lumen.** The smaller the lumen of a blood vessel, the greater its resistance to blood flow. Resistance is inversely proportional to the fourth power of the diameter (d) of the blood vessel's lumen (R ∝ 1/d⁴). The smaller the diameter of the blood vessel, the greater the resistance it offers to blood flow. For example, if the diameter of a blood vessel decreases by one-half, its resistance to blood flow increases 16 times. Vasoconstriction narrows the lumen whereas vasodilation widens it. Normally, moment-to-moment fluctuations in blood flow through a given tissue are due to vasoconstriction and vasodilation of the tissue's arterioles. As arterioles dilate, resistance decreases, and blood pressure falls. As arterioles constrict, resistance increases, and blood pressure rises.

2. **Blood viscosity.** The viscosity (thickness) of blood depends mostly on the ratio of red blood cells to plasma (fluid) volume, and to a smaller extent on the concentration of proteins in plasma. The higher the blood's viscosity, the higher the resistance. Any condition that increases the viscosity of blood, such as dehydration or polycythemia (an unusually high number of red blood cells), thus increases blood pressure. A depletion of

plasma proteins or red blood cells, due to anemia or hemorrhage, decreases viscosity and thus decreases blood pressure.

3. *Total blood vessel length.* Resistance to blood flow through a vessel is directly proportional to the length of the blood vessel. The longer a blood vessel, the greater the resistance. Obese people often have hypertension (elevated blood pressure) due to an increase in total blood vessel length caused by the additional blood vessels in their adipose tissue. An estimated 650 km (about 400 miles) of additional blood vessels develop for each extra kilogram (2.2 lb) of fat.

Systemic vascular resistance (SVR), also known as *total peripheral resistance (TPR),* refers to all the vascular resistances offered by systemic blood vessels. The diameters of arteries and veins are large, and thus their resistance is very small because most of the blood does not come into physical contact with the walls of the blood vessel. The smallest vessels—arterioles, capillaries, and venules—contribute the most resistance. A major function of arterioles is to control SVR—and therefore blood pressure and blood flow to particular tissues—by changing their diameters. Arterioles need vasodilate or vasoconstrict only slightly to have a large effect on SVR. The main center for regulation of SVR is the vasomotor center in the brain stem (described shortly).

Venous Return

Venous return, the volume of blood flowing back to the heart through the systemic veins, occurs due to the pressure generated by contractions of the heart's left ventricle. The pressure difference from venules (averaging about 16 mmHg) to the right ventricle (0 mmHg), although small, normally is sufficient to cause venous return to the heart. If pressure increases in the right atrium or ventricle, venous return will decrease. One cause of increased pressure in the right atrium is an incompetent (leaky) tricuspid valve, which lets blood regurgitate (flow backward) as the ventricles contract. The result is decreased venous return and buildup of blood on the venous side of the systemic circulation.

When you stand, the pressure pushing blood up the veins in your lower limbs is barely enough to overcome the force of gravity pushing it back down. Besides the heart, two other mechanisms "pump" blood from the lower body back to the heart: (1) the skeletal muscle pump, and (2) the respiratory pump. Both pumps depend on the presence of valves in veins.

The **skeletal muscle pump** operates as follows (Figure 21.9):

1 While you are standing at rest, both the venous valve closer to the heart (proximal valve) and the one farther from the heart (distal valve) in this part of the leg are open, and blood flows upward toward the heart.

2 Contraction of leg muscles, such as when you stand on tiptoes or take a step, compresses the vein. The compression pushes blood through the proximal valve, an action called *milking.* At the same time, the distal valve in the uncom-

pressed segment of the vein closes as some blood is pushed against it. People who are immobilized through injury or disease lack these contractions of leg muscles. As a result, their venous return is slower and they may develop circulation problems.

3 Just after muscle relaxation, pressure falls in the previously compressed section of vein, which causes the proximal valve to close. The distal valve now opens because blood pressure in the foot is higher than in the leg, and the vein fills with blood from the foot.

The **respiratory pump** is also based on alternating compression and decompression of veins. During inhalation, the diaphragm moves downward, which causes a decrease in pressure in the thoracic cavity and an increase in pressure in the abdominal cavity. As a result, abdominal veins are compressed, and a greater volume of blood moves from the compressed abdominal veins into the decompressed thoracic veins and then into the right atrium. When the pressures reverse during exhalation, the valves in the veins prevent backflow of blood from the thoracic veins to the abdominal veins.

Figure 21.9 Action of the skeletal muscle pump in returning blood to the heart. **1** At rest, both proximal and distal venous valves are open and blood flows toward the heart. **2** Contraction of leg muscles pushes blood through the proximal valve while closing the distal valve. **3** As the leg muscles relax, the proximal valve closes and the distal valve opens. When the vein fills with blood from the foot, the proximal valve reopens.

🔑 **Milking refers to skeletal muscle contractions that drive venous blood toward the heart.**

What mechanisms, besides cardiac contractions, act as pumps to boost venous return?

Figure 21.10 summarizes the factors that increase blood pressure through increasing cardiac output or systemic vascular resistance.

Velocity of Blood Flow

Earlier we saw that blood flow is the *volume* of blood that flows through any tissue in a given time period (in mL/min). The speed or *velocity* of blood flow (in cm/sec) is inversely related to the cross-sectional area. Velocity is slowest where the total cross-sectional area is greatest (Figure 21.11). Each time an artery branches, the total cross-sectional area of all its branches is greater than the cross-sectional area of the original vessel, and blood flow is slower in the branches. Conversely, when branches combine as venules unite to form veins, the total cross-sectional area becomes smaller and flow becomes faster. In an adult, the cross-sectional area of the aorta is only 3–5 cm², and the average velocity of the blood there is 40 cm/sec. In capillaries, the total cross-sectional area is 4500–6000 cm², and the velocity of blood flow is less than 0.1 cm/sec. In the two venae cavae combined, the cross-sectional area is about 14 cm², and the velocity is about 15 cm/sec. Thus, the velocity of blood flow decreases as blood flows from the aorta to arteries to arterioles to capillaries, and increases as it leaves capillaries and returns to the heart. The relatively slow rate of flow through capillaries aids the exchange of materials between blood and interstitial fluid.

Circulation time is the time required for a drop of blood to pass from the right atrium, through the pulmonary circulation, back to the left atrium, through the systemic circulation down to the foot, and back again to the right atrium. In a resting person, circulation time normally is about 1 minute.

 Syncope

Syncope (SIN-kō-pē), or fainting, is a sudden, temporary loss of consciousness that is not due to head trauma, followed by spontaneous recovery. It is most commonly due to cerebral ischemia, lack of sufficient blood flow to the brain. Syncope may occur for several reasons:

*   *Vasodepressor syncope* is due to sudden emotional stress or real, threatened, or fantasized injury.

Figure 21.10 Summary of factors that increase blood pressure. Changes noted within green boxes increase cardiac output, whereas changes noted within blue boxes increase systemic vascular resistance.

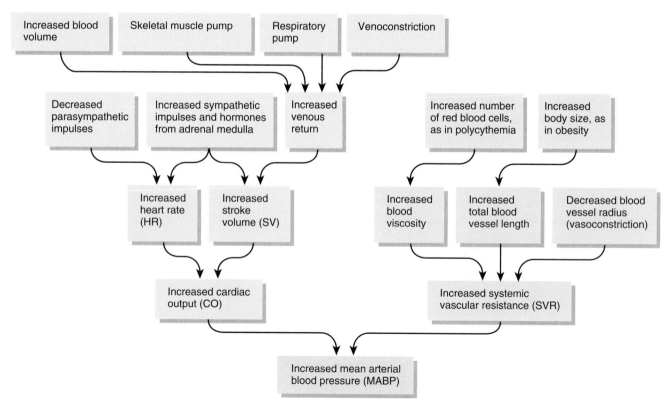

Increases in cardiac output and increases in systemic vascular resistance will increase mean arterial blood pressure.

Which type of blood vessel exerts the major control of systemic vascular resistance, and how does it achieve this?

Figure 21.11 Relationship between velocity (speed) of blood flow and total cross-sectional area in different types of blood vessels.

 Velocity of blood flow is slowest in the capillaries because they have the largest total cross-sectional area.

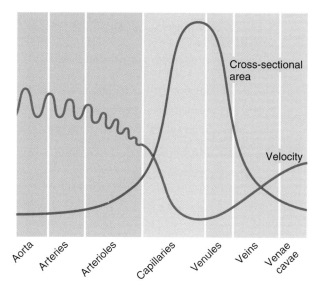

In which blood vessels is the velocity of flow fastest?

- *Situational syncope* is caused by pressure stress associated with urination, defecation, or severe coughing.
- *Drug-induced syncope* may be caused by drugs such as antihypertensives, diuretics, vasodilators, and tranquilizers.
- *Orthostatic hypotension,* an excessive decrease in blood pressure that occurs upon standing up, may cause fainting. ■

▶ CHECKPOINT

9. Explain how blood pressure and resistance determine volume of blood flow.

10. What is systemic vascular resistance and what factors contribute to it?

11. How is the return of venous blood to the heart accomplished?

12. Why is the velocity of blood flow faster in arteries and veins than in capillaries?

CONTROL OF BLOOD PRESSURE AND BLOOD FLOW

▶ OBJECTIVE

- **Describe how blood pressure is regulated.**

Several interconnected negative feedback systems control blood pressure by adjusting heart rate, stroke volume, systemic vascular resistance, and blood volume. Some systems allow rapid

adjustments to cope with sudden changes, such as the drop in blood pressure in the brain that occurs when you get out of bed; others act more slowly to provide long-term regulation of blood pressure. The body may also require adjustments to the distribution of blood flow. During exercise, for example, a greater percentage of the total blood flow is diverted to skeletal muscles.

Role of the Cardiovascular Center

In Chapter 20, we noted how the **cardiovascular (CV) center** in the medulla oblongata helps regulate heart rate and stroke volume. The CV center also controls neural, hormonal, and local negative feedback systems that regulate blood pressure and blood flow to specific tissues. Groups of neurons scattered within the CV center regulate heart rate, contractility (force of contraction) of the ventricles, and blood vessel diameter. Some of the CV center's neurons stimulate the heart (cardiostimulatory center), whereas others inhibit the heart (cardioinhibitory center). Still others control blood vessel diameter (vasomotor center), either by causing constriction (vasoconstrictor center) or dilation (vasodilator center). Because these clusters of neurons communicate with one another, function together, and are not clearly separated anatomically, we discuss them here as a group.

The cardiovascular center receives input both from higher brain regions and from sensory receptors (Figure 21.12). Nerve impulses descend from the cerebral cortex, limbic system, and hypothalamus to affect the cardiovascular center. For example, even before you start to run a race, your heart rate may increase due to nerve impulses conveyed from the limbic system to the CV center. If your body temperature rises during a race, the hypothalamus sends nerve impulses to the CV center. The resulting vasodilation of skin blood vessels allows heat to dissipate more rapidly from the surface of the skin. The three main types of sensory receptors that provide input to the cardiovascular center are proprioceptors, baroreceptors, and chemoreceptors. *Proprioceptors* monitor movements of joints and muscles and provide input to the cardiovascular center during physical activity. Their activity accounts for the rapid increase in heart rate at the beginning of exercise. *Baroreceptors* monitor changes in pressure and stretch in the walls of blood vessels, and *chemoreceptors* monitor the concentration of various chemicals in the blood.

Output from the cardiovascular center flows along sympathetic and parasympathetic neurons of the ANS (Figure 21.12). Sympathetic impulses reach the heart via the **cardiac accelerator nerves.** An increase in sympathetic stimulation increases heart rate and contractility, whereas a decrease in sympathetic stimulation decreases heart rate and contractility. Parasympathetic stimulation, conveyed along the **vagus nerves (cranial nerve X),** decreases heart rate. Thus, opposing sympathetic (stimulatory) and parasympathetic (inhibitory) influences control the heart.

The cardiovascular center also continually sends impulses to smooth muscle in blood vessel walls via **vasomotor nerves.**

Figure 21.12 Location and function of the cardiovascular (CV) center in the medulla oblongata. The CV center receives input from higher brain centers, proprioceptors, baroreceptors, and chemoreceptors. Then, it provides output to the sympathetic and parasympathetic divisions of the autonomic nervous system (ANS).

🔑 **The cardiovascular center is the main region for nervous system regulation of the heart and blood vessels.**

INPUT TO CARDIOVASCULAR CENTER (nerve impulses)

From higher brain centers: cerebral cortex, limbic system, and hypothalamus

From proprioceptors: monitor joint movements

From baroreceptors: monitor blood pressure

From chemoreceptors: monitor blood acidity (H^+), CO_2, and O_2

OUTPUT TO EFFECTORS (increased frequency of nerve impulses)

Heart: decreased rate

Heart: increased rate and contractility

Blood vessels: vasoconstriction

Vagus nerves (parasympathetic)

Cardiac accelerator nerves (sympathetic)

Vasomotor nerves (sympathetic)

Cardiovascular (CV) center

 What types of effector tissues are regulated by the cardiovascular center?

These sympathetic neurons exit the spinal cord through all thoracic and the first one or two lumbar spinal nerves and then pass into the sympathetic trunk ganglia (see Figure 17.2 on page 569). From there, impulses propagate along sympathetic neurons that innervate blood vessels in viscera and peripheral areas. The vasomotor region of the cardiovascular center continually sends impulses over these routes to arterioles throughout the body, but especially to those in the skin and abdominal viscera. The result is a moderate state of tonic contraction or vasoconstriction, called **vasomotor tone,** that sets the resting level of systemic vascular resistance. Sympathetic stimulation of most veins causes constriction that moves blood out of venous blood reservoirs and increases blood pressure.

Neural Regulation of Blood Pressure

The nervous system regulates blood pressure via negative feedback loops that occur as two types of reflexes: baroreceptor reflexes and chemoreceptor reflexes.

Baroreceptor Reflexes

Baroreceptors, pressure-sensitive sensory receptors, are located in the aorta, internal carotid arteries (arteries in the neck that supply blood to the brain), and other large arteries in the neck

and chest. They send impulses to the cardiovascular center to help regulate blood pressure. The two most important **baroreceptor reflexes** are the carotid sinus reflex and the aortic reflex.

Baroreceptors in the wall of the carotid sinuses initiate the **carotid sinus reflex,** which helps regulate blood pressure in the brain. The **carotid sinuses** are small widenings of the right and left internal carotid arteries just above the point where they branch from the common carotid arteries (Figure 21.13). Blood pressure stretches the wall of the carotid sinus, which stimulates the baroreceptors. Nerve impulses propagate from the carotid sinus baroreceptors over sensory axons in the glossopharyngeal nerves (cranial nerve IX) to the cardiovascular center in the medulla oblongata. Baroreceptors in the wall of the ascending aorta and arch of the aorta initiate the **aortic reflex,** which regulates systemic blood pressure. Nerve impulses from aortic baroreceptors reach the cardiovascular center via sensory axons of the vagus nerves (cranial nerve X).

When blood pressure falls, the baroreceptors are stretched less, and they send nerve impulses at a slower rate to the cardiovascular center (Figure 21.14 on page 712). In response, the CV center decreases parasympathetic stimulation of the heart via motor axons of the vagus nerves and increases sympathetic stimulation of the heart via cardiac accelerator nerves. Another consequence of increased sympathetic stimulation is increased epinephrine and

Figure 21.13 ANS innervation of the heart and the baroreceptor reflexes that help regulate blood pressure.

Baroreceptors are pressure-sensitive neurons that monitor stretching.

Cardiovascular (CV) center

Glossopharyngeal nerves (cranial nerve IX)

Medulla oblongata

Vagus nerves (cranial nerve X, parasympathetic)

Baroreceptors in carotid sinus

Baroreceptors in arch of aorta

SA node

AV node

Ventricular myocardium

Spinal cord

Cardiac accelerator nerve (sympathetic)

Sympathetic trunk ganglion

Key:

Sensory (afferent) neurons

Motor (efferent) neurons

Which cranial nerves conduct impulses to the cardiovascular center from baroreceptors in the carotid sinuses and the arch of the aorta?

norepinephrine secretion by the adrenal medulla. As the heart beats faster and more forcefully, and as systemic vascular resistance increases, cardiac output and systemic vascular resistance rise, and blood pressure increases to the normal level.

Conversely, when an increase in pressure is detected, the baroreceptors send impulses at a faster rate. The CV center responds by increasing parasympathetic stimulation and decreasing sympathetic stimulation. The resulting decreases in heart rate and force of contraction reduce the cardiac output. Moreover, the cardiovascular center also slows the rate at which it sends sympathetic impulses along vasomotor neurons that normally cause vasoconstriction. The resulting vasodilation lowers systemic vascular resistance. Decreased cardiac output and decreased systemic vascular resistance both lower systemic arterial blood pressure.

Moving from a prone (lying down) to an erect position decreases blood pressure and blood flow in the head and upper part of the body. The baroreceptor reflexes, however, quickly counteract the drop in pressure. Sometimes these reflexes operate more slowly than normal, especially in the elderly, in which case a person can faint due to reduced brain blood flow upon standing up too quickly.

Carotid Sinus Massage and Carotid Sinus Syncope

Because the carotid sinus is close to the anterior surface of the neck, it is possible to stimulate the baroreceptors there by putting pressure on the neck. Physicians sometimes use **carotid sinus massage,** which involves carefully massaging the neck over the carotid sinus, to slow heart rate in a person who has paroxysmal supraventricular tachycardia, a type of tachycardia that originates in the atria. Anything that stretches or puts pressure on the carotid sinus, such as hyperextension of the head, tight collars, or carrying heavy shoulder loads, may also slow heart rate and can cause **carotid sinus syncope,** fainting due to inappropriate stimulation of the carotid sinus baroreceptors. ■

Chemoreceptor Reflexes

Chemoreceptors, sensory receptors that monitor the chemical composition of blood, are located close to the baroreceptors of the carotid sinus and arch of the aorta in small structures called **carotid bodies** and **aortic bodies,** respectively. These chemoreceptors detect changes in blood level of O_2, CO_2, and H^+. *Hypoxia* (lowered O_2 availability), *acidosis* (an increase in H^+

Figure 21.14   Negative feedback regulation of blood pressure via baroreceptor reflexes.

🔑 **When blood pressure decreases, heart rate increases.**

 Does this negative feedback cycle represent the changes that occur when you lie down or when you stand up?

concentration), or *hypercapnia* (excess CO_2) stimulates the chemoreceptors to send impulses to the cardiovascular center. In response, the CV center increases sympathetic stimulation to arterioles and veins, producing vasoconstriction and an increase in blood pressure. These chemoreceptors also provide input to the respiratory center in the brain stem to adjust the rate of breathing.

Hormonal Regulation of Blood Pressure

Several hormones help regulate blood pressure and blood flow by altering cardiac output, changing systemic vascular resistance, or adjusting the total blood volume:

1. *Renin–angiotensin–aldosterone (RAA) system.* When blood volume falls or blood flow to the kidneys decreases, juxtaglomerular cells in the kidneys secrete **renin** into the bloodstream. In sequence, renin and angiotensin converting enzyme (ACE) act on their substrates to produce the active hormone **angiotensin II,** which raises blood pressure in two ways. First, angiotensin II is a potent vasoconstrictor; it raises blood pressure by increasing systemic vascular resistance. Second, it stimulates secretion of **aldosterone,** which increases reabsorption of sodium ions (Na^+) and water by the kidneys. The water reabsorption increases total blood volume, which increases blood pressure.

2. *Epinephrine and norepinephrine.* In response to sympathetic stimulation, the adrenal medulla releases epinephrine and norepinephrine. These hormones increase cardiac output by increasing the rate and force of heart contractions. They also cause vasoconstriction of arterioles and veins in the skin and abdominal organs and vasodilation of arterioles in cardiac and skeletal muscle, which helps increase blood flow to muscle during exercise.

3. *Antidiuretic hormone (ADH).* ADH is produced by the hypothalamus and released from the posterior pituitary in response to dehydration or decreased blood volume. Among other actions, ADH causes vasoconstriction, which increases blood pressure. For this reason ADH is also called **vasopressin.**

4. *Atrial natriuretic peptide (ANP).* Released by cells in the atria of the heart, ANP lowers blood pressure by causing vasodilation and by promoting the loss of salt and water in the urine, which reduces blood volume.

Table 21.1 summarizes the regulation of blood pressure by hormones.

Autoregulation of Blood Pressure

In each capillary bed, local changes can regulate vasomotion. When vasodilators produce local dilation of arterioles and relaxation of precapillary sphincters, blood flow into capillary networks is increased, which restores O_2 level to normal. Vasoconstrictors have the opposite effect. The ability of a tissue to

Table 21.1 Blood Pressure Regulation by Hormones

Factor Influencing Blood Pressure	Hormone	Effect on Blood Pressure
Cardiac Output		
Increased heart rate and contractility	Norepinephrine Epinephrine	Increase
Systemic Vascular Resistance		
Vasoconstriction	Angiotensin II Antidiuretic hormone (vasopressin) Norepinephrine* Epinephrine*	Increase
Vasodilation	Atrial natriuretic peptide Epinephrine† Nitric oxide	Decrease
Blood Volume		
Blood volume increase	Aldosterone Antidiuretic hormone	Increase
Blood volume decrease	Atrial natriuretic peptide	Decrease

*Acts at α_1 receptors in arterioles of abdomen and skin.
†Acts at β_2 receptors in arterioles of cardiac and skeletal muscle; norepinephrine has a much smaller vasodilating effect.

automatically adjust its blood flow to match its metabolic demands is called **autoregulation.** In tissues such as the heart and skeletal muscle, where the demand for O_2 and nutrients and for the removal of wastes can increase as much as tenfold during physical activity, autoregulation is an important contributor to increased blood flow through the tissue. Autoregulation also controls regional blood flow in the brain; blood distribution to various parts of the brain changes dramatically for different mental and physical activities. During a conversation, for example, blood flow increases to the motor speech areas when one is talking and increases to the auditory areas when one is listening.

Two general types of stimuli cause autoregulatory changes in blood flow:

1. *Physical changes.* Warming promotes vasodilation, whereas cooling causes vasoconstriction. Furthermore, smooth muscle in arteriole walls exhibits a **myogenic response**—it contracts more forcefully when it is stretched and relaxes when stretching lessens. If, for example, blood flow through an arteriole decreases, then stretching of the arteriole walls decreases. As a result, the smooth muscle relaxes and produces vasodilation, which increases blood flow.

2. *Vasodilating and vasoconstricting chemicals.* Several types of cells—including white blood cells, platelets, smooth muscle fibers, macrophages, and endothelial cells—release a wide variety of chemicals that alter blood-vessel diameter. Vasodilating chemicals released by metabolically active tissue cells include K^+, H^+, lactic acid (lactate), and adenosine (from ATP). Another important vasodilator released by endothelial cells is nitric oxide (NO), named endothelium-derived relaxation factor (EDRF) before its chemical identity was known. Tissue trauma or inflammation causes release of vasodilating kinins and histamine. Vasoconstrictors include thromboxane A2, superoxide radicals, serotonin (from platelets), and endothelins (from endothelial cells).

An important difference between the pulmonary and systemic circulations is their autoregulatory response to changes in O_2 level. The walls of blood vessels in the systemic circulation *dilate* in response to low O_2. With vasodilation, O_2 delivery increases, which restores the normal O_2 level. By contrast, the walls of blood vessels in the pulmonary circulation *constrict* in response to low levels of O_2. This response ensures that blood mostly bypasses those alveoli (air sacs) in the lungs that are poorly ventilated by fresh air. Thus, most blood flows to better-ventilated areas of the lung.

▶ CHECKPOINT

13. What are the principal inputs to and outputs from the cardiovascular center?

14. How do the carotid sinus and the aortic reflexes operate?

15. What is the role of chemoreceptors in the regulation of blood pressure?

16. How do hormones regulate blood pressure?

17. What is autoregulation and how does it differ in the systemic and pulmonary circulations?

CHECKING CIRCULATION

▶ OBJECTIVE

• **Define pulse, and define systolic, diastolic, and pulse pressures.**

Pulse

The alternate expansion and recoil of elastic arteries after each systole of the left ventricle create a traveling pressure wave that is called the **pulse.** The pulse is strongest in the arteries closest to the heart, becomes weaker in the arterioles, and disappears altogether in the capillaries. The pulse may be felt in any artery that lies near the surface of the body and runs over a bone or other firm structure. Table 21.2 depicts some common pulse points.

The pulse rate normally is the same as the heart rate, about 70 to 80 beats per minute at rest. **Tachycardia** (tak′-i-KAR-dē-a; *tachy-* = fast) is a rapid resting heart or pulse rate over 100

Table 21.2 Pulse Points

Structure	Location	Structure	Location
Superficial temporal artery	Lateral to orbit of eye.	Femoral artery	Inferior to inguinal ligament.
Facial artery	Mandible (lower jawbone) on a line with the corners of the mouth.	Popliteal artery	Posterior to the knee.
Common carotid artery	Lateral to larynx (voice box).	Radial artery	Distal half of wrist.
Brachial artery	Medial side of biceps brachii muscle.	Dorsal artery of the foot (dorsalis pedis artery)	Superior to instep of foot.

beats/min. **Bradycardia** (brād′-i-KAR-dē-a; *brady-* = slow) is a slow resting heart or pulse rate under 60 beats/min. Endurance-trained athletes normally exhibit bradycardia.

Measuring Blood Pressure

In clinical use, the term **blood pressure** usually refers to the pressure in arteries generated by the left ventricle during systole and the pressure remaining in the arteries when the ventricle is in diastole. Blood pressure is usually measured in the brachial artery in the left arm (see Figure 21.20a). The device used to measure blood pressure is a **sphygmomanometer** (sfig′-mō-ma-NOM-e-ter; *sphygmo-* = pulse; *manometer* = instrument used to measure pressure). When the pressure cuff is inflated above the blood pressure attained during systole, the artery is com-

pressed so that blood flow stops. The technician places a stethoscope below the cuff, and slowly deflates the cuff. When the cuff is deflated enough to allow the artery to open, a spurt of blood passes through, resulting in the first sound heard through the stethoscope. This sound corresponds to **systolic blood pressure (SBP),** the force of blood pressure on arterial walls just after ventricular contraction (Figure 21.15). As the cuff is deflated further, the sounds suddenly become faint. This level, called the **diastolic blood pressure (DBP),** represents the force exerted by the blood remaining in arteries during ventricular relaxation. At pressures below diastolic blood pressure, sounds disappear altogether. The various sounds that are heard while taking blood pressure are called **Korotkoff sounds** (kō-ROT-kof).

The normal blood pressure of a young adult male is 120 mmHg systolic and 80 mmHg diastolic, reported as "120 over

Figure 21.15 Relationship of blood pressure changes to cuff pressure.

As the cuff is deflated, sounds first occur at the systolic blood pressure; the sounds suddenly become faint at the diastolic blood pressure.

If a blood pressure is reported as "142 over 95," what are the diastolic, systolic, and pulse pressures? Does this person have hypertension as defined on page 758?

80" and written as 120/80. In young adult females, the pressures are 8 to 10 mmHg less. People who exercise regularly and are in good physical condition may have even lower blood pressures. Thus, blood pressure slightly lower than 120/80 may be a sign of good health and fitness.

The difference between systolic and diastolic pressure is called **pulse pressure.** This pressure, normally about 40 mmHg, provides information about the condition of the cardiovascular system. For example, conditions such as atherosclerosis and patent (open) ductus arteriosus greatly increase pulse pressure. The normal ratio of systolic pressure to diastolic pressure to pulse pressure is about 3:2:1.

▶ CHECKPOINT

18. Where may the pulse be felt?

19. What do tachycardia and bradycardia mean?

20. How are systolic and diastolic blood pressures measured with a sphygmomanometer?

SHOCK AND HOMEOSTASIS

OBJECTIVE

• **Define shock, and describe the four types of shock.**

Shock is a failure of the cardiovascular system to deliver enough O_2 and nutrients to meet cellular metabolic needs. The causes of shock are many and varied, but all are characterized by inadequate blood flow to body tissues. With inadequate oxygen delivery, cells switch from aerobic to anaerobic production of ATP, and lactic acid accumulates in body fluids. If shock persists,

cells and organs become damaged, and cells may die unless proper treatment begins quickly.

Types of Shock

Shock can be of four different types: (1) **hypovolemic shock** (hī-pō-vō-LĒ-mik; *hypo-* = low; *-volemic* = volume) due to decreased blood volume, (2) **cardiogenic shock** due to poor heart function, (3) **vascular shock** due to inappropriate vasodilation, and (4) **obstructive shock** due to obstruction of blood flow.

A common cause of hypovolemic shock is acute (sudden) hemorrhage. The blood loss may be external as occurs in trauma, or internal, as in rupture of an aortic aneurysm. Loss of body fluids through excessive sweating, diarrhea, or vomiting also can cause hypovolemic shock. Other conditions—for instance, diabetes mellitus—may cause excessive loss of fluid in the urine. Sometimes, hypovolemic shock is due to inadequate intake of fluid. Whatever the cause, when the volume of body fluids falls, venous return to the heart declines, filling of the heart lessens, stroke volume decreases, and cardiac output decreases.

In cardiogenic shock, the heart fails to pump adequately, most often because of a myocardial infarction (heart attack). Other causes of cardiogenic shock include poor perfusion of the heart (ischemia), heart valve problems, excessive preload or afterload, impaired contractility of heart muscle fibers, and arrhythmias.

Even with normal blood volume and cardiac output, shock may occur if blood pressure drops due to a decrease in systemic vascular resistance. A variety of conditions can cause inappropriate dilation of arterioles or venules. In *anaphylactic shock,* a severe allergic reaction—for example, to a bee sting—releases histamine and other mediators that cause vasodilation. In *neurogenic shock,* vasodilation may occur following trauma to the head that causes malfunction of the cardiovascular center in the medulla. Shock stemming from certain bacterial toxins that produce vasodilation is termed *septic shock.* In the United States, septic shock causes more than 100,000 deaths per year and is the most common cause of death in hospital critical care units.

Obstructive shock occurs when blood flow through a portion of the circulation is blocked. The most common cause is *pulmonary embolism,* a blood clot lodged in a blood vessel of the lungs.

Homeostatic Responses to Shock

The major mechanisms of compensation in shock are *negative feedback systems* that work to return cardiac output and arterial blood pressure to normal. When shock is mild, compensation by homeostatic mechanisms prevents serious damage. In an otherwise healthy person, compensatory mechanisms can maintain adequate blood flow and blood pressure despite an acute blood loss of as much as 10% of total volume. Figure 21.16 shows several of the negative feedback systems that respond to hypovolemic shock.

Figure 21.16   **Negative feedback systems that can restore normal blood pressure during hypovolemic shock.**

Homeostatic mechanisms can compensate for an acute blood loss of as much as 10% of total blood volume.

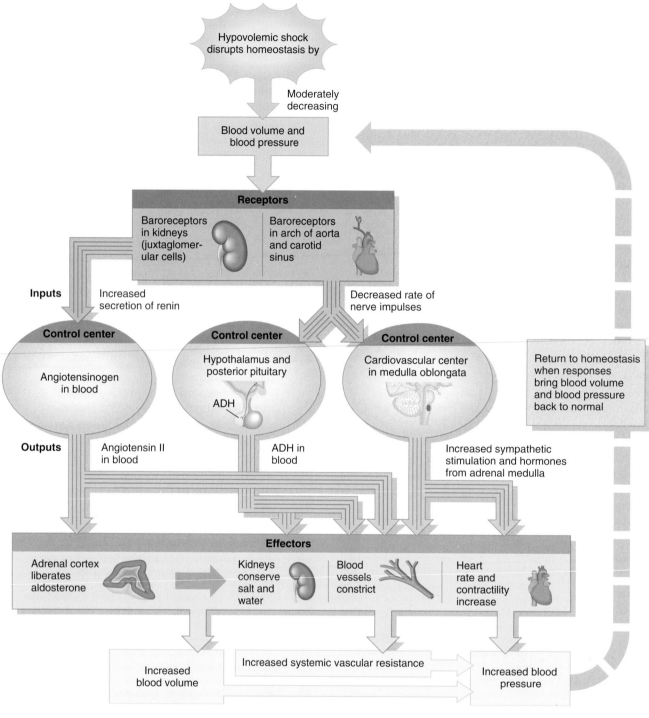

Does almost-normal blood pressure in a person who has lost blood indicate that the patient's tissues are receiving adequate perfusion (blood flow)?

1. ***Activation of the renin–angiotensin–aldosterone system.*** Decreased blood flow to the kidneys causes the kidneys to secrete renin and initiates the renin–angiotensin–aldosterone system (see Figure 18.16 on page 611). Recall that angiotensin II causes vasoconstriction and stimulates the adrenal cortex to secrete aldosterone, a hormone that increases reabsorption of Na^+ and water by the kidneys. The increases in systemic vascular resistance and blood volume help raise blood pressure.

2. ***Secretion of antidiuretic hormone.*** In response to decreased blood pressure, the posterior pituitary releases more antidiuretic hormone (ADH). ADH enhances water reabsorption by the kidneys, which conserves remaining blood volume. It also causes vasoconstriction, which increases systemic vascular resistance.

3. ***Activation of the sympathetic division of the ANS.*** As blood pressure decreases, the aortic and carotid baroreceptors initiate powerful sympathetic responses throughout most of the body. One result is marked vasoconstriction of arterioles and veins of the skin, kidneys, and other abdominal viscera. (Vasoconstriction does not occur in the brain or heart.) The constriction of arterioles increases systemic vascular resistance and the constriction of veins increases venous return. Both effects help maintain an adequate blood pressure. Sympathetic stimulation also increases heart rate and contractility and increases secretion of epinephrine and norepinephrine by the adrenal medulla. These hormones intensify vasoconstriction and increase heart rate and contractility, all of which help raise blood pressure.

4. ***Release of local vasodilators.*** In response to *hypoxia*, cells liberate vasodilators—including K^+, H^+, lactic acid, adenosine, and nitric oxide—that dilate arterioles and relax precapillary sphincters. Such vasodilation increases local blood flow and may restore O_2 level to normal in part of the body. However, vasodilation also has the potentially harmful effect of decreasing systemic vascular resistance and thus lowering the blood pressure.

If blood volume drops more than 10–20%, or if the heart cannot bring blood pressure up sufficiently, compensatory mechanisms may fail to maintain adequate blood flow to tissues. At this point, shock becomes life threatening as damaged cells start to die.

Signs and Symptoms of Shock

Even though the signs and symptoms of shock vary with the severity of the condition, most can be predicted in light of the responses generated by the negative feedback systems that attempt to correct the problem. Among the signs and symptoms of shock are the following:

- Systolic blood pressure is lower than 90 mmHg.
- Resting heart rate is rapid due to sympathetic stimulation and increased blood levels of epinephrine and norepinephrine.
- Pulse is weak and rapid due to reduced cardiac output and fast heart rate.
- Skin is cool, pale, and clammy due to sympathetic constric-

tion of skin blood vessels and sympathetic stimulation of sweating.
- Mental state is altered due to reduced oxygen supply to the brain.
- Urine formation is reduced due to increased levels of aldosterone and antidiuretic hormone (ADH).
- The person is thirsty due to loss of extracellular fluid.
- The pH of blood is low (acidosis) due to buildup of lactic acid.
- The person may have nausea due to impaired blood flow to the digestive organs due to sympathetic vasoconstriction.

▶ CHECKPOINT

21. Which symptoms of hypovolemic shock relate to actual body fluid loss, and which relate to the negative feedback systems that attempt to maintain blood pressure and blood flow?

CIRCULATORY ROUTES

▶ OBJECTIVE

- **Describe and compare the major routes that blood takes through various regions of the body.**

Blood vessels are organized into **circulatory routes** that carry blood to specific organs in the body (Figure 21.17). The routes are parallel—in most cases a portion of the cardiac output flows separately to each tissue of the body. Each organ receives its own supply of freshly oxygenated blood. As noted earlier, the two main circulatory routes are the systemic circulation and pulmonary circulation. The pulmonary and systemic circulations are different in two important ways. First, blood in the pulmonary circulation need not be pumped as far as blood in the systemic circulation. Second, compared to systemic arteries, pulmonary arteries have larger diameters, thinner walls, and less elastic tissue. As a result, the resistance to pulmonary blood flow is very low, which means that less pressure is needed to move blood through the lungs. The peak systolic pressure in the right ventricle is only 20% of that in the left ventricle.

The Systemic Circulation

The **systemic circulation** includes the arteries and arterioles that carry oxygenated blood from the left ventricle to systemic capillaries, plus the veins and venules that return deoxygenated blood to the right atrium. Blood leaving the aorta and flowing through the systemic arteries is a bright red color. As blood flows through capillaries, it loses some of its oxygen and picks up carbon dioxide, becoming a dark red color. All systemic arteries branch from the **aorta.** Completing the circuit, all the veins of the systemic circulation drain into the **superior vena cava,** the **inferior vena cava,** or the **coronary sinus,** which in turn empty into the right atrium. The bronchial arteries, which carry nutrients to the lungs, also are part of the systemic circulation.

Figure 21.17 Circulatory routes. Large black arrows indicate the systemic circulation (detailed in Exhibits 21.3–21.12), small black arrows the pulmonary circulation (detailed in Figure 21.30), and red arrows the hepatic portal circulation (detailed in Figure 21.29). Refer to Figure 20.8 on page 671 for details of the coronary circulation, and to Figure 21.31 for details of the fetal circulation.

🔑 **Blood vessels are organized into various routes that deliver blood to tissues of the body.**

■ = Oxygenated blood
■ = Deoxygenated blood

Systemic capillaries of head, neck, and upper limbs

Left pulmonary artery

Aorta

Pulmonary trunk

Pulmonary capillaries of left lung

Superior vena cava

Left pulmonary veins

Left atrium

Right atrium

Right ventricle

Left ventricle

Inferior vena cava

Celiac trunk

Common hepatic artery

Splenic artery

Hepatic vein

Left gastric artery

Sinusoids of liver

Capillaries of spleen

Capillaries of stomach

Hepatic portal vein

Common iliac vein

Superior mesenteric artery

Systemic capillaries of gastrointestinal tract

Internal iliac vein

Inferior mesenteric artery

External iliac vein

Common iliac artery

Systemic capillaries of pelvis

Internal iliac artery

External iliac artery

Venules

Arterioles

Systemic capillaries of lower limbs

❓ **What are the two main circulatory routes?**

Exhibits 21.1–21.12 and Figures 21.18–21.28 describe the main arteries and veins of the systemic circulation. The blood vessels are organized in the exhibits according to body regions. Figure 21.18a provides an overview of the major arteries, and

Figure 21.24 provides an overview of the major veins. As you study the various blood vessels in the exhibits, refer to these two figures to see the relationships of the blood vessels under consideration to other regions of the body.

718

(text continues on page 751)

Exhibit 21.1 The Aorta and Its Branches (Figure 21.18)

▶ **OBJECTIVES**

• Identify the four principal divisions of the aorta.

• Locate the major arterial branches arising from each division.

The **aorta** (= to lift up) is the largest artery of the body, with a diameter of 2–3 cm (about 1 in.). Its four principal divisions are the ascending aorta, arch of the aorta, thoracic aorta, and abdominal aorta. The portion of the aorta that emerges from the left ventricle posterior to the pulmonary trunk is the **ascending aorta.** The beginning of the aorta contains the aortic valve (see Figure 20.4a on page 665). The ascending aorta gives off two coronary artery branches that supply the myocardium of the heart. Then it turns to the left, forming the **arch of the aorta,** which descends and ends at the level of the intervertebral disc between the fourth and fifth thoracic vertebrae. As the aorta continues to descend, it lies close to the vertebral bodies, passes through the

diaphragm, and divides at the level of the fourth lumbar vertebra into two **common iliac arteries,** which carry blood to the lower limbs. The section of the aorta between the arch of the aorta and the diaphragm is called the **thoracic aorta;** the section between the diaphragm and the common iliac arteries is the **abdominal aorta.** Each division of the aorta gives off arteries that branch into distributing arteries that lead to various organs. Within the organs, the arteries divide into arterioles and then into capillaries that service the systemic tissues (all tissues except the alveoli of the lungs).

▶ **CHECKPOINT**

What general regions do each of the four principal divisions of the aorta supply?

Division and Branches	Region Supplied
Ascending Aorta	
Right and left coronary arteries	Heart.
Arch of the Aorta	
Brachiocephalic trunk (brā′-kē-ō-se-FAL-ik)	
Right common carotid artery (ka-ROT-id)	Right side of head and neck.
Right subclavian artery (sub-KLĀ-vē-an)	Right upper limb.
Left common carotid artery	Left side of head and neck.
Left subclavian artery	Left upper limb.
Thoracic Aorta (*thorac-* = chest)	
Pericardial arteries (per-i-KAR-dē-al)	Pericardium.
Bronchial arteries (BRONG-kē-al)	Bronchi of lungs.
Esophageal arteries (e-sof′-a-JĒ-al)	Esophagus.
Mediastinal arteries (mē′-dē-as-TĪ-nal)	Structures in mediastinum.
Posterior intercostal arteries (in′-ter-KOS-tal)	Intercostal and chest muscles.
Subcostal arteries (sub-KOS-tal)	Same as posterior intercostals.
Superior phrenic arteries (FREN-ik)	Superior and posterior surfaces of diaphragm.
Abdominal Aorta	
Inferior phrenic arteries (FREN-ik)	Inferior surface of diaphragm.
Celiac trunk (SĒ-lē-ak)	
Common hepatic artery (he-PAT-ik)	Liver.
Left gastric artery (GAS-trik)	Stomach and esophagus.
Splenic artery (SPLĒN-ik)	Spleen, pancreas, and stomach.
Superior mesenteric artery (MES-en-ter′-ik)	Small intestine, cecum, ascending and transverse colons, and pancreas.
Suprarenal arteries (su′-pra-RĒ-nal)	Adrenal (suprarenal) glands.
Renal arteries (RĒ-nal)	Kidneys.
Gonadal arteries (gō-NAD-al)	
Testicular arteries (tes-TIK-ū-lar)	Testes (male).
Ovarian arteries (ō-VAR-ē-an)	Ovaries (female).
Inferior mesenteric artery	Transverse, descending, and sigmoid colons; rectum.
Common iliac arteries (IL-ē-ak)	
External iliac arteries	Lower limbs.
Internal iliac arteries	Uterus (female), prostate gland (male), muscles of buttocks, and urinary bladder.

(continues)

Exhibit 21.1 The Aorta and Its Branches (Figure 21.18) *(continued)*

Figure 21.18 Aorta and its principal branches.

All systemic arteries branch from the aorta.

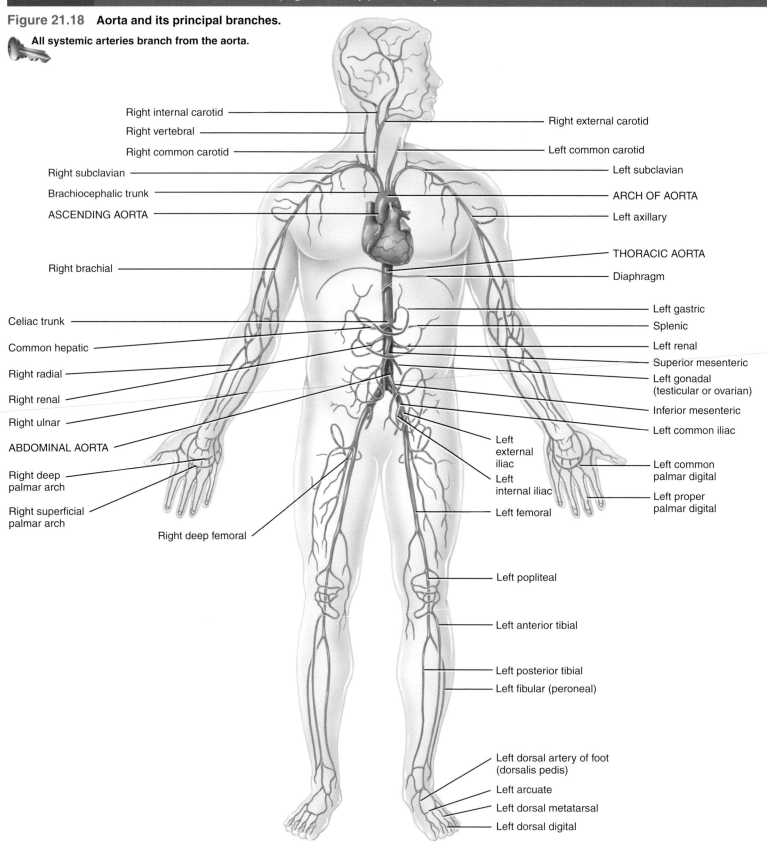

(a) Overall anterior view of the principal branches of the aorta

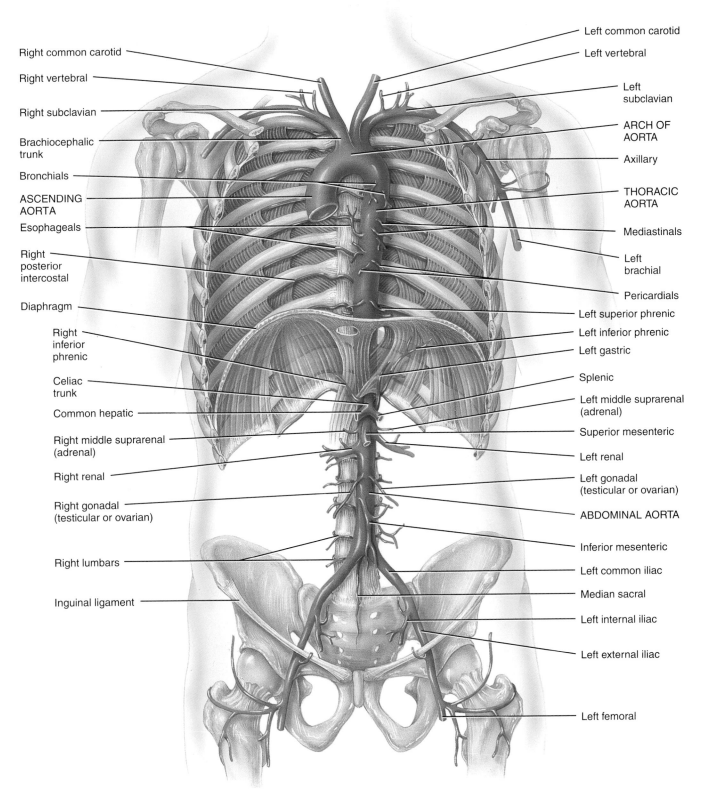

Right common carotid

Right vertebral

Right subclavian

Brachiocephalic trunk

Bronchials

ASCENDING AORTA

Esophageals

Right posterior intercostal

Diaphragm

Right inferior phrenic

Celiac trunk

Common hepatic

Right middle suprarenal (adrenal)

Right renal

Right gonadal (testicular or ovarian)

Right lumbars

Inguinal ligament

Left common carotid

Left vertebral

Left subclavian

ARCH OF AORTA

Axillary

THORACIC AORTA

Mediastinals

Left brachial

Pericardials

Left superior phrenic

Left inferior phrenic

Left gastric

Splenic

Left middle suprarenal (adrenal)

Superior mesenteric

Left renal

Left gonadal (testicular or ovarian)

ABDOMINAL AORTA

Inferior mesenteric

Left common iliac

Median sacral

Left internal iliac

Left external iliac

Left femoral

(b) Detailed anterior view of the principal branches of the aorta

What are the four subdivisions of the aorta?

Exhibit 21.2   Ascending Aorta (Figure 21.19)

▶ **OBJECTIVE**

• Identify the two primary arterial branches of the ascending aorta.

The **ascending aorta** is about 5 cm (2 in.) in length and begins at the aortic valve. It is directed superiorly, slightly anteriorly, and to the right. It ends at the level of the sternal angle, where it becomes the arch of the aorta. The beginning of the ascending aorta is posterior to the pulmonary trunk and right auricle; the right pulmonary artery is posterior to it. At its origin, the ascending aorta contains three dilations called aortic sinuses. Two of these, the right and left sinuses, give rise to the right and left coronary arteries, respectively.

The right and left **coronary arteries** (*coron-* = crown) arise from the ascending aorta just superior to the aortic semilunar valve. They form a crownlike ring around the heart, giving off branches to the atrial and ventricular myocardium. The **posterior interventricular branch** (in-ter-ven-TRIK-ū-lar; *inter-* = between) of the right coronary artery supplies both ventricles, and the **marginal branch** supplies the right ventricle. The **anterior interventricular branch,** also known as the **left anterior descending (LAD) branch,** of the left coronary artery supplies both ventricles, and the **circumflex branch** (SER-kum-flex; *circum-* = around; *-flex* = to bend) supplies the left atrium and left ventricle.

▶ **CHECKPOINT**

Which branches of the coronary arteries supply the left ventricle? Why does the left ventricle have such an extensive arterial blood supply?

SCHEME OF DISTRIBUTION

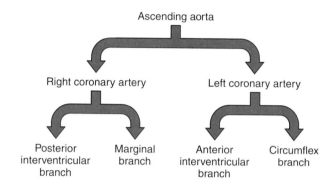

Ascending aorta

Right coronary artery — Left coronary artery

Posterior interventricular branch — Marginal branch

Anterior interventricular branch — Circumflex branch

Figure 21.19   Ascending aorta and its branches.

🔑 **The ascending aorta is the first division of the aorta.**

Left common carotid artery
Brachiocephalic trunk
Superior vena cava
Ascending aorta
Right coronary artery
Right atrium
Marginal branch
Right ventricle
Inferior vena cava

Left subclavian artery
Arch of aorta
Left pulmonary artery
Pulmonary trunk
Left pulmonary veins
Left coronary artery
Circumflex branch
Anterior interventricular branch
Posterior interventricular branch
Left ventricle

 Which arteries arise from the ascending aorta?

Exhibit 21.3   The Arch of the Aorta (Figure 21.20)

▶ **OBJECTIVE**

• Identify the three principal arteries that branch from the arch of the aorta.

The arch of the aorta is 4–5 cm (almost 2 in.) in length and is the continuation of the ascending aorta. It emerges from the pericardium posterior to the sternum at the level of the sternal angle. The arch is directed superiorly and posteriorly to the left and then inferiorly; it ends at the intervertebral disc between the fourth and fifth thoracic vertebrae, where it becomes the thoracic aorta. Three major arteries branch from the superior aspect of the arch of the aorta: the brachiocephalic trunk, the left common carotid, and the left subclavian. The first and largest branch from the arch of the aorta is the **brachiocephalic trunk** (brā′-kē-ō-se-FAL-ik; *brachio-* = arm; *-cephalic* = head). It extends superiorly, bending slightly to the right, and divides at the right sternoclavicular joint

to form the right subclavian artery and right common carotid artery. The second branch from the arch of the aorta is the **left common carotid artery** (ka-ROT-id), which divides into the same branches with the same names as the right common carotid artery. The third branch from the arch of the aorta is the **left subclavian artery** (sub-KLĀ-vē-an), which distributes blood to the left vertebral artery and vessels of the left upper limb. Arteries branching from the left subclavian artery are similar in distribution and name to those branching from the right subclavian artery. The accompanying table focuses on the principal arteries originating from the brachiocephalic trunk.

▶ **CHECKPOINT**

What general regions do the arteries that arise from the arch of the aorta supply?

Branch	Description and Region Supplied
Brachiocephalic trunk	The **brachiocephalic trunk** divides to form the right subclavian artery and right common carotid artery (Figure 21.20a).
Right subclavian artery (sub-KLĀ-vē-an)	The **right subclavian artery** extends from the brachiocephalic trunk to the first rib and then passes into the armpit (axilla). The general distribution of the artery is to the brain and spinal cord, neck, shoulder, thoracic viscera and wall, and scapular muscles.
Axillary artery (AK-sil-ār-ē = armpit)	Continuation of the right subclavian artery into the axilla is called the **axillary artery.** (Note that the right subclavian artery, which passes deep to the clavicle, is a good example of the practice of giving the same vessel different names as it passes through different regions.) Its general distribution is the shoulder, thoracic and scapular muscles, and humerus.
Brachial artery (BRĀ-kē-al = arm)	The **brachial artery** is the continuation of the axillary artery into the arm. The brachial artery provides the main blood supply to the arm and is superficial and palpable along its course. It begins at the tendon of the teres major muscle and ends just distal to the bend of the elbow. At first, the brachial artery is medial to the humerus, but as it descends it gradually curves laterally and passes through the cubital fossa, a triangular depression anterior to the elbow where you can easily detect the pulse of the brachial artery and listen to the various sounds when taking a person's blood pressure. Just distal to the bend in the elbow, the brachial artery divides into the radial artery and ulnar artery.
Radial artery (RĀ-dē-al = radius)	The **radial artery** is the smaller branch and is a direct continuation of the brachial artery. It passes along the lateral (radial) aspect of the forearm and then through the wrist and hand, supplying these structures with blood. At the wrist, the radial artery contacts the distal end of the radius, where it is covered only by fascia and skin. Because of its superficial location at this point, it is a common site for measuring radial pulse.
Ulnar artery (UL-nar = ulna)	The **ulnar artery,** the larger branch of the brachial artery, passes along the medial (ulnar) aspect of the forearm and then into the wrist and hand, supplying these structures with blood. In the palm, branches of the radial and ulnar arteries anastomose to form the superficial palmar arch and the deep palmar arch.
Superficial palmar arch (*palma* = palm)	The **superficial palmar arch** is formed mainly by the ulnar artery, with a contribution from a branch of the radial artery. The arch is superficial to the long flexor tendons of the fingers and extends across the palm at the bases of the metacarpals. It gives rise to **common palmar digital arteries,** which supply the palm. Each divides into a pair of **proper palmar digital arteries,** which supply the fingers.
Deep palmar arch	Mainly the radial artery forms the **deep palmar arch,** with a contribution from a branch of the ulnar artery. The arch is deep to the long flexor tendons of the fingers and extends across the palm, just distal to the bases of the metacarpals. Arising from the deep palmar arch are **palmar metacarpal arteries,** which supply the palm and anastomose with the common palmar digital arteries of the superficial palmar arch.

(continues)

Exhibit 21.3   The Arch of the Aorta (Figure 21.20) *(continued)*

Branch	Description and Region Supplied
Brachiocephalic trunk (continued)	
Vertebral artery (VER-te-bral)	Before passing into the axilla, the right subclavian artery gives off a major branch to the brain called the **right vertebral artery** (Figure 21.20b). The right vertebral artery passes through the foramina of the transverse processes of the sixth through first cervical vertebrae and enters the skull through the foramen magnum to reach the inferior surface of the brain. Here it unites with the left vertebral artery to form the **basilar** (BAS-i-lar) **artery.** The vertebral artery supplies the posterior portion of the brain with blood. The basilar artery passes along the midline of the anterior aspect of the brain stem. It gives off several branches (posterior cerebral and cerebellar arteries) that supply the cerebellum and pons of the brain and the inner ear.
Right common carotid artery	The **right common carotid artery** begins at the bifurcation of the brachiocephalic trunk, posterior to the right sternoclavicular joint, and passes superiorly in the neck to supply structures in the head (Figure 21.20b). At the superior border of the larynx (voice box), it divides into the right external and right internal carotid arteries.
External carotid artery	The **external carotid artery** begins at the superior border of the larynx and terminates near the temporomandibular joint in the substance of the parotid gland, where it divides into two branches: the superficial temporal and maxillary arteries. The carotid pulse can be detected in the external carotid artery just anterior to the sternocleidomastoid muscle at the superior border of the larynx. The general distribution of the external carotid artery is to structures *external* to the skull.
Internal carotid artery	The **internal carotid artery** has no branches in the neck and supplies structures *internal* to the skull. It enters the cranial cavity through the carotid foramen in the temporal bone. The internal carotid artery supplies blood to the eyeball and other orbital structures, ear, most of the cerebrum of the brain, pituitary gland, and external nose. The terminal branches of the internal carotid artery are the **anterior cerebral artery,** which supplies most of the medial surface of the cerebrum and deep masses of gray matter within the cerebrum, and the **middle cerebral artery,** which supplies most of the lateral surface of the cerebrum (Figure 21.20c). Inside the cranium, anastomoses of the left and right internal carotid arteries along with the basilar artery form an arrangement of blood vessels at the base of the brain near the hypophyseal fossa called the **cerebral arterial circle (circle of Willis).** From this circle (Figure 21.20c) arise arteries supplying most of the brain. Essentially, the cerebral arterial circle is formed by the union of the **anterior cerebral arteries** (branches of internal carotids) and **posterior cerebral arteries** (branches of basilar artery). The posterior cerebral arteries supply the inferolateral surface of the temporal lobe and lateral and medial surfaces of the occipital lobe of the cerebrum, deep masses of gray matter within the cerebrum, and midbrain. The posterior cerebral arteries are connected with the internal carotid arteries by the **posterior communicating arteries.** The **anterior communicating artery** connects the anterior cerebral arteries. The **internal carotid arteries** are also considered part of the cerebral arterial circle. The functions of the cerebral arterial circle are to equalize blood pressure to the brain and provide alternate routes for blood flow to the brain, should the arteries become damaged.
Left common carotid artery	See description in the Introduction, on page 723.
Left subclavian artery	See description in the Introduction, on page 723.

SCHEME OF DISTRIBUTION

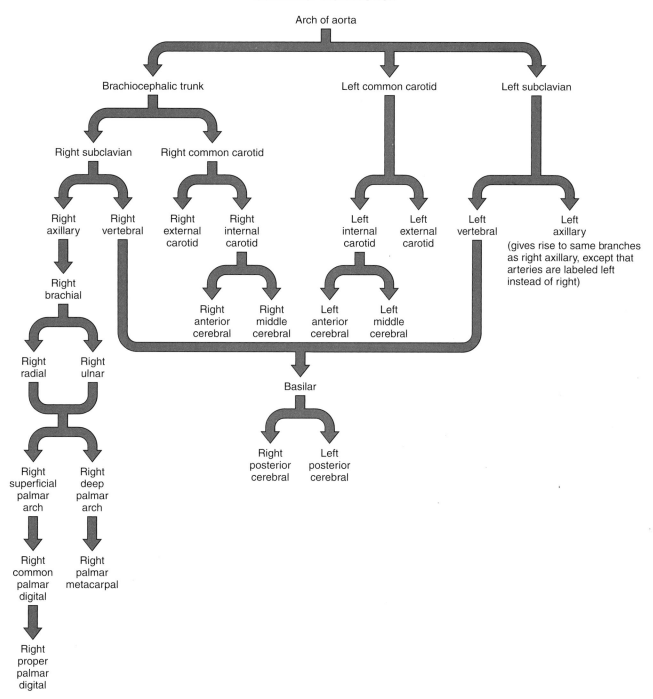

Exhibit 21.3   The Arch of the Aorta (Figure 21.20) *(continued)*

Figure 21.20   Arch of the aorta and its branches.   Note in (c) the arteries that constitute the cerebral arterial circle (circle of Willis).

🔑 **The arch of the aorta ends at the level of the intervertebral disc between the fourth and fifth thoracic vertebrae.**

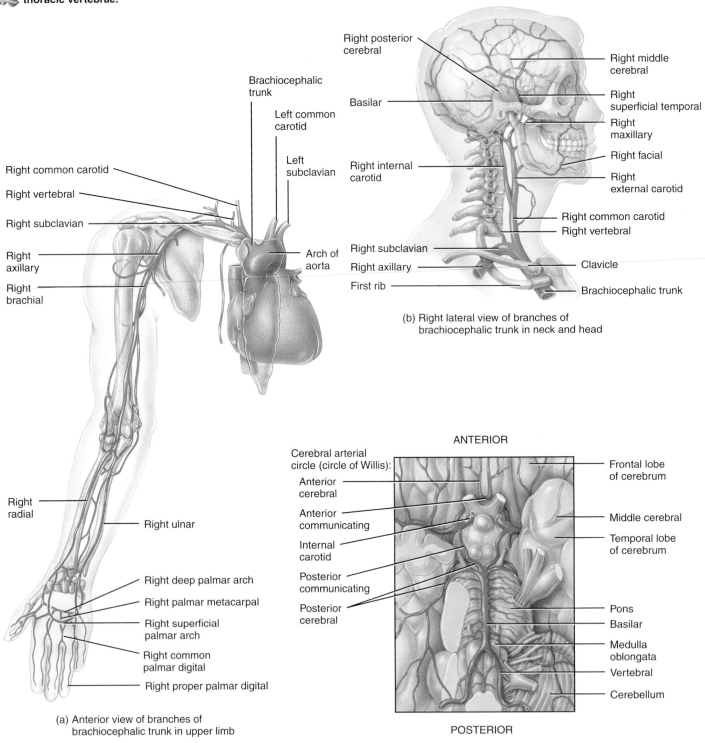

(a) Anterior view of branches of brachiocephalic trunk in upper limb

(b) Right lateral view of branches of brachiocephalic trunk in neck and head

(c) Inferior view of base of brain showing cerebral arterial circle

 What are the three major branches of the arch of the aorta, in order of their origination?

Exhibit 21.4 Thoracic Aorta (Figure 21.21)

▶ OBJECTIVE
• Identify the visceral and parietal branches of the thoracic aorta.

The **thoracic aorta** is about 20 cm (8 in.) long and is a continuation of the arch of the aorta. It begins at the level of the intervertebral disc between the fourth and fifth thoracic vertebrae, where it lies to the left of the vertebral column. As it descends, it moves closer to the midline and extends through an opening in the diaphragm (aortic hiatus), which is located anterior to the vertebral column at the level of the intervertebral disc between the twelfth thoracic and first lumbar vertebrae.

Along its course, the thoracic aorta sends off numerous small arteries, **visceral branches** to viscera and **parietal branches** to body wall structures.

▶ CHECKPOINT
What general regions do the visceral and parietal branches of the thoracic aorta supply?

Branch	Description and Region Supplied
Visceral	
Pericardial arteries (per′-i-KAR-dē-al); *peri-* = around; *cardia-* = heart)	Two or three tiny **pericardial arteries** supply blood to the pericardium.
Bronchial arteries (BRONG-kē-al = windpipe)	One right and two left **bronchial arteries** supply the bronchial tubes, pleurae, bronchial lymph nodes, and esophagus. (Whereas the right bronchial artery arises from the third posterior intercostal artery, the two left bronchial arteries arise from the thoracic aorta.)
Esophageal arteries (e-sof′-a-JĒ-al; *eso-* = to carry; *phage-* = food)	Four or five **esophageal arteries** supply the esophagus.
Mediastinal arteries (mē′-dē-as-TĪ-nal)	Numerous small **mediastinal arteries** supply blood to structures in the mediastinum.
Parietal	
Posterior intercostal arteries (in′-ter-KOS-tal; *inter-* = between; *costa* = rib)	Nine pairs of **posterior intercostal arteries** supply the intercostal, pectoralis major and minor, and serratus anterior muscles; overlying subcutaneous tissue and skin; mammary glands; and vertebrae, meninges, and spinal cord.
Subcostal arteries (sub-KOS-tal; *sub-* = under)	The left and right **subcostal arteries** have a distribution similar to that of the posterior intercostals.
Superior phrenic arteries (FREN-ik = pertaining to the diaphragm)	Small **superior phrenic arteries** supply the superior and posterior surfaces of the diaphragm.

SCHEME OF DISTRIBUTION

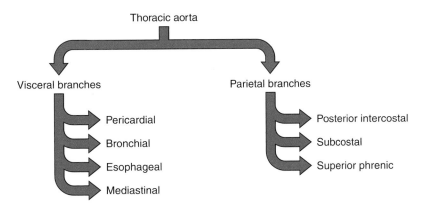

(continues)

Exhibit 21.4   Thoracic Aorta (Figure 21.21) *(continued)*

Figure 21.21   Thoracic aorta and abdominal aorta and their principal branches.

🔑 **The thoracic aorta is the continuation of the ascending aorta.**

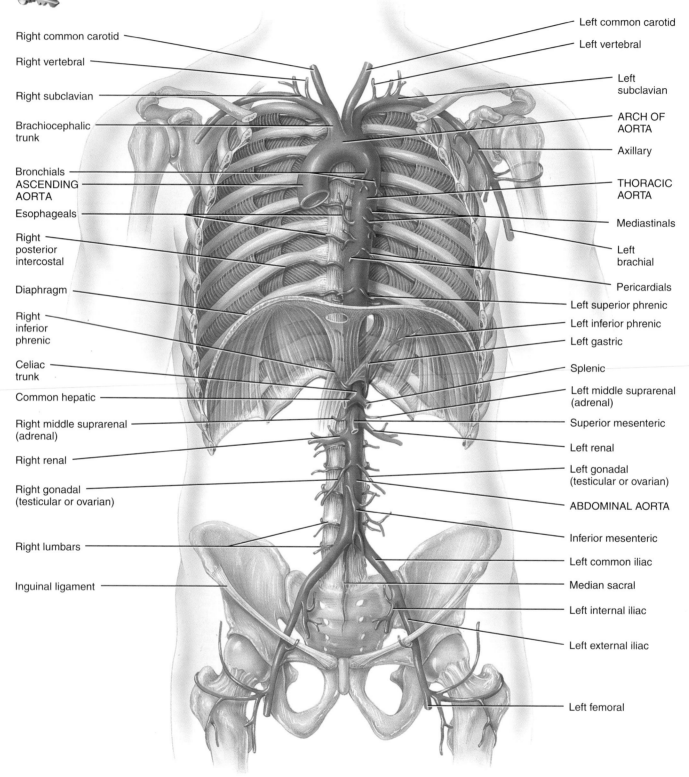

Right common carotid

Right vertebral

Right subclavian

Brachiocephalic trunk

Bronchials

ASCENDING AORTA

Esophageals

Right posterior intercostal

Diaphragm

Right inferior phrenic

Celiac trunk

Common hepatic

Right middle suprarenal (adrenal)

Right renal

Right gonadal (testicular or ovarian)

Right lumbars

Inguinal ligament

Left common carotid

Left vertebral

Left subclavian

ARCH OF AORTA

Axillary

THORACIC AORTA

Mediastinals

Left brachial

Pericardials

Left superior phrenic

Left inferior phrenic

Left gastric

Splenic

Left middle suprarenal (adrenal)

Superior mesenteric

Left renal

Left gonadal (testicular or ovarian)

ABDOMINAL AORTA

Inferior mesenteric

Left common iliac

Median sacral

Left internal iliac

Left external iliac

Left femoral

Anterior view of the principal branches of the aorta

 Where does the thoracic aorta begin?

Exhibit 21.5 Abdominal Aorta (Figure 21.22)

► **OBJECTIVE**

• Identify the visceral and parietal branches of the abdominal aorta.

The **abdominal aorta** is the continuation of the thoracic aorta. It begins at the aortic hiatus in the diaphragm and ends at about the level of the fourth lumbar vertebra, where it divides into the right and left common iliac arteries. The abdominal aorta lies anterior to the vertebral column.

As with the thoracic aorta, the abdominal aorta gives off visceral and parietal branches. The unpaired visceral branches arise from the anterior surface of the aorta and include the **celiac trunk** and the **superior mesenteric** and **inferior mesenteric arteries** (see Figure 21.21).

The paired visceral branches arise from the lateral surfaces of the aorta and include the **suprarenal, renal,** and **gonadal arteries.** The unpaired parietal branch is the **median sacral artery.** The paired parietal branches arise from the posterolateral surfaces of the aorta and include the **inferior phrenic** and **lumbar arteries.**

► CHECKPOINT

Name the paired visceral and parietal branches and the unpaired visceral and parietal branches of the abdominal aorta, and indicate the general regions they supply.

Branch	Description and Region Supplied
Unpaired Visceral Branches	
Celiac trunk (SĒ-lē-ak)	The **celiac trunk (artery)** is the first visceral branch from the aorta inferior to the diaphragm, at about the level of the twelfth thoracic vertebra (Figure 21.22a). Almost immediately, the celiac trunk divides into three branches—the left gastric, splenic, and common hepatic arteries (Figure 21.22a).

1. The **left gastric artery** (GAS-trik = stomach) is the smallest of the three branches. It passes superiorly to the left toward the esophagus and then turns to follow the lesser curvature of the stomach. It supplies the stomach and esophagus.

2. The **splenic artery** (SPLĒN-ik = spleen) is the largest branch of the celiac trunk. It arises from the left side of the celiac trunk distal to the left gastric artery, and passes horizontally to the left along the pancreas. Before reaching the spleen, it gives rise to three arteries:
   • **Pancreatic artery** (pan-krē-AT-ik), which supplies the pancreas.
   • **Left gastroepiploic artery** (gas′-trō-ep′-i-PLŌ-ik; *epiplo-* = omentum), which supplies the stomach and greater omentum.
   • **Short gastric artery,** which supplies the stomach.

3. The **common hepatic artery** (he-PAT-ik = liver) is intermediate in size between the left gastric and splenic arteries. Unlike the other two branches of the celiac trunk, the common hepatic artery arises from the right side. It gives rise to three arteries:
   • **Proper hepatic artery,** which supplies the liver, gallbladder, and stomach.
   • **Right gastric artery,** which supplies the stomach.
   • **Gastroduodenal artery** (gas′-trō-doo′-ō-DĒ-nal), which supplies the stomach, duodenum of the small intestine, pancreas, and greater omentum.

| **Superior mesenteric** (MES-en-ter′-ik; *meso-* = middle; *enteric* = pertaining to the intestines) | The **superior mesenteric artery** (Figure 21.22b) arises from the anterior surface of the abdominal aorta about 1 cm inferior to the celiac trunk at the level of the first lumbar vertebra. It extends inferiorly and anteriorly and between the layers of mesentery, which is a portion of the peritoneum that attaches the small intestine to the posterior abdominal wall. It anastomoses extensively and has five branches: |

1. The **inferior pancreaticoduodenal artery** (pan′-krē-at′-i-kō-doo′-ō-DĒ-nal) supplies the pancreas and duodenum.

2. The **jejunal** (je-JOO-nal) and **ileal arteries** (IL-ē-al) supply the jejunum and ileum of the small intestine, respectively.

3. The **ileocolic artery** (il′ē-ō-KŌL-ik) supplies the ileum and ascending colon of the large intestine.

4. The **right colic artery** (KŌL-ik) supplies the ascending colon.

5. The **middle colic artery** supplies the transverse colon of the large intestine.

| **Inferior mesenteric** | The **inferior mesenteric artery** (Figure 21.22c) arises from the anterior aspect of the abdominal aorta at the level of the third lumbar vertebra and then passes inferiorly to the left of the aorta. It anastomoses extensively and has three branches: |

1. The **left colic artery** supplies the transverse colon and descending colon of the large intestine.

2. The **sigmoid arteries** (SIG-moyd) supply the descending colon and sigmoid colon of the large intestine.

3. The **superior rectal artery** (REK-tal) supplies the rectum of the large intestine.

(continues)

Exhibit 21.5   Abdominal Aorta (Figure 21.22) *(continued)*

Branch	Description and Region Supplied
Paired Visceral Branches	
Suprarenal arteries (soo'-pra-RĒ-nal; *supra-* = above; *ren-* = kidney)	Although there are three pairs of **suprarenal (adrenal) arteries** that supply the adrenal (suprarenal) glands (superior, middle, and inferior), only the middle pair originates directly from the abdominal aorta (see Figure 21.21). The middle suprarenal arteries arise at the level of the first lumbar vertebra at or superior to the renal arteries. The superior suprarenal arteries arise from the inferior phrenic artery, and the inferior suprarenal arteries originate from the renal arteries.
Renal arteries (RĒ-nal = pertaining to the kidney)	The right and left **renal arteries** usually arise from the lateral aspects of the abdominal aorta at the superior border of the second lumbar vertebra, about 1 cm inferior to the superior mesenteric artery (see Figure 21.21). The right renal artery, which is longer than the left, arises slightly lower than the left and passes posterior to the right renal vein and inferior vena cava. The left renal artery is posterior to the left renal vein and is crossed by the inferior mesenteric vein. The renal arteries carry blood to the kidneys, adrenal (suprarenal) glands, and ureters. Their distribution within the kidneys is discussed in Chapter 26.
Gonadal (gō-NAD-al; *gon-* = seed) [**testicular** (test-TIK-ū-lar) or **ovarian arteries** (ō-VAR-ē-an)]	The **gonadal arteries** arise from the abdominal aorta at the level of the second lumbar vertebra just inferior to the renal arteries (see Figure 21.21). In males, the gonadal arteries are specifically referred to as the **testicular arteries.** They pass through the inguinal canal and supply the testes, epididymis, and ureters. In females, the gonadal arteries are called the **ovarian arteries.** They are much shorter than the testicular arteries and supply the ovaries, uterine (Fallopian) tubes, and ureters.
Unpaired Parietal Branch	
Median sacral artery (SĀ-kral = pertaining to the sacrum)	The **median sacral artery** arises from the posterior surface of the abdominal aorta about 1 cm superior to the bifurcation of the aorta into the right and left common iliac arteries (see Figure 21.21). The median sacral artery supplies the sacrum and coccyx.
Paired Parietal Branches	
Inferior phrenic arteries (FREN-ik = pertaining to the diaphragm)	The **inferior phrenic arteries** are the first paired branches of the abdominal aorta, immediately superior to the origin of the celiac trunk (see Figure 21.21). (They may also arise from the renal arteries.) The inferior phrenic arteries are distributed to the inferior surface of the diaphragm and adrenal (suprarenal) glands.
Lumbar arteries (LUM-bar = pertaining to the loin)	The four pairs of **lumbar arteries** arise from the posterolateral surface of the abdominal aorta (see Figure 21.21). They supply the lumbar vertebrae, spinal cord and its meninges, and the muscles and skin of the lumbar region of the back.

SCHEME OF DISTRIBUTION

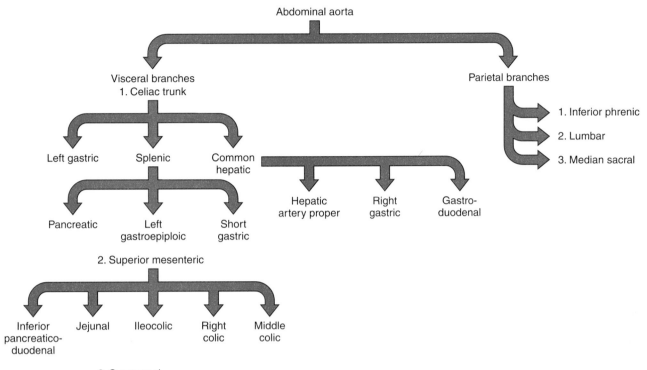

(continues)

Exhibit 21.5   Abdominal Aorta (Figure 21.22) *(continued)*

Figure 21.22   Abdominal aorta and its principal branches.

The abdominal aorta is the continuation of the thoracic aorta.

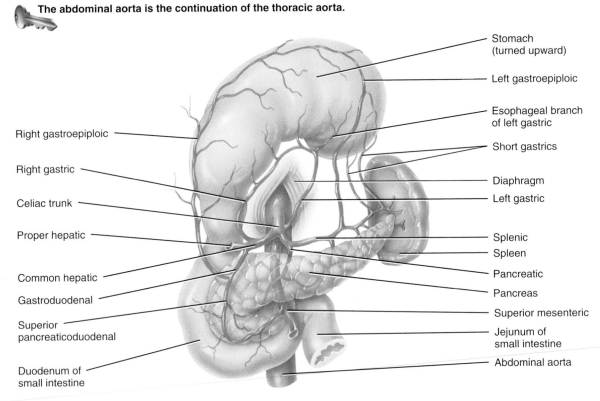

(a) Anterior view of celiac trunk and its branches

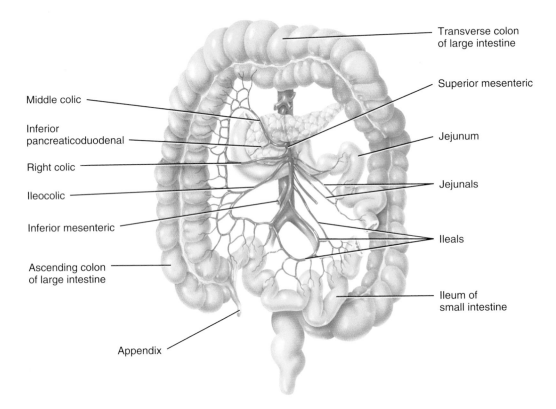

(b) Anterior view of superior mesenteric artery and its branches

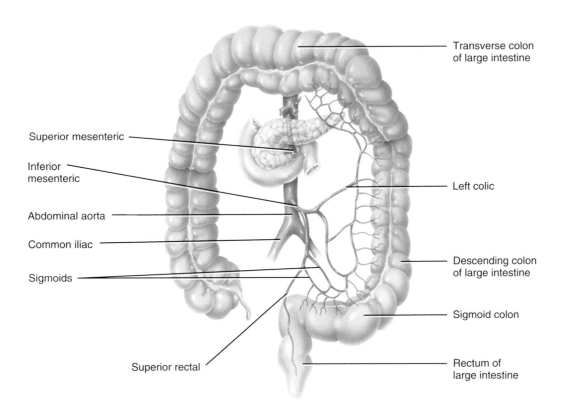

Superior mesenteric

Inferior mesenteric

Abdominal aorta

Common iliac

Sigmoids

Transverse colon of large intestine

Left colic

Descending colon of large intestine

Sigmoid colon

Superior rectal

Rectum of large intestine

(c) Anterior view of inferior mesenteric artery and its branches

 Where does the abdominal aorta begin?

Exhibit 21.6   Arteries of the Pelvis and Lower Limbs (Figure 21.23)

▶ **OBJECTIVE**

• Identify the two major branches of the common iliac arteries.

The abdominal aorta ends by dividing into the right and left **common iliac arteries.** These, in turn, divide into the **internal** and **external iliac arteries.** In sequence, the external iliacs become the **femoral arteries**

in the thighs, the **popliteal arteries** posterior to the knee, and the **anterior** and **posterior tibial arteries** in the legs.

▶ **CHECKPOINT**

What general regions do the internal and external iliac arteries supply?

Branch	Description and Region Supplied
Common iliac arteries (IL-ē-ak = pertaining to the ilium)	At about the level of the fourth lumbar vertebra, the abdominal aorta divides into the right and left **common iliac arteries,** the terminal branches of the abdominal aorta. Each passes inferiorly about 5 cm (2 in.) and gives rise to two branches: internal iliac and external iliac arteries. The general distribution of the common iliac arteries is to the pelvis, external genitals, and lower limbs.
Internal iliac arteries	The **internal iliac (hypogastric) arteries** are the primary arteries of the pelvis. They begin at the bifurcation of the common iliac arteries anterior to the sacroiliac joint at the level of the lumbosacral intervertebral disc. They pass posteromedially as they descend in the pelvis and divide into anterior and posterior divisions. The general distribution of the internal iliac arteries is to the pelvis, buttocks, external genitals, and thigh.
External iliac arteries	The **external iliac arteries** are larger than the internal iliac arteries. Like the internal iliac arteries, they begin at the bifurcation of the common iliac arteries. They descend along the medial border of the psoas major muscles following the pelvic brim, pass posterior to the midportion of the inguinal ligaments, and become the femoral arteries. The general distribution of the external iliac arteries is to the lower limbs. Specifically, branches of the external iliac arteries supply the muscles of the anterior abdominal wall, the cremaster muscle in males and the round ligament of the uterus in females, and the lower limbs.
Femoral arteries (FEM-o-ral = pertaining to the thigh)	The **femoral arteries** descend along the anteromedial aspects of the thighs to the junction of the middle and lower third of the thighs. Here they pass through an opening in the tendon of the adductor magnus muscle, where they emerge posterior to the femurs as the popliteal arteries. A pulse may be felt in the femoral artery just inferior to the inguinal ligament. The general distribution of the femoral arteries is to the lower abdominal wall, groin, external genitals, and muscles of the thigh.
Popliteal arteries (pop′-li-TĒ-al = posterior surface of the knee)	The **popliteal arteries** are the continuation of the femoral arteries through the popliteal fossa (space behind the knee). They descend to the inferior border of the popliteus muscles, where they divide into the anterior and posterior tibial arteries. A pulse may be detected in the popliteal arteries. In addition to supplying the adductor magnus and hamstring muscles and the skin on the posterior aspect of the legs, branches of the popliteal arteries also supply the gastrocnemius, soleus, and plantaris muscles of the calf, knee joint, femur, patella, and fibula.
Anterior tibial arteries (TIB-ē-al = pertaining to the shin bone)	The **anterior tibial arteries** descend from the bifurcation of the popliteal arteries. They are smaller than the posterior tibial arteries. The anterior tibial arteries descend through the anterior muscular compartment of the leg. They pass through the interosseous membrane that connects the tibia and fibula, lateral to the tibia. The anterior tibial arteries supply the knee joints, anterior compartment muscles of the legs, skin over the anterior aspects of the legs, and ankle joints. At the ankles, the anterior tibial arteries become the **dorsal arteries of the foot (dorsalis pedis arteries),** also arteries from which a pulse may be detected. The dorsal arteries of the foot supply the muscles, skin, and joints on the dorsal aspects of the feet. On the dorsum of the feet, the dorsalis pedis arteries give off a transverse branch at the first (medial) cuneiform bone called the **arcuate arteries** (*arcuat-* = bowed) that run laterally over the bases of the metatarsals. From the arcuate arteries branch **dorsal metatarsal arteries,** which supply the feet. The dorsal metatarsal arteries terminate by dividing into the **dorsal digital arteries,** which supply the toes.
Posterior tibial arteries	The **posterior tibial arteries,** the direct continuations of the popliteal arteries, descend from the bifurcation of the popliteal arteries. They pass down the posterior muscular compartment of the legs posterior to the medial malleolus of the tibia. They terminate by dividing into the medial and lateral plantar arteries. Their general distribution is to the muscles, bones, and joints of the leg and foot. Major branches of the posterior tibial arteries are the **fibular (peroneal) arteries,** which supply the fibularis, soleus, tibialis posterior, and flexor hallucis muscles. They also supply the fibula, tarsus, and lateral aspect of the heel. The bifurcation of the posterior tibial arteries into the medial and lateral plantar arteries occurs deep to the flexor retinaculum on the medial side of the feet. The **medial plantar arteries** (PLAN-tar = sole of foot) supply the abductor hallucis and flexor digitorum brevis muscles and the toes. The **lateral plantar arteries** unite with a branch of the dorsal arteries of the foot to form the **plantar arch.** The arch begins at the base of the fifth metatarsal and extends medially across the metacarpals. As the arch crosses the foot, it gives off **plantar metatarsal arteries,** which supply the feet. These terminate by dividing into **plantar digital arteries,** which supply the toes.

SCHEME OF DISTRIBUTION

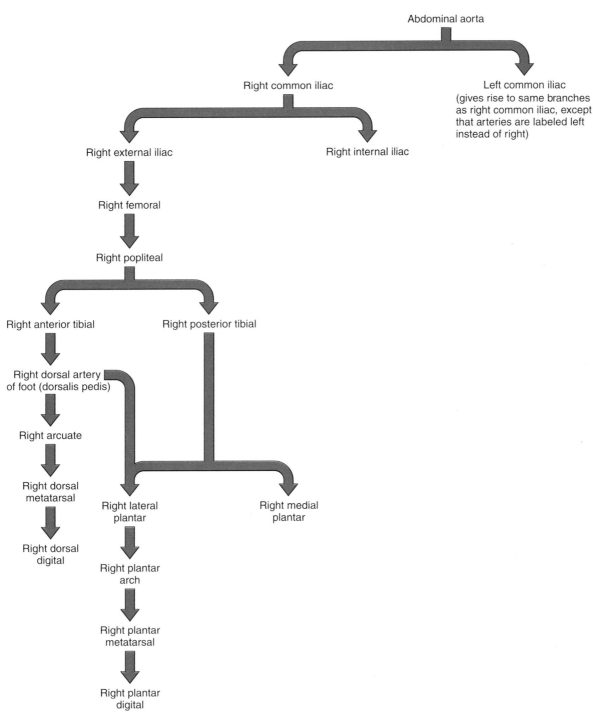

(continues)

Exhibit 21.6   Arteries of the Pelvis and Lower Limbs (Figure 21.23) *(continued)*

Figure 21.23   Arteries of the pelvis and right lower limb.

🔑 The internal iliac arteries carry most of the blood supply to the pelvic viscera and wall.

Right common iliac
Right internal iliac
Right external iliac

Abdominal aorta
Left common iliac

Right femoral

Right popliteal

Right anterior tibial

Right posterior tibial

Right fibular (peroneal)

Right dorsal artery of foot (dorsalis pedis)
Right arcuate
Right dorsal metatarsal
Right dorsal digital

Right lateral plantar
Right medial plantar
Right plantar arch
Right plantar digital
Right plantar metatarsal

(a) Anterior view

(b) Posterior view

 At what point does the abdominal aorta divide into the common iliac arteries?

Exhibit 21.7 Veins of the Systemic Circulation (Figure 21.24)

► OBJECTIVE

- Identify the three systemic veins that return deoxygenated blood to the heart.

Whereas arteries distribute blood to various parts of the body, veins drain blood away from them. For the most part, arteries are deep, whereas veins may be superficial or deep. Superficial veins are located just beneath the skin and can easily be seen. Because there are no large superficial arteries, the names of superficial veins do not correspond to those of arteries. Superficial veins are clinically important as sites for withdrawing blood or giving injections. Deep veins generally travel alongside arteries and usually bear the same name. Arteries usually follow definite pathways; veins are more difficult to follow because they connect in irregular networks in which many tributaries merge to form a large vein. Although only one systemic artery, the aorta, takes oxygenated blood away from the heart (left ventricle), three systemic veins, the **coronary sinus, superior vena cava,** and **inferior vena cava,** return deoxygenated blood to the heart (right atrium). The coronary sinus receives blood from the cardiac veins; the superior vena cava receives blood from other veins superior to the diaphragm, except the air sacs (alveoli) of the lungs; the inferior vena cava receives blood from veins inferior to the diaphragm.

► CHECKPOINT

What are the three tributaries of the coronary sinus?

Vein	Description and Region Drained
Coronary sinus (KOR-ō-nar-ē; *corona* = crown)	The **coronary sinus** is the main vein of the heart; it receives almost all venous blood from the myocardium. It is located in the coronary sulcus (see Figure 20.3c) and opens into the right atrium between the orifice of the inferior vena cava and the tricuspid valve. It is a wide venous channel into which three veins drain. It receives the **great cardiac vein** (in the anterior interventricular sulcus) into its left end, and the **middle cardiac vein** (in the posterior interventricular sulcus) and the **small cardiac vein** into its right end. Several **anterior cardiac veins** drain directly into the right atrium.
Superior vena cava (SVC) (VĒ-na CĀ-va; *vena* = vein; *cava* = cavelike)	The **SVC** is about 7.5 cm. (3 in.) long and 2 cm (1 in.) in diameter and empties its blood into the superior part of the right atrium. It begins posterior to the right first costal cartilage by the union of the right and left brachiocephalic veins and ends at the level of the right third costal cartilage, where it enters the right atrium. The SVC drains the head, neck, chest, and upper limbs.
Inferior vena cava (IVC)	The **IVC** is the largest vein in the body, about 3.5 cm (1.4 in.) in diameter. It begins anterior to the fifth lumbar vertebra by the union of the common iliac veins, ascends behind the peritoneum to the right of the midline, pierces the costal tendon of the diaphragm at the level of the eighth thoracic vertebra, and enters the inferior part of the right atrium. The IVC drains the abdomen, pelvis, and lower limbs. The inferior vena cava is commonly compressed during the later stages of pregnancy by the enlarging uterus, producing edema of the ankles and feet and temporary varicose veins.

(continues)

Exhibit 21.7 Veins of the Systemic Circulation (Figure 21.24) *(continued)*

Figure 21.24 Principal veins.

Deoxygenated blood returns to the heart via the superior and inferior venae cavae and the coronary sinus.

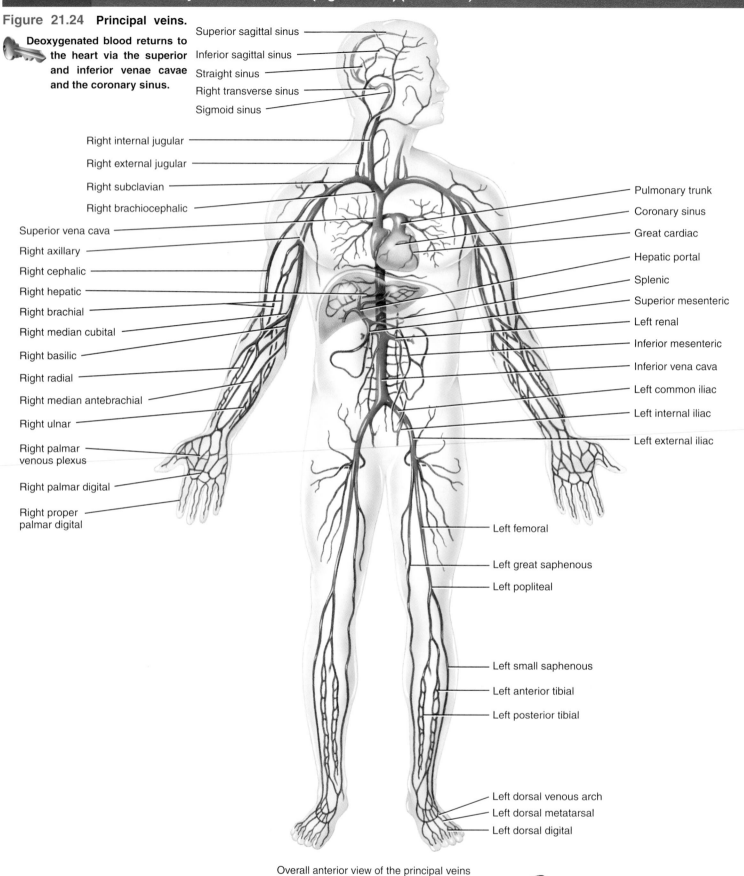

Superior sagittal sinus
Inferior sagittal sinus
Straight sinus
Right transverse sinus
Sigmoid sinus

Right internal jugular
Right external jugular
Right subclavian
Right brachiocephalic
Superior vena cava
Right axillary
Right cephalic
Right hepatic
Right brachial
Right median cubital
Right basilic
Right radial
Right median antebrachial
Right ulnar
Right palmar venous plexus
Right palmar digital
Right proper palmar digital

Pulmonary trunk
Coronary sinus
Great cardiac
Hepatic portal
Splenic
Superior mesenteric
Left renal
Inferior mesenteric
Inferior vena cava
Left common iliac
Left internal iliac
Left external iliac

Left femoral
Left great saphenous
Left popliteal
Left small saphenous
Left anterior tibial
Left posterior tibial
Left dorsal venous arch
Left dorsal metatarsal
Left dorsal digital

Overall anterior view of the principal veins

Which general regions of the body are drained by the superior vena cava and the inferior vena cava?

Exhibit 21.8 Veins of the Head and Neck (Figure 21.25)

▶ **OBJECTIVE**

• Identify the three major veins that drain blood from the head.

Most blood draining from the head passes into three pairs of veins: the **internal jugular, external jugular,** and **vertebral veins.** Within the brain, all veins drain into dural venous sinuses and then into the internal jugular veins. **Dural venous sinuses** are endothelial-lined venous channels between layers of the cranial dura mater.

▶ **CHECKPOINT**

Which general areas are drained by the internal jugular, external jugular, and vertebral veins?

Vein	Description and Region Drained
Internal jugular veins (JUG-ū-lar; *jugular* = throat)	The flow of blood from the dural venous sinuses into the internal jugular veins is as follows (Figure 21.25): The **superior sagittal sinus** (SAJ-i-tal = straight) begins at the frontal bone, where it receives a vein from the nasal cavity, and passes posteriorly to the occipital bone. Along its course, it receives blood from the superior, medial, and lateral aspects of the cerebral hemispheres, meninges, and cranial bones. The superior sagittal sinus usually turns to the right and drains into the right transverse sinus. The **inferior sagittal sinus** is much smaller than the superior sagittal sinus; it begins posterior to the attachment of the falx cerebri and receives the great cerebral vein to become the straight sinus. The great cerebral vein drains the deeper parts of the brain. Along its course the inferior sagittal sinus also receives tributaries from the superior and medial aspects of the cerebral hemispheres. The **straight sinus** runs in the tentorium cerebelli and is formed by the union of the inferior sagittal sinus and the great cerebral vein. The straight sinus also receives blood from the cerebellum and usually drains into the left transverse sinus. The **transverse sinuses** begin near the occipital bone, pass laterally and anteriorly, and become the sigmoid sinuses near the temporal bone. The transverse sinuses receive blood from the cerebrum, cerebellum, and cranial bones. The **sigmoid sinuses** (SIG-moyd = S-shaped) are located along the temporal bone. They pass through the jugular foramina, where they terminate in the internal jugular veins. The sigmoid sinuses drain the transverse sinuses. The **cavernous sinuses** (KAV-er-nus = cavelike) are located on either side of the sphenoid bone. They receive blood from the ophthalmic veins from the orbits, and from the cerebral veins from the cerebral hemispheres. They ultimately empty into the transverse sinuses and internal jugular veins. The cavernous sinuses are unique because they have nerves and a major blood vessel passing through them on their way to the orbit and face. The oculomotor nerve (cranial nerve III), trochlear nerve (cranial nerve IV), and ophthalmic and maxillary branches of the trigeminal nerve (cranial nerve V), as well as the internal carotid arteries pass through the cavernous sinuses. The right and left **internal jugular veins** pass inferiorly on either side of the neck lateral to the internal carotid and common carotid arteries. They then unite with the subclavian veins posterior to the clavicles at the sternoclavicular joints to form the right and left **brachiocephalic veins** (brā-kē-ō-se-FAL-ik; *brachio-* = arm; *cephal-* = head). From here blood flows into the superior vena cava. The general structures drained by the internal jugular veins are the brain (through the dural venous sinuses), face, and neck.
External jugular veins	The right and left **external jugular veins** begin in the parotid glands near the angle of the mandible. They are superficial veins that descend through the neck across the sternocleidomastoid muscles. They terminate at a point opposite the middle of the clavicle, where they empty into the subclavian veins. The general structures drained by the external jugular veins are external to the cranium, such as the scalp and superficial and deep regions of the face.
Vertebral veins (VER-te-bral; *vertebra* = vertebrae)	The right and left **vertebral veins** originate inferior to the occipital condyles. They descend through successive transverse foramina of the first six cervical vertebrae and emerge from the foramina of the skin cervical vertebra to enter the brachiocephalic veins in the root of the neck. The vertebral veins drain deep structures in the neck such as the cervical vertebrae, cervical spinal cord, and some neck muscles.

(continues)

Exhibit 21.8   Veins of the Head and Neck (Figure 21.25) *(continued)*

SCHEME OF DRAINAGE

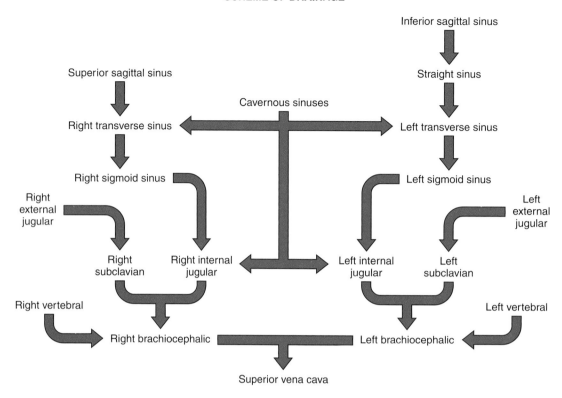

Figure 21.25   **Principal veins of the head and neck.**

Blood draining from the head passes into the internal jugular, external jugular, and vertebral veins.

Right lateral view

 Into which veins in the neck does all venous blood in the brain drain?

Exhibit 21.9 Veins of the Upper Limbs (Figure 21.26)

▶ **O B J E C T I V E**

• Identify the principal veins that drain the upper limbs.

Both superficial and deep veins return blood from the upper limbs to the heart. **Superficial veins** are located just deep to the skin and are often visible. They anastomose extensively with one another and with deep veins, and they do not accompany arteries. Superficial veins are larger than deep veins and return most of the blood from the upper limbs. **Deep veins** are located deep in the body. They usually accompany arteries and have the same names as the corresponding arteries. Both superficial and deep veins have valves, which are more numerous in the deep veins.

▶ C H E C K P O I N T

Where do the cephalic, basilic, median antebrachial, radial, and ulnar veins originate?

Vein	Description and Region Drained
Superficial	
Cephalic veins (se-FAL-ik = pertaining to the head)	The principal superficial veins that drain the upper limbs are the cephalic and basilic veins. They originate in the hand and convey blood from the smaller superficial veins into the axillary veins. The **cephalic veins** begin on the lateral aspect of the **dorsal venous networks of the hands (dorsal venous arches),** networks of veins on the dorsum of the hands formed by the **dorsal metacarpal veins** (Figure 21.26a). These veins, in turn, drain the **dorsal digital veins,** which pass along the sides of the fingers. Following their formation from the dorsal venous networks of the hands, the cephalic veins arch around the radial side of the forearms to the anterior surface and ascend through the entire limbs along the anterolateral surface. The cephalic veins end where they join the axillary veins, just inferior to the clavicles. **Accessory cephalic veins** originate either from a venous plexus on the dorsum of the forearms or from the medial aspects of the dorsal venous networks of the hands and unite with the cephalic veins just inferior to the elbow. The cephalic veins drain blood from the lateral aspect of the upper limbs.
Basilic veins (ba-SIL-ik = royal, of prime importance)	The **basilic veins** begin on the medial aspects of the dorsal venous networks of the hands and ascend along the posteromedial surface of the forearm and anteromedial surface of the arm (Figure 21.26b). They drain blood from the medial aspects of the upper limbs. Anterior to the elbow, the basilic veins are connected to the cephalic veins by the **median cubital veins** (*cubital* = pertaining to the elbow), which drain the forearm. If veins must be punctured for an injection, transfusion, or removal of a blood sample, the medial cubital veins are preferred. After receiving the median cubital veins, the basilic veins continue ascending until they reach the middle of the arm. There they penetrate the tissues deeply and run alongside the brachial arteries until they join the brachial veins. As the basilic and brachial veins merge in the axillary area, they form the axillary veins.
Median antebrachial veins (an′-tē-BRĀ-kē-al; *ante-* = before, in front of; *brachi-* =arm)	The **median antebrachial veins (median veins of the forearm)** begin in the **palmar venous plexuses,** networks of veins on the palms. The plexuses drain the **palmar digital veins** in the fingers. The median antebrachial veins ascend anteriorly in the forearms to join the basilic or median cubital veins, sometimes both. They drain the palms and forearms.
Deep	
Radial veins (RĀ-dē-al = pertaining to the radius)	The paired **radial veins** begin at the **deep palmar venous arches** (Figure 21.26c). These arches drain the **palmar metacarpal veins** in the palms. The radial veins drain the lateral aspects of the forearms and pass alongside the radial arteries. Just inferior to the elbow joint, the radial veins unite with the ulnar veins to form the brachial veins.
Ulnar veins (UL-nar = pertaining to the ulna)	The paired **ulnar veins,** which are larger than the radial veins, begin at the **superficial palmar venous arches.** These arches drain the **common palmar digital veins** and the **proper palmar digital veins** in the fingers. The ulnar veins drain the medial aspect of the forearms, pass alongside the ulnar arteries, and join with the radial veins to form the brachial veins.
Brachial veins (BRĀ-kē-al; *brachi-* = arm)	The paired **brachial veins** accompany the brachial arteries. They drain the forearms, elbow joints, arms, and humerus. They pass superiorly and join with the basilic veins to form the axillary veins.
Axillary veins (AK-sil-ār-ē; *axilla* = armpit)	The **axillary veins** ascend to the outer borders of the first ribs, where they become the subclavian veins. The axillary veins receive tributaries that correspond to the branches of the axillary arteries. The axillary veins drain the arms, axillas, and superolateral chest wall.
Subclavian veins (sub-KLĀ-vē-an; *sub-* = under; *clavian* = pertaining to the clavicle)	The **subclavian veins** are continuations of the axillary veins that terminate at the sternal end of the clavicles, where they unite with the internal jugular veins to form the brachiocephalic veins. The subclavian veins drain the arms, neck, and thoracic wall. The thoracic duct of the lymphatic system delivers lymph into the left subclavian vein at the junction with the internal jugular. The right lymphatic duct delivers lymph into the right subclavian vein at the corresponding junction (see Figure 22.3a).

(continues)

Exhibit 21.9   Veins of the Upper Limbs (Figure 21.26) *(continued)*

SCHEME OF DRAINAGE

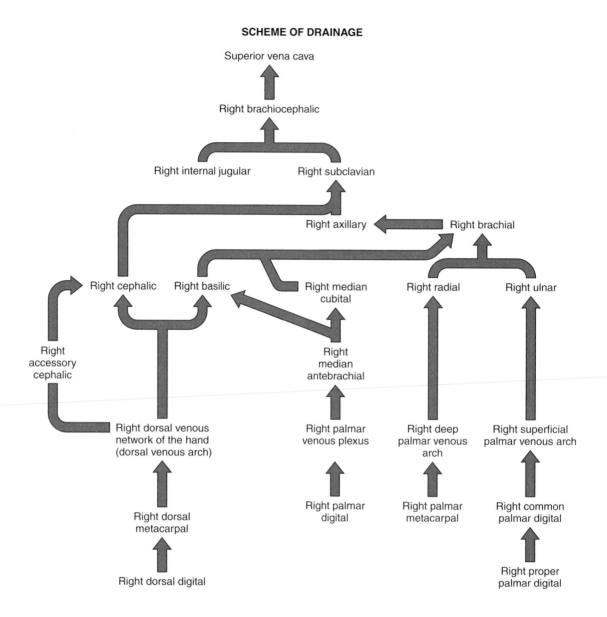

Figure 21.26 Principal veins of the right upper limb.

 Deep veins usually accompany arteries that have similar names.

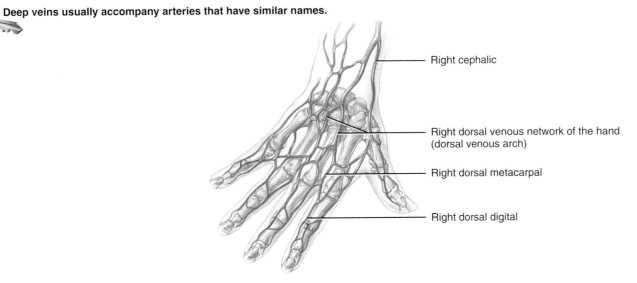

(a) Posterior view of superficial veins of the hand

Right cephalic

Right dorsal venous network of the hand (dorsal venous arch)

Right dorsal metacarpal

Right dorsal digital

(b) Anterior view of superficial veins

Right external jugular

Right subclavian

Right brachiocephalic

Right axillary

Right basilic

Right cephalic

Right internal jugular

Superior vena cava

Sternum

Right median cubital

Right basilic

Right median antebrachial

Right palmar venous plexus

Right palmar digital

(c) Anterior view of deep veins

Right external jugular

Right subclavian

Right brachiocephalic

Right axillary

Right brachial

Right internal jugular

Superior vena cava

Right radials

Right ulnars

Right deep palmar venous arch

Right superficial palmar venous arch

Right common palmar digital

Right palmar metacarpal

Right proper palmar digital

From which vein in the upper limb is a blood sample often taken?

Exhibit 21.10 | *Veins of the Thorax (Figure 21.27)*

► **OBJECTIVE**

• Identify the components of the azygos system of veins.

Although the brachiocephalic veins drain some portions of the thorax, most thoracic structures are drained by a network of veins, called the **azygos system,** that runs on either side of the vertebral column. The system consists of three veins—the **azygos, hemiazygos,** and **accessory hemiazygos veins**—that show considerable variation in origin, course, tributaries, anastomoses, and termination. Ultimately they empty into the superior vena cava.

► **CHECKPOINT**

What is the importance of the azygos system relative to the inferior vena cava?

Vein	Description and Region Drained
Brachiocephalic vein (brā´-kē-ō-se-FAL-ik; *brachio-* = arm; *cephalic* = pertaining to the head)	The right and left **brachiocephalic veins,** formed by the union of the subclavian and internal jugular veins, drain blood from the head, neck, upper limbs, mammary glands, and superior thorax. The brachiocephalic veins unite to form the superior vena cava. Because the superior vena cava is to the right of the body's midline, the left brachiocephalic vein is longer than the right. The right brachiocephalic vein is anterior and to the right of the brachiocephalic trunk. The left brachiocephalic vein is anterior to the brachiocephalic trunk, the left common carotid and left subclavian arteries, the trachea, the left vagus nerve (cranial nerve X), and phrenic nerve.
Azygos system (az-Ī-gos = unpaired)	The **azygos system,** besides collecting blood from the thorax and abdominal wall, may serve as a bypass for the inferior vena cava that drains blood from the lower body. Several small veins directly link the azygos system with the inferior vena cava. Large veins that drain the lower limbs and abdomen conduct blood into the azygos system. If the inferior vena cava or hepatic portal vein becomes obstructed, the azygos system can return blood from the lower body to the superior vena cava.
Azygos vein	The **azygos vein** is anterior to the vertebral column, slightly to the right of the midline. It usually begins at the junction of the right ascending lumbar and right subcostal veins near the diaphragm. At the level of the fourth thoracic vertebra, it arches over the root of the right lung to end in the superior vena cava. Generally, the azygos vein drains the right side of the thoracic wall, thoracic viscera, and abdominal wall. Specifically, the azygos vein receives blood from most of the **right posterior intercostal, hemiazygos, accessory hemiazygos, esophageal, mediastinal, pericardial,** and **bronchial veins.**
Hemiazygos vein (HEM-ē-az-ī-gos; *hemi-* = half)	The **hemiazygos vein** is anterior to the vertebral column and slightly to the left of the midline. It often begins at the junction of the left ascending lumbar and left subcostal veins. It terminates by joining the azygos vein at about the level of the ninth thoracic vertebra. Generally, the hemiazygos vein drains the left side of the thoracic wall, thoracic viscera, and abdominal wall. Specifically, the hemiazygos vein receives blood from the ninth through eleventh **left posterior intercostal, esophageal, mediastinal,** and sometimes the **accessory hemiazygos veins.**
Accessory hemiazygos vein	The **accessory hemiazygos vein** is also anterior to the vertebral column and to the left of the midline. It begins at the fourth or fifth intercostal space and descends from the fifth to the eighth thoracic vertebra or ends in the hemiazygos vein. It terminates by joining the azygos vein at about the level of the eighth thoracic vertebra. The accessory hemiazygos vein drains the left side of the thoracic wall. It receives blood from the fourth through eighth **left posterior intercostal veins** (the first through third left posterior intercostal veins open into the left brachiocephalic vein), **left bronchial,** and **mediastinal veins.**

SCHEME OF DRAINAGE

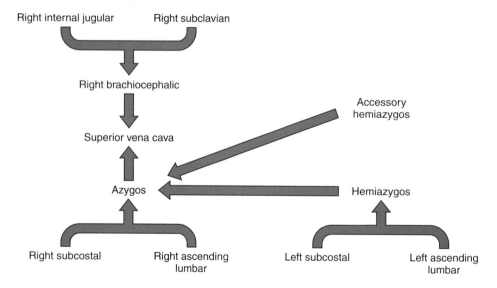

Figure 21.27 Principal veins of the thorax, abdomen, and pelvis.

Most thoracic structures are drained by the azygos system of veins.

Right internal jugular

Right external jugular

Right brachiocephalic

Superior vena cava

Right posterior intercostal

Azygos

Esophageals

Mediastinals

Diaphragm

Hepatics

Right suprarenal

Right subcostal

Right renal

Right ascending lumbar

Right gonadal (testicular or ovarian)

Right lumbar

Right common iliac

Right internal iliac

Right external iliac

Left internal jugular

Left external jugular

Left subclavian

Left brachiocephalic

Left superior intercostal

Left axillary

Left cephalic

Left posterior intercostal

Bronchial

Left brachial

Accessory hemiazygos

Left basilic

Left inferior phrenics

Hemiazygos

Left suprarenal

Left renal

Left ascending lumbar

Left gonadal (testicular or ovarian)

Inferior vena cava

Left common iliac

Inguinal ligament

Middle sacral

Left internal iliac

Left external iliac

Left femoral

Anterior view

 Which vein returns blood from the abdominopelvic viscera to the heart?

Exhibit 21.11 Veins of the Abdomen and Pelvis (Figure 21.27)

▶ **OBJECTIVE**

• Identify the principal veins that drain the abdomen and pelvis.

Blood from the abdominal and pelvic viscera and abdominal wall returns to the heart via the **inferior vena cava.** Many small veins enter the inferior vena cava. Most carry return flow from parietal branches of the abdominal aorta, and their names correspond to the names of the arteries.

The inferior vena cava does not receive veins directly from the gastrointestinal tract, spleen, pancreas, and gallbladder. These organs pass their blood into a common vein, the **hepatic portal vein,** which delivers

the blood to the liver. The superior mesenteric and splenic veins unite to form the hepatic portal vein (see Figure 21.29). This special flow of venous blood is called **hepatic portal circulation,** which is described shortly. After passing through the liver for processing, blood drains into the hepatic veins, which empty into the inferior vena cava.

▶ **CHECKPOINT**

What structures do the lumbar, gonadal, renal, suprarenal, inferior phrenic, and hepatic veins drain?

Vein	Description and Region Drained
Inferior vena cava (VĒ-na CĀ-va; *vena* = vein; *cava* = cavelike)	The two common iliac veins that drain the lower limbs, pelvis, and abdomen unite to form the **inferior vena cava** (see Figure 21.27). The inferior vena cava extends superiorly through the abdomen and thorax to the right atrium.
Common iliac veins (IL-ē-ak = pertaining to the ilium)	The **common iliac veins** are formed by the union of the internal and external iliac veins anterior to the sacroiliac joint and represent the distal continuation of the inferior vena cava at their bifurcation. The right common iliac vein is much shorter than the left and is also more vertical. Generally, the common iliac veins drain the pelvis, external genitals, and lower limbs.
Internal iliac veins	The **internal iliac veins** begin near the superior portion of the greater sciatic notch and run medial to their corresponding arteries. Generally, the veins drain the thigh, buttocks, external genitals, and pelvis.
External iliac veins	The **external iliac veins** are companions of the internal iliac arteries and begin at the inguinal ligaments as continuations of the femoral veins. They end anterior to the sacroiliac joint where they join with the internal iliac veins to form the common iliac veins. The external iliac veins drain the lower limbs, cremaster muscle in males, and the abdominal wall.
Lumbar veins (LUM-bar = pertaining to the loin)	A series of parallel **lumbar veins,** usually four on each side, drain blood from both sides of the posterior abdominal wall, vertebral canal, spinal cord, and meninges. The lumbar veins run horizontally with the lumbar arteries. The lumbar veins connect at right angles with the right and left **ascending lumbar veins,** which form the origin of the corresponding azygos or hemiazygos vein. The lumbar veins drain blood into the ascending lumbars and then run to the inferior vena cava, where they release the remainder of the flow.
Gonadal veins (gō-NAD-al; *gono* = seed) [**testicular** (tes-TIK-ū-lar) or **ovarian** (ō-VAR-ē-an)]	The **gonadal veins** ascend with the gonadal arteries along the posterior abdominal wall. In the male, the gonadal veins are called the testicular veins. The **testicular veins** drain the testes (the left testicular vein empties into the left renal vein, and the right testicular vein drains into the inferior vena cava). In the female, the gonadal veins are called the ovarian veins. The **ovarian veins** drain the ovaries. The left ovarian vein empties into the left renal vein, and the right ovarian vein drains into the inferior vena cava.
Renal veins (RĒ-nal = pertaining to the kidney)	The **renal veins** are large and pass anterior to the renal arteries. The left renal vein is longer than the right renal vein and passes anterior to the abdominal aorta. It receives the left testicular (or ovarian), left inferior phrenic, and usually left suprarenal veins. The right renal vein empties into the inferior vena cava posterior to the duodenum. The renal veins drain the kidneys.
Suprarenal veins (soo′-pra-RĒ-nal; *supra* = above)	The **suprarenal veins** drain the adrenal (suprarenal) glands (the left suprarenal vein empties into the left renal vein, and the right suprarenal vein empties into the inferior vena cava).
Inferior phrenic veins (FREN- ik = pertaining to the diaphragm)	The **inferior phrenic veins** drain the diaphragm (the left inferior phrenic vein usually sends one tributary to the left suprarenal vein, which empties into the left renal vein, and another tributary that empties into the inferior vena cava; the right inferior phrenic vein empties into the inferior vena cava).
Hepatic veins (he-PAT-ik = pertaining to the liver)	The **hepatic veins** drain the liver.

SCHEME OF DRAINAGE

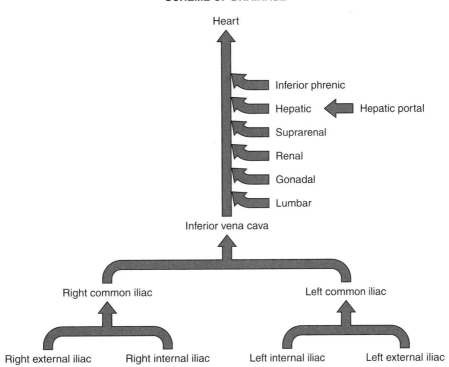

Exhibit 21.12 Veins of the Lower Limbs (Figure 21.28)

► OBJECTIVE

• Identify the principal superficial and deep veins that drain the lower limbs.

As with the upper limbs, blood from the lower limbs is drained by both **superficial** and **deep veins.** The superficial veins often anastomose with each other and with deep veins along their length. Deep veins, for

the most part, have the same names as corresponding arteries. All veins of the lower limbs have valves, which are more numerous than in veins of the upper limbs.

► CHECKPOINT

Why are the great saphenous veins clinically important?

Vein	Description and Region Drained
Superficial veins	
Great saphenous veins (sa-FĒ-nus; *saphen-* = clearly visible)	The **great (long) saphenous veins,** the longest veins in the body, ascend from the foot to the groin in the subcutaneous layer. They begin at the medial end of the dorsal venous arches of the foot. The **dorsal venous arches** (VĒ-nus) are networks of veins on the dorsum of the foot formed by the **dorsal digital veins,** which collect blood from the toes, and then unite in pairs to form the **dorsal metatarsal veins,** which parallel the metatarsals. As the dorsal metatarsal veins approach the foot, they combine to form the dorsal venous arches. The great saphenous veins pass anterior to the medial malleolus of the tibia and then superiorly along the medial aspect of the leg and thigh just deep to the skin. They receive tributaries from superficial tissues and connect with the deep veins as well. They empty into the femoral veins at the groin. Generally, the great saphenous veins drain mainly the medial side of the leg and thigh, the groin, external genitals, and abdominal wall.
	Along their length, the great saphenous veins have from 10 to 20 valves, with more located in the leg than the thigh. These veins are more likely to be subject to varicosities than other veins in the lower limbs because they must support a long column of blood and are not well supported by skeletal muscles.
	The great saphenous veins are often used for prolonged administration of intravenous fluids. This is particularly important in very young children and in patients of any age who are in shock and whose veins are collapsed. The great saphenous veins are also often used as a source of vascular grafts, especially for coronary bypass surgery. In the procedure, the vein is removed and then reversed so that the valves do not obstruct the flow of blood.
Small saphenous veins	The **small (short) saphenous veins** begin at the lateral aspect of the dorsal venous arches of the foot. They pass posterior to the lateral malleolus of the fibula and ascend deep to the skin along the posterior aspect of the leg. They empty into the popliteal veins in the popliteal fossa, posterior to the knee. Along their length, the small saphenous veins have from 9 to 12 valves. The small saphenous veins drain the foot and posterior aspect of the leg. They may communicate with the great saphenous veins in the proximal thigh.
Deep veins	
Posterior tibial veins (TIB-ē-al)	The **plantar digital veins** on the plantar surfaces of the toes unite to form the **plantar metatarsal veins,** which parallel the metatarsals. They unite to form the **deep plantar venous arches.** From each arch emerges the **medial** and **lateral plantar veins.**
	The medial and lateral plantar veins, posterior to the medial malleolus of the tibia, form the paired **posterior tibial veins,** which sometimes merge into a single vessel. They accompany the posterior tibial artery through the leg. They ascend deep to the muscles in the posterior aspect of the leg and drain the foot and posterior compartment muscles. About two-thirds of the way up the leg, the posterior tibial veins drain blood from the **fibular (peroneal) veins,** which drain the lateral and posterior leg muscles. The posterior tibial veins unite with the anterior tibial veins just inferior to the popliteal fossa to form the popliteal veins.
Anterior tibial veins	The paired **anterior tibial veins** arise in the dorsal venous arch and accompany the anterior tibial artery. They ascend in the interosseous membrane between the tibia and fibula and unite with the posterior tibial veins to form the popliteal vein. The anterior tibial veins drain the ankle joint, knee joint, tibiofibular joint, and anterior portion of the leg.
Popliteal veins (pop′-li-TĒ-al = pertaining to the hollow behind knee)	The anterior and posterior tibial veins unite to form the **popliteal veins.** The popliteal veins also receive blood from the small saphenous veins and tributaries that correspond to branches of the popliteal artery. The popliteal veins drain the knee joint and the skin, muscles, and bones of portions of the calf and thigh around the knee joint.
Femoral veins (FEM-o-ral)	The **femoral veins** accompany the femoral arteries and are the continuations of the popliteal veins just superior to the knee. The femoral veins extend up the posterior surface of the thighs and drain the muscles of the thighs, femurs, external genitals, and superficial lymph nodes. The largest tributaries of the femoral veins are the **deep femoral veins.** Just before penetrating the abdominal wall, the femoral veins receive the deep femoral veins and the great saphenous veins. The veins formed from this union penetrate the body wall and enter the pelvic cavity. Here they are known as the **external iliac veins.**

SCHEME OF DRAINAGE

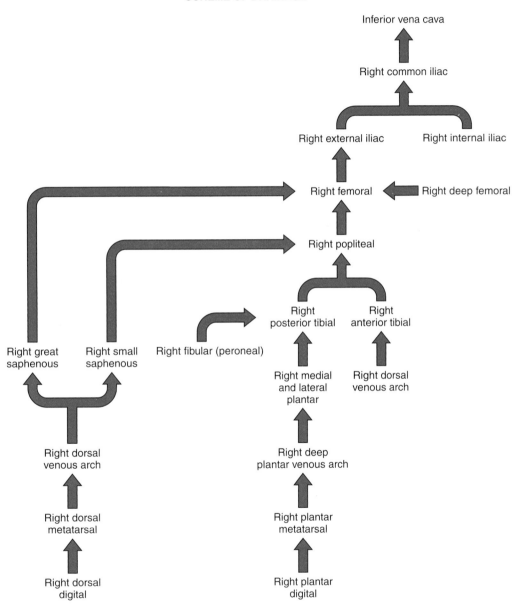

(continues)

Exhibit 21.12   Veins of the Lower Limbs (Figure 21.28) (continued)

Figure 21.28   Principal veins of the pelvis and lower limbs.

Deep veins usually bear the names of their companion arteries.

Inferior vena cava

Right common iliac

Right internal iliac

Right external iliac

Left common iliac

Right deep femoral

Right femoral

Right accessory saphenous

Right great saphenous

Right popliteal

Right small saphenous

Right anterior tibial

Right great saphenous

Right fibular (peroneal)

Right small saphenous

Right posterior tibial

Right dorsal venous arch

Right dorsal metatarsal

Right dorsal digital

Right medial plantar

Right deep plantar venous arch

Right plantar digital

Right lateral plantar

Right plantar metatarsal

(a) Anterior view

(b) Posterior view

 Which veins of the lower limb are superficial?

The Hepatic Portal Circulation

The **hepatic portal circulation** carries venous blood from gastrointestinal organs and spleen to the liver. A vein that carries blood from one capillary network to another is called a **portal vein**. The **hepatic portal vein** (*hepat-* = liver) receives blood from capillaries of gastrointestinal organs and the spleen and delivers it to the sinusoids of the liver (Figure 21.29). After a meal, hepatic portal blood is rich in nutrients absorbed from the gastrointestinal tract. The liver stores some of them and modifies others before they pass into the general circulation. For example, the liver converts glucose into glycogen for storage, thereby reducing blood glucose level shortly after a meal. The liver also detoxifies harmful substances, such as alcohol, that have been absorbed from the gastrointestinal tract and destroys bacteria by phagocytosis. While the liver is receiving nutrient-rich but deoxygenated blood via the hepatic portal vein, it also is receiving oxygenated blood via the hepatic artery, a branch of the celiac trunk. The oxygenated blood mixes with the deoxygenated blood in sinusoids. Eventually, blood leaves the

Figure 21.29 Hepatic portal circulation.   A schematic diagram of blood flow through the liver, including arterial circulation, is shown in (b). As usual, deoxygenated blood is indicated in blue, oxygenated blood in red.

> The hepatic portal circulation delivers venous blood from the organs of the gastrointestinal tract and spleen to the liver.

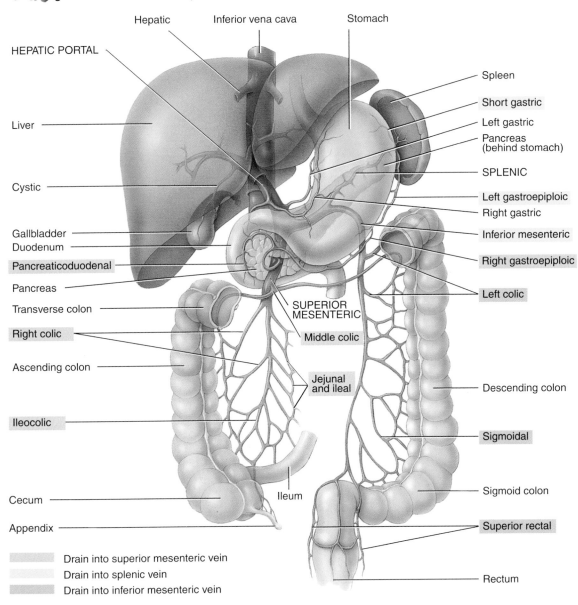

Drain into superior mesenteric vein
Drain into splenic vein
Drain into inferior mesenteric vein

(a) Anterior view of veins draining into the hepatic portal vein

(continues)

Figure 21.29 *(continued)*

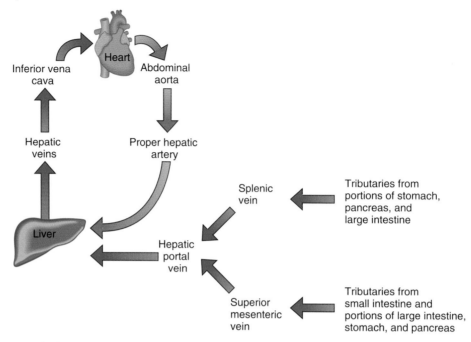

(b) Scheme of principal blood vessels of hepatic portal circulation and arterial
supply and venous drainage of liver

Which veins carry blood away from the liver?

sinusoids of the liver through the **hepatic veins,** which drain into the inferior vena cava.

The superior mesenteric and splenic veins unite to form the hepatic portal vein. The **superior mesenteric vein** drains blood from the small intestine and portions of the large intestine, stomach, and pancreas through the *jejunal, ileal, ileocolic, right colic, middle colic, pancreaticoduodenal,* and *right gastroepiploic veins.* The **splenic vein** drains blood from the stomach, pancreas, and portions of the large intestine through the *short gastric, left gastroepiploic, pancreatic,* and *inferior mesenteric veins.* The inferior mesenteric vein, which passes into the splenic vein, drains portions of the large intestine through the superior *rectal, sigmoidal,* and *left colic veins.* The *right* and *left gastric veins,* which open directly into the hepatic portal vein, drain the stomach. The *cystic vein,* which also opens into the hepatic portal vein, drains the gallbladder.

The Pulmonary Circulation

The **pulmonary circulation** (*pulmo-* = lung) carries deoxygenated blood from the right ventricle to the air sacs (alveoli) within the lungs and returns oxygenated blood from the air sacs to the left atrium (Figure 21.30). The **pulmonary trunk** emerges from the right ventricle and passes superiorly, posteriorly, and to the left. It then divides into two branches: the **right pulmonary artery** to the right lung and the **left pulmonary artery** to the left lung. After birth, the pulmonary arteries are the only arteries that carry deoxygenated blood. On entering the lungs, the

branches divide and subdivide until finally they form capillaries around the air sacs (alveoli) within the lungs. CO_2 passes from the blood into the air sacs and is exhaled. Inhaled O_2 passes from the air within the lungs into the blood. The pulmonary capillaries unite to form venules and eventually **pulmonary veins,** which exit the lungs and carry the oxygenated blood to the left atrium. Two left and two right pulmonary veins enter the left atrium. After birth, the pulmonary veins are the only veins that carry oxygenated blood. Contractions of the left ventricle then eject the oxygenated blood into the systemic circulation.

The Fetal Circulation

The circulatory system of a fetus, called the **fetal circulation,** exists only in the fetus and contains special structures that allow the developing fetus to exchange materials with its mother (Figure 21.31 on pages 754–755). It differs from the postnatal (after birth) circulation because the lungs, kidneys, and gastrointestinal organs do not begin to function until birth. The fetus obtains O_2 and nutrients from and eliminates CO_2 and other wastes into the maternal blood.

The exchange of materials between fetal and maternal circulations occurs through the **placenta** (pla-SEN-ta), which forms inside the mother's uterus and attaches to the umbilicus (navel) of the fetus by the **umbilical cord** (um-BIL-i-kal). The placenta communicates with the mother's cardiovascular system through many small blood vessels that emerge from the uterine wall. The umbilical cord contains blood vessels that branch into capillaries

Figure 21.30 **Pulmonary circulation.**

The pulmonary circulation brings deoxygenated blood from the right ventricle to the lungs and returns oxygenated blood from the lungs to the left atrium.

Superior vena cava
Right pulmonary artery
Pulmonary trunk
Right pulmonary veins
Right lung
Inferior vena cava

Arch of aorta
Ascending aorta
Left pulmonary artery
Left pulmonary veins
Left lung
Diaphragm
Abdominal aorta

(a) Anterior view

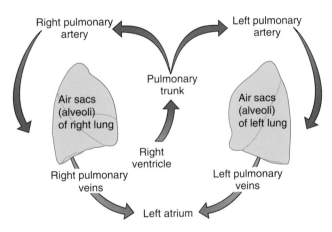

Right pulmonary artery
Left pulmonary artery
Pulmonary trunk
Air sacs (alveoli) of right lung
Air sacs (alveoli) of left lung
Right ventricle
Right pulmonary veins
Left pulmonary veins
Left atrium

(b) Scheme of pulmonary circulation

After birth, which are the only arteries that carry deoxygenated blood?

in the placenta. Wastes from the fetal blood diffuse out of the capillaries, into spaces containing maternal blood (intervillous spaces) in the placenta, and finally into the mother's uterine veins. Nutrients travel the opposite route—from the maternal blood vessels to the intervillous spaces to the fetal capillaries. Normally, there is no direct mixing of maternal and fetal blood because all exchanges occur by diffusion through capillary walls.

Blood passes from the fetus to the placenta via two **umbilical arteries** (Figure 21.31a, c). These branches of the internal iliac (hypogastric) arteries are within the umbilical cord. At the placenta, fetal blood picks up O_2 and nutrients and eliminates CO_2 and wastes. The oxygenated blood returns from the

placenta via a single **umbilical vein.** This vein ascends to the liver of the fetus, where it divides into two branches. Whereas some blood flows through the branch that joins the hepatic portal vein and enters the liver, most of the blood flows into the second branch, the **ductus venosus** (DUK-tus ve-NŌ-sus), which drains into the inferior vena cava.

Deoxygenated blood returning from lower body regions of the fetus mingles with oxygenated blood from the ductus venosus in the inferior vena cava. This mixed blood then enters the right atrium. Deoxygenated blood returning from upper body regions of the fetus enters the superior vena cava and passes into the right atrium.

Figure 21.31 Fetal circulation and changes at birth. The boxes between parts (a) and (b) describe the fate of certain fetal structures once postnatal circulation is established.

🔑 **The lungs and gastrointestinal organs do not begin to function until birth.**

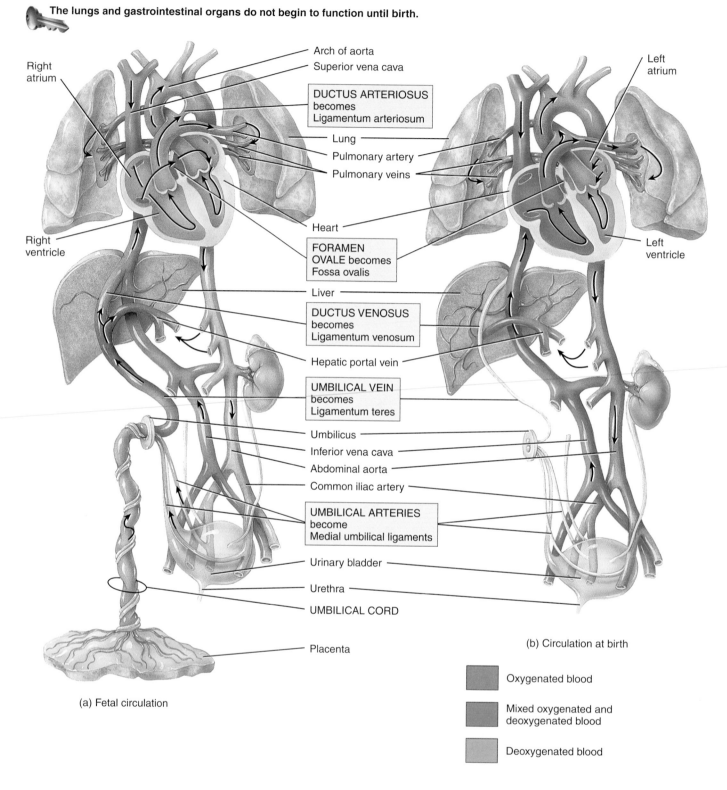

(a) Fetal circulation

(b) Circulation at birth

■ Oxygenated blood

■ Mixed oxygenated and deoxygenated blood

■ Deoxygenated blood

Most of the fetal blood does not pass from the right ventricle to the lungs, as it does in postnatal circulation, because an opening called the **foramen ovale** (fō-RĀ-men ō-VAL-ē) exists in the septum between the right and left atria. About one-third of the blood that enters the right atrium passes through the foramen ovale into the left atrium and joins the systemic circulation. The blood that does pass into the right ventricle is pumped into the pulmonary trunk, but little of this blood reaches the nonfunction-

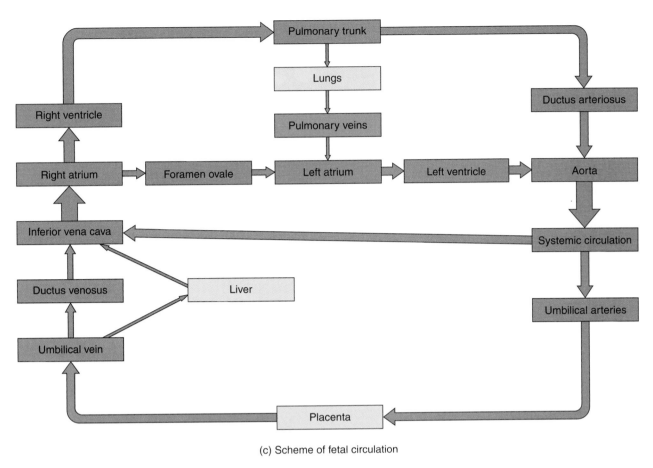

(c) Scheme of fetal circulation

Which structure provides for exchange of materials between mother and fetus?

ing fetal lungs. Instead, most is sent through the **ductus arterio-sus** (ar-tē-rē-Ō-sus), a vessel that connects the pulmonary trunk with the aorta, so that most blood bypasses the fetal lungs. The blood in the aorta is carried to all fetal tissues through the systemic circulation. When the common iliac arteries branch into the external and internal iliacs, part of the blood flows into the internal iliacs, into the umbilical arteries, and back to the placenta for another exchange of materials.

After birth, when pulmonary (lung), renal, and digestive functions begin, the following vascular changes occur (Figure 21.31b):

1. When the umbilical cord is tied off, blood no longer flows through the umbilical arteries, they fill with connective tissue, and the distal portions of the umbilical arteries become fibrous cords called the **medial umbilical ligaments.** Although the arteries are closed functionally only a few minutes after birth, complete obliteration of the lumens may take 2 to 3 months.

2. The umbilical vein collapses but remains as the **ligamentum teres (round ligament),** a structure that attaches the umbilicus to the liver.

3. The ductus venosus collapses but remains as the **ligamentum venosum,** a fibrous cord on the inferior surface of the liver.

4. The placenta is expelled as the **"afterbirth."**

5. The foramen ovale normally closes shortly after birth to become the **fossa ovalis,** a depression in the interatrial septum. When an infant takes its first breath, the lungs expand and blood flow to the lungs increases. Blood returning from the lungs to the heart increases pressure in the left atrium. This closes the foramen ovale by pushing the valve that guards it against the interatrial septum. Permanent closure occurs in about a year.

6. The ductus arteriosus closes by vasoconstriction almost immediately after birth and becomes the **ligamentum arteriosum.** Complete anatomical obliteration of the lumen takes 1 to 3 months.

▶ CHECKPOINT

22. Prepare a diagram to show the hepatic portal circulation. Why is this route important?

23. Prepare a diagram to show the route of the pulmonary circulation.

24. Discuss the anatomy and physiology of the fetal circulation. Indicate the function of the umbilical arteries, umbilical vein, ductus venosus, foramen ovale, and ductus arteriosus.

DEVELOPMENT OF BLOOD VESSELS AND BLOOD

▶ O B J E C T I V E

- **Describe the development of blood vessels and blood.**

The human yolk sac and ovum have little yolk to nourish the developing embryo. Blood and blood vessel formation starts as early as 15–16 days in the **mesoderm** of the yolk sac, chorion, and connecting stalk.

Blood vessels develop from mesenchyme that differentiates into cells called **angioblasts**. The cells aggregate to form isolated masses of cells called **blood islands** (Figure 21.32). Spaces soon appear in the islands and become the lumens of the blood vessels. Some of the angioblasts immediately around the spaces give rise to the *endothelial lining of the blood vessels*. Angioblasts around the endothelium form the *tunics* (interna, media, and externa) of the larger blood vessels. Growth and fusion of blood islands form an extensive network of blood vessels throughout the embryo.

Endothelial cells outside the embryo produce *blood plasma* and *blood cells*, which appear in the blood vessels of the yolk sac, chorion, and allantois at about three weeks. Blood formation in the embryo itself begins at about the fifth week in the liver and the twelfth week in the spleen, red bone marrow, and thymus.

▶ C H E C K P O I N T

25. What are the sites of blood cell production outside the embryo and within the embryo?

AGING AND THE CARDIOVASCULAR SYSTEM

▶ O B J E C T I V E

- **Explain the effects of aging on the cardiovascular system.**

General changes in the cardiovascular system associated with aging include decreased compliance of the aorta and large arteries, reduction in cardiac muscle fiber size, progressive loss of cardiac muscular strength, reduced cardiac output, a decline in maximum heart rate, and an increase in systolic blood pressure. There is an increase in the incidence of coronary artery disease (CAD), the major cause of heart disease and death in older Americans. Congestive heart failure, a set of symptoms associated with impaired pumping of the heart, is also prevalent in older individuals. Atherosclerosis of blood vessels that serve brain tissue reduces brain blood flow and thus decreases nourish-

Figure 21.32   Development of blood vessels and blood cells from blood islands.

🔑 Blood vessel development begins in the embryo on about the 15th or 16th day.

 From which germ cell layer are blood vessels and blood derived?

ment to the brain. As a result, some brain cells malfunction or die. By age 80, cerebral blood flow is 20% less and renal blood flow is 50% less than in the same person at age 30.

• • •

To appreciate the many ways the blood, heart, and blood vessels contribute to homeostasis of other body systems, examine *Focus on Homeostasis: The Cardiovascular System.*

Body System	Contribution of the Cardiovascular System
For all body systems	The heart pumps blood through blood vessels to body tissues, delivering oxygen and nutrients and removing wastes by means of capillary exchange. Circulating blood keeps body tissues at a proper temperature.
Integumantary system	Blood delivers clotting factors and white blood cells that aid in hemostasis when skin is damaged and contribute to repair of injured skin. Changes in skin blood flow contribute to body temperature regulation by adjusting the amount of heat loss via the skin. Blood flowing in skin may give skin a pink hue.
Skeletal system	Blood delivers calcium and phosphate ions that are needed for building bone matrix, hormones that govern building and breakdown of bone matrix, and erythropoietin that stimulates production of red blood cells by red bone marrow.
Muscular system	Blood circulating through exercising muscle removes heat and lactic acid.
Nervous system	Endothelial cells lining choroid plexuses in brain ventricles help produce cerebrospinal fluid (CSF) and contribute to the blood–brain barrier.
Endocrine system	Circulating blood delivers most hormones to their target tissues. Atrial cells secrete atrial natriuretic peptide.
Lymphatic and immune system	Circulating blood distributes lymphocytes, antibodies, and macrophages that carry out immune functions. Lymph forms from excess interstitial fluid, which filters from blood plasma due to blood pressure generated by heart.
Respiratory system	Circulating blood transports oxygen from the lungs to body tissues and carbon dioxide to the lungs for exhalation.
Digestive system	Blood carries newly absorbed nutrients and water to liver. Blood distributes hormones that aid digestion.
Urinary system	Heart and blood vessels deliver 20% of the resting cardiac output to the kidneys, where blood is filtered, needed substances are reabsorbed, and unneeded substances remain as part of urine, which is excreted.
Reproductive systems	Vasodilation of arterioles in penis and clitoris cause erection during sexual intercourse. Blood distributes hormones that regulate reproductive functions.

757

DISORDERS: HOMEOSTATIC IMBALANCES

Hypertension

Hypertension, or persistently high blood pressure, is defined as systolic blood pressure of 140 mmHg or greater and diastolic blood pressure of 90 mmHg or greater. Recall that a blood pressure of 120/80 is normal and desirable in a healthy adult. In industrialized societies, hypertension is the most common disorder affecting the heart and blood vessels; it is a major cause of heart failure, kidney disease, and stroke. Blood pressure values for adults are classified as follows:

Optimal	Systolic less than 120 mmHg; diastolic less than 80 mmHg
Normal	Systolic less than 130 mmHg; diastolic less than 85 mmHg
High-normal	Systolic 130–139 mmHg; diastolic 85–89 mmHg
Hypertension	Systolic 140 mmHg or greater; diastolic 90 mmHg or greater
Stage 1	Systolic 140–159 mmHg; diastolic 90–99 mmHg
Stage 2	Systolic 160–179 mmHg; diastolic 100–109 mmHg
Stage 3	Systolic 180 mmHg or greater; diastolic 110 mmHg or greater

Types and Causes of Hypertension

Between 90% and 95% of all cases of hypertension are **primary hypertension,** which is a persistently elevated blood pressure that cannot be attributed to any identifiable cause. The remaining 5–10% of cases are **secondary hypertension,** which has an identifiable underlying cause. Several disorders cause secondary hypertension:

- *Obstruction of renal blood flow* or disorders that damage renal tissue may cause the kidneys to release excessive amounts of renin into the blood. The resulting high level of angiotensin II causes vasoconstriction, thus increasing systemic vascular resistance.

- *Hypersecretion of aldosterone*—resulting, for instance, from a tumor of the adrenal cortex—stimulates excess reabsorption of salt and water by the kidneys, which increases the volume of body fluids.

- *Hypersecretion of epinephrine and norepinephrine* by a **pheochromocytoma** (fē-ō-krō′-mō-sī-TŌ-ma), a tumor of the adrenal medulla. Epinephrine and norepinephrine increase heart rate and contractility and increase systemic vascular resistance.

Damaging Effects of Untreated Hypertension

High blood pressure is known as the "silent killer" because it can cause considerable damage to the blood vessels, heart, brain, and kidneys before it causes pain or other noticeable symptoms. It is a major risk factor for the number-one (heart disease) and number-three (stroke) causes of death in the United States. In blood vessels, hypertension causes thickening of the tunica media, accelerates development of atherosclerosis and coronary artery disease, and increases systemic vascular resistance. In the heart, hypertension increases the afterload, which forces the ventricles to work harder to eject blood.

The normal response to an increased workload due to vigorous and regular exercise is hypertrophy of the myocardium, especially in the wall of the left ventricle. An increased afterload, however, leads to myocardial hypertrophy that is accompanied by muscle damage and fibrosis (a buildup of collagen fibers between the muscle fibers). As a result, the left ventricle enlarges, weakens, and dilates. Because arteries in the brain are usually less protected by surrounding tissues than are the major arteries in other parts of the body, prolonged hypertension can eventually cause them to rupture, and a stroke occurs due to a brain hemorrhage. Hypertension also damages kidney arterioles, causing them to thicken, which narrows the lumen; because the blood supply to the kidneys is thereby reduced, the kidneys secrete more renin, which elevates the blood pressure even more.

Lifestyle Changes to Reduce Hypertension

Although several categories of drugs (described next) can reduce elevated blood pressure, the following lifestyle changes are also effective in managing hypertension:

- *Lose weight.* This is the best treatment for high blood pressure short of using drugs. Loss of even a few pounds helps reduce blood pressure in overweight hypertensive individuals.

- *Limit alcohol intake.* Drinking in moderation may lower the risk of coronary heart disease, mainly among males over 45 and females over 55. Moderation is defined as no more than one 12-oz beer per day for females and no more than two 12-oz beers per day for males.

- *Exercise.* Becoming more physically fit by engaging in moderate activity (such as brisk walking) several times a week for 30 to 45 minutes can lower systolic blood pressure by about 10 mmHg.

- *Reduce intake of sodium (salt).* Roughly half the people with hypertension are "salt sensitive." For them, a high-salt diet appears to promote hypertension, and a low-salt diet can lower their blood pressure.

- *Maintain recommended dietary intake of potassium, calcium, and magnesium.* Higher levels of potassium, calcium, and magnesium in the diet are associated with a lower risk of hypertension.

- *Don't smoke.* Smoking has devastating effects on the heart and can augment the damaging effects of high blood pressure by promoting vasoconstriction.

- *Manage stress.* Various meditation and biofeedback techniques help some people reduce high blood pressure. These methods may work by decreasing the daily release of epinephrine and norepinephrine by the adrenal medulla.

Drug Treatment of Hypertension

Drugs having several different mechanisms of action are effective in lowering blood pressure. Many people are successfully treated with *diuretics,* agents that decrease blood pressure by decreasing blood volume because they increase elimination of water and salt in the urine. *ACE (angiotensin converting enzyme) inhibitors* block formation of angiotensin II and thereby promote vasodilation and decrease the liberation of aldosterone. *Beta blockers* reduce blood pressure by inhibiting the secretion of renin and by decreasing heart rate and contractility. *Vasodilators* relax the smooth muscle in arterial walls, causing vasodilation and lowering blood pressure by lowering systemic vascular resistance. An important category of vasodilators are the *calcium channel blockers,* which slow the inflow of Ca^{2+} into vascular smooth muscle cells. They reduce the heart's workload by slowing Ca^{2+} entry into pacemaker cells and regular myocardial fibers, thereby decreasing heart rate and the force of myocardial contraction.

MEDICAL TERMINOLOGY

Aneurysm (AN-ū-rizm) A thin, weakened section of the wall of an artery or a vein that bulges outward, forming a balloonlike sac. Common causes are atherosclerosis, syphilis, congenital blood vessel defects, and trauma. If untreated, the aneurysm enlarges and the blood vessel wall becomes so thin that it bursts. The result is massive hemorrhage with shock, severe pain, stroke, or death.

Angiogenesis (an′-jē-ō-JEN-e-sis) Formation of new blood vessels.

Aortography (a′-or-TOG-ra-fē) X-ray examination of the aorta and its main branches after injection of a radiopaque dye.

Arteritis (ar′-te-RĪ-tis; *-itis* = inflammation of) Inflammation of an artery, probably due to an autoimmune response.

Carotid endarterectomy (ka-ROT-id end′-ar-ter-EK-tō-mē) The removal of atherosclerotic plaque from the carotid artery to restore greater blood flow to the brain.

Claudication (klaw′-di-KĀ-shun) Pain and lameness or limping caused by defective circulation of the blood in the vessels of the limbs.

Deep venous thrombosis The presence of a thrombus (blood clot) in a deep vein of the lower limbs. It may lead to (1) pulmonary embolism, if the thrombus dislodges and then lodges within the pulmonary arterial blood flow, and (2) postphlebitic syndrome, which consists of edema, pain, and skin changes due to destruction of venous valves.

Hypotension (hī-pō-TEN-shun) Low blood pressure; most commonly used to describe an acute drop in blood pressure, as occurs during excessive blood loss.

Normotensive (nor′-mō-TEN-siv) Characterized by normal blood pressure.

Occlusion (ō-KLOO-shun) The closure or obstruction of the lumen of a structure such as a blood vessel. An example is an atherosclerotic plaque in an artery.

Orthostatic hypotension (or′-thō-STAT-ik; *ortho-* = straight; *-static* = causing to stand) An excessive lowering of systemic blood pressure when a person assumes an erect or semi-erect posture; it is usually a sign of a disease. May be caused by excessive fluid loss, certain drugs, and cardiovascular or neurogenic factors. Also called **postural hypotension.**

Phlebitis (fle-BĪ-tis; *phleb-* = vein) Inflammation of a vein, often in a leg.

Thrombectomy (throm-BEK-tō-mē; *thrombo-* = clot) An operation to remove a blood clot from a blood vessel.

Thrombophlebitis (throm′-bō-fle-BĪ-tis) Inflammation of a vein involving clot formation. Superficial thrombophlebitis occurs in veins under the skin, especially in the calf.

White coat (office) hypertension A stress-induced syndrome found in patients who have elevated blood pressure when being examined by health-care personnel, but otherwise have normal blood pressure.

STUDY OUTLINE

STRUCTURE AND FUNCTION OF BLOOD VESSELS (p. 697)

1. Arteries carry blood away from the heart. The wall of an artery consists of a tunica interna, a tunica media (which maintains elasticity and contractility), and a tunica externa.
2. Large arteries are termed elastic (conducting) arteries, and medium-sized arteries are called muscular (distributing) arteries.
3. Many arteries anastomose: The distal ends of two or more vessels unite. An alternate blood route from an anastomosis is called collateral circulation. Arteries that do not anastomose are called end arteries.
4. Arterioles are small arteries that deliver blood to capillaries.
5. Through constriction and dilation, arterioles assume a key role in regulating blood flow from arteries into capillaries and in altering arterial blood pressure.
6. Capillaries are microscopic blood vessels through which materials are exchanged between blood and tissue cells; some capillaries are continuous, whereas others are fenestrated.
7. Capillaries branch to form an extensive network throughout a tissue. This network increases surface area, allowing a rapid exchange of large quantities of materials.
8. Precapillary sphincters regulate blood flow through capillaries.
9. Microscopic blood vessels in the liver are called sinusoids.
10. Venules are small vessels that continue from capillaries and merge to form veins.
11. Veins consist of the same three tunics as arteries but have a thinner tunica interna and media. The lumen of a vein is also larger than that of a comparable artery.
12. Veins contain valves to prevent backflow of blood.
13. Weak valves can lead to varicose veins.
14. Vascular (venous) sinuses are veins with very thin walls.
15. Systemic veins are collectively called blood reservoirs because they hold a large volume of blood. If the need arises, this blood can be shifted into other blood vessels through vasoconstriction of veins.
16. The principal blood reservoirs are the veins of the abdominal organs (liver and spleen) and skin.

CAPILLARY EXCHANGE (p. 703)

1. Substances enter and leave capillaries by diffusion, transcytosis, or bulk flow.
2. The movement of water and solutes (except proteins) through capillary walls depends on hydrostatic and osmotic pressures.
3. The near equilibrium between filtration and reabsorption in capillaries is called Starling's law of the capillaries.
4. Edema is an abnormal increase in interstitial fluid.

HEMODYNAMICS: FACTORS AFFECTING BLOOD FLOW (p. 705)

1. The velocity of blood flow is inversely related to the cross-sectional area of blood vessels; blood flows slowest where cross-sectional area is greatest.
2. The velocity of blood flow decreases from the aorta to arteries to capillaries and increases in venules and veins.
3. Blood pressure and resistance determine blood flow.
4. Blood flows from regions of higher to lower pressure. The higher the resistance, however, the lower the blood flow.
5. Cardiac output equals the mean arterial blood pressure divided by total resistance (CO = MABP ÷ R).
6. Blood pressure is the pressure exerted on the walls of a blood vessel.
7. Factors that affect blood pressure are cardiac output, blood volume, viscosity, resistance, and the elasticity of arteries.
8. As blood leaves the aorta and flows through the systemic circulation, its pressure progressively falls to 0 mmHg by the time it reaches the right ventricle.
9. Resistance depends on blood vessel diameter, blood viscosity, and total blood vessel length.
10. Venous return depends on pressure differences between the venules and the right ventricle.
11. Blood return to the heart is maintained by several factors, including skeletal muscular contractions, valves in veins (especially in the limbs), and pressure changes associated with breathing.

CONTROL OF BLOOD PRESSURE AND BLOOD FLOW (p. 709)

1. The cardiovascular (CV) center is a group of neurons in the medulla oblongata that regulates heart rate, contractility, and blood vessel diameter.
2. The cardiovascular center receives input from higher brain regions and sensory receptors (baroreceptors and chemoreceptors).
3. Output from the cardiovascular center flows along sympathetic and parasympathetic axons. Sympathetic impulses propagated along cardioaccelerator nerves increase heart rate and contractility, whereas parasympathetic impulses propagated along vagus nerves decrease heart rate.
4. Baroreceptors monitor blood pressure, and chemoreceptors monitor blood levels of O_2, CO_2, and hydrogen ions. The carotid sinus reflex helps regulate blood pressure in the brain. The aortic reflex regulates general systemic blood pressure.
5. Hormones that help regulate blood pressure are epinephrine, norepinephrine, ADH (vasopressin), angiotensin II, and ANP.
6. Autoregulation refers to local, automatic adjustments of blood flow in a given region to meet a particular tissue's need.
7. O_2 level is the principal stimulus for autoregulation.

CHECKING CIRCULATION (p. 713)

1. Pulse is the alternate expansion and elastic recoil of an artery wall with each heartbeat. It may be felt in any artery that lies near the surface or over a hard tissue.
2. A normal resting pulse (heart) rate is 70–80 beats/min.
3. Blood pressure is the pressure exerted by blood on the wall of an artery when the left ventricle undergoes systole and then diastole. It is measured by the use of a sphygmomanometer.

4. Systolic blood pressure (SBP) is the arterial blood pressure during ventricular contraction. Diastolic blood pressure (DBP) is the arterial blood pressure during ventricular relaxation. Normal blood pressure is 120/80 mmHg.
5. Pulse pressure is the difference between systolic and diastolic blood pressure. It normally is about 40 mmHg.

SHOCK AND HOMEOSTASIS (p. 715)

1. Shock is a failure of the cardiovascular system to deliver enough O_2 and nutrients to meet the metabolic needs of cells.
2. Types of shock include hypovolemic, cardiogenic, vascular, and obstructive.
3. Signs and symptoms of shock include systolic blood pressure less than 90 mmHg; rapid resting heart rate; weak, rapid pulse; clammy, cool, pale skin; sweating; hypotension; altered mental state; decreased urinary output; thirst; and acidosis.

CIRCULATORY ROUTES (p. 717)

1. The two main circulatory routes are the systemic and pulmonary circulations.
2. Among the subdivisions of the systemic circulation are the coronary (cardiac) circulation and the hepatic portal circulation.
3. The systemic circulation carries oxygenated blood from the left ventricle through the aorta to all parts of the body, including some lung tissue, but *not* the air sacs (alveoli) of the lungs, and returns the deoxygenated blood to the right atrium.
4. The aorta is divided into the ascending aorta, the arch of the aorta, and the descending aorta. Each section gives off arteries that branch to supply the whole body.
5. Blood returns to the heart through the systemic veins. All veins of the systemic circulation drain into the superior or inferior venae cavae or the coronary sinus, which, in turn, empty into the right atrium.
6. The major blood vessels of the systemic circulation may be reviewed in Exhibits 21.1–21.12.
7. The hepatic portal circulation directs venous blood from the gastrointestinal organs and spleen into the hepatic portal vein of the liver before it returns to the heart. It enables the liver to utilize nutrients and detoxify harmful substances in the blood.
8. The pulmonary circulation takes deoxygenated blood from the right ventricle to the alveoli within the lungs and returns oxygenated blood from the alveoli to the left atrium.
9. Fetal circulation exists only in the fetus. It involves the exchange of materials between fetus and mother via the placenta.
10. The fetus derives O_2 and nutrients from and eliminates CO_2 and wastes into maternal blood. At birth, when pulmonary (lung), digestive, and liver functions begin, the special structures of fetal circulation are no longer needed.

DEVELOPMENT OF BLOOD VESSELS AND BLOOD (p. 756)

1. Blood vessels develop from angioblasts, which form isolated masses called blood islands.
2. Blood is produced outside the embryo by the endothelium of blood vessels of the yolk sac, chorion, and connecting stalk at about three weeks. Within the embryo, blood is produced by the liver at about the fifth week and in the spleen, red bone marrow, and thymus at about the twelfth week.

AGING AND THE CARDIOVASCULAR SYSTEM (p. 756)

1. General changes associated with aging include reduced compliance of blood vessels, reduction in cardiac muscle size, reduced cardiac output, and increased systolic blood pressure.

2. The incidence of coronary artery disease (CAD), congestive heart failure (CHF), and atherosclerosis increases with age.

SELF-QUIZ QUESTIONS

Fill in the blanks in the following statements.

1. The _____ reflex helps maintain normal blood pressure in the brain; the _____ reflex governs general systemic blood pressure.

2. In addition to the pressure created by contraction of the left ventricle, venous return is aided by the _____ and the _____, both of which depend on the presence of valves in the veins.

Indicate whether the following statements are true or false.

3. The most important method of capillary exchange is simple diffusion.

4. The overall function of the cardiovascular system is to ensure adequate circulation of blood to all body tissues to enable capillary exchange between blood plasma, interstitial fluid, and tissue cells.

Choose the one best answer to the following questions.

5. Which of the following pairs are mismatched? (a) muscular arteries: conducting arteries, (b) arterioles: resistance vessels, (c) elastic arteries: pressure reservoirs, (d) capillaries: exchange vessels, (e) veins: blood reservoirs.

6. The main vessels that permit the exchange of nutrients and wastes between the blood and tissue cells through interstitial fluid are (a) arteries, (b) arterioles, (c) capillaries, (d) venules, (e) veins.

7. Which of the following are causes of edema? (1) increased blood colloid osmotic pressure, (2) increased blood hydrostatic pressure in capillaries, (3) decreased plasma protein concentration, (4) increased capillary permeability, (5) blockage of lymph vessels. (a) 1, 2, and 3, (b) 2, 3, and 4, (c) 3, 4, and 5, (d) 1, 3, 4, and 5, (e) 2, 3, 4, and 5.

8. Systemic vascular resistance depends on which of the following factors? (1) blood viscosity, (2) total blood vessel length, (3) size of lumen, (4) type of blood vessel, (5) oxygen concentration of the blood. (a) 1, 2, and 3, (b) 2, 3, and 4, (c) 3, 4, and 5, (d) 1, 3, and 5, (e) 2, 4, and 5.

9. Which of the following help regulate blood pressure and help control regional blood flow? (1) baroreceptor and chemoreceptor reflexes, (2) hormones, (3) autoregulation, (4) H^+ concentration of blood, (5) oxygen concentration of the blood. (a) 1, 2, and 4, (b) 2, 4, and 5, (c) 1, 4, and 5, (d) 1, 2, 3, 4, and 5, (e) 3, 4, and 5.

10. Which of the following could cause vasoconstriction? (1) atrial natriuretic peptide, (2) angiotensin II, (3) lactic acid, (4) increased sympathetic stimulation to smooth muscle, (5) increase in body temperature. (a) 1, 2, 3, 4, and 5, (b) 1, 2, 3, and 4, (c) 2, 4, and 5, (d) 1, 3, and 4, (e) 2 and 4.

11. Substances enter and leave capillaries by (1) diffusion, (2) bulk flow, (3) vasomotion, (4) transcytosis, (5) filtration. (a) 1, 2, 3, 4, and 5, (b) 1, 4, and 5, (c) 1, 2, 4, and 5, (d) 2, 3, and 4, (e) 2, 4, and 5.

12. Match the following:
_____ (a) pressure generated by the pumping of the heart; pushes fluid out of capillaries
_____ (b) pressure created by proteins present in the interstitial fluid; pulls fluid out of capillaries
_____ (c) balance of pressure; determines whether blood volume and interstitial fluid remain steady or changes
_____ (d) force due to presence of plasma proteins; pulls fluid into capillaries from interstitial spaces
_____ (e) pressure due to fluid in interstitial spaces; pushes fluid back into capillaries

(1) net filtration pressure
(2) blood hydrostatic pressure
(3) interstitial fluid hydrostatic pressure
(4) blood colloid osmotic pressure
(5) interstitial fluid osmotic pressure

13. Match the following:
_____ (a) supplies blood to the kidney
_____ (b) drains blood from the small intestine, portions of the large intestine, stomach, and pancreas
_____ (c) supply and drain blood from the heart muscle
_____ (d) supply blood to the lower limbs
_____ (e) drain oxygenated blood from the lungs and carry it to the left atrium
_____ (f) supplies blood to the stomach, liver, and pancreas
_____ (g) supply blood to the brain
_____ (h) supplies blood to the large intestine
_____ (i) drain blood from the head
_____ (j) detours venous blood from the gastrointestinal organs and spleen through the liver before it returns to the heart
_____ (k) drain most of the thorax and abdominal wall; can serve as a bypass for the inferior vena cava
_____ (l) a part of the venous circulation of the leg; a vessel used in heart bypass surgery

(1) superior mesenteric vein
(2) inferior mesenteric artery
(3) pulmonary veins
(4) coronary vessels
(5) hepatic portal circulation
(6) carotid arteries
(7) jugular veins
(8) celiac trunk
(9) common iliac arteries
(10) azygous veins
(11) renal artery
(12) saphenous vein

14. Match the following:
(1) shock (2) pulse (3) tachycardia (4) bradycardia
(5) systolic blood pressure (6) diastolic blood pressure
_____ (a) a traveling pressure wave created by the alternate expansion and recoil of elastic arteries after each systole of the left ventricle
_____ (b) the lowest blood pressure in arteries during ventricular relaxation
_____ (c) a slow resting heart rate or pulse rate
_____ (d) an inadequate cardiac output that results in a failure of the cardiovascular system to deliver enough oxygen and nutrients to meet the metabolic needs of body cells
_____ (e) a rapid resting heart rate or pulse rate
_____ (f) the highest force with which blood pushes against arterial walls as a result of ventricular contraction

15. Match the following:
(1) ductus venosus (2) ductus arteriosus (3) foramen ovale
(4) umbilical arteries (5) umbilical vein
_____ (a) returns oxygenated blood from the placenta
_____ (b) an opening in the septum between the right and left atria
_____ (c) becomes the ligamentum venosum after birth
_____ (d) passes blood from the fetus to the placenta
_____ (e) bypasses the nonfunctioning lungs; becomes the ligamentum arteriosum at birth
_____ (f) become the medial umbilical ligaments at birth
_____ (g) transports oxygenated blood into the inferior vena cava
_____ (h) becomes the ligamentum teres at birth

CRITICAL THINKING QUESTIONS

1. Which structures that are present in the fetal circulation are absent in the adult circulatory system? Why do these changes occur?
HINT *A mother may want to do everything for her baby after he's born, but she can't breathe for him.*

2. Josef is thinking about the route that his blood follows on its way from his heart to his hands. He realizes that the route is a bit different for blood flowing to the left as compared to the right hand. Trace the pathway of blood from the heart to the right and left hands.
HINT *The lack of symmetry is closer to the heart than to the hand.*

3. Chris spent the week before the exam lying on a beach studying the back of his eyelids. "What was that question about sinuses doing on the cardiovascular system test?" he complained. How would you enlighten Chris?
HINT *In bone, sinuses are filled with air; these aren't.*

ANSWERS TO FIGURE QUESTIONS

21.1 The femoral artery has the thicker wall; the femoral vein has the wider lumen.

21.2 Due to atherosclerosis, less energy is stored in the less-compliant elastic arteries during systole; thus, the heart must pump harder to maintain the same rate of blood flow.

21.3 Metabolically active tissues use O_2 and produce wastes more rapidly than inactive tissues.

21.4 Materials cross capillary walls through intercellular clefts and fenestrations, via transcytosis in pinocytic vesicles, and through the plasma membranes of endothelial cells.

21.5 When you are standing, gravity causes pooling of blood in the veins of the limbs. The valves prevent backflow as the blood proceeds toward the right atrium after each heartbeat. When you are erect, gravity aids the flow of blood in neck veins back toward the heart.

21.6 Blood volume in veins is about 60% of 5 liters, or 3 liters; blood volume in capillaries is about 5% of 5 liters, or 250 mL.

21.7 Blood colloid osmotic pressure is lower than normal in a person with a low level of plasma proteins, and therefore capillary reabsorption is low. The result is edema (see page 705).

21.8 Mean blood pressure in the aorta is closer to diastolic than to systolic pressure; here it is about 93 mmHg.

21.9 The skeletal muscle and respiratory pumps aid venous return.

21.10 Vasodilation and vasoconstriction of arterioles is the main regulator of systemic vascular resistance.

21.11 Velocity of blood flow is fastest in the aorta and arteries.

21.12 The effector tissues regulated by the cardiovascular center are cardiac muscle in the heart and smooth muscle in blood vessel walls.

21.13 Impulses to the cardiovascular center pass from baroreceptors in the carotid sinuses via the glossopharyngeal nerves (cranial nerve IX) and from baroreceptors in the arch of the aorta via the vagus nerves (cranial nerve X).

21.14 It represents a change that occurs when you stand up because gravity causes pooling of blood in leg veins once you are upright, decreasing the blood pressure in your upper body.

21.15 Diastolic blood pressure = 95 mmHg; systolic blood pressure = 142 mmHg; pulse pressure = 47 mmHg. This person has stage I hypertension because the systolic blood pressure is greater than 140 mmHg and the diastolic blood pressure is greater than 90 mmHg.

21.16 Not necessarily; if systemic vascular resistance has increased greatly, tissue perfusion may be inadequate.

21.17 The two main circulatory routes are the systemic and pulmonary circulations.

21.18 The subdivisions of the aorta are the ascending aorta, arch of the aorta, thoracic aorta, and abdominal aorta.

21.19 The coronary arteries arise from the ascending aorta.

21.20 Branches of the arch of aorta are the brachiocephalic trunk, left common carotid artery, and left subclavian artery.

21.21 The thoracic aorta begins at the level of the intervertebral disc between T4 and T5.

21.22 The abdominal aorta begins at the aortic hiatus in the diaphragm.

21.23 The abdominal aorta divides into the common iliac arteries at about the level of L4.

21.24 The superior vena cava drains regions above the diaphragm, and the inferior vena cava drains regions below the diaphragm.

21.25 All venous blood in the brain drains into the internal jugular veins.

21.26 The median cubital vein is often used for withdrawing blood.

21.27 The inferior vena cava returns blood from abdominopelvic viscera to the heart.

21.28 Superficial veins of the lower limbs are the dorsal venous arch and the great saphenous and small saphenous veins.

21.29 The hepatic veins carry blood away from the liver.

21.30 The pulmonary arteries carry deoxygenated blood.

21.31 Exchange of materials between mother and fetus occurs across the placenta.

21.32 Blood vessels and blood are derived from mesoderm.

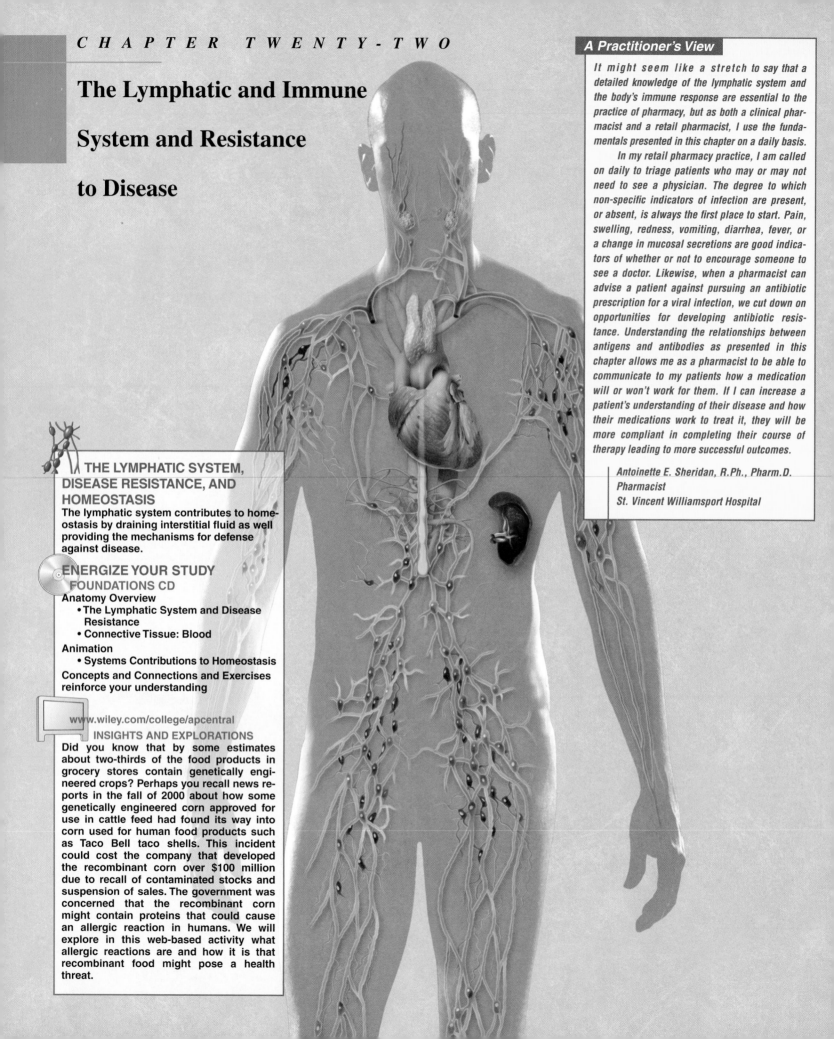

The Lymphatic and Immune System and Resistance to Disease

THE LYMPHATIC SYSTEM, DISEASE RESISTANCE, AND HOMEOSTASIS

The lymphatic system contributes to homeostasis by draining interstitial fluid as well providing the mechanisms for defense against disease.

ENERGIZE YOUR STUDY
FOUNDATIONS CD

Anatomy Overview
- The Lymphatic System and Disease Resistance
- Connective Tissue: Blood

Animation
- Systems Contributions to Homeostasis

Concepts and Connections and Exercises reinforce your understanding

www.wiley.com/college/apcentral

INSIGHTS AND EXPLORATIONS

Did you know that by some estimates about two-thirds of the food products in grocery stores contain genetically engineered crops? Perhaps you recall news reports in the fall of 2000 about how some genetically engineered corn approved for use in cattle feed had found its way into corn used for human food products such as Taco Bell taco shells. This incident could cost the company that developed the recombinant corn over $100 million due to recall of contaminated stocks and suspension of sales. The government was concerned that the recombinant corn might contain proteins that could cause an allergic reaction in humans. We will explore in this web-based activity what allergic reactions are and how it is that recombinant food might pose a health threat.

A Practitioner's View

It might seem like a stretch to say that a detailed knowledge of the lymphatic system and the body's immune response are essential to the practice of pharmacy, but as both a clinical pharmacist and a retail pharmacist, I use the fundamentals presented in this chapter on a daily basis.

In my retail pharmacy practice, I am called on daily to triage patients who may or may not need to see a physician. The degree to which non-specific indicators of infection are present, or absent, is always the first place to start. Pain, swelling, redness, vomiting, diarrhea, fever, or a change in mucosal secretions are good indicators of whether or not to encourage someone to see a doctor. Likewise, when a pharmacist can advise a patient against pursuing an antibiotic prescription for a viral infection, we cut down on opportunities for developing antibiotic resistance. Understanding the relationships between antigens and antibodies as presented in this chapter allows me as a pharmacist to be able to communicate to my patients how a medication will or won't work for them. If I can increase a patient's understanding of their disease and how their medications work to treat it, they will be more compliant in completing their course of therapy leading to more successful outcomes.

Antoinette E. Sheridan, R.Ph., Pharm.D.
Pharmacist
St. Vincent Williamsport Hospital

Maintaining homeostasis in the body requires ways to continually combat harmful agents in our environment. Despite constant exposure to a variety of **pathogens** (PATH-ō-jens), which are disease-producing microbes such as bacteria and viruses, most people remain healthy. The body surface also endures cuts and bumps, exposure to ultraviolet rays in sunlight, chemical toxins, and minor burns with an array of defensive ploys. **Resistance** is the ability to ward off damage or disease through our defenses. Vulnerability or lack of resistance is termed **susceptibility.**

The two general types of resistance are (1) nonspecific resistance or innate defenses and (2) specific resistance or immunity. **Nonspecific resistance** or **innate defenses** are present at birth and include defense mechanisms that provide *immediate* but *general* protection against invasion by a wide range of pathogens. Mechanical and chemical barriers of the skin and mucous membranes provide the first line of defense in nonspecific resistance. The acidity of gastric juice in the stomach, for example, kills many bacteria in food. **Specific resistance** or **immunity** develops in response to contact with a *particular* invader. It occurs more *slowly* than nonspecific resistance mechanisms and involves activation of specific lymphocytes that can combat a specific invader.

The body system responsible for specific resistance (and some aspects of nonspecific resistance) is the lymphatic and immune system. This system is closely allied with the cardiovascular system, and it also functions with the digestive system in the absorption of fatty foods. In this chapter, we will explore the mechanisms that provide defenses against intruders and promote the repair of damaged body tissues.

LYMPHATIC AND IMMUNE SYSTEM

> **OBJECTIVES**

- **Describe the components and major functions of the lymphatic and immune system.**
- **Describe the organization of lymphatic vessels.**
- **Describe the formation and flow of lymph.**
- **Compare the structure and functions of the primary and secondary lymphatic organs and tissues.**

The **lymphatic and immune system** (lim-FAT-ik) consists of a fluid called lymph, vessels called lymphatic vessels to transport the fluid, a number of structures and organs containing lymphatic tissue, and red bone marrow, where stem cells develop into various types of blood cells including lymphocytes (Figure 22.1). It assists in circulating body fluids and helps defend the body against disease-causing agents. As you will see shortly, most components of blood plasma filter through blood capillary walls to form interstitial fluid. After interstitial fluid passes into lymphatic vessels, it is called **lymph** (LIMF = clear fluid). Therefore, interstitial fluid and lymph are very similar; the major difference between the two is location. Whereas interstitial fluid is found between cells, lymph is located within lymphatic vessels and lymphatic tissue.

Lymphatic tissue is a specialized form of reticular connective tissue (see Table 4.3C on page 123) that contains large numbers of lymphocytes. Recall from Chapter 19 that lymphocytes are agranular white blood cells. Two types of lymphocytes participate in immune responses: *B cells* and *T cells.*

Functions of the Lymphatic and Immune System

The lymphatic and immune system has three primary functions:

1. *Draining excess interstitial fluid.* Lymphatic vessels drain excess interstitial fluid from tissue spaces and return it to the blood.

2. *Transporting dietary lipids.* Lymphatic vessels transport the lipids and lipid-soluble vitamins (A, D, E, and K) absorbed by the gastrointestinal tract to the blood.

3. *Carrying out immune responses.* Lymphatic tissue initiates highly specific responses directed against particular microbes or abnormal cells. Lymphocytes, aided by macrophages, recognize foreign cells, microbes, toxins, and cancer cells and respond to them in two basic ways: (1) In cell-mediated immune responses, T cells destroy the intruders by causing them to rupture or by releasing cytotoxic (cell-killing) substances. (2) In antibody-mediated immune responses, B cells differentiate into plasma cells that protect us against disease by producing antibodies, proteins that combine with and cause destruction of specific foreign substances.

Lymphatic Vessels and Lymph Circulation

Lymphatic vessels begin as lymphatic capillaries. These tiny vessels are closed at one end and located in the spaces between cells (Figure 22.2 on page 767). Just as blood capillaries converge to form venules and then veins, lymphatic capillaries unite to form larger lymphatic vessels (see Figure 22.1), which resemble veins in structure but have thinner walls and more valves. At intervals along the lymphatic vessels, lymph flows through lymph nodes, encapsulated masses of B cells and T cells. In the skin, lymphatic vessels lie in the subcutaneous tissue and generally follow veins; lymphatic vessels of the viscera generally follow arteries, forming plexuses (networks) around them. Tissues that lack lymphatic capillaries include avascular tissues (such as cartilage, the epidermis, and the cornea of the eye), the central nervous system, portions of the spleen, and bone marrow.

Figure 22.1 Components of the lymphatic and immune system.

The lymphatic and immune system consists of lymph, lymphatic vessels, lymphatic tissues, and red bone marrow.

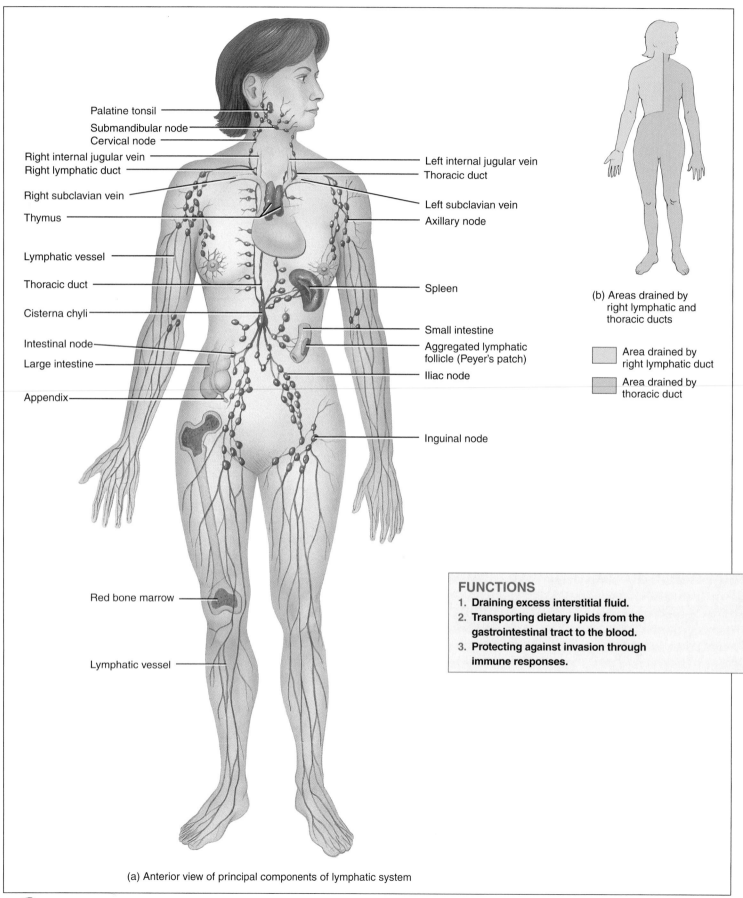

Palatine tonsil

Submandibular node

Cervical node

Right internal jugular vein

Right lymphatic duct

Right subclavian vein

Thymus

Lymphatic vessel

Thoracic duct

Cisterna chyli

Intestinal node

Large intestine

Appendix

Red bone marrow

Lymphatic vessel

Left internal jugular vein

Thoracic duct

Left subclavian vein

Axillary node

Spleen

Small intestine

Aggregated lymphatic follicle (Peyer's patch)

Iliac node

Inguinal node

(b) Areas drained by right lymphatic and thoracic ducts

Area drained by right lymphatic duct

Area drained by thoracic duct

FUNCTIONS

1. **Draining excess interstitial fluid.**
2. **Transporting dietary lipids from the gastrointestinal tract to the blood.**
3. **Protecting against invasion through immune responses.**

(a) Anterior view of principal components of lymphatic system

What tissue contains stem cells that develop into lymphocytes?

Figure 22.2 Lymphatic capillaries.

 Lymphatic capillaries are found throughout the body except in avascular tissues, the central nervous system, portions of the spleen, and bone marrow.

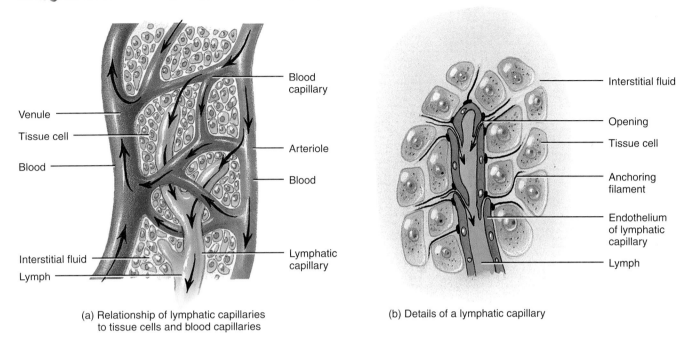

(a) Relationship of lymphatic capillaries to tissue cells and blood capillaries

(b) Details of a lymphatic capillary

Is lymph more similar to blood plasma or to interstitial fluid? Why?

Lymphatic Capillaries

Lymphatic capillaries are slightly larger in diameter than blood capillaries and have a unique structure that permits interstitial fluid to flow into them but not out. The ends of endothelial cells that make up the wall of a lymphatic capillary overlap (Figure 22.2b). When pressure is greater in the interstitial fluid than in lymph, the cells separate slightly, like the opening of a one-way swinging door, and interstitial fluid enters the lymphatic capillary. When pressure is greater inside the lymphatic capillary, the cells adhere more closely, and lymph cannot escape back into interstitial fluid. Attached to the lymphatic capillaries are *anchoring filaments,* which contain elastic fibers. They extend out from the lymphatic capillary, attaching lymphatic endothelial cells to surrounding tissues. When excess interstitial fluid accumulates and causes tissue swelling, the anchoring filaments are pulled, making the openings between cells even larger so that more fluid can flow into the lymphatic capillary.

In the small intestine, specialized lymphatic capillaries called **lacteals** (LAK-tē-als; *lact-* = milky) carry dietary lipids into lymphatic vessels and ultimately into the blood. The presence of these lipids causes the lymph draining the small intestine to appear creamy white; such lymph is referred to as **chyle** (KĪL = juice). Elsewhere, lymph is a clear, pale-yellow fluid.

Lymph Trunks and Ducts

Lymph passes from lymphatic capillaries into lymphatic vessels and then through lymph nodes. Lymphatic vessels exiting lymph nodes pass lymph either toward another node within the same group or on to another group of nodes. From the most proximal group of each chain of nodes, the exiting vessels unite to form **lymph trunks.** The principal trunks are the **lumbar, intestinal, bronchomediastinal, subclavian,** and **jugular trunks** (Figure 22.3). Lymph passes from lymph trunks into two main channels, the thoracic duct and the right lymphatic duct, and then drains into venous blood.

The **thoracic (left lymphatic) duct** is about 38–45 cm (15–18 in.) long and begins as a dilation called the **cisterna chyli** (sis-TER-na KĪ-lē; *cisterna* = cavity or reservoir) anterior to the second lumbar vertebra. The thoracic duct is the main duct for return of lymph to blood. It receives lymph from the left side of the head, neck, and chest, the left upper limb, and the entire body inferior to the ribs (see Figure 22.16). The thoracic duct drains lymph into venous blood via the **left subclavian vein.**

The cisterna chyli receives lymph from the right and left lumbar trunks and from the intestinal trunk. The lumbar trunks drain lymph from the lower limbs, the wall and viscera of the pelvis, the kidneys, the adrenal glands, and the deep lymphatic vessels that drain lymph from most of the abdominal wall. The intestinal trunk drains lymph from the stomach, intestines, pancreas, spleen, and part of the liver.

In the neck, the thoracic duct also receives lymph from the left jugular, left subclavian, and left bronchomediastinal trunks. The left jugular trunk drains lymph from the left side of the head and neck, and the left subclavian trunk drains lymph from the

Figure 22.3   Routes for drainage of lymph from lymph trunks into the thoracic and right lymphatic ducts.

 All lymph returns to the bloodstream through the thoracic (left) lymphatic duct and right lymphatic duct.

Right internal jugular vein

RIGHT JUGULAR TRUNK

RIGHT SUBCLAVIAN TRUNK

RIGHT LYMPHATIC DUCT

Right subclavian vein

Right brachiocephalic vein

RIGHT BRONCHO-MEDIASTINAL TRUNK

Superior vena cava

Rib

Intercostal muscle

Azygos vein

CISTERNA CHYLI

RIGHT LUMBAR TRUNK

Inferior vena cava

Esophagus

Trachea

Left internal jugular vein

LEFT JUGULAR TRUNK

LEFT SUBCLAVIAN TRUNK

THORACIC (LEFT LYMPHATIC) DUCT

Left brachiocephalic vein

Left subclavian vein

First rib

LEFT BRONCHO-MEDIASTINAL TRUNK

Accessory hemiazygos vein

THORACIC (LEFT LYMPHATIC) DUCT

Hemiazygos vein

LEFT LUMBAR TRUNK

INTESTINAL TRUNK

(a) Overall anterior view

Right jugular trunk   Right lymphatic duct

Right subclavian trunk   Left jugular trunk   Left subclavian trunk

Thoracic (left lymphatic) duct

Right broncho-mediastinal trunk

Left broncho-mediastinal trunk

(b) Detailed anterior view

Which lymphatic vessels empty into the cisterna chyli, and which duct receives lymph from the cisterna chyli?

left upper limb. The left bronchomediastinal trunk drains lymph from the left side of the deeper parts of the anterior thoracic wall, the superior part of the anterior abdominal wall, the anterior part of the diaphragm, the left lung, and the left side of the heart.

The **right lymphatic duct** (Figure 22.3) is about 1.2 cm (0.5 in.) long and drains lymph from the upper-right side of the body into venous blood via the **right subclavian vein** (see Figure 22.1b). Three lymphatic trunks drain into the right lymphatic duct: (1) The right jugular trunk drains the right side of the head and neck. (2) The right subclavian trunk drains the right upper

limb. (3) The right bronchomediastinal trunk drains the right side of the thorax, the right lung, the right side of the heart, and part of the liver.

Formation and Flow of Lymph

Most components of blood plasma freely filter through the capillary walls to form interstitial fluid. More fluid filters out of blood capillaries, however, than returns to them by reabsorption (see Figure 21.7 on page 704). The excess filtered fluid—about 3 liters per day—drains into lymphatic vessels and becomes lymph. Because most plasma proteins are too large to leave blood vessels, interstitial fluid contains only a small amount of protein. Proteins that do leave blood plasma, however, cannot return to the blood directly by diffusion because the concentration gradient (high level of proteins inside blood capillaries, low level outside) opposes such movement. The proteins, however, can move readily through the more permeable lymphatic capillaries into lymph. Thus, an important function of lymphatic vessels is to return lost plasma proteins to the bloodstream.

Like veins, lymphatic vessels contain valves, which ensure the one-way movement of lymph. Ultimately, lymph drains into venous blood through the right lymphatic duct and the thoracic duct at the junction of the internal jugular and subclavian veins (Figure 22.3). Thus, the sequence of fluid flow is blood capillaries (blood) → interstitial spaces (interstitial fluid) → lymphatic capillaries (lymph) → lymphatic vessels (lymph) → lymphatic ducts (lymph) → subclavian veins (blood). Figure 22.4 illustrates this sequence, as well as the relationship of the lymphatic and cardiovascular systems.

The same two "pumps" that aid return of venous blood to the heart maintain the flow of lymph.

1. **Skeletal muscle pump.** The "milking action" of skeletal muscle contractions (see Figure 21.9 on page 707) compresses lymphatic vessels (as well as veins) and forces lymph toward the subclavian veins.

2. **Respiratory pump.** Lymph flow is also maintained by pressure changes that occur during inhalation (breathing in). Lymph flows from the abdominal region, where the pressure is higher, toward the thoracic region, where it is lower. When the pressures reverse during exhalation (breathing out), the valves prevent backflow of lymph. In addition, when a lymphatic vessel distends, the smooth muscle in its wall contracts, which helps move lymph from one segment of the vessel to the next.

Edema and Lymph Flow

Edema, an excessive accumulation of interstitial fluid in tissue spaces, may be caused by an obstruction to lymph flow, such as an infected lymph node or a blocked lymphatic vessel. Edema may also result from increased capillary blood pressure, which causes excess interstitial fluid to form faster than it can pass into lymphatic vessels or be reabsorbed back into the capillaries. ■

Figure 22.4 **Schematic diagram showing the relationship of the lymphatic and immune system to the cardiovascular system.**

 The sequence of fluid flow is blood capillaries (blood) → interstitial spaces (interstitial fluid) → lymphatic capillaries (lymph) → lymphatic vessels (lymph) → lymphatic ducts (lymph) → subclavian veins (blood).

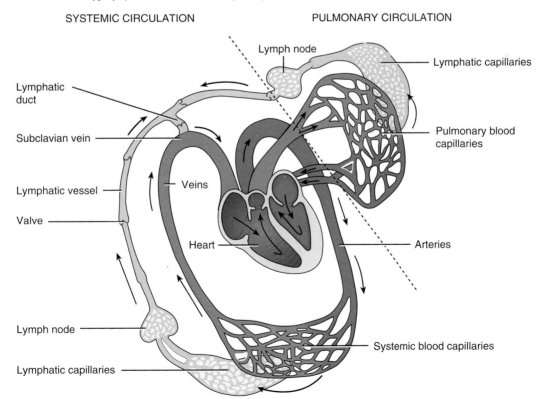

Arrows show direction of flow of lymph and blood

 Does inhalation promote or hinder the flow of lymph?

Lymphatic Organs and Tissues

Lymphatic organs and tissues, which are widely distributed throughout the body, are classified into two groups based on their functions. **Primary lymphatic organs** are the sites where stem cells divide and become **immunocompetent,** that is, capable of mounting an immune response. The primary lymphatic organs are the **red bone marrow** (in flat bones and the epiphyses of long bones of adults) and the **thymus.** Pluripotent stem cells in red bone marrow give rise to mature, immunocompetent B cells and to pre-T cells, which migrate to and become immunocompetent T cells in the thymus. The **secondary lymphatic organs** and **tissues** are the sites where most immune responses occur. They include **lymph nodes,** the **spleen,** and **lymphatic nodules (follicles).** The thymus, lymph nodes, and spleen are considered organs because each is surrounded by a connective tissue capsule; lymphatic nodules, in contrast, are not organs because they lack a capsule.

Thymus

The **thymus** is a bilobed organ located in the mediastinum between the sternum and the aorta (Figure 22.5a). An enveloping layer of connective tissue holds the two lobes closely together, but a connective tissue **capsule** encloses each lobe separately. Extensions of the capsule, called **trabeculae** (tra-BEK-ū-lē = little beams) penetrate inward and divide the lobes into **lobules** (Figure 22.5b).

Each thymic lobule consists of a deeply staining outer cortex and a lighter-staining central medulla (Figure 22.5b). The **cortex** is composed of large numbers of T cells and scattered dendritic cells, epithelial cells, and macrophages. Immature T cells (pre-T cells) migrate from red bone marrow to the cortex of the thymus, where they proliferate and begin to mature. **Dendritic cells** (*dendr-* =a tree), so-named because they have long, branched projections that resemble the dendrites of a neuron, assist the maturation process. As you will see shortly, dendritic

Figure 22.5 Thymus. (See Tortora, *A Photographic Atlas of the Human Body,* Figure 7.2.)

🔑 **The bilobed thymus is largest at puberty and then atrophies with age.**

Blood vessels

Capsule

Lobule:
Cortex

Thymic (Hassall's) corpuscle

Medulla

Trabecula

LM 30x

(b) Thymic lobules

Thyroid gland

Trachea

Right common carotid artery

Superior vena cava

Brachiocephalic veins

Thymus

Parietal pericardium

Right lung

Left lung

Diaphragm

(a) Thymus of adolescent

T cell

Thymic (Hassall's) corpuscle

Epithelial cell

LM 385x

(c) Details of the thymic medulla

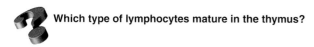

Which type of lymphocytes mature in the thymus?

cells in other parts of the body, such as lymph nodes, play another key role in immune responses. Each of the specialized **epithelial cells** in the cortex has several long processes that surround and serve as a framework for as many as 50 T cells. These epithelial cells help "educate" the pre-T cells in a process known as positive selection (see Figure 22.20). Additionally, they produce thymic hormones that are thought to aid in the maturation of T cells. Only about 2% of developing T cells survive in the cortex. The remaining cells die via apoptosis (programmed cell death). Thymic macrophages help clear out the debris of dead and dying cells. The surviving T cells enter the medulla.

The **medulla** consists of widely scattered, more mature T cells, epithelial cells, dendritic cells, and macrophages (Figure 22.5c). Some of the epithelial cells become arranged into concentric layers of flat cells that degenerate and become filled with keratohyalin granules and keratin. These clusters are called **thymic (Hassall's) corpuscles.** Although their role is uncertain, they may serve as sites of T cell death in the medulla. T cells that leave the thymus via the blood are carried to lymph nodes, the spleen, and other lymphatic tissues where they colonize parts of these organs and tissues.

In infants, the thymus is large, having a mass of about 70 g (2.3 oz). After puberty, adipose and areolar connective tissue begin to replace the thymic tissue. By the time a person reaches maturity, the gland has atrophied considerably, and in old age it may weigh only 3 g (0.1 oz). Before the thymus atrophies, it populates the secondary lymphatic organs and tissues with T cells. However, some T cells continue to proliferate in the thymus throughout an individual's lifetime.

Lymph Nodes

Located along lymphatic vessels are about 600 bean-shaped **lymph nodes.** They are scattered throughout the body, both superficially and deep, and usually occur in groups (see Figure 22.1). Large groups of lymph nodes are present near the mammary glands and in the axillae and groin. Lymph nodes are 1–25 mm (0.04–1 in.) long and are covered by a **capsule** of dense connective tissue that extends into the node (Figure 22.6). The capsular extensions, called **trabeculae,** divide the node into compartments, provide support, and provide a route for blood vessels into the interior of a node. Internal to the capsule is a supporting network of reticular fibers and fibroblasts. The capsule, trabeculae, reticular fibers, and fibroblasts constitute the *stroma,* or framework, of a lymph node.

The *parenchyma* of a lymph node is divided into a superficial cortex and a deep medulla. Within the **outer cortex** are egg-shaped aggregates of B cells called **lymphatic nodules (follicles).** A lymphatic nodule consisting chiefly of B cells is called a *primary lymphatic nodule.* Most lymphatic nodules in the outer cortex are *secondary lymphatic nodules* (Figure 22.6), which form in response to an antigenic challenge and are sites of

plasma cell and memory B cell formation. After B cells in a primary lymphatic nodule recognize an antigen, the primary lymphatic nodule develops into a secondary lymphatic nodule. The center of a secondary lymphatic nodule contains a region of light-staining cells called a *germinal center.* In the germinal center are B cells, follicular dendritic cells (a special type of dendritic cell), and macrophages. When follicular dendritic cells "present" an antigen, B cells proliferate and develop into antibody-producing plasma cells or develop into memory B cells. Memory B cells persist after an immune response and "remember" having encountered a specific antigen. B cells that do not develop properly undergo **apoptosis** (programmed cell death) and are destroyed by macrophages. The region of a secondary lymphatic nodule surrounding the germinal center is composed of dense accumulations of B cells that have migrated away from their site of origin within the nodule.

The **inner cortex,** also called the *paracortex,* does not contain lymphatic nodules. It consists mainly of T cells and dendritic cells that enter a lymph node from other tissues. The dendritic cells present antigens to T cells, causing their proliferation. The newly formed T cells then migrate from the lymph node to areas of the body where there is antigenic activity.

The **medulla** of a lymph node contains B cells, antibody-producing plasma cells that have migrated out of the cortex into the medulla, and macrophages. The various cells are embedded in a network of reticular fibers and reticular cells.

Lymph flows through a node in one direction only (Figure 22.6a). It enters through **afferent lymphatic vessels** (*afferent* = to carry toward), which penetrate the convex surface of the node at several points. The afferent vessels contain valves that open toward the center of the node, such that the lymph is directed *inward.* Within the node, lymph enters **sinuses,** which are a series of irregular channels that contain branching reticular fibers, lymphocytes, and macrophages. From the afferent lymphatic vessels, lymph flows into the **subcapsular sinus,** immediately beneath the capsule. Then lymph flows through **trabecular sinuses,** which extend through the cortex parallel to the trabeculae, and into **medullary sinuses,** which extend through the medulla. The medullary sinuses drain into one or two **efferent lymphatic vessels** (*efferent* = to carry away), which are wider than afferent vessels and fewer in number. They contain valves that open away from the center of the node to convey lymph, antibodies secreted by plasma cells, and activated T cells *out* of the node. Efferent lymphatic vessels emerge from one side of the lymph node at a slight depression called a **hilus** (HĪ-lus). Blood vessels also enter and leave the node at the hilus.

Lymph nodes function to filter lymph. As lymph enters one end of a node, foreign substances are trapped by the reticular fibers within the sinuses of the node. Then macrophages destroy some foreign substances by phagocytosis while lymphocytes destroy others by immune responses. Filtered lymph then leaves the other end of the node.

Figure 22.6   Structure of a lymph node.   Arrows indicate the direction of lymph flow through a lymph node. (See Tortora, *A Photographic Atlas of the Human Body,* Figure 7.4.)

🔑 **Lymph nodes are present throughout the body, usually clustered in groups.**

Outer Cortex

Cells of inner cortex
T cells
Dendritic cells

Cells around germinal center
B cells

Cells in germinal center
B cells
Follicular dendritic cells
Macrophages

Cells of medulla
B cells
Plasma cells
Macrophages

Afferent lymphatic vessel
Valve

Subcapsular sinus
Reticular fiber
Trabecula
Trabecular sinus
Outer cortex:
Germinal center in secondary lymphatic nodule
Cells around germinal center
Inner cortex
Medulla
Medullary sinus
Reticular fiber

Efferent lymphatic vessels

Valve

Hilus

Capsule

Afferent lymphatic vessel

Route of lymph flow through a lymph node:
Afferent lymphatic vessel
↓
Subcapsular sinus
↓
Trabecular sinus
↓
Medullary sinus
↓
Efferent lymphatic vessel

(a) Partially sectioned lymph node showing overall structure

Capsule
Subcapsular sinus
Trabecula
Trabecular sinus
Outer cortex
Germinal center in secondary lymphatic nodule
Inner cortex
Medullary sinus
Medulla

LM 55x

(b) Portion of a lymph node

What happens to foreign substances that enter a lymph node in lymph?

Metastasis Through Lymphatic Vessels

Metastasis (me-TAS-ta-sis; *meta-* = beyond; *stasis* = to stand), the spread of a disease from one part of the body to another, can occur via lymphatic vessels. All malignant tumors eventually exhibit metastasis. Cancer cells may travel in the blood or lymph and establish new tumors where they lodge. When metastasis occurs via lymphatic vessels, secondary tumor sites can be predicted according to the direction of lymph flow from the primary tumor site. Cancerous lymph nodes feel enlarged, firm, nontender, and fixed to underlying structures. By contrast, most lymph nodes that are enlarged due to an infection are not firm, are moveable, and are very tender. ■

Spleen

The oval **spleen** is the largest single mass of lymphatic tissue in the body, measuring about 12 cm (5 in.) in length (Figure 22.7a). It is located in the left hypochondriac region between the stomach and diaphragm. The superior surface of the spleen is smooth and convex and conforms to the concave surface of the

---

Figure 22.7 Structure of the spleen. (See Tortora, *A Photographic Atlas of the Human Body,* Figure 7.3.)

The spleen is the largest single mass of lymphatic tissue in the body.

SUPERIOR

Splenic artery
Gastric impression

POSTERIOR

Splenic vein
Colic impression
Hilus
Renal impression

ANTERIOR

INFERIOR

(a) Visceral surface

Splenic artery
Splenic vein
White pulp
Red pulp:
  Venous sinus
  Splenic cord
Central artery
Trabecula
Capsule

(b) Internal structure

Capsule
Red pulp
White pulp
Central artery
Trabecula

LM 66x

(c) Portion of the spleen

After birth, what are the main functions of the spleen?

diaphragm. Neighboring organs make indentations in the visceral surface of the spleen—the gastric impression (stomach), the renal impression (left kidney), and the colic impression (left flexure of colon). Like lymph nodes, the spleen has a hilus. Through it pass the splenic artery, splenic vein, and efferent lymphatic vessels.

A capsule of dense connective tissue surrounds the spleen and is covered, in turn, by a serous membrane, the visceral peritoneum. Trabeculae extend inward from the capsule. The capsule plus trabeculae, reticular fibers, and fibroblasts constitute the stroma of the spleen; the parenchyma of the spleen consists of two different kinds of tissue called white pulp and red pulp (Figure 22.7b, c). **White pulp** is lymphatic tissue, consisting mostly of lymphocytes and macrophages arranged around branches of the splenic artery called **central arteries.** The **red pulp** consists of blood-filled **venous sinuses** and cords of splenic tissue called **splenic (Billroth's) cords.** Splenic cords consist of red blood cells, macrophages, lymphocytes, plasma cells, and granulocytes. Veins are closely associated with the red pulp.

Blood flowing into the spleen through the splenic artery enters the central arteries of the white pulp. Within the white pulp, B cells and T cells carry out immune functions, similar to lymph nodes, while spleen macrophages destroy blood-borne pathogens by phagocytosis. Within the red pulp, the spleen performs three functions related to blood cells: (1) removal by macrophages of ruptured, worn out, or defective blood cells and platelets; (2) storage of platelets, up to one-third of the body's supply; and (3) production of blood cells (hemopoiesis) during fetal life.

Ruptured Spleen

The spleen is the organ most often damaged in cases of abdominal trauma. Severe blows over the inferior left chest or superior abdomen can fracture the protecting ribs. Such crushing injury may rupture the spleen, which causes severe intraperitoneal hemorrhage and shock. Prompt removal of the spleen, called a **splenectomy,** is needed to prevent death due to bleeding. Other structures, particularly red bone marrow and the liver, can take over some functions normally carried out by the spleen. Immune functions, however, decrease in the absence of a spleen. The spleen's absence also places the patient at higher risk for sepsis (a blood infection) due to loss of the filtering and phagocytic functions of the spleen. To reduce the risk of sepsis, patients with a history of a splenectomy take prophylactic antibiotics before any invasive procedures. ■

Lymphatic Nodules

Lymphatic nodules are egg-shaped masses of lymphatic tissue that are not surrounded by a capsule. Because they are scattered throughout the lamina propria (connective tissue) of mucous membranes lining the gastrointestinal, urinary, and reproductive tracts and the respiratory airways, lymphatic nodules in these areas are also referred to as **mucosa-associated lymphatic tissue (MALT).**

Although many lymphatic nodules are small and solitary, some occur in multiple large aggregations in specific parts of the body. Among these are the tonsils in the pharyngeal region and the aggregated lymphatic follicles (Peyer's patches) in the ileum of the small intestine. Aggregations of lymphatic nodules also occur in the appendix. Usually there are five **tonsils,** which form a ring at the junction of the oral cavity and oropharynx and at the junction of the nasal cavity and nasopharynx (see Figure 23.2b on page 808). The tonsils are strategically positioned to participate in immune responses against inhaled or ingested foreign substances. The single **pharyngeal tonsil** (fa-RIN-jē-al) or **adenoid** is embedded in the posterior wall of the nasopharynx. The two **palatine tonsils** (PAL-a-tīn) lie at the posterior region of the oral cavity, one on either side; these are the tonsils commonly removed in a tonsillectomy. The paired **lingual tonsils** (LIN-gwal), located at the base of the tongue, may also require removal during a tonsillectomy.

► CHECKPOINT

1. How are interstitial fluid and lymph similar, and how do they differ?
2. How do lymphatic vessels differ in structure from veins?
3. Construct a diagram that shows the route of lymph circulation.
4. What is the role of the thymus in immunity?
5. What functions do lymph nodes serve?
6. Describe the functions of the spleen and tonsils.

DEVELOPMENT OF LYMPHATIC TISSUES

► OBJECTIVE

- **Describe the development of lymphatic tissues.**

Lymphatic tissues begin to develop by the end of the fifth week of embryonic life. *Lymphatic vessels* develop from **lymph sacs** that arise from developing veins, which are derived from **mesoderm.**

The first lymph sacs to appear are the paired **jugular lymph sacs** at the junction of the internal jugular and subclavian veins (Figure 22.8). From the jugular lymph sacs, lymphatic capillary plexuses spread to the thorax, upper limbs, neck, and head. Some of the plexuses enlarge and form lymphatic vessels in their respective regions. Each jugular lymph sac retains at least one connection with its jugular vein, the left one developing into the superior portion of the thoracic duct (left lymphatic duct).

The next lymph sac to appear is the unpaired **retroperitoneal lymph sac** at the root of the mesentery of the intestine. It

Figure 22.8   Development of lymphatic tissues.

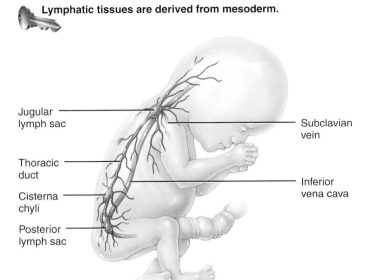

Lymphatic tissues are derived from mesoderm.

Jugular lymph sac

Subclavian vein

Thoracic duct

Cisterna chyli

Inferior vena cava

Posterior lymph sac

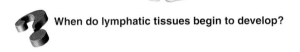

When do lymphatic tissues begin to develop?

develops from the primitive vena cava and mesonephric (primitive kidney) veins. Capillary plexuses and lymphatic vessels spread from the retroperitoneal lymph sac to the abdominal viscera and diaphragm. The sac establishes connections with the cisterna chyli but loses its connections with neighboring veins.

At about the time the retroperitoneal lymph sac is developing, another lymph sac, the **cisterna chyli,** develops inferior to the diaphragm on the posterior abdominal wall. It gives rise to the inferior portion of the *thoracic duct* and the *cisterna chyli* of the thoracic duct. Like the retroperitoneal lymph sac, the cisterna chyli also loses its connections with surrounding veins.

The last of the lymph sacs, the paired **posterior lymph sacs,** develop from the iliac veins. The posterior lymph sacs produce capillary plexuses and lymphatic vessels of the abdominal wall, pelvic region, and lower limbs. The posterior lymph sacs join the cisterna chyli and lose their connections with adjacent veins.

With the exception of the anterior part of the sac from which the cisterna chyli develops, all lymph sacs become invaded by **mesenchymal cells** and are converted into groups of *lymph nodes.*

The *spleen* develops from **mesenchymal cells** between layers of the dorsal mesentery of the stomach. The *thymus* arises as an outgrowth of the **third pharyngeal pouch** (see Figure 18.21a on page 623).

▶ CHECKPOINT

7.   What are the names of the four lymph sacs from which lymphatic vessels develop?

NONSPECIFIC RESISTANCE: INNATE DEFENSES

▶ OBJECTIVE

• Describe the mechanisms of nonspecific resistance to disease.

Although several mechanisms contribute to innate defenses or nonspecific resistance to disease, they all have two things in common. They are present at birth, and they offer immediate protection against a wide variety of pathogens and foreign substances. Nonspecific resistance, as its name suggests, lacks specific responses to specific invaders; instead, its protective mechanisms function the same way, regardless of the type of invader. Innate defense mechanisms include the external physical and chemical barriers provided by the skin and mucous membranes. They also include various internal nonspecific defenses, such as antimicrobial proteins, natural killer cells and phagocytes, inflammation, and fever.

First Line of Defense: Skin and Mucous Membranes

The skin and mucous membranes of the body are the first line of defense against pathogens. Both physical and chemical barriers discourage pathogens and foreign substances from penetrating the body and causing disease.

With its many layers of closely packed, keratinized cells, the outer epithelial layer of the skin—the **epidermis**—provides a formidable physical barrier to the entrance of microbes (see Figure 5.1 on page 141). In addition, periodic shedding of epidermal cells helps remove microbes at the skin surface. Bacteria rarely penetrate the intact surface of healthy epidermis. If this surface is broken by cuts, burns, or punctures, however, pathogens can penetrate the epidermis and invade adjacent tissues or circulate in the blood to other parts of the body.

The epithelial layer of **mucous membranes,** which line body cavities, secretes a fluid called **mucus** that lubricates and moistens the cavity surface. Because mucus is slightly viscous, it traps many microbes and foreign substances. The mucous membrane of the nose has mucus-coated **hairs** that trap and filter microbes, dust, and pollutants from inhaled air. The mucous membrane of the upper respiratory tract contains **cilia,** microscopic hairlike projections on the surface of the epithelial cells. The waving action of cilia propels inhaled dust and microbes that have become trapped in mucus toward the throat. Coughing and sneezing accelerate movement of mucus and its entrapped pathogens out of the body.

Other fluids produced by various organs also help protect epithelial surfaces of the skin and mucous membranes. The **lacrimal apparatus** (LAK-ri-mal) of the eyes (see Figure 16.5 on page 533) manufactures and drains away tears in response to irritants. Blinking spreads tears over the surface of the eyeball, and the continual washing action of tears helps to dilute microbes and keep them from settling on the surface of the eye. **Saliva,** produced by the salivary glands, washes microbes from the surfaces of the teeth and from the mucous membrane of the mouth, much as tears wash the eyes. The flow of saliva reduces colonization of the mouth by microbes.

The cleansing of the urethra by the **flow of urine** retards microbial colonization of the urinary system. **Vaginal secretions** likewise move microbes out of the body in females. **Defecation** and **vomiting** also expel microbes. For example, in response to some microbial toxins, the smooth muscle of the lower gastrointestinal tract contracts vigorously; the resulting diarrhea rapidly expels many of the microbes.

Certain chemicals also contribute to the high degree of resistance of the skin and mucous membranes to microbial invasion. Sebaceous (oil) glands of the skin secrete an oily substance called **sebum** that forms a protective film over the surface of the skin. The unsaturated fatty acids in sebum inhibit the growth of certain pathogenic bacteria and fungi.

The acidity of the skin (pH 3–5) is caused in part by the secretion of fatty acids and lactic acid. **Perspiration** helps flush microbes from the surface of the skin, and it also contains **lysozyme,** an enzyme capable of breaking down the cell walls of certain bacteria. Lysozyme is also present in tears, saliva, nasal secretions, and tissue fluids, where it also exhibits antimicrobial activity. **Gastric juice,** produced by the glands of the stomach, is a mixture of hydrochloric acid, enzymes, and mucus. The strong acidity of gastric juice (pH 1.2–3.0) destroys many bacteria and most bacterial toxins. Vaginal secretions also are slightly acidic, which discourages bacterial growth.

Second Line of Defense: Internal Defenses

When pathogens penetrate the mechanical and chemical barriers of the skin and mucous membranes, they encounter a second line of defense: internal antimicrobial proteins, phagocytes, natural killer cells, inflammation, and fever.

Antimicrobial Proteins

Blood and interstitial fluids contain three main types of **antimicrobial proteins** that discourage microbial growth: interferons, complement, and transferrins.

1. Lymphocytes, macrophages, and fibroblasts infected with viruses produce proteins called **interferons** (in′-ter-FĒR-ons), or **IFNs.** Once released by virus-infected cells, IFNs diffuse to uninfected neighboring cells, where they induce synthesis of antiviral proteins that interfere with viral replication. Although IFNs do not prevent viruses from attaching to and penetrating host cells, they do stop replication. Viruses can cause disease only if they can replicate within body cells. IFNs are an important defense against infection by many different viruses. The three types of interferon are alpha-, beta-, and gamma-IFN.

2. A group of normally inactive proteins in blood plasma and on plasma membranes makes up the **complement system.** When activated, these proteins "complement" or enhance certain immune, allergic, and inflammatory reactions.

3. Iron-binding proteins called **transferrins** inhibit the growth of certain bacteria by reducing the amount of available iron.

Natural Killer Cells and Phagocytes

When microbes penetrate the skin and mucous membranes or bypass the antimicrobial proteins in blood, the next nonspecific defense consists of natural killer cells and phagocytes. About 5–10% of lymphocytes in the blood are **natural killer (NK) cells.** They also are present in the spleen, lymph nodes, and red bone marrow. Although NK cells lack the membrane molecules that identify B cells and T cells, they have the ability to kill a wide variety of infectious microbes and certain tumor cells. NK cells attack cells that display abnormal plasma membrane proteins called major histocompatibility complex (MHC) antigens, which are described on page 782. NK cells cause cellular destruction in at least two ways:

1. NK cells release **perforins,** chemicals that when inserted into the plasma membrane of a microbe make the membrane so leaky that cytolysis occurs.

2. Upon encountering a virus-infected or tumor cell, NK cells release molecules that enter the target cells and cause apoptosis.

Phagocytes (*phago-* = eat; *-cytes* = cells) are specialized cells that perform **phagocytosis** (*-osis* = process), the ingestion of microbes or other particles such as cellular debris (see Figure 3.14 on page 73). The two major types of phagocytes are **neutrophils** and **macrophages** (MAK-rō-fā-jez). When an infection occurs, neutrophils and monocytes migrate to the infected area. During this migration, the monocytes enlarge and develop into actively phagocytic macrophages. **Wandering macrophages** leave the blood and migrate to infected areas. Other macrophages, called **fixed macrophages,** stand guard in specific tissues. Among the fixed macrophages are histiocytes in the skin and subcutaneous layer, stellate reticuloendothelial cells (Kupffer cells) in the liver, alveolar macrophages in the lungs, microglia in the nervous system, and tissue macrophages in the spleen, lymph nodes, and red bone marrow. In addition to being an innate defense mechanism, phagocytosis plays a vital role in immunity, as discussed later in the chapter.

Phagocytosis occurs in five phases: chemotaxis, adherence, ingestion, digestion, and killing (Figure 22.9):

1 Phagocytosis begins with **chemotaxis,** a chemically stimulated movement of phagocytes to a site of damage. Chemicals that attract phagocytes might come from the microbes, white blood cells, damaged tissue cells, or activated complement proteins.

2 Attachment of the phagocyte to the microbe or other foreign material is termed **adherence.** The binding of complement to the invading pathogen enhances adherence.

3 Following adherence, the plasma membrane of the phagocyte extends projections, called **pseudopods,** that engulf the microbe in a process called **ingestion.** When the pseudopods meet, they fuse, surrounding the microorganism with a sac called a **phagosome.**

4 The phagosome enters the cytoplasm and merges with lysosomes to form a single, larger structure called a **phagolysosome.** The lysosome contributes lysozyme, which breaks down microbial cell walls, and other digestive enzymes that degrade carbohydrates, proteins, lipids, and nucleic acids. The phagocyte also forms lethal oxidants, such as superoxide anion (O_2^-), hypochlorite anion (OCl^-), and hydrogen peroxide (H_2O_2), in a process called an **oxidative burst.**

5 The chemical onslaught provided by lysozyme, digestive enzymes, and oxidants within a phagolysosome quickly kills many types of microbes. Any materials that cannot be degraded further remain in structures called **residual bodies.**

Microbial Evasion of Phagocytosis

Some microbes, such as the bacteria that cause pneumonia, have extracellular structures called capsules that prevent adherence. This makes it physically difficult for phagocytes to engulf the microbes. Other microbes, such as the toxin-producing bacteria that cause one kind of food poisoning, may be ingested but not killed; instead, the toxins they produce (leukocidins) may kill the phagocytes by causing the release the phagocyte's own lysosomal enzymes into its cytoplasm. Still other microbes—such as the bacteria that cause tuberculosis—inhibit fusion of phagosomes and lysosomes and thus prevent exposure of the microbes to lysosomal enzymes. These bacteria can also apparently use chemicals in their cell walls to counter the effects of lethal oxidants produced by phagocytes. Subsequent multiplication of the microbes within phagosomes may eventually destroy the phagocyte. ■

Inflammation

Inflammation is a nonspecific, defensive response of the body to tissue damage. Among the conditions that may produce inflammation are pathogens, abrasions, chemical irritations, distortion or disturbances of cells, and extreme temperatures. The four characteristic signs and symptoms of inflammation are **redness, pain, heat,** and **swelling.** Inflammation can also cause the **loss of function** in the injured area, depending on the site and extent of the injury. Inflammation is an attempt to dispose of microbes, toxins, or foreign material at the site of injury, to prevent their spread to other tissues, and to prepare the site for tissue repair. Thus, it helps restore tissue homeostasis.

Because inflammation is one of the body's innate defenses, the response of a tissue to, say, a cut is similar to the response to

Figure 22.9 Phagocytosis of a microbe.

 The major types of phagocytes are neutrophils and macrophages.

(a) Phases of phagocytosis

SEM about 2500x

(b) Phagocyte engulfing a yeast cell

 What chemicals are responsible for killing ingested microbes?

damage caused by burns, radiation, or bacterial or viral invasion. In each case, the inflammatory response has three basic stages: (1) vasodilation and increased permeability of blood vessels, (2) emigration (movement) of phagocytes from the blood into interstitial fluid, and, ultimately, (3) tissue repair.

VASODILATION AND INCREASED PERMEABILITY OF BLOOD VESSELS   Two immediate changes occur in the blood vessels in a region of tissue injury: **vasodilation** (increase in the diameter) of arterioles and **increased permeability** of capillaries (Figure 22.10). Increased permeability means that substances normally retained in blood are permitted to pass from the blood vessels. Vasodilation allows more blood to flow through the damaged area, and increased permeability permits defensive proteins such as antibodies and clotting factors to enter the injured area from the blood. The increased blood flow also helps remove microbial toxins and dead cells.

Among the substances that contribute to vasodilation, increased permeability, and other aspects of the inflammatory response are the following:

- **Histamine.** In response to injury, mast cells in connective tissue and basophils and platelets in blood release histamine. Neutrophils and macrophages attracted to the site of injury also stimulate the release of histamine, which causes vasodilation and increased permeability of blood vessels.

- **Kinins.** These polypeptides, formed in blood from inactive precursors called kininogens, induce vasodilation and increased permeability and serve as chemotactic agents for phagocytes.

- **Prostaglandins (PGs).** These lipids, especially those of the E series, are released by damaged cells and intensify the effects of histamine and kinins. PGs also may stimulate the emigration of phagocytes through capillary walls.

- **Leukotrienes (LTs).** Produced by basophils and mast cells by breakdown of membrane phospholipids, LTs cause increased permeability; they also function in adherence of phagocytes to pathogens and as chemotactic agents that attract phagocytes.

- **Complement.** Different components of the complement system stimulate histamine release, attract neutrophils by chemotaxis, and promote phagocytosis; some components can also destroy bacteria.

Dilation of arterioles and increased permeability of capillaries produce three of the symptoms of inflammation, heat, redness (erythema), and swelling (edema). Heat and redness result from the large amount of blood that accumulates in the damaged area. As the local temperature rises slightly, metabolic reactions proceed more rapidly and release additional heat. Edema results from increased permeability of blood vessels, which permits more fluid to move from blood plasma into tissue spaces.

Pain is a prime symptom of inflammation. It results from injury to neurons and from toxic chemicals released by microbes. Kinins affect some nerve endings, causing much of the pain associated with inflammation. Prostaglandins intensify and prolong the pain associated with inflammation. Pain may also be due to increased pressure from edema.

The increased permeability of capillaries allows leakage of blood-clotting factors into tissues. The clotting cascade is set into motion, and fibrinogen is ultimately converted to an insoluble, thick mesh of fibrin threads that localizes and traps invading microbes and blocks their spread.

EMIGRATION OF PHAGOCYTES   Generally, within an hour after the inflammatory process starts, phagocytes appear on the scene. As large amounts of blood accumulate, neutrophils begin to stick to the inner surface of the endothelium (lining) of blood vessels (Figure 22.10). Then the neutrophils begin to squeeze through the wall of the blood vessel to reach the damaged area. This process, called **emigration,** depends on chemotaxis. Neutrophils attempt to destroy the invading microbes by phagocytosis. A steady stream of neutrophils is ensured by the production and release of additional cells from bone marrow.

Figure 22.10   Inflammation.

 The three stages of inflammation are as follows: (1) vasodilation and increased permeability of blood vessels, (2) phagocyte emigration, and (3) tissue repair.

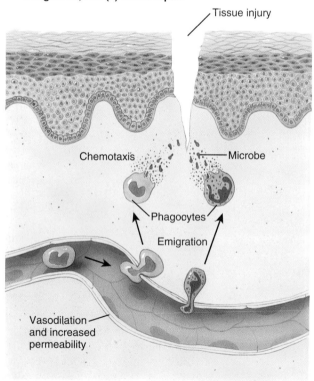

Phagocytes migrate from blood to site of tissue injury

What causes each of the following signs and symptoms of inflammation: redness, pain, heat, and swelling?

Such an increase in white blood cells in the blood is termed **leukocytosis.**

Although neutrophils predominate in the early stages of infection, they die off rapidly. As the inflammatory response continues, monocytes follow the neutrophils into the infected area. Once in the tissue, monocytes transform into wandering macrophages that add to the phagocytic activity of the fixed macrophages already present. True to their name, macrophages are much more potent phagocytes than are neutrophils. They are large enough to engulf damaged tissue, worn-out neutrophils, and invading microbes.

Eventually, macrophages also die. Within a few days, a pocket of dead phagocytes and damaged tissue forms; this collection of dead cells and fluid is called **pus.** Pus formation occurs in most inflammatory responses and usually continues until the infection subsides. At times, pus reaches the surface of the body or drains into an internal cavity and is dispersed; on other occasions the pus remains even after the infection is terminated. In this case, the pus is gradually destroyed over a period of days and is absorbed.

Abscesses and Ulcers

If pus cannot drain out of an inflamed region, the result is an **abscess**—an excessive accumulation of pus in a confined space. Common examples are pimples and boils. When superficial inflamed tissue sloughs off the surface of an organ or tissue, the resulting open sore is called an **ulcer.** People with poor circulation—for instance, diabetics with advanced atherosclerosis—are susceptible to ulcers in the tissues of their legs. These ulcers, which are called stasis ulcers, develop because of poor oxygen and nutrient supply to tissues that then become very susceptible to even a very mild injury or an infection. ■

Fever

Fever is an abnormally high body temperature that occurs because the hypothalamic thermostat is reset. It commonly occurs during infection and inflammation. Many bacterial toxins elevate body temperature, sometimes by triggering release of fever-causing cytokines such as interleukin-1. Elevated body temperature intensifies the effects of interferons, inhibits the growth of some microbes, and speeds up body reactions that aid repair.

Table 22.1 summarizes the components of nonspecific resistance.

▶ CHECKPOINT

8. What physical and chemical factors provide protection from disease in the skin and mucous membranes?

9. What internal defenses provide protection against microbes that penetrate the skin and mucous membranes?

10. How are the activities of natural killer cells and phagocytes similar and different?

11. What are the main signs, symptoms, and stages of inflammation?

Table 22.1	Summary of Nonspecific Resistance (Innate Defenses)
Component	**Functions**
First Line of Defense: Skin and Mucous Membranes	
Physical Factors	
Epidermis of skin	Forms a physical barrier to the entrance of microbes.
Mucous membranes	Inhibit the entrance of many microbes, but not as effective as intact skin.
Mucus	Traps microbes in respiratory and gastrointestinal tracts.
Hairs	Filter out microbes and dust in nose.
Cilia	Together with mucus, trap and remove microbes and dust from upper respiratory tract.
Lacrimal apparatus	Tears dilute and wash away irritating substances and microbes.
Saliva	Washes microbes from surfaces of teeth and mucous membranes of mouth.
Urine	Washes microbes from urethra.
Defecation and vomiting	Expel microbes from body.
Chemical Factors	
Sebum	Forms a protective acidic film over the skin surface that inhibits growth of many microbes.
Lysozyme	Antimicrobial substance in perspiration, tears, saliva, nasal secretions, and tissue fluids.
Gastric juice	Destroys bacteria and most toxins in stomach.
Vaginal secretions	Slight acidity discourages bacterial growth; flush microbes out of vagina.
Second Line of Defense: Internal Defenses	
Antimicrobial Proteins	
Interferons (IFNs)	Protect uninfected host cells from viral infection.
Complement system	Causes cytolysis of microbes, promotes phagocytosis, and contributes to inflammation.
Natural killer (NK) cells	Kill a wide variety of microbes and certain tumor cells.
Phagocytes	Ingest foreign particulate matter.
Inflammation	Confines and destroys microbes and initiates tissue repair.
Fever	Intensifies the effects of interferons, inhibits growth of some microbes, and speeds up body reactions that aid repair.

SPECIFIC RESISTANCE: IMMUNITY

▶ O B J E C T I V E S

- **Define immunity, and describe how T cells and B cells arise.**
- **Explain the relationship between an antigen and an antibody.**
- **Compare the functions of cell-mediated immunity and antibody-mediated immunity.**

The ability of the body to defend itself against specific invading agents such as bacteria, toxins, viruses, and foreign tissues is called **specific resistance** or **immunity.** Substances that are recognized as foreign and provoke immune responses are called **antigens (Ags).** Two properties distinguish immunity from nonspecific defenses: (1) *specificity* for particular foreign molecules (antigens), which also involves distinguishing self from nonself molecules, and (2) *memory* for most previously encountered antigens such that a second encounter prompts an even more rapid and vigorous response. The branch of science that deals with the responses of the body when challenged by antigens is called **immunology** (im′-ū-NOL-ō-jē ; *immuno-* = free from service or exempt; *-logy* = study of). The **immune system** includes the cells and tissues that carry out immune responses.

Maturation of T Cells and B Cells

The cells that develop **immunocompetence,** the ability to carry out immune responses if properly stimulated, are lymphocytes called B cells and T cells. Both develop in primary lymphatic organs (red bone marrow and the thymus) from pluripotent stem cells that originate in red bone marrow (see Figure 19.3 on page 638). B cells complete their development in bone marrow, a process that continues throughout life. T cells develop from pre-T cells that migrate from bone marrow into the thymus, where they mature (Figure 22.11). Most T cells arise before puberty, but they continue to mature and leave the thymus throughout life.

Before T cells leave the thymus or B cells leave red bone marrow, they begin to make several distinctive proteins that are inserted into the plasma membrane. Some of these proteins function as **antigen receptors**—molecules capable of recognizing specific antigens (Figure 22.11). In addition, T cells exit the thymus as either CD4$^+$ or CD8$^+$ cells, which means their plasma membrane includes either a protein called CD4 or one called CD8. As we will see later in this chapter, these two types of T cells have very different functions.

Types of Immune Responses

Immunity consists of two kinds of closely allied responses, both triggered by antigens. In **cell-mediated immune responses,** CD8$^+$ T cells proliferate into cytotoxic T cells that directly attack the invading antigen. In **antibody-mediated immune responses,** B cells transform into plasma cells, which synthesize and secrete specific proteins called **antibodies (Abs)** or **immunoglobulins** (im′-ū-nō-GLOB-ū-lins). A given antibody can bind to and inactivate a specific antigen. Most CD4$^+$ T cells become helper T cells that aid both cell-mediated and antibody-mediated immune responses.

To some extent, each type of immune response specializes in dealing with certain types of invaders. Cell-mediated immunity is particularly effective against (1) intracellular pathogens that reside within host cells (primarily fungi, parasites, and viruses); (2) some cancer cells; and (3) foreign tissue transplants. Thus, cell-mediated immunity always involves cells attacking cells. Antibody-mediated immunity works mainly against (1) antigens present in body fluids and (2) extracellular pathogens that multiply in body fluids but rarely enter body cells (primarily bacteria). However, a given pathogen often provokes both types of immune responses.

Antigens and Antigen Receptors

Antigens have two important characteristics: immunogenicity and reactivity. **Immunogenicity** (im′-ū-nō-je-NIS-i-tē; *-genic* = producing) is the ability to provoke an immune response by stimulating the production of specific antibodies, the proliferation of specific T cells, or both. The term *antigen* derives from its function as an *anti*body *gen*erator. **Reactivity** is the ability of the antigen to react specifically with the antibodies or cells it provoked. Strictly speaking, immunologists define antigens as substances that have reactivity; substances with both immunogenicity and reactivity are considered **complete antigens.** Commonly, however, the term *antigen* implies both immunogenicity and reactivity, and we use the word in this way.

Entire microbes or parts of microbes may act as antigens. Chemical components of bacterial structures such as flagella, capsules, and cell walls are antigenic, as are bacterial toxins. Nonmicrobial examples of antigens include chemical components of pollen, egg white, incompatible blood cells, and transplanted tissues and organs. The huge variety of antigens in the environment provides myriad opportunities for provoking immune responses. Typically, just certain parts of a large antigen molecule act as the triggers for immune responses. The small parts of antigen molecules that initiate immune responses are called **epitopes** (EP-i-tōps), or *antigenic determinants* (Figure 22.12). Most antigens have many epitopes, each of which induces production of a specific antibody or activates a specific T cell.

Antigens that get past the nonspecific defenses generally follow one of three routes into lymphatic tissue: (1) Most antigens that enter the bloodstream (for example, through an injured blood vessel) are trapped as they flow through the spleen. (2) Antigens that penetrate the skin enter lymphatic vessels and lodge in lymph nodes. (3) Antigens that penetrate mucous membranes are entrapped by mucosa-associated lymphatic tissue (MALT).

Chemical Nature of Antigens

Antigens are large, complex molecules. Most often, they are proteins. However, nucleic acids, lipoproteins, glycoproteins, and

Figure 22.11 **B cells and pre-T cells arise from pluripotent stem cells in red bone marrow.**
B cells and T cells develop in primary lymphatic tissues (red bone marrow and the thymus) and are activated in secondary lymphatic tissues (lymph nodes, spleen, and lymphatic nodules).

 The two types of immune responses are cell-mediated immune responses and antibody-mediated immune responses.

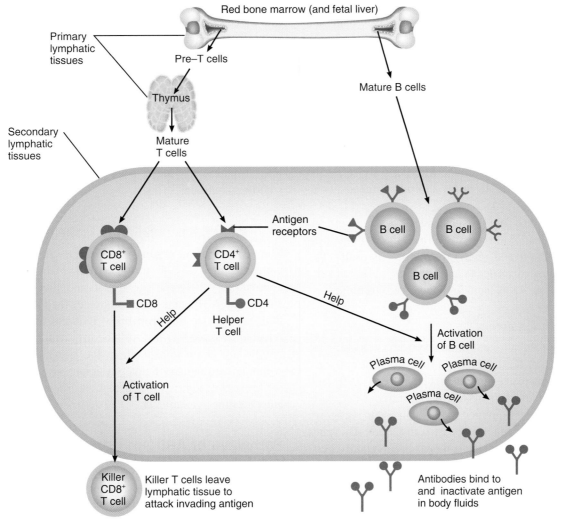

CELL-MEDIATED IMMUNE RESPONSES
Directed against intracellular pathogens, such as viruses, some cancer cells, and tissue transplants

ANTIBODY-MEDIATED IMMUNE RESPONSES
Directed against extracellular pathogens, such as bacteria

Which type of T cell participates in both cell-mediated and antibody-mediated immune responses?

Figure 22.12 **Epitopes (antigenic determinants).**

Most antigens have several epitopes that induce the production of different antibodies or activate different T cells.

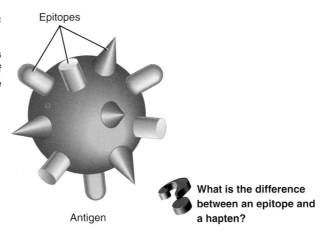

What is the difference between an epitope and a hapten?

certain large polysaccharides may also act as antigens. T cells respond only to antigens that include protein; B cells respond to antigens made of proteins, certain lipids, carbohydrates, and nucleic acids. Complete antigens usually have large molecular weights of 10,000 daltons or more, but large molecules that have simple, repeating subunits—for example, cellulose and most plastics—are not usually antigenic. This is why plastic materials can be used in artificial heart valves or joints.

A smaller substance that has reactivity but lacks immunogenicity is called a **hapten** (HAP-ten = to grasp). A hapten can stimulate an immune response only if it is attached to a larger carrier molecule. An example is the small lipid toxin in poison ivy, which triggers an immune response after combining with a body protein. Likewise, some drugs, such as penicillin, may combine with proteins in the body to form immunogenic complexes. Such hapten-stimulated immune responses are responsible for some allergic reactions to drugs and other substances in the environment (see page 798).

As a rule, antigens are foreign substances; they are not usually part of body tissues. However, sometimes the immune system fails to distinguish "friend" (self) from "foe" (nonself). The result is an autoimmune disorder (see page 799) in which self-molecules or cells are attacked as though they were foreign.

Diversity of Antigen Receptors

An amazing feature of the human immune system is its ability to recognize and bind to at least a billion (10^9) different epitopes. Before a particular antigen ever enters the body, T cells and B cells that can recognize and respond to that intruder are ready and waiting. Cells of the immune system can even recognize artificially made molecules that do not exist in nature. The basis for the ability to recognize so many epitopes is an equally large diversity of antigen receptors. Given that human cells contain only about 35,000 genes, how could a billion or more different antigen receptors possibly be generated?

The answer to this puzzle turned out to be simple in concept. The diversity of antigen receptors in both B cells and T cells is the result of shuffling and rearranging a few hundred versions of several small gene segments. This process is called **genetic recombination.** The gene segments are put together in different combinations as the lymphocytes are developing from stem cells in red bone marrow and the thymus. The situation is similar to shuffling a deck of 52 cards and then dealing out three cards. If you did this over and over, you could generate many more than 52 different sets of three cards. Because of genetic recombination, each B cell or T cell has a unique set of gene segments that codes for its unique antigen receptor. After transcription and translation, the receptor molecules are inserted into the plasma membrane.

Major Histocompatibility Complex Antigens

Located in the plasma membrane of body cells are "self-antigens," the **major histocompatibility complex (MHC)** anti-

gens. These transmembrane glycoproteins are also called *human leukocyte antigens (HLA)* because they were first identified on white blood cells. Unless you have an identical twin, your MHC antigens are unique. Thousands to several hundred thousand MHC molecules mark the surface of each of your body cells except red blood cells. Although MHC antigens are the reason that tissues may be rejected when they are transplanted from one person to another, their normal function is to help T cells recognize that an antigen is foreign, not self. Such recognition is an important first step in any immune response.

The two types of major histocompatibility complex antigens are class I and class II. Class I MHC (MHC-I) molecules are built into the plasma membranes of all body cells except red blood cells. Class II MHC (MHC-II) molecules appear only on the surface of antigen-presenting cells (described in the next section), thymic cells, and T cells that have been activated by exposure to an antigen.

Histocompatibility Testing

The success of an organ or tissue transplant depends on **histocompatibility** (his'-tō-kom-pat-i-BIL-i-tē)—that is, the tissue compatibility between the donor and the recipient. The more similar the MHC antigens, the greater the histocompatibility, and thus the greater the probability that the transplant will not be rejected. **Tissue typing (histocompatibility testing)** is done before any organ transplant. In the United States, a nationwide computerized registry helps physicians select the most histocompatible and needy organ transplant recipients whenever donor organs become available. ■

Pathways of Antigen Processing

For an immune response to occur, B cells and T cells must recognize that a foreign antigen is present. Whereas B cells can recognize and bind to antigens in lymph, interstitial fluid, or blood plasma, T cells only recognize fragments of antigenic proteins that are "presented" together with major histocompatibility complex self-antigens. Presentation occurs in the following way. As proteins inside body cells are being broken down, some peptide fragments associate with a peptide-binding groove of newly synthesized MHC molecules. This association stabilizes the MHC molecule and aids its proper folding, so that the MHC molecule can be inserted into the plasma membrane. When a peptide fragment comes from a *self-protein,* T cells ignore the peptide–MHC complex. When the peptide fragment is from a *foreign protein,* however, a few T cells recognize it as an intruder, and an immune response ensues. Preparation of a foreign antigen for display at the cell surface is termed *processing and presenting* of antigen. It occurs in two ways, depending on whether the antigen is located inside or outside body cells.

Processing of Exogenous Antigens

Foreign antigens that are present in fluids outside body cells are termed *exogenous antigens.* They include intruders such as bacteria and bacterial toxins, worm parasites, inhaled pollen and dust, and viruses that have not yet infected a body cell. A special class of cells called **antigen-presenting cells (APCs)** process and present exogenous antigens. APCs include macrophages, B cells, and dendritic cells. They are strategically located in places where antigens are likely to penetrate nonspecific defenses and enter the body, such as the epidermis and dermis of the skin (Langerhans cells are a type of dendritic cell); mucous membranes that line the respiratory, gastrointestinal, urinary, and reproductive tracts; and lymph nodes. After processing an antigen, APCs migrate from tissues via lymphatic vessels to lymph nodes.

The steps in the processing and presenting of an exogenous antigen by an antigen-presenting cell occur as follows (Figure 22.13):

1 **Ingestion of the antigen.** Antigen-presenting cells ingest antigens by phagocytosis or endocytosis. Ingestion could occur almost anywhere in the body that invaders, such as microbes, have penetrated the nonspecific defenses.

2 **Digestion of antigen into peptide fragments.** Within the phagosome or endosome, digestive enzymes split large antigens into short peptide fragments. At the same time, the antigen-presenting cell is synthesizing MHC-II molecules and packaging them into vesicles. The MCH-II molecules are part of the inner membrane surface of the vesicles.

3 **Fusion of vesicles.** The vesicles containing antigen peptide fragments and MHC-II molecules merge and fuse.

4 **Binding of peptide fragments to MHC-II molecules.** After fusion of the two types of vesicles, antigen peptide fragments bind to MHC-II molecules.

5 **Insertion of antigen–MHC-II complex into the plasma membrane.** The combined vesicle that contains antigen–MHC-II complexes undergoes exocytosis. As a result, the antigen–MHC-II complexes are inserted into the plasma membrane.

After processing an antigen, the antigen-presenting cell migrates to lymphatic tissue to present the antigen to T cells. Within lymphatic tissue, a small number of T cells that have compatibly shaped receptors recognize and bind to the antigen fragment–MHC-II complex, triggering either a cell-mediated or an antibody-mediated immune response. The presentation of exogenous antigen together with MHC-II molecules by antigen-presenting cells informs T cells that intruders are present in the body and that combative action should begin.

Processing of Endogenous Antigens

Foreign antigens that are synthesized within a body cell are termed *endogenous antigens.* Such antigens may be viral proteins produced after a virus infects the cell and takes over the cell's metabolic machinery, or they may be abnormal proteins synthesized by a cancerous cell. Fragments of endogenous antigens associate with major histocompatibility complex-I molecules inside infected cells. The resulting endogenous antigen

Figure 22.13 Processing and presenting of exogenous antigen by an antigen-presenting cell (APC).

Except for identical twins, major histocompatibility complex (MHC) molecules are unique in each person. They help T cells recognize foreign invaders.

APCs present exogenous antigens in association with MHC-II molecules

What types of cells are APCs, and where in the body are they found?

fragment–MHC-I complex then moves to the plasma membrane, where it is displayed at the surface of the cell. Most cells of the body can process and present endogenous antigens. The display of an endogenous antigen bound to an MHC-I molecule signals that a cell has been infected and needs help.

Cytokines

Cytokines are small protein hormones that stimulate or inhibit many normal cell functions, such as cell growth and differentiation. Lymphocytes and antigen-presenting cells secrete cytokines, as do fibroblasts, endothelial cells, monocytes, hepatocytes, and kidney cells. Some cytokines stimulate proliferation of progenitor blood cells in red bone marrow. Others regulate activities of cells involved in nonspecific defenses or immune responses, as described in Table 22.2.

 Cytokine Therapy

Cytokine therapy is the use of cytokines to treat medical conditions. Interferons were the first cytokines shown to be effective against a human cancer. Alpha-interferon (Intron A) is approved in the United States for treating Kaposi (KAP-ō-sē)

sarcoma, a cancer that often occurs in patients infected with HIV, the virus that causes AIDS. Other approved uses for alpha-interferon include treating genital herpes caused by herpes virus; treating hepatitis B and C, caused by the hepatitis B and C viruses; and treating hairy cell leukemia. A form of beta-interferon (Betaseron) slows the progression of multiple sclerosis (MS) and lessens the frequency and severity of MS attacks. Of the interleukins, the one most widely used to fight cancer is interleukin-2. Although this treatment is effective in causing tumor regression in some patients, it also can be very toxic. Among the adverse effects are high fever, severe weakness, difficulty breathing due to pulmonary edema, and hypotension leading to shock. ■

▶ C H E C K P O I N T

12. What is immunocompetence, and which body cells display it?
13. How do the major histocompatibility complex class I and class II self-antigens function?
14. How do antigens arrive at lymphatic tissues?
15. How do antigen-presenting cells process exogenous and endogenous antigens?
16. What are cytokines, what is their origin, and how do they function?

Table 22.2	Summary of Cytokines Participating in Immune Responses
Cytokine	**Origins and Functions**
Interleukin-1 (IL-1)	Produced by monocytes and macrophages; promotes proliferation of helper T cells; acts on hypothalamus to cause fever.
Interleukin-2 (IL-2) (T cell growth factor)	Secreted by helper T cells; costimulates the proliferation of helper T cells, cytotoxic T cells, and B cells; activates NK cells.
Interleukin-4 (IL-4) (B cell stimulating factor)	Produced by activated helper T cells; costimulator for B cells; causes plasma cells to secrete IgE antibodies (see Table 22.3); promotes growth of T cells.
Interleukin-5 (IL-5)	Produced by certain activated CD4+ T cells and activated mast cells; costimulator for B cells; causes plasma cells to secrete IgA antibodies.
Tumor necrosis factor (TNF)	Produced mainly by macrophages; stimulates accumulation of neutrophils and macrophages at sites of inflammation and stimulates their killing of microbes; stimulates macrophages to produce IL-1; induces synthesis of colony-stimulating factors by endothelial cells and fibroblasts; exerts an interferon-like protective effect against viruses; and functions as an endogenous pyrogen to induce fever.
Transforming growth factor beta (TGF-β)	Secreted by T cells and macrophages; has some positive effects but is thought to be important for turning off immune responses; inhibits proliferation of T cells and activation of macrophages.
Gamma-interferon (γ-IFN)	Secreted by helper and cytotoxic T cells and NK cells; strongly stimulates phagocytosis by neutrophils and macrophages; activates NK cells; enhances both cell-mediated and antibody-mediated immune responses.
Alpha- and beta-interferons (α-IFN and β-IFN)	Produced by virus-infected cells to inhibit viral replication in uninfected cells; produced by antigen-stimulated macrophages to stimulate T cell growth; activate NK cells, inhibit cell growth, and suppress formation of some tumors.
Lymphotoxin (LT)	Secreted by cytotoxic T cells; kills cells by causing fragmentation of DNA.
Perforin	Secreted by cytotoxic T cells and perhaps by NK cells; perforates cell membranes of target cells, which causes cytolysis.
Macrophage migration inhibiting factor	Produced by T cells; prevents macrophages from leaving site of infection.

CELL-MEDIATED IMMUNITY

OBJECTIVE

• **Describe the steps of a cell-mediated immune response.**

A cell-mediated immune response begins with *activation* of a small number of T cells by a specific antigen. Once a T cell has been activated, it undergoes *proliferation* and *differentiation* into a clone of **effector cells,** a population of identical cells that can recognize the same antigen and carry out some aspect of the immune attack. Finally, the immune response results in *elimination* of the intruder.

Activation, Proliferation, and Differentiation of T Cells

Antigen receptors on the surface of T cells, called **T-cell receptors (TCRs),** recognize and bind to specific foreign antigen fragments that are presented in antigen–MHC complexes. There are millions of different T cells; each has its own unique TCRs that can recognize a specific antigen–MHC complex. At any given time, most T cells are inactive. When an antigen enters the body, only a few T cells have TCRs that can recognize and bind to the antigen. Antigen recognition by a TCR is the *first signal* in activation of a T cell.

A T cell becomes activated only if it binds to the foreign antigen and at the same time receives a *second signal,* a process known as **costimulation.** Of the more than 20 known costimulators, some are cytokines, such as **interleukin-2.** Other costimulators include pairs of plasma membrane molecules, one on the surface of the T cell and a second on the surface of an antigen-presenting cell, that enable the two cells to adhere to one another for a period of time.

The need for two signals is a little like starting and driving a car: When you insert the correct key (antigen) in the ignition (TCR) and turn it, the car starts (recognition of specific antigen), but it cannot move forward until you move the gear shift into drive (costimulation). The need for costimulation may prevent immune responses from occurring accidentally. Different co-stimulators affect the activated T cell in different ways, just as shifting a car into reverse has a different effect than shifting it into drive. Moreover, recognition (antigen binding to a receptor) without costimulation leads to a prolonged *state of inactivity* called **anergy** in both T cells and B cells. Anergy is rather like a car in neutral with its engine running until it's out of gas!

Once a T cell has received two signals (antigen recognition and costimulation), it is **activated.** An activated T cell enlarges and begins to **proliferate** (divide several times) and to **differentiate** (form more highly specialized cells). The result is a population of identical cells, termed a **clone,** that can recognize the same specific antigen. Before the first exposure to a given antigen, only a few T cells might be able to recognize it, but once an immune response has begun, there are thousands. Activation,

proliferation, and differentiation of T cells occur in the secondary lymphatic organs and tissues. If you have ever noticed swollen tonsils or lymph nodes in your neck, the proliferation of lymphocytes participating in an immune response was likely the cause.

Types of T Cells

The three main types of differentiated T cells are helper T cells, cytotoxic T cells, and memory T cells.

Helper T Cells

Most T cells that display CD4 develop into **helper T (T_H) cells,** also known as **CD4$^+$ T cells.** Resting (inactive) T_H cells recognize antigen fragments associated with major histocompatibility complex class II (MHC-II) molecules at the surface of an antigen-presenting cell (Figure 22.14a). Several different pairs of molecules, one attached to the membrane of the APC and the other attached to the membrane of the helper T cell, interact with each other to hold the two cells tightly together and achieve co-stimulation.

Within hours after costimulation, helper T cells start secreting a variety of cytokines (see Table 22.2). Different subsets of T_H cells specialize in the production of particular cytokines. One very important cytokine produced by T_H cells is interleukin-2 (IL-2), which is needed for virtually all immune responses and is the prime trigger of T cell proliferation. IL-2 can act as a costimulator for resting T_H or cytotoxic T cells, and it enhances activation and proliferation of T cells, B cells, and natural killer cells.

Some actions of interleukin-2 provide a good example of a beneficial positive feedback system. As noted earlier, activation of a helper T cell stimulates it to start secreting IL-2, which then acts in an autocrine manner by binding to IL-2 receptors on the plasma membrane of the cell that secreted it. One effect is stimulation of cell division. As the T_H cells proliferate, a positive feedback effect occurs because they secrete more IL-2, which causes further cell division. IL-2 may also act in a paracrine manner by binding to IL-2 receptors on neighboring T_H cells, cytotoxic T cells, or B cells. If any of these cells have already bound an antigen, IL-2 serves as a costimulator to activate them.

Cytotoxic T Cells

T cells that display CD8 develop into **cytotoxic T (T_C) cells,** also termed **CD8$^+$ cells.** T_C cells recognize foreign antigens combined with major histocompatibility complex class I (MHC-I) molecules on the surfaces of (1) body cells infected by viruses, (2) some tumor cells, and (3) cells of a tissue transplant (Figure 22.14b). However, to become cytolytic (able to lyse cells) T_C cells need costimulation by interleukin-2 or other cytokines produced by helper T cells. (Recall that T_H cells are activated by antigen associated with MHC-II molecules.) Thus, maximal activation of T_C cells requires presentation of antigen associated with both MHC-I and MHC-II molecules.

Figure 22.14 **Activation, proliferation, and differentiation of T cells.**

The binding of CD4 to MHC-II and CD8 to MHC-I helps anchor the T-cell receptor (TCR)–antigen interaction so that antigen recognition can occur.

(a) Helper T (T$_H$) cells or CD4$^+$ cells

(b) Cytotoxic T (T$_C$) cells or CD8$^+$ cells

 What are the first and second signals in activation of a T cell?

Memory T Cells

T cells that remain from a proliferated clone after a cell-mediated immune response are termed **memory T cells.** Should a pathogen bearing the same foreign antigen invade the body later, thousands of memory cells are available to initiate a far swifter reaction than occurred during the first invasion. The second response usually is so fast and so vigorous that the pathogens are destroyed before any signs or symptoms of disease can occur.

Elimination of Invaders

Cytotoxic T cells are the soldiers that march forth to do battle with foreign invaders in cell-mediated immune responses. They leave secondary lymphatic organs and tissues and migrate to the site of invasion, infection, or tumor formation. They recognize and attach to target cells bearing the antigen that stimulated activation and proliferation of their progenitor cells. Then, the cytotoxic T cells deliver a "lethal hit" that kills the target cells without damaging themselves (Figure 22.15). After detaching from a target cell, a cytotoxic T cell can seek out and destroy another invader that displays the same antigen.

Cytotoxic T cells use two killing mechanisms. In the first, granules containing the protein perforin undergo exocytosis from the cytotoxic T cell. **Perforin** forms holes in the plasma membrane of the target cell, which allow extracellular fluid to flow in, and the cell bursts—a process known as **cytolysis.** In the second mechanism, the cytotoxic T cell secretes a toxic molecule known as **lymphotoxin** that activates enzymes within the target cell. These enzymes cause the target cell's DNA to fragment, and the cell dies. In these ways, cytotoxic T cells destroy infected cells. Also, cytotoxic T cells secrete gamma-interferon, which activates phagocytic cells at the scene of the battle. Cytotoxic T cells are especially effective against slowly developing bacterial diseases (such as tuberculosis and brucellosis), some viruses, fungi, cancer cells associated with viral infection, and transplanted cells.

Immunological Surveillance

When a normal cell transforms into a cancerous cell, it often displays novel cell surface components called **tumor antigens.** These molecules are rarely, if ever, displayed on the surface of normal cells. If the immune system recognizes a tumor antigen as nonself, it can destroy the cancer cells carrying them. Such immune responses, called **immunological surveillance,** are carried out by cytotoxic T cells, macrophages, and natural killer cells. Immunological surveillance is most effective in eliminating tumor cells due to cancer-causing viruses. For this reason, transplant recipients who are taking immunosuppressive drugs to prevent transplant rejection have an increased incidence of virus-associated cancers. Their risk for other types of cancer is not increased.

Figure 22.15 Activity of cytotoxic T cells. After delivering a "lethal hit," a cytotoxic T cell can detach and attack another target cell displaying the same antigen.

Cytotoxic T cells kill microbes directly by secreting perforin and lymphotoxin.

Besides cells infected by viruses, what other types of target cells are attacked by cytotoxic T cells?

Graft Rejection

Organ transplantation involves the replacement of an injured or diseased organ, such as the heart, liver, kidney, lungs, or pancreas, with an organ donated by another individual. Usually, the immune system recognizes the proteins in the transplanted organ as foreign and mounts both cell-mediated and antibody-mediated immune responses against them. This phenomenon is known as **graft rejection.** The more closely matched are the major histocompatibility complex proteins of the donor and recipient, the weaker is the graft rejection response. To reduce the risk of graft rejection, organ transplant recipients receive immunosuppressive drugs. One such drug is *cyclosporine,* derived from a fungus, which inhibits secretion of interleukin-2 by helper T cells but has only a minimal effect on B cells. Thus, the risk of rejection is diminished while maintaining resistance to some diseases. ■

▶ CHECKPOINT

17. What are the functions of helper, cytotoxic, and memory T cells?

18. How do cytotoxic T cells kill their targets?

19. How is immunological surveillance useful?

ANTIBODY-MEDIATED IMMUNITY

OBJECTIVES

• **Describe the steps in an antibody-mediated immune response.**

• **Describe the chemical characteristics and actions of antibodies.**

The body contains not only millions of different T cells but also millions of different B cells, each capable of responding to a specific antigen. Whereas cytotoxic T cells leave lymphatic tissues to seek out and destroy a foreign antigen, B cells stay put. In the presence of a foreign antigen, specific B cells in lymph nodes, the spleen, or mucosa-associated lymphatic tissue become activated. They then differentiate into plasma cells that secrete specific antibodies, which in turn circulate in the lymph and blood to reach the site of invasion.

Activation, Proliferation, and Differentiation of B Cells

During activation of a B cell, an antigen binds to **B-cell receptors (BCRs)** (Figure 22.16). B-cell receptors, which are integral transmembrane proteins, are chemically similar to the antibodies that eventually are secreted by plasma cells. Although B cells can respond to an unprocessed antigen present in lymph or interstitial fluid, their response is much more intense when nearby antigen-presenting cells also process and present antigen to them. Some antigen is then taken into the B cell, broken down into peptide fragments and combined with MHC-II self-antigen, and moved to the B cell plasma membrane. Helper T cells recognize the antigen–MHC-II complex and deliver the costimulation needed for B cell proliferation and differentiation. The helper T cell produces interleukin-2 and other cytokines that function as costimulators to activate B cells. Interleukin-4 and interleukin-6, also produced by helper T cells, enhance B cell proliferation, B cell differentiation into plasma cells, and secretion of antibodies by plasma cells.

Some of the activated B cells enlarge, divide, and differentiate into a clone of antibody-secreting **plasma cells.** A few days after exposure to an antigen, a plasma cell is secreting hundreds of millions of antibodies daily and secretion occurs for about 4 or 5 days, until the plasma cell dies. Most antibodies travel in lymph and blood to the invasion site. Some activated B cells do not differentiate into plasma cells but rather remain as **memory B cells** that are ready to respond more rapidly and forcefully should the same antigen reappear at a future time.

Different antigens stimulate different B cells to develop into plasma cells and their accompanying memory B cells. The B cells of a particular clone are capable of secreting only one kind of antibody, which is identical in specificity to the antigen receptor displayed by the B cell that first responded to the antigen. Each specific antigen activates only those B cells that are predestined (by the combination of gene segments they carry) to secrete antibody specific to that antigen. Antibodies produced by a clone of plasma cells enter the circulation and form

Figure 22.16 Activation, proliferation, and differentiation of B cells into plasma cells and memory cells. Plasma cells are actually much larger than B cells.

Plasma cells secrete antibodies.

B cell recognizing unprocessed antigen

Costimulation by several interleukins

B cell displaying processed antigen is recognized by helper T cell, which releases costimulators

Proliferation and differentiation

Plasma cells

Memory cells

Antibodies

Clones of plasma cells secrete antibodies with same specificity as antigen receptor on progenitor (inactive) B cell.

Long-lived memory B cells remain to respond to same antigen when it appears again.

How many different kinds of antibodies will be secreted by the plasma cells in the clone shown here?

antigen–antibody complexes with the antigen that initiated their production.

Antibodies

An **antibody (Ab)** can combine specifically with the epitope on the antigen that triggered its production. The antibody's structure matches its antigen much as a lock fits a specific key. In theory, plasma cells could secrete as many different antibodies as there are different B-cell receptors because the same recombined gene segments code for both the BCR and the antibodies eventually secreted by plasma cells.

Antibody Structure

Antibodies belong to a group of glycoproteins called globulins, and for this reason they are also known as **immunoglobulins (Igs).** Most antibodies contain four polypeptide chains (Figure 22.17). Two of the chains are identical to each other and are called **heavy (H) chains;** each consists of about 450 amino acids. Short carbohydrate chains are attached to each heavy polypeptide chain. The two other polypeptide chains, also identical to each other, are called **light (L) chains,** and each consists of about 220 amino acids. A disulfide bond (S–S) holds each light chain to a heavy chain. Two disulfide bonds also link the midregion of the two heavy chains; this part of the antibody displays considerable flexibility and is called the **hinge region.** Because the antibody "arms" can move somewhat as the hinge region bends, an antibody can assume either a T shape (Figure 22.17a) or a Y shape (Figure 22.17b). Beyond the hinge region, parts of the two heavy chains form the **stem region.**

Within each H and L chain are two distinct regions. The tips of the H and L chains, called the **variable (V) regions,** constitute the **antigen-binding site.** The variable region, which is different for each kind of antibody, is the part of the antibody that recognizes and attaches specifically to a particular antigen. Because most antibodies have two antigen binding sites, they are said to be bivalent. Flexibility at the hinge allows the antibody to simultaneously bind to two epitopes that are some distance apart—for example, on the surface of a microbe.

The remainder of each H and L chain, called the **constant (C) region,** is nearly the same in all antibodies of the same class and is responsible for the type of antigen–antibody reaction that occurs. However, the constant region of the H chain differs from one class of antibody to another, and its structure serves as a basis for distinguishing five different classes, designated IgG, IgA, IgM, IgD, and IgE. Each class has a distinct chemical structure and a specific biological role. Because they appear first and are relatively short-lived, IgM antibodies indicate a recent invasion. In a sick patient, the responsible pathogen may be suggested by the presence of high levels of IgM specific to a particular organism. Resistance of the fetus and newborn baby to infection stems mainly from maternal IgG antibodies that cross the placenta before birth and IgA antibodies in breast milk after birth. Table 22.3 summarizes the structures and functions of the five classes of antibodies.

Antibody Actions

Even though the five classes of immunoglobulins have actions that differ somewhat, all act in some way to disable antigens. Actions of antibodies include:

- **Neutralizing antigen.** The reaction of antibody with antigen blocks or neutralizes some bacterial toxins and prevents attachment of some viruses to body cells.

- **Immobilizing bacteria.** If antibodies form against antigens on the cilia or flagella of motile bacteria, the antigen–antibody reaction may cause the bacteria to lose their motility, which limits their spread into nearby tissues.

- **Agglutinating and precipitating antigen.** Because antibodies have two or more sites for binding to antigen, the antigen–antibody reaction may cross-link pathogens to one another, causing agglutination (clumping together). Phagocytic cells ingest agglutinated microbes more readily.

Figure 22.17 Chemical structure of the immunoglobulin G (IgG) class of antibody. Each molecule is composed of four polypeptide chains (two heavy and two light) plus a short carbohydrate chain attached to each heavy chain. In (a), each circle represents one amino acid. In (b), V_L = variable regions of light chain, C_L = constant region of light chain, V_H = variable region of heavy chain, and C_H = constant region of heavy chain.

 An antibody combines only with the epitope on the antigen that triggered its production.

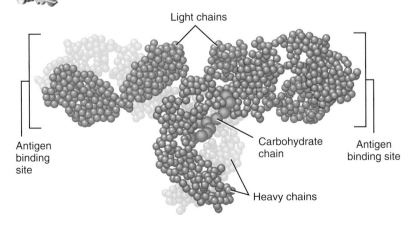

(a) Model of IgG molecule

(b) Diagram of IgG heavy and light chains

What is the function of the variable regions in an antibody molecule?

Table 22.3   Classes of Immunoglobulins (Igs)

Name and Structure	Characteristics and Functions
IgG	Most abundant, about 80% of all antibodies in the blood; found in blood, lymph, and the intestines; monomer (one-unit) structure. Protects against bacteria and viruses by enhancing phagocytosis, neutralizing toxins, and triggering the complement system. It is the only class of antibody to cross the placenta from mother to fetus, thereby conferring considerable immune protection in newborns.
IgA	Found mainly in sweat, tears, saliva, mucus, breast milk and gastrointestinal secretions. Smaller quantities are present in blood and lymph. Makes up 10–15% of all antibodies in the blood; occurs as monomers and dimers (two units). Levels decrease during stress, lowering resistance to infection. Provides localized protection on mucous membranes against bacteria and viruses.
IgM	About 5–10% of all antibodies in the blood; also found in lymph. Occurs as pentamers (five units); first antibody class to be secreted by plasma cells after an initial exposure to any antigen. Activates complement and causes agglutination and lysis of microbes. Also present as monomers on the surfaces of B cells, where they serve as antigen receptors. In blood plasma, the anti-A and anti-B antibodies of the ABO blood group, which bind to A and B antigens during incompatible blood transfusions, are also IgM antibodies (see Figure 19.12 on page 652).
IgD	Mainly found on the surfaces of B cells as antigen receptors, where it occurs as monomers; involved in activation of B cells. About 0.2% of all antibodies in the blood.
IgE	Less than 0.1% of all antibodies in the blood; occurs as monomers; located on mast cells and basophils. Involved in allergic and hypersensitivity reactions; provides protection against parasitic worms.

Likewise, soluble antigens may come out of solution and form a more-easily phagocytized precipitate when cross-linked by antibodies.

- *Activating complement.* Antigen–antibody complexes initiate the classical pathway of the complement system (discussed shortly).

- *Enhancing phagocytosis.* The stem region of an antibody acts as a "flag" that attracts phagocytes once antigens have bound to the antibody's variable region. Antibodies enhance the activity of phagocytes by causing agglutination and precipitation, by activating complement, and by coating microbes so that they are more susceptible to phagocytosis.

Monoclonal Antibodies

The antibodies produced against a given antigen by plasma cells can be harvested from an individual's blood. However, because an antigen typically has many epitopes, several different clones of plasma cells produce different antibodies against the antigen. If a single plasma cell could be isolated and induced to proliferate into a clone of identical plasma cells, then a large quantity of identical antibodies could be produced. Unfortunately, lymphocytes and plasma cells are difficult to grow in culture, so scientists sidestepped this difficulty by fusing B cells with tumor cells that grow easily and proliferate endlessly. The resulting hybrid cell is called a **hybridoma** (hī-bri-DŌ-ma). Hybridomas are long-term sources of large quantities of pure, identical antibodies, called **monoclonal antibodies (MAbs)** because they come from a single clone of identical cells. One clinical use of monoclonal antibodies is for measuring levels of a drug in a patient's blood. Other uses include the diagnosis of strep throat, pregnancy, allergies, and diseases such as hepatitis, rabies, and some sexually transmitted diseases. MAbs have also been used to detect cancer at an early stage and to ascertain the extent of metastasis. They may also be useful in preparing vaccines to counteract the rejection associated with transplants, to treat autoimmune diseases, and perhaps to treat AIDS. ∎

Role of the Complement System in Immunity

The **complement system** is a defensive system consisting of plasma proteins that attack and destroy microbes. The system can be activated via two pathways (classical and alternative), both of which are an ordered sequence or cascade of reactions. Both pathways lead to the same events: inflammation, enhancement of phagocytosis, and bursting of microbes.

The complement system consists of more than 30 different proteins in blood plasma and in plasma membranes. Included are proteins called C1 through C9 (the C stands for complement) and proteins called factors B, D, and P (properdin). The **classical pathway,** which is so-named because it was discovered first, begins by the binding of antibody to antigen (Figure 22.18). The antigen could be a bacterium or other foreign cell. The antigen–antibody attachment activates complement protein C1 through binding to the stem region of the antibody. In the ensuing cascade of enzymatic reactions, inactive complement proteins are split into active fragments termed C2a, C2b, and so forth. The **alternative pathway** does not involve antibodies. It begins by the interaction between polysaccharides on the surface of a microbe and factors B, D, and P (Figure 22.18). This interaction activates complement protein C3 and the cascade begins.

The consequences of activation of the classical and alternative pathways are as follows:

1 *Activation of inflammation.* Some complement proteins (C3a, C4a, and C5a) contribute to the development of inflammation: They dilate arterioles, which increases blood flow to the area, and cause the release of histamine from mast cells, basophils, and platelets. Because histamine increases the permeability of blood capillaries, white blood cells can more easily move into tissues to combat infection or allergy. Other complement proteins serve as chemotactic agents that attract phagocytes to the site of microbial invasion.

2 *Opsonization.* Complement fragment C3b binds to the surface of a microbe and then interacts with receptors on phagocytes, which promotes phagocytosis. Making microbes more susceptible to phagocytosis is known as **opsonization** (op-sō-ni-ZĀ-shun).

3 *Cytolysis.* Several complement proteins (C5b, C6, C7, C8, and C9) come together to form a **membrane attack complex (MAC)** that inserts into the plasma membrane of the microbe, forming large holes. The ensuing leakiness allows fluid to flow into the cell's interior, and the microbe swells and bursts (cytolysis).

Figure 22.18 The classical and alternative pathways of the complement system.

When activated, complement proteins enhance certain immune, allergic, and inflammatory reactions.

Which pathway for activation of complement is linked to the immune system? Explain why.

Immunological Memory

A hallmark of immune responses is memory for specific antigens that have triggered immune responses in the past. Immunological memory is due to the presence of long-lasting antibodies and very long-lived lymphocytes that arise during proliferation and differentiation of antigen-stimulated B cells and T cells.

Immune responses, whether cell-mediated or antibody-mediated, are much quicker and more intense after a second or subsequent exposure to an antigen than after the first exposure. Initially, only a few cells have the correct specificity to respond, and the immune response may take several days to build to maximum intensity. Because thousands of memory cells exist after an initial encounter with an antigen, the next time the same antigen appears they can proliferate and differentiate into plasma cells or cytotoxic T cells within hours.

One measure of immunological memory is the amount of antibody in serum, called the *antibody titer* (TĪ-ter). After an initial contact with an antigen, no antibodies are present for a period of several days. Then, a slow rise in the antibody titer occurs, first IgM and then IgG, followed by a gradual decline in antibody titer (Figure 22.19). This is the **primary response.**

Memory cells may remain for decades. Every new encounter with the same antigen results in a rapid proliferation of memory cells. The antibody titer after subsequent encounters is far greater than during a primary response and consists mainly of IgG antibodies. This accelerated, more intense response is called the **secondary response.** Antibodies produced during a secondary response have an even higher affinity for the antigen than those produced during a primary response, and thus they are more successful in disposing of it.

Primary and secondary responses occur during microbial infection. When you recover from an infection without taking antimicrobial drugs, it is usually because of the primary response. If the same microbe infects you later, the secondary response could be so swift that the microbes are destroyed before you exhibit any signs or symptoms of infection.

Immunological memory provides the basis for immunization by vaccination against certain diseases (for example, polio). When you receive the vaccine, which may contain weakened or killed whole microbes or portions of microbes, your B cells and T cells are activated. Should you subsequently encounter the living pathogen as an infecting microbe, your body initiates a secondary response. Table 22.4 summarizes the various types of antigen encounters that provide naturally and artificially acquired immunity.

▶ CHECKPOINT

20. How do the five classes of antibodies differ in structure and function?

21. How are cell-mediated and antibody-mediated immune responses similar and different?

22. In what ways does the complement system augment antibody-mediated immune responses?

23. How is the secondary response to an antigen different from the primary response?

Figure 22.19 Production of antibodies in the primary (after first exposure) and secondary (after second exposure) responses to a given antigen.

🔑 Immunological memory is the basis for successful immunization by vaccination.

❓ According to this graph, how much more IgG is circulating in the blood in the secondary response than in the primary response? (*Hint:* Notice that each mark on the antibody titer axis represents a 10-fold increase.)

Table 22.4	Types of Immunity
Type of Immunity	**How Acquired**
Naturally acquired active immunity	Antigen recognition by B cells and T cells and costimulation lead to antibody-secreting plasma cells, cytotoxic T cells, and B and T memory cells.
Naturally acquired passive immunity	Transfer of IgG antibodies from mother to fetus across placenta, or of IgA antibodies from mother to baby in milk during breast-feeding.
Artificially acquired active immunity	Antigens introduced during a vaccination stimulate cell-mediated and antibody-mediated immune responses, leading to production of memory cells. The antigens are pretreated to be immunogenic but not pathogenic; that is, they will trigger an immune response but not cause significant illness.
Artificially acquired passive immunity	Intravenous injection of immunoglobulins (antibodies).

SELF-RECOGNITION AND SELF-TOLERANCE

OBJECTIVE

• **Describe how self-recognition and self-tolerance develop.**

To function properly, your T cells must have two traits: (1) They must be able to *recognize* your own major histocompatibility complex (MHC) proteins, a process known as **self-recognition,** and (2) they must *lack reactivity* to peptide fragments from your own proteins, a condition known as **self-tolerance** (Figure 22.20). B cells also display self-tolerance. Loss of self-tolerance leads to the development of autoimmune diseases (see page 799).

Pre-T cells in the thymus develop the capability for self-recognition via **positive selection** (Figure 22.20a). In this process, some pre-T cells express T-cell receptors (TCRs) that interact with self-MHC proteins on epithelial cells in the thymic cortex. Because of this interaction, the T cells can recognize the MHC part of an antigen–MHC complex. These T cells survive. Other immature T cells that fail to interact with thymic epithelial cells are not able to recognize self-MHC proteins. These cells undergo apoptosis.

The development of self-tolerance, in contrast, occurs by a weeding-out process called **negative selection** in which the T cells interact with dendritic cells located at the junction of the cortex and medulla in the thymus. In this process, T cells with receptors that recognize self-peptide fragments or other self-

Figure 22.20   Development of self-recognition and self-tolerance.   MHC = major histocompatibility complex. TCR = T-cell receptor.

Positive selection allows recognition of self-MHC proteins; negative selection provides self-tolerance of your own peptides and other self-antigens.

(a) Positive and negative selection of T cells in the thymus

(b) Selection of T cells after they emerge from the thymus

(c) Selection of B cells

How does deletion differ from anergy?

antigens are eliminated or inactivated (Figure 22.20a). The T cells selected to survive do not respond to self-antigens, the fragments of molecules that are normally present in the body. Negative selection occurs via both deletion and anergy. In **deletion,** self-reactive T cells undergo apoptosis and die, whereas in **anergy** they remain alive but are unresponsive to antigenic stimulation. Only 1–5% of the immature T cells in the thymus receive the proper signals to survive apoptosis during both positive and negative selection and emerge as mature, immunocompetent T cells.

Once T cells have emerged from the thymus, they may still encounter an unfamiliar self-protein; in such cases they may also become anergic if there is no costimulator (Figure 22.20b). Deletion of self-reactive T cells may also occur after they leave the thymus.

B cells also develop tolerance through deletion and anergy (Figure 22.20c). While B cells are developing in bone marrow, those cells exhibiting antigen receptors that recognize common self-antigens (such as MHC proteins or blood group antigens) are deleted. Once B cells are released into the blood, however, anergy appears to be the main mechanism for preventing responses to self-proteins. When B cells encounter an antigen not associated with an antigen-presenting cell, the necessary costimulation signal often is missing. In this case, the B cell is likely to become anergic (inactivated) rather than activated.

Table 22.5 summarizes the activities of cells involved in immune responses.

Cancer Immunology

Although the immune system responds to cancerous cells, for example, in tumor surveillance, often immunity provides inadequate protection, as evidenced by the number of people dying each year from cancer. Over the past 25 years, considerable research has focused on *cancer immunology,* the study of ways to use immune responses for detecting, monitoring, and treating cancer. For example, some tumors of the colon release *carcinoembryonic antigen (CEA)* into the blood and prostate cancer cells release *prostate-specific antigen (PSA).* Detecting these antigens in blood does not provide definitive diagnosis of cancer, however, because both antigens are also released in certain noncancerous conditions. Still, high levels of cancer-related antigens in the blood often do indicate the presence of a malignant tumor.

Finding ways to induce our immune system to mount vigorous attacks against cancerous cells has been an elusive goal. Many different techniques have been tried, with only modest success. In one method, inactive lymphocytes are removed in a blood sample and cultured with interleukin-2. The resulting *lymphokine-activated killer (LAK) cells* are then transfused back into the patient's blood. Although LAK cells have produced dramatic improvement in a few cases, severe complications affect most patients. In another method, lymphocytes procured from a small biopsy sample of a tumor are cultured with interleukin-2. After their proliferation in culture, such *tumor-infiltrating lymphocytes (TILs)* are reinjected. About a quarter of patients with

Table 22.5	Summary of Functions of Cells Participating in Immune Responses
Cell	**Functions**
Antigen-Presenting Cells (APCs)	
Macrophage	Phagocytosis; processing and presentation of foreign antigens to T cells; secretion of interleukin-1, which stimulates secretion of interleukin-2 by helper T cells and induces proliferation of B cells; secretion of interferons that stimulate T cell growth.
Dendritic cell	Processes and presents antigen to T cells and B cells; found in mucous membranes, skin, and lymph nodes.
B cell	Processes and presents antigen to helper T cells.
Lymphocytes	
Cytotoxic T (T_C) cell or CD8$^+$ T cell	Causes lysis and death of foreign cells by releasing perforin and lymphotoxin; releases other cytokines that attract macrophages and increase their phagocytic activity (gamma-IFN) and prevent macrophage migration from site of action (macrophage migration inhibition factor).
Helper T (T_H) cell or CD4$^+$ T cell	Cooperates with B cells to amplify antibody production by plasma cells and secretes interleukin-2, which stimulates proliferation of T cells and B cells. May secrete gamma-IFN and tumor necrosis factor (TNF), which stimulate inflammatory response.
Memory T cell	Remains in lymphatic tissue and recognizes original invading antigens, even years after the first encounter.
B cell	Differentiates into antibody-producing plasma cell.
Plasma cell	Descendant of B cell that produces and secretes antibodies.
Memory B cell	Descendent of B cell that remains after an immune response and is ready to respond rapidly and forcefully should the same antigen enter the body in the future.

malignant melanoma and renal-cell carcinoma who received TIL therapy showed significant improvement. The many studies currently underway provide reason to hope that immune-based methods will eventually lead to cures for cancer. ■

STRESS AND IMMUNITY

▶ O B J E C T I V E

• **Describe the effects of stress on immunity.**

The field of **psychoneuroimmunology (PNI)** deals with communication pathways that link the nervous, endocrine, and immune systems. PNI research appears to justify what people have long observed: Your thoughts, feelings, moods, and beliefs influence your level of health and the course of disease. For example, cortisol, a hormone secreted by the adrenal cortex in association with the stress response, inhibits immune system activity.

If you want to observe the relationship between lifestyle and immune function, visit a college campus. As the semester progresses and the workload accumulates, an increasing number of students can be found in the waiting rooms of student health services. When work and stress pile up, health habits can change. Many people smoke or consume more alcohol when stressed, two habits detrimental to optimal immune function. Under stress, people are less likely to eat well or exercise regularly, two habits that enhance immunity.

People resistant to the negative health effects of stress are more likely to experience a sense of control over the future, a commitment to their work, expectations of generally positive outcomes for themselves, and feelings of social support. To increase your stress resistance, cultivate an optimistic outlook, get involved in your work, and build good relationships with others.

Adequate sleep and relaxation are especially important for a healthy immune system. But when there aren't enough hours in the day, you may be tempted to steal some from the night. While skipping sleep may give you a few more hours of productive time in the short run, in the long run you end up even farther behind, especially if getting sick keeps you out of commission for several days, blurs your concentration, and blocks your creativity.

Even if you make time to get eight hours of sleep, stress can cause insomnia. If you find yourself tossing and turning at night, it's time to improve your stress management and relaxation skills! Be sure to unwind from the day before going to bed.

▶ C H E C K P O I N T

24. What do positive selection, negative selection, and anergy accomplish?
25. Have you ever observed a connection between stress and illness in your own life?

AGING AND THE IMMUNE SYSTEM

▶ O B J E C T I V E

• **Describe the effects of aging on the immune system.**

With advancing age, most people become more susceptible to all types of infections and malignancies. Their response to vaccines is decreased, and they tend to produce more autoantibodies (antibodies against their body's own molecules). In addition, the immune system exhibits lowered levels of function. For example, T cells become less responsive to antigens, and fewer T cells respond to infections. This may result from age-related atrophy of the thymus or decreased production of thymic hormones. Because the T cell population decreases with age, B cells are also less responsive. Consequently, antibody levels do not increase as rapidly in response to a challenge by an antigen, resulting in increased susceptibility to various infections. It is for this key reason that elderly individuals are encouraged to get influenza (flu) vaccinations each year.

• • •

To appreciate the many ways that the lymphatic and immune system contributes to homeostasis of other body systems, examine *Focus on Homeostasis: The Lymphatic and Immune System* on page 796.

Next, in Chapter 23, we will explore the structure and function of the respiratory system and see how its operation is regulated by the nervous system. Most importantly, the respiratory system provides for gas exchange—taking in oxygen and blowing off carbon dioxide. The cardiovascular system aids gas exchange by transporting blood containing the gases between the lungs and tissue cells.

Focus on
Homeostasis:

The
Lymphatic
and
Immune
System

Body System	Contribution of the Lymphatic and Immune System
For all body systems	B cells, T cells, and antibodies protect all body systems from attack by harmful foreign invaders (pathogens), foreign cells, and cancer cells.
Integumentary system	Lymphatic vessels drain excess interstitial fluid and leaked plasma proteins from dermis of skin. Immune system cells (Langerhans cells) in skin help protect skin. Lymphatic tissue provides IgA antibodies in sweat.
Skeletal system	Lymphatic vessels drain excess interstitial fluid and leaked plasma proteins from connective tissue around bones.
Nervous system	Lymphatic vessels drain excess interstitial fluid and leaked plasma proteins from the peripheral nervous system.
Muscular system	Lymphatic vessels drain excess interstitial fluid and leaked plasma proteins from muscles.
Endocrine system	Flow of lymph helps distribute some hormones and cytokines. Lymphatic vessels drain excess interstitial fluid and leaked plasma proteins from endocrine glands.
Cardiovascular system	Lymph returns excess fluid filtered from blood capillaries and leaked plasma proteins to venous blood. Macrophages in spleen destroy aged red blood cells and remove debris in blood.
Respiratory system	Tonsils, alveolar macrophages, and MALT (mucosa-associated lymphatic tissue) help protect airways and lungs from pathogens. Lymphatic vessels drain excess interstitial fluid from lungs.
Digestive system	Tonsils and MALT help defend against toxins and pathogens that penetrate the body from the gastrointestinal tract. Digestive system provides IgA antibodies in saliva and gastrointestinal secretions. Lymphatic vessels pick up absorbed dietary lipids and fat-soluble vitamins from the small intestine and transport them to the blood. Lymphatic vessels drain excess interstitial fluid and leaked plasma proteins from organs of the digestive system.
Urinary system	Lymphatic vessels drain excess interstitial fluid and leaked plasma proteins from organs of the urinary system. MALT helps defend against toxins and pathogens that penetrate the body via the urethra.
Reproductive systems	Lymphatic vessels drain excess interstitial fluid and leaked plasma proteins from organs of the reproductive system. MALT helps defend against toxins and pathogens that penetrate the body via the vagina and penis. In females, sperm deposited in the vagina are not attacked as foreign invaders due to inhibition of immune responses. IgG antibodies can cross the placenta to provide protection to a developing fetus. Lymphatic tissue provides IgA antibodies in milk of nursing mother.

DISORDERS: HOMEOSTATIC IMBALANCES

AIDS: Acquired Immunodeficiency Syndrome

Acquired immunodeficiency syndrome is a condition in which a person experiences a telltale assortment of infections due to the progressive destruction of immune system cells by the **human immunodeficiency virus (HIV).** As we will see, AIDS represents the end stage of infection by HIV. A person who is infected with HIV may be symptom free for many years, even while the virus is actively attacking the immune system. HIV infection is serious and usually fatal because it takes control of and destroys the very cells that the body deploys to attack the virus. In the two decades after the first five cases were reported in 1981, 22 million people died of AIDS. Worldwide, 35 to 40 million people are currently infected with HIV.

Epidemiology

Because HIV is present in the blood and some body fluids, it is most effectively transmitted by actions or practices that involve the exchange of blood or body fluids between people. Within most populations, HIV is transmitted in semen or vaginal fluid during unprotected (without a condom) anal, vaginal, or oral sex. The likelihood of transmission of HIV from an infected person to an uninfected person varies significantly depending on the type of exposure or contact involved. The risk of becoming infected with HIV through unprotected oral sex is lower than that of unprotected anal or vaginal sex. HIV also is transmitted by direct blood-to-blood contact, such as occurs among intravenous drug users who share hypodermic needles. People at high risk are the sexual partners of HIV-infected individuals; those at lesser risk include healthcare professionals who may be accidentally stuck by HIV-contaminated hypodermic needles. In addition, HIV may be transmitted from a mother to her baby during a vaginal delivery or during breast-feeding.

In the United States and Europe before 1985, HIV was unknowingly spread by the transfusion of blood and blood products containing the virus. Effective HIV screening of blood instituted after 1985 has largely eliminated this mode of HIV transmission in the United States and other developed nations. By contrast, in sub-Saharan Africa, where two-thirds of those infected with HIV live, 25% of transfused blood is not screened for HIV.

Most people in economically advantaged, industrial nations who are diagnosed as having AIDS are either homosexual men who have engaged in unprotected anal intercourse or intravenous drug users. The rate of new HIV infections in the United States paints a different and disturbing picture: The greatest increases in new HIV infections are among African-Americans, Hispanics, women, and teenagers. In developing nations, HIV is largely transmitted during unprotected heterosexual intercourse, but the problem is compounded by a contaminated blood supply and unsanitary health care practices.

HIV is a very fragile virus; it cannot survive for long outside the human body. The virus is not transmitted by insect bites. One cannot become infected by casual physical contact with an HIV-infected person, such as by hugging or sharing household items. The virus can be eliminated from personal care items and medical equipment by exposing them to heat (135°F for 10 minutes) or by cleaning them with common disinfectants such as hydrogen peroxide, rubbing alcohol, Lysol® household bleach, or germicidal cleansers such as Betadine or Hibiclens. Standard dishwashing and clothes washing also kills HIV.

The chance of transmitting or of being infected by HIV during vaginal or anal intercourse can be greatly reduced—although not entirely eliminated—by the use of latex condoms. Public health programs aimed at encouraging drug users not to share needles have proved effective at checking the increase in new HIV infections in this population. Also, the prophylactic administration of certain drugs such as AZT (discussed shortly) to pregnant HIV-infected women has proved remarkably effective in reducing the transmission of the virus to their babies.

Pathogenesis of HIV Infection

HIV consists of two identical strands of genetic material surrounded by a protective protein coat. To replicate, a virus must enter a host cell, where it uses the cell's enzymes, ribosomes, and nutrients to reproduce copies of its genetic information. Unlike most viruses, however, HIV is a form of **retrovirus,** a virus whose genetic information is carried in RNA instead of DNA. Once inside the host cell, a viral enzyme called **reverse transcriptase** reads the viral RNA strand and makes a DNA copy.

The coat that surrounds HIV's RNA and reverse transcriptase is composed of many molecules of a protein termed P24. In addition, the coat is wrapped by an envelope composed of a lipid bilayer penetrated by a distinctive assortment of glycoproteins (Figure 22.21). One of these glycoproteins, GP120, functions as the docking protein that binds to CD4 molecules on T cells, macrophages, and dendritic cells. In addition, HIV must simultaneously attach to a coreceptor in the host cell's plasma membrane—a molecule designated CCR5 in antigen-presenting cells

Figure 22.21 Human immunodeficiency virus (HIV), the causative agent of AIDS. The core contains RNA and reverse transcriptase, plus several other enzymes. The protein coat (capsid) around the core consists of multiple copies of a protein called P24. The envelope consists of a lipid bilayer studded with glycoproteins (GP120 and GP41) that assist HIV's binding to and entering target cells.

🔑 **HIV is most effectively transmitted by practices that involve the exchange of body fluids.**

100–140nm

Which cells of the immune system are attacked by HIV?

and CXCR4 in T cells—before it can gain entry into the cell. Another glycoprotein, GP41, helps the viral lipid bilayer fuse with the host cell's lipid bilayer.

The binding of HIV's docking proteins with the receptors and coreceptors in the host cell's plasma membrane causes receptor-mediated endocytosis, which brings the virus into the cell's cytoplasm. Once inside the cell, HIV sheds its protein coat. The reverse transcriptase enzyme makes a DNA copy of the viral RNA, and the viral DNA copy becomes integrated into the cell's DNA. Thus, the viral DNA is duplicated along with the host cell's DNA during normal cell division. In addition, the viral DNA can cause the infected cell to begin producing millions of copies of viral RNA and to assemble new protein coats for each copy. The new HIV copies bud off from the cell's plasma membrane and circulate in the blood to infect other cells.

HIV mainly damages helper ($CD4^+$) T cells, and it does so in various ways. Over 10 billion viral copies may be produced each day. The viruses bud so rapidly from an infected cell's plasma membrane that cell lysis eventually occurs. In addition, the body's defenses attack the infected cells, killing them but not all the viruses they harbor. In most HIV-infected individuals, $CD4^+$ T cells are initially replaced as fast as they are destroyed. After several years, however, the body's ability to replace $CD4^+$ T cells is slowly exhausted, and the number of $CD4^+$ T cells in circulation progressively declines.

Signs, Symptoms, and Diagnosis of HIV Infection

Immediately after being infected with HIV, most people experience a brief flu-like illness. Common signs and symptoms are fever, fatigue, rash, headache, joint pain, sore throat, and swollen lymph nodes. In addition, about 50% of infected people have night sweats. After three to four weeks, plasma cells begin secreting antibodies to components of the HIV protein coat. These antibodies are detectable in blood plasma and form the basis for some of the screening tests for HIV. When people test "HIV-positive," this usually means they have antibodies to HIV epitopes in their bloodstream. If an acute HIV infection is suspected but the antibody test is negative, laboratory tests based on detection of HIV's RNA or P24 coat protein in blood plasma can confirm the presence of HIV.

Progression to AIDS

After a period of 2 to 10 years, the virus destroys enough $CD4^+$ T cells that most infected people begin to experience symptoms of immunodeficiency. HIV-infected people commonly have enlarged lymph nodes and experience persistent fatigue, involuntary weight loss, night sweats, skin rashes, diarrhea, and various lesions of the mouth and gums. In addition, the virus may begin to infect neurons in the brain, affecting the person's memory and producing visual disturbances.

As the immune system slowly collapses, an HIV-infected person becomes susceptible to a host of *opportunistic infections.* These are diseases caused by microorganisms that are normally held in check but now proliferate because of the defective immune system. AIDS is diagnosed when the $CD4^+$ T cell count drops below 200 cells per microliter (= cubic millimeter) of blood or when opportunistic infections arise, whichever occurs first. Typically, opportunistic infections eventually are the cause of death.

Treatment of HIV Infection

At present, infection with HIV cannot be cured, and despite intensive research, no effective vaccine is yet available to provide immunity against HIV. However, two categories of drugs have proved successful in extending the life of many HIV-infected individuals. The first category, *reverse transcriptase inhibitors,* interferes with the action of the reverse transcriptase enzyme that the virus uses to convert its RNA into a DNA copy. Among the drugs in this category are zidovudine (also known as AZT), didanosine (ddI), dideoxycytidine (ddC), and stavudine (d4T). Trizivir, approved in 2000 for treatment of HIV infection, combines three reverse transcriptase inhibitors (abacavir, lamivudine, and zidovudine) in one pill. The second category of drugs is the *protease inhibitors.* These drugs interfere with the action of protease, a viral enzyme that cuts proteins into pieces to assemble the coat of newly produced HIV particles. Drugs in this category include nelfinavir, saquinavir, ritonavir, and indinavir.

In 1996, physicians treating HIV-infected patients widely adopted *highly active antiretroviral therapy (HAART),* also known as *triple therapy*—a combination of two differently acting reverse transcriptase inhibitors and one protease inhibitor. This regimen has been used extensively for treating patients at virtually all stages of HIV infection, regardless of the $CD4^+$ T cell count and viral load (the number of copies of HIV RNA in a milliliter of plasma). Most HIV-infected individuals receiving HAART experience a drastic reduction in viral load and an increase in the number of $CD4^+$ T cells in their blood. Not only does HAART delay the progression of HIV infection to AIDS, but many individuals with AIDS have seen the remission or disappearance of opportunistic infections and an apparent return to health. Unfortunately, triple therapy is very costly (exceeding $10,000 per year), the dosing schedule is grueling, and not all people can tolerate the toxic side effects of these drugs. Data published in 2001 suggests that for some patients HAART could be postponed until the $CD4^+$ T cell count drops to 200 cells per microliter.

Allergic Reactions

A person who is overly reactive to a substance that is tolerated by most other people is said to be **hypersensitive (allergic).** Whenever an allergic reaction takes place, some tissue injury occurs. The antigens that induce an allergic reaction are called **allergens.** Common allergens include certain foods (milk, peanuts, shellfish, eggs), antibiotics (penicillin, tetracycline), vaccines (pertussis, typhoid), venoms (honeybee, wasp, snake), cosmetics, chemicals in plants such as poison ivy, pollens, dust, molds, iodine-containing dyes used in certain x-ray procedures, and even microbes.

There are four basic types of hypersensitivity reactions: type I (anaphylactic), type II (cytotoxic), type III (immune-complex), and type IV (cell-mediated). The first three are antibody-mediated immune responses; the last is a cell-mediated immune response.

Type I (anaphylactic) reactions are the most common and occur within a few minutes after a person sensitized to an allergen is reexposed to it. In response to certain allergens, some people produce IgE antibodies that bind to the surface of mast cells and basophils. The next time the same allergen enters the body, it attaches to the IgE antibodies already present. In response, the mast cells and basophils release histamine, prostaglandins, leukotrienes, and kinins. Collectively, these mediators cause vasodilation, increased blood capillary permeability, increased smooth muscle contraction in the airways of the lungs, and increased mucus secretion. As a result, a person may experience inflammatory responses, difficulty in breathing through the constricted airways, and a runny nose from excess mucus secretion. In **anaphylactic**

shock, which may occur in a susceptible individual who has just received a triggering drug or been stung by a wasp, wheezing and shortness of breath as airways constrict are usually accompanied by shock due to vasodilation and fluid loss from blood. This life-threatening emergency is usually treated by injecting epinephrine to dilate the airways and strengthen the heartbeat.

Type II (cytotoxic) reactions are caused by antibodies (IgG or IgM) directed against antigens on a person's blood cells (red blood cells, lymphocytes, or platelets) or tissue cells. The reaction of antibodies and antigens usually leads to activation of complement. Type II reactions, which may occur in incompatible blood transfusion reactions, damage cells by causing lysis.

Type III (immune-complex) reactions involve antigens, antibodies (IgA or IgM), and complement. When certain ratios of antigen to antibody occur, the immune complexes are small enough to escape phagocytosis, but they become trapped in the basement membrane under the endothelium of blood vessels, where they activate complement and cause inflammation. Glomerulonephritis and rheumatoid arthritis (RA) arise in this way.

Type IV (cell-mediated) reactions or **delayed hypersensitivity reactions** usually appear 12–72 hours after exposure to an allergen. Type IV reactions occur when allergens are taken up by antigen-presenting cells (such as Langerhans cells in the skin) that migrate to lymph nodes and present the allergen to T cells, which then proliferate. Some of the new T cells return to the site of allergen entry into the body, where they produce gamma-interferon, which activates macrophages, and tumor necrosis factor, which stimulates an inflammatory response. Intracellular bacteria such as *Mycobacterium tuberculosis* trigger this type of cell-mediated immune response, as do certain haptens, such as poison ivy toxin. The skin test for tuberculosis also is a delayed hypersensitivity reaction.

Autoimmune Diseases

In an **autoimmune disease** (aw-tō-i-MŪN) or **autoimmunity,** the immune system fails to display self-tolerance and attacks the person's own tissues. Autoimmune diseases usually arise in early adulthood and are common, afflicting an estimated 5% of adults in North America and Europe. Females suffer autoimmune diseases twice as often as males. Recall that self-reactive B cells and T cells normally are deleted or undergo anergy during negative selection (see Figure 22.20). Apparently, this process is not 100% effective. Under the influence of unknown environmental triggers and certain genes that make some people more susceptible, self-tolerance breaks down, leading to activation of self-reactive clones of T cells and B cells. These cells then generate cell-mediated or antibody-mediated immune responses against self-antigens.

A variety of mechanisms produce different autoimmune diseases. Some involve production of **autoantibodies,** antibodies that bind to and stimulate or block self-antigens. For example, autoantibodies that mimic TSH (thyroid-stimulating hormone) are present in Graves disease and stimulate secretion of thyroid hormones (thus producing hyperthyroidism), whereas autoantibodies that bind to and block acetylcholine receptors cause muscle weakness in myasthenia gravis. Other autoimmune diseases involve activation of cytotoxic T cells that destroy certain body cells. Examples include type 1 diabetes mellitus, in which T cells attack the insulin-producing pancreatic beta cells, and multiple sclerosis (MS), in which T cells attack myelin sheaths around

axons of neurons. Inappropriate activation of helper T cells or excessive production of gamma-interferon also occur in certain autoimmune diseases. Other autoimmune disorders include rheumatoid arthritis (RA), systemic lupus erythematosus (SLE), rheumatic fever, hemolytic and pernicious anemias, Addison's disease, Hashimoto's thyroiditis, and ulcerative colitis.

Therapies for various autoimmune diseases include removal of the thymus (thymectomy), injections of interferon beta, immunosuppressive drugs, and plasmapheresis, in which the person's blood plasma is filtered to remove antibodies and antigen–antibody complexes.

Severe Combined Immunodeficiency Disease

Severe combined immunodeficiency disease (SCID) is a rare inherited disorder in which both B cells and T cells are missing or inactive in providing immunity. Scientists have now identified mutations in several genes that are responsible for some types of SCID. One gene codes for an enzyme known as adenosine deaminase (ADA). Deficiency of ADA leads to accumulations of adenosine, ATP (adenosine triphosphate), and deoxy-ATP, which inhibit activity of an enzyme needed for synthesis of DNA. Thus, lymphocytes that lack ADA cannot proliferate to carry out their immune functions. Other mutations affect the enzymes needed to shuffle and rearrange gene segments to generate diverse B-cell and T-cell receptors. Another gene, on the X chromosome, codes for one of the three polypeptide chains (called the gamma chain) that together form the interleukin-2 receptor. This chain also is part of the receptors for Il-4, Il-7, Il-11, and Il-15. In this type of SCID, which afflicts boys exclusively, T cells fail to mature because interleukin signals cannot be received.

The most famous patient with SCID was David, the "bubble boy." He was placed in a sterile see-through chamber shortly after birth in the 1970s to protect him from microbes that his body could not fight. At age 12, David underwent a bone marrow transplant in an effort to correct his disorder. Eighty days after the transplant and still in a germ-free environment, David developed some of the symptoms of infectious mononucleosis, a condition caused by the Epstein–Barr virus. He was brought out of isolation for easier treatment, with the hope that the transplant would provide the same protection as his sterile plastic chamber. It probably was successful but, unfortunately, David died 4 months later from cancer associated with Epstein–Barr virus. David's legacy has continued, however, because some of his cells were used in the experiments that localized the genetic defect to the gene that codes for the gamma chain of the interleukin-2 receptor.

In some cases, an infusion of red bone marrow cells from a sibling having very similar MHC (HLA) antigens can provide normal stem cells that give rise to normal B and T cells. The result can be a complete cure. Less than 30% of afflicted patients, however, have a compatible sibling who could be the donor. Another approach for treatment of SCID is to provide good copies of the defective gene through gene therapy. In 1990 and 1991, two girls with ADA deficiency were the first persons in the United States to receive gene therapy for a life-threatening disorder, and they have continued to improve while receiving infusions of gene-corrected T cells. Because the procedure involves infusion of mature T cells, however, patients must be retreated every couple of months for life.

Infectious Mononucleosis

Infectious mononucleosis (IM) is a contagious disease caused by the *Epstein–Barr virus (EBV)*. "Mono" occurs mainly in children and young adults, and more often in females than in males by a 3:1 ratio. The virus most commonly enters the body through intimate oral contact such as kissing. It then multiplies in lymphatic tissues and spreads into the blood, where it infects and multiplies in B cells. Because of this infection, the B cells become enlarged and abnormal in appearance such that they resemble monocytes, the primary reason for the term *mononucleosis*. Signs and symptoms include an elevated white blood cell count with an abnormally high percentage of lymphocytes, fatigue, headache, dizziness, sore throat, enlarged and tender lymph nodes, and fever. There is no cure for infectious mononucleosis, but the disease usually runs its course in a few weeks.

Lymphomas

Lymphomas (lim-FŌ-mas; *lymph-* = clear water; *-oma* = tumor) are cancers of the lymphatic organs, especially the lymph nodes. Most have no known cause. The two main types of lymphomas are Hodgkin disease and non-Hodgkin lymphoma.

Hodgkin disease (HD) is characterized by a painless, nontender enlargement of one or more lymph nodes, most commonly in the neck, chest, and axilla. If the disease has metastasized from these sites, fever, night sweats, weight loss, and bone pain also occur. HD primarily affects individuals between ages 15 and 35 and those over 60, and it is more common in males. If diagnosed early, HD has a 90–95% cure rate.

Non-Hodgkin lymphoma (NHL), which is more common than HD, occurs in all age groups, the incidence increasing with age to a maximum between ages 45 and 70. NHL may start the same way as HD but may also include an enlarged spleen, anemia, and general malaise. Up to half of all individuals with NHL are cured or survive for a lengthy period. Treatment options for both HD and NHL include radiation therapy, chemotherapy, and bone marrow transplantation.

Systemic Lupus Erythematosus

Systemic lupus erythematosus (er-e'-thēm-a-TŌ-sus), **SLE,** or **lupus** (*lupus* = wolf) is a chronic autoimmune, inflammatory disease that affects multiple body systems. Most cases of SLE occur in women between the ages of 15 and 25, more often in blacks than in whites. Although the cause of SLE is not known, both a genetic predisposition to the disease and environmental factors contribute. Sex hormones appear to influence the development of SLE. Females are nine times more likely than males to suffer from SLE. The disorder often occurs in females who exhibit extremely low levels of androgens.

Signs and symptoms of SLE include joint pain, slight fever, fatigue, oral ulcers, weight loss, enlarged lymph nodes and spleen, photosensitivity, rapid loss of large amounts of scalp hair, and sometimes an eruption across the bridge of the nose and cheeks called a "butterfly rash." The erosive nature of some of the SLE skin lesions was thought to resemble the damage inflicted by the bite of a wolf—thus, the term lupus.

Two immunological features of SLE are excessive activation of B cells and inappropriate production of autoantibodies against DNA (anti-DNA antibodies) and other components of cellular nuclei such as histone proteins. Triggers of B cell activation are thought to include various chemicals and drugs, viral and bacterial antigens, and exposure to sunlight. Circulating complexes of abnormal autoantibodies and their "antigens" cause damage in tissues throughout the body. Kidney damage occurs as the complexes become trapped in the basement membrane of kidney capillaries thereby obstructing blood filtering. Renal failure is the most common cause of death.

MEDICAL TERMINOLOGY

Adenitis (ad'-e-NĪ-tis; *aden-* = gland; *-itis* = inflammation of) Enlarged, tender, and inflamed lymph nodes resulting from an infection.

Allograft (AL-ō-graft; *allo-* = other) A transplant between genetically distinct individuals of the same species. Skin transplants from other people and blood transfusions are allografts.

Autograft (AW-tō-graft; *auto-* = self) A transplant in which one's own tissue is grafted to another part of the body (such as skin grafts for burn treatment or plastic surgery).

Chronic fatigue syndrome (CFS) A disorder, usually occurring in young adults and primarily in females, characterized by (1) extreme fatigue that impairs normal activities for at least 6 months and (2) the absence of other known diseases (cancer, infections, drug abuse, toxicity, or psychiatric disorders) that might produce similar symptoms.

Congenital thymic hypoplasia (DiGeorge syndrome) A disorder in which the thymus is absent or much smaller than normal; caused by faulty development of the third and fourth pharyngeal pouches (see Figure 18.21 on page 623). Thymic deficiency decreases cell-mediated immune responses, but antibody-mediated immune responses are more or less normal. Transplantation of a fetal thymus provides excellent treatment of babies that completely lack a thymus.

Gamma globulin (GLOB-ū-lin) Suspension of immunoglobulins from blood consisting of antibodies that react with a specific pathogen. It is prepared by injecting the pathogen into animals, removing blood from the animals after antibodies have been produced, isolating the antibodies, and injecting them into a human to provide short-term immunity.

Hypersplenism (hī-per-SPLĒN-izm; *hyper-* = over) Abnormal splenic activity due to splenic enlargement and associated with an increased rate of destruction of normal blood cells and platelets.

Lymphadenopathy (lim-fad'-e-NOP-a-thē; *lymph-* = clear fluid; *-pathy* = disease) Enlarged, sometimes tender lymph glands.

Lymphedema (lim'-fe-DĒ-ma; *edema* = swelling) Accumulation of lymph producing subcutaneous tissue swelling.

Splenomegaly (splē'-nō-MEG-a-lē; *mega-* = large) Enlarged spleen.

Tonsillectomy (ton'-si-LEK-tō -mē; *-ectomy* = excision) Removal of a tonsil.

Xenograft (ZEN-ō-graft; *xeno-* = strange or foreign) A transplant between animals of different species. Xenografts from porcine (pig) or bovine (cow) tissue may be used in a human as a physiological dressing for severe burns.

STUDY OUTLINE

INTRODUCTION (p. 765)

1. The ability to ward off disease is called resistance. Lack of resistance is called susceptibility.
2. Nonspecific resistance refers to a wide variety of body responses against a wide range of pathogens; specific resistance or immunity involves activation of specific lymphocytes to combat a particular foreign substance.

THE LYMPHATIC AND IMMUNE SYSTEM (p. 765)

1. The lymphatic and immune system carries out immune responses and consists of lymph, lymphatic vessels, and structures and organs that contain lymphatic tissue (specialized reticular tissue containing many lymphocytes).
2. It drains interstitial fluid, transports dietary lipids, and protects against invasion through immune responses.
3. Lymphatic vessels begin as closed-ended lymphatic capillaries in tissue spaces between cells.
4. Interstitial fluid drains into lymphatic capillaries, thus forming lymph.
5. Lymph capillaries merge to form larger vessels, called lymphatic vessels, which convey lymph into and out of lymph nodes.
6. The route of lymph flow is from lymph capillaries to lymphatic vessels to lymph trunks to the thoracic duct (or right lymphatic duct) to the subclavian veins.
7. Lymph flows because of skeletal muscle contractions and respiratory movements. Valves in lymphatic vessels also aid flow of lymph.
8. The primary lymphatic organs are red bone marrow and the thymus. Secondary lymphatic organs are lymph nodes, spleen, and lymphatic nodules.
9. The thymus lies between the sternum and the large blood vessels above the heart. It is the site of T cell maturation.
10. Lymph nodes are encapsulated, egg-shaped structures located along lymphatic vessels.
11. Lymph enters nodes through afferent lymphatic vessels, is filtered, and exits through efferent lymphatic vessels.
12. Lymph nodes are the site of proliferation of plasma cells and T cells.
13. The spleen is the largest single mass of lymphatic tissue in the body. It is a site of B cell proliferation into plasma cells and phagocytosis of bacteria and worn-out red blood cells.
14. Lymphatic nodules are scattered throughout the mucosa of the gastrointestinal, respiratory, urinary, and reproductive tracts. This lymphatic tissue is termed mucosa-associated lymphatic tissue (MALT).

DEVELOPMENT OF LYMPHATIC TISSUES (p. 774)

1. Lymphatic vessels develop from lymph sacs, which arise from developing veins. Thus, they are derived from mesoderm.
2. Lymph nodes develop from lymph sacs that become invaded by mesenchymal cells.

NONSPECIFIC RESISTANCE: INNATE DEFENSES (p. 775)

1. Mechanisms of nonspecific resistance include mechanical factors, chemical factors, antimicrobial proteins, natural killer cells, phagocytes, inflammation, and fever.
2. The skin and mucous membranes are the first line of defense against entry of pathogens.
3. Antimicrobial proteins include interferons, the complement system, and transferrins.
4. Natural killer cells and phagocytes attack and kill pathogens and defective cells in the body.
5. Inflammation aids disposal of microbes, toxins, or foreign material at the site of an injury, and prepares the site for tissue repair.
6. Fever intensifies the antiviral effects of interferons, inhibits growth of some microbes, and speeds up body reactions that aid repair.
7. Table 22.1 on page 779 summarizes the components of nonspecific resistance.

SPECIFIC RESISTANCE: IMMUNITY (p. 780)

1. Specific resistance to disease involves the production of a specific lymphocyte or antibody against a specific antigen and is called immunity.
2. B cells and T cells derive from stem cells in red bone marrow.
3. T cells complete their maturation and develop immunocompetence in the thymus.
4. In cell-mediated immune responses, cytotoxic T cells directly attack the invading antigen, whereas in antibody-mediated immune responses, plasma cells secrete antibodies.
5. Antigens (Ags) are chemical substances that are recognized as foreign by the immune system.
6. Antigen receptors exhibit great diversity due to genetic recombination.
7. "Self-antigens" termed major histocompatibility complex (MHC) antigens are unique to each person's body cells. All cells except red blood cells display MHC-I molecules; some cells also display MHC-II molecules.
8. Cells called antigen-presenting cells (APCs), which include macrophages, B cells, and dendritic cells, process antigens.
9. Exogenous antigens (formed outside the body) are presented together with MHC-II molecules to T cells, whereas endogenous (formed inside a body cell) antigens are presented together with MHC-I molecules.
10. Cytokines are small protein hormones needed for many normal cell functions. Some of them regulate immune responses (see Table 22.2 on page 784).

CELL-MEDIATED IMMUNITY (p. 785)

1. In a cell-mediated immune response, an antigen is recognized, specific T cells proliferate and differentiate into effector cells, and the antigen is eliminated.
2. T-cell receptors (TCRs) recognize antigen fragments associated with MHC molecules on the surface of a body cell.
3. Proliferation of T cells requires costimulation, either by cytokines such as interleukin-2 or by pairs of plasma membrane molecules.
4. T cells consist of several subpopulations. Helper T cells display CD4 protein, recognize antigen fragments associated with MHC-II molecules, and secrete several cytokines, most importantly interleukin-2, which acts as a costimulator for other helper T cells, cytotoxic T cells, and B cells. Cytotoxic T cells display CD8 protein and recognize antigen fragments associated with MHC-I molecules. Memory T cells remain after a cell-

mediated immune response and initiate a faster response when a pathogen bearing the same foreign antigen invades the body again.

5. Cytotoxic T cells eliminate invaders by secreting perforin, which causes cytolysis, and lymphotoxin, which causes fragmentation of the DNA of a target cell.

6. Cytotoxic T cells, macrophages, and natural killer cells carry out immunological surveillance, recognizing and destroying cancerous cells that display tumor antigens.

ANTIBODY-MEDIATED IMMUNITY (p. 788)

1. B cells can respond to unprocessed antigens, but their response is more intense when dendritic cells present antigen to them. Interleukin-2 and other cytokines secreted by helper T cells provide costimulation for the proliferation of B cells.

2. An activated B cell develops into a clone of antibody-producing plasma cells.

3. An antibody (Ab) is a protein that combines specifically with the antigen that triggered its production.

4. Antibodies consist of heavy and light chains and variable and constant regions.

5. Based on chemistry and structure, antibodies are grouped into five principal classes, (IgG, IgA, IgM, IgD, and IgE), each with specific biological roles.

6. Actions of antibodies include neutralization of antigen, immobilization of bacteria, agglutination and precipitation of antigen, activation of complement, and enhancement of phagocytosis.

7. Complement is a group of proteins whose functions complement immune responses and help clear antigens from the body.

8. Immunization against certain microbes is possible because memory B cells and memory T cells remain after a primary response to an antigen. The secondary response provides protection should the same microbe enter the body again.

SELF-RECOGNITION AND SELF-TOLERANCE (p. 793)

1. T cells undergo positive selection to ensure that they can recognize self-MHC proteins (self-recognition), and negative selection to ensure that they do not react to other self-proteins (self-tolerance). Negative selection involves both deletion and anergy.

2. B cells develop tolerance through deletion and anergy.

STRESS AND IMMUNITY (p. 795)

1. Psychoneuroimmunology (PNI) deals with communication pathways that link the nervous, endocrine, and immune systems. Thoughts, feelings, moods, and beliefs influence health and the course of disease.

2. Under stress, people are less likely to eat well or exercise regularly, two habits that enhance immunity.

AGING AND THE IMMUNE SYSTEM (p. 795)

1. With advancing age, individuals become more susceptible to infections and malignancies, respond less well to vaccines, and produce more autoantibodies.

2. Immune responses also diminish with age.

 SELF-QUIZ QUESTIONS

Fill in the blanks in the following statements.

1. The first line of nonspecific defense against pathogens are the _____ and _____; the second line of nonspecific defense are the _____, _____, and _____.

2. The two distinguishing features of immunity are _____ and _____.

Indicate whether the following statements are true or false.

3. The sequence of fluid flow from blood vessel to blood vessel by way of the lymphatic system is arteries (blood) to blood capillaries (blood) to interstitial spaces (interstitial fluid) to lymphatic capillaries (lymph) to lymphatic vessels (lymph) to lymphatic ducts (lymph) to subclavian veins (blood).

4. A person's T cells must be able to recognize the person's own MHC molecules, a process known as self-recognition, and lack reactivity to peptide fragments from the person's own proteins, a condition known as self-tolerance.

Choose the one best answer to the following questions.

5. Which of the following are functions of the lymphatic system? (1) draining interstitial fluid, (2) draining intracellular fluid, (3) transporting dietary lipids, (4) transporting nucleic acids, (5) protecting against invasion. (a) 1, 2, and 3, (b) 2, 3 and 4, (c) 3, 4, and 5, (d) 1, 2, and 4, (e) 1, 3 and 5.

6. Which of the following statements are *correct*? (1) Lymphatic vessels are found throughout the body, execept in avascular tissues, the CNS, portions of the spleen, and bone marrow. (2) Lymphatic capillaries allow interstitial fluid to flow into them but not out of them. (3) Anchoring filaments attach lymphatic endothelial cells to surrounding tissues. (4) Lymphatic vessels freely receive all the components of blood, including the formed elements. (5) Lymph vessels directly connect to blood vessels by way of the subclavian veins. (a) 1, 3, 4 and 5, (b) 2, 3, 4 and 5, (c) 1, 2, 3 and 4, (d) 1, 2, 4 and 5, (e) 1, 2, 3, and 5.

7. Which of the following are mechanical factors that help fight pathogens and disease? (1) tight junctions of epidermal cells, (2) mucus of mucous membranes, (3) saliva, (4) interferons, (5) complement. (a) 1, 3, and 4, (b) 2, 4, and 5, (c) 1, 4, and 5, (d) 1, 2, and 3, (e) 1, 2, and 4.

8. Which of the following are processes involved in inflammation? (1) vasodilation and increased permeability of blood vessels, (2) phagocyte emigration, (3) tissue repair, (4) opsonization, (5) adherence. (a) 1, 2, and 3, (b) 2, 3, and 4, (c) 3, 4, and 5, (d) 1, 3, and 5, (e) 2, 4, and 5.

9. Antibody-mediated immunity works mainly aganist (a) foregin tissue transplants, (b) intracellular pathogens, (c) extracellular pathogens, (d) cancer cells, (e) viruses.

10. Which of the following are functions of antibodies? (1) neutralization of antigens, (2) immobilization of bacteria, (3) agglutination and precipitation of antigens, (4) activation of complement, (5) enhancement of phagocytosis. (a) 1, 3 and 4, (b) 2, 4, and 5, (c) 1, 2, 3 and 4, (d) 1, 2, 3 and 5, (e) 1, 2, 3, 4, and 5.

11. Which of the following are *true*? (1) Lymphatic vessels resemble arteries. (2) Lymph is very similar to interstitial fluid. (3) Lacteals are specialized lymphatic capillaries responsible for transporting dietary lipids. (4) Lymph is normally a cloudy, pale yellow fluid. (5) The thoracic duct drains lymph from the upper right side of the body. (6) Lymph flow is maintained by skeletal muscle contractions, one-way valves, and breathing movements. (a) 1, 2, 5, and 6, (b) 2, 3, and 6, (c) 2, 3, 4, and 6, (d) 2, 4, and 6, (e) 3, 5, and 6.

12. Place the stages of phagocytosis in the correct order of occurrence. (1) formation of phagolysosome, (2) adherence to microbe, (3) destruction of microbe, (4) ingestion to form a phagosome, (5) chemotaxic attraction of phagocyte. (a) 2, 5, 4, 1, 3, (b) 4, 5, 2, 1, 3, (c) 5, 2, 4, 1, 3, (d) 5, 4, 2, 3, 1, (e) 2, 5, 1, 4, 3.

13. Match the following:
(1) red bone marrow   (2) thymus   (3) lymph nodes
(4) spleen   (5) lymphatic nodules
_____ (a) bean-shaped structures located along the length of lymphatic vessels; contain T cells, macrophages, and follicular dendritic cells; filter lymph
_____ (b) produces pre-T cells and B cells; found in flat bones and epiphyses of long bones
_____ (c) the single largest mass of lymphatic tissue in the body
_____ (d) responsible for the maturation of T cells
_____ (e) nonencapsulated clusters of lymphocytes located in all mucous membranes

14. Match the following:
_____ (a) recognize foreign antigens combined with MHC-1 molecules on the surface of body cells infected by viruses, some turmor cells, and cells of a tissue transplant
_____ (b) are programmed to recognize the reappearance of a previously encountered antigen
_____ (c) differentiate into plasma cells that secrete specific antibodies
_____ (d) process and present exogenous antigens; include macrophages, B cells, and dendritic cells
_____ (e) secrete cytokines as costimulators
_____ (f) ingest microbes or any foreign particulate matter; include neutrophils and macrophages
_____ (g) lymphocytes that have the ability to kill a wide variety of infectious microbes plus certain spontaneously arising tumor cells; lack antigen receptors

(1) helper T cells
(2) cytotoxic T cells
(3) memory T cells
(4) B cells
(5) NK cells
(6) phagocytes
(7) antigen-presenting cells

15. Match the following (answers may be used more than once):
_____ (a) participate in inflammation, opsonization, and cytolysis
_____ (b) stimulate histamine release, attract neutrophills by chemotaxis, promote phagocytosis, and destroy bacteria
_____ (c) glycoproteins that mark the surface of all body cells except for RBCs; distinguish self from nonself
_____ (d) foreign antigens present in fluids outside body cells
_____ (e) foreign antigens synthesized within body cells
_____ (f) small protein hormones that stimulate or inhibit many normal cell functions; serve as costimulators for B cell and T cell activity
_____ (g) a substance that has reactivity but lacks immunogenicity
_____ (h) causes vasodilation and increased permeability of blood vessels; is found in mast cells in connective tissue and in basophills and platelets in blood
_____ (i) polypeptides formed in blood; induce vasodilation and increased permeability of blood vessels; serve as chemotaxic agents for phagocytes
_____ (j) small parts of antigens that initiate immune responses
_____ (k) produced by virus-infected cells; they interfere with viral replication in host cells
_____ (l) chemicals released by NK and cytotoxic T cells that can cause cytolysis in microbes
_____ (m) glycoproteins that contain four polypeptide chains, two of which are identical to each other and two of which are variable and contain the anitgen-binding site

(1) exogenous antigens
(2) endogenous antigens
(3) interferons
(4) hapten
(5) cytokines
(6) perforins
(7) histamine
(8) major histocompatibility complex (MHC) antigens
(9) kinins
(10) antibodies
(11) complement proteins
(12) epitopes

CRITICAL THINKING QUESTIONS

1. Three-year-old Tariq was running barefoot through the grass when he felt a sharp pain and ran crying to his mother about a stepping on a "buzzy bug." Mom pulled an insect stinger out of his foot. Thirty minutes later, the sole of Tariq's foot was swollen around the stinger hole and itchy. What type of immune response is Tariq exhibiting?
   HINT *Ice and topical cortisone ointment relieved the symptoms.*

2. On the second day of Cara's wilderness backpacking trip, she got a splinter stuck in her right thumb. It didn't bother her too much (her feet hurt more than her thumb), so she put a bandage on it and ignored it. At the end of the trip, Cara had red streaks running along her right arm and tender swollen bumps in her right armpit. What happened to her arm?
   HINT *She should have paid more attention to that splinter.*

3. Esperanza watched as her mother got her "flu shot." "Why do you need a shot if you're not sick?" she asked. "So I won't get sick," answered her mom. Explain how the influenza vaccination prevents illness.
   HINT *A flu shot won't protect against measles, chicken pox, or pneumonia; it only works against the flu.*

ANSWERS TO FIGURE QUESTIONS

22.1 Red bone marrow contains stem cells that develop into lymphocytes.

22.2 Lymph is more similar to interstitial fluid than to blood plasma because the protein content of lymph is low.

22.3 The left and right lumbar trunks and the intestinal trunk empty into the cisterna chyli, which then drains into the thoracic duct.

22.4 Inhalation promotes the movement of lymph from abdominal lymphatic vessels toward the thoracic region.

22.5 T cells mature in the thymus.

22.6 Foreign substances that enter a lymph node in lymph may be phagocytized by macrophages or attacked by lymphocytes that mount immune responses.

22.7 White pulp of the spleen functions in immunity; red pulp of the spleen performs functions related to blood cells.

22.8 Lymphatic tissues begin to develop by the end of the fifth week of gestation.

22.9 Lysozyme, digestive enzymes, and oxidants can kill microbes.

22.10 *Redness* results from increased blood flow due to vasodilation; *pain,* from injury of nerve fibers, irritation by microbial toxins, kinins, and prostaglandins, and pressure due to edema; *heat,* from increased blood flow and heat released by locally increased metabolic reactions; *swelling,* from leakage of fluid from capillaries due to increased permeability.

22.11 Helper T cells participate in both cell-mediated and antibody-mediated immune responses.

22.12 Epitopes are small immunogenic parts of a larger antigen; haptens are small molecules that become immunogenic only when they attach to a body protein.

22.13 APCs include macrophages in tissues throughout the body, B cells in blood and lymphatic tissue, and dendritic cells in mucous membranes and the skin.

22.14 The first signal in T cell activation is antigen binding to a TCR; the second signal is a costimulator, such as a cytokine or another pair of plasma membrane molecules.

22.15 Cytotoxic T cells attack some tumor cells and transplanted tissue cells, as well as cells infected by viruses.

22.16 A clone of plasma cells secretes just one kind of antibody.

22.17 The variable regions recognize and bind to a specific antigen.

22.18 The classical pathway for the activation of complement is linked to antibody-mediated immunity because Ag–Ab complexes activate C1.

22.19 At peak secretion, approximately 1000 times more IgG is produced in the secondary response than in the primary response.

22.20 In deletion, self-reactive T cells or B cells die; in anergy, T cells or B cells are alive but are unresponsive to antigenic stimulation.

22.21 HIV attacks helper T cells.

The Respiratory System

A Practitioner's View

I used a previous edition of this textbook while in school for Respiratory Therapy. I have recently graduated from Massage Therapy School and have often referred to that edition. After reviewing this edition, I can see that there have been many improvements made over the years.

The Respiratory chapter begins with a thorough representation of the basic anatomy of the respiratory system, which is essential for any respiratory therapy student. I particularly like the part in the trachea section, which provides important information about tracheotomy and intubation. In the section on the lungs, while explaining the parietal and visceral pleurae, the section goes right into the potential for pathology in these membranes called pleurisy in these membrane. This gives students a look forward into the later months of study. The Pulmonary ventilation section is more than adequate.

The chapter reviews and quizzes are very helpful for students who have pre-test anxieties. The use of more cadaver photos gives students a more realistic impression of human anatomy. The medical terminology section is more informative than many other texts I have referred to. The figures and tables are great. Fig.23.18 and 23.19 are the best representation that I have ever seen.

Giovanni Curra, RRT, CMT
Fayette Community Hospital

THE RESPIRATORY SYSTEM AND HOMEOSTASIS

The respiratory system contributes to homeostasis by providing for the exchange of gases — oxygen and carbon dioxide — between the atmospheric air, blood, and tissues cells. It also helps adjust the pH of body fluids.

ENERGIZE YOUR STUDY
FOUNDATIONS CD
Anatomy Overview
- The Respiratory System

Animations
- Systems Contributions to Homeostasis
- The Case of the Worried Smoker
- The Case of the Coughing Boy

Concepts and Connections and Exercises reinforce your understanding

www.wiley.com/college/apcentral

INSIGHTS AND EXPLORATIONS
The American Lung Association reports that nearly 335,000 Americans die of lung disease every year. One out of every seven deaths is due to lung disease. Perhaps most disturbing is the claim that lung disease and other breathing problems are the number one killer of infants less than one-year old. Come explore in this web-based exercise the many respiratory disorders that affect people of all ages and genders.

C ells continually use oxygen (O_2) for the metabolic reactions that release energy from nutrient molecules and produce ATP. At the same time, these reactions release carbon dioxide (CO_2). Because an excessive amount of CO_2 produces acidity that can be toxic to cells, excess CO_2 must be eliminated quickly and efficiently. The two systems that cooperate to supply O_2 and eliminate CO_2 are the cardiovascular and respiratory systems. The respiratory system provides for gas exchange—intake of O_2 and elimination of CO_2—whereas the cardiovascular system transports blood containing the gases between the lungs and body cells. Failure of either system disrupts homeostasis by causing rapid death of cells from oxygen starvation and buildup of waste products. In addition to functioning in gas exchange, the respiratory system also participates in regulating blood pH, contains receptors for the sense of smell, filters inspired air, produces sounds, and rids the body of some water and heat in exhaled air.

RESPIRATORY SYSTEM ANATOMY

▶ OBJECTIVES

• **Describe the anatomy and histology of the nose, pharynx, larynx, trachea, bronchi, and lungs.**
• **Identify the functions of each respiratory system structure.**

The **respiratory system** consists of the nose, pharynx (throat), larynx (voice box), trachea (windpipe), bronchi, and lungs (Figure 23.1). Structurally, the respiratory system consists of two parts: (1) The **upper respiratory system** includes the nose, pharynx, and associated structures. (2) The **lower respiratory system** includes the larynx, trachea, bronchi, and lungs. Functionally, the respiratory system also consists of two parts: (1) The **conducting portion** consists of a series of interconnecting cavities and tubes both outside and within the lungs—the nose, pharynx, larynx, trachea, bronchi, bronchioles, and terminal bronchioles—that filter, warm, and moisten air and conduct it into the lungs. (2) The **respiratory portion** consists of tissues within the lungs where gas exchange occurs—the respiratory bronchioles, alveolar ducts, alveolar sacs, and alveoli, the main sites of gas exchange between air and blood. The volume of the conducting portion in an adult is about 150 mL; that of the respiratory portion is 5 to 6 liters.

The branch of medicine that deals with the diagnosis and treatment of diseases of the ears, nose, and throat is called **otorhinolaryngology** (ō′-tō-rī′-nō-lar′-in-GOL-ō-jē; *oto-* = ear; *rhino-* = nose; *laryngo-* = voice box; *-logy* = study of). A **pulmonologist** is a specialist in the diagnosis and treatment of diseases of the lungs.

Nose

The **nose** can be divided into external and internal portions. The external nose consists of a supporting framework of bone and hyaline cartilage covered with muscle and skin and lined by a mucous membrane. The frontal bone, nasal bones, and maxillae form the bony framework of the external nose (Figure 23.2a on page 808). The cartilaginous framework of the external nose consists of the **septal cartilage,** which forms the anterior portion of the nasal septum; the **lateral nasal cartilages** inferior to the nasal bones; and the **alar cartilages,** which form a portion of the walls of the nostrils. Because it consists of pliable hyaline cartilage, the cartilaginous framework of the external nose is somewhat flexible. On the undersurface of the external nose are two openings called the **external nares** (NĀ-rez; singular is **naris**) or **nostrils.** Figure 23.3 on page 809 shows the surface anatomy of the nose. The interior structures of the external nose have three functions: (1) warming, moistening, and filtering incoming air; (2) detecting olfactory stimuli; and (3) modifying speech vibrations as they pass through the large, hollow resonating chambers. *Resonance* refers to prolonging, amplifying, or modifying a sound by vibration.

The internal nose is a large cavity in the anterior aspect of the skull that lies inferior to the nasal bone and superior to the mouth; it also includes muscle and mucous membrane. Anteriorly, the internal nose merges with the external nose, and posteriorly it communicates with the pharynx through two openings called the **internal nares** or **choanae** (kō-Ā-nē) (see Figure 23.2b). Ducts from the paranasal sinuses (frontal, sphenoidal, maxillary, and ethmoidal paranasal sinuses) and the nasolacrimal ducts also open into the internal nose. The lateral walls of the internal nose are formed by the ethmoid, maxillae, lacrimal, palatine, and inferior nasal conchae bones (see Figure 7.9 on page 197); the ethmoid also forms the roof. The palatine bones and palatine processes of the maxillae, which together constitute the hard palate, form the floor of the internal nose.

The space within the internal nose is called the **nasal cavity.** The anterior portion of the nasal cavity, just inside the nostrils, is called the **vestibule** and is surrounded by cartilage; the superior part of the nasal cavity is surrounded by bone. A vertical partition, the **nasal septum,** divides the nasal cavity into right and left sides. The anterior portion of the septum consists primarily of hyaline cartilage; the remainder is formed by the vomer, perpendicular plate of the ethmoid, maxillae, and palatine bones (see Figure 7.11 on page 199).

When air enters the nostrils, it passes first through the vestibule, which is lined by skin containing coarse hairs that filter out large dust particles. Three shelves formed by projections of the superior, middle, and inferior nasal conchae extend out of each lateral wall of the cavity. The conchae, almost reaching the septum, subdivide each side of the nasal cavity into a series of groovelike passageways—the **superior, middle,** and **inferior meatuses** (mē-

Figure 23.1 Structures of the respiratory system.

The upper respiratory system includes the nose, pharynx, and associated structures; the lower respiratory system includes the larynx, trachea, bronchi, and lungs.

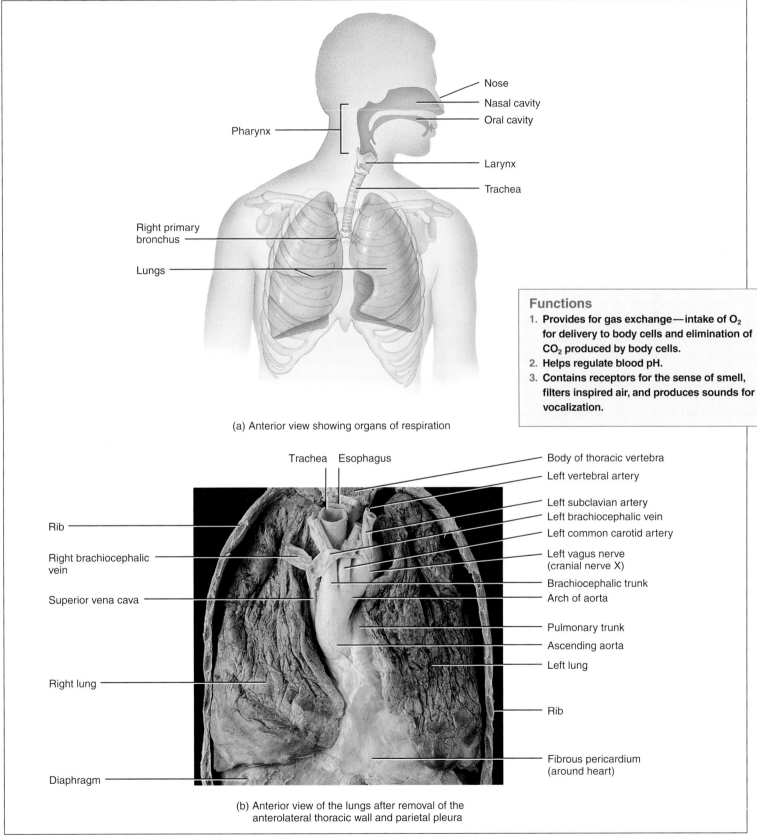

Nose
Nasal cavity
Oral cavity

Pharynx

Larynx

Trachea

Right primary bronchus

Lungs

Functions

1. **Provides for gas exchange—intake of O_2 for delivery to body cells and elimination of CO_2 produced by body cells.**
2. **Helps regulate blood pH.**
3. **Contains receptors for the sense of smell, filters inspired air, and produces sounds for vocalization.**

(a) Anterior view showing organs of respiration

Trachea   Esophagus

Body of thoracic vertebra
Left vertebral artery

Rib

Left subclavian artery
Left brachiocephalic vein
Left common carotid artery

Right brachiocephalic vein

Left vagus nerve (cranial nerve X)

Brachiocephalic trunk

Superior vena cava

Arch of aorta

Pulmonary trunk
Ascending aorta
Left lung

Right lung

Rib

Fibrous pericardium (around heart)

Diaphragm

(b) Anterior view of the lungs after removal of the anterolateral thoracic wall and parietal pleura

Which structures are part of the conducting portion of the respiratory system?

Figure 23.2 Respiratory structures in the head and neck. (See Tortora, *A Photographic Atlas of the Human Body,* Figures 11.2 and 11.3.)

 As air passes through the nose, it is warmed, filtered, and moistened.

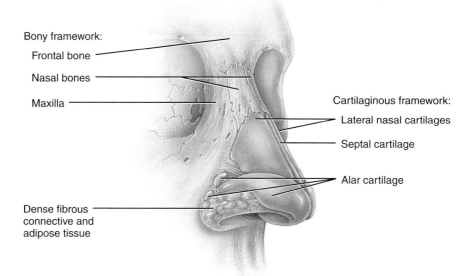

Bony framework:
Frontal bone
Nasal bones
Maxilla

Cartilaginous framework:
Lateral nasal cartilages
Septal cartilage
Alar cartilage

Dense fibrous connective and adipose tissue

(a) Anterolateral view of external portion of nose showing cartilaginous and bony framework

Sagittal plane

Nasal meatuses
Superior
Middle
Inferior

Frontal sinus
Frontal bone
Olfactory epithelium

Sphenoid bone
Sphenoidal sinus
Internal naris
Pharyngeal tonsil
Nasopharynx
Orifice of auditory (Eustachian) tube
Uvula
Palatine tonsil
Fauces
Oropharynx
Epiglottis
Laryngopharynx (hypopharynx)
Esophagus
Trachea

Superior
Middle
Inferior
Nasal conchae (turbinates)

Vestibule
External naris
Maxilla
Oral cavity
Palatine bone
Soft palate
Lingual tonsil
Hyoid bone
Ventricular fold (false vocal cord)
Laryngeal sinus (ventricle)
Vocal fold (true vocal cord)
Larynx
Thyroid cartilage
Cricoid cartilage
Thyroid gland

(b) Sagittal section of the left side of the head and neck showing the location of respiratory structures

 What is the path taken by air molecules into and through the nose?

Figure 23.3 Surface anatomy of the nose.

 The external nose has a cartilaginous and a bony framework.

Anterior view

1. **Root**: Superior attachment of the nose to the frontal bone
2. **Apex**: Tip of nose
3. **Bridge**: Bony framework of nose formed by nasal bones
4. **External naris**: Nostril; external opening into nasal cavity

 What part of the nose is attached to the frontal bone?

Ā-tus-ez = openings or passages). Mucous membrane lines the cavity and its shelves. The arrangement of conchae and meatuses increases surface area in the internal nose and prevents dehydration by acting as a baffle that traps water droplets during exhalation.

The olfactory receptors lie in the membrane lining the superior nasal conchae and adjacent septum. This region is called the **olfactory epithelium.** Inferior to the olfactory epithelium, the mucous membrane contains capillaries and pseudostratified ciliated columnar epithelium with many goblet cells. As inspired air whirls around the conchae and meatuses, it is warmed by blood in the capillaries. Mucus secreted by the goblet cells moistens the air and traps dust particles. Drainage from the nasolacrimal ducts and perhaps secretions from the paranasal sinuses also help moisten the air. The cilia move the mucus and trapped dust particles toward the pharynx, at which point they can be swallowed or spit out, thus removing particles from the respiratory tract.

Rhinoplasty

Rhinoplasty (RĪ-nō-plas′-tē; *-plasty* = to mold or to shape), commonly called a "nose job," is a surgical procedure in which the structure of the external nose is altered. Although rhinoplasty is often done for cosmetic reasons, it is sometimes performed to repair a fractured nose or a deviated nasal septum. In the proce-dure, both a local and general anesthetic are given, and with instruments inserted through the nostrils, the nasal cartilage is reshaped, and the nasal bones are fractured and repositioned, to achieve the desired shape. An internal packing and splint are inserted to keep the nose in the desired position while it heals. ■

Pharynx

The **pharynx** (FAIR-inks), or throat, is a funnel-shaped tube about 13 cm (5 in.) long that starts at the internal nares and extends to the level of the cricoid cartilage, the most inferior cartilage of the larynx (voice box) (Figure 23.4). The pharynx lies just posterior to the nasal and oral cavities, superior to the larynx, and just anterior to the cervical vertebrae. Its wall is composed of skeletal muscles and is lined with a mucous membrane. The pharynx functions as a passageway for air and food, provides a resonating chamber for speech sounds, and houses the tonsils, which participate in immunological reactions against foreign invaders.

The pharynx can be divided into three anatomical regions: (1) nasopharynx, (2) oropharynx, and (3) laryngopharynx. (See the lower orientation diagram in Figure 23.4.) The muscles of the entire pharynx are arranged in two layers, an outer circular layer and an inner longitudinal layer.

The superior portion of the pharynx, called the **nasopharynx,** lies posterior to the nasal cavity and extends to the plane of the soft palate. There are five openings in its wall: two internal nares, two openings that lead into the auditory (pharyngotympanic) tubes (commonly known as the Eustachian tubes), and the opening into the oropharynx. The posterior wall also contains the **pharyngeal tonsil.** Through the internal nares, the nasopharynx receives air from the nasal cavity and receives packages of dust-laden mucus. The nasopharynx is lined with pseudostratified ciliated columnar epithelium, and the cilia move the mucus down toward the most inferior part of the pharynx. The nasopharynx also exchanges small amounts of air with the auditory (Eustachian) tubes to equalize air pressure between the pharynx and the middle ear.

The intermediate portion of the pharynx, the **oropharynx,** lies posterior to the oral cavity and extends from the soft palate inferiorly to the level of the hyoid bone. It has only one opening, the **fauces** (FAW-sēz = throat), the opening from the mouth. This portion of the pharynx has both respiratory and digestive functions because it is a common passageway for air, food, and drink. Because the oropharynx is subject to abrasion by food particles, it is lined with nonkeratinized stratified squamous epithelium. Two pairs of tonsils, the **palatine** and **lingual tonsils,** are found in the oropharynx.

The inferior portion of the pharynx, the **laryngopharynx** (la-rin′-gō-FAIR-inks), or **hypopharynx,** begins at the level of the hyoid bone. It opens into the esophagus (food tube) posteriorly and the larynx (voice box) anteriorly. Like the oropharynx, the laryngopharynx is both a respiratory and a digestive pathway and is lined by nonkeratinized stratified squamous epithelium.

Figure 23.4 Pharynx. (See Tortora, *A Photographic Atlas of the Human Body*, Figure 11.4.)

🔑 The three subdivisions of the pharynx are the (1) nasopharynx, (2) oropharynx, and (3) laryngopharynx.

Sagittal plane

Pharyngeal tonsil
Opening of auditory (Eustachian) tube
NASOPHARYNX
Soft palate
Palatine tonsil
Fauces
OROPHARYNX
Lingual tonsil
Epiglottis
LARYNGOPHARYNX (hypopharynx)
Esophagus

Inferior nasal concha
Hard palate
Oral cavity
Tongue
Mandible
Hyoid bone
Thyroid cartilage (Adam's apple)
Cricoid cartilage
Trachea

Nasopharynx
Oropharynx
Laryngopharynx

Regions of the pharynx

Sagittal section showing the regions of the pharynx

❓ What are the superior and inferior borders of the pharynx?

Larynx

The **larynx** (LAIR-inks), or voice box, is a short passageway that connects the laryngopharynx with the trachea. It lies in the midline of the neck anterior to the fourth through sixth cervical vertebrae (C4–C6).

The wall of the larynx is composed of nine pieces of cartilage (Figure 23.5). Three occur singly (thyroid cartilage, epiglottis, and cricoid cartilage), and three occur in pairs (arytenoid, cuneiform, and corniculate cartilages). Of the paired cartilages, the arytenoid cartilages are the most important because they influence the positions and tensions of the vocal folds (true vocal cords). Whereas the extrinsic muscles of the larynx connect the cartilages to other structures in the throat, the intrinsic muscles connect the cartilages to each other.

The **thyroid cartilage (Adam's apple)** consists of two fused plates of hyaline cartilage that form the anterior wall of the larynx and give it a triangular shape. It is usually larger in males than in females due to the influence of male sex hormones on its growth during puberty. The ligament that connects the thyroid cartilage to the hyoid bone is called the **thyrohyoid membrane.**

The **epiglottis** (*epi-* = over; *glottis* = tongue) is a large, leaf-shaped piece of elastic cartilage that is covered with epithelium (see also Figure 23.4). The "stem" of the epiglottis is the tapered inferior portion that is attached to the anterior rim of the thyroid cartilage and hyoid bone. The broad superior "leaf" portion of the epiglottis is unattached and is free to move up and down like a trap door. During swallowing, the pharynx and larynx rise. Elevation of the pharynx widens it to receive food or drink; elevation of the larynx causes the epiglottis to move down and form a lid over the glottis, closing it off. The **glottis** consists of a pair of folds of mucous membrane, the vocal folds (true vocal cords) in the larynx, and the space between them called the **rima glottidis** (RĪ-ma GLOT-ti-dis). The closing of the larynx in this way during swallowing routes liquids and foods into the esophagus and keeps them out of the larynx and airways inferior to it. When small particles of dust, smoke, food, or liquids pass into the larynx, a cough reflex occurs, usually expelling the material.

The **cricoid cartilage** (KRĪ-koyd = ringlike) is a ring of hyaline cartilage that forms the inferior wall of the larynx. It is attached to the first ring of cartilage of the trachea by the **cricotracheal ligament.** The thyroid cartilage is connected to the cricoid cartilage by the **cricothyroid ligament.** The cricoid cartilage is the landmark for making an emergency airway (a tracheostomy; see page 813).

Figure 23.5   Larynx.   (See Tortora, *A Photographic Atlas of the Human Body,* Figures 11.5 and 11.6.)

 The larynx is composed of nine pieces of cartilage.

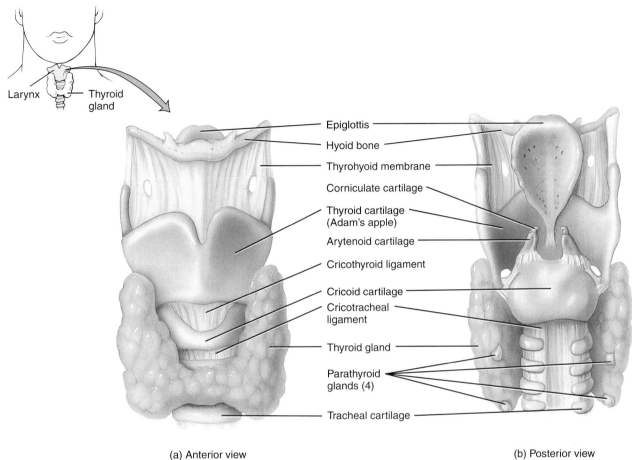

(a) Anterior view   (b) Posterior view

(c) Sagittal section

 How does the epiglottis prevent aspiration of foods and liquids?

The paired **arytenoid cartilages** (ar′-i-TE-noyd = ladle-like) are triangular pieces of mostly hyaline cartilage located at the posterior, superior border of the cricoid cartilage. They attach to the vocal folds and intrinsic pharyngeal muscles. Supported by the arytenoid cartilages, the intrinsic pharyngeal muscles contract and thus move the vocal folds.

The paired **corniculate cartilages** (kor-NIK-ū-lāt = shaped like a small horn), which are horn-shaped pieces of elastic cartilage, are located at the apex of each arytenoid cartilage. The paired **cuneiform cartilages** (KŪ-ne-i-form = wedge-shaped), which are club-shaped elastic cartilages anterior to the corniculate cartilages, support the vocal folds and lateral aspects of the epiglottis.

The lining of the larynx superior to the vocal folds is nonkeratinized stratified squamous epithelium. The lining of the larynx inferior to the vocal folds is pseudostratified ciliated columnar epithelium consisting of ciliated columnar cells, goblet cells, and basal cells. Its mucus helps trap dust not removed in the upper passages. Whereas the cilia in the upper respiratory tract move mucus and trapped particles *down* toward the pharynx, the cilia in the lower respiratory tract move them *up* toward the pharynx.

The Structures of Voice Production

The mucous membrane of the larynx forms two pairs of folds (Figure 23.5c): a superior pair called the **ventricular folds (false vocal cords)** and an inferior pair called simply the **vocal folds (true vocal cords)**. The space between the ventricular folds is known as the **rima vestibuli**. The **laryngeal sinus (ventricle)** is a lateral expansion of the middle portion of the laryngeal cavity between the ventricular folds above and the vocal folds below (see Figure 23.2b).

When the ventricular folds are brought together, they function in holding the breath against pressure in the thoracic cavity, such as might occur when a person strains to lift a heavy object. Deep to the mucous membrane of the vocal folds, which is lined by nonkeratinized stratified squamous epithelium, are bands of elastic ligaments stretched between pieces of rigid cartilage like the strings on a guitar. Intrinsic laryngeal muscles attach to both the rigid cartilage and the vocal folds. When the muscles contract, they pull the elastic ligaments tight and stretch the vocal folds out into the airways so that the rima glottidis is narrowed. If air is directed against the vocal folds, they vibrate and produce sounds (phonation) and set up sound waves in the column of air in the pharynx, nose, and mouth. The greater the pressure of air, the louder the sound.

When the intrinsic muscles of the larynx contract, they pull on the arytenoid cartilages, which causes them to pivot. Contraction of the posterior cricoarytenoid muscles, for example, moves the vocal folds apart (abduction), thereby opening the rima glottidis (Figure 23.6a). By contrast, contraction of the lateral cricoarytenoid muscles moves the vocal folds together (adduction), thereby closing the rima glottidis (Figure 23.6b). Other intrinsic muscles can elongate (and place tension on) or shorten (and relax) the vocal folds.

Pitch is controlled by the tension on the vocal folds. If they are pulled taut by the muscles, they vibrate more rapidly, and a higher pitch results. Decreasing the muscular tension on the vocal folds produces lower-pitch sounds. Due to the influence of androgens (male sex hormones), vocal folds are usually thicker and longer in males than in females, and therefore they vibrate more slowly. Thus, men's voices generally have a lower range of pitch than women's.

Sound originates from the vibration of the vocal folds, but other structures are necessary for converting the sound into recognizable speech. The pharynx, mouth, nasal cavity, and paranasal sinuses all act as resonating chambers that give the voice its human and individual quality. We produce the vowel sounds by constricting and relaxing the muscles in the wall of the pharynx. Muscles of the face, tongue, and lips help us enunciate words.

Whispering is accomplished by closing all but the posterior portion of the rima glottidis. Because the vocal folds do not vibrate during whispering, there is no pitch to this form of speech. However, we can still produce intelligible speech while whispering by changing the shape of the oral cavity as we enunciate. As the size of the oral cavity changes, its resonance qualities change, which imparts a vowel-like pitch to the air as it rushes toward the lips.

Laryngitis and Cancer of the Larynx

Laryngitis is an inflammation of the larynx that is most often caused by a respiratory infection or irritants such as cigarette smoke. Inflammation of the vocal folds causes hoarseness or loss of voice by interfering with the contraction of the folds or by causing them to swell to the point where they cannot vibrate freely. Many long-term smokers acquire a permanent hoarseness from the damage done by chronic inflammation. **Cancer of the larynx** is found almost exclusively in individuals who smoke. The condition is characterized by hoarseness, pain on swallowing, or pain radiating to an ear. Treatment consists of radiation therapy and/or surgery. ■

Trachea

The **trachea** (TRA-ke-a = sturdy), or windpipe, is a tubular passageway for air that is about 12 cm (5 in.) long and 2.5 cm (1 in.) in diameter. It is located anterior to the esophagus (Figure 23.7 on page 814) and extends from the larynx to the superior border of the fifth thoracic vertebra (T5), where it divides into right and left primary bronchi (see Figure 23.8).

The layers of the tracheal wall, from deep to superficial, are (1) mucosa, (2) submucosa, (3) hyaline cartilage, and (4) adventitia, which is composed of areolar connective tissue. The mucosa of the trachea consists of an epithelial layer of pseudostratified ciliated columnar epithelium and an underlying layer

Figure 23.6 Movement of the vocal folds. (See Tortora, *A Photographic Atlas of the Human Body*, Figure 11.7.)

The glottis consists of a pair of folds of mucous membrane in the larynx (the vocal folds) and the space between them (the rima glottidis).

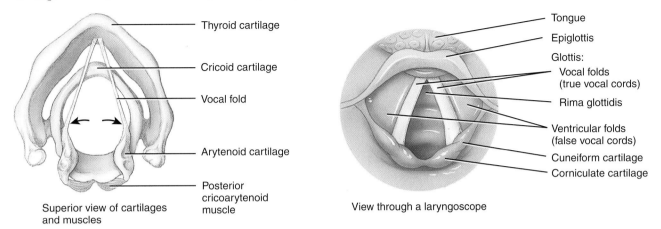

Thyroid cartilage

Cricoid cartilage

Vocal fold

Arytenoid cartilage

Posterior cricoarytenoid muscle

Superior view of cartilages and muscles

Tongue

Epiglottis

Glottis:
Vocal folds (true vocal cords)

Rima glottidis

Ventricular folds (false vocal cords)

Cuneiform cartilage

Corniculate cartilage

View through a laryngoscope

(a) Movement of vocal folds apart (abduction)

Lateral cricoarytenoid muscle

(b) Movement of vocal folds together (adduction)

What is the main function of the vocal folds?

of lamina propria that contains elastic and reticular fibers. The epithelium consists of ciliated columnar cells and goblet cells that reach the luminal surface, plus basal cells that do not (see Table 4.1I on page 113). The epithelium provides the same protection against dust as the membrane lining the nasal cavity and larynx. The submucosa consists of areolar connective tissue that contains seromucous glands and their ducts.

The 16–20 incomplete, horizontal rings of hyaline cartilage resemble the letter C and are stacked one on top of another. They may be felt through the skin inferior to the larynx. The open part of each C-shaped cartilage ring faces the esophagus (Figure 23.7), an arrangement that accommodates slight expansion of the esophagus into the trachea during swallowing. Transverse smooth muscle fibers, called the **trachealis muscle,** and elastic connective tissue stabilize the open ends of the cartilage rings. The solid C-shaped cartilage rings provide a semirigid support so that the tracheal wall does not collapse inward (espe-

cially during inhalation) and obstruct the air passageway. The adventitia of the trachea consists of areolar connective tissue that joins the trachea to surrounding tissues.

Tracheostomy and Intubation

Several conditions may block airflow by obstructing the trachea. For example, the rings of cartilage that support the trachea may collapse due to a crushing injury to the chest, inflammation of the mucous membrane may cause it to swell so much that the airway closes, or vomit or a foreign object may be aspirated into it. Two methods are used to reestablish airflow past a tracheal obstruction. If the obstruction is superior to the level of the larynx, a **tracheostomy** (trā-kē-OS-tō-mē), an operation to make an opening into the trachea, may be performed. In this procedure, a skin incision is followed by a short longitudinal incision into the trachea inferior to the cricoid cartilage. The patient can then breathe through

Figure 23.7 Location of the trachea in relation to the esophagus.

The trachea is anterior to the esophagus and extends from the larynx to the superior border of the fifth thoracic vertebra.

Transverse section of the trachea and esophagus

POSTERIOR

Lumen of esophagus

Esophagus   Trachea

Transverse plane

Trachealis muscle

Lumen of trachea

Tracheal cartilage (hyaline)

LM 2.6x

ANTERIOR

What is the benefit of not having cartilage between the trachea and the esophagus?

a metal or plastic tracheal tube inserted through the incision. The second method is **intubation,** in which a tube is inserted into the mouth or nose and passed inferiorly through the larynx and trachea. The firm wall of the tube pushes aside any flexible obstruction, and the lumen of the tube provides a passageway for air; any mucus clogging the trachea can be suctioned out through the tube. ■

Bronchi

At the superior border of the fifth thoracic vertebra, the trachea divides into a **right primary bronchus** (BRON-kus = windpipe), which goes into the right lung, and a **left primary bronchus,** which goes into the left lung (Figure 23.8). The right primary bronchus is more vertical, shorter, and wider than the left. As a result, an aspirated object is more likely to enter and lodge in the right primary bronchus than the left. Like the trachea, the primary bronchi (BRON-kē) contain incomplete rings of cartilage and are lined by pseudostratified ciliated columnar epithelium.

At the point where the trachea divides into right and left primary bronchi is an internal ridge called the **carina** (ka-RĪ-na = keel of a boat). It is formed by a posterior and somewhat inferior projection of the last tracheal cartilage. The mucous membrane of the carina is one of the most sensitive areas of the entire larynx and trachea for triggering a cough reflex. Widening and distortion of the carina is a serious sign because it usually indicates a carcinoma of the lymph nodes around the region where the trachea divides.

On entering the lungs, the primary bronchi divide to form smaller bronchi—the **secondary (lobar) bronchi,** one for each lobe of the lung. (The right lung has three lobes; the left lung has two.) The secondary bronchi continue to branch, forming still smaller bronchi, called **tertiary (segmental) bronchi,** that divide into **bronchioles.** Bronchioles, in turn, branch repeatedly, and the smallest ones branch into even smaller tubes called **terminal bronchioles.** This extensive branching from the trachea resembles an inverted tree and is commonly referred to as the **bronchial tree.**

As the branching becomes more extensive in the bronchial tree, several structural changes may be noted. First, the mucous membrane in the bronchial tree changes from pseudostratified ciliated columnar epithelium in the primary bronchi, secondary bronchi, and tertiary bronchi to ciliated simple columnar epithelium with some goblet cells in larger bronchioles, to mostly ciliated simple cuboidal epithelium with no goblet cells in smaller bronchioles, to mostly nonciliated simple cuboidal epithelium in terminal bronchioles. (In regions where simple nonciliated cuboidal epithelium is present, inhaled particles are removed by macrophages.) Second, plates of cartilage gradually replace the incomplete rings of cartilage in primary bronchi and finally disappear in the distal bronchioles. Third, as the amount of cartilage decreases, the amount of smooth muscle increases. Smooth

Figure 23.8 **Branching of airways from the trachea: the bronchial tree.**
(See Tortora, *A Photographic Atlas of the Human Body,* Figure 11.8.)

🔑 The bronchial tree begins at the trachea and ends at the terminal bronchioles.

Larynx

Trachea

Visceral pleura

Parietal pleura

Pleural cavity

Right secondary
bronchus

Right primary
bronchus

Right tertiary
bronchus

Right bronchiole

Right terminal
bronchiole

Carina

Left primary bronchus

Left secondary bronchus

Left tertiary bronchus

Left bronchiole

Left terminal bronchiole

Diaphragm

BRANCHING OF
BRONCHIAL TREE

Trachea
↓
Primary bronchi
↓
Secondary bronchi
↓
Tertiary bronchi
↓
Bronchioles
↓
Terminal bronchioles

Anterior view

 How many lobes and secondary bronchi are present in each lung?

muscle encircles the lumen in spiral bands. Because there is no supporting cartilage, however, muscle spasms can close off the airways. This is what happens during an asthma attack, and it can be a life-threatening situation. During exercise, activity in the sympathetic division of the autonomic nervous system (ANS) increases and the adrenal medulla releases the hormones epinephrine and norepinephrine, both of which cause relaxation of smooth muscle in the bronchioles, which dilates the airways. The result is improved lung ventilation because air reaches the alveoli more quickly. The parasympathetic division of the ANS and mediators of allergic reactions such as histamine cause contraction of bronchiolar smooth muscle and result in constriction of distal bronchioles.

▶ CHECKPOINT

1. What functions do the respiratory and cardiovascular systems have in common?

2. What structural and functional features are different in the upper and lower respiratory systems?

3. How do the structure and functions of the external nose and the internal nose compare?

4. What are the three anatomical regions of the pharynx? List the roles of each in respiration.

5. How does the larynx function in respiration and voice production?

6. Describe the location, structure, and function of the trachea.

7. What is the bronchial tree? Describe its structure.

Lungs

The **lungs** (= lightweights, because they float) are paired cone-shaped organs in the thoracic cavity. They are separated from each other by the heart and other structures in the mediastinum, which separates the thoracic cavity into two anatomically distinct chambers. As a result, should trauma cause one lung to collapse, the other may remain expanded. Two layers of serous membrane, collectively called the **pleural membrane** (PLOOR-al; *pleur-* = side), enclose and protect each lung. The superficial layer lines the wall of the thoracic cavity and is called the **parietal pleura;** the deep layer, the **visceral pleura,** covers the lungs themselves (Figure 23.9). Between the visceral and parietal pleurae is a small space, the **pleural cavity,** which contains a small amount of lubricating fluid secreted by the membranes. This fluid reduces friction between the membranes, allowing them to slide easily over one another during breathing. Pleural fluid also causes the two membranes to adhere to one another just as a film of water causes two glass slides to stick together, a phenomenon called surface tension. Separate pleural cavities surround the left and right lungs. Inflammation of the pleural membrane, called **pleurisy** or **pleuritis,** may in its early stages cause pain due to friction between the parietal and visceral layers of the pleura. If the inflammation persists, excess fluid accumulates in the pleural space, a condition known as **pleural effusion.**

The lungs extend from the diaphragm to just slightly superior to the clavicles and lie against the ribs anteriorly and posteriorly (Figure 23.10a). The broad inferior portion of the lung, the **base,** is concave and fits over the convex area of the diaphragm. The narrow superior portion of the lung is the **apex.** The surface of the lung lying against the ribs, the **costal surface,** matches the rounded curvature of the ribs. The **mediastinal (medial) surface** of each lung contains a region, the **hilus,** through which bronchi, pulmonary blood vessels, lymphatic vessels, and nerves enter and exit (Figure 23.10e). These structures are held together by the pleura and connective tissue and constitute the **root** of the lung. Medially, the left lung also contains a concavity, the **cardiac notch,** in which the heart lies. Due to the space occupied by the heart, the left lung is about 10% smaller than the right lung. Although the right lung is thicker and broader, it is

Figure 23.9   Relationship of the pleural membranes to the lungs.   The arrow in the inset indicates the direction from which the lungs are viewed (superior).

The parietal pleura lines the thoracic cavity, whereas the visceral pleura covers the lungs.

Transverse plane

View

ANTERIOR

Sternum   Visceral pericardium   Pericardial cavity

Parietal and fibrous pericardium

Parietal pleura

Pleural cavity

Oblique fissure

Skin

Visceral pleura

Right lung

Rib

Heart

Esophagus

Thoracic aorta

Body of fifth thoracic vertebra

Spinal cord

LATERAL

MEDIAL

POSTERIOR

Inferior view of a transverse section through the thoracic cavity showing the pleural cavity and pleural membranes

What type of membrane is the pleural membrane?

Figure 23.10 **Surface anatomy of the lungs.** (See Tortora, *A Photographic Atlas of the Human Body,* Figures 11.12 and 11.14.)

The oblique fissure divides the left lung into two lobes. The oblique and horizontal fissures divide the right lung into three lobes.

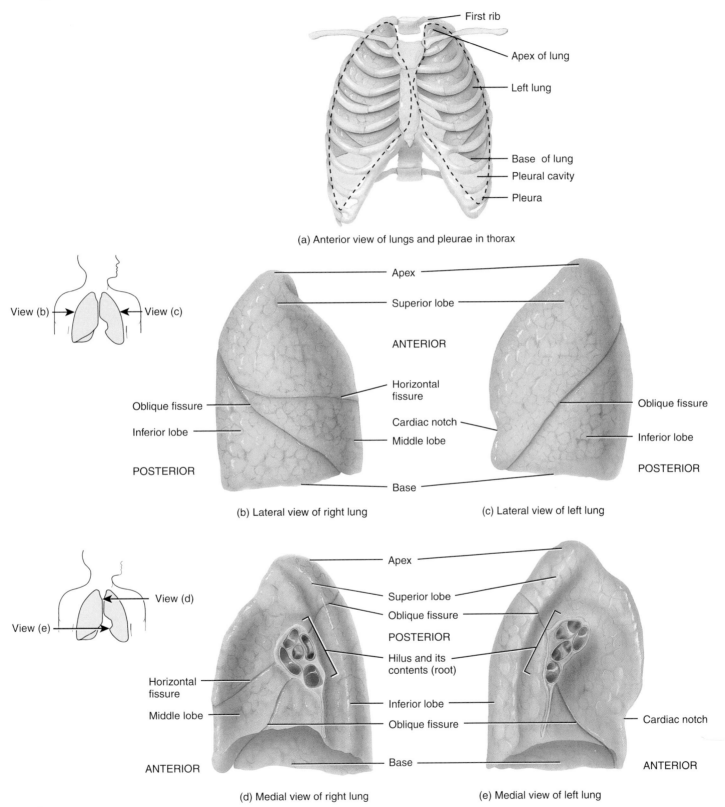

(a) Anterior view of lungs and pleurae in thorax

(b) Lateral view of right lung

(c) Lateral view of left lung

(d) Medial view of right lung

(e) Medial view of left lung

Why are the right and left lungs slightly different in size and shape?

also somewhat shorter than the left lung because the diaphragm is higher on the right side, accommodating the liver that lies inferior to it.

The lungs almost fill the thorax (see Figure 23.10a). The apex of the lungs lies superior to the medial third of the clavicles and is the only area that can be palpated. The anterior, lateral, and posterior surfaces of the lungs lie against the ribs. The base of the lungs extends from the sixth costal cartilage anteriorly to the spinous process of the tenth thoracic vertebra posteriorly. The pleura extends about 5 cm below the base from the sixth costal cartilage anteriorly to the twelfth rib posteriorly. Thus, the lungs do not completely fill the pleural cavity in this area. Removal of excessive fluid in the pleural cavity can be accomplished without injuring lung tissue by inserting the needle posteriorly through the seventh intercostal space, a procedure termed **thoracentesis** (thor'-a-sen-TĒ-sis; -centesis = puncture).

Lobes, Fissures, and Lobules

One or two fissures divide each lung into lobes (Figure 23.10b–e). Both lungs have an **oblique fissure,** which extends inferiorly and anteriorly; the right lung also has a **horizontal fissure.** The oblique fissure in the left lung separates the **superior lobe** from the **inferior lobe.** In the right lung, the superior part of the oblique fissure separates the superior lobe from the inferior lobe, whereas the inferior part of the oblique fissure separates the inferior lobe from the **middle lobe.** The horizontal fissure of the right lung subdivides the superior lobe, thus forming a middle lobe.

Each lobe receives its own secondary (lobar) bronchus. Thus, the right primary bronchus gives rise to three secondary (lobar) bronchi called the **superior, middle,** and **inferior secondary (lobar) bronchi,** whereas the left primary bronchus gives rise to **superior** and **inferior secondary (lobar) bronchi.** Within the substance of the lung, the secondary bronchi give rise to the **tertiary (segmental) bronchi,** which are constant in both origin and distribution—there are ten tertiary bronchi in each lung. The segment of lung tissue that each tertiary bronchus supplies is called a **bronchopulmonary segment.** Bronchial and pulmonary disorders (such as tumors or abscesses) that are localized in a bronchopulmonary segment may be surgically removed without seriously disrupting the surrounding lung tissue.

Each bronchopulmonary segment of the lungs has many small compartments called **lobules,** each of which is wrapped in elastic connective tissue and contains a lymphatic vessel, an arteriole, a venule, and a branch from a terminal bronchiole (Figure 23.11a). Terminal bronchioles subdivide into microscopic branches called **respiratory bronchioles** (Figure 23.11b). As the respiratory bronchioles penetrate more deeply into the lungs, the epithelial lining changes from simple cuboidal to simple squamous. Respiratory bronchioles, in turn, subdivide into several (2–11) **alveolar ducts.** The respiratory passages from the trachea to the alveolar ducts contain about 25 orders of branching; that is, branching—from the trachea into primary bronchi (first-order branching) into secondary bronchi (second-order branching) and so on down to the alveolar ducts—occurs about 25 times.

Alveoli

Around the circumference of the alveolar ducts are numerous alveoli and alveolar sacs. An **alveolus** (al-VĒ-ō-lus) is a cup-shaped outpouching lined by simple squamous epithelium and supported by a thin elastic basement membrane; an **alveolar sac** consists of two or more alveoli that share a common opening (Figure 23.11a, b). The walls of alveoli consist of two types of alveolar epithelial cells (Figure 23.12 on page 820). **Type I alveolar cells,** the predominant cells, are simple squamous epithelial cells that form a nearly continuous lining of the alveolar wall. **Type II alveolar cells,** also called **septal cells,** are fewer in number and are found between type I alveolar cells. The thin type I alveolar cells are the main sites of gas exchange. Type II alveolar cells, which are rounded or cuboidal epithelial cells whose free surfaces contain microvilli, secrete alveolar fluid, which keeps the surface between the cells and the air moist. Included in the alveolar fluid is **surfactant** (sur-FAK-tant), a complex mixture of phospholipids and lipoproteins. Surfactant lowers the surface tension of alveolar fluid, which reduces the tendency of alveoli to collapse. Associated with the alveolar wall are **alveolar macrophages (dust cells),** wandering phagocytes that remove fine dust particles and other debris in the alveolar spaces. Also present are fibroblasts that produce reticular and elastic fibers. Underlying the layer of type I alveolar cells is an elastic basement membrane. On the outer surface of the alveoli, the lobule's arteriole and venule disperse into a network of blood capillaries (see Figure 23.11a) that consist of a single layer of endothelial cells and basement membrane.

The exchange of O_2 and CO_2 between the air spaces in the lungs and the blood takes place by diffusion across the alveolar and capillary walls, which together form the **respiratory membrane.** Extending from the alveolar air space to blood plasma, the respiratory membrane consists of four layers (see Figure 23.12b):

1. A layer of type I and type II alveolar cells and associated alveolar macrophages that constitutes the **alveolar wall.**

2. An **epithelial basement membrane** underlying the alveolar wall.

3. A **capillary basement membrane** that is often fused to the epithelial basement membrane.

4. The **endothelial cells** of the capillary.

Despite having several layers, the respiratory membrane is very thin—only 0.5μm thick, about one-sixteenth the diameter of a red blood cell. This thinness allows rapid diffusion of gases. Moreover, it has been estimated that the lungs contain 300 million alveoli, providing an immense surface area of 70 m^2 (750 ft^2)—about the size of a handball court—for the exchange of gases.

Nebulization

Many respiratory disorders are treated by means of **nebulization** (neb-ū-li-ZA-shun). This procedure consists of administering medication in the form of droplets that are suspended in air into the respiratory tract. The patient inhales the medication as a fine mist. Nebulization therapy can be used with many different types of drugs, such as chemicals that relax the smooth muscle of the airways, chemicals that reduce the thickness of mucus, and antibiotics. ■

Blood Supply to the Lungs

The lungs receive blood via two sets of arteries: pulmonary arteries and bronchial arteries. Deoxygenated blood passes through the pulmonary trunk, which divides into a left pulmonary artery that enters the left lung and a right pulmonary artery that enters the right lung. Return of the oxygenated blood to the heart occurs by way of the four pulmonary veins, which drain into the left atrium (see Figure 21.30 on page 753). A unique feature of pulmonary blood vessels is their constriction in response to localized hypoxia (low O_2 level). In all other body tissues, hypoxia causes dilation of blood vessels, which serves to increase blood flow to a tissue that is not receiving adequate O_2. In the lungs, however, vasoconstriction in response to hypoxia diverts pulmonary blood from poorly ventilated areas to well-ventilated regions of the lungs. This phenomenon is known as **ventilation-perfusion coupling** because the perfusion (blood flow) to each area of the lungs matches the extent of ventilation (airflow) to alveoli in that area.

Figure 23.11   Microscopic anatomy of a lobule of the lungs.

Alveolar sacs consist of two or more alveoli that share a common opening.

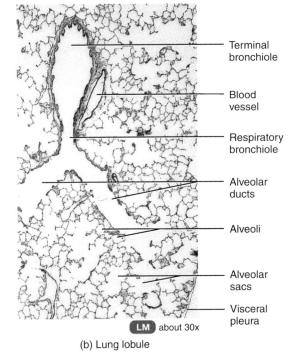

(a) Diagram of a portion of a lobule of the lung

(b) Lung lobule

What types of cells make up the wall of an alveolus?

Figure 23.12 Structural components of an alveolus. The respiratory membrane consists of a layer of type I and type II alveolar cells, an epithelial basement membrane, a capillary basement membrane, and the capillary endothelium.

The exchange of respiratory gases occurs by diffusion across the respiratory membrane.

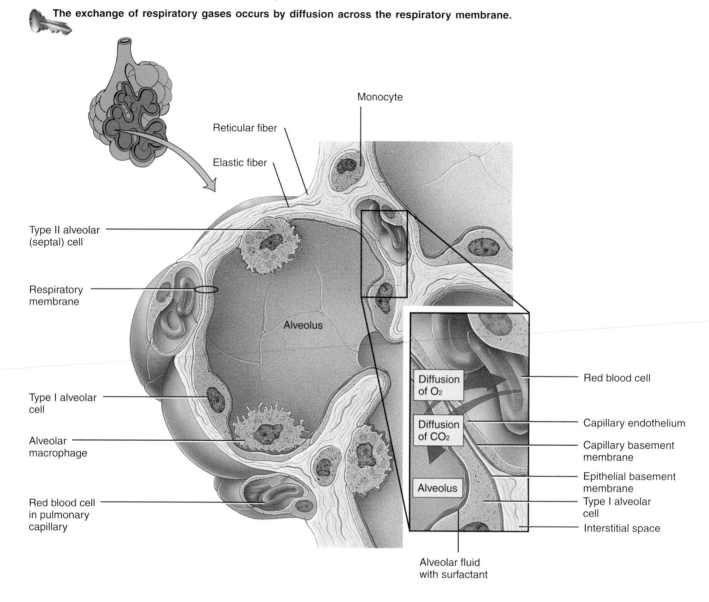

(a) Section through an alveolus
showing its cellular components

(b) Details of respiratory membrane

 How thick is the respiratory membrane?

Bronchial arteries, which branch from the aorta, deliver oxygenated blood to the lungs. This blood mainly perfuses the walls of the bronchi and bronchioles. Connections exist between branches of the bronchial arteries and branches of the pulmonary arteries, however, and most blood returns to the heart via pulmonary veins. Some blood, however, drains into bronchial veins, branches of the azygos system, and returns to the heart via the superior vena cava.

▶ C H E C K P O I N T

8. Where are the lungs located? Distinguish the parietal pleura from the visceral pleura.

9. Define each of the following parts of a lung: base, apex, costal surface, medial surface, hilus, root, cardiac notch, lobe, and lobule.

10. What is a bronchopulmonary segment?

11. Describe the histology and function of the respiratory membrane.

PULMONARY VENTILATION

OBJECTIVE

• **Describe the events that cause inhalation and exhalation.**

The process of gas exchange in the body, called **respiration,** occurs in three basic steps:

1. **Pulmonary ventilation** (*pulmon-* = lung), or breathing, is the mechanical flow of air into (inhalation) and out of (exhalation) the lungs.

2. **External respiration** is the exchange of gases between the air spaces of the lungs and the blood in pulmonary capillaries. In this process, pulmonary capillary blood gains O_2 and loses CO_2.

3. **Internal respiration** is the exchange of gases between blood in systemic capillaries and tissue cells. The blood loses O_2 and gains CO_2. Within cells, the metabolic reactions that consume O_2 and give off CO_2 during the production of ATP are termed *cellular respiration* (discussed in Chapter 25).

Pulmonary ventilation (breathing) is the process by which gases flow between the atmosphere and lung alveoli. Air flows between the atmosphere and the lungs because of alternating pressure differences created by contraction and relaxation of respiratory muscles. The rate of airflow and the amount of effort needed for breathing is also influenced by alveolar surface tension, compliance of the lungs, and airway resistance.

Pressure Changes During Pulmonary Ventilation

Air moves into the lungs when the air pressure inside the lungs is less than the air pressure in the atmosphere and out of the lungs when the pressure inside the lungs is greater than the pressure in the atmosphere.

Inhalation

Breathing in is called **inhalation (inspiration).** Just before each inhalation, the air pressure inside the lungs is equal to the pressure of the atmosphere, which at sea level is about 760 millimeters of mercury (mmHg), or 1 atmosphere (atm). For air to flow into the lungs, the pressure inside the alveoli must become lower than the atmospheric pressure. This condition is achieved by increasing the volume of the lungs.

The pressure of a gas in a closed container is inversely proportional to the volume of the container. This means that if the size of a closed container is increased, the pressure of the gas inside the container decreases, and that if the size of the container is decreased, then the pressure inside it increases. This inverse relationship between volume and pressure, called **Boyle's law,** may be demonstrated as follows (Figure 23.13): Suppose we place a gas in a cylinder that has a movable piston and a pressure gauge, and that the initial pressure created by the

Figure 23.13   Boyle's law.

 The volume of a gas varies inversely with its pressure.

Volume = 1 liter
Pressure = 1 atm

Volume = 1/2 liter
Pressure = 2 atm

If the volume is decreased from 1 liter to 1/4 liter, how does the pressure change?

gas molecules striking the wall of the container is 1 atm. If the piston is pushed down, the gas is compressed into a smaller volume, so that the same number of gas molecules strike less wall area. The gauge shows that the pressure doubles as the gas is compressed to half its original volume. In other words, the same number of molecules in half the volume produces twice the pressure. Conversely, if the piston is raised to increase the volume, the pressure decreases. Thus, the pressure of a gas varies inversely with volume. Boyle's law also applies to everyday activities such as the operation of a bicycle pump or the blowing up of a balloon.

Differences in pressure caused by changes in lung volume force air into our lungs when we inhale and out when we exhale. For inhalation to occur, the lungs must expand, which increases lung volume and thus decreases the pressure in the lungs to below atmospheric pressure. The first step in expanding the lungs during normal quiet inhalation involves contraction of the main muscles of inhalation, the diaphragm and external intercostals (Figure 23.14).

The most important muscle of inhalation is the diaphragm, the dome-shaped skeletal muscle that forms the floor of the thoracic cavity. It is innervated by fibers of the phrenic nerves, which emerge from the spinal cord at cervical levels 3, 4, and 5. Contraction of the diaphragm causes it to flatten and increases the vertical dimension of the thoracic cavity. During normal quiet inhalation, the diaphragm descends about 1 cm (0.4 in.), producing a pressure difference of 1–3 mmHg and the inhalation of about 500 mL of air. In strenuous breathing, the diaphragm may descend 10 cm (4 in.), which produces a pressure difference of 100 mmHg and the inhalation of 2–3 liters of air. Contraction of the diaphragm is responsible for about 75% of the air that enters the lungs during quiet breathing.

The next most important muscles of inhalation are the external intercostals. When these muscles contract, they elevate the

Figure 23.14 Muscles of inhalation and exhalation and their actions. The pectoralis minor muscle (not shown here) is illustrated in Figure 11.14a on page 343.

🔑 **During deep, labored breathing, accessory muscles of inhalation (sternocleidomastoids, scalenes, and pectoralis minors) participate.**

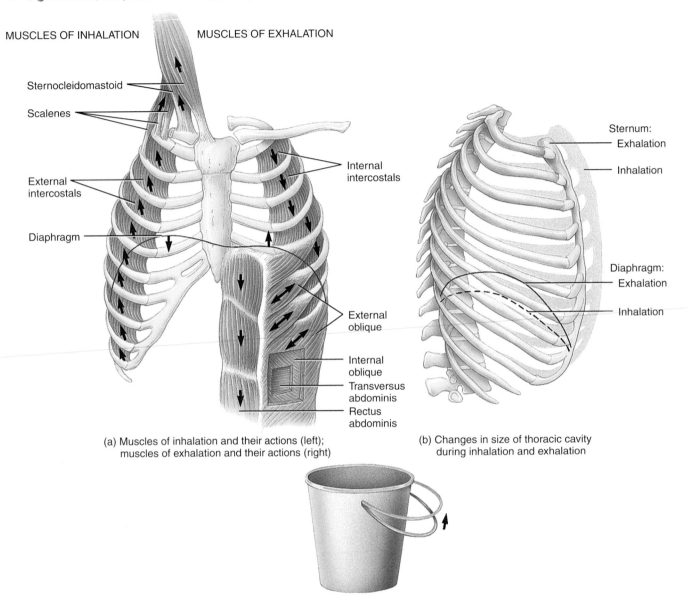

MUSCLES OF INHALATION   MUSCLES OF EXHALATION

Sternocleidomastoid

Scalenes

External intercostals

Diaphragm

Internal intercostals

External oblique

Internal oblique

Transversus abdominis

Rectus abdominis

Sternum:
Exhalation
Inhalation

Diaphragm:
Exhalation
Inhalation

(a) Muscles of inhalation and their actions (left); muscles of exhalation and their actions (right)

(b) Changes in size of thoracic cavity during inhalation and exhalation

(c) During inhalation, the ribs move upward and outward like the handle on a bucket

 Right now, what is the main muscle that powers your breathing?

ribs. As a result, there is an increase in the anteroposterior and lateral diameters of the chest cavity. Contraction of the external intercostals is responsible for about 25% of the air that enters the lungs during normal quiet breathing.

During quiet inhalations, the pressure between the two pleural layers, called **intrapleural pressure,** is always subatmospheric (lower than atmospheric pressure). Just before in-

halation, it is about 4 mmHg less than the atmospheric pressure, or about 756 mmHg if the atmospheric pressure is 760 mmHg (Figure 23.15). As the diaphragm and external intercostals contract and the overall size of the thoracic cavity increases, the volume of the pleural cavity also increases, which causes intrapleural pressure to decrease to about 754 mmHg. During expansion of the thorax, the parietal and visceral pleurae

Figure 23.15 Pressure changes in pulmonary ventilation. During inhalation, the diaphragm contracts, the chest expands, the lungs are pulled outward, and alveolar pressure decreases. During exhalation, the diaphragm relaxes, the lungs recoil inward, and alveolar pressure increases, forcing air out of the lungs.

Air moves into the lungs when alveolar pressure is less than atmospheric pressure, and out of the lungs when alveolar pressure is greater than atmospheric pressure.

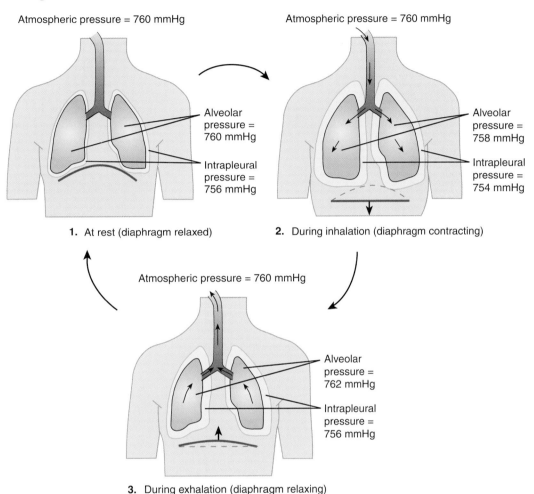

How does the intrapleural pressure change during a normal, quiet breath?

normally adhere tightly because of the subatmospheric pressure between them and because of the surface tension created by their moist adjoining surfaces. As the thoracic cavity expands, the parietal pleura lining the cavity is pulled outward in all directions, and the visceral pleura and lungs are pulled along with it.

As the volume of the lungs increases in this way, the pressure inside the lungs, called the **alveolar (intrapulmonic) pressure,** drops from 760 to 758 mmHg. A pressure difference is thus established between the atmosphere and the alveoli. Because air always flows from a region of higher pressure to a region of lower pressure, inhalation takes place. Air continues to flow into the lungs as long as a pressure difference exists.

During deep, forceful inhalations, accessory muscles of inspiration also participate in increasing the size of the thoracic cavity (see Figure 23.14a). The muscles are so-named because they make little, if any, contribution during normal quiet inhalation, but during exercise or forced ventilation they may contract vigorously. The accessory muscles of inhalation include the sternocleidomastoid muscles, which elevate the sternum; the scalene muscles, which elevate the first two ribs; and the pectoralis minor muscles, which elevate the third through fifth ribs. Because both normal quiet inhalation and inhalation during exercise or forced ventilation involve muscular contraction, the process of inhalation is said to be *active.*

Figure 23.16a summarizes the events of inhalation.

Figure 23.16 Summary of events of inhalation and exhalation.

Inhalation and exhalation are caused by changes in alveolar pressure.

During normal quiet inhalation, the diaphragm and external intercostals contract. During labored inhalation, sternocleidomastoid, scalenes, and pectoralis minor also contract.

Alveolar pressure increases to 762 mmHg

Atmospheric pressure is about 760 mmHg at sea level

Thoracic cavity increases in size and volume of lungs expands

Alveolar pressure decreases to 758 mmHg

During normal quiet exhalation, diaphragm and external intercostals relax. During forceful exhalation, abdominal and internal intercostal muscles contract.

Thoracic cavity decreases in size and lungs recoil

(a) Inhalation

(b) Exhalation

What is normal atmospheric pressure at sea level?

Exhalation

Breathing out, called **exhalation (expiration),** is also due to a pressure gradient, but in this case the gradient is in the opposite direction: The pressure in the lungs is greater than the pressure of the atmosphere. Normal exhalation during quiet breathing, unlike inhalation, is a *passive process* because no muscular contractions are involved. Instead, exhalation results from **elastic recoil** of the chest wall and lungs, both of which have a natural tendency to spring back after they have been stretched. Two inwardly directed forces contribute to elastic recoil: (1) the recoil of elastic fibers that were stretched during inhalation and (2) the inward pull of surface tension due to the film of alveolar fluid.

Exhalation starts when the inspiratory muscles relax. As the diaphragm relaxes, its dome moves superiorly owing to its elasticity. As the external intercostals relax, the ribs are depressed. These movements decrease the vertical, lateral, and anteroposterior dimensions of the thoracic cavity, which decreases lung volume. In turn, the alveolar pressure increases, to about 762 mmHg. Air then flows from the area of higher pressure in the alveoli to the area of lower pressure in the atmosphere (see Figure 23.15).

Exhalation becomes active only during forceful breathing, as occurs while playing a wind instrument or during exercise. During these times, muscles of exhalation—the abdominals and internal intercostals (see Figure 23.14a)—contract, which increases pressure in the abdominal region and thorax. Contraction of the abdominal muscles moves the inferior ribs downward and compresses the abdominal viscera, thereby forcing the diaphragm superiorly. Contraction of the internal intercostals, which extend inferiorly and posteriorly between adjacent ribs, pulls the ribs inferiorly. Although intrapleural pressure is always less than alveolar pressure, it may briefly exceed atmospheric pressure during a forceful exhalation, such as during a cough.

Figure 23.16b summarizes the events of exhalation.

Other Factors Affecting Pulmonary Ventilation

Whereas air pressure differences drive airflow during inhalation and exhalation, three other factors affect the rate of airflow and the ease of pulmonary ventilation: surface tension of the alveolar fluid, compliance of the lungs, and airway resistance.

Surface Tension of Alveolar Fluid

As noted earlier, a thin layer of alveolar fluid coats the luminal surface of alveoli and exerts a force known as **surface tension.** Surface tension arises at all air–water interfaces because the polar water molecules are more strongly attracted to each other than they are to gas molecules in the air. When liquid surrounds a sphere of air, as in an alveolus or a soap bubble, surface tension produces an inwardly directed force. Soap bubbles "burst" because they collapse inward due to surface tension. In the lungs, surface tension causes the alveoli to assume the smallest possible diameter. During breathing, surface tension must be overcome to expand the lungs during each inhalation. Surface tension also accounts for two-thirds of lung elastic recoil, which decreases the size of alveoli during exhalation.

The surfactant present in alveolar fluid reduces its surface tension below the surface tension of pure water. A deficiency of surfactant in premature infants causes *respiratory distress syndrome,* in which the surface tension of alveolar fluid is greatly increased, so that many alveoli collapse at the end of each exhalation. Great effort is then needed at the next inhalation to reopen the collapsed alveoli.

 Pneumothorax

The pleural cavities are sealed off from the outside environment, which prevents them from equalizing their pressure with that of the atmosphere. Injuries of the chest wall that allow air to enter the intrapleural space either from the outside or from the alveoli cause **pneumothorax** (*pneumo-* = air), filling of the pleural cavity with air. Because the intrapleural pressure is then equal to atmospheric pressure rather than subatmospheric, surface tension and recoil of elastic fibers cause the lung to collapse. ∎

Compliance of the Lungs

Compliance refers to how much effort is required to stretch the lungs and chest wall. High compliance means that the lungs and chest wall expand easily; low compliance means that they resist expansion. By analogy, a thin balloon that is easy to inflate has high compliance, whereas a heavy and stiff balloon that takes a lot of effort to inflate has low compliance. In the lungs, compliance is related to two principal factors: elasticity and surface tension. The lungs normally have high compliance and expand easily because elastic fibers in lung tissue are easily stretched and surfactant in alveolar fluid reduces surface tension. Decreased compliance is a common feature in pulmonary conditions that (1) scar lung tissue (for example, tuberculosis), (2) cause lung tissue to become filled with fluid (pulmonary edema), (3) produce a deficiency in surfactant, or (4) impede lung expansion in any way (for example, paralysis of the intercostal muscles). An increase in lung compliance occurs in emphysema due to destruction of elastic fibers in alveolar walls.

Airway Resistance

As is true for blood flow through blood vessels, the rate of airflow through the airways depends on both the pressure difference and the resistance: Airflow equals the pressure difference between the alveoli and the atmosphere divided by the resistance. The walls of the airways, especially the bronchioles, offer some resistance to the normal flow of air into and out of the lungs. As the lungs expand during inhalation, the bronchioles enlarge because their walls are pulled outward in all directions. Larger-diameter airways have decreased resistance. Airway resistance then increases during exhalation as the diameter of bronchioles decreases. Also, the degree of contraction or relaxation of smooth muscle in the walls of the airways regulates airway diameter and thus resistance. Increased signals from the sympathetic division of the autonomic nervous system cause relaxation of this smooth muscle, which results in bronchodilation and decreased resistance.

Any condition that narrows or obstructs the airways increases resistance, and more pressure is required to maintain the same airflow. The hallmark of asthma or chronic obstructive pulmonary disease (COPD)—emphysema or chronic bronchitis—is increased airway resistance due to obstruction or collapse of airways.

Breathing Patterns and Modified Respiratory Movements

The term for the normal pattern of quiet breathing is **eupnea** (ūp-NĒ-a; *eu-* = good, easy, or normal; *-pnea* = breath). Eupnea can consist of shallow, deep, or combined shallow and deep breathing. A pattern of shallow (chest) breathing, called **costal breathing,** consists of an upward and outward movement of the chest due to contraction of the external intercostal muscles. A pattern of deep (abdominal) breathing, called **diaphragmatic breathing,** consists of the outward movement of the abdomen due to the contraction and descent of the diaphragm.

Respirations also provide humans with methods for expressing emotions such as laughing, sighing, and sobbing. Moreover, respiratory air can be used to expel foreign matter from the lower air passages through actions such as sneezing and coughing. Respiratory movements are also modified and controlled during talking and singing. Some of the modified respiratory movements that express emotion or clear the airways are listed in Table 23.1. All these movements are reflexes, but some of them also can be initiated voluntarily.

▶ CHECKPOINT

12. What are the basic differences among pulmonary ventilation, external respiration, and internal respiration?

13. Compare what happens during quiet versus forceful pulmonary ventilation.

14. Describe how alveolar surface tension, compliance, and airway resistance affect pulmonary ventilation.

15. Define the various kinds of modified respiratory movements.

Table 23.1	Modified Respiratory Movements
Movement	**Description**
Coughing	A long-drawn and deep inhalation followed by a complete closure of the rima glottidis, which results in a strong exhalation that suddenly pushes the rima glottidis open and sends a blast of air through the upper respiratory passages. Stimulus for this reflex act may be a foreign body lodged in the larynx, trachea, or epiglottis.
Sneezing	Spasmodic contraction of muscles of exhalation that forcefully expels air through the nose and mouth. Stimulus may be an irritation of the nasal mucosa.
Sighing	A long-drawn and deep inhalation immediately followed by a shorter but forceful exhalation.
Yawning	A deep inhalation through the widely opened mouth producing an exaggerated depression of the mandible. It may be stimulated by drowsiness, fatigue, or someone else's yawning, but precise cause is unknown.
Sobbing	A series of convulsive inhalations followed by a single prolonged exhalation. The rima glottidis closes earlier than normal after each inhalation so only a little air enters the lungs with each inhalation.
Crying	An inhalation followed by many short convulsive exhalations, during which the rima glottidis, remains open and the vocal folds vibrate; accompanied by characteristic facial expressions and tears.
Laughing	The same basic movements as crying, but the rhythm of the movements and the facial expressions usually differ from those of crying. Laughing and crying are sometimes indistinguishable.
Hiccupping	Spasmodic contraction of the diaphragm followed by a spasmodic closure of the rima glottidis, which produces a sharp sound on inhalation. Stimulus is usually irritation of the sensory nerve endings of the gastrointestinal tract.
Valsalva maneuver	Forced exhalation against a closed rima glottidis as may occur during periods of straining while defecating.

LUNG VOLUMES AND CAPACITIES

▶ O B J E C T I V E

• **Define the various lung volumes and capacities.**

While at rest, a healthy adult averages 12 breaths a minute, with each inhalation and exhalation moving about 500 mL of air into and out of the lungs. The volume of one breath is called the **tidal**

volume (V_T). Thus, the **minute ventilation (MV)**—the total volume of air inhaled and exhaled each minute—is respiratory rate multiplied by tidal volume:

$$MV = 12 \text{ breaths/min} \times 500 \text{ mL/breath}$$
$$= 6 \text{ liters/min}$$

A lower-than-normal minute ventilation usually is a sign of pulmonary malfunction. The apparatus commonly used to measure the volume of air exchanged during breathing and the respiratory rate is a **spirometer** (*spiro-* = breathe; *meter* = measuring device) or **respirometer.** The record is called a **spirogram.** Inhalation is recorded as an upward deflection, and exhalation is recorded as a downward deflection (Figure 23.17).

Tidal volume varies considerably from one person to another and in the same person at different times. In a typical adult, about 70% of the tidal volume (350 mL) actually reaches the respiratory portion of the respiratory system—the respiratory bronchioles, alveolar ducts, alveolar sacs, and alveoli—and participates in external respiration. The other 30% (150 mL) remains in the conducting airways of the nose, pharynx, larynx, trachea, bronchi, bronchioles, and terminal bronchioles. Collectively, these conducting airways are known as the **anatomic dead space.** (An easy rule of thumb for determining the volume of your anatomic dead space in milliliters is that it is about the same as your ideal weight in pounds.) Not all of the minute ventilation can be used in gas exchange because some of it remains in the anatomic dead space. The **alveolar ventilation rate** is the volume of air per minute that reaches the alveoli and other respiratory portions. In the example just given, alveolar ventilation rate would be 350 mL/breath × 12 breaths/min = 4200 mL/min.

Several other lung volumes are defined relative to forceful breathing. In general, these volumes are larger in males, taller individuals, and younger adults, and smaller in females, shorter individuals, and the elderly. Various disorders also may be diagnosed by comparison of actual and predicted normal values for one's gender, height, and age. The values given here are averages for young adults.

By taking a very deep breath, you can inhale a good deal more than 500 mL. This additional inhaled air, called the **inspiratory reserve volume,** is about 3100 mL (Figure 23.17). Even more air can be inhaled if inhalation follows forced exhalation. If you inhale normally and then exhale as forcibly as possible, you should be able to push out 1200 mL of air in addition to the 500 mL of tidal volume. The extra 1200 mL is called the **expiratory reserve volume.** The $FEV_{1.0}$ is the **forced expiratory volume in 1 second,** which is the volume of air that can be exhaled from the lungs in 1 second with maximal effort following a maximal inhalation. Typically, chronic obstructive pulmonary disease (COPD) greatly reduces $FEV_{1.0}$. That is because COPD increases airway resistance.

Even after the expiratory reserve volume is exhaled, considerable air remains in the lungs because the subatmospheric intrapleural pressure keeps the alveoli slightly inflated, and some air also remains in the noncollapsible airways. This volume,

Figure 23.17 Spirogram of lung volumes and capacities (average values for a healthy adult).
Note that the spirogram is read from right (start of record) to left (end of record).

 Lung capacities are combinations of various lung volumes.

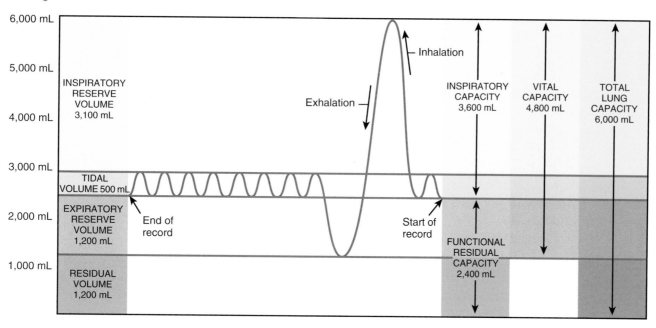

LUNG VOLUMES   LUNG CAPACITIES

If you breathe in as deeply as possible and then exhale as much air as you can, which lung capacity have you demonstrated?

which cannot be measured by spirometry, is called the **residual volume** and amounts to about 1200 mL.

If the thoracic cavity is opened, the intrapleural pressure rises to equal the atmospheric pressure and forces out some of the residual volume. The air remaining is called the **minimal volume.** Minimal volume provides a medical and legal tool for determining whether a baby was born dead or died after birth. The presence of minimal volume can be demonstrated by placing a piece of lung in water and observing if it floats. Fetal lungs contain no air, and so the lung of a stillborn baby will not float in water.

Lung capacities are combinations of specific lung volumes (see Figure 23.17). **Inspiratory capacity** is the sum of tidal volume and inspiratory reserve volume (500 mL + 3100 mL = 3600 mL). **Functional residual capacity** is the sum of residual volume and expiratory reserve volume (1200 mL + 1200 mL = 2400 mL). **Vital capacity** is the sum of inspiratory reserve volume, tidal volume, and expiratory reserve volume (4800 mL). Finally, **total lung capacity** is the sum of vital capacity and residual volume (4800 mL + 1200 mL = 6000 mL).

▶ CHECKPOINT

16. What is a spirometer?

17. What is the difference between a lung volume and a lung capacity?

18. How is minute ventilation calculated?

19. Define alveolar ventilation rate and $FEV_{1.0}$.

EXCHANGE OF OXYGEN AND CARBON DIOXIDE

▶ OBJECTIVES

• **Explain Dalton's law and Henry's law.**

• **Describe the exchange of oxygen and carbon dioxide in external and internal respiration.**

The exchange of oxygen and carbon dioxide between alveolar air and pulmonary blood occurs via passive diffusion, which is governed by the behavior of gases as described by Dalton's law and Henry's law. Whereas Dalton's law is important for understanding how gases move down their pressure differences by diffusion, Henry's law helps explain how the solubility of a gas relates to its diffusion.

Gas Laws: Dalton's Law and Henry's Law

According to **Dalton's law,** each gas in a mixture of gases exerts its own pressure as if no other gases were present. The pressure of a specific gas in a mixture is called its *partial pressure* and is denoted as P_x, where the subscript is the chemical formula of the gas. The total pressure of the mixture is calculated simply by adding all the partial pressures. Atmospheric air is a mixture of gases—nitrogen (N_2), oxygen, water vapor (H_2O), and carbon dioxide, plus other gases present in small quantities. Atmospheric pressure is the sum of the pressures of all these gases:

$$\text{Atmospheric pressure (760 mmHg)} =$$
$$P_{N_2} + P_{O_2} + P_{H_2O} + P_{CO_2} + P_{\text{other gases}}$$

We can determine the partial pressure exerted by each component in the mixture by multiplying the percentage of the gas in the mixture by the total pressure of the mixture. Atmospheric air is 78.6% nitrogen, 20.9% oxygen, 0.04% carbon dioxide, and 0.06% other gases; a variable amount of water vapor is also present, about 0.4% on a cool, dry day. Thus, the partial pressures of the gases in inhaled air are as follows:

$$
\begin{aligned}
P_{N_2} &= 0.786 \times 760 \text{ mmHg} &&= 597.4 \text{ mmHg} \\
P_{O_2} &= 0.209 \times 760 \text{ mmHg} &&= 158.8 \text{ mmHg} \\
P_{H_2O} &= 0.004 \times 760 \text{ mmHg} &&= 3.0 \text{ mmHg} \\
P_{CO_2} &= 0.0004 \times 760 \text{ mmHg} &&= 0.3 \text{ mmHg} \\
P_{\text{other gases}} &= 0.0006 \times 760 \text{ mmHg} &&= \underline{0.5 \text{ mmHg}} \\
&&\text{Total} &= 760 \text{ mmHg}
\end{aligned}
$$

These partial pressures govern the movement of O_2 and CO_2 between the atmosphere and lungs, between the lungs and blood, and between the blood and body cells. When a mixture of gases diffuses across a permeable membrane, each gas diffuses from the area where its partial pressure is greater to the area where its partial pressure is less. The greater the difference in partial pressure, the faster the rate of diffusion. Each gas behaves as if the other gases in the mixture were not present and diffuses at a rate determined by its own partial pressure.

Compared with inhaled air, alveolar air has less O_2 (20.9% versus 13.6%) and more CO_2 (0.04% versus 5.2%) for two reasons. First, gas exchange in the alveoli increases the CO_2 content and decreases the O_2 content of alveolar air. Second, when air is inhaled it becomes humidified as it passes along the moist mucosal linings. As water vapor content of the air increases, the relative percentage that is O_2 decreases. In contrast, exhaled air contains more O_2 than alveolar air (16% versus 13.6%) and less CO_2 (4.5% versus 5.2%) because some of the exhaled air was in the anatomic dead space and did not participate in gas exchange. Exhaled air is a mixture of alveolar air and inhaled air that was in the anatomic dead space.

Henry's law states that the quantity of a gas that will dissolve in a liquid is proportional to the partial pressure of the gas and its solubility coefficient. In body fluids, the ability of a gas to stay in solution is greater when its partial pressure is higher and when it has a high solubility coefficient in water. The higher the partial pressure of a gas over a liquid and the higher the solubility coefficient, the more gas will stay in solution. In comparison to oxygen, much more CO_2 is dissolved in blood plasma because the solubility coefficient of CO_2 is 24 times greater than that of O_2. Even though the air we breathe contains almost 79% N_2, this gas has no known effect on bodily functions, and at sea level pressure very little of it dissolves in blood plasma because its solubility coefficient is very low.

An everyday experience gives a demonstration of Henry's law. You have probably noticed that a soft drink makes a hissing sound when the top of the container is removed, and bubbles rise to the surface for some time afterward. The gas dissolved in carbonated beverages is CO_2. Because the soft drink is bottled or canned under high pressure and capped, the CO_2 remains dissolved so long as the container is unopened. Once you remove the cap, the pressure decreases and the gas begins to bubble out of solution.

Henry's law explains two conditions resulting from changes in the solubility of nitrogen in body fluids. Even though the air we breathe contains about 79% nitrogen, this gas has no known effect on bodily functions, and very little of it dissolves in blood plasma because of its low solubility coefficient at sea level pressure. As the total air pressure increases, the partial pressures of all its gases increase. When a scuba diver breathes air under high pressure, the nitrogen in the mixture can have serious negative effects. Because the partial pressure of nitrogen is higher in a mixture of compressed air than in air at sea level pressure, a considerable amount of nitrogen dissolves in plasma and interstitial fluid. Excessive amounts of dissolved nitrogen may produce giddiness and other symptoms similar to alcohol intoxication. The condition is called **nitrogen narcosis** or "rapture of the depths."

If a diver comes to the surface slowly, the dissolved nitrogen can be eliminated by exhaling it. However, if the ascent is too rapid, nitrogen comes out of solution too quickly and forms gas bubbles in the tissues, resulting in **decompression sickness** (the **bends**). The effects of decompression sickness typically result from bubbles in nervous tissue and can be mild or severe, depending on the number of bubbles formed. Symptoms include joint pain, especially in the arms and legs, dizziness, shortness of breath, extreme fatigue, paralysis, and unconsciousness.

Hyperbaric Oxygenation

A major clinical application of Henry's law is **hyperbaric oxygenation** (*hyper* = over; *baros* = pressure). Using pressure to cause more O_2 to dissolve in the blood is an effective technique in treating patients infected by anaerobic bacteria, such as those that cause tetanus and gangrene. (Anaerobic bacteria cannot live in the presence of free O_2.) A person undergoing hyperbaric oxygenation is placed in a hyperbaric chamber, which contains O_2 at a pressure of 3 to 4 atmospheres (2280–3040 mmHg). As body tissues pick up the O_2, the bacteria are killed. Hyperbaric chambers may also be used for treating certain heart disorders, carbon monoxide poisoning, gas embolisms, crush injuries, cerebral edema, certain hard-to-treat bone infections caused by anaerobic bacteria, smoke inhalation, near-drowning, asphyxia, vascular insufficiencies, and burns. ∎

External and Internal Respiration

External respiration or **pulmonary gas exchange** is the diffusion of O_2 from air in the alveoli of the lungs to blood in pulmonary capillaries and the diffusion of CO_2 in the opposite direction (Figure 23.18a). External respiration in the lungs con-

Figure 23.18 Changes in partial pressures of oxygen and carbon dioxide (in mmHg) during external and internal respiration.

Gases diffuse from areas of higher partial pressure to areas of lower partial pressure.

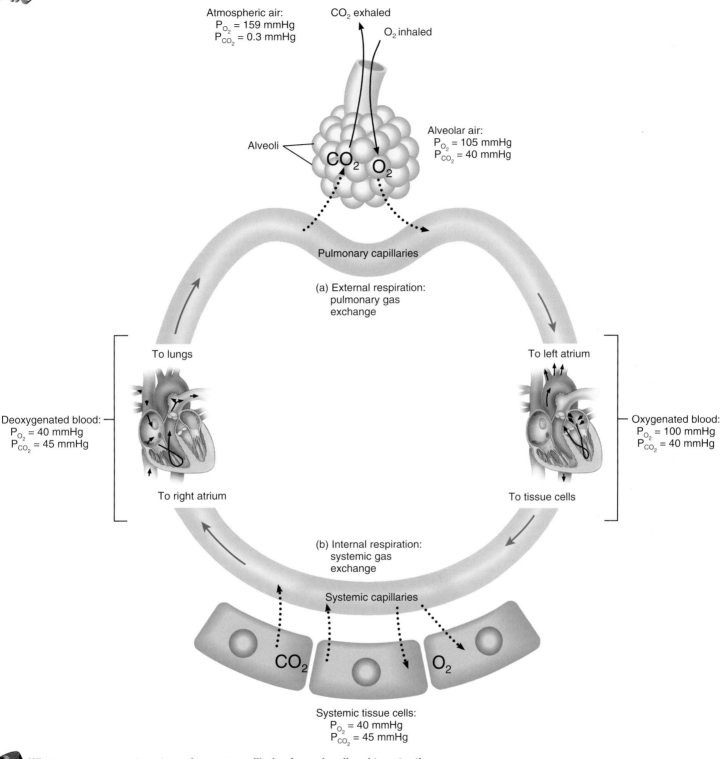

Atmospheric air:
P_{O_2} = 159 mmHg
P_{CO_2} = 0.3 mmHg

CO_2 exhaled

O_2 inhaled

Alveolar air:
P_{O_2} = 105 mmHg
P_{CO_2} = 40 mmHg

Alveoli

CO_2 O_2

Pulmonary capillaries

(a) External respiration: pulmonary gas exchange

To lungs

To left atrium

Deoxygenated blood:
P_{O_2} = 40 mmHg
P_{CO_2} = 45 mmHg

Oxygenated blood:
P_{O_2} = 100 mmHg
P_{CO_2} = 40 mmHg

To right atrium

To tissue cells

(b) Internal respiration: systemic gas exchange

Systemic capillaries

CO_2 O_2

Systemic tissue cells:
P_{O_2} = 40 mmHg
P_{CO_2} = 45 mmHg

What causes oxygen to enter pulmonary capillaries from alveoli and to enter tissue cells from systemic capillaries?

verts **deoxygenated blood** (depleted of some O_2) coming from the right side of the heart into **oxygenated blood** (saturated with O_2) that returns to the left side of the heart (see Figure 21.30 on page 753). As blood flows through the pulmonary capillaries, it picks up O_2 from alveolar air and unloads CO_2 into alveolar air. Although this process is commonly called an "exchange" of gases, each gas diffuses independently from the area where its partial pressure is higher to the area where its partial pressure is lower.

As Figure 23.18a shows, O_2 diffuses from alveolar air, where its partial pressure is 105 mmHg, into the blood in pulmonary capillaries, where P_{O_2} is only 40 mmHg in a resting person. If you have been exercising, the P_{O_2} will be even lower because contracting muscle fibers are using more O_2. Diffusion continues until the P_{O_2} of pulmonary capillary blood increases to match the P_{O_2} of alveolar air, 105 mmHg. Because blood leaving pulmonary capillaries near alveolar air spaces mixes with a small volume of blood that has flowed through conducting portions of the respiratory system, where gas exchange does not occur, the P_{O_2} of blood in the pulmonary veins is slightly less than the P_{O_2} in pulmonary capillaries, about 100 mmHg.

While O_2 is diffusing from alveolar air into deoxygenated blood, CO_2 is diffusing in the opposite direction. The P_{CO_2} of deoxygenated blood is 45 mmHg in a resting person, whereas P_{CO_2} of alveolar air is 40 mmHg. Because of this difference in P_{CO_2}, carbon dioxide diffuses from deoxygenated blood into the alveoli until the P_{CO_2} of the blood decreases to 40 mmHg. Exhalation keeps alveolar P_{CO_2} at 40 mmHg. Oxygenated blood returning to the left side of the heart in the pulmonary veins thus has a P_{CO_2} of 40 mmHg.

The number of capillaries near alveoli in the lungs is very large, and blood flows slowly enough through these capillaries that it picks up a maximal amount of O_2. During vigorous exercise, when cardiac output is increased, blood flows more rapidly through both the systemic and pulmonary circulations. As a result, blood's transit time in the pulmonary capillaries is shorter. Still, the P_{O_2} of blood in the pulmonary veins normally reaches 100 mmHg. In diseases that decrease the rate of gas diffusion, however, the blood may not come into full equilibrium with alveolar air, especially during exercise. When this happens, the P_{O_2} declines and P_{CO_2} rises in systemic arterial blood.

The left ventricle pumps oxygenated blood into the aorta and through the systemic arteries to systemic capillaries. The exchange of O_2 and CO_2 between systemic capillaries and tissue cells is called **internal respiration** or **systemic gas exchange** (Figure 23.18b). As O_2 leaves the bloodstream, oxygenated blood is converted into deoxygenated blood. Unlike external respiration, which occurs only in the lungs, internal respiration occurs in tissues throughout the body.

The P_{O_2} of blood pumped into systemic capillaries is higher (100 mmHg) than the P_{O_2} in tissue cells (40 mmHg at rest) because the cells constantly use O_2 to produce ATP. Due to this pressure difference, oxygen diffuses out of the capillaries into tissue cells and blood P_{O_2} drops to 40 mmHg by the time the blood exits systemic capillaries.

While O_2 diffuses from the systemic capillaries into tissue cells, CO_2 diffuses in the opposite direction. Because tissue cells are constantly producing CO_2, the P_{CO_2} of cells (45 mmHg at rest) is higher than that of systemic capillary blood (40 mmHg). As a result, CO_2 diffuses from tissue cells through interstitial fluid into systemic capillaries until the P_{CO_2} in the blood increases to 45 mmHg. The deoxygenated blood then returns to the heart and is pumped to the lungs for another cycle of external respiration.

In a person at rest, tissue cells, on average, need only 25% of the available O_2 in oxygenated blood and deoxygenated blood retains 75% of its O_2 content. During exercise, more O_2 diffuses from the blood into metabolically active cells, such as contracting skeletal muscle fibers. Active cells use more O_2 for ATP production, and the O_2 content of deoxygenated blood drops below 75%.

The *rate* of pulmonary and systemic gas exchange depends on several factors.

- **Partial pressure difference of the gases.** Alveolar P_{O_2} must be higher than blood P_{O_2} for oxygen to diffuse from alveolar air into the blood. The rate of diffusion is faster when the difference between P_{O_2} in alveolar air and pulmonary capillary blood is larger; diffusion is slower when the difference is smaller. The differences between P_{O_2} and P_{CO_2} in alveolar air versus pulmonary blood increase during exercise. The larger partial pressure differences accelerate the rates of gas diffusion. The partial pressures of O_2 and CO_2 in alveolar air also depend on the rate of airflow into and out of the lungs. Certain drugs (such as morphine) slow ventilation, thereby decreasing the amount of O_2 and CO_2 that can exchanged between alveolar air and blood. With increasing altitude, the total atmospheric pressure decreases, as does the partial pressure of O_2—from 159 mmHg at sea level, to 110 mmHg at 10,000 ft, to 73 mmHg at 20,000 ft. Although O_2 still is 20.9% of the total, the P_{O_2} of inhaled air decreases with increasing altitude. Alveolar P_{O_2} decreases correspondingly, and O_2 diffuses into the blood more slowly. The common signs and symptoms of **high altitude sickness**—shortness of breath, headache, fatigue, insomnia, nausea, and dizziness—are due to a lower level of oxygen in the blood.

- **Surface area available for gas exchange.** The surface area of the alveoli is huge (about 70 m^2 or 750 ft^2). In addition, many capillaries surround each alveolus, so many that as much as 900 mL of blood is able to participate in gas exchange at any instant. Any pulmonary disorder that decreases the functional surface area of the respiratory membranes decreases the rate of external respiration. In emphysema, for example, alveolar walls disintegrate, so surface area is smaller than normal and pulmonary gas exchange is slowed.

- **Diffusion distance.** The respiratory membrane is very thin, so diffusion occurs quickly. Also, the capillaries are so narrow that the red blood cells must pass through them in single file, which minimizes the diffusion distance from an alveolar air space to hemoglobin inside red blood cells. Buildup of interstitial fluid between alveoli, as occurs in pulmonary edema, slows the rate of gas exchange because it increases diffusion distance.

- *Molecular weight and solubility of the gases.* Because O_2 has a lower molecular weight than CO_2, it could be expected to diffuse across the respiratory membrane about 1.2 times faster. However, the solubility of CO_2 in the fluid portions of the respiratory membrane is about 24 times greater than that of O_2. Taking both of these factors into account, net outward CO_2 diffusion occurs 20 times more rapidly than net inward O_2 diffusion. Consequently, when diffusion is slower than normal, for example, in emphysema or pulmonary edema, O_2 insufficiency (hypoxia) typically occurs before there is significant retention of CO_2 (hypercapnia).

▶ CHECKPOINT

20. How does the partial pressure of oxygen change as altitude changes?

21. What are the diffusion paths of oxygen and carbon dioxide during external and internal respiration?

TRANSPORT OF OXYGEN AND CARBON DIOXIDE

▶ OBJECTIVE

- **Describe how the blood transports oxygen and carbon dioxide.**

The blood transports gases between the lungs and body tissues. When O_2 and CO_2 enter the blood, certain chemical reactions occur that aid in gas transport and gas exchange.

Oxygen Transport

Oxygen does not dissolve easily in water, and therefore only about 1.5% of the O_2 is dissolved in blood plasma, which is mostly water. About 98.5% of blood O_2 is bound to hemoglobin in red blood cells (Figure 23.19). Each 100 mL of oxygenated blood contains the equivalent of 20 mL of gaseous O_2. The amount dissolved in the plasma is 0.3 mL and the amount bound to hemoglobin is 19.7 mL.

The heme portion of hemoglobin contains four atoms of iron, each capable of binding to a molecule of O_2 (see Figure 19.4b, c on page 640). Oxygen and hemoglobin bind in an easily reversible reaction to form **oxyhemoglobin:**

$$\text{Hb} \quad + \quad O_2 \quad \underset{\substack{\text{Dissociation} \\ \text{of } O_2}}{\overset{\substack{\text{Binding of } O_2}}{\rightleftharpoons}} \quad \text{Hb}-O_2$$

Reduced hemoglobin   Oxygen   Oxyhemoglobin
(deoxyhemoglobin)

Because 98.5% of the O_2 is bound to hemoglobin and therefore trapped inside RBCs, only the dissolved O_2 (1.5%) can diffuse out of tissue capillaries into tissue cells. Thus, it is important to understand the factors that promote O_2 binding to and dissociation (separation) from hemoglobin.

The Relation Between Hemoglobin and Oxygen Partial Pressure

The most important factor that determines how much O_2 binds to hemoglobin is the P_{O_2}; the higher the P_{O_2}, the more O_2 combines with Hb. When reduced hemoglobin (Hb) is completely converted to oxyhemoglobin (Hb–O_2), the hemoglobin is said to be **fully saturated;** when hemoglobin consists of a mixture of Hb and Hb–O_2, it is **partially saturated.** The **percent saturation of hemoglobin** expresses the average saturation of hemoglobin with oxygen. For instance, if each hemoglobin molecule has bound two O_2 molecules, then the hemoglobin is 50% saturated because each Hb can bind a maximum of four O_2.

The relation between the percent saturation of hemoglobin and P_{O_2} is illustrated in the oxygen–hemoglobin dissociation curve in Figure 23.20 on page 833. Note that when the P_{O_2} is high, hemoglobin binds with large amounts of O_2 and is almost 100% saturated. When P_{O_2} is low, hemoglobin is only partially saturated. In other words, the greater the P_{O_2}, the more O_2 will bind to hemoglobin, until all the available hemoglobin molecules are saturated. Therefore, in pulmonary capillaries, where P_{O_2} is high, a lot of O_2 binds to hemoglobin. In tissue capillaries, where the P_{O_2} is lower, hemoglobin does not hold as much O_2, and the O_2 is unloaded via diffusion into tissue cells. Note that hemoglobin is still 75% saturated with O_2 at a P_{O_2} of 40 mmHg, the average P_{O_2} of tissue cells in a person at rest. This is the basis for the earlier statement that only 25% of the available O_2 unloads from hemoglobin and is used by tissue cells under resting conditions.

When the P_{O_2} is between 60 and 100 mmHg, hemoglobin is 90% or more saturated with O_2 (Figure 23.20). Thus, blood picks up a nearly full load of O_2 from the lungs even when the P_{O_2} of alveolar air is as low as 60 mmHg. The Hb–P_{O_2} curve explains why people can still perform well at high altitudes or when they have certain cardiac and pulmonary diseases, even though P_{O_2} may drop as low as 60 mmHg. Note also in the curve that at a considerably lower P_{O_2} of 40 mmHg, hemoglobin is still 75% saturated with O_2. However, oxygen saturation of Hb drops to 35% at 20 mmHg. Between 40 and 20 mmHg, large amounts of O_2 are released from hemoglobin in response to only small decreases in P_{O_2}. In active tissues such as contracting muscles, P_{O_2} may drop well below 40 mmHg. Then, a large percentage of the O_2 is released from hemoglobin, providing more O_2 to metabolically active tissues.

Other Factors Affecting Hemoglobin's Affinity for Oxygen

Although P_{O_2} is the most important factor that determines the percent O_2 saturation of hemoglobin, several other factors influence the tightness or **affinity** with which hemoglobin binds O_2. In effect, these factors shift the entire curve either to the left (higher affinity) or to the right (lower affinity). The changing affinity of hemoglobin for O_2 is another example of how homeostatic mechanisms adjust body activities to cellular needs. Each one makes sense if you keep in mind that metabolically active

Figure 23.19 Transport of oxygen (O_2) and carbon dioxide (CO_2) in the blood.

Most O_2 is transported by hemoglobin as oxyhemoglobin ($Hb-O_2$) within red blood cells; most CO_2 is transported in blood plasma as bicarbonate ions (HCO_3^-).

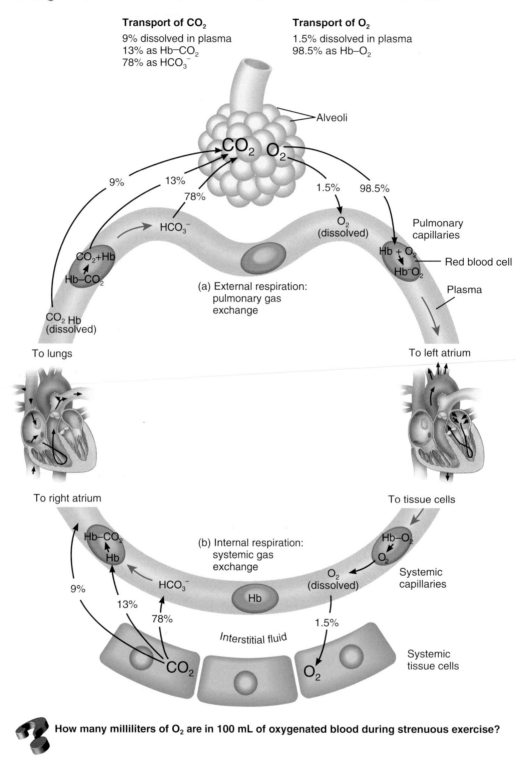

Transport of CO_2
9% dissolved in plasma
13% as $Hb-CO_2$
78% as HCO_3^-

Transport of O_2
1.5% dissolved in plasma
98.5% as $Hb-O_2$

Alveoli

CO_2 O_2

9% 13% 78% 1.5% 98.5%

HCO_3^- O_2 (dissolved)

Pulmonary capillaries

CO_2+Hb $Hb + O_2$ Red blood cell
$Hb-CO_2$ $Hb-O_2$

CO_2 Hb (dissolved)

(a) External respiration: pulmonary gas exchange

Plasma

To lungs

To left atrium

To right atrium

To tissue cells

$Hb-CO_2$ $Hb-O_2$
Hb O_2

(b) Internal respiration: systemic gas exchange

9% HCO_3^- Systemic capillaries

13% 78% Hb O_2 (dissolved) 1.5%

Interstitial fluid

CO_2 O_2

Systemic tissue cells

How many milliliters of O_2 are in 100 mL of oxygenated blood during strenuous exercise?

tissue cells need O_2 and produce acids, CO_2, and heat as wastes. The following four factors affect hemoglobin's affinity for O_2:

1. Acidity (pH). As acidity increases (pH decreases), the affinity of hemoglobin for O_2 decreases, and O_2 dissociates more

readily from hemoglobin (Figure 23.21a). In other words, increasing acidity enhances the unloading of oxygen from hemoglobin. The main acids produced by metabolically active tissues are lactic acid and carbonic acid. When pH decreases, the entire oxygen–hemoglobin dissociation curve shifts to the right; at any

Figure 23.20 Oxygen–hemoglobin dissociation curve showing the relationship between hemoglobin saturation and P_{O_2} at normal body temperature.

As P_{O_2} increases, more O_2 combines with hemoglobin.

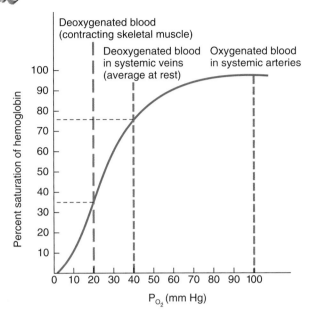

What point on the curve represents blood in your pulmonary veins right now? In your pulmonary veins if you were jogging?

Figure 23.21 Oxygen–hemoglobin dissociation curves showing the relationship of (a) pH and (b) P_{CO_2} to hemoglobin saturation at normal body temperature. As pH increases or P_{CO_2} decreases, O_2 combines more tightly with hemoglobin, so that less is available to tissues. The broken lines emphasize these relationships.

As pH decreases or P_{CO_2} increases, the affinity of hemoglobin for O_2 declines, so less O_2 combines with hemoglobin and more is available to tissues.

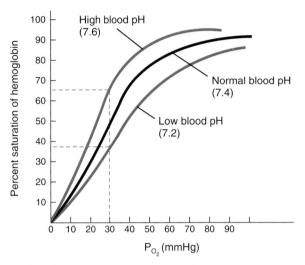

(a) Effect of pH on affinity of hemoglobin for oxygen

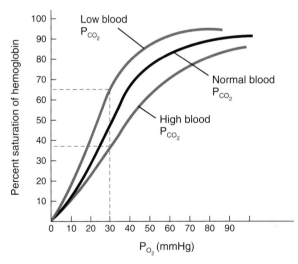

(b) Effect of P_{CO_2} on affinity of hemoglobin for oxygen

In comparison to the value when you are sitting, is the affinity of your hemoglobin for O_2 higher or lower when you are exercising? What is the benefit of this change to you?

given P_{O_2}, Hb is less saturated with O_2, a change termed the **Bohr effect.** The Bohr effect works both ways: An increase in H^+ in blood causes O_2 to unload from hemoglobin, and the binding of O_2 to hemoglobin causes unloading of H^+ from hemoglobin. The explanation for the Bohr effect is that hemoglobin can act as a buffer for hydrogen ions (H^+). But when H^+ bind to amino acids in hemoglobin, they slightly alter its structure and thereby decrease its oxygen-carrying capacity. Thus, lowered pH drives O_2 off hemoglobin, making more O_2 available for tissue cells. By contrast, elevated pH increases the affinity of hemoglobin for O_2 and shifts the oxygen–hemoglobin dissociation curve to the left.

2. *Partial pressure of carbon dioxide.* CO_2 also can bind to hemoglobin, and the effect is similar to that of H^+ (shifting the curve to the right). As P_{CO_2} rises, hemoglobin releases O_2 more readily (Figure 23.21b). P_{CO_2} and pH are related factors because low blood pH (acidity) results from high P_{CO_2}. As CO_2 enters the blood, much of it is temporarily converted to carbonic acid (H_2CO_3), a reaction catalyzed by an enzyme in red blood cells called *carbonic anhydrase (CA):*

$$CO_2 + H_2O \overset{CA}{\rightleftharpoons} H_2CO_3 \rightleftharpoons H^+ + HCO_3^-$$

Carbon dioxide   Water   Carbonic acid   Hydrogen ion   Bicarbonate ion

The carbonic acid thus formed in red blood cells dissociates into hydrogen ions and bicarbonate ions. As the H^+ concentra-

tion increases, pH decreases. Thus, an increased P_{CO_2} produces a more acidic environment, which helps release O_2 from hemoglobin. During exercise, lactic acid—a byproduct of anaerobic metabolism within muscles—also decreases blood pH. Decreased P_{CO_2} (and elevated pH) shifts the saturation curve to the left.

3. **Temperature.** Within limits, as temperature increases, so does the amount of O_2 released from hemoglobin (Figure 23.22). Heat is a byproduct of the metabolic reactions of all cells, and the heat released by contracting muscle fibers tends to raise body temperature. Metabolically active cells require more O_2 and liberate more acids and heat. The acids and heat, in turn, promote release of O_2 from oxyhemoglobin. Fever produces a similar result. In contrast, during hypothermia (lowered body temperature) cellular metabolism slows, the need for O_2 is reduced, and more O_2 remains bound to hemoglobin (a shift to the left in the saturation curve).

4. **BPG.** A substance found in red blood cells called **2,3-bisphosphoglycerate (BPG),** previously called diphosphoglycerate (DPG), decreases the affinity of hemoglobin for O_2 and thus helps unload O_2 from hemoglobin. BPG is formed in red blood cells when they break down glucose to produce ATP in a process called glycolysis (described on page 911). When BPG combines with hemoglobin, by binding to the terminal amino groups of the two beta globin chains, the hemoglobin binds O_2 less tightly at the heme group sites. The greater the level of BPG, the more O_2 is unloaded from hemoglobin. Certain hormones, such as thyroxine, human growth hormone, epinephrine, norepinephrine, and testosterone, increase the formation of BPG. The level of BPG also is higher in people living at higher altitudes.

Oxygen Affinity of Fetal and Adult Hemoglobin

Fetal hemoglobin (Hb-F) differs from **adult hemoglobin (Hb-A)** in structure and in its affinity for O_2. Hb-F has a higher affinity for O_2 because it binds BPG less strongly. Thus, when P_{O_2} is low, Hb-F can carry up to 30% more O_2 than maternal Hb-A (Figure 23.23). As the maternal blood enters the placenta, O_2 is readily transferred to fetal blood. This is very important because the O_2 saturation in maternal blood in the placenta is quite low, and the fetus might suffer hypoxia were it not for the greater affinity of fetal hemoglobin for O_2.

Carbon Monoxide Poisoning

Carbon monoxide (CO) is a colorless and odorless gas found in exhaust fumes from automobiles, gas furnaces, and space heaters and in tobacco smoke. It is a byproduct of the combustion of carbon-containing materials such as coal, gas, and wood. CO binds to the heme group of hemoglobin, just as O_2 does, except that the binding of carbon monoxide to hemoglobin is over 200 times as strong as the binding of O_2 to hemoglobin. Thus, at a concentration as small as 0.1% ($P_{CO} = 0.5\,mmHg$), CO will combine with half the available hemoglobin molecules and reduce the oxygen-carrying capacity of the blood by 50%. Elevated blood levels of CO cause **carbon monoxide poisoning,** one sign of which is a bright, cherry-red color of the lips and oral mucosa. Administering pure oxygen, which speeds up the separation of carbon monoxide from hemoglobin, may rescue the person. ■

Carbon Dioxide Transport

Under normal resting conditions, each 100 mL of deoxygenated blood contains the equivalent of 53 mL of gaseous CO_2, which is transported in the blood in three main forms (see Figure 23.19):

Figure 23.22 Oxygen–hemoglobin dissociation curves showing the effect of temperature changes.

As temperature increases, the affinity of hemoglobin for O_2 decreases.

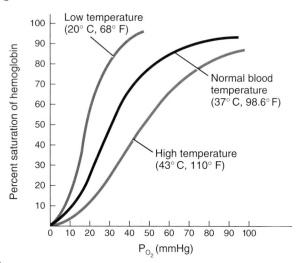

Is O_2 more available or less available to tissue cells when you have a fever? Why?

Figure 23.23 Oxygen–hemoglobin dissociation curves comparing fetal and maternal hemoglobin.

Fetal hemoglobin has a higher affinity for O_2 than does adult hemoglobin.

The P_{O_2} of placental blood is about 40 mmHg. What are the O_2 saturations of maternal and fetal hemoglobin at this P_{O_2}?

1. ***Dissolved CO₂.*** The smallest percentage—about 9%—is dissolved in blood plasma. Upon reaching the lungs, it diffuses into alveolar air and is exhaled.

2. ***Carbamino compounds.*** A somewhat higher percentage, about 13%, combines with the amino groups of amino acids and proteins in blood to form **carbamino compounds.** Because the most prevalent protein in blood is hemoglobin (inside red blood cells), most of the CO_2 transported in this manner is bound to hemoglobin. The main CO_2 binding sites are the terminal amino acids in the two alpha and two beta globin chains. Previous higher estimates of the amount of carbamino compounds formed in blood ignored the competing effect of bis-phosphoglycerate (BPG), which also binds to the terminal amino acids of the two beta chains. Hemoglobin that has bound CO_2 is termed **carbaminohemoglobin (Hb–CO₂):**

$$Hb \quad + \quad CO_2 \quad \rightleftharpoons \quad Hb\text{–}CO_2$$
Hemoglobin   Carbon dioxide   Carbaminohemoglobin

The formation of carbaminohemoglobin is greatly influenced by P_{CO_2}. For example, in tissue capillaries P_{CO_2} is relatively high, which promotes formation of carbaminohemoglobin. But in pulmonary capillaries, P_{CO_2} is relatively low, and the CO_2 readily splits apart from globin and enters the alveoli by diffusion.

3. ***Bicarbonate ions.*** The greatest percentage of CO_2—about 78%—is transported in blood plasma as **bicarbonate ions (HCO₃⁻).** As CO_2 diffuses into systemic capillaries and enters red blood cells, it reacts with water in the presence of the enzyme carbonic anhydrase (CA) to form carbonic acid, which dissociates into H^+ and HCO_3^-:

$$CO_2 \;+\; H_2O \; \overset{CA}{\rightleftharpoons} \; H_2CO_3 \; \rightleftharpoons \; H^+ \;+\; HCO_3^-$$
Carbon dioxide   Water   Carbonic acid   Hydrogen ion   Bicarbonate ion

Thus, as blood picks up CO_2, HCO_3^- accumulates inside RBCs. Some HCO_3^- moves out into the blood plasma, down its concentration gradient. In exchange, chloride ions (Cl^-) move from plasma into the RBCs. This exchange of negative ions, which maintains the electrical balance between blood plasma and RBC cytosol, is known as the **chloride shift** (see Figure 23.24b). The net effect of these reactions is that CO_2 is removed from tissue cells and transported in blood plasma as HCO_3^-. As blood passes through pulmonary capillaries in the lungs, all these reactions reverse as CO_2 is exhaled.

The amount of CO_2 that can be transported in the blood is influenced by the percent saturation of hemoglobin with oxygen. The lower the amount of oxyhemoglobin (Hb–O₂), the higher the CO_2 carrying capacity of the blood, a relationship known as the **Haldane effect.** Two characteristics of deoxyhemoglobin give rise to the Haldane effect: (1) Deoxyhemoglobin binds to and thus transports more CO_2 than does Hb–O₂. (2) Deoxyhemoglobin also buffers more H^+ than does Hb–O₂, thereby removing H^+ from solution and promoting conversion of CO_2 to HCO_3^- via the reaction catalyzed by carbonic anhydrase.

Summary of Gas Exchange and Transport in Lungs and Tissues

Deoxygenated blood returning to the pulmonary capillaries in the lungs (Figure 23.24a) contains CO_2 dissolved in blood plasma, CO_2 combined with globin as carbaminohemoglobin (Hb–CO₂), and CO_2 incorporated into HCO_3^- within RBCs. The RBCs have also picked up H^+, some of which binds to and therefore is buffered by hemoglobin (Hb–H). As blood passes through the pulmonary capillaries, molecules of CO_2 dissolved in blood plasma and CO_2 that dissociates from the globin portion of hemoglobin diffuse into alveolar air and are exhaled. At the same time, inhaled O_2 is diffusing from alveolar air into RBCs and is binding to hemoglobin to form oxyhemoglobin (Hb–O₂). Carbon dioxide also is released from HCO_3^- when H^+ combines with HCO_3^- inside RBCs. The H_2CO_3 formed from this reaction then splits into CO_2, which is exhaled, and H_2O. As the concentration of HCO_3^- declines inside RBCs in pulmonary capillaries, HCO_3^- diffuses in from the blood plasma, in exchange for Cl^-. In sum, oxygenated blood leaving the lungs has increased O_2 content and decreased amounts of CO_2 and H^+. In systemic capillaries, as cells use O_2 and produce CO_2, the chemical reactions reverse (Figure 23.24b).

▶ CHECKPOINT

22. In a resting person, how many O_2 molecules are attached to each hemoglobin molecule, on average, in blood in the pulmonary arteries? In blood in the pulmonary veins?

23. What is the relationship between hemoglobin and P_{O_2}? How do temperature, H^+, P_{CO_2}, and BPG influence the affinity of Hb for O_2?

24. Why can hemoglobin unload more oxygen as blood flows through capillaries of metabolically active tissues, such as skeletal muscle during exercise, than is unloaded at rest?

▌ CONTROL OF RESPIRATION

▶ OBJECTIVE

- **Explain how the nervous system controls breathing and list the factors that can alter the rate and depth of breathing.**

At rest, about 200 mL of O_2 are used each minute by body cells. During strenuous exercise, however, O_2 use typically increases 15- to 20-fold in normal healthy adults, and as much as 30-fold in elite endurance-trained athletes. Several mechanisms help match respiratory effort to metabolic demand.

Respiratory Center

The basic rhythm of respiration is controlled by groups of neurons in the brain stem. The area from which nerve impulses are sent to respiratory muscles is called the **respiratory center** and consists of groups of neurons functionally divided into three

Figure 23.24 Summary of chemical reactions that occur during gas exchange. (a) As carbon dioxide (CO_2) is exhaled, hemoglobin (Hb) inside red blood cells in pulmonary capillaries unloads CO_2 and picks up O_2 from alveolar air. Binding of O_2 to Hb–H releases hydrogen ions (H^+). Bicarbonate ions (HCO_3^-) pass into the RBC and bind to released H^+, forming carbonic acid (H_2CO_3). The H_2CO_3 dissociates into water (H_2O) and CO_2, and the CO_2 diffuses from blood into alveolar air. To maintain electrical balance, a chloride ion (Cl^-) exits the RBC for each HCO_3^- that enters (reverse chloride shift). (b) CO_2 diffuses out of tissue cells that produce it and enters red blood cells, where some of it binds to hemoglobin, forming carbaminohemoglobin (Hb–CO_2). This reaction causes O_2 to dissociate from oxyhemoglobin (Hb–O_2). Other molecules of CO_2 combine with water to produce bicarbonate ions (HCO_3^-) and hydrogen ions (H^+). As Hb buffers H^+, the Hb releases O_2 (Bohr effect). To maintain electrical balance, a chloride ion (Cl^-) enters the RBC for each HCO_3^- that exits (chloride shift).

🔑 **Hemoglobin inside red blood cells transports O_2, CO_2, and H^+.**

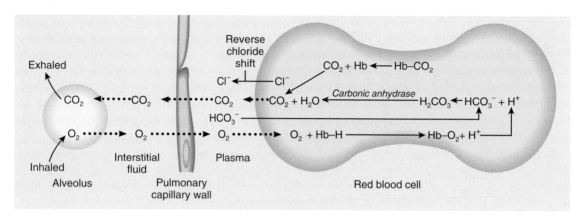

(a) Exchange of O_2 and CO_2 in pulmonary capillaries (external respiration)

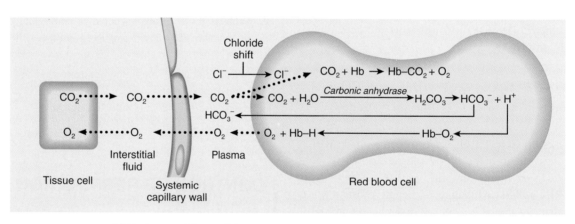

(b) Exchange of O_2 and CO_2 in systemic capillaries (internal respiration)

❓ **Would you expect the concentration of HCO_3^- to be higher in blood plasma taken from a systemic artery or a systemic vein?**

areas: the medullary rhythmicity area, the pneumotaxic area, and the apneustic area (Figure 23.25).

Medullary Rhythmicity Area

The **medullary rhythmicity area** (rith-MIS-i-tē) in the medulla oblongata controls the basic rhythm of respiration. Within the medullary rhythmicity area are both inspiratory and expiratory areas. Figure 23.26 shows the relationships of the inspiratory and expiratory areas during normal quiet breathing and forceful breathing.

During quiet breathing, inhalation lasts for about 2 seconds and exhalation lasts for about 3 seconds. Nerve impulses generated in the **inspiratory area** establish the basic rhythm of breathing. While the inspiratory area is active, it generates nerve

Figure 23.25   Locations of areas of the respiratory center.

The respiratory center is composed of neurons in the medullary rhythmicity area in the medulla oblongata plus the pneumotaxic and apneustic areas in the pons.

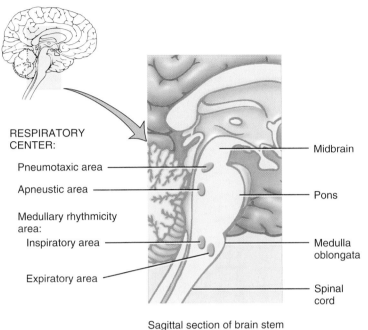

RESPIRATORY CENTER:

Pneumotaxic area

Apneustic area

Medullary rhythmicity area:

   Inspiratory area

   Expiratory area

Midbrain

Pons

Medulla oblongata

Spinal cord

Sagittal section of brain stem

Which area contains neurons that are active and then inactive in a repeating cycle?

impulses for about 2 seconds (Figure 23.26a). The impulses propagate to the external intercostal muscles via intercostal nerves and to the diaphragm via the phrenic nerves. When the nerve impulses reach the diaphragm and external intercostal muscles, the muscles contract and inhalation occurs. Even when all incoming nerve connections to the inspiratory area are cut or blocked, neurons in this area still rhythmically discharge impulses that cause inhalation. At the end of 2 seconds, the inspiratory area becomes inactive and nerve impulses cease. With no impulses arriving, the diaphragm and external intercostal muscles relax for about 3 seconds, allowing passive elastic recoil of the lungs and thoracic wall. Then, the cycle repeats.

The neurons of the **expiratory area** remain inactive during quiet breathing. However, during forceful breathing nerve impulses from the inspiratory area activate the expiratory area (Figure 23.26b). Impulses from the expiratory area cause contraction of the internal intercostal and abdominal muscles, which decreases the size of the thoracic cavity and causes forceful exhalation.

Pneumotaxic Area

Although the medullary rhythmicity area controls the basic rhythm of respiration, other sites in the brain stem help coordinate the transition between inhalation and exhalation. One of these sites is the **pneumotaxic area** (noo-mō-TAK-sik; *pneumo-* = air or breath; *-taxic* = arrangement) in the upper pons (see Figure 23.25), which transmits inhibitory impulses to the inspiratory area. The major effect of these nerve impulses is to help turn off the inspiratory area before the lungs become too

Figure 23.26   Roles of the medullary rhythmicity area in controlling (a) the basic rhythm of respiration and (b) forceful breathing.

During normal, quiet breathing, the expiratory area is inactive; during forceful breathing, the inspiratory area activates the expiratory area.

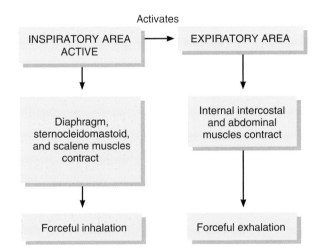

(a) During normal quiet breathing

(b) During forceful breathing

Which nerves convey impulses from the respiratory center to the diaphragm?

full of air. In other words, the impulses shorten the duration of inhalation. When the pneumotaxic area is more active, breathing rate is more rapid.

Apneustic Area

Another part of the brain stem that coordinates the transition between inhalation and exhalation is the **apneustic area** (ap-NOO-stik) in the lower pons (see Figure 23.25). This area sends stimulatory impulses to the inspiratory area that activate it and prolong inhalation. The result is a long, deep inhalation. When the pneumotaxic area is active, it overrides the apneustic area.

Regulation of the Respiratory Center

Although the basic rhythm of respiration is set and coordinated by the inspiratory area, the rhythm can be modified in response to inputs from other brain regions, receptors in the peripheral nervous system, and other factors.

Cortical Influences on Respiration

Because the cerebral cortex has connections with the respiratory center, we can voluntarily alter our pattern of breathing. We can even refuse to breathe at all for a short time. Voluntary control is protective because it enables us to prevent water or irritating gases from entering the lungs. The ability to not breathe, however, is limited by the buildup of CO_2 and H^+ in the body. When P_{CO_2} and H^+ concentrations increase to a certain level, the inspiratory area is strongly stimulated, nerve impulses are sent along the phrenic and intercostal nerves to inspiratory muscles, and breathing resumes, whether the person wants it or not. It is impossible for people to kill themselves by voluntarily holding their breath. Even if breath is held long enough to cause fainting, breathing resumes when consciousness is lost. Nerve impulses from the hypothalamus and limbic system also stimulate the respiratory center, allowing emotional stimuli to alter respirations as, for example, in laughing and crying.

Chemoreceptor Regulation of Respiration

Certain chemical stimuli modulate how quickly and how deeply we breathe. The respiratory system functions to maintain proper levels of CO_2 and O_2 and is very responsive to changes in the levels of either in body fluids. Sensory neurons that are responsive to chemicals are termed **chemoreceptors.** Chemoreceptors in two locations monitor levels of CO_2, H^+, and O_2 and provide input to the respiratory center (Figure 23.27). **Central chemoreceptors** are located in the medulla oblongata in the *central* nervous system. They respond to changes in H^+ concentration or P_{CO_2}, or both, in cerebrospinal fluid. **Peripheral chemoreceptors** are located in the **aortic bodies,** clusters of chemoreceptors located in the wall of the arch of the aorta, and in the **carotid bodies,** which are oval nodules in the wall of the left and right common carotid arteries where they divide into the internal and external carotid arteries. (The chemoreceptors

Figure 23.27 Locations of peripheral chemoreceptors.

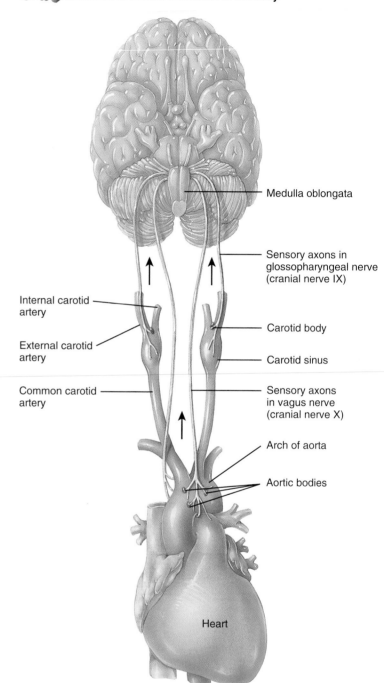

Chemoreceptors are sensory neurons that respond to changes in the levels of certain chemicals in the body.

- Medulla oblongata
- Sensory axons in glossopharyngeal nerve (cranial nerve IX)
- Internal carotid artery
- Carotid body
- External carotid artery
- Carotid sinus
- Common carotid artery
- Sensory axons in vagus nerve (cranial nerve X)
- Arch of aorta
- Aortic bodies
- Heart

 Which chemicals stimulate peripheral chemoreceptors?

of the aortic bodies are located close to the aortic baroreceptors, and the carotid bodies are located close to the carotid sinus baroreceptors. Recall that baroreceptors monitor blood pressure.) These chemoreceptors are part of the *peripheral* nervous system and are sensitive to changes in P_{O_2}, H^+, and P_{CO_2} in the blood. Axons of sensory neurons from the aortic bodies are part of the vagus nerve (cranial nerve X), whereas those from the carotid bodies are part of the right and left glossopharyngeal nerves (cranial nerve IX).

Because CO_2 is lipid soluble, it easily diffuses into cells, where in the presence of carbonic anhydrase it combines with water (H_2O) to form carbonic acid (H_2CO_3). Carbonic acid quickly breaks down into H^+ and HCO_3^-. Thus, an increase in CO_2 in the blood causes an increase in H^+ inside cells, and a decrease in CO_2 causes a decrease in H^+.

Normally, the P_{CO_2} in arterial blood is 40 mmHg. If even a slight increase in P_{CO_2} occurs—a condition called **hypercapnia** or **hypercarbia**—the central chemoreceptors are stimulated and respond vigorously to the resulting increase in H^+ level. The peripheral chemoreceptors also are stimulated by both the high P_{CO_2} and the rise in H^+. In addition, the peripheral chemoreceptors (but not the central chemoreceptors) respond to a deficiency of O_2. When P_{O_2} in arterial blood falls from a normal level of 105 mmHg but is still above 50 mmHg, the peripheral chemoreceptors are stimulated. Severe deficiency of O_2 depresses activity of the central chemoreceptors and inspiratory area, which then do not respond well to any inputs and send fewer impulses to the muscles of inhalation. As the breathing rate decreases or breathing ceases altogether, P_{O_2} falls lower and lower, thereby establishing a positive feedback cycle with a possibly fatal result.

The chemoreceptors participate in a negative feedback system that regulates the levels of CO_2, O_2, and H^+ in the blood (Figure 23.28). As a result of increased P_{CO_2}, decreased pH (increased H^+), or decreased P_{O_2}, input from the central and peripheral chemoreceptors causes the inspiratory area to become highly active, and the rate and depth of breathing increase. Rapid and deep breathing, called **hyperventilation,** allows the inhalation of more O_2 and exhalation of more CO_2 until P_{CO_2} and H^+ are lowered to normal.

If arterial P_{CO_2} is lower than 40 mmHg—a condition called **hypocapnia** or **hypocarbia**—the central and peripheral chemoreceptors are not stimulated, and stimulatory impulses are not sent to the inspiratory area. As a result, the area sets its own moderate pace until CO_2 accumulates and the P_{CO_2} rises to 40 mmHg. The inspiratory center is more strongly stimulated when P_{CO_2} is rising above normal than when P_{O_2} is falling below normal. As a result, people who hyperventilate voluntarily and cause hypocapnia can hold their breath for an unusually long period. Swimmers were once encouraged to hyperventilate just before diving in to compete. However, this practice is risky because the O_2 level may fall dangerously low and cause fainting before the P_{CO_2} rises high enough to stimulate inhalation. A person who faints on land may suffer bumps and bruises, but one who faints in the water may drown.

Figure 23.28 **Regulation of breathing in response to changes in blood P_{CO_2}, P_{O_2}, and pH (H^+ concentration) via negative feedback control.**

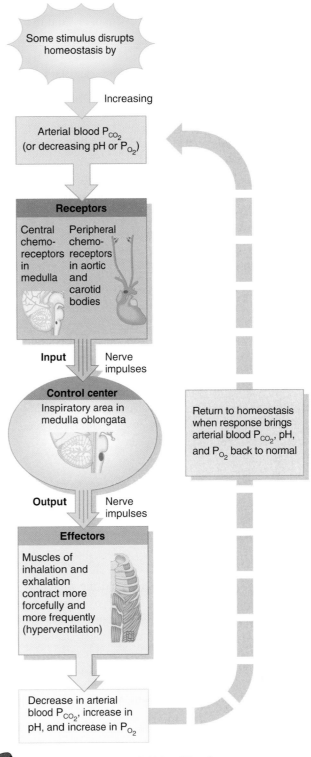

An increase in arterial blood P_{CO_2} stimulates the inspiratory center.

 What is the normal arterial blood P_{CO_2}?

Hypoxia

Hypoxia (hī-POK-sē-a; *hypo-* = under) is a deficiency of O_2 at the tissue level. Based on the cause, we can classify hypoxia into four types, as follows:

1. **Hypoxic hypoxia** is caused by a low P_{O_2} in arterial blood as a result of high altitude, airway obstruction, or fluid in the lungs.

2. In **anemic hypoxia,** too little functioning hemoglobin is present in the blood, which reduces O_2 transport to tissue cells. Among the causes are hemorrhage, anemia, and failure of hemoglobin to carry its normal complement of O_2, as in carbon monoxide poisoning.

3. In **ischemic hypoxia,** blood flow to a tissue is so reduced that too little O_2 is delivered to it, even though P_{O_2} and oxyhemoglobin level are normal.

4. In **histotoxic hypoxia,** the blood delivers adequate O_2 to tissues, but the tissues are unable to use it properly because of the action of some toxic agent. One cause is cyanide poisoning, in which cyanide blocks an enzyme needed for O_2 utilization during ATP synthesis. ■

Proprioceptor Stimulation of Respiration

As soon as you start exercising, your rate and depth of breathing increase, even before changes in P_{O_2}, P_{CO_2}, or H^+ level occur. The main stimulus for these quick changes in respiratory effort is input from proprioceptors, which monitor movement of joints and muscles. Nerve impulses from the proprioceptors stimulate the inspiratory area of the medulla oblongata. At the same time, axon collaterals (branches) of upper motor neurons that originate in the primary motor cortex (precentral gyrus) also feed excitatory impulses into the inspiratory area.

The Inflation Reflex

Located in the walls of bronchi and bronchioles are stretch-sensitive receptors called **baroreceptors** or **stretch receptors.** When these receptors become stretched during overinflation of the lungs, nerve impulses are sent along the cranial nerve X (vagus) to the inspiratory and apneustic areas. In response, the inspiratory area is inhibited directly, and the apneustic area is inhibited from activating the inspiratory area. As a result, exhalation begins. As air leaves the lungs during exhalation, the lungs deflate and the stretch receptors are no longer stimulated. Thus, the inspiratory and apneustic areas are no longer inhibited, and a new inhalation begins. This reflex, referred to as the **inflation (Hering–Breuer) reflex,** is mainly a protective mechanism for preventing excessive inflation of the lungs rather than a key component in the normal regulation of respiration.

Other Influences on Respiration

Other factors that contribute to regulation of respiration include the following:

- **Limbic system stimulation.** Anticipation of activity or emotional anxiety may stimulate the limbic system, which then sends excitatory input to the inspiratory area, increasing the rate and depth of ventilation.

- **Temperature.** An increase in body temperature, as occurs during a fever or vigorous muscular exercise, increases the rate of respiration. A decrease in body temperature decreases respiratory rate. A sudden cold stimulus (such as plunging into cold water) causes temporary **apnea** (AP-nē-a; *a-* = without; *-pnea* = breath), an absence of breathing.

- **Pain.** A sudden, severe pain brings about brief apnea, but a prolonged somatic pain increases respiratory rate. Visceral pain may slow respiratory rate.

- **Stretching the anal sphincter muscle.** This action increases the respiratory rate and is sometimes used to stimulate respiration in a newborn baby or a person who has stopped breathing.

- **Irritation of airways.** Physical or chemical irritation of the pharynx or larynx brings about an immediate cessation of breathing followed by coughing or sneezing.

- **Blood pressure.** The carotid and aortic baroreceptors that detect changes in blood pressure have a small effect on respiration. A sudden rise in blood pressure decreases the rate of respiration, and a drop in blood pressure increases the respiratory rate.

Table 23.2 summarizes the stimuli that affect the rate and depth of ventilation.

▶ CHECKPOINT

25. How does the medullary rhythmicity area function in regulating respiration?

26. How are the apneustic and pneumotaxic areas related to the control of respiration?

27. How do the cerebral cortex, levels of CO_2 and O_2, proprioceptors, inflation reflex, temperature changes, pain, and irritation of the airways modify respiration?

EXERCISE AND THE RESPIRATORY SYSTEM

▶ OBJECTIVE

- **Describe the effects of exercise on the respiratory system.**

During exercise, the respiratory and cardiovascular systems make adjustments in response to both the intensity and duration of the exercise. The effects of exercise on the heart are discussed in Chapter 20. Here we focus on how exercise affects the respiratory system.

Recall that the heart pumps the same amount of blood to the lungs as to all the rest of the body. Thus, as cardiac output rises,

Table 23.2	Summary of Stimuli that Affect Ventilation Rate and Depth
Stimuli that Increase Ventilation Rate and Depth	**Stimuli that Decrease Ventilation Rate and Depth**
Voluntary hyperventilation controlled by cerebral cortex and anticipation of activity by stimulation of the limbic system.	Voluntary hypoventilation controlled by cerebral cortex.
Increase in arterial blood P_{CO_2} above 40 mmHg (causes an increase in H^+) detected by peripheral and central chemoreceptors.	Decrease in arterial blood P_{CO_2} below 40 mmHg (causes a decrease in H^+) detected by peripheral and central chemoreceptors.
Decrease in arterial blood P_{O_2} from 100 mmHg to 50 mmHg.	Decrease in arterial blood P_{O_2} below 50 mmHg.
Increased activity of proprioceptors.	Decreased activity of proprioceptors.
Increase in body temperature.	Decrease in body temperature decreases rate of respiration, and sudden cold stimulus causes apnea.
Prolonged pain.	Severe pain causes apnea.
Decrease in blood pressure.	Increase in blood pressure.
Stretching anal sphincter.	Irritation of pharynx or larynx by touch or chemicals causes brief apnea followed by coughing or sneezing.

the blood flow to the lungs, termed **pulmonary perfusion,** increases as well. In addition, the **O_2 diffusing capacity,** a measure of the rate at which O_2 can diffuse from alveolar air into the blood, may increase threefold during maximal exercise because more pulmonary capillaries become maximally perfused. As a result, there is a greater surface area available for diffusion of O_2 into pulmonary blood capillaries.

When muscles contract during exercise, they consume large amounts of O_2 and produce large amounts of CO_2. During vigorous exercise, O_2 consumption and pulmonary ventilation both increase dramatically. At the onset of exercise, an abrupt increase in pulmonary ventilation is followed by a more gradual increase. With moderate exercise, the increase is due mostly to an increase in the depth of ventilation rather than to increased breathing rate. When exercise is more strenuous, the frequency of breathing also increases.

The abrupt increase in ventilation at the start of exercise is due to *neural* changes that send excitatory impulses to the inspiratory area in the medulla oblongata. These changes include (1) anticipation of the activity, which stimulates the limbic system; (2) sensory impulses from proprioceptors in muscles, tendons, and joints; and (3) motor impulses from the primary motor cortex (precentral gyrus). The more gradual increase in ventila-

tion during moderate exercise is due to *chemical* and *physical* changes in the bloodstream, including (1) slightly decreased P_{O_2}, due to increased O_2 consumption; (2) slightly increased P_{CO_2}, due to increased CO_2 production by contracting muscle fibers; and (3) increased temperature, due to liberation of more heat as more O_2 is utilized. Moreover, during strenuous exercise, HCO_3^- buffers H^+ released by lactic acid in a reaction that liberates CO_2, which further increases P_{CO_2}.

At the end of an exercise session, an abrupt decrease in pulmonary ventilation is followed by a more gradual decline to the resting level. The initial decrease is due mainly to changes in neural factors when movement stops or slows, whereas the more gradual phase reflects the slower return of blood chemistry levels and temperature to the resting state.

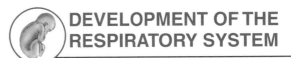

Smoking Lowers Respiratory Efficiency

Smoking may cause a person to become easily "winded" with even moderate exercise because several factors decrease respiratory efficiency in smokers: (1) Nicotine constricts terminal bronchioles, which decreases airflow into and out of the lungs. (2) Carbon monoxide in smoke binds to hemoglobin and reduces its oxygen-carrying capability. (3) Irritants in smoke cause increased mucus secretion by the mucosa of the bronchial tree and swelling of the mucosal lining, both of which impede airflow into and out of the lungs. (4) Irritants in smoke also inhibit the movement of cilia and destroy cilia in the lining of the respiratory system. Thus, excess mucus and foreign debris are not easily removed, which further adds to the difficulty in breathing. (5) With time, smoking leads to destruction of elastic fibers in the lungs and is the prime cause of emphysema (described on page 844). These changes cause collapse of small bronchioles and trapping of air in alveoli at the end of exhalation. The result is less efficient gas exchange. ■

DEVELOPMENT OF THE RESPIRATORY SYSTEM

▶ O B J E C T I V E

• **Describe the development of the respiratory system.**

The development of the mouth and pharynx are discussed in Chapter 24; here we consider the development of the remainder of the respiratory system. At about four weeks of development, the respiratory system begins as an outgrowth of the **endoderm** of the foregut (precursor of some digestive organs) just anterior to the pharynx. This outgrowth is called the **laryngotracheal diverticulum** (see Figure 18.21a on page 623). As the diverticulum grows, it elongates and differentiates into the future epithelial lining of the *larynx* and other structures. Its proximal end maintains a slitlike opening into the pharynx called the *rima* *glottis.* The middle portion of the bud gives rise to the epithelial lining of the *trachea.* The distal portion divides into two **lung buds,** which grow into the epithelial lining of the *bronchi* and *lungs* (Figure 23.29).

As the lung buds develop, they branch repeatedly and give rise to all the *bronchial tubes.* After the sixth month, the closed terminal portions of the tubes dilate and become the *alveoli* of the lungs. **Mesenchymal (mesodermal) cells** form the smooth muscle, cartilage, and connective tissues of the bronchial tubes and the pleural sacs of the lungs.

AGING AND THE RESPIRATORY SYSTEM

▶ OBJECTIVE

• **Describe the effects of aging on the respiratory system.**

With advancing age, the airways and tissues of the respiratory tract, including the alveoli, become less elastic and more rigid; the chest wall becomes more rigid as well. The result is a decrease in lung capacity. In fact, vital capacity (the maximum amount of air that can be expired after maximal inhalation) can decrease as much as 35% by age 70. Moreover, a decrease in blood level of O_2, decreased activity of alveolar macrophages, and diminished ciliary action of the epithelium lining the respiratory tract occur. Owing to all these age-related factors, elderly people are more susceptible to pneumonia, bronchitis, emphysema, and other pulmonary disorders.

▶ CHECKPOINT

28. How does exercise affect the inspiratory area?

29. What structures develop from the laryngotracheal bud?

30. What accounts for the decrease in lung capacity with aging?

• • •

To appreciate the many ways that the respiratory system contributes to homeostasis of other body systems, examine *Focus on Homeostasis: The Respiratory System.* Next, in Chapter 24, we will see how the digestive system makes nutrients available to body cells so that oxygen provided by the respiratory system can be used for ATP production.

Figure 23.29   Development of the bronchial tubes and lungs.

The respiratory system develops from endoderm and mesoderm.

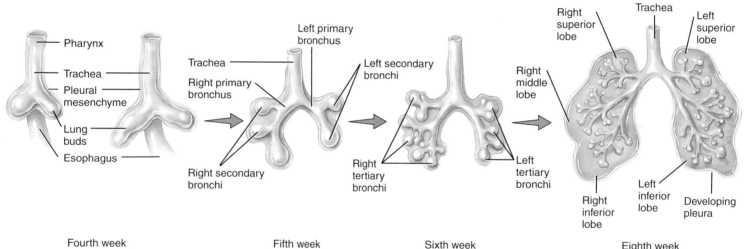

When does the respiratory system begin to develop in an embryo?

Body System	Contribution of the Respiratory System
For all body systems	Provides oxygen and removes carbon dioxide. Helps adjust pH of body fluids through exhalation of carbon dioxide.
Muscular system	Increased rate and depth of breathing support increased activity of skeletal muscles during exercise.
Nervous system	Nose contains receptors for sense of smell (olfaction). Vibrations of air flowing across vocal cords produce sounds for speech.
Endocrine system	Angiotensin converting enzyme (ACE) in lungs catalyzes formation of the hormone angiotensin II from angiotensin I.
Cardiovascular system	During inhalations, respiratory pump aids return of venous blood to the heart.
Lymphatic and immune system	Hairs in nose, cilia and mucus in trachea, bronchi, and smaller airways, and alveolar macrophages contribute to nonspecific resistance to disease. Pharynx (throat) contains lymphatic tissue (tonsils). Respiratory pump (during inhalation) promotes flow of lymph.
Digestive system	Forceful contraction of respiratory muscles can assist in defecation.
Urinary system	Together, respiratory and urinary systems regulate pH of body fluids.
Reproductive systems	Increased rate and depth of breathing support activity during sexual intercourse. Internal respiration provides oxygen to developing fetus.

Focus on Homeostasis:

The Respiratory System

DISORDERS: HOMEOSTATIC IMBALANCES

Asthma

Asthma (AZ-ma = panting) is a disorder characterized by chronic airway inflammation, airway hypersensitivity to a variety of stimuli, and airway obstruction that is at least partially reversible, either spontaneously or with treatment. It affects 3–5% of the U.S. population and is more common in children than in adults. Airway obstruction may be due to smooth muscle spasms in the walls of smaller bronchi and bronchioles, edema of the mucosa of the airways, increased mucus secretion, and damage to the epithelium of the airway.

Individuals with asthma typically react to low concentrations of agents that do not normally cause symptoms in people without asthma. Sometimes the trigger is an allergen such as pollen, house dust mites, molds, or a particular food. Other common triggers of asthma attacks are emotional upset, aspirin, sulfiting agents (used in wine and beer and to keep greens fresh in salad bars), exercise, and breathing cold air or cigarette smoke. In the early phase (acute) response, smooth muscle spasm is accompanied by excessive secretion of mucus that may clog the bronchi and bronchioles and worsen the attack. The late phase (chronic) response is characterized by inflammation, fibrosis, edema, and necrosis (death) of bronchial epithelial cells. A host of mediator chemicals, including leukotrienes, prostaglandins, thromboxane, platelet-activating factor, and histamine, take part.

Symptoms include difficult breathing, coughing, wheezing, chest tightness, tachycardia, fatigue, moist skin, and anxiety. An acute attack is treated by giving an inhaled beta$_2$-adrenergic agonist (albuterol) to help relax smooth muscle in the bronchioles and open up the airways. However, long-term therapy of asthma strives to suppress the underlying inflammation. The anti-inflammatory drugs that are used most often are inhaled corticosteroids (glucocorticoids), cromolyn sodium (Intal), and leukotriene blockers (Accolate).

Chronic Obstructive Pulmonary Disease

Chronic obstructive pulmonary disease (COPD) is a type of respiratory disorder characterized by chronic and recurrent obstruction of airflow, which increases airway resistance. COPD affects about 30 million Americans and is the fourth leading cause of death behind heart disease, cancer, and cerebrovascular disease. The principal types of COPD are emphysema and chronic bronchitis. In most cases, COPD is preventable because its most common cause is cigarette smoking or breathing secondhand smoke. Other causes include air pollution, pulmonary infection, occupational exposure to dusts and gases, and genetic factors. Because men, on average, have more years of exposure to cigarette smoke than women, men are twice as likely as women to suffer from COPD; still, the incidence of COPD in women has risen sixfold in the past 50 years, a reflection of increased smoking among women.

Emphysema

Emphysema (em′-fi-SĒ-ma′ = blown up or full of air) is a disorder characterized by destruction of the walls of the alveoli, producing abnormally large air spaces that remain filled with air during exhalation. With less surface area for gas exchange, O_2 diffusion across the damaged respiratory membrane is reduced. Blood O_2 level is somewhat lowered, and any mild exercise that raises the O_2 requirements of the cells leaves the patient breathless. As increasing numbers of alveolar walls are damaged, lung elastic recoil decreases due to loss of elastic fibers, and an increasing amount of air becomes trapped in the lungs at the end of exhalation. Over several years, added inspiratory exertion increases the size of the chest cage, resulting in a "barrel chest."

Emphysema is generally caused by a long-term irritation; cigarette smoke, air pollution, and occupational exposure to industrial dust are the most common irritants. Some destruction of alveolar sacs may be caused by an imbalance between enzymes called proteases (for example, elastase) and *alpha-1-antitrypsin,* a molecule that inhibits proteases. When decreased production by the liver of the plasma protein alpha-1-antitrypsin occurs, elastase is not inhibited and is free to attack the connective tissue in the walls of alveolar sacs. Cigarette smoke not only deactivates a protein that is apparently crucial in preventing emphysema; it also prevents the repair of affected lung tissue.

Treatment consists of cessation of smoking and removing other environmental irritants, exercise training under careful medical supervision, breathing exercises, use of bronchodilators, and oxygen therapy.

Chronic Bronchitis

Chronic bronchitis is a disorder characterized by excessive secretion of bronchial mucus and accompanied by a productive cough (sputum is raised) that lasts for at least three months of the year for two successive years. Cigarette smoking is the leading cause of chronic bronchitis. Inhaled irritants lead to chronic inflammation with an increase in the size and number of mucous glands and goblet cells in the airway epithelium. The thickened and excessive mucus produced narrows the airway and impairs ciliary function. Thus, inhaled pathogens become embedded in airway secretions and multiply rapidly. Besides a productive cough, symptoms of chronic bronchitis are shortness of breath, wheezing, cyanosis, and pulmonary hypertension. Treatment for chronic bronchitis is similar to that for emphysema.

Lung Cancer

In the United States **lung cancer** is the leading cause of cancer death in both males and females, accounting for 160,000 deaths annually. At the time of diagnosis, lung cancer is usually well advanced, with distant metastases present in about 55% of patients, and regional lymph node involvement in an additional 25%. Most people with lung cancer die within a year of the initial diagnosis; the overall survival rate is only 10–15%. Cigarette smoke is the most common cause of lung cancer. Roughly 85% of lung cancer cases are related to smoking, and the disease is 10 to 30 times more common in smokers than nonsmokers. Exposure to secondhand smoke is also associated with lung cancer and heart disease. In the United States, secondhand smoke causes an estimated 4000 deaths a year from lung cancer, and nearly 40,000 deaths a year from heart disease. Other causes of lung cancer are ionizing radiation and inhaled irritants, such as asbestos and radon gas. Emphysema is a common precursor to the development of lung cancer.

The most common type of lung cancer, **bronchogenic carcinoma,** starts in the epithelium of the bronchial tubes. Bronchogenic tumors are named based on where they arise. For example, *adenocarcinomas* develop in peripheral areas of the lungs from bronchial glands and alveolar cells, *squamous cell carcinomas* develop from the epithelium of larger bronchial tubes, and *small (oat) cell carcinomas* develop from epithelial cells in primary bronchi near the hilus of the lungs and tend to involve the mediastinum early on. Depending on the type of bronchogenic tumors, they may be aggressive, locally invasive, and undergo widespread metastasis. The tumors begin as epithelial lesions that grow to form masses that obstruct the bronchial tubes or invade adjacent lung tissue. Bronchogenic carcinomas metastasize to lymph nodes, the brain, bones, liver, and other organs.

Symptoms of lung cancer are related to the location of the tumor. These may include a chronic cough, spitting blood from the respiratory tract, wheezing, shortness of breath, chest pain, hoarseness, difficulty swallowing, weight loss, anorexia, fatigue, bone pain, confusion, problems with balance, headache, anemia, thrombocytopenia, and jaundice.

Treatment consists of partial or complete surgical removal of a diseased lung (pulmonectomy), radiation therapy, and chemotherapy.

Pneumonia

Pneumonia is an acute infection or inflammation of the alveoli. It is the most common infectious cause of death in the United States, where an estimated 4 million cases occur annually. When certain microbes enter the lungs of susceptible individuals, they release damaging toxins, stimulating inflammation and immune responses that have damaging side effects. The toxins and immune response damage alveoli and bronchial mucous membranes; inflammation and edema cause the alveoli to fill with debris and exudate, interfering with ventilation and gas exchange.

The most common cause of pneumonia is the pneumococcal bacterium *Streptococcus pneumoniae,* but other microbes may also cause pneumonia. Those who are most susceptible to pneumonia are the elderly, infants, immunocompromised individuals (those having AIDS or a malignancy, or those taking immunosuppressive drugs), cigarette smokers, and individuals with an obstructive lung disease. Most cases of pneumonia are preceded by an upper respiratory infection that often is viral. Individuals then develop fever, chills, productive or dry cough, malaise, chest pain, and sometimes dyspnea (difficult breathing) and hemoptysis (spitting blood).

Treatment may involve antibiotics, bronchodilators, oxygen therapy, increased fluid intake, and chest physiotherapy (percussion, vibration, and postural drainage).

Tuberculosis

The bacterium *Mycobacterium tuberculosis* produces an infectious, communicable disease called **tuberculosis (TB)** that most often affects the lungs and the pleurae but may involve other parts of the body. Once the bacteria are inside the lungs, they multiply and cause inflammation, which stimulates neutrophils and macrophages to migrate to the area and engulf the bacteria to prevent their spread. If the immune system is not impaired, the bacteria remain dormant for life, but impaired immunity may enable the bacteria to escape into blood and lymph to infect other organs. In many people, symptoms—fatigue, weight loss, lethargy, anorexia, a low-grade fever, night sweats, cough, dyspnea, chest pain, and hemoptysis—do not develop until the disease is advanced.

During the past several years, the incidence of TB in the United States has risen dramatically. Perhaps the single most important factor related to this increase is the presence of the human immunodeficiency virus (HIV). People infected with HIV are much more likely to develop tuberculosis because their immune systems are impaired. Among the other factors that have contributed to the increased number of cases are homelessness, increased drug abuse, increased immigration from countries with a high prevalence of tuberculosis, increased crowding in housing among the poor, and airborne transmission of tuberculosis in prisons and shelters. In addition, recent outbreaks of tuberculosis involving multi-drug-resistant strains of *Mycobacterium tuberculosis* have occurred because patients fail to complete their antibiotic and other treatment regimens.

Coryza and Influenza

Hundreds of viruses can cause **coryza** (ko-RĪ-za) or the **common cold.** But a group of viruses called *rhinoviruses* is responsible for about 40% of all colds in adults. Typical symptoms include sneezing, excessive nasal secretion, dry cough, and congestion. The uncomplicated common cold is not usually accompanied by a fever. Complications include sinusitis, asthma, bronchitis, ear infections, and laryngitis. Recent investigations suggest an association between emotional stress and the common cold. The higher the stress level, the greater the frequency and duration of colds.

Influenza (flu) is also caused by a virus. Its symptoms include chills, fever (usually higher than 101°F = 39°C), headache, and muscular aches. Coldlike symptoms appear as the fever subsides.

Pulmonary Edema

Pulmonary edema is an abnormal accumulation of fluid in the interstitial spaces and alveoli of the lungs. The edema may arise from increased pulmonary capillary permeability (pulmonary origin) or increased pulmonary capillary pressure (cardiac origin); the latter cause may coincide with congestive heart failure. The most common symptom is dyspnea. Others include wheezing, tachypnea (rapid breathing rate), restlessness, a feeling of suffocation, cyanosis, pallor (paleness), and diaphoresis (excessive perspiration). Treatment consists of administering oxygen, drugs that dilate the bronchioles and lower blood pressure, diuretics to rid the body of excess fluid, and drugs that correct acid–base imbalance; suctioning of airways; and mechanical ventilation.

Cystic Fibrosis

Cystic fibrosis (CF) is an inherited disease of secretory epithelia that affects the airways, liver, pancreas, small intestine, and sweat glands. It is the most common lethal genetic disease in whites: 5% of the population are thought to be genetic carriers. The cause of cystic fibrosis is a genetic mutation affecting a transporter protein that carries chloride ions across the plasma membranes of many epithelial cells. Because dysfunction of sweat glands causes perspiration to contain excessive sodium chloride (salt), measurement of the excess chloride is one index for diagnosing CF. The mutation also disrupts the normal functioning of several organs by causing ducts within them to become obstructed by thick mucus secretions that do not drain easily from the passageways. Buildup of these secretions leads to inflammation and replacement of injured cells with connective tissue that further blocks the ducts. Clogging and infection of the airways leads to difficulty in breathing and eventual destruction of lung tissue. Lung disease accounts for most deaths from CF. Obstruction of small bile ducts in the liver interferes with digestion and disrupts liver function; clogging of pancreatic ducts prevents digestive enzymes from reaching the small intestine. Because pancreatic juice contains the main fat-digesting enzyme, the person fails to absorb fats or fat-soluble vitamins and thus suffers from vitamin A, D, and K deficiency diseases. With respect to the reproductive systems, blockage of the ductus (vas) deferens leads to infertility in males; the formation of dense mucus plugs in the vagina restricts the entry of sperm into the uterus and can lead to infertility in females.

A child suffering from cystic fibrosis is given pancreatic extract and large doses of vitamins A, D, and K. The recommended diet is high in calories, fats, and proteins, with vitamin supplementation and liberal use of salt.

Acute Respiratory Distress Syndrome

Acute respiratory distress syndrome (ARDS) is a form of respiratory failure characterized by excessive leakiness of the respiratory membranes and severe hypoxia. Situations that can cause ARDS include near drowning, aspiration of acidic gastric juice, drug reactions, inhalation of an irritating gas such as ammonia, allergic reactions, various lung infections such as pneumonia or TB, and pulmonary hypertension. ARDS strikes about 250,000 people a year in the United States, and about half of them die despite intensive medical care.

MEDICAL TERMINOLOGY

Abdominal thrust (Heimlich) maneuver (HĪM-lik ma-NOO-ver) First-aid procedure designed to clear the airways of obstructing objects. It is performed by applying a quick upward thrust between the navel and costal margin that causes sudden elevation of the diaphragm and forceful, rapid expulsion of air in the lungs; this action forces air out the trachea to eject the obstructing object. The Heimlich maneuver is also used to expel water from the lungs of near-drowning victims before resuscitation is begun.

Asphyxia (as-FIK-sē-a; *sphyxia* = pulse) Oxygen starvation due to low atmospheric oxygen or interference with ventilation, external respiration, or internal respiration.

Aspiration (as′-pi-RĀ-shun) Inhalation of a foreign substance such as water, food, or a foreign body into the bronchial tree; also, the drawing of a substance in or out by suction.

Atelectasis (at′-ē-LEK-ta-sis; *atel-* = incomplete; *ectasis* = expansion) Incomplete expansion of a lung or a portion of a lung caused by airway obstruction, lung compression, or inadequate pulmonary surfactant.

Bronchiectasis (bron′-kē-EK-ta-sis) A chronic dilation of the bronchi or bronchioles.

Bronchography (bron-KOG-ra-fē) An imaging technique used to visualize the bronchial tree using x rays. After an opaque contrast medium is inhaled through an intratracheal catheter, radiographs of the chest in various positions are taken, and the developed film, a **bronchogram** (BRON-kō-gram), provides a picture of the bronchial tree.

Bronchoscopy (bron-KOS-kō-pē) Visual examination of the bronchi through a **bronchoscope,** an illuminated, flexible tubular instrument that is passed through the mouth (or nose), larynx, and trachea into the bronchi. The examiner can view the interior of the trachea and bronchi to biopsy a tumor, clear an obstructing object or secretions from an airway, take cultures or smears for microscopic examination, stop bleeding, or deliver drugs.

Cheyne–Stokes respiration (CHĀN STŌKS res′-pi-RĀ-shun) A repeated cycle of irregular breathing that begins with shallow breaths that increase in depth and rapidity and then decrease and cease altogether for 15 to 20 seconds. Cheyne–Stokes is normal in infants; it is also often seen just before death from pulmonary, cerebral, cardiac, and kidney disease.

Dyspnea (DISP-nē-a; *dys-* = painful, difficult; *-pnea* = breath) Painful or labored breathing.

Epistaxis (ep′-i-STAK-sis) Loss of blood from the nose due to trauma, infection, allergy, malignant growths, or bleeding disorders. It can be arrested by cautery with silver nitrate, electrocautery, or firm packing. Also called **nosebleed.**

Hemoptysis (hē-MOP-ti-sis; *hemo-* = blood; *-ptysis* = spit) Spitting of blood from the respiratory tract.

Hypoventilation (*hypo-* = below) Slow and shallow breathing.

Rales (RĀLS) Sounds sometimes heard in the lungs that resemble bubbling or rattling. Rales are to the lungs what murmurs are to the heart. Different types are due to the presence of an abnormal type or amount of fluid or mucus within the bronchi or alveoli, or to bronchoconstriction that causes turbulent airflow.

Respirator (RES-pi-rā′-tor) An apparatus fitted to a mask over the nose and mouth, or hooked directly to an endotracheal or tracheotomy tube, that is used to assist or support ventilation or to provide nebulized medication to the air passages.

Respiratory failure A condition in which the respiratory system either cannot supply sufficient O_2 to maintain metabolism or cannot eliminate enough CO_2 to prevent respiratory acidosis (a lower-than-normal pH in extracellular fluid).

Rhinitis (rī-NĪ-tis; *rhin-* = nose) Chronic or acute inflammation of the mucous membrane of the nose.

Sudden infant death syndrome (SIDS) Death of infants between the ages of 1 week and 12 months thought to be due to hypoxia while sleeping in a prone position (on the stomach) and the rebreathing of exhaled air trapped in a depression of the mattress. It is now recommended that normal newborns be placed on their backs for sleeping. Remember: "back to sleep."

Tachypnea (tak′-ip-NĒ-a; *tachy-* = rapid; *-pnea* = breath) Rapid breathing rate.

STUDY OUTLINE

RESPIRATORY SYSTEM ANATOMY (p. 806)

1. The respiratory system consists of the nose, pharynx, larynx, trachea, bronchi, and lungs. They act with the cardiovascular system to supply oxygen (O_2) and remove carbon dioxide (CO_2) from the blood.
2. The external portion of the nose is made of cartilage and skin and is lined with a mucous membrane. Openings to the exterior are the external nares.
3. The internal portion of the nose communicates with the paranasal sinuses and nasopharynx through the internal nares.
4. The nasal cavity is divided by a septum. The anterior portion of the cavity is called the vestibule. The nose warms, moistens, and filters air and functions in olfaction and speech.
5. The pharynx (throat) is a muscular tube lined by a mucous membrane. The anatomic regions are the nasopharynx, oropharynx, and laryngopharynx.
6. The nasopharynx functions in respiration. The oropharynx and laryngopharynx function both in digestion and in respiration.
7. The larynx (voice box) is a passageway that connects the pharynx with the trachea. It contains the thyroid cartilage (Adam's apple); the epiglottis, which prevents food from entering the larynx; the cricoid cartilage, which connects the larynx and trachea; and the paired arytenoid, corniculate, and cuneiform cartilages.
8. The larynx contains vocal folds, which produce sound as they vibrate. Taut folds produce high pitches, and relaxed ones produce low pitches.
9. The trachea (windpipe) extends from the larynx to the primary bronchi. It is composed of C-shaped rings of cartilage and smooth muscle and is lined with pseudostratified ciliated columnar epithelium.
10. The bronchial tree consists of the trachea, primary bronchi, secondary bronchi, tertiary bronchi, bronchioles, and terminal bronchioles. Walls of bronchi contain rings of cartilage; walls of bronchioles contain increasingly smaller plates of cartilage and increasing amounts of smooth muscle.
11. Lungs are paired organs in the thoracic cavity enclosed by the pleural membrane. The parietal pleura is the superficial layer that lines the thoracic cavity; the visceral pleura is the deep layer that covers the lungs.
12. The right lung has three lobes separated by two fissures; the left lung has two lobes separated by one fissure and a depression, the cardiac notch.
13. Secondary bronchi give rise to branches called segmental bronchi, which supply segments of lung tissue called bronchopulmonary segments.
14. Each bronchopulmonary segment consists of lobules, which contain lymphatics, arterioles, venules, terminal bronchioles, respiratory bronchioles, alveolar ducts, alveolar sacs, and alveoli.
15. Alveolar walls consist of type I alveolar cells, type II alveolar cells, and associated alveolar macrophages.
16. Gas exchange occurs across the respiratory membranes.

PULMONARY VENTILATION (p. 821)

1. Pulmonary ventilation, or breathing, consists of inhalation and exhalation.

2. The movement of air into and out of the lungs depends on pressure changes governed in part by Boyle's law, which states that the volume of a gas varies inversely with pressure, assuming that temperature remains constant.
3. Inhalation occurs when alveolar pressure falls below atmospheric pressure. Contraction of the diaphragm and external intercostals increases the size of the thorax, thereby decreasing the intrapleural pressure so that the lungs expand. Expansion of the lungs decreases alveolar pressure so that air moves down a pressure gradient from the atmosphere into the lungs.
4. During forceful inhalation, accessory muscles of inhalation (sternocleidomastoids, scalenes, and pectoralis minors) are also used.
5. Exhalation occurs when alveolar pressure is higher than atmospheric pressure. Relaxation of the diaphragm and external intercostals results in elastic recoil of the chest wall and lungs, which increases intrapleural pressure; lung volume decreases and alveolar pressure increases, so air moves from the lungs to the atmosphere.
6. Forceful exhalation involves contraction of the internal intercostal and abdominal muscles.
7. The surface tension exerted by alveolar fluid is decreased by the presence of surfactant.
8. Compliance is the ease with which the lungs and thoracic wall can expand.
9. The walls of the airways offer some resistance to breathing.
10. Normal quiet breathing is termed eupnea; other patterns are costal breathing and diaphragmatic breathing. Modified respiratory movements, such as coughing, sneezing, sighing, yawning, sobbing, crying, laughing, and hiccupping, are used to express emotions and to clear the airways. See Table 23.1 on page 826.

LUNG VOLUMES AND CAPACITIES (p. 826)

1. Lung volumes exchanged during breathing and the rate of respiration are measured with a spirometer.
2. Lung volumes measured by spirometry include tidal volume, minute ventilation, alveolar ventilation rate, inspiratory reserve volume, expiratory reserve volume, and $FEV_{1.0}$. Other lung volumes are anatomic dead space, residual volume, and minimal volume.
3. Lung capacities, the sum of two or more volumes, include inspiratory, functional residual, vital, and total lung capacities.

EXCHANGE OF OXYGEN AND CARBON DIOXIDE (p. 827)

1. The partial pressure of a gas is the pressure exerted by that gas in a mixture of gases. It is symbolized by P_x, where the subscript is the chemical formula of the gas.
2. According to Dalton's law, each gas in a mixture of gases exerts its own pressure as if all the other gases were not present.
3. Henry's law states that the quantity of a gas that will dissolve in a liquid is proportional to the partial pressure of the gas and its solubility coefficient (given that the temperature remains constant).
4. In internal and external respiration, O_2 and CO_2 diffuse from areas of higher partial pressures to areas of lower partial pressures.
5. External respiration or pulmonary gas exchange is the exchange of gases between alveoli and pulmonary blood capillaries. It depends

on partial pressure differences, a large surface area for gas exchange, a small diffusion distance across the respiratory membrane, and the rate of airflow into and out of the lungs.

6. Internal respiration or systemic gas exchange is the exchange of gases between systemic blood capillaries and tissue cells.

TRANSPORT OF OXYGEN AND CARBON DIOXIDE (p. 831)

1. In each 100 mL of oxygenated blood, 1.5% of the O_2 is dissolved in blood plasma and 98.5% is bound to hemoglobin as oxyhemoglobin ($Hb-O_2$).

2. The binding of O_2 to hemoglobin is affected by P_{O_2}, acidity (pH), P_{CO_2}, temperature, and BPG.

3. Fetal hemoglobin differs from adult hemoglobin in structure and has a higher affinity for O_2.

4. In each 100 mL of deoxygenated blood, 9% of CO_2 is dissolved in blood plasma, 13% combines with hemoglobin as carbaminohemoglobin ($Hb-CO_2$), and 78% is converted to bicarbonate ions (HCO_3^-).

5. In an acidic environment, hemoglobin's affinity for O_2 is lower, and O_2 dissociates more readily from it (Bohr effect).

6. In the presence of O_2, less CO_2 binds to hemoglobin (Haldane effect).

CONTROL OF RESPIRATION (p. 835)

1. The respiratory center consists of a medullary rhythmicity area in the medulla oblongata and a pneumotaxic area and an apneustic area in the pons.

2. The inspiratory area sets the basic rhythm of respiration.

3. The pneumotaxic and apneustic areas coordinate the transition between inhalation and exhalation.

4. Respirations may be modified by a number of factors, including cortical influences; the inflation reflex; chemical stimuli, such as O_2 and CO_2 and H^+ levels; proprioceptor input; blood pressure changes; limbic system stimulation; temperature; pain; and irritation to the airways. (See Table 23.2 on page 841.)

EXERCISE AND THE RESPIRATORY SYSTEM (p. 840)

1. The rate and depth of ventilation change in response to both the intensity and duration of exercise.

2. An increase in pulmonary perfusion and O_2-diffusing capacity occurs during exercise.

3. The abrupt increase in ventilation at the start of exercise is due to neural changes that send excitatory impulses to the inspiratory area in the medulla oblongata. The more gradual increase in ventilation during moderate exercise is due to chemical and physical changes in the bloodstream.

DEVELOPMENT OF THE RESPIRATORY SYSTEM (p. 841)

1. The respiratory system begins as an outgrowth of endoderm called the laryngotracheal bud.

2. Smooth muscle, cartilage, and connective tissue of the bronchial tubes and pleural sacs develop from mesoderm.

AGING AND THE RESPIRATORY SYSTEM (p. 843)

1. Aging results in decreased vital capacity, decreased blood level of O_2, and diminished alveolar macrophage activity.

2. Elderly people are more susceptible to pneumonia, emphysema, bronchitis, and other pulmonary disorders.

Q SELF-QUIZ QUESTIONS

Fill in the blanks in the following statements.

1. The three basic steps of respiration are _____, _____, and _____.

2. For inhalation to occur, air pressure in the alveoli must be _____ than atmospheric pressure; for exhalation to occur, air pressure in the alveoli must be _____ than atmospheric pressure.

3. Oxygen in blood is carried primarily in the form of _____; carbon dioxide is carried as _____, _____, and _____.

4. Write the equation for the chemical reaction that occurs for transport of carbon dioxide as bicarbonate ions in blood:_____.

Indicate whether the following statements are true or false.

5. The carina is the landmark used for an emergency tracheostomy.

6. The most important muscle of inhalation is the diaphragm.

Choose the one best answer to the following questions.

7. Which of the following statements are *correct*? (1) Normal exhalation during quiet breathing is an active process involving intensive muscle contraction. (2) Passive exhalation results from elastic recoil of the chest wall and lungs. (3) Air flow during breathing is due to a pressure gradient between the lungs and the atmospheric air. (4) During normal breathing, the pressure between the two pleural layers (intrapleural pressure) is always subatmospheric. (5) Surface tension of alveolar fluid facilitates inhalation. (a) 1, 2, and 3, (b) 2, 3, and 4, (c) 3, 4, and 5, (d) 1, 3, and 5 (e) 2, 3, and 5.

8. Which of the following factors affect the rate of external respiration? (1) partial pressure differences of the gases, (2) surface area for gas exchange, (3) diffusion distance, (4) solubility and molecular weight of the gases, (5) presence of bis-phosphoglycerate (BPG). (a) 1, 2, and 3, (b) 2, 4, and 5, (c) 1, 2, 4, and 5, (d) 1, 2, 3, and 4 (e) 2, 3, 4, and 5.

9. The most important factor in determining the percent oxygen saturation of hemoglobin is (a) the partial pressure of oxygen, (b) acidity, (c) the partial pressure of carbon dioxide, (d) temperature, (e) BPG.

10. Which gas law that states that each gas in a mixture exerts its own pressure as if all the other gases were not present? (a) Boyle's law, (b) Henry's law, (c) Dalton's law, (d) Haldane's law, (e) Bohr's law.

11. Which of the following statements are *true*? (1) It is impossible for people to kill themselves by holding their breath. (2) The cerebral cortex provides voluntary control over breathing. (3) Emotional stimuli can alter respiration. (4) Certain chemical stimuli can alter the rate and depth of breathing. (5) At the onset of exercise, an abrupt increase in breathing is followed by a more gradual increase in breathing. (a) 1, 2, 4, and 5, (b) 2, 3, and 5, (c) 1, 3, and 5, (d) 2, 3, 4, and 5, (e) 1, 2, ,3, 4, and 5.

12. Match the following:

_____ (a) functions as a passageway for air and food, provides a resonating chamber for speech sounds, and houses the tonsils

_____ (b) site of external respiration

_____ (c) connects the laryngopharynx with the trachea; houses the vocal cords

_____ (d) serous membranes that surround the lungs

_____ (e) functions in warming, moistening, and filtering air; receives olfactory stimuli; is a resonating chamber for sound

_____ (f) form a continuous lining of the alveolar wall; sites of gas exchange

_____ (g) a tubular passageway for air connecting the larynx to the bronchi

_____ (h) secrete alveolar fluid and surfactant

_____ (i) prevents food or fluid from entering the airways

_____ (j) air passageways entering the lungs

(1) nose
(2) pharynx
(3) larynx
(4) epiglottis
(5) trachea
(6) bronchi
(7) pleurae
(8) alveoli
(9) type I alveolar cells
(10) type II alveolar cells

13. Match the following:

_____ (a) a deficiency of oxygen at the tissue level

_____ (b) above-normal partial pressure of carbon dioxide

_____ (c) normal quiet breathing

_____ (d) deep, abdominal breathing

_____ (e) the ease with which the lungs and thoracic wall can be expanded

_____ (f) absence of breathing

_____ (g) rapid and deep breathing

_____ (h) shallow, chest breathing

(1) eupnea
(2) apnea
(3) hyperventilation
(4) costal breathing
(5) diaphragmatic breathing
(6) compliance
(7) hypoxia
(8) hypercapnia

14. Match the following:

_____ (a) total volume of air inhaled and exhaled each minute

_____ (b) tidal volume + inspiratory reserve volume + expiratory reserve volume

_____ (c) additional amount of air inhaled beyond tidal volume when taking a very deep breath

_____ (d) amount of air remaining in lungs after expiratory reserve volume is expelled

_____ (e) tidal volume + inspiratory reserve volume

_____ (f) volume of air in one breath

_____ (g) amount of air exhaled in forced exhalation

_____ (h) provides a medical and legal tool for determining if a baby was born dead or died after birth

(1) tidal volume
(2) residual volume
(3) minute ventilation
(4) expiratory reserve volume
(5) inspiratory reserve volume
(6) minimal volume
(7) inspiratory capacity
(8) vital capacity

15. Match the following:

_____ (a) prevents excessive inflation of the lungs

_____ (b) controls the basic rhythm of respiration

_____ (c) sends stimulatory impulses to the inspiratory area that activate it and prolong inhalation

_____ (d) as acidity increases, the affinity of hemoglobin for oxygen decreases and oxygen dissociates more readily from hemoglobin

_____ (e) transmits inhibitory impulses to turn off the inspiratory area before the lungs become too full of air

_____ (f) the quantity of gas that dissolves in a liquid is proportional to the partial pressure of the gas and its solubility coefficient

_____ (g) relates to the partial pressure of a gas in a mixture of gases

(1) Bohr effect
(2) Dalton's law
(3) medullary rhythmicity area
(4) Henry's law
(5) apneustic area
(6) pneumotaxic area
(7) Hering-Breuer reflex

CRITICAL THINKING QUESTIONS

1. Leung, a member of the college swim team, was measuring his lung volumes during his A&P lab. Although his tidal volume was average for a male of his age and size, his inspiratory reserve volume was the highest in the class. List the normal male values for lung volumes that can be measured with spirometer. Would you expect the average female volumes to be the same? Why or why not?
   HINT   *Total lung capacity includes a volume that cannot be measured by spirometry.*

2. Samir stepped up to the plate and lifted his bat. The pitch was wild, and the ball hit him right in the nose! The x ray revealed a break in both medial bones of the external nose. Describe the structure of the external nose, and identify the location of the break.
   HINT   *Samir may need a rhinoplasty to repair his nose.*

3. Her mother calls Kirsten a "challenging personality," but you have some other adjectives in mind. She's threatening to hold her breath until she, as she puts it, "turns blue, falls down, and dies—and won't you be sorry!" Do you need to worry about the "dies" part?
   HINT   *Only some aspects of respiration are partly under voluntary control.*

ANSWERS TO FIGURE QUESTIONS

23.1 The conducting portion of the respiratory system includes the nose, pharynx, larynx, trachea, bronchi, and bronchioles (except the respiratory bronchioles).

23.2 Air's path is external nares → vestibule → nasal cavity → internal nares.

23.3 The root of the nose attaches it to the frontal bone.

23.4 The superior border of the pharynx is the internal nares; the inferior border of the pharynx is the cricoid cartilage.

23.5 During swallowing, the epiglottis closes over the rima glottidis, the entrance to the trachea.

23.6 The main function of the vocal folds is voice production.

23.7 Because the tissues between the esophagus and trachea are soft, the esophagus can bulge and press against the trachea during swallowing.

23.8 The left lung has two lobes and secondary bronchi; the right lung has three of each.

23.9 The pleural membrane is a serous membrane.

23.10 Because two-thirds of the heart lies to the left of the midline, the left lung contains a cardiac notch to accommodate the presence of the heart.

23.11 The wall of an alveolus is made up of type I alveolar cells, type II alveolar cells, and associated alveolar macrophages.

23.12 The respiratory membrane averages 0.5 μm in thickness.

23.13 The pressure would increase to 4 atm.

23.14 If you are at rest while reading, your diaphragm is responsible for about 75% of each inhalation.

23.15 At the start of inhalation, intrapleural pressure is about 756 mmHg. With contraction of the diaphragm, it decreases to about 754 mmHg as the volume of the space between the two pleural layers expands. With relaxation of the diaphragm, it increases back to 756 mmHg.

23.16 Normal atmospheric pressure at sea level is 760 mmHg.

23.17 Breathing in and then exhaling as much air as possible demonstrates vital capacity.

23.18 In both places, a difference in P_{O_2} promotes oxygen diffusion.

23.19 During strenuous exercise, oxygenated blood contains 20 mL of O_2 per 100 mL of blood (the same as at rest).

23.20 In both cases, hemoglobin in your pulmonary veins would be fully saturated with O_2, a point which is at the upper right of the curve.

23.21 Because lactic acid (lactate) and CO_2 are produced by active skeletal muscles, blood pH decreases slightly and P_{CO_2} increases when you are actively exercising. The result is lowered affinity of hemoglobin for O_2, so more O_2 is available to the working muscles.

23.22 O_2 is more available when you have a fever because the affinity of hemoglobin for O_2 decreases with increasing temperature.

23.23 Fetal Hb is 80% saturated with O_2, whereas maternal Hb is about 75% saturated at a P_{O_2} of 40 mmHg.

23.24 Blood in a systemic vein would have a higher concentration of HCO_3^-.

23.25 The medullary inspiratory area contains autorhythmic neurons.

23.26 The phrenic nerves innervate the diaphragm.

23.27 Peripheral chemoreceptors are responsive to changes in levels of oxygen partial pressure, carbon dioxide partial pressure, and H^+.

23.28 Normal arterial P_{CO_2} is 40 mmHg.

23.29 The respiratory system begins to develop about 4 weeks after fertilization.

The Digestive System

THE DIGESTIVE SYSTEM AND HOMEOSTASIS

The digestive system contributes to homeostasis by breaking down food into forms that can be absorbed and used by body cells. It absorbs water, vitamins, and minerals and eliminates wastes from the body.

ENERGIZE YOUR STUDY

FOUNDATIONS CD

Anatomy Overview
  • The Digestive System

Animations
  • Systems Contributions to Homeostasis
  • The Case of the Coughing Boy
  • The Case of the Girl with the Fruity Breath

Concepts and Connections and Exercises reinforce your understanding

www.wiley.com/college/apcentral

INSIGHTS AND EXPLORATIONS

Most of us are aware that bacteria can cause diarrheal disease. Did you also know that there is a bacterium that can cause peptic ulcers? We used to think that peptic ulcers were the result of too much worry or an inability to handle a stressful life style. Scientists have now shown that peptic ulcers can be caused by the bacterium Helicobacter pylori. In this web-based activity we will explore more about this bacterium, how it can survive the acidity of the stomach, and how a breathalyzer test can be used to diagnose this disease.

The food we eat contains a variety of nutrients, which are used for building new body tissues and repairing damaged tissues. Food is also vital for life because it is our only source of chemical energy. As consumed, however, most food cannot be used either to build tissues or to power the energy-requiring chemical reactions of body cells. Rather, foods must be broken down into molecules that are small enough to enter body cells, a process known as **digestion.** The passage of these smaller molecules through the plasma membranes of cells lining the stomach and intestines and then into the blood and lymph is termed **absorption.** The organs that collectively perform these functions—the **digestive system**—are the focus of this chapter.

The medical specialty that deals with the structure, function, diagnosis, and treatment of diseases of the stomach and intestines is called **gastroenterology** (gas'-trō-en'-ter-OL-ō-jē; *gastro-* = stomach; *entero-* = intestines; *-logy* = study of). The medical specialty that deals with the diagnosis and treatment of disorders of the rectum and anus is called **proctology** (prok-TOL-ō-jē; *proct-* = rectum).

OVERVIEW OF THE DIGESTIVE SYSTEM

▶ O B J E C T I V E S

- **Identify the organs of the digestive system.**
- **Describe the basic processes performed by the digestive system.**

Two groups of organs compose the digestive system (Figure 24.1): the gastrointestinal (GI) tract and the accessory digestive organs. The **gastrointestinal (GI) tract,** or **alimentary canal** (*alimentary* = nourishment), is a continuous tube that extends from the mouth to the anus through the ventral body cavity. Organs of the gastrointestinal tract include the mouth, most of the pharynx, esophagus, stomach, small intestine, and large intestine. The length of the GI tract taken from a cadaver is about 9 m (30 ft). In a living person, it is much shorter because the muscles along the walls of GI tract organs are in a state of tonus (sustained contraction). The **accessory digestive organs** are the teeth, tongue, salivary glands, liver, gallbladder, and pancreas. Teeth aid in the physical breakdown of food, and the tongue assists in chewing and swallowing. The other accessory digestive organs, however, never come into direct contact with food. They produce or store secretions that flow into the GI tract through ducts; the secretions aid in the chemical breakdown of food.

The GI tract contains and processes food from the time it is eaten until it is digested and absorbed or eliminated. In some parts of the GI tract, muscular contractions in the wall physically break down the food by churning it. The contractions also help to dissolve foods by mixing them with fluids secreted into the tract. Enzymes secreted by accessory structures and cells that line the tract break down the food chemically. Wavelike contractions of the smooth muscle in the wall of the GI tract propel the food along the tract, from the esophagus to the anus.

Overall, the digestive system performs six basic processes:

1. **Ingestion.** This process involves taking foods and liquids into the mouth (eating).

2. **Secretion.** Each day, cells within the walls of the GI tract and accessory digestive organs secrete a total of about 7 liters of water, acid, buffers, and enzymes into the lumen of the tract.

3. **Mixing and propulsion.** Alternating contraction and relaxation of smooth muscle in the walls of the GI tract mix food and secretions and propel them toward the anus. This capability of the GI tract to mix and move material along its length is termed **motility.**

4. **Digestion.** Mechanical and chemical processes break down ingested food into small molecules. In **mechanical digestion** the teeth cut and grind food before it is swallowed, and then smooth muscles of the stomach and small intestine churn the food. As a result, food molecules become dissolved and thoroughly mixed with digestive enzymes. In **chemical digestion** the large carbohydrate, lipid, protein, and nucleic acid molecules in food are split into smaller molecules by hydrolysis (see Figure 2.15 on page 44). Digestive enzymes produced by the salivary glands, tongue, stomach, pancreas, and small intestine catalyze these catabolic reactions. A few substances in food can be absorbed without chemical digestion. These include amino acids, cholesterol, glucose, vitamins, minerals, and water.

5. **Absorption.** The entrance of ingested and secreted fluids, ions, and the small molecules that are products of digestion into the epithelial cells lining the lumen of the GI tract is called **absorption.** The absorbed substances pass into blood or lymph and circulate to cells throughout the body.

6. **Defecation.** Wastes, indigestible substances, bacteria, cells sloughed from the lining of the GI tract, and digested materials that were not absorbed leave the body through the anus in a process called **defecation.** The eliminated material is termed **feces.**

▶ C H E C K P O I N T

1. Which components of the digestive system are GI tract organs and which are accessory digestive organs?

2. Which organs of the digestive system come in contact with food, and what are some of their digestive functions?

3. Which kinds of food molecules undergo chemical digestion, and which do not?

Figure 24.1 Organs of the digestive system.

Organs of the gastrointestinal (GI) tract are the mouth, pharynx, esophagus, stomach, small intestine, and large intestine. Accessory digestive organs are the teeth, tongue, salivary glands, liver, gallbladder, and pancreas.

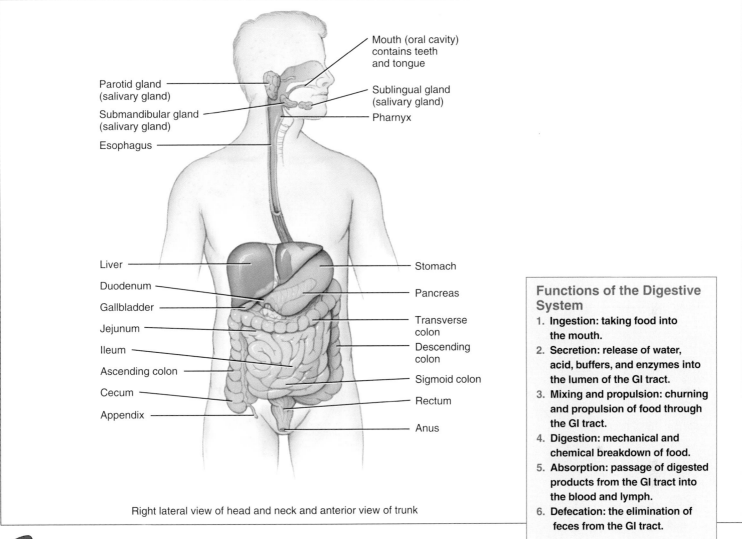

Mouth (oral cavity) contains teeth and tongue

Parotid gland (salivary gland)

Sublingual gland (salivary gland)

Submandibular gland (salivary gland)

Pharnyx

Esophagus

Liver

Duodenum

Gallbladder

Jejunum

Ileum

Ascending colon

Cecum

Appendix

Stomach

Pancreas

Transverse colon

Descending colon

Sigmoid colon

Rectum

Anus

Functions of the Digestive System
1. **Ingestion: taking food into the mouth.**
2. **Secretion: release of water, acid, buffers, and enzymes into the lumen of the GI tract.**
3. **Mixing and propulsion: churning and propulsion of food through the GI tract.**
4. **Digestion: mechanical and chemical breakdown of food.**
5. **Absorption: passage of digested products from the GI tract into the blood and lymph.**
6. **Defecation: the elimination of feces from the GI tract.**

Right lateral view of head and neck and anterior view of trunk

Which structures of the digestive system secrete digestive enzymes?

LAYERS OF THE GI TRACT

OBJECTIVE

• **Describe the layers that form the wall of the gastrointestinal tract.**

The wall of the GI tract from the lower esophagus to the anal canal has the same basic, four-layered arrangement of tissues. The four layers of the tract, from deep to superficial, are the mucosa, submucosa, muscularis, and serosa (Figure 24.2).

Mucosa

The **mucosa,** or inner lining of the GI tract, is a mucous membrane. It is composed of a layer of epithelium in direct contact with the contents of the GI tract, areolar connective tissue, and a thin layer of smooth muscle (muscularis mucosae).

1. The **epithelium** in the mouth, pharynx, esophagus, and anal canal is mainly nonkeratinized stratified squamous epithelium that serves a protective function. Simple columnar epithelium, which functions in secretion and absorption, lines the stomach

Figure 24.2   Layers of the gastrointestinal tract.   Variations in this basic plan may be seen in the esophagus (Figure 24.9), stomach (Figure 24.12), small intestine (Figure 24.23), and large intestine (Figure 24.28).

 The four layers of the GI tract, from deep to superficial, are the mucosa, submucosa, muscularis, and serosa.

What is the function of the enteric plexuses in the wall of the gastrointestinal tract?

and intestines. Neighboring simple columnar epithelial cells are firmly sealed to each other by tight junctions that restrict leakage between the cells. The rate of renewal of GI tract epithelial cells is rapid: Every 5 to 7 days they slough off and are replaced by new cells. Located among the absorptive epithelial cells are exocrine cells that secrete mucus and fluid into the lumen of the tract, and several types of endocrine cells, collectively called **enteroendocrine cells,** that secrete hormones into the bloodstream.

2.   The **lamina propria** (*lamina* = thin, flat plate; *propria* = one's own) is areolar connective tissue containing many blood and lymphatic vessels, which are the routes by which nutrients absorbed into the GI tract reach the other tissues of the body. This layer supports the epithelium and binds it to the muscularis mucosae (discussed next). The lamina propria also contains the

majority of the cells of the **mucosa-associated lymphatic tissue (MALT).** These prominent lymphatic nodules contain immune system cells that protect against disease. MALT is present all along the GI tract, especially in the tonsils, small intestine, appendix, and large intestine, and it contains about as many immune cells as are present in all the rest of the body. The lymphocytes and macrophages in MALT mount immune responses against microbes, such as bacteria, that may penetrate the epithelium.

3.   A thin layer of smooth muscle fibers called the **muscularis mucosae** throws the mucous membrane of the stomach and small intestine into many small folds, which increase the surface area for digestion and absorption. Movements of the muscularis mucosae ensure that all absorptive cells are fully exposed to the contents of the GI tract.

Submucosa

The **submucosa** consists of areolar connective tissue that binds the mucosa to the muscularis. It contains many blood and lymphatic vessels that receive absorbed food molecules. Also located in the submucosa is the **submucosal plexus** or *plexus of Meissner,* an extensive network of neurons. The neurons are part of the **enteric nervous system (ENS),** the "brain of the gut." The ENS consists of about 100 million neurons in two plexuses that extend the entire length of the GI tract. The submucosal plexus contains sensory and motor enteric neurons, plus parasympathetic and sympathetic postganglionic neurons that innervate the mucosa and submucosa. Enteric nerves in the submucosa regulate movements of the mucosa and vasoconstriction of blood vessels. Because its neurons also innervate secretory cells of mucosal and submucosal glands, the ENS is important in controlling secretions by the GI tract. The submucosa may also contain glands and lymphatic tissue.

Muscularis

The **muscularis** of the mouth, pharynx, and superior and middle parts of the esophagus contains *skeletal muscle* that produces voluntary swallowing. Skeletal muscle also forms the external anal sphincter, which permits voluntary control of defecation. Throughout the rest of the tract, the muscularis consists of *smooth muscle* that is generally found in two sheets: an inner sheet of circular fibers and an outer sheet of longitudinal fibers. Involuntary contractions of the smooth muscle help break down food physically, mix it with digestive secretions, and propel it along the tract. Between the layers of the muscularis is a second plexus of the enteric nervous system—the **myenteric plexus** (*my-* = muscle) or *plexus of Auerbach.* The myenteric plexus contains enteric neurons, parasympathetic ganglia, parasympathetic postganglionic neurons, and sympathetic postganglionic neurons that innervate the muscularis. This plexus mostly controls GI tract motility, in particular the frequency and strength of contraction of the muscularis.

Serosa

The **serosa** is the superficial layer of those portions of the GI tract that are suspended in the abdominopelvic cavity. It is a serous membrane composed of areolar connective tissue and simple squamous epithelium. As we will see shortly, the esophagus, which passes through the mediastinum, has a superficial layer called the *adventitia* composed of areolar connective tissue. Inferior to the diaphragm, the serosa is also called the **visceral peritoneum;** it forms a portion of the peritoneum, which we examine in detail next.

▶ CHECKPOINT

4. Where along the GI tract is the muscularis composed of skeletal muscle? Is control of this skeletal muscle voluntary or involuntary?

5. What two plexuses form the enteric nervous system, and where are they located?

PERITONEUM

▶ O B J E C T I V E

• **Describe the peritoneum and its folds.**

The **peritoneum** (per'-i-tō-NĒ-um; *peri-* = around) is the largest serous membrane of the body; it consists of a layer of simple squamous epithelium (mesothelium) with an underlying supporting layer of connective tissue. The peritoneum is divided into the **parietal peritoneum,** which lines the wall of the abdominopelvic cavity, and the **visceral peritoneum,** which covers some of the organs in the cavity and is their serosa (Figure 24.3a). The slim space between the parietal and visceral portions of the peritoneum is called the **peritoneal cavity,** which contains serous fluid. In certain diseases, the peritoneal cavity may become distended by the accumulation of several liters of fluid, a condition called **ascites** (a-SĪ-tēz).

As we will see, some organs lie on the posterior abdominal wall and are covered by peritoneum only on their anterior surfaces. Such organs, including the kidneys and pancreas, are said to be **retroperitoneal** (*retro-* = behind).

Unlike the pericardium and pleurae, which smoothly cover the heart and lungs, the peritoneum contains large folds that weave between the viscera. The folds bind the organs to each other and to the walls of the abdominal cavity. They also contain blood vessels, lymphatic vessels, and nerves that supply the abdominal organs.

The **greater omentum** (ō-MEN-tum = fat skin), the largest peritoneal fold, drapes over the transverse colon and coils of the small intestine like a "fatty apron" (Figure 24.3a, d). Because the greater omentum is a double sheet that folds back upon itself, it is a four-layered structure. From attachments along the stomach and duodenum, the greater omentum extends downward anterior to the small intestine, then turns and extends upward and attaches to the transverse colon. The greater omentum normally contains a considerable amount of adipose tissue. Its adipose tissue content can greatly expand with weight gain, giving rise to the characteristic "beer belly" seen in some overweight individuals. The many lymph nodes of the greater omentum contribute macrophages and antibody-producing plasma cells that help combat and contain infections of the GI tract.

The **falciform ligament** (FAL-si-form; *falc-* = sickle-shaped) attaches the liver to the anterior abdominal wall and diaphragm (Figure 24.3b). The liver is the only digestive organ that is attached to the anterior abdominal wall.

The **lesser omentum** arises as two folds in the serosa of the stomach and duodenum, and it suspends the stomach and duodenum from the liver (Figure 24.3a, c). It contains some lymph nodes.

Another fold of the peritoneum, called the **mesentery** (MEZ-en-ter'-ē; *mes-* = middle), is fan-shaped and binds the small intestine to the posterior abdominal wall (Figure 24.3a, d). It extends from the posterior abdominal wall to wrap around the small intestine and then returns to its origin, forming a double-

Figure 24.3 Relationship of the peritoneal folds to each other and to organs of the digestive system. The size of the peritoneal cavity in (a) is exaggerated for emphasis.

 The peritoneum is the largest serous membrane in the body.

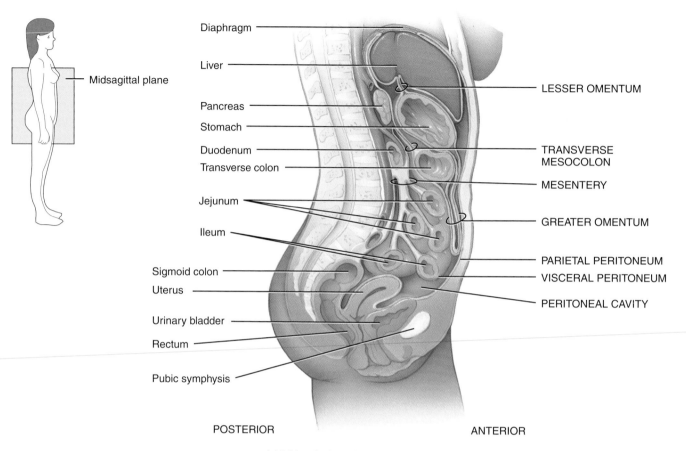

Midsagittal plane

Diaphragm
Liver
Pancreas
Stomach
Duodenum
Transverse colon
Jejunum
Ileum
Sigmoid colon
Uterus
Urinary bladder
Rectum
Pubic symphysis

LESSER OMENTUM
TRANSVERSE MESOCOLON
MESENTERY
GREATER OMENTUM
PARIETAL PERITONEUM
VISCERAL PERITONEUM
PERITONEAL CAVITY

POSTERIOR

ANTERIOR

(a) Midsagittal section showing the peritoneal folds

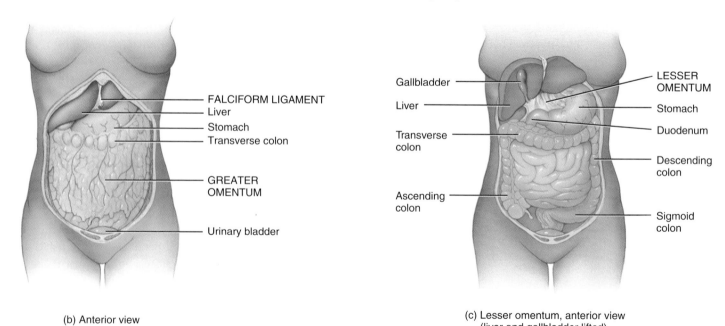

FALCIFORM LIGAMENT
Liver
Stomach
Transverse colon

GREATER OMENTUM

Urinary bladder

(b) Anterior view

Gallbladder
Liver

Transverse colon

Ascending colon

LESSER OMENTUM
Stomach
Duodenum

Descending colon

Sigmoid colon

(c) Lesser omentum, anterior view
(liver and gallbladder lifted)

layered structure. Between the two layers are blood and lymphatic vessels and lymph nodes.

A fold of peritoneum, the **mesocolon** (mez'-ō-KŌ-lon), binds the large intestine to the posterior abdominal wall (Figure 24.3a), It also carries blood and lymphatic vessels to the intestines. Together, the mesentery and mesocolon hold the intestines loosely in place, allowing for a great amount of movement as muscular contractions mix and move the luminal contents along the GI tract.

Peritonitis

Peritonitis is an acute inflammation of the peritoneum. A common cause of the condition is contamination of the peritoneum by infectious microbes, which can result from accidental or surgical wounds in the abdominal wall, or from perforation or rupture of abdominal organs. If, for example, bacteria gain access to the peritoneal cavity through an intestinal perforation or rupture of the appendix, they can produce an acute, life-threatening form of peritonitis. A less serious (but still painful) form of peritonitis can result from the rubbing together of inflamed peritoneal surfaces. ■

▶ CHECKPOINT

6. Where are the visceral peritoneum and parietal peritoneum located?

7. Describe the attachment sites and functions of the mesentery, mesocolon, falciform ligament, lesser omentum, and greater omentum.

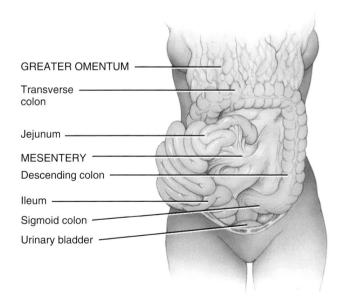

GREATER OMENTUM

Transverse colon

Jejunum

MESENTERY

Descending colon

Ileum

Sigmoid colon

Urinary bladder

(d) Anterior view (greater omentum lifted and small intestine reflected to right side)

Which peritoneal fold binds the small intestine to the posterior abdominal wall?

MOUTH

▶ OBJECTIVES

• **Identify the locations of the salivary glands, and describe the functions of their secretions.**

• **Describe the structure and functions of the tongue.**

• **Identify the parts of a typical tooth, and compare deciduous and permanent dentitions.**

The **mouth,** also referred to as the **oral** or **buccal cavity** (BUK-al; *bucca* = cheeks), is formed by the cheeks, hard and soft palates, and tongue (Figure 24.4). Forming the lateral walls of the oral cavity are the **cheeks**—muscular structures covered externally by skin and internally by nonkeratinized stratified squamous epithelium. The anterior portions of the cheeks end at the lips.

The **lips** or **labia** (= fleshy borders) are fleshy folds surrounding the opening of the mouth. They are covered externally by skin and internally by a mucous membrane. There is a transition zone where the two kinds of covering tissue meet. This portion of the lips is nonkeratinized, and the color of the blood in the underlying blood vessels is visible through the transparent surface layer. The inner surface of each lip is attached to its corresponding gum by a midline fold of mucous membrane called the **labial frenulum** (LĀ-bē-al FREN-ū-lum; *frenulum* = small bridle).

The orbicularis oris muscle and connective tissue lie between the skin and the mucous membrane of the oral cavity. During chewing, contraction of the buccinator muscles in the cheeks and orbicularis oris muscle in the lips helps keep food between the upper and lower teeth. These muscles also assist in speech.

The **vestibule** (= entrance to a canal) of the oral cavity is a space bounded externally by the cheeks and lips and internally by the gums and teeth. The **oral cavity proper** is a space that extends from the gums and teeth to the **fauces** (FAW-sēs = passages), the opening between the oral cavity and the pharynx or throat.

The **hard palate**—the anterior portion of the roof of the mouth—is formed by the maxillae and palatine bones, is covered by mucous membrane, and forms a bony partition between the oral and nasal cavities. The **soft palate,** which forms the posterior portion of the roof of the mouth, is an arch-shaped muscular partition between the oropharynx and nasopharynx that is lined by mucous membrane.

Hanging from the free border of the soft palate is a conical muscular process called the **uvula** (Ū-vū-la = little grape). During swallowing, the soft palate and uvula are drawn superiorly, closing off the nasopharynx and preventing swallowed foods and liquids from entering the nasal cavity. Lateral to the base of the uvula are two muscular folds that run down the lateral sides of the soft palate: Anteriorly, the **palatoglossal arch** extends to the side of the base of the tongue; posteriorly, the **palatopharyngeal arch** (PAL-a-tō-fa-rin'-jē-al) extends to the side of the pharynx. The palatine tonsils are situated between the arches, and the

Figure 24.4   Structures of the mouth (oral cavity).

The cheeks, hard and soft palates, and tongue form the mouth.

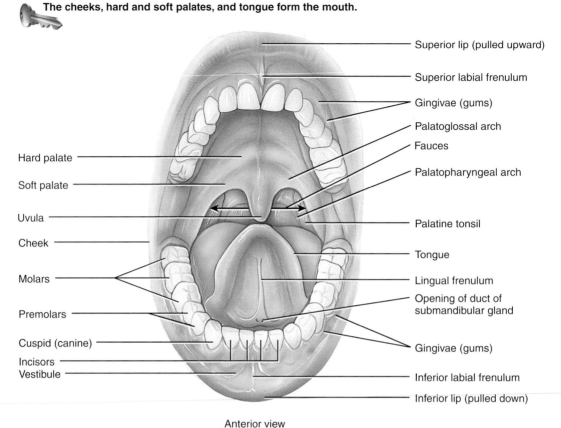

Superior lip (pulled upward)

Superior labial frenulum

Gingivae (gums)

Palatoglossal arch

Fauces

Palatopharyngeal arch

Palatine tonsil

Tongue

Lingual frenulum

Opening of duct of submandibular gland

Gingivae (gums)

Inferior labial frenulum

Inferior lip (pulled down)

Hard palate

Soft palate

Uvula

Cheek

Molars

Premolars

Cuspid (canine)

Incisors

Vestibule

Anterior view

What is the function of the uvula?

lingual tonsils are situated at the base of the tongue. At the posterior border of the soft palate, the mouth opens into the oropharynx through the fauces (Figure 24.4).

Salivary Glands

A **salivary gland** is any cell or organ that releases a secretion called saliva into the oral cavity. Ordinarily, just enough saliva is secreted to keep the mucous membranes of the mouth and pharynx moist and to cleanse the mouth and teeth. When food enters the mouth, however, secretion of saliva increases, and it lubricates, dissolves, and begins the chemical breakdown of the food.

The mucous membrane of the mouth and tongue contains many small salivary glands that open directly, or indirectly via short ducts, to the oral cavity. These glands include *labial, buccal,* and *palatal glands* in the lips, cheeks, and palate, respectively, and *lingual glands* in the tongue, all of which make a small contribution to saliva.

However, most saliva is secreted by the **major salivary glands,** which lie beyond the oral mucosa. Their secretions empty into ducts that lead to the oral cavity. There are three pairs of major salivary glands: the parotid, submandibular, and sublin-

gual glands (Figure 24.5a). The **parotid glands** (*par-* = near; *to-* = ear) are located inferior and anterior to the ears, between the skin and the masseter muscle. Each secretes saliva into the oral cavity via a **parotid duct** that pierces the buccinator muscle to open into the vestibule opposite the second maxillary (upper) molar tooth. The **submandibular glands** are found beneath the base of the tongue in the posterior part of the floor of the mouth. Their ducts, the **submandibular ducts,** run under the mucosa on either side of the midline of the floor of the mouth and enter the oral cavity proper lateral to the lingual frenulum. The **sublingual glands** are superior to the submandibular glands. Their ducts, the **lesser sublingual ducts,** open into the floor of the mouth in the oral cavity proper.

Composition and Functions of Saliva

Chemically, **saliva** is 99.5% water and 0.5% solutes. Among the solutes are ions, including sodium, potassium, chloride, bicarbonate, and phosphate. Also present are some dissolved gases and various organic substances, including urea and uric acid, mucus, immunoglobulin A, the bacteriolytic enzyme lysozyme, and salivary amylase, a digestive enzyme that acts on starch.

Figure 24.5 The three major salivary glands—parotid, sublingual, and submandibular. The submandibular glands, shown in the light micrograph in (b), consist mostly of serous acini (serous-fluid-secreting portions of gland) and a few mucous acini (mucus-secreting portions of gland); the parotid glands consist of serous acini only; and the sublingual glands consist of mostly mucous acini and a few serous acini. (See Tortora, *A Photographic Atlas of the Human Body*, Figure 12.6.)

Saliva lubricates and dissolves foods and begins the chemical breakdown of carbohydrates and lipids.

Parotid duct
Zygomatic arch
PAROTID GLAND
Opening of parotid duct (near second maxillary molar)
Second maxillary molar tooth
Tongue
Lingual frenulum
Submandibular duct
Mylohyoid muscle
SUBMANDIBULAR GLAND
Lesser sublingual duct
SUBLINGUAL GLAND

(a) Location of salivary glands

Mucous acini
Serous acini

LM 350x

(b) Submandibular gland

What is the function of the chloride ions in saliva?

Each major salivary gland supplies different proportions of ingredients to saliva. The parotid glands contain cells that secrete a watery (serous) liquid containing salivary amylase. Because the submandibular glands contain cells similar to those found in the parotid glands, plus some mucous cells, they secrete a fluid that contains amylase but is thickened with mucus. The sublingual glands contain mostly mucous cells, so they secrete a much thicker fluid that contributes only a small amount of amylase to the saliva.

The water in saliva provides a medium for dissolving foods so that they can be tasted and digestive reactions can begin. Chloride ions in the saliva activate salivary amylase. Bicarbonate and phosphate ions buffer acidic foods that enter the mouth; as a result, saliva is only slightly acidic (pH 6.35–6.85). Urea and uric acid are found in saliva because salivary glands (like the sweat glands of the skin) help remove waste molecules from the body. Mucus lubricates the food so it can easily be moved about in the mouth, formed into a ball, and swallowed. Immunoglobulin A (IgA) is a secreted type of antibody that prevents attachment of microbes so they cannot penetrate the epithelium. The enzyme lysozyme kills bacteria. Even though these substances help protect the mucous membrane from infection and the teeth from decay, they are not present in large enough quantities to eliminate all oral bacteria.

Salivation

Secretion of saliva, or **salivation** (sal-i-VĀ-shun), is controlled by the autonomic nervous system. Amounts of saliva secreted daily vary considerably but average 1000–1500 mL (1–1.6 qt). Normally, parasympathetic stimulation promotes continuous secretion of a moderate amount of saliva, which keeps the mucous membranes moist and lubricates the movements of the tongue and lips during speech. The saliva is then swallowed and helps moisten the esophagus. Eventually, most components of saliva are reabsorbed, which prevents fluid loss. Sympathetic stimulation dominates during stress, resulting in dryness of the mouth. During dehydration, the salivary glands stop secreting saliva to conserve water; the resulting dryness in the mouth contributes to the sensation of thirst. Drinking will then not only restore the homeostasis of body water but also moisten the mouth.

The feel and taste of food also are potent stimulators of salivary gland secretions. Chemicals in the food stimulate receptors in taste buds on the tongue, and impulses propagate from the taste buds to the superior and inferior salivatory nuclei in the brain stem. Impulses in parasympathetic neurons of the facial nerve (cranial nerve VII) and glossopharyngeal nerve (cranial nerve IX) then stimulate the secretion of saliva. Saliva continues to be heavily secreted for some time after food is swallowed; this flow of saliva washes out the mouth and dilutes and buffers the remnants of irritating chemicals.

The smell, sight, sound, or thought of food may also stimulate secretion of saliva. These stimuli constitute psychological activation and involve learned behavior. When memories that associate the stimuli with food occur, nerve impulses from the cerebral cortex reach the salivatory nuclei in the brain stem, and the salivary glands are activated. Psychological activation of the glands has some benefit to the body because it allows chemical digestion to start in the mouth as soon as the food is ingested. Salivation also occurs in response to swallowing irritating foods or during nausea due to reflexes originating in the stomach and upper small intestine. This mechanism presumably helps dilute or neutralize the irritating substance.

 Mumps

Although any of the salivary glands may be the target of a nasopharyngeal infection, the mumps virus (a myxovirus) typically attacks the parotid glands. **Mumps** is an inflammation and enlargement of the parotid glands accompanied by moderate fever, malaise (general discomfort), and extreme pain in the throat, especially when swallowing sour foods or acidic juices. Swelling occurs on one or both sides of the face, just anterior to the ramus of the mandible. In about 30% of males past puberty, the testes may also become inflamed; sterility rarely occurs because testicular involvement is usually unilateral (one testis only). Since a vaccine became available for mumps in 1967, the incidence of the disease has declined. ∎

Tongue

The **tongue** is an accessory digestive organ composed of skeletal muscle covered with mucous membrane. Together with its associated muscles, it forms the floor of the oral cavity. The tongue is divided into symmetrical lateral halves by a median septum that extends its entire length, and it is attached inferiorly to the hyoid bone, styloid process of the temporal bone, and mandible. Each half of the tongue consists of an identical complement of extrinsic and intrinsic muscles.

The **extrinsic muscles** of the tongue, which originate outside the tongue (attach to bones in the area) and insert into connective tissues in the tongue, include the hyoglossus, genioglossus, and styloglossus muscles (see Figure 11.7 on page 327). The extrinsic muscles move the tongue from side to side and in and out to maneuver food for chewing, shape the food into a rounded mass, and force the food to the back of the mouth for swallowing. They also form the floor of the mouth and hold the tongue in position. The **intrinsic muscles** originate in and insert into connective tissue within the tongue. They alter the shape and size of the tongue for speech and swallowing. The intrinsic muscles include the longitudinalis superior, longitudinalis inferior, transversus linguae, and verticalis linguae muscles. The **lingual frenulum** (*lingua* = the tongue), a fold of mucous membrane in the midline of the undersurface of the tongue, is attached to the floor of the mouth and aids in limiting the movement of the tongue posteriorly (see Figures 24.4 and 24.5). If a person's lingual frenulum is abnormally short or rigid—a condition called **ankyloglossia** (ang′-kē-lō-GLOSS-ēa)—then eating and speaking are impaired such that the person is said to be "tongue-tied."

The dorsum (upper surface) and lateral surfaces of the tongue are covered with **papillae** (pa-PIL-ē = nipple shaped projections), projections of the lamina propria covered with keratinized epithelium (see Figure 16.2a on page 530). Many papillae contain taste buds, the receptors for gustation (taste). **Fungiform papillae** (FUN-ji-form = mushroomlike) are mushroomlike elevations distributed among the filiform papillae that are more numerous near the tip of the tongue. They appear as red dots on the surface of the tongue, and most of them contain taste buds. **Vallate (circumvallate) papillae** (VAL-āt = wall-like) are arranged in an inverted V shape on the posterior surface of the tongue; all of them contain taste buds. **Foliate papillae** (FŌ-l-ē-āt = leaflike) are located in small trenches on the lateral margins of the tongue but most of their taste buds degenerate in early childhood. **Filiform papillae** (FIL-i-form = threadlike) are pointed, threadlike projections distributed in parallel rows over the anterior two-thirds of the tongue. Although filiform papillae lack taste buds, they contain receptors for touch and increase friction between the tongue and food, making it easier for the tongue to move food in the oral cavity. **Lingual glands** in the lamina propria secrete both mucus and a watery serous fluid that contains the enzyme **lingual lipase,** which acts on triglycerides.

Teeth

The **teeth,** or **dentes** (Figure 24.6), are accessory digestive organs located in sockets of the alveolar processes of the mandible and maxillae. The alveolar processes are covered by the **gingivae** (JIN-ji-vē), or gums, which extend slightly into each socket to form the gingival sulcus. The sockets are lined by the **periodontal ligament** or **membrane** (*odont-* = tooth), which consists of dense fibrous connective tissue and is attached to the socket walls and the cemental surface of the roots. Thus, it anchors the teeth in position and acts as a shock absorber during chewing.

A typical tooth has three major regions. The **crown** is the visible portion above the level of the gums. Embedded in the socket are one to three **roots.** The **neck** is the constricted junction of the crown and root near the gum line.

Teeth are composed primarily of **dentin,** a calcified connective tissue that gives the tooth its basic shape and rigidity. It is harder than bone because of its higher content of calcium salts (70% of dry weight). The dentin encloses a cavity. The enlarged part of the cavity, the **pulp cavity,** lies within the crown and is filled with **pulp,** a connective tissue containing blood vessels, nerves, and lymphatic vessels. Narrow extensions of the pulp cavity, called **root canals,** run through the root of the tooth. Each root canal has an opening at its base, the **apical foramen,** through which blood vessels, lymphatic vessels, and nerves extend.

The dentin of the crown is covered by **enamel** that consists primarily of calcium phosphate and calcium carbonate. Enamel, the hardest substance in the body and the richest in calcium salts (about 95% of dry weight), protects the tooth from the wear and tear of chewing. It also protects against acids that can easily dissolve dentin. The dentin of the root is covered by **cementum,** another bonelike substance, which attaches the root to the periodontal ligament.

The branch of dentistry that is concerned with the prevention, diagnosis, and treatment of diseases that affect the pulp, root, periodontal ligament, and alveolar bone is known as **endodontics** (en′-dō-DON-tiks; *endo-* = within). **Orthodontics** (or′-thō-DON-tiks; *ortho-* = straight) is a branch of dentistry that is concerned with the prevention and correction of abnormally aligned teeth, whereas **periodontics** (per′-ē-ō-DON-tiks) is a branch of dentistry concerned with the treatment of abnormal conditions of the tissues immediately surrounding the teeth.

Humans have two **dentitions,** or sets of teeth: deciduous and permanent. The first of these—the **deciduous teeth** (*decidu-* = falling out), also called **primary teeth, milk teeth,** or **baby teeth**—begin to erupt at about 6 months of age, and one pair of teeth appears at about each month thereafter, until all 20 are present (Figure 24.7a). The incisors, which are closest to the midline, are chisel-shaped and adapted for cutting into food. They are referred to as either **central** or **lateral incisors** based on their position. Next to the incisors, moving posteriorly, are the **cuspids (canines),** which have a pointed surface called a cusp. Cuspids are used to tear and shred food. Incisors and cuspids have only one root apiece. Posterior to them lie the **first** and **second molars,** which have four cusps. Maxillary (upper) molars have three roots; mandibular (lower) molars have two roots. The molars crush and grind food.

All the deciduous teeth are lost—generally between ages 6 and 12 years—and are replaced by the **permanent (secondary) teeth** (Figure 24.7b). The permanent dentition contains 32 teeth that erupt between age 6 and adulthood. The pattern resembles the deciduous dentition, with the following exceptions. The deciduous molars are replaced by the **first** and **second premolars (bicuspids),** which have two cusps and one root (upper first bicuspids have two roots) and are used for crushing and grinding. The permanent molars, which erupt into the mouth posterior to the bicuspids, do not replace any deciduous teeth and erupt as the jaw grows to accommodate them—the **first molars** at age 6, the **second molars** at age 12, the **third molars (wisdom teeth)** after age 17.

Often the human jaw does not have enough room posterior to the second molars to accommodate the eruption of the third molars. In this case, the third molars remain embedded in the

---

Figure 24.6 A typical tooth and surrounding structures.

Teeth are anchored in sockets of the alveolar processes of the mandible and maxillae.

Sagittal plane through tooth

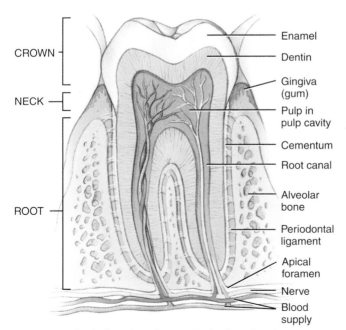

CROWN
— Enamel
— Dentin

NECK
— Gingiva (gum)
— Pulp in pulp cavity
— Cementum
— Root canal

ROOT
— Alveolar bone
— Periodontal ligament
— Apical foramen
— Nerve
— Blood supply

Sagittal section of a mandibular (lower) molar

What type of tissue is the main component of teeth?

Figure 24.7 Dentitions and times of eruptions (indicated in parentheses). A designated letter (deciduous teeth) or number (permanent teeth) uniquely identifies each tooth. Deciduous teeth begin to erupt at 6 months of age, and one pair of teeth appears about each month thereafter, until all 20 are present. (See Tortora, *A Photographic Atlas of the Human Body,* Figure 12.7.)

 There are 20 teeth in a complete deciduous set and 32 teeth in a complete permanent set.

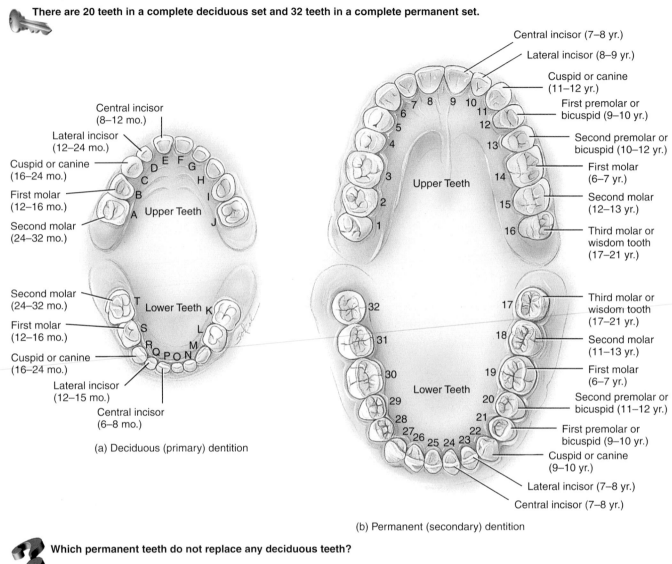

(a) Deciduous (primary) dentition

(b) Permanent (secondary) dentition

Which permanent teeth do not replace any deciduous teeth?

---

alveolar bone and are said to be "impacted." They often cause pressure and pain and must be removed surgically. In some people, third molars may be dwarfed in size or may not develop at all.

Root Canal Therapy

Root canal therapy is a procedure, accomplished in several phases, in which all traces of pulp tissue are removed from the pulp cavity and root canals of a badly diseased tooth. After a hole is made in the tooth, the root canals are filed out and irrigated to remove bacteria. Then, the canals are treated with medication and sealed tightly. The damaged crown is then repaired. ■

Mechanical and Chemical Digestion in the Mouth

Mechanical digestion in the mouth results from chewing, or **mastication** (mas′-ti-KĀ-shun = to chew), in which food is manipulated by the tongue, ground by the teeth, and mixed with saliva. As a result, the food is reduced to a soft, flexible, easily swallowed mass called a **bolus** (= lump). Food molecules begin to dissolve in the water in saliva, an important activity because enzymes can react with food molecules in a liquid medium only.

Two enzymes, salivary amylase and lingual lipase, contribute to chemical digestion in the mouth. **Salivary amylase** initiates the breakdown of starch. Dietary carbohydrates are either monosaccharide and disaccharide sugars or complex polysaccharides such as starches. Most of the carbohydrates we eat are starches, but only monosaccharides can be absorbed into the bloodstream. Thus, ingested disaccharides and starches must be broken down into monosaccharides. The function of salivary amylase is to break certain chemical bonds between glucose units in the starches, which reduces the long-chain polysaccharides to the disaccharide maltose, the trisaccharide maltotriose, and short-chain glucose polymers called α-dextrins. Even though food is usually swallowed too quickly for all the starches to be reduced to disaccharides in the mouth, salivary amylase in the swallowed food continues to act on the starches for about another hour, at which time stomach acids inactivate it. Saliva also contains **lingual lipase,** which is secreted by glands in the tongue. This enzyme becomes activated in the acidic environment of the stomach and thus starts to work after food is swallowed. It breaks down dietary triglycerides into fatty acids and diglycerides.

Table 24.1 summarizes the digestive activities in the mouth.

▶ C H E C K P O I N T

8. What structures form the mouth (oral cavity)?

9. How are the major salivary glands distinguished on the basis of location?

10. How is saliva secretion regulated?

11. What functions do incisors, cuspids, premolars, and molars perform?

12. What is a bolus? How is it formed?

Table 24.1	Summary of Digestive Activities in the Mouth	
Structure	Activity	Result
Cheeks and Lips	Keep food between teeth.	Foods uniformly chewed during mastication.
Salivary glands	Secrete saliva.	Lining of mouth and pharynx moistened and lubricated. Saliva softens, moistens, and dissolves food and cleanses mouth and teeth. Salivary amylase splits starch into smaller fragments.
Tongue		
Extrinsic tongue muscles	Move tongue from side to side and in and out.	Food maneuvered for mastication, shaped into bolus, and maneuvered for swallowing.
Intrinsic tongue muscles	Alter shape of tongue.	Swallowing and speech.
Taste buds	Serve as receptors for gustation (taste) and presence of food in mouth.	Secretion of saliva stimulated by nerve impulses from taste buds to salivary nuclei in brain stem to salivary glands.
Lingual glands	Secrete lingual lipase.	Triglycerides broken down.
Teeth	Cut, tear, and pulverize food.	Solid foods reduced to smaller particles for swallowing.

PHARYNX

▶ O B J E C T I V E

• **Describe the location and function of the pharynx.**

When food is first swallowed, it passes from the mouth into the **pharynx** (= throat), a funnel-shaped tube that extends from the internal nares to the esophagus posteriorly and to the larynx anteriorly (see Figure 23.4 on page 810). The pharynx is composed of skeletal muscle and lined by mucous membrane. Whereas the nasopharynx functions only in respiration, both the oropharynx and laryngopharynx have digestive as well as respiratory functions. Swallowed food passes from the mouth into the oropharynx and laryngopharynx, the muscular contractions of which help propel food into the esophagus and then into the stomach.

The movement of food from the mouth into the stomach is achieved by the act of swallowing, or **deglutition** (dē-gloo-TISH-un) (Figure 24.8). Deglutition is facilitated by saliva and mucus and involves the mouth, pharynx, and esophagus. Swallowing occurs in three stages: (1) the voluntary stage, in which the bolus is passed into the oropharynx; (2) the pharyngeal stage, which is the involuntary passage of the bolus through the

pharynx into the esophagus; and (3) the esophageal stage (described in the section on the esophagus), which is the involuntary passage of the bolus through the esophagus into the stomach.

Swallowing starts when the bolus is forced to the back of the oral cavity and into the oropharynx by the movement of the tongue upward and backward against the palate; these actions constitute the **voluntary stage** of swallowing. With the passage of the bolus into the oropharynx, the involuntary **pharyngeal stage** of swallowing begins (Figure 24.8b). The respiratory passageways close, and breathing is temporarily interrupted. The bolus stimulates receptors in the oropharynx, which send impulses to the **deglutition center** in the medulla oblongata and lower pons of the brain stem. The returning impulses cause the soft palate and uvula to move upward to close off the nasopharynx, and the larynx is pulled forward and upward under the tongue. As the larynx rises, the epiglottis moves backward and downward and seals off the rima glottidis. The movement of the larynx also pulls the vocal cords together, further sealing off the respiratory tract, and widens the opening between the laryngopharynx and esophagus. The bolus passes through the laryngopharynx and enters the esophagus in 1 to 2 seconds. The respiratory passageways then reopen, and breathing resumes.

Figure 24.8   Deglutition (swallowing).   During the pharyngeal stage of deglutition (b), the tongue rises against the palate, the nasopharynx is closed off, the larynx rises, the epiglottis seals off the larynx, and the bolus is passed into the esophagus.

    Deglutition is a mechanism that moves food from the mouth into the stomach.

Nasopharynx

Hard palate

Soft palate

Uvula

Oropharynx

Epiglottis

Laryngopharynx

Larynx

Esophagus

Bolus

Tongue

(a) Position of structures before swallowing

(b) During the pharyngeal stage of swallowing

Is swallowing a voluntary action or an involuntary action?

▶ C H E C K P O I N T

13. What does deglutition mean?

14. How does a bolus pass from the mouth into the stomach?

15. What occurs during the voluntary and pharyngeal stages of swallowing?

ESOPHAGUS

▶ O B J E C T I V E

- **Describe the location, anatomy, histology, and functions of the esophagus.**

The **esophagus** (e-SOF-a-gus = eating gullet) is a collapsible muscular tube that lies posterior to the trachea. It is about 25 cm (10 in.) long. The esophagus begins at the inferior end of the laryngopharynx and passes through the mediastinum anterior to the vertebral column. Then, it pierces the diaphragm through an opening called the **esophageal hiatus,** and ends in the superior portion of the stomach (see Figure 24.1). Sometimes, part of the stomach protrudes above the diaphragm through the esophageal hiatus. This condition is termed **hiatal hernia** (HER-nē-a).

Histology of the Esophagus

The **mucosa** of the esophagus consists of nonkeratinized stratified squamous epithelium, lamina propria (areolar connective

tissue), and a muscularis muscosae (smooth muscle) (Figure 24.9). Near the stomach, the mucosa of the esophagus also contains mucous glands. The stratified squamous epithelium associated with the lips, mouth, tongue, oropharynx, laryngopharynx, and esophagus affords considerable protection against abrasion and wear-and-tear from food particles that are chewed, mixed with secretions, and swallowed. The **submucosa** contains areolar connective tissue, blood vessels, and mucous glands. The **muscularis** of the superior third of the esophagus is skeletal muscle, the intermediate third is skeletal and smooth muscle, and the inferior third is smooth muscle. The superficial layer is known as the **adventitia** (ad-ven-TISH-a), rather than the serosa, because the areolar connective tissue of this layer is not covered by mesothelium and because the connective tissue merges with the connective tissue of surrounding structures of the mediastinum, through which is passes. The adventitia attaches the esophagus to surrounding structures.

Physiology of the Esophagus

The esophagus secretes mucus and transports food into the stomach. It does not produce digestive enzymes, and it does not carry on absorption. The passage of food from the laryngopharynx into the esophagus is regulated at the entrance to the esophagus by a sphincter (a circular band or ring of muscle that is normally contracted) called the **upper esophageal sphincter** (e-sof'-a-JĒ-al). It consists of the cricopharyngeus muscle

Figure 24.9 Histology of the esophagus. A higher magnification view of nonkeratinized, stratified squamous epithelium is shown in Table 4.1E on page 111. (See Tortora, *A Photographic Atlas of the Human Body,* Figure 12.8.)

🔑 **The esophagus secretes mucus and transports food to the stomach.**

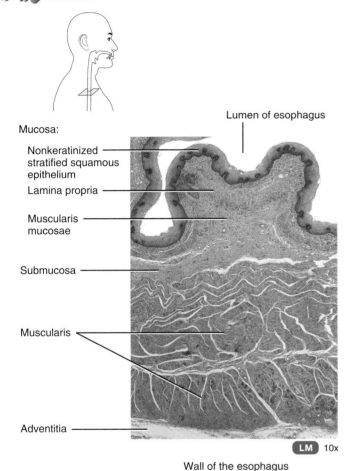

Wall of the esophagus

LM 10x

❓ **In which layers of the esophagus are the glands that secrete lubricating mucus located?**

Figure 24.10 Peristalsis during the esophageal stage of deglutition (swallowing).

🔑 **Peristalsis consists of progressive, wavelike contractions of the muscularis.**

Anterior view of frontal sections of peristalsis in esophagus

❓ **Does peristalsis "push" or "pull" food along the gastrointestinal tract?**

attached to the cricoid cartilage. The elevation of the larynx during the pharyngeal stage of swallowing causes the sphincter to relax, and the bolus enters the esophagus. This sphincter also relaxes during exhalation.

During the **esophageal stage** of swallowing, **peristalsis** (per′-is-STAL-sis; *stalsis* = constriction), a progression of coordinated contractions and relaxations of the circular and longitudinal layers of the muscularis, pushes the food bolus onward (Figure 24.10). (Peristalsis occurs in other tubular structures, including other parts of the GI tract and the ureters, bile ducts, and uterine tubes; in the esophagus it is controlled by the medulla oblongata.) In the section of the esophagus just superior to the bolus, the circular muscle fibers contract, constricting the esophageal wall and squeezing the bolus toward the stomach. Meanwhile, longitudinal fibers inferior to the bolus also con-

tract, which shortens this inferior section and pushes its walls outward so it can receive the bolus. The contractions are repeated in a wave that pushes the food toward the stomach. Mucus secreted by esophageal glands lubricates the bolus and reduces friction. The passage of solid or semisolid food from the mouth to the stomach takes 4 to 8 seconds; very soft foods and liquids pass through in about 1 second.

Just superior to the diaphragm, the esophagus narrows slightly due to a maintained contraction of the muscularis at the lowest part of the esophagus. This physiological sphincter, known as the **lower esophageal sphincter,** relaxes during swallowing and thus allows the bolus to pass from the esophagus into the stomach.

Table 24.2 summarizes the digestive activities of the pharynx and esophagus.

Gastroesophageal Reflux Disease

If the lower esophageal sphincter fails to close adequately after food has entered the stomach, the stomach contents can reflux, or back up, into the inferior portion of the esophagus. This condition is known as **gastroesophageal reflux disease (GERD).** Hydrochloric acid (HCl) from the stomach contents can irritate the esophageal wall, resulting in a burning sensation called **heartburn** because it is experienced in a region very near the heart, even though it is unrelated to any cardiac problem. Drinking alcohol and smoking can cause the sphincter to relax,

Table 24.2	Summary of Digestive Activities in the Pharynx and Esophagus	
Structure	**Activity**	**Result**
Pharynx	Pharyngeal stage of deglutition.	Moves bolus from oropharynx to laryngopharynx and into esophagus; closes air passageways.
Esophagus	Relaxation of upper esophageal sphincter.	Permits entry of bolus from laryngopharynx into esophagus.
	Esophageal stage of deglutition (peristalsis).	Pushes bolus down esophagus.
	Relaxation of lower esophageal sphincter.	Permits entry of bolus into stomach.
	Secretion of mucus.	Lubricates esophagus for smooth passage of bolus.

worsening the problem. The symptoms of GERD often can be controlled by avoiding foods that strongly stimulate stomach acid secretion (coffee, chocolate, tomatoes, fatty foods, orange juice, peppermint, spearmint, and onions). Other acid-reducing strategies include taking over-the-counter histamine-2 (H_2) blockers such as Tagamet HB or Pepcid AC 30 to 60 minutes before eating, and neutralizing already secreted acid with antacids such as Tums or Maalox. Symptoms are less likely to occur if food is eaten in smaller amounts and if the person does not lie down immediately after a meal. GERD may be associated with cancer of the esophagus. ■

▶ C H E C K P O I N T

16. Describe the location and histology of the esophagus. What is its role in digestion?

17. Explain the operation of the upper and lower esophageal sphincters.

STOMACH

▶ OBJECTIVE

- **Describe the location, anatomy, histology, and functions of the stomach.**

The **stomach** is typically a J-shaped enlargement of the GI tract directly inferior to the diaphragm in the epigastric, umbilical, and left hypochondriac regions of the abdomen (see Figure 1.13a on page 20). The stomach connects the esophagus to the duodenum, the first part of the small intestine (Figure 24.11). Because a meal can be eaten much more quickly than the intestines can digest and absorb it, one of the functions of the stomach is to serve as a mixing chamber and holding reservoir. At appropriate intervals after food is ingested, the stomach

forces a small quantity of material into the first portion of the small intestine. The position and size of the stomach vary continually; the diaphragm pushes it inferiorly with each inspiration and pulls it superiorly with each expiration. Empty, it is about the size of a large sausage, but it is the most distensible part of the GI tract and can accommodate a large quantity of food. In the stomach, digestion of starch continues, digestion of proteins and triglycerides begins, the semisolid bolus is converted to a liquid, and certain substances are absorbed.

Anatomy of the Stomach

The stomach has four main regions: the cardia, fundus, body, and pylorus (Figure 24.11). The **cardia** (CAR-dē-a) surrounds the superior opening of the stomach. The rounded portion superior to and to the left of the cardia is the **fundus** (FUN-dus). Inferior to the fundus is the large central portion of the stomach, called the **body.** The region of the stomach that connects to the duodenum is the **pylorus** (pī-LOR-us; *pyl-* = gate; *-orus* = guard); it has two parts, the **pyloric antrum** (AN-trum = cave), which connects to the body of the stomach, and the **pyloric canal,** which leads into the duodenum. When the stomach is empty, the mucosa lies in large folds, called **rugae** (ROO-gē = wrinkles), that can be seen with the unaided eye. The pylorus communicates with the duodenum of the small intestine via a sphincter called the **pyloric sphincter.** The concave medial border of the stomach is called the **lesser curvature,** and the convex lateral border is called the **greater curvature.**

Pylorospasm and Pyloric Stenosis

Two abnormalities of the pyloric sphincter can occur in infants. In **pylorospasm** the muscle fibers of the sphincter fail to relax normally, so food does not pass easily from the stomach to the small intestine, the stomach becomes overly full, and the infant vomits often to relieve the pressure. Pylorospasm is treated by drugs that relax the muscle fibers of the sphincter. **Pyloric stenosis** is a narrowing of the pyloric sphincter, and it must be corrected surgically. The hallmark symptom is projectile vomiting—the spraying of liquid vomitus some distance from the infant. ■

Histology of the Stomach

The stomach wall is composed of the same four basic layers as the rest of the GI tract, with certain modifications. The surface of the **mucosa** is a layer of simple columnar epithelial cells called **surface mucous cells** (Figure 24.12a on page 868). The mucosa contains a **lamina propria** (areolar connective tissue) and a **muscularis mucosae** (smooth muscle). Epithelial cells extend down into the lamina propria, where they form columns of secretory cells called **gastric glands** that line many narrow channels called **gastric pits.** Secretions from several gastric glands flow into each gastric pit and then into the lumen of the stomach.

The four regions of the stomach are the cardia, fundus, body, and pylorus.

Esophagus

FUNDUS

Lower esophageal sphincter

Serosa

Muscularis:

CARDIA

Longitudinal layer

BODY

Circular layer

Lesser curvature

PYLORUS

Oblique layer

Greater curvature

Pyloric sphincter

Rugae of mucosa

PYLORIC CANAL

PYLORIC ANTRUM

Duodenum

Functions of the Stomach

1. Mixes saliva, food, and gastric juice to form chyme.
2. Reservoir for holding food before release into small intestine.
3. Secretes gastric juice, which contains HCl, pepsin, intrinsic factor, and gastric lipase.
4. HCl kills bacteria and denatures proteins. Pepsin begins the digestion of proteins. Intrinsic factor aids absorption of vitamin B_{12}. Gastric lipase aids digestion of triglycerides.
5. Secretes gastrin into blood.

(a) Anterior view of regions of stomach

Esophagus

Lower esophageal sphincter

Pyloric sphincter

CARDIA

FUNDUS

Duodenum

Lesser curvature

PYLORUS

Rugae of mucosa

PYLORIC CANAL

PYLORIC ANTRUM

BODY

Frontal plane

Greater curvature

(b) Frontal section of the internal surface

After a very large meal, does your stomach still have rugae?

Figure 24.12 Histology of the stomach.

Gastric juice is the combined secretions of mucous cells, parietal cells, and chief cells.

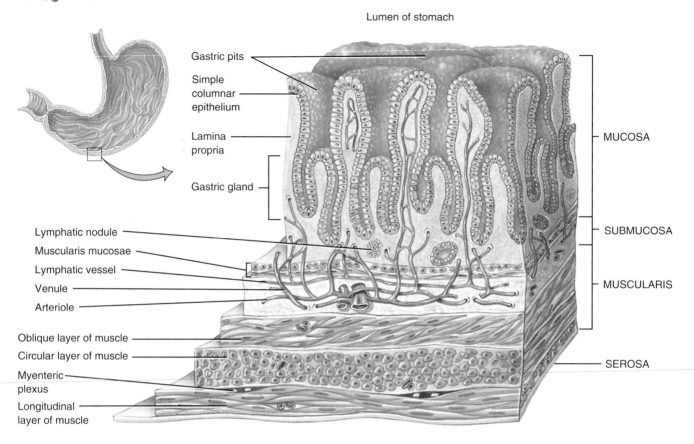

(a) Three dimensional view of layers of the stomach

The gastric glands contain three types of *exocrine gland cells* that secrete their products into the stomach lumen: mucous neck cells, chief cells, and parietal cells. Both surface mucous cells and **mucous neck cells** secrete mucus (Figure 24.12b). **Parietal cells** produce intrinsic factor, which is needed for absorption of vitamin B_{12}, and hydrochloric acid. The **chief cells** secrete pepsinogen and gastric lipase. The secretions of the mucous, parietal, and chief cells form **gastric juice,** which totals 2000–3000 mL (roughly 2–3 qt.) per day. In addition, gastric glands include a type of enteroendocrine cell, the **G cell,** which is located mainly in the pyloric antrum and secretes the hormone gastrin into the bloodstream. As we will see shortly, this hormone stimulates several aspects of gastric activity.

Three additional layers lie deep to the mucosa. The **submucosa** of the stomach is composed of areolar connective tissue. The **muscularis** has three (rather than two) layers of smooth muscle: an outer longitudinal layer, a middle circular layer, and an inner oblique layer. The oblique layer is limited primarily to the body of the stomach. The **serosa** is composed of simple squamous epithelium (mesothelium) and areolar connective tis-

sue, and the portion of the serosa covering the stomach is part of the visceral peritoneum. At the lesser curvature of the stomach, the visceral peritoneum extends upward to the liver as the lesser omentum. At the greater curvature of the stomach, the visceral peritoneum continues downward as the greater omentum and drapes over the intestines.

Mechanical and Chemical Digestion in the Stomach

Several minutes after food enters the stomach, gentle, rippling, peristaltic movements called **mixing waves** pass over the stomach every 15 to 25 seconds. These waves macerate food, mix it with secretions of the gastric glands, and reduce it to a soupy liquid called **chyme** (KĪM = juice). Few mixing waves are observed in the fundus, which primarily has a storage function. As digestion proceeds in the stomach, more vigorous mixing waves begin at the body of the stomach and intensify as they reach the pylorus. The pyloric sphincter normally remains almost, but not completely, closed. As food reaches the pylorus,

Gastric pit

Simple columnar epithelium

Lamina propria

Gastric gland

Muscularis mucosae

Submucosa

Surface mucous cell (secretes mucus)

Mucous neck cell (secretes mucus)

Parietal cell (secretes hydrochloric acid and intrinsic factor)

Chief cell (secretes pepsinogen and gastric lipase)

G cell (secretes the hormone gastrin)

(b) Sectional view of gastric glands and cell types

Gastric pit

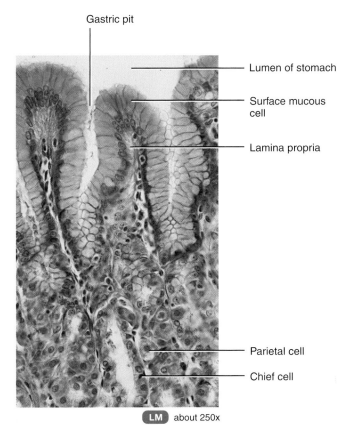

Lumen of stomach

Surface mucous cell

Lamina propria

Parietal cell

Chief cell

LM about 250x

(c) Fundic mucosa

Where is gastrin secreted and what are its functions?

each mixing wave forces several milliliters of chyme into the duodenum through the pyloric sphincter. Most of the chyme is forced back into the body of the stomach, where mixing continues. The next wave pushes the chyme forward again and forces a little more into the duodenum. These forward and backward movements of the gastric contents are responsible for most mixing in the stomach.

Foods may remain in the fundus for about an hour without becoming mixed with gastric juice. During this time, digestion by salivary amylase continues. Soon, however, the churning action mixes chyme with acidic gastric juice, inactivating salivary amylase and activating lingual lipase, which starts to digest triglycerides into fatty acids and diglycerides.

Even though parietal cells secrete hydrogen ions (H^+) and chloride ions (Cl^-) separately into the stomach lumen, the net effect is secretion of hydrochloric acid (HCl). **Proton pumps** powered by H^+/K^+ ATPases actively transport H^+ into the lumen while bringing potassium ions (K^+) into the cell (Figure 24.13). At the same time Cl^- and K^+ diffuse out through Cl^- and K^+ channels in the apical membrane (next to the lumen). The enzyme *carbonic anhydrase,* which is especially plentiful in parietal cells, catalyzes the formation of carbonic acid (H_2CO_3) from water (H_2O) and carbon dioxide (CO_2). As carbonic acid dissociates, it provides a ready source of H^+ for the proton pumps but also generates bicarbonate ions (HCO_3^-). As HCO_3^- builds up in the cytosol, it exits the parietal cell in exchange for Cl^- via Cl^-/HCO_3^- antiporters in the basolateral membrane (next to the lamina propria). HCO_3^- diffuses into nearby blood

Figure 24.13   Secretion of HCl (hydrochloric acid) by parietal cells in the stomach.

Proton pumps, powered by ATP, secrete the H⁺; Cl⁻ diffuses into the stomach lumen through Cl⁻ channels.

Key:

Proton pump (H⁺/K⁺ATPase)

K⁺ (potassium ion) channel

Cl⁻ (chloride ion) channel

CA   Carbonic anhydrase

• • • • ▶   Diffusion

HCO₃⁻ / Cl⁻ antiporter

What molecule is the source of the hydrogen ions that are secreted into gastric juice?

capillaries. This "alkaline tide" of bicarbonate ions entering the bloodstream after a meal may be large enough to slightly elevate blood pH and make urine more alkaline.

The strongly acidic fluid of the stomach kills many microbes in food, and HCl partially denatures (unfolds) proteins in food and stimulates the secretion of hormones that promote the flow of bile and pancreatic juice. Enzymatic digestion of proteins also begins in the stomach. The only proteolytic (protein-digesting) enzyme in the stomach is **pepsin,** which is secreted by chief cells. Because pepsin breaks certain peptide bonds between the amino acids making up proteins, a protein chain of many amino acids is broken down into smaller peptide fragments. Pepsin is most effective in the very acidic environment of the stomach (pH 2); it becomes inactive at higher pH.

What keeps pepsin from digesting the protein in stomach cells along with the food? First, pepsin is secreted in an inactive form called *pepsinogen;* in this form, it cannot digest the proteins in the chief cells that produce it. Pepsinogen is not converted into active pepsin until it comes in contact with active pepsin molecules or hydrochloric acid secreted by parietal cells. Second, the stomach epithelial cells are protected from gastric juices by a 1–3 mm thick layer of alkaline mucus secreted by surface mucous cells and mucous neck cells.

Another enzyme of the stomach is **gastric lipase,** which splits the short-chain triglycerides in fat molecules found in milk into fatty acids and monoglycerides. This enzyme, which has a limited role in the adult stomach, operates best at a pH of 5–6. More important than either lingual lipase or gastric lipase is pancreatic lipase, an enzyme secreted by the pancreas into the small intestine.

Only a small amount of absorption occurs in the stomach because its epithelial cells are impermeable to most materials. However, mucous cells of the stomach absorb some water, ions, and short-chain fatty acids, as well as certain drugs (especially aspirin) and alcohol.

Table 24.3 summarizes the digestive activities of the stomach.

Regulation of Gastric Secretion and Motility

Both neural and hormonal mechanisms control the secretion of gastric juice and the contraction of smooth muscle in the stomach wall. Events in gastric digestion occur in three overlapping phases: the cephalic, gastric, and intestinal phases (Figure 24.14).

Cephalic Phase

The **cephalic phase** of gastric digestion consists of reflexes initiated by sensory receptors in the head. Even before food enters the stomach, the sight, smell, taste, or thought of food initiates this reflex. The cerebral cortex and the feeding center in the hypothalamus send nerve impulses to the medulla oblongata. The medulla then transmits impulses to parasympathetic preganglionic neurons in the vagus nerves (cranial nerve X), which stimulate parasympathetic postganglionic neurons in the submucosal plexus. In turn, impulses from parasympathetic postganglionic neurons stimulate the gastric glands to secrete pepsinogen, hydrochloric acid, and mucus into stomach chyme, and gastrin into the blood. Impulses from parasympathetic neurons also increase stomach motility. Emotions such as anger, fear, and anxiety may slow digestion in the stomach because they stimulate the sympathetic nervous system, which inhibits gastric activity.

Gastric Phase

Once food reaches the stomach, sensory receptors in the stomach initiate both neural and hormonal mechanisms to ensure that gastric secretion and motility continue. This is the **gastric phase** of gastric digestion (Figure 24.14). Food of any kind distends (stretches) the stomach and stimulates stretch receptors in its

Table 24.3 Summary of Digestive Activities in the Stomach

Structure	Activity	Result
Mucosa		
Chief cells	Secrete pepsinogen.	Pepsin, the activated form, breaks certain peptide bonds in proteins.
	Secrete gastric lipase.	Splits short-chain triglycerides into fatty acids and monoglycerides.
Parietal cells	Secrete hydrochloric acid.	Kills microbes in food; denatures proteins; converts pepsinogen into pepsin.
	Secrete intrinsic factor.	Needed for absorption of vitamin B_{12}, which is used in red blood cell formation (erythropoiesis).
Surface mucous cells and **mucous neck cells**	Secrete mucus.	Forms a protective barrier that prevents digestion of stomach wall.
	Absorption.	Small quantity of water, ions, short-chain fatty acids, and some drugs enter the blood-stream.
G cells	Secrete gastrin.	Stimulates parietal cells to secrete HCl and chief cells to secrete pepsinogen; contracts lower esophageal sphincter, increases motility of the stomach, and relaxes pyloric sphincter.
Muscularis	Mixing waves.	Macerate food and mix it with gastric juice, forming chyme.
	Peristalsis.	Forces chyme through pyloric sphincter.
Pyloric sphincter	Opens to permit passage of chyme into duodenum.	Regulates passage of chyme from stomach to duodenum; prevents backflow of chyme from duodenum to stomach.

Figure 24.14 **The cephalic, gastric, and intestinal phases of gastric digestion.** Reflexes initiated by sensory receptors in the head, stomach, and small intestine constitute the cephalic, gastric, and intestinal phases of gastric digestion, respectively.

During the cephalic and gastric phases, digestion in the stomach is stimulated whereas during the intestinal phase, gastric juice secretion and gastric peristalsis are inhibited.

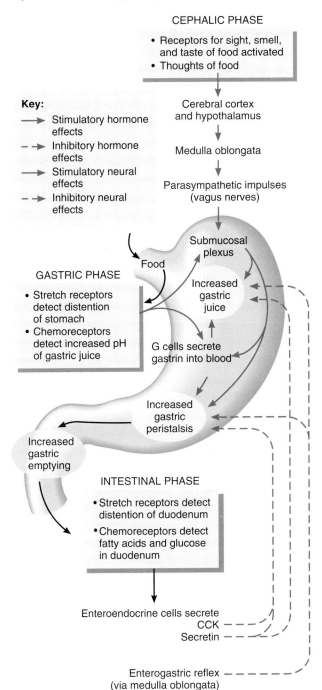

What are the effects on the stomach of increased stimulation of the vagus nerve?

walls. Chemoreceptors in the stomach monitor the pH of the stomach chyme. When the stomach walls are distended or pH increases because proteins have entered the stomach and buffered some of the stomach acid, the stretch receptors and chemoreceptors are activated, and a neural negative feedback loop is set in motion (Figure 24.15). From the stretch receptors and chemoreceptors, nerve impulses propagate to the submucosal plexus, where they activate parasympathetic and enteric neurons. The resulting nerve impulses cause waves of peristalsis and continue to stimulate the flow of gastric juice from parietal cells, chief cells, and mucous cells.

The peristaltic waves mix the food with gastric juice, and when the waves become strong enough, a small quantity of chyme—about 10–15 mL (2–3 teaspoons)—squirts past the pyloric sphincter into the duodenum. As the pH of the stomach chyme becomes more acidic again and the stomach walls are less

Figure 24.15 Neural negative feedback regulation of the pH of gastric juice and gastric motility during the gastric phase of gastric digestion.

Food entering the stomach stimulates secretion of gastric juice and causes vigorous waves of peristalsis.

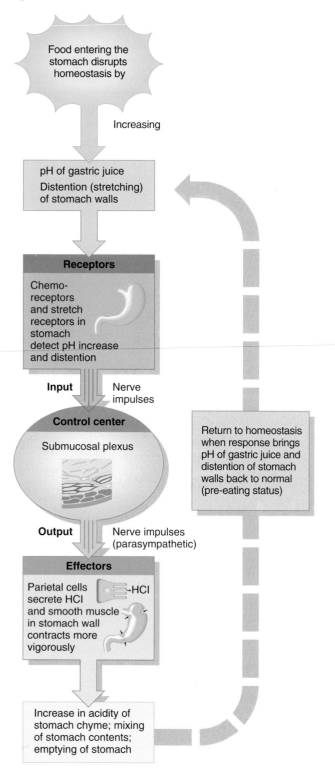

Food entering the stomach disrupts homeostasis by

Increasing

pH of gastric juice
Distention (stretching) of stomach walls

Receptors

Chemo-receptors and stretch receptors in stomach detect pH increase and distention

Input — Nerve impulses

Control center

Submucosal plexus

Return to homeostasis when response brings pH of gastric juice and distention of stomach walls back to normal (pre-eating status)

Output — Nerve impulses (parasympathetic)

Effectors

Parietal cells secrete HCl and smooth muscle in stomach wall contracts more vigorously

Increase in acidity of stomach chyme; mixing of stomach contents; emptying of stomach

Why does food initially cause the pH of the gastric juice to rise?

distended because chyme has passed into the small intestine, this negative feedback cycle suppresses secretion of gastric juice.

Hormonal negative feedback also regulates gastric secretion during the gastric phase (see Figure 24.14). Partially digested proteins buffer H^+, thus increasing pH, and ingested food distends the stomach. Chemoreceptors and stretch receptors monitoring these changes stimulate parasympathetic neurons to release acetylcholine. In turn, acetylcholine stimulates secretion of the hormone **gastrin** by G cells, enteroendocrine cells in the mucosa of the pyloric antrum. (A small amount of gastrin is also secreted by enteroendocrine cells in the small intestine; moreover, some chemicals in food—for example, caffeine—directly stimulate gastrin release.) Gastrin enters the bloodstream and finally reaches its target cells, the gastric glands.

Gastrin stimulates growth of the gastric glands and secretion of large amounts of gastric juice. It also strengthens contraction of the lower esophageal sphincter, increases motility of the stomach, and relaxes the pyloric and ileocecal sphincters (described later). Gastrin secretion is inhibited when the pH of gastric juice drops below 2.0 and is stimulated when the pH rises. This negative feedback mechanism helps provide an optimal low pH for the functioning of pepsin, the killing of microbes, and the denaturing of proteins in the stomach.

Acetylcholine (ACh) released by parasympathetic neurons and gastrin secreted by G cells stimulate parietal cells to secrete more HCl in the presence of histamine. In other words, histamine, a paracrine substance that is released by mast cells in the lamina propria and acts on nearby parietal cells, acts synergistically with acetylcholine and gastrin to enhance their effects. Receptors for all three substances are present in the plasma membrane of parietal cells. The histamine receptors on parietal cells are called H_2 receptors; they mediate different responses than do the H_1 receptors involved in allergic responses.

Intestinal Phase

The **intestinal phase** of gastric digestion is due to activation of receptors in the small intestine. Whereas reflexes initiated during the cephalic and gastric phases stimulate stomach secretory activity and motility, those occurring during the intestinal phase have inhibitory effects (see Figure 24.14) that slow the exit of chyme from the stomach and prevent overloading of the duodenum with more chyme than it can handle. In addition, responses occurring during the intestinal phase promote the continued digestion of foods that have reached the small intestine. When chyme containing fatty acids and glucose leaves the stomach and enters the small intestine, it triggers enteroendocrine cells in the small intestinal mucosa to release into the blood two hormones that affect the stomach—**secretin** (se-KRĒ-tin) and **cholecystokinin** (kō′-lē-sis′-tō-KĪN-in) or **CCK.** With respect to the stomach, secretin mainly decreases gastric secretions, whereas CCK mainly inhibits stomach emptying. Both hormones have other important effects on the pancreas, liver, and gallbladder (explained shortly) that contribute to the regulation of digestive processes.

Regulation of Gastric Emptying

Gastric emptying, the periodic release of chyme from the stomach into the duodenum, is regulated by both neural and hormonal reflexes, as follows (Figure 24.16a):

1. Stimuli such as distention of the stomach and the presence of partially digested proteins, alcohol, and caffeine initiate gastric emptying.

2. These stimuli increase the secretion of gastrin and generate parasympathetic impulses in the vagus nerves.

3. Gastrin and nerve impulses stimulate contraction of the lower esophageal sphincter, increase motility of the stomach, and relax the pyloric sphincter.

4. The net effect of these actions is gastric emptying.

Neural and hormonal reflexes also help ensure that the stomach does not release more chyme than the small intestine can process. The neural reflex called the **enterogastric reflex** and the hormone cholecystokinin inhibit gastric emptying, as follows (Figure 24.16b):

1. Stimuli such as distention of the duodenum and the presence of fatty acids, glucose, and partially digested proteins in the duodenal chyme inhibit gastric emptying.

2. These stimuli then initiate the enterogastric reflex. Nerve impulses propagate from the duodenum to the medulla oblongata, where they inhibit parasympathetic stimulation and stimulate sympathetic activity in the stomach. The same stimuli also increase secretion of cholecystokinin.

3. Increased sympathetic impulses and cholecystokinin both decrease gastric motility.

4. The net effect of these actions is inhibition of gastric emptying.

Within 2 to 4 hours after eating a meal, the stomach has emptied its contents into the duodenum. Foods rich in carbohydrate spend the least time in the stomach; high-protein foods remain somewhat longer, and emptying is slowest after a fat-laden meal containing large amounts of triglycerides. The reason for slower emptying after eating triglycerides is that fatty acids in chyme stimulate release of cholecystokinin, which slows stomach emptying.

Vomiting

Vomiting or *emesis* is the forcible expulsion of the contents of the upper GI tract (stomach and sometimes duodenum) through the mouth. The strongest stimuli for vomiting are irritation and distension of the stomach; other stimuli include unpleasant sights, general anesthesia, dizziness, and certain drugs such as morphine and derivatives of digitalis. Nerve impulses are transmitted to the vomiting center in the medulla oblongata, and returning impulses propagate to the upper GI tract organs, diaphragm, and abdominal muscles. Vomiting involves squeezing the stomach between the diaphragm and abdominal muscles

Figure 24.16 Neural and hormonal regulation of gastric emptying.

🔑 **Foods rich in carbohydrates are the first to leave the stomach.**

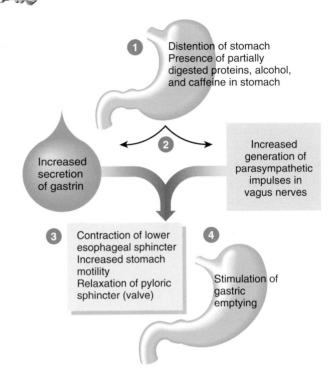

(a) Stimulation of gastric emptying

(b) Inhibition of gastric emptying

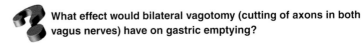

What effect would bilateral vagotomy (cutting of axons in both vagus nerves) have on gastric emptying?

and expelling the contents through open esophageal sphincters. Prolonged vomiting, especially in infants and elderly people, can be serious because the loss of acidic gastric juice can lead to alkalosis (higher than normal blood pH). ■

▶ C H E C K P O I N T

18. Compare the epithelium of the esophagus with that of the stomach. How is each adapted to the function of the organ?

19. What is the importance of rugae, surface mucous cells, mucous neck cells, chief cells, parietal cells, and G cells in the stomach?

20. What is the role of pepsin? Why is it secreted in an inactive form?

21. What are the functions of gastric lipase and lingual lipase in the stomach?

22. Describe the factors that stimulate and inhibit gastric secretion and motility. Why are the three phases of gastric digestion named the cephalic, gastric, and intestinal phases?

PANCREAS

▶ O B J E C T I V E

• **Describe the location, anatomy, histology, and function of the pancreas.**

From the stomach, chyme passes into the small intestine. Because chemical digestion in the small intestine depends on activities of the pancreas, liver, and gallbladder, we first consider the activities of these accessory digestive organs and their contributions to digestion in the small intestine.

Anatomy of the Pancreas

The **pancreas** (*pan-* = all; *-creas* = flesh), a retroperitoneal gland that is about 12–15 cm (5–6 in.) long and 2.5 cm (1 in.) thick, lies posterior to the greater curvature of the stomach. The pancreas consists of a head, a body, and a tail and is connected to the duodenum usually by two ducts (Figure 24.17). The **head** is the expanded portion of the organ near the curve of the duodenum; superior to and to the left of the head are the central **body** and the tapering **tail.**

Pancreatic juices pass from the exocrine cells into small ducts that ultimately unite to form two larger ducts that convey the secretions into the small intestine. The larger of the two ducts is called the **pancreatic duct (duct of Wirsung).** In most people, the pancreatic duct joins the common bile duct from the liver and gallbladder and enters the duodenum as a common duct called the **hepatopancreatic ampulla (ampulla of Vater).** The ampulla opens on an elevation of the duodenal mucosa known as the **major duodenal papilla,** which lies about 10 cm (4 in.) inferior to the pyloric sphincter of the stomach. The smaller of the two ducts, the **accessory duct (duct of Santorini),** leads from the pancreas and empties into the duodenum about 2.5 cm (1 in.) superior to the hepatopancreatic ampulla.

Histology of the Pancreas

The pancreas is made up of small clusters of glandular epithelial cells, about 99% of which are arranged in clusters called **acini** (AS-i-nē) and constitute the *exocrine* portion of the organ (see Figure 18.18b, c on page 615). The cells within acini secrete a mixture of fluid and digestive enzymes called **pancreatic juice.** The remaining 1% of the cells are organized into clusters called **pancreatic islets (islets of Langerhans),** the *endocrine* portion of the pancreas. These cells secrete the hormones glucagon, insulin, somatostatin, and pancreatic polypeptide. The functions of these hormones are discussed in Chapter 18.

Composition and Functions of Pancreatic Juice

Each day the pancreas produces 1200–1500 mL (about 1.2–1.5 qt) of pancreatic juice, which is a clear, colorless liquid consisting mostly of water, some salts, sodium bicarbonate, and several enzymes. The sodium bicarbonate gives pancreatic juice a slightly alkaline pH (7.1–8.2) that buffers acidic gastric juice in chyme, stops the action of pepsin from the stomach, and creates the proper pH for the action of digestive enzymes in the small intestine. The enzymes in pancreatic juice include a carbohydrate-digesting enzyme called **pancreatic amylase;** several protein-digesting enzymes called **trypsin** (TRIP-sin), **chymotrypsin** (kī′-mō-TRIP-sin), **carboxypeptidase** (kar-bok′-sē-PEP-ti-dās), and **elastase** (ē-LAS-tās); the principal triglyceride-digesting enzyme in adults, called **pancreatic lipase;** and nucleic acid–digesting enzymes called **ribonuclease** and **deoxyribonuclease.**

Just as pepsin is produced in the stomach in an inactive form (pepsinogen), so too are the protein-digesting enzymes of the pancreas. Because they are inactive, the enzymes do not digest cells of the pancreas itself. Trypsin is secreted in an inactive form called **trypsinogen** (trip-SIN-ō-jen). Pancreatic acinar cells also secrete a protein called **trypsin inhibitor,** that combines with any trypsin formed accidentally in the pancreas or in pancreatic juice and blocks its enzymatic activity. When trypsinogen reaches the lumen of the small intestine, it encounters an activating brush-border enzyme called **enterokinase** (en′-ter-ō-KĪ-nās), which splits off part of the trypsinogen molecule to form trypsin. In turn, trypsin acts on the inactive precursors (called **chymotrypsinogen, procarboxypeptidase,** and **proelastase**) to produce chymotrypsin, carboxypeptidase, and elastase, respectively.

 Pancreatitis

Inflammation of the pancreas, as may occur in association with alcohol abuse or chronic gallstones, is called **pancreatitis** (pan′-kre-a-TĪ-tis). In a more severe condition known as **acute pancreatitis,** which is associated with heavy alcohol intake or biliary tract obstruction, the pancreatic cells may release either trypsin instead of trypsinogen or insufficient amounts of trypsin

Figure 24.17 Relation of the pancreas to the liver, gallbladder, and duodenum. The inset shows details of the common bile duct and pancreatic duct forming the hepatopancreatic ampulla (ampulla of Vater) and emptying into the duodenum. (See Tortora, *A Photographic Atlas of the Human Body,* Figures 12.10, 12.11.)

🔑 **Pancreatic enzymes digest starches (polysaccharides), proteins, triglycerides, and nucleic acids.**

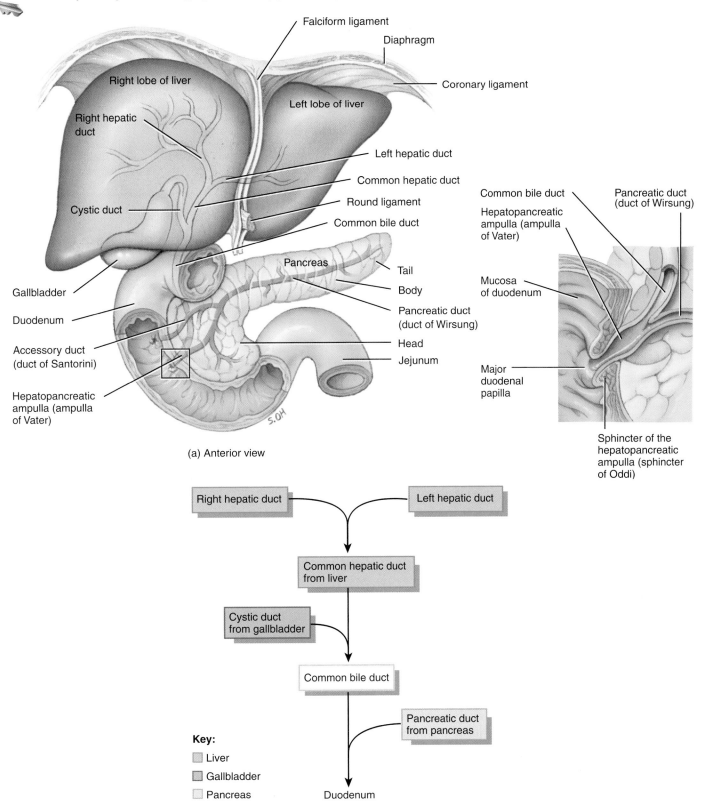

(a) Anterior view

Key:
- ▨ Liver
- ▨ Gallbladder
- ▢ Pancreas

(b) Ducts carrying bile from liver and gallbladder and pancreatic juice from pancreas to the duodenum

❓ **What type of fluid is found in the pancreatic duct? The common bile duct? The hepatopancreatic ampulla?**

inhibitor, and the trypsin begins to digest the pancreatic cells. Patients with acute pancreatitis usually respond to treatment, but recurrent attacks are the rule. ■

Regulation of Pancreatic Secretions

Pancreatic secretion, like gastric secretion, is regulated by both neural and hormonal mechanisms as follows (Figure 24.18).

1 During the cephalic and gastric phases of gastric digestion, parasympathetic impulses are transmitted along the vagus nerves (cranial nerve X) to the pancreas.

2 These parasympathetic nerve impulses stimulate increased secretion of pancreatic enzymes.

3 Acidic chyme containing partially digested fats and proteins enters the small intestine.

4 In response to fatty acids and amino acids, some enteroendocrine cells in the small intestine secrete cholecystokinin (CCK) into the blood. In response to acidic chyme, other enteroendocrine cells in the small intestinal mucosa liberate secretin into the blood.

5 Secretin stimulates the flow of pancreatic juice that is rich in bicarbonate ions.

6 Cholecystokinin (CCK) stimulates a pancreatic secretion that is rich in digestive enzymes.

▶ C H E C K P O I N T

23. Describe the duct system connecting the pancreas to the duodenum.

24. What are pancreatic acini? How do their functions differ from those of the pancreatic islets (islets of Langerhans)?

25. What digestive functions do the components of pancreatic juice have?

26. How is pancreatic juice secretion regulated?

LIVER AND GALLBLADDER

▶ O B J E C T I V E

- **Describe the location, anatomy, histology, and functions of the liver and gallbladder.**

The **liver** is the heaviest gland of the body, weighing about 1.4 kg (about 3 lb) in an average adult, and after the skin it is the second-largest organ of the body. It is inferior to the diaphragm and occupies most of the right hypochondriac and part of the epigastric regions of the abdominopelvic cavity (see Figure 1.13a on page 20).

The **gallbladder** (*gall-* = bile) is a pear-shaped sac that is located in a depression of the posterior surface of the liver. It is 7–10 cm (3–4 in.) long and typically hangs from the anterior inferior margin of the liver (see Figure 24.17).

Figure 24.18   Neural and hormonal stimulation of pancreatic juice secretion.

Parasympathetic stimulation (via the vagus nerves) and acidic chyme in the small intestine stimulate secretin release into the blood. Vagal stimulation and fatty acids and amino acids in chyme within the small intestine stimulate cholecystokinin (CCK) release into the blood.

1 Parasympathetic impulses propagating along vagus nerves

2 Stimulate secretion of pancreatic enzymes

3 Acidic chyme containing partially digested fats (fatty acids) and proteins (amino acids)

CCK
Secretin

4

Secretin stimulates secretion of pancreatic juice rich in bicarbonate ions

CCK stimulates secretion of pancreatic juice rich in digestive enzymes

5   **6**

Blood vessel

Why is secretion of pancreatic juice rich in bicarbonate ions helpful in digestion?

Anatomy of the Liver and Gallbladder

The liver is almost completely covered by visceral peritoneum and is completely covered by a dense irregular connective tissue layer that lies deep to the peritoneum. The liver is divided into two principal lobes—a large **right lobe** and a smaller **left lobe**—by the **falciform ligament** (see Figure 24.17). Even though the right lobe is considered by many anatomists to include an inferior **quadrate lobe** and a posterior **caudate lobe,** based on internal morphology (primarily the distribution of blood vessels), the quadrate and caudate lobes more appropriately belong to the left lobe. The falciform ligament, a fold of the parietal peritoneum, extends from the undersurface of the diaphragm between the two principal lobes of the liver to the superior surface of the liver, helping to suspend the liver. In the free border of the falciform ligament is the **ligamentum teres (round ligament),** a fibrous cord that is a remnant of the umbilical vein of the fetus (see Figure 21.31a, b on page 754); it extends from the liver to the umbilicus. The right and left **coro-**

nary ligaments are narrow reflections of the parietal peritoneum that suspend the liver from the diaphragm.

The parts of the gallbladder are the broad *fundus,* which projects downward beyond the inferior border of the liver; the *body,* the central portion; and the *neck,* the tapered portion. The body and neck project upward.

Histology of the Liver and Gallbladder

The lobes of the liver are made up of many functional units called **lobules** (Figure 24.19). A lobule is typically a six-sided structure (hexagon) that consists of specialized epithelial cells, called **hepatocytes** (*hepat-* = liver; *-cytes* = cells), arranged in irregular, branching, interconnected plates around a **central vein.** Instead of capillaries, the liver has larger, endothelium-lined spaces called **sinusoids,** through which blood passes. Also present in the sinusoids are fixed phagocytes called **stellate reticuloendothelial (Kupffer) cells,** which destroy worn-out white blood cells and red blood cells, bacteria, and other foreign matter in the venous blood draining from the gastrointestinal tract.

Bile, which is secreted by hepatocytes, enters **bile canaliculi** (kan′-a-LIK-ū-lī = small canals), which are narrow intercellular canals that empty into small *bile ductules* (Figure 24.19a). The ductules pass bile into *bile ducts* at the periphery of the lobules. The bile ducts merge and eventually form the larger **right** and **left hepatic ducts,** which unite and exit the liver as the **common hepatic duct** (see Figure 24.17). Farther on, the common hepatic duct joins the **cystic duct** (*cystic* = bladder) from the gallbladder to form the **common bile duct.** Bile enters the cystic duct and is temporarily stored in the gallbladder.

The mucosa of the gallbladder consists of simple columnar epithelium arranged in rugae resembling those of the stomach. The wall of the gallbladder lacks a submucosa. The middle, muscular coat of the wall consists of smooth muscle fibers. Contraction of the smooth muscle fibers ejects the contents of the gallbladder into the **cystic duct.** The gallbladder's outer coat is the visceral peritoneum. The functions of the gallbladder are to store and concentrate bile (up to tenfold) until it is needed in the small intestine. In the concentration process, water and ions are absorbed by the gallbladder mucosa.

Jaundice

Jaundice (JAWN-dis = yellowed) is a yellowish coloration of the sclerae (whites of the eyes), skin, and mucous membranes due to a buildup of a yellow compound called bilirubin. After bilirubin is formed from the breakdown of the heme pigment in aged red blood cells, it is transported to the liver, where it is processed and eventually excreted into bile. The three main categories of jaundice are (1) *prehepatic jaundice,* due to excess production of bilirubin; (2) *hepatic jaundice,* due to congenital liver disease, cirrhosis of the liver, or hepatitis; and (3) *extrahepatic jaundice,* due to blockage of bile drainage by gallstones or cancer of the bowel or the pancreas.

Because the liver of a newborn functions poorly for the first week or so, many babies experience a mild form of jaundice called *neonatal (physiological) jaundice* that disappears as the liver matures. Usually, it is treated by exposing the infant to blue light, which converts bilirubin into substances the kidneys can excrete. ■

Blood Supply of the Liver

The liver receives blood from two sources (Figure 24.20 on page 879). From the hepatic artery it obtains oxygenated blood, and from the hepatic portal vein it receives deoxygenated blood containing newly absorbed nutrients, drugs, and possibly microbes and toxins from the gastrointestinal tract (see Figure 21.29 on page 751). Branches of both the hepatic artery and the hepatic portal vein carry blood into liver sinusoids, where oxygen, most of the nutrients, and certain toxic substances are taken up by the hepatocytes. Products manufactured by the hepatocytes and nutrients needed by other cells are secreted back into the blood, which then drains into the central vein and eventually passes into a hepatic vein. Because blood from the gastrointestinal tract passes through the liver as part of the hepatic portal circulation, the liver is often a site for metastasis of cancer that originates in the GI tract. Branches of the hepatic portal vein, hepatic artery, and bile duct typically accompany each other in their distribution through the liver. Collectively, these three structures are called a **portal triad** (see Figure 24.19). Portal triads are located at the corners of the liver lobules.

Role and Composition of Bile

Each day, hepatocytes secrete 800–1000 mL (about 1 qt) of **bile,** a yellow, brownish, or olive-green liquid. It has a pH of 7.6–8.6 and consists mostly of water and bile acids, bile salts, cholesterol, a phospholipid called lecithin, bile pigments, and several ions.

Bile is partially an excretory product and partially a digestive secretion. Bile salts, which are sodium salts and potassium salts of bile acids (mostly cholic acid and chenodeoxycholic acid), play a role in **emulsification,** the breakdown of large lipid globules into a suspension of droplets about 1 μm in diameter, and in the absorption of lipids following their digestion. The tiny lipid droplets present a very large surface area that allows pancreatic lipase to more rapidly accomplish digestion of triglycerides. Cholesterol is made soluble in bile by bile salts and lecithin.

The principal bile pigment is **conjugated bilirubin.** The phagocytosis of aged red blood cells liberates iron, globin, and bilirubin (derived from heme). The iron and globin are recycled, and some of the bilirubin is converted to **conjugated bilirubin,** in which the bilirubin is attached to glucuronic acid molecules. Conjugated bilirubin is then secreted into the bile and is eventually broken down in the intestine. One of its breakdown products—**stercobilin**—gives feces their normal brown color.

Figure 24.19 Histology of a lobule, the functional unit of the liver.

A lobule consists of hepatocytes arranged around a central vein.

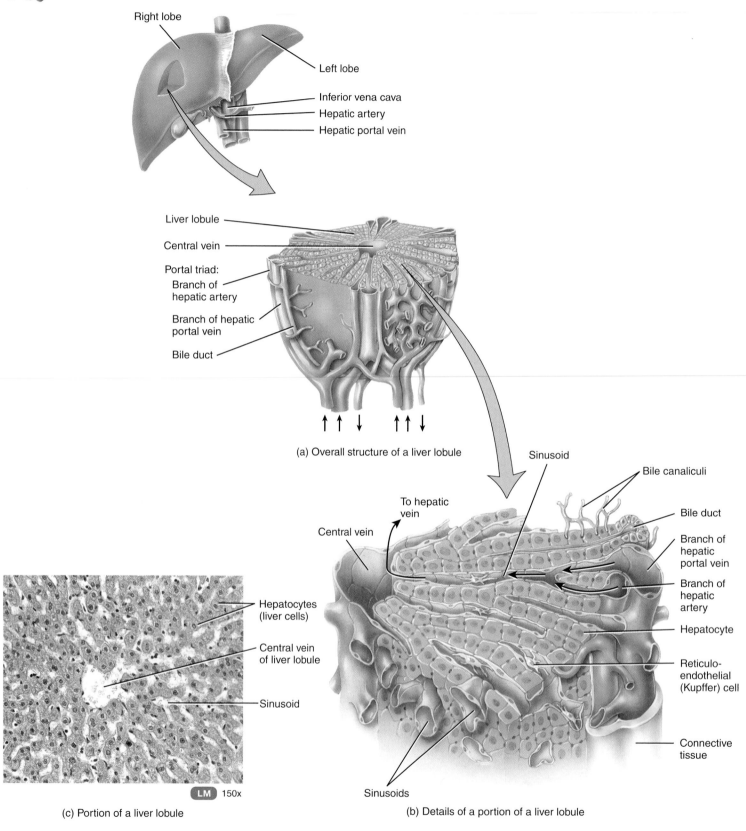

(a) Overall structure of a liver lobule

(c) Portion of a liver lobule

(b) Details of a portion of a liver lobule

Which type of cell in the liver is phagocytic?

Figure 24.20 **Hepatic blood flow: sources, path through the liver, and return to the heart.**

 The liver receives oxygenated blood via the hepatic artery and nutrient-rich deoxygenated blood via the hepatic portal vein.

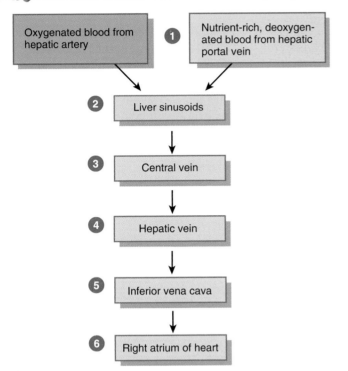

During the first few hours after a meal, how does the chemical composition of blood change as it flows through the liver sinusoids?

Regulation of Bile Secretion

After they have served their function as emulsifying agents, most bile salts are reabsorbed by active transport in the final segment of the small intestine (ileum) and enter portal blood flowing toward the liver. Although hepatocytes continually release bile, they increase production and secretion when the portal blood contains more bile acids; thus, as digestion and absorption continue in the small intestine, bile release increases. Between meals, after most absorption has occurred, bile flows into the gallbladder for storage because the **sphincter of the hepatopancreatic ampulla** (*sphincter of Oddi;* see Figure 24.17) closes off the entrance to the duodenum. After a meal, several neural and hormonal stimuli promote production and release of bile (Figure 24.21):

1. Parasympathetic impulses propagate along axons of the vagus nerve (cranial nerve X) and stimulate the liver to increase bile production.

2. Fatty acids and amino acids in chyme entering the duodenum stimulate some duodenal enteroendocrine cells to secrete the hormone cholecystokinin (CCK) into the blood. Acidic chyme entering the duodenum stimulates other enteroendocrine cells to secrete the hormone secretin into the blood.

3. CCK causes contraction of the wall of the gallbladder, which squeezes stored bile out of the gallbladder into the cystic duct and through the common bile duct. CCK also causes relaxation of the sphincter of the hepatopancreatic ampulla, which allows bile to flow into the duodenum.

Figure 24.21 **Neural and hormonal stimuli that promote production and release of bile.**

Bile salts in bile promote absorption of dietary lipids by emulsifying them.

① Parasympathetic impulses along vagus nerves (cranial nerve X) stimulate bile production by liver

④ Secretin enhances flow of bile rich in HCO_3^- from liver

Liver (site of bile production)

Gallbladder (site of bile storage)

Bile

③ CCK causes contraction of gallbladder

CCK

Chyme

② Fatty acids and amino acids in chyme entering the duodenum stimulate secretion of cholecystokinin (CCK) into blood.

Acidic chyme entering duodenum stimulates secretion of secretin into blood.

Enteroendocrine cells in mucosa of small intestine

Secretin

 In what respect are the effects of secretin on the liver and on the pancreas similar?

④ Secretin, which stimulates secretion of pancreatic juice that is rich in HCO_3^-, also stimulates the secretion of HCO_3^- by hepatocytes into bile.

Functions of the Liver

Besides secreting bile, which is needed for absorption of dietary fats, the liver performs many other vital functions:

- **Carbohydrate metabolism.** The liver is especially important in maintaining a normal blood glucose level. When blood glucose is low, the liver can break down glycogen to glucose and release glucose into the bloodstream. The liver can also convert certain amino acids and lactic acid to glucose, and it can convert other sugars, such as fructose and galactose, into glucose. When blood glucose is high, as occurs just after eating a meal, the liver converts glucose to glycogen and triglycerides for storage.

- **Lipid metabolism.** Hepatocytes store some triglycerides; break down fatty acids to generate ATP; synthesize lipoproteins, which transport fatty acids, triglycerides, and cholesterol to and from body cells; synthesize cholesterol; and use cholesterol to make bile salts.

- **Protein metabolism.** Hepatocytes deaminate (remove the amino group, NH_2, from) amino acids so that the amino acids can be used for ATP production or converted to carbohydrates or fats. The resulting toxic ammonia (NH_3) is then converted into the much less toxic urea, which is excreted in urine. Hepatocytes also synthesize most plasma proteins, such as alpha and beta globulins, albumin, prothrombin, and fibrinogen.

- **Processing of drugs and hormones.** The liver can detoxify substances such as alcohol or excrete drugs such as penicillin, erythromycin, and sulfonamides into bile. It can also chemically alter or excrete thyroid hormones and steroid hormones such as estrogens and aldosterone.

- **Excretion of bilirubin.** As previously noted, bilirubin, derived from the heme of aged red blood cells, is absorbed by the liver from the blood and secreted into bile. Most of the bilirubin in bile is metabolized in the small intestine by bacteria and eliminated in feces.

- **Synthesis of bile salts.** Bile salts are used in the small intestine for the emulsification and absorption of lipids, cholesterol, phospholipids, and lipoproteins.

- **Storage.** In addition to glycogen, the liver is a prime storage site for certain vitamins (A, B_{12}, D, E, and K) and minerals (iron and copper), which are released from the liver when needed elsewhere in the body.

- **Phagocytosis.** The stellate reticuloendothelial (Kupffer) cells of the liver phagocytize aged red blood cells and white blood cells and some bacteria.

- **Activation of vitamin D.** The skin, liver, and kidneys participate in synthesizing the active form of vitamin D.

The liver functions related to metabolism are discussed more fully in Chapter 25.

Gallstones

If bile contains either insufficient bile salts or lecithin or excessive cholesterol, the cholesterol may crystallize to form **gallstones.** As they grow in size and number, gallstones may cause minimal, intermittent, or complete obstruction to the flow of bile from the gallbladder into the duodenum. Treatment consists of using gallstone-dissolving drugs, lithotripsy (shock-wave therapy), or surgery. For people with recurrent gallstones or for whom drugs or lithotripsy is not indicated, *cholecystectomy*—the removal of the gallbladder and its contents—is necessary. More than half a million cholecystectomies are performed each year in the United States. ■

▶ C H E C K P O I N T

27. Draw and label a diagram of a liver lobule.

28. Describe the pathways of blood flow into, through, and out of the liver.

29. How are the liver and gallbladder connected to the duodenum?

30. Once bile has been formed by the liver, how is it collected and transported to the gallbladder for storage?

31. What is the function of bile?

32. How is bile secretion regulated?

SUMMARY: DIGESTIVE HORMONES

▶ O B J E C T I V E

- **Describe the site of secretion and the actions of gastrin, secretin, and cholecystokinin.**

The effects of the three major digestive hormones—gastrin, secretin, and cholecystokinin—and the stimuli that promote release of each are summarized in Table 24.4. All three are secreted into the blood by enteroendocrine cells located in the GI tract mucosa. Gastrin exerts its major effects on the stomach, whereas secretin and cholecystokinin affect the pancreas, liver, and gallbladder most strongly.

Stretching of the stomach as it receives food and the buffering of gastric acids by proteins in food trigger the release of gastrin. Gastrin, in turn, promotes secretion of gastric juice and increases gastric motility so that ingested food becomes well mixed into a thick, soupy chyme. Reflux of the acidic chyme into the esophagus is prevented by tonic contraction of the lower esophageal sphincter, which is enhanced by gastrin.

The major stimulus for secretion of secretin is acidic chyme (high concentration of H^+) entering the small intestine. In turn, secretin promotes secretion of bicarbonate ions (HCO_3^-) into pancreatic juice and bile (see Figures 24.18 and 24.21). HCO_3^- acts to buffer ("soak up") excess H^+. Besides these major

Table 24.4	Major Hormones that Control Digestion	
Hormone	**Stimulus and Site of Secretion**	**Actions**
Gastrin	Distension of stomach, partially digested proteins and caffeine in stomach, and high pH of stomach chyme stimulate gastrin secretion by enteroendocrine G cells, located mainly in the mucosa of pyloric antrum of stomach.	*Major effects:* Promotes secretion of gastric juice, increases gastric motility, and promotes growth of gastric mucosa. *Minor effects:* Constricts lower esophageal sphincter; relaxes pyloric sphincter and ileocecal sphincter.
Secretin	Acidic (high H^+ level) chyme that enters the small intestine stimulates secretion of secretin by enteroendocrine S cells in the mucosa of the duodenum.	*Major effects:* Stimulates secretion of pancreatic juice and bile that are rich in HCO_3^- (bicarbonate ions). *Minor effects:* Inhibits secretion of gastric juice, promotes normal growth and maintenance of the pancreas, and enhances effects of CCK.
Cholecystokinin (CCK)	Partially digested proteins (amino acids), triglycerides and fatty acids that enter the small intestine stimulate secretion of CCK by enteroendocrine CCK cells in the mucosa of the small intestine; CCK is also released in the brain.	*Major effects:* Stimulates secretion of pancreatic juice rich in digestive enzymes, causes ejection of bile from the gallbladder and opening of the sphincter of the hepatopancreatic ampulla (sphincter of Oddi), and induces satiety (feeling full to satisfaction). *Minor effects:* Inhibits gastric emptying, promotes normal growth and maintenance of the pancreas, and enhances effects of secretin.

effects, secretin inhibits secretion of gastric juice, promotes normal growth and maintenance of the pancreas, and enhances the effects of CCK. Overall, secretin causes buffering of acid in chyme that reaches the duodenum and slows production of acid in the stomach.

Amino acids from partially digested proteins and fatty acids from partially digested triglycerides stimulate secretion of cholecystokinin by enteroendocrine cells in the mucosa of the small intestine. CCK stimulates secretion of pancreatic juice that is rich in digestive enzymes (see Figure 24.18) and ejection of bile into the duodenum (see Figure 24.21), slows gastric emptying by promoting contraction of the pyloric sphincter, and produces satiety (feeling full to satisfaction) by acting on the hypothalamus in the brain. Like secretin, CCK promotes normal growth and maintenance of the pancreas; it also enhances the effects of secretin.

Besides these three hormones, at least ten other so-called "gut hormones" are secreted by and have effects on the GI tract. They include *motilin, substance P,* and *bombesin,* which stimulate motility of the intestines; *vasoactive intestinal polypeptide (VIP),* which stimulates secretion of ions and water by the intestines and inhibits gastric acid secretion; *gastrin-releasing peptide,* which stimulates release of gastrin; and *somatostatin,* which inhibits gastrin release. Some of these hormones are thought to act as local hormones (paracrines), whereas others are secreted into the blood or even into the lumen of the GI tract. The physiological roles of these and other gut hormones are still under investigation.

▶ CHECKPOINT

33. What are the stimuli that trigger secretion of gastrin, secretin, and cholecystokinin?

SMALL INTESTINE

▶ OBJECTIVE

• **Describe the location, anatomy, histology, and functions of the small intestine.**

Now that we have examined the sources of the main digestive enzymes and hormones, we can proceed to the next part of the GI tract and pick up the story of digestion and absorption. The major events of digestion and absorption occur in a long tube called the **small intestine.** Because most digestion and absorption of nutrients occur in the small intestine, its structure is specially adapted for these functions. Its length alone provides a large surface area for digestion and absorption, and that area is further increased by circular folds, villi, and microvilli. The small intestine begins at the pyloric sphincter of the stomach, coils through the central and inferior part of the abdominal cavity, and eventually opens into the large intestine. It averages 2.5 cm (1 in.) in diameter; its length is about 3 m (10 ft) in a living person and about 6.5 m (21 ft) in a cadaver due to the loss of smooth muscle tone after death.

Anatomy of the Small Intestine

The small intestine is divided into three regions (Figure 24.22). The **duodenum** (doo-ō-DĒ-num), the shortest region, is retroperitoneal. It starts at the pyloric sphincter of the stomach and extends about 25 cm (10 in.) until it merges with the jejunum. *Duodenum* means "12"; it is so-named because it is about as long as the width of 12 fingers. The **jejunum** (je-JOO-num) is about 1 m (3 ft) long and extends to the ileum. *Jejunum* means "empty," which is how it is found at death. The final and

Figure 24.22   Anatomy of the small intestine.   (a) Regions of the small intestine are the duodenum, jejunum, and ileum. (See Tortora, *A Photographic Atlas of the Human Body,* Figure 12.12.) (b) Circular folds increase the surface area for digestion and absorption in the small intestine.

 Most digestion and absorption occur in the small intestine.

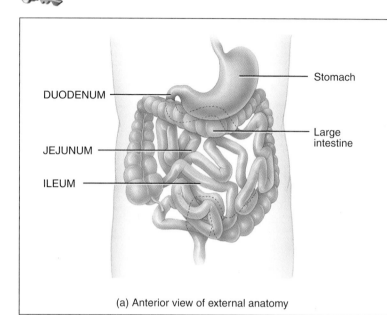

Stomach

DUODENUM

Large intestine

JEJUNUM

ILEUM

Circular folds (plicae circulares)

(a) Anterior view of external anatomy

(b) Internal anatomy of the jejunum

Which portion of the small intestine is the longest?

longest region of the small intestine, the **ileum** (IL-ē-um = twisted), measures about 2 m (6 ft) and joins the large intestine at the **ileocecal sphincter** (il′-ē-ō-SĒ-kal).

Projections called **circular folds** are 10 mm (0.4 in.) permanent ridges in the mucosa (Figure 24.22b). The circular folds begin near the proximal portion of the duodenum and end at about the midportion of the ileum; some extend all the way around the circumference of the intestine, and others extend only part of the way around. They enhance absorption by increasing surface area and causing the chyme to spiral, rather than move in a straight line, as it passes through the small intestine.

Histology of the Small Intestine

Even though the wall of the small intestine is composed of the same four coats that make up most of the GI tract, special features of both the mucosa and the submucosa facilitate the process of digestion and absorption. The mucosa forms a series of fingerlike **villi** (= tufts of hair), projections that are 0.5–1 mm long (Figure 24.23a). The large number of villi (20–40 per square millimeter) vastly increases the surface area of the epithelium available for absorption and digestion and gives the intestinal mucosa a velvety appearance. Each villus (singular form) has a core of lamina propria (areolar connective tissue); embedded in this connective tissue are an arteriole, a venule, a blood capillary network, and a **lacteal** (LAK-tē-al = milky), which is a lymphatic capillary. Nutrients absorbed by

the epithelial cells covering the villus pass through the wall of a capillary or a lacteal to enter blood or lymph, respectively.

The epithelium of the mucosa consists of simple columnar epithelium that contains absorptive cells, goblet cells, enteroendocrine cells, and Paneth cells (Figure 24.23b). The apical (free) membrane of absorptive cells features **microvilli** (mī′-krō-VIL-ī; *micro-* = small); each microvillus is a 1 μm-long cylindrical, membrane-covered projection that contains a bundle of 20–30 actin filaments. In a photomicrograph taken through a light microscope, the microvilli are too small to be seen individually; instead they form a fuzzy line, called the **brush border,** extending into the lumen of the small intestine (see Figure 24.24d). There are an estimated 200 million microvilli per square millimeter of small intestine. Because the microvilli greatly increase the surface area of the plasma membrane, larger amounts of digested nutrients can diffuse into absorptive cells in a given period. The brush border also contains several brush-border enzymes that have digestive functions (discussed shortly). Goblet cells secrete mucus.

The mucosa contains many deep crevices lined with glandular epithelium. Cells lining the crevices form the **intestinal glands (crypts of Lieberkühn)** and secrete intestinal juice. Many of the epithelial cells in the mucosa are goblet cells, which secrete mucus. **Paneth cells,** found in the deepest parts of the intestinal glands, secrete lysozyme, a bactericidal enzyme, and are capable of phagocytosis. They may have a role in regulating the microbial population in the intestines. Three types of

Figure 24.23 Histology of the small intestine.

Circular folds, villi and microvilli increase the surface area of the small intestine for digestion and absorption.

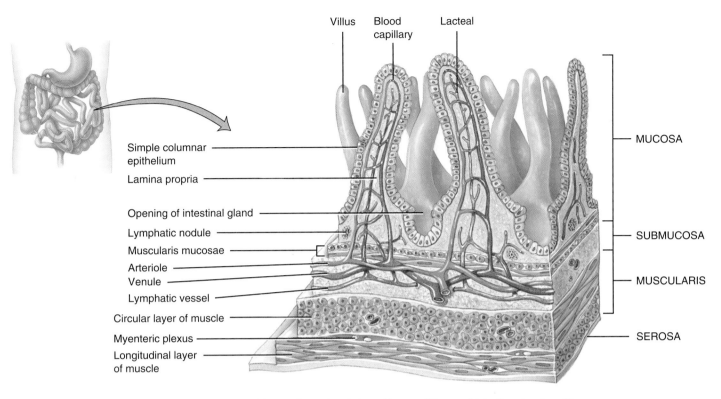

Lumen of small intestine

Villus
Blood capillary
Lacteal

Simple columnar epithelium
Lamina propria

Opening of intestinal gland
Lymphatic nodule
Muscularis mucosae
Arteriole
Venule
Lymphatic vessel
Circular layer of muscle
Myenteric plexus
Longitudinal layer of muscle

MUCOSA

SUBMUCOSA

MUSCULARIS

SEROSA

(a) Three-dimensional view of layers of the small intestine showing villi

Microvilli

Absorptive cell (digests and absorbs nutrients)

Blood capillary
Lacteal
Lamina propria

Mucosa

Goblet cell (secretes mucus)

Intestinal gland

Enteroendocrine cell (secretes the hormones secretin, cholecystokinin, or GIP)

Muscularis mucosae
Arteriole
Venule
Lymphatic vessel

Submucosa

Paneth cell (secretes lysozyme and is capable of phagocytosis)

(b) Enlarged villus showing lacteal, capillaries, intestinal glands, and cell types

What is the functional significance of the blood capillary network and lacteal in the center of each villus?

enteroendocrine cells, also in the deepest part of the intestinal glands, secrete hormones: secretin (by S cells), cholecystokinin (by CCK cells), and glucose-dependent insulinotropic peptide (by K cells). The lamina propria of the small intestine has an abundance of mucosa-associated lymphoid tissue (MALT). **Solitary lymphatic nodules** are most numerous in the distal part of the ileum (see Figure 24.24c). Groups of lymphatic nodules referred to as **aggregated lymphatic follicles (Peyer's patches)** are also present in the ileum. The muscularis mucosae consists of smooth muscle. The submucosa of the duodenum contains **duodenal (Brunner's) glands** (Figure 24.24a), which secrete an alkaline mucus that helps neutralize gastric acid in the chyme. Sometimes the lymphatic tissue of the lamina propria extends through the muscularis mucosae into the submucosa.

The **muscularis** of the small intestine consists of two layers of smooth muscle. The outer, thinner layer contains longitudinal fibers; the inner, thicker layer contains circular fibers. Except for a major portion of the duodenum, the serosa (or visceral peritoneum) completely surrounds the small intestine.

Role of Intestinal Juice and Brush-Border Enzymes

Intestinal juice is a clear yellow fluid secreted in amounts of 1–2 liters (about 1–2 qt) a day. It contains water and mucus and is slightly alkaline (pH 7.6). Together, pancreatic and intestinal juices provide a liquid medium that aids the absorption of substances from chyme as they come in contact with the microvilli. The absorptive epithelial cells synthesize several digestive enzymes, called **brush-border enzymes,** and insert them in the plasma membrane of the microvilli. Thus, some enzymatic digestion occurs at the surface of the epithelial cells that line the villi, rather than in the lumen exclusively, as occurs in other parts of the GI tract. Among the brush-border enzymes are four carbohydrate-digesting enzymes called α-dextrinase, maltase, sucrase, and lactase; protein-digesting enzymes called peptidases (aminopeptidase and dipeptidase); and two types of nucleotide-digesting enzymes, nucleosidases and phosphatases. Also, as cells slough off into the lumen of the small intestine, they break apart and release enzymes that help digest nutrients in the chyme.

Mechanical Digestion in the Small Intestine

The two types of movements of the small intestine—segmentations and a type of peristalsis called migrating motility complexes—are governed mainly by the myenteric plexus. **Segmentations** are localized, mixing contractions that occur in portions of intestine distended by a large volume of chyme. Segmentations mix chyme with the digestive juices and bring the particles of food into contact with the mucosa for absorption; they do not push the intestinal contents along the tract. A seg-

Figure 24.24 Histology of the duodenum and ileum.

Microvilli in the small intestine contain several brush-border enzymes that help digest nutrients.

LM 45x

(a) Wall of the duodenum

mentation starts with the contractions of circular muscle fibers in a portion of the small intestine, an action that constricts the intestine into segments. Next, muscle fibers that encircle the middle of each segment also contract, dividing each segment again. Finally, the fibers that first contracted relax, and each small segment unites with an adjoining small segment so that large segments are formed again. As this sequence of events repeats, the chyme sloshes back and forth. Segmentations occur most rapidly in the duodenum, about 12 times per minute, and progressively slow to about 8 times per minute in the ileum. This movement is similar to alternately squeezing the middle and then the ends of a capped tube of toothpaste.

Lumen of duodenum

Brush border

Simple columnar epithelium

Goblet cell

Absorptive cell

Lamina propria

Intestinal glands (crypts of Lieberkühn)

Muscularis mucosae

Duodenal (Brunner's) gland in submucosa

LM 160x

(b) Three villi from the duodenum of the small intestine

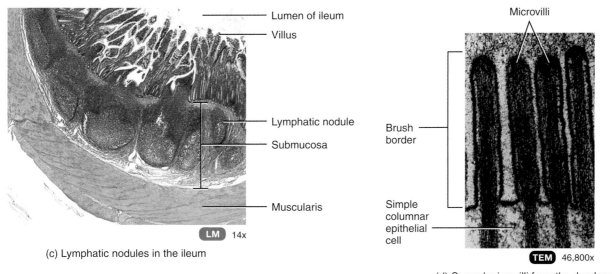

Lumen of ileum

Villus

Lymphatic nodule

Submucosa

Muscularis

LM 14x

(c) Lymphatic nodules in the ileum

Microvilli

Brush border

Simple columnar epithelial cell

TEM 46,800x

(d) Several microvilli from the duodenum

 What is the function of the fluid secreted by duodenal (Brunner's) glands?

After most of a meal has been absorbed, which lessens distention of the wall of the small intestine, segmentation stops and peristalsis begins. The type of peristalsis that occurs in the small intestine, termed a **migrating motility complex (MMC),** begins in the lower portion of the stomach and pushes chyme forward along a short stretch of small intestine before dying out. The MMC slowly migrates down the small intestine, reaching the end of the ileum in 90–120 minutes. Then another MMC begins in the stomach. Altogether, chyme remains in the small intestine for 3–5 hours.

Chemical Digestion in the Small Intestine

In the mouth, salivary amylase converts starch (a polysaccharide) to maltose (a disaccharide), maltotriose (a trisaccharide), and α-dextrins (short-chain, branched fragments of starch with five to ten glucose units). In the stomach, pepsin converts proteins to peptides (small fragments of proteins), and lingual and gastric lipases convert some triglycerides into fatty acids, diglycerides, and monoglycerides. Thus, chyme entering the small intestine contains partially digested carbohydrates, proteins, and lipids. The completion of the digestion of carbohydrates, proteins, and lipids is a collective effort of pancreatic juice, bile, and intestinal juice in the small intestine.

Digestion of Carbohydrates

Even though the action of **salivary amylase** may continue in the stomach for a while, the acidic pH of the stomach destroys salivary amylase and ends its activity. Thus, only a few starches are reduced to maltose by the time chyme leaves the stomach. Those starches not already broken down into maltose, maltotriose, and α-dextrins are cleaved by **pancreatic amylase,** an enzyme in pancreatic juice that acts in the small intestine. Although amylase acts on both glycogen and starches, it does not act on another polysaccharide called cellulose, which is an indigestible plant fiber. After amylase (either salivary or pancreatic) has split starch into smaller fragments, a brush-border enzyme called α-**dextrinase** acts on the resulting α-dextrins, clipping off one glucose unit at a time.

Ingested molecules of sucrose, lactose, and maltose—three disaccharides—are not acted on until they reach the small intestine. Three brush-border enzymes digest the disaccharides into monosaccharides. **Sucrase** breaks sucrose into a molecule of glucose and a molecule of fructose; **lactase** digests lactose into a molecule of glucose and a molecule of galactose; and **maltase** splits maltose and maltotriose into two or three molecules of glucose, respectively. Digestion of carbohydrates ends with the production of monosaccharides, as mechanisms exist for their absorption.

Lactose Intolerance

In some people the mucosal cells of the small intestine fail to produce enough lactase, which is essential for the digestion of lactose. This results in a condition called **lactose intolerance,** in which undigested lactose in chyme retains fluid in the feces, and bacterial fermentation of lactose results in the production of gases. Symptoms of lactose intolerance include diarrhea, gas, bloating, and abdominal cramps after consumption of milk and other dairy products. The severity of symptoms varies from relatively minor to sufficiently serious to require medical attention. Persons with lactose intolerance can take dietary supplements to aid in the digestion of lactose. ■

Digestion of Proteins

Protein digestion starts in the stomach, where proteins are fragmented into peptides by the action of **pepsin.** Enzymes in pancreatic juice—**trypsin, chymotrypsin, carboxypeptidase,** and **elastase**—continue to break down proteins into peptides. Although all these enzymes convert whole proteins into peptides, their actions differ somewhat because each splits peptide bonds between different amino acids. Trypsin, chymotrypsin, and elastase all cleave the peptide bond between a specific amino acid and its neighbor; carboxypeptidase breaks the peptide bond that attaches the terminal amino acid to the carboxyl (acid) end of the peptide. Protein digestion is completed by two **peptidases** in the brush border: aminopeptidase and dipeptidase. **Aminopeptidase** acts on peptides by breaking the peptide bond that attaches the terminal amino acid to the amino end of the peptide. **Dipeptidase** splits dipeptides (two amino acids joined by a peptide bond) into single amino acids.

Digestion of Lipids

The most abundant lipids in the diet are triglycerides, which consist of a molecule of glycerol bonded to three fatty acid molecules (see Figure 2.17 on page 46). Enzymes that split triglycerides and phospholipids are called **lipases.** In adults, most lipid digestion occurs in the small intestine, although some occurs in the stomach through the action of **lingual** and **gastric lipases.** When chyme enters the small intestine, bile salts emulsify the globules of triglycerides into droplets about 1 μm in diameter, which increases the surface area exposed to **pancreatic lipase,** another enzyme in pancreatic juice. This enzyme hydrolyzes triglycerides into fatty acids and monoglycerides, the main end products of triglyceride digestion. Pancreatic and gastric lipases remove two of the three fatty acids from glycerol; the third remains attached to the glycerol, forming a monoglyceride.

Digestion of Nucleic Acids

Pancreatic juice contains two nucleases: **ribonuclease,** which digests RNA, and **deoxyribonuclease,** which digests DNA. The nucleotides that result from the action of the two nucleases are further digested by brush-border enzymes called **nucleosidases** and **phosphatases** into pentoses, phosphates, and nitrogenous bases. These products are absorbed via active transport.

Table 24.5 summarizes the sources, substrates, and products of the digestive enzymes.

Table 24.5   Summary of Digestive Enzymes

Enzyme	Source	Substrates	Products
Saliva			
Salivary amylase	Salivary glands.	Starches (polysaccharides).	Maltose (disaccharide), maltotriose (trisaccharide), and α-dextrins.
Lingual lipase	Glands in the tongue.	Triglycerides (fats and oils) and other lipids.	Fatty acids and diglycerides.
Gastric Juice			
Pepsin (activated from pepsinogen by pepsin and hydrochloric acid)	Stomach chief (zymogenic) cells.	Proteins.	Peptides.
Gastric lipase	Stomach chief (zymogenic) cells.	Short-chain triglycerides (fats and oils) in fat molecules in milk.	Fatty acids and monoglycerides.
Pancreatic Juice			
Pancreatic amylase	Pancreatic acinar cells.	Starches (polysaccharides).	Maltose (disaccharide), maltotriose (trisaccharide), and α-dextrins.
Trypsin (activated from trypsinogen by enterokinase)	Pancreatic acinar cells.	Proteins.	Peptides.
Chymotrypsin (activated from chymotrypsinogen by trypsin)	Pancreatic acinar cells.	Proteins.	Peptides.
Elastase (activated from proelastase by trypsin)	Pancreatic acinar cells.	Proteins.	Peptides.
Carboxypeptidase (activated from procarboxypeptidase by trypsin)	Pancreatic acinar cells.	Terminal amino acid at carboxyl (acid) end of peptides.	Peptides and amino acids.
Pancreatic lipase	Pancreatic acinar cells.	Triglycerides (fats and oils) that have been emulsified by bile salts.	Fatty acids and monoglycerides.
Nucleases			
Ribonuclease	Pancreatic acinar cells.	Ribonucleic acid.	Nucleotides.
Deoxyribonuclease	Pancreatic acinar cells.	Deoxyribonucleic acid.	Nucleotides.
Brush Border			
α-Dextrinase	Small intestine.	α-Dextrins.	Glucose.
Maltase	Small intestine.	Maltose.	Glucose.
Sucrase	Small intestine.	Sucrose.	Glucose and fructose.
Lactase	Small intestine.	Lactose.	Glucose and galactose.
Enterokinase	Small intestine.	Trypsinogen.	Trypsin.
Peptidases			
Aminopeptidase	Small intestine.	Terminal amino acid at amino end of peptides.	Peptides and amino acids.
Dipeptidase	Small intestine.	Dipeptides.	Amino acids.
Nucleosidases and phosphatases	Small intestine.	Nucleotides.	Nitrogenous bases, pentoses, and phosphates.

Regulation of Intestinal Secretion and Motility

The most important mechanisms that regulate small intestinal secretion and motility are enteric reflexes that respond to the presence of chyme; vasoactive intestinal polypeptide (VIP) also stimulates the production of intestinal juice. Segmentation move-ments depend mainly on intestinal distention, which initiates nerve impulses to the enteric plexuses and the central nervous system. Enteric reflexes and returning parasympathetic impulses from the CNS increase motility; sympathetic impulses decrease intestinal motility. Migrating motility complexes strengthen when most nutrients and water have been absorbed—that is, when the walls of the small intestine are less distended. With more vigorous peristalsis, the chyme moves along toward the large intestine as fast as 10 cm/sec. The first remnants of a meal reach the beginning of the large intestine in about 4 hours.

Absorption in the Small Intestine

All the chemical and mechanical phases of digestion from the mouth through the small intestine are directed toward changing food into forms that can pass through the epithelial cells lining the mucosa and into the underlying blood and lymphatic vessels. These forms are monosaccharides (glucose, fructose, and galactose) from carbohydrates; single amino acids, dipeptides, and tripeptides from proteins; and fatty acids, glycerol, and monoglycerides from triglycerides. Passage of these digested nutrients from the gastrointestinal tract into the blood or lymph is called **absorption.**

Absorption of materials occurs via diffusion, facilitated diffusion, osmosis, and active transport. About 90% of all absorption of nutrients occurs in the small intestine; the other 10% occurs in the stomach and large intestine. Any undigested or unabsorbed material left in the small intestine passes on to the large intestine.

Absorption of Monosaccharides

All carbohydrates are absorbed as monosaccharides. The capacity of the small intestine to absorb monosaccharides is huge—an estimated 120 grams per hour. As a result, all dietary carbohydrates that are digested normally are absorbed, leaving only indigestible cellulose and fibers in the feces. Monosaccharides pass from the lumen through the apical membrane via *facilitated diffusion* or *active transport*. Fructose, a monosaccharide found in fruits, is transported via *facilitated diffusion;* glucose and galactose are transported into epithelial cells of the villi via *secondary active transport* that is coupled to the active transport of Na^+ (Figure 24.25a). The transporter has binding sites for one glucose molecule and two sodium ions; unless all three sites are filled, neither substance is transported. Galactose competes with glucose to ride the same transporter. (Because both Na^+ and glucose or galactose move in the same direction, this is a *symporter*. The same type of Na^+–glucose symporter reabsorbs filtered blood glucose in the tubules of the kidneys; see Figure 26.12 on

Figure 24.25 Absorption of digested nutrients in the small intestine. For simplicity, all digested foods are shown in the lumen of the small intestine, even though some nutrients are digested by brush-border enzymes.

🔑 **Long-chain fatty acids and monoglycerides are absorbed into lacteals; other products of digestion enter blood capillaries.**

(a) Mechanisms for movement of nutrients through epithelial cells of the villi page 967.) Monosaccharides then move out of the epithelial cells through their basolateral surfaces via *facilitated diffusion* and enter the capillaries of the villi (see Figure 24.24a, b).

Absorption of Amino Acids, Dipeptides, and Tripeptides

Most proteins are absorbed as amino acids via *active transport* processes that occur mainly in the duodenum and jejunum. About half of the absorbed amino acids are present in food, whereas the other half come from proteins in digestive juices and dead cells that slough off the mucosal surface! Normally, 95–98% of the protein present in the small intestine is digested and absorbed. Several transporters carry different types of amino acids. Some amino acids enter epithelial cells of the villi via Na^+-dependent secondary active transport processes that are similar to the glucose transporter; other amino acids are actively transported by themselves. At least one symporter brings in dipeptides and tripeptides together with H^+; the peptides then are hydrolyzed to single amino acids inside the epithelial cells. Amino acids move out of the epithelial cells via diffusion and enter capillaries of the villus (Figure 24.25a, b). Both monosaccharides and amino acids are transported in the blood to the liver by way of the hepatic portal system. If not removed by hepatocytes, they enter the general circulation.

Absorption of Lipids

All dietary lipids are absorbed via *simple diffusion.* Adults absorb about 95% of the lipids present in the small intestine; due to their lower production of bile, newborn infants absorb only about 85% of lipids. As a result of their emulsification and digestion, triglycerides are broken down into monoglycerides and fatty acids. Recall that lingual and pancreatic lipases remove two of the three fatty acids from glycerol during digestion of a triglyceride; the other fatty acid remains attached to glycerol, thus forming a monoglyceride. The small amount of short-chain fatty acids (having fewer than 10–12 carbon atoms) in the diet passes into the epithelial cells via simple diffusion and follows the same route taken by monosaccharides and amino acids into a blood capillary of a villus (Figure 24.25a, b).

Most dietary fatty acids, however, are long-chain fatty acids. They and monoglycerides reach the bloodstream by a different route and require bile for adequate absorption. Bile salts are amphipathic; they have both polar (hydrophilic) and nonpolar (hydrophobic) portions. Thus, they can form tiny spheres called **micelles** (mī-SELZ = small morsels), which are 2–10 nm in diameter and include 20–50 bile salt molecules. Because they are small and have the polar portions of bile salt molecules at their surface, micelles can dissolve in the water of intestinal fluid. In contrast, partially digested dietary lipids can dissolve in the nonpolar central core of micelles. It is in this form that fatty acids and monoglycerides reach the epithelial cells of the villi.

At the apical surface of the epithelial cells, fatty acids and monoglycerides diffuse into the cells, leaving the micelles behind in chyme. The micelles continually repeat this ferrying function. When chyme reaches the ileum, 90–95% of the bile salts are reabsorbed and returned by the blood to the liver

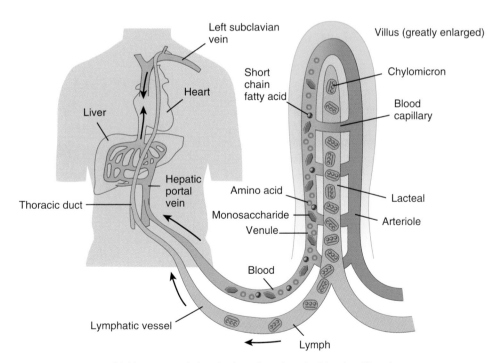

(b) Movement of absorbed nutrients into the blood and lymph

 A monoglyceride may be larger than an amino acid. Why can monoglycerides be absorbed by simple diffusion, whereas amino acids cannot?

through the hepatic portal system for recycling. This cycle of bile salt secretion by hepatocytes into bile, reabsorption by the ileum, and resecretion into bile is called the **enterohepatic circulation.** Insufficient bile salts, due either to obstruction of the bile ducts or removal of the gallbladder, can result in the loss of up to 40% of dietary lipids in feces due to diminished lipid absorption. Moreover, when lipids are not absorbed properly, the fat-soluble vitamins—A, D, E, and K—are not adequately absorbed.

Within the epithelial cells, many monoglycerides are further digested by lipase to glycerol and fatty acids. The fatty acids and glycerol are then recombined to form triglycerides, which aggregate into globules along with phospholipids and cholesterol and become coated with proteins. These large (about 80 nm in diameter) spherical masses are called **chylomicrons.** The hydrophilic protein coat keeps the chylomicrons suspended and prevents them from sticking to each other. Chylomicrons leave the epithelial cell via exocytosis. Because they are so large and bulky, chylomicrons cannot enter blood capillaries in the small intestine; instead, they enter the much leakier lacteals. From there they are transported by way of lymphatic vessels to the thoracic duct and enter the blood at the left subclavian vein (see Figure 24.25b).

Within 10 minutes after their absorption, about half of the chylomicrons have already been removed from the blood as they pass through blood capillaries in the liver and adipose tissue. This removal is accomplished by an enzyme in capillary endothelial cells, called **lipoprotein lipase,** that breaks down triglycerides in chylomicrons and other lipoproteins into fatty acids and glycerol. The fatty acids diffuse into hepatocytes and adipose cells and combine with glycerol during resynthesis of triglycerides. Two or three hours after a meal, few chylomicrons remain in the blood.

Absorption of Electrolytes

Many of the electrolytes absorbed by the small intestine come from gastrointestinal secretions, and some are part of ingested foods and liquids. Sodium ions are actively transported out of intestinal epithelial cells by sodium-potassium pumps (Na^+/K^+ ATPase) after they have moved into epithelial cells via diffusion and secondary active transport. Thus, most of the sodium ions in gastrointestinal secretions are reclaimed and not lost in the feces. Negatively charged bicarbonate, chloride, iodide, and nitrate ions can passively follow Na^+ or be actively transported. Calcium ions also are absorbed actively in a process stimulated by calcitriol. Other electrolytes such as iron, potassium, magnesium, and phosphate ions also are absorbed via active transport mechanisms.

Absorption of Vitamins

The fat-soluble vitamins A, D, E, and K are included with ingested dietary lipids in micelles and are absorbed via simple diffusion. Most water-soluble vitamins, such as most B vitamins and vitamin C, also are absorbed via simple diffusion. Vitamin B_{12}, however, combines with intrinsic factor produced by the stomach, and the combination is absorbed in the ileum via an active transport mechanism.

Absorption of Water

The total volume of fluid that enters the small intestine each day—about 9.3 liters (9.8 qt)—comes from ingestion of liquids (about 2.3 liters) and from various gastrointestinal secretions (about 7.0 liters). Figure 24.26 depicts the amounts of fluid ingested, secreted, absorbed, and excreted by the GI tract. The small intestine absorbs about 8.3 liters of the fluid; the remainder passes into the large intestine, where most of the rest of it—about 0.9 liter—is also absorbed. Only 0.1 liter (100 mL) of water is excreted in the feces each day.

All water absorption in the GI tract occurs via *osmosis* from the lumen of the intestines through epithelial cells and into blood capillaries. Because water can move across the intestinal mucosa in both directions, the absorption of water from the small intes-

Figure 24.26 Daily volumes of fluid ingested, secreted, absorbed, and excreted from the GI tract.

All water absorption in the GI tract occurs via osmosis.

INGESTED AND SECRETED

ABSORBED

Saliva (1 liter)

Ingestion of liquids (2.3 liters)

Gastric juice (2 liters)

Bile (1 liter)

Pancreatic juice (2 liters)

Intestinal juice (1 liter)

Small intestine (8.3 liters)

Large intestine (0.9 liters)

Total ingested and secreted = 9.3 liters

Total absorbed = 9.2 liters

Excreted in feces (0.1 liter)

Fluid balance in GI tract

Which two organs of the digestive system secrete the most fluid?

tine depends on the absorption of electrolytes and nutrients to maintain an osmotic balance with the blood. The absorbed electrolytes, monosaccharides, and amino acids establish a concentration gradient for water that promotes water absorption via osmosis.

Table 24.6 summarizes the digestive activities of the pancreas, liver, gallbladder, and small intestine.

Table 24.6	Summary of Digestive Activities in the Pancreas, Liver, Gallbladder, and Small Intestine
Structure	Activity
Pancreas	Delivers pancreatic juice into the duodenum via the pancreatic duct (see Table 24.5 for pancreatic enzymes and their functions).
Liver	Produces bile (bile salts) necessary for emulsification and absorption of lipids.
Gallbladder	Stores, concentrates, and delivers bile into the duodenum via the common bile duct.
Small Intestine	Major site of digestion and absorption of nutrients and water in the gastrointestinal tract.
Mucosa/ submucosa	
Intestinal glands	Secrete intestinal juice.
Duodenal (Brunner's) glands	Secrete alkaline fluid to buffer stomach acids, and mucus for protection and lubrication.
Microvilli	Microscopic, membrane-covered projections of epithelial cells that contain brush-border enzymes (listed in Table 24.5) and that increase the surface area for digestion and absorption.
Villi	Fingerlike projections of mucosa that are the sites of absorption of digested food and that increase the surface for digestion and absorption.
Circular folds	Folds of mucosa and submucosa that increase the surface area for digestion and absorption.
Muscularis	
Segmentation	Consists of alternating contractions of circular smooth muscle fibers that produce segmentation and resegmentation of sections of the small intestine; mixes chyme with digestive juices and brings food into contact with the mucosa for absorption.
Migrating motility complex (MMC)	A type of peristalsis consisting of waves of contraction and relaxation of circular and longitudinal smooth muscle fibers passing the length of the small intestine; moves chyme toward ileocecal sphincter.

Absorption of Alcohol

The intoxicating and incapacitating effects of alcohol depend on the blood alcohol level. Because it is lipid soluble, alcohol begins to be absorbed in the stomach. However, the surface area available for absorption is much greater in the small intestine than in the stomach, so when alcohol passes into the duodenum, it is absorbed more rapidly. Thus, the longer the alcohol remains in the stomach, the more slowly blood alcohol level rises. Because fatty acids in chyme slow gastric emptying, blood alcohol level will rise more slowly when fat-rich foods, such as pizza, hamburgers, or nachos, are consumed with alcoholic beverages. Also, the enzyme alcohol dehydrogenase, which is present in gastric mucosa cells, breaks down some of the alcohol to acetaldehyde, which is not intoxicating. When the rate of gastric emptying is slower, proportionally more alcohol will be absorbed and converted to acetaldehyde in the stomach, and thus less alcohol will reach the bloodstream. Given identical consumption of alcohol, females often develop higher blood alcohol levels (and therefore experience greater intoxication) than males of comparable size because the activity of gastric alcohol dehydrogenase is up to 60% lower in females than in males. Asian males may also have lower levels of this gastric enzyme. ■

► CHECKPOINT

34. What are the regions of the small intestine?

35. In what ways are the mucosa and submucosa of the small intestine adapted for digestion and absorption?

36. Describe the types of movement that occur in the small intestine.

37. Explain the function of each digestive enzyme.

38. How is small intestinal secretion regulated?

39. Define absorption. How are the end products of carbohydrate and protein digestion absorbed? How are the end products of lipid digestion absorbed?

40. By what routes do absorbed nutrients reach the liver?

41. Describe the absorption of electrolytes, vitamins, and water by the small intestine.

LARGE INTESTINE

► OBJECTIVE

• Describe the anatomy, histology, and functions of the large intestine.

The large intestine is the terminal portion of the GI tract and is divided into four principal regions. The overall functions of the large intestine are the completion of absorption, the production of certain vitamins, the formation of feces, and the expulsion of feces from the body.

Anatomy of the Large Intestine

The **large intestine,** which is about 1.5 m (5 ft) long and 6.5 cm (2.5 in.) in diameter, extends from the ileum to the anus. It is attached to the posterior abdominal wall by its **mesocolon,** which is a double layer of peritoneum. Structurally, the four major regions of the large intestine are the cecum, colon, rectum, and anal canal (Figure 24.27a).

The opening from the ileum into the large intestine is guarded by a fold of mucous membrane called the **ileocecal sphincter (valve),** which allows materials from the small intestine to pass into the large intestine. Hanging inferior to the ileocecal valve is the **cecum,** a blind pouch about 6 cm (2.4 in.) long. Attached to the cecum is a twisted, coiled tube, measuring about 8 cm (3 in.) in length, called the **appendix** or **vermiform appendix** (*vermiform* = worm-shaped; *appendix* = appendage). The mesentery of the appendix, called the **mesoappendix,** attaches the appendix to the inferior part of the mesentery of the ileum.

The open end of the cecum merges with a long tube called the **colon** (= food passage), which is divided into ascending, transverse, descending, and sigmoid portions. Both the ascending and descending colon are retroperitoneal, whereas the trans-

verse and sigmoid colon are not. The **ascending colon** ascends on the right side of the abdomen, reaches the inferior surface of the liver, and turns abruptly to the left to form the **right colic (hepatic) flexure.** The colon continues across the abdomen to the left side as the **transverse colon.** It curves beneath the inferior end of the spleen on the left side as the **left colic (splenic) flexure** and passes inferiorly to the level of the iliac crest as the **descending colon.** The **sigmoid colon** (*sigm-* = S-shaped) begins near the left iliac crest, projects medially to the midline, and terminates as the rectum at about the level of the third sacral vertebra.

The **rectum,** the last 20 cm (8 in.) of the GI tract, lies anterior to the sacrum and coccyx. The terminal 2–3 cm (1 in.) of the rectum is called the **anal canal** (Figure 24.27b). The mucous membrane of the anal canal is arranged in longitudinal folds called **anal columns** that contain a network of arteries and veins. The opening of the anal canal to the exterior, called the **anus,** is guarded by an **internal anal sphincter** of smooth muscle (involuntary) and an

---

Functions of the Large Intestine
1. Haustral churning, peristalsis, and mass peristalsis drive the contents of the colon into the rectum.
2. Bacteria in the large intestine produce some B vitamins and vitamin K.
3. Absorption of some water, ions, and vitamins.
4. Defecation (emptying of the rectum).

Figure 24.27 Anatomy of the large intestine. (See Tortora, *A Photographic Atlas of the Human Body,* Figure 12.13.)

The regions of the large intestine are the cecum, colon, rectum, and anal canal.

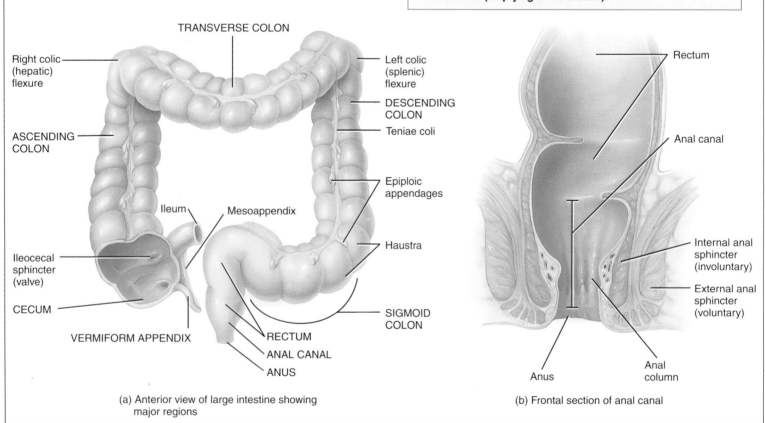

(a) Anterior view of large intestine showing major regions

(b) Frontal section of anal canal

Which portions of the colon are retroperitoneal?

external anal sphincter of skeletal muscle (voluntary). Normally these sphincters keep the anus closed except during the elimination of feces.

Appendicitis

Inflammation of the appendix, termed **appendicitis,** is preceded by obstruction of the lumen of the appendix by chyme, inflammation, a foreign body, a carcinoma of the cecum, stenosis, or kinking of the organ. It is characterized by high fever, elevated white blood cell count, and a neutrophil count higher than 75%. The infection that follows may result in edema and ischemia and may progress to gangrene and perforation within 24 hours. Typically, appendicitis begins with referred pain in the umbilical region of the abdomen, followed by anorexia (loss of appetite), nausea, and vomiting. After several hours the pain localizes in the right lower quadrant (RLQ) and is continuous, dull or severe, and intensified by coughing, sneezing, or body movements. Early appendectomy (removal of the appendix) is recommended because it is safer to operate than to risk rupture, peritonitis, and gangrene. ■

Histology of the Large Intestine

The wall of the large intestine differs from that of the small intestine in several respects. No villi or permanent circular folds are found in the **mucosa,** which consists of simple columnar epithelium, lamina propria (areolar connective tissue), and muscularis mucosae (smooth muscle) (Figure 24.28a). The epithelium contains mostly absorptive and goblet cells (Figure 24.28b and c). The absorptive cells function primarily in water absorption, whereas the goblet cells secrete mucus that lubricates the passage of the colonic contents. Both absorptive and goblet cells are located in long, straight, tubular intestinal glands that extend the full thickness of the mucosa. Solitary lymphatic nodules are also found in the lamina propria of the mucosa and may extend through the muscularis mucosae into the submucosa. The **submucosa** of the large intestine is similar to that found in the rest of the GI tract. The **muscularis** consists of an external layer of longitudinal smooth muscle and an internal layer of circular smooth muscle. Unlike other parts of the GI tract, portions of the longitudinal muscles are thickened, forming three conspicuous longitudinal bands called the **teniae coli** (TĒ-nē-ē KŌ-lī; *teniae* = flat bands), that run most of the length of the large intestine (see Figure 24.27a). The teniae coli are separated by portions of the wall with less or no longitudinal muscle. Tonic contractions of the bands gather the colon into a series of pouches called **haustra** (HAWS-tra = shaped like pouches; singular is **haustrum**), which give the colon a puckered appearance. A single layer of circular smooth muscle lies between teniae coli. The **serosa** of the large intestine is part of the visceral peritoneum. Small pouches of visceral peritoneum filled with fat are attached to teniae coli and are called **epiploic appendages.**

Figure 24.28 **Histology of the large intestine.**

Intestinal glands formed by simple columnar epithelial cells and goblet cells extend the full thickness of the mucosa.

Lumen of large intestine

Openings of intestinal glands

Simple columnar epithelium

Lamina propria

Lymphatic nodule

Muscularis mucosae
Lymphatic vessel
Arteriole
Venule
Circular layer of muscle
Myenteric plexus
Longitudinal layer of muscle

MUCOSA

SUBMUCOSA

MUSCULARIS

SEROSA

(a) Three-dimensional view of layers of the large intestine

(continues)

Figure 24.28   *(continued)*

Openings of intestinal glands

Lamina propria

Microvilli

Absorptive cell (absorbs water)

Intestinal gland

Goblet cell (secretes mucus)

Lymphatic nodule

Muscularis mucosae

Submucosa

(b) Sectional view of intestinal glands and cell types

Lumen of large intestine

Lamina propria

Intestinal gland

Mucosa

Muscularis mucosae

Lymphatic nodule

Submucosa

Muscularis

Serosa

LM 315x

(c) Portion of the wall of the large intestine

Opening of intestinal gland

Lumen of large intestine

Goblet cell

Intestinal gland

Lamina propria

LM 300x

(d) Details of mucosa of large intestine

What is the function of the goblet cells in the large intestine?

Mechanical Digestion in the Large Intestine

The passage of chyme from the ileum into the cecum is regulated by the action of the ileocecal sphincter. Normally, the valve remains partially closed so that the passage of chyme into the cecum usually occurs slowly. Immediately after a meal, a **gastroileal reflex** intensifies ileal peristalsis and forces any chyme in the ileum into the cecum. The hormone gastrin also relaxes the sphincter. Whenever the cecum is distended, the degree of contraction of the ileocecal sphincter intensifies.

Movements of the colon begin when substances pass the ileocecal sphincter. Because chyme moves through the small intestine at a fairly constant rate, the time required for a meal to pass into the colon is determined by gastric emptying time. As food passes through the ileocecal sphincter, it fills the cecum and accumulates in the ascending colon.

One movement characteristic of the large intestine is **haustral churning.** In this process, the haustra remain relaxed and become distended while they fill up. When the distension reaches a certain point, the walls contract and squeeze the contents into the next haustrum. **Peristalsis** also occurs, although at a slower rate (3–12 contractions per minute) than in more proximal portions of the tract. A final type of movement is **mass peristalsis,** a strong peristaltic wave that begins at about the middle of the transverse colon and quickly drives the contents of the colon into the rectum. Because food in the stomach initiates this **gastrocolic reflex** in the colon, mass peristalsis usually takes place three or four times a day, during or immediately after a meal.

Chemical Digestion in the Large Intestine

The final stage of digestion occurs in the colon through the activity of bacteria that inhabit the lumen. Mucus is secreted by the glands of the large intestine, but no enzymes are secreted. Chyme is prepared for elimination by the action of bacteria, which ferment any remaining carbohydrates and release hydrogen, carbon dioxide, and methane gases. These gases contribute to flatus (gas) in the colon, termed *flatulence* when it is excessive. Bacteria also convert any remaining proteins to amino acids and break down the amino acids into simpler substances: indole, skatole, hydrogen sulfide, and fatty acids. Some of the indole and skatole is eliminated in the feces and contributes to their odor; the rest is absorbed and transported to the liver, where these compounds are converted to less toxic compounds and excreted in the urine. Bacteria also decompose bilirubin to simpler pigments, including stercobilin, which give feces their brown color. Several vitamins needed for normal metabolism, including some B vitamins and vitamin K, are bacterial products that are absorbed in the colon.

Absorption and Feces Formation in the Large Intestine

By the time chyme has remained in the large intestine 3–10 hours, it has become solid or semisolid because of water absorption and is now called **feces.** Chemically, feces consist of water, inorganic salts, sloughed-off epithelial cells from the mucosa of the gastrointestinal tract, bacteria, products of bacterial decomposition, unabsorbed digested materials, and indigestible parts of food.

Although 90% of all water absorption occurs in the small intestine, the large intestine absorbs enough to make it an important organ in maintaining the body's water balance. Of the 0.5–1.0 liter of water that enters the large intestine, all but about 100–200 mL is absorbed via osmosis. The large intestine also absorbs ions, including sodium and chloride, and some vitamins.

 Occult Blood

The term **occult blood** refers to blood that is hidden; it is not detectable by the human eye. The main diagnostic value of occult blood testing is to screen for colorectal cancer. Two substances often examined for occult blood are feces and urine. Several types of products are available for at-home testing for hidden blood in feces. The tests are based on color changes when reagents are added to feces. The presence of occult blood in urine may be detected at home by using dip-and-read reagent strips. ∎

The Defecation Reflex

Mass peristaltic movements push fecal material from the sigmoid colon into the rectum. The resulting distention of the rectal wall stimulates stretch receptors, which initiates a **defecation reflex** that empties the rectum. The defecation reflex occurs as follows: In response to distention of the rectal wall, the receptors send sensory nerve impulses to the sacral spinal cord. Motor impulses from the cord travel along parasympathetic nerves back to the descending colon, sigmoid colon, rectum, and anus. The resulting contraction of the longitudinal rectal muscles shortens the rectum, thereby increasing the pressure within it. This pressure, along with voluntary contractions of the diaphragm and abdominal muscles, plus parasympathetic stimulation, open the internal anal sphincter.

The external anal sphincter is voluntarily controlled. If it is voluntarily relaxed, defecation occurs and the feces are expelled through the anus; if it is voluntarily constricted, defecation can be postponed. Voluntary contractions of the diaphragm and abdominal muscles aid defecation by increasing the pressure within the abdomen, which pushes the walls of the sigmoid colon and rectum inward. If defecation does not occur, the feces back up into the sigmoid colon until the next wave of mass peristalsis again stimulates the stretch receptors, further creating the urge to defecate. In infants, the defecation reflex causes automatic emptying of the rectum because voluntary control of the external anal sphincter has not yet developed.

Diarrhea (dī-a-RĒ-a; *dia-* = through; *rrhea* = flow) is an increase in the frequency, volume, and fluid content of the feces caused by increased motility of and decreased absorption by the intestines. When chyme passes too quickly through the small intestine and feces pass too quickly through the large intestine, there is not enough time for absorption. Frequent diarrhea can result in dehydration and electrolyte imbalances. Excessive motility may be caused by lactose intolerance, stress, and microbes that irritate the gastrointestinal mucosa.

Constipation (kon-sti-PĀ-shun; *con-* = together; *stip-* = to press) refers to infrequent or difficult defecation caused by decreased motility of the intestines. Because the feces remain in the colon for prolonged periods, excessive water absorption occurs, and the feces become dry and hard. Constipation may be caused by poor habits (delaying defecation), spasms of the colon, insufficient fiber in the diet, inadequate fluid intake, lack of exercise, emotional stress, and certain drugs. A common treatment is a mild laxative, such as milk of magnesia, which induces defecation. However, many physicians maintain that laxatives are habit-forming, and that adding fiber to the diet, increasing the amount of exercise, and increasing fluid intake are safer ways of controlling this common problem.

Table 24.7 summarizes the digestive activities in the large intestine, and Table 24.8 summarizes the functions of all digestive system organs.

Dietary Fiber

Dietary fiber consists of indigestible plant carbohydrates—such as cellulose, lignin, and pectin—found in fruits, vegetables,

Table 24.7	Summary of Digestive Activities in the Large Intestine	
Structure	**Activity**	**Function(s)**
Lumen	Bacterial activity.	Breaks down undigested carbohydrates, proteins, and amino acids into products that can be expelled in feces or absorbed and detoxified by liver; synthesizes certain B vitamins and vitamin K.
Mucosa	Secretes mucus.	Lubricates colon and protects mucosa.
	Absorption.	Water absorption solidifies feces and contributes to the body's water balance; solutes absorbed include ions and some vitamins.
Muscularis	Haustral churning.	Moves contents from haustrum to haustrum by muscular contractions.
	Peristalsis.	Moves contents along length of colon by contractions of circular and longitudinal muscles.
	Mass peristalsis.	Forces contents into sigmoid colon and rectum.
	Defecation reflex.	Eliminates feces by contractions in sigmoid colon and rectum.

Table 24.8	Summary of Organs of the Digestive System and Their Functions
Organ	**Functions**
Mouth	See the functions of the tongue, salivary glands, and teeth, all of which are in the mouth. Additionally, the lips and cheeks keep food between the teeth during mastication, and buccal glands lining the mouth produce saliva.
Tongue	Maneuvers food for mastication, shapes food into a bolus, maneuvers food for deglutition, detects taste and touch sensations, and initiates digestion of triglycerides.
Salivary glands	Saliva softens, moistens, and dissolves foods; cleanses mouth and teeth; and initiates the digestion of starch.
Teeth	Cut, tear, and pulverize food to reduce solids to smaller particles for swallowing.
Pharynx	Receives a bolus from the oral cavity and passes it into the esophagus.
Esophagus	Receives a bolus from the pharynx and moves it into the stomach. This requires relaxation of the upper esophageal sphincter and secretion of mucus.
Stomach	Mixing waves macerate food, mix it with secretions of gastric glands (gastric juice), and reduce food to chyme. Gastric juice activates pepsin and kills many microbes in food. Intrinsic factor aids absorption of vitamin B_{12}. The stomach serves as a reservoir for food before releasing it into the small intestine.
Pancreas	Pancreatic juice buffers acidic gastric juice in chyme (creating the proper pH for digestion in the small intestine), stops the action of pepsin from the stomach, and contains enzymes that digest carbohydrates, proteins, triglycerides, and nucleic acids.
Liver	Produces bile, which is needed for the emulsification and absorption of lipids in the small intestine.
Gallbladder	Stores and concentrates bile and releases it into the small intestine.
Small intestine	Segmentations mix chyme with digestive juices; migrating motility complexes propel chyme toward the ileocecal sphincter; digestive secretions from the small intestine, pancreas, and liver complete the digestion of carbohydrates, proteins, lipids, and nucleic acids; circular folds, villi, and microvilli increase surface area for absorption; site where about 90% of nutrients and water are absorbed.
Large intestine	Haustral churning, peristalsis, and mass peristalsis drive the contents of the colon into the rectum; bacteria produce some B vitamins and vitamin K; absorption of some water, ions, and vitamins, defecation.

grains, and beans. **Insoluble fiber,** which does not dissolve in water, includes the woody or structural parts of plants such as the skins of fruits and vegetables and the bran coating around wheat and corn kernels. Insoluble fiber passes through the GI tract largely unchanged but speeds up the passage of material through the tract. **Soluble fiber,** which does dissolve in water, forms a gel that slows the passage of material through the tract. It is found in abundance in beans, oats, barley, broccoli, prunes, apples, and citrus fruits.

People who choose a fiber-rich diet may reduce their risk of developing obesity, diabetes, atherosclerosis, gallstones, hemorrhoids, diverticular disease, appendicitis, and colorectal cancer. Soluble fiber also may help lower blood cholesterol because the fiber binds bile salts and retards their reabsorption. As a result, more cholesterol is used to replace the bile salts lost in the feces. ■

▶ CHECKPOINT

42. What are the major regions of the large intestine?

43. How does the muscularis of the large intestine differ from that of the rest of the gastrointestinal tract? What are haustra?

44. Describe the mechanical movements that occur in the large intestine.

45. What is defecation and how does it occur?

46. What activities occur in the large intestine to change its contents into feces?

DEVELOPMENT OF THE DIGESTIVE SYSTEM

▶ OBJECTIVE

• **Describe the development of the digestive system.**

During the fourth week of development, the cells of the **endoderm** form a cavity called the **primitive gut,** the forerunner of the gastrointestinal tract. See Figure 29.12b on page 1078. Soon afterwards the mesoderm forms and splits into two layers (somatic and splanchnic), as shown in Figure 29.9d on page 1074. The splanchnic mesoderm associates with the endoderm of the primitive gut; as a result, the primitive gut has a double-layered wall. The **endodermal layer** gives rise to the *epithelial lining* and *glands* of most of the gastrointestinal tract; the **mesodermal layer** produces the *smooth muscle* and *connective tissue* of the tract.

The primitive gut elongates and differentiates into an anterior **foregut,** an intermediate **midgut,** and a posterior **hindgut** (see Figure 29.12c on page 1078). Until the fifth week of development, the midgut opens into the yolk sac; after that time, the yolk sac constricts and detaches from the midgut, and the midgut seals. In the region of the foregut, a depression consisting of ectoderm, the **stomodeum** (stō-mō-DĒ-um), appears (see Figure 29.12d on page 1078). This develops into the *oral cavity.* The **oropharyngeal membrane** is a depression of fused ectoderm and endoderm on the surface of the embryo that separates

the foregut from the stomodeum. The membrane ruptures during the fourth week of development, so that the foregut is continuous with the outside of the embryo through the oral cavity. Another depression consisting of ectoderm, the **proctodeum** (prok-tō-DĒ-um), forms in the hindgut and goes on to develop into the *anus.* See Figure 29.12d on page 1078. The **cloacal membrane** (klō-Ā-kul) is a fused membrane of ectoderm and endoderm that separates the hindgut from the proctodeum. After it ruptures during the seventh week, the hindgut is continuous with the outside of the embryo through the anus. Thus, the gastrointestinal tract forms a continuous tube from mouth to anus.

The foregut develops into the *pharynx, esophagus, stomach,* and *part of the duodenum.* The midgut is transformed into the *remainder of the duodenum,* the *jejunum,* the *ileum,* and *portions of the large intestine* (cecum, appendix, ascending colon, and most of the transverse colon). The hindgut develops into the *remainder of the large intestine,* except for a portion of the anal canal that is derived from the proctodeum.

As development progresses, the endoderm at various places along the foregut develops into hollow buds that grow into the mesoderm. These buds will develop into the *salivary glands, liver, gallbladder,* and *pancreas.* Each of these organs retains a connection with the gastrointestinal tract through ducts.

▶ CHECKPOINT

47. What structures develop from the foregut, midgut, and hindgut?

AGING AND THE DIGESTIVE SYSTEM

▶ OBJECTIVE

• **Describe the effects of aging on the digestive system.**

Overall changes of the digestive system associated with aging include decreased secretory mechanisms, decreased motility of the digestive organs, loss of strength and tone of the muscular tissue and its supporting structures, changes in neurosensory feedback regarding enzyme and hormone release, and diminished response to pain and internal sensations. In the upper portion of the GI tract, common changes include reduced sensitivity to mouth irritations and sores, loss of taste, periodontal disease, difficulty in swallowing, hiatal hernia, gastritis, and peptic ulcer disease. Changes that may appear in the small intestine include duodenal ulcers, malabsorption, and maldigestion. Other pathologies that increase in incidence with age are appendicitis, gallbladder problems, jaundice, cirrhosis, and acute pancreatitis. Large intestinal changes such as constipation, hemorrhoids, and diverticular disease may also occur. Cancer of the colon or rectum is quite common. • • •

Now that our exploration of the digestive system is completed, you can appreciate the many ways that this system contributes to homeostasis of other body systems by examining *Focus on Homeostasis: The Digestive System* on page 898. Next, in Chapter 25, you will discover how the nutrients absorbed by the GI tract enter into metabolic reactions in the body tissues.

Body System	Contribution of the Digestive System
For all body systems	The digestive system breaks down dietary nutrients into forms that can be absorbed and used by body cells for producing ATP and building body tissues. Absorbs water, minerals, and vitamins needed for growth and function of body tissues; and eliminates wastes from body tissues in feces.
Integumentary system	Small intestine absorbs vitamin D, which skin and kidneys modify to produce the hormone calcitriol. Excess dietary calories are stored as triglycerides in adipose cells in dermis and subcutaneous layer.
Skeletal system	Small intestine absorbs dietary calcium and phosphorus salts needed to build bone matrix.
Muscular system	Liver can convert lactic acid produced by muscles during exercise to glucose.
Nervous system	Gluconeogenesis (synthesis of new glucose molecules) in liver plus digestion and absorption of dietary carbohydrates provide glucose, needed for ATP production by neurons.
Endocrine system	Liver inactivates some hormones, ending their activity. Pancreatic islets release insulin and glucagon. Cells in mucosa of stomach and small intestine release hormones that regulate digestive activities. Liver produces angiotensinogen.
Cardiovascular system	GI tract absorbs water that helps maintain blood volume and iron that is needed for synthesis of hemoglobin in red blood cells. Bilirubin from hemoglobin breakdown is partially excreted in feces. Liver synthesizes most plasma proteins.
Lymphatic and Immune system	Acidity of gastric juice destroys bacteria and most toxins in stomach.
Respiratory system	Pressure of abdominal organs against the diaphragm helps expel air quickly during a forced exhalation.
Urinary system	Absorption of water by GI tract provides water needed to excrete waste products in urine.
Reproductive systems	Digestion and absorption provides adequate nutrients, including fats, for normal development of reproductive structures, for production of gametes (oocytes and sperm), and for fetal growth and development during pregnancy.

DISORDERS: HOMEOSTATIC IMBALANCES

Dental Caries

Dental caries, or tooth decay, involve a gradual demineralization (softening) of the enamel and dentin. If untreated, microorganisms may invade the pulp, causing inflammation and infection, with subsequent death of the pulp and abscess of the alveolar bone surrounding the root's apex. Such teeth are treated by root canal therapy.

Dental caries begin when bacteria, acting on sugars, produce acids that demineralize the enamel. **Dextran,** a sticky polysaccharide produced from sucrose, causes the bacteria to stick to the teeth. Masses of bacterial cells, dextran, and other debris adhering to teeth constitute **dental plaque.** Saliva cannot reach the tooth surface to buffer the acid because the plaque covers the teeth. Brushing the teeth after eating removes the plaque from flat surfaces before the bacteria can produce acids. Dentists also recommend that the plaque between the teeth be removed every 24 hours with dental floss.

Periodontal Disease

Periodontal disease is a collective term for a variety of conditions characterized by inflammation and degeneration of the gingivae, alveolar bone, periodontal ligament, and cementum. One such condition is called **pyorrhea,** the initial symptoms of which are enlargement and inflammation of the soft tissue and bleeding of the gums. Without treatment, the soft tissue may deteriorate and the alveolar bone may be resorbed, causing loosening of the teeth and recession of the gums. Periodontal diseases are often caused by poor oral hygiene; by local ·irritants, such as bacteria, impacted food, and cigarette smoke; or by a poor "bite."

Peptic Ulcer Disease

In the United States, 5–10% of the population develops **peptic ulcer disease (PUD).** An **ulcer** is a craterlike lesion in a membrane; ulcers that develop in areas of the GI tract exposed to acidic gastric juice are called **peptic ulcers.** The most common complication of peptic ulcers is bleeding, which can lead to anemia if enough blood is lost. In acute cases, peptic ulcers can lead to shock and death. Three distinct causes of PUD are recognized: (1) the bacterium *Helicobacter pylori;* (2) nonsteroidal anti-inflammatory drugs (NSAIDs) such as aspirin; and (3) hypersecretion of HCl, as occurs in Zollinger–Ellison syndrome, a gastrin-producing tumor usually of the pancreas.

Helicobacter pylori (previously named *Campylobacter pylori*) is the most frequent cause of PUD. The bacterium produces an enzyme called urease, which splits urea into ammonia and carbon dioxide. While shielding the bacterium from the acidity of the stomach, the ammonia also damages the protective mucous layer of the stomach and the underlying gastric cells. *H. pylori* also produces catalase, an enzyme that may protect the microbe from phagocytosis by neutrophils, plus several adhesion proteins that allow the bacterium to attach itself to gastric cells.

Several therapeutic approaches are helpful in the treatment of PUD. Cigarette smoke, alcohol, caffeine, and NSAIDs should be avoided because they can impair mucosal defensive mechanisms, which increases mucosal susceptibility to the damaging effects of HCl. In cases associated with *H. pylori,* treatment with an antibiotic drug often resolves the problem. Oral antacids such as Tums or Maalox can help temporarily by buffering gastric acid. When hypersecretion of HCl

is the cause of PUD, H_2 blockers (such as Tagamet) or proton pump inhibitors such as omeprazole (Prilosec), which block secretion of H^+ from parietal cells, may be used.

Diverticular Disease

In **diverticular disease,** saclike outpouchings of the wall of the colon, termed **diverticula,** occur in places where the muscularis has weakened and may become inflamed. Development of diverticula is known as **diverticulosis.** Many people who develop diverticulosis have no symptoms and experience no complications. Of those people known to have diverticulosis, 10–25% eventually develop an inflammation known as **diverticulitis.** This condition may be characterized by pain, either constipation or increased frequency of defecation, nausea, vomiting, and low-grade fever. Because diets low in fiber contribute to development of diverticulitis, patients who change to high-fiber diets show marked relief of symptoms. In severe cases, affected portions of the colon may require surgical removal. If diverticula rupture, the release of bacteria into the abdominal cavity can cause peritonitis.

Colorectal Cancer

Colorectal cancer is among the deadliest of malignancies, ranking second to lung cancer in males and third after lung cancer and breast cancer in females. Genetics plays a very important role in that an inherited predisposition contributes to more than half of all cases of colorectal cancer. Intake of alcohol and diets high in animal fat and protein are associated with increased risk of colorectal cancer, whereas dietary fiber, retinoids, calcium, and selenium may be protective. Signs and symptoms of colorectal cancer include diarrhea, constipation, cramping, abdominal pain, and rectal bleeding, either visible or occult. Precancerous growths on the mucosal surface, called **polyps,** also increase the risk of developing colorectal cancer. Screening for colorectal cancer includes testing for blood in the feces, digital rectal examination, sigmoidoscopy, colonoscopy, and barium enema. Tumors may be removed endoscopically or surgically.

Hepatitis

Hepatitis is an inflammation of the liver that can be caused by viruses, drugs, and chemicals, including alcohol. Clinically, several types of viral hepatitis are recognized. **Hepatitis A (infectious hepatitis)** is caused by hepatitis A virus and is spread via fecal contamination of objects such as food, clothing, toys, and eating utensils (fecal–oral route). It is generally a mild disease of children and young adults characterized by loss of appetite, malaise, nausea, diarrhea, fever, and chills. Eventually, jaundice appears. This type of hepatitis does not cause lasting liver damage. Most people recover in 4 to 6 weeks.

Hepatitis B is caused by hepatitis B virus and is spread primarily by sexual contact and contaminated syringes and transfusion equipment. It can also be spread via saliva and tears. Hepatitis B virus can be present for years or even a lifetime, and it can produce cirrhosis and possibly cancer of the liver. Individuals who harbor the active hepatitis B virus also become carriers. Vaccines produced through recombinant DNA technology (for example, Recombivax HB) are available to prevent hepatitis B infection.

Hepatitis C, caused by hepatitis C virus, is clinically similar to hepatitis B. Hepatitis C can cause cirrhosis and possibly liver cancer. In developed nations, donated blood is screened for the presence of hepatitis B and C.

Hepatitis D is caused by hepatitis D virus. It is transmitted like hepatitis B and, in fact, a person must have been co-infected with hepatitis B before contracting hepatitis D. Hepatitis D results in severe liver damage and has a higher fatality rate than infection with hepatitis B virus alone.

Hepatitis E is caused by hepatitis E virus and is spread like hepatitis A. Although it does not cause chronic liver disease, hepatitis E virus is responsible for a very high mortality rate in pregnant women.

Anorexia Nervosa

Anorexia nervosa is a chronic disorder characterized by self-induced weight loss, negative perception of body image, and physiological changes that result from nutritional depletion. Patients with anorexia nervosa have a fixation on weight control and often insist on having a bowel movement every day despite inadequate food intake. They often abuse laxatives, which worsens the fluid and electrolyte imbalances and nutrient deficiencies. The disorder is found predominantly in young, single females, and it may be inherited. Abnormal patterns of menstruation, amenorrhea (absence of menstruation), and a lowered basal metabolic rate reflect the depressant effects of starvation. Individuals may become emaciated and may ultimately die of starvation or one of its complications. Also associated with the disorder are osteoporosis, depression, and brain abnormalities coupled with impaired mental performance. Treatment consists of psychotherapy and dietary regulation.

MEDICAL TERMINOLOGY

Achalasia (ak′-a-LĀ-zē-a; *a-* = without; *chalasis* = relaxation) A condition, caused by malfunction of the myenteric plexus, in which the lower esophageal sphincter fails to relax normally as food approaches. A whole meal may become lodged in the esophagus and enter the stomach very slowly. Distension of the esophagus results in chest pain that is often confused with pain originating from the heart.

Borborygmus (bor′-bō-RIG-mus) A rumbling noise caused by the propulsion of gas through the intestines.

Bulimia (boo-LIM-ē-a; *bu-* = ox; *limia* = hunger) or **binge–purge syndrome** A disorder that typically affects young, single, middle-class, white females, characterized by overeating at least twice a week followed by purging by self-induced vomiting, strict dieting or fasting, vigorous exercise, or use of laxatives or diuretics. It occurs in response to fears of being overweight, or to stress, depression, and physiological disorders such as hypothalamic tumors.

Canker sore (KANG-ker) Painful ulcer on the mucous membrane of the mouth that affects females more often than males, usually between ages 10 and 40. It may be an autoimmune reaction or a food allergy.

Cholecystitis (kō′-lē-sis-TĪ-tis; *chole-* = bile; *cyst-* = bladder; *-itis* = inflammation of) In some cases, an autoimmune inflammation of the gallbladder; other cases are caused by obstruction of the cystic duct by bile stones.

Cirrhosis (sir-RŌ-sis; *cirrh-* = orange-colored) Distorted or scarred liver as a result of chronic inflammation due to hepatitis, chemicals that destroy hepatocytes, parasites that infect the liver, or alcoholism; the hepatocytes are replaced by fibrous or adipose connective tissue. Symptoms include jaundice, edema in the legs, uncontrolled bleeding, and increased sensitivity to drugs.

Colitis (ko-LĪ-tis) Inflammation of the mucosa of the colon and rectum in which absorption of water and salts is reduced, producing watery, bloody feces and, in severe cases, dehydration and salt depletion. Spasms of the irritated muscularis produce cramps. It is thought to be an autoimmune disorder.

Colostomy (kō-LOS-tō-mē; *-stomy* = provide an opening) The diversion of the fecal stream through an opening in the colon, creating a surgical "stoma" (artificial opening) that is affixed to the exterior of the abdominal wall. This opening is a substitute anus through which feces are eliminated into a bag worn on the abdomen.

Dysphagia (dis-FĀ-jē-a; *dys-* = abnormal; *phagia* = to eat) Difficulty in swallowing that may be caused by inflammation, paralysis, obstruction, or trauma.

Enteritis (en′-ter-Ī-tis; *enter-* = intestine) An inflammation of the intestine, particularly the small intestine.

Flatus (FLĀ-tus) Air (gas) in the stomach or intestine, usually expelled through the anus. If the gas is expelled through the mouth, it is called **eructation** or **belching** (burping). Flatus may result from gas released during the breakdown of foods in the stomach or from swallowing air or gas-containing substances such as carbonated drinks.

Gastrectomy (gas-TREK-tō-mē; *gastr-* = stomach; *-ectomy* = to cut out) Removal of all or a portion of the stomach.

Gastroscopy (gas-TROS-kō-pē; *-scopy* = to view with a lighted instrument) Endoscopic examination of the stomach in which the examiner can view the interior of the stomach directly to evaluate an ulcer, tumor, inflammation, or source of bleeding.

Hernia (HER-nē-a) Protrusion of all or part of an organ through a membrane or cavity wall, usually the abdominal cavity. *Diaphragmatic (hiatal) hernia* is the protrusion of the lower esophagus, stomach, or intestine into the thoracic cavity through the esophageal hiatus. *Inguinal hernia* is the protrusion of the hernial sac into the inguinal opening; it may contain a portion of the bowel in an advanced stage and may extend into the scrotal compartment in males, causing strangulation of the herniated part.

Hemorrhoids (HEM-ō-royds = veins likely to bleed) Varicose (enlarged and inflamed) veins in the rectum. Hemorrhoids develop when the veins are put under pressure and become engorged with blood. If the pressure continues, the wall of the vein stretches. Such a distended vessel oozes blood, and bleeding or itching are usually the first signs that a hemorrhoid has developed. Stretching of a vein also favors clot formation, further aggravating swelling and pain. Hemorrhoids may be caused by constipation, which may be brought on by low-fiber diets. Also, repeated straining during defecation forces blood down into the rectal veins, increasing pressure in these veins and possibly causing hemorrhoids. Also called **piles.**

Inflammatory bowel disease (in-FLAM-a-tō′-rē BOW-el) Inflammation of the gastrointestinal tract that exists in two forms. (1) Crohn's disease is an inflammation of any part of the gastrointestinal tract in which the inflammation extends from the mucosa through the submucosa, muscularis, and serosa. (2) Ulcerative colitis is an inflammation of the mucosa of the colon and rectum, usually accompanied by rectal bleeding. Curiously, cigarette smoking increases one's risk for Crohn's disease but decreases the risk of ulcerative colitis.

Irritable bowel syndrome (IBS) Disease of the entire gastrointestinal tract in which a person reacts to stress by developing symptoms (such as cramping and abdominal pain) associated with alternating patterns of diarrhea and constipation. Excessive amounts of mucus may appear in feces, and other symptoms include flatulence, nausea, and loss of appetite. The condition is also known as **irritable colon** or **spastic colitis.**

Malocclusion (mal′-ō-KLOO-zhun; *mal-* = bad; *occlusion* = to fit together) Condition in which the surfaces of the maxillary (upper) and mandibular (lower) teeth fit together poorly.

Traveler's diarrhea Infectious disease of the gastrointestinal tract that results in loose, urgent bowel movements, cramping, abdominal pain, malaise, nausea, and occasionally fever and dehydration. It is acquired through ingestion of food or water contaminated with fecal material that contains bacteria (especially *Escherichia coli*); viruses or protozoan parasites in feces are less common causes.

STUDY OUTLINE

INTRODUCTION (p. 852)

1. The breaking down of larger food molecules into smaller molecules is called digestion; the passage of these smaller molecules into blood and lymph is termed absorption.
2. The organs that collectively perform digestion and absorption constitute the digestive system and are usually composed of two main groups: the gastrointestinal (GI) tract and accessory digestive organs.
3. The GI tract is a continuous tube extending from the mouth to the anus.
4. The accessory digestive organs include the teeth, tongue, salivary glands, liver, gallbladder, and pancreas.

OVERVIEW OF THE DIGESTIVE SYSTEM (p. 852)

1. Digestion includes six basic processes: ingestion, secretion, mixing and propulsion, mechanical and chemical digestion, absorption, and defecation.
2. Mechanical digestion consists of mastication and movements of the gastrointestinal tract that aid chemical digestion.
3. Chemical digestion is a series of hydrolysis reactions that break down large carbohydrates, lipids, proteins, and nucleic acids in foods into smaller molecules that are usable by body cells.

LAYERS OF THE GI TRACT (p. 853)

1. The basic arrangement of layers in most of the gastrointestinal tract, from deep to superficial, is the mucosa, submucosa, muscularis, and serosa.
2. Associated with the lamina propria of the mucosa are extensive patches of lymphatic tissue called mucosa-associated lymphoid tissue (MALT).

PERITONEUM (p. 855)

1. The peritoneum is the largest serous membrane of the body; it lines the wall of the abdominal cavity and covers some abdominal organs.
2. Folds of the peritoneum include the mesentery, mesocolon, falciform ligament, lesser omentum, and greater omentum.

MOUTH (p. 857)

1. The mouth is formed by the cheeks, hard and soft plates, lips, and tongue.
2. The vestibule is the space bounded externally by the cheeks and lips and internally by the teeth and gums.
3. The oral cavity proper extends from the vestibule to the fauces.
4. The tongue, together with its associated muscles, forms the floor of the oral cavity. It is composed of skeletal muscle covered with mucous membrane.
5. The upper surface and sides of the tongue are covered with papillae, some of which contain taste buds.
6. The major portion of saliva is secreted by the major salivary glands, which lie outside the mouth and pour their contents into ducts that empty into the oral cavity.
7. There are three pairs of major salivary glands: parotid, submandibular, and sublingual glands.
8. Saliva lubricates food and starts the chemical digestion of carbohydrates.
9. Salivation is controlled by the nervous system.
10. The teeth (dentes) project into the mouth and are adapted for mechanical digestion.
11. A typical tooth consists of three principal regions: crown, root, and neck.
12. Teeth are composed primarily of dentin and are covered by enamel, the hardest substance in the body.
13. There are two dentitions: deciduous and permanent.
14. Through mastication, food is mixed with saliva and shaped into a soft, flexible mass called a bolus.
15. Salivary amylase begins the digestion of starches, and lingual lipase acts on triglycerides.

PHARYNX (p. 863)

1. Deglution, or swallowing, moves a bolus from the mouth to the stomach.
2. Swallowing consists of a voluntary stage, a pharyngeal stage (involuntary), and an esophageal stage (involuntary).

ESOPHAGUS (p. 864)

1. The esophagus is a collapsible, muscular tube that connects the pharynx to the stomach.
2. It passes a bolus into the stomach by peristalsis.
3. It contains an upper and a lower esophageal sphincter.

STOMACH (p. 866)

1. The stomach connects the esophagus to the duodenum.
2. The principal anatomic regions of the stomach are the cardia, fundus, body, and pylorus.
3. Adaptations of the stomach for digestion include rugae; glands that produce mucus, hydrochloric acid, pepsin, gastric lipase, and intrinsic factor; and a three-layered muscularis.
4. Mechanical digestion consists of mixing waves.
5. Chemical digestion consists mostly of the conversion of proteins into peptides by pepsin.
6. The stomach wall is impermeable to most substances.
7. Among the substances the stomach can absorb are water, certain ions, drugs, and alcohol.
8. Gastric secretion is regulated by neural, paracrine, and hormonal mechanisms.
9. Regulation of gastric secretion occurs in three overlapping phases: the cephalic, gastric, and intestinal phases.
10. During the cephalic and gastric phases, peristalsis is stimulated; during the intestinal phase, motility is inhibited.
11. Gastric emptying is stimulated in response to distension, and gastrin is released in response to the presence of certain types of foods.
12. Gastric emptying is inhibited by the enterogastric reflex and by hormones (CCK and GIP).

PANCREAS (p. 874)

1. The pancreas consists of a head, a body, and a tail and is connected to the duodenum via the pancreatic duct and accessory duct.
2. Endocrine pancreatic islets (islets of Langerhans) secrete hormones, and exocrine acini secrete pancreatic juice.
3. Pancreatic juice contains enzymes that digest starch (pancreatic amylase), proteins (trypsin, chymotrypsin, carboxypeptidase, and elastase), triglycerides (pancreatic lipase), and nucleic acids (ribonuclease and deoxyribonuclease).
4. Pancreatic secretion is regulated by neural and hormonal mechanisms.

LIVER AND GALLBLADDER (p. 876)

1. The liver has left and right lobes; the right lobe includes a quadrate and caudate lobe. The gallbladder is a sac located in a depression on the posterior surface of the liver that stores and concentrates bile.
2. The lobes of the liver are made up of lobules that contain hepato-

cytes (liver cells), sinusoids, stellate reticuloendothelial (Kupffer) cells, and a central vein.
3. Hepatocytes produce bile that is carried by a duct system to the gallbladder for concentration and temporary storage. Cholecystokinin (CCK) stimulates ejection of bile into the common bile duct.
4. Bile's contribution to digestion is the emulsification of dietary lipids.
5. The liver also functions in carbohydrate, lipid, and protein metabolism; processing of drugs and hormones; excretion of bilirubin; synthesis of bile salts; storage of vitamins and minerals; phagocytosis; and activation of vitamin D.
6. Bile secretion is regulated by neural and hormonal mechanisms.

SUMMARY: DIGESTIVE HORMONES (p. 880)

1. Enteroendocrine cells in the mucosa secrete several hormones that regulate digestive processes. The major hormones are gastrin (from the stomach), secretin (from the small intestine), and cholecystokinin (from the small intestine).
2. Gastrin exerts its major effects on the stomach, whereas secretin and cholecystokinin affect the pancreas, liver, and gallbladder most strongly.

SMALL INTESTINE (p. 881)

1. The small intestine extends from the pyloric sphincter to the ileocecal sphincter.
2. It is divided into duodenum, jejunum, and ileum.
3. Its glands secrete fluid and mucus, and the circular folds, villi, and microvilli of its wall provide a large surface area for digestion and absorption.
4. Brush-border enzymes digest α-dextrins, maltose, sucrose, lactose, peptides, and nucleotides at the surface of mucosal epithelial cells.
5. Pancreatic and intestinal brush-border enzymes break down starches into maltose, maltotriose, and α-dextrins (pancreatic amylase), α-dextrins into glucose (α-dextrinase), maltose to glucose (maltase), sucrose to glucose and fructose (sucrase), lactose to glucose and galactose (lactase), and proteins into peptides (trypsin, chymotrypsin, and elastase). Also, enzymes break peptide bonds that attach terminal amino acids to carboxyl ends of peptides (carboxypeptidases) and peptide bonds that attach terminal amino acids to amino ends of peptides (aminopeptidases). Finally, enzymes split dipeptides to amino acids (dipeptidase), triglycerides to fatty acids and monoglycerides (lipases), and nucleotides to pentoses and nitrogenous bases (nucleosidases and phosphatases).
6. Mechanical digestion in the small intestine involves segmentation and migrating motility complexes.
7. The most important regulators of intestinal secretion and motility are enteric reflexes and digestive hormones.
8. Parasympathetic impulses increase motility; sympathetic impulses decrease motility.
9. Absorption occurs via diffusion, facilitated diffusion, osmosis, and active transport; most absorption occurs in the small intestine.
10. Monosaccharides, amino acids, and short-chain fatty acids pass into the blood capillaries.
11. Long-chain fatty acids and monoglycerides are absorbed from

micelles, resynthesized to triglycerides, and formed into chylomicrons.

12. Chylomicrons move into lymph in the lacteal of a villus.

13. The small intestine also absorbs electrolytes, vitamins, and water.

LARGE INTESTINE (p. 891)

1. The large intestine extends from the ileocecal sphincter to the anus.

2. Its regions include the cecum, colon, rectum, and anal canal.

3. The mucosa contains many goblet cells, and the muscularis consists of teniae coli and haustra.

4. Mechanical movements of the large intestine include haustral churning, peristalsis, and mass peristalsis.

5. The last stages of chemical digestion occur in the large intestine through bacterial action. Substances are further broken down, and some vitamins are synthesized.

6. The large intestine absorbs water, ions, and vitamins.

7. Feces consist of water, inorganic salts, epithelial cells, bacteria, and undigested foods.

8. The elimination of feces from the rectum is called defecation.

9. Defecation is a reflex action aided by voluntary contractions of the diaphragm and abdominal muscles and relaxation of the external anal sphincter.

DEVELOPMENT OF THE DIGESTIVE SYSTEM (p. 897)

1. The endoderm of the primitive gut forms the epithelium and glands of most of the gastrointestinal tract.

2. The mesoderm of the primitive gut forms the smooth muscle and connective tissue of the gastrointestinal tract.

AGING AND THE DIGESTIVE SYSTEM (p. 897)

1. General changes include decreased secretory mechanisms, decreased motility, and loss of tone.

2. Specific changes may include loss of taste, pyorrhea, hernias, peptic ulcer disease, constipation, hemorrhoids, and diverticular diseases.

SELF-QUIZ QUESTIONS

Fill in the blanks in the following statements.

1. The four layers of the GI tract, from deep to superficial, are the _____, _____, _____, and _____.

2. The _____ regulates movements of the mucosa and vasoconstriction of blood vessels. The _____ controls GI tract motility, the frequency and strength of contraction of the muscularis.

3. The end-products of chemical digestion of carbohydrates are _____, of proteins are _____, of lipids are _____ and _____, and of nucleic acids are _____, _____, and _____.

Indicate whether the following statements are true or false.

4. All digestive end-products are absorbed from the lumen of the GI tract directly into the blood capillaries of the cardiovascular system and are then carried to the liver for processing.

5. The last stage of digestion occurs in the colon through the activity of the haustra.

Choose the one best answer to the following questions.

6. Which of the following are basic digestive processes? (1) ingestion, (2) secretion, (3) mixing and propulsion, (4) ventilation, (5) absorption, (6) defecation, (7) synthesis, (8) mechanical and chemical digestion. (a) 1, 2, 3, 5, 6, and 8, (b) 2, 3, 4, 6, 7, and 8, (c) 1, 3, 5, and 8, (d) 2, 3, 6, and 8, (e) 1, 3, 6, and 8.

7. Which of the following are functions of the liver? (1) carbohydrate, lipid, and protein metabolism, (2) nucleic acid metabolism, (3) excretion of bilirubin, (4) synthesis of bile salts, (5) activation of vitamin D. (a) 1, 2, 3, and 5, (b) 1, 2, 3, and 4, (c) 1, 3, 4, and 5, (d) 2, 3, 4, and 5, (e) 1, 2, 4, and 5.

8. The part of the wall of the GI tract that contains most of the mucosa-associated lymphatic tissue is the (a) muscularis mucosae, (b) submucosa, (c) muscularis, (d) serosa, (e) lamina propria.

9. Which of the following are *true* concerning the peritoneum? (1) The kidneys and pancreas are retroperitoneal. (2) The greater omentum is the largest of the peritoneal folds. (3) The lesser omentum binds the large intestine to the posterior abdominal wall. (4) Accumulation of fluid in the peritoneal cavity results in a hiatal hernia. (5) The mesentery is associated with the small intestine. (a) 1, 2, 3 and 5, (b) 1, 2, and 5, (c) 2 and 5, (d) 1, 2, 4, and 5, (e) 3, 4, and 5.

10. Which of the following statements regarding the regulation of gastric secretion and motility are *true?* (1) The sight, smell, taste, or thought of food can initate the cephalic phase of gastric activity. (2) The gastric phase begins when food enters the small intestine. (3) Once activated, stretch receptors in the stomach trigger the flow of gastric juice and peristalsis. (4) The intestinal phase reflexes inhibit gastric activity. (5) The enterogastric reflex stimulates gastric emptying. (a) 1, 3, and 4, (b) 2, 4, and 5, (c) 1, 4, and 5, (d) 1, 2, and 5, (e) 1, 2, 3, and 4.

11. Which of the following are *true?* (1) Segmentations in the small intestine help propel chyme through the intestinal tract. (2) The migrating motility complex is a type of peristalis in the small intestine. (3) The large surface area for absorption in the small intestine is due to the presence of circular folds, villi, and microvilli. (4) The mucus-producing cells of the small intestine are Paneth cells. (5) Most long-chain fatty acid and monoglyceride absorption in the small intestine requires the presence of bile salts. (a) 1, 2, and 3, (b) 2, 3, and 5, (c) 1, 2, 3, 4, and 5, (d) 1, 3, and 5, (e) 1, 2, 3, and 5.

12. The release of feces from the large intestine is dependent upon (1) stretching of the rectal walls, (2) voluntary relaxation of the external anal sphincter, (3) involuntary contraction of the diaphragm and abdominal muscles, (4) activity of the intestinal bacteria, (5) sympathetic stimulation of the intestinal sphincter. (a) 2, 4, and 5, (b) 1, 2, and 5, (c) 1, 2, 3, and 5, (d) 1 and 2, (e) 3, 4, and 5.

13. Match the following:

_____ (a) collapsed, muscular tube involved in deglutition and
peristalsis

_____ (b) coiled tube attached to the
cecum

_____ (c) contains Brunner's glands in
the submucosa

_____ (d) produces and secretes bile

_____ (e) contains Peyer's patches in
the submucosa

_____ (f) responsible for ingestion, mastication, and deglutition

_____ (g) responsible for churning, peristalsis, storage, and chemi-
cal digestion with the enzyme
pepsin

_____ (h) storage area for bile

_____ (i) contain acini that release
juices containing several digestive enzymes

_____ (j) passageway for food, fluid,
and air; involved in deglutition

_____ (k) forms a semisolid waste material through haustral churning
and peristalsis

(1) mouth
(2) pharynx
(3) esophagus
(4) stomach
(5) duodenum
(6) ileum
(7) colon
(8) liver
(9) gallbladder
(10) appendix
(11) pancreas

14. Match the following:

_____ (a) an activating brush-border enzyme
that splits off part of the trypsinogen molecule to form trypsin, a
protease

_____ (b) an enzyme that initiates carbohydrate digestion in the mouth

_____ (c) the principal triglyceride-digesting
enzyme in adults

_____ (d) stimulates growth of gastric glands
and secretion of gastric juice

_____ (e) secreted by chief cells in the
stomach; a proteolyic enzyme

_____ (f) stimulates the flow of pancreatic
juice rich in bicarbonates; decreases gastric secretions

_____ (g) a nonenzymatic fat-emulsifying
agent

_____ (h) causes contraction of the gallbladder and stimulates the production of pancreatic juice rich in
digestive enzymes

_____ (i) inhibits gastrin release

_____ (j) stimulates secretion of ions and
water by the intestines and inhibits
gastric acid secretion

(1) gastrin
(2) cholecys-
tokinin
(3) secretin
(4) enterokinase
(5) pepsin
(6) salivary
amylase
(7) pancreatic
lipase
(8) bile
(9) vasoactive
intestinal
polypeptide
(10) somatostatin

15. Match the following:

_____ (a) microvilli of the small intestine that
increase surface area for absorption;
also contains some digestive enzymes

_____ (b) finger-like projections of the mucosa
of the small intestine that increase
surface area

_____ (c) produce hydrochloric acid in the
stomach

_____ (d) secrete lysosome; help regulate microbial population in the intestines

_____ (e) forces the food to the back of the
mouth for swallowing; places food in
contact with the teeth

_____ (f) aids in limiting the movement of the
tongue posteriorly

_____ (g) produce a fluid in the mouth that
helps cleanse the mouth and teeth
and lubricates, dissolves, and begins
the chemical breakdown of food

_____ (h) large mucosal folds in the stomach

_____ (i) secrete pepsinogen and gastric lipase
in the stomach

_____ (j) permanent ridges in the mucosa of
the small intestine; enhance absorption by increasing surface area and
causing chyme to spiral rather than
move in a straight line

_____ (k) stellate reticuloendothelial cells of the
liver; destroy worn-out leukocytes and
red blood cells, bacteria, and other foreign matter in the blood draining the
GI tract

(1) salivary
glands
(2) parietal
cells
(3) chief
cells
(4) brush
border
(5) Kupffer
cells
(6) rugae
(7) tongue
(8) villi
(9) circular
folds
(10) Paneth
cells
(11) lingual
frenulum

CRITICAL THINKING QUESTIONS

1. Katie's first two permanent teeth came in just in time to fill the upper gap in her smile before the start of school. Unfortunately, the day before her class picture, she fell off her bike and re-created the gap in her smile. Use the proper anatomical terms to describe the teeth Katie lost and the ones she has remaining.
   HINT *A complete set of deciduous teeth includes a different number than a complete set of permanent teeth.*

2. Obesity is a health concern for many people. Suppose that a CCK (cholecystokinin) nasal spray is developed as a weight-loss treatment. How would the CCK affect the digestive system and the appetite?
   HINT *CCK is a digestive hormone but it also affects the brain.*

3. The surgeon explained to the woman that he was going to remove the smaller lobe of her liver and transplant it into her child to replace the child's diseased liver. Use proper anatomical terminology to describe the location of this section of the woman's liver?
   HINT *If the hands are positioned incorrectly while performing cardiopulmonary resuscitation (CPR), the xiphoid process may pierce the liver.*

ANSWERS TO FIGURE QUESTIONS

24.1 Digestive enzymes are produced by the salivary glands, tongue, stomach, pancreas, and small intestine.

24.2 The enteric plexuses help regulate secretions and motility of the GI tract.

24.3 Mesentery binds the small intestine to the posterior abdominal wall.

24.4 The uvula helps prevent foods and liquids from entering the nasal cavity during swallowing.

24.5 Chloride ions in saliva activate salivary amylase.

24.6 The main component of teeth is connective tissue, specifically dentin.

24.7 The first, second, and third molars do not replace any deciduous teeth.

24.8 Both. Initiation of swallowing is voluntary and the action is carried out by skeletal muscles. Completion of swallowing—moving a bolus along the esophagus and into the stomach—is involuntary and involves peristalsis by smooth muscle.

24.9 The esophageal mucosa and submucosa contain mucus secreting glands.

24.10 Food is pushed along by contraction of smooth muscle behind the bolus and relaxation of smooth muscle in front of it.

24.11 After a large meal, the rugae stretch and disappear as the stomach fills.

24.12 G cells secrete the hormone gastrin into the blood. Gastrin increases secretion of gastric juice and increases stomach peristalsis.

24.13 Hydrogen ions secreted into gastric juice are derived from carbonic acid (H_2CO_3).

24.14 Stimulation of the vagus nerve (cranial nerve X) causes secretion of gastric juice and increased gastric motility.

24.15 The pH of gastric juice rises due to the buffering action of some amino acids in food proteins.

24.16 A vagotomy would slow the rate of gastric emptying.

24.17 The pancreatic duct contains pancreatic juice (fluid and digestive enzymes); the common bile duct contains bile; the hepatopancreatic ampulla contains pancreatic juice and bile.

24.18 The HCO_3^- in pancreatic juice helps buffer gastric acid and raises the pH of chyme so that pancreatic enzymes can work effectively.

24.19 The phagocytic cell in the liver is the stellate reticuloendothelial (Kupffer) cell.

24.20 While a meal is being absorbed, hepatocytes remove nutrients from blood flowing through liver sinusoids. They also remove O_2 and extract certain toxic substances.

24.21 Secretin promotes liberation of fluid rich in HCO_3^- from both the pancreas and the liver.

24.22 The ileum is the longest part of the small intestine.

24.23 Nutrients being absorbed enter the blood via capillaries or the lymph via lacteals.

24.24 The fluid secreted by Brunner's glands—alkaline mucus—neutralizes gastric acid and protects the mucosal lining of the duodenum.

24.25 Because monoglycerides are nonpolar (hydrophobic) molecules, they can dissolve in and diffuse through the lipid bilayer of the plasma membrane.

24.26 The stomach and pancreas are the two digestive system organs that secrete the largest volumes of fluid.

24.27 The ascending and descending portions of the colon are retroperitoneal.

24.28 Goblet cells secrete mucus to lubricate colonic contents.

Metabolism

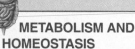

METABOLISM AND HOMEOSTASIS

Metabolic reactions contribute to homeostasis by harvesting chemical energy from nutrients to contribute to the body's growth, repair, and normal functioning.

ENERGIZE YOUR STUDY
FOUNDATIONS CD
Anatomy Overview
- Chemistry Folder: Common Biomolecules

Animations
- Chemistry Folder: Enzymes Functions and ATP; Water and Fluid Flow
- Cell Folder: Cellular Respiration

Concepts and Connections and Exercises reinforce your understanding

www.wiley.com/college/apcentral

INSIGHTS AND EXPLORATIONS

For many of us a diet soda is just another way to enjoy a favorite beverage without the calories. But for people with the metabolic disease phenylketonuria (fen-nel-kee-tone-u-ree-ah) or PKU it is important to know what type of sugar substitute is used. PKU is a genetic disease that is diagnosed at birth with a blood test. For the one infant out of 15,000 born with this disease eating food containing the amino acid phenylalanine can cause mental retardation and other serious medical problems. We will explore in this web-based activity why the development of phenylalanine-free protein from corn by recombinant DNA technology is exciting news for victims of this metabolic disorder.

Whereas plants use the green pigment chlorophyll to trap energy in sunlight, we don't have a similarly functioning pigment in our skin, so the food we eat is our only source of energy for performing biological work. Many molecules needed to maintain cells and tissues can be made from simpler precursors by the body's metabolic reactions; others—the essential amino acids, essential fatty acids, vitamins, and minerals—must be obtained in food because we cannot synthesize them. Carbohydrates, lipids, and proteins in food are prodigested by enzymes in the gastrointestinal tract. The end products of digestion that reach body cells are monosaccharides, fatty acids, glycerol, monoglycerides, and amino acids. Some minerals and many vitamins are part of enzyme systems that catalyze the breakdown and synthesis of carbohydrates, lipids, and proteins. Food molecules absorbed by the gastrointestinal (GI) tract have three main fates:

1. Most food molecules are used to *supply energy* for sustaining life processes, such as active transport, DNA replication, protein synthesis, muscle contraction, maintenance of body temperature, and mitosis.

2. Some food molecules *serve as building blocks* for the synthesis of more complex structural or functional molecules, such as muscle proteins, hormones, and enzymes.

3. Other food molecules are *stored for future use*. For example, glycogen is stored in liver cells, and triglycerides are stored in adipose cells.

In this chapter we discuss how metabolic reactions harvest the chemical energy in foods, how each group of food molecules contributes to the body's growth, repair, and energy needs, and how heat and energy balance is maintained in the body. Finally, we explore some aspects of nutrition.

METABOLIC REACTIONS

OBJECTIVE

- **Explain the role of ATP in anabolism and catabolism.**

Metabolism (me-TAB-ō-lizm; *metabol-* = change) refers to all the chemical reactions of the body. Metabolism is an energy-balancing act between catabolic reactions, which are decomposition reactions, and anabolic reactions, which are synthesis reactions. Overall, catabolic reactions are *exergonic;* they produce more energy than they consume. By contrast, anabolic reactions are *endergonic;* they consume more energy than they produce. The molecule that participates most often in energy exchanges in living cells is **ATP (adenosine triphosphate),** which couples energy-releasing catabolic reactions to energy-requiring anabolic reactions.

The metabolic reactions that occur depend on which enzymes are active in a particular cell at a particular time. Often, catabolic reactions occur in one compartment of a cell, for example, the mitochondria, whereas synthetic reactions take place in another location such as the endoplasmic reticulum.

A molecule synthesized in an anabolic reaction has a limited lifetime. With few exceptions, it will eventually be broken down and its component atoms recycled into other molecules or excreted from the body. Recycling of biological molecules occurs continuously in living tissues, rapidly in some tissues and very slowly in others. Individual cells may be refurbished molecule by molecule, or a whole tissue may be rebuilt cell by cell.

Catabolism and Anabolism Defined

The chemical reactions that break down complex organic molecules into simpler ones are collectively known as **catabolism** (ka-TAB-ō-lizm; *cata-* = downward). Catabolic reactions release the chemical energy stored in organic molecules. Important sets of catabolic reactions occur in glycolysis, the Krebs cycle, and the electron transport chain, each of which will be discussed later in the chapter.

Chemical reactions that combine simple molecules and monomers to form the body's complex structural and functional components are collectively known as **anabolism** (a-NAB-ō-lizm; *ana-* = upward). Examples of anabolic reactions are the formation of peptide bonds between amino acids during protein synthesis, the building of fatty acids into phospholipids that form the plasma membrane bilayer, and the linkage of glucose monomers to form glycogen.

Coupling of Catabolism and Anabolism by ATP

The chemical reactions of living systems depend on efficiently transferring manageable amounts of energy from one molecule to another. The molecule that most often performs this task is ATP, the "energy currency" of a living cell. Like money, it is readily available to "buy" cellular activities; it is spent and remade over and over. A typical cell has about a billion molecules of ATP, each of which typically lasts for less than a minute before being used. Thus, ATP is not a long-term-storage form of currency, like gold in a vault, but rather convenient cash for moment-to-moment transactions.

Recall from Chapter 2 that a molecule of ATP consists of an adenine molecule, a ribose molecule, and three phosphate groups bonded to each other (see Figure 2.25 on page 54). Figure 25.1 shows how ATP links anabolic and catabolic reactions. When the terminal phosphate group is split off ATP in anabolic reactions, adenosine diphosphate (ADP) and a phosphate group (symbolized as Ⓟ) are formed. Some of the energy released is used to drive anabolic reactions such as the formation of glycogen from glucose. Subsequently, energy from complex molecules

Figure 25.1   Role of ATP in linking anabolic and catabolic reactions. When complex molecules and polymers are split apart (catabolism, at left), some of the energy is transferred to form ATP and the rest is given off as heat. When simple molecules and monomers are combined to form complex molecules (anabolism, at right), ATP provides the energy for synthesis, and again some energy is given off as heat.

 The coupling of energy-releasing and energy-requiring reactions is achieved through ATP.

In a pancreatic cell that produces digestive enzymes, does anabolism or catabolism predominate?

is used in catabolic reactions to combine ADP and a phosphate group to resynthesize ATP:

$$\text{ADP} + ⓟ + \text{energy} \longrightarrow \text{ATP}$$

About 40% of the energy released in catabolism is used for cellular functions; the rest is converted to heat, some of which helps maintain normal body temperature. Excess heat is lost to the environment. Compared with machines, which typically convert only 10–20% of energy into work, the 40% efficiency of the body's metabolism is impressive. Still, the body has a continuous need to take in and process external sources of energy so that cells can synthesize enough ATP to sustain life.

▶ CHECKPOINT

1.   What is metabolism? Distinguish between anabolism and catabolism, and give examples of each.

2.   How does ATP provide a link between anabolism and catabolism?

ENERGY TRANSFER

▶ OBJECTIVES

• **Describe oxidation–reduction reactions.**
• **Explain the role of ATP in metabolism.**

Various catabolic reactions transfer energy into the "high-energy" phosphate bonds of ATP. Although the amount of

energy in these bonds is not exceptionally large, it can be released quickly and easily. Before discussing metabolic pathways, we first consider two important aspects of energy transfer: oxidation–reduction reactions and mechanisms of ATP generation.

Oxidation–Reduction Reactions

Oxidation is the *removal of electrons* from an atom or molecule; the result is a *decrease* in the potential energy of the atom or molecule. Because most biological oxidation reactions involve the loss of hydrogen atoms, they are called *dehydrogenation reactions.* An example of an oxidation reaction is the conversion of lactic acid into pyruvic acid:

$$\text{Lactic acid} \xrightarrow[\text{Remove 2H (H}^+ + \text{H}^-)]{\text{Oxidation}} \text{Pyruvic acid}$$

In the preceding reaction, 2H (H$^+$ + H$^-$) means that two neutral hydrogen atoms (2H) are removed as one hydrogen ion (H$^+$) plus one hydride ion (H$^-$).

Reduction is the opposite of oxidation; it is the *addition of electrons* to a molecule. Reduction results in an *increase* in the potential energy of the molecule. An example of a reduction reaction is the conversion of pyruvic acid into lactic acid:

$$\text{Pyruvic acid} \xrightarrow[\text{Add 2 H (H}^+ + \text{H}^-)]{\text{Reduction}} \text{Lactic acid}$$

When a substance is oxidized, the liberated hydrogen atoms do not remain free in the cell but are transferred immediately by coenzymes to another compound. Two coenzymes are commonly used by animal cells to carry hydrogen atoms: **nicotinamide adenine dinucleotide (NAD$^+$)**, a derivative of the B vitamin niacin, and **flavin adenine dinucleotide (FAD),** a derivative of vitamin B$_2$ (riboflavin). The oxidation and reduction states of NAD$^+$ and FAD can be represented as follows:

$$\underset{\text{Oxidized}}{\text{NAD}^+} \underset{-2\,\text{H (H}^+ + \text{H}^-)}{\overset{+2\,\text{H (H}^+ + \text{H}^-)}{\rightleftharpoons}} \underset{\text{Reduced}}{\text{NADH} + \text{H}^+}$$

$$\underset{\text{Oxidized}}{\text{FAD}} \underset{-2\,\text{H (H}^+ + \text{H}^-)}{\overset{+2\,\text{H (H}^+ + \text{H}^-)}{\rightleftharpoons}} \underset{\text{Reduced}}{\text{FADH}_2}$$

When NAD$^+$ is reduced to NADH + H$^+$, the NAD$^+$ gains a hydride ion (H$^-$), and the H$^+$ is released into the surrounding solution. The addition of a hydride ion to NAD$^+$ neutralizes the charge on NAD$^+$ and adds a hydrogen atom so that the reduced form is NADH. When NADH is oxidized to NAD^+, a hydride ion is lost from NADH, which results in one less hydrogen atom and an additional positive charge. Thus, the oxidized form is NAD^+. FAD is reduced to $FADH_2$ when it gains a hydrogen ion and a hydride ion, and $FADH_2$ is oxidized to FAD when it loses the same two ions.

Oxidation and reduction reactions are always coupled; each time one substance is oxidized, another is simultaneously reduced. Such paired reactions are called **oxidation–reduction** or **redox reactions.** For example, when lactic acid is *oxidized* to form pyruvic acid, the two hydrogen atoms removed in the reaction are used to *reduce* NAD^+. This coupled redox reaction may be written as follows:

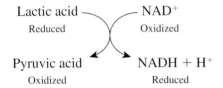

An important point to remember about oxidation–reduction reactions is that oxidation is usually an exergonic (energy-releasing) reaction. Cells use multistep biochemical reactions to release energy from energy-rich, highly reduced compounds (with many hydrogen atoms) to lower-energy, highly oxidized compounds (with many oxygen atoms or multiple bonds). For example, when a cell oxidizes a molecule of glucose ($C_6H_{12}O_6$), the energy in the glucose molecule is removed in a stepwise manner. Ultimately, some of the energy is captured by transferring it to ATP, which then serves as an energy source for energy-requiring reactions within the cell. Compounds that have many hydrogen atoms (for example, glucose) are highly reduced compounds that contain more chemical potential energy than oxidized compounds. For this reason, glucose is a valuable nutrient.

Mechanisms of ATP Generation

Some of the energy released during oxidation reactions is captured within a cell when ATP is formed. Briefly, a phosphate group (P) is added to ADP, with an input of energy, to form ATP. The two high-energy phosphate bonds that can be used to transfer energy are indicated by "squiggles" ($\sim$):

$$\text{Adenosine} - \textcircled{P} \sim \textcircled{P} + \textcircled{P} + \text{energy} \longrightarrow$$
$$\text{ADP}$$
$$\text{Adenosine} - \textcircled{P} \sim \textcircled{P} \sim \textcircled{P}$$
$$\text{ATP}$$

The high-energy phosphate bond that attaches the third phosphate group contains the energy stored in this reaction. The addition of a phosphate group to a molecule, called **phosphorylation** (fos′-for-i-LĀ-shun), increases its potential energy. Organisms use three mechanisms of phosphorylation to generate ATP:

1. **Substrate-level phosphorylation** generates ATP by transferring a high-energy phosphate group from an intermediate phosphorylated metabolic compound—a substrate—directly to ADP. In human cells, this process occurs in the cytosol.

2. **Oxidative phosphorylation** removes electrons from organic compounds and passes them through a series of electron acceptors, called the **electron transport chain,** to molecules of oxygen (O_2). This process occurs in the inner mitochondrial membrane of cells.

3. **Photophosphorylation** occurs only in chlorophyll-containing plant cells or in certain bacteria that contain other light-absorbing pigments.

▶ CHECKPOINT

3. How is a hydride ion different from a hydrogen ion? What is the involvement of both ions in redox reactions?

4. What are three ways that ATP can be generated?

CARBOHYDRATE METABOLISM

▶ OBJECTIVE

• **Describe the fate, metabolism, and functions of carbohydrates.**

During digestion, both polysaccharides and disaccharides are hydrolyzed into the monosaccharides glucose (about 80%), fructose, and galactose. (Some fructose is converted into glucose as it is absorbed through the intestinal epithelial cells.) Hepatocytes (liver cells) convert most of the remaining fructose and practically all the galactose to glucose. Thus, the story of carbohydrate metabolism is really the story of glucose metabolism. Because negative feedback systems maintain blood glucose at about 90 mg/100 mL of plasma (5 mmol/liter), a total of 2–3 g of glucose normally circulates in the blood.

The Fate of Glucose

Because glucose is the body's preferred source for synthesizing ATP, the fate of glucose absorbed from the diet—the use to which it is put—depends on the needs of body cells, which use glucose for:

• *ATP production.* In body cells that require immediate energy, glucose is oxidized to produce ATP. Glucose not needed for immediate ATP production can enter one of several other metabolic pathways.

• *Amino acid synthesis.* Cells throughout the body can use glucose to form several amino acids, which then can be incorporated into proteins.

• *Glycogen synthesis.* Hepatocytes and muscle fibers can perform **glycogenesis** (glī′-kō-JEN-e-sis; *glyco-* = sugar or sweet; *-genesis* = to generate), in which hundreds of glucose monomers are combined to form the polysaccharide glycogen. Total storage capacity of glycogen is about 125 g in the liver and 375 g in skeletal muscles.

• **Triglyceride synthesis.** When the glycogen storage areas are filled up, hepatocytes can transform the glucose to glycerol and fatty acids that can be used for **lipogenesis** (lip′-ō-JEN-ē-sis), the synthesis of triglycerides. Triglycerides then are deposited in adipose tissue, which has virtually unlimited storage capacity.

Glucose Movement into Cells

Before glucose can be used by body cells, it must first pass through the plasma membrane and enter the cytosol. Glucose absorption in the GI tract (and kidney tubules) is accomplished via secondary active transport (Na^+-glucose symporters). Glucose entry into most other body cells occurs via GluT molecules,

a family of facilitated diffusion glucose transporters (see page 68). Insulin increases the insertion of one type of GluT, called GluT4, into the plasma membranes of most body cells, thereby increasing the rate of facilitated diffusion of glucose into cells. In neurons and hepatocytes, however, another type of GluT is always present in the plasma membrane, so glucose entry is always "turned on." Upon entering a cell, glucose becomes phosphorylated. Because GluT cannot transport phosphorylated glucose, this reaction traps glucose within the cell.

Glucose Catabolism

The oxidation of glucose to produce ATP is also known as **cellular respiration,** and it involves four sets of reactions: glycolysis, the formation of acetyl coenzyme A, the Krebs cycle, and the electron transport chain (Figure 25.2).

Figure 25.2 Overview of cellular respiration (oxidation of glucose). A modified version of this figure appears in several places in this chapter to indicate the relationships of particular reactions to the overall process of cellular respiration.

The oxidation of glucose involves glycolysis, the formation of acetyl coenzyme A, the Krebs cycle, and the electron transport chain.

 Which of the four processes shown here is also called anaerobic cellular respiration?

1 *Glycolysis* is a set of reactions in which one glucose molecule is oxidized and two molecules of pyruvic acid are produced. The reactions also produce two molecules of ATP and two energy-containing $NADH + H^+$. Because glycolysis does not require oxygen, it is a way to produce ATP anaerobically (without oxygen) and is known as **anaerobic cellular respiration.**

2 *Formation of acetyl coenzyme A* is a transition step that prepares pyruvic acid for entrance into the Krebs cycle. This step also produces energy-containing $NADH + H^+$ plus carbon dioxide (CO_2).

3 *Krebs cycle reactions* oxidize acetyl coenzyme A and produce CO_2, ATP, energy-containing $NADH + H^+$, and $FADH_2$.

4 *Electron transport chain reactions* oxidize $NADH + H^+$ and $FADH_2$ and transfer their electrons through a series of electron carriers. The Krebs cycle and the electron transport chain together require oxygen to produce ATP and are known as **aerobic cellular respiration.**

Glycolysis

During **glycolysis** (glī-KOL-i-sis; *-lysis* = breakdown), chemical reactions split a six-carbon molecule of glucose into two three-carbon molecules of pyruvic acid (Figure 25.3). Even though glycolysis consumes two ATP molecules, it produces four ATP molecules, for a net gain of two ATP molecules for each glucose molecule that is oxidized.

Figure 25.4 shows the 10 reactions that comprise glycolysis. In the first half of the sequence (reactions **1** through **5**), energy in the form of ATP is "invested" and the 6-carbon glucose is split into two 3-carbon molecules of glyceraldehyde 3-phosphate. *Phosphofructokinase,* the enzyme that catalyzes step **3**, is the key regulator of the rate of glycolysis. The activity of this enzyme is high when ADP concentration is high, in which case ATP is produced rapidly. When the activity of phosphofructokinase is low, most glucose does not enter the reactions of glycolysis but instead undergoes conversion to glycogen for storage. In the second half (reactions **6** through **10**), the two glyceraldehyde 3-phosphate molecules are converted to two pyruvic acid molecules and ATP is generated.

Figure 25.3 Cellular respiration begins with glycolysis.

During glycolysis, each molecule of glucose is converted to two molecules of pyruvic acid.

(a) Cellular respiration

(b) Overview of glycolysis

 For each glucose molecule that undergoes glycolysis, how many ATP molecules are generated?

Figure 25.4 The ten reactions of glycolysis. ① The first reaction phosphorylates glucose, using a phosphate group from an ATP molecule to form glucose 6-phosphate. ② The second reaction converts glucose 6-phosphate to fructose 6-phosphate. ③ The third reaction uses a second ATP to add a second phosphate group to form fructose 1, 6-bisphosphate. ④ and ⑤ Fructose splits into two three-carbon molecules, glyceraldehyde 3-phosphate (G 3-P) and dihydroxyacetone phosphate, each having one phosphate group. ⑥ Oxidation occurs as two molecules of NAD⁺ accept two pairs of electrons and hydrogen ions from two molecules of G 3-P, forming two molecules each of NADH. Many body cells use the two NADH produced in this step to generate four ATPs in the electron transport chain. Hepatocytes, kidney cells, and cardiac muscle fibers, can generate six ATPs from the two NADH. A second phosphate group attaches to G 3-P, forming 1,3-bisphosphoglyceric acid (BPG). ⑦ through ⑩ These reactions generate four molecules of ATP and produce two molecules of pyruvic acid (pyruvate*).

🔑 **Glycolysis results in a net gain of two ATP, two NADH, and two H⁺.**

📞 **Why is the enzyme that catalyzes step ③ called a kinase?**

---

*The carboxyl groups (—COOH) of intermediates in glycolysis and in the citric acid cycle are mostly ionized at the pH of body fluids to —COO⁻. The suffix "-ic acid" indicates the non-ionized form, whereas the ending "-ate" indicates the ionized form. Although the "-ate" names are more correct, we will use the "acid" names because these terms are more familiar.

The Fate of Pyruvic Acid

The fate of pyruvic acid produced during glycolysis depends on the availability of oxygen (Figure 25.5). If oxygen is scarce (anaerobic conditions)—for example, in skeletal muscle fibers during strenuous exercise—then pyruvic acid is reduced via an anaerobic pathway by the addition of two hydrogen atoms to form lactic acid (lactate):

$$2 \text{ Pyruvic acid} + 2 \text{ NADH} + 2 \text{ H}^+ \longrightarrow 2 \text{ Lactic acid} + 2 \text{ NAD}^+$$

Oxidized   Reduced

This reaction regenerates the NAD^+ that was used in the oxidation of glyceraldehyde 3-phosphate (see step ⑥ in Figure 25.4) and thus allows glycolysis to continue. As lactic acid is produced, it rapidly diffuses out of the cell and enters the blood. Hepatocytes remove lactic acid from the blood and convert it back to pyruvic acid.

When oxygen is plentiful (aerobic conditions), most cells convert pyruvic acid to acetyl coenzyme A. This molecule links glycolysis, which occurs in the cytosol, with the Krebs cycle, which occurs in the matrix of mitochondria. Pyruvic acid enters the mitochondrial matrix with the help of a special transporter protein. Because they lack mitochondria, red blood cells can produce ATP through glycolysis only.

Figure 25.5 Fate of pyruvic acid.

🔑 **When oxygen is plentiful, pyruvic acid enters mitochondria, is converted to acetyl coenzyme A, and enters the Krebs cycle (aerobic pathway). When oxygen is scarce, most pyruvic acid is converted to lactic acid via an anaerobic pathway.**

Formation of Acetyl Coenzyme A

Each step in the oxidation of glucose requires a different enzyme, and often a coenzyme as well. The coenzyme used at this point in cellular respiration is **coenzyme A (CoA),** which is derived from pantothenic acid, a B vitamin. During the transitional step between glycolysis and the Krebs cycle, pyruvic acid is prepared for entrance into the cycle. The enzyme *pyruvate dehydrogenase,* which is located exclusively in the mitochondrial matrix, converts pyruvate to a two-carbon fragment by removing a molecule of carbon dioxide (Figure 25.5). The loss of a molecule of CO_2 by a substance is called **decarboxylation** (dē-kar-bok′-si-LĀ-shun). This is the first reaction in cellular respiration that releases CO_2. During this reaction, pyruvic acid is also oxidized. Each pyruvic acid loses two hydrogen atoms in the form of one hydride ion (H^-) plus one hydrogen ion (H^+). The coenzyme NAD^+ is reduced as it picks up the H^- from pyruvic acid; the H^+ is released into the mitochondrial matrix. The reduction of NAD^+ to $NADH + H^+$ is indicated in Figure 25.5 by the curved arrow entering and then leaving the reaction. Recall that the oxidation of one glucose molecule produces two molecules of pyruvic acid, so for each molecule of glucose, two molecules of carbon dioxide are lost and two $NADH + H^+$ are produced. The two-carbon fragment, called an **acetyl group,** attaches to coenzyme A, and the whole complex is called **acetyl coenzyme A (acetyl CoA).**

The Krebs Cycle

Once the pyruvic acid has undergone decarboxylation and the remaining acetyl group has attached to CoA, the resulting compound (acetyl CoA) is ready to enter the Krebs cycle (Figure 25.6). The **Krebs cycle**—named for the biochemist Hans Krebs, who described these reactions in the 1930s—is also known as the **citric acid cycle,** for the first molecule formed when an

In which cellular compartment does glycolysis occur?

Figure 25.6 **After formation of acetyl coenzyme A, the next stage of cellular respiration is the Krebs cycle.**

Reactions of the Krebs cycle occur in the matrix of mitochondria.

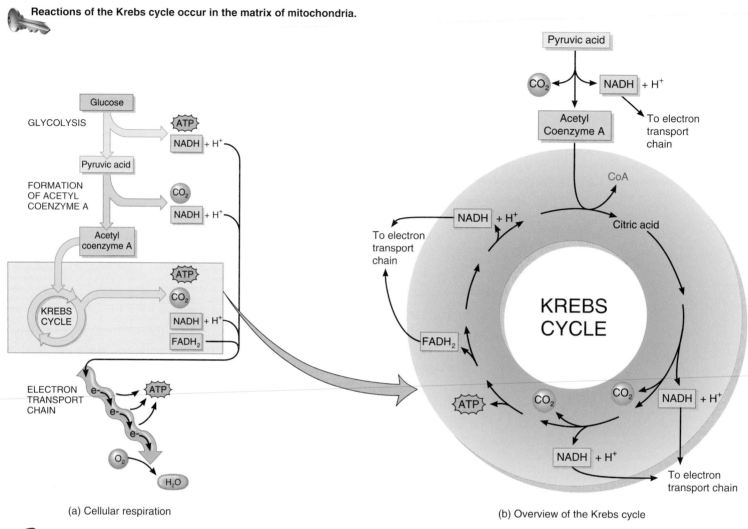

(a) Cellular respiration

(b) Overview of the Krebs cycle

Where in cellular respiration is carbon dioxide given off? What happens to this gas?

acetyl group joins the cycle. The reactions occur in the matrix of mitochondria and consist of a series of oxidation–reduction reactions and decarboxylation reactions that release CO_2. In the Krebs cycle, the oxidation–reduction reactions transfer chemical energy, in the form of electrons, to two coenzymes—NAD^+ and FAD. The pyruvic acid derivatives are oxidized, whereas the coenzymes are reduced. In addition, one step generates ATP. Figure 25.7 shows the reactions of the Krebs cycle in more detail.

The reduced coenzymes (NADH and $FADH_2$) are the most important outcome of the Krebs cycle because they contain the energy originally stored in glucose and then in pyruvic acid. Overall, for every acetyl CoA that enters the Krebs cycle, three NADH, three H^+, and one $FADH_2$ are produced by oxidation–reduction reactions, and one molecule of ATP is generated by substrate-level phosphorylation (see Figure 25.6).

In the electron transport chain, the three NADH + three H^+ will later yield nine ATP molecules, and the $FADH_2$, will later yield two ATP molecules. Thus, each "turn" of the Krebs cycle eventually generates 12 molecules of ATP. Because each glucose provides two acetyl CoA, glucose catabolism via the Krebs cycle and the electron transport chain yields 24 molecules of ATP per glucose molecule.

Liberation of CO_2 occurs as pyruvic acid is converted to acetyl CoA and during the two decarboxylation reactions of the Krebs cycle (see Figure 25.6). Because each molecule of glucose generates two molecules of pyruvic acid, six molecules of CO_2 are liberated from each original glucose molecule catabolized along this pathway. The molecules of CO_2 diffuse out of the mitochondria, through the cytosol and plasma membrane, and then into the blood. Blood transports the CO_2 to the lungs, where it eventually is exhaled.

Figure 25.7 **The eight reactions of the Krebs cycle.** ① *Entry of the acetyl group.* The chemical bond that attaches the acetyl group to coenzyme A (CoA) breaks, and the two-carbon acetyl group attaches to a four-carbon molecule of oxaloacetic acid to form a six-carbon molecule called citric acid. CoA is free to combine with another acetyl group from pyruvic acid and repeat the process. ② *Isomerization.* Citric acid undergoes isomerization to isocitric acid, which has the same molecular formula as citrate. Notice, however, that the hydroxyl group (–OH) is attached to a different carbon. ③ *Oxidative decarboxylation.* Isocitric acid is oxidized and loses a molecule of CO_2, forming alpha-ketoglutaric acid. The H^- from the oxidation is passed on to NAD^+, which is reduced to $NADH + H^+$. ④ *Oxidative decarboxylation.* Alpha-ketoglutaric acid is oxidized, loses a molecule of CO_2, and picks up CoA to form succinyl CoA. ⑤ *Substrate-level phosphorylation.* CoA is displaced by a phosphate group, which is then transferred to guanosine diphosphate (GDP) to form guanosine triphosphate (GTP). GTP can donate a phosphate group to ADP to form ATP. ⑥ *Dehydrogenation.* Succinic acid is oxidized to fumaric acid as two of its hydrogen atoms are transferred to the coenzyme flavin adenine nucleotide (FAD), which is reduced to $FADH_2$. ⑦ *Hydration.* Fumaric acid is converted to malic acid by the addition of a molecule of water. ⑧ *Dehydrogenation.* In the final step in the cycle, malic acid is oxidized to re-form oxaloacetic acid. Two hydrogen atoms are removed and one is transferred to NAD^+, which is reduced to $NADH + H^+$. The regenerated oxaloacetic acid can combine with another molecule of acetyl CoA, beginning a new cycle.

🔑 The three main results of the Krebs cycle are the production of reduced coenzymes ($NADH + H^+$ and $FADH_2$), which contain stored energy; the generation of GTP, a high-energy compound that is used to produce ATP; and the formation of CO_2, which is transported to the lungs and exhaled.

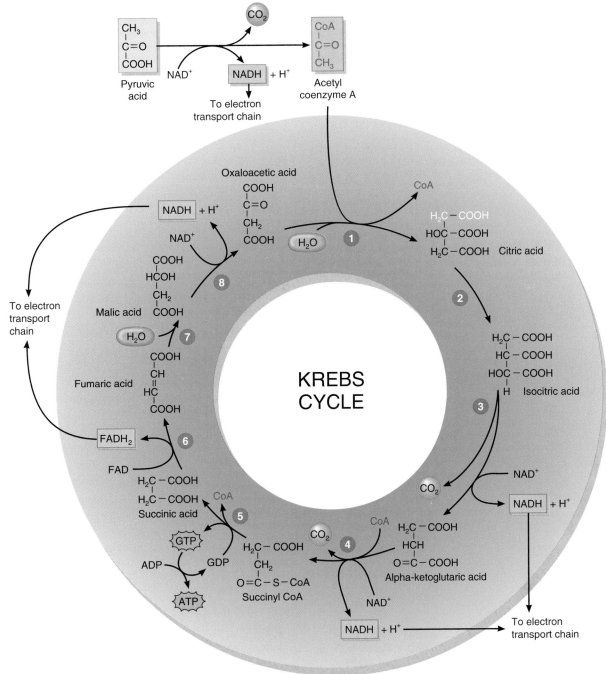

Why is the production of reduced coenzymes important in the Krebs cycle?

915

The Electron Transport Chain

The **electron transport chain** is a series of **electron carriers** that are integral membrane proteins in the inner mitochondrial membrane. This membrane is folded into cristae that increase its surface area, thus accommodating thousands of copies of the transport chain in each mitochondrion. Each carrier in the chain is reduced as it picks up electrons and is oxidized as it gives up electrons. As electrons pass through the chain, exergonic reactions release small amounts of energy in a series of steps, and this energy is used to form ATP. In aerobic cellular respiration, the final electron acceptor of the chain is oxygen. Because this mechanism of ATP generation links chemical reactions (the passage of electrons along the transport chain) with the pumping of hydrogen ions, it is called **chemiosmosis** (kem′-ē-oz-MŌ-sis; *chemi-* = chemical; *-osmosis* = pushing). Briefly, chemiosmosis works as follows (Figure 25.8):

1 Energy from NADH + H⁺ passes along the electron transport chain and is used to pump H⁺ (protons) from the matrix of the mitochondrion into the space between the inner and outer mitochondrial membranes.

2 A high concentration of H⁺ accumulates between the inner and outer mitochondrial membranes.

3 ATP synthesis then occurs as hydrogen ions flow back into the mitochondrial matrix through a special type of H⁺ channel in the inner membrane.

ELECTRON CARRIERS   Several types of molecules and atoms serve as electron carriers:

- **Flavin mononucleotide (FMN),** like FAD (flavin adenine dinucleotide), is a flavoprotein derived from riboflavin (vitamin B₂).

- **Cytochromes** (SĪ-tō-krōmz) are proteins with an iron-containing group (heme) capable of existing alternately in a reduced form (Fe^{2+}) and an oxidized form (Fe^{3+}). The several cytochromes involved in the electron transport chain are cytochromes *b* (cyt *b*), cytochromes c_1 (cyt c_1), cytochrome *c* (cyt *c*), cytochrome *a* (cyt *a*), and cytochrome a_3 (cyt a_3).

- **Iron-sulfur (Fe-S) centers** contain either two or four iron atoms bound to sulfur atoms that form an electron transfer center within a protein.

- **Copper (Cu) atoms** bound to two proteins in the chain also participate in electron transfer.

- **Coenzyme Q,** symbolized **Q,** is a nonprotein, low-molecular-weight carrier that is mobile in the lipid bilayer of the inner membrane.

STEPS IN ELECTRON TRANSPORT AND CHEMIOSMOTIC ATP GENERATION   Within the inner mitochondrial membrane, the carriers of the electron transport chain are clustered into three complexes, each of which acts as a **proton pump** that expels H⁺ from the mitochondrial matrix and helps create an electrochemical gradient of H⁺. Each of the three proton pumps is a complex that transports electrons and pumps H⁺, as shown in Figure 25.9. Notice that oxygen is used to help form water in step **3** . This is the only point in aerobic cellular respiration where O_2 is consumed. Cyanide is a deadly poison because it binds to the cytochrome oxidase complex and blocks this last step in electron transport.

The pumping of H⁺ produces both a concentration gradient of protons and an electrical gradient. The buildup of H⁺ on one side of the inner mitochondrial membrane makes that side positively charged compared with the other side. The resulting electrochemical gradient has potential energy, called the *proton motive force.* Included in the inner mitochondrial membrane are proton channels that allow H⁺ to flow back across the membrane, driven by the proton motive force. As H⁺ flow back, they generate ATP because the H⁺ channels also include an enzyme called **ATP synthase.** The enzyme uses the proton motive force to synthesize ATP from ADP and Ⓟ. This process of chemiosmosis is responsible for most of the ATP produced during cellular respiration.

Figure 25.8   Chemiosmosis.

🔑 **In chemiosmosis, ATP is produced when hydrogen ions diffuse back into the mitochondrial matrix.**

❓ **What is the energy source that powers the proton pumps?**

Summary of Cellular Respiration

The various electron transfers in the electron transport chain generate either 32 or 34 ATP molecules from each molecule of

Figure 25.9 **The actions of the three proton pumps and ATP synthase in the inner membrane of mitochondria.** Each pump is a complex of three or more electron carriers. **❶** The first proton pump is the *NADH dehydrogenase complex,* which contains flavin mononucleotide (FMN) and five or more Fe-S centers. $NADH + H^+$ is oxidized to NAD^+, and FMN is reduced to $FMNH_2$, which in turn is oxidized as it passes electrons to the iron-sulfur centers. Q, which is mobile in the membrane, shuttles electrons to the second pump complex. **❷** The second proton pump is the *cytochrome b-c₁* complex, which contains cytochromes and an iron-sulfur center.

Electrons are passed successively from Q to cyt *b,* to Fe-S, to cyt c_1. The mobile shuttle that passes electrons from the second pump complex to the third is cytochrome *c* (cyt *c*). **❸** The third proton pump is the *cytochrome oxidase complex,* which contains cytochromes *a* and a_3 and two copper atoms. Electrons pass from cyt *c,* to Cu, to cyt *a,* and finally to cyt a_3. Cyt a_3 passes its electrons to one-half of a molecule of oxygen (O_2), which becomes negatively charged and then picks up two H^+ from the surrounding medium to form H_2O.

As the three proton pumps pass electrons from one carrier to the next, they also move protons (H^+) from the matrix into the space between the inner and outer mitochondrial membranes. As protons flow back into the mitochondrial matrix through the H^+ channel in ATP synthase, ATP is synthesized.

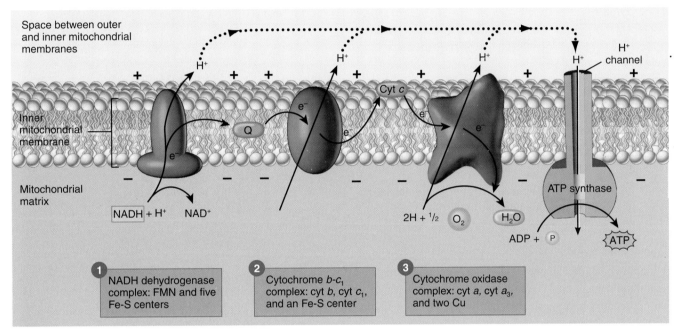

Where is the concentration of H^+ highest?

glucose that is oxidized: either 28 or 30* from the ten molecules of $NADH + H^+$ and two from each of the two molecules of $FADH_2$ (four total). Thus, during cellular respiration, 36 or 38 ATPs can be generated from one molecule of glucose. Note that two of those ATPs come from substrate-level phosphorylation in glycolysis, and two come from substrate-level phosphorylation in the Krebs cycle. The overall reaction is:

$$\underset{\text{Glucose}}{C_6H_{12}O_6} + \underset{\text{Oxygen}}{6\,O_2} + 36 \text{ or } 38 \text{ ADPs} + 36 \text{ or } 38 \,\textcircled{P} \longrightarrow$$

$$6\,CO_2 + 6\,H_2O + 36 \text{ or } 38 \text{ ATPs}$$

$$\underset{\text{dioxide}}{\underset{\text{Carbon}}{}} \quad \underset{\text{Water}}{}$$

Table 25.1 summarizes the ATP yield during cellular respiration. A schematic depiction of the principal reactions of cellular respiration is presented in Figure 25.10. The actual ATP yield may be lower than 36 or 38 ATPs per glucose. One uncertainty is the exact number of H^+ that must be pumped out to generate one ATP during chemiosmosis. Moreover, the ATP generated in mitochondria must be transported out of these organelles into the cytosol for use elsewhere in a cell. Exporting ATP in exchange for the inward movement of ADP that is formed from metabolic reactions in the cytosol uses up part of the proton motive force.

---

* The two NADH produced in the cytosol during glycolysis cannot enter mitochondria. Instead, they donate their electrons to one of two transfer molecules, known as the malate shuttle and the glycerol phosphate shuttle. In cells of the liver, kidneys, and heart, use of the malate shuttle results in three ATPs synthesized for each NADH. In other body cells, such as skeletal muscle fibers and neurons, use of the glycerol phosphate shuttle results in two ATPs synthesized for each NADH.

Table 25.1	Summary of ATP Produced in Cellular Respiration	
Source	**ATP Yield Per Glucose Molecule (Process)**	
Glycolysis		
Oxidation of one glucose molecule to two pyruvic acid molecules	2 ATPs (substrate-level phosphorylation)	
Production of 2 NADH + H⁺	4 or 6 ATPs (oxidative phosphorylation in electron transport chain)	
Formation of Two Molecules of Acetyl Coenzyme A		
2 NADH + 2 H⁺	6 ATPs (oxidative phosphorylation in electron transport chain)	
Krebs Cycle and Electron Transport Chain		
Oxidation of succinyl CoA to succinic acid	2 GTPs that are converted to 2 ATPs (substrate-level phosphorylation)	
Production of 6 NADH + 6 H⁺	18 ATPs (oxidative phosphorylation in electron transport chain)	
Production of 2 FADH₂	4 ATPs (oxidative phosphorylation in electron transport chain)	
Total:	36 or 38 ATPs per glucose molecule (theoretical maximum)	

Glycolysis, the Krebs cycle, and especially the electron transport chain provide all the ATP for cellular activities. Because the Krebs cycle and electron transport chain are aerobic processes, cells cannot carry on their activities for long if oxygen is lacking.

Glucose Anabolism

Even though most of the glucose in the body is catabolized to generate ATP, glucose may take part in or be formed via several anabolic reactions. One is the synthesis of glycogen; another is the synthesis of new glucose molecules from some of the products of protein and lipid breakdown.

Glucose Storage: Glycogenesis

If glucose is not needed immediately for ATP production, it combines with many other molecules of glucose to form **glycogen,** a polysaccharide that is the only stored form of carbohydrate in our bodies. The hormone insulin, from pancreatic beta cells, stimulates hepatocytes and skeletal muscle cells to carry out **glycogenesis,** the synthesis of glycogen (Figure 25.11). The body can store about 500 g (about 1.1 lb) of glycogen, roughly 75% in skeletal muscle fibers and the rest in liver cells. During glycogenesis, glucose is first phosphorylated to glucose 6-phosphate by hexokinase. Glucose 6-phosphate is converted to glucose 1-phosphate, then to uridine diphosphate glucose, and finally to glycogen.

Figure 25.10   Summary of the principal reactions of cellular respiration. ETC = electron transport chain and chemiosmosis.

 Except for glycolysis, which occurs in the cytosol, all other reactions of cellular respiration occur within mitochondria.

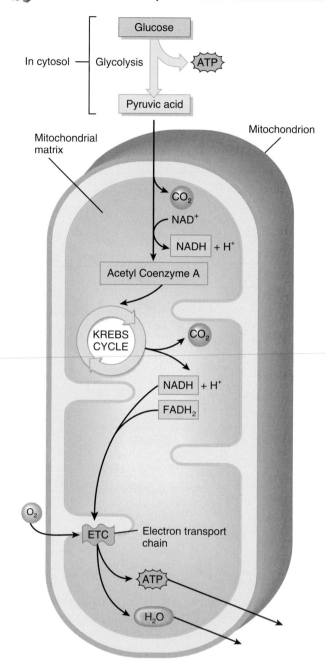

How many molecules of O_2 are used and how many molecules of CO_2 are produced during the complete oxidation of one glucose molecule?

Figure 25.11 Glycogenesis and glycogenolysis.

 The glycogenesis pathway converts glucose into glycogen, whereas the glycogenolysis pathway breaks down glycogen into glucose.

Key:
→ Glycogenesis (stimulated by insulin)
→ Glycogenolysis (stimulated by glucagon and epinephrine)

Besides hepatocytes, which body cells can synthesize glycogen? Why can't they release glucose into the blood?

Glucose Release: Glycogenolysis

When body activities require ATP, glycogen stored in hepatocytes is broken down into glucose and released into the blood to be transported to cells, where it will be catabolized. The process of splitting glycogen into its glucose subunits is called **glycogenolysis** (glī′-kō-je-NOL-e-sis). (Note: Do not confuse *glycogenolysis,* which is the breakdown of glycogen to glucose, with *glycolysis,* the ten reactions that convert glucose to pyruvic acid.)

Glycogenolysis is not a simple reversal of the steps of glycogenesis (Figure 25.11). It begins by splitting glucose molecules off the branched glycogen molecule via phosphorylation to form glucose 1-phosphate. Phosphorylase, the enzyme that catalyzes this reaction, is activated by glucagon from pancreatic alpha cells and epinephrine from the adrenal medulla. Glucose 1-phosphate is then converted to glucose 6-phosphate and finally to glucose, which leaves hepatocytes via glucose transporters (GluT) in the plasma membrane. Phosphorylated glucose molecules cannot ride aboard the GluT transporters, however, and **phosphatase,** the enzyme that converts glucose 6-phosphate into glucose, is absent in skeletal muscle cells. Thus, hepatocytes, which have phosphatase, can release glucose derived from glycogen to the bloodstream, whereas skeletal muscle cells cannot. In skeletal muscle cells, glycogen is broken down into glucose 1-phosphate, which is then catabolized for ATP production via glycolysis and the Krebs cycle. However, the lactic acid pro-

duced by glycolysis in muscle cells can be converted to glucose in the liver. In this way, muscle glycogen can be an indirect source of blood glucose.

Carbohydrate Loading

The amount of glycogen stored in the liver and skeletal muscles varies and can be completely exhausted during long-term athletic endeavors. Thus, many marathon runners and other endurance athletes follow a precise exercise and dietary regimen that includes eating large amounts of complex carbohydrates, such as pasta and potatoes, in the three days before an event. This practice, called **carbohydrate loading,** helps maximize the amount of glycogen available for ATP production in muscles. For athletic events lasting more than an hour, carbohydrate loading has been shown to increase an athlete's endurance. ■

Formation of Glucose from Proteins and Fats: Gluconeogenesis

When your liver runs low on glycogen, it is time to eat. If you don't, your body starts catabolizing triglycerides (fats) and proteins. Actually, the body normally catabolizes some of its triglycerides and proteins, but large-scale triglyceride and protein catabolism does not happen unless you are starving, eating very few carbohydrates, or suffering from an endocrine disorder.

The glycerol part of triglycerides, lactic acid, and certain amino acids can be converted in the liver to glucose (Figure 25.12). The process by which glucose is formed from these noncarbohydrate sources is called **gluconeogenesis** (gloo′-kō-nē′-ō-JEN-e-sis; *neo* = new). An easy way to distinguish this term from glycogenesis or glycogenolysis is to remember that in this

Figure 25.12 Gluconeogenesis, the conversion of noncarbohydrate molecules (amino acids, lactic acid, and glycerol) into glucose.

About 60% of the amino acids in the body can be used for gluconeogenesis.

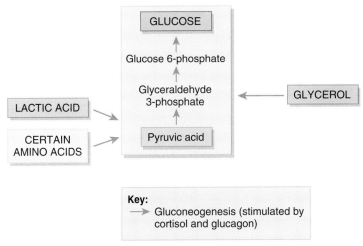

Key:
→ Gluconeogenesis (stimulated by cortisol and glucagon)

What cells can carry out gluconeogenesis and glycogenesis?

case glucose is not converted back from glycogen, but is instead *newly formed*. About 60% of the amino acids in the body can be used for gluconeogenesis. Amino acids such as alanine, cysteine, glycine, serine, and threonine, and lactic acid are converted to pyruvic acid, which then may be synthesized into glucose or enter the Krebs cycle. Glycerol may be converted into glyceraldehyde 3-phosphate, which may form pyruvic acid or be used to synthesize glucose.

Gluconeogenesis is stimulated by cortisol, the main glucocorticoid hormone of the adrenal cortex, and by glucagon from the pancreas. In addition, cortisol stimulates the breakdown of proteins into amino acids, thus expanding the pool of amino acids available for gluconeogenesis. Thyroid hormones (thyroxine and triiodothyronine) also mobilize proteins and may mobilize triglycerides from adipose tissue, thereby making glycerol available for gluconeogenesis.

▶ CHECKPOINT

5. How can glucose move into or out of body cells?

6. What happens during glycolysis?

7. How is acetyl coenzyme A formed?

8. Outline the principal events and outcomes of the Krebs cycle.

9. What happens in the electron transport chain and why is this process called chemiosmosis?

10. Which reactions produce ATP during the complete oxidation of a molecule of glucose?

11. Under what circumstances do glycogenesis and glycogenolysis occur?

12. What is gluconeogenesis, and why is it important?

LIPID METABOLISM

▶ OBJECTIVES

• Describe the lipoproteins that transport lipids in the blood.
• Describe the fate, metabolism, and functions of lipids.

Transport of Lipids by Lipoproteins

Most lipids, such as triglycerides, are nonpolar and therefore very hydrophobic molecules. They do not dissolve in water. To be transported in watery blood, such molecules first must be made more water soluble. This chemical transformation is accomplished by combining them with proteins produced by the liver and intestine. The lipid and protein combinations thus formed are **lipoproteins,** spherical particles with an outer shell of proteins, phospholipids, and cholesterol molecules surrounding an inner core of triglycerides and other lipids (Figure 25.13). The proteins in the outer shell are called **apoproteins (apo)** and are designated by the letters A, B, C, D, and E plus a number. Besides helping to solubilize the lipoprotein in body fluids, each apoprotein also has specific functions.

There are several types of lipoproteins, each having different functions, but all essentially are transport vehicles. They provide

delivery and pickup services so that lipids can be available when cells need them or be removed from circulation when they are not needed. Lipoproteins are categorized and named mainly according to their density, which varies with the ratio of lipids (which have a low density) to proteins (which have a high density). From largest and lightest to smallest and heaviest, the four major classes of lipoproteins are chylomicrons, very low-density lipoproteins (VLDLs), low-density lipoproteins (LDLs), and high-density lipoproteins (HDLs).

Chylomicrons form in mucosal epithelial cells of the small intestine and contain mainly dietary lipids, which they transport to adipose tissue for storage. They contain about 1–2% proteins, 85% triglycerides, 7% phospholipids, and 6–7% cholesterol, plus a small amount of fat-soluble vitamins. Chylomicrons enter lacteals of intestinal villi and are carried by lymph into venous blood and then into the systemic circulation. Their presence gives blood plasma a milky appearance, but they remain in the blood for only a few minutes. As chylomicrons circulate through the capillaries of adipose tissue, one of their apoproteins, **apo C-2,** activates *endothelial lipoprotein lipase,* an enzyme that removes fatty acids from chylomicron triglycerides. The free fatty acids are then taken up by adipocytes for synthesis and storage of triglycerides and by muscle cells for ATP production. Hepatocytes remove chylomicron remnants from the blood via receptor-mediated endocytosis, in which another chylomicron apoprotein, **apo E,** is the docking protein.

Very low-density lipoproteins (VLDLs) form in hepatocytes and contain mainly *endogenous* (made in the body) lipids.

Figure 25.13 A lipoprotein. Shown here is a VLDL.

A single layer of amphipathic phospholipids, cholesterol, and proteins surrounds a core of nonpolar lipids.

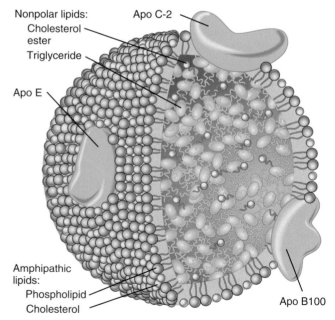

Nonpolar lipids:
Cholesterol ester
Triglyceride
Apo C-2
Apo E
Amphipathic lipids:
Phospholipid
Cholesterol
Apo B100

 Which type of lipoprotein delivers cholesterol to body cells?

VLDLs contain about 10% proteins, 50% triglycerides, 20% phospholipids, and 20% cholesterol. VLDLs transport triglycerides synthesized in hepatocytes to adipocytes for storage. Like chylomicrons, they lose triglycerides as their apo C-2 activates endothelial lipoprotein lipase, and the resulting fatty acids are taken up by adipocytes for storage and by muscle cells for ATP production. As they deposit some of their triglycerides in adipose cells, VLDLs are converted to LDLs.

Low-density lipoproteins (LDLs) contain 25% proteins, 5% triglycerides, 20% phospholipids, and 50% cholesterol. They carry about 75% of the total cholesterol in blood and deliver it to cells throughout the body for use in repair of cell membranes and synthesis of steroid hormones and bile salts. The only apoprotein LDLs contain is **apo B100,** which is the docking protein that binds to LDL receptors for receptor-mediated endocytosis of the LDL into a body cell. Within the cell, the LDL is broken down, and the cholesterol is released to serve the cell's needs. Once a cell has sufficient cholesterol for its activities, a negative feedback system inhibits the cell's synthesis of new LDL receptors.

When present in excessive numbers, LDLs also deposit cholesterol in and around smooth muscle fibers in arteries, forming fatty plaques that increase the risk of coronary artery disease (see page 447). For this reason, the cholesterol in LDLs, called LDL-cholesterol, is known as "bad" cholesterol. Because some people have too few LDL receptors, their body cells remove LDL from the blood less efficiently; as a result, their plasma LDL level is abnormally high, and they are more likely to develop fatty plaques. Furthermore, eating a high-fat diet increases the production of VLDLs, which elevates the LDL level and increases the formation of fatty plaques.

High-density lipoproteins (HDLs), which contain 40–45% proteins, 5–10% triglycerides, 30% phospholipids, and 20% cholesterol, remove excess cholesterol from body cells and the blood and transport it to the liver for elimination. Because HDLs prevent accumulation of cholesterol in the blood, a high HDL level is associated with decreased risk of coronary artery disease. For this reason, HDL-cholesterol is known as "good" cholesterol.

Sources and Significance of Blood Cholesterol

There are two sources of cholesterol in the body. Some is present in foods (eggs, dairy products, organ meats, beef, pork, and processed luncheon meats), but most is synthesized by hepatocytes. Fatty foods that don't contain any cholesterol at all can still dramatically increase blood cholesterol level in two ways. First, a high intake of dietary fats stimulates reabsorption of cholesterol-containing bile back into the blood, so less cholesterol is lost in the feces. Second, when saturated fats are broken down in the body, hepatocytes use some of the breakdown products to make cholesterol.

A lipid profile test usually measures total cholesterol (TC), HDL-cholesterol, and triglycerides (VLDLs). LDL-cholesterol then is calculated by using the following formula: LDL-cholesterol = TC − HDL-cholesterol − (triglycerides/5). In the United States, blood cholesterol is usually measured in milligrams per deciliter (mg/dL); a deciliter is 0.1 liter or 100 mL. For adults, desirable levels of blood cholesterol are total cholesterol under 200 mg/dL, LDL-cholesterol under 130 mg/dL, and HDL-cholesterol over 40 mg/dL. Normally, triglycerides are in the range of 10–190 mg/dL.

As total cholesterol level increases, the risk of coronary artery disease begins to rise. When total cholesterol is above 200 mg/dL (5.2 mmol/liter), the risk of a heart attack doubles with every 50 mg/dL (1.3 mmol/liter) increase in total cholesterol. Total cholesterol of 200–230 mg/dL and LDL of 130–159 mg/dL are borderline-high, whereas total cholesterol above 239 mg/dL and LDL above 159 mg/dL are classified as high blood cholesterol. The ratio of total cholesterol to HDL-cholesterol predicts the risk of developing coronary artery disease. For example, a person with a total cholesterol of 180 mg/dL and HDL of 60 mg/dL has a risk ratio of 3. Ratios above 4 are considered undesirable; the higher the ratio, the greater the risk of developing coronary artery disease.

Among the therapies used to reduce blood cholesterol level are exercise, diet, and drugs. Regular physical activity at aerobic and nearly aerobic levels raises HDL level. Dietary changes are aimed at reducing the intake of total fat, saturated fats, and cholesterol. Drugs used to treat high blood cholesterol levels include cholestyramine (Questran) and colestipol (Colestid), which promote excretion of bile in the feces; nicotinic acid (Liponicin); and the "statin" drugs—atorvastatin (Lipitor), lovastatin (Mevacor), and simvastatin (Zocor), which block the key enzyme (HMG-CoA reductase) needed for cholesterol synthesis.

The Fate of Lipids

Lipids, like carbohydrates, may be oxidized to produce ATP. If the body has no immediate need to use lipids in this way, they are stored in adipose tissue (fat depots) throughout the body and in the liver. A few lipids are used as structural molecules or to synthesize other essential substances. Some examples include phospholipids, which are constituents of plasma membranes; lipoproteins, which are used to transport cholesterol throughout the body; thromboplastin, which is needed for blood clotting; and myelin sheaths, which speed up nerve impulse conduction. Two **essential fatty acids** that the body cannot synthesize are linoleic acid and linolenic acid. Dietary sources include vegetable oils and leafy vegetables. The various functions of lipids in the body may be reviewed in Table 2.7 on page 45.

Triglyceride Storage

A major function of adipose tissue is to remove triglycerides from chylomicrons and VLDLs and store them until they are needed for ATP production in other parts of the body. Triglycerides stored in adipose tissue constitute 98% of all body energy reserves. They are stored more readily than is glycogen, in part because triglycerides are hydrophobic and do not exert osmotic pressure on cell membranes. Adipose tissue also insulates and protects various parts of the body. Adipocytes in the subcutaneous layer contain about 50% of the stored triglycerides. Other adipose tissues account for the other half: about 12% around the kidneys, 10–15% in the omenta, 15% in genital areas, 5–8% between muscles, and 5% behind the eyes, in the sulci of the heart, and attached to the outside of the large intestine. Triglycerides in adipose tissue are continually broken down and resynthesized. Thus, the triglycerides stored in adipose tissue today are not the same molecules that were present last month because they are continually released from storage, transported in the blood, and redeposited in other adipose tissue cells.

Lipid Catabolism: Lipolysis

Muscle, liver, and adipose tissue routinely oxidize fatty acids derived from triglycerides to produce ATP. First, the triglycerides must be split into glycerol and fatty acids, a process called **lipolysis** (li-POL-i-sis) that is catalyzed by enzymes called **lipases.** Two hormones that enhance triglyceride breakdown into

fatty acids and glycerol are epinephrine and norepinephrine, which are released when sympathetic tone increases, as occurs, for example, during exercise. Other lipolytic hormones are cortisol, thyroid hormones, and insulinlike growth factors. By contrast, insulin itself inhibits lipolysis.

The glycerol and fatty acids that result from lipolysis are catabolized via different pathways (Figure 25.14). Glycerol is converted by many cells of the body to glyceraldehyde 3-phosphate, one of the compounds also formed during the catabolism of glucose. If ATP supply in a cell is high, glyceraldehyde 3-phosphate is converted into glucose, an example of gluconeogenesis. If ATP supply in a cell is low, glyceraldehyde 3-phosphate enters the catabolic pathway to pyruvic acid.

Fatty acids are catabolized differently than is glycerol and yield more ATP. The first stage in fatty acid catabolism is a series of reactions, collectively called **beta oxidation,** that occurs in the matrix of mitochondria. Enzymes remove two carbon atoms at a time from the long chain of carbon atoms composing a fatty acid and attach the resulting two-carbon fragment to coenzyme A, forming acetyl CoA. Then, acetyl CoA enters the Krebs cycle (Figure 25.14). A 16-carbon fatty acid such as palmitic acid can yield as many as 129 ATPs upon its complete

Figure 25.14 Pathways of lipid metabolism. Glycerol may be converted to glyceraldehyde 3-phosphate, which can then be converted to glucose or enter the Krebs cycle for oxidation. Fatty acids undergo beta oxidation and enter the Krebs cycle via acetyl coenzyme A. The synthesis of lipids from glucose or amino acids is called lipogenesis.

Glycerol and fatty acids are catabolized in separate pathways.

 What types of cells can carry out lipogenesis, beta oxidation, and lipolysis? What type of cell can carry out ketogenesis?

oxidation via beta oxidation, the Krebs cycle, and the electron transport chain.

As part of normal fatty acid catabolism, hepatocytes can take two acetyl CoA molecules at a time and condense them to form **acetoacetic acid.** This reaction liberates the bulky CoA portion, which cannot diffuse out of cells. Some acetoacetic acid is converted into **beta-hydroxybutyric acid** and **acetone.** The formation of these three substances, collectively known as **ketone bodies,** is called **ketogenesis** (Figure 25.14). Because ketone bodies freely diffuse through plasma membranes, they leave hepatocytes and enter the bloodstream.

Other cells take up acetoacetic acid and attach its four carbons to two coenzyme A molecules to form two acetyl CoA molecules, which can then enter the Krebs cycle for oxidation. Heart muscle and the cortex (outer part) of the kidneys use acetoacetic acid in preference to glucose for generating ATP. Hepatocytes, which make acetoacetic acid, cannot use it for ATP production because they lack the enzyme that transfers acetoacetic acid back to coenzyme A.

Lipid Anabolism: Lipogenesis

Liver cells and adipose cells can synthesize lipids from glucose or amino acids through **lipogenesis** (Figure 25.14), which is stimulated by insulin. Lipogenesis occurs when individuals consume more calories than are needed to satisfy their ATP needs. Excess dietary carbohydrates, proteins, and fats all have the same fate— they are converted into triglycerides. Certain amino acids can undergo the following reactions: amino acids → acetyl CoA → fatty acids → triglycerides. The use of glucose to form lipids takes place via two pathways: (1) glucose → glyceraldehyde 3-phosphate → glycerol and (2) glucose → glyceraldehyde 3-phosphate → acetyl CoA → fatty acids. The resulting glycerol and fatty acids can undergo anabolic reactions to become stored triglycerides, or they can go through a series of anabolic reactions to produce other lipids such as lipoproteins, phospholipids, and cholesterol.

 Ketosis

The level of ketone bodies in the blood normally is very low because other tissues use them for ATP production as fast as they are generated from the breakdown of fatty acids in the liver. During periods of excessive beta oxidation, however, the production of ketone bodies exceeds their uptake and use by body cells. This might occur after a meal rich in triglycerides, or during fasting or starvation, because few carbohydrates are available for catabolism. Excessive beta oxidation may also occur in poorly controlled or untreated diabetes mellitus for two reasons: (1) Because adequate glucose cannot get into cells, triglycerides are used for ATP production, and (2) because insulin normally inhibits lipolysis, a lack of insulin accelerates the pace of lipolysis. When the concentration of ketone bodies in the blood rises above normal—a condition called **ketosis**—the ketone bodies, most of which are acids, must be buffered. If too many accumulate, they decrease the concentration of buffers such as bicarbonate ions, and blood pH falls. Thus, extreme or prolonged ketosis can lead to **acidosis (ketoacidosis),** an abnormally low blood pH. When a diabetic becomes seriously insulin deficient, one of the telltale signs is the sweet smell on the breath from the ketone body acetone. ∎

▶ CHECKPOINT

13. What are the functions of the apoproteins in lipoproteins?
14. Which lipoprotein particles contain "good" and "bad" cholesterol, and why are these terms used?
15. Where are triglycerides stored in the body?
16. Explain the principal events of the catabolism of glycerol and fatty acids.
17. What are ketone bodies? What is ketosis?
18. Define lipogenesis and explain its importance.

PROTEIN METABOLISM

▶ O B J E C T I V E

• **Describe the fate, metabolism, and functions of proteins.**

During digestion, proteins are broken down into amino acids. Unlike carbohydrates and triglycerides, which are stored, proteins are not warehoused for future use. Instead, amino acids are either oxidized to produce ATP or used to synthesize new proteins for body growth and repair. Excess dietary amino acids are not excreted in the urine or feces but instead are converted into glucose (gluconeogenesis) or triglycerides (lipogenesis).

The Fate of Proteins

The active transport of amino acids into body cells is stimulated by insulinlike growth factors (IGFs) and insulin. Almost immediately after digestion, amino acids are reassembled into proteins. Many proteins function as enzymes; others are involved in transportation (hemoglobin) or serve as antibodies, clotting chemicals (fibrinogen), hormones (insulin), or contractile elements in muscle fibers (actin and myosin). Several proteins serve as structural components of the body (collagen, elastin, and keratin). The various functions of proteins in the body may be reviewed in Table 2.8 on page 48.

Protein Catabolism

A certain amount of protein catabolism occurs in the body each day, stimulated mainly by cortisol from the adrenal cortex. Proteins from worn-out cells (such as red blood cells) are broken down into amino acids. Some amino acids are converted into other amino acids, peptide bonds are re-formed, and new proteins are synthesized as part of the recycling process. Hepatocytes convert some amino acids to fatty acids, ketone bodies, or glucose. Also, cells throughout the body oxidize a small amount of amino acids to generate ATP via the Krebs cycle and the electron transport chain. However, before amino acids can be oxidized, they must first be converted to molecules that are part of the Krebs cycle or can enter the Krebs cycle, such as acetyl CoA

(Figure 25.15). Before amino acids can enter the Krebs cycle, their amino group (NH$_2$) must first be removed—a process called **deamination** (dē-am′-i-NĀ-shun). Deamination occurs in hepatocytes and produces ammonia (NH$_3$). The liver cells then convert the highly toxic ammonia to urea, a relatively harmless substance that is excreted in the urine. The conversion of amino acids into glucose (gluconeogenesis) may be reviewed in Figure 25.12; the conversion of amino acids into fatty acids (lipogenesis) or ketone bodies (ketogenesis) is shown in Figure 25.14.

Figure 25.15 **Various points at which amino acids (shown in yellow boxes) enter the Krebs cycle for oxidation.**

Before amino acids can be catabolized, they must first be converted to various substances that can enter the Krebs cycle.

 What group is removed from an amino acid before it can enter the Krebs cycle, and what is this process called?

Protein Anabolism

Protein anabolism, the formation of peptide bonds between amino acids to produce new proteins, is carried out on the ribosomes of almost every cell in the body, directed by the cells' DNA and RNA (see Figure 3.30 on page 89). Insulinlike growth factors, thyroid hormones (T3 and T4), insulin, estrogen, and testosterone stimulate protein synthesis. Because proteins are a main component of most cell structures, adequate dietary protein is especially essential during the growth years, during pregnancy, and when tissue has been damaged by disease or injury. Once dietary intake of protein is adequate, eating more protein will not increase bone or muscle mass; only a regular program of forceful, weight-bearing muscular activity accomplishes that goal.

Of the 20 amino acids in the human body, ten are **essential amino acids:** They must be present in the diet because they cannot be synthesized in the body in adequate amounts. Humans are unable to synthesize seven amino acids (isoleucine, leucine, methionine, phenylalanine, threonine, tryptophan, and valine) and synthesize two others (arginine and histidine) in inadequate amounts, especially in childhood. Because plants synthesize all amino acids, humans can obtain the essential ones by eating plants or animals that eat plants. **Nonessential amino acids** can be synthesized by body cells. They are formed by **transamination,** the transfer of an amino group from an amino acid to pyruvic acid or to an acid in the Krebs cycle. Once the appropriate essential and nonessential amino acids are present in cells, protein synthesis occurs rapidly.

Phenylketonuria

Phenylketonuria (fen'-il-kē'-tō-NOO-rē-a) or **PKU** is a genetic error of protein metabolism characterized by elevated blood levels of the amino acid phenylalanine. Most children with phenylketonuria have a mutation in the gene that codes for the enzyme phenylalanine hydroxylase, the enzyme needed to convert phenylalanine into the amino acid tyrosine, which can enter the Krebs cycle (see Figure 25.15). Because the enzyme is deficient, phenylalanine cannot be metabolized, and what is not used in protein synthesis builds up in the blood. If untreated, the disorder causes vomiting, rashes, seizures, growth deficiency, and severe mental retardation. Newborns are screened for PKU, and mental retardation can be prevented by restricting the child to a diet that supplies only the amount of phenylalanine needed for growth, although learning disabilities may still ensue. Because the artificial sweetener aspartame (NutraSweet) contains phenylalanine, its consumption must be restricted in children with PKU. ■

▶ CHECKPOINT

19. What is deamination and why does it occur?

20. What are the possible fates of the amino acids from protein catabolism?

21. How are essential and nonessential amino acids different?

KEY MOLECULES AT METABOLIC CROSSROADS

▶ O B J E C T I V E

• **Identify the key molecules in metabolism, and describe the reactions and the products they may form.**

Although there are thousands of different chemicals in cells, three molecules—glucose 6-phosphate, pyruvic acid, and acetyl coenzyme A—play pivotal roles in metabolism (Figure 25.16). These molecules stand at "metabolic crossroads;" the reactions that occur depend on the nutritional or activity status of the individual. Reactions ❶ through ❼ in Figure 25.16 occur in the cytosol, reactions ❽ and ❾ occur inside mitochondria, and reactions indicated by ❿ occur on smooth endoplasmic reticulum.

The Role of Glucose 6-Phosphate

Shortly after glucose enters a body cell, a kinase converts it to **glucose 6-phosphate.** Four possible fates await glucose 6-phosphate (see Figure 25.14):

❶ *Synthesis of glycogen.* When glucose is abundant in the bloodstream, such as just after a meal, considerable glucose 6-phosphate is used to synthesize glycogen, the storage form of carbohydrate in animals. Subsequent breakdown of glycogen into glucose 6-phosphate occurs through a slightly different series of reactions. Synthesis and breakdown of glycogen occur mainly in skeletal muscle fibers and hepatocytes.

❷ *Release of glucose into the bloodstream.* If the enzyme glucose 6-phosphatase is present and active, glucose 6-phosphate can be dephosphorylated to glucose. Once glucose is released from the phosphate group, it can leave the cell and enter the bloodstream. Hepatocytes are the main cells that can provide glucose to the bloodstream in this way.

❸ *Synthesis of nucleic acids.* Glucose 6-phosphate is the precursor used by cells throughout the body to make ribose 5-phosphate, a five-carbon sugar that is needed for synthesis of RNA (ribonucleic acid) and DNA (deoxyribonucleic acid). The same sequence of reactions also produces NADPH. This molecule is a hydrogen and electron donor in certain reduction reactions, such as synthesis of fatty acids and steroid hormones.

❹ *Glycolysis.* Some ATP is produced anaerobically via glycolysis, in which glucose 6-phosphate is converted to pyruvic acid, another key molecule in metabolism. Most body cells carry out glycolysis.

The Role of Pyruvic Acid

Each 6-carbon molecule of glucose that undergoes glycolysis yields two 3-carbon molecules of **pyruvic acid.** This molecule

Figure 25.16 Summary of the roles of the key molecules in metabolic pathways. Double-headed arrows indicate that reactions between two molecules may proceed in either direction, if the appropriate enzymes are present and the conditions are favorable; single-headed arrows signify the presence of an irreversible step.

 Three molecules—glucose 6-phosphate, pyruvic acid, and acetyl coenzyme A—stand at "metabolic crossroads." They can undergo different reactions depending on one's nutritional or activity status.

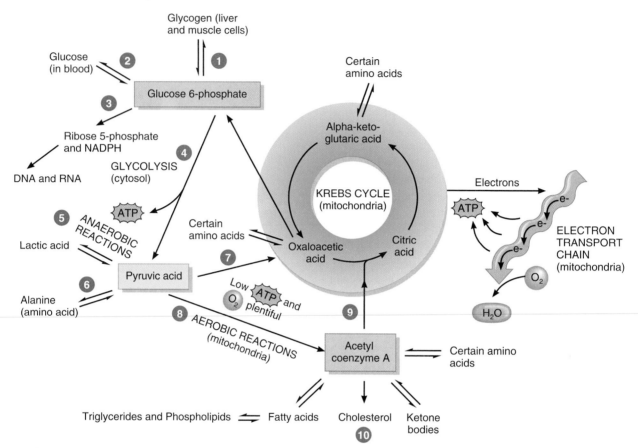

Which substance is the gateway into the Krebs cycle for fuel molecules that are being oxidized to generate ATP?

stands at a metabolic crossroads: Given enough oxygen, the aerobic (oxygen-consuming) reactions of cellular respiration can proceed; if oxygen is in short supply, anaerobic reactions can occur (Figure 25.16):

5 *Production of lactic acid.* When oxygen is in short supply in a tissue, as in actively contracting skeletal or cardiac muscle, some pyruvic acid is changed to lactic acid. The lactic acid then diffuses into the bloodstream and is taken up by hepatocytes, which eventually convert it back to pyruvic acid.

6 *Production of alanine.* One link between carbohydrate and protein metabolism occurs at pyruvic acid. Through transamination, an amino group ($-NH_3$) can either be added to pyruvic acid (a carbohydrate) to produce the amino acid alanine, or be removed from alanine to generate pyruvic acid.

7 *Gluconeogenesis.* Pyruvic acid and certain amino acids also can be converted to oxaloacetic acid, one of the Krebs cycle intermediates, which, in turn, can be used to form glucose 6-phosphate. This sequence of gluconeogenesis reactions bypasses certain one-way reactions of glycolysis.

The Role of Acetyl Coenzyme A

8 When the ATP level in a cell is low but oxygen is plentiful, most pyruvic acid streams toward ATP-producing reactions—the Krebs cycle and electron transport chain—via conversion to **acetyl coenzyme A.**

9 *Entry into the Krebs cycle.* Acetyl CoA is the vehicle for two-carbon acetyl groups to enter the Krebs cycle. Oxidative

Krebs cycle reactions convert acetyl CoA to CO_2 and produce reduced coenzymes (NADH and $FADH_2$) that transfer electrons into the electron transport chain. Oxidative reactions in the electron transport chain, in turn, generate ATP. Most fuel molecules that will be oxidized to generate ATP—glucose, fatty acids, and ketone bodies—are first converted to acetyl CoA.

⑩ Synthesis of lipids. Acetyl CoA also can be used for synthesis of certain lipids, including fatty acids, ketone bodies, and cholesterol. Because pyruvic acid can be converted to acetyl CoA, carbohydrates can be turned into triglycerides; this is a metabolic pathway for storage of some excess carbohydrate calories as fat. Mammals, including humans, cannot reconvert acetyl CoA to pyruvic acid, however, so fatty acids cannot be used to generate glucose or other carbohydrate molecules.

Table 25.2 summarizes carbohydrate, lipid, and protein metabolism.

▶ CHECKPOINT

22. What are the possible fates of glucose 6-phosphate, pyruvic acid, and acetyl coenzyme A in a cell?

METABOLIC ADAPTATIONS

▶ OBJECTIVE

- **Compare metabolism during the absorptive and postabsorptive states.**

Regulation of metabolic reactions depends both on the chemical environment within body cells, such as the levels of ATP and oxygen, and on signals from the nervous and endocrine systems. Some aspects of metabolism depend on how much time has passed since the last meal. During the **absorptive state,** ingested nutrients are entering the bloodstream, and glucose is readily available for ATP production. During the **postabsorptive state,** absorption of nutrients from the GI tract is complete, and energy needs must be met by fuels already in the body. A typical meal requires about 4 hours for complete absorption, and given three meals a day, the absorptive state exists for about 12 hours each day. Assuming no between-meal snacks, the other 12 hours—typically late morning, late afternoon, and most of the night—are spent in the postabsorptive state.

Because the nervous system and red blood cells continue to depend on glucose for ATP production during the postabsorptive

Table 25.2	Summary of Metabolism
Process	**Comments**
Carbohydrates	
Glucose catabolism	Complete oxidation of glucose (cellular respiration) is the chief source of ATP in cells and consists of glycolysis, Krebs cycle, and electron transport chain. Complete oxidation of 1 molecule of glucose yields a maximum of 36 or 38 molecules of ATP.
Glycolysis	Conversion of glucose into pyruvic acid results in the production of some ATP. Reactions do not require oxygen (anaerobic cellular respiration).
Krebs cycle	Cycle includes series of oxidation—reduction reactions in which coenzymes (AND⁺ and FAD) pick up hydrogen ions and hydride ions from oxidized organic acids, and some ATP is produced. CO_2 and H_2O are byproducts. Reactions are aerobic.
Electron transport chain	Third set of reactions in glucose catabolism is another series of oxidation—reduction reactions, in which electrons are passed from one carrier to the next, and most of the ATP is produced. Reactions require oxygen (aerobic cellular respiration).
Glucose anabolism	Some glucose is converted into glycogen (glycogenesis) for storage if not needed immediately for ATP production. Glycogen can be reconverted to glucose (glycogenolysis). The conversion of amino acids, glycerol, and lactic acid into glucose is called gluconeogenesis.
Lipids	
Triglyceride catabolism	Triglycerides are broken down into glycerol and fatty acids. Glycerol may be converted into glucose (gluconeogenesis) or catabolized via glycolysis. Fatty acids are catabolized via beta oxidation into acetyl coenzyme A that can enter the Krebs cycle for ATP production or be converted into ketone bodies (ketogenesis).
Triglyceride anabolism	The synthesis of triglycerides from glucose and fatty acids is called lipogenesis. Triglycerides are stored in adipose tissue.
Proteins	
Protein catabolism	Amino acids are oxidized via the Krebs cycle after deamination. Ammonia resulting from deamination is converted into urea in the liver, passed into blood, and excreted in urine. Amino acids may be converted into glucose (gluconeogenesis), fatty acids, or ketone bodies.
Protein anabolism	Protein synthesis is directed by DNA and utilizes cells' RNA and ribosomes.

state, maintaining a steady blood glucose level is critical during this period. Hormones are the major regulators of metabolism in each state. The effects of insulin dominate in the absorptive state, whereas several other hormones regulate metabolism in the postabsorptive state. During fasting and starvation, many body cells increasingly use ketone bodies for ATP production.

Metabolism During the Absorptive State

Soon after a meal, nutrients start to enter the blood. Recall that ingested food reaches the bloodstream mainly as glucose, amino acids, and triglycerides (in chylomicrons). Two metabolic hall-marks of the absorptive state are the oxidation of glucose for ATP production, which occurs in most body cells, and the storage of excess fuel molecules for future between-meal use, which occurs mainly in hepatocytes, adipocytes, and skeletal muscle fibers.

Absorptive State Reactions

The following reactions dominate during the absorptive state (Figure 25.17):

1 About 50% of the glucose absorbed from a typical meal is oxidized by cells throughout the body to produce ATP via glycolysis, the Krebs cycle, and the electron transport chain.

Figure 25.17   Principal metabolic pathways during the absorptive state.

During the absorptive state, most body cells produce ATP by oxidizing glucose to CO_2 and H_2O.

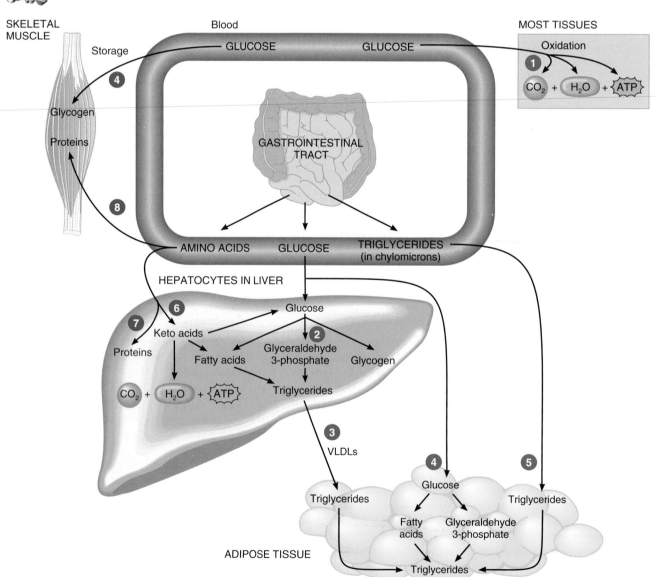

Are the reactions shown in this figure mainly anabolic or catabolic?

2 Most glucose that enters hepatocytes is converted to glycogen. Small amounts may be used for synthesis of fatty acids and glyceraldehyde 3-phosphate.

3 Some fatty acids and triglycerides synthesized in the liver remain there, but hepatocytes package most into VLDLs, which carry lipids to adipose tissue for storage.

4 Adipocytes also take up glucose not picked up by the liver and convert it into triglycerides for storage. Overall, about 40% of the glucose absorbed from a meal is converted to triglycerides, and about 10% is stored as glycogen in skeletal muscles and hepatocytes.

5 Most dietary lipids (mainly triglycerides and fatty acids) are stored in adipose tissue; only a small portion is used for synthesis reactions. Adipocytes obtain the lipids from chylomicrons, from VLDLs, and from their own synthesis reactions.

6 Many absorbed amino acids that enter hepatocytes are deaminated to keto acids, which can either enter the Krebs cycle for ATP production or be used to synthesize glucose or fatty acids.

7 Some amino acids that enter hepatocytes are used to synthesize proteins (for example, plasma proteins).

8 Amino acids not taken up by hepatocytes are used in other body cells (such as muscle cells) for synthesis of proteins or regulatory chemicals such as hormones or enzymes.

Regulation of Metabolism During the Absorptive State

Soon after a meal, glucose-dependent insulinotropic peptide (GIP), plus the rising blood levels of glucose and certain amino acids, stimulate pancreatic beta cells to release insulin. In general, insulin increases the activity of enzymes needed for anabolism and the synthesis of storage molecules; at the same time it decreases the activity of enzymes needed for catabolic or breakdown reactions. Insulin promotes the entry of glucose and amino acids into cells of many tissues, and it stimulates the phosphorylation of glucose in hepatocytes and the conversion of glucose 6-phosphate to glycogen in both liver and muscle cells. In liver and adipose tissue, insulin enhances the synthesis of triglycerides, and in cells throughout the body insulin stimulates protein synthesis. (See page 614 to review the effects of insulin.) Insulinlike growth factors and the thyroid hormones (T3 and T4) also stimulate protein synthesis. Table 25.3 summarizes the hormonal regulation of metabolism in the absorptive state.

Metabolism During the Postabsorptive State

About 4 hours after the last meal, absorption of nutrients from the small intestine is nearly complete, and blood glucose level starts to fall because glucose continues to leave the bloodstream and enter body cells while none is being absorbed from the GI tract. Thus, the main metabolic challenge during the postabsorptive state is to maintain the normal blood glucose level of 70–110 mg/100 mL (3.9–6.1 mmol/liter). Homeostasis of blood

Table 25.3	Hormonal Regulation of Metabolism in the Absorptive State	
Process	**Location(s)**	**Main Stimulating Hormone(s)**
Facilitated diffusion of glucose into cells	Most cells.	Insulin.*
Active transport of amino acids into cells	Most cells.	Insulin.
Glycogenesis (glycogen synthesis)	Hepatocytes and muscle fibers.	Insulin.
Protein synthesis	All body cells.	Insulin, thyroid hormones, and insulinlike growth factors.
Lipogenesis (triglyceride synthesis)	Adipose cells and hepatocytes.	Insulin.

*Facilitated diffusion of glucose into hepatocytes (liver cells) and neurons is always "turned on" and does not require insulin.

glucose concentration is especially important for the nervous system and for red blood cells for the following reasons:

- The dominant fuel molecule for ATP production in the nervous system is glucose, because fatty acids are unable to pass the blood–brain barrier.

- Red blood cells derive all their ATP from glycolysis of glucose because they have no mitochondria, and thus the Krebs cycle and the electron transport chain are not available to them.

Postabsorptive State Reactions

During the postabsorptive state, both *glucose production* and *glucose conservation* help maintain blood glucose level: Hepatocytes produce glucose molecules and export them into the blood, and other body cells switch from glucose to alternative fuels for ATP production to conserve scarce glucose. The major reactions of the postabsorptive state that produce glucose are the following (Figure 25.18):

1 *Breakdown of liver glycogen.* During fasting, a major source of blood glucose is liver glycogen, which can provide about a 4-hour supply of glucose. Liver glycogen is continually being formed and broken down as needed.

2 *Lipolysis.* Glycerol, produced by breakdown of triglycerides in adipose tissue, is also used to form glucose.

3 *Gluconeogenesis using lactic acid.* During exercise, skeletal muscle tissue breaks down stored glycogen (see step **9**) and produces some ATP anaerobically via glycolysis. Some of the pyruvic acid that results is converted to acetyl CoA, and some is converted to lactic acid, which diffuses into the blood. In the liver, lactic acid can be used for gluconeogenesis, and the resulting glucose is released into the blood.

4 *Gluconeogenesis using amino acids.* Modest breakdown of proteins in skeletal muscle and other tissues releases large amounts of amino acids, which then can be converted to glucose by gluconeogenesis in the liver.

Despite these ways to produce glucose, blood glucose level cannot be maintained for very long without further metabolic changes. Thus, a major adjustment must be made during the postabsorptive state to produce ATP while conserving glucose. The following reactions produce ATP without using glucose:

5 *Oxidation of fatty acids.* The fatty acids released by lipolysis of triglycerides cannot be used for glucose production because acetyl CoA cannot readily be converted to pyruvic acid. But most cells can oxidize the fatty acids directly, feed them into the Krebs cycle as acetyl CoA, and produce ATP through the electron transport chain.

6 *Oxidation of lactic acid.* Cardiac muscle can produce ATP aerobically from lactic acid.

7 *Oxidation of amino acids.* In hepatocytes, amino acids may be oxidized directly to produce ATP.

8 *Oxidation of ketone bodies.* Hepatocytes also convert fatty acids to ketone bodies, which can be used by the heart, kidneys, and other tissues for ATP production.

9 *Breakdown of muscle glycogen.* Skeletal muscle cells break down glycogen to glucose 6-phosphate, which undergoes glycolysis and provides ATP for muscle contraction.

Regulation of Metabolism During the Postabsorptive State

Both hormones and the sympathetic division of the autonomic nervous system (ANS) regulate metabolism during the postabsorptive state. The hormones that regulate postabsorptive state metabolism sometimes are called anti-insulin hormones because they counter the effects of insulin that dominate during the absorptive state. As blood glucose level declines, the secretion of insulin falls and the release of anti-insulin hormones rises.

Figure 25.18 Principal metabolic pathways during the postabsorptive state.

The principal function of postabsorptive state reactions is to maintain a normal blood glucose level.

What processes directly elevate blood glucose during this state, and where does each occur?

When blood glucose concentration starts to drop, the pancreatic alpha cells release glucagon at a faster rate, and the beta cells secrete insulin more slowly. The primary target tissue of glucagon is the liver; the major effect is increased release of glucose into the bloodstream due to gluconeogenesis and glycogenolysis.

Low blood glucose also activates the sympathetic branch of the ANS. Glucose-sensitive neurons in the hypothalamus detect low blood glucose and increase sympathetic output. As a result, sympathetic nerve endings release the neurotransmitter norepinephrine, and the adrenal medulla releases two catecholamine hormones—epinephrine and norepinephrine—into the bloodstream. Like glucagon, epinephrine stimulates glycogen breakdown. Epinephrine and norepinephrine are both potent stimulators of lipolysis. These actions of the catecholamines help to increase glucose and free fatty acid levels in the blood. As a result, muscle uses more fatty acids for ATP production, and more glucose is available to the nervous system. Table 25.4 summarizes the hormonal regulation of metabolism in the postabsorptive state.

Metabolism During Fasting and Starvation

The term **fasting** means going without food for many hours or a few days, whereas **starvation** implies weeks or months of food deprivation or inadequate food intake. People can survive without food for two months or more if they drink enough water to prevent dehydration. Although glycogen stores are depleted within a few hours of beginning a fast, catabolism of stored triglycerides and structural proteins can provide energy for several weeks. The amount of adipose tissue the body contains determines the lifespan possible without food.

During fasting and starvation, nervous tissue and RBCs continue to use glucose for ATP production. There is a ready supply of amino acids for gluconeogenesis because lowered insulin and increased cortisol levels slow the pace of protein synthesis and promote protein catabolism. Most cells in the body, especially skeletal muscle cells because they contain so much protein to begin with, can spare a fair amount of protein before their performance is adversely affected. During the first few days of fasting, protein catabolism outpaces protein synthesis by about 75 grams daily as some of the "old" amino acids are being deaminated and used for gluconeogenesis and "new" (that is, dietary) amino acids are lacking.

By the second day of a fast, blood glucose level has stabilized at about 65 mg/100 mL (3.6 mmol/liter), whereas the level of fatty acids in plasma has risen fourfold. Lipolysis of triglycerides in adipose tissue releases glycerol, which is used for gluconeogenesis, and fatty acids. The fatty acids diffuse into muscle fibers and other body cells, where they are used to produce acetyl-CoA, which enters the Krebs cycle. ATP then is synthesized as oxidation proceeds via the Krebs cycle and the electron transport chain.

The most dramatic metabolic change that occurs with fasting and starvation is the increase in the formation of ketone bodies by hepatocytes. During fasting, only small amounts of glucose undergo glycolysis to pyruvic acid, which in turn can be converted to oxaloacetic acid. Acetyl-CoA enters the Krebs cycle by combining with oxaloacetic acid (see Figure 25.16); when oxaloacetic acid is scarce due to fasting, only some of the available acetyl-CoA can enter the Krebs cycle. Surplus acetyl-CoA is used for ketogenesis, mainly in hepatocytes. Ketone body production thus increases as catabolism of fatty acids rises. Lipid-soluble ketone bodies can diffuse through plasma membranes and across the blood–brain barrier and be used as an alternate fuel for ATP production, especially by cardiac and skeletal muscle fibers and neurons. Normally, only a trace of ketone bodies (0.01 mmol/liter) are present in the blood, and thus they are a negligible fuel source. After two days of fasting, however, the level of ketones is 100–300 times higher and supplies roughly a third of the brain's fuel for ATP production. By 40 days of starvation, ketones provide up to two-thirds of the brain's energy needs. The presence of ketones actually reduces the use of glucose for ATP production, which in turn decreases the demand for gluconeogenesis and slows the catabolism of muscle proteins later in starvation to about 20 grams daily.

▶ CHECKPOINT

23. What are the roles of insulin, glucagon, epinephrine, insulinlike growth factors, thyroxine, cortisol, estrogen, and testosterone in regulation of metabolism?

24. Why is ketogenesis more significant during fasting or starvation than during normal absorptive and postabsorptive states?

Table 25.4	Hormonal Regulation of Metabolism in the Postabsorptive State	
Process	**Location(s)**	**Main Stimulating Hormone(s)**
Glycogenolysis (glycogen breakdown)	Hepatocytes and skeletal muscle fibers.	Glucagon and epinephrine.
Lipolysis (triglyceride breakdown)	Adipocytes.	Epinephrine, norepinephrine, cortisol, insulinlike growth factors, thyroid hormones, and others.
Protein breakdown	Most body cells, but especially skeletal muscle fibers.	Cortisol.
Gluconeogenesis (synthesis of glucose from noncarbohydrates)	Hepatocytes and kidney cortex cells.	Glucagon and cortisol.

HEAT AND ENERGY BALANCE

▶ **O B J E C T I V E S**

- **Define basal metabolic rate (BMR), and explain several factors that affect it.**
- **Describe the factors that influence body heat production.**
- **Explain how normal body temperature is maintained by negative feedback loops involving the hypothalamic thermostat.**

The body produces more or less heat depending on the rates of metabolic reactions. Homeostasis of body temperature can be maintained only if the rate of heat loss from the body equals the rate of heat production by metabolism. Thus, it is important to understand the ways in which heat can be lost, gained, or conserved. **Heat** is a form of energy that can be measured as **temperature** and expressed in units called calories. A **calorie (cal)** is defined as the amount of heat required to raise the temperature of 1 gram of water 1°C. Because the calorie is a relatively small unit, the **kilocalorie (kcal)** or **Calorie (Cal),** spelled with an uppercase C, is often used to measure the body's metabolic rate and to express the energy content of foods. A kilocalorie equals 1000 calories. Thus, when we say that a particular food item contains 500 Calories, we are actually referring to kilocalories.

Metabolic Rate

The overall rate at which metabolic reactions use energy is termed the **metabolic rate.** Some of the energy is used to produce ATP, and some is released as heat. Because many factors affect metabolic rate, it is measured under standard conditions, with the body in a quiet, resting, and fasting condition called the **basal state.** The measurement obtained is the **basal metabolic rate (BMR).** The most common way to determine BMR is by measuring the amount of oxygen used per kilocalorie of food metabolized. When the body uses 1 liter of oxygen to oxidize a typical dietary mixture of triglycerides, carbohydrates, and proteins, about 4.8 Cal of energy is released. BMR is 1200–1800 Cal/day in adults, or about 24 Cal/kg of body mass in adult males and 22 Cal/kg in adult females. The added calories needed to support daily activities, such as digestion and walking, range from 500 Cal for a small, relatively sedentary person to over 300 Cal for a person in training for Olympic-level competitions or mountain climbing.

Body Temperature Homeostasis

Despite wide fluctuations in environmental temperature, homeostatic mechanisms can maintain a normal range for internal body temperature. If the rate of body heat production equals the rate of heat loss, the body maintains a constant core temperature near 37°C (98.6°F). **Core temperature** is the temperature in body structures deep to the skin and subcutaneous layer. **Shell temperature** is the temperature near the body surface—in the skin and subcutaneous layer. Depending on the environmental temperature, shell temperature is 1–6°C lower than core temperature. Too high a core temperature kills by denaturing body proteins, whereas too low a core temperature causes cardiac arrhythmias that result in death.

Heat Production

The production of body heat is proportional to metabolic rate. Several factors affect the metabolic rate and thus the rate of heat production:

- *Exercise.* During strenuous exercise, the metabolic rate may increase to as much as 15 times the basal rate. In well-trained athletes, the rate may increase up to 20 times.

- *Hormones.* Thyroid hormones (thyroxine and triiodothyronine) are the main regulators of BMR, which increases as the blood levels of thyroid hormones rise. The response to changing levels of thyroid hormones is slow, however, taking several days to appear. Thyroid hormones increase BMR in part by stimulating aerobic cellular respiration. As cells use more oxygen to produce ATP, more heat is given off, and body temperature rises. Other hormones have minor effects on BMR. Testosterone, insulin, and human growth hormone can increase the metabolic rate by 5–15%.

- *Nervous system.* During exercise or in a stressful situation, the sympathetic division of the autonomic nervous system is stimulated. Its postganglionic neurons release norepinephrine (NE), and it also stimulates release of the hormones epinephrine and norepinephrine by the adrenal medulla. Both epinephrine and norepinephrine increase the metabolic rate of body cells.

- *Body temperature.* The higher the body temperature, the higher the metabolic rate. Each 1°C rise in core temperature increases the rate of biochemical reactions by about 10%. As a result, metabolic rate may be substantially increased during a fever.

- *Ingestion of food.* The ingestion of food raises the metabolic rate 10–20% due to the energy "costs" of digesting, absorbing, and storing nutrients. This effect, *food-induced thermogenesis,* is greatest after eating a high-protein meal and is less after eating carbohydrates and lipids.

- *Age.* The metabolic rate of a child, in relation to its size, is about double that of an elderly person due to the high rates of reactions related to growth.

- *Other factors.* Other factors that affect metabolic rate are gender (lower in females, except during pregnancy and lactation), climate (lower in tropical regions), sleep (lower), and malnutrition (lower).

Mechanisms of Heat Transfer

Maintaining normal body temperature depends on the ability to lose heat to the environment at the same rate as it is produced by metabolic reactions. Heat can be transferred from the body to its surroundings in four ways: via conduction, convection, radiation, and evaporation.

1. **Conduction** is the heat exchange that occurs between molecules of two materials that are in direct contact with each other. At rest, about 3% of body heat is lost via conduction to solid materials in contact with the body, such as a chair, clothing, and jewelry. Heat can also be gained via conduction—for example, while soaking in a hot tub. Because water conducts heat 20 times more effectively than air, heat loss or heat gain via conduction is much greater when the body is submerged in cold or hot water.

2. **Convection** is the transfer of heat by the movement of a fluid (a gas or a liquid) between areas of different temperature. The contact of air or water with your body results in heat transfer by both conduction and convection. When cool air makes contact with the body, it becomes warmed and therefore less dense and is carried away by convection currents created as the less dense air rises. The faster the air moves—for example, by a breeze or a fan—the faster the rate of convection. At rest, about 15% of body heat is lost to the air via conduction and convection.

3. **Radiation** is the transfer of heat in the form of infrared rays between a warmer object and a cooler one without physical contact. Your body loses heat by radiating more infrared waves than it absorbs from cooler objects. If surrounding objects are warmer than you are, you absorb more heat than you lose by radiation. In a room at 21°C (70°F), about 60% of heat loss occurs via radiation in a resting person.

4. **Evaporation** is the conversion of a liquid to a vapor. Every milliliter of evaporating water takes with it a great deal of heat—about 0.58 Cal/mL. Under typical resting conditions, about 22% of heat loss occurs through evaporation of about 700 mL of water per day—300 mL in exhaled air and 400 mL from the skin surface. Because we are not normally aware of this water loss through the skin and mucous membranes of the mouth and respiratory system, it is termed **insensible water loss.** The rate of evaporation is inversely related to relative humidity, the ratio of the actual amount of moisture in the air to the maximum amount it can hold at a given temperature. The higher the relative humidity, the lower the rate of evaporation. At 100% humidity, heat is gained via condensation of water on the skin surface as fast as heat is lost via evaporation. Evaporation provides the main defense against overheating during exercise. Under extreme conditions, a maximum of about 3 liters of sweat can be produced each hour, removing more than 1700 Calories of heat if all of it evaporates. (Note: Sweat that drips off the body rather than evaporating removes very little heat.)

Hypothalamic Thermostat

The control center that functions as the body's thermostat is a group of neurons in the anterior part of the hypothalamus, the **preoptic area.** This area receives impulses from thermoreceptors in the skin and mucous membranes and in the hypothalamus. Neurons of the preoptic area generate nerve impulses at a higher frequency when blood temperature increases, and at a lower frequency when blood temperature decreases.

Nerve impulses from the preoptic area propagate to two other parts of the hypothalamus known as the **heat-losing center** and the **heat-promoting center,** which, when stimulated by the preoptic area, set into operation a series of responses that lower body temperature and raise body temperature, respectively.

Thermoregulation

If core temperature declines, mechanisms that help conserve heat and increase heat production act via several negative feedback loops to raise the body temperature to normal (Figure 25.19). Thermoreceptors in the skin and hypothalamus send nerve impulses to the preoptic area and the heat-promoting center in the hypothalamus, and to hypothalamic neurosecretory cells that produce thyrotropin-releasing hormone (TRH). In response, the hypothalamus discharges nerve impulses and secretes TRH, which in turn stimulates thyrotrophs in the anterior pituitary gland to release thyroid-stimulating hormone (TSH). Nerve impulses from the hypothalamus and TSH then activate several effectors.

Each effector responds in a way that helps increase core temperature to the normal value:

- Nerve impulses from the heat-promoting center stimulate sympathetic nerves that cause blood vessels of the skin to constrict. Vasoconstriction decreases the flow of warm blood, and thus the transfer of heat, from the internal organs to the skin. Slowing the rate of heat loss allows the internal body temperature to increase as metabolic reactions continue to produce heat.

- Nerve impulses in sympathetic nerves leading to the adrenal medulla stimulate the release of epinephrine and norepinephrine into the blood. The hormones, in turn, bring about an increase in cellular metabolism, which increases heat production.

- The heat-promoting center stimulates parts of the brain that increase muscle tone and hence heat production. As muscle tone increases in one muscle (the agonist), the small contractions stretch muscle spindles in its antagonist, initiating a stretch reflex. The resulting contraction in the antagonist stretches muscle spindles in the agonist, and it too develops a stretch reflex. This repetitive cycle—called **shivering**—greatly increases the rate of heat production. During maximal shivering, body heat production can rise to about four times the basal rate in just a few minutes.

- The thyroid gland responds to TSH by releasing more thyroid hormones into the blood. As increased levels of thyroid hormones slowly increase the metabolic rate, body temperature rises.

If core body temperature rises above normal, a negative feedback loop opposite to the one depicted in Figure 25.19 goes into action. The higher temperature of the blood stimulates thermoreceptors that send nerve impulses to the preoptic area, which, in turn, stimulate the heat-losing center and inhibit the heat-promoting center. Nerve impulses from the heat-losing center

Figure 25.19 **Negative feedback mechanisms that conserve heat and increase heat production.**

Core temperature is the temperature in body structures deep to the skin and subcutaneous layer; shell temperature is the temperature near the body surface.

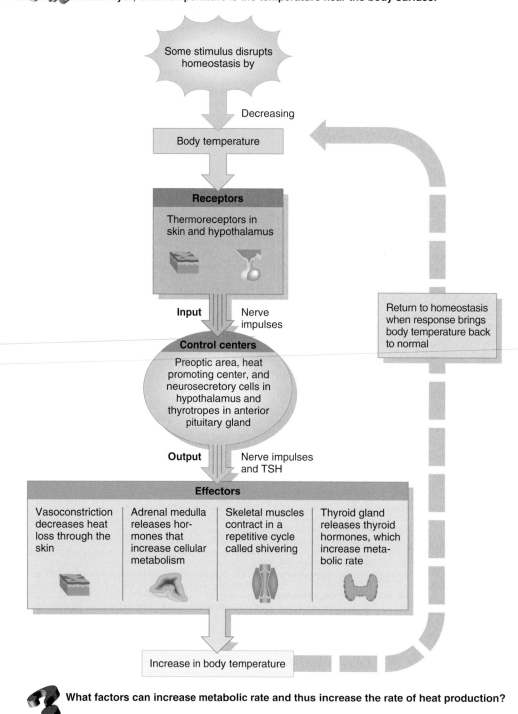

What factors can increase metabolic rate and thus increase the rate of heat production?

cause dilation of blood vessels in the skin. The skin becomes warm, and the excess heat is lost to the environment via radiation and conduction as an increased volume of blood flows from the warmer core of the body into the cooler skin. At the same time, metabolic rate decreases, and shivering does not occur. The high temperature of the blood stimulates sweat glands of the skin via hypothalamic activation of sympathetic nerves. As the water in perspiration evaporates from the surface of the skin, the skin is cooled. All these responses counteract heat-promoting effects and help return body temperature to normal.

Hypothermia

Hypothermia is a lowering of core body temperature to 35°C (95°F) or below. Causes of hypothermia include an overwhelming cold stress (immersion in icy water), metabolic diseases (hypoglycemia, adrenal insufficiency, or hypothyroidism), drugs (alcohol, antidepressants, sedatives, or tranquilizers), burns, and malnutrition. Hypothermia is characterized by the following as core body temperature falls: sensation of cold, shivering, confusion, vasoconstriction, muscle rigidity, bradycardia, acidosis, hypoventilation, hypotension, loss of spontaneous movement, coma, and death (usually caused by cardiac arrhythmias). Because the elderly have reduced metabolic protection against a cold environment coupled with a reduced perception of cold, they are at greater risk for developing hypothermia. ■

Energy Homeostasis and Regulation of Food Intake

Most mature animals and many men and women maintain **energy homeostasis,** the precise matching of energy intake (in food) to energy expenditure over time. When the energy content of food balances the energy used by all the cells of the body, body weight remains constant (unless there is a gain or loss of water). In many people, weight stability persists despite large day-to-day variations in activity and food intake. In the more affluent nations, however, a large fraction of the population is overweight. Easy access to tasty, high-calorie foods and a "couch-potato" lifestyle promote weight gain. Being overweight increases one's risk of dying from a variety of cardiovascular and metabolic disorders, such as diabetes mellitus.

Energy intake depends only on the amount of food consumed (and absorbed), but three components contribute to total energy expenditure.

1. The basal metabolic rate accounts for about 60% of energy expenditure.

2. Physical activity typically adds 30–35% but can be lower in sedentary people. The energy expenditure is partly from voluntary exercise, such as walking, and partly from **nonexercise thermogenesis (NEAT),** the energy costs for maintaining muscle tone, posture while sitting or standing, and involuntary fidgeting movements.

3. **Food-induced thermogenesis,** the heat produced while food is being digested, absorbed, and stored, represents 5–10% of total energy expenditure.

The major site of stored chemical energy in the body is adipose tissue. When energy use exceeds energy input, triglycerides in adipose tissue are catabolized to provide the extra energy, and when energy input exceeds energy expenditure, triglycerides are stored. Over time, the amount of stored triglycerides indicates the excess of energy intake over energy expenditure. Even small differences add up over time. A gain of 20 lb (9 kg) between ages 25 and 55 represents only a tiny imbalance, about 0.3% more energy intake in food than energy expenditure.

Clearly, negative feedback mechanisms are regulating both our energy intake and our energy expenditure. But no sensory receptors exist to monitor one's weight or size. How, then, is food intake regulated? The answer to this question is incomplete, but important advances in understanding regulation of food intake have occurred in the past decade. It depends on many factors, including neural and endocrine signals, levels of certain nutrients in the blood, psychological elements such as stress or depression, signals from the GI tract and the special senses, and neural connections between the hypothalamus and other parts of the brain.

Within the hypothalamus are clusters of neurons that play key roles in regulating food intake. These neurons receive signals that indicate hunger or satiety although they are not precisely organized into "feeding" and "satiety" centers, as was once thought. **Satiety** (sa-TĪ-i-tē) is a feeling of fullness accompanied by lack of desire to eat. Two hypothalamic areas involved in regulation of food intake are the *arcuate nucleus* and the *paraventricular nucleus* (see Figure 14.10 on page 466). In 1994, the first experiments were reported on a mouse gene, named *obese,* that causes overeating and severe obesity when it is mutated. The product of this gene is the hormone **leptin.** In both mice and humans, leptin helps decrease **adiposity,** total body-fat mass. Leptin is synthesized and secreted by adipocytes in proportion to adiposity; as more triglycerides are stored, more leptin is secreted into the bloodstream. Leptin acts on the hypothalamus to inhibit circuits that stimulate eating while also activating circuits that increase energy expenditure. The hormone insulin has a similar, but smaller, effect. Both leptin and insulin pass the blood–brain barrier.

When leptin and insulin levels are *low,* neurons that extend from the arcuate nucleus to the paraventricular nucleus release a neurotransmitter called **neuropeptide Y** that stimulates food intake. Other neurons that extend between the arcuate and paraventricular nuclei release a neurotransmitter called **melanocortin,** which is similar to melanocyte-stimulating hormone (MSH). Leptin stimulates release of melanocortin, which acts to inhibit food intake. Although leptin, neuropeptide Y, and melanocortin are key signaling molecules for maintaining energy homeostasis, several other hormones and neurotransmitters also contribute. An understanding of the brain circuits involved is still far from complete. Other areas of the hypothalamus plus nuclei in the brainstem, limbic system, and cerebral cortex take part.

Achieving energy homeostasis requires regulation of energy intake. Most increases and decreases in food intake are due to changes in meal size rather than changes in number of meals. Many experiments have demonstrated the presence of satiety signals, chemical or neural changes that help terminate eating when "fullness" is attained. For example, an increase in blood glucose level, as occurs after a meal, decreases appetite. Several hormones, such as glucagon, cholecystokinin, estrogens, and epinephrine (acting via beta receptors) act to signal satiety and to increase energy expenditure. Distention of the GI tract, particularly the stomach and duodenum, also contributes to termination of food intake. Other hormones increase appetite and decrease energy expenditure. These include growth hormone-releasing hormone (GHRH), androgens, glucocorticoids, epinephrine (acting via alpha receptors), and progesterone.

Emotional Eating

In addition to keeping us alive, eating serves countless psychological, social, and cultural purposes. We eat to celebrate, punish, comfort, defy, and deny. Eating in response to emotional drives, such as feeling stressed, bored, or tired is called **emotional eating.** Emotional eating is so common that, within limits, it is considered well within the range of normal behavior. Who hasn't at one time or another headed for the refrigerator after a bad day? Problems arise when emotional eating becomes so excessive that it interferes with health. Physical health problems include obesity and associated disorders such as hypertension and heart disease. Psychological health problems include poor self-esteem, an inability to cope effectively with feelings of stress, and in extreme cases, eating disorders, such as anorexia nervosa, bulimia, and obesity.

Eating provides comfort and solace, numbing pain and "feeding the hungry heart." Eating may provide a biochemical "fix" as well. Emotional eaters typically overeat carbohydrate foods (sweets and starches), which may raise brain serotonin levels and lead to feelings of relaxation. Food becomes a way to self-medicate when negative emotions arise. ■

▶ C H E C K P O I N T

25. Define a kilocalorie (kcal). How is the unit used? How does it relate to a calorie?

26. Distinguish between core temperature and shell temperature.

27. In what ways can a person lose heat to or gain heat from the surroundings? How is it possible for a person to lose heat on a sunny beach when the temperature is 40°C (104°F) and the humidity is 85%?

28. What does the term energy homeostasis mean?

29. How is food intake regulated?

▐ NUTRITION

▶ O B J E C T I V E S

• **Describe how to select foods to maintain a healthy diet.**

• **Compare the sources, functions, and importance of minerals and vitamins in metabolism.**

Nutrients are chemical substances in food that body cells use for growth, maintenance, and repair. The six main types of nutrients are water, carbohydrates, lipids, proteins, minerals, and vitamins. Water is the nutrient needed in the largest amount—

about 2–3 liters per day. As the most abundant compound in the body, water provides the medium in which most metabolic reactions occur, and it also participates in some reactions (for example, hydrolysis reactions). The important roles of water in the body can be reviewed on pages 38–39. Three organic nutrients—carbohydrates, lipids, and proteins—provide the energy needed for metabolic reactions and serve as building blocks to make body structures. Some minerals and many vitamins are components of the enzyme systems that catalyze metabolic reactions. *Essential nutrients* are specific nutrient molecules that the body cannot make in sufficient quantity to meet its needs and thus must obtain from the diet. Some amino acids, some fatty acids, vitamins, and minerals are essential nutrients.

Next, we discuss some guidelines for healthy eating and the roles of minerals and vitamins in metabolism.

Guidelines for Healthy Eating

Each gram of protein or carbohydrate in food provides about 4 Calories, whereas a gram of fat (lipids) provides about 9 Calories. On a daily basis, many women and older people need about 1600 Calories; children, teenage girls, active women, and most men need about 2200 Calories; and teenage boys and active men need about 2800 Calories.

We do not know with certainty what levels and types of carbohydrate, fat, and protein are optimal in the diet. Different populations around the world eat radically different diets that are adapted to their particular lifestyles. However, many experts recommend the following distribution of calories: 50–60% from carbohydrates, with less than 15% from simple sugars; less than 30% from fats (triglycerides are the main type of dietary fat), with no more than 10% as saturated fats; and about 12–15% from proteins.

The guidelines for healthy eating are to:

• Eat a variety of foods.

• Maintain a healthy weight.

• Choose foods low in fat, saturated fat, and cholesterol.

• Eat plenty of vegetables, fruits, and grain products.

• Use sugars in moderation only.

• Use salt and sodium in moderation only (less than 6 grams daily).

• If you drink alcoholic beverages, do so in moderation (less than 1 ounce of the equivalent of pure alcohol per day).

To help people achieve a good balance of vitamins, minerals, carbohydrates, fats, and proteins in their food, the U.S. Department of Agriculture developed the food guide pyramid (Figure 25.20). The sections of the pyramid indicate how many servings of each of the five major food groups to eat each day. The smallest number of servings corresponds to a 1600 Cal/day diet, whereas the largest number of servings corresponds to a 2800 Cal/day diet. Because they should be consumed in largest

Figure 25.20 The food guide pyramid. The smallest number of servings corresponds to 1600 Calories per day, whereas the largest number of servings corresponds to 2800 Calories per day. Each example given equals one serving.

The sections of the pyramid show how many servings of five major food groups to eat each day.

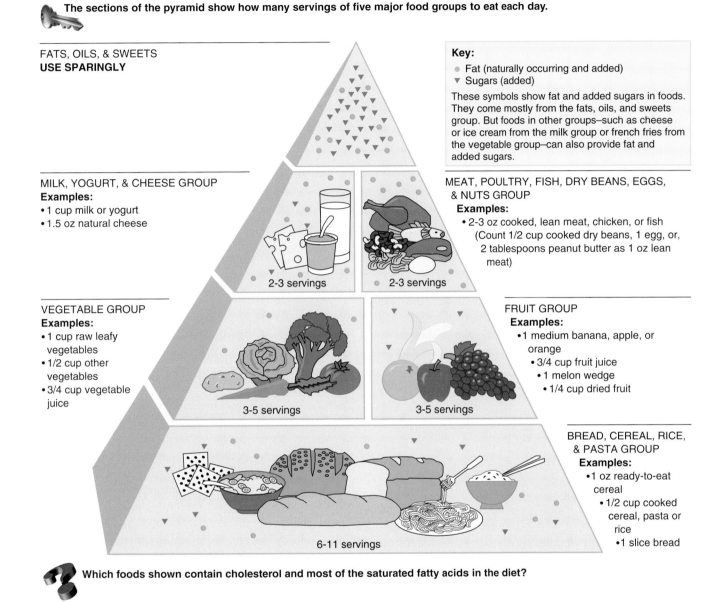

FATS, OILS, & SWEETS
USE SPARINGLY

Key:
- Fat (naturally occurring and added)
- ▼ Sugars (added)

These symbols show fat and added sugars in foods. They come mostly from the fats, oils, and sweets group. But foods in other groups–such as cheese or ice cream from the milk group or french fries from the vegetable group–can also provide fat and added sugars.

MILK, YOGURT, & CHEESE GROUP
Examples:
- 1 cup milk or yogurt
- 1.5 oz natural cheese

2-3 servings

MEAT, POULTRY, FISH, DRY BEANS, EGGS, & NUTS GROUP
Examples:
- 2-3 oz cooked, lean meat, chicken, or fish (Count 1/2 cup cooked dry beans, 1 egg, or, 2 tablespoons peanut butter as 1 oz lean meat)

2-3 servings

VEGETABLE GROUP
Examples:
- 1 cup raw leafy vegetables
- 1/2 cup other vegetables
- 3/4 cup vegetable juice

3-5 servings

FRUIT GROUP
Examples:
- 1 medium banana, apple, or orange
- 3/4 cup fruit juice
- 1 melon wedge
- 1/4 cup dried fruit

3-5 servings

BREAD, CEREAL, RICE, & PASTA GROUP
Examples:
- 1 oz ready-to-eat cereal
- 1/2 cup cooked cereal, pasta or rice
- 1 slice bread

6-11 servings

Which foods shown contain cholesterol and most of the saturated fatty acids in the diet?

quantity, foods rich in complex carbohydrates—the bread, cereal, rice, and pasta group—form the base of the pyramid. Vegetables and fruits form the next level. The health benefits of eating generous amounts of these foods are well documented. To be eaten in smaller quantities are foods on the next level up— the milk, yogurt, and cheese group and the meat, poultry, fish, dry beans, eggs, and nuts group. These two food groups have higher fat and protein content than the food groups below them. To lower your daily intake of fats, choose low-fat foods from these groups—nonfat milk and yogurt, low-fat cheese, fish, and poultry (remove the skin).

The apex of the pyramid is not a food group but rather a caution to use fats, oils, and sweets sparingly. The food guide pyramid does not distinguish among the different types of fatty acids—saturated, polyunsaturated, and monounsaturated—in dietary fats. However, atherosclerosis and coronary artery disease are prevalent in populations that consume large amounts of saturated fats and cholesterol. Populations living around the Mediterranean Sea, in contrast, have low rates of coronary artery disease despite eating a diet that contains up to 40% of the calories as fats. Most of their fat comes from olive oil, which is rich in monounsaturated fatty acids and has no cholesterol. Canola oil, avocados, nuts, and peanut oil are also rich in monounsaturated fatty acids.

Minerals

Minerals are inorganic elements that occur naturally in the Earth's crust. In the body they appear in combination with each other, in combination with organic compounds, or as ions in solution. Minerals constitute about 4% of total body mass and are concentrated most heavily in the skeleton. Minerals with known functions in the body include calcium, phosphorus, potassium, sulfur, sodium, chloride, magnesium, iron, iodide, manganese, copper, cobalt, zinc, fluoride, selenium, and chromium. Table 25.5 describes the vital functions of these minerals. Note that the body generally uses the ions of the minerals rather than the non-ionized form. Some minerals, such as chlorine, are toxic or even fatal if ingested in the non-ionized form. Other minerals—aluminum, boron, silicon, and molybdenum—are present but may have no functions. Typical diets supply adequate amounts of potassium, sodium, chloride, and magnesium. Some attention must be paid to eating foods that provide enough calcium, phosphorus, iron, and iodide. Excess amounts of most minerals are excreted in the urine and feces.

Calcium and phosphorus form part of the matrix of bone. Because minerals do not form long-chain compounds, they are otherwise poor building materials. A major role of minerals is to help regulate enzymatic reactions. Calcium, iron, magnesium, and manganese are constituents of some coenzymes. Magnesium also serves as a catalyst for the conversion of ADP to ATP. Minerals such as sodium and phosphorus work in buffer systems, which help control the pH of body fluids. Sodium also helps regulate the osmosis of water and, along with other ions, is involved in the generation of nerve impulses.

Vitamins

Organic nutrients required in small amounts to maintain growth and normal metabolism are called **vitamins.** Unlike carbohydrates, lipids, or proteins, vitamins do not provide energy or serve as the body's building materials. Most vitamins with known functions are coenzymes.

Most vitamins cannot be synthesized by the body and must be ingested in food. Other vitamins, such as vitamin K, are produced by bacteria in the GI tract and then absorbed. The body can assemble some vitamins if the raw materials, called **provitamins,** are provided. For example, vitamin A is produced by the body from the provitamin beta-carotene, a chemical present in yellow vegetables such as carrots and in dark green vegetables such as spinach. No single food contains all the required vitamins—one of the best reasons to eat a varied diet.

Vitamins are divided into two main groups: fat soluble and water soluble. The **fat-soluble vitamins** are vitamins A, D, E, and K. They are absorbed along with other dietary lipids in the small intestine and packaged into chylomicrons. They cannot be absorbed in adequate quantity unless they are ingested with other lipids. Fat-soluble vitamins may be stored in cells, particu-

larly hepatocytes. The **water-soluble vitamins** include several B vitamins and vitamin C. They are dissolved in body fluids. Excess quantities of these vitamins are not stored but instead are excreted in the urine.

Besides their other functions, three vitamins—C, E, and beta-carotene (a provitamin)—are termed **antioxidant vitamins** because they inactivate oxygen free radicals. Recall that free radicals are highly reactive ions or molecules that carry an unpaired electron in their outermost electron shell (see Figure 2.3 on page 30). Free radicals damage cell membranes, DNA, and other cellular structures and contribute to the formation of artery-narrowing atherosclerotic plaques. Some free radicals arise naturally in the body, and others derive from environmental hazards such as tobacco smoke and radiation. Antioxidant vitamins are thought to play a role in protecting against some kinds of cancer, reducing the buildup of atherosclerotic plaque, delaying some effects of aging, and decreasing the chance of cataract formation in the lens of the eyes. Table 25.6 on pages 940–941 lists the major vitamins, their sources, their functions, and related deficiency disorders.

Vitamin and Mineral Supplements

Most nutritionists recommend eating a balanced diet that includes a variety of foods rather than taking vitamin or mineral supplements, except in special circumstances. Common examples of necessary supplementations include iron for women who have excessive menstrual bleeding; iron and calcium for women who are pregnant or breast-feeding; folic acid (folate) for all women who may become pregnant, to reduce the risk of fetal neural tube defects; calcium for most adults, because they do not receive the recommended amount in their diets; and vitamin B_{12} for strict vegetarians, who eat no meat. Because most North Americans do not ingest in their food the high levels of antioxidant vitamins thought to have beneficial effects, some experts recommend supplementing vitamins C and E. More is not always better, however; larger doses of vitamins or minerals can be very harmful. ■

▶ CHECKPOINT

30. What is a nutrient?

31. Describe the food guide pyramid and give examples of foods from each food group.

32. What is a mineral? Briefly describe the functions of the following minerals: calcium, phosphorus, potassium, sulfur, sodium, chloride, magnesium, iron, iodine, copper, zinc, fluoride, manganese, cobalt, chromium, and selenium.

33. Define a vitamin. Explain how we obtain vitamins. Distinguish between a fat-soluble vitamin and a water-soluble vitamin.

34. For each of the following vitamins, indicate its principal function and the effect(s) of deficiency: A, D, E, K, B_1, B_2, niacin, B_6, B_{12}, pantothenic acid, folic acid, biotin, and C.

Table 25.5	Minerals Vital to the Body	
Mineral	**Comments**	**Importance**
Calcium	Most abundant mineral in body. Appears in combination with phosphates. About 99% is stored in bone and teeth. Blood Ca^{2+} level is controlled by parathyroid hormone(PTH). Calcitriol promotes absorption of dietary calcium. Excess is excreted in feces and urine. Sources are milk, egg yolk, shellfish, and leafy green vegetables.	Formation of bones and teeth, blood clotting, normal muscle and nerve activity, endocytosis and exocytosis, cellular motility, chromosome movement during cell division, glycogen metabolism, and release of neurotransmitters and hormones.
Phosphorus	About 80% is found in bones and teeth as phosphate salts. Blood phosphate level is controlled by parathyroid hormone (PTH). Excess is excreted in urine; small amount is eliminated in feces. Sources are dairy products, meat, fish, poultry, and nuts.	Formation of bones and teeth. Phosphates ($H_2PO_4^-$, HPO_4^-, and PO_4^{3-}) constitute a major buffer system of blood. Plays important role in muscle contraction and nerve activity. Component of many enzymes. Involved in energy transfer (ATP). Component of DNA and RNA.
Potassium	Major cation (K^+) in intracellular fluid. Excess excreted in urine. Present in most foods (meats, fish, poultry, fruits, and nuts).	Needed for generation and conduction of action potentials in neurons and muscle fibers.
Sulfur	Component of many proteins (such as insulin and chrondroitin sulfate), electron carriers in electron transport chain, and some vitamins (thiamine and biotin). Excreted in urine. Sources include beef, liver, lamb, fish, poultry, eggs, cheese, and beans.	As component of hormones and vitamins, regulates various body activities. Needed for ATP production by electron transport chain.
Sodium	Most abundant cation (Na^+) in extracellular fluids; some found in bones. Excreted in urine and perspiration. Normal intake of NaCl (table salt) supplies more than the required amounts.	Strongly affects distribution of water through osmosis. Part of bicarbonate buffer system. Functions in nerve and muscle action potential conduction.
Chloride	Major anion (Cl^-) in extracellular fluid. Excess excreted in urine. Sources include table salt (NaCl), soy sauce, and processed foods.	Plays role in acid–base balance of blood, water balance, and formation of HCl in stomach.
Magnesium	Important cation (Mg^{2+}) in intracellular fluid. Excreted in urine and feces. Widespread in various foods, such as green leafy vegetables, seafood, and whole-grain cereals.	Required for normal functioning of muscle and nervous tissue. Participates in bone formation. Constituent of many coenzymes.
Iron	About 66% found in hemoglobin of blood. Normal losses of iron occur by shedding of hair, epithelial cells, and mucosal cells, and in sweat, urine, feces, bile, and blood lost during menstruation. Sources are meat, liver, shellfish, egg yolk, beans, legumes, dried fruits, nuts, and cereals.	As component of hemoglobin, reversibly binds O_2. Component of cytochromes involved in electron transport chain.
Iodide	Essential component of thyroid hormones. Excreted in urine. Sources are seafood, iodized salt, and vegetables grown in iodine-rich soils.	Required by thyroid gland to synthesize thyroid hormones, which regulate metabolic rate.
Manganese	Some stored in liver and spleen. Most excreted in feces.	Activates several enzymes. Needed for hemoglobin synthesis, urea formation, growth, reproduction, lactation, bone formation, and possibly production and release of insulin, and inhibition of cell damage.
Copper	Some stored in liver and spleen. Most excreted in feces. Sources include eggs, whole-wheat flour, beans, beets, liver, fish, spinach, and asparagus.	Required with iron for synthesis of hemoglobin. Component of coenzymes in electron transport chain and enzyme necessary for melanin formation.
Cobalt	Constituent of vitamin B_{12}.	As part of vitamin B_{12}, required for erythropoiesis.
Zinc	Important component of certain enzymes. Widespread in many foods, especially meats.	As a component of carbonic anhydrase, important in carbon dioxide metabolism. Necessary for normal growth and wound healing, normal taste sensations and appetite, and normal sperm counts in males. As a component of peptidases, it is involved in protein digestion.
Fluoride	Components of bones, teeth, other tissues.	Appears to improve tooth structure and inhibit tooth decay.
Selenium	Important component of certain enzymes. Found in seafood, meat, chicken, tomatoes, egg yolk, milk, mushrooms, and garlic, and cereal grains grown in selenium-rich soil.	Needed for synthesis of thyroid hormones, sperm motility, and proper functioning of the immune system. Also functions as an antioxidant. Prevents chromosome breakage and may play a role in preventing certain birth defects, miscarriage, prostate cancer, and coronary artery disease.
Chromium	Found in high concentrations in brewer's yeast. Also found in wine and some brands of beer.	Needed for normal activity of insulin in carbohydrate and lipid metabolism.

Table 25.6 The Principal Vitamins

Vitamin	Comment and Source	Functions	Deficiency Symptoms and Disorders
Fat-soluble	All require bile salts and some dietary lipids for adequate absorption.		
A	Formed from provitamin beta-carotene (and other provitamins) in GI tract. Stored in liver. Sources of carotene and other provitamins include orange, yellow, and green vegetables; sources of vitamin A include liver and milk.	Maintains general health and vigor of epithelial cells. Beta-carotene acts as an antioxidant to inactivate free radicals.	Deficiency results in atrophy and keratinization of epithelium, leading to dry skin and hair; increased incidence of ear, sinus, respiratory, urinary, and digestive system infections; inability to gain weight; drying of cornea; and skin sores.
		Essential for formation of light-sensitive pigments in photoreceptors of retina.	**Night blindness** or decreased ability for dark adaptation.
		Aids in growth of bones and teeth by helping to regulate activity of osteoblasts and osteoclasts.	Slow and faulty development of bones and teeth.
D	Sunlight converts 7-dehydrocholesterol in the skin to cholecalciferol (vitamin D_3). A liver enzyme then converts cholecalciferol to 25-hydroxycholecalciferol. A second enzyme in the kidneys converts 25-hydroxycholecalciferol to calcitriol (1,25-dihydroxycalciferol), which is the active form of vitamin D. Most is excreted in bile. Dietary sources include fish-liver oils, egg yolk, and fortified milk.	Essential for absorption of calcium and phosphorus from GI tract. Works with parathyroid hormone (PTH) to maintain. Ca^{2+} homeostasis.	Defective utilization of calcium by bones leads to **rickets** in children and **osteomalacia** in adults. Possible loss of muscle tone.
E (tocopherols)	Stored in liver, adipose tissue, and muscles. Sources include fresh nuts and wheat germ, seed oils, and green leafy vegetables.	Inhibits catabolism of certain fatty acids that help form cell structures, especially membranes. Involved in formation of DNA, RNA, and red blood cells. May promote wound healing, contribute to the normal structure and functioning of the nervous system, and prevent scarring. May help protect liver from toxic chemicals such as carbon tetrachloride. Acts as an antioxidant to inactivate free radicals.	May cause oxidation of monounsaturated fats, resulting in abnormal structure and function of mitochondria, lysosomes, and plasma membranes. A possible consequence is hemolytic anemia.
K	Produced by intestinal bacteria. Stored in liver and spleen. Dietary sources include spinach, cauliflower, cabbage, and liver.	Coenzyme essential for synthesis of several clotting factors by liver, including prothrombin.	Delayed clotting time results in excessive bleeding.
Water-soluble	Dissolved in body fluids. Most are not stored in body. Excess intake is eliminated in urine.		
B₁ (thiamine)	Rapidly destroyed by heat. Sources include whole-grain products, eggs, pork, nuts, liver, and yeast.	Acts as coenzyme for many different enzymes that break carbon-to-carbon bonds and are involved in carbohydrate metabolism of pyruvic acid to CO_2 and H_2O. Essential for synthesis of the neurotransmitter acetylcholine.	Improper carbohydrate metabolism leads to buildup of pyruvic and lactic acids and insufficient production of ATP for muscle and nerve cells. Deficiency leads to: (1) **beriberi**, partial paralysis of smooth muscle of GI tract, causing digestive disturbances; skeletal muscle paralysis; and atrophy of limbs; (2) **polyneuritis**, due to degeneration of myelin sheaths; impaired reflexes, impaired sense of touch, stunted growth in children, and poor appetite.

Vitamin	Comment and Source	Functions	Deficiency Symptoms and Disorders
Water-soluble (continued)			
B₂ (riboflavin)	Small amounts supplied by bacteria of GI tract. Dietary sources include yeast, liver, beef, veal, lamb, eggs, whole-grain products, asparagus, peas, beets, and peanuts.	Component of certain coenzymes (for example, FAD and FMN) in carbohydrate and protein metabolism, especially in cells of eye, integument, mucosa of intestine, and blood.	Deficiency may lead to improper utilization of oxygen resulting in blurred vision, cataracts, and corneal ulcerations. Also dermatitis and cracking of skin, lesions of intestinal mucosa, and one type of anemia.
Niacin (nicotinamide)	Derived from amino acid tryptophan. Sources include yeast, meats, liver, fish, whole-grain products, peas, beans, and nuts.	Essential component of NAD and NADP, coenzymes in oxidation–reduction reactions. In lipid metabolism, inhibits production of cholesterol and assists in triglyceride breakdown.	Principal deficiency is **pellagra,** characterized by dermatitis, diarrhea, and psychological disturbances.
B₆ (pyridoxine)	Synthesized by bacteria of GI tract. Stored in liver, muscle, and brain. Other sources include salmon, yeast, tomatoes, yellow corn, spinach, wholegrain products, liver, and yogurt.	Essential coenzyme for normal amino acid metabolism. Assists production of circulating antibodies. May function as coenzyme in triglyceride metabolism.	Most common deficiency symptom is dermatitis of eyes, nose, and mouth. Other symptoms are retarded growth and nausea.
B₁₂ (cyanocobalamin)	Only B vitamin not found in vegetables; only vitamin containing cobalt. Absorption from GI tract depends on intrinsic factor secreted by gastric mucosa. Sources include liver, kidney, milk, eggs, cheese, and meat.	Coenzyme necessary for red blood cell formation, formation of the amino acid methionine, entrance of some amino acids into Krebs cycle, and manufacture of choline (used to synthesize acetylcholine).	Pernicious anemia, neuropsychiatric abnormalities (ataxia, memory loss, weakness, personality and mood changes, and abnormal sensations), and impaired activity of osteoblasts.
Pantothenic acid	Some produced by bacteria of GI tract. Stored primarily in liver and kidneys. Other sources include kidney, liver, yeast, green vegetables, and cereal.	Constituent of coenzyme A, which is essential for transfer of acetyl group from pyruvic acid into Krebs cycle, conversion of lipids and amino acids into glucose, and synthesis of cholesterol and steroid hormones.	Fatigue, muscle spasms, insufficient production of adrenal steroid hormones, vomiting, and insomnia.
Folic acid (folate, folacin)	Synthesized by bacteria of GI tract. Dietary sources include green leafy vegetables, broccoli, asparagus, breads, dried beans, and citrus fruits.	Component of enzyme systems synthesizing nitrogenous bases of DNA and RNA. Essential for normal production of red and white blood cells.	Production of abnormally large red blood cells (macrocytic anemia). Higher risk of neural tube defects in babies born to folate-deficient mothers.
Biotin	Synthesized by bacteria of GI tract. Dietary sources include yeast, liver, egg yolk, and kidneys.	Essential coenzyme for conversion of pyruvic acid to oxaloacetic acid and synthesis of fatty acids and purines.	Mental depression, muscular pain, dermatitis, fatigue, and nausea.
C (ascorbic acid)	Rapidly destroyed by heat. Some stored in glandular tissue and plasma. Sources include citrus fruits, tomatoes, and green vegetables.	Promotes protein synthesis including laying down of collagen in formation of connective tissue. As coenzyme, may combine with poisons, rendering them harmless until excreted. Works with antibodies, promotes wound healing, and functions as an antioxidant.	Scurvy; anemia; many symptoms related to poor collagen formation, including tender swollen gums, loosening of teeth (alveolar processes also deteriorate), poor wound healing, bleeding (vessel walls are fragile because of connective tissue degeneration), and retardation of growth.

DISORDERS: HOMEOSTATIC IMBALANCES

Fever

A **fever** is an elevation of core temperature caused by a resetting of the hypothalamic thermostat. The most common causes of fever are viral or bacterial infections and bacterial toxins; other causes are ovulation, excessive secretion of thyroid hormones, tumors, and reactions to vaccines. When phagocytes ingest certain bacteria, they are stimulated to secrete a **pyrogen** (PĪ-rō-gen; *pyro-* = fire; *-gen* = produce), a fever-producing substance. One pyrogen is interleukin-1. It circulates to the hypothalamus and induces neurons of the preoptic area to secrete prostaglandins. Some prostaglandins can reset the hypothalamic thermostat at a higher temperature, and temperature-regulating reflex mechanisms then act to bring the core body temperature up to this new setting. *Antipyretics* are agents that relieve or reduce fever. Examples include aspirin, acetaminophen (Tylenol), and ibuprofen (Advil), all of which reduce fever by inhibiting synthesis of certain prostaglandins.

Suppose that due to production of pyrogens the thermostat is reset at 39°C (103°F). Now the heat-promoting mechanisms (vasoconstriction, increased metabolism, shivering) are operating at full force. Thus, even though core temperature is climbing higher than normal—say, 38°C (101°F)—the skin remains cold, and shivering occurs. This condition, called a **chill,** is a definite sign that core temperature is rising. After several hours, core temperature reaches the setting of the thermostat, and the chills disappear. But now the body will continue to regulate temperature at 39°C (103°F). When the pyrogens disappear, the thermostat is reset at normal—37.0°C (98.6°F). Because core temperature is high in the beginning, the heat-losing mechanisms (vasodilation and sweating) go into operation to decrease core temperature. The skin becomes warm, and the person begins to sweat. This phase of the fever is called the **crisis,** and it indicates that core temperature is falling.

Although death results if core temperature rises above 44–46°C (112–114°F), up to a point, fever is beneficial. For example, a higher temperature intensifies the effects of interferons and the phagocytic activities of macrophages while hindering replication of some pathogens. Because fever increases heart rate, infection-fighting white blood cells are delivered to sites of infection more rapidly. In addition, antibody production and T cell proliferation increase. Moreover, heat speeds up the rate of chemical reactions, which may help body cells repair themselves more quickly.

Obesity

Obesity is body weight more than 20% above a desirable standard due to an excessive accumulation of adipose tissue. About one-third of the adult population in the United States is obese. (An athlete may be *overweight* due to higher-than-normal amounts of muscle tissue without being obese.) Even moderate obesity is hazardous to health; it is a risk factor in cardiovascular disease, hypertension, pulmonary disease, non-insulin-dependent diabetes mellitus, arthritis, certain cancers (breast, uterus, and colon), varicose veins, and gallbladder disease.

In a few cases, obesity may result from trauma of or tumors in the food-regulating centers in the hypothalamus. In most cases of obesity, no specific cause can be identified. Contributing factors include genetic factors, eating habits taught early in life, overeating to relieve tension, and social customs. Studies indicate that some obese people burn fewer calories during digestion and absorption of a meal, a smaller food-induced thermogenesis effect. Additionally, obese people who lose weight require about 15% fewer calories to maintain normal body weight than do people who have never been obese. Interestingly, people who gain weight easily when deliberately fed excess calories exhibit less NEAT (nonexercise activity thermogenesis, such as occurs with fidgeting) than people who resist weight gains in the face of excess calories. Although leptin suppresses appetite and produces satiety in experimental animals, it is not deficient in most obese people.

Most surplus calories in the diet are converted to triglycerides and stored in adipose cells. Initially, the adipocytes increase in size, but at a maximal size, they divide. As a result, proliferation of adipocytes occurs in extreme obesity. The enzyme endothelial lipoprotein lipase regulates triglyceride storage. The enzyme is very active in abdominal fat but less active in hip fat. Accumulation of fat in the abdomen is associated with higher blood cholesterol level and other cardiac risk factors.

Treatment of obesity is difficult because most people who are successful at losing weight gain it back within two years. Yet, even modest weight loss is associated with health benefits. Treatments for obesity include behavior modification programs, very-low-calorie diets, drugs, and surgery. Behavior modification programs, offered at many hospitals, strive to alter eating behaviors and increase exercise activity. The nutrition program includes a "heart-healthy" diet that includes abundant vegetables but is low in fats, especially saturated fats. A typical exercise program suggests walking for 30 minutes a day, five to seven times a week. Regular exercise enhances both weight loss and weight-loss maintenance. Very-low-calorie (VLC) diets include 400 to 800 kcal/day in a commercially made liquid mixture. The VLC diet is usually prescribed for 12 weeks, under close medical supervision. Two drugs are available to treat obesity. Sibutramine is an appetite suppressant that works by inhibiting reuptake of serotonin and norepinephrine in brain areas that govern eating behavior. Orlistat works by inhibiting the lipases released into the lumen of the GI tract. With less lipase activity, fewer dietary triglycerides are absorbed. For those with extreme obesity who have not responded to other treatments, a surgical procedure may be considered. The two operations most commonly performed—gastric bypass and gastroplasty—both greatly reduce the stomach size so that it can hold just a tiny quantity of food.

MEDICAL TERMINOLOGY

Heat cramps   Cramps that result from profuse sweating. The salt lost in sweat causes painful contractions of muscles; such cramps tend to occur in muscles used while working but do not appear until the person relaxes once the work is done. Drinking salted liquids usually leads to rapid improvement.

Heat exhaustion (heat prostration)   A condition in which the core temperature is generally normal, or a little below, and the skin is cool and moist due to profuse perspiration. Heat exhaustion is usually characterized by loss of fluid and electrolytes, especially salt (NaCl). The salt loss results in muscle cramps, dizziness, vomiting, and fainting; fluid loss may cause low blood pressure. Complete rest, rehydration, and electrolyte replacement are recommended.

Heatstroke (sunstroke) A severe and often fatal disorder caused by exposure to high temperatures, especially when the relative humidity is high, which makes it difficult for the body to lose heat. Blood flow to the skin is decreased, perspiration is greatly reduced, and body temperature rises sharply because of failure of the hypothalamic thermostat. Body temperature may reach 43°C (110°F). Treatment, which must be undertaken immediately, consists of cooling the body by immersing the victim in cool water and by administering fluids and electrolytes.

Kwashiorkor (kwash-ē-OR-kor) A disorder in which protein intake is deficient despite normal or nearly normal caloric intake, character-

ized by edema of the abdomen, enlarged liver, decreased blood pressure, low pulse rate, lower-than-normal body temperature, and sometimes mental retardation. Because the main protein in corn (zein) lacks two essential amino acids, which are needed for growth and tissue repair, many African children whose diet consists largely of cornmeal develop kwashiorkor.

Malnutrition (*mal-* = bad) An imbalance of total caloric intake or intake of specific nutrients, which can be either inadequate or excessive.

Marasmus (mar-AZ-mus) A type of protein–calorie undernutrition that results from inadequate intake of both protein and calories. Its characteristics include retarded growth, low weight, muscle wasting, emaciation, dry skin, and thin, dry, dull hair.

 STUDY OUTLINE

INTRODUCTION (p. 907)

1. Our only source of energy for performing biological work is the food we eat. Food also provides essential substances that we cannot synthesize.
2. Most food molecules absorbed by the gastrointestinal tract are used to supply energy for life processes, serve as building blocks during synthesis of complex molecules, or are stored for future use.

METABOLIC REACTIONS (p. 907)

1. Metabolism refers to all chemical reactions of the body and is of two types: catabolism and anabolism.
2. Catabolism is the term for reactions that break down complex organic compounds into simple ones. Overall, catabolic reactions are exergonic; they produce more energy than they consume.
3. Chemical reactions that combine simple molecules into more complex ones that form the body's structural and functional components are collectively known as anabolism. Overall, anabolic reactions are endergonic; they consume more energy than they produce.
4. The coupling of anabolism and catabolism occurs via ATP.

ENERGY TRANSFER (p. 908)

1. Oxidation is the removal of electrons from a substance; reduction is the addition of electrons to a substance.
2. Two coenzymes that carry hydrogen atoms during coupled oxidation–reduction reactions are nicotinamide adenine dinucleotide (NAD^+) and flavin adenine dinucleotide (FAD).
3. ATP can be generated via substrate-level phosphorylation, oxidative phosphorylation, and photophosphorylation.

CARBOHYDRATE METABOLISM (p. 909)

1. During digestion, polysaccharides and disaccharides are hydro-

lyzed into the monosaccharides glucose (about 80%), fructose, and galactose; the latter two are then converted to glucose.
2. Some glucose is oxidized by cells to provide ATP. Glucose also can be used to synthesize amino acids, glycogen, and triglycerides.
3. Glucose moves into most body cells via facilitated diffusion through glucose transporters (GluT) and becomes phosphorylated to glucose 6-phosphate. In muscle cells, this process is stimulated by insulin. Glucose entry into neurons and hepatocytes is always "turned on."
4. Cellular respiration, the complete oxidation of glucose to CO_2 and H_2O, involves glycolysis, the Krebs cycle, and the electron transport chain.
5. Glycolysis is the breakdown of glucose into two molecules of pyruvic acid; there is a net production of two molecules of ATP.
6. When oxygen is in short supply, pyruvic acid is reduced to lactic acid; under aerobic conditions, pyruvic acid enters the Krebs cycle.
7. Pyruvic acid is prepared for entrance into the Krebs cycle by conversion to a two-carbon acetyl group followed by the addition of coenzyme A to form acetyl coenzyme A.
8. The Krebs cycle involves decarboxylations, oxidations, and reductions of various organic acids.
9. Each molecule of pyruvic acid that is converted to acetyl coenzyme A and then enters the Krebs cycle produces three molecules of CO_2, four molecules of NADH and four H^+, one molecule of $FADH_2$, and one molecule of ATP.
10. The energy originally stored in glucose and then in pyruvic acid is transferred primarily to the reduced coenzymes NADH and $FADH_2$.
11. The electron transport chain involves a series of oxidation–reduction reactions in which the energy in NADH and $FADH_2$ is liberated and transferred to ATP.
12. The electron carriers include FMN, cytochromes, iron–sulfur centers, copper atoms, and coenzyme Q.
13. The electron transport chain yields a maximum of 32 or 34 molecules of ATP and six molecules of H_2O.

14. Table 25.1 on page 918 summarizes the ATP yield during cellular respiration. The complete oxidation of glucose can be represented as follows:

$$C_6H_{12}O_6 + 6\,O_2 + 36 \text{ or } 38 \text{ ADPs} + 36 \text{ or } 38 \text{ ⓟ} \longrightarrow$$
$$6\,CO_2 + 6\,H_2O + 36 \text{ or } 38 \text{ ATPs}$$

15. The conversion of glucose to glycogen for storage in the liver and skeletal muscle is called glycogenesis. It is stimulated by insulin.

16. The conversion of glycogen to glucose is called glycogenolysis. It occurs between meals and is stimulated by glucagon and epinephrine.

17. Gluconeogenesis is the conversion of noncarbohydrate molecules into glucose. It is stimulated by cortisol and glucagon.

LIPID METABOLISM (p. 920)

1. Lipoproteins transport lipids in the bloodstream. Types of lipoproteins include chylomicrons, which carry dietary lipids to adipose tissue; very low-density lipoproteins (VLDLs), which carry triglycerides from the liver to adipose tissue; low-density lipoproteins (LDLs), which deliver cholesterol to body cells; and high-density lipoproteins (HDLs), which remove excess cholesterol from body cells and transport it to the liver for elimination.

2. Cholesterol in the blood comes from two sources: from food and from synthesis by the liver.

3. Lipids may be oxidized to produce ATP or stored as triglycerides in adipose tissue, mostly in the subcutaneous layer.

4. A few lipids are used as structural molecules or to synthesize essential molecules.

5. Adipose tissue contains lipases that catalyze the deposition of triglycerides from chylomicrons and hydrolyze triglycerides into fatty acids and glycerol.

6. In lipolysis, triglycerides are split into fatty acids and glycerol and released from adipose tissue under the influence of epinephrine, norepinephrine, cortisol, thyroid hormones, and insulinlike growth factors.

7. Glycerol can be converted into glucose by conversion into glyceraldehyde 3-phosphate.

8. In beta-oxidation of fatty acids, carbon atoms are removed in pairs from fatty acid chains; the resulting molecules of acetyl coenzyme A enter the Krebs cycle.

9. The conversion of glucose or amino acids into lipids is called lipogenesis; it is stimulated by insulin.

PROTEIN METABOLISM (p. 923)

1. During digestion, proteins are hydrolyzed into amino acids, which enter the liver via the hepatic portal vein.

2. Amino acids, under the influence of insulinlike growth factors and insulin, enter body cells via active transport.

3. Inside cells, amino acids are synthesized into proteins that function as enzymes, hormones, structural elements, and so forth; stored as fat or glycogen; or used for energy.

4. Before amino acids can be catabolized, they must be deaminated and converted to substances that can enter the Krebs cycle.

5. Amino acids may also be converted into glucose, fatty acids, and ketone bodies.

6. Protein synthesis is stimulated by insulinlike growth factors, thyroid hormones, insulin, estrogen, and testosterone.

7. Table 25.2 on page 927 summarizes carbohydrate, lipid, and protein metabolism.

KEY MOLECULES AT METABOLIC CROSSROADS (p. 925)

1. Three molecules play a key role in metabolism: glucose 6-phosphate, pyruvic acid, and acetyl coenzyme A.

2. Glucose 6-phosphate may be converted to glucose, glycogen, ribose 5-phosphate, and pyruvic acid.

3. When ATP is low and oxygen is plentiful, pyruvic acid is converted to acetyl coenzyme A; when oxygen supply is low, pyruvic acid is converted to lactic acid. One link between carbohydrate and protein metabolism occurs via pyruvic acid.

4. Acetyl coenzyme A is the molecule that enters the Krebs cycle; it is also used to synthesize fatty acids, ketone bodies, and cholesterol.

METABOLIC ADAPTATIONS (p. 927)

1. During the absorptive state, ingested nutrients enter the blood and lymph from the GI tract.

2. During the absorptive state, blood glucose is oxidized to form ATP, and glucose transported to the liver is converted to glycogen or triglycerides. Most triglycerides are stored in adipose tissue. Amino acids in hepatocytes are converted to carbohydrates, fats, and proteins. Table 25.3 on page 929 summarizes the hormonal regulation of metabolism during the absorptive state.

3. During the postabsorptive state, absorption is complete and the ATP needs of the body are satisfied by nutrients already present in the body. The major task is to maintain normal blood glucose level by converting glycogen in the liver and skeletal muscle into glucose, converting glycerol into glucose, and converting amino acids into glucose. Fatty acids, ketone bodies, and amino acids are oxidized to supply ATP. Table 25.4 on page 931 summarizes the hormonal regulation of metabolism during the postabsorptive state.

4. Fasting is going without food for a few days; starvation implies weeks or months of inadequate food intake. During fasting and starvation, fatty acids and ketone bodies are increasingly utilized for ATP production.

HEAT AND ENERGY BALANCE (p. 932)

1. Measurement of the metabolic rate under basal conditions is called the basal metabolic rate (BMR).

2. A kilocalorie (kcal) or Calorie is the amount of energy required to raise the temperature of 1000 g of water from 14°C to 15°C.

3. Normal core temperature is maintained by a delicate balance between heat-producing and heat-losing mechanisms.

4. Exercise, hormones, the nervous system, body temperature, ingestion of food, age, gender, climate, sleep, and malnutrition affect metabolic rate.

5. Mechanisms of heat transfer are conduction, convection, radiation, and evaporation. Conduction is the transfer of heat between two substances or objects in contact with each other. Convection is the transfer of heat by a liquid or gas between areas of different temperatures. Radiation is the transfer of heat from a warmer object to a cooler object without physical contact. Evaporation is the conversion of a liquid to a vapor; in the process, heat is lost.

6. The hypothalamic thermostat is in the preoptic area.

7. Responses that produce, conserve, or retain heat when core temperature falls are vasoconstriction; release of epinephrine, norepinephrine, and thyroid hormones; and shivering.

8. Responses that increase heat loss when core temperature increases

include vasodilation, decreased metabolic rate, and evaporation of perspiration.

9. Two nuclei in the hypothalamus that help regulate food intake are the arcuate and paraventricular nuclei. The hormone leptin, released by adipocytes, inhibits release of neuropeptide Y from the arcuate nucleus and thereby decreases food intake. Melanocortin also decreases food intake.

NUTRITION (p. 936)

1. Nutrients include water, carbohydrates, lipids, proteins, minerals, and vitamins.
2. Most teens and adults need between 1600 and 2800 Calories per day.
3. Nutrition experts suggest dietary calories be 50–60% from carbohydrates, 30% or less from fats, and 12–15% from proteins, although the optimal levels of these nutrients may vary.

4. The food guide pyramid indicates how many servings of five food groups are recommended each day to attain the number of calories and variety of nutrients needed for wellness.
5. Minerals known to perform essential functions include calcium, phosphorus, potassium, sulfur, sodium, chloride, magnesium, iron, iodide, manganese, copper, cobalt, zinc, fluoride, selenium, and chromium. Their functions are summarized in Table 25.5 on page 939.
6. Vitamins are organic nutrients that maintain growth and normal metabolism. Many function in enzyme systems.
7. Fat-soluble vitamins are absorbed with fats and include vitamins A, D, E, and K; water-soluble vitamins include the B vitamins and vitamin C.
8. The functions and deficiency disorders of the principal vitamins are summarized in Table 25.6 on pages 940–941.

Q SELF-QUIZ QUESTIONS

Fill in the blanks in the following statements.

1. The thermostat and food intake regulating center of the body is in the _____ of the brain.
2. The three key molecules of metabolism _____, _____, and _____.
3. _____ are the primary regulators of metabolism.

Indicate whether the following statements are true of false.

4. Glycogenesis occurs in the liver and cardiac muscle and is stimulated by insulin.
5. Most of the glucose in the body is used to produce ATP.

Choose the one best answer to the following questions.

6. Which of the following is *not* considered a nutrient? (a) water, (b) carbohydrates, (c) nucleic acids, (d) proteins, (e) lipids.
7. Which of the following factors affect metabolic rate and thus the production of body heat? (1) exercise, (2) hormones, (3) minerals, (4) food intake, (5) waste product removal, (6) age. (a) 1, 2, 3, and 4, (b) 2, 3, 4, and 5, (c) 3, 4, 5, and 6, (d) 1, 2, 4, and 6, (e) 1, 3, 5, and 6.
8. Which of the following are uses for glucose in the body? (1) ATP production, (2) amino acid synthesis, (3) glycogen synthesis, (4) triglyceride synthesis, (5) vitamin synthesis. (a) 1, 2, 3, and 4, (b) 2, 3, 4, and 5, (c) 1, 3, and 5, (d) 2, 4, and 5, (e) 1, 2, 4, and 5.
9. Which of the following is the *correct* sequence for the oxidation of glucose to produce ATP? (a) electron transport chain, Krebs cycle, glycolysis, formation of acetyl CoA, (b) Krebs cycle, formation of acetyl CoA, electron transport chain, glycolysis, (c) glycolysis, electron transport chain, Krebs cycle, formation of acetyl CoA, (d) glycolysis, formation of acetyl CoA, Krebs cycle, electron transport chain, (e) formation of acetyl CoA, Krebs cycle, glycolysis, electron transport chain.
10. Which of the following are absorptive state reactions? (1) aerobic cellular respiration, (2) glycogenesis, (3) glycogenolysis, (4) gluconeogenesis using lactic acid, (5) lipolysis. (a) 1 and 2, (b) 2 and 3, (c) 3 and 4, (d) 4 and 5, (e) 1 and 5.

11. Which of the following would you *not* expect to experience during fasting or starvation? (a) decrease in plasma fatty acid levels, (b) increase in ketone body formation, (c) lipolysis, (d) increased use of ketones for ATP production in the brain, (e) depletion of glycogen.
12. If core body temperature rises above normal, which of the following would occur to cool the body? (1) dilation of vessels in the skin, (2) increased radiation and conduction of heat to the environment, (3) increased metabolic rate, (4) evaporation of perspiration, (5) increased secretion of thyroid hormones. (a) 3, 4, and 5, (b) 1, 2, and 4, (c) 1, 2, and 5, (d) 1, 2, 3, 4, and 5, (e) 1, 2, 4, and 5.
13. Match the following:
    ____ (a) the breakdown of glycogen back to glucose
    ____ (b) the removal of the amino group from an amino acid
    ____ (c) the splitting of a triglyceride into glycerol and fatty acids
    ____ (d) the cleavage of one pair of carbon atoms at a time from a fatty acid
    ____ (e) the formation of ketone bodies
    ____ (f) the synthesis of lipids
    ____ (g) the conversion of glucose into glycogen
    ____ (h) the transfer of an amino group from an amino acid to a substance such as pyruvic acid
    ____ (i) the formation of glucose from noncarbohydrate sources
    ____ (j) the breakdown of glucose into two molecules of pyruvic acid

    (1) glycolysis
    (2) glycogenolysis
    (3) glycogenesis
    (4) gluconeogenesis
    (5) lipolysis
    (6) ketogenesis
    (7) lipogenesis
    (8) deamination
    (9) transamination
    (10) beta oxidation

14. Match the following:

_____ (a) deliver cholesterol to body cells for use in repair of cell membranes and synthesis of steroid hormones

_____ (b) remove excess cholesterol from body cells and transport it to the liver for elimination

_____ (c) organic nutrients required in tiny amounts to maintain growth and normal metabolism

_____ (d) the energy-transfering molecule of the body

_____ (e) nutrient molecules that can be oxidized to produce ATP or stored in adipose tissue

_____ (f) carriers of electrons in the electron transport chain

_____ (g) the body's preferred source for synthesizing ATP

_____ (h) composed of amino acids and are the primary regulatory molecules in the body

_____ (i) acetoacetic acid, betahydroxybutyric acid, and acetone

_____ (j) hormone that acts to decrease total body fat mass

_____ (k) neurotransmitter that stimulates food intake

_____ (j) inorganic substances that perform many vital functions in the body

(1) leptin
(2) minerals
(3) glucose
(4) lipids
(5) proteins
(6) neuropeptide Y
(7) cytochromes
(8) ketone bodies
(9) low-density lipoproteins
(10) ATP
(11) vitamins
(12) high-density lipoproteins

15. Match the following:

_____ (a) the mechanism of ATP generation that links chemical reactions with pumping of hydrogen ions

_____ (b) the removal of electrons from an atom or molecule resulting in a decrease in energy

_____ (c) refers to all the chemical reactions in the body

_____ (d) the oxidation of glucose to produce ATP

_____ (e) the addition of electrons to a molecule resulting in an increase in energy content of the molecule

_____ (f) chemical reactions that break down complex organic molecules into simpler ones

_____ (g) overall rate at which metabolic reactions use energy

_____ (h) removal of CO_2 from a molecule

_____ (i) chemical reactions that combine simple molecules and monomers to make more complex ones

_____ (j) the addition of a phosphate group to a molecule

(1) metabolism
(2) catabolism
(3) anabolism
(4) metabolic rate
(5) oxidation
(6) reduction
(7) phosphorylation
(8) chemiosmosis
(9) decarboxylation
(10) cellular respiration

CRITICAL THINKING QUESTIONS

1. Deb was training for a marathon. With only one day to go before the big event, she had shortened her morning training run and was loading up on bread and pasta for her last meal. Why is Deb eating lots of bread and spaghetti?

HINT *Bread and pasta are high in carbohydrates.*

2. Mr. Hernandez believed that the perfect lawn needed to be cut according to a strict schedule. As he was pushing the lawn mower around the yard on a scorching hot Saturday morning, he began to feel dizzy and nauseous. When he sat down, he was bothered by cramps in his legs, and he nearly fainted. What's wrong with Mr. Hernandez?

HINT *His shirt was soaked with sweat, but his forehead actually felt cool.*

3. Three-year-old Sara's mother was concerned about her child's eating habits. She complained "one day Sara will only eat a few bites, and then the next day she eats like a horse!" The pediatrician told Sara's mom that it sounded like Sara had perfectly normal control over her food intake. How is food intake controlled.

HINT *Sara is too young to count calories.*

ANSWERS TO FIGURE QUESTIONS

25.1 In pancreatic acinar cells, anabolism predominates because they are synthesizing complex molecules (digestive enzymes).

25.2 Glycolysis is also called anaerobic cellular respiration.

25.3 The reactions of glycolysis consume two molecules of ATP but generate four molecules of ATP, for a net of two.

25.4 Kinases are enzymes that phosphorylate (add phosphate to) their substrate.

25.5 Glycolysis occurs in the cytosol.

25.6 CO_2 is given off during the production of acetyl coenzyme A and during the Krebs cycle. It diffuses into the blood, is transported by the blood to the lungs, and is exhaled.

25.7 The production of reduced coenzymes is important in the Krebs cycle because they will subsequently yield ATP in the electron transport chain.

25.8 The energy source that powers the proton pumps is electrons provided by $NADH + H^+$.

25.9 The concentration of H^+ is highest in the space between the inner and outer mitochondrial membranes.

25.10 During the complete oxidation of one glucose molecule, six molecules of O_2 are used and six molecules of CO_2 are produced.

25.11 Skeletal muscle fibers can synthesize glycogen, but they cannot release glucose into the blood because they lack the enzyme phosphatase.

25.12 Hepatocytes can carry out gluconeogenesis and glycogenesis.

25.13 LDLs deliver cholesterol to body cells.

25.14 Hepatocytes and adipose cells carry out lipogenesis, beta oxidation, and lipolysis; hepatocytes carry out ketogenesis.

25.15 An amino group is removed from an amino acid via deamination.

25.16 Acetyl coenzyme A is the gateway into the Krebs cycle.

25.17 Reactions of the absorptive state are mainly anabolic.

25.18 Processes that directly elevate blood glucose during the postabsorptive state include lipolysis (in adipocytes and hepatocytes), gluconeogenesis (in hepatocytes), and glycogenolysis (in hepatocytes).

25.19 Exercise, the sympathetic nervous system, hormones (epinephrine, norepinephrine, thyroxine, testosterone, human growth hormone), elevated body temperature, and ingestion of food increase metabolic rate.

25.20 Foods that contain cholesterol and most of the saturated fatty acids in the diet are milk, yogurt, cheeses, and meats.

The Urinary System

THE URINARY SYSTEM AND HOMEOSTASIS

The urinary system contributes to homeostasis by altering blood composition, forming urine, and regulating body fluid pH.

ENERGIZE YOUR STUDY

FOUNDATIONS CD

Anatomy Overview
• The Urinary System

Animations
• Systems Contributions to Homeostasis
• The Case of the Girl with the Fruity Breath
• The Case of the Man with the Yellow Eyes

Concepts and Connections and Exercises reinforce your understanding

www.wiley.com/college/apcentral

INSIGHTS AND EXPLORATIONS

Fortunately most of us do not have to worry about whether the food we eat, the drugs we take, or the water we drink will be eliminated from our bodies. We have healthy kidneys that do this for us, along with our digestive system. But many people must schedule their weeks around visits to dialysis centers to have waste and extra fluid removed from their blood by a process called dialysis. In this web-based activity we will explore why it is that some people lose the function of their kidneys and how dialysis works.

The **urinary system** consists of two kidneys, two ureters, one urinary bladder, and one urethra (Figure 26.1). After the kidneys filter blood plasma, they return most of the water and solutes to the bloodstream. The remaining water and solutes constitute **urine**, which passes through the ureters and is stored in the urinary bladder until it is excreted from the body through the urethra. **Nephrology** (nef-ROL-ō-jē; *nephro-* = kidney; *-logy* = study of) is the scientific study of the anatomy, physiology, and pathology of the kidneys. The branch of medicine that deals with the male and female urinary systems and the male reproductive system is **urology** (ū-ROL-ō-jē; *uro-* = urine).

OVERVIEW OF KIDNEY FUNCTIONS

OBJECTIVE

• **List the functions of the kidneys.**

The kidneys do the major work of the urinary system. The other parts of the system are mainly passageways and storage areas. Functions of the kidneys include:

• *Regulating blood ionic composition.* The kidneys help regulate the blood levels of several ions, most importantly sodium ions (Na^+), potassium ions (K^+), calcium ions (Ca^{2+}), chloride ions (Cl^-), and phosphate ions (HPO_4^{2-}).

• *Regulating blood pH.* The kidneys excrete a variable amount of hydrogen ions (H^+) into the urine and conserve bicarbonate ions (HCO_3^-), which are an important buffer of H^+ in the blood. Both of these activities help regulate blood pH.

Figure 26.1 **Organs of the urinary system in a female.** (See Tortora, *A Photographic Atlas of the Human Body,* Figure 13.2.)

Urine formed by the kidneys passes first into the ureters, then to the urinary bladder for storage, and finally through the urethra for elimination from the body.

Functions of the Urinary System

1. The kidneys excrete wastes in urine, regulate blood volume and composition, help regulate blood pressure, synthesize glucose, release erythropoietin, and participate in vitamin D synthesis.
2. The ureters transport urine from the kidneys to the urinary bladder.
3. The urinary bladder stores urine.
4. The urethra discharges urine from the body.

Labels: Diaphragm, Esophagus, Left adrenal (suprarenal) gland, Left renal vein, LEFT KIDNEY, Abdominal aorta, Inferior vena cava, LEFT URETER, Rectum, Left ovary, Uterus

RIGHT KIDNEY, Right renal artery, RIGHT URETER, URINARY BLADDER, URETHRA

Anterior view

Which organs constitute the urinary system?

- *Regulating blood volume.* The kidneys adjust blood volume by conserving or eliminating water in the urine. Also, an increase in blood volume increases blood pressure, whereas a decrease in blood volume decreases blood pressure.

- *Regulating blood pressure.* Besides adjusting blood volume, the kidneys help regulate blood pressure by secreting the enzyme renin, which activates the renin–angiotensin–aldosterone pathway (see Figure 18.16 on page 611). Increased renin causes an increase in blood pressure.

- *Maintaining blood osmolarity.* By separately regulating loss of water and loss of solutes in the urine, the kidneys maintain a relatively constant blood osmolarity close to 290 milliosmoles per liter (mOsm/liter).*

- *Producing hormones.* The kidneys produce two hormones. *Calcitriol,* the active form of vitamin D, helps regulate calcium homeostasis (see Figure 18.14 on page 608), and *erythropoietin* stimulates production of red blood cells (see Figure 19.5 on page 641).

- *Regulating blood glucose level.* Like the liver, the kidneys can use the amino acid glutamine in *gluconeogenesis,* the synthesis of new glucose molecules. They can then release glucose into the blood to help maintain a normal blood glucose level.

- *Excreting wastes and foreign substances.* By forming urine, the kidneys help excrete **wastes**—substances that have no useful function in the body. Some wastes excreted in urine result from metabolic reactions in the body. These include ammonia and urea from the deamination of amino acids; bilirubin from the catabolism of hemoglobin; creatinine from the breakdown of creatine phosphate in muscle fibers; and uric acid from the catabolism of nucleic acids. Other wastes excreted in urine are foreign substances from the diet, such as drugs and environmental toxins.

▶ C H E C K P O I N T

1. What are wastes, and how do the kidneys participate in their removal from the body?

---

*The **osmolarity** of a solution is a measure of the total number of dissolved particles per liter of solution. The particles may be molecules, ions, or a mixture of both. To calculate osmolarity, multiply molarity (see page 39) by the number of particles per molecule, once the molecule dissolves. A similar term, *osmolality,* is the number of particles of solute per *kilogram* of water. Because it is easier to measure volumes of solutions than to determine the mass of water they contain, osmolarity is used more commonly than osmolality. Most body fluids and solutions used clinically are dilute, in which case there is less than a 1% difference between the two measures.

ANATOMY AND HISTOLOGY OF THE KIDNEYS

▶ O B J E C T I V E S

- Describe the external and internal gross anatomical features of the kidneys.
- Trace the path of blood flow through the kidneys.
- Describe the structure of renal corpuscles and renal tubules.

The paired **kidneys** are reddish, kidney-bean-shaped organs located just above the waist between the peritoneum and the posterior wall of the abdomen. Because their position is posterior to the peritoneum of the abdominal cavity, they are said to be **retroperitoneal** (re′-trō-per-i-tō-NĒ-al; *retro-* = behind) organs (Figure 26.2). The kidneys are located between the levels of the last thoracic and third lumbar vertebrae, a position where they are partially protected by the eleventh and twelfth pairs of ribs. The right kidney is slightly lower than the left (see Figure 26.1) because the liver occupies considerable space on the right side superior to the kidney.

External Anatomy of the Kidneys

A typical kidney in an adult is 10–12 cm (4–5 in.) long, 5–7 cm (2–3 in.) wide, and 3 cm (1 in.) thick—about the size of a bar of bath soap—and has a mass of 135–150 g (4.5–5 oz). The concave medial border of each kidney faces the vertebral column (see Figure 26.1). Near the center of the concave border is a deep vertical fissure called the **renal hilus** (see Figure 26.3), through which the ureter emerges from the kidney along with blood vessels, lymphatic vessels, and nerves.

Three layers of tissue surround each kidney (Figure 26.2). The deep layer, the **renal capsule** (*ren-* = kidney), is a smooth, transparent sheet of dense irregular connective tissue that is continuous with the outer coat of the ureter. It serves as a barrier against trauma and helps maintain the shape of the kidney. The middle layer, the **adipose capsule,** is a mass of fatty tissue surrounding the renal capsule. It also protects the kidney from trauma and holds it firmly in place within the abdominal cavity. The superficial layer, the **renal fascia,** is another thin layer of dense irregular connective tissue that anchors the kidney to the surrounding structures and to the abdominal wall. On the anterior surface of the kidneys, the renal fascia is deep to the peritoneum.

Nephroptosis (Floating Kidney)

Nephroptosis (nef′-rōp-TŌ-sis; *ptosis* = falling), or **floating kidney,** is an inferior displacement or dropping of the kidney. It occurs when the kidney slips from its normal position because it is not securely held in place by adjacent organs or its covering of fat. Nephroptosis develops most often in very thin people whose adipose capsule or renal fascia is deficient. It is dangerous because the ureter may kink and block urine flow. The resulting backup of urine puts pressure on the kidney, which damages the tissue. Twisting of the ureter also causes pain. ■

Figure 26.2   Position and coverings of the kidneys.   (See Tortora, *A Photographic Atlas of the Human Body,* Figure 13.3.)

The kidneys are surrounded by a renal capsule, adipose capsule, and renal fascia.

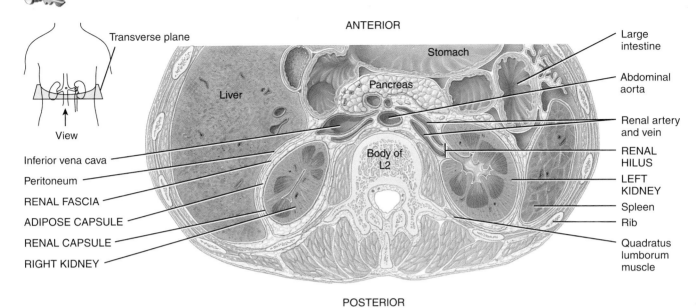

(a) Inferior view of transverse section of abdomen (L2)

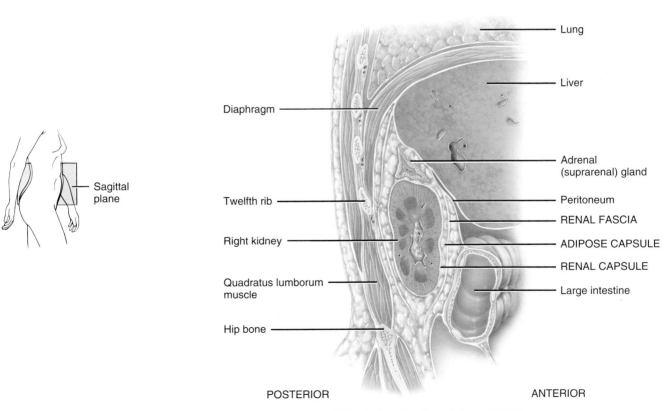

(b) Sagittal section through the right kidney

   Why are the kidneys said to be retroperitoneal?

Internal Anatomy of the Kidneys

A frontal section through the kidney reveals two distinct regions: a superficial, smooth-textured reddish area called the **renal cortex** (*cortex* = rind or bark) and a deep, reddish-brown inner region called the **renal medulla** (*medulla* = inner portion) (Figure 26.3). The medulla consists of 8 to 18 cone-shaped **renal pyramids.** The base (wider end) of each pyramid faces the renal cortex, and its apex (narrower end), called a **renal papilla,** points toward the renal hilus. The renal cortex is the smooth-textured area extending from the renal capsule to the bases of the renal pyramids and into the spaces between them. It is divided into an outer *cortical zone* and an inner *juxtamedullary zone.* Those portions of the renal cortex that extend between renal pyramids are called **renal columns.** A **renal lobe** consists of a renal pyramid, its overlying area of renal cortex, and one-half of each adjacent renal column.

Together, the renal cortex and renal pyramids of the renal medulla constitute the **parenchyma** (functional portion) of the kidney. Within the parenchyma are the functional units of the kidney—about 1 million microscopic structures called **nephrons** (NEF-rons). Urine formed by the nephrons drains into large **papillary ducts,** which extend through the renal papillae of the pyramids. The papillary ducts drain into cuplike structures called **minor** and **major calyces** (KĀ-li-sēz = cups; singular is *calyx*). Each kidney has 8 to 18 minor calyces and 2 to 3 major calyces. A minor calyx receives urine from the papillary ducts of one renal papilla and delivers it to a major calyx. From the major calyces, urine drains into a single large cavity called the **renal pelvis** (*pelv-* = basin) and then out through the ureter to the urinary bladder.

The hilus expands into a cavity within the kidney called the **renal sinus,** which contains part of the renal pelvis, the calyces, and branches of the renal blood vessels and nerves. Adipose tissue helps stabilize the position of these structures in the renal sinus.

Blood and Nerve Supply of the Kidneys

Because the kidneys remove wastes from the blood and regulate its volume and ionic composition, it is not surprising that they are abundantly supplied with blood vessels. Although the kidneys constitute less than 0.5% of total body mass, they receive

Figure 26.3 Internal anatomy of the kidneys. (See Tortora, *A Photographic Atlas of the Human Body,* Figures 13.4, 13.5.)

The two main regions of the kidney parenchyma are the renal cortex and the renal pyramids in the renal medulla.

(a) Frontal section of right kidney

20–25% of the resting cardiac output via the right and left **renal arteries** (Figure 26.4).

Within the kidney, the renal artery divides into several **segmental arteries,** each of which gives off several branches that enter the parenchyma and pass through the renal columns between the renal pyramids as the interlobar arteries. At the bases of the renal pyramids, the **interlobar arteries** arch between the renal medulla and cortex; here they are known as the **arcuate arteries** (AR-kū-āt = shaped like a bow). Divisions of the arcuate arteries produce a series of **interlobular arteries,** which enter the renal cortex and give off branches called **afferent arterioles** (*af-* = toward; *-ferrent* = to carry).

Each nephron receives one afferent arteriole, which divides into a tangled, ball-shaped capillary network called the **glomerulus** (glō-MER-ū-lus = little ball; plural is *glomeruli*). The glomerular capillaries then reunite to form an **efferent arteriole** (*ef-* = out) that carries blood out of the glomerulus. Glomerular capillaries are unique among capillaries in the body because they are positioned between two arterioles, rather than between an arteriole and a venule. Because they are capillary networks and they also play an important role in urine formation, the glomeruli are part of both the cardiovascular and the urinary systems.

The efferent arterioles divide to form the **peritubular capillaries** (*peri-* = around), which surround tubular parts of the

nephron in the renal cortex. Extending from some efferent arterioles are long loop-shaped capillaries called **vasa recta** (VĀ-sa REK-ta; *vasa* = vessels; *recta* = straight) that supply tubular portions of the nephron in the renal medulla (see Figure 26.5b).

The peritubular capillaries eventually reunite to form **peritubular venules** and then **interlobular veins,** which also receive blood from the vasa recta. Then the blood drains through the **arcuate veins** to the **interlobar veins** running between renal pyramids. Blood leaves the kidney through a single **renal vein** that exits at the renal hilus and carries venous blood to the inferior vena cava.

Most renal nerves originate in the *celiac ganglion* and pass through the *renal plexus* into the kidneys along with the renal arteries. Renal nerves are part of the sympathetic division of the autonomic nervous system. Most are vasomotor nerves that regulate the flow of blood through the kidney by causing vasodilation or vasoconstriction of renal arterioles.

The Nephron

Parts of a Nephron

Nephrons are the functional units of the kidneys. Each nephron (Figure 26.5) consists of two parts: a **renal corpuscle** (KOR-pus-sul = tiny body), where blood plasma is filtered, and a **renal tubule** into which the filtered fluid passes. The two components

SUPERIOR

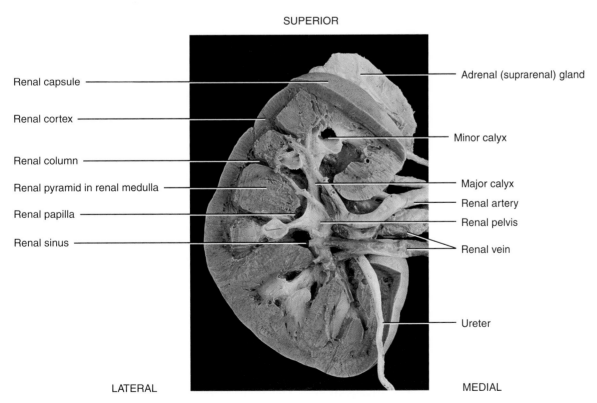

Renal capsule — Adrenal (suprarenal) gland

Renal cortex — Minor calyx

Renal column —

Renal pyramid in renal medulla — Major calyx

Renal papilla — Renal artery

Renal sinus — Renal pelvis

Renal vein

Ureter

LATERAL   MEDIAL

(b) Frontal section of right kidney

What structures pass through the renal hilus?

Figure 26.4   Blood supply of the kidneys.   (See Tortora, *A Photographic Atlas of the Human Body,* Figure 13.6.)

The renal arteries deliver 20–25% of the resting cardiac output to the kidneys.

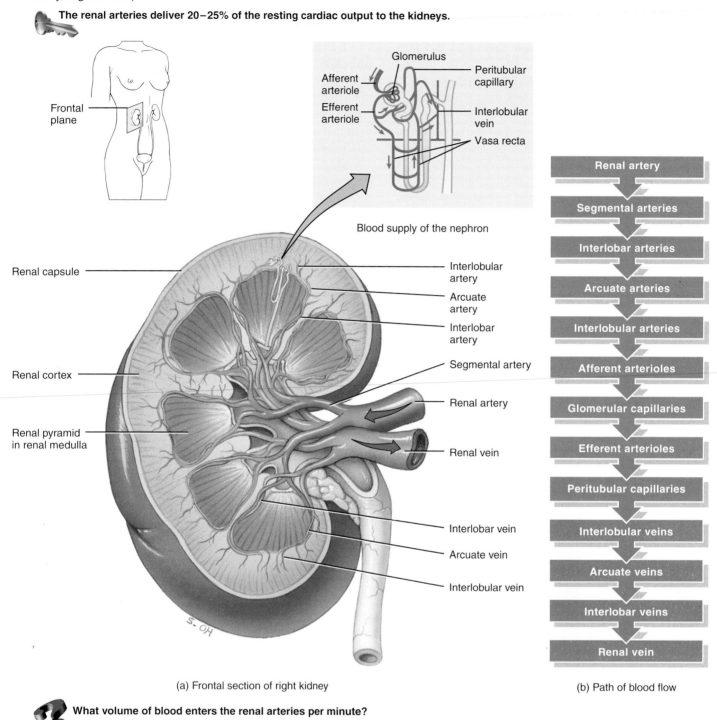

(a) Frontal section of right kidney

(b) Path of blood flow

What volume of blood enters the renal arteries per minute?

---

of a renal corpuscle are the **glomerulus** (capillary network) and the **glomerular (Bowman's) capsule,** a double-walled epithelial cup that surrounds the glomerular capillaries. Blood plasma is first filtered in the glomerular capsule, and then the filtered fluid passes into the renal tubule, which has three main sections. In the order that fluid passes through them, the renal tubule consists of a (1) **proximal convoluted tubule,** (2) **loop of Henle (nephron loop),** and (3) **distal convoluted tubule.** *Proximal* denotes the part of the tubule attached to the glomerular capsule, and *distal* denotes the part that is farther away. *Convoluted* means the tubule is tightly coiled rather than straight. The renal corpuscle and both convoluted tubules lie within the renal cortex, whereas the loop of Henle extends into the renal medulla, makes a hairpin turn, and then returns to the renal cortex.

Figure 26.5 **The structure of nephrons (colored gold) and associated blood vessels.**
(a) A cortical nephron. (b) A juxtamedullary nephron.

Nephrons are the functional units of the kidneys.

FLOW OF FLUID THROUGH A
CORTICAL NEPHRON

Glomerular (Bowman's) capsule
↓
Proximal convoluted tubule
↓
Descending limb of the
loop of Henle
↓
Ascending limb of the
loop of Henle
↓
Distal convoluted tubule
(drains into collecting duct)

(a) Cortical nephron and vascular supply

(continues)

The distal convoluted tubules of several nephrons empty into a single **collecting duct.** Collecting ducts then unite and converge until eventually there are only several hundred large **papillary ducts,** which drain into the minor calyces. The collecting ducts and papillary ducts extend from the renal cortex through the renal medulla to the renal pelvis. Although one kidney has about 1 million nephrons, it has a much smaller number of collecting ducts and even fewer papillary ducts.

In a nephron, the loop of Henle connects the proximal and distal convoluted tubules. The first part of the loop of Henle dips into the renal medulla, where it is called the **descending limb of the loop of Henle** (Figure 26.5). It then makes that hairpin turn and returns to the renal cortex as the **ascending limb of the loop of Henle.** About 80–85% of the nephrons are **cortical nephrons.** Their renal corpuscles lie in the outer portion of the renal cortex, and they have *short* loops of Henle that lie mainly in the cortex and penetrate only into the outer region of the renal medulla (Figure 26.5a). The short loops of Henle receive their blood supply from peritubular capillaries that arise from efferent arterioles. The other 15–20% of the nephrons are

Figure 26.5   *(continued)*

Renal capsule

Proximal convoluted tubule

Peritubular capillary

Distal convoluted tubule

Renal corpuscle:
Glomerular (Bowman's) capsule

Glomerulus

Afferent arteriole

Efferent arteriole

Interlobular artery

Interlobular vein

Arcuate vein

Arcuate artery

Corticomedullary junction

Renal cortex

Renal medulla

Collecting duct

Loop of Henle:
Descending limb

Thick ascending limb

Thin ascending limb

Renal cortex

Renal medulla

Renal papilla

Minor calyx

Kidney

Vasa recta

Papillary duct

FLOW OF FLUID THROUGH A JUXTAMEDULLARY NEPHRON

Glomerular (Bowman's) capsule
↓
Proximal convoluted tubule
↓
Descending limb of the loop of Henle
↓
Thin ascending limb of the loop of Henle
↓
Thick ascending limb of the loop of Henle
↓
Distal convoluted tubule (drains into collecting duct)

Renal papilla

Minor calyx

Urine

(b) Juxtamedullary nephron and vascular supply

What are the basic differences between cortical and juxtamedullary nephrons?

juxtamedullary nephrons (*juxta-* = near to). Their renal corpuscles lie deep in the cortex, close to the medulla, and they have a *long* loop of Henle that extends into the deepest region of the medulla (Figure 26.5b). Long loops of Henle receive their blood supply from peritubular capillaries and from the vasa recta that arise from efferent arterioles. In addition, the ascending limb of the loop of Henle of juxtamedullary nephrons consists of two portions: a **thin ascending limb** followed by a **thick ascending limb** (Figure 26.5b). The lumen of the thin ascending limb is the same as in other areas of the renal tubule; it is only the epithelium that is thinner. Nephrons with long loops of

Henle enable the kidneys to excrete very dilute or very concentrated urine (described on pages 972–973).

Histology of the Nephron and Collecting Duct

A single layer of epithelial cells forms the entire wall of the glomerular capsule, renal tubule, and ducts. Each part, however, has distinctive histological features that reflect its particular functions. In the order that fluid flows through them, the parts are the glomerular capsule, the renal tubule, and the collecting duct.

GLOMERULAR CAPSULE The glomerular (Bowman's) capsule consists of visceral and parietal layers (Figure 26.6a). The visceral layer consists of modified simple squamous epithelial cells called **podocytes** (PŌ-dō-cīts; *podo-* = foot; *-cytes* = cells). The many footlike projections of these cells (pedicels) wrap around the single layer of endothelial cells of the glomerular capillaries and form the inner wall of the capsule. The parietal layer of the glomerular capsule consists of simple squamous epithelium and forms the outer wall of the capsule. Fluid filtered from the glomerular capillaries enters the **capsular (Bowman's)**

Figure 26.6 Histology of a renal corpuscle.

A renal corpuscle consists of a glomerular (Bowman's) capsule and a glomerulus.

Renal corpuscle (external view)

Afferent arteriole

Juxtaglomerular cell

Macula densa

Ascending limb of loop of Henle

Efferent arteriole

Endothelium of glomerulus

Parietal layer of glomerular (Bowman's) capsule

Mesangial cell

Capsular space

Proximal convoluted tubule

Podocyte of visceral layer of glomerular (Bowman's) capsule

Pedicel

(a) Renal corpuscle (internal view)

Glomerulus

Podocytes of visceral layer of glomerular capsule

Capsular space

Simple squamous epithelial cells

Glomerular capsule:
Parietal layer

Visceral layer

Afferent arteriole

Juxtaglomerular cell

Ascending limb of loop of Henle

Macula densa cell

Efferent arteriole

Proximal convoluted tubule

LM 1380x

(b) Renal corpuscle

Is the photomicrograph in (b) from a section through the renal cortex or renal medulla? How can you tell?

space, the space between the two layers of the glomerular capsule. Think of the relationship between the glomerulus and glomerular capsule in the following way. The glomerulus is a fist punched into a limp balloon (the glomerular capsule) until the fist is covered by two layers of balloon (visceral and parietal layers) with a space in between, the capsular space.

RENAL TUBULE AND COLLECTING DUCT Table 26.1 illustrates the histology of the cells that form the renal tubule and collecting duct. In the proximal convoluted tubule, the cells are simple cuboidal epithelial cells with a prominent brush border of microvilli on their apical surface (surface facing the lumen). These microvilli, like those of the small intestine, increase the surface area for reabsorption and secretion. The descending limb of the loop of Henle and the first part of the ascending limb of the loop of Henle (the thin ascending limb) are composed of simple squamous epithelium. (Recall that cortical or short-loop

nephrons lack the thin ascending limb.) The thick ascending limb of the loop of Henle is composed of simple cuboidal to low columnar epithelium.

In each nephron, the final part of the ascending limb of the loop of Henle makes contact with the afferent arteriole serving that renal corpuscle (Figure 26.6a). Because the columnar tubule cells in this region are crowded together, they are known as the **macula densa** (*macula* = spot; *densa* = dense). Alongside the macula densa, the wall of the afferent arteriole (and sometimes the efferent arteriole) contains modified smooth muscle fibers called **juxtaglomerular (JG) cells.** Together with the macula densa, they constitute the **juxtaglomerular apparatus (JGA).** As you will see later, the JGA helps regulate blood pressure within the kidneys. The distal convoluted tubule (DCT) begins a short distance past the macula densa. In the last part of the DCT and continuing into the collecting ducts, two different types of cells are present. Most are **principal cells,** which have receptors

Table 26.1	Histological Features of the Renal Tubule and Collecting Duct	
Region and Histology		**Description**
Proximal convoluted tubule (PCT)	Microvilli Mitochondrion Apical surface	Simple cuboidal epithelial cells with prominent brush borders of microvilli.
Loop of Henle: descending limb and thin ascending limb		Simple squamous epithelial cells.
Loop of Henle: thick ascending limb		Simple cuboidal to low columnar epithelial cells.
Most of distal convoluted tubule (DCT)		Simple cuboidal epithelial cells.
Last part of DCT and all of collecting duct (CD)	Intercalated cell Principal cell	Simple cuboidal epithelium consisting of principal cells and intercalated cells.

for both antidiuretic hormone (ADH) and aldosterone, two hormones that regulate their functions. A smaller number are **intercalated cells,** which play a role in the homeostasis of blood pH. The collecting ducts drain into large papillary ducts, which are lined by simple columnar epithelium.

Number of Nephrons

The number of nephrons is constant from birth. Any increase in kidney size is due solely to the growth of individual nephrons. If nephrons are injured or become diseased, new ones do not form. Signs of kidney dysfunction usually do not become apparent until function declines to less than 25% of normal because the remaining functional nephrons adapt to handle a larger-than-normal load. Surgical removal of one kidney, for example, stimulates hypertrophy (enlargement) of the remaining kidney, which eventually is able to filter blood at 80% of the rate of two normal kidneys. ■

▶ CHECKPOINT

2. Why are the kidneys said to be retroperitoneal?

3. Which branch of the autonomic nervous system innervates renal blood vessels?

4. How do cortical nephrons and juxtamedullary nephrons differ structurally?

5. Where is the juxtaglomerular apparatus (JGA) located and what is its structure?

OVERVIEW OF RENAL PHYSIOLOGY

▶ O B J E C T I V E

• **Identify the three basic functions performed by nephrons and collecting ducts, and indicate where each occurs.**

To produce urine, nephrons and collecting ducts perform three basic processes—glomerular filtration, tubular secretion, and tubular reabsorption (Figure 26.7):

1 *Glomerular filtration.* In the first step of urine production, water and most solutes in blood plasma move across the wall of glomerular capillaries into the glomerular capsule and then into the renal tubule.

2 *Tubular reabsorption.* As filtered fluid flows along the renal tubule and through the collecting duct, tubule cells reabsorb about 99% of the filtered water and many useful solutes. The water and solutes return to the blood as it flows through the peritubular capillaries and vasa recta. Note that the term *reabsorption* refers to the return of substances to the bloodstream. The term *absorption,* by contrast, means entry of new substances into the body, as occurs in the gastrointestinal tract.

3 *Tubular secretion.* As fluid flows along the tubule and through the collecting duct, the tubule and duct cells secrete other materials, such as wastes, drugs, and excess ions, into the fluid. Notice that tubular secretion *removes* a substance from the blood. In other instances of secretion—for

Figure 26.7 Relation of a nephron's structure to its three basic functions: glomerular filtration, tubular reabsorption, and tubular secretion. Excreted substances remain in the urine and subsequently leave the body. For any substance S, excretion rate of S = filtration rate of S − reabsorption rate of S + secretion rate of S.

Glomerular filtration occurs in the renal corpuscle, whereas tubular reabsorption and tubular secretion occur all along the renal tubule and collecting duct.

When cells of the renal tubules secrete the drug penicillin, is the drug being added to or removed from the bloodstream?

instance, secretion of hormones—cells release substances into interstitial fluid and blood.

Solutes in the fluid that drains into the renal pelvis remain in the urine and are excreted. The rate of urinary excretion of any solute is equal to its rate of glomerular filtration, plus its rate of secretion, minus its rate of reabsorption.

By filtering, reabsorbing, and secreting, nephrons help maintain homeostasis of the blood's volume and composition. The situation is somewhat analogous to a recycling center: Garbage trucks dump refuse into an input hopper, where the smaller refuse passes onto a conveyor belt (glomerular filtration of plasma). As the conveyor belt carries the garbage along, workers remove useful items, such as aluminum cans, plastics, and glass containers (reabsorption). Other workers place additional garbage left at the center and larger items onto the conveyor belt (secretion). At the end of the belt, all remaining garbage falls into a truck for transport to the landfill (excretion of wastes in urine).

GLOMERULAR FILTRATION

▶ O B J E C T I V E S

• **Describe the filtration membrane.**
• **Discuss the pressures that promote and oppose glomerular filtration.**

The fluid that enters the capsular space is called the **glomerular filtrate.** The fraction of blood plasma in the afferent arterioles of the kidneys that becomes glomerular filtrate is the **filtration fraction.** Although a filtration fraction of 0.16–0.20 (16–20%) is typical, the value varies considerably in both health and disease. On average, the daily volume of glomerular filtrate in adults is 150 liters in females and 180 liters in males. More than 99% of the glomerular filtrate returns to the bloodstream via tubular reabsorption, however, so only 1–2 liters (about 1–2 qt) are excreted as urine.

The Filtration Membrane

Together, the endothelial cells of glomerular capillaries and the podocytes, which completely encircle the capillaries, form a leaky barrier known as the **filtration membrane.** This sandwichlike assembly permits filtration of water and small solutes but prevents filtration of most plasma proteins, blood cells, and platelets. Substances filtered from the blood cross three barriers—a glomerular endothelial cell, the basal lamina, and a filtration slit formed by a podocyte (Figure 26.8):

❶ Glomerular endothelial cells are quite leaky because they have large **fenestrations** (pores) that measure 70–100 nm (0.07–0.1 μm) in diameter. This size permits all solutes in blood plasma to exit glomerular capillaries but prevents filtration of blood cells and platelets. Located among the glomerular capillaries and in the cleft between afferent and efferent arterioles are **mesangial cells** (*mes-* = in the middle; *-angi* = blood vessel) (see Figure 26.6a). These contractile cells help regulate glomerular filtration.

❷ The **basal lamina,** a layer of acellular material between the endothelium and the podocytes, consists of collagen fibers and proteoglycans in a glycoprotein matrix; it prevents filtration of larger plasma proteins.

❸ Extending from each podocyte are thousands of footlike processes termed **pedicels** (PED-i-sels = little feet) that wrap around glomerular capillaries. The spaces between pedicels are the **filtration slits.** A thin membrane, the **slit membrane,** extends across each filtration slit; it permits the passage of molecules having a diameter smaller than 6–7 nm (0.006–0.007 μm), including water, glucose, vitamins, amino acids, very small plasma proteins, ammonia, urea, and ions. Because albumin, the most plentiful plasma protein, has a diameter of 7.1 nm, less than 1% of it passes the slit membrane.

The principle of filtration—the use of pressure to force fluids and solutes through a membranes—is the same in glomerular capillaries as in capillaries elsewhere in the body (see Starling's law of the capillaries, page 703). However, the volume of fluid filtered by the renal corpuscle is much larger than in other capillaries of the body for three reasons.

1. Glomerular capillaries present a large surface area for filtration because they are long and extensive. The mesangial cells regulate how much of this surface area is available for filtration. When mesangial cells are relaxed, surface area is maximal, and glomerular filtration is very high. Contraction of mesangial cells reduces the available surface area, and glomerular filtration decreases.

2. The filtration membrane is thin and porous. Despite having several layers, the thickness of the filtration membrane is only 0.1 μm. Glomerular capillaries also are about 50 times leakier than capillaries in most other tissues, mainly because of their large fenestrations.

3. Glomerular capillary blood pressure is high. Because the efferent arteriole is smaller in diameter than the afferent arteriole, resistance to the outflow of blood from the glomerulus is high. As a result, blood pressure in glomerular capillaries is considerably higher than in capillaries elsewhere in the body, and a higher pressure produces more filtrate.

Net Filtration Pressure

Glomerular filtration depends on three main pressures. One pressure *promotes* filtration and two pressures *oppose* filtration (Figure 26.9).

Figure 26.8 The filtration membrane. The size of the endothelial fenestrations and filtration slits in (a) have been exaggerated for emphasis.

🔑 **During glomerular filtration, water and solutes pass from blood plasma into the capsular space.**

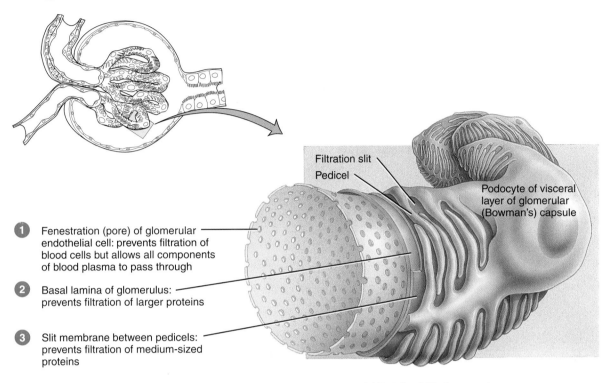

1 Fenestration (pore) of glomerular endothelial cell: prevents filtration of blood cells but allows all components of blood plasma to pass through

2 Basal lamina of glomerulus: prevents filtration of larger proteins

3 Slit membrane between pedicels: prevents filtration of medium-sized proteins

Filtration slit
Pedicel
Podocyte of visceral layer of glomerular (Bowman's) capsule

(a) Details of filtration membrane

Pedicel of podocyte
Filtration slit
Basal lamina
Lumen of glomerulus
Fenestration (pore) of glomerular endothelial cell

TEM 78,000x

(b) Filtration membrane

 Which part of the filtration membrane prevents red blood cells from entering the capsular space?

Figure 26.9   The pressures that drive glomerular filtration.   Taken together, these pressures determine net filtration pressure (NFP).

 Glomerular blood hydrostatic pressure promotes filtration, whereas capsular hydrostatic pressure and blood colloid osmotic pressure oppose filtration.

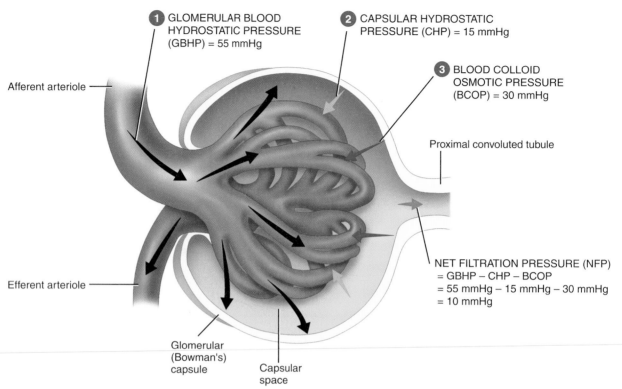

Suppose a tumor is pressing on and obstructing the right ureter. What effect might this have on CHP and thus on NFP in the right kidney? Would the left kidney also be affected?

1 **Glomerular blood hydrostatic pressure (GBHP)** is the blood pressure in glomerular capillaries. Generally, GBHP is about 55 mmHg. It promotes filtration by forcing water and solutes in blood plasma through the filtration membrane.

2 **Capsular hydrostatic pressure (CHP)** is the hydrostatic pressure exerted against the filtration membrane by fluid already in the capsular space and renal tubule. CHP opposes filtration and represents a "back pressure" of about 15 mmHg.

3 **Blood colloid osmotic pressure (BCOP),** which is due to the presence of proteins such as albumin, globulins, and fibrinogen in blood plasma, also opposes filtration. The average BCOP in glomerular capillaries is 30 mmHg.

Net filtration pressure (NFP), the total pressure that promotes filtration, is determined as follows:

Net Filtration Pressure (NFP) = GBHP − CHP − BCOP

By substituting the values just given, normal NFP may be calculated:

$$NFP = 55\,mmHg - 15\,mmHg - 30\,mmHg$$
$$= 10\,mmHg$$

Thus, a pressure of only 10 mmHg causes a normal amount of blood plasma (minus plasma proteins) to filter from the glomerulus into the capsular space.

Loss of Plasma Proteins in Urine Causes Edema

In some kidney diseases, glomerular capillaries are damaged and become so permeable that plasma proteins enter glomerular filtrate. As a result, the filtrate exerts a colloid osmotic pressure that draws water out of the blood. In this situation, the NFP increases, which means more fluid is filtered. At the same time, blood colloid osmotic pressure decreases because plasma proteins are being lost in the urine. Because more fluid filters out of blood capillaries into tissues throughout the body than returns via reabsorption, blood volume decreases and interstitial fluid volume increases. Thus, loss of plasma proteins in urine causes *edema,* an abnormally high volume of interstitial fluid. ■

Glomerular Filtration Rate

The amount of filtrate formed in all the renal corpuscles of both kidneys each minute is the **glomerular filtration rate (GFR).** In adults, the GFR averages 125 mL/min in males and 105 mL/min in females. Homeostasis of body fluids requires that the kidneys maintain a relatively constant GFR. If the GFR is too high, needed substances may pass so quickly through the renal tubules that some are not reabsorbed and are lost in the urine. If the GFR is too low, nearly all the filtrate may be reabsorbed and certain waste products may not be adequately excreted.

GFR is directly related to the pressures that determine net filtration pressure; any change in net filtration pressure will affect GFR. Severe blood loss, for example, reduces mean arterial blood pressure and decreases the glomerular blood hydrostatic pressure. Filtration ceases if glomerular blood hydrostatic pressure drops to 45 mmHg because the opposing pressures add up to 45 mmHg. Amazingly, when systemic blood pressure rises above normal, net filtration pressure and GFR increase very little. GFR is nearly constant when the mean arterial blood pressure is anywhere between 80 and 180 mmHg.

The mechanisms that regulate glomerular filtration rate operate in two main ways: (1) by adjusting blood flow into and out of the glomerulus and (2) by altering the glomerular capillary surface area available for filtration. GFR increases when blood flow into the glomerular capillaries increases. Coordinated control of the diameter of both afferent and efferent arterioles regulates glomerular blood flow. Constriction of the afferent arteriole, for example, decreases blood flow into the glomerulus, whereas dilation of the afferent arteriole increases blood flow into the glomerulus. Three mechanisms control GFR: renal autoregulation, neural regulation, and hormonal regulation.

Renal Autoregulation of GFR

The kidneys themselves help maintain a constant renal blood flow and GFR despite normal, everyday changes in blood pressure (as occur, for instance, during exercise). This capability is called **renal autoregulation** and consists of two mechanisms—the myogenic mechanism and tubuloglomerular feedback. Working together, they can maintain nearly constant GFR over a wide range of systemic blood pressures.

The **myogenic mechanism** (*myo-* = muscle; *-genic* = producing) occurs when stretching triggers contraction of smooth muscle cells in the wall of afferent arterioles. As blood pressure rises, GFR also rises because renal blood flow increases. However, the elevated blood pressure stretches the walls of the afferent arterioles. In response, smooth muscle fibers in the wall of the afferent arteriole contract, which narrows the arteriole's lumen. As a result, renal blood flow decreases, thus reducing GFR to its previous level. Conversely, when arterial blood pressure drops, the smooth muscle cells are stretched less and thus relax. The afferent arterioles dilate, renal blood flow increases, and GFR increases. The myogenic mechanism normalizes renal

blood flow and GFR within seconds after a change in blood pressure.

The second contributor to renal autoregulation, **tubuloglomerular feedback,** is so-named because part of the renal tubules—the macula densa—provides feedback to the glomerulus (Figure 26.10). When GFR is above normal due to elevated systemic blood pressure, filtered fluid flows more rapidly along

Figure 26.10 Tubuloglomerular feedback.

Macula densa cells of the juxtaglomerular apparatus provide negative feedback regulation of glomerular filtration rate.

Why is this process termed autoregulation?

the renal tubules. As a result, the proximal convoluted tubule and loop of Henle have less time to reabsorb Na$^+$, Cl$^-$, and water. Macula densa cells are thought to detect the increased delivery of Na$^+$, Cl$^-$, and water and to inhibit release of nitric oxide (NO) from cells in the juxtaglomerular apparatus (JGA). Because NO causes vasodilation, afferent arterioles constrict when the level of NO declines. As a result, less blood flows into the glomerular capillaries, and GFR decreases. When blood pressure falls and GFR is therefore lower than normal, the opposite sequence of events occurs, although to a lesser degree. Tubuloglomerular feedback operates more slowly than the myogenic mechanism.

Neural Regulation of GFR

Like most blood vessels of the body, those of the kidneys are supplied by sympathetic ANS fibers that release norepinephrine. Norepinephrine causes vasoconstriction through the activation of α_1 receptors, which are particularly plentiful in the smooth muscle fibers of afferent arterioles. At rest, sympathetic stimulation is moderately low, the afferent and efferent arterioles are dilated, and renal autoregulation of GFR prevails. With moderate sympathetic stimulation, both afferent and efferent arterioles constrict to the same degree. Blood flow into and out of the glomerulus is restricted to the same extent, which decreases GFR only slightly. With greater sympathetic stimulation, however, as occurs during exercise or hemorrhage, vasoconstriction of the afferent arterioles predominates. As a result, blood flow into glomerular capillaries is greatly decreased, and GFR drops. This lowering of renal blood flow has two consequences: (1) It reduces urine output, which helps conserve blood volume. (2) It permits greater blood flow to other body tissues.

Hormonal Regulation of GFR

Two hormones contribute to regulation of GFR. Angiotensin II reduces GFR, whereas atrial natriuretic peptide (ANP) increases GFR. **Angiotensin II** is a very potent vasoconstrictor that narrows both afferent and efferent arterioles and reduces renal blood flow, thereby decreasing GFR. Cells in the atria of the heart secrete **atrial natriuretic peptide (ANP).** Stretching of the atria, as occurs when blood volume increases, stimulates secretion of ANP. By causing relaxation of the glomerular mesangial cells, ANP increases the capillary surface area available for filtration. Glomerular filtration rate rises as the surface area increases.

Table 26.2 summarizes the regulation of glomerular filtration rate.

▶ CHECKPOINT

6. If the urinary excretion rate of a drug such as penicillin is greater than the rate at which it is filtered at the glomerulus, how else is it getting into the urine?

7. What is the major chemical difference between blood plasma and glomerular filtrate?

8. Why is there much greater filtration through glomerular capillaries than through capillaries elsewhere in the body?

9. Write the equation for the calculation of net filtration pressure (NFP) and explain the meaning of each term.

10. How is glomerular filtration rate regulated?

Table 26.2	Regulation of Glomerular Filtration Rate (GFR)		
Type of Regulation	**Major Stimulus**	**Mechanism and Site of Action**	**Effect on GFR**
Renal Autoregulation			
Myogenic mechanism	Increased stretching of smooth muscle fibers in afferent arteriole walls due to increased blood pressure.	Stretched smooth muscle fibers contract, thereby narrowing the lumen of the afferent arterioles.	Decrease.
Tubuloglomerular feedback	Rapid delivery of Na$^+$ and Cl$^-$ to the macula densa due to high systemic blood pressure.	Decreased release of nitric oxide (NO) by the juxtaglomerular apparatus causes constriction of afferent arterioles.	Decrease.
Neural Regulation	Increase in level of activity of renal sympathetic nerves releases norepinephrine.	Constriction of afferent arterioles through activation of α_1 receptors and increased release of renin.	Decrease.
Hormonal Regulation			
Angiotensin II	Decreased blood volume or blood pressure stimulates production of angiotensin II.	Constriction of both afferent and efferent arterioles.	Decrease.
Atrial natriuretic peptide (ANP)	Stretching of the atria of the heart stimulates secretion of ANP.	Relaxation of mesangial cells in glomerulus increases capillary surface area available for filtration.	Increase.

TUBULAR REABSORPTION AND TUBULAR SECRETION

▶ **OBJECTIVES**

- **Describe the routes and mechanisms of tubular reabsorption and secretion.**
- **Describe how specific segments of the renal tubule and collecting duct reabsorb water and solutes.**
- **Describe how specific segments of the renal tubule and collecting duct secrete solutes into the urine.**

Principles of Tubular Reabsorption and Secretion

The normal rate of glomerular filtration is so high that the volume of fluid entering the proximal convoluted tubules in half an hour is greater than the total plasma volume. Reabsorption—returning most of the filtered water and many of the filtered solutes to the bloodstream—is the second basic function of the nephron and collecting duct. Normally, about 99% of the filtered water is reabsorbed. Epithelial cells all along the renal tubule and duct carry out reabsorption, but proximal convoluted tubule cells make the largest contribution. Solutes that are reabsorbed by both active and passive processes include glucose, amino acids, urea, and ions such as Na^+ (sodium), K^+ (potassium), Ca^{2+} (calcium), Cl^- (chloride), HCO_3^- (bicarbonate), and HPO_4^{2-} (phosphate). After the proximal convoluted tubule, cells located more distally fine-tune the reabsorption processes to maintain homeostatic balances of water and selected ions. Most small proteins and peptides that pass through the filter also are reabsorbed, usually via pinocytosis. To appreciate the huge extent of tubular reabsorption, look at Table 26.3 and compare the amounts of substances that are filtered, reabsorbed, and excreted in urine.

The third function of nephrons and collecting ducts is tubular secretion, the transfer of materials from the blood and tubule cells into tubular fluid. Secreted substances include hydrogen ions (H^+), K^+, ammonium ions (NH_4^+), creatinine, and certain drugs such as penicillin. Tubular secretion has two important outcomes: (1) The secretion of H^+ helps control blood pH. (2) The secretion of other substances helps eliminate them from the body.

Reabsorption Routes

A substance being reabsorbed from the fluid in the tubule lumen can take one of two routes before entering a peritubular capillary: It can move *between* adjacent tubule cells or *through* an individual tubule cell (Figure 26.11). Along the renal tubule, tight junctions surround and join neighboring cells to one another, much like the plastic rings that hold a six-pack of soda cans together. The **apical membrane** (like the tops of the soda cans) contacts the tubular fluid, and the **basolateral membrane** (like the bottoms and sides of the soda cans) contacts interstitial fluid at the base and sides of the cell.

The tight junctions do not completely seal off the interstitial fluid from the fluid in the tubule lumen. Fluid can leak *between* the cells in a passive process known as **paracellular reabsorption** (*para-* = beside). In some parts of the renal tubule, the paracellular route is thought to account for up to 50% of the reabsorption of certain ions and the water that accompanies them via osmosis. In **transcellular reabsorption** (*trans-* = across), a

Table 26.3	Substances Filtered, Reabsorbed, and Excreted in Urine		
Substance	Filtered* (Enters Glomerular Capsule per Day)	Reabsorbed (Returned to Blood per Day)	Urine (Excreted per Day)
Water	180 liters	178–179 liters	1–2 liters
Proteins	2.0 g	1.9 g	0.1 g
Sodium ions (Na^+)	579 g	575 g	4 g
Chloride ions (Cl^-)	640 g	633.7 g	6.3 g
Bicarbonate ions (HCO_3^-)	275 g	275 g	0.03 g
Glucose	162 g	162 g	0 g
Urea	54 g	24 g	30 g[†]
Potassium ions (K^+)	29.6 g	29.6 g	2.0 g[‡]
Uric acid	8.5 g	7.7 g	0.8 g
Creatinine	1.6 g	0 g	1.7 g

* Assuming GFR is 180 liters per day.
† In addition to being filtered and reabsorbed, urea is secreted.
‡ After virtually all filtered K^+ is reabsorbed in the convoluted tubules and loop of Henle, a variable amount of K^+ is secreted by principal cells in the collecting duct.

Figure 26.11 **Reabsorption routes: paracellular reabsorption and transcellular reabsorption.**

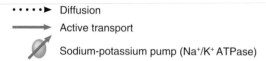 In paracellular reabsorption, water and solutes in tubular fluid return to the bloodstream by moving between tubule cells; in transcellular reabsorption, solutes and water in tubular fluid return to the bloodstream by passing through a tubule cell.

Key:

••••▶ Diffusion

──▶ Active transport

Sodium-potassium pump (Na⁺/K⁺ ATPase)

What is the main function of the tight junctions between tubule cells?

substance passes from the fluid in the tubular lumen through the apical membrane of a tubule cell, across the cytosol, and out into interstitial fluid through the basolateral membrane.

Transport Mechanisms

When renal cells transport solutes out of or into tubular fluid, they move specific substances in one direction only. Not surprisingly, different types of transport proteins are present in the apical and basolateral membranes. Moreover, the tight junctions form a barrier that prevents mixing of proteins in the apical and basolateral membrane compartments. Reabsorption of Na^+ by the renal tubules is especially important because more sodium ions pass the glomerular filters than any other substance except water.

Cells lining the renal tubules, like other cells throughout the body, have a low concentration of Na^+ in their cytosol due to the activity of sodium-potassium pumps (Na^+/K^+ ATPases). These pumps are located in the basolateral membranes and eject Na^+ from the renal tubule cells (Figure 26.11). The absence of sodium-potassium pumps in the apical membrane ensures that

reabsorption of Na^+ is a one-way process. Most sodium ions that cross the apical membrane will be pumped into interstitial fluid at the base and sides of the cell. The amount of ATP used by sodium-potassium pumps in the renal tubules is about 6% of the total ATP consumption at rest. For comparison, this is about the same energy used by the diaphragm as it contracts during quiet breathing.

As we noted in Chapter 3, transport of materials across membranes may be either active or passive. Recall that in **primary active transport** the energy derived from hydrolysis of ATP is used to "pump" a substance across a membrane. Because the sodium-potassium pump uses ATP in this manner, it is a primary active transport pump. In **secondary active transport** the energy stored in an ion's electrochemical gradient, rather than hydrolysis of ATP, drives another substance across a membrane. Secondary active transport couples the movement of an ion down its electrochemical gradient to the "uphill" movement of a second substance against its electrochemical gradient. The membrane proteins that perform secondary active transport are called *symporters* if they move two or more substances in the same direction across a membrane. They are *antiporters* if they move two or more substances in opposite directions across a membrane. Each type of transporter has an upper limit on how fast it can work, just as an escalator has a limit on how many people it can carry from one level to another in a given period. This limit is called the **transport maximum** (T_m) and is measured in mg/min.

Solute reabsorption drives water reabsorption because all water reabsorption occurs via osmosis. About 90% of the reabsorption of water filtered by the kidneys occurs together with the reabsorption of solutes such as Na^+, Cl^-, and glucose. Water reabsorbed together with solutes in tubular fluid is termed **obligatory water reabsorption** (ob-LIG-a-tor'-ē) because the water is "obliged" to follow the solutes when they are reabsorbed. This type of water reabsorption occurs in the proximal convoluted tubule and the descending limb of the loop of Henle because these segments of the nephron are always permeable to water. Reabsorption of the final 10% of the water, a total of 10–20 liters per day, is termed **facultative water reabsorption** (FAK-ul-tā'-tiv). The word *facultative* means "capable of adapting to a need." Facultative water reabsorption is regulated by antidiuretic hormone and occurs mainly in the collecting ducts.

 Glucosuria

When the blood concentration of glucose is above 200 mg/mL, the renal symporters cannot work fast enough to reabsorb all the glucose that enters the glomerular filtrate. As a result, some glucose remains in the urine, a condition called **glucosuria** (gloo'-kō-SOO-rē-a). The most common cause of glucosuria is diabetes mellitus, in which the blood glucose level may rise far above normal because insulin activity is deficient. Rare genetic mutations in the renal Na^+-glucose symporter that greatly reduces its T_m also cause glucosuria. In these cases, glucose appears in the urine even though the blood glucose level is normal. ■

Now that we have discussed the principles of renal transport, we will follow the filtered fluid into the proximal convoluted tubule, loop of Henle, distal convoluted tubule, and collecting ducts. In each segment, we will examine where and how specific substances are reabsorbed and secreted. The filtered fluid becomes *tubular fluid* once it enters the proximal convoluted tubule. Due to reabsorption and secretion, the composition of tubular fluid changes as it flows along the nephron tubule and through the collecting duct. The fluid that drains from papillary ducts into the renal pelvis is *urine*.

Reabsorption and Secretion in the Proximal Convoluted Tubule

The largest amount of solute and water reabsorption from filtered fluid occurs in the proximal convoluted tubules, and most absorptive processes involve Na^+. Two different types of Na^+ transporters are located in the proximal convoluted tubule: (1) Several different Na^+ symporters reabsorb Na^+ together with various other solutes. (2) Na^+/H^+ antiporters reabsorb Na^+ while secreting H^+. These Na^+ transporters normally achieve reabsorption of 100% of most organic solutes, such as glucose and amino acids, from the filtrate. They also reclaim 80–90% of the HCO_3^-; 65% of the water, Na^+, and K^+; 50% of the Cl^-; and a variable amount of Ca^{2+}, Mg^{2+}, and HPO_4^{2-} from the filtered fluid.

Normally, filtered glucose, amino acids, lactic acid, water-soluble vitamins, and other nutrients are not lost in the urine. Rather, they are completely reabsorbed in the first half of the proximal convoluted tubule (PCT) by **Na^+ symporters** located in the apical membrane. Figure 26.12 depicts the operation of Na^+-glucose symporters in the apical membrane of PCT cells. Two Na^+ and a molecule of glucose attach to the symporter protein, which carries them from the tubular fluid into the tubule cell. Glucose molecules brought into proximal convoluted tubule cells by Na^+ symporters exit the basolateral membrane via facilitated diffusion. Then they diffuse into peritubular capillaries. Other Na^+ symporters in the PCT reclaim filtered HPO_4^{2-} (phosphate) and SO_4^{2-} (sulfate) ions, all amino acids, and lactic acid in a similar way.

Another secondary active transport process, the **Na^+/H^+ antiporters,** carry filtered Na^+ down its concentration gradient into a PCT cell as H^+ is moved from the cytosol into the lumen (Figure 26.13a). Thus, Na^+ is reabsorbed into blood and H^+ is secreted into tubular fluid. PCT cells can produce the H^+ needed to keep the antiporters running in the following way. Carbon dioxide (CO_2) diffuses from peritubular blood or tubular fluid or is produced by metabolic reactions within the cells. As also occurs in red blood cells (see Figure 23.24 on page 000), the enzyme *carbonic anhydrase* (CA) catalyzes the reaction of CO_2 with water (H_2O) to form carbonic acid (H_2CO_3), which then dissociates into H^+ and HCO_3^-:

$$CO_2 + H_2O \xrightarrow[\text{anhydrase}]{\text{Carbonic}} H_2CO_3 \longrightarrow H^+ + HCO_3^-$$

This mechanism achieves reabsorption of 80–90% of the filtered bicarbonate ions, thereby safeguarding the body's supply of an important buffer. Figure 26.13b depicts HCO_3^- reabsorption in the proximal convoluted tubule. After H^+ is secreted into the fluid within the lumen of the proximal convoluted tubule, it reacts with filtered HCO_3^- to form H_2CO_3, which readily dissociates into CO_2 and H_2O. Carbon dioxide then diffuses into the tubule cells and joins with H_2O to form H_2CO_3, which dissociates into H^+ and HCO_3^-. As the level of HCO_3^- rises in the cytosol, it exits via facilitated diffusion transporters in the basolateral membrane and diffuses into the blood with Na^+. Thus, every H^+ secreted into the tubular fluid of the proximal convoluted tubule causes reabsorption of a filtered HCO_3^-.

Besides achieving reabsorption of sodium ions, the Na^+ symporters and Na^+/H^+ antiporters promote osmosis of water

Figure 26.12 Reabsorption of glucose by Na^+-glucose symporters in cells of the proximal convoluted tubule (PCT).

Normally, all filtered glucose is reabsorbed in the PCT.

Fluid in tubule lumen

Proximal convoluted tubule cell

Peritubular capillary

ATP

ADP

Na^+ ⋯⋯ Na^+

2 Na^+

Na^+

Glucose

Glucose

Glucose

Tight junction

Interstitial fluid

Brush border (microvilli)

Key:

Na^+-glucose symporter

Glucose facilitated diffusion transporter

⋯⋯► Diffusion

Sodium-potassium pump

How does filtered glucose enter and leave a PCT cell?

Figure 26.13 Actions of Na⁺/H⁺ antiporters in proximal convoluted tubule cells. (a) Reabsorption of sodium ions (Na⁺) and secretion of hydrogen ions (H⁺) via secondary active transport through the apical membrane; (b) reabsorption of bicarbonate ions (HCO₃⁻) via facilitated diffusion through the basolateral membrane. CO_2 = carbon dioxide; H_2CO_3 = carbonic acid; CA = carbonic anhydrase.

🔑 **Na⁺/H⁺ antiporters promote transcellular reabsorption of Na⁺, HCO₃⁻, and water in the proximal convoluted tubule.**

(a) Na⁺ reabsorption and H⁺ secretion

(b) HCO₃⁻ reabsorption

Key:

▬	Na⁺/H⁺ antiporter
⌇	HCO₃⁻ facilitated diffusion transporter
•••••▶	Diffusion
🌀	Sodium-potassium pump

❓ **Which step in Na⁺ movement in part (a) is promoted by the electrochemical gradient?**

and passive reabsorption of other solutes (Figure 26.14). As water leaves the tubular fluid, the concentrations of the remaining filtered solutes increase. In the second half of the PCT, electrochemical gradients for Cl⁻, K⁺, Ca²⁺, Mg²⁺, and urea promote their passive diffusion into peritubular capillaries via both paracellular and transcellular routes. Among these ions, Cl⁻ is present in the highest concentration. Diffusion of negatively charged Cl⁻ into interstitial fluid via the paracellular route makes the interstitial fluid electrically more negative than the tubular fluid. This negativity promotes passive paracellular reabsorption of cations, such as K⁺, Ca²⁺, and Mg²⁺.

Each reabsorbed solute increases the osmolarity, first inside the tubule cell, then in interstitial fluid, and finally in the blood. Water thus moves rapidly from the tubular fluid, via both the paracellular and transcellular routes, into the peritubular capillaries and restores osmotic balance (Figure 26.14). In other words, reabsorption of the solutes creates an osmotic gradient that promotes the reabsorption of water via osmosis. Cells lining the proximal convoluted tubule and the descending limb of the loop of Henle are especially permeable to water because they have many molecules of *aquaporin-1*. This integral protein in the plasma membrane is a water channel that greatly increases the rate of water movement across the apical and basolateral membranes.

Ammonia (NH₃) is a poisonous waste product derived from the deamination (removal of an amino group) of various amino acids, a reaction that occurs mainly in hepatocytes (liver cells). In addition, hepatocytes convert most ammonia to urea, a less-toxic compound. Although tiny amounts of urea and ammonia are present in sweat, most excretion of these nitrogen-containing waste products occurs via the urine. Urea and ammonia in blood are both filtered at the glomerulus and secreted by proximal convoluted tubule cells into the tubular fluid.

Proximal convoluted tubule cells can produce additional NH₃ by deaminating the amino acid glutamine in a reaction that also generates HCO₃⁻. The NH₃ quickly binds H⁺ to become an ammonium ion (NH₄⁺), which can substitute for H⁺ aboard Na⁺/H⁺ antiporters in the apical membrane and be secreted into the tubular fluid. The HCO₃⁻ generated in this reaction moves through the basolateral membrane and then diffuses into the bloodstream, providing additional buffers in blood plasma.

Reabsorption in the Loop of Henle

Because the proximal convoluted tubules reabsorb about 65% of the filtered water (about 80 mL/min), fluid enters the next part of the nephron, the loop of Henle (LOH), at a rate of 40–45 mL/min. The chemical composition of the tubular fluid now is quite different from that of glomerular filtrate because glucose, amino acids, and other nutrients are no longer present. The osmolarity of the tubular fluid is still close to the osmolarity of blood, however, because reabsorption of water by osmosis keeps pace with reabsorption of solutes all along the proximal convoluted tubule.

Figure 26.14 Passive reabsorption of Cl⁻, K⁺, Ca²⁺, Mg²⁺, urea, and water in the second half of the proximal convoluted tubule.

Electrochemical gradients promote passive reabsorption of solutes via both paracellular and transcellular routes.

By what mechanism is water reabsorbed from tubular fluid?

The LOH reabsorbs about 20–30% of the filtered Na⁺, K⁺, Ca²⁺; 10–20% of the filtered HCO_3^-; 35% of the filtered Cl⁻; and 15% of the filtered water. Here, for the first time, reabsorption of water via osmosis is *not* automatically coupled to reabsorption of filtered solutes because part of the loop of Henle is relatively impermeable to water. The loop of Henle thus sets the stage for *independent* regulation of both the *volume* and *osmolarity* of body fluids.

The apical membranes of cells in the thick ascending limb of the LOH have **Na⁺-K⁺-2Cl⁻ symporters** that simultaneously reclaim one Na⁺, one K⁺, and two Cl⁻ from the fluid in the tubular lumen (Figure 26.15). Na⁺ that is actively transported into interstitial fluid at the base and sides of the cell diffuses into the vasa recta. Cl⁻ moves through leakage channels in the basolateral membrane. Because many K⁺ leakage channels are present in the apical membrane, most K⁺ brought in by the symporters moves down its concentration gradient back into the tubular fluid. Thus, the main effect of the Na⁺-K⁺-2Cl⁻ symporters is reabsorption of Na⁺ and Cl⁻.

The movement of positively charged K⁺ into the tubular fluid through the apical membrane channels leaves the interstitial fluid and blood with a negative charge relative to fluid in the ascending limb of the LOH. This relative negativity promotes reabsorption of cations—Na⁺, K⁺, Ca²⁺, and Mg²⁺—via the paracellular route.

Although about 15% of the filtered water is reabsorbed in the *descending* limb of the LOH, little or no water is reabsorbed

Figure 26.15 Na⁺-K⁺-2Cl⁻ symporter in the thick ascending limb of the loop of Henle.

Cells in the thick ascending limb have symporters that simultaneously reabsorb one Na⁺, one K⁺, and two Cl⁻.

Key:

 Na⁺–K⁺–2Cl⁻ symporter

Leakage channels

Sodium-potassium pump

···· ► Diffusion

Why is this process considered secondary active transport? Does water reabsorption accompany ion reabsorption in this region of the nephron?

in the *ascending* limb. In this segment of the tubule, the apical membranes are virtually impermeable to water. Because ions but not water molecules are reabsorbed, the osmolarity of the tubular fluid progressively decreases as fluid flows toward the end of the ascending limb.

Reabsorption in the Distal Convoluted Tubule

Fluid enters the distal convoluted tubules (DCT) at a rate of about 25 mL/min because 80% of the filtered water has now been reabsorbed. As fluid flows along the DCT, reabsorption of

Na$^+$ and Cl$^-$ continues by means of **Na$^+$-Cl$^-$ symporters** in the apical membranes. Sodium-potassium pumps and Cl$^-$ leakage channels in the basolateral membranes then permit reabsorption of Na$^+$ and Cl$^-$ into the peritubular capillaries. The DCT also is the major site where parathyroid hormone (PTH) stimulates reabsorption of Ca^{2+}. Overall, cells of the DCT reabsorb 10–15% of the filtered water.

Reabsorption and Secretion in the Collecting Duct

By the time fluid reaches the end of the distal convoluted tubule, 90–95% of the filtered solutes and water have returned to the bloodstream. Recall that two different types of cells—principal cells and intercalated cells—are present at the end of the distal convoluted tubule and throughout the collecting duct. Whereas principal cells reabsorb Na$^+$ and secrete K$^+$, intercalated cells reabsorb K$^+$ and HCO$_3^-$ and secrete H$^+$.

In contrast to earlier segments of the nephron, Na$^+$ passes through the apical membrane of principal cells via Na$^+$ leakage channels rather than by means of symporters or antiporters (Figure 26.16). The concentration of Na$^+$ in the cytosol remains low, as usual, because the sodium-potassium pumps actively transport Na$^+$ across the basolateral membranes. Then Na$^+$ passively diffuses into the peritubular capillaries from the interstitial spaces around the tubule cells.

Normally, transcellular and paracellular reabsorption in the proximal convoluted tubule and loop of Henle return most filtered K$^+$ to the bloodstream. To adjust for varying dietary intake of potassium and to maintain a stable level of K$^+$ in body fluids, principal cells secrete a variable amount of K$^+$ (Figure 26.16). Because the basolateral sodium-potassium pumps continually bring K$^+$ into principal cells, the intracellular concentration of K$^+$ remains high. K$^+$ leakage channels are present in both the apical and basolateral membranes. Thus, some K$^+$ diffuses down its concentration gradient into the tubular fluid, where the K$^+$ concentration is very low. This secretion mechanism is the main source of K$^+$ excreted in the urine.

Hormonal Regulation of Tubular Reabsorption and Tubular Secretion

Four hormones affect the extent of Na$^+$, Cl$^-$, and water reabsorption as well as K$^+$ secretion by the renal tubules. The most important hormonal regulators of electrolyte reabsorption and secretion are angiotensin II and aldosterone. The major hormone that regulates water reabsorption is antidiuretic hormone. Atrial natriuretic peptide plays a minor role in inhibiting both electrolyte and water reabsorption.

Renin–Angiotensin–Aldosterone System

When blood volume and blood pressure decrease, the walls of the afferent arterioles are stretched less, and the juxtaglomerular

Figure 26.16 Reabsorption of Na$^+$ and secretion of K$^+$ by principal cells in the last part of the distal convoluted tubule and in the collecting duct.

In the apical membrane of principal cells, Na$^+$ leakage channels allow entry of Na$^+$ while K$^+$ leakage channels allow exit of K$^+$ into the tubular fluid.

Key:

• • • • ▶   Diffusion

⊣⊢   Leakage channels

⊗   Sodium-potassium pump

Which hormone stimulates reabsorption and secretion by principal cells, and how does this hormone exert its effect?

cells secrete the enzyme renin into the blood. Sympathetic stimulation also directly stimulates release of renin from juxtaglomerular cells. Renin clips off a 10–amino-acid peptide called angiotensin I from angiotensinogen, which is synthesized by hepatocytes. By clipping off two more amino acids, *angiotensin converting enzyme (ACE)* converts angiotensin I to **angiotensin II,** which is the active hormone.

Angiotensin II affects renal physiology in three main ways:

1. It decreases the glomerular filtration rate by causing vasoconstriction of the afferent arterioles.

2. It enhances reabsorption of Na$^+$, Cl$^-$, and water in the proximal convoluted tubule by stimulating the activity of Na$^+$/H$^+$ antiporters.

3. It stimulates the adrenal cortex to release **aldosterone,** a hormone that in turn stimulates the principal cells in the collecting ducts to reabsorb more Na^+ and Cl^- and secrete more K^+. The osmotic consequence of excreting fewer Na^+ and Cl^- is excreting less water, which increases blood volume.

Antidiuretic Hormone

Antidiuretic hormone (**ADH** or **vasopressin**) is released by the posterior pituitary. It regulates facultative water reabsorption by increasing the water permeability of principal cells in the last part of the distal convoluted tubule and throughout the collecting duct. In the absence of ADH, the apical membranes of principal cells have a very low permeability to water. Within principal cells are tiny vesicles containing many copies of a water channel protein known as **aquaporin-2.*** ADH stimulates insertion of the aquaporin-2–containing vesicles into the apical membranes via exocytosis. As a result, the water permeability of the principal cell's apical membrane increases, and water molecules move more rapidly from the tubular fluid into the cells. Because the basolateral membranes are always relatively permeable to water, water molecules then move rapidly into the blood. The kidneys can produce as little as 400–500 mL of very concentrated urine each day when ADH concentration is maximal, for instance during severe dehydration. When ADH level declines, the aquaporin-2 channels are removed from the apical membrane via endocytosis. The kidneys produce a large volume of dilute urine when ADH level is low.

A negative feedback system involving ADH regulates facultative water reabsorption (Figure 26. 17). When the osmolarity or osmotic pressure of plasma and interstitial fluid increases — that is, when water concentration decreases—by as little as 1%, osmoreceptors in the hypothalamus detect the change. Their nerve impulses stimulate secretion of more ADH into the blood, and the principal cells become more permeable to water. As facultative water reabsorption increases, plasma osmolarity decreases to normal. A second powerful stimulus for ADH secretion is a decrease in blood volume, as occurs in hemorrhaging or severe dehydration. In the pathological absence of ADH activity, a condition known as *diabetes insipidus,* a person may excrete up to 20 liters of very dilute urine daily.

Atrial Natriuretic Peptide

A large increase in blood volume promotes release of atrial natriuretic peptide (ANP) from the heart. Although the importance of ANP in normal regulation of tubular function is unclear, it can inhibit reabsorption of Na^+ and water in the proximal convoluted tubule and collecting duct. ANP also suppresses the secretion of aldosterone and ADH. These effects increase the excretion of Na^+ in urine (natriuresis) and increase urine output (diuresis), which decreases blood volume and blood pressure.

---

* ADH does not govern the previously mentioned water channel (aquaporin-1).

Figure 26.17 Negative feedback regulation of facultative water reabsorption by ADH.

Most water reabsorption (90%) is obligatory; 10% is facultative.

 Besides ADH, which other hormones contribute to the regulation of water reabsorption?

Table 26.4 summarizes hormonal regulation of tubular reabsorption and tubular secretion.

▶ **CHECKPOINT**

11. Draw a diagram that depicts how substances are reabsorbed via the transcellular and paracellular routes. Label the apical membrane and the basolateral membrane Where are the sodium-potassium pumps located?

12. For reabsorption of Na⁺, describe two mechanisms in the PCT, one in the LOH, one in the DCT, and one in the collecting duct. What other solutes are reabsorbed or secreted together with Na⁺ in each case?

13. How do intercalated cells secrete hydrogen ions?

14. Prepare a table that indicates the percentages of filtered water and filtered Na⁺ that are reabsorbed in the PCT, LOH, DCT, and collecting duct. Indicate which hormones, if any, regulate reabsorption in each segment.

PRODUCTION OF DILUTE AND CONCENTRATED URINE

▶ **OBJECTIVE**

- **Describe how the renal tubule and collecting ducts produce dilute and concentrated urine.**

Even though a person's fluid intake can be highly variable, the total volume of fluid in the body normally remains stable. Homeostasis of body fluid volume depends in large part on the ability of the kidneys to regulate the rate of water loss in urine. Normally functioning kidneys produce a large volume of dilute urine when fluid intake is high, and a small volume of concentrated urine when fluid intake is low or fluid loss is large. ADH controls whether dilute urine or concentrated urine is formed. In the absence of ADH, urine is very dilute. However, a high level of ADH stimulates reabsorption of more water into blood, producing a concentrated urine.

Formation of Dilute Urine

Glomerular filtrate has the same ratio of water and solute particles as blood; its osmolarity is about 300 mOsm/liter. As previously noted, fluid leaving the proximal convoluted tubule is still isotonic to plasma. When *dilute* urine is being formed (Figure 26.18), the osmolarity of the fluid in the tubular lumen *increases* as it flows down the descending limb of the loop of Henle, *decreases* as it flows up the ascending limb, and *decreases* still more as it flows through the rest of the nephron and collecting duct. These changes in osmolarity result from the following conditions along the path of tubular fluid:

1. Because the osmolarity of the interstitial fluid of the renal medulla becomes progressively greater, more and more water is reabsorbed by osmosis as tubular fluid flows along the descending limb toward the tip of the loop. (The source of this medullary osmotic gradient is explained shortly.) As a result, the fluid remaining in the lumen becomes progressively more concentrated.

2. Cells lining the thick ascending limb of the loop have symporters that actively reabsorb Na⁺, K⁺, and Cl⁻ from the tubular fluid (see Figure 26.15). The ions pass from the tubular fluid into thick ascending limb cells, then into interstitial fluid, and finally some diffuse into the blood inside the vasa recta.

Table 26.4	Hormonal Regulation of Tubular Reabsorption and Tubular Secretion		
Hormone	**Major Stimuli that Trigger Release**	**Mechanism and Site of Action**	**Effects**
Angiotensin II	Low blood volume or low blood pressure stimulates renin-induced production of angiotensin II.	Stimulates activity of Na⁺/H⁺ antiporters in proximal tubule cells.	Increases reabsorption of Na⁺, other solutes, and water, which increases blood volume.
Aldosterone	Increased angiotensin II level and increased level of plasma K⁺ promote release of aldosterone by adrenal cortex.	Enhances activity of sodium-potassium pumps in basolateral membrane and Na⁺ channels in apical membrane of principal cells in collecting duct.	Increases secretion of K⁺ and reabsorption of Na⁺, Cl⁻; increases reabsorption of water, which increases blood volume.
Antidiuretic hormone (ADH) or vasopressin	Increased osmolarity of extracellular fluid or decreased blood volume promote release of ADH from the posterior pituitary gland.	Stimulates insertion of water-channel proteins (aquaporin-2) into the apical membranes of principal cells.	Increases facultative reabsorption of water, which decreases osmolarity of body fluids.
Atrial natriuretic peptide (ANP)	Stretching of atria of heart stimulates secretion of ANP.	Suppresses reabsorption of Na⁺ and water in proximal tubule and collecting duct; also inhibits secretion of aldosterone and ADH.	Increases excretion of Na⁺ in urine (natriuresis); increases urine output (diuresis) and thus decreases blood volume.

Figure 26.18 Formation of dilute urine. Numbers indicate osmolarity in milliosmoles per liter (mOsm/liter). Heavy brown lines in the ascending limb of the loop of Henle and in the distal convoluted tubule indicate impermeability to water; heavy blue lines indicate the last part of the distal convoluted tubule and the collecting duct, which are impermeable to water in the absence of ADH; light blue areas around the nephron represent interstitial fluid. When ADH is absent, the osmolarity of urine can be as low as 65 mOsm/liter.

🔑 **When ADH level is low, urine is dilute and has an osmolarity less than the osmolarity of blood.**

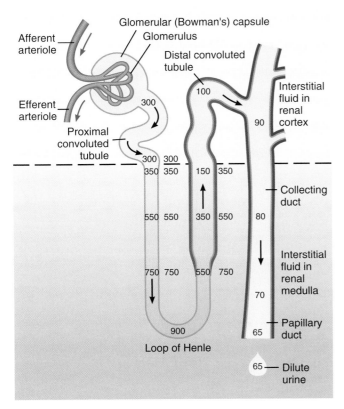

Loop of Henle

❓ **Which portions of the renal tubule and collecting duct reabsorb more solutes than water to produce dilute urine?**

3. Although solutes are being reabsorbed in the thick ascending limb, the water permeability of this portion of the nephron is always quite low, so water cannot follow by osmosis. As solutes—but not water molecules—are leaving the tubular fluid, its osmolarity drops to about 150 mOsm/liter. The fluid entering the distal convoluted tubule is thus more dilute than plasma.

4. While the fluid continues flowing along the distal convoluted tubule, additional solutes but only a few water molecules are reabsorbed. The distal convoluted tubule cells are not very permeable to water and are not regulated by ADH.

5. Finally, the principal cells of the collecting ducts are impermeable to water when the ADH level is very low. Thus, tubular fluid becomes progressively more dilute as it flows onward. By

the time the tubular fluid drains into the renal pelvis, its concentration can be as low as 65–70 mOsm/liter. This is four times more dilute than blood plasma or glomerular filtrate.

Formation of Concentrated Urine

When water intake is low or water loss is high (such as during heavy sweating), the kidneys must conserve water while still eliminating wastes and excess ions. Under the influence of ADH, the kidneys produce a small volume of highly concentrated urine. Urine can be four times more concentrated (up to 1200 mOsm/liter) than blood plasma or glomerular filtrate (300 mOsm/liter).

The ability of ADH to cause excretion of concentrated urine depends on the presence of an **osmotic gradient** of solutes in the interstitial fluid of the renal medulla. Notice in Figure 26.19 that the solute concentration of the interstitial fluid in the kidney increases from about 300 mOsm/liter in the renal cortex to about 1200 mOsm/liter deep in the renal medulla. The three major solutes that contribute to this high osmolarity are Na^+, Cl^-, and urea. Two main factors contribute to building and maintaining this osmotic gradient: (1) differences in solute and water permeability and reabsorption in different sections of the long loops of Henle and the collecting duct, and (2) the countercurrent flow (flow in opposite directions) of fluid in neighboring descending and ascending limbs of the loop of Henle.

Production of concentrated urine occurs as follows (Figure 26.19):

❶ *In long-loop nephrons, symporters in thick ascending limb cells of the loop of Henle establish the osmotic gradient in the renal medulla.* In the thick ascending limb of the LOH, the Na^+-K^+-$2Cl^-$ symporters reabsorb Na^+ and Cl^- from the tubular fluid (Figure 26.19a). Water is not reabsorbed in this segment, however, because the cells are impermeable to water. As a result, the reabsorbed ions become increasingly concentrated in the interstitial fluid of the outer medulla. Cells in the thin ascending limb of the LOH also appear to contribute to building the osmotic gradient in the inner medulla. Those ions that diffuse into the vasa recta are carried deep into the inner medulla by the blood flow (Figure 26.19b). Because blood flow is sluggish in the vasa recta, however, there is time for diffusion of solutes to occur among tubular fluid, interstitial fluid, and blood at each level in the medulla. Thus, fluids in the descending limb, interstitial fluid, and plasma attain the same osmolarity.

❷ *Cells in the collecting ducts reabsorb more water and urea.* When ADH increases the water permeability of the principal cells, water quickly moves via osmosis out of the collecting duct tubular fluid, into the interstitial fluid of the inner medulla, and then into the vasa recta. With loss of water, the urea left behind in the tubular fluid of the collecting duct becomes increasingly concentrated. Because duct cells deep

Figure 26.19 **Mechanism of urine concentration in long-loop juxtamedullary nephrons.** The green line indicates the presence of Na⁺-K⁺-2Cl⁻ symporters that simultaneously reabsorb these ions into the interstitial fluid of the renal medulla; this portion of the nephron is also relatively impermeable to water and urea. All concentrations are in milliosmoles per liter (mOsm/liter).

🔑 **The formation of concentrated urine depends on high concentrations of solutes in interstitial fluid in the renal medulla.**

(a) Reabsorption of Na⁺, Cl⁻ and water in a long-loop juxtamedullary nephron

(b) Recycling of salts and urea in the vasa recta

 Which solutes are the main contributors to the high osmolarity of interstitial fluid in the renal medulla?

in the medulla are permeable to it, urea diffuses from the fluid in the duct into the interstitial fluid of the medulla.

❸ ***Urea recycling causes a buildup of urea in the renal medulla.*** As urea accumulates in the interstitial fluid, some of it diffuses into the tubular fluid in the descending and thin ascending limbs of the long loops of Henle, which also are permeable to urea (Figure 26.19a). However, while the fluid flows through the thick ascending limb, DCT, connecting tubule, and cortical portion of the collecting duct, urea remains in the lumen because cells in these segments are impermeable to it. As fluid flows along the collecting ducts, water reabsorption continues via osmosis because ADH is present. This water reabsorption *further increases* the concentration of urea in the tubular fluid, *more* urea diffuses into the interstitial fluid of the inner renal medulla, and the cycle repeats. The constant transfer of urea between segments of the renal tubule and the interstitial fluid of the medulla is termed *urea recycling.* In this way, reabsorption of water from the tubular fluid of the ducts promotes the buildup of urea in the interstitial fluid of the renal medulla, which in turn promotes water reabsorption. The solutes left behind in the lumen thus become very concentrated, and a small volume of concentrated urine is excreted.

The second contributor to the osmotic gradient in the renal medulla is the **countercurrent mechanism,** which has its basis in the hairpin shape of the long loops of Henle of juxtamedullary nephrons. Note in Figure 26.19a that the descending limb of the loop of Henle carries tubular fluid from the renal cortex deep into the medulla, and the ascending limb carries it in the opposite direction. Thus, fluid flowing in one tube runs counter (opposite) to fluid flowing in a nearby parallel tube, an arrangement called *countercurrent flow.*

The descending limb of the loop of Henle is very permeable to water but impermeable to solutes except urea. Because the osmolarity of the interstitial fluid outside the descending limb is higher than the tubular fluid within it, water moves out of the descending limb via osmosis. This causes the osmolarity of the tubular fluid to increase. As the fluid continues along the descending limb, its osmolarity increases even more: At the hairpin turn of the loop, the osmolarity can be as high as 1200 mOsm/liter in juxtamedullary nephrons.

As previously noted, the ascending limb of the loop is impermeable to water, but its symporters reabsorb Na⁺ and Cl⁻ from the tubular fluid into the interstitial fluid of the renal medulla, so the osmolarity of the tubular fluid progressively decreases as it flows through the ascending limb. At the junction of the medulla and cortex, the osmolarity of the tubular fluid has fallen to about 100 mOsm/liter. Overall, tubular fluid becomes progressively more concentrated as it flows along the descending limb and progressively more dilute as it moves along the ascending limb.

Note in Figure 26.19b that the vasa recta also consists of descending and ascending limbs that are parallel to each other and to the loop of Henle. Just as tubular fluid flows in opposite directions in the loop of Henle, blood flows in opposite directions in the ascending and descending parts of the vasa recta. Blood entering the vasa recta has an osmolarity of about 300 mOsm/liter. As it flows along the descending part into the renal medulla, where the interstitial fluid becomes increasingly concentrated, Na⁺, Cl⁻, and urea diffuse from interstitial fluid into the blood. But after its osmolarity increases, the blood flows into the ascending part of the vasa recta. Here blood flows through a region where the interstitial fluid becomes increasingly less concentrated. As a result, ions and urea diffuse from the blood into interstitial fluid, and reabsorbed water diffuses from interstitial fluid into the vasa recta. The osmolarity of blood leaving the vasa recta is only slightly higher than the osmolarity of blood entering the vasa recta. Thus, the vasa recta provides oxygen and nutrients to the renal medulla without washing out or diminishing the osmotic gradient.

Figure 26.20 summarizes the processes of filtration, reabsorption, and secretion in each segment of the nephron and collecting duct.

Diuretics

Diuretics are substances that slow renal reabsorption of water and thereby cause *diuresis,* an elevated urine flow rate, which in turn reduces blood volume. Diuretic drugs often are prescribed to treat *hypertension* (high blood pressure) because lowering blood volume usually reduces blood pressure. Naturally occurring diuretics include *caffeine* in coffee, tea, and sodas, which inhibits Na⁺ reabsorption, and *alcohol* in beer, wine, and mixed drinks, which inhibits secretion of ADH. Most diuretic drugs act by interfering with a mechanism for reabsorption of filtered Na⁺. For example, loop diuretics, such as furosemide (Lasix), selectively inhibit the Na⁺-K⁺-2Cl⁻ symporters in the thick ascending limb of the loop of Henle (see Figure 26.15). The thiazide diuretics, such as chlorthiazide (Diuril), act in the distal convoluted tubule, where they promote loss of Na⁺ and Cl⁻ in the urine by inhibiting Na⁺-Cl⁻ symporters.

Diuretic drugs that decrease Na⁺ reabsorption in the proximal convoluted tubule or loop of Henle cause a greater-than-normal amount of fluid and solutes to reach more distal parts of the nephron. In response to increased delivery of Na⁺, Cl⁻, and water, the distal convoluted tubule increases its reabsorption of Na⁺ and Cl⁻ and its secretion of K⁺. Although the elevated reabsorption fails to completely make up for the previous diminished reabsorption in the tubule, the enhanced secretion of K⁺ often causes excessive loss of K⁺ in the urine. So-called potassium-sparing diuretics can counteract this effect. They act on principal cells in the collecting duct either to inhibit the action of aldosterone—for example, spironolactone (Aldactone)—or to block the apical membrane Na⁺ leakage channels (for example, amiloride). By slowing reabsorption of Na⁺, other solutes, and water while also inhibiting secretion of K⁺ in principal cells, these drugs promote mild diuresis without excessive urinary loss of K⁺. ■

Figure 26.20   Summary of filtration, reabsorption, and secretion in the nephron and collecting duct.

🔑 Filtration occurs in the renal corpuscle; reabsorption occurs all along the renal tubule and collecting ducts.

PROXIMAL CONVOLUTED TUBULE

Reabsorption (into blood) of filtered:

Water	65% (osmosis)
Na⁺	65% (sodium-potassium pumps, symporters, antiporters)
K⁺	65% (diffusion)
Glucose	100% (symporters and facilitated diffusion)
Amino acids	100% (symporters and facilitated diffusion)
Cl⁻	50% (diffusion)
HCO₃⁻	80–90% (facilitated diffusion)
Urea	50% (diffusion)
Ca²⁺, Mg²⁺	variable (diffusion)

Secretion (into urine) of:

H⁺	variable (antiporters)
NH₄⁺	variable, increases in acidosis (antiporters)
Urea	variable (diffusion)
Creatinine	small amount

At end of PCT, tubular fluid is still isotonic to blood (300 mOsm/liter).

LOOP OF HENLE

Reabsorption (into blood) of:

Water	15% (osmosis in descending limb)
Na⁺	20–30% (symporters in ascending limb)
K⁺	20–30% (symporters in ascending limb)
Cl⁻	35% (symporters in ascending limb)
HCO₃⁻	10–20% (facilitated diffusion)
Ca²⁺, Mg²⁺	variable (diffusion)

Secretion (into urine) of:

Urea	variable (recycling from collecting duct)

At end of loop of Henle, tubular fluid is hypotonic (100–150 mOsm/liter).

RENAL CORPUSCLE

Glomerular filtration rate:
105–125 mL/min of fluid that is isotonic to blood

Filtered substances: water and all solutes present in blood (except proteins) including ions, glucose, amino acids, creatinine, uric acid

DISTAL CONVOLUTED TUBULE

Reabsorption (into blood) of:

Water	10–15% (osmosis)
Na⁺	5% (symporters)
Cl⁻	5% (symporters)
Ca²⁺	variable (stimulated by parathyroid hormone)

PRINCIPAL CELLS IN LATE DISTAL TUBULE AND COLLECTING DUCT

Reabsorption (into blood) of:

Water	5–9% (insertion of water channels stimulated by ADH)
Na⁺	1–4% (sodium-potassium pumps)
Urea	variable (recycling to loop of Henle)

Secretion (into urine) of:

K⁺	variable amount to adjust for dietary intake (leakage channels)

Tubular fluid leaving the collecting duct is dilute when ADH level is low and concentrated when ADH level is high.

INTERCALATED CELLS IN LATE DISTAL TUBULE AND COLLECTING DUCT

Reabsorption (into blood) of:

HCO₃⁻ (new)	varible amount, depends on H⁺ secretion (antiporters)
Urea	variable (recycling to loop of Henle)

Secretion (into urine) of:

H⁺	variable amounts to maintain acid-base homeostasis (H⁺ pumps)

Urine

In which segments of the nephron and collecting duct does secretion occur?

► CHECKPOINT

15. How do symporters in the ascending limb of the LOH and principal cells in the collecting duct contribute to the formation of concentrated urine?

16. How does ADH regulate facultative water reabsorption?

17. What is the countercurrent mechanism? Why is it important?

EVALUATION OF KIDNEY FUNCTION

▶ OBJECTIVES

• **Define urinalysis and describe its importance.**

• **Define renal plasma clearance and describe its importance.**

Routine assessment of kidney function involves evaluating both the quantity and quality of urine and the levels of wastes in the blood.

Urinalysis

An analysis of the volume and physical, chemical, and microscopic properties of urine, called a **urinalysis,** reveals much about the state of the body. Table 26.5 summarizes the major characteristics of normal urine. The volume of urine eliminated per day in a normal adult is 1–2 liters (about 1–2 qt). Fluid intake, blood pressure, blood osmolarity, diet, body temperature,

Table 26.5	Characteristics of Normal Urine
Characteristic	**Description**
Volume	One to two liters in 24 hours but varies considerably.
Color	Yellow or amber but varies with urine concentration and diet. Color is due to urochrome (pigment produced from breakdown of bile) and urobilin (from breakdown of hemoglobin). Concentrated urine is darker in color. Diet (reddish-colored urine from beets), medications, and certain diseases affect color. Kidney stones may produce blood in urine.
Turbidity	Transparent when freshly voided but becomes turbid (cloudy) upon standing.
Odor	Mildly aromatic but becomes ammonia-like upon standing. Some people inherit the ability to form methylmercaptan from digested asparagus that gives urine a characteristic odor. Urine of diabetics has a fruity odor due to presence of ketone bodies.
pH	Ranges between 4.6 and 8.0; average 6.0; varies considerably with diet. High-protein diets increase acidity; vegetarian diets increase alkalinity.
Specific gravity	Specific gravity (density) is the ratio of the weight of a volume of a substance to the weight of an equal volume of distilled water. In urine, it ranges from 1.001 to 1.035. The higher the concentration of solutes, the higher the specific gravity.

diuretics, mental state, and general health influence urine volume. For example, low blood pressure triggers the renin–angiotensin–aldosterone pathway. Aldosterone increases reabsorption of water and salts in the renal tubules and decreases urine volume. By contrast, when blood osmolarity decreases—for example, after drinking a large volume of water—secretion of ADH is inhibited and a larger volume of urine is excreted.

Water accounts for about 95% of the total volume of urine. The remaining 5% consists of electrolytes, solutes derived from cellular metabolism, and exogenous substances such as drugs. Normal urine is virtually protein free. Typical solutes normally present in urine include filtered and secreted electrolytes that are not reabsorbed, urea (from breakdown of proteins), creatinine (from breakdown of creatine phosphate in muscle fibers), uric acid (from breakdown of nucleic acids), urobilinogen (from breakdown of hemoglobin), and small quantities of other substances, such as fatty acids, pigments, enzymes, and hormones.

If disease alters body metabolism or kidney function, traces of substances not normally present may appear in the urine, or normal constituents may appear in abnormal amounts. Table 26.6 lists several abnormal constituents in urine that may be detected as part of a urinalysis. Normal values of urine components and the clinical implications of deviations from normal are listed in Appendix D.

Blood Tests

Two blood-screening tests can provide information about kidney function. One is the **blood urea nitrogen (BUN)** test, which measures the blood nitrogen that is part of the urea resulting from catabolism and deamination of amino acids. When glomerular filtration rate decreases severely, as may occur with renal disease or obstruction of the urinary tract, BUN rises steeply. One strategy in treating such patients is to minimize their protein intake, thereby reducing the rate of urea production.

Another test often used to evaluate kidney function is measurement of **plasma creatinine,** which results from catabolism of creatine phosphate in skeletal muscle. Normally, the blood creatinine level remains steady because the rate of creatinine excretion in the urine equals its discharge from muscle. A creatinine level above 1.5 mg/dL (135 mmol/liter) usually is an indication of poor renal function. Normal values for selected blood tests are listed in Appendix C together with situations that may cause the values to increase or decrease.

Renal Plasma Clearance

Even more useful than BUN and blood creatinine values in the diagnosis of kidney problems is an evaluation of how effectively the kidneys are removing a given substance from blood plasma. **Renal plasma clearance** is the volume of blood that is "cleaned" or cleared of a substance per unit of time, usually expressed in units of *milliliters per minute.* High renal plasma clearance indicates efficient excretion of a substance in the urine; low clearance indicates inefficient excretion. For example,

Table 26.6	Summary of Abnormal Constituents in Urine
Abnormal Constituent	**Comments**
Albumin	A normal constituent of plasma, but it usually appears in only very small amounts in urine because it is too large to pass through capillary fenestrations. The presence of excessive albumin in the urine—**albuminuria** (al′-bū-mi-NOO-rē-a)—indicates an increase in the permeability of filtration membranes due to injury or disease, increased blood pressure, or irritation of kidney cells by substances such as bacterial toxins, ether, or heavy metals.
Glucose	The presence of glucose in the urine is called **glucosuria** (gloo-kō-SOO-rē-a) and usually indicates diabetes mellitus. Occasionally it may be caused by stress, which can cause excessive amounts of epinephrine to be secreted. Epinephrine stimulates the breakdown of glycogen and liberation of glucose from the liver.
Red blood cells (erythrocytes)	The presence of red blood cells in the urine is called **hematuria** (hēm-a-TOO-rē-a) and generally indicates a pathological condition. One cause is acute inflammation of the urinary organs as a result of disease or irritation from kidney stones. Other causes include tumors, trauma, and kidney disease. One should make sure the urine sample was not contaminated with menstrual blood from the vagina.
White blood cells (leukocytes)	The presence of white blood cells and other components of pus in the urine, referred to as **pyuria** (pī-Ū-rē-a), indicates infection in the kidney or other urinary organs.
Ketone bodies	High levels of ketone bodies in the urine, called **ketonuria** (kē-tō-NOO-rē-a), may indicate diabetes mellitus, anorexia, starvation, or simply too little carbohydrate in the diet.
Bilirubin	When red blood cells are destroyed by macrophages, the globin portion of hemoglobin is split off and the heme is converted to biliverdin. Most of the biliverdin is converted to bilirubin, which gives bile its major pigmentation. An above-normal level of bilirubin in urine is called **bilirubinuria** (bil′-ē-roo-bi-NOO-rē-a).
Urobilinogen	The presence of urobilinogen (breakdown product of hemoglobin) in urine is called **urobilinogenuria** (ū′-rō-bi-lin′-ō-je-NOO-rē-a). Traces are normal, but elevated urobilinogen may be due to hemolytic or pernicious anemia, infectious hepatitis, biliary obstruction, jaundice, cirrhosis, congestive heart failure, or infectious mononucleosis.
Casts	**Casts** are tiny masses of material that have hardened and assumed the shape of the lumen of the tubule in which they formed. They are then flushed out of the tubule when filtrate builds up behind them. Casts are named after the cells or substances that compose them or based on their appearance. For example, there are white blood cell casts, red blood cell casts, and epithelial cell casts that contain cells from the walls of the tubules.
Microbes	The number and type of bacteria vary with specific infections in the urinary tract. One of the most common is *E. coli.* The most common fungus to appear in urine is the yeast *Candida albicans,* a cause of vaginitis. The most frequent protozoan seen is *Trichomonas vaginalis,* a cause of vaginitis in females and urethritis in males.

the clearance of glucose normally is zero because it is not excreted at all. Instead, 100% of the filtered glucose is returned to the blood via tubular reabsorption. Knowing a drug's clearance is essential for determining the correct dosage. If clearance is high (one example is penicillin), then the dosage must also be high, and the drug must be given several times a day to maintain an adequate therapeutic level in the blood.

The following equation is used to calculate clearance:

$$\text{Renal plasma clearance of substance S} = \frac{(U \times V)}{P}$$

where U and P are the concentrations of the substance in urine and plasma, respectively (both expressed in the same units, such as mg/mL), and V is the urine flow rate in mL/min.

The clearance of a solute depends on the three basic processes of a nephron: glomerular filtration, tubular reabsorption, and tubular secretion. Consider a substance that is filtered but neither reabsorbed nor secreted. Its clearance equals the glomerular filtration rate because all the molecules that pass the filtration membrane appear in the urine. This is very nearly the situation for creatinine; it easily passes the filter, it is not reabsorbed, and it is secreted only to a very small extent. Measuring the creatinine clearance, which normally is 120–140 mL/min, is

the easiest way to assess glomerular filtration rate. The waste product urea is filtered, reabsorbed, and secreted to varying extents. Its clearance typically is less than the GFR, about 70 mL/min.

 Dialysis

If a person's kidneys are so impaired by disease or injury that they are unable to function adequately, then blood must be cleansed artificially by **dialysis,** the separation of large solutes from smaller ones through use of a selectively permeable membrane. One method of dialysis is the artificial kidney machine, which performs **hemodialysis** (*hemo-* = blood) because it directly filters the patient's blood. As blood flows through tubing made of selectively permeable dialysis membrane, waste products diffuse from the blood into the dialysis solution surrounding the membrane. The dialysis solution is continuously replaced to maintain favorable concentration gradients for diffusion of solutes into and out of the blood. After passing through the dialysis tubing, the cleansed blood flows back into the body.

Continuous ambulatory peritoneal dialysis (CAPD) uses the peritoneal lining of the abdominal cavity as the dialysis membrane. The tip of a catheter is surgically placed in the patient's peritoneal cavity and connected to a sterile dialysis solution. The dialysate flows into the peritoneal cavity from its plastic container by gravity. The solution remains in the cavity until metabolic waste products, excess electrolytes, and extracellular fluid diffuse into the dialysis solution. The solution is then drained from the cavity by gravity into a sterile bag that is discarded. ∎

▶ CHECKPOINT

18. What are the characteristics of normal urine?

19. What chemical substances normally are present in urine?

20. How may kidney function be evaluated?

21. Why are the renal plasma clearances of glucose, urea, and creatinine different? How does each clearance compare to glomerular filtration rate?

URINE TRANSPORTATION, STORAGE, AND ELIMINATION

▶ OBJECTIVE

- Describe the anatomy, histology, and physiology of the ureters, urinary bladder, and urethra.

From collecting ducts, urine drains through papillary ducts into the minor calyces, which join to become major calyces that unite to form the renal pelvis (see Figure 26.3). From the renal pelvis, urine first drains into the ureters and then into the urinary bladder. Urine is then discharged from the body through the single urethra (see Figure 26.1).

Ureters

Each of the two **ureters** (Ū-rē-ters) transports urine from the renal pelvis of one kidney to the urinary bladder. Peristaltic contractions of the muscular walls of the ureters push urine toward the urinary bladder, but hydrostatic pressure and gravity also contribute. Peristaltic waves that pass from the renal pelvis to the urinary bladder vary in frequency from one to five per minute, depending on how fast urine is being formed.

The ureters are 25–30 cm (10–12 in.) long and are thick-walled, narrow tubes that vary in diameter from 1 mm to 10 mm along their course between the renal pelvis and the urinary bladder. Like the kidneys, the ureters are retroperitoneal. At the base of the urinary bladder, the ureters curve medially and pass obliquely through the wall of the posterior aspect of the urinary bladder (Figure 26.21).

Even though there is no anatomical valve at the opening of each ureter into the urinary bladder, a physiological one is quite effective. As the urinary bladder fills with urine, pressure within it compresses the oblique openings into the ureters and prevents the backflow of urine. When this physiological valve is not

Figure 26.21 Ureters, urinary bladder, and urethra in a female. (See Tortora, *A Photographic Atlas of the Human Body,* Figures 13.8, 13.10.)

🔑 Urine is stored in the urinary bladder before being expelled by micturition.

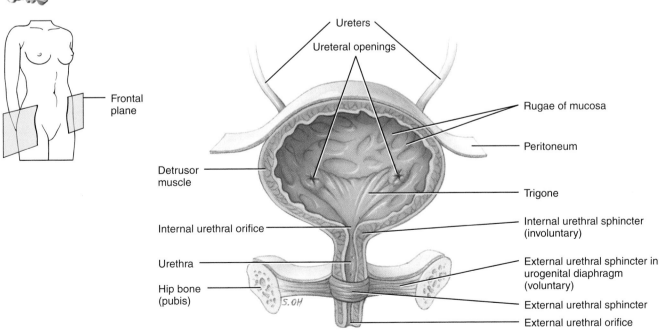

Anterior view of frontal section

 What is a lack of voluntary control over micturition called?

operating properly, it is possible for microbes to travel up the ureters from the urinary bladder to infect one or both kidneys.

Three coats of tissue form the wall of the ureters. The deepest coat, or **mucosa,** is a mucous membrane with **transitional epithelium** (see Table 4.1E on page 111) and an underlying **lamina propria** of areolar connective tissue with considerable collagen, elastic fibers, and lymphatic tissue. Transitional epithelium is able to stretch—a marked advantage for any organ that must accommodate a variable volume of fluid. Mucus secreted by the goblet cells of the mucosa prevents the cells from coming in contact with urine, the solute concentration and pH of which may differ drastically from the cytosol of cells that form the wall of the ureters. Throughout most of the length of the ureters, the intermediate coat, the **muscularis,** is composed of inner longitudinal and outer circular layers of smooth muscle fibers. This arrangement is opposite to that of the gastrointestinal tract, which contains inner circular and outer longitudinal layers. The muscularis of the distal third of the ureters also contains an outer layer of longitudinal muscle fibers. Thus, the muscularis in the distal third of the ureter is inner longitudinal, middle circular, and outer longitudinal. Peristalsis is the major function of the muscularis. The superficial coat of the ureters is the **adventitia,** a layer of areolar connective tissue containing blood vessels, lymphatic vessels, and nerves that serve the muscularis and mucosa. The adventitia blends in with surrounding connective tissue and anchors the ureters in place.

Urinary Bladder

The **urinary bladder** is a hollow, distensible muscular organ situated in the pelvic cavity posterior to the pubic symphysis. In males, it is directly anterior to the rectum; in females, it is anterior to the vagina and inferior to the uterus (see Figure 28.13b on page 1029). Folds of the peritoneum hold the urinary bladder in position. The shape of the urinary bladder depends on how much urine it contains. Empty, it is collapsed; when slightly distended it becomes spherical. As urine volume increases, it becomes pear-shaped and rises into the abdominal cavity. Urinary bladder capacity averages 700–800 mL. It is smaller in females because the uterus occupies the space just superior to the urinary bladder.

Anatomy and Histology of the Urinary Bladder

In the floor of the urinary bladder is a small triangular area called the **trigone** (TRĪ-gon = triangle). The two posterior corners of the trigone contain the two ureteral openings, whereas the opening into the urethra, **the internal urethral orifice,** lies in the anterior corner (Figure 26.21). Because its mucosa is firmly bound to the muscularis, the trigone has a smooth appearance.

Three coats make up the wall of the urinary bladder. The deepest is the **mucosa,** a mucous membrane composed of **tran-**sitional epithelium and an underlying **lamina propria** similar to that of the ureters. Rugae (the folds in the mucosa) are also present to permit expansion of the urinary bladder. Surrounding the mucosa is the intermediate **muscularis,** also called the **detrusor muscle** (de-TROO-ser = to push down), which consists of three layers of smooth muscle fibers: the inner longitudinal, middle circular, and outer longitudinal layers. Around the opening to the urethra the circular fibers form an **internal urethral sphincter;** inferior to it is the **external urethral sphincter,** which is composed of skeletal muscle and is a modification of the urogenital diaphragm muscle (see Figure 11.12 on page 339). The most superficial coat of the urinary bladder on the posterior and inferior surfaces is the **adventitia,** a layer of areolar connective tissue that is continuous with that of the ureters. Over the superior surface of the urinary bladder is the **serosa,** a layer of visceral peritoneum.

The Micturition Reflex

Discharge of urine from the urinary bladder, called **micturition** (mik′-choo-RISH-un; *mictur-* = urinate), is also known as *urination* or *voiding.* Micturition occurs via a combination of involuntary and voluntary muscle contractions. When the volume of urine in the bladder exceeds 200–400 mL, pressure within the bladder increases considerably, and stretch receptors in its wall transmit nerve impulses into the spinal cord. These impulses propagate to the **micturition center** in sacral spinal cord segments S2 and S3 and trigger a spinal reflex called the **micturition reflex.** In this reflex arc, parasympathetic impulses from the micturition center propagate to the urinary bladder wall and internal urethral sphincter. The nerve impulses cause *contraction* of the detrusor muscle and *relaxation* of the internal urethral sphincter muscle. Simultaneously, the micturition center inhibits somatic motor neurons that innervate skeletal muscle in the external urethral sphincter. Upon contraction of the bladder wall and relaxation of the sphincters, urination takes place. Bladder filling causes a sensation of fullness that initiates a conscious desire to urinate before the micturition reflex actually occurs. Although emptying of the urinary bladder is a reflex, in early childhood we learn to initiate it and stop it voluntarily. Through learned control of the external urethral sphincter muscle and certain muscles of the pelvic floor, the cerebral cortex can initiate micturition or delay its occurrence for a limited period.

Urethra

The **urethra** (ū-RĒ-thra) is a small tube leading from the internal urethral orifice in the floor of the urinary bladder to the exterior of the body (see Figure 26.1). In both males and females, the urethra is the terminal portion of the urinary system and the passageway for discharging urine from the body. In males, it discharges semen (fluid that contains sperm) as well.

In females, the urethra lies directly posterior to the pubic symphysis, is directed obliquely inferiorly and anteriorly, and has a length of 4 cm (1.5 in.) (see Figure 28.13a on page 1028). The opening of the urethra to the exterior, the **external urethral orifice,** is located between the clitoris and the vaginal opening (see Figure 28.22 on page 1038). The wall of the female urethra consists of a deep **mucosa** and a superficial **muscularis.** The mucosa is a mucous membrane composed of **epithelium** and **lamina propria** (areolar connective tissue with elastic fibers and a plexus of veins). The muscularis consists of circularly arranged smooth muscle fibers and is continuous with that of the urinary bladder. Near the urinary bladder, the mucosa contains transitional epithelium that is continuous with that of the urinary bladder; near the external urethral orifice, the epithelium is nonkeratinized stratified squamous epithelium for protection. Between these areas, the mucosa contains stratified columnar or pseudostratified columnar epithelium.

In males, the urethra also extends from the internal urethral orifice to the exterior, but its length and passage through the body are considerably different than in females (see Figure 28.3 on page 1016 and Figure 28.12 on page 1028). The male urethra first passes through the prostate, then through the urogenital diaphragm, and finally through the penis, a distance of 15–20 cm (6–8 in.).

The male urethra, which also consists of a deep **mucosa** and a superficial **muscularis,** is subdivided into three anatomical regions: (1) The **prostatic urethra** passes through the prostate. (2) The **membranous urethra,** the shortest portion, passes through the urogenital diaphragm. (3) The **spongy urethra,** the longest portion, passes through the penis. The epithelium of the prostatic urethra is continuous with that of the urinary bladder and consists of transitional epithelium that becomes stratified columnar or pseudostratified columnar epithelium more distally. The mucosa of the membranous urethra contains stratified columnar or pseudostratified columnar epithelium. The epithelium of the spongy urethra is stratified columnar or pseudostratified columnar epithelium, except near the external urethral orifice. There it is nonkeratinized stratified squamous epithelium. The **lamina propria** of the male urethra is areolar connective tissue with elastic fibers and a plexus of veins.

The muscularis of the prostatic urethra is composed of mostly circular smooth muscle fibers superficial to the lamina propria; these circular fibers help form the internal urethral sphincter of the urinary bladder. The muscularis of the membranous urethra consists of circularly arranged skeletal muscle fibers of the urogenital diaphragm that help form the external urethral sphincter of the urinary bladder.

Several glands and other structures associated with reproduction deliver their contents into the male urethra. The prostatic urethra contains the openings from (1) ducts that transport secretions from the **prostate** and from (2) the **seminal vesicles** and **ductus (vas) deferens,** which deliver sperm into the urethra and provide secretions that both neutralize the acidity of the female reproductive tract and contribute to sperm motility and viability. The openings of the ducts of the **bulbourethral (Cowper's) glands** empty into the spongy urethra. They deliver an alkaline substance before ejaculation that neutralizes the acidity of the urethra. The glands also secrete mucus, which lubricates the end of the penis during sexual arousal. Throughout the urethra, but especially in the spongy urethra, the openings of the ducts of **urethral (Littré) glands** discharge mucus during sexual arousal or ejaculation.

 Urinary Incontinence

A lack of voluntary control over micturition is called **urinary incontinence.** In infants and children under 2–3 years old, incontinence is normal because neurons to the external urethral sphincter muscle are not completely developed, and voiding occurs whenever the urinary bladder is sufficiently distended to stimulate the micturition reflex. Urinary incontinence also occurs in adults. Of the more than 10 million U.S. adults who suffer from urinary incontinence, 1–2 million have **stress incontinence.** In this condition, physical stresses that increase abdominal pressure, such as coughing, sneezing, laughing, exercising, pregnancy, or simply walking cause leakage of urine from the urinary bladder. Other causes of incontinence in adults are injury to the nerves controlling the urinary bladder, loss of bladder flexibility with age, disease or irritation of the urinary bladder or urethra, damage to the external urethral sphincter, or certain drugs. Also, compared with nonsmokers, those who smoke have twice the risk of developing incontinence. ■

▶ C H E C K P O I N T

22. What forces help propel urine from the renal pelvis to the urinary bladder?

23. What is micturition? How does the micturition reflex occur?

24. How do the location, length, and histology of the urethra compare in males and females?

WASTE MANAGEMENT IN OTHER BODY SYSTEMS

▶ O B J E C T I V E

• **Describe the ways that body wastes are handled.**

As we have seen, just one of the many functions of the urinary system is to help rid the body of some kinds of waste materials. Besides the kidneys, several other tissues, organs, and processes contribute to the temporary confinement of wastes, the transport of waste materials for disposal, the recycling of materials, and the excretion of excess or toxic substances in the body. These waste management systems include the following:

- **Body buffers.** Buffers bind excess hydrogen ions (H^+), thereby preventing an increase in the acidity of body fluids. Buffers are like wastebaskets in that they have a limited capacity; eventually the H^+, like the paper in a wastebasket, must be eliminated from the body by excretion.

- **Blood.** The bloodstream provides pickup and delivery services for the transport of wastes, in much the same way that garbage trucks and sewer lines serve a community.

- **Liver.** The liver is the primary site for metabolic recycling, as occurs, for example, in the conversion of amino acids into glucose or of glucose into fatty acids. The liver also converts toxic substances into less toxic ones, such as ammonia into urea. These functions of the liver are described in Chapters 24 and 25.

- **Lungs.** With each exhalation, the lungs excrete CO_2, and expel heat and a little water vapor.

- **Sweat (sudoriferous) glands.** Especially during exercise, sweat glands in the skin help eliminate excess heat, water, and CO_2, plus small quantities of salts and urea as well.

- **Gastrointestinal tract.** Through defecation, the gastrointestinal tract excretes solid, undigested foods; wastes; some CO_2; water; salts; and heat.

At about the fifth week, a mesodermal outgrowth, called a **ureteric bud,** develops from the distal portion of the mesonephric duct near the cloaca. The **metanephros,** or ultimate kidney, develops from the ureteric bud and metanephric mesoderm. The ureteric bud forms the *collecting ducts, calyces, renal pelvis,* and *ureter.* The **metanephric mesoderm** forms the *nephrons* of the kidneys. By the third month, the fetal kidneys begin excreting urine into the surrounding amniotic fluid; indeed, fetal urine makes up most of the amniotic fluid.

During development, the cloaca divides into a **urogenital sinus,** into which urinary and genital ducts empty, and a *rectum* that discharges into the anal canal. The *urinary bladder* develops from the urogenital sinus. In females, the *urethra* develops as a result of lengthening of the short duct that extends from the urinary bladder to the urogenital sinus. In males, the urethra is considerably longer and more complicated, but it is also derived from the urogenital sinus.

▶ C H E C K P O I N T

25. Which tissue develops into nephrons?

26. Which tissue gives rise to collecting ducts, calyces, renal pelves, and ureters?

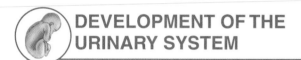

DEVELOPMENT OF THE URINARY SYSTEM

▶ O B J E C T I V E

- **Describe the development of the urinary system.**

Starting in the third week of fetal development, a portion of the mesoderm along the posterior aspect of the embryo, the **intermediate mesoderm,** differentiates into the kidneys. Three pairs of kidneys form within the intermediate mesoderm in succession: the pronephros, the mesonephros, and the metanephros (Figure 26.22). Only the last pair remains as the functional kidneys of the newborn.

The first kidney to form, the **pronephros,** is the most superior of the three and has an associated **pronephric duct.** This duct empties into the **cloaca,** the expanded terminal part of the hindgut, which functions as a common outlet for the urinary, digestive, and reproductive ducts. The pronephros begins to degenerate during the fourth week and is completely gone by the sixth week.

The second kidney, the **mesonephros,** replaces the pronephros. The retained portion of the pronephric duct, which connects to the mesonephros, develops into the **mesonephric duct.** The mesonephros begins to degenerate by the sixth week and is almost gone by the eighth week.

AGING AND THE URINARY SYSTEM

▶ O B J E C T I V E

- **Describe the effects of aging on the urinary system.**

With aging, the kidneys shrink in size, have a decreased blood flow, and filter less blood. The mass of the two kidneys decreases from an average of 260 g in 20-year-olds to less than 200 g by age 80. Likewise, renal blood flow and filtration rate decline by 50% between ages 40 and 70. Kidney diseases that become more common with age include acute and chronic kidney inflammations and renal calculi (kidney stones). Because the sensation of thirst diminishes with age, older individuals also are susceptible to dehydration. Urinary tract infections are more common among the elderly, as are polyuria (excessive urine production), nocturia (excessive urination at night), increased frequency of urination, dysuria (painful urination), urinary retention or incontinence, and hematuria (blood in the urine).

• • •

To appreciate the many ways that the urinary system contributes to homeostasis of other body systems, examine *Focus on Homeostasis: The Urinary System* on page 984. Next, in Chapter 27, we will see how the kidneys and lungs contribute to maintenance of homeostasis of body fluid volume, electrolyte levels in body fluids, and acid–base balance.

Figure 26.22 Development of the urinary system.

Three pairs of kidneys form within intermediate mesoderm in succession: pronephros, mesonephros, and metanephros.

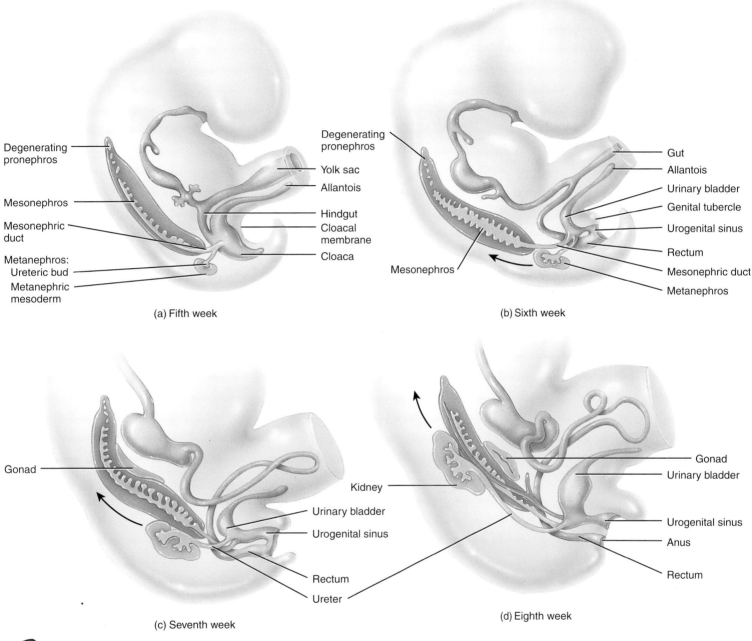

Degenerating pronephros

Mesonephros

Mesonephric duct

Metanephros:
  Ureteric bud
  Metanephric mesoderm

Yolk sac

Allantois

Hindgut

Cloacal membrane

Cloaca

(a) Fifth week

Degenerating pronephros

Mesonephros

Gut

Allantois

Urinary bladder

Genital tubercle

Urogenital sinus

Rectum

Mesonephric duct

Metanephros

(b) Sixth week

Gonad

Urinary bladder

Urogenital sinus

Rectum

Ureter

(c) Seventh week

Kidney

Gonad

Urinary bladder

Urogenital sinus

Anus

Rectum

(d) Eighth week

When do the kidneys begin to develop?

Body System	Contribution of the Urinary System
For all body systems	Kidneys regulate volume, composition, and pH of body fluids by removing wastes and excess substances from blood and excreting them in urine; the ureters transport urine from the kidneys to the urinary bladder, which stores urine until it is eliminated through the urethra.
Integumentary system	Kidneys and skin both contribute to synthesis of calcitriol, the active form of vitamin D.
Skeletal system	Kidneys help adjust levels of blood calcium and phosphates, needed for building bone matrix.
Muscular system	Kidneys help adjust level of blood calcium, needed for contraction of muscle.
Nervous system	Kidneys perform gluconeogenesis, which provides glucose for ATP production in neurons, especially during fasting or starvation.
Endocrine system	Kidneys participate in synthesis of calcitriol, the active form of vitamin D, and release erythropoietin, the hormone that stimulates production of red blood cells.
Cardiovascular system	By increasing or decreasing their reabsorption of water filtered from blood, the kidneys help adjust blood volume and blood pressure; renin released by juxtaglomerular cells in the kidneys raises blood pressure; some bilirubin from hemoglobin breakdown is converted to a yellow pigment (urobilin), which is excreted in urine.
Lymphatic and Immune system	By increasing or decreasing their reabsorption of water filtered from blood, the kidneys help adjust the volume of interstitial fluid and lymph; urine flushes microbes out of urethra.
Respiratory system	Kidneys and lungs cooperate in adjusting pH of body fluids.
Digestive system	Kidneys help synthesize calcitriol, the active form of vitamin D, which is needed for absorption of dietary calcium.
Reproductive systems	In males, the portion of the urethra that extends through the prostate and penis is a passageway for semen as well as urine.

984

DISORDERS: HOMEOSTATIC IMBALANCES

Renal Calculi

The crystals of salts present in urine occasionally precipitate and solidify into insoluble stones called **renal calculi** (*calculi* = pebbles) or **kidney stones.** They commonly contain crystals of calcium oxalate, uric acid, or calcium phosphate. Conditions leading to calculus formation include the ingestion of excessive calcium, low water intake, abnormally alkaline or acidic urine, and overactivity of the parathyroid glands. When a stone lodges in a narrow passage, such as a ureter, the pain can be intense. **Shock-wave lithotripsy** (LITH-ō-trip′-sē; *litho* = stone) offers an alternative to surgical removal of kidney stones. A device, called a lithotripter, delivers brief, high-intensity sound waves through a water-filled cushion. Over a period of 30 to 60 minutes, 1000 or more hydraulic shock waves pulverize the stone, creating fragments that are small enough to wash out in the urine.

Urinary Tract Infections

The term **urinary tract infection (UTI)** is used to describe either an infection of a part of the urinary system or the presence of large numbers of microbes in urine. UTIs are more common in females due to their shorter urethra. Symptoms include painful or burning urination, urgent and frequent urination, low back pain, and bed-wetting. UTIs include *urethritis* (inflammation of the urethra), *cystitis* (inflammation of the urinary bladder), and *pyelonephritis* (inflammation of the kidneys). If pyelonephritis becomes chronic, scar tissue can form in the kidneys and severely impair their function.

Glomerular Diseases

A variety of conditions may damage the kidney glomeruli, either directly or indirectly because of disease elsewhere in the body. Typically, the filtration membrane sustains damage, and its permeability increases.

Glomerulonephritis is an inflammation of the kidney that involves the glomeruli. One of the most common causes is an allergic reaction to the toxins produced by streptococcal bacteria that have recently infected another part of the body, especially the throat. The glomeruli become so inflamed, swollen, and engorged with blood that the filtration membranes allow blood cells and plasma proteins to enter the filtrate. As a result, the urine contains many erythrocytes (hematuria) and a lot of protein. The glomeruli may be permanently damaged, leading to chronic renal failure.

Nephrotic syndrome is a condition characterized by *proteinuria* (protein in the urine) and *hyperlipidemia* (high blood levels of cholesterol, phospholipids, and triglycerides). The proteinuria is due to an increased permeability of the filtration membrane, which permits proteins, especially albumin, to escape from blood into urine. Loss of albumin results in *hypoalbuminemia* (low blood albumin level) once liver production of albumin fails to meet increased urinary losses. Edema, usually seen around the eyes, ankles, feet, and abdomen, occurs in nephrotic syndrome because loss of albumin from the blood decreases blood colloid osmotic pressure. Nephrotic syndrome is associated with several glomerular diseases of unknown cause, as well as with systemic disorders such as diabetes mellitus, systemic lupus erythematosus (SLE), a variety of cancers, and AIDS.

Renal Failure

Renal failure is a decrease or cessation of glomerular filtration. In acute renal failure (ARF), the kidneys abruptly stop working entirely (or almost entirely). The main feature of ARF is the suppression of urine flow, usually characterized either by *oliguria* (olig- =scanty; *-uria* = urine production), which is daily urine output less than 250 mL, or by *anuria,* daily urine output less than 50 mL. Causes include low blood volume (for example, due to hemorrhage), decreased cardiac output, damaged renal tubules, kidney stones, the dyes used to visualize blood vessels in angiograms, nonsteroidal anti-inflammatory drugs, and some antibiotic drugs.

Renal failure causes a multitude of problems. There is edema due to salt and water retention and acidosis due to an inability of the kidneys to excrete acidic substances. In the blood, urea builds up due to impaired renal excretion of metabolic waste products and potassium level rises, which can lead to cardiac arrest. Often, there is anemia because the kidneys no longer produce enough erythropoietin for adequate red blood cell production. Because the kidneys are no longer able to convert vitamin D to calcitriol, which is needed for adequate calcium absorption from the small intestine, osteomalacia also may occur.

Chronic renal failure (CRF) refers to a progressive and usually irreversible decline in glomerular filtration rate (GFR). CRF may result from chronic glomerulonephritis, pyelonephritis, polycystic kidney disease, or traumatic loss of kidney tissue. CRF develops in three stages. In the first stage, *diminished renal reserve,* nephrons are destroyed until about 75% of the functioning nephrons are lost. At this stage, a person may have no signs or symptoms because the remaining nephrons enlarge and take over the function of those that have been lost. Once 75% of the nephrons are lost, the person enters the second stage, called *renal insufficiency,* characterized by a decrease in GFR and increased blood levels of nitrogen-containing wastes and creatinine. Also, the kidneys cannot effectively concentrate or dilute the urine. The final stage, called *end-stage renal failure,* occurs when about 90% of the nephrons have been lost. At this stage, GFR diminishes to 10–15% of normal, oliguria is present, and blood levels of nitrogen-containing wastes and creatinine increase further. People with end-stage renal failure need dialysis therapy and are possible candidates for a kidney transplant operation.

Polycystic Kidney Disease

Polycystic kidney disease (PKD) is one of the most common inherited disorders. In PKD, the kidney tubules become riddled with hundreds or thousands of cysts (fluid-filled cavities). In addition, inappropriate apoptosis (programmed cell death) of cells in noncystic tubules leads to progressive impairment of renal function and eventually to end-stage renal failure.

People with PKD also may have cysts and apoptosis in the liver, pancreas, spleen, and gonads; increased risk of cerebral aneurysms; heart valve defects; and diverticuli in the colon. Typically, symptoms are not noticed until adulthood, when patients may have back pain, urinary tract infections, blood in the urine, hypertension, and large abdominal masses. Using drugs to restore normal blood pressure, restricting protein and salt in the diet, and controlling urinary tract infections may slow progression to renal failure.

Urinary Bladder Cancer

Each year, nearly 12,000 Americans die from **urinary bladder cancer.** It generally strikes people over 50 years of age and is three times more likely to develop in males than females. The disease is typically painless as it develops, but in most cases blood in the urine is a primary sign of the disease. Less often, people experience painful and/or frequent urination.

As long as the disease is identified early and treated promptly, the prognosis is favorable. Fortunately, about 75% of the urinary bladder cancers are confined to the epithelium of the urinary bladder and are easily removed by surgery. The lesions tend to be low grade, meaning that they have only a small potential for metastasis.

Urinary bladder cancer is frequently the result of a carcinogen. About half of all cases occur in people who smoke or have at some time smoked cigarettes. The cancer also tends to develop in people who are exposed to chemicals called aromatic amines. Workers in the leather, dye, rubber, and aluminum industries, as well as painters, are often exposed to these chemicals.

MEDICAL TERMINOLOGY

Azotemia (az-ō-TĒ-mē-a; *azot-* = nitrogen; *-emia* = condition of blood) Presence of urea or other nitrogen-containing substances in the blood.

Cystocele (SIS-tō-sēl; *cysto-* = bladder; *-cele* = hernia or rupture) Hernia of the urinary bladder.

Enuresis (en′-ū-RĒ-sis = to void urine) Involuntary voiding of urine after the age at which voluntary control has typically been attained.

Intravenous pyelogram (in′-tra-VĒ-nus PĪ-el-ō-gram′; *intra-* = within; *veno-* = vein; *pyelo-* = pelvis of kidney; *-gram* =record), or **IVP** Radiograph (x ray) of the kidneys after intravenous injection of a dye.

Nocturnal enuresis (nok-TUR-nal en′-ū-RĒ-sis) Discharge of urine during sleep, resulting in bed-wetting; occurs in about 15% of 5-year-old children and generally resolves spontaneously, afflicting only about 1% of adults. It may have a genetic basis, as bed-wetting occurs more often in identical twins than in fraternal twins and more often in children whose parents or siblings were bedwetters. Possible causes include smaller-than-normal bladder capacity, failure to awaken in response to a full bladder, and above-normal production of urine at night. Also referred to as **nocturia.**

Polyuria (pol′-ē-Ū-rē-a; *poly-* = too much) Excessive formation of urine.

Stricture (STRIK-chur) Narrowing of the lumen of a canal or hollow organ, as may occur in the ureter, urethra, or any other tubular structure in the body.

Urinary retention A failure to completely or normally void urine; may be due to an obstruction in the urethra or neck of the urinary bladder, to nervous contraction of the urethra, or to lack of urge to urinate. In men, an enlarged prostate may constrict the urethra and cause urinary retention.

STUDY OUTLINE

INTRODUCTION (p. 949)

1. The organs of the urinary system are the kidneys, ureters, urinary bladder, and urethra.
2. After the kidneys filter blood and return most water and many solutes to the bloodstream, the remaining water and solutes constitute urine.

OVERVIEW OF KIDNEY FUNCTIONS (p. 949)

1. The kidneys regulate blood ionic composition, blood osmolarity, blood volume, blood pressure, and blood pH.
2. The kidneys also perform gluconeogenesis, release calcitriol and erythropoietin, and excrete wastes and foreign substances.

ANATOMY AND HISTOLOGY OF THE KIDNEYS (p. 950)

1. The kidneys are retroperitoneal organs attached to the posterior abdominal wall.
2. Three layers of tissue surround the kidneys: renal capsule, adipose capsule, and renal fascia.
3. Internally, the kidneys consist of a renal cortex, a renal medulla, renal pyramids, renal papillae, renal columns, calyces, and a renal pelvis.
4. Blood flows into the kidney through the renal artery and successively into segmental, interlobar, arcuate, and interlobular arteries; afferent arterioles; glomerular capillaries; efferent arterioles; peritubular capillaries and vasa recta; and interlobular, arcuate, and interlobar veins before flowing out of the kidney through the renal vein.
5. Vasomotor nerves from the sympathetic division of the autonomic nervous system supply kidney blood vessels; they help regulate flow of blood through the kidney.
6. The nephron is the functional unit of the kidneys. A nephron consists of a renal corpuscle (glomerulus and glomerular or Bowmanms capsule) and a renal tubule.

7. A renal tubule consists of a proximal convoluted tubule, a loop of Henle, and a distal convoluted tubule, which drains into a collecting duct (shared by several nephrons). The loop of Henle consists of a descending limb and an ascending limb.

8. A cortical nephron has a short loop that dips only into the superficial region of the renal medulla; a juxtamedullary nephron has a long loop of Henle that stretches through the renal medulla almost to the renal papilla.

9. The wall of the entire glomerular capsule, renal tubule, and ducts consists of a single layer of epithelial cells. The epithelium has distinctive histological features in different parts of the tubule. Table 26.1 on page 958 summarizes the histological features of the renal tubule and collecting duct.

10. The juxtaglomerular apparatus (JGA) consists of the juxtaglomerular cells of an afferent arteriole and the macula densa of the final portion of the ascending limb of the loop of Henle.

OVERVIEW OF RENAL PHYSIOLOGY (p. 959)

1. Nephrons perform three basic tasks: glomerular filtration, tubular secretion, and tubular reabsorption.

GLOMERULAR FILTRATION (p. 960)

1. Fluid that enters the capsular space is glomerular filtrate.
2. The filtration membrane consists of the glomerular endothelium, basal lamina, and filtration slits between pedicels of podocytes.
3. Most substances in blood plasma easily pass through the glomerular filter. However, blood cells and most proteins normally are not filtered.
4. Glomerular filtrate amounts to up to 180 liters of fluid per day. This large amount of fluid is filtered because the filter is porous and thin, the glomerular capillaries are long, and the capillary blood pressure is high.
5. Glomerular blood hydrostatic pressure (GBHP) promotes filtration, whereas capsular hydrostatic pressure (CHP) and blood colloid osmotic pressure (BCOP) oppose filtration. Net filtration pressure (NFP) = GBHP − CHP − BCOP. NFP is about 10 mmHg.
6. Glomerular filtration rate (GFR) is the amount of filtrate formed in both kidneys per minute; it is normally 105–125 mL/min.
7. Glomerular filtration rate depends on renal autoregulation, neural regulation, and hormonal regulation. Table 26.2 on page 964 summarizes regulation of GFR.

TUBULAR REABSORPTION AND TUBULAR SECRETION (p. 965)

1. Tubular reabsorption is a selective process that reclaims materials from tubular fluid and returns them to the bloodstream. Reabsorbed substances include water, glucose, amino acids, urea, and ions, such as sodium, chloride, potassium, bicarbonate, and phosphate (Table 26.3 on page 965).
2. Some substances not needed by the body are removed from the blood and discharged into the urine via tubular secretion. Included are ions (K^+, H^+, and NH_4^+), urea, creatinine, and certain drugs.
3. Reabsorption routes include both paracellular (between tubule cells) and transcellular (across tubule cells) routes.
4. The maximum amount of a substance that can be reabsorbed per unit time is called the transport maximum (T_m).
5. About 90% of water reabsorption is obligatory; it occurs via osmo-

sis, together with reabsorption of solutes, and is not hormonally regulated. The remaining 10% is facultative water reabsorption, which varies according to body needs and is regulated by ADH.

6. Na^+ are reabsorbed throughout the basolateral membrane via primary active transport.
7. In the proximal convoluted tubule, sodium ions are reabsorbed through the apical membranes via Na^+-glucose symporters and Na^+/H^+ antiporters; water is reabsorbed via osmosis; Cl^-, K^+, Ca^{2+}, Mg^{2+}, and urea are reabsorbed via passive diffusion; and NH_3 and NH_4^+ are secreted.
8. The loop of Henle reabsorbs 20–30% of the filtered Na^+, K^+, Ca^{2+}, and HCO_3^-; 35% of the filtered Cl^-, and 15% of the filtered water.
9. The distal convoluted tubule reabsorbs sodium and chloride ions via Na^+-Cl^- symporters.
10. In the collecting duct, principal cells reabsorb Na^+ and secrete K^+, whereas intercalated cells reabsorb K^+ and HCO_3^- and secrete H^+.
11. Angiotensin II, aldosterone, antidiuretic hormone, and atrial natriuretic peptide regulate solute and water reabsorption, as summarized in Table 26.4 on page 972.

PRODUCTION OF DILUTE AND CONCENTRATED URINE (p. 972)

1. In the absence of ADH, the kidneys produce dilute urine; renal tubules absorb more solutes than water.
2. In the presence of ADH, the kidneys produce concentrated urine; large amounts of water are reabsorbed from the tubular fluid into interstitial fluid, increasing solute concentration of the urine.
3. The countercurrent mechanism establishes an osmotic gradient in the interstitial fluid of the renal medulla that enables production of concentrated urine when ADH is present.

EVALUATION OF KIDNEY FUNCTION (p. 977)

1. A urinalysis is an analysis of the volume and physical, chemical, and microscopic properties of a urine sample. Table 26.5 on page 977 summarizes the principal physical characteristics of urine.
2. Chemically, normal urine contains about 95% water and 5% solutes. The solutes normally include urea, creatinine, uric acid, urobilinogen, and various ions.
3. Table 26.6 on page 978 lists several abnormal components that can be detected in a urinalysis, including albumin, glucose, red and white blood cells, ketone bodies, bilirubin, excessive urobilinogen, casts, and microbes.
4. Renal clearance refers to the ability of the kidneys to clear (remove) a specific substance from blood.

URINE TRANSPORTATION, STORAGE, AND ELIMINATION (p. 979)

1. The ureters are retroperitoneal and consist of a mucosa, muscularis, and adventitia. They transport urine from the renal pelvis to the urinary bladder, primarily via peristalsis.
2. The urinary bladder is located in the pelvic cavity posterior to the pubic symphysis; its function is to store urine before micturition.
3. The urinary bladder consists of a mucosa with rugae, a muscularis (detrusor muscle), and an adventitia (serosa over the superior surface).
4. The micturition reflex discharges urine from the urinary bladder

via parasympathetic impulses that cause contraction of the detrusor muscle and relaxation of the internal urethral sphincter muscle and via inhibition of impulses in somatic motor neurons to the external urethral sphincter.

5.   The urethra is a tube leading from the floor of the urinary bladder to the exterior. Its anatomy and histology differ in females and males. In both sexes, the urethra functions to discharge urine from the body; in males, it discharges semen as well.

WASTE MANAGEMENT IN OTHER BODY SYSTEMS (p. 981)

1.   Besides the kidneys, several other tissues, organs, and processes temporarily confine wastes, transport waste materials for disposal, recycle materials, and excrete excess or toxic substances.
2.   Buffers bind excess H^+, the blood transports wastes, the liver converts toxic substances into less toxic ones, the lungs exhale CO2, sweat glands help eliminate excess heat, and the gastrointestinal tract eliminates solid wastes.

DEVELOPMENT OF THE URINARY SYSTEM (p. 982)

1.   The kidneys develop from intermediate mesoderm.
2.   The kidneys develop in the following sequence: pronephros, mesonephros, and metanephros. Only the metanephros remains and develops into a functional kidney.

AGING AND THE URINARY SYSTEM (p. 982)

1.   With aging, the kidneys shrink in size, have a decreased blood flow, and filter less blood.
2.   Common problems related to aging include urinary tract infections, increased frequency of urination, urinary retention or incontinence, and renal calculi.

Q SELF-QUIZ QUESTIONS

Fill in the blanks in the following statements.

1.   The _____ cells of the ureters and urinary bladder allow distension for urine flow and storage.
2.   Micturition is both a _____ and an _____ reflex.
3.   The three basic processes carried out by nephrons are, _____, _____, and _____.

Indicate whether the following statements are true or false.

4.   When dilute urine is being formed, the osmolarity of the fluid in the tubular lumen increases as it flows down the descending limb of the loop of Henle, decreases as it flows up the ascending limb, and continues to decrease as it flows through the rest of the nephron and collecting duct.
5.   The route of blood flow from the renal artery to the renal vein is renal artery, segmental arteries, interlobar arteries, arcuate arteries, interlobular arteries, afferent arterioles, glomeruli, efferent arterioles, peritubular capillaries, vasa recta, peritubular venules, interlobular veins, arcuate veins, interlobar veins, and renal vein.

Choose the one best answer to the following questions.

6.   Which of the following are features of the renal corpuscle that enhance its filtering capacity? (1) large glomerular capillary surface area, (2) thick, selectively permeable filtration membrane, (3) low glomerular capillary pressure, (4) high glomerular capillary pressure, (5) thin and porous filtration membrane. (a) 1, 2, and 3, (b) 2, 4, and 5, (c) 1, 4, and 5, (d) 2, 3, and 4, (e) 2, 3, and 5.
7.   The main pressure that promotes glomerular filtration is (a) glomerular blood hydrostatic pressure, (b) capsular hydrostatic pressure, (c) blood colloid osmotic pressure, (d) capsular osmotic pressure.
8.   Which of the following statements are *correct*? (1) Glomerular filtration rate (GFR) is directly related to the pressures that determine net filtration pressure. (2) Angiotensin II and atrial natriuretic peptide help regulate GFR. (3) Mechanisms that regulate GFR work by adjusting blood flow into and out of the glomerulus and by altering the glomerular capillary surface area available for fil-

tration. (4) GFR increases when blood flow into glomerular capillaries decreases. (5) Normally, GFR increases very little when systemic blood pressure rises. (a) 1, 2, and 3, (b) 2, 3, and 4, (c) 3, 4, and 5, (d) 1, 2, 3, and 5, (e) 2, 3, 4, and 5.

9.   Which of the following are mechanisms that control GFR? (1) renal autoregulation, (2) neural regulation, (3) hormonal regulation, (4) chemical regulation of ions, (5) presence or absence of a transporter. (a) 1, 2, and 3, (b) 2, 3, and 4, (c) 3, 4, and 5, (d) 1, 3, and 5, (e) 1, 3, and 4.
10.   Which of the following hormones affect Na^+, Cl^-, and water reabsorption and K^+ secretion by the renal tubules? (1) angiotensin II, (2) aldosterone, (3) ADH, (4) atrial natriuretic peptide, (5) thyroid hormone. (a) 1, 3, and 5, (b) 2, 3, and 4, (c) 2, 4, and 5, (d) 1, 2, 4, and 5, (e) 1, 2, 3, and 4.
11.   Which of the following is the correct sequence of tubular fluid flow? (a) proximal convoluted tubule, descending limb of loop of Henle, ascending limb of loop of Henle, collecting duct, proximal convoluted tubule, papillary duct, (b) proximal convoluted tubule, descending limb of loop of Henle, ascending limb of loop of Henle, distal convoluted tubule, collecting duct, papillary duct, (c) papillary duct, distal convoluted tubule, proximal convoluted tubule, collecting duct, ascending limb of loop of Henle, descending limb of loop of Henle, (d) collecting duct, proximal convoluted tubule, ascending limb of loop of Henle, descending limb of loop of Henle, papillary duct, distal convoluted tubule, (e) distal convoluted tubule, ascending limb of loop of Henle, descending limb of loop of Henle, proximal convoluted tubule, papillary duct, collecting duct.
12.   Which of the following are functions of the kidneys? (1) control of blood pH, (2) excretion of wastes, (3) regulation of blood ionic composition, (4) regulation of white blood cell production, (5) regulation of blood volume. (a) 1, 2, 3, and 4, (b) 1, 2, and 5, (c) 1, 2, 3, and 5, (d) 2, 3, and 5, (e) 2, 3, 4, and 5.

13. Match the following:

_____ (a) cells in the last portion of the distal convoluted tubule and in the collecting ducts; regulated by ADH and aldosterone

_____ (b) a capillary network lying in Bowman's capsule and functioning in filtration

_____ (c) the functional unit of the kidney

_____ (d) drains into a collecting duct

_____ (e) combined glomerulus and glomerular capsule; where plasma is filtered

_____ (f) the visceral layer of the glomerular capsule consisting of modified simple squamous epithelial cells

_____ (g) cells of the final portion of the ascending limb of the loop of Henle that make contact with the afferent arteriole

_____ (h) site of obligatory water reabsorption

_____ (i) can secrete H^+ against a concentration gradient

_____ (j) modified smooth muscle cells in the wall of the afferent arteriole

(1) podocytes
(2) glomerulus
(3) renal corpuscle
(4) proximal convoluted tubule
(5) distal convoluted tubule
(6) juxtaglomerular cells
(7) macula densa
(8) principal cells
(9) intercalated cells
(10) nephron

14. Match the following:

_____ (a) measure of blood nitrogen resulting from the catabolism and deamination of amino acids

_____ (b) produced from the catabolism of creatine phosphate in skeletal muscle

_____ (c) volume of blood that is cleared of a substance per unit of time

_____ (d) can result from diabetes mellitus

_____ (e) insoluble stones of crystallized salts

_____ (f) indicates infection in the urinary system

_____ (g) can be caused by damage to the filtration membranes

(1) pyuria
(2) renal calculi
(3) plasma creatinine
(4) BUN test
(5) albuminuria
(6) glucosuria
(7) renal plasma clearance

15. Match the following:

_____ (a) membrane proteins that function as water channels

_____ (b) a secondary active transport process that achieves Na^+ reabsorption, returns filtered HCO_3^- and water to the peritubular capillaries, and secretes H^+

_____ (c) stimulates principal cells to secrete more K^+ into tubular fluid

_____ (d) reduces glomerular filtration rate; increases blood volume and pressure

_____ (e) increases glomerular filtration rate

_____ (f) regulates facultative water reabsorption by increasing the water permeability of principal cells

_____ (g) reabsorb Na^+ together with a variety of other solutes

(1) angiotensin II
(2) atrial natriuretic peptide
(3) Na^+ symporters
(4) Na^+/H^+ antiporters
(5) aquaporins
(6) aldosterone
(7) ADH

 CRITICAL THINKING QUESTIONS

1. Imagine the discovery of a new toxin that blocks renal tubule reabsorption but does not affect filtration. Predict the short-term effects of this toxin.

   HINT *Normally about 99% of glomerular filtrate is reabsorbed.*

2. While traveling down the interstate, little Caitlin told her father to stop the car "Now! I'm so full I'm gonna burst!" Her dad thinks she can make it to the next rest stop. What structures of the urinary bladder will help her "hold it"?

   HINT *Children often visit the emergency room with broken bones, but not with exploding bladders.*

3. Although urinary catheters come in one length only, the number of centimeters of the catheter that must be inserted to reach the bladder and release urine differs significantly in males and females. Why?

   HINT *Why do men and women usually assume different positions when they urinate?*

ANSWERS TO FIGURE QUESTIONS

26.1 The kidneys, ureters, urinary bladder, and urethra are the components of the urinary system.

26.2 The kidneys are retroperitoneal because they are posterior to the peritoneum.

26.3 Blood vessels, lymphatic vessels, nerves, and a ureter pass through the renal hilus.

26.4 About 1200 mL of blood enters the renal arteries each minute.

26.5 Cortical nephrons have glomeruli in the superficial renal cortex, and their short loops of Henle penetrate only into the superficial renal medulla. Juxtamedullary nephrons have glomeruli deep in the renal cortex, and their long loops of Henle extend through the renal medulla nearly to the renal papilla.

26.6 This section must pass through the renal cortex because there are no renal corpuscles in the renal medulla.

26.7 Secreted penicillin is being removed from the bloodstream.

26.8 Endothelial fenestrations (pores) in glomerular capillaries are too small for red blood cells to pass through.

26.9 Obstruction of the right ureter would increase CHP and thus decrease NEP in the right kidney; the obstruction would have no effect on the left kidney.

26.10 *Auto* means self; tubuloglomerular feedback is an example of autoregulation because it takes place entirely within the kidneys.

26.11 The tight junctions form a barrier that prevents diffusion of transporter, channel, and pump proteins between the apical and basolateral membranes.

26.12 Glucose enters a PCT cell via a Na^+-glucose symporter in the apical membrane and leaves via facilitated diffusion through the basolateral membrane.

26.13 The electrochemical gradient promotes movement of Na^+ into the tubule cell through the apical membrane antiporters.

26.14 Reabsorption of the solutes creates an osmotic gradient that promotes the reabsorption of water via osmosis.

26.15 This is secondary active transport because the symporter uses the energy stored in the concentration gradient of Na^+ between extracellular fluid and the cytosol. No water is reabsorbed here because the thick ascending limb of the loop of Henle is virtually impermeable to water.

26.16 In principal cells, aldosterone stimulates secretion of K^+ and reabsorption of Na^+ by increasing the activity of sodium-potassium pumps and number of leakage channels for Na^+ and K^+.

26.17 Aldosterone and atrial natriuretic peptide also influence renal water reabsorption.

26.18 Dilute urine is produced when the thick ascending limb of the loop of Henle, the distal convoluted tubule, and the collecting duct reabsorb more solutes than water.

26.19 The high osmolarity of interstitial fluid in the renal medulla is due mainly to Na^+, Cl^-, and urea.

26.20 Secretion occurs in the proximal convoluted tubule, the loop of Henle, and the collecting duct.

26.21 Lack of voluntary control over micturition is termed urinary incontinence.

26.22 The kidneys start to develop during the third week of development.

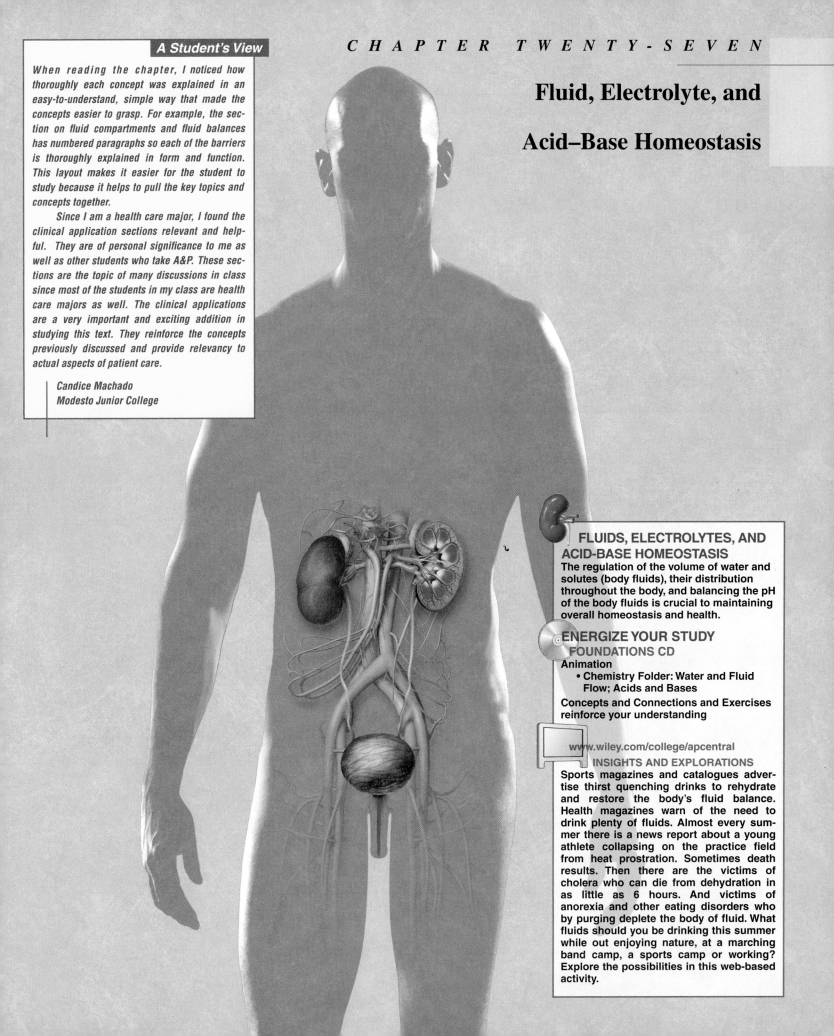

Fluid, Electrolyte, and

Acid–Base Homeostasis

FLUIDS, ELECTROLYTES, AND ACID-BASE HOMEOSTASIS
The regulation of the volume of water and solutes (body fluids), their distribution throughout the body, and balancing the pH of the body fluids is crucial to maintaining overall homeostasis and health.

ENERGIZE YOUR STUDY
FOUNDATIONS CD
Animation
 • Chemistry Folder: Water and Fluid Flow; Acids and Bases
Concepts and Connections and Exercises reinforce your understanding

www.wiley.com/college/apcentral
INSIGHTS AND EXPLORATIONS
Sports magazines and catalogues advertise thirst quenching drinks to rehydrate and restore the body's fluid balance. Health magazines warn of the need to drink plenty of fluids. Almost every summer there is a news report about a young athlete collapsing on the practice field from heat prostration. Sometimes death results. Then there are the victims of cholera who can die from dehydration in as little as 6 hours. And victims of anorexia and other eating disorders who by purging deplete the body of fluid. What fluids should you be drinking this summer while out enjoying nature, at a marching band camp, a sports camp or working? Explore the possibilities in this web-based activity.

n Chapter 26 you learned how the kidneys form urine. One important function of the kidneys is to help maintain fluid balance in the body. The water and dissolved solutes in the body constitute the **body fluids.** Regulatory mechanisms involving the kidneys and other organs normally maintain homeostasis of the body fluids. Malfunction in any or all of them may seriously endanger the functioning of organs throughout the body. In this chapter, we will explore the mechanisms that regulate the volume and distribution of body fluids and examine the factors that determine the concentrations of solutes and the pH of body fluids.

FLUID COMPARTMENTS AND FLUID BALANCE

> OBJECTIVES

- Compare the locations of intracellular fluid (ICF) and extracellular fluid (ECF), and describe the various fluid compartments of the body.
- Describe the sources of water and solute gain and loss, and explain how each is regulated.

In lean adults, body fluids constitute between 55% and 60% of total body mass in females and males, respectively (Figure 27.1). Fluids are present in two main "compartments"—inside cells and outside cells. About two-thirds of body fluid is **intracellular fluid (ICF)** (*intra-* = within) or **cytosol,** the fluid within cells. The other third, called **extracellular fluid (ECF)** (*extra-* = outside), is outside cells and includes all other body fluids. About 80% of the ECF is **interstitial fluid** (*inter-* = between), which occupies the microscopic spaces between tissue cells, and 20% of the ECF is **plasma,** the liquid portion of the blood. Extracellular fluids also include lymph in lymphatic vessels; cerebrospinal fluid in the nervous system; synovial fluid in joints; aqueous humor and vitreous body in the eyes; endolymph and perilymph in the ears; and pleural, pericardial, and peritoneal fluids between serous membranes. Because most of these fluids are similar to interstitial fluid, they are grouped together with it.

Figure 27.1 Body fluid compartments.

The term body fluid refers to body water and its dissolved substances.

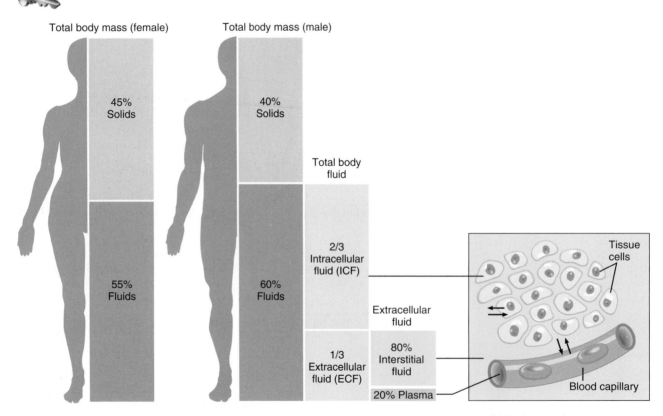

(a) Distribution of body solids and fluids in an average lean, adult female and male

(b) Exchange of water among body fluid compartments

 What is the approximate volume of blood plasma in a lean 60-kg male? In a lean 60-kg female? *(NOTE: One liter of body fluid has a mass of 1 kilogram.)*

Two general "barriers" separate intracellular fluid, interstitial fluid, and blood plasma.

1. The *plasma membrane* of individual cells separates intracellular fluid from the surrounding interstitial fluid. You learned in Chapter 3 that the plasma membrane is a selectively permeable barrier: It allows some substances to cross but blocks the movement of other substances. In addition, active transport pumps work continuously to maintain different concentrations of certain ions in the cytosol and interstitial fluid.

2. *Blood vessel walls* divide the interstitial fluid from blood plasma. Only in capillaries, the smallest blood vessels, are the walls thin enough and leaky enough to permit the exchange of water and solutes between blood plasma and interstitial fluid.

The body is in **fluid balance** when the required amounts of water and solutes are present and are correctly proportioned among the various compartments. **Water** is by far the largest single component of the body, making up 45–75% of total body mass, depending on age and gender. Fatter people have proportionally less water than leaner people because water comprises less than 20% of the mass of adipose tissue. Skeletal muscle tissue, in contrast, is about 65% water. Infants have the highest percentage of water, up to 75% of body mass. The percentage of body mass that is water decreases until about 2 years of age. Until puberty, water accounts for about 60% of body mass in boys and girls. In lean adult males, water still accounts for about 60% of body mass. However, even lean females have more subcutaneous fat than do males. Thus, their percentage of total body water is lower, accounting for about 55% of body mass.

The processes of filtration, reabsorption, diffusion, and osmosis provide for continual exchange of water and solutes among body fluid compartments (Figure 27.1b). Yet, the volume of fluid in each compartment remains remarkably stable. The pressures that promote filtration of fluid from blood capillaries and reabsorption of fluid back into capillaries can be reviewed in Figure 21.7 on page 704. Because osmosis is the primary means of water movement between intracellular fluid and interstitial fluid, the concentration of solutes in these fluids determines the *direction* of water movement. Most solutes in body fluids are **electrolytes,** inorganic compounds that dissociate into ions. Fluid balance depends primarily on electrolyte balance so the two are closely interrelated. Because intake of water and electrolytes rarely occurs in exactly the same proportions as their presence in body fluids, the ability of the kidneys to excrete excess water by producing dilute urine, or to excrete excess electrolytes by producing concentrated urine, is of utmost importance.

Sources of Body Water Gain and Loss

The body can gain water by ingestion and by metabolic synthesis (Figure 27.2). The main sources of body water are ingested

Figure 27.2 Sources of daily water gain and loss under normal conditions. Numbers are average volumes for adults.

 Normally, daily water loss equals daily water gain.

How does each of the following affect fluid balance: Hyperventilation? Vomiting? Fever? Diuretics?

liquids (about 1600 mL) and moist foods (about 700 mL) absorbed from the gastrointestinal (GI) tract, which total about 2300 mL/day. The other source of water is **metabolic water** that is produced in the body mainly when electrons are accepted by oxygen during aerobic cellular respiration (see Figure 25.2 on page 910) and to a smaller extent during dehydration synthesis reactions (see Figure 2.15 on page 44). Metabolic water gain accounts for only 200 mL/day. Daily water gain from these two sources totals about 2500 mL.

Normally, body fluid volume remains constant because water loss equals water gain. Water loss occurs in four ways (Figure 27.2). Each day the kidneys excrete about 1500 mL in urine, the skin evaporates about 600 mL (400 mL through insensible perspiration and 200 mL as sweat), the lungs exhale about 300 mL as water vapor, and the gastrointestinal tract eliminates about 100 mL in feces. In women of reproductive age, additional water is lost in menstrual flow. On average, daily water loss totals about 2500 mL. The amount of water lost by a given route can vary considerably over time. For example, water may literally pour from the skin in the form of sweat during strenuous exertion. In other cases, water may be lost in diarrhea during a GI tract infection.

Regulation of Body Water Gain

The rate of formation of metabolic water is not regulated to maintain homeostasis of body fluids. Rather, the volume of metabolic water formed depends on the level of aerobic cellular respiration, which reflects the demand for ATP in body cells. When more ATP is produced, more water is formed. Regulation of body water gain depends mainly on regulating the volume of water intake, in other words by drinking more or less fluid. An area in the hypothalamus known as the **thirst center** governs the urge to drink.

When water loss is greater than water gain, **dehydration**—a decrease in volume and an increase in osmolarity of body fluids—stimulates thirst (Figure 27.3). When body mass decreases by 2% due to fluid loss, mild dehydration exists. A decrease in blood volume causes blood pressure to fall. This change stimulates the kidneys to release renin, which promotes the formation of angiotensin II. Increased nerve impulses from osmoreceptors in the hypothalamus, triggered by increased blood osmolarity, and increased angiotensin II in the blood both stimulate the thirst center in the hypothalamus. Other signals that stimulate thirst come from (1) neurons in the mouth that detect dryness due to a decreased flow of saliva and (2) baroreceptors that detect lowered blood pressure in the heart and blood vessels. As a result, the sensation of thirst increases, which usually leads to increased fluid intake (if fluids are available) and restoration of normal fluid volume. Overall, fluid gain balances fluid loss. Sometimes, however, either the sensation of thirst does not occur quickly enough or access to fluids is restricted, and significant dehydration ensues. This happens most often in elderly people, in infants, and in those who are in a confused mental state. In situations where heavy sweating or fluid loss from diarrhea or vomiting occurs, it is wise to start replacing body fluids by drinking fluids even before the sensation of thirst appears.

Regulation of Water and Solute Loss

Even though the loss of water and solutes through sweating and exhalation increases during exercise, elimination of *excess* body water or solutes occurs mainly by controlling how much are lost in urine. As you will see, the extent of *urinary salt (NaCl) loss* is the main factor that determines body fluid *volume*. The reason is that in osmosis "water follows solutes," and the two main solutes in extracellular fluid (and in urine) are sodium ions (Na^+) and chloride ions (Cl^-). In a similar way, the main factor that determines body fluid *osmolarity* is the extent of *urinary water loss*.

Because our daily diet contains a highly variable amount of NaCl, urinary excretion of Na^+ and Cl^- must also vary to maintain homeostasis. Hormonal changes regulate the urinary loss of these ions, which in turn affects blood volume. Figure 27.4 depicts the sequence of changes that occur after a salty meal. The increased intake of NaCl produces an increase in plasma levels of Na^+ and Cl^- (the major contributors to osmolarity of extracellular fluid). As a result, the osmolarity of inter-

Figure 27.3 Pathways through which dehydration stimulates thirst.

🔑 Dehydration occurs when water loss is greater than water gain.

```
                        ┌──────────────┐
                        │  Dehydration │
                        └──────────────┘
         ┌─────────────────────┼─────────────────────┐
         ▼                     ▼                       ▼
┌────────────────┐   ┌────────────────┐    ┌────────────────┐
│ Decreased      │   │ Increased blood│    │ Decreased      │
│ flow of saliva │   │ osmolarity     │    │ blood volume   │
└────────────────┘   └────────────────┘    └────────────────┘
         ▼                     ▼                       ▼
┌────────────────┐   ┌────────────────┐    ┌────────────────┐
│ Dry mouth      │   │ Stimulates     │    │ Decreased      │
│ and            │   │ osmoreceptors  │    │ blood          │
│ pharynx        │   │ in hypothalamus│    │ pressure       │
└────────────────┘   └────────────────┘    └────────────────┘
         │                     │                       ▼
         │                     │           ┌────────────────┐
         │                     │           │ Increased      │
         │                     │           │ renin release  │
         │                     │           │ by juxta-      │
         │                     │           │ glomerular cells│
         │                     │           │ of kidneys     │
         │                     │           └────────────────┘
         │                     │                       ▼
         │                     │           ┌────────────────┐
         │                     │           │ Increased      │
         │                     │           │ angiotensin II │
         │                     │           │ formation      │
         │                     │           └────────────────┘
         │                     ▼                       │
         │           ┌────────────────┐                │
         └──────────▶│ Stimulates     │◀───────────────┘
                     │ thirst center in│
                     │ hypothalamus   │
                     └────────────────┘
                             ▼
                     ┌────────────────┐
                     │ Increases thirst│
                     └────────────────┘
                             ▼
                     ┌────────────────┐
                     │ Increases water│
                     │ intake         │
                     └────────────────┘
                             ▼
                     ┌────────────────┐
                     │ Increases body │
                     │ water to normal│
                     │ level and relieves│
                     │ dehydration    │
                     └────────────────┘
```

 Does regulation of these pathways occur via negative or positive feedback? Why?

stitial fluid increases, which causes movement of water from intracellular fluid into interstitial fluid and then into plasma. Such water movement increases blood volume.

The three most important hormones that regulate the extent of renal Na^+ and Cl^- reabsorption (and thus how much is lost in the urine) are **angiotensin II, aldosterone,** and **atrial natriuretic peptide (ANP).** During dehydration angiotensin II and aldosterone promote urinary reabsorption of Na^+ and Cl^- (and water by osmosis with the electrolytes) and thereby conserve the volume of body fluids by reducing urinary loss. An increase in

Figure 27.4 Hormonal changes following increased NaCl intake. The resulting reduced blood volume is achieved by increasing urinary loss of Na⁺ and Cl⁻.

🔑 **The three main hormones that regulate renal Na⁺ and Cl⁻ reabsorption (and thus the amount lost in the urine) are angiotensin II, aldosterone, and atrial natriuretic peptide.**

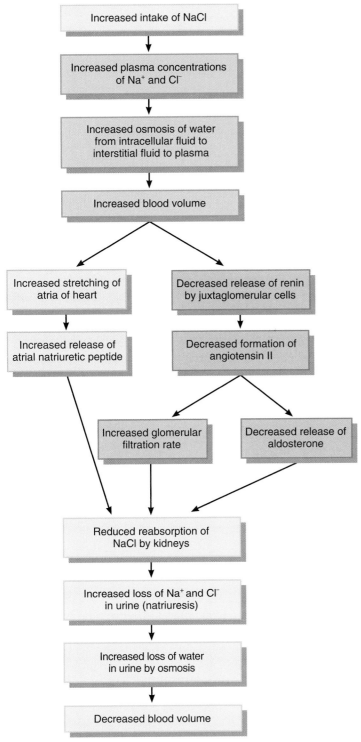

❓ **How does hyperaldosteronism (excessive aldosterone secretion) cause edema?**

blood volume, as might occur after drinking a large volume of fluid, stretches the atria of the heart and promotes release of atrial natriuretic peptide. ANP promotes **natriuresis,** elevated urinary excretion of Na⁺ (and Cl⁻), which decreases blood volume. An increase in blood volume also slows release of renin from juxtaglomerular cells of the kidneys. When renin level declines, less angiotensin II is formed. Decline in angiotensin II from a moderate level to a low level increases glomerular filtration rate and reduces Na⁺, Cl⁻, and water reabsorption in the kidney tubules. In addition, less angiotensin II leads to less aldosterone, which causes reabsorption of filtered Na⁺ and Cl⁻ to slow in the renal collecting ducts. More filtered Na⁺ and Cl⁻ thus remain in the tubular fluid to be excreted in the urine. The osmotic consequence of excreting more Na⁺ and Cl⁻ is loss of more water in urine, which decreases blood volume and blood pressure.

The major hormone that regulates water loss is **antidiuretic hormone (ADH).** This hormone, also known as **vasopressin,** is produced by neurosecretory cells that extend from the hypothalamus to the posterior pituitary. Besides stimulating the thirst mechanism, an increase in the osmolarity of body fluids stimulates release of ADH (see Figure 26.17 on page 971). ADH promotes the insertion of water-channel proteins (aquaporin-2) into the apical membranes of principal cells in the collecting ducts of the kidneys. As a result, the permeability of these cells to water increases. Water molecules move by osmosis from the renal tubular fluid into the cells and then from the cells into the bloodstream. The result is production of a small volume of very concentrated urine. By contrast, intake of plain water decreases the osmolarity of blood and interstitial fluid. Within minutes, ADH secretion shuts down, and soon its blood level is close to zero. When the principal cells are not stimulated by ADH, aquaporin-2 molecules are removed from the apical membrane by endocytosis. As the number of water channels decreases, the water permeability of the principal cells' apical membrane falls, and more water is lost in the urine.

Under some conditions, factors other than blood osmolarity influence ADH secretion. A large decrease in blood volume, which is detected by baroreceptors (sensory neurons that respond to stretching) in the left atrium and in blood vessel walls, also stimulates ADH release. In severe dehydration, glomerular filtration rate decreases because blood pressure falls, and less water is lost in the urine. Conversely, the intake of too much water increases blood pressure, the rate of glomerular filtration rises, and more water is lost in the urine. Hyperventilation (abnormally fast and deep breathing) can increase fluid loss through the exhalation of more water vapor. Vomiting and diarrhea result in fluid loss from the GI tract. Finally, fever, heavy sweating, and destruction of extensive areas of the skin from burns can cause excessive water loss through the skin. In all of these conditions, an increase in ADH secretion will assist the conservation of body fluids.

Table 27.1 summarizes the factors that maintain body water balance.

Table 27.1	Summary of Factors that Maintain Body Water Balance	
Factor	**Mechanism**	**Effect**
Thirst center in hypothalamus	Stimulates desire to drink fluids.	Water gain if thirst is quenched.
Angiotensin II	Stimulates secretion of aldosterone.	Reduces loss of water in urine.
Aldosterone	By promoting urinary reabsorption of Na^+ and Cl^-, increases water reabsorption via osmosis.	Reduces loss of water in urine.
Atrial natriuretic peptide (ANP)	Promotes natriuresis, elevated urinary excretion of Na^+ (and Cl^-), accompanied by water.	Increases loss of water in urine.
Antidiuretic hormone (ADH), also known as **vasopressin**	Promotes insertion of water-channel proteins (aquaporin-2) into the apical membranes of principal cells in the collecting ducts of the kidneys. As a result, the water permeability of these cells increases and more water is reabsorbed.	Reduces loss of water in urine.

Figure 27.5 Series of events in water intoxication.

 Water intoxication is a state in which excessive body water causes cells to become hypotonic and swell.

Excessive blood loss, sweating, vomiting, or diarrhea coupled with intake of plain water

↓

Decreased Na^+ concentration of interstitial fluid and plasma (hyponatremia)

↓

Decreased osmolarity of interstitial fluid and plasma

↓

Osmosis of water from interstitial fluid into intracellular fluid

↓

Water intoxication (cells swell)

↓

Convulsions, coma, and possible death

 Why do solutions used for oral rehydration therapy contain a small amount of table salt (NaCl)?

Movement of Water Between Body Fluid Compartments

Normally, cells neither shrink nor swell because intracellular and interstitial fluids have the same osmolarity. Changes in the osmolarity of interstitial fluid, however, causes fluid imbalances. An increase in the osmolarity of interstitial fluid draws water out of cells, and they shrink slightly. A decrease in the osmolarity of interstitial fluid, by contrast, causes cells to swell. Changes in osmolarity most often result from changes in the concentration of Na^+.

A decrease in the osmolarity of interstitial fluid, as may occur after drinking a large volume of water, inhibits secretion of ADH. Normally, the kidneys then excrete a large volume of dilute urine, which restores the osmotic pressure of body fluids to normal. As a result, body cells swell only slightly, and only for a brief period. But when a person steadily consumes water faster than the kidneys can excrete it (the maximum urine flow rate is about 15 mL/min) or when renal function is poor, the result may be **water intoxication,** a state in which excessive body water causes cells to become hypotonic and to swell dangerously (Figure 27.5). If the body water and Na^+ lost during blood loss or excessive sweating, vomiting, or diarrhea is replaced by drinking plain water, then body fluids become more dilute. This dilution can cause the Na^+ concentration of plasma

and then of interstitial fluid to fall below the normal range (hyponatremia). When the Na^+ concentration of interstitial fluid decreases, its osmolarity also falls. The net result is osmosis of water from interstitial fluid into the cytosol. Water entering the cells causes them to become hypotonic and swell, producing convulsions, coma, and possibly death. To prevent this dire sequence of events in cases of severe electrolyte and water loss, solutions given for intravenous or oral rehydration therapy (ORT) include a small amount of table salt (NaCl).

Enemas and Fluid Balance

An **enema** (EN-e-ma) is the introduction of a solution into the rectum and colon to stimulate bowel activity and evacuate feces. Repeated enemas, especially in young children, increase the risk of fluid and electrolyte imbalances. Because tap water (which is hypotonic) and commercially prepared hypertonic solutions may cause rapid movement of fluid between compartments, an isotonic solution is preferred. ■

▶ CHECKPOINT

1. What is the approximate volume of each of *your* body fluid compartments?

2. Which routes of water gain and loss from the body are regulated?

3. How does thirst help regulate water intake?

4. How do angiotensin II, aldosterone, atrial natriuretic peptide, and antidiuretic hormone regulate the volume and osmolarity of body fluids?

5. What factors govern the movement of water between interstitial fluid and intracellular fluid?

ELECTROLYTES IN BODY FLUIDS

▶ O B J E C T I V E S

- **Compare the electrolyte composition of the three major fluid compartments: plasma, interstitial fluid, and intracellular fluid.**

- **Discuss the functions of sodium, chloride, potassium, bicarbonate, calcium, phosphate, and magnesium ions, and explain how their concentrations are regulated.**

The ions formed when electrolytes dissolve and dissociate serve four general functions in the body: (1) Because they are largely confined to particular fluid compartments and are more numerous than nonelectrolytes, certain ions *control the osmosis of water between fluid compartments.* (2) Ions *help maintain the acid–base balance* required for normal cellular activities. (3) Ions *carry electrical current,* which allows production of action potentials and graded potentials and controls secretion of some hormones and neurotransmitters. (4) Several ions *serve as cofactors* needed for optimal activity of enzymes.

Concentrations of Electrolytes in Body Fluids

To compare the charge carried by ions in different solutions, the concentration of ions is typically expressed in units of **milliequivalents per liter (mEq/liter).** These units give the concentration of cations or anions in a given volume of solution. One equivalent is the positive or negative charge equal to the amount of charge in one mole of H^+; a milliequivalent is one-thousandth of an equivalent. Recall that a mole of a substance is its molecular weight expressed in grams. For ions such as sodium (Na^+), potassium (K^+), and bicarbonate (HCO_3^-), which have a single positive or negative charge, the number of mEq/liter is equal to the number of mmol/liter. For ions such as calcium (Ca^{2+}) or phosphate (HPO_4^{2-}), which have two positive or negative charges, the number of mEq/liter is twice the number of mmol/liter.

Figure 27.6 compares the concentrations of the main electrolytes and protein anions in blood plasma, interstitial fluid, and intracellular fluid. The chief difference between the two extracellular fluids—blood plasma and interstitial fluid—is that

Figure 27.6 Electrolyte and protein anion concentrations in plasma, interstitial fluid, and intracellular fluid. The height of each column represents the milliequivalents per liter (mEq/liter).

The electrolytes present in extracellular fluids are different from those present in intracellular fluid.

Key:
- Blood plasma ⎤ Extracellular
- Interstitial fluid ⎦ fluids
- Intracellular fluid

What are the major cation and the two major anions in ECF and ICF?

blood plasma contains many protein anions, whereas interstitial fluid has very few. Because normal capillary membranes are virtually impermeable to proteins, only a few plasma proteins leak out of blood vessels into the interstitial fluid. This difference in protein concentration is largely responsible for the blood colloid osmotic pressure exerted by blood plasma. In other respects, the two fluids are similar.

The electrolyte content of intracellular fluid, by contrast, differs considerably from that of extracellular fluid. In extracellular fluid, the most abundant cation is Na^+, and the most abundant anion is Cl^-. In intracellular fluid, the most abundant cation is K^+, and the most abundant anions are proteins and HPO_4^{2-}. By actively transporting Na^+ out of cells and K^+ into cells, sodium-potassium pumps (Na^+/K^+ ATPase) play a major role in maintaining the high intracellular concentration of K^+ and high extracellular concentration of Na^+.

Sodium

Sodium ions (Na^+) are the most abundant ions in extracellular fluid, accounting for 90% of the extracellular cations. The normal blood plasma Na^+ concentration is 136–148 mEq/liter. As we have already seen, Na^+ plays a pivotal role in fluid and electrolyte balance because it accounts for almost half of the osmolarity (142 of about 290 mOsm/liter) of extracellular fluid. The flow of Na^+ through voltage-gated channels in the plasma membrane also is necessary for the generation and conduction of action potentials in neurons and muscle fibers. The typical daily intake of Na^+ in North America often far exceeds the body's normal daily requirements due largely to excess dietary salt. The kidneys excrete excess Na^+, but they also can conserve it during periods of shortage.

The Na^+ level in the blood is controlled by aldosterone, antidiuretic hormone (ADH), and atrial natriuretic peptide (ANP). Aldosterone increases renal reabsorption of Na^+. When the blood plasma concentration of Na^+ drops below 135 mEq/liter, a condition called *hyponatremia,* ADH release ceases. The lack of ADH, in turn, permits greater excretion of water in urine and restoration of the normal Na^+ level in ECF. Atrial natriuretic peptide (ANP) increases Na^+ excretion by the kidneys when Na^+ level is above normal, a condition called *hypernatremia.*

Edema and Hypovolemia Indicate Na⁺ Imbalance

If excess sodium ions remain in the body because the kidneys fail to excrete enough of them, water is also osmotically retained. The result is increased blood volume, increased blood pressure, and **edema,** an abnormal accumulation of interstitial fluid. Renal failure and hyperaldosteronism (excessive aldosterone secretion) are two causes of Na^+ retention. Excessive urinary loss of Na^+, by contrast, has the osmotic effect of causing excessive water loss, which results in **hypovolemia,** an abnormally low blood volume. Hypovolemia related to Na^+ loss is most often due to inadequate secretion of aldosterone associated with either adrenal insufficiency or overly vigorous therapy with diuretic drugs. ■

Chloride

Chloride ions (Cl^-) are the most prevalent anions in extracellular fluid. The normal blood plasma Cl^- concentration is 95–105 mEq/liter. Cl^- moves relatively easily between the extracellular and intracellular compartments because most plasma membranes contain many Cl^- leakage channels and antiporters. For this reason, Cl^- can help balance the level of anions in different fluid compartments. One example is the chloride shift that occurs between red blood cells and blood plasma as the blood level of carbon dioxide either increases or decreases (see Figure 23.24 on page 836). In this case, the antiporter exchange of Cl^- for HCO_3^- maintains the correct balance of anions between ECF and ICF. Chloride ions also are part of the hydrochloric acid secreted into gastric juice. ADH helps regulate Cl^- balance in body fluids because it governs the extent of water loss in urine. Processes that increase or decrease renal reabsorption of sodium ions also affect reabsorption of chloride ions. The negatively charged Cl^- follows the positively charged Na^+ due to the electrical attraction of oppositely charged particles.

Potassium

Potassium ions (K^+) are the most abundant cations in intracellular fluid (140 mEq/liter). K^+ plays a key role in establishing the resting membrane potential and in the repolarization phase of action potentials in neurons and muscle fibers; K^+ also helps maintain normal intracellular fluid volume. When K^+ moves into or out of cells, it often is exchanged for H^+ and thereby helps regulate the pH of body fluids.

The normal blood plasma K^+ concentration is 3.5–5.0 mEq/liter and is controlled mainly by aldosterone. When blood plasma K^+ concentration is high, more aldosterone is secreted into the blood. Aldosterone then stimulates principal cells of the renal collecting ducts to secrete more K^+ so excess K^+ is lost in the urine. Conversely, when blood plasma K^+ concentration is low, aldosterone secretion decreases and less K^+ is excreted in urine. Because K^+ is needed during the repolarization phase of action potentials, abnormal K^+ levels can be lethal. For instance, *hyperkalemia* (above normal concentration of K^+ in blood) can cause death due to ventricular fibrillation.

Bicarbonate

Bicarbonate ions (HCO_3^-) are the second most prevalent extracellular anions. Normal blood plasma HCO_3^- concentration is 22–26 mEq/liter in systemic arterial blood and 23–27 mEq/liter in systemic venous blood. HCO_3^- concentration increases as blood flows through systemic capillaries. This happens because the carbon dioxide released by metabolically active cells combines with water to form carbonic acid, which dissociates into H^+ and HCO_3^-. As blood flows through pulmonary capillaries, however, the concentration of HCO_3^- decreases again as carbon dioxide is exhaled. (Figure 23.24 on page 836 shows these reactions.) Intracellular fluid also contains a small amount of HCO_3^-. As previously noted, the exchange of Cl^- for HCO_3^- helps maintain the correct balance of anions in extracellular fluid and intracellular fluid.

The kidneys are the main regulators of blood HCO_3^- concentration. The intercalated cells of the renal tubule can either form HCO_3^- and release it into the blood when the blood level is low (see Figure 27.8) or excrete excess HCO_3^- in the urine when the level in blood is too high. Changes in the blood level of HCO_3^- are considered later in this chapter in the section on acid–base balance.

Calcium

Because such a large amount of calcium is stored in bone, it is the most abundant mineral in the body. About 98% of the calcium in adults is in the skeleton and teeth, where it is combined with phosphates to form a crystal lattice of mineral salts. In body fluids, calcium is mainly an extracellular cation (Ca^{2+}). The normal concentration of free or unattached Ca^{2+} in blood plasma is 4.5–5.5 mEq/liter. In addition, about the same amount of Ca^{2+} is attached to various plasma proteins. Besides contributing to the hardness of bones and teeth, Ca^{2+} plays important roles in blood clotting, neurotransmitter release, maintenance of muscle tone, and excitability of nervous and muscle tissue.

The two main regulators of Ca^{2+} concentration in blood plasma are parathyroid hormone (PTH) and calcitriol (1,25-dihydroxy vitamin D_3), the form of vitamin D that acts as a hormone (see Figure 18.14 on page 608). A low level of Ca^{2+} in blood plasma promotes release of more PTH, which stimulates osteoclasts in bone tissue to release calcium (and phosphate) from bone matrix. Thus, PTH increases bone *resorption*. Parathyroid hormone also enhances *reabsorption* of Ca^{2+} from glomerular filtrate through renal tubule cells and back into blood, and increases production of calcitriol, which in turn increases Ca^{2+} *absorption* from food in the gastrointestinal tract.

Phosphate

About 85% of the phosphate in adults is present as calcium phosphate salts, which are structural components of bone and teeth. The remaining 15% is ionized. Three phosphate ions ($H_2PO_4^-$, HPO_4^{2-}, and PO_4^{3-}) are important intracellular anions. At the normal pH of body fluids, HPO_4^{2-} is the most prevalent form. Phosphates contribute about 100 mEq/liter of anions to intracellular fluid. HPO_4^{2-} is an important buffer of H^+, both in body fluids and in the urine. Although some are "free," most phosphate ions are covalently bound to organic molecules such as lipids (phospholipids), proteins, carbohydrates, nucleic acids (DNA and RNA), and adenosine triphosphate (ATP).

The normal blood plasma concentration of ionized phosphate is only 1.7–2.6 mEq/liter. The same two hormones that govern calcium homeostasis—parathyroid hormone (PTH) and calcitriol—also regulate the level of HPO_4^{2-} in blood plasma. PTH stimulates resorption of bone matrix by osteoclasts, which releases both phosphate and calcium ions into the bloodstream. In the kidneys, however, PTH inhibits reabsorption of phosphate ions while stimulating reabsorption of calcium ions by renal tubular cells. Thus, PTH increases urinary excretion of phosphate and lowers blood phosphate level. Calcitriol promotes absorption of both phosphates and calcium from the digestive tract.

Magnesium

In adults, about 54% of the total body magnesium is part of bone matrix as magnesium salts. The remaining 46% occurs as magnesium ions (Mg^{2+}) in intracellular fluid (45%) and extracellular fluid (1%). Mg^{2+} is the second most common intracellular cation (35 mEq/liter). Functionally, Mg^{2+} is a cofactor for certain enzymes needed for the metabolism of carbohydrates and proteins and for Na^+/K^+ ATPase (the sodium-potassium pump). Mg^{2+} is essential for normal neuromuscular activity, synaptic transmission, and myocardial functioning. In addition, secretion of PTH depends on Mg^{2+}.

Normal blood plasma Mg^{2+} concentration is low, only 1.3–2.1 mEq/liter. Several factors regulate the blood plasma level of Mg^{2+} by varying the rate at which it is excreted in the urine. The kidneys increase urinary excretion of Mg^{2+} in hypercalcemia, hypermagnesemia, increases in extracellular fluid volume, decreases in parathyroid hormone, and acidosis. The opposite conditions decrease renal excretion of Mg^{2+}.

Table 27.2 describes the imbalances that result from the deficiency or excess of several electrolytes.

Those at Risk for Fluid and Electrolyte Imbalances

In a number of situations, people may be at risk for fluid and electrolyte imbalances. Those at risk include people who depend on others for fluid and food, such as infants, the elderly, and the hospitalized; and individuals undergoing medical treatment that involves intravenous infusions, drainages or suctions, and urinary catheters. People who receive diuretics experience excessive fluid losses and require increased fluid intake, or experience fluid retention and have fluid restrictions are also at risk. Finally at risk are postoperative individuals, severe burn or trauma cases, individuals with chronic diseases (congestive heart failure, diabetes, chronic obstructive lung disease, and cancer), people in confinement, and individuals with altered levels of consciousness who may be unable to communicate needs or respond to thirst. ■

▶ CHECKPOINT

6. What are the functions of electrolytes in the body?

7. Name three important extracellular electrolytes and three important intracellular electrolytes.

Table 27.2 Blood Electrolyte Imbalances

Electrolyte*	Deficiency		Excess	
	Name and Causes	Signs and Symptoms	Name and Causes	Signs and Symptoms
Sodium (Na⁺) 136–148 mEq/liter	**Hyponatremia** (hī-pō-na-TRĒ-mē-a) may be due to decreased sodium intake; increased sodium loss through vomiting, diarrhea, aldosterone deficiency, or taking certain diuretics; and excessive water intake.	Muscular weakness; dizziness, headache, and hypotension; tachycardia and shock; mental confusion, stupor, and coma.	**Hypernatremia** may occur with dehydration, water deprivation, or excessive sodium in diet or intravenous fluids; causes hypertonicity of ECF, which pulls water out of body cells into ECF, causing cellular dehydration.	Intense thirst, hypertension, edema, agitation, and convulsions.
Chloride (Cl⁻) 95–105 mEq/liter	**Hypochloremia** (hī-pō-klō-RĒ-mē-a) may be due to excessive vomiting, overhydration, aldosterone deficiency, congestive heart failure, and therapy with certain diuretics such as furosemide (Lasix).	Muscle spasms, metabolic alkalosis, shallow respirations, hypotension, and tetany.	**Hyperchloremia** may result from dehydration due to water loss or water deprivation; excessive chloride intake; or severe renal failure, hyperaldosteronism, certain types of acidosis, and some drugs.	Lethargy, weakness, metabolic acidosis, and rapid, deep breathing.
Potassium (K⁺) 3.5–5.0 mEq/liter	**Hypokalemia** (hī-pō-ka-LĒ-mē-a) may result from excessive loss due to vomiting or diarrhea, decreased potassium intake, hyperaldosteronism, kidney disease, and therapy with some diuretics.	Muscle fatigue, flaccid paralysis, mental confusion, increased urine output, shallow respirations, and changes in the electrocardiogram, including flattening of the T wave.	**Hyperkalemia** may be due to excessive intake, renal failure, aldosterone deficiency, crushing injuries to body tissues, or transfusion of hemolyzed blood.	Irritability, nausea, vomiting, diarrhea, muscular weakness; can cause death by inducing ventricular fibrillation.
Calcium (Ca²⁺) Total = 9–10.5 mg/dL; ionized = 4.5–5.5 mEq/liter	**Hypocalcemia** (hī-pō-kal-SĒ-mē-a) may be due to increased calcium loss, reduced calcium intake, elevated levels of phosphate, or hypoparathyroidism.	Numbness and tingling of the fingers; hyperactive reflexes, muscle cramps, tetany, and convulsions; bone fractures; spasms of laryngeal muscles that can cause death by asphyxiation.	**Hypercalcemia** may result from hyperparathyroidism, some cancers, excessive intake of vitamin D, and Paget's disease of bone.	Lethargy, weakness, anorexia, nausea, vomiting, polyuria, itching, bone pain, depression, confusion, paresthesia, stupor, and coma.
Phosphate (HPO₄²⁻) 1.7–2.6 mEq/liter	**Hypophosphatemia** (hī-pō-fos′-fa-TĒ-mē-a) may occur through increased urinary losses, decreased intestinal absorption, or increased utilization.	Confusion, seizures, coma, chest and muscle pain, numbness and tingling of the fingers, uncoordination, memory loss, and lethargy.	**Hyperphosphatemia** occurs when the kidneys fail to excrete excess phosphate, as happens in renal failure; can also result from increased intake of phosphates or destruction of body cells, which releases phosphates into the blood.	Anorexia, nausea, vomiting, muscular weakness, hyperactive reflexes, tetany, and tachycardia.
Magnesium (Mg²⁺) 1.3–2.1 mEq/liter	**Hypomagnesemia** (hī′-pō-mag′-ne-SĒ-mē-a) may be due to inadequate intake or excessive loss in urine or feces; also occurs in alcoholism, malnutrition, diabetes mellitus, and diuretic therapy.	Weakness, irritability, tetany, delirium, convulsions, confusion, anorexia, nausea, vomiting, paresthesia, and cardiac arrhythmias.	**Hypermagnesemia** occurs in renal failure or due to increased intake of Mg²⁺, such as Mg²⁺-containing antacids; also occurs in aldosterone deficiency and hypothyroidism.	Hypotension, muscular weakness or paralysis, nausea, vomiting, and altered mental functioning.

*Values are normal ranges of blood plasma levels in adults.

ACID–BASE BALANCE

OBJECTIVES

- Compare the roles of buffers, exhalation of carbon dioxide, and kidney excretion of H^+ in maintaining pH of body fluids.
- Define acid–base imbalances, describe their effects on the body, and explain how they are treated.

From our discussion thus far, it is clear that various ions play different roles that help maintain homeostasis. A major homeostatic challenge is keeping the H^+ concentration (pH) of body fluids at an appropriate level. This task—the maintenance of acid–base balance—is of critical importance to normal cellular function. For example, the three-dimensional shape of all body proteins, which enables them to perform specific functions, is very sensitive to pH changes. When the diet contains a large amount of protein, as is typical in North America, cellular metabolism produces more acids than bases, which tends to acidify the blood. (You may wish to review the discussion of acids, bases, and pH on pages 40–41.)

In a healthy person, several mechanisms help maintain the pH of systemic arterial blood between 7.35 and 7.45. (A pH of 7.4 corresponds to a H^+ concentration of 0.00004 mEq/liter = 40 nEq/liter.) Because metabolic reactions often produce a huge excess of H^+, the lack of any mechanism for the disposal of H^+ would cause H^+ level in body fluids to quickly rise to a lethal level. Homeostasis of H^+ concentration within a narrow range is thus essential to survival. The removal of H^+ from body fluids and its subsequent elimination from the body depend on the following three major mechanisms:

1. **Buffer systems.** Buffers act quickly to temporarily bind H^+, removing the highly reactive, excess H^+ from solution. Buffers thus raise pH of body fluids but do not remove H^+ from the body.

2. **Exhalation of carbon dioxide.** By increasing the rate and depth of breathing, more carbon dioxide can be exhaled. Within minutes this reduces the level of carbonic acid in blood, which raises the blood pH (reduces blood H^+ level).

3. **Kidney excretion of H^+.** The slowest mechanism, but the only way to eliminate acids other than carbonic acid, is through their excretion in urine.

We next examine each of these mechanisms in turn.

The Actions of Buffer Systems

Most buffer systems in the body consist of a weak acid and the salt of that acid, which functions as a weak base. Buffers prevent rapid, drastic changes in the pH of body fluids by converting strong acids and bases into weak acids and bases within fractions of a second. Strong acids lower pH more than weak acids because strong acids release H^+ more readily and thus contribute more free hydrogen ions. Similarly, strong bases raise pH more than weak ones. The principal buffer systems of the body fluids are the protein buffer system, the carbonic acid–bicarbonate buffer system, and the phosphate buffer system.

Protein Buffer System

The **protein buffer system** is the most abundant buffer in intracellular fluid and blood plasma. For instance, the protein hemoglobin is an especially good buffer within red blood cells, and albumin is the main protein buffer in blood plasma. Proteins are composed of amino acids, organic molecules that contain at least one carboxyl group ($-COOH$) and at least one amino group ($-NH_2$); these groups are the functional components of the protein buffer system. The free carboxyl group at one end of a protein acts like an acid by releasing H^+ when pH rises; it dissociates as follows:

$$NH_2 - \overset{\overset{\textstyle R}{|}}{\underset{\underset{\textstyle H}{|}}{C}} - COOH \longrightarrow NH_2 - \overset{\overset{\textstyle R}{|}}{\underset{\underset{\textstyle H}{|}}{C}} - COO^- + H^+$$

The H^+ is then able to react with any excess OH^- in the solution to form water. The free amino group at the other end of a protein can act as a base by combining with H^+ when pH falls, as follows:

$$NH_2 - \overset{\overset{\textstyle R}{|}}{\underset{\underset{\textstyle H}{|}}{C}} - COOH + H^+ \longrightarrow {}^+NH_3 - \overset{\overset{\textstyle R}{|}}{\underset{\underset{\textstyle H}{|}}{C}} - COOH$$

Thus, proteins can buffer both acids and bases. Besides the terminal carboxyl and amino groups, side chains that can buffer H^+ are present on seven of the 20 amino acids.

The protein hemoglobin is an important buffer of H^+ in red blood cells (see Figure 23.24 on page 836). As blood flows through the systemic capillaries, carbon dioxide (CO_2) passes from tissue cells into red blood cells, where it combines with water (H_2O) to form carbonic acid (H_2CO_3). Once formed, H_2CO_3 dissociates into H^+ and HCO_3^-. At the same time that CO_2 is entering red blood cells, oxyhemoglobin ($Hb-O_2$) is giving up its oxygen to tissue cells. Reduced hemoglobin (deoxyhemoglobin) is an excellent buffer of H^+, so it picks up most of the H^+. For this reason, reduced hemoglobin usually is written as $Hb-H$. The following reactions summarize these relations:

$$\underset{\text{Water}}{H_2O} + \underset{\substack{\text{Carbon dioxide} \\ \text{(entering RBCs)}}}{CO_2} \longrightarrow \underset{\text{Carbonic acid}}{H_2CO_3}$$

$$\underset{\text{Carbonic acid}}{H_2CO_3} \longrightarrow \underset{\text{Hydrogen ion}}{H^+} + \underset{\text{Bicarbonate ion}}{HCO_3^-}$$

$$\underset{\substack{\text{Oxyhemoglobin} \\ \text{(in RBCs)}}}{Hb-O_2} + \underset{\substack{\text{Hydrogen ion} \\ \text{(from carbonic} \\ \text{acid)}}}{H^+} \longrightarrow \underset{\substack{\text{Reduced} \\ \text{hemoglobin}}}{Hb-H} + \underset{\substack{\text{Oxygen} \\ \text{(released to} \\ \text{tissue cells)}}}{O_2}$$

Carbonic Acid–Bicarbonate Buffer System

The **carbonic acid–bicarbonate buffer system** is based on the *bicarbonate ion* (HCO_3^-), which can act as a weak base, and *carbonic acid* (H_2CO_3), which can act as a weak acid. HCO_3^- is a significant anion in both intracellular and extracellular fluids (Figure 27.6), and its reaction with the CO_2 constantly released during cellular respiration produces H_2CO_3. Because the kidneys also synthesize new HCO_3^- and reabsorb filtered HCO_3^-, this important buffer is not lost in the urine. If there is an excess of H^+, the HCO_3^-, can function as a weak base and remove the excess H^+ as follows:

$$H^+ \quad + \quad HCO_3^- \quad \longrightarrow \quad H_2CO_3$$

Hydrogen ion   Bicarbonate ion   Carbonic acid
                (weak base)

Then, H_2CO_3 dissociates into water and carbon dioxide, and the CO_2 is exhaled from the lungs.

Conversely, if there is a shortage of H^+, the H_2CO_3 can function as a weak acid and provide H^+ as follows:

$$H_2CO_3 \quad \longrightarrow \quad H^+ \quad + \quad HCO_3^-$$

Carbonic acid   Hydrogen ion   Bicarbonate ion
(weak acid)

At a pH of 7.4, HCO_3^- concentration is about 24 mEq/liter, whereas H_2CO_3 concentration is about 1.2 mmol/liter; thus, bicarbonate ions outnumber carbonic acid molecules by 20 to 1. Because CO_2 and H_2O combine to form H_2CO_3, this buffer system cannot protect against pH changes due to respiratory problems in which there is an excess or shortage of CO_2.

Phosphate Buffer System

The **phosphate buffer system** acts via a similar mechanism as the carbonic acid–bicarbonate buffer system. The components of the phosphate buffer system are the ions *dihydrogen phosphate* ($H_2PO_4^-$) and *monohydrogen phosphate* (HPO_4^{2-}). Recall that phosphates are major anions in intracellular fluid and minor ones in extracellular fluids (Figure 27.6). The dihydrogen phosphate ion acts as a weak acid and is capable of buffering strong bases such as OH^-, as follows:

$$OH^- \quad + \quad H_2PO_4^- \quad \longrightarrow \quad H_2O \quad + \quad HPO_4^{2-}$$

Hydroxide ion   Dihydrogen   Water   Monohydrogen
(strong base)   phosphate   phosphate
                 (weak acid)   (weak base)

The monohydrogen phosphate ion, in contrast, acts as a weak base and is capable of buffering the H^+ released by a strong acid such as hydrochloric acid (HCl):

$$H^+ \quad + \quad HPO_4^{2-} \quad \longrightarrow \quad H_2PO_4^-$$

Hydrogen ion   Monohydrogen   Dihydrogen
(strong acid)   phosphate   phosphate
                (weak acid)   (weak acid)

Because the concentration of phosphates is highest in intracellular fluid, the phosphate buffer system is an important regulator of pH in the cytosol. It also acts to a smaller degree in extracellular fluids, and it acts to buffer acids in urine. $H_2PO_4^-$ is formed when excess H^+ in the kidney tubule fluid combine with HPO_4^{2-} (see Figure 27.8). The H^+ that becomes part of the $H_2PO_4^-$ passes into the urine. This reaction is one means whereby the kidneys help maintain blood pH by excreting H^+ in the urine.

Exhalation of Carbon Dioxide

Breathing also plays an important role in maintaining the pH of body fluids. An increase in the carbon dioxide (CO_2) concentration in body fluids increases H^+ concentration and thus lowers the pH (makes body fluids more acidic). Because H_2CO_3 can be eliminated by exhaling CO_2, it is called a **volatile acid.** Conversely, a decrease in the CO_2 concentration of body fluids raises the pH (makes body fluids more alkaline). This chemical interaction is illustrated by the following reversible reactions.

$$CO_2 \; + H_2O \; \rightleftharpoons \; H_2CO_3 \; \rightleftharpoons \; H^+ \; + \; HCO_3^-$$

Carbon   Water   Carbonic   Hydrogen   Bicarbonate
dioxide   acid   ion   ion

Changes in the rate and depth of breathing can alter the pH of body fluids within a couple of minutes. With increased ventilation, more CO_2 is exhaled, the preceding reaction is driven to the left, H^+ concentration falls, and blood pH rises. Doubling the ventilation increases pH by about 0.23 units, from 7.4 to 7.63. If ventilation is slower than normal, less carbon dioxide is exhaled, and the blood pH falls. Reducing ventilation to one-quarter of normal lowers the pH by 0.4 units, from 7.4 to 7.0. These examples show the powerful effect of alterations in breathing on the pH of body fluids.

The pH of body fluids and the rate and depth of breathing interact via a negative feedback loop (Figure 27.7). When the blood acidity increases, the decrease in pH (increase in concentration of H^+) is detected by central chemoreceptors in the medulla oblongata and peripheral chemoreceptors in the aortic and carotid bodies, both of which stimulate the inspiratory area in the medulla oblongata. As a result, the diaphragm and other respiratory muscles contract more forcefully and frequently, so more CO_2 is exhaled. As less H_2CO_3 forms and fewer H^+ are present, blood pH increases. When the response brings blood pH (H^+ concentration) back to normal, there is a return to acid–base homeostasis. The same negative feedback loop operates if the blood level of CO_2 increases. Ventilation increases, which removes more CO_2 from the blood, thus reducing its H^+ concentration so the blood's pH increases.

By contrast, if the pH of the blood increases, the respiratory center is inhibited and the rate and depth of breathing decreases. A decrease in the CO_2 concentration of the blood has the same effect. When breathing decreases, CO_2 accumulates in the blood so its H^+ concentration increases. This respiratory mechanism is a powerful eliminator of acid, but it can only get rid of the sole volatile acid—carbonic acid.

Figure 27.7 Negative feedback regulation of blood pH by the respiratory system.

Exhalation of carbon dioxide lowers the H$^+$ concentration of blood.

Some stimulus disrupts homeostasis by

Decreasing

Blood pH (increase in H$^+$ concentration)

Receptors

Central chemo-receptors in medulla oblongata

Peripheral chemo-receptors in aortic and carotid bodies

Input   Nerve impulses

Control center

Inspiratory area in medulla oblongata

Return to homeostasis when response brings blood pH or H$^+$ concentration back to normal

Output   Nerve impulses

Effectors

Diaphragm contracts more force-fully and frequently so more CO$_2$ is exhaled

As less H$_2$CO$_3$ forms and fewer H$^+$ are present, blood pH increases (H$^+$ concentration decreases)

If you hold your breath for 30 seconds, what is likely to happen to your blood pH?

Kidney Excretion of H$^+$

Metabolic reactions produce **nonvolatile acids** such as sulfuric acid at a rate of about 1 mEq of H$^+$ per day for every kilogram of body mass. The only way to eliminate this huge acid load is to excrete H$^+$ in the urine. Given the magnitude of these contributions to acid–base balance, it's not surprising that renal failure can quickly cause death.

Cells in both the proximal convoluted tubules (PCT) and the collecting ducts of the kidneys secrete hydrogen ions into the tubular fluid. In the PCT, Na$^+$/H$^+$ antiporters secrete H$^+$ as they reabsorb Na$^+$ (see Figure 26.13 on page 968). Even more important for regulation of pH of body fluids, however, are the intercalated cells of the collecting tubule. The *apical* membranes of some intercalated cells include **proton pumps (H$^+$ ATPases)** that secrete H$^+$ into the tubular fluid (Figure 27.8). Intercalated cells can secrete H$^+$ against a concentration gradient so effectively that urine can be up to 1000 times (3 pH units) more acidic than blood. HCO$_3^-$ produced by dissociation of H$_2$CO$_3$ inside intercalated cells crosses the basolateral membrane by means of **Cl$^-$/HCO$_3^-$ antiporters** and then diffuses into peritubular capillaries (Figure 27.8a). The HCO$_3^-$ that enters the blood in this way is *new* (not filtered). For this reason, blood leaving the kidney in the renal vein may have a higher HCO$_3^-$ concentration than blood entering the kidney in the renal artery.

Interestingly, a second type of intercalated cell has proton pumps in its *basolateral* membrane and Cl$^-$/HCO$_3^-$ antiporters in its apical membrane. These intercalated cells secrete HCO$_3^-$ and reabsorb H$^+$. Thus, the two types of intercalated cells help maintain the pH of body fluids in two ways—by excreting excess H$^+$ when pH of body fluids is too low or by excreting excess HCO$_3^-$ when pH is too high.

Some H$^+$ secreted into the tubular fluid of the collecting duct are buffered. By now, most of the filtered HCO$_3^-$ has been reabsorbed; few are left in the tubule lumen to combine with and buffer secreted H$^+$. Two other buffers, however, are present and available to combine with H$^+$ (Figure 27.8b). The most plentiful buffer in the tubular fluid of the collecting duct is HPO$_4^{2-}$ (monohydrogen phosphate ion). In addition, a small amount of NH$_3$ (ammonia) also is present. H$^+$ combines with HPO$_4^{2-}$ to form H$_2$PO$_4^-$ (dihydrogen phosphate ion) and with NH$_3$ to form NH$_4^+$ (ammonium ion). Because these ions cannot diffuse back into tubule cells, they are excreted in the urine.

Table 27.3 summarizes the mechanisms that maintain pH of body fluids.

Acid–Base Imbalances

The normal pH range of systemic arterial blood is between 7.35 (= 45 nEq of H$^+$/liter) and 7.45 (= 35 nEq of H$^+$/liter). **Acidosis** (or **acidemia**) is a condition in which blood pH is below 7.35; **alkalosis** (or **alkalemia**) is a condition in which blood pH is higher than 7.45.

Figure 27.8 Secretion of H⁺ by intercalated cells in the collecting duct. HCO_3^- = bicarbonate ion; CO_2 = carbon dioxide; H_2O = water; H_2CO_3 = carbonic acid; Cl^- = chloride ion; NH_3 = ammonia; NH_4^+ = ammonium ion; HPO_4^{2-} = monohydrogen phosphate ion; $H_2PO_4^-$ = dihydrogen phosphate ion.

 Urine can be up to 1000 times more acidic than blood due to the operation of the proton pumps in the collecting ducts of the kidneys.

(a) Secretion of H⁺

(b) Buffering of H⁺ in urine

Key:

 Proton pump (H⁺ ATPase) in apical membrane

HCO₃⁻/Cl⁻ antiporter in basolateral membrane

$\cdots\blacktriangleright$ Diffusion

 What would be the effects of a drug that blocks the activity of carbonic anhydrase?

Table 27.3	Mechanisms that Maintain pH of Body Fluids
Mechanism	**Comments**
Buffer Systems	Most consist of a weak acid and the salt of that acid, which functions as a weak base. They prevent drastic changes in body fluid pH.
Proteins	The most abundant buffers in body cells and blood. Histidine and cysteine are the two amino acids that contribute most of the buffering capacity of proteins. Hemoglobin inside red blood cells is a good buffer.
Carbonic acid–bicarbonate	Important regulator of blood pH. The most abundant buffers in extracellular fluid (ECF).
Phosphates	Important buffers in intracellular fluid and in urine.
Exhalation of CO₂	With increased exhalation of CO_2, pH rises (fewer H⁺). With decreased exhalation of CO_2, pH falls (more H⁺).
Kidneys	Renal tubules secrete H⁺ into the urine and reabsorb HCO_3^- so it is not lost in the urine.

The major physiological effect of acidosis is depression of the central nervous system through depression of synaptic transmission. If the systemic arterial blood pH falls below 7, depression of the nervous system is so severe that the individual becomes disoriented, then comatose, and may die. Patients with severe acidosis usually die while in a coma. A major physiological effect of alkalosis, by contrast, is overexcitability in both the central nervous system and peripheral nerves. Neurons conduct impulses repetitively, even when not stimulated by normal stimuli; the results are nervousness, muscle spasms, and even convulsions and death.

A change in blood pH that leads to acidosis or alkalosis may be countered by **compensation,** the physiological response to an acid–base imbalance that acts to normalize arterial blood pH. Compensation may be either *complete,* if pH indeed is brought within the normal range, or *partial,* if systemic arterial blood pH is still lower than 7.35 or higher than 7.45. If a person has altered blood pH due to metabolic causes, hyperventilation or hypoventilation can help bring blood pH back toward the normal range; this form of compensation, termed **respiratory compensation,** occurs within minutes and reaches its maximum within hours. If, however, a person has altered blood pH due to respiratory causes, then **renal compensation**—changes in secretion of H⁺ and reabsorption of HCO_3^- by the kidney tubules—can help reverse the change. Renal compensation may begin in minutes, but it takes days to reach maximum effectiveness.

In the discussion that follows, note that both respiratory acidosis and respiratory alkalosis are disorders resulting from changes in the partial pressure of CO_2 (P_{CO_2}) in systemic arterial blood (normal range is 35–45 mmHg). By contrast, both metabolic acidosis and metabolic alkalosis are disorders resulting from changes in HCO_3^- concentration (normal range is 22–26 mEq/liter in systemic arterial blood).

Respiratory Acidosis

The hallmark of **respiratory acidosis** is an abnormally high P_{CO_2} in systemic arterial blood—above 45 mmHg. Inadequate exhalation of CO_2 causes the blood pH to drop. Any condition that decreases the movement of CO_2 from the blood to the alveoli of the lungs to the atmosphere causes a buildup of CO_2, H_2CO_3, and H^+. Such conditions include emphysema, pulmonary edema, injury to the respiratory center of the medulla oblongata, airway obstruction, or disorders of the muscles involved in breathing. If the respiratory problem is not too severe, the kidneys can help raise the blood pH into the normal range by increasing excretion of H^+ and reabsorption of HCO_3^- (renal compensation). The goal in treatment of respiratory acidosis is to increase the exhalation of CO_2, as, for instance, by providing ventilation therapy. In addition, intravenous administration of HCO_3^- may be helpful.

Respiratory Alkalosis

In **respiratory alkalosis,** systemic arterial blood P_{CO_2} falls below 35 mmHg. The cause of the drop in P_{CO_2} and the resulting increase in pH is hyperventilation, which occurs in conditions that stimulate the inspiratory area in the brain stem. Such conditions include oxygen deficiency due to high altitude or pulmonary disease, cerebrovascular accident (stroke), or severe anxiety. Again, renal compensation may bring blood pH into the normal range if the kidneys decrease excretion of H^+ and reabsorption of HCO_3^-. Treatment of respiratory alkalosis is aimed at increasing the level of CO_2 in the body. One simple treatment is to have the person inhale and exhale into a paper bag for a short period; as a result, the person inhales air containing a higher-than-normal concentration of CO_2.

Metabolic Acidosis

In **metabolic acidosis,** the systemic arterial blood HCO_3^- level drops below 22 mEq/liter. Such a decline in this important buffer causes the blood pH to decrease. Three situations may lower the blood level of HCO_3^-: (1) actual loss of HCO_3^-, such as may occur with severe diarrhea or renal dysfunction; (2) accumulation of an acid other than carbonic acid, as may occur in ketosis (described on page 923); or (3) failure of the kidneys to excrete H^+ from metabolism of dietary proteins. If the problem is not too severe, hyperventilation can help bring blood pH into the normal range (respiratory compensation). Treatment of metabolic acidosis consists of administering intravenous solutions of sodium bicarbonate and correcting the cause of the acidosis.

Metabolic Alkalosis

In **metabolic alkalosis,** the systemic arterial blood HCO_3^- concentration is above 26 mEq/liter. A nonrespiratory loss of acid or excessive intake of alkaline drugs causes the blood pH to increase above 7.45. Excessive vomiting of gastric contents, which results in a substantial loss of hydrochloric acid, is probably the most frequent cause of metabolic alkalosis. Other causes include gastric suctioning, use of certain diuretics, endocrine disorders, excessive intake of alkaline drugs (antacids), and severe dehydration. Respiratory compensation through hypoventilation may bring blood pH into the normal range. Treatment of metabolic alkalosis consists of giving fluid solutions to correct Cl^-, K^+, and other electrolyte deficiencies plus correcting the cause of alkalosis.

Table 27.4 summarizes respiratory and metabolic acidosis and alkalosis.

Table 27.4	Summary of Acidosis and Alkalosis		
Condition	**Definition**	**Common Causes**	**Compensatory Mechanism**
Respiratory acidosis	Increased P_{CO_2} (above 45 mmHg) and decreased pH (below 7.35) if there is no compensation.	Hypoventilation due to emphysema, pulmonary edema, trauma to respiratory center, airway obstructions, or dysfunction of muscles of respiration.	*Renal:* increased excretion of H^+; increased reabsorption of HCO_3^-. If compensation is complete, pH will be within the normal range but P_{CO_2} will be high.
Respiratory alkalosis	Decreased P_{CO_2} (below 35 mmHg) and increased pH (above 7.45) if there is no compensation.	Hyperventilation due to oxygen deficiency, pulmonary disease, cerebrovascular accident (CVA), or severe anxiety.	*Renal:* decreased excretion of H^+; decreased reabsorption of HCO_3^-. If compensation is complete, pH will be within the normal range but P_{CO_2} will be low.
Metabolic acidosis	Decreased HCO_3^- (below 22 mEq/liter) and decreased pH (below 7.35) if there is no compensation.	Loss of bicarbonate ions due to diarrhea, accumulation of acid (ketosis), renal dysfunction.	*Respiratory:* hyperventilation, which increases loss of CO_2. If compensation is complete, pH will be within the normal range but HCO_3^- will be low.
Metabolic alkalosis	Increased HCO_3^- (above 26 mEq/liter) and increased pH (above 7.45) if there is no compensation.	Loss of acid due to vomiting, gastric suctioning, or use of certain diuretics; excessive intake of alkaline drugs.	*Respiratory:* hypoventilation, which slows loss of CO_2. If compensation is complete, pH will be within the normal range but HCO_3^- will be high.

Diagnosis of Acid–Base Imbalances

One can often pinpoint the cause of an acid–base imbalance by careful evaluation of three factors in a sample of systemic arterial blood: pH, concentration of HCO_3^-, and P_{CO_2}. These three blood chemistry values are examined in the following four-step sequence:

1. Note whether the pH is high (alkalosis) or low (acidosis).
2. Then decide which value—P_{CO_2} or HCO_3^-—is out of the normal range and could be the *cause* of the pH change. For example, *elevated pH* could be caused by *low P_{CO_2}* or *high HCO_3^-*.
3. If the cause is a *change in P_{CO_2}*, the problem is *respiratory;* if the cause is a *change in HCO_3^-*, the problem is *metabolic*.
4. Now look at the value that doesn't correspond with the observed pH change. If it is within its normal range, there is no compensation. If it is outside the normal range, compensation is occurring and partially correcting the pH imbalance. ■

▶ CHECKPOINT

8. Explain how each of the following buffer systems helps to maintain the pH of body fluids: proteins, carbonic acid–bicarbonate buffers, and phosphates.
9. Define acidosis and alkalosis. Distinguish among respiratory and metabolic acidosis and alkalosis.
10. What are the principal physiological effects of acidosis and alkalosis?

AGING AND FLUID, ELECTROLYTE, AND ACID–BASE BALANCE

▶ OBJECTIVE

• Describe the changes in fluid, electrolyte, and acid–base balance that may occur with aging.

There are significant differences between adults and infants, especially premature infants, with respect to fluid distribution, regulation of fluid and electrolyte balance, and acid–base homeostasis. Accordingly, infants experience more problems than adults in these areas. The differences are related to the following conditions:

• **Proportion and distribution of water.** Whereas a newborn's total body mass is about 75% water (and can be as high as 90% in a premature infant), an adult's is about 55–60% water. (The "adult" percentage is achieved at about 2 years of age.) Moreover, whereas adults have twice as much water in ICF as ECF, the opposite is true in premature infants. Because ECF is subject to more changes than ICF, rapid losses or gains of body water are much more critical in infants. Given that the rate of fluid intake and output is about seven times higher in infants than in adults, the slightest changes in fluid balance can result in severe abnormalities.

• **Metabolic rate.** The metabolic rate of infants is about double that of adults. This results in the production of more metabolic wastes and acids, which can lead to the development of acidosis in infants.

• **Functional development of the kidneys.** In infants, the kidneys are only about half as efficient in concentrating urine as those of adults. (Functional development is not complete until close to the end of the first month after birth.) As a result, the kidneys of newborns can neither concentrate urine nor rid the body of excess acids produced as a result of high metabolic rate as effectively as those of adults.

• **Body surface area.** The ratio of body surface area to body volume of infants is about three times greater than that of adults. This significantly increases water loss through the skin in infants.

• **Breathing rate.** The higher breathing rate of infants (about 30 to 80 times a minute) causes greater water loss from the lungs. Moreover, respiratory alkalosis may occur because greater ventilation eliminates more CO_2 and lowers the P_{CO_2}.

• **Ion concentrations.** Newborns have higher K^+ and Cl^- concentrations than adults. This creates a tendency toward metabolic acidosis.

By comparison with children and younger adults, older adults often have an impaired ability to maintain fluid, electrolyte, and acid–base balance. With increasing age, many people have a decreased volume of intracellular fluid and decreased total body K^+ due to declining skeletal muscle mass and increasing mass of adipose tissue (which contains very little water). Age-related decreases in respiratory and renal functioning may compromise acid–base balance by slowing the exhalation of CO_2 and the excretion of excess acids in urine. Other kidney changes, such as decreased blood flow, decreased glomerular filtration rate, and reduced sensitivity to antidiuretic hormone, have an adverse effect on the ability to maintain fluid and electrolyte balance. Due to a decrease in the number and efficiency of sweat glands, water loss from the skin declines with age. Because of these age-related changes, older adults are susceptible to several fluid and electrolyte disorders:

• *Dehydration* and *hypernatremia* often occur due to inadequate fluid intake or loss of more water than Na^+ in vomit, feces, or urine.

• *Hyponatremia* may occur due to inadequate intake of Na^+; elevated loss of Na^+ in urine, vomit, or diarrhea; or impaired ability of the kidneys to produce dilute urine.

• *Hypokalemia* often occurs in older adults who chronically use laxatives to relieve constipation or who take K^+-depleting diuretic drugs for treatment of hypertension or heart disease.

• *Acidosis* may occur due to impaired ability of the lungs and kidneys to compensate for acid–base imbalances. One cause of acidosis is that renal tubule cells produce less ammonia (NH_3), which then is not available to combine with H^+ and be excreted in urine as NH_4^+; another cause is reduced exhalation of CO_2.

STUDY OUTLINE

FLUID COMPARTMENTS AND FLUID BALANCE (p. 992)

1. Body fluid includes water and dissolved solutes.
2. About two-thirds of the body's fluid is located within cells and is called intracellular fluid (ICF). The other one- third, called extracellular fluid (ECF), includes interstitial fluid; blood plasma and lymph; cerebrospinal fluid; gastrointestinal tract fluids; synovial fluid; fluids of the eyes and ears; pleural, pericardial, and peritoneal fluids; and glomerular filtrate.
3. Fluid balance means that the various body compartments contain the normal amount of water.
4. An inorganic substance that dissociates into ions in solution is called an electrolyte. Fluid balance and electrolyte balance are interrelated.
5. Water is the largest single constituent in the body. It makes up 45–75% of total body mass, depending on age, gender, and the amount of adipose tissue present.
6. Daily water gain and loss are each about 2500 mL. Sources of water gain are ingested liquids and foods, and water produced by cellular respiration and dehydration synthesis reactions (metabolic water). Water is lost from the body via urination, evaporation from the skin surface, exhalation of water vapor, and defecation. In women, menstrual flow is an additional route for loss of body water.
7. The main way to regulate body water gain is by adjusting the volume of water intake, mainly by drinking more or less fluid. The thirst center in the hypothalamus governs the urge to drink.
8. Although increased amounts of water and solutes are lost through sweating and exhalation during exercise, loss of excess body water or excess solutes depends mainly on regulating excretion in the urine. The extent of urinary NaCl loss is the main determinant of body fluid volume, whereas the extent of urinary water loss is the main determinant of body fluid osmolarity.
9. Table 27.1 on page 996 summarizes the factors that regulate water gain and water loss.
10. Angiotensin II and aldosterone reduce urinary loss of Na^+ and Cl^- and thereby increase the volume of body fluids. ANP promotes natriuresis, elevated excretion of Na^+ (and Cl^-), which decreases blood volume.
11. The major hormone that regulates water loss and thus body fluid osmolarity is antidiuretic hormone (ADH).
12. An increase in the osmolarity of interstitial fluid draws water out of cells, and they shrink slightly. A decrease in the osmolarity of interstitial fluid causes cells to swell. Most often a change in osmolarity is due to a change in the concentration of Na^+, the dominant solute in interstitial fluid.
13. When a person consumes water faster than the kidneys can excrete it or when renal function is poor, the result may be water intoxication, in which cells swell dangerously.

ELECTROLYTES IN BODY FLUIDS (p. 997)

1. Ions formed when electrolytes dissolve in body fluids control the osmosis of water between fluid compartments, help maintain acid–base balance, and carry electrical current.
2. The concentrations of cations and anions is expressed in units of milliequivalents/liter (mEq/liter).
3. Blood plasma, interstitial fluid, and intracellular fluid contain varying kinds and amounts of ions.
4. Sodium ions (Na^+) are the most abundant extracellular ions. They are involved in impulse transmission, muscle contraction, and fluid and electrolyte balance. Na^+ level is controlled by aldosterone, antidiuretic hormone, and atrial natriuretic peptide.
5. Chloride ions (Cl^-) are the major extracellular anions. They play a role in regulating osmotic pressure and forming HCl in gastric juice. Cl^- level is controlled indirectly by antidiuretic hormone and by processes that increase or decrease renal reabsorption of Na^+.
6. Potassium ions (K^+) are the most abundant cations in intracellular fluid. They play a key role in the resting membrane potential and action potential of neurons and muscle fibers; help maintain intracellular fluid volume; and contribute to regulation of pH. K^+ level is controlled by aldosterone.
7. Bicarbonate ions (HCO_3^-) are the second most abundant anions in extracellular fluid. They are the most important buffer in blood plasma.
8. Calcium is the most abundant mineral in the body. Calcium salts are structural components of bones and teeth. Ca^{2+}, which are principally extracellular cations, function in blood clotting, neurotransmitter release, and contraction of muscle. Ca^{2+} level is controlled mainly by parathyroid hormone and calcitriol.
9. Phosphate ions ($H_2PO_4^-$, HPO_4^{2-}, and PO_4^{3-}) are principally intracellular anions, and their salts are structural components of bones and teeth. They are also required for the synthesis of nucleic acids and ATP and participate in buffer reactions. Their level is controlled by parathyroid hormone and calcitriol.
10. Magnesium ions (Mg^{2+}) are primarily intracellular cations. They act as cofactors in several enzyme systems.
11. Table 27.2 on page 1000 describes the imbalances that result from deficiency or excess of important body electrolytes.

ACID–BASE BALANCE (p. 1001)

1. The overall acid–base balance of the body is maintained by controlling the H^+ concentration of body fluids, especially extracellular fluid.
2. The normal pH of systemic arterial blood is 7.35–7.45.
3. Homeostasis of pH is maintained by buffer systems, via exhalation of carbon dioxide, and via kidney excretion of H^+ and reabsorption of HCO_3^-.
4. The important buffer systems include proteins, carbonic acid–bicarbonate buffers, and phosphates.
5. An increase in exhalation of carbon dioxide increases blood pH; a decrease of CO_2 in exhalation decreases blood pH.
6. In the proximal convoluted tubules of the kidneys, Na^+/H^+ antiporters secrete H^+ as they reabsorb Na^+. In the collecting ducts of the kidneys, some intercalated cells reabsorb K^+ and HCO_3^- and secrete H^+, whereas other intercalated cells secrete HCO_3^-. In these ways, the kidneys can increase or decrease the pH of body fluids.
7. Table 27.3 on page 1004 summarizes the mechanisms that maintain pH of body fluids.
8. Acidosis is a systemic arterial blood pH below 7.35; its principal

effect is depression of the central nervous system (CNS). Alkalosis is a systemic arterial blood pH above 7.45; its principal effect is overexcitability of the CNS.

9. Respiratory acidosis and alkalosis are disorders due to changes in blood P_{CO_2}, whereas metabolic acidosis and alkalosis are disorders associated with changes in blood HCO_3^- concentration.

10. Metabolic acidosis or alkalosis can be compensated by respiratory mechanisms; respiratory acidosis or alkalosis can be compensated by renal mechanisms.

11. Table 27.4 on page 1005 summarizes respiratory and metabolic acidosis and alkalosis.

12. By examining systemic arterial blood pH, HCO_3^-, and P_{CO_2} values, it is possible to pinpoint the cause of an acid-base imbalance.

AGING AND FLUID, ELECTROLYTE, AND ACID–BASE BALANCE (p. 1006)

1. With increasing age, there is decreased intracellular fluid volume and decreased K^+ due to declining skeletal muscle mass.

2. Decreased kidney function with aging adversely affects fluid and electrolyte balance.

SELF-QUIZ QUESTIONS

Fill in the blanks in the following statements.

1. The two compartments in which water can be found are the _____ and the _____.

2. In the carbonic acid-bicarbonate buffer system, the _____ acts as a weak base, and _____ acts as a weak acid.

3. Most buffers consist of a _____ and a _____.

4. The principal buffer systems of the body are the _____ systems, the _____ systems, and the _____ systems.

Indicate whether the following statements are true or false.

5. The most abundant buffer in intracellular fluid and blood plasma is the protein buffer system.

6. Normally, water loss equals water gain, so body fluid volume is constant.

Choose the one best answer to the following questions.

7. The primary means of regulating body water gain is adjusting (a) the volume of water intake, (b) the rate of cellular respiration, (c) the formation of metabolic water, (d) the volume of metabolic water, (e) the metabolic use of water.

8. Which of the following are ways that dehydration stimulates thirst? (1) It decreases the production of saliva. (2) It increases the production of saliva. (3) It increases osmolarity of body fluids. (4) It decreases osmolarity of body fluids. (5) It decreases blood volume. (6) It increases blood volume. (a) 1, 2, 4, and 6, (b) 1, 3, 5, and 6, (c) 1, 3, and 5, (d) 2, 4, and 6, (e) 1, 4, 5, and 6.

9. Which of the following hormones regulate fluid loss? (1) antidiuretic hormone, (2) aldosterone, (3) atrial natriuretic peptide, (4) thyroxine, (5) cortisol. (a) 1, 3, and 5, (b) 1, 2, and 3, (c) 2, 4, and 5, (d) 2, 3, and 4, (e) 1, 3, and 4.

10. Which of the following are *true* concerning ions in the body? (1) They control osmosis of water between fluid compartments. (2) They help maintain acid–base balance. (3) They carry electrical current. (4) They serve as cofactors for enzyme activity. (5) They serve as neurotransmitters under special circumstances. (a) 1, 3, and 5, (b) 2, 4, and 5, (c) 1, 4, and 5, (d) 1, 2, and 4, (e) 1, 2, 3, and 4.

11. Which of the following statements are *true*? (1) Buffers prevent rapid, drastic changes in pH of a body fluid. (2) Buffers work slowly. (3) Strong acids lower pH more than weak acids because strong acids contribute fewer H^+. (4) Proteins can buffer both acids and bases. (5) Hemoglobin is an important buffer. (a) 1, 2, 3, and 5, (b) 1, 3, 4, and 5, (c) 1, 3, and 5, (d) 1, 4, and 5, (e) 2, 3, and 5.

12. Which of the following statements are *true*? (1) An increase in the carbon dioxide concentration in body fluids increases H^+ concentration and thus lowers pH. (2) Breath holding results in a decline in blood pH. (3) The respiratory buffer mechanism can eliminate a single volatile acid: carbonic acid. (4) The only way to eliminate fixed acids is to excrete H^+ in the urine. (5) When the diet contains a large amount of protein, normal metabolism produces more acids than bases. (a) 1, 2, 3, 4, and 5, (b) 1, 3, 4, and 5, (c) 1, 2, 3, and 4, (d) 1, 2, 4, and 5, (e) 1, 3, and 4.

13. Concerning acid–base imbalances: (1) Acidosis can cause depression of the central nervous system through depression of synaptic transmission. (2) Renal compensation can resolve respiratory alkalosis or acidosis. (3) A major physiological effect of alkalosis is lack of excitability in the central nervous system and peripheral nerves. (4) Resolution of metabolic acidosis and alkalosis occurs through renal compensation. (5) In adjusting blood pH, renal compensation occurs quickly whereas respiratory compensation takes days. (a) 1, 2, and 5, (b) 1 and 2, (c) 2, 3, and 4, (d) 2, 3, and 5, (e) 1, 2, 3, and 5.

14. Match the following:

_____ (a) the most abundant cation in intracellular fluid; plays a key role in establishing the resting membrane potential

_____ (b) the most abundant mineral in the body; plays important roles in blood clotting, neurotransmitter release, maintenance of muscle tone, and excitability of nervous and muscle tissue

_____ (c) second most common intracellular cation; is a cofactor for enzymes involved in carbohydrate, protein, and Na^+/K^+ ATPase metabolism

_____ (d) the most abundant extracellular cation; essential in fluid and electrolyte balance

_____ (e) ions that are mostly combined with lipids, proteins, carbohydrates, nucleic acids, and ATP inside cells

_____ (f) most prevalent extracellular anion; can help balance the level of anions in different fluid compartments

_____ (g) second most prevalent extracellular anion; mainly regulated by the kidneys; important for acid–base balance

_____ (h) substances that act to prevent rapid, drastic changes in the pH of a body fluid

_____ (i) inorganic substances that dissociate into ions when in solution

(1) sodium
(2) chloride
(3) electrolytes
(4) bicarbonate
(5) buffers
(6) phosphate
(7) magnesium
(8) potassium
(9) calcium

15. Match the following:

_____ (a) an abnormal increase in the volume of interstitial fluid

_____ (b) the swelling of cells due to water moving from blood plasma into interstitial fluid and then into cells

_____ (c) occurs when water loss is greater than water gain

_____ (d) condition that can occur as water moves out of blood plasma into interstitial fluid and blood volume decreases

_____ (e) can be caused by emphysema, pulmonary edema, injury to the respiratory center of the medulla oblongata, airway destruction, or disorders of the muscles involved in breathing

_____ (f) can be caused by actual loss of bicarbonate ions, ketosis, or failure of kidneys to excrete H^+

_____ (g) can be caused by excessive vomiting of gastric contents, gastric suctioning, use of certain diuretics, severe dehydration, or excessive intake of alkaline drugs

_____ (h) can be caused by oxygen deficiency at high altitude, stroke, or severe anxiety

(1) respiratory acidosis
(2) respiratory alkalosis
(3) metabolic acidosis
(4) metabolic alkalosis
(5) dehydration
(6) hypovolemia
(7) water intoxication
(8) edema

CRITICAL THINKING QUESTIONS

1. After a late-night party, Gary ended up in the emergency room. According to his friends, Gary had said he was so thirsty that he could drink a tub full of water, so someone dared him to try it. His friends called for an ambulance when Gary suffered a convulsion. What has happened to Gary?

   HINT *Gary was intoxicated, but not with alcohol.*

2. "I'm so mad at Jacques," Aimee told her girl friend. "We're about the same height and the same weight, but when we had our body fat measured in A & P lab, he had less fat than I did!" How do you explain this to Aimee?

   HINT *Fat and water don't mix.*

3. When Carlos ate some of his roommate's leftovers, he thought the food tasted a little odd, but he didn't feel like cooking so he ate it anyway. A while later, Carlos was vomiting the entire contents of his stomach. Carlos then tried to settle his queasy stomach by eating an entire bottle of antacids. What effect will this episode have on Carlos's fluid and electrolyte balance?

   HINT *What is the pH of gastric juice?*

4. Henry is in the intensive care unit because he suffered a severe myocardial infarction three days ago. The lab reports the following values from an arterial blood sample: pH = 7.30, $HCO_3^- = 20$ mEq/liter, $P_{CO_2} = 32$ mmHg. Diagnose Henry's acid–base status and decide whether or not compensation is occurring.

   HINT *Use the four-step method described on page 1006.*

ANSWERS TO FIGURE QUESTIONS

27.1 Plasma volume equals body mass $\times$ percent of body mass that is body fluid, $\times$ proportion of body fluid that is ECF $\times$ proportion of ECF that is plasma $\times$ a conversion factor (1 liter/kg).
For male: blood plasma volume $= 60\text{kg} \times 0.60 \times 1/3 \times 0.20 \times 1$ liter/kg $= 2.4$ liters. For female, blood plasma volume is 2.2 liters.

27.2 Hyperventilation, vomiting, fever, and diuretics all increase fluid loss.

27.3 Negative feedback is in operation because the result (an increase in fluid intake) is opposite to the initiating stimulus (dehydration).

27.4 Elevated aldosterone promotes abnormally high renal reabsorption of NaCl and water, which expands blood volume and increases blood pressure. Increased blood pressure causes more fluid to filter out of capillaries and accumulate in the interstitial fluid; this is edema.

27.5 When both the salt and water are absorbed in the gastrointestinal tract, blood volume increases without a decrease in osmolarity, so water intoxication does not occur.

27.6 In ECF, the major cation is Na^+, and the major anions are Cl^- and HCO_3^-. In ICF, the major cation is K^+, and the major anions are proteins and organic phosphates (for example, ATP).

27.7 Holding one's breath causes blood pH to decrease slightly as CO_2 and H^+ accumulate.

27.8 A carbonic anhydrase inhibitor reduces secretion of H^+ into the urine and reduces reabsorption of Na^+ and HCO_3^- into the blood. It has a diuretic effect and can cause acidosis (lowered pH of the blood) due to loss of HCO_3^- in the urine.

The Reproductive Systems

THE REPRODUCTIVE SYSTEMS AND HOMEOSTASIS

The male and female reproductive organs work together to produce offspring. The female reproductive organs contribute to sustaining the growth of fetuses.

ENERGIZE YOUR STUDY

FOUNDATIONS CD

Anatomy Overview
• The Reproductive Systems

Animation
• Systems Contributions to Homeostasis

Concepts and Connections and Exercises reinforce your understanding

www.wiley.com/college/apcentral

INSIGHTS AND EXPLORATIONS

A young man of 23, planning to start a family, feels a painless lump in one of his testicles. A woman who feels fine is told, after her yearly examination at age 55, that she has ovarian cancer. An African American man of 56, who is without symptoms, is diagnosed with prostate cancer. Young or old, we can fall victim to cancers of the reproductive organs. What are the chances that you will develop one of these cancers? In this web-based activity we will explore who is at risk for developing these cancers and more.

Sexual reproduction is the process by which organisms produce offspring by making germ cells called **gametes** (GAM-ēts = spouses). After the male gamete (sperm cell) unites with the female gamete (secondary oocyte)—an event called **fertilization**—the resulting cell contains one set of chromosomes from each parent. Males and females have anatomically distinct reproductive organs that are adapted for producing gametes, facilitating fertilization, and in females sustaining the growth of the embryo and fetus.

The male and female reproductive organs can be grouped by function. The **gonads**—testes in males and ovaries in females—produce gametes and secrete sex hormones. Various **ducts** then store and transport the gametes, and **accessory sex glands** produce substances that protect the gametes and facilitate their movement. Finally, supporting structures, such as the penis and the uterus, assist the delivery and joining of gametes and, in females, the growth of the fetus during pregnancy.

Gynecology (gī-ne-KOL-ō-jē; *gyneco-* = woman; *-logy* = study of) is the specialized branch of medicine concerned with the diagnosis and treatment of diseases of the female reproductive system. As noted in Chapter 26, **urology** (ū-ROL-ō-jē) is the study of the urinary system. Urologists also diagnose and treat diseases and disorders of the male reproductive system.

THE CELL CYCLE IN THE GONADS

OBJECTIVE

• **Describe the process of meiosis.**

Number of Chromosomes in Somatic Cells and Gametes

Before we can discuss the cell cycle in the gonads, we must first understand the distribution of genetic material in cells. Human somatic cells, such as those in the brain, stomach, and kidneys, contain 23 pairs of chromosomes, or a total of 46 chromosomes. One member of each pair was inherited from each parent. The two chromosomes that make up each pair are called **homologous chromosomes** (hō-MOL-ō-gus; *homo-* = same) or **homologs;** they contain similar genes arranged in the same (or almost the same) order. When examined under a light microscope (see Figure 29.24 on page 1096), homologous chromosomes generally look very similar. The exception to this rule is one pair of chromosomes called the **sex chromosomes,** designated X and Y. In females the homologous pair of sex chromosomes consists of two large X chromosomes; in males the pair consists of an X and a much smaller Y chromosome. The other 22 pairs of chromosomes are called **autosomes.** Because somatic cells contain two sets of chromosomes, they are called **diploid cells** (DIP-loid; *dipl-* = double; *-oid* = form). Geneticists use the symbol n to denote the number of different chromosomes in an organism; in humans, $n = 23$. Diploid cells are $2n$.

In sexual reproduction, each new organism is the result of the union of two different gametes, one produced by each parent. If gametes had the same number of chromosomes as somatic cells, the number of chromosomes would double at fertilization. Because **meiosis** (mī-Ō-sis; *mei-* = lessening; *-osis* = condition of), the reproductive cell division that occurs in the gonads, produces gametes in which the number of chromosomes is reduced by half, gametes contain a single set of 23 chromosomes and thus are **haploid (1n) cells** (HAP-loyd; *hapl-* = single). Fertilization then restores the diploid number of chromosomes.

Meiosis

Meiosis occurs in two successive stages: **meiosis I** and **meiosis II.** During the interphase that precedes meiosis I, the chromosomes of the diploid parent cell replicate. As a result of replication, each chromosome consists of two genetically identical (sister) chromatids, which are attached at their centromeres. This replication of chromosomes is similar to the one that precedes mitosis in somatic cell division.

Meiosis I

Meiosis I, which begins once chromosomal replication is complete, consists of four phases: prophase I, metaphase I, anaphase I, and telophase I (Figure 28.1a). Prophase I is an extended phase in which the chromosomes shorten and thicken, the nuclear envelope and nucleoli disappear, and the mitotic spindle forms. Two events that are not seen in prophase I of mitosis (or in prophase II of meiosis) occur during prophase I of meiosis. First, the two sister chromatids of each pair of homologous chromosomes pair off, an event called **synapsis** (Figure 28.1b). The resulting four chromatids form a structure called a **tetrad.** Second, parts of the chromatids of two homologous chromosomes may be exchanged with one another. Such an exchange between parts of genetically different (nonsister) chromatids is termed **crossing-over** (Figure 28.1b on page 1014). This process, among others, permits an exchange of genes between chromatids of homologous chromosomes. Due to crossing-over, the resulting daughter cells are genetically unlike each other and genetically unlike the parent cell that produced them. Crossing-over results in **genetic recombination**—that is, the formation of new combinations of genes—and accounts for part of the great genetic variation among humans and other organisms that form gametes via meiosis. The random assortment of maternally derived chromosomes and paternally derived chromosomes toward opposite poles is another reason for genetic variation among gametes.

In metaphase I, the homologous pairs of chromosomes line up along the equatorial plane of the cell, with homologous chromosomes side by side (see Figure 28.1a). (No such pairing of

Figure 28.1   Meiosis, reproductive cell division.   Details of events are discussed in the text.

In reproductive cell division, a single diploid parent cell undergoes meiosis I and meiosis II to produce four haploid gametes that are genetically different from the parent cell.

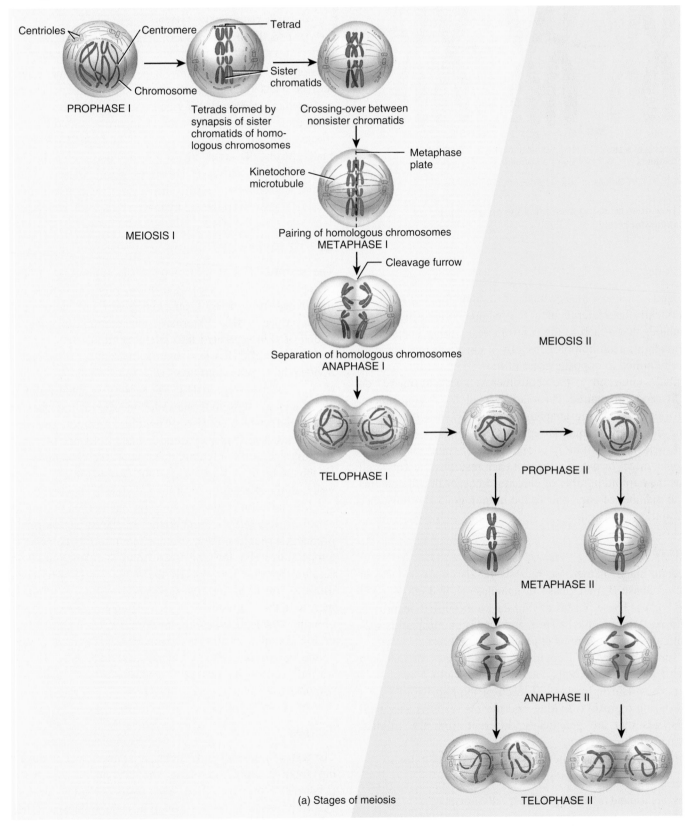

(a) Stages of meiosis

(continues)

Figure 28.1 *(continued)*

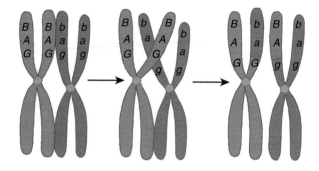

Synapsis of sister   Crossing-over between   Genetic
chromatids   nonsister chromatids   recombination

(b) Details of crossing-over during prophase I

How does crossing-over affect the genetic content of daughter cells?

homologous chromosomes occurs during metaphase of mitosis.) The pericentriolar material of a centrosome forms kinetochore microtubules that attach to the centromeres. During anaphase I, the members of each homologous pair of chromosomes separate as they are pulled to opposite poles of the cell by the kinetochore microtubules attached to the centromeres. The centromeres do not split, and the paired chromatids, held by a centromere, remain together. (During anaphase of mitosis, the centromeres split and the sister chromatids separate.) Telophase I and cytokinesis of meiosis are similar to telophase and cytokinesis of mitosis. The net effect of meiosis I is that each resulting daughter cell contains the haploid number of chromosomes; each cell contains only one member of each pair of the homologous chromosomes present in the parent cell.

Meiosis II

The second stage of meiosis, meiosis II, also consists of four phases: prophase II, metaphase II, anaphase II, and telophase II (Figure 28.1a). These phases are similar to those that occur during mitosis; the centromeres split, and the sister chromatids separate and move toward opposite poles of the cell.

In summary, meiosis I begins with a diploid parent cell and ends with two daughter cells, each with the haploid number of chromosomes. During meiosis II, each of the two haploid cells formed during meiosis I divides, and the net result is four haploid gametes that are genetically different from the original diploid parent cell.

Figure 28.2 compares the events of meiosis and mitosis.

▶ **CHECKPOINT**

1. How are haploid *(n)* and diploid *(2n)* cells different?

2. What are homologous chromosomes?

3. Prepare a table to compare meiosis with mitosis. (*Hint:* To review mitosis in more detail, refer to Figure 3.33 on page 92.)

MALE REPRODUCTIVE SYSTEM

▶ **OBJECTIVES**

• Describe the location, structure, and functions of the organs of the male reproductive system.

• Discuss the process of spermatogenesis in the testes.

The organs of the male reproductive system are the testes, a system of ducts (including the ductus deferens, ejaculatory ducts, and urethra), accessory sex glands (seminal vesicles, prostate, and bulbourethral gland), and several supporting structures, including the scrotum and the penis (Figure 28.3 on pages 1016–1017). The male gonads, the testes, produce sperm and secrete hormones. A system of ducts transports and stores sperm, assists in their maturation, and conveys them to the exterior. Semen contains sperm plus the secretions provided by the accessory sex glands.

Scrotum

The **scrotum** (SKRŌ-tum = bag), the supporting structure for the testes, is a sac consisting of loose skin and superficial fascia that hangs from the root (attached portion) of the penis (Figure 28.3a on page 1018). Externally, the scrotum looks like a single pouch of skin separated into lateral portions by a median ridge called the **raphe** (RĀ-fē = seam). Internally, the scrotal septum divides the scrotum into two sacs, each containing a single testis (Figure 28.4 on page 1018). The septum is made up of superficial fascia and muscle tissue called the **dartos muscle** (DAR-tōs = skinned), which consists of bundles of smooth muscle fibers. The dartos muscle is also found in the subcutaneous tissue of the scrotum and is directly continuous with the subcutaneous tissue of the abdominal wall. When the dartos muscle contracts, it wrinkles the skin of the scrotum and elevates the testes.

The location of the scrotum and the contraction of its muscle fibers regulate the temperature of the testes. Normal sperm production requires a temperature about 2–3°C below core body temperature. This lowered temperature is maintained within the scrotum because it is outside the pelvic cavity. The **cremaster muscle** (krē-MAS-ter = suspender), a small band of skeletal muscle in the spermatic cord that is a continuation of the internal oblique muscle, elevates the testes upon exposure to cold (and during sexual arousal). This action moves the testes closer to the pelvic cavity, where they can absorb body heat. Exposure to warmth reverses the process. The dartos muscle also contracts in response to cold and relaxes in response to warmth.

Testes

The **testes** (TES-tēz), or **testicles,** are paired oval glands measuring about 5 cm (2 in.) long and 2.5 cm (1 in.) in diameter (Figure 28.5 on page 1019). Each **testis** (singular) has a mass of 10–15 grams. The testes develop near the kidneys, in the posterior part of the abdomen, and they usually begin their descent into the scrotum through the inguinal canals (passageways in the anterior abdominal wall; see Figure 28.4) during the latter half

Figure 28.2 **Comparison between mitosis (left) and meiosis (right) in which the parent cell has two pairs of homologous chromosomes.**

The phases of meiosis II and mitosis are similar.

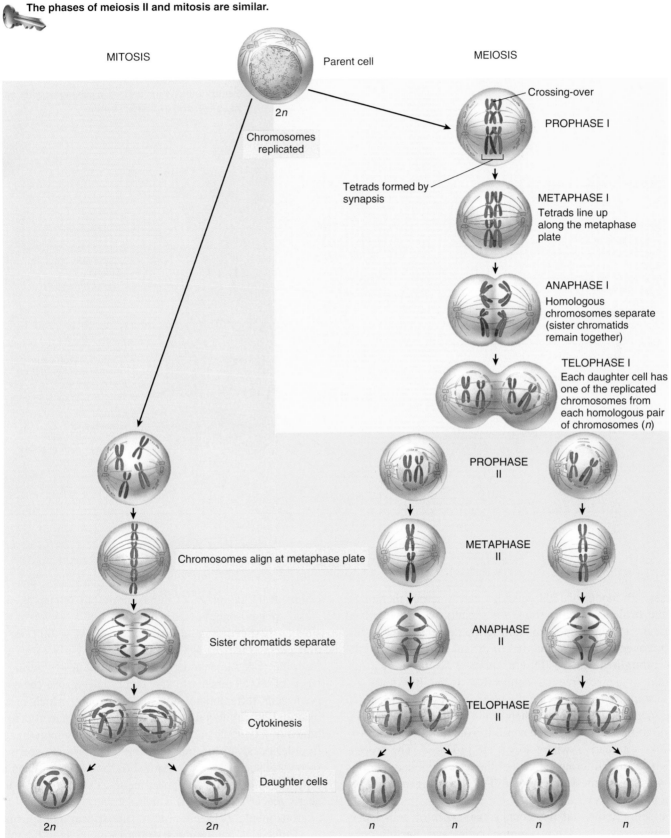

How does anaphase I of meiosis differ from anaphase of mitosis and anaphase II of meiosis?

Reproductive organs are adapted for producing new individuals and passing on genetic material from one generation to the next.

Functions of the Male Reproductive System

1. **Testes:** produce sperm and the male sex hormone testosterone.
2. **Ducts:** transport, store, and assist in maturation of sperm.
3. **Accessory sex glands:** secrete most of the liquid portion of semen.
4. **Penis:** contains the urethra, a passageway for ejaculation of semen and excretion of urine.

Sagittal plane

Sacrum

Seminal vesicle
Vesicorectal pouch
Coccyx
Rectum
Ampulla of ductus (vas) deferens
Ejaculatory duct
Prostatic urethra
Membranous urethra

Anus

Bulb of penis

Epididymis

Testis

Scrotum

Urinary bladder
Ductus (vas) deferens
Suspensory ligament of penis
Pubic symphysis
Prostate
Urogenital diaphragm
Bulbourethral (Cowper's) gland
Corpora cavernosum penis
Spongy (penile) urethra
Penis
Corpus spongiosum penis
Corona
Glans penis
Prepuce (foreskin)
External urethral orifice

(a) Sagittal section

of the seventh month of fetal development. A serous membrane called the **tunica vaginalis** (*tunica* = sheath), which is derived from the peritoneum and forms during the descent of the testes, partially covers the testes. Internal to the tunica vaginalis is the **tunica albuginea** (al′-bū-JIN-ē-a; *albu-* = white), a capsule composed of dense irregular connective tissue. It extends inward, forming septa that divide the testis into a series of internal compartments called **lobules.** Each of the 200–300 lobules contains one to three tightly coiled **seminiferous tubules** (*semin-* = seed; *fer-* = to carry), where sperm are produced. The process by which the seminiferous tubules of the testes produce sperm is called **spermatogenesis** (sper′-ma-tō-JEN-e-sis; *genesis* = beginning process or production).

The seminiferous tubules contain two types of cells: **spermatogenic cells,** the sperm-forming cells, and **Sertoli cells,** which have several functions in supporting spermatogenesis (Figure 28.6 on page 1020). Starting at puberty, sperm production

begins at the periphery of the seminiferous tubules in stem cells called **spermatogonia** (sper′-ma-tō-GŌ-nē-a; *-gonia* = offspring; singular is **spermatogonium).** These cells develop from **primordial germ cells** (*primordi-* = primitive or early form) that arise from the yolk sac endoderm and enter the testes during the fifth week of development. In the embryonic testes, the primordial germ cells differentiate into spermatogonia, which remain dormant during childhood and become active at puberty. Toward the lumen of the tubule are layers of progressively more mature cells. In order of advancing maturity, these are primary spermatocytes, secondary spermatocytes, spermatids, and sperm. When a **sperm cell,** or **spermatozoon** (sper′-ma-tō-ZŌ-on; *-zoon* = life), has nearly reached maturity, it is released into the lumen of the seminiferous tubule. (The plural terms are **sperm** and **spermatozoa.)**

Embedded among the spermatogenic cells in the tubules are large **Sertoli cells** or *sustentacular cells* (sus′-ten-TAK-ū-lar), which extend from the basement membrane to the lumen of the

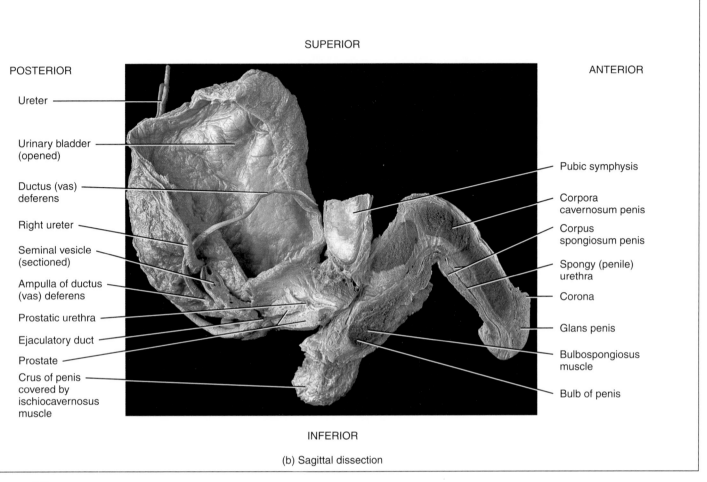

SUPERIOR

POSTERIOR

ANTERIOR

Ureter

Urinary bladder
(opened)

Pubic symphysis

Ductus (vas)
deferens

Corpora
cavernosum penis

Right ureter

Corpus
spongiosum penis

Seminal vesicle
(sectioned)

Spongy (penile)
urethra

Ampulla of ductus
(vas) deferens

Corona

Prostatic urethra

Ejaculatory duct

Glans penis

Prostate

Bulbospongiosus
muscle

Crus of penis
covered by
ischiocavernosus
muscle

Bulb of penis

INFERIOR

(b) Sagittal dissection

 What are the groups of reproductive organs in males, and what are the functions of each group?

tubule. Internal to the basement membrane and spermatogonia, tight junctions join neighboring Sertoli cells to one another. These junctions form an obstruction known as the **blood–testis barrier** because substances must first pass through the Sertoli cells before they can reach the developing sperm. By isolating the developing gametes from the blood, the blood–testis barrier prevents an immune response against the spermatogenic cell's surface antigens, which are recognized as "foreign" by the immune system. The blood–testis barrier does not include spermatogonia.

Sertoli cells support and protect developing spermatogenic cells in several ways. They nourish spermatocytes, spermatids, and sperm; phagocytize excess spermatid cytoplasm as development proceeds; and control movements of spermatogenic cells and the release of sperm into the lumen of the seminiferous tubule. They also produce fluid for sperm transport, secrete androgen-binding protein and the hormone inhibin, and mediate

the effects of testosterone and FSH (follicle-stimulating hormone).

In the spaces between adjacent seminiferous tubules are clusters of cells called **Leydig cells** or *interstitial endocrinocytes* (Figure 28.6a). These cells secrete testosterone, the most prevalent androgen. Although androgens are hormones that promote development of masculine characteristics, they also have other functions, such as promoting libido (sexual desire) in both males and females.

 Cryptorchidism

The condition in which the testes do not descend into the scrotum is called **cryptorchidism** (krip-TOR-ki-dizm; *crypt-* = hidden; *orchid* = testis). It occurs in about 3% of full-term infants and about 30% of premature infants. Untreated bilateral cryptorchidism results in sterility because the cells involved in the

Figure 28.4 The scrotum, the supporting structure for the testes.

 The scrotum consists of loose skin and superficial fascia and supports the testes.

Internal oblique muscle

Aponeurosis of external oblique muscle (cut)

Fundiform ligament of penis

Suspensory ligament of penis

Transverse section of penis:
Corpora cavernosa penis

Spongy (penile) urethra

Corpus spongiosum penis

Scrotal septum

Cremaster muscle

External spermatic fascia

Dartos muscle

Skin of scrotum

Spermatic cord
Superficial inguinal ring
Cremaster muscle
Inguinal canal

Ductus (vas) deferens
Autonomic nerve

Testicular artery

Lymphatic vessel
Pampiniform plexus of testicular veins

Epididymis

Tunica albuginea of testis

Tunica vaginalis (peritoneum)
Internal spermatic fascia

Raphe

Anterior view of scrotum and testes and transverse section of penis

Which muscles help regulate the temperature of the testes?

initial stages of spermatogenesis are destroyed by the higher temperature of the pelvic cavity. The chance of testicular cancer is 30 to 50 times greater in cryptorchid testes, possibly due to abnormal division of germ cells due to the higher temperature of the pelvic cavity. The testes of about 80% of boys with cryptorchidism will descend spontaneously during the first year of life. When the testes remain undescended, the condition can be corrected surgically, ideally before 18 months of age. ■

Spermatogenesis

In humans, spermatogenesis takes 65 to 75 days. It begins in the spermatogonia, which contain the diploid (2*n*) chromosome number (Figure 28.7 on page 1021). Spermatogonia are *stem cells* because after they undergo mitosis, one daughter cell stays near the basement membrane of the seminiferous tubule in an undifferentiated state. As a result, stem cells remain for future cell divisions. The other daughter cell loses contact with the basement membrane, squeezes through the tight junctions of the blood–testis barrier, undergoes developmental changes, and differentiates into a **primary spermatocyte** (SPER-ma-tō-sīt′). Primary spermatocytes, like spermatogonia, are diploid (2*n*); that is, they have 46 chromosomes.

Each primary spermatocyte enlarges and then begins meiosis (Figure 28.7). In meiosis I, DNA replicates, homologous pairs of chromosomes line up at the metaphase plate, and crossing-over occurs. Then, the meiotic spindle forms and pulls one (duplicated) chromosome of each pair to an opposite pole of the dividing cell. The two cells formed by meiosis I are called **secondary spermatocytes.** Each spermatocyte has 23 chromosomes, the haploid number. Each chromosome within a secondary spermatocyte, however, is made up of two chromatids (two copies of the DNA) still attached by a centromere.

In meiosis II, no further replication of DNA occurs. The chromosomes line up in single file along the metaphase plate, and the two chromatids of each chromosome separate. The four

Figure 28.5 Internal and external anatomy of a testis.

The testes are the male gonads, which produce haploid sperm.

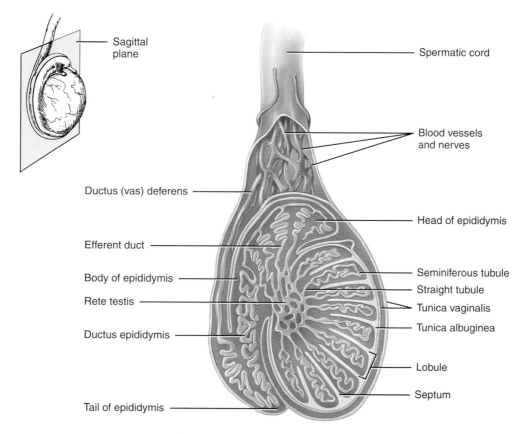

(a) Sagittal section of a testis showing seminiferous tubules

(b) Transverse section

(c) Testis and associated structures (lateral view)

What tissue layers cover and protect the testes?

Figure 28.6 Microscopic anatomy of the seminiferous tubules and stages of sperm production (spermatogenesis). Arrows in (b) indicate the progression of spermatogenic cells from least mature to most mature. The (n) and ($2n$) refer to haploid and diploid chromosome number, respectively.

 Spermatogenesis occurs in the seminiferous tubules of the testes.

Leydig cell (interstitial endocrinocyte)

Basement membrane

Spermatogonium ($2n$) (stem cell)

Primary spermatocyte ($2n$)

Secondary spermatocyte (n)

Spermatid (n)

Sperm cell or spermatozoon (n)

Lumen of seminiferous tubule

Transverse plane

LM 310x

(a) Transverse section of several seminiferous tubules

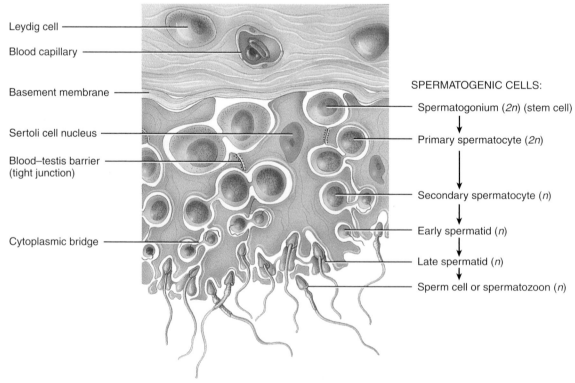

Leydig cell

Blood capillary

Basement membrane

Sertoli cell nucleus

Blood–testis barrier (tight junction)

Cytoplasmic bridge

SPERMATOGENIC CELLS:

Spermatogonium ($2n$) (stem cell)

Primary spermatocyte ($2n$)

Secondary spermatocyte (n)

Early spermatid (n)

Late spermatid (n)

Sperm cell or spermatozoon (n)

(b) Transverse section of a portion of a seminiferous tubule

Which cells produce testosterone?

Figure 28.7 Events in spermatogenesis. Diploid cells (*2n*) have 46 chromosomes; haploid cells (*n*) have 23 chromosomes.

🔑 Spermiogenesis involves the maturation of spermatids into sperm.

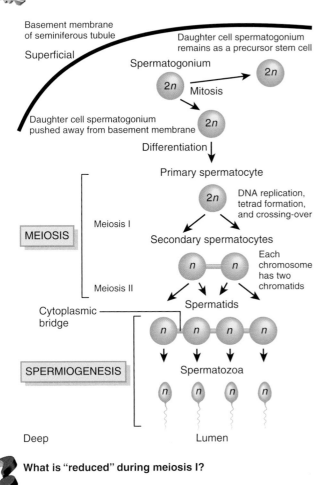

❓ What is "reduced" during meiosis I?

denses and elongates, a flagellum develops, and mitochondria multiply. Sertoli cells dispose of the excess cytoplasm that sloughs off. Finally, sperm are released from their connections to Sertoli cells, an event known as **spermiation.** Sperm then enter the lumen of the seminiferous tubule. Fluid secreted by Sertoli cells pushes sperm along their way, toward the ducts of the testes.

Sperm

Spermatogenesis produces about 300 million sperm per day. Once ejaculated, most do not survive more than 48 hours within the female reproductive tract. A sperm cell consists of three structures highly adapted for reaching and penetrating a secondary oocyte: a head, a midpiece, and a tail (Figure 28.8). The **head** is 4–5 μm long. It contains an **acrosome** (*acro-* = atop; *-some* = body), a lysosomelike vesicle, and a nucleus that has the haploid number of chromosomes (23). Enzymes within the acrosome include hyaluronidase and proteases, which aid penetration of the sperm cell into a secondary oocyte. In the **midpiece** are many mitochondria, which provide ATP for locomotion. The **tail,** a typical flagellum, propels the sperm cell along its way. From head to tip of tail, human sperm are about 70 μm in length.

Hormonal Control of the Testes

Although the initiating factors are not known, at puberty certain hypothalamic neurosecretory cells increase their secretion of **gonadotropin-releasing hormone (GnRH).** This hormone, in

Figure 28.8 Parts of a sperm cell.

🔑 About 300 million sperm mature each day.

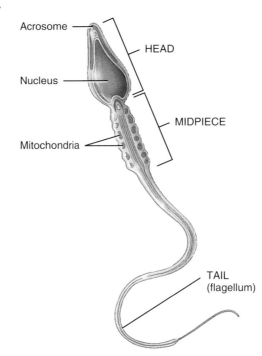

❓ What are the functions of each part of a sperm cell?

haploid cells resulting from meiosis II are called **spermatids.** A single primary spermatocyte therefore produces four spermatids via two rounds of cell division (meiosis I and meiosis II).

A unique process occurs during spermatogenesis. As the sperm cells proliferate, they fail to complete cytoplasmic separation (cytokinesis). The four daughter cells remain in contact via cytoplasmic bridges through their entire development (see Figures 28.6b and 28.7). This pattern of development most likely accounts for the synchronized production of sperm in any given area of seminiferous tubule. It may also have survival value in that half of the sperm contain an X chromosome and half contain a Y chromosome. The larger X chromosome may carry genes needed for spermatogenesis that are lacking on the smaller Y chromosome.

The final stage of spermatogenesis, **spermiogenesis** (sper'-mē-ō-JEN-e-sis), is the maturation of haploid spermatids into sperm. Because no cell division occurs in spermiogenesis, each spermatid develops into a single **sperm cell.** During this process, spherical spermatids transform into elongated, slender sperm. An acrosome (described shortly) forms atop the nucleus, which conturn, stimulates gonadotrophs in the anterior pituitary to increase their secretion of the two gonadotropins, **luteinizing hormone (LH)** and **follicle-stimulating hormone (FSH).** Figure 28.9 shows the hormones and negative feedback loops that control secretion of testosterone and spermatogenesis.

LH stimulates Leydig cells, which are located between seminiferous tubules, to secrete the hormone **testosterone** (tes-TOS-te-rōn). This steroid hormone is synthesized from cholesterol in the testes and is the principal androgen. It is lipid soluble and readily diffuses out of Leydig cells into the interstitial fluid and then into blood. Testosterone acts in a negative feedback manner to suppress secretion of LH by anterior pituitary gonadotrophs and to suppress secretion of GnRH by hypothalamic neurosecretory cells. In some target cells, such as those in the prostate and seminal vesicles, the enzyme 5 alpha-reductase converts testosterone to another androgen called **dihydrotestosterone (DHT)**.

FSH acts indirectly to stimulate spermatogenesis (Figure 28.9). FSH and testosterone act synergistically on the Sertoli cells to stimulate secretion of **androgen-binding protein (ABP)** into the lumen of the seminiferous tubules and into the interstitial fluid around the spermatogenic cells. ABP binds to testosterone, thereby keeping the concentration of testosterone high near the seminiferous tubules. Testosterone stimulates the final steps of spermatogenesis. Once the degree of spermatogenesis required for male reproductive functions has been achieved, Sertoli cells release **inhibin**, a protein hormone named for its inhibition of FSH secretion by the anterior pituitary (Figure 28.9). Inhibin thus inhibits the secretion of hormones needed for spermatogenesis. If spermatogenesis is proceeding too slowly, less inhibin is released, which permits more FSH secretion and an increased rate of spermatogenesis.

Testosterone and dihydrotestosterone both bind to the same androgen receptors, which are found within the nuclei of target cells. The hormone–receptor complex acts to regulate gene expression, turning some genes on and others off. Because of these changes, the androgens produce several effects:

- *Prenatal development.* Before birth, testosterone stimulates the male pattern of development of reproductive system ducts and the descent of the testes. Dihydrotestosterone, by contrast, stimulates development of the external genitals (described on page 1050). Testosterone also is converted in the brain to estrogens (feminizing hormones), which may play a role in the development of certain regions of the brain in males.

- *Development of male sexual characteristics.* At puberty, testosterone and dihydrotestosterone bring about development and enlargement of the male sex organs and the development of masculine secondary sexual characteristics. These include muscular and skeletal growth that results in wide shoulders and narrow hips; pubic, axillary, facial, and chest hair (within hereditary limits); thickening of the skin; increased sebaceous (oil) gland secretion; and enlargement of the larynx and consequent deepening of the voice.

Figure 28.9 Hormonal control of spermatogenesis and actions of testosterone and dihydrotestosterone (DHT). In response to stimulation by FSH and testosterone, Sertoli cells secrete androgen-binding protein (ABP). Dashed red lines indicate negative feedback inhibition.

🔑 Release of FSH is stimulated by GnRH and inhibited by inhibin; release of LH is stimulated by GnRH and inhibited by testosterone.

❓ Which cells secrete inhibin?

- *Development of sexual function.* Androgens contribute to male sexual behavior and spermatogenesis and to sex drive (libido) in both males and females. Recall that the adrenal cortex is the main source of androgens in females.

- **Stimulation of anabolism.** Androgens are anabolic hormones; that is, they stimulate protein synthesis. This effect is obvious in the heavier muscle and bone mass of most men as compared to women.

A negative feedback system regulates testosterone production (Figure 28.10). When testosterone concentration in the blood increases to a certain level, it inhibits the release of GnRH by cells in the hypothalamus. As a result, there is less GnRH in the portal blood that flows from the hypothalamus to the anterior pituitary. Gonadotrophs in the anterior pituitary then release less LH, so the concentration of LH in systemic blood falls. With less stimulation by LH, the Leydig cells in the testes secrete less testosterone, and there is a return to homeostasis. If the testosterone concentration in the blood falls too low, however, GnRH is again released by the hypothalamus and stimulates secretion of LH by the anterior pituitary. LH, in turn, stimulates testosterone production by the testes.

▶ CHECKPOINT

4. How does the scrotum protect the testes?
5. Where are sperm cells produced?
6. What are the functions of Sertoli cells and Leydig cells?
7. What are the principal events of spermatogenesis?
8. Identify the main parts of a sperm cell, and list the functions of each.
9. What are the roles of FSH and LH in the male reproductive system? How is secretion of these hormones controlled?
10. What are the effects of testosterone and inhibin?
11. How is the blood level of testosterone controlled?

Reproductive System Ducts in Males

Ducts of the Testis

Pressure generated by the fluid secreted by Sertoli cells pushes sperm and fluid along the lumen of seminiferous tubules and then into a series of very short ducts called **straight tubules.** The straight tubules lead to a network of ducts in the testis called the **rete testis** (RĒ-tē = network) (see Figure 28.5a). From the rete testis, sperm move into a series of coiled **efferent ducts** in the epididymis that empty into a single tube called the **ductus epididymis.**

Epididymis

The **epididymis** (ep′-i-DID-i-mis; *epi-* = above or over; *-didymis* = testis) is a comma-shaped organ about 4 cm (1.5 in.) long that lies along the posterior border of each testis (see Figure 28.5a). The plural is **epididymides** (ep′-i-did-IM-i-dēs). Each epididymis consists mostly of the tightly coiled **ductus epididymis.** The larger, superior portion of the epididymis, the **head,** is where the efferent ducts from the testis join the ductus epididymis. The **body** is the narrow midportion of the epididymis, and the **tail** is the smaller, inferior portion. At its distal end, the tail of the

Figure 28.10 Negative feedback control of blood level of testosterone.

 Gonadotrophs of the anterior pituitary produce luteinizing hormone (LH).

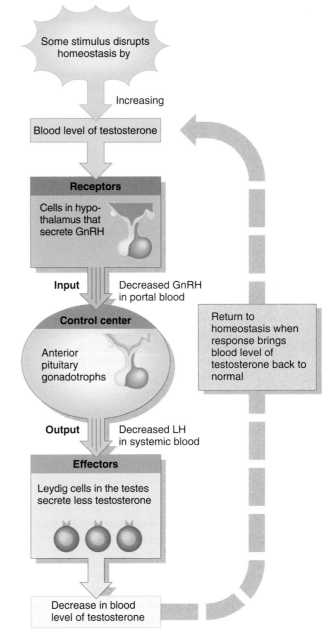

Some stimulus disrupts homeostasis by

Increasing

Blood level of testosterone

Receptors

Cells in hypothalamus that secrete GnRH

Input — Decreased GnRH in portal blood

Control center

Anterior pituitary gonadotrophs

Return to homeostasis when response brings blood level of testosterone back to normal

Output — Decreased LH in systemic blood

Effectors

Leydig cells in the testes secrete less testosterone

Decrease in blood level of testosterone

? Which hormones inhibit secretion of FSH and LH by the anterior pituitary?

epididymis continues as the ductus (vas) deferens (discussed shortly).

The ductus epididymis would measure about 6 m (20 ft) in length if it were straightened out. It is lined with pseudostratified columnar epithelium and encircled by layers of smooth muscle. The free surfaces of the columnar cells contain **stereocilia,** which despite their name are long, branching microvilli that increase surface area for the reabsorption of degenerated sperm.

Functionally, the ductus epididymis is the site where sperm mature, that is, they acquire motility and the ability to fertilize an ovum. This occurs over a 10- to 14-day period. The ductus epididymis also stores sperm and helps propel them by peristaltic contraction of its smooth muscle into the ductus (vas) deferens. Sperm may remain in storage in the ductus epididymis for a month or more.

Ductus Deferens

Within the tail of the epididymis, the ductus epididymis becomes less convoluted, and its diameter increases. Beyond this point, the duct is known as the **ductus deferens** or **vas deferens** (see Figure 28.5a). The ductus deferens, which is about 45 cm (18 in.) long, ascends along the posterior border of the epididymis, passes through the inguinal canal (see Figure 28.4), and enters the pelvic cavity. There it loops over the ureter and passes over the side and down the posterior surface of the urinary bladder (see Figure 28.3a). The dilated terminal portion of the ductus deferens is the **ampulla** (am-PUL-la = little jar; see Figure 28.11). The ductus deferens is lined with pseudostratified columnar epithelium and contains a heavy coat of three layers of smooth muscle. The inner and outer layers are arranged longitudinally, whereas the middle layer encircles the ductus deferens.

Functionally, the ductus deferens stores sperm, which can remain viable here for up to several months. The ductus deferens also conveys sperm from the epididymis toward the urethra by peristaltic contractions of the muscular coat. Sperm that are not ejaculated are eventually reabsorbed.

Spermatic Cord

The **spermatic cord** is a supporting structure of the male reproductive system that ascends out of the scrotum (see Figure 28.4). It consists of the ductus (vas) deferens as it ascends through the scrotum, the testicular artery, autonomic nerves, veins that drain the testes and carry testosterone into circulation (the pampiniform plexus), lymphatic vessels, and the cremaster muscle. The spermatic cord and ilioinguinal nerve pass through the **inguinal canal** (IN-gwinal = groin), an oblique passageway in the anterior abdominal wall just superior and parallel to the medial half of the inguinal ligament. The canal, which is about 4–5 cm (about 2 in.) long, originates at the **deep (abdominal) inguinal ring,** a slitlike opening in the aponeurosis of the transversus abdominis muscle. The canal ends at the **superficial (subcutaneous) inguinal ring** (see Figure 28.4), a triangular opening in the aponeurosis of the external oblique muscle.

Inguinal Hernias

Because the inguinal region is a weak area in the abdominal wall, it is often the site of an **inguinal hernia**—a rupture or separation of a portion of the inguinal area of the abdominal wall resulting in the protrusion of a part of the small intestine. In an *indirect inguinal hernia,* a part of the small intestine protrudes through the deep inguinal ring and enters the scrotum. In a *direct inguinal hernia,* a portion of the small intestine pushes into the posterior wall of the inguinal canal, usually causing a localized bulging in the wall of the canal. Inguinal hernias are much more common in males than in females because the larger inguinal canals in males represent larger weak points in the abdominal wall. ■

Ejaculatory Ducts

Each **ejaculatory duct** (e-JAK-ū-la-tō′-rē; *ejacul-* = to expel) is about 2 cm (1 in.) long and is formed by the union of the duct from the seminal vesicle and the ampulla of the ductus deferens (Figure 28.11). The ejaculatory ducts arise just superior to the base (superior portion) of the prostate and pass inferiorly and anteriorly through the prostate. They terminate in the prostatic urethra, where they eject sperm and seminal vesicle secretions just before **ejaculation,** the powerful propulsion of semen from the urethra to the exterior.

Urethra

In males, the **urethra** is the shared terminal duct of the reproductive and urinary systems; it serves as a passageway for both semen and urine. The urethra, which is about 20 cm (8 in.) long, has three parts that pass through the prostate, the urogenital diaphragm, and the penis (see Figures 28.3a and 28.11). The **prostatic urethra** is 2–3 cm (1 in.) long and passes through the prostate. As this part of the urethra continues inferiorly, it passes through the urogenital diaphragm (a muscular partition between the two ischial and pubic rami; see Figure 11.13 on page 341), where it is known as the **membranous urethra.** The membranous urethra is about 1 cm (0.5 in.) in length. As this duct passes into the corpus spongiosum of the penis, it becomes the **spongy (penile) urethra,** which is about 15–20 cm (6–8 in.) long, and ends at the **external urethral orifice.** The histology of the male urethra may be reviewed on page 981.

Accessory Sex Glands

Whereas the ducts of the male reproductive system store and transport sperm cells, the **accessory sex glands** secrete most of the liquid portion of semen. The accessory sex glands include the seminal vesicles, the prostate, and the bulbourethral glands.

Seminal Vesicles

The paired **seminal vesicles** (VES-i-kuls) or **seminal glands** are convoluted pouchlike structures, about 5 cm (2 in.) in length, lying posterior to the base of the urinary bladder and anterior to the rectum (Figure 28.11). They secrete an alkaline, viscous fluid that contains fructose (a monosaccharide sugar), prostaglandins, and clotting proteins that are different from those in blood. The alkaline nature of the seminal fluid helps to neutralize the acidic environment of the male urethra and female repro-

Figure 28.11 Locations of several accessory reproductive organs in males. The prostate, urethra, and penis have been sectioned to show internal details.

The male urethra has three subdivisions: the prostatic, membranous, and spongy (penile) urethra.

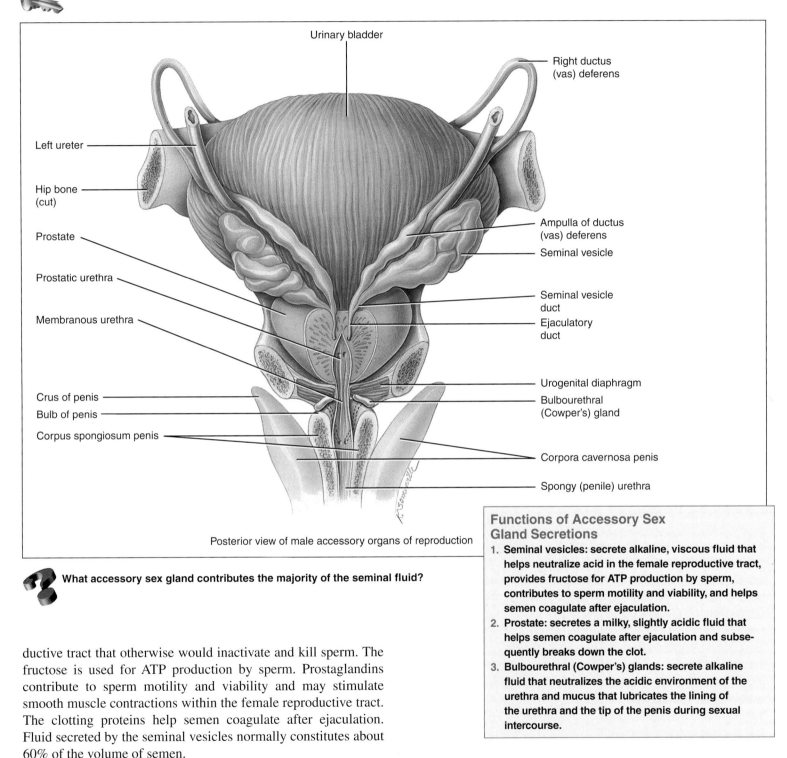

Posterior view of male accessory organs of reproduction

What accessory sex gland contributes the majority of the seminal fluid?

Functions of Accessory Sex Gland Secretions

1. **Seminal vesicles: secrete alkaline, viscous fluid that helps neutralize acid in the female reproductive tract, provides fructose for ATP production by sperm, contributes to sperm motility and viability, and helps semen coagulate after ejaculation.**
2. **Prostate: secretes a milky, slightly acidic fluid that helps semen coagulate after ejaculation and subsequently breaks down the clot.**
3. **Bulbourethral (Cowper's) glands: secrete alkaline fluid that neutralizes the acidic environment of the urethra and mucus that lubricates the lining of the urethra and the tip of the penis during sexual intercourse.**

ductive tract that otherwise would inactivate and kill sperm. The fructose is used for ATP production by sperm. Prostaglandins contribute to sperm motility and viability and may stimulate smooth muscle contractions within the female reproductive tract. The clotting proteins help semen coagulate after ejaculation. Fluid secreted by the seminal vesicles normally constitutes about 60% of the volume of semen.

Prostate

The **prostate** (PROS-tāt) is a single, doughnut-shaped gland about the size of a golf ball. It measures about 4 cm (1.6 in.) from side to side, about 3 cm (1.2 in.) from top to bottom, and about 2 cm (0.8 in.) from front to back. It is inferior to the urinary bladder and surrounds the prostatic urethra (Figure 28.11). The prostate slowly increases in size from birth to puberty, and then it expands rapidly. The size attained by age 30 typically remains stable until about age 45, when further enlargement may occur.

The prostate secretes a milky, slightly acidic fluid (pH about 6.5) that contains several important substances: (1) *Citric acid* in prostatic fluid is used by sperm for ATP production via the Krebs cycle. (2) Several *proteolytic enzymes,* such as *prostate-specific antigen (PSA),* pepsinogen, lysozyme, amylase, and hyaluronidase, eventually break down the clotting proteins from the seminal vesicles. (3) Acid phosphatase is secreted by the prostate, but its function is unknown. Secretions of the prostate enter the prostatic urethra through many prostatic ducts. Prostatic secretions make up about 25% of the volume of semen and contribute to sperm motility and viability.

Bulbourethral Glands

The paired **bulbourethral glands** (bul'-bō-ū-RĒ-thral), or **Cowper's glands,** are about the size of peas. They are located inferior to the prostate on either side of the membranous urethra within the urogenital diaphragm, and their ducts open into the spongy urethra (Figure 28.11). During sexual arousal, the bulbourethral glands secrete an alkaline fluid into the urethra that protects the passing sperm by neutralizing acids from urine in the urethra. They also secrete mucus that lubricates the end of the penis and the lining of the urethra, thereby decreasing the number of sperm damaged during ejaculation.

Semen

Semen (= seed) is a mixture of sperm and **seminal fluid,** a liquid that consists of the secretions of the seminiferous tubules, seminal vesicles, prostate, and bulbourethral glands. The volume of semen in a typical ejaculation is 2.5–5 milliliter (mL), with 50–150 million sperm per mL. When the number falls below 20 million/mL, the male is likely to be infertile. A very large number is required for successful fertilization because only a tiny fraction ever reaches the secondary oocyte.

Despite the slight acidity of prostatic fluid, semen has a slightly alkaline pH of 7.2–7.7 due to the higher pH and larger volume of fluid from the seminal vesicles. The prostatic secretion gives semen a milky appearance, whereas fluids from the seminal vesicles and bulbourethral glands give it a sticky consistency. Seminal fluid provides sperm with a transportation medium and nutrients, and it neutralizes the hostile acidic environment of the male's urethra and the female's vagina. Semen also contains *seminalplasmin*, an antibiotic that can destroy certain bacteria. Seminalplasmin may help decrease the number of naturally occurring bacteria in the semen and in the lower female reproductive tract.

Once ejaculated, liquid semen coagulates within 5 minutes due to the presence of clotting proteins from the seminal vesicles. The functional role of semen coagulation is not known, but the proteins involved are different from those that cause blood coagulation. After about 10 to 20 minutes, semen reliquefies because prostate-specific antigen (PSA) and other proteolytic enzymes produced by the prostate break down the clot. Abnormal or delayed liquefaction of clotted semen may cause complete or partial immobilization of sperm, thereby inhibiting their movement through the cervix of the uterus.

Penis

The **penis** contains the urethra and is a passageway for the ejaculation of semen and the excretion of urine (Figure 28.12). It is cylindrical in shape and consists of a root, body, and glans penis. The **root of the penis** is the attached portion (proximal portion). It consists of the **bulb of the penis,** the expanded portion of the base of the corpus spongiosum penis (described shortly), and the **crura of the penis** (singular is **crus** = resembling a leg), the two separated and tapered portions of the corpora cavernosa penis (Figure 28.12a). The bulb of the penis is attached to the inferior surface of the urogenital diaphragm and is enclosed by the bulbospongiosus muscle. Each crus of the penis is attached to the ischial and inferior pubic rami and is surrounded by the ischiocavernosus muscle (see Figure 11.13 on page 341). Contraction of these skeletal muscles aids ejaculation.

The **body of the penis** is composed of three cylindrical masses of tissue, each surrounded by fibrous tissue called the **tunica albuginea.** The two dorsolateral masses are called the **corpora cavernosa penis** (*corpora* = main bodies; *cavernosa* = hollow). The smaller midventral mass, the **corpus spongiosum penis,** contains the spongy urethra and functions in keeping the spongy urethra open during ejaculation. Fascia and skin enclose all three masses, which consist of erectile tissue permeated by blood sinuses.

Upon sexual stimulation, which may be visual, tactile, auditory, olfactory, or imagined, the arteries supplying the penis dilate, and large quantities of blood enter the blood sinuses. Expansion of these spaces compresses the veins draining the penis, so blood outflow is slowed. These vascular changes, due to local release of nitric oxide and a parasympathetic reflex, result in an **erection,** the enlargement and stiffening of the penis. The penis returns to its flaccid state when the arteries constrict and pressure on the veins is relieved.

Ejaculation is a sympathetic reflex. As part of the reflex, the smooth muscle sphincter at the base of the urinary bladder closes. Therefore, urine is not expelled during ejaculation, and semen does not enter the urinary bladder. Even before ejaculation occurs, peristaltic contractions in the ampulla of the ductus deferens, seminal vesicles, ejaculatory ducts, and prostate propel semen into the penile portion of the urethra (spongy urethra). Typically, this leads to **emission** (ē-MISH-un), the discharge of a small volume of semen before ejaculation. Emission may also occur during sleep (nocturnal emission).

The distal end of the corpus spongiosum penis is a slightly enlarged, acorn-shaped region called the **glans penis;** its margin is the **corona.** The distal urethra enlarges within the glans penis and forms a terminal slitlike opening, the **external urethral orifice.** Covering the glans in an uncircumcised penis is the loosely fitting **prepuce** (PRĒ-poos), or **foreskin.** The weight of the penis is supported by two ligaments that are continuous with the fascia of the penis. (1) The **fundiform ligament** arises

Figure 28.12 Internal structure of the penis. The inset in (b) shows details of the skin and fasciae. (See Tortora, *A Photographic Atlas of the Human Body*, Figure 14.6.)

🔑 **The penis contains the urethra, a pathway for the ejaculation of semen and the excretion of urine.**

Urinary bladder

Internal urethral orifice
Prostatic urethra

Prostate

Orifice of ejaculatory duct

Bulbourethral (Cowper's) gland

Membranous urethra

Urogenital diaphragm

ROOT OF PENIS

Bulb of penis
Crus of penis

Transverse plane

BODY OF PENIS

Corpora cavernosa penis
Corpus spongiosum penis

Deep dorsal vein

Dorsal artery

Skin

Superficial (subcutaneous) dorsal vein

Superficial fascia

Deep fascia

Spongy (penile) urethra

Dorsal

Corpora cavernosa penis

Tunica albuginea of corpora cavernosum

Deep artery of penis

Corona
GLANS PENIS

Frontal plane

Prepuce (foreskin)

Corpus spongiosum penis

Spongy (penile) urethra

Ventral

Tunica albuginea of corpus spongiosum penis

External urethral orifice

(a) Frontal section

(b) Transverse section

❓ **Which tissue masses form the erectile tissue in the penis, and why do they become rigid during sexual arousal?**

from the inferior part of the linea alba. (2) The **suspensory ligament of the penis** arises from the pubic symphysis.

Circumcision

Circumcision (= to cut around) is a surgical procedure in which part or the entire prepuce is removed. It is usually performed just after delivery, 3 to 4 days after birth, or on the eighth day as part of a Jewish religious rite. Although most health-care professionals find no medical justification for circumcision, some feel that it has benefits, such as a lower risk of urinary tract infections, protection against penile cancer, and possibly a lower risk for sexually trans-

mitted diseases. Indeed, studies in several African villages have found lower rates of HIV infection among circumcised men. ∎

▶ CHECKPOINT

12. Which ducts transport sperm within the testes?

13. What structures comprise the spermatic cord?

14. Give the locations of the three subdivisions of the male urethra.

15. Trace the course of sperm through the system of ducts from the seminiferous tubules through the urethra.

16. What is semen? What is its function?

17. How does an erection occur?

FEMALE REPRODUCTIVE SYSTEM

▶ **O B J E C T I V E S**

• **Describe the location, structure, and functions of the organs of the female reproductive system.**

• **Discuss the process of oogenesis in the ovaries.**

The organs of the female reproductive system (Figure 28.13) include the ovaries; the uterine (Fallopian) tubes, or oviducts; the uterus; the vagina; and external organs, which are collectively called the vulva, or pudendum. The mammary glands are part of the integumentary system and also are considered part of the female reproductive system.

Ovaries

The female gonads are the two **ovaries,** one on either side of the uterus, that are the size and shape of unshelled almonds (Figure 28.14 on page 1030). The ovaries produce (1) gametes, secondary oocytes that develop into mature ova (eggs) after fertilization, and (2) hormones, including progesterone and estrogens (the female sex hormones), inhibin, and relaxin. During development, the ovaries arise from the same embryonic gonadal tissue as the testes. When two organs have the same embryonic origin, they are said to be *homologous.*

The ovaries descend to the brim of the superior portion of the pelvic cavity during the third month of development. A

Figure 28.13 Organs of reproduction and surrounding structures in females.

🔑 The organs of reproduction in females include the ovaries, uterine (Fallopian) tubes, uterus, vagina, vulva, and mammary glands.

> **Functions of the Female Reproductive System**
> 1. **Ovaries: produce secondary oocytes and hormones, including progesterone and estrogens (female sex hormones), inhibin, and relaxin.**
> 2. **Uterine tubes: transport a secondary oocyte to the uterus and normally are the sites where fertilization occurs.**
> 3. **Uterus: site of implantation of a fertilized ovum, development of the fetus during pregnancy, and labor.**
> 4. **Vagina: receives the penis during sexual intercourse and is a passageway for childbirth.**
> 5. **Mammary glands: synthesize, secrete, and eject milk for nourishment of the newborn.**

Sagittal plane

Sacrum

Uterosacral ligament

Posterior fornix of vagina

Rectouterine pouch (pouch of Douglas)

Vesicouterine pouch

Coccyx

Rectum

Vagina

Anus

Uterine (Fallopian) tube

Fimbriae

Ovary

Uterus

Round ligament of uterus

Cervix

Urinary bladder

Pubic symphysis

Mons pubis

Clitoris

Urethra

Labium majus

External urethral orifice

Labium minus

(a) Sagittal section

series of ligaments holds them in position (Figure 28.14). The **broad ligament** of the uterus (see also Figure 28.13b), which is itself part of the parietal peritoneum, attaches to the ovaries by a double-layered fold of peritoneum called the **mesovarium.** The **ovarian ligament** anchors the ovaries to the uterus, and the **suspensory ligament** attaches them to the pelvic wall. Each ovary contains a **hilus,** the point of entrance and exit for blood vessels and nerves and along which the mesovarium is attached.

Histology of the Ovary

Each ovary consists of the following parts (Figure 28.15 on page 1031):

- The **germinal epithelium** is a layer of simple epithelium (low cuboidal or squamous) that covers the surface of the ovary. It is continuous with the mesothelium that covers the mesovarium. The term *germinal epithelium* is a misnomer because it does not give rise to ova, although at one time

people believed that it did. Now we know that the progenitors of ova arise from the endoderm of the yolk sac and migrate to the ovaries during embryonic development.

- The **tunica albuginea** is a whitish capsule of dense irregular connective tissue located immediately deep to the germinal epithelium.

- The **ovarian cortex** is a region just deep to the tunica albuginea. It consists of ovarian follicles (described shortly) surrounded by dense irregular connective tissue that contains scattered smooth muscle cells.

- The **ovarian medulla** is deep to the ovarian cortex. The border between the cortex and medulla is indistinct, but the medulla consists of more loosely arranged connective tissue and contains blood vessels, lymphatic vessels, and nerves.

- **Ovarian follicles** (*folliculus* = little bag) are in the cortex and consist of **oocytes** in various stages of development, plus the cells surrounding them. When the surrounding cells form a single layer, they are called **follicular cells;** later in

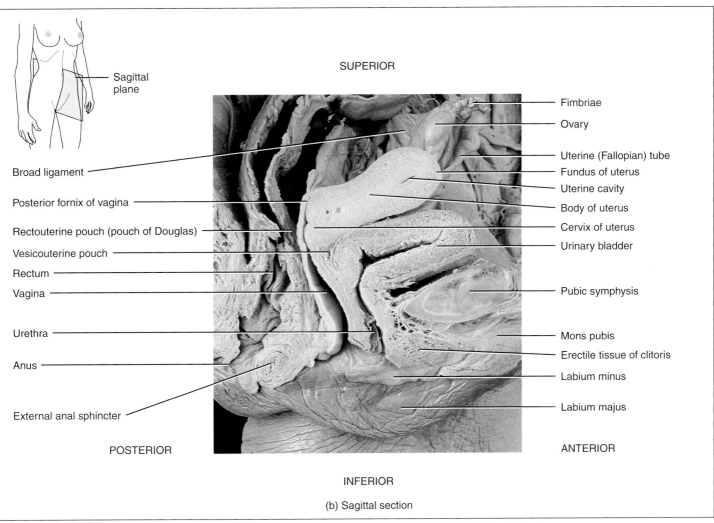

Sagittal plane

SUPERIOR

Fimbriae
Ovary

Broad ligament

Uterine (Fallopian) tube
Fundus of uterus
Uterine cavity

Posterior fornix of vagina

Body of uterus

Rectouterine pouch (pouch of Douglas)

Cervix of uterus

Vesicouterine pouch

Urinary bladder

Rectum

Vagina

Pubic symphysis

Urethra

Mons pubis
Erectile tissue of clitoris

Anus

Labium minus

External anal sphincter

Labium majus

POSTERIOR

ANTERIOR

INFERIOR

(b) Sagittal section

Which structures in males are homologous to the ovaries, the clitoris, the paraurethral glands, and the greater vestibular glands?

Figure 28.14   **Relative positions of the ovaries, the uterus, and the ligaments that support them.**   (See Tortora, *A Photographic Atlas of the Human Body*, Figure 14.9.)

Ligaments holding the ovaries in position are the mesovarium, the ovarian ligament, and the suspensory ligament.

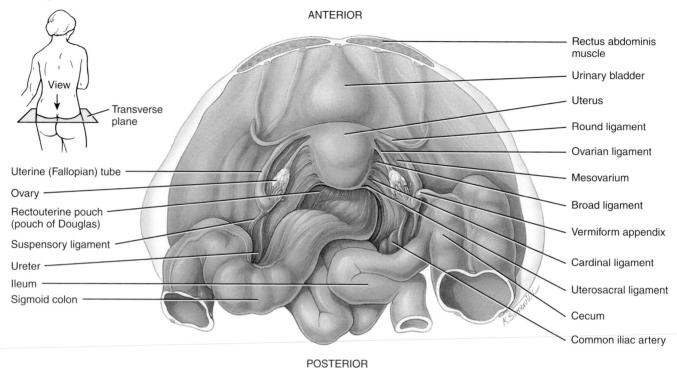

ANTERIOR

Rectus abdominis muscle

Urinary bladder

Uterus

Round ligament

Ovarian ligament

Mesovarium

Broad ligament

Vermiform appendix

Cardinal ligament

Uterosacral ligament

Cecum

Common iliac artery

Uterine (Fallopian) tube

Ovary

Rectouterine pouch (pouch of Douglas)

Suspensory ligament

Ureter

Ileum

Sigmoid colon

POSTERIOR

Superior view of transverse section

To which structures do the mesovarium, ovarian ligament, and suspensory ligament anchor the ovary?

development, when they form several layers, they are referred to as **granulosa cells.** The surrounding cells nourish the developing oocyte and begin to secrete estrogens as the follicle grows larger.

- A **mature (Graafian) follicle** is a large, fluid-filled follicle that is ready to rupture and expel its secondary oocyte, a process known as **ovulation.**

- A **corpus luteum** (= yellow body) contains the remnants of mature follicle after ovulation. The corpus luteum produces progesterone, estrogens, relaxin, and inhibin until it degenerates into fibrous scar tissue called the **corpus albicans** (= white body).

Oogenesis and Follicular Development

The formation of gametes in the ovaries is termed **oogenesis** (ō-ō-JEN-e-sis; *oo-* = egg). Whereas spermatogenesis begins in males at puberty, oogenesis begins in females before they are even born. Oogenesis occurs in essentially the same manner as spermatogenesis, with meiosis taking place and the resulting germ cells undergoing maturation.

During early fetal development, primordial (primitive) germ cells migrate from the endoderm of the yolk sac to the ovaries.

There, germ cells differentiate within the ovaries into **oogonia** (ō'-o-GŌ-nē-a; singular is **oogonium**). Oogonia are diploid (2*n*) stem cells that divide mitotically to produce millions of germ cells. Even before birth, most of these germ cells degenerate in a process known as **atresia** (a-TRĒ-zē-a). A few, however, develop into larger cells called **primary oocytes** (Ō-ō-sītz) that enter prophase of meiosis I during fetal development but do not complete that phase until after puberty. At birth, 200,000 to 2,000,000 oogonia and primary oocytes remain in each ovary. Of these, about 40,000 remain at puberty, and around 400 will mature and ovulate during a woman's reproductive lifetime. The rest undergo atresia.

At the same time oogenesis is occurring, the follicle cells surrounding the oocyte are also undergoing developmental changes. In the beginning, a single layer of follicular cells surrounds each primary oocyte, and the entire structure is called a **primordial follicle** (Figure 28.16a on page 1032). Although the stimulating mechanism is unclear, a few primordial follicles periodically start to grow, even during childhood. They become **primary follicles,** which are surrounded first by one layer of cuboidal follicular cells and then by six to seven layers of cuboidal and low-columnar cells called granulosa cells. As a fol-

Figure 28.15   Histology of the ovary.   The arrows indicate the sequence of developmental stages that occur as part of the maturation of an ovum during the ovarian cycle.

🔑 **The ovaries are the female gonads; they produce haploid oocytes.**

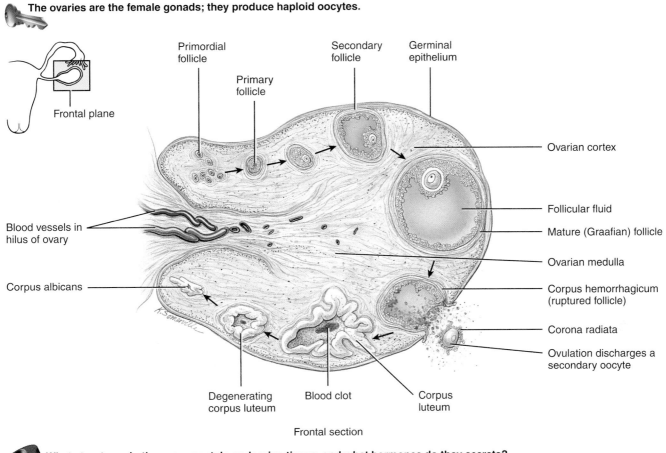

Frontal section

❓ **What structures in the ovary contain endocrine tissue, and what hormones do they secrete?**

licle grows, it forms a clear glycoprotein layer, called the **zona pellucida** (pe-LOO-si-da), between the primary oocyte and the granulosa cells. The innermost layer of granulosa cells becomes firmly attached to the zona pellucida and is called the **corona radiata** (*corona* = crown; *radiata* = radiation) (Figure 28.16b).

The outermost granulosa cells rest on a basement membrane. Encircling the basement membrane is a region called the **theca folliculi.** Although the region of granulosa cells is avascular, many blood capillaries are located in the theca folliculi. As a primary follicle continues to grow, the theca differentiates into two layers: (1) the **theca interna,** a highly vascularized internal layer of cuboidal secretory cells, fibroblasts, and bundles of collagen fibers and (2) the **theca externa,** an outer layer of connective tissue cells, collagen fibers, and smooth muscle fibers. The granulosa cells begin to secrete follicular fluid, which builds up in a cavity called the **antrum** in the center of the follicle, which is now termed a **secondary follicle.**

Each month after puberty, gonadotropins secreted by the anterior pituitary stimulate the resumption of oogenesis (Figure 28.17). Meiosis I resumes in several secondary follicles, although only one will eventually reach the maturity needed for ovulation. The diploid primary oocyte completes meiosis I, producing two haploid cells of unequal size—both with 23 chromosomes (*n*) of two chromatids each. The smaller cell produced by meiosis I, called the **first polar body,** is essentially a packet of discarded nuclear material. The larger cell, known as the **secondary oocyte,** receives most of the cytoplasm. Once a secondary oocyte is formed, it begins meiosis II but then stops in metaphase. The follicle in which these events are taking place—the **mature (Graafian) follicle**—soon ruptures and releases its secondary oocyte. It takes 90 days or longer for a primary follicle to develop into a secondary follicle and then a mature follicle that releases its secondary oocyte.

At ovulation, the secondary oocyte is expelled into the pelvic cavity together with the first polar body and corona radiata. Normally these cells are swept into the uterine tube. If fertilization does not occur, the cells degenerate. If sperm are present in the uterine tube and one penetrates the secondary oocyte, however, meiosis II resumes. The secondary oocyte splits into two haploid (*n*) cells, again of unequal size. The larger cell is the **ovum,** or mature egg; the smaller one is the **second polar body.** The nuclei of the sperm cell and the ovum then unite, forming a

Figure 28.16 Ovarian follicles. (a) Primordial and primary follicles in the ovarian cortex. (b) A secondary follicle.

🔑 **As an ovarian follicle enlarges, follicular fluid accumulates in a cavity called the antrum.**

Primordial follicles   Germinal epithelium   Tunica albuginea   Ovarian cortex

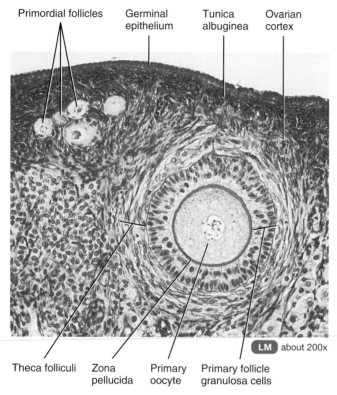

Theca folliculi   Zona pellucida   Primary oocyte   Primary follicle granulosa cells

LM about 200x

(a) Ovarian cortex

Antrum filled with follicular fluid   Corona radiata   Secondary follicle granulosa cells

LM 60x

Theca folliculi   Zona pellucida   Primary oocyte

(b) Secondary follicle

 What happens to most ovarian follicles?

Figure 28.17 Oogenesis. Diploid cells ($2n$) have 46 chromosomes; haploid cells (n) have 23 chromosomes.

🔑 **In a secondary oocyte, meiosis II is completed only if fertilization occurs.**

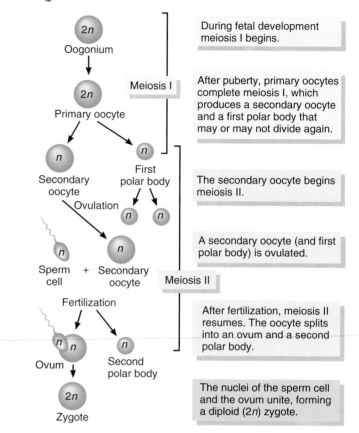

During fetal development meiosis I begins.

After puberty, primary oocytes complete meiosis I, which produces a secondary oocyte and a first polar body that may or may not divide again.

The secondary oocyte begins meiosis II.

A secondary oocyte (and first polar body) is ovulated.

After fertilization, meiosis II resumes. The oocyte splits into an ovum and a second polar body.

The nuclei of the sperm cell and the ovum unite, forming a diploid ($2n$) zygote.

How does the age of a primary oocyte in a female compare with the age of a primary spermatocyte in a male?

diploid ($2n$) **zygote.** If the first polar body undergoes another division to produce two polar bodies, then the primary oocyte ultimately gives rise to three haploid (n) polar bodies, which all degenerate, and a single haploid (n) ovum. Thus, one primary oocyte gives rise to a single gamete (an ovum). By contrast, recall that in males one primary spermatocyte produces four gametes (sperm).

Table 28.1 summarizes the events of oogenesis and follicular development.

Uterine Tubes

Females have two **uterine (Fallopian) tubes,** or **oviducts,** that extend laterally from the uterus (Figure 28.18 on page 1034). The tubes, which measure about 10 cm (4 in.) long and lie between the folds of the broad ligaments of the uterus. They provide a route for sperm to reach an ovum and transport secondary oocytes and fertilized ova from the ovaries to the uterus. The funnel-shaped portion of each tube, called the **infundibulum,** is close to the ovary but is open to the pelvic cavity. It ends in a

Table 28.1	Summary of Oogenesis and Follicular Development

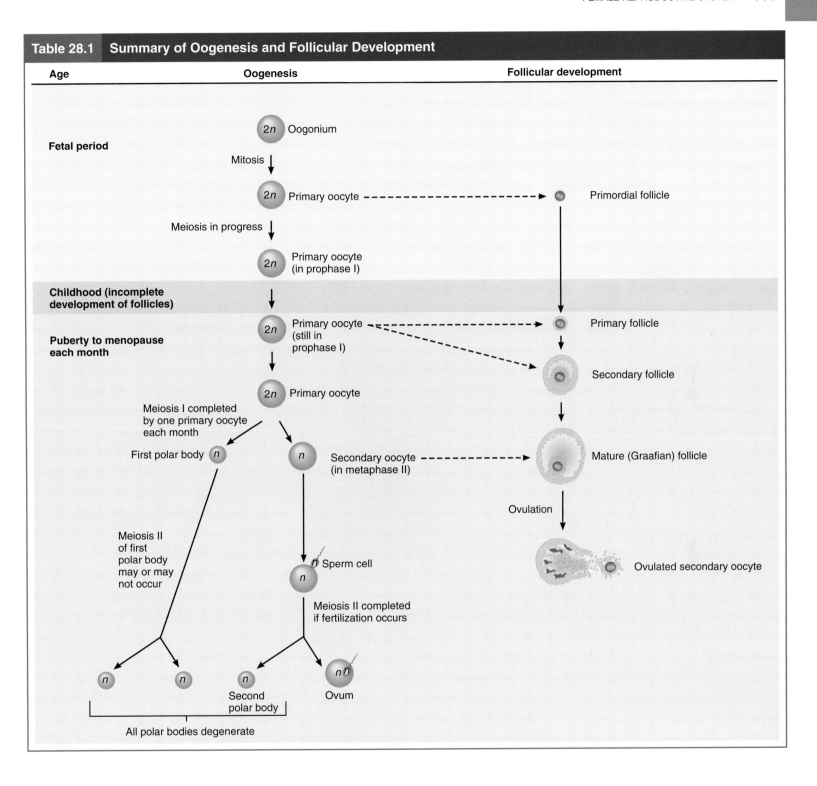

Age	Oogenesis	Follicular development

Fetal period

2n Oogonium

Mitosis

2n Primary oocyte – – – – – – – – → Primordial follicle

Meiosis in progress

2n Primary oocyte (in prophase I)

Childhood (incomplete development of follicles)

2n Primary oocyte (still in prophase I) – – – – – – → Primary follicle

Puberty to menopause each month

Secondary follicle

2n Primary oocyte

Meiosis I completed by one primary oocyte each month

First polar body n   n Secondary oocyte – – – – – – – → Mature (Graafian) follicle (in metaphase II)

Meiosis II of first polar body may or may not occur

Ovulation

Ovulated secondary oocyte

n Sperm cell

Meiosis II completed if fertilization occurs

n   n   n Second polar body   n Ovum

All polar bodies degenerate

fringe of fingerlike projections called **fimbriae** (FIM-brē-ē = fringe), one of which is attached to the lateral end of the ovary. From the infundibulum, the uterine tube extends medially and eventually inferiorly and attaches to the superior lateral angle of the uterus. The **ampulla** of the uterine tube is the widest, longest portion, making up about the lateral two-thirds of its length. The **isthmus** of the uterine tube is the more medial, short, narrow, thick-walled portion that joins the uterus.

Histologically, the uterine tubes are composed of three layers. The internal **mucosa** contains ciliated columnar epithelial cells, which function as a "ciliary conveyor belt" to move the fertilized ovum (or secondary oocyte) along the tube, and secretory cells, which have microvilli and may provide nutrition for the ovum (Figure 28.19 on page 1035). The middle layer, the **muscularis,** is composed of an inner, thick, circular ring of smooth muscle and an outer, thin region of longitudinal smooth

Figure 28.18   Relationship of the uterine (Fallopian) tubes to the ovaries, uterus, and associated structures. In the left side of the drawing the uterine tube and uterus have been sectioned to show internal structures. (See Tortora, *A Photographic Atlas of the Human Body*, Figure 14.9.)

After ovulation, a secondary oocyte and its corona radiata move from the pelvic cavity into the infundibulum of the uterine tube. The uterus is the site of menstruation, implantation of a fertilized ovum, development of the fetus, and labor.

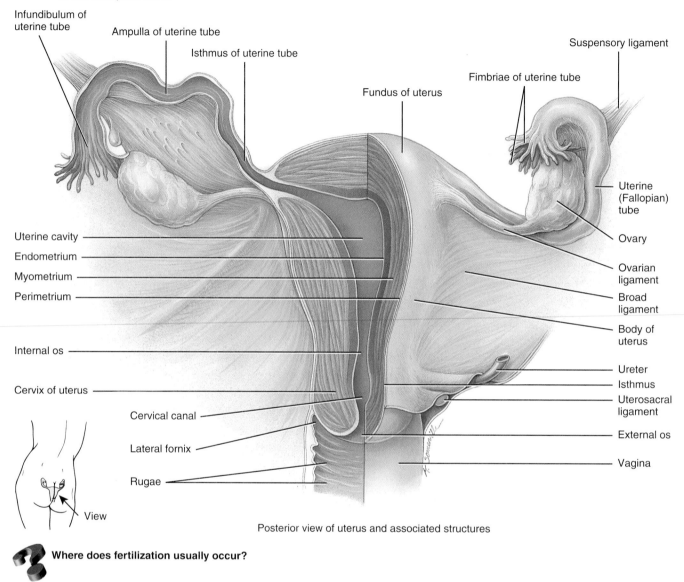

Posterior view of uterus and associated structures

Where does fertilization usually occur?

muscle. Peristaltic contractions of the muscularis and the ciliary action of the mucosal cells help move the oocyte or fertilized ovum toward the uterus. The outer layer of the uterine tubes is a serous membrane, the **serosa.**

Local currents produced by movements of the fimbriae, which surround the ovary during ovulation, sweep the ovulated secondary oocyte from the pelvic cavity into the uterine tube. A sperm cell usually encounters and fertilizes a secondary oocyte in the ampulla of the uterine tube, although fertilization in the pelvic cavity is not uncommon. Fertilization can occur at any time up to about 24 hours after ovulation. Some hours after fertilization, the nuclear materials of the haploid ovum and sperm

unite. The diploid fertilized ovum is now called a **zygote** and begins to undergo cell divisions while moving toward the uterus. It arrives at the uterus 6 to 7 days after ovulation.

Uterus

The **uterus** (womb) serves as part of the pathway for sperm deposited in the vagina to reach the uterine tubes. It is also the site of implantation of a fertilized ovum, development of the fetus during pregnancy, and labor. During reproductive cycles when implantation does not occur, the uterus is the source of menstrual flow.

Figure 28.19   Mucosal cells of the uterine (Fallopian) tube.

 Peristaltic contractions of the muscularis and ciliary action of the mucosa of the uterine tube help move the oocyte or fertilized ovum toward the uterus.

Cilia of ciliated columnar epithelial cell

Secretory cell with microvilli

SEM 4000x

Cilia and secretory cells lining the uterine (Fallopian) tube

 What types of cells line the uterine tubes?

Anatomy of the Uterus

Situated between the urinary bladder and the rectum, the uterus is the size and shape of an inverted pear (see Figure 28.18). In females who have never been pregnant, it is about 7.5 cm (3 in.) long, 5 cm (2 in.) wide, and 2.5 cm (1 in.) thick. The uterus is larger in females who have recently been pregnant, and smaller (atrophied) when sex hormone levels are low, as occurs after menopause.

Anatomical subdivisions of the uterus include: (1) a dome-shaped portion superior to the uterine tubes called the **fundus,** (2) a tapering central portion called the **body,** and (3) an inferior narrow portion called the **cervix** that opens into the vagina. Between the body of the uterus and the cervix is the **isthmus** (IS-mus), a constricted region about 1 cm (0.5 in.) long. The interior of the body of the uterus is called the **uterine cavity,** and the interior of the narrow cervix is called the **cervical canal.** The cervical canal opens into the uterine cavity at the **internal os** (*os* = mouthlike opening) and into the vagina at the **external os.**

Normally, the body of the uterus projects anteriorly and superiorly over the urinary bladder in a position called **anteflexion.** The cervix projects inferiorly and posteriorly and enters the anterior wall of the vagina at nearly a right angle (see Figure 28.13). Several ligaments that are either extensions of the parietal peritoneum or fibromuscular cords maintain the position of the uterus (see Figure 28.14). The paired **broad ligaments** are double folds of peritoneum attaching the uterus to either side of the pelvic cavity. The paired **uterosacral ligaments,** also peritoneal extensions, lie on either side of the rectum and connect the uterus to the sacrum. The **cardinal (lateral cervical) ligaments** are located inferior to the bases of the broad ligaments and extend from the pelvic wall to the cervix and vagina. The **round ligaments** are bands of fibrous connective tissue between the layers of the broad ligament; they extend from a point on the uterus just inferior to the uterine tubes to a portion of the labia majora of the external genitalia. Although the ligaments normally maintain the anteflexed position of the uterus, they also afford the uterine body enough movement such that the uterus may become malpositioned. A posterior tilting of the uterus is called **retroflexion** (*retro-* = backward or behind).

Uterine Prolapse

A condition called **uterine prolapse** (*prolapse* = falling down or downward displacement) may result from weakening of supporting ligaments and pelvic musculature associated with age or disease, traumatic vaginal delivery, chronic straining from coughing or difficult bowel movements, or pelvic tumors. The prolapse may be characterized as *first degree (mild),* in which the cervix remains within the vagina; *second degree (marked),* in which the cervix protrudes to the exterior through the vagina; and *third degree (complete),* in which the entire uterus is outside the vagina. Depending on the degree of prolapse, treatment may involve pelvic exercises, dieting if a patient is overweight, a stool softener to minimize straining during defecation, pessary therapy (placement of a rubber device around the uterine cervix that helps prop up the uterus), or surgery. ■

Histology of the Uterus

Histologically, the uterus consists of three layers of tissue: the perimetrium, myometrium, and endometrium (Figure 28.20). The outer layer—the **perimetrium** (*peri-* = around; *-metrium* = uterus) or serosa—is part of the visceral peritoneum; it is composed of simple squamous epithelium and areolar connective tissue. Laterally, it becomes the broad ligament. Anteriorly, it covers the urinary bladder and forms a shallow pouch, the **vesicouterine pouch** (ves'-i-kō-ū-ter-in; *vesico-* = bladder; see Figure 28.13). Posteriorly, it covers the rectum and forms a deep pouch, the **rectouterine pouch** (rek-tō-ū-ter-in; *recto-* = rectum) or *pouch of Douglas*—the most inferior point in the pelvic cavity.

The middle layer of the uterus, the **myometrium** (*myo-* = muscle), consists of three layers of smooth muscle fibers and is thickest in the fundus and thinnest in the cervix. The thicker middle layer is circular, whereas the inner and outer layers are longitudinal or oblique. During labor and childbirth, coordinated contractions of the myometrium in response to oxytocin from the posterior pituitary help expel the fetus from the uterus.

The inner layer of the uterus, the **endometrium** (*endo-* = within), is highly vascularized and has three components: (1) An

Figure 28.20 Histology of the uterus. The superficial perimetrium (serosa) is not shown.

🔑 **The three layers of the uterus from superficial to deep are the perimetrium (serosa), the myometrium, and the endometrium.**

— Lumen
— Simple columnar epithelium
— Endometrial stroma
— Endometrial gland

Endometrium:
— Stratum functionalis

— Stratum basalis

— Myometrium

LM 25x

Portion of endometrium and myometrium

❓ **What structural features of the endometrium and myometrium contribute to their functions?**

innermost layer of simple columnar epithelium (ciliated and secretory cells) lines the lumen. (2) An underlying endometrial stroma is a very thick region of lamina propria (areolar connective tissue). (3) Endometrial (uterine) glands develop as invaginations of the luminal epithelium and extend almost to the myometrium. The endometrium is divided into two layers. The **stratum functionalis** (*functional layer*) lines the uterine cavity and sloughs off during menstruation. The deeper layer, the **stratum basalis** (*basal layer*), is permanent and gives rise to a new stratum functionalis after each menstruation.

Branches of the internal iliac artery called **uterine arteries** (Figure 28.21) supply blood to the uterus. Uterine arteries give off branches called **arcuate arteries** (= shaped like a bow) that are arranged in a circular fashion in the myometrium. These arteries branch into **radial arteries** that penetrate deeply into the myometrium. Just before the branches enter the endometrium, they divide into two kinds of arterioles: **Straight arterioles** supply the stratum basalis with the materials needed to regenerate the stratum functionalis; **spiral arterioles** supply the stratum

functionalis and change markedly during the menstrual cycle. Blood leaving the uterus is drained by the **uterine veins** into the internal iliac veins. The extensive blood supply of the uterus is essential to support regrowth of a new stratum functionalis after menstruation, implantation of a fertilized ovum, and development of the placenta.

Cervical Mucus

The secretory cells of the mucosa of the cervix produce a secretion called **cervical mucus,** a mixture of water, glycoproteins, lipids, enzymes, and inorganic salts. During their reproductive years, females secrete 20–60 mL of cervical mucus per day. Cervical mucus is more hospitable to sperm at or near the time of ovulation because it is then less viscous and more alkaline (pH 8.5). At other times, viscous mucus forms a cervical plug that physically impedes sperm penetration. Cervical mucus supplements the energy needs of sperm, and both the cervix and cervical mucus protect sperm from phagocytes and the hostile environment of the vagina and uterus. They may also play a role in *capacitation*—a functional change that sperm undergo in the female reproductive tract before they are able to fertilize a secondary oocyte.

 Hysterectomy

Hysterectomy (hiss-ter-EK-tō-mē; *hyster-* = uterus), the surgical removal of the uterus, is the most common gynecological operation. It may be indicated in conditions such as endometriosis, pelvic inflammatory disease, recurrent ovarian cysts, excessive uterine bleeding, and cancer of the cervix, uterus, or ovaries. In a *partial (subtotal) hysterectomy,* the body of the uterus is removed but the cervix is left in place. A *complete hysterectomy* is the removal of both the body and cervix of the uterus. A *radical hysterectomy* includes removal of the body and cervix of the uterus, uterine tubes, possibly the ovaries, the superior portion of the vagina, pelvic lymph nodes, and supporting structures, such as ligaments. ∎

▶ CHECKPOINT

18. How are the ovaries held in position in the pelvic cavity?

19. Describe the principal events of oogenesis.

20. Where are the uterine tubes located? What is their function?

21. Describe the arrangement of ligaments that hold the uterus in its normal position.

22. Describe the histology of the uterus.

23. Discuss the blood supply to the uterus. Why is an abundant blood supply important?

Vagina

The **vagina** (= sheath) is a tubular, 10-cm (4-in.) long fibromuscular canal lined with mucous membrane that extends from the exterior of the body to the uterine cervix (see Figures 28.13 and 28.18). It is the receptacle for the penis during sexual inter-

Figure 28.21 Blood supply of the uterus. The inset shows histological details of the blood vessels of the endometrium.

 Straight arterioles supply the materials needed for regeneration of the stratum functionalis.

Anterior view with left side of uterus partially sectioned

What is the functional significance of the stratum basalis of the endometrium?

course, the outlet for menstrual flow, and the passageway for childbirth. Situated between the urinary bladder and the rectum, the vagina is directed superiorly and posteriorly, where it attaches to the uterus. A recess called the **fornix** (= arch or vault) surrounds the vaginal attachment to the cervix. When properly inserted, a contraceptive diaphragm rests on the fornix, covering the cervix.

The **mucosa** of the vagina is continuous with that of the uterus. Histologically, it consists of nonkeratinized stratified squamous epithelium and areolar connective tissue that lies in a series of transverse folds called **rugae** (ROO-gē). Dendritic cells in the mucosa are antigen-presenting cells (described on page 783). Unfortunately, they also participate in the transmission of viruses—for example, HIV (the virus that causes AIDS)—to a female during intercourse with an infected male. The mucosa of the vagina contains large stores of glycogen, the decomposition of which produces organic acids. The resulting acidic environment retards microbial growth, but it also is harmful to sperm. Alkaline components of semen, mainly from the seminal vesicles, raise the pH of fluid in the vagina and increase viability of the sperm.

The **muscularis** is composed of an outer circular layer and an inner longitudinal layer of smooth muscle that can stretch considerably to accommodate the penis during sexual intercourse and a child during birth.

The **adventitia,** the superficial layer of the vagina, consists of areolar connective tissue. It anchors the vagina to adjacent organs such as the urethra and urinary bladder anteriorly and the rectum and anal canal posteriorly.

A thin fold of vascularized mucous membrane, called the **hymen** (= membrane), forms a border around and partially closes the inferior end of the vaginal opening to the exterior, the **vaginal orifice** (see Figure 28.22). Sometimes the hymen completely covers the orifice, a condition called **imperforate hymen** (im-PER-fō-rāt). Surgery may be needed to open the orifice and permit the discharge of menstrual flow.

Vulva

The term **vulva** (VUL-va = to wrap around), or **pudendum** (pū-DEN-dum), refers to the external genitals of the female (Figure 28.22). The following components comprise the vulva:

Figure 28.22 Components of the vulva (pudendum). (See Tortora, *A Photographic Atlas of the Human Body*, Figure 14.7.)

 The vulva refers to the external genitals of the female.

Inferior view

 What surface structures are anterior to the vaginal opening? Lateral to it?

- Anterior to the vaginal and urethral openings is the **mons pubis** (MONZ PŪ-bis; *mons* = mountain), an elevation of adipose tissue covered by skin and coarse pubic hair that cushions the pubic symphysis.

- From the mons pubis, two longitudinal folds of skin, the **labia majora** (LĀ-bē-a ma-JŌ-ra; *labia* = lips; *majora* = larger), extend inferiorly and posteriorly. The singular term is *labium majus.* The labia majora are covered by pubic hair and contain an abundance of adipose tissue, sebaceous (oil) glands, and apocrine sudoriferous (sweat) glands. They are homologous to the scrotum.

- Medial to the labia majora are two smaller folds of skin called the **labia minora** (mī-NŌ-ra; *minora* = smaller). The singular term is **labium minus.** Unlike the labia majora, the labia minora are devoid of pubic hair and fat and have few sudoriferous glands, but they do contain many sebaceous glands. The labia minora are homologous to the spongy (penile) urethra.

- The **clitoris** (KLI-to-ris) is a small cylindrical mass of erec-

tile tissue and nerves located at the anterior junction of the labia minora. A layer of skin called the **prepuce of the clitoris** is formed at the point where the labia minora unite and covers the body of the clitoris. The exposed portion of the clitoris is the glans. The clitoris is homologous to the glans penis in males. Like the male structure, it is capable of enlargement upon tactile stimulation and has a role in sexual excitement in the female.

- The region between the labia minora is the **vestibule.** Within the vestibule are the hymen (if still present), the vaginal orifice, the external urethral orifice, and the openings of the ducts of several glands. The vestibule is homologous to the membranous urethra of males. The **vaginal orifice,** the opening of the vagina to the exterior, occupies the greater portion of the vestibule and is bordered by the hymen. Anterior to the vaginal orifice and posterior to the clitoris is the **external urethral orifice,** the opening of the urethra to the exterior. On either side of the external urethral orifice are the openings of the ducts of the **paraurethral** *(Skene's)* **glands.** These mucus-secreting glands are embedded in the wall of the urethra. The paraurethral glands are homologous to the prostate. On either side of the vaginal orifice itself are the **greater vestibular** *(Bartholin's)* **glands** (see Figure 28.23), which open by ducts into a groove between the hymen and labia minora. They produce a small quantity of mucus during sexual arousal and intercourse that adds to cervical mucus and also provides lubrication. The greater vestibular glands are homologous to the bulbourethral glands in males. Several **lesser vestibular glands** also open into the vestibule.

- The **bulb of the vestibule** (see Figure 28.23) consists of two elongated masses of erectile tissue just deep to the labia on either side of the vaginal orifice. The bulb of the vestibule becomes engorged with blood during sexual arousal, narrowing the vaginal orifice and placing pressure on the penis during intercourse. The bulb of the vestibule is homologous to the corpus spongiosum and bulb of the penis in males.

Table 28.2 summarizes the homologous structures of the female and male reproductive systems.

Perineum

The **perineum** (per'-i-NĒ-um) is the diamond-shaped area medial to the thighs and buttocks of both males and females

Table 28.2	Summary of Homologous Structures of the Female and Male Reproductive Systems
Female Structures	**Male Structures**
Ovaries	Testes
Ovum	Sperm cell
Labia majora	Scrotum
Labia minora	Spongy (penile) urethra
Vestibule	Membranous urethra
Bulb of vestibule	Corpus spongiosum penis and bulb of penis
Clitoris	Glans penis
Paraurethral glands	Prostate
Greater vestibular glands	Bulbourethral (Cowper's) glands

(Figure 28.23). It contains the external genitals and anus. The perineum is bounded anteriorly by the pubic symphysis, laterally by the ischial tuberosities, and posteriorly by the coccyx. A transverse line drawn between the ischial tuberosities divides the perineum into an anterior **urogenital triangle** (ū'-rō-JEN-i-tal) that contains the external genitalia and a posterior **anal triangle** that contains the anus.

Figure 28.23   Perineum of a female.   (Figure 11.13 on page 341 shows the perineum of a male.)

The perineum is a diamond-shaped area that includes the urogenital triangle and the anal triangle.

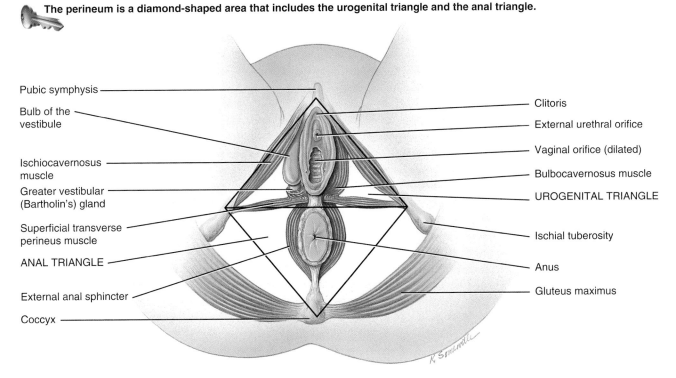

Pubic symphysis
Bulb of the vestibule
Ischiocavernosus muscle
Greater vestibular (Bartholin's) gland
Superficial transverse perineus muscle
ANAL TRIANGLE
External anal sphincter
Coccyx

Clitoris
External urethral orifice
Vaginal orifice (dilated)
Bulbocavernosus muscle
UROGENITAL TRIANGLE
Ischial tuberosity
Anus
Gluteus maximus

Inferior view

 Why is the anterior portion of the perineum called the urogenital triangle?

Episiotomy

During childbirth, the emerging fetus stretches the perineal region. To prevent excessive stretching and even tearing of this region, a physician sometimes performs an **episiotomy** (e-piz-ē-OT-ō-mē; *episi-* = vulva or pubic region; *-otomy* = incision), a perineal cut made with surgical scissors. The cut may be made along the midline or at an approximately 45 degree angle with the midline. The cut enlarges the vaginal opening to make more room for the fetus to pass. In effect, a straight, more easily sutured cut is substituted for a jagged tear. The incision is closed in layers with a suture that is absorbed within a few weeks, so that stitches do not have to be removed. ■

Mammary Glands

The two **mammary glands** (*mamma* = breast) are modified sudoriferous (sweat) glands that produce milk. They lie over the pectoralis major and serratus anterior muscles and are attached to them by a layer of deep fascia composed of dense irregular connective tissue (Figure 28.24).

Each breast has one pigmented projection, the **nipple,** that has a series of closely spaced openings of ducts called **lactifer-ous ducts,** where milk emerges. The circular pigmented area of skin surrounding the nipple is called the **areola** (a-RĒ-ō-la = small space); it appears rough because it contains modified sebaceous (oil) glands. Strands of connective tissue called the **suspensory ligaments of the breast (Cooper's ligaments)** run between the skin and deep fascia and support the breast. These ligaments become looser with age or with excessive strain, as occurs in long-term jogging or high-impact aerobics. Wearing a supportive bra slows the appearance of "Cooper's droop."

Internally, the mammary gland consists of 15 to 20 lobes, or compartments, separated by a variable amount of adipose tissue. In each lobe are several smaller compartments called **lobules,** composed of grapelike clusters of milk-secreting glands termed **alveoli** (= small cavities) embedded in connective tissue. Surrounding the alveoli are **myoepithelial cells,** the contraction of which helps propel milk toward the nipples. When milk is being produced, it passes from the alveoli into a series of **secondary tubules** and then into the **mammary ducts.** Near the nipple, the mammary ducts expand to form sinuses called **lactiferous sinuses** (*lact-* = milk), where some milk may be stored before draining into a lactiferous duct. Each lactiferous duct typically carries milk from one of the lobes to the exterior.

Figure 28.24 Mammary glands. (See Tortora, *A Photographic Atlas of the Human Body*, Figure 14.14.)

The mammary glands function in the synthesis, secretion, and ejection of milk (lactation).

(a) Sagittal section

(b) Anterior view, partially sectioned

What hormones regulate the synthesis and ejection of milk?

The essential functions of the mammary glands are the synthesis, secretion, and ejection of milk; these functions, called **lactation,** are associated with pregnancy and childbirth. Milk production is stimulated largely by the hormone prolactin, from the anterior pituitary, with contributions from progesterone and estrogens. The ejection of milk is stimulated by oxytocin, which is released from the posterior pituitary in response to the sucking of an infant on the mother's nipple (suckling).

Fibrocystic Disease of the Breasts

The breasts of females are highly susceptible to cysts and tumors. In **fibrocystic disease,** the most common cause of breast lumps in females, one or more cysts (fluid-filled sacs) and thickening of alveoli (clusters of milk-secreting cells) develop. The condition, which occurs mainly in females between the ages of 30 and 50, is probably due to a relative excess of estrogens or a deficiency of progesterone in the postovulatory (luteal) phase of the reproductive cycle (discussed shortly). Fibrocystic disease usually causes one or both breasts to become lumpy, swollen, and tender a week or so before menstruation begins. ■

▶ C H E C K P O I N T

24. What is the function of the vagina? Describe its histology.
25. List the parts of the vulva, and explain the functions of each.
26. Describe the structure of the mammary glands. How are they supported?
27. What is the route whereby milk passes from the alveoli of the mammary gland to the nipple?

THE FEMALE REPRODUCTIVE CYCLE

▶ O B J E C T I V E

• **Compare the major events of the ovarian and uterine cycles.**

During their reproductive years, nonpregnant females normally exhibit cyclical changes in the ovaries and uterus. Each cycle takes about a month and involves both oogenesis and preparation of the uterus to receive a fertilized ovum. Hormones secreted by the hypothalamus, anterior pituitary, and ovaries control the main events. The **ovarian cycle** is a series of events in the ovaries that occur during and after the maturation of an oocyte. Steroid hormones released by the ovaries control the **uterine (menstrual) cycle,** a concurrent series of changes in the endometrium of the uterus to prepare it for the arrival of a fertilized ovum that will develop there until birth. If fertilization does not occur, ovarian hormones wane, which causes the stratum functionalis of the endometrium to slough off. The general term **female reproductive cycle** encompasses the ovarian and uterine cycles, the hormonal changes that regulate them, and the related cyclical changes in the breasts and cervix.

Hormonal Regulation of the Female Reproductive Cycle

Gonadotropin-releasing hormone (GnRH) secreted by the hypothalamus controls the ovarian and uterine cycles (Figure 28.25). GnRH stimulates the release of **follicle-stimulating hormone (FSH)** and **luteinizing hormone (LH)** from the anterior pituitary. FSH, in turn, initiates follicular growth and the secretion of estrogens by the growing follicles. LH stimulates the further development of ovarian follicles and their full secretion of estrogens. At midcycle, LH triggers ovulation and then promotes formation of the corpus luteum, the reason for the name luteinizing hormone. Stimulated by LH, the corpus luteum produces and secretes estrogens, progesterone, relaxin, and inhibin.

At least six different estrogens have been isolated from the plasma of human females, but only three are present in significant quantities: *beta (β)-estradiol, estrone,* and *estriol.* In a nonpregnant woman, the most abundant estrogen is *β*-estradiol, which is synthesized from cholesterol in the ovaries.

Estrogens secreted by follicular cells have several important functions (Figure 28.25).

• Estrogens promote the development and maintenance of female reproductive structures, secondary sex characteristics, and the breasts. The secondary sex characteristics include distribution of adipose tissue in the breasts, abdomen, mons pubis, and hips; voice pitch; a broad pelvis; and pattern of hair growth on the head and body.

• Estrogens increase protein anabolism, including building strong bones. In this regard, estrogens are synergistic with human growth hormone (hGH).

• Estrogens lower blood cholesterol level, which is probably the reason that women under age 50 have a much lower risk of coronary artery disease than do men of comparable age.

• Moderate levels of estrogens in the blood inhibit both the release of GnRH by the hypothalamus and secretion of LH and FSH by the anterior pituitary.

Progesterone, secreted mainly by cells of the corpus luteum, acts synergistically with estrogens to prepare and then maintain the endometrium for implantation of a fertilized ovum and to prepare the mammary glands for milk secretion. High levels of progesterone also inhibit secretion of GnRH and LH.

The small quantity of **relaxin,** produced by the corpus luteum during each monthly cycle, relaxes the uterus by inhibiting contractions of the myometrium. Presumably, implantation of a fertilized ovum occurs more readily in a "quiet" uterus. During pregnancy, the placenta produces much more relaxin, and it continues to relax uterine smooth muscle. At the end of pregnancy, relaxin also increases the flexibility of the pubic symphysis and may help dilate the uterine cervix, both of which ease delivery of the baby.

Inhibin is secreted by granulosa cells of growing follicles and by the corpus luteum after ovulation. It inhibits secretion of FSH and, to a lesser extent, LH.

Figure 28.25 Secretion and physiological effects of estrogens, progesterone, relaxin, and inhibin in the female reproductive cycle. Dashed red lines indicate negative feedback inhibition.

The uterine and ovarian cycles are controlled by gonadotropin-releasing hormone (GnRH) and ovarian hormones (estrogens and progesterone).

 Of the several estrogens, which one exerts the major effect?

Phases of the Female Reproductive Cycle

The duration of the female reproductive cycle typically ranges from 24 to 35 days. For this discussion, we assume a duration of 28 days and divide it into four phases: the menstrual phase, the pre-ovulatory phase, ovulation, and the postovulatory phase (Figure 28.26).

Menstrual Phase

The **menstrual phase** (MEN-stroo-al), also called **menstruation** (men'-stroo-Ā-shun) or **menses** (= month), lasts for roughly the first 5 days of the cycle. (By convention, the first day of menstruation is day one of a new cycle.)

EVENTS IN THE OVARIES Beginning about day 25 of the previous cycle and continuing during the menstrual phase, 20 or so small secondary follicles, some in each ovary, begin to enlarge. Follicular fluid, secreted by the granulosa cells and filtered from blood in the capillaries of the theca folliculi, accumulates in the enlarging antrum (space that forms within the follicle) while the oocyte remains near the edge of the follicle (see Figure 28.16b).

EVENTS IN THE UTERUS Menstrual flow from the uterus consists of 50–150 mL of blood, tissue fluid, mucus, and epithelial cells shed from the endometrium. This discharge occurs because the declining level of ovarian hormones, especially progesterone, stimulates release of prostaglandins that cause the uterine spiral arterioles to constrict. As a result, the cells they supply become oxygen-deprived and start to die. Eventually, the entire stratum functionalis sloughs off. At this time the endometrium is very thin, about 2–5mm, because only the stratum basalis remains. The menstrual flow passes from the uterine cavity through the cervix and vagina to the exterior.

Figure 28.26 The female reproductive cycle.   The length of the female reproductive cycle typically is 24 to 36 days; the preovulatory phase is more variable in length than the other phases. (a) Events in the ovarian and uterine cycles and the release of anterior pituitary hormones are correlated with the sequence of the cycle's four phases. In the cycle shown, fertilization and implantation have not occurred. (b) Relative concentrations of anterior pituitary hormones (FSH and LH) and ovarian hormones (estrogens and progesterone) during the phases of a normal female reproductive cycle.

  Estrogens are the primary ovarian hormones before ovulation; after ovulation, both progesterone and estrogens are secreted by the corpus luteum.

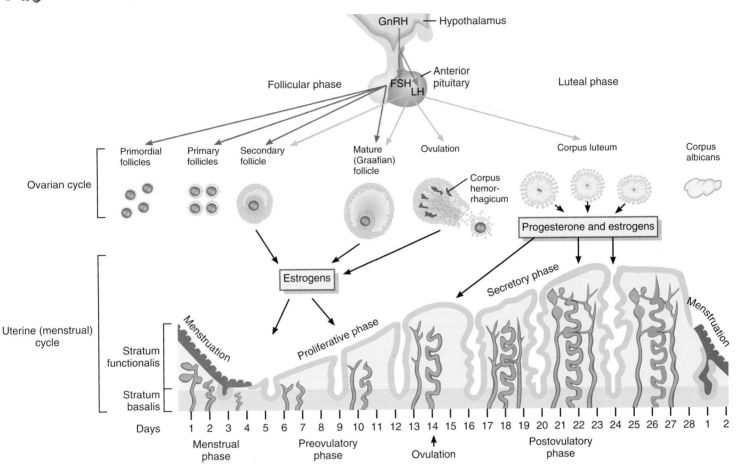

(a) Hormonal regulation of changes in the ovary and uterus

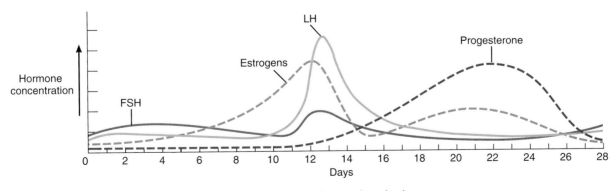

(b) Changes in concentration of anterior pituitary and ovarian hormones

Which hormones are responsible for the proliferative phase of endometrial growth, for ovulation, for growth of the corpus luteum, and for the surge of LH at midcycle?

Preovulatory Phase

The **preovulatory phase** is the time between the end of menstruation and ovulation. The preovulatory phase of the cycle is more variable in length than the other phases and accounts for most of the difference when cycles are shorter or longer than 28 days. It lasts from days 6 to 13 in a 28-day cycle.

EVENTS IN THE OVARIES Under the influence of FSH, the group of about 20 secondary follicles continues to grow and begins to secrete estrogens and inhibin. By about day 6, a single follicle in one of the two ovaries has outgrown all the others to become the **dominant follicle.** Estrogens and inhibin secreted by the dominant follicle decrease the secretion of FSH, which causes other, less well-developed follicles to stop growing and undergo atresia. Fraternal (nonidentical) twins or triplets result when two or three secondary follicles become codominant and later are ovulated and fertilized at about the same time.

Normally, the one dominant follicle becomes the **mature (Graafian) follicle,** which continues to enlarge until it is more than 20 mm in diameter and ready for ovulation (see Figure 28.15). This follicle forms a blisterlike bulge due to the swelling antrum on the surface of the ovary. During the final maturation process, the dominant follicle continues to increase its estrogen production under the influence of an increasing level of LH (Figure 28.26). Although estrogens are the main ovarian hormones before ovulation, small amounts of progesterone are produced by the mature follicle a day or two before ovulation.

With reference to the ovarian cycle, the menstrual and preovulatory phases together are termed the **follicular phase** (fō-LIK-ū-lar) because ovarian follicles are growing and developing.

EVENTS IN THE UTERUS Estrogens liberated into the blood by growing ovarian follicles stimulate the repair of the endometrium; cells of the stratum basalis undergo mitosis and produce a new stratum functionalis. As the endometrium thickens, the short, straight endometrial glands develop, and the arterioles coil and lengthen as they penetrate the stratum functionalis. The thickness of the endometrium approximately doubles, to about 4–10 mm. With reference to the uterine cycle, the preovulatory phase is also termed the **proliferative phase** because the endometrium is proliferating.

Ovulation

Ovulation, the rupture of the mature follicle and the release of the secondary oocyte into the pelvic cavity, usually occurs on day 14 in a 28-day cycle. During ovulation, the secondary oocyte remains surrounded by its zona pellucida and corona radiata. Development of a secondary follicle into a fully mature follicle generally takes a total of about 20 days (spanning the last 6 days of the previous cycle and the first 14 days of the current cycle). During this time the primary oocyte completes meiosis I to become a secondary oocyte, which begins meiosis II but then halts in metaphase.

The *high levels of estrogens* during the last part of the preovulatory phase exert a *positive* feedback effect on the cells that secrete LH and gonadotropin-releasing hormone (GnRH) and cause ovulation, as follows (Figure 28.27):

1 A high concentration of estrogens stimulates more frequent release of GnRH from the hypothalamus. It also directly stimulates gonadotrophs in the anterior pituitary to secrete LH.

2 GnRH promotes the release of FSH and additional LH by the anterior pituitary.

3 LH causes rupture of the dominant follicle and expulsion of a secondary oocyte about 9 hours after the peak of the LH surge. The ovulated oocyte and its corona radiata cells are usually swept into the uterine tube.

From time to time, an oocyte is lost into the pelvic cavity, where it later disintegrates. The small amount of blood that sometimes leaks into the pelvic cavity from the ruptured follicle can cause pain, known as **mittelschmerz** (MIT′-el-shmārtz = pain in the middle), at the time of ovulation.

Figure 28.27 High levels of estrogens exert a positive feedback effect (green arrows) on the hypothalamus and anterior pituitary, thereby increasing secretion of GnRH and LH.

At midcycle, a surge of LH triggers ovulation.

1 High levels of estrogens from almost mature follicle stimulate release of more GnRH and LH

GnRH — Hypothalamus

2 GnRH promotes release of FSH and more LH

LH — Anterior pituitary

3 LH surge brings about ovulation

Ovary

Ovulated secondary oocyte

Almost mature (Graafian) follicle

Corpus hemorrhagicum (ruptured follicle)

What is the effect of rising but still moderate levels of estrogens on the secretion of GnRH, LH, and FSH?

An over-the-counter home test that detects a rising level of LH can be used to predict ovulation a day in advance. FSH also increases at this time, but not as dramatically as LH because FSH is stimulated only by the increase in gonadotropin-releasing hormone. The positive feedback effect of estrogens on the hypothalamus and anterior pituitary does not occur if progesterone is present at the same time.

Postovulatory Phase

The **postovulatory phase** of the female reproductive cycle is the time between ovulation and onset of the next menses. In duration, it is the most constant part of the female reproductive cycle. It lasts for 14 days in a 28-day cycle, from day 15 to day 28 (see Figure 28.26).

EVENTS IN ONE OVARY   After ovulation, the mature follicle collapses, and the basement membrane between the granulosa cells and theca interna breaks down. Once a blood clot forms from minor bleeding of the ruptured follicle, the follicle becomes the **corpus hemorrhagicum** (*hemo-* = blood; *rrhagic-* = bursting forth) (see Figure 28.15). Theca interna cells mix with the granulosa cells as they all become transformed into corpus luteum cells under the influence of LH. Stimulated by LH, the corpus luteum secretes progesterone, estrogen, relaxin, and inhibin. The luteal cells also absorb the blood clot. With reference to the ovarian cycle, this phase is also called the **luteal phase.**

Later events in an ovary that has ovulated an oocyte depend on whether the oocyte is fertilized. If the oocyte *is not fertilized,* the corpus luteum has a lifespan of only 2 weeks. Then, its secretory activity declines, and it degenerates into a corpus albicans (see Figure 28.15). As the levels of progesterone, estrogens, and inhibin decrease, release of GnRH, FSH, and LH rise due to loss of negative feedback suppression by the ovarian hormones. Then, follicular growth resumes and a new ovarian cycle begins.

If the secondary oocyte *is fertilized* and begins to divide, the corpus luteum persists past its normal 2-week life span. It is "rescued" from degeneration by **human chorionic gonadotropin** (kō-rē-ON-ik) **(hCG).** This hormone is produced by the chorion of the embryo beginning about 8 days after fertilization. Like LH, hCG stimulates the secretory activity of the corpus luteum. The presence of hCG in maternal blood or urine is an indicator of pregnancy and is the hormone detected by home pregnancy tests.

EVENTS IN THE UTERUS   Progesterone and estrogens produced by the corpus luteum promote growth and coiling of the endometrial glands, vascularization of the superficial endometrium, and thickening of the endometrium to 12–18 mm (0.48–0.72 in.). Because of the secretory activity of the endometrial glands, which begin to secrete glycogen, this period is termed the **secretory phase** of the uterine cycle. These preparatory changes peak about one week after ovulation, at the time a fertilized ovum might arrive in the uterus. If fertilization does not occur, the level of progesterone declines due to degeneration of the corpus luteum. Withdrawal of progesterone causes menstruation.

Figure 28.28 summarizes the hormonal interactions and cyclical changes in the ovaries and uterus during the ovarian and uterine cycles.

Female Athlete Triad: Disordered Eating, Amenorrhea, and Premature Osteoporosis

The female reproductive cycle can be disrupted by many factors, including weight loss, low body weight, disordered eating, and vigorous physical activity. The observation that three conditions—disordered eating, amenorrhea, and osteoporosis—often occur together in female athletes led researchers to coin the term **female athlete triad** to encompass all three.

Many athletes experience intense pressure from coaches, parents, peers, and themselves to lose weight to improve performance. Hence, they may develop disordered eating behaviors and engage in other harmful weight-loss practices in a struggle to maintain a very low body weight. **Amenorrhea** (ā-men'-ō-RĒ-a; *a-* = without; *men-* = month; *-rrhea* = a flow) is the absence of menstruation. The most common causes of amenorrhea are pregnancy and menopause. In female athletes, amenorrhea results from reduced secretion of gonadotropin-releasing hormone, which decreases the release of LH and FSH. As a result, ovarian follicles fail to develop, ovulation does not occur, synthesis of estrogens and progesterone wanes, and monthly menstrual bleeding ceases. Most cases of the female athlete triad occur in young women whose percentage of body fat is very low. Low levels of the hormone leptin, secreted by adipose cells, may be a contributing factor.

Because estrogens help bones retain calcium and other minerals, chronically low levels of estrogens are associated with loss of bone mineral density. The female athlete triad causes "old bones in young women." In one study, amenorrheic runners in their twenties had low bone mineral densities, similar to those of postmenopausal women 50 to 70 years old! Whereas short periods of amenorrhea in young athletes may cause no lasting harm, long-term cessation of the reproductive cycle may be accompanied by a loss of bone mass, and adolescent athletes may fail to achieve an adequate bone mass. Both situations can lead to premature osteoporosis and irreversible bone damage. ■

▶ CHECKPOINT

28. Describe the function of each of the following hormones in the uterine and ovarian cycles: GnRH, FSH, LH, estrogens, progesterone, and inhibin.

29. Briefly outline the major events of each phase of the uterine cycle, and correlate them with the events of the ovarian cycle.

30. Prepare a labeled diagram of the major hormonal changes that occur during the uterine and ovarian cycles.

Figure 28.28 **Summary of hormonal interactions in the ovarian and uterine cycles.**

Hormones from the anterior pituitary regulate ovarian function, and hormones from the ovaries regulate the changes in the endometrial lining of the uterus.

When declining levels of estrogens and progesterone stimulate secretion of GnRH, is this a positive or a negative feedback effect? Why?

THE HUMAN SEXUAL RESPONSE

▶ O B J E C T I V E

• **Describe the similarities and differences in the sexual responses of males and females.**

During heterosexual **sexual intercourse,** also called **coitus** (KŌ-i-tus = a coming together), sperm are ejaculated from the male urethra into the vagina. The similar sequence of physiological and emotional changes experienced by both males and females

before, during, and after intercourse is termed the **human sexual response.**

Stages of the Human Sexual Response

William Masters and Virginia Johnson, who began their pioneering research on human sexuality in the late 1950s, divided the human sexual response into four stages: excitement, plateau, orgasm, and resolution. During sexual **excitement,** also known as **arousal,** various physical and psychological stimuli trigger parasympathetic reflexes in which nerve impulses propagate from the gray matter of the second, third, and fourth sacral segments of the spinal cord along the pelvic splanchnic nerves. Some of these parasympathetic postganglionic fibers produce relaxation of vascular smooth muscle, which allows **vasocongestion**—engorgement with blood—of genital tissues. Parasympathetic impulses also stimulate the secretion of lubricating fluids, most of which are contributed by the female. Without satisfactory lubrication, sexual intercourse is difficult and painful for both partners and inhibits orgasm.

Other changes include increased heart rate and blood pressure, increased skeletal muscle tone throughout the body, and hyperventilation. Direct physical contact (as in kissing or touching), especially of the glans penis, clitoris, nipples of the breasts, and earlobes is a potent initiator of excitement. However, anticipation or fear, memories, visual, olfactory, and auditory sensations, and fantasies can enhance or diminish the likelihood that excitement will occur. Due to the tactile stimulation of the breast while nursing an infant, feelings of sexual arousal during breastfeeding are fairly common and should not be cause for concern.

The changes that begin during excitement are sustained at an intense level in the **plateau** stage, which may last for only a few seconds or for many minutes. At this stage, many females and some males display a sex flush, a rashlike redness of the face and chest. Generally, the briefest stage is **orgasm** *(climax),* during which the male ejaculates and both sexes experience several rhythmic muscular contractions about 0.8 sec apart, accompanied by intense, pleasurable sensations. The sex flush also is most prominent at this time. During orgasm, bursts of *sympathetic* nerve impulses leave the spinal cord at the first and second lumbar levels and cause rhythmic contractions of smooth muscle in genital organs. At the same time, *somatic motor neurons* from lumbar and sacral segments of the spinal cord trigger powerful, rhythmic contractions of perineal skeletal muscles, particularly the bulbocavernosus muscles, ischiocavernosus muscles, and anal sphincter (see Figure 11.13 on page 341 for males and Figure 28.23 for females).

In males, orgasm usually accompanies ejaculation. Reception of the ejaculate, however, provides little stimulus for a female, especially if she is not already at the plateau stage; this is why a female does not automatically experience orgasm simultaneously with her partner. In both males and females, orgasm is a total body response that may produce milder sensations on some occasions and more intense, explosive sensations at other times. Whereas females may experience two or more orgasms in rapid succession, males enter a **refractory period,** a recovery time during which a second ejaculation and orgasm is physiologically impossible. In some males the refractory period lasts only a few minutes; in others it lasts for several hours.

In the final stage—**resolution,** which begins with a sense of profound relaxation—genital tissues, heart rate, blood pressure, breathing, and muscle tone return to the unaroused state. If sexual excitement has been intense but orgasm has not occurred, resolution takes place more slowly. The four phases of the human sexual response are not always clearly separated from one another and may vary considerably among different people, and even in the same person at different times.

Changes in Males

Most of the time, the penis is flaccid (limp) because sympathetic impulses cause vasoconstriction of its arteries, which limits blood inflow. The first sign of sexual excitement is erection. Parasympathetic impulses cause release of neurotransmitters and local hormones, including the gas nitric oxide, which relaxes vascular smooth muscle in the penis. The arteries of the penis dilate, and blood fills the blood sinuses of the three corpora. Expansion of these erectile tissues compresses the superficial veins that normally drain the penis, resulting in engorgement and rigidity (erection). Parasympathetic impulses also cause the bulbourethral (Cowper's) glands to secrete mucus, which flows through the urethra and provides a small amount of lubrication for intercourse.

During the plateau stage, the head of the penis increases in diameter, and vasocongestion causes the testes to swell. As orgasm begins, rhythmic sympathetic impulses cause peristaltic contractions of smooth muscle in the ducts of each testis, epididymis, and ductus (vas) deferens, as well as in the walls of the seminal vesicles and prostate. These contractions propel sperm and fluid into the urethra (emission). Then, peristaltic contractions in the ducts and urethra combine with rhythmic contractions of skeletal muscles in the perineum and at the base of the penis to propel the semen from the urethra to the exterior.

Changes in Females

The first signs of sexual excitement in females are also due to vasocongestion. Within a few seconds to a minute, parasympathetic impulses stimulate release of fluids that lubricate the walls of the vagina. Although the vaginal mucosa lacks glands, engorgement of its connective tissue with blood during sexual excitement causes lubricating fluid to ooze from capillaries and seep through the epithelial layer via a process called **transudation** (tran′-soo-DĀ-shun; *trans-* = across or through; *sud-* = sweat). Glands within the cervical mucosa and the greater vestibular (Bartholin's) glands contribute a small quantity of lubricating mucus. During the excitement stage, parasympathetic impulses also trigger erection of the clitoris, vasocongestion of the labia, and relaxation of vaginal smooth muscle. Vasocongestion may also cause the breasts to swell and the nipples to become erect.

Late in the plateau stage, pronounced vasocongestion of the distal third of the vagina swells the tissue and narrows the opening. Because of this response, the vagina grips the penis more firmly. If effective sexual stimulation continues, orgasm may occur, associated with 3–15 rhythmic contractions of the vagina, uterus, and perineal muscles.

Erectile Dysfunction

Erectile dysfunction (ED), previously termed *impotence,* is the consistent inability of an adult male to ejaculate or to attain or hold an erection long enough for sexual intercourse. Many cases of impotence are caused by insufficient release of nitric oxide, which relaxes the smooth muscle of the penile arteries. The drug *Viagra* (sildenafil) enhances effects stimulated by nitric oxide in the penis. Other causes of erectile dysfunction include diabetes mellitus, physical abnormalities of the penis, systemic disorders such as syphilis, vascular disturbances (arterial or venous obstructions), neurological disorders, surgery, testosterone deficiency, and drugs (alcohol, antidepressants, antihistamines, antihypertensives, narcotics, nicotine, and tranquilizers). Psychological factors such as anxiety or depression, fear of causing pregnancy, fear of sexually transmitted diseases, religious inhibitions, and emotional immaturity may also cause ED. ■

▶ C H E C K P O I N T

31. What happens during each of the four stages of the human sexual response?

BIRTH CONTROL METHODS AND ABORTION

▶ O B J E C T I V E

• **Compare the various kinds of birth control methods and their effectiveness.**

No single, ideal method of **birth control** exists. The only method of preventing pregnancy that is 100% reliable is total **abstinence,** the avoidance of sexual intercourse. Several other methods are available, each with advantages and disadvantages. These include surgical sterilization, hormonal methods, intrauterine devices, spermicides, barrier methods, periodic abstinence, and coitus interruptus. Table 28.3 gives the failure rates for each method of birth control. We will also discuss induced abortion, the intentional termination of pregnancy.

Surgical Sterilization

Sterilization is a procedure that renders an individual incapable of reproduction. The most common means of sterilization of males is **vasectomy** (vas-EK-tō-mē; -*ectomy* = cut out), in which a portion of each ductus deferens is removed. An incision is made in the posterior side of the scrotum, the ducts are located, each is tied in two places, and the part between the ties is removed. Even though sperm production continues in the testes, sperm can no longer reach the exterior. Instead, they degenerate and are destroyed by phagocytosis. Blood testosterone level is normal, so a vasectomy has no effect on sexual desire or performance. Sterilization in females most often is achieved by performing a **tubal ligation** (lī-GĀ-shun) in which both uterine tubes are tied closed and then cut. As a result, the secondary oocyte cannot pass

Table 28.3	Failure Rates of Several Birth Control Methods	
	Failure Rates*	
Method	**Perfect Use†**	**Typical Use**
None	85%	85%
Complete abstinence	0%	0%
Surgical sterilization		
Vasectomy	0.10%	0.15%
Tubal ligation	0.5%	0.5%
Hormonal methods		
Oral contraceptives	0.1%	3%‡
Norplant	0.3%	0.3%
Depo-provera	0.05%	0.05%
Intrauterine device		
Copper T 380A	0.6%	0.8%
Spermicides	6%	26%‡
Barrier methods		
Male condom	3%	14%‡
Vaginal pouch	5%	21%‡
Diaphragm	6%	20%‡
Periodic abstinence		
Rhythm	9%	25%‡
Sympto-thermal	2%	20%‡
Coitus interruptus	4%	18%‡

*Defined as percentage of women having an unintended pregnancy during the first year of use.
†Failure rate when the method is used correctly and consistently.
‡Includes couples who forgot to use the method.

through the uterine tubes, and sperm cannot reach the oocyte. Tubal ligation reduces the risk of pelvic inflammatory disease in women who are exposed to sexually transmitted infections; it may also reduce the risk of ovarian cancer.

Hormonal Methods

Aside from total abstinence or surgical sterilization, hormonal methods are the most effective means of birth control. Used by 50 million women worldwide, **oral contraceptives** ("the pill") contain various mixtures of synthetic estrogens and progestins (chemicals with actions similar to those of progesterone). They prevent pregnancy mainly by negative feedback inhibition of anterior pituitary secretion of the gonadotropins FSH and LH. The low levels of FSH and LH usually prevent development of a dominant follicle. As a result, estrogen level does not rise, the midcycle LH surge does not occur, and ovulation is not triggered. Thus, there is no secondary oocyte available for fertilization. Even if ovulation does occur, as it does in some cases, oral contraceptives also alter cervical mucus such that it is more hostile to sperm and block implantation in the uterus. If taken properly, the pill is close to 100% effective.

Among the noncontraceptive benefits of oral contraceptives are regulation of the length of menstrual cycles and decreased menstrual flow (and therefore decreased risk of anemia). The pill also provides protection against endometrial and ovarian cancers and reduces the risk of endometriosis. However, oral contraceptives may not be advised for women with a history of blood clotting disorders, cerebral blood vessel damage, migraine headaches, hypertension, liver malfunction, or heart disease. Women who take the pill and smoke face far higher odds of having a heart attack or stroke than do nonsmoking pill users. Smokers should quit smoking or use an alternative method of birth control.

Oral contraceptives also may be used for **emergency contraception (EC),** the so-called "morning-after pill." The relatively high levels of estrogen and progestin in EC pills provide negative feedback inhibition of FSH and LH secretion. Loss of the stimulating effects of these gonadotropic hormones causes the ovaries to cease secretion of their own estrogen and progesterone. In turn, declining levels of estrogen and progesterone induce shedding of the uterine lining, thereby blocking implantation. When two pills are taken within 72 hours after unprotected intercourse, and another two pills are taken 12 hours later, the chance of pregnancy is reduced by 75%.

Other hormonal methods of contraception are Norplant, Depo-provera, and the vaginal ring. **Norplant** consists of six slender hormone-containing capsules that are surgically implanted under the skin of the arm using local anesthesia. They slowly and continually release a progestin, which inhibits ovulation and thickens the cervical mucus. The effects last for 5 years, and Norplant is about as reliable as sterilization. Removing the Norplant capsules restores fertility. **Depo-provera,** which is given as an intramuscular injection once every 3 months, contains progestin, a hormone similar to progesterone, that prevents maturation of the ovum and causes changes in the uterine lining that make pregnancy less likely. The **vaginal ring** is a doughnut-shaped ring that fits in the vagina and releases either a progestin alone or a progestin and an estrogen. It is worn for 3 weeks and removed for 1 week to allow menstruation to occur.

Intrauterine Devices

An **intrauterine device (IUD)** is a small object made of plastic, copper, or stainless steel that is inserted into the cavity of the uterus. IUDs cause changes in the uterine lining that prevent implantation of a fertilized ovum. The IUD most commonly used in the United States today is the Copper T 380A, which is approved for up to 10 years of use and has long-term effectiveness comparable to that of tubal ligation. Some women cannot use IUDs because of expulsion, bleeding, or discomfort.

Spermicides

Various foams, creams, jellies, suppositories, and douches that contain sperm-killing agents, or **spermicides,** make the vagina and cervix unfavorable for sperm survival and are available without prescription. The most widely used spermicide is nonoxynol-9, which kills sperm by disrupting their plasma membrane. It also inactivates the AIDS virus and decreases the incidence of gonorrhea (described on page 1055). A spermicide is more effective when used together with a barrier method such as a diaphragm or a condom.

Barrier Methods

Barrier methods are designed to prevent sperm from gaining access to the uterine cavity and uterine tubes. In addition to preventing pregnancy, barrier methods may also provide some protection against sexually transmitted diseases (STDs) such as AIDS. In contrast, oral contraceptives and IUDs confer no such protection. Among the barrier methods are use of a condom, a vaginal pouch, or a diaphragm.

A **condom** is a nonporous, latex covering placed over the penis that prevents deposition of sperm in the female reproductive tract. A **vaginal pouch,** sometimes called a female condom, is made of two flexible rings connected by a polyurethane sheath. One ring lies inside the sheath and is inserted to fit over the cervix; the other ring remains outside the vagina and covers the female external genitals.

A **diaphragm** is a rubber, dome-shaped structure that fits over the cervix and is used in conjunction with a spermicide. It can be inserted up to 6 hours before intercourse. The diaphragm stops most sperm from passing into the cervix and the spermicide kills most sperm that do get by. Although diaphragm use does decrease the risk of some STDs, it does not fully protect against HIV infection.

Periodic Abstinence

A couple can use their knowledge of the physiological changes that occur during the female reproductive cycle to decide either to abstain from intercourse on those days when pregnancy is a likely result, or to plan intercourse on those days if they wish to conceive a child. In females with normal and regular menstrual cycles, these physiological events help to predict the day on which ovulation is likely to occur.

The first physiologically based method, developed in the 1930s, is known as the **rhythm method.** It takes advantage of the fact that a secondary oocyte is fertilizable for only 24 hours and is available for only 3 to 5 days in each reproductive cycle. During this time (3 days before ovulation, the day of ovulation, and 3 days after ovulation) the couple abstains from intercourse. The effectiveness of the rhythm method for birth control is poor in many women due to their irregular cycles.

Another system is the **sympto-thermal method,** in which couples are instructed to know and understand certain signs of fertility. The signs of ovulation include increased basal body temperature; the production of abundant clear, stretchy cervical mucus; and pain associated with ovulation (mittelschmerz). If a couple abstains from sexual intercourse when the signs of ovulation are present and for 3 days afterward, the chance of pregnancy is decreased. A big problem with this method is that fertilization is very likely if intercourse occurs one or two days *before* ovulation.

Coitus Interruptus

Coitus interruptus is withdrawal of the penis from the vagina just before ejaculation. Failures with this method are due either to failure to withdraw before ejaculation or to preejaculatory emission of sperm-containing fluid from the urethra. In addition, this method offers no protection against transmission of STDs.

Abortion

Abortion refers to the premature expulsion of the products of conception from the uterus, usually before the 20th week of pregnancy. An abortion may be spontaneous (naturally occurring; also called a miscarriage) or induced (intentionally performed). Induced abortions may be performed by vacuum aspiration (suction), infusion of a saline solution, or surgical evacuation (scraping).

Certain drugs, most notably RU 486, can induce a so-called nonsurgical abortion. **RU 486 (mifepristone)** is an antiprogestin; it blocks the action of progesterone by binding to and blocking progesterone receptors. Progesterone prepares the uterine endometrium for implantation and then maintains the uterine lining after implantation. If the level of progesterone falls during pregnancy or if the action of the hormone is blocked, menstruation occurs, and the embryo sloughs off along with the uterine lining. Within 12 hours after taking RU 486, the endometrium starts to degenerate, and within 72 hours, it begins to slough off. A form of prostaglandin E (misoprostol), which stimulates uterine contractions, is given after RU 486 to aid in expulsion of the endometrium. RU 486 can be taken up to 5 weeks after conception. One side effect of the drug is uterine bleeding.

▶ CHECKPOINT

32. How do oral contraceptives reduce the likelihood of pregnancy?

33. Why do some methods of birth control protect against sexually transmitted diseases, whereas others do not?

DEVELOPMENT OF THE REPRODUCTIVE SYSTEMS

▶ OBJECTIVE

• **Describe the development of the male and female reproductive systems.**

The *gonads* develop from the **intermediate mesoderm.** During the fifth week of development, they appear as bulges that protrude into the ventral body cavity (Figure 28.29). Adjacent to the gonads are the **mesonephric (Wolffian) ducts,** which eventually develop into structures of the reproductive system in males. A

second pair of ducts, the **paramesonephric (Müllerian) ducts,** develop lateral to the mesonephric ducts and eventually form structures of the reproductive system in females. Both sets of ducts empty into the urogenital sinus. An early embryo has the potential to follow either the male or the female pattern of development because it contains both sets of ducts and primitive gonads that can differentiate into either testes or ovaries.

Cells of a male embryo have one X chromosome and one Y chromosome. The male pattern of development is initiated by a Y chromosome "master switch" gene named *SRY,* which stands for *S*ex-determining *R*egion of the *Y* chromosome. When the *SRY* gene is expressed during development, its protein product causes the primitive Sertoli cells to begin to differentiate in the gonadal tissues during the seventh week. The developing Sertoli cells secrete a hormone called **Müllerian-inhibiting substance (MIS),** which causes apoptosis of cells within the paramesonephric (Müllerian) ducts. As a result, those cells do not contribute any functional structures to the male reproductive system. Stimulated by human chorionic gonadotropin (hCG), primitive Leydig cells in the gonadal tissue begin to secrete the androgen **testosterone** during the eighth week. Testosterone then stimulates development of the mesonephric duct on each side into the *epididymis, ductus (vas) deferens, ejaculatory duct,* and *seminal vesicle.* The *testes* connect to the mesonephric duct through a series of tubules that eventually become the *seminiferous tubules.* The *prostate* and *bulbourethral glands* are **endodermal** outgrowths of the urethra.

Cells of a female embryo have two X chromosomes and no Y chromosome. Because *SRY* is absent, the gonads develop into *ovaries,* and because MIS is not produced, the paramesonephric ducts flourish. The distal ends of the paramesonephric ducts fuse to form the *uterus* and *vagina* whereas the unfused proximal portions become the *uterine (Fallopian) tubes.* The mesonephric ducts degenerate without contributing any functional structures to the female reproductive system because testosterone is absent. The *greater* and *lesser vestibular glands* develop from **endodermal** outgrowths of the vestibule.

The *external genitals* of both male and female embryos also remain undifferentiated until about the eighth week. Before differentiation, all embryos have an elevated midline swelling called the **genital tubercle** (Figure 28.30). The tubercle consists of the **urethral groove** (opening into the urogenital sinus), paired **urethral folds,** and paired **labioscrotal swellings.**

In male embryos, some testosterone is converted to a second androgen called **dihydrotestosterone (DHT).** DHT stimulates development of the urethra, prostate, and external genitals (scrotum and penis). Part of the genital tubercle elongates and develops into a penis. Fusion of the urethral folds forms the *spongy (penile) urethra* and leaves an opening to the exterior only at the distal end of the penis, the *external urethral orifice.* The labioscrotal swellings develop into the *scrotum.* In the absence of DHT, the genital tubercle gives rise to the *clitoris* in female embryos. The urethral folds remain open as the *labia minora,* and the labioscrotal swellings become the *labia majora.* The urethral groove becomes the *vestibule.* After birth, androgen levels

Figure 28.29   Development of the internal reproductive systems.

The gonads develop from intermediate mesoderm.

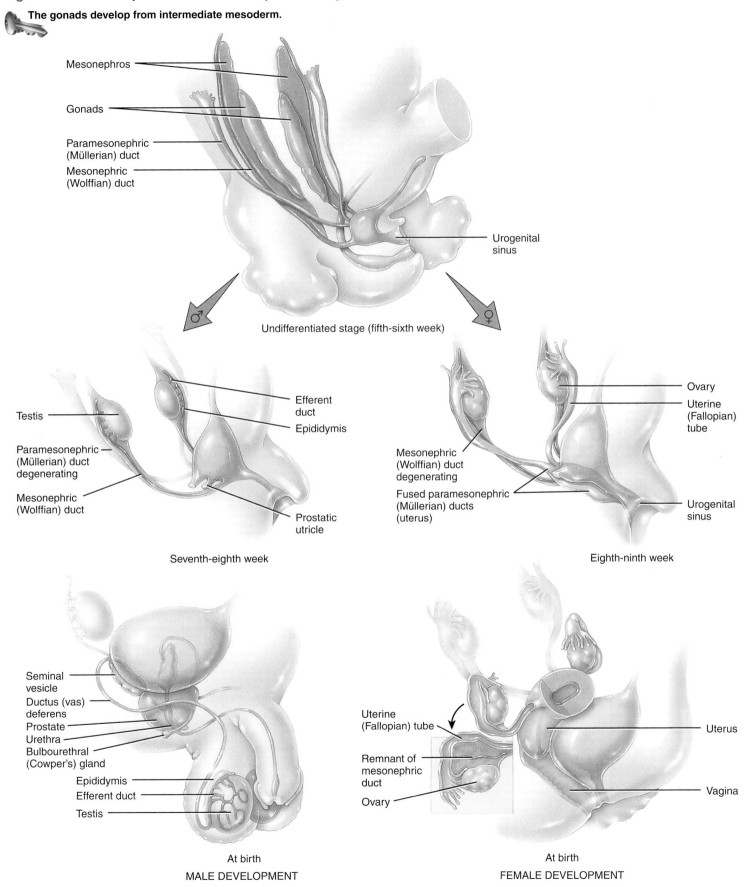

Mesonephros

Gonads

Paramesonephric (Müllerian) duct

Mesonephric (Wolffian) duct

Urogenital sinus

Undifferentiated stage (fifth-sixth week)

Efferent duct

Epididymis

Testis

Paramesonephric (Müllerian) duct degenerating

Mesonephric (Wolffian) duct

Prostatic utricle

Seventh-eighth week

Ovary

Uterine (Fallopian) tube

Mesonephric (Wolffian) duct degenerating

Fused paramesonephric (Müllerian) ducts (uterus)

Urogenital sinus

Eighth-ninth week

Seminal vesicle

Ductus (vas) deferens

Prostate

Urethra

Bulbourethral (Cowper's) gland

Epididymis

Efferent duct

Testis

At birth

MALE DEVELOPMENT

Uterine (Fallopian) tube

Remnant of mesonephric duct

Ovary

Uterus

Vagina

At birth

FEMALE DEVELOPMENT

Which gene is responsible for the development of the gonads into testes?

Figure 28.30   Development of the external genitals.

The external genitals of male and female embryos remain undifferentiated until about the eighth week.

Genital tubercle —

— Glans area
— Urethral folds
— Urethral groove
— Labioscrotal swelling
— Urogenital sinus
— Perineum
— Anus

Undifferentiated stage (about five-week embryo)

♂   ♀

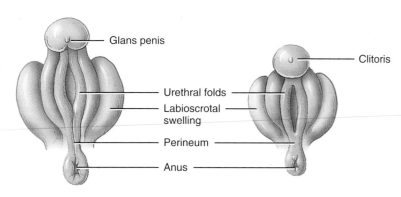

— Glans penis

— Urethral folds
— Labioscrotal swelling
— Perineum
— Anus

— Clitoris

Ten-week embryo

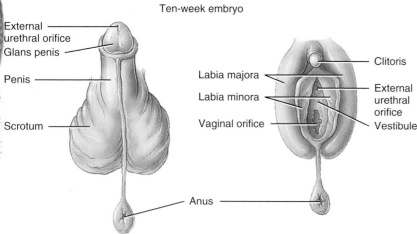

External urethral orifice
Glans penis

Penis —

Scrotum —

— Labia majora
— Labia minora
— Vaginal orifice

— Clitoris
— External urethral orifice
— Vestibule

— Anus

Near birth

MALE DEVELOPMENT   FEMALE DEVELOPMENT

Which hormone is responsible for the differentiation of the external genitals?

---

decline because hCG is no longer present to stimulate secretion of testosterone.

 Deficiency of 5 Alpha-Reductase

A rare genetic mutation leads to a deficiency in 5 alpha-reductase, the enzyme that converts testosterone to dihydrotestosterone. At birth such a baby looks like a female externally because of the absence of dihydrotestosterone during development. At puberty, however, testosterone level rises, masculine characteristics start to appear; and the breasts fail to develop. Internal examination reveals testes and structures that normally develop from the mesonephric duct (epididymis, ductus deferens, seminal vesicle, and ejaculatory duct) rather than ovaries and a uterus. ■

AGING AND THE REPRODUCTIVE SYSTEMS

OBJECTIVE

• **Describe the effects of aging on the reproductive systems.**

During the first decade of life, the reproductive system is in a juvenile state. At about age 10, hormone-directed changes start to occur in both sexes. **Puberty** (PŪ-ber-tē = a ripe age) is the period when secondary sexual characteristics begin to develop and the potential for sexual reproduction is reached. Onset of puberty is marked by pulses or bursts of LH and FSH secretion, each triggered by a pulse of GnRH. Most pulses occur during sleep. As puberty advances, the hormone pulses occur during the day as well as at night. The pulses increase in frequency during a three- to four-year period until the adult pattern is established. The stimuli that cause the GnRH pulses are still unclear, but a role for the hormone leptin is starting to unfold. Just before puberty, leptin levels rise in proportion to adipose tissue mass. Interestingly, leptin receptors are present in both the hypothalamus and anterior pituitary. Mice that lack a functional leptin gene from birth are sterile and remain in a prepubertal state. Giving leptin to such mice elicits secretion of gonadotropins, and they become fertile. Leptin may signal the hypothalamus that long-term energy stores (triglycerides in adipose tissue) are adequate for reproductive functions to begin.

In females, the reproductive cycle normally occurs once each month from **menarche** (me-NAR-kē), the first menses, to **menopause,** the permanent cessation of menses. Thus, the female reproductive system has a time-limited span of fertility between menarche and menopause. For the first 1 to 2 years after menarche, ovulation only occurs in about 10% of the cycles and the luteal phase is short. Gradually, the percentage of ovulatory cycles increases, and the luteal phase reaches its normal duration of 14 days. With age, fertility declines. Between the ages of 40 and 50 the pool of remaining ovarian follicles becomes exhausted. As a result, the ovaries become less responsive to hormonal stimulation. The production of estrogens declines, despite copious secretion of FSH and LH by the anterior pituitary. Many women experience hot flashes and heavy sweating, which coincide with bursts of GnRH release. Other symptoms of menopause are headache, hair loss, muscular pains, vaginal dryness, insomnia, depression, weight gain, and mood swings. Some atrophy of the ovaries, uterine tubes, uterus, vagina, external genitalia, and breasts occurs in postmenopausal women. Due to loss of estrogens, most women experience a decline in bone mineral density after menopause. Sexual desire (libido) does not show a parallel decline; it may be maintained by adrenal sex steroids. The risk of having uterine cancer peaks at about 65 years of age, but cervical cancer is more common in younger women.

In males, declining reproductive function is much more subtle than in females. Healthy men often retain reproductive capacity into their eighties or nineties. At about age 55 a decline in testosterone synthesis leads to reduced muscle strength, fewer viable sperm, and decreased sexual desire. However, abundant sperm may be present even in old age.

Enlargement of the prostate to two to four times its normal size occurs in approximately one-third of all males over age 60. This condition, called **benign prostatic hyperplasia (BPH),** decreases the size of the prostatic urethra and is characterized by frequent urination, nocturia (bed-wetting), hesitancy in urination, decreased force of urinary stream, postvoiding dribbling, and a sensation of incomplete emptying.

▶ C H E C K P O I N T

34. What are the roles of hormones in differentiation of the gonads, the mesonephric ducts, the paramesonephric ducts, and the external genitals?

35. What changes occur in males and females at puberty?

36. What do the terms menarche and menopause mean?

DISORDERS: HOMEOSTATIC IMBALANCES

Reproductive System Disorders in Males

Testicular Cancer

Testicular cancer is the most common cancer in males between the ages of 20 and 35. More than 95% of testicular cancers arise from spermatogenic cells within the seminiferous tubules. An early sign of testicular cancer is a mass in the testis, often associated with a sensation of testicular heaviness or a dull ache in the lower abdomen; pain usually does not occur. To increase the chance for early detection of a testicular cancer, all males should perform regular self-examinations of the testes.

Prostate Disorders

Because the prostate surrounds part of the urethra, any infection, enlargement, or tumor in it can obstruct the flow of urine. Acute and chronic infections of the prostate are common in postpubescent males, often in association with inflammation of the urethra. In **acute prostatitis,** the prostate becomes swollen and tender. **Chronic prostatitis** is one of the most common chronic infections in men of the middle and later years. On examination, the prostate feels enlarged, soft, and very tender, and its surface outline is irregular.

Prostate cancer is the leading cause of death from cancer in men in the United States, having surpassed lung cancer in 1991. Each year it is diagnosed in almost 200,000 U.S. men and causes nearly 40,000 deaths. A blood test can measure the level of prostate-specific antigen (PSA) in the blood. The amount of PSA, which is produced only by prostate epithelial cells, increases with enlargement of the prostate and may indicate infection, benign enlargement, or prostate cancer. Males over the age of 40 should have an annual examination of the prostate gland. In a **digital rectal exam,** a physician palpates the gland through the rectum with the fingers (digits). Many physicians also recommend an annual PSA test for males over age 50. Treatment for prostate cancer may involve surgery, cryotherapy, radiation, hormonal therapy, and chemotherapy. Because many prostate cancers grow very slowly, some urologists recommend "watchful waiting" before treating small tumors in men over age 70.

Reproductive System Disorders in Females

Premenstrual Syndrome and Premenstrual Dysphoric Disorder

Premenstrual syndrome (PMS) is a cyclical disorder of severe physical and emotional distress. It appears during the postovulatory (luteal) phase of the female reproductive cycle and dramatically disappears when menstruation begins. The signs and symptoms are highly variable from one woman to another. They may include edema, weight gain, breast swelling and tenderness, abdominal distension, backache, joint pain, constipation, skin eruptions, fatigue and lethargy, greater need for sleep, depression or anxiety, irritability, mood swings, headache, poor coordination and clumsiness, and cravings for sweet or salty foods. The cause of PMS is unknown. For some women, getting regular exercise; avoiding caffeine, salt, and alcohol; and eating a diet that is high in complex carbohydrates and lean proteins can bring considerable relief.

Premenstrual dysphoric disorder (PMDD) is a more severe syndrome in which PMS-like signs and symptoms do not resolve after the onset of menstruation. Clinical research studies have found that suppression of the reproductive cycle by a drug that interferes with GnRH (leuprolide) decreases symptoms significantly. Because symptoms reappear when estradiol or progesterone is given together with leuprolide, researchers propose that PMDD is caused by abnormal responses to normal levels of these ovarian hormones. **SSRIs** (selective serotonin receptor inhibitors) have shown promise in treating both PMS and PMDD.

Endometriosis

Endometriosis (en-dō-mē-trē-Ō-sis; *endo-* = within; *metri-* = uterus; *osis* = condition) is characterized by the growth of endometrial tissue outside the uterus. The tissue enters the pelvic cavity via the open uterine tubes and may be found in any of several sites—on the ovaries, the rectouterine pouch, the outer surface of the uterus, the sigmoid colon, pelvic and abdominal lymph nodes, the cervix, the abdominal wall, the kidneys, and the urinary bladder. Endometrial tissue responds to hormonal fluctuations, whether it is inside or outside the uterus. With each reproductive cycle, the tissue proliferates and then breaks down and bleeds. When this occurs outside the uterus, it can cause inflammation, pain, scarring, and infertility. Symptoms include premenstrual pain or unusually severe menstrual pain.

Breast Cancer

One in eight women in the United States faces the prospect of **breast cancer.** After lung cancer, it is the second-leading cause of death from cancer in U.S. women. Breast cancer can occur but is rare in males. In females, breast cancer is seldom seen before age 30, and its incidence rises rapidly after menopause. An estimated 5% of the 180,000 cases diagnosed each year in the United States, particularly those that arise in younger women, stem from inherited genetic mutations (changes in the DNA). Researchers have now identified two genes that increase susceptibility to breast cancer: *BRCA 1* (*breast cancer 1*) and *BRCA2*. Mutation of *BRCA1* also confers a high risk for ovarian cancer. In addition, mutations of the *p53* gene increase the risk of breast cancer in both males and females, and mutations of the androgen receptor gene are associated with the occurrence of breast cancer in some males. Despite the fact that breast cancer generally is not painful until it becomes quite advanced, any lump, no matter how small, should be reported to a physician at once. Early detection—by breast self-examination and mammograms—is the best way to increase the chance of survival.

The most effective technique for detecting tumors less than 1 cm (0.4 in.) in diameter is **mammography** (mam-OG-ra-fē; *-graphy* = to record), a type of radiography using very sensitive x-ray film. The image of the breast, called a **mammogram,** is best obtained by compressing the breasts, one at a time, using flat plates. A supplementary procedure for evaluating breast abnormalities is **ultrasound.** Although ultrasound cannot detect tumors smaller than 1 cm in diameter, it can be used to determine whether a lump is a benign, fluid-filled cyst or a solid (and therefore possibly malignant) tumor.

Among the factors that increase the risk of developing breast cancer are (1) a family history of breast cancer, especially in a mother or sister; (2) nulliparity (never having borne a child) or having a first child after age 35; (3) previous cancer in one breast; (4) exposure to ionizing radiation, such as x rays; (5) excessive alcohol intake; and (6) cigarette smoking.

The American Cancer Society recommends the following steps to help diagnose breast cancer as early as possible:

* All women over 20 should develop the habit of monthly breast self-examination.

* A physician should examine the breasts every 3 years when a woman is between the ages of 20 and 40, and every year after age 40.

* A mammogram should be taken in women between the ages of 35 and 39, to be used later for comparison (baseline mammogram).

* Women with no symptoms should have a mammogram every year or two between ages 40 and 49, and every year after age 50.

* Women of any age with a history of breast cancer, a strong family history of the disease, or other risk factors should consult a physician to determine a schedule for mammography.

Treatment for breast cancer may involve hormone therapy, chemotherapy, radiation therapy, **lumpectomy** (removal of the tumor and the immediate surrounding tissue), a modified or radical mastectomy, or a combination of these approaches. A **radical mastectomy** (*mast-* = breast) involves removal of the affected breast along with the underlying pectoral muscles and the axillary lymph nodes. (Lymph nodes are removed because metastasis of cancerous cells usually occurs through lymphatic or blood vessels.) Radiation treatment and chemotherapy may follow the surgery to ensure the destruction of any stray cancer cells. Several types of chemotherapeutic drugs are used to decrease the risk of relapse or disease progression. *Nolvadex (tamoxifen)*, is an estrogen antagonist that binds to and blocks estrogen receptors, thus decreasing the stimulating effect of estrogen itself on breast cancer cells. Used for 20 years now, tamoxifen greatly reduces the risk of cancer recurrence. *Herceptin*, a monoclonal antibody drug, targets an antigen on the surface of breast cancer cells. It is effective in causing regression of tumors and retarding progression of the disease. The early data from clinical trials of two new drugs, *Femara* and *Amimidex*, show relapse rates that are lower than those for tamoxifen. These drugs are inhibitors of aromatase, the enzyme needed for the final step in synthesis of estrogens. Finally, two drugs—tamoxifen and *Evista (raloxifene)*—are being marketed for breast cancer *prevention*. Interestingly, raloxifene blocks estrogen receptors in the breasts and uterus but activates estrogen receptors in bone. Thus, it can be used to treat osteoporosis without increasing a woman's risk of breast or endometrial (uterine) cancer.

Ovarian Cancer

Even though ovarian cancer is the sixth most common form of cancer in females, it is the leading cause of death from all gynecological malignancies (excluding breast cancer) because it is difficult to detect before it metastasizes (spreads) beyond the ovaries. Risk factors associated with ovarian cancer include age (usually over age 50); race (whites are at highest risk); family history of ovarian cancer; more than 40 years of active ovulation; nulliparity or first pregnancy after age 30; a high-fat, low-fiber, vitamin A–deficient diet; and prolonged exposure to asbestos or talc. Early ovarian cancer has no symptoms or only mild ones associated with other common problems, such as abdominal discomfort, heartburn, nausea, loss of appetite, bloating, and flatulence. Later-stage signs and symptoms include an enlarged abdomen, abdominal and/or pelvic pain, persistent gastrointestinal disturbances, urinary complications, menstrual irregularities, and heavy menstrual bleeding.

Cervical Cancer

Cervical cancer, carcinoma of the cervix of the uterus, starts with **cervical dysplasia** (dis-PLĀ-sē-a), a change in the shape, growth, and number of cervical cells. The cells may either return to normal or progress to cancer. In most cases, cervical cancer may be detected in its earliest stages by a Pap test (see page 108). Some evidence links cervical cancer to the virus that causes genital warts, human papillomavirus (HPV). Increased risk is associated with having a large number of sexual partners, having first intercourse at a young age, and smoking cigarettes.

Vulvovaginal Candidiasis

Candida albicans is a yeastlike fungus that commonly grows on mucous membranes of the gastrointestinal and genitourinary tracts. The organism is responsible for **vulvovaginal candidiasis** (vul-vō-VAJ-i-nal can-di-DĪ-a-sis), the most common form of **vaginitis** (vaj′-i′-NĪ-tis), inflammation of the vagina. Candidiasis is characterized by severe itching; a thick, yellow, cheesy discharge; a yeasty odor; and pain. The disorder, experienced at least once by about 75% of females, is usually a result of proliferation of the fungus following antibiotic therapy for another condition. Predisposing conditions include the use of oral contraceptives or cortisone-like medications, pregnancy, and diabetes.

Sexually Transmitted Diseases

A **sexually transmitted disease (STD)** is one that is spread by sexual contact. In most developed countries of the world, such as those of the European Community, Japan, Australia, and New Zealand, the incidence of STDs has declined markedly during the past 25 years. In the United States, by contrast, STDs have been rising to near-epidemic proportions, currently affecting more than 65 million people. AIDS and hepatitis B, which are sexually transmitted diseases that also may be contracted in other ways, are discussed in Chapters 22 and 24, respectively.

Chlamydia

Chlamydia (kla-MID-ē-a) is a sexually transmitted disease caused by the bacterium *Chlamydia trachomatis* (*chlamy-* = cloak). This unusual bacterium cannot reproduce outside body cells; it "cloaks" itself inside cells, where it divides. At present, chlamydia is the most prevalent sexually transmitted disease in the United States. In most cases, the initial infection is asymptomatic and thus difficult to recognize clinically. In males, urethritis is the principal result, causing a clear discharge, burning on urination, frequent urination, and painful urination. Without treatment, the epididymides may also become inflamed, leading to sterility. In 70% of females with chlamydia, symptoms are absent, but chlamydia is the leading cause of pelvic inflammatory disease. Moreover, the uterine tubes may also become inflamed, which increases the risk of ectopic pregnancy (implantation of a fertilized ovum outside the uterus) and infertility due to the formation of scar tissue in the tubes.

Gonorrhea

Gonorrhea (gon-ō-RĒ-a) or **"the clap"** is caused by the bacterium *Neisseria gonorrhoeae*. In the United States, 1–2 million new cases of gonorrhea appear each year, most among individuals aged 15–29 years. Discharges from infected mucus membranes are the source of transmission of the bacteria either during sexual contact or during the passage of a newborn through the birth canal. The infection site can be in the mouth and throat after oral-genital contact, in the vagina and penis after genital intercourse, or in the rectum after recto-genital contact.

Males usually experience urethritis with profuse pus drainage and painful urination. The prostate and epididymis may also become infected. In females, infection typically occurs in the vagina, often with a discharge of pus. Both infected males and females may harbor the disease without any symptoms, however, until it has progressed to a more advanced stage; about 5–10% of males and 50% of females are asymptomatic. In females, the infection and consequent inflammation can proceed from the vagina into the uterus, uterine tubes, and pelvic cavity. An estimated 50,000 to 80,000 women in the United States are made infertile by gonorrhea every year as a result of scar tissue formation that closes the uterine tubes. If bacteria in the birth canal are transmitted to the eyes of a newborn, blindness can result. Administration of a 1% silver nitrate solution in the infant's eyes prevents infection.

Syphilis

Syphilis, caused by the bacterium *Treponema pallidum*, is transmitted through sexual contact or exchange of blood, or through the placenta to a fetus. The disease progresses through several stages. During the *primary stage*, the chief sign is a painless open sore, called a **chancre** (SHANG-ker), at the point of contact. The chancre heals within 1 to 5 weeks. From 6 to 24 weeks later, signs and symptoms such as a skin rash, fever, and aches in the joints and muscles usher in the *secondary state*, which is systemic—the infection spreads to all major body systems. When signs of organ degeneration appear, the disease is said to be in the *tertiary stage*. If the nervous system is involved, the tertiary stage is called **neurosyphilis.** As motor areas become extensively damaged, victims may be unable to control urine and bowel movements. Eventually they may become bedridden and unable even to feed themselves. In addition, damage to the cerebral cortex produces memory loss and personality changes that range from irritability to hallucinations.

Genital Herpes

Genital herpes is an incurable STD. Type II herpes simplex virus (HSV-2) causes genital infections, producing painful blisters on the prepuce, glans penis, and penile shaft in males and on the vulva or sometimes high up in the vagina in females. The blisters disappear and reappear in most patients, but the virus itself remains in the body. A related virus, type I herpes simplex virus (HSV-1), causes cold sores on the mouth and lips. Infected individuals typically experience recurrences of symptoms several times a year.

Genital Warts

Warts are an infectious disease caused by viruses. *Human papillomavirus (HPV)* causes **genital warts,** which is commonly transmitted sexually. Nearly one million people a year develop genital warts in the United States. Patients with a history of genital warts may be at increased risk for cancers of the cervix, vagina, anus, vulva, and penis. There is no cure for genital warts.

MEDICAL TERMINOLOGY

Abnormal uterine bleeding Menstruation of excessive duration or excessive amount, diminished menstrual flow, too frequent menstruation, bleeding between menstrual periods, and postmenopausal bleeding. These abnormalities may be caused by disordered hormonal regulation, emotional factors, fibroid tumors of the uterus, and systemic diseases.

Castration (kas-TRĀ-shun = to prune) Removal, inactivation, or destruction of the gonads; commonly used in reference to removal of the testes only.

Culdoscopy (kul-DOS-kō-pē; *-scopy* = to examine) A procedure in which a culdoscope (endoscope) is inserted through the vagina to view the pelvic cavity.

Dysmenorrhea (dis'-men-oī-Ē-a; *dys-* = difficult or painful) Pain associated with menstruation; the term is usually reserved to describe menstrual symptoms that are severe enough to prevent a woman from functioning normally for one or more days each month. Some cases are caused by uterine tumors, ovarian cysts, pelvic inflammatory disease, or intrauterine devices.

Endocervical curettage   (kū-re-TAHZH; *curette* = scraper) A procedure in which the cervix is dilated and the endometrium of the uterus is scraped with a spoon-shaped instrument called a curette; commonly called a D and C (dilation and curettage).

Hermaphroditism   (her-MAF-rō-dīt-izm) Presence of both ovarian and testicular tissue in one individual.

Hypospadias   (hī′-pō-SPĀ-dē-as; *hypo-* = below) A displaced urethral opening. In males, the displaced opening may be on the underside of the penis, at the penoscrotal junction, between the scrotal folds, or in the perineum; in females, the urethra opens into the vagina.

Leukorrhea   (loo′-kō-RĒ-a; *leuko-* = white) A whitish (nonbloody) vaginal discharge containing mucus and pus cells that may occur at any age and affects most women at some time.

Oophorectomy   (ō′-of-ō-REK-tō-mē; *oophor-* = bearing eggs) Removal of the ovaries.

Ovarian cyst   The most common form of ovarian tumor, in which a fluid-filled follicle or corpus luteum persists and continues growing.

Pelvic inflammatory disease (PID)   A collective term for any extensive bacterial infection of the pelvic organs, especially the uterus, uterine tubes, or ovaries, which is characterized by pelvic soreness, lower back pain, abdominal pain, and urethritis. Often the early symptoms of PID occur just after menstruation. As infection spreads, fever may develop, along with painful abscesses of the reproductive organs.

Salpingectomy   (sal′-pin-JEK-tō-mē; *salpingo* = tube) Removal of a uterine tube (oviduct).

Smegma   (SMEG-ma) The secretion, consisting principally of desquamated epithelial cells, found chiefly around the external genitalia and especially under the foreskin of the male.

STUDY OUTLINE

THE CELL CYCLE IN THE GONADS (p. 1012)

1. Reproduction is the process by which new individuals of a species are produced and the genetic material is passed from generation to generation.
2. The organs of reproduction are grouped as gonads (produce gametes), ducts (transport and store gametes), accessory sex glands (produce materials that support gametes), and supporting structures (have various roles in reproduction).
3. Gametes contain the haploid (*n*) chromosome number (23), and most somatic cells contain the diploid (2*n*) chromosome number (46).
4. Meiosis is the process that produces haploid gametes; it consists of two successive nuclear divisions called meiosis I and meiosis II.
5. During meiosis I, homologous chromosomes undergo synapsis (pairing) and crossing-over; the net result is two haploid daughter cells that are genetically unlike each other and unlike the parent cell that produced them.
6. During meiosis II, the two haploid daughter cells divide to form four haploid cells.

MALE REPRODUCTIVE SYSTEM (p. 1014)

1. The male structures of reproduction include the testes, ductus epididymis, ductus (vas) deferens, ejaculatory duct, urethra, seminal vesicles, prostate, bulbourethral (Cowper's) glands, and penis.
2. The scrotum is a sac that hangs from the root of the penis and consists of loose skin and superficial fascia; it supports the testes.
3. The temperature of the testes is regulated by contraction of the cremaster muscle and dartos muscle, which either elevates them and brings them closer to the pelvic cavity or relaxes and moves them farther from the pelvic cavity.
4. The testes are paired oval glands (gonads) in the scrotum containing seminiferous tubules, in which sperm cells are made; Sertoli cells (sustentacular cells), which nourish sperm cells and secrete

inhibin; and Leydig cells (interstitial endocrinocytes), which produce the male sex hormone testosterone.

5. The testes descend into the scrotum through the inguinal canals during the seventh month of fetal development. Failure of the testes to descend is called cryptorchidism.
6. Secondary oocytes and sperm, both of which are called gametes, are produced in the gonads.
7. Spermatogenesis, which occurs in the testes, is the process whereby immature spermatogonia develop into mature sperm. The spermatogenesis sequence, which includes meiosis I, meiosis II, and spermiogenesis, results in the formation of four haploid sperm (spermatozoa) from each primary spermatocyte.
8. Mature sperm consist of a head, a midpiece, and a tail. Their function is to fertilize a secondary oocyte.
9. At puberty, gonadotropin-releasing hormone (GnRH) stimulates anterior pituitary secretion of FSH and LH. LH stimulates production of testosterone; FSH and testosterone stimulate spermatogenesis. Sertoli cells secrete androgen-binding protein (ABP), which binds to testosterone and keeps its concentration high in the seminiferous tubule.
10. Testosterone controls the growth, development, and maintenance of sex organs; stimulates bone growth, protein anabolism, and sperm maturation; and stimulates development of masculine secondary sex characteristics.
11. Inhibin is produced by Sertoli cells; its inhibition of FSH helps regulate the rate of spermatogenesis.
12. The duct system of the testes includes the seminiferous tubules, straight tubules, and rete testis. Sperm flow out of the testes through the efferent ducts.
13. The ductus epididymis is the site of sperm maturation and storage.
14. The ductus (vas) deferens stores sperm and propels them toward the urethra during ejaculation.
15. Each ejaculatory duct, formed by the union of the duct from the seminal vesicle and ampulla of the ductus (vas) deferens, is the

passageway for ejection of sperm and secretions of the seminal vesicles into the first portion of the urethra, the prostatic urethra.

16. The urethra in males is subdivided into three portions: the prostatic, membranous, and spongy (penile) urethra.

17. The seminal vesicles secrete an alkaline, viscous fluid that contains fructose (used by sperm for ATP production). Seminal fluid constitutes about 60% of the volume of semen and contributes to sperm viability.

18. The prostate secretes a slightly acidic fluid that constitutes about 25% of the volume of semen and contributes to sperm motility.

19. The bulbourethral (Cowper's) glands secrete mucus for lubrication and an alkaline substance that neutralizes acid.

20. Semen is a mixture of sperm and seminal fluid; it provides the fluid in which sperm are transported, supplies nutrients, and neutralizes the acidity of the male urethra and the vagina.

21. The penis consists of a root, a body, and a glans penis.

22. Engorgement of the penile blood sinuses under the influence of sexual excitation is called erection.

FEMALE REPRODUCTIVE SYSTEM (p. 1028)

1. The female organs of reproduction include the ovaries (gonads), uterine (Fallopian) tubes or oviducts, uterus, vagina, and vulva.

2. The mammary glands are part of the integumentary system and also are considered part of the reproductive system in females.

3. The ovaries, the female gonads, are located in the superior portion of the pelvic cavity, lateral to the uterus.

4. Ovaries produce secondary oocytes, discharge secondary oocytes (the process of ovulation), and secrete estrogens, progesterone, relaxin, and inhibin.

5. Oogenesis (the production of haploid secondary oocytes) begins in the ovaries. The oogenesis sequence includes meiosis I and meiosis II, which goes to completion only after an ovulated secondary oocyte is fertilized by a sperm cell.

6. The uterine (Fallopian) tubes transport secondary oocytes from the ovaries to the uterus and are the normal sites of fertilization. Ciliated cells and peristaltic contractions help move a secondary oocyte or fertilized ovum toward the uterus.

7. The uterus is an organ the size and shape of an inverted pear that functions in menstruation, implantation of a fertilized ovum, development of a fetus during pregnancy, and labor. It also is part of the pathway for sperm to reach the uterine tubes to fertilize a secondary oocyte. Normally, the uterus is held in position by a series of ligaments.

8. Histologically, the layers of the uterus are an outer perimetrium (serosa), a middle myometrium, and an inner endometrium.

9. The vagina is a passageway for sperm and the menstrual flow, the receptacle of the penis during sexual intercourse, and the inferior portion of the birth canal. It is capable of considerable distension.

10. The vulva, a collective term for the external genitals of the female, consists of the mons pubis, labia majora, labia minora, clitoris, vestibule, vaginal and urethral orifices, hymen, bulb of the vestibule, and three sets of glands: the paraurethral (Skene's), greater vestibular (Bartholin's), and lesser vestibular glands.

11. The perineum is a diamond-shaped area at the inferior end of the trunk medial to the thighs and buttocks.

12. The mammary glands are modified sweat glands lying superficial to the pectoralis major muscles. Their function is to synthesize, secrete, and eject milk (lactation).

13. Mammary gland development depends on estrogens and progesterone.

14. Milk production is stimulated by prolactin, estrogens, and progesterone; milk ejection is stimulated by oxytocin.

THE FEMALE REPRODUCTIVE CYCLE (p. 1041)

1. The function of the ovarian cycle is to develop a secondary oocyte, whereas that of the uterine (menstrual) cycle is to prepare the endometrium each month to receive a fertilized egg. The female reproductive cycle includes both the ovarian and uterine cycles.

2. The uterine and ovarian cycles are controlled by GnRH from the hypothalamus, which stimulates the release of FSH and LH by the anterior pituitary.

3. FSH stimulates development of secondary follicles and initiates secretion of estrogens by the follicles. LH stimulates further development of the follicles, secretion of estrogens by follicular cells, ovulation, formation of the corpus luteum, and the secretion of progesterone and estrogens by the corpus luteum.

4. Estrogens stimulate the growth, development, and maintenance of female reproductive structures; stimulate the development of secondary sex characteristics; and stimulate protein synthesis.

5. Progesterone works with estrogens to prepare the endometrium for implantation and the mammary glands for milk synthesis.

6. Relaxin relaxes the myometrium at the time of possible implantation. At the end of a pregnancy, relaxin increases the flexibility of the pubic symphysis and helps dilate the uterine cervix to facilitate delivery.

7. During the menstrual phase, the stratum functionalis of the endometrium is shed, discharging blood, tissue fluid, mucus, and epithelial cells.

8. During the preovulatory phase, a group of follicles in the ovaries begins to undergo final maturation. One follicle outgrows the others and becomes dominant while the others degenerate. At the same time, endometrial repair occurs in the uterus. Estrogens are the dominant ovarian hormones during the preovulatory phase.

9. Ovulation is the rupture of the dominant mature (Graafian) follicle and the release of a secondary oocyte into the pelvic cavity. It is brought about by a surge of LH. Signs and symptoms of ovulation include increased basal body temperature; clear, stretchy cervical mucus; changes in the uterine cervix; and abdominal pain.

10. During the postovulatory phase, both progesterone and estrogens are secreted in large quantity by the corpus luteum of the ovary, and the uterine endometrium thickens in readiness for implantation.

11. If fertilization and implantation do not occur, the corpus luteum degenerates, and the resulting low level of progesterone allows discharge of the endometrium followed by the initiation of another reproductive cycle.

12. If fertilization and implantation do occur, the corpus luteum is maintained by placental hCG. The corpus luteum and later the placenta secrete progesterone and estrogens to support pregnancy and breast development for lactation.

THE HUMAN SEXUAL RESPONSE (p. 1046)

1. The similar sequence of changes experienced by both males and females before, during, and after intercourse is termed the human sexual response; it occurs in four stages: excitement (arousal), plateau, orgasm, and resolution.

2. During excitement and plateau, parasympathetic nerve impulses produce genital vasocongestion, engorgement of tissues with blood, and secretion of lubricating fluids. Heart rate, blood pressure, breathing rate, and muscle tone increase.

3. During orgasm, sympathetic and somatic motor nerve impulses cause rhythmical contractions of smooth and skeletal muscles.

4. During resolution, the body relaxes and returns to the unaroused state.

BIRTH CONTROL METHODS AND ABORTION (p. 1048)

1. Birth control methods include surgical sterilization (vasectomy, tubal ligation), hormonal methods, intrauterine devices, spermicides, barrier methods (condom, vaginal pouch, diaphragm), periodic abstinence (rhythm and sympto-thermal methods), coitus interruptus, and induced abortion. See Table 28.2 on page 1039 for failure rates for these methods.

2. Contraceptive pills of the combination type contain estrogens and progestins in concentrations that decrease the secretion of FSH and LH and thereby inhibit development of ovarian follicles and ovulation.

3. An abortion is a spontaneous or induced premature expulsion of the products of conception from the uterus. RU 486 can induce abortion by blocking the action of progesterone.

DEVELOPMENT OF THE REPRODUCTIVE SYSTEMS (p. 1051)

1. The gonads develop from intermediate mesoderm. In the presence of the *SRY* gene, the gonads begin to differentiate into testes during the seventh week. The gonads differentiate into ovaries when the *SRY* gene is absent.

2. In males, testosterone stimulates development of each mesonephric duct into an epididymis, ductus (vas) deferens, ejaculatory duct, and seminal vesicle, and Müllerian-inhibiting substance (MIS) causes the paramesonephric duct cells to die. In females, testosterone and MIS are absent; the paramesonephric ducts develop into the uterine tubes, uterus, and vagina and the mesonephric ducts degenerate.

3. The external genitals develop from the genital tubercle and are stimulated to develop into typical male structures by the hormone dihydrotestosterone (DHT). The external genitals develop into female structures when DHT is not produced, the normal situation in female embryos.

AGING AND THE REPRODUCTIVE SYSTEMS (p. 1052)

1. Puberty is the period when secondary sex characteristics begin to develop and the potential for sexual reproduction is reached.

2. Onset of puberty is marked by pulses or bursts of LH and FSH secretion, each triggered by a pulse of GnRH. The hormone leptin, released by adipose tissue, may signal the hypothalamus that long-term energy stores (triglycerides in adipose tissue) are adequate for reproductive functions to begin.

3. In females, the reproductive cycle normally occurs once each month from menarche, the first menses, to menopause, the permanent cessation of menses.

4. Between the ages of 40 and 50, the pool of remaining ovarian follicles becomes exhausted and levels of progesterone and estrogens decline. Most women experience a decline in bone mineral density after menopause, together with some atrophy of the ovaries, uterine tubes, uterus, vagina, external genitalia, and breasts. Uterine and breast cancer increase in incidence with age.

5. In older males, decreased levels of testosterone are associated with decreased muscle strength, waning sexual desire, and fewer viable sperm; prostate disorders are common.

Q SELF-QUIZ QUESTIONS

Fill in the blanks in the following statements.

1. The period of time when secondary sexual characteristics begin to develop and the potential for sexual reproduction is reached is called _____. The first menses is called _____, and the permanent cessation of menses is called _____.

2. The fluid produced by the male accessory sex glands is called _____.

Indicate whether the following statements are true or false.

3. Spermatogenesis does not occur at normal core body temperature.

4. The stages of the human sexual response are arousal, plateau, orgasm, and resolution.

Choose the one best answer to the following questions.

5. Which of the following are functions of Sertoli cells? (1) protection of developing spermatogenic cells, (2) nourishment of spermatocytes, spermatids, and sperm, (3) phagocytosis of excess sperm cytoplasm as development proceeds, (4) mediation of the effects of testosterone and FSH, (5) control of movements of spermatogenic cells and release of sperm into the lumen of seminiferous tubules. (a) 1, 2, 4, and 5, (b) 1, 2, 3, and 5, (c) 2, 3, 4, and 5, (d) 1, 2, 3, and 4, (e) 1, 2, 3, 4, and 5.

6. Which of the following are true concerning androgens? (1) They stimulate the male pattern of development. (2) They stimulate the male and female patterns of development. (3) They contribute to sex drive in males and females. (4) They contribute to male secondary sex characteristics. (5) They stimulate protein synthesis. (a) 1, 2, and 3, (b) 1, 3, 4, and 5, (c) 1, 2, 4, and 5, (d) 1, 2, 3, and 4, (e) 1, 3, and 5.

7. Which of the following are true concerning estrogens? (1) They promote development and maintenance of female reproductive structures and secondary sex characteristics. (2) They help control fluid and electrolyte balance. (3) They increase protein catabolism. (4) They lower blood cholesterol. (5) In moderate levels, they inhibit the release of GnRH and the secretion of LH and FSH. (a) 1, 4, and 5, (b) 1, 3, 4, and 5, (c) 1, 2, 3, and 5, (d) 1, 2, 3, and 4, (e) 1, 2, 3, 4, and 5.

8. Which of the following statements are correct? (1) A sperm head contains DNA and an acrosome. (2) An acrosome is a specialized lysosome that contains enzymes that enable sperm to produce the ATP needed to propel themselves out of the male reproductive tract. (3) Mitochondria in the midpiece of a sperm produce ATP for sperm motility. (4) A sperm's tail, a flagellum, propels it along its way. (5) Once ejaculated, sperm are viable and normally are able to fertilize a secondary oocyte for 5 days. (a) 1, 2, 3, and 4, (b) 2, 3, 4, and 5, (c) 1, 3, and 4, (d) 2, 4, and 5, (e) 2, 3, and 4.

9. Which of the following statements are correct? (1) Spermatogonia are stem cells because when they undergo mitosis, some of the daughter cells remain to serve as a reservoir of cells for future mitosis. (2) Meiosis I is a division of pairs of chromosomes resulting in daughter cells with only one member of each chromosome pair. (3) Meiosis II separates the chromatids of each chromosome. (4) Spermiogenesis involves the maturation of spermatids into sperm. (5) The process by which the seminiferous tubules produce haploid sperm is called spermatogenesis. (a) 1, 2, 3, and 5, (b) 1, 2, 3, 4, and 5, (c) 1, 3, 4, and 5, (d) 1, 2, 3, and 4, (e) 1, 3, and 5.

10. Which of the following statements are correct? (1) Cells from the endoderm of the yolk sac give rise to oogonia. (2) Ova arise from the germinal epithelium of the ovary. (3) Primary oocytes enter prophase of meiosis I during fetal development but do not complete it until after puberty. (4) Once a secondary oocyte is formed, it proceeds to metaphase of meiosis II and stops at this stage. (5) The secondary oocyte resumes meiosis II and forms the ovum and a polar body only if fertilization occurs. (6) A primary oocyte gives rise to an ovum and four polar bodies. (a) 1, 3, 4, and 5, (b) 1, 3, 4, and 6, (c) 1, 2, 4, and 6, (d) 1, 2, 4, and 5, (e) 1, 2, 5, and 6.

11. Which of the following statements are correct? (1) The female reproductive cycle consists of a menstrual phase, a preovulatory phase, ovulation, and a postovulatory phase. (2) During the menstrual phase, small secondary follicles in the ovary begin to enlarge while the uterus is shedding its lining. (3) During the preovulatory phase, a dominant follicle continues to grow and begins to secrete estrogens and inhibin while the uterine lining begins to rebuild. (4) Ovulation results in the release of an ovum and the shedding of the uterine lining to nourish and support the released ovum. (5) After ovulation, a corpus luteum forms from the ruptured follicle and begins to secrete progesterone and estrogens, which it will continue to do throughout pregnancy if the egg is fertilized. (6) If pregnancy does not occur, then the corpus luteum degenerates into a scar called the corpus albicans, and the uterine lining is prepared to be shed again. (a) 1, 2, 4, and 5, (b) 2, 4, 5, and 6, (c) 1, 4, 5, and 6, (d) 1, 3, 4, and 6, (e) 1, 2, 3, and 6.

12. Match the following:
_____ (a) relaxes the uterus by inhibiting myometrial contractions during monthly cycles; increases flexibility of the pubic symphysis during childbirth
_____ (b) stimulates Leydig cells to secrete testosterone
_____ (c) inhibits production of FSH by the anterior pituitary
_____ (d) stimulates male pattern of development; stimulates protein synthesis; contributes to sex drive
_____ (e) maintains the corpus luteum during the first trimester of pregnancy
_____ (f) promote development of female reproductive structures; lower blood cholesterol
_____ (g) stimulates the initial secretion of estrogens by growing follicles; promotes follicle growth
_____ (h) is secreted by the corpus luteum to maintain the uterine lining during the first trimester of pregnancy

(1) inhibin
(2) LH
(3) FSH
(4) testosterone
(5) estrogens
(6) progesterone
(7) relaxin
(8) human chorionic gonadotropin

13. Match the following:
_____ (a) the process during meiosis when portions of homologous chromosomes may be exchanged with each other
_____ (b) refers to cells containing one-half the chromosome number
_____ (c) the cell produced by the union of an egg and a sperm
_____ (d) all chromosomes except the sex chromosomes
_____ (e) two chromosomes that belong to a pair (one maternal and one paternal)
_____ (f) the degeneration of oogonia before and after birth
_____ (g) a packet of discarded nuclear material from the first or second meiotic division of the egg
_____ (h) refers to cells containing the full chromosome number

(1) autosomes
(2) zygote
(3) homologous chromosomes
(4) haploid
(5) diploid
(6) crossing-over
(7) polar body
(8) atresia

14. Match the following:

_____ (a) site of sperm maturation
_____ (b) the male copulatory organ; a passageway for ejaculation of sperm and excretion of urine
_____ (c) sperm-forming cells
_____ (d) during sexual arousal, produce an alkaline substance that protects sperm by neutralizing acids in the urethra
_____ (e) ejects sperm into the urethra just before ejaculation
_____ (f) the supporting structure for the testes
_____ (g) carries the sperm from the scrotum into the abdominopelvic cavity for release by ejaculation; is cut and tied as a means of sterilization
_____ (h) the shared terminal duct of the reproductive and urinary systems in the male
_____ (i) surrounds the urethra at the base of the urinary bladder; produces secretions that contribute to sperm motility and viability
_____ (j) produce testosterone
_____ (k) support and protect developing spermatogenic cells; secrete inhibin; form the blood-testis barrier
_____ (l) secrete an alkaline fluid to help neutralize acids in the female reproductive tract; secrete fructose for use in ATP production by sperm
_____ (m) site of spermatogenesis

(1) spermatogenic cells
(2) Sertoli cells
(3) Leydig cells
(4) penis
(5) scrotum
(6) epididymis
(7) ductus (vas) deferens
(8) ejaculatory duct
(9) seminiferous tubules
(10) seminal vesicles
(11) prostate
(12) bulbourethral glands
(13) urethra

15. Match the following:

_____ (a) a small, cylindrical mass of erectile tissue and nerves in the female; homologous to the male glans penis
_____ (b) produce mucus in the female during sexual arousal and intercourse; homologous to the male bulbourethral glands
_____ (c) the group of cells that nourish the developing oocyte and begin to secrete estrogens
_____ (d) a pathway for sperm to reach the uterine tubes; the site of menstruation; the site of implantation of a fertilized ovum; the womb
_____ (e) produces progesterone, estrogens, relaxin, and inhibin
_____ (f) draw the ovum into the uterine tube
_____ (g) the opening between the uterus and vagina
_____ (h) mucus-secreting glands in the female that are homologous to the prostate gland
_____ (i) the female copulatory organ; the birth canal
_____ (j) passageway for the ovum to the uterus; usual site of fertilization
_____ (k) refers to the external genitals of the female
_____ (l) the layer of the uterine lining that is partially shed during each monthly cycle

(1) follicle
(2) corpus luteum
(3) uterine tube
(4) fimbriae
(5) uterus
(6) cervix
(7) endometrium
(8) vagina
(9) vulva
(10) clitoris
(11) paraurethral glands
(12) greater vestibular glands

 ## CRITICAL THINKING QUESTIONS

1. Melissa skipped A&P class when the reproductive system was covered because, as she said, "I already know all that stuff." On the exam, she answered "false" to the statement "One target tissue of LH is the testes." Melissa should have attended class. Explain her error.
HINT   *LH is named for its effect in females.*

2. Why is the germinal epithelium of the ovaries superficial to the tunica albuginea, whereas that of the testes is deep? Where is the tunica vaginalis located?
HINT   *Male and female structures develop from common embryonic origins.*

3. If two babies were produced following the fertilization of both a secondary oocyte and a polar body, would the babies be identical twins?
HINT   *Identical twins have an identical chromosomal makeup.*

ANSWERS TO FIGURE QUESTIONS

28.1 The result of crossing-over is that daughter cells are genetically unlike each other and genetically unlike the parent cell that produced them.

28.2 During anaphase I of meiosis, the paired chromatids are held together by a centromere and do not separate. During anaphase II of meiosis, the centromeres do not split and the paired chromatids remain together.

28.3 The gonads (testes) produce gametes (sperm) and hormones; the ducts transport, store, and receive gametes; and the accessory sex glands secrete materials that support gametes.

28.4 The cremaster and dartos muscles help regulate the temperature of the testes.

28.5 The tunica vaginalis and tunica albuginea are tissue layers that cover and protect the testes.

28.6 Leydig cells secrete testosterone.

28.7 During meiosis I, the number of chromosomes in each cell is reduced by half.

28.8 The sperm head contains DNA and enzymes for penetration of a secondary oocyte; the midpiece contains mitochondria for ATP production; the tail consists of a flagellum that provides propulsion.

28.9 Sertoli cells secrete inhibin.

28.10 Testosterone inhibits secretion of LH, and inhibin inhibits secretion of FSH.

28.11 Seminal vesicles contribute the largest volume to seminal fluid.

28.12 During sexual arousal, blood enters the sinuses in the two corpora cavernosa penis faster than it can drain out because the veins become compressed. The trapped blood engorges and stiffens the penis, producing an erection. The corpus spongiosum penis keeps the spongy urethra open so that ejaculation can occur.

28.13 The testes are homologous to the ovaries; the glans penis is homologous to the clitoris; the prostate is homologous to the paraurethral glands; and the bulbourethral gland is homologous to the greater vestibular glands.

28.14 The mesovarium anchors the ovary to the broad ligament of the uterus and the uterine tube; the ovarian ligament anchors it to the uterus; the suspensory ligament anchors it to the pelvic wall.

28.15 Ovarian follicles secrete estrogens; the corpus luteum secretes progesterone, estrogens, relaxin, and inhibin.

28.16 Most ovarian follicles undergo atresia (degeneration).

28.17 Primary oocytes are present in the ovary at birth, so they are as old as the woman is. In males, primary spermatocytes are continually being formed from stem cells (spermatogonia) and thus are only a few days old.

28.18 Fertilization most often occurs in the ampulla of the uterine tube.

28.19 Ciliated columnar epithelial cells and secretory cells with microvilli line the uterine tubes.

28.20 The endometrium is a highly vascularized, secretory epithelium that provides the oxygen and nutrients needed to sustain a fertilized egg; the myometrium is a thick smooth muscle layer that supports the uterine wall during pregnancy and contracts to expel the fetus at birth.

28.21 The stratum basalis of the endometrium provides cells to replace those that are shed (the stratum functionalis) during each menstruation.

28.22 Anterior to the vaginal opening are the mons pubis, clitoris, and prepuce. Lateral to the vaginal opening are the labia minora and labia majora.

28.23 The anterior portion of the perineum is called the urogenital triangle because its borders form a triangle that encloses the urethral (uro-) and vaginal (-genital) orifices.

28.24 Prolactin, estrogens, and progesterone regulate the synthesis of milk. Oxytocin regulates the ejection of milk.

28.25 The principal estrogen is β-estradiol.

28.26 The hormones responsible for the proliferative phase of endometrial growth are estrogens; for ovulation, LH; for growth of the corpus luteum, LH; and for the midcycle surge of LH, estrogens.

28.27 The effect of rising but moderate levels of estrogens is negative feedback inhibition of the secretion of GnRH, LH, and FSH.

28.28 This is negative feedback, because the response is opposite to the stimulus. A reduced amount of negative feedback due to declining levels of estrogens and progesterone stimulates release of GnRH, which in turn increases the production and release of FSH and LH, which ultimately stimulate the secretion of estrogens.

28.29 The *SRY* gene on the Y chromosome is responsible for the development of the gonads into testes.

28.30 The presence of dihydrotestosterone (DHT) stimulates differentiation of the external genitals in males; its absence allows differentiation of the external genitals in females.

Development and Inheritance

DEVELOPMENT, INHERITANCE AND HOMEOSTASIS

The genetic material inherited from parents and normal embryonic and fetal development both play important roles in determining the birth of a healthy child.

ENERGIZE YOUR STUDY

FOUNDATIONS CD
Animation
• **Positive Feedback Control of Labor**
Concepts and Connections and Exercises reinforce your understanding

www.wiley.com/college/apcentral

INSIGHTS AND EXPLORATIONS

Stem cells. What are they? Scientists claim that the study of stem cells holds the promise for new therapies for Parkinson's and Alzheimer's disease, spinal cord injury, stroke, burns, heart disease, diabetes, osteoarthritis, and a host of other diseases. Other scientists believe these cells could be a source of a multitude of cell types to test potential drugs on for safety. Yet there is great controversy and for many an ethical dilemma because some believe the best source for these cells are from human embryonic and fetal tissue from terminated pregnancies. We will explore answers to these questions and issues in this web-based exercise.

D **evelopmental biology** is the study of the sequence of events from the fertilization of a secondary oocyte to the formation of an adult organism. From fertilization through the eighth week of development, the developing human is called an **embryo** (*-bryo* = grow) and this is the period of **embryological development.** **Embryology** (em-brē-OL-ō-jē) is the study of development from the fertilized egg through the eighth week. **Fetal development** begins at week nine and continues until birth. During this time, the developing human is called a **fetus** (FĒ-tus = offspring).

Once sperm and a secondary oocyte have developed through meiosis and maturation, and the sperm have been deposited in the vagina, pregnancy can occur. **Pregnancy** is a sequence of events that begins with fertilization, proceeds to implantation, embryonic development, and fetal development, and normally ends with birth about 38 weeks later, or 40 weeks after the last menstrual period.

Obstetrics (ob-STET-riks; *obstetrix* = midwife) is the branch of medicine that deals with the management of pregnancy, labor, and the **neonatal period,** the first 28 days after birth. **Prenatal development** (prē-NĀ-tal; *pre-* = before; *natal* = birth) is the time from fertilization to birth and includes both embryological and fetal development. It is divided into three periods of three calendar months each, called **trimesters.** Accordingly, obstetricians divide the nine months of pregnancy into three trimesters.

1. The **first trimester** is the most critical stage of development during which the rudiments of all the major organs systems appear, and also during which the developing organism is the most vulnerable to the effects of drugs, radiation, and microbes.

2. The **second trimester** is characterized by the nearly complete development of organ systems. By the end of this stage, the fetus assumes distinctively human features.

3. The **third trimester** represents a period of rapid fetal growth. During the early stages of this period, most of the organ systems are becoming fully functional.

In this chapter, we focus on the developmental sequence from fertilization through implantation, embryonic and fetal development, labor, and birth.

EMBRYONIC PERIOD

OBJECTIVE

- **Explain the major developmental events that occur during the embryonic period.**

First Week of Development

The first week of development is characterized by several significant events including fertilization, cleavage of the zygote, blastocyst formation, and implantation.

Fertilization

During **fertilization** (fer-til-i-ZĀ-shun; *fertil-* = fruitful), the genetic material from a haploid sperm cell (spermatozoon) and a haploid secondary oocyte merges into a single diploid nucleus. Of the about 200 million sperm introduced into the vagina, fewer than 2 million (1%) reach the cervix of the uterus and only about 200 reach the secondary oocyte. Fertilization normally occurs in the uterine (Fallopian) tube within 12 to 24 hours after ovulation. Sperm can remain viable for about 48 hours after deposition in the vagina, although a secondary oocyte is viable for only about 24 hours after ovulation. Thus, pregnancy is *most likely* to occur if intercourse takes place during a 3-day "window"—from 2 days before ovulation to 1 day after ovulation.

The process leading to fertilization begins when sperm swim from the vagina into the cervical canal propelled by the whiplike movements of their tails (flagella). The passage of sperm through the rest of the uterus and then into the uterine tube results mainly from contractions of the walls of these organs. Prostaglandins in semen are believed to stimulate uterine motility at the time of intercourse and to aid in the movement of sperm through the uterus and into the uterine tube. Sperm that reach the vicinity of the oocyte within minutes after ejaculation *are not capable* of fertilizing it until about seven hours later. During this time in the female reproductive tract, mostly in the uterine tube, sperm undergo **capacitation** (ka-pas′-i-TĀ-shun; *capacit-* = capable of), a series of functional changes that cause the sperm's tail to beat even more vigorously and prepare its plasma membrane to fuse with the oocyte's plasma membrane. During capacitation, sperm are acted upon by secretions in the female reproductive tract that result in the removal of cholesterol, glycoproteins, and proteins from the plasma membrane around the acrosome of the sperm.

For fertilization to occur, a sperm cell first must penetrate the **corona radiata** (kō-RŌ-na = crown; rā-dē-A-ta = to shine), the granulosa cells that surround the secondary oocyte, and then the **zona pellucida** (ZŌ-na = zone; pe-LOO-si-da = allowing passage of light), the clear glycoprotein layer between the corona radiata and the oocyte's plasma membrane (Figure 29.1a). One of the glycoproteins in the zona pellucida called ZP3 acts as a sperm receptor. Its binding to specific membrane proteins in the sperm head triggers the **acrosomal**

reaction, the release of the contents of the acrosome. The acrosome is similar to a large lysosome and contains digestive enzymes. The acrosomal enzymes digest a path through the zona pellucida as the lashing sperm tail pushes the sperm cell onward. Many sperm bind to ZP3 molecules and undergo acrosomal reactions, but only the first sperm cell to penetrate the entire zona pellucida and reach the oocyte's plasma membrane fuses with the oocyte.

The fusion of a sperm with a secondary oocyte, called **syngamy** (*syn-* = coming together; *-gamy* = marriage), sets in motion events that block **polyspermy,** fertilization by more than one sperm cell. Within a few seconds, the cell membrane of the oocyte depolarizes, which acts as a *fast block to polyspermy*—a depolarized oocyte cannot fuse with another sperm. Depolarization also triggers the intracellular release of calcium ions, which stimulate exocytosis of secretory vesicles from the oocyte. The molecules released by exocytosis inactivate ZP3 and harden the entire zona pellucida, events that constitute the *slow block to polyspermy.*

Once a sperm cell enters a secondary oocyte, the oocyte first must complete meiosis II. It divides into a larger ovum (mature egg) and a smaller second polar body that fragments and disintegrates (see Figure 28.17 on page 1032). The nucleus in the head of the sperm develops into the **male pronucleus,** and the nucleus of the fertilized ovum develops into the **female pronucleus** (Figure 29.1c). After the male and female pronuclei form, they fuse, producing a single diploid nucleus that contains 23 chromosomes from each pronucleus. Thus, the fusion of the haploid (*n*) pronuclei restores the diploid number (*2n*) of 46 chromosomes. The fertilized ovum now is called a **zygote** (ZĪ-gōt = a joining).

Dizygotic (fraternal) twins are produced from the independent release of two secondary oocytes and the subsequent fertilization of each by different sperm. They are the same age and in the uterus at the same time, but they are genetically as dissimilar as are any other siblings. Dizygotic twins may or may not be the same sex. Because **monozygotic (identical) twins** develop from a single fertilized ovum, they contain exactly the same genetic material and are always the same sex. Monozygotic twins arise from separation of the developing cells into two embryos, which occurs before 8 days after fertilization 99% of the time. Separations that occur later than 8 days are likely to produce **conjoined twins,** a situation in which the twins are joined together and share some body structures.

Cleavage of the Zygote

After fertilization, rapid mitotic cell divisions of the zygote called **cleavage** (KLĒV-ij) take place (Figure 29.2). The first division of the zygote begins about 24 hours after fertilization and is completed about 6 hours later. Each succeeding division takes slightly less time. By the second day after fertilization, the second cleavage is completed and there are four cells (Figure 29.2b). By the end of the third day, there are 16 cells. The progressively smaller cells produced by cleavage are called **blas-**

Figure 29.1 Selected structures and events in fertilization. (a) A sperm cell penetrating the corona radiata and zona pellucida around a secondary oocyte. (b) A sperm cell in contact with a secondary oocyte. (c) Male and female pronuclei.

🔑 **During fertilization, genetic material from a sperm cell and a secondary oocyte merge to form a single diploid nucleus.**

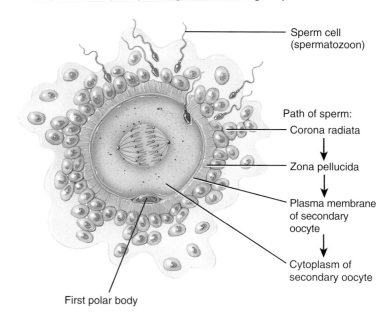

Sperm cell (spermatozoon)

Path of sperm:
- Corona radiata
- Zona pellucida
- Plasma membrane of secondary oocyte
- Cytoplasm of secondary oocyte

First polar body

(a) Sperm cell penetrating a secondary oocyte

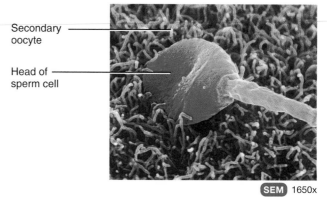

Secondary oocyte

Head of sperm cell

SEM 1650x

(b) Sperm cell in contact with a secondary oocyte

LM 430x

Pronuclei

(c) Male and female pronuclei

What is capacitation?

tomeres (BLAS-tō-mērz; *blasto-* = germ or sprout; *-meres* = parts). Successive cleavages eventually produce a solid sphere of cells called the **morula** (MOR-ū-la; *morula* = mulberry). The morula is still surrounded by the zona pellucida and is about the same size as the original zygote (Figure 29.2c).

Blastocyst Formation

By the end of the fourth day, the number of cells in the morula increases as it continues to move through the uterine tube toward the uterine cavity. When the morula enters the uterine cavity on day 4 or 5, a glycogen-rich secretion from the glands of the endometrium of the uterus passes into the uterine cavity and enters the morula through the zona pellucida. This fluid, called **uterine milk,** along with nutrients stored in the cytoplasm of the blastomeres of the morula, provides nourishment for the developing morula. At the 32-cell stage, the fluid enters the morula, collects between the blastomeres, and reorganizes them around a large fluid-filled cavity called the **blastocyst cavity** (BLAS-tō-sist; *blasto-* = germ or sprout; *-cyst* = bag) (Figure 29.2e). With the formation of this cavity, the developing mass is then called the **blastocyst.** Though it now has hundreds of cells, the blastocyst is still about the same size as the original zygote.

Further rearrangement of the blastomeres results in the formation of two distinct structures: the inner cell mass and trophoblast (Figure 29.2e). The **inner cell mass** is located internally and eventually develops into the embryo. The **trophoblast** (TRŌF-ō-blast; *tropho-* = develop or nourish) is an outer superficial layer of cells that forms the wall of the blastocyst. It will ultimately develop into the fetal portion of the placenta, the site of exchange of nutrients and wastes between the mother and fetus. On about the fifth day after fertilization, the blastocyst hatches from the zona pellucida by digesting a hole in it with an enzyme, and then the blastocyst squeezes through the hole. Shedding of the zona pellucida is necessary to permit implantation.

Stem Cell Research and Therapeutic Cloning

Throughout the book, reference has been made to **stem cells,** the unspecialized cells that have the ability to divide for indefinite periods and give rise to specialized cells. In the context of human development, a zygote (fertilized ovum) is a stem cell. Because it has the potential to form an entire organism, a zygote is known as a *totipotent stem cell* (tō-TIP-ō-tent). Inner cell mass cells, by contrast, can give rise to many (but not all) different types of cells. They are called *pluripotent stem cells* (ploo-RIP-ō-tent). Later, pluripotent stem cells can undergo further specialization into cells that have a specific function. These cells are called *multipotent stem cells* (mul-TIP-ō-tent). Examples that we have already studied include keratinocytes that produce new skin cells, myeloid and lymphoid stem cells that develop into blood cells, and spermatogonia that give rise to sperm.

Figure 29.2 Cleavage and the formation of the morula and blastocyst.

Cleavage refers to the early, rapid mitotic divisions of a zygote.

(a) Cleavage of zygote, two-cell stage (day 1)

Polar bodies
Blastomeres
Zona pellucida

(b) Cleavage, four-cell stage (day 2)

Nucleus
Cytoplasm

(c) Morula (day 4)

(d) Blastocyst, external view (day 5)

(e) Blastocyst, internal view (day 5)

Inner cell mass
Blastocyst cavity
Trophoblast

 What is the histological difference between a morula and a blastocyst?

Pluripotent stem cells currently used in research are derived from (1) the inner cell mass of embryos in the blastocyst stage that were destined to be used for infertility treatments but were not needed and from (2) nonliving fetuses terminated during the first trimester of pregnancy. Because pluripotent stem cells give rise to almost all cell types in the body, they are extremely important in research and health care. For example, they might be used to generate cells and tissues for transplantation to treat conditions such as cancer, Parkinson and Alzheimer disease, spinal cord injury, diabetes, heart disease, stroke, burns, birth defects, osteoarthritis, and rheumatoid arthritis. Pluripotent stem cells may also be used to improve understanding of the underlying processes that occur during development to identify pathways that result in developmental defects and certain diseases. Additionally, pluripotent stem cells may be used to develop and test drugs by first testing them on cells developed from pluripotent stem cells and then using only the most promising drugs for further testing on animals and humans.

On October 13, 2001, researchers reported cloning of the first human embryo to grow cells to treat human diseases. **Therapeutic cloning** is envisioned as a procedure in which the genetic material of a patient with a particular disease is used to create pluripotent stem cells to treat the disease. In the procedure used to produce the first human cloned embryo the genetic material (nucleus) and polar body were removed from an unfertilized secondary oocyte. Then, the researchers injected the genetic material from a granulosa cell from the carona radiata of another secondary oocyte into the unfertilized secondary oocyte. Following injection, the secondary oocyte was exposed to chemicals that induced it to divide. After about 24 hours, the oocyte divided and the resulting cells contained the genetic material only from the injected granulosa cell. Researchers who produced the first human cloned embryos reported that two secondary oocytes divided to form early embryos of two cells each and one secondary oocyte divided to form an early embryo consisting of six cells before growth stopped.

Using the principles of therapeutic cloning, scientists hope to make an embryo clone of a patient, remove the pluripotent stem cells from the embryo, and then use them to grow tissues to treat particular diseases. Presumably, the tissues would not be rejected since they would contain the patient's own genetic material. ■

Implantation

The blastocyst remains free within the uterine cavity for about 2 days before it attaches to the uterine wall. The endometrium is in its secretory phase. About 6 days after fertilization, the blastocyst loosely attaches to the endometrium, in a process called **implantation** (Figure 29.3). As the blastocyst implants, usually in either the posterior portion of the fundus or the body of the uterus, it orients with the inner cell mass toward the endometrium (Figure 29.3b). About seven days after fertilization, the blastocyst attaches to the endometrium more firmly, endome-

Figure 29.3 Relation of a blastocyst to the endometrium of the uterus at the time of implantation.

Implantation, the attachment of a blastocyst to the endometrium, occurs about 6 days after fertilization.

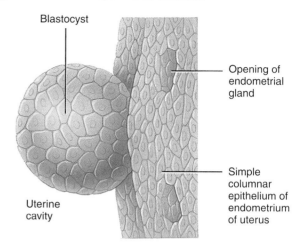

(a) External view of blastocyst, about 6 days after fertilization

trial glands in the vicinity enlarge, and the endometrium becomes more vascularized.

Following implantation, the endometrium is known as the **decidua** (dē-SID-ū-a = falling off). The decidua separates from the endometrium after the fetus is delivered much as it does in normal menstruation. Different regions of the decidua are named based on their positions relative to the site of the implanted blastocyst (Figure 29.4). The **decidua basalis** is the portion of the endometrium between the embryo and the stratum basalis of the uterus; it provides large amounts of glycogen and lipids for the developing embryo and fetus and later becomes the maternal part of the placenta. The **decidua capsularis** is the portion of the endometrium that is located between the embryo and the uterine cavity. The **decidua parietalis** (par-rī-e-TAL-is) is the remaining modified endometrium that lines the noninvolved areas of the rest of the uterus. As the embryo and later the fetus enlarges, the decidua capsularis bulges into the uterine cavity and fuses with the decidua parietalis, thereby obliterating the uterine cavity. By about 27 weeks, the decidua capsularis degenerates and disappears.

The major events associated with the first week of development are summarized in Figure 29.5.

Ectopic Pregnancy

Ectopic pregnancy (ek-TOP-ik; *ec-* = out of; *-topic* = place) is the development of an embryo or fetus outside the uterine cavity. An ectopic pregnancy usually occurs when movement of the fertilized ovum through the uterine tube is impaired. Situations that

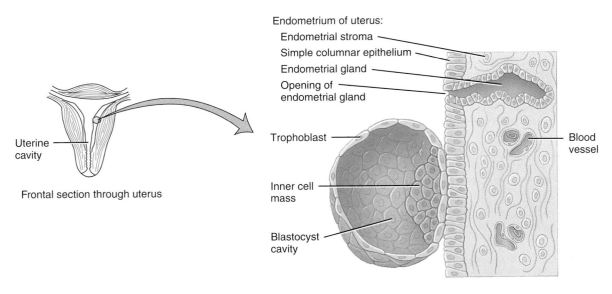

Endometrium of uterus:
- Endometrial stroma
- Simple columnar epithelium
- Endometrial gland
- Opening of endometrial gland

Trophoblast

Inner cell mass

Blastocyst cavity

Blood vessel

Uterine cavity

Frontal section through uterus

(b) Frontal section through endometrium of uterus and blastocyst, about 6 days after fertilization

 How does the blastocyst merge with and burrow into the endometrium?

Figure 29.4   Regions of the decidua.

 The decidua is a modified portion of the endometrium that develops after implantation.

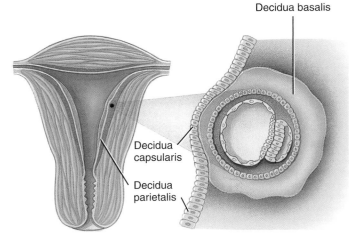

Decidua basalis

Decidua capsularis

Decidua parietalis

Frontal section of uterus   Details of decidua

Which part of the decidua helps form the maternal part of the placenta?

impair movement include scarring due to a prior tubal infection, decreased motility of the uterine tube smooth muscle, or abnormal tubal anatomy. Although the most common sites of ectopic pregnancies are the ampullar and infundibular portions of the uterine tube, ectopic pregnancies may also occur in the ovary, abdominal cavity, or uterine cervix. Compared to nonsmokers, women who smoke and become pregnant are twice as likely to have an ectopic pregnancy because nicotine in cigarette smoke paralyzes the cilia in the lining of the uterine tube (as it does those in the respiratory airways). Scars from pelvic inflammatory disease, previous uterine tube surgery, and previous ectopic pregnancy may also hinder movement of the fertilized ovum.

The signs and symptoms of ectopic pregnancy include one or two missed menstrual cycles followed by bleeding and acute abdominal and pelvic pain. Unless removed, the developing embryo can rupture the uterine tube, often resulting in death of the mother. ■

▶ **CHECKPOINT**
1. Where does fertilization normally occur?
2. How is polyspermy prevented?
3. What is a morula, and how is it formed?
4. Describe the components of a blastocyst.
5. When, where, and how does implantation occur?

Figure 29.5 Summary of events associated with the first week of development.

 Fertilization usually occurs in the uterine tube.

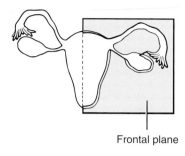

Frontal plane

3. Morula
(3–4 days after fertilization)

2. Cleavage
(first cleavage
completed about 30
hours after fertilization)

4. Blastocyst
(4½–5 days after
fertilization)

Uterine cavity

5. Implantation
(occurs about
6 days after
fertilization)

1. Fertilization
(occurs within
12–24 hours
after ovulation)

Ovary

Ovulation

Uterus:
Endometrium
Myometrium

Frontal section through uterus, uterine tube, and ovary

In which phase of the uterine cycle does implantation occur?

Second Week of Development

Development of the Trophoblast

About 8 days after fertilization, the trophoblast develops into two layers in the region of contact between the blastocyst and endometrium. These are a **syncytiotrophoblast** (sin-sīt′-ē-ō-TRŌF-ō-blast) that contains no distinct cell boundaries, and a **cytotrophoblast** (sī-tō-TRŌF-ō-blast) between the inner cell mass and syncytiotrophoblast that is composed of distinct cells (Figure 29.6a). The two layers of trophoblast become part of the chorion (one of the fetal membranes) as they undergo further growth (see Figure 29.11a inset). During implantation, the syncytiotrophoblast secretes enzymes that enable the blastocyst to penetrate the uterine lining by digesting and liquefying the endometrial cells. Eventually, the blastocyst becomes buried in the endometrium and inner one-third of the myometrium. Another secretion of the trophoblast is human chorionic gonadotropin (hCG), which has actions similar to LH. Human chorionic gonadotropin rescues the corpus luteum from degeneration and sustains its secretion of progesterone and estrogens. These hormones maintain the uterine lining in a secretory state and thereby prevent menstruation. Peak secretion of hGH occurs about the ninth week of pregnancy at which time the placenta is fully developed and produces the progesterone and estrogens that continue to sustain the pregnancy.

Development of the Bilaminar Embryonic Disc

Cells of the inner cell mass also differentiate into two layers around 8 days after fertilization: a hypoblast and epiblast (Figure 29.6a). The **hypoblast (primitive endoderm)** is a layer of columnar cells, whereas the **epiblast (primitive ectoderm)** is a layer of cuboidal cells (two-layered). Cells of the hypoblast and epiblast together form a flat disc referred to as the **bilaminar embryonic disc** (bī-LAM-in-ar = two-layered). In addition, a small cavity appears within the epiblast and eventually enlarges to form the **amniotic cavity** (am-nē-OT-ik; *amnio-* = lamb).

Development of the Amnion

As the amniotic cavity enlarges, a thin protective membrane called the **amnion** (AM-nē-on) develops from the epiblast (Figure 29.6a). Whereas the amnion forms the roof of the amniotic cavity, the epiblast forms the floor. Initially, the amnion overlies only the bilaminar embryonic disc. However, as the embryo grows, the amnion eventually surrounds the entire embryo (see Figure 29.11a inset), creating the amniotic cavity that becomes filled with **amniotic fluid.** Most amniotic fluid is initially derived from a filtrate of maternal blood. Later, the fetus contributes to the fluid by excreting urine into the amniotic cavity. Amniotic fluid serves as a shock absorber for the fetus, helps

Figure 29.6 Principal events of the second week of development.

🔑 About 8 days after fertilization, the trophoblast develops into a syncytiotrophoblast and a cytotrophoblast; the inner cell mass develops into a hypoblast and epiblast (bilaminar embryonic disc).

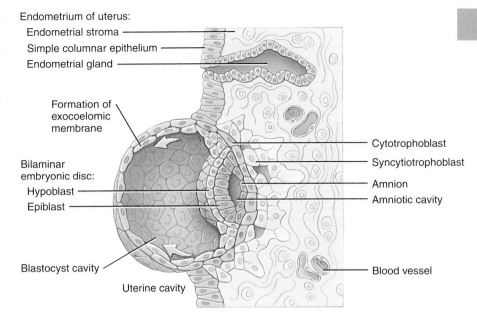

(a) Frontal section through endometrium of uterus showing blastocyst, about 8 days after fertilization

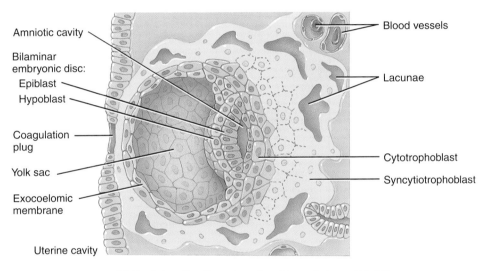

(b) Frontal section through endometrium of uterus showing blastocyst, about 9 days after fertilization

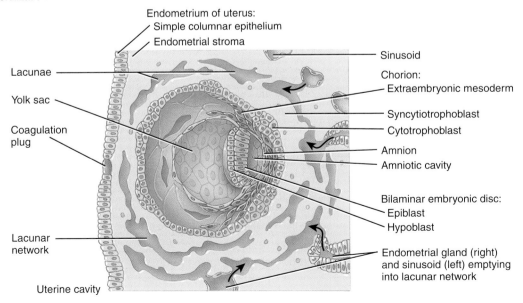

(c) Frontal section through endometrium of uterus showing blastocyst, about 12 days after fertilization

🔑 How is the bilaminar embryonic disc connected to the trophoblast?

regulate fetal body temperature, helps prevent desiccation, and prevents adhesions between the skin of the fetus and surrounding tissues. Embryonic cells are normally sloughed off into amniotic fluid. They can be examined in a procedure called **amniocentesis** (am'-nē-ō-sen-TĒ-sis; *amnio-* = amnion; *-centesis* = puncture to remove fluid), which involves withdrawing some of the amniotic fluid that bathes the developing fetus and analyzing the fetal cells and dissolved substances (see page 1084). The amnion usually ruptures just before birth; it and its fluid constitute the "bag of waters."

Development of the Yolk Sac

Also on the eighth day after fertilization, cells at the edge of the hypoblast migrate and cover the inner surface of the blastocyst wall (Figure 29.6a). The migrating columnar cells become squamous (flat) and form a thin membrane called the **exocoelomic membrane** (ek'-sō-sē-LŌ-mik; *exo-* = outside; *-koilos* = space). On the ninth day after fertilization, the exocoelomic membrane is complete. Together with the hypoblast, the exocoelomic membrane forms the wall of the **yolk sac,** formerly called the blastocyst cavity (Figure 29.6b). As a result, the bilaminar embryonic disc is now positioned between the amniotic cavity and yolk sac. Human embryos receive their nutrients from the endometrium, so the yolk sac is relatively empty, small, and decreases in size as development progresses (see Figure 29.11a inset).

Nevertheless, the yolk sac has several important functions in humans. It transfers nutrients to the embryo during the second and third weeks while uteroplacental circulation (described shortly) is being established. The yolk sac also is the source of blood cells from the third through sixth weeks, at which time the liver takes over. Early in the third week, the yolk sac contains the first cells (primordial germ cells) that will eventually migrate into the developing gonads and there will differentiate into the primitive germ cells (spermatogonia and oogonia). Then during the fourth week, the yolk sac forms part of the gut (gastrointestinal tract). Finally, the yolk sac functions as a shock absorber and helps prevent desiccation of the embryo.

Development of Sinusoids

On the ninth day after fertilization, the blastocyst becomes completely embedded in the endometrium. The site in the endometrial epithelium where the blastocyst implants is sealed by a temporary mass of fibrin called the **coagulation plug** (Figure 29.6b). The ninth day is also when the syncytiotrophoblast expands into the endometrium and around the yolk sac. As the syncytiotrophoblast expands, small spaces called **lacunae** (la-KOO-nē = little lakes) develop within it (Figure 29.6b).

By the twelfth day of development, the lacunae fuse to form larger, interconnecting spaces called **lacunar networks** (Figure 29.6c). Endometrial capillaries around the developing embryo become dilated and are referred to as **sinusoids.** As the syncytiotrophoblast erodes some of the sinusoids and endometrial glands, maternal blood and secretions from the glands enter the lacunar networks and flow through them. Maternal blood is both a rich source of materials for embryonic nutrition and a disposal site for the embryo's wastes.

Development of the Extraembryonic Coelom

About the twelfth day after fertilization, the **extraembryonic mesoderm** develops. These mesodermal cells are derived from the yolk sac and form a connective tissue (mesenchyme) around the amnion and yolk sac (Figure 29.6c). Soon a number of large cavities develop in the extraembryonic mesoderm, which then fuse to form a single, larger cavity called the **extraembryonic coelom** (SĒ-lōm).

Development of the Chorion

The extraembryonic mesoderm, together with the two layers of the trophoblast (the cytotrophoblast and syncytiotrophoblast), constitutes the **chorion** (KOR-ē-on = membrane) (Figure 29.6c). It surrounds the embryo and, later, the fetus (see Figure 29.11a inset). Eventually the chorion becomes the principal embryonic part of the placenta, the structure for exchange of materials between mother and fetus. The chorion protects the embryo and fetus from the immune responses of the mother in two ways: (1) It secretes proteins that block antibody production by the mother. (2) It promotes the production of T lymphocytes that suppress the normal immune response in the uterus. The chorion also produces human chorionic gonadotropin (hCG), an important hormone of pregnancy (see Figure 29.16).

The inner layer of the chorion eventually fuses with the amnion. With the development of the chorion, the extraembryonic coelom is now referred to as the **chorionic cavity.** By the end of the second week of development, the bilaminar embryonic disc becomes connected to the trophoblast by a band of extraembryonic mesoderm called the **connecting (body) stalk** (see Figure 29.7, inset). The connecting stalk is the future umbilical cord.

▶ CHECKPOINT

6. What are the functions of the trophoblast?

7. How is the bilaminar embryonic disc formed?

8. Describe the formation of the amnion, yolk sac, and chorion and explain their functions.

9. Why are sinusoids important during embryonic development?

Third Week of Development

The third week of development begins a six-week period of very rapid embryonic development and differentiation. During the third week, the three primary germ layers are established and lay the groundwork for organ development in weeks four through eight.

Gastrulation

The first major event of the third week of development, **gastrulation** (gas'-troo-LĀ-shun), occurs about 15 days after fertilization. In this process, the two-dimensional bilaminar (two-layered) embryonic disc, consisting of epiblast and hypoblast, transforms into a two-dimensional trilaminar (three-layered) embryonic disc consisting of three primary germ layers, the ectoderm, mesoderm, and endoderm. The primary germ layers are the major embryonic tissues from which the various tissues and organs of the body develop.

Gastrulation involves the rearrangement and migration of cells from the epiblast. The first evidence of gastrulation is the formation of the **primitive streak,** a faint groove on the dorsal surface of the epiblast that elongates from the posterior to the anterior part of the embryo (Figure 29.7a). The primitive streak clearly establishes the head and tail ends of the embryo, as well

Figure 29.7 Gastrulation.

Gastrulation involves the rearrangement and migration of cells from the epiblast.

Amnion

Amniotic cavity

Yolk sac

Cytotrophoblast

Connecting stalk

Bilaminar embryonic disc:
　Epiblast
　Hypoblast

Extraembryonic mesoderm

Uterine cavity

Dorsal surface of bilaminar embryonic disc

Oropharyngeal membrane (future site of mouth)

HEAD END

Yolk sac

Transverse plane

Primitive node

Amnion

Connecting stalk

TAIL END

Primitive streak

Bilaminar embryonic disc:
　Epiblast
　Hypoblast

(a) Dorsal and partial sectional views of embryonic disc, about 15 days after fertilization

Oropharyngeal membrane

Primitive streak

Trilaminar embryonic disc:
　Ectoderm
　Mesoderm
　Endoderm

Yolk sac

(b) Transverse section of trilaminar embryonic disc, about 16 days after fertilization

What is the significance of gastrulation?

as its right and left sides. At the head end of the primitive streak a small group of epiblastic cells forms a rounded structure called the **primitive node.**

Following formation of the primitive streak, cells of the epiblast move inward below the primitive streak and detach from the epiblast (Figure 29.7b). This inward movement is called **invagination** (in-vaj-i-NĀ-shun). Once the cells have invaginated, some of them displace the hypoblast, forming the **endoderm** (*endo-* = inside; *-derm* = skin). Other cells remain between the epiblast and newly formed endoderm to form the **mesoderm** (*meso-* = middle). Cells remaining in the epiblast then form the **ectoderm** (*ecto-* = outside). Whereas the ectoderm and endoderm are epithelia composed of tightly packed cells, the mesoderm is a loosely organized connective tissue (mesenchyme). As the embryo develops, the endoderm ultimately becomes the epithelial lining of the gastrointestinal tract, respiratory tract, and several other organs. The mesoderm gives rise to muscle, bone and other connective tissues, and the peritoneum. The ectoderm develops into the epidermis of the skin and the nervous system. Table 29.1 provides more details about the fates of these primary germ layers.

About 16 days after fertilization, mesodermal cells from the primitive node migrate toward the head end of the embryo and form a hollow tube of cells in the midline called the **notochordal process** (nō-tō-KOR-dal) (Figure 29.8). By days 22–24, the notochordal process becomes a solid cylinder of cells called the **notochord** (nō-tō-KORD; *noto-* = back; *-chord* = cord). This structure plays an extremely important role in **induction,** the process by which one tissue (*inducing tissue*) stimulates the development of an adjacent unspecialized tissue (*responding tissue*) into a specialized one. An inducing tissue usually produces a chemical substance that influences the responding tissue. The notochord induces certain mesodermal cells to develop into the vertebral bodies. It also forms the nucleus pulposus of the intervertebral discs (see Figure 7.24 on page 214).

During the third week of development, two faint depressions appear on the dorsal surface of the embryo. The structure closer to the head end is called the **oropharyngeal membrane** (or-ō-fa-RIN-jē-al; *oro-* = mouth; *-pharyngeal* = pertaining to the pharynx) (Figure 29.8a, b). It breaks down during the fourth week to connect the mouth cavity to the pharynx and the remainder of the gastrointestinal tract. The structure closer to the tail end is called the **cloacal membrane** (klō-Ā-kul = sewer), which degenerates in the seventh week to form the openings of the anus and urinary and reproductive tracts.

When the cloacal membrane appears, the wall of the yolk sac forms a small vascularized outpouching called the **allantois** (a-LAN-tō-is; *allant-* = sausage) that extends into the connecting stalk (Figure 29.8b). In most other mammals, the allantois is used for gas exchange and waste removal. Because of the role of the human placenta in these activities, the allantois is not a prominent structure in humans (see Figure 29.11a, inset). Nevertheless, it does function in early formation of blood and blood vessels and it is associated with the development of the urinary bladder.

Neurulation

In addition to inducing mesodermal cells to develop into vertebral bodies, the notochord also induces ectodermal cells over it to form the **neural plate** (Figure 29.9a on page 1074). (Also see Figure 14.25 on page 490.) By the end of the third week, the lateral edges of the neural plate become more elevated and form the **neural fold** (Figure 29.9b). The depressed midregion is called the **neural groove** (Figure 29.9c). Generally, the neural folds approach each other and fuse, thus converting the neural plate into a **neural tube** (Figure 29.9d). This occurs first near the middle of the embryo and then progresses toward the head and tail ends. Neural tube cells then develop into the brain and spinal cord. The process by which the neural plate, neural folds, and neural tube form is called **neurulation** (noor-oo-LĀ-shun).

Table 29.1	Structures Produced by the Three Primary Germ Layers	
Endoderm	**Mesoderm**	**Ectoderm**
Epithelial lining of gastrointestinal tract (except the oral cavity and anal canal) and the epithelium of its glands.	All skeletal and cardiac muscle tissue and most smooth muscle tissue.	All nervous tissue.
		Epidermis of skin.
Epithelial lining of urinary bladder, gallbladder, and liver.	Cartilage, bone, and other connective tissues.	Hair follicles, arrector pili muscles, nails, epithelium of skin glands (sebaceous and sudoriferous), and mammary glands.
	Blood, red bone marrow, and lymphatic tissue.	
Epithelial lining of pharynx, auditory (Eustachian) tubes, tonsils, larynx, trachea, bronchi, and lungs.	Endothelium of blood vessels and lymphatic vessels.	Lens, cornea, and internal eye muscles.
	Dermis of skin.	Internal and external ear.
Epithelium of thyroid gland, parathyroid glands, pancreas, and thymus.	Fibrous tunic and vascular tunic of eye.	Neuroepithelium of sense organs.
	Middle ear.	Epithelium of oral cavity, nasal cavity, paranasal sinuses, salivary glands, and anal canal.
Epithelial lining of prostate and bulbourethral (Cowper's) glands, vagina, vestibule, urethra, and associated glands such as the greater (Bartholin's) vestibular and lesser vestibular glands.	Mesothelium of ventral body cavity.	
	Epithelium of kidneys and ureters.	Epithelium of pineal gland, pituitary gland, and adrenal medullae.
	Epithelium of adrenal cortex.	
	Epithelium of gonads and genital ducts.	

Figure 29.8 **Development of the notochordal process.**

 The notochordal process develops from the primitive node and later becomes the notochord.

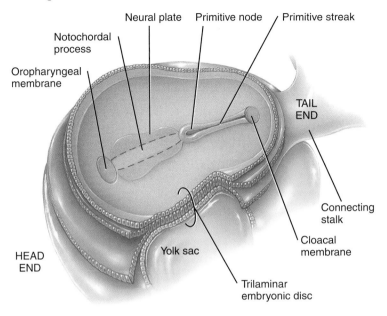

(a) Dorsal and partial sectional views of trilaminar embryonic disc, about 16 days after fertilization

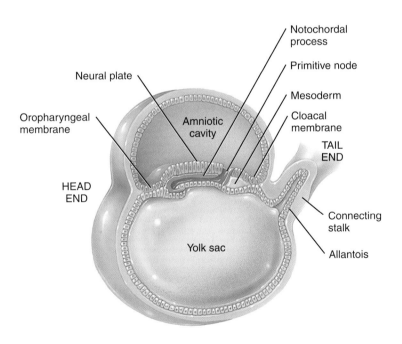

(b) Sagittal section of trilaminar embryonic disc, about 16 days after fertilization

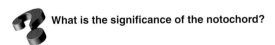 **What is the significance of the notochord?**

As the neural tube forms, some of the ectodermal cells from the tube migrate to form several layers of cells called the **neural crest** (see Figure 14.25b on page 490). Neural crest cells give rise to spinal and cranial nerves and their ganglia, autonomic nervous system ganglia, the meninges of the brain and spinal cord, the adrenal medullae, and several skeletal and muscular components of the head.

At about four weeks after fertilization, the head end of the neural tube develops into three enlarged areas called **primary brain vesicles** (see Figure 14.26 on page 491). These are the **prosencephalon (forebrain), mesencephalon (midbrain),** and **rhombencephalon (hindbrain).** Then, at about five weeks, the prosencephalon develops into **secondary brain vesicles** called the **telencephalon** and **diencephalon** and the rhombencephalon develops into secondary brain vesicles called the **metencephalon** and **myelencephalon.** The areas of the neural tube adjacent to the myelencephalon develop into the spinal cord. The parts of the brain that develop from the various brain vesicles are described on page 490 in Chapter 14.

 Anencephaly

Neural tube defects (NTDs) are caused by arrest of the normal development and closure of the neural tube. These include spina bifida (discussed on page 214) and **anencephaly** (an′-en-SEPH-a-lē; *an-* = without; *encephal* = brain). In anencephaly, the prosencephalon (future cerebral hemispheres, thalamus, and hypothalamus) remains in contact with amniotic fluid and degenerates and the cranial bones fail to form. Usually, the brainstem also is abnormal, so vital functions such as breathing and regulation of the heart may not occur. Infants with anencephaly are stillborn or die within a few days after birth. The condition occurs about once in every 1000 births and is 2 to 4 times more common in females than males. ■

Development of Somites

By about the 17th day after fertilization, the mesoderm adjacent to the notochord and neural tube forms paired longitudinal columns of **paraxial mesoderm** (par-AK-sē-al; *para-* = near) (Figure 29.9b). The mesoderm lateral to the paraxial mesoderm forms paired cylindrical masses called **intermediate mesoderm.** The mesoderm lateral to the intermediate mesoderm consists of a pair of flattened sheets called **lateral plate mesoderm.** The paraxial mesoderm soon segments into a series of paired, cube-shaped structures called **somites** (SŌ-mīts = little bodies). By the end of the fifth week, 42–44 pairs of somites are present. The number of somites that develop over a given period can be correlated to the approximate age of the embryo.

Each somite differentiates into three regions: a **myotome,** a **dermatome,** and a **sclerotome** (see Figure 10.20b on page 301). The myotomes develop into the skeletal muscles of the neck, trunk, and limbs; the dermatomes form connective tissue, including the dermis of the skin; and the sclerotomes give rise to the vertebrae.

Figure 29.9   Neurulation and the development of somites.

Neurulation is the process by which the neural plate, neural folds, and neural tube form.

(a) 17 days

(b) 19 days

(c) 20 days

Dorsal views   (d) 22 days   Transverse sections

Which structures develop from the neural tube and somites?

Development of the Intraembryonic Coelom

In the third week of development, small spaces appear in the lateral plate mesoderm. These spaces soon merge to form a larger cavity called the **intraembryonic coelom.** This cavity splits the lateral plate mesoderm into two parts called the splanchnic mesoderm and somatic mesoderm (Figure 29.9d). During the

second month of development, the intraembryonic coelom divides into the pericardial, pleural, and peritoneal cavities. **Splanchnic mesoderm** (SPLANGK-nik = visceral) forms the heart and the visceral layer of the serous pericardium, blood vessels, the smooth muscle and connective tissues of the respiratory and digestive organs, and the visceral layer of the serous membrane of the pleurae and peritoneum. **Somatic mesoderm** (sō-MAT-ik; *soma-* = body) gives rise to the bones, ligaments, and dermis of the limbs and the parietal layer of the serous membrane of the pericardium, pleurae, and peritoneum.

Development of the Cardiovascular System

At the beginning of the third week, **angiogenesis** (an-jē-ō-JEN-e-sis; *angio-* = vessel; *-genesis* = production), the formation of blood vessels, begins in the extraembryonic mesoderm in the yolk sac, connecting stalk, and chorion. This early development is necessary because there is so little yolk in the yolk sac and ovum to provide adequate nutrition for the rapidly developing embryo. Angiogenesis is initiated when mesodermal cells differentiate into cells called **angioblasts,** which aggregate to form isolated masses of cells referred to as **blood islands** (see Figure 21.32 on page 756). Spaces soon develop in the blood islands and form the lumens of blood vessels. Some angioblasts arrange themselves around each space to form the endothelium of the developing blood vessels. Then the angioblasts form the tunics of the blood vessels around the endothelium. As the blood islands grow and fuse, they soon form an extensive system of blood vessels throughout the embryo.

About 3 weeks after fertilization, blood cells and blood plasma begin to develop *outside* the embryo from endothelial cells in the blood vessels in the walls of the yolk sac, allantois, and chorion. Blood formation begins *within* the embryo at about the fifth week in the liver and the twelfth week in the spleen, red bone marrow, and thymus.

The heart forms from splanchnic mesoderm in the head end of the embryo early in the third week. This region of mesodermal cells is called the **cardiogenic area** (kar-dē-ō-JEN-ik; *cardio-* = heart; *-genic* = producing). In response to induction signals from the underlying endoderm, these mesodermal cells form a pair of **endocardial tubes** (see Figure 20.18 on page 688). The tubes then fuse to form a single **primitive heart tube.** By the end of the third week, the primitive heart tube bends on itself, becomes S-shaped, and begins to beat. It then joins blood vessels in other parts of the embryo, connecting stalk, chorion, and yolk sac to form a primitive cardiovascular system.

Development of the Chorionic Villi and Placenta

By the end of the second week of development, **primary chorionic villi** (ko-rē-ON-ik VIL-ī) develop. These projections of the chorion consist of cytotrophoblast cells that invade and become surrounded by the syncytiotrophoblast of the chorion (Figure 29.10a). Early in the third week, mesenchyme grows into the primary chorionic villi, forming a core of loose connective tissue. At this stage of development, the villi are called **secondary chorionic villi.** By the end of the third week, blood capillaries develop in the villi and the villi become **tertiary chorionic villi** or, simply, **chorionic villi** (Figure 29.10b).

Blood vessels in the chorionic villi connect to the embryonic heart by way of the umbilical arteries and umbilical vein (Figure 29.10c). As a result, maternal and fetal blood vessels are in close

Figure 29.10 Development of chorionic villi.

Blood vessels in chorionic villi connect to the embryonic heart via the umbilical arteries and umbilical vein.

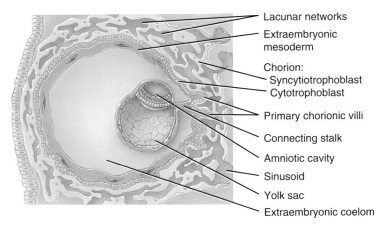

Lacunar networks
Extraembryonic mesoderm
Chorion:
  Syncytiotrophoblast
  Cytotrophoblast
Primary chorionic villi
Connecting stalk
Amniotic cavity
Sinusoid
Yolk sac
Extraembryonic coelom

(a) Frontal section through uterus showing blastocyst, about 13 days after fertilization

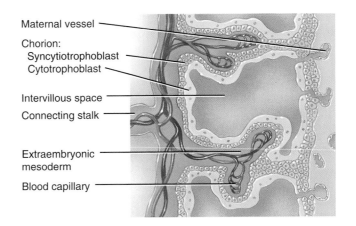

Maternal vessel
Chorion:
  Syncytiotrophoblast
  Cytotrophoblast
Intervillous space
Connecting stalk
Extraembryonic mesoderm
Blood capillary

(b) Details of two chorionic villi, about 21 days after fertilization

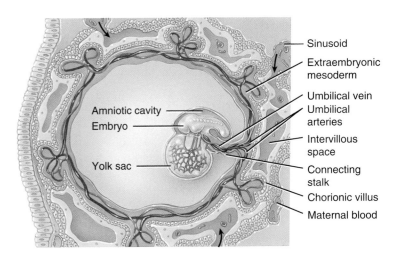

Sinusoid
Extraembryonic mesoderm
Umbilical vein
Umbilical arteries
Intervillous space
Connecting stalk
Chorionic villus
Maternal blood

Amniotic cavity
Embryo
Yolk sac

(c) Frontal section through uterus showing an embryo and its vascular supply, about 21 days after fertilization

 Why is development of chorionic villi important?

proximity. Note, however, that maternal and fetal blood vessels do not join, and the blood they carry does not normally mix. Instead, oxygen and nutrients in the blood of the mother's intervillous spaces, the spaces between chorionic villi, diffuse across the cell membranes into the capillaries of the villi. Waste products such as carbon dioxide diffuse in the opposite direction. From the capillaries of the villi, nutrients and oxygen enter the fetus through the umbilical vein. Wastes leave the fetus through the umbilical arteries, pass into the capillaries of the villi, and diffuse into the maternal blood. A few materials, such as IgG antibodies, pass from the blood of the mother into the capillaries of the villi via transcytosis (described on page 74), in which endocytosis is receptor mediated.

Placentation (plas′-en-TĀ-shun) is the process of forming the **placenta** (pla-SEN-ta = flat cake), the site of exchange of nutrients and wastes between the mother and fetus. The placenta also produces hormones needed to sustain the pregnancy (see Figure 29.16). Recall that chorionic villi develop as outgrowths of the chorion of the fetus by the end of the second week of development (Figure 29.10c) and they grow into the maternally derived intervillous spaces of the decidua basalis. The placenta is unique because it develops from two separate individuals, the mother and the fetus.

By the beginning of the twelfth week, the placenta has two distinct parts: (1) the fetal portion formed by the chorionic villi of the chorion and (2) the maternal portion formed by the decidua basalis of the endometrium (Figure 29.11a). When fully developed, the placenta is shaped like a pancake (Figure 29.11b). Functionally, the placenta allows oxygen and nutrients to diffuse from maternal blood into fetal blood while carbon dioxide and wastes diffuse from fetal blood into maternal blood. The placenta also is a protective barrier because most microorganisms cannot pass through it. However, certain viruses, such as those that cause AIDS, German measles, chickenpox, measles, encephalitis, and poliomyelitis, can cross the placenta. The placenta also stores nutrients such as carbohydrates, proteins, calcium, and iron, which are released into fetal circulation as required, and it produces several hormones that are necessary to maintain pregnancy (discussed later). Many drugs, alcohol, and some substances that can cause birth defects pass freely through the placenta.

The actual connection between the placenta and embryo, and later the fetus, is through the **umbilical cord** (um-BIL-i-kul = navel), which develops from the connecting stalk. The umbilical cord consists of two umbilical arteries that carry deoxygenated fetal blood to the placenta, one umbilical vein that carries oxygenated maternal blood into the fetus, and supporting mucous connective tissue called **Wharton's jelly** derived from the allantois. A layer of amnion surrounds the entire umbilical cord and gives it a shiny appearance (Figure 29.11a).

After the birth of the baby, the placenta detaches from the uterus and is therefore termed the **afterbirth.** At this time, the umbilical cord is tied off and then severed, leaving the baby on its own. The small portion (about an inch) of the cord that remains attached to the infant begins to wither and falls off, usu-

ally within 12 to 15 days after birth. The area where the cord was attached becomes covered by a thin layer of skin, and scar tissue forms. The scar is the **umbilicus** (navel).

Pharmaceutical companies use human placentas as a source of hormones, drugs, and blood; portions of placentas are also used for burn coverage. The placental and umbilical cord veins can also be used in blood vessel grafts, and cord blood can be frozen to provide a future source of pluripotent stem cells, for example, to repopulate red bone marrow following radiotherapy for cancer.

Placenta Previa

In some cases, the entire placenta or part of it may become implanted in the inferior portion of the uterus, near or covering the internal os of the cervix. This condition is called **placenta previa** (PRĒ-vē-a = before or in front of). Although placenta previa may lead to spontaneous abortion, it also occurs in approximately 1 in 250 live births. It is dangerous to the fetus because it may cause premature birth and intrauterine hypoxia due to maternal bleeding. Maternal mortality is increased due to hemorrhage and infection. The most important symptom is sudden, painless, bright-red vaginal bleeding in the third trimester. Cesarean section is the preferred method of delivery in placenta previa. ■

▶ **CHECKPOINT**

10. What is the significance of gastrulation?

11. How do the three primary germ layers form? Why are they important?

12. What is the function of the notochord?

13. Describe how neurulation occurs. Why is it important?

14. What are the functions of somites?

15. How does the cardiovascular system develop?

16. How does the placenta form and what is its importance?

Fourth Week of Development

The fourth through eighth weeks of development are very significant in embryonic development because all major organs appear during this time. The term **organogenesis** (or′-ga-nō-JEN-e-sis) refers to the formation of body organs and systems. By the end of the eighth week, all the major body systems have begun to develop, although their functions for the most part are minimal. Organogenesis requires the presence of blood vessels to supply developing organs with oxygen and other nutrients. However, recent studies suggest that blood vessels play a significant role in organogenesis even before blood begins to flow within them. The endothelial cells of blood vessels apparently provide some type of developmental signal, either a secreted substance or a direct cell-to-cell interaction, that is necessary for organogenesis.

During the fourth week after fertilization, the embryo undergoes dramatic changes in shape and size, nearly tripling its size. It is essentially converted from a flat, two-dimensional trilaminar

Figure 29.11   Placenta and umbilical cord.

🔑 The placenta is formed by the chorionic villi of the embryo and the decidua basalis of the endometrium of the mother.

(a) Details of placenta and umbilical cord

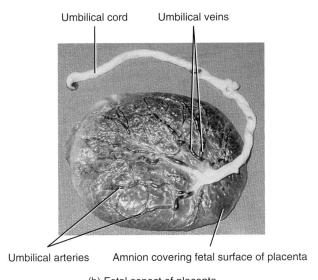

(b) Fetal aspect of placenta

❓ **What is the function of the placenta?**

embryonic disc to a three-dimensional cylinder, a process called **embryonic folding** (Figure 29.12a–d). The cylinder consists of endoderm in the center (gut), ectoderm on the outside (epidermis), and mesoderm in between. The main force responsible for embryonic folding is the different rates of growth of various parts of the embryo, especially the rapid longitudinal growth of the nervous system (neural tube). Folding in the median plane

produces a **head fold** and a **tail fold,** while folding in the horizontal plane results in the two **lateral folds.** Overall, due to the foldings, the embryo curves into a C-shape.

The head fold brings the developing heart and mouth into their eventual adult positions. The tail fold brings the developing anus into its eventual adult position. The lateral folds form as the lateral margins of the trilaminar embryonic disc bend ventrally. As they move toward the midline, the lateral folds incorporate the dorsal part of the yolk sac into the embryo as the **primitive gut,** the forerunner of the gastrointestinal tract (Figure 29.12b). The primitive gut differentiates into an anterior **foregut,** an intermediate **midgut,** and a posterior **hindgut** (Figure 29.12c). The fates of the foregut, midgut, and hindgut are described on page 897. Recall that the oropharyngeal membrane is located in the head end of the embryo (see Figure 29.8). It separates the future pharyngeal region of the foregut from the **stomodeum** (stō-mō-DĒ-um; *stomo-* = mouth), the future oral cavity. Because of head folding, the oropharyngeal membrane moves downward and the foregut and stomodeum move closer to their final positions. When the oropharyngeal membrane ruptures during the fourth week, the pharyngeal region of the pharynx and stomodeum are brought into contact with each other.

In a developing embryo, the last part of the hindgut expands into a cavity called the **cloaca** (see Figure 26.22 on page 983). On the outside of an embryo is a small cavity in the tail region called the **proctodeum** (prok-tō-DĒ-um; *procto-* = anus)

Figure 29.12 Embryonic folding.

Embryonic folding converts the two-dimensional trilaminar embryonic disc into a three-dimensional cylinder.

(a) 22 days

Neural tube, Amniotic cavity, Notochord, Connecting stalk, HEAD END, TAIL END, Oropharyngeal membrane, Yolk sac, Cloacal membrane

Amniotic cavity, Neural tube, Notochord, Intraembryonic coelom, Yolk sac

(b) 24 days

Heart, Primitive gut, Yolk sac

(c) 26 days

Foregut, Tail fold, Head fold, Midgut, Hindgut, Stomodeum, Proctodeum, Yolk sac

Midgut, Intraembryonic coelom, Yolk sac

(d) 28 days

Future pharynx, Oropharyngeal membrane, Stomodeum, Future umbilical cord, Amniotic cavity, Cloacal membrane, Proctodeum

Aorta, Dorsal mesentery, Midgut, Ventral mesentery, Abdominal wall

Sagittal sections   Transverse sections

What are the results of embryonic folding?

(Figure 29.12c). Separating the cloaca from the proctodeum is the **cloacal membrane** (see Figure 29.8). During embryonic development, the cloaca divides into a ventral urogenital sinus and a dorsal anorectal canal. As a result of tail folding, the cloacal membrane moves downward and the urogenital sinus, anorectal canal, and proctodeum move closer to their final positions. When the cloacal membrane ruptures during the seventh week of development, the urogenital and anal openings are created.

Besides embryonic folding, development of somites and the neural tube (previously described) occurs during the fourth week of development. In addition, several **pharyngeal (branchial) arches** develop on each side of the future head and neck regions (Figure 29.13). These paired structures, five in number, begin their development on the 22nd day after fertilization and form swellings on the surface of the embryo. Each pharyngeal arch consists of an outer covering of ectoderm, an inner covering of endoderm, and mesoderm in between. The pharyngeal arches also contain an artery, a cranial nerve, cartilage, and muscle tissue. Also on the outside of the embryo are a series of grooves

Figure 29.13 Development of pharyngeal arches, pharyngeal clefts, and pharyngeal pouches.

The five pairs of pharyngeal pouches consist of ectoderm, mesoderm, and endoderm and contain blood vessels, cranial nerves, cartilage, and muscle tissue.

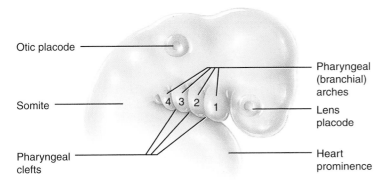

Otic placode

Somite

Pharyngeal clefts

Pharyngeal (branchial) arches

Lens placode

Heart prominence

(a) External view, about 28-day embryo

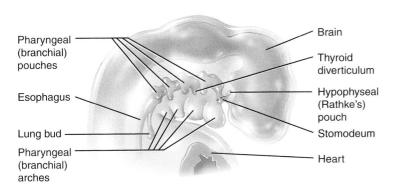

Pharyngeal (branchial) pouches

Esophagus

Lung bud

Pharyngeal (branchial) arches

Brain

Thyroid diverticulum

Hypophyseal (Rathke's) pouch

Stomodeum

Heart

(b) Sagittal section, about 28-day embryo

Why are pharyngeal arches, clefts, and pouches important?

between the pharyngeal arches called **pharyngeal clefts,** which separate the pharyngeal arches (Figure 29.13a). Simultaneous with the development of the pharyngeal arches and clefts, four distinct pairs of **pharyngeal (branchial) pouches** develop inside the embryo (Figure 29.13b). The pharyngeal pouches are endoderm-lined, balloon-like outgrowths of the primitive pharynx, the most cranial part of the foregut.

Together, the pharyngeal arches, clefts, and pouches give rise to the structures of the head and neck. The first sign of a developing ear is a thickened area of ectoderm, the **otic placode** (future internal ear), which can be distinguished about 22 days after fertilization (Figure 29.13a). The eyes also begin their development about 22 days after fertilization. This is evidenced by a thickened area of ectoderm called the **lens placode** (Figure 29.13a).

By the middle of the fourth week, the upper limbs begin their development as outgrowths of mesoderm covered by ectoderm called **upper limb buds** (see Figure 6.13a on page 179). By the end of the fourth week, the **lower limb buds** develop. The heart also forms a distinct projection on the ventral surface of the embryo called the **heart prominence** (see Figure 6.13a). A **tail** is also a distinguishing feature of an embryo at the end of the fourth week (see Figure 6.13a).

Fifth Through Eighth Weeks of Development

During the fifth week of development, there is a very rapid development of the brain and thus growth of the head is considerable. By the end of the sixth week, the head grows even larger relative to the trunk, and the limbs show substantial development (see Figure 6.13b on page 179). In addition, the neck and trunk begin to straighten, and the heart is now four-chambered. By the seventh week, the various regions of the limbs become distinct and the beginnings of digits appear (see Figure 6.13c on page 179). At the start of the eighth week, the final week of the embryonic period, the digits of the hands are short and webbed and the tail is still visible, but shorter. In addition, the eyes are open and the auricles of the ears are visible (see Figure 6.13b on page 179). By the end of the eighth week, all regions of limbs are apparent; the digits are distinct and no longer webbed due to removal of cells via apoptosis. Also, the eyelids come together and may fuse, the tail disappears, and the external genitals begin to differentiate. The embryo now has clearly human characteristics.

▶ C H E C K P O I N T

17. How does embryonic folding occur?

18. How does the primitive gut form and what is its significance?

19. What is the origin of the structures of the head and neck?

20. What are limb buds?

21. What changes occur in the limbs during the second half of the embryonic period?

FETAL PERIOD

▶ O B J E C T I V E

• **Define the fetal period and outline the major events of the fetal period.**

During the fetal period, tissues and organs that developed during the embryonic period grow and differentiate. Very few new structures appear during the fetal period, but the rate of body growth is remarkable, especially between the ninth and sixteenth weeks. For example, by the end of the fetal period, a fetus will have quadrupled its weight at the fourth week. At the beginning of the fetal period, the fetal head is half the length of the fetal body. By the end of the fetal period, the head size is only one-quarter the length of the fetal body. During the same period, the fetal limbs also increase in size from one-eighth to one-half the fetal length. The fetus is also less vulnerable to the damaging effects of drugs, radiation, and microbes than it was as an embryo.

A summary of the major developmental events of the embryonic and fetal period is presented in Table 29.2 and illustrated in Figure 29.14.

Throughout the text we have discussed the developmental biology of the various body systems in their respective chap-

Table 29.2	Summary of Changes During Embryonic and Fetal Development	
Time	**Approximate Size and Weight**	**Representative Changes**
Embryonic period		
1–4 weeks	0.6 cm (3/16 in.)	Primary germ layers and notochord develop. Neurulation occurs. Primary brain vesicles, somites, and intraembryonic coelom develop. Blood vessel formation begins and blood forms in yolk sac, allantois, and chorion. Heart forms and begins to beat. Chorionic villi develop and placental formation begins. The embryo folds. The primitive gut, pharyngeal arches, and limb buds develop. Eyes and ears begin to develop, tail forms, and body systems begin to form.
5–8 weeks	3 cm (1.25 in.) 1 g (1/30 oz)	Primary brain vesicles develop into secondary brain vesicles. Limbs become distinct and digits appear. Heart becomes four-chambered. Eyes are far apart and eyelids are fused. Nose develops and is flat. Face is more human-like. Ossification begins. Blood cells start to form in liver. External genitals begin to differentiate. Tail disappears. Major blood vessels form. Many internal organs continue to develop.
Fetal period		
9–12 weeks	7.5 cm (3 in.) 30 g (1 oz)	Head constitutes about half the length of the fetal body, and fetal length nearly doubles. Brain continues to enlarge. Face is broad, with eyes fully developed, closed, and widely separated. Nose develops a bridge. External ears develop and are low set. Ossification continues. Upper limbs almost reach final relative length but lower limbs are not quite as well developed. Heartbeat can be detected. Gender is distinguishable from external genitals. Urine secreted by fetus is added to amniotic fluid. Red bone marrow, thymus, and spleen participate in blood cell formation. Fetus begins to move, but its movements cannot be felt yet by the mother. Body systems continue to develop.
13–16 weeks	18 cm (6.5–7 in.) 100 g (4 oz)	Head is relatively smaller than rest of body. Eyes move medially to their final positions, and ears move to their final positions on the sides of the head. Lower limbs lengthen. Fetus appears even more human-like. Rapid development of body systems occurs.
17–20 weeks	25–30 cm (10–12 in.) 200–450 g (0.5–1 lb)	Head is more proportionate to rest of body. Eyebrows and head hair are visible. Growth slows but lower limbs continue to lengthen. Vernix caseosa (fatty secretions of sebaceous glands and dead epithelial cells) and lanugo (delicate fetal hair) cover fetus. Brown fat forms and is the site of heat production. Fetal movements are commonly felt by mother (quickening).
21–25 weeks	27–35 cm (11–14 in.) 550–800 g (1.25–1.5 lb)	Head becomes even more proportionate to rest of body. Weight gain is substantial, and skin is pink and wrinkled. By 24 weeks, Type II alveolar cells begin to produce surfactant.
26–29 weeks	32–42 cm (13–17 in.) 110–1350 g (2.5–3 lb)	Head and body are more proportionate and eyes are open. Toenails are visible. Body fat is 3.5% of total body mass and additional subcutaneous fat smoothes out some wrinkles. Testes begin to descend toward scrotum at 28 to 32 weeks. Red bone marrow is major site of blood cell production. Many fetuses born prematurely during this period survive if given intensive care because lungs can provide adequate ventilation and central nervous system is developed enough to control breathing and body temperature.

ters. The following list of these sections is presented here for your review.

▶ Integumentary System (page 153)

▶ Skeletal System (page 178)

▶ Muscular System (page 299)

▶ Nervous System (page 490)

▶ Endocrine System (page 622)

▶ Heart (page 446)

▶ Blood and Blood Vessels (page 756)

▶ Lymphatic System (page 774)

▶ Respiratory System (page 841)

▶ Digestive System (page 897)

▶ Urinary System (page 982)

▶ Reproductive Systems (page 1050)

▶ CHECKPOINT

22. What are the general developmental trends during the fetal period?

23. Using Table 29.2 as a guide, select any one body structure in weeks 9 through 12 and trace its development through the remainder of the fetal period.

Time	Approximate Size and Weight	Representative Changes
Fetal period (continued)		
30–34 weeks	41–45 cm (16.5–18 in.) 2000–2300 g (4.5–5 lb)	Skin is pink and smooth. Fetus assumes upside down position. Pupillary reflex is present by 30 weeks. Body fat is 8% of total body mass. Fetuses 33 weeks and older usually survive if born prematurely.
35–38 weeks	50 cm (20 in.) 3200–3400 g (7–7.5 lb)	By 38 weeks circumference of fetal abdomen is greater than that of head. Skin is usually bluish-pink, and growth slows as birth approaches. Body fat is 16% of total body mass. Testes are usually in scrotum in full-term male infants. Even after birth, an infant is not completely developed; an additional year is required, especially for complete development of the nervous system.

4   8   12   16   20   24   28   32   36   (weeks)

Figure 29.14 Summary of representative developmental events of the embryonic and fetal periods. The embryos and fetuses are not shown at their actual sizes.

🔑 Development during the fetal period is mostly concerned with the growth and differentiation of tissues and organs formed during the embryonic period.

Neural plate

Neural groove

Cut edge of amnion

Somite

Yolk sac

Primitive streak

(a) 20-day embryo

Pharyngeal arches

Lens placode

Heart prominence

Upper limb bud

Tail

Lower limb bud

(c) 32-day embryo

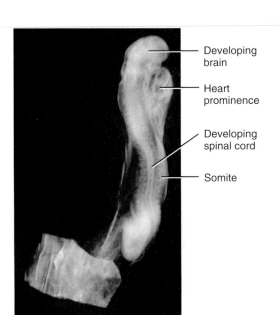

Developing brain

Heart prominence

Developing spinal cord

Somite

(b) 24-day embryo

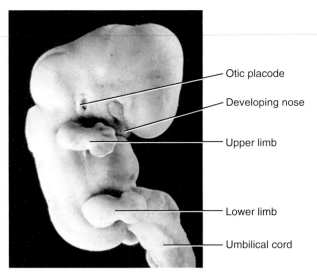

Otic placode

Developing nose

Upper limb

Lower limb

Umbilical cord

(d) 44-day embryo

Ear — Eye
Nose
Upper limb
Lower limb
Umbilical cord

(e) 52-day embryo

Ear
Eye
Nose
Mouth
Upper limb
Umbilical cord
Lower limb

(g) Thirteen-week fetus

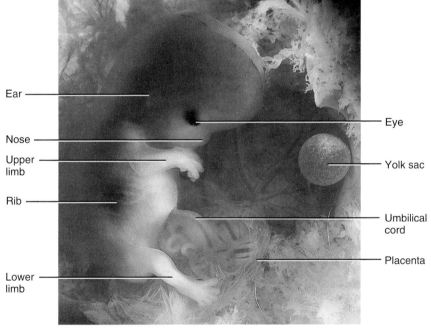

Ear
Nose
Upper limb
Rib
Lower limb
Eye
Yolk sac
Umbilical cord
Placenta

(f) Ten-week fetus

Ear
Eye
Nose
Mouth
Upper limb
Lower limb

(h) Twenty-six-week fetus

 How does mid-fetal weight compare to end-fetal weight?

PRENATAL DIAGNOSTIC TESTS

OBJECTIVE

• **Describe the procedures for fetal ultrasonography, amniocentesis, and chorionic villi sampling.**

Several tests are available to detect genetic disorders and assess fetal well-being. Here we describe fetal ultrasonography, amniocentesis, and chorionic villi sampling (CVS).

Fetal Ultrasonography

If there is a question about the normal progress of a pregnancy, **fetal ultrasonography** (ul′-tra-son-OG-ra-fē) may be performed. By far the most common use of diagnostic ultrasound is to determine a more accurate fetal age when the date of conception is unclear. It is also used to confirm pregnancy, evaluate fetal viability and growth, determine fetal position, identify multiple pregnancies, identify fetal–maternal abnormalities, and serve as an adjunct to special procedures such as amniocentesis. During fetal ultrasonography, a transducer instrument that emits high-frequency sound waves is passed back and forth over the abdomen. The reflected sound waves from the developing fetus are picked up by the transducer and converted to an on-screen image called a **sonogram** (see Table 1.4 on page 21). Because the urinary bladder serves as a landmark during the procedure, the patient needs to drink liquids before the procedure and not void urine to maintain a full bladder.

Amniocentesis

Amniocentesis (am′-nē-ō-sen-TĒ-sis; *amnio-* = amnion; *-centesis* = puncture to remove fluid) involves withdrawing some of the amniotic fluid that bathes the developing fetus and analyzing the fetal cells and dissolved substances. It is used to test for the presence of certain genetic disorders, such as Down syndrome (DS), hemophilia, Tay-Sachs disease, sickle-cell disease, and certain muscular dystrophies. The test is usually done at 14–18 weeks of gestation. All gross chromosomal abnormalities and over 50 biochemical defects can be detected through amniocentesis. It can also reveal the baby's gender, which is important information for the diagnosis of sex-linked disorders. In such disorders, an abnormal gene carried by the mother affects her male offspring only (described on page 1097).

During amniocentesis, the position of the fetus and placenta is first identified using ultrasound and palpation. After the skin is prepared with an antiseptic and a local anesthetic is given, a hypodermic needle is inserted through the mother's abdominal wall and into the amniotic cavity within the uterus. Then, 10 to 30 milliliters of fluid and suspended cells are aspirated (Figure 29.15a) for microscopic examination and biochemical testing. Elevated levels of alpha-fetoprotein (AFP) and acetylcholinesterase may indicate failure of the nervous system to develop properly, as occurs in spina bifida or anencephaly (absence of the cerebrum) or may be due to other developmental or chromosomal problems. Chromosome studies, which require growing the cells for 2–4 weeks in a culture medium, may

Figure 29.15   Amniocentesis and chorionic villi sampling.

To detect genetic abnormalities, amniocentesis is performed at 14–16 weeks of gestation, whereas chorionic villi sampling may be performed as early as 8 weeks of gestation.

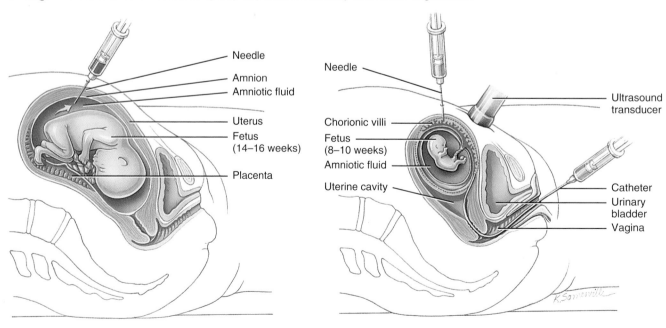

(a) Amniocentesis

(b) Chorionic villi sampling (CVS)

What kinds of information can be provided by amniocentesis?

reveal rearranged, missing, or extra chromosomes. Amniocentesis is performed only when a risk for genetic defects is suspected because there is about a 0.5% chance of spontaneous abortion after the procedure.

Chorionic Villi Sampling

In **chorionic villi sampling** (ko-rē-ON-ik VIL-ī) or **CVS,** a catheter is guided through the vagina and cervix of the uterus and then advanced to the chorionic villi under ultrasound guidance (Figure 29.15b). About 30 milligrams of tissue are suctioned out and prepared for chromosomal analysis. Alternatively, the chorionic villi can be sampled by inserting a needle through the abdominal cavity, as performed in amniocentesis.

CVS can identify the same defects as amniocentesis because chorion cells and fetal cells contain the same genome. Moreover, CVS offers several advantages over amniocentesis: It can be performed as early as eight weeks of gestation, and test results are available in only a few days, permitting an earlier decision on whether or not to continue the pregnancy. In addition, the procedure does not require penetration of the abdomen, uterus, or amniotic cavity by a needle. However, CVS is slightly riskier than amniocentesis; it carries a 1–2% chance of spontaneous abortion after the procedure.

Noninvasive Prenatal Tests

Currently, chorionic villi testing and amniocentesis are the only useful ways to obtain fetal tissue for prenatal testing of gene defects. While these invasive procedures pose relatively little risk when performed by experts, much work has been done to develop **noninvasive prenatal tests,** which do not entail the penetration of any embryonic structure. The goal is to develop accurate, safe, more efficient, and less expensive tests for screening a large population.

The first such test developed was the **maternal alpha-fetoprotein (AFP) test.** In this test, the mother's blood is analyzed for the presence of AFP, a protein synthesized in the fetus that passes into the maternal circulation. The highest levels of AFP normally occur during weeks 12 through 15 of pregnancy. Later, AFP is not produced, and its concentration decreases to a very low level both in the fetus and in maternal blood. A high level of AFP after week 16 usually indicates that the fetus has a neural tube defect, such as spina bifida or anencephaly. Because the test is 95% accurate, it is now recommended that all pregnant women be tested for AFP. A newer test (Quad AFP Plus) probes maternal blood for AFP and three other molecules. The test permits prenatal screening for Down syndrome, trisomy 18, and neural tube defects; it also helps predict the delivery date and may reveal the presence of twins.

▶ CHECKPOINT

24. What conditions can be detected using fetal ultrasonography, amniocentesis, and chorionic villi sampling? What are the advantages of noninvasive prenatal tests?

MATERNAL CHANGES DURING PREGNANCY

▶ OBJECTIVES

- **Describe the sources and functions of the hormones secreted during pregnancy.**
- **Describe the hormonal, anatomical, and physiological changes in the mother during pregnancy.**

Hormones of Pregnancy

During the first 3 to 4 months of pregnancy, the corpus luteum in the ovary continues to secrete **progesterone** and **estrogens,** which maintain the lining of the uterus during pregnancy and prepare the mammary glands to secrete milk. The amounts secreted by the corpus luteum, however, are only slightly more than those produced after ovulation in a normal menstrual cycle. From the third month through the remainder of the pregnancy, the placenta itself provides the high levels of progesterone and estrogens required. As noted previously, the chorion secretes **human chorionic gonadotropin (hCG)** into the blood. In turn, hCG stimulates the corpus luteum to continue production of progesterone and estrogens—an activity required to prevent menstruation and for the continued attachment of the embryo and fetus to the lining of the uterus (Figure 29.16a). By the eighth day after fertilization, hCG can be detected in the blood and urine of a pregnant woman. Peak secretion of hCG occurs at about the ninth week of pregnancy (Figure 29.16b). During the fourth and fifth months the hCG level decreases sharply and then levels off until childbirth.

The chorion begins to secrete estrogens after the first 3 or 4 weeks of pregnancy and progesterone by the sixth week. These hormones are secreted in increasing quantities until the time of birth (Figure 29.16b). By the fourth month, when the placenta is fully established, the secretion of hCG is greatly reduced, and the secretions of the corpus luteum are no longer essential. Thus, from the third month to the ninth month, the placenta supplies the levels of estrogens and progesterone needed to maintain the pregnancy. A high level of progesterone ensures that the uterine myometrium is relaxed and that the cervix is tightly closed. After delivery, estrogens and progesterone in the blood decrease to normal levels.

Relaxin, a hormone produced first by the corpus luteum of the ovary and later by the placenta, increases the flexibility of the pubic symphysis and ligaments of the sacroiliac and sacrococcygeal joints and helps dilate the uterine cervix during labor. Both of these actions ease delivery of the baby.

A third hormone produced by the chorion of the placenta is **human chorionic somatomammotropin (hCS),** also known as **human placental lactogen (hPL).** The rate of secretion of hCS increases in proportion to placental mass, reaching maximum levels after 32 weeks and remaining relatively constant after that. It is thought to help prepare the mammary glands for lactation, enhance maternal growth by increasing protein synthesis, and regulate certain aspects of metabolism in mother and fetus

Figure 29.16   Hormones during pregnancy.

Whereas the corpus luteum produces progesterone and estrogens during the first 3–4 months of pregnancy, the placenta assumes this function from the third month on.

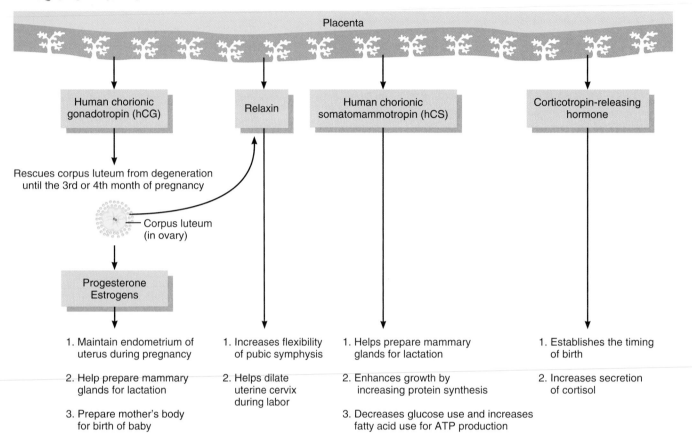

(a) Sources and functions of hormones

(b) Blood levels of hormones during pregnancy

Which hormone is detected by early pregnancy tests?

alike. For example, hCS causes decreased use of glucose by the mother, thus making more available for the fetus. Additionally, hCS promotes the release of fatty acids from adipose tissue, providing an alternative to glucose for the mother's ATP production.

The hormone most recently found to be produced by the placenta is **corticotropin-releasing hormone (CRH),** which in nonpregnant people is secreted only by neurosecretory cells in the hypothalamus. CRH is now thought to be part of the "clock" that establishes the timing of birth. Secretion of CRH by the placenta begins at about 12 weeks and increases enormously toward the end of pregnancy. Women who have higher levels of CRH earlier in pregnancy are more likely to deliver prematurely, whereas those who have low levels are more likely to deliver after their due date. CRH from the placenta has a second important effect: It increases secretion of cortisol, which is needed for maturation of the fetal lungs and the production of surfactant.

Early Pregnancy Tests

Early pregnancy tests detect the tiny amounts of human chorionic gonadotropin (hCG) in the urine that begin to be excreted about 8 days after fertilization. The test kits can detect pregnancy as early as the first day of a missed menstrual period—that is, at about 14 days after fertilization. Chemicals in the kits produce a color change if a reaction occurs between hCG in the urine and hCG antibodies included in the kit.

Several of the test kits available at pharmacies are as sensitive and accurate as test methods used in many hospitals. Still, false-negative and false-positive results can occur. A false-negative result (the test is negative, even though the woman is pregnant) may be due to testing too soon or to an ectopic pregnancy. A false-positive result (the test is positive, but the woman is not pregnant) may be due to excess protein or blood in the urine or to hCG production due to a rare type of uterine cancer. Thiazide diuretics, hormones, steroids, and thyroid drugs may also affect the outcome of an early pregnancy test. ■

Changes During Pregnancy

By about the end of the third month of pregnancy, the uterus occupies most of the pelvic cavity. As the fetus continues to grow, the uterus extends higher and higher into the abdominal cavity. Toward the end of a full-term pregnancy, the uterus fills nearly the entire abdominal cavity, reaching above the costal margin nearly to the xiphoid process of the sternum (Figure 29.17). It pushes the maternal intestines, liver, and stomach superiorly, elevates the diaphragm, and widens the thoracic cavity. Pressure on the stomach may force the stomach contents superiorly into the esophagus, resulting in heartburn. In the pelvic cavity, compression of the ureters and urinary bladder occurs.

Besides the anatomical changes associated with pregnancy, pregnancy-induced physiological changes also occur, including

Figure 29.17 Normal fetal location and position at the end of a full-term pregnancy.

The gestation period is the time interval (about 38 weeks) from fertilization to birth.

Anterior view of position of organs at end of full-term pregnancy

What hormone increases the flexibility of the pubic symphysis and helps dilate the cervix of the uterus to ease delivery of the baby?

weight gain due to the fetus, amniotic fluid, the placenta, uterine enlargement, and increased total body water; increased storage of proteins, triglycerides, and minerals; marked breast enlargement in preparation for lactation; and lower back pain due to lordosis (swayback).

Several changes occur in the maternal cardiovascular system. Stroke volume increases by about 30% and cardiac output rises by 20–30% due to increased maternal blood flow to the placenta and increased metabolism. Heart rate increases 10–15% and blood volume increases 30–50%, mostly during the second half of pregnancy. These increases are necessary to meet the additional demands of the fetus for nutrients and oxygen. When a pregnant woman is lying on her back, the enlarged uterus may compress the aorta, resulting in diminished blood flow to the uterus. Compression of the inferior vena cava also decreases venous return, which leads to edema in the lower limbs and may produce varicose veins. Compression of the renal artery can lead to renal hypertension.

Pulmonary function is also altered during pregnancy to meet the added oxygen demands of the fetus. Tidal volume can increase by 30–40%, expiratory reserve volume can be reduced by up to 40%, functional residual capacity can decline by up to 25%, minute ventilation (the total volume of air inhaled and exhaled each minute) can increase by up to 40%, airway resistance in the bronchial tree can decline by 30–40%, and total body oxygen consumption can increase by about 10–20%. Dyspnea (difficult breathing) also occurs.

With regard to the gastrointestinal tract, pregnant women experience an increase in appetite. A general decrease in GI tract motility can cause constipation, delay gastric emptying time, and produce nausea, vomiting, and heartburn.

Pressure on the urinary bladder by the enlarging uterus can produce urinary symptoms, such as increased frequency and urgency of urination, and stress incontinence. An increase in renal plasma flow up to 35% and an increase in glomerular filtration rate up to 40% increase the renal filtering capacity, which allows faster elimination of the extra wastes produced by the fetus.

Changes in the skin during pregnancy are more apparent in some women than in others. Included are increased pigmentation around the eyes and cheekbones in a masklike pattern (chloasma), in the areolae of the breasts, and in the linea alba of the lower abdomen (linea nigra). Striae (stretch marks) over the abdomen can occur as the uterus enlarges, and hair loss increases.

Changes in the reproductive system include edema and increased vascularity of the vulva and increased pliability and vascularity of the vagina. The uterus increases from its nonpregnant mass of 60–80 g to 900–1200 g at term because of hyperplasia of muscle fibers in the myometrium in early pregnancy and hypertrophy of muscle fibers during the second and third trimesters.

🩺 Pregnancy-Induced Hypertension

About 10–15% of all pregnant women in the United States experience **pregnancy-induced hypertension (PIH),** an elevated blood pressure that is associated with pregnancy. The major cause is **preeclampsia** (prē-ē-KLAMP-sē-a), an abnormal condition of pregnancy characterized by sudden hypertension, large amounts of protein in the urine, and generalized edema that typically appears after the 20th week of pregnancy. Other signs and symptoms are generalized edema, blurred vision, and headaches. Preeclampsia might be related to an autoimmune or allergic reaction resulting from the presence of a fetus. When the condition is also associated with convulsions and coma, it is termed **eclampsia.** ■

▶ C H E C K P O I N T

25. List the hormones involved in pregnancy, and describe the functions of each.

26. What structural and functional changes occur in the mother during pregnancy?

EXERCISE AND PREGNANCY

▶ O B J E C T I V E

• **Explain the effects of pregnancy on exercise and of exercise on pregnancy.**

Only a few changes in early pregnancy affect exercise. A pregnant woman may tire more easily than usual, or morning sickness may interfere with regular exercise. As the pregnancy progresses, weight is gained and posture changes, so more energy is needed to perform activities, and certain maneuvers (sudden stopping, changes in direction, rapid movements) are more difficult to execute. In addition, certain joints, especially the pubic symphysis, become less stable in response to the increased level of the hormone relaxin. As compensation, many mothers-to-be walk with widely spread legs and a shuffling motion.

Although blood shifts from viscera (including the uterus) to the muscles and skin during exercise, there is no evidence of inadequate blood flow to the placenta. The heat generated during exercise may cause dehydration and further increase body temperature. During early pregnancy especially, excessive exercise and heat buildup should be avoided because elevated body temperature has been implicated in neural tube defects. Exercise has no known effect on lactation, provided a woman remains hydrated and wears a bra that provides good support. Overall, moderate physical activity does not endanger the fetus of a healthy woman who has a normal pregnancy.

Among the benefits of exercise to the mother during pregnancy are a greater sense of well-being and fewer minor complaints.

LABOR

▶ O B J E C T I V E

• **Explain the events associated with the three stages of labor.**

Labor is the process by which the fetus is expelled from the uterus through the vagina. **Parturition** (par′-toor-ISH-un; *parturit-* = childbirth) also means giving birth.

The onset of labor is determined by complex interactions of several placental and fetal hormones. Because progesterone inhibits uterine contractions, labor cannot take place until its effects are diminished. Toward the end of gestation, the levels of estrogens in the mother's blood rise sharply, producing changes that overcome the inhibiting effects of progesterone. The rise in estrogens results from increasing secretion by the placenta of corticotropin-releasing hormone, which stimulates the fetal anterior pituitary gland to secrete ACTH (adrenocorticotropic hormone). In turn, ACTH stimulates the fetal adrenal gland to secrete cortisol and dehydroepiandrosterone (DHEA), the major adrenal androgen. The placenta then converts DHEA into an estrogen. High levels of estrogens cause uterine muscle fibers to display receptors for oxytocin and to form gap junctions with one another. Oxytocin released by the posterior pituitary stimulates uterine contractions, and relaxin from the placenta assists by increasing the flexibility of the pubic symphysis and helping dilate the uterine cervix. Estrogen also stimulates the placenta to release prostaglandins, which induce production of enzymes that digest collagen fibers in the cervix, causing it to soften.

Control of labor contractions during parturition occurs via a positive feedback cycle (see Figure 1.4 on page 11). Contractions of the uterine myometrium force the baby's head or body into the cervix, thereby distending (stretching) the cervix. Stretch receptors in the cervix send nerve impulses to neurosecretory cells in the hypothalamus, causing them to release oxytocin into blood capillaries of the posterior pituitary gland. Oxytocin then is carried by the blood to the uterus, where it stimulates the myometrium to contract more forcefully. As the contractions intensify, the baby's body stretches the cervix still more, and the resulting nerve impulses stimulate the secretion of yet more oxytocin. With birth of the infant, the positive feedback cycle is broken because cervical distention suddenly lessens.

Uterine contractions occur in waves (quite similar to peristaltic waves) that start at the top of the uterus and move downward, eventually expelling the fetus. **True labor** begins when uterine contractions occur at regular intervals, usually producing pain. As the interval between contractions shortens, the contractions intensify. Another symptom of true labor in some women is localization of pain in the back that is intensified by walking. The reliable indicator of true labor is dilation of the cervix and the "show," a discharge of a blood-containing mucus that appears in the cervical canal during labor. In **false labor,** pain is felt in the abdomen at irregular intervals, but it does not intensify and walking does not alter it significantly. There is no "show" and no cervical dilation.

True labor can be divided into three stages (Figure 29.18):

❶ *Stage of dilation.* The time from the onset of labor to the complete dilation of the cervix is the **stage of dilation.** This stage, which typically lasts 6–12 hours, features regular contractions of the uterus, usually a rupturing of the amniotic sac, and complete dilation (to 10 cm) of the cervix. If the amniotic sac does not rupture spontaneously, it is ruptured intentionally.

Figure 29.18 Stages of true labor.

🔑 The term *parturition* refers to birth.

❶ Stage of dilation

❷ Stage of expulsion

❸ Placental stage

 What event marks the beginning of the stage of expulsion?

❷ *Stage of expulsion.* The time (10 minutes to several hours) from complete cervical dilation to delivery of the baby is the **stage of expulsion.**

❸ *Placental stage.* The time (5–30 minutes or more) after delivery until the placenta or "afterbirth" is expelled by powerful uterine contractions is the **placental stage.** These contractions also constrict blood vessels that were torn during delivery, thereby reducing the likelihood of hemorrhage.

As a rule, labor lasts longer with first babies, typically about 14 hours. For women who have previously given birth, the average duration of labor is about 8 hours—although the time varies enormously among births. Because the fetus may be squeezed through the birth canal (cervix and vagina) for up to several hours, the fetus is stressed during childbirth: The fetal head is compressed, and the fetus undergoes some degree of intermittent hypoxia due to compression of the umbilical cord and the placenta during uterine contractions. In response to this stress, the fetal adrenal medullae secrete very high levels of epinephrine and norepinephrine, the "fight-or-flight" hormones. Much of the protection against the stresses of parturition, as well as preparation of the infant for surviving extrauterine life, is provided by these hormones. Among other functions, epinephrine and norepinephrine clear the lungs and alter their physiology in readiness for breathing air, mobilize readily usable nutrients for cellular metabolism, and promote an increased flow of blood to the brain and heart.

About 7% of pregnant women do not deliver by 2 weeks after their due date. Such cases carry an increased risk of brain damage to the fetus, and even fetal death, due to inadequate supplies of oxygen and nutrients from an aging placenta. Post-term deliveries may be facilitated by inducing labor, initiated by administration of oxytocin (Pitocin), or by surgical delivery (cesarean section).

Following the delivery of the baby and placenta is a 6-week period during which the maternal reproductive organs and physiology return to the prepregnancy state. This period is called the **puerperium** (pū'-er-PER-ē-um). Through a process of tissue catabolism, the uterus undergoes a remarkable reduction in size, called **involution** (in'-vō-LOO-shun), especially in lactating women. The cervix loses its elasticity and regains its prepregnancy firmness. For 2–4 weeks after delivery, women have a uterine discharge called **lochia** (LŌ-kē-a), which consists initially of blood and later of serous fluid derived from the former site of the placenta.

⚕ Dystocia and Cesarean Section

Dystocia (dis-TŌ-sē-a; *dys-* = painful or difficult; *toc-* = birth), or difficult labor, may result either from an abnormal position (presentation) of the fetus or a birth canal of inadequate size to permit vaginal delivery. In a **breech presentation,** for example, the fetal buttocks or lower limbs, rather than the head, enter the birth canal first; this occurs most often in premature births. If fetal or maternal distress prevents a vaginal birth, the baby may be delivered surgically through an abdominal incision. A low, horizontal cut is made through the abdominal wall and lower portion of the uterus, through which the baby and placenta are removed. Even though it is popularly associated with the birth of Julius Caesar, the true reason this procedure is termed a **cesarean section (C-section)** is because it was described in Roman Law, *lex cesarea,* about 600 years before Julius Caesar was born. Even a history of multiple C-sections need not exclude a pregnant woman from attempting a vaginal delivery. ■

▶ **CHECKPOINT**

27. What hormonal changes induce labor?
28. What is the difference between false labor and true labor?
29. What happens during the stage of dilation, the stage of expulsion, and the placental stage of true labor?

ADJUSTMENTS OF THE INFANT AT BIRTH

▶ **OBJECTIVE**

• **Explain the respiratory and cardiovascular adjustments that occur in an infant at birth.**

During pregnancy, the embryo (and later the fetus) is totally dependent on the mother for its existence. The mother supplies the fetus with oxygen and nutrients, eliminates its carbon dioxide and other wastes, protects it against shocks and temperature changes, and provides antibodies that confer protection against certain harmful microbes. At birth, a physiologically mature baby becomes much more self-supporting, and the newborn's body systems must make various adjustments. Next we examine some of the changes that occur in the respiratory and cardiovascular systems.

Respiratory Adjustments

The fetus depends entirely on the mother for obtaining oxygen and eliminating carbon dioxide. The fetal lungs are either collapsed or partially filled with amniotic fluid, which is absorbed at birth. The production of surfactant begins by the end of the sixth month of development. Because the respiratory system is fairly well developed at least 2 months before birth, premature babies delivered at 7 months are able to breathe and cry. After delivery, the baby's supply of oxygen from the mother ceases and carbon dioxide is no longer being removed. Circulation in the baby continues, but CO_2 builds up in the blood. A rising CO_2 level stimulates the respiratory center in the medulla oblongata, causing the respiratory muscles to contract, and the baby draws its first breath. Because the first inspiration is unusually deep, as the lungs contain no air, the baby exhales vigorously and naturally cries. A full-term baby may breathe 45 times a minute for the first 2 weeks after birth. Breathing rate gradually declines until it approaches a normal rate of 12 breaths per minute.

Cardiovascular Adjustments

After the baby's first inspiration, the cardiovascular system must make several adjustments (see Figure 21.31 on page 754). Closure of the foramen ovale between the atria of the fetal heart, which occurs at the moment of birth, diverts deoxygenated blood to the lungs for the first time. The foramen ovale is closed by two flaps of septal heart tissue that fold together and permanently fuse. The remnant of the foramen ovale is the fossa ovalis.

Once the lungs begin to function, the ductus arteriosus shuts off due to contractions of smooth muscle in its wall, and it becomes the ligamentum arteriosum. The muscle contraction is probably mediated by the polypeptide bradykinin, released from the lungs during their initial inflation. The ductus arteriosus generally does not close completely until about 3 months after birth. Prolonged incomplete closure results in a condition called **patent ductus arteriosus.**

After the umbilical cord is tied off and severed and blood no longer flows through the umbilical arteries, they fill with connective tissue, and their distal portions become the medial umbilical ligaments. The umbilical vein then becomes the ligamentum teres (round ligament) of the liver.

In the fetus, the ductus venosus connects the umbilical vein directly with the inferior vena cava, allowing blood from the placenta to bypass the fetal liver. When the umbilical cord is severed, the ductus venosus collapses, and venous blood from the viscera of the fetus flows into the hepatic portal vein to the liver and then via the hepatic vein to the inferior vena cava. The remnant of the ductus venosus becomes the ligamentum venosum.

At birth, an infant's pulse may range from 120 to 160 beats per minute and may go as high as 180 upon excitation. After birth, oxygen use increases, which stimulates an increase in the rate of red blood cell and hemoglobin production. Moreover, the white blood cell count at birth is very high—sometimes as much as 45,000 cells per microliter—but the count decreases rapidly by the seventh day.

 Premature Infants

Delivery of a physiologically immature baby carries certain risks. A **premature infant** or "preemie" is generally considered a baby who weighs less than 2500 g (5.5 lb) at birth. Poor prenatal care, drug abuse, history of a previous premature delivery, and mother's age below 16 or above 35 increase the chance of premature delivery. The body of a premature infant is not yet ready to sustain some critical functions, and thus its survival is uncertain without medical intervention. The major problem after delivery of an infant under 36 weeks of gestation is respiratory distress syndrome (RDS) of the newborn due to insufficient surfactant. RDS can be eased by use of artificial surfactant and a ventilator that delivers oxygen until the lungs can operate on their own. ■

▶ C H E C K P O I N T

30. What is the importance of surfactant at birth and afterward?

31. What cardiovascular adjustments are made by an infant at birth?

THE PHYSIOLOGY OF LACTATION

▶ O B J E C T I V E

• **Discuss the physiology and hormonal control of lactation.**

Lactation (lak′-TĀ-shun) is the secretion and ejection of milk from the mammary glands. A principal hormone in promoting milk synthesis and secretion is **prolactin (PRL),** which is secreted from the anterior pituitary gland. Even though prolactin levels increase as the pregnancy progresses, no milk secretion occurs because progesterone inhibits the effects of prolactin. After delivery, the levels of estrogens and progesterone in the mother's blood decrease, and the inhibition is removed. The principal stimulus in maintaining prolactin secretion during lactation is the sucking action of the infant. Suckling initiates nerve impulses from stretch receptors in the nipples to the hypothalamus; the impulses decrease hypothalamic release of prolactin-inhibiting hormone (PIH) and increase release of prolactin-releasing hormone (PRH), so more prolactin is released by the anterior pituitary.

Oxytocin causes release of milk into the mammary ducts via the **milk ejection reflex** (Figure 29.19). Milk formed by the glandular cells of the breasts is stored until the baby begins active suckling. Stimulation of touch receptors in the nipple initiates sensory nerve impulses that are relayed to the hypothalamus. In response, secretion of oxytocin from the posterior pituitary increases. Carried by the bloodstream to the mammary glands, oxytocin stimulates contraction of myoepithelial (smooth-muscle-like) cells surrounding the glandular cells and ducts. The resulting compression moves the milk from the alveoli of the mammary glands into the mammary ducts, where it can be suckled. This process is termed **milk ejection (let-down).** Even though the actual ejection of milk does not occur until 30–60 seconds after nursing begins (the latent period), some milk stored in lactiferous sinuses near the nipple is available during the latent period. Stimuli other than suckling, such as hearing a baby's cry or touching the mother's genitals, also can trigger oxytocin release and milk ejection. The suckling stimulation that produces the release of oxytocin also inhibits the release of PIH; this results in increased secretion of prolactin, which maintains lactation.

During late pregnancy and the first few days after birth, the mammary glands secrete a cloudy fluid called **colostrum.** Although it is not as nutritious as milk—it contains less lactose and virtually no fat—colostrum serves adequately until the appearance of true milk on about the fourth day. Colostrum and maternal milk contain important antibodies that protect the infant during the first few months of life.

Following birth of the infant, the prolactin level starts to return to the nonpregnant level. However, each time the mother nurses the infant, nerve impulses from the nipples to the hypothalamus increase the release of PRH (and decrease the release of PIH), resulting in a tenfold increase in prolactin secretion by the anterior pituitary that lasts about an hour. Prolactin acts on the mammary glands to provide milk for the next nursing period.

Figure 29.19 The milk ejection reflex, a positive feedback cycle.

Oxytocin stimulates contraction of myoepithelial cells in the breasts, which squeezes the glandular and duct cells and causes milk ejection.

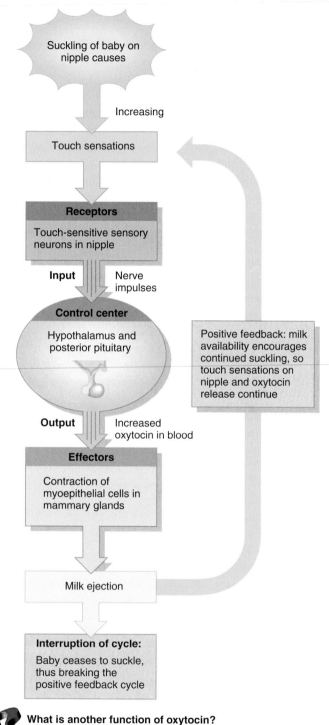

What is another function of oxytocin?

If this surge of prolactin is blocked by injury or disease, or if nursing is discontinued, the mammary glands lose their ability to secrete milk in only a few days. Even though milk secretion normally decreases considerably within 7–9 months after birth, it can continue for several years if nursing continues.

Lactation often blocks ovarian cycles for the first few months following delivery, if the frequency of sucking is about 8–10 times a day. This effect is inconsistent, however, and ovulation commonly precedes the first menstrual period after delivery of a baby. As a result, the mother can never be certain she is not fertile. Breast-feeding is therefore not a very reliable birth control measure. The suppression of ovulation during lactation is believed to occur as follows: During breast-feeding, neural input from the nipple reaches the hypothalamus and causes it to produce neurotransmitters that suppress the release of gonadotropin-releasing hormone (GnRH). As a result, production of LH and FSH decreases, and ovulation is inhibited.

A primary benefit of **breast-feeding** is nutritional: Human milk is a sterile solution that contains amounts of fatty acids, lactose, amino acids, minerals, vitamins, and water that are ideal for the baby's digestion, brain development, and growth. Breast-feeding also benefits infants by providing the following:

- *Beneficial cells.* Several types of white blood cells are present in breast milk. Neutrophils and macrophages serve as phagocytes, ingesting microbes in the baby's gastrointestinal tract. Additionally macrophages produce lysozyme and other immune system components. Plasma cells, which develop from B lymphocytes, produce antibodies against specific microbes, and T lymphocytes kill microbes directly or help mobilize other defenses.

- *Beneficial molecules.* Breast milk also contains an abundance of beneficial molecules. Maternal IgA antibodies in breast milk bind to microbes in the baby's gastrointestinal tract and prevent their migration into other body tissues. Because a mother produces antibodies to whatever disease-causing microbes are present in her environment, her breast milk affords protection against the specific infectious agents to which her baby is also exposed. Additionally, two milk proteins bind to nutrients that many bacteria need to grow and survive, thereby making the nutrients unavailable: B_{12}-binding protein ties up vitamin B_{12}, and lactoferrin ties up iron. Some fatty acids can kill certain viruses by disrupting their membranes, and lysozyme kills bacteria by disrupting their cell walls. Finally, interferons enhance the antimicrobial activity of immune cells.

- *Decreased incidence of diseases later in life.* Breast-feeding provides children with a slight reduction in risk of lymphoma, heart disease, allergies, respiratory and gastrointestinal infections, ear infections, diarrhea, diabetes mellitus, and meningitis. Breast-feeding also protects the mother against osteoporosis and breast cancer.

- *Miscellaneous benefits.* Breast-feeding supports optimal infant growth, enhances intellectual and neurological development, and fosters mother–infant relations by establishing early and prolonged contact between them. Compared to cow's milk, the fats and iron in breast milk are more easily absorbed, the proteins in breast milk are more readily metabolized, and the lower sodium content of breast milk is more suited to an infant's needs. Premature infants benefit from breast-feeding because the milk produced by mothers of premature infants seems to be specially adapted to the infant's needs by having a higher protein content than the milk of mothers of full-term infants. Finally, a baby is less likely to have an allergic reaction to its mother's milk than to milk from another source.

Nursing and Childbirth

Years before oxytocin was discovered, it was common practice in midwifery to let a first-born twin nurse at the mother's breast to speed the birth of the second child. Now we know why this practice is helpful—it stimulates the release of oxytocin. Even after a single birth, nursing promotes expulsion of the placenta (afterbirth) and helps the uterus return to its normal size. Synthetic oxytocin (Pitocin) is often given to induce labor or to increase uterine tone and control hemorrhage just after parturition. ■

▶ **CHECKPOINT**

32. Which hormones contribute to lactation? What is the function of each?

33. What are the benefits of breast-feeding over bottle-feeding?

INHERITANCE

▶ **OBJECTIVE**

- **Define inheritance, and explain the inheritance of dominant, recessive, complex, and sex-linked traits.**

As previously indicated, the genetic material of a father and a mother unite when a sperm cell fuses with a secondary oocyte to form a zygote. Children resemble their parents because they inherit traits passed down from both parents. We now examine some of the principles involved in that process, called inheritance.

Inheritance is the passage of hereditary traits from one generation to the next. It is the process by which you acquired your characteristics from your parents and may transmit some of your traits to your children. The branch of biology that deals with inheritance is called **genetics** (je-NET-iks). The area of health care that offers advice on genetic problems (or potential problems) is called **genetic counseling.**

Genotype and Phenotype

The nuclei of all human cells except gametes contain 23 pairs of chromosomes—the diploid number (2n). One chromosome in each pair came from the mother, and the other came from the father. Each homologue—one of the two chromosomes that make up a pair—contains genes that control the same traits. If a chromosome contains a gene for body hair, for example, its homologue will also contain a gene for body hair in the same position on the chromosome. Such alternative forms of a gene

that code for the same trait and are at the same location on homologous chromosomes are called **alleles** (ah-LĒLZ). For example, one allele of a body hair gene might code for coarse hair, whereas another allele codes for fine hair. A **mutation** (mū-TĀ-shun; *muta-* = change) is a permanent heritable change in an allele that produces a different variant of the same trait.

The relationship of genes to heredity is illustrated by examining the alleles involved in a disorder called **phenylketonuria (PKU).** People with PKU (see page 925) are unable to manufacture the enzyme phenylalanine hydroxylase. The allele that codes for phenylalanine hydroxylase is symbolized as *P* whereas the mutated allele that fails to produce a functional enzyme is symbolized as *p*. The chart in Figure 29.20, which shows the possible combinations of gametes from two parents who each have one *P* and one *p* allele, is called a **Punnett square.** In constructing a Punnett square, the possible paternal alleles in sperm are written at the left side and the possible maternal alleles in ova (or secondary oocytes) are written at the top. The four spaces on the chart show how the alleles can combine in zygotes formed by the union of these sperm and ova to produce the three different genetic makeups, or **genotypes** (JĒ-nō-tīps): *PP*, *Pp*, or *pp*. Notice from the Punnett square that 25% of the offspring

Figure 29.20 Inheritance of phenylketonuria (PKU).

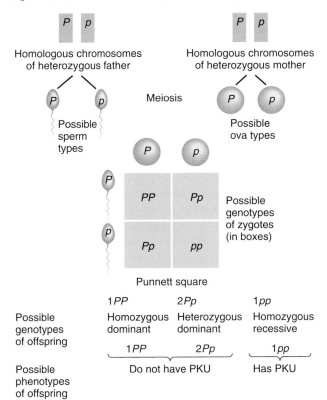

Whereas genotype refers to genetic makeup, phenotype refers to the physical or outward expression of a gene.

 If parents have the genotypes shown here, what is the percent chance that their first child will have PKU? What is the chance of PKU occurring in their second child?

will have the *PP* genotype, 50% will have the *Pp* genotype, and 25% will have the *pp* genotype. People who inherit *PP* or *Pp* genotypes do not have PKU; those with a *pp* genotype suffer from the disorder. Although people with a *Pp* genotype have one PKU allele (*p*), the allele that codes for the normal trait (*P*), is more dominant. An allele that dominates or masks the presence of another allele and is fully expressed (*P* in this example) is said to be a **dominant allele,** and the trait expressed is called a dominant trait. The allele whose presence is completely masked (*p* in this example) is said to be a **recessive allele,** and the trait it controls is called a recessive trait.

By tradition, the symbols for genes are written in italics, with dominant alleles written in capital letters and recessive alleles in lowercase letters. A person with the same alleles on homologous chromosomes (for example, *PP* or *pp*) is said to be **homozygous** for the trait. *PP* is homozygous dominant, and *pp* is homozygous recessive. An individual with different alleles on homologous chromosomes (for example, *Pp*) is said to be **heterozygous** for the trait.

Phenotype (FĒ-nō-tīp; *pheno-* = showing) refers to how the genetic makeup is expressed in the body; it is the physical or outward expression of a gene. A person with *Pp* (a heterozygote) has a different *genotype* from a person with *PP* (a homozygote), but both have the same *phenotype*—normal production of phenylalanine hydroxylase. Heterozygous individuals who carry a recessive gene but do not express it (*Pp*) can pass the gene on to their offspring. Such individuals are called **carriers** of the recessive gene.

Most genes give rise to the same phenotype whether they are inherited from the mother or the father. In a few cases, however, the phenotype is dramatically different, depending on the parental origin. This surprising phenomenon, first appreciated in the 1980s, is called **genomic imprinting.** In humans, the abnormalities most clearly associated with mutation of an imprinted gene are *Angelman syndrome,* which results when the gene for a particular abnormal trait is inherited from the mother, and *Prader-Willi syndrome,* which results when it is inherited from the father.

Alleles that code for normal traits do not always dominate over those that code for abnormal ones, but dominant alleles for severe disorders usually are lethal and cause death of the embryo or fetus. One exception is Huntington disease (HD), which is caused by a dominant allele whose effects do not become manifest until adulthood. Both homozygous dominant and heterozygous people exhibit the disease, whereas homozygous recessive people are normal. HD causes progressive degeneration of the nervous system and eventual death, but because symptoms typically do not appear until after age 30 or 40, many afflicted individuals have already passed the allele for the condition on to their children.

Occasionally an error in cell division, called **nondisjunction,** results in an abnormal number of chromosomes. In this situation, homologous chromosomes or sister chromatids fail to separate properly during anaphase of meiosis or mitosis. A cell that has one or more chromosomes of a set added or deleted is called an **aneuploid** (AN-ū-ployd). A monosomic cell (2*n* − 1) is missing a chromosome; a trisomic cell (2*n* + 1) has an extra chromosome. Most cases of Down syndrome (see page 1099) are aneuploid disorders in which there is trisomy of chromosome 21. Whereas nondisjunction occurs during gametogenesis (meiosis) in most instances, about 2% of Down syndrome cases result from nondisjunction during mitotic divisions in early embryonic development.

Another error in meiosis is a **translocation.** In this case, two chromosomes that are *not* homologous break and interchange portions of their chromosomes. The individual who has a translocation may be perfectly normal if no loss of genetic material took place when the rearrangement occurred. However, when the person makes gametes, some of the gametes may not contain the correct amount and type of genetic material. About 3% of Down syndrome cases result from a translocation of part of chromosome 21 to another chromosome, usually chromosome 14 or 15. The individual who has this translocation is normal and does not even know that he or she is a "carrier." When such a carrier produces gametes, however, some gametes end up with a whole chromosome 21 plus another chromosome with the translocated fragment of chromosome 21. Upon fertilization, the zygote then has three, rather than two, copies of that part of chromosome 21.

Table 29.3 lists some dominant and recessive inherited structural and functional traits in humans.

Variations on Dominant–Recessive Inheritance

Most patterns of inheritance are not simple **dominant–recessive inheritance** in which only dominant and recessive alleles interact. The phenotypic expression of a particular gene may be influenced not only by which alleles are present, but also by other genes and the environment. Moreover, most inherited traits are influenced by more than one gene, and most genes can influence

Table 29.3	Selected Hereditary Traits in Humans
Dominant	**Recessive**
Normal skin pigmentation	Albinism
Near- or farsightedness	Normal vision
PTC taster*	PTC nontaster
Polydactyly (extra digits)	Normal digits
Brachydactyly (short digits)	Normal digits
Syndactylism (webbed digits)	Normal digits
Diabetes insipidus	Normal urine excretion
Huntington disease	Normal nervous system
Widow's peak	Straight hairline
Curved (hyperextended) thumb	Straight thumb
Normal Cl⁻ transport	Cystic fibrosis
Hypercholesterolemia (familial)	Normal cholesterol level

*Ability to taste a chemical compound called phenylthiocarbamide (PTC).

more than a single trait. Next we discuss three variations on dominant–recessive inheritance.

Incomplete Dominance

In **incomplete dominance,** neither member of an allelic pair is dominant over the other, and the heterozygote has a phenotype intermediate between the homozygous dominant and the homozygous recessive phenotypes. An example of incomplete dominance in humans is the inheritance of **sickle-cell disease (SCD)** (Figure 29.21). People with the homozygous dominant genotype Hb^AHb^A form normal hemoglobin; those with the homozygous recessive genotype Hb^SHb^S have sickle-cell disease and severe anemia. Although they are usually healthy, those with the heterozygous genotype Hb^AHb^S have minor problems with anemia because half their hemoglobin is normal and half is not. Heterozygotes are carriers, and they are said to have *sickle-cell trait.*

Multiple-Allele Inheritance

Although a single individual inherits only two alleles for each gene, some genes may have more than two alternate forms, and this is the basis for **multiple-allele inheritance.** One example of multiple-allele inheritance is the inheritance of the ABO blood group. The four blood types (phenotypes) of the ABO group—A, B, AB, and O—result from the inheritance of six combinations of three different alleles of a single gene called the *I* gene: (1) allele I^A produces the A antigen, (2) allele I^B produces the B antigen, and (3) allele i produces neither A nor B antigen. Each person inherits two *I*-gene alleles, one from each parent, that

give rise to the various phenotypes. The six possible genotypes produce four blood types, as follows:

Genotype	Blood type (phenotype)
I^AI^A or I^Ai	A
I^BI^B or I^Bi	B
I^AI^B	AB
ii	O

Notice that both I^A and I^B are inherited as dominant traits, whereas i is inherited as a recessive trait. Because an individual with type AB blood has characteristics of both type A and type B red blood cells expressed in the phenotype, alleles I^A and I^B are said to be **codominant.** In other words, both genes are expressed equally in the heterozygote. Depending on the parental blood types, different offspring may have blood types different from each other. Figure 29.22 shows the blood types offspring could inherit, given the blood types of their parents.

Complex Inheritance

Most inherited traits are not controlled by one gene, but instead by the combined effects of two or more genes, a situation referred to as **polygenic inheritance** (*poly-* = many), or the combined effects of many genes and environmental factors, a situation referred to as **complex inheritance**. Examples of complex traits include skin color, hair color, eye color, height, metabolism rate, and body build. In complex inheritance, one genotype can have many possible phenotypes, depending on the environment, or one phenotype can include many possible genotypes. For

Figure 29.21   Inheritance of sickle-cell disease.

🔑 Sickle-cell disease is an example of incomplete dominance.

Hb^A Hb^S ———— Hb^A Hb^S

Meiosis

Possible sperm types

Possible ova types

Possible genotypes of zygotes (in boxes)

Punnett square

Hb^AHb^A = normal
Hb^AHb^S = carrier of sickle-cell disease
Hb^SHb^S = has sickle-cell disease

❓ What are the distinguishing features of incomplete dominance?

Figure 29.22   The ten possible combinations of parental ABO blood types and the blood types their offspring could inherit. For each possible set of parents, the blue letters represent the blood types their offspring could inherit.

🔑 Inheritance of ABO blood types is an example of multiple-allele inheritance.

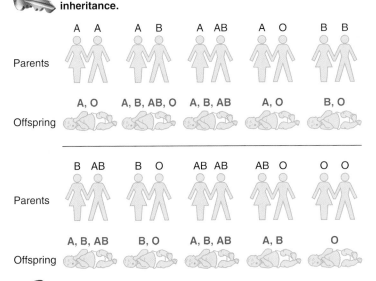

Parents	A A	A B	A AB	A O	B B
Offspring	A, O	A, B, AB, O	A, B, AB	A, O	B, O

Parents	B AB	B O	AB AB	AB O	O O
Offspring	A, B, AB	B, O	A, B, AB	A, B	O

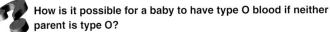

❓ How is it possible for a baby to have type O blood if neither parent is type O?

example, even though a person inherits several genes for tallness, full height potential may not be reached due to environmental factors, such as disease or malnutrition during the growth years. Similarly, the risk of having a child with a neural tube defect is greater in pregnant women who lack adequate folic acid in their diet (an environmental factor). Because neural tube defects are more prevalent in some families than in others, however, one or more genes may also contribute.

Often, a complex trait shows a continuous gradation of small differences between extremes among individuals. Whereas it is relatively easy to predict the risk of passing on an undesirable trait that is due to a single dominant or recessive gene, it is very difficult to make this prediction when the trait is complex. Such traits are difficult to follow in a family because the range of variation is large, the number of different genes involved usually is not known, and the impact of environmental factors may be incompletely understood.

Skin color is a good example of a complex trait. It depends on environmental factors such as sun exposure and nutrition, as well as on several genes. Suppose that skin color is controlled by three separate genes, each having two alleles: *A*, *a*; *B*, *b*; and *C*, *c* (Figure 29.23). Whereas a person with the genotype *AABBCC* is very dark skinned, an individual with the genotype *aabbcc* is

very light skinned, and a person with the genotype *AaBbCc* has an intermediate skin color. Parents having an intermediate skin color may have children with very light, very dark, or intermediate skin color. Note that the **P generation** (parental generation) is the starting generation, the **F₁ generation** (first filial generation) is produced from the P generation, and the **F₂ generation** (second filial generation) is produced from the F₁ generation.

Autosomes, Sex Chromosomes, and Sex Determination

When viewed under a microscope, the 46 human chromosomes in a normal somatic cell can be identified by their size, shape, and staining pattern to be a member of 23 different pairs of chromosomes. In 22 of the pairs, the homologous chromosomes look alike and have the same appearance in both males and females; these 22 pairs are called **autosomes.** The two members of the 23rd pair are termed the **sex chromosomes;** they look different in males and females (Figure 29.24). In females, the pair consists of two chromosomes called X chromosomes. One X chromosome is also present in males, but its mate is a much smaller chromosome called a Y chromosome. The Y chromosome has only 231 genes, less than 10% of the 2968 genes present on chromosome 1, the largest chromosome.

When a spermatocyte undergoes meiosis to reduce its chromosome number, it gives rise to two sperm that contain an X chromosome and two sperm that contain a Y chromosome. Oocytes have no Y chromosomes and produce only X-containing gametes. If the secondary oocyte is fertilized by an X-bearing sperm, the offspring normally is female (XX). Fertilization by a Y-bearing sperm produces a male (XY). Thus, an individual's sex is determined by the father's chromosomes (Figure 29.25).

Both female and male embryos develop identically until about 7 weeks after fertilization. At that point, one or more genes set into motion a cascade of events that leads to the development of a male; in the absence of normal expression of the gene or genes, the female pattern of development occurs. It has been known since 1959 that the Y chromosome is needed to ini-

Figure 29.23 Polygenic inheritance of skin color.

In polygenic inheritance, a trait is controlled by the combined effects of several genes.

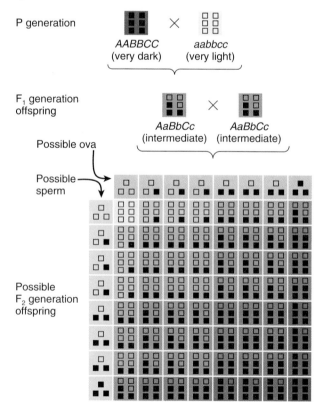

Figure 29.24 Autosomes and sex chromosomes.

Human somatic cells contain 23 different pairs of chromosomes.

 What other traits are transmitted by polygenic inheritance?

What are the two sex chromosomes in females and males?

Figure 29.25   Sex determination.

Sex is determined at the time of fertilization by the presence or absence of a Y chromosome in the sperm.

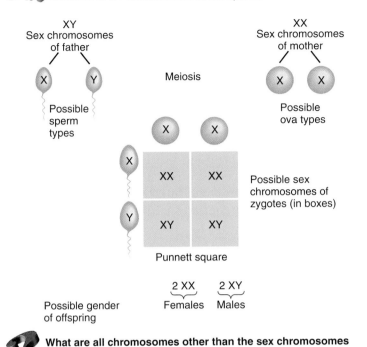

What are all chromosomes other than the sex chromosomes called?

tiate male development. Experiments published in 1991 established that the prime male-determining gene is one called **SRY (sex-determining region of the Y chromosome).** When a small DNA fragment containing this gene was inserted into 11 female mouse embryos, three of them developed as males. (The researchers suspected that the gene failed to be integrated into the genetic material in the other eight.) SRY acts as a molecular switch to turn on the male pattern of development. Only if the SRY gene is present and functional in a fertilized ovum will the fetus develop testes and differentiate into a male; in the absence of SRY, the fetus will develop ovaries and differentiate into a female.

Experiments of nature confirmed the key role of SRY in directing the male pattern of development in humans. In some cases, phenotypic females with an XY genotype were found to have mutated SRY genes. These individuals failed to develop normally as males because their SRY gene was defective. In other cases, phenotypic males with an XX genotype were found to have a small piece of the Y chromosome, including the SRY gene, inserted into one of their X chromosomes.

Sex-Linked Inheritance

The sex chromosomes also are responsible for the transmission of several nonsexual traits. Many of the genes for these traits are present on X chromosomes but are absent from Y chromosomes. This feature produces a pattern of heredity, termed **sex-linked inheritance,** that is different from the patterns already described.

Red–Green Color Blindness

One example of sex-linked inheritance is **red–green color blindness,** the most common type of color blindness. This condition is characterized by a deficiency in either red- or green-sensitive cones, so red and green are seen as the same color (either red or green, depending on which cone is present). The gene for red–green color blindness is a recessive one designated *c*. Normal color vision, designated *C*, dominates. The *C/c* genes are located only on the X chromosome, and thus the ability to see colors depends entirely on the X chromosomes. The possible combinations are as follows:

Genotype	Phenotype
$X^C X^C$	Normal female
$X^C X^c$	Normal female (but a carrier of the recessive gene)
$X^c X^c$	Red–green color-blind female
$X^C Y$	Normal male
$X^c Y$	Red–green color-blind male

Only females who have two X^c genes are red–green color blind. This rare situation can result only from the mating of a color-blind male and a color-blind or carrier female. (In $X^C X^c$ females the trait is masked by the normal, dominant gene.) Because males, in contrast, do not have a second X chromosome that could mask the trait, all males with an X^c gene will be red–green color blind. Figure 29.26 illustrates the inheritance of red–green color blindness in the offspring of a normal male and a carrier female.

Figure 29.26   An example of the inheritance of red–green color blindness.

Red–green color blindness and hemophilia are examples of sex-linked traits.

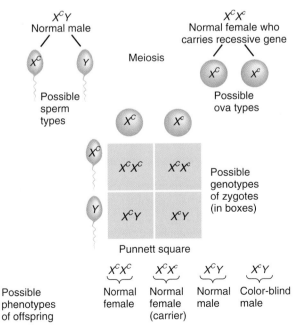

What is the genotype of a red–green color-blind female?

Traits inherited in the manner just described are called **sex-linked traits.** The most common type of **hemophilia**—a condition in which the blood fails to clot or clots very slowly after an injury—is also a sex-linked trait. Like the trait for red–green color blindness, hemophilia is caused by a recessive gene. Other sex-linked traits in humans are fragile X syndrome (described on page 1100), nonfunctional sweat glands, certain forms of diabetes, some types of deafness, uncontrollable rolling of the eyeballs, absence of central incisors, night blindness, one form of cataract, juvenile glaucoma, and juvenile muscular dystrophy.

X-Chromosome Inactivation

Whereas females have two X chromosomes in every cell (except developing oocytes), males have only one. Females thus have a double set of all genes on the X chromosome. A mechanism termed **X-chromosome inactivation (lyonization)** in effect reduces the X-chromosome genes to a single set in females. In each cell of a female's body, one X chromosome is randomly and permanently inactivated early in development, and most of the genes of the inactivated X chromosome are not expressed (transcribed and translated). The nuclei of cells in female mammals contain a dark-staining body, called a **Barr body,** that is not present in the nuclei of cells in males. Geneticist Mary Lyon correctly predicted in 1961 that the Barr body is the inactivated X chromosome. During inactivation, chemical groups that prevent transcription into RNA are added to the X chromosome's DNA. As a result, an inactivated X chromosome reacts differently to histological stains and has a different appearance than the rest of the DNA. In nondividing (interphase) cells, it remains tightly coiled and can be seen as a dark-staining body within the nucleus. In a blood smear, the Barr body of neutrophils is termed a "drumstick" because it looks like a tiny drumstick-shaped projection of the nucleus.

Teratogens

Exposure of a developing embryo or fetus to certain environmental factors can damage the developing organism or even cause death. A **teratogen** (TER-a-tō-jen; *terato-* = monster; *-gen* = creating) is any agent or influence that causes developmental defects in the embryo. In the following sections we briefly discuss several examples.

Chemicals and Drugs

Because the placenta is not an absolute barrier between the maternal and fetal circulations, any drug or chemical that is dangerous to an infant should be considered potentially dangerous to the fetus when given to the mother. Alcohol is by far the number-one fetal teratogen. Intrauterine exposure to even a small amount of alcohol may result in **fetal alcohol syndrome (FAS),** one of the most common causes of mental retardation and the most common preventable cause of birth defects in the United States. The symptoms of FAS may include slow growth before and after birth, characteristic facial features (short palpebral fissures, a thin upper lip, and sunken nasal bridge), defective heart and other organs, malformed limbs, genital abnormalities, and central nervous system damage. Behavioral problems, such as hyperactivity, extreme nervousness, reduced ability to concentrate, and an inability to appreciate cause-and-effect relationships, are common.

Other teratogens include pesticides; defoliants (chemicals that cause plants to shed their leaves prematurely); industrial chemicals; some hormones; antibiotics; oral anticoagulants, anticonvulsants, antitumor agents, thyroid drugs, thalidomide, diethylstilbestrol (DES), and numerous other prescription drugs; LSD; and cocaine. A pregnant woman who uses cocaine, for example, subjects the fetus to higher risk of retarded growth, attention and orientation problems, hyperirritability, a tendency to stop breathing, malformed or missing organs, strokes, and seizures. The risks of spontaneous abortion, premature birth, and still-birth also increase with fetal exposure to cocaine.

Cigarette Smoking

Strong evidence implicates cigarette smoking during pregnancy as a cause of low infant birth weight and there is a strong association between smoking and a higher fetal and infant mortality rate. Women who smoke have a much higher risk of an ectopic pregnancy. Cigarette smoke may be teratogenic and may cause cardiac abnormalities and anencephaly (a developmental defect characterized by the absence of a cerebrum). Maternal smoking also is a significant factor in the development of cleft lip and palate and has been linked with sudden infant death syndrome (SIDS). Infants nursing from smoking mothers have also been found to have an increased incidence of gastrointestinal disturbances. Even a mother's exposure to secondhand cigarette smoke (breathing air containing tobacco smoke) during pregnancy or while nursing predisposes her baby to increased incidence of respiratory problems, including bronchitis and pneumonia, during the first year of life.

Irradiation

Ionizing radiation of various kinds is a potent teratogen. Exposure of pregnant mothers to x rays or radioactive isotopes during the embryo's susceptible period of development may cause microcephaly (small head size relative to the rest of the body), mental retardation, and skeletal malformations. Caution is advised, especially during the first trimester of pregnancy.

▶ CHECKPOINT

34. What do the terms genotype, phenotype, dominant, recessive, homozygous, and heterozygous mean?

35. What are genomic imprinting and nondisjunction?

36. Define incomplete dominance. Give an example.

37. What is multiple-allele inheritance? Give an example.

38. Define complex inheritance and give an example.

39. Why does X-chromosome inactivation occur?

DISORDERS: HOMEOSTATIC IMBALANCES

Infertility

Female infertility, or the inability to conceive, occurs in about 10% of all women of reproductive age in the United States. Female infertility may be caused by ovarian disease, obstruction of the uterine tubes, or conditions in which the uterus is not adequately prepared to receive a fertilized ovum. **Male infertility (sterility)** is an inability to fertilize a secondary oocyte; it does not imply erectile dysfunction (impotence). Male fertility requires production of adequate quantities of viable, normal sperm by the testes, unobstructed transport of sperm though the ducts, and satisfactory deposition in the vagina. The seminiferous tubules of the testes are sensitive to many factors—x rays, infections, toxins, malnutrition, and higher-than-normal scrotal temperatures—that may cause degenerative changes and produce male sterility.

One cause of infertility in females is inadequate body fat. To begin and maintain a normal reproductive cycle, a female must have a minimum amount of body fat. Even a moderate deficiency of fat—10% to 15% below normal weight for height—may delay the onset of menstruation (menarche), inhibit ovulation during the reproductive cycle, or cause amenorrhea (cessation of menstruation). Both dieting and intensive exercise may reduce body fat below the minimum amount and lead to infertility that is reversible, if weight gain or reduction of intensive exercise or both occurs. Studies of very obese women indicate that they, like very lean ones, experience problems with amenorrhea and infertility. Males also experience reproductive problems in response to undernutrition and weight loss. For example, they produce less prostatic fluid and reduced numbers of sperm having decreased motility.

Many fertility-expanding techniques now exist for assisting infertile couples to have a baby. The birth of Louise Joy Brown on July 12, 1978, near Manchester, England, was the first recorded case of **in vitro fertilization (IVF)**—fertilization in a laboratory dish. In the IVF procedure, the mother-to-be is given follicle-stimulating hormone (FSH) soon after menstruation, so that several secondary oocytes, rather than the typical single oocyte, will be produced (superovulation). When several follicles have reached the appropriate size, a small incision is made near the umbilicus, and the secondary oocytes are aspirated from the stimulated follicles and transferred to a solution containing sperm, where the oocytes undergo fertilization. Alternatively, an oocyte may be fertilized in vitro by suctioning a sperm or even a spermatid obtained from the testis into a tiny pipette and then injecting it into the oocyte's cytoplasm. This procedure, termed **intracytoplasmic sperm injection (ICSI),** has been used when infertility is due to impairments in sperm motility or to the failure of spermatids to develop into spermatozoa. When the zygote achieved by IVF reaches the 8-cell or 16-cell stage, it is introduced into the uterus for implantation and subsequent growth.

In **embryo transfer,** a man's semen is used to artificially inseminate a fertile secondary oocyte donor. After fertilization in the donor's uterine tube, the morula or blastocyst is transferred from the donor to the infertile woman, who then carries it (and subsequently the fetus) to term. Embryo transfer is indicated for women who are infertile or who do not want to pass on their own genes because they are carriers of a serious genetic disorder.

In **gamete intrafallopian transfer (GIFT)** the goal is to mimic the normal process of conception by uniting sperm and secondary oocyte in the prospective mother's uterine tubes. It is an attempt to bypass conditions in the female reproductive tract that might prevent fertilization, such as high acidity or inappropriate mucus. In this procedure, a woman is given FSH and LH to stimulate the production of several secondary oocytes, which are aspirated from the mature follicles, mixed outside the body with a solution containing sperm, and then immediately inserted into the uterine tubes.

Down Syndrome

Down syndrome (DS) is a disorder characterized by three, rather than two, copies of at least part of chromosome 21. Overall, one infant in 900 is born with Down syndrome. However, older women are more likely to have a DS baby. The chance of having a baby with this syndrome, which is less than 1 in 3000 for women under age 30, increases to 1 in 300 in the 35–39 age group and to 1 in 9 at age 48.

About 3% of cases are due to translocation of part of chromosome 21 to another chromosome. The rest are due to nondisjunction of chromosome 21. Nondisjunction most often occurs during meiosis, which causes an entire extra chromosome 21 to pass to one of the gametes. The extra chromosome typically comes from the mother, a not too surprising finding given that all her oocytes began meiosis when she herself was a fetus. They may have been exposed to chromosome-damaging chemicals and radiation for years. Undoubtedly, the kinetochore microtubules responsible for pulling sister chromatids to opposite poles of the cell (see Figure 28.1 on page 1013) sustain increasing damage with the passage of time. (Sperm, by contrast, usually are less than 10 weeks old when they fertilize a secondary oocyte.) In 1–2% of cases, nondisjunction of chromosome 21 occurs during an early mitotic division of embryonic development. The resulting individual is a *mosaic*, an organism composed of two genetically different types of cells. Some body cells have 46 chromosomes and other body cells have 47.

Down syndrome is characterized by mental retardation, retarded physical development (short stature and stubby fingers), distinctive facial structures (large tongue, flat profile, broad skull, slanting eyes, and round head), kidney defects, suppressed immune system, and malformations of the heart, ears, hands, and feet. Sexual maturity is rarely attained.

Trinucleotide Repeat Diseases

Repeating triplets of nucleotides in certain genes are known to cause more than a dozen different **trinucleotide repeat diseases.** In each case, a specific sequence of three DNA nucleotides, such as CAG (cytosine-adenine-guanine), that normally repeats several times within a gene becomes greatly expanded during gametogenesis. Sometimes, the number of repeats expands with each succeeding generation, causing a more severe disease that begins at an earlier age.

One trinucleotide repeat disease is **Huntington disease (HD).** Normal people have an average of 18 CAG repeats in each of their two *HD* genes, which codes for a protein named **huntingtin.** Because CAG codes for the amino acid glutamine, huntingtin normally has about 18 glutamines one after the other in a specific part of the protein. When the number of CAG repeats exceeds 35 to 40, the extra glutamines distort the shape of huntingtin, and Huntington disease slowly develops. The normal function of the huntingtin protein is not known, but the abnormal form appears to have a toxic effect on neurons. Because it is a dominant genetic disease, one mutant allele of the *HD* gene is sufficient to cause Huntington disease.

Another trinucleotide repeat disease is **fragile X syndrome,** in which the repeated trinucleotide is in a gene on the X chromosome. When the number of repeats exceeds 230, problems occur with both transcription of the gene and translation of the m-RNA. Effectively, the gene is inactive. Fragile X syndrome is second only to Down syndrome as a cause of inherited mental impairment. It causes learning difficulties, mental retardation, and physical abnormalities such as oversized ears, an elongated forehead, and enlarged testes. The syndrome may also be involved in autism, in which the individual exhibits extreme withdrawal and refusal to communicate. Because males have just one X chromosome and because the affected allele is recessive, fragile X syndrome occurs more often in males than in females. Both Huntington disease and fragile X syndrome can be diagnosed prenatally by measuring the number of repeats present in cells obtained via amniocentesis.

MEDICAL TERMINOLOGY

Breech presentation A malpresentation in which the fetal buttocks or lower limbs present into the maternal pelvis; the most common cause is prematurity.

Emesis gravidarum (EM-e-sis gra-VID-ar-um; *emeo* = to vomit; *gravida* = a pregnant woman) Episodes of nausea and possibly vomiting that are most likely to occur in the morning during the early weeks of pregnancy; also called **morning sickness.** Its cause is unknown, but the high levels of human chorionic gonadotropin (hCG) secreted by the placenta, and of progesterone secreted by the ovaries, have been implicated. In some women the severity of these symptoms requires hospitalization for intravenous feeding, and the condition is then known as **hyperemesis gravidarum.**

Karyotype (KAR-ē-ō-tīp; *karyo-* = nucleus) The chromosomal characteristics of an individual presented as a systematic arrangement of pairs of metaphase chromosomes arrayed in descending order of size and according to the position of the centromere; useful in judging whether chromosomes are normal in number and structure.

Klinefelter's syndrome A sex chromosome aneuploidy, usually due to trisomy XXY, that occurs once in every 500 births. Such individuals are somewhat mentally disadvantaged, sterile males with undeveloped testes, scant body hair, and enlarged breasts.

Lethal gene (LĒ-thal jēn; *lethum* = death) A gene that, when expressed, results in death either in the embryonic state or shortly after birth.

Metafemale syndrome A sex chromosome aneuploidy characterized by at least three X chromosomes (XXX) that occurs about once in every 700 births. These females have underdeveloped genital organs and limited fertility. Generally, they are mentally retarded.

Puerperal fever (pū-ER-per-al; *puer* = child) An infectious disease of childbirth, also called puerperal sepsis and childbed fever. The disease, which results from an infection originating in the birth canal, affects the endometrium. It may spread to other pelvic structures and lead to septicemia.

Turner's syndrome A sex chromosome aneuploidy caused by the presence of a single X chromosome (designated XO); occurs about once in every 5000 births and produces a sterile female with virtually no ovaries and limited development of secondary sex characteristics. Other features include short stature, webbed neck, underdeveloped breasts, and widely spaced nipples. Intelligence usually is normal.

STUDY OUTLINE

EMBRYONIC PERIOD (p. 1063)

1. Pregnancy is a sequence of events that begins with fertilization, proceeds to implantation, embryonic development, and fetal development. It normally ends in birth.
2. During fertilization a sperm cell penetrates a secondary oocyte and their pronuclei unite. Penetration of the zona pellucida is facilitated by enzymes in the sperm's acrosome. The resulting cell is a zygote.
3. Normally, only one sperm cell fertilizes a secondary oocyte because both fast and slow blocks to polyspermy exist.
4. Early rapid cell division of a zygote is called cleavage, and the cells produced by cleavage are called blastomeres. The solid sphere of cells produced by cleavage is a morula.
5. The morula develops into a blastocyst, a hollow ball of cells differentiated into a trophoblast and an inner cell mass.
6. The attachment of a blastocyst to the endometrium is termed implantation; it occurs as a result of enzymatic degradation of the endometrium.
7. After implantation, the endometrium becomes modified and is known as the decidua.
8. The trophoblast develops into the syncytiotrophoblast and cytotrophoblast, both of which become part of the chorion.
9. The inner cell mass differentiates into hypoblast and epiblast, the bilaminar (two-layered) embryonic disc.
10. The amnion is a thin protective membrane that develops from the cytotrophoblast.
11. The exocoelomic membrane and hypoblast form the yolk sac, which transfers nutrients to the embryo, forms blood cells, produces primordial germ cells, and forms part of the gut.
12. Erosion of sinusoids and endometrial glands provides blood and secretions, which enter lacunar networks to supply nutrition to and remove wastes from the embryo.
13. The extraembryonic coelom forms within extraembryonic mesoderm.
14. The extraembryonic mesoderm and trophoblast form the chorion, the principal embryonic part of the placenta.
15. The third week of development is characterized by gastrulation, the conversion of the bilaminar disc into a trilaminar (three-

layered) embryo consisting of ectoderm, mesoderm, and endoderm.

16. The first evidence of gastrulation is formation of the primitive streak and then the primitive node, notochordal process, and notochord.

17. The three primary germ layers form all tissues and organs of the developing organism. Table 29–1 on page 1072 summarizes the structures that develop from the primary germ layers.

18. Also during the third week, the oropharyngeal and cloacal membranes form. The wall of the yolk sac forms a small vascularized outpouching called the allantois, which functions in blood formation and development of the urinary bladder.

19. The process by which the neural plate, neural folds, and neural tube form is called neurulation. The brain and spinal cord develop from the neural tube.

20. Paraxial mesoderm segments to form somites from which skeletal muscles of the neck, trunk, and limbs develop. Somites also form connective tissues and vertebrae.

21. Blood vessel formation, called angiogenesis, begins in mesodermal cells called angioblasts.

22. The heart forms from mesodermal cells called the cardiogenic area. By the end of the third week, the primitive heart beats and circulates blood.

23. Chorionic villi, projections of the chorion, connect to the embryonic heart so that maternal and fetal blood vessels are brought into close proximity. Thus, nutrients and wastes are exchanged between maternal and fetal blood.

24. Placentation refers to formation of the placenta, the site of exchange of nutrients and wastes between the mother and fetus. The placenta also functions as a protective barrier, stores nutrients, and produces several hormones to maintain pregnancy.

25. The actual connection between the placenta and embryo (and later the fetus) is the umbilical cord.

26. Organogenesis refers to the formation of body organs and systems and occurs during the fourth week of development.

27. Conversion of the flat, two-dimensional trilaminar embryonic disc to a three-dimensional cylinder occurs by a process called embryonic folding.

28. Embryo folding brings various organs into their final adult positions and helps form the gastrointestinal tract.

29. Pharyngeal arches, clefts, and pouches give rise to the structures of the head and neck.

30. By the end of the fourth week, upper and lower limb buds develop and by the end of the eighth week the embryo has clearly human features.

FETAL PERIOD (p. 1081)

1. The fetal period is primarily concerned with the growth and differentiation of tissues and organs that developed during the embryonic period.

2. The rate of body growth is remarkable, especially during the ninth and sixteenth weeks.

3. The principal changes associated with embryonic and fetal growth are summarized in Table 29.2 on pages 1081–1082.

PRENATAL DIAGNOSTIC TESTS (p. 1084)

1. Several prenatal diagnostic tests are used to detect genetic disorders and to assess fetal well-being. These include fetal ultrasonography, in which an image of a fetus is displayed on a screen;

amniocentesis, the withdrawal and analysis of amniotic fluid and the fetal cells within it; chorionic villi sampling (CVS), which involves withdrawal of chorionic villi tissue for chromosomal analysis.

2. CVS can be done earlier than amniocentesis, and the results are available more quickly, but it is also slightly riskier than amniocentesis.

3. Noninvasive prenatal tests include the maternal alpha-fetoprotein (AFP) test to detect neural tube defects and the Quad AFP Plus test to detect Down syndrome, trisomy 18, and neural tube defects.

MATERNAL CHANGES DURING PREGNANCY (p. 1085)

1. Pregnancy is maintained by human chorionic gonadotropin (hCG), estrogens, and progesterone.

2. Relaxin increases flexibility of the pubic symphysis and helps dilate the uterine cervix near the end of pregnancy.

3. Human chorionic somatomammotropin (hCS) contributes to breast development, protein anabolism, and catabolism of glucose and fatty acids.

4. Corticotropin-releasing hormone, produced by the placenta, is thought to establish the timing of birth, and stimulates the secretion of cortisol by the fetal adrenal gland.

5. During pregnancy, several anatomical and physiological changes occur in the mother.

EXERCISE AND PREGNANCY (p. 1088)

1. During pregnancy, some joints become less stable, and certain maneuvers are more difficult to execute.

2. Moderate physical activity does not endanger the fetus in a normal pregnancy.

LABOR (p. 1088)

1. Labor is the process by which the fetus is expelled from the uterus through the vagina to the outside. True labor involves dilation of the cervix, expulsion of the fetus, and delivery of the placenta.

2. Oxytocin stimulates uterine contractions via a positive feedback cycle.

ADJUSTMENTS OF THE INFANT AT BIRTH (p. 1090)

1. The fetus depends on the mother for oxygen and nutrients, the removal of wastes, and protection.

2. Following birth, an infant's respiratory and cardiovascular systems undergo changes in adjusting to becoming self-supporting during postnatal life.

THE PHYSIOLOGY OF LACTATION (p. 1091)

1. Lactation refers to the production and ejection of milk by the mammary glands.

2. Milk production is influenced by prolactin (PRL), estrogens, and progesterone.

3. Milk ejection is stimulated by oxytocin.

4. A few of the many benefits of breast-feeding include ideal nutrition for the infant, protection from disease, and decreased likelihood of developing allergies.

INHERITANCE (p. 1093)

1. Inheritance is the passage of hereditary traits from one generation to the next.

2. The genetic makeup of an organism is called its genotype; the traits expressed are called its phenotype.

3. Dominant genes control a particular trait; expression of recessive genes is masked by dominant genes.

4. Many patterns of inheritance do not conform to the simple dominant–recessive patterns.

5. In incomplete dominance, neither member of an allelic pair dominates; phenotypically, the heterozygote is intermediate between the homozygous dominant and the homozygous recessive. An example is sickle-cell disease.

6. In multiple-allele inheritance, genes have more than two alternate forms. An example is the inheritance of ABO blood groups.

7. In complex inheritance, a trait such as skin or eye color is controlled by the combined effects of two or more genes and may be influenced by environmental factors.

8. Each somatic cell has 46 chromosomes—22 pairs of autosomes and 1 pair of sex chromosomes.

9. In females, the sex chromosomes are two X chromosomes; in males, they are one X chromosome and a much smaller Y chromosome, which normally includes the prime male-determining gene, called *SRY*.

10. If the *SRY* gene is present and functional in a fertilized ovum, the fetus will develop testes and differentiate into a male. In the absence of *SRY*, the fetus will develop ovaries and differentiate into a female.

11. Red–green color blindness and hemophilia result from recessive genes located on the X chromosome. They are sex-linked traits that occur primarily in males because of the absence of any counterbalancing dominant genes on the Y chromosome.

12. A mechanism termed X-chromosome inactivation (lyonization) balances the difference in number of X chromosomes between males (one X) and females (two Xs). In each cell of a female's body, one X chromosome is randomly and permanently inactivated early in development and becomes a Barr body.

13. A given phenotype is the result of the interactions of genotype and the environment.

14. Teratogens, which are agents that cause physical defects in developing embryos, include chemicals and drugs, alcohol, nicotine, and ionizing radiation.

 SELF-QUIZ QUESTIONS

Fill in the blanks in the following statements.

1. The three stages of true labor, in order of occurrence, are _____, _____, and _____.

2. Hormones produced by the _____, are responsible for maintaining the pregnancy during the first 3–4 months.

3. Indicate the germ layers responsible for development of the following structures: (a) muscle, bone, and peritoneum: _____; (b) nervous system and epidermis: _____; (c) epithelial linings of respiratory and gastrointestinal tracts: _____.

Indicate whether the following statements are true or false.

4. The principal stimulus in maintaining prolactin secretion during lactation is the sucking action of the infant.

Choose the one best answer to the following questions.

5. The group of cells that secrete enzymes that enable the blastocyst to penetrate the uterine lining is the (a) syncytiotrophoblast, (b) cytotrophoblast, (c) corona radiata, (d) zona pellucida, (e) trophoblast.

6. Which of the following statements are correct? (1) The embryonic period is the first four months of development. (2) The embryonic disc differentiates into the three primary germ layers. (3) The three primary germ layers are the endoderm, mesoderm, and ectoderm. (4) The embryonic membranes are the yolk sac, amnion, chorion, and allantois. (5) Placental development is accomplished during the fourth month of pregnancy. (a) 1, 2, and 3, (b) 2, 3, and 4, (c) 3, 4, and 5, (d) 1, 4, and 5, (e) 2, 3, and 5.

7. The hormone of pregnancy that mimics LH and acts to rescue the corpus luteum from degeneration and to stimulate its continued production of progesterone and estrogens is (a) relaxin, (b) human chorionic somatomammotropin, (c) human placental lactogen, (d) corticotropin-releasing hormone, (e) human chorionic gonadotropin.

8. Match the following:

_____ (a) the penetration of a secondary oocyte by a single sperm cell

_____ (b) fertilization of a secondary oocyte by more than one sperm

_____ (c) the attachment of a blastocyst to the endometrium

_____ (d) the fusion of the genetic material from a haploid sperm and a haploid secondary oocyte into a single diploid nucleus

_____ (e) the induction by the female reproductive tract of functional changes in sperm that allow them to fertilize a secondary oocyte

_____ (f) the examination of embryonic cells sloughed off into the amniotic fluid

_____ (g) an abnormal condition of pregnancy characterized by sudden hypertension, large amounts of protein in urine, and generalized edema

_____ (h) the process of giving birth

_____ (i) the period of time (about 6 weeks) during which the maternal reproductive organs and physiology return to the pre-pregnancy state

(1) fertilization
(2) capacitation
(3) syngamy
(4) polyspermy
(5) implantation
(6) amniocentesis
(7) preeclampsia
(8) parturition
(9) puerperium

9. Which of the following are true concerning fertilization? (1) The sperm first penetrate the zona pellucida and then the corona radiata. (2) The binding of specific membrane proteins in the sperm head to ZP3 causes the release of acrosomal contents. (3) Sperm are able to fertilize the oocyte within minutes after ejaculation. (4) Depolarization of the cell membrane of the secondary oocyte inhibits fertilization by more than one sperm. (5) The oocyte completes meiosis II after fertilization. (a) 1, 2, 4, and 5, (b) 1, 3, and 5, (c) 1, 2, 3, and 4, (d) 1, 4, and 5, (e) 2, 4, and 5.

10. Amniotic fluid (1) is derived entirely from a filtrate of maternal blood, (2) acts as a fetal shock absorber, (3) provides nutrients to the fetus, (4) helps regulate fetal body temperature, (5) prevents adhesions between the skin of the fetus and surrounding tissues. (a) 1, 2, 3, 4, and 5, (b) 2, 4, and 5, (c) 2, 3, 4, and 5, (d) 1, 4, and 5, (e) 1, 2, 4, and 5.

11. Match the following:
_____ (a) the control of inherited traits by the combined effects of many genes
_____ (b) the two alternative forms of a gene that code for the same trait and are at the same location on homologous chromosomes
_____ (c) inheritance based on genes that have more than two alternate forms; an example is the inheritance of blood type
_____ (d) a cell in which one or more chromosomes of a set is added or deleted
_____ (e) refers to an individual with different alleles on homologous chromosomes
_____ (f) traits linked to the X chromosome
_____ (g) neither member of the allelic pair is dominant over the other, and the heterozygote has a phenotype intermediate between the homozygous dominant and homozygous recessive
_____ (h) refers to how the genetic makeup is expressed in the body
_____ (i) a homozygous dominant, homozygous recessive, or heterozygous genetic makeup; the actual gene arrangement
_____ (j) refers to a person with the same alleles on homologous chromosomes
_____ (k) individuals who possess a recessive gene (but do not express it) and can pass the gene on to their offspring
_____ (l) an allele that masks the presence of another allele and is fully expressed

(1) genotype
(2) phenotype
(3) alleles
(4) aneuploid
(5) incomplete dominance
(6) multiple-allele inheritance
(7) polygenic inheritance
(8) sex-linked inheritance
(9) homozygous
(10) heterozygous
(11) carriers
(12) dominant trait

12. Which of the following statements is correct? (a) Normal traits always dominate over abnormal traits. (b) Occasionally an error in meiosis called nondisjunction results in an abnormal number of chromosomes. (c) The mother always determines the sex of the child because she has either an X or a Y gene in her oocytes. (d) Most patterns of inheritance are simple dominant-recessive inheritances. (e) Genes are expressed normally regardless of any outside influence such as chemicals or radiation.

13. Match the following:
_____ (a) the embryonic membrane that entirely surrounds the embryo
_____ (b) functions as an early site of blood formation; contains cells that migrate into the gonads and differentiate into the primitive germ cells
_____ (c) becomes the principal part of the placenta; produces human chorionic gonadotropin
_____ (d) modified endometrium after implantation has occurred; separates from the endometrium after the fetus is delivered
_____ (e) contains the vascular connections between mother and fetus
_____ (f) allows oxygen and nutrients to diffuse from maternal blood into fetal blood
_____ (g) serves as an early site of blood vessel formation; its blood vessels serve as the umbilical connection in the placenta between mother and fetus
_____ (h) finger-like projections of the chorion that are bathed in maternal blood sinuses, thereby bringing maternal and fetal blood vessels into close proximity

(1) decidua
(2) placenta
(3) amnion
(4) chorion
(5) allantois
(6) yolk sac
(7) chorionic villi
(8) umbilical cord

14. Match the following:
_____ (a) a hollow sphere of cells that enters the uterine cavity
_____ (b) cells produced by cleavage
_____ (c) the developing individual from week nine of pregnancy until birth
_____ (d) the outer covering of cells of the blastocyst
_____ (e) membrane derived from trophoblast
_____ (f) early divisions of the zygote
_____ (g) a solid sphere of cells still surrounded by the zona pellucida
_____ (h) event in which differentiation into the three primary germ layers occurs

(1) cleavage
(2) blastomeres
(3) morula
(4) trophoblast
(5) blastocyst
(6) gastrulation
(7) chorion
(8) fetus

15. Which of the following are maternal changes that occur during pregnancy? (1) altered pulmonary function, (2) increased stroke volume, cardiac output, and heart rate, and decreased blood volume, (3) weight gain, (4) increased gastric motility, a delay in gastric emptying time, (5) edema and possible varicose veins. (a) 1, 2, 3, and 4, (b) 2, 3, 4, and 5, (c) 1, 3, 4, and 5, (d) 1, 3, and 5, (e) 2, 4, and 5.

CRITICAL THINKING QUESTIONS

1. The science class was studying genetics at school and Kendra came home in tears. She told her older sister, "We were doing our family tree and when I filled in our traits, I discovered that Mom and Dad can't be my real parents 'cause the traits don't match!" It turns out that Mom and Dad are PTC tasters but Kendra is a nontaster. Could Mom and Dad still be Kendra's parents?

HINT *Everyone says that Kendra is the "spitting image" of her mother.*

2. When pregnant Phoebe learned that Huntington disease (HD) was present in her husband's family, she said, "We won't have to worry about it because no one in my family has ever had it." Explain the inheritance of HD more clearly.

HINT *HD is a trinucleotide repeat disease that often worsens from one generation to the next.*

3. Usually, when you think about arteries, you think about oxygenated blood, but this textbook states that the umbilical arteries carry deoxygenated blood in the fetus. Could the authors, Tortora and Grabowski, be wrong about this? Please explain.

HINT *Think about the pulmonary arteries in an adult.*

ANSWERS TO FIGURE QUESTIONS

29.1 Capacitation entails the functional changes in sperm after they have been deposited in the female reproductive tract that enable them to fertilize a secondary oocyte.

29.2 A morula is a solid ball of cells, whereas a blastocyst consists of a rim of cells (trophoblast) surrounding a cavity (blastocyst cavity) and an inner cell mass.

29.3 The blastocyst secretes digestive enzymes that eat away the endometrial lining at the site of implantation.

29.4 The decidua basalis helps form the maternal part of the placenta.

29.5 Implantation occurs during the secretory phase of the uterine cycle.

29.6 The bilaminar embryonic disc is attached to the trophoblast by the connecting stalk.

29.7 Gastrulation converts a two-dimensional bilaminar embryonic disc into a two-dimensional trilaminar embryonic disc.

29.8 The notochord induces mesodermal cells to develop into vertebral bodies and forms the nucleus pulposus of intervertebral discs.

29.9 The neural tube forms the brain and spinal cord; somites develop into skeletal muscles, connective tissue, and the vertebrae.

29.10 Chorionic villi help to bring the fetal and maternal blood vessels into close proximity.

29.11 The placenta functions in exchange of materials between fetus and mother, as a protective barrier against many microbes, and stores nutrients.

29.12 Because of embryo folding, the embryo curves into a C-shape, various organs are brought into their eventual adult positions, and the primitive gut is formed.

29.13 They give rise to structures of the head and neck.

29.14 During this time, fetal weight quadruples.

29.15 Amniocentesis is used mainly to detect genetic disorders, but it also provides information concerning the maturity (and survivability) of the fetus.

29.16 Early pregnancy tests detect human chorionic gonadotropin (hCG).

29.17 Relaxin increases the flexibility of the pubic symphysis and helps dilate the cervix of the uterus to ease delivery.

29.18 Complete dilation of the cervix marks the onset of the stage of expulsion.

29.19 Oxytocin also stimulates contraction of the uterus during delivery of a baby.

29.20 The odds that a child will have PKU are the same for each child—25%.

29.21 In incomplete dominance, neither member of an allelic pair is dominant; the heterozygote has a phenotype intermediate between the homozygous dominant and the homozygous recessive phenotypes.

29.22 Yes, a baby can have blood type O if each parent has one *i* allele and passes it on to the offspring.

29.23 Hair color, height, and body build, among others, are polygenic traits.

29.24 The female sex chromosomes are XX, and the male sex chromosomes are XY.

29.25 The chromosomes that are not sex chromosomes are called autosomes.

29.26 A red–green color-blind female has an X^cX^c genotype.

Appendix A
Measurements

U.S. Customary System

Parameter	Unit	Relation to Other U.S. Units	SI (Metric) Equivalent
Length	inch	1/12 foot	2.54 centimeters
foot	12 inches	0.305 meter	
yard	36 inches	9.144 meters	
mile	5,280 feet	1.609 kilometers	
Mass	grain	1/1000 pound	64.799 milligrams
dram	1/16 ounce	1.772 grams	
ounce	16 drams	28.350 grams	
pound	16 ounces	453.6 grams	
ton	2,000 pounds	907.18 kilograms	
Volume (Liquid)	ounce	1/16 pint	29.574 milliliters
pint	16 ounces	0.473 liter	
quart	2 pints	0.946 liter	
gallon	4 quarts	3.785 liters	
Volume (Dry)	pint	1/2 quart	0.551 liter
quart	2 pints	1.101 liters	
peck	8 quarts	8.810 liters	
bushel	4 pecks	35.239 liters	

International System (SI)

BASE UNITS

Unit	Quantity	Symbol
meter	length	M
kilogram	mass	Kg
second	time	S
liter	volume	L
mole	amount of matter	Mol

PREFIXES

Prefix	Multiplier	Symbol
tera-	$10^{12} = 1,000,000,000,000$	T
giga-	$10^9 = 1,000,000,000$	G
mega-	$10^6 = 1,000,000$	M
kilo-	$10^3 = 1,000$	k
hecto-	$10^2 = 100$	h
deca-	$10^1 = 10$	da
deci-	$10^{-1} = 0.1$	d
centi-	$10^{-2} = 0.01$	c
milli-	$10^{-3} = 0.001$	m
micro-	$10^{-6} = 0.000,001$	μ
nano-	$10^{-9} = 0.000,000,001$	n
pico-	$10^{-12} = 0.000,000,000,001$	p

Temperature Conversion
Fahrenheit (F) To Celsius (C)
$°C = (°F - 32) \div 1.8$
Celsius (C) To Fahrenheit (F)
$°F = (°C \times 1.8) + 32$

U.S TO SI (Metric) Conversion

When you know	Multiply by	To find
inches	2.54	centimeters
feet	30.48	centimeters
yards	0.91	meters
miles	1.61	kilometers
ounces	28.35	grams
pounds	0.45	kilograms
tons	0.91	metric tons
fluid ounces	29.57	milliliters
pints	0.47	liters
quarts	0.95	liters
gallons	3.79	liters

SI (Metric) TO U.S. Conversion

When you know	Multiply by	To find
millimeters	0.04	inches
centimeters	0.39	inches
meters	3.28	feet
kilometers	0.62	miles
liters	1.06	quarts
cubic meters	35.32	cubic feet
grams	0.035	ounces
kilograms	2.21	pounds

Appendix B
Periodic Table

The periodic table lists the known **chemical elements,** the basic units of matter. The elements in the table are arranged left-to-right in rows in order of their **atomic number,** the number of protons in the nucleus. Each horizontal row, numbered from 1 to 7, is a **period.** All elements in a given period have the same number of electron shells as their period number. For example, an atom of hydrogen or helium each has one electron shell, while an atom of potassium or calcium each has four electron shells. The elements in each column, or **group,** share chemical properties. For example, the elements in column IA are very chemically reactive, whereas the elements in column VIIIA have full electron shells and thus are chemically inert.

Scientists now recognize 113 different elements; 92 occur naturally on Earth, and the rest are produced from the natural elements using particle accelerators or nuclear reactors. Elements are designated by **chemical symbols,** which are the first one or two letters of the element's name in English, Latin, or another language.

Twenty-six of the 92 naturally occurring elements normally are present in your body. Of these, just four elements—oxygen (O), carbon (C), hydrogen (H), and nitrogen (N) (coded blue)—constitute about 96% of the body's mass. Eight others—calcium (Ca), phosphorus (P), potassium (K), sulfur (S), sodium (Na), chlorine (Cl), magnesium (Mg), and iron (Fe) (coded pink)—contribute 3.8% of the body's mass. An additional 14 elements, called **trace elements** because they are present in tiny amounts, account for the remaining 0.2% of the body's mass. The trace elements are aluminum, boron, chromium, cobalt, copper, fluorine, iodine, manganese, molybdenum, selenium, silicon, tin, vanadium, and zinc (coded yellow). Table 2.1 on page 28 provides information about the main chemical elements in the body.

Appendix C
Normal Values For Selected Blood Tests

The system of international (SI) units (Système Internationale d'Unités) is used in most countries and in many medical and scientific journals. Clinical laboratories in the United States, by contrast, usually report values for blood and urine tests in conventional units. The laboratory values in this Appendix give conventional units first, followed by SI equivalents in parentheses. Values listed for various blood tests should be viewed as reference values rather than absolute "normal" values for all well people. Values may vary due to age, gender, diet, and environment of the subject or the equipment, methods, and standards of the lab performing the measurement.

Key To Symbols

g = gram
mg = milligram = 10^{-3} gram
μg = microgram = 10^{-6} gram
U = units
L = liter
dL = deciliter
mL = milliliter
μL = microliter
mEq/L = milliequivalents per liter
mmol/L = millimoles per liter
μmol/L = micromoles per liter
$>$ = greater than; $<$ = less than

Blood Tests

Test (Specimen)	U.S. Reference Values (SI Units)	Values Increase In	Values Decrease In
Aminotransferases (serum)			
Alanine aminotransferase (ALT)	0–35 U/L (same)	Liver disease or liver damage due to toxic drugs.	
Aspartate aminotransferase (AST)	0–35 U/L (same)	Myocardial infarction, liver disease, trauma to skeletal muscles, severe burns.	Beriberi, uncontrolled diabetes mellitus with acidosis, pregnancy.
Ammonia (plasma)	20–120 μg/dL (12–55 μmol/L)	Liver disease, heart failure, emphysema, pneumonia, hemolytic disease of newborn.	Hypertension.
Bilirubin (serum)	Conjugated: <0.5 mg/dL (<5.0 μmol/L)	Conjugated bilirubin: liver dysfunction or gallstones.	
	Unconjugated: 0.2–1.0 mg/dL (18–20 μmol/L) Newborn: 1.0–12.0 mg/dL (<200 μmol/L)	Unconjugated bilirubin: excessive hemolysis of red blood cells.	
Blood urea nitrogen (BUN) (serum)	8–26 mg/dL (2.9–9.3 mmol/L)	Kidney disease, urinary tract obstruction, shock, diabetes, burns, dehydration, myocardial infarction.	Liver failure, malnutrition, overhydration, pregnancy.

Blood Tests

Test (Specimen)	U.S. Reference Values (SI Units)	Values Increase In	Values Decrease In
Carbon dioxide content (bicarbonate + dissolved CO_2) (whole blood)	Arterial: 19–24 mEq/L (19–24 mmol/L) Venous: 22–26 mEq/L (22–26 mmol/L)	Severe diarrhea, severe vomiting, starvation, emphysema, aldosteronism.	Renal failure, diabetic ketoacidosis, shock.
Cholesterol, total (plasma)	<200 mg/dL (<5.2 mmol/L) is desirable	Hypercholesterolemia, uncontrolled diabetes mellitus, hypothyroidism, hypertension, atherosclerosis, nephrosis.	Liver disease, hyperthyroidism, fat malabsorption, pernicious or hemolytic anemia, severe infections.
HDL cholesterol (plasma)	>40 mg/dL (>1.0 mmol/L) is desirable		
LDL cholesterol (plasma)	<130 mg/dL (<3.2 mmol/L) is desirable		
Creatine (serum)	Males: 0.15–0.5 mg/dL (10–40 μmol/L) Females: 0.35–0.9 mg/dL (30–70 μmol/L)	Muscular dystrophy, damage to muscle tissue, electric shock, chronic alcoholism.	
Creatine Kinase (CK), also known as Creatine phosphokinase (CPK) (serum)	0–130 U/L (same)	Myocardial infarction, progressive muscular dystrophy, hypothyroidism, pulmonary edema.	
Creatinine (serum)	0.5–1.2 mg/dL (45–105 μmol/L)	Impaired renal function, urinary tract obstruction, giantism, acromegaly.	Decreased muscle mass, as occurs in muscular dystrophy or myasthenia gravis.
Electrolytes (plasma)	See Table 27.2 on page 1000.		
Gamma-glutamyl transferase (GGT) (serum)	0–30 U/L (same)	Bile duct obstruction, cirrhosis, alcoholism, metastatic liver cancer, congestive heart failure.	
Glucose (plasma)	70–110 mg/dL (3.9–6.1 mmol/L)	Diabetes mellitus, acute stress, hyperthyroidism, chronic liver disease, Cushing's disease.	Addison's disease, hypothyroidism, hyperinsulinism.
Hemoglobin (whole blood)	Males: 14–18 g/100 mL (140–180 g/L) Females: 12–16 g/100 mL (120–160 g/L) Newborns: 14–20 g/100 mL (140–200 g/L)	Polycythemia, congestive heart failure, chronic obstructive pulmonary disease, living at high altitude.	Anemia, severe hemorrhage, cancer, hemolysis, Hodgkin's disease, nutritional deficiency of vitamin B_{12}, systemic lupus erythematosus, kidney disease.
Iron, total (serum)	Males: 80–180 μg/dL (14–32 μmol/L) Females: 60–160 μg/dL (11–29 μmol/L)	Liver disease, hemolytic anemia, iron poisoning.	Iron-deficiency anemia, chronic blood loss, pregnancy (late), chronic heavy menstruation.
Lactic dehydrogenase (LDH) (serum)	71–207 U/L (same)	Myocardial infarction, liver disease, skeletal muscle necrosis, extensive cancer.	
Lipids (serum)		Hyperlipidemia, diabetes mellitus, hypothyroidism.	Fat malabsorption.
Total	400–850 mg/dL (4.0–8.5 g/L)		
Triglycerides	10–190 mg/dL (0.1–1.9 g/L)		
Platelet (thrombocyte) count (whole blood)	150,000–400,000/μL	Cancer, trauma, leukemia, cirrhosis.	Anemias, allergic conditions hemorrhage.
Protein (serum)		Dehydration, shock, chronic infections.	Liver disease, poor protein intake, hemorrhage, diarrhea, malabsorption, chronic renal failure, severe burns.
Total	6–8 g/dL (60–80 g/L)		
Albumin	4–6 g/dL (40–60 g/L)		
Globulin	2.3–3.5 g/dL (23–35 g/L)		

(continues)

Blood Tests

Test (Specimen)	U.S. Reference Values (SI Units)	Values Increase In	Values Decrease In
Red blood cell (erythrocyte) count (whole blood)	Males: 4.5–6.5 million/μL Females: 3.9–5.6 million/μL	Polycythemia, dehydration, living at high altitude.	Hemorrhage, hemolysis, anemias, cancer, overhydration.
Uric acid (urate) (serum)	2.0–7.0 mg/dL (120–420 μmol/L)	Impaired renal function, gout, metastatic cancer, shock, starvation.	
White blood cell (leukocyte) count, total (whole blood)	5,000–10,000/μL (See Table 19.3 on page 646 for relative percentages of different types of WBCs.)	Acute infections, trauma, malignant diseases, cardiovascular diseases. (See also Table 19.2 on page 645.)	Diabetes mellitus, anemia. (See also Table 19.2 on page 645.)

Appendix D
Normal Values For Selected Urine Tests

Urine Tests		
Test (Specimen)	**U.S. Reference Values (SI Units)**	**Clinical Implications**
Amylase (2 hour)	35–260 Somogyi units/hr (6.5–48.1 units/hr)	Values increase in inflammation of the pancreas (pancreatitis) or salivary glands, obstruction of the pancreatic duct, and perforated peptic ulcer.
Bilirubin* (random)	Negative	Values increase in liver disease and obstructive biliary disease.
Blood* (random)	Negative	Values increase in renal disease, extensive burns, transfusion reactions, and hemolytic anemia.
Calcium (Ca^{2+}) (random)	10 mg/dL (2.5 mmol/liter); up to 300 mg/24 hr (7.5 mmol/24 hr)	Amount depends on dietary intake; values increase in hyperparathyroidism, metastatic malignancies, and primary cancer of breasts and lungs; values decrease in hypoparathyroidism and vitamin D deficiency.
Casts (24 hour)		
Epithelial	Occasional	Values increase in nephrosis and heavy metal poisoning.
Granular	Occasional	Values increase in nephritis and pyelonephritis.
Hyaline	Occasional	Values increase in kidney infections.
Red blood cell	Occasional	Values increase in glomerular membrane damage and fever.
White blood cell	Occasional	Values increase in pyelonephritis, kidney stones, and cystitis.
Chloride (Cl$^-$) (24 hour)	140–250 mEq/24 hr (140–250 mmol/24 hr)	Amount depends on dietary salt intake; values increase in Addison's disease, dehydration, and starvation; values decrease in pyloric obstruction, diarrhea, and emphysema.
Color (random)	Yellow, straw, amber	Varies with many disease states, hydration, and diet.
Creatinine (24 hour)	Males: 1.0–2.0 g/24 hr (9–18 mmol/24 hr) Females: 0.8–1.8 g/24 hr (7–16 mmol/24 hr)	Values increase in infections; values decrease in muscular atrophy, anemia, and kidney diseases.
Glucose*	Negative	Values increase in diabetes mellitus, brain injury, and myocardial infarction.
Hydroxycortico-steroids (17-hydroxysteroids) (24 hour)	Males: 5–15 mg/24 hr (13–41 μmol/24 hr) Females: 2–13 mg/24 hr (5–36 μmol/24 hr)	Values increase in Cushing's syndrome, burns, and infections; values decrease in Addison's disease.
Ketone bodies* (random)	Negative	Values increase in diabetic acidosis, fever, anorexia, fasting, and starvation.
17-ketosteroids (24 hour)	Males: 8–25 mg/24 hr (28–87 μmol/24 hr) Females: 5–15 mg/24 hr (17–53 μmol/24 hr)	Values decrease in surgery, burns, infections, adrenogenital syndrome, and Cushing's syndrome.
Odor (random)	Aromatic	Becomes acetonelike in diabetic ketosis.
Osmolality (24 hour)	500–1400 mOsm/kg water (500–1400 mmol/kg water)	Values increase in cirrhosis, congestive heart failure (CHF), and high-protein diets; values decrease in aldosteronism, diabetes insipidus, and hypokalemia.
pH* (random)	4.6–8.0	Values increase in urinary tract infections and severe alkalosis; values decrease in acidosis, emphysema, starvation, and dehydration.

Test (Specimen)	U.S. Reference Values (SI Units)	Clinical Implications
Phenylpyruvic acid (random)	Negative	Values increase in phenylketonuria (PKU).
Potassium (K⁺) (24 hour)	40–80 mEq/24 hr (40–80 mmol/24 hr)	Values increase in chronic renal failure, dehydration, starvation, and Cushing's syndrome; values decrease in diarrhea, malabsorption syndrome, and adrenal cortical insufficiency
Protein* (albumin) (random)	Negative	Values increase in nephritis, fever, severe anemias, trauma, and hyperthyroidism.
Sodium (Na⁺) (24 hour)	75–200 mg/24 hr (75–200 mmol/24 hr)	Amount depends on dietary salt intake; values increase in dehydration, starvation, and diabetic acidosis; values decrease in diarrhea, acute renal failure, emphysema, and Cushing's syndrome.
Specific gravity* (random)	1.001–1.035 (same)	Values increase in diabetes mellitus and excessive water loss; values decrease in absence of antidiuretic hormone (ADH) and severe renal damage.
Urea (random)	25–35 g/24 hr (420–580 mmol/24 hr)	Values increase in response to increased protein intake; values decrease in impaired renal function.
Uric acid (24 hour)	0.4–1.0 g/24 hr (1.5–4.0 mmol/24 hr)	Values increase in gout, leukemia, and liver disease; values decrease in kidney disease.
Urobilinogen* (2 hour)	0.3–1.0 Ehrlich units (1.7–6.0 μmol/24 hr)	Values increase in anemias, hepatitis A (infectious), biliary disease, and cirrhosis; values decrease in cholelithiasis and renal insufficiency.
Volume, total (24 hour)	1000–2000 mL/24 hr (1.0–2.0 liters/24 hr)	Varies with many factors.

*Test often performed using a **dipstick,** a plastic strip impregnated with chemicals that is dipped into a urine specimen to detect particular substances. Certain colors indicate the presence or absence of a substance and sometimes give a rough estimate of the amount(s) present.

Appendix E
Answers

Answers to Self-Quiz Questions

Chapter 1
1. cell **2.** catabolism **3.** homeostasis **4.** false **5.** true **6.** true **7.** d **8.** b **9.** b **10.** e **11.** e **12.** (a) 3, (b) 4, (c) 6, (d) 5, (e) 1, (f) 2 **13.** (a) 6, (b) 1, (c) 11, (d) 5, (e) 10, (f) 8, (g) 7, (h) 9, (i) 4, (j) 3, (k) 2 **14.** (a) 4, (b) 6, (c) 8, (d) 1, (e) 5, (f) 2, (g) 7, (h) 3 **15.** (a) 3, (b) 7, (c) 1, (d) 5, (e) 6, (f) 4, (g) 8, (h) 2

Chapter 2
1. protons, electrons, neutrons **2.** 8 **3.** atomic number **4.** concentration, temperature **5.** true **6.** false **7.** b **8.** d **9.** a **10.** c **11.** b **12.** (a) 1, (b) 2 (c) 1, (d) 4 **13.** a **14.** (a) 8, (b) 1, (c) 7, (d) 2, (e) 5, (f) 3, (g) 4, (h) 6 **15.** (a) 7, (b) 5, (c) 1, (d) 2, (e) 6, (f) 3, (g) 8, (h) 4

Chapter 3
1. plasma membrane, cytoplasm, nucleus **2.** apoptosis, necrosis **3.** cytosol **4.** shortening and loss of protective telomeres on chromosomes, cross-link formation between glucose and proteins, free radical formation **5.** true **6.** true **7.** a **8.** d **9.** b **10.** a **11.** c **12.** (a) 2, (b) 3, (c) 5, (d) 7, (e) 6, (f) 8, (g) 1, (h) 4 **13.** (a) 11, (b) 10, (c) 1, (d) 2, (e) 5, (f) 9, (g) 12, (h) 7, (i) 6, (j) 3, (k) 4, (l) 8 **14.** (a) 2, (b) 9, (c) 3, (d) 5, (e) 11, (f) 8, (g) 1, (h) 6, (i) 10, (j) 7, (k) 4 **15.** (a) 3, (b) 9, (c) 1, (d) 5, (e) 11, (f) 4, (g) 8, (h) 7, (i) 2, (j) 10, (k) 6

Chapter 4
1. tissue **2.** endocrine **3.** cells, ground substance, fibers **4.** true **5.** false **6.** a **7.** c **8.** b **9.** c **10.** b **11.** d **12.** (a) 4, (b) 8, (c) 5, (d) 2, (e) 3, (f) 6, (g) 1, (h) 7 **13.** (a) 3, (b) 5, (c) 8, (d) 12, (e) 9, (f) 7, (g) 11, (h) 6, (i) 2, (j) 4, (k) 10, (l) 1 **14.** (a) 2, (b) 5, (c) 3, (d) 4, (e) 1 **15.** (a) 3, (b) 2, (c) 1

Chapter 5
1. palms, soles, fingertips **2.** strength, extensibility, elasticity **3.** ceruminous **4.** melanin, carotene, hemoglobin **5.** a **6.** b **7.** a **8.** c **9.** c **10.** e **11.** (a) 3, (b) 4, (c) 1, (d) 2 **12.** (a) 5, (b) 13, (c) 3, (d) 7, (e) 4, (f) 2, (g) 6, (h) 8, (i) 10, (j) 12, (k) 11, (l) 1, (m) 9 **13.** (a) 3, (b) 5, (c) 4, (d) 1, (e) 2, (f) 6, (g) 7, (h) 8 **14.** (a) 3, (b) 1, (c) 2, (d) 5, (e) 4 **15.** (a) 4, (b) 3, (c) 2, (d) 1, inflammatory, migratory, proliferative, maturation

Chapter 6
1. hardness, tensile strength **2.** osteons (Haversian systems), trabeculae **3.** hyaline cartilage, loose fibrous connective tissue membranes **4.** true **5.** false **6.** b **7.** b **8.** a **9.** d **10.** a **11.** d **12.** (a) 3, (b) 9, (c) 8, (d) 1, (e) 5, (f) 4, (g) 6, (h) 7, (i) 2, (j) 10 **13.** (a) 10, (b) 2, (c) 7, (d) 1, (e) 8, (f) 3, (g) 6, (h) 4, (i) 9, (j) 5 **14.** (a) 9, (b) 7, (c) 6, (d) 1, (e) 4, (f) 2, (g) 5, (h) 3, (i) 8, (j) 10 **15.** (a) 2, (b) 5, (c) 3, (d) 4, (e) 1

Chapter 7
1. fontanels **2.** thoracic **3.** intervertebral discs **4.** pituitary gland **5.** false **6.** false **7.** a **8.** c **9.** b **10.** e **11.** d **12.** (a) 7, (b) 5, (c) 1, (d) 6, (e) 2, (f) 4, (g) 8, (h) 9, (i) 3, (j) 10, (k) 11, (l) 13, (m) 12 **13.** (a) 2, (b) 3, (c) 5, (d) 6, (e) 4, (f) 1, (g) 5, (h) 4, (i) 2, (j) 4, (k) 3 **14.** (a) 4, (b) 7, (c) 6, (d) 5, (e) 3, (f) 1, (g) 2 **15.** (a) 3, (b) 1, (c) 6, (d) 9, (e) 13, (f) 12, (g) 2, (h) 4, (i) 5, (j) 7, (k) 10, (l) 15, (m) 8, (n) 11, (o) 14

Chapter 8
1. protection, movement **2.** metacarpals **3.** true **4.** true **5.** b **6.** b **7.** d **8.** e **9.** e **10.** c **11.** a **12.** b **13.** b **14.** (a) 4, (b) 3, (c) 3, (d) 6, (e) 7, (f) 1, (g) 3, (h) 2, (i) 5, (j) 9, (k) 8, (l) 2, (m) 4, (n) 6, (o) 7 **15.** (a) 2, (b) 6, (c) 9, (d) 7, (e) 4, (f) 5, (g) 8, (h) 10, (i) 1, (j) 3

Chapter 9
1. joint, articulation, or arthrosis **2.** synovial fluid **3.** osteoarthritis **4.** d **5.** b **6.** a **7.** c **8.** e **9.** e **10.** true **11.** false **12.** (a) 5, (b) 3, (c) 6, (d) 1, (e) 4, (f) 7, (g) 2 **13.** (a) 4, (b) 3, (c) 6, (d) 2, (e) 5, (f) 1 **14.** (a) 6, (b) 4, (c) 5, (d) 1, (e) 3, (f) 2 **15.** (a) 6, (b) 7, (c) 9, (d) 4, (e) 8, (f) 3, (g) 10, (h) 2, (i) 11, (j) 1, (k) 5

Chapter 10
1. electrical excitability, contractility, extensibility, elasticity **2.** actin, myosin **3.** motor unit **4.** false **5.** true **6.** e **7.** a **8.** b **9.** e **10.** d **11.** c **12.** (a) 5, (b) 6, (c) 8, (d) 7, (e) 2, (f) 4, (g) 3, (h) 1 **13.** (a) 5, (b) 8, (c) 7, (d) 10, (e) 6, (f) 9, (g) 4, (h) 1, (i) 2, (j) 3 **14.** (a) 2, (b) 3, (c) 1, (d) 1 and 2 (e) 2 and 3, (f) 2, (g) 1, (h) 3, (i) 1 and 2, (j) 3 **15.** (a) 10, (b) 2, (c) 4, (d) 3, (e) 6, (f) 5, (g) 1, (h) 7, (i) 9, (j) 8

Chapter 11
1. origin, insertion **2.** buccinator **3.** true **4.** true **5.** d **6.** e **7.** a **8.** e **9.** c **10.** a **11.** b **12.** (a) 12, (b) 9, (c) 8, (d) 6, (e) 3, (f) 11, (g) 10, (h) 1, (i) 2, (j) 7, (k) 4, (l) 5 **13.** (a) 3, (b) 1, (c) 2, (d) 1, (e) 2, (f) 3, (g) 3 **14.** (a) 6, (b) 3, (c) 7, (d) 4, (e) 2, (f) 8, (g) 5, (h) 1 **15.** (a) 10, (b) 1, (c) 9, (d) 8, (e) 12, (f) 17, (g) 2, (h) 6, (i) 8, (j) 14, (k) 5, (l) 4, (m) 2, (n) 15, (o) 1, (p) 11, (q) 13, (r) 12, (s) 7, (t) 16, (u) 11, (v) 17, (w) 16, (x) 15, (y) 3

Chapter 12
1. somatic nervous system, autonomic nervous system, enteric nervous system **2.** polarized **3.** true **4.** false **5.** c **6.** b **7.** a **8.** e **9.** c **10.** e **11.** b **12.** a **13.** (a) 6, (b) 11, (c) 1, (d) 2, (e) 9, (f) 13, (g) 4, (h) 8, (i) 7, (j) 12, (k) 5, (l) 3, (m) 10 **14.** (a) 2, (b) 1, (c) 9, (d) 8, (e) 6, (f) 3, (g) 4, (h) 5, (i) 10, (j) 7 **15.** (a) 4, (b) 5, (c) 14, (d) 8, (e) 7, (f) 1, (g) 2, (h) 10, (i) 13, (j) 6, (k) 3, (l) 12, (m) 9, (n) 11

Chapter 13

1. reflexes **2.** mixed **3.** true **4.** true **5.** d **6.** e **7.** d **8.** a **9.** c **10.** (a) 1, (b) 8, (c) 4, (d) 2, (e) 10, (f) 1, (g) 6, (h) 5, (i) 3, (j) 9, (k) 1, (l) 11, (m) 7, (n) 2 **11.** a **12.** b **13.** e **14.** (a) 10, (b) 8, (c) 9, (d) 1, (e) 2, (f) 5, (g) 7, (h) 6, (i) 11, (j) 4, (k) 3 **15.** (a) 2, (b) 1, (c) 3, (d) 4, (e) 1, (f) 2, (g) 4, (h) 3, (i) 5, (j) 1

Chapter 14

1. choroid plexuses **2.** longitudinal fissue **3.** true **4.** false **5.** c **6.** d **7.** e **8.** d **9.** e **10.** a **11.** d **12.** (a) 3, (b) 5, (c) 6, (d) 8, (e) 11, (f) 10, (g) 7, (h) 9, (i) 1, (j) 4, (k) 2, (l) 12, (m) 1, (n) 8, (o) 5, (p) 7, (q) 12, (r) 10, (s) 9 **13.** (a) 5, (b) 9, (c) 11, (d) 6, (e) 3, (f) 1, (g) 10, (h) 8, (i) 2, (j) 4, (k) 7 **14.** (a) 9, (b) 2, (c) 6, (d) 10, (e) 4, (f) 11, (g) 1, (h) 2, (i) 5, (j) 8, (k) 12, (l) 7, (m) 3, (n) 6 and 8, (o) 13, (p) 7, (q) 1 **15.** (a) 9, (b) 2, (c) 6, (d) 8, (e) 7, (f) 5, (g) 3, (h) 10, (i) 12, (j) 11, (k) 4, (l) 1

Chapter 15

1. sensation, perception **2.** adaptation **3.** true **4.** true **5.** d **6.** b **7.** a **8.** d **9.** e **10.** e **11.** d **12.** d **13.** (a) 9, (b) 8, (c) 4, (d) 7, (e) 10, (f) 2, (g) 3, (h) 1, (i) 5, (j) 6, (k) 11 **14.** (a) 10, (b) 8, (c) 7, (d) 1, (e) 4, (f) 3, (g) 5, (h) 6, (i) 9, (j) 2 **15.** (a) 3, (b) 2, (c) 5, (d) 7, (e) 1, (f) 3, (g) 8, (h) 4, (i) 6

Chapter 16

1. sweet, sour, salty, bitter, umami **2.** static, dynamic **3.** true **4.** false **5.** a **6.** d **7.** c **8.** c **9.** c **10.** c **11.** e **12.** a **13.** (a) 1, (b) 5, (c) 7, (d) 6, (e) 1, (f) 8, (g) 2, (h) 4, (i) 3 **14.** (a) 3, (b) 6, (c) 8, (d) 12, (e) 1, (f) 5, (g) 9, (h) 11, (i) 7, (j) 13, (k) 2, (l) 10, (m) 4 **15.** (a) 2, (b) 11, (c) 14, (d) 13, (e) 3, (f) 10, (g) 6, (h) 12, (i) 4, (j) 5, (k) 9, (l) 1, (m) 7, (n) 8

Chapter 17

1. acetylcholine, epinephrine, or norepinephrine **2.** thoracolumbar, craniosacral **3.** true **4.** true **5.** b **6.** c **7.** d **8.** a **9.** a **10.** c **11.** d **12.** e **13.** (a) 2, (b) 5, (c) 6, (d) 1, (e) 3, (f) 4 **14.** (a) 3, (b) 2, (c) 1, (d) 1, (e) 2, (f) 3, (g) 3 **15.** (a) 2, (b) 1, (c) 1, (d) 2, (e) 1, (f) 1, (g) 2, (h) 2

Chapter 18

1. fight-or-flight response, resistance reaction, exhaustion **2.** hypothalamus **3.** exocrine, endocrine **4.** false **5.** true **6.** b **7.** c **8.** d **9.** a **10.** e **11.** a **12.** c **13.** (a) 8, (b) 2, (c) 7, (d) 1, (e) 12, (f) 18, (g) 5, (h) 17, (i) 20, (j) 15, (k) 3, (l) 16, (m) 19, (n) 6, (o) 13, (p) 11, (q) 4, (r) 10, (s) 14, (t) 9 **14.** (a) 10, (b) 8, (c) 2, (d) 4, (e) 1, (f) 6, (g) 9, (h) 7, (i) 5, (j) 3 **15.** (a) 12, (b) 1, (c) 11, (d) 7, (e) 3, (f) 10, (g) 2, (h) 9, (i) 4, (j) 8, (k) 5, (l) 6

Chapter 19

1. blood plasma, formed elements, red blood cells, white blood cells, platelets **2.** serum **3.** platelets, thrombocytes **4.** true **5.** false **6.** e **7.** a **8.** e **9.** a **10.** b **11.** c **12.** d **13.** (a) 6, (b) 8, (c) 3, (d) 10, (e) 2, (f) 4, (g) 1, (h) 7, (i) 9, (j) 5 **14.** (a) 5, (b) 1, (c) 2, (d) 3, (e) 6, (f) 8 (g) 9, (h) 4, (i) 7 **15.** (a) 4, (b) 6, (c) 8, (d) 1, (e) 7, (f) 3, (g) 5, (h) 2

Chapter 20

1. left ventricle **2.** systole, diastole **3.** stroke volume **4.** false **5.** false **6.** true **7.** a **8.** c **9.** b **10.** a **11.** b **12.** (a) 3, (b) 2, (c) 9, (d) 8, (e) 7, (f) 4, (g) 1, (h) 5, (i) 6 **13.** (a) 4, (b) 10, (c) 6, (d) 5, (e) 3, (f) 9, (g) 2, (h) 8, (i) 1, (j) 7 **14.** (a) 3, (b) 7, (c) 2, (d) 5, (e) 1, (f) 6, (g) 4 and 7 **15.** (a) 3, (b) 6, (c) 1, (d) 5, (e) 2, (f) 4

Chapter 21

1. carotid sinus, aortic **2.** skeletal muscle pump, respiratory pump **3.** true **4.** true **5.** a **6.** c **7.** e **8.** a **9.** d **10.** e **11.** c **12.** (a) 2, (b) 5, (c) 1, (d) 4, (e) 3 **13.** (a) 11, (b) 1, (c) 4, (d) 9, (e) 3, (f) 8, (g) 6, (h) 2, (i) 7, (j) 5, (k) 10, (l) 12 **14.** (a) 2, (b) 6, (c) 4, (d) 1, (e) 3, (f) 5 **15.** (a) 5, (b) 3, (c) 1, (d) 4, (e) 2, (f) 4, (g) 1, (h) 5

Chapter 22

1. skin, mucous membranes; antimicrobial proteins, NK cells, phagocytes **2.** specificity, memory **3.** true **4.** true **5.** e **6.** e **7.** d **8.** a **9.** c **10.** e **11.** b **12.** c **13.** (a) 3, (b) 1, (c) 4, (d) 2, (e) 5 **14.** (a) 2, (b) 3, (c) 4, (d) 7, (e) 1, (f) 6, (g) 5 **15.** (a) 11, (b) 11, (c) 8, (d) 1, (e) 2, (f) 5, (g) 4, (h) 7, (i) 9, (j) 12, (k) 3, (l) 6, (m) 10

Chapter 23

1. pulmonary ventilation, external respiration, internal respiration **2.** less, greater **3.** oxyhemoglogin; dissolved CO_2, carbamino compounds, and bicarbonate ion **4.** $CO_2 + H_2O \rightarrow H_2CO_3 \rightarrow H^+ + HCO_3^-$ **5.** false **6.** true **7.** b **8.** d **9.** a **10.** c **11.** e **12.** (a) 2, (b) 8, (c) 3, (d) 7, (e) 1, (f) 9, (g) 5, (h) 10, (i) 4, (j) 6 **13.** (a) 7, (b) 8, (c) 1, (d) 5, (e) 6, (f) 2, (g) 3, (h) 4 **14.** (a) 3, (b) 8, (c) 5, (d) 2, (e) 7, (f) 1, (g) 4, (h) 6 **15.** (a) 7, (b) 3, (c) 5, (d) 1, (e) 6, (f) 4, (g) 2

Chapter 24

1. mucosa, submucosa, muscularis, serosa **2.** submucosal plexus or plexus of Meissner; myenteric plexus or plexus of Auerbach **3.** monosaccharides; amino acids; monoglycerides, fatty acids; pentoses, phosphates, nitrogenous bases **4.** false **5.** false **6.** a **7.** c **8.** e **9.** b **10.** a **11.** b **12.** d **13.** (a) 3, (b) 10, (c) 5, (d) 8, (e) 6, (f) 1, (g) 4, (h) 9, (i) 11, (j) 2, (k) 7 **14.** (a) 4, (b) 6, (c) 7, (d) 1, (e) 5, (f) 3, (g) 8, (h) 2, (i) 10, (j) 9 **15.** (a) 4, (b) 8, (c) 2, (d) 10, (e) 7, (f) 11, (g) 1, (h) 6, (i) 3, (j) 9, (k) 5

Chapter 25

1. hypothalamus **2.** glucose 6-phosphate, pyruvic acid, acetyl coenzyme A **3.** hormones **4.** false **5.** true **6.** c **7.** d **8.** a **9.** d **10.** a **11.** a **12.** b **13.** (a) 2, (b) 8, (c) 5, (d) 10, (e) 6, (f) 7, (g) 3, (h) 9, (i) 4, (j) 1 **14.** (a) 9, (b) 12, (c) 11, (d) 10, (e) 4, (f) 7, (g) 3, (h) 5, (i) 8, (j) 1, (k) 6, (l) 2 **15.** (a) 8, (b) 5, (c) 1, (d) 10, (e) 6, (f) 2, (g) 4, (h) 9, (i) 3, (j) 7

Chapter 26

1. transitional epithelial **2.** voluntary, involuntary **3.** glomerular filtration, tubular reabsorption, tubular secretion **4.** true **5.** true **6.** c **7.** a **8.** d **9.** a **10.** e **11.** b **12.** c **13.** (a) 8, (b) 2, (c) 10, (d) 5, (e) 3, (f) 1, (g) 7, (h) 4, (i) 9, (j) 6 **14.** (a) 4, (b) 3, (c) 7, (d) 6, (e) 2, (f) 1, (g) 5 **15.** (a) 5, (b) 4, (c) 6, (d) 1, (e) 2, (f) 7, (g) 3

Chapter 27

1. intracellular fluid, extracellular fluid **2.** bicarbonate ion, carbonic acid **3.** weak acid, weak base **4.** protein buffer, carbonic acid-bicarbonate, phosphate **5.** true **6.** true **7.** a **8.** c **9.** b **10.** e **11.** d **12.** a **13.** b **14.** (a) 8, (b) 9, (c) 7, (d) 1, (e) 6, (f) 2, (g) 4, (h) 5, (i) 3 **15.** (a) 8, (b) 7, (c) 5, (d) 6, (e) 1, (f) 3, (g) 4, (h) 2

Chapter 28

1. puberty, menarche, menopause **2.** semen **3.** true **4.** true **5.** e **6.** b **7.** a **8.** c **9.** b **10.** a **11.** e **12.** (a) 7, (b) 2, (c) 1, (d) 4, (e) 8, (f) 5, (g) 3, (h) 6 **13.** (a) 6, (b) 4, (c) 2, (d) 1, (e) 3, (f) 8, (g) 7, (h) 5 **14.** (a) 6, (b) 4, (c) 1, (d) 12, (e) 8, (f) 5, (g) 7, (h) 13, (i) 11, (j) 3, (k) 2, (l) 10, (m) 9 **15.** (a) 10, (b) 12, (c) 1, (d) 5, (e) 2, (f) 4, (g) 6, (h) 11, (i) 8, (j) 3, (k) 9, (l) 7

Chapter 29

1. dilation, expulsion, placental **2.** corpus luteum **3.** mesoderm; ectoderm; endoderm **4.** true **5.** a **6.** b **7.** e **8.** (a) 3, (b) 4, (c) 5, (d) 1, (e) 2, (f) 6, (g) 7, (h) 8, (i) 9 **9.** e **10.** b **11.** (a) 7, (b) 3, (c) 6, (d) 4, (e) 10, (f) 8, (g) 5, (h) 2, (i) 1, (j) 9, (k) 11, (l) 12 **12.** b **13.** (a) 3, (b) 6, (c) 4, (d) 1, (e) 8, (f) 2, (g) 5, (h) 7 **14.** (a) 5, (b) 2, (c) 8, (d) 4, (e) 7, (f) 1, (g) 3, (h) 6 **15.** d

Answers to Critical Thinking Questions

Chapter 1

1. Homeostasis is the relative constancy of the body's internal environment. Body temperature should vary within a narrow range around normal body temperature (38°C or 98.6° F).
2. Bilateral means two sides or both arms in this case. The carpal region is the wrist area.
3. A x ray provides a good image of dense tissue such as bones. An MRI is used for imaging soft tissues, not bones. An MRI cannot be used when metal is present due to the magnetic field used.

Chapter 2

1. Fatty acids are in all lipids including plant oils and the phospholipids that compose cell membranes. Sugars (monosaccharides) are needed for ATP production, are a component of nucleotides, and are the basic units of disaccharides and polysaccharides including starch, glycogen, and cellulose.
2. DNA = deoxyribonucleic acid. DNA is composed of 4 different nucleotides. The order of the nucleotides is unique in every person. The 20 different amino acids form proteins.
3. One pH unit equals a tenfold change in H^+ concentration. Pure water has a pH of 7, which equals 1×10^{-7} moles of H^+ per liter. Blood has a pH of 7.4, which equals 0.4×10^{-7} moles H^+ per liter. Thus, blood has less than half as many H^+ as water.

Chapter 3

1. The tissues are destroyed due to autolysis of the cells caused by the release of acids and digestive enzymes from the lysosomes.
2. Synthesis of mucin by the ribosomes on rough endoplasmic reticulum, to transport vesicle, to entry face of Golgi, to transfer vesicle, to medial cisternae, where protein is modified, to transfer vesicle, to exit face, to secretory vesicle, to plasma membrane, where it undergoes exocytosis.
3. The sense strand is composed of DNA codons transcribed into mRNA. Only the exons will be found in the mRNA. The introns are clipped out.

Chapter 4

1. Many possible adaptations including: more adipose tissue for insulation, thicker bone for support, more red blood cells for oxygen transport, and so on.
2. The surface layer of the skin, the epidermis, is keratinized, stratified squamous epithelium. It is avascular. If the pin were stuck straight in, the vascularized connective tissue would be pierced and the finger would bleed.
3. The gelatin portion of the salad represents the ground substance of the connective tissue matrix. The grapes are cells such as fibroblasts. The shredded carrots and coconut are fibers embedded in the ground substance.

Chapter 5

1. The dust particles are mostly keratinocytes shed from the stratum corneum of the skin.
2. It would be neither wise nor feasible to remove the exocrine glands. Essential exocrine glands include the sudoriferous glands (sweat helps control body temperature), sebaceous glands (sebum lubricates the skin), and ceruminous glands (provide protective lubricant for ear canal).
3. The superficial layer of the epidermis of the skin is the stratum corneum. Its cells are full of intermediate filaments of keratin, keratohyalin, and lipids from lamellar granules, making this layer a water-repellent barrier. The epidermis also contains an abundance of desmosomes.

Chapter 6

1. In greenstick fracture, which occurs only in children due to the greater flexibility of their bones, the bone breaks on one side but bends on the other side, resembling what happens when one tries to break a green (not dry) stick. Lynne Marie probably broke her fall with her extended arm, breaking the radius or ulna.
2. At Aunt Edith's age, the production of several hormones (such as estrogens and human growth hormone) necessary for bone remodeling is likely to be decreased. Her decreased height results, in part, from compression of her vertebrae due to flattening of the discs between them. She may also have osteoporosis and an increased susceptibility to fractures resulting in damage to the vertebrae and loss of height.
3. Exercise causes mechanical stress on bones, but because there is effectively zero gravity in space, the pull of gravity on bones is missing. The lack of stress from gravity results in bone demineralization and weakness.

Chapter 7

1. Fontanels, the soft spots between cranial bones, are fibrous connective tissue membranes that will be replaced by bone as the baby matures. They allow the infant's head to be molded during its passage through the birth canal and allow for brain and skull growth in the infant.
2. Barbara probably broke her coccyx or tailbone. The coccygeal vertebrae usually fuse by age 30. The coccyx points inferiorly in females.
3. Infants are born with a single concave curve in the vertebral column. Adults have four curves in their spinal cord-at the cervical (convex), thoracic (concave), lumbar (convex), and sacral regions. Buying a mattress from this company and sleeping on it would result in potentially severe back problems.

Chapter 8

1. Clawfoot is an abnormal elevation of the medial longitudinal arch of the foot.
2. There are 14 phalanges in each hand: two bones in the thumb and three in each of the other fingers. Farmer White has lost five phalanges on his left hand so he has nine remaining on his left and 14 remaining on his right for a total of 23.
3. Snakes don't have appendages, so they don't need shoulders and hips. Pelvic and pectoral girdles connect the upper and lower limbs (appendages) to the axial skeleton, and thus are considered part of the appendicular skeleton.

Chapter 9

1. Katie's vertebral column, head, thigh, lower leg, and lower arm are all flexed. Her lower arm and shoulder are medially rotated.
2. The elbow is a hinge-type synovial joint. The trochlea of the humerus articulates with the trochlear notch of the ulna. The elbow has monaxial (opening and closing) movement similar to a door.
3. Cartilaginous joints can be composed of hyaline cartilage, such as at the epiphyseal plate, or fibrocartilage, such as the intervertebral disc. Sutures are an example of fibrous joints, which are composed of dense fibrous tissue. Synovial (diarthrotic) joints are held together by an articular capsule, which has an inner synovial membrane layer.

Chapter 10

1. Smooth muscle contains both thick and thin filaments as well as intermediate filaments attached to dense bodies. The smooth muscle fiber contracts like a corkscrew turns; the fiber twists like a helix as it contracts and shortens.

2. Ming's muscles performed isometric contractions. Peng used primarily isotonic concentric contraction to lift, isometric to hold, and then isotonic eccentric contractions to lower the barbell.

3. Mr. Klopfer's students have muscle fatigue in their hands. Fatigue has many causes including an increase in lactic acid and ADP, and a decrease in Ca^{2+}, oxygen, creatine phosphate, and glycogen.

Chapter 11

1. Some of the antagonistic pairs for the regions are: upper arm-biceps brachii vs. triceps brachii; upper leg-quadriceps femoris vs. hamstrings; torso-rectus abdominus vs. erector spinae; lower leg-gastrocnemius vs. tibialis anterior.

2. The fulcrum is the knee joint, the load (resistance) is the weight of the upper body and the package, and the effort is the thigh muscles. This is a third-class lever.

3. Lifting the eyebrow uses the frontal belly of the occipitofrontalis; whistling uses the buccinator and orbicularis oris; closing the eyes uses the orbicularis oculi; shaking the head can use several muscles, including sternocleidomastoid, semispinalis capitis, splenius capitis, and longissimus capitis.

Chapter 12

1. The motor neuron is multipolar in structure with an axon and several dendrites projecting from the cell body. The converging circuit has several neurons that converge to form synapses with one common neuron. The bipolar neuron has one dendrite and an axon projecting from the cell body. In a simple circuit, each presynaptic neuron makes a single synapse with one postsynaptic neuron.

2. Gray matter appears gray in color due to the absence of myelin. It is composed of cell bodies, and unmyelinated axons and dendrites. White matter contains many myelinated axons. Lipofuscin, a yellowing pigment, collects in neurons with age.

3. Smelling coffee and hearing alarm are somatic sensory, stretching and yawning are somatic motor, salivating is autonomic (parasympathetic) motor, stomach rumble is enteric motor.

Chapter 13

1. A withdrawal/flexor reflex is ipsilateral and polysynaptic. The route is pain receptor of sensory neuron → spinal cord (integrating center) interneurons → motor neurons → flexor muscles in the leg. Also present is a crossed extensor reflex, which is contralateral and polysynaptic. The route diverges at the spinal cord: Interneurons cross to the other side → motor neurons → extensor muscles in the opposite leg.

2. The spinal cord is anchored in place by the filum terminale and the denticulate ligaments.

3. The needle will pierce the epidermis, the dermis, and the subcutaneous layer and then go between the vertebrae through the epidural space, the dura mater, the subdural space, the arachnoid mater, and into the CSF in the subarachnoid space. CSF is produced in the brain, and the spinal meninges are continuous with cranial meninges.

Chapter 14

1. Movement of the right arm is controlled by the left hemisphere's primary motor area, located in the precentral gyrus. Speech is controlled by Broca's area in the left hemisphere's frontal lobe just superior to the lateral cerebral sulcus.

2. The brain is enclosed by the cranial bones and meninges. The temporal bone houses the middle ear and inner ear, separating these from the brain.

3. The dentist has injected anesthetic into the inferior alveolar nerve, a branch of the mandibular nerve, which numbs the lower teeth and the lower lip. The tongue is numbed by blocking the lingual nerve. The upper teeth and lip are anesthetized by injecting the superior alveolar nerve, a branch of the maxillary nerve.

Chapter 15

1. Chemoreceptors in the nose detect odors. Proprioceptors detect body position and are involved in equilibrium. The receptors for smell are rapidly adapting (phasic), whereas proprioceptors are slowly adapting (tonic).

2. The tickle receptors are free nerve endings in the foot. The impulses travel along the first-order sensory neurons to the posterior gray horn, then along the second-order neurons, crossing to the other side of the spinal cord and up the anterior spinothalamic tract to the thalamus. The third-order neurons extend from the thalamus to the "foot" region of the somatosensory area of the cerebral cortex.

3. Yoshio's perception of feeling in his amputated foot is called phantom limb sensation. Impulses from the remaining proximal section of the sensory neuron are perceived by the brain as still coming from the amputated foot.

Chapter 16

1. Because smell and taste have ties to the cortex and limbic areas, Brenna may be recalling a memory of this or similar food. Pathway: olfactory receptors (cranial nerve I) → olfactory bulbs → olfactory tracts → lateral olfactory area in temporal lobe of cerebral cortex.

2. Fred may have cataracts, a loss of transparency in the lens of the eye. Cataracts are often associated with age, smoking, and exposure to UV light.

3. Auricle → external auditory meatus → tympanic membrane → maleus → incus → stapes → oval window → perilymph of scala vestibuli and tympani → vestibular membrane → endolymph of cochlear duct → basilar membrane → spiral organ.

Chapter 17

1. Digestion and relaxation are controlled by increased stimulation of the parasympathetic division of the ANS. The salivary glands, pancreas, and liver will show increased secretion; the stomach and intestines will have increased activity; the gallbladder will have increased contractions; heart contractions will have decreased force and rate.

2. Stretch receptors in the colon → autonomic sensory neurons → sacral region of spinal cord (integrating center) → parasympathetic preganglionic neuron → terminal ganglion → parasympathetic postganglionic neuron → smooth muscle in colon, rectum, and sphincter (effectors).

3. Nicotine from the cigarette smoke binds to nicotinic receptors on skeletal muscles, mimicking the effect of acetylcholine and causing increased contractions ("twitches"). Nicotine also binds to nicotinic receptors on cells in the adrenal medulla, stimulating release of epinephrine and norepinephrine, which mimics fight-or-flight responses.

Chapter 18

1. The beta cells are one of the cell types in the pancreatic islets of the pancreas. In Type 1 diabetes, only the beta cells are destroyed; the rest of the pancreas is not affected. A successful transplantation of

the beta cells would enable the recipient to produce the hormone insulin and would cure the diabetes.

2. Amanda has an enlarged thyroid gland or goiter. The goiter is probably due to hypothyroidism, which is causing the weight gain, fatigue, mental dullness, and other symptoms.

3. The two adrenal glands are located superior to the two kidneys. The adrenals are about 4 cm high, 2 cm wide, and 1 cm thick. The outer adrenal cortex is composed of three layers: zona glomerulosa, zona fasciculata, and zona reticularis. The inner adrenal medulla is composed of chromaffin cells.

Chapter 19

1. Blood is composed of liquid blood plasma and the formed elements: red blood cells (erythrocytes), white blood cells (leukocytes), and platelets. Blood plasma contains water, proteins, electrolytes, gases, and nutrients. There are about 5 million RBCs, 5000 WBCs, and 250,000 platelets per mL of blood. Leukocytes include neutrophils, eosinophils, basophils, monocytes, and lymphocytes.

2. To determine the blood type, a drop of each of three different antibody solutions (antisera) are added to three separate blood drops. The solutions are anti-A, anti-B, and anti-Rh. In Josef's test, anti-B caused agglutination (clump formation) when added to his blood, indicating the presence of antigen B on the RBCs. Anti-A and anti-Rh did not cause agglutination, indicating the absence of these antigens.

3. Hemostasis occurred. The stages involved are vascular spasm (if an arteriole was cut), platelet plug formation, and coagulation (clotting).

Chapter 20

1. With a normal resting CO of about 5.25 liters/min, and a heart rate of 55 beats per minute, Arian's stroke volume would be 95 mL/beat. His CO during strenuous exercise is 6 times his resting CO, about 31,500 mL/min.

2. Rheumatic fever is caused by inflammation of the bicuspid and the aortic valves following a streptococcal infection. Antibodies produced by the immune system to destroy the bacteria may also attack and damage the heart valves. Heart problems later in life may be related to this damage.

3. Mr. Perkins is suffering from angina pectoris and has several risk factors for coronary artery disease such as smoking, obesity, and male gender. Cardiac angiography involves the use of a cardiac catheter to inject a radiopaque medium into the heart and its vessels. The angiogram may reveal blockages such as atherosclerotic plaques in his coronary arteries.

Chapter 21

1. After birth, the foramen ovale and ductus arteriosus close to establish the pulmonary circulation. The umbilical artery and veins close because the placenta is no longer functioning. The ductus venosus closes so that the liver is no longer bypassed.

2. Right hand: left ventricle → ascending aorta → arch of aorta → brachiocephalic trunk → right subclavian artery → right axillary artery → right brachial artery → right radial and ulnar arteries → right superficial palmar arch. Left hand: left ventricle → ascending aorta → arch of aorta → left subclavian artery → left axillary artery → left brachial artery → left radial and ulnar arteries → left superficial palmar arch.

3. A vascular (venous) sinus is a vein that lacks smooth muscle in its tunica media. Dense connective tissue replaces the tunica media and tunica externa.

Chapter 22

1. Tariq had a hypersensitivity reaction to an insect sting; he was probably allergic to the venom. He had a localized anaphylactic or type I reaction.

2. The site of the embedded splinter had probably become infected with bacteria. The red streaks are the lymphatic vessels that drain the infected area; the swollen tender bumps are the axillary lymph nodes, which are swollen due to the immune response to the infection.

3. Influenza vaccination introduces a weakened or killed virus (which will not cause the disease) to the body. The immune system recognizes the antigen and mounts a primary immune response. Upon exposure to the same flu virus that was in the vaccine, the body will produce a secondary response, which will usually prevent a case of the flu. This is artificially acquired active immunity.

Chapter 23

1. Normal average male volumes are 500 mL for tidal volume during quiet breathing, 3100 mL for inspiratory reserve volume, and 1200 mL for expiratory reserve volume. Average female volumes are less than average male volumes, because females are generally smaller than males.

2. The bones of the external nose are the frontal bone, the two nasal bones (the location of the fracture), and the maxillae. The external nose is also composed of the septal, lateral nasal, and alar cartilages, as well as skin, muscle, and mucous membrane.

3. The cerebral cortex can temporarily allow voluntary breath holding. P_{CO_2} and H^+ levels in the blood and CSF increase with breath holding, strongly stimulating the inspiratory area, which stimulates inspiration to begin again. Breathing will begin again even if the person loses consciousness.

Chapter 24

1. Katie lost the two upper central permanent incisors. The remaining deciduous teeth include the lower central incisors, an upper and lower pair of lateral incisors, an upper and lower pair of cuspids, and upper and lower pair of first molars, and upper and lower pair of second molars.

2. CCK promotes ejection of bile, secretion of pancreatic juice, and contraction of the pyloric sphincter. It also promotes pancreatic growth and enhances the effects of secretin. CCK acts on the hypothalamus to induce satiety (the feeling of fullness) and should therefore decrease the appetite.

3. The smaller left lobe of the liver is separated from the larger right lobe by the falciform ligament. The left lobe is inferior to the diaphragm in the epigastric region of the abdominopelvic cavity.

Chapter 25

1. Deb is eating a diet high in carbohydrates in order to store maximum amounts of glycogen in her skeletal muscles and liver. This practice is called carbohydrate loading. Glycogenolysis, the breakdown of stored glycogen, supplies the muscles with the glucose needed for ATP production via cellular respiration.

2. Mr. Hernandez was suffering from heat exhaustion caused by loss of fluids and electrolytes. Lack of NaCl causes muscle cramps, nausea and vomiting, dizziness, and fainting. Low blood pressure may also result.

3. Sara's normal growth may be maintained despite day-to-day fluctuations in food intake. Many factors govern food intake, including neurons in the hypothalamus, blood glucose level, amount of adipose tissue, CCK, distention of the GI tract, and body temperature.

Chapter 26

1. Without reabsorption, initially 105–125 mL of filtrate would be lost per minute, assuming normal glomerular filtration rate. Fluid loss from the blood would cause a decrease in blood pressure, and therefore a decrease in GBHP. When GBHP dropped below 45mmHg, filtration would stop (assuming normal CHP and BCOP) because NFP would be zero.

2. The urinary bladder can stretch considerably due to the presence of transitional epithelium, rugae, and three layers of smooth muscle in the detrusor muscle.

3. In females the urethra is about 4 cm long; in males the urethra is 15–20 cm long, including its passage through the penis, urogenital diaphragm, and prostate.

Chapter 27

1. Gary has water intoxication. The Na^+ concentration of his plasma and interstitial fluid is below normal. Water moved by osmosis into the cells, resulting in hypotonic intracellular fluid and water intoxication. Decreased plasma volume, due to water movement into the interstitial fluid, caused hypovolemic shock.

2. The average female body contains about 55% water, versus 60% water in the average male. Due to the influence of male and female hormones, the average female has relatively more subcutaneous fat (which contains very little water) and relatively less muscle and other tissues (which have higher water content) than the average male.

3. Excessive vomiting causes a loss of hydrochloric acid in gastric juice, and intake of antacids increases the amount of alkali in the body fluids, resulting in metabolic alkalosis. Vomiting also causes a loss of fluid and may cause dehydration.

4. (Step 1) pH = 7.30 indicates slight acidosis, which could be caused by elevated P_{CO_2} or lowered HCO_3^-. (Step 2) The HCO_3^- is lower than normal (20 mEq/liter), so (step 3) the cause is metabolic. (Step 4) the P_{CO_2} is lower than normal (32 mmHg), so hyperventilation is providing some compensation. Diagnosis: Henry has partially compensated metabolic acidosis. A possible cause is kidney damage that resulted from interruption of blood flow during the heart attack.

Chapter 28

1. LH (luteinizing hormone) is an anterior pituitary hormone that acts on both the male and female reproductive systems. In males, LH stimulates the Leydig cells of the testes to secrete testosterone.

2. The germinal epithelium, located on the surface of the ovaries, is a misnomer. The oocytes are located in the ovarian cortex, deep to the tunica albuginea. The tunica vaginalis covers the testes, superficial to the tunica albuginea. It forms from the peritoneum during the descent of the testes from the abdomen into the scrotum during fetal development.

3. No. The first polar body, formed by meiosis I, would contain half the homologous chromosome pairs, the other half being in the secondary oocyte. The second polar body, formed by meiosis II, would contain chromosomes identical to the ovum; however, fertilization would be by two different sperm, so the babies would still not be identical.

Chapter 29

1. Being able to taste PCT is an autosomal dominant trait. Because Kendra is a PCT nontaster, her genotype is homozygous recessive for this trait. Her parents must both be heterozygous for the trait. They exhibit the dominant phenotype for PCT tasting and are carriers of the recessive trait.

2. Because Huntington disease (HD) is a trinucleotide repeat disease, in which the number of repeats expand with each succeeding generation, it can occur in a child whose parents did not suffer HD. Also, because it is a dominant gene, one mutant allele of the *HD* gene is sufficient to cause HD.

3. All arteries carry blood away from the heart. The umbilical arteries, located in the fetus, carry deoxygenated blood away from the fetal heart towards the placenta where the blood will become oxygenated as it passes through the placenta.

Glossary

Pronunciation Key

1. The most strongly accented syllable appears in capital letters, for example, bilateral (bī-LAT-er-al) and diagnosis (dī-ag-NŌ-sis).
2. If there is a secondary accent, it is noted by a prime ('), for example, constitution (kon'-sti-TOO-shun) and physiology (fiz'-ē-OL-ō-jē). Any additional secondary accents are also noted by a prime, for example, decarboxylation (dē'-kar-bok'-si-LĀ-shun).
3. Vowels marked by a line above the letter are pronounced with the long sound, as in the following common words:

   ā as in māke   ō as in pōle
   ē as in bē   ū as in cute
   ī as in īvy
4. Vowels not marked by a line above the letter are pronounced with the short sound, as in the following words:

   a as in above or at   o as in not
   e as in bet   u as in bud
   i as in sip

5. Other vowel sounds are indicated as follows:

   oy as in oil
   oo as in root
6. Consonant sounds are pronounced as in the following words:

   b as in bat   m as in mother
   ch as in chair   n as in no
   d as in dog   p as in pick
   f as in father   r as in rib
   g as in get   s as in so
   h as in hat   t as in tea
   j as in jump   v as in very
   k as in can   w as in welcome
   ks as in tax   z as in zero
   kw as in quit   zh as in lesion
   l as in let

A

Abdomen (ab-DŌ-men or AB-dō-men) The area between the diaphragm and pelvis.

Abdominal (ab-DŌM-i-nal) **cavity** Superior portion of the abdominopelvic cavity that contains the stomach, spleen, liver, gallbladder, most of the small intestine, and part of the large intestine.

Abdominal thrust maneuver A first-aid procedure for choking. Employs a quick, upward thrust against the diaphragm that forces air out of the lungs with sufficient force to eject any lodged material. Also called the **Heimlich** (HĪM-lik) **maneuver.**

Abdominopelvic (ab-dom'-i-nō-PEL-vic) **cavity** Inferior component of the ventral body cavity that is subdivided into a superior abdominal cavity and an inferior pelvic cavity.

Abduction (ab-DUK-shun) Movement away from the midline of the body.

Abortion (a-BOR-shun) The premature loss (spontaneous) or removal (induced) of the embryo or nonviable fetus; miscarriage due to a failure in the normal process of developing or maturing.

Abscess (AB-ses) A localized collection of pus and liquefied tissue in a cavity.

Absorption (ab-SORP-shun) Intake of fluids or other substances by cells of the skin or mucous membranes; the passage of digested foods from the gastrointestinal tract into blood or lymph.

Absorptive (fed) state Metabolic state during which ingested nutrients are absorbed into the blood or lymph from the gastrointestinal tract.

Accessory duct A duct of the pancreas that empties into the duodenum about 2.5 cm (1 in.) superior to the ampulla of Vater (hepatopancreatic ampulla). Also called the **duct of Santorini** (san'-tō-RĒ-nē).

Accommodation (a-kom-ō-DĀ-shun) An increase in the curvature of the lens of the eye to adjust for near vision.

Acetabulum (as'-e-TAB-ū-lum) The rounded cavity on the external surface of the hip bone that receives the head of the femur.

Acetylcholine (as'-e-til-KŌ-lēn) **(ACh)** A neurotransmitter liberated by many peripheral nervous system neurons and some central nervous system neurons. It is excitatory at neuromuscular junctions but inhibitory at some other synapses (for example, it slows heart rate).

Achalasia (ak'-a-LĀ-zē-a) A condition, caused by malfunction of the myenteric plexus, in which the lower esophageal sphincter fails to relax normally as food approaches. A whole meal may become lodged in the esophagus and enter the stomach very slowly. Distension of the esophagus results in chest pain that is often confused with pain originating from the heart.

Acid (AS-id) A proton donor, or a substance that dissociates into hydrogen ions (H^+) and anions; characterized by an excess of hydrogen ions and a pH less than 7.

Acidosis (as-i-DŌ-sis) A condition in which blood pH is below 7.35. Also known as **acidemia.**

Acini (AS-i-nē) Groups of cells in the pancreas that secrete digestive enzymes.

Acoustic (a-KOOS-tik) Pertaining to sound or the sense of hearing.

Acquired immunodeficiency syndrome (AIDS) A fatal disease caused by the human immunodeficiency virus (HIV). Characterized by a positive HIV-antibody test, low helper T cell count, and certain indicator diseases (for example Kaposi's sarcoma,

Pneumocystis carinii pneumonia, tuberculosis, fungal diseases). Other symptoms include fever or night sweats, coughing, sore throat, fatigue, body aches, weight loss, and enlarged lymph nodes.

Acrosome (AK-rō-sōm) A lysosomelike organelle in the head of a sperm cell containing enzymes that facilitate the penetration of a sperm cell into a secondary oocyte.

Actin (AK-tin) A contractile protein that is part of thin filaments in muscle fibers.

Action potential An electrical signal that propagates along the membrane of a neuron or muscle fiber (cell); a rapid change in membrane potential that involves a depolarization followed by a repolarization. Also called a **nerve action potential** or **nerve impulse** as it relates to a neuron, and a **muscle action potential** as it relates to a muscle fiber.

Activation (ak′-ti-VĀ-shun) **energy** The minimum amount of energy required for a chemical reaction to occur.

Active transport The movement of substances across cell membranes against a concentration gradient, requiring the expenditure of cellular energy (ATP).

Acute (a-KŪT) Having rapid onset, severe symptoms, and a short course; not chronic.

Adaptation (ad′-ap-TĀ-shun) The adjustment of the pupil of the eye to changes in light intensity. The property by which a sensory neuron relays a decreased frequency of action potentials from a receptor, even though the strength of the stimulus remains constant; the decrease in perception of a sensation over time while the stimulus is still present.

Adduction (ad-DUK-shun) Movement toward the midline of the body.

Adenoids (AD-e-noyds) The pharyngeal tonsils.

Adenosine triphosphate (a-DEN-ō-sēn trī-FOS-fāt) **(ATP)** The main energy currency in living cells; used to transfer the chemical energy needed for metabolic reactions. ATP consists of the purine base *adenine* and the five-carbon sugar *ribose*, to which are added, in linear array, three *phosphate* groups.

Adenylate cyclase (a-DEN-i-lāt SĪ-klās) An enzyme that is activated when certain neurotransmitters or hormones bind to their receptors; the enzyme that converts ATP into cyclic AMP, an important second messenger.

Adhesion (ad-HĒ-zhun) Abnormal joining of parts to each other.

Adipocyte (AD-i-pō-sīt) Fat cell, derived from a fibroblast.

Adipose (AD-i-pōz) **tissue** Tissue composed of adipocytes specialized for triglyceride storage and present in the form of soft pads between various organs for support, protection, and insulation.

Adrenal cortex (a-DRĒ-nal KOR-teks) The outer portion of an adrenal gland, divided into three zones; the zona glomerulosa secretes mineralocorticoids, the zona fasciculata secretes glucocorticoids, and the zona reticularis secretes androgens.

Adrenal glands Two glands located superior to each kidney. Also called the **suprarenal** (soo′-pra-RĒ-nal) **glands.**

Adrenal medulla (me-DUL-a) The inner part of an adrenal gland, consisting of cells that secrete epinephrine, norepinephrine, and a small amount of dopamine in response to stimulation by sympathetic preganglionic neurons.

Adrenergic (ad′-ren-ER-jik) **neuron** A neuron that releases epinephrine (adrenaline) or norepinephrine (noradrenaline) as its neurotransmitter.

Adrenocorticotropic (ad-rē′-nō-kor-ti-kō-TRŌP-ik) **hormone (ACTH)** A hormone produced by the anterior pituitary that influences the production and secretion of certain hormones of the adrenal cortex.

Adventitia (ad-ven-TISH-a) The outermost covering of a structure or organ.

Aerobic (air-Ō-bik) Requiring molecular oxygen.

Afferent arteriole (AF-er-ent ar-TĒ-rē-ōl) A blood vessel of a kidney that divides into the capillary network called a glomerulus; there is one afferent arteriole for each glomerulus.

Agglutination (a-gloo′-ti-NĀ-shun) Clumping of microorganisms or blood cells, typically due to an antigen–antibody reaction.

Aggregated lymphatic follicles Clusters of lymph nodules that are most numerous in the ileum. Also called **Peyer's** (PĪ-erz) **patches.**

Albinism (AL-bin-izm) Abnormal, nonpathological, partial, or total absence of pigment in skin, hair, and eyes.

Albumin (al-BŪ-min) The most abundant (60%) and smallest of the plasma proteins; it is the main contributor to blood colloid osmotic pressure (BCOP).

Aldosterone (al-DOS-ter-ōn) A mineralocorticoid produced by the adrenal cortex that promotes sodium and water reabsorption by the kidneys and potassium excretion in urine.

Alkaline (AL-ka-līn) Containing more hydroxide ions (OH^-) than hydrogen ions (H^+); a pH higher than 7.

Alkalosis (al-ka-LŌ-sis) A condition in which blood pH is higher than 7.45. Also known as **alkalemia.**

Allantois (a-LAN-tō-is) A small, vascularized outpouching of the yolk sac that serves as an early site for blood formation and development of the urinary bladder.

Alleles (a-LĒLZ) Alternate forms of a single gene that control the same inherited trait (such as type A blood) and are located at the same position on homologous chromosomes.

Allergen (AL-er-jen) An antigen that evokes a hypersensitivity reaction.

Alpha (AL-fa) **cell** A type of cell in the pancreatic islets (islets of Langerhans) in the pancreas that secretes the hormone glucagon. Also termed an **A cell.**

Alpha receptor A type of receptor for norepinephrine and epinephrine; present on visceral effectors innervated by sympathetic postganglionic neurons.

Alveolar duct Branch of a respiratory bronchiole around which alveoli and alveolar sacs are arranged.

Alveolar macrophage (MAK-rō-fāj) Highly phagocytic cell found in the alveolar walls of the lungs. Also called a **dust cell.**

Alveolar (al-VĒ-ō-lar) **pressure** Air pressure within the lungs. Also called **intrapulmonic pressure.**

Alveolar sac A cluster of alveoli that share a common opening.

Alveolus (al-VĒ-ō-lus) A small hollow or cavity; an air sac in the lungs; milk-secreting portion of a mammary gland. *Plural is* **alveoli** (al-VĒ-ol-ī).

Alzheimer (ALTZ-hī-mer) **disease (AD)** Disabling neurological disorder characterized by dysfunction and death of specific cerebral neurons, resulting in widespread intellectual impairment, personality changes, and fluctuations in alertness.

Amnesia (am-NE-zē-a) A lack or loss of memory.

Amenorrhea (ā-men-ō-RĒ-a) Absence of menstruation.

Amino (a-MĒ-nō) **acid** An organic acid, containing an acidic carboxyl group ($-COOH$) and a basic amino group ($-NH_2$); the monomer used to synthesize polypeptides and proteins.

Amnion (AM-nē-on) A thin, protective fetal membrane that develops from the epiblast; holds the fetus suspended in amniotic fluid. Also called the "**bag of waters.**"

Amniotic (am′-nē-OT-ik) **fluid** Fluid in the amniotic cavity, the space between the developing embryo (or fetus) and amnion; the fluid is initially produced as a filtrate from maternal blood and later includes fetal urine. It functions as a shock absorber, helps regulate fetal body temperature, and helps prevent desiccation.

Amphiarthrosis (am′-fē-ar-THRŌ-sis) A slightly movable joint, in which the articulating bony surfaces are separated by fibrous connective tissue or fibrocartilage to which both are attached; types are syndesmosis and symphysis.

Ampulla (am-PUL-la) A saclike dilation of a canal or duct.

Anabolism (a-NAB-ō-lizm) Synthetic, energy-requiring reactions whereby small molecules are built up into larger ones.

Anaerobic (an-ār-Ō-bik) Not requiring oxygen.

Anal (Ā-nal) **canal** The last 2 or 3 cm (1 in.) of the rectum; opens to the exterior through the anus.

Anal column A longitudinal fold in the mucous membrane of the anal canal that contains a network of arteries and veins.

Anal triangle The subdivision of the female or male perineum that contains the anus.

Analgesia (an-al-JĒ-zē-a) Pain relief; absence of the sensation of pain.

Anaphylaxis (an′-a-fi-LAK-sis) A hypersensitivity (allergic) reaction in which IgE antibodies attach to mast cells and basophils, causing them to produce mediators of anaphylaxis (histamine, leukotrienes, kinins, and prostaglandins) that bring about increased blood permeability, increased smooth muscle contraction, and increased mucus production. Examples are hay fever, hives, and anaphylactic shock.

Anaphase (AN-a-fāz) The third stage of mitosis in which the chromatids that have separated at the centromeres move to opposite poles of the cell.

Anastomosis (a-nas-tō-MŌ-sis) An end-to-end union or joining of blood vessels, lymphatic vessels, or nerves.

Anatomical (an′-a-TOM-i-kal) **position** A position of the body universally used in anatomical descriptions in which the body is erect, the head is level, the eyes face forward, the upper limbs are at the sides, the palms face forward, and the feet are flat on the floor.

Anatomic dead space Spaces of the nose, pharynx, larynx, trachea, bronchi, and bronchioles totaling about 150 mL of the 500 mL in a quiet breath (tidal volume); air in the anatomic dead space does not reach the alveoli to participate in gas exchange.

Anatomy (a-NAT-ō-mē) The structure or study of structure of the body and the relation of its parts to each other.

Androgens (AN-drō-jenz) Masculinizing sex hormones produced by the testes in males and the adrenal cortex in both sexes; responsible for libido (sexual desire); the two main androgens are testosterone and dihydrotestosterone.

Anemia (a-NĒ-mē-a) Condition of the blood in which the number of functional red blood cells or their hemoglobin content is below normal.

Anesthesia (an′-es-THĒ-zē-a) A total or partial loss of feeling or sensation; may be general or local.

Aneuploid (an′-ū-PLOYD) A cell that has one or more chromosomes of a set added or deleted.

Aneurysm (AN-ū-rizm) A saclike enlargement of a blood vessel caused by a weakening of its wall.

Angina pectoris (an-JI-na *or* AN-ji-na PEK-tō-ris) A pain in the chest related to reduced coronary circulation due to coronary artery disease (CAD) or spasms of vascular smooth muscle in coronary arteries.

Angiotensin (an-jē-ō-TEN-sin) Either of two forms of a protein associated with regulation of blood pressure. Angiotensin I is produced by the action of renin on angiotensinogen and is converted by the action of ACE (angiotensin-converting enzyme) into angiotensin II, which stimulates aldosterone secretion by the adrenal cortex, stimulates the sensation of thirst, and causes vasoconstriction with resulting increase in systemic vascular resistance.

Anion (AN-ī-on) A negatively charged ion. An example is the chloride ion (Cl^-).

Ankylosis (ang′-ki-LŌ-sis) Severe or complete loss of movement at a joint as the result of a disease process.

Anoxia (an-OK-sē-a) Deficiency of oxygen.

Antagonist (an-TAG-ō-nist) A muscle that has an action opposite that of the prime mover (agonist) and yields to the movement of the prime mover.

Antagonistic (an-tag-ō-NIST-ik) **effect** A hormonal interaction in which the effect of one hormone on a target cell is opposed by another hormone. For example, calcitonin (CT) lowers blood calcium level, whereas parathyroid hormone (PTH) raises it.

Anterior (an-TĒR-ē-or) Nearer to or at the front of the body. Equivalent to **ventral** in bipeds.

Anterior pituitary Anterior lobe of the pituitary gland. Also called the **adenohypophysis** (ad′-e-nō-hī-POF-i-sis).

Anterior root The structure composed of axons of motor (efferent) neurons that emerges from the anterior aspect of the spinal cord and extends laterally to join a posterior root, forming a spinal nerve. Also called a **ventral root.**

Anterolateral (an′-ter-ō-LAT-er-al) **pathway** Sensory pathway that conveys information related to pain, temperature, crude touch, pressure, tickle, and itch.

Antibody (AN-ti-bod′-ē) A protein produced by plasma cells in response to a specific antigen; the antibody combines with that antigen to neutralize, inhibit, or destroy it. Also called an **immunoglobulin** (im-ū-nō-GLOB-ū-lin) or **Ig.**

Antibody-mediated immunity That component of immunity in which B lymphocytes (B cells) develop into plasma cells that produce antibodies that destroy antigens. Also called **humoral** (HŪ-mor-al) **immunity.**

Anticoagulant (an-tī-cō-AG-ū-lant) A substance that can delay, suppress, or prevent the clotting of blood.

Antidiuretic (an′-ti-dī-ū-RET-ik) Substance that inhibits urine formation.

Antidiuretic hormone (ADH) Hormone produced by neurosecretory cells in the paraventricular and supraoptic nuclei of the hypothalamus that stimulates water reabsorption from kidney tubule cells into the blood and vasoconstriction of arterioles. Also called **vasopressin** (vāz-ō-PRES-in).

Antigen (AN-ti-jen) A substance that has immunogenicity (the ability to provoke an immune response) and reactivity (the ability to react with the antibodies or cells that result from the immune response); contraction of *anti*body *gen*erator. Also termed a **complete antigen.**

Antigen-presenting cell (APC) Special class of migratory cell that processes and presents antigens to T cells during an immune response; APCs include macrophages, B cells, and dendritic cells, which are present in the skin, mucous membranes, and lymph nodes.

Antiporter A transmembrane transporter protein that moves two substances, often Na$^+$ and another substance, in opposite directions across a plasma membrane. Also called a **countertransporter.**

Anulus fibrosus (AN-ū-lus fī-BRŌ-sus) A ring of fibrous tissue and fibrocartilage that encircles the pulpy substance (nucleus pulposus) of an intervertebral disc.

Anuria (an-Ū-rē-a) Absence of urine formation or daily urine output of less than 50 mL.

Anus (Ā-nus) The distal end and outlet of the rectum.

Aorta (ā-OR-ta) The main systemic trunk of the arterial system of the body that emerges from the left ventricle.

Aortic (ā-OR-tik) **body** Cluster of chemoreceptors on or near the arch of the aorta that respond to changes in blood levels of oxygen, carbon dioxide, and hydrogen ions (H$^+$).

Aortic reflex A reflex that helps maintain normal systemic blood pressure; initiated by baroreceptors in the wall of the ascending aorta and arch of the aorta. Nerve impulses from aortic baroreceptors reach the cardiovascular center via sensory axons of the vagus nerves (cranial nerve X).

Aperture (AP-er-chur) An opening or orifice.

Apex (Ā-peks) The pointed end of a conical structure, such as the apex of the heart.

Aphasia (a-FA-zē-a) Loss of ability to express oneself properly through speech or loss of verbal comprehension.

Apnea (AP-nē-a) Temporary cessation of breathing.

Apneustic (ap-NOO-stik) **area** A part of the respiratory center in the pons that sends stimulatory nerve impulses to the inspiratory area that activate and prolong inhalation and inhibit exhalation.

Apocrine (AP-ō-krin) **gland** A type of gland in which the secretory products gather at the free end of the secreting cell and are pinched off, along with some of the cytoplasm, to become the secretion, as in mammary glands.

Aponeurosis (ap′-ō-noo-RŌ-sis) A sheetlike tendon joining one muscle with another or with bone.

Apoptosis (ap′-ō-TŌ-sis *or* ap′-ōp-TŌ-sis) Programmed cell death; a normal type of cell death that removes unneeded cells during embryological development, regulates the number of cells in tissues, and eliminates many potentially dangerous cells such as cancer cells. During apoptosis, the DNA fragments, the nucleus condenses, mitochondria cease to function, and the cytoplasm shrinks, but the plasma membrane remains intact. Phagocytes engulf and digest the apoptotic cells, and an inflammatory response does not occur.

Appositional (ap′-ō-ZISH-o-nal) **growth** Growth due to surface deposition of material, as in the growth in diameter of cartilage and bone. Also called **exogenous** (eks-OJ-e-nus) **growth.**

Aqueous humor (AK-wē-us HŪ-mer) The watery fluid, similar in composition to cerebrospinal fluid, that fills the anterior cavity of the eye.

Arachnoid (a-RAK-noyd) **mater** The middle of the three meninges (coverings) of the brain and spinal cord. Also termed the **arachnoid.**

Arachnoid villus (VIL-us) Berrylike tuft of the arachnoid mater that protrudes into the superior sagittal sinus and through which cerebrospinal fluid is reabsorbed into the bloodstream.

Arbor vitae (AR-bor VĪ-tē) The white matter tracts of the cerebellum, which have a treelike appearance when seen in midsagittal section.

Arch of the aorta The most superior portion of the aorta, lying between the ascending and descending segments of the aorta.

Areola (a-RĒ-ō-la) Any tiny space in a tissue. The pigmented ring around the nipple of the breast.

Arm The part of the upper limb from the shoulder to the elbow.

Arousal (a-ROW-zal) Awakening from sleep, a response due to stimulation of the reticular activating system (RAS).

Arrector pili (a-REK-tor PI-lē) Smooth muscles attached to hairs; contraction pulls the hairs into a vertical position, resulting in "goose bumps."

Arrhythmia (a-RITH-mē-a) An irregular heart rhythm. Also called a **dysrhythmia.**

Arteriole (ar-TĒ-rē-ōl) A small, almost microscopic, artery that delivers blood to a capillary.

Arteriosclerosis (ar-tē-rē-ō-skle-RŌ-sis) Group of diseases characterized by thickening of the walls of arteries and loss of elasticity.

Artery (AR-ter-ē) A blood vessel that carries blood away from the heart.

Arthritis (ar-THRI-tis) Inflammation of a joint.

Arthrology (ar-THROL-ō-jē) The study or description of joints.

Arthroscopy (ar-THROS-co-pē) A procedure for examining the interior of a joint, usually the knee, by inserting an arthroscope into a small incision; used to determine extent of damage, remove torn cartilage, repair cruciate ligaments, and obtain samples for analysis.

Arthrosis (ar-THRŌ-sis) A joint or articulation.

Articular (ar-TIK-ū-lar) **capsule** Sleevelike structure around a synovial joint composed of a fibrous capsule and a synovial membrane.

Articular cartilage (KAR-ti-lij) Hyaline cartilage attached to articular bone surfaces.

Articular disc Fibrocartilage pad between articular surfaces of bones of some synovial joints. Also called a **meniscus** (men-IS-kus).

Articulation (ar-tik′-ū-LĀ-shun) A joint; a point of contact between bones, cartilage and bones, or teeth and bones.

Arytenoid (ar′-i-TĒ-noyd) **cartilages** A pair of small, pyramidal cartilages of the larynx that attach to the vocal folds and intrinsic pharyngeal muscles and can move the vocal folds.

Ascending colon (KŌ-lon) The part of the large intestine that passes superiorly from the cecum to the inferior border of the liver, where it bends at the right colic (hepatic) flexure to become the transverse colon.

Ascites (as-SĪ-tēz) Abnormal accumulation of serous fluid in the peritoneal cavity.

Association areas Large cortical regions on the lateral surfaces of the occipital, parietal, and temporal lobes and on the frontal lobes anterior to the motor areas connected by many motor and sensory axons to other parts of the cortex. The association areas are concerned with motor patterns, memory, concepts of word-hearing and word-seeing, reasoning, will, judgment, and personality traits.

Asthma (AZ-ma) Usually allergic reaction characterized by smooth muscle spasms in bronchi resulting in wheezing and difficult breathing. Also called **bronchial asthma.**

Astigmatism (a-STIG-ma-tizm) An irregularity of the lens or cornea of the eye causing the image to be out of focus and producing faulty vision.

Astrocyte (AS-trō-sīt) A neuroglial cell having a star shape that participates in brain development and the metabolism of neurotransmitters, helps form the blood–brain barrier, helps maintain the proper balance of K$^+$ for generation of nerve impulses, and provides a link between neurons and blood vessels.

Ataxia (a-TAK-sē-a) A lack of muscular coordination, lack of precision.

Atherosclerotic (ath′-er-ō-skle-RO-tic) **plaque** (PLAK) A lesion that results from accumulated cholesterol and smooth muscle fibers (cells) of the tunica media of an artery; may become obstructive.

Atom Unit of matter that makes up a chemical element; consists of a nucleus (containing positively charged protons and uncharged neutrons) and negatively charged electrons that orbit the nucleus.

Atomic mass (weight) Average mass of all stable atoms of an element, reflecting the relative proportion of atoms with different mass numbers.

Atomic number Number of protons in an atom.

Atresia (a-TRĒ-zē-a) Degeneration and reabsorption of an ovarian follicle before it fully matures and ruptures; abnormal closure of a passage, or absence of a normal body opening.

Atrial fibrillation (Ā-trē-al fib-ri-LĀ-shun) Asynchronous contraction of cardiac muscle fibers in the atria that results in the cessation of atrial pumping.

Atrial natriuretic (na′-trē-ū-RET-ik) **peptide** (**ANP**) Peptide hormone, produced by the atria of the heart in response to stretching, that inhibits aldosterone production and thus lowers blood pressure; causes natriuresis, increased urinary excretion of sodium.

Atrioventricular (**AV**) (ā′-trē-ō-ven-TRIK-ū-lar) **bundle** The part of the conduction system of the heart that begins at the atrioventricular (AV) node, passes through the cardiac skeleton separating the atria and the ventricles, then extends a short distance down the interventricular septum before splitting into right and left bundle branches. Also called the **bundle of His** (HISS).

Atrioventricular (**AV**) **node** The part of the conduction system of the heart made up of a compact mass of conducting cells located in the septum between the two atria.

Atrioventricular (**AV**) **valve** A heart valve made up of membranous flaps or cusps that allows blood to flow in one direction only, from an atrium into a ventricle.

Atrium (Ā-trē-um) A superior chamber of the heart.

Atrophy (AT-rō-fē) Wasting away or decrease in size of a part, due to a failure, abnormality of nutrition, or lack of use.

Auditory ossicle (AW-di-tō-rē OS-si-kul) One of the three small bones of the middle ear called the **malleus, incus,** and **stapes.**

Auditory tube The tube that connects the middle ear with the nose and nasopharynx region of the throat. Also called the **Eustachian** (ū-STĀ-shun *or* ū-STĀ-kē-an) **tube** or **pharyngotympanic tube.**

Auscultation (aws-kul-TĀ-shun) Examination by listening to sounds in the body.

Autocrine (AW-tō-krin) Local hormone, such as interleukin-2, that acts on the same cell that secreted it.

Autoimmunity An immunological response against a person's own tissues.

Autolysis (aw-TOL-i-sis) Self-destruction of cells by their own lysosomal digestive enzymes after death or in a pathological process.

Autonomic ganglion (aw′-tō-NOM-ik GANG-lē-on) A cluster of cell bodies of sympathetic or parasympathetic neurons located outside the central nervous system.

Autonomic nervous system (**ANS**) Visceral sensory (afferent) and visceral motor (efferent) neurons. Autonomic motor neurons, both sympathetic and parasympathetic, conduct nerve impulses from the central nervous system to smooth muscle, cardiac muscle, and glands. So named because this part of the nervous system was thought to be self-governing or spontaneous.

Autonomic plexus (PLEK-sus) A network of sympathetic and parasympathetic axons; examples are the cardiac, celiac, and pelvic plexuses, which are located in the thorax, abdomen, and pelvis, respectively.

Autophagy (aw-TOF-a-jē) Process by which worn-out organelles are digested within lysosomes.

Autopsy (AW-top-sē) The examination of the body after death.

Autoregulation (aw-tō-reg-ū-LĀ-shun) A local, automatic adjustment of blood flow in a given region of the body in response to tissue needs.

Autorhythmic cells Cardiac or smooth muscle fibers that are self-excitable (generate impulses without an external stimulus); act as the heart's pacemaker and conduct the pacing impulse through the conduction system of the heart; self-excitable neurons in the central nervous system, as in the inspiratory area of the brain stem.

Autosome (AW-tō-sōm) Any chromosome other than the X and Y chromosomes (sex chromosomes).

Axilla (ak-SIL-a) The small hollow beneath the arm where it joins the body at the shoulders. Also called the **armpit.**

Axon (AK-son) The usually single, long process of a nerve cell that propagates a nerve impulse toward the axon terminals.

Axon terminal Terminal branch of an axon where synaptic vesicles undergo exocytosis to release neurotransmitter molecules.

Azygos (AZ-ī-gos) An anatomical structure that is not paired; occurring singly.

B

B cell A lymphocyte that can develop into a clone of antibody-producing plasma cells or memory cells when properly stimulated by a specific antigen.

Babinski (ba-BIN-skē) **sign** Extension of the great toe, with or without fanning of the other toes, in response to stimulation of the outer margin of the sole; normal up to 18 months of age and indicative of damage to descending motor pathways such as the corticospinal tracts after that.

Back The posterior part of the body; the dorsum.

Ball-and-socket joint A synovial joint in which the rounded surface of one bone moves within a cup-shaped depression or socket of another bone, as in the shoulder or hip joint. Also called a **spheroid** (SFĒ-royd) **joint.**

Baroreceptor (bar′-ō-re-SEP-tor) Neuron capable of responding to changes in blood, air, or fluid pressure. Also called a **pressoreceptor.**

Basal ganglia (GANG-glē-a) Paired clusters of gray matter deep in each cerebral hemisphere including the globus pallidus, putamen, and caudate nucleus. Together, the caudate nucleus and putamen are known as the **corpus striatum.** Nearby structures that are functionally linked to the basal ganglia are the substantia nigra of the midbrain and the subthalamic nuclei of the diencephalon.

Basal metabolic (BĀ-sal met′-a-BOL-ik) **rate** (**BMR**) The rate of metabolism measured under standard or basal conditions (awake, at rest, fasting).

Base A nonacid or a proton acceptor, characterized by excess of hydroxide ions (OH^-) and a pH greater than 7. A ring-shaped, nitrogen-containing organic molecule that is one of the components of a nucleotide, namely, adenine, guanine, cytosine, thymine, and uracil; also known as a **nitrogenous base.**

Basement membrane Thin, extracellular layer between epithelium and connective tissue consisting of a basal lamina and a reticular lamina.

Basilar (BAS-i-lar) **membrane** A membrane in the cochlea of the internal ear that separates the cochlear duct from the scala tympani and on which the spiral organ (organ of Corti) rests.

Basophil (BĀ-sō-fil) A type of white blood cell characterized by a pale nucleus and large granules that stain blue-purple with basic dyes.

Belly The abdomen. The gaster or prominent, fleshy part of a skeletal muscle.

Beta (BĀ-ta) **cell** A type of cell in the pancreatic islets (islets of Langerhans) in the pancreas that secretes the hormone insulin. Also termed a **B cell.**

Beta receptor A type of adrenergic receptor for epinephrine and norepinephrine; found on visceral effectors innervated by sympathetic postganglionic neurons.

Bicuspid (bī-KUS-pid) **valve** Atrioventricular (AV) valve on the left side of the heart. Also called the **mitral valve.**

Bilateral (bī-LAT-er-al) Pertaining to two sides of the body.

Bile (BĪL) A secretion of the liver consisting of water, bile salts, bile pigments, cholesterol, lecithin, and several ions; it emulsifies lipids prior to their digestion.

Bilirubin (bil-ē-ROO-bin) An orange pigment that is one of the end products of hemoglobin breakdown in the hepatocytes and is excreted as a waste material in bile.

Blastocele (BLAS-tō-sēl) The fluid-filled cavity within the blastocyst.

Blastocyst (BLAS-tō-sist) In the development of an embryo, a hollow ball of cells that consists of a blastocele (the internal cavity), trophoblast (outer cells), and inner cell mass.

Blastomere (BLAS-tō-mēr) One of the cells resulting from the cleavage of a fertilized ovum.

Blind spot Area in the retina at the end of the optic nerve (cranial nerve II) in which there are no photoreceptors.

Blood The fluid that circulates through the heart, arteries, capillaries, and veins and that constitutes the chief means of transport within the body.

Blood–brain barrier (BBB) A barrier consisting of specialized brain capillaries and astrocytes that prevents the passage of materials from the blood to the cerebrospinal fluid and brain.

Blood island Isolated mass of mesoderm derived from angioblasts and from which blood vessels develop.

Blood pressure (BP) Force exerted by blood against the walls of blood vessels due to contraction of the heart and influenced by the elasticity of the vessel walls; clinically, a measure of the pressure in arteries during ventricular systole and ventricular diastole. *See also* **mean arterial blood pressure.**

Blood reservoir (REZ-er-vwar) Systemic veins that contain large amounts of blood that can be moved quickly to parts of the body requiring the blood.

Blood–testis barrier (BTB) A barrier formed by Sertoli cells that prevents an immune response against antigens produced by spermatogenic cells by isolating the cells from the blood.

Body cavity A space within the body that contains various internal organs.

Body fluid Body water and its dissolved substances; constitutes about 60% of total body mass.

Bohr (BŌR) **effect** In an acidic environment, oxygen unloads more readily from hemoglobin because when hydrogen ions (H⁺) bind to hemoglobin, they alter the structure of hemoglobin, thereby reducing its oxygen-carrying capacity.

Bolus (BŌ-lus) A soft, rounded mass, usually food, that is swallowed.

Bony labyrinth (LAB-i-rinth) A series of cavities within the petrous portion of the temporal bone forming the vestibule, cochlea, and semicircular canals of the inner ear.

Brachial plexus (BRĀ-kē-al PLEK-sus) A network of nerve axons of the ventral rami of spinal nerves C5, C6, C7, C8, and T1. The nerves that emerge from the brachial plexus supply the upper limb.

Bradycardia (brād′-i-KAR-dē-a) A slow resting heart or pulse rate (under 60 beats per minute).

Brain The part of the central nervous system contained within the cranial cavity.

Brain stem The portion of the brain immediately superior to the spinal cord, made up of the medulla oblongata, pons, and midbrain.

Brain waves Electrical signals that can be recorded from the skin of the head due to electrical activity of brain neurons.

Broad ligament A double fold of parietal peritoneum attaching the uterus to the side of the pelvic cavity.

Broca's (BRŌ-kaz) **area** Motor area of the brain in the frontal lobe that translates thoughts into speech. Also called the **motor speech area.**

Bohr effect Shifting of the oxygen–hemoglobin dissociation curve to the right when pH decreases; at any given partial pressure of oxygen, hemoglobin is less saturated with O_2 at lower pH.

Bronchi (BRONG-kē) Branches of the respiratory passageway including primary bronchi (the two divisions of the trachea), secondary or lobar bronchi (divisions of the primary bronchi that are distributed to the lobes of the lung), and tertiary or segmental bronchi (divisions of the secondary bronchi that are distributed to bronchopulmonary segments of the lung). *Singular is* **bronchus.**

Bronchial tree The trachea, bronchi, and their branching structures up to and including the terminal bronchioles.

Bronchiole (BRONG-kē-ōl) Branch of a tertiary bronchus further dividing into terminal bronchioles (distributed to lobules of the lung), which divide into respiratory bronchioles (distributed to alveolar sacs).

Bronchitis (brong-KĪ-tis) Inflammation of the mucous membrane of the bronchial tree; characterized by hypertrophy and hyperplasia of seromucous glands and goblet cells that line the bronchi and which results in a productive cough.

Bronchopulmonary (brong′-kō-PUL-mō-ner-ē) **segment** One of the smaller divisions of a lobe of a lung supplied by its own branches of a bronchus.

Buccal (BUK-al) Pertaining to the cheek or mouth.

Buffer (BUF-er) **system** A pair of chemicals—one a weak acid and the other the salt of the weak acid, which functions as a weak base—that resists changes in pH.

Bulbourethral (bul′-bō-ū-RĒ-thral) **gland** One of a pair of glands located inferior to the prostate on either side of the urethra that secretes an alkaline fluid into the cavernous urethra. Also called a **Cowper's** (KOW-perz) **gland.**

Bulimia (boo-LIM-ē-a *or* boo-LĒ-mē-a) A disorder characterized by overeating at least twice a week followed by purging by self-induced vomiting, strict dieting or fasting, vigorous exercise, or use of laxatives or diuretics. Also called **binge–purge syndrome.**

Bulk flow The movement of large numbers of ions, molecules, or particles in the same direction due to pressure differences (osmotic, hydrostatic, or air pressure).

Bundle branch One of the two branches of the atrioventricular (AV) bundle made up of specialized muscle fibers (cells) that transmit electrical impulses to the ventricles.

Bursa (BUR-sa) A sac or pouch of synovial fluid located at friction points, especially about joints.

Buttocks (BUT-oks) The two fleshy masses on the posterior aspect of the inferior trunk, formed by the gluteal muscles.

C

Calcaneal (kal-KĀ-nē-al) **tendon** The tendon of the soleus, gastrocnemius, and plantaris muscles at the back of the heel. Also called the **Achilles** (a-KIL-ēz) **tendon.**

Calcification (kal-si-fi-KĀ-shun) Deposition of mineral salts, primarily hydroxyapatite, in a framework formed by collagen fibers in which the tissue hardens. Also called **mineralization** (min′-e-ral-i-ZĀ-shun).

Calcitonin (kal-si-TŌ-nin) **(CT)** A hormone produced by the parafollicular cells of the thyroid gland that can lower the amount of blood calcium and phosphates by inhibiting bone resorption (breakdown of bone matrix) and by accelerating uptake of calcium and phosphates into bone matrix.

Calculus (KAL-kū-lus) A stone, or insoluble mass of crystallized salts or other material, formed within the body, as in the gallbladder, kidney, or urinary bladder.

Callus (KAL-lus) A growth of new bone tissue in and around a fractured area, ultimately replaced by mature bone. An acquired, localized thickening.

Calorie (KAL-ō-rē) A unit of heat. A calorie (cal) is the standard unit and is the amount of heat needed to raise the temperature of 1 g of water from 14°C to 15°C. The **kilocalorie (kcal)** or **Calorie** (spelled with an uppercase C), used to express the caloric value of foods and to measure metabolic rate, is equal to 1000 cal.

Calyx (KĀL-iks) Any cuplike division of the kidney pelvis. *Plural is* **calyces** (KĀ-li-sēz).

Canal (ka-NAL) A narrow tube, channel, or passageway.

Canaliculus (kan′-a-LIK-ū-lus) A small channel or canal, as in bones, where they connect lacunae. *Plural is* **canaliculi** (kan′-a-LIK-ū-lī).

Cancellous (KAN-sel-us) Having a reticular or latticework structure, as in spongy tissue of bone.

Capacitation (ka′-pas-i-TĀ-shun) The functional changes that sperm undergo in the female reproductive tract that allow them to fertilize a secondary oocyte.

Capillary (KAP-i-lar′-ē) A microscopic blood vessel located between an arteriole and venule through which materials are exchanged between blood and interstitial fluid.

Carbohydrate (kar′-bō-HĪ-drāt) An organic compound containing carbon, hydrogen, and oxygen in a particular amount and arrangement and composed of monosaccharide subunits; usually has the general formula $(CH_2O)n$.

Carcinogen (kar-SIN-ō-jen) A chemical substance or radiation that causes cancer.

Cardiac (KAR-dē-ak) **arrest** Cessation of an effective heartbeat in which the heart is completely stopped or in ventricular fibrillation.

Cardiac cycle A complete heartbeat consisting of systole (contraction) and diastole (relaxation) of both atria plus systole and diastole of both ventricles.

Cardiac muscle Striated muscle fibers (cells) that form the wall of the heart; stimulated by an intrinsic conduction system and autonomic motor neurons.

Cardiac notch An angular notch in the anterior border of the left lung into which part of the heart fits.

Cardiac output (CO) The volume of blood pumped from one ventricle of the heart (usually measured from the left ventricle) in 1 min; normally about 5.2 liters/min in an adult at rest.

Cardiac reserve The maximum percentage that cardiac output can increase above normal.

Cardinal ligament A ligament of the uterus, extending laterally from the cervix and vagina as a continuation of the broad ligament.

Cardiology (kar-dē-OL-ō-jē) The study of the heart and diseases associated with it.

Cardiovascular (kar-dē-ō-VAS-kū-lar) **center** Groups of neurons scattered within the medulla oblongata that regulate heart rate, force of contraction, and blood vessel diameter.

Carotene (KAR-o-tēn) Antioxidant precursor of vitamin A, which is needed for synthesis of photopigments; yellow-orange pigment present in the stratum corneum of the epidermis. Accounts for the yellowish coloration of skin. Also termed **beta-carotene.**

Carotid (ka-ROT-id) **body** Cluster of chemoreceptors on or near the carotid sinus that respond to changes in blood levels of oxygen, carbon dioxide, and hydrogen ions.

Carotid sinus A dilated region of the internal carotid artery just above the point where it branches from the common carotid artery; it contains baroreceptors that monitor blood pressure.

Carotid sinus reflex A reflex that helps maintain normal blood pressure in the brain. Nerve impulses propagate from the carotid sinus baroreceptors over sensory axons in the glossopharyngeal nerves (cranial nerve IX) to the cardiovascular center in the medulla oblongata.

Carpal bones The eight bones of the wrist. Also called **carpals.**

Carpus (KAR-pus) A collective term for the eight bones of the wrist.

Cartilage (KAR-ti-lij) A type of connective tissue consisting of chondrocytes in lacunae embedded in a dense network of collagen and elastic fibers and a matrix of chondroitin sulfate.

Cartilaginous (kar′-ti-LAJ-i-nus) **joint** A joint without a synovial (joint) cavity where the articulating bones are held tightly together by cartilage, allowing little or no movement.

Cast A small mass of hardened material formed within a cavity in the body and then discharged from the body; can originate in different areas and can be composed of various materials.

Catabolism (ka-TAB-ō-lizm) Chemical reactions that break down complex organic compounds into simple ones, with the net release of energy.

Catalyst (KAT-a-list) A substance that speeds up a chemical reaction without itself being altered; an enzyme.

Cataract (KAT-a-rakt) Loss of transparency of the lens of the eye or its capsule or both.

Cation (KAT-ī-on) A positively charged ion. An example is a sodium ion (Na^+).

Cauda equina (KAW-da ē-KWĪ-na) A tail-like array of roots of spinal nerves at the inferior end of the spinal cord.

Caudal (KAW-dal) Pertaining to any tail-like structure; inferior in position.

Cecum (SĒ-kum) A blind pouch at the proximal end of the large intestine that attaches to the ileum.

Celiac plexus (PLEK-sus) A large mass of autonomic ganglia and axons located at the level of the superior part of the first lumbar vertebra. Also called the **solar plexus.**

Cell The basic structural and functional unit of all organisms; the smallest structure capable of performing all the activities vital to life.

Cell cycle Growth and division of a single cell into two daughter cells; consists of interphase and cell division.

Cell division Process by which a cell reproduces itself that consists of a nuclear division (mitosis) and a cytoplasmic division (cytokinesis); types include somatic and reproductive cell division.

Cell-mediated immunity That component of immunity in which specially sensitized T lymphocytes (T cells) attach to antigens to destroy them. Also called **cellular immunity.**

Cementum (se-MEN-tum) Calcified tissue covering the root of a tooth.

Center of ossification (os′-i-fi-KĀ-shun) An area in the cartilage model of a future bone where the cartilage cells hypertrophy and then secrete enzymes that result in the calcification of their matrix, resulting in the death of the cartilage cells, followed by the invasion of the area by osteoblasts that then lay down bone.

Central canal A microscopic tube running the length of the spinal cord in the gray commissure. A circular channel running longitudinally in the center of an osteon (Haversian system) of mature compact bone, containing blood and lymphatic vessels and nerves. Also called a **Haversian** (ha-VER-shan) **canal.**

Central fovea (FŌ-vē-a) A depression in the center of the macula lutea of the retina, containing cones only and lacking blood vessels; the area of highest visual acuity (sharpness of vision).

Central nervous system (CNS) That portion of the nervous system that consists of the brain and spinal cord.

Centrioles (SEN-trē-ōlz) Paired, cylindrical structures of a centrosome, each consisting of a ring of microtubules and arranged at right angles to each other.

Centromere (SEN-trō-mēr) The constricted portion of a chromosome where the two chromatids are joined; serves as the point of attachment for the microtubules that pull chromatids during anaphase of cell division.

Centrosome (SEN-trō-sōm) A dense network of small protein fibers near the nucleus of a cell, containing a pair of centrioles and pericentriolar material.

Cephalic (se-FAL-ik) Pertaining to the head; superior in position.

Cerebellar peduncle (ser-e-BEL-ar pe-DUNG-kul) A bundle of nerve axons connecting the cerebellum with the brain stem.

Cerebellum (ser-e-BEL-um) The part of the brain lying posterior to the medulla oblongata and pons; governs balance and coordinates skilled movements.

Cerebral aqueduct (SER-ē-bral AK-we-dukt) A channel through the midbrain connecting the third and fourth ventricles and containing cerebrospinal fluid. Also termed the **aqueduct of Sylvius.**

Cerebral arterial circle A ring of arteries forming an anastomosis at the base of the brain between the internal carotid and basilar arteries and arteries supplying the cerebral cortex. Also called the **circle of Willis.**

Cerebral cortex The surface of the cerebral hemispheres, 2–4 mm thick, consisting of gray matter; arranged in six layers of neuronal cell bodies in most areas.

Cerebral peduncle One of a pair of nerve axon bundles located on the anterior surface of the midbrain, conducting nerve impulses between the pons and the cerebral hemispheres.

Cerebrospinal (se-rē′-brō-SPĪ-nal) **fluid (CSF)** A fluid produced by ependymal cells that cover choroid plexuses in the ventricles of the brain; the fluid circulates in the ventricles, the central canal, and the subarachnoid space around the brain and spinal cord.

Cerebrovascular (se rē′-brō-VAS-kū-lar) **accident (CVA)** Destruction of brain tissue (infarction) resulting from obstruction or rupture of blood vessels that supply the brain. Also called a **stroke** or **brain attack.**

Cerebrum (SER-e-brum *or* se-RĒ-brum) The two hemispheres of the forebrain (derived from the telencephalon), making up the largest part of the brain.

Cerumen (se-ROO-men) Waxlike secretion produced by ceruminous glands in the external auditory meatus (ear canal). Also termed **ear wax.**

Ceruminous (se-ROO-mi-nus) **gland** A modified sudoriferous (sweat) gland in the external auditory meatus that secretes cerumen (ear wax).

Cervical ganglion (SER-vi-kul GANG-glē-on) A cluster of cell bodies of postganglionic sympathetic neurons located in the neck, near the vertebral column.

Cervical plexus (PLEK-sus) A network formed by nerve axons from the ventral rami of the first four cervical nerves and receiving gray rami communicates from the superior cervical ganglion.

Cervix (SER-viks) Neck; any constricted portion of an organ, such as the inferior cylindrical part of the uterus.

Chemical bond Force of attraction in a molecule or compound that holds its atoms together. Examples include ionic and covalent bonds.

Chemical element Unit of matter that cannot be decomposed into a simpler substance by ordinary chemical reactions. Examples include hydrogen (H), carbon (C), and oxygen (O).

Chemically gated channel A channel in a membrane that opens and closes in response to a specific chemical stimulus, such as a neurotransmitter, hormone, or specific type of ion.

Chemical reaction The combination or separation of atoms in which chemical bonds are formed or broken and new products with different properties are produced.

Chemiosmosis (kem′-ē-oz-MŌ-sis) Mechanism for ATP generation that links chemical reactions (electrons passing along the electron transport chain) with pumping of H^+ out of the mitochondrial matrix. ATP synthesis occurs as H^+ diffuse back into the mitochondrial matrix through special H^+ channels in the membrane.

Chemoreceptor (kē′-mō-rē-SEP-tor) Sensory receptor that detects the presence of a specific chemical.

Chemotaxis (kē-mō-TAK-sis) Attraction of phagocytes to microbes by a chemical stimulus.

Chiasm (KĪ-azm) A crossing; especially the crossing of axons in the optic nerve (cranial nerve II).

Chief cell The secreting cell of a gastric gland that produces pepsinogen, the precursor of the enzyme pepsin, and the enzyme gastric lipase. Also called a **zymogenic** (zī′-mō-JEN-ik) **cell.** Cell in the parathyroid glands that secretes parathyroid hormone (PTH). Also called a **principal cell.**

Chiropractic (kī-rō-PRAK-tik) A system of treating disease by using one's hands to manipulate body parts, mostly the vertebral column.

Chloride shift Exchange of bicarbonate ions (HCO_3^-) for chloride ions (Cl^-) between red blood cells and plasma; maintains electrical balance inside red blood cells as bicarbonate ions are produced or eliminated during respiration.

Cholecystectomy (kō′-lē-sis-TEK-tō-mē) Surgical removal of the gallbladder.

Cholecystitis (kō′-lē-sis-TĪ-tis) Inflammation of the gallbladder.

Cholesterol (kō-LES-te-rol) Classified as a lipid, the most abundant steroid in animal tissues; located in cell membranes and used for the synthesis of steroid hormones and bile salts.

Cholinergic (kō′-lin-ER-jik) **neuron** A neuron that liberates acetylcholine as its neurotransmitter.

Chondrocyte (KON-drō-sīt) Cell of mature cartilage.

Chondroitin (kon-DROY-tin) **sulfate** An amorphous matrix material found outside connective tissue cells.

Chordae tendineae (KOR-dē TEN-di-nē-ē) Tendonlike, fibrous cords that connect atrioventricular valves of the heart with papillary muscles.

Chorion (KŌ-rē-on) The most superficial fetal membrane that becomes the principal embryonic portion of the placenta; serves a protective and nutritive function.

Chorionic villi (kō-rē-ON-ik VIL-lī) Fingerlike projections of the chorion that grow into the decidua basalis of the endometrium and contain fetal blood vessels.

Choroid (KŌ-royd) One of the vascular coats of the eyeball.

Choroid plexus (PLEK-sus) A network of capillaries located in the roof of each of the four ventricles of the brain; ependymal cells around choroid plexuses produce cerebrospinal fluid.

Chromaffin (KRŌ-maf-in) **cell** Cell that has an affinity for chrome salts, due in part to the presence of the precursors of the neurotransmitter epinephrine; found, among other places, in the adrenal medulla.

Chromatid (KRŌ-ma-tid) One of a pair of identical connected nucleoprotein strands that are joined at the centromere and separate during cell division, each becoming a chromosome of one of the two daughter cells.

Chromatin (KRŌ-ma-tin) The threadlike mass of genetic material, consisting of DNA and histone proteins, that is present in the nucleus of a nondividing or interphase cell.

Chromatolysis (krō-ma-TOL-i-sis) The breakdown of Nissl bodies into finely granular masses in the cell body of a neuron whose axon has been damaged.

Chromosome (KRŌ-mō-sōm) One of the small, threadlike structures in the nucleus of a cell, normally 46 in a human diploid cell, that bears the genetic material; composed of DNA and proteins (histones) that form a delicate chromatin thread during interphase; becomes packaged into compact rodlike structures that are visible under the light microscope during cell division.

Chronic (KRON-ik) Long term or frequently recurring; applied to a disease that is not acute.

Chronic obstructive pulmonary disease (COPD) A disease, such as bronchitis or emphysema, in which there is some degree of obstruction of airways and consequent increase in airway resistance.

Chyle (KĪL) The milky-appearing fluid found in the lacteals of the small intestine after absorption of lipids in food.

Chylomicron (kī-lō-MĪ-kron) Protein-coated spherical structure that contains triglycerides, phospholipids, and cholesterol and is absorbed into the lacteal of a villus in the small intestine.

Chyme (KĪM) The semifluid mixture of partly digested food and digestive secretions found in the stomach and small intestine during digestion of a meal.

Ciliary (SIL-ē-ar′-ē) **body** One of the three parts of the vascular tunic of the eyeball, the others being the choroid and the iris; includes the ciliary muscle and the ciliary processes.

Ciliary ganglion (GANG-glē-on) A very small parasympathetic ganglion whose preganglionic axons come from the oculomotor nerve (cranial nerve III) and whose postganglionic axons carry nerve impulses to the ciliary muscle and the sphincter muscle of the iris.

Cilium (SIL-ē-um) A hair or hairlike process projecting from a cell that may be used to move the entire cell or to move substances along the surface of the cell. *Plural is* **cilia.**

Circadian (ser-KĀ-dē-an) **rhythm** A cycle of active and nonactive periods in organisms determined by internal mechanisms and repeating about every 24 hours.

Circular folds Permanent, deep, transverse folds in the mucosa and submucosa of the small intestine that increase the surface area for absorption. Also called **plicae circulares** (PLĪ-kē SER-kū-lar-ēs).

Circulation time Time required for blood to pass from the right atrium, through pulmonary circulation, back to the left ventricle, through systemic circulation to the foot, and back again to the right atrium; normally about 1 min.

Circumduction (ser′-kum-DUK-shun) A movement at a synovial joint in which the distal end of a bone moves in a circle while the proximal end remains relatively stable.

Cirrhosis (si-RŌ-sis) A liver disorder in which the parenchymal cells are destroyed and replaced by connective tissue.

Cisterna chyli (sis-TER-na-KĪ-lē) The origin of the thoracic duct.

Cleavage The rapid mitotic divisions following the fertilization of a secondary oocyte, resulting in an increased number of progressively smaller cells, called blastomeres.

Climacteric (klī-mak-TER-ik) Cessation of the reproductive function in the female or diminution of testicular activity in the male.

Climax The peak period or moments of greatest intensity during sexual excitement.

Clitoris (KLI-to-ris) An erectile organ of the female, located at the anterior junction of the labia minora, that is homologous to the male penis.

Clone (KLŌN) A population of identical cells.

Clot The end result of a series of biochemical reactions that changes liquid plasma into a gelatinous mass; specifically, the conversion of fibrinogen into a tangle of polymerized fibrin molecules.

Clot retraction (rē-TRAK-shun) The consolidation of a fibrin clot to pull damaged tissue together.

Clotting Process by which a blood clot is formed. Also known as **coagulation** (cō-ag-ū-LĀ-shun).

Coccyx (KOK-six) The fused bones at the inferior end of the vertebral column.

Cochlea (KŌK-lē-a) A winding, cone-shaped tube forming a portion of the inner ear and containing the spiral organ (organ of Corti).

Cochlear duct The membranous cochlea consisting of a spirally arranged tube enclosed in the bony cochlea and lying along its outer wall. Also called the **scala media** (SCA-la MĒ-dē-a).

Coenzyme A nonprotein organic molecule that is associated with and activates an enzyme; many are derived from vitamins. An example is nicotinamide adenine dinucleotide (NAD), derived from the B vitamin niacin.

Coitus (KŌ-i-tus) Sexual intercourse.

Collagen (KOL-a-jen) A protein that is the main organic constituent of connective tissue.

Collateral circulation The alternate route taken by blood through an anastomosis.

Colliculus (ko-LIK-ū-lus) A small elevation.

Colloid (KOL-loyd) The material that accumulates in the center of thyroid follicles, consisting of thyroglobulin and stored thyroid hormones.

Colon The portion of the large intestine consisting of ascending, transverse, descending, and sigmoid portions.

Colony-stimulating factor (CSF) One of a group of molecules that stimulates development of white blood cells. Examples are macrophage CSF and granulocyte CSF.

Colostrum (kō-LOS-trum) A thin, cloudy fluid secreted by the mammary glands a few days prior to or after delivery before true milk is produced.

Column (KOL-um) Group of white matter tracts in the spinal cord.

Common bile duct A tube formed by the union of the common hepatic duct and the cystic duct that empties bile into the duodenum at the hepatopancreatic ampulla (ampulla of Vater).

Compact (dense) bone tissue Bone tissue that contains few spaces between osteons (Haversian systems); forms the external portion of all bones and the bulk of the diaphysis (shaft) of long bones; is found immediately deep to the periosteum and external to spongy bone.

Complement (KOM-ple-ment) A group of at least 20 normally inactive proteins found in plasma that forms a component of nonspecific resistance and immunity by bringing about cytolysis, inflammation, and opsonization.

Compliance The ease with which the lungs and thoracic wall or blood vessels can be expanded.

Compound A substance that can be broken down into two or more other substances by chemical means.

Concha (KONG-ka) A scroll-like bone found in the skull. *Plural is* **conchae** (KONG-kē).

Concussion (kon-KUSH-un) Traumatic injury to the brain that produces no visible bruising but may result in abrupt, temporary loss of consciousness.

Conduction system A group of autorhythmic cardiac muscle fibers that generates and distributes electrical impulses to stimulate coordinated contraction of the heart chambers; includes the sinoatrial (SA) node, the atrioventricular (AV) node, the atrioventricular (AV) bundle, the right and left bundle branches, and the Purkinje fibers.

Conductivity (kon′-duk-TIV-i-tē) The ability of a cell to propagate (conduct) action potentials along its plasma membrane; characteristic of neurons and muscle fibers (cells).

Condyloid (KON-di-loyd) **joint** A synovial joint structured so that an oval-shaped condyle of one bone fits into an elliptical cavity of another bone, permitting side-to-side and back-and-forth movements, such as the joint at the wrist between the radius and carpals. Also called an **ellipsoidal** (ē-lip-SOYD-al) **joint.**

Cone (KŌN) The type of photoreceptor in the retina that is specialized for highly acute color vision in bright light.

Congenital (kon-JEN-i-tal) Present at the time of birth.

Conjunctiva (kon′-junk-TĪ-va) The delicate membrane covering the eyeball and lining the eyes.

Connective tissue One of the most abundant of the four basic tissue types in the body, performing the functions of binding and supporting; consists of relatively few cells in a generous matrix (the ground substance and fibers between the cells).

Consciousness (KON-shus-nes) A state of wakefulness in which an individual is fully alert, aware, and oriented, partly as a result of feedback between the cerebral cortex and reticular activating system.

Continuous conduction (kon-DUK-shun) Propagation of an action potential (nerve impulse) in a step-by-step depolarization of each adjacent area of an axon membrane.

Contraception (kon′-tra-SEP-shun) The prevention of fertilization or impregnation without destroying fertility.

Contractility (kon′-trak-TIL-i-tē) The ability of cells or parts of cells to actively generate force to undergo shortening for movements. Muscle fibers (cells) exhibit a high degree of contractility.

Contralateral (kon′-tra-LAT-er-al) On the opposite side; affecting the opposite side of the body.

Control center The component of a feedback system, such as the brain, that determines the point at which a controlled condition, such as body temperature, is maintained.

Conus medullaris (KŌ-nus med-ū-LAR-is) The tapered portion of the spinal cord inferior to the lumbar enlargement.

Convergence (con-VER-jens) A synaptic arrangement in which the synaptic end bulbs of several presynaptic neurons terminate on one postsynaptic neuron. The medial movement of the two eyeballs so that both are directed toward a near object being viewed in order to produce a single image.

Convulsion (con-VUL-shun) Violent, involuntary contractions or spasms of an entire group of muscles.

Cornea (KOR-nē-a) The nonvascular, transparent fibrous coat through which the iris of the eye can be seen.

Corona radiata The innermost layer of granulosa cells that is firmly attached to the zona pellucida around a secondary oocyte.

Coronary artery disease (CAD) A condition such as atherosclerosis that causes narrowing of coronary arteries so that blood flow to the heart is reduced. The result is **coronary heart disease (CHD),** in which the heart muscle receives inadequate blood flow due to an interruption of its blood supply.

Coronary circulation The pathway followed by the blood from the ascending aorta through the blood vessels supplying the heart and returning to the right atrium. Also called **cardiac circulation.**

Coronary sinus (SĪ-nus) A wide venous channel on the posterior surface of the heart that collects the blood from the coronary circulation and returns it to the right atrium.

Corpus (KOR-pus) The principal part of any organ; any mass or body.

Corpus albicans (KOR-pus AL-bi-kanz) A white fibrous patch in the ovary that forms after the corpus luteum regresses.

Corpus callosum (kal-LŌ-sum) The great commissure of the brain between the cerebral hemispheres.

Corpus luteum (LOO-tē-um) A yellowish body in the ovary formed when a follicle has discharged its secondary oocyte; secretes estrogens, progesterone, relaxin, and inhibin.

Corpuscle of touch The sensory receptor for the sensation of touch; found in the dermal papillae, especially in palms and soles. Also called a **Meissner** (MĪZ-ner) **corpuscle.**

Cortex (KOR-teks) An outer layer of an organ. The convoluted layer of gray matter covering each cerebral hemisphere.

Costal (KOS-tal) Pertaining to a rib.

Costal cartilage (KAR-ti-lij) Hyaline cartilage that attaches a rib to the sternum.

Countercurrent mechanism One mechanism involved in the ability of the kidneys to produce hypertonic urine.

Cramp A spasmodic, usually painful contraction of a muscle.

Cranial (KRĀ-ne-al) **cavity** A subdivision of the dorsal body cavity formed by the cranial bones and containing the brain.

Cranial nerve One of 12 pairs of nerves that leave the brain; pass through foramina in the skull; and supply sensory and motor neurons to the head, neck, part of the trunk, and viscera of the thorax and abdomen. Each is designated by a Roman numeral and a name.

Craniosacral (krā-nē-ō-SĀ-kral) **outflow** The axons of parasympathetic preganglionic neurons, which have their cell bodies located in nuclei in the brain stem and in the lateral gray matter of the sacral portion of the spinal cord.

Cranium (KRĀ-nē-um) The skeleton of the skull that protects the brain and the organs of sight, hearing, and balance; includes the frontal, parietal, temporal, occipital, sphenoid, and ethmoid bones.

Creatine phosphate (KRĒ-a-tin FOS-fāt) Molecule in striated muscle fibers that contains high-energy phosphate bonds; used to generate ATP rapidly from ADP by transfer of a phosphate group. Also called **phosphocreatine** (fos′-fō-KRĒ-a-tin).

Crenation (krē-NĀ-shun) The shrinkage of red blood cells into knobbed, starry forms when they are placed in a hypertonic solution.

Crista (KRIS-ta) A crest or ridged structure. A small elevation in the ampulla of each semicircular duct that contains receptors for dynamic equilibrium.

Crossed extensor reflex A reflex in which extension of the joints in one limb occurs together with contraction of the flexor muscles of the opposite limb.

Crossing-over The exchange of a portion of one chromatid with another during meiosis. It permits an exchange of genes among chromatids and is one factor that results in genetic variation of progeny.

Crus (KRUS) **of penis** Separated, tapered portion of the corpora cavernosa penis. *Plural is* **crura** (KROO-ra).

Cryptorchidism (krip-TOR-ki-dizm) The condition of undescended testes.

Cuneate (KŪ-nē-āt) **nucleus** A group of neurons in the inferior part of the medulla oblongata in which axons of the cuneate fasciculus terminate.

Cupula (KUP-ū-la) A mass of gelatinous material covering the hair cells of a crista; a sensory receptor in the ampulla of a semicircular canal stimulated when the head moves.

Cushing's syndrome Condition caused by a hypersecretion of glucocorticoids characterized by spindly legs, "moon face," "buffalo hump," pendulous abdomen, flushed facial skin, poor wound healing, hyperglycemia, osteoporosis, hypertension, and increased susceptibility to disease.

Cutaneous (kū-TĀ-nē-us) Pertaining to the skin.

Cyanosis (sī-a-NŌ-sis) A blue or dark purple discoloration, most easily seen in nail beds and mucous membranes, that results from an increased concentration of deoxygenated (reduced) hemoglobin (more than 5 gm/dL).

Cyclic AMP (cyclic adenosine-3′, 5′-monophosphate) Molecule formed from ATP by the action of the enzyme adenylate cyclase; serves as second messenger for some hormones and neurotransmitters.

Cyst (SIST) A sac with a distinct connective tissue wall, containing a fluid or other material.

Cystic (SIS-tik) **duct** The duct that carries bile from the gallbladder to the common bile duct.

Cystitis (sis-TĪ-tis) Inflammation of the urinary bladder.

Cytosol (SĪ-tō-sol) Fluid located within cells. Also called **intracellular** (in′-tra-SEL-ū-lar) **fluid (ICF).**

Cytolysis (sī-TOL-i-sis) The rupture of living cells in which the contents leak out.

Cytochrome (SĪ-tō-krōm) A protein with an iron-containing group (heme) capable of alternating between a reduced form (Fe^{2+}) and an oxidized form (Fe^{3+}).

Cytokines (SĪ-to-kīns) Small protein hormones produced by lymphocytes, fibroblasts, endothelial cells, and antigen-presenting cells that act as autocrine or paracrine substances to stimulate or inhibit cell growth and differentiation, regulate immune responses, or aid nonspecific defenses.

Cytokinesis (sī-tō-ki-NĒ-sis) Distribution of the cytoplasm into two separate cells during cell division; coordinated with nuclear division (mitosis).

Cytoplasm (SĪ-tō-plazm) Cytosol plus all organelles except the nucleus.

Cytoskeleton Complex internal structure of cytoplasm consisting of microfilaments, microtubules, and intermediate filaments.

Cytosol (SĪ-tō-sol) Semifluid portion of cytoplasm in which organelles and inclusions are suspended and solutes are dissolved. Also called **intracellular fluid.**

D

Dartos (DAR-tōs) The contractile tissue deep to the skin of the scrotum.

Decibel (DES-i-bel) **(dB)** A unit for expressing the relative intensity (loudness) of sound.

Decidua (dē-SID-ū-a) That portion of the endometrium of the uterus (all but the deepest layer) that is modified during pregnancy and shed after childbirth.

Deciduous (dē-SID-ū-us) Falling off or being shed seasonally or at a particular stage of development. In the body, referring to the first set of teeth.

Decussation (dē-ku-SĀ-shun) A crossing-over to the opposite (contralateral) side; an example is the crossing of 90% of the axons in the large motor tracts to opposite sides in the medullary pyramids.

Deep Away from the surface of the body or an organ.

Deep fascia (FASH-ē-a) A sheet of connective tissue wrapped around a muscle to hold it in place.

Deep inguinal (IN-gwi-nal) **ring** A slitlike opening in the aponeurosis of the transversus abdominis muscle that represents the origin of the inguinal canal.

Defecation (def-e-KĀ-shun) The discharge of feces from the rectum.

Deglutition (dē-gloo-TISH-un) The act of swallowing.

Dehydration (dē-hī-DRĀ-shun) Excessive loss of water from the body or its parts.

Delta cell A cell in the pancreatic islets (islets of Langerhans) in the pancreas that secretes somatostatin. Also termed a **D cell.**

Demineralization (de-min′-er-al-i-ZĀ-shun) Loss of calcium and phosphorus from bones.

Denaturation (de-nā-chur-Ā-shun) Disruption of the tertiary structure of a protein by heat, changes in pH, or other physical or chemical methods, in which the protein loses its physical properties and biological activity.

Dendrite (DEN-drīt) A neuronal process that carries electrical signals, usually graded potentials, toward the cell body.

Dendritic (den-DRIT-ik) **cell** One type of antigen-presenting cell with long branchlike projections that commonly is present in mucosal linings such as the vagina, in the skin (Langerhans cells in the epidermis), and in lymph nodes (follicular dendritic cells).

Dental caries (KA-rēz) Gradual demineralization of the enamel and dentin of a tooth that may invade the pulp and alveolar bone. Also called **tooth decay.**

Denticulate (den-TIK-ū-lāt) Finely toothed or serrated; characterized by a series of small, pointed projections.

Dentin (DEN-tin) The bony tissues of a tooth enclosing the pulp cavity.

Dentition (den-TI-shun) The eruption of teeth. The number, shape, and arrangement of teeth.

Deoxyribonucleic (dē-ok′-sē-rī′-bō-noo-KLĒ-ik) **acid (DNA)** A nucleic acid constructed of nucleotides consisting of one of four bases (adenine, cytosine, guanine, or thymine), deoxyribose, and a phosphate group; encoded in the nucleotides is genetic information.

Depolarization (dē-pō-lar-i-ZĀ-shun) A reduction of voltage across a plasma membrane; expressed as a change toward less negative (more positive) voltages on the interior surface of the plasma membrane.

Depression (de-PRESH-un) Movement in which a part of the body moves inferiorly.

Dermal papilla (pa-PILL-a) Fingerlike projection of the papillary region of the dermis that may contain blood capillaries or corpuscles of touch (Meissner corpuscles).

Dermatology (der-ma-TOL-ō-jē) The medical specialty dealing with diseases of the skin.

Dermatome (DER-ma-tōm) The cutaneous area developed from one embryonic spinal cord segment and receiving most of its sensory innervation from one spinal nerve. An instrument for incising the skin or cutting thin transplants of skin.

Dermis (DER-mis) A layer of dense irregular connective tissue lying deep to the epidermis.

Descending colon (KŌ-lon) The part of the large intestine descending from the left colic (splenic) flexure to the level of the left iliac crest.

Detritus (de-TRĪ-tus) Particulate matter produced by or remaining after the wearing away or disintegration of a substance or tissue; scales, crusts, or loosened skin.

Detrusor (de-TROO-ser) **muscle** Smooth muscle that forms the wall of the urinary bladder.

Developmental biology The study of development from the fertilized egg to the adult form.

Diagnosis (dī-ag-NŌ-sis) Distinguishing one disease from another or determining the nature of a disease from signs and symptoms by inspection, palpation, laboratory tests, and other means.

Dialysis (dī-AL-i-sis) The removal of waste products from blood by diffusion through a selectively permeable membrane.

Diaphragm (DĪ-a-fram) Any partition that separates one area from another, especially the dome-shaped skeletal muscle between the thoracic and abdominal cavities. Also a dome-shaped device that is placed over the cervix, usually with a spermicide, to prevent conception.

Diaphysis (dī-AF-i-sis) The shaft of a long bone.

Diarrhea (dī-a-RE-a) Frequent defecation of liquid feces caused by increased motility of the intestines.

Diarthrosis (dī-ar-THRŌ-sis) A freely movable joint; types are gliding, hinge, pivot, condyloid, saddle, and ball-and-socket.

Diastole (dī-AS-tō-lē) In the cardiac cycle, the phase of relaxation or dilation of the heart muscle, especially of the ventricles.

Diastolic (dī-as-TOL-ik) **blood pressure** The force exerted by blood on arterial walls during ventricular relaxation; the lowest blood pressure measured in the large arteries, normally about 80 mmHg in a young adult.

Diencephalon (dī′-en-SEF-a-lon) A part of the brain consisting of the thalamus, hypothalamus, epithalamus, and subthalamus.

Diffusion (dif-Ū-zhun) A passive process in which there is a net or greater movement of molecules or ions from a region of high concentration to a region of low concentration until equilibrium is reached.

Digestion (dī-JES-chun) The mechanical and chemical breakdown of food to simple molecules that can be absorbed and used by body cells.

Dilate (DĪ-lāt) To expand or swell.

Diploid (DIP-loyd) Having the number of chromosomes characteristically found in the somatic cells of an organism; having two haploid sets of chromosomes, one each from the mother and father. Symbolized 2*n*.

Direct motor pathways Collections of upper motor neurons with cell bodies in the motor cortex that project axons into the spinal cord, where they synapse with lower motor neurons or interneurons in the anterior horns. Also called the **pyramidal pathways.**

Disease Any change from a state of health.

Dislocation (dis-lō-KĀ-shun) Displacement of a bone from a joint with tearing of ligaments, tendons, and articular capsules. Also called **luxation** (luks-Ā-shun).

Dissect (di-SEKT) To separate tissues and parts of a cadaver or an organ for anatomical study.

Distal (DIS-tal) Farther from the attachment of a limb to the trunk; farther from the point of origin or attachment.

Diuretic (dī-ū-RET-ik) A chemical that increases urine volume by decreasing reabsorption of water, usually by inhibiting sodium reabsorption.

Divergence (dī-VER-jens) A synaptic arrangement in which the synaptic end bulbs of one presynaptic neuron terminate on several postsynaptic neurons.

Diverticulum (dī-ver-TIK-ū-lum) A sac or pouch in the wall of a canal or organ, especially in the colon.

Dominant allele An allele that overrides the influence of an alternate allele on the homologous chromosome; the allele that is expressed.

Dorsal body cavity Cavity near the dorsal (posterior) surface of the body that consists of a cranial cavity and vertebral canal.

Dorsal ramus (RĀ-mus) A branch of a spinal nerve containing motor and sensory axons supplying the muscles, skin, and bones of the posterior part of the head, neck, and trunk.

Dorsiflexion (dor′-si-FLEK-shun) Bending the foot in the direction of the dorsum (upper surface).

Down-regulation Phenomenon in which there is a decrease in the number of receptors in response to an excess of a hormone or neurotransmitter.

Ductus arteriosus (DUK-tus ar-tē-rē-Ō-sus) A small vessel connecting the pulmonary trunk with the aorta; found only in the fetus.

Ductus (vas) deferens (DEF-er-ens) The duct that carries sperm from the epididymis to the ejaculatory duct. Also called the **seminal duct.**

Ductus epididymis (ep′-i-DID-i-mis) A tightly coiled tube inside the epididymis, distinguished into a head, body, and tail, in which sperm undergo maturation.

Ductus venosus (ve-NŌ-sus) A small vessel in the fetus that helps the circulation bypass the liver.

Duodenal (doo-ō-DĒ-nal) **gland** Gland in the submucosa of the duodenum that secretes an alkaline mucus to protect the lining of the small intestine from the action of enzymes and to help neutralize the acid in chyme. Also called **Brunner's** (BRUN-erz) **gland.**

Duodenum (doo′-ō-DĒ-num *or* doo-OD-e-num) The first 25 cm (10 in.) of the small intestine, which connects the stomach and the ileum.

Dura mater (DOO-ra MĀ-ter) The outermost of the three meninges (coverings) of the brain and spinal cord.

Dynamic equilibrium (ē-kwi-LIB-rē-um) The maintenance of body position, mainly the head, in response to sudden movements such as rotation.

Dysfunction (dis-FUNK-shun) Absence of completely normal function.

Dysmenorrhea (dis′-men-ō-RĒ-a) Painful menstruation.

Dysplasia (dis-PLĀ-zē-a) Change in the size, shape, and organization of cells due to chronic irritation or inflammation; may either revert to normal if stress is removed or progress to neoplasia.

Dyspnea (DISP-nē-a) Shortness of breath.

E

Eardrum A thin, semitransparent partition of fibrous connective tissue between the external auditory meatus and the middle ear. Also called the **tympanic membrane.**

Ectoderm The primary germ layer that gives rise to the nervous system and the epidermis of skin and its derivatives.

Ectopic (ek-TOP-ik) Out of the normal location, as in ectopic pregnancy.

Edema (e-DĒ-ma) An abnormal accumulation of interstitial fluid.

Effector (e-FEK-tor) An organ of the body, either a muscle or a gland, that is innervated by somatic or autonomic motor neurons.

Efferent arteriole (EF-er-ent ar-TĒ-rē-ōl) A vessel of the renal vascular system that carries blood from a glomerulus to a peritubular capillary.

Efferent (EF-er-ent) **ducts** A series of coiled tubes that transport sperm from the rete testis to the epididymis.

Eicosanoids (ī-KŌ-sa-noyds) Local hormones derived from a 20-carbon fatty acid (arachidonic acid); two important types are prostaglandins and leukotrienes.

Ejaculation (e-jak-ū-LĀ-shun) The reflex ejection or expulsion of semen from the penis.

Ejaculatory (e-JAK-ū-la-tō-rē) **duct** A tube that transports sperm from the ductus (vas) deferens to the prostatic urethra.

Elasticity (e-las-TIS-i-tē) The ability of tissue to return to its original shape after contraction or extension.

Electrocardiogram (e-lek′-trō-KAR-dē-ō-gram) (**ECG** or **EKG**) A recording of the electrical changes that accompany the cardiac cycle that can be detected at the surface of the body; may be resting, stress, or ambulatory.

Electroencephalogram (e-lek′-trō-en-SEF-a-lō-gram) (**EEG**) A recording of the electrical activity of the brain from the scalp surface; used to diagnose certain diseases (such as epilepsy), furnish information regarding sleep and wakefulness, and confirm brain death.

Electrolyte (ē-LEK-trō-līt) Any compound that separates into ions when dissolved in water and that conducts electricity.

Electromyography (e-lek′-trō-mī-OG-ra-fē) Evaluation of the electrical activity of resting and contracting muscle to ascertain causes of muscular weakness, paralysis, involuntary twitching, and abnormal levels of muscle enzymes; also used as part of biofeedback studies.

Electron transport chain A sequence of electron carrier molecules on the inner mitochondrial membrane that undergo oxidation and reduction as they pump hydrogen ions (H^+) through the membrane. ATP synthesis then occurs as H^+ diffuse back into the mitochondrial matrix through special H^+ channels. *See also* **Chemiosmosis.**

Elevation (el-e-VĀ-shun) Movement in which a part of the body moves superiorly.

Embolism (EM-bō-lizm) Obstruction or closure of a vessel by an embolus.

Embolus (EM-bō-lus) A blood clot, bubble of air or fat from broken bones, mass of bacteria, or other debris or foreign material transported by the blood.

Embryo (EM-brē-ō) The young of any organism in an early stage of development; in humans, the developing organism from fertilization to the end of the eighth week of development.

Embryology (em′-brē-OL-ō-jē) The study of development from the fertilized egg to the end of the eighth week of development.

Emesis (EM-e-sis) Vomiting.

Emigration (em′-e-GRĀ-shun) Process whereby white blood cells (WBCs) leave the bloodstream by rolling along the endothelium, sticking to it, and squeezing between the endothelial cells. Adhesion molecules help WBCs stick to the endothelium. Also known as **migration** or **extravasation.**

Emission (ē-MISH-un) Propulsion of sperm into the urethra due to peristaltic contractions of the ducts of the testes, epididymides, and ductus (vas) deferens as a result of sympathetic stimulation.

Emmetropia (em′-e-TRŌ-pē-a) Normal vision in which light rays are focused exactly on the retina.

Emphysema (em′-fi′-SĒ-ma) A lung disorder in which alveolar walls disintegrate, producing abnormally large air spaces and loss of elasticity in the lungs; typically caused by exposure to cigarette smoke.

Emulsification (ē-mul′-si-fi-KĀ-shun) The dispersion of large lipid globules into smaller, uniformly distributed particles in the presence of bile.

Enamel (e-NAM-el) The hard, white substance covering the crown of a tooth.

End-diastolic (dī-as-TO-lik) **volume (EDV)** The volume of blood, about 130 mL, remaining in a ventricle at the end of its diastole (relaxation).

Endergonic (end′-er-GON-ik) **reaction** Type of chemical reaction in which the energy released as new bonds form is less than the energy needed to break apart old bonds; an energy-requiring reaction.

Endocardium (en-dō-KAR-dē-um) The layer of the heart wall, composed of endothelium and smooth muscle, that lines the inside of the heart and covers the valves and tendons that hold the valves open.

Endochondral ossification (en′-dō-KON-dral os′-i-fi-KĀ-shun) The replacement of cartilage by bone. Also called **intracartilaginous** (in′-tra-kar′-ti-LAJ-i-nus) **ossification.**

Endocrine (EN-dō-krin) **gland** A gland that secretes hormones into interstitial fluid and then the blood; a ductless gland.

Endocrinology (en′-dō-kri-NOL-ō-jē) The science concerned with the structure and functions of endocrine glands and the diagnosis and treatment of disorders of the endocrine system.

Endocytosis (en′-dō-sī-TŌ-sis) The uptake into a cell of large molecules and particles in which a segment of plasma membrane surrounds the substance, encloses it, and brings it in; includes phagocytosis, pinocytosis, and receptor-mediated endocytosis.

Endoderm (EN-dō-derm) A primary germ layer of the developing embryo; gives rise to the gastrointestinal tract, urinary bladder, urethra, and respiratory tract.

Endodontics (en′-dō-DON-tiks) The branch of dentistry concerned with the prevention, diagnosis, and treatment of diseases that affect the pulp, root, periodontal ligament, and alveolar bone.

Endogenous (en-DOJ-e-nus) Growing from or beginning within the organism.

Endolymph (EN-dō-limf′) The fluid within the membranous labyrinth of the internal ear.

Endometrium (en′-dō-MĒ-trē-um) The mucous membrane lining the uterus.

Endomysium (en′-dō-MĪZ-ē-um) Invagination of the perimysium separating each individual muscle fiber (cell).

Endoneurium (en′-dō-NOO-rē-um) Connective tissue wrapping around individual nerve axons (cells).

Endoplasmic reticulum (en′-do-PLAZ-mik re-TIK-ū-lum) **(ER)** A network of channels running through the cytoplasm of a cell that serves in intracellular transportation, support, storage, synthesis, and packaging of molecules. Portions of ER where ribosomes are attached to the outer surface are called **rough ER;** portions that have no ribosomes are called **smooth ER.**

Endorphin (en-DOR-fin) A neuropeptide in the central nervous system that acts as a painkiller.

Endosteum (en-DOS-tē-um) The membrane that lines the medullary (marrow) cavity of bones, consisting of osteogenic cells and scattered osteoclasts.

Endothelium (en′-dō-THĒ-lē-um) The layer of simple squamous epithelium that lines the cavities of the heart, blood vessels, and lymphatic vessels.

End-systolic (sis-TO-lik) **volume (ESV)** The volume of blood, about 60 mL, remaining in a ventricle after its systole (contraction).

Energy The capacity to do work.

Enkephalin (en-KEF-a-lin) A peptide found in the central nervous system that acts as a painkiller.

Enteric (EN-ter-ik) **nervous system** The part of the nervous system that is embedded in the submucosa and muscularis of the gastrointestinal (GI) tract; governs motility and secretions of the GI tract.

Enteroendocrine (en-ter-ō-EN-dō-krin) **cell** A cell of the mucosa of the gastrointestinal tract that secretes a hormone that governs function of the GI tract; hormones secreted include gastrin, cholecystokinin, glucose-dependent insulinotropic peptide (GIP), and secretin.

Enterogastric (en-te-rō-GAS-trik) **reflex** A reflex that inhibits gastric secretion; initiated by food in the small intestine.

Enzyme (EN-zīm) A substance that accelerates chemical reactions; an organic catalyst, usually a protein.

Eosinophil (ē′-ō-SIN-ō-fil) A type of white blood cell characterized by granules that stain red or pink with acid dyes.

Ependymal (e-PEN-de-mal) **cells** Neuroglial cells that cover choroid plexuses and produce cerebrospinal fluid (CSF); they also line the ventricles of the brain and probably assist in the circulation of CSF.

Epicardium (ep′-i-KAR-dē-um) The thin outer layer of the heart wall, composed of serous tissue and mesothelium. Also called the **visceral pericardium.**

Epidemiology (ep′-i-dē-mē-OL-ō-jē) Study of the occurrence and distribution of diseases and disorders in human populations.

Epidermis (ep-i-DERM-is) The superficial, thinner layer of skin, composed of keratinized stratified squamous epithelium.

Epididymis (ep′-i-DID-i-mis) A comma-shaped organ that lies along the posterior border of the testis and contains the ductus epididymis, in which sperm undergo maturation. *Plural is* **epididymides** (ep′-i-DID-i-mi-dēz).

Epidural (ep′-i-DOO-ral) **space** A space between the spinal dura mater and the vertebral canal, containing areolar connective tissue and a plexus of veins.

Epiglottis (ep′-i-GLOT-is) A large, leaf-shaped piece of cartilage lying on top of the larynx, attached to the thyroid cartilage and its unattached portion is free to move up and down to cover the glottis (vocal folds and rima glottidis) during swallowing.

Epimysium (ep′-i-MĪZ-ē-um) Fibrous connective tissue around muscles.

Epinephrine (ep-ē-NEF-rin) Hormone secreted by the adrenal medulla that produces actions similar to those that result from sympathetic stimulation. Also called **adrenaline** (a-DREN-a-lin).

Epineurium (ep′-i-NOO-rē-um) The superficial connective tissue covering around an entire nerve.

Epiphyseal (ep′-i-FIZ-ē-al) **line** The remnant of the epiphyseal plate in the metaphysis of a long bone.

Epiphyseal (ep′-i-FIZ-ē-al) **plate** The hyaline cartilage plate in the metaphysis of a long bone; site of lengthwise growth of long bones.

Epiphysis (ē-PIF-i-sis) The end of a long bone, usually larger in diameter than the shaft (diaphysis).

Episiotomy (e-piz′-ē-OT-ō-mē) A cut made with surgical scissors to avoid tearing of the perineum at the end of the second stage of labor.

Epistaxis (ep′-i-STAK-sis) Loss of blood from the nose due to trauma, infection, allergy, neoplasm, and bleeding disorders. Also called **nosebleed.**

Epithelial (ep′-i-THĒ-lē-al) **tissue** The tissue that forms innermost and outermost surfaces of body structures and forms glands.

Eponychium (ep′-o-NIK-ē-um) Narrow band of stratum corneum at the proximal border of a nail that extends from the margin of the nail wall. Also called the **cuticle.**

Erectile dysfunction Failure to maintain an erection long enough for sexual intercourse. Also known as **impotence** (IM-pō-tens).

Erection (ē-REK-shun) The enlarged and stiff state of the penis or clitoris resulting from the engorgement of the spongy erectile tissue with blood.

Eructation (e-ruk′-TĀ-shun) The forceful expulsion of gas from the stomach. Also called **belching.**

Erythema (er′-i-THĒ-ma) Skin redness usually caused by dilation of the capillaries.

Erythrocyte (e-RITH-rō-sīt) A mature red blood cell.

Erythropoiesis (e-rith′-rō-poy-Ē-sis) The process by which red blood cells are formed.

Erythropoietin (e-rith′-rō-POY-e-tin) A hormone released by the juxtaglomerular cells of the kidneys that stimulates red blood cell production.

Esophagus (e-SOF-a-gus) The hollow muscular tube that connects the pharynx and the stomach.

Essential amino acids Those 10 amino acids that cannot be synthesized by the human body at an adequate rate to meet its needs and therefore must be obtained from the diet.

Estrogens (ES-tro-jenz) Feminizing sex hormones produced by the ovaries; govern development of oocytes, maintenance of female reproductive structures, and appearance of secondary sex characteristics; also affect fluid and electrolyte balance, and protein anabolism. Examples are β-estradiol, estrone, and estriol.

Etiology (ē′-tē-OL-ō-jē) The study of the causes of disease, including theories of the origin and organisms (if any) involved.

Eupnea (ŪP-nē-a) Normal quiet breathing.

Eversion (ē-VER-zhun) The movement of the sole laterally at the ankle joint or of an atrioventricular valve into an atrium during ventricular contraction.

Excitability (ek-sīt′-a-BIL-i-tē) The ability of muscle fibers to receive and respond to stimuli; the ability of neurons to respond to stimuli and generate nerve impulses.

Excitatory postsynaptic potential (EPSP) A small depolarization of a postsynaptic membrane when it is stimulated by an excitatory neurotransmitter. An EPSP is both graded (variable size) and localized (decreases in size away from the point of excitation).

Excrement (EKS-kre-ment) Material eliminated from the body as waste, especially fecal matter.

Excretion (eks-KRĒ-shun) The process of eliminating waste products from the body; also the products excreted.

Exergonic (eks'-er-GON-ik) **reaction** Type of chemical reaction in which the energy released as new bonds form is greater than the energy needed to break apart old bonds; an energy-releasing reaction.

Exocrine (EK-sō-krin) **gland** A gland that secretes its products into ducts that carry the secretions into body cavities, into the lumen of an organ, or to the outer surface of the body.

Exocytosis (ex'-ō-sī-TŌ-sis) A process in which membrane-enclosed secretory vesicles form inside the cell, fuse with the plasma membrane, and release their contents into the interstitial fluid; achieves secretion of materials from a cell.

Exogenous (ex-SOJ-e-nus) Originating outside an organ or part.

Exhalation (eks-ha-LĀ-shun) Breathing out; expelling air from the lungs into the atmosphere. Also called **expiration.**

Exon (EX-on) A region of DNA that codes for synthesis of a protein.

Expiratory (eks-PĪ-ra-tō-rē) **reserve volume** The volume of air in excess of tidal volume that can be exhaled forcibly; about 1200 mL.

Extensibility (ek-sten'-si-BIL-i-tē) The ability of muscle tissue to stretch when it is pulled.

Extension (ek-STEN-shun) An increase in the angle between two bones; restoring a body part to its anatomical position after flexion.

External Located on or near the surface.

External auditory (AW-di-tōr-ē) **canal** or **meatus** (mē-Ā-tus) A curved tube in the temporal bone that leads to the middle ear.

External ear The outer ear, consisting of the pinna, external auditory canal, and tympanic membrane (eardrum).

External nares (NĀ-rez) The external nostrils, or the openings into the nasal cavity on the exterior of the body.

External respiration The exchange of respiratory gases between the lungs and blood. Also called **pulmonary respiration.**

Exteroceptor (eks'-ter-ō-SEP-tor) A sensory receptor adapted for the reception of stimuli from outside the body.

Extracellular fluid (ECF) Fluid outside body cells, such as interstitial fluid and plasma.

Extravasation (eks-trav-a-SĀ-shun) The escape of fluid, especially blood, lymph, or serum, from a vessel into the tissues.

Extrinsic (ek-STRIN-sik) Of external origin.

Extrinsic pathway (of blood clotting) Sequence of reactions leading to blood clotting that is initiated by the release of tissue factor (TF), also known as thromboplastin, that leaks into the blood from damaged cells outside the blood vessels.

Exudate (EKS-oo-dāt) Escaping fluid or semifluid material that oozes from a space and that may contain serum, pus, and cellular debris.

Eyebrow The hairy ridge superior to the eye.

F

Face The anterior aspect of the head.

Facilitated diffusion (fa-SIL-i-tā-ted dif-Ū-zhun) Diffusion in which a substance not soluble by itself in lipids diffuses across a selectively permeable membrane with the help of a transporter protein.

Falciform ligament (FAL-si-form LIG-a-ment) A sheet of parietal peritoneum between the two principal lobes of the liver. The ligamentum teres, or remnant of the umbilical vein, lies within its fold.

Falx cerebelli (FALKS ser'-e-BEL-lē) A small triangular process of the dura mater attached to the occipital bone in the posterior cranial fossa and projecting inward between the two cerebellar hemispheres.

Falx cerebri (FALKS SER-e-brē) A fold of the dura mater extending deep into the longitudinal fissure between the two cerebral hemispheres.

Fascia (FASH-ē-a) A fibrous membrane covering, supporting, and separating muscles.

Fascicle (FAS-i-kul) A small bundle or cluster, especially of nerve or muscle fibers (cells). Also called a **fasciculus** (fa-SIK-ū-lus). *Plural is* **fasciculi** (fa-SIK-yoo-lī).

Fasciculation (fa-sik'-ū-LĀ-shun) Abnormal, spontaneous twitch of all skeletal muscle fibers in one motor unit that is visible at the skin surface; not associated with movement of the affected muscle; present in progressive diseases of motor neurons, for example, poliomyelitis.

Fauces (FAW-sēz) The opening from the mouth into the pharynx.

F cell A cell in the pancreatic islets (islets of Langerhans) that secretes pancreatic polypeptide.

Feces (FĒ-sēz) Material discharged from the rectum and made up of bacteria, excretions, and food residue. Also called **stool.**

Feedback system A sequence of events in which information about the status of a situation is continually reported (fed back) to a control center.

Female reproductive cycle General term for the ovarian and uterine cycles, the hormonal changes that accompany them, and cyclic changes in the breasts and cervix; includes changes in the endometrium of a nonpregnant female that prepares the lining of the uterus to receive a fertilized ovum. Less correctly termed the **menstrual cycle.**

Fertilization (fer'-ti-li-ZĀ-shun) Penetration of a secondary oocyte by a sperm cell, meiotic division of secondary oocyte to form an ovum, and subsequent union of the nuclei of the gametes.

Fetal circulation The cardiovascular system of the fetus, including the placenta and special blood vessels involved in the exchange of materials between fetus and mother.

Fetus (FĒ-tus) In humans, the developing organism *in utero* from the beginning of the third month to birth.

Fever An elevation in body temperature above the normal temperature of 37°C (98.6°F) due to a resetting of the hypothalamic thermostat.

Fibrillation (fi-bri-LĀ-shun) Abnormal, spontaneous twitch of a single skeletal muscle fiber (cell) that can be detected with electromyography but is not visible at the skin surface; not associated with movement of the affected muscle; present in certain disorders of motor neurons, for example, amyotrophic lateral sclerosis (ALS). With reference to cardiac muscle, *see* **Atrial fibrillation** and **Ventricular fibrillation.**

Fibrin (FĪ-brin) An insoluble protein that is essential to blood clotting; formed from fibrinogen by the action of thrombin.

Fibrinogen (fi-BRIN-ō-jen) A clotting factor in blood plasma that by the action of thrombin is converted to fibrin.

Fibrinolysis (fi-bri-NOL-i-sis) Dissolution of a blood clot by the action of a proteolytic enzyme, such as plasmin (fibrinolysin), that dissolves fibrin threads and inactivates fibrinogen and other blood-clotting factors.

Fibroblast (FĪ-brō-blast) A large, flat cell that secretes most of the matrix (extracellular) material of areolar and dense connective tissues.

Fibrous (FĪ-brus) **joint** A joint that allows little or no movement, such as a suture or a syndesmosis.

Fibrous tunic (TOO-nik) The superficial coat of the eyeball, made up of the posterior sclera and the anterior cornea.

Fight-or-flight response The effects produced upon stimulation of the sympathetic division of the autonomic nervous system.

Filiform papilla (FIL-i-form pa-PIL-a) One of the conical projections that are distributed in parallel rows over the anterior two-thirds of the tongue and lack taste buds.

Filtration (fil-TRĀ-shun) The flow of a liquid through a filter (or membrane that acts like a filter) due to a hydrostatic pressure; occurs in capillaries due to blood pressure.

Filtration fraction The percentage of plasma entering the kidneys that becomes glomerular filtrate.

Filtration membrane Site of blood filtration in nephrons of the kidneys, consisting of the endothelium and basement membrane of the glomerulus and the epithelium of the visceral layer of the glomerular (Bowman's) capsule.

Filum terminale (FĪ-lum ter-mi-NAL-ē) Non-nervous fibrous tissue of the spinal cord that extends inferiorly from the conus medullaris to the coccyx.

Fimbriae (FIM-brē-ē) Fingerlike structures, especially the lateral ends of the uterine (Fallopian) tubes.

Fissure (FISH-ur) A groove, fold, or slit that may be normal or abnormal.

Fistula (FIS-tū-la) An abnormal passage between two organs or between an organ cavity and the outside.

Fixator A muscle that stabilizes the origin of the prime mover so that the prime mover can act more efficiently.

Fixed macrophage (MAK-rō-fāj) Stationary phagocytic cell found in the liver, lungs, brain, spleen, lymph nodes, subcutaneous tissue, and red bone marrow. Also called a **histiocyte** (HIS-tē-ō-sīt).

Flaccid (FLAS-sid) Relaxed, flabby, or soft; lacking muscle tone.

Flagellum (fla-JEL-um) A hairlike, motile process on the extremity of a bacterium, protozoan, or sperm cell. *Plural is* **flagella** (fla-JEL-a).

Flatus (FLĀ-tus) Gas in the stomach or intestines, commonly used to denote expulsion of gas through the anus.

Flexion (FLEK-shun) Movement in which there is a decrease in the angle between two bones.

Flexor reflex A protective reflex in which flexor muscles are stimulated while extensor muscles are inhibited.

Follicle (FOL-i-kul) A small secretory sac or cavity; the group of cells that contains a developing oocyte in the ovaries.

Follicle-stimulating hormone (FSH) Hormone secreted by the anterior pituitary that initiates development of ova and stimulates the ovaries to secrete estrogens in females, and initiates sperm production in males.

Fontanel (fon′-ta-NEL) A fibrous connective tissue membrane-filled space where bone formation is not yet complete, especially between the cranial bones of an infant's skull.

Foot The terminal part of the lower limb, from the ankle to the toes.

Foramen (fō-RĀ-men) A passage or opening; a communication between two cavities of an organ, or a hole in a bone for passage of vessels or nerves. *Plural is* **foramina** (fō-RAM-i-na).

Foramen ovale (fō-RĀ-men-ō-VAL-ē) An opening in the fetal heart in the septum between the right and left atria. A hole in the greater wing of the sphenoid bone that transmits the mandibular branch of the trigeminal nerve (cranial nerve V).

Forearm (FOR-arm) The part of the upper limb between the elbow and the wrist.

Fornix (FOR-niks) An arch or fold; a tract in the brain made up of association fibers, connecting the hippocampus with the mammillary bodies; a recess around the cervix of the uterus where it protrudes into the vagina.

Fossa (FOS-a) A furrow or shallow depression.

Fourth ventricle (VEN-tri-kul) A cavity filled with cerebrospinal fluid within the brain lying between the cerebellum and the medulla oblongata and pons.

Fracture (FRAK-choor) Any break in a bone.

Frenulum (FREN-ū-lum) A small fold of mucous membrane that connects two parts and limits movement.

Frontal plane A plane at a right angle to a midsagittal plane that divides the body or organs into anterior and posterior portions. Also called a **coronal** (kō-RŌ-nal) **plane.**

Functional residual (re-ZID-ū-al) **capacity** The sum of residual volume plus expiratory reserve volume; about 2400 mL.

Fundus (FUN-dus) The part of a hollow organ farthest from the opening.

Fungiform papilla (FUN-ji-form pa-PIL-a) A mushroomlike elevation on the upper surface of the tongue appearing as a red dot; most contain taste buds.

G

Gallbladder A small pouch, located inferior to the liver, that stores bile and empties by means of the cystic duct.

Gallstone A solid mass, usually containing cholesterol, in the gallbladder or a bile-containing duct; formed anywhere between bile canaliculi in the liver and the hepatopancreatic ampulla (ampulla of Vater), where bile enters the duodenum. Also called a **biliary calculus.**

Gamete (GAM-ēt) A male or female reproductive cell; a sperm cell or secondary oocyte.

Ganglion (GANG-glē-on) Usually, a group of neuronal cell bodies lying outside the central nervous system (CNS). *Plural is* **ganglia** (GANG-glē-a).

Gastric (GAS-trik) **glands** Glands in the mucosa of the stomach composed of cells that empty their secretions into narrow channels called gastric pits. Types of cells are chief cells (secrete pepsinogen), parietal cells (secrete hydrochloric acid and intrinsic factor), surface mucous and mucous neck cells (secrete mucus), and G cells (secrete gastrin).

Gastroenterology (gas′-trō-en′-ter-OL-ō-jē) The medical specialty that deals with the structure, function, diagnosis, and treatment of diseases of the stomach and intestines.

Gastrointestinal (gas-trō-in-TES-ti-nal) **(GI) tract** A continuous tube running through the ventral body cavity extending from the mouth to the anus. Also called the **alimentary** (al′-i-MEN-tar-ē) **canal.**

Gastrulation (gas′-troo-LĀ-shun) The migration of groups of cells from the epiblast that transform a bilaminar embryonic disc into a trilaminar embryonic disc that consists of the three primary germ layers; transformation of the blastula into the gastrula.

Gene (JĒN) Biological unit of heredity; a segment of DNA located in a definite position on a particular chromosome; a sequence of DNA that codes for a particular mRNA, rRNA, or tRNA.

Generator potential The graded depolarization that results in a change in the resting membrane potential in a receptor (specialized neuronal ending); may trigger a nerve action potential (nerve impulse) if depolarization reaches threshold.

Genetic engineering The manufacture and manipulation of genetic material.

Genetics The study of genes and heredity.

Genitalia (jen′-i-TĀ-lē-a) Reproductive organs.

Genome (JĒ-nōm) The complete set of genes of an organism.

Genotype (JĒ-nō-tīp) The genetic makeup of an individual; the combination of alleles present at one or more chromosomal locations, as distinguished from the appearance, or phenotype, that results from those alleles.

Geriatrics (jer′-ē-AT-riks) The branch of medicine devoted to the medical problems and care of elderly persons.

Gestation (jes-TĀ-shun) The period of development from fertilization to birth.

Gingivae (jin-JI-vē) Gums. They cover the alveolar processes of the mandible and maxilla and extend slightly into each socket.

Gland Specialized epithelial cell or cells that secrete substances; may be exocrine or endocrine.

Glans penis (glanz PĒ-nis) The slightly enlarged region at the distal end of the penis.

Glaucoma (glaw-KŌ-ma) An eye disorder in which there is increased intraocular pressure due to an excess of aqueous humor.

Gliding joint A synovial joint having articulating surfaces that are usually flat, permitting only side-to-side and back-and-forth movements, as between carpal bones, tarsal bones, and the scapula and clavicle. Also called an **arthrodial** (ar-THRŌ-dē-al) **joint.**

Glomerular (glō-MER-ū-lar) **capsule** A double-walled globe at the proximal end of a nephron that encloses the glomerular capillaries. Also called **Bowman's** (BŌ-manz) **capsule.**

Glomerular filtrate (glō-MER-ū-lar FIL-trāt) The fluid produced when blood is filtered by the filtration membrane in the glomeruli of the kidneys.

Glomerular filtration The first step in urine formation in which substances in blood pass through the filtration membrane and the filtrate enters the proximal convoluted tubule of a nephron.

Glomerular filtration rate (GFR) The total volume of fluid that enters all the glomerular (Bowman's) capsules of the kidneys in 1 min; about 100–125 mL/min.

Glomerulus (glō-MER-ū-lus) A rounded mass of nerves or blood vessels, especially the microscopic tuft of capillaries that is surrounded by the glomerular (Bowman's) capsule of each kidney tubule. *Plural is* **glomeruli.**

Glottis (GLOT-is) The vocal folds (true vocal cords) in the larynx plus the space between them (rima glottidis).

Glucagon (GLOO-ka-gon) A hormone produced by the alpha cells of the pancreatic islets (islets of Langerhans) that increases blood glucose level.

Glucocorticoids (gloo-kō-KOR-ti-koyds) Hormones secreted by the cortex of the adrenal gland, especially cortisol, that influence glucose metabolism.

Gluconeogenesis (gloo′-kō-nē-ō-JEN-e-sis) The synthesis of glucose from certain amino acids or lactic acid.

Glucose (GLOO-kōs) A hexose (six-carbon sugar), $C_6H_{12}O_6$, that is a major energy source for the production of ATP by body cells.

Glucosuria (gloo′-kō-SOO-rē-a) The presence of glucose in the urine; may be temporary or pathological. Also called **glycosuria.**

Glycogen (GLĪ-kō-jen) A highly branched polymer of glucose containing thousands of subunits; functions as a compact store of glucose molecules in liver and muscle fibers (cells).

Glycogenesis (glī′-kō-JEN-e-sis) The chemical reactions by which many molecules of glucose are used to synthesize glycogen.

Glycogenolysis (glī-kō-je-NOL-i-sis) The breakdown of glycogen into glucose.

Glycolysis (glī-KOL-i-sis) Series of chemical reactions in the cytosol of a cell in which a molecule of glucose is split into two molecules of pyruvic acid with net production of two ATPs.

Goblet cell A goblet-shaped unicellular gland that secretes mucus; present in epithelium of the airways and intestines.

Goiter (GOY-ter) An enlarged thyroid gland.

Golgi (GOL-jē) **complex** An organelle in the cytoplasm of cells consisting of four to six flattened sacs (cisternae), stacked on one another, with expanded areas at their ends; functions in processing, sorting, packaging, and delivering proteins and lipids to the plasma membrane, lysosomes, and secretory vesicles.

Gomphosis (gom-FŌ-sis) A fibrous joint in which a cone-shaped peg fits into a socket.

Gonad (GŌ-nad) A gland that produces gametes and hormones; the ovary in the female and the testis in the male.

Gonadotropic hormone Anterior pituitary hormone that affects the gonads.

Gracile (GRAS-īl) **nucleus** A group of nerve cells in the inferior part of the medulla oblongata in which axons of the gracile fasciculus terminate.

Gray commissure (KOM-i-shur) A narrow strip of gray matter connecting the two lateral gray masses within the spinal cord.

Gray matter Areas in the central nervous system and ganglia containing neuronal cell bodies, dendrites, unmyelinated axons, axon terminals, and neuroglia; Nissl bodies impart a gray color and there is little or no myelin in gray matter.

Gray ramus communicans (RĀ-mus kō-MŪ-ni-kans) A short nerve containing axons of sympathetic postganglionic neurons; the cell bodies of the neurons are in a sympathetic chain ganglion, and the unmyelinated axons extend via the gray ramus to a spinal nerve and then to the periphery to supply smooth muscle in blood vessels, arrector pili muscles, and sweat glands. *Plural is* **rami communicantes** (RĀ-mē kō-mū-ni-KAN-tēz).

Greater omentum (ō-MEN-tum) A large fold in the serosa of the stomach that hangs down like an apron anterior to the intestines.

Greater vestibular (ves-TIB-ū-lar) **glands** A pair of glands on either side of the vaginal orifice that open by a duct into the space between the hymen and the labia minora. Also called **Bartholin's** (BAR-to-linz) **glands.**

Groin (GROYN) The depression between the thigh and the trunk; the inguinal region.

Gross anatomy The branch of anatomy that deals with structures that can be studied without using a microscope. Also called **macroscopic anatomy.**

Growth An increase in size due to an increase in (1) the number of cells, (2) the size of existing cells as internal components increase in size, or (3) the size of intercellular substances.

Gustatory (GUS-ta-tō′-rē) Pertaining to taste.

Gynecology (gī′-ne-KOL-ō-jē) The branch of medicine dealing with the study and treatment of disorders of the female reproductive system.

Gynecomastia (gīn′-e-kō-MAS-tē-a) Excessive growth (benign) of the male mammary glands due to secretion of estrogens by an adrenal gland tumor (feminizing adenoma).

Gyrus (JĪ-rus) One of the folds of the cerebral cortex of the brain. *Plural is* **gyri** (JĪ-rī). Also called a **convolution.**

H

Hair A threadlike structure produced by hair follicles that develops in the dermis. Also called a **pilus** (PĪ-lus).

Hair follicle (FOL-li-kul) Structure, composed of epithelium and surrounding the root of a hair, from which hair develops.

Hair root plexus (PLEK-sus) A network of dendrites arranged around the root of a hair as free or naked nerve endings that are stimulated when a hair shaft is moved.

Haldane effect Less carbon dioxide binds to hemoglobin when the partial pressure of oxygen is higher.

Hand The terminal portion of an upper limb, including the carpus, metacarpus, and phalanges.

Haploid (HAP-loyd) Having half the number of chromosomes characteristically found in the somatic cells of an organism; characteristic of mature gametes. Symbolized *n.*

Hard palate (PAL-at) The anterior portion of the roof of the mouth, formed by the maxillae and palatine bones and lined by mucous membrane.

Haustra (HAWS-tra) A series of pouches that characterize the colon; caused by tonic contractions of the teniae coli. *Singular is* **haustrum.**

Head The superior part of a human, cephalic to the neck. The superior or proximal part of a structure.

Heart A hollow muscular organ lying slightly to the left of the midline of the chest that pumps the blood through the cardiovascular system.

Heart block An arrhythmia (dysrhythmia) of the heart in which the atria and ventricles contract independently because of a blocking of electrical impulses through the heart at some point in the conduction system.

Heart murmur (MER-mer) An abnormal sound that consists of a flow noise that is heard before, between, or after the normal heart sounds, or that may mask normal heart sounds.

Heat exhaustion Condition characterized by cool, clammy skin, profuse perspiration, and fluid and electrolyte (especially sodium and chloride) loss that results in muscle cramps, dizziness, vomiting, and fainting. Also called **heat prostration.**

Heat stroke Condition produced when the body cannot easily lose heat and characterized by reduced perspiration and elevated body temperature. Also called **sunstroke.**

Hematocrit (hē-MAT-ō-krit) **(Hct)** The percentage of blood made up of red blood cells. Usually measured by centrifuging a blood sample in a graduated tube and then reading the volume of red blood cells and dividing it by the total volume of blood in the sample.

Hematology (hē′-ma-TOL-ō-jē) The study of blood.

Hemiplegia (hem-i-PLĒ-jē-a) Paralysis of the upper limb, trunk, and lower limb on one side of the body.

Hemodynamics (hē-mō-dī-NA-miks) The study of factors and forces that govern the flow of blood through blood vessels.

Hemoglobin (hē′-mō-GLŌ-bin) **(Hb)** A substance in red blood cells consisting of the protein globin and the iron-containing red pigment heme that transports most of the oxygen and some carbon dioxide in blood.

Hemolysis (hē-MOL-i-sis) The escape of hemoglobin from the interior of a red blood cell into the surrounding medium; results from disruption of the cell membrane by toxins or drugs, freezing or thawing, or hypotonic solutions.

Hemolytic disease of the newborn A hemolytic anemia of a newborn child that results from the destruction of the infant's erythrocytes (red blood cells) by antibodies produced by the mother; usually the antibodies are due to an Rh blood type incompatibility. Also called **erythroblastosis fetalis** (e-rith′-rō-blas-TŌ-sis fe-TAL-is).

Hemophilia (hē′-mō-FIL-ē-a) A hereditary blood disorder where there is a deficient production of certain factors involved in blood clotting, resulting in excessive bleeding into joints, deep tissues, and elsewhere.

Hemopoiesis (hē-mō-poy-Ē-sis) Blood cell production, which occurs in red bone marrow after birth. Also called **hematopoiesis** (hem′-a-tō-poy-Ē-sis).

Hemorrhage (HEM-or-rij) Bleeding; the escape of blood from blood vessels, especially when the loss is profuse.

Hemorrhoids (HEM-ō-royds) Dilated or varicosed blood vessels (usually veins) in the anal region. Also called **piles.**

Hemostasis (hē-MŌS-tā-sis) The stoppage of bleeding.

Heparin (HEP-a-rin) An anticoagulant given to slow the conversion of prothrombin to thrombin, thus reducing the risk of blood clot formation; found in basophils, mast cells, and various other tissues, especially the liver and lungs.

Hepatic (he-PAT-ik) Refers to the liver.

Hepatic duct A duct that receives bile from the bile capillaries. Small hepatic ducts merge to form the larger right and left hepatic ducts that unite to leave the liver as the common hepatic duct.

Hepatic portal circulation The flow of blood from the gastrointestinal organs to the liver before returning to the heart.

Hepatocyte (he-PAT-ō-cyte) A liver cell.

Hepatopancreatic (hep′-a-tō-pan′-krē-A-tik) **ampulla** A small, raised area in the duodenum where the combined common bile duct and main pancreatic duct empty into the duodenum. Also called the **ampulla of Vater** (VA-ter).

Hernia (HER-nē-a) The protrusion or projection of an organ or part of an organ through a membrane or cavity wall, usually the abdominal cavity.

Herniated (HER-nē-Ā′-ted) **disc** A rupture of an intervertebral disc so that the nucleus pulposus protrudes into the vertebral cavity. Also called a **slipped disc.**

Heterozygous (he-ter-ō-ZĪ-gus) Possessing different alleles on homologous chromosomes for a particular hereditary trait.

Hiatus (hī-Ā-tus) An opening; a foramen.

Hilus (HĪ-lus) An area, depression, or pit where blood vessels and nerves enter or leave an organ. Also called a **hilum.**

Hinge joint A synovial joint in which a convex surface of one bone fits into a concave surface of another bone, such as the elbow, knee, ankle, and interphalangeal joints. Also called a **ginglymus** (JIN-gli-mus) **joint.**

Hirsutism (HER-soot-izm) An excessive growth of hair in females and children, with a distribution similar to that in adult males, due to the conversion of vellus hairs into large terminal hairs in response to higher-than-normal levels of androgens.

Histamine (HISS-ta-mēn) Substance found in many cells, especially mast cells, basophils, and platelets, released when the cells are injured; results in vasodilation, increased permeability of blood vessels, and constriction of bronchioles.

Histology (hiss-TOL-ō-jē) Microscopic study of the structure of tissues.

Holocrine (HŌL-ō-krin) **gland** A type of gland in which entire secretory cells, along with their accumulated secretions, make up the secretory product of the gland, as in the sebaceous (oil) glands.

Homeostasis (hō′-mē-ō-STĀ-sis) The condition in which the body's internal environment remains relatively constant, within physiological limits.

Homologous chromosomes Two chromosomes that belong to a pair. Also called **homologues.**

Homozygous (hō-mō-ZĪ-gus) Possessing the same alleles on homologous chromosomes for a particular hereditary trait.

Hormone (HOR-mōn) A secretion of endocrine cells that alters the physiological activity of target cells of the body.

Horn An area of gray matter (anterior, lateral, or posterior) in the spinal cord.

Human chorionic gonadotropin (kō-rē-ON-ik gō-nad-ō-TRŌ-pin) **(hCG)** A hormone produced by the developing placenta that maintains the corpus luteum.

Human chorionic somatomammotropin (sō-mat-ō-mam-ō-TRŌ-pin) **(hCS)** Hormone produced by the chorion of the placenta that stimulates breast tissue for lactation, enhances body growth, and regulates metabolism. Also called **human placental lactogen (hPL).**

Human growth hormone (hGH) Hormone secreted by the anterior pituitary that stimulates growth of body tissues, especially skeletal and muscular tissues. Also known as **somatotropin** and **somatotropic hormone (STH).**

Hyaluronic (hī-a-loo-RON-ik) **acid** A viscous, amorphous extracellular material that binds cells together, lubricates joints, and maintains the shape of the eyeballs.

Hyaluronidase (hī-a-loo-RON-i-dās) An enzyme that breaks down hyaluronic acid, increasing the permeability of connective tissues by dissolving the substances that hold body cells together.

Hydride ion (HĪ-drīd Ī-on) A hydrogen nucleus with two orbiting electrons (H⁻), in contrast to the more common **hydrogen ion,** which has no orbiting electrons (H⁺).

Hymen (HĪ-men) A thin fold of vascularized mucous membrane at the vaginal orifice.

Hypercalcemia (hī'-per-kal-SĒ-mē-a) An excess of calcium in the blood.

Hypercapnia (hī'-per-KAP-nē-a) An abnormal increase in the amount of carbon dioxide in the blood.

Hyperextension (hī'-per-ek-STEN-shun) Continuation of extension beyond the anatomical position, as in bending the head backward.

Hyperglycemia (hī'-per-glī-SĒ-mē-a) An elevated blood glucose level.

Hyperkalemia (hī'-per-kā-LĒ-mē-a) An excess of potassium ions in the blood.

Hypermagnesemia (hī'-per-mag'-ne-SĒ-mē-a) An excess of magnesium ions in the blood.

Hypermetropia (hī'-per-mē-TRŌ-pē-a) A condition in which visual images are focused behind the retina, with resulting defective vision of near objects; farsightedness.

Hyperphosphatemia (hī-per-fos'-fa-TĒ-mē-a) An abnormally high level of phosphates in the blood.

Hyperplasia (hī'-per-PLĀ-zē-a) An abnormal increase in the number of normal cells in a tissue or organ, increasing its size.

Hyperpolarization (hī'-per-PŌL-a-ri-zā'-shun) Increase in the internal negativity across a cell membrane, thus increasing the voltage and moving it farther away from the threshold value.

Hypersecretion (hī'-per-se-KRĒ-shun) Overactivity of glands resulting in excessive secretion.

Hypersensitivity (hī'-per-sen-si-TI-vi-tē) Overreaction to an allergen that results in pathological changes in tissues. Also called **allergy.**

Hypertension (hī'-per-TEN-shun) High blood pressure.

Hyperthermia (hī'-per-THERM-ē-a) An elevated body temperature.

Hypertonia (hī'-per-TŌ-nē-a) Increased muscle tone that is expressed as spasticity or rigidity.

Hypertonic (hī'-per-TON-ik) Solution that causes cells to shrink due to loss of water by osmosis.

Hypertrophy (hī-PER-trō-fē) An excessive enlargement or overgrowth of tissue without cell division.

Hyperventilation (hī'-per-ven-ti-LĀ-shun) A rate of respiration higher than that required to maintain a normal partial pressure of carbon dioxide in the blood.

Hypocalcemia (hī'-pō-kal-SĒ-mē-a) A below-normal level of calcium in the blood.

Hypochloremia (hī'-pō-klō-RĒ-mē-a) Deficiency of chloride ions in the blood.

Hypoglycemia (hī'-pō-glī-SĒ-mē-a) An abnormally low concentration of glucose in the blood; can result from excess insulin (injected or secreted).

Hypokalemia (hī'-pō-ka-LĒ-mē-a) Deficiency of potassium ions in the blood.

Hypomagnesemia (hī'-pō-mag'-ne-SĒ-mē-a) Deficiency of magnesium ions in the blood.

Hyponatremia (hī'-pō-na-TRĒ-mē-a) Deficiency of sodium ions in the blood.

Hyponychium (hī'-pō-NIK-ē-um) Free edge of the fingernail.

Hypophosphatemia (hī-pō-fos'-fa-TĒ-mē-a) An abnormally low level of phosphates in the blood.

Hypophyseal fossa (hī'-pō-FIZ-ē-al FOS-a) A depression on the superior surface of the sphenoid bone that houses the pituitary gland.

Hypophyseal (hī'-pō-FIZ-ē-al) **pouch** An outgrowth of ectoderm from the roof of the mouth from which the anterior pituitary develops.

Hypophysis (hī-POF-i-sis) Pituitary gland.

Hyposecretion (hī'-pō-se-KRĒ-shun) Underactivity of glands resulting in diminished secretion.

Hypothalamohypophyseal (hī-pō-tha-lam'-ō-hī-pō-FIZ-ē-al) **tract** A bundle of axons containing secretory vesicles filled with oxytocin or antidiuretic hormone that extend from the hypothalamus to the posterior pituitary.

Hypothalamus (hī'-pō-THAL-a-mus) A portion of the diencephalon, lying beneath the thalamus and forming the floor and part of the wall of the third ventricle.

Hypothermia (hī'-pō-THER-mē-a) Lowering of body temperature below 35°C (95°F); in surgical procedures, it refers to deliberate cooling of the body to slow down metabolism and reduce oxygen needs of tissues.

Hypotonia (hī'-pō-TŌ-nē-a) Decreased or lost muscle tone in which muscles appear flaccid.

Hypotonic (hī'-pō-TON-ik) Solution that causes cells to swell and perhaps rupture due to gain of water by osmosis.

Hypoventilation (hī-pō-ven-ti-LĀ-shun) A rate of respiration lower than that required to maintain a normal partial pressure of carbon dioxide in plasma.

Hypovolemic (hī-pō-vō-LĒ-mik) **shock** A type of shock characterized by decreased blood volume; may be caused by acute hemorrhage or excessive loss of other body fluids, for example, by vomiting, diarrhea, or excessive sweating.

Hypoxia (hī-POKS-ē-a) Lack of adequate oxygen at the tissue level.

Hysterectomy (hiss-te-REK-tō-mē) The surgical removal of the uterus.

I

Ileocecal (il-ē-ō-SĒ-kal) **sphincter** A fold of mucous membrane that guards the opening from the ileum into the large intestine. Also called the **ileocecal valve.**

Ileum (IL-ē-um) The terminal part of the small intestine.

Immunity (im-Ū-ni-tē) The state of being resistant to injury, particularly by poisons, foreign proteins, and invading pathogens.

Immunogenicity (im′-ū-nō-je-NIS-i-tē) Ability of an antigen to provoke an immune response.

Immunoglobulin (im-ū-nō-GLOB-ū-lin) **(Ig)** An antibody synthesized by plasma cells derived from B lymphocytes in response to the introduction of an antigen. Immunoglobulins are divided into five kinds (IgG, IgM, IgA, IgD, IgE).

Immunology (im′-ū-NOL-ō-jē) The study of the responses of the body when challenged by antigens.

Implantation (im-plan-TĀ-shun) The insertion of a tissue or a part into the body. The attachment of the blastocyst to the stratum basalis of the endometrium about 6 days after fertilization.

Incontinence (in-KON-ti-nens) Inability to retain urine, semen, or feces through loss of sphincter control.

Indirect motor pathways Motor tracts that convey information from the brain down the spinal cord for automatic movements, coordination of body movements with visual stimuli, skeletal muscle tone and posture, and balance. Also known as **extrapyramidal pathways.**

Infarction (in-FARK-shun) A localized area of necrotic tissue, produced by inadequate oxygenation of the tissue.

Infection (in-FEK-shun) Invasion and multiplication of microorganisms in body tissues, which may be inapparent or characterized by cellular injury.

Inferior (in-FĒR-ē-or) Away from the head or toward the lower part of a structure. Also called **caudad** (KAW-dad).

Inferior vena cava (VĒ-na CĀ-va) **(IVC)** Large vein that collects blood from parts of the body inferior to the heart and returns it to the right atrium.

Infertility Inability to conceive or to cause conception. Also called **sterility.**

Inflammation (in′-fla-MĀ-shun) Localized, protective response to tissue injury designed to destroy, dilute, or wall off the infecting agent or injured tissue; characterized by redness, pain, heat, swelling, and sometimes loss of function.

Inflation reflex Reflex that prevents overinflation of the lungs. Also called **Hering–Breuer reflex.**

Infundibulum (in′-fun-DIB-ū-lum) The stalklike structure that attaches the pituitary gland to the hypothalamus of the brain. The funnel-shaped, open, distal end of the uterine (Fallopian) tube.

Ingestion (in-JES-chun) The taking in of food, liquids, or drugs, by mouth.

Inguinal (IN-gwi-nal) Pertaining to the groin.

Inguinal canal An oblique passageway in the anterior abdominal wall just superior and parallel to the medial half of the inguinal ligament that transmits the spermatic cord and ilioinguinal nerve in the male and round ligament of the uterus and ilioinguinal nerve in the female.

Inhalation (in-ha-LĀ-shun) The act of drawing air into the lungs. Also termed **inspiration.**

Inheritance The acquisition of body traits by transmission of genetic information from parents to offspring.

Inhibin A hormone secreted by the gonads that inhibits release of follicle-stimulating hormone (FSH) by the anterior pituitary.

Inhibiting hormone Hormone secreted by the hypothalamus that can suppress secretion of hormones by the anterior pituitary.

Inhibitory postsynaptic potential (IPSP) A small hyperpolarization in a neuron; caused by an inhibitory neurotransmitter at a synapse.

Inner cell mass A region of cells of a blastocyst that differentiates into the three primary germ layers—ectoderm, mesoderm, and endoderm—from which all tissues and organs develop; also called an **embryoblast.**

Inorganic (in′-or-GAN-ik) **compound** Compound that usually lacks carbon, usually is small, and often contains ionic bonds. Examples include water and many acids, bases, and salts.

Insertion (in-SER-shun) The attachment of a muscle tendon to a movable bone or the end opposite the origin.

Inspiratory (in-SPĪ-ra-tor-ē) **capacity** Total inspiratory capacity of the lungs; the total of tidal volume plus inspiratory reserve volume; averages 3600 mL.

Inspiratory (in-SPĪ-ra-tor-ē) **reserve volume** Additional inspired air over and above tidal volume; averages 3100 mL.

Insula (IN-soo-la) A triangular area of the cerebral cortex that lies deep within the lateral cerebral fissue, under the parietal, frontal, and temporal lobes.

Insulin (IN-soo-lin) A hormone produced by the beta cells of a pancreatic islet (islet of Langerhans) that decreases the blood glucose level.

Insulinlike growth factor (IGF) Small protein, produced by the liver and other tissues in response to stimulation by human growth hormone (hGH), that mediates most of the effects of human growth hormone. Previously called **somatomedin** (sō′-ma-tō-MĒ-din).

Integrins (IN-te-grinz) A family of transmembrane glycoproteins in plasma membranes that function in cell adhesion; they are present in hemidesmosomes, which anchor cells to a basement membrane, and they mediate adhesion of neutrophils to endothelial cells during emigration.

Integumentary (in-teg′-ū-MEN-tar-e) Relating to the skin.

Intercalated (in-TER-ka-lāt-ed) **disc** An irregular transverse thickening of sarcolemma that contains desmosomes, which hold cardiac muscle fibers (cells) together, and gap junctions, which aid in conduction of muscle action potentials from one fiber to the next.

Intercostal (in′-ter-KOS-tal) **nerve** A nerve supplying a muscle located between the ribs.

Interferons (in′-ter-FĒR-ons) **(IFNs)** Antiviral proteins produced by virus-infected host cells; induce uninfected host cells to synthesize proteins that inhibit viral replication and enhance phagocytic activity of macrophages; types include alpha interferon, beta interferon, and gamma interferon.

Intermediate Between two structures, one of which is medial and one of which is lateral.

Intermediate filament Protein filament, ranging from 8 to 12 nm in diameter, that may provide structural reinforcement, hold organelles in place, and give shape to a cell.

Internal Away from the surface of the body.

Internal capsule A large tract of projection fibers lateral to the thalamus that is the major connection between the cerebral cortex and the brain stem and spinal cord; contains axons of sensory neurons carrying auditory, visual, and somatic sensory signals to the cerebral cortex plus axons of motor neurons descending from the cerebral cortex to the thalamus, subthalamus, brain stem, and spinal cord.

Internal ear The inner ear or labyrinth, lying inside the temporal bone, containing the organs of hearing and balance.

Internal nares (NĀ-rez) The two openings posterior to the nasal cavities opening into the nasopharynx. Also called the **choanae** (kō-Ā-nē).

Internal respiration The exchange of respiratory gases between blood and body cells. Also called **tissue respiration.**

Interneurons (in′-ter-NOO-ronz) Neurons whose axons extend only for a short distance and contact nearby neurons in the brain, spinal cord, or a ganglion; they comprise the vast majority of neurons in the body.

Interoceptor (in′-ter-ō-SEP-tor) Sensory receptor located in blood vessels and viscera that provides information about the body's internal environment.

Interphase (IN-ter-fāz) The period of the cell cycle between cell divisions, consisting of the G_1-(gap or growth) phase, when the cell is engaged in growth, metabolism, and production of substances required for division; S-(synthesis) phase, during which chromosomes are replicated; and G_2-phase.

Interstitial (in′-ter-STISH-al) **fluid** The portion of extracellular fluid that fills the microscopic spaces between the cells of tissues; the internal environment of the body. Also called **intercellular** or **tissue fluid.**

Interstitial growth Growth from within, as in the growth of cartilage. Also called **endogenous** (en-DOJ-e-nus) **growth.**

Interventricular (in′-ter-ven-TRIK-ū-lar) **foramen** A narrow, oval opening through which the lateral ventricles of the brain communicate with the third ventricle. Also called the **foramen of Monro.**

Intervertebral (in′-ter-VER-te-bral) **disc** A pad of fibrocartilage located between the bodies of two vertebrae.

Intestinal gland A gland that opens onto the surface of the intestinal mucosa and secretes digestive enzymes. Also called a **crypt of Lieberkühn** (LĒ-ber-kūn).

Intrafusal (in′-tra-FŪ-sal) **fibers** Three to ten specialized muscle fibers (cells), partially enclosed in a spindle-shaped connective tissue capsule, that make up a muscle spindle.

Intramembranous ossification (in′-tra-MEM-bra-nus os′-i′-fi-KĀ-shun) The method of bone formation in which the bone is formed directly in membranous tissue.

Intraocular (in′-tra-OK-ū-lar) **pressure (IOP)** Pressure in the eyeball, produced mainly by aqueous humor.

Intrapleural pressure Air pressure between the two pleurae of the lungs, usually subatmospheric. Also called **intrathoracic pressure.**

Intrinsic (in-TRIN-sik) Of internal origin.

Intrinsic pathway (of blood clotting) Sequence of reactions leading to blood clotting that is initiated by damage to blood vessel endothelium or platelets; activators of this pathway are contained within blood itself or are in direct contact with blood.

Intrinsic factor (IF) A glycoprotein, synthesized and secreted by the parietal cells of the gastric mucosa, that facilitates vitamin B_{12} absorption in the small intestine.

Intron (IN-tron) A region of DNA that does not code for the synthesis of a protein.

In utero (Ū-ter-ō) Within the uterus.

Invagination (in-vaj′-i-NĀ-shun) The pushing of the wall of a cavity into the cavity itself.

Inversion (in-VER-zhun) The movement of the sole medially at the ankle joint.

In vitro (VĒ-trō) Literally, in glass; outside the living body and in an artificial environment such as a laboratory test tube.

In vivo (VĒ-vō) In the living body.

Ion (Ī-on) Any charged particle or group of particles; usually formed when a substance, such as a salt, dissolves and dissociates.

Ionization (ī′-on-i-ZĀ-shun) Separation of inorganic acids, bases, and salts into ions when dissolved in water. Also called **dissociation.**

Ipsilateral (ip′-si-LAT-er-al) On the same side, affecting the same side of the body.

Iris The colored portion of the vascular tunic of the eyeball seen through the cornea that contains circular and radial smooth muscle; the hole in the center of the iris is the pupil.

Irritable bowel syndrome (IBS) Disease of the entire gastrointestinal tract in which a person reacts to stress by developing symptoms (such as cramping and abdominal pain) associated with alternating patterns of diarrhea and constipation. Excessive amounts of mucus may appear in feces, and other symptoms include flatulence, nausea, and loss of appetite. Also known as **irritable colon** or **spastic colitis.**

Ischemia (is-KĒ-mē-a) A lack of sufficient blood to a body part due to obstruction or constriction of a blood vessel.

Isoantibody A specific antibody in blood plasma that reacts with specific isoantigens and causes the clumping of bacteria, red blood cells, or particles. Also called an **agglutinin.**

Isoantigen A genetically determined antigen located on the surface of red blood cells; basis for the ABO grouping and Rh system of blood classification. Also called an **agglutinogen.**

Isometric contraction A muscle contraction in which tension on the muscle increases, but there is only minimal muscle shortening so that no visible movement is produced.

Isotonic (Ī′-sō-TON-ik) Having equal tension or tone. A solution having the same concentration of impermeable solutes as cytosol.

Isotonic (i′-so-TON-ik) **contraction** Contraction in which the tension remains the same; occurs when a constant load is moved through the range of motions possible at a joint.

Isotopes (Ī-sō-tōps′) Chemical elements that have the same number of protons but different numbers of neutrons. Radioactive isotopes change into other elements with the emission of alpha or beta particles or gamma rays.

Isovolumetric (ī-sō-vol-ū′-MET-rik) **contraction** The period of time, about 0.05 sec, between the start of ventricular systole and the opening of the semilunar valves; the ventricles contract and ventricular pressure rises rapidly, but ventricular volume does not change.

Isovolumetric relaxation The period of time, about 0.05 sec, between the closing of the semilunar valves and the opening of the atrioventricular (AV) valves; there is a drastic decrease in ventricular pressure without a change in ventricular volume.

Isthmus (IS-mus) A narrow strip of tissue or narrow passage connecting two larger parts.

J

Jaundice (JAWN-dis) A condition characterized by yellowness of the skin, the white of the eyes, mucous membranes, and body fluids because of a buildup of bilirubin.

Jejunum (je-JOO-num) The middle part of the small intestine.

Joint kinesthetic (kin′-es-THET-ik) **receptor** A proprioceptive receptor located in a joint, stimulated by joint movement.

Juxtaglomerular (juks-ta-glō-MER-ū-lar) **apparatus (JGA)** Consists of the macula densa (cells of the distal convoluted tubule adjacent to the afferent and efferent arteriole) and juxtaglomerular cells (modified cells of the afferent and sometimes efferent arteriole); secretes renin when blood pressure starts to fall.

K

Keratin (KER-a-tin) An insoluble protein found in the hair, nails, and other keratinized tissues of the epidermis.

Keratinocyte (ke-RAT-in′-ō-sīt) The most numerous of the epidermal cells; produces keratin.

Ketone (KĒ-tōn) **bodies** Substances produced primarily during excessive triglyceride catabolism, such as acetone, acetoacetic acid, and β-hydroxybutyric acid.

Ketosis (kē-TŌ-sis) Abnormal condition marked by excessive production of ketone bodies.

Kidney (KID-nē) One of the paired reddish organs located in the lumbar region that regulates the composition, volume, and pressure of blood and produces urine.

Kidney stone A solid mass, usually consisting of calcium oxalate, uric acid, or calcium phosphate crystals, that may form in any portion of the urinary tract. Also called a **renal calculus.**

Kinesiology (ki-nē′-sē-OL-ō-jē) The study of the movement of body parts.

Kinesthesia (kin-es-THĒ-zē-a) The perception of the extent and direction of movement of body parts; this sense is possible due to nerve impulses generated by proprioceptors.

Korotkoff (kō-ROT-kof) **sounds** The various sounds that are heard while taking blood pressure.

Krebs cycle A series of biochemical reactions that occurs in the matrix of mitochondria in which electrons are transferred to coenzymes and carbon dioxide is formed. The electrons carried by the coenzymes then enter the electron transport chain, which generates a large quantity of ATP. Also called the **citric acid cycle** or **tricarboxylic acid (TCA) cycle.**

Kyphosis (kī-FŌ-sis) An exaggeration of the thoracic curve of the vertebral column, resulting in a "round-shouldered" appearance. Also called **hunchback.**

L

Labia majora (LĀ-bē-a ma-JŌ-ra) Two longitudinal folds of skin extending downward and backward from the mons pubis of the female.

Labia minora (min-OR-a) Two small folds of mucous membrane lying medial to the labia majora of the female.

Labial frenulum (LĀ-bē-al FREN-ū-lum) A medial fold of mucous membrane between the inner surface of the lip and the gums.

Labium (LĀ-bē-um) A lip. A liplike structure. *Plural is* **labia** (LĀ-bē-a).

Labor The process of giving birth in which a fetus is expelled from the uterus through the vagina.

Lacrimal canal A duct, one on each eyelid, beginning at the punctum at the medial margin of an eyelid and conveying tears medially into the nasolacrimal sac.

Lacrimal gland Secretory cells, located at the superior anterolateral portion of each orbit, that secrete tears into excretory ducts that open onto the surface of the conjunctiva.

Lacrimal sac The superior expanded portion of the nasolacrimal duct that receives the tears from a lacrimal canal.

Lactation (lak-TĀ-shun) The secretion and ejection of milk by the mammary glands.

Lacteal (LAK-tē-al) One of many lymphatic vessels in villi of the intestines that absorb triglycerides and other lipids from digested food.

Lacuna (la-KOO-na) A small, hollow space, such as that found in bones in which the osteocytes lie. *Plural is* **lacunae** (la-KOO-nē).

Lambdoid (lam-DOYD) **suture** The joint in the skull between the parietal bones and the occipital bone; sometimes contains sutural (Wormian) bones.

Lamellae (la-MEL-ē) Concentric rings of hard, calcified matrix found in compact bone.

Lamellated corpuscle Oval-shaped pressure receptor located in the dermis or subcutaneous tissue and consisting of concentric layers of connective tissue wrapped around the dendrites of a sensory neuron. Also called a **Pacinian** (pa-SIN-ē-an) **corpuscle.**

Lamina (LAM-i-na) A thin, flat layer or membrane, as the flattened part of either side of the arch of a vertebra. *Plural is* **laminae** (LAM-i-nē).

Lamina propria (PRŌ-prē-a) The connective tissue layer of a mucosa.

Langerhans (LANG-er-hans) **cell** Epidermal dendritic cell that functions as an antigen-presenting cell (APC) during an immune response.

Lanugo (la-NOO-gō) Fine downy hairs that cover the fetus.

Large intestine The portion of the gastrointestinal tract extending from the ileum of the small intestine to the anus, divided structurally into the cecum, colon, rectum, and anal canal.

Laryngopharynx (la-rin′-gō-FAR-inks) The inferior portion of the pharynx, extending downward from the level of the hyoid bone that divides posteriorly into the esophagus and anteriorly into the larynx. Also called the **hypopharynx.**

Laryngotracheal (la-rin′-gō-TRĀ-ke-al) **bud** An outgrowth of endoderm of the foregut from which the respiratory system develops.

Larynx (LAR-inks) The voice box, a short passageway that connects the pharynx with the trachea.

Lateral (LAT-er-al) Farther from the midline of the body or a structure.

Lateral ventricle (VEN-tri-kul) A cavity within a cerebral hemisphere that communicates with the lateral ventricle in the other cerebral hemisphere and with the third ventricle by way of the interventricular foramen.

Leg The part of the lower limb between the knee and the ankle.

Lens A transparent organ constructed of proteins (crystallins) lying posterior to the pupil and iris of the eyeball and anterior to the vitreous body.

Lesion (LĒ-zhun) Any localized, abnormal change in a body tissue.

Lesser omentum (ō-MEN-tum) A fold of the peritoneum that extends from the liver to the lesser curvature of the stomach and the first part of the duodenum.

Lesser vestibular (ves-TIB-ū-lar) **gland** One of the paired mucus-secreting glands with ducts that open on either side of the urethral orifice in the vestibule of the female.

Leukemia (loo-KĒ-mē-a) A malignant disease of the blood-forming tissues characterized by either uncontrolled production and accumulation of immature leukocytes in which many cells fail to reach maturity (acute) or an accumulation of mature leukocytes in the blood because they do not die at the end of their normal life span (chronic).

Leukocyte (LOO-kō-sīt) A white blood cell.

Leukocytosis (loo′-kō-sī-TŌ-sis) An increase in the number of white blood cells, above 10,000 per μL, characteristic of many infections and other disorders.

Leukopenia (loo-kō-PĒ-nē-a) A decrease in the number of white blood cells below 5000 cells per μL.

Leukotriene (loo′-kō-TRĪ-ēn) A type of eicosanoid produced by basophils and mast cells; acts as a local hormone; produces increased vascular permeability and acts as a chemotactic agent for phagocytes in tissue inflammation.

Leydig (LĪ-dig) **cell** A type of cell that secretes testosterone; located in the connective tissue between seminiferous tubules in a mature testis. Also known as **interstitial cell of Leydig** or **interstitial endocrinocyte.**

Libido (li-BĒ-dō) Sexual desire.

Ligament (LIG-a-ment) Dense regular connective tissue that attaches bone to bone.

Ligand (LĪ-gand) A chemical substance that binds to a specific receptor.

Limbic system A part of the forebrain, sometimes termed the visceral brain, concerned with various aspects of emotion and behavior; includes the limbic lobe, dentate gyrus, amygdala, septal nuclei, mammillary bodies, anterior thalamic nucleus, olfactory bulbs, and bundles of myelinated axons.

Lingual frenulum (LIN-gwal FREN-ū-lum) A fold of mucous membrane that connects the tongue to the floor of the mouth.

Lipase An enzyme that splits fatty acids from triglycerides and phospholipids.

Lipid (LIP-id) An organic compound composed of carbon, hydrogen, and oxygen that is usually insoluble in water, but soluble in alcohol, ether, and chloroform; examples include triglycerides (fats and oils), phospholipids, steroids, and eicosanoids.

Lipid bilayer Arrangement of phospholipid, glycolipid, and cholesterol molecules in two parallel sheets in which the hydrophilic "heads" face outward and the hydrophobic "tails" face inward; found in cellular membranes.

Lipogenesis (li-pō-GEN-e-sis) The synthesis of triglycerides.

Lipolysis (lip-OL-i-sis) The splitting of fatty acids from a triglyceride or phospholipid.

Lipoprotein (lip′-ō-PRŌ-tēn) One of several types of particles containing lipids (cholesterol and triglycerides) and proteins that make it water soluble for transport in the blood; high levels of **low-density lipoproteins (LDLs)** are associated with increased risk of atherosclerosis, whereas high levels of **high-density lipoproteins (HDLs)** are associated with decreased risk of atherosclerosis.

Liver Large organ under the diaphragm that occupies most of the right hypochondriac region and part of the epigastric region. Functionally, it produces bile and synthesizes most plasma proteins; interconverts nutrients; detoxifies substances; stores glycogen, iron, and vitamins; carries on phagocytosis of worn-out blood cells and bacteria; and helps synthesize the active form of vitamin D.

Long-term potentiation (LTP) Prolonged, enhanced synaptic transmission that occurs at certain synapses within the hippocampus of the brain; believed to underlie some aspects of memory.

Lordosis (lor-DŌ-sis) An exaggeration of the lumbar curve of the vertebral column. Also called **swayback.**

Lower limb The appendage attached at the pelvic (hip) girdle, consisting of the thigh, knee, leg, ankle, foot, and toes. Also called **lower extremity.**

Lumbar (LUM-bar) Region of the back and side between the ribs and pelvis; loin.

Lumbar plexus (PLEK-sus) A network formed by the anterior (ventral) branches of spinal nerves L1 through L4.

Lumen (LOO-men) The space within an artery, vein, intestine, renal tubule, or other tubular structure.

Lungs Main organs of respiration that lie on either side of the heart in the thoracic cavity.

Lunula (LOO-noo-la) The moon-shaped white area at the base of a nail.

Luteinizing (LOO-tē-in′-īz-ing) **hormone (LH)** A hormone secreted by the anterior pituitary that stimulates ovulation, stimulates progesterone secretion by the corpus luteum, and readies the mammary glands for milk secretion in females; stimulates testosterone secretion by the testes in males.

Lymph (LIMF) Fluid confined in lymphatic vessels and flowing through the lymphatic system until it is returned to the blood.

Lymph node An oval or bean-shaped structure located along lymphatic vessels.

Lymphatic (lim-FAT-ik) **capillary** Closed-ended microscopic lymphatic vessel that begins in spaces between cells and converges with other lymphatic capillaries to form lymphatic vessels.

Lymphatic tissue A specialized form of reticular tissue that contains large numbers of lymphocytes.

Lymphatic vessel A large vessel that collects lymph from lymphatic capillaries and converges with other lymphatic vessels to form the thoracic and right lymphatic ducts.

Lymphocyte (LIM-fō-sīt) A type of white blood cell that helps carry out cell-mediated and antibody-mediated immune responses; found in blood and in lymphatic tissues.

Lysosome (LĪ-sō-sōm) An organelle in the cytoplasm of a cell, enclosed by a single membrane and containing powerful digestive enzymes.

Lysozyme (LĪ-sō-zīm) A bactericidal enzyme found in tears, saliva, and perspiration.

M

Macrophage (MAK-rō-fāj) Phagocytic cell derived from a monocyte; may be fixed or wandering.

Macula (MAK-ū-la) A discolored spot or a colored area. A small, thickened region on the wall of the utricle and saccule that contains receptors for static equilibrium.

Macula lutea (MAK-ū-la LOO-tē-a) The yellow spot in the center of the retina.

Major histocompatibility (MHC) antigens Surface proteins on white blood cells and other nucleated cells that are unique for each person (except for identical siblings); used to type tissues and help prevent rejection of transplanted tissues. Also known as **human leukocyte antigens (HLA).**

Malignant (ma-LIG-nant) Referring to diseases that tend to become worse and cause death, especially the invasion and spreading of cancer.

Mammary (MAM-ar-ē) **gland** Modified sudoriferous (sweat) gland of the female that produces milk for the nourishment of the young.

Mammillary (MAM-i-ler-ē) **bodies** Two small rounded bodies on the inferior aspect of the hypothalamus that are involved in reflexes related to the sense of smell.

Marrow (MAR-ō) Soft, spongelike material in the cavities of bone. Red bone marrow produces blood cells; yellow bone marrow contains adipose tissue that stores triglycerides.

Mast cell A cell found in areolar connective tissue that releases histamine, a dilator of small blood vessels, during inflammation.

Mastication (mas′-ti-KĀ-shun) Chewing.

Matrix (MĀ-triks) The ground substance and fibers between cells in a connective tissue.

Matter Anything that occupies space and has mass.

Mature follicle A large, fluid-filled follicle containing a secondary oocyte and surrounding granulosa cells that secrete estrogens. Also called a **Graafian** (GRAF-ē-an) **follicle.**

Maximal oxygen uptake Maximum rate of oxygen consumption during aerobic catabolism of pyruvic acid that is determined by age, sex, body size, and aerobic training.

Mean arterial blood pressure (MABP) The average force of blood pressure exerted against the walls of arteries; approximately equal to diastolic pressure plus one-third of pulse pressure; for example, 93 mmHg when systolic pressure is 120 mmHg and diastolic pressure is 80 mmHg.

Meatus (mē-Ā-tus) A passage or opening, especially the external portion of a canal.

Mechanoreceptor (me-KAN-ō-rē-sep-tor) Sensory receptor that detects mechanical deformation of the receptor itself or adjacent cells; stimuli so detected include those related to touch, pressure, vibration, proprioception, hearing, equilibrium, and blood pressure.

Medial (MĒ-dē-al) Nearer the midline of the body or a structure.

Medial lemniscus (lem-NIS-kus) A white matter tract that originates in the gracile and cuneate nuclei of the medulla oblongata and extends to the thalamus on the same side; sensory axons in this tract conduct nerve impulses for the sensations of proprioception, fine touch, vibration, hearing, and equilibrium.

Median aperture (AP-er-choor) One of the three openings in the roof of the fourth ventricle through which cerebrospinal fluid enters the subarachnoid space of the brain and cord. Also called the **foramen of Magendie.**

Median plane A vertical plane dividing the body into right and left halves. Situated in the middle.

Mediastinum (mē´-dē-as-TĪ-num) The broad, median partition between the pleurae of the lungs, that extends from the sternum to the vertebral column in the thoracic cavity.

Medulla (me-DUL-la) An inner layer of an organ, such as the medulla of the kidneys.

Medulla oblongata (me-DUL-la ob´-long-GA-ta) The most inferior part of the brain stem. Also termed the **medulla.**

Medullary (MED-ū-lar´-ē) **cavity** The space within the diaphysis of a bone that contains yellow bone marrow. Also called the **marrow cavity.**

Medullary rhythmicity (rith-MIS-i-tē) **area** The neurons of the respiratory center in the medulla oblongata that control the basic rhythm of respiration.

Meiosis (mē-Ō-sis) A type of cell division that occurs during production of gametes, involving two successive nuclear divisions that result in daughter cells with the haploid (*n*) number of chromosomes.

Melanin (MEL-a-nin) A dark black, brown, or yellow pigment found in some parts of the body such as the skin, hair, and pigmented layer of the retina.

Melanocyte (MEL-a-nō-sīt´) A pigmented cell, located between or beneath cells of the deepest layer of the epidermis, that synthesizes melanin.

Melanocyte-stimulating hormone (MSH) A hormone secreted by the anterior pituitary that stimulates the dispersion of melanin granules in melanocytes in amphibians; continued administration produces darkening of skin in humans.

Melatonin (mel-a-TŌN-in) A hormone secreted by the pineal gland that helps set the timing of the body's biological clock.

Membrane A thin, flexible sheet of tissue composed of an epithelial layer and an underlying connective tissue layer, as in an epithelial membrane, or of areolar connective tissue only, as in a synovial membrane.

Membranous labyrinth (mem-BRA-nus LAB-i-rinth) The part of the labyrinth of the internal ear that is located inside the bony labyrinth and separated from it by the perilymph; made up of the semicircular ducts, the saccule and utricle, and the cochlear duct.

Menarche (me-NAR-kē) The first menses (menstrual flow) and beginning of ovarian and uterine cycles.

Meninges (me-NIN-jēz) Three membranes covering the brain and spinal cord, called the dura mater, arachnoid mater, and pia mater. *Singular is* **meninx** (MEN-inks).

Menopause (MEN-ō-pawz) The termination of the menstrual cycles.

Menstruation (men´-stroo-Ā-shun) Periodic discharge of blood, tissue fluid, mucus, and epithelial cells that usually lasts for 5 days; caused by a sudden reduction in estrogens and progesterone. Also called the **menstrual phase** or **menses.**

Merkel (MER-kel) **cell** Type of cell in the epidermis of hairless skin that makes contact with a tactile (Merkel) disc, which functions in touch.

Merocrine (MER-ō-krin) **gland** Gland made up of secretory cells that remain intact throughout the process of formation and discharge of the secretory product, as in the salivary and pancreatic glands.

Mesenchyme (MEZ-en-kīm) An embryonic connective tissue from which all other connective tissues arise.

Mesentery (MEZ-en-ter´-ē) A fold of peritoneum attaching the small intestine to the posterior abdominal wall.

Mesocolon (mez´-ō-KŌ-lon) A fold of peritoneum attaching the colon to the posterior abdominal wall.

Mesoderm The middle primary germ layer that gives rise to connective tissues, blood and blood vessels, and muscles.

Mesothelium (mez´-ō-THĒ-lē-um) The layer of simple squamous epithelium that lines serous membranes.

Mesovarium (mez´-ō-VAR-ē-um) A short fold of peritoneum that attaches an ovary to the broad ligament of the uterus.

Metabolism (me-TAB-ō-lizm) All the biochemical reactions that occur within an organism, including the synthetic (anabolic) reactions and decomposition (catabolic) reactions.

Metacarpus (met´-a-KAR-pus) A collective term for the five bones that make up the palm.

Metaphase (MET-a-phāz) The second stage of mitosis, in which chromatid pairs line up on the metaphase plate of the cell.

Metaphysis (me-TAF-i-sis) Region of a long bone between the diaphysis and epiphysis that contains the epiphyseal plate in a growing bone.

Metarteriole (met´-ar-TĒ-rē-ōl) A blood vessel that emerges from an arteriole, traverses a capillary network, and empties into a venule.

Metastasis (me-TAS-ta-sis) The spread of cancer to surrounding tissues (local) or to other body sites (distant).

Metatarsus (met´-a-TAR-sus) A collective term for the five bones located in the foot between the tarsals and the phalanges.

Micelle (mī-SEL) A spherical aggregate of bile salts that dissolves fatty acids and monoglycerides so that they can be absorbed into small intestinal epithelial cells.

Microfilament (mī-krō-FIL-a-ment) Rodlike protein filament about 6 nm in diameter; constitutes contractile units in muscle fibers (cells) and provides support, shape, and movement in nonmuscle cells.

Microglia (mī-krō-GLĒ-a) Neuroglial cells that carry on phagocytosis.

Microtubule (mī-krō-TOO-būl´) Cylindrical protein filament, ranging in diameter from 18 to 30 nm, consisting of the protein tubulin; provides support, structure, and transportation.

Microvilli (mī´-krō-VIL-ē) Microscopic, fingerlike projections of the plasma membranes of cells that increase surface area for absorption, especially in the small intestine and proximal convoluted tubules of the kidneys.

Micturition (mik′-choo-RISH-un) The act of expelling urine from the urinary bladder. Also called **urination** (yoo-ri-NĀ-shun).

Midbrain The part of the brain between the pons and the diencephalon. Also called the **mesencephalon** (mes′-en-SEF-a-lon).

Middle ear A small, epithelial-lined cavity hollowed out of the temporal bone, separated from the external ear by the eardrum and from the internal ear by a thin bony partition containing the oval and round windows; extending across the middle ear are the three auditory ossicles. Also called the **tympanic** (tim-PAN-ik) **cavity.**

Midline An imaginary vertical line that divides the body into equal left and right sides.

Midsagittal plane A vertical plane through the midline of the body that divides the body or organs into *equal* right and left sides. Also called a **median plane.**

Milk ejection reflex Contraction of alveolar cells to force milk into ducts of mammary glands, stimulated by oxytocin (TO), which is released from the posterior pituitary in response to suckling action. Also called the **milk letdown reflex.**

Mineral Inorganic, homogeneous solid substance that may perform a function vital to life; examples include calcium and phosphorus.

Mineralocorticoids (min′-er-al-ō-KOR-ti-koyds) A group of hormones of the adrenal cortex that help regulate sodium and potassium balance.

Minimal volume The volume of air in the lungs even after the thoracic cavity has been opened, forcing out some of the residual volume.

Minute ventilation (MV) Total volume of air inhaled and exhaled per minute; about 6000 mL.

Miosis Constriction of the pupil of the eye.

Mitochondrion (mī′-tō-KON-drē-on) A double-membraned organelle that plays a central role in the production of ATP; known as the "powerhouse" of the cell.

Mitosis (mī-TŌ-sis) The orderly division of the nucleus of a cell that ensures that each new daughter nucleus has the same number and kind of chromosomes as the original parent nucleus. The process includes the replication of chromosomes and the distribution of the two sets of chromosomes into two separate and equal nuclei.

Mitotic spindle Collective term for a football-shaped assembly of microtubules (nonkinetochore, kinetochore, and aster) that is responsible for the movement of chromosomes during cell division.

Modality (mō-DAL-i-tē) Any of the specific sensory entities, such as vision, smell, taste, or touch.

Modiolus (mō-DĪ-ō′-lus) The central pillar or column of the cochlea.

Mole The weight, in grams, of the combined atomic weights of the atoms that make up a molecule of a substance.

Molecule (MOL-e-kūl) The chemical combination of two or more atoms covalently bonded together.

Monocyte (MON-ō-sīt′) The largest type of white blood cell, characterized by agranular cytoplasm.

Monounsaturated fat A fatty acid that contains one double covalent bond between its carbon atoms; it is not completely saturated with hydrogen atoms. Plentiful in triglycerides of olive and peanut oils.

Mons pubis (monz PŪ-bis) The rounded, fatty prominence over the pubic symphysis, covered by coarse pubic hair.

Morula (MOR-ū-la) A solid sphere of cells produced by successive cleavages of a fertilized ovum about four days after fertilization.

Motor end plate Region of the sarcolemma of a muscle fiber (cell) that includes acetylcholine (ACh) receptors, which bind ACh released by synaptic end bulbs of somatic motor neurons.

Motor neurons (NOO-ronz) Neurons that conduct impulses from the brain toward the spinal cord or out of the brain and spinal cord into cranial or spinal nerves to effectors that may be either muscles or glands. Also called **efferent neurons.**

Motor unit A motor neuron together with the muscle fibers (cells) it stimulates.

Mucin (MŪ-sin) A protein found in mucus.

Mucosa-associated lymphatic tissue (MALT) Lymphatic nodules scattered throughout the lamina propria (connective tissue) of mucous membranes lining the gastrointestinal tract, respiratory airways, urinary tract, and reproductive tract.

Mucous (MŪ-kus) **cell** A unicellular gland that secretes mucus. Two types are mucous neck cells and surface mucous cells in the stomach.

Mucous membrane A membrane that lines a body cavity that opens to the exterior. Also called the **mucosa** (mū-KŌ-sa).

Mucus The thick fluid secretion of goblet cells, mucous cells, mucous glands, and mucous membranes.

Muscarinic (mus′-ka-RIN-ik) **receptor** Receptor for the neurotransmitter acetylcholine found on all effectors innervated by parasympathetic postganglionic axons and on sweat glands innervated by cholinergic sympathetic postganglionic axons; so named because muscarine activates these receptors but does not activate nicotinic receptors for acetylcholine.

Muscle An organ composed of one of three types of muscle tissue (skeletal, cardiac, or smooth), specialized for contraction to produce voluntary or involuntary movement of parts of the body.

Muscle action potential A stimulating impulse that propagates along the sarcolemma and transverse tubules; in skeletal muscle, it is generated by acetylcholine, which increases the permeability of the sarcolemma to cations, especially sodium ions (Na^+).

Muscle fatigue (fa-TĒG) Inability of a muscle to maintain its strength of contraction or tension; may be related to insufficient oxygen, depletion of glycogen, and/or lactic acid buildup.

Muscle spindle An encapsulated proprioceptor in a skeletal muscle, consisting of specialized intrafusal muscle fibers and nerve endings; stimulated by changes in length or tension of muscle fibers.

Muscle tissue A tissue specialized to produce motion in response to muscle action potentials by its qualities of contractility, extensibility, elasticity, and excitability; types include skeletal, cardiac, and smooth.

Muscle tone A sustained, partial contraction of portions of a skeletal or smooth muscle in response to activation of stretch receptors or a baseline level of action potentials in the innervating motor neurons.

Muscular dystrophies (DIS-trō-fēz′) Inherited muscle-destroying diseases, characterized by degeneration of muscle fibers (cells), which causes progressive atrophy of the skeletal muscle.

Muscularis (MUS-kū-la′-ris) A muscular layer (coat or tunic) of an organ.

Muscularis mucosae (mū-KŌ-sē) A thin layer of smooth muscle fibers that underlie the lamina propria of the mucosa of the gastrointestinal tract.

Mutation (mū-TĀ-shun) Any change in the sequence of bases in a DNA molecule resulting in a permanent alteration in some inheritable trait.

Myasthenia (mī-as-THĒ-nē-a) **gravis** Weakness and fatigue of skeletal muscles caused by antibodies directed against acetylcholine receptors.

Myelin (MĪ-e-lin) **sheath** Multilayered lipid and protein covering, formed by Schwann cells and oligodendrocytes, around axons of many peripheral and central nervous system neurons.

Myenteric plexus A network of autonomic axons and postganglionic cell bodies located in the muscularis of the gastrointestinal tract. Also called the **plexus of Auerbach** (OW-er-bak).

Myocardial infarction (mī'-ō-KAR-dē-al in-FARK-shun) **(MI)** Gross necrosis of myocardial tissue due to interrupted blood supply. Also called a **heart attack.**

Myocardium (mī'-ō-KAR-dē-um) The middle layer of the heart wall, made up of cardiac muscle tissue, lying between the epicardium and the endocardium and constituting the bulk of the heart.

Myofibril (mī-ō-FĪ-bril) A threadlike structure, extending longitudinally through a muscle fiber (cell) consisting mainly of thick filaments (myosin) and thin filaments (actin, troponin, and tropomyosin).

Myoglobin (mī-ō-GLŌ-bin) The oxygen-binding, iron-containing protein present in the sarcoplasm of muscle fibers (cells); contributes the red color to muscle.

Myogram (MĪ-ō-gram) The record or tracing produced by a myograph, an apparatus that measures and records the force of muscular contractions.

Myology (mī-OL-ō-jē) The study of muscles.

Myometrium (mī'-ō-MĒ-trē-um) The smooth muscle layer of the uterus.

Myopathy (mī-OP-a-thē) Any abnormal condition or disease of muscle tissue.

Myopia (mī-Ō-pē-a) Defect in vision in which objects can be seen distinctly only when very close to the eyes; nearsightedness.

Myosin (MĪ-ō-sin) The contractile protein that makes up the thick filaments of muscle fibers.

Myotome (MĪ-ō-tōm) A group of muscles innervated by the motor neurons of a single spinal segment. In an embryo, the portion of a somite that develops into some skeletal muscles.

N

Nail A hard plate, composed largely of keratin, that develops from the epidermis of the skin to form a protective covering on the dorsal surface of the distal phalanges of the fingers and toes.

Nail matrix (MĀ-triks) The part of the nail beneath the body and root from which the nail is produced.

Nasal (NĀ-zal) **cavity** A mucosa-lined cavity on either side of the nasal septum that opens onto the face at the external nares and into the nasopharynx at the internal nares.

Nasal septum (SEP-tum) A vertical partition composed of bone (perpendicular plate of ethmoid and vomer) and cartilage, covered with a mucous membrane, separating the nasal cavity into left and right sides.

Nasolacrimal (nā'-zō-LAK-ri-mal) **duct** A canal that transports the lacrimal secretion (tears) from the nasolacrimal sac into the nose.

Nasopharynx (nā'-zō-FAR-inks) The superior portion of the pharynx, lying posterior to the nose and extending inferiorly to the soft palate.

Neck The part of the body connecting the head and the trunk. A constricted portion of an organ such as the neck of the femur or uterus.

Necrosis (ne-KRŌ-sis) A pathological type of cell death that results from disease, injury, or lack of blood supply in which many adjacent cells swell, burst, and spill their contents into the interstitial fluid, triggering an inflammatory response.

Negative feedback The principle governing most control systems; a mechanism of response in which a stimulus initiates actions that reverse or reduce the stimulus.

Neonatal (nē-ō-NĀ-tal) Pertaining to the first four weeks after birth.

Neoplasm (NĒ-ō-plazm) A new growth that may be benign or malignant.

Nephron (NEF-ron) The functional unit of the kidney.

Nerve A cordlike bundle of neuronal axons and/or dendrites and associated connective tissue coursing together outside the central nervous system.

Nerve fiber General term for any process (axon or dendrite) projecting from the cell body of a neuron.

Nerve impulse A wave of depolarization and repolarization that self-propagates along the plasma membrane of a neuron; also called a **nerve action potential.**

Nervous tissue Tissue containing neurons that initiate and conduct nerve impulses to coordinate homeostasis, and neuroglia that provide support and nourishment to neurons.

Net filtration pressure (NFP) Net pressure that promotes fluid outflow at the arterial end of a capillary, and fluid inflow at the venous end of a capillary; net pressure that promotes glomerular filtration in the kidneys.

Neural plate A thickening of ectoderm, induced by the notochord, that forms early in the third week of development and represents the beginning of the development of the nervous system.

Neuralgia (noo-RAL-jē-a) Attacks of pain along the entire course or branch of a peripheral sensory nerve.

Neuritis (noo-RĪ-tis) Inflammation of one or more nerves.

Neuroglia (noo-RŌG-lē-a) Cells of the nervous system that perform various supportive functions. The neuroglia of the central nervous system are the astrocytes, oligodendrocytes, microglia, and ependymal cells; neuroglia of the peripheral nervous system include Schwann cells and satellite cells. Also called **glial** (GLĒ-al) **cells.**

Neurohypophyseal (noo'-rō-hī'-pō-FIZ-ē-al) **bud** An outgrowth of ectoderm located on the floor of the hypothalamus that gives rise to the posterior pituitary.

Neurolemma (noo-rō-LEM-ma) The peripheral, nucleated cytoplasmic layer of the Schwann cell. Also called **sheath of Schwann** (SCHVON).

Neurology (noo-ROL-ō-jē) The study of the normal functioning and disorders of the nervous system.

Neuromuscular (noo-rō-MUS-kū-lar) **junction** A synapse between the axon terminals of a motor neuron and the sarcolemma of a muscle fiber (cell).

Neuron (NOO-ron) A nerve cell, consisting of a cell body, dendrites, and an axon.

Neuropeptide (noo-rō-PEP-tīd) Chain of three to about 40 amino acids that occurs naturally in the nervous system, and that acts primarily to modulate the response of or to a neurotransmitter. Examples are enkephalins and endorphins.

Neurosecretory (noo-rō-SĒC-re-tō-rē) **cell** A neuron that secretes a hypothalamic releasing hormone or inhibiting hormone into blood capillaries of the hypothalmus; a neuron that secretes oxytocin or antidiuretic hormone into blood capillaries of the posterior pituitary.

Neurotransmitter One of a variety of molecules within axon terminals that are released into the synaptic cleft in response to a nerve impulse, and that change the membrane potential of the postsynaptic neuron.

Neutrophil (NOO-trō-fil) A type of white blood cell characterized by granules that stain pale lilac with a combination of acidic and basic dyes.

Nicotinic (nik'-ō-TIN-ik) **receptor** Receptor for the neurotransmitter acetylcholine found on both sympathetic and parasympathetic postganglionic neurons and on skeletal muscle in the motor end plate; so named because nicotine activates these receptors but does not activate muscarinic receptors for acetylcholine.

Nipple A pigmented, wrinkled projection on the surface of the breast that is the location of the openings of the lactiferous ducts for milk release.

Nociceptor (nō'-sē-SEP-tor) A free (naked) nerve ending that detects painful stimuli.

Node of Ranvier (ron-vē-Ā) A space, along a myelinated axon, between the individual Schwann cells that form the myelin sheath and the neurolemma. Also called a **neurofibral node.**

Nondisjunction (non'-dis-JUNGK-shun) Failure of sister chromatids to separate properly during anaphase of mitosis (or meiosis II) or failure of homologous chromosomes to separate properly during meiosis I in which chromatids or chromosomes pass into the same daughter cell; the result is too many copies of that chromosome in the daughter cell or gamete.

Norepinephrine (nor'-ep-ē-NEF-rin) **(NE)** A hormone secreted by the adrenal medulla that produces actions similar to those that result from sympathetic stimulation. Also called **noradrenaline** (nor-a-DREN-a-lin).

Notochord (NŌ-tō-cord) A flexible rod of mesodermal tissue that lies where the future vertebral column will develop and plays a role in induction.

Nuclear medicine The branch of medicine concerned with the use of radioisotopes in the diagnosis and therapy of disease.

Nucleic (noo-KLĒ-ic) **acid** An organic compound that is a long polymer of nucleotides, with each nucleotide containing a pentose sugar, a phosphate group, and one of four possible nitrogenous bases (adenine, cytosine, guanine, and thymine or uracil).

Nucleolus (noo-KLĒ-ō-lus) Spherical body within a cell nucleus composed of protein, DNA, and RNA that is the site of the assembly of small and large ribosomal subunits.

Nucleosome (NOO-klē-ō-sōm) Structural subunit of a chromosome consisting of histones and DNA.

Nucleus (NOO-klē-us) A spherical or oval organelle of a cell that contains the hereditary factors of the cell, called genes. A cluster of unmyelinated nerve cell bodies in the central nervous system. The central part of an atom made up of protons and neutrons.

Nucleus pulposus (pul-PŌ-sus) A soft, pulpy, highly elastic substance in the center of an intervertebral disc; a remnant of the notochord.

Nutrient A chemical substance in food that provides energy, forms new body components, or assists in various body functions.

O

Obesity (ō-BĒS-i-tē) Body weight more than 20% above a desirable standard due to excessive accumulation of fat.

Oblique (ō-BLĒK) **plane** A plane that passes through the body or an organ at an angle between the transverse plane and either the midsagittal, parasagittal, or frontal plane.

Obstetrics (ob-STET-riks) The specialized branch of medicine that deals with pregnancy, labor, and the period of time immediately after delivery (about 6 weeks).

Olfactory (ōl-FAK-tō-rē) Pertaining to smell.

Olfactory bulb A mass of gray matter containing cell bodies of neurons that form synapses with neurons of the olfactory nerve (cranial nerve I), lying inferior to the frontal lobe of the cerebrum on either side of the crista galli of the ethmoid bone.

Olfactory receptor A bipolar neuron with its cell body lying between supporting cells located in the mucous membrane lining the superior portion of each nasal cavity; transduces odors into neural signals.

Olfactory tract A bundle of axons that extends from the olfactory bulb posteriorly to olfactory regions of the cerebral cortex.

Oligodendrocyte (ol'-i-gō-DEN-drō-sīt) A neuroglial cell that supports neurons and produces a myelin sheath around axons of neurons of the central nervous system.

Oligospermia (ol'-i-gō-SPER-mē-a) A deficiency of sperm cells in the semen.

Olive A prominent oval mass on each lateral surface of the superior part of the medulla oblongata.

Oncogenes (ONG-kō-jēnz) Cancer-causing genes; they derive from normal genes, termed **proto-oncogenes,** that encode proteins involved in cell growth or cell regulation but have the ability to transform a normal cell into a cancerous cell when they are mutated or inappropriately activated. One example is *p53.*

Oncology (ong-KOL-ō-jē) The study of tumors.

Oogenesis (ō'-ō-JEN-e-sis) Formation and development of female gametes (oocytes).

Oophorectomy (ō'-of-ō-REK-tō-me) Surgical removal of the ovaries.

Ophthalmic (of-THAL-mik) Pertaining to the eye.

Ophthalmologist (of'-thal-MOL-ō-jist) A physician who specializes in the diagnosis and treatment of eye disorders using drugs, surgery, and corrective lenses.

Ophthalmology (of'-thal-MOL-ō-jē) The study of the structure, function, and diseases of the eye.

Opsin (OP-sin) The glycoprotein portion of a photopigment.

Opsonization (op-sō-ni-ZĀ-shun) The action of some antibodies that renders bacteria and other foreign cells more susceptible to phagocytosis.

Optic (OP-tik) Refers to the eye, vision, or properties of light.

Optic chiasm (KĪ-azm) A crossing point of the optic nerves (cranial nerve II), anterior to the pituitary gland. Also called **optic chiasma.**

Optic disc A small area of the retina containing openings through which the axons of the ganglion cells emerge as the optic nerve (cranial nerve II). Also called the **blind spot.**

Optician (op-TISH-an) A technician who fits, adjusts, and dispenses corrective lenses on prescription of an ophthalmologist or optometrist.

Optic tract A bundle of axons that carry nerve impulses from the retina of the eye between the optic chiasm and the thalamus.

Optometrist (op-TOM-e-trist) Specialist with a doctorate degree in optometry who is licensed to examine and test the eyes and treat visual defects by prescribing corrective lenses.

Ora serrata (Ō-ra ser-RĀ-ta) The irregular margin of the retina lying internal and slightly posterior to the junction of the choroid and ciliary body.

Orbit (OR-bit) The bony, pyramidal-shaped cavity of the skull that holds the eyeball.

Organ A structure composed of two or more different kinds of tissues with a specific function and usually a recognizable shape.

Organelle (or-gan-EL) A permanent structure within a cell with characteristic morphology that is specialized to serve a specific function in cellular activities.

Organic (or-GAN-ik) **compound** Compound that always contains carbon in which the atoms are held together by covalent bonds. Examples include carbohydrates, lipids, proteins, and nucleic acids (DNA and RNA).

Organism (OR-ga-nizm) A total living form; one individual.

Orgasm (OR-gazm) Sensory and motor events involved in ejaculation for the male and involuntary contraction of the perineal muscles in the female at the climax of sexual intercourse.

Orifice (OR-i-fis) Any aperture or opening.

Origin (OR-i-jin) The attachment of a muscle tendon to a stationary bone or the end opposite the insertion.

Oropharynx (or′-ō-FAR-inks) The intermediate portion of the pharynx, lying posterior to the mouth and extending from the soft palate to the hyoid bone.

Orthopedics (or′-thō-PĒ-diks) The branch of medicine that deals with the preservation and restoration of the skeletal system, articulations, and associated structures.

Osmoreceptor (oz′-mō-re-CEP-tor) Receptor in the hypothalamus that is sensitive to changes in blood osmolarity and, in response to high osmolarity (low water concentration), stimulates synthesis and release of antidiuretic hormone (ADH).

Osmosis (os-MŌ-sis) The net movement of water molecules through a selectively permeable membrane from an area of higher water concentration to an area of lower water concentration until equilibrium is reached.

Osmotic pressure The pressure required to prevent the movement of pure water into a solution containing solutes when the solutions are separated by a selectively permeable membrane.

Osseous (OS-ē-us) Bony.

Ossicle (OS-si-kul) One of the small bones of the middle ear (malleus, incus, stapes).

Ossification (os′-i-fi-KĀ-shun) Formation of bone. Also called **osteogenesis.**

Osteoblast (OS-tē-ō-blast) Cell formed from an osteogenic cell that participates in bone formation by secreting some organic components and inorganic salts.

Osteoclast (OS-tē-ō-clast′) A large, multinuclear cell that resorbs (destroys) bone matrix.

Osteocyte (OS-tē-ō-sīt′) A mature bone cell that maintains the daily activities of bone tissue.

Osteogenic (os′-tē-ō-prō-JEN-i-tor) **cell** Stem cell derived from mesenchyme that has mitotic potential and the ability to differentiate into an osteoblast.

Osteogenic (os′-tē-ō-JEN-ik) **layer** The inner layer of the periosteum that contains cells responsible for forming new bone during growth and repair.

Osteology (os′-tē-OL-ō-jē) The study of bones.

Osteon (OS-tē-on) The basic unit of structure in adult compact bone, consisting of a central (Haversian) canal with its concentrically arranged lamellae, lacunae, osteocytes, and canaliculi. Also called a **Haversian** (ha-VER-shan) **system.**

Osteoporosis (os′-tē-ō-pō-RŌ-sis) Age-related disorder characterized by decreased bone mass and increased susceptibility to fractures, often as a result of decreased levels of estrogens.

Otic (Ō-tik) Pertaining to the ear.

Otolith (Ō-tō-lith) A particle of calcium carbonate embedded in the otolithic membrane that functions in maintaining static equilibrium.

Otolithic (ō-tō-LITH-ik) **membrane** Thick, gelatinous, glycoprotein layer located directly over hair cells of the macula in the saccule and utricle of the internal ear.

Otorhinolaryngology (ō′-tō-rī-nō-lar′-in-GOL-ō-jē) The branch of medicine that deals with the diagnosis and treatment of diseases of the ears, nose, and throat.

Oval window A small, membrane-covered opening between the middle ear and inner ear into which the footplate of the stapes fits.

Ovarian (ō-VAR-ē-an) **cycle** A monthly series of events in the ovary associated with the maturation of a secondary oocyte.

Ovarian follicle (FOL-i-kul) A general name for oocytes (immature ova) in any stage of development, along with their surrounding epithelial cells.

Ovarian ligament (LIG-a-ment) A rounded cord of connective tissue that attaches the ovary to the uterus.

Ovary (Ō-var-ē) Female gonad that produces oocytes and the estrogens, progesterone, inhibin, and relaxin hormones.

Ovulation (ov-ū-LĀ-shun) The rupture of a mature ovarian (Graafian) follicle with discharge of a secondary oocyte into the pelvic cavity.

Ovum (Ō-vum) The female reproductive or germ cell; an egg cell; arises through completion of meiosis in a secondary oocyte after penetration by a sperm.

Oxidation (ok-si-DĀ-shun) The removal of electrons from a molecule or, less commonly, the addition of oxygen to a molecule that results in a decrease in the energy content of the molecule. The oxidation of glucose in the body is called **cellular respiration.**

Oxyhemoglobin (ok′-sē-HĒ-mō-glō-bin) **(Hb—O₂)** Hemoglobin combined with oxygen.

Oxytocin (ok′-sē-TŌ-sin) **(OT)** A hormone secreted by neurosecretory cells in the paraventricular and supraoptic nuclei of the hypothalamus that stimulates contraction of smooth muscle in the pregnant uterus and myoepithelial cells around the ducts of mammary glands.

P

P wave The deflection wave of an electrocardiogram that signifies atrial depolarization.

Palate (PAL-at) The horizontal structure separating the oral and the nasal cavities; the roof of the mouth.

Palpate (PAL-pāt) To examine by touch; to feel.

Pancreas (PAN-krē-as) A soft, oblong organ lying along the greater curvature of the stomach and connected by a duct to the duodenum. It is both an exocrine gland (secreting pancreatic juice) and an endocrine gland (secreting insulin, glucagon, somatostatin, and pancreatic polypeptide).

Pancreatic (pan′-krē-AT-ik) **duct** A single large tube that unites with the common bile duct from the liver and gallbladder and drains pancreatic juice into the duodenum at the hepatopancreatic ampulla (ampulla of Vater). Also called the **duct of Wirsung.**

Pancreatic islet A cluster of endocrine gland cells in the pancreas that secretes insulin, glucagon, somatostatin, and pancreatic polypeptide. Also called an **islet of Langerhans** (LANG-er-hanz).

Pancreatic polypeptide Hormone secreted by the F cells of pancreatic islets (islets of Langerhans) that regulates release of pancreatic digestive enzymes.

Papanicolaou (pap′-a-NIK-ō-la-oo) **test** A cytological staining test for the detection and diagnosis of premalignant and malignant conditions of the female genital tract. Cells scraped from the epithelium of the cervix of the uterus are examined microscopically. Also called a **Pap test** or **Pap smear.**

Papilla (pa-PIL-a) A small nipple-shaped projection or elevation.

Paracrine (PAR-a-krin) Local hormone, such as histamine, that acts on neighboring cells without entering the bloodstream.

Paralysis (pa-RAL-a-sis) Loss or impairment of motor function due to a lesion of nervous or muscular origin.

Paranasal sinus (par′-a-NĀ-zal SĪ-nus) A mucus-lined air cavity in a skull bone that communicates with the nasal cavity. Paranasal sinuses are located in the frontal, maxillary, ethmoid, and sphenoid bones.

Paraplegia (par-a-PLĒ-jē-a) Paralysis of both lower limbs.

Parasagittal plane (par-a-SAJ-i-tal) A vertical plane that does not pass through the midline and that divides the body or organs into *unequal* left and right portions.

Parasympathetic (par′-a-sim-pa-THET-ik) **division** One of the two subdivisions of the autonomic nervous system, having cell bodies of preganglionic neurons in nuclei in the brain stem and in the lateral gray horn of the sacral portion of the spinal cord; primarily concerned with activities that conserve and restore body energy.

Parathyroid (par′-a-THĪ-royd) **gland** One of usually four small endocrine glands embedded in the posterior surfaces of the lateral lobes of the thyroid gland.

Parathyroid hormone (PTH) A hormone secreted by the chief (principal) cells of the parathyroid glands that increases blood calcium level and decreases blood phosphate level.

Paraurethral (par′-a-ū-RĒ-thral) **gland** Gland embedded in the wall of the urethra whose duct opens on either side of the urethral orifice and secretes mucus. Also called **Skene's** (SKĒNZ) **gland.**

Parenchyma (par-EN-ki-ma) The functional parts of any organ, as opposed to tissue that forms its stroma or framework.

Parietal (pa-RĪ-e-tal) Pertaining to or forming the outer wall of a body cavity.

Parietal cell A type of secretory cell in gastric glands that produces hydrochloric acid and intrinsic factor. Also called an **oxyntic cell.**

Parkinson disease (PD) Progressive degeneration of the basal ganglia and substantia nigra of the cerebrum resulting in decreased production of dopamine (DA) that leads to tremor, slowing of voluntary movements, and muscle weakness.

Parotid (pa-ROT-id) **gland** One of the paired salivary glands located inferior and anterior to the ears and connected to the oral cavity via a duct (Stensen's) that opens into the inside of the cheek opposite the maxillary (upper) second molar tooth.

Parturition (par′-too-RISH-un) Act of giving birth to young; childbirth, delivery.

Patellar (pa-TELL-ar) **reflex** Extension of the leg by contraction of the quadriceps femoris muscle in response to tapping the patellar ligament. Also called the **knee jerk reflex.**

Patent ductus arteriosus Congenital anatomical heart defect in which the fetal connection between the aorta and pulmonary trunk remains open instead of closing completely after birth.

Pathogen (PATH-ō-jen) A disease-producing microbe.

Pathological (path′-ō-LOJ-i-kal) **anatomy** The study of structural changes caused by disease.

Pectoral (PEK-tō-ral) Pertaining to the chest or breast.

Pediatrician (pē′-dē-a-TRISH-un) A physician who specializes in the care and treatment of children.

Pedicel (PED-i-sel) Footlike structure, as on podocytes of a glomerulus.

Pelvic (PEL-vik) **cavity** Inferior portion of the abdominopelvic cavity that contains the urinary bladder, sigmoid colon, rectum, and internal female and male reproductive structures.

Pelvic splanchnic (PEL-vik SPLANGK-nik) **nerves** Consist of preganglionic parasympathetic axons from the levels of S2, S3, and S4 that supply the urinary bladder, reproductive organs, and the descending and sigmoid colon and rectum.

Pelvis The basinlike structure formed by the two hip bones, the sacrum, and the coccyx. The expanded, proximal portion of the ureter, lying within the kidney and into which the major calyces open.

Penis (PĒ-nis) The organ of urination and copulation in males; used to deposit semen into the female vagina.

Pepsin Protein-digesting enzyme secreted by chief cells of the stomach in the inactive form pepsinogen, which is converted to active pepsin by hydrochloric acid.

Peptic ulcer An ulcer that develops in areas of the gastrointestinal tract exposed to hydrochloric acid; classified as a gastric ulcer if in the lesser curvature of the stomach and as a duodenal ulcer if in the first part of the duodenum.

Percussion (per-KUSH-un) The act of striking (percussing) an underlying part of the body with short, sharp blows as an aid in diagnosing the part by the quality of the sound produced.

Perforating canal A minute passageway by means of which blood vessels and nerves from the periosteum penetrate into compact bone. Also called **Volkmann's** (FŌLK-manz) **canal.**

Pericardial (per′-i-KAR-dē-al) **cavity** Small potential space between the visceral and parietal layers of the serous pericardium that contains pericardial fluid.

Pericardium (per′-i-KAR-dē-um) A loose-fitting membrane that encloses the heart, consisting of a superficial fibrous layer and a deep serous layer.

Perichondrium (per′-i-KON-drē-um) The membrane that covers cartilage.

Perilymph (PER-i-limf) The fluid contained between the bony and membranous labyrinths of the inner ear.

Perimetrium (per′-i-MĒ-trē-um) The serosa of the uterus.

Perimysium (per′-i-MĪZ-ē-um) Invagination of the epimysium that divides muscles into bundles.

Perineum (per′-i-NĒ-um) The pelvic floor; the space between the anus and the scrotum in the male and between the anus and the vulva in the female.

Perineurium (per′-i-NOO-rē-um) Connective tissue wrapping around fascicles in a nerve.

Periodontal (per-ē-ō-DON-tal) **disease** A collective term for conditions characterized by degeneration of gingivae, alveolar bone, periodontal ligament, and cementum.

Periodontal ligament The periosteum lining the alveoli (sockets) for the teeth in the alveolar processes of the mandible and maxillae.

Periosteum (per′-ē-OS-tē-um) The membrane that covers bone and consists of connective tissue, osteogenic cells, and osteoblasts; is essential for bone growth, repair, and nutrition.

Peripheral (pe-RIF-er-al) Located on the outer part or a surface of the body.

Peripheral nervous system (PNS) The part of the nervous system that lies outside the central nervous system, consisting of nerves and ganglia.

Peristalsis (per'-i-STAL-sis) Successive muscular contractions along the wall of a hollow muscular structure.

Peritoneum (per'-i-tō-NĒ-um) The largest serous membrane of the body that lines the abdominal cavity and covers the viscera.

Peritonitis (per'-i-tō-NĪ-tis) Inflammation of the peritoneum.

Permissive (per-MIS-sive) **effect** A hormonal interaction in which the effect of one hormone on a target cell requires previous or simultaneous exposure to another hormone(s) to enhance the response of a target cell or increase the activity of another hormone.

Peroxisome (per-OK-si-sōm) Organelle similar in structure to a lysosome that contains enzymes that use molecular oxygen to oxidize various organic compounds; such reactions produce hydrogen peroxide; abundant in liver cells.

Perspiration Sweat; produced by sudoriferous (sweat) glands and containing water, salts, urea, uric acid, amino acids, ammonia, sugar, lactic acid, and ascorbic acid. Helps maintain body temperature and eliminate wastes.

pH A measure of the concentration of hydrogen ions (H^+) in a solution. The pH scale extends from 0 to 14, with a value of 7 expressing neutrality, values lower than 7 expressing increasing acidity, and values higher than 7 expressing increasing alkalinity.

Phagocytosis (fag'-ō-sī-TŌ-sis) The process by which phagocytes ingest particulate matter; the ingestion and destruction of microbes, cell debris, and other foreign matter.

Phalanx (FĀ-lanks) The bone of a finger or toe. *Plural is* **phalanges** (fa-LAN-jēz).

Pharmacology (far'-ma-KOL-ō-jē) The science of the effects and uses of drugs in the treatment of disease.

Pharynx (FAR-inks) The throat; a tube that starts at the internal nares and runs partway down the neck, where it opens into the esophagus posteriorly and the larynx anteriorly.

Phenotype (FĒ-nō-tīp) The observable expression of genotype; physical characteristics of an organism determined by genetic makeup and influenced by interaction between genes and internal and external environmental factors.

Phlebitis (fle-BĪ-tis) Inflammation of a vein, usually in a lower limb.

Phosphorylation (fos'-for-i-LĀ-shun) The addition of a phosphate group to a chemical compound; types include substrate-level phosphorylation, oxidative phosphorylation, and photo-phosphorylation.

Photopigment A substance that can absorb light and undergo structural changes that can lead to the development of a receptor potential. An example is rhodopsin. In the eye, also called **visual pigment.**

Photoreceptor Receptor that detects light shining on the retina of the eye.

Physiology (fiz'-ē-OL-ō-jē) Science that deals with the functions of an organism or its parts.

Pia mater (PĪ-a-MĀ-ter *or* PĒ-a MA-ter) The innermost of the three meninges (coverings) of the brain and spinal cord.

Pineal (PĪN-ē-al) **gland** A cone-shaped gland located in the roof of the third ventricle that secretes melatonin. Also called the **epiphysis cerebri** (ē-PIF-i-sis se-RĒ-brē).

Pinealocyte (pin-ē-AL-ō-sīt) Secretory cell of the pineal gland that releases melatonin.

Pinna (PIN-na) The projecting part of the external ear composed of elastic cartilage and covered by skin and shaped like the flared end of a trumpet. Also called the **auricle** (OR-i-kul).

Pinocytosis (pi'-nō-sī-TŌ-sis) A process by which most body cells can ingest membrane-surrounded droplets of interstitial fluid.

Pituicyte (pi-TOO-i-sīt) Supporting cell of the posterior pituitary.

Pituitary (pi-TOO-i-tār-ē) **gland** A small endocrine gland occupying the hypophyseal fossa above the sella turcica of the sphenoid bone and attached to the hypothalamus by the infundibulum. Also called the **hypophysis** (hī-POF-i-sis).

Pivot joint A synovial joint in which a rounded, pointed, or conical surface of one bone articulates with a ring formed partly by another bone and partly by a ligament, as in the joint between the atlas and axis and between the proximal ends of the radius and ulna. Also called a **trochoid** (TRŌ-koyd) **joint.**

Placenta (pla-SEN-ta) The special structure through which the exchange of materials between fetal and maternal circulations occurs. Also called the **afterbirth.**

Plantar flexion (PLAN-tar FLEK-shun) Bending the foot in the direction of the plantar surface (sole).

Plaque (PLAK) A layer of dense proteins on the inside of a plasma membrane in adherens junctions and desmosomes. A mass of bacterial cells, dextran (polysaccharide), and other debris that adheres to teeth (dental plaque). See also atherosclerotic plaque.

Plasma (PLAZ-ma) The extracellular fluid found in blood vessels; blood minus the formed elements.

Plasma cell Cell that develops from a B cell (lymphocyte) and produces antibodies.

Plasma (cell) membrane Outer, limiting membrane that separates the cell's internal parts from extracellular fluid or the external environment.

Platelet (PLĀT-let) A fragment of cytoplasm enclosed in a cell membrane and lacking a nucleus; found in the circulating blood; plays a role in hemostasis. Also called a **thrombocyte** (THROM-bō-sīt).

Platelet plug Aggregation of platelets (thrombocytes) at a site where a blood vessel is damaged that helps stop or slow blood loss.

Pleura (PLOOR-a) The serous membrane that covers the lungs and lines the walls of the chest and the diaphragm.

Pleural cavity Small potential space between the visceral and parietal pleurae.

Plexus (PLEK-sus) A network of nerves, veins, or lymphatic vessels.

Pluripotent stem cell Immature stem cell in red bone marrow that gives rise to precursors of all the different mature blood cells.

Pneumotaxic (noo-mō-TAK-sik) **area** A part of the respiratory center in the pons that continually sends inhibitory nerve impulses to the inspiratory area, limiting inhalation and facilitating exhalation.

Podiatry (pō-DĪ-a-trē) The diagnosis and treatment of foot disorders.

Polar body The smaller cell resulting from the unequal division of primary and secondary oocytes during meiosis. The polar body has no function and degenerates.

Polycythemia (pol'-ē-sī-THĒ-mē-a) Disorder characterized by an above-normal hematocrit (above 55%) in which hypertension, thrombosis, and hemorrhage can occur.

Polysaccharide (pol'-ē-SAK-a-rīd) A carbohydrate in which three or more monosaccharides are joined chemically.

Polyunsaturated fat A fatty acid that contains more than one double covalent bond between its carbon atoms; abundant in triglycerides of corn oil, safflower oil, and cottonseed oil.

Polyuria (pol'-ē-Ū-rē-a) An excessive production of urine.

Pons (PONZ) The part of the brain stem that forms a "bridge" between the medulla oblongata and the midbrain, anterior to the cerebellum.

Positive feedback A feedback mechanism in which the response enhances the original stimulus.

Postabsorptive (fasting) state Metabolic state after absorption of food is complete and energy needs of the body must be satisfied using molecules stored previously.

Postcentral gyrus Gyrus of cerebral cortex located immediately posterior to the central sulcus; contains the primary somatosensory area.

Posterior (pos-TĒR-ē-or) Nearer to or at the back of the body. Equivalent to **dorsal** in bipeds.

Posterior column–medial lemniscus pathways Sensory pathways that carry information related to proprioception, fine touch, two-point discrimination, pressure, and vibration. First-order neurons project from the spinal cord to the ipsilateral medulla in the posterior columns (gracile fasciculus and cuneate fasciculus). Second-order neurons project from the medulla to the contralateral thalamus in the medial lemniscus. Third-order neurons project from the thalamus to the somatosensory cortex (postcentral gyrus) on the same side.

Posterior pituitary Posterior lobe of the pituitary gland. Also called the **neurohypophysis** (noo-rō-hī-POF-i-sis).

Posterior root The structure composed of sensory axons lying between a spinal nerve and the dorsolateral aspect of the spinal cord. Also called the **dorsal (sensory) root.**

Posterior root ganglion (GANG-glē-on) A group of cell bodies of sensory neurons and their supporting cells located along the posterior root of a spinal nerve. Also called a **dorsal (sensory) root ganglion.**

Postganglionic neuron (pōst′-gang-lē-ON-ik NOO-ron) The second autonomic motor neuron in an autonomic pathway, having its cell body and dendrites located in an autonomic ganglion and its unmyelinated axon ending at cardiac muscle, smooth muscle, or a gland.

Postsynaptic (pōst-sin-AP-tik) **neuron** The nerve cell that is activated by the release of a neurotransmitter from another neuron and carries nerve impulses away from the synapse.

Precapillary sphincter (SFINGK-ter) A ring of smooth muscle fibers (cells) at the site of origin of true capillaries that regulate blood flow into true capillaries.

Precentral gyrus Gyrus of cerebral cortex located immediately anterior to the central sulcus; contains the primary motor area.

Preganglionic (prē′-gang-lē-ON-ik) **neuron** The first autonomic motor neuron in an autonomic pathway, with its cell body and dendrites in the brain or spinal cord and its myelinated axon ending at an autonomic ganglion, where it synapses with a postganglionic neuron.

Pregnancy Sequence of events that normally includes fertilization, implantation, embryonic growth, and fetal growth and terminates in birth.

Premenstrual syndrome (PMS) Severe physical and emotional stress occurring late in the postovulatory phase of the menstrual cycle and sometimes overlapping with menstruation.

Prepuce (PRĒ-poos) The loose-fitting skin covering the glans of the penis and clitoris. Also called the **foreskin.**

Presbyopia (prez-bē-Ō-pē-a) A loss of elasticity of the lens of the eye due to advancing age with resulting inability to focus clearly on near objects.

Presynaptic (prē-sin-AP-tik) **inhibition** Type of inhibition in which neurotransmitter released by an inhibitory neuron depresses the release of neurotransmitter by a presynaptic neuron.

Presynaptic (prē-sin-AP-tik) **neuron** A neuron that propagates nerve impulses toward a synapse.

Prevertebral ganglion (prē-VER-te-bral GANG-lē-on) A cluster of cell bodies of postganglionic sympathetic neurons anterior to the spinal column and close to large abdominal arteries. Also called a **collateral ganglion.**

Primary germ layer One of three layers of embryonic tissue, called ectoderm, mesoderm, and endoderm, that give rise to all tissues and organs of the body.

Primary motor area A region of the cerebral cortex in the precentral gyrus of the frontal lobe of the cerebrum that controls specific muscles or groups of muscles.

Primary somatosensory area A region of the cerebral cortex posterior to the central sulcus in the postcentral gyrus of the parietal lobe of the cerebrum that localizes exactly the points of the body where somatic sensations originate.

Prime mover The muscle directly responsible for producing a desired motion. Also called an **agonist** (AG-ō-nist).

Primitive gut Embryonic structure formed from the dorsal part of the yolk sac that gives rise to most of the gastrointestinal tract.

Principal cell Cell type in the distal convoluted tubules and collecting ducts of the kidneys that is stimulated by aldosterone and antidiuretic hormone.

Proctology (prok-TOL-ō-jē) The branch of medicine concerned with the rectum and its disorders.

Progeny (PROJ-e-nē) Offspring or descendants.

Progesterone (prō-JES-te-rōn) A female sex hormone produced by the ovaries that helps prepare the endometrium of the uterus for implantation of a fertilized ovum and the mammary glands for milk secretion.

Prognosis (prog-NŌ-sis) A forecast of the probable results of a disorder; the outlook for recovery.

Prolactin (prō-LAK-tin) **(PRL)** A hormone secreted by the anterior pituitary that initiates and maintains milk secretion by the mammary glands.

Prolapse (PRŌ-laps) A dropping or falling down of an organ, especially the uterus or rectum.

Proliferation (prō-lif′-er-Ā-shun) Rapid and repeated reproduction of new parts, especially cells.

Pronation (prō-NĀ-shun) A movement of the forearm in which the palm is turned posteriorly or inferiorly.

Prophase (PRŌ-fāz) The first stage of mitosis during which chromatid pairs are formed and aggregate around the metaphase plate of the cell.

Proprioception (prō-prē-ō-SEP-shun) The perception of the position of body parts, especially the limbs, independent of vision; this sense is possible due to nerve impulses generated by proprioceptors.

Proprioceptor (prō′-prē-ō-SEP-tor) A receptor located in muscles, tendons, joints, or the internal ear (muscle spindles, tendon organs, joint kinesthetic receptors, and hair cells of the vestibular apparatus) that provides information about body position and movements.

Prostaglandin (pros′-ta-GLAN-din) **(PG)** A membrane-associated lipid; released in small quantities and acts as a local hormone.

Prostate (PROS-tāt) A doughnut-shaped gland inferior to the urinary bladder that surrounds the superior portion of the male urethra and secretes a slightly acidic solution that contributes to sperm motility and viability.

Protein An organic compound consisting of carbon, hydrogen, oxygen, nitrogen, and sometimes sulfur and phosphorus; synthesized on ribosomes and made up of amino acids linked by peptide bonds.

Proteasome Tiny cellular organelle in cytosol and nucleus containing proteases that destroy unneeded, damaged, or faulty proteins.

Prothrombin (prō-THROM-bin) An inactive blood-clotting factor synthesized by the liver, released into the blood, and converted to active thrombin in the process of blood clotting by the activated enzyme prothrombinase.

Protraction (prō-TRAK-shun) The movement of the mandible or shoulder girdle forward on a plane parallel with the ground.

Proximal (PROK-si-mal) Nearer the attachment of a limb to the trunk; nearer to the point of origin or attachment.

Pseudopods (SOO-dō-pods) Temporary protrusions of the leading edge of a migrating cell; cellular projections that surround a particle undergoing phagocytosis.

Pterygopalatine ganglion (ter′-i-gō-PAL-a-tīn GANG-glē-on) A cluster of cell bodies of parasympathetic postganglionic neurons ending at the lacrimal and nasal glands.

Ptosis (TŌ-sis) Drooping, as of the eyelid or the kidney.

Puberty (PŪ-ber-tē) The time of life during which the secondary sex characteristics begin to appear and the capability for sexual reproduction is possible; usually occurs between the ages of 10 and 17.

Pubic symphysis A slightly movable cartilaginous joint between the anterior surfaces of the hip bones.

Puerperium (pū′-er-PER-ē-um) The period immediately after childbirth, usually 4–6 weeks.

Pulmonary (PUL-mo-ner′-ē) Concerning or affected by the lungs.

Pulmonary circulation The flow of deoxygenated blood from the right ventricle to the lungs and the return of oxygenated blood from the lungs to the left atrium.

Pulmonary edema (e-DĒ-ma) An abnormal accumulation of interstitial fluid in the tissue spaces and alveoli of the lungs due to increased pulmonary capillary permeability or increased pulmonary capillary pressure.

Pulmonary embolism (EM-bō-lizm) **(PE)** The presence of a blood clot or a foreign substance in a pulmonary arterial blood vessel that obstructs circulation to lung tissue.

Pulmonary ventilation The inflow (inhalation) and outflow (exhalation) of air between the atmosphere and the lungs. Also called **breathing**.

Pulp cavity A cavity within the crown and neck of a tooth, which is filled with pulp, a connective tissue containing blood vessels, nerves, and lymphatic vessels.

Pulse (PULS) The rhythmic expansion and elastic recoil of a systemic artery after each contraction of the left ventricle.

Pulse pressure The difference between the maximum (systolic) and minimum (diastolic) pressures; normally about 40 mmHg.

Pupil The hole in the center of the iris, the area through which light enters the posterior cavity of the eyeball.

Purkinje (pur-KIN-jē) **fiber** Muscle fiber (cell) in the ventricular tissue of the heart specialized for conducting an action potential to the myocardium; part of the conduction system of the heart.

Pus The liquid product of inflammation containing leukocytes or their remains and debris of dead cells.

Pyloric (pī-LOR-ik) **sphincter** A thickened ring of smooth muscle through which the pylorus of the stomach communicates with the duodenum. Also called the **pyloric valve.**

Pyogenesis (pī′-ō-JEN-e-sis) Formation of pus.

Pyramid (PIR-a-mid) A pointed or cone-shaped structure. One of two roughly triangular structures on the anterior aspect of the medulla oblongata composed of the largest motor tracts that run from the cerebral cortex to the spinal cord. A triangular structure in the renal medulla.

Pyramidal (pi-RAM-i-dal) **tracts (pathways)** *See* **Direct motor pathways**.

Q

QRS wave The deflection waves of an electrocardiogram that represent onset of ventricular depolarization.

Quadriplegia (kwod′-ri-PLĒ-jē-a) Paralysis of four limbs: two upper and two lower.

R

Radiographic (rā′-dē-ō-GRAF-ic) **anatomy** Diagnostic branch of anatomy that includes the use of x rays.

Rami communicantes (RĀ-mē kō-mū-ni-KAN-tēz) Branches of a spinal nerve. *Singular is* **ramus communicans** (RĀ-mus kō-MŪ-ni-kans).

Rapid eye movement (REM) sleep Stage of sleep in which dreaming occurs, lasting for 5 to 10 minutes several times during a sleep cycle; characterized by rapid movements of the eyes beneath the eyelids.

Reactivity (rē-ak-TI-vi-tē) Ability of an antigen to react specifically with the antibody whose formation it induced.

Receptor A specialized cell or a distal portion of a neuron that responds to a specific sensory modality, such as touch, pressure, cold, light, or sound, and converts it to an electrical signal (generator or receptor potential). A specific molecule or cluster of molecules that recognizes and binds a particular ligand.

Receptor-mediated endocytosis A highly selective process whereby cells take up specific ligands, which usually are large molecules or particles, by enveloping them within a sac of plasma membrane. Ligands are eventually broken down by enzymes in lysosomes.

Receptor potential Depolarization or hyperpolarization of the plasma membrane of a receptor that alters release of neurotransmitter from the cell; if the neuron that synapses with the receptor cell becomes depolarized to threshold, a nerve impulse is triggered.

Recessive allele An allele whose presence is masked in the presence of a dominant allele on the homologous chromosome.

Reciprocal innervation (re-SIP-rō-kal in-ner-VĀ-shun) The phenomenon by which action potentials stimulate contraction of one muscle and simultaneously inhibit contraction of antagonistic muscles.

Recombinant DNA Synthetic DNA, formed by joining a fragment of DNA from one source to a portion of DNA from another.

Recovery oxygen consumption Elevated oxygen use after exercise ends due to metabolic changes that start during exercise and continue after exercise. Previously called **oxygen debt.**

Recruitment (rē-KROOT-ment) The process of increasing the number of active motor units. Also called **motor unit summation.**

Rectouterine pouch A pocket formed by the parietal peritoneum as it moves posteriorly from the surface of the uterus and is reflected onto the rectum; the most inferior point in the pelvic cavity. Also called the **pouch** or **cul de sac of Douglas.**

Rectum (REK-tum) The last 20 cm (8 in.) of the gastrointestinal tract, from the sigmoid colon to the anus.

Red nucleus A cluster of cell bodies in the midbrain, occupying a large part of the tectum from which axons extend into the rubroreticular and rubrospinal tracts.

Red pulp That portion of the spleen that consists of venous sinuses filled with blood and thin plates of splenic tissue called splenic (Billroth's) cords.

Reduction The addition of electrons to a molecule or, less commonly, the removal of oxygen from a molecule that results in an increase in the energy content of the molecule.

Referred pain Pain that is felt at a site remote from the place of origin.

Reflex Fast response to a change (stimulus) in the internal or external environment that attempts to restore homeostasis.

Reflex arc The most basic conduction pathway through the nervous system, connecting a receptor and an effector and consisting of a receptor, a sensory neuron, an integrating center in the central nervous system, a motor neuron, and an effector.

Refraction (rē-FRAK-shun) The bending of light as it passes from one medium to another.

Refractory (re-FRAK-to-rē) **period** A time period during which an excitable cell (neuron or muscle fiber) cannot respond to a stimulus that is usually adequate to evoke an action potential.

Regional anatomy The division of anatomy dealing with a specific region of the body, such as the head, neck, chest, or abdomen.

Regurgitation (rē-gur'-ji-TĀ-shun) Return of solids or fluids to the mouth from the stomach; backward flow of blood through incompletely closed heart valves.

Relaxin (RLX) A female hormone produced by the ovaries and placenta that increases flexibility of the pubic symphysis and helps dilate the uterine cervix to ease delivery of a baby.

Releasing hormone Hormone secreted by the hypothalamus that can stimulate secretion of hormones of the anterior pituitary.

Remodeling Replacement of old bone by new bone tissue.

Renal (RĒ-nal) Pertaining to the kidneys.

Renal corpuscle (KOR-pus-l) A glomerular (Bowman's) capsule and its enclosed glomerulus.

Renal pelvis A cavity in the center of the kidney formed by the expanded, proximal portion of the ureter, lying within the kidney, and into which the major calyces open.

Renal pyramid A triangular structure in the renal medulla containing the straight segments of renal tubules and the vasa recta.

Renin (RĒ-nin) An enzyme released by the kidney into the plasma, where it converts angiotensinogen into angiotensin I.

Renin–angiotensin–aldosterone (RAA) pathway A mechanism for the control of blood pressure, initiated by the secretion of renin by juxtaglomerular cells of the kidney in response to low blood pressure; renin catalyzes formation of angiotensin I, which is converted to angiotensin II by angiotensin-converting enzyme (ACE), and angiotensin II stimulates secretion of aldosterone.

Repolarization (rē-pō-lar-i-ZĀ-shun) Restoration of a resting membrane potential after depolarization.

Reproduction (rē-prō-DUK-shun) The formation of new cells for growth, repair, or replacement; the production of a new individual.

Reproductive cell division Type of cell division in which gametes (sperm and oocytes) are produced; consists of meiosis and cytokinesis.

Residual (re-ZID-ū-al) **volume** The volume of air still contained in the lungs after a maximal exhalation; about 1200 mL.

Resistance (re-ZIS-tans) Hindrance (impedance) to blood flow as a result of higher viscosity, longer total blood vessel length, and smaller blood vessel radius. Ability to ward off disease. The hindrance encountered by electrical charges as they move from one point to another. The hindrance encountered by air as it moves through the respiratory passageways.

Respiration (res-pi-RĀ-shun) Overall exchange of gases between the atmosphere, blood, and body cells consisting of pulmonary ventilation, external respiration, and internal respiration.

Respiratory center Neurons in the pons and medulla oblongata of the brain stem that regulate the rate and depth of pulmonary ventilation.

Respiratory membrane Structure in the lungs consisting of the alveolar wall and its basement membrane and a capillary endothelium and its basement membrane through which the diffusion of respiratory gases occurs.

Resting membrane potential The voltage difference between the inside and outside of a cell membrane when the cell is not responding to a stimulus; in many neurons and muscle fibers it is -70 to -90 mV, with the inside of the cell negative relative to the outside.

Retention (rē-TEN-shun) A failure to void urine due to obstruction, nervous contraction of the urethra, or absence of sensation of desire to urinate.

Rete (RĒ-tē) **testis** The network of ducts in the testes.

Reticular (re-TIK-ū-lar) **activating system (RAS)** A portion of the reticular formation that has many ascending connections with the cerebral cortex; when this area of the brain stem is active, nerve impulses pass to the thalamus and widespread areas of the cerebral cortex, resulting in generalized alertness or arousal from sleep.

Reticular formation A network of small groups of neuronal cell bodies scattered among bundles of axons (mixed gray and white matter) beginning in the medulla oblongata and extending superiorly through the central part of the brain stem.

Reticulocyte (re-TIK-ū-lō-sīt) An immature red blood cell.

Reticulum (re-TIK-ū-lum) A network.

Retina (RET-i-na) The deep coat of the posterior portion of the eyeball consisting of nervous tissue (where the process of vision begins) and a pigmented layer of epithelial cells that contact the choroid.

Retinal (RE-ti-nal) A derivative of vitamin A that functions as the light-absorbing portion of the photopigment rhodopsin.

Retraction (rē-TRAK-shun) The movement of a protracted part of the body posteriorly on a plane parallel to the ground, as in pulling the lower jaw back in line with the upper jaw.

Retroflexion (re-trō-FLEK-shun) A malposition of the uterus in which it is tilted posteriorly.

Retrograde degeneration (RE-trō-grād dē-jen-er-Ā-shun) Changes that occur in the proximal portion of a damaged axon only as far as the first node of Ranvier; similar to changes that occur during Wallerian degeneration.

Retroperitoneal (re'-trō-per-i-tō-NĒ-al) External to the peritoneal lining of the abdominal cavity.

Rh factor An inherited antigen on the surface of red blood cells in Rh^+ individuals; not present in Rh^- individuals.

Rhinology (rī-NOL-ō-jē) The study of the nose and its disorders.

Rhodopsin (rō-DOP-sin) The photopigment in rods of the retina, consisting of a glycoprotein called opsin and a derivative of vitamin A called retinal.

Ribonucleic (rī-bō-noo-KLĒ-ik) **acid (RNA)** A single-stranded nucleic acid made up of nucleotides, each consisting of a nitrogenous base (adenine, cytosine, guanine, or uracil), ribose, and a phosphate group; three types are messenger RNA (mRNA), transfer RNA (tRNA), and ribosomal RNA (rRNA), each of which has a specific role during protein synthesis.

Ribosome (RĪ-bō-sōm) An organelle in the cytoplasm of cells, composed of a small subunit and a large subunit that contain ribosomal RNA and ribosomal proteins; the site of protein synthesis.

Right heart (atrial) reflex A reflex concerned with maintaining normal venous blood pressure.

Right lymphatic (lim-FAT-ik) **duct** A vessel of the lymphatic system that drains lymph from the upper right side of the body and empties it into the right subclavian vein.

Rigidity (ri-JID-i-tē) Hypertonia characterized by increased muscle tone, but reflexes are not affected.

Rigor mortis State of partial contraction of muscles after death due to lack of ATP; myosin heads (cross bridges) remain attached to actin, thus preventing relaxation.

Rod One of two types of photoreceptor in the retina of the eye; specialized for vision in dim light.

Root canal A narrow extension of the pulp cavity lying within the root of a tooth.

Root of penis Attached portion of penis that consists of the bulb and crura.

Rotation (rō-TĀ-shun) Moving a bone around its own axis, with no other movement.

Round ligament (LIG-a-ment) A band of fibrous connective tissue enclosed between the folds of the broad ligament of the uterus, emerging from the uterus just inferior to the uterine tube, extending laterally along the pelvic wall and through the deep inguinal ring to end in the labia majora.

Round window A small opening between the middle and internal ear, directly inferior to the oval window, covered by the secondary tympanic membrane.

Rugae (ROO-gē) Large folds in the mucosa of an empty hollow organ, such as the stomach and vagina.

S

Saccule (SAK-ūl) The inferior and smaller of the two chambers in the membranous labyrinth inside the vestibule of the internal ear containing a receptor organ for static equilibrium.

Sacral plexus (SĀ-kral PLEK-sus) A network formed by the ventral branches of spinal nerves L4 through S3.

Sacral promontory (PROM-on-tor′-ē) The superior surface of the body of the first sacral vertebra that projects anteriorly into the pelvic cavity; a line from the sacral promontory to the superior border of the pubic symphysis divides the abdominal and pelvic cavities.

Saddle joint A synovial joint in which the articular surface of one bone is saddle shaped and the articular surface of the other bone is shaped like the legs of the rider sitting in the saddle, as in the joint between the trapezium and the metacarpal of the thumb.

Sagittal (SAJ-i-tal) **plane** A plane that divides the body or organs into left and right portions. Such a plane may be **midsagittal (median),** in which the divisions are equal, or **parasagittal,** in which the divisions are unequal.

Saliva (sa-LĪ-va) A clear, alkaline, somewhat viscous secretion produced mostly by the three pairs of salivary glands; contains various salts, mucin, lysozyme, salivary amylase, and lingual lipase (produced by glands in the tongue).

Salivary amylase (SAL-i-ver-ē AM-i-lās) An enzyme in saliva that initiates the chemical breakdown of starch.

Salivary gland One of three pairs of glands that lie external to the mouth and pour their secretory product (saliva) into ducts that empty into the oral cavity; the parotid, submandibular, and sublingual glands.

Salt A substance that, when dissolved in water, ionizes into cations and anions, neither of which are hydrogen ions (H^+) or hydroxide ions (OH^-).

Saltatory (sal-ta-TŌ-rē) **conduction** The propagation of an action potential (nerve impulse) along the exposed parts of a myelinated axon. The action potential appears at successive nodes of Ranvier and therefore seems to leap from node to node.

Sarcolemma (sar′-kō-LEM-ma) The cell membrane of a muscle fiber (cell), especially of a skeletal muscle fiber.

Sarcomere (SAR-kō-mēr) A contractile unit in a striated muscle fiber (cell) extending from one Z disc to the next Z disc.

Sarcoplasm (SAR-kō-plazm) The cytoplasm of a muscle fiber (cell).

Sarcoplasmic reticulum (sar′-kō-PLAZ-mik re-TIK-ū-lum) A network of saccules and tubes surrounding myofibrils of a muscle fiber (cell), comparable to endoplasmic reticulum; functions to reabsorb calcium ions during relaxation and to release them to cause contraction.

Saturated fat A fatty acid that contains only single bonds (no double bonds) between its carbon atoms; all carbon atoms are bonded to the maximum number of hydrogen atoms; prevalent in triglycerides of animal products such as meat, milk, milk products, and eggs.

Scala tympani (SKA-la TIM-pan-ē) The inferior spiral-shaped channel of the bony cochlea, filled with perilymph.

Scala vestibuli (ves-TIB-ū-lē) The superior spiral-shaped channel of the bony cochlea, filled with perilymph.

Schwann (SCHVON) **cell** A neuroglial cell of the peripheral nervous system that forms the myelin sheath and neurolemma around a nerve axon by wrapping around the axon in a jelly-roll fashion.

Sciatica (sī-AT-i-ka) Inflammation and pain along the sciatic nerve; felt along the posterior aspect of the thigh extending down the inside of the leg.

Sclera (SKLE-ra) The white coat of fibrous tissue that forms the superficial protective covering over the eyeball except in the most anterior portion; the posterior portion of the fibrous tunic.

Scleral venous sinus A circular venous sinus located at the junction of the sclera and the cornea through which aqueous humor drains from the anterior chamber of the eyeball into the blood. Also called the **canal of Schlemm** (SHLEM).

Sclerosis (skle-RŌ-sis) A hardening with loss of elasticity of tissues.

Scoliosis (skō′-lē-Ō-sis) An abnormal lateral curvature from the normal vertical line of the backbone.

Scrotum (SKRŌ-tum) A skin-covered pouch that contains the testes and their accessory structures.

Sebaceous (se-BĀ-shus) **gland** An exocrine gland in the dermis of the skin, almost always associated with a hair follicle, that secretes sebum. Also called an **oil gland.**

Sebum (SĒ-bum) Secretion of sebaceous (oil) glands.

Secondary response Accelerated, more intense cell-mediated or antibody-mediated immune response upon a subsequent exposure to an antigen after the primary response.

Secondary sex characteristic A characteristic of the male or female body that develops at puberty under the influence of sex hormones but is not directly involved in sexual reproduction; examples are distribution of body hair, voice pitch, body shape, and muscle development.

Second messenger An intracellular mediator molecule that is produced in response to a first messenger (hormone or neurotransmitter) binding to its receptor in the plasma membrane of a target cell. Initiates a cascade of chemical reactions that produce characteristic effects for that particular target cell.

Secretion (se-KRĒ-shun) Production and release from a cell or a gland of a physiologically active substance.

Selective permeability (per'-mē-a-BIL-i-tē) The property of a membrane by which it permits the passage of certain substances but restricts the passage of others.

Semen (SĒ-men) A fluid discharged at ejaculation by a male that consists of a mixture of sperm and the secretions of the seminiferous tubules, seminal vesicles, prostate, and bulbourethral (Cowper's) glands.

Semicircular canals Three bony channels (anterior, posterior, lateral), filled with perilymph, in which lie the membranous semicircular canals filled with endolymph. They contain receptors for equilibrium.

Semicircular ducts The membranous semicircular canals filled with endolymph and floating in the perilymph of the bony semicircular canals; they contain cristae that are concerned with dynamic equilibrium.

Semilunar (sem'-ē-LOO-nar) **valve** A valve between the aorta or the pulmonary trunk and a ventricle of the heart.

Seminal vesicle (SEM-i-nal VES-i-kul) One of a pair of convoluted, pouchlike structures, lying posterior and inferior to the urinary bladder and anterior to the rectum, that secrete a component of semen into the ejaculatory ducts. Also termed **seminal gland.**

Seminiferous tubule (sem'-i-NI-fer-us TOO-būl) A tightly coiled duct, located in the testis, where sperm are produced.

Senescence (se-NES-ens) The process of growing old.

Sensation A state of awareness of external or internal conditions of the body.

Sensory neurons (NOO-ronz) Neurons that carry sensory information from cranial and spinal nerves into the brain and spinal cord or from a lower to a higher level in the spinal cord and brain. Also called **afferent neurons.**

Septal defect An opening in the septum (interatrial or interventricular) between the left and right sides of the heart.

Septum (SEP-tum) A wall dividing two cavities.

Serous (SIR-us) **membrane** A membrane that lines a body cavity that does not open to the exterior. The external layer of an organ formed by a serous membrane. The membrane that lines the pleural, pericardial, and peritoneal cavities. Also called a **serosa** (se-RŌ-sa).

Sertoli (ser-TŌ-lē) **cell** A supporting cell in the seminiferous tubules that secretes fluid for supplying nutrients to sperm and the hormone inhibin, removes excess cytoplasm from spermatogenic cells, and mediates the effects of FSH and testosterone on spermatogenesis. Also called a **sustentacular** (sus'-ten-TAK-ū-lar) **cell.**

Serum Blood plasma minus its clotting proteins.

Sesamoid (SES-a-moyd) **bones** Small bones usually found in tendons.

Sex chromosomes The twenty-third pair of chromosomes, designated X and Y, which determine the genetic sex of an individual; in males, the pair is XY; in females, XX.

Sexual intercourse The insertion of the erect penis of a male into the vagina of a female. Also called **coitus** (KŌ-i-tus).

Shivering Involuntary contraction of skeletal muscles that generates heat. Also called **involuntary thermogenesis.**

Shock Failure of the cardiovascular system to deliver adequate amounts of oxygen and nutrients to meet the metabolic needs of the body due to inadequate cardiac output. It is characterized by hypotension; clammy, cool, and pale skin; sweating; reduced urine formation; altered mental state; acidosis; tachycardia; weak, rapid pulse; and thirst. Types include hypovolemic, cardiogenic, vascular, and obstructive.

Shoulder joint A synovial joint where the humerus articulates with the scapula.

Sigmoid colon (SIG-moyd KŌ-lon) The S-shaped part of the large intestine that begins at the level of the left iliac crest, projects medially, and terminates at the rectum at about the level of the third sacral vertebra.

Sign Any objective evidence of disease that can be observed or measured such as a lesion, swelling, or fever.

Sinoatrial (si-nō-Ā-trē-al) **(SA) node** A small mass of cardiac muscle fibers (cells) located in the right atrium inferior to the opening of the superior vena cava that spontaneously depolarize and generate a cardiac action potential about 100 times per minute. Also called the **pacemaker.**

Sinus (SĪ-nus) A hollow in a bone (paranasal sinus) or other tissue; a channel for blood (vascular sinus); any cavity having a narrow opening.

Sinusoid (SĪ-nū-soyd) A large, thin-walled, and leaky type of capillary, having large intercellular clefts that may allow proteins and blood cells to pass from a tissue into the bloodstream; present in the liver, spleen, anterior pituitary, parathyroid glands, and red bone marrow.

Skeletal muscle An organ specialized for contraction, composed of striated muscle fibers (cells), supported by connective tissue, attached to a bone by a tendon or an aponeurosis, and stimulated by somatic motor neurons.

Skin The external covering of the body that consists of a superficial, thinner epidermis (epithelial tissue) and a deep, thicker dermis (connective tissue) that is anchored to the subcutaneous layer.

Skull The skeleton of the head consisting of the cranial and facial bones.

Sleep A state of partial unconsciousness from which a person can be aroused; associated with a low level of activity in the reticular activating system.

Sliding-filament mechanism The explanation of how thick and thin filaments slide relative to one another during striated muscle contraction to decrease sarcomere length.

Small intestine A long tube of the gastrointestinal tract that begins at the pyloric sphincter of the stomach, coils through the central and inferior part of the abdominal cavity, and ends at the large intestine; divided into three segments: duodenum, jejunum, and ileum.

Smooth muscle A tissue specialized for contraction, composed of smooth muscle fibers (cells), located in the walls of hollow internal organs, and innervated by autonomic motor neurons.

Sodium-potassium ATPase An active transport pump located in the plasma membrane that transports sodium ions out of the cell and potassium ions into the cell at the expense of cellular ATP. It functions to keep the ionic concentrations of these ions at physiological levels. Also called the **sodium-potassium pump.**

Soft palate (PAL-at) The posterior portion of the roof of the mouth, extending from the palatine bones to the uvula. It is a muscular partition lined with mucous membrane.

Solution A homogeneous molecular or ionic dispersion of one or more substances (solutes) in a dissolving medium (solvent) that is usually liquid.

Somatic (sō-MAT-ik) **cell division** Type of cell division in which a single starting parent cell duplicates itself to produce two identical cells; consists of mitosis and cytokinesis.

Somatic nervous system (SNS) The portion of the peripheral nervous system consisting of somatic sensory (afferent) neurons and somatic motor (efferent) neurons.

Somite (SŌ-mīt) Block of mesodermal cells in a developing embryo that is distinguished into a myotome (which forms most of the skeletal muscles), dermatome (which forms connective tissues), and sclerotome (which forms the vertebrae).

Spasm (SPAZM) A sudden, involuntary contraction of large groups of muscles.

Spasticity (spas-TIS-i-tē) Hypertonia characterized by increased muscle tone, increased tendon reflexes, and pathological reflexes (Babinski sign).

Spermatic (sper-MAT-ik) **cord** A supporting structure of the male reproductive system, extending from a testis to the deep inguinal ring, that includes the ductus (vas) deferens, arteries, veins, lymphatic vessels, nerves, cremaster muscle, and connective tissue.

Spermatogenesis (sper′-ma-tō-JEN-e-sis) The formation and development of sperm in the seminiferous tubules of the testes.

Sperm cell A mature male gamete. Also termed **spermatozoon** (sper′-ma-tō-ZŌ-on).

Spermiogenesis (sper′-mē-ō-JEN-e-sis) The maturation of spermatids into sperm.

Sphincter (SFINGK-ter) A circular muscle that constricts an opening.

Sphincter of the hepatopancreatic ampulla A circular muscle at the opening of the common bile and main pancreatic ducts in the duodenum. Also called the **sphincter of Oddi** (OD-ē).

Sphygmomanometer (sfig′-mō-ma-NOM-e-ter) An instrument for measuring arterial blood pressure.

Spinal (SPĪ-nal) **cord** A mass of nerve tissue located in the vertebral canal from which 31 pairs of spinal nerves originate.

Spinal nerve One of the 31 pairs of nerves that originate on the spinal cord from posterior and anterior roots.

Spinal shock A period from several days to several weeks following transection of the spinal cord and characterized by the abolition of all reflex activity.

Spinothalamic (spī-nō-tha-LAM-ik) **tracts** Sensory (ascending) tracts that convey information up the spinal cord to the thalamus for sensations of pain, temperature, crude touch, and deep pressure.

Spinous (SPĪ-nus) **process** A sharp or thornlike process or projection. Also called a **spine.** A sharp ridge running diagonally across the posterior surface of the scapula.

Spiral organ The organ of hearing, consisting of supporting cells and hair cells that rest on the basilar membrane and extend into the endolymph of the cochlear duct. Also called the **organ of Corti** (KOR-tē).

Spirometer (spī-ROM-e-ter) An apparatus used to measure lung volumes and capacities.

Splanchnic (SPLANK-nik) Pertaining to the viscera.

Spleen (SPLĒN) Large mass of lymphatic tissue between the fundus of the stomach and the diaphragm that functions in formation of blood cells during early fetal development, phagocytosis of ruptured blood cells, and proliferation of B cells during immune responses.

Spongy (cancellous) bone tissue Bone tissue that consists of an irregular latticework of thin plates of bone called trabeculae; spaces between trabeculae of some bones are filled with red bone marrow; found inside short, flat, and irregular bones and in the epiphyses (ends) of long bones.

Sprain Forcible wrenching or twisting of a joint with partial rupture or other injury to its attachments without dislocation.

Squamous (SKWĀ-mus) Flat or scalelike.

Starling's law of the capillaries The movement of fluid between plasma and interstitial fluid is in a state of near equilibrium at the arterial and venous ends of a capillary; that is, filtered fluid and absorbed fluid plus that returned to the lymphatic system are nearly equal.

Starling's law of the heart The force of muscular contraction is determined by the length of the cardiac muscle fibers just before they contract; within limits, the greater the length of stretched fibers, the stronger the contraction.

Starvation (star-VA-shun) The loss of energy stores in the form of glycogen, triglycerides, and proteins due to inadequate intake of nutrients or inability to digest, absorb, or metabolize ingested nutrients.

Stasis (STĀ-sis) Stagnation or halt of normal flow of fluids, as blood or urine, or of the intestinal contents.

Static equilibrium (ē-kwi-LIB-rē-um) The maintenance of posture in response to changes in the orientation of the body, mainly the head, relative to the ground.

Stellate reticuloendothelial (STEL-āt re-tik′-ū-lō-en′-dō-THĒ-lē-al) **cell** Phagocytic cell bordering a sinusoid of the liver. Also called a **Kupffer** (KOOP-fer) **cell.**

Stenosis (sten-Ō-sis) An abnormal narrowing or constriction of a duct or opening.

Stereocilia (ste′-rē-ō-SIL-ē-a) Groups of extremely long, slender, nonmotile microvilli projecting from epithelial cells lining the epididymis.

Stereognosis (ste′-rē-og-NŌ-sis) The ability to recognize the size, shape, and texture of an object by touching it.

Sterile (STE-ril) Free from any living microorganisms. Unable to conceive or produce offspring.

Sterilization (ster′-i-li-ZĀ-shun) Elimination of all living microorganisms. Any procedure that renders an individual incapable of reproduction (for example, castration, vasectomy, hysterectomy, or oophorectomy).

Stimulus Any stress that changes a controlled condition; any change in the internal or external environment that excites a sensory receptor, a neuron, or a muscle fiber.

Stomach The J-shaped enlargement of the gastrointestinal tract directly inferior to the diaphragm in the epigastric, umbilical, and left hypochondriac regions of the abdomen, between the esophagus and small intestine.

Straight tubule (TOO-būl) A duct in a testis leading from a convoluted seminiferous tubule to the rete testis.

Stratum (STRĀ-tum) A layer.

Stratum basalis (ba-SAL-is) The layer of the endometrium next to the myometrium that is maintained during menstruation and gestation and produces a new stratum functionalis following menstruation or parturition.

Stratum functionalis (funk′-shun-AL-is) The layer of the endometrium next to the uterine cavity that is shed during menstruation and that forms the maternal portion of the placenta during gestation.

Stressor A stress that is extreme, unusual, or long-lasting and triggers the stress response.

Stress response Wide-ranging set of bodily changes, triggered by a stressor, that gears the body to meet an emergency. Also known as **general adaptation syndrome (GAS).**

Stretch receptor Receptor in the walls of blood vessels, airways, or organs that monitors the amount of stretching. Also termed **baroreceptor.**

Stretch reflex A monosynaptic reflex triggered by sudden stretching of muscle spindles within a muscle that elicits contraction of that same muscle. Also called a **tendon jerk.**

Stroke volume The volume of blood ejected by either ventricle in one systole; about 70 mL in an adult at rest.

Stroma (STRŌ-ma) The tissue that forms the ground substance, foundation, or framework of an organ, as opposed to its functional parts (parenchyma).

Subarachnoid (sub′-a-RAK-noyd) **space** A space between the arachnoid mater and the pia mater that surrounds the brain and spinal cord and through which cerebrospinal fluid circulates.

Subcutaneous (sub′-kū-TĀ-nē-us) Beneath the skin. Also called **hypodermic** (hi-pō-DER-mik).

Subcutaneous layer A continuous sheet of areolar connective tissue and adipose tissue between the dermis of the skin and the deep fascia of the muscles. Also called the **superficial fascia** (FASH-ē-a).

Subdural (sub-DOO-ral) **space** A space between the dura mater and the arachnoid mater of the brain and spinal cord that contains a small amount of fluid.

Sublingual (sub-LING-gwal) **gland** One of a pair of salivary glands situated in the floor of the mouth deep to the mucous membrane and to the side of the lingual frenulum, with a duct (Rivinus') that opens into the floor of the mouth.

Submandibular (sub′-man-DIB-ū-lar) **gland** One of a pair of salivary glands found inferior to the base of the tongue deep to the mucous membrane in the posterior part of the floor of the mouth, posterior to the sublingual glands, with a duct (Wharton's) situated to the side of the lingual frenulum. Also called the **submaxillary** (sub′-MAK-si-ler-ē) **gland.**

Submucosa (sub-mū-KŌ-sa) A layer of connective tissue located deep to a mucous membrane, as in the gastrointestinal tract or the urinary bladder; the submucosa connects the mucosa to the muscularis layer.

Submucosal plexus A network of autonomic nerve fibers located in the superficial part of the submucous layer of the small intestine. Also called the **plexus of Meissner** (MĪZ-ner).

Substrate A molecule upon which an enzyme acts.

Subthreshold stimulus A stimulus of such weak intensity that it cannot initiate an action potential (nerve impulse).

Sudoriferous (soo′-dor-IF-er-us) **gland** An apocrine or eccrine exocrine gland in the dermis or subcutaneous layer that produces perspiration. Also called a **sweat gland.**

Sulcus (SUL-kus) A groove or depression between parts, especially between the convolutions of the brain. *Plural is* **sulci** (SUL-sī).

Summation (sum-MĀ-shun) The addition of the excitatory and inhibitory effects of many stimuli applied to a neuron. The increased strength of muscle contraction that results when stimuli follow one another in rapid succession.

Superficial (soo′-per-FISH-al) Located on or near the surface of the body or an organ.

Superficial fascia (FASH-ē-a) A continuous sheet of fibrous connective tissue between the dermis of the skin and the deep fascia of the muscles. Also called **subcutaneous** (sub′-kū-TĀ-nē-us) **layer.**

Superficial inguinal (IN-gwi-nal) **ring** A triangular opening in the aponeurosis of the external oblique muscle that represents the termination of the inguinal canal.

Superior (soo-PĒR-ē-or) Toward the head or upper part of a structure. Also called **cephalad** (SEF-a-lad) or **craniad.**

Superior vena cava (VĒ-na CĀ-va) **(SVC)** Large vein that collects blood from parts of the body superior to the heart and returns it to the right atrium.

Supination (soo-pi-NĀ-shun) A movement of the forearm in which the palm is turned anteriorly or superiorly.

Surface anatomy The study of the structures that can be identified from the outside of the body.

Surfactant (sur-FAK-tant) Complex mixture of phospholipids and lipoproteins, produced by type II alveolar (septal) cells in the lungs, that decreases surface tension.

Susceptibility (sus-sep′-ti-BIL-i-tē) Lack of resistance to the damaging effects of an agent such as a pathogen.

Suspensory ligament (sus-PEN-so-rē LIG-a-ment) A fold of peritoneum extending laterally from the surface of the ovary to the pelvic wall.

Sutural (SOO-cher-al) **bone** A small bone located within a suture between certain cranial bones. Also called **Wormian** (WER-mē-an) **bone.**

Suture (SOO-cher) An immovable fibrous joint that joins skull bones.

Sympathetic (sim′-pa-THET-ik) **division** One of the two subdivisions of the autonomic nervous system, having cell bodies of preganglionic neurons in the lateral gray columns of the thoracic segment and the first two or three lumbar segments of the spinal cord; primarily concerned with process involving the expenditure of energy.

Sympathetic trunk ganglion (GANG-glē-on) A cluster of cell bodies of sympathetic postganglionic neurons lateral to the vertebral column, close to the body of a vertebra. These ganglia extend inferiorly through the neck, thorax, and abdomen to the coccyx on both sides of the vertebral column and are connected to one another to form a chain on each side of the vertebral column. Also called **sympathetic chain** or **vertebral chain ganglia.**

Sympathomimetic (sim′-pa-thō-mi-MET-ik) Producing effects that mimic those brought about by the sympathetic division of the autonomic nervous system.

Symphysis (SIM-fi-sis) A line of union. A slightly movable cartilaginous joint such as the pubic symphysis.

Symporter A transmembrane transporter protein that moves two substances, often Na^+ and another substance, in the same direction across a plasma membrane. Also called a **cotransporter.**

Symptom (SIMP-tum) A subjective change in body function not apparent to an observer, such as pain or nausea, that indicates the presence of a disease or disorder of the body.

Synapse (SYN-aps) The functional junction between two neurons or between a neuron and an effector, such as a muscle or gland; may be electrical or chemical.

Synapsis (sin-AP-sis) The pairing of homologous chromosomes during prophase I of meiosis.

Synaptic (sin-AP-tik) **cleft** The narrow gap at a chemical synapse that separates the axon terminal of one neuron from another neuron or muscle fiber (cell) and across which a neurotransmitter diffuses to affect the postsynaptic cell.

Synaptic delay The length of time between the arrival of the action potential at a presynaptic axon terminal and the membrane potential

(IPSP or EPSP) change on the postsynaptic membrane; usually about 0.5 msec.

Synaptic end bulb Expanded distal end of an axon terminal that contains synaptic vesicles. Also called a **synaptic knob.**

Synaptic vesicle Membrane-enclosed sac in a synaptic end bulb that stores neurotransmitters.

Synarthrosis (sin′-ar-THRŌ-sis) An immovable joint such as a suture, gomphosis, and synchondrosis.

Synchondrosis (sin′-kon-DRŌ-sis) A cartilaginous joint in which the connecting material is hyaline cartilage.

Syndesmosis (sin′-dez-MŌ-sis) A slightly movable joint in which articulating bones are united by fibrous connective tissue.

Syndrome (SIN-drōm) A group of signs and symptoms that occur together in a pattern that is characteristic of a particular disease or abnormal condition.

Synergist (SIN-er-jist) A muscle that assists the prime mover by reducing undesired action or unnecessary movement.

Synergistic (syn-er-JIS-tik) **effect** A hormonal interaction in which the effects of two or more hormones acting together is greater or more extensive than the sum of each hormone acting alone.

Synostosis (sin′-os-TŌ-sis) A joint in which the dense fibrous connective tissue that unites bones at a suture has been replaced by bone, resulting in a complete fusion across the suture line.

Synovial (si-NŌ-vē-al) **cavity** The space between the articulating bones of a synovial joint, filled with synovial fluid. Also called a **joint cavity.**

Synovial fluid Secretion of synovial membranes that lubricates joints and nourishes articular cartilage.

Synovial joint A fully movable or diarthrotic joint in which a synovial (joint) cavity is present between the two articulating bones.

Synovial membrane The deeper of the two layers of the articular capsule of a synovial joint, composed of areolar connective tissue that secretes synovial fluid into the synovial (joint) cavity.

System An association of organs that have a common function.

Systemic (sis-TEM-ik) Affecting the whole body; generalized.

Systemic anatomy The anatomic study of particular systems of the body, such as the skeletal, muscular, nervous, cardiovascular, or urinary systems.

Systemic circulation The routes through which oxygenated blood flows from the left ventricle through the aorta to all the organs of the body and deoxygenated blood returns to the right atrium.

Systemic vascular resistance (SVR) All the vascular resistance offered by systemic blood vessels. Also called **total peripheral resistance.**

Systole (SIS-tō-lē) In the cardiac cycle, the phase of contraction of the heart muscle, especially of the ventricles.

Systolic (sis-TOL-ik) **blood pressure** The force exerted by blood on arterial walls during ventricular contraction; the highest pressure measured in the large arteries, about 120 mmHg under normal conditions for a young adult.

T

T cell A lymphocyte that becomes immunocompetent in the thymus and can differentiate into a helper T cell or a cytotoxic T cell, both of which function in cell-mediated immunity.

T wave The deflection wave of an electrocardiogram that represents ventricular repolarization.

Tachycardia (tak′-i-KAR-dē-a) An abnormally rapid resting heartbeat or pulse rate (over 100 beats per minute).

Tactile (TAK-tīl) Pertaining to the sense of touch.

Tactile disc Modified epidermal cell in the stratum basale of hairless skin that functions as a cutaneous receptor for discriminative touch. Also called a **Merkel** (MER-kel) **disc.**

Teniae coli (TĒ-nē-ē KŌ-lī) The three flat bands of thickened, longitudinal smooth muscle running the length of the large intestine, except in the rectum. *Singular is* **tenia coli.**

Target cell A cell whose activity is affected by a particular hormone.

Tarsal bones The seven bones of the ankle. Also called **tarsals.**

Tarsal gland Sebaceous (oil) gland that opens on the edge of each eyelid. Also called a **Meibomian** (mī-BŌ-mē-an) **gland.**

Tarsal plate A thin, elongated sheet of connective tissue, one in each eyelid, giving the eyelid form and support. The aponeurosis of the levator palpebrae superioris is attached to the tarsal plate of the superior eyelid.

Tarsus (TAR-sus) A collective term for the seven bones of the ankle.

Tectorial (tek-TŌ-rē-al) **membrane** A gelatinous membrane projecting over and in contact with the hair cells of the spiral organ (organ of Corti) in the cochlear duct.

Teeth (TĒTH) Accessory structures of digestion, composed of calcified connective tissue and embedded in bony sockets of the mandible and maxilla, that cut, shred, crush, and grind food. Also called **dentes** (DEN-tēz).

Telophase (TEL-ō-fāz) The final stage of mitosis in which the daughter nuclei become established.

Tendon (TEN-don) A white fibrous cord of dense regular connective tissue that attaches muscle to bone.

Tendon organ A proprioceptive receptor, sensitive to changes in muscle tension and force of contraction, found chiefly near the junctions of tendons and muscles. Also called a **Golgi** (GOL-jē) **tendon organ.**

Tendon reflex A polysynaptic, ipsilateral reflex that protects tendons and their associated muscles from damage that might be brought about by excessive tension. The receptors involved are called tendon organs (Golgi tendon organs).

Tentorium cerebelli (ten-TŌ-rē-um ser′-e-BEL-ē) A transverse shelf of dura mater that forms a partition between the occipital lobe of the cerebral hemispheres and the cerebellum and that covers the cerebellum.

Teratogen (TER-a-tō-jen) Any agent or factor that causes physical defects in a developing embryo.

Terminal ganglion (TER-min-al GANG-glē-on) A cluster of cell bodies of parasympathetic postganglionic neurons either lying very close to the visceral effectors or located within the walls of the visceral effectors supplied by the postganglionic neurons.

Testis (TES-tis) Male gonad that produces sperm and the hormones testosterone and inhibin. Also called a **testicle.**

Testosterone (tes-TOS-te-rōn) A male sex hormone (androgen) secreted by interstitial endocrinocytes (Leydig cells) of a mature testis; needed for development of sperm; together with a second androgen termed **dihydrotestosterone (DHT),** controls the growth and development of male reproductive organs, secondary sex characteristics, and body growth.

Tetany (TET-a-nē) Hyperexcitability of neurons and muscle fibers (cells) caused by hypocalcemia and characterized by intermittent or continuous tonic muscular contractions; may be due to hypoparathyroidism.

Thalamus (THAL-a-mus) A large, oval structure located bilaterally on either side of the third ventricle, consisting of two masses of gray matter organized into nuclei; main relay center for sensory impulses ascending to the cerebral cortex.

Thermoreceptor (THER-mō-rē-sep-tor) Sensory receptor that detects changes in temperature.

Thigh The portion of the lower limb between the hip and the knee.

Third ventricle (VEN-tri-kul) A slitlike cavity between the right and left halves of the thalamus and between the lateral ventricles of the brain.

Thirst center A cluster of neurons in the hypothalamus that is sensitive to the osmotic pressure of extracellular fluid and brings about the sensation of thirst.

Thoracic (thō-RAS-ik) **cavity** Superior portion of the ventral body cavity that contains two pleural cavities, the mediastinum, and the pericardial cavity.

Thoracic duct A lymphatic vessel that begins as a dilation called the cisterna chyli, receives lymph from the left side of the head, neck, and chest, the left arm, and the entire body below the ribs, and empties into the left subclavian vein. Also called the **left lymphatic** (lim-FAT-ik) **duct.**

Thoracolumbar (thō′-ra-kō-LUM-bar) **outflow** The axons of sympathetic preganglionic neurons, which have their cell bodies in the lateral gray columns of the thoracic segments and first two or three lumbar segments of the spinal cord.

Thorax (THŌ-raks) The chest.

Threshold potential The membrane voltage that must be reached to trigger an action potential.

Threshold stimulus Any stimulus strong enough to initiate an action potential or activate a sensory receptor.

Thrombin (THROM-bin) The active enzyme formed from prothrombin that converts fibrinogen to fibrin during formation of a blood clot.

Thrombolytic (throm-bō-LIT-ik) **agent** Chemical substance injected into the body to dissolve blood clots and restore circulation; mechanism of action is direct or indirect activation of plasminogen; examples include tissue plasminogen activator (t-PA), streptokinase, and urokinase.

Thrombosis (throm-BŌ-sis) The formation of a clot in an unbroken blood vessel, usually a vein.

Thrombus A stationary clot formed in an unbroken blood vessel, usually a vein.

Thymus (THĪ-mus) A bilobed organ, located in the superior mediastinum posterior to the sternum and between the lungs, in which T cells develop immunocompetence.

Thyroglobulin (thī-rō-GLŌ-bū-lin) **(TGB)** A large glycoprotein molecule produced by follicular cells of the thyroid gland in which some tyrosines are iodinated and coupled to form thyroid hormones.

Thyroid cartilage (THĪ-royd KAR-ti-lij) The largest single cartilage of the larynx, consisting of two fused plates that form the anterior wall of the larynx.

Thyroid follicle (FOL-i-kul) Spherical sac that forms the parenchyma of the thyroid gland and consists of follicular cells that produce thyroxine (T_4) and triiodothyronine (T_3).

Thyroid gland An endocrine gland with right and left lateral lobes on either side of the trachea connected by an isthmus; located anterior to the trachea just inferior to the cricoid cartilage; secretes thyroxine (T_4), triiodothyronine (T_3), and calcitonin.

Thyroid-stimulating hormone (TSH) A hormone secreted by the anterior pituitary that stimulates the synthesis and secretion of thyroxine (T_4) and triiodothyronine (T_3).

Thyroxine (thī-ROK-sēn) **(T_4)** A hormone secreted by the thyroid gland that regulates metabolism, growth and development, and the activity of the nervous system.

Tic Spasmodic, involuntary twitching of muscles that are normally under voluntary control.

Tidal volume The volume of air breathed in and out in any one breath; about 500 mL in quiet, resting conditions.

Tissue A group of similar cells and their intercellular substance joined together to perform a specific function.

Tissue factor (TF) A factor, or collection of factors, whose appearance initiates the blood clotting process. Also called **thromboplastin** (throm-bō-PLAS-tin).

Tissue plasminogen activator (t-PA) An enzyme that dissolves small blood clots by initiating a process that converts plasminogen to plasmin, which degrades the fibrin of a clot.

Tongue A large skeletal muscle covered by a mucous membrane located on the floor of the oral cavity.

Tonicity (tō-NIS-i-tē) A measure of the concentration of impermeable solute particles in a solution relative to cytosol. When cells are bathed in an **isotonic solution,** they neither shrink nor swell.

Tonsil (TON-sil) An aggregation of large lymphatic nodules embedded in the mucous membrane of the throat.

Torn cartilage A tearing of an articular disc (meniscus) in the knee.

Total lung capacity The sum of tidal volume, inspiratory reserve volume, expiratory reserve volume, and residual volume; about 6000 mL in an average adult.

Trabecula (tra-BEK-ū-la) Irregular latticework of thin plates of spongy bone. Fibrous cord of connective tissue serving as supporting fiber by forming a septum extending into an organ from its wall or capsule. *Plural is* **trabeculae** (tra-BEK-ū-lē).

Trabeculae carneae (KAR-nē-ē) Ridges and folds of the myocardium in the ventricles.

Trachea (TRĀ-kē-a) Tubular air passageway extending from the larynx to the fifth thoracic vertebra. Also called the **windpipe.**

Tract A bundle of nerve axons in the central nervous system.

Transcription (trans-KRIP-shun) The first step in the expression of genetic information in which a single strand of DNA serves as a template for the formation of an RNA molecule.

Translation (trans-LĀ-shun) The synthesis of a new protein on a ribosome as dictated by the sequence of codons in messenger RNA.

Transverse colon (trans-VERS KŌ-lon) The portion of the large intestine extending across the abdomen from the right colic (hepatic) flexure to the left colic (splenic) flexure.

Transverse fissure (FISH-er) The deep cleft that separates the cerebrum from the cerebellum.

Transverse plane A plane that divides the body or organs into superior and inferior portions. Also called a **horizontal plane.**

Transverse tubules (TOO-būls) **(T tubules)** Small, cylindrical invaginations of the sarcolemma of striated muscle fibers (cells) that conduct muscle action potentials toward the center of the muscle fiber.

Trauma (TRAW-ma) An injury, either a physical wound or psychic disorder, caused by an external agent or force, such as a physical blow or emotional shock; the agent or force that causes the injury.

Tremor (TREM-or) Rhythmic, involuntary, purposeless contraction of opposing muscle groups.

Triad (TRĪ-ad) A complex of three units in a muscle fiber composed of a transverse tubule and the sarcoplasmic reticulum terminal cisterns on both sides of it.

Tricuspid (trī-KUS-pid) **valve** Atrioventricular (AV) valve on the right side of the heart.

Triglyceride (trī-GLI-cer-īd) A lipid formed from one molecule of glycerol and three molecules of fatty acids that may be either solid (fats) or liquid (oils) at room temperature; the body's most highly concentrated source of chemical potential energy. Found mainly within adipocytes. Also called a **neutral fat** or a **triacylglycerol.**

Trigone (TRĪ-gon) A triangular region at the base of the urinary bladder.

Triiodothyronine (trī-ī-ō-dō-THĪ-rō-nēn) (**T₃**) A hormone produced by the thyroid gland that regulates metabolism, growth and development, and the activity of the nervous system.

Trophoblast (TRŌF-ō-blast) The superficial covering of cells of the blastocyst.

Tropic (TRŌ-pik) **hormone** A hormone whose target is another endocrine gland.

Trunk The part of the body to which the upper and lower limbs are attached.

Tubal ligation (lī-GĀ-shun) A sterilization procedure in which the uterine (Fallopian) tubes are tied and cut.

Tubular reabsorption The movement of filtrate from renal tubules back into blood in response to the body's specific needs.

Tubular secretion The movement of substances in blood into renal tubular fluid in response to the body's specific needs.

Tumor suppressor gene A gene coding for a protein that normally inhibits cell division; loss or alteration of a tumor suppressor gene called *p53* is the most common genetic change in a wide variety of cancer cells.

Tunica albuginea (TOO-ni-ka al′-bū-JIN-ē-a) A dense white fibrous capsule covering a testis or deep to the surface of an ovary.

Tunica externa (eks-TER-na) The superficial coat of an artery or vein, composed mostly of elastic and collagen fibers. Also called the **adventitia.**

Tunica interna (in-TER-na) The deep coat of an artery or vein, consisting of a lining of endothelium, basement membrane, and internal elastic lamina. Also called the **tunica intima** (IN-ti-ma).

Tunica media (MĒ-dē-a) The intermediate coat of an artery or vein, composed of smooth muscle and elastic fibers.

Twitch contraction Brief contraction of all muscle fibers (cells) in a motor unit triggered by a single action potential in its motor neuron.

Tympanic antrum (tim-PAN-ik AN-trum) An air space in the middle ear that leads into the mastoid air cells or sinus.

Type II cutaneous mechanoreceptor A sensory receptor embedded deeply in the dermis and deeper tissues that detects stretching of skin. Also called a **Ruffini corpuscle.**

U

Umbilical cord The long, ropelike structure containing the umbilical arteries and vein that connect the fetus to the placenta.

Umbilicus (um-BIL-i-kus *or* um-bil-Ī-kus) A small scar on the abdomen that marks the former attachment of the umbilical cord to the fetus. Also called the **navel.**

Upper limb The appendage attached at the shoulder girdle, consisting of the arm, forearm, wrist, hand, and fingers. Also called **upper extremity.**

Up-regulation Phenomenon in which there is an increase in the number of receptors in response to a deficiency of a hormone or neurotransmitter.

Uremia (ū-RĒ-mē-a) Accumulation of toxic levels of urea and other nitrogenous waste products in the blood, usually resulting from severe kidney malfunction.

Ureter (Ū-rē-ter) One of two tubes that connect the kidney with the urinary bladder.

Urethra (ū-RĒ-thra) The duct from the urinary bladder to the exterior of the body that conveys urine in females and urine and semen in males.

Urinary (Ū-ri-ner-ē) **bladder** A hollow, muscular organ situated in the pelvic cavity posterior to the pubic symphysis; receives urine via two ureters and stores urine until it is excreted through the urethra.

Urine The fluid produced by the kidneys that contains wastes and excess materials; excreted from the body through the urethra.

Urogenital (ū′-rō-JEN-i-tal) **triangle** The region of the pelvic floor inferior to the pubic symphysis, bounded by the pubic symphysis and the ischial tuberosities, and containing the external genitalia.

Urology (ū-ROL-ō-jē) The specialized branch of medicine that deals with the structure, function, and diseases of the male and female urinary systems and the male reproductive system.

Uterine (Ū-ter-in) **tube** Duct that transports ova from the ovary to the uterus. Also called the **Fallopian** (fal-LŌ-pē-an) **tube** or **oviduct.**

Uterosacral ligament (ū′-ter-ō-SĀ-kral LIG-a-ment) A fibrous band of tissue extending from the cervix of the uterus laterally to the sacrum.

Uterus (Ū-te-rus) The hollow, muscular organ in females that is the site of menstruation, implantation, development of the fetus, and labor. Also called the **womb.**

Utricle (Ū-tri-kul) The larger of the two divisions of the membranous labyrinth located inside the vestibule of the inner ear, containing a receptor organ for static equilibrium.

Uvea (Ū-vē-a) The three structures that together make up the vascular tunic of the eye.

Uvula (Ū-vū-la) A soft, fleshy mass, especially the V-shaped pendant part, descending from the soft palate.

V

Vagina (va-JĪ-na) A muscular, tubular organ that leads from the uterus to the vestibule, situated between the urinary bladder and the rectum of the female.

Vallate papilla (VAL-aī pa-PIL-a) One of the circular projections that is arranged in an inverted V-shaped row at the back of the tongue; the largest of the elevations on the upper surface of the tongue containing taste buds. Also called **circumvallate papilla.**

Valence (VĀ-lens) The combining capacity of an atom; the number of deficit or extra electrons in the outermost electron shell of an atom.

Varicocele (VAR-i-kō-sēl) A twisted vein; especially, the accumulation of blood in the veins of the spermatic cord.

Varicose (VAR-i-kōs) Pertaining to an unnatural swelling, as in the case of a varicose vein.

Vasa recta (VĀ-sa REK-ta) Extensions of the efferent arteriole of a juxtamedullary nephron that run alongside the loop of the nephron (Henle) in the medullary region of the kidney.

Vasa vasorum (va-SŌ-rum) Blood vessels that supply nutrients to the larger arteries and veins.

Vascular (VAS-kū-lar) Pertaining to or containing many blood vessels.

Vascular spasm Contraction of the smooth muscle in the wall of a damaged blood vessel to prevent blood loss.

Vascular (venous) sinus A vein with a thin endothelial wall that lacks a tunica media and externa and is supported by surrounding tissue.

Vascular tunic (TOO-nik) The middle layer of the eyeball, composed of the choroid, ciliary body, and iris. Also called the **uvea** (Ū-ve-a).

Vasectomy (va-SEK-tō-mē) A means of sterilization of males in which a portion of each ductus (vas) deferens is removed.

Vasoconstriction (vāz-ō-kon-STRIK-shun) A decrease in the size of the lumen of a blood vessel caused by contraction of the smooth muscle in the wall of the vessel.

Vasodilation (vāz′-ō-DĪ-lā-shun) An increase in the size of the lumen of a blood vessel caused by relaxation of the smooth muscle in the wall of the vessel.

Vasomotion (vāz-ō-MŌ-shun) Intermittent contraction and relaxation of the smooth muscle of the metarterioles and precapillary sphincters that result in an intermittent blood flow.

Vein A blood vessel that conveys blood from tissues back to the heart.

Vena cava (VĒ-na KĀ-va) One of two large veins that open into the right atrium, returning to the heart all of the deoxygenated blood from the systemic circulation except from the coronary circulation.

Ventral (VEN-tral) Pertaining to the anterior or front side of the body; opposite of dorsal.

Ventral body cavity Cavity near the ventral aspect of the body that contains viscera and consists of a superior thoracic cavity and an inferior abdominopelvic cavity.

Ventral ramus (RĀ-mus) The anterior branch of a spinal nerve, containing sensory and motor fibers to the muscles and skin of the anterior surface of the head, neck, trunk, and the limbs.

Ventricle (VEN-tri-kul) A cavity in the brain filled with cerebrospinal fluid. An inferior chamber of the heart.

Ventricular fibrillation (ven-TRIK-ū-lar fib-ri-LĀ-shun) Asynchronous ventricular contractions; unless reversed by defibrillation, results in heart failure.

Venule (VEN-ūl) A small vein that collects blood from capillaries and delivers it to a vein.

Vermiform appendix (VER-mi-form a-PEN-diks) A twisted, coiled tube attached to the cecum.

Vermilion (ver-MIL-yon) The area of the mouth where the skin on the outside meets the mucous membrane on the inside.

Vermis (VER-mis) The central constricted area of the cerebellum that separates the two cerebellar hemispheres.

Vertebral (VER-te-bral) **canal** A cavity within the vertebral column formed by the vertebral foramina of all the vertebrae and containing the spinal cord. Also called the **spinal canal.**

Vertebral column The 26 vertebrae of an adult and 33 vertebrae of a child; encloses and protects the spinal cord and serves as a point of attachment for the ribs and back muscles. Also called the **backbone, spine,** or **spinal column.**

Vesicle (VES-i-kul) A small bladder or sac containing liquid.

Vesicouterine (ves′-ik-ō-Ū-ter-in) **pouch** A shallow pouch formed by the reflection of the peritoneum from the anterior surface of the uterus, at the junction of the cervix and the body, to the posterior surface of the urinary bladder.

Vestibular (ves-TIB-ū-lar) **apparatus** Collective term for the organs of equilibrium, which includes the saccule, utricle, and semicircular ducts.

Vestibular membrane The membrane that separates the cochlear duct from the scala vestibuli.

Vestibule (VES-ti-būl) A small space or cavity at the beginning of a canal, especially the inner ear, larynx, mouth, nose, and vagina.

Villus (VIL-lus) A projection of the intestinal mucosal cells containing connective tissue, blood vessels, and a lymphatic vessel; functions in the absorption of the end products of digestion. *Plural is* **villi** (VIL-ī).

Viscera (VIS-er-a) The organs inside the ventral body cavity. *Singular is* **viscus** (VIS-kus).

Visceral (VIS-er-al) Pertaining to the organs or to the covering of an organ.

Visceral effectors (e-FEK-torz) Organs of the ventral body cavity that respond to neural stimulation, including cardiac muscle, smooth muscle, and glands.

Vital capacity The sum of inspiratory reserve volume, tidal volume, and expiratory reserve volume; about 4800 mL.

Vital signs Signs necessary to life that include temperature (T), pulse (P), respiratory rate (RR), and blood pressure (BP).

Vitamin An organic molecule necessary in trace amounts that acts as a catalyst in normal metabolic processes in the body.

Vitreous (VIT-rē-us) **body** A soft, jellylike substance that fills the vitreous chamber of the eyeball, lying between the lens and the retina.

Vocal folds Pair of mucous membrane folds below the ventricular folds that function in voice production. Also called **true vocal cords.**

Voltage-gated channel An ion channel in a plasma membrane composed of integral proteins that functions like a gate to permit or restrict the movement of ions across the membrane in response to changes in the voltage.

Vulva (VUL-va) Collective designation for the external genitalia of the female. Also called the **pudendum** (poo-DEN-dum).

W

Wallerian (wal-LE-rē-an) **degeneration** Degeneration of the portion of the axon and myelin sheath of a neuron distal to the site of injury.

Wandering macrophage (MAK-rō-fāj) Phagocytic cell that develops from a monocyte, leaves the blood, and migrates to infected tissues.

Wave summation (sum-MĀ-shun) The increased strength of muscle contraction that results when muscle action potentials occur one after another in rapid succession.

White matter Aggregations or bundles of myelinated and unmyelinated axons located in the brain and spinal cord.

White pulp The regions of the spleen composed of lymphatic tissue, mostly B lymphocytes.

White ramus communicans (RĀ-mus kō-MŪ-ni-kans) The portion of a preganglionic sympathetic axon that branches from the anterior ramus of a spinal nerve to enter the nearest sympathetic trunk ganglion.

X

X-chromosome inactivation The random and permanent inactivation of one X chromosome in each cell of a developing female embryo. Also called **lyonization.**

Xiphoid (ZĪ-foyd) Sword-shaped. The inferior portion of the sternum is the **xiphoid process.**

Y

Yolk sac An extraembryonic membrane composed of the exocoelomic membrane and hypoblast. It transfers nutrients to the embryo, is a source of blood cells, contains primordial germ cells that migrate into the gonads to form primitive germ cells, forms part of the gut, and helps prevent desiccation of the embryo.

Z

Zona fasciculata (ZŌ-na fa-sik′-ū-LA-ta) The middle zone of the adrenal cortex consisting of cells arranged in long, straight cords that secrete glucocorticoid hormones, mainly cortisol.

Zona glomerulosa (glo-mer′-ū-LŌ-sa) The outer zone of the adrenal cortex, directly under the connective tissue covering, consisting of cells arranged in arched loops or round balls that secrete mineralocorticoid hormones, mainly aldosterone.

Zona pellucida (pe-LOO-si-da) Clear glycoprotein layer between a secondary oocyte and the surrounding granulosa cells of the corona radiata.

Zona reticularis (ret-ik′-ū-LAR-is) The inner zone of the adrenal cortex, consisting of cords of branching cells that secrete sex hormones, chiefly androgens.

Zygote (ZĪ-gōt) The single cell resulting from the union of male and female gametes; the fertilized ovum.

Credits

Illustration Credits

Chapter 1 CO art by Keith Kasnot. Table 1.2: Keith Kasnot. 1.1: Tomo Narashima. 1.2–1.4: Jared Schneidman Design. 1.5, 1.6: Kevin Somerville. 1.7: Lynn O'Kelley. 1.8–1.13: Kevin Somerville.

Chapter 2 CO art by Keith Kasnot. 2.1: Imagineering. 2.2, 2.3: Jared Schneidman Design. 2.4: Imagineering. 2.5: Jared Schneidman Design. 2.6–2.11: Imagineering. 2.12–2.15: Jared Schneidman Design. 2.16: Imagineering. 2.17: Jared Schneidman Design. 2.18: Imagineering. 2.19–2.21: Jared Schneidman Design. 2.22: Imagineering. 2.23: Jared Schneidman Design. 2.24: Imagineering. 2.25: Jared Schneidman Design.

Chapter 3 CO art by Keith Kasnot. 3.1, 3.2: Tomo Narashima. 3.3–3.7: Imagineering. 3.8: Jared Schneidman Design. 3.5: Adapted from Bruce Alberts et al., Essential Cell Biology, F12.5, p375 and F12.12, p380 (New York: Garland Publishing Inc., 1998). ©1998 Garland Publishing Inc. 3.9–3.16: Imagineering. 3.17–3.21: Tomo Narashima. 3.22: Imagineering. 3.23–3.25: Tomo Narashima. 3.26–3.33: Imagineering. 3.34: Hilda Muinos.

Chapter 4 CO art by Keith Kasnot. Table 4.1, Table 4.2: Nadine Sokol, Kevin Somerville. Table 4.3: Nadine Sokol, Kevin Somerville, Imagineering. Table 4.4, Table 4.5: Nadine Sokol. 4.1: Kevin Somerville. 4.2: Imagineering. 4.3: Kevin Somerville. 4.4, 4.5: Imagineering.

Chapter 5 CO art by Keith Kasnot. 5.1–5.7: Kevin Somerville. 5.9: Imagineering.

Chapter 6 CO art by Keith Kasnot. 6.1: Leonard Dank. 6.2: Lauren Keswick. 6.3–6.9: Kevin Somerville. 6.10: Leonard Dank. 6.11: Kevin Somerville. 6.12: Jared Schneidman Design. 6.13: Kevin Somerville.

Chapter 7 CO art by Keith Kasnot. Table 7.1: Nadine Sokol. 7.1–7.24: Leonard Dank.

Chapter 8 CO art by Keith Kasnot. Table 8.1, 8.1–8.17: Leonard Dank.

Chapter 9 CO art by Keith Kasnot. 9.1–9.4, 9.11–9.14: Leonard Dank.

Chapter 10 CO art by Keith Kasnot. 10.1, 10.3: Kevin Somerville. 10.4, 10.6: Imagineering. 10.7: Hilda Muinos. 10.8–10.10: Imagineering. 10.11: Kevin Somerville. 10.12: Imagineering. 10.13–10.15: Jared Schneidman Design. 10.16, 10.18: Imagineering. 10.19: Beth Willert. 10.20: Kevin Somerville.

Chapter 11 CO art by Keith Kasnot. Table 11.1: Kevin Somerville. 11.1–11.17: Leonard Dank. 11.18: Leonard Dank, Kevin Somerville. 11.19–11.23: Leonard Dank.

Chapter 12 CO art by Keith Kasnot. Table 12.1: Kevin Somerville. Table 12.3: Jared Schneidman Design. 12.1: Kevin Somerville. 12.2: Jared Schneidman Design. 12.3: Kevin Somerville. 12.4, 12.5: Imagineering. 12.6: Kevin Somerville. 12.7: Sharon Ellis. 12.8–12.12: Imagineering. 12.13: Jared Schneidman Design. 12.14: Imagineering. 12.15: Jared Schneidman Design. 12.16: Nadine Sokol. 12.17: Imagineering.

Chapter 13 CO art by Keith Kasnot. 13.1–13.5: Sharon Ellis. 13.6–13.9: Leonard Dank. 13.10: Kevin Somerville. 13.11: Sharon Ellis. 13.12, 13.13: Steve Oh, Myriam Kirkman-Oh. 13.14: Imagineering. 13.15, 13.16: Steve Oh, Myriam Kirkman-Oh. 13.17: Imagineering.

Chapter 14 CO art by Keith Kasnot. Table 14.1, Table 14.3: Hilda Muinos. 14.1, 14.2: Kevin Somerville. 14.3: Sharon Ellis. 14.4: Kevin Somerville, Imagineering. 14.5–14.8: Sharon Ellis. 14.9: Imagineering. 14.10, 14.11, 14.13–14.15: Sharon Ellis. 14.17: Hilda Muinos. 14.18–14.24: Sharon Ellis. 14.25, 14.26: Kevin Somerville.

Chapter 15 CO art by Keith Kasnot. Table 15.3: Kevin Somerville. 15.1: Imagineering. 15.2, 15.3: Kevin Somerville. 15.4: Leonard Dank. 15.5: Imagineering. 15.6: Kevin Somerville. 15.7: Jared Schneidman Design. 15.8: Kevin Somerville. 15.9, 15.10: Sharon Ellis. 15.11: Imagineering. Adapted from Purves et al., Neuroscience 2e, F26.1 and F26.2, p498 (Sunderland, MA: Sinauer Associates, 1997). ©1997 Sinauer Associates.

Chapter 16 CO art by Keith Kasnot. Table 16.1: Kevin Somerville. Table 16.2: Imagineering. 16.1: Tomo Narashima. 16.2: Molly Borman. 16.3: Imagineering. Adapted from "Making Sense of Taste" by David V. Smith and Robert F. Margolskee, Scientific American, March 2001, page 38. 16.5: Sharon Ellis. 16.6: Tomo Narashima . 16.9: Lynn O'Kelley. 16.10: Tomo Narashima. 16.11, 16.12: Jared Schneidman Design. 16.13: Lynn O'Kelley. 16.14: Jared Schneidman Design. 16.15: Lynn O'Kelley. 16.16: Imagineering. Adapted from Seeley et al., Anatomy and Physiology 4e, F15.22, p480 (New York: WCB McGraw-Hill, 1998) ©1998 The McGraw-Hill Companies. 16.17–16.21: Tomo Narashima. 16.22, 16.23: Tomo Narashima, Sharon Ellis. 16.24, 16.25: Kevin Somerville.

Chapter 17 CO art by Keith Kasnot. 17.1: Jared Schneidman Design. 17.2, 17.3: Imagineering. 17.4: Kevin Somerville. 17.5: Sharon Ellis. 17.6: Imagineering.

Chapter 18 CO art by Keith Kasnot. Table 18.4: Nadine Sokol, Imagineering. Table 18.5–18.8: Nadine Sokol. Table 18.9: Imagineering. Table 18.10: Nadine Sokol. 18.1: Steve Oh. 18.2: Jared Schneidman Design. 18.3, 18.4: Imagineering. 18.5: Lynn O'Kelley. 18.6: Imagineering. 18.7: Jared Schneidman Design. 18.8: Lynn O'Kelley. 18.9: Jared Schneidman Design. 18.10: Molly Borman. 18.11, 18.12: Jared Schneidman Design. 18.13: Molly Borman. 18.14: Jared Schneidman Design. 18.15: Molly Borman. 18.16, 18.17: Jared Schneidman Design. 18.18: Molly Borman. 18.19: Jared Schneidman Design. 18.20: Nadine Sokol. 18.21: Kevin Somerville.

Chapter 19   CO art by Keith Kasnot. Table 19.3: Jared Schneidman Design. 19.1: Hilda Muinos, Nadine Sokol. 19.3, 19.4: Nadine Sokol. 19.5, 19.6, 19.8: Jared Schneidman Design. 19.9: Nadine Sokol. 19.11: Imagineering. 19.12: Jean Jackson. 19.13: Nadine Sokol.

Chapter 20   CO art by Keith Kasnot. 20.1, 20.2: Kevin Somerville, Imagineering. 20.3–20.6: Kevin Somerville. 20.7: Nadine Sokol, Imagineering. 20.8, 20.9: Kevin Somerville. 20.10: Kevin Somerville, Imagineering. 20.11, 20.12: Burmar Technical Corp. 20.13, 20.14: Imagineering. 20.15: Imagineering, Kevin Somerville. 20.16: Hilda Muinos. 20.18: Kevin Somerville. 20.20: Hilda Muinos.

Chapter 21   CO art by Keith Kasnot. Exhibits 21.2–21.12: Keith Ciociola. Table 21.2: Imagineering. 21.1: Kevin Somerville. 21.2: Hilda Muinos. 21.3: Nadine Sokol, Imagineering. 21.4: Kevin Somerville. 21.6, 21.7: Jared Schneidman Design. 21.8: Imagineering. 21.9: Kevin Somerville. 21.10–21.12: Jared Schneidman Design. 21.13: Kevin Somerville. 21.14–21.16: Jared Schneidman Design. 21.17–21.30: Kevin Somerville. 21.31: Kevin Somerville, Keith Ciociola. 21.32: Kevin Somerville.

Chapter 22   CO art by Keith Kasnot. Table 22.3: Jean Jackson. 22.1: Molly Borman. 22.2: Sharon Ellis. 22.3: Molly Borman. 22.4: Nadine Sokol. 22.5: Steve Oh. 22.6: Molly Borman. 22.7: Steve Oh. 22.8: Kevin Somerville. 22.9, 22.10: Molly Borman. 22.11–21.20: Jared Schneidman Design. 22.21: Nadine Sokol, Imagineering.

Chapter 23   CO art by Keith Kasnot. 23.1: Molly Borman, Kevin Somerville. 23.2, 23.4, 23.5: Molly Borman. 23.6: Steve Oh. 23.8, 23.10: Molly Borman. 23.11, 23.12: Kevin Somerville. 23.13: Jared Schneidman Design. 23.14: Kevin Somerville. 23.15: Jared Schneidman Design, Imagineering. 23.16–21.24: Jared Schneidman Design. 23.25: Imagineering. 23.26: Jared Schneidman Design. 23.27: Kevin Somerville. 23.28: Jared Schneidman Design. 23.29: Kevin Somerville.

Chapter 24   CO art by Keith Kasnot. 24.1–24.3: Steve Oh. 24.4: Nadine Sokol. 24.5: Molly Borman. 24.6: Steve Oh. 24.7, 24.8, 24.10: Nadine Sokol. 24.11: Steve Oh. 24.12: Kevin Somerville. 24.13: Imagineering. 24.14–24.16: Jared Schneidman Design. 24.17: Steve Oh, Jared Schneidman Design. 24.18: Jared Schneidman Design. 24.19: Kevin Somerville. 24.20, 24.21: Jared Schneidman Design. 24.23: Kevin Somerville. 24.25, 24.26: Jared Schneidman Design. 24.27: Molly Borman. 24.28: Kevin Somerville.

Chapter 25   CO art by Keith Kasnot. 25.1–25.20: Imagineering.

Chapter 26   CO art by Keith Kasnot. Table 26.1: Nadine Sokol. 26.1, 26.2: Kevin Somerville. 26.3, 26.4: Steve Oh. 26.5: Imagineering. 26.6: Kevin Somerville. 26.7: Nadine Sokol. 26.8: Kevin Somerville. 26.9: Imagineering. 26.10–26.20: Jared Schneidman Design. 26.21: Steve Oh. 26.22: Kevin Somerville.

Chapter 27   CO art by Keith Kasnot. 27.1–27.8: Jared Schneidman Design.

Chapter 28   CO art by Keith Kasnot. Table 28.1, 28.1, 28.2: Imagineering. 28.3–28.6: Kevin Somerville. 28.7: Jared Schneidman Design. 28.8: Kevin Somerville. 28.9, 28.10–28.15, 28.17: Jared Schneidman Design. 28.18, 28.21–28.24: Kevin Somerville. 28.25–28.28: Jared Schneidman Design. 28.29, 28.30: Kevin Somerville.

Chapter 29   CO art by Keith Kasnot. Table 29.2, 29.1–29.15: Kevin Somerville. 29.16: Jared Schneidman Design. 29.17, 29.18: Kevin Somerville. 29.19–29.26: Jared Schneidman Design.

Focus on Homeostasis icons   Imagineering.

Photo Credits

Chapter 1   Figure 1.1: John Wilson White. Figure 1.8a: From Stephen A. Kieffer and E. Robert Heitzman, *An Atlas of Cross–Sectional Anatomy.* Harper & Row, Publishers, New York, 1979. Figure 1.8b: Lester V. Bergman/Project Masters, Inc. Figure 1.8c: Martin Rotker. Figure 1.11: Mark Nielsen. Page 21 (top left): Biophoto Associates/Photo Researchers. Page 21 (top right): Scott Camazine/Photo Researchers. Page 21 (bottom left): Simon Fraser/Photo Researchers. Page 21 (bottom right): Courtesy Andrew Joseph Tortora and Damaris Soler. Page 22: Howard Sochurek/Medical Images, Inc.

Chapter 3   Figure 3.6: Andy Washnik. Figure 3.17c: Courtesy Kent McDonald, UC Berkeley Electron Microscope Laboratory. Figure 3.20b: D. W. Fawcett/Photo Researchers. Figure 3.21b: Biophoto Associates/Photo Researchers. Figure 3.23b: Courtesy Daniel S. Friend, Harvard Medical School. Figure 3.24b: D.W. Fawcett/Photo Researchers. Figure 3.25c: CNRI/Photo Researchers. Figure 3.33: Courtesy Michael Ross, University of Florida.

Chapter 4   Page 109 (top): Biophoto Associates/Photo Researchers. Pages 109 (bottom), 110, 111 (bottom), 112, 113 and 114 (bottom), 121, 122, 123 (top), 124 (bottom), 125, 126 (top), 127 and 130: Courtesy Michael Ross, University of Florida. Page 111 (top): Biophoto Associates/Photo Researchers. Page 114 (top): Lester V. Bergman/The Bergman Collection. Page 123 (bottom): Courtesy Andrew J. Kuntzman. Page 124 (top): Ed Reschke. Page 126 (bottom): John Burbidge/Photo Researchers. Page 131: Biophoto Associates/Photo Researchers. Page 132: © Ed Reschke.

Chapter 5   Figure 5.3b: © L.V. Bergman/Bergman Collection. Figure 5.4b: Science Photo Library/Photo Researchers. Figure 5.8a: Alain Dex/Photo Researchers. Figure 5.8b: Biophoto Associates/Photo Researchers.

Chapter 6   Figure 6.1b: Mark Nielsen. Figure 6.8a: The Bergman Collection. Figure 6.8b: Courtesy Michael Ross, University of Florida. Figure 6.14: P. Motta, Dept. of Anatomy, University La Sapienza, Rome/Photo Researchers.

Chapter 9   Figures 9.5, 9.6, 9.8, 9.7, 9.9 and 9.10: John Wilson White.

Chapter 10   Figure 10.2: Courtesy Fujita. Figure 10.5: Courtesy Denah Appelt and Clara Franzini-Armstrong. Figure 10.17: © John Wiley & Sons. Page 296: Biophoto Associates/Photo Researchers.

Chapter 12   Figure 12.3c: Science VU/Visuals Unlimited. Figure 12.6c: Dennis Kunkel/Phototake. Figure 12.6d: Martin Rotker/Phototake.

Chapter 13   Figure 13.1b: Mark Nielsen. Figure 13.3b: Jean Claude Revy/Phototake. Figure 13.10b: Copyright by Richard Kessel and Randy Kardon, *Tissues and Organs: A Text-Atlas of Scanning Electron Microscopy,* W. H. Freeman and Company, 1979. All rights reserved. Reprinted by permission.

Chapter 14   Figure 14.1b: Mark Nielsen. Figure 14.9e: From Stephen A. Kieffer and E. Robert Heitzman, *An Atlas of Cross-Sectional Anatomy,* Harper and Row, Publishers, 1979. Reproduced with permission. Figure 14.12: From N. Gluhbegovic and T.H. Williams, *The Human Brain: A Photographic Guide,* Harper and Row, Publishers, 1980. Reproduced with permission. Figure 14.16: From *Nature,* November 26, 1992, Vol. 360, page 340. Reproduced with permission from *Nature* and Robert Zatorre, Department of Neuropsychology, McGill University.

Index

COMBINING FORMS, WORD ROOTS, PREFIXES, AND SUFFIXES

Many of the terms used in anatomy and phsiology are compound words; that is, they are made up of word roots and one or more prefixes or suffixes. For example, *leukocyte* is formed from the word roots *leuk-* meaning "white", a connecting vowel (o), and *cyte* meaning "cell." Thus, a leukocyte is a white blood cell. The following list includes some of the most commonly used combining forms, word roots, prefixes, ad suffixes used in the study of anatomy and physiology. Each entry includes a usage example. Learning the meanings of these fundamental word parts will help you remember terms that, at first glance, may seem long or complicated.

COMBINING FORMS AND WORD ROOTS

Acous-, Acu- hearing Acoustics.
Acr- extremity Acromegaly.
Aden- gland Adenoma.
Alg-, Algia- pain Neuralgia.
Angi- vessel Angiocardiography.
Anthr- joint Arthropathy.
Aut-, Auto- self Autolysis.
Audit- hearing Auditory canal.

Bio- life, living Biopsy.
Blast- germ, bud Blastula.
Blephar- eyelid Blepharitis.
Brachi- arm Brachial plexus.
Bronch- trachea, windpipe Bronchoscopy.
Bucc- cheek Buccal.

Capit- head Decapitate.
Carcin- cancer Carcinogenic.
Cardi-, Cardia-, Cardio- heart Cardiogram.
Cephal- head Hydrocephalus.
Cerebro- brain Cerebrospinal fluid.
Chole- bile, gall Cholecystogram.
Chondr-, cartilage Chondrocyte.
Cor-, Coron- heart Coronary.
Cost- rib Costal.
Crani- skull Craniotomy.
Cut- skin Subcutaneous.
Cyst- sac, bladder Cystoscope.

Derma-, Dermato- skin Dermatosis.
Dura- hard Dura mater.

Enter- intestine Enteritis.
Erythr- red Erythrocyte.

Gastr- stomach Gastrointestinal.
Gloss- tongue Hypoglossal.
Glyco- sugar Glycogen.
Gyn-, Gynec- female, woman Gynecology.

Hem-, Hemat- blood Hematoma.
Hepar-, Hepat- liver Hepatitis.
Hist-, Histio- tissue Histology.
Hydr- water Dehydration.
Hyster- uterus Hysterectomy.

Ischi- hip, hip joint Ischium.

Kines- motion Kinesiology.

Labi- lip Labial.
Lacri- tears Lacrimal glands.
Laparo- loin, flank, abdomen Laparoscopy.
Leuko- white Leukocyte.
Lingu- tongue Sublingual glands.
Lip- fat Lipid.
Lumb- lower back, loin Lumbar.

Macul- spot, blotch Macula.
Malign- bad, harmful Malignant.
Mamm-, Mast- breast Mammography, Mastitis.
Meningo- membrane Meningitis.
Myel- marrow, spinal cord Myeloblast.
My-, Myo- muscle Myocardium.

Necro- corpse, dead Necrosis.
Nephro- kidney Nephron.
Neuro- nerve Neurotransmitter.

Ocul- eye Binocular.
Odont- tooth Orthodontic.
Onco- mass, tumor Oncology.
Oo- egg Oocyte.
Opthalm- eye Ophthalmology.
Or- mouth Oral.
Osm- odor, sense of small Anosmia.
Os-, Osseo-, Osteo- bone Osteocyte.
Ot- ear Otitus media.

Palpebr- eyelid Palpebra.
Patho- disease Pathogen.
Pelv- basin Renal pelvis.
Phag- to eat Phagocytosis.
Phleb- vein Phlebitis.
Phren- diaphragm Phrenic.
Pilo- hair Depilatory.
Pneumo- lung, air Pneumothorax.
Pod- foot Podocyte.
Procto- anus, rectum Proctology.
Pulmon- lung Pulmonary.

Ren- kidneys Renal artery.
Rhin- nose Rhinitis.

Scler-, Sclero- hard Atherosclerosis.
Sep-, Spetic- toxic condition due to micoorganisms Septicemia.
Soma-, Somato- body Somatotropin.
Sten- narrow Stenosis.
Stasis-, Stat- stand still Homeostasis.

Tegument- skin, covering Integumentary.
Therm- heat Thermogenesis.
Thromb- clot, lump Thrombus.

Vas- vessel, duct Vasoconstriction.

Zyg- joined Zygote.

PREFIXES

A-, An- without, lack of, deficient Anesthesia.
Ab- away from, from Abnormal.
Ad-, Af- to, toward Adduction, Afferent neuron.
Alb- white Albino.
Alveol- cavity, socket Alveolus.
Andro- male, masculine Androgen.
Ante- before Antebrachial vein.
Anti- against Anticoagulant.

Bas- base, foundation Basal ganglia.
Bi- two, double Biceps.
Brady- slow Bradycardia.

Cata- down, lower, under Catabolism.
Circum- around Circumduction.
Cirrh- yellow Cirrhosis of the liver.
Co-, Con-, Com with, together Congenital.
Contra- against, opposite Contraception.
Crypt- hidden, concealed Cryptorchidism.
Cyano- blue Cyanosis.

De- down, from Deciduous.
Demi-, hemi- half Hemiplegia.
Di-, Diplo- two Diploid.
Dis- separation, apart, away from Dissection.
Dys- painful, difficult Dyspnea.

E-, Ec-, Ef- out from, out of Efferent neuron.
Ecto-, Exo- outside Ectopic pregnancy.
Em-, En- in, on Emmetropia.
End-, Endo- within, inside Endocardium.
Epi- upon, on, above Epidermis.
Eu- good, easy, normal Eupnea.
Ex-, Exo- outside, beyond Exocrine gland.
Extra- outside, beyond, in addition to Extracellular fluid.

Fore- before, in front of Forehead.

Gen- originate, produce, form Genitalia.
Gingiv- gum Gingivitis.

Hemi- half Hemiplegia.
Heter-, Hetero- other, different Heterozygous.
Homeo-, Homo- unchanging, the same, steady Homeostasis.
Hyper- over, above, excessive Hyperglycemia.
Hypo- under, beneath, deficient Hypothalamus.

In-, Im- in, inside, not Incontinent.
Infra- beneath Infraorbital.
Inter- among, between Intercostal.
Intra- within, inside Intracellular fluid.
Ipsi- same Ipsilateral.
Iso- equal, like Isotonic.

Juxta- near to Juxtaglomerular apparatus.

Later- side Lateral.

Macro- large, great Macrophage.
Mal- bad, abnormal Malnutrition.
Medi-, Meso- middle Medial.
Mega-, Megalo- great, large Magakaryocyte.
Melan- black Melanin.
Meta- after, beyond Metacarpus.
Micro- small Microfilament.
Mono- one Monounsaturated fat.

Neo- new Neonatal.

EPONYMS USED IN THIS TEXT

In the life sciences, an eponym is the name of a structure, drug, or disease that is based on the name of a person. For example, you may be more familiar with the Achilles tendon than you are with its more anatomically descriptive term, the calcaneal tendon. Because eponyms remain in frequent use, this listing correlates common eponyms with their anatomical terms.

EPONYM	ANATOMICAL TERM
Achilles tendon	calcaneal tendon
Adam's apple	thyroid cartilage
ampulla of Vater (VA-ter)	hepatopancreatic ampulla
Bartholin's (BAR-tō-linz) gland	greater vestibular gland
Billroth's (BIL-rōtz) cord	splenic cord
Bowman's (BŌ-manz) capsule	glomerular capsule
Bowman's (BŌ-manz) gland	olfactory gland
Broca's (BRŌ-kaz) area	motor speech area
Brunner's (BRUN-erz) gland	duodenal gland
bundle of His (HISS)	artrioventricular (AV) bundle
canal of Schlemm (SHLEM)	scleral venous sinus
circle of Willis (WIL-is)	cerebral arterial circle
Cooper's (KOO-perz) ligament	suspensory ligament of the breast
Cowper's (KOW-perz) gland	bulbourethral gland
crypt of Lieberkühn (LE-ber-kyūn)	intestinal gland
duct of Santorini (san'-tō-RĒ-nē)	accessory duct
duct of Wirsung (VĒR-sung)	pancreatic duct
Eustachian (yoo-STĀ-kē-an)	auditory tube
Fallopian (fal-LŌ-pē-an) tube	uterine tube
gland of Littré (LĒ-tra)	urethral gland
Golgi (GOL-jē) tendon organ	tendon organ
Graafian (GRAF-ē-an) follicle	mature ovarian follicle
Hassall's (HAS-alz) corpuscle	thymic corpuscle
Haversian (ha-VĒR-shun) canal	central canal
Haversian (ha-VĒR-shun) system	osteon
Heimlich (HĪM-lik) maneuver	abdomial thrust maneuver
islet of Langerhans (LANG-er-hanz)	pancreatic islet

EPONYM	ANATOMICAL TERM
Kupffer (KOOP-fer) cell	stellate reticuloendothelial cell
Leydig (LĪ-dig) cell	interstitial endocrinocyte
loop of Henle (HEN-lē)	loop of the nephron
Luschka's (LUSH-kaz) aperture	lateral aperture
Magendie's (ma-JEN-dēz) aperture	median aperture
Meibomian (mi-BŌ-mē-an) gland	tarsal gland
Meissner (MĪS-ner) corpuscle	corpuscle of touch
Merkel (MER-kel) disc	tactile disc
Müllerian (mil-E rē-an) duct	paramesonephric duct
organ of Corti (KOR-tē)	spiral organ
Pacinian (pa-SIN-ē-an) corpuscle	lamellated corpuscle
Peyer's (PĪ-erz) patch	aggregated lymphatic follicle
plexus of Auerbach (OW-er-bak)	myenteric plexus
plexus of Meissner (MĪS-ner)	submucosal plexus
pouch of Douglas	rectouterine pouch
Purkinje (pur-KIN-jē) fiber	conduction myofiber
Rathke's (rath-KĒZ) pouch	hypophyseal pouch
Ruffini (roo-FĒ-nē) corpuscle	type II cutaneous mechanoreceptor
Sertoli (ser-TŌ-lē) cell	sustentacular cell
Skene's (SKĒNZ) gland	paraurethral gland
sphincter of Oddi (OD-dē)	sphincter of the hepatopancreatic ampulla
Volkmann's (FŌLK-manz) canal	perforating canal
Wernicke's (VER-ni-kēz) area	auditory association area
Wharton's (HWAR-tunz) jelly	mucous connective tissue
Wolffian duct	mesonephric duct
Wormian (WER-mē-an) bone	sutural bone